Prepare for Class "Read the Book"

Feature	Description	Benefit	Page
Every chapter begins with....			
Chapter Opening Article & Project	Each chapter begins with a current article and ends with a related project.	The Article describes a real situation. The Project lets you apply what you learned to solve a related problem.	401, 502
Every section begins with....			
Learning Objectives ✓ 1	Each section begins with a list of objectives. Objectives also appear in the text where the objective is covered.	These focus your studying by emphasizing what's most important and where to find it.	423
Most sections contain...			
PREPARING FOR THIS SECTION	Most sections begin with a list of key concepts to review, with page numbers.	Ever forget what you've learned? This feature highlights previously learned material to be used in this section. Review it, and you'll always be prepared to move forward.	423
Now Work the 'Are You Prepared?' Problems	Special problems that support the Preparing for This Section feature.	Not sure you need the Preparing for This Section review? Work the 'Are You Prepared?' problems. If you get one wrong, you'll know exactly what you need to review and where to review it!	423, 432
"Now Work" PROBLEMS	These follow most examples, and direct you to a related exercise.	We learn best by doing. You'll solidify your understanding of examples if you try a similar problem right away, to be sure you understand what you've just read.	431
WARNING	Warnings are provided in the text.	These point out common mistakes and help you to avoid them.	424
Explorations and **Seeing the Concept**	These features often suggest optional graphing utility activities.	You will obtain a deeper and more intuitive understanding of theorems and definitions.	424, 427
In Words	These provide alternative descriptions of select definitions and theorems.	Does math ever look foreign to you? This feature translates math into plain English.	425
Calculus Icon	These appear next to information essential for the study of calculus.	Pay attention– if you spend extra time now, you'll do better later!	429
Step-by-Step, Annotated EXAMPLES	Examples contain detailed, intermediate steps. Many include additional annotations.	Work the examples on your own, uncovering the solution line-by-line as you go. Each line will verify your work. For additional help, consult the blue annotations, which provide the reason why the statement to the left is valid.	417

Practice "Work the Problems"

Feature	Description	Benefit	Page
"Assess Your Understanding" contains a variety of problems at the end of each section.			
'Are You Prepared?' Problems	These assess your retention of the prerequisite material you'll need. Answers are given at the end of the section exercises. This feature is related to the Preparing for This Section feature.	Do you always remember what you've learned? Working these problems is the best way to find out. If you get one wrong, you'll know exactly what you need to review and where to review it!	423, 432
Concepts and Vocabulary	These short-answer questions, mainly Fill-in-the-Blank and True/False items, assess your understanding of key definitions and concepts in the current section.	Learning math is more than memorization – it's about discovering connections. These problems help you understand the 'big ideas' before diving into skill building.	433
Skill Building	Correlated to section examples, these problems provide straightforward practice.	It's important to dig in and develop your skills. These problems provide you with ample practice to do so.	433–435
Applications and Extensions	These problems allow you to apply your skills to real-world problems. These problems also allow you to extend concepts learned in the section.	You will see that the material learned within the section has many uses in everyday life.	435–437
Discussion and Writing	"Discussion and Writing" problems are colored red. These support class discussion, verbalization of mathematical ideas, and writing and research projects.	To verbalize an idea, or to describe it clearly in writing, shows real understanding. These problems nurture that understanding. Many are challenging but you'll get out what you put in.	437
"Now Work" PROBLEMS	Many examples refer you to a related homework problem. These related problems are marked by a pencil and yellow numbers.	If you get stuck while working problems, look for the closest Now Work problem and refer back to the related example to see if it helps.	429
Graphing Calculator	These **optional problems** require the use of a graphing utility, and are marked by a special icon and green numbers.	Your instructor will usually provide guidance on whether or not to do these problems. If so, these problems help to verify and visualize your analytical results.	463

Review "Study for Quizzes and Tests"

Feature	Description	Benefit	Page
Chapter Reviews at the end of each chapter contain. . .			
"Things to Know"	A detailed list of important theorems, formulas, and definitions from the chapter.	Review these and you'll know the most important material in the chapter!	494–495
"You Should be Able to. . . "	Contains a complete list of objectives by section, with corresponding practice exercises.	Do the recommended exercises and you'll have mastery over the key material. If you get something wrong, review the suggested page numbers and try again.	495–496
Review Exercises	These provide comprehensive review and practice of key skills, matched to the Learning Objectives for each section.	Practice makes perfect. These problems combine exercises from all sections, giving you a comprehensive review in one place.	496–500
CHAPTER TEST	About 15–20 problems that can be taken as a Chapter Test. Be sure to take the Chapter Test under test conditions—no notes!	Be prepared. Take the sample practice test. This will get you ready for your instructor's test. If you get a problem wrong, watch the Chapter Test Prep video.	500–501
CUMULATIVE REVIEW	These problem sets appear at the end of each chapter, beginning with Chapter 2. They combine problems from previous chapters, providing an ongoing cumulative review.	These are really important. They will ensure that you are not forgetting anything as you go. These will go a long way toward keeping you constantly primed for quizzes and tests.	501
CHAPTER PROJECTS	The Chapter Project applies what you've learned in the chapter. Additional projects are available on the Instructor's Resource Center (IRC).	The Project gives you an opportunity to apply what you've learned in the chapter to solve a problem related to the opening article. If your instructor allows, these make excellent opportunities to work in a group, which is often the best way of learning math.	502

Eighth Edition

Algebra & Trigonometry

Michael Sullivan

Chicago State University

PEARSON

Prentice
Hall

Upper Saddle River, New Jersey 07458

Library of Congress Cataloging-in-Publication Data
Sullivan, Michael
 Algebra & trigonometry / Michael Sullivan.—8th ed.
 p. cm.
 ISBN 0-13-232903-4 (alk. paper)
 1. Algebra—Textbooks. 2. Trigonometry—Textbooks. I. Title. II. Title:
 Algebra and trigonometry.
 QA154.3.S73 2008
 512'.13—dc22

 2006047007

Editor-in-Chief: Sally Yagan
Acquisitions Editor: Adam Jaworski
Project Manager: Dawn Murrin
Editorial Assistant/Print Supplements Editor:
 Christine Whitlock
Executive Managing Editor: Kathleen Schiaparelli
Senior Managing Editor: Linda Mihatov Behrens
Production Editor: Bob Walters, Prepress
 Management, Inc.
*Assistant Managing Editor, Science and Math
 Supplements:* Karen Bosch
Media Production Editor: Robert Merenoff
Manufacturing Buyer: Maura Zaldivar
Manufacturing Manager: Alexis Heydt-Long
Director of Marketing: Patrice Jones
Marketing Assistant: Kathleen DeChavez
Development Editor: Don Gecewicz
Editor-in-Chief, Development: Carol Trueheart
Managing Editor, Art Management: Abigail Bass
Art Editor: Thomas Benfatti
Art Director: Heather Scott
Cover/Interior Designer: Tamara Newnam

Creative Director: Juan R. López
Director of Creative Services: Paul Belfanti
Director, Image Resource Center: Melinda Patelli
Manager, Rights and Permissions: Zina Arabia
Manager, Visual Research: Beth Brenzel
Cover Image Specialist: Karen Sanatar
Image Permission Coordinator: Joanne Dippel
Photo Researcher: Emily Tietz
Composition: ICC McMillan Inc.
Art Studios: Laserwords and Artworks
 *Senior Managing Editor, Art Production and
 Management:* Patricia Burns
 Manager, Production Technologies: Matthew Haas
 Manager, Illustration Production: Sean Hogan
 Assistant Manager, Illustration Production:
 Ronda C. Whitson
 Illustrations: Stacy Smith, Scott Wieber,
 Nathan Storck, Ryan Currier,
 Mark Landis, Audrey Simonetti, David Lynch
 Quality Assurance: Pamela Taylor,
 Timothy Nguyen, Cathy Shelley

Printed in the United States of America
10 9 8 7 6

ISBN: 0-13-232903-4

Pearson Education LTD., *London*
Pearson Education Australia PTY, Limited, *Sydney*
Pearson Education Singapore, Pte. Ltd
Pearson Education North Asia Ltd. *Hong Kong*
Pearson Education Canada, Ltd., *Toronto*
Pearson Educación de Mexico, S.A. de C.V.
Pearson Education – Japan, *Tokyo*
Pearson Education Malaysia, Pte. Ltd

For the Family

Katy (Murphy) and Pat	Shannon, Patrick, Ryan
Mike and Yola	Michael, Kevin, Marissa
Dan and Sheila	Maeve, Sean
Colleen (O'Hara) and Bill	Kaleigh, Billy, Timmy

For the Family

Kay (Murphy) and Pat Shannon, Patrick, Ryan
Mike and Yola Michael, Kevin, Andrea
Dan and Sheila Maeve, Sean
Coleen O'Hara and Bill Kaleigh, Billy, Henry

Contents

13 Sequences; Induction; the Binomial Theorem 929

14 Counting and Probability 973

Appendix Graphing Utilities A1

Three Distinct Series

Students have different goals, learning styles, and levels of preparation. Instructors have different teaching philosophies, styles, and techniques. Rather than write one series to fit all, the Sullivans have written three distinct series. All share the same goal—to develop a high level of mathematical understanding and an appreciation for the way mathematics can describe the world around us. The manner of reaching that goal, however, differs from series to series.

Contemporary Series, Eighth Edition

The Contemporary Series is the most traditional in approach, yet modern in its treatment of precalculus mathematics. Graphing utility coverage is optional and can be included or excluded at the discretion of the instructor: *College Algebra, College Algebra Essentials, Algebra & Trigonometry, Trigonometry, Precalculus.*

Enhanced with Graphing Utilities Series, Fourth Edition

This series provides a more thorough integration of graphing utilities into topics, allowing students to explore mathematical concepts and foreshadow ideas usually studied in later courses. Using technology, the approach to solving certain problems differs from the Contemporary series, while the emphasis on understanding concepts and building strong skills does not: *College Algebra, College Algebra Essentials, Algebra & Trigonometry, Trigonometry, Precalculus, Precalculus Essentials.*

Concepts through Functions Series, First Edition

This series differs from the others, utilizing a functions approach that serves as the organizing principle tying concepts together. Functions are introduced early in various formats. This approach supports the Rule of Four, which states that functions are represented symbolically, numerically, graphically, and verbally. Each chapter introduces a new type of function and then develops all concepts pertaining to that particular function. The solutions of equations and inequalities, instead of being developed as stand-alone topics, are developed in the context of the underlying functions. Graphing utility coverage is optional and can be included or excluded at the discretion of the instructor: *College Algebra; Precalculus, with a Unit Circle Approach to Trigonometry; Precalculus, with a Right Triangle Approach to Trigonometry.*

Preface to the Instructor

As a professor of mathematics at an urban public university for 35 years, I understand the varied needs of algebra and trigonometry students. Students range from being under-prepared, with little mathematical background and a fear of mathematics, to being highly prepared and motivated. For some, this is their final course in mathematics. For others, it is preparation for future mathematics courses. I have written this text with both groups in mind.

A tremendous benefit of authoring a successful series is the broad-based feedback I receive from teachers and students who have used previous editions. I am sincerely grateful for their support. Virtually every change to this edition is the result of their thoughtful comments and suggestions. I hope that I have been able to take their ideas, and, building upon a successful foundation of the seventh edition, make this series an even better learning and teaching tool for students and teachers.

Features in the Eighth Edition

Rather than provide a list of features here, that information can be found on the foldout in the front cover of this book.

This places the features in their proper context, as building blocks of an overall learning system that has been carefully crafted over the years to help students get the most out of the time they put into studying. Please take the time to review this, and to discuss it with your students at the beginning of your course. My experience has been that when students utilize these features, they are more successful in the course.

New to the Eighth Edition

- **Chapter Tests** are now part of the end of chapter material. These along with the **Cumulative Review** found at the end of each chapter provide ample opportunity for students to prepare for tests and continually review earlier material.

- **Exercise Sets** at the end of each section have been classified according to purpose. Where appropriate, more problems to challenge the better student have been added.

- **Applied problems** have been updated and many new problems involving sourced information as well as data sets have been added to bring relevance and timeliness to these exercises.

- **New Content**

 Section R.3, Geometry Essentials, now contains a discussion of congruent and similar triangles.

 The analysis of the graph of a polynomial function now includes the behavior of the graph near an *x*-intercept.

Examples and exercises that involve a more in-depth analysis of graphing exponential and logarithmic functions is provided.

Examples and exercises that involve graphing a wider variety of inverse trigonometric functions is provided.

- **Chapter 4 Linear and Quadratic Functions** New to this edition, this chapter gives more emphasis to Linear Functions than before, with more applications, and allows for a fuller discussion of the Quadratic Function, its properties, models involving it, and building such functions from data. A full section on quadratic inequalities is also provided.

Organizational Changes in the Eighth Edition

- **Chapter R** Polynomial division is now part of the discussion of polynomials, leaving synthetic division as a stand-alone (optional) section.

- **Chapter 1** Parallel and Perpendicular Lines is now part of the section on Lines. The section on Lines now precedes the section on Circles.

- **Chapter 2** Scatter diagrams; linear curve fitting has been removed and moved to the new chapter on Linear and Quadratic Functions

- **Trigonometric Functions** now contains applications involving right triangles near the beginning of the chapter.

Using the Eighth Edition Effectively with Your Syllabus

To meet the varied needs of diverse syllabi, this book contains more content than is likely to be covered in an *Algebra & Trigonometry* course. As the chart illustrates, this book has been organized with flexibility of use in mind. Within a given chapter, certain sections are optional (see the detail following the flow chart) and can be omitted without loss of continuity.

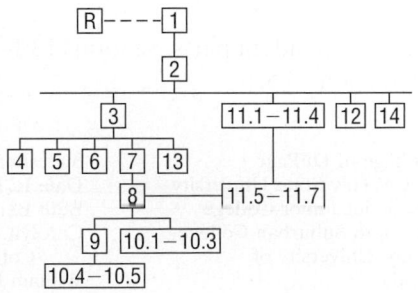

Chapter R Review

This chapter consists of review material. It may be used as the first part of the course or later as a just-in-time review when

the content is required. Specific references to this chapter occur throughout the book to assist in the review process.

Chapter 1 Equations and Inequalities
Primarily a review of Intermediate Algebra topics, this material is prerequisite for later topics. The coverage of complex numbers and quadratic equations with a negative discriminant is optional and may be postponed or skipped entirely without loss of continuity.

Chapter 2 Graphs
This chapter lays the foundation for functions. Section 2.5 is optional.

Chapter 3 Functions and Their Graphs
Perhaps the most important chapter. Section 3.6 is optional.

Chapter 4 Linear and Quadratic Functions
Topic selection depends on your syllabus. Sections 4.2 and 4.4 may be omitted without a loss of continuity.

Chapter 5 Polynomial and Rational Functions
Topic selection depends on your syllabus.

Chapter 6 Exponential and Logarithmic Functions
Sections 6.1–6.6 follow in sequence. Sections 6.7, 6.8, and 6.9 are optional.

Chapter 7 Trigonometric Functions
Section 7.8 may be omitted in a brief course.

Chapter 8 Analytic Trigonometry
Sections 8.2, 8.6, and 8.8 may be omitted in a brief course.

Chapter 9 Applications of Trigonometric Functions
Sections 9.4 and 9.5 may be omitted in a brief course.

Chapter 10 Polar Coordinates; Vectors
Sections 10.1–10.3 and Sections 10.4–10.5 are independent and may be covered separately.

Chapter 11 Analytic Geometry
Sections 11.1–11.4 follow in sequence. Sections 11.5, 11.6, and 11.7 are independent of each other, but each requires Sections 11.1–11.4.

Chapter 12 Systems of Equations and Inequalities
Sections 12.2–12.7 may be covered in any order, but each requires Section 12.1. Section 12.8 requires Section 12.7.

Chapter 13 Sequences; Induction; The Binomial Theorem
There are three independent parts: Sections 13.1–13.3; Section 13.4; and Section 13.5.

Chapter 14 Counting and Probability
The sections follow in sequence.

Acknowledgments

Textbooks are written by authors, but evolve from an idea to final form through the efforts of many people. It was Don Dellen who first suggested this book and series to me. Don is remembered for his extensive contributions to publishing and mathematics.

Thanks are due the following people for their assistance and encouragement in the preparation of this edition:

- From Prentice Hall: Adam Jaworski for his substantial contributions, ideas, and enthusiasm; Patrice Jones, who remains a huge fan and supporter; Dawn Murrin, for her unmatched talent at getting the details right; Bob Walters for his superb organizational skills in directing production; Sally Yagan for her continued support and genuine interest; and the Prentice Hall Sales team, for their continued confidence and personal support of my books.

- Accuracy checkers: Kathleen Miranda, for reading the entire manuscript and providing insightful suggestions; Kevin Bodden and Randy Gallaher, for creating the Solutions Manuals and accuracy checking answers; LaurelTech's editors, Teri Lovelace, Kurt Norlin, George Seki, and Jill McClain-Wardynski, who read the entire manuscript and checked all the answers.

- Reviewers: Debra Kopcso, *Louisiana State University;* Lynn Marecek, *Santa Ana College;* Kathleen Miranda, *SUNY at Old Westbury;* Karla Neal, *Louisiana State University;* Leticia Oropesa, *University of Miami;* Laura Pyzdrowski, *West Virginia University;* Mike Rosenthal, *Florida International University;* Phoebe Rouse, *Louisiana State University;* Brenda Santistevan, *Salt Lake Community College;* Larissa Williamson, *University of Florida.*

Finally, I offer my grateful thanks to the dedicated users and reviewers of my books, whose collective insights form the backbone of each textbook revision.

My list of indebtedness just grows and grows. And, if I've forgotten anyone, please accept my apology. Thank you all.

James Africh, College of DuPage
Steve Agronsky, Cal Poly State University
Grant Alexander, Joliet Junior College
Dave Anderson, South Suburban College
Joby Milo Anthony, University of CentralFlorida
James E. Arnold, University of Wisconsin-Milwaukee
Adel Arshaghi, Center for Educational Merit
Carolyn Autray, University of West Georgia
Agnes Azzolino, Middlesex County College
Wilson P Banks, Illinois State University

Sudeshna Basu, Howard University
Dale R. Bedgood, East Texas State University
Beth Beno, South Suburban College
Carolyn Bernath, Tallahassee Community College
William H. Beyer, University of Akron
Annette Blackwelder, Florida State University
Richelle Blair, Lakeland Community College
Kevin Bodden, Lewis and Clark College
Larry Bouldin, Roane State Community College

Bob Bradshaw, Ohlone College
Trudy Bratten, Grossmont College
Tim Bremer, Broome Community College
Tim Britt, Jackson State Community College
Joanne Brunner, Joliet Junior College
Warren Burch, Brevard Community College
Mary Butler, Lincoln Public Schools
Melanie Butler, West Virginia University
Jim Butterbach, Joliet Junior College
William J. Cable, University of Wisconsin-Stevens Point
Lois Calamia, Brookdale Community College

Jim Campbell, Lincoln Public Schools
Roger Carlsen, Moraine Valley Community College
Elena Catoiu, Joliet Junior College
Mathews Chakkanakuzhi, Palomar College
John Collado, South Suburban College
Nelson Collins, Joliet Junior College
Jim Cooper, Joliet Junior College
Denise Corbett, East Carolina University
Carlos C. Corona, San Antonio College
Theodore C. Coskey, South Seattle Community College
Donna Costello, Plano Senior High School
Paul Crittenden, University of Nebraska at Lincoln
John Davenport, East Texas State University
Faye Dang, Joliet Junior College
Antonio David, Del Mar College
Duane E. Deal, Ball State University
Timothy Deis, University of Wisconsin-Platteville
Vivian Dennis, Eastfield College
Deborah Dillon, R. L. Turner High School
Guesna Dohrman, Tallahassee Community College
Cheryl Doolittle, Iowa State University
Karen R. Dougan, University of Florida
Louise Dyson, Clark College
Paul D. East, Lexington Community College
Don Edmondson, University of Texas-Austin
Erica Egizio, Joliet Junior College
Jason Eltrevoog, Joliet Junior College
Christopher Ennis, University of Minnesota
Kathy Eppler, Salt Lake Community College
Ralph Esparza, Jr., Richland College
Garret J. Etgen, University of Houston
Scott Fallstrom, Shoreline Community College
Pete Falzone, Pensacola Junior College
W.A. Ferguson, University of Illinois-Urbana/Champaign
Iris B. Fetta, Clemson University
Mason Flake, student at Edison Community College
Timothy W. Flood, Pittsburg State University
Merle Friel, Humboldt State University
Richard A. Fritz, Moraine Valley Community College
Dewey Furness, Ricke College
Randy Gallaher, Lewis and Clark College
Tina Garn, University of Arizona
Dawit Getachew, Chicago State University
Wayne Gibson, Rancho Santiago College
Robert Gill, University of Minnesota Duluth
Sudhir Kumar Goel, Valdosta State University
Joan Goliday, Sante Fe Community College
Frederic Gooding, Goucher College
Donald Goral, Northern Virginia Community College
Sue Graupner, Lincoln Public Schools
Jennifer L. Grimsley, University of Charleston
Ken Gurganus, University of North Carolina
James E. Hall, University of Wisconsin-Madison
Judy Hall, West Virginia University
Edward R. Hancock, DeVry Institute of Technology
Julia Hassett, DeVry Institute-Dupage
Christopher Hay-Jahans, University of South Dakota
Michah Heibel, Lincoln Public Schools
LaRae Helliwell, San Jose City College
Celeste Hernandez, Richland College
Brother Herron, Brother Rice High School
Robert Hoburg, Western Connecticut State University

Lynda Hollingsworth, Northwest Missouri State University
Charla Holzbog, Denison High School
Lee Hruby, Naperville North High School
Miles Hubbard, St. Cloud State University
Kim Hughes, California State College-San Bernardino
Ron Jamison, Brigham Young University
Richard A. Jensen, Manatee Community College
Sandra G. Johnson, St. Cloud State University
Tuesday Johnson, New Mexico State University
Moana H. Karsteter, Tallahassee Community College
Donna Katula, Joliet Junior College
Arthur Kaufman, College of Staten Island
Thomas Kearns, North Kentucky University
Jack Keating, Massasoit Community College
Shelia Kellenbarger, Lincoln Public Schools
Rachael Kenney, North Carolina State University
Debra Kopcso, Louisiana State University
Lynne Kowski, Raritan Valley Community College
Keith Kuchar, Manatee Community College
Tor Kwembe, Chicago State University
Linda J. Kyle, Tarrant Country Jr. College
H.E. Lacey, Texas A & M University
Harriet Lamm, Coastal Bend College
James Lapp, Fort Lewis College
Matt Larson, Lincoln Public Schools
Christopher Lattin, Oakton Community College
Julia Ledet, Lousiana State University
Adele LeGere, Oakton Community College
Kevin Leith, University of Houston
JoAnn Lewin, Edison College
Jeff Lewis, Johnson County Community College
Janice C. Lyon, Tallahassee Community College
Jean McArthur, Joliet Junior College
Virginia McCarthy, Iowa State University
Karla McCavit, Albion College
Michael McClendon, University of Central Oklahoma
Tom McCollow, DeVry Institute of Technology
Marilyn McCollum, North Carolina State University
Jill McGowan, Howard University
Will McGowant, Howard University
Angela McNulty, Joliet Junior College
Laurence Maher, North Texas State University
Jay A. Malmstrom, Oklahoma City Community College
Rebecca Mann, Apollo High School
Lynn Marecek, Santa Ana College
Sherry Martina, Naperville North High School
Alec Matheson, Lamar University
Nancy Matthews, University of Oklahoma
James Maxwell, Oklahoma State University-Stillwater
Marsha May, Midwestern State University
Judy Meckley, Joliet Junior College
David Meel, Bowling Green State University
Carolyn Meitler, Concordia University
Samia Metwali, Erie Community College
Rich Meyers, Joliet Junior College
Eldon Miller, University of Mississippi
James Miller, West Virginia University
Michael Miller, Iowa State University
Kathleen Miranda, SUNY at Old Westbury
Thomas Monaghan, Naperville North High School

Maria Montoya, Our Lady of the Lake University
Craig Morse, Naperville North High School
Samad Mortabit, Metropolitan State University
Pat Mower, Washburn University
A. Muhundan, Manatee Community College
Jane Murphy, Middlesex Community College
Richard Nadel, Florida International University
Gabriel Nagy, Kansas State University
Bill Naegele, South Suburban College
Karla Neal, Lousiana State University
Lawrence E. Newman, Holyoke Community College
Dwight Newsome, Pasco-Hernando Community College
James Nymann, University of Texas-El Paso
Mark Omodt, Anoka-Ramsey Community College
Seth F. Oppenheimer, Mississippi State University
Leticia Oropesa, University of Miami
Linda Padilla, Joliet Junior College
E. James Peake, Iowa State University
Kelly Pearson, Murray State University
Philip Pina, Florida Atlantic University
Michael Prophet, University of Northern Iowa
Laura Pyzdrowski, West Virginia University
Neal C. Raber, University of Akron
Thomas Radin, San Joaquin Delta College
Ken A. Rager, Metropolitan State College
Kenneth D. Reeves, San Antonio College
Elsi Reinhardt, Truckee Meadows Community College
Jane Ringwald, Iowa State University
Stephen Rodi, Austin Community College
William Rogge, Lincoln Northeast High School
Howard L. Rolf, Baylor University
Mike Rosenthal, Florida International University
Phoebe Rouse, Lousiana State University
Edward Rozema, University of Tennessee at Chattanooga
Dennis C. Runde, Manatee Community College
Alan Saleski, Loyola University of Chicago
Susan Sandmeyer, Jamestown Community College
Brenda Santistevan, Salt Lake Community College
Linda Schmidt, Greenville Technical College
A.K. Shamma, University of West Florida
Martin Sherry, Lower Columbia College
Tatrana Shubin, San Jose State University
Anita Sikes, Delgado Community College
Timothy Sipka, Alma College
Lori Smellegar, Manatee Community College
Gayle Smith, Loyola Blakefield
John Spellman, Southwest Texas State University
Karen Spike, University of North Carolina
Rajalakshmi Sriram, Okaloosa-Walton Community College
Becky Stamper, Western Kentucky University
Judy Staver, Florida Community College-South
Neil Stephens, Hinsdale South High School
Sonya Stephens, Florida A&M Univeristy
Patrick Stevens, Joliet Junior College
Matthew TenHuisen, University of North Carolina, Wilmington
Christopher Terry, Augusta State University
Diane Tesar, South Suburban College
Tommy Thompson, Brookhaven College

Martha K. Tietze, Shawnee Mission
 Northwest High School
Richard J. Tondra, Iowa State University
Suzanne Topp, Salt Lake Community College
Marvel Townsend, University of Florida
Jim Trudnowski, Carroll College
Robert Tuskey, Joliet Junior College
Richard G. Vinson, University of South
 Alabama
Mary Voxman, University of Idaho

Jennifer Walsh, Daytona Beach Community
 College
Donna Wandke, Naperville North High
 School
Kathryn Wetzel, Amarillo College
Darlene Whitkenack, Northern Illinois
 University
Suzanne Williams, Central Piedmont
 Community College
Larissa Williamson, University of Florida

Christine Wilson, West Virginia University
Brad Wind, Florida International University
Mary Wolyniak, Broome Community
 College
Canton Woods, Auburn University
Tamara S. Worner, Wayne State College
Terri Wright, New Hampshire Community
 Technical College, Manchester
George Zazi, Chicago State University
Steve Zuro, Joliet Junior College

Michael Sullivan
Chicago State University

STUDENT RESOURCES

Student Study-Pack

Everything a student needs to succeed in one place. The Student Study-Pack is available for purchase standalone. Study-Pack contains:

- *Student Solutions Manual*
 Fully worked solutions to odd-numbered exercises.

- *Pearson Tutor Center*
 Tutors provide one-on-one tutoring for any problem with an answer at the back of the book. Students access the Tutor Center via toll-free phone, fax, or email. Available only to college students in the U.S. and Canada.

- *Algebra Review*
 Four chapters of Intermediate Algebra review. Perfect for a slower-paced course or for individual review.

CD Lecture Series

A comprehensive set of CD-ROMs, tied to the textbook, containing short video clips of an instructor working key book examples. The CD Lecture Series can be packaged with the book, or is available for purchase standalone.

Chapter Test Prep Video CD

Packaged at the back of the text, this video CD provides students with step-by-step solutions for each of the exercises in the book's chapter tests.

MathXL®

MathXL® is a powerful online homework, tutorial, and assessment system that accompanies your textbook. Instructors can create, edit, and assign online homework and tests using algorithmically generated exercises correlated at the objective level to the textbook. Student work is tracked in an online gradebook. Students can take chapter tests and receive personalized study plans based on their results. The study plan diagnoses weaknesses and links students to tutorial exercises for objectives they need to study. Students can also access video clips from selected exercises. MathXL® is available to qualified adopters. For more information, visit our website at **www.mathxl.com,** or contact your Prentice Hall sales representative for a demonstration.

MyMathLab

MyMathLab is a text-specific, customizable online course for your textbooks. MyMathLab is powered by CourseCompass™—Pearson Education's online teaching and learning environment—and by MathXL®—our online homework, tutorial, and assessment system. MyMathLab gives you the tools you need to deliver all or a portion of your course online, whether your students are in a lab setting or working from home.

MyMathLab provides a rich and flexible set of course materials, featuring free-response exercises that are algorithmically generated for unlimited practice. Students can use online tools such as video lectures and a multimedia textbook to improve their performance. Instructors can use MyMathLab's homework and test managers to select and assign online exercises correlated to the textbook, and can import TestGen tests for added flexibility. The online gradebook—designed specifically for mathematics—automatically tracks students' homework and test results and gives the instructor control over how to calculate final grades. MyMathLab is available to qualified adopters. For more information, visit our website at **www.mymathlab.com** or contact your Prentice Hall sales representative for a product demonstration.

MathXL® Tutorials on CD

This interactive tutorial CD-ROM provides algorithmically generated practice exercises that are correlated at the objective level to the exercises in the textbook. Every practice exercise is accompanied by an example and a guided solution designed to involve students in the solution process. Selected exercises may also include a video clip to help students visualize concepts. The software provides helpful feedback for incorrect answers and can generate printed summaries of students' progress.

InterAct Math Tutorial Website: www.interactmath.com

Get practice and tutorial help online! This interactive tutorial website provides algorithmically generated practice exercises that correlate directly to the exercises in the textbook. Students can retry an exercise as many times as they like with new values each time for unlimited practice and mastery. Every exercise is accompanied by an interactive guided solution that provides helpful feedback for incorrect answers, and students can also view a worked-out sample problem that steps them through an exercise similar to the one they're working on.

List of Applications

Photo Credits

CHAPTER R Page 1, Michael Newman/PhotoEdit Inc.; Page 31, AP Wide World Photos; Page 38, Whole Sprit Press/Diggins, Julia E., illustrations by Cory don Bell, "String, Straightedge & Shadow, The Story of Geometry;" Page 38, Courtesy of Red River Pess, Inc.; Page 84, Corbis/Bettmann, ©Charless O'Rear/CORBIS.

CHAPTER 1 Pages 85 and 153, Getty Images Inc.—Stone Allstock; Page 96, Bill Aron/PhotoEdit Inc.; Page 108, David Young-Wolff/PhotoEdit Inc.; Page 138, Nancy R. Cohen/Getty Images, Inc.—PhotoDisc; Page 151, Kent Wood/Photo Researchers, Inc.

CHAPTER 2 Pages 155 and 205, Popperfoto/Alamy Images; Page 173, ©Kendra Luck/CORBIS, All Rights Reserved; Page 187, ©Kathy McLaughlin/The Image Works; Page 194, AP Wide World Photos.

CHAPTER 3 Pages 207 and 276, Photollibrary.Com; Page 221, Courtesy of NASA/Jet Propulsion Laboratory/Caltech; Page 229, Jamie Squire/Getty Images; Page 264, Tony Freeman/PhotoEdit Inc.

CHAPTER 4 Pages 277 and 322, Stephen Chernin/Getty Images; Page 312, Erin Garvey/Index Stock Imagery, Inc.;

CHAPTER 5 Pages 323 and 399, Damien Lovegrove/Photo Researchers, Inc.; Page 368, Corbis Digital Stock.

CHAPTER 6 Pages 401 and 502, Kim Karpeles/Alamy Images; Page 456, The Granger Collection; NASA/GFSC/Tom Stack & Associates, Inc.; Page 465, Tony Freeman/Photo Edit Inc.; Page 470, Jim Pickerell/The Image Works; Page 486, ©Theo Allofs/Corbis/Bettmann.

CHAPTER 7 Pages 503 and 599, Texas Automated Buoy System; Page 515, Doug Pensinger/Getty Images; Page 538, © Digital Vision/eStock Photography LLC; Page 573, P. Berndt/Custom Medical Stock Photo, Inc.

CHAPTER 8 Pages 601 and 667, Gemunu Amarasinghe/AP Wide World Photos.

CHAPTER 9 Pages 669 and 711, Getty Images/Photodisc; Page 696, Nelson Hancock/Rough Guides Dorling Kindersley; Page 698, Richard Megna/Fundamental Photographs, NYC.

CHAPTER 10 Pages 713 and 769, Ken Reid/Image Bank/Getty Images; Page 735, Corbis/Bettmann; Page 744, Corbis/Bettmann; Page 754, Library of Congress.

CHAPTER 11 Pages 771 and 834, NASA/Photo Researchers, Inc.; Page 789, Scott McDonald/ Hedrich Blessing.

CHAPTER 12 Pages 835 and 927, AP Wide World Photos; Page 888, Corbis/Bettmann

CHAPTER 13 Pages 929 and 971, NASA/GFSC/Tom Stack & Associates, Inc.; Page 953, The Granger Collection; Page 966, Corbis/Bettmann.

CHAPTER 14 Pages 973 and 1001, Photofest; Page 994, The Granger Collection

To the Student

As you begin, you may feel anxious about the number of theorems, definitions, procedures, and equations. You may wonder if you can learn it all in time. Don't worry, your concerns are normal. This textbook was written with you in mind. If you attend class, work hard, and read and study this book, you will build the knowledge and skills you need to be successful. Here's how you can use the book to your benefit.

Read Carefully

When you get busy, it's easy to skip reading and go right to the problems. Don't. . . the book has a large number of examples and clear explanations to help you break down the mathematics into easy-to-understand steps. Reading will provide you with a clearer understanding, beyond simple memorization. Read before class (not after) so you can ask questions about anything you didn't understand. You'll be amazed at how much more you'll get out of class if you do this.

Use the Features

I use many different methods in the classroom to communicate. Those methods, when incorporated into the book, are called "features". The features serve many purposes, from providing timely review of material you learned before (just when you need it), to providing organized review sessions to help you prepare for quizzes and tests. Take advantage of the features and you will master the material.

To make this easier, I've provided a brief guide to getting the most from this book. Refer to the 'Prepare for Class', 'Practice', and 'Review' pages on the inside front cover of this book. Spend fifteen minutes reviewing the guide and familiarizing yourself with the features by flipping to the page numbers provided. Then, as you read, use them. This is the best way to make the most of your textbook.

Please do not hesitate to contact me, through Prentice Hall, with any questions, suggestions, or comments that would improve this text. I look forward to hearing from you, and good luck with all of your studies.

Best Wishes!

Michael Sullivan

Algebra & Trigonometry

Review

R

A Look Ahead

Chapter R, as the title states, contains review material. Your instructor may choose to cover all or part of it as a regular chapter at the beginning of your course or later as a just-in-time review when the content is required. Regardless, when information in this chapter is needed, a specific reference to this chapter will be made so you can review.

R.1 Real Numbers

PREPARING FOR THIS BOOK *Before getting started, read "To the Student" on page xxvii at the front of this book.*

OBJECTIVES **1** Work with Sets (p. 2)
 2 Classify Numbers (p. 4)
 3 Evaluate Numerical Expressions (p. 8)
 4 Work with Properties of Real Numbers (p. 9)

1 **Work with Sets**

A **set** is a well-defined collection of distinct objects. The objects of a set are called its **elements.** By **well-defined,** we mean that there is a rule that enables us to determine whether a given object is an element of the set. If a set has no elements, it is called the **empty set,** or **null set,** and is denoted by the symbol $\varnothing$.

For example, the set of *digits* consists of the collection of numbers 0, 1, 2, 3, 4, 5, 6, 7, 8, and 9. If we use the symbol D to denote the set of digits, then we can write

$$D = \{0, 1, 2, 3, 4, 5, 6, 7, 8, 9\}$$

In this notation, the braces { } are used to enclose the objects, or **elements,** in the set. This method of denoting a set is called the **roster method.** A second way to denote a set is to use **set-builder notation,** where the set D of digits is written as

$$D = \{ \quad x \quad | \quad x \text{ is a digit}\}$$

Read as "D is the set of all x such that x is a digit."

EXAMPLE 1 | **Using Set-builder Notation and the Roster Method**

(a) $E = \{x | x \text{ is an even digit}\} = \{0, 2, 4, 6, 8\}$
(b) $O = \{x | x \text{ is an odd digit}\} = \{1, 3, 5, 7, 9\}$

Because the elements of a set are distinct, we never repeat elements. For example, we would never write $\{1, 2, 3, 2\}$; the correct listing is $\{1, 2, 3\}$. Because a set is a collection, the order in which the elements are listed is immaterial. $\{1, 2, 3\}$, $\{1, 3, 2\}$, $\{2, 1, 3\}$, and so on, all represent the same set.

If every element of a set A is also an element of a set B, then we say that A is a **subset** of B and write $A \subseteq B$. If two sets A and B have the same elements, then we say that A **equals** B and write $A = B$.

For example, $\{1, 2, 3\} \subseteq \{1, 2, 3, 4, 5\}$ and $\{1, 2, 3\} = \{2, 3, 1\}$.

DEFINITION If A and B are sets, the **intersection** of A with B, denoted $A \cap B$, is the set consisting of elements that belong to both A and B. The **union** of A with B, denoted $A \cup B$, is the set consisting of elements that belong to either A or B, or both.

EXAMPLE 2 | **Finding the Intersection and Union of Sets**

Let $A = \{1, 3, 5, 8\}$, $B = \{3, 5, 7\}$, and $C = \{2, 4, 6, 8\}$. Find:

(a) $A \cap B$ (b) $A \cup B$ (c) $B \cap (A \cup C)$

Solution (a) $A \cap B = \{1, 3, 5, 8\} \cap \{3, 5, 7\} = \{3, 5\}$

(b) $A \cup B = \{1, 3, 5, 8\} \cup \{3, 5, 7\} = \{1, 3, 5, 7, 8\}$

(c) $B \cap (A \cup C) = \{3, 5, 7\} \cap [\{1, 3, 5, 8\} \cup \{2, 4, 6, 8\}]$

$\qquad\qquad\quad = \{3, 5, 7\} \cap \{1, 2, 3, 4, 5, 6, 8\} = \{3, 5\}$

■

Now Work PROBLEM 13

Usually, in working with sets, we designate a **universal set** U, the set consisting of all the elements that we wish to consider. Once a universal set has been designated, we can consider elements of the universal set not found in a given set.

DEFINITION If A is a set, the **complement** of A, denoted $\overline{A}$, is the set consisting of all the elements in the universal set that are not in A.*

EXAMPLE 3 **Finding the Complement of a Set**

If the universal set is $U = \{1, 2, 3, 4, 5, 6, 7, 8, 9\}$ and if $A = \{1, 3, 5, 7, 9\}$, then $\overline{A} = \{2, 4, 6, 8\}$.

■

It follows that $A \cup \overline{A} = U$ and $A \cap \overline{A} = \varnothing$. Do you see why?

Now Work PROBLEM 17

Figure 1

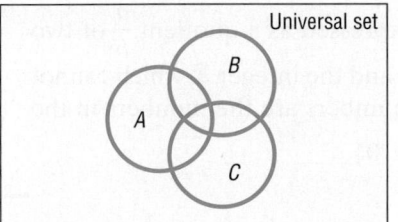
Universal set

It is often helpful to draw pictures of sets. Such pictures, called **Venn diagrams,** represent sets as circles enclosed in a rectangle, which represents the universal set. Such diagrams often help us to visualize various relationships among sets. See Figure 1.

If we know that $A \subseteq B$, we might use the Venn diagram in Figure 2(a). If we know that A and B have no elements in common, that is, if $A \cap B = \varnothing$, we might use the Venn diagram in Figure 2(b). The sets A and B in Figure 2(b) are said to be **disjoint.**

Figure 2

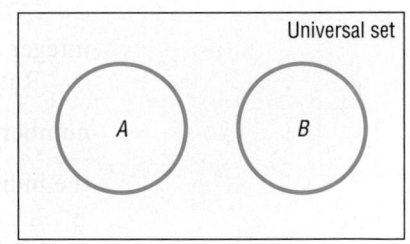

(a) $A \subseteq B$
subset

(b) $A \cap B = \varnothing$
disjoint sets

Figures 3(a), 3(b), and 3(c) use Venn diagrams to illustrate the definitions of intersection, union, and complement, respectively.

Figure 3

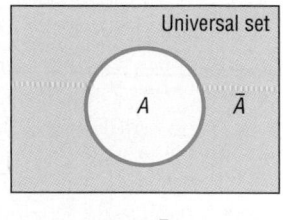

(a) $A \cap B$
intersection

(b) $A \cup B$
union

(c) $\overline{A}$
complement

* Some books use the notation A' for the complement of A.

2 Classify Numbers

It is helpful to classify the various kinds of numbers that we deal with as sets. The **counting numbers,** or **natural numbers,** are the numbers in the set $\{1, 2, 3, 4 \ldots\}$. (The three dots, called an **ellipsis,** indicate that the pattern continues indefinitely.) As their name implies, these numbers are often used to count things. For example, there are 26 letters in our alphabet; there are 100 cents in a dollar. The **whole numbers** are the numbers in the set $\{0, 1, 2, 3, \ldots\}$, that is, the counting numbers together with 0.

DEFINITION The **integers** are the set of numbers $\{\ldots, -3, -2, -1, 0, 1, 2, 3, \ldots\}$.

These numbers are useful in many situations. For example, if your checking account has \$10 in it and you write a check for \$15, you can represent the current balance as $-\$5$.

Notice that the set of counting numbers is a subset of the set of whole numbers. Each time we expand a number system, such as from the whole numbers to the integers, we do so in order to be able to handle new, and usually more complicated, problems. The integers allow us to solve problems requiring both positive and negative counting numbers, such as profit/loss, height above/below sea level, temperature above/below 0°F, and so on.

But integers alone are not sufficient for *all* problems. For example, they do not answer the question "What part of a dollar is 38 cents?" To answer such a question, we enlarge our number system to include *rational numbers.* For example, $\dfrac{38}{100}$ answers the question "What part of a dollar is 38 cents?"

DEFINITION A **rational number** is a number that can be expressed as a quotient $\dfrac{a}{b}$ of two integers. The integer a is called the **numerator,** and the integer b, which cannot be 0, is called the **denominator.** The rational numbers are the numbers in the set $\{x \mid x = \dfrac{a}{b}, \text{ where } a, b \text{ are integers and } b \neq 0\}$.

Examples of rational numbers are $\dfrac{3}{4}, \dfrac{5}{2}, \dfrac{0}{4}, -\dfrac{2}{3}$, and $\dfrac{100}{3}$. Since $\dfrac{a}{1} = a$ for any integer a, it follows that the set of integers is a subset of the set of rational numbers.

Rational numbers may be represented as **decimals.** For example, the rational numbers $\dfrac{3}{4}, \dfrac{5}{2}, -\dfrac{2}{3}$, and $\dfrac{7}{66}$ may be represented as decimals by merely carrying out the indicated division:

$$\frac{3}{4} = 0.75 \qquad \frac{5}{2} = 2.5 \qquad -\frac{2}{3} = -0.666\ldots = -0.\overline{6} \qquad \frac{7}{66} = 0.1060606\ldots = 0.1\overline{06}$$

Notice that the decimal representations of $\dfrac{3}{4}$ and $\dfrac{5}{2}$ terminate, or end. The decimal representations of $-\dfrac{2}{3}$ and $\dfrac{7}{66}$ do not terminate, but they do exhibit a pattern of repetition. For $-\dfrac{2}{3}$, the 6 repeats indefinitely, as indicated by the bar over the 6; for $\dfrac{7}{66}$, the block 06 repeats indefinitely, as indicated by the bar over the 06. It can be shown that every rational number may be represented by a decimal that either terminates or is nonterminating with a repeating block of digits, and vice versa.

On the other hand, some decimals do not fit into either of these categories. Such decimals represent **irrational numbers.** Every irrational number may be represented by a decimal that neither repeats nor terminates. In other words, irrational numbers cannot be written in the form $\dfrac{a}{b}$, where a, b are integers and $b \neq 0$.

Irrational numbers occur naturally. For example, consider the isosceles right triangle whose legs are each of length 1. See Figure 4. The length of the hypotenuse is $\sqrt{2}$, an irrational number.

Also, the number that equals the ratio of the circumference C to the diameter d of any circle, denoted by the symbol π (the Greek letter pi), is an irrational number. See Figure 5.

Figure 4

Figure 5 $\pi = \dfrac{C}{d}$

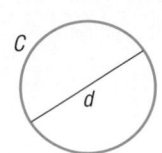

DEFINITION
The set of real numbers is the union of the set of rational numbers with the set of irrational numbers.

Figure 6 shows the relationship of various types of numbers.*

Figure 6

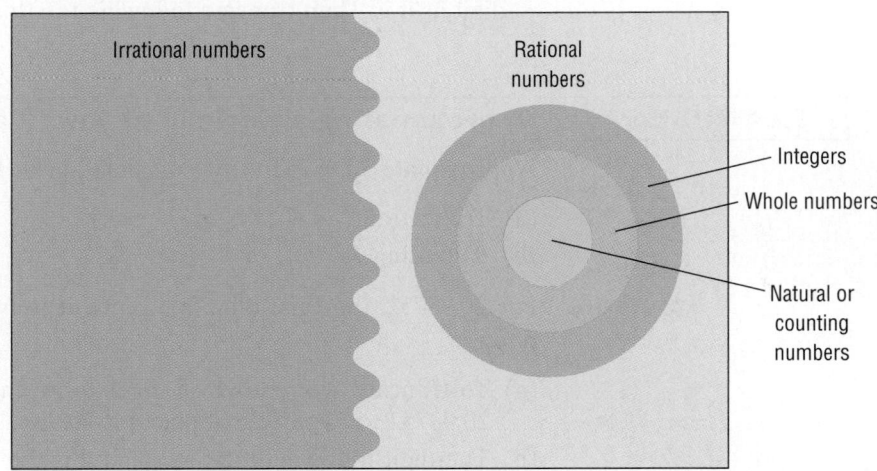

Irrational numbers Rational numbers

Integers

Whole numbers

Natural or counting numbers

Real numbers

EXAMPLE 4

Classifying the Numbers in a Set

List the numbers in the set

$$\left\{-3, \frac{4}{3}, 0.12, \sqrt{2}, \pi, 10, 2.151515\ldots \text{(where the block 15 repeats)}\right\}$$

that are

(a) Natural numbers (b) Integers (c) Rational numbers
(d) Irrational numbers (e) Real numbers

Solution

(a) 10 is the only natural number.

(b) -3 and 10 are integers.

(c) $-3, 10, \dfrac{4}{3}, 0.12$, and $2.151515\ldots$ are rational numbers.

(d) $\sqrt{2}$ and π are irrational numbers.

(e) All the numbers listed are real numbers.

Now Work PROBLEM 23

* The set of real numbers is a subset of the set of complex numbers. We discuss complex numbers in Chapter 1, Section 1.3.

Approximations

Every decimal may be represented by a real number (either rational or irrational), and every real number may be represented by a decimal.

In practice, the decimal representation of an irrational number is given as an approximation. For example, using the symbol $\approx$ (read as "approximately equal to"), we can write

$$\sqrt{2} \approx 1.4142 \qquad \pi \approx 3.1416$$

In approximating decimals, we either *round off* or *truncate* to a given number of decimal places.* The number of places establishes the location of the *final digit* in the decimal approximation.

> **Truncation:** Drop all the digits that follow the specified final digit in the decimal.
>
> **Rounding:** Identify the specified final digit in the decimal. If the next digit is 5 or more, add 1 to the final digit; if the next digit is 4 or less, leave the final digit as it is. Then truncate following the final digit.

EXAMPLE 5 Approximating a Decimal to Two Places

Approximate 20.98752 to two decimal places by

(a) Truncating
(b) Rounding

Solution For 20.98752, the final digit is 8, since it is two decimal places from the decimal point.

(a) To truncate, we remove all digits following the final digit 8. The truncation of 20.98752 to two decimal places is 20.98.
(b) The digit following the final digit 8 is the digit 7. Since 7 is 5 or more, we add 1 to the final digit 8 and truncate. The rounded form of 20.98752 to two decimal places is 20.99.

EXAMPLE 6 Approximating a Decimal to Two and Four Places

Number	Rounded to Two Decimal Places	Rounded to Four Decimal Places	Truncated to Two Decimal Places	Truncated to Four Decimal Places
(a) 3.14159	3.14	3.1416	3.14	3.1415
(b) 0.056128	0.06	0.0561	0.05	0.0561
(c) 893.46125	893.46	893.4613	893.46	893.4612

━━━ **Now Work** PROBLEM 27

Calculators

Calculators are finite machines. As a result, they are incapable of displaying decimals that contain a large number of digits. For example, some calculators are capable of displaying only eight digits. When a number requires more than eight digits,

* Sometimes we say "correct to a given number of decimal places" instead of "truncate."

the calculator either truncates or rounds. To see how your calculator handles decimals, divide 2 by 3. How many digits do you see? Is the last digit a 6 or a 7? If it is a 6, your calculator truncates; if it is a 7, your calculator rounds.

There are different kinds of calculators. An **arithmetic** calculator can only add, subtract, multiply, and divide numbers; therefore, this type is not adequate for this course. **Scientific** calculators have all the capabilities of arithmetic calculators and also contain **function keys** labeled ln, log, sin, cos, tan, x^y, inv, and so on. As you proceed through this text, you will discover how to use many of the function keys. **Graphing** calculators have all the capabilities of scientific calculators and contain a screen on which graphs can be displayed.

For those who have access to a graphing calculator, we have included comments, examples, and exercises marked with a ▦, indicating that a graphing calculator is required. We have also included an appendix that explains some of the capabilities of a graphing calculator. The ▦ comments, examples, and exercises may be omitted without loss of continuity, if so desired.

Operations

In algebra, we use letters such as $x, y, a, b,$ and c to represent numbers. The symbols used in algebra for the operations of addition, subtraction, multiplication, and division are $+, -, \cdot,$ and $/$. The words used to describe the results of these operations are **sum, difference, product,** and **quotient.** Table 1 summarizes these ideas.

Table 1

Operation	Symbol	Words
Addition	$a + b$	Sum: a plus b
Subtraction	$a - b$	Difference: a minus b
Multiplication	$a \cdot b, (a) \cdot b, a \cdot (b), (a) \cdot (b),$ $ab, (a)b, a(b), (a)(b)$	Product: a times b
Division	a/b or $\dfrac{a}{b}$	Quotient: a divided by b

In algebra, we generally avoid using the multiplication sign $\times$ and the division sign $\div$ so familiar in arithmetic. Notice also that when two expressions are placed next to each other without an operation symbol, as in ab, or in parentheses, as in $(a)(b)$, it is understood that the expressions, called **factors,** are to be multiplied.

We also prefer not to use mixed numbers in algebra. When mixed numbers are used, addition is understood; for example, $2\dfrac{3}{4}$ means $2 + \dfrac{3}{4}$. In algebra, use of a mixed number may be confusing because the absence of an operation symbol between two terms is generally taken to mean multiplication. The expression $2\dfrac{3}{4}$ is therefore written instead as 2.75 or as $\dfrac{11}{4}$.

The symbol $=$, called an **equal sign** and read as "equals" or "is," is used to express the idea that the number or expression on the left of the equal sign is equivalent to the number or expression on the right.

EXAMPLE 7 | **Writing Statements Using Symbols**

(a) The sum of 2 and 7 equals 9. In symbols, this statement is written as $2 + 7 = 9$.
(b) The product of 3 and 5 is 15. In symbols, this statement is written as $3 \cdot 5 = 15$.

■

Now Work PROBLEM 39

3 Evaluate Numerical Expressions

Consider the expression $2 + 3 \cdot 6$. It is not clear whether we should add 2 and 3 to get 5, and then multiply by 6 to get 30; or first multiply 3 and 6 to get 18, and then add 2 to get 20. To avoid this ambiguity, we have the following agreement.

> **In Words**
> Multiply first, then add.

We agree that whenever the two operations of addition and multiplication separate three numbers, the multiplication operation always will be performed first, followed by the addition operation.

For $2 + 3 \cdot 6$, we have

$$2 + 3 \cdot 6 = 2 + 18 = 20$$

EXAMPLE 8 **Finding the Value of an Expression**

Evaluate each expression.

(a) $3 + 4 \cdot 5$ (b) $8 \cdot 2 + 1$ (c) $2 + 2 \cdot 2$

Solution (a) $3 + 4 \cdot 5 = 3 + 20 = 23$ (b) $8 \cdot 2 + 1 = 16 + 1 = 17$
 Multiply first Multiply first

(c) $2 + 2 \cdot 2 = 2 + 4 = 6$

✏ **Now Work** PROBLEM 51

To first add 3 and 4 and then multiply the result by 5, we use parentheses and write $(3 + 4) \cdot 5$. Whenever parentheses appear in an expression, it means "perform the operations within the parentheses first!"

EXAMPLE 9 **Finding the Value of an Expression**

(a) $(5 + 3) \cdot 4 = 8 \cdot 4 = 32$
(b) $(4 + 5) \cdot (8 - 2) = 9 \cdot 6 = 54$

When we divide two expressions, as in

$$\frac{2 + 3}{4 + 8}$$

it is understood that the division bar acts like parentheses; that is,

$$\frac{2 + 3}{4 + 8} = \frac{(2 + 3)}{(4 + 8)}$$

The following list gives the rules for the order of operations.

> **Rules for the Order of Operations**
>
> 1. Begin with the innermost parentheses and work outward. Remember that in dividing two expressions the numerator and denominator are treated as if they were enclosed in parentheses.
> 2. Perform multiplications and divisions, working from left to right.
> 3. Perform additions and subtractions, working from left to right.

| EXAMPLE 10 | **Finding the Value of an Expression** |

Evaluate each expression.

(a) $8 \cdot 2 + 3$ (b) $5 \cdot (3 + 4) + 2$

(c) $\dfrac{2 + 5}{2 + 4 \cdot 7}$ (d) $2 + [4 + 2 \cdot (10 + 6)]$

Solution

(a) $8 \cdot 2 + 3 = 16 + 3 = 19$
$\uparrow$
Multiply first

(b) $5 \cdot (3 + 4) + 2 = 5 \cdot 7 + 2 = 35 + 2 = 37$
 $\uparrow$ $\uparrow$
 Parentheses first Multiply before adding

(c) $\dfrac{2 + 5}{2 + 4 \cdot 7} = \dfrac{2 + 5}{2 + 28} = \dfrac{7}{30}$

(d) $2 + [4 + 2 \cdot (10 + 6)] = 2 + [4 + 2 \cdot (16)]$
$$= 2 + [4 + 32] = 2 + [36] = 38$$

Figure 7

Be careful if you use a calculator. For Example 10(c), you need to use parentheses. See Figure 7.* If you don't, the calculator will compute the expression

$$2 + \frac{5}{2} + 4 \cdot 7 = 2 + 2.5 + 28 = 32.5$$

giving a wrong answer.

Now Work PROBLEMS 57 AND 65

4 Work with Properties of Real Numbers

We have used the equal sign to mean that one expression is equivalent to another. Four important properties of equality are listed next. In this list, a, b, and c represent real numbers.

1. The **reflexive property** states that a number always equals itself; that is, $a = a$.
2. The **symmetric property** states that if $a = b$ then $b = a$.
3. The **transitive property** states that if $a = b$ and $b = c$ then $a = c$.
4. The **principle of substitution** states that if $a = b$ then we may substitute b for a in any expression containing a.

Now, let's consider some other properties of real numbers. We begin with an example.

| EXAMPLE 11 | **Commutative Properties** |

(a) $3 + 5 = 8$ (b) $2 \cdot 3 = 6$
 $5 + 3 = 8$ $3 \cdot 2 = 6$
 $3 + 5 = 5 + 3$ $2 \cdot 3 = 3 \cdot 2$

This example illustrates the **commutative property** of real numbers, which states that the order in which addition or multiplication takes place will not affect the final result.

* Notice that we converted the decimal to its fraction form. Consult your manual to see how your calculator does this.

Commutative Properties

$$a + b = b + a \tag{1a}$$
$$a \cdot b = b \cdot a \tag{1b}$$

Here, and in the properties listed next and on pages 11–13, a, b, and c represent real numbers.

EXAMPLE 12 **Associative Properties**

(a) $2 + (3 + 4) = 2 + 7 = 9$
 $(2 + 3) + 4 = 5 + 4 = 9$
 $2 + (3 + 4) = (2 + 3) + 4$

(b) $2 \cdot (3 \cdot 4) = 2 \cdot 12 = 24$
 $(2 \cdot 3) \cdot 4 = 6 \cdot 4 = 24$
 $2 \cdot (3 \cdot 4) = (2 \cdot 3) \cdot 4$ ∎

The way we add or multiply three real numbers will not affect the final result. Expressions such as $2 + 3 + 4$ and $3 \cdot 4 \cdot 5$ present no ambiguity, even though addition and multiplication are performed on one pair of numbers at a time. This property is called the **associative property.**

Associative Properties

$$a + (b + c) = (a + b) + c = a + b + c \tag{2a}$$
$$a \cdot (b \cdot c) = (a \cdot b) \cdot c = a \cdot b \cdot c \tag{2b}$$

The next property is perhaps the most important.

Distributive Property

$$a \cdot (b + c) = a \cdot b + a \cdot c \tag{3a}$$
$$(a + b) \cdot c = a \cdot c + b \cdot c \tag{3b}$$

The **distributive property** may be used in two different ways.

EXAMPLE 13 **Distributive Property**

(a) $2 \cdot (x + 3) = 2 \cdot x + 2 \cdot 3 = 2x + 6$ Use to remove parentheses.
(b) $3x + 5x = (3 + 5)x = 8x$ Use to combine two expressions.
(c) $(x + 2)(x + 3) = x(x + 3) + 2(x + 3) = (x^2 + 3x) + (2x + 6)$
 $= x^2 + (3x + 2x) + 6 = x^2 + 5x + 6$ ∎

‎— **Now Work** PROBLEM 87

The real numbers 0 and 1 have unique properties.

EXAMPLE 14 **Identity Properties**

(a) $4 + 0 = 0 + 4 = 4$ (b) $3 \cdot 1 = 1 \cdot 3 = 3$ ∎

The properties of 0 and 1 illustrated in Example 14 are called the **identity properties.**

Identity Properties

$$0 + a = a + 0 = a \tag{4a}$$
$$a \cdot 1 = 1 \cdot a = a \tag{4b}$$

We call 0 the **additive identity** and 1 the **multiplicative identity.**

For each real number a, there is a real number $-a$, called the **additive inverse** of a, having the following property:

Additive Inverse Property

$$a + (-a) = -a + a = 0 \tag{5a}$$

| EXAMPLE 15 | Finding an Additive Inverse |

(a) The additive inverse of 6 is -6, because $6 + (-6) = 0$.

(b) The additive inverse of -8 is $-(-8) = 8$, because $-8 + 8 = 0$.

The additive inverse of a, that is, $-a$, is often called the *negative* of a or the *opposite* of a. The use of such terms can be dangerous, because they suggest that the additive inverse is a negative number, which may not be the case. For example, the additive inverse of -3, or $-(-3)$, equals 3, a positive number.

For each *nonzero* real number a, there is a real number $\dfrac{1}{a}$, called the **multiplicative inverse** of a, having the following property:

Multiplicative Inverse Property

$$a \cdot \frac{1}{a} = \frac{1}{a} \cdot a = 1 \qquad \text{if } a \neq 0 \tag{5b}$$

The multiplicative inverse $\dfrac{1}{a}$ of a nonzero real number a is also referred to as the **reciprocal** of a.

| EXAMPLE 16 | Finding a Reciprocal |

(a) The reciprocal of 6 is $\dfrac{1}{6}$, because $6 \cdot \dfrac{1}{6} = 1$.

(b) The reciprocal of -3 is $\dfrac{1}{-3}$, because $-3 \cdot \dfrac{1}{-3} = 1$.

(c) The reciprocal of $\dfrac{2}{3}$ is $\dfrac{3}{2}$, because $\dfrac{2}{3} \cdot \dfrac{3}{2} = 1$.

With these properties for adding and multiplying real numbers, we can now define the operations of subtraction and division as follows:

DEFINITION The **difference** $a - b$, also read "a less b" or "a minus b," is defined as

$$a - b = a + (-b) \tag{6}$$

To subtract b from a, add the opposite of b to a.

DEFINITION If b is a nonzero real number, the **quotient** $\dfrac{a}{b}$, also read as "a divided by b" or "the ratio of a to b," is defined as

$$\frac{a}{b} = a \cdot \frac{1}{b} \qquad \text{if } b \neq 0 \tag{7}$$

EXAMPLE 17 **Working with Differences and Quotients**

(a) $8 - 5 = 8 + (-5) = 3$

(b) $4 - 9 = 4 + (-9) = -5$

(c) $\dfrac{5}{8} = 5 \cdot \dfrac{1}{8}$

For any number a, the product of a times 0 is always 0; that is,

In Words
The result of multiplying by zero is zero.

Multiplication by Zero

$$a \cdot 0 = 0 \tag{8}$$

For a nonzero number a,

Division Properties

$$\frac{0}{a} = 0 \qquad \frac{a}{a} = 1 \quad \text{if } a \neq 0 \tag{9}$$

NOTE Division by 0 is not defined. One reason is to avoid the following difficulty: $\dfrac{2}{0} = x$ means to find x such that $0 \cdot x = 2$. But $0 \cdot x$ equals 0 for all x, so there is *no* unique number x such that $\dfrac{2}{0} = x$. ∎

Rules of Signs

$$a(-b) = -(ab) \qquad (-a)b = -(ab) \qquad (-a)(-b) = ab$$
$$-(-a) = a \qquad \frac{a}{-b} = \frac{-a}{b} = -\frac{a}{b} \qquad \frac{-a}{-b} = \frac{a}{b} \tag{10}$$

EXAMPLE 18 **Applying the Rules of Signs**

(a) $2(-3) = -(2 \cdot 3) = -6$ 　　　　(b) $(-3)(-5) = 3 \cdot 5 = 15$

(c) $\dfrac{3}{-2} = \dfrac{-3}{2} = -\dfrac{3}{2}$ 　　　　(d) $\dfrac{-4}{-9} = \dfrac{4}{9}$

(e) $\dfrac{x}{-2} = \dfrac{1}{-2} \cdot x = -\dfrac{1}{2}x$

Cancellation Properties

$$ac = bc \quad \text{implies} \quad a = b \quad \text{if } c \neq 0$$
$$\frac{ac}{bc} = \frac{a}{b} \qquad \qquad \text{if } b \neq 0, c \neq 0 \qquad \textbf{(11)}$$

| EXAMPLE 19 | Using the Cancellation Properties |

(a) If $2x = 6$, then

$$2x = 6$$
$$2x = 2 \cdot 3 \quad \text{Factor 6.}$$
$$x = 3 \quad \text{Cancel the 2's.}$$

NOTE We follow the common practice of using slash marks to indicate cancellations. ■

(b) $\dfrac{18}{12} = \dfrac{3 \cdot \cancel{6}}{2 \cdot \cancel{6}} = \dfrac{3}{2}$
 $\uparrow$ Cancel the 6's.

In Words
If a product equals 0, then one or both of the factors is 0.

Zero-Product Property

$$\text{If } ab = 0, \text{ then } a = 0, \text{ or } b = 0, \text{ or both.} \qquad \textbf{(12)}$$

| EXAMPLE 20 | Using the Zero-Product Property |

If $2x = 0$, then either $2 = 0$ or $x = 0$. Since $2 \neq 0$, it follows that $x = 0$.

Arithmetic of Quotients

$$\frac{a}{b} + \frac{c}{d} = \frac{ad}{bd} + \frac{bc}{bd} = \frac{ad + bc}{bd} \qquad \text{if } b \neq 0, d \neq 0 \qquad \textbf{(13)}$$

$$\frac{a}{b} \cdot \frac{c}{d} = \frac{ac}{bd} \qquad \text{if } b \neq 0, d \neq 0 \qquad \textbf{(14)}$$

$$\frac{\dfrac{a}{b}}{\dfrac{c}{d}} = \frac{a}{b} \cdot \frac{d}{c} = \frac{ad}{bc} \qquad \text{if } b \neq 0, c \neq 0, d \neq 0 \qquad \textbf{(15)}$$

| EXAMPLE 21 | Adding, Subtracting, Multiplying, and Dividing Quotients |

(a) $\dfrac{2}{3} + \dfrac{5}{2} = \dfrac{2 \cdot 2}{3 \cdot 2} + \dfrac{3 \cdot 5}{3 \cdot 2} = \dfrac{2 \cdot 2 + 3 \cdot 5}{3 \cdot 2} = \dfrac{4 + 15}{6} = \dfrac{19}{6}$
 $\uparrow$ By equation (13)

(b) $\dfrac{3}{5} - \dfrac{2}{3} = \dfrac{3}{5} + \left(-\dfrac{2}{3}\right) = \dfrac{3}{5} + \dfrac{2}{3}$
 $\uparrow$ $\qquad\qquad\qquad\qquad$ $\uparrow$
 By equation (6) $\quad$ By equation (10)

 $= \dfrac{3 \cdot 3 + 5 \cdot (-2)}{5 \cdot 3} = \dfrac{9 + (-10)}{15} = \dfrac{-1}{15} = -\dfrac{1}{15}$
 $\uparrow$ By equation (13)

NOTE Slanting the cancellation marks in different directions for different factors, as shown here, is a good practice to follow, since it will help in checking for errors. ■

(c) $\dfrac{8}{3} \cdot \dfrac{15}{4} = \dfrac{8 \cdot 15}{3 \cdot 4} = \dfrac{2 \cdot 4 \cdot 3 \cdot 5}{3 \cdot 4 \cdot 1} = \dfrac{2 \cdot 5}{1} = 10$

$\uparrow$ By equation (14) $\uparrow$ By equation (11)

(d) $\dfrac{\dfrac{3}{5}}{\dfrac{7}{9}} = \dfrac{3}{5} \cdot \dfrac{9}{7} = \dfrac{3 \cdot 9}{5 \cdot 7} = \dfrac{27}{35}$

$\uparrow$ By equation (14)

$\uparrow$ By equation (15)

NOTE In writing quotients, we shall follow the usual convention and write the quotient in lowest terms. That is, we write it so that any common factors of the numerator and the denominator have been removed using the cancellation properties, equation (11). As examples,

$$\frac{90}{24} = \frac{15 \cdot 6}{4 \cdot 6} = \frac{15}{4}$$

$$\frac{24x^2}{18x} = \frac{4 \cdot 6 \cdot x \cdot x}{3 \cdot 6 \cdot x} = \frac{4x}{3} \qquad x \neq 0$$

Now Work PROBLEMS **67, 71, AND 81**

Sometimes it is easier to add two fractions using *least common multiples* (LCM). The LCM of two numbers is the smallest number that each has as a common multiple.

EXAMPLE 22	**Finding the Least Common Multiple of Two Numbers**

Find the least common multiple of 15 and 12.

Solution To find the LCM of 15 and 12, we look at multiples of 15 and 12.

15, 30, 45, 60, 75, 90, 105, 120, ...

12, 24, 36, 48, 60, 72, 84, 96, 108, 120, ...

The *common* multiples are in blue. The *least* common multiple is 60. ■

EXAMPLE 23	**Using the Least Common Multiple to Add Two Fractions**

Find: $\dfrac{8}{15} + \dfrac{5}{12}$

Solution We use the LCM of the denominators of the fractions and rewrite each fraction using the LCM as a common denominator. The LCM of the denominators (12 and 15) is 60. Rewrite each fraction using 60 as the denominator.

$$\frac{8}{15} + \frac{5}{12} = \frac{8}{15} \cdot \frac{4}{4} + \frac{5}{12} \cdot \frac{5}{5}$$

$$= \frac{32}{60} + \frac{25}{60}$$

$$= \frac{32 + 25}{60}$$

$$= \frac{57}{60}$$

$$= \frac{19}{20}$$

■

Now Work PROBLEM **75**

Historical Feature

The real number system has a history that stretches back at least to the ancient Babylonians (1800 BC). It is remarkable how much the ancient Babylonian attitudes resemble our own. As we stated in the text, the fundamental difficulty with irrational numbers is that they cannot be written as quotients of integers or, equivalently, as repeating or terminating decimals. The Babylonians wrote their numbers in a system based on 60 in the same way that we write ours based on 10. They would carry as many places for π as the accuracy of the problem demanded, just as we now use

$$\pi \approx 3\frac{1}{7} \quad \text{or} \quad \pi \approx 3.1416 \quad \text{or} \quad \pi \approx 3.14159$$
$$\text{or} \quad \pi \approx 3.14159265358979$$

depending on how accurate we need to be.

Things were very different for the Greeks, whose number system allowed only rational numbers. When it was discovered that $\sqrt{2}$ was not a rational number, this was regarded as a fundamental flaw in the number concept. So serious was the matter that the Pythagorean Brotherhood (an early mathematical society) is said to have drowned one of its members for revealing this terrible secret. Greek mathematicians then turned away from the number concept, expressing facts about whole numbers in terms of line segments.

In astronomy, however, Babylonian methods, including the Babylonian number system, continued to be used. Simon Stevin (1548–1620), probably using the Babylonian system as a model, invented the decimal system, complete with rules of calculation, in 1585. [Others, for example, al-Kashi of Samarkand (d. 1429), had made some progress in the same direction.] The decimal system so effectively conceals the difficulties that the need for more logical precision began to be felt only in the early 1800s. Around 1880, Georg Cantor (1845–1918) and Richard Dedekind (1831–1916) gave precise definitions of real numbers. Cantor's definition, although more abstract and precise, has its roots in the decimal (and hence Babylonian) numerical system.

Sets and set theory were a spin-off of the research that went into clarifying the foundations of the real number system. Set theory has developed into a large discipline of its own, and many mathematicians regard it as the foundation upon which modern mathematics is built. Cantor's discoveries that infinite sets can also be counted and that there are different sizes of infinite sets are among the most astounding results of modern mathematics.

Historical Problems

The Babylonian number system was based on 60. Thus 2,30 means $2 + \dfrac{30}{60} = 2.5$, and 4,25,14 means

$$4 + \frac{25}{60} + \frac{14}{60^2} = 4 + \frac{1514}{3600} = 4.42055555\ldots$$

1. What are the following numbers in Babylonian notation?

(a) $1\dfrac{1}{3}$ **(b)** $2\dfrac{5}{6}$

2. What are the following Babylonian numbers when written as fractions and as decimals?

(a) 2,20 **(b)** 4,52,30 **(c)** 3,8,29,44

R.1 Assess Your Understanding

Concepts and Vocabulary

1. The numbers in the set $\{x \mid x = \dfrac{a}{b}$, where a, b are integers and $b \neq 0\}$, are called _____ numbers.

2. The value of the expression $4 + 5 \cdot 6 - 3$ is _____.

3. The fact that $2x + 3x = (2 + 3)x$ is a consequence of the _____ Property.

4. "The product of 5 and $x + 3$ equals 6" may be written as _____.

5. *True or False* Rational numbers have decimals that either terminate or are nonterminating with a repeating block of digits.

6. *True or False* The Zero-Product Property states that the product of any number and zero equals zero.

7. *True or False* The least common multiple of 12 and 18 is 6.

8. *True or False* No real number is both rational and irrational.

Skill Building

In Problems 9–20, use U = universal set = $\{0, 1, 2, 3, 4, 5, 6, 7, 8, 9\}$, *A =* $\{1, 3, 4, 5, 9\}$, *B =* $\{2, 4, 6, 7, 8\}$, *and C =* $\{1, 3, 4, 6\}$ *to find each set.*

9. $A \cup B$ **10.** $A \cup C$ **11.** $A \cap B$ **12.** $A \cap C$

13. $(A \cup B) \cap C$ **14.** $(A \cap B) \cup C$ **15.** $\overline{A}$ **16.** $\overline{C}$

17. $\overline{A \cap B}$ **18.** $\overline{B \cup C}$ **19.** $\overline{A} \cup \overline{B}$ **20.** $\overline{B} \cap \overline{C}$

In Problems 21–26, list the numbers in each set that are (a) Natural numbers, (b) Integers, (c) Rational numbers, (d) Irrational numbers, (e) Real numbers.

21. $A = \left\{ -6, \frac{1}{2}, -1.333\ldots \text{(the 3's repeat)}, \pi, 2, 5 \right\}$

22. $B = \left\{ -\frac{5}{3}, 2.060606\ldots \text{(the block 06 repeats)}, 1.25, 0, 1, \sqrt{5} \right\}$

23. $C = \left\{ 0, 1, \frac{1}{2}, \frac{1}{3}, \frac{1}{4} \right\}$

24. $D = \{-1, -1.1, -1.2, -1.3\}$

25. $E = \left\{ \sqrt{2}, \pi, \sqrt{2} + 1, \pi + \frac{1}{2} \right\}$

26. $F = \left\{ -\sqrt{2}, \pi + \sqrt{2}, \frac{1}{2} + 10.3 \right\}$

In Problems 27–38, approximate each number (a) rounded and (b) truncated to three decimal places.

27. 18.9526 **28.** 25.86134 **29.** 28.65319 **30.** 99.05249 **31.** 0.06291 **32.** 0.05388

33. 9.9985 **34.** 1.0006 **35.** $\dfrac{3}{7}$ **36.** $\dfrac{5}{9}$ **37.** $\dfrac{521}{15}$ **38.** $\dfrac{81}{5}$

In Problems 39–48, write each statement using symbols.

39. The sum of 3 and 2 equals 5.

40. The product of 5 and 2 equals 10.

41. The sum of x and 2 is the product of 3 and 4.

42. The sum of 3 and y is the sum of 2 and 2.

43. The product of 3 and y is the sum of 1 and 2.

44. The product of 2 and x is the product of 4 and 6.

45. The difference x less 2 equals 6.

46. The difference 2 less y equals 6.

47. The quotient x divided by 2 is 6.

48. The quotient 2 divided by x is 6.

In Problems 49–86, evaluate each expression.

49. $9 - 4 + 2$ **50.** $6 - 4 + 3$ **51.** $-6 + 4 \cdot 3$ **52.** $8 - 4 \cdot 2$

53. $4 + 5 - 8$ **54.** $8 - 3 - 4$ **55.** $4 + \dfrac{1}{3}$ **56.** $2 - \dfrac{1}{2}$

57. $6 - [3 \cdot 5 + 2 \cdot (3 - 2)]$ **58.** $2 \cdot [8 - 3(4 + 2)] - 3$ **59.** $2 \cdot (3 - 5) + 8 \cdot 2 - 1$ **60.** $1 - (4 \cdot 3 - 2 + 2)$

61. $10 - [6 - 2 \cdot 2 + (8 - 3)] \cdot 2$ **62.** $2 - 5 \cdot 4 - [6 \cdot (3 - 4)]$

63. $(5 - 3)\dfrac{1}{2}$ **64.** $(5 + 4)\dfrac{1}{3}$ **65.** $\dfrac{4 + 8}{5 - 3}$ **66.** $\dfrac{2 - 4}{5 - 3}$

67. $\dfrac{3}{5} \cdot \dfrac{10}{21}$ **68.** $\dfrac{5}{9} \cdot \dfrac{3}{10}$ **69.** $\dfrac{6}{25} \cdot \dfrac{10}{27}$ **70.** $\dfrac{21}{25} \cdot \dfrac{100}{3}$

71. $\dfrac{3}{4} + \dfrac{2}{5}$ **72.** $\dfrac{4}{3} + \dfrac{1}{2}$ **73.** $\dfrac{5}{6} + \dfrac{9}{5}$ **74.** $\dfrac{8}{9} + \dfrac{15}{2}$

75. $\dfrac{5}{18} + \dfrac{1}{12}$ **76.** $\dfrac{2}{15} + \dfrac{8}{9}$ **77.** $\dfrac{1}{30} - \dfrac{7}{18}$ **78.** $\dfrac{3}{14} - \dfrac{2}{21}$

79. $\dfrac{3}{20} - \dfrac{2}{15}$ **80.** $\dfrac{6}{35} - \dfrac{3}{14}$ **81.** $\dfrac{\frac{5}{18}}{\frac{11}{27}}$ **82.** $\dfrac{\frac{5}{21}}{\frac{2}{35}}$

83. $\dfrac{1}{2} \cdot \dfrac{3}{5} + \dfrac{7}{10}$ **84.** $\dfrac{2}{3} + \dfrac{4}{5} \cdot \dfrac{1}{6}$ **85.** $2 \cdot \dfrac{3}{4} + \dfrac{3}{8}$ **86.** $3 \cdot \dfrac{5}{6} - \dfrac{1}{2}$

In Problems 87–98, use the Distributive Property to remove the parentheses.

87. $6(x + 4)$ **88.** $4(2x - 1)$ **89.** $x(x - 4)$ **90.** $4x(x + 3)$

91. $2\left(\dfrac{3}{4}x - \dfrac{1}{2}\right)$ **92.** $3\left(\dfrac{2}{3}x + \dfrac{1}{6}\right)$ **93.** $(x + 2)(x + 4)$ **94.** $(x + 5)(x + 1)$

95. $(x - 2)(x + 1)$ **96.** $(x - 4)(x + 1)$ **97.** $(x - 8)(x - 2)$ **98.** $(x - 4)(x - 2)$

Discussion and Writing

99. Explain to a friend how the Distributive Property is used to justify the fact that $2x + 3x = 5x$.

100. Explain to a friend why $2 + 3 \cdot 4 = 14$, whereas $(2 + 3) \cdot 4 = 20$.

101. Explain why $2(3 \cdot 4)$ is not equal to $(2 \cdot 3) \cdot (2 \cdot 4)$.

102. Explain why $\dfrac{4 + 3}{2 + 5}$ is not equal to $\dfrac{4}{2} + \dfrac{3}{5}$.

103. Is subtraction commutative? Support your conclusion with an example.

104. Is subtraction associative? Support your conclusion with an example.

105. Is division commutative? Support your conclusion with an example.

106. Is division associative? Support your conclusion with an example.

107. If $2 = x$, why does $x = 2$?

108. If $x = 5$, why does $x^2 + x = 30$?

109. Are there any real numbers that are both rational and irrational? Are there any real numbers that are neither? Explain your reasoning.

110. Explain why the sum of a rational number and an irrational number must be irrational.

111. A rational number is defined as the quotient of two integers. When written as a decimal, the decimal will either repeat or terminate. By looking at the denominator of the rational number, there is a way to tell in advance whether its decimal representation will repeat or terminate. Make a list of rational numbers and their decimals. See if you can discover the pattern. Confirm your conclusion by consulting books on number theory at the library. Write a brief essay on your findings.

112. The current time is 12 noon CST. What time (CST) will it be 12,997 hours from now?

113. Both $\dfrac{a}{0} (a \neq 0)$ and $\dfrac{0}{0}$ are undefined, but for different reasons. Write a paragraph or two explaining the different reasons.

R.2 Algebra Essentials

OBJECTIVES
1. Graph Inequalities (p. 18)
2. Find Distance on the Real Number Line (p. 19)
3. Evaluate Algebraic Expressions (p. 20)
4. Determine the Domain of a Variable (p. 21)
5. Use the Laws of Exponents (p. 21)
6. Evaluate Square Roots (p. 23)
7. Use a Calculator to Evaluate Exponents (p. 24)
8. Use Scientific Notation (p. 24)

The Real Number Line

The real numbers can be represented by points on a line called the **real number line.** There is a one-to-one correspondence between real numbers and points on a line. That is, every real number corresponds to a point on the line, and each point on the line has a unique real number associated with it.

Pick a point on the line somewhere in the center, and label it O. This point, called the **origin,** corresponds to the real number 0. See Figure 8. The point 1 unit to the right of O corresponds to the number 1. The distance between 0 and 1 determines the **scale** of the number line. For example, the point associated with the number 2 is twice as far from O as 1. Notice that an arrowhead on the right end of the line indicates the direction in which the numbers increase. Points to the left of the origin correspond to the real numbers -1, -2, and so on. Figure 8 also shows the points associated with the rational numbers $-\dfrac{1}{2}$ and $\dfrac{1}{2}$ and with the irrational numbers $\sqrt{2}$ and π.

Figure 8
Real number line

DEFINITION
The real number associated with a point P is called the **coordinate** of P, and the line whose points have been assigned coordinates is called the **real number line.**

═══➤ **Now Work** PROBLEM 11

The real number line consists of three classes of real numbers, as shown in Figure 9.

Figure 9

1. The **negative real numbers** are the coordinates of points to the left of the origin O.
2. The real number **zero** is the coordinate of the origin O.
3. The **positive real numbers** are the coordinates of points to the right of the origin O.

Negative and positive numbers have the following multiplication properties:

Multiplication Properties of Positive and Negative Numbers

1. The product of two positive numbers is a positive number.
2. The product of two negative numbers is a positive number.
3. The product of a positive number and a negative number is a negative number.

1 Graph Inequalities

Figure 10

(a) $a < b$

(b) $a = b$

(c) $a > b$

An important property of the real number line follows from the fact that, given two numbers (points) a and b, either a is to the left of b, or a is at the same location as b, or a is to the right of b. See Figure 10.

If a is to the left of b, we say that "a is less than b" and write $a < b$. If a is to the right of b, we say that "a is greater than b" and write $a > b$. If a is at the same location as b, then $a = b$. If a is either less than or equal to b, we write $a \leq b$. Similarly, $a \geq b$ means that a is either greater than or equal to b. Collectively, the symbols $<, >, \leq$, and $\geq$ are called **inequality symbols.**

Note that $a < b$ and $b > a$ mean the same thing. It does not matter whether we write $2 < 3$ or $3 > 2$.

Furthermore, if $a < b$ or if $b > a$, then the difference $b - a$ is positive. Do you see why?

EXAMPLE 1 | **Using Inequality Symbols**

(a) $3 < 7$ (b) $-8 > -16$ (c) $-6 < 0$

(d) $-8 < -4$ (e) $4 > -1$ (f) $8 > 0$

In Example 1(a), we conclude that $3 < 7$ either because 3 is to the left of 7 on the real number line or because the difference, $7 - 3 = 4$, is a positive real number.

Similarly, we conclude in Example 1(b) that $-8 > -16$ either because -8 lies to the right of -16 on the real number line or because the difference, $-8 - (-16) = -8 + 16 = 8$, is a positive real number.

Look again at Example 1. Note that the inequality symbol always points in the direction of the smaller number.

Statements of the form $a < b$ or $b > a$ are called **strict inequalities,** whereas statements of the form $a \leq b$ or $b \geq a$ are called **nonstrict inequalities.** An **inequality** is a statement in which two expressions are related by an inequality symbol. The expressions are referred to as the **sides** of the inequality.

Based on the discussion so far, we conclude that

> $a > 0$ is equivalent to a is positive
>
> $a < 0$ is equivalent to a is negative

We sometimes read $a > 0$ by saying that "a is positive." If $a \geq 0$, then either $a > 0$ or $a = 0$, and we may read this as "a is nonnegative."

Now Work PROBLEMS 15 AND 25

We shall find it useful in later work to graph inequalities on the real number line.

EXAMPLE 2	**Graphing Inequalities**

(a) On the real number line, graph all numbers x for which $x > 4$.

(b) On the real number line, graph all numbers x for which $x \leq 5$.

Solution

(a) See Figure 11. Notice that we use a left parenthesis to indicate that the number 4 is *not* part of the graph.

(b) See Figure 12. Notice that we use a right bracket to indicate that the number 5 *is* part of the graph.

Figure 11

```
  ┌─┬─┬─┬─┬─┬─(─┬─┬─┬─►
 -2 -1  0  1  2  3  4  5  6  7
```

Figure 12

```
◄─┬─┬─┬─┬─┬─┬─┬─]─┬─┬─►
 -2 -1  0  1  2  3  4  5  6  7
```

Now Work PROBLEM 31

2 Find Distance on the Real Number Line

Figure 13

```
  ┌──4 units──►├──3 units──┤
◄─┬─●─┬─┬─┬─┬─●─┬─┬─●─┬─►
 -5 -4 -3 -2 -1  0  1  2  3  4
```

The *absolute value* of a number a is the distance from 0 to a on the number line. For example, -4 is 4 units from 0, and 3 is 3 units from 0. See Figure 13. Thus, the absolute value of -4 is 4, and the absolute value of 3 is 3.

A more formal definition of absolute value is given next.

| DEFINITION | The **absolute value** of a real number a, denoted by the symbol $|a|$, is defined by the rules |
| --- | --- |

$$|a| = a \text{ if } a \geq 0 \qquad \text{and} \qquad |a| = -a \text{ if } a < 0$$

For example, since $-4 < 0$, the second rule must be used to get $|-4| = -(-4) = 4$.

EXAMPLE 3	**Computing Absolute Value**

(a) $|8| = 8$ (b) $|0| = 0$ (c) $|-15| = -(-15) = 15$

Look again at Figure 13. The distance from -4 to 3 is 7 units. This distance is the difference $3 - (-4)$, obtained by subtracting the smaller coordinate from the larger. However, since $|3 - (-4)| = |7| = 7$ and $|-4 - 3| = |-7| = 7$, we can use absolute value to calculate the distance between two points without being concerned about which is smaller.

DEFINITION	If P and Q are two points on a real number line with coordinates a and b respectively, the **distance between P and Q**, denoted by $d(P, Q)$, is

$$d(P, Q) = |b - a|$$

Since $|b - a| = |a - b|$, it follows that $d(P, Q) = d(Q, P)$.

EXAMPLE 4

Finding Distance on a Number Line

Let P, Q, and R be points on a real number line with coordinates -5, 7, and -3, respectively. Find the distance

(a) between P and Q (b) between Q and R

Solution See Figure 14.

Figure 14

(a) $d(P, Q) = |7 - (-5)| = |12| = 12$

(b) $d(Q, R) = |-3 - 7| = |-10| = 10$

━━━━━ **Now Work** PROBLEM 37

3 Evaluate Algebraic Expressions

Remember, in algebra we use letters such as x, y, a, b, and c to represent numbers. If the letter used is to represent *any* number from a given set of numbers, it is called a **variable**. A **constant** is either a fixed number, such as 5 or $\sqrt{3}$, or a letter that represents a fixed (possibly unspecified) number.

Constants and variables are combined using the operations of addition, subtraction, multiplication, and division to form *algebraic expressions*. Examples of algebraic expressions include

$$x + 3 \qquad \frac{3}{1 - t} \qquad 7x - 2y$$

To evaluate an algebraic expression, substitute for each variable its numerical value.

EXAMPLE 5

Evaluating an Algebraic Expression

Evaluate each expression if $x = 3$ and $y = -1$.

(a) $x + 3y$ (b) $5xy$ (c) $\dfrac{3y}{2 - 2x}$ (d) $|-4x + y|$

Solution (a) Substitute 3 for x and -1 for y in the expression $x + 3y$.

$$x + 3y = 3 + 3(-1) = 3 + (-3) = 0$$
$$\uparrow$$
$$x = 3, y = -1$$

(b) If $x = 3$ and $y = -1$, then

$$5xy = 5(3)(-1) = -15$$

(c) If $x = 3$ and $y = -1$, then

$$\frac{3y}{2 - 2x} = \frac{3(-1)}{2 - 2(3)} = \frac{-3}{2 - 6} = \frac{-3}{-4} = \frac{3}{4}$$

(d) If $x = 3$ and $y = -1$, then

$$|-4x + y| = |-4(3) + (-1)| = |-12 + (-1)| = |-13| = 13$$

━━━━━ **Now Work** PROBLEMS 39 AND 47

4 Determine the Domain of a Variable

In working with expressions or formulas involving variables, the variables may be allowed to take on values from only a certain set of numbers. For example, in the formula for the area A of a circle of radius r, $A = \pi r^2$, the variable r is necessarily restricted to the positive real numbers. In the expression $\dfrac{1}{x}$, the variable x cannot take on the value 0, since division by 0 is not defined.

DEFINITION

The set of values that a variable may assume is called the **domain of the variable.**

| EXAMPLE 6 | Finding the Domain of a Variable |

The domain of the variable x in the expression

$$\frac{5}{x - 2}$$

is $\{x | x \neq 2\}$, since, if $x = 2$, the denominator becomes 0, which is not defined.

| EXAMPLE 7 | Circumference of a Circle |

In the formula for the circumference C of a circle of radius r,

$$C = 2\pi r$$

the domain of the variable r, representing the radius of the circle, is the set of positive real numbers. The domain of the variable C, representing the circumference of the circle, is also the set of positive real numbers.

In describing the domain of a variable, we may use either set notation or words, whichever is more convenient.

Now Work PROBLEM 57

5 Use the Laws of Exponents

Integer exponents provide a shorthand device for representing repeated multiplications of a real number. For example,

$$3^4 = 3 \cdot 3 \cdot 3 \cdot 3 = 81$$

Additionally, many formulas have exponents. For example,
- The formula for the horsepower rating H of an engine is

$$H = \frac{D^2 N}{2.5}$$

where D is the diameter of a cylinder and N is the number of cylinders.
- A formula for the resistance R of blood flowing in a blood vessel is

$$R = C\frac{L}{r^4}$$

where L is the length of the blood vessel, r is the radius, and C is a positive constant.

DEFINITION

If a is a real number and n is a positive integer, then the symbol a^n represents the product of n factors of a. That is,

$$a^n = \underbrace{a \cdot a \cdot \,...\, \cdot a}_{n\ factors} \tag{1}$$

Here it is understood that $a^1 = a$.

Then $a^2 = a \cdot a$, $a^3 = a \cdot a \cdot a$, and so on. In the expression a^n, a is called the **base** and n is called the **exponent,** or **power.** We read a^n as "a raised to the power n" or as "a to the nth power." We usually read a^2 as "a squared" and a^3 as "a cubed."

In working with exponents, the operation of *raising to a power* is performed before any other operation. As examples,

$$4 \cdot 3^2 = 4 \cdot 9 = 36 \qquad 2^2 + 3^2 = 4 + 9 = 13$$
$$-2^4 = -16 \qquad 5 \cdot 3^2 + 2 \cdot 4 = 5 \cdot 9 + 2 \cdot 4 = 45 + 8 = 53$$

Parentheses are used to indicate operations to be performed first. For example,

$$(-2)^4 = (-2)(-2)(-2)(-2) = 16 \qquad (2 + 3)^2 = 5^2 = 25$$

DEFINITION

If $a \neq 0$, we define

$$a^0 = 1 \quad \text{if } a \neq 0$$

DEFINITION

If $a \neq 0$ and if n is a positive integer, then we define

$$a^{-n} = \frac{1}{a^n} \quad \text{if } a \neq 0$$

Whenever you encounter a negative exponent, think "reciprocal."

EXAMPLE 8 | **Evaluating Expressions Containing Negative Exponents**

(a) $2^{-3} = \dfrac{1}{2^3} = \dfrac{1}{8}$ (b) $x^{-4} = \dfrac{1}{x^4}$ (c) $\left(\dfrac{1}{5}\right)^{-2} = \dfrac{1}{\left(\dfrac{1}{5}\right)^2} = \dfrac{1}{\dfrac{1}{25}} = 25$

▶ **Now Work** PROBLEMS **75** AND **95**

The following properties, called the **Laws of Exponents,** can be proved using the preceding definitions. In the list, a and b are real numbers, and m and n are integers.

THEOREM

Laws of Exponents

$$a^m a^n = a^{m+n} \qquad (a^m)^n = a^{mn} \qquad (ab)^n = a^n b^n$$

$$\frac{a^m}{a^n} = a^{m-n} = \frac{1}{a^{n-m}} \quad \text{if } a \neq 0 \qquad \left(\frac{a}{b}\right)^n = \frac{a^n}{b^n} \quad \text{if } b \neq 0$$

EXAMPLE 9 **Using the Laws of Exponents**

(a) $x^{-3} \cdot x^5 = x^{-3+5} = x^2$ $x \neq 0$

(b) $(x^{-3})^2 = x^{-3 \cdot 2} = x^{-6} = \dfrac{1}{x^6}$ $x \neq 0$

(c) $(2x)^3 = 2^3 \cdot x^3 = 8x^3$

(d) $\left(\dfrac{2}{3}\right)^4 = \dfrac{2^4}{3^4} = \dfrac{16}{81}$

(e) $\dfrac{x^{-2}}{x^{-5}} = x^{-2-(-5)} = x^3$ $x \neq 0$

▬▬▬ **Now Work** PROBLEM 77

EXAMPLE 10 **Using the Laws of Exponents**

Write each expression so that all exponents are positive.

(a) $\dfrac{x^5 y^{-2}}{x^3 y}$ $x \neq 0,\quad y \neq 0$ (b) $\left(\dfrac{x^{-3}}{3y^{-1}}\right)^{-2}$ $x \neq 0,\quad y \neq 0$

Solution (a) $\dfrac{x^5 y^{-2}}{x^3 y} = \dfrac{x^5}{x^3} \cdot \dfrac{y^{-2}}{y} = x^{5-3} \cdot y^{-2-1} = x^2 y^{-3} = x^2 \cdot \dfrac{1}{y^3} = \dfrac{x^2}{y^3}$

(b) $\left(\dfrac{x^{-3}}{3y^{-1}}\right)^{-2} = \dfrac{(x^{-3})^{-2}}{(3y^{-1})^{-2}} = \dfrac{x^6}{3^{-2}(y^{-1})^{-2}} = \dfrac{x^6}{\dfrac{1}{9}y^2} = \dfrac{9x^6}{y^2}$

▬▬▬ **Now Work** PROBLEM 87

6 Evaluate Square Roots

A real number is squared when it is raised to the power 2. The inverse of squaring is finding a **square root.** For example, since $6^2 = 36$ and $(-6)^2 = 36$, the numbers 6 and -6 are square roots of 36.

The symbol $\sqrt{\ }$, called a **radical sign,** is used to denote the **principal,** or non-negative, square root. For example, $\sqrt{36} = 6$.

DEFINITION If a is a nonnegative real number, the nonnegative number b, such that $b^2 = a$ is the **principal square root** of a, is denoted by $b = \sqrt{a}$.

The following comments are noteworthy:

1. Negative numbers do not have square roots (in the real number system), because the square of any real number is *nonnegative.* For example, $\sqrt{-4}$ is not a real number, because there is no real number whose square is -4.

2. The principal square root of 0 is 0, since $0^2 = 0$. That is, $\sqrt{0} = 0$.

3. The principal square root of a positive number is positive.

4. If $c > 0$, then $(\sqrt{c})^2 = c$. For example, $(\sqrt{2})^2 = 2$ and $(\sqrt{3})^2 = 3$.

EXAMPLE 11 **Evaluating Square Roots**

(a) $\sqrt{64} = 8$ (b) $\sqrt{\dfrac{1}{16}} = \dfrac{1}{4}$ (c) $\left(\sqrt{1.4}\right)^2 = 1.4$

Examples 11(a) and (b) are examples of square roots of perfect squares, since $64 = 8^2$ and $\frac{1}{16} = \left(\frac{1}{4}\right)^2$.

Consider the expression $\sqrt{a^2}$. Since $a^2 \geq 0$, the principal square root of a^2 is defined whether $a > 0$ or $a < 0$. However, since the principal square root is non-negative, we need an absolute value to ensure the nonnegative result. That is,

$$\sqrt{a^2} = |a| \qquad a \text{ any real number} \qquad\qquad (2)$$

EXAMPLE 12 **Using Equation (2)**

(a) $\sqrt{(2.3)^2} = |2.3| = 2.3$

(b) $\sqrt{(-2.3)^2} = |-2.3| = 2.3$

(c) $\sqrt{x^2} = |x|$

 Now Work PROBLEM 83

7 Use a Calculator to Evaluate Exponents

Your calculator has either a caret key, $\boxed{\wedge}$ or an $\boxed{x^y}$ key, which is used for computations involving exponents.

EXAMPLE 13 **Exponents on a Graphing Calculator**

Evaluate: $(2.3)^5$

Solution Figure 15 shows the result using a TI-84 graphing calculator.

Figure 15

 Now Work PROBLEM 113

8 Use Scientific Notation

Measurements of physical quantities can range from very small to very large. For example, the mass of a proton is approximately 0.00000000000000000000000000167 kilogram and the mass of Earth is about 5,980,000,000,000,000,000,000,000 kilograms. These numbers obviously are tedious to write down and difficult to read, so we use exponents to rewrite each.

DEFINITION When a number has been written as the product of a number x, where $1 \leq x < 10$, times a power of 10, it is said to be written in **scientific notation.**

In scientific notation,

$$\text{Mass of a proton} = 1.67 \times 10^{-27} \text{ kilogram}$$
$$\text{Mass of Earth} = 5.98 \times 10^{24} \text{ kilograms}$$

> **Converting a Decimal to Scientific Notation**
>
> To change a positive number into scientific notation:
>
> 1. Count the number N of places that the decimal point must be moved to arrive at a number x, where $1 \le x < 10$.
> 2. If the original number is greater than or equal to 1, the scientific notation is $x \times 10^N$. If the original number is between 0 and 1, the scientific notation is $x \times 10^{-N}$.

EXAMPLE 14 **Using Scientific Notation**

Write each number in scientific notation.

(a) 9582 (b) 1.245 (c) 0.285 (d) 0.000561

Solution (a) The decimal point in 9582 follows the 2. We count left from the decimal point

$$9 \;\; 5 \;\; 8 \;\; 2 \;\; .$$
$$ 3 \;\; 2 \;\; 1$$

stopping after three moves, because 9.582 is a number between 1 and 10. Since 9582 is greater than 1, we write

$$9582 = 9.582 \times 10^3$$

(b) The decimal point in 1.245 is between the 1 and 2. Since the number is already between 1 and 10, the scientific notation for it is $1.245 \times 10^0 = 1.245$.

(c) The decimal point in 0.285 is between the 0 and the 2. We count

$$0 \;\; . \;\; 2 \;\; 8 \;\; 5$$
$$ 1$$

stopping after one move, because 2.85 is a number between 1 and 10. Since 0.285 is between 0 and 1, we write

$$0.285 = 2.85 \times 10^{-1}$$

(d) The decimal point in 0.000561 is moved as follows:

$$0 \;\; . \;\; 0 \;\; 0 \;\; 0 \;\; 5 \;\; 6 \;\; 1$$
$$ 1 \;\; 2 \;\; 3 \;\; 4$$

As a result,

$$0.000561 = 5.61 \times 10^{-4}$$

➤ **Now Work** PROBLEM 119

EXAMPLE 15 **Changing from Scientific Notation to Decimals**

Write each number as a decimal.

(a) 2.1×10^4 (b) 3.26×10^{-5} (c) 1×10^{-2}

Solution (a) $2.1 \times 10^4 = 2 \;\; . \;\; 1 \;\; 0 \;\; 0 \;\; 0 \; \times 10^4 = 21{,}000$
$$ 1 \;\; 2 \;\; 3 \;\; 4$$

(b) $3.26 \times 10^{-5} = 0 \;\; 0 \;\; 0 \;\; 0 \;\; 0 \;\; 3 \;\; . \;\; 2 \;\; 6 \times 10^{-5} = 0.0000326$
$$\phantom{3.26 \times 10^{-5} = 0 \;\;} 5 \;\; 4 \;\; 3 \;\; 2 \;\; 1$$

(c) $1 \times 10^{-2} = 0 \;\; 0 \;\; 1 \;\; . \;\; 0 \times 10^{-2} = 0.01$
$$\phantom{1 \times 10^{-2} = 0 \;\;} 2 \;\; 1$$

➤ **Now Work** PROBLEM 127

| EXAMPLE 16 | Using Scientific Notation |

(a) The diameter of the smallest living cell is only about 0.00001 centimeter (cm).* Express this number in scientific notation.

(b) The surface area of Earth is about 1.97×10^8 square miles.† Express the surface area as a whole number.

Solution

(a) $0.00001 \text{ cm} = 1 \times 10^{-5} \text{ cm}$ because the decimal point is moved five places and the number is less than 1.

(b) 1.97×10^8 square miles = 197,000,000 square miles.

━━━━ **Now Work** PROBLEM 153

COMMENT On a calculator, a number such as 3.615×10^{12} is usually displayed as $\boxed{3.615\text{E}12}$. ■

*Powers of Ten, Philip and Phylis Morrison.
†1998 Information Please Almanac.

Historical Feature

The word *algebra* is derived from the Arabic word *al-jabr*. This word is a part of the title of a ninth century work, "Hisâb al-jabr w'al-muqâbalah," written by Mohammed ibn Músâ al-Khowârizmî. The word *al-jabr* means "a restoration," a reference to the fact that, if a number is added to one side of an equation, then it must also be added to the other side in order to "restore" the equality. The title of the work, freely translated, is "The Science of Reduction and Cancellation." Of course, today, algebra has come to mean a great deal more.

R.2 Assess Your Understanding

Concepts and Vocabulary

1. A(n) _____ is a letter used in algebra to represent any number from a given set of numbers.

2. On the real number line, the real number zero is the coordinate of the _____.

3. An inequality of the form $a > b$ is called a(n) _____ inequality.

4. In the expression 2^4, the number 2 is called the _____ and 4 is called the _____.

5. In scientific notation, 1234.5678 = _____.

6. *True or False* The product of two negative real numbers is always greater than zero.

7. *True or False* The distance between two distinct points on the real number line is always greater than zero.

8. *True or False* The absolute value of a real number is always greater than zero.

9. *True or False* When a number is expressed in scientific notation, it is expressed as the product of a number x, $0 \le x < 1$, and a power of 10.

10. *True or False* To multiply two expressions having the same base, retain the base and multiply the exponents.

Skill Building

11. On the real number line, label the points with coordinates 0, 1, -1, $\dfrac{5}{2}$, -2.5, $\dfrac{3}{4}$, and 0.25.

12. Repeat Problem 11 for the coordinates 0, -2, 2, -1.5, $\dfrac{3}{2}$, $\dfrac{1}{3}$, and $\dfrac{2}{3}$.

In Problems 13–22, replace the question mark by $<$, $>$, or $=$, whichever is correct.

13. $\dfrac{1}{2}$? 0

14. 5 ? 6

15. -1 ? -2

16. -3 ? $-\dfrac{5}{2}$

17. π ? 3.14

18. $\sqrt{2}$? 1.41

19. $\dfrac{1}{2}$? 0.5

20. $\dfrac{1}{3}$? 0.33

21. $\dfrac{2}{3}$? 0.67

22. $\dfrac{1}{4}$? 0.25

In Problems 23–28, write each statement as an inequality.

23. x is positive

24. z is negative

25. x is less than 2

26. y is greater than -5

27. x is less than or equal to 1

28. x is greater than or equal to 2

In Problems 29–32, graph the numbers x on the real number line.

29. $x \geq -2$

30. $x < 4$

31. $x > -1$

32. $x \leq 7$

In Problems 33–38, use the given real number line to compute each distance.

33. $d(C, D)$

34. $d(C, A)$

35. $d(D, E)$

36. $d(C, E)$

37. $d(A, E)$

38. $d(D, B)$

In Problems 39–46, evaluate each expression if $x = -2$ and $y = 3$.

39. $x + 2y$

40. $3x + y$

41. $5xy + 2$

42. $-2x + xy$

43. $\dfrac{2x}{x - y}$

44. $\dfrac{x + y}{x - y}$

45. $\dfrac{3x + 2y}{2 + y}$

46. $\dfrac{2x - 3}{y}$

In Problems 47–56, find the value of each expression if $x = 3$ and $y = -2$.

47. $|x + y|$

48. $|x - y|$

49. $|x| + |y|$

50. $|x| - |y|$

51. $\dfrac{|x|}{x}$

52. $\dfrac{|y|}{y}$

53. $|4x - 5y|$

54. $|3x + 2y|$

55. $\big||4x| - |5y|\big|$

56. $3|x| + 2|y|$

In Problems 57–64, determine which of the value(s) (a) through (d), if any, must be excluded from the domain of the variable in each expression:

 (a) $x = 3$ (b) $x = 1$ (c) $x = 0$ (d) $x = -1$

57. $\dfrac{x^2 - 1}{x}$

58. $\dfrac{x^2 + 1}{x}$

59. $\dfrac{x}{x^2 - 9}$

60. $\dfrac{x}{x^2 + 9}$

61. $\dfrac{x^2}{x^2 + 1}$

62. $\dfrac{x^3}{x^2 - 1}$

63. $\dfrac{x^2 + 5x - 10}{x^3 - x}$

64. $\dfrac{-9x^2 - x + 1}{x^3 + x}$

In Problems 65–68, determine the domain of the variable x in each expression.

65. $\dfrac{4}{x - 5}$

66. $\dfrac{-6}{x + 4}$

67. $\dfrac{x}{x + 4}$

68. $\dfrac{x - 2}{x - 6}$

In Problems 69–72, use the formula $C = \dfrac{5}{9}(F - 32)$ for converting degrees Fahrenheit into degrees Celsius to find the Celsius measure of each Fahrenheit temperature.

69. $F = 32°$

70. $F = 212°$

71. $F = 77°$

72. $F = -4°$

In Problems 73–84, simplify each expression.

73. $(-4)^2$

74. -4^2

75. 4^{-2}

76. -4^{-2}

77. $3^{-6} \cdot 3^4$

78. $4^{-2} \cdot 4^3$

79. $(3^{-2})^{-1}$

80. $(2^{-1})^{-3}$

81. $\sqrt{25}$

82. $\sqrt{36}$

83. $\sqrt{(-4)^2}$

84. $\sqrt{(-3)^2}$

In Problems 85–94, simplify each expression. Express the answer so that all exponents are positive. Whenever an exponent is 0 or negative, we assume that the base is not 0.

85. $(8x^3)^2$

86. $(-4x^2)^{-1}$

87. $(x^2 y^{-1})^2$

88. $(x^{-1}y)^3$

89. $\dfrac{x^2 y^3}{xy^4}$

90. $\dfrac{x^{-2}y}{xy^2}$

91. $\dfrac{(-2)^3 x^4 (yz)^2}{3^2 xy^3 z}$

92. $\dfrac{4x^{-2}(yz)^{-1}}{2^3 x^4 y}$

93. $\left(\dfrac{3x^{-1}}{4y^{-1}}\right)^{-2}$

94. $\left(\dfrac{5x^{-2}}{6y^{-2}}\right)^{-3}$

In Problems 95–106, find the value of each expression if $x = 2$ and $y = -1$.

95. $2xy^{-1}$ **96.** $-3x^{-1}y$ **97.** $x^2 + y^2$ **98.** x^2y^2

99. $(xy)^2$ **100.** $(x + y)^2$ **101.** $\sqrt{x^2}$ **102.** $(\sqrt{x})^2$

103. $\sqrt{x^2 + y^2}$ **104.** $\sqrt{x^2} + \sqrt{y^2}$ **105.** x^y **106.** y^x

107. Find the value of the expression $2x^3 - 3x^2 + 5x - 4$ if $x = 2$. What is the value if $x = 1$?

108. Find the value of the expression $4x^3 + 3x^2 - x + 2$ if $x = 1$. What is the value if $x = 2$?

109. What is the value of $\dfrac{(666)^4}{(222)^4}$? **110.** What is the value of $(0.1)^3(20)^3$?

In Problems 111–118, use a calculator to evaluate each expression. Round your answer to three decimal places.

111. $(8.2)^6$ **112.** $(3.7)^5$ **113.** $(6.1)^{-3}$ **114.** $(2.2)^{-5}$

115. $(-2.8)^6$ **116.** $-(2.8)^6$ **117.** $(-8.11)^{-4}$ **118.** $-(8.11)^{-4}$

In Problems 119–126, write each number in scientific notation.

119. 454.2 **120.** 32.14 **121.** 0.013 **122.** 0.00421

123. 32,155 **124.** 21,210 **125.** 0.000423 **126.** 0.0514

In Problems 127–134, write each number as a decimal.

127. 6.15×10^4 **128.** 9.7×10^3 **129.** 1.214×10^{-3} **130.** 9.88×10^{-4}

131. 1.1×10^8 **132.** 4.112×10^2 **133.** 8.1×10^{-2} **134.** 6.453×10^{-1}

Applications and Extensions

In Problems 135–144, express each statement as an equation involving the indicated variables.

135. Area of a Rectangle The area A of a rectangle is the product of its length l and its width w.

136. Perimeter of a Rectangle The perimeter P of a rectangle is twice the sum of its length l and its width w.

137. Circumference of a Circle The circumference C of a circle is the product of π and its diameter d.

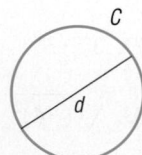

138. Area of a Triangle The area A of a triangle is one-half the product of its base b and its height h.

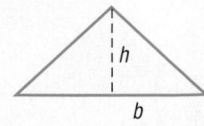

139. Area of an Equilateral Triangle The area A of an equilateral triangle is $\dfrac{\sqrt{3}}{4}$ times the square of the length x of one side.

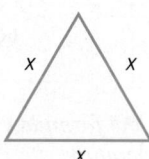

140. Perimeter of an Equilateral Triangle The perimeter P of an equilateral triangle is 3 times the length x of one side.

141. Volume of a Sphere The volume V of a sphere is $\dfrac{4}{3}$ times π times the cube of the radius r.

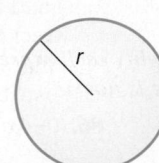

142. Surface Area of a Sphere The surface area S of a sphere is 4 times π times the square of the radius r.

143. **Volume of a Cube** The volume V of a cube is the cube of the length x of a side.

144. **Surface Area of a Cube** The surface area S of a cube is 6 times the square of the length x of a side.

145. **Manufacturing Cost** The weekly production cost C of manufacturing x watches is given by the formula $C = 4000 + 2x$, where the variable C is in dollars.
 (a) What is the cost of producing 1000 watches?
 (b) What is the cost of producing 2000 watches?

146. **Balancing a Checkbook** At the beginning of the month, Mike had a balance of $210 in his checking account. During the next month, he deposited $80, wrote a check for $120, made another deposit of $25, wrote two checks: one for $60 and the other for $32. He was also assessed a monthly service charge of $5. What was his balance at the end of the month?

In Problems 147 and 148, write an inequality using an absolute value to describe each statement.

147. x is at least 6 units from 4.

148. x is more than 5 units from 2.

149. **U.S. Voltage** In the United States, normal household voltage is 110 volts. It is acceptable for the actual voltage x to differ from normal by at most 5 volts. A formula that describes this is
$$|x - 110| \le 5$$
 (a) Show that a voltage of 108 volts is acceptable.
 (b) Show that a voltage of 104 volts is not acceptable.

150. **Foreign Voltage** In other countries, normal household voltage is 220 volts. It is acceptable for the actual voltage x to differ from normal by at most 8 volts. A formula that describes this is
$$|x - 220| \le 8$$
 (a) Show that a voltage of 214 volts is acceptable.
 (b) Show that a voltage of 209 volts is not acceptable.

151. **Making Precision Ball Bearings** The FireBall Company manufactures ball bearings for precision equipment. One of its products is a ball bearing with a stated radius of 3 centimeters (cm). Only ball bearings with a radius within 0.01 cm

of this stated radius are acceptable. If x is the radius of a ball bearing, a formula describing this situation is
$$|x - 3| \le 0.01$$
 (a) Is a ball bearing of radius $x = 2.999$ acceptable?
 (b) Is a ball bearing of radius $x = 2.89$ acceptable?

152. **Body Temperature** Normal human body temperature is 98.6°F. A temperature x that differs from normal by at least 1.5°F is considered unhealthy. A formula that describes this is
$$|x - 98.6| \ge 1.5$$
 (a) Show that a temperature of 97°F is unhealthy.
 (b) Show that a temperature of 100°F is not unhealthy.

153. **Distance from Earth to Its Moon** The distance from Earth to the Moon is about 4×10^8 meters.* Express this distance as a whole number.

154. **Height of Mt. Everest** The height of Mt. Everest is 8872 meters.* Express this height in scientific notation.

155. **Wavelength of Visible Light** The wavelength of visible light is about 5×10^{-7} meter.* Express this wavelength as a decimal.

156. **Diameter of an Atom** The diameter of an atom is about 1×10^{-10} meter.* Express this diameter as a decimal.

157. **Diameter of Copper Wire** The smallest commercial copper wire is about 0.0005 inch in diameter.† Express this diameter using scientific notation.

158. **Smallest Motor** The smallest motor ever made is less than 0.05 centimeter wide.† Express this width using scientific notation.

159. **Astronomy** One light-year is defined by astronomers to be the distance that a beam of light will travel in 1 year (365 days). If the speed of light is 186,000 miles per second, how many miles are in a light-year? Express your answer in scientific notation.

160. **Astronomy** How long does it take a beam of light to reach Earth from the Sun when the Sun is 93,000,000 miles from Earth? Express your answer in seconds, using scientific notation.

161. Does $\frac{1}{3}$ equal 0.333? If not, which is larger? By how much?

162. Does $\frac{2}{3}$ equal 0.666? If not, which is larger? By how much?

Discussion and Writing

163. Is there a positive real number "closest" to 0?

164. **Number game** I'm thinking of a number! It lies between 1 and 10; its square is rational and lies between 1 and 10. The number is larger than π. Correct to two decimal places (that is, truncated to two decimal places) name the number. Now think of your own number, describe it, and challenge a fellow student to name it.

165. Write a brief paragraph that illustrates the similarities and differences between "less than" ($<$) and "less than or equal to" ($\le$).

166. Give a reason why the statement $5 < 8$ is true.

* *Powers of Ten*, Philip and Phylis Morrison.
† *1998 Information Please Almanac.*

R.3 Geometry Essentials

In this section we review some topics studied in geometry that we shall need for our study of algebra.

Figure 16

Hypotenuse
c

b
Leg

$90°$

a
Leg

1 Use the Pythagorean Theorem and Its Converse

The *Pythagorean Theorem* is a statement about *right triangles*. A **right triangle** is one that contains a **right angle,** that is, an angle of 90°. The side of the triangle opposite the 90° angle is called the **hypotenuse;** the remaining two sides are called **legs.** In Figure 16 we have used c to represent the length of the hypotenuse and a and b to represent the lengths of the legs. Notice the use of the symbol ⌐ to show the 90° angle. We now state the Pythagorean Theorem.

PYTHAGOREAN THEOREM

In a right triangle, the square of the length of the hypotenuse is equal to the sum of the squares of the lengths of the legs. That is, in the right triangle shown in Figure 16,

$$c^2 = a^2 + b^2 \qquad (1)$$

A proof of the Pythagorean Theorem is given at the end of this section.

EXAMPLE 1 | **Finding the Hypotenuse of a Right Triangle**

In a right triangle, one leg has length 4 and the other has length 3. What is the length of the hypotenuse?

Solution | Since the triangle is a right triangle, we use the Pythagorean Theorem with $a = 4$ and $b = 3$ to find the length c of the hypotenuse. From equation (1), we have

$$c^2 = a^2 + b^2$$
$$c^2 = 4^2 + 3^2 = 16 + 9 = 25$$
$$c = \sqrt{25} = 5$$

Now Work PROBLEM 13

The converse of the Pythagorean Theorem is also true.

CONVERSE OF THE PYTHAGOREAN THEOREM

In a triangle, if the square of the length of one side equals the sum of the squares of the lengths of the other two sides, the triangle is a right triangle. The 90° angle is opposite the longest side.

A proof is given at the end of this section.

EXAMPLE 2 | **Verifying That a Triangle Is a Right Triangle**

Show that a triangle whose sides are of lengths 5, 12, and 13 is a right triangle. Identify the hypotenuse.

Solution | We square the lengths of the sides.

$$5^2 = 25, \qquad 12^2 = 144, \qquad 13^2 = 169$$

Figure 17

Notice that the sum of the first two squares (25 and 144) equals the third square (169). Hence, the triangle is a right triangle. The longest side, 13, is the hypotenuse. See Figure 17.

Now Work PROBLEM 21

| EXAMPLE 3 | **Applying the Pythagorean Theorem**

Figure 18

Excluding antenna, the tallest inhabited building in the world is Taipei 101 in Taipei, Taiwan. If the indoor observation deck is 1437 feet above ground level, how far can a person standing on the observation deck see (with the aid of a telescope)? Use 3960 miles for the radius of Earth. See Figure 18.

Source: Council on Tall Buildings and Urban Habitat, 2006.

Solution From the center of Earth, draw two radii: one through Taipei 101 and the other to the farthest point a person can see from the observation deck. See Figure 19. Apply the Pythagorean Theorem to the right triangle.

Since 1 mile = 5280 feet, then 1437 feet = $\dfrac{1437}{5280}$ miles. So we have

$$d^2 + (3960)^2 = \left(3960 + \frac{1437}{5280}\right)^2$$

$$d^2 = \left(3960 + \frac{1437}{5280}\right)^2 - (3960)^2 \approx 2155.57$$

$$d \approx 46.43$$

A person can see about 46 miles from the observation tower.

Figure 19

Now Work PROBLEM 53

2 Know Geometry Formulas

Certain formulas from geometry are useful in solving algebra problems. We list some of these next.

For a rectangle of length l and width w,

| Area = lw Perimeter = $2l + 2w$ |

For a triangle with base b and altitude h,

| Area = $\dfrac{1}{2}bh$ |

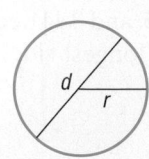

For a circle of radius r (diameter $d = 2r$),

$$\text{Area} = \pi r^2 \qquad \text{Circumference} = 2\pi r = \pi d$$

For a closed rectangular box of length l, width w, and height h,

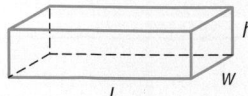

$$\text{Volume} = lwh \qquad \text{Surface area} = 2lh + 2wh + 2lw$$

For a sphere of radius r,

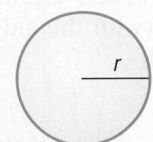

$$\text{Volume} = \frac{4}{3}\pi r^3 \qquad \text{Surface area} = 4\pi r^2$$

For a right circular cylinder of height h and radius r,

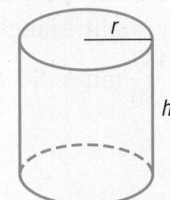

$$\text{Volume} = \pi r^2 h \qquad \text{Surface area} = 2\pi r^2 + 2\pi rh$$

$\rightarrow$ **Now Work** PROBLEM 29

EXAMPLE 4 | **Using Geometry Formulas**

A Christmas tree ornament is in the shape of a semicircle on top of a triangle. How many square centimeters (cm) of copper is required to make the ornament if the height of the triangle is 6 cm and the base is 4 cm?

Solution See Figure 20. The amount of copper required equals the shaded area. This area is the sum of the area of the triangle and the semicircle. The triangle has height $h = 6$ and base $b = 4$. The semicircle has diameter $d = 4$, so its radius is $r = 2$.

Figure 20

$$\text{Area} = \text{Area of triangle} + \text{Area of semicircle}$$

$$= \frac{1}{2}bh + \frac{1}{2}\pi r^2 = \frac{1}{2}(4)(6) + \frac{1}{2}\pi \cdot 2^2 \qquad b = 4; h = 6; r = 2$$

$$= 12 + 2\pi \approx 18.28 \text{ cm}^2$$

About 18.28 cm^2 of copper is required.

$\rightarrow$ **Now Work** PROBLEM 47

3 Understand Congruent Triangles and Similar Triangles

Throughout the text we will make reference to triangles. We begin with a discussion of *congruent* triangles. According to dictionary.com, the word **congruent** means coinciding exactly when superimposed. For example, two angles are congruent if they have the same measure and two line segments are congruent if they have the same length.

In Words
Two triangles are congruent if they have the same size and shape.

DEFINITION

Two triangles are **congruent** if each of the corresponding angles is the same measure and each of the corresponding sides is the same length.

In Figure 21, corresponding angles are equal and the lengths of the corresponding sides are equal: $a = d$, $b = e$, and $c = f$. We conclude that these triangles are congruent.

Figure 21
Congruent triangles

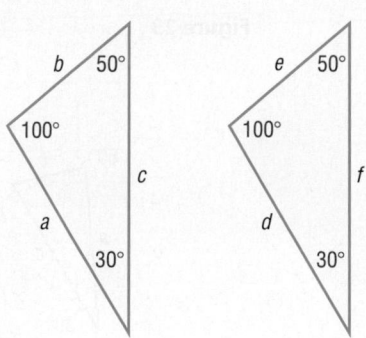

It is not necessary to verify that all three angles and all three sides are the same measure to determine whether two triangles are congruent.

Determining Congruent Triangles

1. **Angle–Side–Angle Case** Two triangles are congruent if two of the angles are equal and the lengths of the corresponding sides between the two angles are equal.

 For example, in Figure 22(a), the two triangles are congruent because two angles and the included side are equal.

2. **Side–Side–Side Case** Two triangles are congruent if the lengths of the corresponding sides of the triangles are equal.

 For example, in Figure 22(b), the two triangles are congruent because the three corresponding sides are all equal.

3. **Side–Angle–Side Case** Two triangles are congruent if the lengths of two corresponding sides are equal and the angles between the two sides are the same.

 For example, in Figure 22(c), the two triangles are congruent because two sides and the included angle are equal.

Figure 22

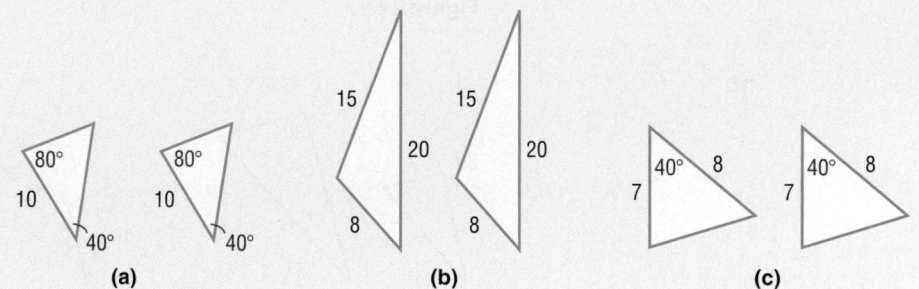

(a) (b) (c)

We contrast congruent triangles with *similar* triangles.

DEFINITION

Two triangles are **similar** if the corresponding angles are equal and the lengths of the corresponding sides are proportional.

For example, the triangles in Figure 23 are similar because the corresponding angles are equal. In addition, the lengths of the corresponding sides are proportional because each side in the triangle on the right is twice as long as each corresponding side in the triangle on the left. That is, the ratio of the corresponding sides is a constant: $\dfrac{d}{a} = \dfrac{e}{b} = \dfrac{f}{c} = 2$.

Figure 23

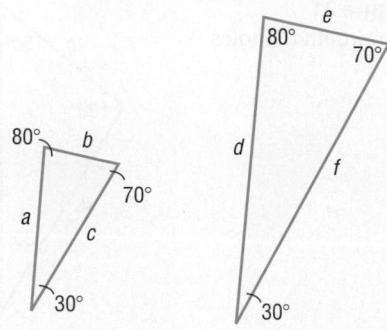

It is not necessary to verify that all three angles are equal and all three sides are proportional to determine whether two triangles are congruent.

Determining Similar Triangles

1. **Angle–Angle Case** Two triangles are similar if two of the corresponding angles are equal.

 For example, in Figure 24(a), the two triangles are similar because two angles are equal.

2. **Side–Side–Side Case** Two triangles are similar if the lengths of all three sides of each triangle are proportional.

 For example, in Figure 24(b), the two triangles are similar because

 $$\frac{10}{30} = \frac{5}{15} = \frac{6}{18} = \frac{1}{3}.$$

3. **Side–Angle–Side Case** Two triangles are similar if two corresponding sides are proportional and the angles between the two sides are equal.

 For example, in Figure 24(c), the two triangles are similar because $\frac{4}{6} = \frac{12}{18} = \frac{2}{3}$ and the angles between the sides are equal.

Figure 24

(a) (b) (c)

| EXAMPLE 5 | **Using Similar Triangles** |

Given that the triangles in Figure 25 are similar, find the missing length x and the angles A, B, and C.

Figure 25

Solution Because the triangles are similar, corresponding angles are equal. So $A = 90°$, $B = 60°$, and $C = 30°$. Also, the corresponding sides are proportional. That is, $\dfrac{3}{5} = \dfrac{6}{x}$. We solve this equation for x.

$$\frac{3}{5} = \frac{6}{x}$$

$$5x \cdot \frac{3}{5} = 5x \cdot \frac{6}{x} \qquad \text{Multiply both sides by } 5x.$$

$$3x = 30 \qquad \text{Simplify.}$$

$$x = 10 \qquad \text{Divide both sides by 3.}$$

The missing length is 10 units.

━━━━━ **Now Work** PROBLEM 41

Proof of the Pythagorean Theorem We begin with a square, each side of length $a + b$. In this square, we can form four right triangles, each having legs equal in length to a and b. See Figure 26. All these triangles are congruent (two sides and their included angle are equal). As a result, the hypotenuse of each is the same, say c, and the pink shading in Figure 26 indicates a square with an area equal to c^2.

Figure 26

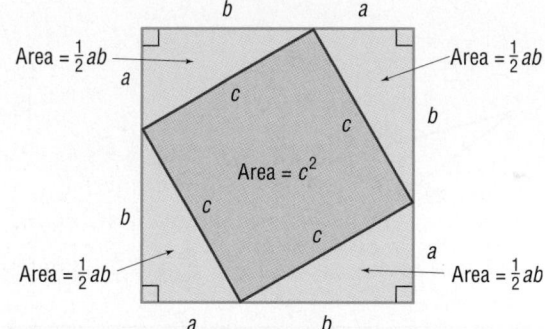

The area of the original square with sides $a + b$ equals the sum of the areas of the four triangles (each of area $\dfrac{1}{2}ab$) plus the area of the square with side c. That is,

$$(a + b)^2 = \frac{1}{2}ab + \frac{1}{2}ab + \frac{1}{2}ab + \frac{1}{2}ab + c^2$$

$$a^2 + 2ab + b^2 = 2ab + c^2$$

$$a^2 + b^2 = c^2$$

The proof is complete. ∎

Figure 27

x

b

a

(a)

c

b

a

(b) $c^2 = a^2 + b^2$

Proof of the Converse of the Pythagorean Theorem We begin with two triangles: one a right triangle with legs a and b and the other a triangle with sides a, b, and c for which $c^2 = a^2 + b^2$. See Figure 27. By the Pythagorean Theorem, the length x of the third side of the first triangle is

$$x^2 = a^2 + b^2$$

But $c^2 = a^2 + b^2$. Hence,

$$x^2 = c^2$$

$$x = c$$

The two triangles have the same sides and are therefore congruent. This means corresponding angles are equal, so the angle opposite side c of the second triangle equals $90°$.
 The proof is complete. ∎

R.3 Assess Your Understanding

Concepts and Vocabulary

1. A(n) _____ triangle is one that contains an angle of 90 degrees. The longest side is called the _____.

2. For a triangle with base b and altitude h, a formula for the area A is _____.

3. The formula for the circumference C of a circle of radius r is _____.

4. Two triangles are_____ if corresponding angles are equal and the lengths of the corresponding sides are proportional.

5. *True or False* In a right triangle, the square of the length of the longest side equals the sum of the squares of the lengths of the other two sides.

6. *True or False* The triangle with sides of length 6, 8, and 10 is a right triangle.

7. *True or False* The volume of a sphere of radius r is $\frac{4}{3}\pi r^2$.

8. *True or False* The triangles shown are congruent.

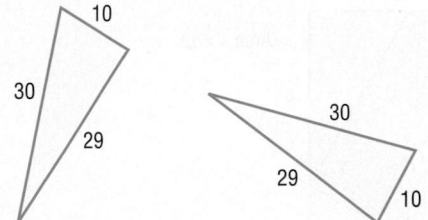

9. *True or False* The triangles shown are similar.

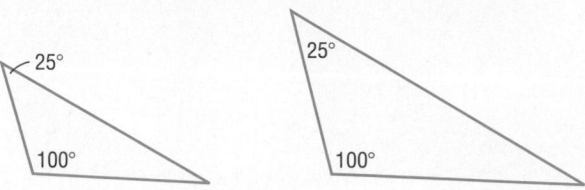

10. *True or False* The triangles shown are similar.

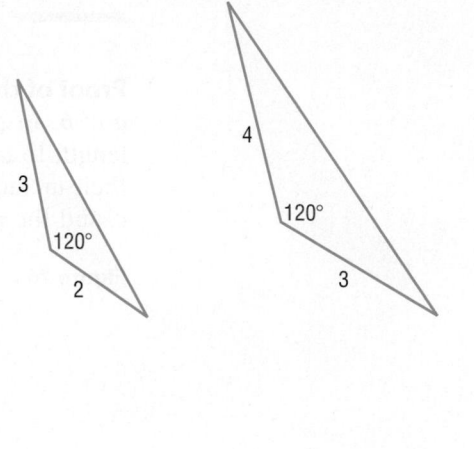

Skill Building

In Problems 11–16, the lengths of the legs of a right triangle are given. Find the hypotenuse.

11. $a = 5$, $b = 12$

12. $a = 6$, $b = 8$

13. $a = 10$, $b = 24$

14. $a = 4$, $b = 3$

15. $a = 7$, $b = 24$

16. $a = 14$, $b = 48$

In Problems 17–24, the lengths of the sides of a triangle are given. Determine which are right triangles. For those that are, identify the hypotenuse.

17. 3, 4, 5

18. 6, 8, 10

19. 4, 5, 6

20. 2, 2, 3

21. 7, 24, 25

22. 10, 24, 26

23. 6, 4, 3

24. 5, 4, 7

25. Find the area A of a rectangle with length 4 inches and width 2 inches.

26. Find the area A of a rectangle with length 9 centimeters and width 4 centimeters.

27. Find the area A of a triangle with height 4 inches and base 2 inches.

28. Find the area A of a triangle with height 9 centimeters and base 4 centimeters.

29. Find the area A and circumference C of a circle of radius 5 meters.

30. Find the area A and circumference C of a circle of radius 2 feet.

31. Find the volume V and surface area S of a rectangular box with length 8 feet, width 4 feet, and height 7 feet.

32. Find the volume V and surface area S of a rectangular box with length 9 inches, width 4 inches, and height 8 inches.

33. Find the volume V and surface area S of a sphere of radius 4 centimeters.

34. Find the volume V and surface area S of a sphere of radius 3 feet.

35. Find the volume V and surface area S of a right circular cylinder with radius 9 inches and height 8 inches.

36. Find the volume V and surface area S of a right circular cylinder with radius 8 inches and height 9 inches.

In Problems 37–40, find the area of the shaded region.

37.

38.

39.

40.

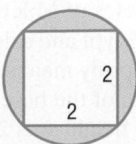

In Problems 41–44, each pair of triangles is similar. Find the missing length x and the missing angles A, B, and C.

41.

42.

43.

44.

Applications and Extensions

45. How many feet does a wheel with a diameter of 16 inches travel after four revolutions?

46. How many revolutions will a circular disk with a diameter of 4 feet have completed after it has rolled 20 feet?

47. In the figure shown, *ABCD* is a square, with each side of length 6 feet. The width of the border (shaded portion) between the outer square *EFGH* and *ABCD* is 2 feet. Find the area of the border.

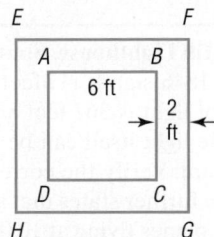

48. Refer to the figure. Square *ABCD* has an area of 100 square feet; square *BEFG* has an area of 16 square feet. What is the area of the triangle *CGF*?

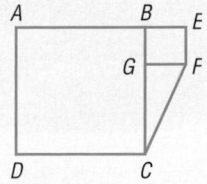

49. **Architecture** A **Norman window** consists of a rectangle surmounted by a semicircle. Find the area of the Norman window shown in the illustration. How much wood frame is needed to enclose the window?

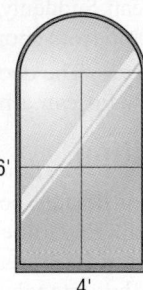

50. **Construction** A circular swimming pool, 20 feet in diameter, is enclosed by a wooden deck that is 3 feet wide. What is the area of the deck? How much fence is required to enclose the deck?

51. How Tall Is the Great Pyramid? The ancient Greek philosopher Thales of Miletus is reported on one occasion to have visited Egypt and calculated the height of the Great Pyramid of Cheops by means of shadow reckoning. Thales knew that each side of the base of the pyramid was 252 paces and that his own height was 2 paces. He measured the length of the pyramid's shadow to be 114 paces and determined the length of his shadow to be 3 paces. See the illustration. Using similar triangles, determine the height of the Great Pyramid in terms of the number of paces.

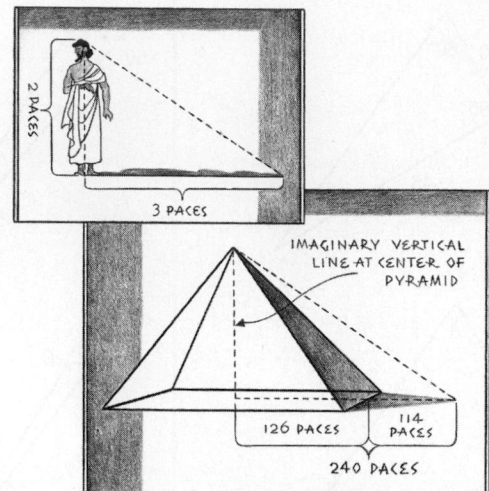

Source: www.anselm.edu/homepage/dbanach/ thales.htm. This site references another source: Selections, from Julia E. Diggins, *String, Straightedge, and Shadow,* Viking Press, New York, 1965, Illustrations by Corydon Bell.

52. The Bermuda Triangle Karen is doing research on the Bermuda Triangle which she defines roughly by Hamilton, Bermuda; San Juan, Puerto Rico; and Fort Lauderdale, Florida. On her atlas Karen measures the straight-line distances from Hamilton to Fort Lauderdale, Fort Lauderdale to San Juan, and San Juan to Hamilton to be approximately 57 millimeters (mm), 58 mm, and 53.5 mm respectively. If the actual distance from Fort Lauderdale to San Juan is 1046 miles, approximate the actual distances from San Juan to Hamilton and from Hamilton to Fort Lauderdale.
Source: www.worldatlas.com

Source: Bermuda Triangle photo reprinted with permission from Red River Press Inc. of Winnipeg, Canada, © *Red River Press, Inc.*

In Problems 53–55, use the facts that the radius of Earth is 3960 miles and 1 mile = 5280 feet.

53. How Far Can You See? The conning tower of the U.S.S. *Silversides,* a World War II submarine now permanently stationed in Muskegon, Michigan, is approximately 20 feet above sea level. How far can you see from the conning tower?

54. How Far Can You See? A person who is 6 feet tall is standing on the beach in Fort Lauderdale, Florida, and looks out onto the Atlantic Ocean. Suddenly, a ship appears on the horizon. How far is the ship from shore?

55. How Far Can You See? The deck of a destroyer is 100 feet above sea level. How far can a person see from the deck?

How far can a person see from the bridge, which is 150 feet above sea level?

56. Suppose that m and n are positive integers with $m > n$. If $a = m^2 - n^2$, $b = 2mn$, and $c = m^2 + n^2$, show that a, b, and c are the lengths of the sides of a right triangle. (This formula can be used to find the sides of a right triangle that are integers, such as 3, 4, 5; 5, 12, 13; and so on. Such triplets of integers are called **Pythagorean triples.**)

Discussion and Writing

57. You have 1000 feet of flexible pool siding and wish to construct a swimming pool. Experiment with rectangular-shaped pools with perimeters of 1000 feet. How do their areas vary? What is the shape of the rectangle with the largest area? Now compute the area enclosed by a circular pool with a perimeter (circumference) of 1000 feet. What would be your choice of shape for the pool? If rectangular, what is your preference for dimensions? Justify your choice. If your only consideration is to have a pool that encloses the most area, what shape should you use?

58. The Gibb's Hill Lighthouse, Southampton, Bermuda, in operation since 1846, stands 117 feet high on a hill 245 feet high, so its beam of light is 362 feet above sea level. A brochure states that the light itself can be seen on the horizon about 26 miles distant. Verify the correctness of this information. The brochure further states that ships 40 miles away can see the light and planes flying at 10,000 feet can see it 120 miles away. Verify the accuracy of these statements. What assumption did the brochure make about the height of the ship?

R.4 Polynomials

We have described algebra as a generalization of arithmetic in which letters are used to represent real numbers. From now on, we shall use the letters at the end of the alphabet, such as x, y, and z, to represent variables and the letters at the beginning of the alphabet, such as a, b, and c, to represent constants. In the expressions $3x + 5$ and $ax + b$, it is understood that x is a variable and that a and b are constants, even though the constants a and b are unspecified. As you will find out, the context usually makes the intended meaning clear.

1 Recognize Monomials

DEFINITION

A **monomial** in one variable is the product of a constant and a variable raised to a nonnegative integer power. A monomial is of the form

NOTE The nonnegative integers are the integers $0, 1, 2, 3, \ldots$. ∎

$$ax^k$$

where a is a constant, x is a variable, and $k \geq 0$ is an integer. The constant a is called the **coefficient** of the monomial. If $a \neq 0$, then k is called the **degree** of the monomial.

EXAMPLE 1 | **Examples of Monomials**

Monomial	Coefficient	Degree	
(a) $6x^2$	6	2	
(b) $-\sqrt{2}x^3$	$-\sqrt{2}$	3	
(c) 3	3	0	Since $3 = 3 \cdot 1 = 3x^0, x \neq 0$
(d) $-5x$	-5	1	Since $-5x = -5x^1$
(e) x^4	1	4	Since $x^4 = 1 \cdot x^4$

Now let's look at some expressions that are not monomials.

EXAMPLE 2 | **Examples of Nonmonomial Expressions**

(a) $3x^{1/2}$ is not a monomial, since the exponent of the variable x is $\dfrac{1}{2}$ and $\dfrac{1}{2}$ is not a nonnegative integer.

(b) $4x^{-3}$ is not a monomial, since the exponent of the variable x is -3 and -3 is not a nonnegative integer.

━━━ **Now Work** PROBLEM 7

2 Recognize Polynomials

Two monomials with the same variable raised to the same power are called **like terms.** For example, $2x^4$ and $-5x^4$ are like terms. In contrast, the monomials $2x^3$ and $2x^5$ are not like terms.

We can add or subtract like terms using the Distributive Property. For example,

$$2x^2 + 5x^2 = (2 + 5)x^2 = 7x^2 \quad \text{and} \quad 8x^3 - 5x^3 = (8 - 5)x^3 = 3x^3$$

The sum or difference of two monomials having different degrees is called a **binomial.** The sum or difference of three monomials with three different degrees is called a **trinomial.** For example,

$x^2 - 2$ is a binomial.

$x^3 - 3x + 5$ is a trinomial.

$2x^2 + 5x^2 + 2 = 7x^2 + 2$ is a binomial.

DEFINITION

A **polynomial** in one variable is an algebraic expression of the form

$$a_n x^n + a_{n-1} x^{n-1} + \cdots + a_1 x + a_0 \qquad (1)$$

In Words
A polynomial is a sum of monomials.

where $a_n, a_{n-1}, \ldots, a_1, a_0$ are constants,* called the **coefficients** of the polynomial, $n \geq 0$ is an integer, and x is a variable. If $a_n \neq 0$, it is called the **leading coefficient,** and n is called the **degree** of the polynomial.

The monomials that make up a polynomial are called its **terms.** If all the coefficients are 0, the polynomial is called the **zero polynomial,** which has no degree.

Polynomials are usually written in **standard form,** beginning with the nonzero term of highest degree and continuing with terms in descending order according to degree. If a power of x is missing, it is because its coefficient is zero.

EXAMPLE 3

Examples of Polynomials

Polynomial	Coefficients	Degree
$-8x^3 + 4x^2 - 6x + 2$	$-8, 4, -6, 2$	3
$3x^2 - 5 = 3x^2 + 0 \cdot x + (-5)$	$3, 0, -5$	2
$8 - 2x + x^2 = 1 \cdot x^2 + (-2)x + 8$	$1, -2, 8$	2
$5x + \sqrt{2} = 5x^1 + \sqrt{2}$	$5, \sqrt{2}$	1
$3 = 3 \cdot 1 = 3 \cdot x^0$	3	0
0	0	No degree

Although we have been using x to represent the variable, letters such as y or z are also commonly used.

$3x^4 - x^2 + 2$ is a polynomial (in x) of degree 4.

$9y^3 - 2y^2 + y - 3$ is a polynomial (in y) of degree 3.

$z^5 + \pi$ is a polynomial (in z) of degree 5.

Algebraic expressions such as

$$\frac{1}{x} \quad \text{and} \quad \frac{x^2 + 1}{x + 5}$$

* The notation a_n is read as "a sub n." The number n is called a **subscript** and should not be confused with an exponent. We use subscripts to distinguish one constant from another when a large or undetermined number of constants is required.

are not polynomials. The first is not a polynomial because $\dfrac{1}{x} = x^{-1}$ has an exponent that is not a nonnegative integer. Although the second expression is the quotient of two polynomials, the polynomial in the denominator has degree greater than 0, so the expression cannot be a polynomial.

──────➤ **Now Work** PROBLEM 17

3 Add and Subtract Polynomials

Polynomials are added and subtracted by combining like terms.

EXAMPLE 4	**Adding Polynomials**

Find the sum of the polynomials:

$$8x^3 - 2x^2 + 6x - 2 \quad \text{and} \quad 3x^4 - 2x^3 + x^2 + x$$

Solution We shall find the sum in two ways.

Horizontal Addition: The idea here is to group the like terms and then combine them.

$$(8x^3 - 2x^2 + 6x - 2) + (3x^4 - 2x^3 + x^2 + x)$$
$$= 3x^4 + (8x^3 - 2x^3) + (-2x^2 + x^2) + (6x + x) - 2$$
$$= 3x^4 + 6x^3 - x^2 + 7x - 2$$

Vertical Addition: The idea here is to vertically line up the like terms in each polynomial and then add the coefficients.

$$
\begin{array}{ccccc}
x^4 & x^3 & x^2 & x^1 & x^0 \\
\end{array}
$$

$$
\begin{array}{r}
8x^3 - 2x^2 + 6x - 2 \\
+\quad 3x^4 - 2x^3 + x^2 + x \\
\hline
3x^4 + 6x^3 - x^2 + 7x - 2
\end{array}
$$

We can subtract two polynomials horizontally or vertically as well.

EXAMPLE 5	**Subtracting Polynomials**

Find the difference: $(3x^4 - 4x^3 + 6x^2 - 1) - (2x^4 - 8x^2 - 6x + 5)$

Solution ***Horizontal Subtraction:***

$$(3x^4 - 4x^3 + 6x^2 - 1) - (2x^4 - 8x^2 - 6x + 5)$$

$$= 3x^4 - 4x^3 + 6x^2 - 1 + \underbrace{(-2x^4 + 8x^2 + 6x - 5)}$$

Be sure to change the sign of each
term in the second polynomial.

$$= (3x^4 - 2x^4) + (-4x^3) + (6x^2 + 8x^2) + 6x + (-1 - 5)$$
↑
Group like terms.

$$= x^4 - 4x^3 + 14x^2 + 6x - 6$$

COMMENT Vertical subtraction will be used when we divide polynomials. ∎

Vertical Subtraction: We line up like terms, change the sign of each coefficient of the second polynomial, and add.

$$
\begin{array}{ccccc}
x^4 & x^3 & x^2 & x^1 & x^0 \\
\end{array}
\qquad
\begin{array}{ccccc}
x^4 & x^3 & x^2 & x^1 & x^0 \\
\end{array}
$$

$$
\begin{array}{ccr}
3x^4 - 4x^3 + 6x^2 \qquad\quad - 1 & = & 3x^4 - 4x^3 + 6x^2 \qquad\quad - 1 \\
- [2x^4 \qquad - 8x^2 - 6x + 5] & = + & -2x^4 \qquad\quad + 8x^2 + 6x - 5 \\
\hline
& & x^4 - 4x^3 + 14x^2 + 6x - 6
\end{array}
$$

The choice of which of these methods to use for adding and subtracting polynomials is left to you. To save space, we shall most often use the horizontal format.

━━➤ **Now Work** PROBLEM 29

4 Multiply Polynomials

Two monomials may be multiplied using the Laws of Exponents and the Commutative and Associative Properties. For example,

$$(2x^3) \cdot (5x^4) = (2 \cdot 5) \cdot (x^3 \cdot x^4) = 10x^{3+4} = 10x^7$$

Products of polynomials are found by repeated use of the Distributive Property and the Laws of Exponents. Again, you have a choice of horizontal or vertical format.

EXAMPLE 6 | **Multiplying Polynomials**

Find the product: $(2x + 5)(x^2 - x + 2)$

Solution | **Horizontal Multiplication:**

$$(2x + 5)(x^2 - x + 2) = 2x(x^2 - x + 2) + 5(x^2 - x + 2)$$
↑
Distributive Property

$$= (2x \cdot x^2 - 2x \cdot x + 2x \cdot 2) + (5 \cdot x^2 - 5 \cdot x + 5 \cdot 2)$$
↑
Distributive Property

$$= (2x^3 - 2x^2 + 4x) + (5x^2 - 5x + 10)$$
↑
Law of Exponents

$$= 2x^3 + 3x^2 - x + 10$$
↑
Combine like terms.

Vertical Multiplication: The idea here is very much like multiplying a two-digit number by a three-digit number.

$$
\begin{array}{rr}
x^2 - x + 2 & \\
2x + 5 & \\
\hline
2x^3 - 2x^2 + 4x & \text{This line is } 2x(x^2 - x + 2). \\
(+) \qquad 5x^2 - 5x + 10 & \text{This line is } 5(x^2 - x + 2). \\
\hline
2x^3 + 3x^2 - x + 10 & \text{Sum of the above two lines.}
\end{array}
$$

━━➤ **Now Work** PROBLEM 45

5 Know Formulas for Special Products

Certain products, which we call **special products,** occur frequently in algebra. We can calculate them easily using the **FOIL** (*First, Outer, Inner, Last*) method of multiplying two binomials.

$$(ax + b)(cx + d) = ax(cx + d) + b(cx + d)$$

$$= \overbrace{ax \cdot cx}^{\text{First}} + \overbrace{ax \cdot d}^{\text{Outer}} + \overbrace{b \cdot cx}^{\text{Inner}} + \overbrace{b \cdot d}^{\text{Last}}$$

$$= acx^2 + adx + bcx + bd$$

$$= acx^2 + (ad + bc)x + bd$$

EXAMPLE 7	Using FOIL

(a) $(x - 3)(x + 3) = x^2 + 3x - 3x - 9 = x^2 - 9$

$\qquad\qquad\qquad\qquad\ \ \text{F}\quad\ \ \text{O}\quad\ \ \text{I}\quad\ \ \text{L}$

(b) $(x + 2)^2 = (x + 2)(x + 2) = x^2 + 2x + 2x + 4 = x^2 + 4x + 4$

(c) $(x - 3)^2 = (x - 3)(x - 3) = x^2 - 3x - 3x + 9 = x^2 - 6x + 9$

(d) $(x + 3)(x + 1) = x^2 + x + 3x + 3 = x^2 + 4x + 3$

(e) $(2x + 1)(3x + 4) = 6x^2 + 8x + 3x + 4 = 6x^2 + 11x + 4$

Now Work PROBLEMS 47 AND 55

Some products have been given special names because of their form. The following special products are based on Examples 7(a), (b), and (c).

Difference of Two Squares

$$(x - a)(x + a) = x^2 - a^2 \qquad\qquad\qquad \textbf{(2)}$$

Squares of Binomials, or Perfect Squares

$$(x + a)^2 = x^2 + 2ax + a^2 \qquad\qquad\qquad \textbf{(3a)}$$

$$(x - a)^2 = x^2 - 2ax + a^2 \qquad\qquad\qquad \textbf{(3b)}$$

EXAMPLE 8	Using Special Product Formulas

(a) $(x - 5)(x + 5) = x^2 - 5^2 = x^2 - 25$ *Difference of two squares*

(b) $(x + 7)^2 = x^2 + 2 \cdot 7 \cdot x + 7^2 = x^2 + 14x + 49$ *Square of a binomial*

(c) $(2x + 1)^2 = (2x)^2 + 2 \cdot 1 \cdot 2x + 1^2 = 4x^2 + 4x + 1$ *Notice that we used 2x in place of x in formula (3a)*

(d) $(3x - 4)^2 = (3x)^2 - 2 \cdot 4 \cdot 3x + 4^2 = 9x^2 - 24x + 16$ *Replace x by 3x in formula (3b).*

Now Work PROBLEMS 65, 67, AND 69

Let's look at some more examples that lead to general formulas.

| EXAMPLE 9 | Cubing a Binomial |

(a) $(x + 2)^3 = (x + 2)(x + 2)^2 = (x + 2)(x^2 + 4x + 4)$ Formula (3a)

$$= (x^3 + 4x^2 + 4x) + (2x^2 + 8x + 8)$$

$$= x^3 + 6x^2 + 12x + 8$$

(b) $(x - 1)^3 = (x - 1)(x - 1)^2 = (x - 1)(x^2 - 2x + 1)$ Formula (3b)

$$= (x^3 - 2x^2 + x) - (x^2 - 2x + 1)$$

$$= x^3 - 3x^2 + 3x - 1$$

Cubes of Binomials, or Perfect Cubes

$$(x + a)^3 = x^3 + 3ax^2 + 3a^2x + a^3 \qquad \text{(4a)}$$

$$(x - a)^3 = x^3 - 3ax^2 + 3a^2x - a^3 \qquad \text{(4b)}$$

Now Work PROBLEM 85

| EXAMPLE 10 | Forming the Difference of Two Cubes |

$$(x - 1)(x^2 + x + 1) = x(x^2 + x + 1) - 1(x^2 + x + 1)$$

$$= x^3 + x^2 + x - x^2 - x - 1$$

$$= x^3 - 1$$

| EXAMPLE 11 | Forming the Sum of Two Cubes |

$$(x + 2)(x^2 - 2x + 4) = x(x^2 - 2x + 4) + 2(x^2 - 2x + 4)$$

$$= x^3 - 2x^2 + 4x + 2x^2 - 4x + 8$$

$$= x^3 + 8$$

Examples 10 and 11 lead to two more special products.

Difference of Two Cubes

$$(x - a)(x^2 + ax + a^2) = x^3 - a^3 \qquad \text{(5)}$$

Sum of Two Cubes

$$(x + a)(x^2 - ax + a^2) = x^3 + a^3 \qquad \text{(6)}$$

6 Divide Polynomials Using Long Division

The procedure for dividing two polynomials is similar to the procedure for dividing two integers.

EXAMPLE 12	Dividing Two Integers

Divide 842 by 15.

Solution

$$
\begin{array}{r}
56 \quad \leftarrow \text{Quotient} \\
\text{Divisor} \rightarrow \quad 15\overline{)842} \quad \leftarrow \text{Dividend} \\
\underline{75} \quad \leftarrow 5\cdot 15 \text{ (subtract)} \\
92 \\
\underline{90} \quad \leftarrow 6\cdot 15 \text{ (subtract)} \\
2 \quad \leftarrow \text{Remainder}
\end{array}
$$

So, $\dfrac{842}{15} = 56 + \dfrac{2}{15}$.

In the long division process detailed in Example 12, the number 15 is called the **divisor,** the number 842 is called the **dividend,** the number 56 is called the **quotient,** and the number 2 is called the **remainder.**

To check the answer obtained in a division problem, multiply the quotient by the divisor and add the remainder. The answer should be the dividend.

$$
\boxed{(\text{Quotient})(\text{Divisor}) + \text{Remainder} = \text{Dividend}}
$$

For example, we can check the results obtained in Example 12 as follows:

$$
(56)(15) + 2 = 840 + 2 = 842
$$

To divide two polynomials, we first must write each polynomial in standard form. The process then follows a pattern similar to that of Example 12. The next example illustrates the procedure.

EXAMPLE 13	Dividing Two Polynomials

Find the quotient and the remainder when

$$
3x^3 + 4x^2 + x + 7 \quad \text{is divided by} \quad x^2 + 1
$$

Solution Each polynomial is in standard form. The dividend is $3x^3 + 4x^2 + x + 7$, and the divisor is $x^2 + 1$.

STEP 1: Divide the leading term of the dividend, $3x^3$, by the leading term of the divisor, x^2. Enter the result, $3x$, over the term $3x^3$, as follows:

$$
\begin{array}{r}
3x \\
x^2 + 1\overline{)3x^3 + 4x^2 + x + 7}
\end{array}
$$

STEP 2: Multiply $3x$ by $x^2 + 1$ and enter the result below the dividend.

$$
\begin{array}{r}
3x \\
x^2 + 1\overline{)3x^3 + 4x^2 + \; x + 7} \\
\underline{3x^3 \qquad\quad + 3x} \quad \leftarrow 3x\cdot(x^2 + 1) = 3x^3 + 3x
\end{array}
$$

Notice that we align the 3x term under the x to make the next step easier.

STEP 3: Subtract and bring down the remaining terms.

$$
\begin{array}{r}
3x \\
x^2 + 1\overline{)3x^3 + 4x^2 + \; x + 7} \\
\underline{3x^3 \qquad\quad + 3x} \quad \leftarrow \text{Subtract (change the signs and add).} \\
4x^2 - 2x + 7 \quad \leftarrow \text{Bring down the } 4x^2 \text{ and the 7.}
\end{array}
$$

STEP 4: Repeat Steps 1–3 using $4x^2 - 2x + 7$ as the dividend.

$$
\begin{array}{r}
3x + 4 \\
x^2 + 1 \overline{)\, 3x^3 + 4x^2 + x + 7 } \\
3x^3 + 3x \\
\hline
4x^2 - 2x + 7 \\
4x^2 + 4 \\
\hline
-2x + 3
\end{array}
$$

← Divide $4x^2$ by x^2 to get 4.

← Multiply $x^2 + 1$ by 4; subtract.

Since x^2 does not divide $-2x$ evenly (that is, the result is not a monomial), the process ends. The quotient is $3x + 4$, and the remainder is $-2x + 3$.

✓ **Check:** (Quotient)(Divisor) + Remainder

$= (3x + 4)(x^2 + 1) + (-2x + 3)$

$= 3x^3 + 3x + 4x^2 + 4 + (-2x + 3)$

$= 3x^3 + 4x^2 + x + 7 =$ Dividend

Then

$$
\frac{3x^3 + 4x^2 + x + 7}{x^2 + 1} = 3x + 4 + \frac{-2x + 3}{x^2 + 1}
$$

The next example combines the steps involved in long division.

EXAMPLE 14 **Dividing Two Polynomials**

Find the quotient and the remainder when

$$x^4 - 3x^3 + 2x - 5 \quad \text{is divided by} \quad x^2 - x + 1$$

Solution In setting up this division problem, it is necessary to leave a space for the missing x^2 term in the dividend.

$$
\begin{array}{r}
x^2 - 2x - 3 \leftarrow \text{Quotient}\\
\text{Divisor} \rightarrow \quad x^2 - x + 1 \overline{)\, x^4 - 3x^3 + 2x - 5 } \leftarrow \text{Dividend}\\
\text{Subtract} \rightarrow \quad x^4 - x^3 + x^2 \\
\hline
-2x^3 - x^2 + 2x - 5\\
\text{Subtract} \rightarrow \quad -2x^3 + 2x^2 - 2x \\
\hline
-3x^2 + 4x - 5\\
\text{Subtract} \rightarrow \quad -3x^2 + 3x - 3\\
\hline
x - 2 \leftarrow \text{Remainder}
\end{array}
$$

✓ **Check:** (Quotient)(Divisor) + Remainder

$= (x^2 - 2x - 3)(x^2 - x + 1) + x - 2$

$= x^4 - x^3 + x^2 - 2x^3 + 2x^2 - 2x - 3x^2 + 3x - 3 + x - 2$

$= x^4 - 3x^3 + 2x - 5 =$ Dividend

As a result,

$$
\frac{x^4 - 3x^3 + 2x - 5}{x^2 - x + 1} = x^2 - 2x - 3 + \frac{x - 2}{x^2 - x + 1}
$$

The process of dividing two polynomials leads to the following result:

THEOREM Let Q be a polynomial of positive degree and let P be a polynomial whose degree is greater than or equal to the degree of Q. The remainder after dividing P by Q is either the zero polynomial or a polynomial whose degree is less than the degree of the divisor Q.

━━━━━ **Now Work** PROBLEM 93

7 Work with Polynomials in Two Variables

A **monomial in two variables** x and y has the form $ax^n y^m$, where a is a constant, x and y are variables, and n and m are nonnegative integers. The **degree** of a monomial is the sum of the powers of the variables.

For example,

$$2xy^3, \quad x^2 y^2, \quad \text{and} \quad x^3 y$$

are monomials, each of which has degree 4.

A **polynomial in two variables** x and y is the sum of one or more monomials in two variables. The **degree of a polynomial** in two variables is the highest degree of all the monomials with nonzero coefficients.

EXAMPLE 15 Examples of Polynomials in Two Variables

$$3x^2 + 2x^3 y + 5 \qquad \pi x^3 - y^2 \qquad x^4 + 4x^3 y - xy^3 + y^4$$

Two variables, Two variables, Two variables,
degree is 4. degree is 3. degree is 4.

Multiplying polynomials in two variables is handled in the same way as polynomials in one variable.

EXAMPLE 16 Using a Special Product Formula

To multiply $(2x - y)^2$, use the Squares of Binomials formula (3b) with $2x$ instead of x and y instead of a.

$$(2x - y)^2 = (2x)^2 - 2 \cdot y \cdot 2x + y^2$$
$$= 4x^2 - 4xy + y^2$$

━━━━━ **Now Work** PROBLEM 79

R.4 Assess Your Understanding

Concepts and Vocabulary

1. The polynomial $3x^4 - 2x^3 + 13x^2 - 5$ is of degree _____. The leading coefficient is _____.

2. $(x^2 - 4)(x^2 + 4) = $ _____.

3. $(x - 2)(x^2 + 2x + 4) = $ _____.

4. *True or False* $4x^{-2}$ is a monomial of degree -2.

5. *True or False* The degree of the product of two nonzero polynomials equals the sum of their degrees.

6. *True or False* $(x + a)(x^2 + ax + a) = x^3 + a^3$.

Skill Building

In Problems 7–16, tell whether the expression is a monomial. If it is, name the variable(s) and the coefficient and give the degree of the monomial. If it is not a monomial, state why not.

7. $2x^3$ **8.** $-4x^2$ **9.** $\dfrac{8}{x}$ **10.** $-2x^{-3}$ **11.** $-2xy^2$

12. $5x^2y^3$ **13.** $\dfrac{8x}{y}$ **14.** $-\dfrac{2x^2}{y^3}$ **15.** $x^2 + y^2$ **16.** $3x^2 + 4$

In Problems 17–26, tell whether the expression is a polynomial. If it is, give its degree. If it is not, state why not.

17. $3x^2 - 5$ **18.** $1 - 4x$ **19.** 5 **20.** $-\pi$ **21.** $3x^2 - \dfrac{5}{x}$

22. $\dfrac{3}{x} + 2$ **23.** $2y^3 - \sqrt{2}$ **24.** $10z^2 + z$ **25.** $\dfrac{x^2 + 5}{x^3 - 1}$ **26.** $\dfrac{3x^3 + 2x - 1}{x^2 + x + 1}$

In Problems 27–46, add, subtract, or multiply, as indicated. Express your answer as a single polynomial in standard form.

27. $(x^2 + 4x + 5) + (3x - 3)$ **28.** $(x^3 + 3x^2 + 2) + (x^2 - 4x + 4)$

29. $(x^3 - 2x^2 + 5x + 10) - (2x^2 - 4x + 3)$ **30.** $(x^2 - 3x - 4) - (x^3 - 3x^2 + x + 5)$

31. $(6x^5 + x^3 + x) + (5x^4 - x^3 + 3x^2)$ **32.** $(10x^5 - 8x^2) + (3x^3 - 2x^2 + 6)$

33. $(x^2 - 3x + 1) + 2(3x^2 + x - 4)$ **34.** $-2(x^2 + x + 1) + (-5x^2 - x + 2)$

35. $6(x^3 + x^2 - 3) - 4(2x^3 - 3x^2)$ **36.** $8(4x^3 - 3x^2 - 1) - 6(4x^3 + 8x - 2)$

37. $(x^2 - x + 2) + (2x^2 - 3x + 5) - (x^2 + 1)$ **38.** $(x^2 + 1) - (4x^2 + 5) + (x^2 + x - 2)$

39. $9(y^2 - 3y + 4) - 6(1 - y^2)$ **40.** $8(1 - y^3) + 4(1 + y + y^2 + y^3)$

41. $x(x^2 + x - 4)$ **42.** $4x^2(x^3 - x + 2)$ **43.** $-2x^2(4x^3 + 5)$

44. $5x^3(3x - 4)$ **45.** $(x + 1)(x^2 + 2x - 4)$ **46.** $(2x - 3)(x^2 + x + 1)$

In Problems 47–64, multiply the polynomials using the FOIL method. Express your answer as a single polynomial in standard form.

47. $(x + 2)(x + 4)$ **48.** $(x + 3)(x + 5)$ **49.** $(2x + 5)(x + 2)$

50. $(3x + 1)(2x + 1)$ **51.** $(x - 4)(x + 2)$ **52.** $(x + 4)(x - 2)$

53. $(x - 3)(x - 2)$ **54.** $(x - 5)(x - 1)$ **55.** $(2x + 3)(x - 2)$

56. $(2x - 4)(3x + 1)$ **57.** $(-2x + 3)(x - 4)$ **58.** $(-3x - 1)(x + 1)$

59. $(-x - 2)(-2x - 4)$ **60.** $(-2x - 3)(3 - x)$ **61.** $(x - 2y)(x + y)$

62. $(2x + 3y)(x - y)$ **63.** $(-2x - 3y)(3x + 2y)$ **64.** $(x - 3y)(-2x + y)$

In Problems 65–88, multiply the polynomials using the special product formulas. Express your answer as a single polynomial in standard form.

65. $(x - 7)(x + 7)$ **66.** $(x - 1)(x + 1)$ **67.** $(2x + 3)(2x - 3)$ **68.** $(3x + 2)(3x - 2)$

69. $(x + 4)^2$ **70.** $(x + 5)^2$ **71.** $(x - 4)^2$ **72.** $(x - 5)^2$

73. $(3x + 4)(3x - 4)$ **74.** $(5x - 3)(5x + 3)$ **75.** $(2x - 3)^2$ **76.** $(3x - 4)^2$

77. $(x + y)(x - y)$ **78.** $(x + 3y)(x - 3y)$ **79.** $(3x + y)(3x - y)$ **80.** $(3x + 4y)(3x - 4y)$

81. $(x + y)^2$ **82.** $(x - y)^2$ **83.** $(x - 2y)^2$ **84.** $(2x + 3y)^2$

85. $(x - 2)^3$ **86.** $(x + 1)^3$ **87.** $(2x + 1)^3$ **88.** $(3x - 2)^3$

In Problems 89–104, find the quotient and the remainder. Check your work by verifying that

$$(Quotient)(Divisor) + Remainder = Dividend$$

89. $4x^3 - 3x^2 + x + 1$ divided by $x + 2$ **90.** $3x^3 - x^2 + x - 2$ divided by $x + 2$

91. $4x^3 - 3x^2 + x + 1$ divided by x^2 **92.** $3x^3 - x^2 + x - 2$ divided by x^2

93. $5x^4 - 3x^2 + x + 1$ divided by $x^2 + 2$ **94.** $5x^4 - x^2 + x - 2$ divided by $x^2 + 2$

95. $4x^5 - 3x^2 + x + 1$ divided by $2x^3 - 1$ **96.** $3x^5 - x^2 + x - 2$ divided by $3x^3 - 1$

97. $2x^4 - 3x^3 + x + 1$ divided by $2x^2 + x + 1$ **98.** $3x^4 - x^3 + x - 2$ divided by $3x^2 + x + 1$

99. $-4x^3 + x^2 - 4$ divided by $x - 1$ **100.** $-3x^4 - 2x - 1$ divided by $x - 1$

101. $1 - x^2 + x^4$ divided by $x^2 + x + 1$ **102.** $1 - x^2 + x^4$ divided by $x^2 - x + 1$

103. $x^3 - a^3$ divided by $x - a$ **104.** $x^5 - a^5$ divided by $x - a$

Discussion and Writing

105. Explain why the degree of the product of two nonzero polynomials equals the sum of their degrees.

106. Explain why the degree of the sum of two polynomials of different degrees equals the larger of their degrees.

107. Give a careful statement about the degree of the sum of two polynomials of the same degree.

108. Do you prefer adding two polynomials using the horizontal method or the vertical method? Write a brief position paper defending your choice.

109. Do you prefer to memorize the rule for the square of a binomial $(x + a)^2$ or to use FOIL to obtain the product? Write a brief position paper defending your choice.

R.5 Factoring Polynomials

> **OBJECTIVES** **1** Factor the Difference of Two Squares and the Sum and the Difference of Two Cubes (p. 50)
> **2** Factor Perfect Squares (p. 51)
> **3** Factor a Second-Degree Polynomial: $x^2 + Bx + C$ (p. 52)
> **4** Factor by Grouping (p. 53)
> **5** Factor a Second-Degree Polynomial: $Ax^2 + Bx + C, A \neq 1$ (p. 54)

Consider the following product:

$$(2x + 3)(x - 4) = 2x^2 - 5x - 12$$

The two polynomials on the left side are called **factors** of the polynomial on the right side. Expressing a given polynomial as a product of other polynomials, that is, finding the factors of a polynomial, is called **factoring.**

 We shall restrict our discussion here to factoring polynomials in one variable into products of polynomials in one variable, where all coefficients are integers. We call this **factoring over the integers.**

COMMENT *Over the real numbers, $3x + 4$ factors into $3(x + \frac{4}{3})$. It is the noninteger $\frac{4}{3}$ that causes $3x + 4$ to be prime over the integers.* ∎

Any polynomial can be written as the product of 1 times itself or as -1 times its additive inverse. If a polynomial cannot be written as the product of two other polynomials (excluding 1 and -1), then the polynomial is said to be **prime.** When a polynomial has been written as a product consisting only of prime factors, it is said to be **factored completely.** Examples of prime polynomials (over the integers) are

$$2, \quad 3, \quad 5, \quad x, \quad x + 1, \quad x - 1, \quad 3x + 4 \quad x^2 + 4$$

The first factor to look for in a factoring problem is a common monomial factor present in each term of the polynomial. If one is present, use the Distributive Property to factor it out.

EXAMPLE 1 **Identifying Common Monomial Factors**

Polynomial	Common Monomial Factor	Remaining Factor	Factored Form
$2x + 4$	2	$x + 2$	$2x + 4 = 2(x + 2)$
$3x - 6$	3	$x - 2$	$3x - 6 = 3(x - 2)$
$2x^2 - 4x + 8$	2	$x^2 - 2x + 4$	$2x^2 - 4x + 8 = 2(x^2 - 2x + 4)$
$8x - 12$	4	$2x - 3$	$8x - 12 = 4(2x - 3)$
$x^2 + x$	x	$x + 1$	$x^2 + x = x(x + 1)$
$x^3 - 3x^2$	x^2	$x - 3$	$x^3 - 3x^2 = x^2(x - 3)$
$6x^2 + 9x$	$3x$	$2x + 3$	$6x^2 + 9x = 3x(2x + 3)$

Notice that, once all common monomial factors have been removed from a polynomial, the remaining factor is either a prime polynomial of degree 1 or a polynomial of degree 2 or higher. (Do you see why?)

➤ **Now Work** PROBLEM 5

1 Factor the Difference of Two Squares and the Sum and Difference of Two Cubes

When you factor a polynomial, first check for common monomial factors. Then see whether you can use one of the special formulas discussed in the previous section.

Difference of Two Squares	$x^2 - a^2 = (x - a)(x + a)$
Perfect Squares	$x^2 + 2ax + a^2 = (x + a)^2$
	$x^2 - 2ax + a^2 = (x - a)^2$
Sum of Two Cubes	$x^3 + a^3 = (x + a)(x^2 - ax + a^2)$
Difference of Two Cubes	$x^3 - a^3 = (x - a)(x^2 + ax + a^2)$

EXAMPLE 2 **Factoring the Difference of Two Squares**

Factor completely: $x^2 - 4$

Solution We notice that $x^2 - 4$ is the difference of two squares, x^2 and 2^2.

$$x^2 - 4 = (x - 2)(x + 2)$$

EXAMPLE 3	**Factoring the Difference of Two Cubes**

Factor completely: $x^3 - 1$

Solution Because $x^3 - 1$ is the difference of two cubes, x^3 and 1^3, we find that

$$x^3 - 1 = (x - 1)(x^2 + x + 1)$$

EXAMPLE 4	**Factoring the Sum of Two Cubes**

Factor completely: $x^3 + 8$

Solution Because $x^3 + 8$ is the sum of two cubes, x^3 and 2^3, we have

$$x^3 + 8 = (x + 2)(x^2 - 2x + 4)$$

EXAMPLE 5	**Factoring the Difference of Two Squares**

Factor completely: $x^4 - 16$

Solution Because $x^4 - 16$ is the difference of two squares, $x^4 = (x^2)^2$ and $16 = 4^2$, we have

$$x^4 - 16 = (x^2 - 4)(x^2 + 4)$$

But $x^2 - 4$ is also the difference of two squares. Thus,

$$x^4 - 16 = (x^2 - 4)(x^2 + 4) = (x - 2)(x + 2)(x^2 + 4)$$

━━━━━ **Now Work** PROBLEMS 15 AND 33

2 Factor Perfect Squares

When the first term and third term of a trinomial are both positive and are perfect squares, such as x^2, $9x^2$, 1, and 4, check to see whether the trinomial is a perfect square.

EXAMPLE 6	**Factoring Perfect Squares**

Factor completely: $x^2 + 6x + 9$

Solution The first term, x^2, and the third term, $9 = 3^2$, are perfect squares. Because the middle term $6x$ is twice the product of x and 3, we have a perfect square.

$$x^2 + 6x + 9 = (x + 3)^2$$

EXAMPLE 7	**Factoring Perfect Squares**

Factor completely: $9x^2 - 6x + 1$

Solution The first term, $9x^2 = (3x)^2$, and the third term, $1 = 1^2$, are perfect squares. Because the middle term, $-6x$, is -2 times the product of $3x$ and 1, we have a perfect square.

$$9x^2 - 6x + 1 = (3x - 1)^2$$

EXAMPLE 8	**Factoring Perfect Squares**

Factor completely: $25x^2 + 30x + 9$

Solution The first term, $25x^2 = (5x)^2$, and the third term, $9 = 3^2$, are perfect squares. Because the middle term, $30x$, is twice the product of $5x$ and 3, we have a perfect square.

$$25x^2 + 30x + 9 = (5x + 3)^2$$

━━━━━ **Now Work** PROBLEMS 25 AND 93

If a trinomial is not a perfect square, it may be possible to factor it using the technique discussed next.

3 Factor a Second-degree Polynomial: $x^2 + Bx + C$

The idea behind factoring a second-degree polynomial like $x^2 + Bx + C$ is to see whether it can be made equal to the product of two, possibly equal, first-degree polynomials.

For example, we know that

$$(x + 3)(x + 4) = x^2 + 7x + 12$$

The factors of $x^2 + 7x + 12$ are $x + 3$ and $x + 4$. Notice the following:

$$x^2 + 7x + 12 = (x + 3)(x + 4)$$

 12 is the product of 3 and 4

 7 is the sum of 3 and 4

In general, if $x^2 + Bx + C = (x + a)(x + b) = x^2 + (a + b)x + ab$, then $ab = C$ and $a + b = B$.

> To factor a second-degree polynomial $x^2 + Bx + C$, find integers whose product is C and whose sum is B. That is, if there are numbers a, b, where $ab = C$ and $a + b = B$, then
> $$x^2 + Bx + C = (x + a)(x + b)$$

EXAMPLE 9 **Factoring Trinomials**

Factor completely: $x^2 + 7x + 10$

Solution First, determine all integers whose product is 10 and then compute their sums.

Integers whose product is 10	1, 10	−1, −10	2, 5	−2, −5
Sum	11	−11	7	−7

The integers 2 and 5 have a product of 10 and add up to 7, the coefficient of the middle term. As a result, we have

$$x^2 + 7x + 10 = (x + 2)(x + 5)$$

EXAMPLE 10 **Factoring Trinomials**

Factor completely: $x^2 - 6x + 8$

Solution First, determine all integers whose product is 8 and then compute each sum.

Integers whose product is 8	1, 8	−1, −8	2, 4	−2, −4
Sum	9	−9	6	−6

Since −6 is the coefficient of the middle term,

$$x^2 - 6x + 8 = (x - 2)(x - 4)$$

EXAMPLE 11 | **Factoring Trinomials**

Factor completely: $x^2 - x - 12$

Solution | First, determine all integers whose product is -12 and then compute each sum.

Integers whose product is -12	$1, -12$	$-1, 12$	$2, -6$	$-2, 6$	$3, -4$	$-3, 4$
Sum	-11	11	-4	4	-1	1

Since -1 is the coefficient of the middle term,

$$x^2 - x - 12 = (x + 3)(x - 4)$$

EXAMPLE 12 | **Factoring Trinomials**

Factor completely: $x^2 + 4x - 12$

Solution | The integers -2 and 6 have a product of -12 and have the sum 4. Thus,

$$x^2 + 4x - 12 = (x - 2)(x + 6)$$

To avoid errors in factoring, always check your answer by multiplying it out to see if the result equals the original expression.

When none of the possibilities works, the polynomial is prime.

EXAMPLE 13 | **Identifying Prime Polynomials**

Show that $x^2 + 9$ is prime.

Solution | First, list the integers whose product is 9 and then compute their sums.

Integers whose product is 9	$1, 9$	$-1, -9$	$3, 3$	$-3, -3$
Sum	10	-10	6	-6

Since the coefficient of the middle term in $x^2 + 9 = x^2 + 0x + 9$ is 0 and none of the sums equals 0, we conclude that $x^2 + 9$ is prime.

Example 13 demonstrates a more general result:

THEOREM | Any polynomial of the form $x^2 + a^2$, a real, is prime.

Now Work PROBLEMS 39 AND 77

4 Factor by Grouping

Sometimes a common factor does not occur in every term of the polynomial, but in each of several groups of terms that together make up the polynomial. When this happens, the common factor can be factored out of each group by means of the Distributive Property. This technique is called **factoring by grouping.**

EXAMPLE 14 | **Factoring by Grouping**

Factor completely by grouping: $(x^2 + 2)x + (x^2 + 2) \cdot 3$

Solution | Notice the common factor $x^2 + 2$. By applying the Distributive Property, we have

$$(x^2 + 2)x + (x^2 + 2) \cdot 3 = (x^2 + 2)(x + 3)$$

Since $x^2 + 2$ and $x + 3$ are prime, the factorization is complete.

△ The next example shows a factoring problem that occurs in calculus.

| EXAMPLE 15 | **Factoring by Grouping** |

Factor completely by grouping: $3(x-1)^2(x+2)^4 + 4(x-1)^3(x+2)^3$

Solution Here, $(x-1)^2(x+2)^3$ is a common factor of $3(x-1)^2(x+2)^4$ and of $4(x-1)^3(x+2)^3$. As a result,

$$3(x-1)^2(x+2)^4 + 4(x-1)^3(x+2)^3 = (x-1)^2(x+2)^3[3(x+2) + 4(x-1)]$$
$$= (x-1)^2(x+2)^3[3x+6+4x-4]$$
$$= (x-1)^2(x+2)^3(7x+2)$$ ∎

| EXAMPLE 16 | **Factoring by Grouping** |

Factor completely by grouping: $x^3 - 4x^2 + 2x - 8$

Solution To see if factoring by grouping will work, group the first two terms and the last two terms. Then look for a common factor in each group. In this example, we can factor x^2 from $x^3 - 4x^2$ and 2 from $2x - 8$. The remaining factor in each case is the same, $x - 4$. This means that factoring by grouping will work, as follows:

$$x^3 - 4x^2 + 2x - 8 = (x^3 - 4x^2) + (2x - 8)$$
$$= x^2(x-4) + 2(x-4)$$
$$= (x-4)(x^2+2)$$

Since $x^2 + 2$ and $x - 4$ are prime, the factorization is complete. ∎

✏ **Now Work** PROBLEMS 51 AND 121

5 Factor a Second-degree Polynomial: $Ax^2 + Bx + C$, $A \neq 1$

To factor a second-degree polynomial $Ax^2 + Bx + C$, when $A \neq 1$ and A, B, and C have no common factors, follow these steps:

> **Steps for Factoring $Ax^2 + Bx + C$,**
> **when $A \neq 1$ and A, B, and C Have No Common Factors**
>
> **STEP 1:** Find the value of AC.
> **STEP 2:** Find integers whose product is AC that add up to B. That is, find a and b so that $ab = AC$ and $a + b = B$.
> **STEP 3:** Write $Ax^2 + Bx + C = Ax^2 + ax + bx + C$.
> **STEP 4:** Factor this last expression by grouping.

| EXAMPLE 17 | **Factoring Trinomials** |

Factor completely: $2x^2 + 5x + 3$

Solution Comparing $2x^2 + 5x + 3$ to $Ax^2 + Bx + C$, we find that $A = 2$, $B = 5$, and $C = 3$.

STEP 1: The value of AC is $2 \cdot 3 = 6$.
STEP 2: Determine the integers whose product is $AC = 6$ and compute their sums.

Integers whose product is 6	1, 6	−1, −6	2, 3	−2, −3
Sum	7	−7	5	−5

STEP 3: The integers whose product is 6 that add up to $B = 5$ are 2 and 3.

$$2x^2 + 5x + 3 = 2x^2 + 2x + 3x + 3$$

STEP 4: Factor by grouping.

$$2x^2 + 2x + 3x + 3 = (2x^2 + 2x) + (3x + 3)$$
$$= 2x(x + 1) + 3(x + 1)$$
$$= (x + 1)(2x + 3)$$

As a result,

$$2x^2 + 5x + 3 = (x + 1)(2x + 3)$$

EXAMPLE 18 | **Factoring Trinomials**

Factor completely: $2x^2 - x - 6$

Solution Comparing $2x^2 - x - 6$ to $Ax^2 + Bx + C$, we find that $A = 2, B = -1$, and $C = -6$.

STEP 1: The value of AC is $2 \cdot (-6) = -12$.
STEP 2: Determine the integers whose product is $AC = -12$ and compute their sums.

Integers whose product is −12	1, −12	−1, 12	2, −6	−2, 6	3, −4	−3, 4
Sum	−11	11	−4	4	−1	1

STEP 3: The integers whose product is −12 that add up to $B = -1$ are −4 and 3.

$$2x^2 - x - 6 = 2x^2 - 4x + 3x - 6$$

STEP 4: Factor by grouping.

$$2x^2 - 4x + 3x - 6 = (2x^2 - 4x) + (3x - 6)$$
$$= 2x(x - 2) + 3(x - 2)$$
$$= (x - 2)(2x + 3)$$

As a result,

$$2x^2 - x - 6 = (x - 2)(2x + 3)$$

Now Work PROBLEM 57

SUMMARY

Type of Polynomial	Method	Example
Any polynomial	Look for common monomial factors. (Always do this first!)	$6x^2 + 9x = 3x(2x + 3)$
Binomials of degree 2 or higher	Check for a special product: Difference of two squares, $x^2 - a^2$ Difference of two cubes, $x^3 - a^3$ Sum of two cubes, $x^3 + a^3$	$x^2 - 16 = (x - 4)(x + 4)$ $x^3 - 64 = (x - 4)(x^2 + 4x + 16)$ $x^3 + 27 = (x + 3)(x^2 - 3x + 9)$
Trinomials of degree 2	Check for a perfect square, $(x \pm a)^2$ Factoring $x^2 + Bx + C$ (p. 52) Factoring $Ax^2 + Bx + C$ (p. 54)	$x^2 + 8x + 16 = (x + 4)^2$ $x^2 - 10x + 25 = (x - 5)^2$ $x^2 - x - 2 = (x - 2)(x + 1)$ $6x^2 + x - 1 = (2x + 1)(3x - 1)$
Three or more terms	Grouping	$2x^3 - 3x^2 + 4x - 6 = (2x - 3)(x^2 + 2)$

R.5 Assess Your Understanding

Concepts and Vocabulary

1. If factored completely, $3x^3 - 12x =$ _____.

2. If a polynomial cannot be written as the product of two other polynomials (excluding 1 and −1), then the polynomial is said to be _____.

3. **True or False** The polynomial $x^2 + 4$ is prime.

4. **True or False** $3x^3 - 2x^2 - 6x + 4 = (3x - 2)(x^2 + 2)$.

Skill Building

In Problems 5–14, factor each polynomial by removing the common monomial factor.

5. $3x + 6$

6. $7x - 14$

7. $ax^2 + a$

8. $ax - a$

9. $x^3 + x^2 + x$

10. $x^3 - x^2 + x$

11. $2x^2 - 2x$

12. $3x^2 - 3x$

13. $3x^2y - 6xy^2 + 12xy$

14. $60x^2y - 48xy^2 + 72x^3y$

In Problems 15–22, factor the difference of two squares.

15. $x^2 - 1$

16. $x^2 - 4$

17. $4x^2 - 1$

18. $9x^2 - 1$

19. $x^2 - 16$

20. $x^2 - 25$

21. $25x^2 - 4$

22. $36x^2 - 9$

In Problems 23–32, factor the perfect squares.

23. $x^2 + 2x + 1$

24. $x^2 - 4x + 4$

25. $x^2 + 4x + 4$

26. $x^2 - 2x + 1$

27. $x^2 - 10x + 25$

28. $x^2 + 10x + 25$

29. $4x^2 + 4x + 1$

30. $9x^2 + 6x + 1$

31. $16x^2 + 8x + 1$

32. $25x^2 + 10x + 1$

In Problems 33–38, factor the sum or difference of two cubes.

33. $x^3 - 27$

34. $x^3 + 125$

35. $x^3 + 27$

36. $27 - 8x^3$

37. $8x^3 + 27$

38. $64 - 27x^3$

In Problems 39–50, factor each polynomial.

39. $x^2 + 5x + 6$

40. $x^2 + 6x + 8$

41. $x^2 + 7x + 6$

42. $x^2 + 9x + 8$

43. $x^2 + 7x + 10$

44. $x^2 + 11x + 10$

45. $x^2 - 10x + 16$

46. $x^2 - 17x + 16$

47. $x^2 - 7x - 8$

48. $x^2 - 2x - 8$

49. $x^2 + 7x - 8$

50. $x^2 + 2x - 8$

In Problems 51–56, factor by grouping.

51. $2x^2 + 4x + 3x + 6$

52. $3x^2 - 3x + 2x - 2$

53. $2x^2 - 4x + x - 2$

54. $3x^2 + 6x - x - 2$

55. $6x^2 + 9x + 4x + 6$

56. $9x^2 - 6x + 3x - 2$

In Problems 57–68, factor each polynomial.

57. $3x^2 + 4x + 1$

58. $2x^2 + 3x + 1$

59. $2z^2 + 5z + 3$

60. $6z^2 + 5z + 1$

61. $3x^2 + 2x - 8$

62. $3x^2 + 10x + 8$

63. $3x^2 - 2x - 8$

64. $3x^2 - 10x + 8$

65. $3x^2 + 14x + 8$

66. $3x^2 - 14x + 8$

67. $3x^2 + 10x - 8$

68. $3x^2 - 10x - 8$

In Problems 69–116, factor completely each polynomial. If the polynomial cannot be factored, say it is prime.

69. $x^2 - 36$

70. $x^2 - 9$

71. $2 - 8x^2$

72. $3 - 27x^2$

73. $x^2 + 11x + 10$

74. $x^2 + 5x + 4$

75. $x^2 - 10x + 21$

76. $x^2 - 6x + 8$

77. $4x^2 - 8x + 32$

78. $3x^2 - 12x + 15$

79. $x^2 + 4x + 16$

80. $x^2 + 12x + 36$

81. $15 + 2x - x^2$

82. $14 + 6x - x^2$

83. $3x^2 - 12x - 36$

84. $x^3 + 8x^2 - 20x$

85. $y^4 + 11y^3 + 30y^2$

86. $3y^3 - 18y^2 - 48y$

87. $4x^2 + 12x + 9$

88. $9x^2 - 12x + 4$

89. $6x^2 + 8x + 2$

90. $8x^2 + 6x - 2$

91. $x^4 - 81$

92. $x^4 - 1$

93. $x^6 - 2x^3 + 1$

94. $x^6 + 2x^3 + 1$

95. $x^7 - x^5$

96. $x^8 - x^5$

97. $16x^2 + 24x + 9$

98. $9x^2 - 24x + 16$

99. $5 + 16x - 16x^2$

100. $5 + 11x - 16x^2$

101. $4y^2 - 16y + 15$ **102.** $9y^2 + 9y - 4$ **103.** $1 - 8x^2 - 9x^4$ **104.** $4 - 14x^2 - 8x^4$

105. $x(x + 3) - 6(x + 3)$ **106.** $5(3x - 7) + x(3x - 7)$ **107.** $(x + 2)^2 - 5(x + 2)$ **108.** $(x - 1)^2 - 2(x - 1)$

109. $(3x - 2)^3 - 27$ **110.** $(5x + 1)^3 - 1$ **111.** $3(x^2 + 10x + 25) - 4(x + 5)$

112. $7(x^2 - 6x + 9) + 5(x - 3)$ **113.** $x^3 + 2x^2 - x - 2$ **114.** $x^3 - 3x^2 - x + 3$

115. $x^4 - x^3 + x - 1$ **116.** $x^4 + x^3 + x + 1$

Applications and Extensions

In Problems 117–126, expressions that occur in calculus are given. Factor completely each expression.

117. $2(3x + 4)^2 + (2x + 3) \cdot 2(3x + 4) \cdot 3$ **118.** $5(2x + 1)^2 + (5x - 6) \cdot 2(2x + 1) \cdot 2$

119. $2x(2x + 5) + x^2 \cdot 2$ **120.** $3x^2(8x - 3) + x^3 \cdot 8$

121. $2(x + 3)(x - 2)^3 + (x + 3)^2 \cdot 3(x - 2)^2$ **122.** $4(x + 5)^3(x - 1)^2 + (x + 5)^4 \cdot 2(x - 1)$

123. $(4x - 3)^2 + x \cdot 2(4x - 3) \cdot 4$ **124.** $3x^2(3x + 4)^2 + x^3 \cdot 2(3x + 4) \cdot 3$

125. $2(3x - 5) \cdot 3(2x + 1)^3 + (3x - 5)^2 \cdot 3(2x + 1)^2 \cdot 2$ **126.** $3(4x + 5)^2 \cdot 4(5x + 1)^2 + (4x + 5)^3 \cdot 2(5x + 1) \cdot 5$

127. Show that $x^2 + 4$ is prime. **128.** Show that $x^2 + x + 1$ is prime.

Discussion and Writing

129. Make up a polynomial that factors into a perfect square.

130. Explain to a fellow student what you look for first when presented with a factoring problem. What do you do next?

R.6 Synthetic Division

OBJECTIVE 1 Divide Polynomials Using Synthetic Division (p. 57)

1 Divide Polynomials Using Synthetic Division

To find the quotient as well as the remainder when a polynomial of degree 1 or higher is divided by $x - c$, a shortened version of long division, called **synthetic division**, makes the task simpler.

To see how synthetic division works, we will use long division to divide the polynomial $2x^3 - x^2 + 3$ by $x - 3$.

$$\begin{array}{r} 2x^2 + 5x + 15 \quad \leftarrow \text{Quotient} \\ x - 3 \overline{) 2x^3 - x^2 \qquad + 3} \\ \underline{2x^3 - 6x^2} \\ 5x^2 \\ \underline{5x^2 - 15x} \\ 15x + 3 \\ \underline{15x - 45} \\ 48 \quad \leftarrow \text{Remainder} \end{array}$$

CHECK: (Divisor) · (Quotient) + Remainder

$$= (x - 3)(2x^2 + 5x + 15) + 48$$
$$= 2x^3 + 5x^2 + 15x - 6x^2 - 15x - 45 + 48$$
$$= 2x^3 - x^2 + 3$$

The process of synthetic division arises from rewriting the long division in a more compact form, using simpler notation. For example, in the long division above, the terms in blue are not really necessary because they are identical to the terms directly above them. With these terms removed, we have

$$
\begin{array}{r}
2x^2 + 5x\ \ + 15 \\
x - 3\overline{)2x^3 -\ \ x^2\qquad\quad + 3} \\
-6x^2\qquad\qquad \\
\hline
5x^2\qquad\qquad \\
-15x\quad \\
\hline
15x\quad \\
-45 \\
\hline
48
\end{array}
$$

Most of the x's that appear in this process can also be removed, provided that we are careful about positioning each coefficient. In this regard, we will need to use 0 as the coefficient of x in the dividend, because that power of x is missing. Now we have

$$
\begin{array}{r}
2x^2 + 5x + 15 \\
x - 3\overline{)2\ \ -1\qquad 0\qquad 3} \\
-6\qquad\qquad \\
\hline
5\qquad\qquad \\
-15\quad \\
\hline
15\quad \\
-45 \\
\hline
48
\end{array}
$$

We can make this display more compact by moving the lines up until the numbers in blue align horizontally.

$$
\begin{array}{rl}
2x^2 + 5x + 15 & \text{Row 1} \\
x - 3\overline{)2\ \ -1\qquad 0\qquad 3} & \text{Row 2} \\
-6\ -15 -45 & \text{Row 3} \\
\hline
\bigcirc\quad 5\qquad 15\qquad 48 & \text{Row 4}
\end{array}
$$

Because the leading coefficient of the divisor is always 1, we know that the leading coefficient of the dividend will also be the leading coefficient of the quotient. So we place the leading coefficient of the quotient, 2, in the circled position. Now, the first three numbers in row 4 are precisely the coefficients of the quotient, and the last number in row 4 is the remainder. Thus, row 1 is not really needed, so we can compress the process to three rows, where the bottom row contains both the coefficients of the quotient and the remainder.

$$
\begin{array}{rl}
x - 3\overline{)2\ \ -1\qquad 0\qquad 3} & \text{Row 1} \\
-6\ -15 -45 & \text{Row 2 (subtract)} \\
\hline
2\qquad 5\qquad 15\qquad 48 & \text{Row 3}
\end{array}
$$

Recall that the entries in row 3 are obtained by subtracting the entries in row 2 from those in row 1. Rather than subtracting the entries in row 2, we can change the sign of each entry and add. With this modification, our display will look like this:

$$
\begin{array}{rl}
x - 3\overline{)2\ \ -1\qquad 0\qquad 3} & \text{Row 1} \\
6\qquad 15\qquad 45 & \text{Row 2 (add)} \\
\hline
2\qquad 5\qquad 15\qquad 48 & \text{Row 3}
\end{array}
$$

Notice that the entries in row 2 are three times the prior entries in row 3. Our last modification to the display replaces the $x - 3$ by 3. The entries in row 3 give the quotient and the remainder, as shown next.

$$\begin{array}{r|rrr} 3 & 2 & -1 & 0 & 3 \\ & & 6 & 15 & 45 \\ \hline & 2 & 5 & 15 & 48 \end{array}$$

Row 1
Row 2 (add)
Row 3

$$\underbrace{2x^2 + 5x + 15}_{\text{Quotient}} \quad \underbrace{48}_{\text{Remainder}}$$

Let's go through an example step by step.

EXAMPLE 1 **Using Synthetic Division to Find the Quotient and Remainder**

Use synthetic division to find the quotient and remainder when

$$x^3 - 4x^2 - 5 \quad \text{is divided by} \quad x - 3$$

Solution **STEP 1:** Write the dividend in descending powers of x. Then copy the coefficients, remembering to insert a 0 for any missing powers of x.

$$1 \quad -4 \quad 0 \quad -5 \quad \text{Row 1}$$

STEP 2: Insert the usual division symbol. In synthetic division, the divisor is of the form $x - c$, and c is the number placed to the left of the division symbol. Here, since the divisor is $x - 3$, we insert 3 to the left of the division symbol.

$$3 \overline{)1 \quad -4 \quad 0 \quad -5} \quad \text{Row 1}$$

STEP 3: Bring the 1 down two rows, and enter it in row 3.

$$\begin{array}{r|rrrr} 3 & 1 & -4 & 0 & -5 \\ & \downarrow & & & \\ \hline & 1 & & & \end{array}$$

Row 1
Row 2
Row 3

STEP 4: Multiply the latest entry in row 3 by 3, and place the result in row 2, one column over to the right.

$$\begin{array}{r|rrrr} 3 & 1 & -4 & 0 & -5 \\ & & 3 & & \\ \hline & 1 & & & \end{array}$$

Row 1
Row 2
Row 3

STEP 5: Add the entry in row 2 to the entry above it in row 1, and enter the sum in row 3.

$$\begin{array}{r|rrrr} 3 & 1 & -4 & 0 & -5 \\ & & 3 & & \\ \hline & 1 & -1 & & \end{array}$$

Row 1
Row 2
Row 3

STEP 6: Repeat Steps 4 and 5 until no more entries are available in row 1.

$$\begin{array}{r|rrrr} 3 & 1 & -4 & 0 & -5 \\ & & 3 & -3 & -9 \\ \hline & 1 & -1 & -3 & -14 \end{array}$$

Row 1
Row 2
Row 3

STEP 7: The final entry in row 3, the -14, is the remainder; the other entries in row 3, the 1, -1, and -3, are the coefficients (in descending order) of a polynomial whose degree is 1 less than that of the dividend. This is the quotient. Thus,

$$\text{Quotient} = x^2 - x - 3 \qquad \text{Remainder} = -14$$

✓ **Check:** (Divisor)(Quotient) + Remainder

$$= (x - 3)(x^2 - x - 3) + (-14)$$
$$= (x^3 - x^2 - 3x - 3x^2 + 3x + 9) + (-14)$$
$$= x^3 - 4x^2 - 5 = \text{Dividend}$$

Let's do an example in which all seven steps are combined.

| EXAMPLE 2 | Using Synthetic Division to Verify a Factor |

Use synthetic division to show that $x + 3$ is a factor of

$$2x^5 + 5x^4 - 2x^3 + 2x^2 - 2x + 3$$

Solution The divisor is $x + 3 = x - (-3)$, so we place -3 to the left of the division symbol. Then the row 3 entries will be multiplied by -3, entered in row 2, and added to row 1.

$$
\begin{array}{r|rrrrrr}
-3) & 2 & 5 & -2 & 2 & -2 & 3 \quad \text{Row 1}\\
 & & -6 & 3 & -3 & 3 & -3 \quad \text{Row 2}\\
\hline
 & 2 & -1 & 1 & -1 & 1 & 0 \quad \text{Row 3}
\end{array}
$$

Because the remainder is 0, we have

(Divisor)(Quotient) + Remainder

$= (x + 3)(2x^4 - x^3 + x^2 - x + 1) = 2x^5 + 5x^4 - 2x^3 + 2x^2 - 2x + 3$

As we see, $x + 3$ is a factor of $2x^5 + 5x^4 - 2x^3 + 2x^2 - 2x + 3$. ∎

As Example 2 illustrates, the remainder after division gives information about whether the divisor is, or is not, a factor. We shall have more to say about this in Chapter 5.

Now Work PROBLEMS 7 AND 17

R.6 Assess Your Understanding

Concepts and Vocabulary

1. To check division, the expression that is being divided, the dividend, should equal the product of the _____ and the _____ plus the _____.

2. To divide $2x^3 - 5x + 1$ by $x + 3$ using synthetic division, the first step is to write _____) _____.

3. *True or False* In using synthetic division, the divisor is always a polynomial of degree 1, whose leading coefficient is 1.

4. *True or False*
$$
\begin{array}{r|rrrr}
-2) & 5 & 3 & 2 & 1\\
 & & -10 & 14 & -32\\
\hline
 & 5 & -7 & 16 & -31
\end{array}
$$
means $\dfrac{5x^3 + 3x^2 + 2x + 1}{x + 2} = 5x^2 - 7x + 16 + \dfrac{-31}{x + 2}$.

Skill Building

In Problems 5–16, use synthetic division to find the quotient and remainder when:

5. $x^3 - x^2 + 2x + 4$ is divided by $x - 2$

6. $x^3 + 2x^2 - 3x + 1$ is divided by $x + 1$

7. $3x^3 + 2x^2 - x + 3$ is divided by $x - 3$

8. $-4x^3 + 2x^2 - x + 1$ is divided by $x + 2$

9. $x^5 - 4x^3 + x$ is divided by $x + 3$

10. $x^4 + x^2 + 2$ is divided by $x - 2$

11. $4x^6 - 3x^4 + x^2 + 5$ is divided by $x - 1$

12. $x^5 + 5x^3 - 10$ is divided by $x + 1$

13. $0.1x^3 + 0.2x$ is divided by $x + 1.1$

14. $0.1x^2 - 0.2$ is divided by $x + 2.1$

15. $x^5 - 1$ is divided by $x - 1$

16. $x^5 + 1$ is divided by $x + 1$

In Problems 17–26, use synthetic division to determine whether $x - c$ is a factor of the given polynomial.

17. $4x^3 - 3x^2 - 8x + 4$; $x - 2$

18. $-4x^3 + 5x^2 + 8$; $x + 3$

19. $3x^4 - 6x^3 - 5x + 10$; $x - 2$

20. $4x^4 - 15x^2 - 4$; $x - 2$

21. $3x^6 + 82x^3 + 27$; $x + 3$

22. $2x^6 - 18x^4 + x^2 - 9$; $x + 3$

23. $4x^6 - 64x^4 + x^2 - 15; \quad x + 4$

24. $x^6 - 16x^4 + x^2 - 16; \quad x + 4$

25. $2x^4 - x^3 + 2x - 1; \quad x - \dfrac{1}{2}$

26. $3x^4 + x^3 - 3x + 1; \quad x + \dfrac{1}{3}$

Applications and Extensions

27. Find the sum of $a, b, c,$ and d if

$$\frac{x^3 - 2x^2 + 3x + 5}{x + 2} = ax^2 + bx + c + \frac{d}{x + 2}$$

Discussion and Writing

28. When dividing a polynomial by $x - c$, do you prefer to use long division or synthetic division? Does the value of c make a difference to you in choosing? Give reasons.

R.7 Rational Expressions

> **OBJECTIVES** **1** Reduce a Rational Expression to Lowest Terms (p. 61)
>
> **2** Multiply and Divide Rational Expressions (p. 62)
>
> **3** Add and Subtract Rational Expressions (p. 63)
>
> **4** Use the Least Common Multiple Method (p. 65)
>
> **5** Simplify Complex Rational Expressions (p. 67)

1 Reduce a Rational Expression to Lowest Terms

If we form the quotient of two polynomials, the result is called a **rational expression.** Some examples of rational expressions are

(a) $\dfrac{x^3 + 1}{x}$ (b) $\dfrac{3x^2 + x - 2}{x^2 + 5}$ (c) $\dfrac{x}{x^2 - 1}$ (d) $\dfrac{xy^2}{(x - y)^2}$

Expressions (a), (b), and (c) are rational expressions in one variable, x, whereas (d) is a rational expression in two variables, x and y.

Rational expressions are described in the same manner as rational numbers. In expression (a), the polynomial $x^3 + 1$ is called the **numerator,** and x is called the **denominator.** When the numerator and denominator of a rational expression contain no common factors (except 1 and −1), we say that the rational expression is **reduced to lowest terms,** or **simplified.**

The polynomial in the denominator of a rational expression cannot be equal to 0 because division by 0 is not defined. For example, for the expression $\dfrac{x^3 + 1}{x}$, x cannot take on the value 0. The domain of the variable x is $\{x \mid x \neq 0\}$.

A rational expression is reduced to lowest terms by factoring completely the numerator and the denominator and canceling any common factors by using the Cancellation Property:

$$\frac{a\cancel{c}}{b\cancel{c}} = \frac{a}{b} \qquad \text{if } b \neq 0, c \neq 0 \tag{1}$$

| EXAMPLE 1 | Reducing a Rational Expression to Lowest Terms |

Reduce to lowest terms: $\dfrac{x^2 + 4x + 4}{x^2 + 3x + 2}$

Solution We begin by factoring the numerator and the denominator.

$$x^2 + 4x + 4 = (x + 2)(x + 2)$$
$$x^2 + 3x + 2 = (x + 2)(x + 1)$$

WARNING Apply the Cancellation Property only to rational expressions written in factored form. Be sure to cancel only common factors! ∎

Since a common factor, $x + 2$, appears, the original expression is not in lowest terms. To reduce it to lowest terms, we use the Cancellation Property:

$$\frac{x^2 + 4x + 4}{x^2 + 3x + 2} = \frac{\cancel{(x + 2)}(x + 2)}{\cancel{(x + 2)}(x + 1)} = \frac{x + 2}{x + 1} \qquad x \neq -2, -1$$

∎

| EXAMPLE 2 | Reducing Rational Expressions to Lowest Terms |

Reduce each rational expression to lowest terms.

(a) $\dfrac{x^3 - 8}{x^3 - 2x^2}$ 　　　　　　　　　　　(b) $\dfrac{8 - 2x}{x^2 - x - 12}$

Solution 　(a) $\dfrac{x^3 - 8}{x^3 - 2x^2} = \dfrac{\cancel{(x - 2)}(x^2 + 2x + 4)}{x^2 \cancel{(x - 2)}} = \dfrac{x^2 + 2x + 4}{x^2} \qquad x \neq 0, 2$

(b) $\dfrac{8 - 2x}{x^2 - x - 12} = \dfrac{2(4 - x)}{(x - 4)(x + 3)} = \dfrac{2(-1)\cancel{(x - 4)}}{\cancel{(x - 4)}(x + 3)} = \dfrac{-2}{x + 3} \qquad x \neq -3, 4$

∎

▬▬▬━ **Now Work** PROBLEM 5

2 Multiply and Divide Rational Expressions

The rules for multiplying and dividing rational expressions are the same as the rules for multiplying and dividing rational numbers. If $\dfrac{a}{b}$ and $\dfrac{c}{d}$, $b \neq 0$, $d \neq 0$, are two rational expressions, then

$$\frac{a}{b} \cdot \frac{c}{d} = \frac{ac}{bd} \qquad \text{if } b \neq 0, d \neq 0 \tag{2}$$

$$\frac{\dfrac{a}{b}}{\dfrac{c}{d}} = \frac{a}{b} \cdot \frac{d}{c} = \frac{ad}{bc} \qquad \text{if } b \neq 0, c \neq 0, d \neq 0 \tag{3}$$

In using equations (2) and (3) with rational expressions, be sure first to factor each polynomial completely so that common factors can be canceled. Leave your answer in factored form.

| EXAMPLE 3 | Multiplying and Dividing Rational Expressions |

Perform the indicated operation and simplify the result. Leave your answer in factored form.

(a) $\dfrac{x^2 - 2x + 1}{x^3 + x} \cdot \dfrac{4x^2 + 4}{x^2 + x - 2}$

(b) $\dfrac{\dfrac{x + 3}{x^2 - 4}}{\dfrac{x^2 - x - 12}{x^3 - 8}}$

Solution

(a) $\dfrac{x^2 - 2x + 1}{x^3 + x} \cdot \dfrac{4x^2 + 4}{x^2 + x - 2} = \dfrac{(x-1)^2}{x(x^2+1)} \cdot \dfrac{4(x^2+1)}{(x+2)(x-1)}$

$$= \dfrac{(x-1)^{\cancel{2}}(4)\cancel{(x^2+1)}}{x\cancel{(x^2+1)}(x+2)\cancel{(x-1)}}$$

$$= \dfrac{4(x-1)}{x(x+2)} \qquad x \neq -2, 0, 1$$

(b) $\dfrac{\dfrac{x+3}{x^2-4}}{\dfrac{x^2-x-12}{x^3-8}} = \dfrac{x+3}{x^2-4} \cdot \dfrac{x^3-8}{x^2-x-12}$

$$= \dfrac{x+3}{(x-2)(x+2)} \cdot \dfrac{(x-2)(x^2+2x+4)}{(x-4)(x+3)}$$

$$= \dfrac{\cancel{(x+3)}\,\cancel{(x-2)}(x^2+2x+4)}{\cancel{(x-2)}(x+2)(x-4)\cancel{(x+3)}}$$

$$= \dfrac{x^2+2x+4}{(x+2)(x-4)} \qquad x \neq -3, -2, 2, 4$$

Now Work PROBLEMS **17** AND **25**

3 Add and Subtract Rational Expressions

The rules for adding and subtracting rational expressions are the same as the rules for adding and subtracting rational numbers. So, if the denominators of two rational expressions to be added (or subtracted) are equal, we add (or subtract) the numerators and keep the common denominator.

In Words

To add (or subtract) two rational expressions with the same denominator, keep the common denominator and add (or subtract) the numerators.

If $\dfrac{a}{b}$ and $\dfrac{c}{b}$ are two rational expressions, then

$$\dfrac{a}{b} + \dfrac{c}{b} = \dfrac{a+c}{b} \qquad \dfrac{a}{b} - \dfrac{c}{b} = \dfrac{a-c}{b} \qquad \text{if } b \neq 0 \qquad \text{(4)}$$

EXAMPLE 4

Adding and Subtracting Rational Expressions with Equal Denominators

Perform the indicated operation and simplify the result. Leave your answer in factored form.

(a) $\dfrac{2x^2-4}{2x+5} + \dfrac{x+3}{2x+5} \qquad x \neq -\dfrac{5}{2}$

(b) $\dfrac{x}{x-3} - \dfrac{3x+2}{x-3} \qquad x \neq 3$

Solution

(a) $\dfrac{2x^2-4}{2x+5} + \dfrac{x+3}{2x+5} = \dfrac{(2x^2-4)+(x+3)}{2x+5}$

$$= \dfrac{2x^2+x-1}{2x+5} = \dfrac{(2x-1)(x+1)}{2x+5}$$

(b) $\dfrac{x}{x-3} - \dfrac{3x+2}{x-3} = \dfrac{x-(3x+2)}{x-3} = \dfrac{x-3x-2}{x-3}$

$$= \dfrac{-2x-2}{x-3} = \dfrac{-2(x+1)}{x-3}$$

| EXAMPLE 5 | **Adding Rational Expressions Whose Denominators Are Additive Inverses of Each Other** |

Perform the indicated operation and simplify the result. Leave your answer in factored form.

$$\frac{2x}{x-3} + \frac{5}{3-x} \qquad x \neq 3$$

Solution Notice that the denominators of the two rational expressions are different. However, the denominator of the second expression is the additive inverse of the denominator of the first. That is,

$$3 - x = -x + 3 = -1 \cdot (x - 3) = -(x - 3)$$

Then

$$\frac{2x}{x-3} + \frac{5}{3-x} = \frac{2x}{x-3} + \frac{5}{-(x-3)} = \frac{2x}{x-3} + \frac{-5}{x-3}$$

$$\uparrow \qquad\qquad\qquad \uparrow$$
$$3 - x = -(x-3) \qquad\qquad \frac{a}{-b} = \frac{-a}{b}$$

$$= \frac{2x + (-5)}{x-3} = \frac{2x - 5}{x-3} \quad \blacksquare$$

─────── **Now Work** PROBLEMS 37 AND 43

If the denominators of two rational expressions to be added or subtracted are not equal, we can use the general formulas for adding and subtracting quotients.

$$\frac{a}{b} + \frac{c}{d} = \frac{a \cdot d}{b \cdot d} + \frac{b \cdot c}{b \cdot d} = \frac{ad + bc}{bd} \qquad \text{if } b \neq 0, d \neq 0 \qquad \textbf{(5a)}$$

$$\frac{a}{b} - \frac{c}{d} = \frac{a \cdot d}{b \cdot d} - \frac{b \cdot c}{b \cdot d} = \frac{ad - bc}{bd} \qquad \text{if } b \neq 0, d \neq 0 \qquad \textbf{(5b)}$$

| EXAMPLE 6 | **Adding and Subtracting Rational Expressions with Unequal Denominators** |

Perform the indicated operation and simplify the result. Leave your answer in factored form.

(a) $\dfrac{x-3}{x+4} + \dfrac{x}{x-2} \qquad x \neq -4, 2$

(b) $\dfrac{x^2}{x^2-4} - \dfrac{1}{x} \qquad x \neq -2, 0, 2$

Solution (a) $\dfrac{x-3}{x+4} + \dfrac{x}{x-2} = \dfrac{x-3}{x+4} \cdot \dfrac{x-2}{x-2} + \dfrac{x+4}{x+4} \cdot \dfrac{x}{x-2}$

$$\uparrow$$
$$(5a)$$

$$= \frac{(x-3)(x-2) + (x+4)(x)}{(x+4)(x-2)}$$

$$= \frac{x^2 - 5x + 6 + x^2 + 4x}{(x+4)(x-2)} = \frac{2x^2 - x + 6}{(x+4)(x-2)}$$

(b) $\dfrac{x^2}{x^2-4}-\dfrac{1}{x}=\dfrac{x^2}{x^2-4}\cdot\dfrac{x}{x}-\dfrac{x^2-4}{x^2-4}\cdot\dfrac{1}{x}=\dfrac{x^2(x)-(x^2-4)(1)}{(x^2-4)(x)}$

$(5b)$

$$=\dfrac{x^3-x^2+4}{(x-2)(x+2)(x)}$$

Now Work PROBLEM 47

4 Use the Least Common Multiple Method

If the denominators of two rational expressions to be added (or subtracted) have common factors, we usually do not use the general rules given by equations (5a) and (5b). Just as with fractions, we apply the **least common multiple (LCM) method.** The LCM method uses the polynomial of least degree that has each denominator polynomial as a factor.

The LCM Method for Adding or Subtracting Rational Expressions

The Least Common Multiple (LCM) Method requires four steps:

STEP 1: Factor completely the polynomial in the denominator of each rational expression.

STEP 2: The LCM of the denominator is the product of each of these factors raised to a power equal to the greatest number of times that the factor occurs in the polynomials.

STEP 3: Write each rational expression using the LCM as the common denominator.

STEP 4: Add or subtract the rational expressions using equation (4).

Let's work an example that only requires Steps 1 and 2.

EXAMPLE 7	Finding the Least Common Multiple

Find the least common multiple of the following pair of polynomials:

$$x(x-1)^2(x+1) \quad \text{and} \quad 4(x-1)(x+1)^3$$

Solution **STEP 1:** The polynomials are already factored completely as

$$x(x-1)^2(x+1) \quad \text{and} \quad 4(x-1)(x+1)^3$$

STEP 2: Start by writing the factors of the left-hand polynomial. (Or you could start with the one on the right.)

$$x(x-1)^2(x+1)$$

Now look at the right-hand polynomial. Its first factor, 4, does not appear in our list, so we insert it.

$$4x(x-1)^2(x+1)$$

The next factor, $x-1$, is already in our list, so no change is necessary. The final factor is $(x+1)^3$. Since our list has $x+1$ to the first power only, we replace $x+1$ in the list by $(x+1)^3$. The LCM is

$$4x(x-1)^2(x+1)^3$$

Notice that the LCM is, in fact, the polynomial of least degree that contains $x(x - 1)^2(x + 1)$ and $4(x - 1)(x + 1)^3$ as factors.

━━━━➤ **Now Work** PROBLEM 53

| EXAMPLE 8 | **Using the Least Common Multiple to Add Rational Expressions** |

Perform the indicated operation and simplify the result. Leave your answer in factored form.

$$\frac{x}{x^2 + 3x + 2} + \frac{2x - 3}{x^2 - 1} \qquad x \neq -2, -1, 1$$

Solution **STEP 1:** Factor completely the polynomials in the denominators.

$$x^2 + 3x + 2 = (x + 2)(x + 1)$$
$$x^2 - 1 = (x - 1)(x + 1)$$

STEP 2: The LCM is $(x + 2)(x + 1)(x - 1)$. Do you see why?
STEP 3: Write each rational expression using the LCM as the denominator.

$$\frac{x}{x^2 + 3x + 2} = \frac{x}{(x + 2)(x + 1)} = \frac{x}{(x + 2)(x + 1)} \cdot \frac{x - 1}{x - 1} = \frac{x(x - 1)}{(x + 2)(x + 1)(x - 1)}$$

⬆ Multiply numerator and
denominator by x − 1 to get
the LCM in the denominator.

$$\frac{2x - 3}{x^2 - 1} = \frac{2x - 3}{(x - 1)(x + 1)} = \frac{2x - 3}{(x - 1)(x + 1)} \cdot \frac{x + 2}{x + 2} = \frac{(2x - 3)(x + 2)}{(x - 1)(x + 1)(x + 2)}$$

⬆ Multiply numerator and
denominator by x + 2 to get
the LCM in the denominator.

STEP 4: Now we can add by using equation (4).

$$\frac{x}{x^2 + 3x + 2} + \frac{2x - 3}{x^2 - 1} = \frac{x(x - 1)}{(x + 2)(x + 1)(x - 1)} + \frac{(2x - 3)(x + 2)}{(x + 2)(x + 1)(x - 1)}$$

$$= \frac{(x^2 - x) + (2x^2 + x - 6)}{(x + 2)(x + 1)(x - 1)}$$

$$= \frac{3x^2 - 6}{(x + 2)(x + 1)(x - 1)} = \frac{3(x^2 - 2)}{(x + 2)(x + 1)(x - 1)} \quad ∎$$

| EXAMPLE 9 | **Using the Least Common Multiple to Subtract Rational Expressions** |

Perform the indicated operations and simplify the result. Leave your answer in factored form.

$$\frac{3}{x^2 + x} - \frac{x + 4}{x^2 + 2x + 1} \qquad x \neq -1, 0$$

Solution **STEP 1:** Factor completely the polynomials in the denominators.

$$x^2 + x = x(x + 1)$$
$$x^2 + 2x + 1 = (x + 1)^2$$

STEP 2: The LCM is $x(x + 1)^2$.

STEP 3: Write each rational expression using the LCM as the denominator.

$$\frac{3}{x^2 + x} = \frac{3}{x(x+1)} = \frac{3}{x(x+1)} \cdot \frac{x+1}{x+1} = \frac{3(x+1)}{x(x+1)^2}$$

$$\frac{x+4}{x^2 + 2x + 1} = \frac{x+4}{(x+1)^2} = \frac{x+4}{(x+1)^2} \cdot \frac{x}{x} = \frac{x(x+4)}{x(x+1)^2}$$

STEP 4: Subtract, using equation (4).

$$\frac{3}{x^2 + x} - \frac{x+4}{x^2 + 2x + 1} = \frac{3(x+1)}{x(x+1)^2} - \frac{x(x+4)}{x(x+1)^2}$$

$$= \frac{3(x+1) - x(x+4)}{x(x+1)^2}$$

$$= \frac{3x + 3 - x^2 - 4x}{x(x+1)^2}$$

$$= \frac{-x^2 - x + 3}{x(x+1)^2}$$

Now Work PROBLEM 63

5 Simplify Complex Rational Expressions

When sums and/or differences of rational expressions appear as the numerator and/or denominator of a quotient, the quotient is called a **complex rational expression.*** For example,

$$\frac{1 + \dfrac{1}{x}}{1 - \dfrac{1}{x}} \quad \text{and} \quad \frac{\dfrac{x^2}{x^2 - 4} - 3}{\dfrac{x - 3}{x + 2} - 1}$$

are complex rational expressions. To **simplify** a complex rational expression means to write it as a rational expression reduced to lowest terms. This can be accomplished in either of two ways.

Simplifying a Complex Rational Expression

METHOD 1: Treat the numerator and denominator of the complex rational expression separately, performing whatever operations are indicated and simplifying the results. Follow this by simplifying the resulting rational expression.

METHOD 2: Find the LCM of the denominators of all rational expressions that appear in the complex rational expression. Multiply the numerator and denominator of the complex rational expression by the LCM and simplify the result.

We will use both methods in the next example. By carefully studying each method, you can discover situations in which one method may be easier to use than the other.

EXAMPLE 10 Simplifying a Complex Rational Expression

Simplify: $\dfrac{\dfrac{1}{2} + \dfrac{3}{x}}{\dfrac{x+3}{4}}$ $x \neq -3, 0$

* Some texts use the term **complex fraction.**

Solution **Method 1:** First, we perform the indicated operation in the numerator, and then we divide.

$$\frac{\dfrac{1}{2}+\dfrac{3}{x}}{\dfrac{x+3}{4}} = \frac{\dfrac{1\cdot x + 2\cdot 3}{2\cdot x}}{\dfrac{x+3}{4}} = \frac{\dfrac{x+6}{2x}}{\dfrac{x+3}{4}} = \frac{x+6}{2x}\cdot\frac{4}{x+3}$$

↑ Rule for adding quotients Rule for dividing quotients ↑

$$= \frac{(x+6)\cdot 4}{2\cdot x\cdot(x+3)} = \frac{2\cdot 2\cdot(x+6)}{2\cdot x\cdot(x+3)} = \frac{2(x+6)}{x(x+3)}$$

↑ Rule for multiplying quotients

Method 2: The rational expressions that appear in the complex rational expression are

$$\frac{1}{2},\ \frac{3}{x},\ \frac{x+3}{4}$$

The LCM of their denominators is $4x$. We multiply the numerator and denominator of the complex rational expression by $4x$ and then simplify.

$$\frac{\dfrac{1}{2}+\dfrac{3}{x}}{\dfrac{x+3}{4}} = \frac{4x\cdot\left(\dfrac{1}{2}+\dfrac{3}{x}\right)}{4x\cdot\left(\dfrac{x+3}{4}\right)} = \frac{4x\cdot\dfrac{1}{2}+4x\cdot\dfrac{3}{x}}{\dfrac{4x\cdot(x+3)}{4}}$$

↑ Multiply the numerator and denominator by $4x$. ↑ Use the Distributive Property in the numerator.

$$= \frac{2\cdot 2x\cdot\dfrac{1}{2}+4x\cdot\dfrac{3}{x}}{\dfrac{4x\cdot(x+3)}{4}} = \frac{2x+12}{x(x+3)} = \frac{2(x+6)}{x(x+3)}$$

↑ Simplify ↑ Factor

EXAMPLE 11 **Simplifying a Complex Rational Expression**

Simplify: $\dfrac{\dfrac{x^2}{x-4}+2}{\dfrac{2x-2}{x}-1}$ $x\neq 0,2,4$

Solution We will use Method 1.

$$\frac{\dfrac{x^2}{x-4}+2}{\dfrac{2x-2}{x}-1} = \frac{\dfrac{x^2}{x-4}+\dfrac{2(x-4)}{x-4}}{\dfrac{2x-2}{x}-\dfrac{x}{x}} = \frac{\dfrac{x^2+2x-8}{x-4}}{\dfrac{2x-2-x}{x}}$$

$$= \frac{\dfrac{(x+4)(x-2)}{x-4}}{\dfrac{x-2}{x}} = \frac{(x+4)(x-2)}{x-4}\cdot\frac{x}{x-2}$$

$$= \frac{(x+4)\cdot x}{x-4}$$

━━━► **Now Work** PROBLEM 73

Application

| EXAMPLE 12 | Solving an Application in Electricity |

An electrical circuit contains two resistors connected in parallel, as shown in Figure 28. If the resistance of each is R_1 and R_2 ohms, respectively, their combined resistance R is given by the formula

Figure 28

$$R = \dfrac{1}{\dfrac{1}{R_1} + \dfrac{1}{R_2}}$$

Express R as a rational expression; that is, simplify the right-hand side of this formula. Evaluate the rational expression if $R_1 = 6$ ohms and $R_2 = 10$ ohms.

Solution We will use Method 2. If we consider 1 as the fraction $\dfrac{1}{1}$, then the rational expressions in the complex rational expression are

$$\dfrac{1}{1}, \quad \dfrac{1}{R_1}, \quad \dfrac{1}{R_2}$$

The LCM of the denominators is $R_1 R_2$. We multiply the numerator and denominator of the complex rational expression by $R_1 R_2$ and simplify.

$$\dfrac{1}{\dfrac{1}{R_1} + \dfrac{1}{R_2}} = \dfrac{1 \cdot R_1 R_2}{\left(\dfrac{1}{R_1} + \dfrac{1}{R_2}\right) \cdot R_1 R_2} = \dfrac{R_1 R_2}{\dfrac{1}{R_1} \cdot R_1 R_2 + \dfrac{1}{R_2} \cdot R_1 R_2} = \dfrac{R_1 R_2}{R_2 + R_1}$$

Thus,

$$R = \dfrac{R_1 R_2}{R_2 + R_1}$$

If $R_1 = 6$ and $R_2 = 10$, then

$$R = \dfrac{6 \cdot 10}{10 + 6} = \dfrac{60}{16} = \dfrac{15}{4} \quad \text{ohms}$$

R.7 Assess Your Understanding

Concepts and Vocabulary

1. When the numerator and denominator of a rational expression contain no common factors (except 1 and −1), the rational expression is in _____ _____.

2. LCM is an abbreviation for _____ _____ _____.

3. **True or False** The rational expression $\dfrac{2x^3 - 4x}{x - 2}$ is reduced to lowest terms.

4. **True or False** The LCM of $2x^3 + 6x^2$ and $6x^4 + 4x^3$ is $4x^3(x + 1)$.

Skill Building

In Problems 5–16, reduce each rational expression to lowest terms.

5. $\dfrac{3x + 9}{x^2 - 9}$

6. $\dfrac{4x^2 + 8x}{12x + 24}$

7. $\dfrac{x^2 - 2x}{3x - 6}$

8. $\dfrac{15x^2 + 24x}{3x^2}$

9. $\dfrac{24x^2}{12x^2 - 6x}$

10. $\dfrac{x^2 + 4x + 4}{x^2 - 4}$

11. $\dfrac{y^2 - 25}{2y^2 - 8y - 10}$

12. $\dfrac{3y^2 - y - 2}{3y^2 + 5y + 2}$

13. $\dfrac{x^2 + 4x - 5}{x^2 - 2x + 1}$

14. $\dfrac{x - x^2}{x^2 + x - 2}$

15. $\dfrac{x^2 + 5x - 14}{2 - x}$

16. $\dfrac{2x^2 + 5x - 3}{1 - 2x}$

In Problems 17–34, perform the indicated operation and simplify the result. Leave your answer in factored form.

17. $\dfrac{3x+6}{5x^2} \cdot \dfrac{x}{x^2-4}$

18. $\dfrac{3}{2x} \cdot \dfrac{x^2}{6x+10}$

19. $\dfrac{4x^2}{x^2-16} \cdot \dfrac{x^3-64}{2x}$

20. $\dfrac{12}{x^2+x} \cdot \dfrac{x^3+1}{4x-2}$

21. $\dfrac{4x-8}{-3x} \cdot \dfrac{12}{12-6x}$

22. $\dfrac{6x-27}{5x} \cdot \dfrac{2}{4x-18}$

23. $\dfrac{x^2-3x-10}{x^2+2x-35} \cdot \dfrac{x^2+4x-21}{x^2+9x+14}$

24. $\dfrac{x^2+x-6}{x^2+4x-5} \cdot \dfrac{x^2-25}{x^2+2x-15}$

25. $\dfrac{\dfrac{6x}{x^2-4}}{\dfrac{3x-9}{2x+4}}$

26. $\dfrac{\dfrac{12x}{5x+20}}{\dfrac{4x^2}{x^2-16}}$

27. $\dfrac{\dfrac{8x}{x^2-1}}{\dfrac{10x}{x+1}}$

28. $\dfrac{\dfrac{x-2}{4x}}{\dfrac{x^2-4x+4}{12x}}$

29. $\dfrac{\dfrac{4-x}{4+x}}{\dfrac{4x}{x^2-16}}$

30. $\dfrac{\dfrac{3+x}{3-x}}{\dfrac{x^2-9}{9x^3}}$

31. $\dfrac{\dfrac{x^2+7x+12}{x^2-7x+12}}{\dfrac{x^2+x-12}{x^2-x-12}}$

32. $\dfrac{\dfrac{x^2+7x+6}{x^2+x-6}}{\dfrac{x^2+5x-6}{x^2+5x+6}}$

33. $\dfrac{\dfrac{2x^2-x-28}{3x^2-x-2}}{\dfrac{4x^2+16x+7}{3x^2+11x+6}}$

34. $\dfrac{\dfrac{9x^2+3x-2}{12x^2+5x-2}}{\dfrac{9x^2-6x+1}{8x^2-10x-3}}$

In Problems 35–52, perform the indicated operation and simplify the result. Leave your answer in factored form.

35. $\dfrac{x}{2} + \dfrac{5}{2}$

36. $\dfrac{3}{x} - \dfrac{6}{x}$

37. $\dfrac{x^2}{2x-3} - \dfrac{4}{2x-3}$

38. $\dfrac{3x^2}{2x-1} - \dfrac{9}{2x-1}$

39. $\dfrac{x+1}{x-3} + \dfrac{2x-3}{x-3}$

40. $\dfrac{2x-5}{3x+2} + \dfrac{x+4}{3x+2}$

41. $\dfrac{3x+5}{2x-1} - \dfrac{2x-4}{2x-1}$

42. $\dfrac{5x-4}{3x+4} - \dfrac{x+1}{3x+4}$

43. $\dfrac{4}{x-2} + \dfrac{x}{2-x}$

44. $\dfrac{6}{x-1} - \dfrac{x}{1-x}$

45. $\dfrac{4}{x-1} - \dfrac{2}{x+2}$

46. $\dfrac{2}{x+5} - \dfrac{5}{x-5}$

47. $\dfrac{x}{x+1} + \dfrac{2x-3}{x-1}$

48. $\dfrac{3x}{x-4} + \dfrac{2x}{x+3}$

49. $\dfrac{x-3}{x+2} - \dfrac{x+4}{x-2}$

50. $\dfrac{2x-3}{x-1} - \dfrac{2x+1}{x+1}$

51. $\dfrac{x}{x^2-4} + \dfrac{1}{x}$

52. $\dfrac{x-1}{x^3} + \dfrac{x}{x^2+1}$

In Problems 53–60, find the LCM of the given polynomials.

53. $x^2-4, \quad x^2-x-2$

54. $x^2-x-12, \quad x^2-8x+16$

55. $x^3-x, \quad x^2-x$

56. $3x^2-27, \quad 2x^2-x-15$

57. $4x^3-4x^2+x, \quad 2x^3-x^2, \quad x^3$

58. $x-3, \quad x^2+3x, \quad x^3-9x$

59. $x^3-x, \quad x^3-2x^2+x, \quad x^3-1$

60. $x^2+4x+4, \quad x^3+2x^2, \quad (x+2)^3$

In Problems 61–72, perform the indicated operations and simplify the result. Leave your answer in factored form.

61. $\dfrac{x}{x^2-7x+6} - \dfrac{x}{x^2-2x-24}$

62. $\dfrac{x}{x-3} - \dfrac{x+1}{x^2+5x-24}$

63. $\dfrac{4x}{x^2-4} - \dfrac{2}{x^2+x-6}$

64. $\dfrac{3x}{x-1} - \dfrac{x-4}{x^2-2x+1}$

65. $\dfrac{3}{(x-1)^2(x+1)} + \dfrac{2}{(x-1)(x+1)^2}$

66. $\dfrac{2}{(x+2)^2(x-1)} - \dfrac{6}{(x+2)(x-1)^2}$

67. $\dfrac{x+4}{x^2-x-2} - \dfrac{2x+3}{x^2+2x-8}$

68. $\dfrac{2x-3}{x^2+8x+7} - \dfrac{x-2}{(x+1)^2}$

69. $\dfrac{1}{x} - \dfrac{2}{x^2+x} + \dfrac{3}{x^3-x^2}$

70. $\dfrac{x}{(x-1)^2} + \dfrac{2}{x} - \dfrac{x+1}{x^3-x^2}$

71. $\dfrac{1}{h}\left(\dfrac{1}{x+h} - \dfrac{1}{x}\right)$

72. $\dfrac{1}{h}\left[\dfrac{1}{(x+h)^2} - \dfrac{1}{x^2}\right]$

In Problems 73–84, perform the indicated operations and simplify the result. Leave your answer in factored form.

73. $\dfrac{1 + \dfrac{1}{x}}{1 - \dfrac{1}{x}}$

74. $\dfrac{4 + \dfrac{1}{x^2}}{3 - \dfrac{1}{x^2}}$

75. $\dfrac{2 - \dfrac{x+1}{x}}{3 + \dfrac{x-1}{x+1}}$

76. $\dfrac{1 - \dfrac{x}{x+1}}{2 - \dfrac{x-1}{x}}$

77. $\dfrac{\dfrac{x+4}{x-2} - \dfrac{x-3}{x+1}}{x+1}$

78. $\dfrac{\dfrac{x-2}{x+1} - \dfrac{x}{x-2}}{x+3}$

79. $\dfrac{\dfrac{x-2}{x+2} + \dfrac{x-1}{x+1}}{\dfrac{x}{x+1} - \dfrac{2x-3}{x}}$

80. $\dfrac{\dfrac{2x+5}{x} - \dfrac{x}{x-3}}{\dfrac{x^2}{x-3} - \dfrac{(x+1)^2}{x+3}}$

81. $1 - \dfrac{1}{1 - \dfrac{1}{x}}$

82. $1 - \dfrac{1}{1 - \dfrac{1}{1-x}}$

83. $\dfrac{2(x-1)^{-1} + 3}{3(x-1)^{-1} + 2}$

84. $\dfrac{4(x+2)^{-1} - 3}{3(x+2)^{-1} - 1}$

In Problems 85–92, expressions that occur in calculus are given. Reduce each expression to lowest terms.

85. $\dfrac{(2x+3)\cdot 3 - (3x-5)\cdot 2}{(3x-5)^2}$

86. $\dfrac{(4x+1)\cdot 5 - (5x-2)\cdot 4}{(5x-2)^2}$

87. $\dfrac{x\cdot 2x - (x^2+1)\cdot 1}{(x^2+1)^2}$

88. $\dfrac{x\cdot 2x - (x^2-4)\cdot 1}{(x^2-4)^2}$

89. $\dfrac{(3x+1)\cdot 2x - x^2\cdot 3}{(3x+1)^2}$

90. $\dfrac{(2x-5)\cdot 3x^2 - x^3\cdot 2}{(2x-5)^2}$

91. $\dfrac{(x^2+1)\cdot 3 - (3x+4)\cdot 2x}{(x^2+1)^2}$

92. $\dfrac{(x^2+9)\cdot 2 - (2x-5)\cdot 2x}{(x^2+9)^2}$

Applications and Extensions

93. The Lensmaker's Equation The focal length f of a lens with index of refraction n is

$$\frac{1}{f} = (n-1)\left[\frac{1}{R_1} + \frac{1}{R_2}\right]$$

where R_1 and R_2 are the radii of curvature of the front and back surfaces of the lens. Express f as a rational expression. Evaluate the rational expression for $n = 1.5$, $R_1 = 0.1$ meter, and $R_2 = 0.2$ meter.

94. Electrical Circuits An electrical circuit contains three resistors connected in parallel. If the resistance of each is R_1, R_2, and R_3 ohms, respectively, their combined resistance R is given by the formula

$$\frac{1}{R} = \frac{1}{R_1} + \frac{1}{R_2} + \frac{1}{R_3}$$

Express R as a rational expression. Evaluate R for $R_1 = 5$ ohms, $R_2 = 4$ ohms, and $R_3 = 10$ ohms.

Discussion and Writing

95. The following expressions are called **continued fractions**:

$$1 + \frac{1}{x}, \quad 1 + \frac{1}{1 + \dfrac{1}{x}}, \quad 1 + \frac{1}{1 + \dfrac{1}{1 + \dfrac{1}{x}}}, \quad 1 + \frac{1}{1 + \dfrac{1}{1 + \dfrac{1}{1 + \dfrac{1}{x}}}}, \quad \dots$$

Each simplifies to an expression of the form

$$\frac{ax + b}{bx + c}$$

Trace the successive values of a, b, and c as you "continue" the fraction. Can you discover the patterns that these values follow? Go to the library and research Fibonacci numbers. Write a report on your findings.

96. Explain to a fellow student when you would use the LCM method to add two rational expressions. Give two examples of adding two rational expressions, one in which you use the LCM and the other in which you do not.

97. Which of the two methods given in the text for simplifying complex rational expressions do you prefer? Write a brief paragraph stating the reasons for your choice.

R.8 *n*th Roots; Rational Exponents

PREPARING FOR THIS SECTION *Before getting started, review the following:*

- Exponents, Square Roots (Section R.2, pp. 21–24)

Now Work the 'Are You Prepared?' problems on page 77.

OBJECTIVES 1 Work with *n*th Roots (p. 72)
2 Simplify Radicals (p. 73)
3 Rationalize Denominators (p. 74)
4 Simplify Expressions with Rational Exponents (p. 75)

1 Work with *n*th Roots

DEFINITION

The **principal *n*th root of a real number *a*,** $n \geq 2$ an integer, symbolized by $\sqrt[n]{a}$, is defined as follows:

$$\sqrt[n]{a} = b \quad \text{means} \quad a = b^n$$

where $a \geq 0$ and $b \geq 0$ if *n* is even and *a*, *b* are any real numbers if *n* is odd.

Notice that if *a* is negative and *n* is even then $\sqrt[n]{a}$ is not defined. When it is defined, the principal *n*th root of a number is unique.

The symbol $\sqrt[n]{a}$ for the principal *n*th root of *a* is called a **radical;** the integer *n* is called the **index,** and *a* is called the **radicand.** If the index of a radical is 2, we call $\sqrt[2]{a}$ the **square root** of *a* and omit the index 2 by simply writing $\sqrt{a}$. If the index is 3, we call $\sqrt[3]{a}$ the **cube root** of *a*.

EXAMPLE 1 **Simplifying Principal *n*th Roots**

(a) $\sqrt[3]{8} = \sqrt[3]{2^3} = 2$ (b) $\sqrt[3]{-64} = \sqrt[3]{(-4)^3} = -4$

(c) $\sqrt[4]{\dfrac{1}{16}} = \sqrt[4]{\left(\dfrac{1}{2}\right)^4} = \dfrac{1}{2}$ (d) $\sqrt[6]{(-2)^6} = |-2| = 2$

These are examples of **perfect roots,** since each simplifies to a rational number. Notice the absolute value in Example 1(d). If *n* is even, the principal *n*th root must be nonnegative.

In general, if $n \geq 2$ is an integer and *a* is a real number, we have

$$\sqrt[n]{a^n} = a \qquad \text{if } n \geq 3 \text{ is odd} \tag{1a}$$
$$\sqrt[n]{a^n} = |a| \qquad \text{if } n \geq 2 \text{ is even} \tag{1b}$$

Now Work PROBLEM 7

Radicals provide a way of representing many irrational real numbers. For example, there is no rational number whose square is 2. Using radicals, we can say that $\sqrt{2}$ is the positive number whose square is 2.

EXAMPLE 2	**Using a Calculator to Approximate Roots**

Use a calculator to approximate $\sqrt[5]{16}$.

Solution Figure 29 shows the result using a TI-84 plus graphing calculator.

Figure 29

Now Work PROBLEM 101

2 Simplify Radicals

Let $n \geq 2$ and $m \geq 2$ denote positive integers, and let a and b represent real numbers. Assuming that all radicals are defined, we have the following properties:

Properties of Radicals

$$\sqrt[n]{ab} = \sqrt[n]{a}\,\sqrt[n]{b} \qquad \text{(2a)}$$

$$\sqrt[n]{\frac{a}{b}} = \frac{\sqrt[n]{a}}{\sqrt[n]{b}} \qquad \text{(2b)}$$

$$\sqrt[n]{a^m} = (\sqrt[n]{a})^m \qquad \text{(2c)}$$

When used in reference to radicals, the direction to "simplify" will mean to remove from the radicals any perfect roots that occur as factors. Let's look at some examples of how the preceding rules are applied to simplify radicals.

EXAMPLE 3	**Simplifying Radicals**

(a) $\sqrt{32} = \sqrt{16 \cdot 2} = \sqrt{16} \cdot \sqrt{2} = 4\sqrt{2}$

 Factor out 16, (2a)
 a perfect square.

(b) $\sqrt[3]{16} = \sqrt[3]{8 \cdot 2} = \sqrt[3]{8} \cdot \sqrt[3]{2} = \sqrt[3]{2^3} \cdot \sqrt[3]{2} = 2\sqrt[3]{2}$

 Factor out 8, (2a)
 a perfect cube.

(c) $\sqrt[3]{-16x^4} = \sqrt[3]{-8 \cdot 2 \cdot x^3 \cdot x} = \sqrt[3]{(-8x^3)(2x)}$

 Factor perfect Group perfect
 cubes inside radical. cubes.

$$= \sqrt[3]{(-2x)^3 \cdot 2x} = \sqrt[3]{(-2x)^3} \cdot \sqrt[3]{2x} = -2x\sqrt[3]{2x}$$

 (2a)

(d) $\sqrt[4]{\dfrac{16x^5}{81}} = \sqrt[4]{\dfrac{2^4 x^4 x}{3^4}} = \sqrt[4]{\left(\dfrac{2x}{3}\right)^4 \cdot x} = \sqrt[4]{\left(\dfrac{2x}{3}\right)^4} \cdot \sqrt[4]{x} = \left|\dfrac{2x}{3}\right| \sqrt[4]{x}$

Now Work PROBLEMS 11 AND 17

Two or more radicals can be combined, provided that they have the same index and the same radicand. Such radicals are called **like radicals.**

EXAMPLE 4	**Combining Like Radicals**

(a) $-8\sqrt{12} + \sqrt{3} = -8\sqrt{4 \cdot 3} + \sqrt{3}$

$$= -8 \cdot \sqrt{4}\,\sqrt{3} + \sqrt{3}$$

$$= -16\sqrt{3} + \sqrt{3} = -15\sqrt{3}$$

(b) $\sqrt[3]{8x^4} + \sqrt[3]{-x} + 4\sqrt[3]{27x} = \sqrt[3]{2^3 x^3 x} + \sqrt[3]{-1 \cdot x} + 4\sqrt[3]{3^3 x}$

$$= \sqrt[3]{(2x)^3} \cdot \sqrt[3]{x} + \sqrt[3]{-1} \cdot \sqrt[3]{x} + 4\sqrt[3]{3^3} \cdot \sqrt[3]{x}$$

$$= 2x\sqrt[3]{x} - 1 \cdot \sqrt[3]{x} + 12\sqrt[3]{x}$$

$$= (2x + 11)\sqrt[3]{x}$$

Now Work PROBLEM 33

3 Rationalize Denominators

When radicals occur in quotients, it is customary to rewrite the quotient so that the new denominator contains no radicals. This process is referred to as **rationalizing the denominator.**

The idea is to multiply by an appropriate expression so that the new denominator contains no radicals. For example:

If a Denominator Contains the Factor	Multiply by	To Obtain a Denominator Free of Radicals
$\sqrt{3}$	$\sqrt{3}$	$(\sqrt{3})^2 = 3$
$\sqrt{3} + 1$	$\sqrt{3} - 1$	$(\sqrt{3})^2 - 1^2 = 3 - 1 = 2$
$\sqrt{2} - 3$	$\sqrt{2} + 3$	$(\sqrt{2})^2 - 3^2 = 2 - 9 = -7$
$\sqrt{5} - \sqrt{3}$	$\sqrt{5} + \sqrt{3}$	$(\sqrt{5})^2 - (\sqrt{3})^2 = 5 - 3 = 2$
$\sqrt[3]{4}$	$\sqrt[3]{2}$	$\sqrt[3]{4} \cdot \sqrt[3]{2} = \sqrt[3]{8} = 2$

In rationalizing the denominator of a quotient, be sure to multiply both the numerator and the denominator by the expression.

EXAMPLE 5	**Rationalizing Denominators**

Rationalize the denominator of each expression:

(a) $\dfrac{1}{\sqrt{3}}$ (b) $\dfrac{5}{4\sqrt{2}}$ (c) $\dfrac{\sqrt{2}}{\sqrt{3} - 3\sqrt{2}}$

Solution (a) The denominator contains the factor $\sqrt{3}$, so we multiply the numerator and denominator by $\sqrt{3}$ to obtain

$$\frac{1}{\sqrt{3}} = \frac{1}{\sqrt{3}} \cdot \frac{\sqrt{3}}{\sqrt{3}} = \frac{\sqrt{3}}{(\sqrt{3})^2} = \frac{\sqrt{3}}{3}$$

(b) The denominator contains the factor $\sqrt{2}$, so we multiply the numerator and denominator by $\sqrt{2}$ to obtain

$$\frac{5}{4\sqrt{2}} = \frac{5}{4\sqrt{2}} \cdot \frac{\sqrt{2}}{\sqrt{2}} = \frac{5\sqrt{2}}{4(\sqrt{2})^2} = \frac{5\sqrt{2}}{4 \cdot 2} = \frac{5\sqrt{2}}{8}$$

(c) The denominator contains the factor $\sqrt{3} - 3\sqrt{2}$, so we multiply the numerator and denominator by $\sqrt{3} + 3\sqrt{2}$ to obtain

$$\frac{\sqrt{2}}{\sqrt{3} - 3\sqrt{2}} = \frac{\sqrt{2}}{\sqrt{3} - 3\sqrt{2}} \cdot \frac{\sqrt{3} + 3\sqrt{2}}{\sqrt{3} + 3\sqrt{2}} = \frac{\sqrt{2}(\sqrt{3} + 3\sqrt{2})}{(\sqrt{3})^2 - (3\sqrt{2})^2}$$

$$= \frac{\sqrt{2}\sqrt{3} + 3(\sqrt{2})^2}{3 - 18} = \frac{\sqrt{6} + 6}{-15} = -\frac{6 + \sqrt{6}}{15}$$

Now Work PROBLEM 47

4 Simplify Expressions with Rational Exponents

Radicals are used to define rational exponents.

DEFINITION

If a is a real number and $n \geq 2$ is an integer, then

$$a^{1/n} = \sqrt[n]{a} \tag{3}$$

provided that $\sqrt[n]{a}$ exists.

Note that if n is even and $a < 0$ then $\sqrt[n]{a}$ and $a^{1/n}$ do not exist.

EXAMPLE 6 | **Writing Expressions Containing Fractional Exponents as Radicals**

(a) $4^{1/2} = \sqrt{4} = 2$ (b) $8^{1/2} = \sqrt{8} = 2\sqrt{2}$

(c) $(-27)^{1/3} = \sqrt[3]{-27} = -3$ (d) $16^{1/3} = \sqrt[3]{16} = 2\sqrt[3]{2}$

DEFINITION

If a is a real number and m and n are integers containing no common factors, with $n \geq 2$, then

$$a^{m/n} = \sqrt[n]{a^m} = \left(\sqrt[n]{a}\right)^m \tag{4}$$

provided that $\sqrt[n]{a}$ exists.

We have two comments about equation (4):

1. The exponent $\dfrac{m}{n}$ must be in lowest terms and n must be positive.

2. In simplifying the rational expression $a^{m/n}$, either $\sqrt[n]{a^m}$ or $\left(\sqrt[n]{a}\right)^m$ may be used, the choice depending on which is easier to simplify. Generally, taking the root first, as in $\left(\sqrt[n]{a}\right)^m$, is easier.

EXAMPLE 7 | **Using Equation (4)**

(a) $4^{3/2} = \left(\sqrt{4}\right)^3 = 2^3 = 8$ (b) $(-8)^{4/3} = \left(\sqrt[3]{-8}\right)^4 = (-2)^4 = 16$

(c) $(32)^{-2/5} = \left(\sqrt[5]{32}\right)^{-2} = 2^{-2} = \dfrac{1}{4}$ (d) $25^{6/4} = 25^{3/2} = \left(\sqrt{25}\right)^3 = 5^3 = 125$

━━━ **Now Work** PROBLEM 55

It can be shown that the Laws of Exponents hold for rational exponents. The next example illustrates using the Law of Exponents to simplify.

EXAMPLE 8 | **Simplifying Expressions Containing Rational Exponents**

Simplify each expression. Express your answer so that only positive exponents occur. Assume that the variables are positive.

(a) $(x^{2/3}y)(x^{-2}y)^{1/2}$ (b) $\left(\dfrac{2x^{1/3}}{y^{2/3}}\right)^{-3}$ (c) $\left(\dfrac{9x^2y^{1/3}}{x^{1/3}y}\right)^{1/2}$

Solution
(a) $(x^{2/3}y)(x^{-2}y)^{1/2} = (x^{2/3}y)[(x^{-2})^{1/2}y^{1/2}]$

$= x^{2/3}yx^{-1}y^{1/2}$

$= (x^{2/3} \cdot x^{-1})(y \cdot y^{1/2})$

$= x^{-1/3}y^{3/2}$

$= \dfrac{y^{3/2}}{x^{1/3}}$

(b) $\left(\dfrac{2x^{1/3}}{y^{2/3}}\right)^{-3} = \left(\dfrac{y^{2/3}}{2x^{1/3}}\right)^{3} = \dfrac{(y^{2/3})^{3}}{(2x^{1/3})^{3}} = \dfrac{y^{2}}{2^{3}(x^{1/3})^{3}} = \dfrac{y^{2}}{8x}$

(c) $\left(\dfrac{9x^{2}y^{1/3}}{x^{1/3}y}\right)^{1/2} = \left(\dfrac{9x^{2-(1/3)}}{y^{1-(1/3)}}\right)^{1/2} = \left(\dfrac{9x^{5/3}}{y^{2/3}}\right)^{1/2} = \dfrac{9^{1/2}(x^{5/3})^{1/2}}{(y^{2/3})^{1/2}} = \dfrac{3x^{5/6}}{y^{1/3}}$ ∎

━━ **Now Work** PROBLEM 71

 The next two examples illustrate some algebra that you will need to know for certain calculus problems.

EXAMPLE 9	**Writing an Expression as a Single Quotient**

Write the following expression as a single quotient in which only positive exponents appear.

$$(x^{2} + 1)^{1/2} + x \cdot \frac{1}{2}(x^{2} + 1)^{-1/2} \cdot 2x$$

Solution
$(x^{2} + 1)^{1/2} + x \cdot \dfrac{1}{2}(x^{2} + 1)^{-1/2} \cdot 2x = (x^{2} + 1)^{1/2} + \dfrac{x^{2}}{(x^{2} + 1)^{1/2}}$

$= \dfrac{(x^{2} + 1)^{1/2}(x^{2} + 1)^{1/2} + x^{2}}{(x^{2} + 1)^{1/2}}$

$= \dfrac{(x^{2} + 1) + x^{2}}{(x^{2} + 1)^{1/2}}$

$= \dfrac{2x^{2} + 1}{(x^{2} + 1)^{1/2}}$ ∎

━━ **Now Work** PROBLEM 77

EXAMPLE 10	**Factoring an Expression Containing Rational Exponents**

Factor: $\dfrac{4}{3}x^{1/3}(2x + 1) + 2x^{4/3}$

Solution
We begin by writing $2x^{4/3}$ as a fraction with 3 as denominator.

$\dfrac{4}{3}x^{1/3}(2x + 1) + 2x^{4/3} = \dfrac{4x^{1/3}(2x + 1)}{3} + \dfrac{6x^{4/3}}{3} = \dfrac{4x^{1/3}(2x + 1) + 6x^{4/3}}{3}$

↑
Add the two fractions

$= \dfrac{2x^{1/3}[2(2x + 1) + 3x]}{3} = \dfrac{2x^{1/3}(7x + 2)}{3}$

↑
2 and $x^{1/3}$ are common factors ↑
Simplify

∎

━━ **Now Work** PROBLEM 89

Historical Feature

The radical sign, $\sqrt{}$, was first used in print by Christoff Rudolff in 1525. It is thought to be the manuscript form of the letter *r* (for the Latin word *radix* = *root*), although this is not quite conclusively confirmed. It took a long time for $\sqrt{}$ to become the standard symbol for a square root and much longer to standardize $\sqrt[3]{}$, $\sqrt[4]{}$, $\sqrt[5]{}$, and so on. The indexes of the root were placed in every conceivable position, with

$$\sqrt[3]{8}, \quad \sqrt{}③8, \quad \text{and} \quad \underset{3}{\sqrt{}}8$$

all being variants for $\sqrt[3]{8}$. The notation $\sqrt{}\sqrt{}16$ was popular for $\sqrt[4]{16}$. By the 1700s, the index had settled where we now put it.

The bar on top of the present radical symbol, as follows,

$$\sqrt{a^2 + 2ab + b^2}$$

is the last survivor of the **vinculum,** a bar placed atop an expression to indicate what we would now indicate with parentheses. For example,

$$\overline{ab + c} = a(b + c)$$

R.8 Assess Your Understanding

'Are You Prepared?' *Answers are given at the end of these exercises. If you get a wrong answer, read the pages in red.*

1. $(-3)^2 =$ _____; $-3^2 =$ _____ (pp. 21–24)

2. $\sqrt{16} =$ _____; $\sqrt{(-4)^2} =$ _____ (pp. 21–24)

Concepts and Vocabulary

3. In the symbol $\sqrt[n]{a}$, the integer *n* is called the _____.

5. We call $\sqrt[3]{a}$ the _____ _____ of *a*.

4. *True or False* $\sqrt[5]{-32} = -2$

6. *True or False* $\sqrt[4]{(-3)^4} = -3$

Skill Building

In Problems 7–42, simplify each expression. Assume that all variables are positive when they appear.

7. $\sqrt[3]{27}$

8. $\sqrt[4]{16}$

9. $\sqrt[3]{-8}$

10. $\sqrt[3]{-1}$

11. $\sqrt{8}$

12. $\sqrt[3]{54}$

13. $\sqrt[3]{-8x^4}$

14. $\sqrt[4]{48x^5}$

15. $\sqrt[4]{x^{12}y^8}$

16. $\sqrt[5]{x^{10}y^5}$

17. $\sqrt[4]{\dfrac{x^9 y^7}{xy^3}}$

18. $\sqrt[3]{\dfrac{3xy^2}{81x^4y^2}}$

19. $\sqrt{36x}$

20. $\sqrt{9x^5}$

21. $\sqrt{3x^2}\sqrt{12x}$

22. $\sqrt{5x}\sqrt{20x^3}$

23. $\left(\sqrt{5}\sqrt[3]{9}\right)^2$

24. $\left(\sqrt[3]{3}\sqrt{10}\right)^4$

25. $\left(3\sqrt{6}\right)\left(2\sqrt{2}\right)$

26. $\left(5\sqrt{8}\right)\left(-3\sqrt{3}\right)$

27. $3\sqrt{2} + 4\sqrt{2}$

28. $6\sqrt{5} - 4\sqrt{5}$

29. $-\sqrt{18} + 2\sqrt{8}$

30. $2\sqrt{12} - 3\sqrt{27}$

31. $\left(\sqrt{3} + 3\right)\left(\sqrt{3} - 1\right)$

32. $\left(\sqrt{5} - 2\right)\left(\sqrt{5} + 3\right)$

33. $5\sqrt[3]{2} - 2\sqrt[3]{54}$

34. $9\sqrt[3]{24} - \sqrt[3]{81}$

35. $\left(\sqrt{x} - 1\right)^2$

36. $\left(\sqrt{x} + \sqrt{5}\right)^2$

37. $\sqrt[3]{16x^4} - \sqrt[3]{2x}$

38. $\sqrt[4]{32x} + \sqrt[4]{2x^5}$

39. $\sqrt{8x^3} - 3\sqrt{50x}$

40. $3x\sqrt{9y} + 4\sqrt{25y}$

41. $\sqrt[3]{16x^4y} - 3x\sqrt[3]{2xy} + 5\sqrt[3]{-2xy^4}$

42. $8xy - \sqrt{25x^2y^2} + \sqrt[3]{8x^3y^3}$

In Problems 43–54, rationalize the denominator of each expression. Assume that all variables are positive when they appear.

43. $\dfrac{1}{\sqrt{2}}$

44. $\dfrac{2}{\sqrt{3}}$

45. $\dfrac{-\sqrt{3}}{\sqrt{5}}$

46. $\dfrac{-\sqrt{3}}{\sqrt{8}}$

47. $\dfrac{\sqrt{3}}{5 - \sqrt{2}}$

48. $\dfrac{\sqrt{2}}{\sqrt{7} + 2}$

49. $\dfrac{2 - \sqrt{5}}{2 + 3\sqrt{5}}$

50. $\dfrac{\sqrt{3} - 1}{2\sqrt{3} + 3}$

51. $\dfrac{5}{\sqrt[3]{2}}$ **52.** $\dfrac{-2}{\sqrt[3]{9}}$ **53.** $\dfrac{\sqrt{x+h}-\sqrt{x}}{\sqrt{x+h}+\sqrt{x}}$ **54.** $\dfrac{\sqrt{x+h}+\sqrt{x-h}}{\sqrt{x+h}-\sqrt{x-h}}$

In Problems 55–66, simplify each expression.

55. $8^{2/3}$ **56.** $4^{3/2}$ **57.** $(-27)^{1/3}$ **58.** $16^{3/4}$ **59.** $16^{3/2}$ **60.** $25^{3/2}$

61. $9^{-3/2}$ **62.** $16^{-3/2}$ **63.** $\left(\dfrac{9}{8}\right)^{3/2}$ **64.** $\left(\dfrac{27}{8}\right)^{2/3}$ **65.** $\left(\dfrac{8}{9}\right)^{-3/2}$ **66.** $\left(\dfrac{8}{27}\right)^{-2/3}$

In Problems 67–74, simplify each expression. Express your answer so that only positive exponents occur. Assume that the variables are positive.

67. $x^{3/4}x^{1/3}x^{-1/2}$ **68.** $x^{2/3}x^{1/2}x^{-1/4}$ **69.** $(x^3y^6)^{1/3}$ **70.** $(x^4y^8)^{3/4}$

71. $\dfrac{(x^2y)^{1/3}(xy^2)^{2/3}}{x^{2/3}y^{2/3}}$ **72.** $\dfrac{(xy)^{1/4}(x^2y^2)^{1/2}}{(x^2y)^{3/4}}$ **73.** $\dfrac{(16x^2y^{-1/3})^{3/4}}{(xy^2)^{1/4}}$ **74.** $\dfrac{(4x^{-1}y^{1/3})^{3/2}}{(xy)^{3/2}}$

Applications and Extensions

In Problems 75–88, expressions that occur in calculus are given. Write each expression as a single quotient in which only positive exponents and/or radicals appear.

75. $\dfrac{x}{(1+x)^{1/2}}+2(1+x)^{1/2}$ $x>-1$

76. $\dfrac{1+x}{2x^{1/2}}+x^{1/2}$ $x>0$

77. $2x(x^2+1)^{1/2}+x^2\cdot\dfrac{1}{2}(x^2+1)^{-1/2}\cdot 2x$

78. $(x+1)^{1/3}+x\cdot\dfrac{1}{3}(x+1)^{-2/3}$ $x\neq -1$

79. $\sqrt{4x+3}\cdot\dfrac{1}{2\sqrt{x-5}}+\sqrt{x-5}\cdot\dfrac{1}{5\sqrt{4x+3}}$ $x>5$

80. $\dfrac{\sqrt[3]{8x+1}}{3\sqrt[3]{(x-2)^2}}+\dfrac{\sqrt[3]{x-2}}{24\sqrt[3]{(8x+1)^2}}$ $x\neq 2, x\neq -\dfrac{1}{8}$

81. $\dfrac{\sqrt{1+x}-x\cdot\dfrac{1}{2\sqrt{1+x}}}{1+x}$ $x>-1$

82. $\dfrac{\sqrt{x^2+1}-x\cdot\dfrac{2x}{2\sqrt{x^2+1}}}{x^2+1}$

83. $\dfrac{(x+4)^{1/2}-2x(x+4)^{-1/2}}{x+4}$ $x>-4$

84. $\dfrac{(9-x^2)^{1/2}+x^2(9-x^2)^{-1/2}}{9-x^2}$ $-3<x<3$

85. $\dfrac{\dfrac{x^2}{(x^2-1)^{1/2}}-(x^2-1)^{1/2}}{x^2}$ $x<-1$ or $x>1$

86. $\dfrac{(x^2+4)^{1/2}-x^2(x^2+4)^{-1/2}}{x^2+4}$

87. $\dfrac{\dfrac{1+x^2}{2\sqrt{x}}-2x\sqrt{x}}{(1+x^2)^2}$ $x>0$

88. $\dfrac{2x(1-x^2)^{1/3}+\dfrac{2}{3}x^3(1-x^2)^{-2/3}}{(1-x^2)^{2/3}}$ $x\neq -1, x\neq 1$

In Problems 89–98, expressions that occur in calculus are given. Factor each expression. Express your answer so that only positive exponents occur.

89. $(x+1)^{3/2}+x\cdot\dfrac{3}{2}(x+1)^{1/2}$ $x\geq -1$

90. $(x^2+4)^{4/3}+x\cdot\dfrac{4}{3}(x^2+4)^{1/3}\cdot 2x$

91. $6x^{1/2}(x^2 + x) - 8x^{3/2} - 8x^{1/2}$ $x \geq 0$

92. $6x^{1/2}(2x + 3) + x^{3/2} \cdot 8$ $x \geq 0$

93. $3(x^2 + 4)^{4/3} + x \cdot 4(x^2 + 4)^{1/3} \cdot 2x$

94. $2x(3x + 4)^{4/3} + x^2 \cdot 4(3x + 4)^{1/3}$

95. $4(3x + 5)^{1/3}(2x + 3)^{3/2} + 3(3x + 5)^{4/3}(2x + 3)^{1/2}$ $x \geq -\dfrac{3}{2}$

96. $6(6x + 1)^{1/3}(4x - 3)^{3/2} + 6(6x + 1)^{4/3}(4x - 3)^{1/2}$ $x \geq \dfrac{3}{4}$

97. $3x^{-1/2} + \dfrac{3}{2}x^{1/2}$ $x > 0$

98. $8x^{1/3} - 4x^{-2/3}$ $x \neq 0$

In Problems 99–106, use a calculator to approximate each radical. Round your answer to two decimal places.

99. $\sqrt{2}$ **100.** $\sqrt{7}$ **101.** $\sqrt[3]{4}$ **102.** $\sqrt[3]{-5}$

103. $\dfrac{2 + \sqrt{3}}{3 - \sqrt{5}}$ **104.** $\dfrac{\sqrt{5} - 2}{\sqrt{2} + 4}$ **105.** $\dfrac{3\sqrt[3]{5} - \sqrt{2}}{\sqrt{3}}$ **106.** $\dfrac{2\sqrt{3} - \sqrt[3]{4}}{\sqrt{2}}$

Applications and Extensions

107. Calculating the Amount of Gasoline in a Tank A Shell station stores its gasoline in underground tanks that are right circular cylinders lying on their sides. See the illustration. The volume *V* of gasoline in the tank (in gallons) is given by the formula

$$V = 40h^2\sqrt{\dfrac{96}{h} - 0.608}$$

where *h* is the height of the gasoline (in inches) as measured on a depth stick.
(a) If $h = 12$ inches, how many gallons of gasoline are in the tank?
(b) If $h = 1$ inch, how many gallons of gasoline are in the tank?

108. Inclined Planes The final velocity *v* of an object in feet per second (ft/sec) after it slides down a frictionless inclined plane of height *h* feet is

$$v = \sqrt{64h + v_0^2}$$

where v_0 is the initial velocity (in ft/sec) of the object.
(a) What is the final velocity *v* of an object that slides down a frictionless inclined plane of height 4 feet? Assume that the initial velocity is 0.
(b) What is the final velocity *v* of an object that slides down a frictionless inclined plane of height 16 feet? Assume that the initial velocity is 0.
(c) What is the final velocity *v* of an object that slides down a frictionless inclined plane of height 2 feet with an initial velocity of 4 ft/sec?

Problems 109–112 require the following information.

Period of a Pendulum *The period T, in seconds, of a pendulum of length l, in feet, may be approximated using the formula*

$$T = 2\pi\sqrt{\dfrac{l}{32}}$$

In Problems 109–112, express your answer both as a square root and as a decimal.
109. Find the period *T* of a pendulum whose length is 64 feet.
110. Find the period *T* of a pendulum whose length is 16 feet.
111. Find the period *T* of a pendulum whose length is 8 inches.
112. Find the period *T* of a pendulum whose length is 4 inches.

Discussion and Writing

113. Give an example to show that $\sqrt{a^2}$ is not equal to *a*. Use it to explain why $\sqrt{a^2} = |a|$.

'Are you Prepared?' Answers

 1. $9; -9$ **2.** $4; 4$

CHAPTER REVIEW

Things to Know

Set (pp. 2–3) Well-defined collection of distinct objects, called elements

Empty set or null set	$\varnothing$	Set that has no elements
Equality	$A = B$	A and B have the same elements.
Subset	$A \subseteq B$	Each element of A is also an element of B.
Intersection	$A \cap B$	Set consisting of elements that belong to both A and B
Union	$A \cup B$	Set consisting of elements that belong to either A or B, or both
Universal set	U	Set consisting of all the elements that we wish to consider
Complement	$\overline{A}$	Set consisting of elements of the universal set that are not in A

Classify numbers (pp. 4–5)

Counting numbers	$1, 2, 3 \ldots$
Whole numbers	$0, 1, 2, 3 \ldots$
Integers	$\ldots, -2, -1, 0, 1, 2, \ldots$
Rational numbers	Quotients of two integers (denominator not equal to 0); terminating or repeating decimals
Irrational numbers	Nonrepeating nonterminating decimals
Real numbers	Rational or irrational numbers

Work with properties of real numbers (pp. 9–14)

Commutative properties	$a + b = b + a, \quad a \cdot b = b \cdot a$
Associative properties	$a + (b + c) = (a + b) + c, \quad a \cdot (b \cdot c) = (a \cdot b) \cdot c$
Distributive property	$a \cdot (b + c) = a \cdot b + a \cdot c$
Identity properties	$a + 0 = a \quad a \cdot 1 = a$
Inverse properties	$a + (-a) = 0 \quad a \cdot \dfrac{1}{a} = 1 \quad$ where $a \neq 0$
Zero-Product Property	If $ab = 0$, then $a = 0$ or $b = 0$ or both.

Absolute value (p. 19) $|a| = a$ if $a \geq 0 \quad |a| = -a$ if $a < 0$

Exponents (p. 22)

$a^n = \underbrace{a \cdot a \cdot \ldots \cdot a}_{n \text{ factors}} \quad n$ a positive integer $\quad\quad a^0 = 1 \quad a \neq 0 \quad\quad a^{-n} = \dfrac{1}{a^n} \quad a \neq 0, \quad n$ a positive integer

Laws of exponents (p. 22)

$a^m \cdot a^n = a^{m+n} \quad\quad (a^m)^n = a^{mn} \quad\quad (a \cdot b)^n = a^n \cdot b^n \quad\quad \dfrac{a^m}{a^n} = a^{m-n} = \dfrac{1}{a^{n-m}} \quad\quad \left(\dfrac{a}{b}\right)^n = \dfrac{a^n}{b^n}$

Pythagorean Theorem (p. 30)

In a right triangle, the square of the length of the hypotenuse is equal to the sum of the squares of the lengths of the legs.

Converse of the Pythagorean Theorem (p. 30)

In a triangle, if the square of the length of one side equals the sum of the squares of the lengths of the other two sides, then the triangle is a right triangle.

Polynomial (p. 40)

Algebraic expression of the form $a_n x^n + a_{n-1} x^{n-1} + \cdots + a_1 x + a_0$, n a nonnegative integer

Special products/factoring formulas (pp. 43–44)

$(x - a)(x + a) = x^2 - a^2$

$(x + a)^2 = x^2 + 2ax + a^2 \quad\quad (x - a)^2 = x^2 - 2ax + a^2$

$(x + a)(x + b) = x^2 + (a + b)x + ab$

$(ax + b)(cx + d) = acx^2 + (ad + bc)x + bd$

$(x + a)^3 = x^3 + 3ax^2 + 3a^2 x + a^3 \quad\quad (x - a)^3 = x^3 - 3ax^2 + 3a^2 x - a^3$

$(x - a)(x^2 + ax + a^2) = x^3 - a^3 \quad\quad (x + a)(x^2 - ax + a^2) = x^3 + a^3$

Radicals; Rational exponents (pp. 72 and 75)

$\sqrt[n]{a} = b$ means $a = b^n$ where $a \geq 0$ and $b \geq 0$ if $n \geq 2$ is even and a, b are any real numbers if $n \geq 3$ is odd.

$a^{m/n} = \sqrt[n]{a^m} = (\sqrt[n]{a})^m$ $m, n \geq 2$ m/n in lowest terms

Objectives

Section		You should be able to . . .	Review Exercises
R.1	1	Work with sets (p. 2)	1–8
	2	Classify numbers (p. 4)	9, 10
	3	Evaluate numerical expressions (p. 8)	11–14, 109
	4	Work with properties of real numbers (p. 9)	17, 18
R.2	1	Graph inequalities (p. 18)	19, 20
	2	Find distance on the real number line (p. 19)	21, 22
	3	Evaluate algebraic expressions (p. 20)	23–27, 108
	4	Determine the domain of a variable (p. 21)	29, 30
	5	Use the Laws of Exponents (p. 21)	83, 84
	6	Evaluate square roots (p. 23)	15, 27
	7	Use a calculator to evaluate exponents (p. 24)	33
	8	Use scientific notation (p. 24)	36, 105
R.3	1	Use the Pythagorean Theorem and its converse (p. 30)	106, 107, 112
	2	Know geometry formulas (p. 31)	110, 111
	3	Understand congruent triangles and similar triangles (p. 32)	31, 32
R.4	1	Recognize monomials (p. 39)	34
	2	Recognize polynomials (p. 40)	35
	3	Add and subtract polynomials (p. 41)	37, 38
	4	Multiply polynomials (p. 42)	39–44
	5	Know formulas for special products (p. 43)	41, 42
	6	Divide polynomials using long division (p. 44)	45–50
	7	Work with polynomials in two variables (p. 47)	39–40
R.5	1	Factor the difference of two squares and the sum and difference of two cubes (p. 50)	57, 58, 61, 62
	2	Factor perfect squares (p. 51)	65, 66
	3	Factor a second-degree polynomial: $x^2 + Bx + C$ (p. 52)	51, 52, 64
	4	Factor by grouping (p. 53)	59, 60
	5	Factor a second-degree polynomial: $Ax^2 + Bx + C, A \neq 1$ (p. 54)	53–56, 63
R.6	1	Divide polynomials using synthetic division (p. 57)	45, 46, 49, 50
R.7	1	Reduce a rational expression to lowest terms (p. 61)	67, 68
	2	Multiply and divide rational expressions (p. 62)	69, 70, 75
	3	Add and subtract rational expressions (p. 63)	71–74
	4	Use the least common multiple method (p. 65)	71–74
	5	Simplify complex rational expressions (p. 67)	76
R.8	1	Work with nth roots (p. 72)	28, 87–90
	2	Simplify radicals (p. 73)	77–82, 87–90, 101, 102
	3	Rationalize denominators (p. 74)	91–96
	4	Simplify expressions with rational exponents (p. 75)	85, 86, 97–100, 103, 104

Review Exercises

In Problems 1–8, use U = universal set = $\{1, 2, 3, 4, 5, 6, 7, 8, 9\}$, $A = \{1, 3, 5, 7\}$, $B = \{3, 5, 6, 7, 8\}$, *and* $C = \{2, 3, 7, 8, 9\}$ *to find each set.*

1. $A \cup B$
2. $B \cup C$
3. $A \cap C$
4. $A \cap B$

5. $\overline{A}$
6. $\overline{B}$
7. $\overline{B \cap C}$
8. $\overline{A \cup B}$

In Problems 9 and 10, list the numbers in the set that are (a) Natural numbers, (b) Integers, (c) Rational numbers, (d) Irrational numbers, (e) Real numbers.

9. $A = \left\{-10, 0.65, 1.343434\ldots, \sqrt{7}, \dfrac{1}{9}\right\}$

10. $B = \left\{0, -5, \dfrac{1}{3}, 0.59, 1.333\ldots, 2\sqrt{2}, \dfrac{\pi}{2}\right\}$

In Problems 11–16, evaluate each expression.

11. $-6 + 4 \cdot (8 - 3)$

12. $\dfrac{5}{18} + \dfrac{1}{12}$

13. $\dfrac{4}{3} \cdot \dfrac{9}{16}$

14. $\dfrac{\frac{5}{18}}{\frac{11}{27}}$

15. $\sqrt{(-3)^2}$

16. $(-3)^5$

In Problems 17 and 18, use the Distributive Property to remove the parentheses.

17. $4(x - 3)$

18. $(x - 2)(3x + 1)$

In Problems 19 and 20, graph the numbers x on the real number line.

19. $x > 3$

20. $x \le 5$

In Problems 21 and 22, let P, Q, and R be points on a real number line with coordinates −2, 3, and 9, respectively.

21. Find the distance between P and Q.

22. Find the distance between Q and R.

In Problems 23–28, evaluate each expression if $x = -5$ and $y = 7$.

23. $\dfrac{4x}{x + y}$

24. $|2x - 3y|$

25. $5x^{-1}y^2$

26. $2x^2y^{-2}$

27. $\sqrt{x^2}$

28. $\sqrt[3]{x^3}$

In Problems 29 and 30, determine the domain of the variable.

29. $\dfrac{3}{x - 6}$

30. $\dfrac{x + 1}{x + 5}$

In Problems 31 and 32, the triangles are similar. Find the missing length x and the missing angles A, B, and C.

31.

32.

33. Use a calculator to evaluate $(1.5)^4$.

34. Identify the coefficient and degree of the monomial $-5x^3$.

35. Identify the coefficients and degree of the polynomial $3x^5 + 4x^4 - 2x^3 + 5x - 12$.

36. Write 3.275×10^5 as an integer.

In Problems 37–44, perform the indicated operation.

37. $(2x^4 - 8x^3 + 5x - 1) + (6x^3 + x^2 + 4)$

38. $(x^3 + 8x^2 - 3x + 4) - (4x^3 - 7x^2 - 2x + 3)$

39. $(2x + y)(3x - 5y)$

40. $(2x - 5y)(3x + 2y)$

41. $(4x + 1)(4x - 1)$

42. $(5x + 2)^2$

43. $(x + 1)(x + 2)(x - 3)$

44. $(x + 1)(x + 3)(x - 5)$

In Problems 45–50, find the quotient and remainder. Check your work by verifying that

$$(Quotient)(Divisor) + Remainder = Dividend$$

45. $3x^3 - x^2 + x + 4$ divided by $x - 3$

46. $2x^3 - 3x^2 + x + 1$ divided by $x - 2$

47. $-3x^4 + x^2 + 2$ divided by $x^2 + 1$

48. $-4x^3 + x^2 - 2$ divided by $x^2 - 1$

49. $x^5 + 1$ divided by $x + 1$

50. $x^5 - 1$ divided by $x - 1$

In Problems 51–66, factor each polynomial completely (over the integers). If the polynomial cannot be factored, say it is prime.

51. $x^2 + 5x - 14$

52. $x^2 - 9x + 14$

53. $6x^2 - 5x - 6$

54. $6x^2 + x - 2$

55. $3x^2 - 15x - 42$

56. $2x^3 + 18x^2 + 28x$

57. $8x^3 + 1$

58. $27x^3 - 8$

59. $2x^3 + 3x^2 - 2x - 3$

60. $2x^3 + 3x^2 + 2x + 3$

61. $25x^2 - 4$

62. $16x^2 - 1$

63. $9x^2 + 1$

64. $x^2 - x + 1$

65. $x^2 + 8x + 16$

66. $4x^2 + 12x + 9$

In Problems 67–76, perform the indicated operation and simplify. Leave your answer in factored form.

67. $\dfrac{2x^2 + 11x + 14}{x^2 - 4}$

68. $\dfrac{x^2 - 5x - 14}{4 - x^2}$

69. $\dfrac{9x^2 - 1}{x^2 - 9} \cdot \dfrac{3x - 9}{9x^2 + 6x + 1}$

70. $\dfrac{x^2 - 25}{x^3 - 4x^2 - 5x} \cdot \dfrac{x^2 + x}{1 - x^2}$

71. $\dfrac{x + 1}{x - 1} - \dfrac{x - 1}{x + 1}$

72. $\dfrac{x}{x + 1} - \dfrac{2x}{x + 2}$

73. $\dfrac{3x + 4}{x^2 - 4} - \dfrac{2x - 3}{x^2 + 4x + 4}$

74. $\dfrac{x^2}{2x^2 + 5x - 3} + \dfrac{x^2}{2x^2 - 5x + 2}$

75. $\dfrac{\dfrac{x^2 - 1}{x^2 - 5x + 6}}{\dfrac{x + 1}{x - 2}}$

76. $\dfrac{\dfrac{x + 4}{3}}{\dfrac{1}{4} + \dfrac{2}{x}}$

In Problems 77–86, simplify each expression. All exponents must be positive. Assume that $x > 0$ and $y > 0$, when they appear.

77. $\sqrt{32}$

78. $\sqrt{75}$

79. $\sqrt[3]{-16}$

80. $\sqrt[5]{64}$

81. $5\sqrt{8} - 2\sqrt{32}$

82. $4\sqrt{12} + 5\sqrt{27}$

83. $\dfrac{(x^2 y)^{-4}}{(xy)^{-3}}$

84. $\left(\dfrac{x^2 y^2}{x^{-1}}\right)^2$

85. $(25x^{-4/3} y^{-2/3})^{3/2}$

86. $(27x^{-3/2} y^{5/2})^{2/3}$

In Problems 87–90, simplify each expression.

87. $\sqrt{\dfrac{9x^2}{25y^4}}$ $x \geq 0, y > 0$

88. $\dfrac{\sqrt{2x}\,\sqrt{5xy^3}}{\sqrt{10y}}$ $x \geq 0, y > 0$

89. $\sqrt[3]{27x^4 y^{12}}$

90. $\dfrac{\sqrt[4]{243x^5 y}}{\sqrt[4]{3xy^9}}$ $x > 0, y > 0$

In Problems 91–96, rationalize the denominator of each expression.

91. $\dfrac{4}{\sqrt{5}}$

92. $\dfrac{-2}{\sqrt{3}}$

93. $\dfrac{2}{1 - \sqrt{2}}$

94. $\dfrac{-4}{1 + \sqrt{3}}$

95. $\dfrac{1 + \sqrt{5}}{1 - \sqrt{5}}$

96. $\dfrac{4\sqrt{3} + 2}{2\sqrt{3} + 1}$

In Problems 97–102, write each expression as a single quotient in which only positive exponents and/or radicals appear.

97. $(2 + x^2)^{1/2} + x \cdot \dfrac{1}{2}(2 + x^2)^{-1/2} \cdot 2x$

98. $(x^2 + 4)^{2/3} + x \cdot \dfrac{2}{3}(x^2 + 4)^{-1/3} \cdot 2x$

99. $\dfrac{(x + 4)^{1/2} \cdot 2x - x^2 \cdot \dfrac{1}{2}(x + 4)^{-1/2}}{x + 4}$ $x > -4$

100. $\dfrac{(x^2 + 4)^{1/2} \cdot 2x - x^2 \cdot \dfrac{1}{2}(x^2 + 4)^{-1/2} \cdot 2x}{x^2 + 4}$

101. $\dfrac{\dfrac{x^2}{\sqrt{x^2 - 1}} - \sqrt{x^2 - 1}}{x^2}$

102. $\dfrac{\dfrac{4 + x^2}{2\sqrt{x}} - 2x\sqrt{x}}{(4 + x^2)^2}$

In Problems 103 and 104, factor each expression.

103. $3(x^2 + 4)^{4/3} + x \cdot 4(x^2 + 4)^{1/3} \cdot 2x$

104. $2x(3x + 5)^{5/3} + x^2 \cdot 4(3x + 5)^{2/3}$

105. U.S. Population According to the U.S. Census Bureau, the U.S. population in 2000 was 281,421,906. Write the U.S. population in scientific notation.

106. Find the hypotenuse of a right triangle whose legs are of lengths 5 and 8.

107. The lengths of the sides of a triangle are 12, 16, and 20. Is this a right triangle?

108. Manufacturing Cost The weekly production cost C of manufacturing x calculators is given by the formula

$$C = 3000 + 6x - \dfrac{x^2}{1000}$$

(a) What is the cost of producing 1000 calculators?
(b) What is the cost of producing 3000 calculators?

109. Quarterly Corporate Earnings In the first quarter of its fiscal year, a company posted earnings of $1.20 per share. During the second and third quarters, it posted losses of $0.75 per share and $0.30 per share, respectively. In the fourth quarter, it earned a modest $0.20 per share. What were the annual earnings per share of this company?

110. Design A window consists of a rectangle surmounted by a triangle. Find the area of the window shown in the illustration. How much wood frame is needed to enclose the window?

4 ft 4 ft

6 ft

5 ft

111. Construction A statue with a circular base of radius 3 feet is enclosed by a circular pond as shown in the illustration. What is the surface area of the pond? How much fence is required to enclose the pond?

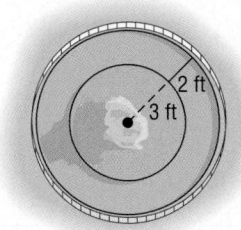

2 ft
3 ft

112. How far can a pilot see? On a recent flight to San Francisco, the pilot announced that we were 139 miles from the city, flying at an altitude of 35,000 feet. The pilot claimed that he could see the Golden Gate Bridge and beyond. Was he telling the truth? How far could he see?

113. Use the material in this chapter to create a problem that uses each of the following words:
(a) Simplify (b) Factor (c) Reduce

CHAPTER TEST

1. List the numbers in the set that are (a) Natural numbers, (b) Integers, (c) Rational numbers, (d) Irrational numbers, (e) Real numbers.

$$\left\{0, 1.2, \sqrt{2}, 7, \frac{1}{2}, \pi\right\}$$

2. Evaluate each expression if $x = -3$ and $y = 4$.
(a) $3x^{-1}y^2$ (b) $|2x - 3y|$
(c) $\sqrt{x^2 + y^2}$ (d) $5x^3 - 3x^2 + 2x - 6$

3. The triangles below are similar. Find the missing length x and the missing angles A, B, and C.

2 95° 3
40° 45°

B
1 x
A C

4. Perform the indicated operation.
(a) $(-2x^3 + 4x^2 - 6x + 10) - (6x^3 - 7x^2 + 8x - 1)$
(b) $(2x - 3)(-5x + 2)$

5. Factor each polynomial.
(a) $x^2 - 6x + 8$
(b) $4x^2 - 25$
(c) $6x^2 - 19x - 7$

6. Find the quotient and remainder if $x^3 - 3x^2 + 8x - 10$ is divided by $x - 2$.

7. Simplify each expression.
(a) $\sqrt{27}$ (b) $\sqrt[3]{-8}$ (c) $\dfrac{(xy^2)^{-3}}{(x^2y)^{-2}}$ (d) $(16x^{4/3}y^{-2/3})^{3/2}$

8. Rationalize the denominator of $\dfrac{3}{5 + \sqrt{3}}$.

9. Graph the numbers x on the real number line: $x > -3$.

10. Fill in the blanks The Zero-Product Property states that if $a \cdot b = 0$ then either ____ or ____, or both.

11. Construction A rectangular swimming pool, 20 feet long and 10 feet wide, is enclosed by a wooden deck that is 3 feet wide. What is the area of the deck? How much fence is required to enclose the deck?

20 ft 3 ft
10 ft
3 ft

Equations and Inequalities

1

A Look Ahead

Chapter 1, Equations and Inequalities, reviews many topics covered in Intermediate Algebra. If your instructor decides to exclude complex numbers from the course, don't be alarmed. The book has been designed so that the topic of complex numbers can be included or excluded without any confusion later on.

Outline

1.1 Linear Equations

PREPARING FOR THIS SECTION *Before getting started, review the following:*

- Properties of Real Numbers (Section R.1, pp. 9–14)
- Domain of a Variable (Section R.2, p. 21)

Now Work the 'Are You Prepared?' problems on page 94.

OBJECTIVES 1 Solve a Linear Equation (p. 88)
 2 Solve Equations That Lead to Linear Equations (p. 90)
 3 Solve Applied Problems Involving Linear Equations (p. 92)

An **equation in one variable** is a statement in which two expressions, at least one containing the variable, are equal. The expressions are called the **sides** of the equation. Since an equation is a statement, it may be true or false, depending on the value of the variable. Unless otherwise restricted, the admissible values of the variable are those in the domain of the variable. These admissible values of the variable, if any, that result in a true statement are called **solutions,** or **roots,** of the equation. To **solve an equation** means to find all the solutions of the equation.

For example, the following are all equations in one variable, x:

$$x + 5 = 9 \qquad x^2 + 5x = 2x - 2 \qquad \frac{x^2 - 4}{x + 1} = 0 \qquad \sqrt{x^2 + 9} = 5$$

The first of these statements, $x + 5 = 9$, is true when $x = 4$ and false for any other choice of x. That is, 4 is a solution of the equation $x + 5 = 9$. We also say that 4 **satisfies** the equation $x + 5 = 9$, because, when we substitute 4 for x, a true statement results.

Sometimes an equation will have more than one solution. For example, the equation

$$\frac{x^2 - 4}{x + 1} = 0$$

has $x = -2$ and $x = 2$ as solutions.

Usually, we will write the solution of an equation in set notation. This set is called the **solution set** of the equation. For example, the solution set of the equation $x^2 - 9 = 0$ is $\{-3, 3\}$.

Some equations have no real solution. For example, $x^2 + 9 = 5$ has no real solution, because there is no real number whose square when added to 9 equals 5.

An equation that is satisfied for every value of the variable for which both sides are defined is called an **identity.** For example, the equation

$$3x + 5 = x + 3 + 2x + 2$$

is an identity, because this statement is true for any real number x.

> One method for solving an equation is to replace the original equation by a succession of equivalent equations until an equation with an obvious solution is obtained.

For example, all the following equations are equivalent.

$$2x + 3 = 13$$
$$2x = 10$$
$$x = 5$$

We conclude that the solution set of the original equation is $\{5\}$.

How do we obtain equivalent equations? In general, there are five ways.

Procedures That Result in Equivalent Equations

1. Interchange the two sides of the equation:

$$\text{Replace} \quad 3 = x \quad \text{by} \quad x = 3$$

2. Simplify the sides of the equation by combining like terms, eliminating parentheses, and so on:

$$\text{Replace} \quad (x + 2) + 6 = 2x + (x + 1)$$
$$\text{by} \quad x + 8 = 3x + 1$$

3. Add or subtract the same expression on both sides of the equation:

$$\text{Replace} \quad 3x - 5 = 4$$
$$\text{by} \quad (3x - 5) + 5 = 4 + 5$$

4. Multiply or divide both sides of the equation by the same nonzero expression:

$$\text{Replace} \quad \frac{3x}{x - 1} = \frac{6}{x - 1} \quad x \neq 1$$
$$\text{by} \quad \frac{3x}{x - 1} \cdot (x - 1) = \frac{6}{x - 1} \cdot (x - 1)$$

5. If one side of the equation is 0 and the other side can be factored, then we may use the Zero-Product Property[*] and set each factor equal to 0:

$$\text{Replace} \quad x(x - 3) = 0$$
$$\text{by} \quad x = 0 \quad \text{or} \quad x - 3 = 0$$

WARNING Squaring both sides of an equation does not necessarily lead to an equivalent equation. ∎

Whenever it is possible to solve an equation in your head, do so. For example,

The solution of $2x = 8$ is $x = 4$.

The solution of $3x - 15 = 0$ is $x = 5$.

Now Work PROBLEM 9

Often, though, some rearrangement is necessary.

EXAMPLE 1	**Solving an Equation**

Solve the equation: $3x - 5 = 4$

Solution We replace the original equation by a succession of equivalent equations.

$$3x - 5 = 4$$
$$(3x - 5) + 5 = 4 + 5 \qquad \text{Add 5 to both sides.}$$
$$3x = 9 \qquad \text{Simplify.}$$
$$\frac{3x}{3} = \frac{9}{3} \qquad \text{Divide both sides by 3.}$$
$$x = 3 \qquad \text{Simplify.}$$

The last equation, $x = 3$, has the single solution 3. All these equations are equivalent, so 3 is the only solution of the original equation, $3x - 5 = 4$.

[*] The Zero-Product Property says that if $ab = 0$ then $a = 0$ or $b = 0$ or both equal 0.

✓ **Check:** It is a good practice to check the solution by substituting 3 for x in the original equation.

$$3x - 5 = 4$$
$$3(3) - 5 \overset{?}{=} 4$$
$$9 - 5 \overset{?}{=} 4$$
$$4 = 4$$

The solution checks. The solution set is $\{3\}$. ∎

━━━━ **Now Work** PROBLEM 23

Steps for Solving Equations

STEP 1: List any restrictions on the domain of the variable.
STEP 2: Simplify the equation by replacing the original equation by a succession of equivalent equations following the procedures listed earlier.
STEP 3: If the result of Step 2 is a product of factors equal to 0, use the Zero-Product Property and set each factor equal to 0 (procedure 5).
STEP 4: Check your solution(s).

1 Solve a Linear Equation

Linear equations are equations such as

$$3x + 12 = 0 \qquad -2x + 5 = 0 \qquad \frac{1}{2}x - \sqrt{3} = 0$$

A general definition is given next.

DEFINITION A **linear equation in one variable** is equivalent to an equation of the form

$$ax + b = 0$$

where a and b are real numbers and $a \neq 0$. ⌐

Sometimes, a linear equation is called a **first-degree equation,** because the left side is a polynomial in x of degree 1.

It is relatively easy to solve a linear equation. The idea is to *isolate* the variable:

$$ax + b = 0 \qquad a \neq 0$$
$$ax = -b \qquad \text{Subtract } b \text{ from both sides.}$$
$$x = \frac{-b}{a} \qquad \text{Divide both sides by } a, \, a \neq 0.$$

The linear equation $ax + b = 0$, $a \neq 0$, has the single solution given by the formula $x = -\dfrac{b}{a}$.

EXAMPLE 2	**Solving a Linear Equation**

Solve the equation: $\dfrac{1}{2}(x + 5) - 4 = \dfrac{1}{3}(2x - 1)$

Solution To clear the equation of fractions, we multiply both sides by 6, the least common multiple of the denominators of the fractions $\frac{1}{2}$ and $\frac{1}{3}$.

$$\frac{1}{2}(x + 5) - 4 = \frac{1}{3}(2x - 1)$$

$$6\left[\frac{1}{2}(x + 5) - 4\right] = 6\left[\frac{1}{3}(2x - 1)\right] \qquad \text{Multiply both sides by 6, the LCM of 2 and 3.}$$

$$3(x + 5) - 6 \cdot 4 = 2(2x - 1) \qquad \begin{array}{l}\text{Use the Distributive Property on the left}\\ \text{and the Associative Property on the right.}\end{array}$$

$$3x + 15 - 24 = 4x - 2 \qquad \text{Use the Distributive Property.}$$

$$3x - 9 = 4x - 2 \qquad \text{Combine like terms.}$$

$$3x - 9 + 9 = 4x - 2 + 9 \qquad \text{Add 9 to each side.}$$

$$3x = 4x + 7 \qquad \text{Simplify.}$$

$$3x - 4x = 4x + 7 - 4x \qquad \text{Subtract } 4x \text{ from each side.}$$

$$-x = 7 \qquad \text{Simplify.}$$

$$x = -7 \qquad \text{Multiply both sides by } -1.$$

✓ Check: $\frac{1}{2}(x + 5) - 4 = \frac{1}{2}(-7 + 5) - 4 = \frac{1}{2}(-2) - 4 = -1 - 4 = -5$

$\frac{1}{3}(2x - 1) = \frac{1}{3}[2(-7) - 1] = \frac{1}{3}(-14 - 1) = \frac{1}{3}(-15) = -5$

Since the two expressions are equal, the solution $x = -7$ checks and the solution set is $\{-7\}$.

━━━━ **Now Work** PROBLEM 33

EXAMPLE 3 **Solving a Linear Equation Using a Calculator**

Solve the equation: $2.78x + \dfrac{2}{17.931} = 54.06$

Round the answer to two decimal places.

Solution To avoid rounding errors, we solve for x before using the calculator.

$$2.78x + \frac{2}{17.931} = 54.06$$

$$2.78x = 54.06 - \frac{2}{17.931} \qquad \text{Subtract } \frac{2}{17.931} \text{ from each side.}$$

$$x = \frac{54.06 - \dfrac{2}{17.931}}{2.78} \qquad \text{Divide each side by 2.78.}$$

Now use your calculator. The solution, rounded to two decimal places, is 19.41.

✓ Check: We store the unrounded solution 19.40592134 in memory and proceed to evaluate $2.78x + \dfrac{2}{17.931}$.

$$(2.78)(19.40592134) + \frac{2}{17.931} = 54.06$$

━━━━ **Now Work** PROBLEM 65

2 Solve Equations That Lead to Linear Equations

The next three examples illustrate equations that lead to linear equations upon simplification.

EXAMPLE 4	Solving Equations

Solve the equation: $(2y + 1)(y - 1) = (y + 5)(2y - 5)$

Solution

$$(2y + 1)(y - 1) = (y + 5)(2y - 5)$$

$2y^2 - y - 1 = 2y^2 + 5y - 25$ Multiply and combine like terms.

$-y - 1 = 5y - 25$ Subtract $2y^2$ from each side.

$-y = 5y - 24$ Add 1 to each side.

$-6y = -24$ Subtract $5y$ from each side.

$y = 4$ Divide both sides by -6

✓ Check: $(2y + 1)(y - 1) = [2(4) + 1](4 - 1) = (8 + 1)(3) = (9)(3) = 27$

$(y + 5)(2y - 5) = (4 + 5)[2(4) - 5] = (9)(8 - 5) = (9)(3) = 27$

Since the two expressions are equal, the solution $y = 4$ checks.
The solution set is $\{4\}$.

EXAMPLE 5	Solving Equations

Solve the equation: $\dfrac{3}{x - 2} = \dfrac{1}{x - 1} + \dfrac{7}{(x - 1)(x - 2)}$

Solution

First, we note that the domain of the variable is $\{x \mid x \neq 1, x \neq 2\}$. We clear the equation of fractions by multiplying both sides by the least common multiple of the denominators of the three fractions, $(x - 1)(x - 2)$.

$$\frac{3}{x - 2} = \frac{1}{x - 1} + \frac{7}{(x - 1)(x - 2)}$$

$(x - 1)(x - 2)\dfrac{3}{x - 2} = (x - 1)(x - 2)\left[\dfrac{1}{x - 1} + \dfrac{7}{(x - 1)(x - 2)}\right]$ Multiply both sides by $(x - 1)(x - 2)$. Cancel on the left.

$3x - 3 = (x - 1)(x - 2)\dfrac{1}{x - 1} + (x - 1)(x - 2)\dfrac{7}{(x - 1)(x - 2)}$ Use the Distributive Property on each side; cancel on the right.

$3x - 3 = (x - 2) + 7$

$3x - 3 = x + 5$ Combine like terms.

$2x = 8$ Add 3 to each side. Subtract x from each side.

$x = 4$ Divide by 2.

✓ Check: $\dfrac{3}{x - 2} = \dfrac{3}{4 - 2} = \dfrac{3}{2}$

$\dfrac{1}{x - 1} + \dfrac{7}{(x - 1)(x - 2)} = \dfrac{1}{4 - 1} + \dfrac{7}{(4 - 1)(4 - 2)} = \dfrac{1}{3} + \dfrac{7}{3 \cdot 2} = \dfrac{2}{6} + \dfrac{7}{6} = \dfrac{9}{6} = \dfrac{3}{2}$

Since the two expressions are equal, the solution $x = 4$ checks.
The solution set is $\{4\}$.

Now Work PROBLEM 59

The next example is of an equation that has no solution.

| EXAMPLE 6 | An Equation with No Solution |

Solve the equation: $\dfrac{3x}{x-1} + 2 = \dfrac{3}{x-1}$

Solution First, we note that the domain of the variable is $\{x \mid x \neq 1\}$. Since the two quotients in the equation have the same denominator, $x - 1$, we can simplify by multiplying both sides by $x - 1$. The resulting equation is equivalent to the original equation, since we are multiplying by $x - 1$, which is not 0. (Remember, $x \neq 1$).

$$\frac{3x}{x-1} + 2 = \frac{3}{x-1}$$

$$\left(\frac{3x}{x-1} + 2\right) \cdot (x-1) = \frac{3}{x-1} \cdot (x-1) \qquad \text{Multiply both sides by } x-1; \text{ cancel on the right.}$$

$$\frac{3x}{x-1} \cdot (x-1) + 2 \cdot (x-1) = 3 \qquad \text{Use the Distributive Property on the left side; cancel on the left.}$$

$$3x + 2x - 2 = 3 \qquad \text{Simplify.}$$

$$5x - 2 = 3 \qquad \text{Combine like terms.}$$

$$5x = 5 \qquad \text{Add 2 to each side.}$$

$$x = 1 \qquad \text{Divide both sides by 5.}$$

The solution appears to be 1. But recall that $x = 1$ is not in the domain of the variable. The equation has no solution.

■

━━ **Now Work** PROBLEM 49

| EXAMPLE 7 | Converting to Fahrenheit from Celsius |

In the United States we measure temperature in both degrees Fahrenheit (°F) and degrees Celsius (°C), which are related by the formula $C = \dfrac{5}{9}(F - 32)$. What are the Fahrenheit temperatures corresponding to Celsius temperatures of 0°, 10°, 20°, and 30°C?

Solution We could solve four equations for F by replacing C each time by 0, 10, 20, and 30. Instead, it is much easier and faster first to solve the equation $C = \dfrac{5}{9}(F - 32)$ for F and then substitute in the values of C.

$$C = \frac{5}{9}(F - 32)$$

$$9C = 5(F - 32) \qquad \text{Multiply both sides by 9.}$$

$$9C = 5F - 160 \qquad \text{Use the Distributive Property.}$$

$$5F - 160 = 9C \qquad \text{Interchange sides.}$$

$$5F = 9C + 160 \qquad \text{Add 160 to each side.}$$

$$F = \frac{9}{5}C + 32 \qquad \text{Divide both sides by 5.}$$

We can now do the required arithmetic.

$$0°\text{C:} \quad F = \frac{9}{5}(0) + 32 = 32°\text{F}$$

$$10°\text{C:} \quad F = \frac{9}{5}(10) + 32 = 50°\text{F}$$

$$20°\text{C:} \quad F = \frac{9}{5}(20) + 32 = 68°\text{F}$$

$$30°\text{C:} \quad F = \frac{9}{5}(30) + 32 = 86°\text{F}$$

■

3 Solve Applied Problems Involving Linear Equations

Although each situation has its unique features, we can provide an outline of the steps to follow to solve applied problems.

Steps for Solving Applied Problems

STEP 1: Read the problem carefully, perhaps two or three times. Pay particular attention to the question being asked in order to identify what you are looking for. If you can, determine realistic possibilities for the answer.

STEP 2: Assign a letter (variable) to represent what you are looking for, and, if necessary, express any remaining unknown quantities in terms of this variable.

STEP 3: Make a list of all the known facts, and translate them into mathematical expressions. These may take the form of an equation (or, later, an inequality) involving the variable. If possible, draw an appropriately labeled diagram to assist you. Sometimes a table or chart helps.

STEP 4: Solve the equation for the variable, and then answer the question, usually using a complete sentence.

STEP 5: Check the answer with the facts in the problem. If it agrees, congratulations! If it does not agree, try again.

Let's look at two examples.

EXAMPLE 8 | **Investments**

A total of $18,000 is invested, some in stocks and some in bonds. If the amount invested in bonds is half that invested in stocks, how much is invested in each category?

Solution

STEP 1: We are being asked to find the amount of two investments. These amounts must total $18,000. (Do you see why?)

STEP 2: If we let x equal the amounts invested in stocks, then the rest of the money, $18,000 - x$, is the amount invested in bonds. Do you see why? Look at Step 3.

STEP 3: We set up a table:

Amount in Stocks	Amount in Bonds	Reason
x	$18,000 - x$	Total invested is $18,000

We also know that

Total amount invested in bonds	is	one-half that in stocks
$18,000 - x$	$=$	$\frac{1}{2}(x)$

STEP 4:
$$18,000 - x = \frac{1}{2}x$$

$$18,000 = x + \frac{1}{2}x \qquad \text{Add x to both sides.}$$

$$18,000 = \frac{3}{2}x \qquad \text{Simplify.}$$

$$\left(\frac{2}{3}\right)18,000 = \left(\frac{2}{3}\right)\left(\frac{3}{2}x\right) \qquad \text{Multiply both sides by } \frac{2}{3}.$$

$$12,000 = x \qquad \text{Simplify.}$$

So, $12,000 is invested in stocks and $18,000 - $12,000 = $6000 is invested in bonds.

STEP 5: The total invested is $12,000 + $6000 = $18,000, and the amount in bonds, $6000, is half that in stocks, $12,000.

━━━━━━━ **Now Work** PROBLEM 83

EXAMPLE 9 **Determining an Hourly Wage**

Shannon grossed $435 one week by working 52 hours. Her employer pays time-and-a-half for all hours worked in excess of 40 hours. With this information, can you determine Shannon's regular hourly wage?

Solution STEP 1: We are looking for an hourly wage. Our answer will be in dollars per hour.

STEP 2: Let x represent the regular hourly wage; x is measured in dollars per hour.

STEP 3: We set up a table:

	Hours Worked	Hourly Wage	Salary
Regular	40	x	$40x$
Overtime	12	$1.5x$	$12(1.5x) = 18x$

The sum of regular salary plus overtime salary will equal $435. From the table, $40x + 18x = 435$.

STEP 4:
$$40x + 18x = 435$$
$$58x = 435$$
$$x = 7.50$$

Shannon's regular hourly wage is $7.50 per hour.

STEP 5: Forty hours yields a salary of $40(7.50) = 300, and 12 hours of overtime yields a salary of $12(1.5)(7.50) = 135, for a total of $435.

━━━━━━━ **Now Work** PROBLEM 87

SUMMARY Steps for Solving a Linear Equation

To solve a linear equation, follow these steps:

STEP 1: List any restrictions on the variable.

STEP 2: If necessary, clear the equation of fractions by multiplying both sides by the least common multiple (LCM) of the denominators of all the fractions.

STEP 3: Remove all parentheses and simplify.

STEP 4: Collect all terms containing the variable on one side and all remaining terms on the other side.

STEP 5: Simplify and solve.

STEP 6: Check your solution(s).

Historical Feature

Solving equations is among the oldest of mathematical activities, and efforts to systematize this activity determined much of the shape of modern mathematics.

Consider the following problem and its solution using only words: Solve the problem of how many apples Jim has, given that

"Bob's five apples and Jim's apples together make twelve apples" by thinking,

"Jim's apples are all twelve apples less Bob's five apples" and then concluding,

"Jim has seven apples."

The mental steps translated into algebra are

$$5 + x = 12$$
$$x = 12 - 5$$
$$= 7$$

The solution of this problem using only words is the earliest form of algebra. Such problems were solved exactly this way in Babylonia in 1800 BC. We know almost nothing of mathematical work before this date, although most authorities believe the sophistication of the earliest known texts indicates that a long period of previous development must have occurred. The method of writing out equations in words persisted for thousands of years, and although it now seems extremely cumbersome, it was used very effectively by many generations of mathematicians. The Arabs developed a good deal of the theory of cubic equations while writing out all the equations in words. About AD 1500, the tendency to abbreviate words in the written equations began to lead in the direction of modern notation; for example, the Latin word *et* (meaning *and*) developed into the plus sign, $+$. Although the occasional use of letters to represent variables dates back to AD 1200, the practice did not become common until about AD 1600. Development thereafter was rapid, and by 1635 algebraic notation did not differ essentially from what we use now.

1.1 Assess Your Understanding

'Are You Prepared?' *Answers are given at the end of these exercises. If you get a wrong answer, read the pages listed in red.*

1. The fact that $2(x + 3) = 2x + 6$ is because of the _____ Property. (p. 10)

2. The fact that $3x = 0$ implies that $x = 0$ is a result of the _____ Property. (p. 13)

3. The domain of the variable in the expression $\dfrac{x}{x - 4}$ is _____. (p. 21)

Concepts and Vocabulary

4. *True or False* Multiplying both sides of an equation by any number results in an equivalent equation.

5. An equation that is satisfied for every value of the variable for which both sides are defined is called a(n) _____.

6. An equation of the form $ax + b = 0$ is called a(n) _____ equation or a(n) _____ equation.

7. *True or False* The solution of the equation $3x - 8 = 0$ is $\dfrac{3}{8}$.

8. *True or False* Some equations have no solution.

Skill Building

In Problems 9–16, mentally solve each equation.

9. $7x = 21$

10. $6x = -24$

11. $3x + 15 = 0$

12. $6x + 18 = 0$

13. $2x - 3 = 0$

14. $3x + 4 = 0$

15. $\dfrac{1}{3}x = \dfrac{5}{12}$

16. $\dfrac{2}{3}x = \dfrac{9}{2}$

In Problems 17–64, solve each equation.

17. $3x + 4 = x$

18. $2x + 9 = 5x$

19. $2t - 6 = 3 - t$

20. $5y + 6 = -18 - y$

21. $6 - x = 2x + 9$

22. $3 - 2x = 2 - x$

23. $3 + 2n = 4n + 7$

24. $6 - 2m = 3m + 1$

25. $2(3 + 2x) = 3(x - 4)$

26. $3(2 - x) = 2x - 1$

27. $8x - (3x + 2) = 3x - 10$

28. $7 - (2x - 1) = 10$

29. $\dfrac{3}{2}x + 2 = \dfrac{1}{2} - \dfrac{1}{2}x$

30. $\dfrac{1}{3}x = 2 - \dfrac{2}{3}x$

31. $\dfrac{1}{2}x - 5 = \dfrac{3}{4}x$

32. $1 - \dfrac{1}{2}x = 6$

33. $\dfrac{2}{3}p = \dfrac{1}{2}p + \dfrac{1}{3}$

34. $\dfrac{1}{2} - \dfrac{1}{3}p = \dfrac{4}{3}$

35. $0.9t = 0.4 + 0.1t$

36. $0.9t = 1 + t$

37. $\dfrac{x+1}{3} + \dfrac{x+2}{7} = 2$

38. $\dfrac{2x+1}{3} + 16 = 3x$

39. $\dfrac{2}{y} + \dfrac{4}{y} = 3$

40. $\dfrac{4}{y} - 5 = \dfrac{5}{2y}$

41. $\dfrac{1}{2} + \dfrac{2}{x} = \dfrac{3}{4}$

42. $\dfrac{3}{x} - \dfrac{1}{3} = \dfrac{1}{6}$

43. $(x+7)(x-1) = (x+1)^2$

44. $(x+2)(x-3) = (x+3)^2$

45. $x(2x-3) = (2x+1)(x-4)$

46. $x(1+2x) = (2x-1)(x-2)$

47. $z(z^2+1) = 3 + z^3$

48. $w(4 - w^2) = 8 - w^3$

49. $\dfrac{x}{x-2} + 3 = \dfrac{2}{x-2}$

50. $\dfrac{2x}{x+3} = \dfrac{-6}{x+3} - 2$

51. $\dfrac{2x}{x^2-4} = \dfrac{4}{x^2-4} - \dfrac{3}{x+2}$

52. $\dfrac{x}{x^2-9} + \dfrac{4}{x+3} = \dfrac{3}{x^2-9}$

53. $\dfrac{x}{x+2} = \dfrac{3}{2}$

54. $\dfrac{3x}{x-1} = 2$

55. $\dfrac{5}{2x-3} = \dfrac{3}{x+5}$

56. $\dfrac{-4}{x+4} = \dfrac{-3}{x+6}$

57. $\dfrac{6t+7}{4t-1} = \dfrac{3t+8}{2t-4}$

58. $\dfrac{8w+5}{10w-7} = \dfrac{4w-3}{5w+7}$

59. $\dfrac{4}{x-2} = \dfrac{-3}{x+5} + \dfrac{7}{(x+5)(x-2)}$

60. $\dfrac{-4}{2x+3} + \dfrac{1}{x-1} = \dfrac{1}{(2x+3)(x-1)}$

61. $\dfrac{2}{y+3} + \dfrac{3}{y-4} = \dfrac{5}{y+6}$

62. $\dfrac{5}{5z-11} + \dfrac{4}{2z-3} = \dfrac{-3}{5-z}$

63. $\dfrac{x}{x^2-1} - \dfrac{x+3}{x^2-x} = \dfrac{-3}{x^2+x}$

64. $\dfrac{x+1}{x^2+2x} - \dfrac{x+4}{x^2+x} = \dfrac{-3}{x^2+3x+2}$

In Problems 65–68, use a calculator to solve each equation. Round the solution to two decimal places.

65. $3.2x + \dfrac{21.3}{65.871} = 19.23$

66. $6.2x - \dfrac{19.1}{83.72} = 0.195$

67. $14.72 - 21.58x = \dfrac{18}{2.11}x + 2.4$

68. $18.63x - \dfrac{21.2}{2.6} = \dfrac{14}{2.32}x - 20$

Applications and Extensions

In Problems 69–74, solve each equation. The letters a, b, and c are constants.

69. $ax - b = c, \quad a \neq 0$

70. $1 - ax = b, \quad a \neq 0$

71. $\dfrac{x}{a} + \dfrac{x}{b} = c, \quad a \neq 0, b \neq 0, a \neq -b$

72. $\dfrac{a}{x} + \dfrac{b}{x} = c, \quad c \neq 0$

73. $\dfrac{1}{x-a} + \dfrac{1}{x+a} = \dfrac{2}{x-1}$

74. $\dfrac{b+c}{x+a} = \dfrac{b-c}{x-a}, \quad c \neq 0, a \neq 0$

75. Find the number a for which $x = 4$ is a solution of the equation

$$x + 2a = 16 + ax - 6a$$

76. Find the number b for which $x = 2$ is a solution of the equation

$$x + 2b = x - 4 + 2bx$$

Problems 77–82 list some formulas that occur in applications. Solve each formula for the indicated variable.

77. Electricity $\dfrac{1}{R} = \dfrac{1}{R_1} + \dfrac{1}{R_2}$ for R

78. Finance $A = P(1 + rt)$ for r

79. Mechanics $F = \dfrac{mv^2}{R}$ for R

80. Chemistry $PV = nRT$ for T

81. Mathematics $S = \dfrac{a}{1-r}$ for r

82. Mechanics $v = -gt + v_0$ for t

83. Finance A total of $20,000 is to be invested, some in bonds and some in certificates of deposit (CDs). If the amount invested in bonds is to exceed that in CDs by $3000, how much will be invested in each type of investment?

84. Finance A total of $10,000 is to be divided between Sean and George, with George to receive $3000 less than Sean. How much will each receive?

85. Internet Searches In November 2005, Google and Yahoo! search engines were used to conduct a total of 3.57 billion online searches. Google was used to conduct 0.53 billion more searches than Yahoo!. How many searches were conducted on each search engine?

Source: comScore Networks

86. Sharing the Cost of a Pizza Judy and Tom agree to share the cost of an $18 pizza based on how much each ate. If Tom ate $\frac{2}{3}$ the amount that Judy ate, how much should each pay? [**Hint:** Some pizza may be left.]

Tom's portion

Judy's portion

87. Computing Hourly Wages Sandra, who is paid time-and-a-half for hours worked in excess of 40 hours, had gross weekly wages of $442 for 48 hours worked. What is her regular hourly rate?

88. Computing Hourly Wages Leigh is paid time-and-a-half for hours worked in excess of 40 hours and double-time for hours worked on Sunday. If Leigh had gross weekly wages of $342 for working 50 hours, 4 of which were on Sunday, what is her regular hourly rate?

89. Computing Grades Going into the final exam, which will count as two tests, Brooke has test scores of 80, 83, 71, 61, and 95. What score does Brooke need on the final in order to have an average score of 80?

90. Computing Grades Going into the final exam, which will count as two-thirds of the final grade, Mike has test scores of 86, 80, 84, and 90. What score does Mike need on the final in order to earn a B, which requires an average score of 80? What does he need to earn an A, which requires an average of 90?

91. Business: Discount Pricing A builder of tract homes reduced the price of a model by 15%. If the new price is $425,000, what was its original price? How much can be saved by purchasing the model?

92. Business: Discount Pricing A car dealer, at a year-end clearance, reduces the list price of last year's models by 15%. If a certain four-door model has a discounted price of $8000, what was its list price? How much can be saved by purchasing last year's model?

93. Business: Marking up the Price of Books A college book store marks up the price that it pays the publisher for a book by 35%. If the selling price of a book is $92.00, how much did the bookstore pay for this book?

94. Personal Finance: Cost of a Car The suggested list price of a new car is $18,000. The dealer's cost is 85% of list. How much will you pay if the dealer is willing to accept $100 over cost for the car?

95. Business: Theater Attendance The manager of the Coral Theater wants to know whether the majority of its patrons are adults or children. One day in July, 5200 tickets were sold and the receipts totaled $29,961. The adult admission is $7.50, and the children's admission is $4.50. How many adult patrons were there?

96. Business: Discount Pricing A wool suit, discounted by 30% for a clearance sale, has a price tag of $399. What was the suit's original price?

97. Geometry The perimeter of a rectangle is 60 feet. Find its length and width if the length is 8 feet longer than the width.

98. Geometry The perimeter of a rectangle is 42 meters. Find its length and width if the length is twice the width.

99. Internet Users In March 2006, 152 million people in the United States were Internet users, which accounted for 21.9% of the world's online audience. How many people worldwide were Internet users in March 2006?
Source: comScore Networks

Discussion and Writing

100. One step in the following list contains an error. Identify it and explain what is wrong.

$$x = 2 \tag{1}$$
$$3x - 2x = 2 \tag{2}$$
$$3x = 2x + 2 \tag{3}$$
$$x^2 + 3x = x^2 + 2x + 2 \tag{4}$$
$$x^2 + 3x - 10 = x^2 + 2x - 8 \tag{5}$$
$$(x - 2)(x + 5) = (x - 2)(x + 4) \tag{6}$$
$$x + 5 = x + 4 \tag{7}$$
$$1 = 0 \tag{8}$$

101. The equation

$$\frac{5}{x + 3} + 3 = \frac{8 + x}{x + 3}$$

has no solution, yet when we go through the process of solving it we obtain $x = -3$. Write a brief paragraph to explain what causes this to happen.

102. Make up an equation that has no solution and give it to a fellow student to solve. Ask the fellow student to write a critique of your equation.

'Are You Prepared?' Answers

1. Distributive **2.** Zero-Product **3.** $\{x \mid x \neq 4\}$

1.2 Quadratic Equations

PREPARING FOR THIS SECTION *Before getting started, review the following:*

- Factoring Polynomials (Section R.5, pp. 49–55)
- Zero-Product Property (Section R.1, p. 13)
- Square Roots (Section R.2, pp. 23–24)

Now Work the 'Are You Prepared?' problems on page 106.

OBJECTIVES **1** Solve a Quadratic Equation by Factoring (p. 97)
2 Know How to Complete the Square (p. 99)
3 Solve a Quadratic Equation by Completing the Square (p. 100)
4 Solve a Quadratic Equation Using the Quadratic Formula (p. 102)
5 Solve Applied Problems Involving Quadratic Equations (p. 105)

Quadratic equations are equations such as

$$2x^2 + x + 8 = 0$$
$$3x^2 - 5x + 6 = 0$$
$$x^2 - 9 = 0$$

A general definition is given next.

DEFINITION A **quadratic equation** is an equation equivalent to one of the form

$$ax^2 + bx + c = 0 \qquad (1)$$

where a, b, and c are real numbers and $a \neq 0$.

A quadratic equation written in the form $ax^2 + bx + c = 0$ is said to be in **standard form.**

Sometimes, a quadratic equation is called a **second-degree equation,** because the left side is a polynomial of degree 2. We shall discuss three ways of solving quadratic equations: by factoring, by completing the square, and by using the quadratic formula.

1 Solve a Quadratic Equation by Factoring

When a quadratic equation is written in standard form $ax^2 + bx + c = 0$, it may be possible to factor the expression on the left side into the product of two first-degree polynomials. Then, by using the Zero-Product Property and setting each factor equal to 0, we can solve the resulting linear equations and obtain the solutions of the quadratic equation.

Let's look at an example.

EXAMPLE 1 **Solving a Quadratic Equation by Factoring**

Solve the equation:

(a) $x^2 + 6x = 0$ (b) $2x^2 = x + 3$

Solution (a) The equation is in the standard form specified in equation (1). The left side may be factored as

$$x^2 + 6x = 0$$
$$x(x + 6) = 0 \quad \text{Factor.}$$

Using the Zero-Product Property, we set each factor equal to 0 and then solve the resulting first-degree equations.

$$x = 0 \quad \text{or} \quad x + 6 = 0 \qquad \text{Zero-Product Property}$$
$$x = 0 \quad \text{or} \qquad x = -6 \qquad \text{Solve.}$$

The solution set is $\{0, -6\}$.

(b) We put the equation $2x^2 = x + 3$ in standard form by adding $-x - 3$ to both sides.

$$2x^2 = x + 3$$
$$2x^2 - x - 3 = 0 \qquad \text{Add } -x - 3 \text{ to both sides.}$$

The left side may now be factored as

$$(2x - 3)(x + 1) = 0 \qquad \text{Factor.}$$

so that

$$2x - 3 = 0 \quad \text{or} \quad x + 1 = 0 \qquad \text{Zero-Product Property}$$
$$x = \frac{3}{2} \qquad\qquad x = -1 \qquad \text{Solve.}$$

The solution set is $\left\{-1, \dfrac{3}{2}\right\}$.

When the left side factors into two linear equations with the same solution, the quadratic equation is said to have a **repeated solution.** We also call this solution a **root of multiplicity 2,** or a **double root.**

EXAMPLE 2	**Solving a Quadratic Equation by Factoring**

Solve the equation: $9x^2 - 6x + 1 = 0$

Solution This equation is already in standard form, and the left side can be factored.

$$9x^2 - 6x + 1 = 0$$
$$(3x - 1)(3x - 1) = 0$$

so

$$x = \frac{1}{3} \quad \text{or} \quad x = \frac{1}{3}$$

This equation has only the repeated solution $\dfrac{1}{3}$. The solution set is $\left\{\dfrac{1}{3}\right\}$.

➜ **Now Work** PROBLEMS 11 AND 21

The Square Root Method

Suppose that we wish to solve the quadratic equation

$$x^2 = p \tag{2}$$

where $p \geq 0$ is a nonnegative number. We proceed as in the earlier examples.

$$x^2 - p = 0 \qquad \text{Put in standard form.}$$
$$(x - \sqrt{p})(x + \sqrt{p}) = 0 \qquad \text{Factor (over the real numbers).}$$
$$x = \sqrt{p} \quad \text{or} \quad x = -\sqrt{p} \qquad \text{Solve.}$$

We have the following result:

$$\boxed{\text{If } x^2 = p \text{ and } p \geq 0, \text{ then } x = \sqrt{p} \text{ or } x = -\sqrt{p}. \tag{3}}$$

When statement (3) is used, it is called the **Square Root Method.** In statement (3), note that if $p > 0$ the equation $x^2 = p$ has two solutions, $x = \sqrt{p}$ and $x = -\sqrt{p}$. We usually abbreviate these solutions as $x = \pm\sqrt{p}$, read as "x equals plus or minus the square root of p."

For example, the two solutions of the equation

$$x^2 = 4$$

are

$$x = \pm\sqrt{4} \qquad \text{Use the Square Root Method.}$$

and, since $\sqrt{4} = 2$, we have

$$x = \pm 2$$

The solution set is $\{-2, 2\}$.

| EXAMPLE 3 | **Solving a Quadratic Equation Using the Square Root Method** |

Solve each equation.

(a) $x^2 = 5$ (b) $(x - 2)^2 = 16$

Solution (a) We use the Square Root Method to get

$$x^2 = 5$$
$$x = \pm\sqrt{5} \qquad \text{Use the Square Root Method.}$$
$$x = \sqrt{5} \quad \text{or} \quad x = -\sqrt{5}$$

The solution set is $\{-\sqrt{5}, \sqrt{5}\}$.

(b) We use the Square Root Method to get

$$(x - 2)^2 = 16$$
$$x - 2 = \pm\sqrt{16} \qquad \text{Use the Square Root Method.}$$
$$x - 2 = \pm 4$$
$$x - 2 = 4 \quad \text{or} \quad x - 2 = -4$$
$$x = 6 \quad \text{or} \qquad x = -2$$

The solution set is $\{-2, 6\}$.

Now Work PROBLEM 31

2 Know How to Complete the Square

We now introduce the method of **completing the square.** The idea behind this method is to *adjust* the left side of a quadratic equation, $ax^2 + bx + c = 0$, so that it becomes a perfect square, that is, the square of a first-degree polynomial. For example, $x^2 + 6x + 9$ and $x^2 - 4x + 4$ are perfect squares because

$$x^2 + 6x + 9 = (x + 3)^2 \quad \text{and} \quad x^2 - 4x + 4 = (x - 2)^2$$

How do we adjust the left side? We do it by adding the appropriate number to the left side to create a perfect square. For example, to make $x^2 + 6x$ a perfect square, we add 9.

Let's look at several examples of completing the square when the coefficient of x^2 is 1:

Start	Add	Result
$x^2 + 8x$	16	$x^2 + 8x + 16 = (x + 4)^2$
$x^2 + 2x$	1	$x^2 + 2x + 1 = (x + 1)^2$
$x^2 - 6x$	9	$x^2 - 6x + 9 = (x - 3)^2$
$x^2 + x$	$\dfrac{1}{4}$	$x^2 + x + \dfrac{1}{4} = \left(x + \dfrac{1}{2}\right)^2$

Do you see the pattern? Provided that the coefficient of x^2 is 1, we complete the square by adding the square of $\frac{1}{2}$ of the coefficient of x.

Procedure for Completing a Square

Start	Add	Result
$x^2 + mx$	$\left(\dfrac{m}{2}\right)^2$	$x^2 + mx + \left(\dfrac{m}{2}\right)^2 = \left(x + \dfrac{m}{2}\right)^2$

EXAMPLE 4 Completing the Square

Determine the number that must be added to each expression to complete the square. Then factor the expression.

Start	Add	Result	Factored Form
$y^2 + 8y$	$\left(\dfrac{1}{2} \cdot 8\right)^2 = 16$	$y^2 + 8y + 16$	$(y + 4)^2$
$x^2 + 12x$	$\left(\dfrac{1}{2} \cdot 12\right)^2 = 36$	$x^2 + 12x + 36$	$(x + 6)^2$
$a^2 - 20a$	$\left(\dfrac{1}{2} \cdot (-20)\right)^2 = 100$	$a^2 - 20a + 100$	$(a - 10)^2$
$p^2 - 5p$	$\left(\dfrac{1}{2} \cdot (-5)\right)^2 = \dfrac{25}{4}$	$p^2 - 5p + \dfrac{25}{4}$	$\left(p - \dfrac{5}{2}\right)^2$

Notice that the factored form of a perfect square is either

$$x^2 + bx + \left(\frac{b}{2}\right)^2 = \left(x + \frac{b}{2}\right)^2 \quad \text{or} \quad x^2 - bx + \left(\frac{b}{2}\right)^2 = \left(x - \frac{b}{2}\right)^2$$

Now Work PROBLEM 35

Are you wondering why we call making an expression a perfect square "completing the square"? Look at the square in Figure 1. Its area is $(y + 4)^2$. The yellow area is y^2 and each orange area is $4y$ (for a total area of $8y$). The sum of these areas is $y^2 + 8y$. To complete the square we need to add the area of the green region: $4 \cdot 4 = 16$. As a result, $y^2 + 8y + 16 = (y + 4)^2$.

Figure 1

3 Solve a Quadratic Equation by Completing the Square

EXAMPLE 5 Solving a Quadratic Equation by Completing the Square

Solve by completing the square: $x^2 + 5x + 4 = 0$

Solution We always begin this procedure by rearranging the equation so that the constant is on the right side.

$$x^2 + 5x + 4 = 0$$
$$x^2 + 5x = -4$$

Since the coefficient of x^2 is 1, we can complete the square on the left side by adding $\left(\dfrac{1}{2} \cdot 5\right)^2 = \dfrac{25}{4}$. Of course, in an equation, whatever we add to the left side also must be added to the right side. So we add $\dfrac{25}{4}$ to *both* sides.

$$x^2 + 5x + \frac{25}{4} = -4 + \frac{25}{4} \qquad \text{Add } \frac{25}{4} \text{ to both sides.}$$

$$\left(x + \frac{5}{2}\right)^2 = \frac{9}{4} \qquad \text{Factor.}$$

$$x + \frac{5}{2} = \pm\sqrt{\frac{9}{4}} \qquad \text{Use the Square Root Method.}$$

$$x + \frac{5}{2} = \pm\frac{3}{2}$$

$$x = -\frac{5}{2} \pm \frac{3}{2}$$

$$x = -\frac{5}{2} + \frac{3}{2} = -1 \quad \text{or} \quad x = -\frac{5}{2} - \frac{3}{2} = -4$$

The solution set is $\{-4, -1\}$.

THE SOLUTION OF THE EQUATION IN EXAMPLE 5 ALSO CAN BE OBTAINED BY FACTORING. REWORK EXAMPLE 5 USING THIS TECHNIQUE

The next example illustrates an equation that cannot be solved by factoring.

| EXAMPLE 6 | **Solving a Quadratic Equation by Completing the Square** |

Solve by completing the square: $2x^2 - 8x - 5 = 0$

Solution First, we rewrite the equation.

$$2x^2 - 8x - 5 = 0$$
$$2x^2 - 8x = 5$$

Next, we divide both sides by 2 so that the coefficient of x^2 is 1. (This enables us to complete the square at the next step.)

$$x^2 - 4x = \frac{5}{2}$$

Finally, we complete the square by adding 4 to both sides.

$$x^2 - 4x + 4 = \frac{5}{2} + 4$$

$$(x - 2)^2 = \frac{13}{2}$$

$$x - 2 = \pm\sqrt{\frac{13}{2}} \qquad \text{Use the Square Root Method.}$$

$$x - 2 = \pm\frac{\sqrt{26}}{2} \qquad \sqrt{\frac{13}{2}} = \frac{\sqrt{13}}{\sqrt{2}} = \frac{\sqrt{13}}{\sqrt{2}} \cdot \frac{\sqrt{2}}{\sqrt{2}} = \frac{\sqrt{26}}{2}$$

$$x = 2 \pm \frac{\sqrt{26}}{2}$$

NOTE If we wanted an approximation, say rounded to two decimal places, of these solutions, we would use a calculator to get $\{-0.55, 4.55\}$. ∎

The solution set is $\left\{2 - \dfrac{\sqrt{26}}{2}, 2 + \dfrac{\sqrt{26}}{2}\right\}$

Now Work PROBLEM 41

4 Solve a Quadratic Equation Using the Quadratic Formula

We can use the method of completing the square to obtain a general formula for solving any quadratic equation

$$ax^2 + bx + c = 0 \qquad a \neq 0$$

NOTE There is no loss in generality to assume that $a > 0$, since if $a < 0$ we can multiply by -1 to obtain an equivalent equation with a positive leading coefficient. ∎

As in Examples 5 and 6, we rearrange the terms as

$$ax^2 + bx = -c \quad a > 0$$

Since $a > 0$, we can divide both sides by a to get

$$x^2 + \frac{b}{a}x = -\frac{c}{a}$$

Now the coefficient of x^2 is 1. To complete the square on the left side, add the square of $\frac{1}{2}$ of the coefficient of x; that is, add

$$\left(\frac{1}{2} \cdot \frac{b}{a}\right)^2 = \frac{b^2}{4a^2}$$

to both sides. Then

$$x^2 + \frac{b}{a}x + \frac{b^2}{4a^2} = \frac{b^2}{4a^2} - \frac{c}{a}$$

$$\left(x + \frac{b}{2a}\right)^2 = \frac{b^2 - 4ac}{4a^2} \qquad \frac{b^2}{4a^2} - \frac{c}{a} = \frac{b^2}{4a^2} - \frac{4ac}{4a^2} = \frac{b^2 - 4ac}{4a^2} \qquad \textbf{(4)}$$

Provided that $b^2 - 4ac \geq 0$, we now can use the Square Root Method to get

$$x + \frac{b}{2a} = \pm\sqrt{\frac{b^2 - 4ac}{4a^2}}$$

$$x + \frac{b}{2a} = \frac{\pm\sqrt{b^2 - 4ac}}{2a} \qquad \text{\small The square root of a quotient equals the quotient of the square roots.}$$

$$\text{\small Also, } \sqrt{4a^2} = 2a \text{ since } a > 0.$$

$$x = -\frac{b}{2a} \pm \frac{\sqrt{b^2 - 4ac}}{2a} \qquad \text{\small Add } -\frac{b}{2a} \text{ to both sides.}$$

$$= \frac{-b \pm \sqrt{b^2 - 4ac}}{2a} \qquad \text{\small Combine the quotients on the right.}$$

What if $b^2 - 4ac$ is negative? Then equation (4) states that the left expression (a real number squared) equals the right expression (a negative number). Since this occurrence is impossible for real numbers, we conclude that if $b^2 - 4ac < 0$ the quadratic equation has no *real* solution. (We discuss quadratic equations for which the quantity $b^2 - 4ac < 0$ in detail in the next section.)

We now state the *quadratic formula*.

THEOREM

Consider the quadratic equation

$$ax^2 + bx + c = 0 \qquad a \neq 0$$

If $b^2 - 4ac < 0$, this equation has no real solution.

If $b^2 - 4ac \geq 0$, the real solution(s) of this equation is (are) given by the **quadratic formula.**

Quadratic Formula

$$x = \frac{-b \pm \sqrt{b^2 - 4ac}}{2a} \qquad \textbf{(5)}$$

The quantity $b^2 - 4ac$ is called the **discriminant** of the quadratic equation, because its value tells us whether the equation has real solutions. In fact, it also tells us how many solutions to expect.

> **Discriminant of a Quadratic Equation**
>
> For a quadratic equation $ax^2 + bx + c = 0$:
>
> **1.** If $b^2 - 4ac > 0$, there are two unequal real solutions.
> **2.** If $b^2 - 4ac = 0$, there is a repeated solution, a root of multiplicity 2.
> **3.** If $b^2 - 4ac < 0$, there is no real solution.

When asked to find the real solutions, if any, of a quadratic equation, always evaluate the discriminant first to see how many real solutions there are.

EXAMPLE 7 | **Solving a Quadratic Equation Using the Quadratic Formula**

Use the quadratic formula to find the real solutions, if any, of the equation

$$3x^2 - 5x + 1 = 0$$

Solution The equation is in standard form, so we compare it to $ax^2 + bx + c = 0$ to find a, b, and c.

$$3x^2 - 5x + 1 = 0$$
$$ax^2 + bx + c = 0 \quad a = 3, b = -5, c = 1$$

With $a = 3$, $b = -5$, and $c = 1$, we evaluate the discriminant $b^2 - 4ac$.

$$b^2 - 4ac = (-5)^2 - 4(3)(1) = 25 - 12 = 13$$

Since $b^2 - 4ac > 0$, there are two real solutions, which can be found using the quadratic formula.

$$x = \frac{-b \pm \sqrt{b^2 - 4ac}}{2a} = \frac{-(-5) \pm \sqrt{13}}{2(3)} = \frac{5 \pm \sqrt{13}}{6}$$

The solution set is $\left\{ \dfrac{5 - \sqrt{13}}{6}, \dfrac{5 + \sqrt{13}}{6} \right\}$.

EXAMPLE 8 | **Solving a Quadratic Equation Using the Quadratic Formula**

Use the quadratic formula to find the real solutions, if any, of the equation

$$\frac{25}{2}x^2 - 30x + 18 = 0$$

Solution The equation is given in standard form. However, to simplify the arithmetic, we clear the fractions.

$$\frac{25}{2}x^2 - 30x + 18 = 0$$
$$25x^2 - 60x + 36 = 0 \quad \text{Clear fractions; multiply by 2.}$$
$$ax^2 + bx + c = 0 \quad \text{Compare to standard form.}$$

With $a = 25$, $b = -60$, and $c = 36$, we evaluate the discriminant.

$$b^2 - 4ac = (-60)^2 - 4(25)(36) = 3600 - 3600 = 0$$

The equation has a repeated solution, which we find by using the quadratic formula.

$$x = \frac{-b \pm \sqrt{b^2 - 4ac}}{2a} = \frac{60 \pm \sqrt{0}}{50} = \frac{60}{50} = \frac{6}{5}$$

The solution set is $\left\{\dfrac{6}{5}\right\}$.

| EXAMPLE 9 | Solving a Quadratic Equation Using the Quadratic Formula |

Use the quadratic formula to find the real solutions, if any, of the equation

$$3x^2 + 2 = 4x$$

Solution The equation, as given, is not in standard form.

$$3x^2 + 2 = 4x$$
$$3x^2 - 4x + 2 = 0 \quad \text{Put in standard form.}$$
$$ax^2 + bx + c = 0 \quad \text{Compare to standard form.}$$

With $a = 3$, $b = -4$, and $c = 2$, we find

$$b^2 - 4ac = (-4)^2 - 4(3)(2) = 16 - 24 = -8$$

Since $b^2 - 4ac < 0$, the equation has no real solution.

> **Now Work** PROBLEMS 51 AND 61

Sometimes a given equation can be transformed into a quadratic equation so that it can be solved using the quadratic formula.

| EXAMPLE 10 | Solving a Quadratic Equation Using the Quadratic Formula |

Find the real solutions, if any, of the equation: $9 + \dfrac{3}{x} - \dfrac{2}{x^2} = 0, \; x \neq 0$

Solution In its present form, the equation

$$9 + \frac{3}{x} - \frac{2}{x^2} = 0$$

is not a quadratic equation. However, it can be transformed into one by multiplying each side by x^2. The result is

$$9x^2 + 3x - 2 = 0$$

Although we multiplied each side by x^2, we know that $x^2 \neq 0$ (do you see why?), so this quadratic equation is equivalent to the original equation.

Using $a = 9$, $b = 3$, and $c = -2$, the discriminant is

$$b^2 - 4ac = 3^2 - 4(9)(-2) = 9 + 72 = 81$$

Since $b^2 - 4ac > 0$, the new equation has two real solutions.

$$x = \frac{-b \pm \sqrt{b^2 - 4ac}}{2a} = \frac{-3 \pm \sqrt{81}}{2(9)} = \frac{-3 \pm 9}{18}$$

$$x = \frac{-3 + 9}{18} = \frac{6}{18} = \frac{1}{3} \quad \text{or} \quad x = \frac{-3 - 9}{18} = \frac{-12}{18} = -\frac{2}{3}$$

The solution set is $\left\{-\dfrac{2}{3}, \dfrac{1}{3}\right\}$.

SUMMARY Procedure for Solving a Quadratic Equation

To solve a quadratic equation, first put it in standard form:

$$ax^2 + bx + c = 0$$

Then:

STEP 1: Identify a, b, and c.

STEP 2: Evaluate the discriminant, $b^2 - 4ac$.

STEP 3: (a) If the discriminant is negative, the equation has no real solution.
(b) If the discriminant is zero, the equation has one real solution, a repeated root.
(c) If the discriminant is positive, the equation has two distinct real solutions.

If you can easily spot factors, use the factoring method to solve the equation. Otherwise, use the quadratic formula or the method of completing the square.

5 Solve Applied Problems Involving Quadratic Equations

 Many applied problems require the solution of a quadratic equation. Let's look at one that you will probably see again in a slightly different form if you study calculus.

| **EXAMPLE 11** | **Constructing a Box** |

From each corner of a square piece of sheet metal, remove a square of side 9 centimeters. Turn up the edges to form an open box. If the box is to hold 144 cubic centimeters (cm^3), what should be the dimensions of the piece of sheet metal?

Solution We use Figure 2 as a guide. We have labeled by x the length of a side of the square piece of sheet metal. The box will be of height 9 centimeters, and its square base will measure $x - 18$ on each side. The volume V (Length $\times$ Width $\times$ Height) of the box is therefore

$$V = (x - 18)(x - 18) \cdot 9 = 9(x - 18)^2$$

Figure 2

Since the volume of the box is to be 144 cm^3, we have

$$9(x - 18)^2 = 144 \qquad \text{\small{$V = 144$}}$$
$$(x - 18)^2 = 16 \qquad \text{\small{Divide each side by 9.}}$$
$$x - 18 = \pm 4 \qquad \text{\small{Use the Square Root Method.}}$$
$$x = 18 + 4$$
$$x = 22 \quad \text{or} \quad x = 14$$

We discard the solution $x = 14$ (do you see why?) and conclude that the sheet metal should be 22 centimeters by 22 centimeters.

✓ **Check:** If we begin with a piece of sheet metal 22 centimeters by 22 centimeters, cut out a 9 centimeter square from each corner, and fold up the edges, we get a box whose dimensions are 9 by 4 by 4, with volume $9 \times 4 \times 4 = 144 \text{ cm}^3$, as required.

━━ **Now Work** PROBLEM 105

Historical Feature

P roblems using quadratic equations are found in the oldest known mathematical literature. Babylonians and Egyptians were solving such problems before 1800 BC. Euclid solved quadratic equations geometrically in his *Data* (300 BC), and the Hindus and Arabs gave rules for solving any quadratic equation with real roots. Because negative numbers were not freely used before AD 1500, there were several different types of quadratic equations, each with its own rule.

Thomas Harriot (1560–1621) introduced the method of factoring to obtain solutions, and François Viète (1540–1603) introduced a method that is essentially completing the square.

Until modern times it was usual to neglect the negative roots (if there were any), and equations involving square roots of negative quantities were regarded as unsolvable until the 1500s.

Historical Problems

1. *One of al-Khwǎiizmi solutions* We solve $x^2 + 12x = 85$ by drawing the square shown. The area of the four white rectangles and the yellow square is $x^2 + 12x$. We then set this expression equal to 85 to get the equation $x^2 + 12x = 85$. If we add the four blue squares, we will have a larger square of known area. Complete the solution.

2. *Viète's method* We solve $x^2 + 12x - 85 = 0$ by letting $x = u + z$. Then

$$(u + z)^2 + 12(u + z) - 85 = 0$$
$$u^2 + (2z + 12)u + (z^2 + 12z - 85) = 0$$

Now select z so that $2z + 12 = 0$ and finish the solution.

3. *Another method to get the quadratic formula* Look at equation (4) on page 102. Rewrite the right side as $\left(\dfrac{\sqrt{b^2 - 4ac}}{2a}\right)^2$ and then subtract it from each side. The right side is now 0 and the left side

is a difference of two squares. If you factor this difference of two squares, you will easily be able to get the quadratic formula, and, moreover, the quadratic expression is factored, which is sometimes useful.

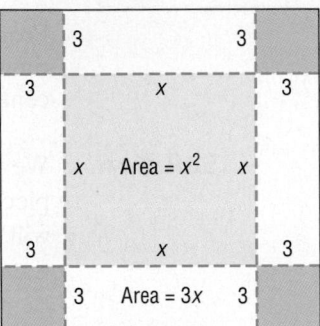

1.2 Assess Your Understanding

'Are You Prepared?' *Answers are given at the end of these exercises. If you get a wrong answer, read the pages listed in red.*

1. Factor: $x^2 - 5x - 6$ (pp. 49–55)

2. Factor: $2x^2 - x - 3$ (pp. 49–55)

3. The solution set of the equation $(x - 3)(3x + 5) = 0$ is _____. (p. 13)

4. *True or False* $\sqrt{x^2} = |x|$. (pp. 23–24)

Concepts and Vocabulary

5. To complete the square of the expression $x^2 + 5x$, you would _____ the number _____.

6. The quantity $b^2 - 4ac$ is called the _____ of a quadratic equation. If it is _____, the equation has no real solution.

7. *True or False* Quadratic equations always have two real solutions.

8. *True or False* If the discriminant of a quadratic equation is positive, then the equation has two solutions that are negatives of one another.

Skill Building

In Problems 9–28, solve each equation by factoring.

9. $x^2 - 9x = 0$

10. $x^2 + 4x = 0$

11. $x^2 - 25 = 0$

12. $x^2 - 9 = 0$

13. $z^2 + z - 6 = 0$

14. $v^2 + 7v + 6 = 0$

15. $2x^2 - 5x - 3 = 0$

16. $3x^2 + 5x + 2 = 0$

17. $3t^2 - 48 = 0$ **18.** $2y^2 - 50 = 0$ **19.** $x(x - 8) + 12 = 0$ **20.** $x(x + 4) = 12$

21. $4x^2 + 9 = 12x$ **22.** $25x^2 + 16 = 40x$ **23.** $6(p^2 - 1) = 5p$ **24.** $2(2u^2 - 4u) + 3 = 0$

25. $6x - 5 = \dfrac{6}{x}$ **26.** $x + \dfrac{12}{x} = 7$ **27.** $\dfrac{4(x - 2)}{x - 3} + \dfrac{3}{x} = \dfrac{-3}{x(x - 3)}$ **28.** $\dfrac{5}{x + 4} = 4 + \dfrac{3}{x - 2}$

In Problems 29–34, solve each equation by the Square Root Method.

29. $x^2 = 25$ **30.** $x^2 = 36$ **31.** $(x - 1)^2 = 4$

32. $(x + 2)^2 = 1$ **33.** $(2y + 3)^2 = 9$ **34.** $(3z - 2)^2 = 4$

In Problems 35–40, what number should be added to complete the square of each expression?

35. $x^2 - 8x$ **36.** $x^2 - 4x$ **37.** $x^2 + \dfrac{1}{2}x$

38. $x^2 - \dfrac{1}{3}x$ **39.** $y^2 - \dfrac{2}{3}y$ **40.** $z^2 - \dfrac{2}{5}z$

In Problems 41–46, solve each equation by completing the square.

41. $x^2 + 4x = 21$ **42.** $x^2 - 6x = 13$ **43.** $x^2 - \dfrac{1}{2}x - \dfrac{3}{16} = 0$

44. $x^2 + \dfrac{2}{3}x - \dfrac{1}{3} = 0$ **45.** $3x^2 + x - \dfrac{1}{2} = 0$ **46.** $2x^2 - 3x - 1 = 0$

In Problems 47–70, find the real solutions, if any, of each equation. Use the quadratic formula.

47. $x^2 - 4x + 2 = 0$ **48.** $x^2 + 4x + 2 = 0$ **49.** $x^2 - 4x - 1 = 0$ **50.** $x^2 + 6x + 1 = 0$

51. $2x^2 - 5x + 3 = 0$ **52.** $2x^2 + 5x + 3 = 0$ **53.** $4y^2 - y + 2 = 0$ **54.** $4t^2 + t + 1 = 0$

55. $4x^2 = 1 - 2x$ **56.** $2x^2 = 1 - 2x$ **57.** $4x^2 = 9x$ **58.** $5x = 4x^2$

59. $9t^2 - 6t + 1 = 0$ **60.** $4u^2 - 6u + 9 = 0$ **61.** $\dfrac{3}{4}x^2 - \dfrac{1}{4}x - \dfrac{1}{2} = 0$ **62.** $\dfrac{2}{3}x^2 - x - 3 = 0$

63. $\dfrac{5}{3}x^2 - x = \dfrac{1}{3}$ **64.** $\dfrac{3}{5}x^2 - x = \dfrac{1}{5}$ **65.** $2x(x + 2) = 3$ **66.** $3x(x + 2) = 1$

67. $4 - \dfrac{1}{x} - \dfrac{2}{x^2} = 0$ **68.** $4 + \dfrac{1}{x} - \dfrac{1}{x^2} = 0$ **69.** $\dfrac{3x}{x - 2} + \dfrac{1}{x} = 4$ **70.** $\dfrac{2x}{x - 3} + \dfrac{1}{x} = 4$

In Problems 71–78, find the real solutions, if any, of each equation. Use the quadratic formula and a calculator. Express any solutions rounded to two decimal places.

71. $x^2 - 4.1x + 2.2 = 0$ **72.** $x^2 + 3.9x + 1.8 = 0$ **73.** $x^2 + \sqrt{3}x - 3 = 0$

74. $x^2 + \sqrt{2}x - 2 = 0$ **75.** $\pi x^2 - x - \pi = 0$ **76.** $\pi x^2 + \pi x - 2 = 0$

77. $3x^2 + 8\pi x + \sqrt{29} = 0$ **78.** $\pi x^2 - 15\sqrt{2}x + 20 = 0$

In Problems 79–92, find the real solutions, if any, of each equation. Use any method.

79. $x^2 - 5 = 0$ **80.** $x^2 - 6 = 0$ **81.** $16x^2 - 8x + 1 = 0$

82. $9x^2 - 12x + 4 = 0$ **83.** $10x^2 - 19x - 15 = 0$ **84.** $6x^2 + 7x - 20 = 0$

85. $2 + z = 6z^2$ **86.** $2 = y + 6y^2$ **87.** $x^2 + \sqrt{2}x = \dfrac{1}{2}$

88. $\dfrac{1}{2}x^2 = \sqrt{2}x + 1$ **89.** $x^2 + x = 4$ **90.** $x^2 + x = 1$

91. $\dfrac{x}{x - 2} + \dfrac{2}{x + 1} = \dfrac{7x + 1}{x^2 - x - 2}$ **92.** $\dfrac{3x}{x + 2} + \dfrac{1}{x - 1} = \dfrac{4 - 7x}{x^2 + x - 2}$

In Problems 93–98, use the discriminant to determine whether each quadratic equation has two unequal real solutions, a repeated real solution, or no real solution, without solving the equation.

93. $2x^2 - 6x + 7 = 0$ **94.** $x^2 + 4x + 7 = 0$ **95.** $9x^2 - 30x + 25 = 0$

96. $25x^2 - 20x + 4 = 0$ **97.** $3x^2 + 5x - 8 = 0$ **98.** $2x^2 - 3x - 7 = 0$

Applications and Extensions

99. College Tuition and Fees The average annual published undergraduate tuition-and-fee charges C, in dollars, for public four-year institutions from academic years 2000–2001 through 2005–2006 can be estimated by the equation $C = 20.2x^2 + 314.5x + 3467.6$, where x is the number of years afer the 2000–2001 academic year. Assuming the model will remain valid beyond 2005–2006, in what academic year will average annual tuition-and-fee charges be $8000?

Source: College Board, *Trends in College Pricing 2005*

100. Women's Weekly Earnings The median weekly earnings E, in dollars, for full-time women wage and salary workers ages 16 years and older from 2000 through 2004 can be estimated by the equation $E = 0.14x^2 + 7.8x + 540$, where x is the number of years after 2000. Assuming the model will remain valid beyond 2004, in what year will the median weekly earnings be $632?

Source: U.S. Department of Labor, *Highlights of Women's Earnings in 2004,* September 2005

101. Dimensions of a Window The area of the opening of a rectangular window is to be 143 square feet. If the length is to be 2 feet more than the width, what are the dimensions?

102. Dimensions of a Window The area of a rectangular window is to be 306 square centimeters. If the length exceeds the width by 1 centimeter, what are the dimensions?

103. Geometry Find the dimensions of a rectangle whose perimeter is 26 meters and whose area is 40 square meters.

104. Watering a Field An adjustable water sprinkler that sprays water in a circular pattern is placed at the center of a square field whose area is 1250 square feet (see the figure). What is the shortest radius setting that can be used if the field is to be completely enclosed within the circle?

105. Constructing a Box An open box is to be constructed from a square piece of sheet metal by removing a square of side 1 foot from each corner and turning up the edges. If the box is to hold 4 cubic feet, what should be the dimensions of the sheet metal?

106. Constructing a Box Rework Problem 105 if the piece of sheet metal is a rectangle whose length is twice its width.

107. Physics A ball is thrown vertically upward from the top of a building 96 feet tall with an initial velocity of 80 feet per second. The distance s (in feet) of the ball from the ground after t seconds is $s = 96 + 80t - 16t^2$.

(a) After how many seconds does the ball strike the ground?

(b) After how many seconds will the ball pass the top of the building on its way down?

108. Physics An object is propelled vertically upward with an initial velocity of 20 meters per second. The distance s (in meters) of the object from the ground after t seconds is $s = -4.9t^2 + 20t$.

(a) When will the object be 15 meters above the ground?

(b) When will it strike the ground?

(c) Will the object reach a height of 100 meters?

109. Reducing the Size of a Candy Bar A jumbo chocolate bar with a rectangular shape measures 12 centimeters in length, 7 centimeters in width, and 3 centimeters in thickness. Due to escalating costs of cocoa, management decides to reduce the volume of the bar by 10%. To accomplish this reduction, management decides that the new bar should have the same 3 centimeter thickness, but the length and width of each should be reduced an equal number of centimeters. What should be the dimensions of the new candy bar?

110. Reducing the Size of a Candy Bar Rework Problem 109 if the reduction is to be 20%.

111. Constructing a Border around a Pool A circular pool measures 10 feet across. One cubic yard of concrete is to be used to create a circular border of uniform width around the pool. If the border is to have a depth of 3 inches, how wide will the border be? (1 cubic yard = 27 cubic feet) See the illustration.

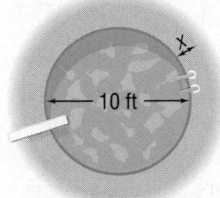

112. Constructing a Border around a Pool Rework Problem 111 if the depth of the border is 4 inches.

113. Constructing a Border around a Garden A landscaper, who just completed a rectangular flower garden measuring 6 feet by 10 feet, orders 1 cubic yard of premixed cement, all of which is to be used to create a border of uniform width around the garden. If the border is to have a depth of 3 inches, how wide will the border be? (1 cubic yard = 27 cubic feet)

114. Dimensions of a Patio A contractor orders 8 cubic yards of premixed cement, all of which is to be used to pour a patio that will be 4 inches thick. If the length of the patio is specified to be twice the width, what will be the patio dimensions? (1 cubic yard = 27 cubic feet)

115. Comparing TVs The screen size of a television is determined by the length of the diagonal of the rectangular screen.

traditional
4:3

LCD
16:9

Traditional TVs come in a 4 : 3 format, meaning the ratio of the length to the width of the rectangular is 4 to 3. What is the area of a 37-inch traditional TV screen? What is the area of a 37-inch LCD TV whose screen is in a 16 : 9 format? Which screen is larger?

[**Hint:** If x is the length of a 4 : 3 format screen, then $\frac{3}{4}x$ is the width.]

116. **Comparing TVs** Refer to Problem 115. Find the screen area of a traditional 50-inch TV and compare it with a 50-inch Plasma TV whose screen is in a 16 : 9 format. Which screen is larger?

117. The sum of the consecutive integers $1, 2, 3, \ldots, n$ is given by the formula $\frac{1}{2}n(n + 1)$. How many consecutive integers, starting with 1, must be added to get a sum of 666?

118. **Geometry** If a polygon of n sides has $\frac{1}{2}n(n - 3)$ diagonals, how many sides will a polygon with 65 diagonals have? Is there a polygon with 80 diagonals?

119. Show that the sum of the roots of a quadratic equation is $-\dfrac{b}{a}$.

120. Show that the product of the roots of a quadratic equation is $\dfrac{c}{a}$.

121. Find k such that the equation $kx^2 + x + k = 0$ has a repeated real solution.

122. Find k such that the equation $x^2 - kx + 4 = 0$ has a repeated real solution.

123. Show that the real solutions of the equation $ax^2 + bx + c = 0$ are the negatives of the real solutions of the equation $ax^2 - bx + c = 0$. Assume that $b^2 - 4ac \geq 0$.

124. Show that the real solutions of the equation $ax^2 + bx + c = 0$ are the reciprocals of the real solutions of the equation $cx^2 + bx + a = 0$. Assume that $b^2 - 4ac \geq 0$.

Discussion and Writing

125. Which of the following pairs of equations are equivalent? Explain.
 (a) $x^2 = 9;\quad x = 3$
 (b) $x = \sqrt{9};\quad x = 3$
 (c) $(x - 1)(x - 2) = (x - 1)^2;\quad x - 2 = x - 1$

126. Describe three ways that you might solve a quadratic equation. State your preferred method; explain why you chose it.

127. Explain the benefits of evaluating the discriminant of a quadratic equation before attempting to solve it.

128. Create three quadratic equations: one having two distinct solutions, one having no real solution, and one having exactly one real solution.

129. The word *quadratic* seems to imply four (*quad*), yet a quadratic equation is an equation that involves a polynomial of degree 2. Investigate the origin of the term *quadratic* as it is used in the expression *quadratic equation*. Write a brief essay on your findings.

'Are You Prepared?' Answers

1. $(x - 6)(x + 1)$ **2.** $(2x - 3)(x + 1)$ **3.** $\left\{-\dfrac{5}{3}, 3\right\}$ **4.** True

1.3 Complex Numbers; Quadratic Equations in the Complex Number System*

PREPARING FOR THIS SECTION *Before getting started, review the following:*

- Classification of Numbers (Section R.1, pp. 4–5)
- Rationalizing Denominators (Section R.8, p. 74)

Now Work the 'Are You Prepared?' problems on page 116.

OBJECTIVES 1 Add, Subtract, Multiply, and Divide Complex Numbers (p. 110)
2 Solve Quadratic Equations in the Complex Number System (p. 114)

Complex Numbers

One property of a real number is that its square is nonnegative. For example, there is no real number x for which

$$x^2 = -1$$

To remedy this situation, we introduce a new number called the *imaginary unit*.

* This section may be omitted without any loss of continuity.

DEFINITION | The **imaginary unit,** which we denote by *i*, is the number whose square is −1. That is,

$$i^2 = -1$$

This should not surprise you. If our universe were to consist only of integers, there would be no number *x* for which $2x = 1$. This unfortunate circumstance was remedied by introducing numbers such as $\frac{1}{2}$ and $\frac{2}{3}$, the *rational numbers*. If our universe were to consist only of rational numbers, there would be no *x* whose square equals 2. That is, there would be no number *x* for which $x^2 = 2$. To remedy this, we introduced numbers such as $\sqrt{2}$ and $\sqrt[3]{5}$, the *irrational numbers*. The *real numbers,* you will recall, consist of the rational numbers and the irrational numbers. Now, if our universe were to consist only of real numbers, then there would be no number *x* whose square is −1. To remedy this, we introduce a number *i*, whose square is −1.

In the progression outlined, each time we encountered a situation that was unsuitable, we introduced a new number system to remedy this situation. And each new number system contained the earlier number system as a subset. The number system that results from introducing the number *i* is called the **complex number system.**

DEFINITION | **Complex numbers** are numbers of the form $a + bi$, where *a* and *b* are real numbers. The real number *a* is called the **real part** of the number $a + bi$; the real number *b* is called the **imaginary part** of $a + bi$; and *i* is the imaginary unit, so $i^2 = -1$.

For example, the complex number $-5 + 6i$ has the real part −5 and the imaginary part 6.

When a complex number is written in the form $a + bi$, where *a* and *b* are real numbers, we say it is in **standard form.** However, if the imaginary part of a complex number is negative, such as in the complex number $3 + (-2)i$, we agree to write it instead in the form $3 - 2i$.

Also, the complex number $a + 0i$ is usually written merely as *a*. This serves to remind us that the real numbers are a subset of the complex numbers. The complex number $0 + bi$ is usually written as bi. Sometimes the complex number bi is called a **pure imaginary number.**

1 Add, Subtract, Multiply, and Divide Complex Numbers

Equality, addition, subtraction, and multiplication of complex numbers are defined so as to preserve the familiar rules of algebra for real numbers. Two complex numbers are equal if and only if their real parts are equal and their imaginary parts are equal. That is,

Equality of Complex Numbers

$$a + bi = c + di \quad \text{if and only if } a = c \text{ and } b = d \tag{1}$$

Two complex numbers are added by forming the complex number whose real part is the sum of the real parts and whose imaginary part is the sum of the imaginary parts. That is,

Sum of Complex Numbers

$$(a + bi) + (c + di) = (a + c) + (b + d)i \tag{2}$$

To subtract two complex numbers, we use this rule:

Difference of Complex Numbers

$$(a + bi) - (c + di) = (a - c) + (b - d)i \qquad \textbf{(3)}$$

EXAMPLE 1 **Adding and Subtracting Complex Numbers**

(a) $(3 + 5i) + (-2 + 3i) = [3 + (-2)] + (5 + 3)i = 1 + 8i$

(b) $(6 + 4i) - (3 + 6i) = (6 - 3) + (4 - 6)i = 3 + (-2)i = 3 - 2i$

━━━━━ **Now Work** PROBLEM 13

Products of complex numbers are calculated as illustrated in Example 2.

EXAMPLE 2 **Multiplying Complex Numbers**

$$(5 + 3i) \cdot (2 + 7i) = 5 \cdot (2 + 7i) + 3i(2 + 7i) = 10 + 35i + 6i + 21i^2$$

$$\underset{\text{Distributive Property}}{\uparrow} \qquad\qquad \underset{\text{Distributive Property}}{\uparrow}$$

$$= 10 + 41i + 21(-1)$$

$$\underset{i^2 = -1}{\uparrow}$$

$$= -11 + 41i$$

Based on the procedure of Example 2, we define the **product** of two complex numbers as follows:

Product of Complex Numbers

$$(a + bi) \cdot (c + di) = (ac - bd) + (ad + bc)i \qquad \textbf{(4)}$$

Do not bother to memorize formula (4). Instead, whenever it is necessary to multiply two complex numbers, follow the usual rules for multiplying two binomials, as in Example 2, remembering that $i^2 = -1$. For example,

$$(2i)(2i) = 4i^2 = -4$$

$$(2 + i)(1 - i) = 2 - 2i + i - i^2 = 3 - i$$

━━━━━ **Now Work** PROBLEM 19

Algebraic properties for addition and multiplication, such as the commutative, associative, and distributive properties, hold for complex numbers. The property that every nonzero complex number has a multiplicative inverse, or reciprocal, requires a closer look.

DEFINITION If $z = a + bi$ is a complex number, then its **conjugate,** denoted by $\bar{z}$, is defined as

$$\bar{z} = \overline{a + bi} = a - bi$$

For example, $\overline{2 + 3i} = 2 - 3i$ and $\overline{-6 - 2i} = -6 + 2i$.

| EXAMPLE 3 | **Multiplying a Complex Number by Its Conjugate** |

Find the product of the complex number $z = 3 + 4i$ and its conjugate $\bar{z}$.

Solution Since $\bar{z} = 3 - 4i$, we have

$$z\bar{z} = (3 + 4i)(3 - 4i) = 9 - 12i + 12i - 16i^2 = 9 + 16 = 25$$

The result obtained in Example 3 has an important generalization.

THEOREM

The product of a complex number and its conjugate is a nonnegative real number. That is, if $z = a + bi$, then

$$z\bar{z} = a^2 + b^2 \qquad\qquad \textbf{(5)}$$

Proof If $z = a + bi$, then

$$z\bar{z} = (a + bi)(a - bi) = a^2 - (bi)^2 = a^2 - b^2i^2 = a^2 + b^2 \qquad \blacksquare$$

To express the reciprocal of a nonzero complex number z in standard form, multiply the numerator and denominator of $\dfrac{1}{z}$ by $\bar{z}$. That is, if $z = a + bi$ is a nonzero complex number, then

$$\frac{1}{a + bi} = \frac{1}{z} = \frac{1}{z} \cdot \frac{\bar{z}}{\bar{z}} = \frac{\bar{z}}{z\bar{z}} = \frac{a - bi}{a^2 + b^2}$$

$$\uparrow$$
$$\text{Use (5).}$$

$$= \frac{a}{a^2 + b^2} - \frac{b}{a^2 + b^2}i$$

| EXAMPLE 4 | **Writing the Reciprocal of a Complex Number in Standard Form** |

Write $\dfrac{1}{3 + 4i}$ in standard form $a + bi$; that is, find the reciprocal of $3 + 4i$.

Solution The idea is to multiply the numerator and denominator by the conjugate of $3 + 4i$, that is, by the complex number $3 - 4i$. The result is

$$\frac{1}{3 + 4i} = \frac{1}{3 + 4i} \cdot \frac{3 - 4i}{3 - 4i} = \frac{3 - 4i}{9 + 16} = \frac{3}{25} - \frac{4}{25}i$$

To express the quotient of two complex numbers in standard form, we multiply the numerator and denominator of the quotient by the conjugate of the denominator.

| EXAMPLE 5 | **Writing the Quotient of a Complex Number in Standard Form** |

Write each of the following in standard form.

(a) $\dfrac{1 + 4i}{5 - 12i}$ (b) $\dfrac{2 - 3i}{4 - 3i}$

Solution

(a) $\dfrac{1+4i}{5-12i} = \dfrac{1+4i}{5-12i} \cdot \dfrac{5+12i}{5+12i} = \dfrac{5+12i+20i+48i^2}{25+144}$

$= \dfrac{-43+32i}{169} = -\dfrac{43}{169} + \dfrac{32}{169}i$

(b) $\dfrac{2-3i}{4-3i} = \dfrac{2-3i}{4-3i} \cdot \dfrac{4+3i}{4+3i} = \dfrac{8+6i-12i-9i^2}{16+9}$

$= \dfrac{17-6i}{25} = \dfrac{17}{25} - \dfrac{6}{25}i$

■

━━━ **Now Work** PROBLEM 27

EXAMPLE 6	**Writing Other Expressions in Standard Form**

If $z = 2 - 3i$ and $w = 5 + 2i$, write each of the following expressions in standard form.

(a) $\dfrac{z}{w}$ (b) $\overline{z+w}$ (c) $z + \overline{z}$

Solution

(a) $\dfrac{z}{w} = \dfrac{z \cdot \overline{w}}{w \cdot \overline{w}} = \dfrac{(2-3i)(5-2i)}{(5+2i)(5-2i)} = \dfrac{10-4i-15i+6i^2}{25+4}$

$= \dfrac{4-19i}{29} = \dfrac{4}{29} - \dfrac{19}{29}i$

(b) $\overline{z+w} = \overline{(2-3i)+(5+2i)} = \overline{7-i} = 7+i$

(c) $z + \overline{z} = (2-3i) + (2+3i) = 4$

■

The conjugate of a complex number has certain general properties that we shall find useful later.

For a real number $a = a + 0i$, the conjugate is $\overline{a} = \overline{a+0i} = a - 0i = a$. That is,

THEOREM	The conjugate of a real number is the real number itself.

Other properties of the conjugate that are direct consequences of the definition are given next. In each statement, z and w represent complex numbers.

THEOREM

The conjugate of the conjugate of a complex number is the complex number itself.

$$(\overline{\overline{z}}) = z \tag{6}$$

The conjugate of the sum of two complex numbers equals the sum of their conjugates.

$$\overline{z+w} = \overline{z} + \overline{w} \tag{7}$$

The conjugate of the product of two complex numbers equals the product of their conjugates.

$$\overline{z \cdot w} = \overline{z} \cdot \overline{w} \tag{8}$$

We leave the proofs of equations (6), (7), and (8) as exercises.

Powers of i

The **powers of i** follow a pattern that is useful to know.

$$i^1 = i$$
$$i^2 = -1$$
$$i^3 = i^2 \cdot i = -1 \cdot i = -i$$
$$i^4 = i^2 \cdot i^2 = (-1)(-1) = 1$$

$$i^5 = i^4 \cdot i = 1 \cdot i = i$$
$$i^6 = i^4 \cdot i^2 = -1$$
$$i^7 = i^4 \cdot i^3 = -i$$
$$i^8 = i^4 \cdot i^4 = 1$$

And so on. The powers of i repeat with every fourth power.

EXAMPLE 7 **Evaluating Powers of i**

(a) $i^{27} = i^{24} \cdot i^3 = (i^4)^6 \cdot i^3 = 1^6 \cdot i^3 = -i$

(b) $i^{101} = i^{100} \cdot i^1 = (i^4)^{25} \cdot i = 1^{25} \cdot i = i$

EXAMPLE 8 **Writing the Power of a Complex Number in Standard Form**

Write $(2 + i)^3$ in standard form.

Solution We use the special product formula for $(x + a)^3$.

$$(x + a)^3 = x^3 + 3ax^2 + 3a^2x + a^3$$

Using this special product formula,

$$(2 + i)^3 = 2^3 + 3 \cdot i \cdot 2^2 + 3 \cdot i^2 \cdot 2 + i^3$$
$$= 8 + 12i + 6(-1) + (-i)$$
$$= 2 + 11i.$$

Now Work PROBLEM 41

2 Solve Quadratic Equations in the Complex Number System

Quadratic equations with a negative discriminant have no real number solution. However, if we extend our number system to allow complex numbers, quadratic equations will always have a solution. Since the solution to a quadratic equation involves the square root of the discriminant, we begin with a discussion of square roots of negative numbers.

DEFINITION If N is a positive real number, we define the **principal square root of $-N$,** denoted by $\sqrt{-N}$, as

$$\sqrt{-N} = \sqrt{N}\,i$$

where i is the imaginary unit and $i^2 = -1$.

WARNING In writing $\sqrt{-N} = \sqrt{N}i$ be sure to place i outside the $\sqrt{}$ symbol. ∎

EXAMPLE 9 **Evaluating the Square Root of a Negative Number**

(a) $\sqrt{-1} = \sqrt{1}\,i = i$ (b) $\sqrt{-4} = \sqrt{4}\,i = 2i$

(c) $\sqrt{-8} = \sqrt{8}\,i = 2\sqrt{2}\,i$

EXAMPLE 10 | **Solving Equations**

Solve each equation in the complex number system.

(a) $x^2 = 4$ (b) $x^2 = -9$

Solution (a) $x^2 = 4$

$x = \pm\sqrt{4} = \pm 2$

The equation has two solutions, -2 and 2. The solution set is $\{-2, 2\}$.

(b) $x^2 = -9$

$x = \pm\sqrt{-9} = \pm\sqrt{9}\,i = \pm 3i$

The equation has two solutions, $-3i$ and $3i$. The solution set is $\{-3i, 3i\}$.

- **Now Work** PROBLEMS 49 AND 53

WARNING When working with square roots of negative numbers, do not set the square root of a product equal to the product of the square roots (which can be done with positive numbers). To see why, look at this calculation: We know that $\sqrt{100} = 10$. However, it is also true that $100 = (-25)(-4)$, so

$$10 = \sqrt{100} = \sqrt{(-25)(-4)} \neq \sqrt{-25}\,\sqrt{-4} = \left(\sqrt{25}\,i\right)\left(\sqrt{4}\,i\right) = (5i)(2i) = 10i^2 = -10$$

$\uparrow$
Here is the error. ∎

Because we have defined the square root of a negative number, we can now restate the quadratic formula without restriction.

THEOREM | **Quadratic Formula**

In the complex number system, the solutions of the quadratic equation $ax^2 + bx + c = 0$, where a, b, and c are real numbers and $a \neq 0$, are given by the formula

$$x = \frac{-b \pm \sqrt{b^2 - 4ac}}{2a} \tag{9}$$

EXAMPLE 11 | **Solving Quadratic Equations in the Complex Number System**

Solve the equation $x^2 - 4x + 8 = 0$ in the complex number system.

Solution Here $a = 1$, $b = -4$, $c = 8$, and $b^2 - 4ac = 16 - 4(1)(8) = -16$. Using equation (9), we find that

$$x = \frac{-(-4) \pm \sqrt{-16}}{2(1)} = \frac{4 \pm \sqrt{16}\,i}{2} = \frac{4 \pm 4i}{2} = 2 \pm 2i$$

The equation has two solutions $2 - 2i$ and $2 + 2i$. The solution set is $\{2 - 2i, 2 + 2i\}$.

✓ **Check:** $2 + 2i$: $(2 + 2i)^2 - 4(2 + 2i) + 8 = 4 + 8i + 4i^2 - 8 - 8i + 8$
$$= 4 - 4 = 0$$

$2 - 2i$: $(2 - 2i)^2 - 4(2 - 2i) + 8 = 4 - 8i + 4i^2 - 8 + 8i + 8$
$$= 4 - 4 = 0$$
∎

- **Now Work** PROBLEM 59

The discriminant $b^2 - 4ac$ of a quadratic equation still serves as a way to determine the character of the solutions.

Character of the Solutions of a Quadratic Equation

In the complex number system, consider a quadratic equation $ax^2 + bx + c = 0$ with real coefficients.

1. If $b^2 - 4ac > 0$, the equation has two unequal real solutions.
2. If $b^2 - 4ac = 0$, the equation has a repeated real solution, a double root.
3. If $b^2 - 4ac < 0$, the equation has two complex solutions that are not real. The solutions are conjugates of each other.

The third conclusion in the display is a consequence of the fact that if $b^2 - 4ac = -N < 0$ then, by the quadratic formula, the solutions are

$$x = \frac{-b + \sqrt{b^2 - 4ac}}{2a} = \frac{-b + \sqrt{-N}}{2a} = \frac{-b + \sqrt{N}\,i}{2a} = \frac{-b}{2a} + \frac{\sqrt{N}}{2a}i$$

and

$$x = \frac{-b - \sqrt{b^2 - 4ac}}{2a} = \frac{-b - \sqrt{-N}}{2a} = \frac{-b - \sqrt{N}\,i}{2a} = \frac{-b}{2a} - \frac{\sqrt{N}}{2a}i$$

which are conjugates of each other.

EXAMPLE 12 | **Determining the Character of the Solution of a Quadratic Equation**

Without solving, determine the character of the solution of each equation.

(a) $3x^2 + 4x + 5 = 0$ (b) $2x^2 + 4x + 1 = 0$ (c) $9x^2 - 6x + 1 = 0$

Solution (a) Here $a = 3, b = 4,$ and $c = 5,$ so $b^2 - 4ac = 16 - 4(3)(5) = -44.$ The solutions are two complex numbers that are not real and are conjugates of each other.

(b) Here $a = 2, b = 4,$ and $c = 1,$ so $b^2 - 4ac = 16 - 8 = 8.$ The solutions are two unequal real numbers.

(c) Here $a = 9, b = -6,$ and $c = 1,$ so $b^2 - 4ac = 36 - 4(9)(1) = 0.$ The solution is a repeated real number, that is, a double root.

■

Now Work PROBLEM 73

1.3 Assess Your Understanding

'Are You Prepared?' *Answers are given at the end of these exercises. If you get a wrong answer, read the pages listed in red.*

1. Name the integers and the rational numbers in the set $\left\{-3, 0, \sqrt{2}, \dfrac{6}{5}, \pi\right\}$. (pp. 4–5)

2. *True or False* Rational numbers and irrational numbers are in the set of real numbers. (pp. 4–5)

3. Rationalize the denominator of $\dfrac{3}{2 + \sqrt{3}}$. (p. 74)

Concepts and Vocabulary

4. In the complex number $5 + 2i$, the number 5 is called the _____ part; the number 2 is called the _____ part; the number i is called the _____ _____.

5. The equation $x^2 = -4$ has the solution set _____.

6. *True or False* The conjugate of $2 + 5i$ is $-2 - 5i$.

7. *True or False* All real numbers are complex numbers.

8. *True or False* If $2 - 3i$ is a solution of a quadratic equation with real coefficients, then $-2 + 3i$ is also a solution.

Skill Building

In Problems 9–46, write each expression in the standard form $a + bi$.

9. $(2 - 3i) + (6 + 8i)$ 10. $(4 + 5i) + (-8 + 2i)$ 11. $(-3 + 2i) - (4 - 4i)$ 12. $(3 - 4i) - (-3 - 4i)$

13. $(2 - 5i) - (8 + 6i)$ 14. $(-8 + 4i) - (2 - 2i)$ 15. $3(2 - 6i)$ 16. $-4(2 + 8i)$

17. $2i(2 - 3i)$ **18.** $3i(-3 + 4i)$ **19.** $(3 - 4i)(2 + i)$ **20.** $(5 + 3i)(2 - i)$

21. $(-6 + i)(-6 - i)$ **22.** $(-3 + i)(3 + i)$ **23.** $\dfrac{10}{3 - 4i}$ **24.** $\dfrac{13}{5 - 12i}$

25. $\dfrac{2 + i}{i}$ **26.** $\dfrac{2 - i}{-2i}$ **27.** $\dfrac{6 - i}{1 + i}$ **28.** $\dfrac{2 + 3i}{1 - i}$

29. $\left(\dfrac{1}{2} + \dfrac{\sqrt{3}}{2}i\right)^2$ **30.** $\left(\dfrac{\sqrt{3}}{2} - \dfrac{1}{2}i\right)^2$ **31.** $(1 + i)^2$ **32.** $(1 - i)^2$

33. i^{23} **34.** i^{14} **35.** i^{-15} **36.** i^{-23}

37. $i^6 - 5$ **38.** $4 + i^3$ **39.** $6i^3 - 4i^5$ **40.** $4i^3 - 2i^2 + 1$

41. $(1 + i)^3$ **42.** $(3i)^4 + 1$ **43.** $i^7(1 + i^2)$ **44.** $2i^4(1 + i^2)$

45. $i^6 + i^4 + i^2 + 1$ **46.** $i^7 + i^5 + i^3 + i$

In Problems 47–52, perform the indicated operations and express your answer in the form a + bi.

47. $\sqrt{-4}$ **48.** $\sqrt{-9}$ **49.** $\sqrt{-25}$

50. $\sqrt{-64}$ **51.** $\sqrt{(3 + 4i)(4i - 3)}$ **52.** $\sqrt{(4 + 3i)(3i - 4)}$

In Problems 53–72, solve each equation in the complex number system.

53. $x^2 + 4 = 0$ **54.** $x^2 - 4 = 0$ **55.** $x^2 - 16 = 0$ **56.** $x^2 + 25 = 0$

57. $x^2 - 6x + 13 = 0$ **58.** $x^2 + 4x + 8 = 0$ **59.** $x^2 - 6x + 10 = 0$ **60.** $x^2 - 2x + 5 = 0$

61. $8x^2 - 4x + 1 = 0$ **62.** $10x^2 + 6x + 1 = 0$ **63.** $5x^2 + 1 = 2x$ **64.** $13x^2 + 1 = 6x$

65. $x^2 + x + 1 = 0$ **66.** $x^2 - x + 1 = 0$ **67.** $x^3 - 8 = 0$ **68.** $x^3 + 27 = 0$

69. $x^4 = 16$ **70.** $x^4 = 1$ **71.** $x^4 + 13x^2 + 36 = 0$ **72.** $x^4 + 3x^2 - 4 = 0$

In Problems 73–78, without solving, determine the character of the solutions of each equation in the complex number system.

73. $3x^2 - 3x + 4 = 0$ **74.** $2x^2 - 4x + 1 = 0$ **75.** $2x^2 + 3x = 4$

76. $x^2 + 6 = 2x$ **77.** $9x^2 - 12x + 4 = 0$ **78.** $4x^2 + 12x + 9 = 0$

79. $2 + 3i$ is a solution of a quadratic equation with real coefficients. Find the other solution.

80. $4 - i$ is a solution of a quadratic equation with real coefficients. Find the other solution.

In Problems 81–84, z = 3 − 4i and w = 8 + 3i. Write each expression in the standard form a + bi.

81. $z + \overline{z}$ **82.** $w - \overline{w}$ **83.** $z\overline{z}$ **84.** $\overline{z - w}$

Applications and Extensions

85. Electrical Circuits The impedance Z, in ohms, of a circuit element is defined as the ratio of the phasor voltage V, in volts, across the element to the phasor current I, in amperes, through the elements. That is, $Z = \dfrac{V}{I}$. If the voltage across a circuit element is $18 + i$ volts and the current through the element is $3 - 4i$ amperes, determine the impedance.

86. Parallel Circuits In an ac circuit with two parallel pathways, the total impedance Z, in ohms, satisfies the formula $\dfrac{1}{Z} = \dfrac{1}{Z_1} + \dfrac{1}{Z_2}$, where Z_1 is the impedance of the first pathway and Z_2 is the impedance of the second pathway. Determine the total impedance if the impedances of the two pathways are $Z_1 = 2 + i$ ohms and $Z_2 = 4 - 3i$ ohms.

87. Use $z = a + bi$ to show that $z + \overline{z} = 2a$ and $z - \overline{z} = 2bi$.

88. Use $z = a + bi$ to show that $\overline{\overline{z}} = z$.

89. Use $z = a + bi$ and $w = c + di$ to show that $\overline{z + w} = \overline{z} + \overline{w}$.

90. Use $z = a + bi$ and $w = c + di$ to show that $\overline{z \cdot w} = \overline{z} \cdot \overline{w}$.

Discussion and Writing

91. Explain to a friend how you would add two complex numbers and how you would multiply two complex numbers. Explain any differences in the two explanations.

92. Write a brief paragraph that compares the method used to rationalize the denominator of a radical expression and the method used to write the quotient of two complex numbers in standard form.

'Are You Prepared?' Answers

1. Integers: $\{-3, 0\}$; rational numbers: $\left\{-3, 0, \dfrac{6}{5}\right\}$ **2.** True **3.** $3(2 - \sqrt{3})$

1.4 Radical Equations; Equations Quadratic in Form; Factorable Equations

PREPARING FOR THIS SECTION *Before getting started, review the following:*

- Square Roots (Section R.2, pp. 23–24)
- Factoring Polynomials (Section R.5, pp. 49–55)
- *n*th Roots; Rational Exponents (Section R.8, pp. 72–76)

Now Work the 'Are You Prepared?' problems on page 122.

OBJECTIVES 1 Solve Radical Equations (p. 118)
2 Solve Equations Quadratic in Form (p. 119)
3 Solve Equations by Factoring (p. 121)

1 Solve Radical Equations

When the variable in an equation occurs in a square root, cube root, and so on, that is, when it occurs in a radical, the equation is called a **radical equation.** Sometimes a suitable operation will change a radical equation to one that is linear or quadratic. A commonly used procedure is to isolate the most complicated radical on one side of the equation and then eliminate it by raising each side to a power equal to the index of the radical. Care must be taken, however, because apparent solutions that are not, in fact, solutions of the original equation may result. These are called **extraneous solutions.** Therefore, we need to check all answers when working with radical equations.

EXAMPLE 1 **Solving a Radical Equation**

Find the real solutions of the equation: $\sqrt[3]{2x - 4} - 2 = 0$

Solution The equation contains a radical whose index is 3. We isolate it on the left side.

$$\sqrt[3]{2x - 4} - 2 = 0$$
$$\sqrt[3]{2x - 4} = 2$$

Now raise each side to the third power (the index of the radical is 3) and solve.

$$\left(\sqrt[3]{2x - 4}\right)^3 = 2^3 \quad \text{Raise each side to the power 3.}$$
$$2x - 4 = 8 \quad \text{Simplify.}$$
$$2x = 12 \quad \text{Add 4 to both sides.}$$
$$x = 6 \quad \text{Divide both sides by 2.}$$

✓ Check: $\sqrt[3]{2(6) - 4} - 2 = \sqrt[3]{12 - 4} - 2 = \sqrt[3]{8} - 2 = 2 - 2 = 0.$

The solution set is $\{6\}$.

Now Work PROBLEM 7

EXAMPLE 2 **Solving a Radical Equation**

Find the real solutions of the equation: $\sqrt{x - 1} = x - 7$

Solution We square both sides since the index of a square root is 2.

$$\sqrt{x - 1} = x - 7$$
$$\left(\sqrt{x - 1}\right)^2 = (x - 7)^2 \quad \text{Square both sides.}$$
$$x - 1 = x^2 - 14x + 49 \quad \text{Remove parentheses.}$$
$$x^2 - 15x + 50 = 0 \quad \text{Put in standard form.}$$

$$(x - 10)(x - 5) = 0 \qquad \text{Factor.}$$
$$x = 10 \quad \text{or} \quad x = 5 \qquad \text{Apply the Zero-Product Property and solve.}$$

✓ **Check:**

$$x = 10: \quad \sqrt{x - 1} = \sqrt{10 - 1} = \sqrt{9} = 3 \text{ and } x - 7 = 10 - 7 = 3$$
$$x = 5: \quad \sqrt{x - 1} = \sqrt{5 - 1} = \sqrt{4} = 2 \text{ and } x - 7 = 5 - 7 = -2$$

The solution $x = 5$ is extraneous; the only solution of the equation is $x = 10$. The solution set is $\{10\}$.

━━━ **Now Work** PROBLEM 19

Sometimes we need to raise each side to a power more than once in order to solve a radical equation.

EXAMPLE 3	**Solving a Radical Equation**

Find the real solutions of the equation: $\sqrt{2x + 3} - \sqrt{x + 2} = 2$

Solution First, we choose to isolate the more complicated radical expression (in this case, $\sqrt{2x + 3}$) on the left side.

$$\sqrt{2x + 3} = \sqrt{x + 2} + 2$$

Now square both sides (the index of the radical on the left is 2).

$$\left(\sqrt{2x + 3}\right)^2 = \left(\sqrt{x + 2} + 2\right)^2 \qquad \text{Square both sides.}$$
$$2x + 3 = \left(\sqrt{x + 2}\right)^2 + 4\sqrt{x + 2} + 4 \qquad \text{Remove parentheses.}$$
$$2x + 3 = x + 2 + 4\sqrt{x + 2} + 4 \qquad \text{Simplify.}$$
$$2x + 3 = x + 6 + 4\sqrt{x + 2} \qquad \text{Combine like terms.}$$

Because the equation still contains a radical, we isolate the remaining radical on the right side and again square both sides.

$$x - 3 = 4\sqrt{x + 2} \qquad \text{Isolate the radical on the right side.}$$
$$(x - 3)^2 = \left(4\sqrt{x + 2}\right)^2 \qquad \text{Square both sides.}$$
$$x^2 - 6x + 9 = 16x + 32 \qquad \text{Remove parentheses.}$$
$$x^2 - 22x - 23 = 0 \qquad \text{Put in standard form.}$$
$$(x - 23)(x + 1) = 0 \qquad \text{Factor.}$$
$$x = 23 \quad \text{or} \quad x = -1$$

The original equation appears to have the solution set $\{-1, 23\}$. However, we have not yet checked.

✓ **Check:**

$$x = 23: \quad \sqrt{2x + 3} - \sqrt{x + 2} = \sqrt{2(23) + 3} - \sqrt{23 + 2} = \sqrt{49} - \sqrt{25} = 7 - 5 = 2$$
$$x = -1: \quad \sqrt{2x + 3} - \sqrt{x + 2} = \sqrt{2(-1) + 3} - \sqrt{-1 + 2} = \sqrt{1} - \sqrt{1} = 1 - 1 = 0$$

The equation has only one solution, 23; the solution -1 is extraneous. The solution set is $\{23\}$.

━━━ **Now Work** PROBLEM 29

2 Solve Equations Quadratic in Form

The equation $x^4 + x^2 - 12 = 0$ is not quadratic in x, but it is quadratic in x^2. That is, if we let $u = x^2$, we get $u^2 + u - 12 = 0$, a quadratic equation. This equation can be solved for u and, in turn, by using $u = x^2$, we can find the solutions x of the original equation.

In general, if an appropriate substitution u transforms an equation into one of the form

$$au^2 + bu + c = 0 \qquad a \neq 0$$

then the original equation is called an **equation of the quadratic type** or an **equation quadratic in form.**

The difficulty of solving such an equation lies in the determination that the equation is, in fact, quadratic in form. After you are told an equation is quadratic in form, it is easy enough to see it, but some practice is needed to enable you to recognize such equations on your own.

EXAMPLE 4 **Solving Equations That Are Quadratic in Form**

Find the real solutions of the equation: $(x + 2)^2 + 11(x + 2) - 12 = 0$

Solution For this equation, let $u = x + 2$. Then $u^2 = (x + 2)^2$, and the original equation,

$$(x + 2)^2 + 11(x + 2) - 12 = 0$$

becomes

$$u^2 + 11u - 12 = 0 \quad \text{Let } u = x + 2. \text{ Then } u^2 = (x + 2)^2.$$
$$(u + 12)(u - 1) = 0 \quad \text{Factor.}$$
$$u = -12 \quad \text{or} \quad u = 1 \quad \text{Solve.}$$

But we want to solve for x. Because $u = x + 2$, we have

$$x + 2 = -12 \quad \text{or} \quad x + 2 = 1$$
$$x = -14 \qquad\qquad x = -1$$

✓ Check: $x = -14$: $(-14 + 2)^2 + 11(-14 + 2) - 12$
$$= (-12)^2 + 11(-12) - 12 = 144 - 132 - 12 = 0$$
$x = -1$: $(-1 + 2)^2 + 11(-1 + 2) - 12 = 1 + 11 - 12 = 0$

The original equation has the solution set $\{-14, -1\}$.

EXAMPLE 5 **Solving Equations That Are Quadratic in Form**

Find the real solutions of the equation: $(x^2 - 1)^2 + (x^2 - 1) - 12 = 0$

Solution For the equation $(x^2 - 1)^2 + (x^2 - 1) - 12 = 0$, we let $u = x^2 - 1$ so that $u^2 = (x^2 - 1)^2$. Then the original equation,

$$(x^2 - 1)^2 + (x^2 - 1) - 12 = 0$$

becomes

$$u^2 + u - 12 = 0 \quad \text{Let } u = x^2 - 1. \text{ Then } u^2 = (x^2 - 1)^2.$$
$$(u + 4)(u - 3) = 0 \quad \text{Factor.}$$
$$u = -4 \quad \text{or} \quad u = 3 \quad \text{Solve.}$$

But remember that we want to solve for x. Because $u = x^2 - 1$, we have

$$x^2 - 1 = -4 \quad \text{or} \quad x^2 - 1 = 3$$
$$x^2 = -3 \qquad\qquad x^2 = 4$$

The first of these has no real solution; the second has the solution set $\{-2, 2\}$.

✓ Check: $x = -2$: $(4 - 1)^2 + (4 - 1) - 12 = 9 + 3 - 12 = 0$
$x = 2$: $(4 - 1)^2 + (4 - 1) - 12 = 9 + 3 - 12 = 0$

The original equation has the solution set $\{-2, 2\}$.

| EXAMPLE 6 | **Solving Equations That Are Quadratic in Form** |

Find the real solutions of the equation: $x + 2\sqrt{x} - 3 = 0$

Solution For the equation $x + 2\sqrt{x} - 3 = 0$, let $u = \sqrt{x}$. Then $u^2 = x$, and the original equation,

$$x + 2\sqrt{x} - 3 = 0$$

becomes

$$u^2 + 2u - 3 = 0 \quad \text{Let } u = \sqrt{x}. \text{ Then } u^2 = x.$$
$$(u + 3)(u - 1) = 0 \quad \text{Factor.}$$
$$u = -3 \quad \text{or} \quad u = 1 \quad \text{Solve.}$$

Since $u = \sqrt{x}$, we have $\sqrt{x} = -3$ or $\sqrt{x} = 1$. The first of these, $\sqrt{x} = -3$, has no real solution, since the square root of a real number is never negative. The second, $\sqrt{x} = 1$, has the solution $x = 1$.

✓ Check: $1 + 2\sqrt{1} - 3 = 1 + 2 - 3 = 0$

The original equation has the solution set {1}.

✎ ANOTHER METHOD FOR SOLVING EXAMPLE 6 WOULD BE TO TREAT IT AS A RADICAL EQUATION. SOLVE IT THIS WAY FOR PRACTICE.

The idea should now be clear. If an equation contains an expression and that same expression squared, make a substitution for the expression. You may get a quadratic equation.

✎ **Now Work** PROBLEM 51

3 Solve Equations by Factoring

We have already solved certain quadratic equations using factoring. Let's look at examples of other kinds of equations that can be solved by factoring.

| EXAMPLE 7 | **Solving Equations by Factoring** |

Solve the equation: $x^4 = 4x^2$

Solution We begin by collecting all terms on one side. This results in 0 on one side and an expression to be factored on the other.

$$x^4 = 4x^2$$
$$x^4 - 4x^2 = 0$$
$$x^2(x^2 - 4) = 0 \quad \text{Factor.}$$
$$x^2 = 0 \quad \text{or} \quad x^2 - 4 = 0 \quad \text{Apply the Zero-Product Property.}$$
$$x^2 = 4$$
$$x = 0 \quad \text{or} \quad x = -2 \quad \text{or} \quad x = 2$$

The solution set is $\{-2, 0, 2\}$.

✓ Check: $x = -2$: $(-2)^4 = 16$ and $4(-2)^2 = 16$ So -2 is a solution.
 $x = 0$: $0^4 = 0$ and $4 \cdot 0^2 = 0$ So 0 is a solution.
 $x = 2$: $2^4 = 16$ and $4 \cdot 2^2 = 16$ So 2 is a solution.

| EXAMPLE 8 | Solving Equations by Factoring |

Solve the equation: $x^3 - x^2 - 4x + 4 = 0$

Solution Do you recall the method of factoring by grouping? (If not, review pp. 53–54.) We group the terms of $x^3 - x^2 - 4x + 4 = 0$ as follows:

$$(x^3 - x^2) - (4x - 4) = 0$$

Factor out x^2 from the first grouping and 4 from the second.

$$x^2(x - 1) - 4(x - 1) = 0$$

This reveals the common factor $(x - 1)$, so we have

$$(x^2 - 4)(x - 1) = 0$$
$$(x - 2)(x + 2)(x - 1) = 0 \qquad \text{Factor again.}$$
$$x - 2 = 0 \quad \text{or} \quad x + 2 = 0 \quad \text{or} \quad x - 1 = 0 \quad \text{Set each factor equal to 0.}$$
$$x = 2 \qquad\qquad x = -2 \qquad\qquad x = 1 \quad \text{Solve.}$$

The solution set is $\{-2, 1, 2\}$.

✓ **Check:**

$x = -2$: $(-2)^3 - (-2)^2 - 4(-2) + 4 = -8 - 4 + 8 + 4 = 0$ −2 is a solution.
$x = 1$: $1^3 - 1^2 - 4(1) + 4 = 1 - 1 - 4 + 4 = 0$ 1 is a solution.
$x = 2$: $2^3 - 2^2 - 4(2) + 4 = 8 - 4 - 8 + 4 = 0$ 2 is a solution.

━ **Now Work** PROBLEM 79

1.4 Assess Your Understanding

'Are You Prepared?' *Answers are given at the end of these exercises. If you get a wrong answer, read the pages listed in red.*

1. *True or False* The principal square root of any nonnegative real number is always nonnegative. (pp. 23–24)

2. $\sqrt[3]{-8} = $ _____ (pp. 72–76)

3. Factor $6x^3 - 2x^2$ (pp. 49–55)

Concepts and Vocabulary

4. When an apparent solution does not satisfy the original equation, it is called a(n) _____ solution.

5. If u is an expression involving x, the equation $au^2 + bu + c = 0, a \neq 0$, is called a(n) equation _____ _____ _____.

6. *True or False* Radical equations sometimes have extraneous solutions.

Skill Building

In Problems 7–40, find the real solutions of each equation.

7. $\sqrt{2t - 1} = 1$

8. $\sqrt{3t + 4} = 2$

9. $\sqrt{3t + 4} = -6$

10. $\sqrt{5t + 3} = -2$

11. $\sqrt[3]{1 - 2x} - 3 = 0$

12. $\sqrt[3]{1 - 2x} - 1 = 0$

13. $\sqrt[4]{5x - 4} = 2$

14. $\sqrt[5]{2x - 3} = -1$

15. $\sqrt[5]{x^2 + 2x} = -1$

16. $\sqrt[4]{x^2 + 16} = \sqrt{5}$

17. $x = 8\sqrt{x}$

18. $x = 3\sqrt{x}$

19. $\sqrt{15 - 2x} = x$

20. $\sqrt{12 - x} = x$

21. $x = 2\sqrt{x - 1}$

22. $x = 2\sqrt{-x - 1}$

23. $\sqrt{x^2 - x - 4} = x + 2$

24. $\sqrt{3 - x + x^2} = x - 2$

25. $3 + \sqrt{3x + 1} = x$

26. $2 + \sqrt{12 - 2x} = x$

27. $\sqrt{2x + 3} - \sqrt{x + 1} = 1$

28. $\sqrt{3x + 7} + \sqrt{x + 2} = 1$

29. $\sqrt{3x + 1} - \sqrt{x - 1} = 2$

30. $\sqrt{3x - 5} - \sqrt{x + 7} = 2$

31. $\sqrt{3 - 2\sqrt{x}} = \sqrt{x}$

32. $\sqrt{10 + 3\sqrt{x}} = \sqrt{x}$

33. $(3x + 1)^{1/2} = 4$

34. $(3x - 5)^{1/2} = 2$

35. $(5x - 2)^{1/3} = 2$

36. $(2x + 1)^{1/3} = -1$

37. $(x^2 + 9)^{1/2} = 5$

38. $(x^2 - 16)^{1/2} = 9$

39. $x^{3/2} - 3x^{1/2} = 0$

40. $x^{3/4} - 9x^{1/4} = 0$

In Problems 41–72, find the real solutions of each equation.

41. $x^4 - 5x^2 + 4 = 0$

42. $x^4 - 10x^2 + 25 = 0$

43. $3x^4 - 2x^2 - 1 = 0$

44. $2x^4 - 5x^2 - 12 = 0$

45. $x^6 + 7x^3 - 8 = 0$

46. $x^6 - 7x^3 - 8 = 0$

47. $(x + 2)^2 + 7(x + 2) + 12 = 0$

48. $(2x + 5)^2 - (2x + 5) - 6 = 0$

49. $(3x + 4)^2 - 6(3x + 4) + 9 = 0$

50. $(2 - x)^2 + (2 - x) - 20 = 0$

51. $2(s + 1)^2 - 5(s + 1) = 3$

52. $3(1 - y)^2 + 5(1 - y) + 2 = 0$

53. $x - 4x\sqrt{x} = 0$

54. $x + 8\sqrt{x} = 0$

55. $x + \sqrt{x} = 20$

56. $x + \sqrt{x} = 6$

57. $t^{1/2} - 2t^{1/4} + 1 = 0$

58. $z^{1/2} - 4z^{1/4} + 4 = 0$

59. $4x^{1/2} - 9x^{1/4} + 4 = 0$

60. $x^{1/2} - 3x^{1/4} + 2 = 0$

61. $\sqrt[4]{5x^2 - 6} = x$

62. $\sqrt[4]{4 - 5x^2} = x$

63. $x^2 + 3x + \sqrt{x^2 + 3x} = 6$

64. $x^2 - 3x - \sqrt{x^2 - 3x} = 2$

65. $\dfrac{1}{(x + 1)^2} = \dfrac{1}{x + 1} + 2$

66. $\dfrac{1}{(x - 1)^2} + \dfrac{1}{x - 1} = 12$

67. $3x^{-2} - 7x^{-1} - 6 = 0$

68. $2x^{-2} - 3x^{-1} - 4 = 0$

69. $2x^{2/3} - 5x^{1/3} - 3 = 0$

70. $3x^{4/3} + 5x^{2/3} - 2 = 0$

71. $\left(\dfrac{v}{v + 1}\right)^2 + \dfrac{2v}{v + 1} = 8$

72. $\left(\dfrac{y}{y - 1}\right)^2 = 6\left(\dfrac{y}{y - 1}\right) + 7$

In Problems 73–88, find the real solutions of each equation by factoring.

73. $x^3 - 9x = 0$

74. $x^4 - x^2 = 0$

75. $4x^3 = 3x^2$

76. $x^5 = 4x^3$

77. $x^3 + x^2 - 20x = 0$

78. $x^3 + 6x^2 - 7x = 0$

79. $x^3 + x^2 - x - 1 = 0$

80. $x^3 + 4x^2 - x - 4 = 0$

81. $x^3 - 3x^2 - 4x + 12 = 0$

82. $x^3 - 3x^2 - x + 3 = 0$

83. $2x^3 + 4 = x^2 + 8x$

84. $3x^3 + 4x^2 = 27x + 36$

85. $5x^3 + 45x = 2x^2 + 18$

86. $3x^3 + 12x = 5x^2 + 20$

87. $x(x^2 - 3x)^{1/3} + 2(x^2 - 3x)^{4/3} = 0$

88. $3x(x^2 + 2x)^{1/2} - 2(x^2 + 2x)^{3/2} = 0$

In Problems 89–94, find the real solutions of each equation. Use a calculator to express any solutions rounded to two decimal places.

89. $x - 4x^{1/2} + 2 = 0$

90. $x^{2/3} + 4x^{1/3} + 2 = 0$

91. $x^4 + \sqrt{3}x^2 - 3 = 0$

92. $x^4 + \sqrt{2}x^2 - 2 = 0$

93. $\pi(1 + t)^2 = \pi + 1 + t$

94. $\pi(1 + r)^2 = 2 + \pi(1 + r)$

95. If $k = \dfrac{x + 3}{x - 3}$ and $k^2 - k = 12$, find x.

96. If $k = \dfrac{x + 3}{x - 4}$ and $k^2 - 3k = 28$, find x.

Applications

97. Physics: Using Sound to Measure Distance The distance to the surface of the water in a well can sometimes be found by dropping an object into the well and measuring the time elapsed until a sound is heard. If t_1 is the time (measured in seconds) that it takes for the object to strike the water, then t_1 will obey the equation $s = 16t_1^2$, where s is the distance (measured in feet). It follows that $t_1 = \dfrac{\sqrt{s}}{4}$. Suppose that t_2 is the time that it takes for the sound of the impact to reach your ears. Because sound waves are known to travel at a speed of approximately 1100 feet per second, the time t_2 to travel the distance s will be $t_2 = \dfrac{s}{1100}$. See the illustration.

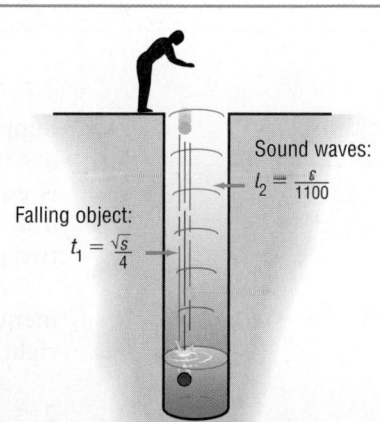

Sound waves: $t_2 = \dfrac{s}{1100}$

Falling object: $t_1 = \dfrac{\sqrt{s}}{4}$

Now $t_1 + t_2$ is the total time that elapses from the moment that the object is dropped to the moment that a sound is heard. We have the equation

$$\text{Total time elapsed} = \frac{\sqrt{s}}{4} + \frac{s}{1100}$$

Find the distance to the water's surface if the total time elapsed from dropping a rock to hearing it hit water is 4 seconds.

98. Crushing Load A civil engineer relates the thickness T, in inches, and height H, in feet, of a square wooden pillar to its crushing load L, in tons, using the model $T = \sqrt[4]{\dfrac{LH^2}{25}}$. If a square wooden pillar is 4 inches thick and 10 feet high, what is its crushing load?

99. Foucault's Pendulum The period of a pendulum is the time it takes the pendulum to make one full swing back and forth. The period T, in seconds, is given by the formula $T = 2\pi\sqrt{\dfrac{l}{32}}$, where l is the length, in feet, of the pendulum. In 1851, Jean-Bernard-Leon Foucault demonstrated the axial rotation of Earth using a large pendulum that he hung in the Pantheon in Paris. The period of Foucault's pendulum was approximately 16.5 seconds. What was its length?

Discussion and Writing

100. Make up a radical equation that has no solution.

101. Make up a radical equation that has an extraneous solution.

102. Discuss the step in the solving process for radical equations that leads to the possibility of extraneous solutions. Why is there no such possibility for linear and quadratic equations?

'Are You Prepared?' Answers

1. True **2.** -2 **3.** $2x^2(3x - 1)$

1.5 Solving Inequalities

PREPARING FOR THIS SECTION *Before getting started, review the following:*

• Algebra Essentials (Section R.2, pp. 17–19)

Now work the 'Are You Prepared?' problems on page 132.

OBJECTIVES	1 Use Interval Notation (p. 125)
	2 Use Properties of Inequalities (p. 126)
	3 Solve Inequalities (p. 128)
	4 Solve Combined Inequalities (p. 129)

Suppose that a and b are two real numbers and $a < b$. We shall use the notation $a < x < b$ to mean that x is a number *between* a and b. The expression $a < x < b$ is equivalent to the two inequalities $a < x$ and $x < b$. Similarly, the expression $a \leq x \leq b$ is equivalent to the two inequalities $a \leq x$ and $x \leq b$. The remaining two possibilities, $a \leq x < b$ and $a < x \leq b$, are defined similarly.

Although it is acceptable to write $3 \geq x \geq 2$, it is preferable to reverse the inequality symbols and write instead $2 \leq x \leq 3$ so that, as you read from left to right, the values go from smaller to larger.

A statement such as $2 \leq x \leq 1$ is false because there is no number x for which $2 \leq x$ and $x \leq 1$. Finally, we never mix inequality symbols, as in $2 \leq x \geq 3$.

1 Use Interval Notation

Let a and b represent two real numbers with $a < b$.

DEFINITION

A **closed interval,** denoted by **[a, b],** consists of all real numbers x for which $a \leq x \leq b$.

An **open interval,** denoted by **(a, b),** consists of all real numbers x for which $a < x < b$.

The **half-open,** or **half-closed, intervals** are **(a, b],** consisting of all real numbers x for which $a < x \leq b$, and **[a, b),** consisting of all real numbers x for which $a \leq x < b$.

In each of these definitions, a is called the **left endpoint** and b the **right endpoint** of the interval.

The symbol ∞ (read as "infinity") is not a real number, but a notational device used to indicate unboundedness in the positive direction. The symbol $-\infty$ (read as "negative infinity") also is not a real number, but a notational device used to indicate unboundedness in the negative direction. Using the symbols ∞ and $-\infty$, we can define five other kinds of intervals:

$[a, \infty)$	Consists of all real numbers x for which $x \geq a$
(a, ∞)	Consists of all real numbers x for which $x > a$
$(-\infty, a]$	Consists of all real numbers x for which $x \leq a$
$(-\infty, a)$	Consists of all real numbers x for which $x < a$
$(-\infty, \infty)$	Consists of all real numbers x

Note that ∞ and $-\infty$ are never included as endpoints, since neither is a real number.

Table 1 summarizes interval notation, corresponding inequality notation, and their graphs.

Table 1

Interval	Inequality	Graph
The open interval (a, b)	$a < x < b$	
The closed interval $[a, b]$	$a \leq x \leq b$	
The half-open interval $[a, b)$	$a \leq x < b$	
The half-open interval $(a, b]$	$a < x \leq b$	
The interval $[a, \infty)$	$x \geq a$	
The interval (a, ∞)	$x > a$	
The interval $(-\infty, a]$	$x \leq a$	
The interval $(-\infty, a)$	$x < a$	
The interval $(-\infty, \infty)$	All real numbers	

EXAMPLE 1 **Writing Inequalities Using Interval Notation**

Write each inequality using interval notation.

(a) $1 \leq x \leq 3$ (b) $-4 < x < 0$ (c) $x > 5$ (d) $x \leq 1$

Solution

(a) $1 \leq x \leq 3$ describes all numbers x between 1 and 3, inclusive. In interval notation, we write $[1, 3]$.

(b) In interval notation, $-4 < x < 0$ is written $(-4, 0)$.

(c) $x > 5$ consists of all numbers x greater than 5. In interval notation, we write $(5, \infty)$.

(d) In interval notation, $x \leq 1$ is written $(-\infty, 1]$.

| EXAMPLE 2 | **Writing Intervals Using Inequality Notation** |

Write each interval as an inequality involving x.

(a) $[1, 4)$ (b) $(2, \infty)$ (c) $[2, 3]$ (d) $(-\infty, -3]$

Solution

(a) $[1, 4)$ consists of all numbers x for which $1 \leq x < 4$.

(b) $(2, \infty)$ consists of all numbers x for which $x > 2$.

(c) $[2, 3]$ consists of all numbers x for which $2 \leq x \leq 3$.

(d) $(-\infty, -3]$ consists of all numbers x for which $x \leq -3$.

=== **Now Work** PROBLEMS 11, 23, AND 31

2 Use Properties of Inequalities

The product of two positive real numbers is positive, the product of two negative real numbers is positive, and the product of 0 and 0 is 0. For any real number a, the value of a^2 is 0 or positive; that is, a^2 is nonnegative. This is called the **nonnegative property.**

> **In Words**
> The square of a real number is never negative.

Nonnegative Property

For any real numbers a,

$$a^2 \geq 0 \tag{1}$$

If we add the same number to both sides of an inequality, we obtain an equivalent inequality. For example, since $3 < 5$, then $3 + 4 < 5 + 4$ or $7 < 9$. This is called the **addition property** of inequalities.

Addition Property of Inequalities

For real numbers a, b, and c,

$$\text{If } a < b, \text{ then } a + c < b + c. \tag{2a}$$

$$\text{If } a > b, \text{ then } a + c > b + c. \tag{2b}$$

The addition property states that the sense, or direction, of an inequality remains unchanged if the same number is added to each side. Figure 3 illustrates the addition property (2a). In Figure 3(a), we see that a lies to the left of b. If c is positive, then $a + c$ and $b + c$ each lie c units to the right of a and b, respectively. Consequently, $a + c$ must lie to the left of $b + c$; that is, $a + c < b + c$. Figure 3(b) illustrates the situation if c is negative.

Figure 3

(a) If $a < b$ and $c > 0$, then $a + c < b + c$.

(b) If $a < b$ and $c < 0$, then $a + c < b + c$.

=== DRAW AN ILLUSTRATION SIMILAR TO FIGURE 3
THAT ILLUSTRATES THE ADDITION PROPERTY (2b)

EXAMPLE 3 | **Addition Property of Inequalities**

(a) If $x < -5$, then $x + 5 < -5 + 5$ or $x + 5 < 0$.
(b) If $x > 2$, then $x + (-2) > 2 + (-2)$ or $x - 2 > 0$.

■

===== **Now Work** PROBLEM 39

We will use two examples to arrive at our next property.

EXAMPLE 4 | **Multiplying an Inequality by a Positive Number**

Express as an inequality the result of multiplying each side of the inequality $3 < 7$ by 2.

Solution We begin with

$$3 < 7$$

Multiplying each side by 2 yields the numbers 6 and 14, so we have

$$6 < 14$$

■

EXAMPLE 5 | **Multiplying an Inequality by a Negative Number**

Express as an inequality the result of multiplying each side of the inequality $9 > 2$ by -4.

Solution We begin with

$$9 > 2$$

Multiplying each side by -4 yields the numbers -36 and -8, so we have

$$-36 < -8$$

■

In Words
Multiplying by a negative number reverses the inequality.

Note that the effect of multiplying both sides of $9 > 2$ by the negative number -4 is that the direction of the inequality symbol is reversed.

Examples 4 and 5 illustrate the following general **multiplication properties** for inequalities:

Multiplication Properties for Inequalities

For real numbers a, b, and c,

> If $a < b$ and if $c > 0$, then $ac < bc$.
> If $a < b$ and if $c < 0$, then $ac > bc$. **(3a)**
>
> If $a > b$ and if $c > 0$, then $ac > bc$.
> If $a > b$ and if $c < 0$, then $ac < bc$. **(3b)**

The multiplication properties state that the sense, or direction, of an inequality *remains the same* if each side is multiplied by a *positive* real number, whereas the direction is *reversed* if each side is multiplied by a *negative* real number.

EXAMPLE 6 | **Multiplication Property of Inequalities**

(a) If $2x < 6$, then $\frac{1}{2}(2x) < \frac{1}{2}(6)$ or $x < 3$.

(b) If $\frac{x}{-3} > 12$, then $-3\left(\frac{x}{-3}\right) < -3(12)$ or $x < -36$.

(c) If $-4x < -8$, then $\dfrac{-4x}{-4} > \dfrac{-8}{-4}$ or $x > 2$.

(d) If $-x > 8$, then $(-1)(-x) < (-1)(8)$ or $x < -8$.

━━━━━ **Now Work** PROBLEM 45

The **reciprocal property** states that the reciprocal of a positive real number is positive and that the reciprocal of a negative real number is negative.

Reciprocal Property for Inequalities

If $a > 0$, then $\dfrac{1}{a} > 0$ $\qquad$ If $\dfrac{1}{a} > 0$, then $a > 0$	**(4a)**
If $a < 0$, then $\dfrac{1}{a} < 0$ $\qquad$ If $\dfrac{1}{a} < 0$, then $a < 0$	**(4b)**

3 Solve Inequalities

An **inequality in one variable** is a statement involving two expressions, at least one containing the variable, separated by one of the inequality symbols $<$, $\le$, $>$, or $\ge$. To **solve an inequality** means to find all values of the variable for which the statement is true. These values are called **solutions** of the inequality.

For example, the following are all inequalities involving one variable x:

$$x + 5 < 8 \qquad 2x - 3 \ge 4 \qquad x^2 - 1 \le 3 \qquad \frac{x + 1}{x - 2} > 0$$

As with equations, one method for solving an inequality is to replace it by a series of equivalent inequalities until an inequality with an obvious solution, such as $x < 3$, is obtained. We obtain equivalent inequalities by applying some of the same properties as those used to find equivalent equations. The addition property and the multiplication properties form the bases for the following procedures.

Procedures That Leave the Inequality Symbol Unchanged

1. Simplify both sides of the inequality by combining like terms and eliminating parentheses:

 Replace $\qquad (x + 2) + 6 > 2x + 5(x + 1)$

 by $\qquad\qquad x + 8 > 7x + 5$

2. Add or subtract the same expression on both sides of the inequality:

 Replace $\qquad\qquad 3x - 5 < 4$

 by $\qquad (3x - 5) + 5 < 4 + 5$

3. Multiply or divide both sides of the inequality by the same positive expression:

 Replace $\qquad 4x > 16$ by $\dfrac{4x}{4} > \dfrac{16}{4}$

Procedures That Reverse the Sense or Direction of the Inequality Symbol

1. Interchange the two sides of the inequality:

 Replace $\qquad 3 < x$ by $x > 3$

2. Multiply or divide both sides of the inequality by the same *negative* expression:

 Replace $\qquad -2x > 6$ by $\dfrac{-2x}{-2} < \dfrac{6}{-2}$

As the examples that follow illustrate, we solve inequalities using many of the same steps that we would use to solve equations. In writing the solution of an inequality, we may use either set notation or interval notation, whichever is more convenient.

| EXAMPLE 7 | **Solving an Inequality** |

Solve the inequality: $3 - 2x < 5$
Graph the solution set.

Solution

$$3 - 2x < 5$$
$$3 - 2x - 3 < 5 - 3 \qquad \text{Subtract 3 from both sides.}$$
$$-2x < 2 \qquad \text{Simplify.}$$
$$\frac{-2x}{-2} > \frac{2}{-2} \qquad \text{Divide both sides by } -2. \text{ (The sense of the inequality symbol is reversed.)}$$
$$x > -1 \qquad \text{Simplify.}$$

Figure 4

The solution set is $\{x \mid x > -1\}$ or, using interval notation, all numbers in the interval $(-1, \infty)$. See Figure 4 for the graph. ∎

| EXAMPLE 8 | **Solving an Inequality** |

Solve the inequality: $4x + 7 \geq 2x - 3$
Graph the solution set.

Solution

$$4x + 7 \geq 2x - 3$$
$$4x + 7 - 7 \geq 2x - 3 - 7 \qquad \text{Subtract 7 from both sides.}$$
$$4x \geq 2x - 10 \qquad \text{Simplify.}$$
$$4x - 2x \geq 2x - 10 - 2x \qquad \text{Subtract 2x from both sides.}$$
$$2x \geq -10 \qquad \text{Simplify.}$$
$$\frac{2x}{2} \geq \frac{-10}{2} \qquad \text{Divide both sides by 2. (The direction of the inequality symbol is unchanged.)}$$
$$x \geq -5 \qquad \text{Simplify.}$$

Figure 5

The solution set is $\{x \mid x \geq -5\}$ or, using interval notation, all numbers in the interval $[-5, \infty)$. See Figure 5 for the graph. ∎

 Now Work PROBLEM 53

4 Solve Combined Inequalities

| EXAMPLE 9 | **Solving Combined Inequalities** |

Solve the inequality: $-5 < 3x - 2 < 1$
Graph the solution set.

Solution

Recall that the inequality

$$-5 < 3x - 2 < 1$$

is equivalent to the two inequalities

$$-5 < 3x - 2 \quad \text{and} \quad 3x - 2 < 1$$

We will solve each of these inequalities separately.

$$-5 < 3x - 2 \qquad\qquad\qquad 3x - 2 < 1$$

$-5 + 2 < 3x - 2 + 2$ Add 2 to both sides. $3x - 2 + 2 < 1 + 2$

$$-3 < 3x \qquad\qquad \text{Simplify.} \qquad\qquad 3x < 3$$

$$\frac{-3}{3} < \frac{3x}{3} \qquad \text{Divide both sides by 3.} \qquad \frac{3x}{3} < \frac{3}{3}$$

$$-1 < x \qquad\qquad \text{Simplify.} \qquad\qquad x < 1$$

The solution set of the original pair of inequalities consists of all x for which

$$-1 < x \quad \text{and} \quad x < 1$$

Figure 6

This may be written more compactly as $\{x | -1 < x < 1\}$. In interval notation, the solution is $(-1, 1)$. See Figure 6 for the graph.

We observe in the preceding process that the two inequalities we solved required exactly the same steps. A shortcut to solving the original inequality algebraically is to deal with the two inequalities at the same time, as follows:

$$-5 < \quad 3x - 2 \quad < 1$$

$-5 + 2 < 3x - 2 + 2 < 1 + 2$ Add 2 to each part.

$$-3 < \quad 3x \quad < 3 \qquad \text{Simplify.}$$

$$\frac{-3}{3} < \quad \frac{3x}{3} \quad < \frac{3}{3} \qquad \text{Divide each part by 3.}$$

$$-1 < \quad x \quad < 1 \qquad \text{Simplify.}$$

We use this shortcut in the next example.

| EXAMPLE 10 | **Solving Combined Inequalities** |

Solve the inequality: $-1 \le \dfrac{3 - 5x}{2} \le 9$

Graph the solution set.

Solution

$$-1 \le \quad \frac{3 - 5x}{2} \quad \le 9$$

$$2(-1) \le 2\left(\frac{3 - 5x}{2}\right) \le 2(9) \qquad \text{Multiply each part by 2 to remove the denominator.}$$

$$-2 \le \quad 3 - 5x \quad \le 18 \qquad \text{Simplify.}$$

$$-2 - 3 \le 3 - 5x - 3 \le 18 - 3 \qquad \begin{array}{l}\text{Subtract 3 from each part to isolate}\\ \text{the term containing x.}\end{array}$$

$$-5 \le \quad -5x \quad \le 15 \qquad \text{Simplify.}$$

$$\frac{-5}{-5} \ge \quad \frac{-5x}{-5} \quad \ge \frac{15}{-5} \qquad \begin{array}{l}\text{Divide each part by } -5 \text{ (reverse the sense}\\ \text{of each inequality symbol).}\end{array}$$

$$1 \ge \quad x \quad \ge -3 \qquad \text{Simplify.}$$

$$-3 \le \quad x \quad \le 1 \qquad \begin{array}{l}\text{Reverse the order so that the numbers}\\ \text{get larger as you read from left to right.}\end{array}$$

Figure 7

The solution set is $\{x | -3 \le x \le 1\}$, that is, all x in the interval $[-3, 1]$. Figure 7 illustrates the graph.

Now Work PROBLEM 73

EXAMPLE 11	Using the Reciprocal Property to Solve an Inequality

Solve the inequality: $(4x - 1)^{-1} > 0$
Graph the solution set.

Solution Since $(4x - 1)^{-1} = \dfrac{1}{4x - 1}$ and since the Reciprocal Property states that when $\dfrac{1}{a} > 0$ then $a > 0$, we have

$$(4x - 1)^{-1} > 0$$
$$\frac{1}{4x - 1} > 0$$
$$4x - 1 > 0 \quad \text{Reciprocal Property}$$
$$4x > 1$$
$$x > \frac{1}{4}$$

Figure 8

The solution set is $\left\{ x \middle| x > \dfrac{1}{4} \right\}$, that is, all x in the interval $\left(\dfrac{1}{4}, \infty \right)$. Figure 8 illustrates the graph.

Now Work PROBLEM 83

EXAMPLE 12	Creating Equivalent Inequalities

If $-1 < x < 4$, find a and b so that $a < 2x + 1 < b$.

Solution The idea here is to change the middle part of the combined inequality from x to $2x + 1$, using properties of inequalities.

$$
\begin{aligned}
-1 <\ &x\ < 4 \\
-2 <\ &2x\ < 8 \quad \text{Multiply each part by 2.} \\
-1 <\ &2x + 1 < 9 \quad \text{Add 1 to each part.}
\end{aligned}
$$

Now we see that $a = -1$ and $b = 9$.

Now Work PROBLEM 91

Application

Let's look at an applied problem involving inequalities.

EXAMPLE 13	Physics: Ohm's Law

In electricity, Ohm's law states that $E = IR$, where E is the voltage (in volts), I is the current (in amperes), and R is the resistance (in ohms). An air-conditioning unit is rated at a resistance of 10 ohms. If the voltage varies from 110 to 120 volts, inclusive, what corresponding range of current will the air conditioner draw?

Solution The voltage lies between 110 and 120, inclusive, so

$$
\begin{aligned}
110 \le\ &E\ \le 120 \\
110 \le\ &IR\ \le 120 \quad \text{Ohm's law, } E = IR \\
110 \le\ &I(10) \le 120 \quad R = 10 \\
\frac{110}{10} \le\ &\frac{I(10)}{10} \le \frac{120}{10} \quad \text{Divide each part by 10.} \\
11 \le\ &I\ \le 12 \quad \text{Simplify.}
\end{aligned}
$$

The air conditioner will draw between 11 and 12 amperes of current, inclusive.

1.5 Assess Your Understanding

'Are You Prepared?' *Answers are given at the end of these exercises. If you get a wrong answer, read the pages listed in red.*

1. Graph the inequality: $x \geq -2$. (pp. 17–19)

2. *True or False* $-5 > -3$ (pp. 17–19)

Concepts and Vocabulary

3. If each side of an inequality is multiplied by a(n) _____ number, then the sense of the inequality symbol is reversed.

4. A(n) _____ _____, denoted $[a, b]$, consists of all real numbers x for which $a \leq x \leq b$.

5. The _____ _____ state that the sense, or direction, of an inequality remains the same if each side is multiplied by a positive number, while the direction is reversed if each side is multiplied by a negative number.

In Problems 6–9, assume that $a < b$ and $c < 0$.

6. *True or False* $a + c < b + c$

7. *True or False* $a - c < b - c$

8. *True or False* $ac > bc$

9. *True or False* $\dfrac{a}{c} < \dfrac{b}{c}$

10. *True or False* The square of any real number is always nonnegative.

Skill Building

In Problems 11–16, express the graph shown in blue using interval notation. Also express each as an inequality involving x.

11.

12.

13.

14.

15.

16.

In Problems 17–22, an inequality is given. Write the inequality obtained by:
(a) *Adding 3 to each side of the given inequality.*
(b) *Subtracting 5 from each side of the given inequality.*
(c) *Multiplying each side of the given inequality by 3.*
(d) *Multiplying each side of the given inequality by −2.*

17. $3 < 5$ **18.** $2 > 1$ **19.** $4 > -3$ **20.** $-3 > -5$ **21.** $2x + 1 < 2$ **22.** $1 - 2x > 5$

In Problems 23–30, write each inequality using interval notation, and illustrate each inequality using the real number line.

23. $0 \leq x \leq 4$ **24.** $-1 < x < 5$ **25.** $4 \leq x < 6$ **26.** $-2 < x < 0$

27. $x \geq 4$ **28.** $x \leq 5$ **29.** $x < -4$ **30.** $x > 1$

In Problems 31–38, write each interval as an inequality involving x, and illustrate each inequality using the real number line.

31. $[2, 5]$ **32.** $(1, 2)$ **33.** $(-3, -2)$ **34.** $[0, 1)$

35. $[4, \infty)$ **36.** $(-\infty, 2]$ **37.** $(-\infty, -3)$ **38.** $(-8, \infty)$

In Problems 39–52, fill in the blank with the correct inequality symbol.

39. If $x < 5$, then $x - 5$ _____ 0.

40. If $x < -4$, then $x + 4$ _____ 0.

41. If $x > -4$, then $x + 4$ _____ 0.

42. If $x > 6$, then $x - 6$ _____ 0.

43. If $x \geq -4$, then $3x$ _____ -12.

44. If $x \leq 3$, then $2x$ _____ 6.

45. If $x > 6$, then $-2x$ _____ -12.

46. If $x > -2$, then $-4x$ _____ 8.

47. If $x \geq 5$, then $-4x$ _____ -20.

48. If $x \leq -4$, then $-3x$ _____ 12.

49. If $2x > 6$, then x _____ 3.

50. If $3x \leq 12$, then x _____ 4.

51. If $-\dfrac{1}{2}x \leq 3$, then x _____ -6.

52. If $-\dfrac{1}{4}x > 1$, then x _____ -4.

In Problems 53–88, solve each inequality. Express your answer using set notation or interval notation. Graph the solution set.

53. $x + 1 < 5$

54. $x - 6 < 1$

55. $1 - 2x \leq 3$

56. $2 - 3x \leq 5$

57. $3x - 7 > 2$

58. $2x + 5 > 1$

59. $3x - 1 \geq 3 + x$

60. $2x - 2 \geq 3 + x$

61. $-2(x + 3) < 8$

62. $-3(1 - x) < 12$

63. $4 - 3(1 - x) \leq 3$

64. $8 - 4(2 - x) \leq -2x$

65. $\frac{1}{2}(x - 4) > x + 8$

66. $3x + 4 > \frac{1}{3}(x - 2)$

67. $\frac{x}{2} \geq 1 - \frac{x}{4}$

68. $\frac{x}{3} \geq 2 + \frac{x}{6}$

69. $0 \leq 2x - 6 \leq 4$

70. $4 \leq 2x + 2 \leq 10$

71. $-5 \leq 4 - 3x \leq 2$

72. $-3 < 3 - 2x \leq 9$

73. $-3 < \frac{2x - 1}{4} < 0$

74. $0 < \frac{3x + 2}{2} < 4$

75. $1 < 1 - \frac{1}{2}x < 4$

76. $0 < 1 - \frac{1}{3}x < 1$

77. $(x + 2)(x - 3) > (x - 1)(x + 1)$

78. $(x - 1)(x + 1) > (x - 3)(x + 4)$

79. $x(4x + 3) \leq (2x + 1)^2$

80. $x(9x - 5) \leq (3x - 1)^2$

81. $\frac{1}{2} \leq \frac{x + 1}{3} < \frac{3}{4}$

82. $\frac{1}{3} < \frac{x + 1}{2} \leq \frac{2}{3}$

83. $(4x + 2)^{-1} < 0$

84. $(2x - 1)^{-1} > 0$

85. $0 < \frac{2}{x} < \frac{3}{5}$

86. $0 < \frac{4}{x} < \frac{2}{3}$

87. $0 < (2x - 4)^{-1} < \frac{1}{2}$

88. $0 < (3x + 6)^{-1} < \frac{1}{3}$

Applications and Extensions

In Problems 89–98, find a and b.

89. If $-1 < x < 1$, then $a < x + 4 < b$.

90. If $-3 < x < 2$, then $a < x - 6 < b$.

91. If $2 < x < 3$, then $a < -4x < b$.

92. If $-4 < x < 0$, then $a < \frac{1}{2}x < b$.

93. If $0 < x < 4$, then $a < 2x + 3 < b$.

94. If $-3 < x < 3$, then $a < 1 - 2x < b$.

95. If $-3 < x < 0$, then $a < \frac{1}{x + 4} < b$.

96. If $2 < x < 4$, then $a < \frac{1}{x - 6} < b$.

97. If $6 < 3x < 12$, then $a < x^2 < b$.

98. If $0 < 2x < 6$, then $a < x^2 < b$.

99. What is the domain of the variable in the expression $\sqrt{3x + 6}$?

100. What is the domain of the variable in the expression $\sqrt{8 + 2x}$?

101. A young adult may be defined as someone older than 21, but less than 30 years of age. Express this statement using inequalities.

102. Middle-aged may be defined as being 40 or more and less than 60. Express this statement using inequalities.

103. **Life Expectancy** The Social Security Administration determined that an average 30-year-old male in 2005 could

expect to live at least 49.66 more years and an average 30-year-old female in 2005 could expect to live at least 53.58 more years.

(a) To what age can an average 30-year-old male expect to live? Express your answer as an inequality.

(b) To what age can an average 30-year-old female expect to live? Express your answer as an inequality.

(c) Who can expect to live longer, a male or a female? By how many years?

Source: Actuarial Study No. 120, August 2005

104. **General Chemistry** For a certain ideal gas, the volume V (in cubic centimeters) equals 20 times the temperature T (in degrees Celsius). If the temperature varies from 80° to 120° C inclusive, what is the corresponding range of the volume of the gas?

105. **Real Estate** A real estate agent agrees to sell an apartment complex according to the following commission schedule: $45,000 plus 25% of the selling price in excess of $900,000. Assuming that the complex will sell at some price between

$900,000 and $1,100,000 inclusive, over what range does the agent's commission vary? How does the commission vary as a percent of selling price?

106. Sales Commission A used car salesperson is paid a commission of $25 plus 40% of the selling price in excess of owner's cost. The owner claims that used cars typically sell for at least owner's cost plus $200 and at most owner's cost plus $3000. For each sale made, over what range can the salesperson expect the commission to vary?

107. Federal Tax Withholding The percentage method of withholding for federal income tax (2006) states that a single person whose weekly wages, after subtracting withholding allowances, are over $620, but not over $1409, shall have $78.30 plus 25% of the excess over $620 withheld. Over what range does the amount withheld vary if the weekly wages vary from $700 to $900 inclusive?

Source: Employer's Tax Guide. Department of the Treasury, Internal Revenue Service, Publication 2006.

108. Exercising Sue wants to lose weight. For healthy weight loss, the American College of Sports Medicine (ACSM) recommends 200 to 300 minutes of exercise per week. For the first six days of the week, Sue exercised 40, 45, 0, 50, 25, and 35 minutes. How long should Sue exercise on the seventh day in order to stay within the ACSM guidelines?

109. Electricity Rates Commonwealth Edison Company's charge for electricity in May 2006 is 8.275¢ per kilowatt-hour. In addition, each monthly bill contains a customer charge of $7.58. If last year's bills ranged from a low of $63.47 to a high of $214.53, over what range did usage vary (in kilowatt-hours)?

Source: Commonwealth Edison Co., Chicago, Illinois, 2006.

110. Water Bills The Village of Oak Lawn charges homeowners $28.84 per quarter-year plus $2.28 per 1000 gallons for water usage in excess of 12,000 gallons. In 2006 one homeowner's quarterly bill ranged from a high of $74.44 to a low of $42.52. Over what range did water usage vary?

Source: Village of Oak Lawn, Illinois, April 2006.

111. Markup of a New Car The markup over dealer's cost of a new car ranges from 12% to 18%. If the sticker price is $18,000, over what range will the dealer's cost vary?

112. IQ Tests A standard intelligence test has an average score of 100. According to statistical theory, of the people who take the test, the 2.5% with the highest scores will have scores of more than 1.96σ above the average, where σ (sigma, a number called the **standard deviation**) depends on the nature of the test. If $\sigma = 12$ for this test and there is (in principle) no upper limit to the score possible on the test, write the interval of possible test scores of the people in the top 2.5%.

113. Computing Grades In your Economics 101 class, you have scores of 68, 82, 87, and 89 on the first four of five tests. To get a grade of B, the average of the first five test scores must be greater than or equal to 80 and less than 90.

(a) Solve an inequality to find the range of the score that you need on the last test to get a B.

(b) What score do you need if the fifth test counts double?

What do I need to get a B?

114. "Light" Foods For food products to be labeled "light," the U.S. Food and Drug Administration requires that the altered product must either contain one-third or fewer calories than the regular product or it must contain one-half or less fat than the regular product. If a serving of Miracle Whip® Light contains 20 calories and 1.5 grams of fat, then what must be true about either the number of calories or the grams of fat in a serving of regular Miracle Whip®?

115. Arithmetic Mean If $a < b$, show that $a < \dfrac{a+b}{2} < b$. The number $\dfrac{a+b}{2}$ is called the **arithmetic mean** of a and b.

116. Refer to Problem 115. Show that the arithmetic mean of a and b is equidistant from a and b.

117. Geometric Mean If $0 < a < b$, show that $a < \sqrt{ab} < b$. The number $\sqrt{ab}$ is called the **geometric mean** of a and b.

118. Refer to Problems 115 and 117. Show that the geometric mean of a and b is less than the arithmetic mean of a and b.

119. Harmonic Mean For $0 < a < b$, let h be defined by

$$\frac{1}{h} = \frac{1}{2}\left(\frac{1}{a} + \frac{1}{b}\right)$$

Show that $a < h < b$. The number h is called the **harmonic mean** of a and b.

120. Refer to Problems 115, 117, and 119. Show that the harmonic mean of a and b equals the geometric mean squared divided by the arithmetic mean.

121. Another Reciprocal Property Prove that if $0 < a < b$, then $0 < \dfrac{1}{b} < \dfrac{1}{a}$.

Discussion and Writing

122. Make up an inequality that has no solution. Make up one that has exactly one solution.

123. The inequality $x^2 + 1 < -5$ has no real solution. Explain why.

124. Do you prefer to use inequality notation or interval notation to express the solution to an inequality? Give your reasons. Are there particular circumstances when you prefer one to the other? Cite examples.

125. How would you explain to a fellow student the underlying reason for the multiplication properties for inequalities (page 127), that is, the sense or direction of an inequality remains the same if each side is multiplied by a positive real number, whereas the direction is reversed if each side is multiplied by a negative real number.

'Are You Prepared?' Answers

1. (number line showing points at -4, -2, 0) **2.** False

1.6 Equations and Inequalities Involving Absolute Value

PREPARING FOR THIS SECTION *Before getting started, review the following:*

- Algebra Essentials (Chapter R, Section R.2, pp. 17–26)

Now Work the 'Are You Prepared?' problems on page 137.

OBJECTIVES **1** Solve Equations Involving Absolute Value (p. 135)
 2 Solve Inequalities Involving Absolute Value (p. 135)

1 Solve Equations Involving Absolute Value

Recall that, on the real number line, the absolute value of a equals the distance from the origin to the point whose coordinate is a. For example, there are two points whose distance from the origin is 5 units, -5 and 5. So the equation $|x| = 5$ will have the solution set $\{-5, 5\}$. This leads to the following result:

THEOREM If a is a positive real number and if u is any algebraic expression, then

$$|u| = a \quad \text{is equivalent to} \quad u = a \quad \text{or} \quad u = -a \qquad \textbf{(1)}$$

EXAMPLE 1 **Solving an Equation Involving Absolute Value**

Solve the equations:

(a) $|x + 4| = 13$ (b) $|2x - 3| + 2 = 7$

Solution (a) This follows the form of equation (1), where $u = x + 4$. There are two possibilities.

$$x + 4 = 13 \quad \text{or} \quad x + 4 = -13$$
$$x = 9 \quad \text{or} \quad x = -17$$

The solution set is $\{-17, 9\}$.

(b) The equation $|2x - 3| + 2 = 7$ is not in the form of equation (1). We proceed as follows:

$$|2x - 3| + 2 = 7$$
$$|2x - 3| = 5 \qquad \text{Subtract 2 from each side.}$$
$$2x - 3 = 5 \quad \text{or} \quad 2x - 3 = -5 \qquad \text{Apply (1).}$$
$$2x = 8 \quad \text{or} \quad 2x = -2$$
$$x = 4 \quad \text{or} \quad x = -1$$

The solution set is $\{-1, 4\}$.

 Now Work PROBLEM 9

2 Solve Inequalities Involving Absolute Value

EXAMPLE 2 **Solving an Inequality Involving Absolute Value**

Solve the inequality: $|x| < 4$

Solution We are looking for all points whose coordinate x is a distance less than 4 units from the origin. See Figure 9 for an illustration. Because any number x between -4 and 4 satisfies the condition $|x| < 4$, the solution set consists of all numbers x for which $-4 < x < 4$, that is, all x in the interval $(-4, 4)$.

Figure 9

Less than 4 units
from origin 0

0

-5 -4 -3 -2 -1 0 1 2 3 4

We are led to the following results:

THEOREM If a is a positive number and if u is an algebraic expression, then

$\lvert u \rvert < a$ is equivalent to $-a < u < a$	**(2)**
$\lvert u \rvert \leq a$ is equivalent to $-a \leq u \leq a$	**(3)**

In other words, $\lvert u \rvert < a$ is equivalent to $-a < u$ and $u < a$.

Figure 10

See Figure 10 for an illustration of statement (3).

EXAMPLE 3 **Solving an Inequality Involving Absolute Value**

Solve the inequality: $\lvert 2x + 4 \rvert \leq 3$
Graph the solution set.

Solution

$$\lvert 2x + 4 \rvert \leq 3$$ This follows the form of statement (3); the expression $u = 2x + 4$ is inside the absolute value bars.

$$-3 \leq \quad 2x + 4 \quad \leq 3$$ Apply statement (3).

$$-3 - 4 \leq 2x + 4 - 4 \leq 3 - 4$$ Subtract 4 from each part.

$$-7 \leq \quad\quad 2x \quad\quad \leq -1$$ Simplify.

$$\frac{-7}{2} \leq \quad \frac{2x}{2} \quad \leq \frac{-1}{2}$$ Divide each part by 2.

$$-\frac{7}{2} \leq \quad\quad x \quad\quad \leq -\frac{1}{2}$$ Simplify.

Figure 11

The solution set is $\left\{ x \middle| -\dfrac{7}{2} \leq x \leq -\dfrac{1}{2} \right\}$, that is, all x in the interval $\left[-\dfrac{7}{2}, -\dfrac{1}{2} \right]$. See Figure 11 for the graph of the solution set.

EXAMPLE 4 **Solving an Inequality Involving Absolute Value**

Solve the inequality: $\lvert 1 - 4x \rvert < 5$
Graph the solution set.

Solution

$$\lvert 1 - 4x \rvert < 5$$ This expression follows the form of statement (2); the expression $u = 1 - 4x$ is inside the absolute value bars.

$$-5 < \quad 1 - 4x \quad < 5$$ Apply statement (2).

$$-5 - 1 < 1 - 4x - 1 < 5 - 1$$ Subtract 1 from each part.

$$-6 < \quad\quad -4x \quad\quad < 4$$ Simplify.

$$\frac{-6}{-4} > \quad \frac{-4x}{-4} \quad > \frac{4}{-4}$$ Divide each part by -4, which reverses the sense of the inequality symbols.

$$\frac{3}{2} > \quad\quad x \quad\quad > -1$$ Simplify.

$$-1 < \quad\quad x \quad\quad < \frac{3}{2}$$ Rearrange the ordering.

Figure 12

The solution set is $\left\{ x \middle| -1 < x < \dfrac{3}{2} \right\}$, that is, all x in the interval $\left(-1, \dfrac{3}{2} \right)$. See Figure 12 for the graph of the solution set.

➤ **Now Work** PROBLEM 39

| EXAMPLE 5 | **Solving an Inequality Involving Absolute Value** |

Solve the inequality: $|x| > 3$
Graph the solution set.

Figure 13

Solution We are looking for all points whose coordinate x is a distance greater than 3 units from the origin. Figure 13 illustrates the situation. We conclude that any number x less than -3 or greater than 3 satisfies the condition $|x| > 3$. Consequently, the solution set consists of all numbers x for which $x < -3$ or $x > 3$, that is, all x in $(-\infty, -3) \cup (3, \infty)$.*

THEOREM If a is a positive number and u is an algebraic expression, then

$$|u| > a \quad \text{is equivalent to} \quad u < -a \quad \text{or} \quad u > a \qquad \textbf{(4)}$$

$$|u| \geq a \quad \text{is equivalent to} \quad u \leq -a \quad \text{or} \quad u \geq a \qquad \textbf{(5)}$$

Figure 14

$|u| \leq a, a > 0$

See Figure 14 for an illustration of statement (5).

| EXAMPLE 6 | **Solving an Inequality Involving Absolute Value** |

Solve the inequality: $|2x - 5| > 3$
Graph the solution set.

Solution

$|2x - 5| > 3$ This follows the form of statement (4); the expression $u = 2x - 5$ is inside the absolute value bars.

$2x - 5 < -3$ or $2x - 5 > 3$ Apply statement (4).

$2x - 5 + 5 < -3 + 5$ or $2x - 5 + 5 > 3 + 5$ Add 5 to each part.

$2x < 2$ or $2x > 8$ Simplify.

$\dfrac{2x}{2} < \dfrac{2}{2}$ or $\dfrac{2x}{2} > \dfrac{8}{2}$ Divide each part by 2.

$x < 1$ or $x > 4$ Simplify.

Figure 15

The solution set is $\{x \mid x < 1 \text{ or } x > 4\}$, that is, all x in $(-\infty, 1) \cup (4, \infty)$. See Figure 15 for the graph of the solution set.

WARNING A common error to be avoided is to attempt to write the solution $x < 1$ or $x > 4$ as the combined inequality $1 > x > 4$, which is incorrect, since there are no numbers x for which $1 > x$ and $x > 4$. ■

━━━ **Now Work** PROBLEM 43

*Recall that the symbol $\cup$ stands for the union of two sets. Refer to p. 2 if necessary.

1.6 Assess Your Understanding

'Are You Prepared?' *Answers are given at the end of these exercises. If you get a wrong answer, read the pages listed in red.*

1. $|-2| = $ _____ (p. 19)

2. *True or False* $|x| \geq 0$ for any real number x. (p. 19)

Concepts and Vocabulary

3. The solution set of the equation $|x| = 5$ is {_____}.

4. The solution set of the inequality $|x| < 5$ is $\{x \mid$ _____$\}$.

5. *True or False* The equation $|x| = -2$ has no solution.

6. *True or False* The inequality $|x| \geq -2$ has the set of real numbers as solution set.

Skill Building

In Problems 7–34, solve each equation.

7. $|2x| = 6$

8. $|3x| = 12$

9. $|2x + 3| = 5$

10. $|3x - 1| = 2$

11. $|1 - 4t| + 8 = 13$

12. $|1 - 2z| + 6 = 9$

13. $|-2x| = |8|$

14. $|-x| = |1|$

15. $|-2|x = 4$

16. $|3|x = 9$

17. $\frac{2}{3}|x| = 9$

18. $\frac{3}{4}|x| = 9$

19. $\left|\frac{x}{3} + \frac{2}{5}\right| = 2$

20. $\left|\frac{x}{2} - \frac{1}{3}\right| = 1$

21. $|u - 2| = -\frac{1}{2}$

22. $|2 - v| = -1$

23. $4 - |2x| = 3$

24. $5 - \left|\frac{1}{2}x\right| = 3$

25. $|x^2 - 9| = 0$

26. $|x^2 - 16| = 0$

27. $|x^2 - 2x| = 3$

28. $|x^2 + x| = 12$

29. $|x^2 + x - 1| = 1$

30. $|x^2 + 3x - 2| = 2$

31. $\left|\frac{3x - 2}{2x - 3}\right| = 2$

32. $\left|\frac{2x + 1}{3x + 4}\right| = 1$

33. $|x^2 + 3x| = |x^2 - 2x|$

34. $|x^2 - 2x| = |x^2 + 6x|$

In Problems 35–62, solve each inequality. Express your answer using set notation or interval notation. Graph the solution set.

35. $|2x| < 8$

36. $|3x| < 15$

37. $|3x| > 12$

38. $|2x| > 6$

39. $|x - 2| + 2 < 3$

40. $|x + 4| + 3 < 5$

41. $|3t - 2| \le 4$

42. $|2u + 5| \le 7$

43. $|2x - 3| \ge 2$

44. $|3x + 4| \ge 2$

45. $|1 - 4x| - 7 < -2$

46. $|1 - 2x| - 4 < -1$

47. $|1 - 2x| > 3$

48. $|2 - 3x| > 1$

49. $|-4x| + |-5| \le 1$

50. $|-x| - |4| \le 2$

51. $|-2x| > |-3|$

52. $|-x - 2| \ge 1$

53. $-|2x - 1| \ge -3$

54. $-|1 - 2x| \ge -3$

55. $|2x| < -1$

56. $|3x| \ge 0$

57. $|5x| \ge -1$

58. $|6x| < -2$

59. $\left|\frac{2x + 3}{3} - \frac{1}{2}\right| < 1$

60. $3 - |x + 1| < \frac{1}{2}$

61. $5 + |x - 1| > \frac{1}{2}$

62. $\left|\frac{2x - 3}{2} + \frac{1}{3}\right| > 1$

Applications and Extensions

63. Body Temperature "Normal" human body temperature is 98.6°F. If a temperature x that differs from normal by at least 1.5° is considered unhealthy, write the condition for an unhealthy temperature x as an inequality involving an absolute value, and solve for x.

64. Household Voltage In the United States, normal household voltage is 110 volts. However, it is not uncommon for actual voltage to differ from normal voltage by at most 5 volts. Express this situation as an inequality involving an absolute value. Use x as the actual voltage and solve for x.

65. Reading Books A Gallup poll conducted May 20–22, 2005, found that Americans read an average of 13.4 books per year. Gallup is 99% confident that the result from this poll is off by fewer than 1.35 books from the actual average x. Express this situation as an inequality involving absolute value, and solve

the inequality for x to determine the interval in which the actual average is likely to fall.

Note: In statistics, this interval is called a 99% **confidence interval.**

66. Speed of Sound According to data from the Hill Aerospace Museum (Hill Air Force Base, Utah), the speed of sound varies depending on altitude, barometric pressure, and temperature. For example, at 20,000 feet, 13.75 inches of mercury, and −12.3°F, the speed of sound is about 707 miles per hour, but the speed can vary from this result by as much as 55 miles per hour as conditions change.

(a) Express this situation as an inequality involving an absolute value.

(b) Using x for the speed of sound, solve for x to find an interval for the speed of sound.

67. Express the fact that x differs from 3 by less than $\frac{1}{2}$ as an inequality involving an absolute value. Solve for x.

68. Express the fact that x differs from −4 by less than 1 as an inequality involving an absolute value. Solve for x.

69. Express the fact that x differs from −3 by more than 2 as an inequality involving an absolute value. Solve for x.

70. Express the fact that x differs from 2 by more than 3 as an inequality involving an absolute value. Solve for x.

In Problems 71–76, find a and b.

71. If $|x - 1| < 3$, then $a < x + 4 < b$.

72. If $|x + 2| < 5$, then $a < x - 2 < b$.

73. If $|x + 4| \leq 2$, then $a \leq 2x - 3 \leq b$.

74. If $|x - 3| \leq 1$, then $a \leq 3x + 1 \leq b$.

75. If $|x - 2| \leq 7$, then $a \leq \dfrac{1}{x - 10} \leq b$.

76. If $|x + 1| \leq 3$, then $a \leq \dfrac{1}{x + 5} \leq b$.

77. Show that if $a > 0, b > 0$, and $\sqrt{a} < \sqrt{b}$, then $a < b$.

[**Hint:** $b - a = (\sqrt{b} - \sqrt{a})(\sqrt{b} + \sqrt{a})$.]

78. Show that $a \leq |a|$.

79. Prove the *triangle inequality* $|a + b| \leq |a| + |b|$.

[**Hint:** Expand $|a + b|^2 = (a + b)^2$, and use the result of Problem 78.]

80. Prove that $|a - b| \geq |a| - |b|$.

[**Hint:** Apply the triangle inequality from Problem 79 to $|a| = |(a - b) + b|$.]

81. If $a > 0$, show that the solution set of the inequality

$$x^2 < a$$

consists of all numbers x for which

$$-\sqrt{a} < x < \sqrt{a}$$

82. If $a > 0$, show that the solution set of the inequality

$$x^2 > a$$

consists of all numbers x for which

$$x < -\sqrt{a} \quad \text{or} \quad x > \sqrt{a}$$

In Problems 83–90, use the results found in Problems 81 and 82 to solve each inequality.

83. $x^2 < 1$

84. $x^2 < 4$

85. $x^2 \geq 9$

86. $x^2 \geq 1$

87. $x^2 \leq 16$

88. $x^2 \leq 9$

89. $x^2 > 4$

90. $x^2 > 16$

91. Solve $|3x - |2x + 1|| = 4$. **92.** Solve $|x + |3x - 2|| = 2$.

Discussion and Writing

93. The equation $|x| = -2$ has no solution. Explain why.

94. The inequality $|x| > -0.5$ has all real numbers as the solution. Explain why.

95. The inequality $|x| > 0$ has as solution set $\{x | x \neq 0\}$. Explain why.

'Are You Prepared?' Answers

1. 2 **2.** True

1.7 Problem Solving: Interest, Mixture, Uniform Motion, Constant Rate Job Applications

OBJECTIVES **1** Translate Verbal Descriptions into Mathematical Expressions (p. 140)

2 Solve Interest Problems (p. 141)

3 Solve Mixture Problems (p. 142)

4 Solve Uniform Motion Problems (p. 143)

5 Solve Constant Rate Job Problems (p. 145)

Applied (word) problems do not come in the form "Solve the equation...." Instead, they supply information using words, a verbal description of the real problem. So, to solve applied problems, we must be able to translate the verbal description into the language of mathematics. We do this by using variables to represent unknown quantities and then finding relationships (such as equations) that involve these variables. The process of doing all this is called **mathematical modeling.**

Any solution to the mathematical problem must be checked against the mathematical problem, the verbal description, and the real problem. See Figure 16 on page 140 for an illustration of the **modeling process.**

Figure 16

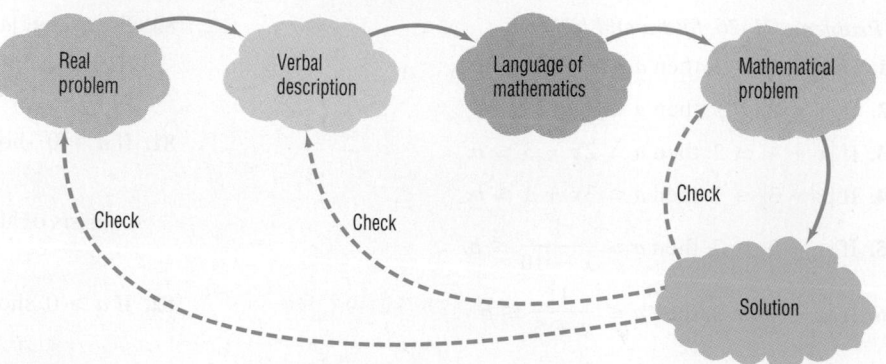

1 Translate Verbal Descriptions into Mathematical Expressions

Let's look at a few examples that will help you to translate certain words into mathematical symbols.

| EXAMPLE 1 | **Translating Verbal Descriptions into Mathematical Expressions** |

(a) For uniform motion, the velocity of an object equals the distance traveled divided by the time required.

Translation: If v is the velocity, s the distance, and t the time, then $v = \dfrac{s}{t}$.

(b) Let x denote a number.

The number 5 times as large as x is $5x$.

The number 3 less than x is $x - 3$.

The number that exceeds x by 4 is $x + 4$.

The number that, when added to x, gives 5 is $5 - x$.

Now Work PROBLEM 7

Always check the units used to measure the variables of an applied problem. In Example 1(a), if v is measured in miles per hour, then the distance s must be expressed in miles and the time t must be expressed in hours. It is a good practice to check units to be sure that they are consistent and make sense.

The steps to follow for setting up applied problems, given earlier, are repeated next:

Steps for Setting up Applied Problems

STEP 1: Read the problem carefully, perhaps two or three times. Pay particular attention to the question being asked in order to identify what you are looking for. If you can, determine realistic possibilities for the answer.

STEP 2: Assign a letter (variable) to represent what you are looking for, and, if necessary, express any remaining unknown quantities in terms of this variable.

STEP 3: Make a list of all the known facts, and translate them into mathematical expressions. These may take the form of an equation or an inequality involving the variable. If possible, draw an appropriately labeled diagram to assist you. Sometimes a table or chart helps.

STEP 4: Solve the equation for the variable, and then answer the question.

STEP 5: Check the answer with the facts in the problem. If it agrees, congratulations! If it does not agree, try again.

2 Solve Interest Problems

Interest is money paid for the use of money. The total amount borrowed (whether by an individual from a bank in the form of a loan or by a bank from an individual in the form of a savings account) is called the **principal.** The **rate of interest,** expressed as a percent, is the amount charged for the use of the principal for a given period of time, usually on a yearly (that is, on a per annum) basis.

Simple Interest Formula

If a principal of P dollars is borrowed for a period of t years at a per annum interest rate r, expressed as a decimal, the interest I charged is

$$I = Prt \qquad (1)$$

Interest charged according to formula (1) is called **simple interest.** When using formula (1), be sure to express r as a decimal.

| EXAMPLE 2 | Finance: Computing Interest on a Loan |

Suppose that Juanita borrows $500 for 6 months at the simple interest rate of 9% per annum. What is the interest that Juanita will be charged on the loan? How much does Juanita owe after 6 months?

Solution The rate of interest is given per annum, so the actual time that the money is borrowed must be expressed in years. The interest charged would be the principal, $500, times the rate of interest (9% = 0.09) times the time in years, $\frac{1}{2}$:

$$\text{Interest charged} = I = Prt = (500)(0.09)\left(\frac{1}{2}\right) = \$22.50$$

After 6 months, Juanita will owe what she borrowed plus the interest:

$$\$500 + \$22.50 = \$522.50$$

| EXAMPLE 3 | Financial Planning |

Candy has $70,000 to invest and requires an overall rate of return of 9%. She can invest in a safe, government-insured certificate of deposit, but it only pays 8%. To obtain 9%, she agrees to invest some of her money in noninsured corporate bonds paying 12%. How much should be placed in each investment to achieve her goal?

Solution **STEP 1:** The question is asking for two dollar amounts: the principal to invest in the corporate bonds and the principal to invest in the certificate of deposit.

STEP 2: We let x represent the amount (in dollars) to be invested in the bonds. Then $70,000 - x$ is the amount that will be invested in the certificate. (Do you see why?)

STEP 3: We set up a table.

	Principal ($)	Rate	Time (yr)	Interest ($)
Bonds	x	12% = 0.12	1	0.12x
Certificate	70,000 − x	8% = 0.08	1	0.08(70,000 − x)
Total	70,000	9% = 0.09	1	0.09(70,000) = 6300

Since the total interest from the investments is equal to $0.09(70,000) = 6300$, we must have the equation

$$0.12x + 0.08(70,000 - x) = 6300$$

(Note that the units are consistent: the unit is dollars on each side.)

STEP 4: $0.12x + 5600 - 0.08x = 6300$

$$0.04x = 700$$

$$x = 17,500$$

Candy should place $17,500 in the bonds and $70,000 − $17,500 = $52,500 in the certificate.

STEP 5: The interest on the bonds after 1 year is $0.12(\$17,500) = \2100; the interest on the certificate after 1 year is $0.08(\$52,500) = \4200. The total annual interest is $6300, the required amount.

━━━━━━━ **Now Work** PROBLEM 17

3 Solve Mixture Problems

Oil refineries sometimes produce gasoline that is a blend of two or more types of fuel; bakeries occasionally blend two or more types of flour for their bread. These problems are referred to as **mixture problems** because they combine two or more quantities to form a mixture.

| EXAMPLE 4 | **Blending Coffees** |

The manager of a Starbucks store decides to experiment with a new blend of coffee. She will mix some B grade Colombian coffee that sells for $5 per pound with some A grade Arabica coffee that sells for $10 per pound to get 100 pounds of the new blend. The selling price of the new blend is to be $7 per pound, and there is to be no difference in revenue from selling the new blend versus selling the other types. How many pounds of the B grade Colombian and A grade Arabica coffees are required?

Solution Let x represent the number of pounds of the B grade Colombian coffee. Then $100 - x$ equals the number of pounds of the A grade Arabica coffee. See Figure 17.

Figure 17

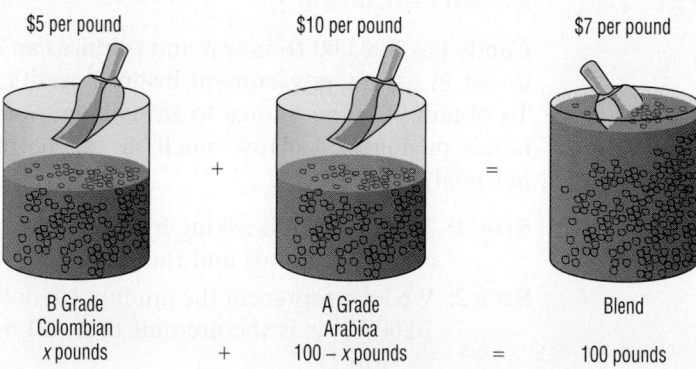

Since there is to be no difference in revenue between selling the A and B grades separately versus the blend, we have

$$\left\{\begin{matrix}\text{Price per pound}\\ \text{of B grade}\end{matrix}\right\}\left\{\begin{matrix}\text{\# Pounds}\\ \text{B grade}\end{matrix}\right\} + \left\{\begin{matrix}\text{Price per pound}\\ \text{of A grade}\end{matrix}\right\}\left\{\begin{matrix}\text{\# Pounds}\\ \text{A grade}\end{matrix}\right\} = \left\{\begin{matrix}\text{Price per pound}\\ \text{of blend}\end{matrix}\right\}\left\{\begin{matrix}\text{\# Pounds}\\ \text{blend}\end{matrix}\right\}$$

$$\$5 \quad\cdot\quad x \quad + \quad \$10 \quad\cdot (100 - x) = \quad \$7 \quad\cdot\quad 100$$

We have the equation

$$5x + 10(100 - x) = 700$$
$$5x + 1000 - 10x = 700$$
$$-5x = -300$$
$$x = 60$$

The manager should blend 60 pounds of B grade Colombian coffee with $100 - 60 = 40$ pounds of A grade Arabica coffee to get the desired blend.

✓ Check: The 60 pounds of B grade coffee would sell for $(\$5)(60) = \300, and the 40 pounds of A grade coffee would sell for $(\$10)(40) = \400; the total revenue, $700, equals the revenue obtained from selling the blend, as desired. ∎

━━━━ **Now Work** PROBLEM 21

4 Solve Uniform Motion Problems

Objects that move at a constant velocity (or speed) are said to be in **uniform motion.**[*] When the average velocity of an object is known, it can be interpreted as its constant velocity. For example, a bicyclist traveling at an average velocity of 25 miles per hour is in uniform motion.

Uniform Motion Formula

If an object moves at an average velocity v, the distance s covered in time t is given by the formula

$$s = vt \qquad \qquad \textbf{(2)}$$

That is, Distance = Velocity · Time.

EXAMPLE 5 **Physics: Uniform Motion**

Tanya, who is a long-distance runner, runs at an average velocity of 8 miles per hour (mi/hr). Two hours after Tanya leaves your house, you leave in your Honda and follow the same route. If your average velocity is 40 mi/hr, how long will it be before you catch up to Tanya? How far will each of you be from your home?

Solution Refer to Figure 18. We use t to represent the time (in hours) that it takes the Honda to catch up to Tanya. When this occurs, the total time elapsed for Tanya is $t + 2$ hours.

Figure 18

Set up the following table:

	Velocity mi/hr	Time hr	Distance mi
Tanya	8	$t + 2$	$8(t + 2)$
Honda	40	t	$40t$

[*] Technically, velocity has both magnitude (speed) and direction. Since all velocities in this section are in the positive direction, we make no distinction between speed and velocity.

Since the distance traveled is the same, we are led to the following equation:

$$8(t + 2) = 40t$$
$$8t + 16 = 40t$$
$$32t = 16$$
$$t = \frac{1}{2} \text{ hour}$$

It will take the Honda $\frac{1}{2}$ hour to catch up to Tanya. Each of you will have gone 20 miles.

✓ **Check:** In 2.5 hours, Tanya travels a distance of $(2.5)(8) = 20$ miles. In $\frac{1}{2}$ hour, the Honda travels a distance of $\left(\frac{1}{2}\right)(40) = 20$ miles. ∎

EXAMPLE 6 | **Physics: Uniform Motion**

A motorboat heads upstream a distance of 24 miles on a river whose current is running at 3 miles per hour (mi/hr). The trip up and back takes 6 hours. Assuming that the motorboat maintained a constant speed relative to the water, what was its speed?

Solution

See Figure 19. We use v to represent the constant speed of the motorboat relative to the water. Then the true speed going upstream is $v - 3$ mi/hr, and the true speed going downstream is $v + 3$ mi/hr. Since Distance = Velocity × Time, then Time = $\dfrac{\text{Distance}}{\text{Velocity}}$. We set up a table.

Figure 19

	Velocity mi/hr	Distance mi	Time = $\dfrac{\text{Distance}}{\text{Velocity}}$ hr
Upstream	$v - 3$	24	$\dfrac{24}{v - 3}$
Downstream	$v + 3$	24	$\dfrac{24}{v + 3}$

Since the total time up and back is 6 hours, we have

$$\frac{24}{v - 3} + \frac{24}{v + 3} = 6$$

$$\frac{24(v + 3) + 24(v - 3)}{(v - 3)(v + 3)} = 6 \qquad \text{Add the quotients on the left.}$$

$$\frac{48v}{v^2 - 9} = 6 \qquad \text{Simplify.}$$

$$48v = 6(v^2 - 9) \qquad \text{Multiply both sides by } v^2 - 9.$$

$$6v^2 - 48v - 54 = 0 \qquad \text{Place in standard form.}$$

$$v^2 - 8v - 9 = 0 \qquad \text{Divide by 6.}$$

$$(v - 9)(v + 1) = 0 \qquad \text{Factor.}$$

$$v = 9 \quad \text{or} \quad v = -1 \qquad \text{Apply the Zero-Product Property and solve.}$$

We discard the solution $v = -1$ mi/hr, so the speed of the motorboat relative to the water is 9 mi/hr. ∎

Now Work PROBLEM 27

5 Solve Constant Rate Job Problems

This section involves jobs that are performed at a **constant rate.** Our assumption is that, if a job can be done in t units of time, then $\frac{1}{t}$ of the job is done in 1 unit of time. Let's look at an example.

| EXAMPLE 7 | Working Together to Do a Job |

At 10 AM Danny is asked by his father to weed the garden. From past experience, Danny knows that this will take him 4 hours, working alone. His older brother, Mike, when it is his turn to do this job, requires 6 hours. Since Mike wants to go golfing with Danny and has a reservation for 1 PM, he agrees to help Danny. Assuming no gain or loss of efficiency, when will they finish if they work together? Can they make the golf date?

Solution We set up Table 2. In 1 hour, Danny does $\frac{1}{4}$ of the job, and in 1 hour, Mike does $\frac{1}{6}$ of the job. Let t be the time (in hours) that it takes them to do the job together. In 1 hour, then, $\frac{1}{t}$ of the job is completed. We reason as follows:

Table 2

	Hours to Do Job	Part of Job Done in 1 Hour
Danny	4	$\frac{1}{4}$
Mike	6	$\frac{1}{6}$
Together	t	$\frac{1}{t}$

$$\left(\begin{array}{c}\text{Part done by Danny}\\\text{in 1 hour}\end{array}\right) + \left(\begin{array}{c}\text{Part done by Mike}\\\text{in 1 hour}\end{array}\right) = \left(\begin{array}{c}\text{Part done together}\\\text{in 1 hour}\end{array}\right)$$

From Table 2,

$$\frac{1}{4} + \frac{1}{6} = \frac{1}{t}$$
$$\frac{3}{12} + \frac{2}{12} = \frac{1}{t}$$
$$\frac{5}{12} = \frac{1}{t}$$
$$5t = 12$$
$$t = \frac{12}{5}$$

Working together, the job can be done in $\frac{12}{5}$ hours, or 2 hours, 24 minutes. They should make the golf date, since they will finish at 12:24 PM. ∎

═══► **Now Work** PROBLEM 33

1.7 Assess Your Understanding

Concepts and Vocabulary

1. The process of using variables to represent unknown quantities and then finding relationships that involve these variables is referred to as _____ _____.

2. The money paid for the use of money is _____.

3. Objects that move at a constant velocity are said to be in _____ _____.

4. *True or False* The amount charged for the use of principal for a given period of time is called the rate of interest.

5. *True or False* If an object moves at an average velocity v, the distance s covered in time t is given by the formula $s = vt$.

6. Suppose that you want to mix two coffees in order to obtain 100 pounds of the blend. If x represents the number of pounds of coffee A, write an algebraic expression that represents the number of pounds of coffee B.

Applications and Extensions

In Problems 7–16, translate each sentence into a mathematical equation. Be sure to identify the meaning of all symbols.

7. **Geometry** The area of a circle is the product of the number π and the square of the radius.

8. **Geometry** The circumference of a circle is the product of the number π and twice the radius.

9. **Geometry** The area of a square is the square of the length of a side.

10. **Geometry** The perimeter of a square is four times the length of a side.

11. **Physics** Force equals the product of mass and acceleration.

12. **Physics** Pressure is force per unit area.

13. **Physics** Work equals force times distance.

14. **Physics** Kinetic energy is one-half the product of the mass and the square of the velocity.

15. **Business** The total variable cost of manufacturing x dishwashers is $150 per dishwasher times the number of dishwashers manufactured.

16. **Business** The total revenue derived from selling x dishwashers is $250 per dishwasher times the number of dishwashers sold.

17. **Financial Planning** Betsy, a recent retiree, requires $6000 per year in extra income. She has $50,000 to invest and can invest in B-rated bonds paying 15% per year or in a certificate of deposit (CD) paying 7% per year. How much money should be invested in each to realize exactly $6000 in interest per year?

18. **Financial Planning** After 2 years, Betsy (see Problem 17) finds that she will now require $7000 per year. Assuming that the remaining information is the same, how should the money be reinvested?

19. **Banking** A bank loaned out $12,000, part of it at the rate of 8% per year and the rest at the rate of 18% per year. If the interest received totaled $1000, how much was loaned at 8%?

20. **Banking** Wendy, a loan officer at a bank, has $1,000,000 to lend and is required to obtain an average return of 18% per year. If she can lend at the rate of 19% or at the rate of 16%, how much can she lend at the 16% rate and still meet her requirement?

21. **Blending Teas** The manager of a store that specializes in selling tea decides to experiment with a new blend. She will mix some Earl Grey tea that sells for $5 per pound with some Orange Pekoe tea that sells for $3 per pound to get 100 pounds of the new blend. The selling price of the new blend is to be $4.50 per pound, and there is to be no difference in revenue from selling the new blend versus selling the other types. How many pounds of the Earl Grey tea and Orange Pekoe tea are required?

22. **Business: Blending Coffee** A coffee manufacturer wants to market a new blend of coffee that sells for $3.90 per pound by mixing two coffees that sell for $2.75 and $5 per pound, respectively. What amounts of each coffee should be blended to obtain the desired mixture?

 [**Hint:** Assume that the total weight of the desired blend is 100 pounds.]

23. **Business: Mixing Nuts** A nut store normally sells cashews for $9.00 per pound and almonds for $3.50 per pound. But at the end of the month the almonds had not sold well, so, in order to sell 60 pounds of almonds, the manager decided to mix the 60 pounds of almonds with some cashews and sell the mixture for $7.50 per pound. How many pounds of cashews should be mixed with the almonds to ensure no change in the profit?

24. **Business: Mixing Candy** A candy store sells boxes of candy containing caramels and cremes. Each box sells for $12.50 and holds 30 pieces of candy (all pieces are the same size). If the caramels cost $0.25 to produce and the cremes cost $0.45 to produce, how many of each should be in a box to make a profit of $3?

25. **Physics: Uniform Motion** A motorboat can maintain a constant speed of 16 miles per hour relative to the water. The boat makes a trip upstream to a certain point in 20 minutes; the return trip takes 15 minutes. What is the speed of the current? See the figure.

26. **Physics: Uniform Motion** A motorboat heads upstream on a river that has a current of 3 miles per hour. The trip upstream takes 5 hours, and the return trip takes 2.5 hours. What is the speed of the motorboat? (Assume that the motorboat maintains a constant speed relative to the water.)

27. **Physics: Uniform Motion** A motorboat maintained a constant speed of 15 miles per hour relative to the water in going 10 miles upstream and then returning. The total time for the trip was 1.5 hours. Use this information to find the speed of the current.

28. **Physics: Uniform Motion** Two cars enter the Florida Turnpike at Commercial Boulevard at 8:00 AM, each heading for Wildwood. One car's average speed is 10 miles per hour more than the other's. The faster car arrives at Wildwood at 11:00 AM, $\frac{1}{2}$ hour before the other car. What was the average speed of each car? How far did each travel?

29. **Moving Walkways** The speed of a moving walkway is typically about 2.5 feet per second. Walking on such a moving walkway, it takes Karen a total of 40 seconds to travel 50 feet with the movement of the walkway and then back again against the movement of the walkway. What is Karen's normal walking speed?

 Source: Answers.com

30. **Moving Walkways** The Gare Montparnasse train station in Paris has a high-speed version of a moving walkway. If he walks while riding this moving walkway, Jean Claude can travel 200 meters in 30 seconds less time than if he stands still on the moving walkway. If Jean Claude walks at a normal rate of 1.5 meters per second, what is the speed of the Gare Montparnasse walkway?

 Source: Answers.com

31. **Tennis Anyone?** A regulation doubles tennis court has an area of 2808 square feet. If it is 6 feet longer than twice its width, determine the dimensions of the court.

 Source: United States Tennis Association

32. Laser Printers It takes an HP LaserJet 1300 laser printer 10 minutes longer to complete a 600-page print job by itself than it takes an HP LaserJet 2420 to complete the same job by itself. Together the two printers can complete the job in 12 minutes. How long does it take each printer to complete the print job alone? What is the speed of each printer?
Source: Hewlett-Packard

33. Working Together on a Job Trent can deliver his newspapers in 30 minutes. It takes Lois 20 minutes to do the same route. How long would it take them to deliver the newspapers if they work together?

34. Working Together on a Job Patrice, by himself, can paint four rooms in 10 hours. If he hires April to help, they can do the same job together in 6 hours. If he lets April work alone, how long will it take her to paint four rooms?

35. Enclosing a Garden A gardener has 46 feet of fencing to be used to enclose a rectangular garden that has a border 2 feet wide surrounding it. See the figure.

(a) If the length of the garden is to be twice its width, what will be the dimensions of the garden?

(b) What is the area of the garden?

(c) If the length and width of the garden are to be the same, what would be the dimensions of the garden?

(d) What would be the area of the square garden?

36. Construction A pond is enclosed by a wooden deck that is 3 feet wide. The fence surrounding the deck is 100 feet long.

(a) If the pond is square, what are its dimensions?

(b) If the pond is rectangular and the length of the pond is to be three times its width, what are its dimensions?

(c) If the pond is circular, what is its diameter?

(d) Which pond has the most area?

37. Football A tight end can run the 100-yard dash in 12 seconds. A defensive back can do it in 10 seconds. The tight end catches a pass at his own 20-yard line with the defensive back at the 15-yard line. (See the figure.) If no other players are nearby, at what yard line will the defensive back catch up to the tight end?
[**Hint:** At time $t = 0$, the defensive back is 5 yards behind the tight end.]

38. Computing Business Expense Therese, an outside salesperson, uses her car for both business and pleasure. Last year, she traveled 30,000 miles, using 900 gallons of gasoline. Her car gets 40 miles per gallon on the highway and 25 in the city. She can deduct all highway travel, but no city travel, on her taxes. How many miles should Therese be allowed as a business expense?

39. Mixing Water and Antifreeze How much water should be added to 1 gallon of pure antifreeze to obtain a solution that is 60% antifreeze?

40. Mixing Water and Antifreeze The cooling system of a certain foreign-made car has a capacity of 15 liters. If the system is filled with a mixture that is 40% antifreeze, how much of this mixture should be drained and replaced by pure antifreeze so that the system is filled with a solution that is 60% antifreeze?

41. Chemistry: Salt Solutions How much water must be evaporated from 32 ounces of a 4% salt solution to make a 6% salt solution?

42. Chemistry: Salt Solutions How much water must be evaporated from 240 gallons of a 3% salt solution to produce a 5% salt solution?

43. Purity of Gold The purity of gold is measured in karats, with pure gold being 24 karats. Other purities of gold are expressed as proportional parts of pure gold. Thus, 18-karat gold is $\dfrac{18}{24}$, or 75% pure gold; 12-karat gold is $\dfrac{12}{24}$, or 50% pure gold; and so on. How much 12-karat gold should be mixed with pure gold to obtain 60 grams of 16-karat gold?

44. Chemistry: Sugar Molecules A sugar molecule has twice as many atoms of hydrogen as it does oxygen and one more atom of carbon than oxygen. If a sugar molecule has a total of 45 atoms, how many are oxygen? How many are hydrogen?

45. Running a Race Mike can run the mile in 6 minutes, and Dan can run the mile in 9 minutes. If Mike gives Dan a head start of 1 minute, how far from the start will Mike pass Dan? How long does it take? See the figure.

46. Range of an Airplane An air rescue plane averages 300 miles per hour in still air. It carries enough fuel for 5 hours of flying time. If, upon takeoff, it encounters a head wind of 30 mi/hr, how far can it fly and return safely? (Assume that the wind remains constant.)

47. Emptying Oil Tankers An oil tanker can be emptied by the main pump in 4 hours. An auxiliary pump can empty the tanker in 9 hours. If the main pump is started at 9 AM, when should the auxiliary pump be started so that the tanker is emptied by noon?

48. Cement Mix A 20-pound bag of Economy brand cement mix contains 25% cement and 75% sand. How much pure cement must be added to produce a cement mix that is 40% cement?

49. Emptying a Tub A bathroom tub will fill in 15 minutes with both faucets open and the stopper in place. With both faucets closed and the stopper removed, the tub will empty in 20 minutes. How long will it take for the tub to fill if both faucets are open and the stopper is removed?

50. Using Two Pumps A 5-horsepower (hp) pump can empty a pool in 5 hours. A smaller, 2-hp pump empties the same pool in 8 hours. The pumps are used together to begin emptying this pool. After two hours, the 2-hp pump breaks down. How long will it take the larger pump to empty the pool?

51. A Biathlon Suppose that you have entered an 87-mile biathlon that consists of a run and a bicycle race. During your run, your average velocity is 6 miles per hour, and during your bicycle race, your average velocity is 25 miles per hour. You finish the race in 5 hours. What is the distance of the run? What is the distance of the bicycle race?

52. Cyclists Two cyclists leave a city at the same time, one going east and the other going west. The westbound cyclist bikes 5 mph faster than the eastbound cyclist. After 6 hours they are 246 miles apart. How fast is each cyclist riding?

53. Comparing Olympic Heroes In the 1984 Olympics, C. Lewis of the United States won the gold medal in the 100-meter race with a time of 9.99 seconds. In the 1896 Olympics, Thomas Burke, also of the United States, won the gold medal in the 100-meter race in 12.0 seconds. If they ran in the same race repeating their respective times, by how many meters would Lewis beat Burke?

54. Constructing a Coffee Can A 39-ounce can of Hills Bros.® coffee requires 188.5 square inches of aluminum. If its height is 7 inches, what is its radius? [**Hint:** The surface area S of a right cylinder is $S = 2\pi r^2 + 2\pi rh$, where r is the radius and h is the height.]

7 in.

Discussion and Writing

55. Critical Thinking You are the manager of a clothing store and have just purchased 100 dress shirts for $20.00 each. After 1 month of selling the shirts at the regular price, you plan to have a sale giving 40% off the original selling price. However, you still want to make a profit of $4 on each shirt at the sale price. What should you price the shirts at initially to ensure this? If, instead of 40% off at the sale, you give 50% off, by how much is your profit reduced?

56. Critical Thinking Make up a word problem that requires solving a linear equation as part of its solution. Exchange problems with a friend. Write a critique of your friend's problem.

57. Critical Thinking Without solving, explain what is wrong with the following mixture problem: How many liters of 25% ethanol should be added to 20 liters of 48% ethanol to obtain a solution of 58% ethanol? Now go through an algebraic solution. What happens?

58. Computing Average Speed In going from Chicago to Atlanta, a car averages 45 miles per hour, and in going from Atlanta to Miami, it averages 55 miles per hour. If Atlanta is halfway between Chicago and Miami, what is the average speed from Chicago to Miami? Discuss an intuitive solution. Write a paragraph defending your intuitive solution. Then solve the problem algebraically. Is your intuitive solution the same as the algebraic one? If not, find the flaw.

59. Speed of a Plane On a recent flight from Phoenix to Kansas City, a distance of 919 nautical miles, the plane arrived 20 minutes early. On leaving the aircraft, I asked the captain, "What was our tail wind?" He replied, "I don't know, but our ground speed was 550 knots." How can you determine if enough information is provided to find the tail wind? If possible, find the tail wind. (1 knot = 1 nautical mile per hour)

CHAPTER REVIEW

Things to Know

Quadratic formula (pp. 102 and 115)

If $ax^2 + bx + c = 0$, $a \neq 0$, then $x = \dfrac{-b \pm \sqrt{b^2 - 4ac}}{2a}$

If $b^2 - 4ac < 0$, there are no real solutions.

Discriminant (pp. 103 and 116)

If $b^2 - 4ac > 0$, there are two distinct real solutions.

If $b^2 - 4ac = 0$, there is one repeated real solution.

If $b^2 - 4ac < 0$, there are no real solutions, but there are two distinct complex solutions that are not real; the complex solutions are conjugates of each other.

Interval notation (p. 125)

$[a, b]$	$\{x \mid a \le x \le b\}$	$[a, \infty)$	$\{x \mid x \ge a\}$
$[a, b)$	$\{x \mid a \le x < b\}$	(a, ∞)	$\{x \mid x > a\}$
$(a, b]$	$\{x \mid a < x \le b\}$	$(-\infty, a]$	$\{x \mid x \le a\}$
(a, b)	$\{x \mid a < x < b\}$	$(-\infty, a)$	$\{x \mid x < a\}$
		$(-\infty, \infty)$	All real numbers

Properties of inequalities

Addition property (p. 126) If $a < b$, then $a + c < b + c$

 If $a > b$, then $a + c > b + c$

Multiplication properties (p. 127) (a) If $a < b$ and if $c > 0$, then $ac < bc$ (b) If $a > b$ and if $c > 0$, then $ac > bc$

 If $a < b$ and if $c < 0$, then $ac > bc$ If $a > b$ and if $c < 0$, then $ac < bc$

Reciprocal properties (p. 128) If $a > 0$, then $\dfrac{1}{a} > 0$ If $a < 0$, then $\dfrac{1}{a} < 0$

 If $\dfrac{1}{a} > 0$, then $a > 0$ If $\dfrac{1}{a} < 0$, then $a < 0$

Absolute value

If $|u| = a, a > 0$, then $u = -a$ or $u = a$ (p. 135)

If $|u| \le a, a > 0$, then $-a \le u \le a$ (p. 136)

If $|u| \ge a, a > 0$, then $u \le -a$ or $u \ge a$ (p. 137)

Objectives

Section		You should be able to . . .	Review Exercises
1.1	1	Solve a linear equation (p. 88)	1–6, 11–12
	2	Solve equations that lead to linear equations (p. 90)	7, 8
	3	Solve applied problems involving linear equations (p. 92)	86, 105
1.2	1	Solve a quadratic equation by factoring (p. 97)	10, 13, 14, 33–36
	2	Know how to complete the square (p. 99)	61–64
	3	Solve a quadratic equation by completing the square (p. 100)	9, 10, 13–16, 19, 20, 33–36
	4	Solve a quadratic equation using the quadratic formula (p. 102)	9, 10, 13–16, 19, 20, 33–36
	5	Solve applied problems involving quadratic equations (p. 105)	88, 94, 99, 100, 104, 106
1.3	1	Add, subtract, multiply, and divide complex numbers (p. 110)	65–74
	2	Solve quadratic equations in the complex number system (p. 114)	75–82
1.4	1	Solve radical equations (p. 118)	17, 18, 23–30, 37, 38
	2	Solve equations quadratic in form (p. 119)	21, 22, 31, 32
	3	Solve equations by factoring (p. 121)	43–46
1.5	1	Use interval notation (p. 125)	47–60
	2	Use properties of inequalities (p. 126)	47–60
	3	Solve inequalities (p. 128)	47, 48
	4	Solve combined inequalities (p. 129)	49–52
1.6	1	Solve equations involving absolute value (p. 135)	39–42
	2	Solve inequalities involving absolute value (p. 135)	53–60
1.7	1	Translate verbal descriptions into mathematical expressions (p. 140)	83, 84
	2	Solve interest problems (p. 141)	85, 86
	3	Solve mixture problems (p. 142)	86, 97, 98, 101
	4	Solve uniform motion problems (p. 143)	87, 89–93, 109
	5	Solve constant rate job problems (p. 145)	95, 96, 103, 107

Review Exercises

In Problems 1–46, find all the real solutions, if any, of each equation. (Where they appear, a, b, m, and n are positive constants.)

1. $2 - \dfrac{x}{3} = 8$

2. $\dfrac{x}{4} - 2 = 4$

3. $-2(5 - 3x) + 8 = 4 + 5x$

4. $(6 - 3x) - 2(1 + x) = 6x$

5. $\dfrac{3x}{4} - \dfrac{x}{3} = \dfrac{1}{12}$

6. $\dfrac{4 - 2x}{3} + \dfrac{1}{6} = 2x$

7. $\dfrac{x}{x - 1} = \dfrac{6}{5} \quad x \neq 1$

8. $\dfrac{4x - 5}{3 - 7x} = 2 \quad x \neq \dfrac{3}{7}$

9. $x(1 - x) = 6$

10. $x(1 + x) = 6$

11. $\dfrac{1}{2}\left(x - \dfrac{1}{3}\right) = \dfrac{3}{4} - \dfrac{x}{6}$

12. $\dfrac{1 - 3x}{4} = \dfrac{x + 6}{3} + \dfrac{1}{2}$

13. $(x - 1)(2x + 3) = 3$

14. $x(2 - x) = 3(x - 4)$

15. $2x + 3 = 4x^2$

16. $1 + 6x = 4x^2$

17. $\sqrt[3]{x^2 - 1} = 2$

18. $\sqrt{1 + x^3} = 3$

19. $x(x + 1) + 2 = 0$

20. $3x^2 - x + 1 = 0$

21. $x^4 - 5x^2 + 4 = 0$

22. $3x^4 + 4x^2 + 1 = 0$

23. $\sqrt{2x - 3} + x = 3$

24. $\sqrt{2x - 1} = x - 2$

25. $\sqrt[4]{2x + 3} = 2$

26. $\sqrt[5]{3x + 1} = -1$

27. $\sqrt{x + 1} + \sqrt{x - 1} = \sqrt{2x + 1}$

28. $\sqrt{2x - 1} - \sqrt{x - 5} = 3$

29. $2x^{1/2} - 3 = 0$

30. $3x^{1/4} - 2 = 0$

31. $x^{-6} - 7x^{-3} - 8 = 0$

32. $6x^{-1} - 5x^{-1/2} + 1 = 0$

33. $x^2 + m^2 = 2mx + (nx)^2 \quad n \neq 1, n \neq -1$

34. $b^2x^2 + 2ax = x^2 + a^2 \quad b \neq 1, b \neq -1$

35. $10a^2x^2 - 2abx - 36b^2 = 0$

36. $\dfrac{1}{x - m} + \dfrac{1}{x - n} = \dfrac{2}{x} \quad x \neq 0, x \neq m, x \neq n$

37. $\sqrt{x^2 + 3x + 7} - \sqrt{x^2 - 3x + 9} + 2 = 0$

38. $\sqrt{x^2 + 3x + 7} - \sqrt{x^2 + 3x + 9} = 2$

39. $|2x + 3| = 7$

40. $|3x - 1| = 5$

41. $|2 - 3x| + 2 = 9$

42. $|1 - 2x| + 1 = 4$

43. $2x^3 = 3x^2$

44. $5x^4 = 9x^3$

45. $2x^3 + 5x^2 - 8x - 20 = 0$

46. $3x^3 + 5x^2 - 3x - 5 = 0$

In Problems 47–60, solve each inequality. Express your answer using set notation or interval notation. Graph the solution set.

47. $\dfrac{2x - 3}{5} + 2 \leq \dfrac{x}{2}$

48. $\dfrac{5 - x}{3} \leq 6x - 4$

49. $-9 \leq \dfrac{2x + 3}{-4} \leq 7$

50. $-4 < \dfrac{2x - 2}{3} < 6$

51. $2 < \dfrac{3 - 3x}{12} < 6$

52. $-3 \leq \dfrac{5 - 3x}{2} \leq 6$

53. $|3x + 4| < \dfrac{1}{2}$

54. $|1 - 2x| < \dfrac{1}{3}$

55. $|2x - 5| \geq 9$

56. $|3x + 1| \geq 10$

57. $2 + |2 - 3x| \leq 4$

58. $\dfrac{1}{2} + \left|\dfrac{2x - 1}{3}\right| \leq 1$

59. $1 - |2 - 3x| < -4$

60. $1 - \left|\dfrac{2x - 1}{3}\right| < -2$

In Problems 61–64, what number should be added to complete the square of each expression?

61. $x^2 + 6x$

62. $x^2 - 10x$

63. $x^2 - \dfrac{4}{3}x$

64. $x^2 + \dfrac{4}{5}x$

In Problems 65–74, use the complex number system and write each expression in the standard form a + bi.

65. $(6 + 3i) - (2 - 4i)$

66. $(8 - 3i) + (-6 + 2i)$

67. $4(3 - i) + 3(-5 + 2i)$

68. $2(1 + i) - 3(2 - 3i)$

69. $\dfrac{3}{3 + i}$

70. $\dfrac{4}{2 - i}$

71. i^{50}

72. i^{29}

73. $(2 + 3i)^3$

74. $(3 - 2i)^3$

In Problems 75–82, solve each equation in the complex number system.

75. $x^2 + x + 1 = 0$

76. $x^2 - x + 1 = 0$

77. $2x^2 + x - 2 = 0$

78. $3x^2 - 2x - 1 = 0$

79. $x^2 + 3 = x$

80. $2x^2 + 1 = 2x$

81. $x(1 - x) = 6$

82. $x(1 + x) = 2$

83. Translate the following statement into a mathematical expression: The perimeter p of a rectangle is the sum of two times the length l and two times the width w.

84. Translate the following statement into a mathematical expression: The total cost C of manufacturing x bicycles in one day is $50,000 plus $95 times the number of bicycles manufactured.

85. Banking A bank lends out $9000 at 7% simple interest. At the end of 1 year, how much interest is owed on the loan?

86. Financial Planning Steve, a recent retiree, requires $5000 per year in extra income. He has $70,000 to invest and can invest in A-rated bonds paying 8% per year or in a certificate of deposit (CD) paying 5% per year. How much money should be invested in each to realize exactly $5000 in interest per year?

87. Lightning and Thunder A flash of lightning is seen, and the resulting thunderclap is heard 3 seconds later. If the speed of sound averages 1100 feet per second, how far away is the storm?

88. Physics: Intensity of Light The intensity I (in candlepower) of a certain light source obeys the equation $I = \dfrac{900}{x^2}$, where x is the distance (in meters) from the light. Over what range of distances can an object be placed from this light source so that the range of intensity of light is from 1600 to 3600 candlepower, inclusive?

89. Extent of Search and Rescue A search plane has a cruising speed of 250 miles per hour and carries enough fuel for at most 5 hours of flying. If there is a wind that averages 30 miles per hour and the direction of the search is with the wind one way and against it the other, how far can the search plane travel before it has to turn back?

90. Extent of Search and Rescue If the search plane described in Problem 89 is able to add a supplementary fuel tank that allows for an additional 2 hours of flying, how much farther can the plane extend its search?

91. Rescue at Sea A life raft, set adrift from a sinking ship 150 miles offshore, travels directly toward a Coast Guard station at the rate of 5 miles per hour. At the time that the raft is set adrift, a rescue helicopter is dispatched from the Coast Guard station. If the helicopter's average speed is 90 miles per hour, how long will it take the helicopter to reach the life raft?

92. Physics: Uniform Motion Two bees leave two locations 150 meters apart and fly, without stopping, back and forth between these two locations at average speeds of 3 meters

per second and 5 meters per second, respectively. How long is it until the bees meet for the first time? How long is it until they meet for the second time?

93. Physics: Uniform Motion A Metra commuter train leaves Union Station in Chicago at 12 noon. Two hours later, an Amtrak train leaves on the same track, traveling at an average speed that is 50 miles per hour faster than the Metra train. At 3 PM the Amtrak train is 10 miles behind the commuter train. How fast is each going?

94. Physics An object is thrown down from the top of a building 1280 feet tall with an initial velocity of 32 feet per second. The distance s (in feet) of the object from the ground after t seconds is $s = 1280 - 32t - 16t^2$.

(a) When will the object strike ground?

(b) What is the height of the object after 4 seconds?

95. Working Together to Get a Job Done Clarissa and Shawna, working together, can paint the exterior of a house in 6 days. Clarissa by herself can complete this job in 5 days less than Shawna. How long will it take Clarissa to complete the job by herself?

96. Emptying a Tank Two pumps of different sizes, working together, can empty a fuel tank in 5 hours. The larger pump can empty this tank in 4 hours less than the smaller. If the larger pump is out of order, how long will it take the smaller one to do the job alone?

97. Chemistry: Salt Solutions How much water should be added to 64 ounces of a 10% salt solution to make a 2% salt solution?

98. Chemistry: Salt Solutions How much water must be evaporated from 64 ounces of a 2% salt solution to make a 10% salt solution?

99. Geometry The hypotenuse of a right triangle measures 13 centimeters. Find the lengths of the legs if their sum is 17 centimeters.

100. Geometry The diagonal of a rectangle measures 10 inches. If the length is 2 inches more than the width, find the dimensions of the rectangle.

101. Chemistry: Mixing Acids A laboratory has 60 cubic centimeters (cm^3) of a solution that is 40% HCl acid. How many cubic centimeters of a 15% solution of HCl acid should be mixed with the 60 cm^3 of 40% acid to obtain a solution of 25% HCl? How much of the 25% solution is there?

102. Framing a Painting An artist has 50 inches of oak trim to frame a painting. The frame is to have a border 3 inches wide surrounding the painting.

(a) If the painting is square, what are its dimensions? What are the dimensions of the frame?

(b) If the painting is rectangular with a length twice its width, what are the dimensions of the painting? What are the dimensions of the frame?

103. Using Two Pumps An 8-horsepower (hp) pump can fill a tank in 8 hours. A smaller, 3-hp pump fills the same tank in 12 hours. The pumps are used together to begin filling this tank. After four hours, the 8-hp pump breaks down. How long will it take the smaller pump to fill the tank?

104. Pleasing Proportion One formula stating the relationship between the length l and width w of a rectangle of "pleasing proportion" is $l^2 = w(l + w)$. How should a 4 foot by 8 foot sheet of plasterboard be cut so that the result is a rectangle of "pleasing proportion" with a width of 4 feet?

105. Finance An inheritance of $900,000 is to be divided among Scott, Alice, and Tricia in the following manner: Alice is to receive $\frac{3}{4}$ of what Scott gets, while Tricia gets $\frac{1}{2}$ of what Scott gets. How much does each receive?

106. Business: Determining the Cost of a Charter A group of 20 senior citizens can charter a bus for a one-day excursion trip for $15 per person. The charter company agrees to reduce the price of each ticket by 10¢ for each additional passenger in excess of 20 who goes on the trip, up to a maximum of 44 passengers (the capacity of the bus). If the final bill from the charter company was $482.40, how many seniors went on the trip, and how much did each pay?

107. Utilizing Copying Machines A new copying machine can do a certain job in 1 hour less than an older copier. Together they can do this job in 72 minutes. How long would it take the older copier by itself to do the job?

108. In a 100-meter race, Todd crosses the finish line 5 meters ahead of Scott. To even things up, Todd suggests to Scott that they race again, this time with Todd lining up 5 meters behind the start.

(a) Assuming that Todd and Scott run at the same pace as before, does the second race end in a tie?

(b) If not, who wins?

(c) By how many meters does he win?

(d) How far back should Todd start so that the race ends in a tie?

After running the race a second time, Scott, to even things up, suggests to Todd that he (Scott) line up 5 meters in front of the start.

(e) Assuming again that they run at the same pace as in the first race, does the third race result in a tie?

(f) If not, who wins?

(g) By how many meters?

(h) How far ahead should Scott start so that the race ends in a tie?

109. Physics: Uniform Motion A man is walking at an average speed of 4 miles per hour alongside a railroad track. A freight train, going in the same direction at an average speed of 30 miles per hour, requires 5 seconds to pass the man. How long is the freight train? Give your answer in feet.

CHAPTER TEST

In Problems 1–7, find all the real solutions, if any, of each equation.

1. $\dfrac{2x}{3} - \dfrac{x}{2} = \dfrac{5}{12}$

2. $x(x - 1) = 6$

3. $x^4 - 3x^2 - 4 = 0$

4. $\sqrt{2x - 5} + 2 = 4$

5. $|2x - 3| + 7 = 10$

6. $3x^3 + 2x^2 - 12x - 8 = 0$

7. $3x^2 - x + 1 = 0$

In Problems 8–10, solve each inequality. Express your answer using interval notation. Graph the solution set.

8. $-3 \le \dfrac{3x - 4}{2} \le 6$

9. $|3x + 4| < 8$

10. $2 + |2x - 5| \ge 9$

11. Write $\dfrac{-2}{3 - i}$ in the standard form $a + bi$.

12. Solve the equation $4x^2 - 4x + 5 = 0$ in the complex number system.

13. Blending Coffee A coffee house has 20 pounds of a coffee that sells for $4 per pound. How many pounds of a coffee that sells for $8 per pound should be mixed with the 20 pounds of $4-per-pound coffee to obtain a blend that will sell for $5 per pound? How much of the $5-per-pound coffee is there to sell?

CHAPTER PROJECTS

I. Home Mortgages While you may not be in the market for a home right now, it is probably an event that will occur for you within the next few years. The formula below gives the monthly payment P required to pay off a loan L at an annual rate of interest r, expressed as a decimal, but usually given as a percent. The time t, measured in months, is the duration of the loan, so a 15-year mortgage requires $t = 12 \times 15 = 180$ monthly payments.

$$P = L \left[\dfrac{\dfrac{r}{12}}{1 - \left(1 + \dfrac{r}{12}\right)^{-t}} \right] \quad \begin{array}{l} P = \text{monthly payment} \\ L = \text{loan amount} \\ r = \text{annual rate of interest,} \\ \quad \text{expressed as a decimal} \\ t = \text{length of loan, in months} \end{array} \quad \textbf{(1)}$$

On May 12, 2006, average interest rates were:

(a) 6.58% on 30-year mortgages

(b) 6.17% on 15-year mortgages

1. For each of these loans, calculate the monthly payment for a loan of $200,000.

2. Then compute the total amount paid over the term of the loan.

3. Finally, calculate the interest paid.

On May 5, 2006, average interest rates were:

(a) 6.59% on 30-year mortgages

(b) 6.22% on 15-year mortgages

4. For each of these loans, calculate the monthly payment for a loan of $200,000.

5. Then compute the total amount paid over the term of the loan.

6. Finally, calculate the interest paid.

7. Solve equation (1) for L.

8. If you can afford to pay $1000 per month for a mortgage payment, calculate the amount you can borrow on May 5, 2006.

 (a) On a 30-year mortgage

 (b) On a 15-year mortgage

9. Repeat Problem 8 for the rates on May 12, 2006.

10. Check with your local lending institution for current rates for 30-year and 15-year mortgages. Calculate how much you can borrow with a $1000 per month payment.

11. Repeat Problem 10 if you can afford a payment of $1300 per month.

12. Do you think that the interest rate plays an important role in determining how much you can afford to pay for a house?

13. Comment on the two types of mortgages: 30-year and 15-year. Which would you take? Why?

The following project is also available on the Instructor's Resource Center (IRC):

II. Project at Motorola How Many Cellular Phones Can I Make? An industrial engineer uses a model involving equations to be sure production levels meet customer demand.

Graphs

The First Modern Olympics: Athens, 1896

The birth of the modern Olympic Games

By John Gettings—"I hereby proclaim the opening of the first International Olympic Games at Athens." With these words on April 6, 1896, King George I of Greece welcomed the crowd that had gathered in the newly reconstructed Panathenean Stadium to the modern-day Olympic Summer Games.

The event was the idea of Baron Pierre de Coubertin of France who traveled the world to gather support for his dream to have nations come together and overcome national disputes, all in the name of sport.

The program for the Games included track and field, fencing, weightlifting, rifle and pistol shooting, tennis, cycling, swimming, gymnastics, and wrestling. Although 14 nations participated, most of the athletes were Greek.

The Games reached their high point on Day 11 with the first modern-day marathon. The idea to hold an event to commemorate the Ancient Olympic games was suggested by a friend of de Coubertin and was met with great anticipation. The race was run from Marathon to Athens (estimated at 22–26 miles), watched by more than 100,000 people and won by a Greek runner, Spiridon Louis.

Source: Fact Monster™ Database, © 2006 Pearson Education, Inc.

—See the Chapter Project—

Source: Adapted from John Gettings, "The First Modern Olympics: Athens, 1896," accessed at www.infoplease.com. ©2006 Pearson Education, Inc. Reprinted with permission.

A Look Back

In Chapter R we reviewed algebra essentials and geometry essentials. In Chapter 1 we studied equations in one variable.

A Look Ahead

Here we connect algebra and geometry using the rectangular coordinate system and use it to graph equations in two variables. The idea of using a system of rectangular coordinates dates back to ancient times, when such a system was used for surveying and city planning. Apollonius of Perga, in 200 BC, used a form of rectangular coordinates in his work on conics, although this use does not stand out as clearly as it does in modern treatments. Sporadic use of rectangular coordinates continued until the 1600s. By that time, algebra has developed sufficiently so that René Descartes (1596–1650) and Pierre de Fermat (1601–1665) could take the crucial step, which was the use of rectangular coordinates to translate geometry problems into algebra problems, and vice versa. This step was important for two reasons. First, it allowed both geometers and algebraists to gain new insights into their subjects, which previously had been regarded as separate, but now were seen to be connected in many important ways. Second, these insights made the development of calculus possible, which greatly enlarged the number of areas in which mathematics could be applied and made possible a much deeper understanding of these areas.

Outline

2.1 The Distance and Midpoint Formulas

PREPARING FOR THIS SECTION *Before getting started, review the following:*

- Algebra Essentials (Chapter R, Section R.2, pp. 17–26)
- Geometry Essentials (Chapter R, Section R.3, pp. 30–35)

Now Work the 'Are You Prepared?' problems on page 160.

OBJECTIVES 1 Use the Distance Formula (p. 157)
2 Use the Midpoint Formula (p. 159)

Rectangular Coordinates

Figure 1

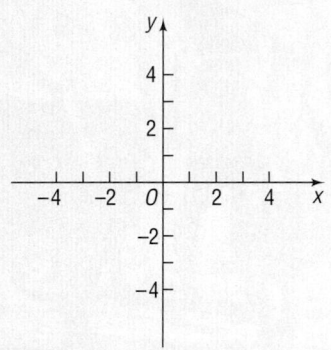

We locate a point on the real number line by assigning it a single real number, called the *coordinate of the point.* For work in a two-dimensional plane, we locate points by using two numbers.

We begin with two real number lines located in the same plane: one horizontal and the other vertical. We call the horizontal line the **x-axis,** the vertical line the **y-axis,** and the point of intersection the **origin O.** See Figure 1. We assign coordinates to every point on these number lines using a convenient scale. We usually use the same scale on each axis. In applications, however, different scales appropriate to the application may be used.

The origin O has a value of 0 on both the x-axis and y-axis. Points on the x-axis to the right of O are associated with positive real numbers, and those to the left of O are associated with negative real numbers. Points on the y-axis above O are associated with positive real numbers, and those below O are associated with negative real numbers. In Figure 1, the x-axis and y-axis are labeled as x and y, respectively, and we have used an arrow at the end of each axis to denote the positive direction.

The coordinate system described here is called a **rectangular** or **Cartesian* coordinate system.** The plane formed by the x-axis and y-axis is sometimes called the **xy-plane,** and the x-axis and y-axis are referred to as the **coordinate axes.**

Any point P in the xy-plane can be located by using an **ordered pair** (x, y) of real numbers. Let x denote the signed distance of P from the y-axis (*signed* means that, if P is to the right of the y-axis, then $x > 0$, and if P is to the left of the y-axis, then $x < 0$); and let y denote the signed distance of P from the x-axis. The ordered pair (x, y), also called the **coordinates** of P, then gives us enough information to locate the point P in the plane.

Figure 2

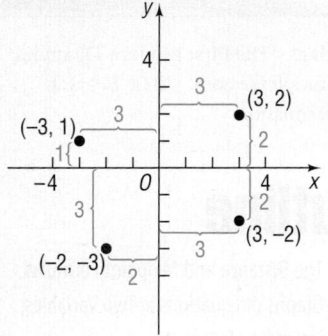

For example, to locate the point whose coordinates are $(-3, 1)$, go 3 units along the x-axis to the left of O and then go straight up 1 unit. We **plot** this point by placing a dot at this location. See Figure 2, in which the points with coordinates $(-3, 1)$, $(-2, -3)$, $(3, -2)$, and $(3, 2)$ are plotted.

The origin has coordinates $(0, 0)$. Any point on the x-axis has coordinates of the form $(x, 0)$, and any point on the y-axis has coordinates of the form $(0, y)$.

If (x, y) are the coordinates of a point P, then x is called the **x-coordinate,** or **abscissa,** of P and y is the **y-coordinate,** or **ordinate,** of P. We identify the point P by its coordinates (x, y) by writing $P = (x, y)$. Usually, we will simply say "the point (x, y)" rather than "the point whose coordinates are (x, y)."

The coordinate axes divide the xy-plane into four sections called **quadrants,** as shown in Figure 3. In quadrant I, both the x-coordinate and the y-coordinate of all points are positive; in quadrant II, x is negative and y is positive; in quadrant III, both x and y are negative; and in quadrant IV, x is positive and y is negative. Points on the coordinate axes belong to no quadrant.

Figure 3

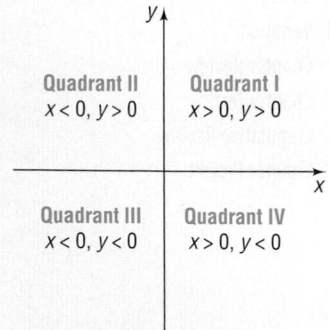

━━ **Now Work** PROBLEM 11

* Named after René Descartes (1596–1650), a French mathematician, philosopher, and theologian.

COMMENT On a graphing calculator, you can set the scale on each axis. Once this has been done, you obtain the **viewing rectangle.** See Figure 4 for a typical viewing rectangle. You should now read Section 1, *The Viewing Rectangle*, in the Appendix.

Figure 4

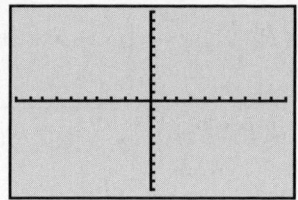

1 Use the Distance Formula

If the same units of measurement, such as inches, centimeters, and so on, are used for both the *x*-axis and *y*-axis, then all distances in the *xy*-plane can be measured using this unit of measurement.

| EXAMPLE 1 | **Finding the Distance between Two Points** |

Find the distance *d* between the points $(1, 3)$ and $(5, 6)$.

Solution First we plot the points $(1, 3)$ and $(5, 6)$ and connect them with a straight line. See Figure 5(a). We are looking for the length *d*. We begin by drawing a horizontal line from $(1, 3)$ to $(5, 3)$ and a vertical line from $(5, 3)$ to $(5, 6)$, forming a right triangle, as shown in Figure 5(b). One leg of the triangle is of length 4 (since $|5 - 1| = 4$), and the other is of length 3 (since $|6 - 3| = 3$). By the Pythagorean Theorem, the square of the distance *d* that we seek is

$$d^2 = 4^2 + 3^2 = 16 + 9 = 25$$
$$d = \sqrt{25} = 5$$

Figure 5

(a) (b)

The **distance formula** provides a straightforward method for computing the distance between two points.

THEOREM **Distance Formula**

The distance between two points $P_1 = (x_1, y_1)$ and $P_2 = (x_2, y_2)$, denoted by $d(P_1, P_2)$, is

$$d(P_1, P_2) = \sqrt{(x_2 - x_1)^2 + (y_2 - y_1)^2} \qquad \textbf{(1)}$$

In Words
To compute the distance between two points, find the difference of the x-coordinates, square it, and add this to the square of the difference of the y-coordinates. The square root of this sum is the distance.

Proof of the Distance Formula Let (x_1, y_1) denote the coordinates of point P_1 and let (x_2, y_2) denote the coordinates of point P_2. Assume that the line joining P_1 and P_2 is neither horizontal nor vertical. Refer to Figure 6(a). The coordinates of P_3 are (x_2, y_1). The horizontal distance from P_1 to P_3 is the absolute value of

the difference of the x-coordinates, $|x_2 - x_1|$. The vertical distance from P_3 to P_2 is the absolute value of the difference of the y-coordinates, $|y_2 - y_1|$. See Figure 6(b). The distance $d(P_1, P_2)$ that we seek is the length of the hypotenuse of the right triangle, so, by the Pythagorean Theorem, it follows that

$$[d(P_1, P_2)]^2 = |x_2 - x_1|^2 + |y_2 - y_1|^2$$
$$= (x_2 - x_1)^2 + (y_2 - y_1)^2$$
$$d(P_1, P_2) = \sqrt{(x_2 - x_1)^2 + (y_2 - y_1)^2}$$

Figure 6

(a)

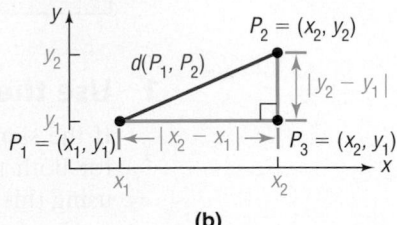

(b)

Now, if the line joining P_1 and P_2 is horizontal, then the y-coordinate of P_1 equals the y-coordinate of P_2; that is, $y_1 = y_2$. Refer to Figure 7(a). In this case, the distance formula (1) still works, because, for $y_1 = y_2$, it reduces to

$$d(P_1, P_2) = \sqrt{(x_2 - x_1)^2 + 0^2} = \sqrt{(x_2 - x_1)^2} = |x_2 - x_1|$$

Figure 7

(a)

(b)

A similar argument holds if the line joining P_1 and P_2 is vertical. See Figure 7(b). The distance formula is valid in all cases. ∎

EXAMPLE 2

Using the Distance Formula

Find the distance d between the points $(-4, 5)$ and $(3, 2)$.

Solution Using the distance formula, equation (1), the solution is obtained as follows:

$$d = \sqrt{[3 - (-4)]^2 + (2 - 5)^2} = \sqrt{7^2 + (-3)^2}$$
$$= \sqrt{49 + 9} = \sqrt{58} \approx 7.62$$

Now Work PROBLEMS **15 AND 19**

The distance between two points $P_1 = (x_1, y_1)$ and $P_2 = (x_2, y_2)$ is never a negative number. Furthermore, the distance between two points is 0 only when the points are identical, that is, when $x_1 = x_2$ and $y_1 = y_2$. Also, because $(x_2 - x_1)^2 = (x_1 - x_2)^2$ and $(y_2 - y_1)^2 = (y_1 - y_2)^2$, it makes no difference whether the distance is computed from P_1 to P_2 or from P_2 to P_1; that is, $d(P_1, P_2) = d(P_2, P_1)$.

The introduction to this chapter mentioned that rectangular coordinates enable us to translate geometry problems into algebra problems, and vice versa. The next example shows how algebra (the distance formula) can be used to solve geometry problems.

EXAMPLE 3	**Using Algebra to Solve Geometry Problems**

Consider the three points $A = (-2, 1)$, $B = (2, 3)$, and $C = (3, 1)$.

(a) Plot each point and form the triangle ABC.

(b) Find the length of each side of the triangle.

(c) Verify that the triangle is a right triangle.

(d) Find the area of the triangle.

Solution

(a) Points A, B, C and triangle ABC are plotted in Figure 8.

(b) We use the distance formula, equation (1).

Figure 8

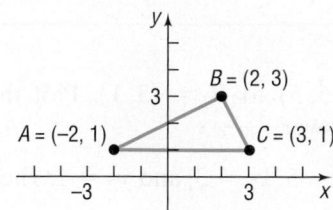

$$d(A, B) = \sqrt{[2 - (-2)]^2 + (3 - 1)^2} = \sqrt{16 + 4} = \sqrt{20} = 2\sqrt{5}$$

$$d(B, C) = \sqrt{(3 - 2)^2 + (1 - 3)^2} = \sqrt{1 + 4} = \sqrt{5}$$

$$d(A, C) = \sqrt{[3 - (-2)]^2 + (1 - 1)^2} = \sqrt{25 + 0} = 5$$

(c) To show that the triangle is a right triangle, we need to show that the sum of the squares of the lengths of two of the sides equals the square of the length of the third side. (Why is this sufficient?) Looking at Figure 8, it seems reasonable to conjecture that the right angle is at vertex B. We shall check to see whether

$$[d(A, B)]^2 + [d(B, C)]^2 = [d(A, C)]^2$$

We find that

$$[d(A, B)]^2 + [d(B, C)]^2 = \left(2\sqrt{5}\right)^2 + \left(\sqrt{5}\right)^2$$

$$= 20 + 5 = 25 = [d(A, C)]^2$$

so it follows from the converse of the Pythagorean Theorem that triangle ABC is a right triangle.

(d) Because the right angle is at vertex B, the sides AB and BC form the base and height of the triangle. Its area is

$$\text{Area} = \frac{1}{2}(\text{Base})(\text{Height}) = \frac{1}{2}\left(2\sqrt{5}\right)\left(\sqrt{5}\right) = 5 \text{ square units}$$

Now Work PROBLEM 29

2 Use the Midpoint Formula

Figure 9

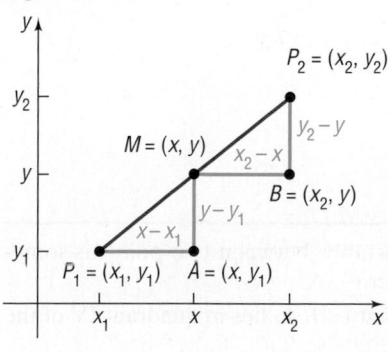

We now derive a formula for the coordinates of the **midpoint of a line segment.** Let $P_1 = (x_1, y_1)$ and $P_2 = (x_2, y_2)$ be the endpoints of a line segment, and let $M = (x, y)$ be the point on the line segment that is the same distance from P_1 as it is from P_2. See Figure 9. The triangles P_1AM and MBP_2 are congruent. Do you see why? $d(P_1, M) = d(M, P_2)$ is given; $\angle AP_1M = \angle BMP_2$,* and $\angle P_1MA = \angle MP_2B$. So, we have angle–side–angle. Because triangles P_1AM and MBP_2 are congruent, corresponding sides are equal in length. That is,

$$x - x_1 = x_2 - x \qquad \text{and} \qquad y - y_1 = y_2 - y$$

$$2x = x_1 + x_2 \qquad\qquad\qquad 2y = y_1 + y_2$$

$$x = \frac{x_1 + x_2}{2} \qquad\qquad\qquad y = \frac{y_1 + y_2}{2}$$

* A postulate from geometry states that the transversal $\overline{P_1P_2}$ forms congruent corresponding angles with the parallel line segments $\overline{P_1A}$ and $\overline{MB}$.

THEOREM	**Midpoint Formula**

The midpoint $M = (x, y)$ of the line segment from $P_1 = (x_1, y_1)$ to $P_2 = (x_2, y_2)$ is

In Words

To find the midpoint of a line segment, average the x-coordinates and average the y-coordinates of the endpoints.

$$M = (x, y) = \left(\frac{x_1 + x_2}{2}, \frac{y_1 + y_2}{2} \right) \qquad (2)$$

EXAMPLE 4	**Finding the Midpoint of a Line Segment**

Find the midpoint of a line segment from $P_1 = (-5, 5)$ to $P_2 = (3, 1)$. Plot the points P_1 and P_2 and their midpoint. Check your answer.

Solution We apply the midpoint formula (2) using $x_1 = -5$, $y_1 = 5$, $x_2 = 3$, and $y_2 = 1$. Then the coordinates (x, y) of the midpoint M are

Figure 10

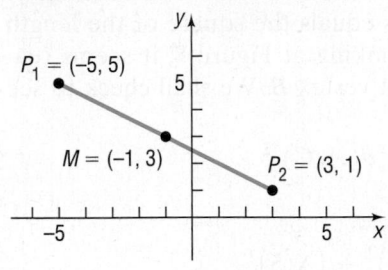

$$x = \frac{x_1 + x_2}{2} = \frac{-5 + 3}{2} = -1 \quad \text{and} \quad y = \frac{y_1 + y_2}{2} = \frac{5 + 1}{2} = 3$$

That is, $M = (-1, 3)$. See Figure 10.

✔ **Check:** Because M is the midpoint, we check the answer by verifying that $d(P_1, M) = d(M, P_2)$:

$$d(P_1, M) = \sqrt{[-1 - (-5)]^2 + (3 - 5)^2} = \sqrt{16 + 4} = \sqrt{20}$$

$$d(M, P_2) = \sqrt{[3 - (-1)]^2 + (1 - 3)^2} = \sqrt{16 + 4} = \sqrt{20}$$

■

➡ **Now Work** PROBLEM 35

2.1 Assess Your Understanding

'Are You Prepared?' *Answers are given at the end of these exercises. If you get a wrong answer, read the pages listed in red.*

1. On the real number line the origin is assigned the number _____. (p. 17–26)

2. If -3 and 5 are the coordinates of two points on the real number line, the distance between these points is _____. (p. 17–26)

3. If 3 and 4 are the legs of a right triangle, the hypotenuse is _____. (pp. 30–35)

4. Use the converse of the Pythagorean Theorem to show that a triangle whose sides are of lengths 11, 60, and 61 is a right triangle. (p. 30–35)

Concepts and Vocabulary

5. If (x, y) are the coordinates of a point P in the xy-plane, then x is called the _____ of P and y is the _____ of P.

6. The coordinate axes divide the xy-plane into four sections called _____.

7. If three distinct points P, Q, and R all lie on a line and if $d(P, Q) = d(Q, R)$, then Q is called the _____ of the line segment from P to R.

8. **True or False** The distance between two points is sometimes a negative number.

9. **True or False** The point $(-1, 4)$ lies in quadrant IV of the Cartesian plane.

10. **True or False** The midpoint of a line segment is found by averaging the x-coordinates and averaging the y-coordinates of the endpoints.

Skill Building

In Problems 11 and 12, plot each point in the xy-plane. Tell in which quadrant or on what coordinate axis each point lies.

11. (a) $A = (-3, 2)$ (d) $D = (6, 5)$
 (b) $B = (6, 0)$ (e) $E = (0, -3)$
 (c) $C = (-2, -2)$ (f) $F = (6, -3)$

12. (a) $A = (1, 4)$ (d) $D = (4, 1)$
 (b) $B = (-3, -4)$ (e) $E = (0, 1)$
 (c) $C = (-3, 4)$ (f) $F = (-3, 0)$

13. Plot the points $(2, 0)$, $(2, -3)$, $(2, 4)$, $(2, 1)$, and $(2, -1)$. Describe the set of all points of the form $(2, y)$, where y is a real number.

14. Plot the points $(0, 3)$, $(1, 3)$, $(-2, 3)$, $(5, 3)$, and $(-4, 3)$. Describe the set of all points of the form $(x, 3)$, where x is a real number.

In Problems 15–28, find the distance $d(P_1, P_2)$ between the points P_1 and P_2.

15.
16.
17.
18.

19. $P_1 = (3, -4)$; $P_2 = (5, 4)$
20. $P_1 = (-1, 0)$; $P_2 = (2, 4)$

21. $P_1 = (-3, 2)$; $P_2 = (6, 0)$
22. $P_1 = (2, -3)$; $P_2 = (4, 2)$

23. $P_1 = (4, -3)$; $P_2 = (6, 4)$
24. $P_1 = (-4, -3)$; $P_2 = (6, 2)$

25. $P_1 = (-0.2, 0.3)$; $P_2 = (2.3, 1.1)$
26. $P_1 = (1.2, 2.3)$; $P_2 = (-0.3, 1.1)$

27. $P_1 = (a, b)$; $P_2 = (0, 0)$
28. $P_1 = (a, a)$; $P_2 = (0, 0)$

In Problems 29–34, plot each point and form the triangle ABC. Verify that the triangle is a right triangle. Find its area.

29. $A = (-2, 5)$; $B = (1, 3)$; $C = (-1, 0)$
30. $A = (-2, 5)$; $B = (12, 3)$; $C = (10, -11)$

31. $A = (-5, 3)$; $B = (6, 0)$; $C = (5, 5)$
32. $A = (-6, 3)$; $B = (3, -5)$; $C = (-1, 5)$

33. $A = (4, -3)$; $B = (0, -3)$; $C = (4, 2)$
34. $A = (4, -3)$; $B = (4, 1)$; $C = (2, 1)$

In Problems 35–44, find the midpoint of the line segment joining the points P_1 and P_2.

35. $P_1 = (3, -4)$; $P_2 = (5, 4)$
36. $P_1 = (-2, 0)$; $P_2 = (2, 4)$

37. $P_1 = (-3, 2)$; $P_2 = (6, 0)$
38. $P_1 = (2, -3)$; $P_2 = (4, 2)$

39. $P_1 = (4, -3)$; $P_2 = (6, 1)$
40. $P_1 = (-4, -3)$; $P_2 = (2, 2)$

41. $P_1 = (-0.2, 0.3)$; $P_2 = (2.3, 1.1)$
42. $P_1 = (1.2, 2.3)$; $P_2 = (-0.3, 1.1)$

43. $P_1 = (a, b)$; $P_2 = (0, 0)$
44. $P_1 = (a, a)$; $P_2 = (0, 0)$

Applications and Extensions

45. Find all points having an x-coordinate of 2 whose distance from the point $(-2, -1)$ is 5.

46. Find all points having a y-coordinate of -3 whose distance from the point $(1, 2)$ is 13.

47. Find all points on the x-axis that are 5 units from the point $(4, -3)$.

48. Find all points on the y-axis that are 5 units from the point $(4, 4)$.

49. The midpoint of the line segment from P_1 to P_2 is $(-1, 4)$. If $P_1 = (-3, 6)$, what is P_2?

50. The midpoint of the line segment from P_1 to P_2 is $(5, -4)$. If $P_2 = (7, -2)$, what is P_1?

51. **Geometry** The **medians** of a triangle are the line segments from each vertex to the midpoint of the opposite side (see the figure). Find the lengths of the medians of the triangle with vertices at $A = (0, 0)$, $B = (6, 0)$, and $C = (4, 4)$.

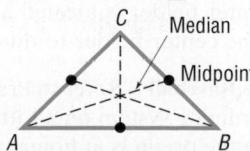

52. Geometry An **equilateral triangle** is one in which all three sides are of equal length. If two vertices of an equilateral triangle are $(0, 4)$ and $(0, 0)$, find the third vertex. How many of these triangles are possible?

53. Geometry Find the midpoint of each diagonal of a square with side of length s. Draw the conclusion that the diagonals of a square intersect at their midpoints.

[**Hint:** Use $(0, 0)$, $(0, s)$, $(s, 0)$, and (s, s) as the vertices of the square.]

54. Geometry Verify that the points $(0, 0)$, $(a, 0)$, and $\left(\dfrac{a}{2}, \dfrac{\sqrt{3}\,a}{2}\right)$ are the vertices of an equilateral triangle. Then show that the midpoints of the three sides are the vertices of a second equilateral triangle (refer to Problem 52).

*In Problems 55–58, find the length of each side of the triangle determined by the three points P_1, P_2, and P_3. State whether the triangle is an isosceles triangle, a right triangle, neither of these, or both. (An **isosceles triangle** is one in which at least two of the sides are of equal length.)*

55. $P_1 = (2, 1)$; $\quad P_2 = (-4, 1)$; $\quad P_3 = (-4, -3)$

56. $P_1 = (-1, 4)$; $\quad P_2 = (6, 2)$; $\quad P_3 = (4, -5)$

57. $P_1 = (-2, -1)$; $\quad P_2 = (0, 7)$; $\quad P_3 = (3, 2)$

58. $P_1 = (7, 2)$; $\quad P_2 = (-4, 0)$; $\quad P_3 = (4, 6)$

59. Baseball A major league baseball "diamond" is actually a square, 90 feet on a side (see the figure). What is the distance directly from home plate to second base (the diagonal of the square)?

60. Little League Baseball The layout of a Little League playing field is a square, 60 feet on a side.* How far is it directly from home plate to second base (the diagonal of the square)?

Source: Little League Baseball, Official Regulations and Playing Rules, 2006.

61. Baseball Refer to Problem 59. Overlay a rectangular coordinate system on a major league baseball diamond so that the origin is at home plate, the positive x-axis lies in the direction from home plate to first base, and the positive y-axis lies in the direction from home plate to third base.

(a) What are the coordinates of first base, second base, and third base? Use feet as the unit of measurement.

(b) If the right fielder is located at $(310, 15)$, how far is it from the right fielder to second base?

(c) If the center fielder is located at $(300, 300)$, how far is it from the center fielder to third base?

62. Little League Baseball Refer to Problem 60. Overlay a rectangular coordinate system on a Little League baseball diamond so that the origin is at home plate, the positive x-axis

lies in the direction from home plate to first base, and the positive y-axis lies in the direction from home plate to third base.

(a) What are the coordinates of first base, second base, and third base? Use feet as the unit of measurement.

(b) If the right fielder is located at $(180, 20)$, how far is it from the right fielder to second base?

(c) If the center fielder is located at $(220, 220)$, how far is it from the center fielder to third base?

63. Distance between Moving Objects A Dodge Neon and a Mack truck leave an intersection at the same time. The Neon heads east at an average speed of 30 miles per hour, while the truck heads south at an average speed of 40 miles per hour. Find an expression for their distance apart d (in miles) at the end of t hours.

64. Distance of a Moving Object from a Fixed Point A hot-air balloon, headed due east at an average speed of 15 miles per hour and at a constant altitude of 100 feet, passes over an intersection (see the figure). Find an expression for the distance d (measured in feet) from the balloon to the intersection t seconds later.

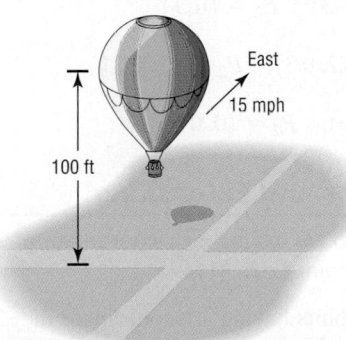

65. Drafting Error When a draftsman draws three lines that are to intersect at one point, the lines may not intersect as intended and subsequently will form an **error triangle.** If this error triangle is long and thin, one estimate for the location of the desired point is the midpoint of the shortest side. The figure shows one such error triangle.

Source: www.uwgb.edu/dutchs/STRUCTGE/sl00.htm

Inc., in 2004. How does your result compare to the reported value of $256 billion?

Source: Wal-Mart Stores, Inc., 2006 Annual Report

(a) Find an estimate for the desired intersection point.

(b) Find the length of the median for the midpoint found in part (a). See Problem 51.

66. Net Sales The figure illustrates how net sales of Wal-Mart Stores, Inc., have grown from 2002 through 2006. Use the midpoint formula to estimate the net sales of Wal-Mart Stores,

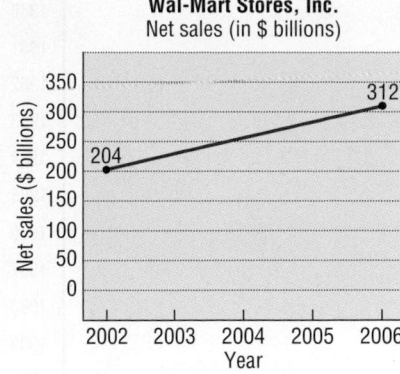

Wal-Mart Stores, Inc.
Net sales (in $ billions)

'Are You Prepared?' Answers

1. 0 **2.** 8 **3.** 5 **4.** $11^2 + 60^2 = 121 + 3600 = 3721 = 61^2$

2.2 Graphs of Equations in Two Variables; Intercepts; Symmetry

PREPARING FOR THIS SECTION *Before getting started, review the following:*

• Solving Equations (Section 1.1, pp. 86–93)

Now Work the 'Are You Prepared' problems on page 171.

OBJECTIVES 1 Graph Equations by Plotting Points (p. 163)

2 Find Intercepts from a Graph (p. 165)

3 Find Intercepts from an Equation (p. 166)

4 Test an Equation for Symmetry with Respect to the *x*-Axis, the *y*-Axis, and the Origin (p. 167)

5 Know How to Graph Key Equations (p. 169)

1 Graph Equations by Plotting Points

An **equation in two variables,** say *x* and *y*, is a statement in which two expressions involving *x* and *y* are equal. The expressions are called the **sides** of the equation. Since an equation is a statement, it may be true or false, depending on the value of the variables. Any values of *x* and *y* that result in a true statement are said to **satisfy** the equation.

For example, the following are all equations in two variables *x* and *y*:

$$x^2 + y^2 = 5 \qquad 2x - y = 6 \qquad y = 2x + 5 \qquad x^2 = y$$

The first of these, $x^2 + y^2 = 5$, is satisfied for $x = 1, y = 2$, since $1^2 + 2^2 = 1 + 4 = 5$. Other choices of *x* and *y*, such as $x = -1, y = -2$, also satisfy this equation. It is not satisfied for $x = 2$ and $y = 3$, since $2^2 + 3^2 = 4 + 9 = 13 \neq 5$.

The **graph of an equation in two variables** *x* and *y* consists of the set of points in the *xy*-plane whose coordinates (x, y) satisfy the equation.

Graphs play an important role in helping us visualize the relationships that exist between two variable quantities. Table 1 shows the average price of regular

Table 1
Adjusted average price (in dollars) of gasoline in California, 1978–2005.

Year	Price	Year	Price
1978	1.5651	1992	1.3969
1979	1.9829	1993	1.5189
1980	2.4929	1994	1.4709
1981	2.4977	1995	1.4669
1982	2.1795	1996	1.5397
1983	1.8782	1997	1.5329
1984	1.8310	1998	1.3246
1985	1.7540	1999	1.5270
1986	1.3459	2000	1.8249
1987	1.3274	2001	1.7536
1988	1.3111	2002	1.5955
1989	1.3589	2003	1.8950
1990	1.4656	2004	2.1521
1991	1.4973	2005	2.4730

Figure 11

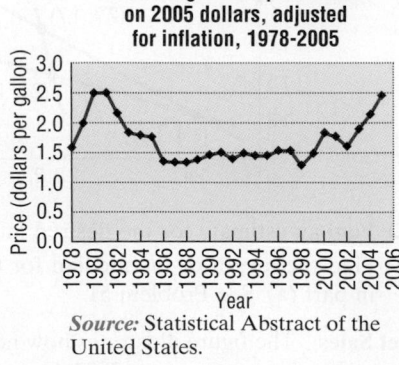

California gasoline prices based on 2005 dollars, adjusted for inflation, 1978-2005

Source: Statistical Abstract of the United States.

gasoline in California based on 2005 dollars (adjusted for inflation) for the years 1978–2005. In Figure 11, we graph the data from the table, using the year along the x-axis and the price along the y-axis. From the graph, we can see the price was highest in 1980, 1981, and 2005 at about $2.50 per gallon.

EXAMPLE 1 | **Determining Whether a Point Is on the Graph of an Equation**

Determine if the following points are on the graph of the equation $2x - y = 6$.

(a) $(2, 3)$ (b) $(2, -2)$

Solution

(a) For the point $(2, 3)$, we check to see if $x = 2, y = 3$ satisfies the equation $2x - y = 6$.

$$2x - y = 2(2) - 3 = 4 - 3 = 1 \neq 6$$

The equation is not satisfied, so the point $(2, 3)$ is not on the graph of $2x - y = 6$.

(b) For the point $(2, -2)$, we have

$$2x - y = 2(2) - (-2) = 4 + 2 = 6$$

The equation is satisfied, so the point $(2, -2)$ is on the graph of $2x - y = 6$.

Now Work PROBLEM 11

EXAMPLE 2 | **Graphing an Equation by Plotting Points**

Graph the equation: $y = 2x + 5$

Solution

We want to find all points (x, y) that satisfy the equation. To locate some of these points (and get an idea of the pattern of the graph), we assign some numbers to x and find corresponding values for y.

Figure 12

If	Then	Point on Graph
$x = 0$	$y = 2(0) + 5 = 5$	$(0, 5)$
$x = 1$	$y = 2(1) + 5 = 7$	$(1, 7)$
$x = -5$	$y = 2(-5) + 5 = -5$	$(-5, -5)$
$x = 10$	$y = 2(10) + 5 = 25$	$(10, 25)$

By plotting these points and then connecting them, we obtain the graph of the equation (a *line*), as shown in Figure 12.

| EXAMPLE 3 | Graphing an Equation by Plotting Points |

Graph the equation: $y = x^2$

Solution Table 2 provides several points on the graph. In Figure 13 we plot these points and connect them with a smooth curve to obtain the graph (a *parabola*).

Table 2

x	$y = x^2$	(x, y)
−4	16	(−4, 16)
−3	9	(−3, 9)
−2	4	(−2, 4)
−1	1	(−1, 1)
0	0	(0, 0)
1	1	(1, 1)
2	4	(2, 4)
3	9	(3, 9)
4	16	(4, 16)

Figure 13

The graphs of the equations shown in Figures 12 and 13 do not show all points. For example, in Figure 12, the point (20, 45) is a part of the graph of $y = 2x + 5$, but it is not shown. Since the graph of $y = 2x + 5$ could be extended out as far as we please, we use arrows to indicate that the pattern shown continues. It is important when illustrating a graph to present enough of the graph so that any viewer of the illustration will "see" the rest of it as an obvious continuation of what is actually there. This is referred to as a **complete graph.**

One way to obtain a complete graph of an equation is to plot a sufficient number of points on the graph until a pattern becomes evident. Then these points are connected with a smooth curve following the suggested pattern. But how many points are sufficient? Sometimes knowledge about the equation tells us. For example, we will learn in the next section that, if an equation is of the form $y = mx + b$, then its graph is a line. In this case, only two points are needed to obtain the graph.

One purpose of this book is to investigate the properties of equations in order to decide whether a graph is complete. Sometimes we shall graph equations by plotting points. Shortly, we shall investigate various techniques that will enable us to graph an equation without plotting so many points.

 COMMENT Another way to obtain the graph of an equation is to use a graphing utility. Read Section 2, *Using a Graphing Utility to Graph Equations*, in the Appendix. ∎

Two techniques that sometimes reduce the number of points required to graph an equation involve finding *intercepts* and checking for *symmetry*.

2 Find Intercepts from a Graph

The points, if any, at which a graph crosses or touches the coordinate axes are called the **intercepts.** See Figure 14. The x-coordinate of a point at which the graph crosses or touches the x-axis is an **x-intercept,** and the y-coordinate of a point at which the graph crosses or touches the y-axis is a **y-intercept.** For a graph to be complete, all its intercepts must be displayed.

Figure 14

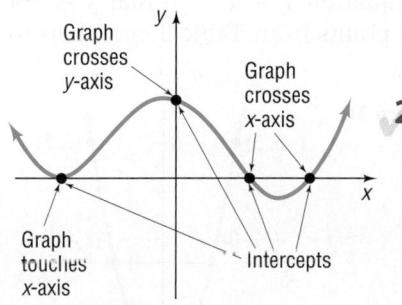

| EXAMPLE 4 | Finding Intercepts from a Graph |

Find the intercepts of the graph in Figure 15 shown on p. 166. What are its x-intercepts? What are its y-intercepts?

Figure 15

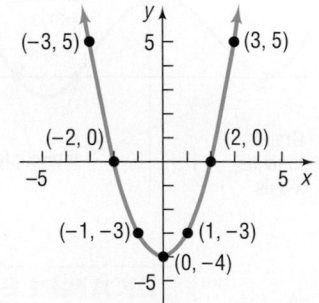

Solution

The intercepts of the graph are the points

$$(-3, 0), \quad (0, 3), \quad \left(\frac{3}{2}, 0\right), \quad \left(0, -\frac{4}{3}\right), \quad (0, -3.5), \quad (4.5, 0)$$

The x-intercepts are -3, $\frac{3}{2}$, and 4.5; the y-intercepts are -3.5, $-\frac{4}{3}$, and 3.

In Example 4, you should notice the following usage: If we do not specify the type of intercept (x- versus y-), then we report the intercept as an ordered pair. However, if we specify the type of intercept, then we only report the coordinate of the intercept. For x-intercepts, we report the x-coordinate of the intercept; for y-intercepts, we report the y-coordinate of the intercept.

➤ **Now Work** PROBLEM 39(a)

3 Find Intercepts from an Equation

The intercepts of a graph can be found from its equation by using the fact that points on the x-axis have y-coordinates equal to 0 and points on the y-axis have x-coordinates equal to 0.

 COMMENT For many equations, finding intercepts may not be so easy. In such cases, a graphing utility can be used. Read the first part of Section 3, *Using a Graphing Utility to Locate Intercepts and Check for Symmetry*, in the Appendix, to find out how a graphing utility locates intercepts. ■

Procedure for Finding Intercepts

1. To find the x-intercept(s), if any, of the graph of an equation, let $y = 0$ in the equation and solve for x, where x is a real number.
2. To find the y-intercept(s), if any, of the graph of an equation, let $x = 0$ in the equation and solve for y, where y is a real number.

EXAMPLE 5

Finding Intercepts from an Equation

Find the x-intercept(s) and the y-intercept(s) of the graph of $y = x^2 - 4$. Then graph $y = x^2 - 4$ by plotting points.

Solution

To find the x-intercept(s), we let $y = 0$ and obtain the equation

$$\begin{array}{ll} x^2 - 4 = 0 & y = x^2 - 4 \text{ with } y = 0 \\ (x + 2)(x - 2) = 0 & \text{Factor.} \\ x + 2 = 0 \quad \text{or} \quad x - 2 = 0 & \text{Zero-Product Property} \\ x = -2 \quad \text{or} \quad x = 2 & \text{Solve.} \end{array}$$

The equation has two solutions, -2 and 2. The x-intercepts are -2 and 2.

To find the y-intercept(s), we let $x = 0$ in the equation.

$$y = x^2 - 4$$
$$= 0^2 - 4 = -4$$

The y-intercept is -4.

Since $x^2 \geq 0$ for all x, we deduce from the equation $y = x^2 - 4$ that $y \geq -4$ for all x. This information, the intercepts, and the points from Table 3 enable us to graph $y = x^2 - 4$. See Figure 16.

Table 3

x	$y = x^2 - 4$	(x, y)
-3	5	$(-3, 5)$
-1	-3	$(-1, -3)$
1	-3	$(1, -3)$
3	5	$(3, 5)$

Figure 16

➤ **Now Work** PROBLEM 21

4 Test an Equation for Symmetry with Respect to the *x*-Axis, the *y*-Axis, and the Origin

We just saw the role that intercepts play in obtaining key points on the graph of an equation. Another helpful tool for graphing equations involves symmetry, particularly symmetry with respect to the *x*-axis, the *y*-axis, and the origin.

DEFINITION A graph is said to be **symmetric with respect to the *x*-axis** if, for every point (x, y) on the graph, the point $(x, -y)$ is also on the graph.

Figure 17 illustrates the definition. When a graph is symmetric with respect to the *x*-axis, notice that the part of the graph above the *x*-axis is a reflection or mirror image of the part below it, and vice versa.

Figure 17
Symmetry with respect to the *x*-axis

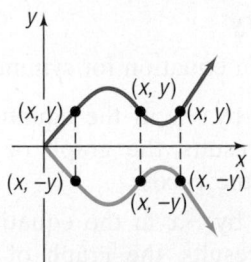

EXAMPLE 6 **Points Symmetric with Respect to the *x*-Axis**

If a graph is symmetric with respect to the *x*-axis and the point $(3, 2)$ is on the graph, then the point $(3, -2)$ is also on the graph.

DEFINITION A graph is said to be **symmetric with respect to the *y*-axis** if, for every point (x, y) on the graph, the point $(-x, y)$ is also on the graph.

Figure 18
Symmetry with respect to the *y*-axis

Figure 18 illustrates the definition. When a graph is symmetric with respect to the *y*-axis, notice that the part of the graph to the right of the *y*-axis is a reflection of the part to the left of it, and vice versa.

EXAMPLE 7 **Points Symmetric with Respect to the *y*-Axis**

If a graph is symmetric with respect to the *y*-axis and the point $(5, 8)$ is on the graph, then the point $(-5, 8)$ is also on the graph.

DEFINITION A graph is said to be **symmetric with respect to the origin** if, for every point (x, y) on the graph, the point $(-x, -y)$ is also on the graph.

Figure 19
Symmetry with respect to the origin

Figure 19 illustrates the definition. Notice that symmetry with respect to the origin may be viewed in three ways:

1. As a reflection about the *y*-axis, followed by a reflection about the *x*-axis

2. As a projection along a line through the origin so that the distances from the origin are equal

3. As half of a complete revolution about the origin

EXAMPLE 8	Points Symmetric with Respect to the Origin

If a graph is symmetric with respect to the origin and the point $(4, 2)$ is on the graph, then the point $(-4, -2)$ is also on the graph.

Now Work PROBLEMS 29 AND 39(b)

When the graph of an equation is symmetric with respect to a coordinate axis or the origin, the number of points that you need to plot in order to see the pattern is reduced. For example, if the graph of an equation is symmetric with respect to the y-axis, then, once points to the right of the y-axis are plotted, an equal number of points on the graph can be obtained by reflecting them about the y-axis. Because of this, before we graph an equation, we first want to determine whether it has any symmetry. The following tests are used for this purpose.

Tests for Symmetry

To test the graph of an equation for symmetry with respect to the

x-Axis Replace y by $-y$ in the equation and simplify. If an equivalent equation results, the graph of the equation is symmetric with respect to the x-axis.

y-Axis Replace x by $-x$ in the equation and simplify. If an equivalent equation results, the graph of the equation is symmetric with respect to the y-axis.

Origin Replace x by $-x$ and y by $-y$ in the equation and simplify. If an equivalent equation results, the graph of the equation is symmetric with respect to the origin.

EXAMPLE 9	Testing an Equation for Symmetry

Test $y = \dfrac{4x^2}{x^2 + 1}$ for symmetry.

Solution *x-Axis:* To test for symmetry with respect to the x-axis, replace y by $-y$. Since $-y = \dfrac{4x^2}{x^2 + 1}$ is not equivalent to $y = \dfrac{4x^2}{x^2 + 1}$, the graph of the equation is not symmetric with respect to the x-axis.

y-Axis: To test for symmetry with respect to the y-axis, replace x by $-x$. Since $y = \dfrac{4(-x)^2}{(-x)^2 + 1} = \dfrac{4x^2}{x^2 + 1}$ is equivalent to $y = \dfrac{4x^2}{x^2 + 1}$, the graph of the equation is symmetric with respect to the y-axis.

Origin: To test for symmetry with respect to the origin, replace x by $-x$ and y by $-y$.

$$-y = \frac{4(-x)^2}{(-x)^2 + 1} \qquad \text{Replace } x \text{ by } -x \text{ and } y \text{ by } -y.$$

$$-y = \frac{4x^2}{x^2 + 1} \qquad \text{Simplify.}$$

$$y = -\frac{4x^2}{x^2 + 1} \qquad \text{Multiply both sides by } -1.$$

Since the result is not equivalent to the original equation, the graph of the equation $y = \dfrac{4x^2}{x^2 + 1}$ is not symmetric with respect to the origin.

Seeing the Concept

Figure 20 shows the graph of $y = \dfrac{4x^2}{x^2 + 1}$ using a graphing utility. Do you see the symmetry with respect to the y-axis?

Figure 20

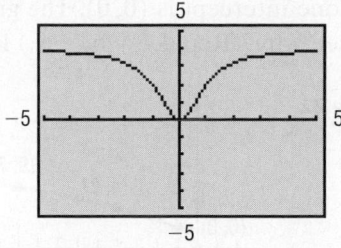

━━━ **Now Work** PROBLEM 59

5 Know How to Graph Key Equations

The next three examples use intercepts, symmetry, and point plotting to obtain the graphs of key equations. It is important to know the graphs of these key equations because we use them later. The first of these is $y = x^3$.

EXAMPLE 10

Graphing the Equation $y = x^3$ by Finding Intercepts, Checking for Symmetry, and Plotting Points

Graph the equation $y = x^3$ by plotting points. Find any intercepts and check for symmetry first.

Solution

First, we seek the intercepts. When $x = 0$, then $y = 0$; and when $y = 0$, then $x = 0$. The origin $(0, 0)$ is the only intercept. Now we test for symmetry.

x-Axis: Replace y by $-y$. Since $-y = x^3$ is not equivalent to $y = x^3$, the graph is not symmetric with respect to the x-axis.

y-Axis: Replace x by $-x$. Since $y = (-x)^3 = -x^3$ is not equivalent to $y = x^3$, the graph is not symmetric with respect to the y-axis.

Origin: Replace x by $-x$ and y by $-y$. Since $-y = (-x)^3 = -x^3$ is equivalent to $y = x^3$ (multiply both sides by -1), the graph is symmetric with respect to the origin.

To graph $y = x^3$, we use the equation to obtain several points on the graph. Because of the symmetry, we only need to locate points on the graph for which $x \geq 0$. See Table 4. Since $(1, 1)$ is on the graph, and the graph is symmetric with respect to the origin, the point $(-1, -1)$ is also on the graph. Plot the points from Table 4. Figure 21 shows the graph.

Table 4

x	$y = x^3$	(x, y)
0	0	$(0, 0)$
1	1	$(1, 1)$
2	8	$(2, 8)$
3	27	$(3, 27)$

Figure 21

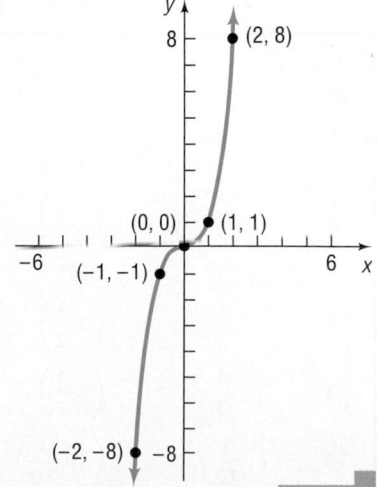

Graphing the Equation $x = y^2$

Graph the equation $x = y^2$. Find any intercepts and check for symmetry first.

Solution

The lone intercept is $(0, 0)$. The graph is symmetric with respect to the x-axis. (Do you see why? Replace y by $-y$.) Figure 22 shows the graph.

Figure 22
$x = y^2$

Figure 23
$y = \sqrt{x}$

If we restrict y so that $y \geq 0$, the equation $x = y^2$, $y \geq 0$, may be written equivalently as $y = \sqrt{x}$. The portion of the graph of $x = y^2$ in quadrant I is therefore the graph of $y = \sqrt{x}$. See Figure 23.

COMMENT To see the graph of the equation $x = y^2$ on a graphing calculator, you will need to graph two equations: $Y_1 = \sqrt{x}$ and $Y_2 = -\sqrt{x}$. We discuss why in Chapter 3. See Figure 24.

Figure 24

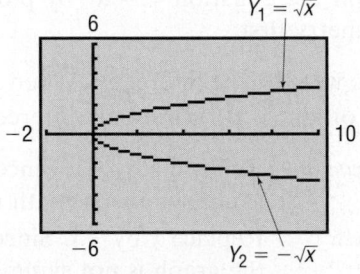

Graphing the Equation $y = \dfrac{1}{x}$

Graph the equation $y = \dfrac{1}{x}$. Find any intercepts and check for symmetry first.

Solution

We check for intercepts first. If we let $x = 0$, we obtain 0 in the denominator, which makes y undefined. We conclude that there is no y-intercept. If we let $y = 0$, we get the equation $\dfrac{1}{x} = 0$, which has no solution. We conclude that there is no x-intercept. The graph of $y = \dfrac{1}{x}$ does not cross or touch the coordinate axes.

Next we check for symmetry:

x-Axis: Replacing y by $-y$ yields $-y = \dfrac{1}{x}$, which is not equivalent to $y = \dfrac{1}{x}$.

y-Axis: Replacing x by $-x$ yields $y = \dfrac{1}{-x} = -\dfrac{1}{x}$, which is not equivalent to $y = \dfrac{1}{x}$.

Origin: Replacing x by $-x$ and y by $-y$ yields $-y = -\dfrac{1}{x}$, which is equivalent to $y = \dfrac{1}{x}$. The graph is symmetric with respect to the origin.

Now, we set up Table 5, listing several points on the graph. Because of the symmetry with respect to the origin, we use only positive values of x. From Table 5 we

Table 5

x	$y = \dfrac{1}{x}$	(x, y)
$\dfrac{1}{10}$	10	$\left(\dfrac{1}{10}, 10\right)$
$\dfrac{1}{3}$	3	$\left(\dfrac{1}{3}, 3\right)$
$\dfrac{1}{2}$	2	$\left(\dfrac{1}{2}, 2\right)$
1	1	$(1, 1)$
2	$\dfrac{1}{2}$	$\left(2, \dfrac{1}{2}\right)$
3	$\dfrac{1}{3}$	$\left(3, \dfrac{1}{3}\right)$
10	$\dfrac{1}{10}$	$\left(10, \dfrac{1}{10}\right)$

Figure 25

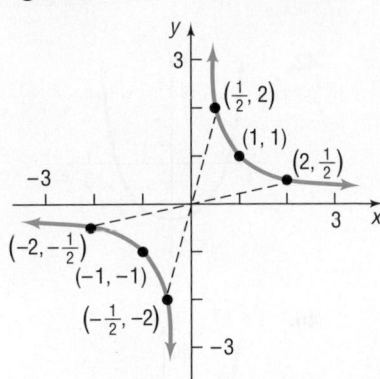

infer that if x is a large and positive number, then $y = \dfrac{1}{x}$ is a positive number close to 0. We also infer that if x is a positive number close to 0, then $y = \dfrac{1}{x}$ is a large and positive number. Armed with this information, we can graph the equation.

Figure 25 illustrates some of these points and the graph of $y = \dfrac{1}{x}$. Observe how the absence of intercepts and the existence of symmetry with respect to the origin were utilized.

COMMENT Refer to Example 3 in the Appendix, Section 3, for the graph of $y = \dfrac{1}{x}$ using a graphing utility.

2.2 Assess Your Understanding

'Are You Prepared?' *Answers are given at the end of these exercises. If you get a wrong answer, read the pages listed in red.*

1. Solve the equation $2(x + 3) - 1 = -7$. (pp. 86–93)

2. Solve the equation $x^2 - 9 = 0$. (p. 86–93)

Concepts and Vocabulary

3. The points, if any, at which a graph crosses or touches the co-ordinate axes are called _____ .

4. The x-intercepts of the graph of an equation are those x-values for which _____ .

5. If for every point (x, y) on the graph of an equation the point $(-x, y)$ is also on the graph, then the graph is symmetric with respect to the _____ .

6. If the graph of an equation is symmetric with respect to the y-axis and -4 is an x-intercept of this graph, then _____ is also an x-intercept.

7. If the graph of an equation is symmetric with respect to the origin and $(3, -4)$ is a point on the graph, then _____ is also a point on the graph.

8. *True or False* To find the y-intercepts of the graph of an equation, let $x = 0$ and solve for y.

9. *True or False* The y-coordinate of a point at which the graph crosses or touches the x-axis is an x-intercept.

10. *True or False* If a graph is symmetric with respect to the x-axis, then it cannot be symmetric with respect to the y-axis.

Skill Building

In Problems 11–16, determine which of the given points are on the graph of the equation.

11. Equation: $y = x^4 - \sqrt{x}$
 Points: $(0, 0); (1, 1); (-1, 0)$

12. Equation: $y = x^3 - 2\sqrt{x}$
 Points: $(0, 0); (1, 1); (1, -1)$

13. Equation: $y^2 = x^2 + 9$
 Points: $(0, 3); (3, 0); (-3, 0)$

14. Equation: $y^3 = x + 1$
 Points: $(1, 2); (0, 1); (-1, 0)$

15. Equation: $x^2 + y^2 = 4$
 Points: $(0, 2); (-2, 2); \left(\sqrt{2}, \sqrt{2}\right)$

16. Equation: $x^2 + 4y^2 = 4$
 Points: $(0, 1); (2, 0); \left(2, \dfrac{1}{2}\right)$

In Problems 17–28, find the intercepts and graph each equation by plotting points. Be sure to label the intercepts.

17. $y = x + 2$

18. $y = x - 6$

19. $y = 2x + 8$

20. $y = 3x - 9$

21. $y = x^2 - 1$

22. $y = x^2 - 9$

23. $y = -x^2 + 4$

24. $y = -x^2 + 1$

25. $2x + 3y = 6$

26. $5x + 2y = 10$

27. $9x^2 + 4y = 36$

28. $4x^2 + y = 4$

In Problems 29–38, plot each point. Then plot the point that is symmetric to it with respect to (a) the x-axis; (b) the y-axis; (c) the origin.

29. $(3, 4)$

30. $(5, 3)$

31. $(-2, 1)$

32. $(4, -2)$

33. $(5, -2)$

34. $(-1, -1)$

35. $(-3, -4)$

36. $(4, 0)$

37. $(0, -3)$

38. $(-3, 0)$

In Problems 39–50, the graph of an equation is given. (a) Find the intercepts. (b) Indicate whether the graph is symmetric with respect to the x-axis, the y-axis, or the origin.

39.

40.

41.

42.

43.

44.

45.

46.

47.

48.

49.

50.

In Problems 51–54, draw a complete graph so that it has the type of symmetry indicated.

51. *y*-axis

52. *x*-axis

53. Origin

54. *y*-axis

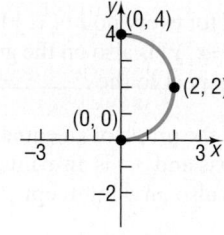

In Problems 55–70, list the intercepts and test for symmetry.

55. $y^2 = x + 4$

56. $y^2 = x + 9$

57. $y = \sqrt[3]{x}$

58. $y = \sqrt[5]{x}$

59. $x^2 + y - 9 = 0$

60. $x^2 - y - 4 = 0$

61. $9x^2 + 4y^2 = 36$

62. $4x^2 + y^2 = 4$

63. $y = x^3 - 27$

64. $y = x^4 - 1$

65. $y = x^2 - 3x - 4$

66. $y = x^2 + 4$

67. $y = \dfrac{3x}{x^2 + 9}$

68. $y = \dfrac{x^2 - 4}{2x}$

69. $y = \dfrac{-x^3}{x^2 - 9}$

70. $y = \dfrac{x^4 + 1}{2x^5}$

In Problems 71–74, draw a quick sketch of each equation.

71. $y = x^3$

72. $x = y^2$

73. $y = \sqrt{x}$

74. $y = \dfrac{1}{x}$

75. If $(3, b)$ is a point on the graph of $y = 4x + 1$, what is b?

76. If $(-2, b)$ is a point on the graph of $2x + 3y = 2$, what is b?

77. If $(a, 4)$ is a point on the graph of $y = x^2 + 3x$, what is a?

78. If $(a, -5)$ is a point on the graph of $y = x^2 + 6x$, what is a?

Applications and Extensions

79. Given that the point $(1, 2)$ is on the graph of an equation that is symmetric with respect to the origin, what other point is on the graph?

80. If the graph of an equation is symmetric with respect to the y-axis and 6 is an x-intercept of this graph, name another x-intercept.

81. If the graph of an equation is symmetric with respect to the origin and -4 is an x-intercept of this graph, name another x-intercept.

82. If the graph of an equation is symmetric with respect to the x-axis and 2 is a y-intercept, name another y-intercept.

83. Microphones In studios and on stages, cardioid microphones are often preferred for the richness they add to voices and for their ability to reduce the level of sound from the sides and rear of the microphone. Suppose one such cardiod pattern is given by the equation $(x^2 + y^2 - x)^2 = x^2 + y^2$.

 (a) Find the intercepts of the graph of the equation.

 (b) Test for symmetry with respect to the x-axis, y-axis, and origin.

 Source: www.notaviva.com

84. Solar Energy The solar electric generating systems at Kramer Junction, California, use parabolic troughs to heat a heat-transfer fluid to a high temperature. This fluid is used to generate steam that drives a power conversion system to produce electricity. For troughs 7.5 feet wide, an equation for the cross-section is $16y^2 = 120x - 225$.

 (a) Find the intercepts of the graph of the equation.

 (b) Test for symmetry with respect to the x-axis, y-axis, and origin.

 Source: U.S. Department of Energy

Discussion and Writing

In Problem 85, you may use a graphing utility, but it is not required.

85. (a) Graph $y = \sqrt{x^2}$, $y = x$, $y = |x|$, and $y = (\sqrt{x})^2$, noting which graphs are the same.

 (b) Explain why the graphs of $y = \sqrt{x^2}$ and $y = |x|$ are the same.

 (c) Explain why the graphs of $y = x$ and $y = (\sqrt{x})^2$ are not the same.

 (d) Explain why the graphs of $y = \sqrt{x^2}$ and $y = x$ are not the same.

86. Explain what is meant by a complete graph.

87. Draw a graph of an equation that contains two x-intercepts; at one the graph crosses the x-axis, and at the other the graph touches the x-axis.

88. Make up an equation with the intercepts $(2, 0)$, $(4, 0)$, and $(0, 1)$. Compare your equation with a friend's equation. Comment on any similarities.

89. Draw a graph that contains the points $(-2, -1)$, $(0, 1)$, $(1, 3)$, and $(3, 5)$. Compare your graph with those of other students. Are most of the graphs almost straight lines? How many are "curved"? Discuss the various ways that these points might be connected.

90. An equation is being tested for symmetry with respect to the x-axis, the y-axis, and the origin. Explain why, if two of these symmetries are present, the remaining one must also be present.

'Are You Prepared?' Answers

1. $\{-6\}$ **2.** $\{-3, 3\}$

2.3 Lines

OBJECTIVES 1 Calculate and Interpret the Slope of a Line (p. 174)

2 Graph Lines Given a Point and the Slope (p. 176)

3 Find the Equation of a Vertical Line (p. 177)

4 Use the Point–Slope Form of a Line; Identify Horizontal Lines (p. 178)

5 Find the Equation of a Line Given Two Points (p. 179)

6 Write the Equation of a Line in Slope–Intercept Form (p. 179)

7 Identify the Slope and y-Intercept of a Line from Its Equation (p. 180)

8 Graph Lines Written in General Form Using Intercepts (p. 181)

9 Find Equations of Parallel Lines (p. 182)

10 Find Equations of Perpendicular Lines (p. 183)

In this section we study a certain type of equation that contains two variables, called a *linear equation*, and its graph, a *line*.

1 Calculate and Interpret the Slope of a Line

Figure 26

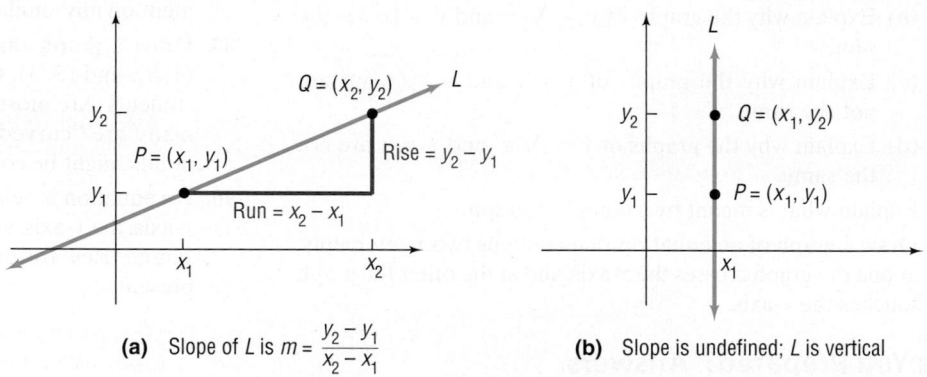

Consider the staircase illustrated in Figure 26. Each step contains exactly the same horizontal **run** and the same vertical **rise.** The ratio of the rise to the run, called the *slope*, is a numerical measure of the steepness of the staircase. For example, if the run is increased and the rise remains the same, the staircase becomes less steep. If the run is kept the same, but the rise is increased, the staircase becomes more steep. This important characteristic of a line is best defined using rectangular coordinates.

DEFINITION

Let $P = (x_1, y_1)$ and $Q = (x_2, y_2)$ be two distinct points. If $x_1 \neq x_2$, the **slope** m of the nonvertical line L containing P and Q is defined by the formula

$$m = \frac{y_2 - y_1}{x_2 - x_1} \qquad x_1 \neq x_2 \qquad \text{(1)}$$

If $x_1 = x_2$, L is a **vertical line** and the slope m of L is **undefined** (since this results in division by 0).

Figure 27(a) provides an illustration of the slope of a nonvertical line; Figure 27(b) illustrates a vertical line.

Figure 27

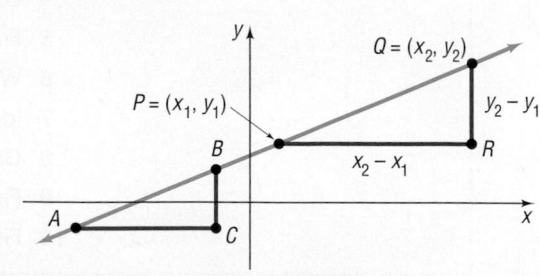

(a) Slope of L is $m = \dfrac{y_2 - y_1}{x_2 - x_1}$

(b) Slope is undefined; L is vertical

As Figure 27(a) illustrates, the slope m of a nonvertical line may be viewed as

$$m = \frac{y_2 - y_1}{x_2 - x_1} = \frac{\text{Rise}}{\text{Run}}$$

Two comments about computing the slope of a nonvertical line may prove helpful:

1. Any two distinct points on the line can be used to compute the slope of the line. (See Figure 28 for justification.)

Figure 28
Triangles *ABC* and *PQR* are similar (equal angles), so ratios of corresponding sides are proportional. Then

Slope using *P* and *Q* $= \dfrac{y_2 - y_1}{x_2 - x_1} =$

$\dfrac{d(B, C)}{d(A, C)} =$ Slope using *A* and *B*

2. The slope of a line may be computed from $P = (x_1, y_1)$ to $Q = (x_2, y_2)$ or from Q to P because

$$\frac{y_2 - y_1}{x_2 - x_1} = \frac{y_1 - y_2}{x_1 - x_2}$$

Finally, we can also express the slope m of a nonvertical line as

$$m = \frac{y_2 - y_1}{x_2 - x_1} = \frac{\text{Change in } y}{\text{Change in } x} = \frac{\Delta y}{\Delta x}$$

That is, the slope m of a nonvertical line L measures the amount that y changes as x changes from x_1 to x_2. The expression $\dfrac{\Delta y}{\Delta x}$ is called the **average rate of change** of y with respect to x.

Since any two distinct points can be used to compute the slope of a line, the average rate of change of a line is always the same number.

EXAMPLE 1	**Finding and Interpreting the Slope of a Line Given Two Points**

The slope m of the line containing the points $(1, 2)$ and $(5, -3)$ may be computed as

$$m = \frac{-3 - 2}{5 - 1} = \frac{-5}{4} = -\frac{5}{4} \quad \text{or as} \quad m = \frac{2 - (-3)}{1 - 5} = \frac{5}{-4} = -\frac{5}{4}$$

For every 4-unit change in x, y will change by -5 units. That is, if x increases by 4 units, then y will decrease by 5 units. The average rate of change of y with respect to x is $-\dfrac{5}{4}$.

◼

Now Work PROBLEMS 11 AND 17

To get a better idea of the meaning of the slope m of a line L, consider the following example.

EXAMPLE 2	**Finding the Slopes of Various Lines Containing the Same Point (2, 3)**

Compute the slopes of the lines L_1, L_2, L_3, and L_4 containing the following pairs of points. Graph all four lines on the same set of coordinate axes.

$$L_1: \quad P = (2, 3) \qquad Q_1 = (-1, -2)$$
$$L_2: \quad P = (2, 3) \qquad Q_2 = (3, -1)$$
$$L_3: \quad P = (2, 3) \qquad Q_3 = (5, 3)$$
$$L_4: \quad P = (2, 3) \qquad Q_4 = (2, 5)$$

Figure 29

$m_1 = \frac{5}{3}$ m_4 undefined

Solution Let $m_1, m_2, m_3,$ and m_4 denote the slopes of the lines L_1, L_2, L_3, and L_4, respectively. Then

$$m_1 = \frac{-2 - 3}{-1 - 2} = \frac{-5}{-3} = \frac{5}{3} \quad \text{A rise of 5 divided by a run of 3}$$

$$m_2 = \frac{-1 - 3}{3 - 2} = \frac{-4}{1} = -4$$

$$m_3 = \frac{3 - 3}{5 - 2} = \frac{0}{3} = 0$$

m_4 is undefined because $x_1 = x_2 = 2$

The graphs of these lines are given in Figure 29.

◼

Figure 29 illustrates the following facts:

1. When the slope of a line is positive, the line slants upward from left to right (L_1).
2. When the slope of a line is negative, the line slants downward from left to right (L_2).
3. When the slope is 0, the line is horizontal (L_3).
4. When the slope is undefined, the line is vertical (L_4).

Figure 30

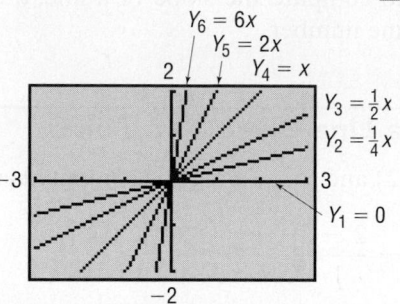

Seeing the Concept

On the same screen, graph the following equations:

$Y_1 = 0$ *Slope of line is 0.*

$Y_2 = \dfrac{1}{4}x$ *Slope of line is $\dfrac{1}{4}$.*

$Y_3 = \dfrac{1}{2}x$ *Slope of line is $\dfrac{1}{2}$.*

$Y_4 = x$ *Slope of line is 1.*

$Y_5 = 2x$ *Slope of line is 2.*

$Y_6 = 6x$ *Slope of line is 6.*

See Figure 30.

Figure 31

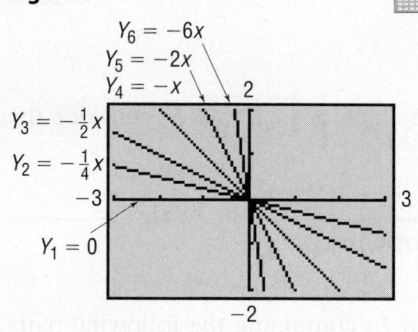

Seeing the Concept

On the same screen, graph the following equations:

$Y_1 = 0$ *Slope of line is 0.*

$Y_2 = -\dfrac{1}{4}x$ *Slope of line is $-\dfrac{1}{4}$.*

$Y_3 = -\dfrac{1}{2}x$ *Slope of line is $-\dfrac{1}{2}$.*

$Y_4 = -x$ *Slope of line is -1.*

$Y_5 = -2x$ *Slope of line is -2.*

$Y_6 = -6x$ *Slope of line is -6.*

See Figure 31.

Figures 30 and 31 illustrate that the closer the line is to the vertical position, the greater the magnitude of the slope.

2 Graph Lines Given a Point and the Slope

The next example illustrates how the slope of a line can be used to graph the line.

EXAMPLE 3 **Graphing a Line Given a Point and a Slope**

Draw a graph of the line that contains the point $(3, 2)$ and has a slope of:

(a) $\dfrac{3}{4}$ (b) $-\dfrac{4}{5}$

Solution (a) Slope $= \dfrac{\text{Rise}}{\text{Run}}$. The fact that the slope is $\dfrac{3}{4}$ means that for every horizontal movement (run) of 4 units to the right there will be a vertical movement (rise) of 3 units. Look at Figure 32. If we start at the given point $(3, 2)$ and move 4 units to the right and 3 units up, we reach the point $(7, 5)$. By drawing the line through this point and the point $(3, 2)$, we have the graph.

Figure 32

(b) The fact that the slope is

$$-\frac{4}{5} = \frac{-4}{5} = \frac{\text{Rise}}{\text{Run}}$$

means that for every horizontal movement of 5 units to the right there will be a corresponding vertical movement of -4 units (a downward movement). If we start at the given point $(3, 2)$ and move 5 units to the right and then 4 units down, we arrive at the point $(8, -2)$. By drawing the line through these points, we have the graph. See Figure 33.

Alternatively, we can set

$$-\frac{4}{5} = \frac{4}{-5} = \frac{\text{Rise}}{\text{Run}}$$

Figure 33

so that for every horizontal movement of -5 units (a movement to the left) there will be a corresponding vertical movement of 4 units (upward). This approach brings us to the point $(-2, 6)$, which is also on the graph shown in Figure 33.

━━━ **Now Work** PROBLEM 23

3 Find the Equation of a Vertical Line

Now that we have discussed the slope of a line, we are ready to derive equations of lines. As we shall see, there are several forms of the equation of a line. Let's start with an example.

| EXAMPLE 4 | **Graphing a Line** |

Graph the equation: $x = 3$

Solution To graph $x = 3$, recall that we are looking for all points (x, y) in the plane for which $x = 3$. No matter what y-coordinate is used, the corresponding x-coordinate always equals 3. Consequently, the graph of the equation $x = 3$ is a vertical line with x-intercept 3 and undefined slope. See Figure 34.

Figure 34

As suggested by Example 4, we have the following result:

| THEOREM | **Equation of a Vertical Line** |

A vertical line is given by an equation of the form

$$x = a$$

where a is the x-intercept.

COMMENT To graph an equation using a graphing utility, we need to express the equation in the form y = {expression in x}. But x = 3 cannot be put in this form. To overcome this, most graphing utilities have special commands for drawing vertical lines. DRAW, LINE, PLOT, and VERT are among the more common ones. Consult your manual to determine the correct methodology for your graphing utility. ∎

Figure 35

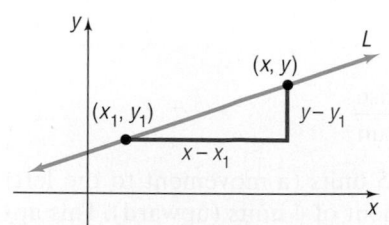

4 Use the Point–Slope Form of a Line; Identify Horizontal Lines

Now let L be a nonvertical line with slope m and containing the point (x_1, y_1). See Figure 35. For any other point (x, y) on L, we have

$$m = \frac{y - y_1}{x - x_1} \quad \text{or} \quad y - y_1 = m(x - x_1)$$

THEOREM

Point–Slope Form of an Equation of a Line

An equation of a nonvertical line with slope m that contains the point (x_1, y_1) is

$$y - y_1 = m(x - x_1) \tag{2}$$

EXAMPLE 5

Using the Point–Slope Form of a Line

Figure 36

An equation of the line with slope 4 and containing the point $(1, 2)$ can be found by using the point–slope form with $m = 4$, $x_1 = 1$, and $y_1 = 2$.

$$y - y_1 = m(x - x_1)$$
$$y - 2 = 4(x - 1) \qquad \text{$m = 4$, $x_1 = 1$, $y_1 = 2$}$$
$$y = 4x - 2 \qquad \text{Solve for y.}$$

See Figure 36 for the graph. ∎

━━━ **Now Work** PROBLEM 45

EXAMPLE 6

Finding the Equation of a Horizontal Line

Find an equation of the horizontal line containing the point $(3, 2)$.

Solution

Because all the y-values are equal on a horizontal line, the slope of a horizontal line is 0. To get an equation, we use the point–slope form with $m = 0$, $x_1 = 3$, and $y_1 = 2$.

Figure 37

$$y - y_1 = m(x - x_1)$$
$$y - 2 = 0 \cdot (x - 3) \qquad \text{$m = 0$, $x_1 = 3$, and $y_1 = 2$}$$
$$y - 2 = 0$$
$$y = 2$$

See Figure 37 for the graph. ∎

As suggested by Example 6, we have the following result:

THEOREM

Equation of a Horizontal Line

A horizontal line is given by an equation of the form

$$y = b$$

where b is the y-intercept.

5 Find the Equation of a Line Given Two Points

We use the slope formula and the point–slope form of a line to find the equation of a line given two points.

EXAMPLE 7 Finding an Equation of a Line Given Two Points

Find an equation of the line containing the points $(2, 3)$ and $(-4, 5)$. Graph the line.

Solution We first compute the slope of the line.

$$m = \frac{5 - 3}{-4 - 2} = \frac{2}{-6} = -\frac{1}{3}$$

Figure 38

We use the point $(2, 3)$ and the slope $m = -\dfrac{1}{3}$ to get the point–slope form of the equation of the line.

$$y - 3 = -\frac{1}{3}(x - 2)$$

See Figure 38 for the graph.

In the solution to Example 7, we could have used the other point, $(-4, 5)$, instead of the point $(2, 3)$. The equation that results, although it looks different, is equivalent to the equation that we obtained in the example. (Try it for yourself.)

Now Work PROBLEM 37

6 Write the Equation of a Line in Slope–Intercept Form

Another useful equation of a line is obtained when the slope m and y-intercept b are known. In this event, we know both the slope m of the line and a point $(0, b)$ on the line; then we may use the point–slope form, equation (2), to obtain the following equation:

$$y - b = m(x - 0) \quad \text{or} \quad y = mx + b$$

THEOREM

Slope–Intercept Form of an Equation of a Line

An equation of a line with slope m and y-intercept b is

$$y = mx + b \tag{3}$$

Now Work PROBLEM 51 (EXPRESS ANSWER
IN SLOPE–INTERCEPT FORM)

Figure 39 $y = mx + 2$

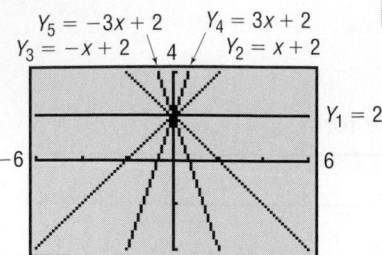

$Y_5 = -3x + 2$ $Y_4 = 3x + 2$
$Y_3 = -x + 2$ $Y_2 = x + 2$
$Y_1 = 2$

Seeing the Concept

To see the role that the slope m plays, graph the following lines on the same screen.

$$Y_1 = 2$$
$$Y_2 = x + 2$$
$$Y_3 = -x + 2$$
$$Y_4 = 3x + 2$$
$$Y_5 = -3x + 2$$

See Figure 39. What do you conclude about the lines $y = mx + 2$?

Figure 40 $y = 2x + b$

$Y_2 = 2x + 1$
$Y_1 = 2x$
$Y_4 = 2x + 4$
$Y_3 = 2x - 1$
$Y_5 = 2x - 4$

Seeing the Concept

To see the role of the y-intercept b, graph the following lines on the same screen.

$$Y_1 = 2x$$
$$Y_2 = 2x + 1$$
$$Y_3 = 2x - 1$$
$$Y_4 = 2x + 4$$
$$Y_5 = 2x - 4$$

See Figure 40. What do you conclude about the lines $y = 2x + b$?

7 Identify the Slope and *y*-Intercept of a Line from Its Equation

When the equation of a line is written in slope–intercept form, it is easy to find the slope m and y-intercept b of the line. For example, suppose that the equation of a line is

$$y = -2x + 7$$

Compare it to $y = mx + b$.

$$y = -2x + 7$$
$$y = \quad mx + b$$

The slope of this line is -2 and its y-intercept is 7.

Now Work PROBLEM 71

EXAMPLE 8 | **Finding the Slope and *y*-Intercept**

Find the slope m and y-intercept b of the equation $2x + 4y = 8$. Graph the equation.

Solution

To obtain the slope and y-intercept, we write the equation in slope–intercept form by solving for y.

$$2x + 4y = 8$$
$$4y = -2x + 8$$
$$y = -\frac{1}{2}x + 2 \quad y = mx + b$$

Figure 41

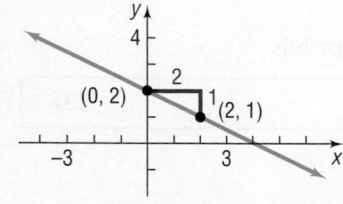

The coefficient of x, $-\dfrac{1}{2}$, is the slope, and the y-intercept is 2. We can graph the line using the fact that the y-intercept is 2 and the slope is $-\dfrac{1}{2}$. Then, starting at the point $(0, 2)$, go to the right 2 units and then down 1 unit to the point $(2, 1)$. See Figure 41.

Now Work PROBLEM 77

8 Graph Lines Written in General Form Using Intercepts

Refer to Example 8. The form of the equation of the line $2x + 4y = 8$, is called the *general form*.

DEFINITION

The equation of a line is in **general form*** when it is written as

$$Ax + By = C \qquad (4)$$

where A, B, and C are real numbers and A and B are not both 0.

When we want to graph an equation that is written in general form, we can solve the equation for y and write the equation in slope–intercept form as we did in Example 8. Another approach to graphing the equation would be to find its intercepts. Remember, the intercepts of the graph of an equation are the points where the graph crosses or touches a coordinate axis.

EXAMPLE 9 | **Graphing an Equation in General Form Using Its Intercepts**

Graph the equation $2x + 4y = 8$ by finding its intercepts.

Solution
To obtain the x-intercept, let $y = 0$ in the equation and solve for x.

$$2x + 4y = 8$$
$$2x + 4(0) = 8 \quad \text{Let } y = 0.$$
$$2x = 8$$
$$x = 4 \quad \text{Divide both sides by 2.}$$

The x-intercept is 4 and the point $(4, 0)$ is on the graph of the equation.
To obtain the y-intercept, let $x = 0$ in the equation and solve for y.

$$2x + 4y = 8$$
$$2(0) + 4y = 8 \quad \text{Let } x = 0.$$
$$4y = 8$$
$$y = 2 \quad \text{Divide both sides by 4.}$$

Figure 42

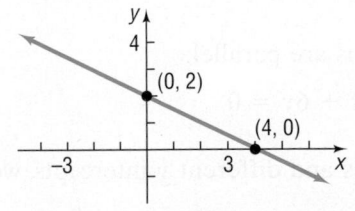

The y-intercept is 2 and the point $(0, 2)$ is on the graph of the equation.
We plot the points $(4, 0)$ and $(0, 2)$ on a Cartesian plane and draw a line through the points. See Figure 42.

Now Work PROBLEM 91

Every line has an equation that is equivalent to an equation written in general form. For example, a vertical line whose equation is

$$x = a$$

can be written in the general form

$$1 \cdot x + 0 \cdot y = a \quad A = 1, B = 0, C = a$$

A horizontal line whose equation is

$$y = b$$

can be written in the general form

$$0 \cdot x + 1 \cdot y = b \quad A = 0, B = 1, C = b$$

*Some books use the term **standard form**.

Lines that are neither vertical nor horizontal have general equations of the form

$$Ax + By = C \quad A \neq 0 \text{ and } B \neq 0$$

Because the equation of every line can be written in general form, any equation equivalent to equation (4) is called a **linear equation.**

Figure 43

9 Find Equations of Parallel Lines

When two lines (in the plane) do not intersect (that is, they have no points in common), they are said to be **parallel.** Look at Figure 43. There we have drawn two parallel lines and have constructed two right triangles by drawing sides parallel to the coordinate axes. The right triangles are similar. (Do you see why? Two angles are equal.) Because the triangles are similar, the ratios of corresponding sides are equal.

This suggests the following result:

THEOREM

Criterion for Parallel Lines

Two nonvertical lines are parallel if and only if their slopes are equal and they have different y-intercepts.

The use of the words "if and only if" in the preceding theorem means that actually two statements are being made, one the converse of the other.

If two nonvertical lines are parallel, then their slopes are equal and they have different y-intercepts.

If two nonvertical lines have equal slopes and they have different y-intercepts, then they are parallel.

EXAMPLE 10

Showing That Two Lines Are Parallel

Show that the lines given by the following equations are parallel:

$$L_1: \ 2x + 3y = 6, \qquad L_2: \ 4x + 6y = 0$$

Solution

To determine whether these lines have equal slopes and different y-intercepts, we write each equation in slope–intercept form:

Figure 44

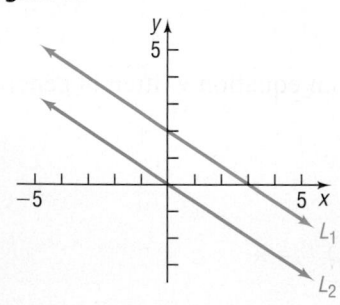

$$
\begin{array}{ll}
L_1: \ 2x + 3y = 6 & L_2: \ 4x + 6y = 0 \\
\quad 3y = -2x + 6 & \quad 6y = -4x \\
\quad y = -\dfrac{2}{3}x + 2 & \quad y = -\dfrac{2}{3}x
\end{array}
$$

$$
\text{Slope} = -\frac{2}{3}; \ y\text{-intercept} = 2 \qquad \text{Slope} = -\frac{2}{3}; \ y\text{-intercept} = 0
$$

Because these lines have the same slope, $-\dfrac{2}{3}$, but different y-intercepts, the lines are parallel. See Figure 44.

■

EXAMPLE 11

Finding a Line That Is Parallel to a Given Line

Find an equation for the line that contains the point $(2, -3)$ and is parallel to the line $2x + y = 6$.

Solution Since the two lines are to be parallel, the slope of the line that we seek equals the slope of the line $2x + y = 6$. We begin by writing the equation of the line $2x + y = 6$ in slope–intercept form.

$$2x + y = 6$$
$$y = -2x + 6$$

The slope is -2. Since the line that we seek also has slope -2 and contains the point $(2, -3)$, we use the point–slope form to obtain its equation.

$$y - y_1 = m(x - x_1) \qquad \text{Point–slope form}$$
$$y - (-3) = -2(x - 2) \qquad m = -2, x_1 = 2, y_1 = -3$$
$$y + 3 = -2x + 4 \qquad \text{Simplify.}$$
$$y = -2x + 1 \qquad \text{Slope–intercept form}$$
$$2x + y = 1 \qquad \text{General form}$$

This line is parallel to the line $2x + y = 6$ and contains the point $(2, -3)$. See Figure 45.

Figure 45

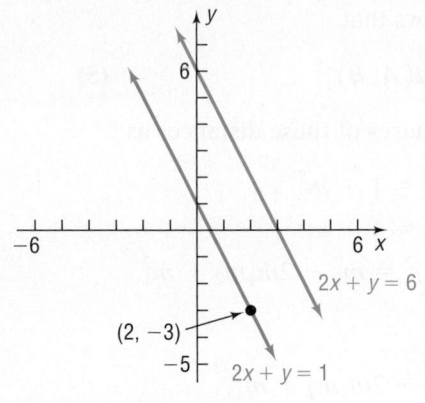

Now Work PROBLEM 59

10 Find Equations of Perpendicular Lines

When two lines intersect at a right angle (90°), they are said to be **perpendicular.** See Figure 46.

Figure 46

The following result gives a condition, in terms of their slopes, for two lines to be perpendicular.

THEOREM **Criterion for Perpendicular Lines**

Two nonvertical lines are perpendicular if and only if the product of their slopes is -1.

Here we shall prove the "only if" part of the statement:

If two nonvertical lines are perpendicular, then the product of their slopes is -1.

In Problem 128 you are asked to prove the "if" part of the theorem; that is:

If two nonvertical lines have slopes whose product is -1, then the lines are perpendicular.

Figure 47

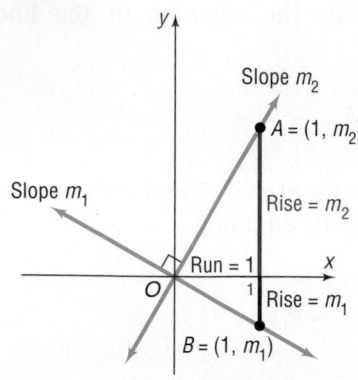

Proof Let m_1 and m_2 denote the slopes of the two lines. There is no loss in generality (that is, neither the angle nor the slopes are affected) if we situate the lines so that they meet at the origin. See Figure 47. The point $A = (1, m_2)$ is on the line having slope m_2, and the point $B = (1, m_1)$ is on the line having slope m_1. (Do you see why this must be true?)

Suppose that the lines are perpendicular. Then triangle OAB is a right triangle. As a result of the Pythagorean Theorem, it follows that

$$[d(O, A)]^2 + [d(O, B)]^2 = [d(A, B)]^2 \qquad \textbf{(5)}$$

Using the distance formula, we can write the squares of these distances as

$$[d(O, A)]^2 = (1 - 0)^2 + (m_2 - 0)^2 = 1 + m_2^2$$
$$[d(O, B)]^2 = (1 - 0)^2 + (m_1 - 0)^2 = 1 + m_1^2$$
$$[d(A, B)]^2 = (1 - 1)^2 + (m_2 - m_1)^2 = m_2^2 - 2m_1m_2 + m_1^2$$

Using these facts in equation (5), we get

$$(1 + m_2^2) + (1 + m_1^2) = m_2^2 - 2m_1m_2 + m_1^2$$

which, upon simplification, can be written as

$$m_1m_2 = -1$$

If the lines are perpendicular, the product of their slopes is -1. ∎

You may find it easier to remember the condition for two nonvertical lines to be perpendicular by observing that the equality $m_1m_2 = -1$ means that m_1 and m_2 are negative reciprocals of each other; that is, either $m_1 = -\dfrac{1}{m_2}$ or $m_2 = -\dfrac{1}{m_1}$.

EXAMPLE 12 **Finding the Slope of a Line Perpendicular to Another Line**

If a line has slope $\dfrac{3}{2}$, any line having slope $-\dfrac{2}{3}$ is perpendicular to it.

EXAMPLE 13 **Finding the Equation of a Line Perpendicular to a Given Line**

Find an equation of the line that contains the point $(1, -2)$ and is perpendicular to the line $x + 3y = 6$. Graph the two lines.

Solution We first write the equation of the given line in slope–intercept form to find its slope.

$$x + 3y = 6$$
$$3y = -x + 6 \qquad \textit{Proceed to solve for y.}$$
$$y = -\frac{1}{3}x + 2 \qquad \textit{Place in the form } y = mx + b.$$

The given line has slope $-\dfrac{1}{3}$. Any line perpendicular to this line will have slope 3.

Because we require the point $(1, -2)$ to be on this line with slope 3, we use the point–slope form of the equation of a line.

$$y - y_1 = m(x - x_1) \qquad \textit{Point–slope form}$$
$$y - (-2) = 3(x - 1) \qquad \textit{m = 3, } x_1 = 1, y_1 = -2$$

Figure 48

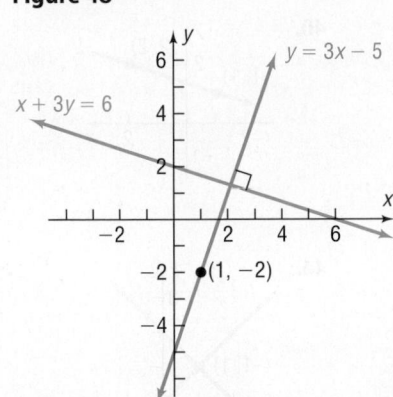

To obtain other forms of the equation, we proceed as follows:

$$y + 2 = 3(x - 1)$$
$$y + 2 = 3x - 3 \qquad \text{Simplify.}$$
$$y = 3x - 5 \qquad \text{Slope–intercept form}$$
$$3x - y = 5 \qquad \text{General form}$$

Figure 48 shows the graphs.

— **Now Work** PROBLEM 65

WARNING *Be sure to use a square screen when you graph perpendicular lines. Otherwise, the angle between the two lines will appear distorted. A discussion of square screens is given in section 5 of the Appendix.*

2.3 Assess Your Understanding

Concepts and Vocabulary

1. The slope of a vertical line is _____; the slope of a horizontal line is _____.

2. For the line $2x + 3y = 6$, the x-intercept is _____ and the y-intercept is _____.

3. A horizontal line is given by an equation of the form _____, where b is the _____.

4. *True or False* Vertical lines have an undefined slope.

5. *True or False* The slope of the line $2y = 3x + 5$ is 3.

6. *True or False* The point $(1, 2)$ is on the line $2x + y = 4$.

7. Two nonvertical lines have slopes m_1 and m_2, respectively. The lines are parallel if _____ and the _____ are unequal; the lines are perpendicular if _____.

8. The lines $y = 2x + 3$ and $y = ax + 5$ are parallel if $a =$ _____.

9. The lines $y = 2x - 1$ and $y = ax + 2$ are perpendicular if $a =$ _____.

10. *True or False* Perpendicular lines have slopes that are reciprocals of one another.

Skill Building

In Problems 11–14, (a) find the slope of the line and (b) interpret the slope.

11.

12.

13.

14.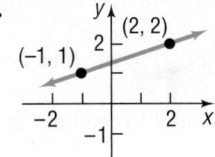

In Problems 15–22, plot each pair of points and determine the slope of the line containing them. Graph the line.

15. $(2, 3); (4, 0)$

16. $(4, 2); (3, 4)$

17. $(-2, 3); (2, 1)$

18. $(-1, 1); (2, 3)$

19. $(-3, -1); (2, -1)$

20. $(4, 2); (-5, 2)$

21. $(-1, 2); (-1, -2)$

22. $(2, 0); (2, 2)$

In Problems 23–30, graph the line containing the point P and having slope m.

23. $P = (1, 2); m = 3$

24. $P = (2, 1); m = 4$

25. $P = (2, 4); m = -\dfrac{3}{4}$

26. $P = (1, 3); m = -\dfrac{2}{5}$

27. $P = (-1, 3); m = 0$

28. $P = (2, -4); m = 0$

29. $P = (0, 3);$ slope undefined

30. $P = (-2, 0);$ slope undefined

In Problems 31–36, the slope and a point on a line are given. Use this information to locate three additional points on the line. Answers may vary.

[**Hint:** *It is not necessary to find the equation of the line. See Example 3.*]

31. Slope 4; point $(1, 2)$

32. Slope 2; point $(-2, 3)$

33. Slope $-\dfrac{3}{2}$; point $(2, -4)$

34. Slope $\dfrac{4}{3}$; point $(-3, 2)$

35. Slope -2; point $(-2, -3)$

36. Slope -1; point $(4, 1)$

In Problems 37–44, find an equation of the line L.

37.

38.

39.

40.

41.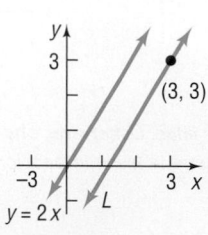

L is parallel to y = 2x

42.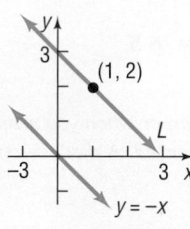

L is parallel to y = −x

43.

L is perpendicular to y = 2x

44.

L is perpendicular to y = −x

In Problems 45–70, find an equation for the line with the given properties. Express your answer using either the general form or the slope–intercept form of the equation of a line, whichever you prefer.

45. Slope = 3; containing the point $(-2, 3)$

46. Slope = 2; containing the point $(4, -3)$

47. Slope $= -\dfrac{2}{3}$; containing the point $(1, -1)$

48. Slope $= \dfrac{1}{2}$; containing the point $(3, 1)$

49. Containing the points $(1, 3)$ and $(-1, 2)$

50. Containing the points $(-3, 4)$ and $(2, 5)$

51. Slope = −3; y-intercept = 3

52. Slope = −2; y-intercept = −2

53. x-intercept = 2; y-intercept = −1

54. x-intercept = −4; y-intercept = 4

55. Slope undefined; containing the point $(2, 4)$

56. Slope undefined; containing the point $(3, 8)$

57. Horizontal; containing the point $(-3, 2)$

58. Vertical; containing the point $(4, -5)$

59. Parallel to the line $y = 2x$; containing the point $(-1, 2)$

60. Parallel to the line $y = -3x$; containing the point $(-1, 2)$

61. Parallel to the line $2x - y = -2$; containing the point $(0, 0)$

62. Parallel to the line $x - 2y = -5$; containing the point $(0, 0)$

63. Parallel to the line $x = 5$; containing the point $(4, 2)$

64. Parallel to the line $y = 5$; containing the point $(4, 2)$

65. Perpendicular to the line $y = \dfrac{1}{2}x + 4$; containing the point $(1, -2)$

66. Perpendicular to the line $y = 2x - 3$; containing the point $(1, -2)$

67. Perpendicular to the line $2x + y = 2$; containing the point $(-3, 0)$

68. Perpendicular to the line $x - 2y = -5$; containing the point $(0, 4)$

69. Perpendicular to the line $x = 8$; containing the point $(3, 4)$

70. Perpendicular to the line $y = 8$; containing the point $(3, 4)$

In Problems 71–90, find the slope and y-intercept of each line. Graph the line.

71. $y = 2x + 3$

72. $y = -3x + 4$

73. $\dfrac{1}{2}y = x - 1$

74. $\dfrac{1}{3}x + y = 2$

75. $y = \dfrac{1}{2}x + 2$

76. $y = 2x + \dfrac{1}{2}$

77. $x + 2y = 4$

78. $-x + 3y = 6$

79. $2x - 3y = 6$

80. $3x + 2y = 6$

81. $x + y = 1$

82. $x - y = 2$

83. $x = -4$

84. $y = -1$

85. $y = 5$

86. $x = 2$

87. $y - x = 0$

88. $x + y = 0$

89. $2y - 3x = 0$

90. $3x + 2y = 0$

In Problems 91–100, (a) find the intercepts of the graph of each equation and (b) graph the equation.

91. $2x + 3y = 6$

92. $3x - 2y = 6$

93. $-4x + 5y = 40$

94. $6x - 4y = 24$

95. $7x + 2y = 21$

96. $5x + 3y = 18$

97. $\frac{1}{2}x + \frac{1}{3}y = 1$

98. $x - \frac{2}{3}y = 4$

99. $0.2x - 0.5y = 1$

100. $-0.3x + 0.4y = 1.2$

101. Find an equation of the *x*-axis.

102. Find an equation of the *y*-axis.

In Problems 103–106, the equations of two lines are given. Determine if the lines are parallel, perpendicular, or neither.

103. $y = 2x - 3$
$y = 2x + 4$

104. $y = \frac{1}{2}x - 3$
$y = -2x + 4$

105. $y = 4x + 5$
$y = -4x + 2$

106. $y = -2x + 3$
$y = -\frac{1}{2}x + 2$

In Problems 107–110, write an equation of each line. Express your answer using either the general form or the slope–intercept form of the equation of a line, whichever you prefer.

107.

108.

109.

110.
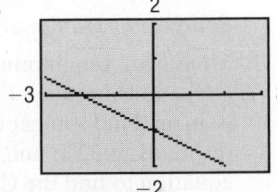

Applications and Extensions

111. Geometry Use slopes to show that the triangle whose vertices are $(-2, 5)$, $(1, 3)$, and $(-1, 0)$ is a right triangle.

112. Geometry Use slopes to show that the quadrilateral whose vertices are $(1, -1)$, $(4, 1)$, $(2, 2)$, and $(5, 4)$ is a parallelogram.

113. Geometry Use slopes to show that the quadrilateral whose vertices are $(-1, 0)$, $(2, 3)$, $(1, -2)$, and $(4, 1)$ is a rectangle.

114. Geometry Use slopes and the distance formula to show that the quadrilateral whose vertices are $(0, 0)$, $(1, 3)$, $(4, 2)$, and $(3, -1)$ is a square.

115. Truck Rentals A truck rental company rents a moving truck for one day by charging $29 plus $0.20 per mile. Write a linear equation that relates the cost C, in dollars, of renting the truck to the number x of miles driven. What is the cost of renting the truck if the truck is driven 110 miles? 230 miles?

116. Cost Equation The **fixed costs** of operating a business are the costs incurred regardless of the level of production. Fixed costs include rent, fixed salaries, and costs of leasing machinery. The **variable costs** of operating a business are the costs that change with the level of output. Variable costs include raw materials, hourly wages, and electricity. Suppose that a manufacturer of jeans has fixed daily costs of $500 and variable costs of $8 for each pair of jeans manufactured. Write a linear equation that relates the daily cost C, in dollars, of manufacturing the jeans to the number x of jeans manufactured. What is the cost of manufacturing 400 pairs of jeans? 740 pairs?

117. Cost of Sunday Home Delivery The cost to the *Chicago Tribune* for Sunday home delivery is approximately $0.53 per newspaper with fixed costs of $1,070,000. Write a linear

equation that relates the cost C and the number x of copies delivered.

Source: Chicago Tribune, 2002.

118. Wages of a Car Salesperson Dan receives $375 per week for selling new and used cars at a car dealership in Oak Lawn, Illinois. In addition, he receives 5% of the profit on any sales that he generates. Write a linear equation that relates Dan's weekly salary S when he has sales that generate a profit of x dollars.

119. Electricity Rates in Illinois Commonwealth Edison Company supplies electricity to residential customers for a monthly customer charge of $7.58 plus 8.275 cents per kilowatt-hour for up to 400 kilowatt-hours.

(a) Write a linear equation that relates the monthly charge C, in dollars, to the number x of kilowatt-hours used in a month, $0 \leq x \leq 400$.

(b) Graph this equation.

(c) What is the monthly charge for using 100 kilowatt-hours?

(d) What is the monthly charge for using 300 kilowatt-hours?

(e) Interpret the slope of the line.

Source: Commonwealth Edison Company, April, 2006.

120. **Electricity Rates in Florida** Florida Power & Light Company supplies electricity to residential customers for a monthly customer charge of $5.17 plus 10.07 cents per kilowatt-hour for up to 1000 kilowatt-hours.

(a) Write a linear equation that relates the monthly charge C, in dollars, to the number x of kilowatt-hours used in a month, $0 \le x \le 1000$.

(b) Graph this equation.

(c) What is the monthly charge for using 200 kilowatt-hours?

(d) What is the monthly charge for using 500 kilowatt-hours?

(e) Interpret the slope of the line.

Source: Florida Power & Light Company, April, 2006.

121. **Measuring Temperature** The relationship between Celsius (°C) and Fahrenheit (°F) degrees of measuring temperature is linear. Find a linear equation relating °C and °F if 0°C corresponds to 32°F and 100°C corresponds to 212°F. Use the equation to find the Celsius measure of 70°F.

122. **Measuring Temperature** The Kelvin (K) scale for measuring temperature is obtained by adding 273 to the Celsius temperature.

(a) Write a linear equation relating K and °C.

(b) Write a linear equation relating K and °F (see Problem 121).

123. **Access Ramp** A wooden access ramp is being built to reach a platform that sits 30 inches above the floor. The ramp drops 2 inches for every 25-inch run.

(a) Write a linear equation that relates the height y above the floor to the horizontal distance x from the platform.

(b) Find and interpret the x-intercept of the graph of your equation.

(c) Design requirements stipulate that the maximum run be 30 feet and that the maximum slope be a drop of 1 inch for each 12 inches of run. Will this ramp meet the requirements? Explain.

(d) What slopes could be used to obtain the 30-inch rise and still meet design requirements?

Source: www.adaptiveaccess.com/wood_ramps.php

124. **Cigarette Use** A study by the *Partnership for a Drug-Free America* indicated that, in 1998, 42% of teens in grades 7 through 12 had recently used cigarettes. A similar study in 2005 indicated that 22% of such teens had recently used cigarettes.

(a) Write a linear equation that relates the percent of teens y to the number of years after 1998, x.

(b) Find the intercepts of the graph of your equation.

(c) Do the intercepts have any meaningful interpretation?

(d) Use your equation to predict the percent for the year 2019. Is this result reasonable?

Source: www.drugfree.org

125. **Product Promotion** A cereal company finds that the number of people who will buy one of its products in the first month that it is introduced is linearly related to the amount of money it spends on advertising. If it spends $40,000 on advertising, then 100,000 boxes of cereal will be sold, and if it spends $60,000, then 200,000 boxes will be sold.

(a) Write a linear equation that relates the amount A spent on advertising to the number x of boxes the company aims to sell.

(b) How much advertising is needed to sell 300,000 boxes of cereal?

(c) Interpret the slope.

126. Show that the line containing the points (a, b) and (b, a), $a \ne b$, is perpendicular to the line $y = x$. Also show that the midpoint of (a, b) and (b, a) lies on the line $y = x$.

127. The equation $2x - y = C$ defines a **family of lines,** one line for each value of C. On one set of coordinate axes, graph the members of the family when $C = -4, C = 0,$ and $C = 2$. Can you draw a conclusion from the graph about each member of the family?

128. Prove that if two nonvertical lines have slopes whose product is -1 then the lines are perpendicular. [**Hint:** Refer to Figure 47 and use the converse of the Pythagorean Theorem.]

Discussion and Writing

129. Which of the following equations might have the graph shown? (More than one answer is possible.)

(a) $2x + 3y = 6$

(b) $-2x + 3y = 6$

(c) $3x - 4y = -12$

(d) $x - y = 1$

(e) $x - y = -1$

(f) $y = 3x - 5$

(g) $y = 2x + 3$

(h) $y = -3x + 3$

130. Which of the following equations might have the graph shown? (More than one answer is possible.)

(a) $2x + 3y = 6$

(b) $2x - 3y = 6$

(c) $3x + 4y = 12$

(d) $x - y = 1$

(e) $x - y = -1$

(f) $y = -2x - 1$

(g) $y = -\dfrac{1}{2}x + 10$

(h) $y = x + 4$

131. The figure shows the graph of two parallel lines. Which of the following pairs of equations might have such a graph?

(a) $x - 2y = 3$
 $x + 2y = 7$
(b) $x + y = 2$
 $x + y = -1$
(c) $x - y = -2$
 $x - y = 1$
(d) $x - y = -2$
 $2x - 2y = -4$
(e) $x + 2y = 2$
 $x + 2y = -1$

132. The figure shows the graph of two perpendicular lines. Which of the following pairs of equations might have such a graph?

(a) $y - 2x = 2$
 $y + 2x = -1$
(b) $y - 2x = 0$
 $2y + x = 0$
(c) $2y - x = 2$
 $2y + x = -2$
(d) $y - 2x = 2$
 $x + 2y = -1$
(e) $2x + y = -2$
 $2y + x = -2$

133. **m is for Slope** The accepted symbol used to denote the slope of a line is the letter *m*. Investigate the origin of this symbolism. Begin by consulting a French dictionary and looking up the French word *monter*. Write a brief essay on your findings.

134. **Grade of a Road** The term *grade* is used to describe the inclination of a road. How does this term relate to the notion of slope of a line? Is a 4% grade very steep? Investigate the grades of some mountainous roads and determine their slopes. Write a brief essay on your findings.

135. **Carpentry** Carpenters use the term *pitch* to describe the steepness of staircases and roofs. How does pitch relate to slope? Investigate typical pitches used for stairs and for roofs. Write a brief essay on your findings.

136. Can the equation of every line be written in slope–intercept form? Why?

137. Does every line have exactly one *x*-intercept and one *y*-intercept? Are there any lines that have no intercepts?

138. What can you say about two lines that have equal slopes and equal *y*-intercepts?

139. What can you say about two lines with the same *x*-intercept and the same *y*-intercept? Assume that the *x*-intercept is not 0.

140. If two distinct lines have the same slope, but different *x*-intercepts, can they have the same *y*-intercept?

141. If two distinct lines have the same *y*-intercept, but different slopes, can they have the same *x*-intercept?

142. Which form of the equation of a line do you prefer to use? Justify your position with an example that shows that your choice is better than another. Have reasons.

2.4 Circles

PREPARING FOR THIS SECTION *Before getting started, review the following:*

• Completing the Square (Section 1.2, pp. 99–100) • Square Root Method (Section 1.2, pp. 98–99)

 Now Work the 'Are You Prepared?' problems on page 193.

OBJECTIVES 1 Write the Standard Form of the Equation of a Circle (p. 189)
 2 Graph a Circle (p. 191)
 3 Work with the General Form of the Equation of a Circle (p. 192)

1 Write the Standard Form of the Equation of a Circle

One advantage of a coordinate system is that it enables us to translate a geometric statement into an algebraic statement, and vice versa. Consider, for example, the following geometric statement that defines a circle.

DEFINITION A **circle** is a set of points in the *xy*-plane that are a fixed distance *r* from a fixed point (h, k). The fixed distance *r* is called the **radius,** and the fixed point (h, k) is called the **center** of the circle.

Figure 49

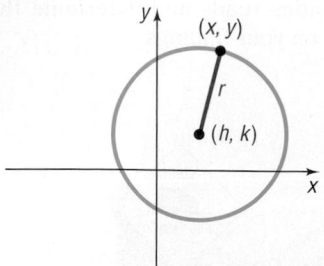

Figure 49 shows the graph of a circle. To find the equation, we let (x, y) represent the coordinates of any point on a circle with radius r and center (h, k). Then the distance between the points (x, y) and (h, k) must always equal r. That is, by the distance formula

$$\sqrt{(x - h)^2 + (y - k)^2} = r$$

or, equivalently,

$$(x - h)^2 + (y - k)^2 = r^2$$

DEFINITION

The **standard form of an equation of a circle** with radius r and center (h, k) is

$$(x - h)^2 + (y - k)^2 = r^2 \qquad (1)$$

THEOREM

The standard form of an equation of a circle of radius r with center at the origin $(0, 0)$ is

$$x^2 + y^2 = r^2$$

DEFINITION

If the radius $r = 1$, the circle whose center is at the origin is called the **unit circle** and has the equation

$$x^2 + y^2 = 1$$

See Figure 50. Notice that the graph of the unit circle is symmetric with respect to the x-axis, the y-axis, and the origin.

Figure 50
Unit circle $x^2 + y^2 = 1$

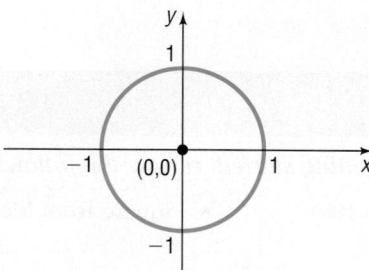

| EXAMPLE 1 | **Writing the Standard Form of the Equation of a Circle** |

Write the standard form of the equation of the circle with radius 5 and center $(-3, 6)$.

Solution
Using the form of equation (1) and substituting the values $r = 5$, $h = -3$, and $k = 6$, we have

$$(x - h)^2 + (y - k)^2 = r^2$$
$$(x + 3)^2 + (y - 6)^2 = 25$$

Now Work PROBLEM 7

2 Graph a Circle

The graph of any equation in the form of equation (1) is that of a circle with radius r and center (h, k).

EXAMPLE 2	**Graphing a Circle**

Graph the equation: $(x + 3)^2 + (y - 2)^2 = 16$

Solution
Since the equation is in the form of equation (1), its graph is a circle. To graph the equation, we first compare the given equation to the standard form of the equation of a circle. The comparison yields information about the circle.

Figure 51

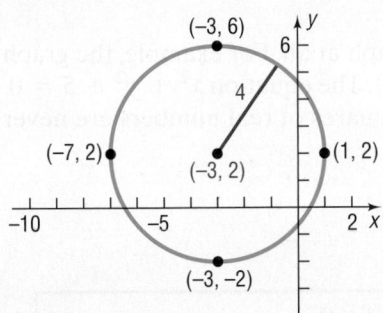

$$(x + 3)^2 + (y - 2)^2 = 16$$
$$(x - (-3))^2 + (y - 2)^2 = 4^2$$
$$(x - h)^2 + (y - k)^2 = r^2$$

We see that $h = -3$, $k = 2$, and $r = 4$. The circle has center $(-3, 2)$ and a radius of 4 units. To graph this circle, we first plot the center $(-3, 2)$. Since the radius is 4, we can locate four points on the circle by plotting points 4 units to the left, to the right, up, and down from the center. These four points can then be used as guides to obtain the graph. See Figure 51.

Now Work PROBLEMS 23(a) AND (b)

EXAMPLE 3	**Finding the Intercepts of a Circle**

For the circle $(x + 3)^2 + (y - 2)^2 = 16$, find the intercepts, if any, of its graph.

Solution
This is the equation discussed and graphed in Example 2. To find the x-intercepts, if any, let $y = 0$. Then

$$(x + 3)^2 + (y - 2)^2 = 16$$
$$(x + 3)^2 + (0 - 2)^2 = 16 \qquad y = 0$$
$$(x + 3)^2 + 4 = 16 \qquad \text{Simplify.}$$
$$(x + 3)^2 = 12 \qquad \text{Simplify.}$$
$$x + 3 = \pm\sqrt{12} \qquad \text{Apply the Square Root Method.}$$
$$x = -3 \pm 2\sqrt{3} \qquad \text{Solve for } x.$$

The x-intercepts are $-3 - 2\sqrt{3} \approx -6.46$ and $-3 + 2\sqrt{3} \approx 0.46$. To find the y-intercepts, if any, we let $x = 0$. Then

$$(x + 3)^2 + (y - 2)^2 = 16$$
$$(0 + 3)^2 + (y - 2)^2 = 16$$
$$9 + (y - 2)^2 = 16$$
$$(y - 2)^2 = 7$$
$$y - 2 = \pm\sqrt{7}$$
$$y = 2 \pm \sqrt{7}$$

The y-intercepts are $2 - \sqrt{7} \approx -0.65$ and $2 + \sqrt{7} \approx 4.65$. Look back at Figure 51 to verify the approximate locations of the intercepts.

Now Work PROBLEM 23(c)

3 Work with the General Form of the Equation of a Circle

If we eliminate the parentheses from the standard form of the equation of the circle given in Example 2, we get

$$(x + 3)^2 + (y - 2)^2 = 16$$
$$x^2 + 6x + 9 + y^2 - 4y + 4 = 16$$

which we find, upon simplifying, is equivalent to

$$x^2 + y^2 + 6x - 4y - 3 = 0 \qquad \textbf{(2)}$$

It can be shown that any equation of the form

$$x^2 + y^2 + ax + by + c = 0$$

has a graph that is a circle, or a point, or has no graph at all. For example, the graph of the equation $x^2 + y^2 = 0$ is the single point $(0, 0)$. The equation $x^2 + y^2 + 5 = 0$, or $x^2 + y^2 = -5$, has no graph, because sums of squares of real numbers are never negative.

DEFINITION

When its graph is a circle, the equation

$$x^2 + y^2 + ax + by + c = 0$$

is referred to as the **general form of the equation of a circle.**

───── **Now Work** PROBLEM 13

If an equation of a circle is in the general form, we use the method of completing the square to put the equation in standard form so that we can identify its center and radius.

────────────────────────────────

EXAMPLE 4 **Graphing a Circle Whose Equation Is in General Form**

Graph the equation $x^2 + y^2 + 4x - 6y + 12 = 0$

Solution We complete the square in both x and y to put the equation in standard form. Group the expression involving x, group the expression involving y, and put the constant on the right side of the equation. The result is

$$(x^2 + 4x) + (y^2 - 6y) = -12$$

Next, complete the square of each expression in parentheses. Remember that any number added on the left side of the equation must also be added on the right.

$$(x^2 + 4x + 4) + (y^2 - 6y + 9) = -12 + 4 + 9$$

$$\left(\frac{4}{2}\right)^2 = 4 \qquad \left(\frac{-6}{2}\right)^2 = 9$$

$$(x + 2)^2 + (y - 3)^2 = 1 \quad \text{Factor.}$$

We recognize this equation as the standard form of the equation of a circle with radius 1 and center $(-2, 3)$.

To graph the equation use the center $(-2, 3)$ and the radius 1. See Figure 52.

Figure 52

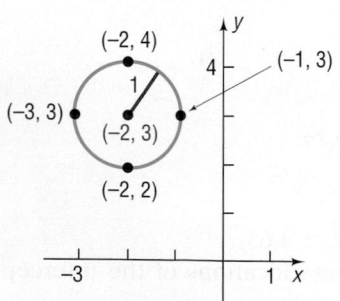

───── **Now Work** PROBLEM 27

| EXAMPLE 5 | Using a Graphing Utility to Graph a Circle |

Graph the equation: $x^2 + y^2 = 4$

Solution This is the equation of a circle with center at the origin and radius 2. To graph this equation, we must first solve for y.

Figure 53

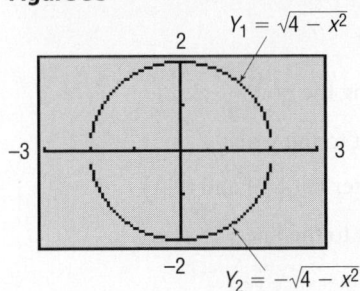

$$x^2 + y^2 = 4$$
$$y^2 = 4 - x^2 \qquad \text{Subtract } x^2 \text{ from each side.}$$
$$y = \pm\sqrt{4 - x^2} \qquad \begin{array}{l}\text{Apply the Square Root}\\ \text{Method to solve for } y\end{array}$$

There are two equations to graph: first, we graph $Y_1 = \sqrt{4 - x^2}$ and then $Y_2 = -\sqrt{4 - x^2}$ on the same square screen. (Your circle will appear oval if you do not use a square screen.) See Figure 53.

2.4 Assess Your Understanding

Are You Prepared? *Answers are given at the end of these exercises. If you get a wrong answer, read the pages listed in red.*

1. To complete the square of $x^2 + 10x$, you would (*add/subtract*) the number _____. (pp. 99–100)

2. Use the Square Root Method to solve the equation $(x - 2)^2 = 9$. (pp. 98–99)

Concepts and Vocabulary

3. *True or False* Every equation of the form

$$x^2 + y^2 + ax + by + c = 0$$

has a circle as its graph.

4. For a circle, the _____ is the distance from the center to any point on the circle.

5. *True or False* The radius of the circle $x^2 + y^2 = 9$ is 3.

6. *True or False* The center of the circle

$$(x + 3)^2 + (y - 2)^2 = 13$$

is $(3, -2)$.

Skill Building

In Problems 7–10, find the center and radius of each circle. Write the standard form of the equation.

7.

8.

9.

10.
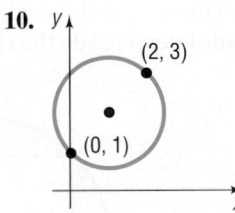

In Problems 11–20, write the standard form of the equation and the general form of the equation of each circle of radius r and center (h, k). Graph each circle.

11. $r = 2$; $(h, k) = (0, 0)$

12. $r = 3$; $(h, k) = (0, 0)$

13. $r = 2$; $(h, k) = (0, 2)$

14. $r = 3$; $(h, k) = (1, 0)$

15. $r = 5$; $(h, k) = (4, -3)$

16. $r = 4$; $(h, k) = (2, -3)$

17. $r = 4$; $(h, k) = (-2, 1)$

18. $r = 7$; $(h, k) = (-5, -2)$

19. $r = \dfrac{1}{2}$; $(h, k) = \left(\dfrac{1}{2}, 0\right)$

20. $r = \dfrac{1}{2}$; $(h, k) = \left(0, -\dfrac{1}{2}\right)$

In Problems 21–34, (a) find the center (h, k) and radius r of each circle; (b) graph each circle; (c) find the intercepts, if any.

21. $x^2 + y^2 = 4$

22. $x^2 + (y - 1)^2 = 1$

23. $2(x - 3)^2 + 2y^2 = 8$

24. $3(x + 1)^2 + 3(y - 1)^2 = 6$

25. $x^2 + y^2 - 2x - 4y - 4 = 0$

26. $x^2 + y^2 + 4x + 2y - 20 = 0$

27. $x^2 + y^2 + 4x - 4y - 1 = 0$

28. $x^2 + y^2 - 6x + 2y + 9 = 0$

29. $x^2 + y^2 - x + 2y + 1 = 0$

30. $x^2 + y^2 + x + y - \dfrac{1}{2} = 0$

31. $2x^2 + 2y^2 - 12x + 8y - 24 = 0$

32. $2x^2 + 2y^2 + 8x + 7 = 0$

33. $2x^2 + 8x + 2y^2 = 0$

34. $3x^2 + 3y^2 - 12y = 0$

In Problems 35–42, find the standard form of the equation of each circle.

35. Center at the origin and containing the point $(-2, 3)$

36. Center $(1, 0)$ and containing the point $(-3, 2)$

37. Center $(2, 3)$ and tangent to the *x*-axis

38. Center $(-3, 1)$ and tangent to the *y*-axis

39. With endpoints of a diameter at $(1, 4)$ and $(-3, 2)$

40. With endpoints of a diameter at $(4, 3)$ and $(0, 1)$

41. Center $(-1, 3)$ and tangent to the line $y = 2$

42. Center $(4, -2)$ and tangent to the line $x = 1$

In Problems 43–46, match each graph with the correct equation.

(a) $(x - 3)^2 + (y + 3)^2 = 9$ (b) $(x + 1)^2 + (y - 2)^2 = 4$ (c) $(x - 1)^2 + (y + 2)^2 = 4$ (d) $(x + 3)^2 + (y - 3)^2 = 9$

43.

44.

45.

46.
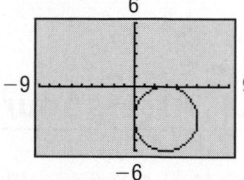

Applications and Extensions

47. Find the area of the square in the figure.

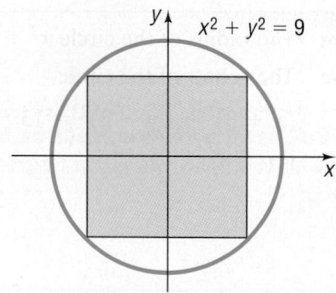

48. Find the area of the shaded region in the figure assuming the quadrilateral inside the circle is a square.

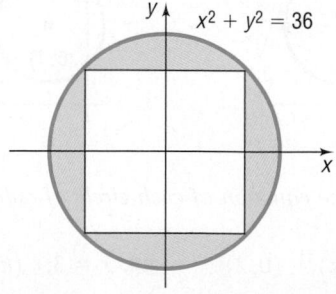

49. Ferris Wheel The original Ferris wheel was built in 1893 by Pittsburg, Pennsylvania, bridge builder George W. Ferris. The Ferris wheel was originally built for the 1893 World's Fair in Chicago, but was also later reconstructed for the 1904 World's Fair in St. Louis. It had a maximum height of 264 feet and a wheel diameter of 250 feet. Find an equation for the wheel if the center of the wheel is on the *y*-axis.

Source: inventors.about.com

50. Ferris Wheel In 2006, the star of Nanchang (in the Jiangxi province) opened as the world's largest Ferris wheel. It has a maximum height of 160 meters and a diameter of 153 meters, with one full rotation taking approximately 30 minutes. Find an equation for the wheel if the center of the wheel is on the *y*-axis.

Source: AsiaOne Travel

Note: Two other even larger Ferris wheels are reportedly to be completed in Asia by 2008 in time for the 2008 summer Olympics

51. Weather Satellites Earth is represented on a map of a portion of the solar system so that its surface is the circle with equation $x^2 + y^2 + 2x + 4y - 4091 = 0$. A weather

satellite circles 0.6 unit above Earth with the center of its circular orbit at the center of Earth. Find the equation for the orbit of the satellite on this map.

52. The **tangent line** to a circle may be defined as the line that intersects the circle in a single point, called the **point of tangency**. See the figure.

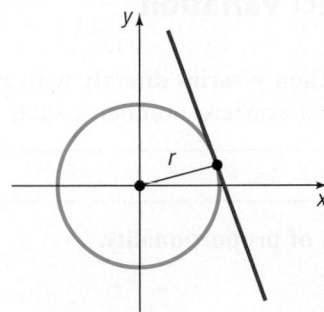

If the equation of the circle is $x^2 + y^2 = r^2$ and the equation of the tangent line is $y = mx + b$, show that:
(a) $r^2(1 + m^2) = b^2$

[**Hint:** The quadratic equation $x^2 + (mx + b)^2 = r^2$ has exactly one solution.]

(b) The point of tangency is $\left(\dfrac{-r^2 m}{b}, \dfrac{r^2}{b}\right)$.

(c) The tangent line is perpendicular to the line containing the center of the circle and the point of tangency.

53. **The Greek Method** The Greek method for finding the equation of the tangent line to a circle uses the fact that at any point on a circle the lines containing the center and the tangent line are perpendicular (see Problem 52). Use this method to find an equation of the tangent line to the circle $x^2 + y^2 = 9$ at the point $(1, 2\sqrt{2})$.

54. Use the Greek method described in Problem 53 to find an equation of the tangent line to the circle $x^2 + y^2 - 4x + 6y + 4 = 0$ at the point $(3, 2\sqrt{2} - 3)$.

55. Refer to Problem 52. The line $x - 2y + 4 = 0$ is tangent to a circle at $(0, 2)$. The line $y = 2x - 7$ is tangent to the same circle at $(3, -1)$. Find the center of the circle.

56. Find an equation of the line containing the centers of the two circles

$$x^2 + y^2 - 4x + 6y + 4 = 0$$

and

$$x^2 + y^2 + 6x + 4y + 9 = 0$$

57. If a circle of radius 2 is made to roll along the x-axis, what is an equation for the path of the center of the circle?

58. If the circumference of a circle is 6π, what is its radius?

Discussion and Writing

59. Which of the following equations might have the graph shown? (More than one answer is possible.)
(a) $(x - 2)^2 + (y + 3)^2 = 13$
(b) $(x - 2)^2 + (y - 2)^2 = 8$
(c) $(x - 2)^2 + (y - 3)^2 = 13$
(d) $(x + 2)^2 + (y - 2)^2 = 8$
(e) $x^2 + y^2 - 4x - 9y = 0$
(f) $x^2 + y^2 + 4x - 2y = 0$
(g) $x^2 + y^2 - 9x - 4y = 0$
(h) $x^2 + y^2 - 4x - 4y = 4$

60. Which of the following equations might have the graph shown? (More than one answer is possible.)
(a) $(x - 2)^2 + y^2 = 3$
(b) $(x + 2)^2 + y^2 = 3$
(c) $x^2 + (y - 2)^2 = 3$
(d) $(x + 2)^2 + y^2 = 4$
(e) $x^2 + y^2 + 10x + 16 = 0$
(f) $x^2 + y^2 + 10x - 2y = 1$
(g) $x^2 + y^2 + 9x + 10 = 0$
(h) $x^2 + y^2 - 9x - 10 = 0$

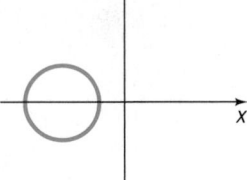

61. Explain how the center and radius of a circle can be used to graph a circle.

Are You Prepared? Answers

1. add; 25

2. $\{-1, 5\}$

2.5 Variation

OBJECTIVES 1 Construct a Model Using Direct Variation (p. 196)

2 Construct a Model Using Inverse Variation (p. 197)

3 Construct a Model Using Joint or Combined Variation (p. 197)

When a mathematical model is developed for a real-world problem, it often involves relationships between quantities that are expressed in terms of proportionality:

Force is proportional to acceleration.

When an ideal gas is held at a constant temperature, pressure and volume are inversely proportional.

The force of attraction between two heavenly bodies is inversely proportional to the square of the distance between them.

Revenue is directly proportional to sales.

Each of these statements illustrates the idea of **variation,** or how one quantity varies in relation to another quantity. Quantities may vary *directly, inversely,* or *jointly.*

1 Construct a Model Using Direct Variation

DEFINITION Let x and y denote two quantities. Then y **varies directly** with x, or y is **directly proportional to** x, if there is a nonzero number k such that

$$y = kx$$

The number k is called the **constant of proportionality.**

Figure 54
$y = kx; k > 0, x \geq 0$

The graph in Figure 54 illustrates the relationship between y and x if y varies directly with x and $k > 0$, $x \geq 0$. Note that the constant of proportionality is, in fact, the slope of the line.

If we know that two quantities vary directly, then knowing the value of each quantity in one instance enables us to write a formula that is true in all cases.

EXAMPLE 1 **Mortgage Payments**

The monthly payment p on a mortgage varies directly with the amount borrowed B. If the monthly payment on a 30-year mortgage is $6.65 for every $1000 borrowed, find a formula that relates the monthly payment p to the amount borrowed B for a mortgage with these terms. Then find the monthly payment p when the amount borrowed B is $120,000.

Solution Because p varies directly with B, we know that

$$p = kB$$

for some constant k. Because $p = 6.65$ when $B = 1000$, it follows that

$$6.65 = k(1000)$$
$$k = 0.00665 \quad \text{Solve for } k.$$

Figure 55

Since $p = kB$, we have

$$p = 0.00665B$$

In particular, when $B = \$120,000$, we find that

$$p = 0.00665(\$120,000) = \$798$$

Figure 55 illustrates the relationship between the monthly payment p and the amount borrowed B.

Now Work PROBLEMS 3 AND 21

2 Construct a Model Using Inverse Variation

Figure 56

$y = \dfrac{k}{x}; k > 0, x > 0$

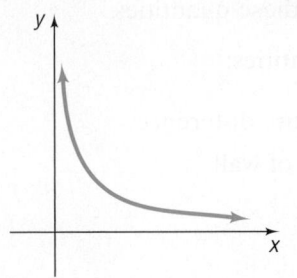

DEFINITION Let x and y denote two quantities. Then y **varies inversely** with x, or y is **inversely proportional to** x, if there is a nonzero constant k such that

$$y = \frac{k}{x}$$

The graph in Figure 56 illustrates the relationship between y and x if y varies inversely with x and $k > 0$, $x > 0$.

EXAMPLE 2

Maximum Weight That Can Be Supported by a Piece of Pine

Figure 57

See Figure 57. The maximum weight W that can be safely supported by a 2-inch by 4-inch piece of pine varies inversely with its length l. Experiments indicate that the maximum weight that a 10-foot-long 2-by-4 piece of pine can support is 500 pounds. Write a general formula relating the maximum weight W (in pounds) to length l (in feet). Find the maximum weight W that can be safely supported by a length of 25 feet.

Solution Because W varies inversely with l, we know that

$$W = \frac{k}{l}$$

for some constant k. Because $W = 500$ when $l = 10$, we have

$$500 = \frac{k}{10}$$
$$k = 5000$$

Since $W = \dfrac{k}{l}$, we have

$$W = \frac{5000}{l}$$

In particular, the maximum weight W that can be safely supported by a piece of pine 25 feet in length is

$$W = \frac{5000}{25} = 200 \text{ pounds}$$

Figure 58 illustrates the relationship between the weight W and the length l.

Figure 58

 Now Work PROBLEM 31

3 Construct a Model Using Joint Variation or Combined Variation

When a variable quantity Q is proportional to the product of two or more other variables, we say that Q **varies jointly** with these quantities. Finally, combinations of direct and/or inverse variation may occur. This is usually referred to as **combined variation.**

Let's look at an example.

EXAMPLE 3	Loss of Heat Through a Wall

The loss of heat through a wall varies jointly with the area of the wall and the difference between the inside and outside temperatures and varies inversely with the thickness of the wall. Write an equation that relates these quantities.

Solution We begin by assigning symbols to represent the quantities:

$$L = \text{Heat loss} \qquad T = \text{Temperature difference}$$
$$A = \text{Area of wall} \qquad d = \text{Thickness of wall}$$

Then

$$L = k\frac{AT}{d}$$

where k is the constant of proportionality.

In direct or inverse variation, the quantities that vary may be raised to powers. For example, in the early seventeenth century, Johannes Kepler (1571–1630) discovered that the square of the period of revolution T around the Sun varies directly with the cube of its mean distance a from the Sun. That is, $T^2 = ka^3$, where k is the constant of proportionality.

EXAMPLE 4	Force of the Wind on a Window

The force F of the wind on a flat surface positioned at a right angle to the direction of the wind varies jointly with the area A of the surface and the square of the speed v of the wind. A wind of 30 miles per hour blowing on a window measuring 4 feet by 5 feet has a force of 150 pounds. See Figure 59. What is the force on a window measuring 3 feet by 4 feet caused by a wind of 50 miles per hour?

Solution Since F varies jointly with A and v^2, we have

$$F = kAv^2$$

Figure 59

Wind

where k is the constant of proportionality. We are told that $F = 150$ when $A = 4 \cdot 5 = 20$ and $v = 30$. Then we have

$$150 = k(20)(900) \qquad F = kAv^2, F = 150, A = 20, v = 30$$

$$k = \frac{1}{120}$$

Since $F = kAv^2$; we have

$$F = \frac{1}{120}Av^2$$

For a wind of 50 miles per hour blowing on a window whose area is $A = 3 \cdot 4 = 12$ square feet, the force F is

$$F = \frac{1}{120}(12)(2500) = 250 \text{ pounds}$$

Now Work PROBLEM 39

2.5 Assess Your Understanding

Concepts and Vocabulary

1. If x and y are two quantities, then y is directly proportional to x if there is a nonzero number k such that _____.

2. *True or False* If y varies directly with x, then $y = \dfrac{k}{x}$, where k is a constant.

Skill Building

In Problems 3–14, write a general formula to describe each variation.

3. y varies directly with x; $y = 2$ when $x = 10$

4. v varies directly with t; $v = 16$ when $t = 2$

5. A varies directly with x^2; $A = 4\pi$ when $x = 2$

6. V varies directly with x^3; $V = 36\pi$ when $x = 3$

7. F varies inversely with d^2; $F = 10$ when $d = 5$

8. y varies inversely with $\sqrt{x}$; $y = 4$ when $x = 9$

9. z varies directly with the sum of the squares of x and y; $z = 5$ when $x = 3$ and $y = 4$

10. T varies jointly with the cube root of x and the square of d; $T = 18$ when $x = 8$ and $d = 3$

11. M varies directly with the square of d and inversely with the square root of x; $M = 24$ when $x = 9$ and $d = 4$

12. z varies directly with the sum of the cube of x and the square of y; $z = 1$ when $x = 2$ and $y = 3$

13. The square of T varies directly with the cube of a and inversely with the square of d; $T = 2$ when $a = 2$ and $d = 4$

14. The cube of z varies directly with the sum of the squares of x and y; $z = 2$ when $x = 9$ and $y = 4$

Applications and Extensions

In Problems 15–20, write an equation that relates the quantities.

15. **Geometry** The volume V of a sphere varies directly with the cube of its radius r. The constant of proportionality is $\dfrac{4\pi}{3}$.

16. **Geometry** The square of the length of the hypotenuse c of a right triangle varies jointly with the sum of the squares of the lengths of its legs a and b. The constant of proportionality is 1.

17. **Geometry** The area A of a triangle varies jointly with the lengths of the base b and the height h. The constant of proportionality is $\dfrac{1}{2}$.

18. **Geometry** The perimeter p of a rectangle varies jointly with the sum of the lengths of its sides l and w. The constant of proportionality is 2.

19. **Physics: Newton's Law** The force F (in newtons) of attraction between two bodies varies jointly with their masses m and M (in kilograms) and inversely with the square of the distance d (in meters) between them. The constant of proportionality is $G = 6.67 \times 10^{-11}$.

20. **Physics: Simple Pendulum** The **period** of a pendulum is the time required for one oscillation; the pendulum is usually referred to as **simple** when the angle made to the vertical is less than 5°. The period T of a simple pendulum (in seconds) varies directly with the square root of its length l (in feet). The constant of proportionality is $\dfrac{2\pi}{\sqrt{32}}$.

21. **Mortgage Payments** The monthly payment p on a mortgage varies directly with the amount borrowed B. If the monthly payment on a 30-year mortgage is $6.49 for every

$1000 borrowed, find a linear equation that relates the monthly payment p to the amount borrowed B for a mortgage with the same terms. Then find the monthly payment p when the amount borrowed B is $145,000.

22. **Mortgage Payments** The monthly payment p on a mortgage varies directly with the amount borrowed B. If the monthly payment on a 15-year mortgage is $8.99 for every $1000 borrowed, find a linear equation that relates the monthly payment p to the amount borrowed B for a mortgage with the same terms. Then find the monthly payment p when the amount borrowed B is $175,000.

23. **Physics: Falling Objects** The distance s that an object falls is directly proportional to the square of the time t of the fall. If an object falls 16 feet in 1 second, how far will it fall in 3 seconds? How long will it take an object to fall 64 feet?

24. **Physics: Falling Objects** The velocity v of a falling object is directly proportional to the time t of the fall. If, after 2 seconds, the velocity of the object is 64 feet per second, what will its velocity be after 3 seconds?

25. **Physics: Stretching a Spring** The elongation E of a spring balance varies directly with the applied weight W (see the figure). If $E = 3$ when $W = 20$, find E when $W = 15$.

26. **Physics: Vibrating String** The rate of vibration of a string under constant tension varies inversely with the length of the string. If a string is 48 inches long and vibrates 256 times per second, what is the length of a string that vibrates 576 times per second?

27. **Revenue Equation** At the corner Shell station, the revenue R varies directly with the number g of gallons of gasoline sold. If the revenue is $47.40 when the number of gallons sold is 12, find a linear equation that relates revenue R to the number g of gallons of gasoline. Then find the revenue R when the number of gallons of gasoline sold is 10.5.

28. **Cost Equation** The cost C of chocolate-covered almonds varies directly with the number A of pounds of almonds purchased. If the cost is $23.75 when the number of pounds of chocolate-covered almonds purchased is 5, find a linear equation that relates the cost C to the number A of pounds of almonds purchased. Then find the cost C when the number of pounds of almonds purchased is 3.5.

29. **Demand** Suppose that the demand D for candy at the movie theater is inversely related to the price p.
 (a) When the price of candy is $2.75 per bag, the theater sells 156 bags of candy. Express the demand for candy in terms of its price.
 (b) Determine the number of bags of candy that will be sold if the price is raised to $3 a bag.

30. **Driving to School** The time t that it takes to get to school varies inversely with your average speed s.
 (a) Suppose that it takes you 40 minutes to get to school when your average speed is 30 miles per hour. Express the driving time to school in terms of average speed.
 (b) Suppose that your average speed to school is 40 miles per hour. How long will it take you to get to school?

31. **Pressure** The volume of a gas V held at a constant temperature in a closed container varies inversely with its pressure P. If the volume of a gas is 600 cubic centimeters (cm^3) when the pressure is 150 millimeters of mercury (mm Hg), find the volume when the pressure is 200 mm Hg.

32. **Resistance** The current i in a circuit is inversely proportional to its resistance Z measured in ohms. Suppose that when the current in a circuit is 30 amperes the resistance is 8 ohms. Find the current in the same circuit when the resistance is 10 ohms.

33. **Weight** The weight of an object above the surface of Earth varies inversely with the square of the distance from the center of Earth. If Maria weighs 125 pounds when she is on the surface of Earth (3960 miles from the center), determine Maria's weight if she is at the top of Mount McKinley (3.8 miles from the surface of Earth).

34. **Intensity of Light** The intensity I of light (measured in foot-candles) varies inversely with the square of the distance from the bulb. Suppose that the intensity of a 100-watt light bulb at a distance of 2 meters is 0.075 foot-candle. Determine the intensity of the bulb at a distance of 5 meters.

35. **Geometry** The volume V of a right circular cylinder varies jointly with the square of its radius r and its height h. The constant of proportionality is π. See the figure. Write an equation for V.

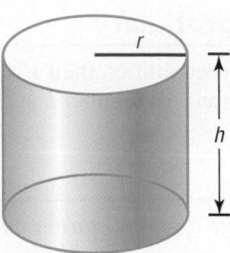

36. **Geometry** The volume V of a right circular cone varies jointly with the square of its radius r and its height h. The constant of proportionality is $\dfrac{\pi}{3}$. See the figure. Write an equation for V.

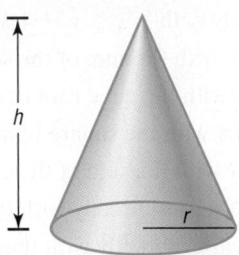

37. **Weight of a Body** The weight of a body above the surface of Earth varies inversely with the square of the distance from the center of Earth. If a certain body weighs 55 pounds when it is 3960 miles from the center of Earth, how much will it weigh when it is 3965 miles from the center?

38. **Force of the Wind on a Window** The force exerted by the wind on a plane surface varies jointly with the area of the surface and the square of the velocity of the wind. If the force on an area of 20 square feet is 11 pounds when the wind velocity is 22 miles per hour, find the force on a surface area of 47.125 square feet when the wind velocity is 36.5 miles per hour.

39. **Horsepower** The horsepower (hp) that a shaft can safely transmit varies jointly with its speed (in revolutions per minute, rpm) and the cube of its diameter. If a shaft of a certain material 2 inches in diameter can transmit 36 hp at 75 rpm, what diameter must the shaft have in order to transmit 45 hp at 125 rpm?

40. **Chemistry: Gas Laws** The volume V of an ideal gas varies directly with the temperature T and inversely with the pressure P. Write an equation relating V, T, and P using k as the constant of proportionality. If a cylinder contains oxygen at a temperature of 300 K and a pressure of 15 atmospheres in a volume of 100 liters, what is the constant of proportionality k? If a piston is lowered into the cylinder, decreasing the volume occupied by the gas to 80 liters and raising the temperature to 310 K, what is the gas pressure?

41. **Physics: Kinetic Energy** The kinetic energy K of a moving object varies jointly with its mass m and the square of its velocity v. If an object weighing 25 kilograms and moving with a velocity of 10 meters per second has a kinetic energy of 1250 joules, find its kinetic energy when the velocity is 15 meters per second.

42. Electrical Resistance of a Wire The electrical resistance of a wire varies directly with the length of the wire and inversely with the square of the diameter of the wire. If a wire 432 feet long and 4 millimeters in diameter has a resistance of 1.24 ohms, find the length of a wire of the same material whose resistance is 1.44 ohms and whose diameter is 3 millimeters.

43. Measuring the Stress of Materials The stress in the material of a pipe subject to internal pressure varies jointly with the internal pressure and the internal diameter of the pipe and inversely with the thickness of the pipe. The stress is 100 pounds per square inch when the diameter is 5 inches,

the thickness is 0.75 inch, and the internal pressure is 25 pounds per square inch. Find the stress when the internal pressure is 40 pounds per square inch if the diameter is 8 inches and the thickness is 0.50 inch.

44. Safe Load for a Beam The maximum safe load for a horizontal rectangular beam varies jointly with the width of the beam and the square of the thickness of the beam and inversely with its length. If an 8-foot beam will support up to 750 pounds when the beam is 4 inches wide and 2 inches thick, what is the maximum safe load in a similar beam 10 feet long, 6 inches wide, and 2 inches thick?

Discussion and Writing

45. In the early 17th century, Johannes Kepler discovered that the square of the period T varies directly with the cube of its mean distance a from the Sun. Go to the library and research this law and Kepler's other two laws. Write a brief paper about these laws and Kepler's place in history.

46. Using a situation that has not been discussed in the text, write a real-world problem that you think involves two variables that vary directly. Exchange your problem with another student's to solve and critique.

47. Using a situation that has not been discussed in the text, write a real-world problem that you think involves two variables that vary inversely. Exchange your problem with another student's to solve and critique.

48. Using a situation that has not been discussed in the text, write a real-world problem that you think involves three variables that vary jointly. Exchange your problem with another student's to solve and critique.

CHAPTER REVIEW

Things to Know

Formulas

Distance formula (p. 157)	$d = \sqrt{(x_2 - x_1)^2 + (y_2 - y_1)^2}$
Midpoint formula (p. 160)	$(x, y) = \left(\dfrac{x_1 + x_2}{2}, \dfrac{y_1 + y_2}{2} \right)$
Slope (p. 174)	$m = \dfrac{y_2 - y_1}{x_2 - x_1}$ if $x_1 \neq x_2$; undefined if $x_1 = x_2$
Parallel lines (p. 182)	Equal slopes ($m_1 = m_2$) and different y-intercepts ($b_1 \neq b_2$)
Perpendicular lines (p. 183)	Product of slopes is -1 ($m_1 \cdot m_2 = -1$)
Direct variation (p. 196)	$y = kx$
Inverse variation (p. 197)	$y = \dfrac{k}{x}$

Equations of Lines and Circles

Vertical line (p. 177)	$x = a$; a is the x-intercept
Horizontal line (p. 179)	$y = b$; b is the y-intercept
Point–slope form of the equation of a line (p. 178)	$y - y_1 = m(x - x_1)$; m is the slope of the line, (x_1, y_1) is a point on the line
Slope–intercept form of the equation of a line (p. 179)	$y = mx + b$; m is the slope of the line, b is the y-intercept
General form of the equation of a line (p. 181)	$Ax + By = C$; A, B not both 0
Standard form of the equation of a circle (p. 190)	$(x - h)^2 + (y - k)^2 = r^2$; r is the radius of the circle, (h, k) is the center of the circle
Equation of the unit circle (p. 190)	$x^2 + y^2 = 1$
General form of the equation of a circle (p. 192)	$x^2 + y^2 + ax + by + c = 0$, with restrictions on a, b, and c

Objectives

Section	You should be able to ...	Review Exercises
2.1	**1** Use the distance formula (p. 157)	1(a)–6(a), 48, 49(a), 50
	2 Use the midpoint formula (p. 159)	1(b)–6(b), 50

(continued)

Section	You should be able to ...	Review Exercises
2.2	**1** Graph equations by plotting points (p. 163)	7
	2 Find intercepts from a graph (p. 165)	8
	3 Find intercepts from an equation (p. 166)	9–16
	4 Test an equation for symmetry with respect to the x-axis, the y-axis, and the origin (p. 167)	9–16
	5 Know how to graph key equations (p. 169)	45–46
2.3	**1** Calculate and interpret the slope of a line (p. 174)	1(c)–6(c), 1(d)–6(d), 49(b), 51
	2 Graph lines given a point and the slope (p. 176)	47
	3 Find the equation of a vertical line (p. 177)	29
	4 Use the point–slope form of a line; identify horizontal lines (p. 178)	27, 28
	5 Find the equation of a line given two points (p. 179)	30–32
	6 Write the equation of a line in slope–intercept form (p. 179)	27, 28, 30–36
	7 Identify the slope and y-intercept of a line from its equation (p. 180)	37–40
	8 Graph lines written in general form using intercepts (p. 181)	41–44
	9 Find equations of parallel lines (p. 182)	33, 34
	10 Find equations of perpendicular lines (p. 183)	35, 36
2.4	**1** Write the standard form of the equation of a circle (p. 189)	17–20
	2 Graph a circle (p. 191)	21–26
	3 Work with the general form of the equation of a circle (p. 192)	23–26
2.5	**1** Construct a model using direct variation (p. 196)	52, 53, 55
	2 Construct a model using inverse variation (p. 197)	54
	3 Construct a model using joint or combined variation (p. 197)	55

Review Exercises

In Problems 1–6, find the following for each pair of points:
(a) The distance between the points
(b) The midpoint of the line segment connecting the points
(c) The slope of the line containing the points
(d) Interpret the slope found in part (c)

1. $(0, 0); (4, 2)$ **2.** $(0, 0); (-4, 6)$

3. $(1, -1); (-2, 3)$ **4.** $(-2, 2); (1, 4)$

5. $(4, -4); (4, 8)$ **6.** $(-3, 4); (2, 4)$

7. Graph $y = x^2 + 4$ by plotting points.

8. List the intercepts of the given graph.

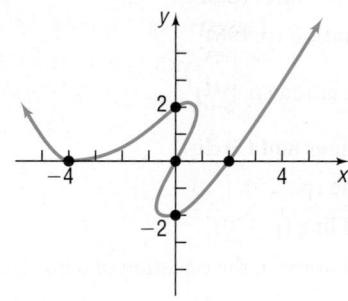

In Problems 9–16, list the intercepts and test for symmetry with respect to the x-axis, the y-axis, and the origin.

9. $2x = 3y^2$ **10.** $y = 5x$ **11.** $x^2 + 4y^2 = 16$ **12.** $9x^2 - y^2 = 9$

13. $y = x^4 + 2x^2 + 1$ **14.** $y = x^3 - x$ **15.** $x^2 + x + y^2 + 2y = 0$ **16.** $x^2 + 4x + y^2 - 2y = 0$

In Problems 17–20, find the standard form of the equation of the circle whose center and radius are given.

17. $(h, k) = (-2, 3); r = 4$ **18.** $(h, k) = (3, 4); r = 4$ **19.** $(h, k) = (-1, -2); r = 1$ **20.** $(h, k) = (2, -4); r = 3$

In Problems 21–26, find the center and radius of each circle. Graph each circle. Find the intercepts, if any, of each circle.

21. $x^2 + (y - 1)^2 = 4$ **22.** $(x + 2)^2 + y^2 = 9$ **23.** $x^2 + y^2 - 2x + 4y - 4 = 0$

24. $x^2 + y^2 + 4x - 4y - 1 = 0$ **25.** $3x^2 + 3y^2 - 6x + 12y = 0$ **26.** $2x^2 + 2y^2 - 4x = 0$

In Problems 27–36, find an equation of the line having the given characteristics. Express your answer using either the general form or the slope–intercept form of the equation of a line, whichever you prefer.

27. Slope $= -2$; containing the point $(3, -1)$

28. Slope $= 0$; containing the point $(-5, 4)$

29. Vertical; containing the point $(-3, 4)$

30. x-intercept $= 2$; containing the point $(4, -5)$

31. y-intercept $= -2$; containing the point $(5, -3)$

32. Containing the points $(3, -4)$ and $(2, 1)$

33. Parallel to the line $2x - 3y = -4$; containing the point $(-5, 3)$

34. Parallel to the line $x + y = 2$; containing the point $(1, -3)$

35. Perpendicular to the line $x + y = 2$; containing the point $(4, -3)$

36. Perpendicular to the line $3x - y = -4$; containing the point $(-2, 4)$

In Problems 37–40, find the slope and y-intercept of each line. Graph the line, labeling any intercepts.

37. $4x - 5y = -20$

38. $3x + 4y = 12$

39. $\dfrac{1}{2}x - \dfrac{1}{3}y = -\dfrac{1}{6}$

40. $-\dfrac{3}{4}x + \dfrac{1}{2}y = 0$

In Problems 41–44, find the intercepts and graph each line.

41. $2x - 3y = 12$

42. $x - 2y = 8$

43. $\dfrac{1}{2}x + \dfrac{1}{3}y = 2$

44. $\dfrac{1}{3}x - \dfrac{1}{4}y = 1$

45. Sketch a graph of $y = x^3$.

46. Sketch a graph of $y = \sqrt{x}$.

47. Graph the line with slope $\dfrac{2}{3}$ containing the point $(1, 2)$.

48. Show that the points $A = (3, 4)$, $B = (1, 1)$, and $C = (-2, 3)$ are the vertices of an isosceles triangle.

49. Show that the points $A = (-2, 0)$, $B = (-4, 4)$, and $C = (8, 5)$ are the vertices of a right triangle in two ways:
(a) By using the converse of the Pythagorean Theorem
(b) By using the slopes of the lines joining the vertices

50. The endpoints of the diameter of a circle are $(-3, 2)$ and $(5, -6)$. Find the center and radius of the circle. Write the standard equation of this circle.

51. Show that the points $A = (2, 5)$, $B = (6, 1)$, and $C = (8, -1)$ lie on a line by using slopes.

52. Mortgage Payments The monthly payment p on a mortgage varies directly with the amount borrowed B. If the monthly payment on a 30-year mortgage is \$854.00 when \$130,000 is borrowed, find an equation that relates the monthly payment p to the amount borrowed B for a mortgage with the same terms. Then find the monthly payment p when the amount borrowed B is \$165,000.

53. Revenue Function At the corner Esso station, the revenue R varies directly with the number g of gallons of gasoline sold. If the revenue is \$46.67 when the number of gallons sold is 13, find an equation that relates revenue R to the number g of gallons of gasoline. Then find the revenue R when the number of gallons of gasoline sold is 11.2.

54. Weight of a Body The weight of a body varies inversely with the square of its distance from the center of Earth. Assuming that the radius of Earth is 3960 miles, how much would a man weigh at an altitude of 1 mile above Earth's surface if he weighs 200 pounds on Earth's surface?

55. Kepler's Third Law of Planetary Motion Kepler's Third Law of Planetary Motion states that the square of the period of revolution T of a planet varies directly with the cube of its mean distance a from the Sun. If the mean distance of Earth from the Sun is 93 million miles, what is the mean distance of the planet Mercury from the Sun, given that Mercury has a "year" of 88 days?

56. Create four problems that you might be asked to do given the two points $(-3, 4)$ and $(6, 1)$. Each problem should involve a different concept. Be sure that your directions are clearly stated.

57. Describe each of the following graphs in the xy-plane. Give justification.
 (a) $x = 0$ (b) $y = 0$ (c) $x + y = 0$ (d) $xy = 0$ (e) $x^2 + y^2 = 0$

58. Suppose that you have a rectangular field that requires watering. Your watering system consists of an arm of variable length that rotates so that the watering pattern is a circle. Decide where to position the arm and what length it should be so that the entire field is watered most efficiently. When does it become desirable to use more than one arm?

 [**Hint:** Use a rectangular coordinate system positioned as shown in the figures. Write equations for the circle(s) swept out by the watering arm(s).]

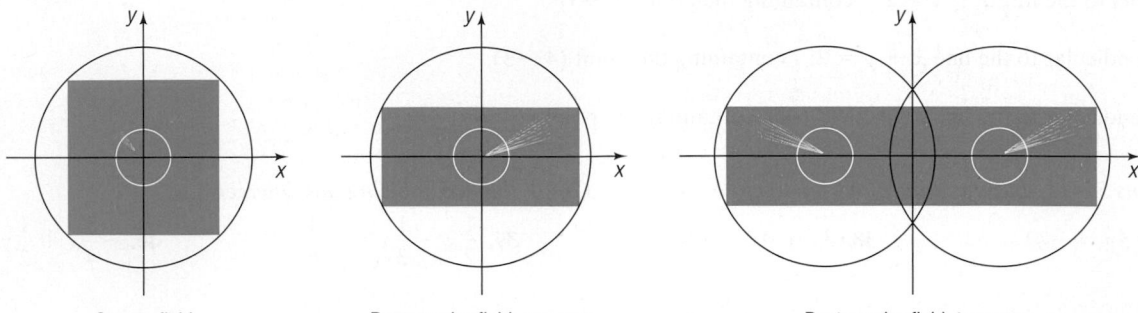

Square field Rectangular field, one arm Rectangular field, two arms

CHAPTER TEST

In Problems 1–3, use $P_1 = (-1, 3)$ and $P_2 = (5, -1)$.

1. Find the distance from P_1 to P_2.

2. Find the midpoint of the line segment joining P_1 and P_2.

3. (a) Find the slope of the line containing P_1 and P_2.
 (b) Interpret this slope.

4. Graph $y = x^2 - 9$ by plotting points.

5. Sketch the graph of $y^2 = x$.

6. List the intercepts and test for symmetry: $x^2 + y = 9$.

7. Write the slope–intercept form of the line with slope -2 containing the point $(3, -4)$. Graph the line.

8. Write the general form of the circle with center $(4, -3)$ and radius 5.

9. Find the center and radius of the circle $x^2 + y^2 + 4x - 2y - 4 = 0$. Graph this circle.

10. For the line $2x + 3y = 6$, find a line parallel to it containing the point $(1, -1)$. Also find a line perpendicular to it containing the point $(0, 3)$.

11. Resistance due to a Conductor The resistance (in ohms) of a circular conductor varies directly with the length of the conductor and inversely with the square of the radius of the conductor. If 50 feet of wire with a radius of 6×10^{-3} inch has a resistance of 10 ohms, what would be the resistance of 100 feet of the same wire if the radius is increased to 7×10^{-3} inch?

CUMULATIVE REVIEW

In Problems 1–8, find the real solution(s) of each equation.

1. $3x - 5 = 0$

2. $x^2 - x - 12 = 0$

3. $2x^2 - 5x - 3 = 0$

4. $x^2 - 2x - 2 = 0$

5. $x^2 + 2x + 5 = 0$

6. $\sqrt{2x + 1} = 3$

7. $|x - 2| = 1$

8. $\sqrt{x^2 + 4x} = 2$

In Problems 9 and 10, solve each equation in the complex number system.

9. $x^2 = -9$

10. $x^2 - 2x + 5 = 0$

In Problems 11–14, solve each inequality. Graph the solution set.

11. $2x - 3 \le 7$

12. $-1 < x + 4 < 5$

13. $|x - 2| \le 1$

14. $|2 + x| > 3$

15. Find the distance between the points $P = (-1, 3)$ and $Q = (4, -2)$. Find the midpoint of the line segment from P to Q.

16. Which of the following points are on the graph of $y = x^3 - 3x + 1$?
 (a) $(-2, -1)$ (b) $(2, 3)$ (c) $(3, 1)$

17. Sketch the graph of $y = x^3$.

18. Find the equation of the line containing the points $(-1, 4)$ and $(2, -2)$. Express your answer in slope–intercept form.

19. Find the equation of the line perpendicular to the line $y = 2x + 1$ and containing the point $(3, 5)$. Express your answer in slope–intercept form and graph the line.

20. Graph the equation $x^2 + y^2 - 4x + 8y - 5 = 0$.

CHAPTER PROJECT

Predicting Olympic Performance Measurements of human performance over time sometimes follow a strong linear relationship for reasonably short periods. In 2004 the Summer Olympic Games returned to Greece, the home of both the ancient Olympics and the first modern Olympics. The following data represent the winning times (in hours) for men and women in the Olympic marathon.

Year	Men	Women
1984	2.16	2.41
1988	2.18	2.43
1992	2.22	2.54
1996	2.21	2.43
2000	2.17	2.39

Source: www.hickoksports.com/history/olmtandf.shtml

1. Treating year as the independent variable and the winning value as the dependent variable, find linear equations relating these variables (separately for men and women) using the data for the years 1992 and 1996. Compare the equations and comment on any similarities or differences.

2. Interpret the slopes in your equations from part 1. Do the y-intercepts have a reasonable interpretation? Why or why not?

3. Use your lines to predict the winning time in the 2004 Olympics. Compare your predictions to the actual results (2.18 hours for men and 2.44 hours for women). How well did your equations do in predicting the winning times?

4. Repeat parts 1 to 3 using the data for the years 1996 and 2000. How do your results compare?

5. Would your equations be useful in predicting the winning marathon times in the 2104 Summer Olympics? Why or why not?

6. Pick your favorite Winter Olympics event and find the winning value (that is distance, time, or the like) in two Winter Olympics prior to 2006. Repeat parts 1 to 3 using your selected event and years and compare to the actual results of the 2006 Winter Olympics in Torino, Italy.

Functions and Their Graphs

3

—See the Chapter Project—

U.S. Consumers Make Wireless Top Choice

July 11, 2005—WASHINGTON, D.C., CTIA-The Wireless Association President and CEO Steve Largent highlights the Federal Communications Commission (FCC) report on local telephone competition, which indicates for the first time there are more wireless subscribers in the United States than wireline service lines. According to the FCC report, as of December 31, 2004 there were 181.1 million wireless subscribers in the U.S., compared to a combined 178 million incumbent local exchange carrier switched access lines and competitive local exchange carrier switched access lines. Also according to the report, there was a 15% increase in wireless subscribers for the 12-month period ending December 31, 2004.

"This significant milestone reflects consumers' approval of a vibrant, ultracompetitive industry that lets us communicate how we want, where we want and with whom we want," said Largent.

"Wireless is so popular because it lets people communicate on their terms. And it's not just about talking anymore. You can browse the web, take pictures and video, download music, play games or conduct business. Wireless can satisfy a lot of different communication needs and desires and that's extremely popular to millions of consumers."

Source: www.CTIA.org

A Look Back

So far, we have developed techniques for graphing equations containing two variables.

A Look Ahead

In this chapter, we look at a special type of equation involving two variables called a *function*. This chapter deals with what a function is, how to graph functions, properties of functions, and how functions are used in applications. The word function apparently was introduced by René Descartes in 1637. For him, a function simply meant any positive integral power of a variable *x*. Gottfried Wilhelm Leibniz (1646–1716), who always emphasized the geometric side of mathematics, used the word function to denote any quantity associated with a curve, such as the coordinates of a point on the curve. Leonhard Euler (1707–1783) employed the word to mean any equation or formula involving variables and constants. His idea of a function is similar to the one most often seen in courses that precede calculus. Later, the use of functions in investigating heat flow equations led to a very broad definition, due to Lejeune Dirichlet (1805–1859), which describes a function as a correspondence between two sets. It is his definition that we use here.

Outline

3.1 Functions

PREPARING FOR THIS SECTION *Before getting started, review the following:*

- Intervals (Section 1.5, pp. 125–126)
- Evaluating Algebraic Expressions, Domain of a Variable (Chapter R, Section R.2, pp. 20–21)

- Solving Inequalities (Section 1.5, pp. 128–131)

\ **Now Work** the 'Are You Prepared?' problems on page 219.

OBJECTIVES
1 Determine Whether a Relation Represents a Function (p. 208)
2 Find the Value of a Function (p. 212)
3 Find the Domain of a Function (p. 215)
4 Form the Sum, Difference, Product, and Quotient of Two Functions (p. 217)

1 Determine Whether a Relation Represents a Function

We often see situations where one variable is somehow linked to the value of another variable. For example, an individual's level of education is linked to annual income. Engine size is linked to gas mileage. When the value of one variable is related to the value of a second variable, we have a *relation*. A **relation** is a correspondence between two sets. If x and y are two elements in these sets and if a relation exists between x and y, then we say that x **corresponds** to y or that y **depends on** x, and we write $x \rightarrow y$.

We have a number of ways to express relations between two sets. For example, the equation $y = 3x - 1$ shows a relation between x and y. It says that if we take some number x, multiply it by 3, and then subtract 1 we obtain the corresponding value of y. In this sense, x serves as the **input** to the relation and y is the **output** of the relation. We can also express this relation as a graph as shown in Figure 1.

Not only can a relation be expressed through an equation or graph, but we can also express a relation through a technique called *mapping*. A **map** illustrates a relation by using a set of inputs and drawing arrows to the corresponding element in the set of outputs. Ordered pairs can be used to represent $x \rightarrow y$ as (x, y). We illustrate these two concepts in Example 1.

Figure 1

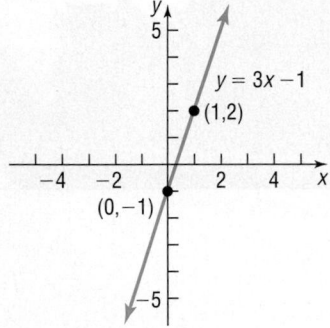

$y = 3x - 1$
$(1, 2)$
$(0, -1)$

EXAMPLE 1 **Maps and Ordered Pairs as Relations**

Figure 2 shows a relation between states and the number of representatives each state has in the House of Representatives. The relation might be named "number of representatives."

Figure 2

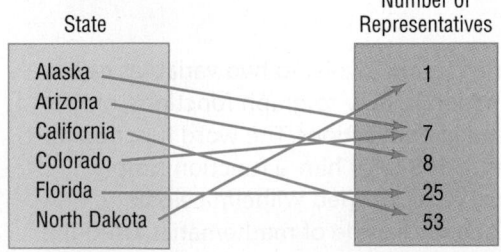

State / Number of Representatives

Alaska, Arizona, California, Colorado, Florida, North Dakota → 1, 7, 8, 25, 53

In this relation, Alaska corresponds to 7, Arizona corresponds to 8, and so on. Using ordered pairs, this relation would be expressed as

{(Alaska, 7), (Arizona, 8), (California, 53), (Colorado, 7), (Florida, 25), (North Dakota, 1)}

We now present one of the most important concepts in algebra—the *function*. A function is a special type of relation. To understand the idea behind a function,

let's revisit the relation presented in Example 1. If we were to ask, "How many representatives does Alaska have?," you would respond 7. In other words, each input *state* corresponds to a single output *number of representatives*.

Let's consider a second relation where we have a correspondence between four people and their phone numbers. See Figure 3. Notice that Colleen has two telephone numbers. If asked, "What is Colleen's phone number?", you cannot assign a single number to her.

Figure 3

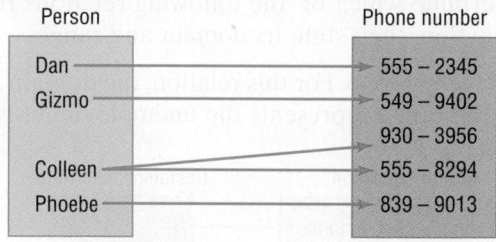

Let's look at one more relation. Figure 4 is a relation that shows a correspondence between *animals* and life *expectancy*. If asked to determine the life expectancy of a dog, we would all respond "11 years." If asked to determine the life expectancy of a rabbit, we would all respond "7 years."

Figure 4

Notice that the relations presented in Figures 2 and 4 have something in common. What is it? The common link between these two relations is that each input corresponds to exactly one output. This leads to the definition of a *function*.

DEFINITION

Let X and Y be two nonempty sets.* A **function** from X into Y is a relation that associates with each element of X exactly one element of Y.

The set X is called the **domain** of the function. For each element x in X, the corresponding element y in Y is called the **value** of the function at x, or the **image** of x. The set of all images of the elements in the domain is called the **range** of the function. See Figure 5.

Figure 5

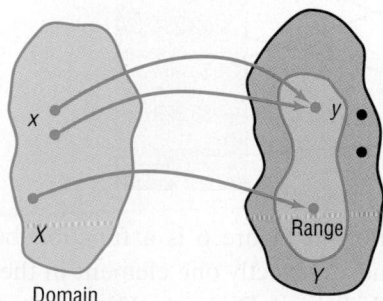

* The sets X and Y will usually be sets of real numbers, in which case a (real) function results. The two sets can also be sets of complex numbers, and then we have defined a complex function. In the broad definition (due to Lejeune Dirichlet), X and Y can be any two sets.

Since there may be some elements in Y that are not the image of some x in X, it follows that the range of a function may be a subset of Y, as shown in Figure 5.

Not all relations between two sets are functions. The next example shows how to determine whether a relation is a function or not.

EXAMPLE 2 **Determining Whether a Relation Represents a Function**

Determine which of the following relations represent a function. If the relation is a function, then state its domain and range.

(a) See Figure 6. For this relation, the domain represents the level of education and the range represents the unemployment rate.

Figure 6

SOURCE: *Statistical Abstract of the United States, 2006*

(b) See Figure 7. For this relation, the domain represents the number of calories in a sandwich from a fast-food restaurant and the range represents the fat content (in grams).

Figure 7

SOURCE: *Each company's Web site*

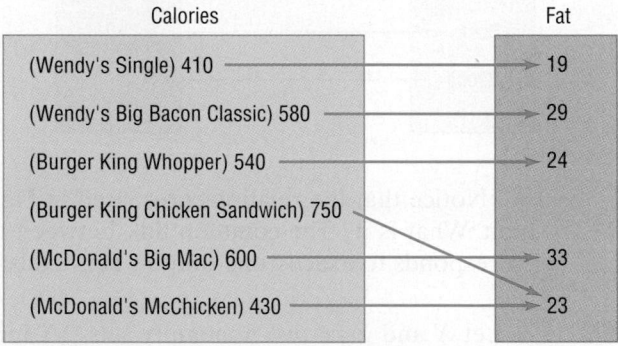

(c) See Figure 8. For this relation, the domain represents the weight (in carats) of pear-cut diamonds and the range represents the price (in dollars).

Figure 8

SOURCE: *diamonds.com*

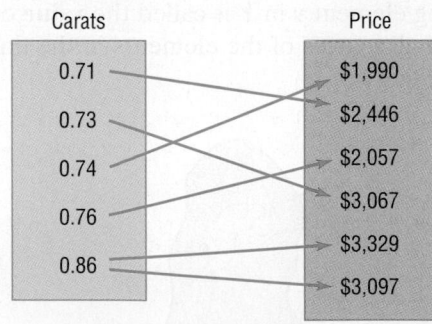

Solution

(a) The relation in Figure 6 is a function because each element in the domain corresponds to exactly one element in the range. The domain of the relation is {No High School Diploma, High School Diploma, Some College, College Graduate} and the range of the relation is {8.5%, 5.0%, 4.2%, 2.7%}.

(b) The relation in Figure 7 is a function because each element in the domain corresponds to exactly one element in the range. The domain of the relation is {410, 580, 540, 750, 600, 430}. The range of the relation is {19, 29, 24, 33, 23}.

Notice that it is okay for more than one element in the domain to correspond to the same element in the range (McDonald's and Burger King's chicken sandwich both have 23 grams of fat).

(c) The relation in Figure 8 is not a function because each element in the domain does not correspond to exactly one element in the range. If a 0.86-carat diamond is chosen from the domain, a single price cannot be assigned to it.

━━━➤ **Now Work** PROBLEM 15

The idea behind a function is its predictability. If the input is known, we can use the function to determine the output. With "nonfunctions," we don't have this predictability. Look back at Figure 7. The inputs are $\{410, 580, 540, 750, 600, 430\}$. The correspondence is *number of fat grams*, and the outputs are $\{19, 29, 24, 33, 23\}$. If asked, "How many grams of fat are in a 410-calorie sandwich?", we can use the correspondence to answer "19." Now consider Figure 8. If asked, "What is the price of a 0.86-carat diamond?", we could not give a single response because two outputs result from the single input "0.86". For this reason, the relation in Figure 8 is not a function.

We may also think of a function as a set of ordered pairs (x, y) in which no ordered pairs have the same first element and different second elements. The set of all first elements x is the domain of the function, and the set of all second elements y is its range. Each element x in the domain corresponds to exactly one element y in the range.

EXAMPLE 3 | **Determining Whether a Relation Represents a Function**

Determine whether each relation represents a function. If it is a function, state the domain and range.

(a) $\{(1, 4), (2, 5), (3, 6), (4, 7)\}$

(b) $\{(1, 4), (2, 4), (3, 5), (6, 10)\}$

(c) $\{(-3, 9), (-2, 4), (0, 0), (1, 1), (-3, 8)\}$

Solution

(a) This relation is a function because there are no ordered pairs with the same first element and different second elements. The domain of this function is $\{1, 2, 3, 4\}$, and its range is $\{4, 5, 6, 7\}$.

(b) This relation is a function because there are no ordered pairs with the same first element and different second elements. The domain of this function is $\{1, 2, 3, 6\}$, and its range is $\{4, 5, 10\}$.

(c) This relation is not a function because there are two ordered pairs, $(-3, 9)$ and $(-3, 8)$, that have the same first element and different second elements.

In Example 3(b), notice that 1 and 2 in the domain each have the same image in the range. This does not violate the definition of a function; two different first elements can have the same second element. A violation of the definition occurs when two ordered pairs have the same first element and different second elements, as in Example 3(c).

━━━➤ **Now Work** PROBLEM 19

Up to now we have shown how to identify when a relation is a function for relations defined by mappings (Example 2) and ordered pairs (Example 3). We know that relations can also be expressed as equations. We discuss next the circumstances under which equations are functions.

To determine whether an equation, where y depends on x, is a function, it is often easiest to solve the equation for y. If any value of x in the domain corresponds to more than one y, the equation does not define a function; otherwise, it does define a function.

EXAMPLE 4	**Determining Whether an Equation Is a Function**

Determine if the equation $y = 2x - 5$ defines y as a function of x.

Solution The equation tells us to take an input x, multiply it by 2, and then subtract 5. For any input x, these operations yield only one output y. For example, if $x = 1$, then $y = 2(1) - 5 = -3$. If $x = 3$, then $y = 2(3) - 5 = 1$. For this reason, the equation is a function. ∎

EXAMPLE 5	**Determining Whether an Equation Is a Function**

Determine if the equation $x^2 + y^2 = 1$ defines y as a function of x.

Solution To determine whether the equation $x^2 + y^2 = 1$, which defines the unit circle, is a function, we need to solve the equation for y.

$$x^2 + y^2 = 1$$
$$y^2 = 1 - x^2$$
$$y = \pm\sqrt{1 - x^2}$$

For values of x between -1 and 1, two values of y result. For example, if $x = 0$, then $y = \pm 1$, so two different outputs result from the same input. This means that the equation $x^2 + y^2 = 1$ does not define a function. ∎

➤ **Now Work** PROBLEM 33

2 Find the Value of a Function

Functions are often denoted by letters such as f, F, g, G, and others. If f is a function, then for each number x in its domain the corresponding image in the range is designated by the symbol $f(x)$, read as "f of x" or as "f at x." We refer to $f(x)$ as the **value of f at the number x**; $f(x)$ is the number that results when x is given and the function f is applied; $f(x)$ is the output corresponding to x or the image of x; $f(x)$ does *not* mean "f times x." For example, the function given in Example 4 may be written as $y = f(x) = 2x - 5$. Then $f\left(\dfrac{3}{2}\right) = -2$.

Figure 9 illustrates some other functions. Notice that, in every function, for each x in the domain there is one value in the range.

Figure 9

(a) $f(x) = x^2$

(b) $F(x) = \dfrac{1}{x}$

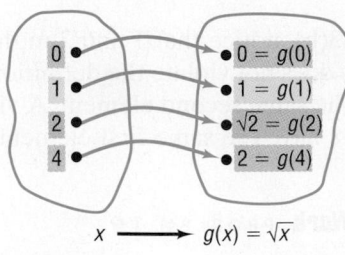

(c) $g(x) = \sqrt{x}$

(d) $G(x) = 3$

Figure 10

Input x

Output
$y = f(x)$

Sometimes it is helpful to think of a function f as a machine that receives as input a number from the domain, manipulates it, and outputs the value. See Figure 10.

The restrictions on this input/output machine are as follows:

1. It only accepts numbers from the domain of the function.

2. For each input, there is exactly one output (which may be repeated for different inputs).

For a function $y = f(x)$, the variable x is called the **independent variable,** because it can be assigned any of the permissible numbers from the domain. The variable y is called the **dependent variable,** because its value depends on x.

Any symbol can be used to represent the independent and dependent variables. For example, if f is the *cube function,* then f can be given by $f(x) = x^3$ or $f(t) = t^3$ or $f(z) = z^3$. All three functions are the same. Each tells us to cube the independent variable to get the output. In practice, the symbols used for the independent and dependent variables are based on common usage, such as using C for cost in business.

The independent variable is also called the **argument** of the function. Thinking of the independent variable as an argument can sometimes make it easier to find the value of a function. For example, if f is the function defined by $f(x) = x^3$, then f tells us to cube the argument. Thus, $f(2)$ means to cube 2, $f(a)$ means to cube the number a, and $f(x + h)$ means to cube the quantity $x + h$.

EXAMPLE 6	Finding Values of a Function

For the function f defined by $f(x) = 2x^2 - 3x$, evaluate

(a) $f(3)$ (b) $f(x) + f(3)$ (c) $3f(x)$

(d) $f(-x)$ (e) $-f(x)$ (f) $f(3x)$

(g) $f(x + 3)$ (h) $\dfrac{f(x + h) - f(x)}{h}$ $h \neq 0$

Solution

(a) We substitute 3 for x in the equation for f, $f(x) = 2x^2 - 3x$, to get

$$f(3) = 2(3)^2 - 3(3) = 18 - 9 = 9$$

(b) $f(x) + f(3) = (2x^2 - 3x) + (9) = 2x^2 - 3x + 9$

(c) We multiply the equation for f by 3.

$$3f(x) = 3(2x^2 - 3x) = 6x^2 - 9x$$

(d) We substitute $-x$ for x in the equation for f and simplify.

$$f(-x) = 2(-x)^2 - 3(-x) = 2x^2 + 3x$$

(e) $-f(x) = -(2x^2 - 3x) = -2x^2 + 3x$

(f) We substitute $3x$ for x in the equation for f and simplify.

$$f(3x) = 2(3x)^2 - 3(3x) = 2(9x^2) - 9x = 18x^2 - 9x$$

(g) We substitute $x + 3$ for x in the equation for f and simplify.

$$f(x + 3) = 2(x + 3)^2 - 3(x + 3) \qquad \text{Notice the use of parentheses here.}$$
$$= 2(x^2 + 6x + 9) - 3x - 9$$
$$= 2x^2 + 12x + 18 - 3x - 9$$
$$= 2x^2 + 9x + 9$$

(h) $\dfrac{f(x+h)-f(x)}{h} = \dfrac{[2(x+h)^2 - 3(x+h)] - [2x^2 - 3x]}{h}$

$\uparrow$
$f(x+h) = 2(x+h)^2 - 3(x+h)$

$\quad = \dfrac{2(x^2 + 2xh + h^2) - 3x - 3h - 2x^2 + 3x}{h}$ Simplify.

$\quad = \dfrac{2x^2 + 4xh + 2h^2 - 3h - 2x^2}{h}$ Distribute and combine like terms.

$\quad = \dfrac{4xh + 2h^2 - 3h}{h}$ Combine like terms.

$\quad = \dfrac{h(4x + 2h - 3)}{h}$ Factor out h.

$\quad = 4x + 2h - 3$ Divide out the h's.

Notice in this example that $f(x+3) \neq f(x) + f(3)$, $f(-x) \neq -f(x)$, and $3f(x) \neq f(3x)$.

 The expression in part (h) is called the **difference quotient** of f, an important expression in calculus.

Now Work PROBLEMS **39** AND **73**

Most calculators have special keys that enable you to find the value of certain commonly used functions. For example, you should be able to find the square function $f(x) = x^2$, the square root function $f(x) = \sqrt{x}$, the reciprocal function $f(x) = \dfrac{1}{x} = x^{-1}$, and many others that will be discussed later in this book (such as $\ln x$ and $\log x$). Verify the results of Example 7, which follows, on your calculator.

EXAMPLE 7 **Finding Values of a Function on a Calculator**

(a) $f(x) = x^2$ $f(1.234) = 1.234^2 = 1.522756$

(b) $F(x) = \dfrac{1}{x}$ $F(1.234) = \dfrac{1}{1.234} \approx 0.8103727715$

(c) $g(x) = \sqrt{x}$ $g(1.234) = \sqrt{1.234} \approx 1.110855526$

COMMENT Graphing calculators can be used to evaluate any function that you wish. Figure 11 shows the result obtained in Example 6(a) on a TI-84 Plus graphing calculator with the function to be evaluated, f(x) = 2x² − 3x, in Y₁.

Figure 11

Implicit Form of a Function

In general, when a function f is defined by an equation in x and y, we say that the function f is given **implicitly**. If it is possible to solve the equation for y in terms of x, then we write $y = f(x)$ and say that the function is given **explicitly**. For example,

Implicit Form	Explicit Form
$3x + y = 5$	$y = f(x) = -3x + 5$
$x^2 - y = 6$	$y = f(x) = x^2 - 6$
$xy = 4$	$y = f(x) = \dfrac{4}{x}$

COMMENT The explicit form of a function is the form required by a graphing calculator. ∎

We list next a summary of some important facts to remember about a function f.

SUMMARY Important Facts about Functions

(a) For each x in a domain of a function f, there is exactly one image $f(x)$ in the range; however, an element in the range can result from more than one x in the domain.

(b) f is the symbol that we use to denote the function. It is symbolic of the equation that we use to get from an x in the domain to $f(x)$ in the range.

(c) If $y = f(x)$, then x is called the independent variable or argument of f, and y is called the dependent variable or the value of f at x.

3 Find the Domain of a Function

Often the domain of a function f is not specified; instead, only the equation defining the function is given. In such cases, we agree that the **domain of f** is the largest set of real numbers for which the value $f(x)$ is a real number. The domain of a function f is the same as the domain of the variable x in the expression $f(x)$.

EXAMPLE 8 Finding the Domain of a Function

Find the domain of each of the following functions:

(a) $f(x) = x^2 + 5x$ (b) $g(x) = \dfrac{3x}{x^2 - 4}$ (c) $h(t) = \sqrt{4 - 3t}$

Solution

(a) The function tells us to square a number and then add five times the number. Since these operations can be performed on any real number, we conclude that the domain of f is the set of all real numbers.

(b) The function g tells us to divide $3x$ by $x^2 - 4$. Since division by 0 is not defined, the denominator $x^2 - 4$ can never be 0, so x can never equal -2 or 2. The domain of the function g is $\{x \mid x \neq -2, x \neq 2\}$.

(c) The function h tells us to take the square root of $4 - 3t$. But only nonnegative numbers have real square roots, so the expression under the square root must be nonnegative (greater than or equal to zero). This requires that

$$4 - 3t \geq 0$$
$$-3t \geq -4$$
$$t \leq \frac{4}{3}$$

The domain of h is $\left\{ t \mid t \leq \dfrac{4}{3} \right\}$ or the interval $\left(-\infty, \dfrac{4}{3} \right]$. ∎

For the functions that we will encounter in this book, the following steps may prove helpful for finding the domain of a function whose domain is a subset of real numbers and is defined by an equation.

Start with the set of real numbers

1. If the equation has a denominator, exclude any numbers that give a zero denominator.
2. If the equation has a radical of even index, exclude any numbers that cause the expression inside the radical to be negative.

━━━ **Now Work** PROBLEM 51

If x is in the domain of a function f, we shall say that **f is defined at x**, or **$f(x)$ exists.** If x is not in the domain of f, we say that **f is not defined at x**, or **$f(x)$ does not exist.** For example, if $f(x) = \dfrac{x}{x^2 - 1}$, then $f(0)$ exists, but $f(1)$ and $f(-1)$ do not exist. (Do you see why?)

We have not said much about finding the range of a function. The reason is that when a function is defined by an equation it is often difficult to find the range.* Therefore, we shall usually be content to find just the domain of a function when the function is defined by an equation. We shall express the domain of a function using inequalities, interval notation, set notation, or words, whichever is most convenient.

When we use functions in applications, the domain may be restricted by physical or geometric considerations. For example, the domain of the function f defined by $f(x) = x^2$ is the set of all real numbers. However, if f is used to obtain the area of a square when the length x of a side is known, then we must restrict the domain of f to the positive real numbers, since the length of a side can never be 0 or negative.

| **EXAMPLE 9** | **Finding the Domain in an Application** |

Express the area of a circle as a function of its radius. Find the domain.

Solution

Figure 12

See Figure 12. We know that the formula for the area A of a circle of radius r is $A = \pi r^2$. If we use r to represent the independent variable and A to represent the dependent variable, the function expressing this relationship is

$$A(r) = \pi r^2$$

In this setting, the domain is $\{r \mid r > 0\}$. (Do you see why?)

Observe in the solution to Example 9 that we used the symbol A in two ways: It is used to name the function, and it is used to symbolize the dependent variable. This double use is common in applications and should not cause any difficulty.

━━━ **Now Work** PROBLEM 87

* In Section 6.2 we discuss a way to find the range for a special class of functions.

4 Form the Sum, Difference, Product, and Quotient of Two Functions

Next we introduce some operations on functions. We shall see that functions, like numbers, can be added, subtracted, multiplied, and divided. For example, if $f(x) = x^2 + 9$ and $g(x) = 3x + 5$, then

$$f(x) + g(x) = (x^2 + 9) + (3x + 5) = x^2 + 3x + 14$$

The new function $y = x^2 + 3x + 14$ is called the *sum function* $f + g$. Similarly,

$$f(x) \cdot g(x) = (x^2 + 9)(3x + 5) = 3x^3 + 5x^2 + 27x + 45$$

The new function $y = 3x^3 + 5x^2 + 27x + 45$ is called the *product function* $f \cdot g$. The general definitions are given next.

DEFINITION

If f and g are functions:

The **sum** $f + g$ is the function defined by

$$(f + g)(x) = f(x) + g(x)$$

The domain of $f + g$ consists of the numbers x that are in the domains of both f and g. That is, domain of $f + g$ = domain of $f \cap$ domain of g.

DEFINITION

The **difference** $f - g$ is the function defined by

$$(f - g)(x) = f(x) - g(x)$$

The domain of $f - g$ consists of the numbers x that are in the domains of both f and g. That is, domain of $f - g$ = domain of $f \cap$ domain of g.

DEFINITION

The **product** $f \cdot g$ is the function defined by

$$(f \cdot g)(x) = f(x) \cdot g(x)$$

The domain of $f \cdot g$ consists of the numbers x that are in the domains of both f and g. That is, domain of $f \cdot g$ = domain of $f \cap$ domain of g.

DEFINITION

The **quotient** $\dfrac{f}{g}$ is the function defined by

$$\left(\frac{f}{g}\right)(x) = \frac{f(x)}{g(x)} \qquad g(x) \neq 0$$

The domain of $\dfrac{f}{g}$ consists of the numbers x for which $g(x) \neq 0$ that are in the domains of both f and g. That is,

$$\text{domain of } \frac{f}{g} = \{x \mid g(x) \neq 0\} \cap \text{domain of } f \cap \text{domain of } g.$$

| EXAMPLE 10 | Operations on Functions |

Let f and g be two functions defined as

$$f(x) = \frac{1}{x+2} \quad \text{and} \quad g(x) = \frac{x}{x-1}$$

Find the following, and determine the domain in each case.

(a) $(f + g)(x)$ (b) $(f - g)(x)$

(c) $(f \cdot g)(x)$ (d) $\left(\dfrac{f}{g}\right)(x)$

Solution The domain of f is $\{x \mid x \neq -2\}$ and the domain of g is $\{x \mid x \neq 1\}$.

(a) $(f + g)(x) = f(x) + g(x) = \dfrac{1}{x+2} + \dfrac{x}{x-1}$

$$= \frac{x-1}{(x+2)(x-1)} + \frac{x(x+2)}{(x+2)(x-1)} = \frac{x^2 + 3x - 1}{(x+2)(x-1)}$$

The domain of $f + g$ consists of those numbers x that are in the domains of both f and g. Therefore, the domain of $f + g$ is $\{x \mid x \neq -2, x \neq 1\}$.

(b) $(f - g)(x) = f(x) - g(x) = \dfrac{1}{x+2} - \dfrac{x}{x-1}$

$$= \frac{x-1}{(x+2)(x-1)} - \frac{x(x+2)}{(x+2)(x-1)} = \frac{-(x^2 + x + 1)}{(x+2)(x-1)}$$

The domain of $f - g$ consists of those numbers x that are in the domains of both f and g. Therefore, the domain of $f - g$ is $\{x \mid x \neq -2, x \neq 1\}$.

(c) $(f \cdot g)(x) = f(x) \cdot g(x) = \dfrac{1}{x+2} \cdot \dfrac{x}{x-1} = \dfrac{x}{(x+2)(x-1)}$

The domain of $f \cdot g$ consists of those numbers x that are in the domains of both f and g. Therefore, the domain of $f \cdot g$ is $\{x \mid x \neq -2, x \neq 1\}$.

(d) $\left(\dfrac{f}{g}\right)(x) = \dfrac{f(x)}{g(x)} = \dfrac{\dfrac{1}{x+2}}{\dfrac{x}{x-1}} = \dfrac{1}{x+2} \cdot \dfrac{x-1}{x} = \dfrac{x-1}{x(x+2)}$

The domain of $\dfrac{f}{g}$ consists of the numbers x for which $g(x) \neq 0$ that are in the domains of both f and g. Since $g(x) = 0$ when $x = 0$, we exclude 0 as well as -2 and 1 from the domain. The domain of $\dfrac{f}{g}$ is $\{x \mid x \neq -2, x \neq 0, x \neq 1\}$.

— **Now Work** PROBLEM 61

In calculus, it is sometimes helpful to view a complicated function as the sum, difference, product, or quotient of simpler functions. For example,

$F(x) = x^2 + \sqrt{x}$ is the sum of $f(x) = x^2$ and $g(x) = \sqrt{x}$.

$H(x) = \dfrac{x^2 - 1}{x^2 + 1}$ is the quotient of $f(x) = x^2 - 1$ and $g(x) = x^2 + 1$.

SUMMARY

We list here some of the important vocabulary introduced in this section, with a brief description of each term.

Function	A relation between two sets of real numbers so that each number x in the first set, the domain, has corresponding to it exactly one number y in the second set.
	A set of ordered pairs (x, y) or $(x, f(x))$ in which no first element is paired with two different second elements.
	The range is the set of y values of the function that are the images of the x values in the domain.
	A function f may be defined implicitly by an equation involving x and y or explicitly by writing $y = f(x)$.
	Numbers for which $f(x) = 0$ are the zeros of f.
Unspecified domain	If a function f is defined by an equation and no domain is specified, then the domain will be taken to be the largest set of real numbers for which the equation defines a real number.
Function notation	$y = f(x)$
	f is a symbol for the function.
	x is the independent variable or argument.
	y is the dependent variable.
	$f(x)$ is the value of the function at x, or the image of x.

3.1 Assess Your Understanding

'Are You Prepared?' *Answers are given at the end of these exercises. If you get a wrong answer, read the pages listed in red.*

1. The inequality $-1 < x < 3$ can be written in interval notation as _____. (pp. 125–126)

2. If $x = -2$, the value of the expression $3x^2 - 5x + \dfrac{1}{x}$ is _____. (pp. 20–21)

3. The domain of the variable in the expression $\dfrac{x-3}{x+4}$ is _____. (pp. 20–21)

4. Solve the inequality: $3 - 2x > 5$. Graph the solution set. (pp. 128–131)

Concepts and Vocabulary

5. If f is a function defined by the equation $y = f(x)$, then x is called the _____ variable and y is the _____ variable.

6. The set of all images of the elements in the domain of a function is called the _____.

7. If the domain of f is all real numbers in the interval $[0, 7]$ and the domain of g is all real numbers in the interval $[-2, 5]$, the domain of $f + g$ is all real numbers in the interval _____.

8. The domain of $\dfrac{f}{g}$ consists of numbers x for which $g(x)$ _____ 0 that are in the domains of both _____ and _____.

9. If $f(x) = x + 1$ and $g(x) = x^3$, then _____ $= x^3 - (x + 1)$.

10. **True or False** Every relation is a function.

11. **True or False** The domain of $(f \cdot g)(x)$ consists of the numbers x that are in the domains of both f and g.

12. **True or False** The independent variable is sometimes referred to as the argument of the function.

13. **True or False** If no domain is specified for a function f, then the domain of f is taken to be the set of real numbers.

14. **True or False** The domain of the function $f(x) = \dfrac{x^2 - 4}{x}$ is $\{x | x \neq \pm 2\}$.

Skill Building

In Problems 15–26, determine whether each relation represents a function. For each function, state the domain and range.

15.

16.

17.

18.

19. $\{(2, 6), (-3, 6), (4, 9), (2, 10)\}$

20. $\{(-2, 5), (-1, 3), (3, 7), (4, 12)\}$

21. $\{(1, 3), (2, 3), (3, 3), (4, 3)\}$

22. $\{(0, -2), (1, 3), (2, 3), (3, 7)\}$

23. $\{(-2, 4), (-2, 6), (0, 3), (3, 7)\}$

24. $\{(-4, 4), (-3, 3), (-2, 2), (-1, 1), (-4, 0)\}$

25. $\{(-2, 4), (-1, 1), (0, 0), (1, 1)\}$

26. $\{(-2, 16), (-1, 4), (0, 3), (1, 4)\}$

In Problems 27–38, determine whether the equation defines y as a function of x.

27. $y = x^2$

28. $y = x^3$

29. $y = \dfrac{1}{x}$

30. $y = |x|$

31. $y^2 = 4 - x^2$

32. $y = \pm\sqrt{1 - 2x}$

33. $x = y^2$

34. $x + y^2 = 1$

35. $y = 2x^2 - 3x + 4$

36. $y = \dfrac{3x - 1}{x + 2}$

37. $2x^2 + 3y^2 = 1$

38. $x^2 - 4y^2 = 1$

In Problems 39–46, find the following for each function:

(a) $f(0)$ (b) $f(1)$ (c) $f(-1)$ (d) $f(-x)$ (e) $-f(x)$ (f) $f(x + 1)$ (g) $f(2x)$ (h) $f(x + h)$

39. $f(x) = 3x^2 + 2x - 4$

40. $f(x) = -2x^2 + x - 1$

41. $f(x) = \dfrac{x}{x^2 + 1}$

42. $f(x) = \dfrac{x^2 - 1}{x + 4}$

43. $f(x) = |x| + 4$

44. $f(x) = \sqrt{x^2 + x}$

45. $f(x) = \dfrac{2x + 1}{3x - 5}$

46. $f(x) = 1 - \dfrac{1}{(x + 2)^2}$

In Problems 47–60, find the domain of each function.

47. $f(x) = -5x + 4$

48. $f(x) = x^2 + 2$

49. $f(x) = \dfrac{x}{x^2 + 1}$

50. $f(x) = \dfrac{x^2}{x^2 + 1}$

51. $g(x) = \dfrac{x}{x^2 - 16}$

52. $h(x) = \dfrac{2x}{x^2 - 4}$

53. $F(x) = \dfrac{x - 2}{x^3 + x}$

54. $G(x) = \dfrac{x + 4}{x^3 - 4x}$

55. $h(x) = \sqrt{3x - 12}$

56. $G(x) = \sqrt{1 - x}$

57. $f(x) = \dfrac{4}{\sqrt{x - 9}}$

58. $f(x) = \dfrac{x}{\sqrt{x - 4}}$

59. $p(x) = \sqrt{\dfrac{2}{x - 1}}$

60. $q(x) = \sqrt{-x - 2}$

In Problems 61–70, for the given functions f and g, find the following. For parts (a)–(d), also find the domain.

(a) $(f + g)(x)$ (b) $(f - g)(x)$ (c) $(f \cdot g)(x)$ (d) $\left(\dfrac{f}{g}\right)(x)$

(e) $(f + g)(3)$ (f) $(f - g)(4)$ (g) $(f \cdot g)(2)$ (h) $\left(\dfrac{f}{g}\right)(1)$

61. $f(x) = 3x + 4; \quad g(x) = 2x - 3$

62. $f(x) = 2x + 1; \quad g(x) = 3x - 2$

63. $f(x) = x - 1; \quad g(x) = 2x^2$

64. $f(x) = 2x^2 + 3; \quad g(x) = 4x^3 + 1$

65. $f(x) = \sqrt{x}; \quad g(x) = 3x - 5$

66. $f(x) = |x|; \quad g(x) = x$

67. $f(x) = 1 + \dfrac{1}{x}; \quad g(x) = \dfrac{1}{x}$

68. $f(x) = \sqrt{x - 1}; \quad g(x) = \sqrt{4 - x}$

69. $f(x) = \dfrac{2x + 3}{3x - 2}; \quad g(x) = \dfrac{4x}{3x - 2}$

70. $f(x) = \sqrt{x + 1}; \quad g(x) = \dfrac{2}{x}$

71. Given $f(x) = 3x + 1$ and $(f + g)(x) = 6 - \dfrac{1}{2}x$, find the function g.

72. Given $f(x) = \dfrac{1}{x}$ and $\left(\dfrac{f}{g}\right)(x) = \dfrac{x + 1}{x^2 - x}$, find the function g.

⚠ In Problems 73–80, find the difference quotient of f; that is, find $\dfrac{f(x+h)-f(x)}{h}$, $h \neq 0$, for each function. Be sure to simplify.

73. $f(x) = 4x + 3$

74. $f(x) = -3x + 1$

75. $f(x) = x^2 - x + 4$

76. $f(x) = x^2 + 5x - 1$

77. $f(x) = 3x^2 - 2x + 6$

78. $f(x) = 4x^2 + 5x - 7$

79. $f(x) = x^3 - 2$

80. $f(x) = \dfrac{1}{x+3}$

Applications and Extensions

81. If $f(x) = 2x^3 + Ax^2 + 4x - 5$ and $f(2) = 5$, what is the value of A?

82. If $f(x) = 3x^2 - Bx + 4$ and $f(-1) = 12$, what is the value of B?

83. If $f(x) = \dfrac{3x + 8}{2x - A}$ and $f(0) = 2$, what is the value of A?

84. If $f(x) = \dfrac{2x - B}{3x + 4}$ and $f(2) = \dfrac{1}{2}$, what is the value of B?

85. If $f(x) = \dfrac{2x - A}{x - 3}$ and $f(4) = 0$, what is the value of A? Where is f not defined?

86. If $f(x) = \dfrac{x - B}{x - A}$, $f(2) = 0$ and $f(1)$ is undefined, what are the values of A and B?

87. Geometry Express the area A of a rectangle as a function of the length x if the length of the rectangle is twice its width.

88. Geometry Express the area A of an isosceles right triangle as a function of the length x of one of the two equal sides.

89. Constructing Functions Express the gross salary G of a person who earns \$10 per hour as a function of the number x of hours worked.

90. Constructing Functions Tiffany, a commissioned salesperson, earns \$100 base pay plus \$10 per item sold. Express her gross salary G as a function of the number x of items sold.

91. Population as a Function of Age The function

$$P(a) = 0.015a^2 - 4.962a + 290.580$$

represents the population P (in millions) of Americans in 2005 that are a years of age or older.
Source: U.S. Census Bureau

(a) Identify the dependent and independent variable.

(b) Evaluate $P(20)$. Provide a verbal explanation of the meaning of $P(20)$.

(c) Evaluate $P(0)$. Provide a verbal explanation of the meaning of $P(0)$.

92. Number of Rooms The function

$$N(r) = -1.44r^2 + 14.52r - 14.96$$

represents the number N of housing units (in millions) in 2005 that have r rooms, where r is an integer and $2 \leq r \leq 9$.
Source: U.S. Census Bureau

(a) Identify the dependent and independent variable.

(b) Evaluate $N(3)$. Provide a verbal explanation of the meaning of $N(3)$.

93. Effect of Gravity on Earth If a rock falls from a height of 20 meters on Earth, the height H (in meters) after x seconds is approximately

$$H(x) = 20 - 4.9x^2$$

(a) What is the height of the rock when $x = 1$ second? $x = 1.1$ seconds? $x = 1.2$ seconds? $x = 1.3$ seconds?

(b) When is the height of the rock 15 meters? When is it 10 meters? When is it 5 meters?

(c) When does the rock strike the ground?

94. Effect of Gravity on Jupiter If a rock falls from a height of 20 meters on the planet Jupiter, its height H (in meters) after x seconds is approximately

$$H(x) = 20 - 13x^2$$

(a) What is the height of the rock when $x = 1$ second? $x = 1.1$ seconds? $x = 1.2$ seconds?

(b) When is the height of the rock 15 meters? When is it 10 meters? When is it 5 meters?

(c) When does the rock strike the ground?

95. Cost of Trans-Atlantic Travel A Boeing 747 crosses the Atlantic Ocean (3000 miles) with an airspeed of 500 miles per hour. The cost C (in dollars) per passenger is given by

$$C(x) = 100 + \dfrac{x}{10} + \dfrac{36,000}{x}$$

where x is the ground speed (airspeed ± wind).

(a) What is the cost per passenger for quiescent (no wind) conditions?

(b) What is the cost per passenger with a head wind of 50 miles per hour?

(c) What is the cost per passenger with a tail wind of 100 miles per hour?

(d) What is the cost per passenger with a head wind of 100 miles per hour?

96. Cross-sectional Area The cross-sectional area of a beam cut from a log with radius 1 foot is given by the function $A(x) = 4x\sqrt{1 - x^2}$, where x represents the length, in feet, of half the base of the beam. See the figure. Determine the cross-sectional area of the beam if the length of half the base of the beam is as follows:

(a) One-third of a foot

(b) One-half of a foot

(c) Two-thirds of a foot

$A(x) = 4x\sqrt{1 - x^2}$

97. Economics The **participation rate** is the number of people in the labor force divided by the civilian population (excludes military). Let $L(x)$ represent the size of the labor force in year x and $P(x)$ represent the civilian population in year x. Determine a function that represents the participation rate R as a function of x.

98. Crimes Suppose that $V(x)$ represents the number of violent crimes committed in year x and $P(x)$ represents the number of property crimes committed in year x. Determine a function T that represents the combined total of violent crimes and property crimes in year x.

99. Health Care Suppose that $P(x)$ represents the percentage of income spent on health care in year x and $I(x)$ represents income in year x. Determine a function H that represents total health care expenditures in year x.

100. Income Tax Suppose that $I(x)$ represents the income of an individual in year x before taxes and $T(x)$ represents the individual's tax bill in year x. Determine a function N that represents the individual's net income (income after taxes) in year x.

101. Profit Function Suppose that the revenue R, in dollars, from selling x cell phones, in hundreds, is $R(x) = -1.2x^2 + 220x$. The cost C, in dollars, of selling x cell phones is $C(x) = 0.05x^3 - 2x^2 + 65x + 500$.

(a) Find the profit function, $P(x) = R(x) - C(x)$.

(b) Find the profit if $x = 15$ hundred cell phones are sold.

(c) Interpret $P(15)$.

102. Profit Function Suppose that the revenue R, in dollars, from selling x clocks is $R(x) = 30x$. The cost C, in dollars, of selling x clocks is $C(x) = 0.1x^2 + 7x + 400$.

(a) Find the profit function, $P(x) = R(x) - C(x)$.

(b) Find the profit if $x = 30$ clocks are sold.

(c) Interpret $P(30)$.

103. Some functions f have the property that $f(a + b) = f(a) + f(b)$ for all real numbers a and b. Which of the following functions have this property?

(a) $h(x) = 2x$ (b) $g(x) = x^2$

(c) $F(x) = 5x - 2$ (d) $G(x) = \dfrac{1}{x}$

Discussion and Writing

104. Are the functions $f(x) = x - 1$ and $g(x) = \dfrac{x^2 - 1}{x + 1}$ the same? Explain.

105. Investigate when, historically, the use of the function notation $y = f(x)$ first appeared.

'Are You Prepared?' Answers

1. $(-1, 3)$ **2.** 21.5 **3.** $\{x \mid x \neq -4\}$ **4.** $\{x \mid x < -1\}$

3.2 The Graph of a Function

PREPARING FOR THIS SECTION *Before getting started, review the following:*

- Graphs of Equations (Section 2.2, pp. 163–165)
- Intercepts (Section 2.2, pp. 165–167)

Now Work the 'Are You Prepared?' problems on page 226.

OBJECTIVES 1 Identify the Graph of a Function (p. 223)
 2 Obtain Information from or about the Graph of a Function (p. 224)

In applications, a graph often demonstrates more clearly the relationship between two variables than, say, an equation or table would. For example, Table 1 shows the average price of gasoline in California, adjusted for inflation (based on 2005 dollars) for the years 1978–2005. If we plot these data and then connect the points, we obtain Figure 13.

Table 1

Year	Price	Year	Price	Year	Price	Year	Price
1978	1.5651	1985	1.7540	1992	1.3969	1999	1.5270
1979	1.9829	1986	1.3459	1993	1.5189	2000	1.8249
1980	2.4929	1987	1.3274	1994	1.4709	2001	1.7536
1981	2.4977	1988	1.3111	1995	1.4669	2002	1.5955
1982	2.1795	1989	1.3589	1996	1.5397	2003	1.8950
1983	1.8782	1990	1.4656	1997	1.5329	2004	2.1521
1984	1.8310	1991	1.4973	1998	1.3246	2005	2.4730

Source: Statistical Abstract of the United States.

Figure 13

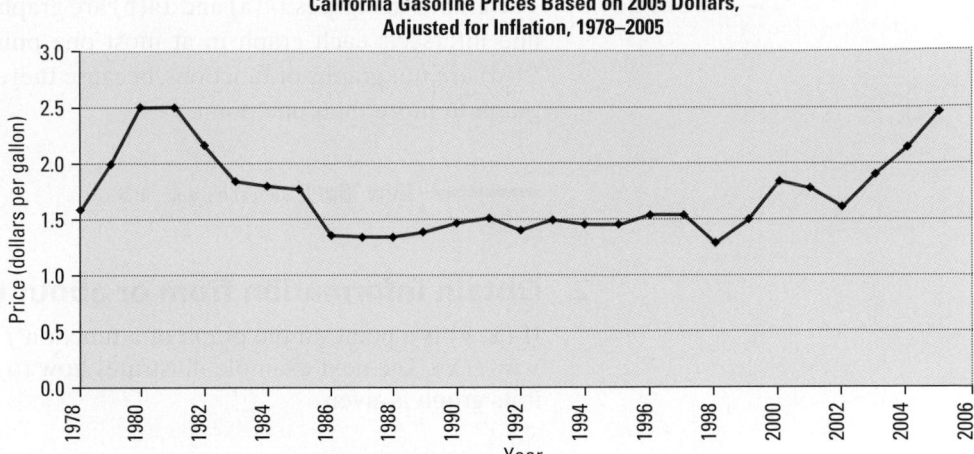

California Gasoline Prices Based on 2005 Dollars,
Adjusted for Inflation, 1978–2005

SOURCE: Statistical Abstact of the United States.

COMMENT When we select a viewing window to graph a function on a calculator, the values of Xmin, Xmax give the domain that we wish to view, while Ymin, Ymax give the range that we wish to view. These settings usually do not represent the actual domain and range of the function. ∎

We can see from the graph that the price of gasoline (adjusted for inflation) rose rapidly from 1978 to 1981 and was falling from 2000 to 2002. The graph also shows that the lowest price occurred in 1998. To learn information such as this from an equation requires that some calculations be made.

Look again at Figure 13. The graph shows that for each date on the horizontal axis there is only one price on the vertical axis. The graph represents a function, although the exact rule for getting from date to price is not given.

When a function is defined by an equation in x and y, the **graph of the function** is the graph of the equation, that is, the set of points (x, y) in the xy-plane that satisfies the equation.

1 Identify the Graph of a Function

Not every collection of points in the xy-plane represents the graph of a function. Remember, for a function, each number x in the domain has exactly one image y in the range. This means that the graph of a function cannot contain two points with the same x-coordinate and different y-coordinates. Therefore, the graph of a function must satisfy the following **vertical-line test.**

THEOREM

Vertical-line Test

A set of points in the xy-plane is the graph of a function if and only if every vertical line intersects the graph in at most one point.

In other words, if any vertical line intersects a graph at more than one point, the graph is not the graph of a function.

EXAMPLE 1

Identifying the Graph of a Function

Which of the graphs in Figure 14 are graphs of functions?

Figure 14

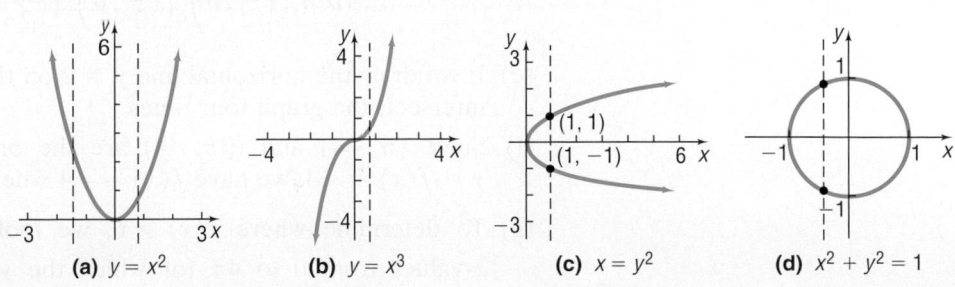

(a) $y = x^2$ (b) $y = x^3$ (c) $x = y^2$ (d) $x^2 + y^2 = 1$

Solution The graphs in Figures 14(a) and 14(b) are graphs of functions, because every vertical line intersects each graph in at most one point. The graphs in Figures 14(c) and 14(d) are not graphs of functions, because there is a vertical line that intersects each graph in more than one point.

→ **Now Work** PROBLEM 15

2 Obtain Information from or about the Graph of a Function

If (x, y) is a point on the graph of a function f, then y is the value of f at x; that is, $y = f(x)$. The next example illustrates how to obtain information about a function if its graph is given.

EXAMPLE 2 | **Obtaining Information from the Graph of a Function**

Let f be the function whose graph is given in Figure 15. (The graph of f might represent the distance that the bob of a pendulum is from its *at-rest* position. Negative values of y mean that the pendulum is to the left of the at-rest position, and positive values of y mean that the pendulum is to the right of the at-rest position.)

Figure 15

(a) What are $f(0)$, $f\left(\dfrac{3\pi}{2}\right)$, and $f(3\pi)$?

(b) What is the domain of f?

(c) What is the range of f?

(d) List the intercepts. (Recall that these are the points, if any, where the graph crosses or touches the coordinate axes.)

(e) How often does the line $y = 2$ intersect the graph?

(f) For what values of x does $f(x) = -4$?

(g) For what values of x is $f(x) > 0$?

Solution

(a) Since $(0, 4)$ is on the graph of f, the y-coordinate 4 is the value of f at the x-coordinate 0; that is, $f(0) = 4$. In a similar way, we find that when $x = \dfrac{3\pi}{2}$, then $y = 0$, so $f\left(\dfrac{3\pi}{2}\right) = 0$. When $x = 3\pi$, then $y = -4$, so $f(3\pi) = -4$.

(b) To determine the domain of f, we notice that the points on the graph of f have x-coordinates between 0 and 4π, inclusive; and for each number x between 0 and 4π, there is a point $(x, f(x))$ on the graph. The domain of f is $\{x | 0 \le x \le 4\pi\}$ or the interval $[0, 4\pi]$.

(c) The points on the graph all have y-coordinates between -4 and 4, inclusive; and for each such number y, there is at least one number x in the domain. The range of f is $\{y | -4 \le y \le 4\}$ or the interval $[-4, 4]$.

(d) The intercepts are

$$(0, 4), \left(\frac{\pi}{2}, 0\right), \left(\frac{3\pi}{2}, 0\right), \left(\frac{5\pi}{2}, 0\right), \quad \text{and} \quad \left(\frac{7\pi}{2}, 0\right)$$

(e) If we draw the horizontal line $y = 2$ on the graph in Figure 15, we find that it intersects the graph four times.

(f) Since $(\pi, -4)$ and $(3\pi, -4)$ are the only points on the graph for which $y = f(x) = -4$, we have $f(x) = -4$ when $x = \pi$ and $x = 3\pi$.

(g) To determine where $f(x) > 0$, we look at Figure 15 and determine the x-values from 0 to 4π for which the y-coordinate is positive. This occurs

on $\left[0, \dfrac{\pi}{2}\right) \cup \left(\dfrac{3\pi}{2}, \dfrac{5\pi}{2}\right) \cup \left(\dfrac{7\pi}{2}, 4\pi\right]$. Using inequality notation, $f(x) > 0$

for $0 \le x < \dfrac{\pi}{2}$ or $\dfrac{3\pi}{2} < x < \dfrac{5\pi}{2}$ or $\dfrac{7\pi}{2} < x \le 4\pi$.

When the graph of a function is given, its domain may be viewed as the shadow created by the graph on the x-axis by vertical beams of light. Its range can be viewed as the shadow created by the graph on the y-axis by horizontal beams of light. Try this technique with the graph given in Figure 15.

Now Work PROBLEMS 9 AND 13

EXAMPLE 3

Obtaining Information about the Graph of a Function

Consider the function: $f(x) = \dfrac{x + 1}{x + 2}$

(a) Is the point $\left(1, \dfrac{1}{2}\right)$ on the graph of f?

(b) If $x = 2$, what is $f(x)$? What point is on the graph of f?

(c) If $f(x) = 2$, what is x? What point is on the graph of f?

(d) What are the x-intercepts of the graph of f (if any)? What point(s) are on the graph of f?

Solution

(a) When $x = 1$, then

$$f(x) = \dfrac{x + 1}{x + 2}$$

$$f(1) = \dfrac{1 + 1}{1 + 2} = \dfrac{2}{3}$$

The point $\left(1, \dfrac{2}{3}\right)$ is on the graph of f; the point $\left(1, \dfrac{1}{2}\right)$ is not.

(b) If $x = 2$, then

$$f(x) = \dfrac{x + 1}{x + 2}$$

$$f(2) = \dfrac{2 + 1}{2 + 2} = \dfrac{3}{4}$$

The point $\left(2, \dfrac{3}{4}\right)$ is on the graph of f.

(c) If $f(x) = 2$, then

$$f(x) = 2$$

$$\dfrac{x + 1}{x + 2} = 2$$

$x + 1 = 2(x + 2)$ Multiply both sides by x + 2.

$x + 1 = 2x + 4$ Remove parentheses.

$x = -3$ Solve for x.

If $f(x) = 2$, then $x = -3$. The point $(-3, 2)$ is on the graph of f.

(d) The x-intercepts of the graph of f are the real solutions of the equation $f(x) = 0$. The only real solution of the equation $f(x) = \dfrac{x + 1}{x + 2} = 0$, is $x = -1$, so -1 is the only x-intercept. Since $f(-1) = 0$, the point $(-1, 0)$ is on the graph of f.

Now Work PROBLEM 25

EXAMPLE 4	Average Cost Function

The average cost $\overline{C}$ of manufacturing x computers per day is given by the function

$$\overline{C}(x) = 0.56x^2 - 34.39x + 1212.57 + \frac{20,000}{x}$$

Determine the average cost of manufacturing:

(a) 30 computers in a day

(b) 40 computers in a day

(c) 50 computers in a day

 (d) Graph the function $\overline{C} = \overline{C}(x), 0 < x \leq 80$.

(e) Create a TABLE with TblStart = 1 and ΔTbl = 1. Which value of x minimizes the average cost?

Solution (a) The average cost of manufacturing $x = 30$ computers is

$$\overline{C}(30) = 0.56(30)^2 - 34.39(30) + 1212.57 + \frac{20,000}{30} = \$1351.54$$

(b) The average cost of manufacturing $x = 40$ computers is

$$\overline{C}(40) = 0.56(40)^2 - 34.39(40) + 1212.57 + \frac{20,000}{40} = \$1232.97$$

(c) The average cost of manufacturing $x = 50$ computers is

$$\overline{C}(50) = 0.56(50)^2 - 34.39(50) + 1212.57 + \frac{20,000}{50} = \$1293.07$$

(d) See Figure 16 for the graph of $\overline{C} = \overline{C}(x)$.

(e) With the function $\overline{C} = \overline{C}(x)$ in Y_1, we create Table 2. We scroll down until we find a value of x for which Y_1 is smallest. Table 3 shows that manufacturing $x = 41$ computers minimizes the average cost at $1231.74 per computer.

Figure 16

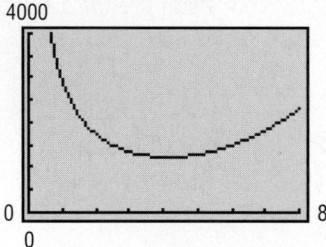

Table 2

X	Y1
1	21179
2	11146
3	7781.1
4	6084
5	5054.6
6	4359.7
7	3856.4

Y1=.56X²-34.39X...

Table 3

X	Y1
38	1240.7
39	1235.9
40	1233
41	1231.7
42	1232.2
43	1234.4
44	1238.1

Y1=1231.74487805

▬▬▬▶ **Now Work** PROBLEM 31

SUMMARY

Graph of a function The collection of points (x, y) that satisfies the equation $y = f(x)$.

Vertical Line Test A collection of points is the graph of a function provided that every vertical line intersects the graph in at most one point.

3.2 Assess Your Understanding

'Are You Prepared?' *Answers are given at the end of these exercises. If you get a wrong answer, read the pages listed in red.*

1. The intercepts of the equation $x^2 + 4y^2 = 16$ are _____. (pp. 165–167)

2. *True or False* The point $(-2, -6)$ is on the graph of the equation $x = 2y - 2$. (pp. 163–165)

Concepts and Vocabulary

3. A set of points in the xy-plane is the graph of a function if and only if every _____ line intersects the graph in at most one point.

4. If the point $(5, -3)$ is a point on the graph of f, then $f(___) = ___$.

5. Find a so that the point $(-1, 2)$ is on the graph of $f(x) = ax^2 + 4$.

6. *True or False* A function can have more than one y-intercept.

7. *True or False* The graph of a function $y = f(x)$ always crosses the y-axis.

8. *True or False* The y-intercept of the graph of the function $y = f(x)$, whose domain is all real numbers, is $f(0)$.

Skill Building

9. Use the given graph of the function f to answer parts (a)–(n).

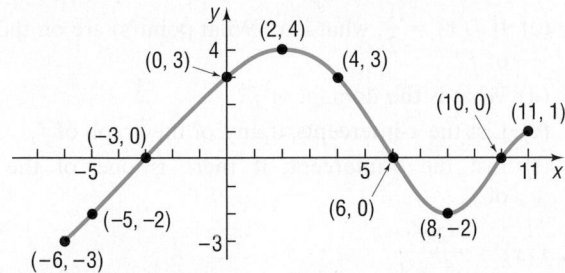

(a) Find $f(0)$ and $f(-6)$.
(b) Find $f(6)$ and $f(11)$.
(c) Is $f(3)$ positive or negative?
(d) Is $f(-4)$ positive or negative?
(e) For what values of x is $f(x) = 0$?
(f) For what values of x is $f(x) > 0$?
(g) What is the domain of f?
(h) What is the range of f?
(i) What are the x-intercepts?
(j) What is the y-intercept?

(k) How often does the line $y = \dfrac{1}{2}$ intersect the graph?

(l) How often does the line $x = 5$ intersect the graph?
(m) For what values of x does $f(x) = 3$?
(n) For what values of x does $f(x) = -2$?

10. Use the given graph of the function f to answer parts (a)–(n).

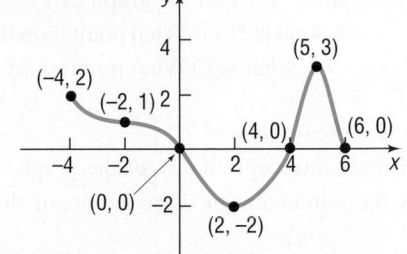

(a) Find $f(0)$ and $f(6)$.
(b) Find $f(2)$ and $f(-2)$.
(c) Is $f(3)$ positive or negative?
(d) Is $f(-1)$ positive or negative?
(e) For what values of x is $f(x) = 0$?
(f) For what values of x is $f(x) < 0$?
(g) What is the domain of f?
(h) What is the range of f?
(i) What are the x-intercepts?
(j) What is the y-intercept?
(k) How often does the line $y = -1$ intersect the graph?
(l) How often does the line $x = 1$ intersect the graph?
(m) For what value of x does $f(x) = 3$?
(n) For what value of x does $f(x) = -2$?

In Problems 11–22, determine whether the graph is that of a function by using the vertical-line test. If it is, use the graph to find:

(a) *The domain and range*
(b) *The intercepts, if any*
(c) *Any symmetry with respect to the x-axis, the y-axis, or the origin*

11.

12.

13.

14.

15.

16.

17.

18.

19. **20.** **21.** **22.**

In Problems 23–28, answer the questions about the given function.

23. $f(x) = 2x^2 - x - 1$

(a) Is the point $(-1, 2)$ on the graph of f?

(b) If $x = -2$, what is $f(x)$? What point is on the graph of f?

(c) If $f(x) = -1$, what is x? What point(s) are on the graph of f?

(d) What is the domain of f?

(e) List the x-intercepts, if any, of the graph of f.

(f) List the y-intercept, if there is one, of the graph of f.

24. $f(x) = -3x^2 + 5x$

(a) Is the point $(-1, 2)$ on the graph of f?

(b) If $x = -2$, what is $f(x)$? What point is on the graph of f?

(c) If $f(x) = -2$, what is x? What point(s) are on the graph of f?

(d) What is the domain of f?

(e) List the x-intercepts, if any, of the graph of f.

(f) List the y-intercept, if there is one, of the graph of f.

25. $f(x) = \dfrac{x + 2}{x - 6}$

(a) Is the point $(3, 14)$ on the graph of f?

(b) If $x = 4$, what is $f(x)$? What point is on the graph of f?

(c) If $f(x) = 2$, what is x? What point(s) are on the graph of f?

(d) What is the domain of f?

(e) List the x-intercepts, if any, of the graph of f.

(f) List the y-intercept, if there is one, of the graph of f.

26. $f(x) = \dfrac{x^2 + 2}{x + 4}$

(a) Is the point $\left(1, \dfrac{3}{5}\right)$ on the graph of f?

(b) If $x = 0$, what is $f(x)$? What point is on the graph of f?

(c) If $f(x) = \dfrac{1}{2}$, what is x? What point(s) are on the graph of f?

(d) What is the domain of f?

(e) List the x-intercepts, if any, of the graph of f.

(f) List the y-intercept, if there is one, of the graph of f.

27. $f(x) = \dfrac{2x^2}{x^4 + 1}$

(a) Is the point $(-1, 1)$ on the graph of f?

(b) If $x = 2$, what is $f(x)$? What point is on the graph of f?

(c) If $f(x) = 1$, what is x? What point(s) are on the graph of f?

(d) What is the domain of f?

(e) List the x-intercepts, if any, of the graph of f.

(f) List the y-intercept, if there is one, of the graph of f.

28. $f(x) = \dfrac{2x}{x - 2}$

(a) Is the point $\left(\dfrac{1}{2}, -\dfrac{2}{3}\right)$ on the graph of f?

(b) If $x = 4$, what is $f(x)$? What point is on the graph of f?

(c) If $f(x) = 1$, what is x? What point(s) are on the graph of f?

(d) What is the domain of f?

(e) List the x-intercepts, if any, of the graph of f.

(f) List the y-intercept, if there is one, of the graph of f.

Applications and Extensions

29. Free-throw Shots According to physicist Peter Brancazio, the key to a successful foul shot in basketball lies in the arc of the shot. Brancazio determined the optimal angle of the arc from the free-throw line to be 45 degrees. The arc also depends on the velocity with which the ball is shot. If a player shoots a foul shot, releasing the ball at a 45-degree angle from a position 6 feet above the floor, then the path of the ball can be modeled by the quadratic function

$$h(x) = -\frac{44x^2}{v^2} + x + 6$$

where h is the height of the ball above the floor, x is the forward distance of the ball in front of the foul line, and v is the initial velocity with which the ball is shot in feet per second.

Suppose a player shoots a ball with an initial velocity of 28 feet per second.

(a) Determine the height of the ball after it has traveled 8 feet in front of the foul line.

(b) Determine the height of the ball after it has traveled 12 feet in front of the foul line.

(c) Find additional points and graph the path of the basketball.

(d) The center of the hoop is 10 feet above the floor and 15 feet in front of the foul line. Will the ball go through the hoop? Why or why not? If not, with what initial velocity must the ball be shot in order for the ball to go through the hoop?

Source: The Physics of Foul Shots, Discover, *Vol. 21, No. 10, October 2000*

30. **Granny Shots** The last player in the NBA to use an underhand foul shot (a "granny" shot) was Hall of Fame forward Rick Barry who retired in 1980. Barry believes that current NBA players could increase their free-throw percentage if they were to use an underhand shot. Since underhand shots are released from a lower position, the angle of the shot must be increased. If a player shoots an underhand foul shot, releasing the ball at a 70-degree angle from a position 3.5 feet above the floor, then the path of the ball can be modeled by the quadratic function $h(x) = -\dfrac{136x^2}{v^2} + 2.7x + 3.5$, where h is the height of the ball above the floor, x is the forward distance of the ball in front of the foul line, and v is the initial velocity with which the ball is shot in feet per second.
 - (a) The center of the hoop is 10 feet above the floor and 15 feet in front of the foul line. Determine the initial velocity with which the ball must be shot in order for the ball to go through the hoop.
 - (b) Write the function for the path of the ball using the velocity found in part (a).
 - (c) Determine the height of the ball after it has traveled 9 feet in front of the foul line.
 - (d) Find additional points and graph the path of the basketball.

 Source: The Physics of Foul Shots, Discover, Vol. 21, No. 10, October 2000

31. **Motion of a Golf Ball** A golf ball is hit with an initial velocity of 130 feet per second at an inclination of 45° to the horizontal. In physics, it is established that the height h of the golf ball is given by the function
$$h(x) = \frac{-32x^2}{130^2} + x$$
where x is the horizontal distance that the golf ball has traveled.

- (a) Determine the height of the golf ball after it has traveled 100 feet.
- (b) What is the height after it has traveled 300 feet?
- (c) What is the height after it has traveled 500 feet?
- (d) How far was the golf ball hit?
- (e) Use a graphing utility to graph the function $h = h(x)$.
- (f) Use a graphing utility to determine the distance that the ball has traveled when the height of the ball is 90 feet.
- (g) Create a TABLE with TblStart = 0 and ΔTbl = 25. To the nearest 25 feet, how far does the ball travel before it reaches a maximum height? What is the maximum height?
- (h) Adjust the value of ΔTbl until you determine the distance, to within 1 foot, that the ball travels before it reaches a maximum height.

32. **Cross-sectional Area** The cross-sectional area of a beam cut from a log with radius 1 foot is given by the function $A(x) = 4x\sqrt{1 - x^2}$, where x represents the length, in feet, of half the base of the beam. See the figure.

$$A(x) = 4x\sqrt{1 - x^2}$$

- (a) Find the domain of A.
- (b) Use a graphing utility to graph the function $A = A(x)$.
- (c) Create a TABLE with TblStart = 0 and ΔTbl = 0.1 for $0 \le x \le 1$. Which value of x maximizes the cross-sectional area? What should be the length of the base of the beam to maximize the cross-sectional area?

33. **Cost of Trans-Atlantic Travel** A Boeing 747 crosses the Atlantic Ocean (3000 miles) with an airspeed of 500 miles per hour. The cost C (in dollars) per passenger is given by
$$C(x) = 100 + \frac{x}{10} + \frac{36{,}000}{x}$$
where x is the ground speed (airspeed ± wind).
- (a) Use a graphing utility to graph the function $C = C(x)$.
- (b) Create a TABLE with TblStart = 0 and ΔTbl = 50.
- (c) To the nearest 50 miles per hour, what ground speed minimizes the cost per passenger?

34. **Effect of Elevation on Weight** If an object weighs m pounds at sea level, then its weight W (in pounds) at a height of h miles above sea level is given approximately by
$$W(h) = m\left(\frac{4000}{4000 + h}\right)^2$$
- (a) If Amy weighs 120 pounds at sea level, how much will she weigh on Pike's Peak, which is 14,110 feet above sea level?
- (b) Use a graphing utility to graph the function $W = W(h)$. Use $m = 120$ pounds.
- (c) Create a Table with TblStart = 0 and ΔTbl = 0.5 to see how the weight W varies as h changes from 0 to 5 miles.
- (d) At what height will Amy weigh 119.95 pounds?
- (e) Does your answer to part (d) seem reasonable? Explain.

35. The graph of two functions, f and g, is illustrated. Use the graph to answer parts (a)–(f).

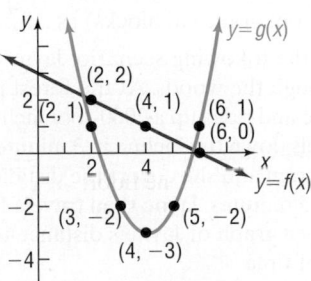

- (a) $(f + g)(2)$
- (b) $(f + g)(4)$
- (c) $(f - g)(6)$
- (d) $(g - f)(6)$
- (e) $(f \cdot g)(2)$
- (f) $\left(\dfrac{f}{g}\right)(4)$

Discussion and Writing

36. Describe how you would proceed to find the domain and range of a function if you were given its graph. How would your strategy change if you were given the equation defining the function instead of its graph?

37. How many *x*-intercepts can the graph of a function have? How many *y*-intercepts can the graph of a function have?

38. Is a graph that consists of a single point the graph of a function? Can you write the equation of such a function?

39. Match each of the following functions with the graph that best describes the situation.
 (a) The cost of building a house as a function of its square footage
 (b) The height of an egg dropped from a 300-foot building as a function of time
 (c) The height of a human as a function of time
 (d) The demand for Big Macs as a function of price
 (e) The height of a child on a swing as a function of time

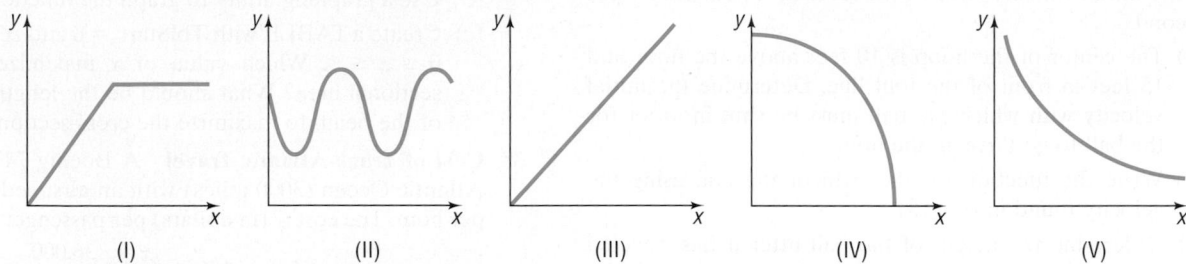

(I) (II) (III) (IV) (V)

40. Match each of the following functions with the graph that best describes the situation.
 (a) The temperature of a bowl of soup as a function of time
 (b) The number of hours of daylight per day over a 2-year period
 (c) The population of Florida as a function of time
 (d) The distance travelled by a car going at a constant velocity as a function of time
 (e) The height of a golf ball hit with a 7-iron as a function of time

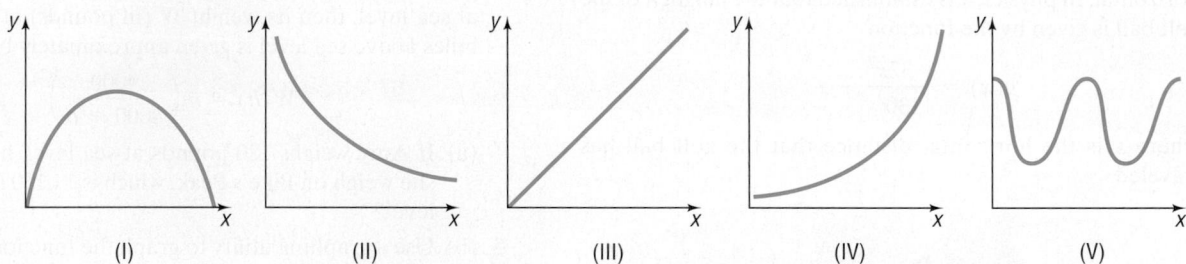

(I) (II) (III) (IV) (V)

41. Consider the following scenario: Barbara decides to take a walk. She leaves home, walks 2 blocks in 5 minutes at a constant speed, and realizes that she forgot to lock the door. So Barbara runs home in 1 minute. While at her doorstep, it takes her 1 minute to find her keys and lock the door. Barbara walks 5 blocks in 15 minutes and then decides to jog home. It takes her 7 minutes to get home. Draw a graph of Barbara's distance from home (in blocks) as a function of time.

42. Consider the following scenario: Jayne enjoys riding her bicycle through the woods. At the forest preserve, she gets on her bicycle and rides up a 2000-foot incline in 10 minutes. She then travels down the incline in 3 minutes. The next 5000 feet is level terrain and she covers the distance in 20 minutes. She rests for 15 minutes. Jayne then travels 10,000 feet in 30 minutes. Draw a graph of Jayne's distance traveled (in feet) as a function of time.

43. The following sketch represents the distance *d* (in miles) that Kevin is from home as a function of time *t* (in hours). Answer the questions based on the graph. In parts (a)–(g), how many hours elapsed and how far was Kevin from home during this time?

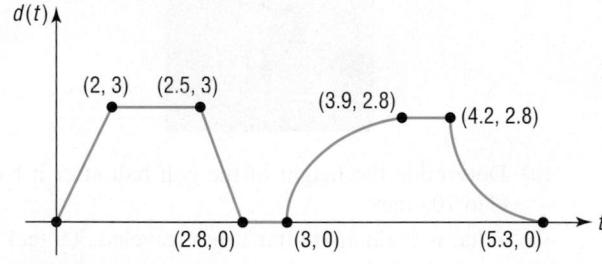

(a) From *t* = 0 to *t* = 2
(b) From *t* = 2 to *t* = 2.5
(c) From *t* = 2.5 to *t* = 2.8
(d) From *t* = 2.8 to *t* = 3
(e) From *t* = 3 to *t* = 3.9
(f) From *t* = 3.9 to *t* = 4.2
(g) From *t* = 4.2 to *t* = 5.3
(h) What is the farthest distance that Kevin is from home?
(i) How many times did Kevin return home?

44. The following sketch represents the speed v (in miles per hour) of Michael's car as a function of time t (in minutes).

(a) Over what interval of time is Michael traveling fastest?

(b) Over what interval(s) of time is Michael's speed zero?

(c) What is Michael's speed between 0 and 2 minutes?

(d) What is Michael's speed between 4.2 and 6 minutes?

(e) What is Michael's speed between 7 and 7.4 minutes?

(f) When is Michael's speed constant?

45. Draw the graph of a function whose domain is $\{x \mid -3 \le x \le 8, \ x \ne 5\}$ and whose range is $\{y \mid -1 \le y \le 2, \ y \ne 0\}$. What point(s) in the rectangle $-3 \le x \le 8, -1 \le y \le 2$ cannot be on the graph? Compare your graph with those of other students. What differences do you see?

46. Is there a function whose graph is symmetric with respect to the x-axis? Explain.

'Are You Prepared?' Answers

1. $(-4, 0), (4, 0), (0, -2), (0, 2)$ **2.** False

3.3 Properties of Functions

PREPARING FOR THIS SECTION *Before getting started, review the following:*

- Intervals (Section 1.5, pp. 125–126)
- Intercepts (Section 2.2, pp. 165–167)
- Slope of a Line (Section 2.3, pp. 174–176)

- Point–Slope Form of a Line (Section 2.3, pp. 176–179)
- Symmetry (Section 2.2, pp. 167–168)

Now Work the 'Are You Prepared?' problems on page 238.

OBJECTIVES **1** Determine Even and Odd Functions from a Graph (p. 231)

2 Identify Even and Odd Functions from the Equation (p. 232)

3 Use a Graph to Determine Where a Function Is Increasing, Decreasing, or Constant (p. 233)

4 Use a Graph to Locate Local Maxima and Local Minima (p. 234)

5 Use a Graphing Utility to Approximate Local Maxima and Local Minima and to Determine Where a Function Is Increasing or Decreasing (p. 235)

6 Find the Average Rate of Change of a Function (p. 236)

It is easiest to obtain the graph of a function $y = f(x)$ by knowing certain properties that the function has and the impact of these properties on the way that the graph will look.

1 Determine Even and Odd Functions from a Graph

The words *even* and *odd,* when applied to a function f, describe the symmetry that exists for the graph of the function.

A function f is even, if and only if, whenever the point (x, y) is on the graph of f then the point $(-x, y)$ is also on the graph. Using function notation, we define an even function as follows:

DEFINITION A function f is **even** if, for every number x in its domain, the number $-x$ is also in the domain and

$$f(-x) = f(x)$$

A function f is odd, if and only if, whenever the point (x, y) is on the graph of f then the point $(-x, -y)$ is also on the graph. Using function notation, we define an odd function as follows:

DEFINITION A function f is **odd** if, for every number x in its domain, the number $-x$ is also in the domain and

$$f(-x) = -f(x)$$

Refer to page 168, where the tests for symmetry are listed. The following results are then evident.

THEOREM A function is even if and only if its graph is symmetric with respect to the y-axis. A function is odd if and only if its graph is symmetric with respect to the origin.

EXAMPLE 1 **Determining Even and Odd Functions from the Graph**

Determine whether each graph given in Figure 17 is the graph of an even function, an odd function, or a function that is neither even nor odd.

Figure 17

(a) (b) (c)

Solution (a) The graph in Figure 17(a) is that of an even function, because the graph is symmetric with respect to the y-axis.

(b) The function whose graph is given in Figure 17(b) is neither even nor odd, because the graph is neither symmetric with respect to the y-axis nor symmetric with respect to the origin.

(c) The function whose graph is given in Figure 17(c) is odd, because its graph is symmetric with respect to the origin.

■

Now Work PROBLEMS 21(a), (b), AND (d)

2 Identify Even and Odd Functions from the Equation

In the next example, we use algebraic techniques to verify whether a given function is even, odd, or neither.

EXAMPLE 2 **Identifying Even and Odd Functions Algebraically**

Determine whether each of the following functions is even, odd, or neither. Then determine whether the graph is symmetric with respect to the y-axis, the origin, or neither.

(a) $f(x) = x^2 - 5$ (b) $g(x) = x^3 - 1$

(c) $h(x) = 5x^3 - x$ (d) $F(x) = |x|$

Solution (a) To determine whether f is even, odd, or neither, we replace x by $-x$ in $f(x) = x^2 - 5$. Then

$$f(-x) = (-x)^2 - 5 = x^2 - 5 = f(x)$$

Since $f(-x) = f(x)$, we conclude that f is an even function, and the graph is symmetric with respect to the y-axis.

(b) We replace x by $-x$ in $g(x) = x^3 - 1$. Then

$$g(-x) = (-x)^3 - 1 = -x^3 - 1$$

Since $g(-x) \neq g(x)$ and $g(-x) \neq -g(x) = -(x^3 - 1) = -x^3 + 1$, we conclude that g is neither even nor odd. The graph is not symmetric with respect to the y-axis nor is it symmetric with respect to the origin.

(c) We replace x by $-x$ in $h(x) = 5x^3 - x$. Then

$$h(-x) = 5(-x)^3 - (-x) = -5x^3 + x = -(5x^3 - x) = -h(x)$$

Since $h(-x) = -h(x)$, h is an odd function, and the graph of h is symmetric with respect to the origin.

(d) We replace x by $-x$ in $F(x) = |x|$. Then

$$F(-x) = |-x| = |-1| \cdot |x| = |x| = F(x)$$

Since $F(-x) = F(x)$, F is an even function, and the graph of F is symmetric with respect to the y-axis.

■

━━━ **Now Work** PROBLEM 33

3 Use a Graph to Determine Where a Function Is Increasing, Decreasing, or Constant

 Consider the graph given in Figure 18. If you look from left to right along the graph of the function, you will notice that parts of the graph are rising, parts are falling, and parts are horizontal. In such cases, the function is described as *increasing, decreasing,* or *constant,* respectively.

Figure 18

| EXAMPLE 3 | **Determining Where a Function Is Increasing, Decreasing, or Constant from Its Graph**

Where is the function in Figure 18 increasing? Where is it decreasing? Where is it constant?

Solution To answer the question of where a function is increasing, where it is decreasing, and where it is constant, we use strict inequalities involving the independent variable x, or we use open intervals* of x-coordinates. The graph in Figure 18 is rising (increasing) from the point $(-4, -2)$ to the point $(0, 4)$, so we conclude that it is increasing on the open interval $(-4, 0)$ or for $-4 < x < 0$. The graph is falling (decreasing) from the point $(-6, 0)$ to the point $(-4, -2)$ and from the point $(3, 4)$ to the point $(6, 1)$. We conclude that the graph is decreasing on the open intervals

WARNING We describe the behavior of a graph in terms of its x-values. Do not say the graph in Figure 18 is increasing from the point $(-4, -2)$ to $(0, 4)$. Rather, say it is increasing on the interval $(-4, 0)$.

■ * The open interval (a, b) consists of all real numbers x for which $a < x < b$.

$(-6, -4)$ and $(3, 6)$ or for $-6 < x < -4$ and $3 < x < 6$. The graph is constant on the open interval $(0, 3)$ or for $0 < x < 3$.

More precise definitions follow:

DEFINITIONS

A function f is **increasing** on an open interval I if, for any choice of x_1 and x_2 in I, with $x_1 < x_2$, we have $f(x_1) < f(x_2)$.

A function f is **decreasing** on an open interval I if, for any choice of x_1 and x_2 in I, with $x_1 < x_2$, we have $f(x_1) > f(x_2)$.

A function f is **constant** on an open interval I if, for all choices of x in I, the values $f(x)$ are equal.

Figure 19 illustrates the definitions. The graph of an increasing function goes up from left to right, the graph of a decreasing function goes down from left to right, and the graph of a constant function remains at a fixed height.

Figure 19

(a) For $x_1 < x_2$ in I,
$f(x_1) < f(x_2)$;
f is increasing on I

(b) For $x_1 < x_2$ in I,
$f(x_1) > f(x_2)$;
f is decreasing on I

(c) For all x in I, the values of
f are equal; f is constant on I

Now Work PROBLEMS **1 1 , 1 3 , 1 5 , AND 2 1 (c)**

4 Use a Graph to Locate Local Maxima and Local Minima

When the graph of a function is increasing to the left of $x = c$ and decreasing to the right of $x = c$, then at c the value of f is largest. This value is called a *local maximum* of f. See Figure 20(a).

When the graph of a function is decreasing to the left of $x = c$ and is increasing to the right of $x = c$, then at c the value of f is the smallest. This value is called a *local minimum* of f. See Figure 20(b).

Figure 20

increasing decreasing

The local maximum
is $f(c)$ and occurs
at $x = c$.

(a)

decreasing increasing

The local minimum
is $f(c)$ and occurs
at $x = c$.

(b)

DEFINITIONS

A function f has a **local maximum** at c if there is an open interval I containing c so that, for all x not equal to c in I, $f(x) < f(c)$. We call $f(c)$ a **local maximum of f**.

A function f has a **local minimum** at c if there is an open interval I containing c so that, for all x not equal to c in I, $f(x) > f(c)$. We call $f(c)$ a **local minimum of f**.

If f has a local maximum at c, then the value of f at c is greater than the values of f near c. If f has a local minimum at c, then the value of f at c is less than the values of f near c. The word *local* is used to suggest that it is only near c that the value $f(c)$ is largest or smallest.

EXAMPLE 4

Finding Local Maxima and Local Minima from the Graph of a Function and Determining Where the Function Is Increasing, Decreasing, or Constant

Figure 21

Figure 21 shows the graph of a function f.

(a) At what number(s), if any, does f have a local maximum?
(b) What are the local maxima?
(c) At what number(s), if any, does f have a local minimum?
(d) What are the local minima?
(e) List the intervals on which f is increasing. List the intervals on which f is decreasing.

Solution

The domain of f is the set of real numbers.

(a) f has a local maximum at 1, since for all x close to 1, $x \neq 1$, we have $f(x) < f(1)$.
(b) The local maximum is $f(1) = 2$.
(c) f has a local minimum at -1 and at 3.
(d) The local minima are $f(-1) = 1$ and $f(3) = 0$.
(e) The function whose graph is given in Figure 21 is increasing for all values of x between -1 and 1 and for all values of x greater than 3. That is, the function is increasing on the intervals $(-1, 1)$ and $(3, \infty)$ or for $-1 < x < 1$ and $x > 3$. The function is decreasing for all values of x less than -1 and for all values of x between 1 and 3. That is, the function is decreasing on the intervals $(-\infty, -1)$ and $(1, 3)$ or for $x < -1$ and $1 < x < 3$. ∎

WARNING The y-value is the local maximum or local minimum and it occurs at some x-value. For Figure 21, we say the local maximum is 2 and that the local maximum occurs at x = 1. ∎

 Now Work PROBLEMS **17** AND **19**

 5 Use a Graphing Utility to Approximate Local Maxima and Local Minima and to Determine Where a Function Is Increasing or Decreasing

To locate the exact value at which a function f has a local maximum or a local minimum usually requires calculus. However, a graphing utility may be used to approximate these values by using the MAXIMUM and MINIMUM features.

EXAMPLE 5

Using a Graphing Utility to Approximate Local Maxima and Minima and to Determine Where a Function Is Increasing or Decreasing

(a) Use a graphing utility to graph $f(x) = 6x^3 - 12x + 5$ for $-2 < x < 2$. Approximate where f has a local maximum and where f has a local minimum.
(b) Determine where f is increasing and where it is decreasing.

Solution

(a) Graphing utilities have a feature that finds the maximum or minimum point of a graph within a given interval. Graph the function f for $-2 < x < 2$. See Figure 22(a). Using MAXIMUM, we find that the local maximum is 11.53 and it occurs at $x = -0.82$, rounded to two decimal places. See Figure 22(b). Using MINIMUM, we find that the local minimum is -1.53 and it occurs at $x = 0.82$, rounded to two decimal places.

Figure 22

(a) (b)

(b) Looking at Figures 22(a) and (b), we see that the graph of f is increasing from $x = -2$ to $x = -0.82$ and from $x = 0.82$ to $x = 2$, so f is increasing on the intervals $(-2, -0.82)$ and $(0.82, 2)$ or for $-2 < x < -0.82$ and $0.82 < x < 2$. The graph is decreasing from $x = -0.82$ to $x = 0.82$, so f is decreasing on the interval $(-0.82, 0.82)$ or for $-0.82 < x < 0.82$.

◀───── **Now Work** PROBLEM 45

6 Find the Average Rate of Change of a Function

In Section 2.3, we said that the slope of a line could be interpreted as the average rate of change of the line. To find the average rate of change of a function between any two points on its graph, we calculate the slope of the line containing the two points.

DEFINITION

If a and b, $a \neq b$, are in the domain of a function $y = f(x)$, the **average rate of change of f** from a to b is defined as

$$\text{Average rate of change} = \frac{\Delta y}{\Delta x} = \frac{f(b) - f(a)}{b - a} \qquad a \neq b \qquad \textbf{(1)}$$

The symbol Δy in (1) is the "change in y" and Δx is the "change in x." The average rate of change of f is the change in y divided by the change in x.

EXAMPLE 6 **Finding the Average Rate of Change**

Find the average rate of change of $f(x) = 3x^2$:

(a) From 1 to 3 (b) From 1 to 5 (c) From 1 to 7

Solution (a) The average rate of change of $f(x) = 3x^2$ from 1 to 3 is

$$\frac{\Delta y}{\Delta x} = \frac{f(3) - f(1)}{3 - 1} = \frac{27 - 3}{3 - 1} = \frac{24}{2} = 12$$

(b) The average rate of change of $f(x) = 3x^2$ from 1 to 5 is

$$\frac{\Delta y}{\Delta x} = \frac{f(5) - f(1)}{5 - 1} = \frac{75 - 3}{5 - 1} = \frac{72}{4} = 18$$

Figure 23

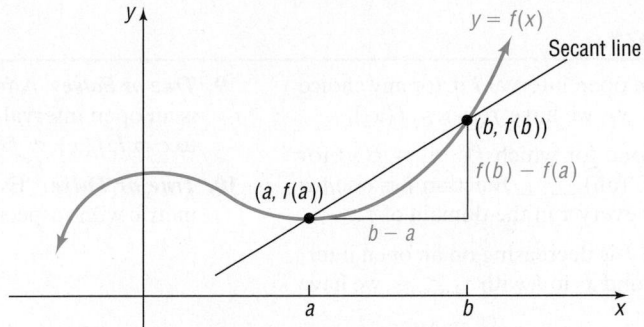

(c) The average rate of change of $f(x) = 3x^2$ from 1 to 7 is

$$\frac{\Delta y}{\Delta x} = \frac{f(7) - f(1)}{7 - 1}$$

$$= \frac{147 - 3}{7 - 1} = \frac{144}{6} = 24$$

See Figure 23 for a graph of $f(x) = 3x^2$. The function f is increasing for $x > 0$. The fact that the average rates of change are getting larger indicates that the graph is getting steeper; that is, it is increasing at an increasing rate.

Now Work PROBLEM 53

The Secant Line

The average rate of change of a function has an important geometric interpretation. Look at the graph of $y = f(x)$ in Figure 24. We have labeled two points on the graph: $(a, f(a))$ and $(b, f(b))$. The line containing these two points is called the **secant line**; its slope is

$$m_{\text{sec}} = \frac{f(b) - f(a)}{b - a}$$

Figure 24

THEOREM

Slope of the Secant Line

The average rate of change of a function from a to b equals the slope of the secant line containing the two points $(a, f(a))$ and $(b, f(b))$ on its graph.

EXAMPLE 7

Finding the Equation of a Secant Line

Suppose that $g(x) = 3x^2 - 2x + 3$.

(a) Find the average rate of change of g from -2 to 1.

(b) Find an equation of the secant line containing $(-2, g(-2))$ and $(1, g(1))$.

Solution

(a) The average rate of change of $g(x) = 3x^2 - 2x + 3$ from -2 to 1 is

$$\text{Average rate of change} = \frac{g(1) - g(-2)}{1 - (-2)}$$

$$= \frac{4 - 19}{3} \qquad \begin{aligned} g(1) &= 3(1^2) - 2(1) + 3 = 4 \\ g(-2) &= 3(-2)^2 - 2(-2) + 3 = 19 \end{aligned}$$

$$= -\frac{15}{3} = -5$$

(b) The slope of the secant line containing $(-2, g(-2)) = (-2, 19)$ and $(1, g(1)) = (1, 4)$ is $m_{sec} = -5$. We use the point–slope form to find an equation of the secant line.

$$y - y_1 = m_{sec}(x - x_1) \qquad \text{Point–slope form of the secant line}$$
$$y - 19 = -5(x - (-2)) \qquad x_1 = -2, \, y_1 = g(-2) = 19, \, m_{sec} = -5$$
$$y - 19 = -5x - 10 \qquad \text{Simplify.}$$
$$y = -5x + 9 \qquad \text{Slope–intercept form of the secant line}$$

━━━━▶ **Now Work** PROBLEM 59

3.3 Assess Your Understanding

'Are You Prepared?' *Answers are given at the end of these exercises. If you get a wrong answer, read the pages listed in red.*

1. The interval $(2, 5)$ can be written as the inequality _____. (pp. 125–126)

2. The slope of the line containing the points $(-2, 3)$ and $(3, 8)$ is _____. (pp. 174–176)

3. Test the equation $y = 5x^2 - 1$ for symmetry with respect to the x-axis, the y-axis, and the origin. (pp. 167–168)

4. Write the point–slope form of the line with slope 5 containing the point $(3, -2)$. (pp. 176–179)

5. The intercepts of the equation $y = x^2 - 9$ are _____. (pp. 165–167)

Concepts and Vocabulary

6. A function f is _____ on an open interval I if, for any choice of x_1 and x_2 in I, with $x_1 < x_2$, we have $f(x_1) < f(x_2)$.

7. A(n) _____ function f is one for which $f(-x) = f(x)$ for every x in the domain of f; (an) _____ function f is one for which $f(-x) = -f(x)$ for every x in the domain of f.

8. **True or False** A function f is decreasing on an open interval I if, for any choice of x_1 and x_2 in I, with $x_1 < x_2$, we have $f(x_1) > f(x_2)$.

9. **True or False** A function f has a local maximum at c if there is an open interval I containing c so that, for all x not equal to c in I, $f(x) < f(c)$.

10. **True or False** Even functions have graphs that are symmetric with respect to the origin.

Skill Building

In Problems 11–20, use the graph of the function f given.

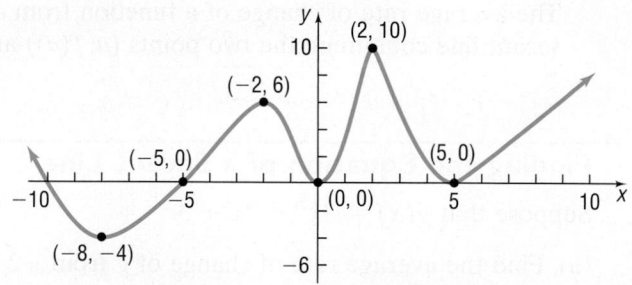

11. Is f increasing on the interval $(-8, -2)$?

12. Is f decreasing on the interval $(-8, -4)$?

13. Is f increasing on the interval $(2, 10)$?

14. Is f decreasing on the interval $(2, 5)$?

15. List the interval(s) on which f is increasing.

16. List the interval(s) on which f is decreasing.

17. Is there a local maximum at 2? If yes, what is it?

18. Is there a local maximum at 5? If yes, what is it?

19. List the numbers at which f has a local maximum. What are these local maxima?

20. List the numbers at which f has a local minimum. What are these local minima?

In Problems 21–28, the graph of a function is given. Use the graph to find:

(a) The intercepts, if any
(b) The domain and range
(c) The intervals on which it is increasing, decreasing, or constant
(d) Whether it is even, odd, or neither

21.

22.

23.

24.

25.

26.

27.

28.
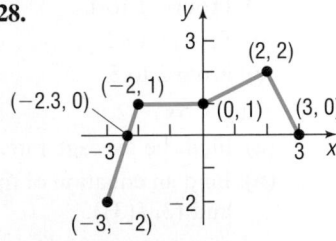

In Problems 29–32, the graph of a function f is given. Use the graph to find:

(a) The numbers, if any, at which f has a local maximum. What are these local maxima?
(b) The numbers, if any, at which f has a local minimum. What are these local minima?

29.

30.

31.

32.
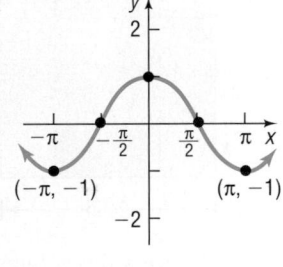

In Problems 33–44, determine algebraically whether each function is even, odd, or neither.

33. $f(x) = 4x^3$

34. $f(x) = 2x^4 - x^2$

35. $g(x) = -3x^2 - 5$

36. $h(x) = 3x^3 + 5$

37. $F(x) = \sqrt[3]{x}$

38. $G(x) = \sqrt{x}$

39. $f(x) = x + |x|$

40. $f(x) = \sqrt[3]{2x^2 + 1}$

41. $g(x) = \dfrac{1}{x^2}$

42. $h(x) = \dfrac{x}{x^2 - 1}$

43. $h(x) = \dfrac{-x^3}{3x^2 - 9}$

44. $F(x) = \dfrac{2x}{|x|}$

In Problems 45–52, use a graphing utility to graph each function over the indicated interval and approximate any local maxima and local minima. Determine where the function is increasing and where it is decreasing. Round answers to two decimal places.

45. $f(x) = x^3 - 3x + 2$ $(-2, 2)$

46. $f(x) = x^3 - 3x^2 + 5$ $(-1, 3)$

47. $f(x) = x^5 - x^3$ $(-2, 2)$

48. $f(x) = x^4 - x^2$ $(-2, 2)$

49. $f(x) = -0.2x^3 - 0.6x^2 + 4x - 6$ $(-6, 4)$

50. $f(x) = -0.4x^3 + 0.6x^2 + 3x - 2$ $(-4, 5)$

51. $f(x) = 0.25x^4 + 0.3x^3 - 0.9x^2 + 3$ $(-3, 2)$

52. $f(x) = -0.4x^4 - 0.5x^3 + 0.8x^2 - 2$ $(-3, 2)$

53. Find the average rate of change of $f(x) = -2x^2 + 4$
 (a) From 0 to 2
 (b) From 1 to 3
 (c) From 1 to 4

54. Find the average rate of change of $f(x) = -x^3 + 1$
 (a) From 0 to 2
 (b) From 1 to 3
 (c) From -1 to 1

55. Find the average rate of change of $g(x) = x^3 - 2x + 1$
 (a) From -3 to -2
 (b) From -1 to 1
 (c) From 1 to 3

56. Find the average rate of change of $h(x) = x^2 - 2x + 3$
 (a) From -1 to 1
 (b) From 0 to 2
 (c) From 2 to 5

57. $f(x) = 5x - 2$
 (a) Find the average rate of change from 1 to 3.
 (b) Find an equation of the secant line containing $(1, f(1))$ and $(3, f(3))$.

58. $f(x) = -4x + 1$
 (a) Find the average rate of change from 2 to 5.
 (b) Find an equation of the secant line containing $(2, f(2))$ and $(5, f(5))$.

59. $g(x) = x^2 - 2$
 (a) Find the average rate of change from -2 to 1.
 (b) Find an equation of the secant line containing $(-2, g(-2))$ and $(1, g(1))$.

60. $g(x) = x^2 + 1$
 (a) Find the average rate of change from -1 to 2.
 (b) Find an equation of the secant line containing $(-1, g(-1))$ and $(2, g(2))$.

61. $h(x) = x^2 - 2x$
 (a) Find the average rate of change from 2 to 4.
 (b) Find an equation of the secant line containing $(2, h(2))$ and $(4, h(4))$.

62. $h(x) = -2x^2 + x$
 (a) Find the average rate of change from 0 to 3.
 (b) Find an equation of the secant line containing $(0, h(0))$ and $(3, h(3))$.

Applications and Extensions

63. Constructing an Open Box An open box with a square base is to be made from a square piece of cardboard 24 inches on a side by cutting out a square from each corner and turning up the sides. See the figure.

 (a) Express the volume V of the box as a function of the length x of the side of the square cut from each corner.
 (b) What is the volume if a 3-inch square is cut out?
 (c) What is the volume if a 10-inch square is cut out?
 (d) Graph $V = V(x)$. For what value of x is V largest?

64. Constructing an Open Box An open box with a square base is required to have a volume of 10 cubic feet.
 (a) Express the amount A of material used to make such a box as a function of the length x of a side of the square base.
 (b) How much material is required for a base 1 foot by 1 foot?
 (c) How much material is required for a base 2 feet by 2 feet?
 (d) Use a graphing utility to graph $A = A(x)$. For what value of x is A smallest?

65. Maximum Height of a Ball The height s of a ball (in feet) thrown with an initial velocity of 80 feet per second from an initial height of 6 feet is given as a function of the time t (in seconds) by

$$s(t) = -16t^2 + 80t + 6$$

 (a) Use a graphing utility to graph $s = s(t)$.
 (b) Determine the time at which height is maximum.
 (c) What is the maximum height?

66. Maximum Height of a Ball On July 1, 2004, the Cassini probe became the first spacecraft to orbit the planet Saturn. Although Saturn is about 764 times the size of Earth, it has a very similar gravitational force. The height s of an object thrown upward from Saturn's surface with an initial velocity of 100 feet per second is given as a function of time t (in seconds) by $s(t) = -17.28t^2 + 100t$.
 (a) Use a graphing utility to graph $s = s(t)$.
 (b) Determine the time at which height is a maximum.
 (c) What is the maximum height?
 (d) The same object thrown from the surface of Earth would have a height given by $s(t) = -16t^2 + 100t$. Determine the maximum height of the object on Earth and compare this to your result from part (c).

67. Minimum Average Cost The average cost per hour in dollars of producing x riding lawn mowers is given by

$$\overline{C}(x) = 0.3x^2 + 21x - 251 + \frac{2500}{x}$$

 (a) Use a graphing utility to graph $\overline{C} = \overline{C}(x)$
 (b) Determine the number of riding lawn mowers to produce in order to minimize average cost.
 (c) What is the minimum average cost?

68. Medicine Concentration The concentration C of a medication in the bloodstream t hours after being administered is given by

$$C(t) = -0.002x^4 + 0.039t^3 - 0.285t^2 + 0.766t + 0.085$$

 (a) After how many hours will the concentration be highest?
 (b) A woman nursing a child must wait until the concentration is below 0.5 before she can feed him. After taking the medication, how long must she wait before feeding her child?

69. E-coli Growth A strain of E-coli Beu 397-recA441 is placed into a nutrient broth at 30° Celsius and allowed to grow. The data shown below are collected. The population is measured in grams and the time in hours. Since population P depends on time t and each input corresponds to exactly one output, we can say that population is a function of time; so $P(t)$ represents the population at time t.

(a) Find the average rate of change of the population from 0 to 2.5 hours.

(b) Find the average rate of change of the population from 4.5 to 6 hours.

(c) What is happening to the average rate of change as time passes?

Time (hours), t	Population (grams), P
0	0.09
2.5	0.18
3.5	0.26
4.5	0.35
6	0.50

70. e-Filing Tax Returns The Internal Revenue Service Restructuring and Reform Act (RRA) was signed into law by President Bill Clinton in 1998. A major objective of the RRA was to promote electronic filing of tax returns. The data in the table show the percentage of individual income tax returns filed electronically for filing years 1998–2006. Since the percentage P of returns filed electronically depends on the filing year y and each input corresponds to exactly one output, the percentage of returns filed electronically is a function of the filing year; so $P(y)$ represents the percentage of returns filed electronically for filing year y.

(a) Find the average rate of change of the percentage of e-filed returns from 1998 to 2000.

(b) Find the average rate of change of the percentage of e-filed returns from 2001 to 2003.

(c) Find the average rate of change of the percentage of e-filed returns from 2004 to 2006.

(d) What is happening to the average rate of change as time passes?

Year	Percentage of returns e-filed
1998	20.7
1999	23.5
2000	27.6
2001	30.7
2002	35.6
2003	40.2
2004	46.5
2005	51.1
2006	57.1

SOURCE: Internal Revenue Service

71. For the function $f(x) = x^2$, compute each average rate of change:

(a) From 0 to 1 (d) From 0 to 0.01

(b) From 0 to 0.5 (e) From 0 to 0.001

(c) From 0 to 0.1

(f) Use a graphing utility to graph each of the secant lines along with f.

(g) What do you think is happening to the secant lines?

(h) What is happening to the slopes of the secant lines? Is there some number that they are getting closer to? What is that number?

72. For the function $f(x) = x^2$, compute each average rate of change:

(a) From 1 to 2 (d) From 1 to 1.01

(b) From 1 to 1.5 (e) From 1 to 1.001

(c) From 1 to 1.1

(f) Use a graphing utility to graph each of the secant lines along with f.

(g) What do you think is happening to the secant lines?

(h) What is happening to the slopes of the secant lines? Is there some number that they are getting closer to? What is that number?

Problems 73–80 require the following discussion of a secant line. The slope of the secant line containing the two points $(x, f(x))$ and $(x + h, f(x + h))$ on the graph of a function $y = f(x)$ may be given as

$$m_{sec} = \frac{f(x + h) - f(x)}{(x + h) - x} = \frac{f(x + h) - f(x)}{h} \qquad h \neq 0$$

*In calculus, this expression is called the **difference quotient of f**.*

(a) *Express the slope of the secant line of each function in terms of x and h. Be sure to simplify your answer.*

(b) *Find m_{sec} for $h = 0.5, 0.1$, and 0.01 at $x = 1$. What value does m_{sec} approach as h approaches 0?*

(c) *Find the equation for the secant line at $x = 1$ with $h = 0.01$.*

(d) *Use a graphing utility to graph f and the secant line found in part (c) on the same viewing window.*

73. $f(x) = 2x + 5$

74. $f(x) = -3x + 2$

75. $f(x) = x^2 + 2x$

76. $f(x) = 2x^2 + x$

77. $f(x) = 2x^2 - 3x + 1$

78. $f(x) = -x^2 + 3x - 2$

79. $f(x) = \dfrac{1}{x}$

80. $f(x) = \dfrac{1}{x^2}$

Discussion and Writing

81. Draw the graph of a function that has the following properties: domain: all real numbers; range: all real numbers; intercepts: $(0, -3)$ and $(3, 0)$; a local maximum of -2 is at -1; a local minimum of -6 is at 2. Compare your graph with others. Comment on any differences.

82. Redo Problem 81 with the following additional information: increasing on $(-\infty, -1), (2, \infty)$; decreasing on $(-1, 2)$. Again compare your graph with others and comment on any differences.

83. How many x-intercepts can a function defined on an interval have if it is increasing on that interval? Explain.

84. Suppose that a friend of yours does not understand the idea of increasing and decreasing functions. Provide an explanation, complete with graphs, that clarifies the idea.

85. Can a function be both even and odd? Explain.

86. Using a graphing utility, graph $y = 5$ on the interval $(-3, 3)$. Use MAXIMUM to find the local maxima on $(-3, 3)$. Comment on the result provided by the calculator.

'Are You Prepared?' Answers

1. $2 < x < 5$ **2.** 1 **3.** symmetric with respect to the y-axis **4.** $y + 2 = 5(x - 3)$ **5.** $(-3, 0), (3, 0), (0, -9)$

3.4 Library of Functions; Piecewise-defined Functions

PREPARING FOR THIS SECTION *Before getting started, review the following:*

- Intercepts (Section 2.2, pp. 165–167)

- Graphs of Key Equations (Section 2.2: Example 3, p. 165; Example 10, p. 169; Example 11, p. 170; Example 12, p. 170)

\ **Now Work** the 'Are You Prepared?' problems on page 248.

OBJECTIVES 1 Graph the Functions Listed in the Library of Functions (p. 244)
2 Graph Piecewise-defined Functions (p. 247)

Figure 25

We now introduce a few more functions to add to our list of important functions. We begin with the *square root function*.

On page 170, we graphed the equation $y = \sqrt{x}$. Figure 25 shows a graph of the function $f(x) = \sqrt{x}$. Based on the graph, we have the following properties:

Properties of $f(x) = \sqrt{x}$

1. The domain and the range are the set of nonnegative real numbers.

2. The x-intercept of the graph of $f(x) = \sqrt{x}$ is 0. The y-intercept of the graph of $f(x) = \sqrt{x}$ is also 0.

3. The function is neither even nor odd.

4. It is increasing on the interval $(0, \infty)$.

5. It has a minimum value of 0 at $x = 0$.

EXAMPLE 1 | **Graphing the Cube Root Function**

(a) Determine whether $f(x) = \sqrt[3]{x}$ is even, odd, or neither. State whether the graph of f is symmetric with respect to the y-axis or symmetric with respect to the origin.

(b) Determine the intercepts, if any, of the graph of $f(x) = \sqrt[3]{x}$.

(c) Graph $f(x) = \sqrt[3]{x}$.

Solution (a) Because

$$f(-x) = \sqrt[3]{-x} = -\sqrt[3]{x} = -f(x)$$

the function is odd. The graph of f is symmetric with respect to the origin.

(b) The y-intercept is $f(0) = \sqrt[3]{0} = 0$. The x-intercept is found by solving the equation $f(x) = 0$.

$$f(x) = 0$$
$$\sqrt[3]{x} = 0 \qquad f(x) = \sqrt[3]{x}$$
$$x = 0 \qquad \text{Cube both sides of the equation.}$$

The x-intercept is also 0.

(c) We use the function to form Table 4 and obtain some points on the graph. Because of the symmetry with respect to the origin, we need to find only points (x, y) for which $x \geq 0$. Figure 26 shows the graph of $f(x) = \sqrt[3]{x}$.

Table 4

x	$y = f(x) = \sqrt[3]{x}$	(x, y)
0	0	$(0, 0)$
$\dfrac{1}{8}$	$\dfrac{1}{2}$	$\left(\dfrac{1}{8}, \dfrac{1}{2}\right)$
1	1	$(1, 1)$
2	$\sqrt[3]{2} \approx 1.26$	$\left(2, \sqrt[3]{2}\right)$
8	2	$(8, 2)$

Figure 26

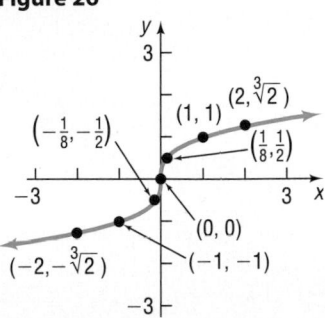

From the results of Example 1 and Figure 26, we have the following properties of the cube root function.

Properties of $f(x) = \sqrt[3]{x}$

1. The domain and the range are the set of all real numbers.
2. The x-intercept of the graph of $f(x) = \sqrt[3]{x}$ is 0. The y-intercept of the graph of $f(x) = \sqrt[3]{x}$ is also 0.
3. The graph is symmetric with respect to the origin. The function is odd.
4. It is increasing on the interval $(-\infty, \infty)$.
5. It does not have a local minimum or a local maximum.

EXAMPLE 2

Graphing the Absolute Value Function

(a) Determine whether $f(x) = |x|$ is even, odd, or neither. State whether the graph of f is symmetric with respect to the y-axis or symmetric with respect to the origin.
(b) Determine the intercepts, if any, of the graph of $f(x) = |x|$.
(c) Graph $f(x) = |x|$.

Solution

(a) Because

$$f(-x) = |-x|$$
$$= |x| = f(x)$$

the function is even. The graph of f is symmetric with respect to the y-axis.

(b) The y-intercept is $f(0) = |0| = 0$. The x-intercept is found by solving the equation $f(x) = 0$ or $|x| = 0$. So the x-intercept is 0.

(c) We use the function to form Table 5 and obtain some points on the graph. Because of the symmetry with respect to the y-axis, we need to find only points (x, y) for which $x \geq 0$. Figure 27 shows the graph of $f(x) = |x|$.

Table 5

x	y = f(x) = \|x\|	(x, y)
0	0	(0, 0)
1	1	(1, 1)
2	2	(2, 2)
3	3	(3, 3)

Figure 27

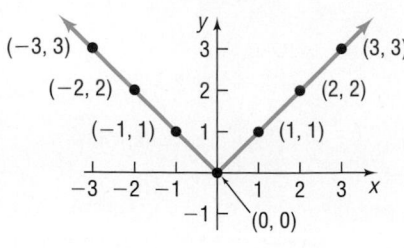

From the results of Example 2 and Figure 27, we have the following properties of the absolute value function.

Properties of $f(x) = |x|$

1. The domain is the set of all real numbers. The range of f is $\{y | y \geq 0\}$.
2. The x-intercept of the graph of $f(x) = |x|$ is 0. The y-intercept of the graph of $f(x) = |x|$ is also 0.
3. The graph is symmetric with respect to the y-axis. The function is even.
4. It is decreasing on the interval $(-\infty, 0)$. It is increasing on the interval $(0, \infty)$.
5. It has a local minimum of 0 at $x = 0$.

 Seeing the Concept

Graph $y = |x|$ on a square screen and compare what you see with Figure 27. Note that some graphing calculators use abs(x) for absolute value.

1 Graph the Functions Listed in the Library of Functions

We now provide a summary of the key functions that we have encountered. In going through this list, pay special attention to the properties of each function, particularly to the shape of each graph. Knowing these graphs will lay the foundation for later graphing techniques.

Constant Function

$$f(x) = b \qquad b \text{ is a real number}$$

Figure 28
Constant Function

See Figure 28.

The domain of a **constant function** is the set of all real numbers; its range is the set consisting of a single number b. Its graph is a horizontal line whose y-intercept is b. The constant function is an even function whose graph is constant over its domain.

Identity Function

$$f(x) = x$$

Figure 29
Identity Function

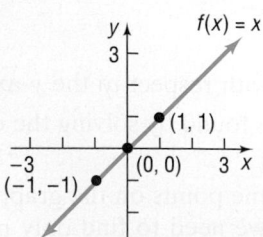

See Figure 29.

The domain and the range of the **identity function** are the set of all real numbers. Its graph is a line whose slope is $m = 1$ and whose y-intercept is 0. The line consists of all points for which the x-coordinate equals the y-coordinate. The identity function is an odd function that is increasing over its domain. Note that the graph bisects quadrants I and III.

Figure 30
Square Function

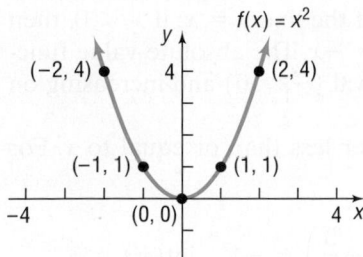

Square Function

$$f(x) = x^2$$

See Figure 30.

The domain of the **square function** f is the set of all real numbers; its range is the set of nonnegative real numbers. The graph of this function is a parabola whose intercept is at $(0, 0)$. The square function is an even function that is decreasing on the interval $(-\infty, 0)$ and increasing on the interval $(0, \infty)$.

Figure 31
Cube Function

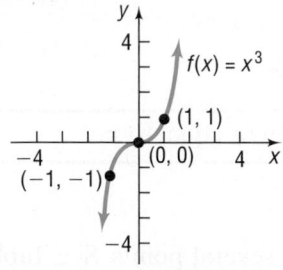

Cube Function

$$f(x) = x^3$$

See Figure 31.

The domain and the range of the **cube function** are the set of all real numbers. The intercept of the graph is at $(0, 0)$. The cube function is odd and is increasing on the interval $(-\infty, \infty)$.

Square Root Function

$$f(x) = \sqrt{x}$$

Figure 32
Square Root Function

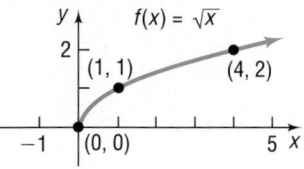

See Figure 32.

The domain and the range of the **square root function** are the set of nonnegative real numbers. The intercept of the graph is at $(0, 0)$. The square root function is neither even nor odd and is increasing on the interval $(0, \infty)$.

Cube Root Function

$$f(x) = \sqrt[3]{x}$$

Figure 33
Cube Root Function

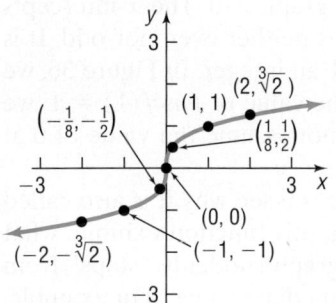

See Figure 33.

The domain and the range of the **cube root function** are the set of all real numbers. The intercept of the graph is at $(0, 0)$. The cube root function is an odd function that is increasing on the interval $(-\infty, \infty)$.

Reciprocal Function

$$f(x) = \frac{1}{x}$$

Refer to Example 12, page 170, for a discussion of the equation $y = \frac{1}{x}$. See Figure 34.

The domain and the range of the **reciprocal function** are the set of all nonzero real numbers. The graph has no intercepts. The reciprocal function is decreasing on the intervals $(-\infty, 0)$ and $(0, \infty)$ and is an odd function.

Figure 34
Reciprocal Function

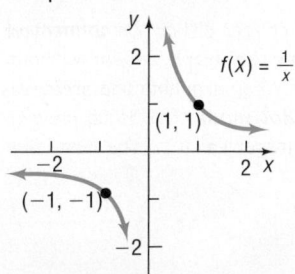

Absolute Value Function

$$f(x) = |x|$$

See Figure 35 on page 246.

Figure 35
Absolute Value Function

The domain of the **absolute value function** is the set of all real numbers; its range is the set of nonnegative real numbers. The intercept of the graph is at $(0, 0)$. If $x \geq 0$, then $f(x) = x$, and the graph of f is part of the line $y = x$; if $x < 0$, then $f(x) = -x$, and the graph of f is part of the line $y = -x$. The absolute value function is an even function; it is decreasing on the interval $(-\infty, 0)$ and increasing on the interval $(0, \infty)$.

The notation $\text{int}(x)$ stands for the largest integer less than or equal to x. For example,

$$\text{int}(1) = 1, \quad \text{int}(2.5) = 2, \quad \text{int}\left(\frac{1}{2}\right) = 0, \quad \text{int}\left(-\frac{3}{4}\right) = -1, \quad \text{int}(\pi) = 3$$

This type of correspondence occurs frequently enough in mathematics that we give it a name.

DEFINITION

Greatest Integer Function

$$f(x) = \text{int}(x)^* = \text{greatest integer less than or equal to } x$$

Table 6

x	$y = f(x)$ $= \text{int}(x)$	(x, y)
-1	-1	$(-1, -1)$
$-\dfrac{1}{2}$	-1	$\left(-\dfrac{1}{2}, -1\right)$
$-\dfrac{1}{4}$	-1	$\left(-\dfrac{1}{4}, -1\right)$
0	0	$(0, 0)$
$\dfrac{1}{4}$	0	$\left(\dfrac{1}{4}, 0\right)$
$\dfrac{1}{2}$	0	$\left(\dfrac{1}{2}, 0\right)$
$\dfrac{3}{4}$	0	$\left(\dfrac{3}{4}, 0\right)$

We obtain the graph of $f(x) = \text{int}(x)$ by plotting several points. See Table 6. For values of x, $-1 \leq x < 0$, the value of $f(x) = \text{int}(x)$ is -1; for values of x, $0 \leq x < 1$, the value of f is 0. See Figure 36 for the graph.

Figure 36
Greatest Integer Function

The domain of the **greatest integer function** is the set of all real numbers; its range is the set of integers. The y-intercept of the graph is 0. The x-intercepts lie in the interval $[0, 1)$. The greatest integer function is neither even nor odd. It is constant on every interval of the form $[k, k + 1)$, for k an integer. In Figure 36, we use a solid dot to indicate, for example, that at $x = 1$ the value of f is $f(1) = 1$; we use an open circle to illustrate that the function does not assume the value of 0 at $x = 1$.

From the graph of the greatest integer function, we can see why it is also called a **step function.** At $x = 0$, $x = \pm 1$, $x = \pm 2$, and so on, this function exhibits what is called a *discontinuity;* that is, at integer values, the graph suddenly "steps" from one value to another without taking on any of the intermediate values. For example, to the immediate left of $x = 3$, the y-coordinates of the points on the graph are 2, and to the immediate right of $x = 3$, the y-coordinates of the points on the graph are 3.

 COMMENT When graphing a function using a graphing utility, you can choose either the **connected mode,** in which points plotted on the screen are connected, making the graph appear without any breaks, or the **dot mode,** in which only the points plotted appear. When graphing the greatest integer function with a graphing utility, it is necessary to be in the **dot mode.** This is to prevent the utility from "connecting the dots" when f(x) changes from one integer value to the next. See Figure 37.

* Some books use the notation $f(x) = [x]$ instead of $\text{int}(x)$.

Figure 37
$f(x) = \text{int}(x)$

(a) Connected mode

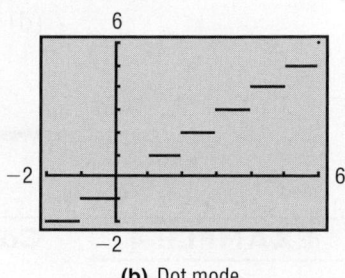

(b) Dot mode

The functions that we have discussed so far are basic. Whenever you encounter one of them, you should see a mental picture of its graph. For example, if you encounter the function $f(x) = x^2$, you should see in your mind's eye a picture like Figure 30.

Now Work PROBLEMS 9 THROUGH 16

2 Graph Piecewise-defined Functions

Sometimes a function is defined differently on different parts of its domain. For example, the absolute value function $f(x) = |x|$ is actually defined by two equations: $f(x) = x$ if $x \geq 0$ and $f(x) = -x$ if $x < 0$. For convenience, we generally combine these equations into one expression as

$$f(x) = |x| = \begin{cases} x & \text{if } x \geq 0 \\ -x & \text{if } x < 0 \end{cases}$$

When functions are defined by more than one equation, they are called **piecewise-defined** functions.

Let's look at another example of a piecewise-defined function.

| EXAMPLE 3 | Analyzing a Piecewise-defined Function |

The function f is defined as

$$f(x) = \begin{cases} -x + 1 & \text{if } -1 \leq x < 1 \\ 2 & \text{if } x = 1 \\ x^2 & \text{if } x > 1 \end{cases}$$

(a) Find $f(0)$, $f(1)$, and $f(2)$. (b) Determine the domain of f.
(c) Graph f. (d) Use the graph to find the range of f.

Solution

(a) To find $f(0)$, we observe that when $x = 0$ the equation for f is given by $f(x) = -x + 1$. So we have

$$f(0) = -0 + 1 = 1$$

When $x = 1$, the equation for f is $f(x) = 2$. Thus,

$$f(1) = 2$$

When $x = 2$, the equation for f is $f(x) = x^2$. So

$$f(2) = 2^2 = 4$$

(b) To find the domain of f, we look at its definition. We conclude that the domain of f is $\{x | x \geq -1\}$, or the interval $[-1, \infty)$.

(c) To graph f, we graph "each piece." First we graph the line $y = -x + 1$ and keep only the part for which $-1 \leq x < 1$. Then we plot the point $(1, 2)$ because, when $x = 1$, $f(x) = 2$. Finally, we graph the parabola $y = x^2$ and keep only the part for which $x > 1$. See Figure 38.

Figure 38

(d) From the graph, we conclude that the range of f is $\{y | y > 0\}$, or the interval $(0, \infty)$.

━━━━━ **Now Work** PROBLEM 29

| EXAMPLE 4 | Cost of Electricity |

In May 2006, Commonwealth Edison Company supplied electricity to residences for a monthly customer charge of $7.58 plus 8.275¢ per kilowatt-hour (kWhr) for the first 400 kWhr supplied in the month and 6.208¢ per kWhr for all usage over 400 kWhr in the month.

(a) What is the charge for using 300 kWhr in a month?
(b) What is the charge for using 700 kWhr in a month?
(c) If C is the monthly charge for x kWhr, express C as a function of x.

Source: Commonwealth Edison Co., Chicago, Illinois, 2006.

Solution (a) For 300 kWhr, the charge is $7.58 plus 8.275¢ = $0.08275 per kWhr. That is,

$$\text{Charge} = \$7.58 + \$0.08275(300) = \$32.41$$

(b) For 700 kWhr, the charge is $7.58 plus 8.275¢ per kWhr for the first 400 kWhr plus 6.208¢ per kWhr for the 300 kWhr in excess of 400. That is,

$$\text{Charge} = \$7.58 + \$0.08275(400) + \$0.06208(300) = \$59.30$$

(c) If $0 \le x \le 400$, the monthly charge C (in dollars) can be found by multiplying x times $0.08275 and adding the monthly customer charge of $7.58. So, if $0 \le x \le 400$, then $C(x) = 0.08275x + 7.58$.
For $x > 400$, the charge is $0.08275(400) + 7.58 + 0.06208(x - 400)$, since $x - 400$ equals the usage in excess of 400 kWhr, which costs $0.06208 per kWhr. That is, if $x > 400$, then

$$
\begin{aligned}
C(x) &= 0.08275(400) + 7.58 + 0.06208(x - 400) \\
&= 40.68 + 0.06208(x - 400) \\
&= 0.06208x + 15.848
\end{aligned}
$$

Figure 39

The rule for computing C follows two equations:

$$
C(x) = \begin{cases} 0.08275x + 7.58 & \text{if } 0 \le x \le 400 \\ 0.06208x + 15.848 & \text{if } x > 400 \end{cases}
$$

See Figure 39 for the graph.

━━━━━

3.4 Assess Your Understanding

'Are You Prepared?' *Answers are given at the end of these exercises. If you get a wrong answer, read the pages listed in red.*

1. Sketch the graph of $y = \sqrt{x}$. (p. 170)

2. Sketch the graph of $y = \dfrac{1}{x}$. (p. 170)

3. List the intercepts of the equation $y = x^3 - 8$. (pp. 165–167)

Concepts and Vocabulary

4. The function $f(x) = x^2$ is decreasing on the interval _____.

5. When functions are defined by more than one equation, they are called _____ functions.

6. *True or False* The cube function is odd and is increasing on the interval $(-\infty, \infty)$.

7. *True or False* The cube root function is odd and is decreasing on the interval $(-\infty, \infty)$.

8. *True or False* The domain and the range of the reciprocal function are the set of all real numbers.

Skill Building

In Problems 9–16, match each graph to its function.

A. *Constant function*
B. *Identity function*
C. *Square function*
D. *Cube function*
E. *Square root function*
F. *Reciprocal function*
G. *Absolute value function*
H. *Cube root function*

9.

10.

11.

12.

13.

14.

15.

16.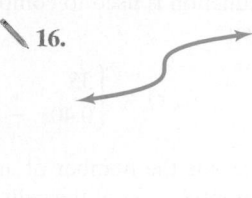

In Problems 17–24, sketch the graph of each function. Be sure to label three points on the graph.

17. $f(x) = x$

18. $f(x) = x^2$

19. $f(x) = x^3$

20. $f(x) = \sqrt{x}$

21. $f(x) = \dfrac{1}{x}$

22. $f(x) = |x|$

23. $f(x) = \sqrt[3]{x}$

24. $f(x) = 3$

25. If $f(x) = \begin{cases} x^2 & \text{if } x < 0 \\ 2 & \text{if } x = 0 \\ 2x + 1 & \text{if } x > 0 \end{cases}$

 find: (a) $f(-2)$ (b) $f(0)$ (c) $f(2)$

26. If $f(x) = \begin{cases} -3x & \text{if } x < -1 \\ 0 & \text{if } x = -1 \\ 2x^2 + 1 & \text{if } x > -1 \end{cases}$

 find: (a) $f(-2)$ (b) $f(-1)$ (c) $f(0)$

27. If $f(x) = \begin{cases} 2x - 4 & \text{if } -1 \le x \le 2 \\ x^3 - 2 & \text{if } 2 < x \le 3 \end{cases}$

 find: (a) $f(0)$ (b) $f(1)$ (c) $f(2)$ (d) $f(3)$

28. If $f(x) = \begin{cases} x^3 & \text{if } -2 \le x < 1 \\ 3x + 2 & \text{if } 1 \le x \le 4 \end{cases}$

 find: (a) $f(-1)$ (b) $f(0)$ (c) $f(1)$ (d) $f(3)$

In Problems 29–40:

(a) *Find the domain of each function.*
(b) *Locate any intercepts.*
(c) *Graph each function.*
(d) *Based on the graph, find the range.*

29. $f(x) = \begin{cases} 2x & \text{if } x \ne 0 \\ 1 & \text{if } x = 0 \end{cases}$

30. $f(x) = \begin{cases} 3x & \text{if } x \ne 0 \\ 4 & \text{if } x = 0 \end{cases}$

31. $f(x) = \begin{cases} -2x + 3 & x < 1 \\ 3x - 2 & x \ge 1 \end{cases}$

32. $f(x) = \begin{cases} x + 3 & \text{if } x < -2 \\ -2x - 3 & \text{if } x \ge -2 \end{cases}$

33. $f(x) = \begin{cases} x + 3 & \text{if } -2 \le x < 1 \\ 5 & \text{if } x = 1 \\ -x + 2 & \text{if } x > 1 \end{cases}$

34. $f(x) = \begin{cases} 2x + 5 & \text{if } -3 \le x < 0 \\ -3 & \text{if } x = 0 \\ -5x & \text{if } x > 0 \end{cases}$

35. $f(x) = \begin{cases} 1 + x & \text{if } x < 0 \\ x^2 & \text{if } x \ge 0 \end{cases}$

36. $f(x) = \begin{cases} \dfrac{1}{x} & \text{if } x < 0 \\ \sqrt[3]{x} & \text{if } x \ge 0 \end{cases}$

37. $f(x) = \begin{cases} |x| & \text{if } -2 \le x < 0 \\ x^3 & \text{if } x > 0 \end{cases}$

38. $f(x) = \begin{cases} 2 - x & \text{if } -3 \le x < 1 \\ \sqrt{x} & \text{if } x > 1 \end{cases}$

39. $f(x) = 2\,\text{int}(x)$

40. $f(x) = \text{int}(2x)$

In Problems 41–44, the graph of a piecewise-defined function is given. Write a definition for each function.

41.

42.

43.

44.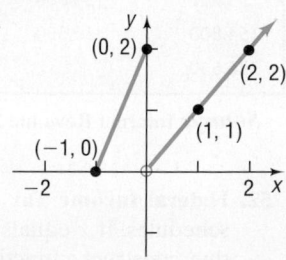

45. If $f(x) = \text{int}(2x)$, find (a) $f(1.2)$, (b) $f(1.6)$, (c) $f(-1.8)$.

46. If $f(x) = \text{int}\left(\dfrac{x}{2}\right)$, find (a) $f(1.2)$, (b) $f(1.6)$, (c) $f(-1.8)$.

Applications and Extensions

47. Cell Phone Service Sprint PCS offers a monthly cellular phone plan for $35. It includes 300 anytime minutes and charges $0.40 per minute for additional minutes. The following function is used to compute the monthly cost for a subscriber:

$$C(x) = \begin{cases} 35 & \text{if } 0 < x \le 300 \\ 0.40x - 85 & \text{if } x > 300 \end{cases}$$

where x is the number of anytime minutes used. Compute the monthly cost of the cellular phone for use of the following anytime minutes:

(a) 200 (b) 365 (c) 301

Source: Sprint PCS

48. Parking at O'Hare International Airport The short-term (no more than 24 hours) parking fee F (in dollars) for parking x hours at O'Hare International Airport's main parking garage can be modeled by the function

$$F(x) = \begin{cases} 3 & \text{if } 0 < x \le 3 \\ 5 \, \text{int}(x + 1) + 1 & \text{if } 3 < x < 9 \\ 50 & \text{if } 9 \le x \le 24 \end{cases}$$

Determine the fee for parking in the short-term parking garage for

(a) 2 hours

(b) 7 hours

(c) 15 hours

(d) 8 hours and 24 minutes

Source: O'Hare International Airport

49. Cost of Natural Gas In May 2006, Peoples Energy had the following rate schedule for natural gas usage in single-family residences:

Monthly service charge	$9.45
Per therm service charge	
1st 50 therms	$0.36375/therm
Over 50 therms	$0.11445/therm
Gas charge	$0.7958/therm

(a) What is the charge for using 50 therms in a month?

(b) What is the charge for using 500 therms in a month?

(c) Construct a function that relates the monthly charge C for x therms of gas.

(d) Graph this function.

Source: Peoples Energy, Chicago, Illinois, 2006

50. Cost of Natural Gas In May 2006, Nicor Gas had the following rate schedule for natural gas usage in single-family residences:

Monthly customer charge	$8.85
Distribution charge	
1st 20 therms	$0.1557/therm
Next 30 therms	$0.0663/therm
Over 50 therms	$0.0519/therm
Gas supply charge	$0.66/therm

(a) What is the charge for using 40 therms in a month?

(b) What is the charge for using 202 therms in a month?

(c) Construct a function that gives the monthly charge C for x therms of gas.

(d) Graph this function.

Source: Nicor Gas, Aurora, Illinois, 2006

51. Federal Income Tax Two 2006 Tax Rate Schedules are given in the accompanying table. If x equals taxable income and y equals the tax due, construct a function $y = f(x)$ for Schedule X.

REVISED 2006 TAX RATE SCHEDULES											
Schedule X— Single						**Schedule Y-1—Married Filing jointly or qualifying Widow(er)**					
If Taxable Income Is over	But Not over	The Tax Is This Amount		Plus This %	Of the Excess over	If Taxable Income Is over	But Not over	The Tax Is This Amount		Plus This %	Of the Excess over
0	7550	0.00	+	10%	$0	$0	15,100	0.00	+	10%	$0
7550	30,650	755.00	+	15%	7550	15,100	61,300	1510.00	+	15%	15,100
30,650	74,200	4220.00	+	25%	30,650	61,300	123,700	8440.00	+	25%	61,300
74,200	154,800	15,107.50	+	28%	74,200	123,700	188,450	24,040.00	+	28%	123,700
154,800	336,550	37,675.50	+	33%	154,800	188,450	336,550	42,176.00	+	33%	188,450
336,550		97,653.00	+	35%	336,550	336,550	—	91,043.00	+	35%	336,550

Source: Internal Revenue Service

52. Federal Income Tax Refer to the revised 2006 tax rate schedules. If x equals taxable income and y equals the tax due, construct a function $y = f(x)$ for Schedule Y-1.

53. Cost of Transporting Goods A trucking company transports goods between Chicago and New York, a distance of 960 miles. The company's policy is to charge, for each pound,

$0.50 per mile for the first 100 miles, $0.40 per mile for the next 300 miles, $0.25 per mile for the next 400 miles, and no charge for the remaining 160 miles.

(a) Graph the relationship between the cost of transportation in dollars and mileage over the entire 960-mile route.

(b) Find the cost as a function of mileage for hauls between 100 and 400 miles from Chicago.

(c) Find the cost as a function of mileage for hauls between 400 and 800 miles from Chicago.

54. **Car Rental Costs** An economy car rented in Florida from National Car Rental® on a weekly basis costs $95 per week. Extra days cost $24 per day until the day rate exceeds the weekly rate, in which case the weekly rate applies. Also, any part of a day used counts as a full day. Find the cost C of renting an economy car as a piecewise-defined function of the number x of days used, where $7 \leq x \leq 14$. Graph this function.

55. **Minimum Payments for Credit Cards** Holders of credit cards issued by banks, department stores, oil companies, and so on, receive bills each month that state minimum amounts that must be paid by a certain due date. The minimum due depends on the total amount owed. One such credit card company uses the following rules: For a bill of less than $10, the entire amount is due. For a bill of at least $10 but less than $500, the minimum due is $10. A minimum of $30 is due on a bill of at least $500 but less than $1000, a minimum of $50 is due on a bill of at least $1000 but less than $1500, and a minimum of $70 is due on bills of $1500 or more. Find the function f that describes the minimum payment due on a bill of x dollars. Graph f.

56. **Interest Payments for Credit Cards** Refer to Problem 55. The card holder may pay any amount between the minimum due and the total owed. The organization issuing the card charges the card holder interest of 1.5% per month for the first $1000 owed and 1% per month on any unpaid balance over $1000. Find the function g that gives the amount of interest charged per month on a balance of x dollars. Graph g.

57. **Wind Chill** The wind chill factor represents the equivalent air temperature at a standard wind speed that would produce the same heat loss as the given temperature and wind speed. One formula for computing the equivalent temperature is

$$W = \begin{cases} t & 0 \leq v < 1.79 \\ 33 - \dfrac{(10.45 + 10\sqrt{v} - v)(33 - t)}{22.04} & 1.79 \leq v \leq 20 \\ 33 - 1.5958(33 - t) & v > 20 \end{cases}$$

where v represents the wind speed (in meters per second) and t represents the air temperature (°C). Compute the wind chill for the following:

(a) An air temperature of 10°C and a wind speed of 1 meter per second (m/sec)

(b) An air temperature of 10°C and a wind speed of 5 m/sec

(c) An air temperature of 10°C and a wind speed of 15 m/sec

(d) An air temperature of 10°C and a wind speed of 25 m/sec

(e) Explain the physical meaning of the equation corresponding to $0 \leq v < 1.79$.

(f) Explain the physical meaning of the equation corresponding to $v > 20$.

58. **Wind Chill** Redo Problem 57(a)–(d) for an air temperature of −10°C.

59. **First-class Mail** In 2006 the U.S. Postal Service charged $0.39 postage for a first-class letter weighing up to 1 ounce, plus $0.24 for each additional ounce up to 13 ounces. First-class rates do not apply to letters weighing more than 13 ounces. Write a piecewise-defined function C for the first-class postage charged for a letter weighing x ounces. Graph the function.

Source: United States Postal Service

Discussion and Writing

In Problems 60–67, use a graphing utility.

60. **Exploration** Graph $y = x^2$. Then on the same screen graph $y = x^2 + 2$, followed by $y = x^2 + 4$, followed by $y = x^2 - 2$. What pattern do you observe? Can you predict the graph of $y = x^2 - 4$? Of $y = x^2 + 5$?

61. **Exploration** Graph $y = x^2$. Then on the same screen graph $y = (x - 2)^2$, followed by $y = (x - 4)^2$, followed by $y = (x + 2)^2$. What pattern do you observe? Can you predict the graph of $y = (x + 4)^2$? Of $y = (x - 5)^2$?

62. **Exploration** Graph $y = |x|$. Then on the same screen graph $y = 2|x|$, followed by $y = 4|x|$, followed by $y = \frac{1}{2}|x|$. What pattern do you observe? Can you predict the graph of $y = \frac{1}{4}|x|$? Of $y = 5|x|$?

63. **Exploration** Graph $y = x^2$. Then on the same screen graph $y = -x^2$. What pattern do you observe? Now try $y = |x|$ and $y = -|x|$. What do you conclude?

64. **Exploration** Graph $y = \sqrt{x}$. Then on the same screen graph $y = \sqrt{-x}$. What pattern do you observe? Now try $y = 2x + 1$ and $y = 2(-x) + 1$. What do you conclude?

65. **Exploration** Graph $y = x^3$. Then on the same screen graph $y = (x - 1)^3 + 2$. Could you have predicted the result?

66. **Exploration** Graph $y = x^2$, $y = x^4$, and $y = x^6$ on the same screen. What do you notice is the same about each graph? What do you notice that is different?

67. **Exploration** Graph $y = x^3$, $y = x^5$, and $y = x^7$ on the same screen. What do you notice is the same about each graph? What do you notice that is different?

68. **Consider the equation**

$$y = \begin{cases} 1 & \text{if } x \text{ is rational} \\ 0 & \text{if } x \text{ is irrational} \end{cases}$$

Is this a function? What is its domain? What is its range? What is its y-intercept, if any? What are its x-intercepts, if any? Is it even, odd, or neither? How would you describe its graph?

69. Define some functions that pass through $(0, 0)$ and $(1, 1)$ and are increasing for $x \geq 0$. Begin your list with $y = \sqrt{x}$, $y = x$, and $y = x^2$. Can you propose a general result about such functions?

'Are You Prepared?' Answers

1.

2.

3. $(0, -8), (2, 0)$

3.5 Graphing Techniques: Transformations

OBJECTIVES
1 Graph Functions Using Vertical and Horizontal Shifts (p. 252)
2 Graph Functions Using Compressions and Stretches (p. 255)
3 Graph Functions Using Reflections about the x-Axis and the y-Axis (p. 258)

At this stage, if you were asked to graph any of the functions defined by $y = x$, $y = x^2$, $y = x^3$, $y = \sqrt{x}$, $y = \sqrt[3]{x}$, $y = |x|$, or $y = \dfrac{1}{x}$, your response should be, "Yes, I recognize these functions and know the general shapes of their graphs." (If this is not your answer, review the previous section, Figures 29 through 35.)

Sometimes we are asked to graph a function that is "almost" like one that we already know how to graph. In this section, we look at some of these functions and develop techniques for graphing them. Collectively, these techniques are referred to as **transformations.**

1 Graph Functions Using Vertical and Horizontal Shifts

EXAMPLE 1	**Vertical Shift Up**

Use the graph of $f(x) = x^2$ to obtain the graph of $g(x) = x^2 + 3$.

Solution We begin by obtaining some points on the graphs of f and g. For example, when $x = 0$, then $y = f(0) = 0$ and $y = g(0) = 3$. When $x = 1$, then $y = f(1) = 1$ and $y = g(1) = 4$. Table 7 lists these and a few other points on each graph. Notice that each y-coordinate of a point on the graph of g is 3 units larger than the y-coordinate of the corresponding point on the graph of f. We conclude that the graph of g is identical to that of f, except that it is shifted vertically up 3 units. See Figure 40.

Table 7

x	$y = f(x)$ $= x^2$	$y = g(x)$ $= x^2 + 3$
−2	4	7
−1	1	4
0	0	3
1	1	4
2	4	7

Figure 40

Exploration

Figure 41

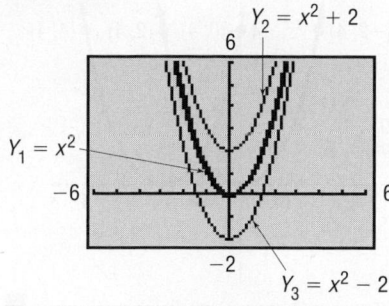

On the same screen, graph each of the following functions:

$$Y_1 = x^2$$
$$Y_2 = x^2 + 2$$
$$Y_3 = x^2 - 2$$

Result Figure 41 illustrates the graphs. You should have observed a general pattern. With $Y_1 = x^2$ on the screen, the graph of $Y_2 = x^2 + 2$ is identical to that of $Y_1 = x^2$, except that it is shifted vertically up 2 units. The graph of $Y_3 = x^2 - 2$ is identical to that of $Y_1 = x^2$, except that it is shifted vertically down 2 units.

We are led to the following conclusions:

> If a positive real number k is added to the right side of a function $y = f(x)$, the graph of the new function $y = f(x) + k$ is the graph of f **shifted vertically up** k units.

> If a positive real number k is subtracted from the right side of a function $y = f(x)$, the graph of the new function $y = f(x) - k$ is the graph of f **shifted vertically down** k units.

Let's look at another example.

| EXAMPLE 2 | **Vertical Shift Down** |

Use the graph of $f(x) = x^2$ to obtain the graph of $g(x) = x^2 - 4$.

Solution Table 8 lists some points on the graphs of f and g. Notice that each y-coordinate of g is 4 units less than the corresponding y-coordinate of f. The graph of g is identical to that of f, except that it is shifted down 4 units. See Figure 42.

Table 8

x	y = f(x) = x²	y = g(x) = x² − 4
−2	4	0
−1	1	−3
0	0	−4
1	1	−3
2	4	0

Figure 42

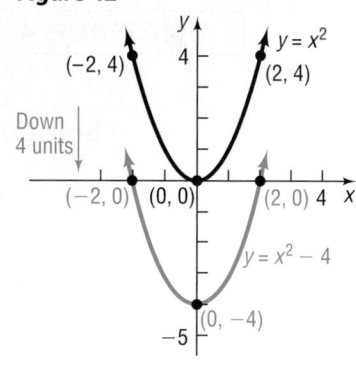

Now Work PROBLEM 35

| EXAMPLE 3 | **Horizontal Shift to the Right** |

Use the graph of $f(x) = x^2$ to obtain the graph of $g(x) = (x - 2)^2$.

Solution The function $g(x) = (x - 2)^2$ is basically a square function. Table 9 lists some points on the graphs of f and g. Note that when $f(x) = 0$ then $x = 0$, and when $g(x) = 0$, then $x = 2$. Also, when $f(x) = 4$, then $x = -2$ or 2, and when $g(x) = 4$, then $x = 0$ or 4. We conclude that the graph of g is identical to that of f, except that it is shifted 2 units to the right. See Figure 43.

Table 9

x	$y = f(x)$ $= x^2$	$y = g(x)$ $= (x - 2)^2$
−2	4	16
0	0	4
2	4	0
4	16	4

Figure 43

Exploration

Figure 44

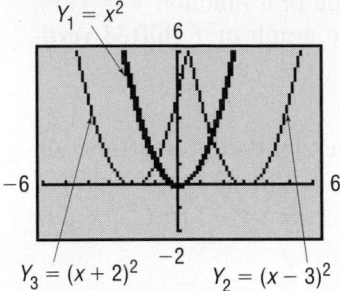

On the same screen, graph each of the following functions:

$$Y_1 = x^2$$
$$Y_2 = (x - 3)^2$$
$$Y_3 = (x + 2)^2$$

Result Figure 44 illustrates the graphs. You should observe the following pattern. With the graph of $Y_1 = x^2$ on the screen, the graph of $Y_2 = (x - 3)^2$ is identical to that of $Y_1 = x^2$, except that it is shifted horizontally to the right 3 units. The graph of $Y_3 = (x + 2)^2$ is identical to that of $Y_1 = x^2$, except that it is shifted horizontally to the left 2 units.

We are led to the following conclusion.

> If the argument x of a function f is replaced by $x - h, h > 0$, the graph of the new function $y = f(x - h)$ is the graph of f **shifted horizontally right** h units. If the argument x of a function f is replaced by $x + h, h > 0$, the graph of the new function $y = f(x + h)$ is the graph of f **shifted horizontally left** h units.

EXAMPLE 4	**Horizontal Shift to the Left**

Use the graph of $f(x) = x^2$ to obtain the graph of $g(x) = (x + 4)^2$.

Solution Again, the function $g(x) = (x + 4)^2$ is basically a square function. Its graph is the same as that of f, except that it is shifted horizontally 4 units to the left. See Figure 45.

Figure 45

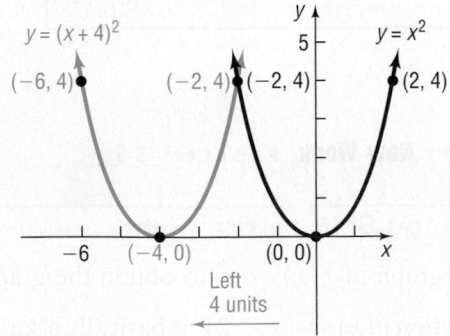

═══ **Now Work** PROBLEM 39

Vertical and horizontal shifts are sometimes combined.

EXAMPLE 5	**Combining Vertical and Horizontal Shifts**

Graph the function: $f(x) = (x + 3)^2 - 5$

Solution We graph f in steps. First, we note that the rule for f is basically a square function, so we begin with the graph of $y = x^2$ as shown in Figure 46(a). Next, to get the graph of $y = (x + 3)^2$, we shift the graph of $y = x^2$ horizontally 3 units to the left. See Figure 46(b). Finally, to get the graph of $y = (x + 3)^2 - 5$, we shift the graph of $y = (x + 3)^2$ vertically down 5 units. See Figure 46(c). Note the points plotted on each graph. Using key points can be helpful in keeping track of the transformation that has taken place.

Figure 46

(a) (b) (c)

 Check: Graph $Y_1 = f(x) = (x + 3)^2 - 5$ and compare the graph to Figure 46(c).

In Example 5, if the vertical shift had been done first, followed by the horizontal shift, the final graph would have been the same. Try it for yourself.

Now Work PROBLEM 41

2 Graph Functions Using Compressions and Stretches

EXAMPLE 6	**Vertical Stretch**

Use the graph of $f(x) = |x|$ to obtain the graph of $g(x) = 2|x|$.

Solution To see the relationship between the graphs of f and g, we form Table 10, listing points on each graph. For each x, the y-coordinate of a point on the graph of g is 2 times as large as the corresponding y-coordinate on the graph of f. The graph of $f(x) = |x|$ is vertically stretched by a factor of 2 to obtain the graph of $g(x) = 2|x|$ [for example, $(1, 1)$ is on the graph of f, but $(1, 2)$ is on the graph of g]. See Figure 47.

Table 10

	$y = f(x)$	$y = g(x)$
x	$= \lvert x \rvert$	$= 2\lvert x \rvert$
-2	2	4
-1	1	2
0	0	0
1	1	2
2	2	4

Figure 47

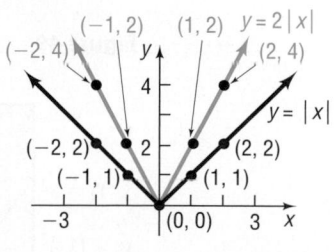

| EXAMPLE 7 | Vertical Compression |

Use the graph of $f(x) = |x|$ to obtain the graph of $g(x) = \dfrac{1}{2}|x|$.

Solution For each x, the y-coordinate of a point on the graph of g is $\dfrac{1}{2}$ as large as the corresponding y-coordinate on the graph of f. The graph of $f(x) = |x|$ is vertically compressed by a factor of $\dfrac{1}{2}$ to obtain the graph of $g(x) = \dfrac{1}{2}|x|$ [for example, $(2, 2)$ is on the graph of f, but $(2, 1)$ is on the graph of g.]. See Table 11 and Figure 48.

Table 11

| x | $y = f(x)$ $= |x|$ | $y = g(x)$ $= \dfrac{1}{2}|x|$ |
|---|---|---|
| -2 | 2 | 1 |
| -1 | 1 | $\dfrac{1}{2}$ |
| 0 | 0 | 0 |
| 1 | 1 | $\dfrac{1}{2}$ |
| 2 | 2 | 1 |

Figure 48

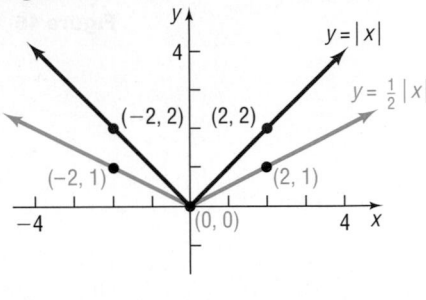

When the right side of a function $y = f(x)$ is multiplied by a positive number a, the graph of the new function $y = af(x)$ is obtained by multiplying each y-coordinate on the graph of $y = f(x)$ by a. The new graph is a **vertically compressed** (if $0 < a < 1$) or a **vertically stretched** (if $a > 1$) version of the graph of $y = f(x)$.

 Now Work PROBLEM 43

What happens if the argument x of a function $y = f(x)$ is multiplied by a positive number a, creating a new function $y = f(ax)$? To find the answer, we look at the following Exploration.

Exploration On the same screen, graph each of the following functions:

$$Y_1 = f(x) = x^2$$
$$Y_2 = f(2x) = (2x)^2$$
$$Y_3 = f\left(\dfrac{1}{2}x\right) = \left(\dfrac{1}{2}x\right)^2$$

Result You should have obtained the graphs shown in Figure 49. The graph of $Y_2 = (2x)^2$ is the graph of $Y_1 = x^2$ compressed horizontally. Look also at Table 12(a).

Figure 49

Table 12

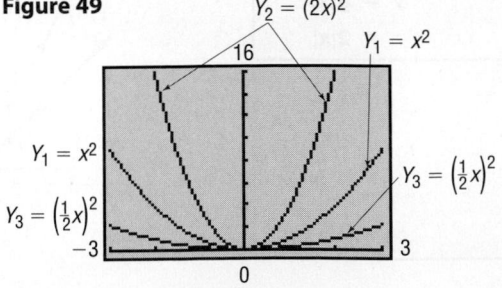

(a) (b)

Notice in Table 12(a) that (1, 1), (2, 4), (4, 16), and (16, 256) are points on the graph of $Y_1 = x^2$. Also, (0.5, 1), (1, 4), (2, 16), and (8, 256) are points on the graph of $Y_2 = (2x)^2$. For equal y-coordinates, the x-coordinate on the graph of Y_2 is $\frac{1}{2}$ the x-coordinate of Y_1. The graph of $Y_2 = (2x)^2$ is obtained by multiplying the x-coordinate of each point on the graph of $Y_1 = x^2$ by $\frac{1}{2}$.

The graph of $Y_3 = \left(\frac{1}{2}x\right)^2$ is the graph of $Y_1 = x^2$ stretched horizontally. Look at Table 12(b). Notice that (0.5, 0.25), (1, 1), (2, 4), and (4, 16) are points on the graph of $Y_1 = x^2$. Also, (1, 0.25), (2, 1), (4, 4), and (8, 16) are points on the graph of $Y_3 = \left(\frac{1}{2}x\right)^2$. For equal y-coordinates, the x-coordinate on the graph of Y_3 is 2 times the x-coordinate on Y_1. The graph of $Y_3 = \left(\frac{1}{2}x\right)^2$ is obtained by multiplying the x-coordinate of each point on the graph of $Y_1 = x^2$ by a factor of 2.

If the argument x of a function $y = f(x)$ is multiplied by a positive number a, the graph of the new function $y = f(ax)$ is obtained by multiplying each x-coordinate of $y = f(x)$ by $\frac{1}{a}$. A **horizontal compression** results if $a > 1$, and a **horizontal stretch** occurs if $0 < a < 1$.

Let's look at an example.

EXAMPLE 8 **Graphing Using Stretches and Compressions**

The graph of $y = f(x)$ is given in Figure 50. Use this graph to find the graphs of:
(a) $y = 3f(x)$ (b) $y = f(3x)$

Figure 50

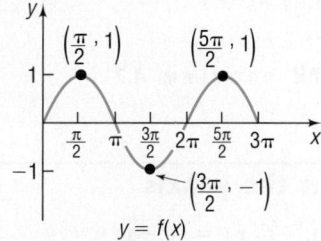

Solution (a) The graph of $y = 3f(x)$ is obtained by multiplying each y-coordinate of $y = f(x)$ by a factor of 3. See Figure 51(a).

(b) The graph of $y = f(3x)$ is obtained from the graph of $y = f(x)$ by multiplying each x-coordinate of $y = f(x)$ by a factor of $\frac{1}{3}$. See Figure 51(b).

Figure 51

Now Work PROBLEMS 65(e) AND (g)

3 Graph Functions Using Reflections about the *x*-Axis and the *y*-Axis

| EXAMPLE 9 | **Reflection about the *x*-Axis** |

Graph the function: $f(x) = -x^2$

Solution We begin with the graph of $y = x^2$, as shown in black in Figure 52. For each point (x, y) on the graph of $y = x^2$, the point $(x, -y)$ is on the graph of $y = -x^2$, as indicated in Table 13. We can draw the graph of $y = -x^2$ by reflecting the graph of $y = x^2$ about the *x*-axis. See Figure 52.

Figure 52

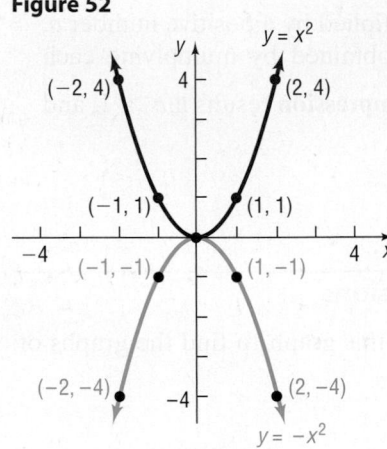

Table 13

x	$y = x^2$	$y = -x^2$
−2	4	−4
−1	1	−1
0	0	0
1	1	−1
2	4	−4

When the right side of the function $y = f(x)$ is multiplied by −1, the graph of the new function $y = -f(x)$ is the **reflection about the *x*-axis** of the graph of the function $y = f(x)$.

═══ **Now Work** PROBLEM **47**

| EXAMPLE 10 | **Reflection about the *y*-Axis** |

Graph the function: $f(x) = \sqrt{-x}$

Solution First, notice that the domain of f consists of all real numbers x for which $-x \geq 0$ or, equivalently, $x \leq 0$. To get the graph of $f(x) = \sqrt{-x}$, we begin with the graph of $y = \sqrt{x}$, as shown in Figure 53. For each point (x, y) on the graph of $y = \sqrt{x}$, the point $(-x, y)$ is on the graph of $y = \sqrt{-x}$. We obtain the graph of $y = \sqrt{-x}$ by reflecting the graph of $y = \sqrt{x}$ about the *y*-axis. See Figure 53.

Figure 53

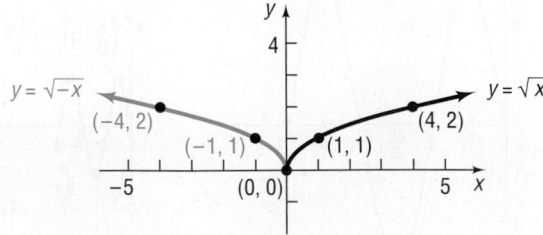

When the graph of the function $y = f(x)$ is known, the graph of the new function $y = f(-x)$ is the **reflection about the *y*-axis** of the graph of the function $y = f(x)$.

SUMMARY OF GRAPHING TECHNIQUES

To Graph:	Draw the Graph of f and:	Functional Change to $f(x)$
Vertical shifts		
$y = f(x) + k, \quad k > 0$	Raise the graph of f by k units.	Add k to $f(x)$.
$y = f(x) - k, \quad k > 0$	Lower the graph of f by k units.	Subtract k from $f(x)$.
Horizontal shifts		
$y = f(x + h), \quad h > 0$	Shift the graph of f to the left h units.	Replace x by $x + h$.
$y = f(x - h), \quad h > 0$	Shift the graph of f to the right h units.	Replace x by $x - h$.
Compressing or stretching		
$y = af(x), \quad a > 0$	Multiply each y-coordinate of $y = f(x)$ by a. Stretch the graph of f vertically if $a > 1$. Compress the graph of f vertically if $0 < a < 1$.	Multiply $f(x)$ by a.
$y = f(ax), \quad a > 0$	Multiply each x-coordinate of $y = f(x)$ by $\dfrac{1}{a}$. Stretch the graph of f horizontally if $0 < a < 1$. Compress the graph of f horizontally if $a > 1$.	Replace x by ax.
Reflection about the x-axis		
$y = -f(x)$	Reflect the graph of f about the x-axis.	Multiply $f(x)$ by -1.
Reflection about the y-axis		
$y = f(-x)$	Reflect the graph of f about the y-axis.	Replace x by $-x$.

The examples that follow combine some of the procedures outlined in this section to get the required graph.

EXAMPLE 11 **Determining the Function Obtained from a Series of Transformations**

Find the function that is finally graphed after the following three transformations are applied to the graph of $y = |x|$.

1. Shift left 2 units.
2. Shift up 3 units.
3. Reflect about the y-axis.

Solution
1. Shift left 2 units: Replace x by $x + 2$. $y = |x + 2|$
2. Shift up 3 units: Add 3. $y = |x + 2| + 3$
3. Reflect about the y-axis: Replace x by $-x$. $y = |-x + 2| + 3$

➤ **Now Work** PROBLEM 27

EXAMPLE 12 **Combining Graphing Procedures**

Graph the function: $f(x) = \dfrac{3}{x - 2} + 1$

Solution We use the following steps to obtain the graph of f:

STEP 1: $y = \dfrac{1}{x}$ Reciprocal function.

STEP 2: $y = \dfrac{3}{x}$ Multiply by 3; vertical stretch of the graph of $y = \dfrac{1}{x}$ by a factor of 3.

STEP 3: $y = \dfrac{3}{x - 2}$ Replace x by x − 2; horizontal shift to the right 2 units.

STEP 4: $y = \dfrac{3}{x - 2} + 1$ Add 1; vertical shift up 1 unit.

See Figure 54.

Figure 54

(a) $y = \dfrac{1}{x}$ Multiply by 3; Vertical stretch **(b)** $y = \dfrac{3}{x}$ Replace x by x − 2; Horizontal shift right 2 units **(c)** $y = \dfrac{3}{x-2}$ Add 1; Vertical shift up 1 unit **(d)** $y = \dfrac{3}{x-2} + 1$

Other ordering of the steps shown in Example 12 would also result in the graph of f. For example, try this one:

STEP 1: $y = \dfrac{1}{x}$ Reciprocal function

STEP 2: $y = \dfrac{1}{x - 2}$ Replace x by x − 2; horizontal shift to the right 2 units.

STEP 3: $y = \dfrac{3}{x - 2}$ Multiply by 3; vertical stretch of the graph of $y = \dfrac{1}{x - 2}$ by a factor of 3.

STEP 4: $y = \dfrac{3}{x - 2} + 1$ Add 1; vertical shift up 1 unit.

EXAMPLE 13 **Combining Graphing Procedures**

Graph the function: $f(x) = \sqrt{1 - x} + 2$

Solution Because horizontal shifts require the form $x - h$, we begin by rewriting $f(x)$ as $f(x) = \sqrt{1 - x} + 2 = \sqrt{-(x - 1)} + 2$. Now use the following steps:

STEP 1: $y = \sqrt{x}$ Square root function

STEP 2: $y = \sqrt{-x}$ Replace x by −x; reflect about the y-axis.

STEP 3: $y = \sqrt{-(x - 1)} = \sqrt{1 - x}$ Replace x by x − 1; horizontal shift to the right 1 unit.

STEP 4: $y = \sqrt{1 - x} + 2$ Add 2; vertical shift up 2 units.

See Figure 55.

Figure 55

(a) $y = \sqrt{x}$ Replace x by −x; Reflect about y-axis **(b)** $y = \sqrt{-x}$ Replace x by x − 1; Horizontal shift right 1 unit **(c)** $y = \sqrt{-(x - 1)}$
$= \sqrt{-x + 1}$
$= \sqrt{1 - x}$ Add 2; Vertical shift up 2 units **(d)** $y = \sqrt{1 - x} + 2$

Now Work PROBLEM **57**

3.5 Assess Your Understanding

Concepts and Vocabulary

1. Suppose that the graph of a function f is known. Then the graph of $y = f(x - 2)$ may be obtained by a(n) _____ shift of the graph of f to the _____ a distance of 2 units.

2. Suppose that the graph of a function f is known. Then the graph of $y = f(-x)$ may be obtained by a reflection about the _____-axis of the graph of the function $y = f(x)$.

3. Suppose that the x-intercepts of the graph of $y = f(x)$ are $-2, 1,$ and 5. The x-intercepts of $y = f(x + 3)$ are _____, _____, and _____.

4. *True or False* The graph of $y = -f(x)$ is the reflection about the x-axis of the graph of $y = f(x)$.

5. *True or False* To obtain the graph of $y = f(x + 2) - 3$, shift the graph of $y = f(x)$ horizontally to the right 2 units and vertically down 3 units.

6. *True or False* Suppose that the x-intercepts of the graph of $y = f(x)$ are -3 and 2. Then the x-intercepts of the graph of $y = 2f(x)$ are -3 and 2.

Skill Building

In Problems 7–18, match each graph to one of the following functions:

A. $y = x^2 + 2$

B. $y = -x^2 + 2$

C. $y = |x| + 2$

D. $y = -|x| + 2$

E. $y = (x - 2)^2$

F. $y = -(x + 2)^2$

G. $y = |x - 2|$

H. $y = -|x + 2|$

I. $y = 2x^2$

J. $y = -2x^2$

K. $y = 2|x|$

L. $y = -2|x|$

7.

8.

9.

10.

11.

12.

13.

14.

15.

16.

17.

18.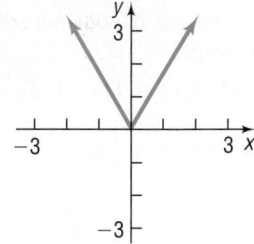

In Problems 19–26, write the function whose graph is the graph of $y = x^3$, but is:

19. Shifted to the right 4 units

20. Shifted to the left 4 units

21. Shifted up 4 units

22. Shifted down 4 units

23. Reflected about the y-axis

24. Reflected about the x-axis

25. Vertically stretched by a factor of 4

26. Horizontally stretched by a factor of 4

In Problems 27–30, find the function that is finally graphed after the following transformations are applied to the graph of $y = \sqrt{x}$.

27. (1) Shift up 2 units
 (2) Reflect about the x-axis
 (3) Reflect about the y-axis

28. (1) Reflect about the x-axis
 (2) Shift right 3 units
 (3) Shift down 2 units

29. (1) Reflect about the x-axis
 (2) Shift up 2 units
 (3) Shift left 3 units

30. (1) Shift up 2 units
 (2) Reflect about the y-axis
 (3) Shift left 3 units

31. If $(3, 0)$ is a point on the graph of $y = f(x)$, which of the following points must be on the graph of $y = -f(x)$?
(a) $(0, 3)$ (b) $(0, -3)$
(c) $(3, 0)$ (d) $(-3, 0)$

32. If $(3, 0)$ is a point on the graph of $y = f(x)$, which of the following points must be on the graph of $y = f(-x)$?
(a) $(0, 3)$ (b) $(0, -3)$
(c) $(3, 0)$ (d) $(-3, 0)$

33. If $(0, 3)$ is a point on the graph of $y = f(x)$, which of the following points must be on the graph of $y = 2f(x)$?
(a) $(0, 3)$ (b) $(0, 2)$
(c) $(0, 6)$ (d) $(6, 0)$

34. If $(3, 0)$ is a point on the graph of $y = f(x)$, which of the following points must be on the graph of $y = \dfrac{1}{2}f(x)$?
(a) $(3, 0)$ (b) $\left(\dfrac{3}{2}, 0\right)$
(c) $\left(0, \dfrac{3}{2}\right)$ (d) $\left(\dfrac{1}{2}, 0\right)$

In Problems 35–64, graph each function using the techniques of shifting, compressing, stretching, and/or reflecting. Start with the graph of the basic function (for example, $y = x^2$) and show all stages.

35. $f(x) = x^2 - 1$ **36.** $f(x) = x^2 + 4$ **37.** $g(x) = x^3 + 1$

38. $g(x) = x^3 - 1$ **39.** $h(x) = \sqrt{x - 2}$ **40.** $h(x) = \sqrt{x + 1}$

41. $f(x) = (x - 1)^3 + 2$ **42.** $f(x) = (x + 2)^3 - 3$ **43.** $g(x) = 4\sqrt{x}$

44. $g(x) = \dfrac{1}{2}\sqrt{x}$ **45.** $h(x) = \dfrac{1}{2x}$ **46.** $h(x) = 3\sqrt[3]{x}$

47. $f(x) = -\sqrt[3]{x}$ **48.** $f(x) = -\sqrt{x}$ **49.** $g(x) = \sqrt[3]{-x}$

50. $g(x) = -\dfrac{1}{x}$ **51.** $h(x) = -x^3 + 2$ **52.** $h(x) = \dfrac{1}{-x} + 2$

53. $f(x) = 2(x + 1)^2 - 3$ **54.** $f(x) = 3(x - 2)^2 + 1$ **55.** $g(x) = \sqrt{x - 2} + 1$

56. $g(x) = |x + 1| - 3$ **57.** $h(x) = \sqrt{-x} - 2$ **58.** $h(x) = \dfrac{4}{x} + 2$

59. $f(x) = -(x + 1)^3 - 1$ **60.** $f(x) = -4\sqrt{x - 1}$ **61.** $g(x) = 2|1 - x|$

62. $g(x) = 4\sqrt{2 - x}$ **63.** $h(x) = 2\,\text{int}(x - 1)$ **64.** $h(x) = \text{int}(-x)$

In Problems 65–68, the graph of a function f is illustrated. Use the graph of f as the first step toward graphing each of the following functions:

 (a) $F(x) = f(x) + 3$ (b) $G(x) = f(x + 2)$ (c) $P(x) = -f(x)$ (d) $H(x) = f(x + 1) - 2$

 (e) $Q(x) = \dfrac{1}{2}f(x)$ (f) $g(x) = f(-x)$ (g) $h(x) = f(2x)$

65.

66.

67.

68.

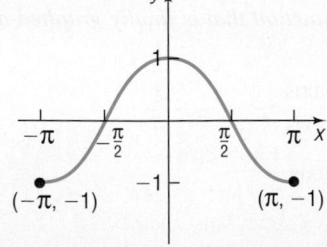

Applications and Extensions

69. Suppose that the x-intercepts of the graph of $y = f(x)$ are -5 and 3.

(a) What are the x-intercepts of the graph of $y = f(x + 2)$?

(b) What are the x-intercepts of the graph of $y = f(x - 2)$?

(c) What are the x-intercepts of the graph of $y = 4f(x)$?

(d) What are the x-intercepts of the graph of $y = f(-x)$?

70. Suppose that the x-intercepts of the graph of $y = f(x)$ are -8 and 1.

(a) What are the x-intercepts of the graph of $y = f(x + 4)$?

(b) What are the x-intercepts of the graph of $y = f(x - 3)$?

(c) What are the x-intercepts of the graph of $y = 2f(x)$?

(d) What are the x-intercepts of the graph of $y = f(-x)$?

71. Suppose that the function $y = f(x)$ is increasing on the interval $(-1, 5)$.

(a) Over what interval is the graph of $y = f(x + 2)$ increasing?

(b) Over what interval is the graph of $y = f(x - 5)$ increasing?

(c) What can be said about the graph of $y = -f(x)$?

(d) What can be said about the graph of $y = f(-x)$?

72. Suppose that the function $y = f(x)$ is decreasing on the interval $(-2, 7)$.

(a) Over what interval is the graph of $y = f(x + 2)$ decreasing?

(b) Over what interval is the graph of $y = f(x - 5)$ decreasing?

(c) What can be said about the graph of $y = -f(x)$?

(d) What can be said about the graph of $y = f(-x)$?

73. The graph of a function f is illustrated in the figure.

(a) Draw the graph of $y = |f(x)|$.

(b) Draw the graph of $y = f(|x|)$.

74. The graph of a function f is illustrated in the figure.

(a) Draw the graph of $y = |f(x)|$.

(b) Draw the graph of $y = f(|x|)$.

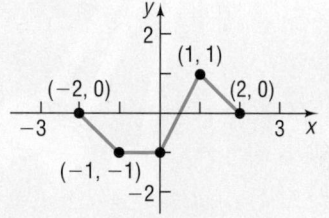

In Problems 75–84, complete the square of each quadratic expression. Then graph each function using the technique of shifting. (If necessary, refer to Section 1.2 to review completing the square.)

75. $f(x) = x^2 + 2x$ **76.** $f(x) = x^2 - 6x$ **77.** $f(x) = x^2 - 8x + 1$ **78.** $f(x) = x^2 + 4x + 2$

79. $f(x) = x^2 + x + 1$ **80.** $f(x) = x^2 - x + 1$ **81.** $f(x) = 2x^2 - 12x + 19$ **82.** $f(x) = 3x^2 + 6x + 1$

83. $f(x) = -3x^2 - 12x - 17$ **84.** $f(x) = -2x^2 - 12x - 13$

85. The equation $y = (x - c)^2$ defines a *family of parabolas,* one parabola for each value of c. On one set of coordinate axes, graph the members of the family for $c = 0$, $c = 3$, and $c = -2$.

86. Repeat Problem 85 for the family of parabolas $y = x^2 + c$.

87. Thermostat Control Energy conservation experts estimate that homeowners can save 5 to 10 percent on winter heating bills by programming their thermostats 5 to 10 degrees lower while sleeping. In the given graph, the temperature T (in degrees Fahrenheit) of a home is given as a function of time t (in hours after midnight) over a 24-hour period.

Time (hours after midnight)

(a) At what temperature is the thermostat set during day-time hours? At what temperature is the thermostat set overnight?

(b) The homeowner reprograms the thermostat to $y = T(t) - 2$. Explain how this affects the temperature in the house. Graph this new function.

(c) The homeowner reprograms the thermostat to $y = T(t + 1)$. Explain how this affects the temperature in the house. Graph this new function.

Source: Roger Albright, 547 Ways to be Fuel Smart, 2000.

88. Digital Music Revenues The total projected worldwide digital music revenues R, in millions of dollars, for the years 2005 through 2010 can be estimated by the function

$$R(x) = 170.7x^2 + 1373x + 1080$$

where x is the number of years after 2005.

(a) Find $R(0)$, $R(3)$, and $R(5)$ and explain what each value represents.

(b) Find $r = R(x - 5)$

(c) Find $r(5)$, $r(8)$, and $r(10)$ and explain what each value represents.

(d) In the model r, what does x represent?

(e) Would there be an advantage in using the model r when estimating the projected revenues for a given year instead of the model R?

Source: eMarketer.com, May 2006

89. Temperature Measurements The relationship between the Celsius (°C) and Fahrenheit (°F) scales for measuring temperature is given by the equation

$$F = \frac{9}{5}C + 32$$

The relationship between the Celsius (°C) and Kelvin (K) scales is $K = C + 273$. Graph the equation $F = \frac{9}{5}C + 32$ using degrees Fahrenheit on the y-axis and degrees Celsius on the x-axis. Use the techniques introduced in this section to obtain the graph showing the relationship between Kelvin and Fahrenheit temperatures.

90. Period of a Pendulum The period T (in seconds) of a simple pendulum is a function of its length l (in feet) defined by the equation

$$T = 2\pi\sqrt{\frac{l}{g}}$$

where $g \approx 32.2$ feet per second per second is the acceleration of gravity.
(a) Use a graphing utility to graph the function $T = T(l)$.
(b) Now graph the functions $T = T(l + 1), T = T(l + 2)$, and $T = T(l + 3)$.
(c) Discuss how adding to the length l changes the period T.
(d) Now graph the functions $T = T(2l), T = T(3l)$, and $T = T(4l)$.
(e) Discuss how multiplying the length l by factors of 2, 3, and 4 changes the period T.

91. Cigar Company Profits The daily profits of a cigar company from selling x cigars are given by

$$p(x) = -0.05x^2 + 100x - 2000$$

The government wishes to impose a tax on cigars (sometimes called a *sin tax*) that gives the company the option of either paying a flat tax of $10,000 per day or a tax of 10% on profits. As chief financial officer (CFO) of the company, you need to decide which tax is the better option for the company.
(a) On the same screen, graph $Y_1 = p(x) - 10,000$ and $Y_2 = (1 - 0.10)p(x)$.
(b) Based on the graph, which option would you select? Why?
(c) Using the terminology learned in this section, describe each graph in terms of the graph of $p(x)$.
(d) Suppose that the government offered the options of a flat tax of $4800 or a tax of 10% on profits. Which would you select? Why?

Discussion and Writing

92. Suppose that the graph of a function f is known. Explain how the graph of $y = 4f(x)$ differs from the graph of $y = f(4x)$.

3.6 Mathematical Models: Building Functions

> **OBJECTIVE** 1 Build and Analyze Functions (p. 264)

1 Build and Analyze Functions

Real-world problems often result in mathematical models that involve functions. These functions need to be constructed or built based on the information given. In building functions, we must be able to translate the verbal description into the language of mathematics. We do this by assigning symbols to represent the independent and dependent variables and then by finding the function or rule that relates these variables.

| EXAMPLE 1 | **Finding the Distance from the Origin to a Point on a Graph** |

Let $P = (x, y)$ be a point on the graph of $y = x^2 - 1$.

(a) Express the distance d from P to the origin O as a function of x.
(b) What is d if $x = 0$?
(c) What is d if $x = 1$?
(d) What is d if $x = \frac{\sqrt{2}}{2}$?

 (e) Use a graphing utility to graph the function $d = d(x), x \geq 0$. Rounded to two decimal places, find the value(s) of x at which d has a local minimum. [This gives the point(s) on the graph of $y = x^2 - 1$ closest to the origin.]

Solution

Figure 56

(a) Figure 56 illustrates the graph of $y = x^2 - 1$. The distance d from P to O is

$$d = \sqrt{(x - 0)^2 + (y - 0)^2} = \sqrt{x^2 + y^2}$$

Since P is a point on the graph of $y = x^2 - 1$, we substitute $x^2 - 1$ for y. Then

$$d(x) = \sqrt{x^2 + (x^2 - 1)^2} = \sqrt{x^4 - x^2 + 1}$$

We have expressed the distance d as a function of x.

(b) If $x = 0$, the distance d is

$$d(0) = \sqrt{1} = 1$$

(c) If $x = 1$, the distance d is

$$d(1) = \sqrt{1 - 1 + 1} = 1$$

(d) If $x = \dfrac{\sqrt{2}}{2}$, the distance d is

$$d\left(\frac{\sqrt{2}}{2}\right) = \sqrt{\left(\frac{\sqrt{2}}{2}\right)^4 - \left(\frac{\sqrt{2}}{2}\right)^2 + 1} = \sqrt{\frac{1}{4} - \frac{1}{2} + 1} = \frac{\sqrt{3}}{2}$$

Figure 57

(e) Figure 57 shows the graph of $Y_1 = \sqrt{x^4 - x^2 + 1}$. Using the MINIMUM feature on a graphing utility, we find that when $x \approx 0.71$ the value of d is smallest. The local minimum is $d \approx 0.87$ rounded to two decimal places. [Since $d(x)$ is even, by symmetry, it follows that when $x \approx -0.71$ the value of d is also a local minimum.]

━━━━ **Now Work** PROBLEM 1

EXAMPLE 2 **Area of a Rectangle**

A rectangle has one corner in quadrant I on the graph of $y = 25 - x^2$, another at the origin, a third on the positive y-axis, and the fourth on the positive x-axis. See Figure 58.

Figure 58

(a) Express the area A of the rectangle as a function of x.

(b) What is the domain of A?

 (c) Graph $A = A(x)$.

(d) For what value of x is the area largest?

Solution

(a) The area A of the rectangle is $A = xy$, where $y = 25 - x^2$. Substituting this expression for y, we obtain $A(x) = x(25 - x^2) = 25x - x^3$.

(b) Since (x, y) is in quadrant I, we have $x > 0$. Also, $y = 25 - x^2 > 0$, which implies that $x^2 < 25$, so $-5 < x < 5$. Combining these restrictions, we have the domain of A as $\{x \mid 0 < x < 5\}$, or $(0, 5)$ using interval notation.

(c) See Figure 59 for the graph of $A = A(x)$.

(d) Using MAXIMUM, we find that the maximum area is 48.11 at $x = 2.89$, each rounded to two decimal places. See Figure 60.

Figure 59

Figure 60

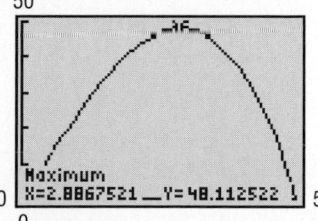

━━━━ **Now Work** PROBLEM 7

| EXAMPLE 3 | Making a Playpen* |

A manufacturer of children's playpens makes a square model that can be opened at one corner and attached at right angles to a wall or, perhaps, the side of a house. If each side is 3 feet in length, the open configuration doubles the available area in which the child can play from 9 square feet to 18 square feet. See Figure 61.

Now suppose that we place hinges at the outer corners to allow for a configuration like the one shown in Figure 62.

Figure 61

Figure 62

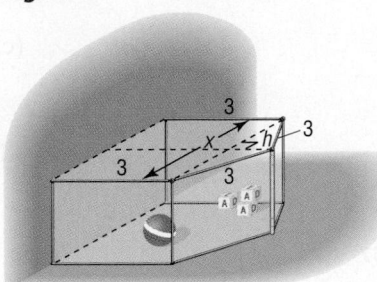

(a) Express the area A of the configuration shown in Figure 62 as a function of the distance x between the two parallel sides.

(b) Find the domain of A.

(c) Find A if $x = 5$.

(d) Graph $A = A(x)$. For what value of x is the area largest? What is the maximum area?

Solution

(a) Refer to Figure 62. The area A that we seek consists of the area of a rectangle (with width 3 and length x) and the area of an isosceles triangle (with base x and two equal sides of length 3). The height h of the triangle may be found using the Pythagorean Theorem.

$$h^2 + \left(\frac{x}{2}\right)^2 = 3^2$$

$$h^2 = 3^2 - \left(\frac{x}{2}\right)^2 = 9 - \frac{x^2}{4} = \frac{36 - x^2}{4}$$

$$h = \frac{1}{2}\sqrt{36 - x^2}$$

The area A enclosed by the playpen is

$$A = \text{area of rectangle} + \text{area of triangle} = 3x + \frac{1}{2}x\left(\frac{1}{2}\sqrt{36 - x^2}\right)$$

$$A(x) = 3x + \frac{x\sqrt{36 - x^2}}{4}$$

Now the area A is expressed as a function of x.

(b) To find the domain of A, we note first that $x > 0$, since x is a length. Also, the expression under the square root must be positive, so

$$36 - x^2 > 0$$

$$x^2 < 36$$

$$-6 < x < 6$$

Combining these restrictions, we find that the domain of A is $0 < x < 6$, or $(0, 6)$ using interval notation.

* Adapted from *Proceedings, Summer Conference for College Teachers on Applied Mathematics* (University of Missouri, Rolla), 1971.

Figure 63

Maximum
X=5.5836276 Y=19.816513

(c) If $x = 5$, the area is

$$A(5) = 3(5) + \frac{5}{4}\sqrt{36 - (5)^2} \approx 19.15 \text{ square feet}$$

If the length of the playpen is 5 feet, its area is 19.15 square feet.

(d) See Figure 63. The maximum area is about 19.82 square feet, obtained when x is about 5.58 feet.

3.6 Assess Your Understanding

Applications and Extensions

1. Let $P = (x, y)$ be a point on the graph of $y = x^2 - 8$.
 (a) Express the distance d from P to the origin as a function of x.
 (b) What is d if $x = 0$?
 (c) What is d if $x = 1$?
 (d) Use a graphing utility to graph $d = d(x)$.
 (e) For what values of x is d smallest?

2. Let $P = (x, y)$ be a point on the graph of $y = x^2 - 8$.
 (a) Express the distance d from P to the point $(0, -1)$ as a function of x.
 (b) What is d if $x = 0$?
 (c) What is d if $x = -1$?
 (d) Use a graphing utility to graph $d = d(x)$.
 (e) For what values of x is d smallest?

3. Let $P = (x, y)$ be a point on the graph of $y = \sqrt{x}$.
 (a) Express the distance d from P to the point $(1, 0)$ as a function of x.
 (b) Use a graphing utility to graph $d = d(x)$.
 (c) For what values of x is d smallest?

4. Let $P = (x, y)$ be a point on the graph of $y = \dfrac{1}{x}$.
 (a) Express the distance d from P to the origin as a function of x.
 (b) Use a graphing utility to graph $d = d(x)$.
 (c) For what values of x is d smallest?

5. A right triangle has one vertex on the graph of $y = x^3$, $x > 0$, at (x, y), another at the origin, and the third on the positive y-axis at $(0, y)$, as shown in the figure. Express the area A of the triangle as a function of x.

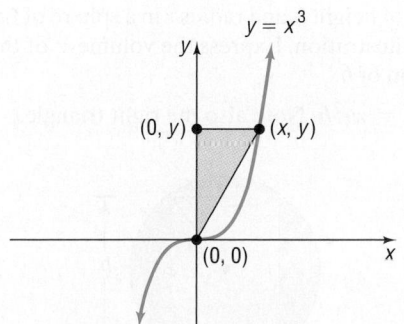

6. A right triangle has one vertex on the graph of $y = 9 - x^2$, $x > 0$, at (x, y), another at the origin, and the

third on the positive x-axis at $(x, 0)$. Express the area A of the triangle as a function of x.

7. A rectangle has one corner in quadrant I on the graph of $y = 16 - x^2$, another at the origin, a third on the positive y-axis, and the fourth on the positive x-axis. See the figure.

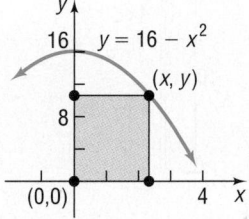

 (a) Express the area A of the rectangle as a function of x.
 (b) What is the domain of A?
 (c) Graph $A = A(x)$. For what value of x is A largest?

8. A rectangle is inscribed in a semicircle of radius 2. See the figure. Let $P = (x, y)$ be the point in quadrant I that is a vertex of the rectangle and is on the circle.

 (a) Express the area A of the rectangle as a function of x.
 (b) Express the perimeter p of the rectangle as a function of x.
 (c) Graph $A = A(x)$. For what value of x is A largest?
 (d) Graph $p = p(x)$. For what value of x is p largest?

9. A rectangle is inscribed in a circle of radius 2. See the figure. Let $P = (x, y)$ be the point in quadrant I that is a vertex of the rectangle and is on the circle.

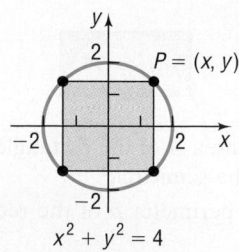

(a) Express the area A of the rectangle as a function of x.

(b) Express the perimeter p of the rectangle as a function of x.

(c) Graph $A = A(x)$. For what value of x is A largest?

(d) Graph $p = p(x)$. For what value of x is p largest?

10. A circle of radius r is inscribed in a square. See the figure.

(a) Express the area A of the square as a function of the radius r of the circle.

(b) Express the perimeter p of the square as a function of r.

11. A wire 10 meters long is to be cut into two pieces. One piece will be shaped as a square, and the other piece will be shaped as a circle. See the figure.

(a) Express the total area A enclosed by the pieces of wire as a function of the length x of a side of the square.

(b) What is the domain of A?

(c) Graph $A = A(x)$. For what value of x is A smallest?

12. A wire 10 meters long is to be cut into two pieces. One piece will be shaped as an equilateral triangle, and the other piece will be shaped as a circle.

(a) Express the total area A enclosed by the pieces of wire as a function of the length x of a side of the equilateral triangle.

(b) What is the domain of A?

(c) Graph $A = A(x)$. For what value of x is A smallest?

13. A wire of length x is bent into the shape of a circle.

(a) Express the circumference of the circle as a function of x.

(b) Express the area of the circle as a function of x.

14. A wire of length x is bent into the shape of a square.

(a) Express the perimeter of the square as a function of x.

(b) Express the area of the square as a function of x.

15. A semicircle of radius r is inscribed in a rectangle so that the diameter of the semicircle is the length of the rectangle. See the figure.

(a) Express the area A of the rectangle as a function of the radius r of the semicircle.

(b) Express the perimeter p of the rectangle as a function of r.

16. An equilateral triangle is inscribed in a circle of radius r. See the figure. Express the circumference C of the circle as a function of the length x of a side of the triangle.

[**Hint:** First show that $r^2 = \dfrac{x^2}{3}$.]

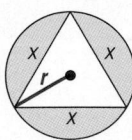

17. An equilateral triangle is inscribed in a circle of radius r. See the figure in Problem 16. Express the area A within the circle, but outside the triangle, as a function of the length x of a side of the triangle.

18. Two cars leave an intersection at the same time. One is headed south at a constant speed of 30 miles per hour, and the other is headed west at a constant speed of 40 miles per hour (see the figure). Express the distance d between the cars as a function of the time t.

[**Hint:** At $t = 0$, the cars leave the intersection.]

19. Two cars are approaching an intersection. One is 2 miles south of the intersection and is moving at a constant speed of 30 miles per hour. At the same time, the other car is 3 miles east of the intersection and is moving at a constant speed of 40 miles per hour.

(a) Express the distance d between the cars as a function of time t.

[**Hint:** At $t = 0$, the cars are 2 miles south and 3 miles east of the intersection, respectively.]

(b) Use a graphing utility to graph $d = d(t)$. For what value of t is d smallest?

20. Inscribing a Cylinder in a Sphere Inscribe a right circular cylinder of height h and radius r in a sphere of fixed radius R. See the illustration. Express the volume V of the cylinder as a function of h.

[**Hint:** $V = \pi r^2 h$. Note also the right triangle.]

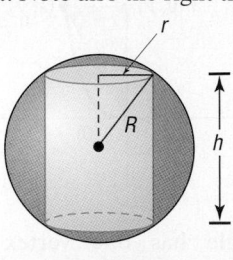

Sphere

21. Inscribing a Cylinder in a Cone Inscribe a right circular cylinder of height h and radius r in a cone of fixed radius R and fixed height H. See the illustration. Express the volume V of the cylinder as a function of r.

[**Hint:** $V = \pi r^2 h$. Note also the similar triangles.]

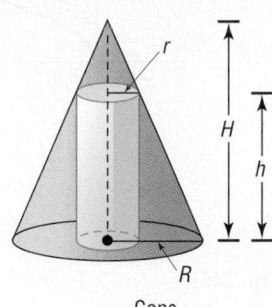

Cone

22. Installing Cable TV MetroMedia Cable is asked to provide service to a customer whose house is located 2 miles from the road along which the cable is buried. The nearest connection box for the cable is located 5 miles down the road. See the figure.

(a) If the installation cost is $500 per mile along the road and $700 per mile off the road, express the total cost C of installation as a function of the distance x (in miles) from the connection box to the point where the cable installation turns off the road. Give the domain.

(b) Compute the cost if $x = 1$ mile.

(c) Compute the cost if $x = 3$ miles.

(d) Graph the function $C = C(x)$. Use TRACE to see how the cost C varies as x changes from 0 to 5.

(e) What value of x results in the least cost?

23. Time Required to Go from an Island to a Town An island is 2 miles from the nearest point P on a straight shoreline. A town is 12 miles down the shore from P. See the illustration.

(a) If a person can row a boat at an average speed of 3 miles per hour and the same person can walk 5 miles per hour, express the time T that it takes to go from the island to town as a function of the distance x from P to where the person lands the boat.

(b) What is the domain of T?

(c) How long will it take to travel from the island to town if the person lands the boat 4 miles from P?

(d) How long will it take if the person lands the boat 8 miles from P?

24. Filling a Conical Tank Water is poured into a container in the shape of a right circular cone with radius 4 feet and height 16 feet. See the figure. Express the volume V of the water in the cone as a function of the height h of the water.

[**Hint:** The volume V of a cone of radius r and height h is $V = \dfrac{1}{3}\pi r^2 h$.]

CHAPTER REVIEW

Library of Functions

Constant function (p. 244)

$$f(x) = b$$

The graph is a horizontal line with y-intercept b.

Identity function (p. 244)

$$f(x) = x$$

The graph is a line with slope 1 and y-intercept 0.

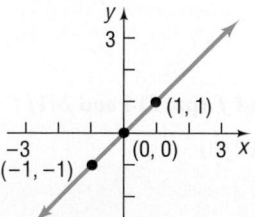

Square function (p. 245)

$f(x) = x^2$

The graph is a parabola with intercept at $(0, 0)$.

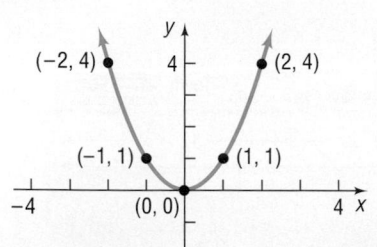

Cube function (p. 245)

$f(x) = x^3$

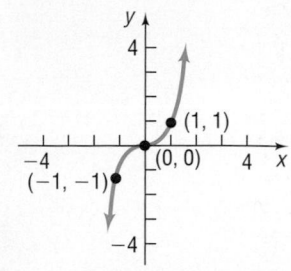

Square root function (p. 245)

$f(x) = \sqrt{x}$

Cube root function (p. 245)

$f(x) = \sqrt[3]{x}$

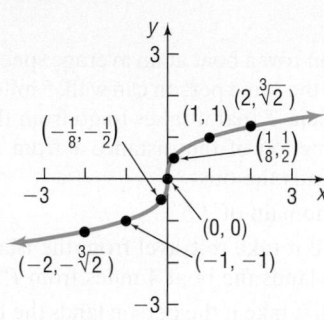

Reciprocal function (p. 245)

$f(x) = \dfrac{1}{x}$

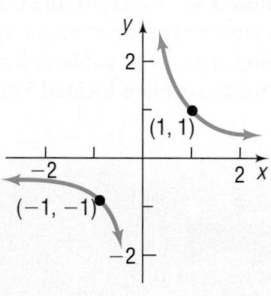

Absolute value function (pp. 245–246)

$f(x) = |x|$

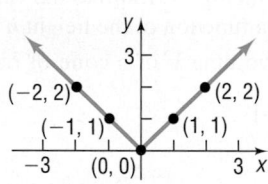

Greatest integer function (p. 246)

$f(x) = \text{int}(x)$

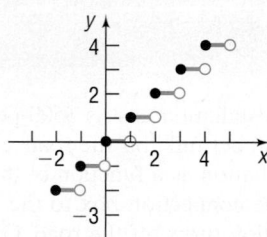

Things to Know

Function (pp. 208–212)

A relation between two sets so that each element x in the first set, the domain, has corresponding to it exactly one element y in the second set. The range is the set of y values of the function for the x values in the domain.

A function can also be characterized as a set of ordered pairs (x, y) in which no first element is paired with two different second elements.

Function notation (pp. 212–215)

$y = f(x)$

f is a symbol for the function.

x is the argument, or independent variable.

y is the dependent variable.

$f(x)$ is the value of the function at x, or the image of x.

A function f may be defined implicitly by an equation involving x and y or explicitly by writing $y = f(x)$.

Difference quotient of f (pp. 214 and 241)

$$\dfrac{f(x + h) - f(x)}{h} \quad h \neq 0$$

Domain (pp. 209 and 215)

If unspecified, the domain of a function f is the largest set of real numbers for which $f(x)$ is a real number.

Vertical-line test (p. 223)	A set of points in the plane is the graph of a function if and only if every vertical line intersects the graph in at most one point.
Even function f (p. 231)	$f(-x) = f(x)$ for every x in the domain ($-x$ must also be in the domain).
Odd function f (p. 232)	$f(-x) = -f(x)$ for every x in the domain ($-x$ must also be in the domain).
Increasing function (p. 234)	A function f is increasing on an open interval I if, for any choice of x_1 and x_2 in I, with $x_1 < x_2$, we have $f(x_1) < f(x_2)$.
Decreasing function (p. 234)	A function f is decreasing on an open interval I if, for any choice of x_1 and x_2 in I, with $x_1 < x_2$, we have $f(x_1) > f(x_2)$.
Constant function (p. 234)	A function f is constant on an open interval I if, for all choices of x in I, the values of $f(x)$ are equal.
Local maximum (p. 235)	A function f has a local maximum at c if there is an open interval I containing c so that, for all x not equal to c in I, $f(x) < f(c)$.
Local minimum (p. 235)	A function f has a local minimum at c if there is an open interval I containing c so that, for all x not equal to c in I, $f(x) > f(c)$.
Average rate of change of a function (p. 236)	The average rate of change of f from a to b is

$$\frac{\Delta y}{\Delta x} = \frac{f(b) - f(a)}{b - a} \qquad a \neq b$$

Objectives

Section	You should be able to . . .	Review Exercises
3.1	**1** Determine whether a relation represents a function (p. 208)	1, 2
	2 Find the value of a function (p. 212)	3–8, 23, 24, 71, 72, 73
	3 Find the domain of a function (p. 215)	9–16
	4 Form the sum, difference, product, and quotient of two functions (p. 217)	17–22
3.2	**1** Identify the graph of a function (p. 223)	47–50
	2 Obtain information from or about the graph of a function (p. 224)	25(a)–(e), 26(a)–(e), 27(a), 27(d) 27(f), 28(a), 28(d), 28(f)
3.3	**1** Determine even and odd functions from a graph (p. 231)	27(e), 28(e)
	2 Identify even and odd functions from the equation (p. 232)	29–36
	3 Use a graph to determine where a function is increasing, decreasing, or constant (p. 233)	27(b), 28(b)
	4 Use a graph to locate local maxima and local minima (p. 234)	27(c), 28(c)
	5 Use a graphing utility to approximate local maxima and local minima and to determine where a function is increasing or decreasing (p. 235)	37–40, 75(e), 76(b)
	6 Find the average rate of change of a function (p. 236)	41–46
3.4	**1** Graph the functions listed in the library of functions (p. 244)	51–54
	2 Graph piecewise-defined functions (p. 247)	67–70
3.5	**1** Graph functions using vertical and horizontal shifts (p. 252)	25(f), 26(f), 26(g) 55, 56, 59–66
	2 Graph functions using compressions and stretches (p. 255)	25(g), 26(h), 57, 58, 65, 66
	3 Graph functions using reflections about the x-axis and the y-axis (p. 258)	25(h), 57, 61, 62, 66
3.6	**1** Build and analyze functions (p. 264)	74–76

Review Exercises

In Problems 1 and 2, determine whether each relation represents a function. For each function, state the domain and range.

1. $\{(-1, 0), (2, 3), (4, 0)\}$

2. $\{(4, -1), (2, 1), (4, 2)\}$

In Problems 3–8, find the following for each function:

(a) $f(2)$ (b) $f(-2)$ (c) $f(-x)$ (d) $-f(x)$ (e) $f(x - 2)$ (f) $f(2x)$

3. $f(x) = \dfrac{3x}{x^2 - 1}$

4. $f(x) = \dfrac{x^2}{x + 1}$

5. $f(x) = \sqrt{x^2 - 4}$

6. $f(x) = |x^2 - 4|$

7. $f(x) = \dfrac{x^2 - 4}{x^2}$

8. $f(x) = \dfrac{x^3}{x^2 - 9}$

In Problems 9–16, find the domain of each function.

9. $f(x) = \dfrac{x}{x^2 - 9}$

10. $f(x) = \dfrac{3x^2}{x - 2}$

11. $f(x) = \sqrt{2 - x}$

12. $f(x) = \sqrt{x + 2}$

13. $h(x) = \dfrac{\sqrt{x}}{|x|}$

14. $g(x) = \dfrac{|x|}{x}$

15. $f(x) = \dfrac{x}{x^2 + 2x - 3}$

16. $F(x) = \dfrac{1}{x^2 - 3x - 4}$

In Problems 17–22, find $f + g$, $f - g$, $f \cdot g$, and $\dfrac{f}{g}$ for each pair of functions. State the domain of each one.

17. $f(x) = 2 - x$; $\ g(x) = 3x + 1$

18. $f(x) = 2x - 1$; $\ g(x) = 2x + 1$

19. $f(x) = 3x^2 + x + 1$; $\ g(x) = 3x$

20. $f(x) = 3x$; $\ g(x) = 1 + x + x^2$

21. $f(x) = \dfrac{x + 1}{x - 1}$; $\ g(x) = \dfrac{1}{x}$

22. $f(x) = \dfrac{1}{x - 3}$; $\ g(x) = \dfrac{3}{x}$

In Problems 23 and 24, find the difference quotient of each function f; that is, find

$$\dfrac{f(x + h) - f(x)}{h} \qquad h \neq 0$$

23. $f(x) = -2x^2 + x + 1$

24. $f(x) = 3x^2 - 2x + 4$

25. Using the graph of the function f shown:

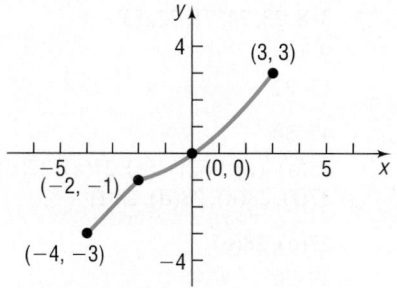

(a) Find the domain and the range of f.
(b) List the intercepts.
(c) Find $f(-2)$.
(d) For what value of x does $f(x) = -3$?
(e) Solve $f(x) > 0$.
(f) Graph $y = f(x - 3)$.
(g) Graph $y = f\left(\dfrac{1}{2}x\right)$.
(h) Graph $y = -f(x)$.

26. Using the graph of the function g shown:

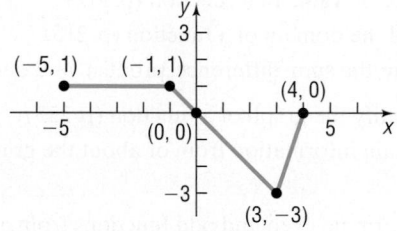

(a) Find the domain and the range of g.
(b) Find $g(-1)$.
(c) List the intercepts.
(d) For what value of x does $g(x) = -3$?
(e) Solve $g(x) > 0$.
(f) Graph $y = g(x - 2)$.
(g) Graph $y = g(x) + 1$.
(h) Graph $y = 2g(x)$.

In Problems 27 and 28, use the graph of the function f to find:
(a) The domain and the range of f.
(b) The intervals on which f is increasing, decreasing, or constant.
(c) The local minima and local maxima.
(d) Whether the graph is symmetric with respect to the x-axis, the y-axis, or the origin.
(e) Whether the function is even, odd, or neither.
(f) The intercepts, if any.

27.

28.

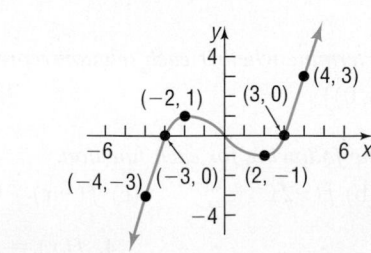

In Problems 29–36, determine (algebraically) whether the given function is even, odd, or neither.

29. $f(x) = x^3 - 4x$

30. $g(x) = \dfrac{4 + x^2}{1 + x^4}$

31. $h(x) = \dfrac{1}{x^4} + \dfrac{1}{x^2} + 1$

32. $F(x) = \sqrt{1 - x^3}$

33. $G(x) = 1 - x + x^3$

34. $H(x) = 1 + x + x^2$

35. $f(x) = \dfrac{x}{1 + x^2}$

36. $g(x) = \dfrac{1 + x^2}{x^3}$

In Problems 37–40, use a graphing utility to graph each function over the indicated interval. Approximate any local maxima and local minima. Determine where the function is increasing and where it is decreasing.

37. $f(x) = 2x^3 - 5x + 1 \quad (-3, 3)$

38. $f(x) = -x^3 + 3x - 5 \quad (-3, 3)$

39. $f(x) = 2x^4 - 5x^3 + 2x + 1 \quad (-2, 3)$

40. $f(x) = -x^4 + 3x^3 - 4x + 3 \quad (-2, 3)$

In Problems 41 and 42, find the average rate of change of f:

(a) From 1 to 2 (b) From 0 to 1 (c) From 2 to 4

41. $f(x) = 8x^2 - x$

42. $f(x) = 2x^3 + x$

In Problems 43–46, find the average rate of change from 2 to 3 for each function f. Be sure to simplify.

43. $f(x) = 2 - 5x$

44. $f(x) = 2x^2 + 7$

45. $f(x) = 3x - 4x^2$

46. $f(x) = x^2 - 3x + 2$

In Problems 47–50, is the graph shown the graph of a function?

47.

48.

49.

50.

In Problems 51–54, sketch the graph of each function. Be sure to label at least three points.

51. $f(x) = |x|$

52. $f(x) = \sqrt[3]{x}$

53. $f(x) = \sqrt{x}$

54. $f(x) = \dfrac{1}{x}$

In Problems 55–66, graph each function using the techniques of shifting, compressing or stretching, and reflections. Identify any intercepts on the graph. State the domain and, based on the graph, find the range.

55. $F(x) = |x| - 4$

56. $f(x) = |x| + 4$

57. $g(x) = -2|x|$

58. $g(x) = \dfrac{1}{2}|x|$

59. $h(x) = \sqrt{x - 1}$

60. $h(x) = \sqrt{x} - 1$

61. $f(x) = \sqrt{1 - x}$

62. $f(x) = -\sqrt{x + 3}$

63. $h(x) = (x - 1)^2 + 2$

64. $h(x) = (x + 2)^2 - 3$

65. $g(x) = 3(x - 1)^3 + 1$

66. $g(x) = -2(x + 2)^3 - 8$

In Problems 67–70,

(a) Find the domain of each function. (b) Locate any intercepts.

(c) Graph each function. (d) Based on the graph, find the range.

67. $f(x) = \begin{cases} 3x & \text{if } -2 < x \le 1 \\ x + 1 & \text{if } x > 1 \end{cases}$

68. $f(x) = \begin{cases} x - 1 & \text{if } -3 < x < 0 \\ 3x - 1 & \text{if } x \ge 0 \end{cases}$

69. $f(x) = \begin{cases} x & \text{if } -4 \le x < 0 \\ 1 & \text{if } x = 0 \\ 3x & \text{if } x > 0 \end{cases}$

70. $f(x) = \begin{cases} x^2 & \text{if } -2 \le x \le 2 \\ 2x - 1 & \text{if } x > 2 \end{cases}$

71. A function f is defined by

$$f(x) = \frac{Ax + 5}{6x - 2}$$

If $f(1) = 4$, find A.

72. A function g is defined by

$$g(x) = \frac{A}{x} + \frac{8}{x^2}$$

If $g(-1) = 0$, find A.

73. Spheres The volume V of a sphere of radius r is $V = \frac{4}{3}\pi r^3$; the surface area S of this sphere is $S = 4\pi r^2$. If the radius doubles, how does the volume change? How does the surface area change?

74. Page Design A page with dimensions of $8\frac{1}{2}$ inches by 11 inches has a border of uniform width x surrounding the printed matter of the page, as shown in the figure.

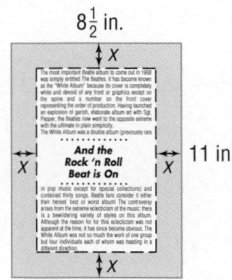

$8\frac{1}{2}$ in.

11 in.

(a) Express the area A of the printed part of the page as a function of the width x of the border.
(b) Give the domain and the range of A.
(c) Find the area of printed page for borders of widths 1 inch, 1.2 inches, and 1.5 inches.

(d) Graph the function $A = A(x)$.
(e) Use TRACE to determine what margin should be used to obtain areas of 70 square inches and 50 square inches.

75. Material Needed to Make a Drum A steel drum in the shape of a right circular cylinder is required to have a volume of 100 cubic feet.
(a) Express the amount A of material required to make the drum as a function of the radius r of the cylinder.
(b) How much material is required if the drum is of radius 3 feet?
(c) How much material is required if the drum is of radius 4 feet?
(d) How much material is required if the drum is of radius 5 feet?
(e) Graph $A = A(r)$. For what value of r is A smallest?

Hint: The volume V and surface area S of a right circular cylinder are

$$V = \pi r^2 h \qquad S = 2\pi r^2 + 2\pi rh$$

where h is the height and r is the radius

76. A rectangle has one vertex in quadrant I on the graph of $y = 10 - x^2$, another at the origin, one on the positive x-axis, and one on the positive y-axis.
(a) Express the area A of the rectangle as a function of x.
(b) Find the largest area A that can be enclosed by the rectangle.

CHAPTER TEST

1. Determine whether each relation represents a function. For each function, state the domain and range.
(a) $\{(2, 5), (4, 6), (6, 7), (8, 8)\}$
(b) $\{(1, 3), (4, -2), (-3, 5), (1, 7)\}$
(c)

(d)

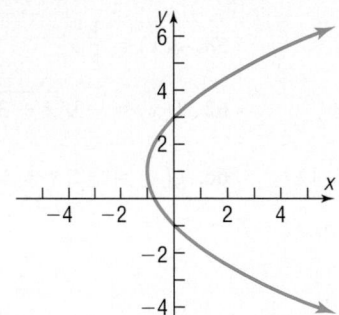

In Problems 2–4, find the domain of each function and evaluate each function at $x = -1$.

2. $f(x) = \sqrt{4 - 5x}$

3. $g(x) = \dfrac{x + 2}{|x + 2|}$

4. $h(x) = \dfrac{x - 4}{x^2 + 5x - 36}$

5. Using the graph of the function f:

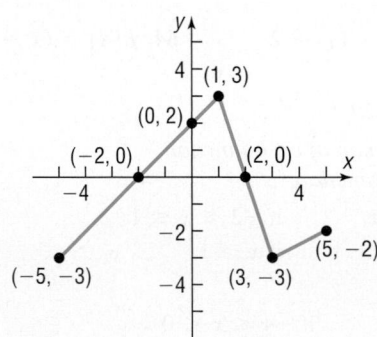

(a) Find the domain and the range of f.
(b) List the intercepts.
(c) Find $f(1)$.
(d) For what value(s) of x does $f(x) = -3$?
(e) Solve $f(x) < 0$.

6. Use a graphing utility to graph the function $f(x) = -x^4 + 2x^3 + 4x^2 - 2$ on the interval $(-5, 5)$. Approximate any local maxima and local minima rounded to two decimal places. Determine where the function is increasing and where it is decreasing.

7. Consider the function $g(x) = \begin{cases} 2x + 1 & \text{if } x < -1 \\ x - 4 & \text{if } x \geq -1 \end{cases}$

 (a) Graph the function. (b) List the intercepts.

 (c) Find $g(-5)$. (d) Find $g(2)$.

8. For the function $f(x) = 3x^2 - 2x + 4$, find the average rate of change of f from 3 to 4.

9. For the functions $f(x) = 2x^2 + 1$ and $g(x) = 3x - 2$, find the following and simplify:

 (a) $f - g$

 (b) $f \cdot g$

 (c) $f(x + h) - f(x)$

10. Graph each function using the techniques of shifting, compressing or stretching, and reflections. Start with the graph of the basic function and show all stages.

 (a) $h(x) = -2(x + 1)^3 + 3$

 (b) $g(x) = |x + 4| + 2$

11. The variable interest rate on a student loan changes each July 1 based on the bank prime loan rate. For the years 1992–2004, this rate can be approximated by the model

$r(x) = -0.115x^2 + 1.183x + 5.623$, where x is the number of years since 1992 and r is the interest rate as a percent.

 (a) Use a graphing utility to estimate the highest rate during this time period. During which year was the interest rate the highest?

 (b) Use the model to estimate the rate in 2010. Does this value seem reasonable?

Source: U.S. Federal Reserve

12. A community skating rink is in the shape of a rectangle with semicircles attached at the ends. The length of the rectangle is 20 feet less than twice the width. The thickness of the ice is 0.75 inch.

 (a) Write the ice volume, V, as a function of the width, x.

 (b) How much ice is in the rink if the width is 90 feet?

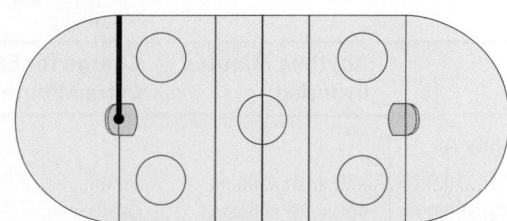

CUMULATIVE REVIEW

In Problems 1–6, find the real solutions of each equation.

1. $3x - 8 = 10$

2. $3x^2 - x = 0$

3. $x^2 - 8x - 9 = 0$

4. $6x^2 - 5x + 1 = 0$

5. $|2x + 3| = 4$

6. $\sqrt{2x + 3} = 2$

In Problems 7–9, solve each inequality. Graph the solution set.

7. $2 - 3x > 6$

8. $|2x - 5| < 3$

9. $|4x + 1| \geq 7$

10. (a) Find the distance from $P_1 = (-2, -3)$ to $P_2 = (3, -5)$.

 (b) What is the midpoint of the line segment from P_1 to P_2?

 (c) What is the slope of the line containing the points P_1 to P_2?

In Problems 11–14, graph each equation.

11. $3x - 2y = 12$

12. $x = y^2$

13. $x^2 + (y - 3)^2 = 16$

14. $y = \sqrt{x}$

15. For the equation $3x^2 - 4y = 12$, find the intercepts and check for symmetry.

16. Find the slope–intercept form of the equation of the line containing the points $(-2, 4)$ and $(6, 8)$.

In Problems 17–19, graph each function.

17. $f(x) = (x + 2)^2 - 3$

18. $f(x) = \dfrac{1}{x}$

19. $f(x) = \begin{cases} 2 - x & \text{if } x \leq 2 \\ |x| & \text{if } x > 2 \end{cases}$

CHAPTER PROJECTS

I. **Cell Phone Service** This month, as you were paying your bills, you noticed that your cell phone service contract had expired. Since cell phone numbers are now portable, you decided to investigate different companies. Since you regularly travel outside your local calling area, you decide to look only at plans with no roaming charges. Here are your findings:

	Anytime Minutes Included	Charge for Each Extra Minute	Mobile-to-Mobile Minutes	National Long Distance	Nights and Weekends
Company A:					
Plan A1: $39.99	450 with rollover	$0.45	Unlimited	Included	5000 (9 PM–6 AM)
Plan A2: $59.99	900 with rollover	$0.40	Unlimited	Included	Unlimited (9 PM–6 AM)
Company B:					
Plan B1: $39.99	600	$0.40	None	Included	Unlimited (9 PM–7 AM)
Plan B2: $49.99	1000	$0.40	None	Included	Unlimited (9 PM–7 AM)
Company C:					
Plan C1: $59.99	550	551–1050, $5 for each 50 minutes; Above 1050, $0.10 each minute	Unlimited	Included	Unlimited (7 PM–7 AM)
Plan C2: $69.99	800	801–1300, $5 for each 50 minutes; Above 1300, $0.10 each minute	Unlimited	Included	Unlimited (7 PM–7 AM)

Source: Based on rates from the Web sites of the companies Cingular, T Mobile and Sprint PCS for the zip code 76201 on May 20, 2006. (*www.cingular.com, www.t-mobile.com,www.sprintpcs.com*)

Each plan requires a 2-year contract.

1. Determine the total cost of each plan for the life of the contract, assuming that you stay within the allotted anytime minutes provided by each contract.

2. If you expect to use 400 anytime, 200 mobile-to-mobile, and 4500 night and weekend minutes per month, which plan provides the best deal? If you expect to use 400 anytime, 200 mobile-to-mobile, and 5500 night and weekend minutes per month, which plan provides the best deal? If you expect to use 500 anytime, and 1000 mobile-to-mobile, and 2000 night and weekend minutes per month, which plan provides the best deal?

3. Ignoring any mobile-to-mobile and night and weekend usage, if you expect to use 850 anytime minutes each month, which option provides the best deal? What if you use 1050 anytime minutes per month?

4. Each monthly charge includes a specific number of peak time minutes in the monthly fee. Write a function for each option, where C is the monthly cost and x is the number of anytime minutes used.

5. Graph each function from part 4.

6. For each of the companies A, B, and C, determine the average price per minute for each plan, based on no extra minutes used. For each company, which plan is better?

7. Looking at the three plans that you found to be the best for companies A, B, and C, in part 6, which of these three seems to be the best deal?

8. Based on your own cell phone usage, which plan would be the best for you?

The following projects are available on the Instructor's Resource Center (IRC):

II. **Project at Motorola: *Wireless Internet Service*** Use functions and their graphs to analyze the total cost of various wireless Internet service plans.

III. **Cost of Cable** When government regulations and customer preference influence the path of a new cable line, the Pythagorean Theorem can be used to assess the cost of installation.

IV. **Oil Spill** Functions are used to analyze the size and spread of an oil spill from a leaking tanker.

Linear and Quadratic Functions

4

Valuing Hewlett-Packard

17 MAY 2006—With the market focused on Hewlett-Packard Co. (HPQ) taking market share from Dell Inc. (DELL), the casual observer might be fooled into mistaking it as a computer hardware company. It's not; while the Compaq acquisition of 2002 expanded the company's hardware business, 53.4 percent of HPQ's profits still come from its imaging and printing division.

When we value stocks, we use the capital asset pricing model to price shares, which examines the volatility of the stock through beta and then projects how much earnings growth would be needed to justify the current stock price. Historically, HPQ has a beta of 2.05. The computer peripherals industry average, though, is 1.98. But the industry average beta for the computer hardware companies is 1.58.

A weighted average beta for the two industries seems appropriate for a company that walks the line between them. In this case, a beta weighted between the computer peripherals and computer hardware industries by the percentages of HPQ's profits yields an expected beta of 1.8. If we use this beta in a capital-asset-pricing model, assuming that Hewlett-Packard will maintain a price-to-earnings-to-growth ratio of 1.3 and that full fiscal year 2006 earnings per share will be at the low end of management's guidance, $2.04, then a new position taken at the current price would require an annualized, per-share earnings growth rate of 9.9% over the next five years to break even. The average long-term growth rate among the 12 analysts who provide one is 12.3% for Hewlett-Packard; this is well above what our analysis indicates that we require, implying a strong undervaluation even after the price spike.

Source: Adapted from *Paul DeMartino, "Valuing Hewlett-Packard," May 17, 2006, www.reuters.com. Adapted with permission.*

—See the Chapter Project—

A Look Back

Up to now, our discussion has focused on graphs of equations and the idea of a function. We have learned how to graph equations using the point-plotting method, intercepts, and the tests for symmetry. In addition, we have learned what a function is and how to identify whether a relation represents a function. We also introduced properties of functions, such as domain/range, increasing/decreasing, even/odd, and average rate of change.

A Look Ahead

Going forward, we will look at classes of functions. In this chapter, we focus on linear and quadratic functions and their properties and applications.

Outline

4.1 Linear Functions and Their Properties

PREPARING FOR THIS SECTION *Before getting started, review the following:*

- Lines (Section 2.3, pp. 173–185)
- Graphs of Equations in Two Variables; Intercepts; Symmetry (Section 2.2, pp. 163–171)
- Linear Equations (Section 1.1, pp. 86–93)

- Functions (Section 3.1, pp. 208–219)
- The Graph of a Function (Section 3.2, pp. 222–226)
- Properties of Functions (Section 3.3, pp. 231–238)

Now Work the 'Are You Prepared?' problems on page 284.

OBJECTIVES 1 Graph Linear Functions (p. 278)
2 Use Average Rate of Change to Identify Linear Functions (p. 278)
3 Determine Whether a Linear Function is Increasing, Decreasing, or Constant (p. 281)
4 Work with Applications of Linear Functions (p. 282)

1 Graph Linear Functions

In Section 2.3 we discussed lines. In particular, we developed the slope–intercept form of the equation of a line $y = mx + b$. When we write the slope–intercept form of a line using function notation, we have a *linear function*.

DEFINITION A **linear function** is a function of the form

$$f(x) = mx + b$$

The graph of a linear function is a line with slope m and y-intercept b. Its domain is the set of all real numbers.

EXAMPLE 1 **Graphing a Linear Function**

Figure 1

Graph the linear function: $f(x) = -3x + 7$

Solution This is a linear function with slope $m = -3$ and y-intercept $b = 7$. To graph this function, we start by plotting the point $(0, 7)$, the y-intercept, and use the slope to find an additional point by moving right 1 unit and down 3 units. See Figure 1.

Alternatively, we could have found an additional point by evaluating the function at some $x \neq 0$. For $x = 1$, we find $f(1) = -3(1) + 7 = 4$ and obtain the point $(1, 4)$ on the graph. ∎

Now Work PROBLEMS 13(a) AND (b)

2 Use Average Rate of Change to Identify Linear Functions

Look at Table 1, which shows certain values of the independent variable x and corresponding values of the dependent variable y for the function $f(x) = -3x + 7$. Notice that as the value of the independent variable, x, increases by 1 the value of the dependent variable y decreases by 3. That is, the average rate of change of y with respect to x is a constant, -3.

Table 1

It is not a coincidence that the average rate of change of the linear function $f(x) = -3x + 7$ is the slope of the linear function. That is, $\dfrac{\Delta y}{\Delta x} = m = -3$. The following theorem states this fact.

THEOREM

Average Rate of Change of a Linear Function

Linear functions have a constant average rate of change. That is, the average rate of change of a linear function $f(x) = mx + b$ is

$$\frac{\Delta y}{\Delta x} = m$$

Proof The average rate of change of $f(x) = mx + b$ from x_1 to x_2, $x_1 \neq x_2$, is

$$\frac{\Delta y}{\Delta x} = \frac{f(x_2) - f(x_1)}{x_2 - x_1} = \frac{(mx_2 + b) - (mx_1 + b)}{x_2 - x_1}$$

$$= \frac{mx_2 - mx_1}{x_2 - x_1} = \frac{m(x_2 - x_1)}{x_2 - x_1} = m \qquad \blacksquare$$

Based on the theorem just proved, the average rate of change of the function $g(x) = -\dfrac{2}{5}x + 5$ is $-\dfrac{2}{5}$.

Now Work PROBLEM 13(C)

As it turns out, only linear functions have a constant average rate of change. Because of this, we can use the average rate of change to determine whether a particular function is linear or not. Functions that are not linear are said to be **nonlinear.**

EXAMPLE 2

Using the Average Rate of Change to Identify Linear Functions

(a) A strain of E-coli Beu 397-rccA441 is placed into a Petri dish at 30° Celsius and allowed to grow. The data shown in Table 2 are collected. The population is measured in grams and the time in hours. Plot the ordered pairs (x, y) in the Cartesian plane and use the average rate of change to determine whether the function is linear.

(b) The data in Table 3 represent the maximum number of heartbeats that a healthy individual should have during a 15-second interval of time while exercising for different ages. Plot the ordered pairs (x, y) in the Cartesian plane and use the average rate of change to determine whether the function is linear.

Table 2

Time (hours), x	Population (grams), y	(x, y)
0	0.09	(0, 0.09)
1	0.12	(1, 0.12)
2	0.16	(2, 0.16)
3	0.22	(3, 0.22)
4	0.29	(4, 0.29)
5	0.39	(5, 0.39)

Table 3

Age, x	Maximum Number of Heart Beats, y	(x, y)
20	50	(20, 50)
30	47.5	(30, 47.5)
40	45	(40, 45)
50	42.5	(50, 42.5)
60	40	(60, 40)
70	37.5	(70, 37.5)

Source: American Heart Association

Solution

We compute the average rate of change of each function. If the average rate of change is constant, the function is linear. If the average rate of change is not constant, the function is nonlinear.

(a) Figure 2 shows the points listed in Table 2 plotted in the Cartesian plane. Notice that it is impossible to draw a straight line that contains all the points. Table 4 displays the average rate of change of population.

Figure 2

y axis: Population (grams), y — values 0.1, 0.2, 0.3, 0.4
x axis: Time (hours), x — values 0, 1, 2, 3, 4, 5

Table 4

Time (hours), x	Population (grams), y	Average Rate of Change $= \dfrac{\Delta y}{\Delta x}$
0	0.09	
		$\dfrac{0.12 - 0.09}{1 - 0} = 0.03$
1	0.12	
		0.04
2	0.16	
		0.06
3	0.22	
		0.07
4	0.29	
		0.10
5	0.39	

Because the average rate of change is not constant, we know that the function is not linear. In fact, because the average rate of change is increasing as the value of the independent variable increases, we say that the function is increasing at an increasing rate. So not only is the population increasing over time, but it is also growing more rapidly as time passes.

(b) Figure 3 shows the points listed in Table 3 plotted in the Cartesian plane. We can see that the data in Figure 3 lie on a straight line. Table 5 contains the average rate of change of the maximum number of heartbeats. The average rate of change of the heartbeat data is constant, -0.25 beats per year, so the function is linear.

Figure 3

Table 5

Age, x	Maximum Number of Heartbeats, y	Average Rate of Change = $\dfrac{\Delta y}{\Delta x}$
20	50	
		$\dfrac{47.5 - 50}{30 - 20} = -0.25$
30	47.5	
		-0.25
40	45	
		-0.25
50	42.5	
		-0.25
60	40	
		-0.25
70	37.5	

➤ **Now Work** PROBLEM 21

3 Determine Whether a Linear Function Is Increasing, Decreasing, or Constant

Look back at the Seeing the Concept on page 176. We know that when the slope m of a linear function is positive ($m > 0$) the line slants upward from left to right. When the slope m of a linear function is negative ($m < 0$), the line slants downward from left to right. When the slope m of a linear function is zero ($m = 0$), the line is horizontal.

Based on these results, we have the following theorem.

THEOREM

Increasing, Decreasing, and Constant Linear Functions

A linear function $f(x) = mx + b$ is increasing over its domain if its slope, m, is positive. It is decreasing over its domain if its slope, m, is negative. It is constant over its domain if its slope, m, is zero.

EXAMPLE 3

Determining Whether a Linear Function Is Increasing, Decreasing, or Constant

Determine whether the following linear functions are increasing, decreasing, or constant.

(a) $f(x) = 5x - 2$ (b) $g(x) = -2x + 8$

(c) $s(t) = \dfrac{3}{4}t - 4$ (d) $h(z) = 7$

Solution

(a) For the linear function $f(x) = 5x - 2$, the slope is 5, which is positive. The function f is increasing on the interval $(-\infty, \infty)$.

(b) For the linear function $g(x) = -2x + 8$, the slope is -2, which is negative. The function g is decreasing on the interval $(-\infty, \infty)$.

(c) For the linear function $s(t) = \dfrac{3}{4}t - 4$, the slope is $\dfrac{3}{4}$, which is positive. The function s is increasing on the interval $(-\infty, \infty)$.

(d) We can write the linear function h as $h(z) = 0z + 7$. Because the slope is 0, the function h is constant on the interval $(-\infty, \infty)$.

➤ **Now Work** PROBLEM 13(d)

4 Work with Applications of Linear Functions

There are many applications of linear functions. Let's look at one from accounting.

| EXAMPLE 4 | Straight-line Depreciation |

Book value is the value of an asset that a company uses to create its balance sheet. Some companies depreciate their assets using straight-line depreciation so that the value of the asset declines by a fixed amount each year. The amount of the decline depends on the useful life that the company places on the asset. Suppose that a company just purchased a fleet of new cars for its sales force at a cost of $28,000 per car. The company chooses to depreciate each vehicle using the straight-line method over 7 years. This means that each car will depreciate by $\dfrac{\$28{,}000}{7} = \4000 per year.

(a) Write a linear function that expresses the book value V of each car as a function of its age, x.

(b) Graph the linear function.

(c) What is the book value of each car after 3 years?

(d) Interpret the slope.

(e) When will the book value of each car be $8000?
 [**Hint:** Solve the equation $V(x) = 8000$.]

Solution

(a) If we let $V(x)$ represent the value of each car after x years, then $V(0)$ represents the original value of each car, so $V(0) = \$28{,}000$. The y-intercept of the linear function is $28,000. Because each car depreciates by $4000 per year, the slope of the linear function is -4000. The linear function that represents the book value V of each car after x years is

$$V(x) = -4000x + 28{,}000$$

(b) Figure 4 shows the graph of V.

(c) The book value of each car after 3 years is

$$V(3) = -4000(3) + 28{,}000$$
$$= \$16{,}000$$

(d) Since the slope of $V(x) = -4000x + 28{,}000$ is -4000, the average rate of change of book value is $-\$4000$/year. So for each additional year that passes the book value of the car decreases by $4000.

(e) To find when the book value is $8000, we solve the equation

$$V(x) = 8000$$
$$-4000x + 28{,}000 = 8000$$
$$-4000x = -20{,}000 \qquad \text{Subtract 28,000 from each side.}$$
$$x = \frac{-20{,}000}{-4000} = 5 \qquad \text{Divide by } -4000.$$

The car will have a book value of $8000 when it is 5 years old.

Figure 4

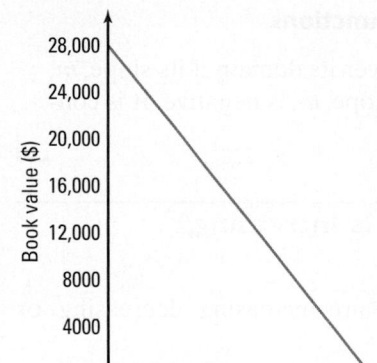

Age of vehicle (years)

Now Work PROBLEM 47

Next we look at an application from economics.

| EXAMPLE 5 | Supply and Demand |

The **quantity supplied** of a good is the amount of a product that a company is willing to make available for sale at a given price. The **quantity demanded** of a good is the amount of a product that consumers are willing to purchase at a given price.

Suppose that the quantity supplied, S, and quantity demanded, D, of cellular telephones each month are given by the following functions:

$$S(p) = 60p - 900$$

$$D(p) = -15p + 2850$$

where p is the price (in dollars) of the telephone.

(a) The **equilibrium price** of a product is defined as the price at which quantity supplied equals quantity demanded. That is, the equilibrium price is the price at which $S(p) = D(p)$. Find the equilibrium price of cellular telephones. What is the **equilibrium quantity,** the amount demanded (or supplied) at the equilibrium price?

(b) Determine the prices for which quantity supplied is greater than quantity demanded. That is, solve the inequality $S(p) > D(p)$.

(c) Graph $S = S(p)$, $D = D(p)$ and label the equilibrium price.

Solution (a) To find the equilibrium price, we solve the equation $S(p) = D(p)$.

$$60p - 900 = -15p + 2850 \quad S(p) = 60p - 900;$$
$$ D(p) = -15p + 2850$$

$$60p = -15p + 3750 \quad \text{Add 900 to each side.}$$

$$75p = 3750 \quad \text{Add 15p to each side.}$$

$$p = 50 \quad \text{Divide each side by 75.}$$

The equilibrium price is $50 per cellular phone. To find the equilibrium quantity, we evaluate either $S(p)$ or $D(p)$ at $p = 50$.

$$S(50) = 60(50) - 900 = 2100$$

The equilibrium quantity is 2100 cellular phones. At a price of $50 per phone, the company will produce and sell 2100 phones each month and have no shortages or excess inventory.

(b) We solve the inequality $S(p) > D(p)$.

$$60p - 900 > -15p + 2850 \quad S(p) > D(p)$$

$$60p > -15p + 3750 \quad \text{Add 900 to each side.}$$

$$75p > 3750 \quad \text{Add 15p to each side.}$$

$$p > 50 \quad \text{Divide each side by 75.}$$

If the company charges more than $50 per phone, quantity supplied will exceed quantity demanded. In this case the company will have excess phones in inventory.

(c) Figure 5 shows the graphs of $S = S(p)$ and $D = D(p)$ with the equilibrium point labeled.

Figure 5

Now Work PROBLEM 41

4.1 Assess Your Understanding

'Are You Prepared?' *Answers are given at the end of these exercises. If you get a wrong answer, read the pages listed in red.*

1. Graph $y = 2x - 3$. (pp. 173–185)

2. Find the slope of the line joining the points $(2, 5)$ and $(-1, 3)$. (pp. 173–174)

3. Find the average rate of change of $f(x) = 3x^2 - 2$, from 2 to 4. (pp. 231–238)

4. Solve: $60x - 900 = -15x + 2850$. (pp. 86–93)

5. If $f(x) = x^2 - 4$, find $f(-2)$. (pp. 208–219)

6. *True or False* The graph of the function $f(x) = x^2$ is increasing on the interval $(0, \infty)$. (pp. 231–238)

Concepts and Vocabulary

7. For the graph of the linear function $f(x) = mx + b$, m is the _____ and b is the _____.

8. For the graph of the linear function $H(z) = -4z + 3$, the slope is _____ and the y-intercept is _____.

9. If the slope m of the graph of a linear function is _____, the function is increasing over its domain.

10. *True or False* The slope of a line is the average rate of change of the linear function.

11. *True or False* If the average rate of change of a linear function is $\frac{2}{3}$, then if y increases by 3, x will increase by 2.

12. *True or False* The average rate of change of $f(x) = 2x + 8$ is 8.

Skill Building

In Problems 13–20, a linear function is given.
 (a) Determine the slope and y-intercept of each function.
 (b) Use the slope and y-intercept to graph the linear function.
 (c) Determine the average rate of change of each function.
 (d) Determine whether the linear function is increasing, decreasing, or constant.

13. $f(x) = 2x + 3$

14. $g(x) = 5x - 4$

15. $h(x) = -3x + 4$

16. $p(x) = -x + 6$

17. $f(x) = \frac{1}{4}x - 3$

18. $h(x) = -\frac{2}{3}x + 4$

19. $F(x) = 4$

20. $G(x) = -2$

In Problems 21–28, determine whether the given function is linear or nonlinear. If it is linear, determine the slope.

21.

x	y = f(x)
−2	4
−1	1
0	−2
1	−5
2	−8

22.

x	y = f(x)
−2	1/4
−1	1/2
0	1
1	2
2	4

23.

x	y = f(x)
−2	−8
−1	−3
0	0
1	1
2	0

24.

x	y = f(x)
−2	−4
−1	0
0	4
1	8
2	12

25.

x	y = f(x)
−2	−26
−1	−4
0	2
1	−2
2	−10

26.

x	y = f(x)
−2	−4
−1	−3.5
0	−3
1	−2.5
2	−2

27.

x	y = f(x)
−2	8
−1	8
0	8
1	8
2	8

28.

x	y = f(x)
−2	0
−1	1
0	4
1	9
2	16

Applications and Extensions

29. Suppose that $f(x) = 4x - 1$ and $g(x) = -2x + 5$.
 (a) Solve $f(x) = 0$. (b) Solve $f(x) > 0$.
 (c) Solve $f(x) = g(x)$. (d) Solve $f(x) \leq g(x)$.
 (e) Graph $y = f(x)$ and $y = g(x)$ and label the point that represents the solution to the equation $f(x) = g(x)$.

30. Suppose that $f(x) = 3x + 5$ and $g(x) = -2x + 15$.
 (a) Solve $f(x) = 0$. (b) Solve $f(x) < 0$.
 (c) Solve $f(x) = g(x)$. (d) Solve $f(x) \geq g(x)$.
 (e) Graph $y = f(x)$ and $y = g(x)$ and label the point that represents the solution to the equation $f(x) = g(x)$.

31. In parts (a)–(f), use the following figure.

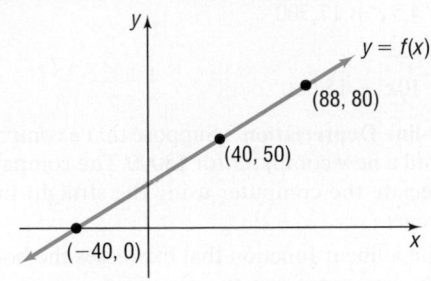

(a) Solve $f(x) = 50$. (b) Solve $f(x) = 80$.

(c) Solve $f(x) = 0$. (d) Solve $f(x) > 50$.

(e) Solve $f(x) \le 80$. (f) Solve $0 < f(x) < 80$.

32. In parts (a)–(f), use the following figure.

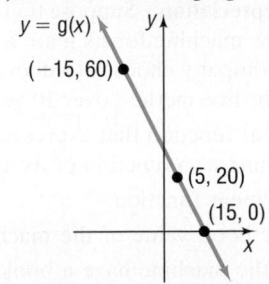

(a) Solve $g(x) = 20$. (b) Solve $g(x) = 60$.

(c) Solve $g(x) = 0$. (d) Solve $g(x) > 20$.

(e) Solve $g(x) \le 60$. (f) Solve $0 < g(x) < 60$.

33. In parts (a) and (b) use the following figure.

(a) Solve the equation: $f(x) = g(x)$.

(b) Solve the inequality: $f(x) > g(x)$.

34. In parts (a) and (b), use the following figure.

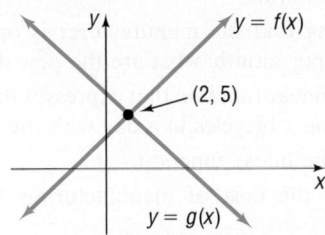

(a) Solve the equation: $f(x) = g(x)$.

(b) Solve the inequality: $f(x) \le g(x)$.

35. In parts (a) and (b), use the following figure.

(a) Solve the equation: $f(x) = g(x)$.

(b) Solve the inequality: $g(x) \le f(x) < h(x)$.

36. In parts (a) and (b), use the following figure.

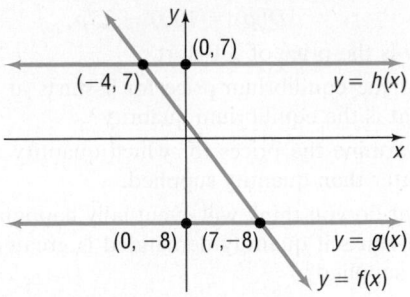

(a) Solve the equation: $f(x) = g(x)$.

(b) Solve the inequality: $g(x) < f(x) \le h(x)$.

37. Car Rentals The cost C, in dollars, of renting a moving truck for a day is given by the function $C(x) = 0.25x + 35$, where x is the number of miles driven.

(a) What is the cost if you drive $x = 40$ miles?

(b) If the cost of renting the moving truck is $80, how many miles did you drive?

(c) Suppose that you want the cost to be no more than $100. What is the maximum number of miles that you can drive?

38. Phone Charges The monthly cost C, in dollars, for international calls on a certain cellular phone plan is given by the function $C(x) = 0.38x + 5$, where x is the number of minutes used.

(a) What is the cost if you talk on the phone for $x = 50$ minutes?

(b) Suppose that your monthly bill is $29.32. How many minutes did you use the phone?

(c) Suppose that you budget yourself $60 per month for the phone. What is the maximum number of minutes that you can talk?

39. Disability Benefits The average monthly benefit B, in dollars, for individuals on disability is given by the function

$$B(t) = 19.25t + 585.72$$

where t is the number of years since January 1, 1990.

(a) What was the average monthly benefit in 2000 ($t = 10$)?

(b) In what year will the average monthly benefit be $893.72? [**Hint:** use a table with $\triangle t = 1$.]

(c) In what year will the average monthly benefit exceed $1000?

40. Health Expenditures The total private health expenditures H, in billions of dollars, is given by the function

$$H(t) = 26t + 411$$

where t is the number of years since January 1, 1990.

(a) What were total private health expenditures in 2000 ($t = 10$)?

(b) In what year will total private health expenditures be $879 billion? [**Hint:** use a table with $\triangle t = 1$.]

(c) In what year will total private health expenditures exceed $1 trillion ($1,000 billion)?

41. Supply and Demand Suppose that the quantity supplied S and quantity demanded D of T-shirts at a concert are given by the following functions:

$$S(p) = -200 + 50p$$
$$D(p) = 1000 - 25p$$

where p is the price of a T-shirt.
(a) Find the equilibrium price for T-shirts at this concert. What is the equilibrium quantity?
(b) Determine the prices for which quantity demanded is greater than quantity supplied.
(c) What do you think will eventually happen to the price of T-shirts if quantity demanded is greater than quantity supplied?

42. Supply and Demand Suppose that the quantity supplied S and quantity demanded D of hot dogs at a baseball game are given by the following functions:

$$S(p) = -2000 + 3000p$$
$$D(p) = 10,000 - 1000p$$

where p is the price of a hot dog.
(a) Find the equilibrium price for hot dogs at the baseball game. What is the equilibrium quantity?
(b) Determine the prices for which quantity demanded is less than quantity supplied.
(c) What do you think will eventually happen to the price of hot dogs if quantity demanded is less than quantity supplied?

43. Taxes The function $T(x) = 0.15(x - 7300) + 730$ represents the tax bill T of a single person whose adjusted gross income is x dollars for income between $7300 and $29,700, inclusive in 2005.

Source: Internal Revenue Service

(a) What is the domain of this linear function?
(b) What is a single filer's tax bill if adjusted gross income is $18,000?
(c) Which variable is independent and which is dependent?
(d) Graph the linear function over the domain specified in part (a).
(e) What is a single filer's adjusted gross income if the tax bill is $2860?

44. Luxury Tax In 2002, major league baseball signed a labor agreement with the players. In this agreement, any team whose payroll exceeds $128 million in 2005 will have to pay a luxury tax of 22.5% (for first-time offenses). The linear function $T(p) = 0.225(p - 128)$ describes the luxury tax T of a team whose payroll is p (in millions of dollars).
(a) What is the implied domain of this linear function?
(b) What is the luxury tax for a team whose payroll is $160 million?
(c) Graph the linear function.
(d) What is the payroll of a team that pays a luxury tax of $11.7 million?

*The point at which a company's profits equal zero is called the company's **break-even point**. For Problems 45 and 46, let R represent a company's revenue, let C represent the company's costs, and let x represent the number of units produced and sold each day.*
(a) Find the firm's break-even point; that is, find x so that R = C.
(b) Find the values of x such that R(x) > C(x). This represents the number of units that the company must sell to earn a profit.

45. $R(x) = 8x$
$C(x) = 4.5x + 17,500$

46. $R(x) = 12x$
$C(x) = 10x + 15,000$

47. Straight-line Depreciation Suppose that a company has just purchased a new computer for $3000. The company chooses to depreciate the computer using the straight-line method over 3 years.
(a) Write a linear function that expresses the book value V of the computer as a function of its age x.
(b) Graph the linear function.
(c) What is the book value of the computer after 2 years?
(d) When will the computer have a book value of $2000?

48. Straight-line Depreciation Suppose that a company has just purchased a new machine for its manufacturing facility for $120,000. The company chooses to depreciate the machine using the straight-line method over 10 years.
(a) Write a linear function that expresses the book value V of the machine as a function of its age x.
(b) Graph the linear function.
(c) What is the book value of the machine after 4 years?
(d) When will the machine have a book value of $72,000?

49. Cost Function The simplest cost function is the linear cost function, $C(x) = mx + b$, where the y-intercept b represents the fixed costs of operating a business and the slope m represents the cost of each item produced. Suppose that a small bicycle manufacturer has daily fixed costs of $1800 and each bicycle costs $90 to manufacture.
(a) Write a linear function that expresses the cost C of manufacturing x bicycles in a day.
(b) Graph the linear function.
(c) What is the cost of manufacturing 14 bicycles in a day?
(d) How many bicycles could be manufactured for $3780?

50. Cost Function Refer to Problem 49. Suppose that the landlord of the building increases the bicycle manufacturer's rent by $100 per month.
(a) Assuming that the manufacturer is open for business 20 days per month, what are the new daily fixed costs?
(b) Write a linear function that expresses the cost C of manufacturing x bicycles in a day with the higher rent.
(c) Graph the linear function.
(d) What is the cost of manufacturing 14 bicycles in a day?
(e) How many bicycles could be manufactured for $3780?

51. Truck Rentals A truck rental company rents a truck for one day by charging $29 plus $0.07 per mile.
(a) Write a linear function that relates the cost C, in dollars, of renting the truck to the number x of miles driven.
(b) What is the cost of renting the truck if the truck is driven 110 miles? 230 miles?

52. Long Distance A phone company offers a domestic long distance package by charging $5 plus $0.05 per minute.
(a) Write a linear function that relates the cost C, in dollars, of talking x minutes.
(b) What is the cost of talking 105 minutes? 180 minutes?

Discussion and Writing

53. Which of the following functions might have the graph shown? (More than one answer is possible.)

(a) $f(x) = 2x - 7$

(b) $g(x) = -3x + 4$

(c) $H(x) = 5$

(d) $F(x) = 3x + 4$

(e) $G(x) = \frac{1}{2}x + 2$

54. Which of the following functions might have the graph shown? (More than one answer is possible.)

(a) $f(x) = 3x + 1$

(b) $g(x) = -2x + 3$

(c) $H(x) = 3$

(d) $F(x) = -4x - 1$

(e) $G(x) = -\frac{2}{3}x + 3$

55. Under what circumstances is a linear function $f(x) = mx + b$ odd? Can a linear function ever be even?

'Are You Prepared?' Answers

1.

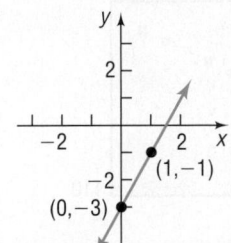

2. $\frac{2}{3}$ **3.** 36 **4.** $\{50\}$ **5.** 0 **6.** True

4.2 Building Linear Functions from Data

OBJECTIVES **1** Draw and Interpret Scatter Diagrams (p. 287)

2 Distinguish between Linear and Nonlinear Relations (p. 288)

3 Use a Graphing Utility to Find the Line of Best Fit (p. 289)

1 Draw and Interpret Scatter Diagrams

In Section 4.1, we built linear functions from verbal descriptions. Linear functions can also be constructed by fitting a linear function to data. The first step is to plot the ordered pairs using rectangular coordinates. The resulting graph is called a **scatter diagram.**

EXAMPLE 1 **Drawing and Interpreting a Scatter Diagram**

The data listed in Table 6 represent the apparent temperature versus the relative humidity in a room whose actual temperature is 72° Fahrenheit.

Table 6

Relative Humidity (%), x	Apparent Temperature °F, y	(x, y)
0	64	(0, 64)
10	65	(10, 65)
20	67	(20, 67)
30	68	(30, 68)
40	70	(40, 70)
50	71	(50, 71)
60	72	(60, 72)
70	73	(70, 73)
80	74	(80, 74)
90	75	(90, 75)
100	76	(100, 76)

(a) Draw a scatter diagram by hand treating relative humidity as the independent variable.

(b) Use a graphing utility to draw a scatter diagram.*

(c) Describe what happens to the apparent temperature as the relative humidity increases.

Solution

(a) To draw a scatter diagram by hand, we plot the ordered pairs listed in Table 6, with relative humidity as the *x*-coordinate and apparent temperature as the *y*-coordinate. See Figure 6(a). Notice that the points in a scatter diagram are not connected.

(b) Figure 6(b) shows a scatter diagram using a graphing utility.

Figure 6

(a)

(b)

(c) We see from the scatter diagrams that, as the relative humidity increases, the apparent temperature increases.

━━━ Now Work PROBLEM 9(a)

2 Distinguish between Linear and Nonlinear Relations

Scatter diagrams are used to help us see the type of relation that exists between two variables. In this text, we will discuss a variety of different relations that may exist between two variables. For now, we concentrate on distinguishing between linear and nonlinear relations. See Figure 7.

Figure 7

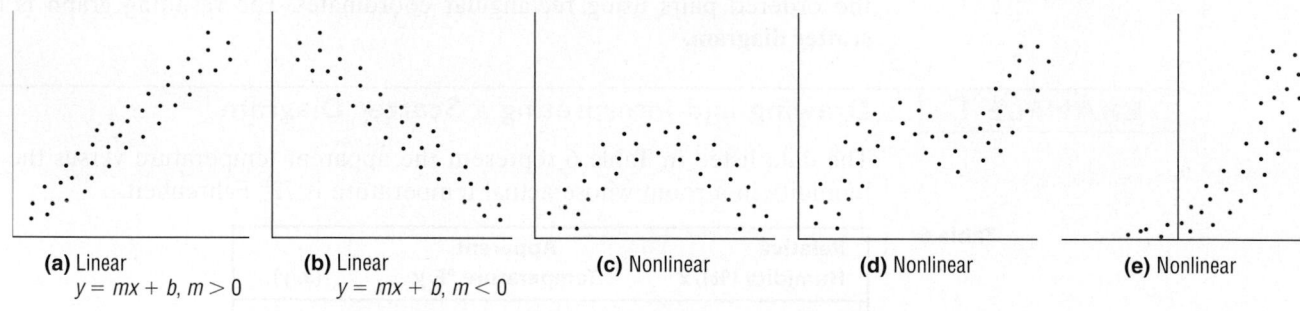

(a) Linear
$y = mx + b, m > 0$

(b) Linear
$y = mx + b, m < 0$

(c) Nonlinear

(d) Nonlinear

(e) Nonlinear

| EXAMPLE 2 | Distinguishing between Linear and Nonlinear Relations |

Determine whether the relation between the two variables in Figure 8 is linear or nonlinear.

Figure 8

(a)

(b)

(c)

(d)

* Consult your owner's manual for the proper keystrokes.

Solution (a) Linear (b) Nonlinear (c) Nonlinear (d) Nonlinear

 Now Work PROBLEM 3

In this section we will study data whose scatter diagrams imply that a linear relation exists between the two variables.

Suppose that the scatter diagram of a set of data appears to be linearly related as in Figure 7(a) or (b). We might wish to find an equation of a line that relates the two variables. One way to obtain an equation for such data is to draw a line through two points on the scatter diagram and determine the equation of the line.

EXAMPLE 3 **Finding an Equation for Linearly Related Data**

Using the data in Table 6 from Example 1,

(a) Select two points and find an equation of the line containing the points.
(b) Graph the line on the scatter diagram obtained in Example 1(a).

Solution (a) Select two points, say $(10, 65)$ and $(70, 73)$. The slope of the line joining the points $(10, 65)$ and $(70, 73)$ is

$$m = \frac{73 - 65}{70 - 10} = \frac{8}{60} = \frac{2}{15}$$

The equation of the line with slope $\frac{2}{15}$ and passing through $(10, 65)$ is found using the point–slope form with $m = \frac{2}{15}$, $x_1 = 10$, and $y_1 = 65$.

$$y - y_1 = m(x - x_1) \quad \textit{Point–slope form}$$
$$y - 65 = \frac{2}{15}(x - 10) \quad m = \tfrac{2}{15}, \ x_1 = 10, \ y_1 = 65$$
$$y = \frac{2}{15}x + \frac{191}{3} \quad y = mx + b$$

(b) Figure 9 shows the scatter diagram from Example 1(a) with the graph of the line.

Figure 9

Apparent Temperature (vertical axis: 62, 64, 66, 68, 70, 72, 74, 76, 78, 80)

Relative Humidity (horizontal axis: 0, 10, 20, 30, 40, 50, 60, 70, 80, 90, 100)

 Select two other points and complete the solution. Add the line obtained to Figure 9.

 Now Work PROBLEMS 9(b) AND (c)

3 Use a Graphing Utility to Find the Line of Best Fit

The line obtained in Example 3 depends on the selection of points, which will vary from person to person. So the line that we found might be different from the line you found. Although the line we found in Example 3 appears to fit the data well, there may be a line that "fits it better." Do you think your line fits the data better? Is there a *line of best fit*? As it turns out, there is a method for finding the line that best fits linearly related data (called the **line of best fit**).*

EXAMPLE 4 **Finding the Line of Best Fit**

Using the data in Table 6 from Example 1:

(a) Find the line of best fit using a graphing utility.
(b) Graph the line of best fit on the scatter diagram obtained in Example 1(b).

* We shall not discuss the underlying mathematics of lines of best fit in this book.

(c) Interpret the slope of the line of best fit.

(d) Use the line of best fit to predict the apparent temperature of a room whose actual temperature is 72°F and relative humidity is 45%.

Solution

(a) Graphing utilities contain built-in programs that find the line of best fit for a collection of points in a scatter diagram. (Look in your owner's manual under Linear Regression or Line of Best Fit for details on how to execute the program.) Upon executing the LINear REGression program, we obtain the results shown in Figure 10. The output that the utility provides shows us the equation $y = ax + b$, where a is the slope of the line and b is the y-intercept. The line of best fit that relates relative humidity to apparent temperature may be expressed as the line $y = 0.121x + 64.409$.

(b) Figure 11 shows the graph of the line of best fit, along with the scatter diagram.

(c) The slope of the line of best fit is 0.121, which means that, for every 1% increase in the relative humidity, apparent room temperature increases 0.121°F.

(d) Letting $x = 45$ in the equation of the line of best fit, we obtain $y = 0.121(45) + 64.409 \approx 70°F$, which is the apparent temperature in the room.

Figure 10

Figure 11

✏ **Now Work** PROBLEMS 9(d) AND (e)

Does the line of best fit appear to be a good fit? In other words, does the line appear to accurately describe the relation between temperature and relative humidity?

And just how "good" is this line of best fit? The answers are given by what is called the *correlation coefficient*. Look again at Figure 10. The last line of output is $r = 0.994$. This number, called the **correlation coefficient,** r, $-1 \le r \le 1$, is a measure of the strength of the *linear relation* that exists between two variables. The closer that $|r|$ is to 1, the more perfect the linear relationship is. If r is close to 0, there is little or no *linear* relationship between the variables. A negative value of $r, r < 0$, indicates that as x increases y decreases; a positive value of $r, r > 0$, indicates that as x increases y does also. The data given in Example 1, having a correlation coefficient of 0.994, are indicative of a strong linear relationship with positive slope.

4.2 Assess Your Understanding

Concepts and Vocabulary

1. A(n) _____ is used to help us to see the type of relation, if any, that may exist between two variables.

2. *True or False* The correlation coefficient is a measure of the strength of a linear relation between two variables and must lie between −1 and 1, inclusive.

Skill Building

In Problems 3–8, examine the scatter diagram and determine whether the type of relation that may exist is linear or nonlinear.

3.

4.

5.

6.

7.

8.

In Problems 9–14
 (a) *Draw a scatter diagram.*
 (b) *Select two points from the scatter diagram and find the equation of the line containing the points selected.**
 (c) *Graph the line found in part (b) on the scatter diagram.*
 (d) *Use a graphing utility to find the line of best fit.*
 (e) *Use a graphing utility to draw the scatter diagram and graph the line of best fit on it.*

9.

x	3	4	5	6	7	8	9
y	4	6	7	10	12	14	16

10.

x	3	5	7	9	11	13
y	0	2	3	6	9	11

11.

x	-2	-1	0	1	2
y	-4	0	1	4	5

12.

x	-2	-1	0	1	2
y	7	6	3	2	0

13.

x	-20	-17	-15	-14	-10
y	100	120	118	130	140

14.

x	-30	-27	-25	-20	-14
y	10	12	13	13	18

Applications and Extensions

15. Consumption and Disposable Income An economist wants to estimate a line that relates personal consumption expenditures C and disposable income I. Both C and I are in thousands of dollars. She interviews eight heads of households for families of size 3 and obtains the data shown below. Let I represent the independent variable and C the dependent variable.

 (a) Draw a scatter diagram.
 (b) Find a line that fits the data.*
 (c) Interpret the slope. The slope of this line is called the **marginal propensity to consume.**
 (d) Predict the consumption of a family whose disposable income is $42,000.
 (e) Use a graphing utility to find the line of best fit to the data. What is the correlation coefficient?

I (000)	C (000)
20	16
20	18
18	13
27	21
36	27
37	26
45	36
50	39

16. Marginal Propensity to save The same economist as in Problem 15 wants to estimate a line that relates savings S and disposable income I. Let $S = I - C$ be the dependent variable and I the independent variable.

 (a) Draw a scatter diagram.
 (b) Find a line that fits the data.*
 (c) Interpret the slope. The slope of this line is called the **marginal propensity to save.**
 (d) Predict the savings of a family whose income is $42,000.
 (e) Use a graphing utility to find the line of best fit. What is the correlation coefficient?

17. Candy The following data represent the weight (in grams) of various candy bars and the corresponding number of calories.

 (a) Draw a scatter diagram of the data treating weight as the independent variable.
 (b) What type of relation appears to exist between the weight of a candy bar and the number of calories?

Candy Bar	Weight, x	Calories, y
Hershey's Milk Chocolate®	44.28	230
Nestle's Crunch®	44.84	230
Butterfinger®	61.30	270
Baby Ruth®	66.45	280
Almond Joy®	47.33	220
Twix® (with Caramel)	58.00	280
Snickers®	61.12	280
Heath®	39.52	210

Source: Megan Pocius, Student at Joliet Junior College

 (c) Select two points and find an equation of the line containing the points.*
 (d) Graph the line on the scatter diagram drawn in part (a).
 (e) Predict the number of calories in a candy bar that weighs 62.3 grams.
 (f) Interpret the slope of the line found in part (c).

18. Raisins The following data represent the weight w (in grams) of a box of raisins and the number N of raisins in the box.

Weight (in grams), w	Number of Raisins, N
42.3	87
42.7	91
42.8	93
42.4	87
42.6	89
42.4	90
42.3	82
42.5	86
42.7	86
42.5	86

Source: Jennifer Maxwell, Student at Joliet Junior College

 (a) Does the relation defined by the set of ordered pairs (w, N) represent a function?
 (b) Draw a scatter diagram of the data treating weight as the independent variable.
 (c) Select two points and find the equation of the line containing the points.*

*Answers will vary. We will use the first and the last data points in the answer section.

(d) Graph the line on the scatter diagram drawn in part (b).

(e) Express the relationship found in part (c) using function notation.

(f) Predict the number of raisins in a box that weighs 42.5 grams.

(g) Interpret the slope of the line found in part (c).

19. Height versus Head Circumference A pediatrician wanted to estimate a linear function that relates a child's height, H, to his or her head circumference, C. She randomly selects nine children from her practice, measures their height and head circumference, and obtains the data shown below. Let H represent the independent variable and C the dependent variable.

(a) Use a graphing utility to draw a scatter diagram.

(b) Use a graphing utility to find the line of best fit to the data. Express the solution using function notation.

(c) Interpret the slope.

(d) Predict the head circumference of a child that is 26 inches tall.

(e) Predict the height of a child whose head circumference is 17.4 inches.

Height, *H* (inches)	Head Circumference, *C* (inches)
25.25	16.4
25.75	16.9
25	16.9
27.75	17.6
26.5	17.3
27	17.5
26.75	17.3
26.75	17.5
27.5	17.5

Source: Denise Slucki, Student at Joliet Junior College

20. Gestation Period versus Life Expectancy A researcher would like to estimate the linear function relating the gestation period of an animal, G, and its life expectancy, L. She collects the following data.

Animal	Gestation (or incubation) Period, *G* (days)	Life Expectancy, *L* (years)
Cat	63	11
Chicken	22	7.5
Dog	63	11
Duck	28	10
Goat	151	12
Lion	108	10
Parakeet	18	8
Pig	115	10
Rabbit	31	7
Squirrel	44	9

Source: Time Almanac 2000

Let G represent the independent variable and L the dependent variable.

(a) Use a graphing utility to draw a scatter diagram.

(b) Use a graphing utility to find the line of best fit to the data. Express the solution using function notation.

(c) Interpret the slope.

(d) Predict the life expectancy of an animal whose gestation period is 89 days.

21. Demand for Jeans The marketing manager at Levi-Strauss wishes to find a function that relates the demand D for men's jeans and p, the price of the jeans. The following data were obtained based on a price history of the jeans.

Price ($/Pair), *p*	Demand (Pairs of Jeans Sold per Day), *D*
20	60
22	57
23	56
23	53
27	52
29	49
30	44

(a) Does the relation defined by the set of ordered pairs (p, D) represent a function?

(b) Draw a scatter diagram of the data.

(c) Using a graphing utility, find the line of best fit relating price and quantity demanded. What is the correlation coefficient?

(d) Interpret the slope.

(e) Express the relationship found in part (c) using function notation.

(f) What is the domain of the function?

(g) How many jeans will be demanded if the price is $28 a pair?

22. Advertising and Sales Revenue A marketing firm wishes to find a function that relates the sales S of a product and A, the amount spent on advertising the product. The data are obtained from past experience. Advertising and sales are measured in thousands of dollars.

Advertising Expenditures, *A*	Sales, *S*
20	335
22	339
22.5	338
24	343
24	341
27	350
28.3	351

(a) Does the relation defined by the set of ordered pairs (A, S) represent a function?

(b) Draw a scatter diagram of the data.

(c) Using a graphing utility, find the line of best fit relating advertising expenditures and sales. What is the correlation coefficient?

(d) Interpret the slope.

(e) Express the relationship found in part (c) using function notation.

(f) What is the domain of the function?

(g) Predict sales if advertising expenditures are $25,000.

Discussion and Writing

23. **Maternal Age versus Down Syndrome** A biologist would like to know how the age of the mother affects the incidence rate of Down syndrome. The following data represent the age of the mother and the incidence rate of Down syndrome per 1000 pregnancies.

Draw a scatter diagram treating age of the mother as the independent variable. Explain why it would not make sense to find the line of best fit for these data.

24. Find the line of best fit for the ordered pairs $(1, 5)$ and $(3, 8)$. What is the correlation coefficient for these data? Why is this result reasonable?

25. What does a correlation coefficient of 0 imply?

Age of Mother, x	Incidence of Down Syndrome, y
33	2.4
34	3.1
35	4
36	5
37	6.7
38	8.3
39	10
40	13.3
41	16.7
42	22.2
43	28.6
44	33.3
45	50

Source: E. B. Hook, P. K. Cross, and D. M. Schreinemachers, "Chromosomal abnormality rates at amniocentesis and in live-born infants," *Journal of the American Medical Association*, 249(15), 1983, pp. 2034-2038.

4.3 Quadratic Functions and Their Properties

PREPARING FOR THIS SECTION *Before getting started, review the following:*

- Intercepts (Section 2.2, pp. 165–166)
- Graphing Techniques: Transformations (Section 3.5, pp. 252–260)
- Completing the Square (Section 1.2, pp. 99–100)
- Quadratic Equations (Section 1.2, pp. 100–106)

Now Work the 'Are You Prepared?' problems on page 302.

OBJECTIVES 1 Graph a Quadratic Function Using Transformations (p. 295)
2 Identify the Vertex and Axis of Symmetry of a Quadratic Function (p. 297)
3 Graph a Quadratic Function Using Its Vertex, Axis, and Intercepts (p. 297)
4 Find the Maximum or Minimum Value of a Quadratic Function (p. 301)

Quadratic Functions

Here are some examples of quadratic functions.

$$F(x) = 3x^2 - 5x + 1 \qquad g(x) = -6x^2 + 1 \qquad H(x) = \frac{1}{2}x^2 + \frac{2}{3}x$$

DEFINITION

A **quadratic function** is a function of the form

$$f(x) = ax^2 + bx + c$$

where a, b, and c are real numbers and $a \neq 0$. The domain of a quadratic function is the set of all real numbers.

We may also say that a quadratic function is a function defined by a second-degree polynomial in one variable.

Many applications require a knowledge of quadratic functions. For example, suppose that Texas Instruments collects the data shown in Table 7, which relate the number of calculators sold at the price p (in dollars) per calculator. Since the price of a product determines the quantity that will be purchased, we treat price as the independent variable. The relationship between the number x of calculators sold and the price p per calculator is given by the linear equation

$$x = 21,000 - 150p$$

Table 7

Price per Calculator, p (Dollars)	Number of Calculators, x
60	12,000
65	11,250
70	10,500
75	9,750
80	9,000
85	8,250
90	7,500

Then the revenue R derived from selling x calculators at the price p per calculator is equal to the unit selling price p of the calculator times the number x of units actually sold. That is,

$$R = xp$$
$$R(p) = (21,000 - 150p)p \quad \text{\small x = 21,000 - 150p}$$
$$= -150p^2 + 21,000p$$

So the revenue R is a quadratic function of the price p. Figure 12 illustrates the graph of this revenue function, whose domain is $0 \leq p \leq 140$, since both x and p must be nonnegative. In the next section we shall determine the price p that maximizes revenue.

Figure 12

Figure 13
Path of a cannonball

A second situation in which a quadratic function appears involves the motion of a projectile. Based on Newton's Second Law of Motion (force equals mass times acceleration, $F = ma$), it can be shown that, ignoring air resistance, the path of a projectile propelled upward at an inclination to the horizontal is the graph of a quadratic function. See Figure 13 for an illustration. In the next section we shall analyze the path of a projectile.

1 Graph a Quadratic Function Using Transformations

We know how to graph the square function $f(x) = x^2$. Figure 14 shows the graph of three functions of the form $f(x) = ax^2, a > 0$, for $a = 1, a = \dfrac{1}{2}$, and $a = 3$.

Notice that the larger the value of a, the "narrower" the graph is, and the smaller the value of a, the "wider" the graph is.

Figure 15 shows the graphs of $f(x) = ax^2$ for $a < 0$. Notice that these graphs are reflections about the x-axis of the graphs in Figure 14. Based on the results of these two figures, we can draw some general conclusions about the graph of $f(x) = ax^2$. First, as $|a|$ increases, the graph becomes "taller" (a vertical stretch), and as $|a|$ gets closer to zero, the graph gets "shorter" (a vertical compression). Second, if a is positive, the graph opens "up," and if a is negative, the graph opens "down."

Figure 14

Figure 15

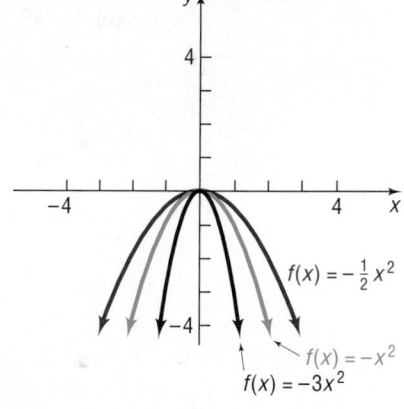

Figure 16

Graphs of a quadratic function,
$f(x) = ax^2 + bx + c, a \neq 0$

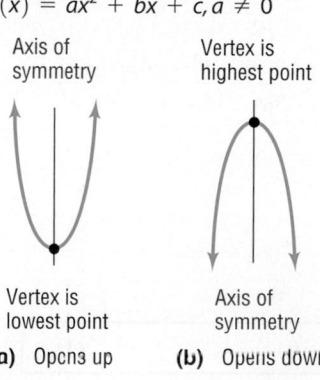

Axis of symmetry

Vertex is highest point

Vertex is lowest point

Axis of symmetry

(a) Opens up
$a > 0$

(b) Opens down
$a < 0$

The graphs in Figures 14 and 15 are typical of the graphs of all quadratic functions, which we call **parabolas.*** Refer to Figure 16, where two parabolas are pictured. The one on the left **opens up** and has a lowest point; the one on the right **opens down** and has a highest point. The lowest or highest point of a parabola is called the **vertex.** The vertical line passing through the vertex in each parabola in Figure 16 is called the **axis of symmetry** (usually abbreviated to **axis**) of the parabola. Because the parabola is symmetric about its axis, the axis of symmetry of a parabola can be used to find additional points on the parabola.

The parabolas shown in Figure 16 are the graphs of a quadratic function $f(x) = ax^2 + bx + c, a \neq 0$. Notice that the coordinate axes are not included in the figure. Depending on the values of a, b, and c, the axes could be placed anywhere. The important fact is that the shape of the graph of a quadratic function will look like one of the parabolas in Figure 16.

In the following example, we use techniques from Section 3.5 to graph a quadratic function $f(x) = ax^2 + bx + c, a \neq 0$. In so doing, we shall complete the square and write the function f in the form $f(x) = a(x - h)^2 + k$.

EXAMPLE 1	**Graphing a Quadratic Function Using Transformations**

Graph the function $f(x) = 2x^2 + 8x + 5$. Find the vertex and axis of symmetry.

* We shall study parabolas using a geometric definition later in this book.

Solution We begin by completing the square on the right side.

$$f(x) = 2x^2 + 8x + 5$$

$$= 2(x^2 + 4x) + 5 \qquad \text{Factor out the 2 from } 2x^2 + 8x.$$

$$= 2(x^2 + 4x + 4) + 5 - 8 \qquad \text{Complete the square of } x^2 + 4x \text{ by adding 4.}$$
$$\qquad\qquad\qquad\qquad\qquad\quad \text{Notice that the factor of 2 requires that 8 be}$$
$$= 2(x + 2)^2 - 3 \qquad\qquad\quad \text{added and subtracted.}$$

The graph of f can be obtained in three stages, as shown in Figure 17. Now compare this graph to the graph in Figure 16(a). The graph of $f(x) = 2x^2 + 8x + 5$ is a parabola that opens up and has its vertex (lowest point) at $(-2, -3)$. Its axis of symmetry is the line $x = -2$.

Figure 17

(a) $y = x^2$ — Multiply by 2; Vertical stretch

(b) $y = 2x^2$ — Replace x by $x + 2$; Shift left 2 units

(c) $y = 2(x + 2)^2$ — Subtract 3; Shift down 3 units

(d) $y = 2(x + 2)^2 - 3$ — Vertex $(-2, -3)$

Now Work PROBLEM 27

The method used in Example 1 can be used to graph any quadratic function $f(x) = ax^2 + bx + c, a \neq 0$, as follows:

$$f(x) = ax^2 + bx + c$$

$$= a\left(x^2 + \frac{b}{a}x\right) + c \qquad \text{Factor out } a \text{ from } ax^2 + bx.$$

$$= a\left(x^2 + \frac{b}{a}x + \frac{b^2}{4a^2}\right) + c - a\left(\frac{b^2}{4a^2}\right) \qquad \text{Complete the square by adding } \frac{b^2}{4a^2}.$$
$$\qquad\qquad\qquad\qquad\qquad\qquad\qquad\qquad\quad \text{Look closely at this step!}$$

$$= a\left(x + \frac{b}{2a}\right)^2 + c - \frac{b^2}{4a} \qquad \text{Factor.}$$

$$= a\left(x + \frac{b}{2a}\right)^2 + \frac{4ac - b^2}{4a} \qquad c - \frac{b^2}{4a} = c \cdot \frac{4a}{4a} - \frac{b^2}{4a} = \frac{4ac - b^2}{4a}$$

Based on these results, we conclude the following:

If $h = -\dfrac{b}{2a}$ and $k = \dfrac{4ac - b^2}{4a}$, then

$$\boxed{f(x) = ax^2 + bx + c = a(x - h)^2 + k \qquad\qquad\qquad \textbf{(1)}}$$

The graph of $f(x) = a(x - h)^2 + k$ is the parabola $y = ax^2$ shifted horizontally h units (replace x by $x - h$) and vertically k units (add k). As a result, the vertex is at (h, k), and the graph opens up if $a > 0$ and down if $a < 0$. The axis of symmetry is the vertical line $x = h$.

For example, compare equation (1) with the solution given in Example 1.

$$f(x) = 2(x + 2)^2 - 3$$
$$= 2(x - (-2))^2 + (-3)$$
$$= a(x - h)^2 + k$$

We conclude that $a = 2$, so the graph opens up. Also, we find that $h = -2$ and $k = -3$, so its vertex is at $(-2, -3)$.

2 Identify the Vertex and Axis of Symmetry of a Quadratic Function

We do not need to complete the square to obtain the vertex. In almost every case, it is easier to obtain the vertex of a quadratic function f by remembering that its x-coordinate is $h = -\dfrac{b}{2a}$. The y-coordinate k can then be found by evaluating f at $-\dfrac{b}{2a}$. That is, $k = f\left(-\dfrac{b}{2a}\right)$.

We summarize these remarks as follows:

> **Properties of the Graph of a Quadratic Function**
>
> $$f(x) = ax^2 + bx + c \qquad a \neq 0$$
>
> $$\text{Vertex} = \left(-\frac{b}{2a}, f\left(-\frac{b}{2a}\right)\right) \quad \text{Axis of symmetry: the line } x = -\frac{b}{2a} \qquad \textbf{(2)}$$
>
> Parabola opens up if $a > 0$; the vertex is a minimum point.
> Parabola opens down if $a < 0$; the vertex is a maximum point.

EXAMPLE 2 Locating the Vertex without Graphing

Without graphing, locate the vertex and axis of symmetry of the parabola defined by $f(x) = -3x^2 + 6x + 1$. Does it open up or down?

Solution For this quadratic function, $a = -3, b = 6$, and $c = 1$. The x-coordinate of the vertex is

$$h = -\frac{b}{2a} = -\frac{6}{-6} = 1$$

The y-coordinate of the vertex is

$$k = f\left(-\frac{b}{2a}\right) = f(1) = -3 + 6 + 1 = 4$$

The vertex is located at the point $(1, 4)$. The axis of symmetry is the line $x = 1$. Because $a = -3 < 0$, the parabola opens down.

$\blacksquare$

3 Graph a Quadratic Function Using Its Vertex, Axis, and Intercepts

The information we gathered in Example 2, together with the location of the intercepts, usually provides enough information to graph $f(x) = ax^2 + bx + c, a \neq 0$.

The y-intercept is the value of f at $x = 0$; that is, the y-intercept is $f(0) = c$.

The x-intercepts, if there are any, are found by solving the quadratic equation

$$ax^2 + bx + c = 0$$

This equation has two, one, or no real solutions, depending on whether the discriminant $b^2 - 4ac$ is positive, 0, or negative. Depending on the value of the discriminant, the graph of f has x-intercepts, as follows:

The x-Intercepts of a Quadratic Function

1. If the discriminant $b^2 - 4ac > 0$, the graph of $f(x) = ax^2 + bx + c$ has two distinct x-intercepts so it crosses the x-axis in two places.
2. If the discriminant $b^2 - 4ac = 0$, the graph of $f(x) = ax^2 + bx + c$ has one x-intercept so it touches the x-axis at its vertex.
3. If the discriminant $b^2 - 4ac < 0$, the graph of $f(x) = ax^2 + bx + c$ has no x-intercept so it does not cross or touch the x-axis.

Figure 18 illustrates these possibilities for parabolas that open up.

Figure 18
$f(x) = ax^2 + bx + c,\ a > 0$

(a) $b^2 - 4ac > 0$
Two x-intercepts

(b) $b^2 - 4ac = 0$
One x-intercept

(c) $b^2 - 4ac < 0$
No x-intercepts

EXAMPLE 3 **Graphing a Quadratic Function Using Its Vertex, Axis, and Intercepts**

(a) Use the information from Example 2 and the locations of the intercepts to graph $f(x) = -3x^2 + 6x + 1$.
(b) Determine the domain and the range of f.
(c) Determine where f is increasing and where it is decreasing.

Solution

(a) In Example 2, we found the vertex to be at $(1, 4)$ and the axis of symmetry to be $x = 1$. The y-intercept is found by letting $x = 0$. The y-intercept is $f(0) = 1$. The x-intercepts are found by solving the equation $f(x) = 0$. This results in the equation

$$-3x^2 + 6x + 1 = 0 \quad a = -3, b = 6, c = 1$$

The discriminant $b^2 - 4ac = (6)^2 - 4(-3)(1) = 36 + 12 = 48 > 0$, so the equation has two real solutions and the graph has two x-intercepts. Using the quadratic formula, we find that

$$x = \frac{-b + \sqrt{b^2 - 4ac}}{2a} = \frac{-6 + \sqrt{48}}{-6} = \frac{-6 + 4\sqrt{3}}{-6} \approx -0.15$$

and

$$x = \frac{-b - \sqrt{b^2 - 4ac}}{2a} = \frac{-6 - \sqrt{48}}{-6} = \frac{-6 - 4\sqrt{3}}{-6} \approx 2.15$$

The x-intercepts are approximately -0.15 and 2.15.

The graph is illustrated in Figure 19. Notice how we used the y-intercept and the axis of symmetry, $x = 1$, to obtain the additional point $(2, 1)$ on the graph.

Figure 19

(b) The domain of f is the set of all real numbers. Based on the graph, the range of f is the interval $(-\infty, 4]$.

(c) The function f is increasing on the interval $(-\infty, 1)$ and decreasing on the interval $(1, \infty)$.

 Graph the function in Example 3 by completing the square and using transformations. Which method do you prefer?

Now Work PROBLEM 35

If the graph of a quadratic function has only one x-intercept or no x-intercepts, it is usually necessary to plot an additional point to obtain the graph.

EXAMPLE 4 | **Graphing a Quadratic Function Using Its Vertex, Axis, and Intercepts**

(a) Graph $f(x) = x^2 - 6x + 9$ by determining whether the graph opens up or down and by finding its vertex, axis of symmetry, y-intercept, and x-intercepts, if any.

(b) Determine the domain and the range of f.

(c) Determine where f is increasing and where it is decreasing.

Solution

(a) For $f(x) = x^2 - 6x + 9$, we have $a = 1, b = -6$, and $c = 9$. Since $a = 1 > 0$, the parabola opens up. The x-coordinate of the vertex is

$$h = -\frac{b}{2a} = -\frac{-6}{2(1)} = 3$$

The y-coordinate of the vertex is

$$k = f(3) = (3)^2 - 6(3) + 9 = 0$$

So the vertex is at $(3, 0)$. The axis of symmetry is the line $x = 3$. The y-intercept is $f(0) = 9$. Since the vertex $(3, 0)$ lies on the x-axis, the graph touches the x-axis at the x-intercept. By using the axis of symmetry and the y-intercept at $(0, 9)$, we can locate the additional point $(6, 9)$ on the graph. See Figure 20.

(b) The domain of f is the set of all real numbers. Based on the graph, the range of f is the interval $[0, \infty)$.

(c) The function f is decreasing on the interval $(-\infty, 3)$ and increasing on the interval $(3, \infty)$.

Figure 20

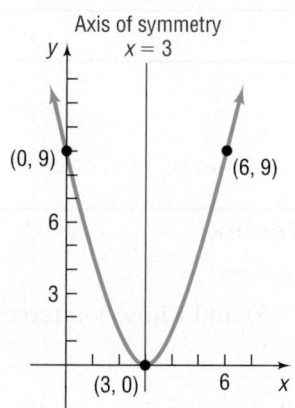

Now Work PROBLEM 43

EXAMPLE 5 | **Graphing a Quadratic Function Using Its Vertex, Axis, and Intercepts**

(a) Graph $f(x) = 2x^2 + x + 1$ by determining whether the graph opens up or down and by finding its vertex, axis of symmetry, y-intercept, and x-intercepts, if any.

(b) Determine the domain and the range of f.

(c) Determine where f is increasing and where it is decreasing.

Solution

(a) For $f(x) = 2x^2 + x + 1$, we have $a = 2, b = 1$, and $c = 1$. Since $a = 2 > 0$, the parabola opens up. The x-coordinate of the vertex is

$$h = -\frac{b}{2a} = -\frac{1}{4}$$

The y-coordinate of the vertex is

$$k = f\left(-\frac{1}{4}\right) = 2\left(\frac{1}{16}\right) + \left(-\frac{1}{4}\right) + 1 = \frac{7}{8}$$

NOTE In Example 5, since the vertex is above the x-axis and the parabola opens up, we can conclude that the graph of the quadratic function will have no x-intercepts. ∎

So the vertex is at $\left(-\frac{1}{4}, \frac{7}{8}\right)$. The axis of symmetry is the line $x = -\frac{1}{4}$. The y-intercept is $f(0) = 1$. The x-intercept(s), if any, obey the equation $2x^2 + x + 1 = 0$. Since the discriminant $b^2 - 4ac = (1)^2 - 4(2)(1) = -7 < 0$, this equation has no real solutions, and therefore the graph has no x-intercepts.

We use the point $(0, 1)$ and the axis of symmetry $x = -\frac{1}{4}$ to locate the additional point $\left(-\frac{1}{2}, 1\right)$ on the graph. See Figure 21.

Figure 21

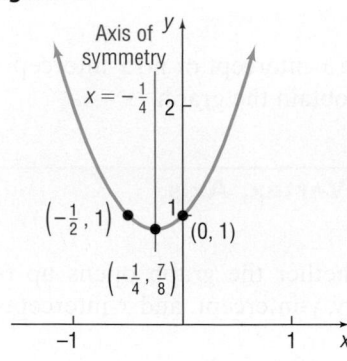

(b) The domain of f is the set of all real numbers. Based on the graph, the range of f is the interval $\left[\frac{7}{8}, \infty\right)$.

(c) The function f is decreasing on the interval $\left(-\infty, -\frac{1}{4}\right)$ and is increasing on the interval $\left(-\frac{1}{4}, \infty\right)$.

∎

━ **Now Work** PROBLEM **47**

Given the vertex (h, k) and one additional point on the graph of a quadratic function $f(x) = ax^2 + bx + c, a \neq 0$, we can use

$$f(x) = a(x - h)^2 + k \qquad \textbf{(3)}$$

to obtain the quadratic function.

| EXAMPLE 6 | **Finding the Quadratic Function Given Its Vertex and One Other Point** |

Determine the quadratic function whose vertex is $(1, -5)$ and whose y-intercept is -3. The graph of the parabola is shown in Figure 22.

Solution The vertex is $(1, -5)$, so $h = 1$ and $k = -5$. Substitute these values into equation (3).

Figure 22

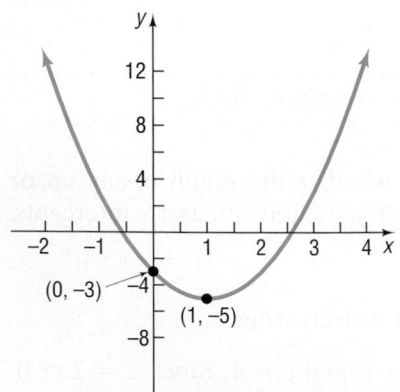

$$f(x) = a(x - h)^2 + k \quad \text{Equation (3)}$$
$$f(x) = a(x - 1)^2 - 5 \quad h = 1, k = -5$$

To determine the value of a, we use the fact that $f(0) = -3$ (the y-intercept).

$$f(x) = a(x - 1)^2 - 5$$
$$-3 = a(0 - 1)^2 - 5 \quad x = 0, y = f(0) = -3$$
$$-3 = a - 5$$
$$a = 2$$

The quadratic function whose graph is shown in Figure 22 is

$$f(x) = a(x - h)^2 + k = 2(x - 1)^2 - 5 = 2x^2 - 4x - 3$$

∎

━ **Now Work** PROBLEM **53**

SUMMARY Steps for Graphing a Quadratic Function $f(x) = ax^2 + bx + c, a \neq 0$

Option 1

STEP 1: Complete the square in x to write the quadratic function in the form $f(x) = a(x - h)^2 + k$.

STEP 2: Graph the function in stages using transformations.

Option 2

STEP 1: Determine whether the parabola open up $(a > 0)$ or down $(a < 0)$.

STEP 2: Determine the vertex $\left(-\dfrac{b}{2a}, f\left(-\dfrac{b}{2a} \right) \right)$.

STEP 3: Determine the axis of symmetry, $x = -\dfrac{b}{2a}$.

STEP 4: Determine the y-intercept, $f(0)$, and the x-intercepts, if any.
 (a) If $b^2 - 4ac > 0$, the graph of the quadratic function has two x-intercepts, which are found by solving the equation $ax^2 + bx + c = 0$.
 (b) If $b^2 - 4ac = 0$, the vertex is the x-intercept.
 (c) If $b^2 - 4ac < 0$, there are no x-intercepts.

STEP 5: Determine an additional point by using the y-intercept and the axis of symmetry.

STEP 6: Plot the points and draw the graph.

4 Find the Maximum or Minimum Value of a Quadratic Function

The graph of a quadratic function

$$f(x) = ax^2 + bx + c \qquad a \neq 0$$

is a parabola with vertex at $\left(-\dfrac{b}{2a}, f\left(-\dfrac{b}{2a} \right) \right)$. This vertex is the highest point on the graph if $a < 0$ and the lowest point on the graph if $a > 0$. If the vertex is the highest point $(a < 0)$, then $f\left(-\dfrac{b}{2a} \right)$ is the **maximum value** of f. If the vertex is the lowest point $(a > 0)$, then $f\left(-\dfrac{b}{2a} \right)$ is the **minimum value** of f.

EXAMPLE 7 **Finding the Maximum or Minimum Value of a Quadratic Function**

Determine whether the quadratic function

$$f(x) = x^2 - 4x - 5$$

has a maximum or minimum value. Then find the maximum or minimum value.

Solution We compare $f(x) = x^2 - 4x - 5$ to $f(x) = ax^2 + bx + c$. We conclude that $a = 1, b = -4$, and $c = -5$. Since $a > 0$, the graph of f opens up, so the vertex is a minimum point. The minimum value occurs at

$$x = -\frac{b}{2a} = -\frac{-4}{2(1)} = \frac{4}{2} = 2$$

$$\underset{\uparrow}{}$$

$$a = 1, b = -4$$

The minimum value is

$$f\left(-\frac{b}{2a} \right) = f(2) = 2^2 - 4(2) - 5 = 4 - 8 - 5 = -9$$

Now Work PROBLEM 61

4.3 Assess Your Understanding

'Are You Prepared?' *Answers are given at the end of these exercises. If you get a wrong answer, read the pages listed in red.*

1. List the intercepts of the equation $y = x^2 - 9$. (pp. 165–166)
2. Find the real solutions of the equation $2x^2 + 7x - 4 = 0$. (pp. 100–106)
3. To complete the square of $x^2 - 5x$, you add the number _____. (pp. 99–100)
4. To graph $y = (x - 4)^2$, you shift the graph of $y = x^2$ to the _____ a distance of _____ units. (pp. 252–260)

Concepts and Vocabulary

5. The graph of a quadratic function is called a(n) _____.
6. The vertical line passing through the vertex of a parabola is called the _____.
7. The x-coordinate of the vertex of $f(x) = ax^2 + bx + c$, $a \neq 0$, is _____.
8. **True or False** The graph of $f(x) = 2x^2 + 3x - 4$ opens up.
9. **True or False** The y-coordinate of the vertex of $f(x) = -x^2 + 4x + 5$ is $f(2)$.
10. **True or False** If the discriminant $b^2 - 4ac = 0$, the graph of $f(x) = ax^2 + bx + c, a \neq 0$, will touch the x-axis at its vertex.

Skill Building

In Problems 11–18, match each graph to one the following functions.

11. $f(x) = x^2 - 1$
12. $f(x) = -x^2 - 1$
13. $f(x) = x^2 - 2x + 1$
14. $f(x) = x^2 + 2x + 1$
15. $f(x) = x^2 - 2x + 2$
16. $f(x) = x^2 + 2x$
17. $f(x) = x^2 - 2x$
18. $f(x) = x^2 + 2x + 2$

A

B

C

D

E

F

G

H
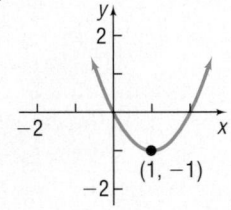

In Problems 19–34, graph the function f by starting with the graph of $y = x^2$ and using transformations (shifting, compressing, stretching, and/or reflection).

[Hint: If necessary, write f in the form $f(x) = a(x - h)^2 + k$.]

19. $f(x) = \frac{1}{4}x^2$
20. $f(x) = 2x^2$
21. $f(x) = \frac{1}{4}x^2 - 2$
22. $f(x) = 2x^2 - 3$

23. $f(x) = \frac{1}{4}x^2 + 2$
24. $f(x) = 2x^2 + 4$
25. $f(x) = \frac{1}{4}x^2 + 1$
26. $f(x) = -2x^2 - 2$

27. $f(x) = x^2 + 4x + 2$
28. $f(x) = x^2 - 6x - 1$
29. $f(x) = 2x^2 - 4x + 1$
30. $f(x) = 3x^2 + 6x$

31. $f(x) = -x^2 - 2x$
32. $f(x) = -2x^2 + 6x + 2$
33. $f(x) = \frac{1}{2}x^2 + x - 1$
34. $f(x) = \frac{2}{3}x^2 + \frac{4}{3}x - 1$

In Problems 35–52, (a) graph each quadratic function by determining whether its graph opens up or down and by finding its vertex, axis of symmetry, y-intercept, and x-intercepts, if any. (b) Determine the domain and the range of the function. (c) Determine where the function is increasing and where it is decreasing.

35. $f(x) = x^2 + 2x$
36. $f(x) = x^2 - 4x$
37. $f(x) = -x^2 - 6x$
38. $f(x) = -x^2 + 4x$
39. $f(x) = 2x^2 - 8x$
40. $f(x) = 3x^2 + 18x$
41. $f(x) = x^2 + 2x - 8$
42. $f(x) = x^2 - 2x - 3$
43. $f(x) = x^2 + 2x + 1$
44. $f(x) = x^2 + 6x + 9$
45. $f(x) = 2x^2 - x + 2$
46. $f(x) = 4x^2 - 2x + 1$

47. $f(x) = -2x^2 + 2x - 3$

48. $f(x) = -3x^2 + 3x - 2$

49. $f(x) = 3x^2 + 6x + 2$

50. $f(x) = 2x^2 + 5x + 3$

51. $f(x) = -4x^2 - 6x + 2$

52. $f(x) = 3x^2 - 8x + 2$

In Problems 53–58, determine the quadratic function whose graph is given.

53.

54.

55.

56.

57.

58.

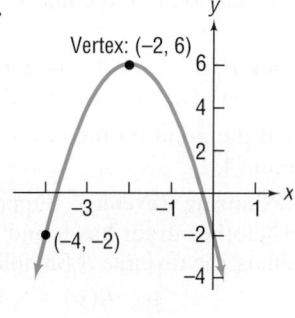

In Problems 59–66, determine, without graphing, whether the given quadratic function has a maximum value or a minimum value and then find the value.

59. $f(x) = 2x^2 + 12x$

60. $f(x) = -2x^2 + 12x$

61. $f(x) = 2x^2 + 12x - 3$

62. $f(x) = 4x^2 - 8x + 3$

63. $f(x) = -x^2 + 10x - 4$

64. $f(x) = -2x^2 + 8x + 3$

65. $f(x) = -3x^2 + 12x + 1$

66. $f(x) = 4x^2 - 4x$

Applications and Extensions

67. The graph of the function $f(x) = ax^2 + bx + c$ has vertex at $(0, 2)$ and passes through the point $(1, 8)$. Find $a, b,$ and c.

68. The graph of the function $f(x) = ax^2 + bx + c$ has vertex at $(1, 4)$ and passes through the point $(-1, -8)$. Find $a, b,$ and c.

In Problems 69–74, for the given functions f and g,
(a) Graph f and g on the same Cartesian plane.
(b) Solve $f(x) = g(x)$.
(c) Use the result of part (b) to label the points of intersection of the graphs of f and g.
(d) Shade the region for which $f(x) > g(x)$, that is, the region below f and above g.

69. $f(x) = 2x - 1; \quad g(x) = x^2 - 4$

70. $f(x) = -2x - 1; \quad g(x) = x^2 - 9$

71. $f(x) = -x^2 + 4; \quad g(x) = -2x + 1$

72. $f(x) = -x^2 + 9; \quad g(x) = 2x + 1$

73. $f(x) = -x^2 + 5x; \quad g(x) = x^2 + 3x - 4$

74. $f(x) = -x^2 + 7x - 6; \quad g(x) = x^2 + x - 6$

Answer Problems 75 and 76 using the following: A quadratic function of the form $f(x) = ax^2 + bx + c$ with $b^2 - 4ac > 0$ may also be written in the form $f(x) = a(x - r_1)(x - r_2)$, where r_1 and r_2 are the x-intercepts of the graph of the quadratic function.

75. (a) Find a quadratic function whose x-intercepts are -3 and 1 with $a = 1; a = 2; a = -2; a = 5$.

(b) How does the value of a affect the intercepts?

(c) How does the value of a affect the axis of symmetry?

(d) How does the value of a affect the vertex?

(e) Compare the x-coordinate of the vertex with the midpoint of the x-intercepts. What might you conclude?

76. (a) Find a quadratic function whose x-intercepts are -5 and 3 with $a = 1; a = 2; a = -2; a = 5$.

(b) How does the value of a affect the intercepts?

(c) How does the value of a affect the axis of symmetry?

(d) How does the value of a affect the vertex?

(e) Compare the x-coordinate of the vertex with the midpoint of the x-intercepts. What might you conclude?

77. Suppose that $f(x) = x^2 + 4x - 21$
 (a) What is the vertex of f?
 (b) What are the x-intercepts of the graph of f?
 (c) Solve $f(x) = -21$ for x. What points are on the graph of f?
 (d) Use the information obtained in parts (a)–(c) to graph $f(x) = x^2 + 4x - 21$.

78. Suppose that $f(x) = x^2 + 2x - 8$
 (a) What is the vertex of f?
 (b) What are the x-intercepts of the graph of f?
 (c) Solve $f(x) = -8$ for x. What points are on the graph of f?
 (d) Use the information obtained in parts (a)–(c) to graph $f(x) = x^2 + 2x - 8$.

79. Find the point on the line $y = x$ that is closest to the point $(3, 1)$.
 [**Hint:** Express the distance d from the point to the line as a function of x, and then find the minimum value of $[d(x)]^2$.

80. Find the point on the line $y = x + 1$ that is closest to the point $(4, 1)$.

81. Maximizing Revenue Suppose that the manufacturer of a gas clothes dryer has found that, when the unit price is p dollars, the revenue R (in dollars) is
$$R(p) = -4p^2 + 4000p$$
What unit price should be established for the dryer to maximize revenue? What is the maximum revenue?

82. Maximizing Revenue The John Deere company has found that the revenue, in dollars, from sales of heavy-duty tractors is a function of the unit price p, in dollars, that it charges. If the revenue R is
$$R(p) = -\frac{1}{2}p^2 + 1900p$$
what unit price p should be charged to maximize revenue? What is the maximum revenue?

83. Minimizing Marginal Cost The **marginal cost** of a product can be thought of as the cost of producing one additional unit of output. For example, if the marginal cost of producing the 50th product is \$6.20, it cost \$6.20 to increase production from 49 to 50 units of output. Suppose the marginal cost C (in dollars) to produce x thousand mp3 players is given by the function
$$C(x) = x^2 - 140x + 7400$$
 (a) How many players should be produced to minimize the marginal cost?
 (b) What is the minimum marginal cost?

84. Minimizing Marginal Cost (See Problem 83.) The marginal cost C (in dollars) of manufacturing x cell phones (in thousands) is given by
$$C(x) = 5x^2 - 200x + 4000.$$
 (a) How many cell phones should be manufactured to minimize the marginal cost?
 (b) What is the minimum marginal cost?

85. Hunting The function
$$H(x) = -3.24x^2 + 242.1x - 738.4$$
models the number of individuals whose age is x and engage in hunting activities.
 (a) What is the age at which there are the most hunters? Approximately how many hunters are this age?

 (b) Are the number of hunters increasing or decreasing for individuals who are between 40 and 45 years of age?
 Source: National Sporting Goods Association

86. Advanced Degrees The function
$$P(x) = -0.008x^2 + 0.815x - 9.983$$
models the percentage of the U.S. population whose age is x that have earned an advanced degree (more than a bachelor's degree) in March 2003.
 (a) What is the age for which the highest percentage of Americans have earned an advanced degree? What is the highest percentage?
 (b) Is the percentage of Americans who have earned an advanced degree increasing or decreasing for individuals between the ages of 40 and 50?
 Source: U.S. Census Bureau

87. Male Murder Victims The function
$$M(x) = 1.00x^2 - 136.74x + 4764.89$$
models the number of male murder victims who are x years of age ($20 \le x < 90$).
 (a) Use the model to approximate the number of male murder victims who are $x = 23$ years of age.
 (b) At what age is the number of male murder victims 1456?
 (c) Describe what happens to the number of male murder victims as age increases from 20 to 65.
 Source: Federal Bureau of Investigation

88. Health Care Expenditures The function
$$H(x) = 0.004x^2 - 0.197x + 5.406$$
models the percentage of total income that an individual who is x years of age spends on health care.
 (a) Use the model to approximate the percentage of total income an individual who is $x = 45$ years of age spends on health care.
 (b) At what age is the percentage of income spent on health care 10%?
 (c) Using a graphing utility, graph $H = H(x)$.
 (d) Based on the graph drawn in part (c), describe what happens to the percentage of income spent on health care as individuals age.
 Source: Bureau of Labor Statistics

89. Business The monthly revenue R achieved by selling x wristwatches is figured to be $R(x) = 75x - 0.2x^2$. The monthly cost C of selling x wristwatches is $C(x) = 32x + 1750$.
 (a) How many wristwatches must the firm sell to maximize revenue? What is the maximum revenue?
 (b) Profit is given as $P(x) = R(x) - C(x)$. What is the profit function?
 (c) How many wristwatches must the firm sell to maximize profit? What is the maximum profit?
 (d) Provide a reasonable explanation as to why the answers found in parts (a) and (c) differ. Explain why a quadratic function is a reasonable model for revenue.

90. Business The daily revenue R achieved by selling x boxes of candy is figured to be $R(x) = 9.5x - 0.04x^2$. The daily cost C of selling x boxes of candy is $C(x) = 1.25x + 250$.

(a) How many boxes of candy must the firm sell to maximize revenue? What is the maximum revenue?

(b) Profit is given as $P(x) = R(x) - C(x)$. What is the profit function?

(c) How many boxes of candy must the firm sell to maximize profit? What is the maximum profit?

(d) Provide a reasonable explanation as to why the answers found in parts (a) and (c) differ. Explain why a quadratic function is a reasonable model for revenue.

91. Let $f(x) = ax^2 + bx + c$, where $a, b,$ and c are odd integers. If x is an integer, show that $f(x)$ must be an odd integer.

[**Hint:** x is either an even integer or an odd integer.]

Discussion and Writing

92. Make up a quadratic function that opens down and has only one x-intercept. Compare yours with others in the class. What are the similarities? What are the differences?

93. On one set of coordinate axes, graph the family of parabolas $f(x) = x^2 + 2x + c$ for $c = -3, c = 0,$ and $c = 1$. Describe the characteristics of a member of this family.

94. On one set of coordinate axes, graph the family of parabolas $f(x) = x^2 + bx + 1$ for $b = -4, b = 0,$ and $b = 4$. Describe the general characteristics of this family.

95. State the circumstances that cause the graph of a quadratic function $f(x) = ax^2 + bx + c$ to have no x-intercepts.

96. Why does the graph of a quadratic function open up if $a > 0$ and down if $a < 0$?

97. Can a quadratic function have a range of $(-\infty, \infty)$? Justify your answer.

98. What are the possibilities for the number of times the graphs of two different quadratic functions intersect?

'Are You Prepared?' Answers

1. $(0, -9), (-3, 0), (3, 0)$ **2.** $\left\{-4, \dfrac{1}{2}\right\}$ **3.** $\dfrac{25}{4}$ **4.** right; 4

4.4 Quadratic Models; Building Quadratic Functions from Data

PREPARING FOR THIS SECTION *Before getting started, review the following:*

- Problem Solving (Section 1.7, pp. 139–145)
- Building Linear Functions from Data (Section 4.2, pp. 287–290)

Now Work the 'Are You Prepared?' problems on page 310.

> **OBJECTIVES** **1** Solve Applied Problems Involving Quadratic Functions (p. 305)
> **2** Use a Graphing Utility to Find the Quadratic Function of Best Fit (p. 309)

In this section we will first discuss models that lead to a quadratic function when a verbal description of the problem is given. We end the section by fitting a quadratic function to data, which is another form of modeling.

When a mathematical model leads to a quadratic function, the properties of the graph of the quadratic function can provide important information about the model. In particular, we can use the quadratic function to determine the maximum or minimum value of the function. The fact that the graph of a quadratic function has a maximum or minimum value enables us to answer questions involving **optimization,** that is, finding the maximum or minimum values in models involving quadratic functions.

1 Solve Applied Problems Involving Quadratic Functions

In economics, revenue R, in dollars, is defined as the amount of money received from the sale of an item and is equal to the unit selling price p, in dollars, of the item times the number x of units actually sold. That is,

$$R = xp$$

In economics, the Law of Demand states that p and x are related: As one increases, the other decreases. The equation that relates p and x is called the **demand equation.**

| EXAMPLE 1 | Maximizing Revenue |

The marketing department at Texas Instruments has found that, when certain calculators are sold at a price of p dollars per unit, the number x of calculators sold is given by the demand equation

$$x = 21{,}000 - 150p$$

(a) Express the revenue R as a function of the price p.
(b) What unit price should be used to maximize revenue?
(c) If this price is charged, what is the maximum revenue?
(d) How many units are sold at this price?
(e) Graph R.

Solution (a) The revenue R is $R = xp$, where $x = 21{,}000 - 150p$.

$$R = xp = (21{,}000 - 150p)p = -150p^2 + 21{,}000p$$

(b) The function R is a quadratic function with $a = -150, b = 21{,}000$, and $c = 0$. Because $a < 0$, the vertex is the highest point on the parabola. The revenue R is therefore a maximum when the price p is

$$p = -\frac{b}{2a} = -\frac{21{,}000}{2(-150)} = -\frac{21{,}000}{-300} = \$70.00$$
$$\uparrow$$
$$a = -150, b = 21{,}000$$

(c) The maximum revenue R is

$$R(70) = -150(70)^2 + 21{,}000(70) = \$735{,}000$$

(d) The number of calculators sold is given by the demand equation $x = 21{,}000 - 150p$. At a price of $p = \$70$,

$$x = 21{,}000 - 150(70) = 10{,}500$$

calculators are sold.

(e) See Figure 23 for the graph of R.

Figure 23

Now Work PROBLEM 3

| EXAMPLE 2 | Maximizing the Area Enclosed by a Fence |

A farmer has 2000 yards of fence to enclose a rectangular field. What are the dimensions of the rectangle that encloses the most area?

Solution Figure 24 illustrates the situation. The available fence represents the perimeter of the rectangle. If x is the length and w is the width, then

$$2x + 2w = 2000 \qquad\qquad \textbf{(1)}$$

The area A of the rectangle is

$$A = xw$$

Figure 24

To express A in terms of a single variable, we solve equation (1) for w and substitute the result in $A = xw$. Then A involves only the variable x. [You could also solve equation (1) for x and express A in terms of w alone. Try it!]

$$2x + 2w = 2000 \qquad \text{Equation (1)}$$
$$2w = 2000 - 2x \qquad \text{Solve for } w.$$
$$w = \frac{2000 - 2x}{2} = 1000 - x$$

Then the area A is

$$A = xw = x(1000 - x) = -x^2 + 1000x$$

Now, A is a quadratic function of x.

$$A(x) = -x^2 + 1000x \qquad a = -1, b = 1000, c = 0$$

Figure 25

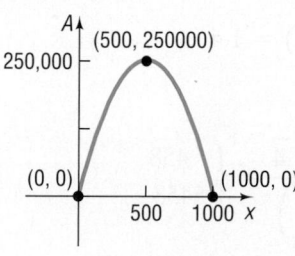

Figure 25 shows a graph of $A(x) = -x^2 + 1000x$. Since $a < 0$, the vertex is a maximum point on the graph of A. The maximum value occurs at

$$x = -\frac{b}{2a} = -\frac{1000}{2(-1)} = 500$$

The maximum value of A is

$$A\left(-\frac{b}{2a}\right) = A(500) = -500^2 + 1000(500) = -250{,}000 + 500{,}000 = 250{,}000$$

The largest rectangle that can be enclosed by 2000 yards of fence has an area of 250,000 square yards. Its dimensions are 500 yards by 500 yards.

━━━ **Now Work** PROBLEM 7

EXAMPLE 3	**Analyzing the Motion of a Projectile**

A projectile is fired from a cliff 500 feet above the water at an inclination of 45° to the horizontal, with a muzzle velocity of 400 feet per second. In physics, it is established that the height h of the projectile above the water is given by

$$h(x) = \frac{-32x^2}{(400)^2} + x + 500$$

where x is the horizontal distance of the projectile from the base of the cliff. See Figure 26.

Figure 26

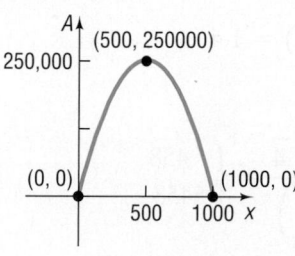

(a) Find the maximum height of the projectile.
(b) How far from the base of the cliff will the projectile strike the water?

Solution (a) The height of the projectile is given by a quadratic function.

$$h(x) = \frac{-32x^2}{(400)^2} + x + 500 = \frac{-1}{5000}x^2 + x + 500$$

We are looking for the maximum value of h. Since $a < 0$, the maximum value is obtained at the vertex. We compute

$$x = -\frac{b}{2a} = -\frac{1}{2\left(-\dfrac{1}{5000}\right)} = \frac{5000}{2} = 2500$$

The maximum height of the projectile is

$$h(2500) = \frac{-1}{5000}(2500)^2 + 2500 + 500 = -1250 + 2500 + 500 = 1750 \text{ ft}$$

(b) The projectile will strike the water when the height is zero. To find the distance x traveled, we need to solve the equation

$$h(x) = \frac{-1}{5000}x^2 + x + 500 = 0$$

We find the discriminant first.

$$b^2 - 4ac = 1^2 - 4\left(\frac{-1}{5000}\right)(500) = 1.4$$

Then

$$x = \frac{-b \pm \sqrt{b^2 - 4ac}}{2a} = \frac{-1 \pm \sqrt{1.4}}{2\left(-\dfrac{1}{5000}\right)} \approx \begin{cases} -458 \\ 5458 \end{cases}$$

We discard the negative solution and find that the projectile will strike the water at a distance of about 5458 feet from the base of the cliff.

— **Now Work** PROBLEM 11

Seeing the Concept

Graph

$$h(x) = \frac{-1}{5000}x^2 + x + 500$$

$0 \le x \le 5500$

Use MAXIMUM to find the maximum height of the projectile, and use ROOT or ZERO to find the distance from the base of the cliff to where it strikes the water. Compare your results with those obtained in Example 3.

EXAMPLE 4 **The Golden Gate Bridge**

The Golden Gate Bridge, a suspension bridge, spans the entrance to San Francisco Bay. Its 746-foot-tall towers are 4200 feet apart. The bridge is suspended from two huge cables more than 3 feet in diameter; the 90-foot-wide roadway is 220 feet above the water. The cables are parabolic in shape* and touch the road surface at the center of the bridge. Find the height of the cable above the road at a distance of 1000 feet from the center.

Solution See Figure 27. We begin by choosing the placement of the coordinate axes so that the x-axis coincides with the road surface and the origin coincides with the center of the bridge. As a result, the twin towers will be vertical (height $746 - 220 = 526$ feet above the road) and located 2100 feet from the center. Also, the cable, which has the shape of a parabola, will extend from the towers, open up, and have its vertex at $(0, 0)$. The choice of placement of the axes enables us to identify the equation of the parabola as $y = ax^2, a > 0$. We can also see that the points $(-2100, 526)$ and $(2100, 526)$ are on the graph.

Figure 27

* A cable suspended from two towers is in the shape of a **catenary,** but when a horizontal roadway is suspended from the cable, the cable takes the shape of a parabola.

Based on these facts, we can find the value of a in $y = ax^2$.

$$y = ax^2$$

$$526 = a(2100)^2 \qquad x = 2100, y = 526$$

$$a = \frac{526}{(2100)^2}$$

The equation of the parabola is therefore

$$y = \frac{526}{(2100)^2}x^2$$

The height of the cable when $x = 1000$ is

$$y = \frac{526}{(2100)^2}(1000)^2 \approx 119.3 \text{ feet}$$

The cable is 119.3 feet above the road at a distance of 1000 feet from the center of the bridge.

 Now Work PROBLEM 13

2 Use a Graphing Utility to Find the Quadratic Function of Best Fit

In Section 4.2 we found the line of best fit for data that appeared to be linearly related. It was noted that data may also follow a nonlinear relation. Figures 28(a) and (b) show scatter diagrams of data that follow a quadratic relation.

Figure 28

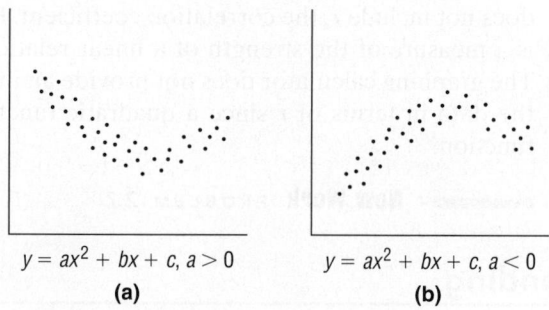

$y = ax^2 + bx + c,\ a > 0$ $y = ax^2 + bx + c,\ a < 0$

(a) (b)

EXAMPLE 5 **Fitting a Quadratic Function to Data**

A farmer collected the data given in Table 8 (on page 310), which shows crop yields Y for various amounts x of fertilizer used.

(a) Draw a scatter diagram of the data. Comment on the type of relation that may exist between the two variables.

(b) Use a graphing utility to find the quadratic function of best fit to these data.

(c) Use the function found in part (b) to determine the optimal amount of fertilizer to apply.

(d) Use the function found in part (b) to predict crop yield when the optimal amount of fertilizer is applied.

(e) Draw the quadratic function of best fit on the scatter diagram.

Solution (a) Figure 29 shows the scatter diagram, from which it appears that the data follow a quadratic relation, with $a < 0$.

Table 8

Plot	Fertilizer, x (Pounds/100 ft²)	Yield, Y (Bushels)
1	0	4
2	0	6
3	5	10
4	5	7
5	10	12
6	10	10
7	15	15
8	15	17
9	20	18
10	20	21
11	25	20
12	25	21
13	30	21
14	30	22
15	35	21
16	35	20
17	40	19
18	40	19

Figure 31

(b) Upon executing the QUADratic REGression program, we obtain the results shown in Figure 30. The output that the utility provides shows us the equation $y = ax^2 + bx + c$. The quadratic function of best fit is

$$Y(x) = -0.0171x^2 + 1.0765x + 3.8939$$

Figure 29

Figure 30

where x represents the amount of fertilizer used and Y represents crop yield.

(c) Based on the quadratic function of best fit, the optimal amount of fertilizer to apply is

$$x = -\frac{b}{2a} = -\frac{1.0765}{2(-0.0171)} \approx 31.5 \text{ pounds of fertilizer per 100 square feet}$$

(d) We evaluate the function $Y(x)$ for $x = 31.5$.

$$Y(31.5) = -0.0171(31.5)^2 + 1.0765(31.5) + 3.8939 \approx 20.8 \text{ bushels}$$

If we apply 31.5 pounds of fertilizer per 100 square feet, the crop yield will be 20.8 bushels according to the quadratic function of best fit.

(e) Figure 31 shows the graph of the quadratic function found in part (b) drawn on the scatter diagram.

Look again at Figure 30. Notice that the output given by the graphing calculator does not include r, the correlation coefficient. Recall that the correlation coefficient is a measure of the strength of a linear relation that exists between two variables. The graphing calculator does not provide an indication of how well the function fits the data in terms of r since a quadratic function cannot be expressed as a linear function.

Now Work PROBLEM 27

4.4 Assess Your Understanding

1. Translate the following sentence to a mathematical equation: The total revenue R from selling x hot dogs is $3 times the number of hot dogs sold. (pp. 139–145)

2. Use a graphing utility to find the line of best fit for the following data: (pp. 287–290)

x	3	5	5	6	7	8
y	10	13	12	15	16	19

Applications and Extensions

3. **Demand Equation** The price p (in dollars) and the quantity x sold of a certain product obey the demand equation

$$p = -\frac{1}{6}x + 100 \qquad 0 \le x \le 600$$

(a) Express the revenue R as a function of x. (Remember, $R = xp$.)

(b) What is the revenue if 200 units are sold?

(c) What quantity x maximizes revenue? What is the maximum revenue?

(d) What price should the company charge to maximize revenue?

4. **Demand Equation** The price p (in dollars) and the quantity x sold of a certain product obey the demand equation

$$p = -\frac{1}{3}x + 100 \qquad 0 \le x \le 300$$

(a) Express the revenue R as a function of x.

(b) What is the revenue if 100 units are sold?

(c) What quantity x maximizes revenue? What is the maximum revenue?

(d) What price should the company charge to maximize revenue?

5. **Demand Equation** The price p (in dollars) and the quantity x sold of a certain product obey the demand equation

$$x = -5p + 100, \qquad 0 \le p \le 20$$

(a) Express the revenue R as a function of x.

(b) What is the revenue if 15 units are sold?

(c) What quantity x maximizes revenue? What is the maximum revenue?

(d) What price should the company charge to maximize revenue?

6. **Demand Equation** The price p (in dollars) and the quantity x sold of a certain product obey the demand equation

$$x = -20p + 500, \qquad 0 \le p \le 25$$

(a) Express the revenue R as a function of x.

(b) What is the revenue if 20 units are sold?

(c) What quantity x maximizes revenue? What is the maximum revenue?

(d) What price should the company charge to maximize revenue?

7. **Enclosing a Rectangular Field** David has 400 yards of fencing and wishes to enclose a rectangular area.

(a) Express the area A of the rectangle as a function of the width w of the rectangle.

(b) For what value of w is the area largest?

(c) What is the maximum area?

8. **Enclosing a Rectangular Field** Beth has 3000 feet of fencing available to enclose a rectangular field.

(a) Express the area A of the rectangle as a function of x, where x is the length of the rectangle.

(b) For what value of x is the area largest?

(c) What is the maximum area?

9. **Enclosing the Most Area with a Fence** A farmer with 4000 meters of fencing wants to enclose a rectangular plot that borders on a river. If the farmer does not fence the side along the river, what is the largest area that can be enclosed? (See the figure.)

$4000 - 2x$

10. **Enclosing the Most Area with a Fence** A farmer with 2000 meters of fencing wants to enclose a rectangular plot that borders on a straight highway. If the farmer does not fence the side along the highway, what is the largest area that can be enclosed?

11. **Analyzing the Motion of a Projectile** A projectile is fired from a cliff 200 feet above the water at an inclination of 45° to the horizontal, with a muzzle velocity of 50 feet per second. The height h of the projectile above the water is given by

$$h(x) = \frac{-32x^2}{(50)^2} + x + 200$$

where x is the horizontal distance of the projectile from the face of the cliff.

(a) At what horizontal distance from the face of the cliff is the height of the projectile a maximum?

(b) Find the maximum height of the projectile.

(c) At what horizontal distance from the face of the cliff will the projectile strike the water?

(d) Using a graphing utility, graph the function h, $0 \le x \le 200$.

(e) Use a graphing utility to verify the solutions found in parts (b) and (c).

(f) When the height of the projectile is 100 feet above the water, how far is it from the cliff?

12. **Analyzing the Motion of a Projectile** A projectile is fired at an inclination of 45° to the horizontal, with a muzzle velocity of 100 feet per second. The height h of the projectile is given by

$$h(x) = \frac{-32x^2}{(100)^2} + x$$

where x is the horizontal distance of the projectile from the firing point.

(a) At what horizontal distance from the firing point is the height of the projectile a maximum?

(b) Find the maximum height of the projectile.

(c) At what horizontal distance from the firing point will the projectile strike the ground?

(d) Using a graphing utility, graph the function h, $0 \le x \le 350$.

(e) Use a graphing utility to verify the results obtained in parts (b) and (c).

(f) When the height of the projectile is 50 feet above the ground, how far has it traveled horizontally?

13. **Suspension Bridge** A suspension bridge with weight uniformly distributed along its length has twin towers that extend 75 meters above the road surface and are 400 meters apart. The cables are parabolic in shape and are suspended from the tops of the towers. The cables touch the road surface at the center of the bridge. Find the height of the cables at a point 100 meters from the center. (Assume that the road is level.)

14. **Architecture** A parabolic arch has a span of 120 feet and a maximum height of 25 feet. Choose suitable rectangular coordinate axes and find the equation of the parabola. Then calculate the height of the arch at points 10 feet, 20 feet, and 40 feet from the center.

15. **Constructing Rain Gutters** A rain gutter is to be made of aluminum sheets that are 12 inches wide by turning up the edges 90°. See the illustration. What depth will provide maximum cross-sectional area and hence allow the most water to flow?

16. Norman Windows A **Norman window** has the shape of a rectangle surmounted by a semicircle of diameter equal to the width of the rectangle. See the figure. If the perimeter of the window is 20 feet, what dimensions will admit the most light (maximize the area)?

[**Hint:** Circumference of a circle $= 2\pi r$; area of a circle $= \pi r^2$, where r is the radius of the circle.]

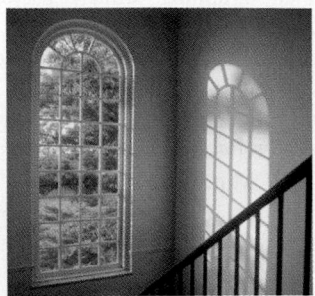

17. Constructing a Stadium A track and field playing area is in the shape of a rectangle with semicircles at each end. See the figure. The inside perimeter of the track is to be 1500 meters. What should the dimensions of the rectangle be so that the area of the rectangle is a maximum?

18. Architecture A special window has the shape of a rectangle surmounted by an equilateral triangle. See the figure. If the perimeter of the window is 16 feet, what dimensions will admit the most light?

[**Hint:** Area of an equilateral triangle $= \left(\dfrac{\sqrt{3}}{4}\right)x^2$, where x is the length of a side of the triangle.]

19. Stopping Distance An accepted relationship between *stopping distance, d* (in feet), and the *speed* of a car, v (in mph), is $d = 1.1v + 0.06v^2$ on dry, level concrete.

(a) How many feet will it take a car traveling 45 mph to stop on dry, level concrete?

(b) If an accident occurs 200 feet ahead of you, what is the maximum speed you should be traveling to avoid being involved?

(c) What might the term $1.1v$ represent?

Source: www2.nsta.org/Energy/fn_braking.html

20. Insurance Claims The years 1999 to 2005 were particularly costly for insurance companies, with 7 of the 10 most costly catastrophes in U.S. history (as of 2005). For the United States Automobile Association (USAA) and its affiliates, the total cost of claims for catastrophic losses, in millions, can be approximated by $C(x) = 34.87x^2 - 98.1x + 258.3$, where $x = 0$ for 1999, $x = 1$ for 2000, $x = 2$ for 2001, and so on.

(a) Estimate the total cost of claims for the year 2003.

(b) According to the model, during which year were catastrophic loss claims at a minimum?

(c) Would $C(x)$ be useful for predicting total catastrophic loss claims for the year 2015? Why or why not?

Source: USAA Report to Members 2005

21. Chemical Reactions A self-catalytic chemical reaction results in the formation of a compound that causes the formation ratio to increase. If the reaction rate V is given by

$$V(x) = kx(a - x), \qquad 0 \le x \le a$$

where k is a positive constant, a is the initial amount of the compound, and x is the variable amount of the compound, for what value of x is the reaction rate a maximum?

22. Calculus: Simpson's Rule The figure shows the graph of $y = ax^2 + bx + c$. Suppose that the points $(-h, y_0)$, $(0, y_1)$, and (h, y_2) are on the graph. It can be shown that the area enclosed by the parabola, the x-axis, and the lines $x = -h$ and $x = h$ is

$$\text{Area} = \frac{h}{3}(2ah^2 + 6c)$$

Show that this area may also be given by

$$\text{Area} = \frac{h}{3}(y_0 + 4y_1 + y_2)$$

23. Use the result obtained in Problem 22 to find the area enclosed by $f(x) = -5x^2 + 8$, the x-axis, and the lines $x = -1$ and $x = 1$.

24. Use the result obtained in Problem 22 to find the area enclosed by $f(x) = 2x^2 + 8$, the x-axis, and the lines $x = -2$ and $x = 2$.

25. Use the result obtained in Problem 22 to find the area enclosed by $f(x) = x^2 + 3x + 5$, the x-axis, and the lines $x = -4$ and $x = 4$.

26. Use the result obtained in Problem 22 to find the area enclosed by $f(x) = -x^2 + x + 4$, the x-axis, and the lines $x = -1$ and $x = 1$.

27. Life Cycle Hypothesis An individual's income varies with his or her age. The following table shows the median income I of individuals of different age groups within the United States for 2003. For each age group, let the class midpoint represent the independent variable x. For the class

"65 years and older," we will assume that the class midpoint is 69.5.

Age (years)	Class Midpoint, x	Median Income ($), I
15–24	19.5	8,614
25–34	29.5	26,212
35–44	39.5	30,914
45–54	49.5	32,583
55–64	59.5	28,068
65 and older	69.5	14,664

Source: U.S. Census Bureau, 2003 Annual Social and Economic Supplement

(a) Draw a scatter diagram of the data. Comment on the type of relation that may exist between the two variables.

(b) The quadratic function of best fit to these data is

$$I(x) = -34.3x^2 + 3157x - 39,115$$

Use this function to determine the age at which an individual can expect to earn the most income.

(c) Use the function to predict the peak income earned.

(d) Use a graphing utility to verify that the function given in part (b) is the quadratic function of best fit.

(e) With a graphing utility, draw a scatter diagram of the data and then graph the quadratic function of best fit on the scatter diagram.

28. **Life Cycle Hypothesis** An individual's income varies with his or her age. The following table shows the median income *I* of individuals of different age groups within the United States for 2004. For each age group, let the class midpoint represent the independent variable *x*. For the class "65 years and older," we will assume that the class midpoint is 69.5.

Age (years)	Class Midpoint, x	Median Income ($), I
15–24	19.5	8,782
25–34	29.5	26,642
35–44	39.5	31,629
45–54	49.5	32,908
55–64	59.5	28,518
65 and older	69.5	15,193

Source: U.S. Census Bureau, 2004 Annual Social and Economic Supplement

(a) Draw a scatter diagram of the data. Comment on the type of relation that may exist between the two variables.

(b) The quadratic function of best fit to these data is

$$I(x) = -34.5x^2 + 3186x - 39,335$$

Use this function to determine the age at which an individual can expect to earn the most income.

(c) Use the function to predict the peak income earned.

(d) Use a graphing utility to verify that the function given in part (b) is the quadratic function of best fit.

(e) With a graphing utility, draw a scatter diagram of the data and then graph the quadratic function of best fit on the scatter diagram.

29. **Height of a Ball** A shot-putter throws a ball at an inclination of 45° to the horizontal. The following data represent the height *h* of the ball at the instant that it has traveled *x* feet horizontally.

Distance, x	Height, h
20	25
40	40
60	55
80	65
100	71
120	77
140	77
160	75
180	71
200	64

(a) Draw a scatter diagram of the data. Comment on the type of relation that may exist between the two variables.

(b) The quadratic function of best fit to these data is

$$h(x) = -0.0037x^2 + 1.03x + 5.7$$

Use this function to determine the horizontal distance the ball will travel before it reaches its maximum height.

(c) Use the function to find the maximum height of the ball.

(d) Use a graphing utility to verify that the function given in part (b) is the quadratic function of best fit.

(e) With a graphing utility, draw a scatter diagram of the data and then graph the quadratic function of best fit on the scatter diagram.

30. **Miles per Gallon** An engineer collects data showing the speed *s* of a Ford Taurus and its average miles per gallon, *M*. See the table.

Speed, s	Miles per Gallon, M
30	18
35	20
40	23
40	25
45	25
50	28
55	30
60	29
65	26
65	25
70	25

(a) Draw a scatter diagram of the data. Comment on the type of relation that may exist between the two variables.

(b) The quadratic function of best fit to these data is

$$M(s) = -0.017s^2 + 1.93s - 25.34$$

Use this function to determine the speed that maximizes miles per gallon.

(c) Use the function to predict miles per gallon for a speed of 63 miles per hour.

(d) Use a graphing utility to verify that the function given in part (b) is the quadratic function of best fit.

(e) With a graphing utility, draw a scatter diagram of the data and then graph the quadratic function of best fit on the scatter diagram.

Discussion and Writing

31. Refer to Example 1 on page 306. Notice that if the price charged for the calculators is $0 or $140 the revenue is $0. It is easy to explain why revenue would be $0 if the price charged is $0, but how can revenue be $0 if the price charged is $140?

'Are You Prepared?' Answers

1. $R = 3x$ **2.** $y = 1.7826x + 4.0652$

4.5 Inequalities Involving Quadratic Functions

PREPARING FOR THIS SECTION *Before getting started, review the following:*

• Solve Inequalities (Section 1.5, p. 131) • Use Interval Notation (Section 1.5, pp. 125–126)

Now Work the 'Are You Prepared?' problem on page 316.

> **OBJECTIVE 1** Solve Inequalities Involving a Quadratic Function (p. 314)

1 Solve Inequalities Involving a Quadratic Function

In this section we solve inequalities that involve quadratic functions. We will accomplish this by using their graphs. For example, to solve the inequality

$$ax^2 + bx + c > 0 \qquad a \neq 0$$

we graph the function $f(x) = ax^2 + bx + c$ and, from the graph, determine where it is above the x-axis, that is, where $f(x) > 0$. To solve the inequality $ax^2 + bx + c < 0, a \neq 0$, we graph the function $f(x) = ax^2 + bx + c$ and determine where the graph is below the x-axis. If the inequality is not strict, we include the x-intercepts in the solution.

Let's look at an example.

EXAMPLE 1 **Solving an Inequality**

Solve the inequality $x^2 - 4x - 12 \leq 0$ and graph the solution set.

Solution We graph the function $f(x) = x^2 - 4x - 12$. The intercepts are

y-intercept:	$f(0) = -12$	Evaluate f at 0.
x-intercepts (if any):	$x^2 - 4x - 12 = 0$	Solve f(x) = 0.
	$(x - 6)(x + 2) = 0$	Factor.
	$x - 6 = 0$ or $x + 2 = 0$	Apply the Zero-Product Property.
	$x = 6$ or $x = -2$	

The y-intercept is -12; the x-intercepts are -2 and 6.

The vertex is at $x = -\dfrac{b}{2a} = -\dfrac{-4}{2} = 2$. Since $f(2) = -16$, the vertex is $(2, -16)$.

See Figure 32 for the graph.

Figure 32

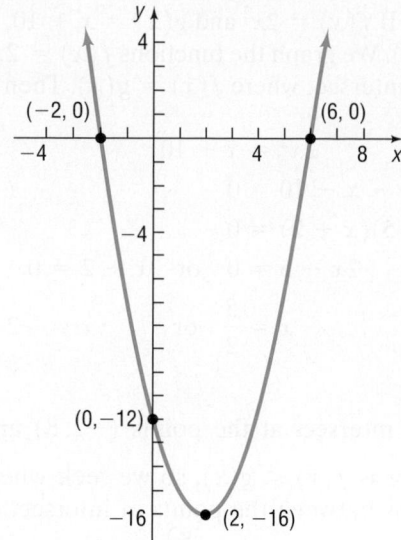

The graph is below the x-axis for $-2 < x < 6$. Since the original inequality is not strict, we include the x-intercepts. The solution set is $\{x \mid -2 \le x \le 6\}$ or, using interval notation, $[-2, 6]$. See Figure 33 for the graph of the solution set.

Figure 33

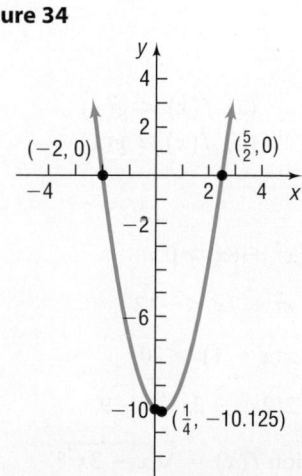

Now Work PROBLEM 9

EXAMPLE 2	Solving an Inequality

Solve the inequality $2x^2 < x + 10$ and graph the solution set.

Solution

Method 1 We rearrange the inequality so that 0 is on the right side.

$$2x^2 < x + 10$$
$$2x^2 - x - 10 < 0 \qquad \text{Subtract } x + 10 \text{ from both sides.}$$

This inequality is equivalent to the one that we wish to solve.

Next we graph the function $f(x) = 2x^2 - x - 10$. The intercepts are

y-intercept: $f(0) = -10$ Evaluate f at 0.

x-intercepts (if any): $2x^2 - x - 10 = 0$ Solve $f(x) = 0$.

$$(2x - 5)(x + 2) = 0 \qquad \text{Factor.}$$
$$2x - 5 = 0 \quad \text{or} \quad x + 2 = 0 \qquad \text{Apply the Zero-Product Property.}$$
$$x = \frac{5}{2} \quad \text{or} \quad x = -2$$

The y-intercept is -10; the x-intercepts are -2 and $\dfrac{5}{2}$.

Figure 34

The vertex is at $x = \dfrac{b}{2a} = \dfrac{-1}{4} = \dfrac{1}{4}$. Since $f\left(\dfrac{1}{4}\right) = 10.125$, the vertex is $\left(\dfrac{1}{4}, -10.125\right)$. See Figure 34 for the graph.

The graph is below the x-axis between $x = -2$ and $x = \dfrac{5}{2}$. Since the inequality is strict, the solution set is $\left\{x \mid -2 < x < \dfrac{5}{2}\right\}$ or, using interval notation, $\left(-2, \dfrac{5}{2}\right)$.

Method 2 If $f(x) = 2x^2$ and $g(x) = x + 10$, the inequality that we wish to solve is $f(x) < g(x)$. We graph the functions $f(x) = 2x^2$ and $g(x) = x + 10$. See Figure 35. The graphs intersect where $f(x) = g(x)$. Then

Figure 35

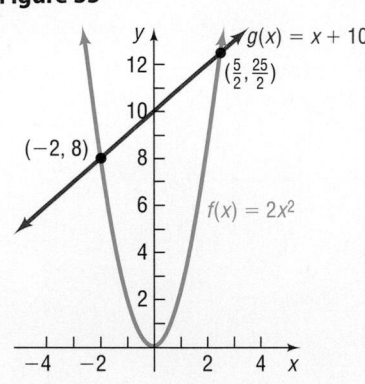

$$2x^2 = x + 10 \qquad f(x) = g(x)$$
$$2x^2 - x - 10 = 0$$
$$(2x - 5)(x + 2) = 0 \qquad \text{Factor.}$$
$$2x - 5 = 0 \quad \text{or} \quad x + 2 = 0 \qquad \text{Apply the Zero-Product Property.}$$
$$x = \frac{5}{2} \quad \text{or} \qquad x = -2$$

The graphs intersect at the points $(-2, 8)$ and $\left(\frac{5}{2}, \frac{25}{2}\right)$. The inequality that we wish to solve is $f(x) < g(x)$, so we seek where the graph of f is below that of g. This happens between the points of intersection. Since the inequality is strict, the solution set is $\left\{x \mid -2 < x < \frac{5}{2}\right\}$ or, using interval notation, $\left(-2, \frac{5}{2}\right)$.

Figure 36

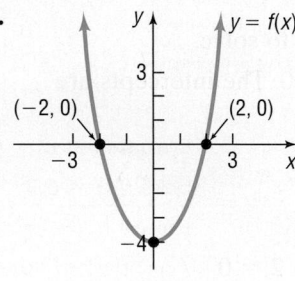

See Figure 36 for the graph of the solution set.

══ **Now Work** PROBLEMS 5 AND 13

4.5 Assess Your Understanding

'Are You Prepared?' *Answers are given at the end of these exercises. If you get a wrong answer, read the pages listed in red.*

1. Solve the inequality $-3x - 2 < 7$ (p. 131)

2. Write $(-2, 7]$ using inequality notation. (pp. 125–126)

Skill Building

In Problems 3–6, use the figure to solve each inequality.

3.

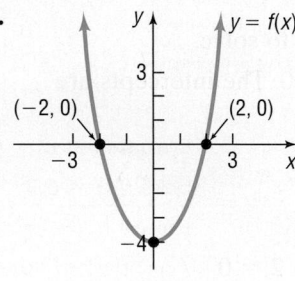

(a) $f(x) > 0$
(b) $f(x) \leq 0$

4.

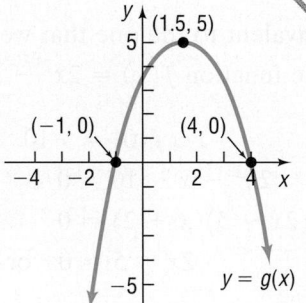

(a) $g(x) < 0$
(b) $g(x) \geq 0$

5.

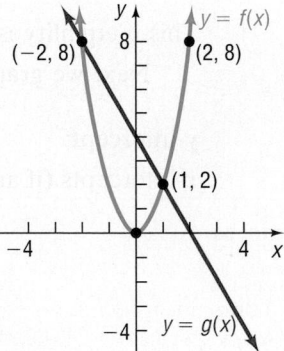

(a) $g(x) \geq f(x)$
(b) $f(x) > g(x)$

6.

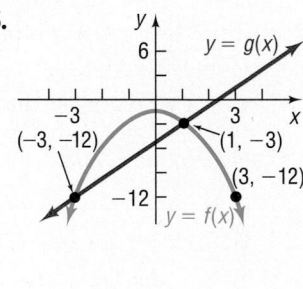

(a) $f(x) < g(x)$
(b) $f(x) \geq g(x)$

In Problems 7–22, solve each inequality.

7. $x^2 - 3x - 10 < 0$

8. $x^2 + 3x - 10 > 0$

9. $x^2 - 4x > 0$

10. $x^2 + 8x > 0$

11. $x^2 - 9 < 0$

12. $x^2 - 1 < 0$

13. $x^2 + x > 12$

14. $x^2 + 7x < -12$

15. $2x^2 < 5x + 3$

16. $6x^2 < 6 + 5x$

17. $x(x - 7) > 8$

18. $x(x + 1) > 20$

19. $4x^2 + 9 < 6x$

20. $25x^2 + 16 < 40x$

21. $6(x^2 - 1) > 5x$

22. $2(2x^2 - 3x) > -9$

23. What is the domain of the function $f(x) = \sqrt{x^2 - 16}$?

24. What is the domain of the function $f(x) = \sqrt{x - 3x^2}$?

In Problems 25–32, use the given functions f and g.

(a) *Solve* $f(x) = 0$.　　(b) *Solve* $g(x) = 0$.　　(c) *Solve* $f(x) = g(x)$.　　(d) *Solve* $f(x) > 0$.

(e) *Solve* $g(x) \le 0$.　　(f) *Solve* $f(x) > g(x)$.　　(g) *Solve* $f(x) \ge 1$.

25. $f(x) = x^2 - 1$
$g(x) = 3x + 3$

26. $f(x) = -x^2 + 3$
$g(x) = -3x + 3$

27. $f(x) = -x^2 + 1$
$g(x) = 4x + 1$

28. $f(x) = -x^2 + 4$
$g(x) = -x - 2$

29. $f(x) = x^2 - 4$
$g(x) = -x^2 + 4$

30. $f(x) = x^2 - 2x + 1$
$g(x) = -x^2 + 1$

31. $f(x) = x^2 - x - 2$
$g(x) = x^2 + x - 2$

32. $f(x) = -x^2 - x + 1$
$g(x) = -x^2 + x + 6$

Applications and Extensions

33. Physics A ball is thrown vertically upward with an initial velocity of 80 feet per second. The distance s (in feet) of the ball from the ground after t seconds is $s(t) = 80t - 16t^2$.

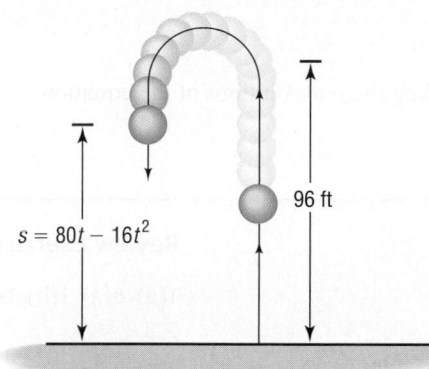

$s = 80t - 16t^2$

96 ft

(a) At what time t will the ball strike the ground?

(b) For what time t is the ball more than 96 feet above the ground?

34. Physics A ball is thrown vertically upward with an initial velocity of 96 feet per second. The distance s (in feet) of the ball from the ground after t seconds is $s(t) = 96t - 16t^2$.

(a) At what time t will the ball strike the ground?

(b) For what time t is the ball more than 128 feet above the ground?

35. Revenue Suppose that the manufacturer of a gas clothes dryer has found that, when the unit price is p dollars, the revenue R (in dollars) is

$$R(p) = -4p^2 + 4000p$$

(a) At what prices p is revenue zero?

(b) For what range of prices will revenue exceed $800,000?

36. Revenue The John Deere company has found that the revenue from sales of heavy-duty tractors is a function of the unit price p, in dollars, that it charges. If the revenue R, in dollars, is

$$R(p) = -\frac{1}{2}p^2 + 1900p$$

(a) At what prices p is revenue zero?

(b) For what range of prices will revenue exceed $1,200,000?

37. Artillery A projectile fired from the point $(0, 0)$ at an angle to the positive x-axis has a trajectory given by

$$y = cx - (1 + c^2)\left(\frac{g}{2}\right)\left(\frac{x}{v}\right)^2$$

where

$x =$ horizontal distance in meters

$y =$ height in meters

$v =$ initial muzzle velocity in meters per second (m/sec)

$g =$ acceleration due to gravity = 9.81 meters per second squared (m/sec^2)

$c > 0$ is a constant determined by the angle of elevation.

A howitzer fires an artillery round with a muzzle velocity of 897 m/sec.

(a) If the round must clear a hill 200 meters high at a distance of 2000 meters in front of the howitzer, what c values are permitted in the trajectory equation?

(b) If the goal in part (a) is to hit a target on the ground 75 kilometers away, is it possible to do so? If so, for what values of c? If not, what is the maximum distance the round will travel?

Source: www.answers.com

38. Runaway Car Using Hooke's Law, we can show that the *work* done in compressing a spring a distance of x feet from its at-rest position is $W = \frac{1}{2}kx^2$, where k is a stiffness constant depending on the spring. It can also be shown that the work done by a body in motion before it comes to rest is given by $\widetilde{W} = \frac{w}{2g}v^2$, where $w =$ weight of the object (lb), $g =$ acceleration due to gravity (32.2 ft/sec^2), and $v =$ object's velocity (in ft/sec). A parking garage has a spring shock absorber at the end of a ramp to stop runaway cars. The spring has a stiffness constant $k = 9450$ lb/ft and must be able to stop a 4000-lb car traveling at 25 mph. What is the least compression required of the spring? Express your answer using feet to the nearest tenth. **[Hint:** Solve $W > \widetilde{W}, x \ge 0$.]

Source: www.sciforums.com

Discussion and Writing

39. Show that the inequality $(x - 4)^2 \le 0$ has exactly one solution.

40. Show that the inequality $(x - 2)^2 > 0$ has one real number that is not a solution.

41. Explain why the inequality $x^2 + x + 1 > 0$ has all real numbers as the solution set.

42. Explain why the inequality $x^2 - x + 1 < 0$ has the empty set as solution set.

'Are You Prepared?' Answers

1. $\{x | x > -3\}$ or $(-3, \infty)$　　**2.** $-2 < x \le 7$

CHAPTER REVIEW

Things to Know

Linear function (p. 278)

$f(x) = mx + b$

Average rate of change $= m$

The graph is a line with slope m and y-intercept b.

Quadratic function (pp. 293–301)

$f(x) = ax^2 + bx + c, a \neq 0$

The graph is a parabola that opens up if $a > 0$ and opens down if $a < 0$.

Vertex: $\left(-\dfrac{b}{2a}, f\left(-\dfrac{b}{2a} \right) \right)$

Axis of symmetry: $x = -\dfrac{b}{2a}$

y-intercept: $f(0) = c$

x-intercept(s): If any, found by finding the real solutions of the equation $ax^2 + bx + c = 0$

Objectives

Section		You should be able to . . .	Review Exercises
4.1	1	Graph linear functions (p. 278)	1(a)–6(a), 1(b)–6(b)
	2	Use average rate of change to identify linear functions (p. 278)	7, 8
	3	Determine whether a linear function is increasing, decreasing, or constant (p. 281)	1(c)–6(c)
	4	Work with applications of linear functions (p. 282)	35, 36
4.2	1	Draw and interpret scatter diagrams (p. 287)	45(a), 46(a)
	2	Distinguish between linear and nonlinear relations (p. 288)	45(b), 46(a)
	3	Use a graphing utility to find the line of best fit (p. 289)	45(c)
4.3	1	Graph a quadratic function using transformations (p. 295)	9–14
	2	Identify the vertex and axis of symmetry of a quadratic function (p. 297)	15–24
	3	Graph a quadratic function using its vertex, axis, and intercepts (p. 297)	15–24
	4	Find the maximum or minimum value of a quadratic function (p. 301)	25–30, 37–42
4.4	1	Solve applied problems involving quadratic functions (p. 305)	37–44
	2	Use a graphing utility to find the quadratic function of best fit (p. 309)	46
4.5	1	Solve inequalities involving a quadratic function (p. 314)	31–34

Review Exercises

In Problems 1–6:

(a) Determine the slope and y-intercept of each linear function.

(b) Graph each function. Label the intercepts.

(c) Determine whether the function is increasing, decreasing, or constant.

1. $f(x) = 2x - 5$

2. $g(x) = -4x + 7$

3. $h(x) = \dfrac{4}{5}x - 6$

4. $F(x) = -\dfrac{1}{3}x + 1$

5. $G(x) = 4$

6. $H(x) = -3$

In Problems 7 and 8, determine whether the function is linear or nonlinear. If the function is linear, state its slope.

7.

x	y = f(x)
−1	−2
0	3
1	8
2	13
3	18

8.

x	y = g(x)
−1	−3
0	4
1	7
2	6
3	1

In Problems 9–14, graph each quadratic function using transformations (shifting, compressing, stretching, and/or reflecting).

9. $f(x) = (x - 2)^2 + 2$

10. $f(x) = (x + 1)^2 - 4$

11. $f(x) = -(x - 4)^2$

12. $f(x) = (x - 1)^2 - 3$

13. $f(x) = 2(x + 1)^2 + 4$

14. $f(x) = -3(x + 2)^2 + 1$

In Problems 15–24, (a) graph each quadratic function by determining whether its graph opens up or down and by finding its vertex, axis of symmetry, y-intercept, and x-intercepts, if any. (b) Determine the domain and the range of the function. (c) Determine where the function is increasing and where it is decreasing.

15. $f(x) = (x - 2)^2 + 2$

16. $f(x) = (x + 1)^2 - 4$

17. $f(x) = \frac{1}{4}x^2 - 16$

18. $f(x) = -\frac{1}{2}x^2 + 2$

19. $f(x) = -4x^2 + 4x$

20. $f(x) = 9x^2 - 6x + 3$

21. $f(x) = \frac{9}{2}x^2 + 3x + 1$

22. $f(x) = -x^2 + x + \frac{1}{2}$

23. $f(x) = 3x^2 + 4x - 1$

24. $f(x) = -2x^2 - x + 4$

In Problems 25–30, determine whether the given quadratic function has a maximum value or a minimum value, and then find the value.

25. $f(x) = 3x^2 - 6x + 4$

26. $f(x) = 2x^2 + 8x + 5$

27. $f(x) = -x^2 + 8x - 4$

28. $f(x) = -x^2 - 10x - 3$

29. $f(x) = -3x^2 + 12x + 4$

30. $f(x) = -2x^2 + 4$

In Problems 31–34, solve each quadratic inequality.

31. $x^2 + 6x - 16 < 0$

32. $3x^2 - 2x - 1 \geq 0$

33. $3x^2 \geq 14x + 5$

34. $4x^2 < 13x - 3$

35. Comparing Phone Companies Marissa must decide between one of two companies as her long-distance phone provider. Company A charges a monthly fee of $7.00 plus $0.06 per minute, while Company B does not have a monthly fee, but charges $0.08 per minute.
(a) Find a linear function that relates cost, C, to total minutes on the phone, x, for each company.
(b) Determine the number of minutes x for which the bill from Company A will equal the bill from Company B.
(c) Over what interval of minutes x will the bill from Company B be less than the bill from Company A?

36. Sales Commissions Bill was just offered a sales position for a computer company. His salary would be $15,000 per year plus 1% of his total annual sales.
(a) Find a linear function that relates Bill's annual salary, S, to his total annual sales, x.
(b) In 2005, Bill had total annual sales of $1,000,000. What was Bill's salary?
(c) What would Bill have to sell to earn $100,000?
(d) Determine the sales required of Bill for his salary to exceed $150,000.

37. Demand Equation The price p (in dollars) and the quantity x sold of a certain product obey the demand equation

$$p = -\frac{1}{10}x + 150 \qquad 0 \leq x \leq 1500$$

(a) Express the revenue R as a function of x.
(b) What is the revenue if 100 units are sold?
(c) What quantity x maximizes revenue? What is the maximum revenue?
(d) What price should the company charge to maximize revenue?

38. Landscaping A landscape engineer has 200 feet of border to enclose a rectangular pond. What dimensions will result in the largest pond?

39. Enclosing the Most Area with a Fence A farmer with 10,000 meters of fencing wants to enclose a rectangular field and then divide it into two plots with a fence parallel to one of the

sides. See the figure. What is the largest area that can be enclosed?

40. Architecture A special window in the shape of a rectangle with semicircles at each end is to be constructed so that the outside dimensions are 100 feet in length. See the illustration. Find the dimensions of the rectangle that maximizes its area.

41. Minimizing Marginal Cost Callaway Golf Company has determined that the marginal cost C of manufacturing x Big Bertha golf clubs may be expressed by the quadratic function

$$C(x) = 4.9x^2 - 617.4x + 19,600$$

(a) How many clubs should be manufactured to minimize the marginal cost?
(b) At this level of production, what is the marginal cost?

42. Violent Crimes The function

$$V(t) = 5.0t^2 - 87.3t + 1761.1$$

models the number V (in thousands) of violent crimes committed in the United States t years after 1995, based on data obtained from the Federal Bureau of Investigation. So $t = 0$ represents 1995, $t = 1$ represents 1996, and so on.

(a) Determine the year in which the fewest violent crimes were committed.
(b) Approximately how many violent crimes were committed during this year?
(c) Using a graphing utility, graph $V = V(t)$. Were the number of violent crimes increasing or decreasing during the years 1995 to 2003?

43. A rectangle has one vertex on the line $y = 10 - x$, $x > 0$, another at the origin, one on the positive x-axis, and one on the positive y-axis. Express the area A of the rectangle as a function of x. Find the largest area A that can be enclosed by the rectangle.

44. Parabolic Arch Bridge A horizontal bridge is in the shape of a parabolic arch. Given the information shown in the figure, what is the height h of the arch 2 feet from shore?

10 ft
h
2 ft
20 ft

45. Bone Length Research performed at NASA, led by Dr. Emily R. Morey-Holton, measured the lengths of the right humerus and right tibia in 11 rats that were sent to space on Spacelab Life Sciences 2. The following data were collected.

Right Humerus (mm), x	Right Tibia (mm), y
24.80	36.05
24.59	35.57
24.59	35.57
24.29	34.58
23.81	34.20
24.87	34.73
25.90	37.38
26.11	37.96
26.63	37.46
26.31	37.75
26.84	38.50

Source: NASA Life Sciences Data Archive

(a) Draw a scatter diagram of the data treating length of the right humerus as the independent variable.
(b) Based on the scatter diagram, do you think that there is a linear relation between the length of the right humerus and the length of the right tibia?
(c) Use a graphing utility to find the line of best fit relating length of the right humerus and length of the right tibia.
(d) Predict the length of the right tibia on a rat whose right humerus is 26.5 millimeters (mm).

46. Advertising A small manufacturing firm collected the following data on advertising expenditures A (in thousands of dollars) and total revenue R (in thousands of dollars).

Advertising	Total Revenue
20	$6101
22	$6222
25	$6350
25	$6378
27	$6453
28	$6423
29	$6360
31	$6231

ADVERTISE HERE

(a) Draw a scatter diagram of the data. Comment on the type of relation that may exist between the two variables.
(b) The quadratic function of best fit to these data is

$$R(A) = -7.76A^2 + 411.88A + 942.72$$

Use this function to determine the optimal level of advertising.
(c) Use the function to predict the total revenue when the optimal level of advertising is spent.
(d) Use a graphing utility to verify that the function given in part (b) is the quadratic function of best fit.
(e) Use a graphing utility to draw a scatter diagram of the data and then graph the quadratic function of best fit on the scatter diagram.

CHAPTER TEST

1. For the linear function $f(x) = -4x + 3$,
(a) Find the slope and y-intercept.
(b) Determine whether f is increasing, decreasing, or constant.
(c) Graph f.

In Problems 2 and 3, find the intercepts of each quadratic function.

2. $f(x) = 3x^2 - 2x - 8$

3. $G(x) = -2x^2 + 4x + 1$

4. Given that $f(x) = x^2 + 3x$ and $g(x) = 5x + 3$, solve $f(x) = g(x)$. Graph each function and label the points of intersection.

5. Graph $f(x) = (x - 3)^2 - 2$ using transformations.

6. For the quadratic function $f(x) = 3x^2 - 12x + 4$,
 (a) Determine whether the graph opens up or down.
 (b) Determine the vertex.
 (c) Determine the axis of symmetry.
 (d) Determine the intercepts.
 (e) Use the information from parts (a)–(d) to graph f.

7. Determine whether $f(x) = -2x^2 + 12x + 3$ has a maximum or minimum value. Then find the maximum or minimum value.

8. Solve $x^2 - 10x + 24 \geq 0$.

9. **RV Rental** The weekly rental cost of a 20-foot recreational vehicle is $129.50 plus $0.15 per mile.
 (a) Find a linear function that expresses the cost C as a function of miles driven m.
 (b) What is the rental cost if 860 miles are driven?
 (c) How many miles were driven if the rental cost is $213.80?

10. The variable interest rate on a student loan changes each July 1 based on the bank prime loan rate. For the years 1992 to 2004, this rate can be approximated by the model

$$r(x) = -0.115x^2 + 1.183x + 5.623$$

where x is the number of years since 1992 and r is the interest rate as a percent.
Source: U.S. Federal Reserve
 (a) During which year was the interest rate the highest? Determine the highest rate during this time period.
 (b) Use the model to estimate the rate in 2010. Does this value seem reasonable?

CUMULATIVE REVIEW

1. Find the distance between the points $P = (-1, 3)$ and $Q = (4, -2)$. Find the midpoint of the line segment P to Q.

2. Which of the following points are on the graph of $y = x^3 - 3x + 1$?
 (a) $(-2, -1)$ (b) $(2, 3)$ (c) $(3, 1)$

3. Solve the inequality $5x + 3 \geq 0$ and graph the solution set.

4. Find the equation of the line containing the points $(-1, 4)$ and $(2, -2)$. Express your answer in slope–intercept form and graph the line.

5. Find the equation of the line perpendicular to the line $y = 2x + 1$ and containing the point $(3, 5)$. Express your answer in slope–intercept form and graph the line.

6. Graph the equation $x^2 + y^2 - 4x + 8y - 5 = 0$.

7. Does the following relation represent a function?
 $\{(-3, 8), (1, 3), (2, 5), (3, 8)\}$.

8. For the function f defined by $f(x) = x^2 - 4x + 1$, find:
 (a) $f(2)$
 (b) $f(x) + f(2)$
 (c) $f(-x)$
 (d) $-f(x)$
 (e) $f(x + 2)$
 (f) $\dfrac{f(x + h) - f(x)}{h}$ $h \neq 0$

9. Find the domain of $h(z) = \dfrac{3z - 1}{6z - 7}$.

10. Is the following graph the graph of a function?

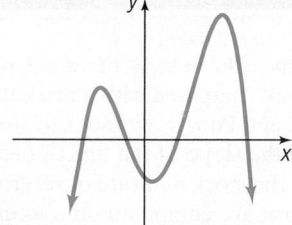

11. Consider the function $f(x) = \dfrac{x}{x + 4}$.

 (a) Is the point $\left(1, \dfrac{1}{4}\right)$ on the graph of f?
 (b) If $x = -2$, what is $f(x)$? What point is on the graph of f?
 (c) If $f(x) = 2$, what is x? What point is on the graph of f?

12. Is the function $f(x) = \dfrac{x^2}{2x + 1}$ even, odd, or neither?

13. Approximate the local maxima and local minima of $f(x) = x^3 - 5x + 1$ on $(-4, 4)$. Determine where the function is increasing and where it is decreasing.

14. If $f(x) = 3x + 5$ and $g(x) = 2x + 1$,
 (a) Solve $f(x) = g(x)$.
 (b) Solve $f(x) > g(x)$.

15. For the graph of the function *f*,

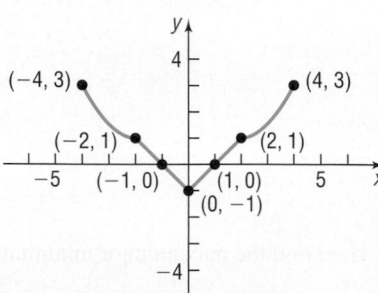

(a) Find the domain and the range of *f*.
(b) Find the intercepts.
(c) Is the graph of *f* symmetric with respect to the *x*-axis, the *y*-axis, or the origin?
(d) Find $f(2)$.
(e) For what value(s) of *x* is $f(x) = 3$?
(f) Solve $f(x) < 0$.
(g) Graph $y = f(x) + 2$.
(h) Graph $y = f(-x)$.
(i) Graph $y = 2f(x)$.
(j) Is *f* even, odd, or neither?
(k) Find the interval(s) on which *f* is increasing.

CHAPTER PROJECTS

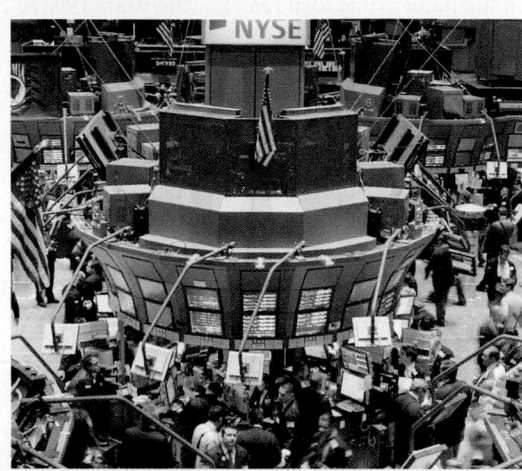

I. Analyzing a Stock The beta, β, of a stock represents the relative risk of a stock compared with a market basket of stocks, such as Standard and Poor's 500 Index of stocks. Beta is computed by finding the slope of the line of best fit between the rate of return of the stock and rate of return of the S&P 500. The rates of return are computed on a weekly basis.

1. Find the weekly closing price of your favorite stock and the S&P 500 for 20 weeks. One good source is on the Internet at */finance.yahoo.com.*

2. Compute the rate of return by computing the weekly percentage change in the closing price of your stock and the weekly percentage change in the S&P 500 using the following formula:

$$\text{Weekly \% change} = \frac{P_2 - P_1}{P_1}$$

where P_1 is last week's price and P_2 is this week's price.

3. Using a graphing utility, find the line of best fit, treating the weekly rate of return of the S&P 500 as the independent variable and the weekly rate of return of your stock as the dependent variable.

4. What is the beta of your stock?

5. Compare your result with that of the Value Line Investment Survey found in your library. What might account for any differences?

The following projects are available on the Instructor's Resource Center (IRC):

II. Cannons A battery commander uses the weight of a missile, its initial velocity, and the position of its gun to determine where the missile will travel.

III. First and Second Differences Finite differences provide a numerical method that is used to estimate the graph of an unknown function.

IV. CBL Experiment Computer simulation is used to study the physical properties of a bouncing ball.

Polynomial and Rational Functions

<div style="text-align: right; font-size: 4em; font-weight: bold;">5</div>

We Don't Want to Be Number One Again—in Tree Pollen

LEXINGTON, Ky. (March 29, 2006)—Were your allergies especially bothersome last spring? It's no wonder. In the spring of 2005, Lexington, Ky., was the tree pollen capital of the U.S. No allergy sufferer is rooting for the Bluegrass to earn that title again in 2006. Either way, allergy sufferers should be prepared. The spring allergy season is here.

If you can't stop sneezing, blowing your nose, and rubbing your itchy eyes, you may have allergic rhinitis, as do 40 million other Americans. The pollens of trees such as maple, oak and walnut trigger much of the allergic rhinitis, commonly referred to as hay fever, in the early spring. Pollinating grasses and weeds are responsible for summer hay fever. Perennial allergic rhinitis can be triggered any time of year by indoor allergens, such as animal dander, mold or dust mites.

"Spring can be the start of misery for many allergy sufferers because they have multiple pollen allergies," said Dr. Beth Miller, director of the UK Asthma, Allergy and Sinus Clinics.

Source: Adapted from *Melanie Jackson, "We Don't Want to Be Number One Again—In Tree Pollen," University of Kentucky News, March 29, 2006, http://news.uky.edu.*

—See the Chapter Project—

A Look Back

In Chapter 3, we began our discussion of functions. We defined domain and range and independent and dependent variables; we found the value of a function and graphed functions. We continued our study of functions by listing the properties that a function might have, like being even or odd, and we created a library of functions, naming key functions and listing their properties, including their graphs.

In Chapter 4, we discussed linear functions and quadratic functions, which belong to the class of *polynomial functions*.

A Look Ahead

In this chapter, we look at two general classes of functions, polynomial functions and rational functions, and examine their properties. Polynomial functions are arguably the simplest expressions in algebra. For this reason, they are often used to approximate other, more complicated functions. Rational functions are ratios of polynomial functions.

Outline

5.1 Polynomial Functions and Models

PREPARING FOR THIS SECTION *Before getting started, review the following:*

- Polynomials (Chapter R, Section R.4, pp. 39–47)
- Using a Graphing Utility to Approximate Local Maxima and Local Minima (Section 3.3, pp. 235–236)
- Intercepts of a Function (Section 3.2, pp. 224–225)

- Graphing Techniques: Transformations (Section 3.5, pp. 252–260)
- Intercepts (Section 2.2, pp. 165–166)

Now Work the 'Are You Prepared?' problems on page 339.

OBJECTIVES 1 Identify Polynomial Functions and Their Degree (p. 324)

2 Graph Polynomial Functions Using Transformations (p. 328)

3 Identify the Real Zeros of a Polynomial Function and Their Multiplicity (p. 329)

4 Analyze the Graph of a Polynomial Function (p. 336)

1 Identify Polynomial Functions and Their Degree

In Chapter 4 we studied the linear function $f(x) = mx + b$, which we can write as

$$f(x) = a_1x + a_0$$

and we studied the quadratic function $f(x) = ax^2 + bx + c$, $a \neq 0$, which we can write as

$$f(x) = a_2x^2 + a_1x + a_0, \ a_2 \neq 0$$

Each of these functions is an example of a *polynomial function*.

DEFINITION A **polynomial function** is a function of the form

$$f(x) = a_nx^n + a_{n-1}x^{n-1} + \cdots + a_1x + a_0 \qquad \textbf{(1)}$$

where $a_n, a_{n-1}, \ldots, a_1, a_0$ are real numbers and n is a nonnegative integer. The domain of a polynomial function is the set of all real numbers.

A polynomial function is a function whose rule is given by a polynomial in one variable. The **degree** of a polynomial function is the largest power of x that appears. The zero polynomial function $f(x) = 0$ is not assigned a degree.

Polynomial functions are among the simplest expressions in algebra. They are easy to evaluate: only addition and repeated multiplication are required. Because of this, they are often used to approximate other, more complicated functions. In this section, we investigate properties of this important class of functions.

EXAMPLE 1 **Identifying Polynomial Functions**

Determine which of the following are polynomial functions. For those that are, state the degree; for those that are not, tell why not.

(a) $f(x) = 2 - 3x^4$ (b) $g(x) = \sqrt{x}$ (c) $h(x) = \dfrac{x^2 - 2}{x^3 - 1}$

(d) $F(x) = 0$ (e) $G(x) = 8$ (f) $H(x) = -2x^3(x - 1)^2$

Solution (a) f is a polynomial function of degree 4.

(b) g is not a polynomial function because $g(x) = \sqrt{x} = x^{\frac{1}{2}}$, so the variable x is raised to the $\dfrac{1}{2}$ power, which is not a nonnegative integer.

(c) h is not a polynomial function. It is the ratio of two polynomials, and the polynomial in the denominator is of positive degree.

(d) F is the zero polynomial function; it is not assigned a degree.

(e) G is a nonzero constant function. It is a polynomial function of degree 0 since $G(x) = 8 = 8x^0$.

(f) $H(x) = -2x^3(x - 1)^2 = -2x^3(x^2 - 2x + 1) = -2x^5 + 4x^4 - 2x^3$. So H is a polynomial function of degree 5. Do you see a way to find the degree of H without multiplying out?

Now Work PROBLEMS 11 AND 15

We have already discussed in detail polynomial functions of degrees 0, 1, and 2. See Table 1 for a summary of the properties of the graphs of these polynomial functions.

Table 1

Degree	Form	Name	Graph
No degree	$f(x) = 0$	Zero function	The x-axis
0	$f(x) = a_0, \quad a_0 \neq 0$	Constant function	Horizontal line with y-intercept a_0
1	$f(x) = a_1x + a_0, \quad a_1 \neq 0$	Linear function	Nonvertical, nonhorizontal line with slope a_1 and y-intercept a_0
2	$f(x) = a_2x^2 + a_1x + a_0, \quad a_2 \neq 0$	Quadratic function	Parabola: Graph opens up if $a_2 > 0$; graph opens down if $a_2 < 0$

One objective of this section is to analyze the graph of a polynomial function. If you take a course in calculus, you will learn that the graph of every polynomial function is both smooth and continuous. By **smooth,** we mean that the graph contains no sharp corners or cusps; by **continuous,** we mean that the graph has no gaps or holes and can be drawn without lifting pencil from paper. See Figures 1(a) and (b).

Figure 1

(a) Graph of a polynomial function: smooth, continuous

(b) Cannot be the graph of a polynomial function

Power Functions

We begin the analysis of the graph of a polynomial function by discussing *power functions,* a special kind of polynomial function.

DEFINITION

A **power function of degree** n is a monomial of the form

$$f(x) = ax^n \qquad (2)$$

where a is a real number, $a \neq 0$, and $n > 0$ is an integer.

Examples of power functions are

$$f(x) = 3x \qquad f(x) = -5x^2 \qquad f(x) = 8x^3 \qquad f(x) = -5x^4$$
$$\text{degree 1} \qquad \text{degree 2} \qquad \text{degree 3} \qquad \text{degree 4}$$

The graph of a power function of degree 1, $f(x) = ax$, is a straight line, with slope a, that passes through the origin. The graph of a power function of degree 2, $f(x) = ax^2$, is a parabola, with vertex at the origin, that opens up if $a > 0$ and down if $a < 0$.

If we know how to graph a power function of the form $f(x) = x^n$, a compression or stretch and, perhaps, a reflection about the x-axis will enable us to obtain the graph of $g(x) = ax^n$. Consequently, we shall concentrate on graphing power functions of the form $f(x) = x^n$.

We begin with power functions of even degree of the form $f(x) = x^n$, $n \geq 2$ and n even. The domain of f is the set of all real numbers, and the range is the set of nonnegative real numbers. Such a power function is an even function (do you see why?), so its graph is symmetric with respect to the y-axis. Its graph always contains the origin and the points $(-1, 1)$ and $(1, 1)$.

If $n = 2$, the graph is the familiar parabola $y = x^2$ that opens up, with vertex at the origin. If $n \geq 4$, the graph of $f(x) = x^n$, n even, will be closer to the x-axis than the parabola $y = x^2$ if $-1 < x < 1$ and farther from the x-axis than the parabola $y = x^2$ if $x < -1$ or if $x > 1$. Figure 2(a) illustrates this conclusion. Figure 2(b) shows the graphs of $y = x^4$ and $y = x^8$ for comparison.

Figure 2

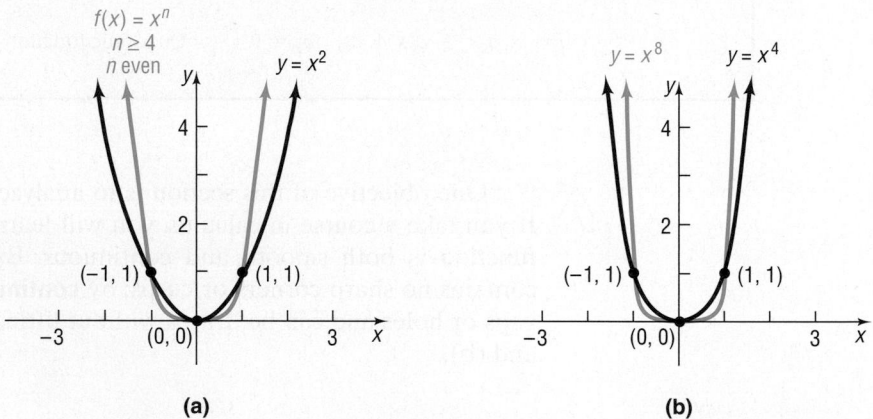

(a) (b)

From Figure 2, we can see that as n increases the graph of $f(x) = x^n$, $n \geq 2$ and n even, tends to flatten out near the origin and to increase very rapidly when x is far from 0. For large n, it may appear that the graph coincides with the x-axis near the origin, but it does not; the graph actually touches the x-axis only at the origin (see Table 2). Also, for large n, it may appear that for $x < -1$ or for $x > 1$ the graph is vertical, but it is not; it is only increasing very rapidly in these intervals. If the graphs were enlarged many times, these distinctions would be clear.

Table 2

	$x = 0.1$	$x = 0.3$	$x = 0.5$
$f(x) = x^8$	10^{-8}	0.0000656	0.0039063
$f(x) = x^{20}$	10^{-20}	$3.487 \cdot 10^{-11}$	0.000001
$f(x) = x^{40}$	10^{-40}	$1.216 \cdot 10^{-21}$	$9.095 \cdot 10^{-13}$

 Seeing the Concept

Graph $Y_1 = x^4$, $Y_2 = x^8$, and $Y_3 = x^{12}$ using the viewing rectangle $-2 \leq x \leq 2, -4 \leq y \leq 16$. Then graph each again using the viewing rectangle $-1 \leq x \leq 1, 0 \leq y \leq 1$. See Figure 3. TRACE along one of the graphs to confirm that for x close to 0 the graph is above the x-axis and that for $x > 0$ the graph is increasing.

Figure 3

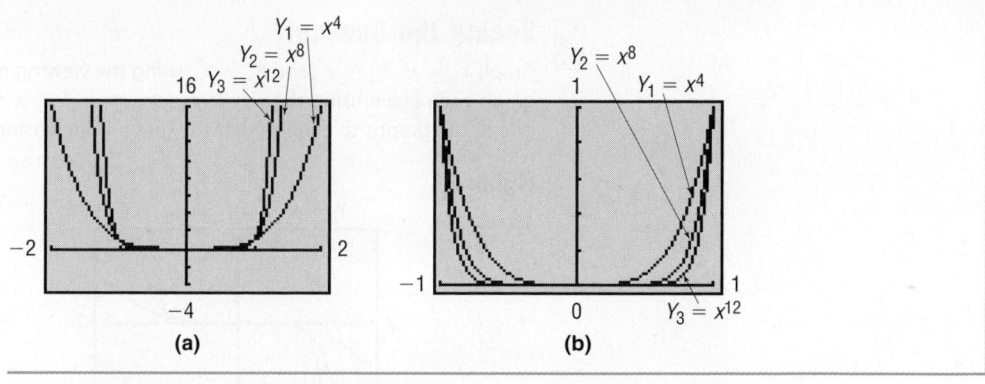

(a) (b)

Properties of Power Functions, $f(x) = x^n$, n Is an Even Integer

1. f is an even function, so its graph is symmetric with respect to the y-axis.
2. The domain is the set of all real numbers. The range is the set of non-negative real numbers.
3. The graph always contains the points $(-1, 1)$, $(0, 0)$, and $(1, 1)$.
4. As the exponent n increases in magnitude, the graph becomes more vertical when $x < -1$ or $x > 1$; but for x near the origin, the graph tends to flatten out and lie closer to the x-axis.

Now we consider power functions of odd degree of the form $f(x) = x^n$, $n \geq 3$ and n odd. The domain and the range of f are the set of real numbers. Such a power function is an odd function (do you see why?), so its graph is symmetric with respect to the origin. Its graph always contains the origin and the points $(-1, -1)$ and $(1, 1)$.

The graph of $f(x) = x^n$ when $n = 3$ has been shown several times and is repeated in Figure 4. If $n \geq 5$, the graph of $f(x) = x^n$, n odd, will be closer to the x-axis than that of $y = x^3$ if $-1 < x < 1$ and farther from the x-axis than that of $y = x^3$ if $x < -1$ or if $x > 1$. Figure 4 also illustrates this conclusion.

Figure 5 shows the graph of $y = x^5$ and the graph of $y = x^9$ for further comparison.

Figure 4 **Figure 5**

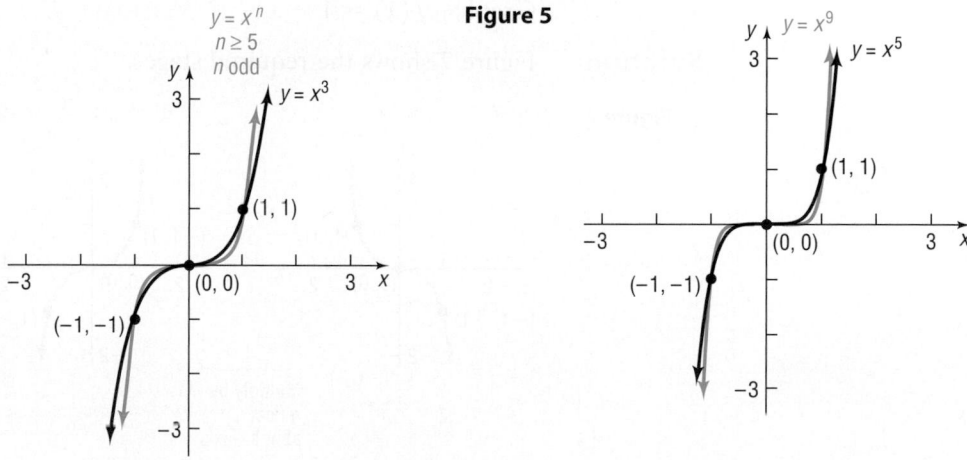

It appears that each graph coincides with the x-axis near the origin, but it does not; each graph actually crosses the x-axis at the origin. Also, it appears that as x increases the graph becomes vertical, but it does not; each graph is increasing very rapidly.

Seeing the Concept

Graph $Y_1 = x^3$, $Y_2 = x^7$, and $Y_3 = x^{11}$ using the viewing rectangle $-2 \leq x \leq 2$, $-16 \leq y \leq 16$. Then graph each again using the viewing rectangle $-1 \leq x \leq 1$, $-1 \leq y \leq 1$. See Figure 6. TRACE along one of the graphs to confirm that the graph is increasing and crosses the x-axis at the origin.

Figure 6

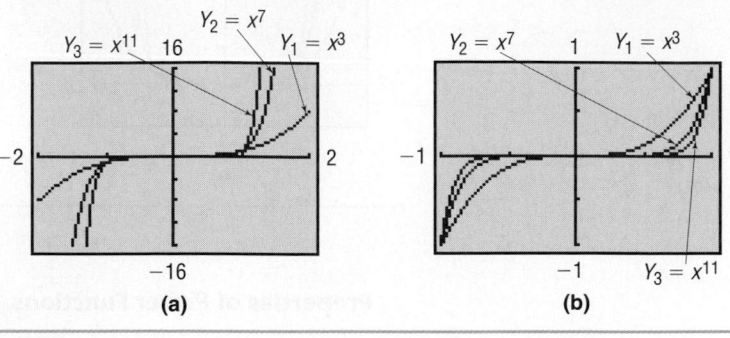

(a)

(b)

To summarize:

Properties of Power Functions, $f(x) = x^n$, n Is an Odd Integer

1. f is an odd function, so its graph is symmetric with respect to the origin.
2. The domain and the range are the set of all real numbers.
3. The graph always contains the points $(-1, -1)$, $(0, 0)$, and $(1, 1)$.
4. As the exponent n increases in magnitude, the graph becomes more vertical when $x < -1$ or $x > 1$; but for x near the origin, the graph tends to flatten out and lie closer to the x-axis.

2 Graph Polynomial Functions Using Transformations

The methods of shifting, compression, stretching, and reflection studied in Section 3.5, when used with the facts just presented, will enable us to graph polynomial functions that are transformations of power functions.

EXAMPLE 2 Graphing Polynomial Functions Using Transformations

Graph: $f(x) = 1 - x^5$

Solution Figure 7 shows the required stages.

Figure 7

(a) $y = x^5$ Multiply by -1; reflect about x-axis (b) $y = -x^5$ Add 1; shift up 1 unit (c) $y = -x^5 + 1 = 1 - x^5$

EXAMPLE 3 Graphing Polynomial Functions Using Transformations

Graph: $f(x) = \dfrac{1}{2}(x - 1)^4$

Solution Figure 8 shows the required stages.

Figure 8

Replace x by $x - 1$;
shift right
1 unit

Multiply by $\frac{1}{2}$;
compression by
a factor of $\frac{1}{2}$

(a) $y = x^4$ (b) $y = (x - 1)^4$ (c) $y = \frac{1}{2}(x - 1)^4$

─── **Now Work** PROBLEMS **23** AND **29**

3 Identify the Real Zeros of a Polynomial Function and Their Multiplicity

Figure 9 shows the graph of a polynomial function with four x-intercepts. Notice that at the x-intercepts the graph must either cross the x-axis or touch the x-axis. Consequently, between consecutive x-intercepts the graph is either above the x-axis or below the x-axis. We will make use of this property of the graph of a polynomial function shortly.

Figure 9

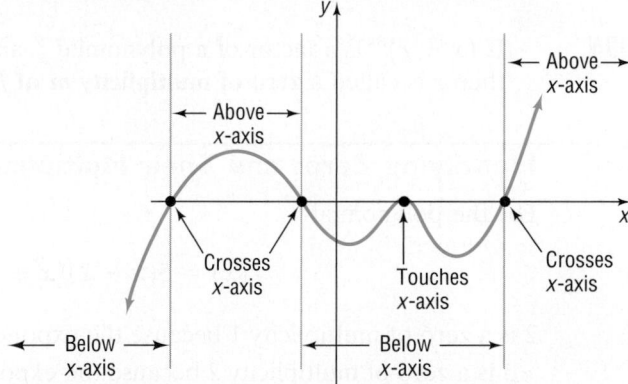

Above
x-axis

Above
x-axis

Crosses
x-axis

Touches
x-axis

Crosses
x-axis

Below
x-axis

Below
x-axis

If a polynomial function f is factored completely, it is easy to solve the equation $f(x) = 0$ using the Zero-Product Property and locate the x-intercepts of the graph. For example, if $f(x) = (x - 1)^2(x + 3)$, then the solutions of the equation

$$f(x) = (x - 1)^2(x + 3) = 0$$

are identified as 1 and -3. Based on this result, we make the following observations:

DEFINITION If f is a function and r is a real number for which $f(r) = 0$, then r is called a **real zero of f**.

As a consequence of this definition, the following statements are equivalent.

1. r is a real zero of a polynomial function f
2. r is an x-intercept of the graph of f
3. $x - r$ is a factor of f

So the real zeros of a polynomial function are the x-intercepts of its graph, and they are found by solving the equation $f(x) = 0$.

| EXAMPLE 4 | Finding a Polynomial from Its Zeros |

(a) Find a polynomial of degree 3 whose zeros are $-3, 2$, and 5.

 (b) Use a graphing utility to graph the polynomial found in part (a) to verify your result.

Solution (a) If r is a real zero of a polynomial f, then $x - r$ is a factor of f. This means that $x - (-3) = x + 3$, $x - 2$, and $x - 5$ are factors of f. As a result, any polynomial of the form

Figure 10

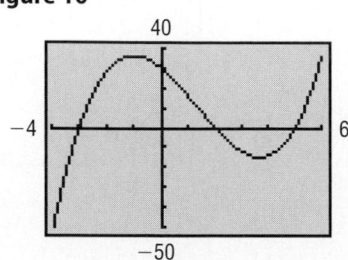

$$f(x) = a(x + 3)(x - 2)(x - 5)$$

where a is any nonzero real number, qualifies. The value of a causes a stretch, compression, or reflection, but does not affect the x-intercepts. Do you know why?

(b) We choose to graph f with $a = 1$. Then

$$f(x) = (x + 3)(x - 2)(x - 5) = x^3 - 4x^2 - 11x + 30$$

Figure 10 shows the graph of f. Notice that the x-intercepts are $-3, 2$, and 5.

Seeing the Concept

Graph the function found in Example 4 for $a = 2$ and $a = -1$. Does the value of a affect the zeros of f? How does the value of a affect the graph of f?

 Now Work PROBLEM 37

If the same factor $x - r$ occurs more than once, r is called a **repeated,** or **multiple, zero of f.** More precisely, we have the following definition.

DEFINITION If $(x - r)^m$ is a factor of a polynomial f and $(x - r)^{m+1}$ is not a factor of f, then r is called a **zero of multiplicity m of f.***

| EXAMPLE 5 | Identifying Zeros and Their Multiplicities |

For the polynomial

$$f(x) = 5(x - 2)(x + 3)^2\left(x - \frac{1}{2}\right)^4$$

2 is a zero of multiplicity 1 because the exponent on the factor $x - 2$ is 1.
-3 is a zero of multiplicity 2 because the exponent on the factor $x + 3$ is 2.
$\frac{1}{2}$ is a zero of multiplicity 4 because the exponent on the factor $x - \frac{1}{2}$ is 4.

 Now Work PROBLEM 45(a)

Suppose that it is possible to factor completely a polynomial function and, as a result, locate all the x-intercepts of its graph (the real zeros of the function). These x-intercepts then divide the x-axis into open intervals and, on each such interval, the graph of the polynomial will be either above or below the x-axis. Let's look at an example.

| EXAMPLE 6 | Graphing a Polynomial Using Its x-Intercepts |

For the polynomial: $f(x) = x^2(x - 2)$

(a) Find the x- and y-intercepts of the graph of f.

(b) Use the x-intercepts to find the intervals on which the graph of f is above the x-axis and the intervals on which the graph of f is below the x-axis.

* Some books use the terms **multiple root** and **root of multiplicity m.**

(c) Locate other points on the graph and connect all the points plotted with a smooth, continuous curve.

Solution (a) The y-intercept is $f(0) = 0^2(0 - 2) = 0$. The x-intercepts satisfy the equation

$$f(x) = x^2(x - 2) = 0$$

from which we find

$$x^2 = 0 \quad \text{or} \quad x - 2 = 0$$
$$x = 0 \quad \text{or} \quad x = 2$$

The x-intercepts are 0 and 2.

(b) The two x-intercepts divide the x-axis into three intervals:

$$(-\infty, 0) \qquad (0, 2) \qquad (2, \infty)$$

Since the graph of f crosses or touches the x-axis only at $x = 0$ and $x = 2$, it follows that the graph of f is either above the x-axis $[f(x) > 0]$ or below the x-axis $[f(x) < 0]$ on each of these three intervals. To see where the graph lies, we only need to pick a number in each interval, evaluate f there, and see whether the value is positive (above the x-axis) or negative (below the x-axis). See Table 3.

(c) In constructing Table 3, we obtained three additional points on the graph: $(-1, -3)$, $(1, -1)$, and $(3, 9)$. Figure 11 illustrates these points, the intercepts, and a smooth, continuous curve (the graph of f) connecting them.

Table 3

	0	2	
Interval	$(-\infty, 0)$	$(0, 2)$	$(2, \infty)$
Number chosen	-1	1	3
Value of f	$f(-1) = -3$	$f(1) = -1$	$f(3) = 9$
Location of graph	Below x-axis	Below x-axis	Above x-axis
Point on graph	$(-1, -3)$	$(1, -1)$	$(3, 9)$

Figure 11

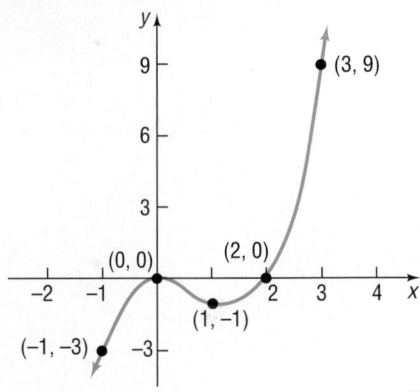

Look again at Table 3. Since the graph of $f(x) = x^2(x - 2)$ is below the x-axis on both sides of 0, the graph of f *touches* the x-axis at $x = 0$, a *zero of multiplicity 2*. Since the graph of f is below the x-axis for $x < 2$ and above the x-axis for $x > 2$, the graph of f *crosses* the x-axis at $x = 2$, a *zero of multiplicity 1*.

This suggests the following results:

If r Is a Zero of Even Multiplicity

The sign of $f(x)$ does not change from one side to the other side of r.

The graph of f **touches** the x-axis at r.

If r Is a Zero of Odd Multiplicity

The sign of $f(x)$ changes from one side to the other side of r.

The graph of f **crosses** the x-axis at r.

Now Work PROBLEM 45(b)

Behavior Near a Zero

We have just learned how the multiplicity of a zero can be used to determine whether the graph of a function touches or crosses the x-axis at the zero. However, we can learn more about the behavior of the graph near its zeros than just whether the graph crosses or touches the x-axis. Consider the function $f(x) = x^2(x - 2)$ whose graph is drawn in Figure 11. The zeros of f are 0 and 2. Table 4 shows the values of $f(x) = x^2(x - 2)$ and $y = -2x^2$ for x near 0. Figure 12 shows the points $(-0.1, -0.021), (-0.05, -0.0051)$, and so on, that are on the graph of $f(x) = x^2(x - 2)$ along with the graph of $y = -2x^2$ on the same Cartesian plane. From the table and graph, we can see that the points on the graph of $f(x) = x^2(x - 2)$ and the points on the graph of $y = -2x^2$ are indistinguishable near $x = 0$. So $y = -2x^2$ describes the behavior of the graph of $f(x) = x^2(x - 2)$ near $x = 0$.

Table 4

x	$f(x) = x^2(x - 2)$	$y = -2x^2$
-0.1	-0.021	-0.02
-0.05	-0.005125	-0.005
-0.03	-0.001827	-0.0018
-0.01	-0.000201	-0.0002
0	0	0
0.01	-0.000199	-0.0002
0.03	-0.001773	-0.0018
0.05	-0.004875	-0.005
0.1	-0.019	-0.02

Figure 12

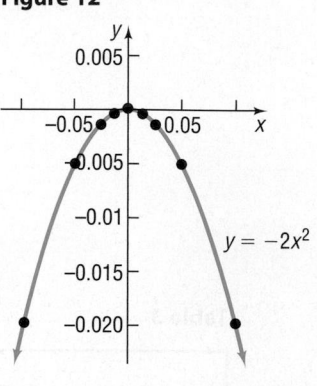

But how did we know that the function $f(x) = x^2(x - 2)$ behaves like $y = -2x^2$ when x is close to 0? In other words, where did $y = -2x^2$ come from? Because the zero, 0, comes from the factor x^2, we evaluate all factors in the function f at 0 with the exception of x^2.

$$f(x) = x^2(x - 2)$$
$$\approx x^2(0 - 2)$$
$$= -2x^2$$

The factor x^2 gives rise to the zero, so we keep the factor x^2 and let $x = 0$ in the remaining factors to find the behavior near 0.

This tells us that the graph of $f(x) = x^2(x - 2)$ will behave like the graph of $y = -2x^2$ near $x = 0$.

Now let's discuss the behavior of $f(x) = x^2(x - 2)$ near $x = 2$, the other zero. Because the zero, 2, comes from the factor $x - 2$, we evaluate all factors of the function f at 2 with the exception of $x - 2$.

$$f(x) = x^2(x - 2)$$
$$\approx 2^2(x - 2)$$
$$= 4(x - 2)$$

The factor $x - 2$ gives rise to the zero, so we keep the factor $x - 2$ and let $x = 2$ in the remaining factors to find the behavior near 2.

So the graph of $f(x) = x^2(x - 2)$ will behave like the graph of $y = 4(x - 2)$ near $x = 2$. Table 5 verifies that $f(x) = x^2(x - 2)$ and $y = 4(x - 2)$ have similar values for x near 2. Figure 13 shows the points $(1.9, -0.361), (1.99, -0.0396)$, and so on, that are on the graph of $f(x) = x^2(x - 2)$ along with the graph of $y = 4(x - 2)$ on the same Cartesian plane. We can see that the points on the graph of $f(x) = x^2(x - 2)$ and the points on the graph of $y = 4(x - 2)$ are indistinguishable near $x = 2$. So $y = 4(x - 2)$, a line with slope 4, describes the behavior of the graph of $f(x) = x^2(x - 2)$ near $x = 2$.

Table 5

x	$f(x) = x^2(x - 2)$	$y = 4(x - 2)$
1.9	−0.361	−0.4
1.99	−0.0396	−0.04
1.999	−0.003996	−0.004
2	0	0
2.001	0.004004	0.004
2.01	0.0404	0.04
2.1	0.441	0.4

Figure 13

Figure 14

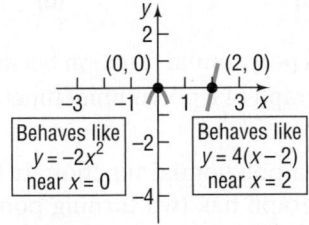

Figure 14 illustrates how we would use this information to begin to graph $f(x) = x^2(x - 2)$.

By determining the multiplicity of a real zero, we determine whether the graph crosses or touches the x-axis at the zero. By determining the behavior of the graph near the real zero, we determine how the graph touches or crosses the x-axis.

 Now Work PROBLEM 45(c)

Turning Points

Look again at Figure 11. We cannot be sure just how low the graph actually goes between $x = 0$ and $x = 2$. But we do know that somewhere in the interval $(0, 2)$ the graph of f must change direction (from decreasing to increasing). The points at which a graph changes direction are called **turning points.** In calculus, such points are called **local maxima** or **local minima,** and techniques for locating them are given. So we shall not ask for the location of turning points in our graphs. Instead, we will use the following result from calculus, which tells us the maximum number of turning points that the graph of a polynomial function can have.

THEOREM **Turning Points**

If f is a polynomial function of degree n, then f has at most $n - 1$ turning points.

If the graph of a polynomial function f has $n - 1$ turning points, the degree of f is at least n.

For example, the graph of $f(x) = x^2(x - 2)$ shown in Figure 11 is the graph of a polynomial of degree 3 and has $3 - 1 = 2$ turning points: one at $(0, 0)$ and the other somewhere between $x = 0$ and $x = 2$.

Exploration

A graphing utility can be used to locate the turning points of a graph. Graph $Y_1 = x^2(x - 2)$. Use MINIMUM to find the location of the turning point for $0 < x < 2$. See Figure 15.

Figure 15

Now Work PROBLEM 45(d)

EXAMPLE 7	**Identifying the Graph of a Polynomial Function**

Which of the graphs in Figure 16 could be the graph of a polynomial function? For those that could, list the real zeros and state the least degree the polynomial can have. For those that could not, say why not.

Figure 16

(a) (b) (c) (d)

Solution

(a) The graph in Figure 16(a) cannot be the graph of a polynomial function because of the gap that occurs at $x = -1$. Remember, the graph of a polynomial function is continuous—no gaps or holes.

(b) The graph in Figure 16(b) could be the graph of a polynomial function. It has three real zeros, at -2, at 1, and at 2. Since the graph has two turning points, the degree of the polynomial function must be at least 3.

(c) The graph in Figure 16(c) cannot be the graph of a polynomial function because of the cusp at $x = 1$. Remember, the graph of a polynomial function is smooth.

(d) The graph in Figure 16(d) could be the graph of a polynomial function. It has two real zeros, at -2 and at 1. Since the graph has three turning points, the degree of the polynomial function is at least 4.

━━━━━━━━━━ **Now Work** PROBLEM 57

End Behavior

One last remark about Figure 11. For very large values of x, either positive or negative, the graph of $f(x) = x^2(x - 2)$ looks like the graph of $y = x^3$. To see why, we write f in the form

$$f(x) = x^2(x - 2) = x^3 - 2x^2 = x^3\left(1 - \frac{2}{x}\right)$$

Now, for large values of x, either positive or negative, the term $\frac{2}{x}$ is close to 0, so for large values of x

$$f(x) = x^3 - 2x^2 = x^3\left(1 - \frac{2}{x}\right) \approx x^3$$

The behavior of the graph of a function for large values of x, either positive or negative, is referred to as its **end behavior.**

THEOREM

End Behavior

For large values of x, either positive or negative, the graph of the polynomial

$$f(x) = a_n x^n + a_{n-1}x^{n-1} + \cdots + a_1 x + a_0$$

resembles the graph of the power function

$$y = a_n x^n$$

Look back at Figures 2 and 4. Based on the above theorem and the previous discussion on power functions, the end behavior of a polynomial can only be of four types. See Figure 17.

Figure 17
End Behavior of $f(x) = a_n x^n + a_{n-1}x^{n-1} + \cdots + a_1 x + a_0$

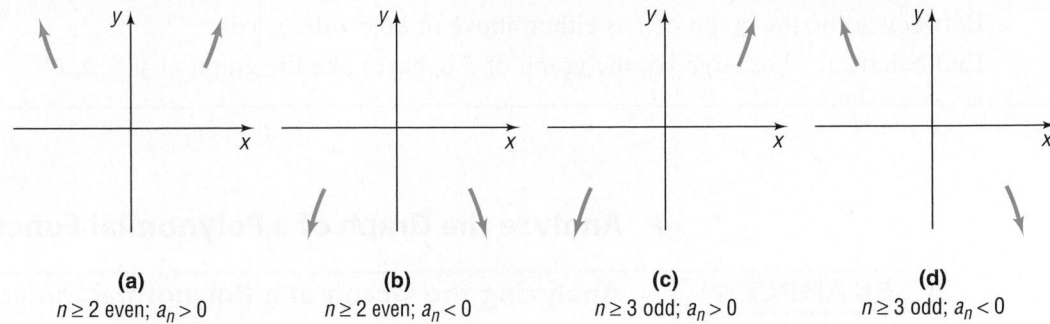

(a)	(b)	(c)	(d)
$n \geq 2$ even; $a_n > 0$	$n \geq 2$ even; $a_n < 0$	$n \geq 3$ odd; $a_n > 0$	$n \geq 3$ odd; $a_n < 0$

For example, if $f(x) = -2x^4 + x^3 + 4x^2 - 7x + 1$, the graph of f will resemble the graph of the power function $y = -2x^4$ for large $|x|$. The graph of f will look like Figure 17(b) for large $|x|$.

━━━ Now Work PROBLEM 45(e)

EXAMPLE 8 **Identifying the Graph of a Polynomial Function**

Which of the graphs in Figure 18 could be the graph of

$$f(x) = x^4 + 5x^3 + 5x^2 - 5x - 6?$$

Figure 18

| (a) | (b) | (c) | (d) |

Solution The y-intercept of f is $f(0) = -6$. We can eliminate the graph in Figure 18(a), whose y-intercept is positive.

We don't have any methods for finding the x-intercepts of f, so we move on to investigate the turning points of each graph. Since f is of degree 4, the graph of f has at most 3 turning points. We eliminate the graph in Figure 18(c) since that graph has 5 turning points.

Now we look at end behavior. For large values of x, the graph of f will behave like the graph of $y = x^4$. This eliminates the graph in Figure 18(d), whose end behavior is like the graph of $y = -x^4$.

Only the graph in Figure 18(b) could be (and, in fact, is) the graph of $f(x) = x^4 + 5x^3 + x^2 - 5x - 6$.

━━━ Now Work PROBLEM 61

SUMMARY Graph of a Polynomial Function $f(x) = a_n x^n + a_{n-1} x^{n-1} + \cdots + a_1 x + a_0 \quad a_n \neq 0$

Degree of the polynomial f: n

Maximum number of turning points: $n - 1$

At a zero of even multiplicity: The graph of f touches the x-axis.

At a zero of odd multiplicity: The graph of f crosses the x-axis.

Between zeros, the graph of f is either above or below the x-axis.

End behavior: For large $|x|$, the graph of f behaves like the graph of $y = a_n x^n$.

4 Analyze the Graph of a Polynomial Function

| EXAMPLE 9 | Analyzing the Graph of a Polynomial Function |

For the polynomial $f(x) = x^3 + x^2 - 12x$:

(a) Find the x- and y-intercepts of the graph of f.

(b) Determine whether the graph crosses or touches the x-axis at each x-intercept.

(c) End behavior: find the power function that the graph of f resembles for large values of $|x|$.

(d) Determine the maximum number of turning points on the graph of f.

(e) Determine the behavior of the graph of f near each x-intercept.

(f) Put all the information together to obtain the graph of f.

Solution

(a) The y-intercept is $f(0) = 0$.
To find the x-intercepts, if any, we first factor f.

$$f(x) = x^3 + x^2 - 12x$$
$$= x(x^2 + x - 12)$$
$$= x(x + 4)(x - 3)$$

We find the x-intercepts by solving the equation

$$f(x) = x(x + 4)(x - 3) = 0$$
$$x = 0 \quad \text{or} \quad x + 4 = 0 \quad \text{or} \quad x - 3 = 0 \quad \text{Zero-Product Property}$$
$$x = 0 \quad \text{or} \quad x = -4 \quad \text{or} \quad x = 3 \quad \text{Solve.}$$

The x-intercepts, or real zeros of f, are -4, 0, and 3.

(b) Since each real zero of f is of multiplicity 1, the graph of f will cross the x-axis at each x-intercept.

(c) End behavior: the graph of f resembles that of the power function $y = x^3$ for large values of $|x|$.

(d) The graph of f will contain at most two turning points.

(e) The three x-intercepts are -4, 0, and 3.

Near -4: $f(x) = x(x + 4)(x - 3) \approx -4(x + 4)(-4 - 3) = 28(x + 4)$ A line with slope 28

Near 0: $f(x) = x(x + 4)(x - 3) \approx x(0 + 4)(0 - 3) = -12x$ A line with slope -12

Near 3: $f(x) = x(x + 4)(x - 3) \approx 3(3 + 4)(x - 3) = 21(x - 3)$ A line with slope 21

(f) Figure 19(a) illustrates the information obtained from parts (a), (b), (c), and (e). The graph of f is given in Figure 19(b). Notice that we evaluated f at -5, -2, 1, and 4 to help establish the scale on the y-axis.

Figure 19

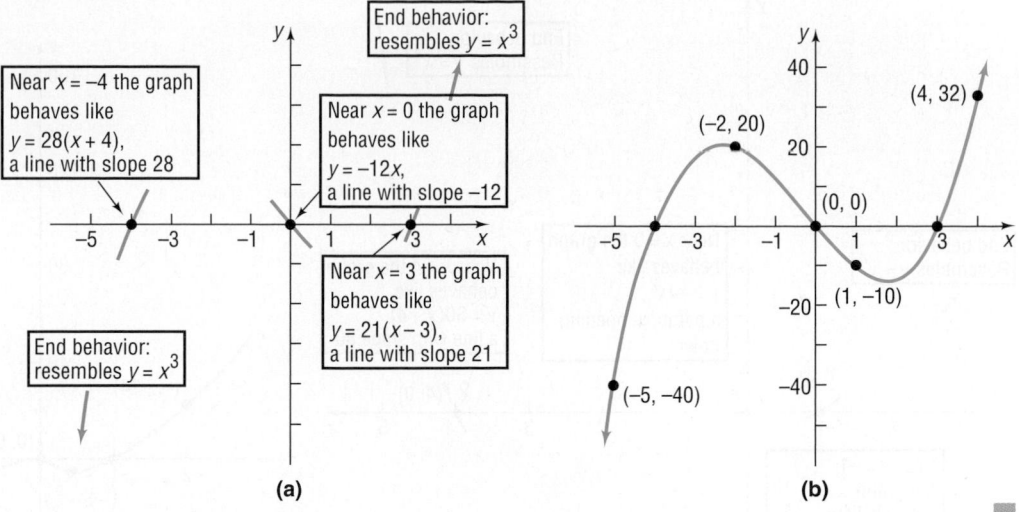

(a) (b)

<image type="calculator_icon">
Exploration

Graph $Y_1 = x^3 + x^2 - 12x$. Compare what you see with Figure 19(b). Use MAXIMUM/MINIMUM to locate the two turning points.
</image>

━━━ **Now Work** PROBLEM 67

| EXAMPLE 10 | **Analyzing the Graph of a Polynomial Function** |

Follow the instructions of Example 9 for the following polynomial:

$$f(x) = x^2(x - 4)(x + 1)$$

Solution (a) The y-intercept is $f(0) = 0$. The x-intercepts satisfy the equation

$$f(x) = x^2(x - 4)(x + 1) = 0$$

So

$$x^2 = 0 \quad \text{or} \quad x - 4 = 0 \quad \text{or} \quad x + 1 = 0$$
$$x = 0 \quad \text{or} \qquad x = 4 \quad \text{or} \qquad x = -1$$

The x-intercepts are -1, 0, and 4.

(b) The intercept 0 is a zero of multiplicity 2, so the graph of f will touch the x-axis at 0; 4 and -1 are zeros of multiplicity 1, so the graph of f will cross the x-axis at 4 and -1.

(c) End behavior: the graph of f resembles that of the power function $y = x^4$ for large values of $|x|$.

(d) The graph of f will contain at most three turning points.

(e) The three x-intercepts are -1, 0, and 4.

Near -1: $f(x) = x^2(x - 4)(x + 1) \approx (-1)^2(-1 - 4)(x + 1) = -5(x + 1)$ *A line with slope −5*

Near 0: $f(x) = x^2(x - 4)(x + 1) \approx x^2(0 - 4)(0 + 1) = -4x^2$ *A parabola opening down*

Near 4: $f(x) = x^2(x - 4)(x + 1) \approx 4^2(x - 4)(4 + 1) = 80(x - 4)$ *A line with slope 80*

(f) Figure 20(a) illustrates the information obtained from parts (a), (b), (c), and (e).

The graph of f is given in Figure 20(b). Notice that we evaluated f at $-2, -\dfrac{1}{2},$ $2,$ and 5 to help establish the scale on the y-axis.

Figure 20

(a) (b)

Exploration

Graph $Y_1 = x^2(x - 4)(x + 1)$. Compare what you see with Figure 20(b). Use MAXIMUM/MINIMUM to locate the two turning points besides (0, 0).

SUMMARY **Steps for Analyzing the Graph of a Polynomial**

To analyze the graph of a polynomial function $y = f(x)$, follow these steps:

STEP 1: (a) Find the y-intercept by letting $x = 0$ and finding the value of $f(0)$.
 (b) Find the x-intercepts, if any, by solving the equation $f(x) = 0$.

STEP 2: Determine whether the graph of f crosses or touches the x-axis at each x-intercept.

STEP 3: End behavior: find the power function that the graph of f resembles for large values of $|x|$.

STEP 4: Determine the maximum number of turning points on the graph of f.

STEP 5: Determine the behavior of the graph of f near each x-intercept.

STEP 6: Put all the information together to obtain the graph of f. This usually requires finding additional points on the graph.

Now Work PROBLEM 77

For polynomial functions that have noninteger coefficients and for polynomials that are not easily factored, we utilize the graphing utility early in the analysis of the graph. This is because the amount of information that can be obtained from algebraic analysis is limited.

EXAMPLE 11 **Using a Graphing Utility to Analyze the Graph of a Polynomial Function**

For the polynomial $f(x) = x^3 + 2.48x^2 - 4.3155x + 2.484406$:

(a) Find the degree of the polynomial. Determine the end behavior; that is, find the power function that the graph of f resembles for large values of $|x|$.

(b) Graph f using a graphing utility.

(c) Find the x- and y-intercepts of the graph.

(d) Use a TABLE to find points on the graph around each x-intercept. Determine on which intervals the graph is above and below the x-axis.

(e) Determine the local maxima and local minima, if any exist, rounded to two decimal places. That is, locate any turning points.

(f) Use the information obtained in parts (a) to (e) to draw a complete graph of f by hand. Be sure to label the intercepts, turning points, and the points obtained in part (d).

(g) Find the domain of f. Use the graph to find the range of f.

(h) Use the graph to determine where f is increasing and decreasing.

Solution

Figure 21

Table 6

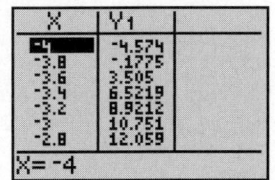

(a) The degree of the polynomial is 3. End behavior: the graph of f resembles that of the power function $y = x^3$ for large values of $|x|$.

(b) See Figure 21 for the graph of f.

(c) The y-intercept is $f(0) = 2.484406$. In Examples 9 and 10 we could easily factor $f(x)$ to find the x-intercepts. However, it is not readily apparent how $f(x)$ factors in this example. Therefore, we use a graphing utility's ZERO (or ROOT) feature and find the lone x-intercept to be -3.79, rounded to two decimal places.

(d) Table 6 shows values of x around the x-intercept. The points $(-4, -4.57)$ and $(-3, 10.75)$ are on the graph. The graph is below the x-axis on the interval $(-\infty, -3.79)$ and above the x-axis on the interval $(-3.79, \infty)$.

(e) From the graph we see that it has two turning points: one between -3 and -2, the other between 0 and 1. Rounded to two decimal places, the local maximum is 13.36 and occurs at $x = -2.28$; the local minimum is 1 and occurs at $x = 0.63$. The turning points are $(-2.28, 13.36)$ and $(0.63, 1)$.

(f) Figure 22 shows a graph of f drawn by hand using the information obtained in parts (a) to (e).

Figure 22

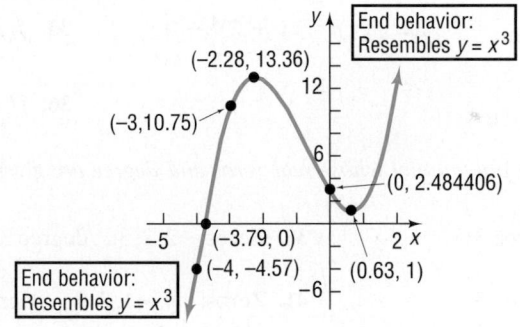

(g) The domain and the range of f are the set of all real numbers.

(h) Based on the graph, f is decreasing on the interval $(-2.28, 0.63)$ and is increasing on the intervals $(-\infty, -2.28)$ and $(0.63, \infty)$.

Now Work PROBLEM 89

5.1 Assess Your Understanding

'Are You Prepared?' *Answers are given at the end of these exercises. If you get a wrong answer, read the pages listed in red.*

1. The intercepts of the equation $9x^2 + 4y = 36$ are _____. (pp. 165–166)

2. *True or False* The expression $4x^3 - 3.6x^2 - \sqrt{2}$ is a polynomial. (pp. 39–47)

3. To graph $y = x^2 - 4$, you would shift the graph of $y = x^2$ _____ a distance of _____ units. (pp. 252–260)

4. *True or False* The x-intercepts of the graph of a function $y = f(x)$ are the real solutions of the equation $f(x) = 0$. (pp. 224–225)

Concepts and Vocabulary

5. The graph of every polynomial function is both _____ and _____ .

6. A real number r for which $f(r) = 0$ is called a(n) _____ of the function f.

7. If r is a real zero of even multiplicity of a function f, the graph of f _____ the x-axis at r.

8. *True or False* The graph of $f(x) = x^2(x - 3)(x + 4)$ has exactly three x-intercepts.

9. *True or False* The x-intercepts of the graph of a polynomial function are called turning points.

10. *True or False* End behavior: the graph of the function $f(x) = 3x^4 - 6x^2 + 2x + 5$ resembles $y = x^4$ for large values of $|x|$.

Skill Building

In Problems 11–22, determine which functions are polynomial functions. For those that are, state the degree. For those that are not, tell why not.

11. $f(x) = 4x + x^3$

12. $f(x) = 5x^2 + 4x^4$

13. $g(x) = \dfrac{1 - x^2}{2}$

14. $h(x) = 3 - \dfrac{1}{2}x$

15. $f(x) = 1 - \dfrac{1}{x}$

16. $f(x) = x(x - 1)$

17. $g(x) = x^{3/2} - x^2 + 2$

18. $h(x) = \sqrt{x}(\sqrt{x} - 1)$

19. $F(x) = 5x^4 - \pi x^3 + \dfrac{1}{2}$

20. $F(x) = \dfrac{x^2 - 5}{x^3}$

21. $G(x) = 2(x - 1)^2(x^2 + 1)$

22. $G(x) = -3x^2(x + 2)^3$

In Problems 23–36, use transformations of the graph of $y = x^4$ or $y = x^5$ to graph each function.

23. $f(x) = (x + 1)^4$

24. $f(x) = (x - 2)^5$

25. $f(x) = x^5 - 3$

26. $f(x) = x^4 + 2$

27. $f(x) = \dfrac{1}{2}x^4$

28. $f(x) = 3x^5$

29. $f(x) = -x^5$

30. $f(x) = -x^4$

31. $f(x) = (x - 1)^5 + 2$

32. $f(x) = (x + 2)^4 - 3$

33. $f(x) = 2(x + 1)^4 + 1$

34. $f(x) = \dfrac{1}{2}(x - 1)^5 - 2$

35. $f(x) = 4 - (x - 2)^5$

36. $f(x) = 3 - (x + 2)^4$

In Problems 37–44, form a polynomial whose real zeros and degree are given. Answers will vary depending on the choice of a leading coefficient.

37. Zeros: $-1, 1, 3$; degree 3

38. Zeros: $-2, 2, 3$; degree 3

39. Zeros: $-3, 0, 4$; degree 3

40. Zeros: $-4, 0, 2$; degree 3

41. Zeros: $-4, -1, 2, 3$; degree 4

42. Zeros: $-3, -1, 2, 5$; degree 4

43. Zeros: -1, multiplicity 1; 3, multiplicity 2; degree 3

44. Zeros: -2, multiplicity 2; 4, multiplicity 1; degree 3

In Problems 45–56, for each polynomial function:

(a) List each real zero and its multiplicity.
(b) Determine whether the graph crosses or touches the x-axis at each x-intercept.
(c) Determine the behavior of the graph near each x-intercept.
(d) Determine the maximum number of turning points on the graph.
(e) Determine the end behavior; that is, find the power function that the graph of f resembles for large values of $|x|$.

45. $f(x) = 3(x - 7)(x + 3)^2$

46. $f(x) = 4(x + 4)(x + 3)^3$

47. $f(x) = 4(x^2 + 1)(x - 2)^3$

48. $f(x) = 2(x - 3)(x + 4)^3$

49. $f(x) = -2\left(x + \dfrac{1}{2}\right)^2(x^2 + 4)^2$

50. $f(x) = \left(x - \dfrac{1}{3}\right)^2(x - 1)^3$

51. $f(x) = (x - 5)^3(x + 4)^2$

52. $f(x) = (x + \sqrt{3})^2(x - 2)^4$

53. $f(x) = 3(x^2 + 8)(x^2 + 9)^2$

54. $f(x) = -2(x^2 + 3)^3$

55. $f(x) = -2x^2(x^2 - 2)$

56. $f(x) = 4x(x^2 - 3)$

In Problems 57–60, identify which of the graphs could be the graph of a polynomial function. For those that could, list the real zeros and state the least degree the polynomial can have. For those that could not, say why not.

57.

58.

59.

60.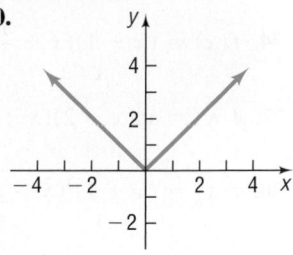

In Problems 61–64, decide which of the polynomial functions in the list might have the given graph. (More than one answer may be possible.)

61.

62.

(a) $y = -4x(x - 1)(x - 2)$ (d) $y = x(x - 1)^2(x - 2)^2$
(b) $y = x^2(x - 1)^2(x - 2)$ (e) $y = x^3(x - 1)(x - 2)$
(c) $y = 3x(x - 1)(x - 2)$ (f) $y = -x(1 - x)(x - 2)$

(a) $y = 2x^3(x - 1)(x - 2)^2$ (d) $y = x^2(x - 1)^2(x - 2)^2$
(b) $y = x^2(x - 1)(x - 2)$ (e) $y = 5x(x - 1)^2(x - 2)$
(c) $y = x^3(x - 1)^2(x - 2)$ (f) $y = -2x(x - 1)^2(2 - x)$

63.

64.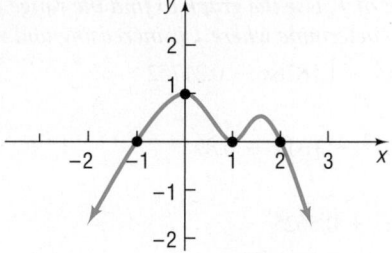

(a) $y = \dfrac{1}{2}(x^2 - 1)(x - 2)$

(b) $y = -\dfrac{1}{2}(x^2 + 1)(x - 2)$

(c) $y = (x^2 - 1)\left(1 - \dfrac{x}{2}\right)$

(d) $y = -\dfrac{1}{2}(x^2 - 1)^2(x - 2)$

(e) $y = \left(x^2 + \dfrac{1}{2}\right)(x^2 - 1)(2 - x)$

(f) $y = -(x - 1)(x - 2)(x + 1)$

(a) $y = -\dfrac{1}{2}(x^2 - 1)(x - 2)(x + 1)$

(b) $y = -\dfrac{1}{2}(x^2 + 1)(x - 2)(x + 1)$

(c) $y = -\dfrac{1}{2}(x + 1)^2(x - 1)(x - 2)$

(d) $y = (x - 1)^2(x + 1)\left(1 - \dfrac{x}{2}\right)$

(e) $y = -(x - 1)^2(x - 2)(x + 1)$

(f) $y = -\left(x^2 + \dfrac{1}{2}\right)(x - 1)^2(x + 1)(x - 2)$

In Problems 65–88:

(a) *Find the x- and y-intercepts of each polynomial function f.*
(b) *Determine whether the graph of f crosses or touches the x-axis at each x-intercept.*
(c) *End behavior: find the power function that the graph of f resembles for large values of |x|.*
(d) *Determine the maximum number of turning points on the graph of f.*
(e) *Determine the behavior of the graph of f near each x-intercept.*
(f) *Put all the information together to obtain the graph of f. (You may need to locate additional points on the graph.)*

65. $f(x) = (x - 1)^2$ **66.** $f(x) = (x - 2)^3$ **67.** $f(x) = x^2(x - 3)$

68. $f(x) = x(x + 2)^2$ **69.** $f(x) = 6x^3(x + 4)$ **70.** $f(x) = 5x(x - 1)^3$

71. $f(x) = -4x^2(x + 2)$

72. $f(x) = -\frac{1}{2}x^3(x + 4)$

73. $f(x) = (x - 1)(x - 2)(x + 4)$

74. $f(x) = (x + 1)(x + 4)(x - 3)$

75. $f(x) = 4x - x^3$

76. $f(x) = x - x^3$

77. $f(x) = x^2(x - 2)(x + 2)$

78. $f(x) = x^2(x - 3)(x + 4)$

79. $f(x) = (x + 2)^2(x - 2)^2$

80. $f(x) = (x + 1)^3(x - 3)$

81. $f(x) = (x - 1)^2(x - 3)(x + 1)$

82. $f(x) = (x + 1)^2(x - 3)(x - 1)$

83. $f(x) = (x + 2)^2(x - 4)^2$

84. $f(x) = (x - 2)^2(x + 2)(x + 4)$

85. $f(x) = x^2(x - 2)(x^2 + 3)$

86. $f(x) = x^2(x^2 + 1)(x + 4)$

87. $f(x) = -x^2(x^2 - 1)(x + 1)$

88. $f(x) = -x^2(x^2 - 4)(x - 5)$

In Problems 89–98, for each polynomial function f:

(a) *Find the degree of the polynomial. Determine the end behavior: that is, find the power function that the graph of f resembles for large values of $|x|$.*

(b) *Graph f using a graphing utility.*

(c) *Find the x- and y-intercepts of the graph, rounded to two decimal places.*

(d) *Use TABLE to find points on the graph around each x-intercept. Determine on which intervals the graph is above and below the x-axis.*

(e) *Determine the local maxima and local minima, if any exist, rounded to two decimal places. That is, locate any turning points.*

(f) *Use the information obtained in parts (a) to (e) to draw a complete graph of f by hand. Be sure to label the intercepts, turning points, and the points obtained in part (d).*

(g) *Find the domain of f. Use the graph to find the range of f.*

(h) *Use the graph to determine where f is increasing and where f is decreasing.*

89. $f(x) = x^3 + 0.2x^2 - 1.5876x - 0.31752$

90. $f(x) = x^3 - 0.8x^2 - 4.6656x + 3.73248$

91. $f(x) = x^3 + 2.56x^2 - 3.31x + 0.89$

92. $f(x) = x^3 - 2.91x^2 - 7.668x - 3.8151$

93. $f(x) = x^4 - 2.5x^2 + 0.5625$

94. $f(x) = x^4 - 18.5x^2 + 50.2619$

95. $f(x) = 2x^4 - \pi x^3 + \sqrt{5}x - 4$

96. $f(x) = -1.2x^4 + 0.5x^2 - \sqrt{3}x + 2$

97. $f(x) = -2x^5 - \sqrt{2}x^2 - x - \sqrt{2}$

98. $f(x) = \pi x^5 + \pi x^4 + \sqrt{3}x + 1$

99. Hurricanes In 2005, hurricane Katrina struck the Gulf Coast of the United States, killing 1289 people and causing an estimated $200 billion in damage. The following data represent the number of major hurricane strikes in the U.S. (category 3, 4, or 5) each decade from 1921 to 2000.

Decade, x	Major Hurricanes Striking United States, H
1921–1930, 1	5
1931–1940, 2	8
1941–1950, 3	10
1951–1960, 4	8
1961–1970, 5	6
1971–1980, 6	4
1981–1990, 7	5
1991–2000, 8	5

Source: National Oceanic & Atmospheric Administration

(a) Draw a scatter diagram of the data. Comment on the type of relation that may exist between the two variables.

(b) The cubic function of best fit to these data is

$$H(x) = 0.16x^3 - 2.32x^2 + 9.33x - 2.21$$

Use this function to predict the number of major hurricanes that struck the United States between 1961 and 1970.

(c) Use a graphing utility to verify that the function given in part (b) is the cubic function of best fit.

(d) With a graphing utility, draw a scatter diagram of the data and then graph the cubic function of best fit on the scatter diagram.

(e) Concern has risen about the increase in the number and intensity of hurricanes, but some scientists believe this is just a natural fluctuation that could last another decade or two. Use your model to predict the number of major hurricanes that will strike the United States between 2001 and 2010. Does your result appear to agree with what these scientists believe?

(f) From 2001–2005, five major hurricanes had struck the United States. Does this support or contradict your prediction in part (e)?

100. Active Duty The following data represent the number of active duty military personnel (in millions) in the four major branches of the U.S. military for the years 1998–2005, where 1 represents 1998, 2 represents 1999, and so on.

Year, x	Active Duty Personnel, N
1998, 1	1.41
1999, 2	1.39
2000, 3	1.38
2001, 4	1.39
2002, 5	1.41
2003, 6	1.42
2004, 7	1.41
2005, 8	1.39

Source: infoplease.com

(a) Draw a scatter diagram of the data. Comment on the type of relation that may exist between the two variables.

(b) The cubic function of best fit to these data is

$$N(x) = -0.00146x^3 + 0.0198x^2 - 0.0742x + 1.4671$$

Use this function to predict the number of active duty military personnel in 2002.

(c) Use a graphing utility to verify that the function given in part (b) is the cubic function of best fit.

(d) With a graphing utility, draw a scatter diagram of the data and then graph the cubic function of best fit on the scatter diagram.

(e) Do you think that the function given in part (b) will be useful in predicting the number of active duty military personnel in 2010? Explain.

101. Temperature The following data represent the temperature T (°Fahrenheit) in Kansas City, Missouri, x hours after midnight on May 15, 2005.

Hours after Midnight, x	Temperature (°F), T
3	45.0
6	44.1
9	51.1
12	57.9
15	63.0
18	63.0
21	59.0
24	54.0

Source: The Weather Underground

(a) Draw a scatter diagram of the data. Comment on the type of relation that may exist between the two variables.

(b) Find the average rate of change in temperature from 9 AM to 12 noon.

(c) What is the average rate of change in temperature from 3 PM to 6 PM?

(d) The cubic function of best fit to these data is

$$T(x) = -0.0103x^3 + 0.32x^2 - 1.37x + 45.39$$

Use this function to predict the temperature at 5 PM.

(e) Use a graphing utility to verify that the function given in part (b) is the cubic function of best fit.

(f) With a graphing utility, draw a scatter diagram of the data and then graph the cubic function of best fit on the scatter diagram.

(g) Interpret the y-intercept.

Discussion and Writing

102. Can the graph of a polynomial function have no y-intercept? Can it have no x-intercepts? Explain.

103. Write a few paragraphs that provide a general strategy for graphing a polynomial function. Be sure to mention the following: degree, intercepts, end behavior, and turning points.

104. Make up a polynomial that has the following characteristics: crosses the x-axis at -1 and 4, touches the x-axis at 0 and 2, and is above the x-axis between 0 and 2. Give your polynomial to a fellow classmate and ask for a written critique of your polynomial.

105. Make up two polynomials, not of the same degree, with the following characteristics: crosses the x-axis at -2, touches the x-axis at 1, and is above the x-axis between -2 and 1. Give your polynomials to a fellow classmate and ask for a written critique of your polynomials.

106. The graph of a polynomial function is always smooth and continuous. Name a function studied earlier that is smooth and not continuous. Name one that is continuous, but not smooth.

107. Which of the following statements are true regarding the graph of the cubic polynomial $f(x) = x^3 + bx^2 + cx + d$? (Give reasons for your conclusions.)

(a) It intersects the y-axis in one and only one point.

(b) It intersects the x-axis in at most three points.

(c) It intersects the x-axis at least once.

(d) For $|x|$ very large, it behaves like the graph of $y = x^3$.

(e) It is symmetric with respect to the origin.

(f) It passes through the origin.

108. The illustration shows the graph of a polynomial function.

(a) Is the degree of the polynomial even or odd?

(b) Is the leading coefficient positive or negative?

(c) Is the function even, odd, or neither?

(d) Why is x^2 necessarily a factor of the polynomial?

(e) What is the minimum degree of the polynomial?

(f) Formulate five different polynomials whose graphs could look like the one shown. Compare yours to those of other students. What similarities do you see? What differences?

109. Design a polynomial function with the following characteristics: degree 6; four distinct real zeros, one of multiplicity 3; y-intercept 3; behaves like $y = -5x^6$ for large values of $|x|$. Is this polynomial unique? Compare your polynomial with those of other students. What terms will be the same as everyone else's? Add some more characteristics, such as symmetry or naming the real zeros. How does this modify the polynomial?

'Are You Prepared?' Answers

1. $(-2, 0), (2, 0), (0, 9)$ **2.** True **3.** down; 4 **4.** True

5.2 Properties of Rational Functions

PREPARING FOR THIS SECTION *Before getting started, review the following:*

- Rational Expressions (Chapter R, Section R.7, pp. 61–68)
- Polynomial Division (Chapter R, Section R.4, pp. 44–47)
- Graph of $f(x) = \dfrac{1}{x}$ (Section 2.2, Example 12, pp. 170–171)
- Graphing Techniques: Transformations (Section 3.5, pp. 252–260)

Now Work the 'Are You Prepared?' problems on page 352.

OBJECTIVES **1** Find the Domain of a Rational Function (p. 344)

 2 Find the Vertical Asymptotes of a Rational Function (p. 347)

 3 Find the Horizontal or Oblique Asymptotes of a Rational Function (p. 348)

Ratios of integers are called *rational numbers*. Similarly, ratios of polynomial functions are called *rational functions*. Examples of rational functions are

$$R(x) = \frac{x^2 - 4}{x^2 + x + 1} \qquad F(x) = \frac{x^3}{x^2 - 4} \qquad G(x) = \frac{3x^2}{x^4 - 1}$$

DEFINITION

A **rational function** is a function of the form

$$R(x) = \frac{p(x)}{q(x)}$$

where p and q are polynomial functions and q is not the zero polynomial. The domain of a rational function is the set of all real numbers except those for which the denominator q is 0.

1 Find the Domain of a Rational Function

EXAMPLE 1 Finding the Domain of a Rational Function

(a) The domain of $R(x) = \dfrac{2x^2 - 4}{x + 5}$ is the set of all real numbers x except -5; that is, the domain is $\{x \mid x \neq -5\}$.

(b) The domain of $R(x) = \dfrac{1}{x^2 - 4}$ is the set of all real numbers x except -2 and 2; that is, the domain is $\{x \mid x \neq -2, x \neq 2\}$.

(c) The domain of $R(x) = \dfrac{x^3}{x^2 + 1}$ is the set of all real numbers.

(d) The domain of $R(x) = \dfrac{3}{x^2 + 2}$ is the set of all real numbers.

(e) The domain of $R(x) = \dfrac{x^2 - 1}{x - 1}$ is the set of all real numbers x except 1; that is, the domain is $\{x \mid x \neq 1\}$.

It is important to observe that the functions

$$R(x) = \frac{x^2 - 1}{x - 1} \quad \text{and} \quad f(x) = x + 1$$

are not equal, since the domain of R is $\{x \mid x \neq 1\}$ and the domain of f is the set of all real numbers.

Now Work PROBLEM 13

If $R(x) = \dfrac{p(x)}{q(x)}$ is a rational function and if p and q have no common factors, then the rational function R is said to be in **lowest terms.** For a rational function $R(x) = \dfrac{p(x)}{q(x)}$ in lowest terms, the real zeros, if any, of the numerator are the x-intercepts of the graph of R and so will play a major role in the graph of R. The real zeros of the denominator of R [that is, the numbers x, if any, for which $q(x) = 0$], although not in the domain of R, also play a major role in the graph of R. We will discuss this role shortly.

We have already discussed the properties of the rational function $f(x) = \dfrac{1}{x}$.

(Refer to Example 12, pages 170–171). The next rational function that we take up is $H(x) = \dfrac{1}{x^2}$.

EXAMPLE 2

Graphing $y = \dfrac{1}{x^2}$

Analyze the graph of: $H(x) = \dfrac{1}{x^2}$

Solution

The domain of $H(x) = \dfrac{1}{x^2}$ is the set of all real numbers x except 0. The graph has no y-intercept, because x can never equal 0. The graph has no x-intercept because the equation $H(x) = 0$ has no solution. Therefore, the graph of H will not cross either of the coordinate axes. Because

$$H(-x) = \frac{1}{(-x)^2} = \frac{1}{x^2} = H(x)$$

H is an even function, so its graph is symmetric with respect to the y-axis.

Table 7 shows the behavior of $H(x) = \dfrac{1}{x^2}$ for selected positive numbers x. (We will use symmetry to obtain the graph of H when $x < 0$). From the first three rows of Table 7, we see that, as the values of x approach (get closer to) 0, the values of $H(x)$ become larger and larger positive numbers. When this happens, we say that H is **unbounded in the positive direction.** We symbolize this by writing $H \to \infty$ (read as "H **approaches infinity**"). In calculus, **limits** are used to convey these ideas. There we use the symbolism $\lim\limits_{x \to 0} H(x) = \infty$, read "the limit of $H(x)$ as x approaches zero equals infinity," to mean that $H(x) \to \infty$ as $x \to 0$.

Look at the last four rows of Table 7. As $x \to \infty$, the values of $H(x)$ approach 0 (the end behavior of the graph). In calculus, this is symbolized by writing

Table 7

x	$H(x) = \dfrac{1}{x^2}$
$\dfrac{1}{2}$	4
$\dfrac{1}{100}$	10,000
$\dfrac{1}{10,000}$	100,000,000
1	1
2	$\dfrac{1}{4}$
100	$\dfrac{1}{10,000}$
10,000	$\dfrac{1}{100,000,000}$

$\lim\limits_{x \to \infty} H(x) = 0$. Figure 23 shows the graph. Notice the use of red dashed lines to convey the ideas discussed above.

Figure 23

$H(x) = \dfrac{1}{x^2}$

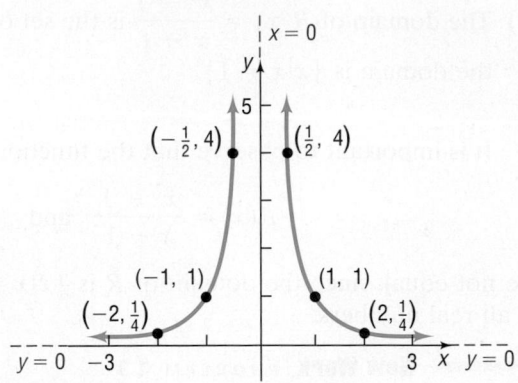

Sometimes transformations (shifting, compressing, stretching, and reflection) can be used to graph a rational function.

EXAMPLE 3	**Using Transformations to Graph a Rational Function**

Graph the rational function: $R(x) = \dfrac{1}{(x-2)^2} + 1$

Solution First, we take note of the fact that the domain of R is the set of all real numbers except $x = 2$. To graph R, we start with the graph of $y = \dfrac{1}{x^2}$. See Figure 24 for the steps.

Figure 24

Replace x by $x - 2$; shift right 2 units

Add 1; shift up 1 unit

(a) $y = \dfrac{1}{x^2}$

(b) $y = \dfrac{1}{(x-2)^2}$

(c) $y = \dfrac{1}{(x-2)^2} + 1$

Now Work PROBLEM 31

Asymptotes

In Figure 24(c), notice that as the values of x become more negative, that is, as x becomes **unbounded in the negative direction** ($x \to -\infty$, read as "x **approaches negative infinity**"), the values $R(x)$ approach 1. In fact, we can conclude the following from Figure 24(c):

1. As $x \to -\infty$, the values $R(x)$ approach 1. $[\lim\limits_{x \to -\infty} R(x) = 1]$
2. As x approaches 2, the values $R(x) \to \infty$. $[\lim\limits_{x \to 2} R(x) = \infty]$
3. As $x \to \infty$, the values $R(x)$ approach 1. $[\lim\limits_{x \to \infty} R(x) = 1]$

This behavior of the graph is depicted by the vertical line $x = 2$ and the horizontal line $y = 1$. These lines are called *asymptotes* of the graph, which we define as follows:

DEFINITION

Let R denote a function.

If, as $x \to -\infty$ or as $x \to \infty$, the values of $R(x)$ approach some fixed number L, then the line $y = L$ is a **horizontal asymptote** of the graph of R. [Refer to Figures 25(a) and (b).]

If, as x approaches some number c, the values $|R(x)| \to \infty$, then the line $x = c$ is a **vertical asymptote** of the graph of R. The graph of R never intersects a vertical asymptote. [Refer to Figures 25(c) and (d).]

Figure 25

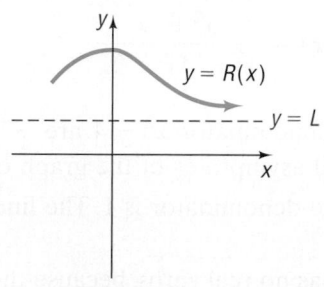

(a) End behavior:
As $x \to \infty$, the values of $R(x)$ approach L
[symbolized by $\lim\limits_{x \to \infty} R(x) = L$].
That is, the points on the graph of R are getting closer to the line $y = L$; $y = L$ is a horizontal asymptote.

(b) End behavior:
As $x \to -\infty$, the values of $R(x)$ approach L
[symbolized by $\lim\limits_{x \to -\infty} R(x) = L$].
That is, the points on the graph of R are getting closer to the line $y = L$; $y = L$ is a horizontal asymptote.

(c) As x approaches c, the values of $|R(x)| \to \infty$
[for $x < c$, this is symbolized by $\lim\limits_{x \to c^-} R(x) = \infty$;
for $x > c$, this is symbolized by $\lim\limits_{x \to c^+} R(x) = \infty$]. That is, the points on the graph of R are getting closer to the line $x = c$; $x = c$ is a vertical asymptote.

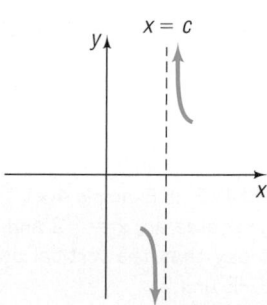

(d) As x approaches c, the values of $|R(x)| \to \infty$
[for $x < c$, this is symbolized by $\lim\limits_{x \to c^-} R(x) = -\infty$;
for $x > c$, this is symbolized by $\lim\limits_{x \to c^+} R(x) = \infty$]. That is, the points on the graph of R are getting closer to the line $x = c$; $x = c$ is a vertical asymptote.

A horizontal asymptote, when it occurs, describes the end behavior of the graph as $x \to \infty$ or as $x \to -\infty$. **The graph of a function may intersect a horizontal asymptote.**

A vertical asymptote, when it occurs, describes the behavior of the graph when x is close to some number c. **The graph of a function will never intersect a vertical asymptote.**

There is a third possibility. If, as $x \to -\infty$ or as $x \to \infty$, the value of a rational function $R(x)$ approaches a linear expression $ax + b$, $a \neq 0$, then the line $y = ax + b$, $a \neq 0$, is an **oblique asymptote** of R. Figure 26 shows an oblique asymptote. An oblique asymptote, when it occurs, describes the end behavior of the graph. **The graph of a function may intersect an oblique asymptote.**

Figure 26

 Now Work PROBLEM 23

2 Find the Vertical Asymptotes of a Rational Function

The vertical asymptotes of a rational function $R(x) = \dfrac{p(x)}{q(x)}$, in lowest terms, are located at the real zeros of the denominator of $q(x)$. Suppose that r is a real zero of q, so $x - r$ is a factor of q. As x approaches r, symbolized as $x \to r$, the values of $x - r$ approach 0, causing the ratio to become unbounded, that is, $|R(x)| \to \infty$. Based on the definition, we conclude that the line $x = r$ is a vertical asymptote.

THEOREM

Locating Vertical Asymptotes

A rational function $R(x) = \dfrac{p(x)}{q(x)}$, *in lowest terms*, will have a vertical asymptote $x = r$ if r is a real zero of the *denominator q*. That is, if $x - r$ is a factor of the denominator q of a rational function $R(x) = \dfrac{p(x)}{q(x)}$, in lowest terms, R will have the vertical asymptote $x = r$.

WARNING If a rational function is not in lowest terms, an application of this theorem will result in an incorrect listing of vertical asymptotes. ∎

| EXAMPLE 4 | **Finding Vertical Asymptotes** |

Find the vertical asymptotes, if any, of the graph of each rational function.

(a) $R(x) = \dfrac{x}{x^2 - 4}$ (b) $F(x) = \dfrac{x + 3}{x - 1}$

(c) $H(x) = \dfrac{x^2}{x^2 + 1}$ (d) $G(x) = \dfrac{x^2 - 9}{x^2 + 4x - 21}$

Solution

(a) R is in lowest terms and the zeros of the denominator $x^2 - 4$ are -2 and 2. The lines $x = -2$ and $x = 2$ are the vertical asymptotes of the graph of R.

WARNING In Example 4(a), the vertical asymptotes are $x = -2$ and $x = 2$. Do not say that the vertical asymptotes are -2 and 2. ∎

(b) F is in lowest terms and the only zero of the denominator is 1. The line $x = 1$ is the vertical asymptote of the graph of F.

(c) H is in lowest terms and the denominator has no real zeros, because the equation $x^2 + 1$ has no real solutions. The graph of H has no vertical asymptotes.

(d) Factor $G(x)$ to determine if it is in lowest terms.

$$G(x) = \frac{x^2 - 9}{x^2 + 4x - 21} = \frac{(x + 3)(x - 3)}{(x + 7)(x - 3)} = \frac{x + 3}{x + 7} \qquad x \neq 3$$

The only zero of the denominator of $G(x)$ in lowest terms is -7. The line $x = -7$ is the only vertical asymptote of the graph of G. ∎

As Example 4 points out, rational functions can have no vertical asymptotes, one vertical asymptote, or more than one vertical asymptote. However, the graph of a rational function will never intersect any of its vertical asymptotes. (Do you know why?)

 Exploration

Graph each of the following rational functions:

$$R(x) = \frac{1}{x - 1} \qquad R(x) = \frac{1}{(x - 1)^2} \qquad R(x) = \frac{1}{(x - 1)^3} \qquad R(x) = \frac{1}{(x - 1)^4}$$

Each has the vertical asymptote $x = 1$. What happens to the value of $R(x)$ as x approaches 1 from the right side of the vertical asymptote; that is, what is $\lim\limits_{x \to 1^+} R(x)$? What happens to the value of $R(x)$ as x approaches 1 from the left side of the vertical asymptote; that is, what is $\lim\limits_{x \to 1^-} R(x)$? How does the multiplicity of the zero in the denominator affect the graph of R?

Now Work PROBLEM 45 (FIND THE VERTICAL ASYMPTOTES, IF ANY.)

3 Find the Horizontal or Oblique Asymptotes of a Rational Function

The procedure for finding horizontal and oblique asymptotes is somewhat more involved. To find such asymptotes, we need to know how the values of a function behave as $x \to -\infty$ or as $x \to \infty$.

If a rational function $R(x)$ is **proper,** that is, if the degree of the numerator is less than the degree of the denominator, then as $x \to -\infty$ or as $x \to \infty$ the value of $R(x)$ approaches 0. Consequently, the line $y = 0$ (the x-axis) is a horizontal asymptote of the graph.

THEOREM

If a rational function is proper, the line $y = 0$ is a horizontal asymptote of its graph.

| EXAMPLE 5 | **Finding Horizontal Asymptotes** |

Find the horizontal asymptotes, if any, of the graph of

$$R(x) = \frac{x - 12}{4x^2 + x + 1}$$

Solution Since the degree of the numerator, 1, is less than the degree of the denominator, 2, the rational function R is proper. We conclude that the line $y = 0$ is a horizontal asymptote of the graph of R.

To see why $y = 0$ is a horizontal asymptote of the function R in Example 5, we need to investigate the behavior of R as $x \to -\infty$ and $x \to \infty$. When $|x|$ is unbounded, the numerator of R, which is $x - 12$, can be approximated by the power function $y = x$, while the denominator of R, which is $4x^2 + x + 1$, can be approximated by the power function $y = 4x^2$. Applying these ideas to $R(x)$, we find

$$R(x) = \frac{x - 12}{4x^2 + x + 1} \underset{\uparrow}{\approx} \frac{x}{4x^2} = \frac{1}{4x} \underset{\uparrow}{\to} 0$$

<div align="center">For $|x|$ unbounded As $x \to -\infty$ or $x \to \infty$</div>

This shows that the line $y = 0$ is a horizontal asymptote of the graph of R.

If a rational function $R(x) = \dfrac{p(x)}{q(x)}$ is **improper,** that is, if the degree of the numerator is greater than or equal to the degree of the denominator, we must use long division to write the rational function as the sum of a polynomial $f(x)$ plus a proper rational function $\dfrac{r(x)}{q(x)}$. That is, we write

$$R(x) = \frac{p(x)}{q(x)} = f(x) + \frac{r(x)}{q(x)}$$

where $f(x)$ is a polynomial and $\dfrac{r(x)}{q(x)}$ is a proper rational function. Since $\dfrac{r(x)}{q(x)}$ is proper, $\dfrac{r(x)}{q(x)} \to 0$ as $x \to -\infty$ or as $x \to \infty$. As a result,

$$R(x) = \frac{p(x)}{q(x)} \to f(x) \qquad \text{as } x \to -\infty \text{ or as } x \to \infty$$

The possibilities are listed next.

1. If $f(x) = b$, a constant, the line $y = b$ is a horizontal asymptote of the graph of R.

2. If $f(x) = ax + b, a \neq 0$, the line $y = ax + b$ is an oblique asymptote of the graph of R.

3. In all other cases, the graph of R approaches the graph of f, and there are no horizontal or oblique asymptotes.

The following examples demonstrate these conclusions.

| EXAMPLE 6 | Finding Horizontal or Oblique Asymptotes |

Find the horizontal or oblique asymptotes, if any, of the graph of

$$H(x) = \frac{3x^4 - x^2}{x^3 - x^2 + 1}$$

Solution Since the degree of the numerator, 4, is larger than the degree of the denominator, 3, the rational function H is improper. To find any horizontal or oblique asymptotes, we use long division.

$$
\begin{array}{r}
3x + 3 \\
x^3 - x^2 + 1 \overline{)3x^4 \quad\quad - x^2} \\
\underline{3x^4 - 3x^3 \quad\quad + 3x} \\
3x^3 - x^2 - 3x \\
\underline{3x^3 - 3x^2 \quad\quad + 3} \\
2x^2 - 3x - 3
\end{array}
$$

As a result,

$$H(x) = \frac{3x^4 - x^2}{x^3 - x^2 + 1} = 3x + 3 + \frac{2x^2 - 3x - 3}{x^3 - x^2 + 1}$$

As $x \to -\infty$ or as $x \to \infty$,

$$\frac{2x^2 - 3x - 3}{x^3 - x^2 + 1} \approx \frac{2x^2}{x^3} = \frac{2}{x} \to 0$$

As $x \to -\infty$ or as $x \to \infty$, we have $H(x) \to 3x + 3$. We conclude that the graph of the rational function H has an oblique asymptote $y = 3x + 3$.

■

| EXAMPLE 7 | Finding Horizontal or Oblique Asymptotes |

Find the horizontal or oblique asymptotes, if any, of the graph of

$$R(x) = \frac{8x^2 - x + 2}{4x^2 - 1}$$

Solution Since the degree of the numerator, 2, equals the degree of the denominator, 2, the rational function R is improper. To find any horizontal or oblique asymptotes, we use long division.

$$
\begin{array}{r}
2 \\
4x^2 - 1 \overline{)8x^2 - x + 2} \\
\underline{8x^2 \quad\quad - 2} \\
- x + 4
\end{array}
$$

As a result,

$$R(x) = \frac{8x^2 - x + 2}{4x^2 - 1} = 2 + \frac{-x + 4}{4x^2 - 1}$$

Then, as $x \to -\infty$ or as $x \to \infty$,

$$\frac{-x + 4}{4x^2 - 1} \approx \frac{-x}{4x^2} = \frac{-1}{4x} \to 0$$

As $x \to -\infty$ or as $x \to \infty$, we have $R(x) \to 2$. We conclude that $y = 2$ is a horizontal asymptote of the graph.

■

In Example 7, we note that the quotient 2 obtained by long division is the quotient of the leading coefficients of the numerator polynomial and the denominator polynomial $\left(\dfrac{8}{4}\right)$. This means that we can avoid the long division process for rational functions whose numerator and denominator *are of the same degree* and conclude that the quotient of the leading coefficients will give us the horizontal asymptote.

━━━━━━━━ **Now Work** PROBLEMS 41 AND 43

EXAMPLE 8	**Finding Horizontal or Oblique Asymptotes**

Find the horizontal or oblique asymptotes, if any, of the graph of

$$G(x) = \frac{2x^5 - x^3 + 2}{x^3 - 1}$$

Solution Since the degree of the numerator, 5, is larger than the degree of the denominator, 3, the rational function G is improper. To find any horizontal or oblique asymptotes, we use long division.

$$
\begin{array}{r}
2x^2 - 1 \\
x^3 - 1 \overline{\smash{)}2x^5 - x^3 \qquad\quad + 2} \\
\underline{2x^5 \qquad\quad - 2x^2} \\
-x^3 + 2x^2 + 2 \\
\underline{-x^3 \qquad\quad + 1} \\
2x^2 + 1
\end{array}
$$

As a result,

$$G(x) = \frac{2x^5 - x^3 + 2}{x^3 - 1} = 2x^2 - 1 + \frac{2x^2 + 1}{x^3 - 1}$$

Then, as $x \to -\infty$ or as $x \to \infty$,

$$\frac{2x^2 + 1}{x^3 - 1} \approx \frac{2x^2}{x^3} = \frac{2}{x} \to 0$$

As $x \to -\infty$ or as $x \to \infty$, we have $G(x) \to 2x^2 - 1$. We conclude that, for large values of $|x|$, the graph of G approaches the graph of $y = 2x^2 - 1$. That is, the graph of G will look like the graph of $y = 2x^2 - 1$ as $x \to -\infty$ or $x \to \infty$. Since $y = 2x^2 - 1$ is not a linear function, G has no horizontal or oblique asymptotes.

━━━━━━━

We now summarize the procedure for finding horizontal and oblique asymptotes.

SUMMARY Finding Horizontal and Oblique Asymptotes of a Rational Function R

Consider the rational function

$$R(x) = \frac{p(x)}{q(x)} = \frac{a_n x^n + a_{n-1}x^{n-1} + \cdots + a_1 x + a_0}{b_m x^m + b_{m-1}x^{m-1} + \cdots + b_1 x + b_0}$$

in which the degree of the numerator is n and the degree of the denominator is m.

1. If the degree of the numerator is less than the degree of the denominator, R is a proper rational function, and the graph of R will have the horizontal asymptote $y = 0$ (the x-axis).

2. If the degree of the numerator is greater than or equal to the degree of the denominator, then R is improper. Here long division is used.

 (a) If the degree of the numerator equals the degree of the denominator, the quotient obtained will be the number $\dfrac{a_n}{b_m}$, and the line $y = \dfrac{a_n}{b_m}$ is a horizontal asymptote.

 (b) If the degree of the numerator is one more than the degree of the denominator, the quotient obtained is of the form $ax + b$ (a polynomial of degree 1), and the line $y = ax + b$ is an oblique asymptote.

 (c) If the degree of the numerator is two or more than the degree of the denominator, the quotient obtained is a polynomial of degree 2 or higher, and R has neither a horizontal nor an oblique asymptote. In this case, for $|x|$ unbounded, the graph of R will behave like the graph of the quotient.

 Note: The graph of a rational function either has one horizontal or one oblique asymptote or else has no horizontal and no oblique asymptote. It cannot have both a horizontal asymptote and an oblique asymptote.

5.2 Assess Your Understanding

'Are You Prepared?' *Answers are given at the end of these exercises. If you get a wrong answer, read the pages listed in red.*

1. ***True or False*** The quotient of two polynomial expressions is a rational expression. (pp. 61–68)

2. What is the quotient and remainder when $3x^4 - x^2$ is divided by $x^3 - x^2 + 1$. (pp. 44–47)

3. Graph $y = \dfrac{1}{x}$. (pp. 170–171)

4. Graph $y = 2(x + 1)^2 - 3$ using transformations. (pp. 252–260)

Concepts and Vocabulary

5. The line _____ is a horizontal asymptote of $R(x) = \dfrac{x^3 - 1}{x^3 + 1}$.

6. The line _____ is a vertical asymptote of $R(x) = \dfrac{x - 1}{x + 1}$.

7. For a rational function R, if the degree of the numerator is less than the degree of the denominator, then R is _____.

8. ***True or False*** The domain of every rational function is the set of all real numbers.

9. ***True or False*** If an asymptote is neither horizontal nor vertical, it is called oblique.

10. ***True or False*** If the degree of the numerator of a rational function equals the degree of the denominator, then the ratio of the leading coefficients gives rise to the horizontal asymptote.

Skill Building

In Problems 11–22, find the domain of each rational function.

11. $R(x) = \dfrac{4x}{x - 3}$

12. $R(x) = \dfrac{5x^2}{3 + x}$

13. $H(x) = \dfrac{-4x^2}{(x - 2)(x + 4)}$

14. $G(x) = \dfrac{6}{(x + 3)(4 - x)}$

15. $F(x) = \dfrac{3x(x - 1)}{2x^2 - 5x - 3}$

16. $Q(x) = \dfrac{-x(1 - x)}{3x^2 + 5x - 2}$

17. $R(x) = \dfrac{x}{x^3 - 8}$

18. $R(x) = \dfrac{x}{x^4 - 1}$

19. $H(x) = \dfrac{3x^2 + x}{x^2 + 4}$

20. $G(x) = \dfrac{x - 3}{x^4 + 1}$

21. $R(x) = \dfrac{3(x^2 - x - 6)}{4(x^2 - 9)}$

22. $F(x) = \dfrac{-2(x^2 - 4)}{3(x^2 + 4x + 4)}$

In Problems 23–28, use the graph shown to find:

(a) *The domain and range of each function* (b) *The intercepts, if any* (c) *Horizontal asymptotes, if any*
(d) *Vertical asymptotes, if any* (e) *Oblique asymptotes, if any*

23.

24.

25.

26.

27.

28.

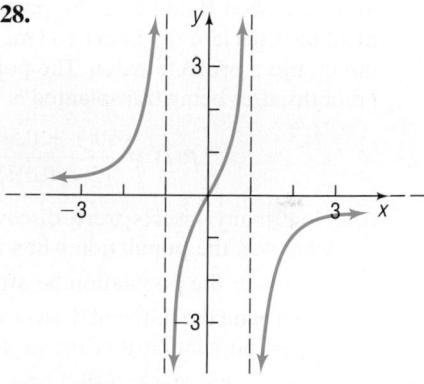

In Problems 29–40, graph each rational function using transformations.

29. $F(x) = 2 + \dfrac{1}{x}$

30. $Q(x) = 3 + \dfrac{1}{x^2}$

31. $R(x) = \dfrac{1}{(x-1)^2}$

32. $R(x) = \dfrac{3}{x}$

33. $H(x) = \dfrac{-2}{x+1}$

34. $G(x) = \dfrac{2}{(x+2)^2}$

35. $R(x) = \dfrac{-1}{x^2 + 4x + 4}$

36. $R(x) = \dfrac{1}{x-1} + 1$

37. $G(x) = 1 + \dfrac{2}{(x-3)^2}$

38. $F(x) = 2 - \dfrac{1}{x+1}$

39. $R(x) = \dfrac{x^2 - 4}{x^2}$

40. $R(x) = \dfrac{x-4}{x}$

In Problems 41–52, find the vertical, horizontal, and oblique asymptotes, if any, of each rational function.

41. $R(x) = \dfrac{3x}{x+4}$

42. $R(x) = \dfrac{3x+5}{x-6}$

43. $H(x) = \dfrac{x^3 - 8}{x^2 - 5x + 6}$

44. $G(x) = \dfrac{-x^2 + 1}{x^2 - 5x + 6}$

45. $T(x) = \dfrac{x^3}{x^4 - 1}$

46. $P(x) = \dfrac{4x^5}{x^3 - 1}$

47. $Q(x) = \dfrac{5 - x^2}{3x^4}$

48. $F(x) = \dfrac{-2x^2 + 1}{2x^3 + 4x^2}$

49. $R(x) = \dfrac{3x^4 + 4}{x^3 + 3x}$

50. $R(x) = \dfrac{6x^2 + x + 12}{3x^2 - 5x - 2}$

51. $G(x) = \dfrac{x^3 - 1}{x - x^2}$

52. $F(x) = \dfrac{x-1}{x - x^3}$

Applications and Extensions

53. Gravity In physics, it is established that the acceleration due to gravity, g (in meters/sec^2), at a height h meters above sea level is given by

$$g(h) = \frac{3.99 \times 10^{14}}{(6.374 \times 10^6 + h)^2}$$

where 6.374×10^6 is the radius of Earth in meters.

(a) What is the acceleration due to gravity at sea level?

(b) The Sears Tower in Chicago, Illinois, is 443 meters tall. What is the acceleration due to gravity at the top of the Sears Tower?

(c) The peak of Mount Everest is 8848 meters above sea level. What is the acceleration due to gravity on the peak of Mount Everest?

(d) Find the horizontal asymptote of $g(h)$.

(e) Solve $g(h) = 0$. How do you interpret your answer?

54. Population Model A rare species of insect was discovered in the Amazon Rain Forest. To protect the species, environmentalists declare the insect endangered and transplant the insect into a protected area. The population P of the insect t months after being transplanted is

$$P(t) = \frac{50(1 + 0.5t)}{(2 + 0.01t)}$$

(a) How many insects were discovered? In other words, what was the population when $t = 0$?

(b) What will the population be after 5 years?

(c) Determine the horizontal asymptote of $P(t)$. What is the largest population that the protected area can sustain?

55. Resistance in Parallel Circuits From Ohm's law for circuits, it follows that the total resistance R_{tot} of two components hooked in parallel is given by the equation

$$R_{\text{tot}} = \frac{R_1 R_2}{R_1 + R_2}$$

where R_1 and R_2 are the individual resistances.

(a) Let $R_1 = 10$ ohms, and graph R_{tot} as a function of R_2.

(b) Find and interpret any asymptotes of the graph obtained in part (a).

(c) If $R_2 = 2\sqrt{R_1}$, what value of R_1 will yield an R_{tot} of 17 ohms?

Source: en.wikipedia.org/wiki/Series_and_parallel_circuits

56. Newton's Method In calculus you will learn that, if

$$p(x) = a_n x^n + a_{n-1} x^{n-1} + \cdots + a_1 x + a_0$$

is a polynomial, then the *derivative* of $p(x)$ is

$$p'(x) = n a_n x^{n-1} + (n-1) a_{n-1} x^{n-2} + \cdots + 2 a_2 x + a_1$$

Newton's Method is an efficient method for finding the x-intercepts (or real zeros) of a function, such as $p(x)$. The steps below outline Newton's Method

STEP 1: Select an initial value x_0 that is somewhat close to the x-intercept being sought.

STEP 2: Find values for x using the relation

$$x_{n+1} = x_n - \frac{p(x_n)}{p'(x_n)} \quad n = 1, 2, \ldots$$

until you get two consecutive values x_k and x_{k+1} that agree to whatever decimal place accuracy you desire.

STEP 3: The approximate zero will be x_{k+1}.

Consider the polynomial $p(x) = x^3 - 7x - 40$.

(a) Evaluate $p(5)$ and $p(-3)$.

(b) What might we conclude about a zero of p. Explain.

(c) Use Newton's Method to approximate an x-intercept, r, $-3 < r < 5$, of $p(x)$ to four decimal places.

(d) Use a graphing utility to graph $p(x)$ and verify your answer in part (c).

(e) Using a graphing utility, evaluate $p(r)$ to verify your result.

Discussion and Writing

57. If the graph of a rational function R has the vertical asymptote $x = 4$, the factor $x - 4$ must be present in the denominator of R. Explain why.

58. If the graph of a rational function R has the horizontal asymptote $y = 2$, the degree of the numerator of R equals the degree of the denominator of R. Explain why.

59. Can the graph of a rational function have both a horizontal and an oblique asymptote? Explain.

60. Make up a rational function that has $y = 2x + 1$ as an oblique asymptote. Explain the methodology that you used.

'Are You Prepared?' Answers

1. True **2.** Quotient: $3x + 3$; Remainder: $2x^2 - 3x - 3$ **3.**

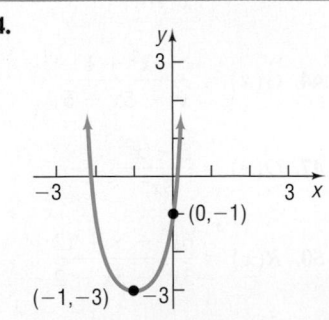

5.3 The Graph of a Rational Function

PREPARING FOR THIS SECTION *Before getting started, review the following:*

- Intercepts (Section 2.2, pp. 165–166)

Now Work the 'Are You Prepared?' problems on page 366.

OBJECTIVES 1 Analyze the Graph of a Rational Function (p. 355)
2 Solve Applied Problems Involving Rational Functions (p. 365)

1 Analyze the Graph of a Rational Function

We commented earlier that calculus provides the tools required to graph a polynomial function accurately. The same holds true for rational functions. However, we can gather together quite a bit of information about their graphs to get an idea of the general shape and position of the graph.

In the examples that follow, we will analyze the graph of a rational function by applying the following steps:

Analyzing the Graph of a Rational Function R

STEP 1: Factor the numerator and denominator of R and find its domain. If 0 is in the domain, find the y-intercept, $R(0)$, and plot it.

STEP 2: Write R in lowest terms as $\dfrac{p(x)}{q(x)}$ and find the real zeros of the numerator; that is, find the real solutions of the equation $p(x) = 0$, if any. These are the x-intercepts of the graph. Determine the behavior of the graph of R near each x-intercept, using the same procedure as for polynomial functions. Plot each x-intercept and indicate the behavior of the graph near it.

STEP 3: With R written in lowest terms as $\dfrac{p(x)}{q(x)}$, find the real zeros of the denominator; that is, find the real solutions of the equation $q(x) = 0$, if any. These determine the vertical asymptotes of the graph. Graph each vertical asymptote using a dashed line.

STEP 4: Locate any horizontal or oblique asymptotes using the procedure given in the previous section. Graph the asymptotes using a dashed line. Determine the points, if any, at which the graph of R intersects these asymptotes. Plot any such points.

STEP 5: Using the real zeros of the numerator and the denominator of the given equation for R, divide the x-axis into intervals and determine where the graph is above the x-axis and where it is below the x-axis by choosing a number in each interval and evaluating R there. Plot the points found.

STEP 6: Analyze the behavior of the graph of R near each asymptote and indicate this behavior on the graph.

STEP 7: Put all the information together to obtain the graph of R.

EXAMPLE 1 Analyzing the Graph of a Rational Function

Analyze the graph of the rational function: $R(x) = \dfrac{x - 1}{x^2 - 4}$

Solution **Step 1:** We factor the numerator and denominator of R, obtaining

$$R(x) = \frac{x - 1}{(x + 2)(x - 2)}$$

The domain of R is $\{x \mid x \neq -2, x \neq 2\}$. The y-intercept is

$$R(0) = \frac{-1}{-4} = \frac{1}{4}$$

Plot the point $\left(0, \frac{1}{4}\right)$.

Step 2: R is in lowest terms. The real zero of the numerator satisfies the equation $x - 1 = 0$. The only x-intercept is 1.

$$\text{Near 1:} \quad R(x) = \frac{x - 1}{(x + 2)(x - 2)} \approx \frac{x - 1}{(1 + 2)(1 - 2)} = -\frac{1}{3}(x - 1)$$

Plot the point $(1, 0)$ and indicate a line with slope $-\frac{1}{3}$ there.

Step 3: R is in lowest terms. The real zeros of the denominator are the real solutions of the equation $(x + 2)(x - 2) = 0$, that is, -2 and 2. The graph of R has two vertical asymptotes: the lines $x = -2$ and $x = 2$. Graph each of these asymptotes using dashed lines.

Step 4: The degree of the numerator is less than the degree of the denominator, so R is proper and the line $y = 0$ (the x-axis) is a horizontal asymptote of the graph. Indicate this line by graphing $y = 0$ using a dashed line.

To determine if the graph of R intersects the horizontal asymptote, we solve the equation $R(x) = 0$.

$$\frac{x - 1}{x^2 - 4} = 0$$

$$x - 1 = 0 \quad \text{Multiply both sides by } x^2 - 4.$$

$$x = 1$$

The only solution is $x = 1$, so the graph of R intersects the horizontal asymptote at $(1, 0)$. We have already plotted this point.

Step 5: The zero of the numerator, 1, and the zeros of the denominator, -2 and 2, divide the x-axis into four intervals:

$$(-\infty, -2) \quad (-2, 1) \quad (1, 2) \quad (2, \infty)$$

Now construct Table 8.

Table 8

Interval	$(-\infty, -2)$	$(-2, 1)$	$(1, 2)$	$(2, \infty)$
Number chosen	-3	0	$\dfrac{3}{2}$	3
Value of R	$R(-3) = -0.8$	$R(0) = \dfrac{1}{4}$	$R\left(\dfrac{3}{2}\right) = -\dfrac{2}{7}$	$R(3) = 0.4$
Location of graph	Below x-axis	Above x-axis	Below x-axis	Above x-axis
Point on graph	$(-3, -0.8)$	$\left(0, \dfrac{1}{4}\right)$	$\left(\dfrac{3}{2}, -\dfrac{2}{7}\right)$	$(3, 0.4)$

Plot the points from Table 8. You should now have Figure 27(a).

Step 6: Next, we determine the behavior of the graph near the asymptotes.

• Since the x-axis is a horizontal asymptote and the graph lies below the x-axis for $x < -2$, we can sketch a portion of the graph by placing a small arrow to the far left and under the x-axis.

- Since the line $x = -2$ is a vertical asymptote and the graph lies below the x-axis for $x < -2$, we continue by placing an arrow well below the x-axis and approaching the line $x = -2$ on the left.

- Since the graph is above the x-axis for $-2 < x < 1$ and $x = -2$ is a vertical asymptote, the graph will continue on the right of $x = -2$ at the top. Similar explanations account for the other arrows shown in Figure 27(b).

STEP 7: Figure 27(c) shows the complete graph.

Figure 27

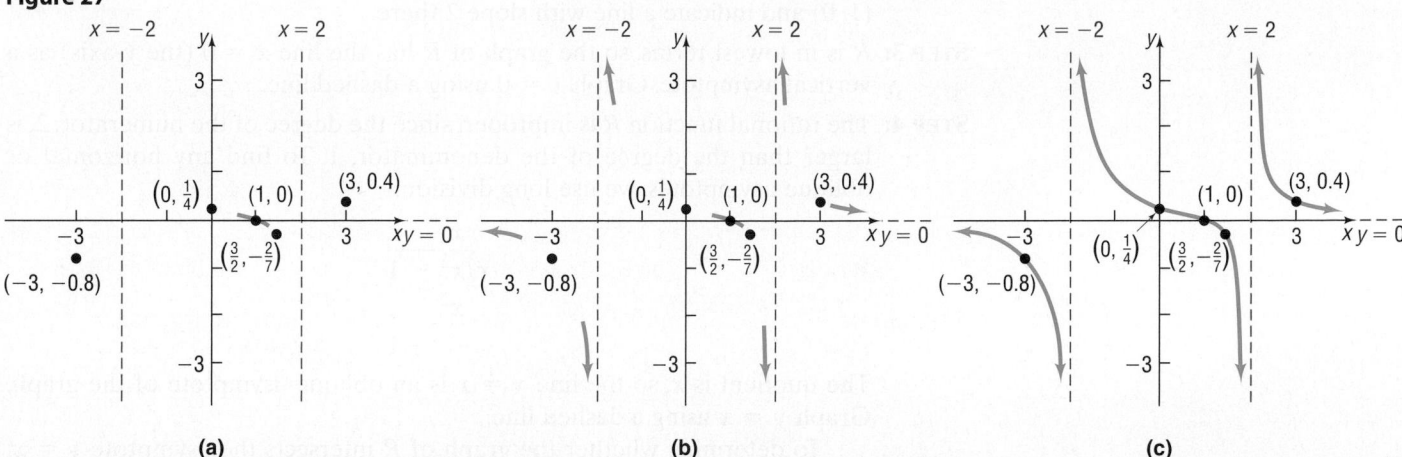

(a) (b) (c)

Exploration

Graph $R(x) = \dfrac{x-1}{x^2-4}$

Result The analysis just completed in Example 1 helps us to set the viewing rectangle to obtain a complete graph. Figure 28(a) shows the graph of $R(x) = \dfrac{x-1}{x^2-4}$ in connected mode, and Figure 28(b) shows it in dot mode. Notice in Figure 28(a) that the graph has vertical lines at $x = -2$ and $x = 2$. This is due to the fact that, when the graphing utility is in connected mode, it will connect the dots between consecutive pixels. We know that the graph of R does not cross the lines $x = -2$ and $x = 2$, since R is not defined at $x = -2$ or $x = 2$. So, when graphing rational functions, dot mode should be used to avoid extraneous vertical lines that are not part of the graph. See Figure 28(b).

Figure 28

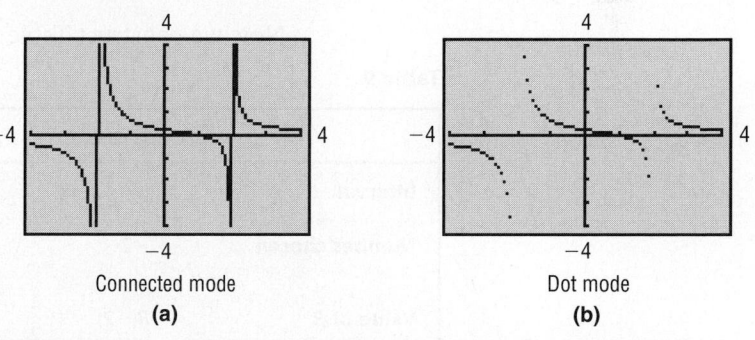

Connected mode Dot mode
(a) (b)

Now Work PROBLEM 7

EXAMPLE 2	Analyzing the Graph of a Rational Function

Analyze the graph of the rational function: $R(x) = \dfrac{x^2 - 1}{x}$

Solution **STEP 1:** The domain of R is $\{x \mid x \neq 0\}$. Because x cannot equal 0, there is no y-intercept. Now factor R to obtain

$$R(x) = \frac{(x+1)(x-1)}{x}$$

STEP 2: R is in lowest terms.

Solving the equation $R(x) = 0$, we find the graph has two x-intercepts: -1 and 1.

$$\text{Near } -1: \quad R(x) = \frac{(x+1)(x-1)}{x} \approx \frac{(x+1)(-1-1)}{-1} = 2(x+1)$$

$$\text{Near } 1: \quad R(x) = \frac{(x+1)(x-1)}{x} \approx \frac{(1+1)(x-1)}{1} = 2(x-1)$$

Plot the point $(-1, 0)$ and indicate a line with slope 2 there. Plot the point $(1, 0)$ and indicate a line with slope 2 there.

STEP 3: R is in lowest terms, so the graph of R has the line $x = 0$ (the y-axis) as a vertical asymptote. Graph $x = 0$ using a dashed line.

STEP 4: The rational function R is improper, since the degree of the numerator, 2, is larger than the degree of the denominator, 1. To find any horizontal or oblique asymptotes, we use long division.

$$\begin{array}{r} x \\ x\overline{)x^2 - 1} \\ \underline{x^2 } \\ -1 \end{array}$$

The quotient is x, so the line $y = x$ is an oblique asymptote of the graph. Graph $y = x$ using a dashed line.

To determine whether the graph of R intersects the asymptote $y = x$, we solve the equation $R(x) = x$.

$$R(x) = \frac{x^2 - 1}{x} = x$$
$$x^2 - 1 = x^2$$
$$-1 = 0 \quad \text{Impossible}$$

We conclude that the equation $\dfrac{x^2 - 1}{x} = x$ has no solution, so the graph of R does not intersect the line $y = x$.

STEP 5: The zeros of the numerator are -1 and 1; the zero of the denominator is 0. We use these values to divide the x-axis into four intervals:

$$(-\infty, -1) \quad (-1, 0) \quad (0, 1) \quad (1, \infty)$$

Now we construct Table 9.

Table 9

	-1 0 1 $\rightarrow x$			
Interval	$(-\infty, -1)$	$(-1, 0)$	$(0, 1)$	$(1, \infty)$
Number chosen	-2	$-\dfrac{1}{2}$	$\dfrac{1}{2}$	2
Value of R	$R(-2) = -\dfrac{3}{2}$	$R\left(-\dfrac{1}{2}\right) = \dfrac{3}{2}$	$R\left(\dfrac{1}{2}\right) = -\dfrac{3}{2}$	$R(2) = \dfrac{3}{2}$
Location of graph	Below x-axis	Above x-axis	Below x-axis	Above x-axis
Point on graph	$\left(-2, -\dfrac{3}{2}\right)$	$\left(-\dfrac{1}{2}, \dfrac{3}{2}\right)$	$\left(\dfrac{1}{2}, -\dfrac{3}{2}\right)$	$\left(2, \dfrac{3}{2}\right)$

Plot the points from Table 9. You should now have Figure 29(a).

STEP 6: • Since the graph of R is below the x-axis for $x < -1$ and is above the x-axis for $x > 1$, and since the graph of R does not intersect the oblique asymptote $y = x$, the graph of R will approach the line $y = x$ as shown in Figure 29(b).

• Since the graph of R is above the x-axis for $-1 < x < 0$, the graph of R will approach the vertical asymptote $x = 0$ at the top to the left of $x = 0$.

Figure 29

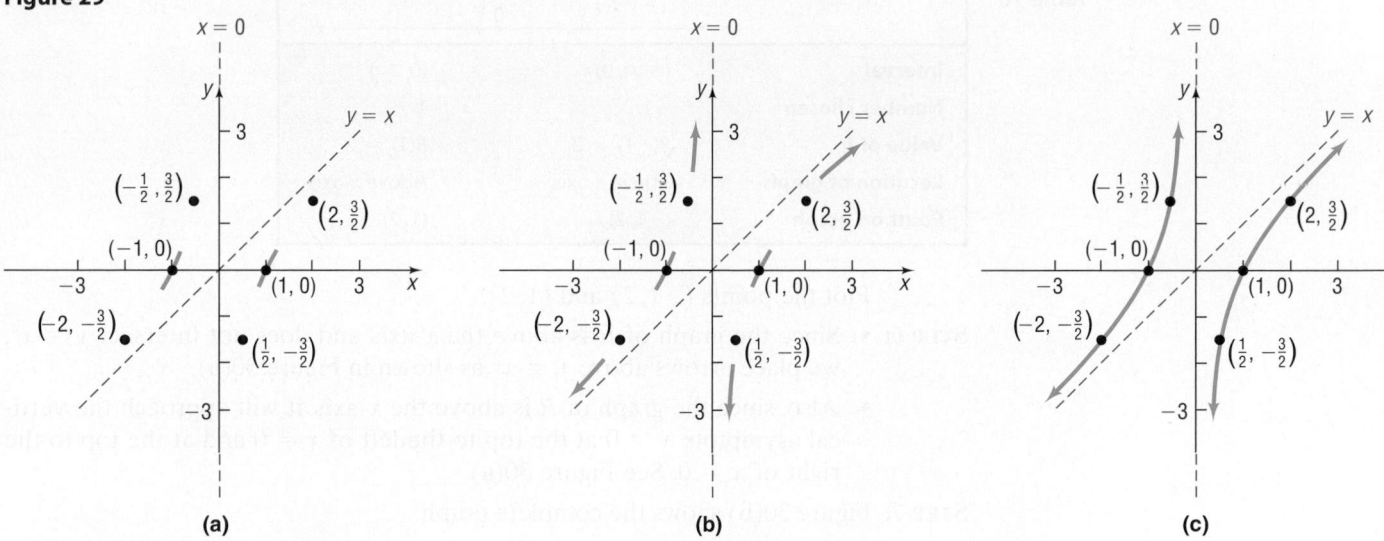

(a) (b) (c)

Seeing the Concept

Graph $R(x) = \dfrac{x^2 - 1}{x}$ and compare what you see with Figure 29(c). Could you have predicted from the graph that $y = x$ is an oblique asymptote? Graph $y = x$ and ZOOM-OUT. What do you observe?

- Since the graph of R is below the x-axis for $0 < x < 1$, the graph of R will approach the vertical asymptote $x = 0$ at the bottom to the right of $x = 0$. See Figure 29(b).

STEP 7: The complete graph is given in Figure 29(c).

Now Work PROBLEM 15

EXAMPLE 3	**Analyzing the Graph of a Rational Function**

Analyze the graph of the rational function: $R(x) = \dfrac{x^4 + 1}{x^2}$

Solution

STEP 1: R is completely factored. The domain of R is $\{x \mid x \neq 0\}$. There is no y-intercept.

STEP 2: R is in lowest terms. Since $x^4 + 1 = 0$ has no real solutions, there are no x-intercepts.

STEP 3: R is in lowest terms, so $x = 0$ (the y-axis) is a vertical asymptote of R. Graph the line $x = 0$ using dashes.

STEP 4: The rational function R is improper. To find any horizontal or oblique asymptotes, we use long division.

$$
\begin{array}{r}
x^2 \\
x^2 \overline{)x^4 + 1} \\
\underline{x^4 } \\
1
\end{array}
$$

The quotient is x^2, so the graph has no horizontal or oblique asymptotes. However, the graph of R will approach the graph of $y = x^2$ as $x \to -\infty$ and as $x \to \infty$. The graph of R does not intersect $y = x^2$. Graph $y = x^2$ using dashes.

STEP 5: The numerator has no zeros, and the denominator has one zero at 0. We divide the x-axis into the two intervals

$$(-\infty, 0) \qquad (0, \infty)$$

and construct Table 10.

Table 10

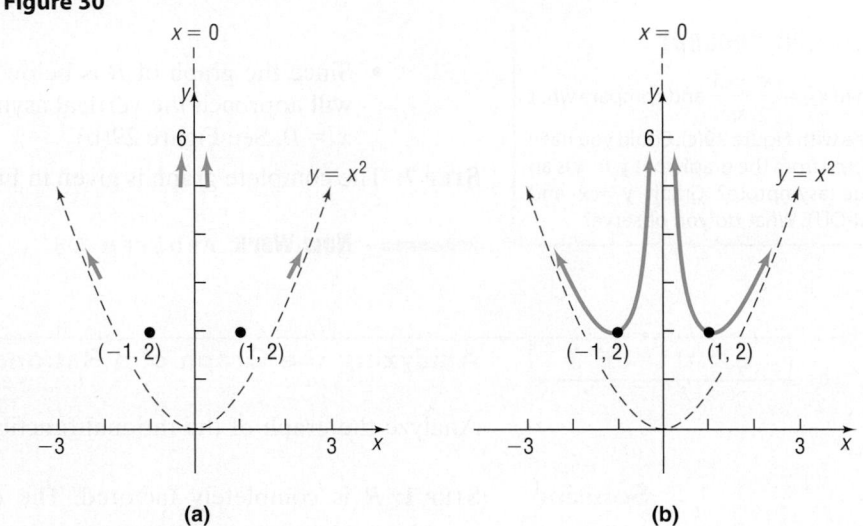

	$(-\infty, 0)$	$(0, \infty)$
Interval	$(-\infty, 0)$	$(0, \infty)$
Number chosen	-1	1
Value of R	$R(-1) = 2$	$R(1) = 2$
Location of graph	Above x-axis	Above x-axis
Point on graph	$(-1, 2)$	$(1, 2)$

Plot the points $(-1, 2)$ and $(1, 2)$.

STEP 6:
- Since the graph of R is above the x-axis and does not intersect $y = x^2$, we place arrows above $y = x^2$ as shown in Figure 30(a).
- Also, since the graph of R is above the x-axis, it will approach the vertical asymptote $x = 0$ at the top to the left of $x = 0$ and at the top to the right of $x = 0$. See Figure 30(a).

STEP 7: Figure 30(b) shows the complete graph.

Figure 30

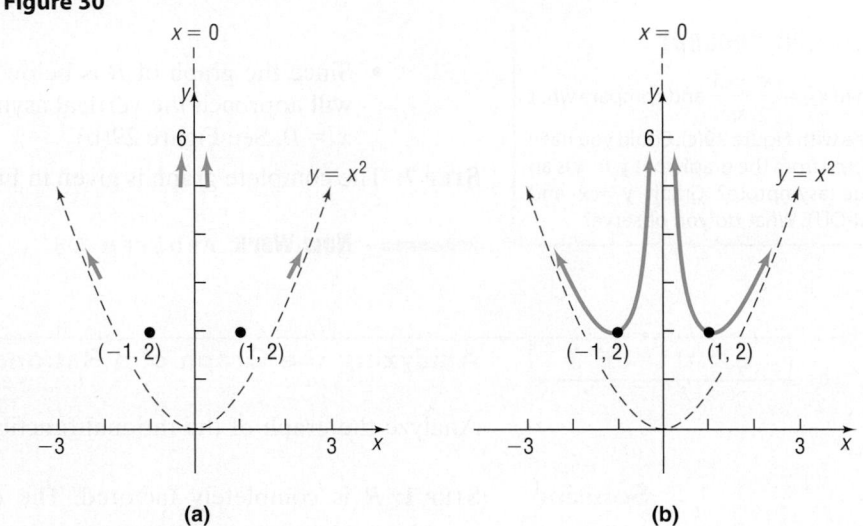

(a) (b)

Seeing the Concept

Graph $R(x) = \dfrac{x^4 + 1}{x^2}$ and compare what you see with Figure 30(b). Use MINIMUM to find the two turning points. Enter $Y_2 = x^2$ and ZOOM-OUT. What do you see?

—— **Now Work** PROBLEM 13

EXAMPLE 4 **Analyzing the Graph of a Rational Function**

Analyze the graph of the rational function: $R(x) = \dfrac{3x^2 - 3x}{x^2 + x - 12}$

Solution **STEP 1:** We factor R to get

$$R(x) = \frac{3x(x - 1)}{(x + 4)(x - 3)}$$

The domain of R is $\{x \mid x \neq -4, x \neq 3\}$. The y-intercept is $R(0) = 0$. Plot the point $(0, 0)$.

STEP 2: R is in lowest terms. Since the real solutions of the equation $3x(x - 1) = 0$ are $x = 0$ and $x = 1$, the graph has two x-intercepts, 0 and 1. We determine the behavior of the graph of R near each x-intercept.

Near 0: $R(x) = \dfrac{3x(x - 1)}{(x + 4)(x - 3)} \approx \dfrac{3x(0 - 1)}{(0 + 4)(0 - 3)} = \dfrac{1}{4}x$

Near 1: $R(x) = \dfrac{3x(x - 1)}{(x + 4)(x - 3)} \approx \dfrac{3(1)(x - 1)}{(1 + 4)(1 - 3)} = -\dfrac{3}{10}(x - 1)$

Plot the point $(0, 0)$ and show a line with slope $\dfrac{1}{4}$ there.

Plot the point $(1, 0)$ and show a line with slope $-\dfrac{3}{10}$ there.

STEP 3: R is in lowest terms. Since the real solutions of the equation $(x + 4)(x - 3) = 0$ are $x = -4$ and $x = 3$, the graph of R has two vertical asymptotes, the lines $x = -4$ and $x = 3$. Plot these lines using dashes.

STEP 4: Since the degree of the numerator equals the degree of the denominator, the graph has a horizontal asymptote. To find it, we either use long division or form the quotient of the leading coefficient of the numerator, 3, and the leading coefficient of the denominator, 1. The graph of R has the horizontal asymptote $y = 3$.

To find out whether the graph of R intersects the asymptote, we solve the equation $R(x) = 3$.

$$R(x) = \frac{3x^2 - 3x}{x^2 + x - 12} = 3$$
$$3x^2 - 3x = 3x^2 + 3x - 36$$
$$-6x = -36$$
$$x = 6$$

The graph intersects the line $y = 3$ only at $x = 6$, and $(6, 3)$ is a point on the graph of R. Plot the point $(6, 3)$ and the line $y = 3$ using dashes.

STEP 5: The zeros of the numerator, 0 and 1, and the zeros of the denominator, -4 and 3, divide the x-axis into five intervals:

$$(-\infty, -4) \quad (-4, 0) \quad (0, 1) \quad (1, 3) \quad (3, \infty)$$

Construct Table 11.

Table 11

Interval	$(-\infty, -4)$	$(-4, 0)$	$(0, 1)$	$(1, 3)$	$(3, \infty)$
Number chosen	-5	-2	$\dfrac{1}{2}$	2	4
Value of R	$R(-5) = 11.25$	$R(-2) = -1.8$	$R\left(\dfrac{1}{2}\right) = \dfrac{1}{15}$	$R(2) = -1$	$R(4) = 4.5$
Location of graph	Above x-axis	Below x-axis	Above x-axis	Below x-axis	Above x-axis
Point on graph	$(-5, 11.25)$	$(-2, -1.8)$	$\left(\dfrac{1}{2}, \dfrac{1}{15}\right)$	$(2, -1)$	$(4, 4.5)$

Plot the points from Table 11. Figure 31(a) shows the graph we have so far.

STEP 6:
- Since the graph of R is above the x-axis for $x < -4$ and only crosses the line $y = 3$ at $(6, 3)$, as x approaches $-\infty$ the graph of R will approach the horizontal asymptote $y = 3$ from above.

- The graph of R will approach the vertical asymptote $x = -4$ at the top to the left of $x = -4$ and at the bottom to the right of $x = -4$.

- The graph of R will approach the vertical asymptote $x = 3$ at the bottom to the left of $x = 3$ and at the top to the right of $x = 3$.

- We do not know whether the graph of R crosses or touches the line $y = 3$ at $(6, 3)$. To see whether the graph, in fact, crosses or touches the line $y = 3$, we plot an additional point to the right of $(6, 3)$. We use $x = 7$ to find $R(7) = \dfrac{63}{22} < 3$. The graph crosses $y = 3$ at $x = 6$. Because $(6, 3)$ is the only point where the graph of R intersects the asymptote $y = 3$, the graph must approach the line $y = 3$ from below as $x \to \infty$. See Figure 31(b).

STEP 7: The complete graph is shown in Figure 31(c).

Figure 31

(a)

(b)

(c)

Exploration

Graph $R(x) = \dfrac{3x^2 - 3x}{x^2 + x - 12}$.

Result Figure 32 shows the graph in connected mode, and Figure 33(a) shows it in dot mode. Neither graph displays clearly the behavior between the two x-intercepts, 0 and 1. Nor do they clearly display the fact that the graph crosses the horizontal asymptote at $(6, 3)$. To see these parts better, we graph R for $-1 \leq x \leq 2$ and $-1 \leq y \leq 0.5$ [Figure 33(b)] and for $4 \leq x \leq 60$ and $2.5 \leq y \leq 3.5$. [Figure 34(b)].

Figure 32

Connected mode

Figure 33

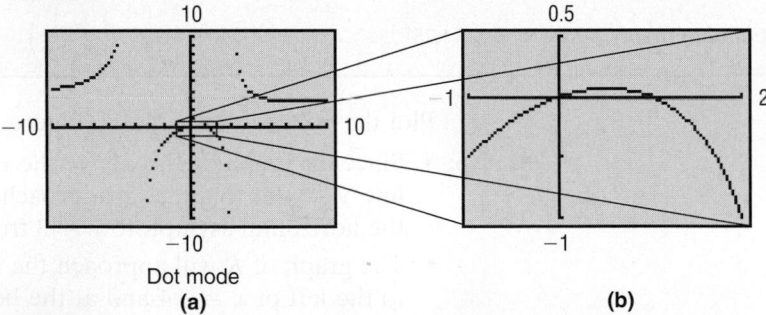

Dot mode
(a)

(b)

Figure 34

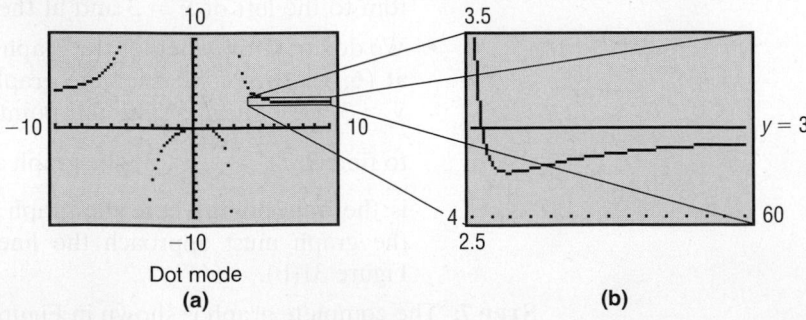

Dot mode
(a)

(b)

The new graphs reflect the behavior produced by the analysis. Furthermore, we observe two turning points, one between 0 and 1 and the other to the right of 4. Rounded to two decimal places, these turning points are (0.52, 0.07) and (11.48, 2.75).

| EXAMPLE 5 | **Analyzing the Graph of a Rational Function with a Hole** |

Analyze the graph of the rational function: $R(x) = \dfrac{2x^2 - 5x + 2}{x^2 - 4}$

Solution **STEP 1:** We factor R and obtain

$$R(x) = \frac{(2x - 1)(x - 2)}{(x + 2)(x - 2)}$$

The domain of R is $\{x \mid x \neq -2, x \neq 2\}$. The y-intercept is $R(0) = -\dfrac{1}{2}$.

Plot the point $\left(0, -\dfrac{1}{2}\right)$.

STEP 2: In lowest terms,

$$R(x) = \frac{2x - 1}{x + 2} \qquad x \neq -2$$

The graph has one x-intercept: $\dfrac{1}{2}$.

Near $\dfrac{1}{2}$: $R(x) = \dfrac{2x - 1}{x + 2} \approx \dfrac{2x - 1}{\dfrac{1}{2} + 2} = \dfrac{2}{5}(2x - 1)$

Plot the point $\left(\dfrac{1}{2}, 0\right)$ showing a line with slope $\dfrac{4}{5}$.

STEP 3: Look at R in lowest terms. The graph has one vertical asymptote, $x = -2$, since $x + 2$ is the only factor of the denominator of $R(x)$ *in lowest terms*. Remember though, the rational function is undefined at both $x = 2$ and $x = -2$. Graph the line $x = -2$ using dashes.

STEP 4: Since the degree of the numerator equals the degree of the denominator, the graph has a horizontal asymptote. To find it, we either use long division or form the quotient of the leading coefficient of the numerator, 2, and the leading coefficient of the denominator, 1. The graph of R has the horizontal asymptote $y = 2$. Graph the line $y = 2$ using dashes.

To find out whether the graph of R intersects the horizontal asymptote $y = 2$, we solve the equation $R(x) = 2$.

$$R(x) = \frac{2x - 1}{x + 2} = 2$$
$$2x - 1 = 2(x + 2)$$
$$2x - 1 = 2x + 4$$
$$-1 = 4 \qquad \text{Impossible}$$

The graph does not intersect the line $y = 2$.

STEP 5: Look at the given expression for R. The zeros of the numerator and denominator, -2, $\dfrac{1}{2}$, and 2, divide the x-axis into four intervals:

$$(-\infty, -2) \qquad \left(-2, \frac{1}{2}\right) \qquad \left(\frac{1}{2}, 2\right) \qquad (2, \infty)$$

Construct Table 12.

Table 12

	$(-\infty, -2)$	$\left(-2, \frac{1}{2}\right)$	$\left(\frac{1}{2}, 2\right)$	$(2, \infty)$
Interval	$(-\infty, -2)$	$\left(-2, \frac{1}{2}\right)$	$\left(\frac{1}{2}, 2\right)$	$(2, \infty)$
Number chosen	-3	-1	1	3
Value of R	$R(-3) = 7$	$R(-1) = -3$	$R(1) = \frac{1}{3}$	$R(3) = 1$
Location of graph	Above x-axis	Below x-axis	Above x-axis	Above x-axis
Point on graph	$(-3, 7)$	$(-1, -3)$	$\left(1, \frac{1}{3}\right)$	$(3, 1)$

Plot these points.

STEP 6: • From Table 12 we know the graph of R is above the x-axis for $x < -2$. From Step 4 we know the graph of R does not intersect the asymptote $y = 2$. Therefore, the graph of R will approach $y = 2$ from above as $x \to -\infty$ and will approach the vertical asymptote $x = -2$ at the top from the left.

• Since the graph of R is below the x-axis for $-2 < x < \frac{1}{2}$, the graph of R will approach $x = -2$ at the bottom from the right.

• Finally, since the graph of R is above the x-axis for $x > \frac{1}{2}$ and does not intersect the horizontal asymptote $y = 2$, the graph of R will approach $y = 2$ from below as $x \to \infty$. See Figure 35(a).

STEP 7: See Figure 35(b) for the complete graph. Since R is not defined at 2, there is a hole at the point $\left(2, \frac{3}{4}\right)$.

Figure 35

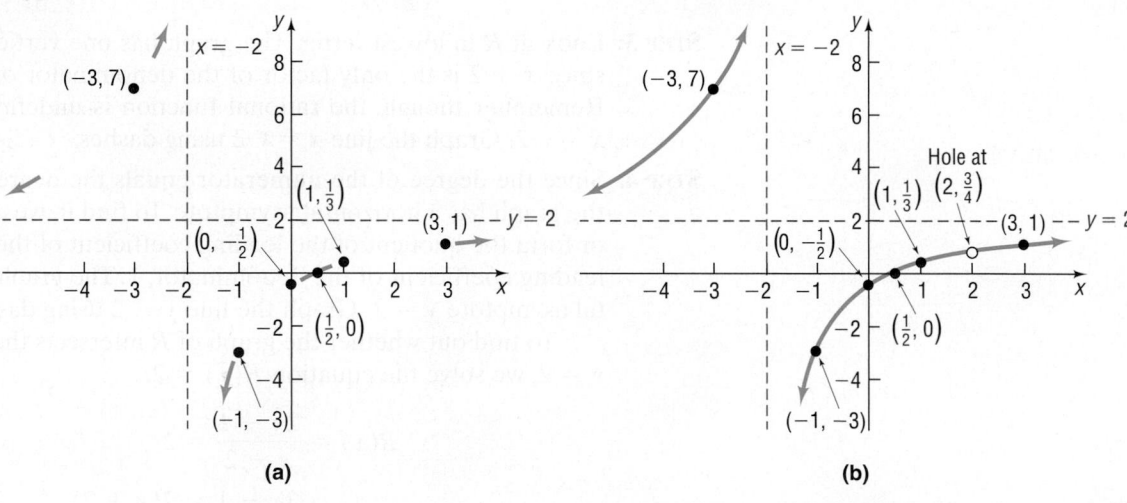

(a) (b)

Exploration

Graph $R(x) = \dfrac{2x^2 - 5x + 2}{x^2 - 4}$. Do you see the hole at $\left(2, \frac{3}{4}\right)$? TRACE along the graph. Did you obtain an ERROR at $x = 2$? Are you convinced that an algebraic analysis of a rational function is required in order to accurately interpret the graph obtained with a graphing utility?

NOTE The coordinates of the hole were obtained by evaluating R in lowest terms at 2. R in lowest terms is $\dfrac{2x - 1}{x + 2}$, which, at $x = 2$, is $\dfrac{2(2) - 1}{2 + 2} = \dfrac{3}{4}$. ∎

As Example 5 shows, **the real zeros of the denominator of a rational function give rise to either vertical asymptotes or holes on the graph.**

Now Work PROBLEM 33

We now discuss the problem of finding a rational function from its graph.

| EXAMPLE 6 | **Constructing a Rational Function from Its Graph** |

Find a rational function that might have the graph shown in Figure 36.

Figure 36

Solution The numerator of a rational function $R(x) = \dfrac{p(x)}{q(x)}$ in lowest terms determines the

x-intercepts of its graph. The graph shown in Figure 36 has x-intercepts -2 (even multiplicity; graph touches the x-axis) and 5 (odd multiplicity; graph crosses the x-axis). So one possibility for the numerator is $p(x) = (x + 2)^2(x - 5)$.

The denominator of a rational function in lowest terms determines the vertical asymptotes of its graph. The vertical asymptotes of the graph are $x = -5$ and $x = 2$. Since $R(x)$ approaches ∞ to the left of $x = -5$ and $R(x)$ approaches $-\infty$ to the right of $x = -5$, then $x + 5$ is a factor of odd multiplicity in $q(x)$. Also, because $R(x)$ approaches $-\infty$ on both sides of $x = 2$, then $x - 2$ is a factor of even multiplicity in $q(x)$. A possibility for the denominator is $q(x) = (x + 5)(x - 2)^2$. So far we have $R(x) = \dfrac{(x + 2)^2(x - 5)}{(x + 5)(x - 2)^2}$.

The horizontal asymptote of the graph given in Figure 36 is $y = 2$, so we know that the degree of the numerator must equal the degree of the denominator and the quotient of leading coefficients must be $\dfrac{2}{1}$. This leads to

$$R(x) = \frac{2(x + 2)^2(x - 5)}{(x + 5)(x - 2)^2}$$

Figure 37

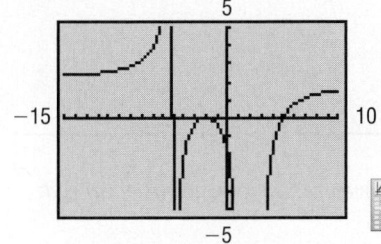

✓ **Check:** Figure 37 shows the graph of R on a graphing utility. Since Figure 37 looks similar to Figure 36, we have found a rational function R for the graph in Figure 36.

━━━━ **Now Work** PROBLEM 45

2 Solve Applied Problems Involving Rational Functions

| EXAMPLE 7 | **Finding the Least Cost of a Can** |

Reynolds Metal Company manufactures aluminum cans in the shape of a right circular cylinder with a capacity of 500 cubic centimeters $\left(\dfrac{1}{2} \text{ liter}\right)$. The top and bottom of the can are made of a special aluminum alloy that costs 0.05¢ per square centimeter. The sides of the can are made of material that costs 0.02¢ per square centimeter.

(a) Express the cost C of material for the can as a function of the radius r of the can.

(b) Find any vertical asymptotes. Discuss the cost of the can near any vertical asymptotes.

 (c) Use a graphing utility to graph the function $C = C(r)$.

(d) What value of r will result in the least cost?

(e) What is this least cost?

Solution

Figure 38

(a) Figure 38 illustrates the components of a can in the shape of a right circular cylinder. Notice that the material required to produce a cylindrical can of height h and radius r consists of a rectangle of area $2\pi rh$ and two circles, each of area πr^2. The total cost C (in cents) of manufacturing the can is therefore

$$C = \text{Cost of the top and bottom} + \text{Cost of the side}$$

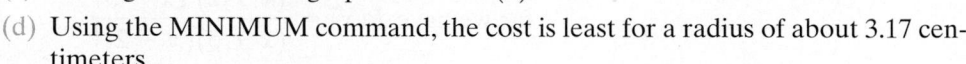

$$= \underbrace{2(\pi r^2)}_{\substack{\text{Total area} \\ \text{of top and} \\ \text{bottom}}} \underbrace{(0.05)}_{\substack{\text{Cost/unit} \\ \text{area}}} + \underbrace{(2\pi rh)}_{\substack{\text{Total} \\ \text{area of} \\ \text{side}}} \underbrace{(0.02)}_{\substack{\text{Cost/unit} \\ \text{area}}}$$

$$= 0.10\pi r^2 + 0.04\pi rh$$

But we have the additional restriction that the height h and radius r must be chosen so that the volume V of the can is 500 cubic centimeters. Since $V = \pi r^2 h$, we have

$$500 = \pi r^2 h \quad \text{so} \quad h = \frac{500}{\pi r^2}$$

Substituting this expression for h, the cost C, in cents, as a function of the radius r is

$$C(r) = 0.10\pi r^2 + 0.04\pi r \frac{500}{\pi r^2} = 0.10\pi r^2 + \frac{20}{r} = \frac{0.10\pi r^3 + 20}{r}$$

Figure 39

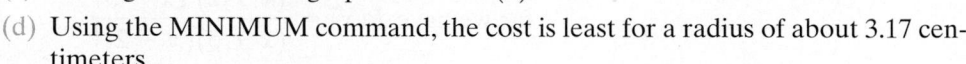

Minimum
X=3.1692012 Y=9.4661027

(b) The only vertical asymptote is $r = 0$. As the radius r of the can gets closer to 0, the cost C of the can gets higher.

(c) See Figure 39 for the graph of $C = C(r)$.

(d) Using the MINIMUM command, the cost is least for a radius of about 3.17 centimeters.

(e) The least cost is $C(3.17) \approx 9.47$¢.

━━━ **Now Work** PROBLEM 55

5.3 Assess Your Understanding

'Are You Prepared?' *The answer is given at the end of these exercises. If you get a wrong answer, read the pages listed in red.*

1. The intercepts of the equation $y = \dfrac{x-1}{x^2-4}$ are _____. (pp. 165–166)

Concepts and Vocabulary

2. If the numerator and the denominator of a rational function have no common factors, the rational function is _____.

3. *True or False* The graph of a polynomial function sometimes has a hole.

4. *True or False* The graph of a rational function never intersects a horizontal asymptote.

5. *True or False* The graph of a rational function sometimes intersects a vertical asymptote.

6. *True or False* The graph of a rational function sometimes has a hole.

Skill Building

In Problems 7–44, follow Steps 1 through 7 on page 355 to analyze the graph of each function.

7. $R(x) = \dfrac{x+1}{x(x+4)}$

8. $R(x) = \dfrac{x}{(x-1)(x+2)}$

9. $R(x) = \dfrac{3x+3}{2x+4}$

10. $R(x) = \dfrac{2x + 4}{x - 1}$

11. $R(x) = \dfrac{3}{x^2 - 4}$

12. $R(x) = \dfrac{6}{x^2 - x - 6}$

13. $P(x) = \dfrac{x^4 + x^2 + 1}{x^2 - 1}$

14. $Q(x) = \dfrac{x^4 - 1}{x^2 - 4}$

15. $H(x) = \dfrac{x^3 - 1}{x^2 - 9}$

16. $G(x) = \dfrac{x^3 + 1}{x^2 + 2x}$

17. $R(x) = \dfrac{x^2}{x^2 + x - 6}$

18. $R(x) = \dfrac{x^2 + x - 12}{x^2 - 4}$

19. $G(x) = \dfrac{x}{x^2 - 4}$

20. $G(x) = \dfrac{3x}{x^2 - 1}$

21. $R(x) = \dfrac{3}{(x - 1)(x^2 - 4)}$

22. $R(x) = \dfrac{-4}{(x + 1)(x^2 - 9)}$

23. $H(x) = \dfrac{x^2 - 1}{x^4 - 16}$

24. $H(x) = \dfrac{x^2 + 4}{x^4 - 1}$

25. $F(x) = \dfrac{x^2 - 3x - 4}{x + 2}$

26. $F(x) = \dfrac{x^2 + 3x + 2}{x - 1}$

27. $R(x) = \dfrac{x^2 + x - 12}{x - 4}$

28. $R(x) = \dfrac{x^2 - x - 12}{x + 5}$

29. $F(x) = \dfrac{x^2 + x - 12}{x + 2}$

30. $G(x) = \dfrac{x^2 - x - 12}{x + 1}$

31. $R(x) = \dfrac{x(x - 1)^2}{(x + 3)^3}$

32. $R(x) = \dfrac{(x - 1)(x + 2)(x - 3)}{x(x - 4)^2}$

33. $R(x) = \dfrac{x^2 + x - 12}{x^2 - x - 6}$

34. $R(x) = \dfrac{x^2 + 3x - 10}{x^2 + 8x + 15}$

35. $R(x) = \dfrac{6x^2 - 7x - 3}{2x^2 - 7x + 6}$

36. $R(x) = \dfrac{8x^2 + 26x + 15}{2x^2 - x - 15}$

37. $R(x) = \dfrac{x^2 + 5x + 6}{x + 3}$

38. $R(x) = \dfrac{x^2 + x - 30}{x + 6}$

39. $f(x) = x + \dfrac{1}{x}$

40. $f(x) = 2x + \dfrac{9}{x}$

41. $f(x) = x^2 + \dfrac{1}{x}$

42. $f(x) = 2x^2 + \dfrac{16}{x}$

43. $f(x) = x + \dfrac{1}{x^3}$

44. $f(x) = 2x + \dfrac{9}{x^3}$

In Problems 45–48, find a rational function that might have the given graph. (More than one answer might be possible.)

45.

46.

47.

48.

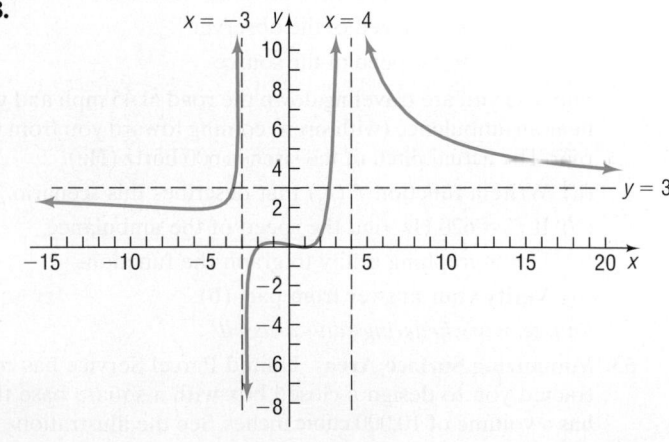

Applications and Extensions

49. Drug Concentration The concentration C of a certain drug in a patient's bloodstream t hours after injection is given by

$$C(t) = \frac{t}{2t^2 + 1}$$

(a) Find the horizontal asymptote of $C(t)$. What happens to the concentration of the drug as t increases?

(b) Using your graphing utility, graph $C(t)$.

(c) Determine the time at which the concentration is highest.

50. Drug Concentration The concentration C of a certain drug in a patient's bloodstream t minutes after injection is given by

$$C(t) = \frac{50t}{t^2 + 25}$$

(a) Find the horizontal asymptote of $C(t)$. What happens to the concentration of the drug as t increases?

(b) Using your graphing utility, graph $C(t)$.

(c) Determine the time at which the concentration is highest.

51. Minimum Cost A rectangular area adjacent to a river is to be fenced in; no fence is needed on the river side. The enclosed area is to be 1000 square feet. Fencing for the side parallel to the river is $5 per linear foot, and fencing for the other two sides is $8 per linear foot; the four corner posts are $25 apiece. Let x be the length of one of the sides perpendicular to the river.

(a) Write a function $C(x)$ that describes the cost of the project.

(b) What is the domain of C?

(c) Use a graphing utility to graph $C(x)$.

(d) Find the dimensions of the cheapest enclosure.

Source: www.uncwil.edu/courses/math111hb/PandR/rational/ rational.html

52. Doppler Effect The Doppler effect (named after Christian Doppler) is the change in the pitch (frequency) of the sound from a source (s) as heard by an observer (o) when one or both are in motion. If we assume both the source and the observer are moving in the same direction, the relationship is

$$f' = f_a\left(\frac{v - v_o}{v - v_s}\right)$$

where
f' = perceived pitch by the observer
f_a = actual pitch of the source
v = speed of sound in air (assume 772.4 mph)
v_o = speed of the observer
v_s = speed of the source

Suppose you are traveling down the road at 45 mph and you hear an ambulance (with siren) coming toward you from the rear. The actual pitch of the siren is 600 hertz (Hz).

(a) Write a function $f'(v_s)$ that describes this scenario.

(b) If $f' = 620$ Hz, find the speed of the ambulance.

(c) Use a graphing utility to graph the function.

(d) Verify your answer from part (b).

Source: www.kettering.edu/~drussell/

53. Minimizing Surface Area United Parcel Service has contracted you to design a closed box with a square base that has a volume of 10,000 cubic inches. See the illustration.

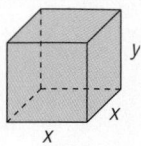

(a) Express the surface area S of the box as a function of x.

(b) Using a graphing utility, graph the function found in part (a).

(c) What is the minimum amount of cardboard that can be used to construct the box?

(d) What are the dimensions of the box that minimize the surface area?

(e) Why might UPS be interested in designing a box that minimizes the surface area?

54. Minimizing Surface Area United Parcel Service has contracted you to design an open box with a square base that has a volume of 5000 cubic inches. See the illustration.

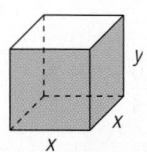

(a) Express the surface area S of the box as a function of x.

(b) Using a graphing utility, graph the function found in part (a).

(c) What is the minimum amount of cardboard that can be used to construct the box?

(d) What are the dimensions of the box that minimize the surface area?

(e) Why might UPS be interested in designing a box that minimizes the surface area?

55. Cost of a Can A can in the shape of a right circular cylinder is required to have a volume of 500 cubic centimeters. The top and bottom are made of material that costs 6¢ per square centimeter, while the sides are made of material that costs 4¢ per square centimeter.

(a) Express the total cost C of the material as a function of the radius r of the cylinder. (Refer to Figure 38.)

(b) Graph $C = C(r)$. For what value of r is the cost C a minimum?

56. Material Needed to Make a Drum A steel drum in the shape of a right circular cylinder is required to have a volume of 100 cubic feet.

(a) Express the amount A of material required to make the drum as a function of the radius r of the cylinder.

(b) How much material is required if the drum's radius is 3 feet?

(c) How much material is required if the drum's radius is 4 feet?

(d) How much material is required if the drum's radius is 5 feet?

(e) Graph $A = A(r)$. For what value of r is A smallest?

Discussion and Writing

57. Graph each of the following functions:

$$y = \frac{x^2 - 1}{x - 1} \qquad y = \frac{x^3 - 1}{x - 1}$$

$$y = \frac{x^4 - 1}{x - 1} \qquad y = \frac{x^5 - 1}{x - 1}$$

Is $x = 1$ a vertical asymptote? Why not? What is happening for $x = 1$? What do you conjecture about $y = \frac{x^n - 1}{x - 1}$, $n \geq 1$ an integer, for $x = 1$?

58. Graph each of the following functions:

$$y = \frac{x^2}{x - 1} \qquad y = \frac{x^4}{x - 1} \qquad y = \frac{x^6}{x - 1} \qquad y = \frac{x^8}{x - 1}$$

What similarities do you see? What differences?

59. Write a few paragraphs that provide a general strategy for graphing a rational function. Be sure to mention the following: proper, improper, intercepts, and asymptotes.

60. Create a rational function that has the following characteristics: crosses the x-axis at 2; touches the x-axis at -1; one vertical asymptote at $x = -5$ and another at $x = 6$; and one horizontal asymptote, $y = 3$. Compare your function to a fellow classmate's. How do they differ? What are their similarities?

61. Create a rational function that has the following characteristics: crosses the x-axis at 3; touches the x-axis at -2; one vertical asymptote, $x = 1$; and one horizontal asymptote, $y = 2$. Give your rational function to a fellow classmate and ask for a written critique of your rational function.

62. Create a rational function with the following characteristics: three real zeros, one of multiplicity 2; y-intercept 1; vertical asymptotes $x = -2$ and $x = 3$; oblique asymptote $y = 2x + 1$. Is this rational function unique? Compare your function with those of other students. What will be the same as everyone else's? Add some more characteristics, such as symmetry or naming the real zeros. How does this modify the rational function?

'Are You Prepared?' Answers

1. $\left(0, \dfrac{1}{4}\right), (1, 0)$

5.4 Polynomial and Rational Inequalities

PREPARING FOR THIS SECTION *Before getting started, review the following:*

- Solving Inequalities (Section 1.5, pp. 124–131)

Now Work the 'Are You Prepared?' problem on page 373.

OBJECTIVES 1 Solve Polynomial Inequalities (p. 370)

2 Solve Rational Inequalities (p. 372)

In this section we solve inequalities that involve polynomials of degree 3 and higher, as well as some that involve rational expressions. To solve such inequalities, we use the information obtained in the previous three sections about the graph of polynomial and rational functions. The general idea follows:

Suppose that the polynomial or rational inequality is in one of the forms

$$f(x) < 0 \qquad f(x) > 0 \qquad f(x) \leq 0 \qquad f(x) \geq 0$$

Locate the zeros of f if f is a polynomial function, and locate the zeros of the numerator and the denominator if f is a rational function. If we use these zeros to divide the real number line into intervals, we know that on each interval the graph of f is either above the x-axis [$f(x) > 0$] or below the x-axis [$f(x) < 0$]. In other words, we have found the solution of the inequality.

The following steps provide more detail.

Steps for Solving Polynomial and Rational Inequalities

STEP 1: Write the inequality so that a polynomial or rational expression f is on the left side and zero is on the right side in one of the following forms:

$$f(x) > 0 \qquad f(x) \geq 0 \qquad f(x) < 0 \qquad f(x) \leq 0$$

For rational expressions, be sure that the left side is written as a single quotient and find its domain.

STEP 2: Determine the real numbers at which the expression f on the left side equals zero and, if the expression is rational, the real numbers at which the expression f on the left side is undefined.

STEP 3: Use the numbers found in Step 2 to separate the real number line into intervals.

STEP 4: Select a number in each interval and evaluate f at the number.
 (a) If the value of f is positive, then $f(x) > 0$ for all numbers x in the interval.
 (b) If the value of f is negative, then $f(x) < 0$ for all numbers x in the interval.

If the inequality is not strict, include the solutions of $f(x) = 0$ in the solution set.

1 Solve Polynomial Inequalities

| EXAMPLE 1 | Solving a Polynomial Inequality |

Solve the inequality $x^4 \leq 4x^2$, and graph the solution set.

Solution **STEP 1:** Rearrange the inequality so that 0 is on the right side.

$$x^4 \leq 4x^2$$
$$x^4 - 4x^2 \leq 0 \qquad \text{Subtract } 4x^2 \text{ from both sides of the inequality.}$$

This inequality is equivalent to the one that we wish to solve.

STEP 2: Find the real zeros of $f(x) = x^4 - 4x^2$ by solving the equation $x^4 - 4x^2 = 0$.

$$x^4 - 4x^2 = 0$$
$$x^2(x^2 - 4) = 0 \qquad \text{Factor.}$$
$$x^2(x + 2)(x - 2) = 0 \qquad \text{Factor.}$$
$$x = 0 \quad \text{or} \quad x = -2 \quad \text{or} \quad x = 2 \qquad \text{Use the Zero-Product Property and solve.}$$

STEP 3: We use these zeros of f to separate the real number line into four intervals:

$$(-\infty, -2) \qquad (-2, 0) \qquad (0, 2) \qquad (2, \infty)$$

STEP 4: Select a number in each interval and evaluate $f(x) = x^4 - 4x^2$ to determine if $f(x)$ is positive or negative. See Table 13.

Table 13

	-2	0	2	x
Interval	$(-\infty, -2)$	$(-2, 0)$	$(0, 2)$	$(2, \infty)$
Number chosen	-3	-1	1	3
Value of f	$f(-3) = 45$	$f(-1) = -3$	$f(1) = -3$	$f(3) = 45$
Conclusion	Positive	Negative	Negative	Positive

Based on Table 13, we know that $f(x) < 0$ for all x in the intervals $(-2, 0) \cup (0, 2)$, that is, for all x such that $-2 < x < 0$ or $0 < x < 2$. However, because the original inequality is not strict, numbers x that satisfy the equation $f(x) = x^4 - 4x^2 = 0$ are also solutions of the inequality $x^4 \leq 4x^2$. Thus, we include -2, 0, and 2. The solution set of the given inequality is $\{x | -2 \leq x \leq 2\}$ or, using interval notation, $[-2, 2]$.

Figure 40 shows the graph of the solution set.

Figure 40

━━━━ **Now Work** PROBLEMS 5 AND 13

━━

| EXAMPLE 2 | **Solving a Polynomial Inequality** |

Solve the inequality $x^4 > x$, and graph the solution set.

Solution **STEP 1:** Rearrange the inequality so that 0 is on the right side.

$$x^4 > x$$
$$x^4 - x > 0 \quad \text{Subtract x from both sides of the inequality.}$$

This inequality is equivalent to the one that we wish to solve.

STEP 2: Find the real zeros of $f(x) = x^4 - x$ by solving $x^4 - x = 0$.

$$x^4 - x = 0$$
$$x(x^3 - 1) = 0 \qquad \text{Factor out x.}$$
$$x(x - 1)(x^2 + x + 1) = 0 \qquad \text{Factor the difference of two cubes.}$$
$$x = 0 \quad \text{or} \quad x - 1 = 0 \quad \text{or} \quad x^2 + x + 1 = 0 \quad \text{Set each factor equal to zero}$$
$$x = 0 \quad \text{or} \qquad x = 1 \qquad\qquad\qquad\qquad \text{and solve. The equation}$$

and solve. The equation $x^2 + x + 1 = 0$ has no real solutions. (Do you see why?)

STEP 3: We use these zeros 0 and 1 to separate the real number line into three intervals:

$$(-\infty, 0) \qquad (0, 1) \qquad (1, \infty)$$

STEP 4: We choose a number in each interval and evaluate $f(x) = x^4 - x$ to determine if $f(x)$ is positive or negative. See Table 14.

Table 14

	0	1	
Interval	$(-\infty, 0)$	$(0, 1)$	$(1, \infty)$
Number chosen	-1	$\dfrac{1}{2}$	2
Value of f	$f(-1) = 2$	$f\left(\dfrac{1}{2}\right) = -\dfrac{7}{16}$	$f(2) = 14$
Conclusion	Positive	Negative	Positive

Based on Table 14, we know that $f(x) > 0$ for all x in $(-\infty, 0) \cup (1, \infty)$, that is, for all numbers x for which $x < 0$ or $x > 1$. Because the original inequality is strict, the solution set of the given inequality is $\{x | x < 0 \text{ or } x > 1\}$ or, using interval notation, $(-\infty, 0) \cup (1, \infty)$. Figure 41 shows the graph of the solution set.

Figure 41

━━━━ **Now Work** PROBLEM 17

2 Solve Rational Inequalities

| EXAMPLE 3 | **Solving a Rational Inequality** |

Solve the inequality $\dfrac{(x + 3)(2 - x)}{(x - 1)^2} > 0$, and graph the solution set.

Solution **STEP 1:** The domain of the variable x is $\{x \mid x \neq 1\}$. The inequality is already in a form with 0 on the right side.

STEP 2: Let $f(x) = \dfrac{(x + 3)(2 - x)}{(x - 1)^2}$. The real zeros of the numerator of f are -3 and 2; the real zero of the denominator is 1.

STEP 3: We use the zeros -3, 1, and 2, found in Step 2, to separate the real number line into four intervals:

$$(-\infty, -3) \qquad (-3, 1) \qquad (1, 2) \qquad (2, \infty)$$

STEP 4: Select a number in each interval and evaluate $f(x) = \dfrac{(x + 3)(2 - x)}{(x - 1)^2}$ to determine if $f(x)$ is positive or negative. See Table 15.

Table 15

Interval	$(-\infty, -3)$	$(-3, 1)$	$(1, 2)$	$(2, \infty)$
Number chosen	-4	0	$\dfrac{3}{2}$	3
Value of f	$f(-4) = -\dfrac{6}{25}$	$f(0) = 6$	$f\left(\dfrac{3}{2}\right) = 9$	$f(3) = -\dfrac{3}{2}$
Conclusion	Negative	Positive	Positive	Negative

Based on Table 15, we know that $f(x) > 0$ for all x in $(-3, 1) \cup (1, 2)$, that is, for all x such that $-3 < x < 1$ or $1 < x < 2$. Because the original inequality is strict, the solution set of the given inequality is $\{x \mid -3 < x < 2, x \neq 1\}$ or, using interval notation, $(-3, 1) \cup (1, 2)$. Figure 42 shows the graph of the solution set. Notice the hole at $x = 1$ to indicate that 1 is to be excluded.

Figure 42

━━━ **Now Work** PROBLEM 21

| EXAMPLE 4 | **Solving a Rational Inequality** |

Solve the inequality $\dfrac{4x + 5}{x + 2} \geq 3$, and graph the solution set.

Solution **STEP 1:** The domain of the variable x is $\{x \mid x \neq -2\}$. Rearrange the inequality so that 0 is on the right side. Then express the left side as a single quotient.

$$\frac{4x + 5}{x + 2} - 3 \geq 0 \quad \text{Subtract 3 from both sides of the inequality.}$$

$$\frac{4x + 5}{x + 2} - 3\left(\frac{x + 2}{x + 2}\right) \geq 0 \quad \text{Multiply } -3 \text{ by } \frac{x + 2}{x + 2}.$$

$$\frac{4x + 5 - 3x - 6}{x + 2} \geq 0 \quad \text{Write as a single quotient.}$$

$$\frac{x - 1}{x + 2} \geq 0 \quad \text{Combine like terms.}$$

The domain of the variable is $\{x \mid x \neq -2\}$.

STEP 2: Let $f(x) = \dfrac{x-1}{x+2}$. The zero of the numerator of f is 1, and the zero of the denominator is -2.

STEP 3: We use the zeros found in Step 2, -2 and 1, to separate the real number line into three intervals:

$$(-\infty, -2) \quad (-2, 1) \quad (1, \infty)$$

STEP 4: Select a number in each interval and evaluate $f(x) = \dfrac{4x+5}{x+2} - 3$ to determine if $f(x)$ is positive or negative. See Table 16.

Table 16

Interval	$(-\infty, -2)$	$(-2, 1)$	$(1, \infty)$
Number chosen	-3	0	2
Value of f	$f(-3) = 4$	$f(0) = -\dfrac{1}{2}$	$f(2) = \dfrac{1}{4}$
Conclusion	Positive	Negative	Positive

Based on Table 16, we know that $f(x) > 0$ for all x in $(-\infty, -2) \cup (1, \infty)$, that is, for all x such that $x < -2$ or $x > 1$. Because the original inequality is not strict, numbers x that satisfy the equation $f(x) = \dfrac{x-1}{x+2} = 0$ are also solutions of the inequality. Since $\dfrac{x-1}{x+2} = 0$ only if $x = 1$, we conclude that the solution set is $\{x \mid x < -2 \text{ or } x \geq 1\}$ or, using interval notation, $(-\infty, -2) \cup [1, \infty)$.

Figure 43

Figure 43 shows the graph of the solution set.

— **Now Work** PROBLEM 29

5.4 Assess Your Understanding

'Are You Prepared?' *The answer is given at the end of these exercises. If you get a wrong answer, read the pages listed in red.*

1. Solve the inequality: $3 - 4x > 5$. Graph the solution set. (pp. 124–131)

Concepts and Vocabulary

2. *True or False* The first step in solving the inequality $x^2 + 4x \geq -4$ is to factor the expression $x^2 + 4x$.

Skill Building

In Problems 3–40, solve each inequality.

3. $(x-5)^2(x+2) < 0$

4. $(x-5)(x+2)^2 > 0$

5. $x^3 - 4x^2 > 0$

6. $x^3 + 8x^2 < 0$

7. $x^3 - 9x \leq 0$

8. $x^3 - x \geq 0$

9. $2x^3 > -8x^2$

10. $3x^3 < -15x^2$

11. $(x-1)(x^2+x+4) \geq 0$

12. $(x+2)(x^2-x+1) > 0$

13. $(x-1)(x-2)(x-3) \leq 0$

14. $(x+1)(x+2)(x+3) \leq 0$

15. $x^3 - 2x^2 - 3x > 0$

16. $x^3 + 2x^2 - 3x > 0$

17. $x^4 > x^2$

18. $x^4 < 9x^2$

19. $x^4 > 1$

20. $x^3 > 1$

21. $\dfrac{x+1}{x-1} > 0$

22. $\dfrac{x-3}{x+1} > 0$

23. $\dfrac{(x-1)(x+1)}{x} \leq 0$

24. $\dfrac{(x-3)(x+2)}{x-1} \leq 0$

25. $\dfrac{(x-2)^2}{x^2-1} \geq 0$

26. $\dfrac{(x+5)^2}{x^2-4} \geq 0$

27. $6x - 5 < \dfrac{6}{x}$

28. $x + \dfrac{12}{x} < 7$

29. $\dfrac{x+4}{x-2} \le 1$ **30.** $\dfrac{x+2}{x-4} \ge 1$ **31.** $\dfrac{3x-5}{x+2} \le 2$ **32.** $\dfrac{x-4}{2x+4} \ge 1$

33. $\dfrac{1}{x-2} < \dfrac{2}{3x-9}$ **34.** $\dfrac{5}{x-3} > \dfrac{3}{x+1}$ **35.** $\dfrac{2x+5}{x+1} > \dfrac{x+1}{x-1}$ **36.** $\dfrac{1}{x+2} > \dfrac{3}{x+1}$

37. $\dfrac{x^2(3+x)(x+4)}{(x+5)(x-1)} \ge 0$ **38.** $\dfrac{x(x^2+1)(x-2)}{(x-1)(x+1)} \ge 0$ **39.** $\dfrac{(3-x)^3(2x+1)}{x^3-1} < 0$ **40.** $\dfrac{(2-x)^3(3x-2)}{x^3+1} < 0$

Application and Extensions

41. For what positive numbers will the cube of a number exceed four times its square?

42. For what positive numbers will the cube of a number be less than the number?

43. What is the domain of the function $f(x) = \sqrt{x^4 - 16}$?

44. What is the domain of the function $f(x) = \sqrt{x^3 - 3x^2}$?

45. What is the domain of the function $f(x) = \sqrt{\dfrac{x-2}{x+4}}$?

46. What is the domain of the function $f(x) = \sqrt{\dfrac{x-1}{x+4}}$?

In Problems 47–50, determine where the graph of f is below the graph of g by solving the inequality $f(x) \le g(x)$. Graph f and g together.

47. $f(x) = x^4 - 1$
$g(x) = -2x^2 + 2$

48. $f(x) = x^4 - 1$
$g(x) = x - 1$

49. $f(x) = x^4 - 4$
$g(x) = 3x^2$

50. $f(x) = x^4$
$g(x) = 2 - x^2$

51. Average Cost Suppose that the daily cost C of manufacturing bicycles is given by $C(x) = 80x + 5000$. Then the average daily cost $\overline{C}$ is given by $\overline{C}(x) = \dfrac{80x + 5000}{x}$. How many bicycles must be produced each day for the average cost to be no more than \$100?

52. Average Cost See Problem 51. Suppose that the government imposes a \$1000 per day tax on the bicycle manufacturer so that the daily cost C of manufacturing x bicycles is now given by $C(x) = 80x + 6000$. Now the average daily cost $\overline{C}$ is given by $\overline{C}(x) = \dfrac{80x + 6000}{x}$. How many bicycles must be produced each day for the average cost to be no more than \$100?

53. Bungee Jumping Originating on Pentecost Island in the Pacific, the practice of a person jumping from a high place harnessed to a flexible attachment was introduced to Western culture in 1979 by the Oxford University Dangerous Sport

Club. One important parameter to know before attempting a bungee jump is the amount the cord will stretch at the bottom of the fall. The stiffness of the cord is related to the amount of stretch by the equation

$$K = \dfrac{2W(S + L)}{S^2}$$

where W = weight of the jumper (pounds)
K = cord's stiffness (pounds per foot)
L = free length of the cord (feet)
S = stretch (feet)

(a) A 150-pound person plans to jump off a ledge attached to a cord of length 42 feet. If the stiffness of the cord is no less than 16 pounds per foot, how much will the cord stretch?

(b) If safety requirements will not permit the jumper to get any closer than 3 feet to the ground, what is the minimum height required for the ledge in part (a)?

Source: American Institute of Physics, Physics News Update, No. 150, November 5, 1993

54. Gravitational Force According to Newton's Law of universal gravitation, the attractive force F between two bodies is given by

$$F = G\dfrac{m_1 m_2}{r^2}$$

where m_1, m_2 are the masses of the two bodies
r = distance between the Two bodies
G = gravitational constant 6.6742×10^{-11} newtons meter2 kilogram^{-2}

Suppose an object is traveling directly from Earth to the moon. The mass of the Earth is 5.9742×10^{24} kilograms, the mass of the moon is 7.349×10^{22} kilograms and the mean distance from Earth to the moon is 384,400 kilometers. For an object between Earth and the moon, how far from Earth is the force on the object due to the moon greater than the force on the object due to Earth?

Source: www.solarviews.com;en.wikipedia.org

Discussion and Writing

55. Make up an inequality that has no solution. Make up one that has exactly one solution.

56. The inequality $x^4 + 1 < -5$ has no solution. Explain why.

57. A student attempted to solve the inequality $\dfrac{x+4}{x-3} \le 0$ by multiplying both sides of the inequality by $x - 3$ to get

$x + 4 \le 0$. This led to a solution of $\{x | x \le -4\}$. Is the student correct? Explain.

58. Write a rational inequality whose solution set is $\{x | -3 < x \le 5\}$.

'Are You Prepared?'

1. $\left\{ x \Big| x < -\dfrac{1}{2} \right\}$

5.5 The Real Zeros of a Polynomial Function

PREPARING FOR THIS SECTION *Before getting started, review the following:*

- Evaluating Functions (Section 3.1, pp. 212–214)
- Factoring Polynomials (Chapter R, Section R.5, pp. 49–55)
- Synthetic Division (Chapter R, Section R.6, pp. 57–60)
- Polynomial Division (Chapter R, Section R.4, pp. 44–47)
- Quadratic Formula (Section 1.2, pp. 102–104)

Now Work the 'Are You Prepared?' problems on page 386.

OBJECTIVES 1 Use the Remainder and Factor Theorems (p. 375)

2 Use Descartes' Rule of Signs to Determine the Number of Positive and the Number of Negative Real Zeros of a Polynomial Function (p. 378)

3 Use the Rational Zeros Theorem to List the Potential Rational Zeros of a Polynomial Function (p. 379)

4 Find the Real Zeros of a Polynomial Function (p. 381)

5 Solve Polynomial Equations (p. 382)

6 Use the Theorem for Bounds on Zeros (p. 383)

7 Use the Intermediate Value Theorem (p. 384)

In this section, we discuss techniques that can be used to find the real zeros of a polynomial function. Recall that if r is a real zero of a polynomial function f then $f(r) = 0$, r is an x-intercept of the graph of f, and r is a solution of the equation $f(x) = 0$.

For polynomial and rational functions, we have seen the importance of the real zeros for graphing. In most cases, however, the real zeros of a polynomial function are difficult to find using algebraic methods. No nice formulas like the quadratic formula are available to help us find zeros for polynomials of degree 3 or higher. Formulas do exist for solving any third- or fourth-degree polynomial equation, but they are somewhat complicated. No general formulas exist for polynomial equations of degree 5 or higher. Refer to the Historical Feature at the end of this section for more information.

1 Use the Remainder and Factor Theorems

When we divide one polynomial (the dividend) by another (the divisor), we obtain a quotient polynomial and a remainder, the remainder being either the zero polynomial or a polynomial whose degree is less than the degree of the divisor. To check our work, we verify that

$$(Quotient)(Divisor) + Remainder = Dividend$$

This checking routine is the basis for a famous theorem called the **division algorithm*** **for polynomials,** which we now state without proof.

THEOREM **Division Algorithm for Polynomials**

If $f(x)$ and $g(x)$ denote polynomial functions and if $g(x)$ is not the zero polynomial, there are unique polynomial functions $q(x)$ and $r(x)$ such that

$$\frac{f(x)}{g(x)} = q(x) + \frac{r(x)}{g(x)} \quad \text{or} \quad f(x) = q(x)g(x) + r(x) \quad \text{(1)}$$

<center>dividend quotient divisor remainder</center>

where $r(x)$ is either the zero polynomial or a polynomial of degree less than that of $g(x)$.

* A systematic process in which certain steps are repeated a finite number of times is called an **algorithm.** For example, long division is an algorithm.

In equation (1), $f(x)$ is the **dividend,** $g(x)$ is the **divisor,** $q(x)$ is the **quotient,** and $r(x)$ is the **remainder.**

If the divisor $g(x)$ is a first-degree polynomial of the form

$$g(x) = x - c, \quad c \text{ a real number}$$

then the remainder $r(x)$ is either the zero polynomial or a polynomial of degree 0. As a result, for such divisors, the remainder is some number, say R, and we may write

$$f(x) = (x - c)q(x) + R \qquad (2)$$

This equation is an identity in x and is true for all real numbers x. Suppose that $x = c$. Then equation (2) becomes

$$f(c) = (c - c)q(c) + R$$
$$f(c) = R$$

Substitute $f(c)$ for R in equation (2) to obtain

$$f(x) = (x - c)q(x) + f(c) \qquad (3)$$

We have now proved the **Remainder Theorem.**

REMAINDER THEOREM
Let f be a polynomial function. If $f(x)$ is divided by $x - c$, then the remainder is $f(c)$.

| EXAMPLE 1 | **Using the Remainder Theorem** |

Find the remainder if $f(x) = x^3 - 4x^2 - 5$ is divided by

(a) $x - 3$ \qquad (b) $x + 2$

Solution (a) We could use long division or synthetic division, but it is easier to use the Remainder Theorem, which says that the remainder is $f(3)$.

$$f(3) = (3)^3 - 4(3)^2 - 5 = 27 - 36 - 5 = -14$$

The remainder is -14.

(b) To find the remainder when $f(x)$ is divided by $x + 2 = x - (-2)$, we evaluate $f(-2)$.

$$f(-2) = (-2)^3 - 4(-2)^2 - 5 = -8 - 16 - 5 = -29$$

The remainder is -29.

Compare the method used in Example 1(a) with the method used in Example 1 of Chapter R, Section R.6. Which method do you prefer? Give reasons.

 COMMENT A graphing utility provides another way to find the value of a function using the eVALUEate feature. Consult your manual for details. Then check the results of Example 1. ▪

An important and useful consequence of the Remainder Theorem is the **Factor Theorem.**

FACTOR THEOREM
Let f be a polynomial function. Then $x - c$ is a factor of $f(x)$ if and only if $f(c) = 0$.

The Factor Theorem actually consists of two separate statements:

1. If $f(c) = 0$, then $x - c$ is a factor of $f(x)$.
2. If $x - c$ is a factor of $f(x)$, then $f(c) = 0$.

The proof requires two parts.

Proof

1. Suppose that $f(c) = 0$. Then, by equation (3), we have
$$f(x) = (x - c)q(x)$$
for some polynomial $q(x)$. That is, $x - c$ is a factor of $f(x)$.

2. Suppose that $x - c$ is a factor of $f(x)$. Then there is a polynomial function q such that
$$f(x) = (x - c)q(x)$$
Replacing x by c, we find that
$$f(c) = (c - c)q(c) = 0 \cdot q(c) = 0$$
This completes the proof. ∎

One use of the Factor Theorem is to determine whether a polynomial has a particular factor.

EXAMPLE 2	**Using the Factor Theorem**

Use the Factor Theorem to determine whether the function
$$f(x) = 2x^3 - x^2 + 2x - 3$$
has the factor

(a) $x - 1$ (b) $x + 3$

Solution The Factor Theorem states that if $f(c) = 0$ then $x - c$ is a factor.

(a) Because $x - 1$ is of the form $x - c$ with $c = 1$, we find the value of $f(1)$. We choose to use substitution.
$$f(1) = 2(1)^3 - (1)^2 + 2(1) - 3 = 2 - 1 + 2 - 3 = 0$$
By the Factor Theorem, $x - 1$ is a factor of $f(x)$.

(b) To test the factor $x + 3$, we first need to write it in the form $x - c$. Since $x + 3 = x - (-3)$, we find the value of $f(-3)$. We choose to use synthetic division.

$$
\begin{array}{r|rrrr}
-3 & 2 & -1 & 2 & -3 \\
 & & -6 & 21 & -69 \\
\hline
 & 2 & -7 & 23 & -72
\end{array}
$$

Because $f(-3) = -72 \neq 0$, we conclude from the Factor Theorem that $x - (-3) = x + 3$ is not a factor of $f(x)$.

━━━▶ **Now Work** PROBLEM 11

In Example 2(a), we found that $x - 1$ was a factor of f. To write f in factored form, we use long division or synthetic division. Using synthetic division, we find that

$$
\begin{array}{r|rrrr}
1 & 2 & -1 & 2 & -3 \\
 & & 2 & 1 & 3 \\
\hline
 & 2 & 1 & 3 & 0
\end{array}
$$

The quotient is $q(x) = 2x^2 + x + 3$ with a remainder of 0, as expected. We can write f in factored form as
$$f(x) = 2x^3 - x^2 + 2x - 3 = (x - 1)(2x^2 + x + 3)$$

The next theorem concerns the number of real zeros that a polynomial function may have. In counting the zeros of a polynomial, we count each zero as many times as its multiplicity.

THEOREM **Number of Real Zeros**

A polynomial function cannot have more real zeros than its degree.

Proof The proof is based on the Factor Theorem. If r is a real zero of a polynomial function f, then $f(r) = 0$ and, hence, $x - r$ is a factor of $f(x)$. Each real zero corresponds to a factor of degree 1. Because f cannot have more first-degree factors than its degree, the result follows. ∎

2 Use Descartes' Rule of Signs to Determine the Number of Positive and the Number of Negative Real Zeros of a Polynomial Function

Descartes' Rule of Signs provides information about the number and location of the real zeros of a polynomial function written in standard form (descending powers of x). It requires that we count the number of variations in the sign of the coefficients of $f(x)$ and $f(-x)$.

For example, the following polynomial function has two variations in the signs of the coefficients.

$$f(x) = -3x^7 + 4x^4 + 3x^2 - 2x - 1$$
$$= \underbrace{-3x^7 + 0x^6 + 0x^5 + 4x^4}_{- \text{ to } +} + 0x^3 + \underbrace{3x^2 - 2x}_{+ \text{ to } -} - 1$$

Notice that we ignored the zero coefficients in $0x^6, 0x^5$, and $0x^3$ in counting the number of variations in the sign of $f(x)$. Replacing x by $-x$, we get

$$f(-x) = -3(-x)^7 + 4(-x)^4 + 3(-x)^2 - 2(-x) - 1$$
$$= 3x^7 + 4x^4 + 3x^2 + \underbrace{2x - 1}_{+ \text{ to } -}$$

which has one variation in sign.

THEOREM **Descartes' Rule of Signs**

Let f denote a polynomial function written in standard form.

The number of positive real zeros of f either equals the number of variations in the sign of the nonzero coefficients of $f(x)$ or else equals that number less an even integer.

The number of negative real zeros of f either equals the number of variations in the sign of the nonzero coefficients of $f(-x)$ or else equals that number less an even integer.

We shall not prove Descartes' Rule of Signs. Let's see how it is used.

EXAMPLE 3 **Using the Number of Real Zeros Theorem and Descartes' Rule of Signs**

Discuss the real zeros of $f(x) = 3x^6 - 4x^4 + 3x^3 + 2x^2 - x - 3$.

Solution Because the polynomial is of degree 6, by the Number of Real Zeros Theorem there are at most six real zeros. Since there are three variations in the sign of the nonzero

coefficients of $f(x)$, by Descartes' Rule of Signs we expect either three or one positive real zeros. To continue, we look at $f(-x)$.

$$f(-x) = 3x^6 - 4x^4 - 3x^3 + 2x^2 + x - 3$$

There are three variations in sign, so we expect either three or one negative real zeros. Equivalently, we now know that the graph of f has either three or one positive x-intercepts and three or one negative x-intercepts. ∎

─── **Now Work** PROBLEM 21

3 Use the Rational Zeros Theorem to List the Potential Rational Zeros of a Polynomial Function

The next result, called the **Rational Zeros Theorem,** provides information about the rational zeros of a polynomial *with integer coefficients*.

THEOREM

Rational Zeros Theorem

Let f be a polynomial function of degree 1 or higher of the form

$$f(x) = a_n x^n + a_{n-1} x^{n-1} + \cdots + a_1 x + a_0 \qquad a_n \neq 0, \quad a_0 \neq 0$$

where each coefficient is an integer. If $\dfrac{p}{q}$, in lowest terms, is a rational zero of f, then p must be a factor of a_0, and q must be a factor of a_n. ⌐

EXAMPLE 4	**Listing Potential Rational Zeros**

List the potential rational zeros of

$$f(x) = 2x^3 + 11x^2 - 7x - 6$$

Solution

Because f has integer coefficients, we may use the Rational Zeros Theorem. First, we list all the integers p that are factors of the constant term $a_0 = -6$ and all the integers q that are factors of the leading coefficient $a_3 = 2$.

$$p: \quad \pm 1, \pm 2, \pm 3, \pm 6 \quad \text{Factors of } -6$$
$$q: \quad \pm 1, \pm 2 \qquad\qquad \text{Factors of } 2$$

Now we form all possible ratios $\dfrac{p}{q}$.

$$\frac{p}{q}: \quad \pm 1, \pm 2, \pm 3, \pm 6, \pm \frac{1}{2}, \pm \frac{3}{2}$$

If f has a rational zero, it will be found in this list, which contains 12 possibilities. ∎

─── **Now Work** PROBLEM 33

> **In Words**
> For the polynomial function $f(x) = 2x^3 + 11x^2 - 7x - 6$, we know 5 is not a zero, because 5 is not in the list of potential rational zeros. However, -1 may or may not be a zero.

Be sure that you understand what the Rational Zeros Theorem says: For a polynomial with integer coefficients, *if* there is a rational zero, it is one of those listed. It may be the case that the function does not have any rational zeros.

Long division, synthetic division, or substitution can be used to test each potential rational zero to determine whether it is indeed a zero. To make the work easier, integers are usually tested first. Let's continue this example.

| EXAMPLE 5 | Finding the Rational Zeros of a Polynomial Function |

Continue working with Example 4 to find the rational zeros of

$$f(x) = 2x^3 + 11x^2 - 7x - 6$$

Write f in factored form.

Solution We gather all the information that we can about the zeros.

STEP 1: There are at most three real zeros.

STEP 2: By Descartes' Rule of Signs, there is one positive real zero. Also, because

$$f(-x) = -2x^3 + 11x^2 + 7x - 6$$

there are two negative zeros or no negative zeros.

STEP 3: Now we use the list of potential rational zeros obtained in Example 4:

$$\pm 1, \pm 2, \pm 3, \pm 6, \pm \frac{1}{2}, \pm \frac{3}{2}.$$

We choose to test the potential rational zero 1 using substitution.

$$f(1) = 2(1)^3 + 11(1)^2 - 7(1) - 6 = 2 + 11 - 7 - 6 = 0$$

Since $f(1) = 0$, 1 is a zero and $x - 1$ is a factor of f. We can use long division or synthetic division to factor f.

$$f(x) = 2x^3 + 11x^2 - 7x - 6$$

$$= (x - 1)(2x^2 + 13x + 6)$$

Now any solution of the equation $2x^2 + 13x + 6 = 0$ will be a zero of f. Because of this, we call the equation $2x^2 + 13x + 6 = 0$ a **depressed equation** of f. Since the degree of the depressed equation of f is less than that of the original polynomial, it is easier to work with the depressed equation to find the zeros of f.

STEP 4: The depressed equation $2x^2 + 13x + 6 = 0$ is a quadratic equation with discriminant $b^2 - 4ac = 169 - 48 = 121 > 0$. The equation has two real solutions, which can be found by factoring.

$$2x^2 + 13x + 6 = (2x + 1)(x + 6) = 0$$

$$2x + 1 = 0 \quad \text{or} \quad x + 6 = 0$$

$$x = -\frac{1}{2} \qquad \qquad x = -6$$

The zeros of f are -6, $-\dfrac{1}{2}$, and 1.

We know from Step 3 that $f(x) = 2x^3 + 11x^2 - 7x - 6$ can be written as $f(x) = (x - 1)(2x^2 + 13x + 6)$. Because $2x^2 + 13x + 6 = (2x + 1)(x + 6)$, we write f in factored form as

$$f(x) = (x - 1)(2x + 1)(x + 6)$$

Notice that the three zeros of f found in this example are among those given in the list of potential rational zeros in Example 4.

4 Find the Real Zeros of a Polynomial Function

To obtain information about the real zeros of a polynomial function, follow these steps:

> **Steps for Finding the Real Zeros of a Polynomial Function**
>
> **STEP 1:** Use the degree of the polynomial to determine the maximum number of real zeros.
>
> **STEP 2:** Use Descartes' Rule of Signs to determine the possible number of positive zeros and negative zeros.
>
> **STEP 3:** (a) If the polynomial has integer coefficients, use the Rational Zeros Theorem to identify those rational numbers that potentially could be zeros.
>
> (b) Use substitution, synthetic division, or long division to test each potential rational zero. Each time that a zero (and thus a factor) is found, repeat Step 3 on the depressed equation.
>
> **STEP 4:** In attempting to find the zeros, remember to use (if possible) the factoring techniques that you already know (special products, factoring by grouping, and so on).

EXAMPLE 6 | **Finding the Real Zeros of a Polynomial Function**

Find the real zeros of $f(x) = x^5 - 5x^4 + 12x^3 - 24x^2 + 32x - 16$. Write f in factored form.

Solution We gather all the information that we can about the zeros.

STEP 1: There are at most five real zeros.

STEP 2: By Descartes' Rule of Signs, there are five, three, or one positive zeros. There are no negative zeros because

$$f(-x) = -x^5 - 5x^4 - 12x^3 - 24x^2 - 32x - 16$$

has no variation in sign.

STEP 3: Because the leading coefficient $a_5 = 1$ and there are no negative zeros, the potential rational zeros are the integers $1, 2, 4, 8,$ and 16, the positive factors of the constant term, 16. We test the potential rational zero 1 first, using synthetic division.

$$
\begin{array}{r|rrrrr}
1) & 1 & -5 & 12 & -24 & 32 & -16 \\
 & & 1 & -4 & 8 & -16 & 16 \\
\hline
 & 1 & -4 & 8 & -16 & 16 & 0
\end{array}
$$

The remainder is $f(1) = 0$, so 1 is a zero and $x - 1$ is a factor of f. Using the entries in the bottom row of the synthetic division, we can begin to factor f.

$$f(x) = x^5 - 5x^4 + 12x^3 - 24x^2 + 32x - 16$$
$$= (x - 1)(x^4 - 4x^3 + 8x^2 - 16x + 16)$$

We now work with the first depressed equation:

$$q_1(x) = x^4 - 4x^3 + 8x^2 - 16x + 16 = 0$$

REPEAT STEP 3: The potential rational zeros of q_1 are still $1, 2, 4, 8,$ and 16. We test 1 first, since it may be a repeated zero of f.

$$
\begin{array}{r|rrrrr}
1) & 1 & -4 & 8 & -16 & 16 \\
 & & 1 & -3 & 5 & -11 \\
\hline
 & 1 & -3 & 5 & -11 & 5
\end{array}
$$

Since the remainder is 5, 1 is not a repeated zero. We try 2 next.

$$
\begin{array}{r|rrrrr}
2) & 1 & -4 & 8 & -16 & 16 \\
 & & 2 & -4 & 8 & -16 \\
\hline
 & 1 & -2 & 4 & -8 & 0
\end{array}
$$

The remainder is $f(2) = 0$, so 2 is a zero and $x - 2$ is a factor of f. Again using the bottom row, we find

$$f(x) = x^5 - 5x^4 + 12x^3 - 24x^2 + 32x - 16$$
$$= (x - 1)(x - 2)(x^3 - 2x^2 + 4x - 8)$$

The remaining zeros satisfy the new depressed equation

$$q_2(x) = x^3 - 2x^2 + 4x - 8 = 0$$

Notice that $q_2(x)$ can be factored using grouping. (Alternatively, you could repeat Step 3 and check the potential rational zero 2.) Then

$$x^3 - 2x^2 + 4x - 8 = 0$$
$$x^2(x - 2) + 4(x - 2) = 0$$
$$(x^2 + 4)(x - 2) = 0$$
$$x^2 + 4 = 0 \quad \text{or} \quad x - 2 = 0$$
$$x = 2$$

Since $x^2 + 4 = 0$ has no real solutions, the real zeros of f are 1 and 2, the latter being a zero of multiplicity 2. The factored form of f is

$$f(x) = x^5 - 5x^4 + 12x^3 - 24x^2 + 32x - 16$$
$$= (x - 1)(x - 2)^2(x^2 + 4)$$

━━━━ **Now Work** PROBLEM 45

5 Solve Polynomial Equations

EXAMPLE 7 | **Solving a Polynomial Equation**

Find the real solutions of the equation: $x^5 - 5x^4 + 12x^3 - 24x^2 + 32x - 16 = 0$

Solution The real solutions of this equation are the real zeros of the polynomial function

$$f(x) = x^5 - 5x^4 + 12x^3 - 24x^2 + 32x - 16$$

Using the result of Example 6, the real zeros of f are 1 and 2. These are the real solutions of the equation

$$x^5 - 5x^4 + 12x^3 - 24x^2 + 32x - 16 = 0$$

━━━━ **Now Work** PROBLEM 57

In Example 6, the quadratic factor $x^2 + 4$ that appears in the factored form of f is called *irreducible,* because the polynomial $x^2 + 4$ cannot be factored over the real numbers. In general, we say that a quadratic factor $ax^2 + bx + c$ is **irreducible** if it cannot be factored over the real numbers, that is, if it is prime over the real numbers.

Refer to Examples 5 and 6. The polynomial function of Example 5 has three real zeros, and its factored form contains three linear factors. The polynomial function of Example 6 has two distinct real zeros, and its factored form contains two distinct linear factors and one irreducible quadratic factor.

THEOREM Every polynomial function with real coefficients can be uniquely factored into a product of linear factors and/or irreducible quadratic factors.

We shall prove this result in Section 5.6, and, in fact, we shall draw several additional conclusions about the zeros of a polynomial function. One conclusion is worth noting now. If a polynomial with real coefficients is of odd degree, it must contain at least one linear factor. (Do you see why?) This means that it must have at least one real zero.

THEOREM A polynomial function with real coefficients of odd degree has at least one real zero.

6 Use the Theorem for Bounds on Zeros

The search for the real zeros of a polynomial function can be reduced somewhat if *bounds* on the zeros are found. A number M is a **bound** on the zeros of a polynomial if every zero lies between $-M$ and M, inclusive. That is, M is a bound on the zeros of a polynomial f if

$$-M \leq \text{any real zero of } f \leq M$$

THEOREM **Bounds on Zeros**

Let f denote a polynomial function whose leading coefficient is 1.

$$f(x) = x^n + a_{n-1}x^{n-1} + \cdots + a_1 x + a_0$$

A bound M on the real zeros of f is the smaller of the two numbers

$$\boxed{\text{Max}\{1, |a_0| + |a_1| + \cdots + |a_{n-1}|\}, \quad 1 + \text{Max}\{|a_0|, |a_1|, \ldots, |a_{n-1}|\} \qquad \textbf{(4)}}$$

where Max{ } means "choose the largest entry in { }."

An example will help to make the theorem clear.

EXAMPLE 8 **Using the Theorem for Finding Bounds on Zeros**

Find a bound on the real zeros of each polynomial.

(a) $f(x) = x^5 + 3x^3 - 9x^2 + 5$

(b) $g(x) = 4x^5 - 2x^3 + 2x^2 + 1$

Solution (a) The leading coefficient of f is 1.

$$f(x) = x^5 + 3x^3 - 9x^2 + 5 \quad a_4 = 0, a_3 = 3, a_2 = -9, a_1 = 0, a_0 = 5$$

We evaluate the two expressions in (4).

$$\text{Max}\{1, |a_0| + |a_1| + \cdots + |a_{n-1}|\} = \text{Max}\{1, |5| + |0| + |-9| + |3| + |0|\}$$
$$= \text{Max}\{1, 17\} = 17$$
$$1 + \text{Max}\{|a_0|, |a_1|, \ldots, |a_{n-1}|\} = 1 + \text{Max}\{|5|, |0|, |-9|, |3|, |0|\}$$
$$= 1 + 9 = 10$$

The smaller of the two numbers, 10, is the bound. Every real zero of f lies between -10 and 10.

(b) First we write g so that it is the product of a constant times a polynomial whose leading coefficient is 1.

$$g(x) = 4x^5 - 2x^3 + 2x^2 + 1 = 4\left(x^5 - \frac{1}{2}x^3 + \frac{1}{2}x^2 + \frac{1}{4} \right)$$

Next we evaluate the two expressions in (4) with $a_4 = 0$, $a_3 = -\frac{1}{2}$, $a_2 = \frac{1}{2}$, $a_1 = 0$, and $a_0 = \frac{1}{4}$.

$$\text{Max}\{1, |a_0| + |a_1| + \cdots + |a_{n-1}|\} = \text{Max}\left\{ 1, \left|\frac{1}{4}\right| + |0| + \left|\frac{1}{2}\right| + \left|-\frac{1}{2}\right| + |0| \right\}$$

$$= \text{Max}\left\{ 1, \frac{5}{4} \right\} = \frac{5}{4}$$

$$1 + \text{Max}\{|a_0|, |a_1|, \ldots, |a_{n-1}|\} = 1 + \text{Max}\left\{ \left|\frac{1}{4}\right|, |0|, \left|\frac{1}{2}\right|, \left|-\frac{1}{2}\right|, |0| \right\}$$

$$= 1 + \frac{1}{2} = \frac{3}{2}$$

The smaller of the two numbers, $\frac{5}{4}$, is the bound. Every real zero of g lies between $-\frac{5}{4}$ and $\frac{5}{4}$.

COMMENT The bounds on the zeros of a polynomial provide good choices for setting Xmin and Xmax of the viewing-rectangle. With these choices, all the x-intercepts of the graph can be seen. ∎

━━━━ **Now Work** PROBLEM 81

7 Use the Intermediate Value Theorem

The next result, called the **Intermediate Value Theorem**, is based on the fact that the graph of a polynomial function is continuous; that is, it contains no "holes" or "gaps."

THEOREM

Intermediate Value Theorem

Let f denote a polynomial function. If $a < b$ and if $f(a)$ and $f(b)$ are of opposite sign, there is at least one real zero of f between a and b.

Although the proof of this result requires advanced methods in calculus, it is easy to "see" why the result is true. Look at Figure 44.

Figure 44
If $f(a) < 0$ and $f(b) > 0$, there is a zero between a and b.

EXAMPLE 9

Using the Intermediate Value Theorem to Locate Real Zeros

Show that $f(x) = x^5 - x^3 - 1$ has a zero between 1 and 2.

Solution We evaluate f at 1 and at 2.

$$f(1) = -1 \quad \text{and} \quad f(2) = 23$$

Because $f(1) < 0$ and $f(2) > 0$, it follows from the Intermediate Value Theorem that f has a zero between 1 and 2. ∎

━━━━━━▶ **Now Work** PROBLEM 89

Let's look at the polynomial f of Example 9 more closely. Based on Descartes' Rule of Signs, f has exactly one positive real zero. Based on the Rational Zeros Theorem, 1 is the only potential positive rational zero. Since $f(1) \neq 0$, we conclude that the zero between 1 and 2 is irrational. We can use the Intermediate Value Theorem to approximate it. The steps to use follow:

Approximating the Real Zeros of a Polynomial Function

STEP 1: Find two consecutive integers a and $a + 1$ such that f has a zero between them.

STEP 2: Divide the interval $[a, a + 1]$ into 10 equal subintervals.

STEP 3: Evaluate f at each endpoint of the subintervals until the Intermediate Value Theorem applies; this interval then contains a zero.

STEP 4: Repeat the process starting at Step 2 until the desired accuracy is achieved.

EXAMPLE 10	**Approximating the Real Zeros of a Polynomial Function**

Find the positive zero of $f(x) = x^5 - x^3 - 1$ correct to two decimal places.

Solution

Exploration We examine the polynomial f given in Example 10. The Theorem on Bounds of Zeros tells us that every zero is between -2 and 2. If we graph f using $-2 \leq x \leq 2$ (see Figure 45), we see that f has exactly one x-intercept. Using ZERO or ROOT, we find this zero to be 1.24 rounded to two decimal places. Correct to two decimal places, the zero is 1.23.

Figure 45

From Example 9 we know that the positive zero is between 1 and 2. We divide the interval $[1, 2]$ into 10 equal subintervals: $[1, 1.1]$, $[1.1, 1.2]$, $[1.2, 1.3]$, $[1.3, 1.4]$, $[1.4, 1.5]$, $[1.5, 1.6]$, $[1.6, 1.7]$, $[1.7, 1.8]$, $[1.8, 1.9]$, $[1.9, 2]$. Now we find the value of f at each endpoint until the Intermediate Value Theorem applies.

$$f(x) = x^5 - x^3 - 1$$

$$f(1.0) = -1 \qquad\qquad f(1.2) = -0.23968$$

$$f(1.1) = -0.72049 \qquad f(1.3) = 0.51593$$

We can stop here and conclude that the zero is between 1.2 and 1.3. Now we divide the interval $[1.2, 1.3]$ into 10 equal subintervals and proceed to evaluate f at each endpoint.

$$f(1.20) = -0.23968 \qquad f(1.23) \approx -0.0455613$$

$$f(1.21) \approx -0.1778185 \quad f(1.24) \approx 0.025001$$

$$f(1.22) \approx -0.1131398$$

We conclude that the zero lies between 1.23 and 1.24, and so, correct to two decimal places, the zero is 1.23. ∎

━━━━━━▶ **Now Work** PROBLEM 101

COMMENT The TABLE feature of a graphing calculator makes the computations in the solution to Example 10 a lot easier. ∎

There are many other numerical techniques for approximating the zeros of a polynomial. The one outlined in Example 10 (a variation of the *bisection method*) has the advantages that it will always work, that it can be programmed rather easily on a computer, and each time it is used another decimal place of accuracy is achieved. See Problem 115 for the bisection method, which places the zero in a succession of intervals, with each new interval being half the length of the preceding one.

Historical Feature

Formulas for the solution of third- and fourth-degree polynomial equations exist, and, while not very practical, they do have an interesting history.

In the 1500s in Italy, mathematical contests were a popular pastime, and persons possessing methods for solving problems kept them secret. (Solutions that were published were already common knowledge.) Niccolo of Brescia (1499–1557), commonly referred to as Tartaglia ("the stammerer"), had the secret for solving cubic (third-degree) equations, which gave him a decided advantage in the contests. Girolamo Cardano (1501–1576) found out that Tartaglia had the secret, and, being interested in cubics, he requested it from Tartaglia. The reluctant Tartaglia hesitated for some time, but finally, swearing Cardano to secrecy with midnight oaths by candlelight, told him the secret. Cardano then pub-

lished the solution in his book *Ars Magna* (1545), giving Tartaglia the credit but rather compromising the secrecy. Tartaglia exploded into bitter recriminations, and each wrote pamphlets that reflected on the other's mathematics, moral character, and ancestry.

The quartic (fourth-degree) equation was solved by Cardano's student Lodovico Ferrari, and this solution also was included, with credit and this time with permission, in the *Ars Magna*.

Attempts were made to solve the fifth-degree equation in similar ways, all of which failed. In the early 1800s, P. Ruffini, Niels Abel, and Evariste Galois all found ways to show that it is not possible to solve fifth-degree equations by formula, but the proofs required the introduction of new methods. Galois's methods eventually developed into a large part of modern algebra.

Historical Problems

Problems 1–8 develop the Tartaglia–Cardano solution of the cubic equation and show why it is not altogether practical.

1. Show that the general cubic equation $y^3 + by^2 + cy + d = 0$ can be transformed into an equation of the form $x^3 + px + q = 0$ by using the substitution $y = x - \dfrac{b}{3}$.

2. In the equation $x^3 + px + q = 0$, replace x by $H + K$. Let $3HK = -p$, and show that $H^3 + K^3 = -q$.

3. Based on Problem 2, we have the two equations

$$3HK = -p \quad \text{and} \quad H^3 + K^3 = -q$$

Solve for K in $3HK = -p$ and substitute into $H^3 + K^3 = -q$. Then show that

$$H = \sqrt[3]{\dfrac{-q}{2} + \sqrt{\dfrac{q^2}{4} + \dfrac{p^3}{27}}}$$

[**Hint:** Look for an equation that is quadratic in form.]

4. Use the solution for H from Problem 3 and the equation $H^3 + K^3 = -q$ to show that

$$K = \sqrt[3]{\dfrac{-q}{2} - \sqrt{\dfrac{q^2}{4} + \dfrac{p^3}{27}}}$$

5. Use the results from Problems 2 to 4 to show that the solution of $x^3 + px + q = 0$ is

$$x = \sqrt[3]{\dfrac{-q}{2} + \sqrt{\dfrac{q^2}{4} + \dfrac{p^3}{27}}} + \sqrt[3]{\dfrac{-q}{2} - \sqrt{\dfrac{q^2}{4} + \dfrac{p^3}{27}}}$$

6. Use the result of Problem 5 to solve the equation $x^3 - 6x - 9 = 0$.

7. Use a calculator and the result of Problem 5 to solve the equation $x^3 + 3x - 14 = 0$.

8. Use the methods of this chapter to solve the equation $x^3 + 3x - 14 = 0$.

5.5 Assess Your Understanding

'Are You Prepared?' *Answers are given at the end of these exercises. If you get a wrong answer, read the pages listed in red.*

1. Find $f(-1)$ if $f(x) = 2x^2 - x$. (pp. 210–214)
2. Factor the expression $6x^2 + x - 2$. (pp. 49–55)
3. Find the quotient and remainder if $3x^4 - 5x^3 + 7x - 4$ is divided by $x - 3$. (pp. 44–47 or 57–60)
4. Solve the equation $x^2 + x - 3 = 0$. (pp. 102–104)

Concepts and Vocabulary

5. In the process of polynomial division, (Divisor)(Quotient) + _____ = _____.

6. When a polynomial function f is divided by $x - c$, the remainder is _____.

7. If a function f, whose domain is all real numbers, is even and if 4 is a zero of f, then _____ is also a zero.

8. *True or False* Every polynomial function of degree 3 with real coefficients has exactly three real zeros.

9. *True or False* The only potential rational zeros of $f(x) = 2x^5 - x^3 + x^2 - x + 1$ are $\pm 1, \pm 2$.

10. *True or False* If f is a polynomial function of degree 4 and if $f(2) = 5$, then

$$\dfrac{f(x)}{x - 2} = p(x) + \dfrac{5}{x - 2}$$

where $p(x)$ is a polynomial of degree 3.

Skill Building

In Problems 11–20, use the Remainder Theorem to find the remainder when $f(x)$ is divided by $x - c$. Then use the Factor Theorem to determine whether $x - c$ is a factor of $f(x)$.

11. $f(x) = 4x^3 - 3x^2 - 8x + 4;\quad x - 2$
12. $f(x) = -4x^3 + 5x^2 + 8;\quad x + 3$

13. $f(x) = 3x^4 - 6x^3 - 5x + 10; \quad x - 2$

14. $f(x) = 4x^4 - 15x^2 - 4; \quad x - 2$

15. $f(x) = 3x^6 + 82x^3 + 27; \quad x + 3$

16. $f(x) = 2x^6 - 18x^4 + x^2 - 9; \quad x + 3$

17. $f(x) = 4x^6 - 64x^4 + x^2 - 15; \quad x + 4$

18. $f(x) = x^6 - 16x^4 + x^2 - 16; \quad x + 4$

19. $f(x) = 2x^4 - x^3 + 2x - 1; \quad x - \dfrac{1}{2}$

20. $f(x) = 3x^4 + x^3 - 3x + 1; \quad x + \dfrac{1}{3}$

In Problems 21–32, tell the maximum number of real zeros that each polynomial function may have. Then use Descartes' Rule of Signs to determine how many positive and how many negative zeros each polynomial function may have. Do not attempt to find the zeros.

21. $f(x) = -4x^7 + x^3 - x^2 + 2$

22. $f(x) = 5x^4 + 2x^2 - 6x - 5$

23. $f(x) = 2x^6 - 3x^2 - x + 1$

24. $f(x) = -3x^5 + 4x^4 + 2$

25. $f(x) = 3x^3 - 2x^2 + x + 2$

26. $f(x) = -x^3 - x^2 + x + 1$

27. $f(x) = -x^4 + x^2 - 1$

28. $f(x) = x^4 + 5x^3 - 2$

29. $f(x) = x^5 + x^4 + x^2 + x + 1$

30. $f(x) = x^5 - x^4 + x^3 - x^2 + x - 1$

31. $f(x) = x^6 - 1$

32. $f(x) = x^6 + 1$

In Problems 33–44, list the potential rational zeros of each polynomial function. Do not attempt to find the zeros.

33. $f(x) = 3x^4 - 3x^3 + x^2 - x + 1$

34. $f(x) = x^5 - x^4 + 2x^2 + 3$

35. $f(x) = x^5 - 6x^2 + 9x - 3$

36. $f(x) = 2x^5 - x^4 - x^2 + 1$

37. $f(x) = -4x^3 - x^2 + x + 2$

38. $f(x) = 6x^4 - x^2 + 2$

39. $f(x) = 6x^4 - x^2 + 9$

40. $f(x) = -4x^3 + x^2 + x + 6$

41. $f(x) = 2x^5 - x^3 + 2x^2 + 12$

42. $f(x) = 3x^5 - x^2 + 2x + 18$

43. $f(x) = 6x^4 + 2x^3 - x^2 + 20$

44. $f(x) = -6x^3 - x^2 + x + 10$

In Problems 45–56, use the Rational Zeros Theorem to find all the real zeros of each polynomial function. Use the zeros to factor f over the real numbers.

45. $f(x) = x^3 + 2x^2 - 5x - 6$

46. $f(x) = x^3 + 8x^2 + 11x - 20$

47. $f(x) = 2x^3 - x^2 + 2x - 1$

48. $f(x) = 2x^3 + x^2 + 2x + 1$

49. $f(x) = 2x^3 - 4x^2 - 10x + 20$

50. $f(x) = 3x^3 + 6x^2 - 15x - 30$

51. $f(x) = 2x^4 + x^3 - 7x^2 - 3x + 3$

52. $f(x) = 2x^4 - x^3 - 5x^2 + 2x + 2$

53. $f(x) = x^4 + x^3 - 3x^2 - x + 2$

54. $f(x) = x^4 - x^3 - 6x^2 + 4x + 8$

55. $f(x) = 4x^4 + 5x^3 + 9x^2 + 10x + 2$

56. $f(x) = 3x^4 + 4x^3 + 7x^2 + 8x + 2$

In Problems 57–68, solve each equation in the real number system.

57. $x^4 - x^3 + 2x^2 - 4x - 8 = 0$

58. $2x^3 + 3x^2 + 2x + 3 = 0$

59. $3x^3 + 4x^2 - 7x + 2 = 0$

60. $2x^3 - 3x^2 - 3x - 5 = 0$

61. $3x^3 - x^2 - 15x + 5 = 0$

62. $2x^3 - 11x^2 + 10x + 8 = 0$

63. $x^4 + 4x^3 + 2x^2 - x + 6 = 0$

64. $x^4 - 2x^3 + 10x^2 - 18x + 9 = 0$

65. $x^3 - \dfrac{2}{3}x^2 + \dfrac{8}{3}x + 1 = 0$

66. $x^3 + \dfrac{3}{2}x^2 + 3x - 2 = 0$

67. $2x^4 - 19x^3 + 57x^2 - 64x + 20 = 0$

68. $2x^4 + x^3 - 24x^2 + 20x + 16 = 0$

In Problems 69–80, graph each polynomial function.

69. $f(x) = x^3 + 2x^2 - 5x - 6$

70. $f(x) = x^3 + 8x^2 + 11x - 20$

71. $f(x) = 2x^3 - x^2 + 2x - 1$

72. $f(x) = 2x^3 + x^2 + 2x + 1$

73. $f(x) = x^4 + x^2 - 2$

74. $f(x) = x^4 - 3x^2 - 4$

75. $f(x) = 4x^4 + 7x^2 - 2$

76. $f(x) = 4x^4 + 15x^2 - 4$

77. $f(x) = x^4 + x^3 - 3x^2 - x + 2$

78. $f(x) = x^4 - x^3 - 6x^2 + 4x + 8$

79. $f(x) = 4x^5 - 8x^4 - x + 2$

80. $f(x) = 4x^5 + 12x^4 - x - 3$

In Problems 81–88, find bounds on the real zeros of each polynomial function.

81. $f(x) = x^4 - 3x^2 - 4$

82. $f(x) = x^4 - 5x^2 - 36$

83. $f(x) = x^4 + x^3 - x - 1$

84. $f(x) = x^4 - x^3 + x - 1$

85. $f(x) = 3x^4 + 3x^3 - x^2 - 12x - 12$

86. $f(x) = 3x^4 - 3x^3 - 5x^2 + 27x - 36$

87. $f(x) = 4x^5 - x^4 + 2x^3 - 2x^2 + x - 1$

88. $f(x) = 4x^5 + x^4 + x^3 + x^2 - 2x - 2$

In Problems 89–94, use the Intermediate Value Theorem to show that each polynomial function has a zero in the given interval.

89. $f(x) = 8x^4 - 2x^2 + 5x - 1$; $[0, 1]$

90. $f(x) = x^4 + 8x^3 - x^2 + 2$; $[-1, 0]$

91. $f(x) = 2x^3 + 6x^2 - 8x + 2$; $[-5, -4]$

92. $f(x) = 3x^3 - 10x + 9$; $[-3, -2]$

93. $f(x) = x^5 - x^4 + 7x^3 - 7x^2 - 18x + 18$; $[1.4, 1.5]$

94. $f(x) = x^5 - 3x^4 - 2x^3 + 6x^2 + x + 2$; $[1.7, 1.8]$

In Problems 95–98, each equation has a solution r in the interval indicated. Use the method of Example 10 to approximate this solution correct to two decimal places.

95. $8x^4 - 2x^2 + 5x - 1 = 0$; $0 \le r \le 1$

96. $x^4 + 8x^3 - x^2 + 2 = 0$; $-1 \le r \le 0$

97. $2x^3 + 6x^2 - 8x + 2 = 0$; $-5 \le r \le -4$

98. $3x^3 - 10x + 9 = 0$; $-3 \le r \le -2$

In Problems 99–102, each polynomial function has exactly one positive zero. Use the method of Example 10 to approximate the zero correct to two decimal places.

99. $f(x) = x^3 + x^2 + x - 4$

100. $f(x) = 2x^4 + x^2 - 1$

101. $f(x) = 2x^4 - 3x^3 - 4x^2 - 8$

102. $f(x) = 3x^3 - 2x^2 - 20$

Applications and Extensions

103. Find k such that $f(x) = x^3 - kx^2 + kx + 2$ has the factor $x - 2$.

104. Find k such that $f(x) = x^4 - kx^3 + kx^2 + 1$ has the factor $x + 2$.

105. What is the remainder when $f(x) = 2x^{20} - 8x^{10} + x - 2$ is divided by $x - 1$?

106. What is the remainder when $f(x) = -3x^{17} + x^9 - x^5 + 2x$ is divided by $x + 1$?

107. Use the Factor Theorem to prove that $x - c$ is a factor of $x^n - c^n$ for any positive integer n.

108. Use the Factor Theorem to prove that $x + c$ is a factor of $x^n + c^n$ if $n \ge 1$ is an odd integer.

109. One solution of the equation $x^3 - 8x^2 + 16x - 3 = 0$ is 3. Find the sum of the remaining solutions.

110. One solution of the equation $x^3 + 5x^2 + 5x - 2 = 0$ is -2. Find the sum of the remaining solutions.

111. Geometry What is the length of the edge of a cube if, after a slice 1 inch thick is cut from one side, the volume remaining is 294 cubic inches?

112. Geometry What is the length of the edge of a cube if its volume could be doubled by an increase of 6 centimeters in one edge, an increase of 12 centimeters in a second edge, and a decrease of 4 centimeters in the third edge?

113. Let $f(x)$ be a polynomial function whose coefficients are integers. Suppose that r is a real zero of f and that the leading coefficient of f is 1. Use the Rational Zeros Theorem to show that r is either an integer or an irrational number.

114. Prove the Rational Zeros Theorem.

[Hint: Let $\dfrac{p}{q}$, where p and q have no common factors except 1 and -1, be a zero of the polynomial function

$$f(x) = a_n x^n + a_{n-1} x^{n-1} + \cdots + a_1 x + a_0$$

whose coefficients are all integers. Show that

$$a_n p^n + a_{n-1} p^{n-1} q + \cdots + a_1 p q^{n-1} + a_0 q^n = 0$$

Now, because p is a factor of the first n terms of this equation, p must also be a factor of the term $a_0 q^n$. Since p is not a factor of q (why?), p must be a factor of a_0. Similarly, q must be a factor of a_n.**]**

115. Bisection Method for Approximating Zeros of a Function f We begin with two consecutive integers, a and $a + 1$, such that $f(a)$ and $f(a + 1)$ are of opposite sign. Evaluate f at the midpoint m_1 of a and $a + 1$. If $f(m_1) = 0$, then m_1 is the zero of f, and we are finished. Otherwise, $f(m_1)$ is of opposite sign to either $f(a)$ or $f(a + 1)$. Suppose that it is $f(a)$ and $f(m_1)$ that are of opposite sign. Now evaluate f at the midpoint m_2 of a and m_1. Repeat this process until the

desired degree of accuracy is obtained. Note that each iteration places the zero in an interval whose length is half that of the previous interval. Use the bisection method to solve Problems 95–102.

[**Hint:** The process ends when both endpoints agree to the desired number of decimal places.]

Discussion and Writing

116. Is $\dfrac{1}{3}$ a zero of $f(x) = 2x^3 + 3x^2 - 6x + 7$? Explain.

117. Is $\dfrac{1}{3}$ a zero of $f(x) = 4x^3 - 5x^2 - 3x + 1$? Explain.

118. Is $\dfrac{3}{5}$ a zero of $f(x) = 2x^6 - 5x^4 + x^3 - x + 1$? Explain.

119. Is $\dfrac{2}{3}$ a zero of $f(x) = x^7 + 6x^5 - x^4 + x + 2$? Explain.

'Are You Prepared?' Answers

1. 3 **2.** $(3x + 2)(2x - 1)$ **3.** Quotient: $3x^3 + 4x^2 + 12x + 43$; Remainder: 125 **4.** $\left\{ \dfrac{-1 - \sqrt{13}}{2}, \dfrac{-1 + \sqrt{13}}{2} \right\}$

5.6 Complex Zeros; Fundamental Theorem of Algebra

PREPARING FOR THIS SECTION *Before getting started, review the following:*

- Complex Numbers (Section 1.3, pp. 109–114)
- Quadratic Equations in the Complex Number System (Section 1.3, pp. 114–116)

Now Work the 'Are You Prepared?' problems on page 393.

OBJECTIVES **1** Use the Conjugate Pairs Theorem (p. 390)
2 Find a Polynomial Function with Specified Zeros (p. 391)
3 Find the Complex Zeros of a Polynomial (p. 392)

In Section 1.2, we found the real solutions of a quadratic equation. That is, we found the real zeros of a polynomial function of degree 2. Then, in Section 1.3 we found the solutions of a quadratic equation in the complex number system. These are referred to as the *complex zeros* of a polynomial function of degree 2.

In Section 5.5 we found the real zeros of polynomial functions of degree 3 or higher. In this section we will find the *complex zeros* of polynomial functions of degree 3 or higher.

DEFINITION A variable in the complex number system is referred to as a **complex variable**. A **complex polynomial function** f of degree n is a function of the form

$$f(x) = a_n x^n + a_{n-1} x^{n-1} + \cdots + a_1 x + a_0 \qquad \textbf{(1)}$$

where $a_n, a_{n-1}, \ldots, a_1, a_0$ are complex numbers, $a_n \neq 0$, n is a nonnegative integer, and x is a complex variable. As before, a_n is called the **leading coefficient** of f. A complex number r is called a **complex zero** of f if $f(r) = 0$.

In most of our work the coefficients in (1) will be real numbers.

We have learned that some quadratic equations have no real solutions, but that in the complex number system every quadratic equation has a solution, either real or complex. The next result, proved by Karl Friedrich Gauss (1777–1855) when he was 22 years old,* gives an extension to complex polynomials. In fact, this result is so important and useful that it has become known as the **Fundamental Theorem of Algebra.**

* In all, Gauss gave four different proofs of this theorem, the first one in 1799 being the subject of his doctoral dissertation.

FUNDAMENTAL THEOREM OF ALGEBRA

Every complex polynomial function $f(x)$ of degree $n \geq 1$ has at least one complex zero.

We shall not prove this result, as the proof is beyond the scope of this book. However, using the Fundamental Theorem of Algebra and the Factor Theorem, we can prove the following result:

THEOREM

Every complex polynomial function $f(x)$ of degree $n \geq 1$ can be factored into n linear factors (not necessarily distinct) of the form

$$f(x) = a_n(x - r_1)(x - r_2) \cdot \cdots \cdot (x - r_n) \qquad (2)$$

where $a_n, r_1, r_2, \ldots, r_n$ are complex numbers. That is, every complex polynomial function of degree $n \geq 1$ has exactly n complex zeros, some of which may repeat.

Proof Let

$$f(x) = a_n x^n + a_{n-1} x^{n-1} + \cdots + a_1 x + a_0$$

By the Fundamental Theorem of Algebra, f has at least one zero, say r_1. Then, by the Factor Theorem, $x - r_1$ is a factor, and

$$f(x) = (x - r_1)q_1(x)$$

where $q_1(x)$ is a complex polynomial of degree $n - 1$ whose leading coefficient is a_n. Repeating this argument n times, we arrive at

$$f(x) = (x - r_1)(x - r_2) \cdot \cdots \cdot (x - r_n)q_n(x)$$

where $q_n(x)$ is a complex polynomial of degree $n - n = 0$ whose leading coefficient is a_n. That is, $q_n(x) = a_n x^0 = a_n$, and so

$$f(x) = a_n(x - r_1)(x - r_2) \cdot \cdots \cdot (x - r_n)$$

We conclude that every complex polynomial function $f(x)$ of degree $n \geq 1$ has exactly n (not necessarily distinct) zeros. ∎

1 Use the Conjugate Pairs Theorem

We can use the Fundamental Theorem of Algebra to obtain valuable information about the complex zeros of polynomials whose coefficients are real numbers.

CONJUGATE PAIRS THEOREM

Let $f(x)$ be a polynomial whose coefficients are real numbers. If $r = a + bi$ is a zero of f, the complex conjugate $\bar{r} = a - bi$ is also a zero of f.

In other words, for polynomials whose coefficients are real numbers, the complex zeros occur in conjugate pairs. This result should not be all that surprising since the complex zeros of a quadratic function occurred in conjugate pairs.

Proof Let

$$f(x) = a_n x^n + a_{n-1} x^{n-1} + \cdots + a_1 x + a_0$$

where $a_n, a_{n-1}, \ldots, a_1, a_0$ are real numbers and $a_n \neq 0$. If $r = a + bi$ is a zero of f, then $f(r) = f(a + bi) = 0$, so

$$a_n r^n + a_{n-1} r^{n-1} + \cdots + a_1 r + a_0 = 0$$

We take the conjugate of both sides to get

$$\overline{a_n r^n + a_{n-1} r^{n-1} + \cdots + a_1 r + a_0} = \overline{0}$$

$$\overline{a_n r^n} + \overline{a_{n-1} r^{n-1}} + \cdots + \overline{a_1 r} + \overline{a_0} = \overline{0} \quad \text{\textit{The conjugate of a sum equals the sum}}$$
of the conjugates (see Section 1.3).

$$\overline{a_n}(\overline{r})^n + \overline{a_{n-1}}(\overline{r})^{n-1} + \cdots + \overline{a_1}\overline{r} + \overline{a_0} = \overline{0} \quad \text{\textit{The conjugate of a product equals the}}$$
product of the conjugates.

$$a_n(\overline{r})^n + a_{n-1}(\overline{r})^{n-1} + \cdots + a_1\overline{r} + a_0 = 0 \quad \text{\textit{The conjugate of a real number equals the}}$$
real number.

This last equation states that $f(\overline{r}) = 0$; that is, $\overline{r} = a - bi$ is a zero of f. ∎

The importance of this result should be clear. Once we know that, say, $3 + 4i$ is a zero of a polynomial with real coefficients, then we know that $3 - 4i$ is also a zero. This result has an important corollary.

COROLLARY A polynomial f of odd degree with real coefficients has at least one real zero.

Proof Because complex zeros occur as conjugate pairs in a polynomial with real coefficients, there will always be an even number of zeros that are not real numbers. Consequently, since f is of odd degree, one of its zeros has to be a real number. ∎

For example, the polynomial $f(x) = x^5 - 3x^4 + 4x^3 - 5$ has at least one zero that is a real number, since f is of degree 5 (odd) and has real coefficients.

EXAMPLE 1 Using the Conjugate Pairs Theorem

A polynomial f of degree 5 whose coefficients are real numbers has the zeros 1, $5i$, and $1 + i$. Find the remaining two zeros.

Solution Since f has coefficients that are real numbers, complex zeros appear as conjugate pairs. It follows that $-5i$, the conjugate of $5i$, and $1 - i$, the conjugate of $1 + i$, are the two remaining zeros. ∎

━━━━━▶ **Now Work** PROBLEM 7

2 Find a Polynomial Function with Specified Zeros

EXAMPLE 2 Finding a Polynomial Function Whose Zeros Are Given

Find a polynomial f of degree 4 whose coefficients are real numbers that has the zeros 1, 1, and $-4 + i$.

Solution Since $-4 + i$ is a zero, by the Conjugate Pairs Theorem, $-4 - i$ must also be a zero of f. Because of the Factor Theorem, if $f(c) = 0$, then $x - c$ is a factor of $f(x)$. So we can now write f as

$$f(x) = a(x - 1)(x - 1)[x - (-4 + i)][x - (-4 - i)]$$

where a is any real number. Then

$$f(x) = a(x - 1)(x - 1)[x - (-4 + i)][x - (-4 - i)]$$
$$= a(x^2 - 2x + 1)[x^2 - (-4 + i)x - (-4 - i)x + (-4 + i)(-4 - i)]$$
$$= a(x^2 - 2x + 1)(x^2 + 4x - ix + 4x + ix + 16 + 4i - 4i - i^2)$$
$$= a(x^2 - 2x + 1)(x^2 + 8x + 17)$$
$$= a(x^4 + 8x^3 + 17x^2 - 2x^3 - 16x^2 - 34x + x^2 + 8x + 17)$$
$$= a(x^4 + 6x^3 + 2x^2 - 26x + 17)$$

Exploration

Graph the function f found in Example 2 for $a = 1$. Does the value of a affect the zeros of f? How does the value of a affect the graph of f? What information about f is sufficient to uniquely determine a?

Figure 46

Result A quick analysis of the polynomial f tells us what to expect:

At most three turning points.

For large $|x|$, the graph will behave like $y = x^4$.

A repeated real zero at 1, so the graph will touch the x-axis at 1.

The only x-intercept is at 1; the y-intercept is 17.

Figure 46 shows the complete graph. (Do you see why? The graph has exactly three turning points.) The value of a causes a stretch or compression; a reflection also occurs if $a < 0$. The zeros are not affected.

If any point other than an x-intercept on the graph of f is known, then a can be determined. For example, if $(2, 3)$ is on the graph, then $f(2) = 3 = a(37)$, so $a = 3/37$. Why won't an x-intercept work?

Now we can prove the theorem we conjectured in Section 5.5.

THEOREM

Every polynomial function with real coefficients can be uniquely factored over the real numbers into a product of linear factors and/or irreducible quadratic factors.

Proof Every complex polynomial f of degree n has exactly n zeros and can be factored into a product of n linear factors. If its coefficients are real, those zeros that are complex numbers will always occur as conjugate pairs. As a result, if $r = a + bi$ is a complex zero, then so is $\bar{r} = a - bi$. Consequently, when the linear factors $x - r$ and $x - \bar{r}$ of f are multiplied, we have

$$(x - r)(x - \bar{r}) = x^2 - (r + \bar{r})x + r\bar{r} = x^2 - 2ax + a^2 + b^2$$

This second-degree polynomial has real coefficients and is irreducible (over the real numbers). Thus, the factors of f are either linear or irreducible quadratic factors. ∎

Now Work PROBLEM 17

3 Find the Complex Zeros of a Polynomial

EXAMPLE 3

Finding the Complex Zeros of a Polynomial

Find the complex zeros of the polynomial function

$$f(x) = 3x^4 + 5x^3 + 25x^2 + 45x - 18$$

Write f in factored form.

Solution

STEP 1: The degree of f is 4. So f will have four complex zeros.

STEP 2: Descartes' Rule of Signs provides information about the real zeros. For this polynomial, there is one positive real zero. There are three or one negative real zeros because

$$f(-x) = 3x^4 - 5x^3 + 25x^2 - 45x - 18$$

has three variations in sign.

STEP 3: The Rational Zeros Theorem provides information about the potential rational zeros of polynomials with integer coefficients. For this polynomial (which has integer coefficients), the potential rational zeros are

$$\pm \frac{1}{3}, \quad \pm \frac{2}{3}, \quad \pm 1, \quad \pm 2, \quad \pm 3, \quad \pm 6, \quad \pm 9, \quad \pm 18$$

We test 1 first:

$$
\begin{array}{r|rrrr}
1) & 3 & 5 & 25 & 45 & -18 \\
 & & 3 & 8 & 33 & 78 \\
\hline
 & 3 & 8 & 33 & 78 & 60 \\
\end{array}
$$
1 is not a zero.

We test -1:

$$
\begin{array}{r|rrrr}
-1) & 3 & 5 & 25 & 45 & -18 \\
 & & -3 & -2 & -23 & -22 \\
\hline
 & 3 & 2 & 23 & 22 & -40 \\
\end{array}
$$
-1 is not a zero.

We test 2:

$$
\begin{array}{r|rrrr}
2) & 3 & 5 & 25 & 45 & -18 \\
 & & 6 & 22 & 94 & 278 \\
\hline
 & 3 & 11 & 47 & 139 & 260 \\
\end{array}
$$
2 is not a zero.

We test -2:

$$
\begin{array}{r|rrrr}
-2) & 3 & 5 & 25 & 45 & -18 \\
 & & -6 & 2 & -54 & 18 \\
\hline
 & 3 & -1 & 27 & -9 & 0 \\
\end{array}
$$

Since $f(-2) = 0$, then -2 is a zero and $x + 2$ is a factor of f. The depressed equation is

$$3x^3 - x^2 + 27x - 9 = 0$$

STEP 4: We factor the depressed equation by grouping.

$$3x^3 - x^2 + 27x - 9 = 0$$
$$x^2(3x - 1) + 9(3x - 1) = 0 \quad \text{Factor } x^2 \text{ from } 3x^3 - x^2 \text{ and 9 from } 27x - 9.$$
$$(x^2 + 9)(3x - 1) = 0 \quad \text{Factor out the common factor } 3x - 1.$$
$$x^2 + 9 = 0 \quad \text{or} \quad 3x - 1 = 0 \quad \text{Apply the Zero-Product Property.}$$
$$x^2 = -9 \quad \text{or} \quad x = \frac{1}{3}$$
$$x = -3i, \quad x = 3i \quad \text{or} \quad x = \frac{1}{3}$$

The four complex zeros of f are $\left\{ -3i, 3i, -2, \dfrac{1}{3} \right\}$.

The factored form of f is

$$f(x) = 3x^4 + 5x^3 + 25x^2 + 45x - 18$$
$$= 3(x + 3i)(x - 3i)(x + 2)\left(x - \frac{1}{3} \right)$$

━━━━ **Now Work** PROBLEM 33

5.6 Assess Your Understanding

'Are You Prepared?' *Answers are given at the end of these exercises. If you get the wrong answer, read the pages listed in red.*

1. Find the sum and the product of the complex numbers $3 - 2i$ and $-3 + 5i$. (pp. 109–114)

2. In the complex number system, solve $x^2 + 2x + 2 = 0$. (pp. 114–116)

Concepts and Vocabulary

3. Every polynomial function of odd degree with real coefficients will have at least _____ real zero(s).

4. If $3 + 4i$ is a zero of a polynomial function of degree 5 with real coefficients, then so is _____.

5. *True or False* A polynomial function of degree n with real coefficients has exactly n complex zeros. At most n of them are real zeros.

6. *True or False* A polynomial function of degree 4 with real coefficients could have $-3, 2 + i, 2 - i$, and $-3 + 5i$ as its zeros.

Skill Building

In Problems 7–16, information is given about a polynomial $f(x)$ whose coefficients are real numbers. Find the remaining zeros of f.

7. Degree 3; zeros: $3, 4 - i$

8. Degree 3; zeros: $4, 3 + i$

9. Degree 4; zeros: $i, 1 + i$

10. Degree 4; zeros: $1, 2, 2 + i$

11. Degree 5; zeros: $1, i, 2i$

12. Degree 5; zeros: $0, 1, 2, i$

13. Degree 4; zeros: $i, 2, -2$

14. Degree 4; zeros: $2 - i, -i$

15. Degree 6; zeros: $2, 2 + i, -3 - i, 0$

16. Degree 6; zeros: $i, 3 - 2i, -2 + i$

In Problems 17–22, form a polynomial $f(x)$ with real coefficients having the given degree and zeros. Answers will vary depending on the choice of leading coefficient.

17. Degree 4; zeros: $3 + 2i$; 4, multiplicity 2

18. Degree 4; zeros: $i, 1 + 2i$

19. Degree 5; zeros: 2; $-i$; $1 + i$

20. Degree 6; zeros: $i, 4 - i$; $2 + i$

21. Degree 4; zeros: 3, multiplicity 2; $-i$

22. Degree 5; zeros: 1, multiplicity 3; $1 + i$

In Problems 23–30, use the given zero to find the remaining zeros of each function.

23. $f(x) = x^3 - 4x^2 + 4x - 16$; zero: $2i$

24. $g(x) = x^3 + 3x^2 + 25x + 75$; zero: $-5i$

25. $f(x) = 2x^4 + 5x^3 + 5x^2 + 20x - 12$; zero: $-2i$

26. $h(x) = 3x^4 + 5x^3 + 25x^2 + 45x - 18$; zero: $3i$

27. $h(x) = x^4 - 9x^3 + 21x^2 + 21x - 130$; zero: $3 - 2i$

28. $f(x) = x^4 - 7x^3 + 14x^2 - 38x - 60$; zero: $1 + 3i$

29. $h(x) = 3x^5 + 2x^4 + 15x^3 + 10x^2 - 528x - 352$; zero: $-4i$

30. $g(x) = 2x^5 - 3x^4 - 5x^3 - 15x^2 - 207x + 108$; zero: $3i$

In Problems 31–40, find the complex zeros of each polynomial function. Write f in factored form.

31. $f(x) = x^3 - 1$

32. $f(x) = x^4 - 1$

33. $f(x) = x^3 - 8x^2 + 25x - 26$

34. $f(x) = x^3 + 13x^2 + 57x + 85$

35. $f(x) = x^4 + 5x^2 + 4$

36. $f(x) = x^4 + 13x^2 + 36$

37. $f(x) = x^4 + 2x^3 + 22x^2 + 50x - 75$

38. $f(x) = x^4 + 3x^3 - 19x^2 + 27x - 252$

39. $f(x) = 3x^4 - x^3 - 9x^2 + 159x - 52$

40. $f(x) = 2x^4 + x^3 - 35x^2 - 113x + 65$

Discussion and Writing

In Problems 41 and 42, explain why the facts given are contradictory.

41. $f(x)$ is a polynomial of degree 3 whose coefficients are real numbers; its zeros are $4 + i, 4 - i$, and $2 + i$.

42. $f(x)$ is a polynomial of degree 3 whose coefficients are real numbers; its zeros are $2, i$, and $3 + i$.

43. $f(x)$ is a polynomial of degree 4 whose coefficients are real numbers; three of its zeros are $2, 1 + 2i$, and $1 - 2i$. Explain why the remaining zero must be a real number.

44. $f(x)$ is a polynomial of degree 4 whose coefficients are real numbers; two of its zeros are -3 and $4 - i$. Explain why one of the remaining zeros must be a real number. Write down one of the missing zeros.

'Are You Prepared?' Answers

1. Sum: $3i$; product: $1 + 21i$

2. $-1 - i, -1 + i$

CHAPTER REVIEW

Things to Know

Power function (pp. 325–328)

$f(x) = x^n$, $n \geq 2$ even

Domain: all real numbers Range: nonnegative real numbers

Passes through $(-1, 1), (0, 0), (1, 1)$

Even function

Decreasing on $(-\infty, 0)$, increasing on $(0, \infty)$

$f(x) = x^n, \quad n \geq 3$ odd

Domain: all real numbers Range: all real numbers

Passes through $(-1, -1), (0, 0), (1, 1)$

Odd function

Increasing on $(-\infty, \infty)$

Polynomial function (pp. 324, 328–339)

$f(x) = a_n x^n + a_{n-1} x^{n-1}$
$+ \cdots + a_1 x + a_0, \quad a_n \neq 0$

Domain: all real numbers

At most $n - 1$ turning points

End behavior: Behaves like $y = a_n x^n$ for large $|x|$

Real Zeros of a polynomial function f (p. 329)

Real Numbers for which $f(x) = 0$; the real zeros of f are the x-intercepts of the graph of f.

Rational function (p. 344–352)

$R(x) = \dfrac{p(x)}{q(x)}$

p, q are polynomial functions.

Domain: $\{x | q(x) \neq 0\}$

Vertical asymptotes: With $R(x)$ in lowest terms, if $q(r) = 0$ for some real number, then $x = r$ is a vertical asymptote.

Horizontal or oblique asymptotes: See the summary on pages 351–352.

Remainder Theorem (p. 376)

If a polynomial function $f(x)$ is divided by $x - c$, then the remainder is $f(c)$.

Factor Theorem (p. 376)

$x - c$ is a factor of a polynomial function $f(x)$ if and only if $f(c) = 0$.

Descartes' Rule of Signs (p. 378)

Let f denote a polynomial function written in standard form. The number of positive zeros of f either equals the number of variations in sign of the nonzero coefficients of $f(x)$ or else equals that number less some even integer. The number of negative zeros of f either equals the number of variations in sign of the nonzero coefficients of $f(-x)$ or else equals that number less some even integer.

Rational Zeros Theorem (p. 379)

Let f be a polynomial function of degree 1 or higher of the form

$$f(x) = a_n x^n + a_{n-1} x^{n-1} + \cdots + a_1 x + a_0 \quad a_n \neq 0, a_0 \neq 0$$

where each coefficient is an integer. If $\dfrac{p}{q}$, in lowest terms, is a rational zero of f, then p must be a factor of a_0, and q must be a factor of a_n.

Intermediate Value Theorem (p. 384)

Let f be a polynomial function. If $a < b$ and $f(a)$ and $f(b)$ are of opposite sign, then there is at least one real zero of f between a and b.

Fundamental Theorem of Algebra (p. 390)

Every complex polynomial function $f(x)$ of degree $n \geq 1$ has at least one complex zero.

Conjugate Pairs Theorem (p. 390)

Let $f(x)$ be a polynomial whose coefficients are real numbers. If $r = a + bi$ is a zero of f, then its complex conjugate $\bar{r} = a - bi$ is also a zero of f.

Objectives

Section	You should be able to . . .	Review Exercises
5.1	1 Identify polynomial functions and their degree (p. 324)	1–4
	2 Graph polynomial functions using transformations (p. 328)	5–10
	3 Identify the real zeros of a polynomial function and their multiplicity (p. 329)	11–18 (a) and (b)
	4 Analyze the graph of a polynomial function (p. 336)	11–18
5.2	1 Find the domain of a rational function (p. 344)	19–22
	2 Find the vertical asymptotes of a rational function (p. 347)	19–22
	3 Find the horizontal or oblique asymptotes of a rational function (p. 348)	19–22
5.3	1 Analyze the graph of a rational function (p. 355)	23–34
	2 Solve applied problems involving rational functions (p. 365)	89
5.4	1 Solve polynomial inequalities (p. 370)	35, 36
	2 Solve rational inequalities (p. 372)	37–44

Review Exercises

In Problems 1–4, determine which functions are polynomial functions. For those that are, state the degree. For those that are not, tell why not.

1. $f(x) = 4x^5 - 3x^2 + 5x - 2$ **2.** $f(x) = \dfrac{3x^5}{2x+1}$ **3.** $f(x) = 3x^2 + 5x^{1/2} - 1$ **4.** $f(x) = 3$

In Problems 5–10, graph each function using transformations (shifting, compressing, stretching, and reflection). Show all the stages.

5. $f(x) = (x+2)^3$ **6.** $f(x) = -x^3 + 3$ **7.** $f(x) = -(x-1)^4$

8. $f(x) = (x-1)^4 - 2$ **9.** $f(x) = (x-1)^4 + 2$ **10.** $f(x) = (1-x)^3$

In Problems 11–18:
(a) Find the x- and y-intercepts of each polynomial function f.
(b) Determine whether the graph of f touches or crosses the x-axis at each x-intercept.
(c) End behavior: find the power function that the graph of f resembles for large values of $|x|$.
(d) Determine the maximum number of turning points of the graph of f.
(e) Determine the behavior of the graph of f near each x-intercept.
(f) Put all the information together to obtain the graph of f. (You may need to locate additional points on the graph.)

11. $f(x) = x(x+2)(x+4)$ **12.** $f(x) = x(x-2)(x-4)$ **13.** $f(x) = (x-2)^2(x+4)$

14. $f(x) = (x-2)(x+4)^2$ **15.** $f(x) = -2x^3 + 4x^2$ **16.** $f(x) = -4x^3 + 4x$

17. $f(x) = (x-1)^2(x+3)(x+1)$ **18.** $f(x) = (x-4)(x+2)^2(x-2)$

In Problems 19–22, find the domain of each rational function. Find any horizontal, vertical, or oblique asymptotes.

19. $R(x) = \dfrac{x+2}{x^2-9}$ **20.** $R(x) = \dfrac{x^2+4}{x-2}$ **21.** $R(x) = \dfrac{x^2+3x+2}{(x+2)^2}$ **22.** $R(x) = \dfrac{x^3}{x^3-1}$

In Problems 23–34, discuss each rational function following the seven steps given on page 355.

23. $R(x) = \dfrac{2x-6}{x}$ **24.** $R(x) = \dfrac{4-x}{x}$ **25.** $H(x) = \dfrac{x+2}{x(x-2)}$ **26.** $H(x) = \dfrac{x}{x^2-1}$

27. $R(x) = \dfrac{x^2+x-6}{x^2-x-6}$ **28.** $R(x) = \dfrac{x^2-6x+9}{x^2}$ **29.** $F(x) = \dfrac{x^3}{x^2-4}$ **30.** $F(x) = \dfrac{3x^3}{(x-1)^2}$

31. $R(x) = \dfrac{2x^4}{(x-1)^2}$ **32.** $R(x) = \dfrac{x^4}{x^2-9}$ **33.** $G(x) = \dfrac{x^2-4}{x^2-x-2}$ **34.** $F(x) = \dfrac{(x-1)^2}{x^2-1}$

In Problems 35–44, solve each inequality. Graph the solution set.

35. $x^3 + x^2 < 4x + 4$ **36.** $x^3 + 4x^2 \geq x + 4$ **37.** $\dfrac{6}{x+3} \geq 1$ **38.** $\dfrac{-2}{1-3x} < 1$

39. $\dfrac{2x-6}{1-x} < 2$ **40.** $\dfrac{3-2x}{2x+5} \geq 2$ **41.** $\dfrac{(x-2)(x-1)}{x-3} \geq 0$ **42.** $\dfrac{x+1}{x(x-5)} \leq 0$

43. $\dfrac{x^2-8x+12}{x^2-16} > 0$ **44.** $\dfrac{x(x^2+x-2)}{x^2+9x+20} \leq 0$

In Problems 45–48, find the remainder R when f(x) is divided by g(x). Is g a factor of f?

45. $f(x) = 8x^3 - 3x^2 + x + 4$; $g(x) = x - 1$

46. $f(x) = 2x^3 + 8x^2 - 5x + 5$; $g(x) = x - 2$

47. $f(x) = x^4 - 2x^3 + 15x - 2$; $g(x) = x + 2$

48. $f(x) = x^4 - x^2 + 2x + 2$; $g(x) = x + 1$

49. Find the value of $f(x) = 12x^6 - 8x^4 + 1$ at $x = 4$.

50. Find the value of $f(x) = -16x^3 + 18x^2 - x + 2$ at $x = -2$.

In Problems 51 and 52, use Descartes' Rule of Signs to determine how many positive and negative zeros each polynomial function may have. Do not attempt to find the zeros.

51. $f(x) = 12x^8 - x^7 + 8x^4 - 2x^3 + x + 3$

52. $f(x) = -6x^5 + x^4 + 5x^3 + x + 1$

53. List all the potential rational zeros of $f(x) = 12x^8 - x^7 + 6x^4 - x^3 + x - 3$.

54. List all the potential rational zeros of $f(x) = -6x^5 + x^4 + 2x^3 - x + 1$.

In Problems 55–60, use the Rational Zeros Theorem to find all the real zeros of each polynomial function. Use the zeros to factor f over the real numbers.

55. $f(x) = x^3 - 3x^2 - 6x + 8$

56. $f(x) = x^3 - x^2 - 10x - 8$

57. $f(x) = 4x^3 + 4x^2 - 7x + 2$

58. $f(x) = 4x^3 - 4x^2 - 7x - 2$

59. $f(x) = x^4 - 4x^3 + 9x^2 - 20x + 20$

60. $f(x) = x^4 + 6x^3 + 11x^2 + 12x + 18$

In Problems 61–64, solve each equation in the real number system.

61. $2x^4 + 2x^3 - 11x^2 + x - 6 = 0$

62. $3x^4 + 3x^3 - 17x^2 + x - 6 = 0$

63. $2x^4 + 7x^3 + x^2 - 7x - 3 = 0$

64. $2x^4 + 7x^3 - 5x^2 - 28x - 12 = 0$

In Problems 65–68, find bounds to the real zeros of each polynomial function.

65. $f(x) = x^3 - x^2 - 4x + 2$

66. $f(x) = x^3 + x^2 - 10x - 5$

67. $f(x) = 2x^3 - 7x^2 - 10x + 35$

68. $f(x) = 3x^3 - 7x^2 - 6x + 14$

In Problems 69–72, use the Intermediate Value Theorem to show that each polynomial has a zero in the given interval.

69. $f(x) = 3x^3 - x - 1$; $[0, 1]$

70. $f(x) = 2x^3 - x^2 - 3$; $[1, 2]$

71. $f(x) = 8x^4 - 4x^3 - 2x - 1$; $[0, 1]$

72. $f(x) = 3x^4 + 4x^3 - 8x - 2$; $[1, 2]$

In Problems 73–76, each polynomial has exactly one positive zero. Approximate the zero correct to two decimal places.

73. $f(x) = x^3 - x - 2$

74. $f(x) = 2x^3 - x^2 - 3$

75. $f(x) = 8x^4 - 4x^3 - 2x - 1$

76. $f(x) = 3x^4 + 4x^3 - 8x - 2$

In Problems 77–80, information is given about a complex polynomial f(x) whose coefficients are real numbers. Find the remaining zeros of f. Then find a polynomial with real coefficients that has the zeros.

77. Degree 3; zeros: $4 + i, 6$

78. Degree 3; zeros: $3 + 4i, 5$

79. Degree 4; zeros: $i, 1 + i$

80. Degree 4; zeros: $1, 2, 1 + i$

In Problems 81–88, find the complex zeros of each polynomial function f(x). Write f in factored form.

81. $f(x) = x^3 - 3x^2 - 6x + 8$

82. $f(x) = x^3 - x^2 - 10x - 8$

83. $f(x) = 4x^3 + 4x^2 - 7x + 2$

84. $f(x) = 4x^3 - 4x^2 - 7x - 2$

85. $f(x) = x^4 - 4x^3 + 9x^2 - 20x + 20$

86. $f(x) = x^4 + 6x^3 + 11x^2 + 12x + 18$

87. $f(x) = 2x^4 + 2x^3 - 11x^2 + x - 6$

88. $f(x) = 3x^4 + 3x^3 - 17x^2 + x - 6$

89. Making a Can A can in the shape of a right circular cylinder is required to have a volume of 250 cubic centimeters.
 (a) Express the amount A of material to make the can as a function of the radius r of the cylinder.
 (b) How much material is required if the can is of radius 3 centimeters?
 (c) How much material is required if the can is of radius 5 centimeters?
 (d) Graph $A = A(r)$. For what value of r is A smallest?

CHAPTER TEST

1. Graph $f(x) = (x - 3)^4 - 2$ using transformations.

2. For the polynomial function $g(x) = 2x^3 + 5x^2 - 28x - 15$,
 (a) Determine the maximum number of real zeros that the function may have.
 (b) Find bounds to the zeros of the function.
 (c) List the potential rational zeros.
 (d) Determine the real zeros of g. Factor g over the reals.
 (e) Find the x- and y-intercepts of the graph of g.
 (f) Determine whether the graph crosses or touches the x-axis at each x-intercept.
 (g) Find the power function that the graph of g resembles for large values of $|x|$.
 (h) Determine the behavior of the graph of g near each x-intercept.
 (i) Put all the information together to obtain the graph of g.

3. Find the complex zeros of $f(x) = x^3 - 4x^2 + 25x - 100$.

4. Solve $3x^3 + 2x - 1 = 8x^2 - 4$ in the complex number system.

In Problems 5 and 6, find the domain of each function. Find any horizontal, vertical, or oblique asymptotes.

5. $g(x) = \dfrac{2x^2 - 14x + 24}{x^2 + 6x - 40}$ 6. $r(x) = \dfrac{x^2 + 2x - 3}{x + 1}$

7. Sketch the graph of the function in Problem 6. Label all intercepts, vertical asymptotes, horizontal asymptotes, and oblique asymptotes.

In Problems 8 and 9, write a function that meets the given conditions.

8. Fourth-degree polynomial with real coefficients; zeros: $-2, 0, 3 + i$

9. Rational function; asymptotes: $y = 2$, $x = 4$; domain: $\{x \mid x \neq 4, x \neq 9\}$

10. Use the Intermediate Value Theorem to show that the function $f(x) = -2x^2 - 3x + 8$ has at least one real zero on the interval $[0, 4]$.

11. Solve: $\dfrac{x + 2}{x - 3} < 2$

CUMULATIVE REVIEW

1. Find the distance between the points $P = (1, 3)$ and $Q = (-4, 2)$.

2. Solve the inequality $x^2 \geq x$ and graph the solution set.

3. Solve the inequality $x^2 - 3x < 4$ and graph the solution set.

4. Find a linear function with slope -3 that contains the point $(-1, 4)$. Graph the function.

5. Find the equation of the line parallel to the line $y = 2x + 1$ and containing the point $(3, 5)$. Express your answer in slope–intercept form and graph the line.

6. Graph the equation $y = x^3$.

7. Does the relation $\{(3, 6), (1, 3), (2, 5), (3, 8)\}$ represent a function? Why or why not?

8. Solve the equation $x^3 - 6x^2 + 8x = 0$.

9. Solve the inequality $3x + 2 \leq 5x - 1$ and graph the solution set.

10. Find the center and radius of the circle $x^2 + 4x + y^2 - 2y - 4 = 0$. Graph the circle.

11. For the equation $y = x^3 - 9x$, determine the intercepts and test for symmetry.

12. Find an equation of the line perpendicular to $3x - 2y = 7$ that contains the point $(1, 5)$.

13. Is the following graph the graph of a function? Why or why not?

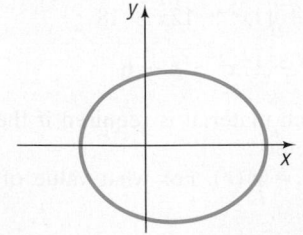

14. For the function $f(x) = x^2 + 5x - 2$, find
 (a) $f(3)$ (b) $f(-x)$
 (c) $-f(x)$ (d) $f(3x)$
 (e) $\dfrac{f(x + h) - f(x)}{h}$ $h \neq 0$

15. Answer the following questions regarding the function
$$f(x) = \frac{x + 5}{x - 1}$$
 (a) What is the domain of f?
 (b) Is the point $(2, 6)$ on the graph of f?
 (c) If $x = 3$, what is $f(x)$? What point is on the graph of f?
 (d) If $f(x) = 9$, what is x? What point is on the graph of f?

16. Graph the function $f(x) = -3x + 7$.

17. Graph $f(x) = 2x^2 - 4x + 1$ by determining whether its graph opens up or down and by finding its vertex, axis of symmetry, y-intercept, and x-intercepts, if any.

18. Find the average rate of change of $f(x) = x^2 + 3x + 1$ from 1 to 2. Use this result to find the equation of the secant line containing $(1, f(1))$ and $(2, f(2))$.

19. In parts (a) to (f), use the following graph.

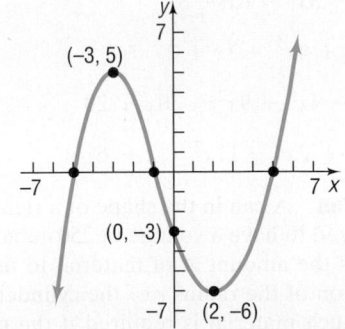

(a) Determine the intercepts.

(b) Based on the graph, tell whether the graph is symmetric with respect to the *x*-axis, the *y*-axis, and/or the origin.

(c) Based on the graph, tell whether the function is even, odd, or neither.

(d) List the intervals on which *f* is increasing. List the intervals on which *f* is decreasing.

(e) List the numbers, if any, at which *f* has a local maximum. What are these local maxima?

(f) List the numbers, if any, at which *f* has a local minimum. What are these local minima?

20. Determine algebraically whether the function

$$f(x) = \frac{5x}{x^2 - 9}$$

is even, odd, or neither.

21. For the function $f(x) = \begin{cases} 2x + 1 & \text{if } -3 < x < 2 \\ -3x + 4 & \text{if } x \ge 2 \end{cases}$

(a) Find the domain of *f*.

(b) Locate any intercepts.

(c) Graph the function.

(d) Based on the graph, find the range.

22. Graph the function $f(x) = -3(x + 1)^2 + 5$ using transformations.

23. Suppose that $f(x) = x^2 - 5x + 1$ and $g(x) = -4x - 7$.

(a) Find $f + g$ and state its domain.

(b) Find $\dfrac{f}{g}$ and state its domain.

24. Demand Equation The price *p* (in dollars) and the quantity *x* sold of a certain product obey the demand equation

$$p = -\frac{1}{10}x + 150, \qquad 0 \le x \le 1500$$

(a) Express the revenue *R* as a function of *x*.

(b) What is the revenue if 100 units are sold?

(c) What quantity *x* maximizes revenue? What is the maximum revenue?

(d) What price should the company charge to maximize revenue?

CHAPTER PROJECTS

I. Weed Pollen The table shows the tree pollen count for Lexington, Kentucky, from April 1, 2006, to May 16, 2006, during the prime season for tree pollen.

Date	Count	Date	Count	Date	Count	Date	Count	Date	Count
4/1	76	4/10	19	4/19	1162	4/29	359	5/8	63
4/2	43	4/11	207	4/20	924	4/30	164	5/9	91
4/3	114	4/12	903	4/21	1030	5/1	65	5/10	166
4/4	73	4/13	1058	4/22	82	5/2	69	5/11	80
4/5	140	4/14	981	4/23	1514	5/3	30	5/12	153
4/6	138	4/15	1832	4/25	724	5/4	181	5/13	71
4/7	153	4/16	2058	4/26	102	5/5	95	5/14	63
4/8	264	4/17	1387	4/27	110	5/6	300	5/15	13
4/9	203	4/18	1188	4/28	320	5/7	91	5/16	9

Source: National Allergy Bureau at the American Academy of Allergy, Asthma and Immunology website. Used with permission.

1. Let the independent variable D represent the date, where $D = 1$ on 4/1, $D = 2$ on 4/2, $D = 3$ on 4/3, and $D = 46$ on 5/16. Let the dependent variable P represent the number of grains per cubic meter of weed pollen. Draw a scatter diagram of the data using your graphing utility and by hand on graph paper.

2. In the hand-drawn scatter diagram, sketch a smooth curve that fits the data. Try to have the curve pass through as many of the points as closely as possible. How many turning points are there? What does this tell you about the degree of the polynomial that could be fit to the data? If the ends of the graph on the left and on the right are extended downward, how many real zeros are indicated by the graph? How many complex? Why?

3. Use a graphing utility to determine the quartic function of best fit. Graph it on your graphing utility. Does it look like what you expected from your hand-drawn graph? Explain.

4. Use a graphing utility to determine the cubic function of best fit. Graph it on your graphing utility. Does it look like what you expected from your hand-drawn graph? Explain.

5. Use a graphing utility to determine the quadratic function of best fit. Graph it on your graphing utility. Does it look like what you expected from your hand-drawn graph? Explain.

6. Which of the three functions seems to fit the best? Explain your reasoning.

7. Investigate weed pollen counts at other times of the year. (Go to the National Allergy Bureau at www.aaaai.org.) Would the polynomial you found in part (3) work for other 6-week periods? Why or why not? What about for the same period of time in other years?

The following project is available at the Instructor's Resource Center (IRC):

II. Theory of Equations The coefficients of a polynomial function can be found if its zeros are known, an advantage of using polynomials in modeling.

Exponential and Logarithmic Functions

The McDonald's Scalding Coffee Case

April 3, 1996—There is a lot of hype about the McDonald's scalding coffee case. No one is in favor of frivolous cases or outlandish results; however, it is important to understand some points that were not reported in most of the stories about the case. McDonald's coffee was not only hot, it was scalding, capable of almost instantaneous destruction of skin, flesh, and muscle.

Plaintiff's expert, a scholar in thermodynamics applied to human skin burns, testified that liquids at 180 degrees will cause a full thickness burn to human skin in two to seven seconds. Other testimony showed that, as the temperature decreases toward 155 degrees, the extent of the burn relative to that temperature decreases exponentially. Thus, if (the) spill had involved coffee at 155 degrees, the liquid would have cooled and given her time to avoid a serious burn.

Source: ATLA fact sheet © 1995, 1996 Consumer Attorneys of CA. Used with permission of the Association of Trial Lawyers of America.

—See the Chapter Project—

A Look Back

Until now, our study of functions has concentrated on polynomial and rational functions. These functions belong to the class of **algebraic functions,** that is, functions that can be expressed in terms of sums, differences, products, quotients, powers, or roots of polynomials. Functions that are not algebraic are termed **transcendental** (they transcend, or go beyond, algebraic functions).

A Look Ahead

In this chapter, we study two transcendental functions: the exponential function and the logarithmic function. These functions occur frequently in a wide variety of applications, such as biology, chemistry, economics, and psychology.

The chapter begins with a discussion of composite, one-to-one, and inverse functions, concepts needed to see the relationship between exponential and logarithmic functions.

Outline

6.1 Composite Functions

PREPARING FOR THIS SECTION *Before getting started, review the following:*

- Finding Values of a Function (Section 3.1, pp. 212–214)
- Domain of a Function (Section 3.1, pp. 209–211 and 215–216)

Now Work the 'Are You Prepared?' problems on page 407.

> **OBJECTIVES** 1 Form a Composite Function (p. 402)
>
> 2 Find the Domain of a Composite Function (p. 403)

1 Form a Composite Function

Figure 1

Suppose that an oil tanker is leaking oil and you want to determine the area of the circular oil patch around the ship. See Figure 1. It is determined that the oil is leaking from the tanker in such a way that the radius of the circular patch of oil around the ship is increasing at a rate of 3 feet per minute. Therefore, the radius r of the oil patch at any time t, in minutes, is given by $r(t) = 3t$. So after 20 minutes the radius of the oil patch is $r(20) = 3(20) = 60$ feet.

The area A of a circle as a function of the radius r is given by $A(r) = \pi r^2$. The area of the circular patch of oil after 20 minutes is $A(60) = \pi(60)^2 = 3600\pi$ square feet. Notice that $60 = r(20)$, so $A(60) = A(r(20))$. The argument of the function A is itself a function!

In general, we can find the area of the oil patch as a function of time t by evaluating $A(r(t))$. The function $A(r(t))$ is a special type of function called a *composite function*.

As another example, consider the function $y = (2x + 3)^2$. If we write $y = f(u) = u^2$ and $u = g(x) = 2x + 3$, then, by a substitution process, we can obtain the original function: $y = f(u) = f(g(x)) = (2x + 3)^2$.

In general, suppose that f and g are two functions and that x is a number in the domain of g. By evaluating g at x, we get $g(x)$. If $g(x)$ is in the domain of f, then we may evaluate f at $g(x)$ and obtain the expression $f(g(x))$. The correspondence from x to $f(g(x))$ is called a *composite function* $f \circ g$.

DEFINITION

Given two functions f and g, the **composite function,** denoted by $f \circ g$ (read as "f composed with g"), is defined by

$$(f \circ g)(x) = f(g(x))$$

The domain of $f \circ g$ is the set of all numbers x in the domain of g such that $g(x)$ is in the domain of f.

Look carefully at Figure 2. Only those x's in the domain of g for which $g(x)$ is in the domain of f can be in the domain of $f \circ g$. The reason is that if $g(x)$ is not in the domain of f then $f(g(x))$ is not defined. Because of this, the domain of $f \circ g$ is a subset of the domain of g; the range of $f \circ g$ is a subset of the range of f.

Figure 2

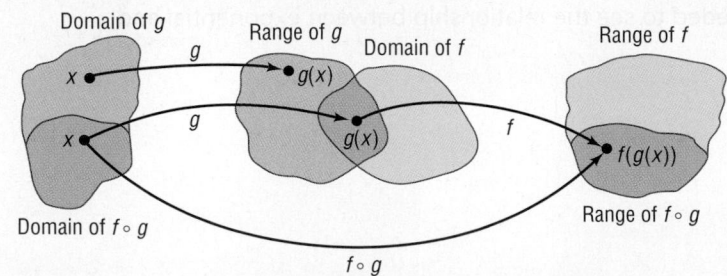

Domain of g Range of g Domain of f Range of f

$x \quad g \quad g(x)$

$x \quad g \quad g(x) \quad f \quad f(g(x))$

Domain of $f \circ g$ Range of $f \circ g$

$f \circ g$

Figure 3 provides a second illustration of the definition. Here x is the input to the function g, yielding $g(x)$. Then $g(x)$ is the input to the function f, yielding $f(g(x))$. Notice that the "inside" function g in $f(g(x))$ is done first.

Figure 3

INPUT x $\longrightarrow$ $\quad g \quad$ $\xrightarrow{g(x)}$ $\quad f \quad$ $\longrightarrow$ OUTPUT $f(g(x))$

Let's look at some examples.

EXAMPLE 1 | **Evaluating a Composite Function**

Suppose that $f(x) = 2x^2 - 3$ and $g(x) = 4x$. Find:

(a) $(f \circ g)(1)$ (b) $(g \circ f)(1)$ (c) $(f \circ f)(-2)$ (d) $(g \circ g)(-1)$

Solution (a) $(f \circ g)(1) = f(g(1)) = f(4) = 2 \cdot 4^2 - 3 = 29$
$$\underset{\substack{\uparrow \\ g(x) = 4x \\ g(1) = 4}}{} \quad \underset{\substack{\uparrow \\ f(x) = 2x^2 - 3}}{}$$

(b) $(g \circ f)(1) = g(f(1)) = g(-1) = 4 \cdot (-1) = -4$
$$\underset{\substack{\uparrow \\ f(x) = 2x^2 - 3 \\ f(1) = -1}}{} \quad \underset{\substack{\uparrow \\ g(x) = 4x}}{}$$

(c) $(f \circ f)(-2) = f(f(-2)) = f(5) = 2 \cdot 5^2 - 3 = 47$
$$\underset{\substack{\uparrow \\ f(-2) = 2(-2)^2 - 3 = 5}}{}$$

(d) $(g \circ g)(-1) = g(g(-1)) = g(-4) = 4 \cdot (-4) = -16$
$$\underset{\substack{\uparrow \\ g(-1) = -4}}{}$$

Figure 4

 COMMENT Graphing calculators can be used to evaluate composite functions.* Let $Y_1 = f(x) = 2x^2 - 3$ and $Y_2 = g(x) = 4x$. Then, using a TI-84 Plus graphing calculator, $(f \circ g)(1)$ is found as shown in Figure 4. Notice that this is the result obtained in Example 1(a). ∎

━━━ **Now Work** PROBLEM 11

2 Find the Domain of a Composite Function

EXAMPLE 2 | **Finding a Composite Function and Its Domain**

Suppose that $f(x) = x^2 + 3x - 1$ and $g(x) = 2x + 3$.

Find: (a) $f \circ g$ (b) $g \circ f$

Then find the domain of each composite function.

Solution The domain of f and the domain of g are the set of all real numbers.

(a) $(f \circ g)(x) = f(g(x)) = f(2x + 3) = (2x + 3)^2 + 3(2x + 3) - 1$
$$\underset{\substack{\uparrow \\ f(x) = x^2 + 3x - 1}}{}$$

$$= 4x^2 + 12x + 9 + 6x + 9 - 1 = 4x^2 + 18x + 17$$

Since the domains of both f and g are the set of all real numbers, the domain of $f \circ g$ is the set of all real numbers.

* Consult your owner's manual for the appropriate keystrokes.

(b) $(g \circ f)(x) = g(f(x)) = g(x^2 + 3x - 1) = 2(x^2 + 3x - 1) + 3$

$$\underset{\substack{\big\uparrow \\ g(x) = 2x + 3}}{}$$

$$= 2x^2 + 6x - 2 + 3 = 2x^2 + 6x + 1$$

Since the domains of both f and g are the set of all real numbers, the domain of $f \circ g$ is the set of all real numbers.

Look back at Figure 2. In determining the domain of the composite function $(f \circ g)(x) = f(g(x))$, keep the following two thoughts in mind about the input x.

1. $g(x)$ must be defined so that any x not in the domain of g must be excluded.

2. $f(g(x))$ must be defined so that any x for which $g(x)$ is not in the domain of f must be excluded.

| EXAMPLE 3 | **Finding the Domain of $f \circ g$** |

Find the domain of $(f \circ g)(x)$ if $f(x) = \dfrac{1}{x + 2}$ and $g(x) = \dfrac{4}{x - 1}$.

Solution For $(f \circ g)(x) = f(g(x))$, we first note that the domain of g is $\{x | x \neq 1\}$, so we exclude 1 from the domain of $f \circ g$. Next we note that the domain of f is $\{x | x \neq -2\}$, which means that $g(x)$ cannot equal -2. We solve the equation $g(x) = -2$ to determine what values of x to exclude.

$$\dfrac{4}{x - 1} = -2 \qquad g(x) = -2$$

$$4 = -2(x - 1)$$

$$4 = -2x + 2$$

$$2x = -2$$

$$x = -1$$

We also exclude -1 from the domain of $f \circ g$. The domain of $f \circ g$ is $\{x | x \neq -1, x \neq 1\}$.

✓ **Check:** For $x = 1$, $g(x) = \dfrac{4}{x - 1}$ is not defined, so $(f \circ g)(x) = f(g(x))$ is not defined.

For $x = -1$, $g(-1) = \dfrac{4}{-2} = -2$, and $(f \circ g)(-1) = f(g(-1)) = f(-2)$ is not defined.

Now Work PROBLEM 21

| EXAMPLE 4 | **Finding a Composite Function and Its Domain** |

Suppose that $f(x) = \dfrac{1}{x + 2}$ and $g(x) = \dfrac{4}{x - 1}$

Find: (a) $f \circ g$ (b) $f \circ f$

Then find the domain of each composite function.

Solution The domain of f is $\{x | x \neq -2\}$ and the domain of g is $\{x | x \neq 1\}$.

(a) $(f \circ g)(x) = f(g(x)) = f\left(\dfrac{4}{x - 1}\right) = \dfrac{1}{\dfrac{4}{x - 1} + 2} = \dfrac{x - 1}{4 + 2(x - 1)} = \dfrac{x - 1}{2x + 2} = \dfrac{x - 1}{2(x + 1)}$

$$\underset{\substack{\big\uparrow \\ f(x) = \frac{1}{x+2}}}{} \qquad \underset{\substack{\big\uparrow \\ \text{Multiply by } \frac{x-1}{x-1}.}}{}$$

In Example 3, we found the domain of $f \circ g$ to be $\{x | x \neq -1, x \neq 1\}$.

We could also find the domain of $f \circ g$ by first looking at the domain of g: $\{x|x \neq 1\}$. We exclude 1 from the domain of $f \circ g$ as a result. Then we look at $f \circ g$ and notice that x cannot equal -1, since $x = -1$ results in division by 0. So we also exclude -1 from the domain of $f \circ g$. Therefore, the domain of $f \circ g$ is $\{x|x \neq -1, x \neq 1\}$.

(b) $(f \circ f)(x) = f(f(x)) = f\left(\dfrac{1}{x+2}\right) = \dfrac{1}{\underset{\uparrow}{\dfrac{1}{x+2}} + 2} = \dfrac{x+2}{\underset{\uparrow}{1 + 2(x+2)}} = \dfrac{x+2}{2x+5}$

$\qquad f(x) = \dfrac{1}{x+2} \qquad\qquad$ Multiply by $\dfrac{x+2}{x+2}$.

The domain of $f \circ f$ consists of those x in the domain of f, $\{x|x \neq -2\}$, for which

$$f(x) = \frac{1}{x+2} \neq -2 \qquad \frac{1}{x+2} = -2$$
$$1 = -2(x+2)$$
$$1 = -2x - 4$$
$$2x = -5$$
$$x = -\frac{5}{2}$$

or, equivalently,

$$x \neq -\frac{5}{2}$$

The domain of $f \circ f$ is $\left\{x \middle| x \neq -\dfrac{5}{2}, x \neq -2\right\}$.

We could also find the domain of $f \circ f$ by recognizing that -2 is not in the domain of f and so should be excluded from the domain of $f \circ f$. Then, looking at $f \circ f$, we see that x cannot equal $-\dfrac{5}{2}$. Do you see why? Therefore, the domain of $f \circ f$ is $\left\{x \middle| x \neq -\dfrac{5}{2}, x \neq -2\right\}$.

━━━━━ **Now Work** PROBLEMS **33** AND **35**

Look back at Example 2, which illustrates that, in general, $f \circ g \neq g \circ f$. Sometimes $f \circ g$ does equal $g \circ f$, as shown in the next example.

| **EXAMPLE 5** | **Showing That Two Composite Functions Are Equal** |

If $f(x) = 3x - 4$ and $g(x) = \dfrac{1}{3}(x + 4)$, show that

$$(f \circ g)(x) = (g \circ f)(x) = x$$

for every x in the domain of $f \circ g$ and $g \circ f$.

Solution $(f \circ g)(x) = f(g(x))$

$\qquad = f\left(\dfrac{x+4}{3}\right) \qquad g(x) = \dfrac{1}{3}(x+4) = \dfrac{x+4}{3}$

$\qquad = 3\left(\dfrac{x+4}{3}\right) - 4 \qquad$ Substitute $g(x)$ into the rule for f, $f(x) = 3x - 4$.

$\qquad = x + 4 - 4 = x$

Using a graphing calculator, let

$$Y_1 = f(x) = 3x - 4$$

$$Y_2 = g(x) = \frac{1}{3}(x + 4)$$

$$Y_3 = f \circ g, \; Y_4 = g \circ f.$$

Using the viewing window $-3 \le x \le 3$, $-2 \le y \le 2$, graph only Y_3 and Y_4. What do you see? TRACE to verify that $Y_3 = Y_4$.

$$(g \circ f)(x) = g(f(x))$$
$$= g(3x - 4) \qquad f(x) = 3x - 4$$
$$= \frac{1}{3}[(3x - 4) + 4] \quad \text{Substitute } f(x) \text{ into the rule for } g, \; g(x) = \frac{1}{3}(x + 4).$$
$$= \frac{1}{3}(3x) = x$$

We conclude that $(f \circ g)(x) = (g \circ f)(x) = x$.

In Section 6.2, we shall see that there is an important relationship between functions f and g for which $(f \circ g)(x) = (g \circ f)(x) = x$.

Now Work PROBLEM 45

Calculus Application

 Some techniques in calculus require that we be able to determine the components of a composite function. For example, the function $H(x) = \sqrt{x + 1}$ is the composition of the functions f and g, where $f(x) = \sqrt{x}$ and $g(x) = x + 1$, because $H(x) = (f \circ g)(x) = f(g(x)) = f(x + 1) = \sqrt{x + 1}$.

EXAMPLE 6 | **Finding the Components of a Composite Function**

Find functions f and g such that $f \circ g = H$ if $H(x) = (x^2 + 1)^{50}$.

Solution The function H takes $x^2 + 1$ and raises it to the power 50. A natural way to decompose H is to raise the function $g(x) = x^2 + 1$ to the power 50. If we let $f(x) = x^{50}$ and $g(x) = x^2 + 1$, then

Figure 5

$$(f \circ g)(x) = f(g(x))$$
$$= f(x^2 + 1)$$
$$= (x^2 + 1)^{50} = H(x)$$

See Figure 5.

Other functions f and g may be found for which $f \circ g = H$ in Example 6. For example, if $f(x) = x^2$ and $g(x) = (x^2 + 1)^{25}$, then

$$(f \circ g)(x) = f(g(x)) = f((x^2 + 1)^{25}) = [(x^2 + 1)^{25}]^2 = (x^2 + 1)^{50}$$

Although the functions f and g found as a solution to Example 6 are not unique, there is usually a "natural" selection for f and g that comes to mind first.

EXAMPLE 7 | **Finding the Components of a Composite Function**

Find functions f and g such that $f \circ g = H$ if $H(x) = \dfrac{1}{x + 1}$.

Solution Here H is the reciprocal of $g(x) = x + 1$. If we let $f(x) = \dfrac{1}{x}$ and $g(x) = x + 1$, we find that

$$(f \circ g)(x) = f(g(x)) = f(x + 1) = \frac{1}{x + 1} = H(x)$$

Now Work PROBLEM 53

6.1 Assess Your Understanding

'Are You Prepared?' *Answers are given at the end of these exercises. If you get a wrong answer, read the pages listed in red.*

1. Find $f(3)$ if $f(x) = -4x^2 + 5x$. (pp. 212–214)
2. Find $f(3x)$ if $f(x) = 4 - 2x^2$. (pp. 212–214)
3. Find the domain of the function $f(x) = \dfrac{x^2 - 1}{x^2 - 25}$. (pp. 215–216)

Concepts and Vocabulary

4. If $f(x) = x + 1$ and $g(x) = x^3$, then _____ $= (x + 1)^3$.
5. *True or False* $f(g(x)) = f(x) \cdot g(x)$.
6. *True or False* The domain of the composite function $(f \circ g)(x)$ is the same as the domain of $g(x)$.

Skill Building

In Problems 7 and 8, evaluate each expression using the values given in the table.

7.

x	−3	−2	−1	0	1	2	3
f(x)	−7	−5	−3	−1	3	5	5
g(x)	8	3	0	−1	0	3	8

(a) $(f \circ g)(1)$ (b) $(f \circ g)(-1)$ (c) $(g \circ f)(-1)$ (d) $(g \circ f)(0)$ (e) $(g \circ g)(-2)$ (f) $(f \circ f)(-1)$

8.

x	−3	−2	−1	0	1	2	3
f(x)	11	9	7	5	3	1	−1
g(x)	−8	−3	0	1	0	−3	−8

(a) $(f \circ g)(1)$ (b) $(f \circ g)(2)$ (c) $(g \circ f)(2)$ (d) $(g \circ f)(3)$ (e) $(g \circ g)(1)$ (f) $(f \circ f)(3)$

In Problems 9 and 10, evaluate each expression using the graphs of $y = f(x)$ and $y = g(x)$ shown below.

9. (a) $g(f(-1))$ (b) $g(f(0))$
 (c) $f(g(-1))$ (d) $f(g(4))$

10. (a) $g(f(1))$ (b) $g(f(5))$
 (c) $f(g(0))$ (d) $f(g(2))$

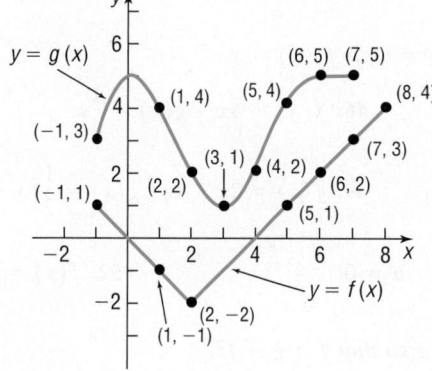

In Problems 11–20, for the given functions f and g, find:

(a) $(f \circ g)(4)$ (b) $(g \circ f)(2)$ (c) $(f \circ f)(1)$ (d) $(g \circ g)(0)$

11. $f(x) = 2x$; $g(x) = 3x^2 + 1$
12. $f(x) = 3x + 2$; $g(x) = 2x^2 - 1$

13. $f(x) = 4x^2 - 3$; $g(x) = 3 - \dfrac{1}{2}x^2$
14. $f(x) = 2x^2$; $g(x) = 1 - 3x^2$

15. $f(x) = \sqrt{x}$; $g(x) = 2x$
16. $f(x) = \sqrt{x + 1}$; $g(x) = 3x$

17. $f(x) = |x|$; $g(x) = \dfrac{1}{x^2 + 1}$
18. $f(x) = |x - 2|$; $g(x) = \dfrac{3}{x^2 + 2}$

19. $f(x) = \dfrac{3}{x + 1}$; $g(x) = \sqrt[3]{x}$
20. $f(x) = x^{3/2}$; $g(x) = \dfrac{2}{x + 1}$

In Problems 21–28, find the domain of the composite function f ∘ g.

21. $f(x) = \dfrac{3}{x-1}$; $g(x) = \dfrac{2}{x}$

22. $f(x) = \dfrac{1}{x+3}$; $g(x) = -\dfrac{2}{x}$

23. $f(x) = \dfrac{x}{x-1}$; $g(x) = -\dfrac{4}{x}$

24. $f(x) = \dfrac{x}{x+3}$; $g(x) = \dfrac{2}{x}$

25. $f(x) = \sqrt{x}$; $g(x) = 2x + 3$

26. $f(x) = x - 2$; $g(x) = \sqrt{1-x}$

27. $f(x) = x^2 + 1$; $g(x) = \sqrt{x-1}$

28. $f(x) = x^2 + 4$; $g(x) = \sqrt{x-2}$

In Problems 29–44, for the given functions f and g, find:

 (a) $f \circ g$ (b) $g \circ f$ (c) $f \circ f$ (d) $g \circ g$

State the domain of each composite function.

29. $f(x) = 2x + 3$; $g(x) = 3x$

30. $f(x) = -x$; $g(x) = 2x - 4$

31. $f(x) = 3x + 1$; $g(x) = x^2$

32. $f(x) = x + 1$; $g(x) = x^2 + 4$

33. $f(x) = x^2$; $g(x) = x^2 + 4$

34. $f(x) = x^2 + 1$; $g(x) = 2x^2 + 3$

35. $f(x) = \dfrac{3}{x-1}$; $g(x) = \dfrac{2}{x}$

36. $f(x) = \dfrac{1}{x+3}$; $g(x) = -\dfrac{2}{x}$

37. $f(x) = \dfrac{x}{x-1}$; $g(x) = -\dfrac{4}{x}$

38. $f(x) = \dfrac{x}{x+3}$; $g(x) = \dfrac{2}{x}$

39. $f(x) = \sqrt{x}$; $g(x) = 2x + 3$

40. $f(x) = \sqrt{x-2}$; $g(x) = 1 - 2x$

41. $f(x) = x^2 + 1$; $g(x) = \sqrt{x-1}$

42. $f(x) = x^2 + 4$; $g(x) = \sqrt{x-2}$

43. $f(x) = \dfrac{x-5}{x+1}$; $g(x) = \dfrac{x+2}{x-3}$

44. $f(x) = \dfrac{2x-1}{x-2}$; $g(x) = \dfrac{x+4}{2x-5}$

In Problems 45–52, show that $(f \circ g)(x) = (g \circ f)(x) = x$.

45. $f(x) = 2x$; $g(x) = \dfrac{1}{2}x$

46. $f(x) = 4x$; $g(x) = \dfrac{1}{4}x$

47. $f(x) = x^3$; $g(x) = \sqrt[3]{x}$

48. $f(x) = x + 5$; $g(x) = x - 5$

49. $f(x) = 2x - 6$; $g(x) = \dfrac{1}{2}(x + 6)$

50. $f(x) = 4 - 3x$; $g(x) = \dfrac{1}{3}(4 - x)$

51. $f(x) = ax + b$; $g(x) = \dfrac{1}{a}(x - b)$ $a \neq 0$

52. $f(x) = \dfrac{1}{x}$; $g(x) = \dfrac{1}{x}$

In Problems 53–58, find functions f and g so that $f \circ g = H$.

53. $H(x) = (2x + 3)^4$

54. $H(x) = (1 + x^2)^3$

55. $H(x) = \sqrt{x^2 + 1}$

56. $H(x) = \sqrt{1 - x^2}$

57. $H(x) = |2x + 1|$

58. $H(x) = |2x^2 + 3|$

Applications and Extensions

59. If $f(x) = 2x^3 - 3x^2 + 4x - 1$ and $g(x) = 2$, find $(f \circ g)(x)$ and $(g \circ f)(x)$.

60. If $f(x) = \dfrac{x+1}{x-1}$, find $(f \circ f)(x)$.

61. If $f(x) = 2x^2 + 5$ and $g(x) = 3x + a$, find a so that the graph of $f \circ g$ crosses the y-axis at 23.

62. If $f(x) = 3x^2 - 7$ and $g(x) = 2x + a$, find a so that the graph of $f \circ g$ crosses the y-axis at 68.

In Problems 63 and 64, use the functions f and g to find:

 (a) $f \circ g$ (b) $g \circ f$

 (c) the domain of $f \circ g$ and of $g \circ f$ (d) the conditions for which $f \circ g = g \circ f$

63. $f(x) = ax + b$; $g(x) = cx + d$

64. $f(x) = \dfrac{ax+b}{cx+d}$; $g(x) = mx$

65. Surface Area of a Balloon The surface area S (in square meters) of a hot-air balloon is given by

$$S(r) = 4\pi r^2$$

where r is the radius of the balloon (in meters). If the radius r is increasing with time t (in seconds) according to the formula $r(t) = \frac{2}{3}t^3, t \geq 0$, find the surface area S of the balloon as a function of the time t.

66. Volume of a Balloon The volume V (in cubic meters) of the hot-air balloon described in Problem 65 is given by $V(r) = \frac{4}{3}\pi r^3$. If the radius r is the same function of t as in Problem 65, find the volume V as a function of the time t.

67. Automobile Production The number N of cars produced at a certain factory in 1 day after t hours of operation is given by $N(t) = 100t - 5t^2, 0 \leq t \leq 10$. If the cost C (in dollars) of producing N cars is $C(N) = 15,000 + 8000N$, find the cost C as a function of the time t of operation of the factory.

68. Environmental Concerns The spread of oil leaking from a tanker is in the shape of a circle. If the radius r (in feet) of the spread after t hours is $r(t) = 200\sqrt{t}$, find the area A of the oil slick as a function of the time t.

69. Production Cost The price p, in dollars, of a certain product and the quantity x sold obey the demand equation

$$p = -\frac{1}{4}x + 100 \quad 0 \leq x \leq 400$$

Suppose that the cost C, in dollars, of producing x units is

$$C = \frac{\sqrt{x}}{25} + 600$$

Assuming that all items produced are sold, find the cost C as a function of the price p.

[**Hint:** Solve for x in the demand equation and then form the composite.]

70. Cost of a Commodity The price p, in dollars, of a certain commodity and the quantity x sold obey the demand equation

$$p = -\frac{1}{5}x + 200 \quad 0 \leq x \leq 1000$$

Suppose that the cost C, in dollars, of producing x units is

$$C = \frac{\sqrt{x}}{10} + 400$$

Assuming that all items produced are sold, find the cost C as a function of the price p.

71. Volume of a Cylinder The volume V of a right circular cylinder of height h and radius r is $V = \pi r^2 h$. If the height is twice the radius, express the volume V as a function of r.

72. Volume of a Cone The volume V of a right circular cone is $V = \frac{1}{3}\pi r^2 h$. If the height is twice the radius, express the volume V as a function of r.

73. Foreign Exchange Traders often buy foreign currency in hope of making money when the currency's value changes. For example, on February 17, 2006, one U.S. dollar could purchase 0.8382 Euros, and one Euro could purchase 140.9687 yen. Let $f(x)$ represent the number of Euros you can buy with x dollars, and let $g(x)$ represent the number of yen you can buy with x Euros.

(a) Find a function that relates dollars to Euros.

(b) Find a function that relates Euros to yen.

(c) Use the results of parts (a) and (b) to find a function that relates dollars to yen. That is, find $(g \circ f)(x) = g(f(x))$.

(d) What is $g(f(1000))$?

74. Temperature Conversion The function $C(F) = \frac{5}{9}(F - 32)$ converts a temperature in degrees Fahrenheit, F, to a temperature in degrees Celsius, C. The function $K(C) = C + 273$, converts a temperature in degrees Celsius to a temperature in kelvins, K.

(a) Find a function that converts a temperature in degrees Fahrenheit to a temperature in kelvins.

(b) Determine 80 degrees Fahrenheit in kelvins.

75. If f and g are odd functions, show that the composite function $f \circ g$ is also odd.

76. If f is an odd function and g is an even function, show that the composite functions $f \circ g$ and $g \circ f$ are both even.

'Are You Prepared?' Answers

1. -21 **2.** $4 - 18x^2$ **3.** $\{x \mid x \neq -5, x \neq 5\}$

6.2 One-to-One Functions; Inverse Functions

PREPARING FOR THIS SECTION *Before getting started, review the following:*

- Functions (Section 3.1, pp. 208–219)
- Increasing/Decreasing Functions (Section 3.3, pp. 233–234)

Now Work the 'Are You Prepared?' problems on page 419.

OBJECTIVES **1** Determine Whether a Function Is One-to-One (p. 410)

2 Determine the Inverse of a Function Defined by a Map or a Set of Ordered Pairs (p. 412)

3 Obtain the Graph of the Inverse Function from the Graph of the Function (p. 414)

4 Find the Inverse of a Function Defined by an Equation (p. 416)

1 Determine Whether a Function Is One-to-One

In Section 3.1, we presented four different ways to represent a function: as (1) a map, (2) a set of ordered pairs, (3) a graph, and (4) an equation. For example, Figures 6 and 7 illustrate two different functions represented as mappings. The function in Figure 6 shows the correspondence between states and their population (in millions). The function in Figure 7 shows a correspondence between animals and life expectancy (in years).

Figure 6

State	Population (in millions)
Indiana	6.2
Washington	6.1
South Dakota	0.8
North Carolina	8.3
Tennessee	5.8

Figure 7

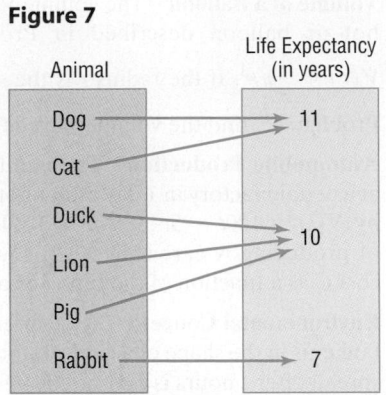

Animal	Life Expectancy (in years)
Dog	11
Cat	
Duck	10
Lion	
Pig	
Rabbit	7

Suppose we asked a group of people to name the state that has a population of 0.8 million based on the function in Figure 6. Everyone in the group would respond South Dakota. Now, if we asked the same group of people to name the animal whose life expectancy is 11 years based on the function in Figure 7, some would respond dog, while others would respond cat. What is the difference between the functions in Figures 6 and 7? In Figure 6, we can see that each element in the domain corresponds to exactly one element in the range. In Figure 7, this is not the case: two different elements in the domain correspond to the same element in the range. We give functions such as the one in Figure 6 a special name.

DEFINITION

A function is **one-to-one** if any two different inputs in the domain correspond to two different outputs in the range. That is, if x_1 and x_2 are two different inputs of a function f, then $f(x_1) \neq f(x_2)$.

> **In Words**
> A function is not one-to-one if two different inputs correspond to the same output.

Put another way, a function f is one-to-one if no y in the range is the image of more than one x in the domain. A function is not one-to-one if two different elements in the domain correspond to the same element in the range. So the function in Figure 7 is not one-to-one because two different elements in the domain, *dog* and *cat*, both correspond to 11. Figure 8 illustrates the distinction of one-to-one functions, non–one-to-one functions, and nonfunctions.

Figure 8

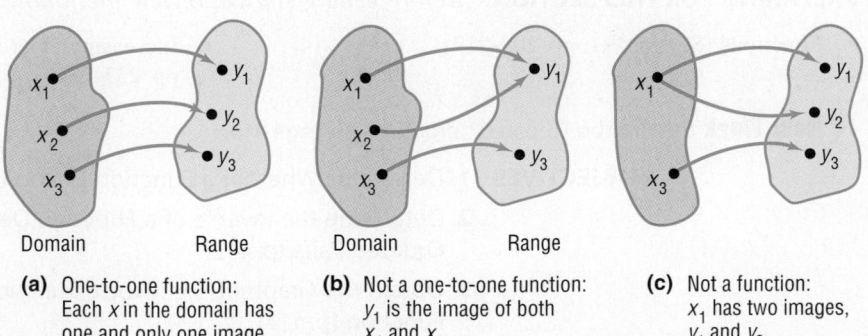

(a) One-to-one function: Each *x* in the domain has one and only one image in the range

(b) Not a one-to-one function: y_1 is the image of both x_1 and x_2.

(c) Not a function: x_1 has two images, y_1 and y_2.

| EXAMPLE 1 | Determining Whether a Function Is One-to-One |

Determine whether the following functions are one-to-one.

(a) For the following function, the domain represents the age of five males and the range represents their HDL (good) cholesterol (mg/dL).

Age HDL Cholesterol

(b) $\{(-2, 6), (-1, 3), (0, 2), (1, 5), (2, 8)\}$

Solution

(a) The function is not one-to-one because there are two different inputs, 55 and 61, that correspond to the same output, 38.

(b) The function is one-to-one because there are no two distinct inputs that correspond to the same output.

━━━ **Now Work** PROBLEMS 9 AND 13

For functions defined by an equation $y = f(x)$ and for which the graph of f is known, there is a simple test, called the **horizontal-line test,** to determine whether f is one-to-one.

Figure 9

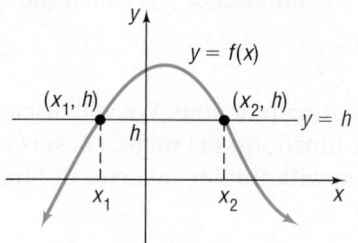

$f(x_1) = f(x_2) = h$ and $x_1 \neq x_2$; f is not a one-to-one function.

| THEOREM | Horizontal-line Test |

If every horizontal line intersects the graph of a function f in at most one point, then f is one-to-one.

The reason that this test works can be seen in Figure 9, where the horizontal line $y = h$ intersects the graph at two distinct points, (x_1, h) and (x_2, h). Since h is the image of both x_1 and x_2, $x_1 \neq x_2$, f is not one-to-one. Based on Figure 9, we can state the horizontal-line test in another way: If the graph of any horizontal line intersects the graph of a function f at more than one point, then f is not one-to-one.

| EXAMPLE 2 | Using the Horizontal-line Test |

For each function, use its graph to determine whether the function is one-to-one.

(a) $f(x) = x^2$ (b) $g(x) = x^3$

Solution

(a) Figure 10(a) on page 412 illustrates the horizontal-line test for $f(x) = x^2$. The horizontal line $y = 1$ intersects the graph of f twice, at $(1, 1)$ and at $(-1, 1)$, so f is not one-to-one.

(b) Figure 10(b) illustrates the horizontal-line test for $g(x) = x^3$. Because every horizontal line intersects the graph of g exactly once, it follows that g is one-to-one.

Figure 10

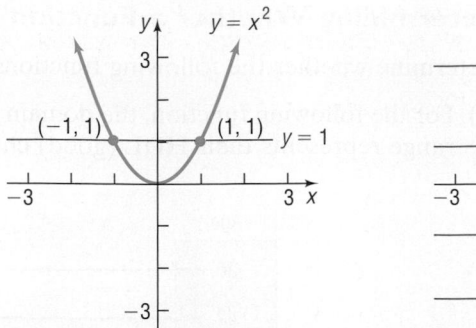

(a) A horizontal line intersects the graph twice; f is not one-to-one

(b) Every horizontal line intersects the graph exactly once; g is one-to-one

━━━━━━ **Now Work** PROBLEM 17

Let's look more closely at the one-to-one function $g(x) = x^3$. This function is an increasing function. Because an increasing (or decreasing) function will always have different y-values for unequal x-values, it follows that a function that is increasing (or decreasing) over its domain is also a one-to-one function.

THEOREM

A function that is increasing on an interval I is a one-to-one function on I.

A function that is decreasing on an interval I is a one-to-one function on I.

2 Determine the Inverse of a Function Defined by a Map or a Set of Ordered Pairs

We begin with the definition.

DEFINITION

Suppose f is a one-to-one function. Then, to each x in the domain of f, there is exactly one y in the range (because f is a function); and to each y in the range of f, there is exactly one x in the domain (because f is one-to-one). The correspondence from the range of f back to the domain of f is called the **inverse function of f**.

Recall that we have a variety of ways of representing functions. We will discuss how to find inverses for all four representations of functions: (1) maps, (2) sets of ordered pairs (3) graphs, and (4) equations. We begin with finding inverses of functions represented by maps or sets of ordered pairs.

EXAMPLE 3 Finding the Inverse of a Function Defined by a Map

Find the inverse of the following function. Let the domain of the function represent certain states, and let the range represent the state's population (in millions). State the domain and the range of the inverse function.

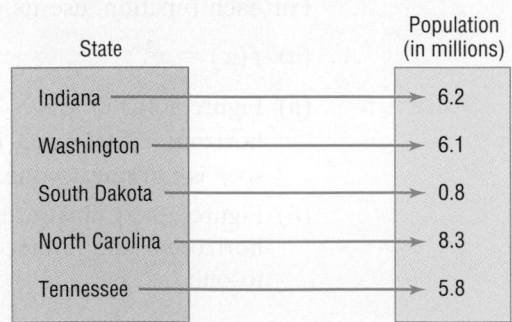

Solution The function is one-to-one. To find the inverse function, we interchange the elements in the domain with the elements in the range. For example, the function receives as input Indiana and outputs 6.2 million. So the inverse receives as input 6.2 million and outputs Indiana. The inverse function is shown next.

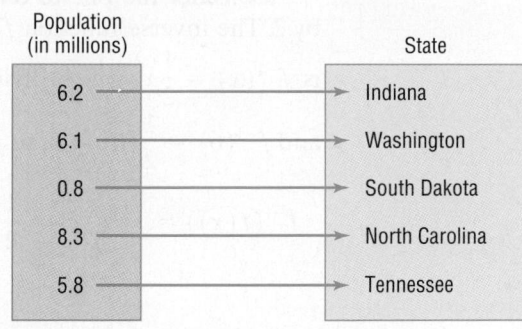

The domain of the inverse function is $\{6.2, 6.1, 0.8, 8.3, 5.8\}$. The range of the inverse function is $\{$Indiana, Washington, South Dakota, North Carolina, Tennessee$\}$.

If the function f is a set of ordered pairs (x, y), then the inverse of f is the set of ordered pairs (y, x).

| **EXAMPLE 4** | **Finding the Inverse of a Function Defined by a Set of Ordered Pairs** |

Find the inverse of the following one-to-one function:

$$\{(-3, -27), (-2, -8), (-1, -1), (0, 0), (1, 1), (2, 8), (3, 27)\}$$

Solution The inverse of the given function is found by interchanging the entries in each ordered pair and so is given by

$$\{(-27, -3), (-8, -2), (-1, -1), (0, 0), (1, 1), (8, 2), (27, 3)\}$$

 ➤ **Now Work** PROBLEMS **23** AND **27**

Figure 11

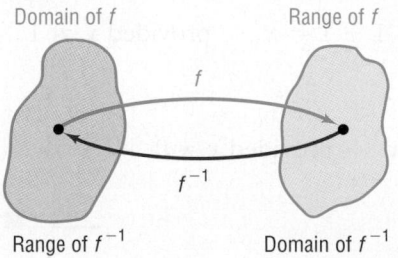

Domain of f Range of f

f

f^{-1}

Range of f^{-1} Domain of f^{-1}

Remember, if f is a one-to-one function, it has an inverse function. We use the symbol f^{-1} to denote the inverse of f. Figure 11 illustrates this definition.

Based on Figure 11, two facts are now apparent about a one-to-one function f and its inverse f^{-1}.

| Domain of f = Range of f^{-1} Range of f = Domain of f^{-1} |

Look again at Figure 11 to visualize the relationship. If we start with x, apply f, and then apply f^{-1}, we get x back again. If we start with x, apply f^{-1}, and then apply f, we get the number x back again. To put it simply, what f does, f^{-1} undoes, and vice versa. See the illustration that follows.

WARNING Be careful! f^{-1} is a symbol for the inverse function of f. The -1 used in f^{-1} is not an exponent. That is, f^{-1} does not mean the reciprocal of f; $f^{-1}(x)$ is not equal to $\dfrac{1}{f(x)}$. ∎

$$\boxed{\text{Input } x} \xrightarrow{\text{Apply } f} \boxed{f(x)} \xrightarrow{\text{Apply } f^{-1}} \boxed{f^{-1}(f(x)) = x}$$

$$\boxed{\text{Input } x} \xrightarrow{\text{Apply } f^{-1}} \boxed{f^{-1}(x)} \xrightarrow{\text{Apply } f} \boxed{f(f^{-1}(x)) = x}$$

In other words,

$$f^{-1}(f(x)) = x \text{ where } x \text{ is in the domain of } f$$
$$f(f^{-1}(x)) = x \text{ where } x \text{ is in the domain of } f^{-1}$$

Consider the one-to-one function $f(x) = 2x$, which multiplies the argument x by 2. The inverse function f^{-1} undoes whatever f does. So the inverse function of f is $f^{-1}(x) = \dfrac{1}{2}x$, which divides the argument by 2. For example, $f(3) = 2(3) = 6$ and $f^{-1}(6) = \dfrac{1}{2}(6) = 3$, so f^{-1} undoes what f did. We can verify this by showing that

Figure 12

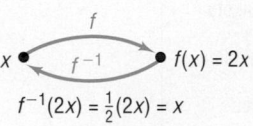

$$f^{-1}(2x) = \tfrac{1}{2}(2x) = x$$

$$f^{-1}(f(x)) = f^{-1}(2x) = \frac{1}{2}(2x) = x \quad \text{and} \quad f(f^{-1}(x)) = f\left(\frac{1}{2}x\right) = 2\left(\frac{1}{2}x\right) = x$$

$$\underset{f^{-1}(x) = \frac{1}{2}x}{\uparrow} \qquad\qquad\qquad \underset{f(x) = 2x}{\uparrow}$$

See Figure 12.

EXAMPLE 5 **Verifying Inverse Functions**

(a) We verify that the inverse of $g(x) = x^3$ is $g^{-1}(x) = \sqrt[3]{x}$ by showing that

$$g^{-1}(g(x)) = g^{-1}(x^3) = \sqrt[3]{x^3} = x \quad \text{for all } x \text{ in the domain of } g$$
$$g(g^{-1}(x)) = g(\sqrt[3]{x}) = (\sqrt[3]{x})^3 = x \quad \text{for all } x \text{ in the domain of } g^{-1}$$

(b) We verify that the inverse of $f(x) = 2x + 3$ is $f^{-1}(x) = \dfrac{1}{2}(x - 3)$ by showing that

$$f^{-1}(f(x)) = f^{-1}(2x + 3) = \frac{1}{2}[(2x + 3) - 3] = \frac{1}{2}(2x) = x \quad \begin{array}{l}\text{for all } x \text{ in the} \\ \text{domain of } f\end{array}$$
$$f(f^{-1}(x)) = f\left(\frac{1}{2}(x - 3)\right) = 2\left[\frac{1}{2}(x - 3)\right] + 3 = (x - 3) + 3 = x \quad \begin{array}{l}\text{for all } x \text{ in the} \\ \text{domain of } f^{-1}\end{array}$$

EXAMPLE 6 **Verifying Inverse Functions**

Verify that the inverse of $f(x) = \dfrac{1}{x - 1}$ is $f^{-1}(x) = \dfrac{1}{x} + 1$. For what values of x is $f^{-1}(f(x)) = x$? For what values of x is $f(f^{-1}(x)) = x$?

Solution The domain of f is $\{x | x \neq 1\}$ and the domain of f^{-1} is $\{x | x \neq 0\}$. Now

$$f^{-1}(f(x)) = f^{-1}\left(\frac{1}{x - 1}\right) = \frac{1}{\dfrac{1}{x - 1}} + 1 = x - 1 + 1 = x \qquad \text{provided } x \neq 1$$

$$f(f^{-1}(x)) = f\left(\frac{1}{x} + 1\right) = \frac{1}{\dfrac{1}{x} + 1 - 1} = \frac{1}{\dfrac{1}{x}} = x \qquad \text{provided } x \neq 0$$

Now Work PROBLEM 31

Figure 13

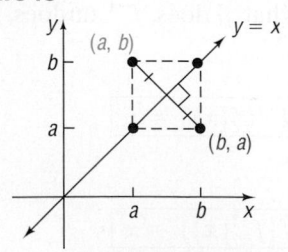

3 Obtain the Graph of the Inverse Function from the Graph of the Function

Suppose that (a, b) is a point on the graph of a one-to-one function f defined by $y = f(x)$. Then $b = f(a)$. This means that $a = f^{-1}(b)$, so (b, a) is a point on the graph of the inverse function f^{-1}. The relationship between the point (a, b) on f and the point (b, a) on f^{-1} is shown in Figure 13. The line segment containing (a, b) and

(b, a) is perpendicular to the line $y = x$ and is bisected by the line $y = x$. (Do you see why?) It follows that the point (b, a) on f^{-1} is the reflection about the line $y = x$ of the point (a, b) on f.

THEOREM

The graph of a function f and the graph of its inverse f^{-1} are symmetric with respect to the line $y = x$.

Figure 14 illustrates this result. Notice that, once the graph of f is known, the graph of f^{-1} may be obtained by reflecting the graph of f about the line $y = x$.

Figure 14

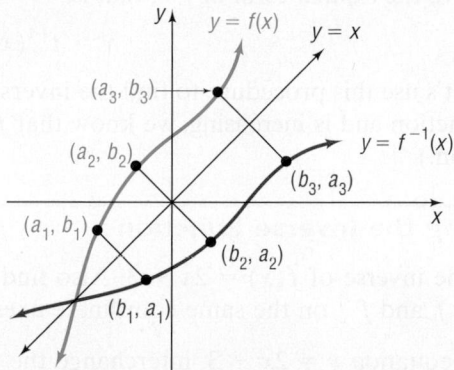

Seeing the Concept

Simultaneously graph $Y_1 = x$, $Y_2 = x^3$, and $Y_3 = \sqrt[3]{x}$ on a square screen with $-3 \leq x \leq 3$. What do you observe about the graphs of $Y_2 = x^3$, its inverse $Y_3 = \sqrt[3]{x}$, and the line $Y_1 = x$? Repeat this experiment by simultaneously graphing $Y_1 = x$, $Y_2 = 2x + 3$, and $Y_3 = \dfrac{1}{2}(x - 3)$ on a square screen with $-6 \leq x \leq 3$. Do you see the symmetry of the graph of Y_2 and its inverse Y_3 with respect to the line $Y_1 = x$?

EXAMPLE 7

Graphing the Inverse Function

The graph in Figure 15(a) is that of a one-to-one function $y = f(x)$. Draw the graph of its inverse.

Solution

We begin by adding the graph of $y = x$ to Figure 15(a). Since the points $(-2, -1), (-1, 0),$ and $(2, 1)$ are on the graph of f, we know that the points $(-1, -2), (0, -1),$ and $(1, 2)$ must be on the graph of f^{-1}. Keeping in mind that the graph of f^{-1} is the reflection about the line $y = x$ of the graph of f, we can draw f^{-1}. See Figure 15(b).

Figure 15

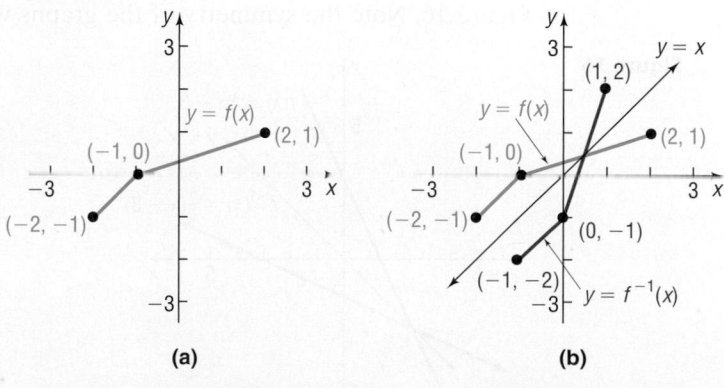

(a) (b)

⟵ **Now Work** PROBLEM 41

4 Find the Inverse of a Function Defined by an Equation

The fact that the graphs of a one-to-one function f and its inverse function f^{-1} are symmetric with respect to the line $y = x$ tells us more. It says that we can obtain f^{-1} by interchanging the roles of x and y in f. Look again at Figure 14. If f is defined by the equation

$$y = f(x)$$

then f^{-1} is defined by the equation

$$x = f(y)$$

The equation $x = f(y)$ defines f^{-1} *implicitly*. If we can solve this equation for y, we will have the *explicit* form of f^{-1}, that is,

$$y = f^{-1}(x)$$

Let's use this procedure to find the inverse of $f(x) = 2x + 3$. (Since f is a linear function and is increasing, we know that f is one-to-one and so has an inverse function.)

EXAMPLE 8	**Finding the Inverse Function**

Find the inverse of $f(x) = 2x + 3$. Also find the domain and range of f and f^{-1}. Graph f and f^{-1} on the same coordinate axes.

Solution In the equation $y = 2x + 3$, interchange the variables x and y. The result,

$$x = 2y + 3$$

is an equation that defines the inverse f^{-1} implicitly. To find the explicit form, we solve for y.

$$2y + 3 = x$$
$$2y = x - 3$$
$$y = \frac{1}{2}(x - 3)$$

The explicit form of the inverse f^{-1} is therefore

$$f^{-1}(x) = \frac{1}{2}(x - 3)$$

which we verified in Example 5(b).

Next we find

$$\text{Domain of } f = \text{Range of } f^{-1} = (-\infty, \infty)$$
$$\text{Range of } f = \text{Domain of } f^{-1} = (-\infty, \infty)$$

The graphs of $f(x) = 2x + 3$ and its inverse $f^{-1}(x) = \frac{1}{2}(x - 3)$ are shown in Figure 16. Note the symmetry of the graphs with respect to the line $y = x$.

Figure 16

We now outline the steps to follow for finding the inverse of a one-to-one function.

Procedure for Finding the Inverse of a One-to-One Function

STEP 1: In $y = f(x)$, interchange the variables x and y to obtain

$$x = f(y)$$

This equation defines the inverse function f^{-1} implicitly.

STEP 2: If possible, solve the implicit equation for y in terms of x to obtain the explicit form of f^{-1}:

$$y = f^{-1}(x)$$

STEP 3: Check the result by showing that

$$f^{-1}(f(x)) = x \quad \text{and} \quad f(f^{-1}(x)) = x$$

EXAMPLE 9 **Finding the Inverse Function**

The function

$$f(x) = \frac{2x + 1}{x - 1} \qquad x \neq 1$$

is one-to-one. Find its inverse and check the result.

Solution **STEP 1:** Interchange the variables x and y in

$$y = \frac{2x + 1}{x - 1}$$

to obtain

$$x = \frac{2y + 1}{y - 1}$$

STEP 2: Solve for y.

$$x = \frac{2y + 1}{y - 1}$$

$x(y - 1) = 2y + 1$ Multiply both sides by y − 1.

$xy - x = 2y + 1$ Apply the Distributive Property.

$xy - 2y = x + 1$ Subtract 2y from both sides; add x to both sides.

$(x - 2)y = x + 1$ Factor.

$y = \dfrac{x + 1}{x - 2}$ Divide by x − 2.

The inverse is

$$f^{-1}(x) = \frac{x + 1}{x - 2} \qquad x \neq 2 \quad \text{Replace y by } f^{-1}(x).$$

STEP 3: ✓ Check:

$$f^{-1}(f(x)) = f^{-1}\left(\frac{2x + 1}{x - 1}\right) = \frac{\dfrac{2x + 1}{x - 1} + 1}{\dfrac{2x + 1}{x - 1} - 2} = \frac{2x + 1 + x - 1}{2x + 1 - 2(x - 1)} = \frac{3x}{3} = x \quad x \neq 1$$

$$f(f^{-1}(x)) = f\left(\frac{x + 1}{x - 2}\right) = \frac{2\left(\dfrac{x + 1}{x - 2}\right) + 1}{\dfrac{x + 1}{x - 2} - 1} = \frac{2(x + 1) + x - 2}{x + 1 - (x - 2)} = \frac{3x}{3} = x \quad x \neq 2$$

—➤ **Now Work** PROBLEM 49

We said in Chapter 3 that finding the range of a function f is not easy. However, if f is one-to-one, we can find its range by finding the domain of the inverse function f^{-1}.

EXAMPLE 10	**Finding the Range of a Function**

Find the domain and range of

$$f(x) = \frac{2x + 1}{x - 1}$$

Solution The domain of f is $\{x \mid x \neq 1\}$. To find the range of f, we first find the inverse f^{-1}. Based on Example 9, we have

$$f^{-1}(x) = \frac{x + 1}{x - 2}$$

The domain of f^{-1} is $\{x \mid x \neq 2\}$, so the range of f is $\{y \mid y \neq 2\}$. ∎

—➤ **Now Work** PROBLEM 63

If a function is not one-to-one, it has no inverse function. Sometimes, though, an appropriate restriction on the domain of such a function will yield a new function that is one-to-one. Then the function defined on the restricted domain has an inverse function. Let's look at an example of this common practice.

EXAMPLE 11	**Finding the Inverse of a Domain-restricted Function**

Find the inverse of $y = f(x) = x^2$ if $x \geq 0$.

Solution The function $y = x^2$ is not one-to-one. [Refer to Example 2(a).] However, if we restrict the domain of this function to $x \geq 0$, as indicated, we have a new function that is increasing and therefore is one-to-one. As a result, the function defined by $y = f(x) = x^2$, $x \geq 0$, has an inverse function, f^{-1}.

We follow the steps given previously to find f^{-1}.

STEP 1: In the equation $y = x^2$, $x \geq 0$, interchange the variables x and y. The result is

$$x = y^2 \qquad y \geq 0$$

This equation defines (implicitly) the inverse function.

STEP 2: We solve for y to get the explicit form of the inverse. Since $y \geq 0$, only one solution for y is obtained: $y = \sqrt{x}$. So $f^{-1}(x) = \sqrt{x}$.

STEP 3: ✓ Check: $f^{-1}(f(x)) = f^{-1}(x^2) = \sqrt{x^2} = |x| = x$ since $x \geq 0$
$$f(f^{-1}(x)) = f(\sqrt{x}) = (\sqrt{x})^2 = x$$

∎

Figure 17

Figure 17 illustrates the graphs of $f(x) = x^2$, $x \geq 0$, and $f^{-1}(x) = \sqrt{x}$.

SUMMARY

1. If a function f is one-to-one, then it has an inverse function f^{-1}.
2. Domain of f = Range of f^{-1}; Range of f = Domain of f^{-1}.
3. To verify that f^{-1} is the inverse of f, show that $f^{-1}(f(x)) = x$ for every x in the domain of f and $f(f^{-1}(x)) = x$ for every x in the domain of f^{-1}.
4. The graphs of f and f^{-1} are symmetric with respect to the line $y = x$.
5. To find the range of a one-to-one function f, find the domain of its inverse function f^{-1}.

6.2 Assess Your Understanding

'Are You Prepared?' *Answers are given at the end of these exercises. If you get a wrong answer, read the pages listed in red.*

1. Is the set of ordered pairs $\{(1, 3), (2, 3), (-1, 2)\}$ a function? Why or why not? (pp. 208–210)
2. Where is the function $f(x) = x^2$ increasing? Where is it decreasing? (pp. 233–234)
3. What is the domain of $f(x) = \dfrac{x + 5}{x^2 + 3x - 18}$? (pp. 215–216)

Concepts and Vocabulary

4. If every horizontal line intersects the graph of a function f at no more than one point, f is a(n) _____ function.
5. If f^{-1} denotes the inverse of a function f, then the graphs of f and f^{-1} are symmetric with respect to the line _____.
6. If the domain of a one-to-one function f is $[4, \infty)$, the range of its inverse, f^{-1}, is _____.
7. **True or False** If f and g are inverse functions, the domain of f is the same as the domain of g.
8. **True or False** If f and g are inverse functions, their graphs are symmetric with respect to the line $y = x$.

Skill Building

In Problems 9–16, determine whether the function is one-to-one.

9.

10.

11.

12.
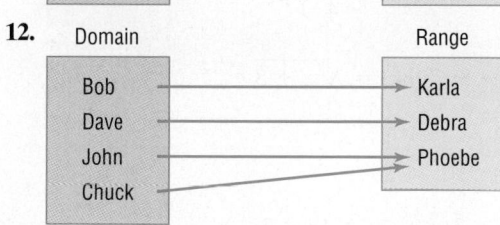

13. $\{(2, 6), (-3, 6), (4, 9), (1, 10)\}$
14. $\{(-2, 5), (-1, 3), (3, 7), (4, 12)\}$
15. $\{(0, 0), (1, 1), (2, 16), (3, 81)\}$
16. $\{(1, 2), (2, 8), (3, 18), (4, 32)\}$

In Problems 17–22, the graph of a function f is given. Use the horizontal-line test to determine whether f is one-to-one.

17.

18.

19.

20. **21.** **22.**

In Problems 23–30, find the inverse of each one-to-one function. State the domain and the range of each inverse function.

23.

Location	Annual Rainfall (inches)
Mt Waialeale, Hawaii	460.00
Monrovia, Liberia	202.01
Pago Pago, American Samoa	196.46
Moulmein, Burma	191.02
Lae, Papua New Guinea	182.87

Source: Information Please Almanac

24.

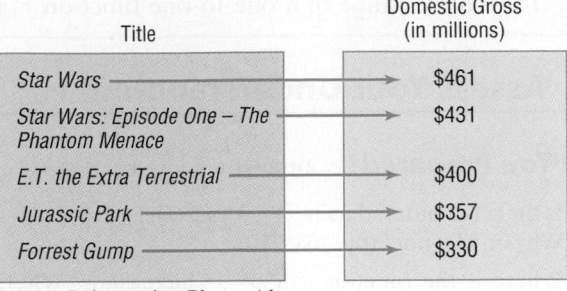

Source: Information Please Almanac

25.

Source: eterm.com

26.

Source: United States Statistical Abstract

27. $\{(-3, 5), (-2, 9), (-1, 2), (0, 11), (1, -5)\}$

28. $\{(-2, 2), (-1, 6), (0, 8), (1, -3), (2, 9)\}$

29. $\{(-2, 1), (-3, 2), (-10, 0), (1, 9), (2, 4)\}$

30. $\{(-2, -8), (-1, -1), (0, 0), (1, 1), (2, 8)\}$

In Problems 31–40, verify that the functions f and g are inverses of each other by showing that $f(g(x)) = x$ and $g(f(x)) = x$. Give any values of x that need to be excluded.

31. $f(x) = 3x + 4; \quad g(x) = \frac{1}{3}(x - 4)$

32. $f(x) = 3 - 2x; \quad g(x) = -\frac{1}{2}(x - 3)$

33. $f(x) = 4x - 8; \quad g(x) = \frac{x}{4} + 2$

34. $f(x) = 2x + 6; \quad g(x) = \frac{1}{2}x - 3$

35. $f(x) = x^3 - 8; \quad g(x) = \sqrt[3]{x + 8}$

36. $f(x) = (x - 2)^2, x \geq 2; \quad g(x) = \sqrt{x} + 2$

37. $f(x) = \frac{1}{x}; \quad g(x) = \frac{1}{x}$

38. $f(x) = x; \quad g(x) = x$

39. $f(x) = \frac{2x + 3}{x + 4}; \quad g(x) = \frac{4x - 3}{2 - x}$

40. $f(x) = \frac{x - 5}{2x + 3}; \quad g(x) = \frac{3x + 5}{1 - 2x}$

In Problems 41–46, the graph of a one-to-one function f is given. Draw the graph of the inverse function f^{-1}. For convenience (and as a hint), the graph of $y = x$ is also given.

41. **42.** **43.**

44.

45.

46.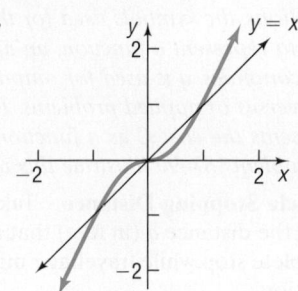

In Problems 47–58, the function f is one-to-one. Find its inverse and check your answer. State the domain and the range of f and f^{-1}. Graph f, f^{-1}, and y = x on the same coordinate axes.

47. $f(x) = 3x$

48. $f(x) = -4x$

49. $f(x) = 4x + 2$

50. $f(x) = 1 - 3x$

51. $f(x) = x^3 - 1$

52. $f(x) = x^3 + 1$

53. $f(x) = x^2 + 4 \quad x \geq 0$

54. $f(x) = x^2 + 9 \quad x \geq 0$

55. $f(x) = \dfrac{4}{x}$

56. $f(x) = -\dfrac{3}{x}$

57. $f(x) = \dfrac{1}{x - 2}$

58. $f(x) = \dfrac{4}{x + 2}$

In Problems 59–70, the function f is one-to-one. Find its inverse and check your answer. State the domain of f and find its range using f^{-1}.

59. $f(x) = \dfrac{2}{3 + x}$

60. $f(x) = \dfrac{4}{2 - x}$

61. $f(x) = \dfrac{3x}{x + 2}$

62. $f(x) = -\dfrac{2x}{x - 1}$

63. $f(x) = \dfrac{2x}{3x - 1}$

64. $f(x) = -\dfrac{3x + 1}{x}$

65. $f(x) = \dfrac{3x + 4}{2x - 3}$

66. $f(x) = \dfrac{2x - 3}{x + 4}$

67. $f(x) = \dfrac{2x + 3}{x + 2}$

68. $f(x) = \dfrac{-3x - 4}{x - 2}$

69. $f(x) = \dfrac{x^2 - 4}{2x^2} \quad x > 0$

70. $f(x) = \dfrac{x^2 + 3}{3x^2} \quad x > 0$

Applications and Extensions

71. Use the graph of $y = f(x)$ given in Problem 41 to evaluate the following:
 (a) $f(-1)$
 (b) $f(1)$
 (c) $f^{-1}(1)$
 (d) $f^{-1}(2)$

72. Use the graph of $y = f(x)$ given in Problem 42 to evaluate the following:
 (a) $f(2)$
 (b) $f(1)$
 (c) $f^{-1}(0)$
 (d) $f^{-1}(-1)$

73. If $f(7) = 13$ and f is one-to-one, what is $f^{-1}(13)$?

74. If $g(-5) = 3$ and g is one-to-one, what is $g^{-1}(3)$?

75. The domain of a one-to-one function f is $[5, \infty)$, and its range is $[-2, \infty)$. State the domain and the range of f^{-1}.

76. The domain of a one-to-one function f is $[0, \infty)$, and its range is $[5, \infty)$. State the domain and the range of f^{-1}.

77. The domain of a one-to-one function g is the set of all real numbers, and its range is $[0, \infty)$. State the domain and the range of g^{-1}.

78. The domain of a one-to-one function g is $[0, 15]$, and its range is $(0, 8)$. State the domain and the range of g^{-1}.

79. A function $y = f(x)$ is increasing on the interval $(0, 5)$. What conclusions can you draw about the graph of $y = f^{-1}(x)$?

80. A function $y = f(x)$ is decreasing on the interval $(0, 5)$. What conclusions can you draw about the graph of $y = f^{-1}(x)$?

81. Find the inverse of the linear function
$$f(x) = mx + b \quad m \neq 0$$

82. Find the inverse of the function
$$f(x) = \sqrt{r^2 - x^2} \quad 0 \leq x \leq r$$

83. A function f has an inverse function. If the graph of f lies in quadrant I, in which quadrant does the graph of f^{-1} lie?

84. A function f has an inverse function. If the graph of f lies in quadrant II, in which quadrant does the graph of f^{-1} lie?

85. The function $f(x) = |x|$ is not one-to-one. Find a suitable restriction on the domain of f so that the new function that results is one-to-one. Then find the inverse of f.

86. The function $f(x) = x^4$ is not one-to-one. Find a suitable restriction on the domain of f so that the new function that results is one-to-one. Then find the inverse of f.

In applications, the symbols used for the independent and dependent variables are often based on common usage. So, rather than using $y = f(x)$ to represent a function, an applied problem might use $C = C(q)$ to represent the cost C of manufacturing q units of a good since, in economics, q is used for output. Because of this, the inverse notation f^{-1} used in a pure mathematics problem is not used when finding inverses of applied problems. Rather, the inverse of a function such as $C = C(q)$ will be $q = q(C)$. So $C = C(q)$ is a function that represents the cost C as a function of the output q, while $q = q(C)$ is a function that represents the output q as a function of the cost C. Problems 87–90 illustrate this idea.

87. Vehicle Stopping Distance Taking into account reaction time, the distance d (in feet) that a car requires to come to a complete stop while traveling r miles per hour is given by the function

$$d(r) = 6.97r - 90.39$$

(a) Express the speed r at which the car is traveling as a function of the distance d required to come to a complete stop.

(b) Verify that $r = r(d)$ is the inverse of $d = d(r)$ by showing that $r(d(r)) = r$ and $d(r(d)) = d$.

(c) Predict the speed that a car was traveling if the distance required to stop was 300 feet.

88. Height and Head Circumference The head circumference C of a child is related to the height H of the child (both in inches) through the function

$$H(C) = 2.15C - 10.53$$

(a) Express the head circumference C as a function of height H.

(b) Verify that $C = C(H)$ is the inverse of $H = H(C)$ by showing that $H(C(H)) = H$ and $C(H(C)) = C$.

(c) Predict the head circumference of a child who is 26 inches tall.

89. Ideal Body Weight The ideal body weight W for men (in kilograms) as a function of height h (in inches) is given by the function

$$W(h) = 50 + 2.3(h - 60)$$

(a) What is the ideal weight of a 6-foot male?

(b) Express the height h as a function of weight W.

(c) Verify that $h = h(W)$ is the inverse of $W = W(h)$ by showing that $h(W(h)) = h$ and $W(h(W)) = W$.

(d) What is the height of a male who is at his ideal weight of 80 kilograms?

Note: The ideal body weight W for women (in kilograms) as a function of height h (in inches) is given by $W(h) = 45.5 + 2.3(h - 60)$.

90. Temperature Conversion The function $F(C) = \dfrac{9}{5}C + 32$ converts a temperature from C degrees Celsius to F degrees Fahrenheit.

(a) Express the temperature in degrees Celsius C as a function of the temperature in degrees Fahrenheit F.

(b) Verify that $C = C(F)$ is the inverse of $F = F(C)$ by showing that $C(F(C)) = C$ and $F(C(F)) = F$.

(c) What is the temperature in degrees Celsius if it is 70 degrees Fahrenheit?

91. Income Taxes The function

$$T(g) = 4220 + 0.25(g - 30,650)$$

represents the 2006 federal income tax T (in dollars) due for a "single" filer whose adjusted gross income is g dollars, where $30,650 \le g \le 74,200$.

(a) What is the domain of the function T?

(b) Given that the tax due T is an increasing linear function of adjusted gross income g, find the range of the function T.

(c) Find adjusted gross income g as a function of federal income tax T. What are the domain and the range of this function?

92. Income Taxes The function

$$T(g) = 1510 + 0.15(g - 15,100)$$

represents the 2006 federal income tax T (in dollars) due for a "married filing jointly" filer whose adjusted gross income is g dollars, where $15,100 \le g \le 61,300$.

(a) What is the domain of the function T?

(b) Given that the tax due T is an increasing linear function of adjusted gross income g, find the range of the function T.

(c) Find adjusted gross income g as a function of federal income tax T. What are the domain and the range of this function?

93. Gravity on Earth If a rock falls from a height of 100 meters on Earth, the height H (in meters) after t seconds is approximately

$$H(t) = 100 - 4.9t^2$$

(a) In general, quadratic functions are not one-to-one. However, the function $H(t)$ is one-to-one. Why?

(b) Find the inverse of H and verify your result.

(c) How long will it take a rock to fall 80 meters?

94. Period of a Pendulum The period T (in seconds) of a simple pendulum as a function of its length l (in feet) is given by

$$T(l) = 2\pi\sqrt{\dfrac{l}{32.2}}$$

(a) Express the length l as a function of the period T.

(b) How long is a pendulum whose period is 3 seconds?

95. Given

$$f(x) = \dfrac{ax + b}{cx + d}$$

find $f^{-1}(x)$. If $c \ne 0$, under what conditions on a, b, c, and d is $f = f^{-1}$?

Discussion and Writing

96. Can a one-to-one function and its inverse be equal? What must be true about the graph of f for this to happen? Give some examples to support your conclusion.

97. Draw the graph of a one-to-one function that contains the points $(-2, -3)$, $(0, 0)$, and $(1, 5)$. Now draw the graph of its inverse. Compare your graph to those of other students. Discuss any similarities. What differences do you see?

98. Give an example of a function whose domain is the set of real numbers and that is neither increasing nor decreasing on its domain, but is one-to-one.

[**Hint:** Use a piecewise-defined function.]

99. Is every odd function one-to-one? Explain.

100. Suppose $C(g)$ represents the cost C, in dollars, of manufacturing g cars. Explain what $C^{-1}(800{,}000)$ represents.

'Are You Prepared?' Answers

1. Yes; for each input x there is one output y.

2. Increasing on $(0, \infty)$; decreasing on $(-\infty, 0)$

3. $\{x \mid x \neq -6, x \neq 3\}$

6.3 Exponential Functions

PREPARING FOR THIS SECTION *Before getting started, review the following:*

- Exponents (Section R.2, pp. 21–24, and Section R.8, pp. 75–76)
- Graphing Techniques: Transformations (Section 3.5, pp. 252–260)

- Average Rate of Change (Section 3.3, pp. 236–238)
- Solving Equations (Section 1.1, pp. 86–92)
- Horizontal Asymptotes (Section 5.2, pp. 346–352)

Now Work the 'Are You Prepared?' problems on page 432.

OBJECTIVES **1** Evaluate Exponential Functions (p. 423)
 2 Graph Exponential Functions (p. 425)
 3 Define the Number e (p. 429)
 4 Solve Exponential Equations (p. 431)

1 Evaluate Exponential Functions

In Chapter R, Section R.8, we give a definition for raising a real number a to a rational power. Based on that discussion, we gave meaning to expressions of the form

$$a^r$$

where the base a is a positive real number and the exponent r is a rational number.

 But what is the meaning of a^x, where the base a is a positive real number and the exponent x is an irrational number? Although a rigorous definition requires methods discussed in calculus, the basis for the definition is easy to follow: Select a rational number r that is formed by truncating (removing) all but a finite number of digits from the irrational number x. Then it is reasonable to expect that

$$a^x \approx a^r$$

For example, take the irrational number $\pi = 3.14159\ldots$. Then an approximation to a^π is

$$a^\pi \approx a^{3.14}$$

where the digits after the hundredths position have been removed from the value for π. A better approximation would be

$$a^\pi \approx a^{3.14159}$$

where the digits after the hundred-thousandths position have been removed. Continuing in this way, we can obtain approximations to a^π to any desired degree of accuracy.

Most calculators have an $\boxed{x^y}$ key or a caret key $\boxed{\wedge}$ for working with exponents. To evaluate expressions of the form a^x, enter the base a, then press the $\boxed{x^y}$ key (or the $\boxed{\wedge}$ key), enter the exponent x, and press $\boxed{=}$ (or $\boxed{\text{enter}}$).

| EXAMPLE 1 | **Using a Calculator to Evaluate Powers of 2** |

Using a calculator, evaluate:

(a) $2^{1.4}$ (b) $2^{1.41}$ (c) $2^{1.414}$ (d) $2^{1.4142}$ (e) $2^{\sqrt{2}}$

Solution

(a) $2^{1.4} \approx 2.639015822$ (b) $2^{1.41} \approx 2.657371628$

(c) $2^{1.414} \approx 2.66474965$ (d) $2^{1.4142} \approx 2.665119089$

(e) $2^{\sqrt{2}} \approx 2.665144143$

➤ **Now Work** PROBLEM 11

It can be shown that the familiar laws for rational exponents hold for real exponents.

THEOREM

Laws of Exponents

If s, t, a, and b are real numbers with $a > 0$ and $b > 0$, then

$$a^s \cdot a^t = a^{s+t} \qquad (a^s)^t = a^{st} \qquad (ab)^s = a^s \cdot b^s$$

$$1^s = 1 \qquad a^{-s} = \frac{1}{a^s} = \left(\frac{1}{a}\right)^s \qquad a^0 = 1 \qquad \textbf{(1)}$$

We are now ready for the following definition:

DEFINITION

An **exponential function** is a function of the form

$$f(x) = a^x$$

where a is a positive real number ($a > 0$) and $a \neq 1$. The domain of f is the set of all real numbers.

WARNING It is important to distinguish a power function, $g(x) = x^n$, $n \geq 2$, an integer, from an exponential function, $f(x) = a^x$, $a \neq 1$, $a > 0$. In a power function, the base is a variable and the exponent is a constant. In an exponential function, the base is a constant and the exponent is a variable. ∎

We exclude the base $a = 1$ because this function is simply the constant function $f(x) = 1^x = 1$. We also need to exclude bases that are negative; otherwise, we would have to exclude many values of x from the domain, such as $x = \frac{1}{2}$ and $x = \frac{3}{4}$. [Recall that $(-2)^{1/2} = \sqrt{-2}$, $(-3)^{3/4} = \sqrt[4]{(-3)^3} = \sqrt[4]{-27}$, and so on, are not defined in the set of real numbers.]

Some examples of exponential functions are

$$f(x) = 2^x \qquad F(x) = \left(\frac{1}{3}\right)^x$$

Notice for each function that the base is a constant and the exponent is a variable.

You may wonder what role the base a plays in the exponential function $f(x) = a^x$. We use the following Exploration to find out.

Exploration

(a) Evaluate $f(x) = 2^x$ at $x = -2, -1, 0, 1, 2$, and 3.

(b) Evaluate $g(x) = 3x + 2$ at $x = -2, -1, 0, 1, 2$, and 3.

(c) Comment on the pattern that exists in the values of f and g.

Result (a) Table 1 shows the values of $f(x) = 2^x$ for $x = -2, -1, 0, 1, 2$, and 3.

(b) Table 2 shows the values of $g(x) = 3x + 2$ for $x = -2, -1, 0, 1, 2$, and 3.

Table 1

x	f(x) = 2ˣ
−2	$f(-2) = 2^{-2} = \dfrac{1}{2^2} = \dfrac{1}{4}$
−1	$\dfrac{1}{2}$
0	1
1	2
2	4
3	8

Table 2

x	g(x) = 3x + 2
−2	$g(-2) = 3(-2) + 2 = -4$
−1	−1
0	2
1	5
2	8
3	11

(c) In Table 1 we notice that each value of the exponential function $f(x) = a^x = 2^x$ could be found by multiplying the previous value of the function by the base, $a = 2$. For example,

$$f(-1) = 2 \cdot f(-2) = 2 \cdot \frac{1}{4} = \frac{1}{2}, \quad f(0) = 2 \cdot f(-1) = 2 \cdot \frac{1}{2} = 1, \quad f(1) = 2 \cdot f(0) = 2 \cdot 1 = 2$$

and so on.

Put another way, the ratio of consecutive outputs is constant for unit increases in the input. The constant equals the value of the base a of the exponential function. For example, for the function $f(x) = 2^x$, we notice that

$$\frac{f(-1)}{f(-2)} = \frac{\frac{1}{2}}{\frac{1}{4}} = 2, \quad \frac{f(1)}{f(0)} = \frac{2}{1} = 2, \quad \frac{f(x + 1)}{f(x)} = \frac{2^{x+1}}{2^x} = 2$$

and so on.

From Table 2 we see that the function $g(x) = 3x + 2$ does not have the ratio of consecutive outputs that are constant because it is not exponential. For example,

$$\frac{g(-1)}{g(-2)} = \frac{-1}{-4} = \frac{1}{4} \neq \frac{g(1)}{g(0)} = \frac{5}{2}$$

Instead, because $g(x) = 3x + 2$ is a linear function, for unit increases in the input, the outputs increase by a fixed amount equal to the value of the slope, 3.

The results of the Exploration lead to the following result.

THEOREM

For an exponential function $f(x) = a^x, a > 0, a \neq 1$, if x is any real number, then

$$\frac{f(x + 1)}{f(x)} = a$$

In Words
For an exponential function $f(x) = a^x$, for 1-unit changes in the input x, the ratio of consecutive outputs is the constant a.

Proof

$$\frac{f(x + 1)}{f(x)} = \frac{a^{x+1}}{a^x} = a^{x+1-x} = a^1 = a \qquad \blacksquare$$

Now Work PROBLEM 21

2 Graph Exponential Functions

First, we graph the exponential function $f(x) = 2^x$.

EXAMPLE 2 Graphing an Exponential Function

Graph the exponential function: $f(x) = 2^x$

Solution The domain of $f(x) = 2^x$ consists of all real numbers. We begin by locating some points on the graph of $f(x) = 2^x$, as listed in Table 3.

Table 3

x	$f(x) = 2^x$
-10	$2^{-10} \approx 0.00098$
-3	$2^{-3} = \dfrac{1}{8}$
-2	$2^{-2} = \dfrac{1}{4}$
-1	$2^{-1} = \dfrac{1}{2}$
0	$2^0 = 1$
1	$2^1 = 2$
2	$2^2 = 4$
3	$2^3 = 8$
10	$2^{10} = 1024$

Figure 18

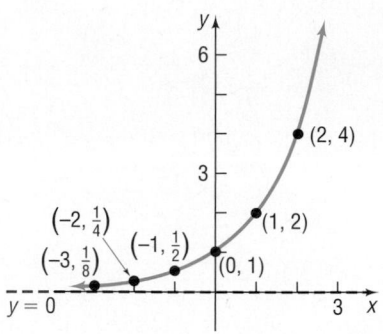

Since $2^x > 0$ for all x, the range of f is the interval $(0, \infty)$. From this, we conclude that the graph has no x-intercepts, and, in fact, the graph will lie above the x-axis. As Table 3 indicates, the y-intercept is 1. Table 3 also indicates that as $x \to -\infty$ the value of $f(x) = 2^x$ gets closer and closer to 0. We conclude that the x-axis is a horizontal asymptote to the graph as $x \to -\infty$. This gives us the end behavior of the graph for x large and negative.

To determine the end behavior for x large and positive, look again at Table 3. As $x \to \infty$, $f(x) = 2^x$ grows very quickly, causing the graph of $f(x) = 2^x$ to rise very rapidly. It is apparent that f is an increasing function and so is one-to-one.

Using all this information, we plot some of the points from Table 3 and connect them with a smooth, continuous curve, as shown in Figure 18. ∎

As we shall see, graphs that look like the one in Figure 18 occur very frequently in a variety of situations. For example, look at the graph in Figure 19, which illustrates the number of cellular telephone subscribers at the end of each year from 1985 to 2005. We might conclude from this graph that the number of cellular telephone subscribers is growing *exponentially*.

Figure 19

Number of Cellular Phone Subscribers at Year End

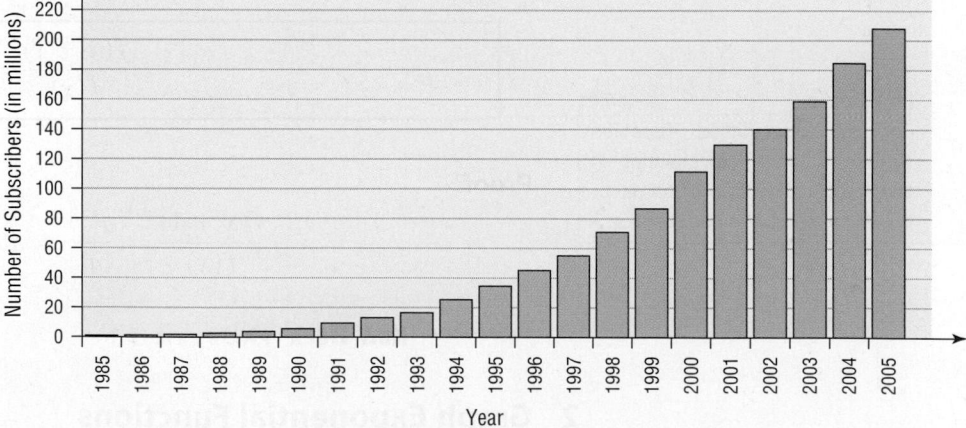

Source: ©2006 CTIA–The Wireless Association®. All Rights Reserved.

We shall have more to say about situations that lead to exponential growth later in this chapter. For now, we continue to seek properties of the exponential functions.

The graph of $f(x) = 2^x$ in Figure 18 is typical of all exponential functions that have a base larger than 1. Such functions are increasing functions and hence are one-to-one. Their graphs lie above the x-axis, pass through the point $(0, 1)$, and thereafter rise rapidly as $x \to \infty$. As $x \to -\infty$, the x-axis $(y = 0)$ is a horizontal

Figure 20

asymptote. There are no vertical asymptotes. Finally, the graphs are smooth and continuous, with no corners or gaps.

Figure 20 illustrates the graphs of two more exponential functions whose bases are larger than 1. Notice that for the larger base the graph is steeper when $x > 0$ and is closer to the x-axis when $x < 0$.

Seeing the Concept

Graph $Y_1 = 2^x$ and compare what you see to Figure 18. Clear the screen and graph $Y_1 = 3^x$ and $Y_2 = 6^x$ and compare what you see to Figure 20. Clear the screen and graph $Y_1 = 10^x$ and $Y_2 = 100^x$. What viewing rectangle seems to work best?

The following display summarizes the information that we have about $f(x) = a^x, a > 1$.

Figure 21
$f(x) = a^x, a > 1$

Properties of the Exponential Function $f(x) = a^x$, $a > 1$

1. The domain is the set of all real numbers; the range is the set of positive real numbers.
2. There are no x-intercepts; the y-intercept is 1.
3. The x-axis ($y = 0$) is a horizontal asymptote as $x \to -\infty$.
4. $f(x) = a^x, a > 1$, is an increasing function and is one-to-one.
5. The graph of f contains the points $\left(-1, \dfrac{1}{a}\right)$, $(0, 1)$, and $(1, a)$.
6. The graph of f is smooth and continuous, with no corners or gaps. See Figure 21.

Now we consider $f(x) = a^x$ when $0 < a < 1$.

EXAMPLE 3 **Graphing an Exponential Function**

Graph the exponential function: $f(x) = \left(\dfrac{1}{2}\right)^x$

Table 4

x	$f(x) = \left(\dfrac{1}{2}\right)^x$
-10	$\left(\dfrac{1}{2}\right)^{-10} = 1024$
-3	$\left(\dfrac{1}{2}\right)^{-3} = 8$
-2	$\left(\dfrac{1}{2}\right)^{-2} = 4$
-1	$\left(\dfrac{1}{2}\right)^{-1} = 2$
0	$\left(\dfrac{1}{2}\right)^{0} = 1$
1	$\left(\dfrac{1}{2}\right)^{1} = \dfrac{1}{2}$
2	$\left(\dfrac{1}{2}\right)^{2} = \dfrac{1}{4}$
3	$\left(\dfrac{1}{2}\right)^{3} = \dfrac{1}{8}$
10	$\left(\dfrac{1}{2}\right)^{10} \approx 0.00098$

Solution

The domain of $f(x) = \left(\dfrac{1}{2}\right)^x$ consists of all real numbers. As before, we locate some points on the graph by creating Table 4. Since $\left(\dfrac{1}{2}\right)^x > 0$ for all x, the range of f is the interval $(0, \infty)$. The graph lies above the x-axis and so has no x-intercepts. The y-intercept is 1. As $x \to -\infty$, $f(x) = \left(\dfrac{1}{2}\right)^x$ grows very quickly. As $x \to \infty$, the values of $f(x)$ approach 0. The x-axis ($y = 0$) is a horizontal asymptote as $x \to \infty$. It is apparent that f is a decreasing function and so is one-to-one. Figure 22 illustrates the graph.

Figure 22

We could have obtained the graph of $y = \left(\dfrac{1}{2}\right)^x$ from the graph of $y = 2^x$ using transformations. The graph of $y = \left(\dfrac{1}{2}\right)^x = 2^{-x}$ is a reflection about the y-axis of the graph of $y = 2^x$ (replace x by $-x$). See Figures 23(a) and 23(b).

Figure 23

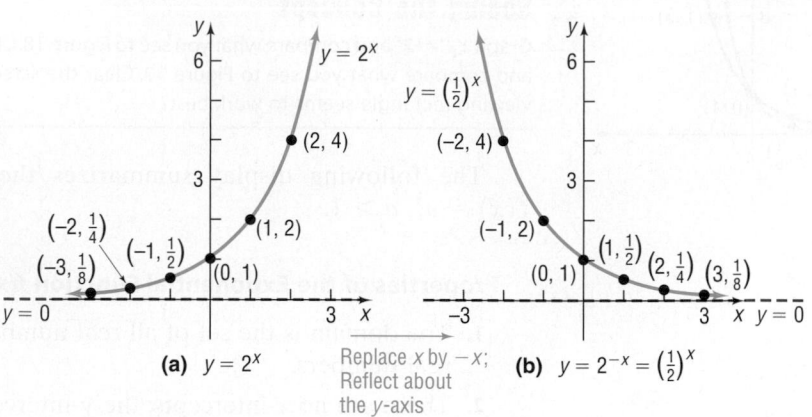

(a) $y = 2^x$

Replace x by $-x$;
Reflect about
the y-axis

(b) $y = 2^{-x} = \left(\dfrac{1}{2}\right)^x$

Seeing the Concept

Using a graphing utility, simultaneously graph:

(a) $Y_1 = 3^x,\, Y_2 = \left(\dfrac{1}{3}\right)^x$ (b) $Y_1 = 6^x,\, Y_2 = \left(\dfrac{1}{6}\right)^x$

Conclude that the graph of $Y_2 = \left(\dfrac{1}{a}\right)^x$, for $a > 0$, is the reflection about the y-axis of the graph of $Y_1 = a^x$.

Figure 24

The graph of $f(x) = \left(\dfrac{1}{2}\right)^x$ in Figure 22 is typical of all exponential functions that have a base between 0 and 1. Such functions are decreasing and one-to-one. Their graphs lie above the x-axis and pass through the point $(0, 1)$. The graphs rise rapidly as $x \to -\infty$. As $x \to \infty$, the x-axis $(y = 0)$ is a horizontal asymptote. There are no vertical asymptotes. Finally, the graphs are smooth and continuous, with no corners or gaps.

Figure 24 illustrates the graphs of two more exponential functions whose bases are between 0 and 1. Notice that the smaller base results in a graph that is steeper when $x < 0$. When $x > 0$, the graph of the equation with the smaller base is closer to the x-axis.

The following display summarizes the information that we have about the function $f(x) = a^x, 0 < a < 1$.

Figure 25

$f(x) = a^x, 0 < a < 1$

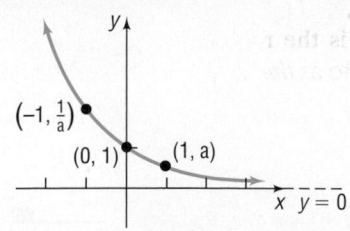

Properties of the Exponential Function $f(x) = a^x$, $0 < a < 1$

1. The domain is the set of all real numbers; the range is the set of positive real numbers.

2. There are no x-intercepts; the y-intercept is 1.

3. The x-axis $(y = 0)$ is a horizontal asymptote as $x \to \infty$.

4. $f(x) = a^x, 0 < a < 1$, is a decreasing function and is one-to-one.

5. The graph of f contains the points $\left(-1, \dfrac{1}{a}\right)$, $(0, 1)$, and $(1, a)$.

6. The graph of f is smooth and continuous, with no corners or gaps. See Figure 25.

| EXAMPLE 4 | **Graphing Exponential Functions Using Transformations** |

Graph $f(x) = 2^{-x} - 3$ and determine the domain, range, and horizontal asymptote of f.

Solution We begin with the graph of $y = 2^x$. Figure 26 shows the stages.

Figure 26

Replace x by $-x$;
reflect
(a) $y = 2^x$ about y-axis

Subtract 3;
shift down
(b) $y = 2^{-x}$ 3 units

(c) $y = 2^{-x} - 3$

As Figure 26(c) illustrates, the domain of $f(x) = 2^{-x} - 3$ is the interval $(-\infty, \infty)$ and the range is the interval $(-3, \infty)$. The horizontal asymptote of f is the line $y = -3$.

━━━ **Now Work** PROBLEM 37

3 Define the Number e

As we shall see shortly, many problems that occur in nature require the use of an exponential function whose base is a certain irrational number, symbolized by the letter e.

Let's look at one way of arriving at this important number e.

DEFINITION The **number e** is defined as the number that the expression

$$\left(1 + \frac{1}{n}\right)^n \qquad (2)$$

 approaches as $n \to \infty$. In calculus, this is expressed using limit notation as

$$e = \lim_{n \to \infty}\left(1 + \frac{1}{n}\right)^n$$

Table 5, on page 430, illustrates what happens to the defining expression (2) as n takes on increasingly large values. The last number in the right column in the table is correct to nine decimal places and is the same as the entry given for e on your calculator (if expressed correctly to nine decimal places).

The exponential function $f(x) = e^x$, whose base is the number e, occurs with such frequency in applications that it is usually referred to as *the* exponential function. Indeed, most calculators have the key $\boxed{e^x}$ or $\boxed{\exp(x)}$, which may be used to evaluate the exponential function for a given value of x.*

* If your calculator does not have one of these keys, refer to your Owner's Manual.

Table 5

n	$\dfrac{1}{n}$	$1 + \dfrac{1}{n}$	$\left(1 + \dfrac{1}{n}\right)^n$
1	1	2	2
2	0.5	1.5	2.25
5	0.2	1.2	2.48832
10	0.1	1.1	2.59374246
100	0.01	1.01	2.704813829
1,000	0.001	1.001	2.716923932
10,000	0.0001	1.0001	2.718145927
100,000	0.00001	1.00001	2.718268237
1,000,000	0.000001	1.000001	2.718280469
1,000,000,000	10^{-9}	$1 + 10^{-9}$	2.718281827

Table 6

x	e^x
-2	0.14
-1	0.37
0	1
1	2.72
2	7.39

Figure 27
$y = e^x$

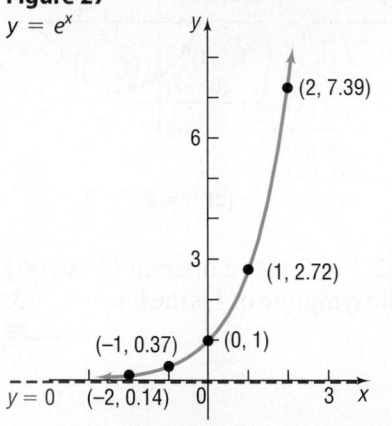

Now use your calculator to approximate e^x for $x = -2$, $x = -1$, $x = 0$, $x = 1$, and $x = 2$, as we have done to create Table 6. The graph of the exponential function $f(x) = e^x$ is given in Figure 27. Since $2 < e < 3$, the graph of $y = e^x$ lies between the graphs of $y = 2^x$ and $y = 3^x$. Do you see why? (Refer to Figures 18 and 20.)

 Seeing the Concept

Graph $Y_1 = e^x$ and compare what you see to Figure 27. Use eVALUEate or TABLE to verify the points on the graph shown in Figure 27. Now graph $Y_2 = 2^x$ and $Y_3 = 3^x$ on the same screen as $Y_1 = e^x$. Notice that the graph of $Y_1 = e^x$ lies between these two graphs.

EXAMPLE 5 **Graphing Exponential Functions Using Transformations**

Graph $f(x) = -e^{x-3}$ and determine the domain, range, and horizontal asymptote of f.

Solution We begin with the graph of $y = e^x$. Figure 28 shows the stages.

Figure 28

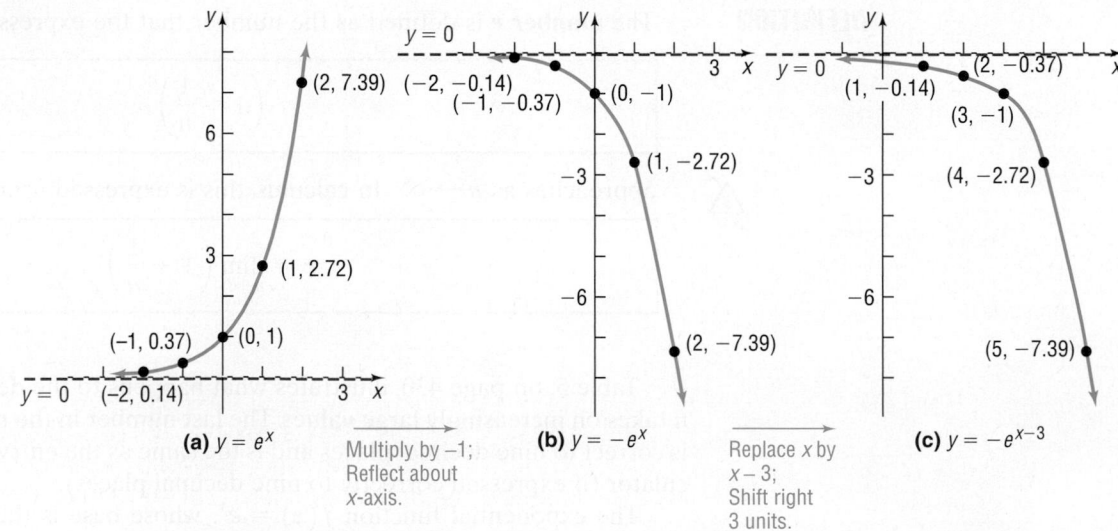

As Figure 28(c) illustrates, the domain of $f(x) = -e^{x-3}$ is the interval $(-\infty, \infty)$, and the range is the interval $(-\infty, 0)$. The horizontal asymptote is the line $y = 0$.

✏️ **Now Work** PROBLEM 49

4 Solve Exponential Equations

Equations that involve terms of the form a^x, $a > 0$, $a \neq 1$, are often referred to as **exponential equations.** Such equations can sometimes be solved by appropriately applying the Laws of Exponents and property (3):

$$\text{If} \quad a^u = a^v, \quad \text{then} \quad u = v \tag{3}$$

Property (3) is a consequence of the fact that exponential functions are one-to-one. To use property (3), each side of the equality must be written with the same base.

EXAMPLE 6 | **Solving an Exponential Equation**

Solve: $3^{x+1} = 81$

Solution Since $81 = 3^4$, we can write the equation as

$$3^{x+1} = 81 = 3^4$$

Now we have the same base, 3, on each side, so we can set the exponents equal to each other to obtain

$$x + 1 = 4$$
$$x = 3$$

The solution set is $\{3\}$.

➤ **Now Work** PROBLEM 59

EXAMPLE 7 | **Solving an Exponential Equation**

Solve: $e^{-x^2} = (e^x)^2 \cdot \dfrac{1}{e^3}$

Solution We use the Laws of Exponents first to get the base e on the right side.

$$(e^x)^2 \cdot \frac{1}{e^3} = e^{2x} \cdot e^{-3} = e^{2x-3}$$

As a result,

$$e^{-x^2} = e^{2x-3}$$
$$-x^2 = 2x - 3 \quad \text{Apply Property (3).}$$
$$x^2 + 2x - 3 = 0 \quad \text{Place the quadratic equation in standard form.}$$
$$(x + 3)(x - 1) = 0 \quad \text{Factor.}$$
$$x = -3 \quad \text{or} \quad x = 1 \quad \text{Use the Zero-Product Property.}$$

The solution set is $\{-3, 1\}$.

Many applications involve the exponential function. Let's look at one.

EXAMPLE 8 | **Exponential Probability**

Between 9:00 PM and 10:00 PM cars arrive at Burger King's drive-thru at the rate of 12 cars per hour (0.2 car per minute). The following formula from statistics can be used to determine the probability that a car will arrive within t minutes of 9:00 PM.

$$F(t) = 1 - e^{-0.2t}$$

(a) Determine the probability that a car will arrive within 5 minutes of 9 PM (that is, before 9:05 PM).

(b) Determine the probability that a car will arrive within 30 minutes of 9 PM (before 9:30 PM).

(c) What value does F approach as t becomes unbounded in the positive direction?

(d) Graph F using your graphing utility.

Solution (a) The probability that a car will arrive within 5 minutes is found by evaluating $F(t)$ at $t = 5$.

$$F(5) = 1 - e^{-0.2(5)} \approx 0.63212$$
$$\uparrow$$
Use a calculator.

We conclude that there is a 63% probability that a car will arrive within 5 minutes.

(b) The probability that a car will arrive within 30 minutes is found by evaluating $F(t)$ at $t = 30$.

$$F(30) = 1 - e^{-0.2(30)} \approx 0.9975$$
$$\uparrow$$
Use a calculator.

There is a 99.75% probability that a car will arrive within 30 minutes.

(c) As time passes, the probability that a car will arrive increases. The value that F approaches can be found by letting $t \to \infty$. Since $e^{-0.2t} = \dfrac{1}{e^{0.2t}}$, it follows that $e^{-0.2t} \to 0$ as $t \to \infty$. We conclude that F approaches 1 as t gets large.

(d) See Figure 29 for the graph of F.

Figure 29

Now **Work** PROBLEM 101

SUMMARY Properties of the Exponential Function

$f(x) = a^x, \quad a > 1$	Domain: the interval $(-\infty, \infty)$; range: the interval $(0, \infty)$
	x-intercepts: none; y-intercept: 1
	Horizontal asymptote: x-axis $(y = 0)$ as $x \to -\infty$
	Increasing; one-to-one; smooth; continuous
	See Figure 21 for a typical graph.
$f(x) = a^x, \quad 0 < a < 1$	Domain: the interval $(-\infty, \infty)$; Range: the interval $(0, \infty)$.
	x-intercepts: none; y-intercept: 1
	Horizontal asymptote: x-axis $(y = 0)$ as $x \to \infty$
	Decreasing; one-to-one; smooth; continuous
	See Figure 25 for a typical graph.

If $a^u = a^v$, then $u = v$.

6.3 Assess Your Understanding

'Are You Prepared?' *Answers are given at the end of these exercises. If you get a wrong answer, read the pages listed in red.*

1. $4^3 = $ _____; $8^{2/3} = $ _____; $3^{-2} = $ _____. (pp. 21–24 and pp. 75–76)

2. Solve: $5x - 2 = 3$ (pp. 86–92)

3. *True or False* To graph $y = (x - 2)^3$, shift the graph of $y = x^3$ to the left 2 units. (pp. 252–260)

4. Find the average rate of change of $f(x) = 3x - 5$ from $x = 0$ to $x = 4$. (pp. 236–238)

5. *True or False* The function $f(x) = \dfrac{2x}{x - 3}$ has $y = 2$ as a horizontal asymptote. (pp. 346–352)

Concepts and Vocabulary

6. The graph of every exponential function $f(x) = a^x$, $a > 0, a \neq 1$, passes through three points: _____, _____, and _____.

7. If the graph of the exponential function $f(x) = a^x$, $a > 0, a \neq 1$, is decreasing, then a must be less than _____.

8. If $3^x = 3^4$, then $x = $ _____.

9. *True or False* The graphs of $y = 3^x$ and $y = \left(\dfrac{1}{3}\right)^x$ are identical.

10. *True or False* The range of the exponential function $f(x) = a^x, a > 0, a \neq 1$, is the set of all real numbers.

Skill Building

In Problems 11–20, approximate each number using a calculator. Express your answer rounded to three decimal places.

11. (a) $3^{2.2}$ (b) $3^{2.23}$ (c) $3^{2.236}$ (d) $3^{\sqrt{5}}$

12. (a) $5^{1.7}$ (b) $5^{1.73}$ (c) $5^{1.732}$ (d) $5^{\sqrt{3}}$

13. (a) $2^{3.14}$ (b) $2^{3.141}$ (c) $2^{3.1415}$ (d) 2^{π}

14. (a) $2^{2.7}$ (b) $2^{2.71}$ (c) $2^{2.718}$ (d) 2^{e}

15. (a) $3.1^{2.7}$ (b) $3.14^{2.71}$ (c) $3.141^{2.718}$ (d) π^{e}

16. (a) $2.7^{3.1}$ (b) $2.71^{3.14}$ (c) $2.718^{3.141}$ (d) e^{π}

17. $e^{1.2}$

18. $e^{-1.3}$

19. $e^{-0.85}$

20. $e^{2.1}$

*In Problems 21–28, determine whether the given function is exponential or not. For those that are exponential functions, identify the value of the base. [**Hint:** Look at the ratio of consecutive values.]*

21.

x	f(x)
−1	3
0	6
1	12
2	18
3	30

22.

x	g(x)
−1	2
0	5
1	8
2	11
3	14

23.

x	H(x)
−1	$\dfrac{1}{4}$
0	1
1	4
2	16
3	64

24.

x	F(x)
−1	$\dfrac{2}{3}$
0	1
1	$\dfrac{3}{2}$
2	$\dfrac{9}{4}$
3	$\dfrac{27}{8}$

25.

x	f(x)
−1	$\dfrac{3}{2}$
0	3
1	6
2	12
3	24

26.

x	g(x)
−1	6
0	1
1	0
2	3
3	10

27.

x	H(x)
−1	2
0	4
1	6
2	8
3	10

28.

x	F(x)
−1	$\dfrac{1}{2}$
0	$\dfrac{1}{4}$
1	$\dfrac{1}{8}$
2	$\dfrac{1}{16}$
3	$\dfrac{1}{32}$

In Problems 29–36, the graph of an exponential function is given. Match each graph to one of the following functions.

A. $y = 3^x$

B. $y = 3^{-x}$

C. $y = -3^x$

D. $y = -3^{-x}$

E. $y = 3^x - 1$

F. $y = 3^{x-1}$

G. $y = 3^{1-x}$

H. $y = 1 - 3^x$

29.

30.

31.

32.

33. **34.** **35.** **36.**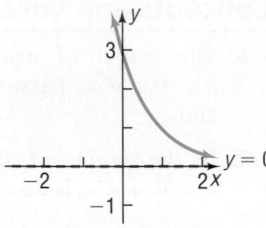

In Problems 37–48, use transformations to graph each function. Determine the domain, range, and horizontal asymptote of each function.

37. $f(x) = 2^x + 1$

38. $f(x) = 3^x - 2$

39. $f(x) = 3^{x-1}$

40. $f(x) = 2^{x+2}$

41. $f(x) = 3 \cdot \left(\frac{1}{2}\right)^x$

42. $f(x) = 4 \cdot \left(\frac{1}{3}\right)^x$

43. $f(x) = 3^{-x} - 2$

44. $f(x) = -3^x + 1$

45. $f(x) = 2 + 4^{x-1}$

46. $f(x) = 1 - 2^{x+3}$

47. $f(x) = 2 + 3^{x/2}$

48. $f(x) = 1 - 2^{-x/3}$

In Problems 49–56, begin with the graph of $y = e^x$ [Figure 27] and use transformations to graph each function. Determine the domain, range, and horizontal asymptote of each function.

49. $f(x) = e^{-x}$

50. $f(x) = -e^x$

51. $f(x) = e^{x+2}$

52. $f(x) = e^x - 1$

53. $f(x) = 5 - e^{-x}$

54. $f(x) = 9 - 3e^{-x}$

55. $f(x) = 2 - e^{-x/2}$

56. $f(x) = 7 - 3e^{2x}$

In Problems 57–76, solve each equation.

57. $7^x = 7^3$

58. $5^x = 5^{-6}$

59. $2^{-x} = 16$

60. $3^{-x} = 81$

61. $\left(\frac{1}{5}\right)^x = \frac{1}{25}$

62. $\left(\frac{1}{4}\right)^x = \frac{1}{64}$

63. $2^{2x-1} = 4$

64. $5^{x+3} = \frac{1}{5}$

65. $3^{x^3} = 9^x$

66. $4^{x^2} = 2^x$

67. $8^{-x+14} = 16^x$

68. $9^{-x+15} = 27^x$

69. $3^{x^2-7} = 27^{2x}$

70. $5^{x^2+8} = 125^{2x}$

71. $4^x \cdot 2^{x^2} = 16^2$

72. $9^{2x} \cdot 27^{x^2} = 3^{-1}$

73. $e^x = e^{3x+8}$

74. $e^{3x} = e^{2-x}$

75. $e^{x^2} = e^{3x} \cdot \frac{1}{e^2}$

76. $(e^4)^x \cdot e^{x^2} = e^{12}$

77. Suppose that $f(x) = 2^x$.

(a) What is $f(4)$? What point is on the graph of f?

(b) If $f(x) = \frac{1}{16}$, what is x? What point is on the graph of f?

78. Suppose that $f(x) = 3^x$.

(a) What is $f(4)$? What point is on the graph of f?

(b) If $f(x) = \frac{1}{9}$, what is x? What point is on the graph of f?

79. Suppose that $g(x) = 4^x + 2$.

(a) What is $g(-1)$? What point is on the graph of g?

(b) If $g(x) = 66$, what is x? What point is on the graph of g?

80. Suppose that $g(x) = 5^x - 3$.

(a) What is $g(-1)$? What point is on the graph of g?

(b) If $g(x) = 122$, what is x? What point is on the graph of g?

81. Suppose that $H(x) = 3 \cdot \left(\frac{1}{2}\right)^x - 2$.

(a) What is $H(-2)$? What point is on the graph of H?

(b) If $H(x) = -\frac{13}{8}$, what is x? What point is on the graph of H?

82. Suppose that $F(x) = -2 \cdot \left(\frac{1}{3}\right)^x + 1$.

(a) What is $F(-1)$? What point is on the graph of F?

(b) If $F(x) = -53$, what is x? What point is on the graph of F?

83. If $4^x = 7$, what does 4^{-2x} equal?

84. If $2^x = 3$, what does 4^{-x} equal?

85. If $3^{-x} = 2$, what does 3^{2x} equal?

86. If $5^{-x} = 3$, what does 5^{3x} equal?

In Problems 87–90, determine the exponential function whose graph is given.

87.

88.

89.

90.

In Problems 91–94, graph each function. Based on the graph, state the domain and the range and find any intercepts.

91. $f(x) = \begin{cases} e^{-x} & \text{if } x < 0 \\ e^x & \text{if } x \geq 0 \end{cases}$

92. $f(x) = \begin{cases} e^x & \text{if } x < 0 \\ e^{-x} & \text{if } x \geq 0 \end{cases}$

93. $f(x) = \begin{cases} -e^x & \text{if } x < 0 \\ -e^{-x} & \text{if } x \geq 0 \end{cases}$

94. $f(x) = \begin{cases} -e^{-x} & \text{if } x < 0 \\ -e^x & \text{if } x \geq 0 \end{cases}$

Applications and Extensions

95. Optics If a single pane of glass obliterates 3% of the light passing through it, the percent p of light that passes through n successive panes is given approximately by the function

$$p(n) = 100(0.97)^n$$

(a) What percent of light will pass through 10 panes?

(b) What percent of light will pass through 25 panes?

96. Atmospheric Pressure The atmospheric pressure p on a balloon or plane decreases with increasing height. This pressure, measured in millimeters of mercury, is related to the height h (in kilometers) above sea level by the function

$$p(h) = 760e^{-0.145h}$$

(a) Find the atmospheric pressure at a height of 2 kilometers (over a mile).

(b) What is it at a height of 10 kilometers (over 30,000 feet)?

97. Depreciation The price p, in dollars, of a Honda Civic DX Sedan that is x years old is given by

$$p(x) = 16,630(0.90)^x$$

(a) How much does a 3-year-old Civic DX Sedan cost?

(b) How much does a 9-year-old Civic DX Sedan cost?

98. Healing of Wounds The normal healing of wounds can be modeled by an exponential function. If A_0 represents the original area of the wound and if A equals the area of the wound, then the function

$$A(n) = A_0 e^{-0.35n}$$

describes the area of a wound after n days following an injury when no infection is present to retard the healing. Suppose that a wound initially had an area of 100 square millimeters.

(a) If healing is taking place, how large will the area of the wound be after 3 days?

(b) How large will it be after 10 days?

99. Drug Medication The function

$$D(h) = 5e^{-0.4h}$$

can be used to find the number of milligrams D of a certain drug that is in a patient's bloodstream h hours after the drug has been administered. How many milligrams will be present after 1 hour? After 6 hours?

100. Spreading of Rumors A model for the number N of people in a college community who have heard a certain rumor is

$$N = P(1 - e^{-0.15d})$$

where P is the total population of the community and d is the number of days that have elapsed since the rumor began. In a community of 1000 students, how many students will have heard the rumor after 3 days?

101. Exponential Probability Between 12:00 PM and 1:00 PM, cars arrive at Citibank's drive-thru at the rate of 6 cars per hour (0.1 car per minute). The following formula from probability can be used to determine the probability that a car will arrive within t minutes of 12:00 PM:

$$F(t) = 1 - e^{-0.1t}$$

(a) Determine the probability that a car will arrive within 10 minutes of 12:00 PM (that is, before 12:10 PM).

(b) Determine the probability that a car will arrive within 40 minutes of 12:00 PM (before 12:40 PM).

(c) What value does F approach as t becomes unbounded in the positive direction?

(d) Graph F using a graphing utility.

(e) Using TRACE, determine how many minutes are needed for the probability to reach 50%.

102. Exponential Probability Between 5:00 PM and 6:00 PM, cars arrive at Jiffy Lube at the rate of 9 cars per hour (0.15 car per minute). The following formula from probability can be used to determine the probability that a car will arrive within t minutes of 5:00 PM:

$$F(t) = 1 - e^{-0.15t}$$

(a) Determine the probability that a car will arrive within 15 minutes of 5:00 PM (that is, before 5:15 PM).

(b) Determine the probability that a car will arrive within 30 minutes of 5:00 PM (before 5:30 PM).

(c) What value does F approach as t becomes unbounded in the positive direction?

(d) Graph F using a graphing utility.

(e) Using TRACE, determine how many minutes are needed for the probability to reach 60%.

103. Poisson Probability Between 5:00 PM and 6:00 PM, cars arrive at McDonald's drive-thru at the rate of 20 cars per hour. The following formula from probability can be used to determine the probability that x cars will arrive between 5:00 PM and 6:00 PM.

$$P(x) = \frac{20^x e^{-20}}{x!}$$

where

$$x! = x \cdot (x - 1) \cdot (x - 2) \cdots \cdot 3 \cdot 2 \cdot 1$$

(a) Determine the probability that $x = 15$ cars will arrive between 5:00 PM and 6:00 PM.

(b) Determine the probability that $x = 20$ cars will arrive between 5:00 PM and 6:00 PM.

104. Poisson Probability People enter a line for the *Demon Roller Coaster* at the rate of 4 per minute. The following formula from probability can be used to determine the probability that x people will arrive within the next minute.

$$P(x) = \frac{4^x e^{-4}}{x!}$$

where

$$x! = x \cdot (x - 1) \cdot (x - 2) \cdots \cdot 3 \cdot 2 \cdot 1$$

(a) Determine the probability that $x = 5$ people will arrive within the next minute.

(b) Determine the probability that $x = 8$ people will arrive within the next minute.

105. Relative Humidity The relative humidity is the ratio (expressed as a percent) of the amount of water vapor in the air to the maximum amount that it can hold at a specific temperature. The relative humidity, R, is found using the following formula:

$$R = 10^{\left(\frac{4221}{T+459.4} - \frac{4221}{D+459.4} + 2\right)}$$

where T is the air temperature (in °F) and D is the dew point temperature (in °F).

(a) Determine the relative humidity if the air temperature is 50° Fahrenheit and the dew point temperature is 41° Fahrenheit.

(b) Determine the relative humidity if the air temperature is 68° Fahrenheit and the dew point temperature is 59° Fahrenheit.

(c) What is the relative humidity if the air temperature and the dew point temperature are the same?

106. Learning Curve Suppose that a student has 500 vocabulary words to learn. If the student learns 15 words after 5 minutes, the function

$$L(t) = 500(1 - e^{-0.0061t})$$

approximates the number of words L that the student will learn after t minutes.

(a) How many words will the student learn after 30 minutes?

(b) How many words will the student learn after 60 minutes?

107. Current in a *RL* Circuit The equation governing the amount of current I (in amperes) after time t (in seconds) in a single *RL* circuit consisting of a resistance R (in ohms), an inductance L (in henrys), and an electromotive force E (in volts) is

$$I = \frac{E}{R}\left[1 - e^{-(R/L)t}\right]$$

(a) If $E = 120$ volts, $R = 10$ ohms, and $L = 5$ henrys, how much current I_1 is flowing after 0.3 second? After 0.5 second? After 1 second?

(b) What is the maximum current?

(c) Graph this function $I = I_1(t)$, measuring I along the y-axis and t along the x-axis.

(d) If $E = 120$ volts, $R = 5$ ohms, and $L = 10$ henrys, how much current I_2 is flowing after 0.3 second? After 0.5 second? After 1 second?

(e) What is the maximum current?

(f) Graph this function $I = I_2(t)$ on the same coordinate axes as $I_1(t)$.

108. Current in a *RC* Circuit The equation governing the amount of current I (in amperes) after time t (in microseconds) in a single *RC* circuit consisting of a resistance R (in ohms), a capacitance C (in microfarads), and an electromotive force E (in volts) is

$$I = \frac{E}{R} e^{-t/(RC)}$$

(a) If $E = 120$ volts, $R = 2000$ ohms, and $C = 1.0$ microfarad, how much current I_1 is flowing initially $(t = 0)$? After 1000 microseconds? After 3000 microseconds?

(b) What is the maximum current?

(c) Graph this function $I = I_1(t)$, measuring I along the y-axis and t along the x-axis.

(d) If $E = 120$ volts, $R = 1000$ ohms, and $C = 2.0$ microfarads, how much current I_2 is flowing initially? After 1000 microseconds? After 3000 microseconds?

(e) What is the maximum current?

(f) Graph this function $I = I_2(t)$ on the same coordinate axes as $I_1(t)$.

109. Another Formula for *e* Use a calculator to compute the values of

$$2 + \frac{1}{2!} + \frac{1}{3!} + \cdots + \frac{1}{n!}$$

for $n = 4, 6, 8,$ and 10. Compare each result with e.

[**Hint:** $1! = 1, 2! = 2 \cdot 1, 3! = 3 \cdot 2 \cdot 1,$
$n! = n(n - 1) \cdots \cdot (3)(2)(1).$]

110. Another Formula for e Use a calculator to compute the various values of the expression. Compare the values to e.

$$2 + \cfrac{1}{1 + \cfrac{1}{2 + \cfrac{2}{3 + \cfrac{3}{4 + 4}}}}$$
etc.

111. Difference Quotient If $f(x) = a^x$, show that

$$\frac{f(x+h) - f(x)}{h} = a^x \cdot \frac{a^h - 1}{h} \quad h \neq 0$$

112. If $f(x) = a^x$, show that $f(A + B) = f(A) \cdot f(B)$.

113. If $f(x) = a^x$, show that $f(-x) = \dfrac{1}{f(x)}$.

114. If $f(x) = a^x$, show that $f(\alpha x) = [f(x)]^\alpha$.

Problems 115 and 116 provide definitions for two other transcendental functions.

115. The **hyperbolic sine function,** designated by sinh x, is defined as

$$\sinh x = \frac{1}{2}(e^x - e^{-x})$$

(a) Show that $f(x) = \sinh x$ is an odd function.

(b) Graph $f(x) = \sinh x$ using a graphing utility.

116. The **hyperbolic cosine function,** designated by cosh x, is defined as

$$\cosh x = \frac{1}{2}(e^x + e^{-x})$$

(a) Show that $f(x) = \cosh x$ is an even function.

(b) Graph $f(x) = \cosh x$ using a graphing utility.

(c) Refer to Problem 115. Show that, for every x,

$$(\cosh x)^2 - (\sinh x)^2 = 1$$

117. Historical Problem Pierre de Fermat (1601–1665) conjectured that the function

$$f(x) = 2^{(2^x)} + 1$$

for $x = 1, 2, 3, \ldots$, would always have a value equal to a prime number. But Leonhard Euler (1707–1783) showed that this formula fails for $x = 5$. Use a calculator to determine the prime numbers produced by f for $x = 1, 2, 3, 4$. Then show that $f(5) = 641 \times 6,700,417$, which is not prime.

Discussion and Writing

118. The bacteria in a 4-liter container double every minute. After 60 minutes the container is full. How long did it take to fill half the container?

119. Explain in your own words what the number e is. Provide at least two applications that use this number.

120. Do you think that there is a power function that increases more rapidly than an exponential function whose base is greater than 1? Explain.

121. As the base a of an exponential function $f(x) = a^x, a > 1$, increases, what happens to the behavior of its graph for $x > 0$? What happens to the behavior of its graph for $x < 0$?

122. The graphs of $y = a^{-x}$ and $y = \left(\dfrac{1}{a}\right)^x$ are identical. Why?

'Are You Prepared?' Answers

1. 64; 4; $\dfrac{1}{9}$ **2.** {1} **3.** False **4.** 3 **5.** True

6.4 Logarithmic Functions

PREPARING FOR THIS SECTION *Before getting started, review the following:*

- Solving Inequalities (Section 1.5, pp. 128–131)
- Quadratic Inequalities (Section 4.5, pp. 314–316)
- Polynomial and Rational Inequalities (Section 5.4, pp. 369–373)

Now Work the 'Are You Prepared?' problems on page 446.

OBJECTIVES **1** Change Exponential Expressions to Logarithmic Expressions and Logarithmic Expressions to Exponential Expressions (p. 438)
 2 Evaluate Logarithmic Expressions (p. 438)
 3 Determine the Domain of a Logarithmic Function (p. 439)
 4 Graph Logarithmic Functions (p. 440)
 5 Solve Logarithmic Equations (p. 443)

Recall that a one-to-one function $y = f(x)$ has an inverse function that is defined (implicitly) by the equation $x = f(y)$. In particular, the exponential function $y = f(x) = a^x$, $a > 0$, $a \neq 1$, is one-to-one and hence has an inverse function that is defined implicitly by the equation

$$x = a^y, \qquad a > 0, \qquad a \neq 1$$

This inverse function is so important that it is given a name, the *logarithmic function*.

DEFINITION

The **logarithmic function to the base a,** where $a > 0$ and $a \neq 1$, is denoted by $y = \log_a x$ (read as "y is the logarithm to the base a of x") and is defined by

$$y = \log_a x \quad \text{if and only if} \quad x = a^y$$

The domain of the logarithmic function $y = \log_a x$ is $x > 0$.

As this definition illustrates, a *logarithm* is a name for a certain exponent.

EXAMPLE 1

Relating Logarithms to Exponents

(a) If $y = \log_3 x$, then $x = 3^y$. For example, $4 = \log_3 81$ is equivalent to $81 = 3^4$.

(b) If $y = \log_5 x$, then $x = 5^y$. For example, $-1 = \log_5\left(\dfrac{1}{5}\right)$ is equivalent to $\dfrac{1}{5} = 5^{-1}$.

1 Change Exponential Expressions to Logarithmic Expressions and Logarithmic Expressions to Exponential Expressions

We can use the definition of a logarithm to convert from exponential form to logarithmic form, and vice versa, as the following two examples illustrate.

EXAMPLE 2

Changing Exponential Expressions to Logarithmic Expressions

Change each exponential expression to an equivalent expression involving a logarithm.

(a) $1.2^3 = m$ (b) $e^b = 9$ (c) $a^4 = 24$

Solution

We use the fact that $y = \log_a x$ and $x = a^y$, $a > 0$, $a \neq 1$, are equivalent.

(a) If $1.2^3 = m$, then $3 = \log_{1.2} m$. (b) If $e^b = 9$, then $b = \log_e 9$.

(c) If $a^4 = 24$, then $4 = \log_a 24$.

Now Work PROBLEM 9

EXAMPLE 3

Changing Logarithmic Expressions to Exponential Expressions

Change each logarithmic expression to an equivalent expression involving an exponent.

(a) $\log_a 4 = 5$ (b) $\log_e b = -3$ (c) $\log_3 5 = c$

Solution

(a) If $\log_a 4 = 5$, then $a^5 = 4$. (b) If $\log_e b = -3$, then $e^{-3} = b$.

(c) If $\log_3 5 = c$, then $3^c = 5$.

Now Work PROBLEM 17

2 Evaluate Logarithmic Expressions

To find the exact value of a logarithm, we write the logarithm in exponential notation and use the fact that if $a^u = a^v$ then $u = v$.

| EXAMPLE 4 | Finding the Exact Value of a Logarithmic Expression |

Find the exact value of:

(a) $\log_2 16$

(b) $\log_3 \dfrac{1}{27}$

Solution

(a) $y = \log_2 16$

$2^y = 16$ *Change to exponential form.*

$2^y = 2^4$ $16 = 2^4$

$y = 4$ *Equate exponents.*

Therefore, $\log_2 16 = 4$.

(b) $y = \log_3 \dfrac{1}{27}$

$3^y = \dfrac{1}{27}$ *Change to exponential form.*

$3^y = 3^{-3}$ $\dfrac{1}{27} = \dfrac{1}{3^3} = 3^{-3}$

$y = -3$ *Equate exponents.*

Therefore, $\log_3 \dfrac{1}{27} = -3$.

═══ **Now Work** PROBLEM 25

3 Determine the Domain of a Logarithmic Function

The logarithmic function $y = \log_a x$ has been defined as the inverse of the exponential function $y = a^x$. That is, if $f(x) = a^x$, then $f^{-1}(x) = \log_a x$. Based on the discussion given in Section 6.2 on inverse functions, for a function f and its inverse f^{-1}, we have

Domain of f^{-1} = Range of f and Range of f^{-1} = Domain of f

Consequently, it follows that

> Domain of the logarithmic function = Range of the exponential function = $(0, \infty)$
>
> Range of the logarithmic function = Domain of the exponential function = $(-\infty, \infty)$

In the next box, we summarize some properties of the logarithmic function:

> $y = \log_a x$ (defining equation: $x = a^y$)
>
> Domain: $0 < x < \infty$ Range: $-\infty < y < \infty$

The domain of a logarithmic function consists of the *positive* real numbers, so the argument of a logarithmic function must be greater than zero.

| EXAMPLE 5 | Finding the Domain of a Logarithmic Function |

Find the domain of each logarithmic function.

(a) $F(x) = \log_2(x + 3)$ (b) $g(x) = \log_5\left(\dfrac{1 + x}{1 - x}\right)$ (c) $h(x) = \log_{1/2}|x|$

Solution

(a) The domain of F consists of all x for which $x + 3 > 0$, that is, $x > -3$. Using interval notation, the domain of f is $(-3, \infty)$.

(b) The domain of g is restricted to

$$\frac{1 + x}{1 - x} > 0$$

Solving this inequality, we find that the domain of g consists of all x between -1 and 1, that is, $-1 < x < 1$ or, using interval notation, $(-1, 1)$.

(c) Since $|x| > 0$, provided that $x \neq 0$, the domain of h consists of all real numbers except zero or, using interval notation, $(-\infty, 0) \cup (0, \infty)$.

═══ **Now Work** PROBLEMS 39 AND 45

4 Graph Logarithmic Functions

Since exponential functions and logarithmic functions are inverses of each other, the graph of the logarithmic function $y = \log_a x$ is the reflection about the line $y = x$ of the graph of the exponential function $y = a^x$, as shown in Figure 30.

Figure 30

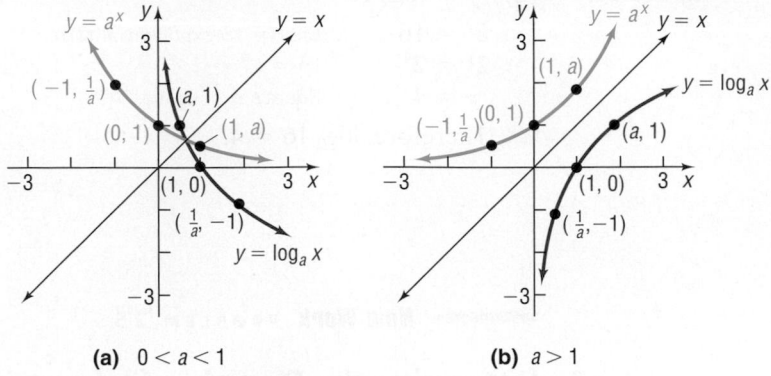

(a) $0 < a < 1$ **(b)** $a > 1$

For example, to graph $y = \log_2 x$, graph $y = 2^x$ and reflect it about the line $y = x$. See Figure 31. To graph $y = \log_{1/3} x$, graph $y = \left(\dfrac{1}{3}\right)^x$ and reflect it about the line $y = x$. See Figure 32.

Figure 31

Figure 32

━━━ **Now Work** PROBLEM 59

Properties of the Logarithmic Function $f(x) = \log_a x$

1. The domain is the set of positive real numbers; the range is the set of all real numbers.

2. The x-intercept of the graph is 1. There is no y-intercept.

3. The y-axis ($x = 0$) is a vertical asymptote of the graph.

4. A logarithmic function is decreasing if $0 < a < 1$ and increasing if $a > 1$.

5. The graph of f contains the points $(1, 0)$, $(a, 1)$, and $\left(\dfrac{1}{a}, -1\right)$.

6. The graph is smooth and continuous, with no corners or gaps.

If the base of a logarithmic function is the number e, then we have the **natural logarithm function.** This function occurs so frequently in applications that it is given a special symbol, **ln** (from the Latin, *logarithmus naturalis*). That is,

$$y = \ln x \quad \text{if and only if} \quad x = e^y \qquad \textbf{(1)}$$

Since $y = \ln x$ and the exponential function $y = e^x$ are inverse functions, we can obtain the graph of $y = \ln x$ by reflecting the graph of $y = e^x$ about the line $y = x$. See Figure 33.

Using a calculator with an $\boxed{\ln}$ key, we can obtain other points on the graph of $f(x) = \ln x$. See Table 7.

Figure 33

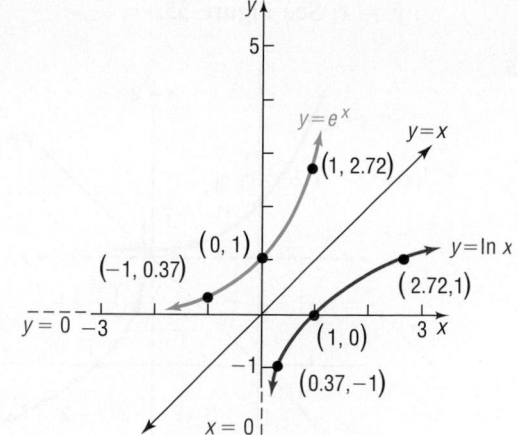

Table 7

x	ln x
$\frac{1}{2}$	−0.69
2	0.69
3	1.10

Seeing the Concept

Graph $Y_1 = e^x$ and $Y_2 = \ln x$ on the same square screen. Use eVALUEate to verify the points on the graph given in Figure 33. Do you see the symmetry of the two graphs with respect to the line $y = x$?

EXAMPLE 6 **Graphing a Logarithmic Function and Its Inverse**

(a) Find the domain of the logarithmic function $f(x) = -\ln(x - 2)$.

(b) Graph f.

(c) From the graph, determine the range and vertical asymptote of f.

(d) Find f^{-1}, the inverse of f.

(e) Use f^{-1} to find the range of f.

(f) Graph f^{-1}.

Solution (a) The domain of f consists of all x for which $x - 2 > 0$ or, equivalently, $x > 2$.

(b) To obtain the graph of $y = -\ln(x - 2)$, we begin with the graph of $y = \ln x$ and use transformations. See Figure 34.

Figure 34

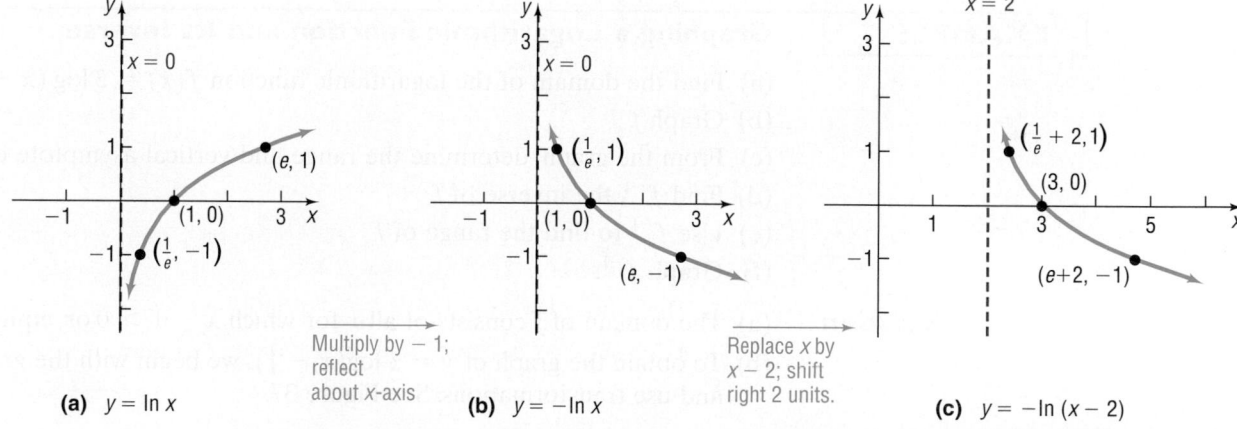

(a) $y = \ln x$ Multiply by -1; reflect about x-axis (b) $y = -\ln x$ Replace x by $x - 2$; shift right 2 units. (c) $y = -\ln(x - 2)$

(c) The range of $f(x) = -\ln(x - 2)$ is the set of all real numbers. The vertical asymptote is $x = 2$. [Do you see why? The original asymptote $(x = 0)$ is shifted to the right 2 units.]

(d) We begin with $y = -\ln(x - 2)$. The inverse function is defined (implicitly) by the equation

$$x = -\ln(y - 2)$$

We proceed to solve for y.

$$-x = \ln(y - 2) \qquad \text{Isolate the logarithm.}$$

$$e^{-x} = y - 2 \qquad \text{Change to an exponential expression.}$$

$$y = e^{-x} + 2 \qquad \text{Solve for y.}$$

The inverse of f is $f^{-1}(x) = e^{-x} + 2$.

(e) The range of f is the domain of f^{-1}, the set of all real numbers, confirming what we found from the graph of f.

(f) To graph f^{-1}, we use the graph of f in Figure 34 (c) and reflect it about the line $y = x$. See Figure 35.

Figure 35

Figure 36

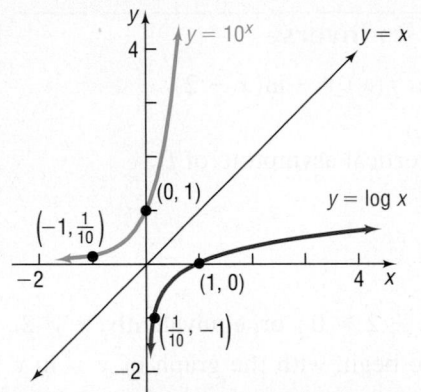

━━━ **Now Work** PROBLEM 71

If the base of a logarithmic function is the number 10, then we have the **common logarithm function.** If the base a of the logarithmic function is not indicated, it is understood to be 10. That is,

$$y = \log x \quad \text{if and only if} \quad x = 10^y$$

Since $y = \log x$ and the exponential function $y = 10^x$ are inverse functions, we can obtain the graph of $y = \log x$ by reflecting the graph of $y = 10^x$ about the line $y = x$. See Figure 36.

EXAMPLE 7 **Graphing a Logarithmic Function and Its Inverse**

(a) Find the domain of the logarithmic function $f(x) = 3 \log (x - 1)$.

(b) Graph f.

(c) From the graph, determine the range and vertical asymptote of f.

(d) Find f^{-1}, the inverse of f.

(e) Use f^{-1} to find the range of f.

(f) Graph f^{-1}.

Solution

(a) The domain of f consists of all x for which $x - 1 > 0$ or, equivalently, $x > 1$.

(b) To obtain the graph of $y = 3 \log(x - 1)$, we begin with the graph of $y = \log x$ and use transformations. See Figure 37.

Figure 37

(a) $y = \log x$ **(b)** $y = \log (x - 1)$ **(c)** $y = 3 \log (x - 1)$

(c) The range of $f(x) = 3 \log(x - 1)$ is the set of all real numbers. The vertical asymptote is $x = 1$.

(d) We begin with $y = 3 \log(x - 1)$. The inverse function is defined (implicitly) by the equation

$$x = 3 \log(y - 1)$$

We proceed to solve for y.

$$\frac{x}{3} = \log(y - 1) \qquad \text{Isolate the logarithm.}$$

$$10^{x/3} = y - 1 \qquad \text{Change to an exponential expression.}$$

$$y = 10^{x/3} + 1 \qquad \text{Solve for y.}$$

The inverse of f is $f^{-1}(x) = 10^{x/3} + 1$.

(e) The range of f is the domain of f^{-1}, the set of all real numbers, confirming what we found from the graph of f.

(f) To graph f^{-1}, we use the graph of f in Figure 37(c) and reflect it about the line $y = x$. See Figure 38.

Figure 38

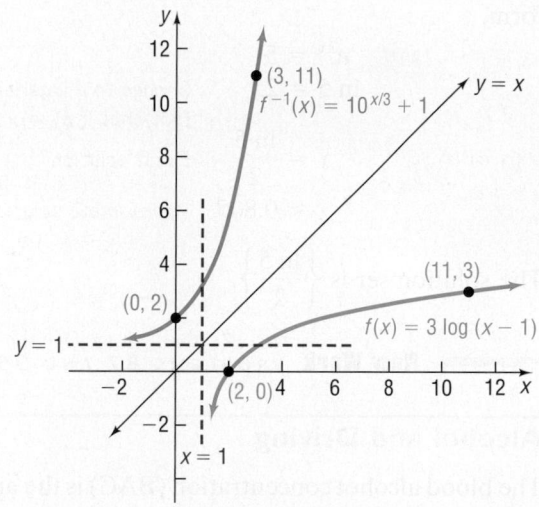

Now Work PROBLEM 79

5 Solve Logarithmic Equations

Equations that contain logarithms are called **logarithmic equations.** Care must be taken when solving logarithmic equations algebraically. In the expression $\log_a M$, remember that a and M are positive and $a \neq 1$. Be sure to check each apparent solution in the original equation and discard any that are extraneous.

Some logarithmic equations can be solved by changing from a logarithmic expression to an exponential expression.

EXAMPLE 8 **Solving a Logarithmic Equation**

Solve: (a) $\log_3(4x - 7) = 2$ (b) $\log_x 64 = 2$

Solution (a) We can obtain an exact solution by changing the logarithm to exponential form.

$$\log_3(4x - 7) = 2$$

$$4x - 7 = 3^2 \qquad \text{Change to an exponential expression.}$$

$$4x - 7 = 9$$

$$4x = 16$$

$$x = 4$$

✓Check: $\log_3(4x - 7) = \log_3(16 - 7) = \log_3 9 = 2$ $3^2 = 9$

The solution set is {4}.

(b) We can obtain an exact solution by changing the logarithm to exponential form.

$$\log_x 64 = 2$$
$$x^2 = 64 \qquad \text{Change to an exponential expression.}$$
$$x = \pm\sqrt{64} = \pm 8 \quad \text{Square Root Method}$$

The base of a logarithm is always positive. As a result, we discard -8. We check the solution 8.

✓Check: $\log_8 64 = 2$ $8^2 = 64$

The solution set is {8}.

EXAMPLE 9	**Using Logarithms to Solve Exponential Equations**

Solve: $e^{2x} = 5$

Solution We can obtain an exact solution by changing the exponential equation to logarithmic form.

$$e^{2x} = 5$$
$$\ln 5 = 2x \qquad \text{Change to a logarithmic expression using the}$$
$$\qquad\qquad\qquad \text{fact that if } e^y = x \text{ then } y = \ln x.$$
$$x = \frac{\ln 5}{2} \qquad \text{Exact solution}$$
$$\approx 0.805 \quad \text{Approximate solution}$$

The solution set is $\left\{\dfrac{\ln 5}{2}\right\}$.

Now Work PROBLEMS 87 AND 99

EXAMPLE 10	**Alcohol and Driving**

The blood alcohol concentration (BAC) is the amount of alcohol in a person's bloodstream. A BAC of 0.04% means that a person has 4 parts alcohol per 10,000 parts blood in the body. Relative risk is defined as the likelihood of one event occurring divided by the likelihood of a second event occurring. For example, if an individual with a BAC of 0.02% is 1.4 times as likely to have a car accident as an individual that has not been drinking, the relative risk of an accident with a BAC of 0.02% is 1.4. Recent medical research suggests that the relative risk R of having an accident while driving a car can be modeled by the equation

$$R = e^{kx}$$

where x is the percent of concentration of alcohol in the bloodstream and k is a constant.

(a) Medical research indicates that the relative risk of a person having an accident with a BAC of 0.02% is 1.4. Find the constant k in the equation.

(b) Using this value of k, what is the relative risk if the concentration is 0.17%?

(c) Using this same value of k, what BAC corresponds to a relative risk of 100?

(d) If the law asserts that anyone with a relative risk of 5 or more should not have driving privileges, at what concentration of alcohol in the bloodstream should a driver be arrested and charged with a DUI (driving under the influence)?

Solution (a) For a concentration of alcohol in the blood of 0.02% and a relative risk of 1.4, we let $x = 0.02$ and $R = 1.4$ in the equation and solve for k.

$$R = e^{kx}$$
$$1.4 = e^{k(0.02)} \qquad R = 1.4; x = 0.02$$

$$0.02k = \ln 1.4 \qquad \text{Change to a logarithmic expression.}$$

$$k = \frac{\ln 1.4}{0.02} \approx 16.82 \quad \text{Solve for } k.$$

(b) For a concentration of 0.17%, we have $x = 0.17$. Using $k = 16.82$ in the equation, we find the relative risk R to be

$$R = e^{kx} = e^{(16.82)(0.17)} = 17.5$$

For a concentration of alcohol in the blood of 0.17%, the relative risk of an accident is about 17.5. That is, a person with a BAC of 0.17% is 17.5 times as likely to have a car accident as a person with no alcohol in the bloodstream.

(c) For a relative risk of 100, we have $R = 100$. Using $k = 16.82$ in the equation $R = e^{kx}$, we find the concentration x of alcohol in the blood obeys

$$100 = e^{16.82x} \qquad R = e^{kx}, R = 100; k = 16.82$$

$$16.82x = \ln 100 \qquad \text{Change to a logarithmic expression.}$$

$$x = \frac{\ln 100}{16.82} \approx 0.27 \quad \text{Solve for } x.$$

NOTE A BAC of 0.30 results in a loss of consciousness in most people. ■

For a concentration of alcohol in the blood of 0.27% the relative risk of an accident is 100.

(d) For a relative risk of 5, we have $R = 5$. Using $k = 16.82$ in the equation $R = e^{kx}$, we find the concentration x of alcohol in the bloodstream obeys

$$5 = e^{16.82x}$$

$$16.82x = \ln 5$$

$$x = \frac{\ln 5}{16.82} \approx 0.096$$

NOTE Most states use 0.08 or 0.10 as the blood alcohol content at which a DUI citation is given. ■

A driver with a BAC of 0.096% or more should be arrested and charged with DUI. ■

SUMMARY Properties of the Logarithmic Function

$f(x) = \log_a x, \quad a > 1$
$(y = \log_a x \text{ means } x = a^y)$

Domain: the interval $(0, \infty)$; range: the interval $(-\infty, \infty)$

x-intercept: 1; y-intercept: none; vertical asymptote: $x = 0$ (y-axis); increasing; one-to-one

See Figure 39(a) for a typical graph.

$f(x) = \log_a x, \quad 0 < a < 1$
$(y = \log_a x \text{ means } x = a^y)$

Domain: the interval $(0, \infty)$; range: the interval $(-\infty, \infty)$

x-intercept: 1; y-intercept: none; vertical asymptote: $x = 0$ (y-axis); decreasing; one-to-one

See Figure 39(b) for a typical graph.

Figure 39

(a) $a > 1$

(b) $0 < a < 1$

6.4 Assess Your Understanding

'Are You Prepared?' *Answers are given at the end of these exercises. If you get a wrong answer, read the pages listed in red.*

1. Solve the inequality: $3x - 7 \leq 8 - 2x$ (pp. 128–131)

2. Solve the inequality: $x^2 - x - 6 > 0$ (pp. 314–316)

3. Solve the inequality: $\dfrac{x-1}{x+4} > 0$ (pp. 369–373)

Concepts and Vocabulary

4. The domain of the logarithmic function $f(x) = \log_a x$ is _____.

5. The graph of every logarithmic function $f(x) = \log_a x$, $a > 0, a \neq 1$, passes through three points: _____, _____, and _____.

6. If the graph of a logarithmic function $f(x) = \log_a x$, $a > 0, a \neq 1$, is increasing, then its base must be larger than _____.

7. **True or False** If $y = \log_a x$, then $y = a^x$.

8. **True or False** The graph of $f(x) = \log_a x, a > 0, a \neq 1$, has an x-intercept equal to 1 and no y-intercept.

Skill Building

In Problems 9–16, change each exponential expression to an equivalent expression involving a logarithm.

9. $9 = 3^2$

10. $16 = 4^2$

11. $a^2 = 1.6$

12. $a^3 = 2.1$

13. $2^x = 7.2$

14. $3^x = 4.6$

15. $e^x = 8$

16. $e^{2.2} = M$

In Problems 17–24, change each logarithmic expression to an equivalent expression involving an exponent.

17. $\log_2 8 = 3$

18. $\log_3\left(\dfrac{1}{9}\right) = -2$

19. $\log_a 3 = 6$

20. $\log_b 4 = 2$

21. $\log_3 2 = x$

22. $\log_2 6 = x$

23. $\ln 4 = x$

24. $\ln x = 4$

In Problems 25–36, find the exact value of each logarithm without using a calculator.

25. $\log_2 1$

26. $\log_8 8$

27. $\log_5 25$

28. $\log_3\left(\dfrac{1}{9}\right)$

29. $\log_{1/2} 16$

30. $\log_{1/3} 9$

31. $\log_{10}\sqrt{10}$

32. $\log_5 \sqrt[3]{25}$

33. $\log_{\sqrt{2}} 4$

34. $\log_{\sqrt{3}} 9$

35. $\ln \sqrt{e}$

36. $\ln e^3$

In Problems 37–48, find the domain of each function.

37. $f(x) = \ln(x - 3)$

38. $g(x) = \ln(x - 1)$

39. $F(x) = \log_2 x^2$

40. $H(x) = \log_5 x^3$

41. $f(x) = 3 - 2\log_4\left[\dfrac{x}{2} - 5\right]$

42. $g(x) = 8 + 5\ln(2x + 3)$

43. $f(x) = \ln\left(\dfrac{1}{x+1}\right)$

44. $g(x) = \ln\left(\dfrac{1}{x-5}\right)$

45. $g(x) = \log_5\left(\dfrac{x+1}{x}\right)$

46. $h(x) = \log_3\left(\dfrac{x}{x-1}\right)$

47. $f(x) = \sqrt{\ln x}$

48. $g(x) = \dfrac{1}{\ln x}$

In Problems 49–56, use a calculator to evaluate each expression. Round your answer to three decimal places.

49. $\ln\dfrac{5}{3}$

50. $\dfrac{\ln 5}{3}$

51. $\dfrac{\ln\dfrac{10}{3}}{0.04}$

52. $\dfrac{\ln\dfrac{2}{3}}{-0.1}$

53. $\dfrac{\ln 4 + \ln 2}{\log 4 + \log 2}$

54. $\dfrac{\log 15 + \log 20}{\ln 15 + \ln 20}$

55. $\dfrac{2\ln 5 + \log 50}{\log 4 - \ln 2}$

56. $\dfrac{3\log 80 - \ln 5}{\log 5 + \ln 20}$

57. Find a so that the graph of $f(x) = \log_a x$ contains the point $(2, 2)$.

58. Find a so that the graph of $f(x) = \log_a x$ contains the point $\left(\dfrac{1}{2}, -4\right)$.

In Problems 59–62, graph each function and its inverse on the same Cartesian plane.

59. $f(x) = 3^x; f^{-1}(x) = \log_3 x$

60. $f(x) = 4^x; f^{-1}(x) = \log_4 x$

61. $f(x) = \left(\dfrac{1}{2}\right)^x; f^{-1}(x) = \log_{\frac{1}{2}} x$

62. $f(x) = \left(\dfrac{1}{3}\right)^x; f^{-1}(x) = \log_{\frac{1}{3}} x$

In Problems 63–70, the graph of a logarithmic function is given. Match each graph to one of the following functions:

A. $y = \log_3 x$ B. $y = \log_3(-x)$ C. $y = -\log_3 x$ D. $y = -\log_3(-x)$

E. $y = \log_3 x - 1$ F. $y = \log_3(x - 1)$ G. $y = \log_3(1 - x)$ H. $y = 1 - \log_3 x$

63.

64.

65.

66.

67.

68.

69.

70.

In Problems 71–86, use the given function f to:
(a) *Find the domain of f.*
(b) *Graph f.*
(c) *From the graph, determine the range and any asymptotes of f.*
(d) *Find* f^{-1}, *the inverse of f.*
(e) *Use* f^{-1} *to find the range of f.*
(f) *Graph* f^{-1}.

71. $f(x) = \ln(x + 4)$ **72.** $f(x) = \ln(x - 3)$ **73.** $f(x) = 2 + \ln x$ **74.** $f(x) = -\ln(-x)$

75. $f(x) = \ln(2x) - 3$ **76.** $f(x) = -2\ln(x + 1)$ **77.** $f(x) = \log(x - 4) + 2$ **78.** $f(x) = \dfrac{1}{2}\log x - 5$

79. $f(x) = \dfrac{1}{2}\log(2x)$ **80.** $f(x) = \log(-2x)$ **81.** $f(x) = 3 + \log_3(x + 2)$ **82.** $f(x) = 2 - \log_3(x + 1)$

83. $f(x) = e^{x+2} - 3$ **84.** $f(x) = 3e^x + 2$ **85.** $f(x) = 2^{x/3} + 4$ **86.** $f(x) = -3^{x+1}$

In Problems 87–110, solve each equation.

87. $\log_3 x = 2$ **88.** $\log_5 x = 3$ **89.** $\log_2(2x + 1) = 3$ **90.** $\log_3(3x - 2) = 2$

91. $\log_x 4 = 2$ **92.** $\log_x\left(\dfrac{1}{8}\right) = 3$ **93.** $\ln e^x = 5$ **94.** $\ln e^{-2x} = 8$

95. $\log_4 64 = x$ **96.** $\log_5 625 = x$ **97.** $\log_3 243 = 2x + 1$ **98.** $\log_6 36 = 5x + 3$

99. $e^{3x} = 10$ **100.** $e^{-2x} = \dfrac{1}{3}$ **101.** $e^{2x+5} = 8$ **102.** $e^{-2x+1} = 13$

103. $\log_3(x^2 + 1) = 2$ **104.** $\log_5(x^2 + x + 4) = 2$ **105.** $\log_2 8^x = -3$ **106.** $\log_3 3^x = -1$

107. $5e^{0.2x} = 7$ **108.** $8 \cdot 10^{2x-7} = 3$ **109.** $2 \cdot 10^{2-x} = 5$ **110.** $4e^{x+1} = 5$

111. Suppose that $G(x) = \log_3(2x + 1)$.
 (a) What is the domain of G?
 (b) What is $G(40)$? What point is on the graph of G?
 (c) If $G(x) = 2$, what is x? What point is on the graph of G?

112. Suppose that $F(x) = \log_2(x + 1) - 3$.
 (a) What is the domain of F?
 (b) What is $F(7)$? What point is on the graph of f?
 (c) If $F(x) = -1$, what is x? What point is on the graph of F?

In Problems 113–116, graph each function. Based on the graph, state the domain and the range and find any intercepts.

113. $f(x) = \begin{cases} \ln(-x) & \text{if } x < 0 \\ \ln x & \text{if } x > 0 \end{cases}$

114. $f(x) = \begin{cases} \ln(-x) & \text{if } x \leq -1 \\ -\ln(-x) & \text{if } -1 < x < 0 \end{cases}$

115. $f(x) = \begin{cases} -\ln x & \text{if } 0 < x < 1 \\ \ln x & \text{if } x \geq 1 \end{cases}$

116. $f(x) = \begin{cases} \ln x & \text{if } 0 < x < 1 \\ -\ln x & \text{if } x \geq 1 \end{cases}$

Applications and Extensions

117. Chemistry The pH of a chemical solution is given by the formula

$$\text{pH} = -\log_{10}[\text{H}^+]$$

where $[\text{H}^+]$ is the concentration of hydrogen ions in moles per liter. Values of pH range from 0 (acidic) to 14 (alkaline).
 (a) What is the pH of a solution for which $[\text{H}^+]$ is 0.1?
 (b) What is the pH of a solution for which $[\text{H}^+]$ is 0.01?
 (c) What is the pH of a solution for which $[\text{H}^+]$ is 0.001?
 (d) What happens to pH as the hydrogen ion concentration decreases?
 (e) Determine the hydrogen ion concentration of an orange (pH = 3.5).
 (f) Determine the hydrogen ion concentration of human blood (pH = 7.4).

118. Diversity Index **Shannon's diversity index** is a measure of the diversity of a population. The diversity index is given by the formula

$$H = -(p_1 \log p_1 + p_2 \log p_2 + \cdots + p_n \log p_n)$$

where p_1 is the proportion of the population that is species 1, p_2 is the proportion of the population that is species 2, and so on.
 (a) According to the U.S. Census Bureau, the distribution of race in the United States in 2000 was as follows:

Race	Proportion
American Indian or Native Alaskan	0.014
Asian	0.041
Black or African American	0.128
Hispanic	0.124
Native Hawaiian or Pacific Islander	0.003
White	0.690

Source: U.S. Census Bureau

 Compute the diversity index of the United States in 2000.
 (b) The largest value of the diversity index is given by $H_{\max} = \log(S)$, where S is the number of categories of race. Compute $H_{\max}$.
 (c) The **evenness ratio** is given by $E_H = \dfrac{H}{H_{\max}}$, where $0 \leq E_H \leq 1$. If $E_H = 1$, there is complete evenness. Compute the evenness ratio for the United States.
 (d) Obtain the distribution of race for the United States in 1990 from the Census Bureau. Compute Shannon's diversity index. Is the United States becoming more diverse? Why?

119. Atmospheric Pressure The atmospheric pressure p on a balloon or an aircraft decreases with increasing height. This pressure, measured in millimeters of mercury, is related to the height h (in kilometers) above sea level by the formula

$$p = 760e^{-0.145h}$$

 (a) Find the height of an aircraft if the atmospheric pressure is 320 millimeters of mercury.
 (b) Find the height of a mountain if the atmospheric pressure is 667 millimeters of mercury.

120. Healing of Wounds The normal healing of wounds can be modeled by an exponential function. If A_0 represents the original area of the wound and if A equals the area of the wound, then the formula

$$A = A_0 e^{-0.35n}$$

describes the area of a wound after n days following an injury when no infection is present to retard the healing. Suppose that a wound initially had an area of 100 square millimeters.
 (a) If healing is taking place, after how many days will the wound be one-half its original size?
 (b) How long before the wound is 10% of its original size?

121. Exponential Probability Between 12:00 PM and 1:00 PM, cars arrive at Citibank's drive-thru at the rate of 6 cars per hour (0.1 car per minute). The following formula from statistics can be used to determine the probability that a car will arrive within t minutes of 12:00 PM.

$$F(t) = 1 - e^{-0.1t}$$

 (a) Determine how many minutes are needed for the probability to reach 50%.
 (b) Determine how many minutes are needed for the probability to reach 80%.
 (c) Is it possible for the probability to equal 100%? Explain.

122. Exponential Probability Between 5:00 PM and 6:00 PM, cars arrive at Jiffy Lube at the rate of 9 cars per hour (0.15 car per minute). The following formula from statistics can be used to determine the probability that a car will arrive within t minutes of 5:00 PM.

$$F(t) = 1 - e^{-0.15t}$$

 (a) Determine how many minutes are needed for the probability to reach 50%.
 (b) Determine how many minutes are needed for the probability to reach 80%.

123. Drug Medication The formula

$$D = 5e^{-0.4h}$$

can be used to find the number of milligrams D of a certain drug that is in a patient's bloodstream h hours after the drug was administered. When the number of milligrams reaches 2, the drug is to be administered again. What is the time between injections?

124. Spreading of Rumors A model for the number N of people in a college community who have heard a certain rumor is

$$N = P(1 - e^{-0.15d})$$

where P is the total population of the community and d is the number of days that have elapsed since the rumor began. In a community of 1000 students, how many days will elapse before 450 students have heard the rumor?

125. Current in a RL Circuit The equation governing the amount of current I (in amperes) after time t (in seconds) in a simple RL circuit consisting of a resistance R (in ohms), an inductance L (in henrys), and an electromotive force E (in volts) is

$$I = \frac{E}{R}[1 - e^{-(R/L)t}]$$

If $E = 12$ volts, $R = 10$ ohms, and $L = 5$ henrys, how long does it take to obtain a current of 0.5 ampere? Of 1.0 ampere? Graph the equation.

126. Learning Curve Psychologists sometimes use the function

$$L(t) = A(1 - e^{-kt})$$

to measure the amount L learned at time t. The number A represents the amount to be learned, and the number k measures the rate of learning. Suppose that a student has an amount A of 200 vocabulary words to learn. A psychologist determines that the student learned 20 vocabulary words after 5 minutes.

(a) Determine the rate of learning k.

(b) Approximately how many words will the student have learned after 10 minutes?

(c) After 15 minutes?

(d) How long does it take for the student to learn 180 words?

Loudness of Sound *Problems 127–130 use the following discussion: The **loudness** $L(x)$, measured in decibels, of a sound of intensity x, measured in watts per square meter, is defined as $L(x) = 10 \log \dfrac{x}{I_0}$, where $I_0 = 10^{-12}$ watt per square meter is the least intense sound that a human ear can detect. Determine the loudness, in decibels, of each of the following sounds.*

127. Normal conversation: intensity of $x = 10^{-7}$ watt per square meter.

128. Amplified rock music: intensity of 10^{-1} watt per square meter.

129. Heavy city traffic: intensity of $x = 10^{-3}$ watt per square meter.

130. Diesel truck traveling 40 miles per hour 50 feet away: intensity 10 times that of a passenger car traveling 50 miles per hour 50 feet away whose loudness is 70 decibels.

The Richter Scale *Problems 131 and 132 use the following discussion: The **Richter scale** is one way of converting seismographic readings into numbers the Richter scale that provide an easy reference for measuring the magnitude M of an earthquake. All earthquakes are compared to a **zero-level earthquake** whose seismographic reading measures 0.001 millimeter at a distance of 100 kilometers from the epicenter. An earthquake whose seismographic reading measures x millimeters has **magnitude** $M(x)$, given by*

$$M(x) = \log\left(\frac{x}{x_0}\right)$$

where $x_0 = 10^{-3}$ is the reading of a zero-level earthquake the same distance from its epicenter. In Problems 131 and 132, determine the magnitude of each earthquake.

131. Magnitude of an Earthquake Mexico City in 1985: seismographic reading of 125,892 millimeters 100 kilometers from the center.

132. Magnitude of an Earthquake San Francisco in 1906: seismographic reading of 7943 millimeters 100 kilometers from the center.

133. Alcohol and Driving The concentration of alcohol in a person's bloodstream is measurable. Suppose that the relative risk R of having an accident while driving a car can be modeled by the equation

$$R = e^{kx}$$

where x is the percent of concentration of alcohol in the bloodstream and k is a constant.

(a) Suppose that a concentration of alcohol in the bloodstream of 0.03 percent results in a relative risk of an accident of 1.4. Find the constant k in the equation.

(b) Using this value of k, what is the relative risk if the concentration is 0.17 percent?

(c) Using the same value of k, what concentration of alcohol corresponds to a relative risk of 100?

(d) If the law asserts that anyone with a relative risk of having an accident of 5 or more should not have driving privileges, at what concentration of alcohol in the bloodstream should a driver be arrested and charged with a DUI?

(e) Compare this situation with that of Example 10. If you were a lawmaker, which situation would you support? Give your reasons.

Discussion and Writing

134. Is there any function of the form $y = x^{\alpha}, 0 < \alpha < 1$, that increases more slowly than a logarithmic function whose base is greater than 1? Explain.

135. In the definition of the logarithmic function, the base a is not allowed to equal 1. Why?

136. Critical Thinking In buying a new car, one consideration might be how well the price of the car holds up over time. Different makes of cars have different depreciation rates. One way to compute a depreciation rate for a car is given here. Suppose that the current prices of a certain Mercedes automobile are as follows:

		Age in Years			
New	**1**	**2**	**3**	**4**	**5**
$38,000	$36,600	$32,400	$28,750	$25,400	$21,200

Use the formula New = Old(e^{Rt}) to find R, the annual depreciation rate, for a specific time t. When might be the best time to trade in the car? Consult the NADA ("blue") book and compare two like models that you are interested in. Which has the better depreciation rate?

'Are You Prepared?' Answers

1. $x \leq 3$ **2.** $x < -2$ or $x > 3$ **3.** $x < -4$ or $x > 1$

6.5 Properties of Logarithms

OBJECTIVES 1 Work with the Properties of Logarithms (p. 450)

2 Write a Logarithmic Expression as a Sum or Difference of Logarithms (p. 452)

3 Write a Logarithmic Expression as a Single Logarithm (p. 453)

4 Evaluate Logarithms Whose Base Is Neither 10 nor e (p. 454)

1 Work with the Properties of Logarithms

Logarithms have some very useful properties that can be derived directly from the definition and the laws of exponents.

EXAMPLE 1 **Establishing Properties of Logarithms**

(a) Show that $\log_a 1 = 0$. (b) Show that $\log_a a = 1$.

Solution (a) This fact was established when we graphed $y = \log_a x$ (see Figure 30). To show the result algebraically, let $y = \log_a 1$. Then

$$y = \log_a 1$$
$$a^y = 1 \qquad \text{Change to an exponential expression.}$$
$$a^y = a^0 \qquad a^0 = 1$$
$$y = 0 \qquad \text{Solve for } y.$$
$$\log_a 1 = 0 \qquad y = \log_a 1$$

(b) Let $y = \log_a a$. Then

$$y = \log_a a$$
$$a^y = a \qquad \text{Change to an exponential expression.}$$
$$a^y = a^1 \qquad a = a^1$$
$$y = 1 \qquad \text{Solve for } y.$$
$$\log_a a = 1 \qquad y = \log_a a$$

To summarize:

$$\log_a 1 = 0 \qquad \log_a a = 1$$

THEOREM **Properties of Logarithms**

In the properties given next, M and a are positive real numbers, $a \neq 1$, and r is any real number.

The number $\log_a M$ is the exponent to which a must be raised to obtain M. That is,

$$a^{\log_a M} = M \tag{1}$$

The logarithm to the base a of a raised to a power equals that power. That is,

$$\log_a a^r = r \tag{2}$$

The proof uses the fact that $y = a^x$ and $y = \log_a x$ are inverses.

Proof of Property (1) For inverse functions,

$$f(f^{-1}(x)) = x \quad \text{for all } x \text{ in the domain of } f^{-1}$$

Using $f(x) = a^x$ and $f^{-1}(x) = \log_a x$, we find

$$f(f^{-1}(x)) = a^{\log_a x} = x \quad \text{for } x > 0$$

Now let $x = M$ to obtain $a^{\log_a M} = M$, where $M > 0$. ∎

Proof of Property (2) For inverse functions,

$$f^{-1}(f(x)) = x \quad \text{for all } x \text{ in the domain of } f$$

Using $f(x) = a^x$ and $f^{-1}(x) = \log_a x$, we find

$$f^{-1}(f(x)) = \log_a a^x = x \quad \text{for all real numbers } x$$

Now let $x = r$ to obtain $\log_a a^r = r$, where r is any real number. ∎

EXAMPLE 2 **Using Properties (1) and (2)**

(a) $2^{\log_2 \pi} = \pi$ (b) $\log_{0.2} 0.2^{-\sqrt{2}} = -\sqrt{2}$ (c) $\ln e^{kt} = kt$

━━━ **Now Work** PROBLEM 9

Other useful properties of logarithms are given next.

THEOREM **Properties of Logarithms**

In the following properties, M, N, and a are positive real numbers, $a \neq 1$, and r is any real number.

The Log of a Product Equals the Sum of the Logs

$$\log_a(MN) = \log_a M + \log_a N \tag{3}$$

The Log of a Quotient Equals the Difference of the Logs

$$\log_a\left(\frac{M}{N}\right) = \log_a M - \log_a N \tag{4}$$

The Log of a Power Equals the Product of the Power and the Log

$$\log_a M^r = r \log_a M \tag{5}$$

We shall derive properties (3) and (5) and leave the derivation of property (4) as an exercise (see Problem 103).

Proof of Property (3) Let $A = \log_a M$ and let $B = \log_a N$. These expressions are equivalent to the exponential expressions

$$a^A = M \quad \text{and} \quad a^B = N$$

Now

$$
\begin{aligned}
\log_a(MN) = \log_a(a^A a^B) &= \log_a a^{A+B} && \text{Law of Exponents}\\
&= A + B && \text{Property (2) of logarithms}\\
&= \log_a M + \log_a N
\end{aligned}
$$
∎

Proof of Property (5) Let $A = \log_a M$. This expression is equivalent to

$$a^A = M$$

Now

$$
\begin{aligned}
\log_a M^r = \log_a(a^A)^r &= \log_a a^{rA} && \text{Law of Exponents}\\
&= rA && \text{Property (2) of logarithms}\\
&= r \log_a M
\end{aligned}
$$
∎

── **Now Work** PROBLEM **13**

2 Write a Logarithmic Expression as a Sum or Difference of Logarithms

 Logarithms can be used to transform products into sums, quotients into differences, and powers into factors. Such transformations prove useful in certain types of calculus problems.

EXAMPLE 3 | **Writing a Logarithmic Expression as a Sum of Logarithms**

Write $\log_a\!\left(x\sqrt{x^2 + 1}\right)$, $x > 0$, as a sum of logarithms. Express all powers as factors.

Solution
$$
\begin{aligned}
\log_a\!\left(x\sqrt{x^2 + 1}\right) &= \log_a x + \log_a \sqrt{x^2 + 1} && \log_a(M \cdot N) = \log_a M + \log_a N\\
&= \log_a x + \log_a(x^2 + 1)^{1/2}\\
&= \log_a x + \frac{1}{2}\log_a(x^2 + 1) && \log_a M^r = r \log_a M
\end{aligned}
$$
∎

EXAMPLE 4 | **Writing a Logarithmic Expression as a Difference of Logarithms**

Write

$$\ln \frac{x^2}{(x - 1)^3} \qquad x > 1$$

as a difference of logarithms. Express all powers as factors.

Solution

$$\ln \frac{x^2}{(x-1)^3} = \ln x^2 - \ln(x-1)^3 = 2\ln x - 3\ln(x-1)$$

$$\underset{\uparrow}{\log_a\left(\frac{M}{N}\right) = \log_a M - \log_a N} \qquad \underset{\uparrow}{\log_a M^r = r\log_a M}$$

EXAMPLE 5 | **Writing a Logarithmic Expression as a Sum and Difference of Logarithms**

Write

$$\log_a \frac{\sqrt{x^2+1}}{x^3(x+1)^4} \quad x > 0$$

as a sum and difference of logarithms. Express all powers as factors.

Solution

WARNING In using properties (3) through (5), be careful about the values that the variable may assume. For example, the domain of the variable for $\log_a x$ is $x > 0$ and for $\log_a(x-1)$ it is $x > 1$. If we add these functions, the domain is $x > 1$. That is, the equality

$$\log_a x + \log_a(x-1) = \log_a[x(x-1)]$$

is true only for $x > 1$. ■

$$\log_a \frac{\sqrt{x^2+1}}{x^3(x+1)^4} = \log_a \sqrt{x^2+1} - \log_a[x^3(x+1)^4] \qquad \text{Property (4)}$$

$$= \log_a \sqrt{x^2+1} - [\log_a x^3 + \log_a(x+1)^4] \qquad \text{Property (3)}$$

$$= \log_a(x^2+1)^{1/2} - \log_a x^3 - \log_a(x+1)^4$$

$$= \frac{1}{2}\log_a(x^2+1) - 3\log_a x - 4\log_a(x+1) \qquad \text{Property (5)}$$

■

✏ **Now Work** PROBLEM 45

3 Write a Logarithmic Expression as a Single Logarithm

Another use of properties (3) through (5) is to write sums and/or differences of logarithms with the same base as a single logarithm. This skill will be needed to solve certain logarithmic equations discussed in the next section.

EXAMPLE 6 | **Writing Expressions as a Single Logarithm**

Write each of the following as a single logarithm.

(a) $\log_a 7 + 4\log_a 3$

(b) $\frac{2}{3}\ln 8 - \ln(3^4 - 8)$

(c) $\log_a x + \log_a 9 + \log_a(x^2+1) - \log_a 5$

Solution

(a) $\log_a 7 + 4\log_a 3 = \log_a 7 + \log_a 3^4 \qquad r\log_a M = \log_a M^r$

$$= \log_a 7 + \log_a 81$$

$$= \log_a(7 \cdot 81) \qquad \log_a M + \log_a N = \log_a(M \cdot N)$$

$$= \log_a 567$$

(b) $\frac{2}{3}\ln 8 - \ln(3^4 - 8) = \ln 8^{2/3} - \ln(81 - 8) \qquad r\log_a M = \log_a M^r$

$$= \ln 4 - \ln 73$$

$$= \ln\left(\frac{4}{73}\right) \qquad \log_a M - \log_a N = \log_a\left(\frac{M}{N}\right)$$

(c) $\log_a x + \log_a 9 + \log_a(x^2 + 1) - \log_a 5 = \log_a(9x) + \log_a(x^2 + 1) - \log_a 5$

$$= \log_a[9x(x^2 + 1)] - \log_a 5$$

$$= \log_a\left[\frac{9x(x^2 + 1)}{5}\right]$$

WARNING A common error made by some students is to express the logarithm of a sum as the sum of logarithms.

$$\log_a(M + N) \quad \text{is not equal to} \quad \log_a M + \log_a N$$

Correct statement $\quad \log_a(MN) = \log_a M + \log_a N \qquad$ Property (3)

Another common error is to express the difference of logarithms as the quotient of logarithms.

$$\log_a M - \log_a N \quad \text{is not equal to} \quad \frac{\log_a M}{\log_a N}$$

Correct statement $\quad \log_a M - \log_a N = \log_a\left(\dfrac{M}{N}\right) \qquad$ Property (4)

A third common error is to express a logarithm raised to a power as the product of the power times the logarithm.

$$(\log_a M)^r \quad \text{is not equal to} \quad r\log_a M$$

Correct statement $\quad \log_a M^r = r\log_a M \qquad$ Property (5) ∎

─── **Now Work** PROBLEM 51

Two other properties of logarithms that we need to know are consequences of the fact that the logarithmic function $y = \log_a x$ is a one-to-one function.

THEOREM **Properties of Logarithms**

In the following properties, M, N, and a are positive real numbers, $a \neq 1$.

> If $M = N$, then $\log_a M = \log_a N$. $\qquad$ **(6)**
>
> If $\log_a M = \log_a N$, then $M = N$. $\qquad$ **(7)**

When property (6) is used, we start with the equation $M = N$ and say "take the logarithm of both sides" to obtain $\log_a M = \log_a N$.

Properties (6) and (7) are useful for solving *exponential and logarithmic equations,* a topic discussed in the next section.

4 Evaluate Logarithms Whose Base Is Neither 10 nor *e*

Logarithms to the base 10, common logarithms, were used to facilitate arithmetic computations before the widespread use of calculators. (See the Historical Feature at the end of this section.) Natural logarithms, that is, logarithms whose base is the number *e*, remain very important because they arise frequently in the study of natural phenomena.

Common logarithms are usually abbreviated by writing **log,** with the base understood to be 10, just as natural logarithms are abbreviated by **ln,** with the base understood to be *e*.

Most calculators have both ⬚ log ⬚ and ⬚ ln ⬚ keys to calculate the common logarithm and natural logarithm of a number. Let's look at an example to see how to approximate logarithms having a base other than 10 or *e*.

| EXAMPLE 7 | **Approximating a Logarithm Whose Base Is Neither 10 nor e** |

Approximate $\log_2 7$. Round the answer to four decimal places.

Solution Let $y = \log_2 7$. Then $2^y = 7$, so

$$2^y = 7$$

$$\ln 2^y = \ln 7 \qquad \text{Property (6)}$$

$$y \ln 2 = \ln 7 \qquad \text{Property (5)}$$

$$y = \frac{\ln 7}{\ln 2} \qquad \text{Exact value}$$

$$y \approx 2.8074 \qquad \text{Approximate value rounded to four decimal places}$$

Example 7 shows how to approximate a logarithm whose base is 2 by changing to logarithms involving the base e. In general, we use the **Change-of-Base Formula.**

THEOREM **Change-of-Base Formula**

If $a \neq 1$, $b \neq 1$, and M are positive real numbers, then

$$\log_a M = \frac{\log_b M}{\log_b a} \qquad \qquad (8)$$

Proof We derive this formula as follows: Let $y = \log_a M$. Then

$$a^y = M$$

$$\log_b a^y = \log_b M \qquad \text{Property (6)}$$

$$y \log_b a = \log_b M \qquad \text{Property (5)}$$

$$y = \frac{\log_b M}{\log_b a} \qquad \text{Solve for } y.$$

$$\log_a M = \frac{\log_b M}{\log_b a} \qquad y = \log_a M$$

Since calculators have keys only for $\boxed{\log}$ and $\boxed{\ln}$, in practice, the Change-of-Base Formula uses either $b = 10$ or $b = e$. That is,

$$\log_a M = \frac{\log M}{\log a} \quad \text{and} \quad \log_a M = \frac{\ln M}{\ln a} \qquad \qquad (9)$$

| EXAMPLE 8 | **Using the Change-of-Base Formula** |

Approximate: (a) $\log_5 89$

(b) $\log_{\sqrt{2}} \sqrt{5}$

Round answers to four decimal places.

Solution (a) $\log_5 89 = \dfrac{\log 89}{\log 5} \approx \dfrac{1.949390007}{0.6989700043} \approx 2.7889$

or

$$\log_5 89 = \frac{\ln 89}{\ln 5} \approx \frac{4.48863637}{1.609437912} \approx 2.7889$$

$$\text{(b) } \log_{\sqrt{2}} \sqrt{5} = \frac{\log \sqrt{5}}{\log \sqrt{2}} = \frac{\frac{1}{2} \log 5}{\frac{1}{2} \log 2} = \frac{\log 5}{\log 2} \approx 2.3219$$

or

$$\log_{\sqrt{2}} \sqrt{5} = \frac{\ln \sqrt{5}}{\ln \sqrt{2}} = \frac{\frac{1}{2} \ln 5}{\frac{1}{2} \ln 2} = \frac{\ln 5}{\ln 2} \approx 2.3219$$

──── **Now Work** PROBLEMS 17 AND 65

 COMMENT To graph logarithmic functions when the base is different from e or 10 requires the Change-of-Base Formula. For example, to graph $y = \log_2 x$, we would instead graph $y = \dfrac{\ln x}{\ln 2}$. Try it.

──── **Now Work** PROBLEM 73

SUMMARY Properties of Logarithms

In the list that follows, a, b, M, N, and r are real numbers. Also, $a > 0$, $a \neq 1$, $b > 0$, $b \neq 1$, $M > 0$ and $N > 0$.

Definition $y = \log_a x$ means $x = a^y$

Properties of logarithms $\log_a 1 = 0;$ $\log_a a = 1$ $\log_a M^r = r \log_a M$

$a^{\log_a M} = M;$ $\log_a a^r = r$ If $M = N$, then $\log_a M = \log_a N$.

$\log_a(MN) = \log_a M + \log_a N$ If $\log_a M = \log_a N$, then $M = N$.

$\log_a\left(\dfrac{M}{N}\right) = \log_a M - \log_a N$

Change-of-Base Formula $\log_a M = \dfrac{\log_b M}{\log_b a}$

Historical Feature

John Napier
(1550–1617)

ogarithms were invented about 1590 by John Napier (1550–1617) and Joost Bürgi (1552–1632), working independently. Napier, whose work had the greater influence, was a Scottish lord, a secretive man whose neighbors were inclined to believe him to be in league with the devil. His approach to logarithms was very different from ours; it was based on the relationship between arithmetic and geometric sequences, discussed in a later chapter, and not on the inverse function relationship of logarithms to exponential functions (described in Sec-

tion 6.4). Napier's tables, published in 1614, listed what would now be called *natural logarithms* of sines and were rather difficult to use. A London professor, Henry Briggs, became interested in the tables and visited Napier. In their conversations, they developed the idea of common logarithms, which were published in 1617. Their importance for calculation was immediately recognized, and by 1650 they were being printed as far away as China. They remained an important calculation tool until the advent of the inexpensive handheld calculator about 1972, which has decreased their calculational, but not their theoretical, importance.

A side effect of the invention of logarithms was the popularization of the decimal system of notation for real numbers.

6.5 Assess Your Understanding

Concepts and Vocabulary

1. The logarithm of a product equals the _____ of the logarithms.

2. If $\log_8 M = \dfrac{\log_5 7}{\log_5 8}$, then $M =$ _____.

3. $\log_a M^r =$ _____.

4. True or False $\ln(x + 3) - \ln(2x) = \dfrac{\ln(x + 3)}{\ln(2x)}$

5. True or False $\log_2(3x^4) = 4 \log_2(3x)$

6. True or False $\log_2 16 = \dfrac{\ln 16}{\ln 2}$

Skill Building

In Problems 7–22, use properties of logarithms to find the exact value of each expression. Do not use a calculator.

7. $\log_3 3^{71}$

8. $\log_2 2^{-13}$

9. $\ln e^{-4}$

10. $\ln e^{\sqrt{2}}$

11. $2^{\log_2 7}$

12. $e^{\ln 8}$

13. $\log_8 2 + \log_8 4$

14. $\log_6 9 + \log_6 4$

15. $\log_6 18 - \log_6 3$

16. $\log_8 16 - \log_8 2$

17. $\log_2 6 \cdot \log_6 4$

18. $\log_3 8 \cdot \log_8 9$

19. $3^{\log_3 5 - \log_3 4}$

20. $5^{\log_5 6 + \log_5 7}$

21. $e^{\log_{e^2} 16}$

22. $e^{\log_{e^2} 9}$

In Problems 23–30, suppose that $\ln 2 = a$ and $\ln 3 = b$. Use properties of logarithms to write each logarithm in terms of a and b.

23. $\ln 6$

24. $\ln \dfrac{2}{3}$

25. $\ln 1.5$

26. $\ln 0.5$

27. $\ln 8$

28. $\ln 27$

29. $\ln \sqrt[5]{6}$

30. $\ln \sqrt[4]{\dfrac{2}{3}}$

In Problems 31–50, write each expression as a sum and/or difference of logarithms. Express powers as factors.

31. $\log_5(25x)$

32. $\log_3 \dfrac{x}{9}$

33. $\log_2 z^3$

34. $\log_7(x^5)$

35. $\ln(ex)$

36. $\ln \dfrac{e}{x}$

37. $\ln(xe^x)$

38. $\ln \dfrac{x}{e^x}$

39. $\log_a(u^2 v^3)$ $u > 0, v > 0$

40. $\log_2\left(\dfrac{a}{b^2}\right)$ $a > 0, b > 0$

41. $\ln\left(x^2\sqrt{1 - x}\right)$ $0 < x < 1$

42. $\ln\left(x\sqrt{1 + x^2}\right)$ $x > 0$

43. $\log_2\left(\dfrac{x^3}{x - 3}\right)$ $x > 3$

44. $\log_5\left(\dfrac{\sqrt[3]{x^2 + 1}}{x^2 - 1}\right)$ $x > 1$

45. $\log\left[\dfrac{x(x + 2)}{(x + 3)^2}\right]$ $x > 0$

46. $\log\left[\dfrac{x^3\sqrt{x + 1}}{(x - 2)^2}\right]$ $x > 2$

47. $\ln\left[\dfrac{x^2 - x - 2}{(x + 4)^2}\right]^{1/3}$ $x > 2$

48. $\ln\left[\dfrac{(x - 4)^2}{x^2 - 1}\right]^{2/3}$ $x > 4$

49. $\ln \dfrac{5x\sqrt{1 + 3x}}{(x - 4)^3}$ $x > 4$

50. $\ln\left[\dfrac{5x^2\sqrt[3]{1 - x}}{4(x + 1)^2}\right]$ $0 < x < 1$

In Problems 51–64, write each expression as a single logarithm.

51. $3 \log_5 u + 4 \log_5 v$

52. $2 \log_3 u - \log_3 v$

53. $\log_3 \sqrt{x} - \log_3 x^3$

54. $\log_2\left(\dfrac{1}{x}\right) + \log_2\left(\dfrac{1}{x^2}\right)$

55. $\log_4(x^2 - 1) - 5 \log_4(x + 1)$

56. $\log(x^2 + 3x + 2) - 2 \log(x + 1)$

57. $\ln\left(\dfrac{x}{x - 1}\right) + \ln\left(\dfrac{x + 1}{x}\right) - \ln(x^2 - 1)$

58. $\log\left(\dfrac{x^2 + 2x - 3}{x^2 - 4}\right) - \log\left(\dfrac{x^2 + 7x + 6}{x + 2}\right)$

59. $8 \log_2 \sqrt{3x - 2} - \log_2\left(\dfrac{4}{x}\right) + \log_2 4$

60. $21 \log_3 \sqrt[3]{x} + \log_3(9x^2) - \log_3 9$

61. $2 \log_a(5x^3) - \dfrac{1}{2} \log_a(2x + 3)$

62. $\dfrac{1}{3} \log(x^3 + 1) + \dfrac{1}{2} \log(x^2 + 1)$

63. $2 \log_2(x + 1) - \log_2(x + 3) - \log_2(x - 1)$

64. $3 \log_5(3x + 1) - 2 \log_5(2x - 1) - \log_5 x$

In Problems 65–72, use the Change-of-Base Formula and a calculator to evaluate each logarithm. Round your answer to three decimal places.

65. $\log_3 21$ **66.** $\log_5 18$ **67.** $\log_{1/3} 71$ **68.** $\log_{1/2} 15$

69. $\log_{\sqrt{2}} 7$ **70.** $\log_{\sqrt{5}} 8$ **71.** $\log_\pi e$ **72.** $\log_\pi \sqrt{2}$

In Problems 73–78, graph each function using a graphing utility and the Change-of-Base Formula.

73. $y = \log_4 x$ **74.** $y = \log_5 x$ **75.** $y = \log_2(x + 2)$ **76.** $y = \log_4(x - 3)$

77. $y = \log_{x-1}(x + 1)$ **78.** $y = \log_{x+2}(x - 2)$

Applications and Extensions

79. If $f(x) = \ln x$, $g(x) = e^x$, and $h(x) = x^2$, find:
(a) $(f \circ g)(x)$. What is the domain of $f \circ g$?
(b) $(g \circ f)(x)$. What is the domain of $g \circ f$?
(c) $(f \circ g)(5)$
(d) $(f \circ h)(x)$. What is the domain of $f \circ h$?
(e) $(f \circ h)(e)$

80. If $f(x) = \log_2 x$, $g(x) = 2^x$, and $h(x) = 4x$, find:
(a) $(f \circ g)(x)$. What is the domain of $f \circ g$?
(b) $(g \circ f)(x)$. What is the domain of $g \circ f$?
(c) $(f \circ g)(3)$
(d) $(f \circ h)(x)$. What is the domain of $f \circ h$?
(e) $(f \circ h)(8)$

In Problems 81–90, express y as a function of x. The constant C is a positive number.

81. $\ln y = \ln x + \ln C$

82. $\ln y = \ln(x + C)$

83. $\ln y = \ln x + \ln(x + 1) + \ln C$

84. $\ln y = 2 \ln x - \ln(x + 1) + \ln C$

85. $\ln y = 3x + \ln C$

86. $\ln y = -2x + \ln C$

87. $\ln(y - 3) = -4x + \ln C$

88. $\ln(y + 4) = 5x + \ln C$

89. $3 \ln y = \dfrac{1}{2}\ln(2x + 1) - \dfrac{1}{3}\ln(x + 4) + \ln C$

90. $2 \ln y = -\dfrac{1}{2}\ln x + \dfrac{1}{3}\ln(x^2 + 1) + \ln C$

91. Find the value of $\log_2 3 \cdot \log_3 4 \cdot \log_4 5 \cdot \log_5 6 \cdot \log_6 7 \cdot \log_7 8$

92. Find the value of $\log_2 4 \cdot \log_4 6 \cdot \log_6 8$

93. Find the value of $\log_2 3 \cdot \log_3 4 \cdot \cdots \cdot \log_n(n + 1) \cdot \log_{n+1} 2$

94. Find the value of $\log_2 2 \cdot \log_2 4 \cdot \cdots \cdot \log_2 2^n$

95. Show that $\log_a\left(x + \sqrt{x^2 - 1}\right) + \log_a\left(x - \sqrt{x^2 - 1}\right) = 0$.

96. Show that $\log_a\left(\sqrt{x} + \sqrt{x - 1}\right) + \log_a\left(\sqrt{x} - \sqrt{x - 1}\right) = 0$.

97. Show that $\ln(1 + e^{2x}) = 2x + \ln(1 + e^{-2x})$.

98. Difference Quotient If $f(x) = \log_a x$, show that $\dfrac{f(x + h) - f(x)}{h} = \log_a\left(1 + \dfrac{h}{x}\right)^{1/h}$, $h \neq 0$.

99. If $f(x) = \log_a x$, show that $-f(x) = \log_{1/a} x$.

100. If $f(x) = \log_a x$, show that $f(AB) = f(A) + f(B)$.

101. If $f(x) = \log_a x$, show that $f\left(\dfrac{1}{x}\right) = -f(x)$.

102. If $f(x) = \log_a x$, show that $f(x^\alpha) = \alpha f(x)$.

103. Show that $\log_a\left(\dfrac{M}{N}\right) = \log_a M - \log_a N$, where a, M, and N are positive real numbers and $a \neq 1$.

104. Show that $\log_a\left(\dfrac{1}{N}\right) = -\log_a N$, where a and N are positive real numbers and $a \neq 1$.

Discussion and Writing

105. Graph $Y_1 = \log(x^2)$ and $Y_2 = 2\log(x)$ using a graphing utility. Are they equivalent? What might account for any differences in the two functions?

106. Write an example that illustrates why $(\log_a x)^r \neq r \log_a x$.

107. Write an example that illustrates why $\log_2(x + y) \neq \log_2 x + \log_2 y$.

108. Does $3^{\log_3(-5)} = -5$? Why or why not?

6.6 Logarithmic and Exponential Equations

PREPARING FOR THIS SECTION *Before getting started, review the following:*

- Solving Equations Using a Graphing Utility (Appendix, Section 4, pp. A8–A10)
- Equations Quadratic in Form (Section 1.4, pp. 119–121)
- Solving Quadratic Equations (Section 1.2, pp. 97–106)

Now Work the 'Are You Prepared?' problems on page 463.

OBJECTIVES **1** Solve Logarithmic Equations (p. 459)

2 Solve Exponential Equations (p. 461)

3 Solve Logarithmic and Exponential Equations Using a Graphing Utility (p. 462)

1 Solve Logarithmic Equations

In Section 6.4 we solved logarithmic equations by changing a logarithmic expression to an exponential expression. That is, we used the definition of a logarithm:

$$y = \log_a x \quad \text{is equivalent to} \quad x = a^y \qquad a > 0, a \neq 1$$

For example, to solve the equation $\log_2(1 - 2x) = 3$, we use the equivalent exponential expression $1 - 2x = 2^3$ and solve for x.

$$
\begin{aligned}
\log_2(1 - 2x) &= 3 \\
1 - 2x &= 2^3 \qquad \text{\small Change to an exponential expression.} \\
-2x &= 7 \qquad \text{\small Simplify.} \\
x &= -\frac{7}{2} \qquad \text{\small Solve.}
\end{aligned}
$$

You should check this solution for yourself.

For most logarithmic equations, some manipulation of the equation (usually using properties of logarithms) is required to obtain a solution. Also, to avoid extraneous solutions with logarithmic equations, we determine the domain of the variable first.

We begin with an example of a logarithmic equation that requires using the fact that a logarithmic function is a one-to-one function.

$$\text{If } \log_a M = \log_a N, \text{ then } M = N \qquad M, N, \text{ and } a \text{ are positive and } a \neq 1$$

EXAMPLE 1 **Solving a Logarithmic Equation**

Solve: $2 \log_5 x = \log_5 9$

Solution The domain of the variable in this equation is $x > 0$. Because each logarithm is to the same base, 5, we can obtain an exact solution as follows:

$$
\begin{aligned}
2 \log_5 x &= \log_5 9 \\
\log_5 x^2 &= \log_5 9 \qquad \text{\small } \log_a M^r = r \log_a M \\
x^2 &= 9 \qquad \text{\small If } \log_a M = \log_a N, \text{ then } M = N. \\
x = 3 \quad \text{or} \quad &\cancel{x = -3} \qquad \text{\small Recall that the domain of the variable is } x > 0. \\
&\qquad\qquad\qquad\quad \text{\small Therefore, } -3 \text{ is extraneous and we discard it.}
\end{aligned}
$$

✓Check: $2 \log_5 3 \overset{?}{=} \log_5 9$

$\log_5 3^2 \overset{?}{=} \log_5 9$ $r \log_a M = \log_a M^r$

$\log_5 9 = \log_5 9$

The solution set is $\{3\}$.

■

━━━ **Now Work** PROBLEM 13

In the next example we employ the log of a product property to solve a logarithmic equation.

EXAMPLE 2	**Solving a Logarithmic Equation**

Solve: $\log_5(x + 6) + \log_5(x + 2) = 1$

Solution The domain of the variable requires that $x + 6 > 0$ and $x + 2 > 0$ so $x > -6$ and $x > -2$. This means any solution must satisfy $x > -2$. To obtain an exact solution, we need to express the left side as a single logarithm. Then we will change the expression to exponential form.

$\log_5(x + 6) + \log_5(x + 2) = 1$

$\log_5[(x + 6)(x + 2)] = 1$ $\log_a M + \log_a N = \log_a(MN)$

$(x + 6)(x + 2) = 5^1 = 5$ Change to an exponential expression.

$x^2 + 8x + 12 = 5$ Simplify.

$x^2 + 8x + 7 = 0$ Place the quadratic equation in standard form.

$(x + 7)(x + 1) = 0$ Factor.

$x = -7$ or $x = -1$ Zero-Product Property

WARNING A negative solution is not automatically extraneous. You must determine whether the potential solution causes the argument of any logarithmic expression in the equation to be negative. ■

Only $x = -1$ satisfies the restriction that $x > -2$, so $x = -7$ is extraneous. The solution set is $\{-1\}$, which you should check.

■

━━━ **Now Work** PROBLEM 21

EXAMPLE 3	**Solving a Logarithmic Equation**

Solve: $\ln x + \ln(x - 4) = \ln(x + 6)$

Solution The domain of the variable requires that $x > 0$, $x - 4 > 0$, and $x + 6 > 0$. As a result, the domain of the variable here is $x > 4$. We begin the solution using the log of a product property.

$\ln x + \ln(x - 4) = \ln(x + 6)$

$\ln[x(x - 4)] = \ln(x + 6)$ $\ln M + \ln N = \ln(MN)$

$x(x - 4) = x + 6$ If $\ln M = \ln N$, then $M = N$.

$x^2 - 4x = x + 6$ Simplify.

$x^2 - 5x - 6 = 0$ Place the quadratic equation in standard form.

$(x - 6)(x + 1) = 0$ Factor.

$x = 6$ or $x = -1$ Zero-Product Property

Since the domain of the variable is $x > 4$, we discard -1 as extraneous. The solution set is $\{6\}$, which you should check.

■

━━━ **Now Work** PROBLEM 31

2 Solve Exponential Equations

In Sections 6.3 and 6.4, we solved exponential equations algebraically by expressing each side of the equation using the same base. That is, we used the one-to-one property of the exponential function:

$$\text{If } a^u = a^v, \quad \text{then } u = v \qquad a > 0, a \neq 1$$

For example, to solve the exponential equation $4^{2x+1} = 16$, we notice that $16 = 4^2$ and apply the property above to obtain $2x + 1 = 2$, from which we find $x = \dfrac{1}{2}$.

For most exponential equations, we cannot express each side of the equation using the same base. In such cases, algebraic techniques can sometimes be used to obtain exact solutions. When algebraic techniques cannot be used, we use a graphing utility to obtain approximate solutions. You should pay particular attention to the form of equations for which exact solutions are obtained.

In the next example we solve two exponential equations by changing the exponential expression to a logarithmic expression.

EXAMPLE 4	Solving an Exponential Equation

Solve: (a) $2^x = 5$ (b) $8 \cdot 3^x = 5$

Solution (a) Since 5 cannot be written as an integral power of 2, we write the exponential equation as the equivalent logarithmic equation.

$$2^x = 5$$
$$x = \log_2 5$$

Now use the Change of Base Formula (9) on page 455.

$$x = \frac{\ln 5}{\ln 2} \qquad \text{Exact solution}$$
$$\approx 2.322 \quad \text{Approximate solution}$$

The solution set is $\left\{ \dfrac{\ln 5}{\ln 2} \right\}$.

(b) $8 \cdot 3^x = 5$
$$3^x = \frac{5}{8} \qquad \text{Solve for } 3^x.$$

$$x = \log_3\left(\frac{5}{8}\right) = \frac{\ln\left(\dfrac{5}{8}\right)}{\ln 3} \qquad \text{Exact solution}$$

$$\approx -0.428 \quad \text{Approximate solution}$$

The solution set is $\left\{ \dfrac{\ln\left(\dfrac{5}{8}\right)}{\ln 3} \right\}$.

━━━━▶ **Now Work** PROBLEM 35

EXAMPLE 5	Solving an Exponential Equation

Solve: $5^{x-2} = 3^{3x+2}$

Solution Because the bases are different, we first apply Property (6), Section 6.5 (take the natural logarithm of each side), and then use a property of logarithms. The result is an equation in x that we can solve.

$$5^{x-2} = 3^{3x+2}$$

$$\ln 5^{x-2} = \ln 3^{3x+2} \qquad \text{If } M = N, \ln M = \ln N.$$

$$(x - 2) \ln 5 = (3x + 2) \ln 3 \qquad \ln M^r = r \ln M$$

$$(\ln 5)x - 2 \ln 5 = (3 \ln 3)x + 2 \ln 3 \qquad \text{Distribute.}$$

$$(\ln 5)x - (3 \ln 3)x = 2 \ln 3 + 2 \ln 5 \qquad \text{Place terms involving } x \text{ on the left.}$$

$$(\ln 5 - 3 \ln 3)x = 2(\ln 3 + \ln 5) \qquad \text{Factor.}$$

$$x = \frac{2(\ln 3 + \ln 5)}{\ln 5 - 3 \ln 3} \qquad \text{Exact solution}$$

$$\approx -3.212 \qquad \text{Approximate solution}$$

The solution set is $\left\{ \dfrac{2(\ln 3 + \ln 5)}{\ln 5 - 3 \ln 3} \right\}$.

━━━ **Now Work** PROBLEM 45

The next example deals with an exponential equation that is quadratic in form.

EXAMPLE 6 **Solving an Exponential Equation That Is Quadratic in Form**

Solve: $4^x - 2^x - 12 = 0$

Solution We note that $4^x = (2^2)^x = 2^{(2x)} = (2^x)^2$, so the equation is quadratic in form, and we can rewrite it as

$$(2^x)^2 - 2^x - 12 = 0 \quad \text{Let } u = 2^x; \text{ then } u^2 - u - 12 = 0.$$

Now we can factor as usual.

$$(2^x - 4)(2^x + 3) = 0 \qquad (u - 4)(u + 3) = 0$$

$$2^x - 4 = 0 \quad \text{or} \quad 2^x + 3 = 0 \qquad u - 4 = 0 \quad \text{or} \quad u + 3 = 0$$

$$2^x = 4 \qquad\qquad 2^x = -3 \qquad u = 2^x = 4 \qquad u = 2^x = -3$$

The equation on the left has the solution $x = 2$, since $2^x = 4 = 2^2$; the equation on the right has no solution, since $2^x > 0$ for all x. The only solution is 2. The solution set is $\{2\}$.

━━━ **Now Work** PROBLEM 53

3 Solve Logarithmic and Exponential Equations Using a Graphing Utility

The algebraic techniques introduced in this section to obtain exact solutions apply only to certain types of logarithmic and exponential equations. Solutions for other types are usually studied in calculus, using numerical methods. For such types, we can use a graphing utility to approximate the solution.

EXAMPLE 7 **Solving Equations Using a Graphing Utility**

Solve: $x + e^x = 2$

Express the solution(s) rounded to two decimal places.

Figure 40

Solution The solution is found by graphing $Y_1 = x + e^x$ and $Y_2 = 2$. Since Y_1 is an increasing function (do you know why?), there is only one point of intersection for Y_1 and Y_2. Figure 40 shows the graphs of Y_1 and Y_2. Using the INTERSECT command, the solution is 0.44 rounded to two decimal places.

━━━━━━ **Now Work** PROBLEM 63

6.6 Assess Your Understanding

'Are You Prepared?' *Answers are given at the end of these exercises. If you get a wrong answer, read the pages listed in red.*

1. Solve $x^2 - 7x - 30 = 0$. (pp. 97–106)

2. Solve $(x + 3)^2 - 4(x + 3) + 3 = 0$. (pp. 119–121)

3. Approximate the solution(s) to $x^3 = x^2 - 5$ using a graphing utility. (pp. A8–A10)

4. Approximate the solution(s) to $x^3 - 2x + 2 = 0$ using a graphing utility. (pp. A8–A10)

Skill Building

In Problems 5–32, solve each logarithmic equation. Express irrational solutions in exact form and as a decimal rounded to 3 decimal places.

5. $\log_4 x = 2$

6. $\log(x + 6) = 1$

7. $\log_2(5x) = 4$

8. $\log_3(3x - 1) = 2$

9. $\log_4(x + 2) = \log_4 8$

10. $\log_5(2x + 3) = \log_5 3$

11. $\frac{1}{2}\log_3 x = 2\log_3 2$

12. $-2\log_4 x = \log_4 9$

13. $3\log_2 x = -\log_2 27$

14. $2\log_5 x = 3\log_5 4$

15. $3\log_2(x - 1) + \log_2 4 = 5$

16. $2\log_3(x + 4) - \log_3 9 = 2$

17. $\log x + \log(x + 15) = 2$

18. $\log x + \log(x - 21) = 2$

19. $\log(2x + 1) = 1 + \log(x - 2)$

20. $\log(2x) - \log(x - 3) = 1$

21. $\log_2(x + 7) + \log_2(x + 8) = 1$

22. $\log_6(x + 4) + \log_6(x + 3) = 1$

23. $\log_8(x + 6) = 1 - \log_8(x + 4)$

24. $\log_5(x + 3) = 1 - \log_5(x - 1)$

25. $\ln x + \ln(x + 2) = 4$

26. $\ln(x + 1) - \ln x = 2$

27. $\log_3(x + 1) + \log_3(x + 4) = 2$

28. $\log_2(x + 1) + \log_2(x + 7) = 3$

29. $\log_{1/3}(x^2 + x) - \log_{1/3}(x^2 - x) = -1$

30. $\log_4(x^2 - 9) - \log_4(x + 3) = 3$

31. $\log_a(x - 1) - \log_a(x + 6) = \log_a(x - 2) - \log_a(x + 3)$

32. $\log_a x + \log_a(x - 2) = \log_a(x + 4)$

In Problems 33–60, solve each exponential equation. Express irrational solutions in exact form and as a decimal rounded to 3 decimal places.

33. $2^{x-5} = 8$

34. $5^{-x} = 25$

35. $2^x = 10$

36. $3^x = 14$

37. $8^{-x} = 1.2$

38. $2^{-x} = 1.5$

39. $5(2^{3x}) = 8$

40. $0.3(4^{0.2x}) = 0.2$

41. $3^{1-2x} = 4^x$

42. $2^{x+1} = 5^{1-2x}$

43. $\left(\frac{3}{5}\right)^x = 7^{1-x}$

44. $\left(\frac{4}{3}\right)^{1-x} = 5^x$

45. $1.2^x = (0.5)^{-x}$

46. $0.3^{1+x} = 1.7^{2x-1}$

47. $\pi^{1-x} = e^x$

48. $e^{x+3} = \pi^x$

49. $2^{2x} + 2^x - 12 = 0$

50. $3^{2x} + 3^x - 2 = 0$

51. $3^{2x} + 3^{x+1} - 4 = 0$

52. $2^{2x} + 2^{x+2} - 12 = 0$

53. $16^x + 4^{x+1} - 3 = 0$

54. $9^x - 3^{x+1} + 1 = 0$

55. $25^x - 8 \cdot 5^x = -16$

56. $36^x - 6 \cdot 6^x = -9$

57. $3 \cdot 4^x + 4 \cdot 2^x + 8 = 0$

58. $2 \cdot 49^x + 11 \cdot 7^x + 5 = 0$

59. $4^x - 10 \cdot 4^{-x} = 3$

60. $3^x - 14 \cdot 3^{-x} = 5$

In Problems 61–74, use a graphing utility to solve each equation. Express your answer rounded to two decimal places.

61. $\log_5(x + 1) - \log_4(x - 2) = 1$

62. $\log_2(x - 1) - \log_6(x + 2) = 2$

63. $e^x = -x$

64. $e^{2x} = x + 2$

65. $e^x = x^2$

66. $e^x = x^3$

67. $\ln x = -x$

68. $\ln(2x) = -x + 2$

69. $\ln x = x^3 - 1$

70. $\ln x = -x^2$

71. $e^x + \ln x = 4$

72. $e^x - \ln x = 4$

73. $e^{-x} = \ln x$

74. $e^{-x} = -\ln x$

Applications and Extensions

In Problems 75–86, solve each equation. Express irrational solutions in exact form and as a decimal rounded to 3 decimal places.

75. $\log_2(x + 1) - \log_4 x = 1$
[**Hint:** Change $\log_4 x$ to base 2.]

76. $\log_2(3x + 2) - \log_4 x = 3$

77. $\log_{16} x + \log_4 x + \log_2 x = 7$

78. $\log_9 x + 3\log_3 x = 14$

79. $\left(\sqrt[3]{2}\right)^{2-x} = 2^{x^2}$

80. $\log_2 x^{\log_2 x} = 4$

81. $\dfrac{e^x + e^{-x}}{2} = 1$
[**Hint:** Multiply each side by e^x.]

82. $\dfrac{e^x + e^{-x}}{2} = 3$

83. $\dfrac{e^x - e^{-x}}{2} = 2$

84. $\dfrac{e^x - e^{-x}}{2} = -2$

85. $\log_5 x + \log_3 x = 1$
[**Hint:** Use the Change-of-Base Formula and factor out $\ln x$.]

86. $\log_2 x + \log_6 x = 3$

87. $f(x) = \log_2(x + 3)$ and $g(x) = \log_2(3x + 1)$.
(a) Solve $f(x) = 3$. What point is on the graph of f?
(b) Solve $g(x) = 4$. What point is on the graph of g?
(c) Solve $f(x) = g(x)$. Do the graphs of f and g intersect? If so, where?
(d) Solve $f(x) + g(x) = 7$.
(e) Solve $f(x) - g(x) = 2$.

88. $f(x) = \log_3(x + 5)$ and $g(x) = \log_3(x - 1)$.
(a) Solve $f(x) = 2$. What point is on the graph of f?
(b) Solve $g(x) = 3$. What point is on the graph of g?
(c) Solve $f(x) = g(x)$. Do the graphs of f and g intersect? If so, where?
(d) Solve $f(x) + g(x) = 3$.
(e) Solve $f(x) - g(x) = 2$.

89. (a) If $f(x) = 3^{x+1}$ and $g(x) = 2^{x+2}$, graph f and g on the same Cartesian plane.
(b) Find the point(s) of intersection of the graphs of f and g by solving $f(x) = g(x)$. Round answers to three decimal places. Label this point on the graph drawn in part (a).
(c) Based on the graph, solve $f(x) > g(x)$.

90. (a) If $f(x) = 5^{x-1}$ and $g(x) = 2^{x+1}$, graph f and g on the same Cartesian plane.
(b) Find the point(s) of intersection of the graphs of f and g by solving $f(x) = g(x)$. Label this point on the graph drawn in part (a).
(c) Based on the graph, solve $f(x) > g(x)$.

91. (a) Graph $f(x) = 3^x$ and $g(x) = 10$ on the same Cartesian plane.
(b) Shade the region bounded by the y-axis, $f(x) = 3^x$, and $g(x) = 10$ on the graph drawn in part (a).
(c) Solve $f(x) = g(x)$ and label the point of intersection on the graph drawn in part (b).

92. (a) Graph $f(x) = 2^x$ and $g(x) = 12$ on the same Cartesian plane.
(b) Shade the region bounded by the y-axis, $f(x) = 2^x$, and $g(x) = 12$ on the graph drawn in part (a).
(c) Solve $f(x) = g(x)$ and label the point of intersection on the graph drawn in part (b).

93. (a) Graph $f(x) = 2^{x+1}$ and $g(x) = 2^{-x+2}$ on the same Cartesian plane.

(b) Shade the region bounded by the y-axis, $f(x) = 2^{x+1}$, and $g(x) = 2^{-x+2}$ on the graph draw in part (a).
(c) Solve $f(x) = g(x)$ and label the point of intersection on the graph drawn in part (b).

94. (a) Graph $f(x) = 3^{-x+1}$ and $g(x) = 3^{x-2}$ on the same Cartesian plane.
(b) Shade the region bounded by the y-axis, $f(x) = 3^{-x+1}$, and $g(x) = 3^{x-2}$ on the graph draw in part (a).
(c) Solve $f(x) = g(x)$ and label the point of intersection on the graph drawn in part (b).

95. (a) Graph $f(x) = 2^x - 4$.
(b) Based on the graph, solve $f(x) < 0$.

96. (a) Graph $g(x) = 3^x - 9$.
(b) Based on the graph, solve $g(x) > 0$.

97. A Population Model The resident population of the United States in 2006 was 298 million people and was growing at a rate of 0.9% per year. Assuming that this growth rate continues, the model $P(t) = 298(1.009)^{t-2006}$ represents the population P (in millions of people) in year t.
(a) According to this model, when will the population of the United States be 310 million people?
(b) According to this model, when will the population of the United States be 360 million people?

Source: Statistical Abstract of the United States, 125th ed., 2006.

98. A Population Model The population of the world in 2006 was 6.53 billion people and was growing at a rate of 1.14% per year. Assuming that this growth rate continues, the model $P(t) = 6.53(1.0114)^{t-2006}$ represents the population P (in billions of people) in year t.

(a) According to this model, when will the population of the world be 9.25 billion people?

(b) According to this model, when will the population of the world be 11.75 billion people?

Source: U.S. Census Bureau.

99. Depreciation The value V of a Chevy Cobalt that is t years old can be modeled by $V(t) = 14,512(0.82)^t$.

(a) According to the model, when will the car be worth $9000?

(b) According to the model, when will the car be worth $4000?

(c) According to the model, when will the car be worth $2000?

Source: Kelley Blue Book

100. Depreciation The value V of a Dodge Stratus that is t years old can be modeled by $V(t) = 19,282(0.84)^t$.

(a) According to the model, when will the car be worth $15,000?

(b) According to the model, when will the car be worth $8000?

(c) According to the model, when will the car be worth $2000?

Source: Kelley Blue Book

Discussion and Writing

101. Fill in reasons for each step in the following two solutions.

Solve: $\log_3(x-1)^2 = 2$

Solution A

$\log_3(x-1)^2 = 2$

$(x-1)^2 = 3^2 = 9$ _____

$(x-1) = \pm 3$ _____

$x - 1 = -3$ or $x - 1 = 3$ _____

$x = -2$ or $x = 4$ _____

Solution B

$\log_3(x-1)^2 = 2$

$2\log_3(x-1) = 2$ _____

$\log_3(x-1) = 1$ _____

$x - 1 = 3^1 = 3$ _____

$x = 4$ _____

Both solutions given in Solution A check. Explain what caused the solution $x = -2$ to be lost in Solution B.

'Are You Prepared?' Answers

1. $\{-3, 10\}$ **2.** $\{-2, 0\}$ **3.** $\{-1.43\}$ **4.** $\{-1.77\}$

6.7 Compound Interest

PREPARING FOR THIS SECTION *Before getting started, review the following:*

• Simple Interest (Section 1.7, pp. 141–142)

Now Work the 'Are You Prepared?' problems on page 472.

OBJECTIVES **1** Determine the Future Value of a Lump Sum of Money (p. 465)

 2 Calculate Effective Rates of Return (p. 469)

 3 Determine the Present Value of a Lump Sum of Money (p. 470)

 4 Determine the Rate of Interest or Time Required to Double a Lump Sum of Money (p. 471)

1 Determine the Future Value of a Lump Sum of Money

Interest is money paid for the use of money. The total amount borrowed (whether by an individual from a bank in the form of a loan or by a bank from an individual in the form of a savings account) is called the **principal.** The **rate of interest,**

expressed as a percent, is the amount charged for the use of the principal for a given period of time, usually on a yearly (that is, per annum) basis.

THEOREM

Simple Interest Formula

If a principal of P dollars is borrowed for a period of t years at a per annum interest rate r, expressed as a decimal, the interest I charged is

$$I = Prt \tag{1}$$

Interest charged according to formula (1) is called **simple interest.**

In working with problems involving interest, we define the term **payment period** as follows:

Annually:	Once per year	**Monthly:**	12 times per year
Semiannually:	Twice per year	**Daily:**	365 times per year*
Quarterly:	Four times per year		

When the interest due at the end of a payment period is added to the principal so that the interest computed at the end of the next payment period is based on this new principal amount (old principal + interest), the interest is said to have been **compounded. Compound interest** is interest paid on principal and previously earned interest.

EXAMPLE 1

Computing Compound Interest

A credit union pays interest of 8% per annum compounded quarterly on a certain savings plan. If $1000 is deposited in such a plan and the interest is left to accumulate, how much is in the account after 1 year?

Solution

We use the simple interest formula, $I = Prt$. The principal P is $1000 and the rate of interest is 8% = 0.08. After the first quarter of a year, the time t is $\frac{1}{4}$ year, so the interest earned is

$$I = Prt = (\$1000)(0.08)\left(\frac{1}{4}\right) = \$20$$

The new principal is $P + I = \$1000 + \$20 = \$1020$. At the end of the second quarter, the interest on this principal is

$$I = (\$1020)(0.08)\left(\frac{1}{4}\right) = \$20.40$$

At the end of the third quarter, the interest on the new principal of $1020 + $20.40 = $1040.40 is

$$I = (\$1040.40)(0.08)\left(\frac{1}{4}\right) = \$20.81$$

Finally, after the fourth quarter, the interest is

$$I = (\$1061.21)(0.08)\left(\frac{1}{4}\right) = \$21.22$$

After 1 year the account contains $1061.21 + $21.22 = $1082.43.

*Most banks use a 360-day "year." Why do you think they do?

The pattern of the calculations performed in Example 1 leads to a general formula for compound interest. To fix our ideas, let P represent the principal to be invested at a per annum interest rate r that is compounded n times per year, so the time of each compounding period is $\frac{1}{n}$ years. (For computing purposes, r is expressed as a decimal.) The interest earned after each compounding period is given by formula (1).

$$\text{Interest} = \text{principal} \times \text{rate} \times \text{time} = P \cdot r \cdot \frac{1}{n} = P \cdot \left(\frac{r}{n}\right)$$

The amount A after one compounding period is

$$A = P + P \cdot \left(\frac{r}{n}\right) = P \cdot \left(1 + \frac{r}{n}\right)$$

After two compounding periods, the amount A, based on the new principal $P \cdot \left(1 + \frac{r}{n}\right)$, is

$$A = \underbrace{P \cdot \left(1 + \frac{r}{n}\right)}_{\substack{\text{New} \\ \text{principal}}} + \underbrace{P \cdot \left(1 + \frac{r}{n}\right)\left(\frac{r}{n}\right)}_{\substack{\text{Interest on} \\ \text{new principal}}} = P \cdot \left(1 + \frac{r}{n}\right)\left(1 + \frac{r}{n}\right) = P \cdot \left(1 + \frac{r}{n}\right)^2$$

Factor out $P \cdot \left(1 + \frac{r}{n}\right)$

After three compounding periods, the amount A is

$$A = P \cdot \left(1 + \frac{r}{n}\right)^2 + P \cdot \left(1 + \frac{r}{n}\right)^2\left(\frac{r}{n}\right) = P \cdot \left(1 + \frac{r}{n}\right)^2 \cdot \left(1 + \frac{r}{n}\right) = P \cdot \left(1 + \frac{r}{n}\right)^3$$

Continuing this way, after n compounding periods (1 year), the amount A is

$$A = P \cdot \left(1 + \frac{r}{n}\right)^n$$

Because t years will contain $n \cdot t$ compounding periods, after t years we have

$$A = P \cdot \left(1 + \frac{r}{n}\right)^{nt}$$

THEOREM

Compound Interest Formula

The amount A after t years due to a principal P invested at an annual interest rate r compounded n times per year is

$$A = P \cdot \left(1 + \frac{r}{n}\right)^{nt} \tag{2}$$

Exploration

To see the effects of compounding interest monthly on an initial deposit of \$1, graph $Y_1 = \left(1 + \frac{r}{12}\right)^{12x}$ with $r = 0.06$ and $r = 0.12$ for $0 \le x \le 30$. What is the future value of \$1 in 30 years when the interest rate per annum is $r = 0.06$ (6%)? What is the future value of \$1 in 30 years when the interest rate per annum is $r = 0.12$ (12%)? Does doubling the interest rate double the future value?

For example, to rework Example 1, we would use $P = \$1000, r = 0.08, n = 4$ (quarterly compounding), and $t = 1$ year to obtain

$$A = P \cdot \left(1 + \frac{r}{n}\right)^{nt} = 1000\left(1 + \frac{0.08}{4}\right)^{4 \cdot 1} = \$1082.43$$

In equation (2), the amount A is typically referred to as the **future value** of the account, while P is called the **present value.**

━━➤ **Now Work** PROBLEM 3

| EXAMPLE 2 | Comparing Investments Using Different Compounding Periods |

Investing $1000 at an annual rate of 10% compounded annually, semiannually, quarterly, monthly, and daily will yield the following amounts after 1 year:

Annual compounding ($n = 1$): $A = P \cdot (1 + r)$

$$= (\$1000)(1 + 0.10) = \$1100.00$$

Semiannual compounding ($n = 2$): $A = P \cdot \left(1 + \dfrac{r}{2}\right)^2$

$$= (\$1000)(1 + 0.05)^2 = \$1102.50$$

Quarterly compounding ($n = 4$): $A = P \cdot \left(1 + \dfrac{r}{4}\right)^4$

$$= (\$1000)(1 + 0.025)^4 = \$1103.81$$

Monthly compounding ($n = 12$): $A = P \cdot \left(1 + \dfrac{r}{12}\right)^{12}$

$$= (\$1000)\left(1 + \dfrac{0.10}{12}\right)^{12} = \$1104.71$$

Daily compounding ($n = 365$): $A = P \cdot \left(1 + \dfrac{r}{365}\right)^{365}$

$$= (\$1000)\left(1 + \dfrac{0.10}{365}\right)^{365} = \$1105.16$$

From Example 2, we can see that the effect of compounding more frequently is that the amount after 1 year is higher: $1000 compounded 4 times a year at 10% results in $1103.81, $1000 compounded 12 times a year at 10% results in $1104.71, and $1000 compounded 365 times a year at 10% results in $1105.16. This leads to the following question: What would happen to the amount after 1 year if the number of times that the interest is compounded were increased without bound?

Let's find the answer. Suppose that P is the principal, r is the per annum interest rate, and n is the number of times that the interest is compounded each year. The amount after 1 year is

$$A = P \cdot \left(1 + \frac{r}{n}\right)^n$$

Rewrite this expression as follows:

$$A = P \cdot \left(1 + \frac{r}{n}\right)^n = P \cdot \left(1 + \frac{1}{\frac{n}{r}}\right)^n = P \cdot \left[\left(1 + \frac{1}{\frac{n}{r}}\right)^{n/r}\right]^r \underset{\underset{h = \frac{n}{r}}{\uparrow}}{=} P \cdot \left[\left(1 + \frac{1}{h}\right)^h\right]^r \quad \textbf{(3)}$$

Now suppose that the number n of times that the interest is compounded per year gets larger and larger; that is, suppose that $n \to \infty$. Then $h = \dfrac{n}{r} \to \infty$, and the expression in brackets in Equation (3) equals e. That is, $A \to Pe^r$.

Table 8 compares $\left(1 + \dfrac{r}{n}\right)^n$, for large values of n, to e^r for $r = 0.05$, $r = 0.10$, $r = 0.15$, and $r = 1$. The larger that n gets, the closer $\left(1 + \dfrac{r}{n}\right)^n$ gets to e^r.

No matter how frequent the compounding, the amount after 1 year has the definite ceiling Pe^r.

Table 8

	$\left(1 + \frac{r}{n}\right)^n$			
	$n = 100$	$n = 1000$	$n = 10{,}000$	e^r
$r = 0.05$	1.0512580	1.0512698	1.051271	1.0512711
$r = 0.10$	1.1051157	1.1051654	1.1051704	1.1051709
$r = 0.15$	1.1617037	1.1618212	1.1618329	1.1618342
$r = 1$	2.7048138	2.7169239	2.7181459	2.7182818

When interest is compounded so that the amount after 1 year is Pe^r, we say the interest is **compounded continuously.**

THEOREM

Continuous Compounding

The amount A after t years due to a principal P invested at an annual interest rate r compounded continuously is

$$A = Pe^{rt} \qquad\qquad (4)$$

EXAMPLE 3

Using Continuous Compounding

The amount A that results from investing a principal P of $1000 at an annual rate r of 10% compounded continuously for a time t of 1 year is

$$A = \$1000e^{0.10} = (\$1000)(1.10517) = \$1105.17$$

══ **Now Work** PROBLEM 11

2 Calculate Effective Rates of Return

The **effective rate of interest** is the equivalent annual simple rate of interest that would yield the same amount as compounding after 1 year. For example, based on Example 3, a principal of $1000 will result in $1105.17 at a rate of 10% compounded continuously. To get this same amount using a simple rate of interest would require that interest of $1105.17 − $1000.00 = $105.17 be earned on the principal. Since $105.17 is 10.517% of $1000, a simple rate of interest of 10.517% is needed to equal 10% compounded continuously. The effective rate of interest of 10% compounded continuously is 10.517%.

Based on the results of Examples 2 and 3, we find the following comparisons:

	Annual Rate	Effective Rate
Annual compounding	10%	10%
Semiannual compounding	10%	10.25%
Quarterly compounding	10%	10.381%
Monthly compounding	10%	10.471%
Daily compounding	10%	10.516%
Continuous compounding	10%	10.517%

EXAMPLE 4

Computing the Effective Rate of Interest

On January 2, 2007, $2000 is placed in an Individual Retirement Account (IRA) that will pay interest of 7% per annum compounded continuously.

(a) What will the IRA be worth on January 1, 2027?

(b) What is the effective annual rate of interest?

Solution (a) On January, 1, 2027, the initial principal of $2000 will have earned interest of 7% compounded continuously for 20 years. The amount A after 20 years is

$$A = Pe^{rt} = \$2000e^{(0.07)(20)} = \$8110.40$$

(b) First, we compute the interest earned on $2000 at $r = 7\%$ compounded continuously for 1 year.

$$A = \$2000e^{0.07(1)} = \$2145.02$$

So the interest earned is $2145.02 − $2000.00 = $145.02. To find the effective rate of interest R, we use the simple interest formula $I = PRt$, with $I = \$145.02$, $P = \$2000$, and $t = 1$.

$$\$145.02 = \$2000 \cdot R \cdot 1 \qquad I = PRt$$

$$R = \frac{\$145.02}{\$2000} = 0.07251 \qquad \text{Solve for } s.$$

The effective annual rate of interest R is 7.251%.

Now Work PROBLEM 27

Exploration
For the IRA described in Example 4, how long will it be until $A = \$4000$? 6000? [**Hint:** Graph $Y_1 = 2000e^{0.07x}$ and $Y_2 = 4000$. Use INTERSECT to find x.]

3 Determine the Present Value of a Lump Sum of Money

When people engaged in finance speak of the "time value of money," they are usually referring to the *present value* of money. The **present value** of A dollars to be received at a future date is the principal that you would need to invest now so that it will grow to A dollars in the specified time period. The present value of money to be received at a future date is always less than the amount to be received, since the amount to be received will equal the present value (money invested now) *plus* the interest accrued over the time period.

We use the compound interest formula (2) to get a formula for present value. If P is the present value of A dollars to be received after t years at a per annum interest rate r compounded n times per year, then, by formula (2),

$$A = P \cdot \left(1 + \frac{r}{n}\right)^{nt}$$

To solve for P, we divide both sides by $\left(1 + \frac{r}{n}\right)^{nt}$. The result is

$$\frac{A}{\left(1 + \frac{r}{n}\right)^{nt}} = P \quad \text{or} \quad P = A \cdot \left(1 + \frac{r}{n}\right)^{-nt}$$

THEOREM | **Present Value Formulas**

The present value P of A dollars to be received after t years, assuming a per annum interest rate r compounded n times per year, is

$$P = A \cdot \left(1 + \frac{r}{n}\right)^{-nt} \qquad (5)$$

If the interest is compounded continuously,

$$P = Ae^{-rt} \qquad (6)$$

To prove (6), solve formula (4) for P.

| EXAMPLE 5 | Computing the Value of a Zero-coupon Bond |

A zero-coupon (noninterest-bearing) bond can be redeemed in 10 years for $1000. How much should you be willing to pay for it now if you want a return of

(a) 8% compounded monthly?

(b) 7% compounded continuously?

Solution (a) We are seeking the present value of $1000. We use formula (5) with $A = \$1000$, $n = 12, r = 0.08$, and $t = 10$.

$$P = A \cdot \left(1 + \frac{r}{n}\right)^{-nt} = \$1000\left(1 + \frac{0.08}{12}\right)^{-12(10)} = \$450.52$$

For a return of 8% compounded monthly, you should pay $450.52 for the bond.

(b) Here we use formula (6) with $A = \$1000, r = 0.07$, and $t = 10$.

$$P = Ae^{-rt} = \$1000e^{-(0.07)(10)} = \$496.59$$

For a return of 7% compounded continuously, you should pay $496.59 for the bond.

━━━━━━➤ **Now Work** PROBLEM 13

4 Determine the Rate of Interest or Time Required to Double a Lump Sum of Money

| EXAMPLE 6 | Rate of Interest Required to Double an Investment |

What annual rate of interest compounded annually should you seek if you want to double your investment in 5 years?

Solution If P is the principal and we want P to double, the amount A will be $2P$. We use the compound interest formula with $n = 1$ and $t = 5$ to find r.

$$A = P \cdot \left(1 + \frac{r}{n}\right)^{nt}$$

$$2P = P \cdot (1 + r)^5 \qquad \text{\textit{A = 2P, n = 1, t = 5}}$$

$$2 = (1 + r)^5 \qquad \text{\textit{Cancel the P's.}}$$

$$1 + r = \sqrt[5]{2} \qquad \text{\textit{Take the fifth root of each side.}}$$

$$r = \sqrt[5]{2} - 1 \approx 1.148698 - 1 = 0.148698$$

The annual rate of interest needed to double the principal in 5 years is 14.87%.

━━━━━━➤ **Now Work** PROBLEM 31

| EXAMPLE 7 | Time Required to Double or Triple an Investment |

(a) How long will it take for an investment to double in value if it earns 5% compounded continuously?

(b) How long will it take to triple at this rate?

Solution (a) If P is the initial investment and we want P to double, the amount A will be $2P$. We use formula (4) for continuously compounded interest with $r = 0.05$.

Then

$$A = Pe^{rt}$$
$$2P = Pe^{0.05t} \qquad \text{\textit{A} = 2\textit{P}, \textit{r} = 0.05}$$
$$2 = e^{0.05t} \qquad \text{Cancel the \textit{P}'s.}$$
$$0.05t = \ln 2 \qquad \text{Rewrite as a logarithm.}$$
$$t = \frac{\ln 2}{0.05} \approx 13.86 \qquad \text{Solve for \textit{t}.}$$

It will take about 14 years to double the investment.

(b) To triple the investment, we set $A = 3P$ in formula (4).

$$A = Pe^{rt}$$
$$3P = Pe^{0.05t} \qquad \text{\textit{A} = 3\textit{P}, \textit{r} = 0.05}$$
$$3 = e^{0.05t} \qquad \text{Cancel the \textit{P}'s.}$$
$$0.05t = \ln 3 \qquad \text{Rewrite as a logarithm.}$$
$$t = \frac{\ln 3}{0.05} \approx 21.97 \qquad \text{Solve for \textit{t}.}$$

It will take about 22 years to triple the investment.

Now Work PROBLEM 35

6.7 Assess Your Understanding

'Are You Prepared?' *Answers are given at the end of these exercises. If you get a wrong answer, read the pages listed in red.*

1. What is the interest due if $500 is borrowed for 6 months at a simple interest rate of 6% per annum? (pp. 141–142)

2. If you borrow $5000 and, after 9 months, pay off the loan in the amount of $5500, what per annum rate of interest was charged? (pp. 141–142)

Skill Building

In Problems 3–12, find the amount that results from each investment.

3. $100 invested at 4% compounded quarterly after a period of 2 years

4. $50 invested at 6% compounded monthly after a period of 3 years

5. $500 invested at 8% compounded quarterly after a period of $2\frac{1}{2}$ years

6. $300 invested at 12% compounded monthly after a period of $1\frac{1}{2}$ years

7. $600 invested at 5% compounded daily after a period of 3 years

8. $700 invested at 6% compounded daily after a period of 2 years

9. $10 invested at 11% compounded continuously after a period of 2 years

10. $40 invested at 7% compounded continuously after a period of 3 years

11. $100 invested at 10% compounded continuously after a period of $2\frac{1}{4}$ years

12. $100 invested at 12% compounded continuously after a period of $3\frac{3}{4}$ years

In Problems 13–22, find the principal needed now to get each amount; that is, find the present value.

13. To get $100 after 2 years at 6% compounded monthly

14. To get $75 after 3 years at 8% compounded quarterly

15. To get $1000 after $2\frac{1}{2}$ years at 6% compounded daily

16. To get $800 after $3\frac{1}{2}$ years at 7% compounded monthly

17. To get $600 after 2 years at 4% compounded quarterly

18. To get $300 after 4 years at 3% compounded daily

19. To get $80 after $3\frac{1}{4}$ years at 9% compounded continuously

20. To get $800 after $2\frac{1}{2}$ years at 8% compounded continuously

21. To get $400 after 1 year at 10% compounded continuously

22. To get $1000 after 1 year at 12% compounded continuously

In Problems 23–26, which of the two rates would yield the larger amount in 1 year?
[**Hint:** *Start with a principal of $10,000 in each instance.*]

23. 6% compounded quarterly or $6\frac{1}{4}$% compounded annually

24. 9% compounded quarterly or $9\frac{1}{4}$% compounded annually

25. 9% compounded monthly or 8.8% compounded daily

26. 8% compounded semiannually or 7.9% compounded daily

In Problems 27–30, find the effective rate of interest.

27. For 5% compounded quarterly

28. For 6% compounded monthly

29. For 5% compounded continuously

30. For 6% compounded continuously

31. What rate of interest compounded annually is required to double an investment in 3 years?

32. What rate of interest compounded annually is required to double an investment in 6 years?

33. What rate of interest compounded annually is required to triple an investment in 5 years?

34. What rate of interest compounded annually is required to triple an investment in 10 years?

35. (a) How long does it take for an investment to double in value if it is invested at 8% compounded monthly?

 (b) How long does it take if the interest is compounded continuously?

36. (a) How long does it take for an investment to triple in value if it is invested at 6% compounded monthly?

 (b) How long does it take if the interest is compounded continuously?

37. What rate of interest compounded quarterly will yield an effective interest rate of 7%?

38. What rate of interest compounded continuously will yield an effective interest rate of 6%?

Applications and Extensions

39. Time Required to Reach a Goal If Tanisha has $100 to invest at 8% per annum compounded monthly, how long will it be before she has $150? If the compounding is continuous, how long will it be?

40. Time Required to Reach a Goal If Angela has $100 to invest at 10% per annum compounded monthly, how long will it be before she has $175? If the compounding is continuous, how long will it be?

41. Time Required to Reach a Goal How many years will it take for an initial investment of $10,000 to grow to $25,000? Assume a rate of interest of 6% compounded continuously.

42. Time Required to Reach a Goal How many years will it take for an initial investment of $25,000 to grow to $80,000? Assume a rate of interest of 7% compounded continuously.

43. Price Appreciation of Homes What will a $90,000 house cost 5 years from now if the price appreciation for homes over that period averages 3% compounded annually?

44. Credit Card Interest Sears charges 1.25% per month on the unpaid balance for customers with charge accounts (interest is compounded monthly). A customer charges $200 and does not pay her bill for 6 months. What is the bill at that time?

45. Saving for a Car Jerome will be buying a used car for $15,000 in 3 years. How much money should he ask his parents for now so that, if he invests it at 5% compounded continuously, he will have enough to buy the car?

46. Paying off a Loan John will require $3000 in 6 months to pay off a loan that has no prepayment privileges. If he has the $3000 now, how much of it should he save in an account paying 3% compounded monthly so that in 6 months he will have exactly $3000?

47. Return on a Stock George is contemplating the purchase of 100 shares of a stock selling for $15 per share. The stock pays no dividends. The history of the stock indicates that it should grow at an annual rate of 15% per year. How much will the 100 shares of stock be worth in 5 years?

48. Return on an Investment A business purchased for $650,000 in 2001 is sold in 2004 for $850,000. What is the annual rate of return for this investment?

49. Comparing Savings Plans Jim places $1000 in a bank account that pays 5.6% compounded continuously. After 1 year, will he have enough money to buy a computer system that costs $1060? If another bank will pay Jim 5.9% compounded monthly, is this a better deal?

50. Savings Plans On January 1, Kim places $1000 in a certificate of deposit that pays 6.8% compounded continuously and matures in 3 months. Then Kim places the $1000 and the interest in a passbook account that pays 5.25% compounded monthly. How much does Kim have in the passbook account on May 1?

51. Comparing IRA Investments Will invests $2000 in his IRA in a bond trust that pays 9% interest compounded semiannually. His friend Henry invests $2000 in his IRA in a certificate of deposit that pays $8\frac{1}{2}$% compounded continuously. Who has more money after 20 years, Will or Henry?

52. Comparing Two Alternatives Suppose that April has access to an investment that will pay 10% interest compounded continuously. Which is better: to be given $1000 now so that she can take advantage of this investment opportunity or to be given $1325 after 3 years?

53. College Costs The average cost of college at 4-year private colleges was $29,026 in 2005. This was a 5.5% increase from the previous year.

Source: Trends in College Pricing 2005, College Board

(a) If the cost of college increases by 5.5% each year, what will be the average cost of college at a 4-year private college in 2015?

(b) College savings plans, such as a 529 plan, allow individuals to put money aside now to help pay for college later. If one such plan offers a rate of 4% compounded continuously, how much should be put in a college savings plan in 2005 to pay for 1 year of the cost of college at a 4-year private college in 2015?

54. Analyzing Interest Rates on a Mortgage Colleen and Bill have just purchased a house for $650,000, with the seller holding a second mortgage of $100,000. They promise to pay the seller $100,000 plus all accrued interest 5 years from now. The seller offers them three interest options on the second mortgage:

(a) Simple interest at 12% per annum

(b) $11\frac{1}{2}$% interest compounded monthly

(c) $11\frac{1}{4}$% interest compounded continuously

Which option is best; that is, which results in the least interest on the loan?

55. Federal Deficit At the end of fiscal year 2005, the federal budget deficit was $319 billion. At that time, 20-year Series EE bonds had a fixed rate of 3.2% compounded semiannually. If the federal government financed this deficit through EE bonds, how much would it have to pay back in 2025?

Source: U.S. Treasury Department

56. Federal Deficit On February 6, 2006, President Bush proposed the fiscal year 2007 federal budget. The proposal projected a fiscal year 2006 deficit of $423 billion and a fiscal year 2007 deficit of $354 billion. Assuming the deficit decreases at the same rate each year, when will the deficit be cut to $100 billion?

Source: Office of Management and Budget

Inflation *Problems 57–62 require the following discussion.* **Inflation** *is a term used to describe the erosion of the purchasing power of money. For example, suppose the annual inflation rate is 3%. Then $1000 worth of purchasing power now will have only $970 worth of purchasing power in one year because 3% of the original $1000 (0.03 × 1000 = 30) has been eroded due to inflation. In general, if the rate of inflation averages r% over n years, the amount A that $P will purchase after n years is*

$$A = P \cdot (1 - r)^n$$

where r is expressed as a decimal.

57. Inflation If the inflation rate averages 3%, how much will $1000 purchase in 2 years?

58. Inflation If the inflation rate averages 2%, how much will $1000 purchase in 3 years?

59. Inflation If the amount that $1000 will purchase is only $950 after 2 years, what was the average inflation rate?

60. Inflation If the amount that $1000 will purchase is only $930 after 2 years, what was the average inflation rate?

61. Inflation If the average inflation rate is 2%, how long is it until purchasing power is cut in half?

62. Inflation If the average inflation rate is 4%, how long is it until purchasing power is cut in half?

Problems 63–66 involve zero-coupon bonds. A **zero-coupon bond** *is a bond that is sold now at a discount and will pay its face value at the time when it matures; no interest payments are made.*

63. Zero-Coupon Bonds A zero-coupon bond can be redeemed in 20 years for $10,000. How much should you be willing to pay for it now if you want a return of:

(a) 10% compounded monthly?

(b) 10% compounded continuously?

64. Zero-Coupon Bonds A child's grandparents are considering buying a $40,000 face value zero-coupon bond at birth so that she will have enough money for her college education 17 years later. If they want a rate of return of 8% compounded annually, what should they pay for the bond?

65. Zero-Coupon Bonds How much should a $10,000 face value zero-coupon bond, maturing in 10 years, be sold for now if its rate of return is to be 8% compounded annually?

66. Zero-Coupon Bonds If Pat pays $12,485.52 for a $25,000 face value zero-coupon bond that matures in 8 years, what is his annual rate of return?

67. Time to Double or Triple an Investment The formula

$$t = \frac{\ln m}{n \ln\left(1 + \dfrac{r}{n}\right)}$$

can be used to find the number of years t required to multiply an investment m times when r is the per annum interest rate compounded n times a year.

(a) How many years will it take to double the value of an IRA that compounds annually at the rate of 12%?

(b) How many years will it take to triple the value of a savings account that compounds quarterly at an annual rate of 6%?

(c) Give a derivation of this formula.

68. Time to Reach an Investment Goal The formula

$$t = \frac{\ln A - \ln P}{r}$$

can be used to find the number of years t required for an investment P to grow to a value A when compounded continuously at an annual rate r.

(a) How long will it take to increase an initial investment of $1000 to $8000 at an annual rate of 10%?

(b) What annual rate is required to increase the value of a $2000 IRA to $30,000 in 35 years?

(c) Give a derivation of this formula.

*Problems 69–72, require the following discussion. The **Consumer Price Index (CPI)** indicates the relative change in price over time for a fixed basket of goods and services. It is a cost of living index that helps measure the effect of inflation on the cost of goods and services. The CPI uses the base period 1982–1984 for comparison (the CPI for this period is 100). The CPI for January 2006 was 198.3. This means that $100 in the period 1982–1984 had the same purchasing power as $198.30 in January 2006. In general, if the rate of inflation averages r% over n years, then the CPI index after n years is*

$$\text{CPI} = \text{CPI}_0\left(1 + \frac{r}{100}\right)^n$$

*where CPI_0 is the CPI index at the beginning of the n-year period. **Source:** U.S. Bureau of Labor Statistics*

69. Consumer Price Index

(a) The CPI was 152.4 for 1995 and 195.3 for 2005. Assuming that annual inflation remained constant for this time period, determine the average annual inflation rate.

(b) Using the inflation rate from part (a), in what year will the CPI reach 300?

70. Consumer Price Index If the current CPI is 234.2 and the average annual inflation rate is 2.8%, what will be the CPI in 5 years?

71. Consumer Price Index If the average annual inflation rate is 3.1%, how long will it take for the CPI index to double?

72. Consumer Price Index The base period for the CPI changed in 1998. Under the previous weight and item structure, the CPI for 1995 was 456.5. If the average annual inflation rate was 5.57%, what year was used as the base period for the CPI?

Discussion and Writing

73. Explain in your own words what the term *compound interest* means. What does *continuous compounding* mean?

74. Explain in your own words the meaning of *present value*.

75. Critical Thinking You have just contracted to buy a house and will seek financing in the amount of $100,000. You go to several banks. Bank 1 will lend you $100,000 at the rate of 8.75% amortized over 30 years with a loan origination fee of 1.75%. Bank 2 will lend you $100,000 at the rate of 8.375% amortized over 15 years with a loan origination fee of 1.5%. Bank 3 will lend you $100,000 at the rate of 9.125% amortized over 30 years with no loan origination fee. Bank 4 will lend you $100,000 at the rate of 8.625% amortized over 15 years with no loan origination fee. Which loan would you take? Why? Be sure to have sound reasons for your choice.

Use the information in the table to assist you. If the amount of the monthly payment does not matter to you, which loan would you take? Again, have sound reasons for your choice. Compare your final decision with others in the class. Discuss.

	Monthly Payment	Loan Origination Fee
Bank 1	$786.70	$1,750.00
Bank 2	$977.42	$1,500.00
Bank 3	$813.63	$0.00
Bank 4	$990.68	$0.00

'Are You Prepared?' Answers

1. $15

2. $13\frac{1}{3}\%$

6.8 Exponential Growth and Decay Models; Newton's Law; Logistic Growth and Decay Models

OBJECTIVES
1 Find Equations of Populations That Obey the Law of Uninhibited Growth (p. 475)
2 Find Equations of Populations That Obey the Law of Decay (p. 478)
3 Use Newton's Law of Cooling (p. 479)
4 Use Logistic Models (p. 481)

1 Find Equations of Populations That Obey the Law of Uninhibited Growth

Many natural phenomena have been found to follow the law that an amount A varies with time t according to the function

$$A(t) = A_0 e^{kt} \tag{1}$$

Here A_0 is the original amount ($t = 0$) and $k \neq 0$ is a constant.

If $k > 0$, then equation (1) states that the amount A is increasing over time; if $k < 0$, the amount A is decreasing over time. In either case, when an amount A varies over time according to equation (1), it is said to follow the **exponential law** or the **law of uninhibited growth** $(k > 0)$ **or decay** $(k < 0)$. See Figure 41.

Figure 41

(a) $A(t) = A_0 e^{kt}, k > 0$ **(b)** $A(t) = A_0 e^{kt}, k < 0$

For example, we saw in Section 6.7 that continuously compounded interest follows the law of uninhibited growth. In this section we shall look at three additional phenomena that follow the exponential law.

Cell division is the growth process of many living organisms, such as amoebas, plants, and human skin cells. Based on an ideal situation in which no cells die and no by-products are produced, the number of cells present at a given time follows the law of uninhibited growth. Actually, however, after enough time has passed, growth at an exponential rate will cease due to the influence of factors such as lack of living space and dwindling food supply. The law of uninhibited growth accurately reflects only the early stages of the cell division process.

The cell division process begins with a culture containing N_0 cells. Each cell in the culture grows for a certain period of time and then divides into two identical cells. We assume that the time needed for each cell to divide in two is constant and does not change as the number of cells increases. These new cells then grow, and eventually each divides in two, and so on.

Uninhibited Growth of Cells

A model that gives the number N of cells in a culture after a time t has passed (in the early stages of growth) is

$$N(t) = N_0 e^{kt} \qquad k > 0 \qquad \qquad \textbf{(2)}$$

where N_0 is the initial number of cells and k is a positive constant that represents the growth rate of the cells.

In using formula (2) to model the growth of cells, we are using a function that yields positive real numbers, even though we are counting the number of cells, which must be an integer. This is a common practice in many applications.

EXAMPLE 1	Bacterial Growth

A colony of bacteria that grows according to the law of uninhibited growth is modeled by the function $N(t) = 100e^{0.045t}$, where N is measured in grams and t is measured in days.

(a) Determine the initial amount of bacteria.

(b) What is the growth rate of the bacteria?

(c) What is the population after 5 days?

(d) How long will it take for the population to reach 140 grams?

(e) What is the doubling time for the population?

Solution (a) The initial amount of bacteria, N_0, is obtained when $t = 0$, so

$$N_0 = N(0) = 100e^{0.045(0)} = 100 \text{ grams}$$

(b) Compare $N(t) = 100e^{0.045t}$ to $N(t) = 100e^{kt}$. The value of k, 0.045, indicates a growth rate of 4.5%.

(c) The population after 5 days is $N(5) = 100e^{0.045(5)} = 125.2$ grams.

(d) To find how long it takes for the population to reach 140 grams, we solve the equation $N(t) = 140$.

$$100e^{0.045t} = 140$$

$e^{0.045t} = 1.4$ Divide both sides of the equation by 100.

$0.045t = \ln 1.4$ Rewrite as a logarithm.

$t = \dfrac{\ln 1.4}{0.045}$ Divide both sides of the equation by 0.045.

≈ 7.5 days

(e) The population doubles when $N(t) = 200$ grams, so we find the doubling time by solving the equation $200 = 100e^{0.045t}$ for t.

$$200 = 100e^{0.045t}$$

$2 = e^{0.045t}$ Divide both sides of the equation by 100.

$\ln 2 = 0.045t$ Rewrite as a logarithm.

$t = \dfrac{\ln 2}{0.045}$ Divide both sides of the equation by 0.045.

≈ 15.4 days

The population doubles approximately every 15.4 days.

──── **Now Work** PROBLEM 1

| EXAMPLE 2 | **Bacterial Growth** |

A colony of bacteria increases according to the law of uninhibited growth.

(a) If N is the number of cells and t is the time in hours, express N as a function of t.

(b) If the number of bacteria doubles in 3 hours, find the function that gives the number of cells in the culture.

(c) How long will it take for the size of the colony to triple?

(d) How long will it take for the population to double a second time (that is, increase four times)?

Solution (a) Using formula (2), the number N of cells at a time t is

$$N(t) = N_0 e^{kt}$$

where N_0 is the initial number of bacteria present and k is a positive number.

(b) We seek the number k. The number of cells doubles in 3 hours, so we have

$$N(3) = 2N_0$$

But $N(3) = N_0 e^{k(3)}$, so

$N_0 e^{k(3)} = 2N_0$

$e^{3k} = 2$ Divide both sides by N_0.

$3k = \ln 2$ Write the exponential equation as a logarithm.

$k = \dfrac{1}{3} \ln 2$

The function that models this growth process is therefore

$$N(t) = N_0 e^{\left(\frac{1}{3}\ln 2\right)t}$$

(c) The time t needed for the size of the colony to triple requires that $N = 3N_0$. We substitute $3N_0$ for N to get

$$3N_0 = N_0 e^{\left(\frac{1}{3}\ln 2\right)t}$$

$$3 = e^{\left(\frac{1}{3}\ln 2\right)t}$$

$$\left(\frac{1}{3}\ln 2\right)t = \ln 3$$

$$t = \frac{3\ln 3}{\ln 2} \approx 4.755 \text{ hours}$$

It will take about 4.755 hours or 4 hours, 45 minutes for the size of the colony to triple.

(d) If a population doubles in 3 hours, it will double a second time in 3 more hours, for a total time of 6 hours.

∎

2 Find Equations of Populations That Obey the Law of Decay

Radioactive materials follow the law of uninhibited decay.

Uninhibited Radioactive Decay

The amount A of a radioactive material present at time t is given by

$$A(t) = A_0 e^{kt} \qquad k < 0 \tag{3}$$

where A_0 is the original amount of radioactive material and k is a negative number that represents the rate of decay.

All radioactive substances have a specific **half-life,** which is the time required for half of the radioactive substance to decay. In **carbon dating,** we use the fact that all living organisms contain two kinds of carbon, carbon 12 (a stable carbon) and carbon 14 (a radioactive carbon with a half-life of 5600 years). While an organism is living, the ratio of carbon 12 to carbon 14 is constant. But when an organism dies, the original amount of carbon 12 present remains unchanged, whereas the amount of carbon 14 begins to decrease. This change in the amount of carbon 14 present relative to the amount of carbon 12 present makes it possible to calculate when the organism died.

EXAMPLE 3	**Estimating the Age of Ancient Tools**

Traces of burned wood along with ancient stone tools in an archeological dig in Chile were found to contain approximately 1.67% of the original amount of carbon 14. If the half-life of carbon 14 is 5600 years, approximately when was the tree cut and burned?

Solution Using formula (3), the amount A of carbon 14 present at time t is

$$A(t) = A_0 e^{kt}$$

where A_0 is the original amount of carbon 14 present and k is a negative number. We first seek the number k. To find it, we use the fact that after 5600 years half of

the original amount of carbon 14 remains, so $A(5600) = \dfrac{1}{2}A_0$. Then

$$\dfrac{1}{2}A_0 = A_0 e^{k(5600)}$$

$$\dfrac{1}{2} = e^{5600k} \qquad \text{Divide both sides of the equation by } A_0.$$

$$5600k = \ln \dfrac{1}{2} \qquad \text{Rewrite as a logarithm.}$$

$$k = \dfrac{1}{5600} \ln \dfrac{1}{2}$$

Formula (3) therefore becomes

$$A(t) = A_0 e^{\frac{\ln \frac{1}{2}}{5600} t}$$

If the amount A of carbon 14 now present is 1.67% of the original amount, it follows that

$$0.0167 A_0 = A_0 e^{\frac{\ln \frac{1}{2}}{5600} t}$$

$$0.0167 = e^{\frac{\ln \frac{1}{2}}{5600} t} \qquad \text{Divide both sides of the equation by } A_0.$$

$$\dfrac{\ln \dfrac{1}{2}}{5600} t = \ln 0.0167 \qquad \text{Rewrite as a logarithm.}$$

$$t = \dfrac{5600}{\ln \dfrac{1}{2}} \ln 0.0167 \approx 33{,}062 \text{ years}$$

The tree was cut and burned about 33,062 years ago. Some archeologists use this conclusion to argue that humans lived in the Americas 33,000 years ago, much earlier than is generally accepted.

━━━━━ **Now Work** PROBLEM 3

3 Use Newton's Law of Cooling

Newton's Law of Cooling* states that the temperature of a heated object decreases exponentially over time toward the temperature of the surrounding medium.

Newton's Law of Cooling

The temperature u of a heated object at a given time t can be modeled by the following function:

$$u(t) = T + (u_0 - T)e^{kt} \qquad k < 0 \tag{4}$$

where T is the constant temperature of the surrounding medium, u_0 is the initial temperature of the heated object, and k is a negative constant.

* Named after Sir Isaac Newton (1643–1727), one of the cofounders of calculus.

| EXAMPLE 4 | **Using Newton's Law of Cooling** |

An object is heated to 100°C (degrees Celsius) and is then allowed to cool in a room whose air temperature is 30°C.

(a) If the temperature of the object is 80°C after 5 minutes, when will its temperature be 50°C?

(b) Determine the elapsed time before the temperature of the object is 35°C.

(c) What do you notice about the temperature as time passes?

Solution (a) Using formula (4) with $T = 30$ and $u_0 = 100$, the temperature $u(t)$ (in degrees Celsius) of the object at time t (in minutes) is

$$u(t) = 30 + (100 - 30)e^{kt} = 30 + 70e^{kt}$$

where k is a negative constant. To find k, we use the fact that $u = 80$ when $t = 5$. Then

$$u(t) = 30 + 70e^{kt}$$
$$80 = 30 + 70e^{k(5)} \qquad u(5) = 80$$
$$50 = 70e^{5k} \qquad \text{Simplify}$$
$$e^{5k} = \frac{50}{70} \qquad \text{Solve for } e^{5k}$$
$$5k = \ln\frac{5}{7} \qquad \text{Take ln of both sides}$$
$$k = \frac{1}{5}\ln\frac{5}{7} \approx -0.0673 \quad \text{Solve for } k$$

Formula (4) therefore becomes

$$u(t) = 30 + 70e^{\frac{\ln\frac{5}{7}}{5}t} \qquad\qquad\qquad \textbf{(5)}$$

We want to find t when $u = 50°C$, so

$$50 = 30 + 70e^{\frac{\ln\frac{5}{7}}{5}t}$$
$$20 = 70e^{\frac{\ln\frac{5}{7}}{5}t} \qquad\qquad \text{Simplify}$$
$$e^{\frac{\ln\frac{5}{7}}{5}t} = \frac{20}{70}$$
$$\frac{\ln\frac{5}{7}}{5}t = \ln\frac{2}{7} \qquad\qquad \text{Take ln of both sides}$$
$$t = \frac{5}{\ln\frac{5}{7}}\ln\frac{2}{7} \approx 18.6 \text{ minutes} \quad \text{Solve for } t$$

The temperature of the object will be 50°C after about 18.6 minutes or 18 minutes, 36 seconds.

(b) If $u = 35°C$, then, based on equation (5), we have

$$35 = 30 + 70e^{\frac{\ln\frac{5}{7}}{5}t}$$
$$5 = 70e^{\frac{\ln\frac{5}{7}}{5}t} \qquad\qquad \text{Simplify}$$

$$e^{\frac{\ln\frac{5}{7}}{5}t} = \frac{5}{70}$$

$$\frac{\ln\frac{5}{7}}{5}t = \ln\frac{5}{70} \qquad \text{Take ln of both sides}$$

$$t = \frac{\ln\frac{5}{70}}{\ln\frac{5}{7}} \approx 39.2 \text{ minutes} \quad \text{Solve for } t$$

The object will reach a temperature of 35°C after about 39.2 minutes.

(c) Look at equation (5). As t increases, the exponent $\dfrac{\ln\frac{5}{7}}{5}t$ becomes unbounded in the negative direction, since $\ln\dfrac{5}{7} < 0$. As a result, the value of $e^{\frac{\ln\frac{5}{7}}{5}t}$ approaches zero, so the value of u, the temperature of the object, approaches 30°C, the air temperature of the room.

◾

Now Work PROBLEM 13

4 Use Logistic Models

The exponential growth model $A(t) = A_0 e^{kt}$, $k > 0$, assumes uninhibited growth, meaning that the value of the function grows without limit. Recall that we stated that cell division could be modeled using this function, assuming that no cells die and no by-products are produced. However, cell division eventually is limited by factors such as living space and food supply. The **logistic model**, given next, can describe situations where the growth or decay of the dependent variable is limited.

Logistic Model

In a logistic growth model, the population P after time t is given by the function

$$P(t) = \frac{c}{1 + ae^{-bt}} \qquad (6)$$

where a, b, and c are constants with $c > 0$. The model is a growth model if $b > 0$; the model is a decay model if $b < 0$.

The number c is called the **carrying capacity** (for growth models) because the value $P(t)$ approaches c as t approaches infinity; that is, $\lim\limits_{t\to\infty} P(t) = c$. The number $|b|$ is the growth rate for $b > 0$ and the decay rate for $b < 0$. Figure 42(a) shows the graph of a typical logistic growth function, and Figure 42(b) shows the graph of a typical logistic decay function.

Figure 42

(a) (b)

Based on the figures, we have the following properties of logistic growth functions.

Properties of the Logistic Growth Function, Equation (6)

1. The domain is the set of all real numbers. The range is the interval $(0, c)$, where c is the carrying capacity.
2. There are no x-intercepts; the y-intercept is $P(0)$.
3. There are two horizontal asymptotes: $y = 0$ and $y = c$.
4. $P(t)$ is an increasing function if $b > 0$ and a decreasing function if $b < 0$.
5. There is an **inflection point** where $P(t)$ equals $\dfrac{1}{2}$ of the carrying capacity. The inflection point is the point on the graph where the graph changes from being curved upward to curved downward for growth functions and the point where the graph changes from being curved downward to curved upward for decay functions.
6. The graph is smooth and continuous, with no corners or gaps.

EXAMPLE 5	**Fruit Fly Population**

Fruit flies are placed in a half-pint milk bottle with a banana (for food) and yeast plants (for food and to provide a stimulus to lay eggs). Suppose that the fruit fly population after t days is given by

$$P(t) = \frac{230}{1 + 56.5e^{-0.37t}}$$

(a) State the carrying capacity and the growth rate.
(b) Determine the initial population.
(c) What is the population after 5 days?
(d) How long does it take for the population to reach 180?
 (e) Use a graphing utility to determine how long it takes for the population to reach one-half of the carrying capacity by graphing $Y_1 = P(t)$ and $Y_2 = 115$ and using INTERSECT.

Solution (a) As $t \to \infty$, $e^{-0.37t} \to 0$ and $P(t) \to \dfrac{230}{1}$. The carrying capacity of the half-pint bottle is 230 fruit flies. The growth rate is $|b| = |0.37| = 37\%$ per day.

(b) To find the initial number of fruit flies in the half-pint bottle, we evaluate $P(0)$.

$$P(0) = \frac{230}{1 + 56.5e^{-0.37(0)}}$$
$$= \frac{230}{1 + 56.5}$$
$$= 4$$

So, initially, there were 4 fruit flies in the half-pint bottle.

(c) To find the number of fruit flies in the half-pint bottle after 5 days, we evaluate $P(5)$.

$$P(5) = \frac{230}{1 + 56.5e^{-0.37(5)}} \approx 23 \text{ fruit flies}$$

After 5 days, there are approximately 23 fruit flies in the bottle.

(d) To determine when the population of fruit flies will be 180, we solve the equation $P(t) = 180$.

$$\frac{230}{1 + 56.5e^{-0.37t}} = 180$$

$$230 = 180(1 + 56.5e^{-0.37t})$$

$1.2778 = 1 + 56.5e^{-0.37t}$ Divide both sides by 180.

$0.2778 = 56.5e^{-0.37t}$ Subtract 1 from both sides.

$0.0049 = e^{-0.37t}$ Divide both sides by 56.5.

$\ln(0.0049) = -0.37t$ Rewrite as a logarithmic expression.

$t \approx 14.4$ days Divide both sides by −0.37.

It will take approximately 14.4 days (14 days, 10 hours) for the population to reach 180 fruit flies.

(e) One-half of the carrying capacity is 115 fruit flies. We solve $P(t) = 115$ by graphing $Y_1 = \dfrac{230}{1 + 56.5e^{-0.37t}}$ and $Y_2 = 115$ and using INTERSECT. See Figure 43.

The population will reach one-half of the carrying capacity in about 10.9 days (10 days, 22 hours).

Figure 43

$Y_1 = \dfrac{230}{1 + 56.5e^{-0.37t}}$

$Y_2 = 115$

Intersection
X=10.903353 Y=115

 Look back at Figure 43. Notice the point where the graph reaches 115 fruit flies (one-half of the carrying capacity): the graph changes from being curved upward to being curved downward. Using the language of calculus, we say the graph changes from increasing at an increasing rate to increasing at a decreasing rate. For any logistic growth function, when the population reaches one-half the carrying capacity, the population growth starts to slow down.

Now Work PROBLEM 2 3

 Exploration

On the same viewing rectangle, graph

$$Y_1 = \frac{500}{1 + 24e^{-0.03t}} \text{ and } Y_2 = \frac{500}{1 + 24e^{-0.08t}}$$

What effect does the growth rate $|b|$ have on the logistic growth function?

EXAMPLE 6 **Wood Products**

The EFISCEN wood product model classifies wood products according to their life-span. There are four classifications: short (1 year), medium short (4 years), medium long (16 years), and long (50 years). Based on data obtained from the European Forest Institute, the percentage of remaining wood products after t years for wood products with long life-spans (such as those used in the building industry) is given by

$$P(t) = \frac{100.3952}{1 + 0.0316e^{0.0581t}}$$

(a) What is the decay rate?

(b) What is the percentage of remaining wood products after 10 years?

(c) How long does it take for the percentage of remaining wood products to reach 50%?

(d) Explain why the numerator given in the model is reasonable.

Solution (a) The decay rate is $|b| = |-0.0581| = 5.81\%$.

(b) Evaluate $P(10)$.

$$P(10) = \frac{100.3952}{1 + 0.0316e^{0.0581(10)}} \approx 95.0$$

So 95% of long-life-span wood products remains after 10 years.

(c) Solve the equation $P(t) = 50$.

$$\frac{100.3952}{1 + 0.0316e^{0.0581t}} = 50$$

$$100.3952 = 50(1 + 0.0316e^{0.0581t})$$

$$2.0079 = 1 + 0.0316e^{0.0581t} \qquad \text{Divide both sides by 50.}$$

$$1.0079 = 0.0316e^{0.0581t} \qquad \text{Subtract 1 from both sides.}$$

$$31.8957 = e^{0.0581t} \qquad \text{Divide both sides by 0.0316.}$$

$$\ln(31.8957) = 0.0581t \qquad \text{Rewrite as a logarithmic expression.}$$

$$t \approx 59.6 \text{ years} \qquad \text{Divide both sides by 0.0581.}$$

It will take approximately 59.6 years for the percentage of long-life-span wood products remaining to reach 50%.

(d) The numerator of 100.3952 is reasonable because the maximum percentage of wood products remaining that is possible is 100%.

$\blacksquare$

6.8 Assess Your Understanding

Applications and Extensions

1. **Growth of an Insect Population** The size P of a certain insect population at time t (in days) obeys the function $P(t) = 500e^{0.02t}$.
 (a) Determine the number of insects at $t = 0$ days.
 (b) What is the growth rate of the insect population?
 (c) What is the population after 10 days?
 (d) When will the insect population reach 800?
 (e) When will the insect population double?

2. **Growth of Bacteria** The number N of bacteria present in a culture at time t (in hours) obeys the law of uninhibited growth $N(t) = 1000e^{0.01t}$.
 (a) Determine the number of bacteria at $t = 0$ hours.
 (b) What is the growth rate of the bacteria?
 (c) What is the population after 4 hours?
 (d) When will the number of bacteria reach 1700?
 (e) When will the number of bacteria double?

3. **Radioactive Decay** Strontium 90 is a radioactive material that decays according to the function $A(t) = A_0e^{-0.0244t}$, where A_0 is the initial amount present and A is the amount present at time t (in years). Assume that a scientist has a sample of 500 grams of strontium 90.
 (a) What is the decay rate of strontium 90?
 (b) How much strontium 90 is left after 10 years?
 (c) When will 400 grams of strontium 90 be left?
 (d) What is the half-life of strontium 90?

4. **Radioactive Decay** Iodine 131 is a radioactive material that decays according to the function $A(t) = A_0e^{-0.087t}$, where A_0 is the initial amount present and A is the amount present at time t (in days). Assume that a scientist has a sample of 100 grams of iodine 131.
 (a) What is the decay rate of iodine 131?
 (b) How much iodine 131 is left after 9 days?
 (c) When will 70 grams of iodine 131 be left?
 (d) What is the half-life of iodine 131?

5. **Growth of a Colony of Mosquitoes** The population of a colony of mosquitoes obeys the law of uninhibited growth.
 (a) If N is the population of the colony and t is the time in days, express N as a function of t.
 (b) If there are 1000 mosquitoes initially and there are 1800 after 1 day, what is the size of the colony after 3 days?
 (c) How long is it until there are 10,000 mosquitoes?

6. **Bacterial Growth** A culture of bacteria obeys the law of uninhibited growth.
 (a) If N is the number of bacteria in the culture and t is the time in hours, express N as a function of t.
 (b) If 500 bacteria are present initially and there are 800 after 1 hour, how many will be present in the culture after 5 hours?
 (c) How long is it until there are 20,000 bacteria?

7. **Population Growth** The population of a southern city follows the exponential law.
 (a) If N is the population of the city and t is the time in years, express N as a function of t.
 (b) If the population doubled in size over an 18-month period and the current population is 10,000, what will the population be 2 years from now?

8. **Population Decline** The population of a midwestern city follows the exponential law.
 (a) If N is the population of the city and t is the time in years, express N as a function of t.
 (b) If the population decreased from 900,000 to 800,000 from 2003 to 2005, what will the population be in 2007?

9. **Radioactive Decay** The half-life of radium is 1690 years. If 10 grams are present now, how much will be present in 50 years?

10. **Radioactive Decay** The half-life of radioactive potassium is 1.3 billion years. If 10 grams are present now, how much will be present in 100 years? In 1000 years?

11. **Estimating the Age of a Tree** A piece of charcoal is found to contain 30% of the carbon 14 that it originally had. When did the tree die from which the charcoal came? Use 5600 years as the half-life of carbon 14.

12. **Estimating the Age of a Fossil** A fossilized leaf contains 70% of its normal amount of carbon 14. How old is the fossil?

13. **Cooling Time of a Pizza Pan** A pizza pan is removed at 5:00 PM from an oven whose temperature is fixed at 450°F into a room that is a constant 70°F. After 5 minutes, the pan is at 300°F.
 (a) At what time is the temperature of the pan 135°F?
 (b) Determine the time that needs to elapse before the pan is 160°F.
 (c) What do you notice about the temperature as time passes?

14. **Newton's Law of Cooling** A thermometer reading 72°F is placed in a refrigerator where the temperature is a constant 38°F.
 (a) If the thermometer reads 60°F after 2 minutes, what will it read after 7 minutes?
 (b) How long will it take before the thermometer reads 39°F?
 (c) Determine the time needed to elapse before the thermometer reads 45°F.
 (d) What do you notice about the temperature as time passes?

15. **Newton's Law of Heating** A thermometer reading 8°C is brought into a room with a constant temperature of 35°C. If the thermometer reads 15°C after 3 minutes, what will it read after being in the room for 5 minutes? For 10 minutes?
 [**Hint:** You need to construct a formula similar to equation (4).]

16. **Warming Time of a Beerstein** A beerstein has a temperature of 28°F. It is placed in a room with a constant temperature of 70°F. After 10 minutes, the temperature of the stein has risen to 35°F. What will the temperature of the stein be after 30 minutes? How long will it take the stein to reach a temperature of 45°F? (See the hint given for Problem 15.)

17. **Decomposition of Chlorine in a Pool** Under certain water conditions, the free chlorine (hypochlorous acid, $HOCl$) in a swimming pool decomposes according to the law of uninhibited decay. After shocking his pool, Ben tested the water and found the amount of free chlorine to be 2.5 parts per million (ppm). Twenty-four hours later, Ben tested the water again and found the amount of free chlorine to be 2.2 ppm. What will be the reading after 3 days (that is, 72 hours)? When the chlorine level reaches 1.0 ppm, Ben must shock the pool again. How long can Ben go before he must shock the pool again?

18. **Decomposition of Dinitrogen Pentoxide** At 45°C, dinitrogen pentoxide (N_2O_5) decomposes into nitrous dioxide (NO_2) and oxygen (O_2) according to the law of uninhibited decay. An initial amount of 0.25 M of dinitrogen pentoxide decomposes to 0.15 M in 17 minutes. How much dinitrogen pentoxide will remain after 30 minutes? How long will it take until 0.01 M of dinitrogen pentoxide remains?

19. **Decomposition of Sucrose** Reacting with water in an acidic solution at 35°C, sucrose ($C_{12}H_{22}O_{11}$) decomposes into glucose ($C_6H_{12}O_6$) and fructose ($C_6H_{12}O_6$)* according to the law of uninhibited decay. An initial amount of 0.40 M of sucrose decomposes to 0.36 M in 30 minutes. How much sucrose will remain after 2 hours? How long will it take until 0.10 M of sucrose remains?

20. **Decomposition of Salt in Water** Salt ($NaCl$) decomposes in water into sodium (Na^+) and chloride (Cl^-) ions according to the law of uninhibited decay. If the initial amount of salt is 25 kilograms and, after 10 hours, 15 kilograms of salt is left, how much salt is left after 1 day? How long does it take until $\dfrac{1}{2}$ kilogram of salt is left?

21. **Radioactivity from Chernobyl** After the release of radioactive material into the atmosphere from a nuclear power plant at Chernobyl (Ukraine) in 1986, the hay in Austria was contaminated by iodine 131 (half-life 8 days). If it is safe to feed the hay to cows when 10% of the iodine 131 remains, how long did the farmers need to wait to use this hay?

22. **Pig Roasts** The hotel Bora-Bora is having a pig roast. At noon, the chef put the pig in a large earthen oven. The pig's original temperature was 75°F. At 2:00 PM the chef checked the pig's temperature and was upset because it had reached

* Author's Note: Surprisingly, the chemical formulas for glucose and fructose are the same. This is not a typo.

only 100°F. If the oven's temperature remains a constant 325°F, at what time may the hotel serve its guests, assuming that pork is done when it reaches 175°F?

23. **Proportion of the Population That Owns a DVD Player** The logistic growth model

$$P(t) = \frac{0.9}{1 + 6e^{-0.32t}}$$

relates the proportion of U.S. households that own a DVD player to the year. Let $t = 0$ represent 2000, $t = 1$ represent 2001, and so on.

(a) Determine the maximum proportion of households that will own a DVD player.

(b) What proportion of households owned a DVD player in 2000 ($t = 0$)?

(c) What proportion of households owned a DVD player in 2005 ($t = 5$)?

(d) When will 0.8 (80%) of U.S. households own a DVD player?

24. **Market Penetration of Intel's Coprocessor** The logistic growth model

$$P(t) = \frac{0.90}{1 + 3.5e^{-0.339t}}$$

relates the proportion of new personal computers (PCs) sold at Best Buy that have Intel's latest coprocessor t months after it has been introduced.

(a) Determine the maximum proportion of PCs sold at Best Buy that will have Intel's latest coprocessor.

(b) What proportion of computers sold at Best Buy will have Intel's latest coprocessor when it is first introduced ($t = 0$)?

(c) What proportion of PCs sold will have Intel's latest coprocessor $t = 4$ months after it is introduced?

(d) When will 0.75 (75%) of PCs sold at Best Buy have Intel's latest coprocessor?

(e) How long will it be before 0.45 (45%) of the PCs sold by Best Buy have Intel's latest coprocessor?

25. **Population of a Bacteria Culture** The logistic growth model

$$P(t) = \frac{1000}{1 + 32.33e^{-0.439t}}$$

represents the population (in grams) of a bacterium after t hours.

(a) Determine the carrying capacity of the environment.

(b) What is the growth rate of the bacteria?

(c) Determine the initial population size.

(d) What is the population after 9 hours?

(e) When will the population be 700 grams?

(f) How long does it take for the population to reach one-half the carrying capacity?

26. **Population of an Endangered Species** Often environmentalists capture an endangered species and transport the species to a controlled environment where the species can produce offspring and regenerate its population. Suppose that six American bald eagles are captured, transported to Montana, and set free. Based on experience, the environmentalists

expect the population to grow according to the model

$$P(t) = \frac{500}{1 + 83.33e^{-0.162t}}$$

where t is measured in years.

(a) Determine the carrying capacity of the environment.

(b) What is the growth rate of the bald eagle?

(c) What is the population after 3 years?

(d) When will the population be 300 eagles?

(e) How long does it take for the population to reach one-half of the carrying capacity?

27. **The *Challenger* Disaster** After the *Challenger* disaster in 1986, a study was made of the 23 launches that preceded the fatal flight. A mathematical model was developed involving the relationship between the Fahrenheit temperature x around the O-rings and the number y of eroded or leaky primary O-rings. The model stated that

$$y = \frac{6}{1 + e^{-(5.085 - 0.1156x)}}$$

where the number 6 indicates the 6 primary O-rings on the spacecraft.

(a) What is the predicted number of eroded or leaky primary O-rings at a temperature of 100°F?

(b) What is the predicted number of eroded or leaky primary O-rings at a temperature of 60°F?

(c) What is the predicted number of eroded or leaky primary O-rings at a temperature of 30°F?

(d) Graph the equation using a graphing utility. At what temperature is the predicted number of eroded or leaky O-rings 1? 3? 5?

Source: Linda Tappin, "Analyzing Data Relating to the *Challenger* Disaster," *Mathematics Teacher,* Vol. 87, No. 6, September 1994, pp. 423–426.

6.9 Building Exponential, Logarithmic, and Logistic Models from Data

PREPARING FOR THIS SECTION *Before getting started, review the following:*

- Building Linear Functions from Data (Section 4.2, pp. 287–290)
- Building Quadratic Functions from Data (Section 4.4, pp. 309–310)

OBJECTIVES **1** Use a Graphing Utility to Fit an Exponential Function to Data (p. 487)
 2 Use a Graphing Utility to Fit a Logarithmic Function to Data (p. 489)
 3 Use a Graphing Utility to Fit a Logistic Function to Data (p. 490)

In Section 4.2, we discussed how to find the linear function of best fit ($y = ax + b$), and in Section 4.4, we discussed how to find the quadratic function of best fit ($y = ax^2 + bx + c$).

In this section we will discuss how to use a graphing utility to find equations of best fit that describe the relation between two variables when the relation is thought to be exponential ($y = ab^x$), logarithmic ($y = a + b \ln x$), or logistic $\left(y = \dfrac{c}{1 + ae^{-bx}} \right)$. As before, we draw a scatter diagram of the data to help to determine the appropriate model to use.

Figure 44 shows scatter diagrams that will typically be observed for the three models. Below each scatter diagram are any restrictions on the values of the parameters.

Figure 44

$y = ab^x, a > 0, b > 1$	$y = ab^x, 0 < b < 1, a > 0$	$y = a + b \ln x, a > 0, b < 0$	$y = a + b \ln x, a > 0, b > 0$	$y = \dfrac{c}{1 + ae^{-bx}}, a > 0, b > 0, c > 0$
Exponential	Exponential	Logarithmic	Logarithmic	Logistic

Most graphing utilities have REGression options that fit data to a specific type of curve. Once the data have been entered and a scatter diagram obtained, the type of curve that you want to fit to the data is selected. Then that REGression option is used to obtain the curve of *best fit* of the type selected.

The correlation coefficient r will appear only if the model can be written as a linear expression. As it turns out, r will appear for the linear, power, exponential, and logarithmic models, since these models can be written as a linear expression. Remember, the closer $|r|$ is to 1, the better the fit.

Let's look at some examples.

 1 Use a Graphing Utility to Fit an Exponential Function to Data

We saw in Section 6.7 that the future value of money behaves exponentially, and we saw in Section 6.8 that growth and decay models also behave exponentially. The next example shows how data can lead to an exponential model.

| EXAMPLE 1 | Fitting an Exponential Function to Data |

Kathleen is interested in finding a function that explains the growth of cell phone usage in the United States. She gathers data on the number (in millions) of U.S. cell phone subscribers from 1985 through 2005. The data are shown in Table 9.

Table 9

Year, x	Number of Subscribers (in millions), y
1985 (x = 1)	0.34
1986 (x = 2)	0.68
1987 (x = 3)	1.23
1988 (x = 4)	2.07
1989 (x = 5)	3.51
1990 (x = 6)	5.28
1991 (x = 7)	7.56
1992 (x = 8)	11.03
1993 (x = 9)	16.01
1994 (x = 10)	24.13
1995 (x = 11)	33.76
1996 (x = 12)	44.04
1997 (x = 13)	55.31
1998 (x = 14)	69.21
1999 (x = 15)	86.05
2000 (x = 16)	109.48
2001 (x = 17)	128.37
2002 (x = 18)	140.77
2003 (x = 19)	158.72
2004 (x = 20)	182.14
2005 (x = 21)	207.90

Source: ©2006 CTIA–The Wireless Association®. All Rights Reserved.

(a) Using a graphing utility, draw a scatter diagram with year as the independent variable.
(b) Using a graphing utility, fit an exponential function to the data.
(c) Express the function found in part (b) in the form $A = A_0 e^{kt}$.
(d) Graph the exponential function found in part (b) or (c) on the scatter diagram.
(e) Using the solution to part (b) or (c), predict the number of U.S. cell phone subscribers in 2009.
(f) Interpret the value of k found in part (c).

Solution

(a) Enter the data into the graphing utility, letting 1 represent 1985, 2 represent 1986, and so on. We obtain the scatter diagram shown in Figure 45.

(b) A graphing utility fits the data in Figure 45 to an exponential function of the form $y = ab^x$ using the EXPonential REGression option. From Figure 46 we find that $y = ab^x = 0.66703(1.3621)^x$. Notice that $|r|$ is close to 1, indicating a good fit.

Figure 45

```
243

0                    23
0
```

Figure 46

```
ExpReg
y=a*b^x
a=.6670278215
b=1.362123394
r²=.9482687614
r=.9737909228
```

(c) To express $y = ab^x$ in the form $A = A_0 e^{kt}$, where $x = t$ and $y = A$, we proceed as follows:

$$ab^x = A_0 e^{kt} \quad x = t$$

When $x = t = 0$, we find that $a = A_0$. This leads to

$$a = A_0, \quad b^x = e^{kt}$$
$$b^x = (e^k)^t$$
$$b = e^k \quad x = t$$

Since $y = ab^x = 0.66703(1.3621)^x$, we find that $a = 0.66703$ and $b = 1.3621$:

$$a = A_0 = 0.66703 \quad \text{and} \quad b = e^k = 1.3621$$

We want to find k, so we rewrite $e^k = 1.3621$ as a logarithm and obtain

$$k = \ln(1.3621) \approx 0.3090$$

As a result, $A = A_0 e^{kt} = 0.66703 e^{0.3090t}$.

(d) See Figure 47 for the graph of the exponential function of best fit.

(e) Let $t = 25$ (end of 2009) in the function found in part (c). The predicted number (in millions) of cell phone subscribers in the United States in 2009 is

$$A_0 e^{kt} = 0.66703 e^{0.3090(25)} = 1512.01$$

This prediction (1,512 millons) far exceeds what the U.S. population will be in 2009 (currently the U.S. population is about 300 million). See the answer in part (f).

Figure 47

```
243

0                    23
0
```

(f) The value of $k = 0.3090$ represents the growth rate of the number of cell phone subscribers in the United States. Over the period 1985 through 2005, the number of cell phone subscribers has grown at an annual rate of 30.9% compounded continuously. This growth rate is not sustainable and caused the answer to part (e). See Problem 12.

Now Work PROBLEM 1

2 Use a Graphing Utility to Fit a Logarithmic Function to Data

Many relations between variables do not follow an exponential model; instead, the independent variable is related to the dependent variable using a logarithmic model.

| EXAMPLE 2 | Fitting a Logarithmic Function to Data |

Jodi, a meteorologist, is interested in finding a function that explains the relation between the height of a weather balloon (in kilometers) and the atmospheric pressure (measured in millimeters of mercury) on the balloon. She collects the data shown in Table 10.

(a) Using a graphing utility, draw a scatter diagram of the data with atmospheric pressure as the independent variable.

(b) It is known that the relation between atmospheric pressure and height follows a logarithmic model. Using a graphing utility, fit a logarithmic function to the data.

(c) Draw the logarithmic function found in part (b) on the scatter diagram.

(d) Use the function found in part (b) to predict the height of the weather balloon if the atmospheric pressure is 560 millimeters of mercury.

Table 10

Atmospheric Pressure, p	Height, h
760	0
740	0.184
725	0.328
700	0.565
650	1.079
630	1.291
600	1.634
580	1.862
550	2.235

Solution

(a) After entering the data into the graphing utility, we obtain the scatter diagram shown in Figure 48.

(b) A graphing utility fits the data in Figure 48 to a logarithmic function of the form $y = a + b \ln x$ by using the LOGarithm REGression option. See Figure 49. The logarithmic function of best fit to the data is

$$h(p) = 45.7863 - 6.9025 \ln p$$

where h is the height of the weather balloon and p is the atmospheric pressure. Notice that $|r|$ is close to 1, indicating a good fit.

(c) Figure 50 shows the graph of $h(p) = 45.7863 - 6.9025 \ln p$ on the scatter diagram.

Figure 48

Figure 49

Figure 50

(d) Using the function found in part (b), Jodi predicts the height of the weather balloon when the atmospheric pressure is 560 to be

$$h(560) = 45.7863 - 6.9025 \ln 560$$
$$\approx 2.108 \text{ kilometers}$$

Now Work PROBLEM 7

3 Use a Graphing Utility to Fit a Logistic Function to Data

Logistic growth models can be used to model situations for which the value of the dependent variable is limited. Many real-world situations conform to this scenario. For example, the population of the human race is limited by the availability of natural resources such as food and shelter. When the value of the dependent variable is limited, a logistic growth model is often appropriate.

EXAMPLE 3 **Fitting a Logistic Function to Data**

The data in Table 11 represent the amount of yeast biomass in a culture after t hours.

Table 11

Time (in hours)	Yeast Biomass	Time (in hours)	Yeast Biomass
0	9.6	10	513.3
1	18.3	11	559.7
2	29.0	12	594.8
3	47.2	13	629.4
4	71.1	14	640.8
5	119.1	15	651.1
6	174.6	16	655.9
7	257.3	17	659.6
8	350.7	18	661.8
9	441.0		

Source: Tor Carlson (Über Geschwindigkeit und Grösse der Hefevermehrung in Würze, Biochemische Zeitschrift, Bd. 57, pp. 313–334, 1913)

(a) Using a graphing utility, draw a scatter diagram of the data with time as the independent variable.
(b) Using a graphing utility, fit a logistic function to the data.
(c) Using a graphing utility, graph the function found in part (b) on the scatter diagram.
(d) What is the predicted carrying capacity of the culture?
(e) Use the function found in part (b) to predict the population of the culture at $t = 19$ hours.

Solution (a) See Figure 51 for a scatter diagram of the data.

Figure 51

(b) A graphing utility fits a logistic growth model of the form $y = \dfrac{c}{1 + ae^{-bx}}$ by using the LOGISTIC regression option. See Figure 52. The logistic function of best fit to the data is

$$y = \frac{663.0}{1 + 71.6e^{-0.5470x}}$$

where y is the amount of yeast biomass in the culture and x is the time.

(c) See Figure 53 for the graph of the logistic function of best fit.

Figure 52

Figure 53

(d) Based on the logistic growth function found in part (b), the carrying capacity of the culture is 663.

(e) Using the logistic growth function found in part (b), the predicted amount of yeast biomass at $t = 19$ hours is

$$y = \frac{663.0}{1 + 71.6e^{-0.5470(19)}} = 661.6$$

Now Work PROBLEM 9

6.9 Assess Your Understanding

Applications and Extensions

1. **Biology** A strain of E-coli Beu 397-recA441 is placed into a nutrient broth at 30°Celsius and allowed to grow. The following data are collected. Theory states that the number of bacteria in the petri dish will initially grow according to the law of uninhibited growth. The population is measured using an optical device in which the amount of light that passes through the petri dish is measured.

Time (hours), x	Population, y
0	0.09
2.5	0.18
3.5	0.26
4.5	0.35
6	0.50

Source: Dr. Polly Lavery, Joliet Junior College

(a) Draw a scatter diagram treating time as the independent variable.

(b) Using a graphing utility, fit an exponential function to the data.

(c) Express the function found in part (b) in the form $N(t) = N_0e^{kt}$.

(d) Graph the exponential function found in part (b) or (c) on the scatter diagram.

(e) Use the exponential function from part (b) or (c) to predict the population at $x = 7$ hours.

(f) Use the exponential function from part (b) or (c) to predict when the population will reach 0.75.

2. **Biology** A strain of E-coli SC18del-recA718 is placed into a nutrient broth at 30°Celsius and allowed to grow. The

following data are collected. Theory states that the number of bacteria in the petri dish will initially grow according to the law of uninhibited growth. The population is measured using an optical device in which the amount of light that passes through the petri dish is measured.

Time (hours), x	Population, y
2.5	0.175
3.5	0.38
4.5	0.63
4.75	0.76
5.25	1.20

Source: Dr. Polly Lavery, Joliet Junior College

(a) Draw a scatter diagram treating time as the independent variable.

(b) Using a graphing utility, fit an exponential function to the data.

(c) Express the function found in part (b) in the form $N(t) = N_0e^{kt}$.

(d) Graph the exponential function found in part (b) or (c) on the scatter diagram.

(e) Use the exponential function from part (b) or (c) to predict the population at $x = 6$ hours.

(f) Use the exponential function from part (b) or (c) to predict when the population will reach 2.1.

3. **Chemistry** A chemist has a 100-gram sample of a radioactive material. He records the amount of radioactive

material every week for 6 weeks and obtains the following data:

Week	Weight (in Grams)
0	100.0
1	88.3
2	75.9
3	69.4
4	59.1
5	51.8
6	45.5

(a) Using a graphing utility, draw a scatter diagram with week as the independent variable.

(b) Using a graphing utility, fit an exponential function to the data.

(c) Express the function found in part (b) in the form $A(t) = A_0 e^{kt}$.

(d) Graph the exponential function found in part (b) or (c) on the scatter diagram.

(e) From the result found in part (b), determine the half-life of the radioactive material.

(f) How much radioactive material will be left after 50 weeks?

(g) When will there be 20 grams of radioactive material?

4. **Chemistry** A chemist has a 1000-gram sample of a radioactive material. She records the amount of radioactive material remaining in the sample every day for a week and obtains the following data:

Day	Weight (in Grams)
0	1000.0
1	897.1
2	802.5
3	719.8
4	651.1
5	583.4
6	521.7
7	468.3

(a) Using a graphing utility, draw a scatter diagram with day as the independent variable.

(b) Using a graphing utility, fit an exponential function to the data.

(c) Express the function found in part (b) in the form $A(t) = A_0 e^{kt}$.

(d) Graph the exponential function found in part (b) or (c) on the scatter diagram.

(e) From the result found in part (b), find the half-life of the radioactive material.

(f) How much radioactive material will be left after 20 days?

(g) When will there be 200 grams of radioactive material?

5. **Cigarette Production** The following data represent the number of cigarettes (in billions) produced in the United States by year.

Year	Cigarette Production (in billions of pieces)
1995	747
1998	680
1999	607
2000	565
2001	562
2002	532
2003	499
2004	493

Source: Statistical Abstract of the United States, 2006

(a) Let t = the number of years since 1995. Using a graphing utility, draw a scatter diagram of the data using t as the independent variable and number of cigarettes as the dependent variable.

(b) Using a graphing utility, fit an exponential function to the data.

(c) Express the function found in part (b) in the form $A(t) = A_0 e^{kt}$.

(d) Graph the exponential function found in part (b) or (c) on the scatter diagram.

(e) Use the exponential function from part (b) or (c) to predict the number of cigarettes that will be produced in the United States in 2010.

(f) Use the exponential function from part (b) or (c) to predict when the number of cigarettes produced in the United States will decrease to 230 billion.

6. **Cigarette Exports** The following data represent the number of cigarettes (in billions) exported from the United States by year.

Year	Cigarette Exports (in billions of pieces)
1995	231.1
1998	201.3
1999	151.4
2000	147.9
2001	133.9
2002	127.4
2003	121.5
2004	118.7

Source: Statistical Abstract of the United States, 2006

(a) Let t = the number of years since 1995. Using a graphing utility, draw a scatter diagram of the data using t as the independent variable and number of cigarettes as the dependent variable.

(b) Using a graphing utility, fit an exponential function to the data.

(c) Express the function found in part (b) in the form $A(t) = A_0 e^{kt}$.

(d) Graph the exponential function found in part (b) or (c) on the scatter diagram.

(e) Use the exponential function from part (b) or (c) to predict the number of cigarettes that will be exported from the United States in 2010.

(f) Use the exponential function from part (b) or (c) to predict when the number of cigarettes exported from the United States will decrease to 50 billion.

7. Economics and Marketing The following data represent the price and quantity demanded in 2005 for IBM personal computers.

Price ($/Computer)	Quantity Demanded
2300	152
2000	159
1700	164
1500	171
1300	176
1200	180
1000	189

(a) Using a graphing utility, draw a scatter diagram of the data with price as the dependent variable.

(b) Using a graphing utility, fit a logarithmic function to the data.

(c) Using a graphing utility, draw the logarithmic function found in part (b) on the scatter diagram.

(d) Use the function found in part (b) to predict the number of IBM personal computers that will be demanded if the price is $1650.

8. Economics and Marketing The following data represent the price and quantity supplied in 2005 for IBM personal computers.

Price ($/Computer)	Quantity Supplied
2300	180
2000	173
1700	160
1500	150
1300	137
1200	130
1000	113

(a) Using a graphing utility, draw a scatter diagram of the data with price as the dependent variable.

(b) Using a graphing utility, fit a logarithmic function to the data.

(c) Using a graphing utility, draw the logarithmic function found in part (b) on the scatter diagram.

(d) Use the function found in part (b) to predict the number of IBM personal computers that will be supplied if the price is $1650.

9. Population Model The following data represent the population of the United States. An ecologist is interested in finding a function that describes the population of the United States.

Year	Population
1900	76,212,168
1910	92,228,496
1920	106,021,537
1930	123,202,624
1940	132,164,569
1950	151,325,798
1960	179,323,175
1970	203,302,031
1980	226,542,203
1990	248,709,873
2000	281,421,906

Source: U.S. Census Bureau

(a) Using a graphing utility, draw a scatter diagram of the data using the year as the independent variable and population as the dependent variable.

(b) Using a graphing utility, fit a logistic function to the data.

(c) Using a graphing utility, draw the function found in part (b) on the scatter diagram.

(d) Based on the function found in part (b), what is the carrying capacity of the United States?

(e) Use the function found in part (b) to predict the population of the United States in 2004.

(f) When will the United States population be 300,000,000?

(g) Compare actual U.S. Census figures to the prediction found in part (e). Discuss any differences.

10. Population Model The following data represent the world population. An ecologist is interested in finding a function that describes the world population.

Year	Population (in Billions)
1993	5.531
1994	5.611
1995	5.691
1996	5.769
1997	5.847
1998	5.925
1999	6.003
2000	6.080
2001	6.157

Source: U.S. Census Bureau

(a) Using a graphing utility, draw a scatter diagram of the data using year as the independent variable and population as the dependent variable.

(b) Using a graphing utility, fit a logistic function to the data.

(c) Using a graphing utility, draw the function found in part (b) on the scatter diagram.

(d) Based on the function found in part (b), what is the carrying capacity of the world?

(e) Use the function found in part (b) to predict the population of the world in 2004.

(f) When will world population be 7 billion?

(g) Compare actual U.S. Census figures to the prediction found in part (e). Discuss any differences.

11. Cable Subscribers The following data represent the number of basic cable TV subscribers in the United States. A market

Year	Subscribers (1,000)
1975 ($t = 5$)	9,800
1980 ($t = 10$)	17,500
1985 ($t = 15$)	35,440
1990 ($t = 20$)	50,520
1992 ($t = 22$)	54,300
1994 ($t = 24$)	58,373
1996 ($t = 26$)	62,300
1998 ($t = 28$)	64,650
2000 ($t = 30$)	66,250
2002 ($t = 32$)	66,472
2004 ($t = 34$)	65,853

Source: Statistical Abstract of the United States, 2006

researcher believes that external factors, such as satellite TV, have affected the growth of cable subscribers. She is interested in finding a function that can be used to describe the number of cable TV subscribers in the United States.

(a) Using a graphing utility, draw a scatter diagram of the data using the number of years after 1970, t, as the independent variable and number of subscribers as the dependent variable.

(b) Using a graphing utility, fit a logistic function to the data.

(c) Using a graphing utility, draw the function found in part (b) on the scatter diagram.

(d) Based on the function found in part (b), what is the carrying capacity of the cable TV market in the United States?

(e) Use the function found in part (b) to predict the number of cable TV subscribers in the United States in 2015.

12. Cell Phone Users Refer to the data in Table 9 (p. 488).

(a) Using a graphing utility, fit a logistic function to the data.

(b) Graph the logistic function found in part (b) on a scatter diagram of the data.

(c) What is the predicted carrying capacity of U.S. cell phone subscribers?

(d) Use the function found in part (b) to predict the number of U.S. cell phone subscribers at the end of 2009.

(e) Compare the answer to part (d) above with the answer to Example 1, part (e). How do you explain the different predictions?

CHAPTER REVIEW

Things to Know

Composite function (p. 402)

$(f \circ g)(x) = f(g(x))$ The domain of $f \circ g$ is the set of all numbers x in the domain of g for which $g(x)$ is in the domain of f.

One-to-one function f (p. 410)

A function for which any two different inputs in the domain correspond to two different output in the range.

For any choice of elements x_1, x_2 in the domain of f, if $x_1 \neq x_2$, then $f(x_1) \neq f(x_2)$.

Horizontal-line test (p. 411)

If every horizontal line intersects the graph of a function f in at most one point, f is one-to-one.

Inverse function f^{-1} of f (pp. 412–415)

Domain of f = range of f^{-1}; range of f = domain of f^{-1}

$f^{-1}(f(x)) = x$ for all x in the domain of f, and $f(f^{-1}(x)) = x$ for all x in the domain of f^{-1}.

The graphs of f and f^{-1} are symmetric with respect to the line $y = x$.

Properties of the exponential function (pp. 427 and 428)

$f(x) = a^x, \quad a > 1$

Domain: the interval $(-\infty, \infty)$

Range: the interval $(0, \infty)$

x-intercepts: none; y-intercept: 1

Horizontal asymptote: x-axis ($y = 0$) as $x \to -\infty$

Increasing; one-to-one; smooth; continuous

See Figure 21 for a typical graph.

$f(x) = a^x, \quad 0 < a < 1$

Domain: the interval $(-\infty, \infty)$

Range: the interval $(0, \infty)$

x-intercepts: none; y-intercept: 1

Horizontal asymptote: x-axis ($y = 0$) as $x \to \infty$

Decreasing; one-to-one; smooth; continuous

See Figure 25 for a typical graph.

Number _e_ (p. 429) Value approached by the expression $\left(1 + \frac{1}{n}\right)^n$ as $n \to \infty$; that is, $\lim\limits_{n\to\infty}\left(1 + \frac{1}{n}\right)^n = e$

Property of exponents (p. 431) If $a^u = a^v$, then $u = v$.

Properties of the logarithmic function (p. 440)

$f(x) = \log_a x, \quad a > 1$ Domain: the interval $(0, \infty)$
($y = \log_a x$ means $x = a^y$) Range: the interval $(-\infty, \infty)$
 x-intercept: 1; y-intercept; none
 Vertical asymptote: $x = 0$ (y-axis)
 Increasing; one-to-one; smooth; continuous
 See Figure 30(b) for a typical graph.

$f(x) = \log_a x, \quad 0 < a < 1$ Domain: the interval $(0, \infty)$
($y = \log_a x$ means $x = a^y$) Range: the interval $(-\infty, \infty)$
 x-intercept: 1; y-intercept; none
 Vertical asymptote: $x = 0$ (y-axis)
 Decreasing; one-to-one; smooth; continuous
 See Figure 30(a) for a typical graph.

Natural logarithm (p. 440) $y = \ln x$ means $x = e^y$.

Properties of logarithms (pp. 450–452, 454)

$\log_a 1 = 0 \quad \log_a a = 1 \quad a^{\log_a M} = M \quad \log_a a^r = r$

$\log_a(MN) = \log_a M + \log_a N \quad \log_a\left(\frac{M}{N}\right) = \log_a M - \log_a N$

$\log_a M^r = r \log_a M$
If $M = N$, then $\log_a M = \log_a N$.
If $\log_a M = \log_a N$, then $M = N$.

Formulas

Change-of-Base Formula (p. 455) $\log_a M = \dfrac{\log_b M}{\log_b a}$

Compound Interest Formula (p. 463) $A = P \cdot \left(1 + \dfrac{r}{n}\right)^{nt}$

Continuous compounding (p. 469) $A = Pe^{rt}$

Present Value Formulas (p. 470) $P = A \cdot \left(1 + \dfrac{r}{n}\right)^{-nt}$ or $P = Ae^{-rt}$

Growth and decay (p. 475) $A(t) = A_0 e^{kt}$

Newton's Law of Cooling (p. 479) $u(t) = T + (u_0 - T)e^{kt} \quad k < 0$

Logistic model (p. 481) $P(t) = \dfrac{c}{1 + ae^{-bt}}$

Objectives

Section		You should be able to . . .	Review Exercises
6.1	1	Form a composite function (p. 402)	1–12
	2	Find the domain of a composite function (p. 403)	7–12
6.2	1	Determine whether a function is one-to-one (p. 410)	13(a), 14(a), 15, 16
	2	Determine the inverse of a function defined by a map on or a set of ordered pairs (p. 412)	13(b), 14(b)
	3	Obtain the graph of the inverse function from the graph of the function (p. 414)	15, 16
	4	Find the inverse of a function defined by an equation (p. 416)	17–22
6.3	1	Evaluate exponential functions (p. 423)	23(a), (c), 24(a), (c), 87(a)
	2	Graph exponential functions (p. 425)	55–60
	3	Define the number e (p. 429)	59, 60
	4	Solve exponential equations (p. 431)	63–66, 71, 72, 74–76, 87(b)

Review Exercises

In Problems 1–6, for the given functions f and g find:

(a) $(f \circ g)(2)$ (b) $(g \circ f)(-2)$ (c) $(f \circ f)(4)$ (d) $(g \circ g)(-1)$

1. $f(x) = 3x - 5;$ $g(x) = 1 - 2x^2$

2. $f(x) = 4 - x;$ $g(x) = 1 + x^2$

3. $f(x) = \sqrt{x + 2};$ $g(x) = 2x^2 + 1$

4. $f(x) = 1 - 3x^2;$ $g(x) = \sqrt{4 - x}$

5. $f(x) = e^x;$ $g(x) = 3x - 2$

6. $f(x) = \dfrac{2}{1 + 2x^2};$ $g(x) = 3x$

In Problems 7–12, find $f \circ g, g \circ f, f \circ f,$ and $g \circ g$ for each pair of functions. State the domain of each composite function.

7. $f(x) = 2 - x;$ $g(x) = 3x + 1$

8. $f(x) = 2x - 1;$ $g(x) = 2x + 1$

9. $f(x) = 3x^2 + x + 1;$ $g(x) = |3x|$

10. $f(x) = \sqrt{3x};$ $g(x) = 1 + x + x^2$

11. $f(x) = \dfrac{x + 1}{x - 1};$ $g(x) = \dfrac{1}{x}$

12. $f(x) = \sqrt{x - 3};$ $g(x) = \dfrac{3}{x}$

In Problems 13 and 14, (a) verify that the function is one-to-one, and (b) find the inverse of the given function.

13. $\{(1, 2), (3, 5), (5, 8), (6, 10)\}$

14. $\{(-1, 4), (0, 2), (1, 5), (3, 7)\}$

In Problems 15 and 16, state why the graph of the function is one-to-one. Then draw the graph of the inverse function f^{-1}. For convenience (and as a hint), the graph of $y = x$ is also given.

15.

16.

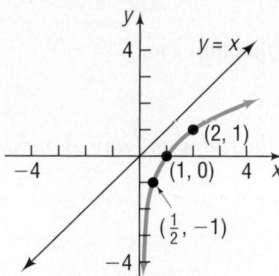

In Problems 17–22, the function f is one-to-one. Find the inverse of each function and check your answer. Find the domain and the range of f and f^{-1}.

17. $f(x) = \dfrac{2x + 3}{5x - 2}$

18. $f(x) = \dfrac{2 - x}{3 + x}$

19. $f(x) = \dfrac{1}{x - 1}$

20. $f(x) = \sqrt{x - 2}$

21. $f(x) = \dfrac{3}{x^{1/3}}$

22. $f(x) = x^{1/3} + 1$

In Problems 23 and 24, $f(x) = 3^x$ and $g(x) = \log_3 x$.

23. Evaluate: (a) $f(4)$ (b) $g(9)$ (c) $f(-2)$ (d) $g\!\left(\dfrac{1}{27}\right)$

24. Evaluate: (a) $f(1)$ (b) $g(81)$ (c) $f(-4)$ (d) $g\!\left(\dfrac{1}{243}\right)$

In Problems 25 and 26, convert each exponential expression to an equivalent expression involving a logarithm. In Problems 27 and 28, convert each logarithmic expression to an equivalent expression involving an exponent.

25. $5^2 = z$

26. $a^5 = m$

27. $\log_5 u = 13$

28. $\log_a 4 = 3$

In Problems 29–32, find the domain of each logarithmic function.

29. $f(x) = \log(3x - 2)$

30. $F(x) = \log_5(2x + 1)$

31. $H(x) = \log_2(x^2 - 3x + 2)$

32. $F(x) = \ln(x^2 - 9)$

In Problems 33–38, evaluate each expression. Do not use a calculator.

33. $\log_2\!\left(\dfrac{1}{8}\right)$

34. $\log_3 81$

35. $\ln e^{\sqrt{2}}$

36. $e^{\ln 0.1}$

37. $2^{\log_2 0.4}$

38. $\log_2 2^{\sqrt{3}}$

In Problems 39–44, write each expression as the sum and/or difference of logarithms. Express powers as factors.

39. $\log_3\!\left(\dfrac{uv^2}{w}\right)$, $u > 0, v > 0, w > 0$

40. $\log_2\!\left(a^2\sqrt{b}\right)^4$, $a > 0, b > 0$

41. $\log\!\left(x^2\sqrt{x^3 + 1}\right)$, $x > 0$

42. $\log_5\!\left(\dfrac{x^2 + 2x + 1}{x^2}\right)$, $x > 0$

43. $\ln\!\left(\dfrac{x\sqrt[3]{x^2 + 1}}{x - 3}\right)$, $x > 3$

44. $\ln\!\left(\dfrac{2x + 3}{x^2 - 3x + 2}\right)^2$, $x > 2$

In Problems 45–50, write each expression as a single logarithm.

45. $3\log_4 x^2 + \dfrac{1}{2}\log_4 \sqrt{x}$

46. $-2\log_3\!\left(\dfrac{1}{x}\right) + \dfrac{1}{3}\log_3 \sqrt{x}$

47. $\ln\!\left(\dfrac{x - 1}{x}\right) + \ln\!\left(\dfrac{x}{x + 1}\right) - \ln(x^2 - 1)$

48. $\log(x^2 - 9) - \log(x^2 + 7x + 12)$

49. $2\log 2 + 3\log x - \dfrac{1}{2}[\log(x + 3) + \log(x - 2)]$

50. $\dfrac{1}{2}\ln(x^2 + 1) - 4\ln\dfrac{1}{2} - \dfrac{1}{2}[\ln(x - 4) + \ln x]$

In Problems 51 and 52, use the Change-of-Base Formula and a calculator to evaluate each logarithm. Round your answer to three decimal places.

51. $\log_4 19$

52. $\log_2 21$

In Problems 53 and 54, graph each function using a graphing utility and the Change-of-Base Formula.

53. $y = \log_3 x$ **54.** $y = \log_7 x$

In Problems 55–62, use the given function f to:
 (a) *Find the domain of f.*
 (b) *Graph f.*
 (c) *From the graph, determine the range and any asymptotes of f.*
 (d) *Find f^{-1}, the inverse of f.*
 (e) *Use f^{-1} to find the range of f.*
 (f) *Graph f^{-1}.*

55. $f(x) = 2^{x-3}$ **56.** $f(x) = -2^x + 3$ **57.** $f(x) = \frac{1}{2}(3^{-x})$ **58.** $f(x) = 1 + 3^{-x}$

59. $f(x) = 1 - e^{-x}$ **60.** $f(x) = 3e^{x-2}$ **61.** $f(x) = \frac{1}{2}\ln(x+3)$ **62.** $f(x) = 3 + \ln(2x)$

In Problems 63–82, solve each equation. Express irrational solutions in exact form and as a decimal rounded to 3 decimal places.

63. $4^{1-2x} = 2$ **64.** $8^{6+3x} = 4$ **65.** $3^{x^2+x} = \sqrt{3}$ **66.** $4^{x-x^2} = \frac{1}{2}$

67. $\log_x 64 = -3$ **68.** $\log_{\sqrt{2}} x = -6$ **69.** $5^x = 3^{x+2}$ **70.** $5^{x+2} = 7^{x-2}$

71. $9^{2x} = 27^{3x-4}$ **72.** $25^{2x} = 5^{x^2-12}$ **73.** $\log_3 \sqrt{x-2} = 2$ **74.** $2^{x+1} \cdot 8^{-x} = 4$

75. $8 = 4^{x^2} \cdot 2^{5x}$ **76.** $2^x \cdot 5 = 10^x$ **77.** $\log_6(x+3) + \log_6(x+4) = 1$

78. $\log(7x - 12) = 2\log x$ **79.** $e^{1-x} = 5$ **80.** $e^{1-2x} = 4$

81. $9^x + 4 \cdot 3^x - 3 = 0$ **82.** $4^x - 14 \cdot 4^{-x} = 5$

83. Suppose that $f(x) = \log_2(x-2) + 1$.
 (a) Graph f.
 (b) What is $f(6)$? What point is on the graph of f?
 (c) Solve $f(x) = 4$. What point is on the graph of f?
 (d) Based on the graph drawn in part (a), solve $f(x) > 0$.
 (e) Find $f^{-1}(x)$. Graph f^{-1} on the same Cartesian plane as f.

84. Suppose that $f(x) = \log_3(x+1) - 4$.
 (a) Graph f.
 (b) What is $f(8)$? What point is on the graph of f?
 (c) Solve $f(x) = -3$. What point is on the graph of f?
 (d) Based on the graph drawn in part (a), solve $f(x) < 0$.
 (e) Find $f^{-1}(x)$. Graph f^{-1} on the same Cartesian plane as f.

In Problems 85 and 86, use the following result: If x is the atmospheric pressure (measured in millimeters of mercury), then the formula for the altitude h(x) (measured in meters above sea level) is

$$h(x) = (30T + 8000)\log\left(\frac{P_0}{x}\right)$$

where T is the temperature (in degrees Celsius) and P_0 is the atmospheric pressure at sea level, which is approximately 760 millimeters of mercury.

85. Finding the Altitude of an Airplane At what height is a Piper Cub whose instruments record an outside temperature of 0°C and a barometric pressure of 300 millimeters of mercury?

86. Finding the Height of a Mountain How high is a mountain if instruments placed on its peak record a temperature of 5°C and a barometric pressure of 500 millimeters of mercury?

87. Amplifying Sound An amplifier's power output P (in watts) is related to its decibel voltage gain d by the formula

$$P = 25e^{0.1d}.$$

 (a) Find the power output for a decibel voltage gain of 4 decibels.
 (b) For a power output of 50 watts, what is the decibel voltage gain?

88. Limiting Magnitude of a Telescope A telescope is limited in its usefulness by the brightness of the star that it is aimed at and by the diameter of its lens. One measure of a star's brightness is its *magnitude;* the dimmer the star, the larger its magnitude. A formula for the limiting magnitude L of a telescope,

that is, the magnitude of the dimmest star that it can be used to view, is given by

$$L = 9 + 5.1 \log d$$

where d is the diameter (in inches) of the lens.
(a) What is the limiting magnitude of a 3.5-inch telescope?
(b) What diameter is required to view a star of magnitude 14?

89. **Salvage Value** The number of years n for a piece of machinery to depreciate to a known salvage value can be found using the formula

$$n = \frac{\log s - \log i}{\log(1 - d)}$$

where s is the salvage value of the machinery, i is its initial value, and d is the annual rate of depreciation.
(a) How many years will it take for a piece of machinery to decline in value from $90,000 to $10,000 if the annual rate of depreciation is 0.20 (20%)?
(b) How many years will it take for a piece of machinery to lose half of its value if the annual rate of depreciation is 15%?

90. **Funding a College Education** A child's grandparents purchase a $10,000 bond fund that matures in 18 years to be used for her college education. The bond fund pays 4% interest compounded semiannually. How much will the bond fund be worth at maturity? What is the effective rate of interest? How long will it take the bond to double in value under these terms?

91. **Funding a College Education** A child's grandparents wish to purchase a bond that matures in 18 years to be used for her college education. The bond pays 4% interest compounded semiannually. How much should they pay so that the bond will be worth $85,000 at maturity?

92. **Funding an IRA** First Colonial Bankshares Corporation advertised the following IRA investment plans.

Target IRA Plans

For each $5000 Maturity Value Desired

Deposit:	At a Term of:
$620.17	20 Years
$1045.02	15 Years
$1760.92	10 Years
$2967.26	5 Years

(a) Assuming continuous compounding, what annual rate of interest did they offer?
(b) First Colonial Bankshares claims that $4000 invested today will have a value of over $32,000 in 20 years. Use the answer found in part (a) to find the actual value of $4000 in 20 years. Assume continuous compounding.

93. **Estimating the Date That a Prehistoric Man Died** The bones of a prehistoric man found in the desert of New Mexico contain approximately 5% of the original amount of carbon 14. If the half-life of carbon 14 is 5600 years, approximately how long ago did the man die?

94. **Temperature of a Skillet** A skillet is removed from an oven whose temperature is 450°F and placed in a room whose temperature is 70°F. After 5 minutes, the temperature of the skillet is 400°F. How long will it be until its temperature is 150°F?

95. **World Population** The annual growth rate of the world's population in 2005 was $k = 1.15\% = 0.0115$. The population of the world in 2005 was 6,451,058,790. Letting $t = 0$ represent 2005, use the uninhibited growth model to predict the world's population in the year 2015.
Source: U.S. Census Bureau

96. **Radioactive Decay** The half-life of radioactive cobalt is 5.27 years. If 100 grams of radioactive cobalt is present now, how much will be present in 20 years? In 40 years?

97. **Federal Deficit** In fiscal year 2005, the federal deficit was $319 billion. At that time, 10-year treasury notes were paying 4.25% interest per annum. If the federal government financed this deficit through 10-year notes, how much would if have to pay back in 2015?
Source: U.S. Treasury Department

98. **Logistic Growth** The logistic growth model

$$P(t) = \frac{0.8}{1 + 1.67e^{-0.16t}}$$

represents the proportion of new cars with a global positioning system (GPS). Let $t = 0$ represent 2006, $t = 1$ represent 2007, and so on.
(a) What proportion of new cars in 2006 had a GPS?
(b) Determine the maximum proportion of new cars that have a GPS.
(c) Using a graphing utility, graph $P(t)$.
(d) When will 75% of new cars have a GPS?

99. **CBL Experiment** The following data were collected by placing a temperature probe in a portable heater, removing the probe, and then recording temperature over time.

Time (sec.)	Temperature (°F)
0	165.07
1	164.77
2	163.99
3	163.22
4	162.82
5	161.96
6	161.20
7	160.45
8	159.35
9	158.61
10	157.89
11	156.83
12	156.11
13	155.08
14	154.40
15	153.72

According to Newton's Law of Cooling, these data should follow an exponential model.
(a) Using a graphing utility, draw a scatter diagram for the data.
(b) Using a graphing utility, fit an exponential function to the data.

(c) Graph the exponential function found in part (b) on the scatter diagram.

(d) Predict how long it will take for the probe to reach a temperature of 110°F.

100. **Wind Chill Factor** The following data represent the wind speed (mph) and wind chill factor at an air temperature of 15°F.

Wind Speed (mph)	Wind Chill Factor (°F)
5	7
10	3
15	0
20	−2
25	−4
30	−5
35	−7

Source: U.S. National Weather Service

(a) Using a graphing utility, draw a scatter diagram with wind speed as the independent variable.

(b) Using a graphing utility, fit a logarithmic function to the data.

(c) Using a graphing utility, draw the logarithmic function found in part (b) on the scatter diagram.

(d) Use the function found in part (b) to predict the wind chill factor if the air temperature is 15°F and the wind speed is 23 mph.

101. **Spreading of a Disease** Jack and Diane live in a small town of 50 people. Unfortunately, both Jack and Diane have a cold.

Those who come in contact with someone who has this cold will themselves catch the cold. The following data represent the number of people in the small town who have caught the cold after *t* days.

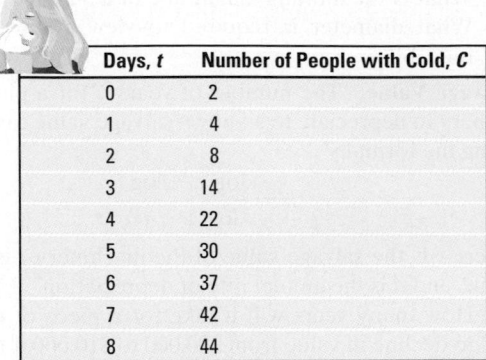

Days, *t*	Number of People with Cold, *C*
0	2
1	4
2	8
3	14
4	22
5	30
6	37
7	42
8	44

(a) Using a graphing utility, draw a scatter diagram of the data. Comment on the type of relation that appears to exist between the days and number of people with a cold.

(b) Using a graphing utility, fit a logistic function to the data.

(c) Graph the function found in part (b) on the scatter diagram.

(d) According to the function found in part (b), what is the maximum number of people who will catch the cold? In reality, what is the maximum number of people who could catch the cold?

(e) Sometime between the second and third day, 10 people in the town had a cold. According to the model found in part (b), when did 10 people have a cold?

(f) How long will it take for 46 people to catch the cold?

CHAPTER TEST

1. Given $f(x) = \dfrac{x+2}{x-2}$ and $g(x) = 2x + 5$, find:

 (a) $f \circ g$ and state its domain
 (b) $(g \circ f)(-2)$
 (c) $(f \circ g)(-2)$

2. Determine whether the function is one-to-one.

 (a) $y = 4x^2 + 3$
 (b) $y = \sqrt{x+3} - 5$

3. Find the inverse of $f(x) = \dfrac{2}{3x-5}$ and check your answer. State the domain and the range of f and f^{-1}.

4. If the point $(3, -5)$ is on the graph of a one-to-one function f, what point must be on the graph of f^{-1}?

In Problems 5–7, find the unknown value without using a calculator.

5. $3^x = 243$ 6. $\log_b 16 = 2$ 7. $\log_5 x = 4$

In Problems 8–11, use a calculator to evaluate each expression. Round your answer to three decimal places.

8. $e^3 + 2$ 9. $\log 20$
10. $\log_3 21$ 11. $\ln 133$

In Problems 12 and 13, use the given function f to:

 (a) *Find the domain of f.*
 (b) *Graph f.*
 (c) *From the graph, determine the range and any asymptotes of f.*
 (d) *Find f^{-1}, the inverse of f.*
 (e) *Use f^{-1} to find the range of f.*
 (f) *Graph f^{-1}.*

12. $f(x) = 4^{x+1} - 2$

13. $f(x) = 1 - \log_5(x - 2)$

In Problems 14–19, solve each equation.

14. $5^{x+2} = 125$ 15. $\log(x + 9) = 2$

16. $8 - 2e^{-x} = 4$ 17. $\log(x^2 + 3) = \log(x + 6)$

18. $7^{x+3} = e^x$ 19. $\log_2(x - 4) + \log_2(x + 4) = 3$

20. Write $\log_2\left(\dfrac{4x^3}{x^2 - 3x - 18}\right)$ as the sum and/or difference of logarithms. Express powers as factors.

21. A 50-mg sample of a radioactive substance decays to 34 mg after 30 days. How long will it take for there to be 2 mg remaining?

22. (a) If $1000 is invested at 5% compounded monthly, how much is there after 8 months?

(b) If you want to have $1000 in 9 months, how much do you need now to place in a savings account now that pays 5% compounded quarterly?

(c) How long does it take to double your money if you can invest it at 6% compounded annually?

23. The decibel level, D, of sound is given by the equation

$$D = 10 \log\left(\frac{I}{I_0}\right),$$ where I is the intensity of the sound and

$I_0 = 10^{-12}$ watt per square meter.

(a) If the shout of a single person measures 80 decibels, how loud will the sound be if two people shout at the same time? That is, how loud would the sound be if the intensity doubled?

(b) The pain threshold for sound is 125 decibels. If the Athens Olympic Stadium 2004 (Olympiako Stadio Athinas 'Spyros Louis') can seat 74,400 people, how many people in the crowd need to shout at the same time for the resulting sound level to meet or exceed the pain threshold? (Ignore any possible sound dampening.)

CUMULATIVE REVIEW

1. Is the following graph the graph of a function? If it is, is the function one-to-one?

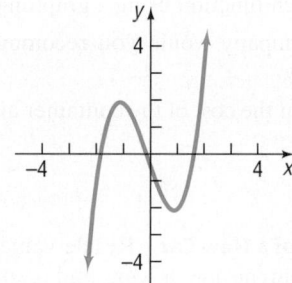

2. For the function $f(x) = 2x^2 - 3x + 1$, find the following:

(a) $f(3)$ (b) $f(-x)$ (c) $f(x + h)$

3. Determine which of the following points are on the graph of $x^2 + y^2 = 1$.

(a) $\left(\frac{1}{2}, \frac{1}{2}\right)$ (b) $\left(\frac{1}{2}, \frac{\sqrt{3}}{2}\right)$

4. Solve the equation $3(x - 2) = 4(x + 5)$.

5. Graph the line $2x - 4y = 16$.

6. (a) Graph the quadratic function $f(x) = -x^2 + 2x - 3$ by determining whether its graph opens up or down and by finding its vertex, axis of symmetry, y-intercept, and x-intercept(s), if any.

(b) Solve $f(x) \le 0$.

7. Determine the quadratic function whose graph is given in the figure.

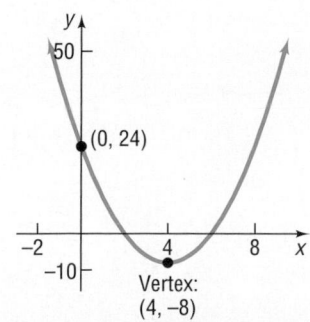

8. Graph $f(x) = 3(x + 1)^3 - 2$ using transformations.

9. Given that $f(x) = x^2 + 2$ and $g(x) = \dfrac{2}{x - 3}$, find $f(g(x))$ and state its domain. What is $f(g(5))$?

10. For the polynomial function $f(x) = 4x^3 + 9x^2 - 30x - 8$:

(a) Find the real zeros of f.

(b) Determine the intercepts of the graph of f.

(c) Use a graphing utility to approximate the local maxima and local minima.

(d) Draw a complete graph of f. Be sure to label the intercepts and turning points.

11. For the function $g(x) = 3^x + 2$:

(a) Graph g using transformations. State the domain, range, and horizontal asymptote of g.

(b) Determine the inverse of g. State the domain, range, and vertical asymptote of g^{-1}.

(c) On the same graph as g, graph g^{-1}.

12. Solve the equation $4^{x-3} = 8^{2x}$.

13. Solve the equation: $\log_3(x + 1) + \log_3(2x - 3) = \log_9 9$

14. Suppose that $f(x) = \log_3(x + 2)$. Solve:

(a) $f(x) = 0$. (b) $f(x) > 0$. (c) $f(x) = 3$.

15. **Data Analysis** The following data represent the percent of all drivers by age that have been stopped by the police for any reason within the past year. The median age represents the midpoint of the upper and lower limit for the age range.

Age Range	Median Age, x	Percentage Stopped, y
16–19	17.5	18.2
20–29	24.5	16.8
30–39	34.5	11.3
40–49	44.5	9.4
50–59	54.5	7.7
≥60	69.5	3.8

(a) Using your graphing utility, draw a scatter diagram of the data treating median age, x, as the independent variable.

(b) Determine a model that you feel best describes the relation between median age and percentage stopped. You may choose from among linear, quadratic, cubic, power, exponential, logarithmic, or logistic models.

(c) Provide a justification for the model that you selected in part (b).

CHAPTER PROJECTS

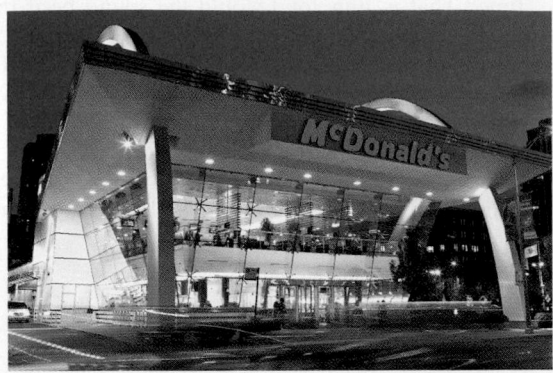

I. **Hot Coffee** A fast-food restaurant wants a special container to hold coffee. The restaurant wishes the container to quickly cool the coffee from 200° to 130°F and keep the liquid between 110° and 130°F as long as possible. The restaurant has three containers to select from.

1. The CentiKeeper Company has a container that reduces the temperature of a liquid from 200° to 100°F in 30 minutes by maintaining a constant temperature of 70°F.

2. The TempControl Company has a container that reduces the temperature of a liquid from 200° to 110°F in 25 minutes by maintaining a constant temperature of 60°F.

3. The Hot'n'Cold Company has a container that reduces the temperature of a liquid from 200° to 120°F in 20 minutes by maintaining a constant temperature of 65°F.

You need to recommend which container the restaurant should purchase.

(a) Use Newton's Law of Cooling to find a function relating the temperature of the liquid over time for each container.

(b) How long does it take each container to lower the coffee temperature from 200° to 130°F?

(c) How long will the coffee temperature remain between 110° and 130°F? This temperature is considered the optimal drinking temperature.

(d) Graph each function using a graphing utility.

(e) Which company would you recommend to the restaurant? Why?

(f) How might the cost of the container affect your decision?

The following projects are available on the Instructor's Resource Center (IRC):

II. **Project at Motorola** *Thermal Fatigue of Solder Connections* Product reliability is a major concern of a manufacturer. Here a logarithmic transformation is used to simplify the analysis of a cell phone's ability to withstand temperature change.

III. **Depreciation of a New Car** Resale value is a factor to consider when purchasing a car, and exponential functions provide a way to compare the depreciation rates of different makes and models.

Trigonometric Functions

<div style="font-size:3em;text-align:right">7</div>

Surf's Up: Using Models To Predict Huge Waves

GALVESTON, Feb. 15, 2005—If you are a ship captain and there might be 50-foot waves headed your way, you would appreciate some information about them, right?

That's the idea behind a wave model system a Texas A&M University at Galveston professor has developed. His detailed wave prediction system is currently in use in the Gulf of Mexico and the Gulf of Maine.

Vijay Panchang, head of the Department of Maritime Systems Engineering, doesn't make waves; he predicts what they will do, when they will do it, and how high they will get.

Using data provided daily from NOAA and his own complex mathematical models, Panchang and research engineer Doncheng Li provide daily wave model predictions for much of the Texas coast, the Gulf of Mexico, and the Gulf of Maine. Their simulations, updated every 12 hours, provide a forecast for two days ahead.

Because the models use wind data, tsunamis that are created by undersea earthquakes cannot be predicted. But that is not to say his modeling system does not come up with some big waves.

His wave model predicted big waves in November 2003 in the Gulf of Maine, and it was accurate; waves as high as 30 feet were recorded during one storm even in coastal regions.

Last summer during Hurricane Ivan, a buoy located 60 miles south of the Alabama coast recorded a whopping 60-foot wave. "There may have been higher waves because right after recording the 60-foot wave, the buoy snapped and stopped functioning," he says.

Source: Science Daily www.sciencedaily.com/releases/2005/02/050222193810.htm. Posted February 23, 2005.

—See the Chapter Project—

A Look Back

In Chapter 3, we began our discussion of functions. We defined domain and range and independent and dependent variables; we found the value of a function and graphed functions. We continued our study of functions by listing properties that a function might have, like being even or odd, and we created a library of functions, naming key functions and listing their properties, including the graph.

A Look Ahead

In this chapter we define the trigonometric functions, six functions that have wide application. We shall talk about their domain and range, see how to find values, graph them, and develop a list of their properties.

There are two widely accepted approaches to the development of the trigonometric functions: one uses right triangles; the other uses circles, especially the unit circle. In this book, we develop the trigonometric functions using right triangles. In Section 7.5, we introduce trigonometric functions using the unit circle and show that this approach leads to the definition using right triangles.

Outline

7.1 Angles and Their Measure

PREPARING FOR THIS SECTION *Before getting started, review the following:*

- Circumference and Area of a Circle (Chapter R, Review, Section R.3, pp. 31–32)

Now Work the 'Are You Prepared?' problems on page 513.

OBJECTIVES 1 Convert between Decimals and Degrees, Minutes, Seconds Forms for Angles (p. 506)
2 Find the Arc Length of a Circle (p. 507)
3 Convert from Degrees to Radians and from Radians to Degrees (p. 508)
4 Find the Area of a Sector of a Circle (p. 511)
5 Find the Linear Speed of an Object Traveling in Circular Motion (p. 512)

Figure 1

A **ray,** or **half-line,** is that portion of a line that starts at a point V on the line and extends indefinitely in one direction. The starting point V of a ray is called its **vertex.** See Figure 1.

If two rays are drawn with a common vertex, they form an **angle.** We call one of the rays of an angle the **initial side** and the other the **terminal side.** The angle formed is identified by showing the direction and amount of rotation from the initial side to the terminal side. If the rotation is in the counterclockwise direction, the angle is **positive;** if the rotation is clockwise, the angle is **negative.** See Figure 2.

Figure 2

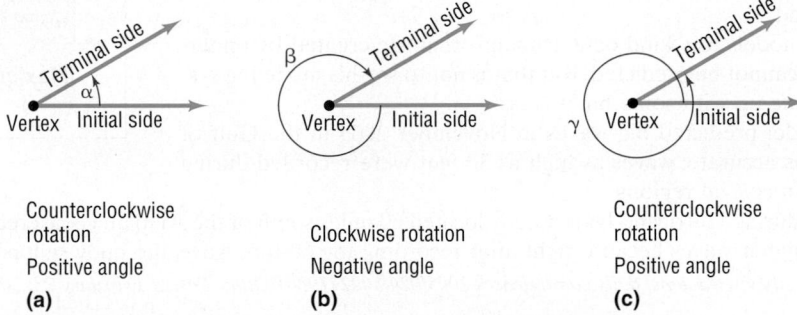

Lowercase Greek letters, such as α (alpha), β (beta), γ (gamma), and θ (theta), will often be used to denote angles. Notice in Figure 2(a) that the angle α is positive because the direction of the rotation from the initial side to the terminal side is counterclockwise. The angle β in Figure 2(b) is negative because the rotation is clockwise. The angle γ in Figure 2(c) is positive. Notice that the angle α in Figure 2(a) and the angle γ in Figure 2(c) have the same initial side and the same terminal side. However, α and γ are unequal, because the amount of rotation required to go from the initial side to the terminal side is greater for angle γ than for angle α.

An angle θ is said to be in **standard position** if its vertex is at the origin of a rectangular coordinate system and its initial side coincides with the positive x-axis. See Figure 3.

Figure 3

(a) θ is in standard position; θ is positive

(b) θ is in standard position; θ is negative

When an angle θ is in standard position, the terminal side will lie either in a quadrant, in which case we say that θ **lies in that quadrant,** or θ will lie on the x-axis or the y-axis, in which case we say that θ is a **quadrantal angle.** For example, the angle θ in Figure 4(a) lies in quadrant II, the angle θ in Figure 4(b) lies in quadrant IV, and the angle θ in Figure 4(c) is a quadrantal angle.

Figure 4

 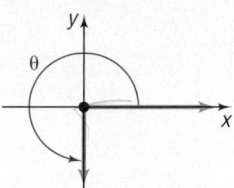

(a) θ lies in quadrant II **(b)** θ lies in quadrant IV **(c)** θ is a quadrantal angle

We measure angles by determining the amount of rotation needed for the initial side to become coincident with the terminal side. The two commonly used measures for angles are *degrees* and *radians*.

Degrees

HISTORICAL NOTE One counterclockwise rotation is 360° due to the Babylonian year, which had 360 days in it. ■

The angle formed by rotating the initial side exactly once in the counterclockwise direction until it coincides with itself (1 revolution) is said to measure 360 degrees, abbreviated 360°. **One degree, 1°,** is $\frac{1}{360}$ revolution. A **right angle** is an angle that measures 90°, or $\frac{1}{4}$ revolution; a **straight angle** is an angle that measures 180°, or $\frac{1}{2}$ revolution. See Figure 5. As Figure 5(b) shows, it is customary to indicate a right angle by using the symbol ∟.

Figure 5

(a) 1 revolution counterclockwise, 360° **(b)** right angle, $\frac{1}{4}$ revolution counter-clockwise, 90° **(c)** straight angle, $\frac{1}{2}$ revolution counter-clockwise, 180°

It is also customary to refer to an angle that measures θ degrees as an angle *of* θ degrees.

| EXAMPLE 1 | **Drawing an Angle** |

Draw each angle.

(a) 45° (b) −90° (c) 225° (d) 405°

Solution

(a) An angle of 45° is $\frac{1}{2}$ of a right angle. See Figure 6.

(b) An angle of −90° is $\frac{1}{4}$ revolution in the clockwise direction. See Figure 7.

Figure 6

Figure 7

(c) An angle of 225° consists of a rotation through 180° followed by a rotation through 45°. See Figure 8.

(d) An angle of 405° consists of 1 revolution (360°) followed by a rotation through 45°. See Figure 9.

Figure 8

Figure 9

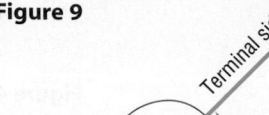

Now Work PROBLEM 11

1 Convert between Decimals and Degrees, Minutes, Seconds Forms for Angles

Although subdivisions of a degree may be obtained by using decimals, we also may use the notion of *minutes* and *seconds*. **One minute,** denoted by **1′**, is defined as $\frac{1}{60}$ degree.

One second, denoted by **1″**, is defined as $\frac{1}{60}$ minute, or equivalently, $\frac{1}{3600}$ degree.

An angle of, say, 30 degrees, 40 minutes, 10 seconds is written compactly as 30°40′10″. To summarize:

$$
\begin{array}{c}
1 \text{ counterclockwise revolution} = 360° \\
1° = 60' \qquad 1' = 60''
\end{array}
\tag{1}
$$

It is sometimes necessary to convert from the degree, minute, second notation (D°M′S″) to a decimal form, and vice versa. Check your calculator; it should be capable of doing the conversion for you.

Before getting started, though, you must set the mode to degrees because there are two common ways to measure angles: degree mode and radian mode. (We will define radians shortly.) Usually, a menu is used to change from one mode to another. Check your owner's manual to find out how your particular calculator works.

Now let's see how to convert from the degree, minute, second notation (D°M′S″) to a decimal form, and vice versa, by looking at some examples:

$$15°30' = 15.5° \quad \text{because} \quad 30' = 30 \cdot 1' = 30 \cdot \left(\frac{1}{60}\right)° = 0.5°$$

$$1' = \left(\frac{1}{60}\right)°$$

$$32.25° = 32°15' \quad \text{because} \quad 0.25° = \left(\frac{1}{4}\right)° = \frac{1}{4} \cdot 1° = \frac{1}{4}(60') = 15'$$

$$1° = 60'$$

EXAMPLE 2 **Converting between Degrees, Minutes, Seconds Form and Decimal Form**

(a) Convert 50°6′21″ to a decimal in degrees. Round the answer to four decimal places.

(b) Convert 21.256° to the D°M′S″ form. Round the answer to the nearest second.

Solution (a) Because $1' = \left(\dfrac{1}{60}\right)^{\circ}$ and $1'' = \left(\dfrac{1}{60}\right)' = \left(\dfrac{1}{60} \cdot \dfrac{1}{60}\right)^{\circ}$, we convert as follows:

$$50°6'21'' = 50° + 6' + 21''$$

$$= 50° + 6 \cdot \left(\frac{1}{60}\right)^{\circ} + 21 \cdot \left(\frac{1}{60} \cdot \frac{1}{60}\right)^{\circ}$$

$$\approx 50° + 0.1° + 0.0058°$$

$$= 50.1058°$$

(b) We proceed as follows:

$$21.256° = 21° + 0.256°$$

$$ = 21° + (0.256)(60') \qquad \text{Convert fraction of degree to minutes; } 1° = 60'.$$

$$ = 21° + 15.36'$$

$$ = 21° + 15' + 0.36'$$

$$ = 21° + 15' + (0.36)(60'') \qquad \text{Convert fraction of minute to seconds; } 1' = 60''.$$

$$ = 21° + 15' + 21.6''$$

$$ \approx 21°15'22'' \qquad \text{Round to the nearest second.}$$

Now Work PROBLEMS **23** AND **29**

In many applications, such as describing the exact location of a star or the precise position of a ship at sea, angles measured in degrees, minutes, and even seconds are used. For calculation purposes, these are transformed to decimal form. In other applications, especially those in calculus, angles are measured using *radians*.

Radians

A **central angle** is a positive angle whose vertex is at the center of a circle. The rays of a central angle subtend (intersect) an arc on the circle. If the radius of the circle is r and the length of the arc subtended by the central angle is also r, then the measure of the angle is **1 radian.** See Figure 10(a).

For a circle of radius 1, the rays of a central angle with measure 1 radian would subtend an arc of length 1. For a circle of radius 3, the rays of a central angle with measure 1 radian would subtend an arc of length 3. See Figure 10(b).

Figure 10

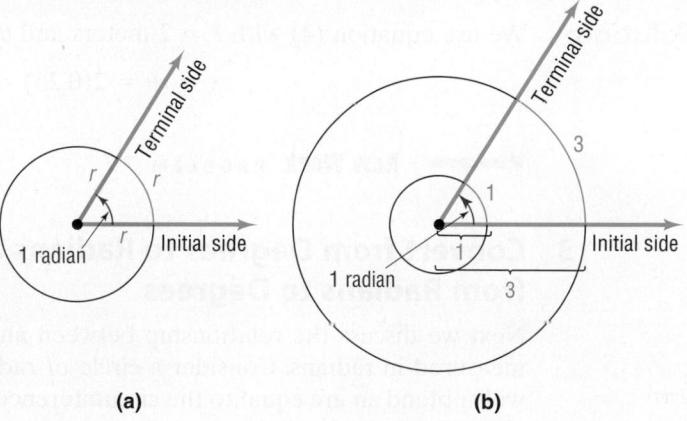

(a) (b)

2 Find the Arc Length of a Circle

Now consider a circle of radius r and two central angles, θ and θ_1, measured in radians. Suppose that these central angles subtend arcs of lengths s and s_1, respectively,

Figure 11 $\dfrac{\theta}{\theta_1} = \dfrac{s}{s_1}$

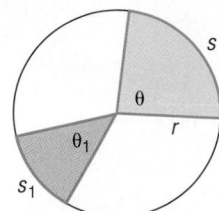

as shown in Figure 11. From geometry, we know that the ratio of the measures of the angles equals the ratio of the corresponding lengths of the arcs subtended by these angles; that is,

$$\frac{\theta}{\theta_1} = \frac{s}{s_1} \tag{2}$$

Suppose that $\theta_1 = 1$ radian. Refer again to Figure 10(a). The length s_1 of the arc subtended by the central angle $\theta_1 = 1$ radian equals the radius r of the circle. Then $s_1 = r$, so equation (2) reduces to

$$\frac{\theta}{1} = \frac{s}{r} \quad \text{or} \quad s = r\theta \tag{3}$$

THEOREM

Arc Length

For a circle of radius r, a central angle of θ radians subtends an arc whose length s is

$$s = r\theta \tag{4}$$

NOTE Formulas must be consistent with regard to the units used. In equation (4), we write

$$s = r\theta$$

To see the units, however, we must go back to equation (3) and write

$$\frac{\theta \text{ radians}}{1 \text{ radian}} = \frac{s \text{ length units}}{r \text{ length units}}$$

$$s \text{ length units} = r \text{ length units} \frac{\theta \text{ radians}}{1 \text{ radian}}$$

Since the radians cancel, we are left with

$$s \text{ length units} = (r \text{ length units})\theta \quad s = r\theta$$

where θ appears to be "dimensionless" but, in fact, is measured in radians. So, in using the formula $s = r\theta$, the dimension for θ is radians, and any convenient unit of length (such as inches or meters) may be used for s and r. ∎

EXAMPLE 3

Finding the Length of an Arc of a Circle

Find the length of the arc of a circle of radius 2 meters subtended by a central angle of 0.25 radian.

Solution

We use equation (4) with $r = 2$ meters and $\theta = 0.25$. The length s of the arc is

$$s = r\theta = 2(0.25) = 0.5 \text{ meter}$$

Now Work PROBLEM 71

3 Convert from Degrees to Radians and from Radians to Degrees

Figure 12
1 revolution = 2π radians

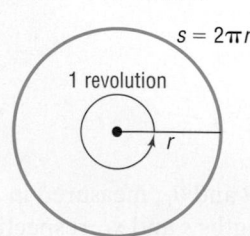

Next we discuss the relationship between angles measured in degrees and angles measured in radians. Consider a circle of radius r. A central angle of 1 revolution will subtend an arc equal to the circumference of the circle (Figure 12). Because the circumference of a circle equals $2\pi r$, we use $s = 2\pi r$ in equation (4) to find that, for an angle θ of 1 revolution,

$$s = r\theta$$
$$2\pi r = r\theta \qquad \theta = 1 \text{ revolution; } s = 2\pi r$$
$$\theta = 2\pi \text{ radians} \qquad \text{Solve for } \theta.$$

From this we have,

$$1 \text{ revolution} = 2\pi \text{ radians} \qquad \textbf{(5)}$$

Since 1 revolution = 360°, we have

$$360° = 2\pi \text{ radians}$$

Dividing both sides by 2 yields

$$180° = \pi \text{ radians} \qquad \textbf{(6)}$$

Divide both sides of equation (6) by 180. Then

$$1 \text{ degree} = \frac{\pi}{180} \text{ radian}$$

Divide both sides of (6) by π. Then

$$\frac{180}{\pi} \text{ degrees} = 1 \text{ radian}$$

We have the following two conversion formulas:

$$1 \text{ degree} = \frac{\pi}{180} \text{ radian} \qquad 1 \text{ radian} = \frac{180}{\pi} \text{ degrees} \qquad \textbf{(7)}$$

EXAMPLE 4	**Converting from Degrees to Radians**

Convert each angle in degrees to radians.

(a) 60° (b) 150° (c) −45° (d) 90° (e) 107°

Solution

(a) $60° = 60 \cdot 1 \text{ degree} = 60 \cdot \dfrac{\pi}{180} \text{ radian} = \dfrac{\pi}{3} \text{ radians}$

(b) $150° = 150 \cdot 1° = 150 \cdot \dfrac{\pi}{180} \text{ radian} = \dfrac{5\pi}{6} \text{ radians}$

(c) $-45° = -45 \cdot \dfrac{\pi}{180} \text{ radian} = -\dfrac{\pi}{4} \text{ radian}$

(d) $90° = 90 \cdot \dfrac{\pi}{180} \text{ radian} = \dfrac{\pi}{2} \text{ radians}$

(e) $107° = 107 \cdot \dfrac{\pi}{180} \text{ radian} \approx 1.868 \text{ radians}$

Example 4, parts (a)–(d), illustrates that angles that are "nice" fractions of a revolution are expressed in radian measure as fractional multiples of π, rather than as decimals. For example, a right angle, as in Example 4(d), is left in the form $\dfrac{\pi}{2}$ radians, which is exact, rather than using the approximation $\dfrac{\pi}{2} \approx \dfrac{3.1416}{2} = 1.5708$ radians. When the fractions are not "nice," we use the decimal approximation of the angle, as in Example 4(e).

═══ **Now Work** PROBLEMS 3 5 AND 6 1

EXAMPLE 5 **Converting Radians to Degrees**

Convert each angle in radians to degrees.

(a) $\dfrac{\pi}{6}$ radian (b) $\dfrac{3\pi}{2}$ radians (c) $-\dfrac{3\pi}{4}$ radians

(d) $\dfrac{7\pi}{3}$ radians (e) 3 radians

Solution (a) $\dfrac{\pi}{6}$ radian $= \dfrac{\pi}{6} \cdot 1 \text{ radian} = \dfrac{\pi}{6} \cdot \dfrac{180}{\pi}$ degrees $= 30°$

(b) $\dfrac{3\pi}{2}$ radians $= \dfrac{3\pi}{2} \cdot \dfrac{180}{\pi}$ degrees $= 270°$

(c) $-\dfrac{3\pi}{4}$ radians $= -\dfrac{3\pi}{4} \cdot \dfrac{180}{\pi}$ degrees $= -135°$

(d) $\dfrac{7\pi}{3}$ radians $= \dfrac{7\pi}{3} \cdot \dfrac{180}{\pi}$ degrees $= 420°$

(e) 3 radians $= 3 \cdot \dfrac{180}{\pi}$ degrees $\approx 171.89°$

═══ **Now Work** PROBLEM 47

Table 1 lists the degree and radian measures of some commonly encountered angles. You should learn to feel equally comfortable using degree or radian measure for these angles.

Table 1

Degrees	0°	30°	45°	60°	90°	120°	135°	150°	180°
Radians	0	$\dfrac{\pi}{6}$	$\dfrac{\pi}{4}$	$\dfrac{\pi}{3}$	$\dfrac{\pi}{2}$	$\dfrac{2\pi}{3}$	$\dfrac{3\pi}{4}$	$\dfrac{5\pi}{6}$	π
Degrees		210°	225°	240°	270°	300°	315°	330°	360°
Radians		$\dfrac{7\pi}{6}$	$\dfrac{5\pi}{4}$	$\dfrac{4\pi}{3}$	$\dfrac{3\pi}{2}$	$\dfrac{5\pi}{3}$	$\dfrac{7\pi}{4}$	$\dfrac{11\pi}{6}$	2π

EXAMPLE 6 **Finding the Distance between Two Cities**

See Figure 13(a). The latitude of a location L is the angle formed by a ray drawn from the center of Earth to the Equator and a ray drawn from the center of Earth to L. See Figure 13(b). Glasgow, Montana, is due north of Albuquerque, New Mexico. Find the distance between Glasgow (48°9′ north latitude) and Albuquerque (35°5′ north latitude). Assume that the radius of Earth is 3960 miles.

Figure 13

(a) (b)

Solution The measure of the central angle between the two cities is $48°9' - 35°5' = 13°4'$. We use equation (4), $s = r\theta$, but first we must convert the angle of $13°4'$ to radians.

$$\theta = 13°4' \approx 13.0667° = 13.0667 \cdot \frac{\pi}{180} \text{ radian} \approx 0.228 \text{ radian}$$

$$\underset{\substack{\uparrow \\ 4' = \frac{4°}{60}}}{}$$

We use $\theta = 0.228$ radian and $r = 3960$ miles in equation (4). The distance between the two cities is

$$s = r\theta = 3960 \cdot 0.228 \approx 903 \text{ miles}$$

NOTE If the measure of an angle is given as 5, it is understood to mean 5 radians; if the measure of an angle is given as 5°, it means 5 degrees. ∎

When an angle is measured in degrees, the degree symbol will always be shown. However, when an angle is measured in radians, we will follow the usual practice and omit the word *radians*. So, if the measure of an angle is given as $\frac{\pi}{6}$, it is understood to mean $\frac{\pi}{6}$ radian.

➡ **Now Work** PROBLEM 101

4 Find the Area of a Sector of a Circle

Figure 14

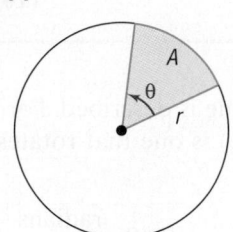

Consider a circle of radius r. Suppose that θ, measured in radians, is a central angle of this circle. See Figure 14. We seek a formula for the area A of the sector (shown in blue) formed by the angle θ.

Now consider a circle of radius r and two central angles θ and θ_1, both measured in radians. See Figure 15. From geometry, we know the ratio of the measures of the angles equals the ratio of the corresponding areas of the sectors formed by these angles. That is,

$$\frac{\theta}{\theta_1} = \frac{A}{A_1}$$

Figure 15 $\dfrac{\theta}{\theta_1} = \dfrac{A}{A_1}$

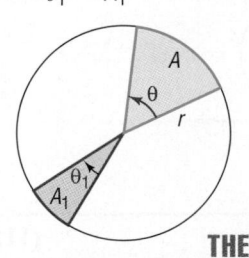

Suppose that $\theta_1 = 2\pi$ radians. Then $A_1 = $ area of the circle $ = \pi r^2$. Solving for A, we find

$$A = A_1 \frac{\theta}{\theta_1} = \pi r^2 \frac{\theta}{2\pi} = \frac{1}{2} r^2 \theta$$

$$\underset{\substack{\uparrow \\ A_1 = \pi r^2 \\ \theta_1 = 2\pi}}{}$$

THEOREM **Area of a Sector**

The area A of the sector of a circle of radius r formed by a central angle of θ radians is

$$A = \frac{1}{2} r^2 \theta \qquad \qquad \textbf{(8)}$$

EXAMPLE 7 **Finding the Area of a Sector of a Circle**

Find the area of the sector of a circle of radius 2 feet formed by an angle of 30°. Round the answer to two decimal places.

Solution We use equation (8) with $r = 2$ feet and $\theta = 30° = \frac{\pi}{6}$ radian. [Remember, in equation (8), θ must be in radians.]

$$A = \frac{1}{2} r^2 \theta = \frac{1}{2} (2)^2 \frac{\pi}{6} = \frac{\pi}{3}$$

The area A of the sector is 1.05 square feet, rounded to two decimal places. ∎

➡ **Now Work** PROBLEM 79

5 Find the Linear Speed of an Object Traveling in Circular Motion

We have already defined the average speed of an object as the distance traveled divided by the elapsed time.

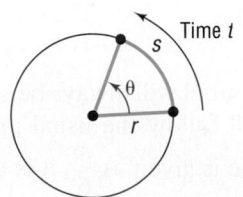

Figure 16 $v = \dfrac{s}{t}$

DEFINITION

Suppose that an object moves around a circle of radius r at a constant speed. If s is the distance traveled in time t around this circle, then the **linear speed** v of the object is defined as

$$v = \frac{s}{t} \qquad (9)$$

As this object travels around the circle, suppose that θ (measured in radians) is the central angle swept out in time t. See Figure 16.

DEFINITION

The **angular speed** ω (the Greek letter omega) of this object is the angle θ (measured in radians) swept out, divided by the elapsed time t; that is,

$$\omega = \frac{\theta}{t} \qquad (10)$$

Angular speed is the way the turning rate of an engine is described. For example, an engine idling at 900 rpm (revolutions per minute) is one that rotates at an angular speed of

$$900\,\frac{\text{revolutions}}{\text{minute}} = 900\,\frac{\text{revolutions}}{\text{minute}} \cdot 2\pi\,\frac{\text{radians}}{\text{revolution}} = 1800\pi\,\frac{\text{radians}}{\text{minute}}$$

There is an important relationship between linear speed and angular speed:

$$\text{linear speed} = v = \underset{\substack{\uparrow \\ (9)}}{\frac{s}{t}} = \underset{\substack{\uparrow \\ s\,=\,r\theta}}{\frac{r\theta}{t}} = r\left(\frac{\theta}{t}\right) = \underset{\substack{\uparrow \\ (10)}}{r \cdot \omega}$$

So,

$$v = r\omega \qquad (11)$$

where ω is measured in radians per unit time.

When using equation (11), remember that $v = \dfrac{s}{t}$ (the linear speed) has the dimensions of length per unit of time (such as feet per second or miles per hour), r (the radius of the circular motion) has the same length dimension as s, and ω (the angular speed) has the dimensions of radians per unit of time. If the angular speed is given in terms of *revolutions* per unit of time (as is often the case), be sure to convert it to *radians* per unit of time before attempting to use equation (11). Remember, 1 revolution = 2π radians.

EXAMPLE 8 | **Finding Linear Speed**

A child is spinning a rock at the end of a 2-foot rope at the rate of 180 revolutions per minute (rpm). Find the linear speed of the rock when it is released.

Solution

Look at Figure 17. The rock is moving around a circle of radius $r = 2$ feet. The angular speed ω of the rock is

$$\omega = 180\,\frac{\text{revolutions}}{\text{minute}} = 180\,\frac{\text{revolutions}}{\text{minute}} \cdot 2\pi\,\frac{\text{radians}}{\text{revolution}} = 360\pi\,\frac{\text{radians}}{\text{minute}}$$

Figure 17

From equation (11), the linear speed v of the rock is

$$v = r\omega = 2 \text{ feet} \cdot 360\pi \frac{\text{radians}}{\text{minute}} = 720\pi \frac{\text{feet}}{\text{minute}} \approx 2262 \frac{\text{feet}}{\text{minute}}$$

The linear speed of the rock when it is released is 2262 ft/min ≈ 25.7 mi/hr.

Now Work PROBLEM 97

Historical Feature

Trigonometry was developed by Greek astronomers, who regarded the sky as the inside of a sphere, so it was natural that triangles on a sphere were investigated early (by Menelaus of Alexandria about AD 100) and that triangles in the plane were studied much later. The first book containing a systematic treatment of plane and spherical trigonometry was written by the Persian astronomer Nasir Eddin (about AD 1250).

Regiomontanus (1436–1476) is the person most responsible for moving trigonometry from astronomy into mathematics. His work was improved by Copernicus (1473–1543) and Copernicus's student

Rhaeticus (1514–1576). Rhaeticus's book was the first to define the six trigonometric functions as ratios of sides of triangles, although he did not give the functions their present names. Credit for this is due to Thomas Finck (1583), but Finck's notation was by no means universally accepted at the time. The notation was finally stabilized by the textbooks of Leonhard Euler (1707–1783).

Trigonometry has since evolved from its use by surveyors, navigators, and engineers to present applications involving ocean tides, the rise and fall of food supplies in certain ecologies, brain wave patterns, and many other phenomena.

7.1 Assess Your Understanding

'Are You Prepared?' *Answers are given at the end of these exercises. If you get a wrong answer, read the pages listed in red.*

1. What is the formula for the circumference C of a circle of radius r? (pp. 31–32)

2. What is the formula for the area A of a circle of radius r? (pp. 31–32)

Concepts and Vocabulary

3. An angle θ is in _____ _____ if its vertex is at the origin of a rectangular coordinate system and its initial side coincides with the positive x-axis.

4. On a circle of radius r, a central angle of θ radians subtends an arc of length $s =$ _____; the area of the sector formed by this angle θ is $A =$ _____.

5. An object travels around a circle of radius r with constant speed. If s is the distance traveled in time t around the circle and θ is the central angle (in radians) swept out in time t, then the linear speed of the object is $v =$ _____ and the angular speed of the object is $\omega =$ _____.

6. *True or False* $\pi = 180$.

7. *True or False* $180° = \pi$ radians.

8. *True or False* On the unit circle, if s is the length of the arc subtended by a central angle θ, measured in radians, then $s = \theta$.

9. *True or False* The area A of the sector of a circle of radius r formed by a central angle of θ degrees is $A = \frac{1}{2}r^2\theta$.

10. *True or False* For circular motion on a circle of radius r, linear speed equals angular speed divided by r.

Skill Building

In Problems 11–22, draw each angle.

11. 30° 12. 60° 13. 135° 14. −120° 15. 450° 16. 540°

17. $\dfrac{3\pi}{4}$ 18. $\dfrac{4\pi}{3}$ 19. $-\dfrac{\pi}{6}$ 20. $-\dfrac{2\pi}{3}$ 21. $\dfrac{16\pi}{3}$ 22. $\dfrac{21\pi}{4}$

In Problems 23–28, convert each angle to a decimal in degrees. Round your answer to two decimal places.

23. 40°10′25″ 24. 61°42′21″ 25. 1°2′3″ 26. 73°40′40″ 27. 9°9′9″ 28. 98°22′45″

In Problems 29–34, convert each angle to D°M′S″ form. Round your answer to the nearest second.

29. 40.32° 30. 61.24° 31. 18.255° 32. 29.411° 33. 19.99° 34. 44.01°

In Problems 35–46, convert each angle in degrees to radians. Express your answer as a multiple of π.

35. 30° **36.** 120° **37.** 240° **38.** 330° **39.** −60° **40.** −30°

41. 180° **42.** 270° **43.** −135° **44.** −225° **45.** −90° **46.** −180°

In Problems 47–58, convert each angle in radians to degrees.

47. $\dfrac{\pi}{3}$ **48.** $\dfrac{5\pi}{6}$ **49.** $-\dfrac{5\pi}{4}$ **50.** $-\dfrac{2\pi}{3}$ **51.** $\dfrac{\pi}{2}$ **52.** 4π

53. $\dfrac{\pi}{12}$ **54.** $\dfrac{5\pi}{12}$ **55.** $-\dfrac{\pi}{2}$ **56.** $-\pi$ **57.** $-\dfrac{\pi}{6}$ **58.** $-\dfrac{3\pi}{4}$

In Problems 59–64, convert each angle in degrees to radians. Express your answer in decimal form, rounded to two decimal places.

59. 17° **60.** 73° **61.** −40° **62.** −51° **63.** 125° **64.** 350°

In Problems 65–70, convert each angle in radians to degrees. Express your answer in decimal form, rounded to two decimal places.

65. 3.14 **66.** 0.75 **67.** 2 **68.** 3 **69.** 6.32 **70.** $\sqrt{2}$

In Problems 71–78, s denotes the length of the arc of a circle of radius r subtended by the central angle θ. Find the missing quantity. Round answers to three decimal places.

71. $r = 10$ meters, $\theta = \dfrac{1}{2}$ radian, $s = ?$

72. $r = 6$ feet, $\theta = 2$ radians, $s = ?$

73. $\theta = \dfrac{1}{3}$ radian, $s = 2$ feet, $r = ?$

74. $\theta = \dfrac{1}{4}$ radian, $s = 6$ centimeters, $r = ?$

75. $r = 5$ miles, $s = 3$ miles, $\theta = ?$

76. $r = 6$ meters, $s = 8$ meters, $\theta = ?$

77. $r = 2$ inches, $\theta = 30°$, $s = ?$

78. $r = 3$ meters, $\theta = 120°$, $s = ?$

In Problems 79–86, A denotes the area of the sector of a circle of radius r formed by the central angle θ. Find the missing quantity. Round answers to three decimal places.

79. $r = 10$ meters, $\theta = \dfrac{1}{2}$ radian, $A = ?$

80. $r = 6$ feet, $\theta = 2$ radians, $A = ?$

81. $\theta = \dfrac{1}{3}$ radian, $A = 2$ square feet, $r = ?$

82. $\theta = \dfrac{1}{4}$ radian, $A = 6$ square centimeters, $r = ?$

83. $r = 5$ miles, $A = 3$ square miles, $\theta = ?$

84. $r = 6$ meters, $A = 8$ square meters, $\theta = ?$

85. $r = 2$ inches, $\theta = 30°$, $A = ?$

86. $r = 3$ meters, $\theta = 120°$, $A = ?$

In Problems 87–90, find the length s and area A. Round answers to three decimal places.

87. **88.** **89.** **90.**

Applications and Extensions

91. Movement of a Minute Hand The minute hand of a clock is 6 inches long. How far does the tip of the minute hand move in 15 minutes? How far does it move in 25 minutes? Round answers to two decimal places.

92. Movement of a Pendulum A pendulum swings through an angle of 20° each second. If the pendulum is 40 inches long, how far does its tip move each second? Round answers to two decimal places.

93. Area of a Sector Find the area of the sector of a circle of radius 4 meters formed by an angle of 45°. Round the answer to two decimal places.

94. Area of a Sector Find the area of the sector of a circle of radius 3 centimeters formed by an angle of 60°. Round the answer to two decimal places.

95. Watering a Lawn A water sprinkler sprays water over a distance of 30 feet while rotating through an angle of 135°. What area of lawn receives water?

135°
30 ft

96. Designing a Water Sprinkler An engineer is asked to design a water sprinkler that will cover a field of 100 square yards that is in the shape of a sector of a circle of radius 50 yards. Through what angle should the sprinkler rotate?

97. Motion on a Circle An object is traveling around a circle with a radius of 5 centimeters. If in 20 seconds a central angle of $\frac{1}{3}$ radian is swept out, what is the angular speed of the object? What is its linear speed?

98. Motion on a Circle An object is traveling around a circle with a radius of 2 meters. If in 20 seconds the object travels 5 meters, what is its angular speed? What is its linear speed?

99. Bicycle Wheels The diameter of each wheel of a bicycle is 26 inches. If you are traveling at a speed of 35 miles per hour on this bicycle, through how many revolutions per minute are the wheels turning?

100. Car Wheels The radius of each wheel of a car is 15 inches. If the wheels are turning at the rate of 3 revolutions per second, how fast is the car moving? Express your answer in inches per second and in miles per hour.

In Problems 101–104, the latitude of a location L is the angle formed by a ray drawn from the center of Earth to the Equator and a ray drawn from the center of Earth to L. See the figure.

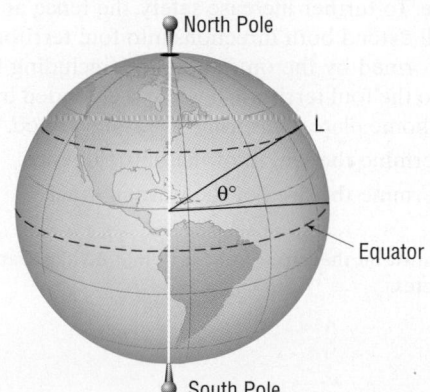

North Pole

L

θ°

Equator

South Pole

101. Distance between Cities Memphis, Tennessee, is due north of New Orleans, Louisiana. Find the distance between Memphis (35°9′ north latitude) and New Orleans (29°57′ north latitude). Assume that the radius of Earth is 3960 miles.

102. Distance between Cities Charleston, West Virginia, is due north of Jacksonville, Florida. Find the distance between Charleston (38°21′ north latitude) and Jacksonville (30°20′ north latitude). Assume that the radius of Earth is 3960 miles.

103. Linear Speed on Earth Earth rotates on an axis through its poles. The distance from the axis to a location on Earth 30° north latitude is about 3429.5 miles. Therefore, a location on Earth at 30° north latitude is spinning on a circle of radius 3429.5 miles. Compute the linear speed on the surface of Earth at 30° north latitude.

104. Linear Speed on Earth Earth rotates on an axis through its poles. The distance from the axis to a location on Earth 40° north latitude is about 3033.5 miles. Therefore, a location on Earth at 40° north latitude is spinning on a circle of radius 3033.5 miles. Compute the linear speed on the surface of Earth at 40° north latitude.

105. Speed of the Moon The mean distance of the Moon from Earth is 2.39×10^5 miles. Assuming that the orbit of the Moon around Earth is circular and that 1 revolution takes 27.3 days, find the linear speed of the Moon. Express your answer in miles per hour.

106. Speed of Earth The mean distance of Earth from the Sun is 9.29×10^7 miles. Assuming that the orbit of Earth around the Sun is circular and that 1 revolution takes 365 days, find the linear speed of Earth. Express your answer in miles per hour.

107. Pulleys Two pulleys, one with radius 2 inches and the other with radius 8 inches, are connected by a belt. (See the figure.) If the 2-inch pulley is caused to rotate at 3 revolutions per minute, determine the revolutions per minute of the 8-inch pulley.

[*Hint:* The linear speeds of the pulleys are the same; both equal the speed of the belt.]

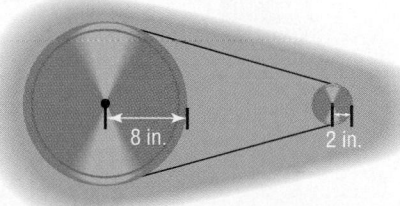

8 in. 2 in.

108. Ferris Wheels A neighborhood carnival has a Ferris wheel whose radius is 30 feet. You measure the time it takes for one revolution to be 70 seconds. What is the linear speed (in feet per second) of this Ferris wheel? What is the angular speed in radians per second?

109. Computing the Speed of a River Current To approximate the speed of the current of a river, a circular paddle wheel with radius 4 feet is lowered into the water. If the current causes the wheel to rotate at a speed of 10 revolutions

per minute, what is the speed of the current? Express your answer in miles per hour.

110. **Spin Balancing Tires** A spin balancer rotates the wheel of a car at 480 revolutions per minute. If the diameter of the wheel is 26 inches, what road speed is being tested? Express your answer in miles per hour. At how many revolutions per minute should the balancer be set to test a road speed of 80 miles per hour?

111. **The Cable Cars of San Francisco** At the Cable Car Museum you can see the four cable lines that are used to pull cable cars up and down the hills of San Francisco. Each cable travels at a speed of 9.55 miles per hour, caused by a rotating wheel whose diameter is 8.5 feet. How fast is the wheel rotating? Express your answer in revolutions per minute.

112. **Difference in Time of Sunrise** Naples, Florida, is approximately 90 miles due west of Ft. Lauderdale. How much sooner would a person in Ft. Lauderdale first see the rising Sun than a person in Naples? See the hint.

[**Hint:** Consult the figure. When a person at Q sees the first rays of the Sun, a person at P is still in the dark. The person at P sees the first rays after Earth has rotated so that P is at the location Q. Now use the fact that at the latitude of Ft. Lauderdale in 24 hours a length of arc of $2\pi(3559)$ miles is subtended.]

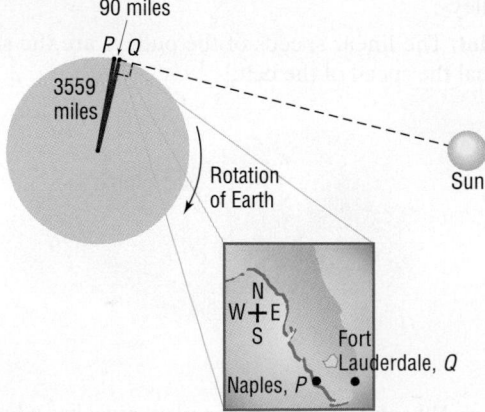

113. **Keeping Up with the Sun** How fast would you have to travel on the surface of Earth at the equator to keep up with the Sun (that is, so that the Sun would appear to remain in the same position in the sky)?

114. **Nautical Miles** A **nautical mile** equals the length of arc subtended by a central angle of 1 minute on a great circle* on the surface of Earth. (See the figure.) If the radius of Earth is taken as 3960 miles, express 1 nautical mile in terms of ordinary, or **statute,** miles.

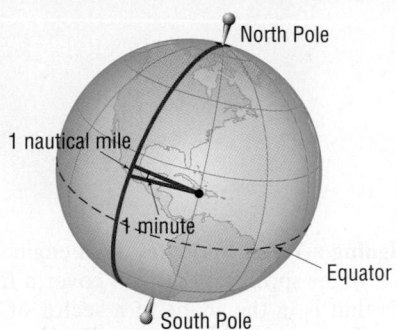

115. **Approximating the Circumference of Earth** Eratosthenes of Cyrene (276–194 BC) was a Greek scholar who lived and worked in Cyrene and Alexandria. One day while visiting in Syene he noticed that the Sun's rays shone directly down a well. On this date 1 year later, in Alexandria, which is 500 miles due north of Syene he measured the angle of the Sun to be about 7.2 degrees. See the figure. Use this information to approximate the radius and circumference of Earth.

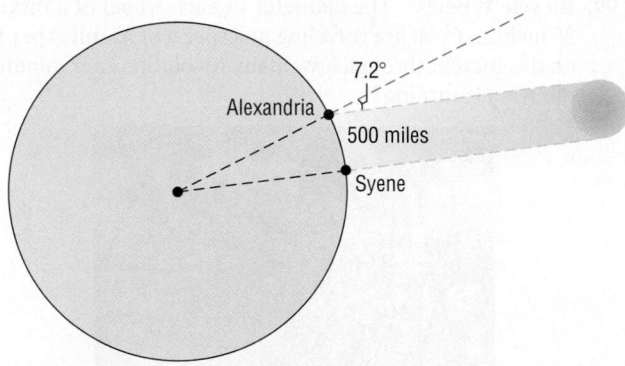

116. **Designing a Little League Field** For a 60-foot Little League Baseball field, the distance from home base to the nearest fence (or other obstruction) on fair territory should be a minimum of 200 feet. The commissioner of parks and recreation is making plans for a new 60-foot field. Because of limited ground availability, he will use the minimum required distance to the outfield fence. To increase safety, however, he plans to include a 10-foot wide warning track on the inside of the fence. To further increase safety, the fence and warning track will extend both directions into foul territory. In total the arc formed by the outfield fence (including the extensions into the foul territories) will be subtended by a central angle at home plate measuring 96°, as illustrated.

(a) Determine the length of the outfield fence.

(b) Determine the area of the warning track.

* Any circle drawn on the surface of Earth that divides Earth into two equal hemispheres.

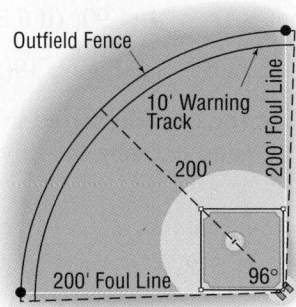

Source: www.littleleague.org

[**Note:** There is a 90° angle between the two foul lines. Then there are two 3° angles between the foul lines and the dotted lines shown. The angle between the two dotted lines outside the 200 foot foul lines is 96°.]

117. Pulleys Two pulleys, one with radius r_1 and the other with radius r_2, are connected by a belt. The pulley with radius r_1 rotates at ω_1 revolutions per minute, whereas the pulley with radius r_2 rotates at ω_2 revolutions per minute. Show that

$$\frac{r_1}{r_2} = \frac{\omega_2}{\omega_1}.$$

Discussion and Writing

118. Do you prefer to measure angles using degrees or radians? Provide justification and a rationale for your choice.

119. What is 1 radian?

120. Which angle has the larger measure: 1 degree or 1 radian? Or are they equal?

121. Explain the difference between linear speed and angular speed.

122. For a circle of radius r, a central angle of θ degrees subtends an arc whose length s is $s = \dfrac{\pi}{180} r\theta$. Discuss whether this is a true or false statement. Give reasons to defend your position.

123. Discuss why ships and airplanes use nautical miles to measure distance. Explain the difference between a nautical mile and a statute mile.

124. Investigate the way that speed bicycles work. In particular, explain the differences and similarities between 5-speed and 9-speed derailleurs. Be sure to include a discussion of linear speed and angular speed.

125. In Example 6, we found that the distance between Albuquerque, New Mexico, and Glasgow, Montana, is approximately 903 miles. According to *mapquest.com*, the distance is approximately 1300 miles. What might account for the difference?

'Are You Prepared?' Answers

1. $C = 2\pi r$ **2.** $A = \pi r^2$

7.2 Right Triangle Trigonometry

PREPARING FOR THIS SECTION *Before getting started, review the following:*

- Geometry Essentials (Chapter R, Review, Section R.3, pp. 30–35)
- Functions (Section 3.1, pp. 208–218)

Now Work the 'Are You Prepared?' problems on page 525.

OBJECTIVES 1 Find the Values of Trigonometric Functions of Acute Angles (p. 517)
 2 Use the Fundamental Identities (p. 519)
 3 Find the Values of the Remaining Trigonometric Functions, Given the Value of One of Them (p. 521)
 4 Use the Complementary Angle Theorem (p. 523)

1 Find the Values of Trigonometric Functions of Acute Angles

A triangle in which one angle is a right angle (90°) is called a **right triangle.** Recall that the side opposite the right angle is called the **hypotenuse,** and the remaining two sides are called the **legs** of the triangle. In Figure 18 we have labeled the hypotenuse as c to indicate that its length is c units, and, in a like manner, we have labeled the legs as a and b. Because the triangle is a right triangle, the Pythagorean Theorem tells us that

$$c^2 = a^2 + b^2$$

Figure 18

Now, suppose that θ is an **acute angle;** that is, $0° < \theta < 90°$ (if θ is measured in degrees) and $0 < \theta < \dfrac{\pi}{2}$ (if θ is measured in radians). See Figure 19(a). Using this acute angle θ, we can form a right triangle, like the one illustrated in Figure 19(b), with hypotenuse of length c and legs of lengths a and b. Using the three sides of this triangle, we can form exactly six ratios:

$$\frac{b}{c}, \quad \frac{a}{c}, \quad \frac{b}{a}, \quad \frac{c}{b}, \quad \frac{c}{a}, \quad \frac{a}{b}$$

Figure 19

(a) Acute angle (b) Right triangle (c) Similar triangles

In fact, these ratios depend only on the size of the angle θ and not on the triangle formed. To see why, look at Figure 19(c). Any two right triangles formed using the angle θ will be similar and, hence, corresponding ratios will be equal. As a result,

$$\frac{b}{c} = \frac{b'}{c'} \quad \frac{a}{c} = \frac{a'}{c'} \quad \frac{b}{a} = \frac{b'}{a'} \quad \frac{c}{b} = \frac{c'}{b'} \quad \frac{c}{a} = \frac{c'}{a'} \quad \frac{a}{b} = \frac{a'}{b'}$$

Because the ratios depend only on the angle θ and not on the triangle itself, we give each ratio a name that involves θ: sine of θ, cosine of θ, tangent of θ, cosecant of θ, secant of θ, and cotangent of θ.

DEFINITION

The six ratios of a right triangle are called **trigonometric functions of acute angles** and are defined as follows:

Function Name	Abbreviation	Value
sine of θ	$\sin \theta$	$\dfrac{b}{c}$
cosine of θ	$\cos \theta$	$\dfrac{a}{c}$
tangent of θ	$\tan \theta$	$\dfrac{b}{a}$
cosecant of θ	$\csc \theta$	$\dfrac{c}{b}$
secant of θ	$\sec \theta$	$\dfrac{c}{a}$
cotangent of θ	$\cot \theta$	$\dfrac{a}{b}$

Figure 20

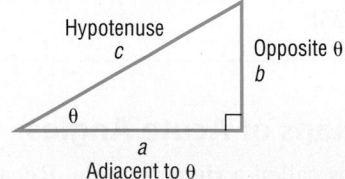

As an aid to remembering these definitions, it may be helpful to refer to the lengths of the sides of the triangle by the names *hypotenuse (c), opposite (b),* and *adjacent (a)*. See Figure 20. In terms of these names, we have the following ratios:

$$\sin \theta = \frac{\text{opposite}}{\text{hypotenuse}} = \frac{b}{c} \qquad \cos \theta = \frac{\text{adjacent}}{\text{hypotenuse}} = \frac{a}{c} \qquad \tan \theta = \frac{\text{opposite}}{\text{adjacent}} = \frac{b}{a}$$

$$\csc \theta = \frac{\text{hypotenuse}}{\text{opposite}} = \frac{c}{b} \qquad \sec \theta = \frac{\text{hypotenuse}}{\text{adjacent}} = \frac{c}{a} \qquad \cot \theta = \frac{\text{adjacent}}{\text{opposite}} = \frac{a}{b} \tag{1}$$

Since a, b, and c are positive, each of the trigonometric functions of an acute angle θ is positive.

| EXAMPLE 1 | **Finding the Value of Trigonometric Functions** |

Find the value of each of the six trigonometric functions of the angle θ in Figure 21.

Solution We see in Figure 21 that the two given sides of the triangle are

$$c = \text{hypotenuse} = 5 \qquad a = \text{adjacent} = 3$$

Figure 21

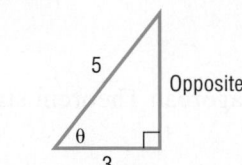

To find the length of the opposite side, we use the Pythagorean Theorem.

$$(\text{adjacent})^2 + (\text{opposite})^2 = (\text{hypotenuse})^2$$
$$3^2 + (\text{opposite})^2 = 5^2$$
$$(\text{opposite})^2 = 25 - 9 = 16$$
$$\text{opposite} = 4$$

WARNING When writing the values of the trigonometric functions, do not forget the argument of the function

$$\sin \theta = \frac{4}{5} \quad \text{correct}$$

$$\sin = \frac{4}{5} \quad \text{incorrect} \qquad ▪$$

Now that we know the lengths of the three sides, we use the ratios in (1) to find the value of each of the six trigonometric functions:

$$\sin \theta = \frac{\text{opposite}}{\text{hypotenuse}} = \frac{4}{5} \qquad \cos \theta = \frac{\text{adjacent}}{\text{hypotenuse}} = \frac{3}{5} \qquad \tan \theta = \frac{\text{opposite}}{\text{adjacent}} = \frac{4}{3}$$

$$\csc \theta = \frac{\text{hypotenuse}}{\text{opposite}} = \frac{5}{4} \qquad \sec \theta = \frac{\text{hypotenuse}}{\text{adjacent}} = \frac{5}{3} \qquad \cot \theta = \frac{\text{adjacent}}{\text{opposite}} = \frac{3}{4}$$

— **Now Work** PROBLEM 11

2 Use the Fundamental Identities

You may have observed some relationships that exist among the six trigonometric functions of acute angles. For example, the **reciprocal identities** are

Reciprocal Identities

$$\csc \theta = \frac{1}{\sin \theta} \qquad \sec \theta = \frac{1}{\cos \theta} \qquad \cot \theta = \frac{1}{\tan \theta} \qquad \textbf{(2)}$$

Two other fundamental identities that are easy to see are the **quotient identities.**

Quotient Identities

$$\tan \theta = \frac{\sin \theta}{\cos \theta} \qquad \cot \theta = \frac{\cos \theta}{\sin \theta} \qquad \textbf{(3)}$$

If $\sin \theta$ and $\cos \theta$ are known, formulas (2) and (3) make it easy to find the values of the remaining trigonometric functions.

| EXAMPLE 2 | **Finding the Values of the Remaining Trigonometric Functions, Given $\sin \theta$ and $\cos \theta$** |

Given $\sin \theta = \dfrac{\sqrt{5}}{5}$ and $\cos \theta = \dfrac{2\sqrt{5}}{5}$, find the value of each of the four remaining trigonometric functions of θ.

Solution Based on formula (3), we have

$$\tan \theta = \frac{\sin \theta}{\cos \theta} = \frac{\dfrac{\sqrt{5}}{5}}{\dfrac{2\sqrt{5}}{5}} = \frac{1}{2}$$

Then we use the reciprocal identities from formula (2) to get

$$\csc \theta = \frac{1}{\sin \theta} = \frac{1}{\frac{\sqrt{5}}{5}} = \frac{5}{\sqrt{5}} = \sqrt{5} \qquad \sec \theta = \frac{1}{\cos \theta} = \frac{1}{\frac{2\sqrt{5}}{5}} = \frac{5}{2\sqrt{5}} = \frac{\sqrt{5}}{2} \qquad \cot \theta = \frac{1}{\tan \theta} = \frac{1}{\frac{1}{2}} = 2$$

══════ **Now Work** PROBLEM 21

Figure 22

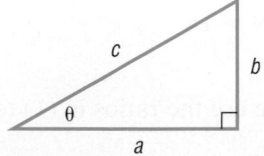

Refer now to the right triangle in Figure 22. The Pythagorean Theorem states that $a^2 + b^2 = c^2$, which we can write as

$$b^2 + a^2 = c^2$$

Dividing each side by c^2, we get

$$\frac{b^2}{c^2} + \frac{a^2}{c^2} = 1 \quad \text{or} \quad \left(\frac{b}{c}\right)^2 + \left(\frac{a}{c}\right)^2 = 1$$

In terms of trigonometric functions of the angle θ, this equation states that

$$(\sin \theta)^2 + (\cos \theta)^2 = 1 \qquad \textbf{(4)}$$

Equation (4) is, in fact, an identity, since the equation is true for any acute angle θ.

It is customary to write $\sin^2 \theta$ instead of $(\sin \theta)^2$, $\cos^2 \theta$ instead of $(\cos \theta)^2$, and so on. With this notation, we can rewrite equation (4) as

$$\sin^2 \theta + \cos^2 \theta = 1 \qquad \textbf{(5)}$$

Another identity can be obtained from equation (5) by dividing each side by $\cos^2 \theta$.

$$\frac{\sin^2 \theta}{\cos^2 \theta} + 1 = \frac{1}{\cos^2 \theta}$$

Now use formulas (2) and (3) to get

$$\tan^2 \theta + 1 = \sec^2 \theta \qquad \textbf{(6)}$$

Similarly, by dividing each side of equation (5) by $\sin^2 \theta$, we get $1 + \cot^2 \theta = \csc^2 \theta$, which we write as

$$\cot^2 \theta + 1 = \csc^2 \theta \qquad \textbf{(7)}$$

Collectively, the identities in equations (5), (6), and (7) are referred to as the **Pythagorean Identities.**

Let's pause here to summarize the fundamental identities.

Fundamental Identities

$$\tan \theta = \frac{\sin \theta}{\cos \theta} \qquad \cot \theta = \frac{\cos \theta}{\sin \theta}$$

$$\csc \theta = \frac{1}{\sin \theta} \qquad \sec \theta = \frac{1}{\cos \theta} \qquad \cot \theta = \frac{1}{\tan \theta}$$

$$\sin^2 \theta + \cos^2 \theta = 1 \qquad \tan^2 \theta + 1 = \sec^2 \theta \qquad \cot^2 \theta + 1 = \csc^2 \theta$$

EXAMPLE 3	Finding the Exact Value of a Trigonometric Expression Using Identities

Find the exact value of each expression. Do not use a calculator.

(a) $\tan 20° - \dfrac{\sin 20°}{\cos 20°}$

(b) $\sin^2 \dfrac{\pi}{12} + \dfrac{1}{\sec^2 \dfrac{\pi}{12}}$

Solution

(a) $\tan 20° - \dfrac{\sin 20°}{\cos 20°} = \tan 20° - \tan 20° = 0$

$$\uparrow$$
$$\frac{\sin\theta}{\cos\theta} = \tan\theta$$

(b) $\sin^2 \dfrac{\pi}{12} + \dfrac{1}{\sec^2 \dfrac{\pi}{12}} = \sin^2 \dfrac{\pi}{12} + \cos^2 \dfrac{\pi}{12} = 1$

$$\qquad\qquad\uparrow\qquad\qquad\qquad\qquad\uparrow$$
$$\cos\theta = \frac{1}{\sec\theta} \qquad \sin^2\theta + \cos^2\theta = 1$$

 Now Work PROBLEM 39

3 Find the Values of the Remaining Trigonometric Functions, Given the Value of One of Them

Once the value of one trigonometric function is known, it is possible to find the value of each of the remaining five trigonometric functions.

EXAMPLE 4	Finding the Values of the Remaining Trigonometric Functions, Given $\sin \theta$, θ Acute

Given that $\sin \theta = \dfrac{1}{3}$ and θ is an acute angle, find the exact value of each of the remaining five trigonometric functions of θ.

Solution

We solve this problem in two ways: The first way uses the definition of the trigonometric functions; the second method uses the fundamental identities.

Solution 1 Using the Definition

We draw a right triangle with acute angle θ, opposite side of length $b = 1$, and hypotenuse of length $c = 3$ $\left(\text{because } \sin \theta = \dfrac{1}{3} = \dfrac{b}{c}\right)$. See Figure 23. The adjacent side a can be found by using the Pythagorean Theorem.

Figure 23

```
    c = 3
  ╱────────┐ b = 1
 ╱  θ      │
╱──────────┘
     a
```

$$a^2 + 1^2 = 3^2 \qquad a^2 + b^2 = c^2; b = 1, c = 3$$
$$a^2 + 1 = 9$$
$$a^2 = 8$$
$$a = 2\sqrt{2}$$

Now the definitions given in equation (1) can be used to find the value of each of the remaining five trigonometric functions. (Refer back to the method used in Example 1.) Using $a = 2\sqrt{2}$, $b = 1$, and $c = 3$, we have

$$\cos \theta = \frac{a}{c} = \frac{2\sqrt{2}}{3} \qquad\qquad \tan \theta = \frac{b}{a} = \frac{1}{2\sqrt{2}} = \frac{\sqrt{2}}{4}$$

$$\csc \theta = \frac{c}{b} = \frac{3}{1} = 3 \quad \sec \theta = \frac{c}{a} = \frac{3}{2\sqrt{2}} = \frac{3\sqrt{2}}{4} \quad \cot \theta = \frac{a}{b} = \frac{2\sqrt{2}}{1} = 2\sqrt{2}$$

Solution 2
Using Identities

We begin by seeking $\cos\theta$, which can be found by using the Pythagorean Identity from equation (5).

$$\sin^2\theta + \cos^2\theta = 1 \qquad \text{Formula (5)}$$

$$\frac{1}{9} + \cos^2\theta = 1 \qquad \sin\theta = \frac{1}{3}$$

$$\cos^2\theta = 1 - \frac{1}{9} = \frac{8}{9}$$

Recall that the trigonometric functions of an acute angle are positive. In particular, $\cos\theta > 0$ for an acute angle θ, so we have

$$\cos\theta = \sqrt{\frac{8}{9}} = \frac{2\sqrt{2}}{3}$$

Now we know that $\sin\theta = \dfrac{1}{3}$ and $\cos\theta = \dfrac{2\sqrt{2}}{3}$, so we can proceed as we did in Example 2.

$$\tan\theta = \frac{\sin\theta}{\cos\theta} = \frac{\frac{1}{3}}{\frac{2\sqrt{2}}{3}} = \frac{1}{2\sqrt{2}} = \frac{\sqrt{2}}{4} \qquad \cot\theta = \frac{1}{\tan\theta} = \frac{1}{\frac{\sqrt{2}}{4}} = \frac{4}{\sqrt{2}} = 2\sqrt{2}$$

$$\sec\theta = \frac{1}{\cos\theta} = \frac{1}{\frac{2\sqrt{2}}{3}} = \frac{3}{2\sqrt{2}} = \frac{3\sqrt{2}}{4} \qquad \csc\theta = \frac{1}{\sin\theta} = \frac{1}{\frac{1}{3}} = 3$$

Finding the Values of the Trigonometric Functions When One Is Known

Given the value of one trigonometric function of an acute angle θ, the exact value of each of the remaining five trigonometric functions of θ can be found in either of two ways.

Method 1 Using the Definition

STEP 1: Draw a right triangle showing the acute angle θ.
STEP 2: Two of the sides can then be assigned values based on the value of the given trigonometric function.
STEP 3: Find the length of the third side by using the Pythagorean Theorem.
STEP 4: Use the definitions in equation (1) to find the value of each of the remaining trigonometric functions.

Method 2 Using Identities

Use appropriately selected identities to find the value of each of the remaining trigonometric functions.

EXAMPLE 5

Given One Value of a Trigonometric Function, Find the Values of the Remaining Ones

Given $\tan\theta = \dfrac{1}{2}$, θ an acute angle, find the exact value of each of the remaining five trigonometric functions of θ.

Solution 1
Using the Definition

Figure 24 shows a right triangle with acute angle θ, where

$$\tan\theta = \frac{1}{2} = \frac{\text{opposite}}{\text{adjacent}} = \frac{b}{a}$$

Figure 24

$$\tan \theta = \frac{1}{2}$$

With $b = 1$ and $a = 2$, the hypotenuse c can be found by using the Pythagorean Theorem.

$$c^2 = a^2 + b^2 = 2^2 + 1^2 = 5$$
$$c = \sqrt{5}$$

Now apply the definitions using $a = 2$, $b = 1$, and $c = \sqrt{5}$.

$$\sin \theta = \frac{b}{c} = \frac{1}{\sqrt{5}} = \frac{\sqrt{5}}{5} \qquad \cos \theta = \frac{a}{c} = \frac{2}{\sqrt{5}} = \frac{2\sqrt{5}}{5}$$

$$\csc \theta = \frac{c}{b} = \frac{\sqrt{5}}{1} = \sqrt{5} \qquad \sec \theta = \frac{c}{a} = \frac{\sqrt{5}}{2} \qquad \cot \theta = \frac{a}{b} = \frac{2}{1} = 2$$

Solution 2
Using Identities

Because we know the value of $\tan \theta$, we use the Pythagorean Identity that involves $\tan \theta$:

$$\tan^2 \theta + 1 = \sec^2 \theta \qquad \text{Formula (6)}$$

$$\frac{1}{4} + 1 = \sec^2 \theta \qquad \tan \theta = \frac{1}{2}$$

$$\sec^2 \theta = \frac{1}{4} + 1 = \frac{5}{4} \qquad \text{Proceed to solve for sec } \theta.$$

$$\sec \theta = \frac{\sqrt{5}}{2}$$

Now we know $\tan \theta = \frac{1}{2}$ and $\sec \theta = \frac{\sqrt{5}}{2}$. Using reciprocal identities, we find

$$\cos \theta = \frac{1}{\sec \theta} = \frac{1}{\dfrac{\sqrt{5}}{2}} = \frac{2}{\sqrt{5}} = \frac{2\sqrt{5}}{5}$$

$$\cot \theta = \frac{1}{\tan \theta} = \frac{1}{\dfrac{1}{2}} = 2$$

To find $\sin \theta$, we use the following reasoning:

$$\tan \theta = \frac{\sin \theta}{\cos \theta}, \quad \text{so} \quad \sin \theta = (\tan \theta)(\cos \theta) = \frac{1}{2} \cdot \frac{2\sqrt{5}}{5} = \frac{\sqrt{5}}{5}$$

$$\csc \theta = \frac{1}{\sin \theta} = \frac{1}{\dfrac{\sqrt{5}}{5}} = \sqrt{5}$$

━━ **Now Work** PROBLEM 25

4 Use the Complementary Angle Theorem

Two acute angles are called **complementary** if their sum is a right angle. Because the sum of the angles of any triangle is $180°$, it follows that, for a right triangle, the two acute angles are complementary.

Refer now to Figure 25; we have labeled the angle opposite side b as B and the angle opposite side a as A. Notice that side a is adjacent to angle B and is opposite angle A. Similarly, side b is opposite angle B and is adjacent to angle A. As a result,

Figure 25

Adjacent to A
opposite B

Adjacent to B
opposite A

$$\sin B = \frac{b}{c} = \cos A \qquad \cos B = \frac{a}{c} = \sin A \qquad \tan B = \frac{b}{a} = \cot A$$

$$\csc B = \frac{c}{b} = \sec A \qquad \sec B = \frac{c}{a} = \csc A \qquad \cot B = \frac{a}{b} = \tan A$$

$$(8)$$

Because of these relationships, the functions sine and cosine, tangent and cotangent, and secant and cosecant are called **cofunctions** of each other. The identities (8) may be expressed in words as follows:

THEOREM

Complementary Angle Theorem

Cofunctions of complementary angles are equal.

Here are examples of this theorem.

Complementary angles	Complementary angles	Complementary angles
$\sin 30° = \cos 60°$	$\tan 40° = \cot 50°$	$\sec 80° = \csc 10°$
Cofunctions	Cofunctions	Cofunctions

If an angle θ is measured in degrees, we will use the degree symbol when writing a trigonometric function of θ, as, for example, in $\sin 30°$ and $\tan 45°$. If an angle θ is measured in radians, then no symbol is used when writing a trigonometric function of θ, as, for example, in $\cos \pi$ and $\sec \dfrac{\pi}{3}$.

If θ is an acute angle measured in degrees, the angle $90° - \theta$ (or $\dfrac{\pi}{2} - \theta$, if θ is in radians) is the angle complementary to θ. Table 2 restates the preceding theorem on cofunctions.

Table 2

θ (Degrees)	θ (Radians)
$\sin \theta = \cos(90° - \theta)$	$\sin \theta = \cos\left(\dfrac{\pi}{2} - \theta\right)$
$\cos \theta = \sin(90° - \theta)$	$\cos \theta = \sin\left(\dfrac{\pi}{2} - \theta\right)$
$\tan \theta = \cot(90° - \theta)$	$\tan \theta = \cot\left(\dfrac{\pi}{2} - \theta\right)$
$\csc \theta = \sec(90° - \theta)$	$\csc \theta = \sec\left(\dfrac{\pi}{2} - \theta\right)$
$\sec \theta = \csc(90° - \theta)$	$\sec \theta = \csc\left(\dfrac{\pi}{2} - \theta\right)$
$\cot \theta = \tan(90° - \theta)$	$\cot \theta = \tan\left(\dfrac{\pi}{2} - \theta\right)$

The angle θ in Table 2 is acute. We will see later (Section 8.4) that these results are valid for any angle θ.

EXAMPLE 6	**Using the Complementary Angle Theorem**

(a) $\sin 62° = \cos(90° - 62°) = \cos 28°$

(b) $\tan \dfrac{\pi}{12} = \cot\left(\dfrac{\pi}{2} - \dfrac{\pi}{12}\right) = \cot \dfrac{5\pi}{12}$

(c) $\cos \dfrac{\pi}{4} = \sin\left(\dfrac{\pi}{2} - \dfrac{\pi}{4}\right) = \sin \dfrac{\pi}{4}$

(d) $\csc \dfrac{\pi}{6} = \sec\left(\dfrac{\pi}{2} - \dfrac{\pi}{6}\right) = \sec \dfrac{\pi}{3}$

EXAMPLE 7	Using the Complementary Angle Theorem

Find the exact value of each expression. Do not use a calculator.

(a) $\sec 28° - \csc 62°$

(b) $\dfrac{\sin 35°}{\cos 55°}$

Solution

(a) $\sec 28° - \csc 62° = \csc(90° - 28°) - \csc 62°$

$\qquad\qquad\qquad\;\; = \csc 62° - \csc 62° = 0$

(b) $\dfrac{\sin 35°}{\cos 55°} = \dfrac{\cos(90° - 35°)}{\cos 55°} = \dfrac{\cos 55°}{\cos 55°} = 1$

▬▬▬ **Now Work** PROBLEM 43

Historical Feature

The name *sine* for the sine function is due to a medieval confusion. The name comes from the Sanskrit word *jīva*, (meaning *chord*), first used in India by Araybhata the Elder (AD 510). He really meant half-chord, but abbreviated it. This was brought into Arabic as *jība*, which was meaningless. Because the proper Arabic word *jaib* would be written the same way (short vowels are not written out in Arabic), *jība*, was pronounced as *jaib*, which meant bosom or hollow, and *jaib* remains as the Arabic word for sine to this day. Scholars translating the Arabic works into Latin found that the word *sinus* also meant bosom or hollow, and from *sinus* we get the word *sine*.

The name *tangent*, due to Thomas Finck (1583), can be understood by looking at Figure 26. The line segment $\overline{DC}$ is tangent to the circle at C. If $d(O, B) = d(O, C) = 1$, then the length of the line segment $\overline{DC}$ is

$$d(D, C) = \frac{d(D, C)}{1} = \frac{d(D, C)}{d(O, C)} = \tan \alpha$$

The old name for the tangent is *umbra versa* (meaning turned shadow), referring to the use of the tangent in solving height problems with shadows.

Figure 26

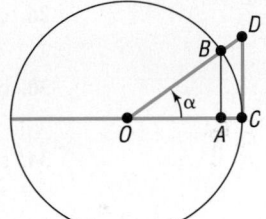

The names of the cofunctions came about as follows. If A and B are complementary angles, then $\cos A = \sin B$. Because B is the complement of A, it was natural to write the cosine of A as *sin co A*. Probably for reasons involving ease of pronunciation, the *co* migrated to the front, and then cosine received a three-letter abbreviation to match sin, sec, and tan. The two other cofunctions were similarly treated, except that the long forms *cotan* and *cosec* survive to this day in some countries.

7.2 Assess Your Understanding

'Are You Prepared?' *Answers are given at the end of these exercises. If you get a wrong answer, read the pages listed in red.*

1. In a right triangle with legs $a = 6$ and $b = 10$, the Pythagorean Theorem tells us that the hypotenuse $c = $ _____. (pp. 30–35)

2. The value of the function $f(x) = 3x - 7$ at 5 is _____. (pp. 208–218)

Concepts and Vocabulary

3. Two acute angles whose sum is a right angle are called _____.

4. The sine and _____ functions are cofunctions.

5. $\tan 28° = \cot$ _____.

6. For any angle θ, $\sin^2 \theta + \cos^2 \theta = $ _____,

7. *True or False* $\tan \theta = \dfrac{\sin \theta}{\cos \theta}$.

8. *True or False* $1 + \tan^2 \theta = \csc^2 \theta$.

9. *True or False* If θ is an acute angle and $\sec \theta = 3$, then $\cos \theta = \dfrac{1}{3}$.

10. *True or False* $\tan \dfrac{\pi}{5} = \cot \dfrac{4\pi}{5}$.

Skill Building

In Problems 11–20, find the value of the six trigonometric functions of the angle θ in each figure.

11.

12.

13.

14.

15.

16.

17.

18.

19.

20.

In Problems 21–24, use identities to find the exact value of each of the four remaining trigonometric functions of the acute angle θ.

21. $\sin \theta = \dfrac{1}{2}$, $\cos \theta = \dfrac{\sqrt{3}}{2}$

22. $\sin \theta = \dfrac{\sqrt{3}}{2}$, $\cos \theta = \dfrac{1}{2}$

23. $\sin \theta = \dfrac{2}{3}$, $\cos \theta = \dfrac{\sqrt{5}}{3}$

24. $\sin \theta = \dfrac{1}{3}$, $\cos \theta = \dfrac{2\sqrt{2}}{3}$

In Problems 25–36, use the definition or identities to find the exact value of each of the remaining five trigonometric functions of the acute angle θ.

25. $\sin \theta = \dfrac{\sqrt{2}}{2}$

26. $\cos \theta = \dfrac{\sqrt{2}}{2}$

27. $\cos \theta = \dfrac{1}{3}$

28. $\sin \theta = \dfrac{\sqrt{3}}{4}$

29. $\tan \theta = \dfrac{1}{2}$

30. $\cot \theta = \dfrac{1}{2}$

31. $\sec \theta = 3$

32. $\csc \theta = 5$

33. $\tan \theta = \sqrt{2}$

34. $\sec \theta = \dfrac{5}{3}$

35. $\csc \theta = 2$

36. $\cot \theta = 2$

In Problems 37–54, use Fundamental Identities and/or the Complementary Angle Theorem to find the exact value of each expression. Do not use a calculator.

37. $\sin^2 20° + \cos^2 20°$

38. $\sec^2 28° - \tan^2 28°$

39. $\sin 80° \csc 80°$

40. $\tan 10° \cot 10°$

41. $\tan 50° - \dfrac{\sin 50°}{\cos 50°}$

42. $\cot 25° - \dfrac{\cos 25°}{\sin 25°}$

43. $\sin 38° - \cos 52°$

44. $\tan 12° - \cot 78°$

45. $\dfrac{\cos 10°}{\sin 80°}$

46. $\dfrac{\cos 40°}{\sin 50°}$

47. $1 - \cos^2 20° - \cos^2 70°$

48. $1 + \tan^2 5° - \csc^2 85°$

49. $\tan 20° - \dfrac{\cos 70°}{\cos 20°}$

50. $\cot 40° - \dfrac{\sin 50°}{\sin 40°}$

51. $\tan 35° \cdot \sec 55° \cdot \cos 35°$

52. $\cot 25° \cdot \csc 65° \cdot \sin 25°$

53. $\cos 35° \sin 55° + \cos 55° \sin 35°$

54. $\sec 35° \csc 55° - \tan 35° \cot 55°$

55. Given $\sin 30° = \dfrac{1}{2}$, use trigonometric identities to find the exact value of

(a) $\cos 60°$

(b) $\cos^2 30°$

(c) $\csc \dfrac{\pi}{6}$

(d) $\sec \dfrac{\pi}{3}$

56. Given $\sin 60° = \dfrac{\sqrt{3}}{2}$, use trigonometric identities to find the exact value of

(a) $\cos 30°$

(b) $\cos^2 60°$

(c) $\sec \dfrac{\pi}{6}$

(d) $\csc \dfrac{\pi}{3}$

57. Given $\tan \theta = 4$, use trigonometric identities to find the exact value of

(a) $\sec^2 \theta$

(b) $\cot \theta$

(c) $\cot \left(\dfrac{\pi}{2} - \theta \right)$

(d) $\csc^2 \theta$

58. Given $\sec \theta = 3$, use trigonometric identities to find the exact value of

(a) $\cos \theta$ (b) $\tan^2 \theta$ (c) $\csc(90° - \theta)$ (d) $\sin^2 \theta$

59. Given $\csc \theta = 4$, use trigonometric identities to find the exact value of

(a) $\sin \theta$ (b) $\cot^2 \theta$ (c) $\sec(90° - \theta)$ (d) $\sec^2 \theta$

60. Given $\cot \theta = 2$, use trigonometric identities to find the exact value of

(a) $\tan \theta$

(b) $\csc^2 \theta$

(c) $\tan \left(\dfrac{\pi}{2} - \theta \right)$

(d) $\sec^2 \theta$

61. Given the approximation $\sin 38° \approx 0.62$, use trigonometric identities to find the approximate value of

(a) $\cos 38°$ (b) $\tan 38°$

(c) $\cot 38°$ (d) $\sec 38°$

(e) $\csc 38°$ (f) $\sin 52°$

(g) $\cos 52°$ (h) $\tan 52°$

62. Given the approximation $\cos 21° \approx 0.93$, use trigonometric identities to find the approximate value of

(a) $\sin 21°$ (b) $\tan 21°$

(c) $\cot 21°$ (d) $\sec 21°$

(e) $\csc 21°$ (f) $\sin 69°$

(g) $\cos 69°$ (h) $\tan 69°$

63. If $\sin \theta = 0.3$, find the exact value of $\sin \theta + \cos\left(\dfrac{\pi}{2} - \theta\right)$.

64. If $\tan \theta = 4$, find the exact value of $\tan \theta + \tan\left(\dfrac{\pi}{2} - \theta\right)$.

65. Find an acute angle θ that satisfies the equation

$$\sin \theta = \cos(2\theta + 30°)$$

66. Find an acute angle θ that satisfies the equation

$$\tan \theta = \cot(\theta + 45°)$$

Applications and Extensions

67. Calculating the Time of a Trip From a parking lot you want to walk to a house on the ocean. The house is located 1500 feet down a paved path that parallels the beach, which is 500 feet wide. Along the path you can walk 300 feet per minute, but in the sand on the beach you can only walk 100 feet per minute. See the illustration.

(a) Calculate the time T if you walk 1500 feet along the paved path and then 500 feet in the sand to the house.

(b) Calculate the time T if you walk in the sand directly toward the ocean for 500 feet and then turn left and walk along the beach for 1500 feet to the house.

(c) Express the time T to get from the parking lot to the beachhouse as a function of the angle θ shown in the illustration.

(d) Calculate the time T if you walk directly from the parking lot to the house.

$$\left[\textbf{Hint: } \tan \theta = \frac{500}{1500}\right]$$

(e) Calculate the time T if you walk 1000 feet along the paved path and then walk directly to the house.

(f) Graph $T = T(\theta)$. For what angle θ is T least? What is x for this angle? What is the minimum time?

(g) Explain why $\tan \theta = \dfrac{1}{3}$ gives the smallest angle θ that is possible.

68. Carrying a Ladder around a Corner Two hallways, one of width 3 feet, the other of width 4 feet, meet at a right angle. See the illustration.

(a) Express the length L of the line segment shown as a function of the angle θ.

(b) Discuss why the length of the longest ladder that can be carried around the corner is equal to the smallest value of L.

69. Electrical Engineering A resistor and an inductor connected in a series network impede the flow of an alternating current. This impedance Z is determined by the reactance X of the inductor and the resistance R of the resistor. The three quantities, all measured in ohms, can be represented by the sides of a right triangle as illustrated, so $Z^2 = X^2 + R^2$. The angle ϕ is called the **phase angle**. Suppose a series network has an inductive reactance of $X = 400$ ohms and a resistance of $R = 600$ ohms.

(a) Find the impedance Z.

(b) Find the values of the six trigonometric functions of the phase angle ϕ.

70. Electrical Engineering Refer to Problem 69. A series network has a resistance of $R = 588$ ohms. The phase angle ϕ is such that $\tan \phi = \dfrac{5}{12}$.

(a) Determine the inductive reactance X and the impedance Z.

(b) Determine the values of the remaining five trigonometric functions of the phase angle ϕ.

71. Geometry Suppose that the angle θ is a central angle of a circle of radius 1 (see the figure). Show that

(a) Angle $OAC = \dfrac{\theta}{2}$

(b) $|CD| = \sin \theta$ and $|OD| = \cos \theta$

(c) $\tan \dfrac{\theta}{2} = \dfrac{\sin \theta}{1 + \cos \theta}$

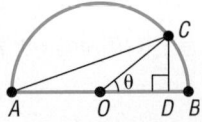

72. Geometry Show that the area A of an isosceles triangle is $A = a^2 \sin \theta \cos \theta$, where a is the length of one of the two equal sides and θ is the measure of one of the two equal angles (see the figure).

73. Geometry Let $n \geq 1$ be any real number and let θ be any angle for which $0 < n\theta < \dfrac{\pi}{2}$. Then we can draw a triangle with the angles θ and $n\theta$ and included side of length 1 (do you see why?) and place it on the unit circle as illustrated.

Now, drop the perpendicular from C to $D = (x, 0)$ and show that

$$x = \frac{\tan(n\theta)}{\tan \theta + \tan(n\theta)}$$

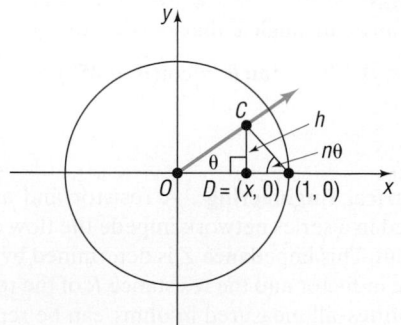

74. Geometry Refer to the figure. The smaller circle, whose radius is a, is tangent to the larger circle, whose radius is b. The ray $\overrightarrow{OA}$ contains a diameter of each circle, and the ray $\overrightarrow{OB}$ is tangent to each circle. Show that

$$\cos \theta = \frac{\sqrt{ab}}{\dfrac{a + b}{2}}$$

(This shows that $\cos \theta$ equals the ratio of the geometric mean of a and b to the arithmetic mean of a and b.)

[**Hint:** First show that $\sin \theta = (b - a)/(b + a)$.]

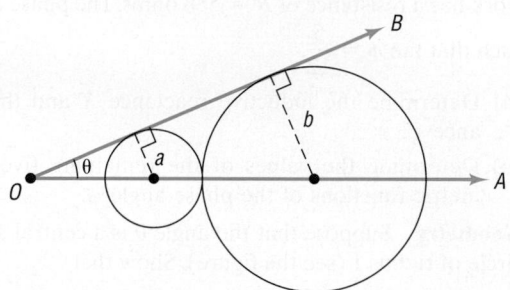

75. Geometry Refer to the figure. If $|OA| = 1$, show that

(a) Area $\triangle OAC = \dfrac{1}{2} \sin \alpha \cos \alpha$

(b) Area $\triangle OCB = \dfrac{1}{2} |OB|^2 \sin \beta \cos \beta$

(c) Area $\triangle OAB = \dfrac{1}{2} |OB| \sin(\alpha + \beta)$

(d) $|OB| = \dfrac{\cos \alpha}{\cos \beta}$

(e) $\sin(\alpha + \beta) = \sin \alpha \cos \beta + \cos \alpha \sin \beta$

[**Hint:** Area $\triangle OAB$ = Area $\triangle OAC$ + Area $\triangle OCB$]

76. Geometry Refer to the figure, where a unit circle is drawn. The line segment DB is tangent to the circle.

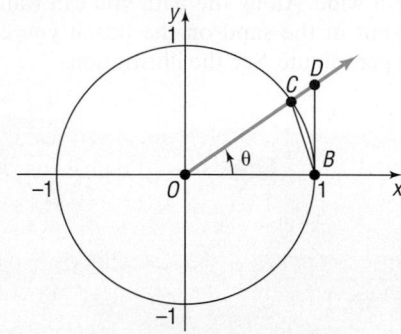

(a) Express the area of $\triangle OBC$ in terms of $\sin \theta$ and $\cos \theta$.
 [**Hint:** Use the altitude from C to the base $\overline{OB} = 1$.]

(b) Express the area of $\triangle OBD$ in terms of $\sin \theta$ and $\cos \theta$.

(c) The area of the sector OBC of the circle is $\dfrac{1}{2} \theta$, where θ is measured in radians. Use the results of parts (a) and (b) and the fact that

Area $\triangle OBC$ < Area of sector OBC < Area $\triangle OBD$

to show that

$$1 < \frac{\theta}{\sin \theta} < \frac{1}{\cos \theta}$$

77. If $\cos \alpha = \tan \beta$ and $\cos \beta = \tan \alpha$, where α and β are acute angles, show that

$$\sin \alpha = \sin \beta = \sqrt{\frac{3 - \sqrt{5}}{2}}$$

78. If θ is an acute angle and $\tan \theta = x$, $x \neq 0$, express the remaining five trigonometric functions in terms of x.

Discussion and Writing

79. If θ is an acute angle, explain why $\sec \theta > 1$.

80. If θ is an acute angle, explain why $0 < \sin \theta < 1$.

81. How would you explain the meaning of the sine function to a fellow student who has just completed college algebra?

82. Look back at Example 5. Which of the two solutions do you prefer? Explain your reasoning.

'Are You Prepared?' Answers

1. $2\sqrt{34}$ **2.** $f(5) = 8$

7.3 Computing the Values of Trigonometric Functions of Acute Angles

OBJECTIVES 1 Find the Exact Values of the Trigonometric Functions of $\frac{\pi}{4} = 45°$ (p. 529)

2 Find the Exact Values of the Trigonometric Functions of $\frac{\pi}{6} = 30°$ and $\frac{\pi}{3} = 60°$ (p. 530)

3 Use a Calculator to Approximate the Values of the Trigonometric Functions of Acute Angles (p. 532)

4 Model and Solve Applied Problems Involving Right Triangles (p. 532)

In the previous section, we developed ways to find the value of each trigonometric function of an acute angle when one of the functions is known. In this section, we discuss the problem of finding the value of each trigonometric function of an acute angle when the angle is given.

For three special acute angles, we can use some results from plane geometry to find the exact value of each of the six trigonometric functions.

1 Find the Exact Values of the Trigonometric Functions of $\frac{\pi}{4} = 45°$

EXAMPLE 1 **Finding the Exact Values of the Trigonometric Functions of $\frac{\pi}{4} = 45°$**

Find the exact values of the six trigonometric functions of $\frac{\pi}{4} = 45°$.

Solution Using the right triangle in Figure 27(a), in which one of the angles is $\frac{\pi}{4} = 45°$, it follows that the other acute angle is also $\frac{\pi}{4} = 45°$, and hence the triangle is isosceles. As a result, side a and side b are equal in length. Since the values of the trigonometric functions of an angle depend only on the angle and not on the size of the triangle, we may assign any values to a and b for which $a = b > 0$. We decide to use the triangle for which

$$a = b = 1$$

Then, by the Pythagorean Theorem,

$$c^2 = a^2 + b^2 = 1 + 1 = 2$$
$$c = \sqrt{2}$$

As a result, we have the triangle in Figure 27(b), from which we find

$$\sin \frac{\pi}{4} = \sin 45° = \frac{b}{c} = \frac{1}{\sqrt{2}} = \frac{\sqrt{2}}{2} \qquad \cos \frac{\pi}{4} = \cos 45° = \frac{a}{c} = \frac{1}{\sqrt{2}} = \frac{\sqrt{2}}{2}$$

Figure 27

(a)

(b)

Using Quotient and Reciprocal Identities, we find

$$\tan\frac{\pi}{4} = \tan 45° = \frac{\sin 45°}{\cos 45°} = \frac{\dfrac{\sqrt{2}}{2}}{\dfrac{\sqrt{2}}{2}} = 1 \qquad \cot\frac{\pi}{4} = \cot 45° = \frac{1}{\tan 45°} = \frac{1}{1} = 1$$

$$\sec\frac{\pi}{4} = \sec 45° = \frac{1}{\cos 45°} = \frac{1}{\dfrac{1}{\sqrt{2}}} = \sqrt{2} \qquad \csc\frac{\pi}{4} = \csc 45° = \frac{1}{\sin 45°} = \frac{1}{\dfrac{1}{\sqrt{2}}} = \sqrt{2}$$

EXAMPLE 2 **Finding the Exact Value of a Trigonometric Expression**

Find the exact value of each expression.

(a) $(\sin 45°)(\tan 45°)$

(b) $\left(\sec\dfrac{\pi}{4}\right)\left(\cot\dfrac{\pi}{4}\right)$

Solution We use the results obtained in Example 1.

(a) $(\sin 45°)(\tan 45°) = \dfrac{\sqrt{2}}{2} \cdot 1 = \dfrac{\sqrt{2}}{2}$

(b) $\left(\sec\dfrac{\pi}{4}\right)\left(\cot\dfrac{\pi}{4}\right) = \sqrt{2} \cdot 1 = \sqrt{2}$

Now Work PROBLEMS 5 AND 17

2 Find the Exact Values of the Trigonometric Functions of $\dfrac{\pi}{6} = 30°$ and $\dfrac{\pi}{3} = 60°$

EXAMPLE 3 **Finding the Exact Values of the Trigonometric Functions of $\dfrac{\pi}{6} = 30°$ and $\dfrac{\pi}{3} = 60°$**

Find the exact values of the six trigonometric functions of $\dfrac{\pi}{6} = 30°$ and $\dfrac{\pi}{3} = 60°$.

Solution Form a right triangle in which one of the angles is $\dfrac{\pi}{6} = 30°$. It then follows that the third angle is $\dfrac{\pi}{3} = 60°$. Figure 28(a) illustrates such a triangle with hypotenuse of length 2. Our problem is to determine a and b.

We begin by placing next to the triangle in Figure 28(a) another triangle congruent to the first, as shown in Figure 28(b). Notice that we now have a triangle whose angles are each 60°. This triangle is therefore equilateral, so each side is of length 2. In particular, the base is $2a = 2$, so $a = 1$. By the Pythagorean Theorem, b satisfies the equation $a^2 + b^2 = c^2$, so we have

$$a^2 + b^2 = c^2$$
$$1^2 + b^2 = 2^2 \qquad a = 1, c = 2$$
$$b^2 = 4 - 1 = 3$$
$$b = \sqrt{3}$$

Figure 28

(a)

(b)

(c)

Using the triangle in Figure 28(c) and the fact that $\frac{\pi}{6} = 30°$ and $\frac{\pi}{3} = 60°$ are complementary angles, we find

$$\sin \frac{\pi}{6} = \sin 30° = \frac{\text{opposite}}{\text{hypotenuse}} = \frac{1}{2} \qquad\qquad \cos \frac{\pi}{3} = \cos 60° = \frac{1}{2}$$

$$\cos \frac{\pi}{6} = \cos 30° = \frac{\text{adjacent}}{\text{hypotenuse}} = \frac{\sqrt{3}}{2} \qquad\qquad \sin \frac{\pi}{3} = \sin 60° = \frac{\sqrt{3}}{2}$$

$$\tan \frac{\pi}{6} = \tan 30° = \frac{\sin 30°}{\cos 30°} = \frac{\frac{1}{2}}{\frac{\sqrt{3}}{2}} = \frac{1}{\sqrt{3}} = \frac{\sqrt{3}}{3} \qquad \cot \frac{\pi}{3} = \cot 60° = \frac{\sqrt{3}}{3}$$

$$\csc \frac{\pi}{6} = \csc 30° = \frac{1}{\sin 30°} = \frac{1}{\frac{1}{2}} = 2 \qquad\qquad \sec \frac{\pi}{3} = \sec 60° = 2$$

$$\sec \frac{\pi}{6} = \sec 30° = \frac{1}{\cos 30°} = \frac{1}{\frac{\sqrt{3}}{2}} = \frac{2}{\sqrt{3}} = \frac{2\sqrt{3}}{3} \qquad \csc \frac{\pi}{3} = \csc 60° = \frac{2\sqrt{3}}{3}$$

$$\cot \frac{\pi}{6} = \cot 30° = \frac{1}{\tan 30°} = \frac{1}{\frac{\sqrt{3}}{3}} = \frac{3}{\sqrt{3}} = \sqrt{3} \qquad \tan \frac{\pi}{3} = \tan 60° = \sqrt{3}$$

Table 3 summarizes the information just derived for the angles $\frac{\pi}{6} = 30°$, $\frac{\pi}{4} = 45°$, and $\frac{\pi}{3} = 60°$. Rather than memorize the entries in Table 3, you can draw the appropriate triangle to determine the values given in the table.

Table 3

θ (Radians)	θ (Degrees)	$\sin \theta$	$\cos \theta$	$\tan \theta$	$\csc \theta$	$\sec \theta$	$\cot \theta$
$\frac{\pi}{6}$	30°	$\frac{1}{2}$	$\frac{\sqrt{3}}{2}$	$\frac{\sqrt{3}}{3}$	2	$\frac{2\sqrt{3}}{3}$	$\sqrt{3}$
$\frac{\pi}{4}$	45°	$\frac{\sqrt{2}}{2}$	$\frac{\sqrt{2}}{2}$	1	$\sqrt{2}$	$\sqrt{2}$	1
$\frac{\pi}{3}$	60°	$\frac{\sqrt{3}}{2}$	$\frac{1}{2}$	$\sqrt{3}$	$\frac{2\sqrt{3}}{3}$	2	$\frac{\sqrt{3}}{3}$

EXAMPLE 4 **Finding the Exact Value of a Trigonometric Expression**

Find the exact value of each expression.

(a) $\sin 45° \cos 30°$ (b) $\tan \frac{\pi}{4} - \sin \frac{\pi}{3}$ (c) $\tan^2 \frac{\pi}{6} + \sin^2 \frac{\pi}{4}$

Solution (a) $\sin 45° \cos 30° = \frac{\sqrt{2}}{2} \cdot \frac{\sqrt{3}}{2} = \frac{\sqrt{6}}{4}$

(b) $\tan \frac{\pi}{4} - \sin \frac{\pi}{3} = 1 - \frac{\sqrt{3}}{2} = \frac{2 - \sqrt{3}}{2}$

(c) $\tan^2 \frac{\pi}{6} + \sin^2 \frac{\pi}{4} = \left(\frac{\sqrt{3}}{3}\right)^2 + \left(\frac{\sqrt{2}}{2}\right)^2 = \frac{1}{3} + \frac{1}{2} = \frac{5}{6}$

━━ **Now Work** PROBLEMS 9 AND 19

The exact values of the trigonometric functions for the angles $\frac{\pi}{6} = 30°$, $\frac{\pi}{4} = 45°$, and $\frac{\pi}{3} = 60°$ are relatively easy to calculate, because the triangles that contain such angles have "nice" geometric features. For most other angles, we can only approximate the value of each trigonometric function. To do this, we will need a calculator.

3 Use a Calculator to Approximate the Values of the Trigonometric Functions of Acute Angles

Before getting started, you must first decide whether to enter the angle in the calculator using radians or degrees and then set the calculator to the correct MODE. (Check your instruction manual to find out how your calculator handles degrees and radians.) Your calculator has the keys marked $\boxed{\text{sin}}$, $\boxed{\text{cos}}$, and $\boxed{\text{tan}}$. To find the values of the remaining three trigonometric functions (secant, cosecant, and cotangent), we use the reciprocal identities.

$$\sec \theta = \frac{1}{\cos \theta} \qquad \csc \theta = \frac{1}{\sin \theta} \qquad \cot \theta = \frac{1}{\tan \theta}$$

| EXAMPLE 5 | **Using a Calculator to Approximate the Value of Trigonometric Functions** |

Use a calculator to find the approximate value of:

(a) $\cos 48°$ (b) $\csc 21°$ (c) $\tan \dfrac{\pi}{12}$

Express your answer rounded to two decimal places.

Solution (a) First, we set the MODE to receive degrees. Rounded to two decimal places,

$$\cos 48° = 0.67$$

(b) Most calculators do not have a csc key. The manufacturers assume the user knows some trigonometry. To find the value of csc 21°, we use the fact that $\csc 21° = \dfrac{1}{\sin 21°}$. Rounded to two decimal places, csc 21° = 2.79.

Figure 29

tan(π/12)
 .2679491924

(c) Set the MODE to receive radians. Figure 29 shows the solution using a TI-84 Plus graphing calculator. Rounded to two decimal places,

$$\tan \frac{\pi}{12} = 0.27$$

Now Work PROBLEM 29

4 Model and Solve Applied Problems Involving Right Triangles

Right triangles can be used to model many types of situations, such as the optimal design of a rain gutter.*

* In applied problems, it is important that answers be reported with both justifiable accuracy and appropriate significant figures. We shall assume that the problem data are accurate to the number of significant digits, resulting in sides being rounded to two decimal places and angles being rounded to one decimal place.

EXAMPLE 6	**Constructing a Rain Gutter**

Figure 30

A rain gutter is to be constructed of aluminum sheets 12 inches wide. After marking off a length of 4 inches from each edge, this length is bent up at an angle θ. See Figure 30.

(a) Express the area A of the opening as a function of θ.

 [**Hint:** Let b denote the vertical height of the bend.]

(b) Find the area A of the opening for $\theta = 30°$, $\theta = 45°$, $\theta = 60°$, and $\theta = 75°$.

(c) Graph $A = A(\theta)$. Find the angle θ that makes A largest. (This bend will allow the most water to flow through the gutter.)

Solution

(a) Look again at Figure 30. The area A of the opening is the sum of the areas of two congruent right triangles and one rectangle. Look at Figure 31, showing one of the triangles in Figure 30 redrawn. We see that

Figure 31

$$\cos \theta = \frac{a}{4} \quad \text{so} \quad a = 4 \cos \theta \qquad \sin \theta = \frac{b}{4} \quad \text{so} \quad b = 4 \sin \theta$$

The area of the triangle is

$$\text{area of triangle} = \frac{1}{2}(\text{base})(\text{height}) = \frac{1}{2}ab = \frac{1}{2}(4 \cos \theta)(4 \sin \theta) = 8 \sin \theta \cos \theta$$

So the area of the two congruent triangles is $16 \sin \theta \cos \theta$.

The rectangle has length 4 and height b, so its area is

$$\text{area of rectangle} = 4b = 4(4 \sin \theta) = 16 \sin \theta$$

The area A of the opening is

$$A = \text{area of the two triangles} + \text{area of the rectangle}$$
$$A(\theta) = 16 \sin \theta \cos \theta + 16 \sin \theta = 16 \sin \theta(\cos \theta + 1)$$

(b) For $\theta = 30°$: $\quad A(30°) = 16 \sin 30°(\cos 30° + 1)$

$$= 16\left(\frac{1}{2}\right)\left(\frac{\sqrt{3}}{2} + 1\right) = 4\sqrt{3} + 8 \approx 14.9$$

The area of the opening for $\theta = 30°$ is about 14.9 square inches.

For $\theta = 45°$: $\quad A(45°) = 16 \sin 45°(\cos 45° + 1)$

$$= 16\left(\frac{\sqrt{2}}{2}\right)\left(\frac{\sqrt{2}}{2} + 1\right) = 8 + 8\sqrt{2} \approx 19.3$$

The area of the opening for $\theta = 45°$ is about 19.3 square inches.

For $\theta = 60°$: $\quad A(60°) = 16 \sin 60°(\cos 60° + 1)$

$$= 16\left(\frac{\sqrt{3}}{2}\right)\left(\frac{1}{2} + 1\right) = 12\sqrt{3} \approx 20.8$$

Figure 32

The area of the opening for $\theta = 60°$ is about 20.8 square inches.

For $\theta = 75°$: $\quad A(75°) = 16 \sin 75°(\cos 75° + 1) \approx 19.5$

The area of the opening for $\theta = 75°$ is about 19.5 square inches.

(c) Figure 32 shows the graph of $A = A(\theta)$. Using MAXIMUM, the angle θ that makes A largest is 60°.

━━━━➤ **Now Work** PROBLEM 51

In addition to developing models using right triangles, we can use right triangle trigonometry to measure heights and distances that are either awkward or impossible to measure by ordinary means. When using right triangles to solve these problems, pay attention to the known measures. This will indicate what trigonometric function to use. For example, if we know the measure of an angle and the length of the side adjacent to the angle, and wish to find the length of the opposite side, we would use the tangent function. Do you know why?

| EXAMPLE 7 | Finding the Width of a River |

Figure 33

A surveyor can measure the width of a river by setting up a transit* at a point C on one side of the river and taking a sighting of a point A on the other side. Refer to Figure 33. After turning through an angle of 90° at C, the surveyor walks a distance of 200 meters to point B. Using the transit at B, the angle θ is measured and found to be 20°. What is the width of the river rounded to the nearest meter?

Solution We seek the length of side b. We know a and θ. So we use the fact that b is opposite θ and a is adjacent to θ and write

$$\tan \theta = \frac{b}{a}$$

which leads to

$$\tan 20° = \frac{b}{200}$$

$$b = 200 \tan 20° \approx 72.79 \text{ meters}$$

The width of the river is 73 meters, rounded to the nearest meter.

━━━━ **Now Work** PROBLEM **59**

Vertical heights can sometimes be measured using either the *angle of elevation* or the *angle of depression*. If a person is looking up at an object, the acute angle measured from the horizontal to a line of sight to the object is called the **angle of elevation.** See Figure 34(a).

Figure 34

(a)

(b)

* An instrument used in surveying to measure angles.

If a person is standing on a cliff looking down at an object, the acute angle made by the line of sight to the object and the horizontal is called the **angle of depression**. See Figure 34(b).

| EXAMPLE 8 | Finding the Height of a Cloud |

Meteorologists find the height of a cloud using an instrument called a **ceilometer**. A ceilometer consists of a **light projector** that directs a vertical light beam up to the cloud base and a **light detector** that scans the cloud to detect the light beam. See Figure 35(a). On December 1, 2006, at Midway Airport in Chicago, a ceilometer was employed to find the height of the cloud cover. It was set up with its light detector 300 feet from its light projector. If the angle of elevation from the light detector to the base of the cloud is 75°, what is the height of the cloud cover?

Figure 35

(a) (b)

Solution Figure 35(b) illustrates the situation. To find the height h, we use the fact that $\tan 75° = \dfrac{h}{300}$, so

$$h = 300 \tan 75° \approx 1120 \text{ feet}$$

The ceiling (height to the base of the cloud cover) is approximately 1120 feet.

Now Work PROBLEM 61

The idea behind Example 8 can also be used to find the height of an object with a base that is not accessible to the horizontal.

| EXAMPLE 9 | Finding the Height of a Statue on a Building |

Adorning the top of the Board of Trade building in Chicago is a statue of Ceres, the Roman goddess of wheat. From street level, two observations are taken 400 feet from the center of the building. The angle of elevation to the base of the statue is found to be 55.1° and the angle of elevation to the top of the statue is 56.5°. See Figure 36(a). What is the height of the statue?

Figure 36

(a) (b)

Solution Figure 36(b) shows two triangles that replicate Figure 36(a). The height of the statue of Ceres will be $b' - b$. To find b and b', we refer to Figure 36(b).

$$\tan 55.1° = \frac{b}{400} \qquad\qquad \tan 56.5° = \frac{b'}{400}$$

$$b = 400 \tan 55.1° \approx 573.39 \qquad b' = 400 \tan 56.5° \approx 604.33$$

The height of the statue is approximately $604.33 - 573.39 = 30.94$ feet ≈ 31 feet. ∎

Now Work PROBLEM 67

7.3 Assess Your Understanding

Concepts and Vocabulary

1. $\tan \dfrac{\pi}{4} + \sin 30° =$ _____.

2. Using a calculator, $\sin 2 =$ _____, rounded to two decimal places.

3. **True or False** Exact values can be found for the trigonometric functions of 60°.

4. **True or False** Exact values can be found for the sine of any angle.

Skill Building

5. Write down the exact value of each of the six trigonometric functions of 45°.

6. Write down the exact value of each of the six trigonometric functions of 30° and of 60°.

In Problems 7–16, $f(\theta) = \sin \theta$ and $g(\theta) = \cos \theta$. Find the exact value of each expression if $\theta = 60°$. Do not use a calculator.

7. $f(\theta)$

8. $g(\theta)$

9. $f\left(\dfrac{\theta}{2}\right)$

10. $g\left(\dfrac{\theta}{2}\right)$

11. $[f(\theta)]^2$

12. $[g(\theta)]^2$

13. $2f(\theta)$

14. $2g(\theta)$

15. $\dfrac{f(\theta)}{2}$

16. $\dfrac{g(\theta)}{2}$

In Problems 17–28, find the exact value of each expression. Do not use a calculator.

17. $4 \cos 45° - 2 \sin 45°$

18. $2 \sin 45° + 4 \cos 30°$

19. $6 \tan 45° - 8 \cos 60°$

20. $\sin 30° \cdot \tan 60°$

21. $\sec \dfrac{\pi}{4} + 2 \csc \dfrac{\pi}{3}$

22. $\tan \dfrac{\pi}{4} + \cot \dfrac{\pi}{4}$

23. $\sec^2 \dfrac{\pi}{6} - 4$

24. $4 + \tan^2 \dfrac{\pi}{3}$

25. $\sin^2 30° + \cos^2 60°$

26. $\sec^2 60° - \tan^2 45°$

27. $1 - \cos^2 30° - \cos^2 60°$

28. $1 + \tan^2 30° - \csc^2 45°$

In Problems 29–46, use a calculator to find the approximate value of each expression. Round the answer to two decimal places.

29. $\sin 28°$

30. $\cos 14°$

31. $\tan 21°$

32. $\cot 70°$

33. $\sec 41°$

34. $\csc 55°$

35. $\sin \dfrac{\pi}{10}$

36. $\cos \dfrac{\pi}{8}$

37. $\tan \dfrac{5\pi}{12}$

38. $\cot \dfrac{\pi}{18}$

39. $\sec \dfrac{\pi}{12}$

40. $\csc \dfrac{5\pi}{13}$

41. $\sin 1$

42. $\tan 1$

43. $\sin 1°$

44. $\tan 1°$

45. $\tan 0.3$

46. $\tan 0.1$

Applications and Extensions

Problems 47–51 require the following discussion.

Projectile Motion The path of a projectile fired at an inclination θ to the horizontal with initial speed v_0 is a parabola. See the figure. The **range R** of the projectile, that is, the horizontal distance that the projectile travels, is found by using the function

$$R(\theta) = \frac{2v_0^2 \sin \theta \cos \theta}{g}$$

where $g \approx 32.2$ feet per second per second ≈ 9.8 meters per second per second is the acceleration due to gravity. The maximum height H of the projectile is given by the function

$$H(\theta) = \frac{v_0^2 \sin^2 \theta}{2g}$$

v_0 = Initial speed

Height, H

Range, R

In Problems 47–50, find the range R and maximum height H of the projectile. Round answers to two decimal places.

47. The projectile is fired at an angle of 45° to the horizontal with an initial speed of 100 feet per second.

48. The projectile is fired at an angle of 30° to the horizontal with an initial speed of 150 meters per second.

49. The projectile is fired at an angle of 25° to the horizontal with an initial speed of 500 meters per second.

50. The projectile is fired at an angle of 50° to the horizontal with an initial speed of 200 feet per second.

51. Inclined Plane See the illustration. If friction is ignored, the time *t* (in seconds) required for a block to slide down an inclined plane is given by the function

$$t(\theta) = \sqrt{\frac{2a}{g \sin\theta \cos\theta}}$$

where *a* is the length (in feet) of the base and $g \approx 32$ feet per second per second is the acceleration due to gravity. How long does it take a block to slide down an inclined plane with base $a = 10$ feet when

(a) $\theta = 30°$? (b) $\theta = 45°$? (c) $\theta = 60°$?

52. Piston Engines See the illustration. In a certain piston engine, the distance *x* (in inches) from the center of the drive shaft to the head of the piston is given by the function

$$x(\theta) = \cos\theta + \sqrt{16 + 0.5(2\cos^2\theta - 1)}$$

where θ is the angle between the crank and the path of the piston head. Find *x* when $\theta = 30°$ and when $\theta = 45°$.

53. Calculating the Time of a Trip Two oceanfront homes are located 8 miles apart on a straight stretch of beach, each a distance of 1 mile from a paved road that parallels the ocean. Sally can jog 8 miles per hour along the paved road, but only 3 miles per hour in the sand on the beach. Because a river flows between the two houses, it is necessary to jog on the sand to the road, continue on the road, and then jog on the sand to get from one house to the other. See the illustration.

(a) Express the time *T* to get from one house to the other as a function of the angle θ shown in the illustration.

(b) Calculate the time *T* for $\theta = 30°$. How long is Sally on the paved road?

(c) Calculate the time *T* for $\theta = 45°$. How long is Sally on the paved road?

(d) Calculate the time *T* for $\theta = 60°$. How long is Sally on the paved road?

(e) Calculate the time *T* for $\theta = 90°$. Describe the path taken.

(f) Calculate the time *T* for $\tan\theta = \dfrac{1}{4}$. Describe the path taken. Explain why θ must be larger than 14°.

(g) Graph $T = T(\theta)$. What angle θ results in the least time? What is the least time? How long is Sally on the paved road?

54. Designing Fine Decorative Pieces A designer of decorative art plans to market solid gold spheres encased in clear crystal cones. Each sphere is of fixed radius *R* and will be enclosed in a cone of height *h* and radius *r*. See the illustration. Many cones can be used to enclose the sphere, each having a different slant angle θ.

(a) Express the volume *V* of the cone as a function of the slant angle θ of the cone.

[**Hint:** The volume *V* of a cone of height *h* and radius *r* is $V = \dfrac{1}{3}\pi r^2 h$.]

(b) What volume *V* is required to enclose a sphere of radius 2 centimeters in a cone whose slant angle θ is 30°? 45°? 60°?

(c) What slant angle θ should be used for the volume *V* of the cone to be a minimum? (This choice minimizes the amount of crystal required and gives maximum emphasis to the gold sphere.)

55. Geometry A right triangle has a hypotenuse of length 8 inches. If one angle is 35°, find the length of each leg.

56. Geometry A right triangle has a hypotenuse of length 10 centimeters. If one angle is 40°, find the length of each leg.

57. Geometry A right triangle contains a 25° angle.
 (a) If one leg is of length 5 inches, what is the length of the hypotenuse?
 (b) There are two answers. How is this possible?

58. Geometry A right triangle contains an angle of $\dfrac{\pi}{8}$ radian.
 (a) If one leg is of length 3 meters, what is the length of the hypotenuse?
 (b) There are two answers. How is this possible?

59. Finding the Width of a Gorge Find the distance from A to C across the gorge illustrated in the figure.

60. Finding the Distance across a Pond Find the distance from A to C across the pond illustrated in the figure.

61. The Eiffel Tower The tallest tower built before the era of television masts, the Eiffel Tower was completed on March 31, 1889. Find the height of the Eiffel Tower (before a television mast was added to the top) using the information given in the illustration.

62. Finding the Distance of a Ship from Shore A person in a small boat, offshore from a vertical cliff known to be 100 feet in height, takes a sighting of the top of the cliff. If the angle of elevation is found to be 25°, how far offshore is the ship?

63. Finding the Distance to a Plateau Suppose that you are headed toward a plateau 50 meters high. If the angle of elevation to the top of the plateau is 20°, how far are you from the base of the plateau?

64. Finding the Reach of a Ladder A 22-foot extension ladder leaning against a building makes a 70° angle with the ground. How far up the building does the ladder touch?

65. Finding the Distance between Two Objects A blimp, suspended in the air at a height of 500 feet, lies directly over a line from Soldier Field to the Adler Planetarium on Lake Michigan (see the figure). If the angle of depression from the blimp to the stadium is 32° and from the blimp to the planetarium is 23°, find the distance between Soldier Field and the Adler Planetarium.

66. Hot-air Balloon While taking a ride in a hot-air balloon in Napa Valley, Francisco wonders how high he is. To find out, he chooses a landmark that is to the east of the balloon and measures the angle of depression to be 54°. A few minutes later, after traveling 100 feet east, the angle of depression to the same landmark is determined to be 61°. Use this information to determine the height of the balloon.

67. Mt. Rushmore To measure the height of Lincoln's caricature on Mt. Rushmore, two sightings 800 feet from the base of the mountain are taken. If the angle of elevation to the bottom of Lincoln's face is 32° and the angle of elevation to the top is 35°, what is the height of Lincoln's face?

68. The CN Tower The CN Tower, located in Toronto, Canada, is the tallest structure in the world. While visiting Toronto, a tourist wondered what the height of the tower above the top of the Sky Pod is. While standing 4000 feet from the tower, she measured the angle to the top of the Sky Pod to be 20.1°. At this same distance, the angle of elevation to the top of the tower was found to be 24.4°. Use this information to determine the height of the tower above the Sky Pod.

69. Finding the Length of a Guy Wire A radio transmission tower is 200 feet high. How long should a guy wire be if it is to be attached to the tower 10 feet from the top and is to make an angle of 69° with the ground?

70. Finding the Height of a Tower A guy wire 80 feet long is attached to the top of a radio transmission tower, making an angle of 65° with the ground. How high is the tower?

71. Washington Monument The angle of elevation of the Sun is 35.1° at the instant the shadow cast by the Washington

Monument is 789 feet long. Use this information to calculate the height of the monument.

72. Finding the Length of a Mountain Trail A straight trail with an inclination of 17° leads from a hotel at an elevation of 9000 feet to a mountain lake at an elevation of 11,200 feet. What is the length of the trail?

73. Constructing a Highway A highway whose primary directions are north–south is being constructed along the west coast of Florida. Near Naples, a bay obstructs the straight path of the road. Since the cost of a bridge is prohibitive, engineers decide to go around the bay. The illustration shows the path that they decide on and the measurements taken. What is the length of highway needed to go around the bay?

74. Photography A camera is mounted on a tripod 4 feet high at a distance of 10 feet from George, who is 6 feet tall. See the illustration. If the camera lens has angles of depression and elevation of 20°, will George's feet and head be seen by the lens? If not, how far back will the camera need to be moved to include George's feet and head?

75. Calculating Pool Shots A Pool player located at **X** wants to shoot the white ball off the top cushion and hit the red ball dead center. He knows from physics that the white ball will come off a cushion at the same angle as it hits a cushion. Where on the top cushion should he hit the white ball?

76. The Freedom Tower The Freedom Tower is to be the centerpiece of the rebuilding of the World Trade Center in New York City. The tower will be 1776 feet tall (not including a broadcast antenna). The angle of elevation from the base of an office building to the top of the tower is 34°. The angle of elevation from the helipad on the roof of the office building to the top of the tower is 20°.

(a) How far away is the office building from the Freedom Tower? Assume the side of the tower is vertical. Round to the nearest foot.

(b) How tall is the office building? Round to the nearest foot.

77. Use a calculator set in radian mode to complete the following table. What can you conclude about the value of $f(\theta) = \dfrac{\sin \theta}{\theta}$ as θ approaches 0?

θ	0.5	0.4	0.2	0.1	0.01	0.001	0.0001	0.00001
$f(\theta) = \dfrac{\sin \theta}{\theta}$								

78. Use a calculator set in radian mode to complete the following table. What can you conclude about the value of $g(\theta) = \dfrac{\cos \theta - 1}{\theta}$ as θ approaches 0?

θ	0.5	0.4	0.2	0.1	0.01	0.001	0.0001	0.00001
$g(\theta) = \dfrac{\cos \theta - 1}{\theta}$								

79. Find the exact value of $\tan 1° \cdot \tan 2° \cdot \tan 3° \cdot \ldots \cdot \tan 89°$.

80. Find the exact value of $\cot 1° \cdot \cot 2° \cdot \cot 3° \cdot \ldots \cdot \cot 89°$.

81. Find the exact value of $\cos 1° \cdot \cos 2° \cdot \ldots \cdot \cos 45° \cdot \csc 46° \cdot \ldots \cdot \csc 89°$.

82. Find the exact value of $\sin 1° \cdot \sin 2° \cdot \ldots \cdot \sin 45° \cdot \sec 46° \cdot \ldots \cdot \sec 89°$.

Discussion and Writing

83. Write a brief paragraph that explains how to quickly compute the trigonometric functions of 30°, 45°, and 60°.

84. Explain how you would measure the width of the Grand Canyon from a point on its ridge.

85. Explain how you would measure the height of a TV tower that is on the roof of a tall building.

7.4 Trigonometric Functions of General Angles

OBJECTIVES
1. Find the Exact Values of the Trigonometric Functions for General Angles (p. 540)
2. Use Coterminal Angles to Find the Exact Value of a Trigonometric Function (p. 542)
3. Determine the Signs of the Trigonometric Functions of an Angle in a Given Quadrant (p. 544)
4. Find the Reference Angle of a General Angle (p. 545)
5. Use a Reference Angle to Find the Exact Value of a Trigonometric Function (p. 546)
6. Find the Exact Values of Trigonometric Functions of an Angle, Given Information about the Functions (p. 547)

Figure 37

1 Find the Exact Values of the Trigonometric Functions for General Angles

To extend the definition of the trigonometric functions to include angles that are not acute, we employ a rectangular coordinate system and place the angle in the standard position so that its vertex is at the origin and its initial side is along the positive x-axis. See Figure 37.

DEFINITION

Let θ be any angle in standard position, and let (a, b) denote the coordinates of any point, except the origin $(0, 0)$, on the terminal side of θ. If $r = \sqrt{a^2 + b^2}$ denotes the distance from $(0, 0)$ to (a, b), then the **six trigonometric functions of θ** are defined as the ratios

$$\sin \theta = \frac{b}{r} \qquad \cos \theta = \frac{a}{r} \qquad \tan \theta = \frac{b}{a}$$

$$\csc \theta = \frac{r}{b} \qquad \sec \theta = \frac{r}{a} \qquad \cot \theta = \frac{a}{b}$$

provided no denominator equals 0. If a denominator equals 0, that trigonometric function of the angle θ is not defined.

Notice in the preceding definitions that if $a = 0$, that is, if the point (a, b) is on the y-axis, then the tangent function and the secant function are undefined. Also, if $b = 0$, that is, if the point (a, b) is on the x-axis, then the cosecant function and the cotangent function are undefined.

By constructing similar triangles, you should be convinced that the values of the six trigonometric functions of an angle θ do not depend on the selection of the point (a, b) on the terminal side of θ, but rather depend only on the angle θ itself. See Figure 38 for an illustration of this when θ lies in quadrant II. Since the triangles are similar, the ratio $\dfrac{b}{r}$ equals the ratio $\dfrac{b'}{r'}$, which equals $\sin \theta$. Also, the ratio $\dfrac{|a|}{r}$ equals the ratio $\dfrac{|a'|}{r'}$, so $\dfrac{a}{r} = \dfrac{a'}{r'}$, which equals $\cos \theta$. And so on.

Figure 38

Also, observe that if θ is acute these definitions reduce to the right triangle definitions given in Section 7.2, as illustrated in Figure 39.

Figure 39

Finally, from the definition of the six trigonometric functions of a general angle, we see that the Quotient and Reciprocal Identities hold. Also, using $r^2 = a^2 + b^2$ and dividing each side by r^2, we can derive the Pythagorean Identities for general angles.

EXAMPLE 1 | **Finding the Exact Values of the Six Trigonometric Functions of θ, Given a Point on the Terminal Side**

Find the exact value of each of the six trigonometric functions of a positive angle θ if $(4, -3)$ is a point on its terminal side.

Solution

Figure 40

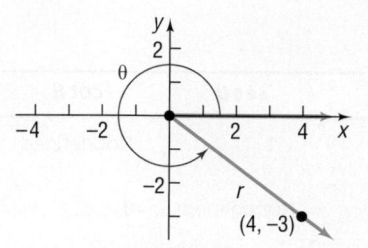

Figure 40 illustrates the situation. For the point $(a, b) = (4, -3)$, we have $a = 4$ and $b = -3$. Then $r = \sqrt{a^2 + b^2} = \sqrt{16 + 9} = 5$, so that

$$\sin \theta = \frac{b}{r} = -\frac{3}{5} \qquad \cos \theta = \frac{a}{r} = \frac{4}{5} \qquad \tan \theta = \frac{b}{a} = -\frac{3}{4}$$

$$\csc \theta = \frac{r}{b} = -\frac{5}{3} \qquad \sec \theta = \frac{r}{a} = \frac{5}{4} \qquad \cot \theta = \frac{a}{b} = -\frac{4}{3}$$

■

Now Work PROBLEM 11

In the next example, we find the exact value of each of the six trigonometric functions at the quadrantal angles $0, \dfrac{\pi}{2}, \pi,$ and $\dfrac{3\pi}{2}$.

EXAMPLE 2 | **Finding the Exact Values of the Six Trigonometric Functions of Quadrantal Angles**

Find the exact values of each of the six trigonometric functions of

(a) $\theta = 0 = 0°$ (b) $\theta = \dfrac{\pi}{2} = 90°$ (c) $\theta = \pi = 180°$ (d) $\theta = \dfrac{3\pi}{2} = 270°$

Solution

Figure 41
$\theta = 0 = 0°$

(a) The point $P = (1, 0)$ is on the terminal side of $\theta = 0 = 0°$ and is a distance of 1 unit from the origin. See Figure 41. Then

$$\sin 0 = \sin 0° = \frac{0}{1} = 0 \qquad \cos 0 = \cos 0° = \frac{1}{1} = 1$$

$$\tan 0 = \tan 0° = \frac{0}{1} = 0 \qquad \sec 0 = \sec 0° = \frac{1}{1} = 1$$

Since the y-coordinate of P is 0, $\csc 0$ and $\cot 0$ are not defined.

Figure 42
$\theta = \dfrac{\pi}{2} = 90°$

(b) The point $P = (0, 1)$ is on the terminal side of $\theta = \dfrac{\pi}{2} = 90°$ and is a distance of 1 unit from the origin. See Figure 42. Then

$$\sin \frac{\pi}{2} = \sin 90° = \frac{1}{1} = 1 \qquad \cos \frac{\pi}{2} = \cos 90° = \frac{0}{1} = 0$$

$$\csc \frac{\pi}{2} = \csc 90° = \frac{1}{1} = 1 \qquad \cot \frac{\pi}{2} = \cot 90° = \frac{0}{1} = 0$$

Since the x-coordinate of P is 0, $\tan \dfrac{\pi}{2}$ and $\sec \dfrac{\pi}{2}$ are not defined.

Figure 43
$\theta = \pi = 180°$

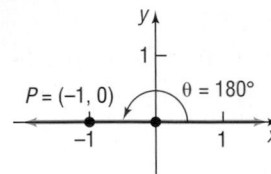

(c) The point $P = (-1, 0)$ is on the terminal side of $\theta = \pi = 180°$ and is a distance of 1 unit from the origin. See Figure 43. Then

$$\sin \pi = \sin 180° = \frac{0}{1} = 0 \qquad \cos \pi = \cos 180° = \frac{-1}{1} = -1$$

$$\tan \pi = \tan 180° = \frac{0}{-1} = 0 \qquad \sec \pi = \sec 180° = \frac{1}{-1} = -1$$

Since the y-coordinate of P is 0, $\csc \pi$ and $\cot \pi$ are not defined.

Figure 44
$\theta = \dfrac{3\pi}{2} = 270°$

(d) The point $P = (0, -1)$ is on the terminal side of $\theta = \dfrac{3\pi}{2} = 270°$ and is a distance of 1 unit from the origin. See Figure 44. Then

$$\sin \frac{3\pi}{2} = \sin 270° = \frac{-1}{1} = -1 \qquad \cos \frac{3\pi}{2} = \cos 270° = \frac{0}{1} = 0$$

$$\csc \frac{3\pi}{2} = \csc 270° = \frac{1}{-1} = -1 \qquad \cot \frac{3\pi}{2} = \cot 270° = \frac{0}{-1} = 0$$

Since the x-coordinate of P is 0, $\tan \dfrac{3\pi}{2}$ and $\sec \dfrac{3\pi}{2}$ are not defined. ∎

Table 4 summarizes the values of the trigonometric functions found in Example 2.

Table 4

θ (Radians)	θ (Degrees)	$\sin \theta$	$\cos \theta$	$\tan \theta$	$\csc \theta$	$\sec \theta$	$\cot \theta$
0	0°	0	1	0	Not defined	1	Not defined
$\dfrac{\pi}{2}$	90°	1	0	Not defined	1	Not defined	0
π	180°	0	−1	0	Not defined	−1	Not defined
$\dfrac{3\pi}{2}$	270°	−1	0	Not defined	−1	Not defined	0

There is no need to memorize Table 4. To find the value of a trigonometric function of a quadrantal angle, draw the angle and apply the definition, as we did in Example 2.

2 Use Coterminal Angles to Find the Exact Value of a Trigonometric Function

DEFINITION

Two angles in standard position are said to be **coterminal** if they have the same terminal side.

See Figure 45.

Figure 45

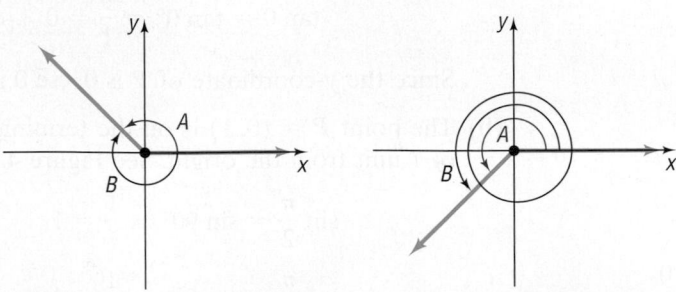

(a) *A* and *B* are coterminal **(b)** *A* and *B* are coterminal

For example, the angles 60° and 420° are coterminal, as are the angles −40° and 320°.

In general, if θ is an angle measured in degrees, then $\theta + 360°k$, where k is any integer, is an angle coterminal with θ. If θ is measured in radians, then $\theta + 2\pi k$, where k is any integer, is an angle coterminal with θ.

Because coterminal angles have the same terminal side, it follows that the values of the trigonometric functions of coterminal angles are equal. We use this fact in the next example.

| **EXAMPLE 3** | **Using a Coterminal Angle to Find the Exact Value of a Trigonometric Function** |

Find the exact value of each of the following:

Figure 46

(a) $\sin 390°$ (b) $\cos 420°$ (c) $\tan \dfrac{9\pi}{4}$ (d) $\sec\left(-\dfrac{7\pi}{4}\right)$ (e) $\csc(-270°)$

Solution (a) It is best to sketch the angle first. See Figure 46. The angle $390°$ is coterminal with $30°$.

$$\sin 390° = \sin(30° + 360°)$$
$$= \sin 30° = \frac{1}{2}$$

Figure 47

(b) See Figure 47. The angle $420°$ is coterminal with $60°$.

$$\cos 420° = \cos(60° + 360°)$$
$$= \cos 60° = \frac{1}{2}$$

(c) See Figure 48. The angle $\dfrac{9\pi}{4}$ is coterminal with $\dfrac{\pi}{4}$.

Figure 48

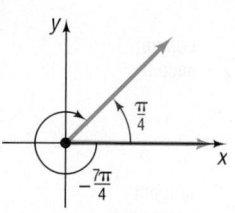

$$\tan \frac{9\pi}{4} = \tan\left(\frac{\pi}{4} + 2\pi\right)$$
$$= \tan \frac{\pi}{4} = 1$$

(d) See Figure 49. The angle $-\dfrac{7\pi}{4}$ is coterminal with $\dfrac{\pi}{4}$.

$$\sec\left(-\frac{7\pi}{4}\right) = \sec\left(\frac{\pi}{4} + 2\pi(-1)\right) = \sec \frac{\pi}{4} = \sqrt{2}$$

Figure 49

(e) See Figure 50. The angle $-270°$ is coterminal with $90°$.

$$\csc(-270°) = \csc(90° + 360°(-1))$$
$$= \csc 90° = 1$$

As Example 3 illustrates, the value of a trigonometric function of any angle is equal to the value of the same trigonometric function of an angle θ coterminal to it, where $0° \leq \theta < 360°$ (or $0 \leq \theta < 2\pi$). Because the angles θ and $\theta + 360°k$ (or $\theta + 2\pi k$), where k is any integer, are coterminal, and because the values of the trigonometric functions are equal for coterminal angles, it follows that

Figure 50

θ **degrees**	θ **radians**
$\sin(\theta + 360°k) = \sin \theta$	$\sin(\theta + 2\pi k) = \sin \theta$
$\cos(\theta + 360°k) = \cos \theta$	$\cos(\theta + 2\pi k) = \cos \theta$
$\tan(\theta + 360°k) = \tan \theta$	$\tan(\theta + 2\pi k) = \tan \theta$
$\csc(\theta + 360°k) = \csc \theta$	$\csc(\theta + 2\pi k) = \csc \theta$
$\sec(\theta + 360°k) = \sec \theta$	$\sec(\theta + 2\pi k) = \sec \theta$
$\cot(\theta + 360°k) = \cot \theta$	$\cot(\theta + 2\pi k) = \cot \theta$ **(1)**

where k is any integer.

These formulas show that the values of the trigonometric functions repeat themselves every 360° (or 2π radians).

Now Work PROBLEM 21

3 Determine the Signs of the Trigonometric Functions of an Angle in a Given Quadrant

If θ is not a quadrantal angle, then it will lie in a particular quadrant. In such a case, the signs of the x-coordinate and y-coordinate of a point (a, b) on the terminal side of θ are known. Because $r = \sqrt{a^2 + b^2} > 0$, it follows that the signs of the trigonometric functions of an angle θ can be found if we know in which quadrant θ lies.

For example, if θ lies in quadrant II, as shown in Figure 51, then a point (a, b) on the terminal side of θ has a negative x-coordinate and a positive y-coordinate. Then,

Figure 51
θ in quadrant II, $a < 0, b > 0, r > 0$

$$\sin \theta = \frac{b}{r} > 0 \qquad \cos \theta = \frac{a}{r} < 0 \qquad \tan \theta = \frac{b}{a} < 0$$

$$\csc \theta = \frac{r}{b} > 0 \qquad \sec \theta = \frac{r}{a} < 0 \qquad \cot \theta = \frac{a}{b} < 0$$

Table 5 lists the signs of the six trigonometric functions for each quadrant. Figure 52 provides two illustrations.

Table 5

Quadrant of θ	$\sin \theta, \csc \theta$	$\cos \theta, \sec \theta$	$\tan \theta, \cot \theta$
I	Positive	Positive	Positive
II	Positive	Negative	Negative
III	Negative	Negative	Positive
IV	Negative	Positive	Negative

Figure 52

(a)　　　　　　(b)

EXAMPLE 4　　**Finding the Quadrant in Which an Angle Lies**

If $\sin \theta < 0$ and $\cos \theta < 0$, name the quadrant in which the angle θ lies.

Solution　　If $\sin \theta < 0$, then θ lies in quadrant III or IV. If $\cos \theta < 0$, then θ lies in quadrant II or III. Therefore, θ lies in quadrant III.

Now Work PROBLEM 33

4 Find the Reference Angle of a General Angle

Once we know in which quadrant an angle lies, we know the sign of each trigono-metric function of this angle. This information, along with the *reference angle* will allow us to evaluate the trigonometric functions of such an angle.

DEFINITION

Let θ denote an angle that lies in a quadrant. The acute angle formed by the terminal side of θ and the *x*-axis is called the **reference angle** for θ.

Figure 53 illustrates the reference angle for some general angles θ. Note that a reference angle is always an acute angle. That is, a reference angle has a measure between 0° and 90°.

Figure 53

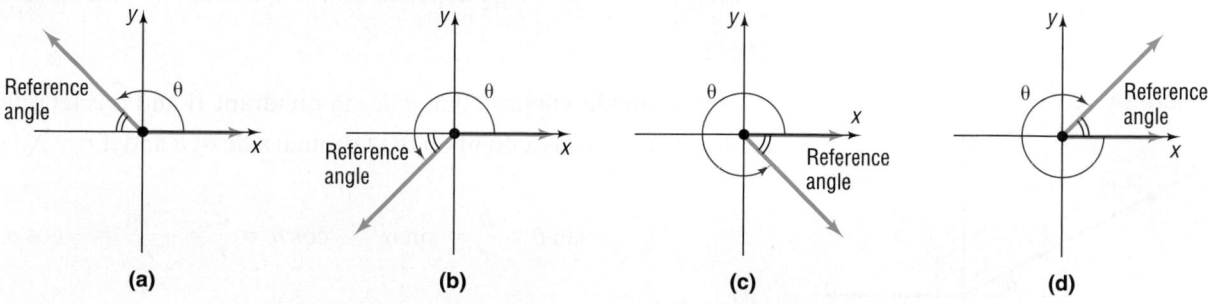

(a) (b) (c) (d)

Although formulas can be given for calculating reference angles, usually it is easier to find the reference angle for a given angle by making a quick sketch of the angle.

EXAMPLE 5

Finding Reference Angles

Find the reference angle for each of the following angles:

(a) 150° (b) −45° (c) $\dfrac{9\pi}{4}$ (d) $-\dfrac{5\pi}{6}$

Solution

(a) Refer to Figure 54. The refer-ence angle for 150° is 30°.

Figure 54

(b) Refer to Figure 55. The refer-ence angle for −45° is 45°.

Figure 55

(c) Refer to Figure 56. The refer-ence angle for $\dfrac{9\pi}{4}$ is $\dfrac{\pi}{4}$.

Figure 56

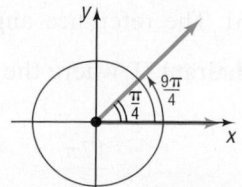

(d) Refer to Figure 57. The refer-ence angle for $-\dfrac{5\pi}{6}$ is $\dfrac{\pi}{6}$.

Figure 57

Now Work PROBLEM 41

5 Use a Reference Angle to Find the Exact Value of a Trigonometric Function

The advantage of using reference angles is that, except for the correct sign, the values of the trigonometric functions of a general angle θ equal the values of the trigonometric functions of its reference angle.

THEOREM

Reference Angles

If θ is an angle that lies in a quadrant and if α is its reference angle, then

$$\sin \theta = \pm\sin \alpha \qquad \cos \theta = \pm\cos \alpha \qquad \tan \theta = \pm\tan \alpha$$
$$\csc \theta = \pm\csc \alpha \qquad \sec \theta = \pm\sec \alpha \qquad \cot \theta = \pm\cot \alpha \qquad \textbf{(2)}$$

where the $+$ or $-$ sign depends on the quadrant in which θ lies.

Figure 58

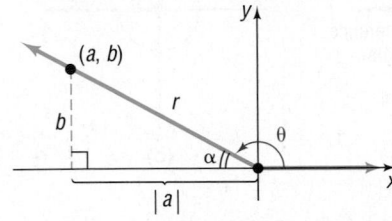

For example, suppose that θ lies in quadrant II and α is its reference angle. See Figure 58. If (a, b) is a point on the terminal side of θ and if $r = \sqrt{a^2 + b^2}$, we have

$$\sin \theta = \frac{b}{r} = \sin \alpha \qquad \cos \theta = \frac{a}{r} = \frac{-|a|}{r} = -\cos \alpha$$

and so on.

The next example illustrates how the theorem on reference angles is used.

EXAMPLE 6

Using the Reference Angle to Find the Exact Value of a Trigonometric Function

Find the exact value of each of the following trigonometric functions using reference angles.

(a) $\sin 135°$ (b) $\cos 600°$ (c) $\cos \dfrac{17\pi}{6}$ (d) $\tan\left(-\dfrac{\pi}{3}\right)$

Figure 59

Solution (a) Refer to Figure 59. The reference angle for $135°$ is $45°$ and $\sin 45° = \dfrac{\sqrt{2}}{2}$. The angle $135°$ is in quadrant II, where the sine function is positive, so

$$\sin 135° = \sin 45° = \frac{\sqrt{2}}{2}$$

Figure 60

(b) Refer to Figure 60. The reference angle for $600°$ is $60°$ and $\cos 60° = \dfrac{1}{2}$. The angle $600°$ is in quadrant III, where the cosine function is negative, so

$$\cos 600° = -\cos 60° = -\frac{1}{2}$$

Figure 61

(c) Refer to Figure 61. The reference angle for $\dfrac{17\pi}{6}$ is $\dfrac{\pi}{6}$ and $\cos \dfrac{\pi}{6} = \dfrac{\sqrt{3}}{2}$. The angle $\dfrac{17\pi}{6}$ is in quadrant II, where the cosine function is negative, so

$$\cos \frac{17\pi}{6} = -\cos \frac{\pi}{6} = -\frac{\sqrt{3}}{2}$$

Figure 62

(d) Refer to Figure 62. The reference angle for $-\dfrac{\pi}{3}$ is $\dfrac{\pi}{3}$ and $\tan\dfrac{\pi}{3} = \sqrt{3}$. The angle $-\dfrac{\pi}{3}$ is in quadrant IV, where the tangent function is negative, so

$$\tan\left(-\frac{\pi}{3}\right) = -\tan\frac{\pi}{3} = -\sqrt{3}$$

Finding the Values of the Trigonometric Functions of a General Angle

If the angle θ is a quadrantal angle, draw the angle, pick a point on its terminal side, and apply the definition of the trigonometric functions.

If the angle θ lies in a quadrant:

1. Find the reference angle α of θ.
2. Find the value of the trigonometric function at α.
3. Adjust the sign ($+$ or $-$) according to the sign of the trigonometric function in the quadrant where θ lies.

═══ **Now Work** PROBLEMS 59 AND 63

6 Find the Exact Values of Trigonometric Functions of an Angle, Given Information about the Functions

EXAMPLE 7 **Finding the Exact Values of Trigonometric Functions**

Given that $\cos\theta = -\dfrac{2}{3}$, $\dfrac{\pi}{2} < \theta < \pi$, find the exact value of each of the remaining trigonometric functions.

Solution The angle θ lies in quadrant II, so we know that $\sin\theta$ and $\csc\theta$ are positive and the other four trigonometric functions are negative. If α is the reference angle for θ, then $\cos\alpha = \dfrac{2}{3} = \dfrac{\text{adjacent}}{\text{hypotenuse}}$. The values of the remaining trigonometric functions of the reference angle α can be found by drawing the appropriate triangle and using the Pythagorean Theorem. We use Figure 63 to obtain

Figure 63

$\cos\alpha = \dfrac{2}{3}$

$$\sin\alpha = \frac{\sqrt{5}}{3} \qquad\qquad \cos\alpha = \frac{2}{3} \qquad \tan\alpha = \frac{\sqrt{5}}{2}$$

$$\csc\alpha = \frac{3}{\sqrt{5}} = \frac{3\sqrt{5}}{5} \qquad \sec\alpha = \frac{3}{2} \qquad \cot\alpha = \frac{2}{\sqrt{5}} = \frac{2\sqrt{5}}{5}$$

Now we assign the appropriate signs to each of these values to find the values of the trigonometric functions of θ.

$$\sin\theta = \frac{\sqrt{5}}{3} \qquad\qquad \cos\theta = -\frac{2}{3} \qquad \tan\theta = -\frac{\sqrt{5}}{2}$$

$$\csc\theta = \frac{3\sqrt{5}}{5} \qquad\qquad \sec\theta = -\frac{3}{2} \qquad \cot\theta = -\frac{2\sqrt{5}}{5}$$

═══ **Now Work** PROBLEM 89

EXAMPLE 8	**Finding the Exact Values of Trigonometric Functions**

If $\tan \theta = -4$ and $\sin \theta < 0$, find the exact value of each of the remaining trigonometric functions of θ.

Solution Since $\tan \theta = -4 < 0$ and $\sin \theta < 0$, it follows that θ lies in quadrant IV. If α is the reference angle for θ, then $\tan \alpha = 4 = \dfrac{4}{1} = \dfrac{b}{a}$. With $a = 1$ and $b = 4$, we find $r = \sqrt{1^2 + 4^2} = \sqrt{17}$. See Figure 64. Then

Figure 64
$\tan \alpha = 4$

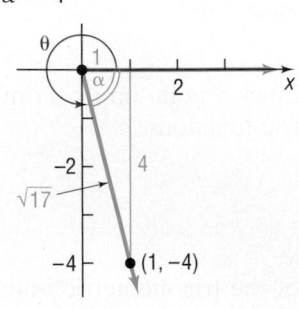

$$\sin \alpha = \frac{4}{\sqrt{17}} = \frac{4\sqrt{17}}{17} \qquad \cos \alpha = \frac{1}{\sqrt{17}} = \frac{\sqrt{17}}{17} \qquad \tan \alpha = \frac{4}{1} = 4$$

$$\csc \alpha = \frac{\sqrt{17}}{4} \qquad\qquad \sec \alpha = \frac{\sqrt{17}}{1} = \sqrt{17} \qquad \cot \alpha = \frac{1}{4}$$

We assign the appropriate sign to each of these to obtain the values of the trigonometric functions of θ.

$$\sin \theta = -\frac{4\sqrt{17}}{17} \qquad\qquad \cos \theta = \frac{\sqrt{17}}{17} \qquad\qquad \tan \theta = -4$$

$$\csc \theta = -\frac{\sqrt{17}}{4} \qquad\qquad \sec \theta = \sqrt{17} \qquad\qquad \cot \theta = -\frac{1}{4}$$

═══ **Now Work** PROBLEM 99

7.4 Assess Your Understanding

Concepts and Vocabulary

1. For an angle θ that lies in quadrant III, the trigonometric functions _____ and _____ are positive.

2. Two angles in standard position that have the same terminal side are _____.

3. The reference angle of $240°$ is _____.

4. *True or False* $\sin 182° = \cos 2°$.

5. *True or False* $\tan \dfrac{\pi}{2}$ is not defined.

6. *True or False* The reference angle is always an acute angle.

7. What is the reference angle of $600°$?

8. In which quadrants is the cosine function positive?

9. If $0 \le \theta < 2\pi$, for what angles θ, if any, is $\tan \theta$ undefined?

10. What is the reference angle of $\dfrac{13\pi}{3}$?

Skill Building

In Problems 11–20, a point on the terminal side of an angle θ is given. Find the exact value of each of the six trigonometric functions of θ.

11. $(-3, 4)$ **12.** $(5, -12)$ **13.** $(2, -3)$ **14.** $(-1, -2)$ **15.** $(-3, -3)$

16. $(2, -2)$ **17.** $\left(\dfrac{\sqrt{3}}{2}, \dfrac{1}{2}\right)$ **18.** $\left(-\dfrac{1}{2}, \dfrac{\sqrt{3}}{2}\right)$ **19.** $\left(\dfrac{\sqrt{2}}{2}, -\dfrac{\sqrt{2}}{2}\right)$ **20.** $\left(-\dfrac{\sqrt{2}}{2}, -\dfrac{\sqrt{2}}{2}\right)$

In Problems 21–32, use a coterminal angle to find the exact value of each expression. Do not use a calculator.

21. $\sin 405°$ **22.** $\cos 420°$ **23.** $\tan 405°$ **24.** $\sin 390°$ **25.** $\csc 450°$ **26.** $\sec 540°$

27. $\cot 390°$ **28.** $\sec 420°$ **29.** $\cos \dfrac{33\pi}{4}$ **30.** $\sin \dfrac{9\pi}{4}$ **31.** $\tan(21\pi)$ **32.** $\csc \dfrac{9\pi}{2}$

In Problems 33–40, name the quadrant in which the angle θ lies.

33. $\sin \theta > 0, \quad \cos \theta < 0$ **34.** $\sin \theta < 0, \quad \cos \theta > 0$ **35.** $\sin \theta < 0, \quad \tan \theta < 0$

36. $\cos \theta > 0, \quad \tan \theta > 0$ **37.** $\cos \theta > 0, \quad \cot \theta < 0$ **38.** $\sin \theta < 0, \quad \cot \theta > 0$

39. $\sec \theta < 0, \quad \tan \theta > 0$ **40.** $\csc \theta > 0, \quad \cot \theta < 0$

In Problems 41–58, find the reference angle of each angle.

41. $-30°$ **42.** $60°$ **43.** $120°$ **44.** $300°$ **45.** $210°$ **46.** $330°$

47. $\dfrac{5\pi}{4}$ **48.** $\dfrac{5\pi}{6}$ **49.** $\dfrac{8\pi}{3}$ **50.** $\dfrac{7\pi}{4}$ **51.** $-135°$ **52.** $-240°$

53. $-\dfrac{2\pi}{3}$ **54.** $-\dfrac{7\pi}{6}$ **55.** $440°$ **56.** $490°$ **57.** $\dfrac{15\pi}{4}$ **58.** $\dfrac{19\pi}{6}$

In Problems 59–88, use the reference angle to find the exact value of each expression. Do not use a calculator.

59. $\sin 150°$ **60.** $\cos 210°$ **61.** $\cos 315°$ **62.** $\sin 120°$ **63.** $\sin 510°$ **64.** $\cos 600°$

65. $\cos(-45°)$ **66.** $\sin(-240°)$ **67.** $\sec 240°$ **68.** $\csc 300°$ **69.** $\cot 330°$ **70.** $\tan 225°$

71. $\sin \dfrac{3\pi}{4}$ **72.** $\cos \dfrac{2\pi}{3}$ **73.** $\cot \dfrac{7\pi}{6}$ **74.** $\csc \dfrac{7\pi}{4}$ **75.** $\cos \dfrac{13\pi}{4}$ **76.** $\tan \dfrac{8\pi}{3}$

77. $\sin\left(-\dfrac{2\pi}{3}\right)$ **78.** $\cot\left(-\dfrac{\pi}{6}\right)$ **79.** $\tan \dfrac{14\pi}{3}$ **80.** $\sec \dfrac{11\pi}{4}$ **81.** $\csc(-315°)$ **82.** $\sec(-225°)$

83. $\sin(8\pi)$ **84.** $\cos(-2\pi)$ **85.** $\tan(7\pi)$ **86.** $\cot(5\pi)$ **87.** $\sec(-3\pi)$ **88.** $\csc\left(-\dfrac{5\pi}{2}\right)$

In Problems 89–106, find the exact value of each of the remaining trigonometric functions of θ.

89. $\sin \theta = \dfrac{12}{13}, \quad \theta$ in Quadrant II **90.** $\cos \theta = \dfrac{3}{5}, \quad \theta$ in Quadrant IV **91.** $\cos \theta = -\dfrac{4}{5}, \quad \theta$ in Quadrant III

92. $\sin \theta = -\dfrac{5}{13}, \quad \theta$ in Quadrant III **93.** $\sin \theta = \dfrac{5}{13}, \quad 90° < \theta < 180°$ **94.** $\cos \theta = \dfrac{4}{5}, \quad 270° < \theta < 360°$

95. $\cos \theta = -\dfrac{1}{3}, \quad 180° < \theta < 270°$ **96.** $\sin \theta = -\dfrac{2}{3}, \quad 180° < \theta < 270°$ **97.** $\sin \theta = \dfrac{2}{3}, \quad \tan \theta < 0$

98. $\cos \theta = -\dfrac{1}{4}, \quad \tan \theta > 0$ **99.** $\sec \theta = 2, \quad \sin \theta < 0$ **100.** $\csc \theta = 3, \quad \cot \theta < 0$

101. $\tan \theta = \dfrac{3}{4}, \quad \sin \theta < 0$ **102.** $\cot \theta = \dfrac{4}{3}, \quad \cos \theta < 0$ **103.** $\tan \theta = -\dfrac{1}{3}, \quad \sin \theta > 0$

104. $\sec \theta = -2, \quad \tan \theta > 0$ **105.** $\csc \theta = -2, \quad \tan \theta > 0$ **106.** $\cot \theta = -2, \quad \sec \theta > 0$

107. Find the exact value of $\sin 40° + \sin 130° + \sin 220° + \sin 310°$.

108. Find the exact value of $\tan 40° + \tan 140°$.

Applications and Extensions

109. If $f(\theta) = \sin \theta = 0.2$, find $f(\theta + \pi)$. **110.** If $g(\theta) = \cos \theta = 0.4$, find $g(\theta + \pi)$. **111.** If $F(\theta) = \tan \theta = 3$, find $F(\theta + \pi)$.

112. If $G(\theta) = \cot \theta = -2$, find $G(\theta + \pi)$. **113.** If $\sin \theta = \dfrac{1}{5}$, find $\csc(\theta + \pi)$. **114.** If $\cos \theta = \dfrac{2}{3}$, find $\sec(\theta + \pi)$.

115. Find the exact value of
$\sin 1° + \sin 2° + \sin 3° + \cdots + \sin 358° + \sin 359°$

116. Find the exact value of
$\cos 1° + \cos 2° + \cos 3° + \cdots + \cos 358° + \cos 359°$

117. Projectile Motion An object is propelled upward at an angle θ, $45° < \theta < 90°$, to the horizontal with an initial velocity of v_0 feet per second from the base of a plane that makes an angle of $45°$ with the horizontal. See the illustration. If air resistance is ignored, the distance R that it travels up the inclined plane is given by the function

$$R(\theta) = \frac{v_0^2 \sqrt{2}}{32}[\sin(2\theta) - \cos(2\theta) - 1]$$

(a) Find the distance R that the object travels along the inclined plane if the initial velocity is 32 feet per second and $\theta = 60°$.

(b) Graph $R = R(\theta)$ if the initial velocity is 32 feet per second.

(c) What value of θ makes R largest?

Discussion and Writing

118. Give three examples that demonstrate how to use the theorem on reference angles.

119. Write a brief paragraph that explains how to quickly compute the value of the trigonometric functions of $0°, 90°, 180°$, and $270°$.

7.5 Unit Circle Approach; Properties of the Trigonometric Functions

PREPARING FOR THIS SECTION *Before getting started, review the following:*

- Unit Circle (Section 2.4, p. 190)
- Functions (Section 3.1, pp. 208–218)
- Even and Odd Functions (Section 3.3, pp. 231–233)

Now Work the 'Are You Prepared?' problems on page 557.

OBJECTIVES 1 Find the Exact Values of the Trigonometric Functions Using the Unit Circle (p. 550)
2 Know the Domain and Range of the Trigonometric Functions (p. 553)
3 Use the Periodic Properties to Find the Exact Values of the Trigonometric Functions (p. 555)
4 Use Even–Odd Properties to Find the Exact Values of the Trigonometric Functions (p. 556)

In this section, we develop important properties of the trigonometric functions. We begin by introducing the trigonometric functions using the unit circle. This approach will lead to the definition given earlier of the trigonometric functions of a general angle.

1 Find the Exact Values of the Trigonometric Functions Using the Unit Circle

Recall that the unit circle is a circle whose radius is 1 and whose center is at the origin of a rectangular coordinate system. Also recall that any circle of radius r has circumference of length $2\pi r$. Therefore, the unit circle (radius = 1) has a circumference of length 2π. In other words, for 1 revolution around the unit circle the length of the arc is 2π units.

The following discussion sets the stage for defining the trigonometric functions using the unit circle.

Let $t \geq 0$ be any real number and let s be the distance from the origin to t on the real number line. See the red portion of Figure 65(a). Now look at the unit circle in Figure 65(a). Beginning at the point $(1, 0)$ on the unit circle, travel $s = t$ units in the counterclockwise direction along the circle to arrive at the point $P = (a, b)$. In this sense, the length $s = t$ units is being **wrapped** around the unit circle.

If $t < 0$, we begin at the point $(1, 0)$ on the unit circle and travel $s = |t|$ units in the clockwise direction to arrive at the point $P = (a, b)$. See Figure 65(b).

If $t > 2\pi$ or if $t < -2\pi$, it will be necessary to travel around the unit circle more than once before arriving at point P. Do you see why?

Figure 65

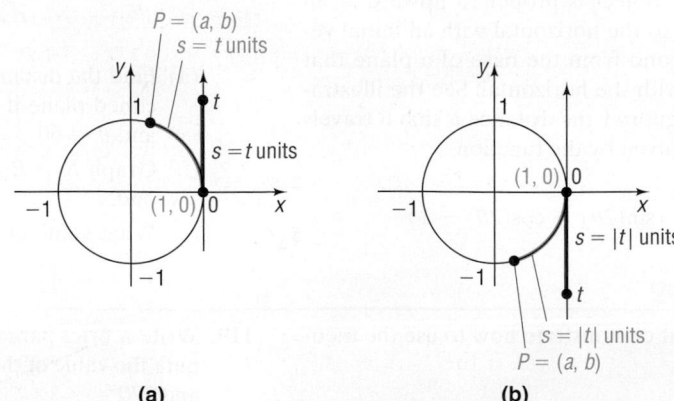

(a) (b)

Let's describe this process another way. Picture a string of length $s = |t|$ units being wrapped around a circle of radius 1 unit. We start wrapping the string around the circle at the point $(1, 0)$. If $t \geq 0$, we wrap the string in the counterclockwise direction; if $t < 0$, we wrap the string in the clockwise direction. The point $P = (a, b)$ is the point where the string ends.

This discussion tells us that, for any real number t, we can locate a unique point $P = (a, b)$ on the unit circle. We call this point **the point P on the unit circle that corresponds to t.** This is the important idea here. No matter what real number t is chosen, there is a unique point P on the unit circle corresponding to it. We use the coordinates of the point $P = (a, b)$ on the unit circle corresponding to the real number t to define the **six trigonometric functions of t.**

DEFINITION

> **In Words**
>
> The sine function takes as input a real number t that corresponds to a point $P = (a, b)$ on the unit circle and outputs the y-coordinate, b. The cosine function takes as input a real number t that corresponds to a point $P = (a, b)$ on the unit circle and outputs the x-coordinate, a.

Let t be a real number and let $P = (a, b)$ be the point on the unit circle that corresponds to t.

The **sine function** associates with t the y-coordinate of P and is denoted by

$$\sin t = b$$

The **cosine function** associates with t the x-coordinate of P and is denoted by

$$\cos t = a$$

If $a \neq 0$, the **tangent function** is defined as

$$\tan t = \frac{b}{a}$$

If $b \neq 0$, the **cosecant function** is defined as

$$\csc t = \frac{1}{b}$$

If $a \neq 0$, the **secant function** is defined as

$$\sec t = \frac{1}{a}$$

If $b \neq 0$, the **cotangent function** is defined as

$$\cot t = \frac{a}{b}$$

Once again, notice in these definitions that if $a = 0$ (that is, if the point P is on the y-axis) the tangent function and the secant function are undefined. Also, if $b = 0$ (that is, if the point P is on the x-axis), the cosecant function and the cotangent function are undefined.

Because we use the unit circle in these definitions of the trigonometric functions, they are also sometimes referred to as **circular functions.**

| **EXAMPLE 1** | **Finding the Values of the Trigonometric Functions Using a Point on the Unit Circle** |

Find the values of $\sin t$, $\cos t$, $\tan t$, $\csc t$, $\sec t$, and $\cot t$ if $P = \left(-\dfrac{1}{2}, \dfrac{\sqrt{3}}{2} \right)$ is the point on the unit circle that corresponds to the real number t.

Figure 66

$P = \left(-\frac{1}{2}, \frac{\sqrt{3}}{2}\right)$

Solution

See Figure 66. We follow the definition of the six trigonometric functions using $P = \left(-\frac{1}{2}, \frac{\sqrt{3}}{2}\right) = (a, b)$. Then, with $a = -\frac{1}{2}$ and $b = \frac{\sqrt{3}}{2}$, we have

$$\sin t = b = \frac{\sqrt{3}}{2} \qquad \cos t = a = -\frac{1}{2} \qquad \tan t = \frac{b}{a} = \frac{\frac{\sqrt{3}}{2}}{-\frac{1}{2}} = -\sqrt{3}$$

$$\csc t = \frac{1}{b} = \frac{1}{\frac{\sqrt{3}}{2}} = \frac{2}{\sqrt{3}} = \frac{2\sqrt{3}}{3} \qquad \sec t = \frac{1}{a} = \frac{1}{-\frac{1}{2}} = -2 \qquad \cot t = \frac{a}{b} = \frac{-\frac{1}{2}}{\frac{\sqrt{3}}{2}} = -\frac{1}{\sqrt{3}} = -\frac{\sqrt{3}}{3}$$

Now Work PROBLEM 9

Trigonometric Functions of Angles

Let $P = (a, b)$ be the point on the unit circle corresponding to the real number t. See Figure 67(a). Let θ be the angle in standard position, measured in radians, whose terminal side is the ray from the origin through P. See Figure 67(b). Since the unit circle has radius 1 unit, if $s = |t|$ units, then from the arc length formula $s = r\theta$, we have $\theta = t$ radians. See Figures 67(c) and (d).

Figure 67

(a)

(b)

(c)

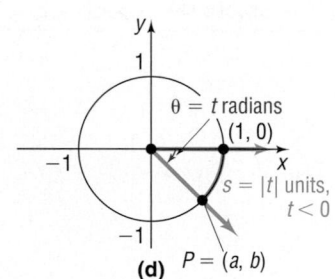

(d)

The point $P = (a, b)$ on the unit circle that corresponds to the real number t is the point P on the terminal side of the angle $\theta = t$ radians. As a result, we can say that

$$\sin t = \sin \theta$$

Real number $\theta = t$ radians

and so on. We can now define the trigonometric functions of the angle θ.

DEFINITION

If $\theta = t$ radians, the six **trigonometric functions of the angle θ** are defined as

$\sin \theta = \sin t$	$\cos \theta = \cos t$	$\tan \theta = \tan t$
$\csc \theta = \csc t$	$\sec \theta = \sec t$	$\cot \theta = \cot t$

Even though the trigonometric functions can be viewed both as functions of real numbers and as functions of angles, it is customary to refer to trigonometric functions of real numbers and trigonometric functions of angles collectively as *the trigonometric functions*. We will follow this practice from now on.

Since the values of the trigonometric functions of an angle θ are determined by the coordinates of the point $P = (a, b)$ on the unit circle corresponding to θ, the units used to measure the angle θ are irrelevant. For example, it does not matter

whether we write $\theta = \dfrac{\pi}{2}$ radians or $\theta = 90°$. In either case, the point on the unit circle corresponding to this angle is $P = (0, 1)$. As a result,

$$\sin \frac{\pi}{2} = \sin 90° = 1 \quad \text{and} \quad \cos \frac{\pi}{2} = \cos 90° = 0$$

To find the exact value of a trigonometric function of an angle θ requires that we locate the corresponding point $P^* = (a^*, b^*)$ on the unit circle. In fact, though, any circle whose center is at the origin can be used.

Let θ be any nonquadrantal angle placed in standard position. Let $P = (a, b)$ be the point on the circle $x^2 + y^2 = r^2$ that corresponds to θ and let $P^* = (a^*, b^*)$ be the point on the unit circle that corresponds to θ. See Figure 68.

Notice that the triangles OA^*P^* and OAP are similar so ratios of corresponding sides are equal.

Figure 68

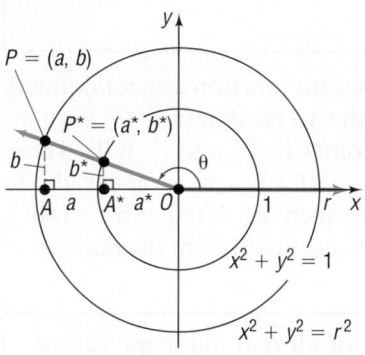

$$\frac{b^*}{1} = \frac{b}{r} \qquad \frac{a^*}{1} = \frac{a}{r} \qquad \frac{b^*}{a^*} = \frac{b}{a}$$

$$\frac{1}{b^*} = \frac{r}{b} \qquad \frac{1}{a^*} = \frac{r}{a} \qquad \frac{a^*}{b^*} = \frac{a}{b}$$

These results lead us to formulate the following theorem:

THEOREM

For an angle θ in standard position, let $P = (a, b)$ be any point on the terminal side of θ that is also on the circle $x^2 + y^2 = r^2$. Then

$$\sin \theta = \frac{b}{r} \qquad \cos \theta = \frac{a}{r} \qquad \tan \theta = \frac{b}{a} \quad a \neq 0$$

$$\csc \theta = \frac{r}{b} \quad b \neq 0 \qquad \sec \theta = \frac{r}{a} \quad a \neq 0 \qquad \cot \theta = \frac{a}{b} \quad b \neq 0$$

This result coincides with the definition given in Section 7.4 for the six trigonometric functions of a general angle θ.

Now Work PROBLEM 15

2 Know the Domain and Range of the Trigonometric Functions

Let θ be an angle in standard position, and let $P = (a, b)$ be the point on the unit circle that corresponds to θ. See Figure 69. Then, by the definition given earlier:

Figure 69

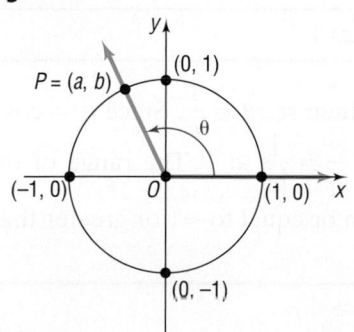

$$\sin \theta = b \qquad \cos \theta = a \qquad \tan \theta = \frac{b}{a} \quad a \neq 0$$

$$\csc \theta = \frac{1}{b} \quad b \neq 0 \qquad \sec \theta = \frac{1}{a} \quad a \neq 0 \qquad \cot \theta = \frac{a}{b} \quad b \neq 0$$

For $\sin \theta$ and $\cos \theta$, θ can be any angle, so it follows that the domain of the sine function and cosine function is the set of all real numbers.

The domain of the sine function is the set of all real numbers.

The domain of the cosine function is the set of all real numbers.

If $a = 0$, then the tangent function and the secant function are not defined. That is, for the tangent function and secant function, the x-coordinate of $P = (a, b)$ cannot be 0. On the unit circle, there are two such points, $(0, 1)$ and $(0, -1)$. These two points correspond to the angles $\dfrac{\pi}{2}(90°)$ and $\dfrac{3\pi}{2}(270°)$ or, more generally, to any angle that is an odd integer multiple of $\dfrac{\pi}{2}(90°)$, such as $\pm\dfrac{\pi}{2}(\pm90°)$, $\pm\dfrac{3\pi}{2}(\pm270°)$,

and $\pm\dfrac{5\pi}{2}(\pm450°)$. Such angles must therefore be excluded from the domain of the tangent function and secant function.

> The domain of the tangent function is the set of all real numbers, except odd integer multiples of $\dfrac{\pi}{2}(90°)$.
>
> The domain of the secant function is the set of all real numbers, except odd integer multiples of $\dfrac{\pi}{2}(90°)$.

If $b = 0$, then the cotangent function and the cosecant function are not defined. For the cotangent function and cosecant function, the y-coordinate of $P = (a, b)$ cannot be 0. On the unit circle, there are two such points, $(1, 0)$ and $(-1, 0)$. These two points correspond to the angles $0 (0°)$ and $\pi (180°)$ or, more generally, to any angle that is an integer multiple of $\pi (180°)$, such as $0(0°)$, $\pm\pi (\pm180°)$, $\pm2\pi (\pm360°)$, and $\pm3\pi (\pm540°)$. Such angles must be excluded from the domain of the cotangent function and cosecant function.

> The domain of the cotangent function is the set of all real numbers, except integer multiples of $\pi (180°)$.
>
> The domain of the cosecant function is the set of all real numbers, except integer multiples of $\pi (180°)$.

Next, we determine the range of each of the six trigonometric functions. Refer again to Figure 69. Let $P = (a, b)$ be the point on the unit circle that corresponds to the angle θ. It follows that $-1 \le a \le 1$ and $-1 \le b \le 1$. Since $\sin \theta = b$ and $\cos \theta = a$, we have

$$-1 \le \sin \theta \le 1 \quad \text{and} \quad -1 \le \cos \theta \le 1$$

The range of both the sine function and the cosine function consists of all real numbers between -1 and 1, inclusive. Using absolute value notation, we have $|\sin \theta| \le 1$ and $|\cos \theta| \le 1$.

If θ is not an integer multiple of $\pi (180°)$, then $\csc \theta = \dfrac{1}{b}$. Since $b = \sin \theta$ and $|b| = |\sin \theta| \le 1$, it follows that $|\csc \theta| = \dfrac{1}{|\sin \theta|} = \dfrac{1}{|b|} \ge 1$. The range of the cosecant function consists of all real numbers less than or equal to -1 or greater than or equal to 1. That is,

$$\csc \theta \le -1 \quad \text{or} \quad \csc \theta \ge 1$$

If θ is not an odd integer multiple of $\dfrac{\pi}{2}(90°)$, then $\sec \theta = \dfrac{1}{a}$. Since $a = \cos \theta$ and $|a| = |\cos \theta| \le 1$, it follows that $|\sec \theta| = \dfrac{1}{|\cos \theta|} = \dfrac{1}{|a|} \ge 1$. The range of the secant function consists of all real numbers less than or equal to -1 or greater than or equal to 1. That is,

$$\sec \theta \le -1 \quad \text{or} \quad \sec \theta \ge 1$$

The range of both the tangent function and the cotangent function consists of all real numbers. That is,

$$-\infty < \tan \theta < \infty \quad \text{and} \quad -\infty < \cot \theta < \infty$$

You are asked to prove this in Problems 93 and 94.
Table 6 summarizes these results.

Table 6

Function	Symbol	Domain	Range
sine	$f(\theta) = \sin\theta$	All real numbers	All real numbers from -1 to 1, inclusive
cosine	$f(\theta) = \cos\theta$	All real numbers	All real numbers from -1 to 1, inclusive
tangent	$f(\theta) = \tan\theta$	All real numbers, except odd integer multiples of $\dfrac{\pi}{2}(90°)$	All real numbers
cosecant	$f(\theta) = \csc\theta$	All real numbers, except integer multiples of π (180°)	All real numbers greater than or equal to 1 or less than or equal to -1
secant	$f(\theta) = \sec\theta$	All real numbers, except odd integer multiples of $\dfrac{\pi}{2}(90°)$	All real numbers greater than or equal to 1 or less than or equal to -1
cotangent	$f(\theta) = \cot\theta$	All real numbers, except integer multiples of π (180°)	All real numbers

—————— **Now Work** PROBLEMS 61 AND 65

3 Use the Periodic Properties to Find the Exact Values of the Trigonometric Functions

Figure 70

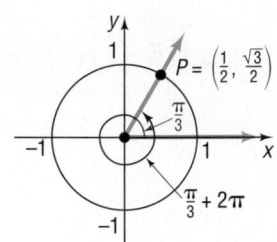

Look at Figure 70. This figure shows that for an angle of $\dfrac{\pi}{3}$ radians the corresponding point P on the unit circle is $\left(\dfrac{1}{2}, \dfrac{\sqrt{3}}{2}\right)$. Notice that for an angle of $\dfrac{\pi}{3} + 2\pi$ radians, the corresponding point P on the unit circle is also $\left(\dfrac{1}{2}, \dfrac{\sqrt{3}}{2}\right)$. As a result,

$$\sin\frac{\pi}{3} = \frac{\sqrt{3}}{2} \quad \text{and} \quad \sin\left(\frac{\pi}{3} + 2\pi\right) = \frac{\sqrt{3}}{2}$$

$$\cos\frac{\pi}{3} = \frac{1}{2} \quad \text{and} \quad \cos\left(\frac{\pi}{3} + 2\pi\right) = \frac{1}{2}$$

Figure 71

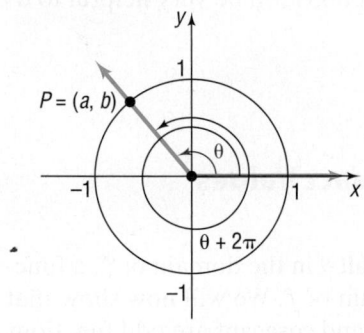

This example illustrates a more general situation. For a given angle θ, measured in radians, suppose that we know the corresponding point $P = (a, b)$ on the unit circle. Now add 2π to θ. The point on the unit circle corresponding to $\theta + 2\pi$ is identical to the point P on the unit circle corresponding to θ. See Figure 71. The values of the trigonometric functions of $\theta + 2\pi$ are equal to the values of the corresponding trigonometric functions of θ.

If we add (or subtract) integer multiples of 2π to θ, the values of the sine and cosine function remain unchanged. That is, for all θ

$$\sin(\theta + 2\pi k) = \sin\theta \qquad \cos(\theta + 2\pi k) = \cos\theta$$

$$\text{where } k \text{ is any integer} \tag{1}$$

Functions that exhibit this kind of behavior are called *periodic functions*.

DEFINITION

A function f is called **periodic** if there is a positive number p such that, whenever θ is in the domain of f, so is $\theta + p$, and

$$f(\theta + p) = f(\theta)$$

If there is a smallest such number p, this smallest value is called the **(fundamental) period** of f.

Based on equation (1), the sine and cosine functions are periodic. In fact, the sine, cosine, secant, and cosecant functions have period 2π. You are asked to prove this in Problems 93 through 96.

Figure 72

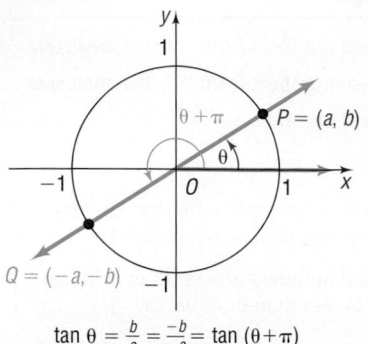

$$\tan \theta = \frac{b}{a} = \frac{-b}{-a} = \tan(\theta + \pi)$$

The tangent and cotangent functions are periodic with period π. See Figure 72 for a partial justification. You are asked to prove this statement in Problems 97 and 98.

These facts are summarized as follows:

Periodic Properties

$$
\begin{array}{lll}
\sin(\theta + 2\pi) = \sin\theta & \cos(\theta + 2\pi) = \cos\theta & \tan(\theta + \pi) = \tan\theta \\
\csc(\theta + 2\pi) = \csc\theta & \sec(\theta + 2\pi) = \sec\theta & \cot(\theta + \pi) = \cot\theta
\end{array}
$$

In Words

Tangent and cotangent have period π; the others have period 2π.

Because the sine, cosine, secant, and cosecant functions have period 2π, once we know their values for $0 \le \theta < 2\pi$, we know all their values; similarly, since the tangent and cotangent functions have period π, once we know their values for $0 \le \theta < \pi$, we know all their values.

EXAMPLE 2

Using Periodic Properties to Find Exact Values

Find the exact value of:

(a) $\sin 420°$

(b) $\tan \dfrac{5\pi}{4}$

(c) $\cos \dfrac{11\pi}{4}$

Solution

(a) $\sin 420° = \sin(60° + 360°) = \sin 60° = \dfrac{\sqrt{3}}{2}$

(b) $\tan \dfrac{5\pi}{4} = \tan\left(\dfrac{\pi}{4} + \pi\right) = \tan \dfrac{\pi}{4} = 1$

(c) $\cos \dfrac{11\pi}{4} = \cos\left(\dfrac{3\pi}{4} + \dfrac{8\pi}{4}\right) = \cos\left(\dfrac{3\pi}{4} + 2\pi\right) = \cos \dfrac{3\pi}{4} = -\dfrac{\sqrt{2}}{2}$

The periodic properties of the trigonometric functions will be very helpful to us when we study their graphs in the next section.

Now Work PROBLEMS 21 AND 79

4 Use Even–Odd Properties to Find the Exact Values of the Trigonometric Functions

Recall that a function f is even if $f(-\theta) = f(\theta)$ for all θ in the domain of f; a function f is odd if $f(-\theta) = -f(\theta)$ for all θ in the domain of f. We will now show that the trigonometric functions sine, tangent, cotangent, and cosecant are odd functions and the functions cosine and secant are even functions.

THEOREM

Even–Odd Properties

$$
\begin{array}{lll}
\sin(-\theta) = -\sin\theta & \cos(-\theta) = \cos\theta & \tan(-\theta) = -\tan\theta \\
\csc(-\theta) = -\csc\theta & \sec(-\theta) = \sec\theta & \cot(-\theta) = -\cot\theta
\end{array}
$$

Figure 73

Proof Let $P = (a, b)$ be the point on the unit circle that corresponds to the angle θ. See Figure 73. The point Q on the unit circle that corresponds to the angle $-\theta$ will have coordinates $(a, -b)$. Using the definition for the trigonometric functions, we have

$$\sin\theta = b \qquad \sin(-\theta) = -b \qquad \cos\theta = a \qquad \cos(-\theta) = a$$

so
$$\sin(-\theta) = -\sin\theta \qquad \cos(-\theta) = \cos\theta$$

Now, using these results and some of the Fundamental Identities, we have

$$\tan(-\theta) = \frac{\sin(-\theta)}{\cos(-\theta)} = \frac{-\sin\theta}{\cos\theta} = -\tan\theta$$

$$\cot(-\theta) = \frac{1}{\tan(-\theta)} = \frac{1}{-\tan\theta} = -\cot\theta$$

$$\sec(-\theta) = \frac{1}{\cos(-\theta)} = \frac{1}{\cos\theta} = \sec\theta$$

$$\csc(-\theta) = \frac{1}{\sin(-\theta)} = \frac{1}{-\sin\theta} = -\csc\theta$$

> **In Words**
> Cosine and secant are even functions; the others are odd functions.

| EXAMPLE 3 | **Finding Exact Values Using Even–Odd Properties** |

Find the exact value of:

(a) $\sin(-45°)$ (b) $\cos(-\pi)$ (c) $\cot\left(-\dfrac{3\pi}{2}\right)$ (d) $\tan\left(-\dfrac{37\pi}{4}\right)$

Solution

(a) $\sin(-45°) = -\sin 45° = -\dfrac{\sqrt{2}}{2}$
 ↑
 Odd function

(b) $\cos(-\pi) = \cos\pi = -1$
 ↑
 Even function

(c) $\cot\left(-\dfrac{3\pi}{2}\right) = -\cot\dfrac{3\pi}{2} = 0$
 ↑
 Odd function

(d) $\tan\left(-\dfrac{37\pi}{4}\right) = -\tan\dfrac{37\pi}{4} = -\tan\left(\dfrac{\pi}{4} + 9\pi\right) = -\tan\dfrac{\pi}{4} = -1$
 ↑ ↑
 Odd function Period is π.

Now Work PROBLEMS 37 AND 73

7.5 Assess Your Understanding

'Are You Prepared?' *Answers are given at the end of these exercises. If you get a wrong answer, read the pages listed in red.*

1. What is the equation of the unit circle? (p. 190)

2. The domain of the function $f(x) = \dfrac{3x-6}{x-4}$ is _____. (pp. 208–218)

3. A function for which $f(x) = f(-x)$ for all x in the domain of f is called a(n) _____ function. (pp. 231–233)

Concepts and Vocabulary

4. The sine, cosine, cosecant, and secant functions have period _____; the tangent and cotangent functions have period _____.

5. The domain of the tangent function is _____.

6. The range of the sine function is _____.

7. If $\sin\theta = 0.2$, then $\sin(-\theta) =$ _____ and $\sin(\theta + 2\pi) =$ _____.

8. *True or False* The only even trigonometric functions are the cosine and secant functions.

Skill Building

In Problems 9–14, the point P on the unit circle that corresponds to a real number t is given. Find sin t, cos t, tan t, csc t, sec t, and cot t.

9. $\left(\dfrac{\sqrt{3}}{2}, -\dfrac{1}{2}\right)$ **10.** $\left(-\dfrac{\sqrt{3}}{2}, -\dfrac{1}{2}\right)$ **11.** $\left(-\dfrac{\sqrt{2}}{2}, -\dfrac{\sqrt{2}}{2}\right)$ **12.** $\left(\dfrac{\sqrt{2}}{2}, -\dfrac{\sqrt{2}}{2}\right)$ **13.** $\left(\dfrac{\sqrt{5}}{3}, \dfrac{2}{3}\right)$ **14.** $\left(-\dfrac{\sqrt{5}}{5}, \dfrac{2\sqrt{5}}{5}\right)$

In Problems 15–20, the point P on the circle $x^2 + y^2 = r^2$ that is also on the terminal side of an angle θ in standard position is given. Find $\sin\theta$, $\cos\theta$, $\tan\theta$, $\csc\theta$, $\sec\theta$, and $\cot\theta$.

15. $(3, -4)$ **16.** $(4, -3)$ **17.** $(-2, 3)$ **18.** $(2, -4)$ **19.** $(-1, -1)$ **20.** $(-3, 1)$

In Problems 21–36, use the fact that the trigonometric functions are periodic to find the exact value of each expression. Do not use a calculator.

21. $\sin 405°$ **22.** $\cos 420°$ **23.** $\tan 405°$ **24.** $\sin 390°$

25. $\csc 450°$ **26.** $\sec 540°$ **27.** $\cot 390°$ **28.** $\sec 420°$

29. $\cos \dfrac{33\pi}{4}$ **30.** $\sin \dfrac{9\pi}{4}$ **31.** $\tan(21\pi)$ **32.** $\csc \dfrac{9\pi}{2}$

33. $\sec \dfrac{17\pi}{4}$ **34.** $\cot \dfrac{17\pi}{4}$ **35.** $\tan \dfrac{19\pi}{6}$ **36.** $\sec \dfrac{25\pi}{6}$

In Problems 37–54, use the even–odd properties to find the exact value of each expression. Do not use a calculator.

37. $\sin(-60°)$ **38.** $\cos(-30°)$ **39.** $\tan(-30°)$ **40.** $\sin(-135°)$ **41.** $\sec(-60°)$

42. $\csc(-30°)$ **43.** $\sin(-90°)$ **44.** $\cos(-270°)$ **45.** $\tan\left(-\dfrac{\pi}{4}\right)$ **46.** $\sin(-\pi)$

47. $\cos\left(-\dfrac{\pi}{4}\right)$ **48.** $\sin\left(-\dfrac{\pi}{3}\right)$ **49.** $\tan(-\pi)$ **50.** $\sin\left(-\dfrac{3\pi}{2}\right)$

51. $\csc\left(-\dfrac{\pi}{4}\right)$ **52.** $\sec(-\pi)$ **53.** $\sec\left(-\dfrac{\pi}{6}\right)$ **54.** $\csc\left(-\dfrac{\pi}{3}\right)$

In Problems 55–60, find the exact value of each expression. Do not use a calculator.

55. $\sin(-\pi) + \cos(5\pi)$ **56.** $\tan\left(-\dfrac{5\pi}{6}\right) - \cot \dfrac{7\pi}{2}$ **57.** $\sec(-\pi) + \csc\left(-\dfrac{\pi}{2}\right)$

58. $\tan(-6\pi) + \cos \dfrac{9\pi}{4}$ **59.** $\sin\left(-\dfrac{9\pi}{4}\right) - \tan\left(-\dfrac{9\pi}{4}\right)$ **60.** $\cos\left(-\dfrac{17\pi}{4}\right) - \sin\left(-\dfrac{3\pi}{2}\right)$

61. What is the domain of the sine function?

62. What is the domain of the cosine function?

63. For what numbers θ is $f(\theta) = \tan\theta$ not defined?

64. For what numbers θ is $f(\theta) = \cot\theta$ not defined?

65. For what numbers θ is $f(\theta) = \sec\theta$ not defined?

66. For what numbers θ is $f(\theta) = \csc\theta$ not defined?

67. What is the range of the sine function?

68. What is the range of the cosine function?

69. What is the range of the tangent function?

70. What is the range of the cotangent function?

71. What is the range of the secant function?

72. What is the range of the cosecant function?

73. Is the sine function even, odd, or neither? Is its graph symmetric? With respect to what?

74. Is the cosine function even, odd, or neither? Is its graph symmetric? With respect to what?

75. Is the tangent function even, odd, or neither? Is its graph symmetric? With respect to what?

76. Is the cotangent function even, odd, or neither? Is its graph symmetric? With respect to what?

77. Is the secant function even, odd, or neither? Is its graph symmetric? With respect to what?

78. Is the cosecant function even, odd, or neither? Is its graph symmetric? With respect to what?

79. If $\sin\theta = 0.3$, find the value of:
$$\sin\theta + \sin(\theta + 2\pi) + \sin(\theta + 4\pi)$$

80. If $\cos\theta = 0.2$, find the value of:
$$\cos\theta + \cos(\theta + 2\pi) + \cos(\theta + 4\pi)$$

81. If $\tan \theta = 3$, find the value of:
$$\tan \theta + \tan(\theta + \pi) + \tan(\theta + 2\pi)$$

82. If $\cot \theta = -2$, find the value of:
$$\cot \theta + \cot(\theta - \pi) + \cot(\theta - 2\pi)$$

In Problems 83–88, use the periodic and even–odd properties.

83. If $f(x) = \sin x$ and $f(a) = \dfrac{1}{3}$, find the exact value of:
 (a) $f(-a)$ (b) $f(a) + f(a + 2\pi) + f(a + 4\pi)$

84. If $f(x) = \cos x$ and $f(a) = \dfrac{1}{4}$, find the exact value of:
 (a) $f(-a)$ (b) $f(a) + f(a + 2\pi) + f(a - 2\pi)$

85. If $f(x) = \tan x$ and $f(a) = 2$, find the exact value of:
 (a) $f(-a)$ (b) $f(a) + f(a + \pi) + f(a + 2\pi)$

86. If $f(x) = \cot x$ and $f(a) = -3$, find the exact value of:
 (a) $f(-a)$ (b) $f(a) + f(a + \pi) + f(a + 4\pi)$

87. If $f(x) = \sec x$ and $f(a) = -4$, find the exact value of:
 (a) $f(-a)$ (b) $f(a) + f(a + 2\pi) + f(a + 4\pi)$

88. If $f(x) = \csc x$ and $f(a) = 2$, find the exact value of:
 (a) $f(-a)$ (b) $f(a) + f(a + 2\pi) + f(a + 4\pi)$

Applications and Extensions

In Problems 89–92, use the figure to approximate the value of the six trigonometric functions at t to the nearest tenth. Then use a calculator to approximate each of the six trigonometric functions at t.

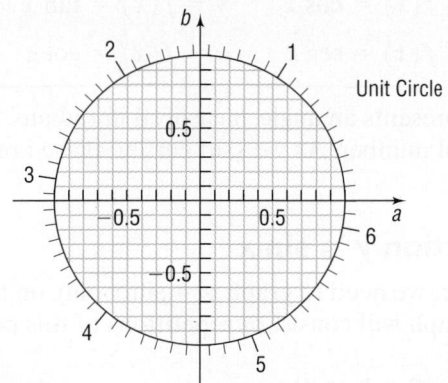

89. (a) $t = 1$ (b) $t = 5.1$

90. (a) $t = 2$ (b) $t = 4$

91. Show that the range of the tangent function is the set of all real numbers.

92. Show that the range of the cotangent function is the set of all real numbers.

93. Show that the period of $f(\theta) = \sin \theta$ is 2π.
[**Hint:** Assume that $0 < p < 2\pi$ exists so that $\sin(\theta + p) = \sin \theta$ for all θ. Let $\theta = 0$ to find p. Then let $\theta = \dfrac{\pi}{2}$ to obtain a contradiction.]

94. Show that the period of $f(\theta) = \cos \theta$ is 2π.

95. Show that the period of $f(\theta) = \sec \theta$ is 2π.

96. Show that the period of $f(\theta) = \csc \theta$ is 2π.

97. Show that the period of $f(\theta) = \tan \theta$ is π.

98. Show that the period of $f(\theta) = \cot \theta$ is π.

99. If θ, $0 < \theta < \pi$, is the angle between a horizontal ray directed to the right (say, the positive x-axis) and a nonhorizontal, nonvertical line L, show that the slope m of L equals $\tan \theta$. The angle θ is called the **inclination** of L.

[**Hint:** See the illustration, where we have drawn the line M parallel to L and passing through the origin. Use the fact that M intersects the unit circle at the point $(\cos \theta, \sin \theta)$.]

Discussion and Writing

100. Explain how you would find the value of $\sin 390°$ using periodic properties.

101. Explain how you would find the value of $\cos(-45°)$ using even–odd properties.

102. Write down five properties of the tangent function. Explain the meaning of each.

103. Describe your understanding of the meaning of a periodic function.

'Are You Prepared?' Answers

1. $x^2 + y^2 = 1$ 2. $\{x \mid x \neq 4\}$ 3. even

7.6 Graphs of the Sine and Cosine Functions*

PREPARING FOR THIS SECTION *Before getting started, review the following:*

- Graphing Techniques: Transformations (Section 3.5, pp. 252–260)

Now Work the 'Are You Prepared?' problems on page 570.

OBJECTIVES 1 Graph Functions of the Form $y = A \sin(\omega x)$ Using Transformations (p. 561)
2 Graph Functions of the Form $y = A \cos(\omega x)$ Using Transformations (p. 563)
3 Determine the Amplitude and Period of Sinusoidal Functions (p. 564)
4 Graph Sinusoidal Functions Using Key Points (p. 565)
5 Find an Equation for a Sinusoidal Graph (p. 569)

Since we want to graph the trigonometric functions in the xy-plane, we shall use the traditional symbols x for the independent variable (or argument) and y for the dependent variable (or value at x) for each function. So we write the six trigonometric functions as

$$y = f(x) = \sin x \qquad y = f(x) = \cos x \qquad y = f(x) = \tan x$$
$$y = f(x) = \csc x \qquad y = f(x) = \sec x \qquad y = f(x) = \cot x$$

 Here the independent variable x represents an angle, measured in radians. In calculus, x will usually be treated as a real number. As we said earlier, these are equivalent ways of viewing x.

The Graph of the Sine Function $y = \sin x$

Since the sine function has period 2π, we need to graph $y = \sin x$ only on the interval $[0, 2\pi]$. The remainder of the graph will consist of repetitions of this portion of the graph.

We begin by constructing Table 7, which lists some points on the graph of $y = \sin x$, $0 \le x \le 2\pi$. As the table shows, the graph of $y = \sin x$, $0 \le x \le 2\pi$, begins at the origin. As x increases from 0 to $\frac{\pi}{2}$, the value of $y = \sin x$ increases from 0 to 1; as x increases from $\frac{\pi}{2}$ to π to $\frac{3\pi}{2}$, the value of y decreases from 1 to 0 to -1; as x increases from $\frac{3\pi}{2}$ to 2π, the value of y increases from -1 to 0. If we plot the points listed in Table 7 and connect them with a smooth curve, we obtain the graph shown in Figure 74.

Table 7

x	$y = \sin x$	(x, y)
0	0	$(0, 0)$
$\dfrac{\pi}{6}$	$\dfrac{1}{2}$	$\left(\dfrac{\pi}{6}, \dfrac{1}{2}\right)$
$\dfrac{\pi}{2}$	1	$\left(\dfrac{\pi}{2}, 1\right)$
$\dfrac{5\pi}{6}$	$\dfrac{1}{2}$	$\left(\dfrac{5\pi}{6}, \dfrac{1}{2}\right)$
π	0	$(\pi, 0)$
$\dfrac{7\pi}{6}$	$-\dfrac{1}{2}$	$\left(\dfrac{7\pi}{6}, -\dfrac{1}{2}\right)$
$\dfrac{3\pi}{2}$	-1	$\left(\dfrac{3\pi}{2}, -1\right)$
$\dfrac{11\pi}{6}$	$-\dfrac{1}{2}$	$\left(\dfrac{11\pi}{6}, -\dfrac{1}{2}\right)$
2π	0	$(2\pi, 0)$

Figure 74
$y = \sin x, 0 \le x \le 2\pi$

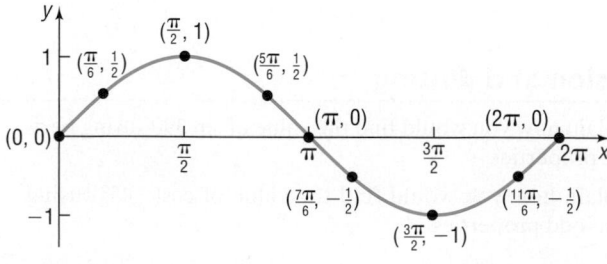

The graph in Figure 74 is one period, or **cycle,** of the graph of $y = \sin x$. To obtain a more complete graph of $y = \sin x$, we continue the graph in each direction, as shown in Figure 75.

* For those who wish to include phase shifts here, Section 7.8 can be covered immediately after Section 7.6 without loss of continuity.

Figure 75
$y = \sin x, \; -\infty < x < \infty$

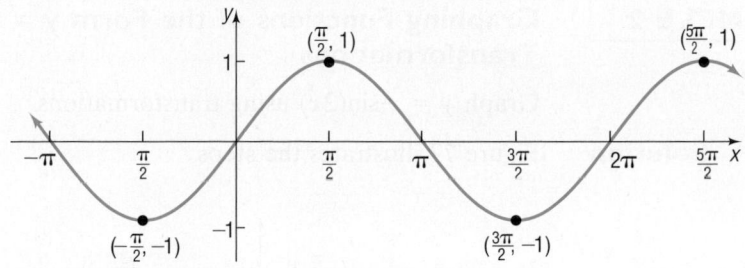

The graph of $y = \sin x$ illustrates some of the facts that we already know about the sine function.

Properties of the Sine Function $y = \sin x$

1. The domain is the set of all real numbers.
2. The range consists of all real numbers from -1 to 1, inclusive.
3. The sine function is an odd function, as the symmetry of the graph with respect to the origin indicates.
4. The sine function is periodic, with period 2π.
5. The x-intercepts are $\ldots, -2\pi, -\pi, 0, \pi, 2\pi, 3\pi, \ldots$; the y-intercept is 0.
6. The maximum value is 1 and occurs at $x = \ldots, -\dfrac{3\pi}{2}, \dfrac{\pi}{2}, \dfrac{5\pi}{2}, \dfrac{9\pi}{2}, \ldots$;

 the minimum value is -1 and occurs at $x = \ldots, -\dfrac{\pi}{2}, \dfrac{3\pi}{2}, \dfrac{7\pi}{2}, \dfrac{11\pi}{2}, \ldots$.

━━━━━➤ **Now Work** PROBLEMS **9, 11, AND 13**

1 Graph Functions of the Form $y = A\sin(\omega x)$ Using Transformations

| EXAMPLE 1 | **Graphing Functions of the Form $y = A\sin(\omega x)$ Using Transformations** |

Graph $y = 3\sin x$ using transformations.

Solution Figure 76 illustrates the steps.

Figure 76

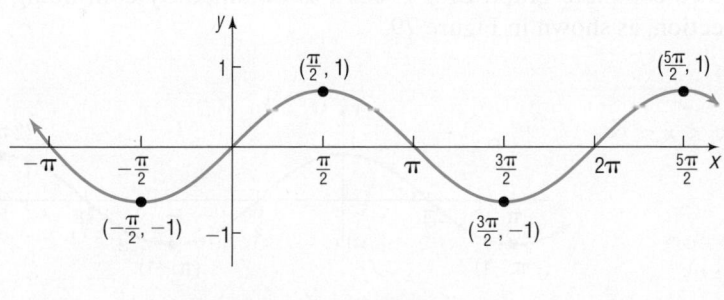

(a) $y = \sin x$

Multiply by 3
vertical strech
by a factor of 3

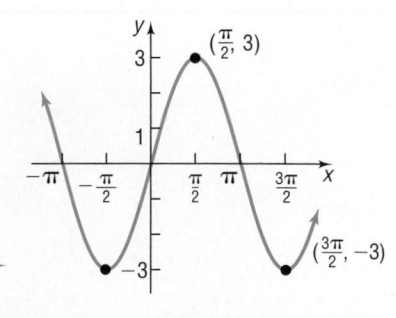

(b) $y = 3\sin x$

| EXAMPLE 2 | Graphing Functions of the Form $y = A \sin(\omega x)$ Using Transformations |

Graph $y = -\sin(2x)$ using transformations.

Solution Figure 77 illustrates the steps.

Figure 77

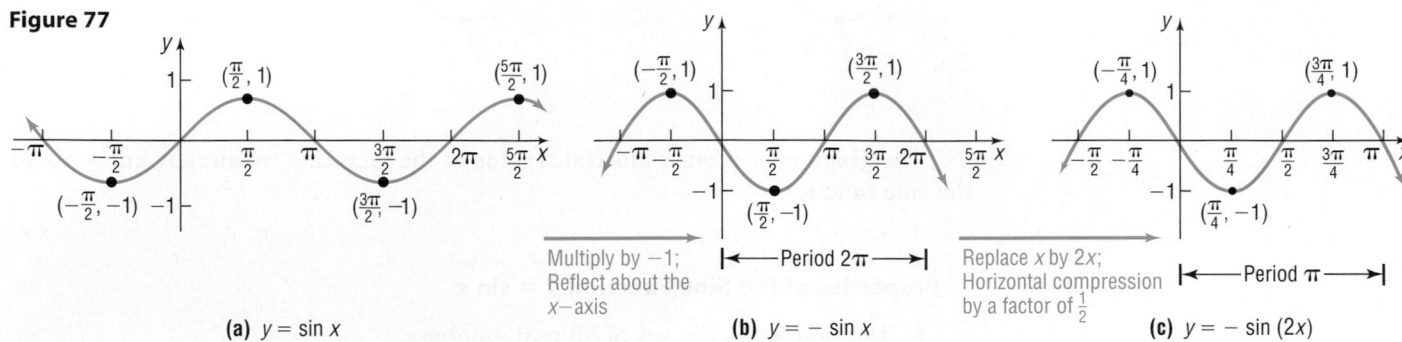

Multiply by -1; Reflect about the x–axis

|←——Period 2π——→|

Replace x by $2x$; Horizontal compression by a factor of $\frac{1}{2}$

|←——Period π——→|

(a) $y = \sin x$ **(b)** $y = -\sin x$ **(c)** $y = -\sin(2x)$

Notice in Figure 77(c) that the period of the function $y = -\sin(2x)$ is π due to the horizontal compression of the original period 2π by a factor of $\frac{1}{2}$.

Now Work PROBLEM 45 USING TRANSFORMATIONS

Table 8

x	$y = \cos x$	(x, y)
0	1	$(0, 1)$
$\dfrac{\pi}{3}$	$\dfrac{1}{2}$	$\left(\dfrac{\pi}{3}, \dfrac{1}{2}\right)$
$\dfrac{\pi}{2}$	0	$\left(\dfrac{\pi}{2}, 0\right)$
$\dfrac{2\pi}{3}$	$-\dfrac{1}{2}$	$\left(\dfrac{2\pi}{3}, -\dfrac{1}{2}\right)$
π	-1	$(\pi, -1)$
$\dfrac{4\pi}{3}$	$-\dfrac{1}{2}$	$\left(\dfrac{4\pi}{3}, -\dfrac{1}{2}\right)$
$\dfrac{3\pi}{2}$	0	$\left(\dfrac{3\pi}{2}, 0\right)$
$\dfrac{5\pi}{3}$	$\dfrac{1}{2}$	$\left(\dfrac{5\pi}{3}, \dfrac{1}{2}\right)$
2π	1	$(2\pi, 1)$

The Graph of the Cosine Function

The cosine function also has period 2π. We proceed as we did with the sine function by constructing Table 8, which lists some points on the graph of $y = \cos x$, $0 \le x \le 2\pi$. As the table shows, the graph of $y = \cos x$, $0 \le x \le 2\pi$, begins at the point $(0, 1)$. As x increases from 0 to $\dfrac{\pi}{2}$ to π, the value of y decreases from 1 to 0 to -1; as x increases from π to $\dfrac{3\pi}{2}$ to 2π, the value of y increases from -1 to 0 to 1. As before, we plot the points in Table 8 to get one period or cycle of the graph. See Figure 78.

Figure 78
$y = \cos x$, $0 \le x \le 2\pi$

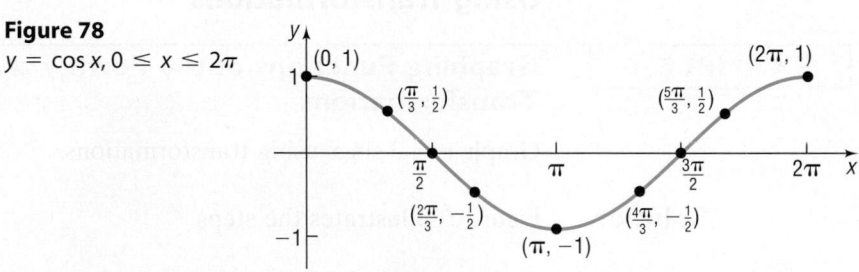

A more complete graph of $y = \cos x$ is obtained by continuing the graph in each direction, as shown in Figure 79.

Figure 79
$y = \cos x$, $-\infty < x < \infty$

The graph of $y = \cos x$ illustrates some of the facts that we already know about the cosine function.

Properties of the Cosine Function

1. The domain is the set of all real numbers.
2. The range consists of all real numbers from -1 to 1, inclusive.
3. The cosine function is an even function, as the symmetry of the graph with respect to the y-axis indicates.
4. The cosine function is periodic, with period 2π.
5. The x-intercepts are $\ldots, -\dfrac{3\pi}{2}, -\dfrac{\pi}{2}, \dfrac{\pi}{2}, \dfrac{3\pi}{2}, \dfrac{5\pi}{2}, \ldots$; the y-intercept is 1.
6. The maximum value is 1 and occurs at $x = \ldots, -2\pi, 0, 2\pi, 4\pi, 6\pi, \ldots$; the minimum value is -1 and occurs at $x = \ldots, -\pi, \pi, 3\pi, 5\pi, \ldots$.

2 Graph Functions of the Form $y = A\cos(\omega x)$ Using Transformations

EXAMPLE 3 Graphing Functions of the Form $y = A\cos(\omega x)$ Using Transformations

Graph $y = 2\cos(3x)$ using transformations.

Solution Figure 80 shows the steps.

Figure 80

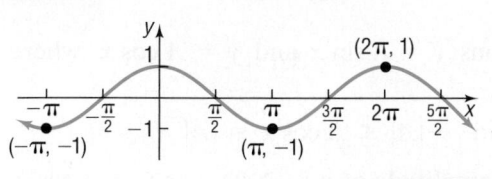

(a) $y = \cos x$

Multiply by 2;
Vertical stretch
by a factor of 2

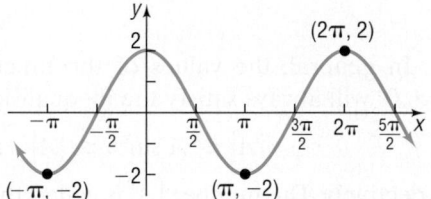

(b) $y = 2\cos x$

Replace x by 3x;
Horizontal
compression by
a factor of $\frac{1}{3}$

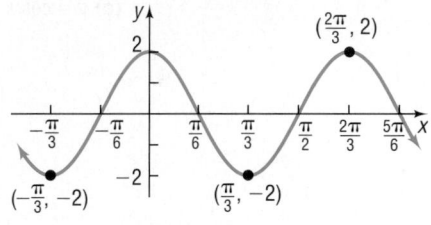

(c) $y = 2\cos(3x)$

Notice in Figure 80(c) that the period of the function $y = 2\cos(3x)$ is $\dfrac{2\pi}{3}$ due to the compression of the original period 2π by a factor of $\dfrac{1}{3}$.

➤ **Now Work** PROBLEM 53 USING TRANSFORMATIONS

Sinusoidal Graphs

Shift the graph of $y = \cos x$ to the right $\dfrac{\pi}{2}$ units to obtain the graph of $y = \cos\left(x - \dfrac{\pi}{2}\right)$. See Figure 81(a). Now look at the graph of $y = \sin x$ in Figure 81(b). We see that the graph of $y = \sin x$ is the same as the graph of $y = \cos\left(x - \dfrac{\pi}{2}\right)$.

Figure 81

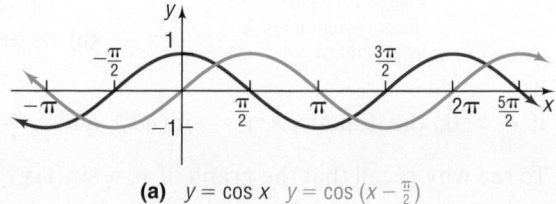

(a) $y = \cos x$ $y = \cos\left(x - \frac{\pi}{2}\right)$

(b) $y = \sin x$

Based on Figure 81, we conjecture that

$$\sin x = \cos\left(x - \frac{\pi}{2}\right)$$

(We shall prove this fact in Chapter 8.) Because of this relationship, the graphs of functions of the form $y = A \sin(\omega x)$ or $y = A \cos(\omega x)$ are referred to as **sinusoidal graphs.**

Let's look at some general properties of sinusoidal graphs.

3 Determine the Amplitude and Period of Sinusoidal Functions

In Figure 82(b) we show the graph of $y = 2 \cos x$. Notice that the values of $y = 2 \cos x$ lie between -2 and 2, inclusive.

Figure 82

(a) $y = \cos x$ Multiply by 2; Vertical stretch by a factor of 2 **(b)** $y = 2 \cos x$

In general, the values of the functions $y = A \sin x$ and $y = A \cos x$, where $A \neq 0$, will always satisfy the inequalities

$$-|A| \leq A \sin x \leq |A| \quad \text{and} \quad -|A| \leq A \cos x \leq |A|$$

respectively. The number $|A|$ is called the **amplitude** of $y = A \sin x$ or $y = A \cos x$. See Figure 83.

Figure 83

$y = A \sin x, A > 0$
Period $= 2\pi$

In Figure 84(b), we show the graph of $y = \cos(3x)$. Notice that the period of this function is $\dfrac{2\pi}{3}$, due to the horizontal compression of the original period 2π by a factor of $\dfrac{1}{3}$.

Figure 84

(a) $y = \cos x$ Replace x by $3x$; Horizontal compression by a factor of $\frac{1}{3}$ **(b)** $y = \cos(3x)$

In general, if $\omega > 0$, the functions $y = \sin(\omega x)$ and $y = \cos(\omega x)$ will have period $T = \dfrac{2\pi}{\omega}$. To see why, recall that the graph of $y = \sin(\omega x)$ is obtained from the

graph of $y = \sin x$ by performing a horizontal compression or stretch by a factor $\dfrac{1}{\omega}$.

This horizontal compression replaces the interval $[0, 2\pi]$, which contains one period of the graph of $y = \sin x$, by the interval $\left[0, \dfrac{2\pi}{\omega}\right]$, which contains one period of the graph of $y = \sin(\omega x)$. The period of the functions $y = \sin(\omega x)$ and $y = \cos(\omega x)$, $\omega > 0$, is $\dfrac{2\pi}{\omega}$.

For example, for the function $y = \cos(3x)$, graphed in Figure 84(b), $\omega = 3$, so the period is $\dfrac{2\pi}{\omega} = \dfrac{2\pi}{3}$.

One period of the graph of $y = \sin(\omega x)$ or $y = \cos(\omega x)$ is called a **cycle.** Figure 85 illustrates the general situation. The blue portion of the graph is one cycle.

Figure 85

$y = A \sin(\omega x),\ A > 0,\ \omega > 0$
Period $= \frac{2\pi}{\omega}$

If $\omega < 0$ in $y = \sin(\omega x)$ or $y = \cos(\omega x)$, we use the Even–Odd Properties of the sine and cosine functions as follows:

$$\sin(-\omega x) = -\sin(\omega x) \quad \text{and} \quad \cos(-\omega x) = \cos(\omega x)$$

This gives us an equivalent form in which the coefficient of x in the argument is positive. For example,

$$\sin(-2x) = -\sin(2x) \quad \text{and} \quad \cos(-\pi x) = \cos(\pi x)$$

Because of this, we can assume $\omega > 0$.

THEOREM

If $\omega > 0$, the amplitude and period of $y = A \sin(\omega x)$ and $y = A \cos(\omega x)$ are given by

$$\text{Amplitude} = |A| \qquad \text{Period} = T = \dfrac{2\pi}{\omega} \qquad \text{(1)}$$

EXAMPLE 4 **Finding the Amplitude and Period of a Sinusoidal Function**

Determine the amplitude and period of $y = 3 \sin(4x)$.

Solution Comparing $y = 3 \sin(4x)$ to $y = A \sin(\omega x)$, we find that $A = 3$ and $\omega = 4$. From equation (1),

$$\text{Amplitude} = |A| = 3 \qquad \text{Period} = T = \dfrac{2\pi}{\omega} = \dfrac{2\pi}{4} = \dfrac{\pi}{2}$$

══════ **Now Work** PROBLEM 23

4 Graph Sinusoidal Functions Using Key Points

So far, we have graphed functions of the form $y = A \sin(\omega x)$ or $y = A \cos(\omega x)$ using transformations. We now introduce another method that can be used to graph these functions.

Figure 86 shows one cycle of the graphs of $y = \sin x$ and $y = \cos x$ on the interval $[0, 2\pi]$. Notice that each graph consists of four parts corresponding to the four subintervals:

$$\left[0, \frac{\pi}{2}\right], \quad \left[\frac{\pi}{2}, \pi\right], \quad \left[\pi, \frac{3\pi}{2}\right], \quad \left[\frac{3\pi}{2}, 2\pi\right]$$

Each subinterval is of length $\frac{\pi}{2}$ (the period 2π divided by 4, the number of parts), and the endpoints of these intervals $x = 0$, $x = \frac{\pi}{2}$, $x = \pi$, $x = \frac{3\pi}{2}$, $x = 2\pi$ give rise to five key points on each graph:

For $y = \sin x$: $(0, 0)$, $\left(\frac{\pi}{2}, 1\right)$, $(\pi, 0)$, $\left(\frac{3\pi}{2}, -1\right)$, $(2\pi, 0)$

For $y = \cos x$: $(0, 1)$, $\left(\frac{\pi}{2}, 0\right)$, $(\pi, -1)$, $\left(\frac{3\pi}{2}, 0\right)$, $(2\pi, 1)$

Look again at Figure 86.

Figure 86

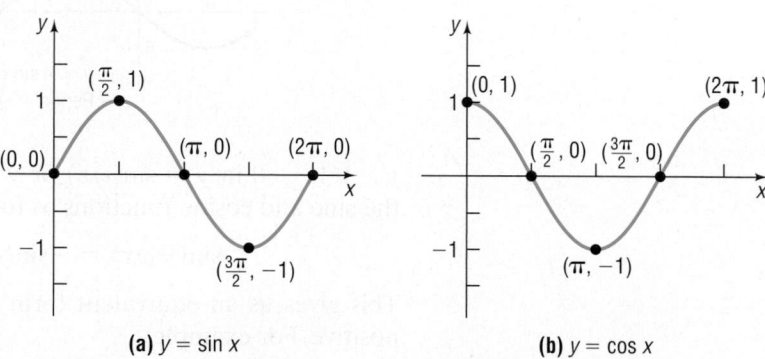

(a) $y = \sin x$ **(b)** $y = \cos x$

Steps for Graphing a Sinusoidal Function of the Form $y = A \sin(\omega x)$ or $y = A \cos(\omega x)$ Using Key Points

STEP 1: Use the amplitude A to determine the maximum and minimum values of the function. This sets the scale for the y-axis.

STEP 2: Use the period $\frac{2\pi}{\omega}$ and divide the interval $\left[0, \frac{2\pi}{\omega}\right]$ into four subintervals of the same length.

STEP 3: Use the endpoints of these subintervals to obtain five key points on the graph.

STEP 4: Connect these points with a sinusoidal graph to obtain the graph of one cycle and extend the graph in each direction to make it complete.

EXAMPLE 5

Graphing a Sinusoidal Function Using Key Points

Graph: $y = 3 \sin(4x)$

Solution

Refer to Example 4. For $y = 3 \sin(4x)$, the amplitude is 3 and the period is $\frac{\pi}{2}$. Because the amplitude is 3, the graph of $y = 3 \sin(4x)$ will lie between -3 and 3 on the y-axis. Because the period is $\frac{\pi}{2}$, one cycle will begin at $x = 0$ and end at $x = \frac{\pi}{2}$.

We divide the interval $\left[0, \frac{\pi}{2}\right]$ into four subintervals, each of length $\frac{\pi}{2} \div 4 = \frac{\pi}{8}$, by finding the following values:

$$0 \qquad \underset{\text{2nd value}}{0 + \frac{\pi}{8} = \frac{\pi}{8}} \qquad \underset{\text{3rd value}}{\frac{\pi}{8} + \frac{\pi}{8} = \frac{\pi}{4}} \qquad \underset{\text{4th value}}{\frac{\pi}{4} + \frac{\pi}{8} = \frac{3\pi}{8}} \qquad \underset{\text{final value}}{\frac{3\pi}{8} + \frac{\pi}{8} = \frac{\pi}{2}}$$

initial value

These values of x determine the x-coordinates of the five key points on the graph. To obtain the y-coordinates of the five key points of $y = 3\sin(4x)$, we multiply the y-coordinates of the five key points for $y = \sin x$ in Figure 86(a) by $A = 3$. The five key points are

$$(0, 0), \quad \left(\frac{\pi}{8}, 3\right), \quad \left(\frac{\pi}{4}, 0\right), \quad \left(\frac{3\pi}{8}, -3\right), \quad \left(\frac{\pi}{2}, 0\right)$$

NOTE We could also obtain the five key points by evaluating $y = 3\sin(4x)$ at each value of x. ■

We plot these five points and fill in the graph of the sine curve as shown in Figure 87(a). We extend the graph in either direction to obtain the complete graph shown in Figure 87(b).

Figure 87

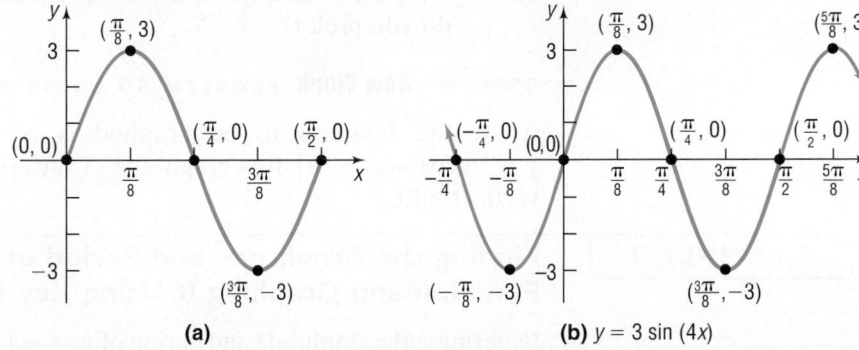

(a)

(b) $y = 3\sin(4x)$

COMMENT To graph a sinusoidal function of the form $y = A\sin(\omega x)$ or $y = A\cos(\omega x)$ using a graphing utility, we use the amplitude to set Ymin and Ymax and use the period to set Xmin and Xmax. ■

✓ **Check:** Graph $y = 3\sin(4x)$ using transformations. Which graphing method do you prefer?

━━━ **Now Work** PROBLEM 45 USING KEY POINTS

| EXAMPLE 6 | **Finding the Amplitude and Period of a Sinusoidal Function and Graphing It Using Key Points** |

Determine the amplitude and period of $y = 2\sin\left(-\dfrac{\pi}{2}x\right)$, and graph the function.

Solution Since the sine function is odd, we can use the equivalent form:

$$y = -2\sin\left(\frac{\pi}{2}x\right)$$

Comparing $y = -2\sin\left(\dfrac{\pi}{2}x\right)$ to $y = A\sin(\omega x)$, we find that $A = -2$ and $\omega = \dfrac{\pi}{2}$.

The amplitude is $|A| = 2$, and the period is $T = \dfrac{2\pi}{\omega} = \dfrac{2\pi}{\dfrac{\pi}{2}} = 4$.

The graph of $y = -2\sin\left(\dfrac{\pi}{2}x\right)$ will lie between -2 and 2 on the y-axis. One cycle will begin at $x = 0$ and end at $x = 4$. We divide the interval $[0, 4]$ into four subintervals, each of length $4 \div 4 = 1$, by finding the following values:

0	0 + 1 = 1	1 + 1 = 2	2 + 1 = 3	3 + 1 = 4
initial value	1st value	2nd value	3rd value	final value

Since $y = -2\sin\left(\dfrac{\pi}{2}x\right)$, we multiply the y-coordinates of the five key points in Figure 86(a) by $A = -2$. The five key points on the graph are

$$(0, 0), \quad (1, -2), \quad (2, 0), \quad (3, 2), \quad (4, 0)$$

We plot these five points and fill in the graph of the sine function as shown in Figure 88(a) on page 568. Extending the graph in each direction, we obtain Figure 88(b).

Figure 88

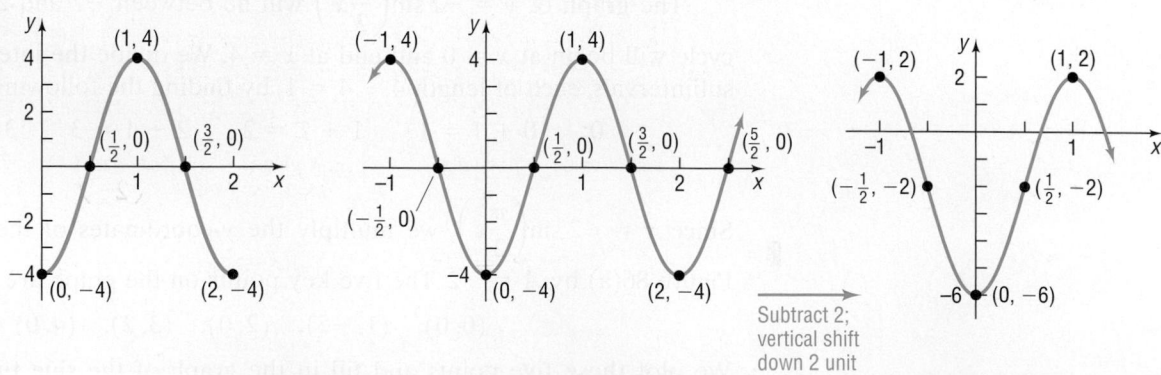

(a)

(b) $y = 2 \sin\left(-\frac{\pi}{2}x\right)$

✓ **Check:** Graph $y = 2 \sin\left(-\frac{\pi}{2}x\right)$ using transformations. Which graphing method do you prefer?

Now Work PROBLEM **49** USING KEY POINTS

If the function to be graphed is of the form $y = A \sin(\omega x) + B$ [or $y = A \cos(\omega x) + B$], first graph $y = A \sin(\omega x)$ [or $y = A \cos(\omega x)$] and then use a vertical shift.

EXAMPLE 7 **Finding the Amplitude and Period of a Sinusoidal Function and Graphing It Using Key Points**

Determine the amplitude and period of $y = -4 \cos(\pi x) - 2$, and graph the function.

Solution We begin by graphing the function $y = -4 \cos(\pi x)$. Comparing $y = -4 \cos(\pi x)$ with $y = A \cos(\omega x)$, we find that $A = -4$ and $\omega = \pi$. The amplitude is $|A| = |-4| = 4$, and the period is $T = \dfrac{2\pi}{\omega} = \dfrac{2\pi}{\pi} = 2$.

The graph of $y = -4 \cos(\pi x)$ will lie between -4 and 4 on the y-axis. One cycle will begin at $x = 0$ and end at $x = 2$. We divide the interval $[0, 2]$ into four subintervals, each of length $2 \div 4 = \dfrac{1}{2}$, by finding the following values:

$$\underset{\text{initial value}}{0} \qquad \underset{\text{1st value}}{0 + \frac{1}{2} = \frac{1}{2}} \qquad \underset{\text{2nd value}}{\frac{1}{2} + \frac{1}{2} = 1} \qquad \underset{\text{3rd value}}{1 + \frac{1}{2} = \frac{3}{2}} \qquad \underset{\text{final value}}{\frac{3}{2} + \frac{1}{2} = 2}$$

Since $y = -4 \cos(\pi x)$, we multiply the y-coordinates of the five key points of $y = \cos x$ shown in Figure 86(b) by $A = -4$ to obtain the five key points on the graph of $y = -4 \cos(\pi x)$:

$$(0, -4), \quad \left(\frac{1}{2}, 0\right), \quad (1, 4), \quad \left(\frac{3}{2}, 0\right), \quad (2, -4)$$

We plot these five points and fill in the graph of the cosine function as shown in Figure 89(a). Extending the graph in each direction, we obtain Figure 89(b), the graph of $y = -4 \cos(\pi x)$. A vertical shift down 2 units gives the graph of $y = -4 \cos(\pi x) - 2$, as shown in Figure 89(c).

Figure 89

(a)

(b) $y = -4 \cos(\pi x)$

Subtract 2;
vertical shift
down 2 unit

(c) $y = -4 \cos(\pi x) - 2$

Now Work PROBLEM **59**

5 Find an Equation for a Sinusoidal Graph

We can also use the ideas of amplitude and period to identify a sinusoidal function when its graph is given.

EXAMPLE 8	**Finding an Equation for a Sinusoidal Graph**

Find an equation for the graph shown in Figure 90.

Figure 90

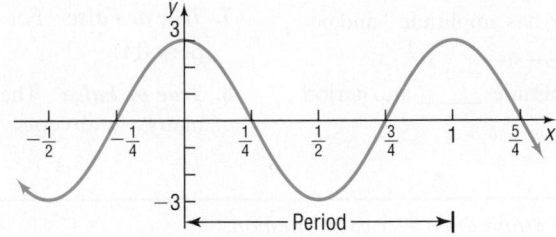

Solution The graph has the characteristics of a cosine function. Do you see why? So we view the equation as a cosine function $y = A\cos(\omega x)$ with $A = 3$ and period $T = 1$. Then $\dfrac{2\pi}{\omega} = 1$, so $\omega = 2\pi$. The cosine function whose graph is given in Figure 90 is

$$y = A\cos(\omega x) = 3\cos(2\pi x)$$

 ✓**Check:** Graph $Y_1 = 3\cos(2\pi x)$ and compare the result with Figure 90.

EXAMPLE 9	**Finding an Equation for a Sinusoidal Graph**

Find an equation for the graph shown in Figure 91.

Figure 91

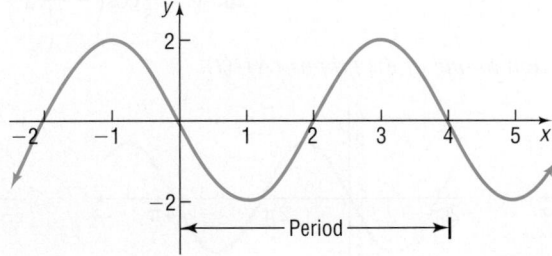

Solution The graph is sinusoidal, with amplitude $|A| = 2$. The period is 4, so $\dfrac{2\pi}{\omega} = 4$ or $\omega = \dfrac{\pi}{2}$. Since the graph passes through the origin, it is easiest to view the equation as a sine function,* but notice that the graph is actually the reflection of a sine function about the x-axis (since the graph is decreasing near the origin). This requires that $A = -2$. The sine function whose graph is given in Figure 91 is

$$y = A\sin(\omega x) = -2\sin\left(\frac{\pi}{2}x\right)$$

 ✓**Check:** Graph $Y_1 = -2\sin\left(\dfrac{\pi}{2}x\right)$ and compare the result with Figure 91.

━━━━━ **Now Work** PROBLEMS **67** AND **71**

* The equation could also be viewed as a cosine function with a horizontal shift, but viewing it as a sine function is easier.

7.6 Assess Your Understanding

'Are You Prepared?' *Answers are given at the end of these exercises. If you get a wrong answer, read the pages listed in red.*

 1. Use transformations to graph $y = 3x^2$ (pp. 252–260) **2.** Use transformations to graph $y = -x^2$. (pp. 252–260)

Concepts and Vocabulary

 3. The maximum value of $y = \sin x, 0 \le x \le 2\pi$, is _____ and occurs at $x = $ _____.

 4. The function $y = A \sin(\omega x), A > 0$, has amplitude 3 and period 2; then $A = $ _____ and $\omega = $ _____.

 5. The function $y = 3 \cos(6x)$ has amplitude _____ and period _____.

 6. *True or False* The graphs of $y = \sin x$ and $y = \cos x$ are identical except for a horizontal shift.

 7. *True or False* For $y = 2 \sin(\pi x)$, the amplitude is 2 and the period is $\dfrac{\pi}{2}$.

 8. *True or False* The graph of the sine function has infinitely many x-intercepts.

Skill Building

In Problems 9–18, if necessary, refer to a graph to answer each question.

 9. What is the y-intercept of $y = \sin x$?

 10. What is the y-intercept of $y = \cos x$?

 11. For what numbers $x, -\pi \le x \le \pi$, is the graph of $y = \sin x$ increasing?

 12. For what numbers $x, -\pi \le x \le \pi$, is the graph of $y = \cos x$ decreasing?

 13. What is the largest value of $y = \sin x$?

 14. What is the smallest value of $y = \cos x$?

 15. For what numbers $x, 0 \le x \le 2\pi$, does $\sin x = 0$?

 16. For what numbers $x, 0 \le x \le 2\pi$, does $\cos x = 0$?

 17. For what numbers $x, -2\pi \le x \le 2\pi$, does $\sin x = 1$? Where does $\sin x = -1$?

 18. For what numbers $x, -2\pi \le x \le 2\pi$, does $\cos x = 1$? Where does $\cos x = -1$?

In Problems 19–28, determine the amplitude and period of each function without graphing.

 19. $y = 2 \sin x$

 20. $y = 3 \cos x$

 21. $y = -4 \cos(2x)$

 22. $y = -\sin\left(\dfrac{1}{2}x\right)$

 23. $y = 6 \sin(\pi x)$

 24. $y = -3 \cos(3x)$

 25. $y = -\dfrac{1}{2}\cos\left(\dfrac{3}{2}x\right)$

 26. $y = \dfrac{4}{3}\sin\left(\dfrac{2}{3}x\right)$

 27. $y = \dfrac{5}{3}\sin\left(-\dfrac{2\pi}{3}x\right)$

 28. $y = \dfrac{9}{5}\cos\left(-\dfrac{3\pi}{2}x\right)$

In Problems 29–38, match the given function to one of the graphs (A)–(J).

(A)

(B)

(C)

(D)

(E)

(F)

(G)

(H)

(I)

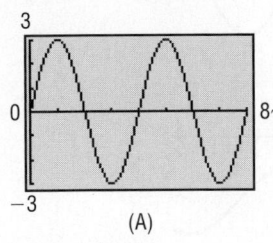

(J)

29. $y = 2 \sin\left(\dfrac{\pi}{2} x\right)$

30. $y = 2 \cos\left(\dfrac{\pi}{2} x\right)$

31. $y = 2 \cos\left(\dfrac{1}{2} x\right)$

32. $y = 3 \cos(2x)$

33. $y = -3 \sin(2x)$

34. $y = 2 \sin\left(\dfrac{1}{2} x\right)$

35. $y = -2 \cos\left(\dfrac{1}{2} x\right)$

36. $y = -2 \cos\left(\dfrac{\pi}{2} x\right)$

37. $y = 3 \sin(2x)$

38. $y = -2 \sin\left(\dfrac{1}{2} x\right)$

In Problems 39–42, match the given function to one of the graphs (A)–(D).

(A)

(B)

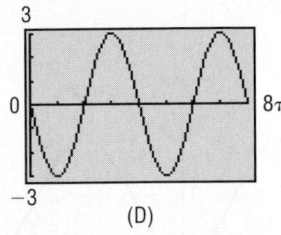

(C)

(D)

39. $y = 3 \sin\left(\dfrac{1}{2} x\right)$

40. $y = -3 \sin(2x)$

41. $y = 3 \sin(2x)$

42. $y = -3 \sin\left(\dfrac{1}{2} x\right)$

In Problems 43–66, graph each function. Be sure to label key points and show at least two cycles.

43. $y = 4 \cos x$

44. $y = 3 \sin x$

45. $y = -4 \sin x$

46. $y = -3 \cos x$

47. $y = \cos(4x)$

48. $y = \sin(3x)$

49. $y = \sin(-2x)$

50. $y = \cos(-2x)$

51. $y = 2 \sin\left(\dfrac{1}{2} x\right)$

52. $y = 2 \cos\left(\dfrac{1}{4} x\right)$

53. $y = -\dfrac{1}{2} \cos(2x)$

54. $y = -4 \sin\left(\dfrac{1}{8} x\right)$

55. $y = 2 \sin x + 3$

56. $y = 3 \cos x + 2$

57. $y = 5 \cos(\pi x) - 3$

58. $y = 4 \sin\left(\dfrac{\pi}{2} x\right) - 2$

59. $y = -6 \sin\left(\dfrac{\pi}{3} x\right) + 4$

60. $y = -3 \cos\left(\dfrac{\pi}{4} x\right) + 2$

61. $y = 5 - 3 \sin(2x)$

62. $y = 2 - 4 \cos(3x)$

63. $y = \dfrac{5}{3} \sin\left(-\dfrac{2\pi}{3} x\right)$

64. $y = \dfrac{9}{5} \cos\left(-\dfrac{3\pi}{2} x\right)$

65. $y = -\dfrac{3}{2} \cos\left(\dfrac{\pi}{4} x\right) + \dfrac{1}{2}$

66. $y = -\dfrac{1}{2} \sin\left(\dfrac{\pi}{8} x\right) + \dfrac{3}{2}$

In Problems 67–70, write the equation of a sine function that has the given characteristics.

67. Amplitude: 3
Period: π

68. Amplitude: 2
Period: 4π

69. Amplitude: 3
Period: 2

70. Amplitude: 4
Period: 1

In Problems 71–84, find an equation for each graph.

71.

72.

73.

74.

75.

76.

77.

78.

79.

80.

81.

82.

83.

84.

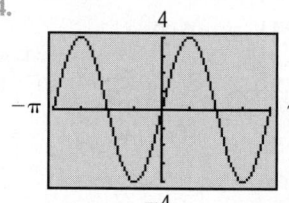

Applications and Extensions

In Problems 85–88, find the average rate of change of f from 0 to $\dfrac{\pi}{2}$.

85. $f(x) = \sin x$

86. $f(x) = \cos x$

87. $f(x) = \sin\left(\dfrac{x}{2}\right)$

88. $f(x) = \cos(2x)$

In Problems 89–92, find $(f \circ g)(x)$ and $(g \circ f)(x)$ and graph each of these functions.

89. $f(x) = \sin x$
$g(x) = 4x$

90. $f(x) = \cos x$
$g(x) = \dfrac{1}{2}x$

91. $f(x) = -2x$
$g(x) = \cos x$

92. $f(x) = -3x$
$g(x) = \sin x$

93. Alternating Current (ac) Circuits The current I, in amperes, flowing through an ac (alternating current) circuit at time t in seconds, is

$$I(t) = 220 \sin(60\pi t) \qquad t \geq 0$$

What is the period? What is the amplitude? Graph this function over two periods.

94. Alternating Current (ac) Circuits The current I, in amperes, flowing through an ac (alternating current) circuit at time t in seconds, is

$$I(t) = 120 \sin(30\pi t) \qquad t \geq 0$$

What is the period? What is the amplitude? Graph this function over two periods.

95. Alternating Current (ac) Generators The voltage V, in volts, produced by an ac generator at time t, in seconds, is

$$V(t) = 220 \sin(120\pi t)$$

(a) What is the amplitude? What is the period?
(b) Graph V over two periods, beginning at $t = 0$.

(c) If a resistance of $R = 10$ ohms is present, what is the current I?
 [**Hint:** Use Ohm's Law, $V = IR$.]
(d) What is the amplitude and period of the current I?
(e) Graph I over two periods, beginning at $t = 0$.

96. Alternating Current (ac) Generators The voltage V, in volts, produced by an ac generator at time t, in seconds, is

$$V(t) = 120 \sin(120\pi t)$$

(a) What is the amplitude? What is the period?
(b) Graph V over two periods, beginning at $t = 0$.
(c) If a resistance of $R = 20$ ohms is present, what is the current I?
 [**Hint:** Use Ohm's Law, $V = IR$.]
(d) What is the amplitude and period of the current I?
(e) Graph I over two periods, beginning at $t = 0$.

97. Alternating Current (ac) Generators The voltage V produced by an ac generator is sinusoidal. As a function of time, the voltage V is

$$V(t) = V_0 \sin(2\pi f t)$$

where f is the **frequency,** the number of complete oscillations (cycles) per second. [In the United States and Canada, f is 60 hertz (Hz).] The **power** P delivered to a resistance R at any time t is defined as

$$P(t) = \frac{[V(t)]^2}{R}$$

(a) Show that $P(t) = \dfrac{V_0^2}{R} \sin^2(2\pi f t)$.

(b) The graph of P is shown in the figure. Express P as a sinusoidal function.

Power in an ac generator

(c) Deduce that

$$\sin^2(2\pi f t) = \frac{1}{2}[1 - \cos(4\pi f t)]$$

98. Bridge Clearance A one-lane highway runs through a tunnel in the shape of one-half a sine curve cycle. The opening is 28 feet wide at road level and is 15 feet tall at its highest point.

(a) Find an equation for the sine curve that fits the opening. Place the origin at the left end of the sine curve.

(b) If the road is 14 feet wide with 7 foot shoulders on each side, what is the height of the tunnel at the edge of the road.

Sources: en.wikipedia.org/wiki/Interstate_Highway_standards and *Ohio Revised Code*

99. Biorhythms In the theory of biorhythms, a sine function of the form

$$P(t) = 50 \sin(\omega t) + 50$$

is used to measure the percent P of a person's potential at time t, where t is measured in days and $t = 0$ is the person's birthday. Three characteristics are commonly measured:

Physical potential: period of 23 days

Emotional potential: period of 28 days

Intellectual potential: period of 33 days

(a) Find ω for each characteristic.

(b) Using a graphing utility, graph all three functions on the same screen.

(c) Is there a time t when all three characteristics have 100% potential? When is it?

(d) Suppose that you are 20 years old today ($t = 7305$ days). Describe your physical, emotional, and intellectual potential for the next 30 days.

100. Graph $y = |\cos x|$, $-2\pi \le x \le 2\pi$.

101. Graph $y = |\sin x|$, $-2\pi \le x \le 2\pi$.

Discussion and Writing

102. Explain how you would scale the x-axis and y-axis before graphing $y = 3 \cos(\pi x)$.

103. Explain the term *amplitude* as it relates to the graph of a sinusoidal function.

104. Explain how the amplitude and period of a sinusoidal graph are used to establish the scale on each coordinate axis.

105. Find an application in your major field that leads to a sinusoidal graph. Write a paper about your findings.

'Are You Prepared?' Answers

1. Vertical stretch by a factor of 3

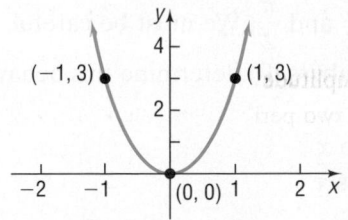

2. Reflection about the x-axis

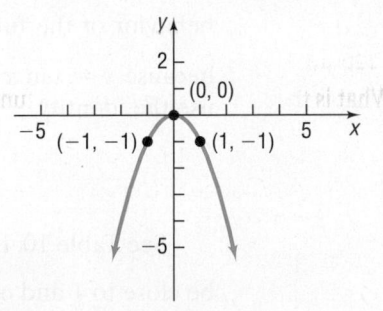

7.7 Graphs of the Tangent, Cotangent, Cosecant, and Secant Functions

PREPARING FOR THIS SECTION *Before getting started, review the following:*

- Vertical Asymptotes (Section 5.2, pp. 346–348)

Now Work the 'Are You Prepared?' problems on page 579.

OBJECTIVES 1 Graph Functions of the Form $y = A \tan(\omega x) + B$
and $y = A \cot(\omega x) + B$ (p. 576)

2 Graph Functions of the Form $y = A \csc(\omega x) + B$
and $y = A \sec(\omega x) + B$ (p. 578)

The Graph of the Tangent Function

Because the tangent function has period π, we only need to determine the graph over some interval of length π. The rest of the graph will consist of repetitions of that graph. Because the tangent function is not defined at $\ldots, -\dfrac{3\pi}{2}, -\dfrac{\pi}{2}, \dfrac{\pi}{2}, \dfrac{3\pi}{2}, \ldots$, we will concentrate on the interval $\left(-\dfrac{\pi}{2}, \dfrac{\pi}{2}\right)$, of length π, and construct Table 9, which lists some points on the graph of $y = \tan x, -\dfrac{\pi}{2} < x < \dfrac{\pi}{2}$. We plot the points in the table and connect them with a smooth curve. See Figure 92 for a partial graph of $y = \tan x$, where $-\dfrac{\pi}{3} \le x \le \dfrac{\pi}{3}$.

Table 9

x	$y = \tan x$	(x, y)
$-\dfrac{\pi}{3}$	$-\sqrt{3} \approx -1.73$	$\left(-\dfrac{\pi}{3}, -\sqrt{3}\right)$
$-\dfrac{\pi}{4}$	-1	$\left(-\dfrac{\pi}{4}, -1\right)$
$-\dfrac{\pi}{6}$	$-\dfrac{\sqrt{3}}{3} \approx -0.58$	$\left(-\dfrac{\pi}{6}, -\dfrac{\sqrt{3}}{3}\right)$
0	0	$(0, 0)$
$\dfrac{\pi}{6}$	$\dfrac{\sqrt{3}}{3} \approx 0.58$	$\left(\dfrac{\pi}{6}, \dfrac{\sqrt{3}}{3}\right)$
$\dfrac{\pi}{4}$	1	$\left(\dfrac{\pi}{4}, 1\right)$
$\dfrac{\pi}{3}$	$\sqrt{3} \approx 1.73$	$\left(\dfrac{\pi}{3}, \sqrt{3}\right)$

Figure 92

$y = \tan x, -\dfrac{\pi}{3} \le x \le \dfrac{\pi}{3}$

To complete one period of the graph of $y = \tan x$, we need to investigate the behavior of the function as x approaches $-\dfrac{\pi}{2}$ and $\dfrac{\pi}{2}$. We must be careful, though, because $y = \tan x$ is not defined at these numbers. To determine this behavior, we use the identity

$$\tan x = \frac{\sin x}{\cos x}$$

See Table 10. If x is close to $\dfrac{\pi}{2} \approx 1.5708$, but remains less than $\dfrac{\pi}{2}$, then $\sin x$ will be close to 1 and $\cos x$ will be positive and close to 0. (To see this, refer back to the

graphs of the sine function and the cosine function.) So the ratio $\dfrac{\sin x}{\cos x}$ will be positive and large. In fact, the closer x gets to $\dfrac{\pi}{2}$, the closer $\sin x$ gets to 1 and $\cos x$ gets to 0, so $\tan x$ approaches $\infty \left(\lim\limits_{x \to \frac{\pi}{2}^-} \tan x = \infty \right)$. In other words, the vertical line $x = \dfrac{\pi}{2}$ is a vertical asymptote to the graph of $y = \tan x$.

Table 10

x	$\sin x$	$\cos x$	$y = \tan x$
$\dfrac{\pi}{3} \approx 1.05$	$\dfrac{\sqrt{3}}{2}$	$\dfrac{1}{2}$	$\sqrt{3} \approx 1.73$
1.5	0.9975	0.0707	14.1
1.57	0.9999	$7.96E^{-4}$	1255.8
1.5707	0.9999	$9.6E^{-5}$	10,381
$\dfrac{\pi}{2} \approx 1.5708$	1	0	Undefined

If x is close to $-\dfrac{\pi}{2}$, but remains greater than $-\dfrac{\pi}{2}$, then $\sin x$ will be close to -1 and $\cos x$ will be positive and close to 0. The ratio $\dfrac{\sin x}{\cos x}$ approaches $-\infty \left(\lim\limits_{x \to \frac{\pi}{2}^+} \tan x = -\infty \right)$. In other words, the vertical line $x = -\dfrac{\pi}{2}$ is also a vertical asymptote to the graph.

With these observations, we can complete one period of the graph. We obtain the complete graph of $y = \tan x$ by repeating this period, as shown in Figure 93.

Figure 93
$y = \tan x$, $-\infty < x < \infty$, x not equal to odd multiples of $\dfrac{\pi}{2}$, $-\infty < y < \infty$

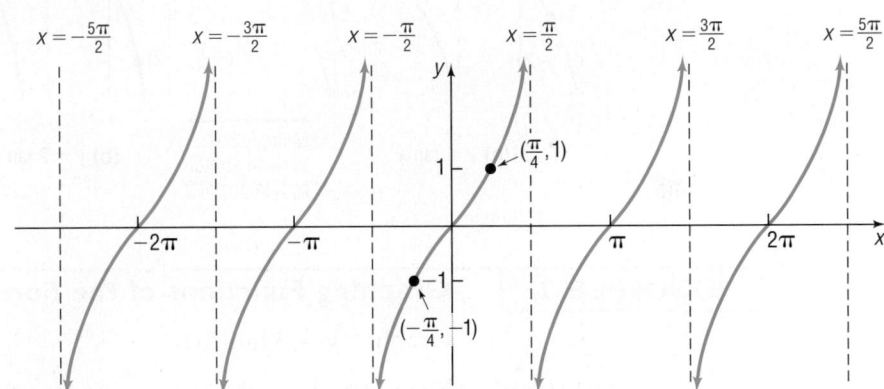

Check: Graph $Y_1 = \tan x$ and compare the result with Figure 93. Use TRACE to see what happens as x gets close to $\dfrac{\pi}{2}$, but is less than $\dfrac{\pi}{2}$. Be sure to set the WINDOW accordingly and to use DOT mode.

The graph of $y = \tan x$ in Figure 93 illustrates the following properties.

Properties of the Tangent Function

1. The domain is the set of all real numbers, except odd multiples of $\dfrac{\pi}{2}$.
2. The range is the set of all real numbers.
3. The tangent function is an odd function, as the symmetry of the graph with respect to the origin indicates.

(continued)

4. The tangent function is periodic, with period π.

5. The x-intercepts are $\ldots, -2\pi, -\pi, 0, \pi, 2\pi, 3\pi, \ldots$; the y-intercept is 0.

6. Vertical asymptotes occur at $x = \ldots, -\dfrac{3\pi}{2}, -\dfrac{\pi}{2}, \dfrac{\pi}{2}, \dfrac{3\pi}{2}, \ldots$.

Now Work PROBLEMS 7 AND 15

1 Graph Functions of the Form $y = A \tan(\omega x) + B$

For tangent functions, there is no concept of amplitude since the range of the tangent function is $(-\infty, \infty)$. The role of A in $y = A \tan(\omega x) + B$ is to provide the magnitude of the vertical stretch. The period of $y = \tan x$ is π, so the period of $y = A \tan(\omega x) + B$ is $\dfrac{\pi}{\omega}$, caused by the horizontal compression of the graph by a factor of $\dfrac{1}{\omega}$. Finally, the presence of B indicates a vertical shift is required.

EXAMPLE 1	**Graphing Functions of the Form $y = A \tan(\omega x) + B$**

Graph: $y = 2 \tan x - 1$

Solution Figure 94 shows the steps using transformations.

Figure 94

(a) $y = \tan x$ Multiply by 2; Vertical stretch by a factor of 2 (b) $y = 2 \tan x$ Subtract 1; vertically shift down one unit (c) $y = 2 \tan x - 1$

EXAMPLE 2	**Graphing Functions of the Form $y = A \tan(\omega x) + B$**

Graph: $y = 3 \tan(2x)$

Solution Figure 95 shows the steps using transformations.

Figure 95

(a) $y = \tan x$ Multiply by 3; Vertical strech by a factor of 3 (b) $y = 3 \tan x$ Replace x by $2x$; Horizontal compression by a factor of $\frac{1}{2}$ (c) $y = 3 \tan(2x)$

Notice in Figure 95(c) that the period of $y = 3\tan(2x)$ is $\dfrac{\pi}{2}$ due to the compression of the original period π by a factor of $\dfrac{1}{2}$. Also notice that the asymptotes are $x = -\dfrac{\pi}{4}$, $x = \dfrac{\pi}{4}$, $x = \dfrac{3\pi}{4}$, and so on, also due to the compression.

—— **Now Work** PROBLEM 21

The Graph of the Cotangent Function

We obtain the graph of $y = \cot x$ as we did the graph of $y = \tan x$. The period of $y = \cot x$ is π. Because the cotangent function is not defined for integer multiples of π, we will concentrate on the interval $(0, \pi)$. Table 11 lists some points on the graph of $y = \cot x$, $0 < x < \pi$. As x approaches 0, but remains greater than 0, the value of $\cos x$ will be close to 1 and the value of $\sin x$ will be positive and close to 0. Hence, the ratio $\dfrac{\cos x}{\sin x} = \cot x$ will be positive and large; so as x approaches 0, with $x > 0$, $\cot x$ approaches ∞ ($\lim\limits_{x \to 0^+} \cot x = \infty$). Similarly, as x approaches π, but remains less than π, the value of $\cos x$ will be close to -1, and the value of $\sin x$ will be positive and close to 0. So the ratio $\dfrac{\cos x}{\sin x} = \cot x$ will be negative and will approach $-\infty$ as x approaches π ($\lim\limits_{x \to \pi^-} \cot x = -\infty$). Figure 96 shows the graph.

Table 11

x	$y = \cot x$	(x, y)
$\dfrac{\pi}{6}$	$\sqrt{3}$	$\left(\dfrac{\pi}{6}, \sqrt{3}\right)$
$\dfrac{\pi}{4}$	1	$\left(\dfrac{\pi}{4}, 1\right)$
$\dfrac{\pi}{3}$	$\dfrac{\sqrt{3}}{3}$	$\left(\dfrac{\pi}{3}, \dfrac{\sqrt{3}}{3}\right)$
$\dfrac{\pi}{2}$	0	$\left(\dfrac{\pi}{2}, 0\right)$
$\dfrac{2\pi}{3}$	$-\dfrac{\sqrt{3}}{3}$	$\left(\dfrac{2\pi}{3}, -\dfrac{\sqrt{3}}{3}\right)$
$\dfrac{3\pi}{4}$	-1	$\left(\dfrac{3\pi}{4}, -1\right)$
$\dfrac{5\pi}{6}$	$-\sqrt{3}$	$\left(\dfrac{5\pi}{6}, -\sqrt{3}\right)$

Figure 96
$y = \cot x$, $-\infty < x < \infty$, x not equal to integral multiples of π, $-\infty < y < \infty$.

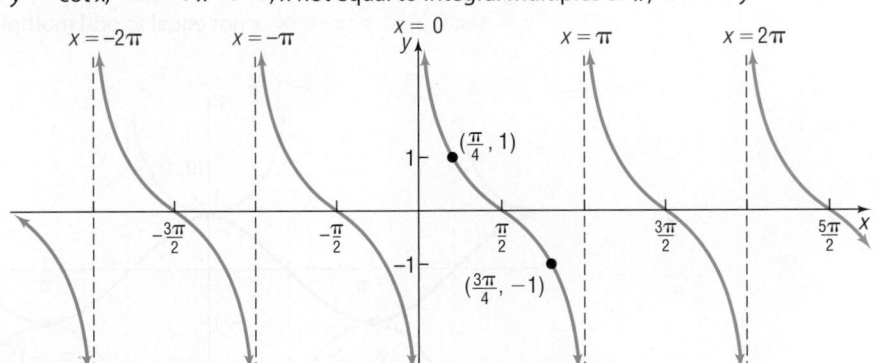

Check: Graph $Y_1 = \cot x$ and compare the result with Figure 96. Use TRACE to see what happens when x is close to 0.

The graph of $y = A\cot(\omega x) + B$ has similar characteristics to those of the tangent function. The cotangent function $y = A\cot(\omega x) + B$ has period $\dfrac{\pi}{\omega}$. The role of A is to provide the magnitude of the vertical stretch; the presence of B indicates a vertical shift is required.

—— **Now Work** PROBLEM 23

The Graph of the Cosecant Function and the Secant Function

The cosecant and secant functions, sometimes referred to as **reciprocal functions,** are graphed by making use of the reciprocal identities

$$\csc x = \frac{1}{\sin x} \quad \text{and} \quad \sec x = \frac{1}{\cos x}$$

For example, the value of the cosecant function $y = \csc x$ at a given number x equals the reciprocal of the corresponding value of the sine function, provided that the value of the sine function is not 0. If the value of $\sin x$ is 0, then x is an integer multiple of π. At such numbers, the cosecant function is not defined. In fact, the graph of the cosecant function has vertical asymptotes at integer multiples of π. Figure 97 shows the graph.

Figure 97

$y = \csc x$, $-\infty < x < \infty$, x not equal to integer multiples of π, $|y| \geq 1$

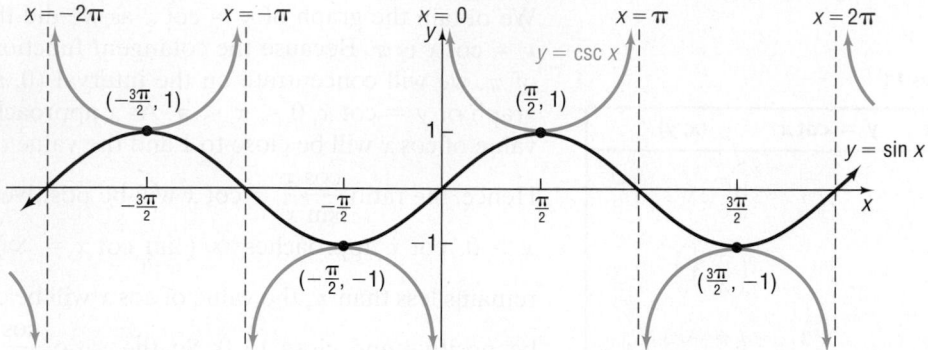

Using the idea of reciprocals, we can similarly obtain the graph of $y = \sec x$. See Figure 98.

Figure 98

$y = \sec x$, $-\infty < x < \infty$, x not equal to odd multiples of $\dfrac{\pi}{2}$, $|y| \geq 1$

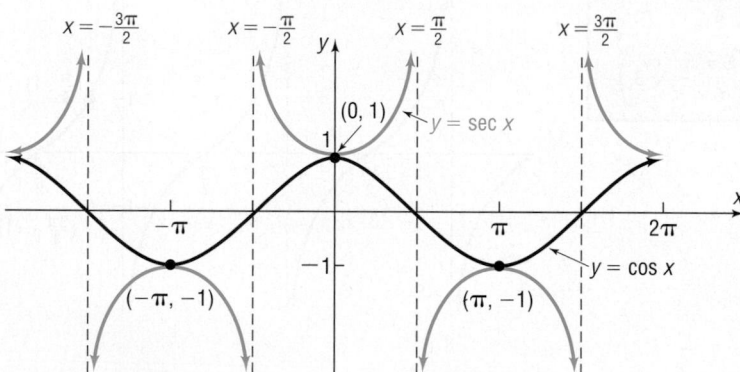

2 Graph Functions of the Form $y = A\,\csc(\omega x) + B$ and $y = A\,\sec(\omega x) + B$

The role of A in these functions is to set the range. The range of $y = \csc x$ is $\{y \mid |y| \geq 1\}$; the range of $y = A\,\csc x$ is $\{y \mid |y| \geq |A|\}$, due to the vertical stretch of the graph by a factor of $|A|$. Just as with the sine and cosine functions, the period of $y = \csc(\omega x)$ and $y = \sec(\omega x)$ becomes $\dfrac{2\pi}{\omega}$, due to the horizontal compression of the graph by a factor of $\dfrac{1}{\omega}$. The presence of B indicates a vertical shift is required.

We shall graph these functions in two ways: using transformations and using the reciprocal function.

| EXAMPLE 3 | Graphing Functions of the Form $y = A\,\csc(\omega x) + B$ |

Graph: $y = 2\,\csc x - 1$

Solution
Using Transformations Figure 99 shows the required steps. .

Figure 99

(a) $y = \csc x$

Multiply by 2;
Vertical stretch
by a factor of 2

(b) $y = 2 \csc x$

Subtract 1;
Vertical shift
down 1 unit

(c) $y = 2 \csc x - 1$

Solution
Using the Reciprocal
Function We graph $y = 2 \csc x - 1$ by first graphing the reciprocal function $y = 2 \sin x - 1$ and then filling in the graph of $y = 2 \csc x - 1$, using the idea of reciprocals. See Figure 100.

Figure 100

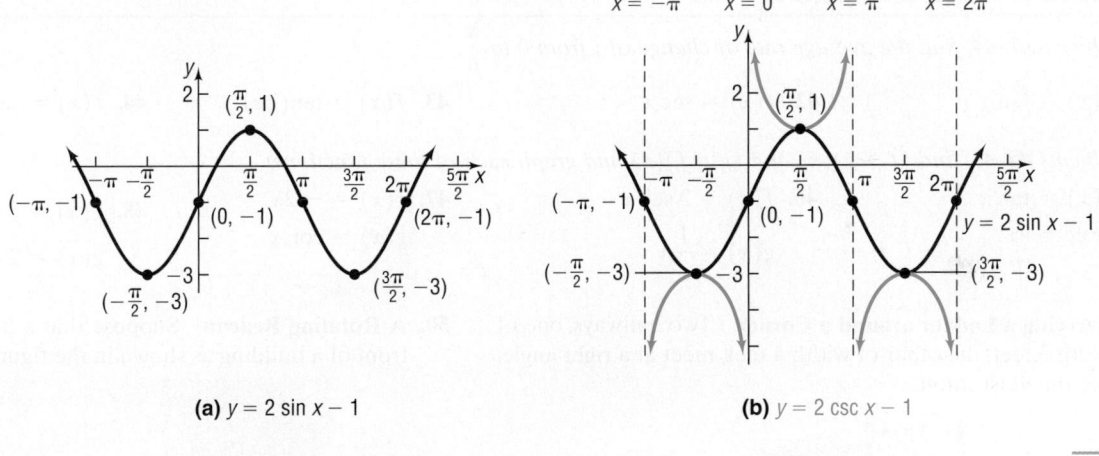

(a) $y = 2 \sin x - 1$

(b) $y = 2 \csc x - 1$

➤ **Now Work** PROBLEM **29**

7.7 Assess Your Understanding

'Are You Prepared?' *Answers are given at the end of these exercises. If you get a wrong answer, read the pages listed in red.*

1. The graph of $y = \dfrac{3x - 6}{x - 4}$ has a vertical asymptote. What is it? (pp. 346–348)

2. *True or False* A function f has at most one vertical asymptote. (pp. 346–348)

Concepts and Vocabulary

3. The graph of $y = \tan x$ is symmetric with respect to the _____ and has vertical asymptotes at _____.

4. The graph of $y = \sec x$ is symmetric with respect to the _____ and has vertical asymptotes at _____.

5. It is easiest to graph $y = \sec x$ by first sketching the graph of _____.

6. *True or False* The graphs of $y = \tan x$, $y = \cot x$, $y = \sec x$, and $y = \csc x$ each have infinitely many vertical asymptotes.

Skill Building

In Problems 7–16, if necessary, refer to the graphs to answer each question.

7. What is the y-intercept of $y = \tan x$?

8. What is the y-intercept of $y = \cot x$?

9. What is the y-intercept of $y = \sec x$?

10. What is the y-intercept of $y = \csc x$?

11. For what numbers x, $-2\pi \le x \le 2\pi$, does $\sec x = 1$? For what numbers x does $\sec x = -1$?

12. For what numbers x, $-2\pi \le x \le 2\pi$, does $\csc x = 1$? For what numbers x does $\csc x = -1$?

13. For what numbers x, $-2\pi \le x \le 2\pi$, does the graph of $y = \sec x$ have vertical asymptotes?

14. For what numbers x, $-2\pi \le x \le 2\pi$, does the graph of $y = \csc x$ have vertical asymptotes?

15. For what numbers x, $-2\pi \le x \le 2\pi$, does the graph of $y = \tan x$ have vertical asymptotes?

16. For what numbers x, $-2\pi \le x \le 2\pi$, does the graph of $y = \cot x$ have vertical asymptotes?

In Problems 17–40, graph each function. Be sure to label key points and show at least two cycles.

17. $y = 3 \tan x$

18. $y = -2 \tan x$

19. $y = 4 \cot x$

20. $y = -3 \cot x$

21. $y = \tan\left(\dfrac{\pi}{2}x\right)$

22. $y = \tan\left(\dfrac{1}{2}x\right)$

23. $y = \cot\left(\dfrac{1}{4}x\right)$

24. $y = \cot\left(\dfrac{\pi}{4}x\right)$

25. $y = 2 \sec x$

26. $y = \dfrac{1}{2} \csc x$

27. $y = -3 \csc x$

28. $y = -4 \sec x$

29. $y = 4 \sec\left(\dfrac{1}{2}x\right)$

30. $y = \dfrac{1}{2} \csc(2x)$

31. $y = -2 \csc(\pi x)$

32. $y = -3 \sec\left(\dfrac{\pi}{2}x\right)$

33. $y = \tan\left(\dfrac{1}{4}x\right) + 1$

34. $y = 2 \cot x - 1$

35. $y = \sec\left(\dfrac{2\pi}{3}x\right) + 2$

36. $y = \csc\left(\dfrac{3\pi}{2}x\right)$

37. $y = \dfrac{1}{2} \tan\left(\dfrac{1}{4}x\right) - 2$

38. $y = 3 \cot\left(\dfrac{1}{2}x\right) - 2$

39. $y = 2 \csc\left(\dfrac{1}{3}x\right) - 1$

40. $y = 3 \sec\left(\dfrac{1}{4}x\right) + 1$

Applications and Extensions

In Problems 41–44, find the average rate of change of f from 0 to $\dfrac{\pi}{6}$.

41. $f(x) = \tan x$

42. $f(x) = \sec x$

43. $f(x) = \tan(2x)$

44. $f(x) = \sec(2x)$

In Problems 45–48, find $(f \circ g)(x)$ and $(g \circ f)(x)$ and graph each of these functions.

45. $f(x) = \tan x$
$g(x) = 4x$

46. $f(x) = 2 \sec x$
$g(x) = \dfrac{1}{2}x$

47. $f(x) = -2x$
$g(x) = \cot x$

48. $f(x) = \dfrac{1}{2}x$
$g(x) = 2 \csc x$

49. Carrying a Ladder around a Corner Two hallways, one of width 3 feet, the other of width 4 feet, meet at a right angle. See the illustration.

(a) Show that the length L of the line segment shown as a function of the angle θ is

$$L(\theta) = 3 \sec \theta + 4 \csc \theta$$

(b) Graph $L = L(\theta)$, $0 < \theta < \dfrac{\pi}{2}$.

(c) For what value of θ is L the least?

(d) What is the length of the longest ladder that can be carried around the corner? Why is this also the least value of L?

50. A Rotating Beacon Suppose that a fire truck is parked in front of a building as shown in the figure.

The beacon light on top of the fire truck is located 10 feet from the wall and has a light on each side. If the beacon light rotates 1 revolution every 2 seconds, then a model for determining the distance d that the beacon of light is from point A on the wall after t seconds is given by

$$d(t) = |10 \tan(\pi t)|$$

(a) Graph $d(t) = |10 \tan(\pi t)|$ for $0 \le t \le 2$.

(b) For what values of t is the function undefined? Explain what this means in terms of the beam of light on the wall.

(c) Fill in the following table.

t	0	0.1	0.2	0.3	0.4
$d(t) = 10\tan(\pi t)$					

(d) Compute $\dfrac{d(0.1) - d(0)}{0.1 - 0}$, $\dfrac{d(0.2) - d(0.1)}{0.2 - 0.1}$, and so on, for each consecutive value of t. These are called **first differences**.

(e) Interpret the first differences found in part (d). What is happening to the speed of the beam of light as d increases?

51. Exploration Graph

$$y = \tan x \quad \text{and} \quad y = -\cot\left(x + \frac{\pi}{2}\right)$$

Do you think that $\tan x = -\cot\left(x + \dfrac{\pi}{2}\right)$?

'Are You Prepared?' Answers

1. $x = 4$ **2.** False

7.8 Phase Shift; Sinusoidal Curve Fitting

> **OBJECTIVES** **1** Graph Sinusoidal Functions of the Form $y = A\sin(\omega x - \phi) + B$ (p. 581)
>
> **2** Find a Sinusoidal Function from Data (p. 585)

1 Graph Sinusoidal Functions of the Form $y = A\sin(\omega x - \phi) + B$

Figure 101
One cycle $y = A\sin(\omega x)$, $A > 0$, $\omega > 0$

We have seen that the graph of $y = A\sin(\omega x)$, $\omega > 0$, has amplitude $|A|$ and period $T = \dfrac{2\pi}{\omega}$. One cycle can be drawn as x varies from 0 to $\dfrac{2\pi}{\omega}$ or, equivalently, as ωx varies from 0 to 2π. See Figure 101.

We now want to discuss the graph of

$$y = A\sin(\omega x - \phi)$$

which may also be written as

$$y = A\sin\left[\omega\left(x - \frac{\phi}{\omega}\right)\right]$$

where $\omega > 0$ and ϕ (the Greek letter phi) are real numbers. The graph will be a sine curve with amplitude $|A|$. As $\omega x - \phi$ varies from 0 to 2π, one period will be traced out. This period will begin when

$$\omega x - \phi = 0 \quad \text{or} \quad x = \frac{\phi}{\omega}$$

and will end when

$$\omega x - \phi = 2\pi \quad \text{or} \quad x = \frac{\phi}{\omega} + \frac{2\pi}{\omega}$$

See Figure 102.

NOTE We can also find the beginning and end of the period by solving the inequality:

$$0 \le \omega x - \phi \le 2\pi$$
$$\phi \le \omega x \le 2\pi + \phi$$
$$\frac{\phi}{\omega} \le x \le \frac{2\pi}{\omega} + \frac{\phi}{\omega}$$ ∎

Figure 102
One cycle $y = A\sin(\omega x - \phi)$, $A > 0$, $\omega > 0$, $\phi > 0$

We see that the graph of $y = A\sin(\omega x - \phi) = A\sin\left[\omega\left(x - \dfrac{\phi}{\omega}\right)\right]$ is the same as the graph of $y = A\sin(\omega x)$, except that it has been shifted $\dfrac{\phi}{\omega}$ units (to the right if $\phi > 0$ and to the left if $\phi < 0$). This number $\dfrac{\phi}{\omega}$ is called the **phase shift** of the graph of $y = A\sin(\omega x - \phi)$.

For the graphs of $y = A\sin(\omega x - \phi)$ or $y = A\cos(\omega x - \phi)$, $\omega > 0$,

$$\text{Amplitude} = |A| \qquad \text{Period} = T = \frac{2\pi}{\omega} \qquad \text{Phase shift} = \frac{\phi}{\omega}$$

The phase shift is to the left if $\phi < 0$ and to the right if $\phi > 0$.

EXAMPLE 1

Finding the Amplitude, Period, and Phase Shift of a Sinusoidal Function and Graphing It

Find the amplitude, period, and phase shift of $y = 3\sin(2x - \pi)$ and graph the function.

Solution Comparing

$$y = 3\sin(2x - \pi) = 3\sin\left[2\left(x - \frac{\pi}{2}\right)\right]$$

to

$$y = A\sin(\omega x - \phi) = A\sin\left[\omega\left(x - \frac{\phi}{\omega}\right)\right]$$

we find that $A = 3$, $\omega = 2$, and $\phi = \pi$. The graph is a sine curve with amplitude $|A| = 3$, period $T = \dfrac{2\pi}{\omega} = \dfrac{2\pi}{2} = \pi$, and phase shift $= \dfrac{\phi}{\omega} = \dfrac{\pi}{2}$.

NOTE We also can find the interval defining one cycle by solving the inequality
$$0 \le 2x - \pi \le 2\pi.$$
Then
$$\pi \le 2x \le 3\pi$$
$$\frac{\pi}{2} \le x \le \frac{3\pi}{2}$$ ∎

The graph of $y = 3\sin(2x - \pi)$ will lie between -3 and 3 on the y-axis. One cycle will begin at $x = \dfrac{\phi}{\omega} = \dfrac{\pi}{2}$ and end at $x = \dfrac{\phi}{\omega} + \dfrac{2\pi}{\omega} = \dfrac{\pi}{2} + \pi = \dfrac{3\pi}{2}$. To find the five key points, we divide the interval $\left[\dfrac{\pi}{2}, \dfrac{3\pi}{2}\right]$ into four subintervals, each of length $\pi \div 4 = \dfrac{\pi}{4}$, by finding the following values of x:

$$\underset{\text{initial value}}{\frac{\pi}{2}} \qquad \underset{\text{2nd value}}{\frac{\pi}{2} + \frac{\pi}{4} = \frac{3\pi}{4}} \qquad \underset{\text{3rd value}}{\frac{3\pi}{4} + \frac{\pi}{4} = \pi} \qquad \underset{\text{4th value}}{\pi + \frac{\pi}{4} = \frac{5\pi}{4}} \qquad \underset{\text{final value}}{\frac{5\pi}{4} + \frac{\pi}{4} = \frac{3\pi}{2}}$$

Use these values of x to determine the five key points on the graph:

$$\left(\frac{\pi}{2}, 0\right), \quad \left(\frac{3\pi}{4}, 3\right), \quad (\pi, 0), \quad \left(\frac{5\pi}{4}, -3\right), \quad \left(\frac{3\pi}{2}, 0\right)$$

We plot these five points and fill in the graph of the sine function as shown in Figure 103(a). Extending the graph in each direction, we obtain Figure 103(b).

Figure 103

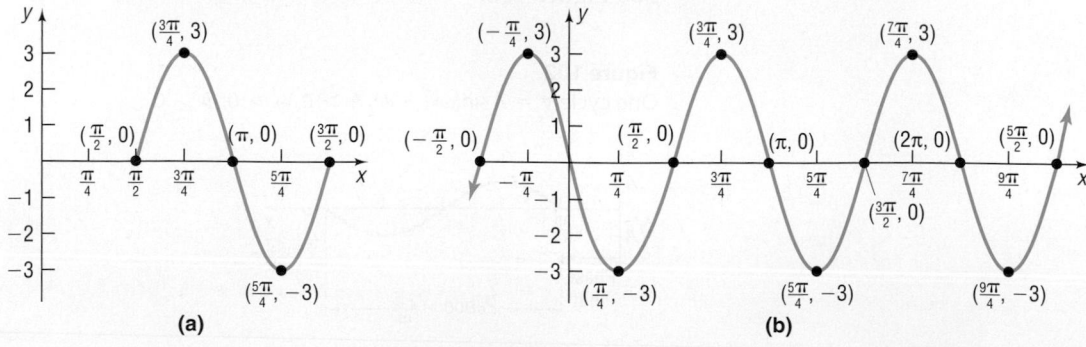

(a)

(b)

The graph of $y = 3 \sin(2x - \pi) = 3 \sin\left[2\left(x - \dfrac{\pi}{2}\right)\right]$ may also be obtained using transformations. See Figure 104.

Figure 104

Multiply by 3;
Vertical stretch
by a factor of 3

Replace x by $2x$;
Horizontal compression
by a factor of $\frac{1}{2}$

Replace x by $x - \frac{\pi}{2}$;
Shift right
$\frac{\pi}{2}$ units

(a) $y = \sin x$

(b) $y = 3 \sin x$

(c) $y = 3 \sin (2x)$

(d) $y = 3 \sin\left[2\left(x - \frac{\pi}{2}\right)\right]$
$= 3 \sin (2x - \pi)$

To graph a sinusoidal function of the form $y = A \sin(\omega x - \phi) + B$, we first graph the function $y = A \sin(\omega x - \phi)$ and then apply a vertical shift.

EXAMPLE 2	**Finding the Amplitude, Period, and Phase Shift of a Sinusoidal Function and Graphing It**

Find the amplitude, period, and phase shift of $y = 2 \cos(4x + 3\pi) + 1$ and graph the function.

Solution We begin by graphing $y = 2 \cos(4x + \pi)$. Comparing

$$y = 2 \cos(4x + 3\pi) = 2 \cos\left[4\left(x + \dfrac{3\pi}{4}\right)\right]$$

to

$$y = A \cos(\omega x - \phi) = A \cos\left[\omega\left(x - \dfrac{\phi}{\omega}\right)\right]$$

NOTE We can also find the interval defining one cycle by solving the inequality

$$0 \le 4x + 3\pi \le 2\pi$$

Then

$$-3\pi \le 4x \le -\pi$$

$$-\dfrac{3\pi}{4} \le x \le -\dfrac{\pi}{4}$$

we see that $A = 2$, $\omega = 4$, and $\phi = -3\pi$. The graph is a cosine curve with amplitude $|A| = 2$, period $T = \dfrac{2\pi}{\omega} = \dfrac{2\pi}{4} = \dfrac{\pi}{2}$, and phase shift $= \dfrac{\phi}{\omega} = -\dfrac{3\pi}{4}$.

The graph of $y = 2 \cos(4x + 3\pi)$ will lie between -2 and 2 on the y-axis. One cycle will begin at $x = \dfrac{\phi}{\omega} = -\dfrac{3\pi}{4}$ and end at $x = \dfrac{\phi}{\omega} + \dfrac{2\pi}{\omega} = -\dfrac{3\pi}{4} + \dfrac{\pi}{2} = -\dfrac{\pi}{4}$.

To find the five key points, we divide the interval $\left[-\dfrac{3\pi}{4}, -\dfrac{\pi}{4}\right]$ into four subintervals, each of the length $\dfrac{\pi}{2} \div 4 = \dfrac{\pi}{8}$, by finding the following values.

$-\dfrac{3\pi}{4}$ $\quad -\dfrac{3\pi}{4} + \dfrac{\pi}{8} = -\dfrac{5\pi}{8}$ $\quad -\dfrac{5\pi}{8} + \dfrac{\pi}{8} = -\dfrac{\pi}{2}$ $\quad -\dfrac{\pi}{2} + \dfrac{\pi}{8} = -\dfrac{3\pi}{8}$ $\quad -\dfrac{3\pi}{8} + \dfrac{\pi}{8} = -\dfrac{\pi}{4}$

initial value 2nd value 3rd value 4th value final value

The five key points on the graph of $y = 2 \cos(4x + \pi)$ are

$$\left(-\dfrac{3\pi}{4}, 2\right), \quad \left(-\dfrac{5\pi}{8}, 0\right), \quad \left(-\dfrac{\pi}{2}, -2\right), \quad \left(-\dfrac{3\pi}{8}, 0\right), \quad \left(-\dfrac{\pi}{4}, 2\right)$$

We plot these five points and fill in the graph of the cosine function as shown in Figure 105(a). Extending the graph in each direction, we obtain Figure 105(b), the graph of $y = 2 \cos(4x + \pi)$. A vertical shift up 1 unit gives the final graph. See Figure 105(c).

Figure 105

(a)

(b) $y = 2\cos(4x + 3\pi)$

Add 1; Vertical shift up 1 unit

(c) $y = 2\cos(4x + 3\pi) + 1$

The graph of $y = 2\cos(4x + 3\pi) + 1 = 2\cos\left[4\left(x + \dfrac{3\pi}{4}\right)\right] + 1$ may also be obtained using transformations. See Figure 106.

Figure 106

(a) $y = 2\cos x$

Replace x by $4x$; Horizontal compression by a factor of $\frac{1}{4}$

(b) $y = 2\cos(4x)$

Replace x by $x + \frac{3\pi}{4}$; Shift left $\frac{3\pi}{4}$ units

(c) $y = 2\cos\left[4\left(x + \frac{3\pi}{4}\right)\right]$ $= 2\cos(4x + 3\pi)$

Add 1; Vertical shift up 1 unit

(d) $y = 2\cos(4x + 3\pi) + 1$

Now Work PROBLEM 3

SUMMARY Steps for Graphing Sinusoidal Functions $y = A\sin(\omega x - \phi) + B$ or $y = A\cos(\omega x - \phi) + B$

STEP 1: Determine the amplitude $|A|$ and period $T = \dfrac{2\pi}{\omega}$.

STEP 2: Determine the starting point of one cycle of the graph, $\dfrac{\phi}{\omega}$.

STEP 3: Determine the ending point of one cycle of the graph, $\dfrac{\phi}{\omega} + \dfrac{2\pi}{\omega}$.

STEP 4: Divide the interval $\left[\dfrac{\phi}{\omega}, \dfrac{\phi}{\omega} + \dfrac{2\pi}{\omega}\right]$ into four subintervals, each of length $\dfrac{2\pi}{\omega} \div 4$.

STEP 5: Use the endpoints of the subintervals to find the five key points on the graph.

STEP 6: Fill in one cycle of the graph.

STEP 7: Extend the graph in each direction to make it complete.

STEP 8: If $B \neq 0$, apply a vertical shift.

2 Find a Sinusoidal Function from Data

Scatter diagrams of data sometimes take the form of a sinusoidal function. Let's look at an example.

The data given in Table 12 represent the average monthly temperatures in Denver, Colorado. Since the data represent *average* monthly temperatures collected over many years, the data will not vary much from year to year and so will essentially repeat each year. In other words, the data are periodic. Figure 107 shows the scatter diagram of these data repeated over 2 years, where $x = 1$ represents January, $x = 2$ represents February, and so on.

Table 12

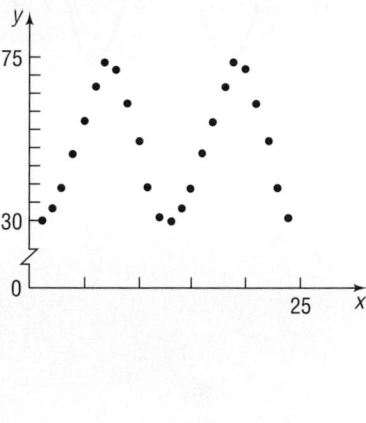

Month, x	Average Monthly Temperature, °F
January, 1	29.7
February, 2	33.4
March, 3	39.0
April, 4	48.2
May, 5	57.2
June, 6	66.9
July, 7	73.5
August, 8	71.4
September, 9	62.3
October, 10	51.4
November, 11	39.0
December, 12	31.0

SOURCE: U.S. National Oceanic and Atmospheric Administration

Figure 107

Notice that the scatter diagram looks like the graph of a sinusoidal function. We choose to fit the data to a sine function of the form

$$y = A \sin(\omega x - \phi) + B$$

where A, B, ω, and ϕ are constants.

EXAMPLE 3 **Finding a Sinusoidal Function from Temperature Data**

Fit a sine function to the data in Table 12.

Solution We begin with a scatter diagram of the data for one year. See Figure 108. The data will be fitted to a sine function of the form

$$y = A \sin(\omega x - \phi) + B$$

STEP 1: To find the amplitude A, we compute

$$\text{Amplitude} = \frac{\text{largest data value} - \text{smallest data value}}{2}$$

$$= \frac{73.5 - 29.7}{2} = 21.9$$

To see the remaining steps in this process, we superimpose the graph of the function $y = 21.9 \sin x$, where x represents months, on the scatter diagram.

Figure 108

Figure 109

Figure 109 shows the two graphs. To fit the data, the graph needs to be shifted vertically, shifted horizontally, and stretched horizontally.

STEP 2: We determine the vertical shift by finding the average of the highest and lowest data values.

$$\text{Vertical shift} = \frac{73.5 + 29.7}{2} = 51.6$$

Now we superimpose the graph of $y = 21.9 \sin x + 51.6$ on the scatter diagram. See Figure 110.

We see that the graph needs to be shifted horizontally and stretched horizontally.

STEP 3: It is easier to find the horizontal stretch factor first. Since the temperatures repeat every 12 months, the period of the function is $T = 12$. Since $T = \dfrac{2\pi}{\omega} = 12$, we find

$$\omega = \frac{2\pi}{12} = \frac{\pi}{6}$$

Now we superimpose the graph of $y = 21.9 \sin\left(\dfrac{\pi}{6}x\right) + 51.6$ on the scatter diagram. See Figure 111. We see that the graph still needs to be shifted horizontally.

Figure 110

Figure 111

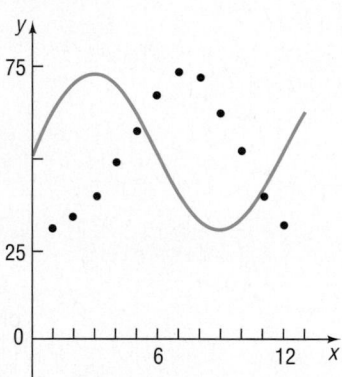

STEP 4: To determine the horizontal shift, we use the period $T = 12$ and divide the interval $[0, 12]$ into four subintervals of length $12 \div 4 = 3$:

$$[0, 3], \quad [3, 6], \quad [6, 9], \quad [9, 12]$$

The sine curve is increasing on the interval $(0, 3)$ and is decreasing on the interval $(3, 9)$, so a local maximum occurs at $x = 3$. The data indicate that a maximum occurs at $x = 7$ (corresponding to July's temperature), so we must shift the graph of the function 4 units to the right by replacing x by $x - 4$. Doing this, we obtain

$$y = 21.9 \sin\left(\frac{\pi}{6}(x - 4)\right) + 51.6$$

Multiplying out, we find that a sine function of the form $y = A \sin(\omega x - \phi) + B$ that fits the data is

$$y = 21.9 \sin\left(\frac{\pi}{6}x - \frac{2\pi}{3}\right) + 51.6$$

Figure 112

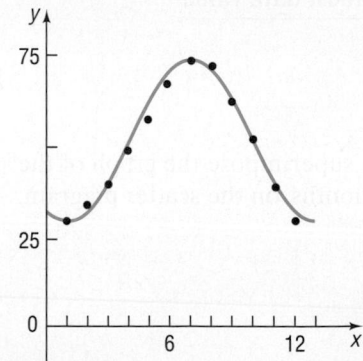

The graph of $y = 21.9 \sin\left(\dfrac{\pi}{6}x - \dfrac{2\pi}{3}\right) + 51.6$ and the scatter diagram of the data are shown in Figure 112.

The steps to fit a sine function

$$y = A \sin(\omega x - \phi) + B$$

to sinusoidal data follow:

Steps for Fitting Data to a Sine Function $y = A \sin(\omega x - \phi) + B$

STEP 1: Determine A, the amplitude of the function.

$$\text{Amplitude} = \frac{\text{largest data value} - \text{smallest data value}}{2}$$

STEP 2: Determine B, the vertical shift of the function.

$$\text{Vertical shift} = \frac{\text{largest data value} + \text{smallest data value}}{2}$$

STEP 3: Determine ω. Since the period T, the time it takes for the data to repeat, is $T = \dfrac{2\pi}{\omega}$, we have

$$\omega = \frac{2\pi}{T}$$

STEP 4: Determine the horizontal shift of the function by using the period of the data. Divide the period into four subintervals of equal length. Determine the x-coordinate for the maximum of the sine function and the x-coordinate for the maximum value of the data. Use this information to determine the value of the phase shift, $\dfrac{\phi}{\omega}$.

Now Work PROBLEM 29(a)–(c)

Let's look at another example. Since the number of hours of sunlight in a day cycles annually, the number of hours of sunlight in a day for a given location can be modeled by a sinusoidal function.

The longest day of the year (in terms of hours of sunlight) occurs on the day of the summer solstice. For locations in the northern hemisphere, the summer solstice is the time when the sun is farthest north. In 2005, the summer solstice occurred on June 21 (the 172nd day of the year) at 2:46 AM EDT. The shortest day of the year occurs on the day of the winter solstice. The winter solstice is the time when the Sun is farthest south (again, for locations in the northern hemisphere). In 2005, the winter solstice occurred on December 21 (the 355th day of the year) at 1:35 PM (EST).

EXAMPLE 4

Finding a Sinusoidal Function for Hours of Daylight

According to the *Old Farmer's Almanac*, the number of hours of sunlight in Boston on the summer solstice is 15.30 and the number of hours of sunlight on the winter solstice is 9.08.

(a) Find a sinusoidal function of the form $y = A \sin(\omega x - \phi) + B$ that fits the data.

(b) Use the function found in part (a) to predict the number of hours of sunlight on April 1, the 91st day of the year.

(c) Draw a graph of the function found in part (a).

(d) Look up the number of hours of sunlight for April 1 in the *Old Farmer's Almanac* and compare it to the results found in part (b).

Source: The *Old Farmer's Almanac*, www.almanac.com/rise

Solution

(a) **STEP 1:** $\text{Amplitude} = \dfrac{\text{largest data value} - \text{smallest data value}}{2}$

$$= \frac{15.30 - 9.08}{2} = 3.11$$

STEP 2: Vertical shift = $\dfrac{\text{largest data value} + \text{smallest data value}}{2}$

$$= \dfrac{15.30 + 9.08}{2} = 12.19$$

STEP 3: The data repeat every 365 days. Since $T = \dfrac{2\pi}{\omega} = 365$, we find

$$\omega = \dfrac{2\pi}{365}$$

So far, we have $y = 3.11 \sin\left(\dfrac{2\pi}{365}x - \phi\right) + 12.19$.

STEP 4: To determine the horizontal shift, we use the period $T = 365$ and divide the interval $[0, 365]$ into four subintervals of length $365 \div 4 = 91.25$:

$$[0, 91.25], \quad [91.25, 182.5], \quad [182.5, 273.75], \quad [273.75, 365]$$

The sine curve is increasing on the interval $(0, 91.25)$ and is decreasing on the interval $(91.25, 273.75)$, so a local maximum occurs at $x = 91.25$. Since the maximum occurs on the summer solstice at $x = 172$, we must shift the graph of the function $172 - 91.25 = 80.75$ units to the right by replacing x by $x - 80.75$. Doing this, we obtain

$$y = 3.11 \sin\left(\dfrac{2\pi}{365}(x - 80.75)\right) + 12.19$$

Multiplying out, we find that a sine function of the form $y = A \sin(\omega x - \phi) + B$ that fits the data is

$$y = 3.11 \sin\left(\dfrac{2\pi}{365}x - \dfrac{323\pi}{730}\right) + 12.19$$

(b) To predict the number of hours of daylight on April 1, we let $x = 91$ in the function found in part (a) and obtain

$$y = 3.11 \sin\left(\dfrac{2\pi}{365} \cdot 91 - \dfrac{323}{730}\pi\right) + 12.19$$

$$\approx 12.74$$

So we predict that there will be about 12.74 hours = 12 hours, 44 minutes of sunlight on April 1 in Boston.

(c) The graph of the function found in part (a) is given in Figure 113.

(d) According to the *Old Farmer's Almanac*, there will be 12 hours 45 minutes of sunlight on April 1 in Boston.

Figure 113

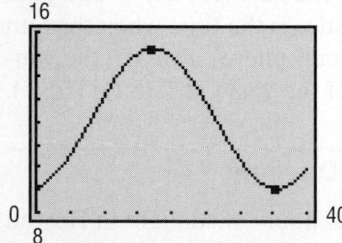

Now Work PROBLEM 35

Certain graphing utilities (such as a TI-83, TI-84 Plus, and TI-86) have the capability of finding the sine function of best fit for sinusoidal data. At least four data points are required for this process.

EXAMPLE 5	Finding the Sine Function of Best Fit

Use a graphing utility to find the sine function of best fit for the data in Table 12. Graph this function with the scatter diagram of the data.

Solution Enter the data from Table 12 and execute the SINe REGression program. The result is shown in Figure 114.

The output that the utility provides shows the equation

$$y = a \sin(bx + c) + d$$

The sinusoidal function of best fit is

$$y = 21.15 \sin(0.55x - 2.35) + 51.19$$

where x represents the month and y represents the average temperature.
Figure 115 shows the graph of the sinusoidal function of best fit on the scatter diagram.

Figure 114

Figure 115

Now Work PROBLEMS 29(d) AND (e)

7.8 Assess Your Understanding

Concepts and Vocabulary

1. For the graph of $y = A \sin(\omega x - \phi)$, the number $\dfrac{\phi}{\omega}$ is called the _____.

2. **True or False** Only two data points are required by a graphing utility to find the sine function of best fit.

Skill Building

In Problems 3–14, find the amplitude, period, and phase shift of each function. Graph each function. Be sure to label key points. Show at least two periods.

3. $y = 4 \sin(2x - \pi)$

4. $y = 3 \sin(3x - \pi)$

5. $y = 2 \cos\left(3x + \dfrac{\pi}{2}\right)$

6. $y = 3 \cos(2x + \pi)$

7. $y = -3 \sin\left(2x + \dfrac{\pi}{2}\right)$

8. $y = -2 \cos\left(2x - \dfrac{\pi}{2}\right)$

9. $y = 4 \sin(\pi x + 2) - 5$

10. $y = 2 \cos(2\pi x + 4) + 4$

11. $y = 3 \cos(\pi x - 2) + 5$

12. $y = 2 \cos(2\pi x - 4) - 1$

13. $y = -3 \sin\left(-2x + \dfrac{\pi}{2}\right)$

14. $y = -3 \cos\left(-2x + \dfrac{\pi}{2}\right)$

In Problems 15–18, write the equation of a sine function that has the given characteristics.

15. Amplitude: 2
Period: π
Phase shift: $\dfrac{1}{2}$

16. Amplitude: 3
Period: $\dfrac{\pi}{2}$
Phase shift: 2

17. Amplitude: 3
Period: 3π
Phase shift: $-\dfrac{1}{3}$

18. Amplitude: 2
Period: π
Phase shift: -2

Applications and Extensions

In Problems 19–26, apply the methods of this and the previous section to graph each function. Be sure to label key points and show at least two periods.

19. $y = 2 \tan(4x - \pi)$

20. $y = \dfrac{1}{2} \cot(2x - \pi)$

21. $y = 3 \csc\left(2x - \dfrac{\pi}{4}\right)$

22. $y = \dfrac{1}{2} \sec(3x - \pi)$

23. $y = -\cot\left(2x + \dfrac{\pi}{2}\right)$

24. $y = -\tan\left(3x + \dfrac{\pi}{2}\right)$

25. $y = -\sec(2\pi x + \pi)$

26. $y = -\csc\left(-\dfrac{1}{2}\pi x + \dfrac{\pi}{4}\right)$

27. Alternating Current (ac) Circuits The current I, in amperes, flowing through an ac (alternating current) circuit at time t, in seconds, is

$$I(t) = 120 \sin\left(30\pi t - \frac{\pi}{3}\right), \qquad t \geq 0$$

What is the period? What is the amplitude? What is the phase shift? Graph this function over two periods.

28. Alternating Current (ac) Circuits The current I, in amperes, flowing through an ac (alternating current) circuit at time t, in seconds, is

$$I(t) = 220 \sin\left(60\pi t - \frac{\pi}{6}\right), \qquad t \geq 0$$

What is the period? What is the amplitude? What is the phase shift? Graph this function over two periods.

29. Monthly Temperature The following data represent the average monthly temperatures for Juneau, Alaska.

Month, x	Average Monthly Temperature, °F
January, 1	24.2
February, 2	28.4
March, 3	32.7
April, 4	39.7
May, 5	47.0
June, 6	53.0
July, 7	56.0
August, 8	55.0
September, 9	49.4
October, 10	42.2
November, 11	32.0
December, 12	27.1

Source: U.S. National Oceanic and Atmospheric Administration

(a) Draw a scatter diagram of the data for one period.
(b) Find a sinusoidal function of the form
 $y = A \sin(\omega x - \phi) + B$ that fits the data.
(c) Draw the sinusoidal function found in part (b) on the scatter diagram.
(d) Use a graphing utility to find the sinusoidal function of best fit.
(e) Draw the sinusoidal function of best fit on a scatter diagram of the data.

30. Monthly Temperature The following data represent the average monthly temperatures for Washington, D.C.

(a) Draw a scatter diagram of the data for one period.
(b) Find a sinusoidal function of the form
 $y = A \sin(\omega x - \phi) + B$ that fits the data.
(c) Draw the sinusoidal function found in part (b) on the scatter diagram.
(d) Use a graphing utility to find the sinusoidal function of best fit.
(e) Graph the sinusoidal function of best fit on a scatter diagram of the data.

Month, x	Average Monthly Temperature, °F
January, 1	34.6
February, 2	37.5
March, 3	47.2
April, 4	56.5
May, 5	66.4
June, 6	75.6
July, 7	80.0
August, 8	78.5
September, 9	71.3
October, 10	59.7
November, 11	49.8
December, 12	39.4

Source: U.S. National Oceanic and Atmospheric Administration

31. Monthly Temperature The following data represent the average monthly temperatures for Indianapolis, Indiana.

Month, x	Average Monthly Temperature, °F
January, 1	25.5
February, 2	29.6
March, 3	41.4
April, 4	52.4
May, 5	62.8
June, 6	71.9
July, 7	75.4
August, 8	73.2
September, 9	66.6
October, 10	54.7
November, 11	43.0
December, 12	30.9

Source: U.S. National Oceanic and Atmospheric Administration

(a) Draw a scatter diagram of the data for one period.
(b) Find a sinusoidal function of the form
 $y = A \sin(\omega x - \phi) + B$ that fits the data.
(c) Draw the sinusoidal function found in part (b) on the scatter diagram.
(d) Use a graphing utility to find the sinusoidal function of best fit.
(e) Graph the sinusoidal function of best fit on a scatter diagram of the data.

32. Monthly Temperature The following data represent the average monthly temperatures for Baltimore, Maryland.

(a) Draw a scatter diagram of the data for one period.

Month, x	Average Monthly Temperature, °F
January, 1	31.8
February, 2	34.8
March, 3	44.1
April, 4	53.4
May, 5	63.4
June, 6	72.5
July, 7	77.0
August, 8	75.6
September, 9	68.5
October, 10	56.6
November, 11	46.8
December, 12	36.7

SOURCE: U.S. National Oceanic and Atmospheric Administration

(b) Find a sinusoidal function of the form
$y = A \sin(\omega x - \phi) + B$ that fits the data.

(c) Draw the sinusoidal function found in part (b) on the scatter diagram.

(d) Use a graphing utility to find the sinusoidal function of best fit.

(e) Graph the sinusoidal function of best fit on a scatter diagram of the data.

33. Tides Suppose that the length of time between consecutive high tides is approximately 12.5 hours. According to the National Oceanic and Atmospheric Administration, on Saturday, August 7, 2004, in Savannah, Georgia, high tide occurred at 3:38 AM (3.6333 hours) and low tide occurred at 10:08 AM (10.1333 hours). Water heights are measured as the amounts above or below the mean lower low water. The height of the water at high tide was 8.2 feet, and the height of the water at low tide was −0.6 foot.

(a) Approximately when will the next high tide occur?

(b) Find a sinusoidal function of the form
$y = A \sin(\omega x - \phi) + B$ that fits the data.

(c) Draw a graph of the function found in part (b).

(d) Use the function found in part (b) to predict the height of the water at the next high tide.

34. Tides Suppose that the length of time between consecutive high tides is approximately 12.5 hours. According to the National Oceanic and Atmospheric Administration, on Saturday, August 7, 2004, in Juneau, Alaska, high tide occurred at 8:11 AM (8.1833 hours) and low tide occurred at 2:14 PM (14.2333 hours). Water heights are measured as the amounts above or below the mean lower low water. The height of the water at high tide was 13.2 feet, and the height of the water at low tide was 2.2 feet.

(a) Approximately when will the next high tide occur?

(b) Find a sinusoidal function of the form
$y = A \sin(\omega x - \phi) + B$ that fits the data.

(c) Draw a graph of the function found in part (b).

(d) Use the function found in part (b) to predict the height of the water at the next high tide.

35. Hours of Daylight According to the *Old Farmer's Almanac*, in Miami, Florida, the number of hours of sunlight on the summer solstice of 2005 is 13.75 and the number of hours of sunlight on the winter solstice is 10.53.

(a) Find a sinusoidal function of the form
$y = A \sin(\omega x - \phi) + B$ that fits the data.

(b) Use the function found in part (a) to predict the number of hours of sunlight on April 1, the 91st day of the year.

(c) Draw a graph of the function found in part (a).

(d) Look up the number of hours of sunlight for April 1 in the *Old Farmer's Almanac*, and compare the actual hours of daylight to the results found in part (c).

36. Hours of Daylight According to the *Old Farmer's Almanac*, in Detroit, Michigan, the number of hours of sunlight on the summer solstice of 2005 is 15.30 and the number of hours of sunlight on the winter solstice is 9.08.

(a) Find a sinusoidal function of the form
$y = A \sin(\omega x - \phi) + B$ that fits the data.

(b) Use the function found in part (a) to predict the number of hours of sunlight on April 1, the 91st day of the year.

(c) Draw a graph of the function found in part (a).

(d) Look up the number of hours of sunlight for April 1 in the *Old Farmer's Almanac*, and compare the actual hours of daylight to the results found in part (c).

37. Hours of Daylight According to the *Old Farmer's Almanac*, in Anchorage, Alaska, the number of hours of sunlight on the summer solstice of 2005 is 19.42 and the number of hours of sunlight on the winter solstice is 5.47.

(a) Find a sinusoidal function of the form
$y = A \sin(\omega x - \phi) + B$ that fits the data.

(b) Use the function found in part (a) to predict the number of hours of sunlight on April 1, the 91st day of the year.

(c) Draw a graph of the function found in part (a).

(d) Look up the number of hours of sunlight for April 1 in the *Old Farmer's Almanac*, and compare the actual hours of daylight to the results found in part (c).

38. Hours of Daylight According to the *Old Farmer's Almanac*, in Honolulu, Hawaii, the number of hours of sunlight on the summer solstice of 2005 is 13.43 and the number of hours of sunlight on the winter solstice is 10.85.

(a) Find a sinusoidal function of the form
$y = A \sin(\omega x - \phi) + B$ that fits the data.

(b) Use the function found in part (a) to predict the number of hours of sunlight on April 1, the 91st day of the year.

(c) Draw a graph of the function found in part (a).

(d) Look up the number of hours of sunlight for April 1 in the *Old Farmer's Almanac*, and compare the actual hours of daylight to the results found in part (c).

Discussion and Writing

39. Explain how the amplitude and period of a sinusoidal graph are used to establish the scale on each coordinate axis.

40. Find an application in your major field that leads to a sinusoidal graph. Write a paper about your findings.

CHAPTER REVIEW

Things to Know

Definitions

Angle in standard position (p. 504)	Vertex is at the origin; initial side is along the positive x-axis.
1 Degree (1°) (p. 505)	$1° = \dfrac{1}{360}$ revolution
1 Radian (p. 507)	The measure of a central angle of a circle whose rays subtend an arc whose length is the radius of the circle
Acute angle (p. 518)	An angle θ whose measure is $0° < \theta < 90°$ or $0 < \theta < \dfrac{\pi}{2}$, θ in radians
Complementary angles (p. 523)	Two acute angles whose sum is $90°$ $\left(\dfrac{\pi}{2} \text{ radians}\right)$
Cofunction (p. 524)	The following pairs of functions are cofunctions of each other: sine and cosine; tangent and cotangent; secant and cosecant.
Trigonometric functions of a general angle (p. 540)	$P = (a, b)$ is the point on the terminal side of θ a distance r from the origin:

$$\sin \theta = \frac{b}{r} \qquad \cos \theta = \frac{a}{r} \qquad \tan \theta = \frac{b}{a}, \quad a \neq 0$$

$$\csc \theta = \frac{r}{b}, \quad b \neq 0 \qquad \sec \theta = \frac{r}{a}, \quad a \neq 0 \qquad \cot \theta = \frac{a}{b}, \quad b \neq 0$$

Reference angle of θ (p. 545)	The acute angle formed by the terminal side of θ and either the positive or negative x-axis
Periodic function (p. 555)	$f(\theta + p) = f(\theta)$, for all θ, $p > 0$, where the smallest such p is the fundamental period

Formulas

1 revolution = 360° (p. 505)	
$\qquad\qquad\;\; = 2\pi$ radians (p. 509)	
$s = r\theta$ (p. 508)	θ is measured in radians; s is the length of the arc subtended by the central angle θ of the circle of radius r;
$A = \dfrac{1}{2}r^2\,\theta$ (p. 511)	A is the area of the sector of a circle of radius r formed by a central angle of θ radians.
$v = r\omega$ (p. 512)	v is the linear speed along the circle of radius r; ω is the angular speed (measured in radians per unit time).

Table of Values

θ (Radians)	θ (Degrees)	$\sin \theta$	$\cos \theta$	$\tan \theta$	$\csc \theta$	$\sec \theta$	$\cot \theta$
0	0°	0	1	0	Not defined	1	Not defined
$\dfrac{\pi}{6}$	30°	$\dfrac{1}{2}$	$\dfrac{\sqrt{3}}{2}$	$\dfrac{\sqrt{3}}{3}$	2	$\dfrac{2\sqrt{3}}{3}$	$\sqrt{3}$
$\dfrac{\pi}{4}$	45°	$\dfrac{\sqrt{2}}{2}$	$\dfrac{\sqrt{2}}{2}$	1	$\sqrt{2}$	$\sqrt{2}$	1
$\dfrac{\pi}{3}$	60°	$\dfrac{\sqrt{3}}{2}$	$\dfrac{1}{2}$	$\sqrt{3}$	$\dfrac{2\sqrt{3}}{3}$	2	$\dfrac{\sqrt{3}}{3}$
$\dfrac{\pi}{2}$	90°	1	0	Not defined	1	Not defined	0
π	180°	0	-1	0	Not defined	-1	Not defined
$\dfrac{3\pi}{2}$	270°	-1	0	Not defined	-1	Not defined	0

Fundamental Identities (p. 520)

$$\tan\theta = \frac{\sin\theta}{\cos\theta} \qquad \cot\theta = \frac{\cos\theta}{\sin\theta}$$

$$\csc\theta = \frac{1}{\sin\theta} \qquad \sec\theta = \frac{1}{\cos\theta} \qquad \cot\theta = \frac{1}{\tan\theta}$$

$$\sin^2\theta + \cos^2\theta = 1 \qquad \tan^2\theta + 1 = \sec^2\theta \qquad \cot^2\theta + 1 = \csc^2\theta$$

Properties of the Trigonometric Functions

$y = \sin x$ (p. 561)
Domain: $-\infty < x < \infty$
Range: $-1 \le y \le 1$
Periodic: period $= 2\pi$ (360°)
Odd function

$y = \cos x$ (pp. 562–563)
Domain: $-\infty < x < \infty$
Range: $-1 \le y \le 1$
Periodic: period $= 2\pi$ (360°)
Even function

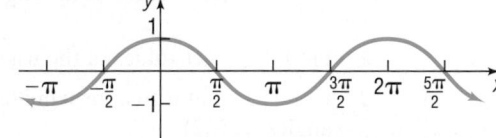

$y = \tan x$ (pp. 575–576)
Domain: $-\infty < x < \infty$, except odd integer multiples of $\frac{\pi}{2}$ (90°)
Range: $-\infty < y < \infty$
Periodic: period $= \pi$ (180°)
Odd function
Vertical asymptotes at odd integer multiples of $\frac{\pi}{2}$

$y = \cot x$ (pp. 577)
Domain: $-\infty < x < \infty$, except integer multiples of π (180°)
Range: $-\infty < y < \infty$
Periodic: period $= \pi$ (180°)
Odd function
Vertical asymptotes at integer multiples of π

$y = \csc x$ (p. 578)
Domain: $-\infty < x < \infty$, except integer multiples of π (180°)
Range: $|y| \ge 1$
Periodic: period $= 2\pi$ (360°)
Odd function
Vertical asymptotes at integer multiples of π

$y = \sec x$ (p. 578)
Domain: $-\infty < x < \infty$, except odd integer multiples of $\frac{\pi}{2}$ (90°)
Range: $|y| \ge 1$
Periodic: period $= 2\pi$ (360°)
Even function
Vertical asymptotes at odd integer multiples of $\frac{\pi}{2}$

Sinusoidal Graphs

$y = A\sin(\omega x) + B, \quad \omega > 0$

$y = A\cos(\omega x) + B, \quad \omega > 0$

$y = A\sin(\omega x - \phi) + B = A\sin\left[\omega\left(x - \frac{\phi}{\omega}\right)\right] + B$

$y = A\cos(\omega x - \phi) + B = A\cos\left[\omega\left(x - \frac{\phi}{\omega}\right)\right] + B$

Period $= \dfrac{2\pi}{\omega}$ (pp. 565, 582)

Amplitude $= |A|$ (pp. 565, 582)

Phase shift $= \dfrac{\phi}{\omega}$ (p. 582)

Objectives

Review Exercises

In Problems 1–4, convert each angle in degrees to radians. Express your answer as a multiple of π.

1. $135°$ **2.** $210°$ **3.** $18°$ **4.** $15°$

In Problems 5–8, convert each angle in radians to degrees.

5. $\dfrac{3\pi}{4}$ **6.** $\dfrac{2\pi}{3}$ **7.** $-\dfrac{5\pi}{2}$ **8.** $-\dfrac{3\pi}{2}$

In Problems 9–30, find the exact value of each expression. Do not use a calculator.

9. $\tan \dfrac{\pi}{4} - \sin \dfrac{\pi}{6}$

10. $\cos \dfrac{\pi}{3} + \sin \dfrac{\pi}{2}$

11. $3 \sin 45° - 4 \tan \dfrac{\pi}{6}$

12. $4 \cos 60° + 3 \tan \dfrac{\pi}{3}$

13. $6 \cos \dfrac{3\pi}{4} + 2 \tan\left(-\dfrac{\pi}{3}\right)$

14. $3 \sin \dfrac{2\pi}{3} - 4 \cos \dfrac{5\pi}{2}$

15. $\sec\left(-\dfrac{\pi}{3}\right) - \cot\left(-\dfrac{5\pi}{4}\right)$

16. $4 \csc \dfrac{3\pi}{4} - \cot\left(-\dfrac{\pi}{4}\right)$

17. $\tan \pi + \sin \pi$

18. $\cos \dfrac{\pi}{2} - \csc\left(-\dfrac{\pi}{2}\right)$

19. $\cos 540° - \tan(-405°)$

20. $\sin 270° + \cos(-180°)$

21. $\sin^2 20° + \dfrac{1}{\sec^2 20°}$

22. $\dfrac{1}{\cos^2 40°} - \dfrac{1}{\cot^2 40°}$

23. $\sec 50° \cos 50°$

24. $\tan 10° \cot 10°$

25. $\dfrac{\sin 50°}{\cos 40°}$

26. $\dfrac{\tan 20°}{\cot 70°}$

27. $\dfrac{\sin(-40°)}{\cos 50°}$

28. $\tan(-20°) \cot 20°$

29. $\sin 400° \sec(-50°)$

30. $\cot 200° \cot(-70°)$

In Problems 31–46, find the exact value of each of the remaining trigonometric functions.

31. $\sin \theta = \dfrac{4}{5}$, $\quad \theta$ is acute

32. $\tan \theta = \dfrac{1}{4}$, $\quad \theta$ is acute

33. $\tan \theta = \dfrac{12}{5}$, $\quad \sin \theta < 0$

34. $\cot \theta = \dfrac{12}{5}$, $\quad \cos \theta < 0$

35. $\sec \theta = -\dfrac{5}{4}$, $\quad \tan \theta < 0$

36. $\csc \theta = -\dfrac{5}{3}$, $\quad \cot \theta < 0$

37. $\sin \theta = \dfrac{12}{13}$, $\quad \theta$ in quadrant II

38. $\cos \theta = -\dfrac{3}{5}$, $\quad \theta$ in quadrant III

39. $\sin \theta = -\dfrac{5}{13}$, $\quad \dfrac{3\pi}{2} < \theta < 2\pi$

40. $\cos \theta = \dfrac{12}{13}$, $\quad \dfrac{3\pi}{2} < \theta < 2\pi$

41. $\tan \theta = \dfrac{1}{3}$, $\quad 180° < \theta < 270°$

42. $\tan \theta = -\dfrac{2}{3}$, $\quad 90° < \theta < 180°$

43. $\sec \theta = 3$, $\quad \dfrac{3\pi}{2} < \theta < 2\pi$

44. $\csc \theta = -4$, $\quad \pi < \theta < \dfrac{3\pi}{2}$

45. $\cot \theta = -2$, $\quad \dfrac{\pi}{2} < \theta < \pi$

46. $\tan \theta = -2$, $\quad \dfrac{3\pi}{2} < \theta < 2\pi$

In Problems 47–62, graph each function. Each graph should contain at least two periods.

47. $y = 2 \sin(4x)$

48. $y = -3 \cos(2x)$

49. $y = -2 \cos\left(x + \dfrac{\pi}{2}\right)$

50. $y = 3 \sin(x - \pi)$

51. $y = \tan(x + \pi)$

52. $y = -\tan\left(x - \dfrac{\pi}{2}\right)$

53. $y = -2 \tan(3x)$

54. $y = 4 \tan(2x)$

55. $y = \cot\left(x + \dfrac{\pi}{4}\right)$

56. $y = -4 \cot(2x)$

57. $y = 4 \sec(2x)$

58. $y = \csc\left(x + \dfrac{\pi}{4}\right)$

59. $y = 4 \sin(2x + 4) - 2$

60. $y = 3 \cos(4x + 2) + 1$

61. $y = 4 \tan\left(\dfrac{x}{2} + \dfrac{\pi}{4}\right)$

62. $y = 5 \cot\left(\dfrac{x}{3} - \dfrac{\pi}{4}\right)$

In Problems 63–66, determine the amplitude and period of each function without graphing.

63. $y = 4 \cos x$

64. $y = \sin(2x)$

65. $y = -8 \sin\left(\dfrac{\pi}{2} x\right)$

66. $y = -2 \cos(3\pi x)$

In Problems 67–74, find the amplitude, period, and phase shift of each function. Graph each function. Show at least two periods.

67. $y = 4 \sin(3x)$

68. $y = 2 \cos\left(\dfrac{1}{3} x\right)$

69. $y = 2 \sin(2x - \pi)$

70. $y = -\cos\left(\dfrac{1}{2} x + \dfrac{\pi}{2}\right)$

71. $y = \dfrac{1}{2} \sin\left(\dfrac{3}{2} x - \pi\right)$

72. $y = \dfrac{3}{2} \cos(6x + 3\pi)$

73. $y = -\dfrac{2}{3} \cos(\pi x - 6)$

74. $y = -7 \sin\left(\dfrac{\pi}{3} x + \dfrac{4}{3}\right)$

In Problems 75–78, find a function whose graph is given.

75.

76.

77.

78.

79. Find the value of each of the six trigonometric functions of the angle θ in the illustration.

80. Use a calculator to approximate sec $10°$. Round the answer to two decimal places.

81. Find the exact value of each of the six trigonometric functions of an angle θ if $(3, -4)$ is a point on the terminal side of θ.

82. Name the quadrant θ lies in if $\cos \theta > 0$ and $\tan \theta < 0$.

83. Find the reference angle of $-\dfrac{4\pi}{5}$.

84. Find the exact value of $\sin t, \cos t,$ and $\tan t$ if $P = \left(-\dfrac{3}{5}, \dfrac{4}{5}\right)$ is the point on the unit circle that corresponds to t.

85. What is the domain and the range of the secant function?

86. (a) Convert the angle $32°20'35''$ to a decimal in degrees. Round the answer to two decimal places.
(b) Convert the angle $63.18°$ to D°M'S'' form. Express the answer to the nearest second.

87. Find the length of the arc subtended by a central angle of $30°$ on a circle of radius 2 feet. What is the area of the sector?

88. The minute hand of a clock is 8 inches long. How far does the tip of the minute hand move in 30 minutes? How far does it move in 20 minutes?

89. Angular Speed of a Race Car A race car is driven around a circular track at a constant speed of 180 miles per hour. If the diameter of the track is $\dfrac{1}{2}$ mile, what is the angular speed of the car? Express your answer in revolutions per hour (which is equivalent to laps per hour).

90. Merry-Go-Rounds A neighborhood carnival has a merry-go-round whose radius is 25 feet. If the time for one revolution is 30 seconds, how fast is the merry-go-round going? Give the linear speed and the angular speed.

91. Lighthouse Beacons The Montauk Point Lighthouse on Long Island has dual beams (two light sources opposite each other). Ships at sea observe a blinking light every 5 seconds. What rotation speed is required to do this?

92. Spin Balancing Tires The radius of each wheel of a car is 16 inches. At how many revolutions per minute should a spin balancer be set to balance the tires at a speed of 90 miles per hour? Is the setting different for a wheel of radius 14 inches? If so, what is this setting?

93. Measuring the Length of a Lake From a stationary hot-air balloon 500 feet above the ground, two sightings of a lake are made (see the figure). How long is the lake?

94. Finding the Speed of a Glider From a glider 200 feet above the ground, two sightings of a stationary object directly in front are taken 1 minute apart (see the figure). What is the speed of the glider?

95. Finding the Width of a River Find the distance from A to C across the river illustrated in the figure.

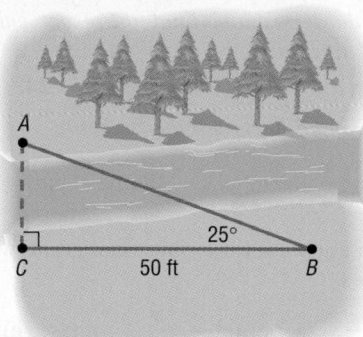

96. Finding the Height of a Building Find the height of the building shown in the figure.

97. Finding the Distance to Shore The Sears Tower in Chicago is 1454 feet tall and is situated about 1 mile inland from the shore of Lake Michigan, as indicated in the figure. An observer in a pleasure boat on the lake directly in front of the Sears Tower looks at the top of the tower and measures the angle of elevation as 5°. How far offshore is the boat?

98. Alternating Voltage The electromotive force E, in volts, in a certain ac circuit obeys the equation

$$E = 120 \sin(120\pi t), \qquad t \geq 0$$

where t is measured in seconds.
(a) What is the maximum value of E?
(b) What is the period?
(c) Graph this function over two periods.

99. Alternating Current The current I, in amperes, flowing through an ac (alternating current) circuit at time t is

$$I = 220 \sin\left(30\pi t + \frac{\pi}{6}\right), \qquad t \geq 0$$

(a) What is the period?
(b) What is the amplitude?
(c) What is the phase shift?
(d) Graph this function over two periods.

100. Monthly Temperature The following data represent the average monthly temperatures for Phoenix, Arizona.

Month, m	Average Monthly Temperature, T
January, 1	51
February, 2	55
March, 3	63
April, 4	67
May, 5	77
June, 6	86
July, 7	90
August, 8	90
September, 9	84
October, 10	71
November, 11	59
December, 12	52

Source: U.S. National Oceanic and Atmospheric Administration

(a) Draw a scatter diagram of the data for one period.
(b) Find a sinusoidal function of the form $y = A \sin(\omega x - \phi) + B$ that fits the data.
(c) Draw the sinusoidal function found in part (b) on the scatter diagram.
(d) Use a graphing utility to find the sinusoidal function of best fit.
(e) Graph the sinusoidal function of best fit on the scatter diagram.

101. Hours of Daylight According to the *Old Farmer's Almanac*, in Las Vegas, Nevada, the number of hours of sunlight on the summer solstice is 14.63 and the number of hours of sunlight on the winter solstice is 9.72.
(a) Find a sinusoidal function of the form $y = A \sin(\omega x - \phi) + B$ that fits the data.
(b) Use the function found in part (a) to predict the number of hours of sunlight on April 1, the 91st day of the year.
(c) Draw a graph of the function found in part (a).
(d) Look up the number of hours of sunlight for April 1 in the *Old Farmer's Almanac* and compare the actual hours of daylight to the results found in part (c).

CHAPTER TEST

In Problems 1–3, convert each angle in degrees to radians. Express your answer as a multiple of π.

1. $260°$ **2.** $-400°$ **3.** $13°$

In Problems 4–6 convert each angle in radians to degrees.

4. $-\dfrac{\pi}{8}$ **5.** $\dfrac{9\pi}{2}$ **6.** $\dfrac{3\pi}{4}$

In Problems 7–12, find the exact value of each expression.

7. $\sin\dfrac{\pi}{6}$ **8.** $\cos\left(-\dfrac{5\pi}{4}\right) - \cos\dfrac{3\pi}{4}$

9. $\cos(-120°)$ **10.** $\tan 330°$

11. $\sin\dfrac{\pi}{2} - \tan\dfrac{19\pi}{4}$ **12.** $2\sin^2 60° - 3\cos 45°$

In Problems 13–16, use a calculator to evaluate each expression. Round your answer to three decimal places.

13. $\sin 17°$ **14.** $\cos\dfrac{2\pi}{5}$ **15.** $\sec 229°$ **16.** $\cot\dfrac{28\pi}{9}$

17. Fill in each table entry with the sign of each function.

	$\sin\theta$	$\cos\theta$	$\tan\theta$	$\sec\theta$	$\csc\theta$	$\cot\theta$
θ in QI						
θ in QII						
θ in QIII						
θ in QIV						

18. If $f(x) = \sin x$ and $f(a) = \dfrac{3}{5}$, find $f(-a)$.

In Problems 19–21 find the value of the remaining five trigonometric functions of θ.

19. $\sin\theta = \dfrac{5}{7}$, θ in quadrant II **20.** $\cos\theta = \dfrac{2}{3}$, $\dfrac{3\pi}{2} < \theta < 2\pi$

21. $\tan\theta = -\dfrac{12}{5}$, $\dfrac{\pi}{2} < \theta < \pi$

In Problems 22–24, the point (x, y) is on the terminal side of angle θ in standard position. Find the exact value of the given trigonometric function.

22. $(2, 7)$, $\sin\theta$ **23.** $(-5, 11)$, $\cos\theta$ **24.** $(6, -3)$, $\tan\theta$

In Problems 25 and 26, graph the function.

25. $y = 2\sin\left(\dfrac{x}{3} - \dfrac{\pi}{6}\right)$ **26.** $y = \tan\left(-x + \dfrac{\pi}{4}\right) + 2$

27. Write an equation for a sinusoidal graph with the following properties:

$$A = -3 \qquad \text{period} = \dfrac{2\pi}{3} \qquad \text{phase shift} = -\dfrac{\pi}{4}$$

28. Logan has a garden in the shape of a sector of a circle; the outer rim of the garden is 25 feet long and the central angle of the sector is 50°. She wants to add a 3 foot-wide walk to the outer rim; how many square feet of paving blocks will she need to build the walk?

29. Hungarian Adrian Annus won the gold medal for the hammer throw at the 2004 Olympics in Athens with a winning distance of 83.19 meters.* The event consists of swinging a 16 pound weight attached to a wire 190 centimeters long—in a circle and then releasing it. Assuming his release is at a 45° angle to the ground, the hammer will travel a distance of $\dfrac{v_0^2}{g}$ meters, where $g = 9.8$ meters/second2 and v_0 is the linear speed of the hammer when released. At what rate (rpm) was he swinging the hammer upon release?

30. A ship is just offshore of New York City. A sighting is taken of the Statue of Liberty, which is about 305 feet tall. If the angle of elevation to the top of the statue is 20°, how far is the ship from the base of the statue?

31. To measure the height of a building, two sightings are taken a distance of 50 feet apart. If the first angle of elevation is 40° and the second is 32°, what is the height of the building?

* Annus was stripped of his medal after refusing to cooperate with postmedal drug testing.

CUMULATIVE REVIEW

1. Find the real solutions, if any, of the equation $2x^2 + x - 1 = 0$.

2. Find an equation for the line with slope -3 containing the point $(-2, 5)$.

3. Find an equation for a circle of radius 4 and center at the point $(0, -2)$.

4. Discuss the equation $2x - 3y = 12$. Graph it.

5. Discuss the equation $x^2 + y^2 - 2x + 4y - 4 = 0$. Graph it.

6. Use transformations to graph the function $y = (x - 3)^2 + 2$.

7. Sketch a graph of each of the following functions. Label at least three points on each graph.
(a) $y = x^2$ (b) $y = x^3$ (c) $y = e^x$
(d) $y = \ln x$ (e) $y = \sin x$ (f) $y = \tan x$

8. Find the inverse function of $f(x) = 3x - 2$.

9. Find the exact value of $(\sin 14°)^2 + (\cos 14°)^2 - 3$.

10. Graph $y = 3\sin(2x)$.

11. Find the exact value of $\tan\dfrac{\pi}{4} - 3\cos\dfrac{\pi}{6} + \csc\dfrac{\pi}{6}$.

12. Find an exponential function for the following graph. Express your answer in the form $y = Ab^x$.

13. Find a sinusoidal function for the following graph.

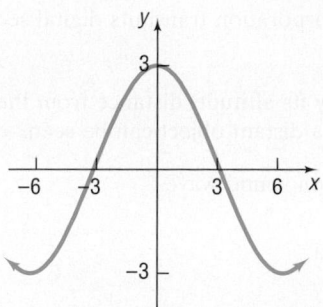

14. (a) Find a linear function that contains the points $(-2, 3)$ and $(1, -6)$. What is the slope? What are the intercepts of the function? Graph the function. Be sure to label the intercepts.

(b) Find a quadratic function that contains the point $(-2, 3)$ with vertex $(1, -6)$. What are the intercepts of the function? Graph the function.

(c) Show that there is no exponential function of the form $f(x) = ae^x$ that contains the points $(-2, 3)$ and $(1, -6)$.

15. (a) Find a polynomial function of degree 3 whose y-intercept is 5 and whose x-intercepts are $-2, 3$, and 5. Graph the function.

(b) Find a rational function whose y-intercept is 5 and whose x-intercepts are $-2, 3$, and 5 that has the line $x = 2$ as a vertical asymptote. Graph the function.

CHAPTER PROJECTS

I. Tides The given table is a partial tide table for November 2006 for the Sabine Bank Lighthouse, a shoal located offshore from Texas where the Sabine River empties into the Gulf of Mexico.

1. On November 15, when was the tide high? This is called *high tide*. On November 19, when was the tide low? This is called *low tide*. Most days will have two high tides and two low tides.

2. Why do you think there is a negative height for the low tide on November 20? What is the height measured against?

3. On your graphing utility, draw a scatter diagram for the data in the table. Let t (time) be the independent variable, with $t = 0$ being 12:00 AM on November 14, $t = 24$ being 12:00 AM on November 15, and so on. Let h be the height in feet. Remember that there are 60 minutes in an hour. Also, make sure your graphing utility is in radian mode.

4. What shape does the data take? What is the period of the data? What is the amplitude? Is the amplitude constant? Explain.

5. Fit a sine curve to the data. Is there a vertical shift? Is there a phase shift?

6. Using your graphing utility, find the sinusoidal function of best fit. How does this function compare to your equation?

7. Using the equation found in part (5) and the sinusoidal equation of best fit found in part (6), predict the high tides and the low tides on November 21.

8. Looking at the times of day that the low tides occur, what do you think causes the low tides to vary so much each day? Explain. Does this seem to have the same type of effect on the high tides? Explain.

Nov	Low Tide Time	Low Tide Ht (ft)	Low Tide Time	Low Tide Ht (ft)	High Tide Time	High Tide Ht (ft)	High Tide Time	High Tide Ht (ft)	Sun/Moon Phase Sunrise/set	Sun/Moon Phase Moonrise/set
14	6:26a	2.0	4:38p	1.4	9:29a	2.2	11:14p	2.8	6:40a/5:20p	1:05a/2:02p
15	6:22a	1.6	5:34p	1.8	11:18a	2.4	11:15p	2.6	6:41a/5:20p	1:58a/2:27p
16	6:28a	1.2	6:25p	2.0	12:37p	2.6	11:16p	2.6	6:41a/5:19p	2:50a/2:52p
17	6:40a	0.8	7:12p	2.4	1:38p	2.8	11:16p	2.6	6:42a/5:19p	3:43a/3:19p
18	6:56a	0.4	7:57p	2.6	2:27p	3.0	11:14p	2.8	6:43a/5:19p	4:38a/3:47p
19	7:17a	0.0	8:38p	2.6	3:10p	3.2	11:05p	2.8	6:44a/5:18p	5:35a/4:20p
20	7:43a	−0.2			3:52p	3.4			6:45a/5:18p	6:34a/4:57p

[**Note:** a, AM; p, PM.]

Sources: National Oceanic and Atmospheric Administration (http://tidesandcurrents.noaa.gov) and U.S. Naval Observatory (http://aa.usno.navy.mil)

The following projects are available at the Instructor's Resource Center (IRC):

II. Project at Motorola *Digital Transmission over the Air* Learn how Motorola Corporation transmits digital sequences by modulating the phase of the carrier waves.

III. Identifying Mountain Peaks in Hawaii The visibility of a mountain is affected by its altitude, distance from the viewer, and the curvature of the earth's surface. Trigonometry can be used to determine whether a distant object can be seen.

IV. CBL Experiment Technology is used to model and study the effects of damping on sound waves.

Analytic Trigonometry

In an Instant, the Land Merged with the Sea

John Lancaster and Peter S. Goodman—Washington Post Foreign Service —Sunday, January 2, 2005—Although the tsunami would have been barely perceptible to passengers on a ship at sea, it rolled outward from the fault zone at about 420 miles per hour, roughly the speed of a commercial jetliner. Even at that speed, however, the surge took about two hours to travel to the Indian mainland.

The tsunami hit coastal areas in different ways. In some places, the first event was the recession of water; in others, it was a flood. The difference was determined at the origin of the tsunami: Along the fault line, the sea floor rose in some places and fell in others. Some coastal areas felt the effect of the sinking first and saw the water recede before it returned, often with a vengeance; elsewhere, the opposite happened. Places farther from the origin of the tsunami, such as Sri Lanka, experienced more waves than did closer areas, such as Sumatra.

As the tsunami approached shallower water near land, its behavior began to change. Its velocity slowed and the distance between its crests, known as wavelength, shortened. The stored energy in the wave was partly transformed into an increase in wave height. What happened next

varied widely according to local bottom contours. In some areas, the tsunami broke like a massive beach wave; in others, it caused a sudden, rapid increase in sea level, like water pouring over the edge of a tub.

"The ocean turns into a river," explained Steven N. Ward, a geophysicist and tsunami expert at the University of California at Santa Cruz. "It comes in for about 10 minutes, and it flows back out for 10 minutes. You can't duck under it, you can't stand up in it, and you can't outrun it."

Source: Excerpt from *"In an Instant, the Land Merged with the Sea,"* by John Lancaster and Peter S. Goodman. Washington Post, January 2, 2005, p. A1. © 2005, The Washington Post. Reprinted with Permission.

—See the Chapter Project—

A Look Back

In Chapter 6, we defined inverse functions and developed their properties, particularly the relationship between the domain and range of a function and its inverse. We learned that the graph of a function and its inverse are symmetric with respect to the line $y = x$.

We continued in Chapter 6 by defining the exponential function and the inverse of the exponential function, the logarithmic function. In Chapter 7, we defined the six trigonometric functions.

A Look Ahead

In the first two sections of this chapter, we define the six inverse trigonometric functions and investigate their properties. In Sections 8.3 through 8.6 of this chapter, we continue the derivation of identities. These identities play an important role in calculus, the physical and life sciences, and economics, where they are used to simplify complicated expressions. The last two sections of this chapter deal with equations that contain trigonometric functions.

Outline

8.1 The Inverse Sine, Cosine, and Tangent Functions

PREPARING FOR THIS SECTION *Before getting started, review the following:*

- Inverse Functions (Section 6.2, pp. 409–418)
- Values of the Trigonometric Functions (Section 7.3, pp. 529–532, and Section 7.4, pp. 540–547)
- Properties of the Sine, Cosine, and Tangent Functions (Section 7.5, pp. 553–557)
- Graphs of the Sine, Cosine, and Tangent Functions (Section 7.6, pp. 560–568, and Section 7.7, pp. 574–576)

Now Work the 'Are You Prepared?' problems on page 611.

OBJECTIVES 1 Find the Exact Value of Inverse Sine, Cosine, and Tangent Functions (p. 603)

2 Find an Approximate Value of Inverse Sine, Cosine, and Tangent Functions (p. 604)

3 Use Properties of Inverse Functions to Find Exact Values of Certain Composite Functions (p. 605)

4 Find the Inverse Function of a Trigonometric Function (p. 610)

5 Solve Equations Involving Inverse Trigonometric Functions (p. 611)

In Section 6.2 we discussed inverse functions, and we noted that if a function is one-to-one it will have an inverse function. We also observed that if a function is not one-to-one it may be possible to restrict its domain in some suitable manner so that the restricted function is one-to-one.

Other properties of a one-to-one function f and its inverse function f^{-1} we discussed in Section 6.2 are summarized below.

1. $f^{-1}(f(x)) = x$ for every x in the domain of f and $f(f^{-1}(x)) = x$ for every x in the domain of f^{-1}.
2. Domain of f = range of f^{-1} and range of f = domain of f^{-1}.
3. The graph of f and the graph of f^{-1} are symmetric with respect to the line $y = x$.
4. If a function $y = f(x)$ has an inverse function, the implicit equation of the inverse function is $x = f(y)$. If we solve this equation for y, we obtain the explicit equation $y = f^{-1}(x)$.

The Inverse Sine Function

In Figure 1, we show the graph of $y = \sin x$. Because every horizontal line $y = b$, where b is between -1 and 1, intersects the graph of $y = \sin x$ infinitely many times, it follows from the horizontal-line test that the function $y = \sin x$ is not one-to-one.

Figure 1
$y = \sin x, -\infty < x < \infty, -1 \leq y \leq 1$

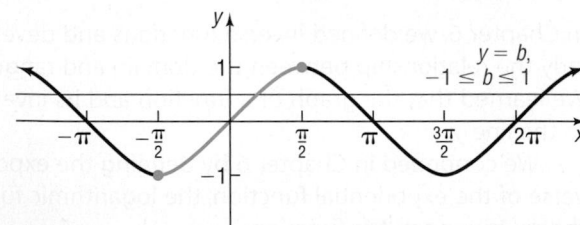

Figure 2
$y = \sin x, -\dfrac{\pi}{2} \leq x \leq \dfrac{\pi}{2}, -1 \leq y \leq 1$

However, if we restrict the domain of $y = \sin x$ to the interval $\left[-\dfrac{\pi}{2}, \dfrac{\pi}{2}\right]$, the restricted function

$$y = \sin x \qquad -\dfrac{\pi}{2} \leq x \leq \dfrac{\pi}{2}$$

is one-to-one and so will have an inverse function.* See Figure 2.

* Although there are many other ways to restrict the domain and obtain a one-to-one function, mathematicians have agreed to use the interval $\left[-\dfrac{\pi}{2}, \dfrac{\pi}{2}\right]$ to define the inverse of $y = \sin x$.

An equation for the inverse of $y = f(x) = \sin x$ is obtained by interchanging x and y. The implicit form of the inverse function is $x = \sin y, -\frac{\pi}{2} \le y \le \frac{\pi}{2}$. The explicit form is called the **inverse sine** of x and is symbolized by $y = f^{-1}(x) = \sin^{-1} x$.

DEFINITION

$$y = \sin^{-1} x \quad \text{means} \quad x = \sin y$$
$$\text{where} \quad -1 \le x \le 1 \quad \text{and} \quad -\frac{\pi}{2} \le y \le \frac{\pi}{2} \tag{1}$$

Because $y = \sin^{-1} x$ means $x = \sin y$, we read $y = \sin^{-1} x$ as "y is the angle or real number whose sine equals x." Alternatively, we can say that "y is the inverse sine of x." Be careful about the notation used. The superscript -1 that appears in $y = \sin^{-1} x$ is not an exponent, but is the symbolism used to denote the inverse function f^{-1} of f. (To avoid this notation, some books use the notation $y = \text{Arcsin } x$ instead of $y = \sin^{-1} x$.)

The inverse of a function f receives as input an element from the range of f and returns as output an element in the domain of f. The restricted sine function, $y = f(x) = \sin x$, receives as input an angle or real number x in the interval $\left[-\frac{\pi}{2}, \frac{\pi}{2} \right]$ and outputs a real number in the interval $[-1, 1]$. Therefore, the inverse sine function $y = \sin^{-1} x$ receives as input a real number in the interval $[-1, 1]$ or $-1 \le x \le 1$, its domain, and outputs an angle or real number in the interval $\left[-\frac{\pi}{2}, \frac{\pi}{2} \right]$ or $-\frac{\pi}{2} \le y \le \frac{\pi}{2}$, its range.

Figure 3
$y = \sin^{-1} x, -1 \le x \le 1, -\frac{\pi}{2} \le y \le \frac{\pi}{2}$

The graph of the inverse sine function can be obtained by reflecting the restricted portion of the graph of $y = f(x) = \sin x$ about the line $y = x$, as shown in Figure 3.

Check: Graph $Y_1 = \sin x$ and $Y_2 = \sin^{-1} x$. Compare the result with Figure 3.

1 Find the Exact Value of an Inverse Sine Function

For some numbers x, it is possible to find the exact value of $y = \sin^{-1} x$.

EXAMPLE 1 **Finding the Exact Value of an Inverse Sine Function**

Find the exact value of: $\sin^{-1} 1$

Solution Let $\theta = \sin^{-1} 1$. We seek the angle $\theta, -\frac{\pi}{2} \le \theta \le \frac{\pi}{2}$, whose sine equals 1.

$$\theta = \sin^{-1} 1 \qquad -\frac{\pi}{2} \le \theta \le \frac{\pi}{2}$$
$$\sin \theta = 1 \qquad -\frac{\pi}{2} \le \theta \le \frac{\pi}{2} \quad \textit{By definition of } y = \sin^{-1} x$$

Now look at Table 1 below and Figure 4 on page 604.

Table 1

θ	$-\frac{\pi}{2}$	$-\frac{\pi}{3}$	$-\frac{\pi}{4}$	$-\frac{\pi}{6}$	0	$\frac{\pi}{6}$	$\frac{\pi}{4}$	$\frac{\pi}{3}$	$\frac{\pi}{2}$
$\sin \theta$	-1	$-\frac{\sqrt{3}}{2}$	$-\frac{\sqrt{2}}{2}$	$-\frac{1}{2}$	0	$\frac{1}{2}$	$\frac{\sqrt{2}}{2}$	$\frac{\sqrt{3}}{2}$	1

We see that the only angle θ within the interval $\left[-\frac{\pi}{2}, \frac{\pi}{2} \right]$ whose sine is 1 is $\frac{\pi}{2}$. (Note that $\sin \frac{5\pi}{2}$ also equals 1, but $\frac{5\pi}{2}$ lies outside the interval $\left[-\frac{\pi}{2}, \frac{\pi}{2} \right]$ and hence

Figure 4

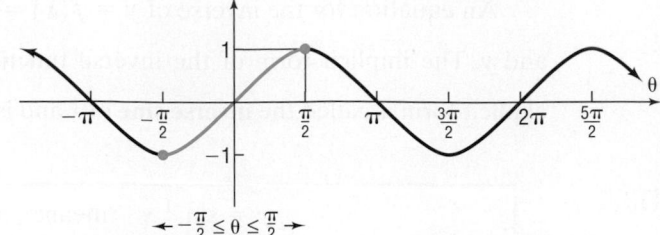

$$-\frac{\pi}{2} \le \theta \le \frac{\pi}{2}$$

is not admissible.) So, since $\sin \frac{\pi}{2} = 1$ and $\frac{\pi}{2}$ is in $\left[-\frac{\pi}{2}, \frac{\pi}{2}\right]$, we conclude that

$$\sin^{-1} 1 = \frac{\pi}{2}$$

━━━ **Now Work** PROBLEM 13

EXAMPLE 2	**Finding the Exact Value of an Inverse Sine Function**

Find the exact value of: $\sin^{-1}\left(-\frac{1}{2}\right)$

Solution Let $\theta = \sin^{-1}\left(-\frac{1}{2}\right)$. We seek the angle $\theta, -\frac{\pi}{2} \le \theta \le \frac{\pi}{2}$, whose sine equals $-\frac{1}{2}$.

$$\theta = \sin^{-1}\left(-\frac{1}{2}\right) \qquad -\frac{\pi}{2} \le \theta \le \frac{\pi}{2}$$

$$\sin \theta = -\frac{1}{2} \qquad -\frac{\pi}{2} \le \theta \le \frac{\pi}{2}$$

(Refer to Table 1 and Figure 4, if necessary.) The only angle within the interval $\left[-\frac{\pi}{2}, \frac{\pi}{2}\right]$ whose sine is $-\frac{1}{2}$ is $-\frac{\pi}{6}$. So, since $\sin\left(-\frac{\pi}{6}\right) = -\frac{1}{2}$ and $-\frac{\pi}{6}$ is in the interval $\left[-\frac{\pi}{2}, \frac{\pi}{2}\right]$, we conclude that

$$\sin^{-1}\left(-\frac{1}{2}\right) = -\frac{\pi}{6}$$

━━━ **Now Work** PROBLEM 19

2 Find an Approximate Value of an Inverse Sine Function

For most numbers x, the value $y = \sin^{-1} x$ must be approximated.

EXAMPLE 3	**Finding an Approximate Value of an Inverse Sine Function**

Find an approximate value of:

(a) $\sin^{-1}\frac{1}{3}$　　　　　　(b) $\sin^{-1}\left(-\frac{1}{4}\right)$

Express the answer in radians rounded to two decimal places.

Solution (a) Because we want the angle measured in radians, we first set the mode of our calculator to radians.* Rounded to two decimal places, we have

$$\sin^{-1}\frac{1}{3} = 0.34$$

* On most calculators, the inverse sine is obtained by pressing │SHIFT│ or │2nd│, followed by │sin│. On some calculators, │sin⁻¹│ is pressed first, then 1/3 is entered; on others, this sequence is reversed. Consult your owner's manual for the correct sequence.

Figure 5

```
sin⁻¹(-1/4)
     -.2526802551
```

(b) Figure 5 shows the solution using a TI-84 Plus graphing calculator in radian mode. We have $\sin^{-1}\left(-\dfrac{1}{4}\right) = -0.25$, rounded to two decimal places.

━━━━ **Now Work** PROBLEM 25

3 Use Properties of Inverse Functions to Find Exact Values of Certain Composite Functions

When we discussed functions and their inverses in Section 6.2, we found that $f^{-1}(f(x)) = x$ for all x in the domain of f and $f(f^{-1}(x)) = x$ for all x in the domain of f^{-1}. In terms of the sine function and its inverse, these properties are of the form

$$f^{-1}(f(x)) = \sin^{-1}(\sin x) = x \qquad \text{where } -\frac{\pi}{2} \le x \le \frac{\pi}{2} \qquad \textbf{(2a)}$$

$$f(f^{-1}(x)) = \sin(\sin^{-1} x) = x \qquad \text{where } -1 \le x \le 1 \qquad \textbf{(2b)}$$

EXAMPLE 4 | **Finding the Exact Value of Certain Composite Functions**

Find the exact value of each of the following composite functions:

(a) $\sin^{-1}\left(\sin \dfrac{\pi}{8}\right)$ (b) $\sin^{-1}\left(\sin \dfrac{5\pi}{8}\right)$

Solution
(a) The composite function $\sin^{-1}\left(\sin \dfrac{\pi}{8}\right)$ follows the form of equation (2a).

Because $\dfrac{\pi}{8}$ is in the interval $\left[-\dfrac{\pi}{2}, \dfrac{\pi}{2}\right]$, we can use (2a). Then

$$\sin^{-1}\left(\sin \frac{\pi}{8}\right) = \frac{\pi}{8}$$

(b) The composite function $\sin^{-1}\left(\sin \dfrac{5\pi}{8}\right)$ follows the form of equation (2a), but

Figure 6

$\dfrac{5\pi}{8}$ is not in the interval $\left[-\dfrac{\pi}{2}, \dfrac{\pi}{2}\right]$. To use (2a) we need to find an angle θ in the interval $\left[-\dfrac{\pi}{2}, \dfrac{\pi}{2}\right]$ for which $\sin \theta = \sin \dfrac{5\pi}{8}$. Then, using (2a), $\sin^{-1}\left(\sin \dfrac{5\pi}{8}\right) = \sin^{-1}(\sin \theta) = \theta$, and we are finished.

Look at Figure 6. The angle $\dfrac{5\pi}{8}$ is in quadrant II. The reference angle of $\dfrac{5\pi}{8}$ is $\dfrac{3\pi}{8}$ and $\sin \dfrac{5\pi}{8} = \dfrac{b}{r} = \sin \dfrac{3\pi}{8}$. Since $\dfrac{3\pi}{8}$ is in the interval $\left[-\dfrac{\pi}{2}, \dfrac{\pi}{2}\right]$, we have

$$\sin^{-1}\left(\sin \frac{5\pi}{8}\right) = \sin^{-1}\left(\sin \frac{3\pi}{8}\right) = \frac{3\pi}{8}$$
$$\underset{\text{Apply (2a).}}{\uparrow}$$

━━━━ **Now Work** PROBLEM 41

EXAMPLE 5 | **Finding the Exact Value of Certain Composite Functions**

Find the exact value, if any, of each composite function.

(a) $\sin(\sin^{-1} 0.5)$ (b) $\sin(\sin^{-1} 1.8)$

Solution

(a) The composite function $\sin(\sin^{-1} 0.5)$ follows the form of equation (2b) and 0.5 is in the interval $[-1, 1]$. So we use (2b):

$$\sin(\sin^{-1} 0.5) = 0.5$$

(b) The composite function $\sin(\sin^{-1} 1.8)$ follows the form of equation (2b), but 1.8 is not in the domain of the inverse sine function. This composite function is not defined.

 Now Work PROBLEM 45

The Inverse Cosine Function

In Figure 7 we show the graph of $y = \cos x$. Because every horizontal line $y = b$, where b is between -1 and 1, intersects the graph of $y = \cos x$ infinitely many times, it follows that the cosine function is not one-to-one.

Figure 7
$y = \cos x, -\infty < x < \infty, -1 \leq y \leq 1$

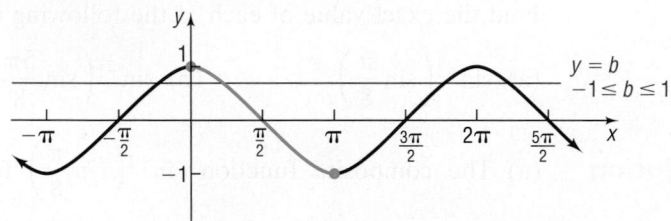

Figure 8
$y = \cos x, 0 \leq x \leq \pi, -1 \leq y \leq 1$

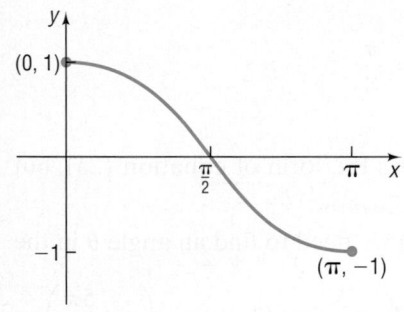

However, if we restrict the domain of $y = \cos x$ to the interval $[0, \pi]$, the restricted function

$$y = \cos x \qquad 0 \leq x \leq \pi$$

is one-to-one and hence will have an inverse function.* See Figure 8.

An equation for the inverse of $y = f(x) = \cos x$ is obtained by interchanging x and y. The implicit form of the inverse function is $x = \cos y, 0 \leq y \leq \pi$. The explicit form is called the **inverse cosine** of x and is symbolized by $y = f^{-1}(x) = \cos^{-1} x$ (or by $y = \text{Arccos } x$).

DEFINITION

$$y = \cos^{-1} x \quad \text{means} \quad x = \cos y$$
$$\text{where} \quad -1 \leq x \leq 1 \quad \text{and} \quad 0 \leq y \leq \pi \qquad (3)$$

Here y is the angle whose cosine is x. Because the range of the cosine function, $y = \cos x$, is $-1 \leq y \leq 1$, the domain of the inverse function $y = \cos^{-1} x$ is $-1 \leq x \leq 1$. Because the restricted domain of the cosine function, $y = \cos x$, is $0 \leq x \leq \pi$, the range of the inverse function $y = \cos^{-1} x$ is $0 \leq y \leq \pi$.

The graph of $y = \cos^{-1} x$ can be obtained by reflecting the restricted portion of the graph of $y = \cos x$ about the line $y = x$, as shown in Figure 9.

* This is the generally accepted restriction to define the inverse cosine function.

Figure 9
$y = \cos^{-1} x, -1 \le x \le 1, 0 \le y \le \pi$

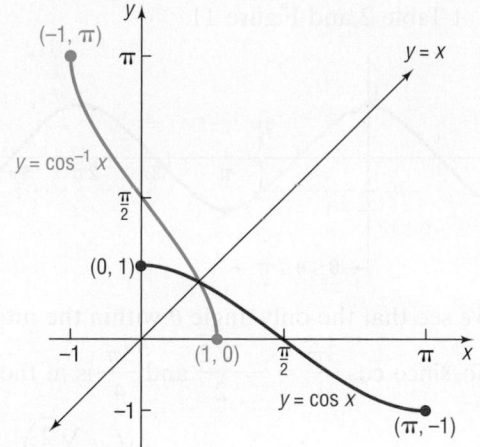

✓ **Check:** Graph $Y_1 = \cos x$ and $Y_2 = \cos^{-1} x$. Compare the result with Figure 9.

| EXAMPLE 6 | **Finding the Exact Value of an Inverse Cosine Function** |

Find the exact value of: $\cos^{-1} 0$

Solution Let $\theta = \cos^{-1} 0$. We seek the angle $\theta, 0 \le \theta \le \pi$, whose cosine equals 0.

$$\theta = \cos^{-1} 0 \qquad 0 \le \theta \le \pi$$
$$\cos \theta = 0 \qquad 0 \le \theta \le \pi$$

Look at Table 2 and Figure 10.

Table 2

θ	$\cos \theta$
0	1
$\dfrac{\pi}{6}$	$\dfrac{\sqrt{3}}{2}$
$\dfrac{\pi}{4}$	$\dfrac{\sqrt{2}}{2}$
$\dfrac{\pi}{3}$	$\dfrac{1}{2}$
$\dfrac{\pi}{2}$	0
$\dfrac{2\pi}{3}$	$-\dfrac{1}{2}$
$\dfrac{3\pi}{4}$	$-\dfrac{\sqrt{2}}{2}$
$\dfrac{5\pi}{6}$	$-\dfrac{\sqrt{3}}{2}$
π	-1

Figure 10

We see that the only angle θ within the interval $[0, \pi]$ whose cosine is 0 is $\dfrac{\pi}{2}$. (Note that $\cos \dfrac{3\pi}{2}$ and $\cos\left(-\dfrac{\pi}{2}\right)$ also equal 0, but they lie outside the interval $[0, \pi]$ and hence are not admissible.) Since $\cos \dfrac{\pi}{2} = 0$ and $\dfrac{\pi}{2}$ is in the interval $[0, \pi]$, we conclude that

$$\cos^{-1} 0 = \frac{\pi}{2}$$

| EXAMPLE 7 | **Finding the Exact Value of an Inverse Cosine Function** |

Find the exact value of: $\cos^{-1}\left(-\dfrac{\sqrt{2}}{2}\right)$

Solution Let $\theta = \cos^{-1}\left(-\dfrac{\sqrt{2}}{2}\right)$. We seek the angle $\theta, 0 \le \theta \le \pi$, whose cosine equals $-\dfrac{\sqrt{2}}{2}$.

$$\theta = \cos^{-1}\left(-\frac{\sqrt{2}}{2}\right) \qquad 0 \le \theta \le \pi$$
$$\cos \theta = -\frac{\sqrt{2}}{2} \qquad 0 \le \theta \le \pi$$

Look at Table 2 and Figure 11.

Figure 11

$$\leftarrow 0 \le \theta \le \pi \rightarrow$$

We see that the only angle θ within the interval $[0, \pi]$ whose cosine is $-\dfrac{\sqrt{2}}{2}$ is $\dfrac{3\pi}{4}$. So, since $\cos\dfrac{3\pi}{4} = -\dfrac{\sqrt{2}}{2}$ and $\dfrac{3\pi}{4}$ is in the interval $[0, \pi]$, we conclude that

$$\cos^{-1}\left(-\frac{\sqrt{2}}{2}\right) = \frac{3\pi}{4}$$

━━━━▶ **Now Work** PROBLEM 23

For the cosine function and its inverse, the following properties hold:

$$f^{-1}(f(x)) = \cos^{-1}(\cos x) = x \qquad \text{where } 0 \le x \le \pi \qquad \textbf{(4a)}$$
$$f(f^{-1}(x)) = \cos(\cos^{-1} x) = x \qquad \text{where } -1 \le x \le 1 \qquad \textbf{(4b)}$$

EXAMPLE 8	**Using Properties of Inverse Functions to Find the Exact Value of Certain Composite Functions**

Find the exact value of:

(a) $\cos^{-1}\left(\cos\dfrac{\pi}{12}\right)$ (b) $\cos[\cos^{-1}(-0.4)]$ (c) $\cos^{-1}\left[\cos\left(-\dfrac{2\pi}{3}\right)\right]$ (d) $\cos(\cos^{-1}\pi)$

Solution (a) $\cos^{-1}\left(\cos\dfrac{\pi}{12}\right) = \dfrac{\pi}{12}$ *$\dfrac{\pi}{12}$ is in the interval $[0, \pi]$; use Property (4a).*

(b) $\cos[\cos^{-1}(-0.4)] = -0.4$ *-0.4 is in the interval $[-1, 1]$; use Property (4b).*

(c) The angle $-\dfrac{2\pi}{3}$ is not in the interval $[0, \pi]$ so we cannot use (4a). However, because the cosine function is even, $\cos\left(-\dfrac{2\pi}{3}\right) = \cos\dfrac{2\pi}{3}$. Since $\dfrac{2\pi}{3}$ is in the inerval $[0, \pi]$ we have

$$\cos^{-1}\left[\cos\left(-\frac{2\pi}{3}\right)\right] = \cos^{-1}\left(\cos\frac{2\pi}{3}\right) = \frac{2\pi}{3} \qquad \text{\textit{$\dfrac{2\pi}{3}$ is in the interval $[0, \pi]$; apply (4a).}}$$

(d) Because π is not in the domain of the inverse cosine function, the interval $[-1, 1]$, $\cos^{-1}\pi$ is not defined. This means the composite function $\cos(\cos^{-1}\pi)$ is also not defined.

━━━━▶ **Now Work** PROBLEMS 37 AND 49

The Inverse Tangent Function

In Figure 12 we show the graph of $y = \tan x$. Because every horizontal line intersects the graph infinitely many times, it follows that the tangent function is not one-to-one.

However, if we restrict the domain of $y = \tan x$ to the interval $\left(-\dfrac{\pi}{2}, \dfrac{\pi}{2}\right)$, the restricted function

$$y = \tan x \qquad -\frac{\pi}{2} < x < \frac{\pi}{2}$$

is one-to-one and hence has an inverse function.* See Figure 13.

* This is the generally accepted restriction.

Figure 12
$y = \tan x$, $-\infty < x < \infty$, x not equal to odd multiples of $\dfrac{\pi}{2}$, $-\infty < y < \infty$

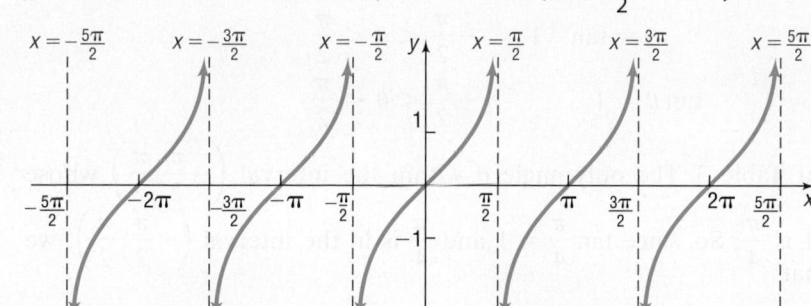

Figure 13
$y = \tan x$, $-\dfrac{\pi}{2} < x < \dfrac{\pi}{2}$, $-\infty < y < \infty$

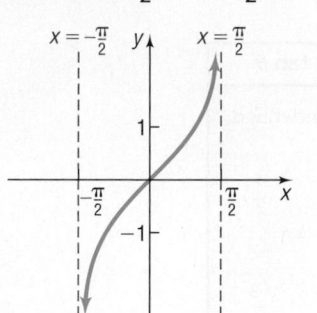

An equation for the inverse of $y = f(x) = \tan x$ is obtained by interchanging x and y. The implicit form of the inverse function is $x = \tan y$, $-\dfrac{\pi}{2} < y < \dfrac{\pi}{2}$. The explicit form is called the **inverse tangent** of x and is symbolized by $y = f^{-1}(x) = \tan^{-1}x$ (or by $y = \text{Arctan}\,x$).

DEFINITION

> $y = \tan^{-1}x$ means $x = \tan y$
>
> where $-\infty < x < \infty$ and $-\dfrac{\pi}{2} < y < \dfrac{\pi}{2}$ **(5)**

Here y is the angle whose tangent is x. The domain of the function $y = \tan^{-1}x$ is $-\infty < x < \infty$, and its range is $-\dfrac{\pi}{2} < y < \dfrac{\pi}{2}$. The graph of $y = \tan^{-1}x$ can be obtained by reflecting the restricted portion of the graph of $y = \tan x$ about the line $y = x$, as shown in Figure 14.

Figure 14
$y = \tan^{-1}x$ $-\infty < x < \infty$ $-\dfrac{\pi}{2} < y < \dfrac{\pi}{2}$

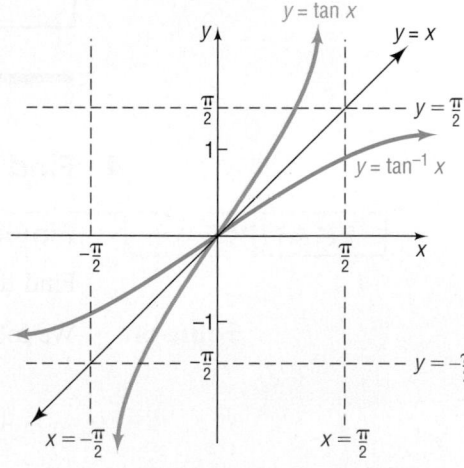

✓ **Check:** Graph $Y_1 = \tan x$ and $Y_2 = \tan^{-1}x$. Compare the result with Figure 14.

| EXAMPLE 9 | **Finding the Exact Value of an Inverse Tangent Function** |

Find the exact value of:

(a) $\tan^{-1} 1$ (b) $\tan^{-1}\left(-\sqrt{3}\right)$

Solution (a) Let $\theta = \tan^{-1} 1$. We seek the angle θ, $-\dfrac{\pi}{2} < \theta < \dfrac{\pi}{2}$, whose tangent equals 1.

$$\theta = \tan^{-1} 1 \qquad -\frac{\pi}{2} < \theta < \frac{\pi}{2}$$

$$\tan \theta = 1 \qquad -\frac{\pi}{2} < \theta < \frac{\pi}{2}$$

Look at Table 3. The only angle θ within the interval $\left(-\dfrac{\pi}{2}, \dfrac{\pi}{2}\right)$ whose tangent is 1 is $\dfrac{\pi}{4}$. So, since $\tan \dfrac{\pi}{4} = 1$ and $\dfrac{\pi}{4}$ is in the interval $\left(-\dfrac{\pi}{2}, \dfrac{\pi}{2}\right)$, we conclude that

$$\tan^{-1} 1 = \frac{\pi}{4}$$

(b) Let $\theta = \tan^{-1}\left(-\sqrt{3}\right)$. We seek the angle θ, $-\dfrac{\pi}{2} < \theta < \dfrac{\pi}{2}$, whose tangent equals $-\sqrt{3}$.

$$\theta = \tan^{-1}\left(-\sqrt{3}\right) \qquad -\frac{\pi}{2} < \theta < \frac{\pi}{2}$$

$$\tan \theta = -\sqrt{3} \qquad -\frac{\pi}{2} < \theta < \frac{\pi}{2}$$

Look at Table 3 or Figure 13 if necessary. The only angle θ within the interval $\left(-\dfrac{\pi}{2}, \dfrac{\pi}{2}\right)$ whose tangent is $-\sqrt{3}$ is $-\dfrac{\pi}{3}$. So, since $\tan\left(-\dfrac{\pi}{3}\right) = -\sqrt{3}$ and $-\dfrac{\pi}{3}$ is in the interval $\left(-\dfrac{\pi}{2}, \dfrac{\pi}{2}\right)$, we conclude that

$$\tan^{-1}\left(-\sqrt{3}\right) = -\frac{\pi}{3}$$

Table 3

θ	$\tan \theta$
$-\dfrac{\pi}{2}$	Undefined
$-\dfrac{\pi}{3}$	$-\sqrt{3}$
$-\dfrac{\pi}{4}$	-1
$-\dfrac{\pi}{6}$	$-\dfrac{\sqrt{3}}{3}$
0	0
$\dfrac{\pi}{6}$	$\dfrac{\sqrt{3}}{3}$
$\dfrac{\pi}{4}$	1
$\dfrac{\pi}{3}$	$\sqrt{3}$
$\dfrac{\pi}{2}$	Undefined

 Now Work PROBLEM 17

For the tangent function and its inverse, the following properties hold:

$$f^{-1}(f(x)) = \tan^{-1}(\tan x) = x \qquad \text{where } -\frac{\pi}{2} < x < \frac{\pi}{2}$$

$$f(f^{-1}(x)) = \tan(\tan^{-1} x) = x \qquad \text{where } -\infty < x < \infty$$

 Now Work PROBLEM 43

4 Find the Inverse Function of a Trigonometric Function

EXAMPLE 10 **Finding the Inverse Function of a Trigonometric Function**

Find the inverse function f^{-1} of $f(x) = 2 \sin x - 1$. State the domain of f and f^{-1}.

Solution We follow the steps on page 417 for finding the inverse function.

$$y = 2 \sin x - 1$$
$$x = 2 \sin y - 1 \qquad \text{\small Interchange } x \text{ and } y.$$
$$x + 1 = 2 \sin y \qquad \text{\small Proceed to solve for } y.$$
$$\sin y = \frac{x + 1}{2}$$
$$y = \sin^{-1} \frac{x + 1}{2} \qquad \text{\small Apply the definition (1).}$$

The inverse function is $f^{-1}(x) = \sin^{-1} \dfrac{x + 1}{2}$.

The domain of $f(x) = 2 \sin x - 1$ is the set of all real numbers. To find the domain of f^{-1}, we note that the argument of the inverse sine function is $\dfrac{x+1}{2}$ and that it must lie in the interval $[-1, 1]$. That is,

$$-1 \le \frac{x+1}{2} \le 1$$

$$-2 \le x + 1 \le 2 \quad \text{Multiply each part by 2.}$$

$$-3 \le x \le 1 \quad \text{Add } -1 \text{ to each part.}$$

The domain of f^{-1} is $\{x \mid -3 \le x \le 1\}$.

▬▬▬ **Now Work** PROBLEM 55

5 Solve Equations Involving Inverse Trigonometric Functions

Equations that contain inverse trigonometric functions are called **inverse trigonometric equations.**

EXAMPLE 11	**Solving an Equation Involving an Inverse Trigonometric Function**

Solve the equation: $3 \sin^{-1}x = \pi$

Solution To solve an equation involving a single inverse trigonometric function, first solve the equation for the inverse trigonometric function.

$$3 \sin^{-1} x = \pi$$

$$\sin^{-1}x = \frac{\pi}{3} \quad \text{Divide both sides by 3.}$$

$$x = \sin \frac{\pi}{3} \quad y = \sin^{-1}x \text{ means } x = \sin y.$$

$$x = \frac{\sqrt{3}}{2}$$

The solution set is $\left\{ \dfrac{\sqrt{3}}{2} \right\}$.

▬▬▬ **Now Work** PROBLEM 61

8.1 Assess Your Understanding

'Are You Prepared?' *Answers are given at the end of these exercises. If you get a wrong answer, read the pages listed in red.*

1. What is the domain and the range of $y = \sin x$? (pp. 553–557)

2. A suitable restriction on the domain of the function $f(x) = (x-1)^2$ to make it one to one would be _____. (pp. 409–418)

3. If the domain of a one-to-one function is $[3, \infty)$, the range of its inverse is _____. (pp. 409–418)

4. **True or False** The graph of $y = \cos x$ is decreasing on the interval $[0, \pi]$. (pp. 560–563)

5. $\tan \dfrac{\pi}{4} = $ _____; $\sin \dfrac{\pi}{3} = $ _____ (pp. 529–532)

6. $\sin\left(-\dfrac{\pi}{6}\right) = $ _____; $\cos \pi = $ _____ (pp. 540–547)

Concepts and Vocabulary

7. $y = \sin^{-1} x$ means _____, where $-1 \le x \le 1$ and $-\dfrac{\pi}{2} \le y \le \dfrac{\pi}{2}$.

8. The value of $\sin^{-1}\left[\sin \dfrac{\pi}{2}\right]$ is _____.

9. $\cos^{-1}\left[\cos\dfrac{\pi}{5}\right] = $ _____.

10. *True or False* The domain of $y = \sin^{-1} x$ is $-\dfrac{\pi}{2} \le x \le \dfrac{\pi}{2}$.

11. *True or False* $\sin(\sin^{-1} 0) = 0$ and $\cos(\cos^{-1} 0) = 0$.

12. *True or False* $y = \tan^{-1} x$ means $x = \tan y$, where $-\infty < x < \infty$ and $-\dfrac{\pi}{2} < y < \dfrac{\pi}{2}$.

Skill Building

In Problems 13–24, find the exact value of each expression.

13. $\sin^{-1} 0$

14. $\cos^{-1} 1$

15. $\sin^{-1}(-1)$

16. $\cos^{-1}(-1)$

17. $\tan^{-1} 0$

18. $\tan^{-1}(-1)$

19. $\sin^{-1}\dfrac{\sqrt{2}}{2}$

20. $\tan^{-1}\dfrac{\sqrt{3}}{3}$

21. $\tan^{-1}\sqrt{3}$

22. $\sin^{-1}\left(-\dfrac{\sqrt{3}}{2}\right)$

23. $\cos^{-1}\left(-\dfrac{\sqrt{3}}{2}\right)$

24. $\sin^{-1}\left(-\dfrac{\sqrt{2}}{2}\right)$

In Problems 25–36, use a calculator to find the value of each expression rounded to two decimal places.

25. $\sin^{-1} 0.1$

26. $\cos^{-1} 0.6$

27. $\tan^{-1} 5$

28. $\tan^{-1} 0.2$

29. $\cos^{-1}\dfrac{7}{8}$

30. $\sin^{-1}\dfrac{1}{8}$

31. $\tan^{-1}(-0.4)$

32. $\tan^{-1}(-3)$

33. $\sin^{-1}(-0.12)$

34. $\cos^{-1}(-0.44)$

35. $\cos^{-1}\dfrac{\sqrt{2}}{3}$

36. $\sin^{-1}\dfrac{\sqrt{3}}{5}$

In Problems 37–44, find the exact value of each expression. Do not use a calculator.

37. $\cos^{-1}\left(\cos\dfrac{4\pi}{5}\right)$

38. $\sin^{-1}\left[\sin\left(-\dfrac{\pi}{10}\right)\right]$

39. $\tan^{-1}\left[\tan\left(-\dfrac{3\pi}{8}\right)\right]$

40. $\sin^{-1}\left[\sin\left(-\dfrac{3\pi}{7}\right)\right]$

41. $\sin^{-1}\left(\sin\dfrac{9\pi}{8}\right)$

42. $\cos^{-1}\left[\cos\left(-\dfrac{5\pi}{3}\right)\right]$

43. $\tan^{-1}\left(\tan\dfrac{4\pi}{5}\right)$

44. $\tan^{-1}\left[\tan\left(-\dfrac{2\pi}{3}\right)\right]$

In Problems 45–52, find the exact value, if any, of each composite function. If there is no value, say it is "not defined." Do not use a calculator.

45. $\sin\left(\sin^{-1}\dfrac{1}{4}\right)$

46. $\cos\left[\cos^{-1}\left(-\dfrac{2}{3}\right)\right]$

47. $\tan\left(\tan^{-1} 4\right)$

48. $\tan\left[\tan^{-1}(-2)\right]$

49. $\cos\left(\cos^{-1} 1.2\right)$

50. $\sin\left[\sin^{-1}(-2)\right]$

51. $\tan\left(\tan^{-1}\pi\right)$

52. $\sin\left[\sin^{-1}(-1.5)\right]$

In Problems 53–60, find the inverse function f^{-1} of each function f. State the domain of f and f^{-1}.

53. $f(x) = 5\sin x + 2$

54. $f(x) = 2\tan x - 3$

55. $f(x) = -2\cos(3x)$

56. $f(x) = 3\sin(2x)$

57. $f(x) = -\tan(x + 1) - 3$

58. $f(x) = \cos(x + 2) + 1$

59. $f(x) = 3\sin(2x + 1)$

60. $f(x) = 2\cos(3x + 2)$

In Problems 61–68, find the exact solution of each equation.

61. $4\sin^{-1}(x) = \pi$

62. $2\cos^{-1} x = \pi$

63. $3\cos^{-1}(2x) = 2\pi$

64. $-6\sin^{-1}(3x) = \pi$

65. $3\tan^{-1} x = \pi$

66. $-4\tan^{-1} x = \pi$

67. $4\cos^{-1} x - 2\pi = 2\cos^{-1} x$

68. $5\sin^{-1} x - 2\pi = 2\sin^{-1} x - 3\pi$

Applications and Extensions

In Problems 69–74, use the following discussion. The formula

$$D = 24\left[1 - \dfrac{\cos^{-1}(\tan i \tan \theta)}{\pi}\right]$$

can be used to approximate the number of hours of daylight D when the declination of the Sun is $i°$ at a location $\theta°$ north latitude for any date between the vernal equinox and autumnal equinox. The declination of the Sun is defined as the angle i between the equatorial plane and any ray of light from the Sun. The latitude of a location is

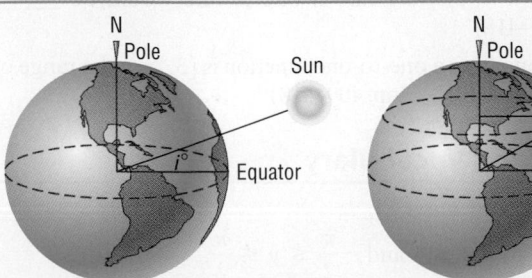

the angle θ between the Equator and the location on the surface of Earth, with the vertex of the angle located at the center of Earth. See the figure. To use the formula, $\cos^{-1}(\tan i \tan \theta)$ must be expressed in radians.

69. Approximate the number of hours of daylight in Houston, Texas (29°45′ north latitude), for the following dates:
 (a) Summer solstice ($i = 23.5°$)
 (b) Vernal equinox ($i = 0°$)
 (c) July 4 ($i = 22°48′$)

70. Approximate the number of hours of daylight in New York, New York (40°45′ north latitude), for the following dates:
 (a) Summer solstice ($i = 23.5°$)
 (b) Vernal equinox ($i = 0°$)
 (c) July 4 ($i = 22°48′$)

71. Approximate the number of hours of daylight in Honolulu, Hawaii (21°18′ north latitude), for the following dates:
 (a) Summer solstice ($i = 23.5°$)
 (b) Vernal equinox ($i = 0°$)
 (c) July 4 ($i = 22°48′$)

72. Approximate the number of hours of daylight in Anchorage, Alaska (61°10′ north latitude), for the following dates:
 (a) Summer solstice ($i = 23.5°$)
 (b) Vernal equinox ($i = 0°$)
 (c) July 4 ($i = 22°48′$)

73. Approximate the number of hours of daylight at the Equator (0° north latitude) for the following dates:
 (a) Summer solstice ($i = 23.5°$)
 (b) Vernal equinox ($i = 0°$)
 (c) July 4 ($i = 22°48′$)
 (d) What do you conclude about the number of hours of daylight throughout the year for a location at the Equator?

74. Approximate the number of hours of daylight for any location that is 66°30′ north latitude for the following dates:
 (a) Summer solstice ($i = 23.5°$)
 (b) Vernal equinox ($i = 0°$)
 (c) July 4 ($i = 22°48′$)
 (d) The number of hours of daylight on the winter solstice may be found by computing the number of hours of daylight on the summer solstice and subtracting this result from 24 hours, due to the symmetry of the orbital path of Earth around the Sun. Compute the number of hours of daylight for this location on the winter solstice. What do you conclude about daylight for a location at 66°30′ north latitude?

75. Being the First to See the Rising Sun Cadillac Mountain, elevation 1530 feet, is located in Acadia National Park, Maine, and is the highest peak on the east coast of the United States. It is said that a person standing on the summit will be the first person in the United States to see the rays of the rising Sun.

How much sooner would a person atop Cadillac Mountain see the first rays than a person standing below, at sea level?
[**Hint:** Consult the figure. When the person at D sees the first rays of the Sun, the person at P does not. The person at P sees the first rays of the Sun only after Earth has rotated so that P is at location Q. Compute the length of the arc subtended by the central angle θ. Then use the fact that, at the latitude of Cadillac Mountain, in 24 hours a length of $2\pi(2710) \approx 17027.4$ miles is subtended, and find the time that it takes to subtend this length.]

76. Movie Theater Screens Suppose that a movie theater has a screen that is 28 feet tall. When you sit down, the bottom of the screen is 6 feet above your eye level. The angle formed by drawing a line from your eye to the bottom of the screen and your eye and the top of the screen is called the **viewing angle.** In the figure θ is the viewing angle. Suppose that you sit x feet from the screen. The viewing angle θ is given by the function

$$\theta(x) = \tan^{-1}\left(\frac{34}{x}\right) - \tan^{-1}\left(\frac{6}{x}\right).$$

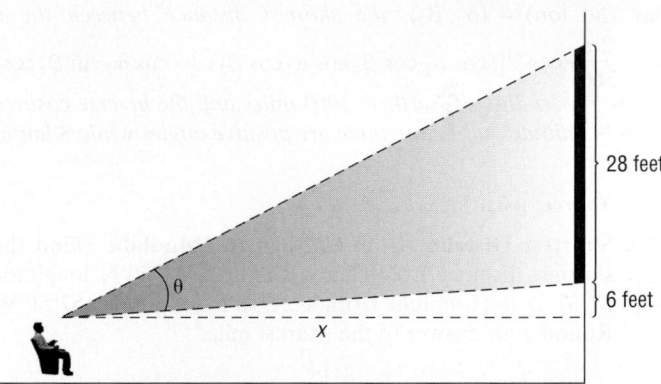

 (a) What is your viewing angle if you sit 10 feet from the screen? 15 feet? 20 feet?
 (b) If there is 5 feet between the screen and the first row of seats and there is 3 feet between each row, which row results in the largest viewing angle?
 (c) Using a graphing utility, graph

$$\theta(x) = \tan^{-1}\left(\frac{34}{x}\right) - \tan^{-1}\left(\frac{6}{x}\right)$$

What value of x results in the largest viewing angle?

77. Area under a Curve The area under the graph of $y = \dfrac{1}{1 + x^2}$ and above the x-axis between $x = a$ and $x = b$ is given by

$$\tan^{-1} b - \tan^{-1} a$$

See the figure.

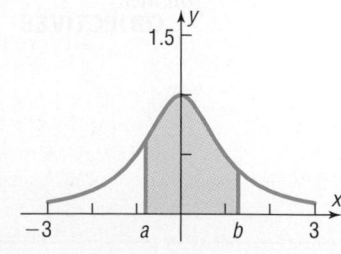

(a) Find the exact area under the graph of $y = \dfrac{1}{1 + x^2}$ and above the x-axis between $x = 0$ and $x = \sqrt{3}$.

(b) Find the exact area under the graph of $y = \dfrac{1}{1 + x^2}$ and above the x-axis between $x = -\dfrac{\sqrt{3}}{3}$ and $x = 1$.

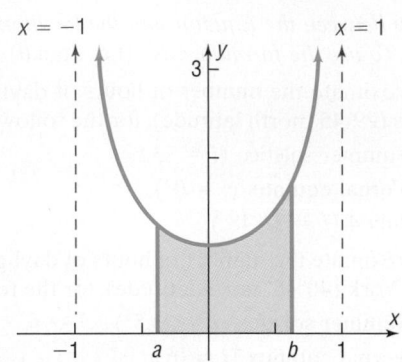

78. Area under a Curve The area under the graph of $y = \dfrac{1}{\sqrt{1 - x^2}}$ and above the x-axis between $x = a$ and $x = b$ is given by

$$\sin^{-1} b - \sin^{-1} a$$

See the figure.

(a) Find the exact area under the graph of $y = \dfrac{1}{\sqrt{1 - x^2}}$ and above the x-axis between $x = 0$ and $x = \dfrac{\sqrt{3}}{2}$.

(b) Find the exact area under the graph of $y = \dfrac{1}{\sqrt{1 - x^2}}$ and above the x-axis between $x = -\dfrac{1}{2}$ and $x = \dfrac{1}{2}$.

Problems 79 and 80 require the following discussion:
The shortest distance between two points on Earth's surface can be determined from the latitude and longitude of the two locations. For example, if location 1 has (lat, lon) $= (\alpha_1, \beta_1)$ *and location 2 has* (lat, lon) $= (\alpha_2, \beta_2)$, *the shortest distance between the two locations is approximately*

$$d = \frac{2\pi}{360} r \cos^{-1}[(\cos \alpha_1 \cos \beta_1 \cos \alpha_2 \cos \beta_2) + (\cos \alpha_1 \sin \beta_1 \cos \alpha_2 \sin \beta_2) + (\sin \alpha_1 \sin \alpha_2)]$$

where r = radius of Earth ≈ 3960 miles and the inverse cosine function is expressed in radians. Also N latitude and E longitude are positive angles while S latitude and W longitude are negative angles.

City	Latitude	Longitude
Chicago, IL	41°50'N	87°37'W
Honolulu, HI	21°18'N	157°50'W
Melbourne, Australia	37°47'S	144°58'E

Source: www.infoplease.com

79. Shortest Distance from Chicago to Honolulu Find the shortest distance from Chicago, latitude 41°50'N, longitude 87°37'W to Honolulu, latitude 21°18'N, longitude 157°50'W. Round your answer to the nearest mile.

80. Shortest Distance from Honolulu to Melbourne, Australia Find the shortest distance from Honolulu to Melbourne, Australia, latitude 37°47'S, longitude 144°58'E. Round your answer to the nearest mile.

'Are You Prepared?' Answers

1. domain: the set of all real numbers; range: $-1 \leq y \leq 1$

2. Two answers are possible: $x \leq 1$ or $x \geq 1$

3. $[3, \infty)$

4. True

5. $1; \dfrac{\sqrt{3}}{2}$

6. $-\dfrac{1}{2}; -1$

8.2 The Inverse Trigonometric Functions (Continued)

PREPARING FOR THIS SECTION *Before getting started, review the following concepts:*

- Finding Exact Values Given the Value of a Trigonometric Function and the Quadrant of the Angle (Section 7.4, pp. 547–548)
- Graphs of the Secant, Cosecant, and Cotangent Functions (Section 7.7, pp. 577–579)
- Domain and Range of the Secant, Cosecant, and Cotangent Functions (Section 7.5, pp. 553–555)

Now Work the 'Are You Prepared?' problems on page 618.

OBJECTIVES 1 Find the Exact Value of Expressions Involving the Inverse Sine, Cosine, and Tangent Functions (p. 615)

2 Know the Definition of the Inverse Secant, Cosecant, and Cotangent Functions (p. 616)

3 Use a Calculator to Evaluate $\sec^{-1} x$, $\csc^{-1} x$, and $\cot^{-1} x$ (p. 616)

4 Write a Trigonometric Expression as an Algebraic Expression (p. 617)

1 Find the Exact Value of Expressions Involving the Inverse Sine, Cosine, and Tangent Functions

EXAMPLE 1	**Finding the Exact Value of Expressions Involving Inverse Trigonometric Functions**

Find the exact value of: $\sin\left(\tan^{-1}\dfrac{1}{2}\right)$

Solution Let $\theta = \tan^{-1}\dfrac{1}{2}$. Then $\tan\theta = \dfrac{1}{2}$, where $-\dfrac{\pi}{2} < \theta < \dfrac{\pi}{2}$. We seek $\sin\theta$. Because $\tan\theta > 0$, it follows that $0 < \theta < \dfrac{\pi}{2}$, so θ lies in quadrant I. Now, in Figure 15 we draw a triangle in quadrant I. Because $\tan\theta = \dfrac{1}{2}$, the side opposite θ is 1 and the side adjacent to θ is 2. The hypotenuse of this triangle is found to be $\sqrt{5}$. Then $\sin\theta = \dfrac{1}{\sqrt{5}}$, and

Figure 15

$\tan\theta = \dfrac{1}{2}$

$$\sin\left(\tan^{-1}\dfrac{1}{2}\right) = \sin\theta = \dfrac{1}{\sqrt{5}} = \dfrac{\sqrt{5}}{5}$$

EXAMPLE 2	**Finding the Exact Value of Expressions Involving Inverse Trigonometric Functions**

Find the exact value of: $\cos\left[\sin^{-1}\left(-\dfrac{1}{3}\right)\right]$

Solution Let $\theta = \sin^{-1}\left(-\dfrac{1}{3}\right)$. Then $\sin\theta = -\dfrac{1}{3}$ and $-\dfrac{\pi}{2} \le \theta \le \dfrac{\pi}{2}$. We seek $\cos\theta$. Because $\sin\theta < 0$, it follows that $-\dfrac{\pi}{2} \le \theta < 0$, so θ lies in quadrant IV. Figure 16 illustrates $\sin\theta = -\dfrac{1}{3}$ for θ in quadrant IV. Then

Figure 16

$\sin\theta = -\dfrac{1}{3}$

$$\cos\left[\sin^{-1}\left(-\dfrac{1}{3}\right)\right] = \cos\theta = \dfrac{2\sqrt{2}}{3}$$

EXAMPLE 3	**Finding the Exact Value of Expressions Involving Inverse Trigonometric Functions**

Find the exact value of: $\tan\left[\cos^{-1}\left(-\dfrac{1}{3}\right)\right]$

Solution Let $\theta = \cos^{-1}\left(-\dfrac{1}{3}\right)$. Then $\cos\theta = -\dfrac{1}{3}$ and $0 \le \theta \le \pi$. We seek $\tan\theta$. Because $\cos\theta < 0$, it follows that $\dfrac{\pi}{2} < \theta \le \pi$, so θ lies in quadrant II. Figure 17 illustrates $\cos\theta = -\dfrac{1}{3}$ for θ in quadrant II. Then

Figure 17

$\cos\theta = -\dfrac{1}{3}$

$$\tan\left[\cos^{-1}\left(-\dfrac{1}{3}\right)\right] = \tan\theta = \dfrac{2\sqrt{2}}{-1} = -2\sqrt{2}$$

━━━━ **Now Work** PROBLEMS **9** AND **27**

2 Know the Definition of the Inverse Secant, Cosecant, and Cotangent Functions

The inverse secant, inverse cosecant, and inverse cotangent functions are defined as follows:

DEFINITION

$$y = \sec^{-1} x \quad \text{means} \quad x = \sec y \qquad\qquad \textbf{(1)}$$

$$\text{where} \quad |x| \geq 1 \quad \text{and} \quad 0 \leq y \leq \pi, \quad y \neq \frac{\pi}{2}{}^{*}$$

$$y = \csc^{-1} x \quad \text{means} \quad x = \csc y \qquad\qquad \textbf{(2)}$$

$$\text{where} \quad |x| \geq 1 \quad \text{and} \quad -\frac{\pi}{2} \leq y \leq \frac{\pi}{2}, \quad y \neq 0^{\dagger}$$

$$y = \cot^{-1} x \quad \text{means} \quad x = \cot y \qquad\qquad \textbf{(3)}$$

$$\text{where} \quad -\infty < x < \infty \quad \text{and} \quad 0 < y < \pi$$

You are encouraged to review the graphs of the cotangent, cosecant, and secant functions in Figures 96, 97, and 98 in Section 7.7 to help you to see the basis for these definitions.

EXAMPLE 4 **Finding the Exact Value of an Inverse Cosecant Function**

Find the exact value of: $\csc^{-1} 2$

Solution Let $\theta = \csc^{-1} 2$. We seek the angle θ, $-\dfrac{\pi}{2} \leq \theta \leq \dfrac{\pi}{2}, \theta \neq 0$, whose cosecant equals 2 $\left(\text{or, equivalently, whose sine equals } \dfrac{1}{2}\right)$.

$$\theta = \csc^{-1} 2 \qquad -\frac{\pi}{2} \leq \theta \leq \frac{\pi}{2}, \quad \theta \neq 0$$

$$\csc \theta = 2 \qquad -\frac{\pi}{2} \leq \theta \leq \frac{\pi}{2}, \quad \theta \neq 0 \qquad {\scriptstyle \sin \theta = \frac{1}{2}}$$

The only angle θ in the interval $-\dfrac{\pi}{2} \leq \theta \leq \dfrac{\pi}{2}, \theta \neq 0$, whose cosecant is 2 $\left[\sin \theta = \dfrac{1}{2}\right]$ is $\dfrac{\pi}{6}$, so $\csc^{-1} 2 = \dfrac{\pi}{6}$.

Now Work PROBLEM 39

3 Use a Calculator to Evaluate $\sec^{-1} x$, $\csc^{-1} x$, and $\cot^{-1} x$

Most calculators do not have keys for evaluating the inverse cotangent, cosecant, and secant functions. The easiest way to evaluate them is to convert to an inverse trigonometric function whose range is the same as the one to be evaluated. In this regard, notice that $y = \cot^{-1} x$ and $y = \sec^{-1} x$, except where undefined, each have the same range as $y = \cos^{-1} x$; $y = \csc^{-1} x$, except where undefined, has the same range as $y = \sin^{-1} x$.

* Most books use this definition. A few use the restriction $0 \leq y < \dfrac{\pi}{2}, \pi \leq y < \dfrac{3\pi}{2}$.

† Most books use this definition. A few use the restriction $-\pi < y \leq -\dfrac{\pi}{2}, 0 < y \leq \dfrac{\pi}{2}$.

| EXAMPLE 5 | **Approximating the Value of Inverse Trigonometric Functions** |

Use a calculator to approximate each expression in radians rounded to two decimal places.

(a) $\sec^{-1} 3$ (b) $\csc^{-1}(-4)$

(c) $\cot^{-1} \dfrac{1}{2}$ (d) $\cot^{-1}(-2)$

Solution First, set your calculator to radian mode.

(a) Let $\theta = \sec^{-1} 3$. Then $\sec \theta = 3$ and $0 \le \theta \le \pi, \theta \ne \dfrac{\pi}{2}$. We seek $\cos \theta$ because $y = \cos^{-1} x$ has the same range as $y = \sec^{-1} x$, except where undefined. Since $\sec \theta = \dfrac{1}{\cos \theta} = 3$, we have $\cos \theta = \dfrac{1}{3}$. Then $\theta = \cos^{-1} \dfrac{1}{3}$ and

$$\sec^{-1} 3 = \theta = \cos^{-1} \dfrac{1}{3} \underset{\underset{\text{Use a calculator.}}{\uparrow}}{\approx} 1.23$$

(b) Let $\theta = \csc^{-1}(-4)$. Then $\csc \theta = -4, -\dfrac{\pi}{2} \le \theta \le \dfrac{\pi}{2}, \theta \ne 0$. We seek $\sin \theta$ because $y = \sin^{-1} x$ has the same range as $y = \csc^{-1} x$, except where undefined. Since $\csc \theta = \dfrac{1}{\sin \theta} = -4$, we have $\sin \theta = -\dfrac{1}{4}$. Then $\theta = \sin^{-1}\left(-\dfrac{1}{4}\right)$, and

$$\csc^{-1}(-4) = \theta = \sin^{-1}\left(-\dfrac{1}{4}\right) \approx -0.25$$

Figure 18

$\cot \theta = \dfrac{1}{2}, 0 < \theta < \pi$

(c) We proceed as before. Let $\theta = \cot^{-1} \dfrac{1}{2}$. Then $\cot \theta = \dfrac{1}{2}, 0 < \theta < \pi$. From these facts we know that θ lies in quadrant I. We seek $\cos \theta$ because $y = \cos^{-1} x$ has the same range as $y = \cot^{-1} x$, except where undefined. To find $\cos \theta$ we use Figure 18. Then $\cos \theta = \dfrac{1}{\sqrt{5}}, 0 < \theta < \dfrac{\pi}{2}, \theta = \cos^{-1}\left(\dfrac{1}{\sqrt{5}}\right)$, and

$$\cot^{-1} \dfrac{1}{2} = \theta = \cos^{-1}\left(\dfrac{1}{\sqrt{5}}\right) \approx 1.11$$

Figure 19

$\cot \theta = -2, 0 < \theta < \pi$

(d) Let $\theta = \cot^{-1}(-2)$. Then $\cot \theta = -2, 0 < \theta < \pi$. From these facts we know that θ lies in quadrant II. We seek $\cos \theta$. To find it we use Figure 19. Then $\cos \theta = -\dfrac{2}{\sqrt{5}}, \dfrac{\pi}{2} < \theta < \pi, \theta = \cos^{-1}\left(-\dfrac{2}{\sqrt{5}}\right)$, and

$$\cot^{-1}(-2) = \theta = \cos^{-1}\left(-\dfrac{2}{\sqrt{5}}\right) \approx 2.68$$

■

➡ **Now Work** PROBLEM 45

4 Write a Trigonometric Expression as an Algebraic Expression

| EXAMPLE 6 | **Writing a Trigonometric Expression as an Algebraic Expression** |

Write $\sin(\tan^{-1} u)$ as an algebraic expression containing u.

Solution Let $\theta = \tan^{-1}u$ so that $\tan \theta = u, -\dfrac{\pi}{2} < \theta < \dfrac{\pi}{2}, -\infty < u < \infty$. As a result, we know that $\sec \theta > 0$. Then

$$\sin(\tan^{-1} u) = \sin \theta = \sin \theta \cdot \frac{\cos \theta}{\cos \theta} = \tan \theta \cos \theta = \frac{\tan \theta}{\sec \theta} = \frac{\tan \theta}{\sqrt{1 + \tan^2 \theta}} = \frac{u}{\sqrt{1 + u^2}}$$

$$\underset{\text{Multiply by 1: } \frac{\cos \theta}{\cos \theta}.}{\uparrow} \qquad \underset{\frac{\sin \theta}{\cos \theta} = \tan \theta}{\uparrow} \qquad \underset{\substack{\sec^2 \theta = 1 + \tan^2 \theta \\ \sec \theta > 0}}{\uparrow}$$

■

━━ **Now Work** PROBLEM 57

8.2 Assess Your Understanding

'Are You Prepared?' *Answers are given at the end of these exercises. If you get a wrong answer, read the pages listed in red.*

1. What is the domain and the range of $y = \sec x$? (pp. 553–555)

2. *True or False* The graph of $y = \sec x$ is increasing on the interval $\left[0, \dfrac{\pi}{2}\right)$ and on the interval $\left(\dfrac{\pi}{2}, \pi\right]$. (pp. 577–579)

3. If $\cot \theta = -2$ and $0 < \theta < \pi$, then $\cos \theta =$ _____. (pp. 547–548)

Concepts and Vocabulary

4. $y = \sec^{-1} x$ means _____, where $|x|$ _____ and _____ $\le y \le$ _____, $y \ne \dfrac{\pi}{2}$.

5. $\cos(\tan^{-1} 1) =$ ____.

6. *True or False* It is impossible to obtain exact values for the inverse secant function.

7. *True or False* $\csc^{-1} 0.5$ is not defined.

8. *True or False* The domain of the inverse cotangent function is the set of real numbers.

Skill Building

In Problems 9–36, find the exact value of each expression.

9. $\cos\left(\sin^{-1} \dfrac{\sqrt{2}}{2}\right)$

10. $\sin\left(\cos^{-1} \dfrac{1}{2}\right)$

11. $\tan\left[\cos^{-1}\left(-\dfrac{\sqrt{3}}{2}\right)\right]$

12. $\tan\left[\sin^{-1}\left(-\dfrac{1}{2}\right)\right]$

13. $\sec\left(\cos^{-1} \dfrac{1}{2}\right)$

14. $\cot\left[\sin^{-1}\left(-\dfrac{1}{2}\right)\right]$

15. $\csc(\tan^{-1} 1)$

16. $\sec\left(\tan^{-1} \sqrt{3}\right)$

17. $\sin[\tan^{-1}(-1)]$

18. $\cos\left[\sin^{-1}\left(-\dfrac{\sqrt{3}}{2}\right)\right]$

19. $\sec\left[\sin^{-1}\left(-\dfrac{1}{2}\right)\right]$

20. $\csc\left[\cos^{-1}\left(-\dfrac{\sqrt{3}}{2}\right)\right]$

21. $\cos^{-1}\left(\cos \dfrac{5\pi}{4}\right)$

22. $\tan^{-1}\left(\tan \dfrac{2\pi}{3}\right)$

23. $\sin^{-1}\left[\sin\left(-\dfrac{7\pi}{6}\right)\right]$

24. $\cos^{-1}\left[\cos\left(-\dfrac{\pi}{3}\right)\right]$

25. $\tan\left(\sin^{-1} \dfrac{1}{3}\right)$

26. $\tan\left(\cos^{-1} \dfrac{1}{3}\right)$

27. $\sec\left(\tan^{-1} \dfrac{1}{2}\right)$

28. $\cos\left(\sin^{-1} \dfrac{\sqrt{2}}{3}\right)$

29. $\cot\left[\sin^{-1}\left(-\dfrac{\sqrt{2}}{3}\right)\right]$

30. $\csc[\tan^{-1}(-2)]$

31. $\sin[\tan^{-1}(-3)]$

32. $\cot\left[\cos^{-1}\left(-\dfrac{\sqrt{3}}{3}\right)\right]$

33. $\sec\left(\sin^{-1} \dfrac{2\sqrt{5}}{5}\right)$

34. $\csc\left(\tan^{-1} \dfrac{1}{2}\right)$

35. $\sin^{-1}\left(\cos \dfrac{3\pi}{4}\right)$

36. $\cos^{-1}\left(\sin \dfrac{7\pi}{6}\right)$

In Problems 37–44, find the exact value of each expression.

37. $\cot^{-1} \sqrt{3}$

38. $\cot^{-1} 1$

39. $\csc^{-1}(-1)$

40. $\csc^{-1} \sqrt{2}$

41. $\sec^{-1} \dfrac{2\sqrt{3}}{3}$

42. $\sec^{-1}(-2)$

43. $\cot^{-1}\left(-\dfrac{\sqrt{3}}{3}\right)$

44. $\csc^{-1}\left(-\dfrac{2\sqrt{3}}{3}\right)$

In Problems 45–56, use a calculator to find the value of each expression rounded to two decimal places.

45. $\sec^{-1} 4$

46. $\csc^{-1} 5$

47. $\cot^{-1} 2$

48. $\sec^{-1}(-3)$

49. $\csc^{-1}(-3)$ **50.** $\cot^{-1}\left(-\dfrac{1}{2}\right)$ **51.** $\cot^{-1}\left(-\sqrt{5}\right)$ **52.** $\cot^{-1}(-8.1)$

53. $\csc^{-1}\left(-\dfrac{3}{2}\right)$ **54.** $\sec^{-1}\left(-\dfrac{4}{3}\right)$ **55.** $\cot^{-1}\left(-\dfrac{3}{2}\right)$ **56.** $\cot^{-1}\left(-\sqrt{10}\right)$

In Problems 57–66, write each trigonometric expression as an algebraic expression in u.

57. $\cos\left(\tan^{-1}u\right)$ **58.** $\sin\left(\cos^{-1}u\right)$ **59.** $\tan\left(\sin^{-1}u\right)$ **60.** $\tan\left(\cos^{-1}u\right)$ **61.** $\sin\left(\sec^{-1}u\right)$

62. $\sin\left(\cot^{-1}u\right)$ **63.** $\cos\left(\csc^{-1}u\right)$ **64.** $\cos\left(\sec^{-1}u\right)$ **65.** $\tan\left(\cot^{-1}u\right)$ **66.** $\tan\left(\sec^{-1}u\right)$

Applications and Extensions

In Problems 67–78, $f(x) = \sin x$, $g(x) = \cos x$, and $h(x) = \tan x$. Find the exact value of each composite function.

67. $g\left(f^{-1}\left(\dfrac{12}{13}\right)\right)$ **68.** $f\left(g^{-1}\left(\dfrac{5}{13}\right)\right)$ **69.** $g^{-1}\left(f\left(\dfrac{7\pi}{4}\right)\right)$ **70.** $f^{-1}\left(g\left(\dfrac{5\pi}{6}\right)\right)$

71. $h\left(f^{-1}\left(-\dfrac{3}{5}\right)\right)$ **72.** $h\left(g^{-1}\left(-\dfrac{4}{5}\right)\right)$ **73.** $g\left(h^{-1}\left(\dfrac{12}{5}\right)\right)$ **74.** $f\left(h^{-1}\left(\dfrac{5}{12}\right)\right)$

75. $g^{-1}\left(f\left(-\dfrac{4\pi}{3}\right)\right)$ **76.** $g^{-1}\left(f\left(-\dfrac{5\pi}{6}\right)\right)$ **77.** $h\left(g^{-1}\left(-\dfrac{1}{4}\right)\right)$ **78.** $h\left(f^{-1}\left(-\dfrac{2}{5}\right)\right)$

*Problems 79 and 80 require the following discussion: When granular materials are allowed to fall freely, they form conical (cone-shaped) piles. The naturally occurring angle of slope, measured from the horizontal, at which the loose material comes to rest is called the **angle of repose** and varies for different materials. The angle of repose θ is related to the height h and base radius r of the conical pile by the equation $\theta = \cot^{-1}\dfrac{r}{h}$. See the illustration.*

79. Angle of Repose: Deicing Salt Due to potential transportation issues (for example, frozen waterways) deicing salt used by highway departments in the Midwest must be ordered early and stored for future use. When deicing salt is stored in a pile 14 feet high, the diameter of the base of the pile is 45 feet.

 (a) Find the angle of repose for deicing salt.

 (b) What is the base diameter of a pile that is 17 feet high?

 (c) What is the height of a pile that has a base diameter of approximately 122 feet?

Source: Salt Institute, The Salt Storage Handbook, 2006

80. Angle of Repose: Bunker Sand The steepness of sand bunkers on a golf course is affected by the angle of repose of the sand (a larger angle of repose allows for steeper bunkers). A freestanding pile of loose sand from a United States Golf Association (USGA) bunker had a height of 4 feet and a base diameter of approximately 6.68 feet.

 (a) Find the angle of repose for USGA bunker sand.

 (b) What is the height of such a pile if the diameter of the base is 8 feet?

 (c) A 6-foot high pile of loose Tour Grade 50/50 sand has a base diameter of approximately 8.44 feet. Which type of sand (USGA or Tour Grade 50/50) would be better suited for steep bunkers?

Source: 2004 Annual Report, Purdue University Turfgrass Science Program

81. Artillery A projectile fired into the first quadrant from the origin of a coordinate system will pass through the point (x, y) at time t according to the relationship $\cot\theta = \dfrac{2x}{2y + gt^2}$ where θ = the angle of elevation of the launcher and g = the acceleration due to gravity = 32.2 feet/second². An artilleryman is firing at an enemy bunker located 2450 feet up the side of a hill that is 6175 feet away. He fires a round, and exactly 2.27 seconds later he scores a direct hit.

 (a) What angle of elevation did he use?

 (b) If the angle of elevation is also given by $\sec\theta = \dfrac{v_0 t}{x}$, where v_0 as the muzzle velocity of the weapon, find the muzzle velocity of the artillery piece he used.

Source: www.egwald.com/geometry/projectile3d.php

82. Using a graphing utility, graph $y = \cot^{-1}x$.

83. Using a graphing utility, graph $y = \sec^{-1}x$.

84. Using a graphing utility, graph $y = \csc^{-1}x$.

Discussion and Writing

85. Explain in your own words how you would use your calculator to find the value of $\cot^{-1}10$.

86. Consult three books on calculus and write down the definition in each of $y = \sec^{-1}x$ and $y = \csc^{-1}x$. Compare these with the definitions given in this book.

'Are You Prepared?' Answers

1. Domain: $\left\{x \mid x \neq \text{odd integer multiples of } \dfrac{\pi}{2}\right\}$; range: $\{y \leq -1 \text{ or } y \geq 1\}$ **2.** True **3.** $\dfrac{-2\sqrt{5}}{5}$

8.3 Trigonometric Identities

PREPARING FOR THIS SECTION *Before getting started, review the following:*

- Fundamental Identities (Section 7.2, p. 520)
- Even–Odd Properties (Section 7.5, p. 556)

Now Work the 'Are You Prepared?' problems on page 624.

OBJECTIVES 1 Use Algebra to Simplify Trigonometric Expressions (p. 621)
2 Establish Identities (p. 622)

We saw in the previous chapter that the trigonometric functions lend themselves to a wide variety of identities. Before establishing some additional identities, let's review the definition of an *identity*.

DEFINITION

Two functions f and g are said to be **identically equal** if

$$f(x) = g(x)$$

for every value of x for which both functions are defined. Such an equation is referred to as an **identity**. An equation that is not an identity is called a **conditional equation.**

For example, the following are identities:

$$(x + 1)^2 = x^2 + 2x + 1 \qquad \sin^2 x + \cos^2 x = 1 \qquad \csc x = \frac{1}{\sin x}$$

The following are conditional equations:

$$2x + 5 = 0 \qquad \text{True only if } x = -\frac{5}{2}$$

$$\sin x = 0 \qquad \text{True only if } x = k\pi, k \text{ an integer}$$

$$\sin x = \cos x \qquad \text{True only if } x = \frac{\pi}{4} + 2k\pi \text{ or } x = \frac{5\pi}{4} + 2k\pi, k \text{ an integer}$$

The following boxes summarize the trigonometric identities that we have established thus far.

Quotient Identities

$$\tan\theta = \frac{\sin\theta}{\cos\theta} \qquad \cot\theta = \frac{\cos\theta}{\sin\theta}$$

Reciprocal Identities

$$\csc\theta = \frac{1}{\sin\theta} \qquad \sec\theta = \frac{1}{\cos\theta} \qquad \cot\theta = \frac{1}{\tan\theta}$$

Pythagorean Identities

$$\sin^2\theta + \cos^2\theta = 1 \qquad \tan^2\theta + 1 = \sec^2\theta$$
$$\cot^2\theta + 1 = \csc^2\theta$$

Even–Odd Identities

$$\sin(-\theta) = -\sin\theta \qquad \cos(-\theta) = \cos\theta \qquad \tan(-\theta) = -\tan\theta$$
$$\csc(-\theta) = -\csc\theta \qquad \sec(-\theta) = \sec\theta \qquad \cot(-\theta) = -\cot\theta$$

This list of identities comprises what we shall refer to as the **basic trigonometric identities.** These identities should not merely be memorized, but should be *known* (just as you know your name rather than have it memorized). In fact, minor variations of a basic identity are often used. For example, we might want to use

$$\sin^2 \theta = 1 - \cos^2 \theta \quad \text{or} \quad \cos^2 \theta = 1 - \sin^2 \theta$$

instead of $\sin^2 \theta + \cos^2 \theta = 1$. For this reason, among others, you need to know these relationships and be quite comfortable with variations of them.

1 Use Algebra to Simplify Trigonometric Expressions

The ability to use algebra to manipulate trigonometric expressions is a key skill that one must have to establish identities. Some of the techniques that are used in establishing identities are multiplying by a "well-chosen 1," writing a trigonometric expression over a common denominator, rewriting a trigonometric expression in terms of sine and cosine only, and factoring.

| EXAMPLE 1 | **Using Algebraic Techniques to Simplify Trigonometric Expressions** |

(a) Simplify $\dfrac{\cot \theta}{\csc \theta}$ by rewriting each trigonometric function in terms of sine and cosine functions.

(b) Show that $\dfrac{\cos \theta}{1 + \sin \theta} = \dfrac{1 - \sin \theta}{\cos \theta}$ by multiplying the numerator and denominator by $1 - \sin \theta$.

(c) Simplify $\dfrac{1 + \sin u}{\sin u} + \dfrac{\cot u - \cos u}{\cos u}$ by rewriting the expression over a common denominator.

(d) Simplify $\dfrac{\sin^2 v - 1}{\tan v \sin v - \tan v}$ by factoring.

Solution　(a) $\dfrac{\cot \theta}{\csc \theta} = \dfrac{\dfrac{\cos \theta}{\sin \theta}}{\dfrac{1}{\sin \theta}} = \dfrac{\cos \theta}{\sin \theta} \cdot \dfrac{\sin \theta}{1} = \cos \theta$

(b) $\dfrac{\cos \theta}{1 + \sin \theta} = \dfrac{\cos \theta}{1 + \sin \theta} \cdot \dfrac{1 - \sin \theta}{1 - \sin \theta} = \dfrac{\cos \theta (1 - \sin \theta)}{1 - \sin^2 \theta}$

　　　　　　$\uparrow$ Multiply by a well-chosen 1: $\dfrac{1 - \sin \theta}{1 - \sin \theta}$

　　　　$= \dfrac{\cos \theta (1 - \sin \theta)}{\cos^2 \theta} = \dfrac{1 - \sin \theta}{\cos \theta}$

(c) $\dfrac{1 + \sin u}{\sin u} + \dfrac{\cot u - \cos u}{\cos u} = \dfrac{1 + \sin u}{\sin u} \cdot \dfrac{\cos u}{\cos u} + \dfrac{\cot u - \cos u}{\cos u} \cdot \dfrac{\sin u}{\sin u}$

　　$= \dfrac{\cos u + \sin u \cos u + \cot u \sin u - \cos u \sin u}{\sin u \cos u} = \dfrac{\cos u + \dfrac{\cos u}{\sin u} \cdot \sin u}{\sin u \cos u}$

　　　　　　　　　　　　　　　　　　　　　$\uparrow \cot u = \dfrac{\cos u}{\sin u}$

　　$= \dfrac{\cos u + \cos u}{\sin u \cos u} = \dfrac{2 \cos u}{\sin u \cos u} = \dfrac{2}{\sin u}$

(d) $\dfrac{\sin^2 v - 1}{\tan v \sin v - \tan v} = \dfrac{(\sin v + 1)(\sin v - 1)}{\tan v (\sin v - 1)} = \dfrac{\sin v + 1}{\tan v}$

Now Work PROBLEMS **9**, **11**, AND **13**

2 Establish Identities

In the examples that follow, the directions will read "Establish the identity. . . . " As you will see, this is accomplished by starting with one side of the given equation (usually the one containing the more complicated expression) and, using appropriate basic identities and algebraic manipulations, arriving at the other side. The selection of appropriate basic identities to obtain the desired result is learned only through experience and lots of practice.

EXAMPLE 2 **Establishing an Identity**

Establish the identity: $\csc\theta \cdot \tan\theta = \sec\theta$

Solution We start with the left side, because it contains the more complicated expression, and apply a reciprocal identity and a quotient identity.

$$\csc\theta \cdot \tan\theta = \frac{1}{\sin\theta} \cdot \frac{\sin\theta}{\cos\theta} = \frac{1}{\cos\theta} = \sec\theta$$

Having arrived at the right side, the identity is established. ∎

Now Work PROBLEM 19

EXAMPLE 3 **Establishing an Identity**

Establish the identity: $\sin^2(-\theta) + \cos^2(-\theta) = 1$

Solution We begin with the left side and, because the arguments are $-\theta$, apply Even–Odd Identities.

$$\begin{aligned}
\sin^2(-\theta) + \cos^2(-\theta) &= [\sin(-\theta)]^2 + [\cos(-\theta)]^2 \\
&= (-\sin\theta)^2 + (\cos\theta)^2 \qquad \text{Even–Odd Identities} \\
&= (\sin\theta)^2 + (\cos\theta)^2 \\
&= 1 \qquad\qquad\qquad\qquad \text{Pythagorean Identity}
\end{aligned}$$

∎

EXAMPLE 4 **Establishing an Identity**

Establish the identity: $\dfrac{\sin^2(-\theta) - \cos^2(-\theta)}{\sin(-\theta) - \cos(-\theta)} = \cos\theta - \sin\theta$

Solution We begin with two observations: The left side contains the more complicated expression. Also, the left side contains expressions with the argument $-\theta$, whereas the right side contains expressions with the argument θ. We decide, therefore, to start with the left side and apply Even–Odd Identities.

$$\begin{aligned}
\frac{\sin^2(-\theta) - \cos^2(-\theta)}{\sin(-\theta) - \cos(-\theta)} &= \frac{[\sin(-\theta)]^2 - [\cos(-\theta)]^2}{\sin(-\theta) - \cos(-\theta)} \\[2mm]
&= \frac{(-\sin\theta)^2 - (\cos\theta)^2}{-\sin\theta - \cos\theta} \qquad \text{Even–Odd Identities} \\[2mm]
&= \frac{(\sin\theta)^2 - (\cos\theta)^2}{-\sin\theta - \cos\theta} \qquad\quad \text{Simplify.} \\[2mm]
&= \frac{(\sin\theta - \cos\theta)(\sin\theta + \cos\theta)}{-(\sin\theta + \cos\theta)} \qquad \text{Factor.} \\[2mm]
&= \cos\theta - \sin\theta \qquad\qquad\qquad\quad \text{Cancel and simplify.}
\end{aligned}$$

∎

| EXAMPLE 5 | **Establishing an Identity** |

Establish the identity: $\dfrac{1 + \tan u}{1 + \cot u} = \tan u$

Solution

$$\frac{1 + \tan u}{1 + \cot u} = \frac{1 + \tan u}{1 + \dfrac{1}{\tan u}} = \frac{1 + \tan u}{\dfrac{\tan u + 1}{\tan u}}$$

$$= \frac{\tan u(1 + \tan u)}{\tan u + 1} = \tan u$$

■

—— **Now Work** PROBLEMS 23 AND 27

When sums or differences of quotients appear, it is usually best to rewrite them as a single quotient, especially if the other side of the identity consists of only one term.

| EXAMPLE 6 | **Establishing an Identity** |

Establish the identity: $\dfrac{\sin \theta}{1 + \cos \theta} + \dfrac{1 + \cos \theta}{\sin \theta} = 2 \csc \theta$

Solution

The left side is more complicated, so we start with it and proceed to add.

$$\frac{\sin \theta}{1 + \cos \theta} + \frac{1 + \cos \theta}{\sin \theta} = \frac{\sin^2 \theta + (1 + \cos \theta)^2}{(1 + \cos \theta)(\sin \theta)} \qquad \text{Add the quotients.}$$

$$= \frac{\sin^2 \theta + 1 + 2 \cos \theta + \cos^2 \theta}{(1 + \cos \theta)(\sin \theta)} \qquad \begin{array}{l}\text{Remove parentheses}\\ \text{in the numerator.}\end{array}$$

$$= \frac{(\sin^2 \theta + \cos^2 \theta) + 1 + 2 \cos \theta}{(1 + \cos \theta)(\sin \theta)} \qquad \text{Regroup.}$$

$$= \frac{2 + 2 \cos \theta}{(1 + \cos \theta)(\sin \theta)} \qquad \text{Pythagorean Identity}$$

$$= \frac{2(1 + \cos \theta)}{(1 + \cos \theta)(\sin \theta)} \qquad \text{Factor and cancel.}$$

$$= \frac{2}{\sin \theta}$$

$$= 2 \csc \theta \qquad \text{Reciprocal Identity}$$

■

—— **Now Work** PROBLEM 49

Sometimes it helps to write one side in terms of sine and cosine functions only.

| EXAMPLE 7 | **Establishing an Identity** |

Establish the identity: $\dfrac{\tan v + \cot v}{\sec v \csc v} = 1$

Solution

$$\frac{\tan v + \cot v}{\sec v \csc v} = \frac{\dfrac{\sin v}{\cos v} + \dfrac{\cos v}{\sin v}}{\dfrac{1}{\cos v} \cdot \dfrac{1}{\sin v}} = \frac{\dfrac{\sin^2 v + \cos^2 v}{\cos v \sin v}}{\dfrac{1}{\cos v \sin v}}$$

 ↑ Change to sines and cosines. ↑ Add the quotients in the numerator.

$$= \frac{1}{\cos v \sin v} \cdot \frac{\cos v \sin v}{1} = 1$$

 ↑ Divide the quotients; $\sin^2 v + \cos^2 v = 1$.

■

—————▶ **Now Work** PROBLEM 69

Sometimes, multiplying the numerator and denominator by an appropriate factor will result in a simplification.

EXAMPLE 8	**Establishing an Identity**

Establish the identity: $\dfrac{1 - \sin\theta}{\cos\theta} = \dfrac{\cos\theta}{1 + \sin\theta}$

Solution We start with the left side and multiply the numerator and the denominator by $1 + \sin\theta$. (Alternatively, we could multiply the numerator and denominator of the right side by $1 - \sin\theta$.)

$$\frac{1 - \sin\theta}{\cos\theta} = \frac{1 - \sin\theta}{\cos\theta} \cdot \frac{1 + \sin\theta}{1 + \sin\theta} \quad \text{Multiply the numerator and denominator by } 1 + \sin\theta.$$

$$= \frac{1 - \sin^2\theta}{\cos\theta(1 + \sin\theta)}$$

$$= \frac{\cos^2\theta}{\cos\theta(1 + \sin\theta)} \quad 1 - \sin^2\theta = \cos^2\theta$$

$$= \frac{\cos\theta}{1 + \sin\theta} \quad \text{Cancel.}$$

■

—————▶ **Now Work** PROBLEM 53

Although a lot of practice is the only real way to learn how to establish identities, the following guidelines should prove helpful.

WARNING Be careful not to handle identities to be established as if they were conditional equations. You cannot establish an identity by such methods as adding the same expression to each side and obtaining a true statement. This practice is not allowed, because the original statement is precisely the one that you are trying to establish. You do not know until it has been established that it is, in fact, true. ■

Guidelines for Establishing Identities

1. It is almost always preferable to start with the side containing the more complicated expression.
2. Rewrite sums or differences of quotients as a single quotient.
3. Sometimes rewriting one side in terms of sine and cosine functions only will help.
4. Always keep your goal in mind. As you manipulate one side of the expression, you must keep in mind the form of the expression on the other side.

8.3 Asssess Your Understanding

'Are You Prepared?' *Answers are given at the end of these exercises. If you get a wrong answer, read the pages listed in red.*

1. *True or False* $\sin^2\theta = 1 - \cos^2\theta$. (p. 520)

2. *True or False* $\sin(-\theta) + \cos(-\theta) = \cos\theta - \sin\theta$. (p. 556)

Concepts and Vocabulary

3. Suppose that f and g are two functions with the same domain. If $f(x) = g(x)$ for every x in the domain, the equation is called a(n) _____. Otherwise, it is called a(n) _____ equation.

4. $\tan^2 \theta - \sec^2 \theta =$ _____.

5. $\cos(-\theta) - \cos \theta =$ _____.

6. *True or False* $\sin(-\theta) + \sin \theta = 0$ for any value of θ.

7. *True or False* In establishing an identity, it is often easiest to just multiply both sides by a well-chosen nonzero expression involving the variable.

8. *True or False* $\tan \theta \cdot \cos \theta = \sin \theta$ for any $\theta \neq (2k + 1)\dfrac{\pi}{2}$.

Skill Building

In Problems 9–18, simplify each trigonometric expression by following the indicated direction.

9. Rewrite in terms of sine and cosine functions: $\tan \theta \cdot \csc \theta$.

10. Rewrite in terms of sine and cosine functions: $\cot \theta \cdot \sec \theta$.

11. Multiply $\dfrac{\cos \theta}{1 - \sin \theta}$ by $\dfrac{1 + \sin \theta}{1 + \sin \theta}$.

12. Multiply $\dfrac{\sin \theta}{1 + \cos \theta}$ by $\dfrac{1 - \cos \theta}{1 - \cos \theta}$.

13. Rewrite over a common denominator:
$$\frac{\sin \theta + \cos \theta}{\cos \theta} + \frac{\cos \theta - \sin \theta}{\sin \theta}$$

14. Rewrite over a common denominator:
$$\frac{1}{1 - \cos v} + \frac{1}{1 + \cos v}$$

15. Multiply and simplify: $\dfrac{(\sin \theta + \cos \theta)(\sin \theta + \cos \theta) - 1}{\sin \theta \cos \theta}$

16. Multiply and simplify: $\dfrac{(\tan \theta + 1)(\tan \theta + 1) - \sec^2 \theta}{\tan \theta}$

17. Factor and simplify: $\dfrac{3 \sin^2 \theta + 4 \sin \theta + 1}{\sin^2 \theta + 2 \sin \theta + 1}$

18. Factor and simplify: $\dfrac{\cos^2 \theta - 1}{\cos^2 \theta - \cos \theta}$

In Problems 19–98, establish each identity.

19. $\csc \theta \cdot \cos \theta = \cot \theta$

20. $\sec \theta \cdot \sin \theta = \tan \theta$

21. $1 + \tan^2(-\theta) = \sec^2 \theta$

22. $1 + \cot^2(-\theta) = \csc^2 \theta$

23. $\cos \theta(\tan \theta + \cot \theta) = \csc \theta$

24. $\sin \theta(\cot \theta + \tan \theta) = \sec \theta$

25. $\tan u \cot u - \cos^2 u = \sin^2 u$

26. $\sin u \csc u - \cos^2 u = \sin^2 u$

27. $(\sec \theta - 1)(\sec \theta + 1) = \tan^2 \theta$

28. $(\csc \theta - 1)(\csc \theta + 1) = \cot^2 \theta$

29. $(\sec \theta + \tan \theta)(\sec \theta - \tan \theta) = 1$

30. $(\csc \theta + \cot \theta)(\csc \theta - \cot \theta) = 1$

31. $\cos^2 \theta(1 + \tan^2 \theta) = 1$

32. $(1 - \cos^2 \theta)(1 + \cot^2 \theta) = 1$

33. $(\sin \theta + \cos \theta)^2 + (\sin \theta - \cos \theta)^2 = 2$

34. $\tan^2 \theta \cos^2 \theta + \cot^2 \theta \sin^2 \theta = 1$

35. $\sec^4 \theta - \sec^2 \theta = \tan^4 \theta + \tan^2 \theta$

36. $\csc^4 \theta - \csc^2 \theta = \cot^4 \theta + \cot^2 \theta$

37. $\sec u - \tan u = \dfrac{\cos u}{1 + \sin u}$

38. $\csc u - \cot u = \dfrac{\sin u}{1 + \cos u}$

39. $3 \sin^2 \theta + 4 \cos^2 \theta = 3 + \cos^2 \theta$

40. $9 \sec^2 \theta - 5 \tan^2 \theta = 5 + 4 \sec^2 \theta$

41. $1 - \dfrac{\cos^2 \theta}{1 + \sin \theta} = \sin \theta$

42. $1 - \dfrac{\sin^2 \theta}{1 - \cos \theta} = -\cos \theta$

43. $\dfrac{1 + \tan v}{1 - \tan v} = \dfrac{\cot v + 1}{\cot v - 1}$

44. $\dfrac{\csc v - 1}{\csc v + 1} = \dfrac{1 - \sin v}{1 + \sin v}$

45. $\dfrac{\sec \theta}{\csc \theta} + \dfrac{\sin \theta}{\cos \theta} = 2 \tan \theta$

46. $\dfrac{\csc \theta - 1}{\cot \theta} = \dfrac{\cot \theta}{\csc \theta + 1}$

47. $\dfrac{1 + \sin \theta}{1 - \sin \theta} = \dfrac{\csc \theta + 1}{\csc \theta - 1}$

48. $\dfrac{\cos \theta + 1}{\cos \theta - 1} = \dfrac{1 + \sec \theta}{1 - \sec \theta}$

49. $\dfrac{1 - \sin v}{\cos v} + \dfrac{\cos v}{1 - \sin v} = 2 \sec v$

50. $\dfrac{\cos v}{1 + \sin v} + \dfrac{1 + \sin v}{\cos v} = 2 \sec v$

51. $\dfrac{\sin \theta}{\sin \theta - \cos \theta} = \dfrac{1}{1 - \cot \theta}$

52. $1 - \dfrac{\sin^2 \theta}{1 + \cos \theta} = \cos \theta$

53. $\dfrac{1 - \sin \theta}{1 + \sin \theta} = (\sec \theta - \tan \theta)^2$

54. $\dfrac{1 - \cos \theta}{1 + \cos \theta} = (\csc \theta - \cot \theta)^2$

55. $\dfrac{\cos \theta}{1 - \tan \theta} + \dfrac{\sin \theta}{1 - \cot \theta} = \sin \theta + \cos \theta$

56. $\dfrac{\cot \theta}{1 - \tan \theta} + \dfrac{\tan \theta}{1 - \cot \theta} = 1 + \tan \theta + \cot \theta$

57. $\tan \theta + \dfrac{\cos \theta}{1 + \sin \theta} = \sec \theta$

58. $\dfrac{\sin \theta \cos \theta}{\cos^2 \theta - \sin^2 \theta} = \dfrac{\tan \theta}{1 - \tan^2 \theta}$

59. $\dfrac{\tan \theta + \sec \theta - 1}{\tan \theta - \sec \theta + 1} = \tan \theta + \sec \theta$

60. $\dfrac{\sin \theta - \cos \theta + 1}{\sin \theta + \cos \theta - 1} = \dfrac{\sin \theta + 1}{\cos \theta}$

61. $\dfrac{\tan \theta - \cot \theta}{\tan \theta + \cot \theta} = \sin^2 \theta - \cos^2 \theta$

62. $\dfrac{\sec \theta - \cos \theta}{\sec \theta + \cos \theta} = \dfrac{\sin^2 \theta}{1 + \cos^2 \theta}$

63. $\dfrac{\tan u - \cot u}{\tan u + \cot u} + 1 = 2 \sin^2 u$

64. $\dfrac{\tan u - \cot u}{\tan u + \cot u} + 2 \cos^2 u = 1$

65. $\dfrac{\sec \theta + \tan \theta}{\cot \theta + \cos \theta} = \tan \theta \sec \theta$

66. $\dfrac{\sec \theta}{1 + \sec \theta} = \dfrac{1 - \cos \theta}{\sin^2 \theta}$

67. $\dfrac{1 - \tan^2 \theta}{1 + \tan^2 \theta} + 1 = 2 \cos^2 \theta$

68. $\dfrac{1 - \cot^2 \theta}{1 + \cot^2 \theta} + 2 \cos^2 \theta = 1$

69. $\dfrac{\sec \theta - \csc \theta}{\sec \theta \csc \theta} = \sin \theta - \cos \theta$

70. $\dfrac{\sin^2 \theta - \tan \theta}{\cos^2 \theta - \cot \theta} = \tan^2 \theta$

71. $\sec \theta - \cos \theta = \sin \theta \tan \theta$

72. $\tan \theta + \cot \theta = \sec \theta \csc \theta$

73. $\dfrac{1}{1 - \sin \theta} + \dfrac{1}{1 + \sin \theta} = 2 \sec^2 \theta$

74. $\dfrac{1 + \sin \theta}{1 - \sin \theta} - \dfrac{1 - \sin \theta}{1 + \sin \theta} = 4 \tan \theta \sec \theta$

75. $\dfrac{\sec \theta}{1 - \sin \theta} = \dfrac{1 + \sin \theta}{\cos^3 \theta}$

76. $\dfrac{1 + \sin \theta}{1 - \sin \theta} = (\sec \theta + \tan \theta)^2$

77. $\dfrac{(\sec v - \tan v)^2 + 1}{\csc v (\sec v - \tan v)} = 2 \tan v$

78. $\dfrac{\sec^2 v - \tan^2 v + \tan v}{\sec v} = \sin v + \cos v$

79. $\dfrac{\sin \theta + \cos \theta}{\cos \theta} - \dfrac{\sin \theta - \cos \theta}{\sin \theta} = \sec \theta \csc \theta$

80. $\dfrac{\sin \theta + \cos \theta}{\sin \theta} - \dfrac{\cos \theta - \sin \theta}{\cos \theta} = \sec \theta \csc \theta$

81. $\dfrac{\sin^3 \theta + \cos^3 \theta}{\sin \theta + \cos \theta} = 1 - \sin \theta \cos \theta$

82. $\dfrac{\sin^3 \theta + \cos^3 \theta}{1 - 2 \cos^2 \theta} = \dfrac{\sec \theta - \sin \theta}{\tan \theta - 1}$

83. $\dfrac{\cos^2 \theta - \sin^2 \theta}{1 - \tan^2 \theta} = \cos^2 \theta$

84. $\dfrac{\cos \theta + \sin \theta - \sin^3 \theta}{\sin \theta} = \cot \theta + \cos^2 \theta$

85. $\dfrac{(2 \cos^2 \theta - 1)^2}{\cos^4 \theta - \sin^4 \theta} = 1 - 2 \sin^2 \theta$

86. $\dfrac{1 - 2 \cos^2 \theta}{\sin \theta \cos \theta} = \tan \theta - \cot \theta$

87. $\dfrac{1 + \sin \theta + \cos \theta}{1 + \sin \theta - \cos \theta} = \dfrac{1 + \cos \theta}{\sin \theta}$

88. $\dfrac{1 + \cos \theta + \sin \theta}{1 + \cos \theta - \sin \theta} = \sec \theta + \tan \theta$

89. $(a \sin \theta + b \cos \theta)^2 + (a \cos \theta - b \sin \theta)^2 = a^2 + b^2$

90. $(2a \sin \theta \cos \theta)^2 + a^2 (\cos^2 \theta - \sin^2 \theta)^2 = a^2$

91. $\dfrac{\tan \alpha + \tan \beta}{\cot \alpha + \cot \beta} = \tan \alpha \tan \beta$

92. $(\tan \alpha + \tan \beta)(1 - \cot \alpha \cot \beta) + (\cot \alpha + \cot \beta)(1 - \tan \alpha \tan \beta) = 0$

93. $(\sin \alpha + \cos \beta)^2 + (\cos \beta + \sin \alpha)(\cos \beta - \sin \alpha) = 2 \cos \beta (\sin \alpha + \cos \beta)$

94. $(\sin \alpha - \cos \beta)^2 + (\cos \beta + \sin \alpha)(\cos \beta - \sin \alpha) = -2 \cos \beta (\sin \alpha - \cos \beta)$

95. $\ln |\sec \theta| = -\ln |\cos \theta|$

96. $\ln |\tan \theta| = \ln |\sin \theta| - \ln |\cos \theta|$

97. $\ln |1 + \cos \theta| + \ln |1 - \cos \theta| = 2 \ln |\sin \theta|$

98. $\ln |\sec \theta + \tan \theta| + \ln |\sec \theta - \tan \theta| = 0$

In Problems 99–102, show that the functions f and g are identically equal.

99. $f(x) = \sin x \cdot \tan x$ $\quad g(x) = \sec x - \cos x$

100. $f(x) = \cos x \cdot \cot x$ $\quad g(x) = \csc x - \sin x$

101. $f(\theta) = \dfrac{1 - \sin \theta}{\cos \theta} - \dfrac{\cos \theta}{1 + \sin \theta}$ $\quad g(\theta) = 0$

102. $f(\theta) = \tan \theta + \sec \theta$ $\quad g(\theta) = \dfrac{\cos \theta}{1 - \sin \theta}$

Applications and Extensions

103. Searchlights A searchlight at the grand opening of a new car dealership casts a spot of light on a wall located 75 meters from the searchlight. The acceleration $\ddot{r}$ of the spot of light is found to be $\ddot{r} = 1200 \sec \theta (2\sec^2 \theta - 1)$. Show that this is equivalent to $\ddot{r} = 1200 \left(\dfrac{1 + \sin^2 \theta}{\cos^3 \theta} \right)$.

Source: Adapted from Hibbeler, *Engineering Mechanics: Dynamics,* 10th ed. © 2004.

104. Optical Measurement Optical methods of measurement often rely on the interference of two light waves. If two light waves, identical except for a phase lag, are mixed together, the resulting intensity, or irradiance, is given by $I_t = 4A^2 \dfrac{(\csc \theta - 1)(\sec \theta + \tan \theta)}{\csc \theta \sec \theta}$. Show that this is equivalent to $I_t = (2A \cos \theta)^2$.

Source: Experimental Techniques, July/August 2002

Discussion and Writing

105. Write a few paragraphs outlining your strategy for establishing identities.

106. Write down the three Pythagorean Identities.

107. Why do you think it is usually preferable to start with the side containing the more complicated expression when establishing an identity?

108. Make up an identity that is not a Fundamental Identity.

'Are You Prepared?' Answers

1. True **2.** True

8.4 Sum and Difference Formulas

PREPARING FOR THIS SECTION *Before getting started, review the following:*

- Distance Formula (Section 2.1, p. 157)
- Values of the Trigonometric Functions (Section 7.3, pp. 529–532 and Section 7.4, pp. 540–547)
- Finding Exact Values Given the Value of a Trigonometric Function and the Quadrant of the Angle (Section 7.4, pp. 546–548)

Now Work the 'Are You Prepared?' problems on page 634.

OBJECTIVES 1 Use Sum and Difference Formulas to Find Exact Values (p. 628)

2 Use Sum and Difference Formulas to Establish Identities (p. 629)

3 Use Sum and Difference Formulas Involving Inverse Trigonometric Functions (p. 633)

In this section, we continue our derivation of trigonometric identities by obtaining formulas that involve the sum or difference of two angles, such as $\cos(\alpha + \beta)$, $\cos(\alpha - \beta)$, or $\sin(\alpha + \beta)$. These formulas are referred to as the **sum and difference formulas.** We begin with the formulas for $\cos(\alpha + \beta)$ and $\cos(\alpha - \beta)$.

THEOREM

Sum and Difference Formulas for the Cosine Function

$$\cos(\alpha + \beta) = \cos \alpha \cos \beta - \sin \alpha \sin \beta \qquad \textbf{(1)}$$

$$\cos(\alpha - \beta) = \cos \alpha \cos \beta + \sin \alpha \sin \beta \qquad \textbf{(2)}$$

In Words

Formula (1) states that the cosine of the sum of two angles equals the cosine of the first angle times the cosine of the second angle minus the sine of the first angle times the sine of the second angle.

Proof We will prove formula (2) first. Although this formula is true for all numbers α and β, we shall assume in our proof that $0 < \beta < \alpha < 2\pi$. We begin with the unit circle and place the angles α and β in standard position, as shown in Figure 20(a). The point P_1 lies on the terminal side of β, so its coordinates are $(\cos \beta, \sin \beta)$; and the point P_2 lies on the terminal side of α, so its coordinates are $(\cos \alpha, \sin \alpha)$.

Figure 20

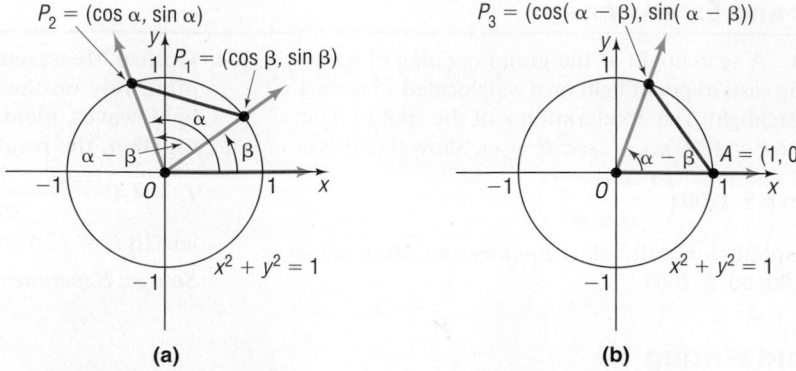

(a) (b)

Now place the angle $\alpha - \beta$ in standard position, as shown in Figure 20(b). The point A has coordinates $(1, 0)$, and the point P_3 is on the terminal side of the angle $\alpha - \beta$, so its coordinates are $(\cos(\alpha - \beta), \sin(\alpha - \beta))$.

Looking at triangle OP_1P_2 in Figure 20(a) and triangle OAP_3 in Figure 20(b), we see that these triangles are congruent. (Do you see why? We have SAS: two sides and the included angle, $\alpha - \beta$, are equal.) As a result, the unknown side of each triangle must be equal; that is,

$$d(A, P_3) = d(P_1, P_2)$$

Using the distance formula, we find that

$$\sqrt{[\cos(\alpha - \beta) - 1]^2 + [\sin(\alpha - \beta) - 0]^2} = \sqrt{(\cos\alpha - \cos\beta)^2 + (\sin\alpha - \sin\beta)^2} \quad \text{\small $d(A, P_3) = d(P_1, P_2)$}$$

$$[\cos(\alpha - \beta) - 1]^2 + \sin^2(\alpha - \beta) = (\cos\alpha - \cos\beta)^2 + (\sin\alpha - \sin\beta)^2 \quad \text{\small Square both sides.}$$

$$\cos^2(\alpha - \beta) - 2\cos(\alpha - \beta) + 1 + \sin^2(\alpha - \beta) = \cos^2\alpha - 2\cos\alpha\cos\beta + \cos^2\beta \quad \text{\small Multiply out the squared terms.}$$
$$+ \sin^2\alpha - 2\sin\alpha\sin\beta + \sin^2\beta$$

$$2 - 2\cos(\alpha - \beta) = 2 - 2\cos\alpha\cos\beta - 2\sin\alpha\sin\beta \quad \text{\small Apply a Pythagorean Identity (3 times).}$$

$$-2\cos(\alpha - \beta) = -2\cos\alpha\cos\beta - 2\sin\alpha\sin\beta \quad \text{\small Subtract 2 from each side.}$$

$$\cos(\alpha - \beta) = \cos\alpha\cos\beta + \sin\alpha\sin\beta \quad \text{\small Divide each side by −2.}$$

This is formula (2). ■

The proof of formula (1) follows from formula (2) and the Even–Odd Identities. We use the fact that $\alpha + \beta = \alpha - (-\beta)$. Then

$$\cos(\alpha + \beta) = \cos[\alpha - (-\beta)]$$

$$= \cos\alpha\cos(-\beta) + \sin\alpha\sin(-\beta) \quad \text{\small Use formula (2).}$$

$$= \cos\alpha\cos\beta - \sin\alpha\sin\beta \quad \text{\small Even–Odd Identities}$$

1 Use Sum and Difference Formulas to Find Exact Values

One use of formulas (1) and (2) is to obtain the exact value of the cosine of an angle that can be expressed as the sum or difference of angles whose sine and cosine are known exactly.

| EXAMPLE 1 | **Using the Sum Formula to Find Exact Values** |

Find the exact value of $\cos 75°$.

Solution Since $75° = 45° + 30°$, we use formula (1) to obtain

$$\cos 75° = \cos(45° + 30°) = \cos 45°\cos 30° - \sin 45°\sin 30°$$
$$\uparrow$$
$$\text{\small Formula (1)}$$

$$= \frac{\sqrt{2}}{2} \cdot \frac{\sqrt{3}}{2} - \frac{\sqrt{2}}{2} \cdot \frac{1}{2} = \frac{1}{4}(\sqrt{6} - \sqrt{2})$$

| EXAMPLE 2 | **Using the Difference Formula to Find Exact Values** |

Find the exact value of $\cos \dfrac{\pi}{12}$.

Solution
$$\cos \frac{\pi}{12} = \cos\left(\frac{3\pi}{12} - \frac{2\pi}{12}\right) = \cos\left(\frac{\pi}{4} - \frac{\pi}{6}\right)$$

$$= \cos \frac{\pi}{4} \cos \frac{\pi}{6} + \sin \frac{\pi}{4} \sin \frac{\pi}{6} \qquad \text{Use formula (2).}$$

$$= \frac{\sqrt{2}}{2} \cdot \frac{\sqrt{3}}{2} + \frac{\sqrt{2}}{2} \cdot \frac{1}{2} = \frac{1}{4}\left(\sqrt{6} + \sqrt{2}\right)$$

■

═══► **Now Work** PROBLEM 11

2 Use Sum and Difference Formulas to Establish Identities

Another use of formulas (1) and (2) is to establish other identities. Two important identities we conjectured earlier in Section 7.6 are given next.

$$\cos\left(\frac{\pi}{2} - \theta\right) = \sin \theta \qquad \textbf{(3a)}$$

$$\sin\left(\frac{\pi}{2} - \theta\right) = \cos \theta \qquad \textbf{(3b)}$$

Proof To prove formula (3a), we use the formula for $\cos(\alpha - \beta)$ with $\alpha = \dfrac{\pi}{2}$ and $\beta = \theta$.

$$\cos\left(\frac{\pi}{2} - \theta\right) = \cos \frac{\pi}{2} \cos \theta + \sin \frac{\pi}{2} \sin \theta$$

$$= 0 \cdot \cos \theta + 1 \cdot \sin \theta$$

$$= \sin \theta$$

To prove formula (3b), we make use of the identity (3a) just established.

$$\sin\left(\frac{\pi}{2} - \theta\right) = \cos\left[\frac{\pi}{2} - \left(\frac{\pi}{2} - \theta\right)\right] = \cos \theta$$
$$\underset{\text{Use (3a).}}{\uparrow}$$

■

Formulas (3a) and (3b) should look familiar. They are the basis for the theorem stated in Chapter 7: Cofunctions of complementary angles are equal. Also, since

$$\cos\left(\frac{\pi}{2} - \theta\right) = \cos\left[-\left(\theta - \frac{\pi}{2}\right)\right] = \cos\left(\theta - \frac{\pi}{2}\right)$$
$$\underset{\text{Even Property of Cosine}}{\uparrow}$$

and since

$$\cos\left(\frac{\pi}{2} - \theta\right) = \sin \theta$$
$$\underset{3(a)}{\uparrow}$$

it follows that $\cos\left(\theta - \dfrac{\pi}{2}\right) = \sin \theta$. The graphs of $y = \cos\left(\theta - \dfrac{\pi}{2}\right)$ and $y = \sin \theta$ are identical.

Having established the identities in formulas (3a) and (3b), we now can derive the sum and difference formulas for $\sin(\alpha + \beta)$ and $\sin(\alpha - \beta)$.

Proof $\sin(\alpha + \beta) = \cos\left[\dfrac{\pi}{2} - (\alpha + \beta)\right]$ Formula (3a)

$$= \cos\left[\left(\dfrac{\pi}{2} - \alpha\right) - \beta\right]$$

$$= \cos\left(\dfrac{\pi}{2} - \alpha\right)\cos \beta + \sin\left(\dfrac{\pi}{2} - \alpha\right)\sin \beta \quad \text{Formula (2)}$$

$$= \sin \alpha \cos \beta + \cos \alpha \sin \beta \quad \text{Formulas (3a) and (3b)}$$

$$\sin(\alpha - \beta) = \sin[\alpha + (-\beta)]$$

$$= \sin \alpha \cos(-\beta) + \cos \alpha \sin(-\beta) \quad \begin{array}{l}\text{Use the sum formula for}\\ \text{sine just obtained.}\end{array}$$

$$= \sin \alpha \cos \beta + \cos \alpha(-\sin \beta) \quad \text{Even–Odd Identities.}$$

$$= \sin \alpha \cos \beta - \cos \alpha \sin \beta \quad \blacksquare$$

> **In Words**
> Formula (4) states that the sine of the sum of two angles equals the sine of the first angle times the cosine of the second angle plus the cosine of the first angle times the sine of the second angle.

THEOREM **Sum and Difference Formulas for the Sine Function**

$$\sin(\alpha + \beta) = \sin \alpha \cos \beta + \cos \alpha \sin \beta \quad \textbf{(4)}$$

$$\sin(\alpha - \beta) = \sin \alpha \cos \beta - \cos \alpha \sin \beta \quad \textbf{(5)}$$

EXAMPLE 3 **Using the Sum Formula to Find Exact Values**

Find the exact value of $\sin \dfrac{7\pi}{12}$.

Solution $\sin \dfrac{7\pi}{12} = \sin\left(\dfrac{3\pi}{12} + \dfrac{4\pi}{12}\right) = \sin\left(\dfrac{\pi}{4} + \dfrac{\pi}{3}\right)$

$$= \sin \dfrac{\pi}{4}\cos \dfrac{\pi}{3} + \cos \dfrac{\pi}{4}\sin \dfrac{\pi}{3} \quad \text{Formula (4)}$$

$$= \dfrac{\sqrt{2}}{2}\cdot\dfrac{1}{2} + \dfrac{\sqrt{2}}{2}\cdot\dfrac{\sqrt{3}}{2} = \dfrac{1}{4}\left(\sqrt{2} + \sqrt{6}\right)$$

➔ **Now Work** PROBLEM 17

EXAMPLE 4 **Using the Difference Formula to Find Exact Values**

Find the exact value of $\sin 80° \cos 20° - \cos 80° \sin 20°$.

Solution The form of the expression $\sin 80° \cos 20° - \cos 80° \sin 20°$ is that of the right side of the formula (5) for $\sin(\alpha - \beta)$ with $\alpha = 80°$ and $\beta = 20°$. That is,

$$\sin 80° \cos 20° - \cos 80° \sin 20° = \sin(80° - 20°) = \sin 60° = \dfrac{\sqrt{3}}{2}$$

➔ **Now Work** PROBLEMS 23 AND 27

EXAMPLE 5 **Finding Exact Values**

If it is known that $\sin \alpha = \dfrac{4}{5}$, $\dfrac{\pi}{2} < \alpha < \pi$, and that $\sin \beta = -\dfrac{2}{\sqrt{5}} = -\dfrac{2\sqrt{5}}{5}$, $\pi < \beta < \dfrac{3\pi}{2}$, find the exact value of

(a) $\cos \alpha$ (b) $\cos \beta$ (c) $\cos(\alpha + \beta)$ (d) $\sin(\alpha + \beta)$

Solution

(a) Since $\sin \alpha = \dfrac{4}{5} = \dfrac{b}{r}$ and $\dfrac{\pi}{2} < \alpha < \pi$, we let $b = 4$ and $r = 5$ and place α in quadrant II. See Figure 21. Since $(a, 4)$ is in quadrant II, we have $a < 0$. The distance from $(a, 4)$ to $(0, 0)$ is 5, so

$$a^2 + 4^2 = 5^2$$
$$a^2 + 16 = 25$$
$$a^2 = 25 - 16 = 9$$
$$a = -3 \qquad a < 0$$

Then

$$\cos \alpha = \frac{a}{r} = -\frac{3}{5}$$

Alternatively, we can find $\cos \alpha$ using identities, as follows:

$$\cos \alpha = -\sqrt{1 - \sin^2 \alpha} = -\sqrt{1 - \frac{16}{25}} = -\sqrt{\frac{9}{25}} = -\frac{3}{5}$$

$\uparrow$
α in quadrant II,
$\cos \alpha < 0$

Figure 21

$\sin \alpha = \dfrac{4}{5}, \dfrac{\pi}{2} < \alpha < \pi$

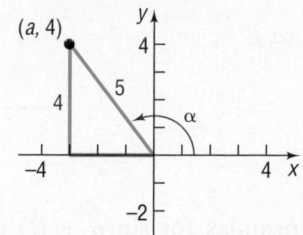

Figure 22

$\sin \beta = \dfrac{-2}{\sqrt{5}}, \pi < \beta < \dfrac{3\pi}{2}$

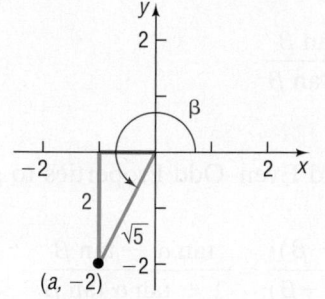

(b) Since $\sin \beta = \dfrac{-2}{\sqrt{5}} = \dfrac{b}{r}$ and $\pi < \beta < \dfrac{3\pi}{2}$, we let $b = -2$ and $r = \sqrt{5}$ and place β in quadrant III. See Figure 22. Since $(a, -2)$ is in quadrant III, we have $a < 0$. The distance from $(a, -2)$ to $(0, 0)$ is $\sqrt{5}$, so

$$a^2 + 2^2 = \left(\sqrt{5}\right)^2$$
$$a^2 = 5 - 4 = 1$$
$$a = -1 \qquad a < 0$$

Then

$$\cos \beta = \frac{a}{r} = \frac{-1}{\sqrt{5}} = -\frac{\sqrt{5}}{5}$$

Alternatively, we can find $\cos \beta$ using identities, as follows:

$$\cos \beta = -\sqrt{1 - \sin^2 \beta} = -\sqrt{1 - \frac{4}{5}} = -\sqrt{\frac{1}{5}} = -\frac{\sqrt{5}}{5}$$

(c) Using the results found in parts (a) and (b) and formula (1), we have

$$\cos(\alpha + \beta) = \cos \alpha \cos \beta - \sin \alpha \sin \beta$$

$$= -\frac{3}{5}\left(-\frac{\sqrt{5}}{5}\right) - \frac{4}{5}\left(-\frac{2\sqrt{5}}{5}\right) = \frac{11\sqrt{5}}{25}$$

(d) $\sin(\alpha + \beta) = \sin \alpha \cos \beta + \cos \alpha \sin \beta$

$$= \frac{4}{5}\left(-\frac{\sqrt{5}}{5}\right) + \left(-\frac{3}{5}\right)\left(-\frac{2\sqrt{5}}{5}\right) = \frac{2\sqrt{5}}{25}$$

— **Now Work** PROBLEMS 31(a), (b), AND (c)

EXAMPLE 6 **Establishing an Identity**

Establish the identity: $\dfrac{\cos(\alpha - \beta)}{\sin \alpha \sin \beta} = \cot \alpha \cot \beta + 1$

Solution
$$\frac{\cos(\alpha - \beta)}{\sin \alpha \sin \beta} = \frac{\cos \alpha \cos \beta + \sin \alpha \sin \beta}{\sin \alpha \sin \beta}$$

$$= \frac{\cos \alpha \cos \beta}{\sin \alpha \sin \beta} + \frac{\sin \alpha \sin \beta}{\sin \alpha \sin \beta}$$

$$= \frac{\cos \alpha}{\sin \alpha} \cdot \frac{\cos \beta}{\sin \beta} + 1$$

$$= \cot \alpha \cot \beta + 1$$

━━━ **Now Work** PROBLEMS 45 AND 57

We use the identity $\tan \theta = \dfrac{\sin \theta}{\cos \theta}$ and the sum formulas for $\sin(\alpha + \beta)$ and $\cos(\alpha + \beta)$ to derive a formula for $\tan(\alpha + \beta)$.

Proof $\tan(\alpha + \beta) = \dfrac{\sin(\alpha + \beta)}{\cos(\alpha + \beta)} = \dfrac{\sin \alpha \cos \beta + \cos \alpha \sin \beta}{\cos \alpha \cos \beta - \sin \alpha \sin \beta}$

Now we divide the numerator and denominator by $\cos \alpha \cos \beta$.

$$\tan(\alpha + \beta) = \frac{\dfrac{\sin \alpha \cos \beta + \cos \alpha \sin \beta}{\cos \alpha \cos \beta}}{\dfrac{\cos \alpha \cos \beta - \sin \alpha \sin \beta}{\cos \alpha \cos \beta}} = \frac{\dfrac{\sin \alpha \cos \beta}{\cos \alpha \cos \beta} + \dfrac{\cos \alpha \sin \beta}{\cos \alpha \cos \beta}}{\dfrac{\cos \alpha \cos \beta}{\cos \alpha \cos \beta} - \dfrac{\sin \alpha \sin \beta}{\cos \alpha \cos \beta}}$$

$$= \frac{\dfrac{\sin \alpha}{\cos \alpha} + \dfrac{\sin \beta}{\cos \beta}}{1 - \dfrac{\sin \alpha}{\cos \alpha} \cdot \dfrac{\sin \beta}{\cos \beta}} = \frac{\tan \alpha + \tan \beta}{1 - \tan \alpha \tan \beta}$$

Proof We use the sum formula for $\tan(\alpha + \beta)$ and Even–Odd Properties to get the difference formula.

$$\tan(\alpha - \beta) = \tan[\alpha + (-\beta)] = \frac{\tan \alpha + \tan(-\beta)}{1 - \tan \alpha \tan(-\beta)} = \frac{\tan \alpha - \tan \beta}{1 + \tan \alpha \tan \beta}$$

We have proved the following results:

THEOREM

Sum and Difference Formulas for the Tangent Function

$$\tan(\alpha + \beta) = \frac{\tan \alpha + \tan \beta}{1 - \tan \alpha \tan \beta} \quad \text{(6)}$$

$$\tan(\alpha - \beta) = \frac{\tan \alpha - \tan \beta}{1 + \tan \alpha \tan \beta} \quad \text{(7)}$$

━━━ **Now Work** PROBLEM 31(d)

EXAMPLE 7 **Establishing an Identity**

Prove the identity: $\tan(\theta + \pi) = \tan \theta$

Solution $\tan(\theta + \pi) = \dfrac{\tan \theta + \tan \pi}{1 - \tan \theta \tan \pi} = \dfrac{\tan \theta + 0}{1 - \tan \theta \cdot 0} = \tan \theta$

The result obtained in Example 7 verifies that the tangent function is periodic with period π, a fact that we discussed earlier.

| EXAMPLE 8 | Establishing an Identity |

Prove the identity: $\tan\left(\theta + \dfrac{\pi}{2}\right) = -\cot\theta$

Solution We cannot use formula (6), since $\tan\dfrac{\pi}{2}$ is not defined. Instead, we proceed as follows:

WARNING Be careful when using formulas (6) and (7). These formulas can be used only for angles α and β for which $\tan\alpha$ and $\tan\beta$ are defined, that is, all angles except odd integer multiples of $\dfrac{\pi}{2}$. ∎

$$\tan\left(\theta + \dfrac{\pi}{2}\right) = \dfrac{\sin\left(\theta + \dfrac{\pi}{2}\right)}{\cos\left(\theta + \dfrac{\pi}{2}\right)} = \dfrac{\sin\theta\cos\dfrac{\pi}{2} + \cos\theta\sin\dfrac{\pi}{2}}{\cos\theta\cos\dfrac{\pi}{2} - \sin\theta\sin\dfrac{\pi}{2}}$$

$$= \dfrac{(\sin\theta)(0) + (\cos\theta)(1)}{(\cos\theta)(0) - (\sin\theta)(1)} = \dfrac{\cos\theta}{-\sin\theta} = -\cot\theta$$ ∎

3 Use Sum and Difference Formulas Involving Inverse Trigonometric Functions

| EXAMPLE 9 | Finding the Exact Value of an Expression Involving Inverse Trigonometric Functions |

Find the exact value of: $\sin\left(\cos^{-1}\dfrac{1}{2} + \sin^{-1}\dfrac{3}{5}\right)$

Solution We seek the sine of the sum of two angles, $\alpha = \cos^{-1}\dfrac{1}{2}$ and $\beta = \sin^{-1}\dfrac{3}{5}$. Then

$$\cos\alpha = \dfrac{1}{2} \quad 0 \le \alpha \le \pi \quad \text{and} \quad \sin\beta = \dfrac{3}{5} \quad -\dfrac{\pi}{2} \le \beta \le \dfrac{\pi}{2}$$

NOTE In Example 9, we could also find $\sin\alpha$ by using $\cos\alpha = \dfrac{1}{2} = \dfrac{a}{r}$, so $a = 1$ and $r = 2$. Then $b = \sqrt{3}$ and $\sin\alpha = \dfrac{b}{r} = \dfrac{\sqrt{3}}{2}$. We could find $\cos\beta$ in a similar fashion. ∎

We use Pythagorean Identities to obtain $\sin\alpha$ and $\cos\beta$. Since $\sin\alpha \ge 0$ and $\cos\beta \ge 0$ (do you know why?), we find

$$\sin\alpha = \sqrt{1 - \cos^2\alpha} = \sqrt{1 - \dfrac{1}{4}} = \sqrt{\dfrac{3}{4}} = \dfrac{\sqrt{3}}{2}$$

$$\cos\beta = \sqrt{1 - \sin^2\beta} = \sqrt{1 - \dfrac{9}{25}} = \sqrt{\dfrac{16}{25}} = \dfrac{4}{5}$$

As a result,

$$\sin\left(\cos^{-1}\dfrac{1}{2} + \sin^{-1}\dfrac{3}{5}\right) = \sin(\alpha + \beta) = \sin\alpha\cos\beta + \cos\alpha\sin\beta$$

$$= \dfrac{\sqrt{3}}{2}\cdot\dfrac{4}{5} + \dfrac{1}{2}\cdot\dfrac{3}{5} = \dfrac{4\sqrt{3} + 3}{10}$$ ∎

═══ **Now Work** PROBLEM 73

| EXAMPLE 10 | Writing a Trigonometric Expression as an Algebraic Expression |

Write $\sin(\sin^{-1}u + \cos^{-1}v)$ as an algebraic expression containing u and v (that is, without any trigonometric functions). Give the restrictions on u and v.

Solution First, for $\sin^{-1}u$, we have $-1 \le u \le 1$ and for $\cos^{-1}v$, we have $-1 \le v \le 1$. Now let $\alpha = \sin^{-1}u$ and $\beta = \cos^{-1}v$. Then

$$\sin\alpha = u \quad -\dfrac{\pi}{2} \le \alpha \le \dfrac{\pi}{2} \quad -1 \le u \le 1$$

$$\cos\beta = v \quad 0 \le \beta \le \pi \quad -1 \le v \le 1$$

Since $-\dfrac{\pi}{2} \le \alpha \le \dfrac{\pi}{2}$, we know that $\cos \alpha \ge 0$. As a result,

$$\cos \alpha = \sqrt{1 - \sin^2 \alpha} = \sqrt{1 - u^2}$$

Similarly, since $0 \le \beta \le \pi$, we know that $\sin \beta \ge 0$. Then

$$\sin \beta = \sqrt{1 - \cos^2 \beta} = \sqrt{1 - v^2}$$

As a result,

$$\sin(\sin^{-1} u + \cos^{-1} v) = \sin(\alpha + \beta) = \sin \alpha \cos \beta + \cos \alpha \sin \beta$$
$$= uv + \sqrt{1 - u^2}\,\sqrt{1 - v^2} \quad \rule{1cm}{0.5mm}\blacksquare$$

✏ **Now Work** PROBLEM 83

SUMMARY Sum and Difference Formulas

$$\cos(\alpha + \beta) = \cos \alpha \cos \beta - \sin \alpha \sin \beta \qquad \cos(\alpha - \beta) = \cos \alpha \cos \beta + \sin \alpha \sin \beta$$

$$\sin(\alpha + \beta) = \sin \alpha \cos \beta + \cos \alpha \sin \beta \qquad \sin(\alpha - \beta) = \sin \alpha \cos \beta - \cos \alpha \sin \beta$$

$$\tan(\alpha + \beta) = \frac{\tan \alpha + \tan \beta}{1 - \tan \alpha \tan \beta} \qquad\qquad \tan(\alpha - \beta) = \frac{\tan \alpha - \tan \beta}{1 + \tan \alpha \tan \beta}$$

8.4 Assess Your Understanding

'Are You Prepared?' *Answers are given at the end of these exercises. If you get a wrong answer, read the pages listed in red.*

1. The distance d from the point $(2, -3)$ to the point $(5, 1)$ is _____. (p. 157)

2. If $\sin \theta = \dfrac{4}{5}$ and θ is in quadrant II, then $\cos \theta = $ _____.
(pp. 546–548)

3. (a) $\sin \dfrac{\pi}{4} \cdot \cos \dfrac{\pi}{3} = $ _____. (pp. 529–532)

 (b) $\tan \dfrac{\pi}{4} - \sin \dfrac{\pi}{6} = $ _____. (pp. 529–532)

Concepts and Vocabulary

4. $\cos(\alpha + \beta) = \cos \alpha \cos \beta $ _____ $ \sin \alpha \sin \beta$.

5. $\sin(\alpha - \beta) = \sin \alpha \cos \beta $ _____ $ \cos \alpha \sin \beta$.

6. *True or False* $\sin(\alpha + \beta) = \sin \alpha + \sin \beta + 2 \sin \alpha \sin \beta$

7. *True or False* $\tan 75° = \tan 30° + \tan 45°$

8. *True or False* $\cos\left(\dfrac{\pi}{2} - \theta\right) = \cos \theta$

Skill Building

In Problems 9–20, find the exact value of each expression.

9. $\sin \dfrac{5\pi}{12}$

10. $\sin \dfrac{\pi}{12}$

11. $\cos \dfrac{7\pi}{12}$

12. $\tan \dfrac{7\pi}{12}$

13. $\cos 165°$

14. $\sin 105°$

15. $\tan 15°$

16. $\tan 195°$

17. $\sin \dfrac{17\pi}{12}$

18. $\tan \dfrac{19\pi}{12}$

19. $\sec\left(-\dfrac{\pi}{12}\right)$

20. $\cot\left(-\dfrac{5\pi}{12}\right)$

In Problems 21–30, find the exact value of each expression.

21. $\sin 20° \cos 10° + \cos 20° \sin 10°$

22. $\sin 20° \cos 80° - \cos 20° \sin 80°$

23. $\cos 70° \cos 20° - \sin 70° \sin 20°$

24. $\cos 40° \cos 10° + \sin 40° \sin 10°$

25. $\dfrac{\tan 20° + \tan 25°}{1 - \tan 20° \tan 25°}$

26. $\dfrac{\tan 40° - \tan 10°}{1 + \tan 40° \tan 10°}$

27. $\sin \dfrac{\pi}{12} \cos \dfrac{7\pi}{12} - \cos \dfrac{\pi}{12} \sin \dfrac{7\pi}{12}$

28. $\cos \dfrac{5\pi}{12} \cos \dfrac{7\pi}{12} - \sin \dfrac{5\pi}{12} \sin \dfrac{7\pi}{12}$

29. $\cos \dfrac{\pi}{12} \cos \dfrac{5\pi}{12} + \sin \dfrac{5\pi}{12} \sin \dfrac{\pi}{12}$

30. $\sin \dfrac{\pi}{18} \cos \dfrac{5\pi}{18} + \cos \dfrac{\pi}{18} \sin \dfrac{5\pi}{18}$

In Problems 31–36, find the exact value of each of the following under the given conditions:

(a) $\sin(\alpha + \beta)$ (b) $\cos(\alpha + \beta)$ (c) $\sin(\alpha - \beta)$ (d) $\tan(\alpha - \beta)$

31. $\sin \alpha = \dfrac{3}{5}, 0 < \alpha < \dfrac{\pi}{2}$; $\cos \beta = \dfrac{2\sqrt{5}}{5}, -\dfrac{\pi}{2} < \beta < 0$ **32.** $\cos \alpha = \dfrac{\sqrt{5}}{5}, 0 < \alpha < \dfrac{\pi}{2}$; $\sin \beta = -\dfrac{4}{5}, -\dfrac{\pi}{2} < \beta < 0$

33. $\tan \alpha = -\dfrac{4}{3}, \dfrac{\pi}{2} < \alpha < \pi$; $\cos \beta = \dfrac{1}{2}, 0 < \beta < \dfrac{\pi}{2}$ **34.** $\tan \alpha = \dfrac{5}{12}, \pi < \alpha < \dfrac{3\pi}{2}$; $\sin \beta = -\dfrac{1}{2}, \pi < \beta < \dfrac{3\pi}{2}$

35. $\sin \alpha = \dfrac{5}{13}, -\dfrac{3\pi}{2} < \alpha < -\pi$; $\tan \beta = -\sqrt{3}, \dfrac{\pi}{2} < \beta < \pi$ **36.** $\cos \alpha = \dfrac{1}{2}, -\dfrac{\pi}{2} < \alpha < 0$; $\sin \beta = \dfrac{1}{3}, 0 < \beta < \dfrac{\pi}{2}$

37. If $\sin \theta = \dfrac{1}{3}$, θ in quadrant II, find the exact value of:

 (a) $\cos \theta$ (b) $\sin\left(\theta + \dfrac{\pi}{6}\right)$ (c) $\cos\left(\theta - \dfrac{\pi}{3}\right)$ (d) $\tan\left(\theta + \dfrac{\pi}{4}\right)$

38. If $\cos \theta = \dfrac{1}{4}$, θ in quadrant IV, find the exact value of:

 (a) $\sin \theta$ (b) $\sin\left(\theta - \dfrac{\pi}{6}\right)$ (c) $\cos\left(\theta + \dfrac{\pi}{3}\right)$ (d) $\tan\left(\theta - \dfrac{\pi}{4}\right)$

In Problems 39–44, use the figures to evaluate each function if $f(x) = \sin x$, $g(x) = \cos x$, and $h(x) = \tan x$.

39. $f(\alpha + \beta)$ **40.** $g(\alpha + \beta)$ **41.** $g(\alpha - \beta)$

42. $f(\alpha - \beta)$ **43.** $h(\alpha + \beta)$ **44.** $h(\alpha - \beta)$

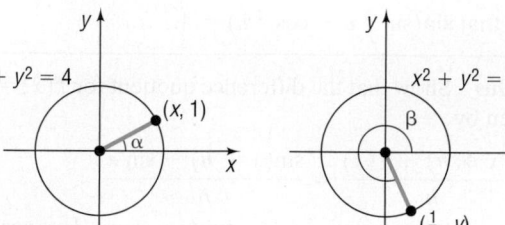

In Problems 45–70, establish each identity.

45. $\sin\left(\dfrac{\pi}{2} + \theta\right) = \cos \theta$ **46.** $\cos\left(\dfrac{\pi}{2} + \theta\right) = -\sin \theta$ **47.** $\sin(\pi - \theta) = \sin \theta$

48. $\cos(\pi - \theta) = -\cos \theta$ **49.** $\sin(\pi + \theta) = -\sin \theta$ **50.** $\cos(\pi + \theta) = -\cos \theta$

51. $\tan(\pi - \theta) = -\tan \theta$ **52.** $\tan(2\pi - \theta) = -\tan \theta$ **53.** $\sin\left(\dfrac{3\pi}{2} + \theta\right) = -\cos \theta$

54. $\cos\left(\dfrac{3\pi}{2} + \theta\right) = \sin \theta$ **55.** $\sin(\alpha + \beta) + \sin(\alpha - \beta) = 2 \sin \alpha \cos \beta$

56. $\cos(\alpha + \beta) + \cos(\alpha - \beta) = 2 \cos \alpha \cos \beta$ **57.** $\dfrac{\sin(\alpha + \beta)}{\sin \alpha \cos \beta} = 1 + \cot \alpha \tan \beta$

58. $\dfrac{\sin(\alpha + \beta)}{\cos \alpha \cos \beta} = \tan \alpha + \tan \beta$ **59.** $\dfrac{\cos(\alpha + \beta)}{\cos \alpha \cos \beta} = 1 - \tan \alpha \tan \beta$

60. $\dfrac{\cos(\alpha - \beta)}{\sin \alpha \cos \beta} = \cot \alpha + \tan \beta$ **61.** $\dfrac{\sin(\alpha + \beta)}{\sin(\alpha - \beta)} = \dfrac{\tan \alpha + \tan \beta}{\tan \alpha - \tan \beta}$

62. $\dfrac{\cos(\alpha + \beta)}{\cos(\alpha - \beta)} = \dfrac{1 - \tan \alpha \tan \beta}{1 + \tan \alpha \tan \beta}$ **63.** $\cot(\alpha + \beta) = \dfrac{\cot \alpha \cot \beta - 1}{\cot \beta + \cot \alpha}$

64. $\cot(\alpha - \beta) = \dfrac{\cot \alpha \cot \beta + 1}{\cot \beta - \cot \alpha}$ **65.** $\sec(\alpha + \beta) = \dfrac{\csc \alpha \csc \beta}{\cot \alpha \cot \beta - 1}$

66. $\sec(\alpha - \beta) = \dfrac{\sec \alpha \sec \beta}{1 + \tan \alpha \tan \beta}$ **67.** $\sin(\alpha - \beta) \sin(\alpha + \beta) = \sin^2 \alpha - \sin^2 \beta$

68. $\cos(\alpha - \beta) \cos(\alpha + \beta) = \cos^2 \alpha - \sin^2 \beta$ **69.** $\sin(\theta + k\pi) = (-1)^k \sin \theta$, k any integer

70. $\cos(\theta + k\pi) = (-1)^k \cos \theta$, k any integer

In Problems 71–82, find the exact value of each expression.

71. $\sin\left(\sin^{-1}\dfrac{1}{2} + \cos^{-1} 0\right)$ **72.** $\sin\left(\sin^{-1}\dfrac{\sqrt{3}}{2} + \cos^{-1} 1\right)$ **73.** $\sin\left[\sin^{-1}\dfrac{3}{5} - \cos^{-1}\left(-\dfrac{4}{5}\right)\right]$

74. $\sin\left[\sin^{-1}\left(-\dfrac{4}{5}\right) - \tan^{-1}\dfrac{3}{4}\right]$

75. $\cos\left(\tan^{-1}\dfrac{4}{3} + \cos^{-1}\dfrac{5}{13}\right)$

76. $\cos\left[\tan^{-1}\dfrac{5}{12} - \sin^{-1}\left(-\dfrac{3}{5}\right)\right]$

77. $\cos\left(\sin^{-1}\dfrac{5}{13} - \tan^{-1}\dfrac{3}{4}\right)$

78. $\cos\left(\tan^{-1}\dfrac{4}{3} + \cos^{-1}\dfrac{12}{13}\right)$

79. $\tan\left(\sin^{-1}\dfrac{3}{5} + \dfrac{\pi}{6}\right)$

80. $\tan\left(\dfrac{\pi}{4} - \cos^{-1}\dfrac{3}{5}\right)$

81. $\tan\left(\sin^{-1}\dfrac{4}{5} + \cos^{-1}1\right)$

82. $\tan\left(\cos^{-1}\dfrac{4}{5} + \sin^{-1}1\right)$

In Problems 83–88, write each trigonometric expression as an algebraic expression containing u and v. Give the restrictions required on u and v.

83. $\cos(\cos^{-1}u + \sin^{-1}v)$

84. $\sin(\sin^{-1}u - \cos^{-1}v)$

85. $\sin(\tan^{-1}u - \sin^{-1}v)$

86. $\cos(\tan^{-1}u + \tan^{-1}v)$

87. $\tan(\sin^{-1}u - \cos^{-1}v)$

88. $\sec(\tan^{-1}u + \cos^{-1}v)$

Applications and Extensions

89. Show that $\sin^{-1}v + \cos^{-1}v = \dfrac{\pi}{2}$.

90. Show that $\tan^{-1}v + \cot^{-1}v = \dfrac{\pi}{2}$.

91. Show that $\tan^{-1}\left(\dfrac{1}{v}\right) = \dfrac{\pi}{2} - \tan^{-1}v$, if $v > 0$.

92. Show that $\cot^{-1}e^v = \tan^{-1}e^{-v}$.

93. Show that $\sin(\sin^{-1}v + \cos^{-1}v) = 1$.

94. Show that $\cos(\sin^{-1}v + \cos^{-1}v) = 0$.

95. Calculus Show that the difference quotient for $f(x) = \sin x$ is given by

$$\frac{f(x+h) - f(x)}{h} = \frac{\sin(x+h) - \sin x}{h}$$

$$= \cos x \cdot \frac{\sin h}{h} - \sin x \cdot \frac{1 - \cos h}{h}$$

96. Calculus Show that the difference quotient for $f(x) = \cos x$ is given by

$$\frac{f(x+h) - f(x)}{h} = \frac{\cos(x+h) - \cos x}{h}$$

$$= -\sin x \cdot \frac{\sin h}{h} - \cos x \cdot \frac{1 - \cos h}{h}$$

97. One, Two, Three

(a) Show that $\tan(\tan^{-1}1 + \tan^{-1}2 + \tan^{-1}3) = 0$.

(b) Conclude from part (a) that

$$\tan^{-1}1 + \tan^{-1}2 + \tan^{-1}3 = \pi.$$

Source: College Mathematics Journal, Vol. 37, No. 3, May 2006

98. Electric Power In an alternating current (ac) circuit, the instantaneous power p at time t is given by

$$p(t) = V_m I_m \cos\phi \sin^2(\omega t) - V_m I_m \sin\phi \sin(\omega t)\cos(\omega t)$$

Show that this is equivalent to

$$p(t) = V_m I_m \sin(\omega t)\sin(\omega t - \phi)$$

Source: HyperPhysics, hosted by Georgia State University

99. Geometry: Angle Between Two Lines Let L_1 and L_2 denote two nonvertical intersecting lines, and let θ denote the acute angle between L_1 and L_2 (see the figure). Show that

$$\tan\theta = \frac{m_2 - m_1}{1 + m_1 m_2}$$

where m_1 and m_2 are the slopes of L_1 and L_2, respectively.

[**Hint:** Use the facts that $\tan\theta_1 = m_1$ and $\tan\theta_2 = m_2$.]

100. If $\alpha + \beta + \gamma = 180°$ and

$$\cot\theta = \cot\alpha + \cot\beta + \cot\gamma, \quad 0 < \theta < 90°$$

show that

$$\sin^3\theta = \sin(\alpha - \theta)\sin(\beta - \theta)\sin(\gamma - \theta)$$

101. If $\tan\alpha = x + 1$ and $\tan\beta = x - 1$, show that

$$2\cot(\alpha - \beta) = x^2$$

Discussion and Writing

102. Discuss the following derivation:

$$\tan\left(\theta + \frac{\pi}{2}\right) = \frac{\tan\theta + \tan\dfrac{\pi}{2}}{1 - \tan\theta\tan\dfrac{\pi}{2}} = \frac{\dfrac{\tan\theta}{\tan\dfrac{\pi}{2}} + 1}{\dfrac{1}{\tan\dfrac{\pi}{2}} - \tan\theta} = \frac{0 + 1}{0 - \tan\theta} = \frac{1}{-\tan\theta} = -\cot\theta$$

Can you justify each step?

103. Explain why formula (7) cannot be used to show that

$$\tan\left(\frac{\pi}{2} - \theta\right) = \cot\theta$$

Establish this identity by using formulas (3a) and (3b).

'Are You Prepared?' Answers

1. 5 **2.** $-\dfrac{3}{5}$ **3.** (a) $\dfrac{\sqrt{2}}{4}$ (b) $\dfrac{1}{2}$

8.5 Double-angle and Half-angle Formulas

OBJECTIVES **1** Use Double-angle Formulas to Find Exact Values (p. 637)
2 Use Double-angle Formulas to Establish Identities (p. 638)
3 Use Half-angle Formulas to Find Exact Values (p. 640)

In this section we derive formulas for $\sin(2\theta)$, $\cos(2\theta)$, $\sin\left(\dfrac{1}{2}\theta\right)$, and $\cos\left(\dfrac{1}{2}\theta\right)$ in terms of $\sin\theta$ and $\cos\theta$. They are derived using the sum formulas.

In the sum formulas for $\sin(\alpha + \beta)$ and $\cos(\alpha + \beta)$, let $\alpha = \beta = \theta$. Then

$$\sin(\alpha + \beta) = \sin\alpha\cos\beta + \cos\alpha\sin\beta$$
$$\sin(\theta + \theta) = \sin\theta\cos\theta + \cos\theta\sin\theta$$
$$\sin(2\theta) = 2\sin\theta\cos\theta$$

and

$$\cos(\alpha + \beta) = \cos\alpha\cos\beta - \sin\alpha\sin\beta$$
$$\cos(\theta + \theta) = \cos\theta\cos\theta - \sin\theta\sin\theta$$
$$\cos(2\theta) = \cos^2\theta - \sin^2\theta$$

An application of the Pythagorean Identity $\sin^2\theta + \cos^2\theta = 1$ results in two other ways to express $\cos(2\theta)$.

$$\cos(2\theta) = \cos^2\theta - \sin^2\theta = (1 - \sin^2\theta) - \sin^2\theta = 1 - 2\sin^2\theta$$

and

$$\cos(2\theta) = \cos^2\theta - \sin^2\theta = \cos^2\theta - (1 - \cos^2\theta) = 2\cos^2\theta - 1$$

We have established the following **Double-angle Formulas:**

THEOREM

Double-angle Formulas

$$\sin(2\theta) = 2\sin\theta\cos\theta \tag{1}$$
$$\cos(2\theta) = \cos^2\theta - \sin^2\theta \tag{2}$$
$$\cos(2\theta) = 1 - 2\sin^2\theta \tag{3}$$
$$\cos(2\theta) = 2\cos^2\theta - 1 \tag{4}$$

1 Use Double-angle Formulas to Find Exact Values

EXAMPLE 1 **Finding Exact Values Using the Double-angle Formula**

If $\sin\theta = \dfrac{3}{5}$, $\dfrac{\pi}{2} < \theta < \pi$, find the exact value of:

(a) $\sin(2\theta)$ (b) $\cos(2\theta)$

Solution (a) Because $\sin(2\theta) = 2\sin\theta\cos\theta$ and we already know that $\sin\theta = \dfrac{3}{5}$, we only need to find $\cos\theta$. Since $\sin\theta = \dfrac{3}{5} = \dfrac{b}{r}$, $\dfrac{\pi}{2} < \theta < \pi$, we let $b = 3$ and $r = 5$ and place θ in quadrant II. See Figure 23. The point $(a, 3)$ is in quadrant II, so $a < 0$. The distance from $(a, 3)$ to $(0, 0)$ is 5, so

Figure 23

$$a^2 + 3^2 = 5^2, \qquad a < 0$$
$$a^2 + 9 = 25$$
$$a^2 = 25 - 9 = 16$$
$$a = -4$$

We find that $\cos\theta = \dfrac{a}{r} = -\dfrac{4}{5}$. Now we use formula (1) to obtain

$$\sin(2\theta) = 2\sin\theta\cos\theta = 2\left(\dfrac{3}{5}\right)\left(-\dfrac{4}{5}\right) = -\dfrac{24}{25}$$

(b) Because we are given $\sin\theta = \dfrac{3}{5}$, it is easiest to use formula (3) to get $\cos(2\theta)$.

$$\cos(2\theta) = 1 - 2\sin^2\theta = 1 - 2\left(\dfrac{9}{25}\right) = 1 - \dfrac{18}{25} = \dfrac{7}{25}$$

WARNING In finding $\cos(2\theta)$ in Example 1(b), we chose to use a version of the Double-angle Formula, formula (3). Note that we are unable to use the Pythagorean Identity $\cos(2\theta) = \pm\sqrt{1 - \sin^2(2\theta)}$, with $\sin(2\theta) = -\dfrac{24}{25}$, because we have no way of knowing which sign to choose. ∎

Now Work PROBLEMS 7(a) AND (b)

2 Use Double-angle Formulas to Establish Identities

| EXAMPLE 2 | **Establishing Identities** |

(a) Develop a formula for $\tan(2\theta)$ in terms of $\tan\theta$.
(b) Develop a formula for $\sin(3\theta)$ in terms of $\sin\theta$ and $\cos\theta$.

Solution (a) In the sum formula for $\tan(\alpha + \beta)$, let $\alpha = \beta = \theta$. Then

$$\tan(\alpha + \beta) = \dfrac{\tan\alpha + \tan\beta}{1 - \tan\alpha\tan\beta}$$

$$\tan(\theta + \theta) = \dfrac{\tan\theta + \tan\theta}{1 - \tan\theta\tan\theta}$$

$$\tan(2\theta) = \dfrac{2\tan\theta}{1 - \tan^2\theta} \tag{5}$$

(b) To get a formula for $\sin(3\theta)$, we use the sum formula and write 3θ as $2\theta + \theta$.

$$\sin(3\theta) = \sin(2\theta + \theta) = \sin(2\theta)\cos\theta + \cos(2\theta)\sin\theta$$

Now use the Double-angle Formulas to get

$$\sin(3\theta) = (2\sin\theta\cos\theta)(\cos\theta) + (\cos^2\theta - \sin^2\theta)(\sin\theta)$$
$$= 2\sin\theta\cos^2\theta + \sin\theta\cos^2\theta - \sin^3\theta$$
$$= 3\sin\theta\cos^2\theta - \sin^3\theta$$

The formula obtained in Example 2(b) can also be written as

$$\sin(3\theta) = 3\sin\theta\cos^2\theta - \sin^3\theta = 3\sin\theta(1 - \sin^2\theta) - \sin^3\theta$$
$$= 3\sin\theta - 4\sin^3\theta$$

That is, $\sin(3\theta)$ is a third-degree polynomial in the variable $\sin\theta$. In fact, $\sin(n\theta)$, n a positive odd integer, can always be written as a polynomial of degree n in the variable $\sin\theta$.*

━━━ **Now Work** PROBLEM 65

By rearranging the Double-angle Formulas (3) and (4), we obtain other formulas that we will use later in this section.

We begin with formula (3) and proceed to solve for $\sin^2\theta$.

$$\cos(2\theta) = 1 - 2\sin^2\theta$$
$$2\sin^2\theta = 1 - \cos(2\theta)$$

$$\sin^2\theta = \frac{1 - \cos(2\theta)}{2} \qquad (6)$$

Similarly, using formula (4), we proceed to solve for $\cos^2\theta$.

$$\cos(2\theta) = 2\cos^2\theta - 1$$
$$2\cos^2\theta = 1 + \cos(2\theta)$$

$$\cos^2\theta = \frac{1 + \cos(2\theta)}{2} \qquad (7)$$

Formulas (6) and (7) can be used to develop a formula for $\tan^2\theta$.

$$\tan^2\theta = \frac{\sin^2\theta}{\cos^2\theta} = \frac{\dfrac{1 - \cos(2\theta)}{2}}{\dfrac{1 + \cos(2\theta)}{2}}$$

$$\tan^2\theta = \frac{1 - \cos(2\theta)}{1 + \cos(2\theta)} \qquad (8)$$

Formulas (6) through (8) do not have to be memorized since their derivations are so straightforward.

 Formulas (6) and (7) are important in calculus. The next example illustrates a problem that arises in calculus requiring the use of formula (7).

EXAMPLE 3	**Establishing an Identity**

Write an equivalent expression for $\cos^4\theta$ that does not involve any powers of sine or cosine greater than 1.

*Due to the work done by P. L. Chebyshëv, these polynomials are sometimes called *Chebyshëv polynomials*.

Solution The idea here is to apply formula (7) twice.

$$\cos^4 \theta = (\cos^2 \theta)^2 = \left(\frac{1 + \cos(2\theta)}{2}\right)^2 \qquad \text{Formula (7)}$$

$$= \frac{1}{4}[1 + 2\cos(2\theta) + \cos^2(2\theta)]$$

$$= \frac{1}{4} + \frac{1}{2}\cos(2\theta) + \frac{1}{4}\cos^2(2\theta)$$

$$= \frac{1}{4} + \frac{1}{2}\cos(2\theta) + \frac{1}{4}\left\{\frac{1 + \cos[2(2\theta)]}{2}\right\} \qquad \text{Formula (7)}$$

$$= \frac{1}{4} + \frac{1}{2}\cos(2\theta) + \frac{1}{8}[1 + \cos(4\theta)]$$

$$= \frac{3}{8} + \frac{1}{2}\cos(2\theta) + \frac{1}{8}\cos(4\theta)$$

━━━━ **Now Work** PROBLEM 41

Identities, such as the Double-angle Formulas, can sometimes be used to rewrite expressions in a more suitable form. Let's look at an example.

| EXAMPLE 4 | **Projectile Motion** |

Figure 24

An object is propelled upward at an angle θ to the horizontal with an initial velocity of v_0 feet per second. See Figure 24. If air resistance is ignored, the **range** R, the horizontal distance that the object travels, is given by the function

$$R(\theta) = \frac{1}{16} v_0^2 \sin\theta \cos\theta$$

(a) Show that $R(\theta) = \frac{1}{32} v_0^2 \sin(2\theta)$.

(b) Find the angle θ for which R is a maximum.

Solution (a) We rewrite the given expression for the range using the Double-angle Formula $\sin(2\theta) = 2\sin\theta\cos\theta$. Then

$$R(\theta) = \frac{1}{16} v_0^2 \sin\theta\cos\theta = \frac{1}{16} v_0^2 \frac{2\sin\theta\cos\theta}{2} = \frac{1}{32} v_0^2 \sin(2\theta)$$

(b) In this form, the largest value for the range R can be found. For a fixed initial speed v_0, the angle θ of inclination to the horizontal determines the value of R. Since the largest value of a sine function is 1, occurring when the argument 2θ is 90°, it follows that for maximum R we must have

$$2\theta = 90°$$
$$\theta = 45°$$

An inclination to the horizontal of 45° results in the maximum range.

3 Use Half-angle Formulas to Find Exact Values

Another important use of formulas (6) through (8) is to prove the *Half-angle Formulas*. In formulas (6) through (8), let $\theta = \dfrac{\alpha}{2}$. Then

$$\boxed{\sin^2\frac{\alpha}{2} = \frac{1 - \cos\alpha}{2} \qquad \cos^2\frac{\alpha}{2} = \frac{1 + \cos\alpha}{2} \qquad \tan^2\frac{\alpha}{2} = \frac{1 - \cos\alpha}{1 + \cos\alpha}} \quad \text{(9)}$$

 The identities in box (9) will prove useful in integral calculus.

If we solve for the trigonometric functions on the left sides of equations (9), we obtain the Half-angle Formulas.

THEOREM

Half-angle Formulas

$$\sin\frac{\alpha}{2} = \pm\sqrt{\frac{1 - \cos\alpha}{2}} \qquad \text{(10a)}$$

$$\cos\frac{\alpha}{2} = \pm\sqrt{\frac{1 + \cos\alpha}{2}} \qquad \text{(10b)}$$

$$\tan\frac{\alpha}{2} = \pm\sqrt{\frac{1 - \cos\alpha}{1 + \cos\alpha}} \qquad \text{(10c)}$$

where the + or − sign is determined by the quadrant of the angle $\dfrac{\alpha}{2}$.

EXAMPLE 5 **Finding Exact Values Using Half-angle Formulas**

Use a Half-angle Formula to find the exact value of:

(a) $\cos 15°$ (b) $\sin(-15°)$

Solution (a) Because $15° = \dfrac{30°}{2}$, we can use the Half-angle Formula for $\cos\dfrac{\alpha}{2}$ with $\alpha = 30°$.

Also, because $15°$ is in quadrant I, $\cos 15° > 0$, we choose the + sign in using formula (10b):

$$\cos 15° = \cos\frac{30°}{2} = \sqrt{\frac{1 + \cos 30°}{2}}$$

$$= \sqrt{\frac{1 + \sqrt{3}/2}{2}} = \sqrt{\frac{2 + \sqrt{3}}{4}} = \frac{\sqrt{2 + \sqrt{3}}}{2}$$

(b) We use the fact that $\sin(-15°) = -\sin 15°$ and then apply formula (10a).

$$\sin(-15°) = -\sin\frac{30°}{2} = -\sqrt{\frac{1 - \cos 30°}{2}}$$

$$= -\sqrt{\frac{1 - \sqrt{3}/2}{2}} = -\sqrt{\frac{2 - \sqrt{3}}{4}} = -\frac{\sqrt{2 - \sqrt{3}}}{2}$$

It is interesting to compare the answer found in Example 5(a) with the answer to Example 2 of Section 8.4. There we calculated

$$\cos\frac{\pi}{12} = \cos 15° = \frac{1}{4}\left(\sqrt{6} + \sqrt{2}\right)$$

Based on this and the result of Example 5(a), we conclude that

$$\frac{1}{4}\left(\sqrt{6} + \sqrt{2}\right) \quad \text{and} \quad \frac{\sqrt{2 + \sqrt{3}}}{2}$$

are equal. (Since each expression is positive, you can verify this equality by squaring each expression.) Two very different looking, yet correct, answers can be obtained, depending on the approach taken to solve a problem.

Now Work PROBLEM 19

| EXAMPLE 6 | **Finding Exact Values Using Half-angle Formulas** |

If $\cos \alpha = -\dfrac{3}{5}$, $\pi < \alpha < \dfrac{3\pi}{2}$, find the exact value of:

(a) $\sin \dfrac{\alpha}{2}$ (b) $\cos \dfrac{\alpha}{2}$ (c) $\tan \dfrac{\alpha}{2}$

Solution First, we observe that if $\pi < \alpha < \dfrac{3\pi}{2}$ then $\dfrac{\pi}{2} < \dfrac{\alpha}{2} < \dfrac{3\pi}{4}$. As a result, $\dfrac{\alpha}{2}$ lies in quadrant II.

(a) Because $\dfrac{\alpha}{2}$ lies in quadrant II, $\sin \dfrac{\alpha}{2} > 0$, so we use the $+$ sign in formula (10a) to get

$$\sin \frac{\alpha}{2} = \sqrt{\frac{1 - \cos \alpha}{2}} = \sqrt{\frac{1 - \left(-\dfrac{3}{5}\right)}{2}}$$

$$= \sqrt{\frac{\dfrac{8}{5}}{2}} = \sqrt{\frac{4}{5}} = \frac{2}{\sqrt{5}} = \frac{2\sqrt{5}}{5}$$

(b) Because $\dfrac{\alpha}{2}$ lies in quadrant II, $\cos \dfrac{\alpha}{2} < 0$, so we use the $-$ sign in formula (10b) to get

$$\cos \frac{\alpha}{2} = -\sqrt{\frac{1 + \cos \alpha}{2}} = -\sqrt{\frac{1 + \left(-\dfrac{3}{5}\right)}{2}}$$

$$= -\sqrt{\frac{\dfrac{2}{5}}{2}} = -\frac{1}{\sqrt{5}} = -\frac{\sqrt{5}}{5}$$

(c) Because $\dfrac{\alpha}{2}$ lies in quadrant II, $\tan \dfrac{\alpha}{2} < 0$, so we use the $-$ sign in formula (10c) to get

$$\tan \frac{\alpha}{2} = -\sqrt{\frac{1 - \cos \alpha}{1 + \cos \alpha}} = -\sqrt{\frac{1 - \left(-\dfrac{3}{5}\right)}{1 + \left(-\dfrac{3}{5}\right)}} = -\sqrt{\frac{\dfrac{8}{5}}{\dfrac{2}{5}}} = -2$$

Another way to solve Example 6(c) is to use the solutions found in parts (a) and (b).

$$\tan \frac{\alpha}{2} = \frac{\sin \dfrac{\alpha}{2}}{\cos \dfrac{\alpha}{2}} = \frac{\dfrac{2\sqrt{5}}{5}}{-\dfrac{\sqrt{5}}{5}} = -2$$

Now Work PROBLEMS 7(c) AND (d)

There is a formula for $\tan \dfrac{\alpha}{2}$ that does not contain $+$ and $-$ signs, making it more useful than Formula 10(c). To derive it, we use the formulas

$$1 - \cos \alpha = 2 \sin^2 \frac{\alpha}{2} \quad \text{Formula (9)}$$

and

$$\sin \alpha = \sin\left[2\left(\frac{\alpha}{2}\right)\right] = 2\sin\frac{\alpha}{2}\cos\frac{\alpha}{2} \quad \text{Double-angle Formula}$$

Then

$$\frac{1 - \cos\alpha}{\sin\alpha} = \frac{2\sin^2\frac{\alpha}{2}}{2\sin\frac{\alpha}{2}\cos\frac{\alpha}{2}} = \frac{\sin\frac{\alpha}{2}}{\cos\frac{\alpha}{2}} = \tan\frac{\alpha}{2}$$

Since it also can be shown that

$$\frac{1 - \cos\alpha}{\sin\alpha} = \frac{\sin\alpha}{1 + \cos\alpha}$$

we have the following two Half-angle Formulas:

Half-angle Formulas for $\tan\dfrac{\alpha}{2}$

$$\tan\frac{\alpha}{2} = \frac{1 - \cos\alpha}{\sin\alpha} = \frac{\sin\alpha}{1 + \cos\alpha} \tag{11}$$

With this formula, the solution to Example 6(c) can be obtained as follows:

$$\cos\alpha = -\frac{3}{5} \quad \pi < \alpha < \frac{3\pi}{2}$$

$$\sin\alpha = -\sqrt{1 - \cos^2\alpha} = -\sqrt{1 - \frac{9}{25}} = -\sqrt{\frac{16}{25}} = -\frac{4}{5}$$

Then, by equation (11),

$$\tan\frac{\alpha}{2} = \frac{1 - \cos\alpha}{\sin\alpha} = \frac{1 - \left(-\dfrac{3}{5}\right)}{-\dfrac{4}{5}} = \frac{\dfrac{8}{5}}{-\dfrac{4}{5}} = -2$$

8.5 Assess Your Understanding

Concepts and Vocabulary

1. $\cos(2\theta) = \cos^2\theta - \underline{\quad} = \underline{\quad} - 1 = 1 - \underline{\quad}$.

2. $\sin^2\dfrac{\theta}{2} = \dfrac{\underline{\quad}}{2}$.

3. $\tan\dfrac{\theta}{2} = \dfrac{1 - \cos\theta}{\underline{\quad}}$.

4. *True or False* $\cos(2\theta)$ has three equivalent forms:
$$\cos^2\theta - \sin^2\theta, \quad 1 - 2\sin^2\theta, \quad 2\cos^2\theta - 1$$

5. *True or False* $\sin(2\theta)$ has two equivalent forms:
$$2\sin\theta\cos\theta \quad \text{and} \quad \sin^2\theta - \cos^2\theta$$

6. *True or False* $\tan(2\theta) + \tan(2\theta) = \tan(4\theta)$

Skill Building

In Problems 7–18, use the information given about the angle θ, $0 \le \theta < 2\pi$, to find the exact value of

(a) $\sin(2\theta)$ (b) $\cos(2\theta)$ (c) $\sin\dfrac{\theta}{2}$ (d) $\cos\dfrac{\theta}{2}$

7. $\sin\theta = \dfrac{3}{5}$ $0 < \theta < \dfrac{\pi}{2}$

8. $\cos\theta = \dfrac{3}{5}$ $0 < \theta < \dfrac{\pi}{2}$

9. $\tan\theta = \dfrac{4}{3}$ $\pi < \theta < \dfrac{3\pi}{2}$

10. $\tan\theta = \dfrac{1}{2}$ $\pi < \theta < \dfrac{3\pi}{2}$

11. $\cos\theta = -\dfrac{\sqrt{6}}{3}$ $\dfrac{\pi}{2} < \theta < \pi$

12. $\sin\theta = -\dfrac{\sqrt{3}}{3}$ $\dfrac{3\pi}{2} < \theta < 2\pi$

13. $\sec \theta = 3 \quad \sin \theta > 0$

14. $\csc \theta = -\sqrt{5} \quad \cos \theta < 0$

15. $\cot \theta = -2 \quad \sec \theta < 0$

16. $\sec \theta = 2 \quad \csc \theta < 0$

17. $\tan \theta = -3 \quad \sin \theta < 0$

18. $\cot \theta = 3 \quad \cos \theta < 0$

In Problems 19–28, use the Half-angle Formulas to find the exact value of each expression.

19. $\sin 22.5°$

20. $\cos 22.5°$

21. $\tan \dfrac{7\pi}{8}$

22. $\tan \dfrac{9\pi}{8}$

23. $\cos 165°$

24. $\sin 195°$

25. $\sec \dfrac{15\pi}{8}$

26. $\csc \dfrac{7\pi}{8}$

27. $\sin\left(-\dfrac{\pi}{8}\right)$

28. $\cos\left(-\dfrac{3\pi}{8}\right)$

In Problems 29–40, use the figures to evaluate each function given that $f(x) = \sin x$, $g(x) = \cos x$, and $h(x) = \tan x$.

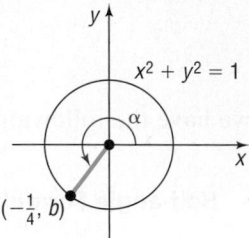

29. $f(2\theta)$

30. $g(2\theta)$

31. $g\left(\dfrac{\theta}{2}\right)$

32. $f\left(\dfrac{\theta}{2}\right)$

33. $h(2\theta)$

34. $h\left(\dfrac{\theta}{2}\right)$

35. $g(2\alpha)$

36. $f(2\alpha)$

37. $f\left(\dfrac{\alpha}{2}\right)$

38. $g\left(\dfrac{\alpha}{2}\right)$

39. $h\left(\dfrac{\alpha}{2}\right)$

40. $h(2\alpha)$

41. Show that $\sin^4 \theta = \dfrac{3}{8} - \dfrac{1}{2}\cos(2\theta) + \dfrac{1}{8}\cos(4\theta)$.

42. Show that $\sin(4\theta) = (\cos \theta)(4 \sin \theta - 8 \sin^3 \theta)$.

43. Develop a formula for $\cos(3\theta)$ as a third-degree polynomial in the variable $\cos \theta$.

44. Develop a formula for $\cos(4\theta)$ as a fourth-degree polynomial in the variable $\cos \theta$.

45. Find an expression for $\sin(5\theta)$ as a fifth-degree polynomial in the variable $\sin \theta$.

46. Find an expression for $\cos(5\theta)$ as a fifth-degree polynomial in the variable $\cos \theta$.

In Problems 47–68, establish each identity.

47. $\cos^4 \theta - \sin^4 \theta = \cos(2\theta)$

48. $\dfrac{\cot \theta - \tan \theta}{\cot \theta + \tan \theta} = \cos(2\theta)$

49. $\cot(2\theta) = \dfrac{\cot^2 \theta - 1}{2 \cot \theta}$

50. $\cot(2\theta) = \dfrac{1}{2}(\cot \theta - \tan \theta)$

51. $\sec(2\theta) = \dfrac{\sec^2 \theta}{2 - \sec^2 \theta}$

52. $\csc(2\theta) = \dfrac{1}{2}\sec \theta \csc \theta$

53. $\cos^2(2u) - \sin^2(2u) = \cos(4u)$

54. $(4 \sin u \cos u)(1 - 2 \sin^2 u) = \sin(4u)$

55. $\dfrac{\cos(2\theta)}{1 + \sin(2\theta)} = \dfrac{\cot \theta - 1}{\cot \theta + 1}$

56. $\sin^2 \theta \cos^2 \theta = \dfrac{1}{8}[1 - \cos(4\theta)]$

57. $\sec^2 \dfrac{\theta}{2} = \dfrac{2}{1 + \cos \theta}$

58. $\csc^2 \dfrac{\theta}{2} = \dfrac{2}{1 - \cos \theta}$

59. $\cot^2 \dfrac{v}{2} = \dfrac{\sec v + 1}{\sec v - 1}$

60. $\tan \dfrac{v}{2} = \csc v - \cot v$

61. $\cos \theta = \dfrac{1 - \tan^2 \dfrac{\theta}{2}}{1 + \tan^2 \dfrac{\theta}{2}}$

62. $1 - \dfrac{1}{2}\sin(2\theta) = \dfrac{\sin^3 \theta + \cos^3 \theta}{\sin \theta + \cos \theta}$

63. $\dfrac{\sin(3\theta)}{\sin \theta} - \dfrac{\cos(3\theta)}{\cos \theta} = 2$

64. $\dfrac{\cos \theta + \sin \theta}{\cos \theta - \sin \theta} - \dfrac{\cos \theta - \sin \theta}{\cos \theta + \sin \theta} = 2 \tan(2\theta)$

65. $\tan(3\theta) = \dfrac{3 \tan \theta - \tan^3 \theta}{1 - 3 \tan^2 \theta}$

66. $\tan \theta + \tan(\theta + 120°) + \tan(\theta + 240°) = 3 \tan(3\theta)$

67. $\ln|\sin \theta| = \dfrac{1}{2}(\ln|1 - \cos(2\theta)| - \ln 2)$

68. $\ln|\cos \theta| = \dfrac{1}{2}(\ln|1 + \cos(2\theta)| - \ln 2)$

In Problems 69–80, find the exact value of each expression.

69. $\sin\left(2\sin^{-1}\dfrac{1}{2}\right)$

70. $\sin\left[2\sin^{-1}\dfrac{\sqrt{3}}{2}\right]$

71. $\cos\left(2\sin^{-1}\dfrac{3}{5}\right)$

72. $\cos\left(2\cos^{-1}\dfrac{4}{5}\right)$

73. $\tan\left[2\cos^{-1}\left(-\dfrac{3}{5}\right)\right]$

74. $\tan\left(2\tan^{-1}\dfrac{3}{4}\right)$

75. $\sin\left(2\cos^{-1}\dfrac{4}{5}\right)$

76. $\cos\left[2\tan^{-1}\left(-\dfrac{4}{3}\right)\right]$

77. $\sin^2\left(\dfrac{1}{2}\cos^{-1}\dfrac{3}{5}\right)$

78. $\cos^2\left(\dfrac{1}{2}\sin^{-1}\dfrac{3}{5}\right)$

79. $\sec\left(2\tan^{-1}\dfrac{3}{4}\right)$

80. $\csc\left[2\sin^{-1}\left(-\dfrac{3}{5}\right)\right]$

Applications and Extensions

81. Laser Projection In a laser projection system, the **optical** or **scanning angle** θ is related to the throw distance D from the scanner to the screen and the projected image width W by the equation $D = \dfrac{\dfrac{1}{2}W}{\csc\theta - \cot\theta}$.

(a) Show that the projected image width is given by
$$W = 2D\tan\dfrac{\theta}{2}.$$

(b) Find the optical angle if the throw distance is 15 feet and the projected image width is 6.5 feet.

Source: Pangolin Laser Systems, Inc.

82. Product of Inertia The **product of inertia** for an area about inclined axes is given by the formula
$$I_{uv} = I_x\sin\theta\cos\theta - I_y\sin\theta\cos\theta + I_{xy}(\cos^2\theta - \sin^2\theta).$$
Show that this is equivalent to
$$I_{uv} = \dfrac{I_x - I_y}{2}\sin 2\theta + I_{xy}\cos 2\theta.$$

Source: Adapted from Hibbeler, *Engineering Mechanics: Statics*, 10th ed., Prentice Hall © 2004.

83. Projectile Motion An object is propelled upward at an angle θ, $45° < \theta < 90°$, to the horizontal with an initial velocity of v_0 feet per second from the base of a plane that makes an angle of 45°; with the horizontal. See the illustration. If air resistance is ignored, the distance R that it travels up the inclined plane is given by the function
$$R(\theta) = \dfrac{v_0^2\sqrt{2}}{16}\cos\theta(\sin\theta - \cos\theta)$$

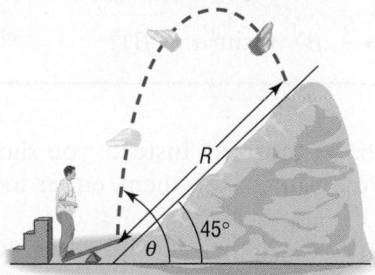

(a) Show that
$$R(\theta) = \dfrac{v_0^2\sqrt{2}}{32}[\sin(2\theta) - \cos(2\theta) - 1]$$

(b) Graph $R = R(\theta)$. (Use $v_0 = 32$ feet per second.)

(c) What value of θ makes R the largest? (Use $v_0 = 32$ feet per second.)

84. Sawtooth Curve An oscilloscope often displays a sawtooth curve. This curve can be approximated by sinusoidal curves of varying periods and amplitudes. A first approximation to the sawtooth curve is given by
$$y = \dfrac{1}{2}\sin(2\pi x) + \dfrac{1}{4}\sin(4\pi x)$$
Show that $y = \sin(2\pi x)\cos^2(\pi x)$.

85. Area of an Isosceles Triangle Show that the area A of an isosceles triangle whose equal sides are of length s and θ is the angle between them is
$$\dfrac{1}{2}s^2\sin\theta$$

[**Hint:** See the illustration. The height h bisects the angle θ and is the perpendicular bisector of the base.]

86. Geometry A rectangle is inscribed in a semicircle of radius 1. See the illustration.

(a) Express the area A of the rectangle as a function of the angle θ shown in the illustration.

(b) Show that $A(\theta) = \sin(2\theta)$.

(c) Find the angle θ that results in the largest area A.

(d) Find the dimensions of this largest rectangle.

87. If $x = 2\tan\theta$, express $\sin(2\theta)$ as a function of x.

88. If $x = 2\tan\theta$, express $\cos(2\theta)$ as a function of x.

89. Find the value of the number C:
$$\frac{1}{2}\sin^2 x + C = -\frac{1}{4}\cos(2x)$$

90. Find the value of the number C:
$$\frac{1}{2}\cos^2 x + C = \frac{1}{4}\cos(2x)$$

91. If $z = \tan\dfrac{\alpha}{2}$, show that $\sin\alpha = \dfrac{2z}{1+z^2}$.

92. If $z = \tan\dfrac{\alpha}{2}$, show that $\cos\alpha = \dfrac{1-z^2}{1+z^2}$.

93. Graph $f(x) = \sin^2 x = \dfrac{1-\cos(2x)}{2}$ for $0 \le x \le 2\pi$ by using transformations.

94. Repeat Problem 93 for $g(x) = \cos^2 x$.

95. Use the fact that
$$\cos\frac{\pi}{12} = \frac{1}{4}\left(\sqrt{6}+\sqrt{2}\right)$$
to find $\sin\dfrac{\pi}{24}$ and $\cos\dfrac{\pi}{24}$.

96. Show that
$$\cos\frac{\pi}{8} = \frac{\sqrt{2+\sqrt{2}}}{2}$$
and use it to find $\sin\dfrac{\pi}{16}$ and $\cos\dfrac{\pi}{16}$.

97. Show that
$$\sin^3\theta + \sin^3(\theta + 120°) + \sin^3(\theta + 240°) = -\frac{3}{4}\sin(3\theta)$$

98. If $\tan\theta = a\tan\dfrac{\theta}{3}$, express $\tan\dfrac{\theta}{3}$ in terms of a.

Discussion and Writing

99. Go to the library and research Chebyshëv polynomials. Write a report on your findings.

8.6 Product-to-Sum and Sum-to-Product Formulas

OBJECTIVES 1 Express Products as Sums (p. 646)

2 Express Sums as Products (p. 647)

1 Express Products as Sums

Sum and difference formulas can be used to derive formulas for writing the products of sines and/or cosines as sums or differences. These identities are usually called the **Product-to-Sum Formulas.**

THEOREM **Product-to-Sum Formulas**

$$\sin\alpha\sin\beta = \frac{1}{2}[\cos(\alpha-\beta) - \cos(\alpha+\beta)] \tag{1}$$

$$\cos\alpha\cos\beta = \frac{1}{2}[\cos(\alpha-\beta) + \cos(\alpha+\beta)] \tag{2}$$

$$\sin\alpha\cos\beta = \frac{1}{2}[\sin(\alpha+\beta) + \sin(\alpha-\beta)] \tag{3}$$

These formulas do not have to be memorized. Instead, you should remember how they are derived. Then, when you want to use them, either look them up or derive them, as needed.

To derive formulas (1) and (2), write down the sum and difference formulas for the cosine:

$$\cos(\alpha-\beta) = \cos\alpha\cos\beta + \sin\alpha\sin\beta \tag{4}$$

$$\cos(\alpha+\beta) = \cos\alpha\cos\beta - \sin\alpha\sin\beta \tag{5}$$

Subtract equation (5) from equation (4) to get

$$\cos(\alpha-\beta) - \cos(\alpha+\beta) = 2\sin\alpha\sin\beta$$

from which

$$\sin\alpha\sin\beta = \frac{1}{2}[\cos(\alpha-\beta) - \cos(\alpha+\beta)]$$

Now add equations (4) and (5) to get

$$\cos(\alpha - \beta) + \cos(\alpha + \beta) = 2 \cos \alpha \cos \beta$$

from which

$$\cos \alpha \cos \beta = \frac{1}{2}[\cos(\alpha - \beta) + \cos(\alpha + \beta)]$$

To derive Product-to-Sum Formula (3), use the sum and difference formulas for sine in a similar way. (You are asked to do this in Problem 43.)

EXAMPLE 1 **Expressing Products as Sums**

Express each of the following products as a sum containing only sines or only cosines.

(a) $\sin(6\theta) \sin(4\theta)$ (b) $\cos(3\theta) \cos \theta$ (c) $\sin(3\theta) \cos(5\theta)$

Solution (a) We use formula (1) to get

$$\sin(6\theta) \sin(4\theta) = \frac{1}{2}[\cos(6\theta - 4\theta) - \cos(6\theta + 4\theta)]$$

$$= \frac{1}{2}[\cos(2\theta) - \cos(10\theta)]$$

(b) We use formula (2) to get

$$\cos(3\theta) \cos \theta = \frac{1}{2}[\cos(3\theta - \theta) + \cos(3\theta + \theta)]$$

$$= \frac{1}{2}[\cos(2\theta) + \cos(4\theta)]$$

(c) We use formula (3) to get

$$\sin(3\theta) \cos(5\theta) = \frac{1}{2}[\sin(3\theta + 5\theta) + \sin(3\theta - 5\theta)]$$

$$= \frac{1}{2}[\sin(8\theta) + \sin(-2\theta)] = \frac{1}{2}[\sin(8\theta) - \sin(2\theta)]$$

Now Work PROBLEM 1

2 Express Sums as Products

The **Sum-to-Product Formulas** are given next.

THEOREM **Sum-to-Product Formulas**

$$\sin \alpha + \sin \beta = 2 \sin \frac{\alpha + \beta}{2} \cos \frac{\alpha - \beta}{2} \qquad (6)$$

$$\sin \alpha - \sin \beta = 2 \sin \frac{\alpha - \beta}{2} \cos \frac{\alpha + \beta}{2} \qquad (7)$$

$$\cos \alpha + \cos \beta = 2 \cos \frac{\alpha + \beta}{2} \cos \frac{\alpha - \beta}{2} \qquad (8)$$

$$\cos \alpha - \cos \beta = -2 \sin \frac{\alpha + \beta}{2} \sin \frac{\alpha - \beta}{2} \qquad (9)$$

We will derive formula (6) and leave the derivations of formulas (7) through (9) as exercises (see Problems 44 through 46).

Proof

$$2 \sin \frac{\alpha + \beta}{2} \cos \frac{\alpha - \beta}{2} = 2 \cdot \frac{1}{2}\left[\sin\left(\frac{\alpha + \beta}{2} + \frac{\alpha - \beta}{2}\right) + \sin\left(\frac{\alpha + \beta}{2} - \frac{\alpha - \beta}{2}\right)\right]$$

$$\uparrow$$
Product-to-Sum Formula (3)

$$= \sin \frac{2\alpha}{2} + \sin \frac{2\beta}{2} = \sin \alpha + \sin \beta \qquad \blacksquare$$

EXAMPLE 2	**Expressing Sums (or Differences) as a Product**

Express each sum or difference as a product of sines and/or cosines.

(a) $\sin(5\theta) - \sin(3\theta)$ (b) $\cos(3\theta) + \cos(2\theta)$

Solution (a) We use formula (7) to get

$$\sin(5\theta) - \sin(3\theta) = 2 \sin \frac{5\theta - 3\theta}{2} \cos \frac{5\theta + 3\theta}{2}$$

$$= 2 \sin \theta \cos(4\theta)$$

(b) $\cos(3\theta) + \cos(2\theta) = 2 \cos \dfrac{3\theta + 2\theta}{2} \cos \dfrac{3\theta - 2\theta}{2}$ Formula (8)

$$= 2 \cos \frac{5\theta}{2} \cos \frac{\theta}{2}$$

Now Work PROBLEM 11

8.6 Assess Your Understanding

Skill Building

In Problems 1–10, express each product as a sum containing only sines or only cosines.

1. $\sin(4\theta) \sin(2\theta)$ 2. $\cos(4\theta) \cos(2\theta)$ 3. $\sin(4\theta) \cos(2\theta)$ 4. $\sin(3\theta) \sin(5\theta)$ 5. $\cos(3\theta) \cos(5\theta)$

6. $\sin(4\theta) \cos(6\theta)$ 7. $\sin \theta \sin(2\theta)$ 8. $\cos(3\theta) \cos(4\theta)$ 9. $\sin \dfrac{3\theta}{2} \cos \dfrac{\theta}{2}$ 10. $\sin \dfrac{\theta}{2} \cos \dfrac{5\theta}{2}$

In Problems 11–18, express each sum or difference as a product of sines and/or cosines.

11. $\sin(4\theta) - \sin(2\theta)$ 12. $\sin(4\theta) + \sin(2\theta)$ 13. $\cos(2\theta) + \cos(4\theta)$ 14. $\cos(5\theta) - \cos(3\theta)$

15. $\sin \theta + \sin(3\theta)$ 16. $\cos \theta + \cos(3\theta)$ 17. $\cos \dfrac{\theta}{2} - \cos \dfrac{3\theta}{2}$ 18. $\sin \dfrac{\theta}{2} - \sin \dfrac{3\theta}{2}$

In Problems 19–36, establish each identity.

19. $\dfrac{\sin \theta + \sin(3\theta)}{2 \sin(2\theta)} = \cos \theta$ 20. $\dfrac{\cos \theta + \cos(3\theta)}{2 \cos(2\theta)} = \cos \theta$ 21. $\dfrac{\sin(4\theta) + \sin(2\theta)}{\cos(4\theta) + \cos(2\theta)} = \tan(3\theta)$

22. $\dfrac{\cos \theta - \cos(3\theta)}{\sin(3\theta) - \sin \theta} = \tan(2\theta)$ 23. $\dfrac{\cos \theta - \cos(3\theta)}{\sin \theta + \sin(3\theta)} = \tan \theta$ 24. $\dfrac{\cos \theta - \cos(5\theta)}{\sin \theta + \sin(5\theta)} = \tan(2\theta)$

25. $\sin \theta[\sin \theta + \sin(3\theta)] = \cos \theta[\cos \theta - \cos(3\theta)]$ 26. $\sin \theta[\sin(3\theta) + \sin(5\theta)] = \cos \theta[\cos(3\theta) - \cos(5\theta)]$

27. $\dfrac{\sin(4\theta) + \sin(8\theta)}{\cos(4\theta) + \cos(8\theta)} = \tan(6\theta)$ 28. $\dfrac{\sin(4\theta) - \sin(8\theta)}{\cos(4\theta) - \cos(8\theta)} = -\cot(6\theta)$

29. $\dfrac{\sin(4\theta) + \sin(8\theta)}{\sin(4\theta) - \sin(8\theta)} = -\dfrac{\tan(6\theta)}{\tan(2\theta)}$ 30. $\dfrac{\cos(4\theta) - \cos(8\theta)}{\cos(4\theta) + \cos(8\theta)} = \tan(2\theta) \tan(6\theta)$

31. $\dfrac{\sin \alpha + \sin \beta}{\sin \alpha - \sin \beta} = \tan \dfrac{\alpha + \beta}{2} \cot \dfrac{\alpha - \beta}{2}$ 32. $\dfrac{\cos \alpha + \cos \beta}{\cos \alpha - \cos \beta} = -\cot \dfrac{\alpha + \beta}{2} \cot \dfrac{\alpha - \beta}{2}$

33. $\dfrac{\sin \alpha + \sin \beta}{\cos \alpha + \cos \beta} = \tan \dfrac{\alpha + \beta}{2}$

34. $\dfrac{\sin \alpha - \sin \beta}{\cos \alpha - \cos \beta} = -\cot \dfrac{\alpha + \beta}{2}$

35. $1 + \cos(2\theta) + \cos(4\theta) + \cos(6\theta) = 4 \cos \theta \cos(2\theta) \cos(3\theta)$

36. $1 - \cos(2\theta) + \cos(4\theta) - \cos(6\theta) = 4 \sin \theta \cos(2\theta) \sin(3\theta)$

Applications and Extensions

37. Touch-Tone Phones On a Touch-Tone phone, each button produces a unique sound. The sound produced is the sum of two tones, given by

$$y = \sin(2\pi l t) \quad \text{and} \quad y = \sin(2\pi h t)$$

where l and h are the low and high frequencies (cycles per second) shown on the illustration. For example, if you touch 7, the low frequency is $l = 852$ cycles per second and the high frequency is $h = 1209$ cycles per second. The sound emitted by touching 7 is

$$y = \sin[2\pi(852)t] + \sin[2\pi(1209)t]$$

Touch-Tone phone

697 cycles/sec
770 cycles/sec
852 cycles/sec
941 cycles/sec

1209 cycles/sec 1477 cycles/sec

1336 cycles/sec

(a) Write this sound as a product of sines and/or cosines.
(b) Determine the maximum value of y.
(c) Graph the sound emitted by touching 7.

38. Touch-Tone Phones
(a) Write the sound emitted by touching the # key as a product of sines and/or cosines.
(b) Determine the maximum value of y.
(c) Graph the sound emitted by touching the # key.

39. Moment of Inertia The moment of inertia I of an object is a measure of how easy it is to rotate the object about some fixed point. In engineering mechanics, it is sometimes necessary to compute moments of inertia with respect to a set of rotated axes. These moments are given by the equations

$$I_u = I_x \cos^2 \theta + I_y \sin^2 \theta - 2I_{xy} \sin \theta \cos \theta$$
$$I_v = I_x \sin^2 \theta + I_y \cos^2 \theta + 2I_{xy} \sin \theta \cos \theta$$

Use product-to-sum formulas to show that

$$I_u = \frac{I_x + I_y}{2} + \frac{I_x - I_y}{2} \cos 2\theta - I_{xy} \sin 2\theta$$

and

$$I_v = \frac{I_x + I_y}{2} - \frac{I_x - I_y}{2} \cos 2\theta + I_{xy} \sin 2\theta$$

Source: Adapted from Hibbeler, *Engineering Mechanics: Statics,* 10th ed., Prentice Hall © 2004.

40. Projectile Motion The range of a projectile propelled downward from the top of an inclined plane at an angle θ to the inclined plane is given by

$$R(\theta) = \frac{2v_0^2 \sin \theta \cos(\theta - \phi)}{g \cos^2 \phi}$$

when the projectile hits the inclined plane. Here v_0 is the initial velocity of the projectile, ϕ is the angle the plane makes with respect to the horizontal, and g is acceleration due to gravity.

(a) Show that for fixed v_0 and ϕ the maximum range down the incline is given by $R_{\max} = \dfrac{v_0^2}{g(1 - \sin \phi)}$.

(b) Determine the maximum range if the projectile has an initial velocity of 50 meters/second, the angle of the plane is $\phi = 35°$, and $g = 9.8$ meters/second2.

41. If $\alpha + \beta + \gamma = \pi$, show that

$$\sin(2\alpha) + \sin(2\beta) + \sin(2\gamma) = 4 \sin \alpha \sin \beta \sin \gamma$$

42. If $\alpha + \beta + \gamma = \pi$, show that

$$\tan \alpha + \tan \beta + \tan \gamma = \tan \alpha \tan \beta \tan \gamma$$

43. Derive formula (3).

44. Derive formula (7).

45. Derive formula (8).

46. Derive formula (9).

8.7 Trigonometric Equations (I)

PREPARING FOR THIS SECTION *Before getting started, review the following:*

- Solving Equations (Section 1.1, pp. 86–92)
- Values of the Trigonometric Functions (Section 7.3, pp. 529–532 and Section 7.4, pp. 540–548)

\\ **Now Work** the 'Are You Prepared?' problems on page 653.

OBJECTIVE 1 Solve Equations Involving a Single Trigonometric Function (p. 650)

1 Solve Equations Involving a Single Trigonometric Function

The previous four sections of this chapter were devoted to trigonometric identities, that is, equations involving trigonometric functions that are satisfied by every value in the domain of the variable. In the remaining two sections, we discuss **trigonometric equations,** that is, equations involving trigonometric functions that are satisfied only by some values of the variable (or, possibly, are not satisfied by any values of the variable). The values that satisfy the equation are called **solutions** of the equation.

EXAMPLE 1 **Checking Whether a Given Number Is a Solution of a Trigonometric Equation**

Determine whether $\theta = \dfrac{\pi}{4}$ is a solution of the equation $\sin \theta = \dfrac{1}{2}$. Is $\theta = \dfrac{\pi}{6}$ a solution?

Solution Replace θ by $\dfrac{\pi}{4}$ in the given equation. The result is

$$\sin \frac{\pi}{4} = \frac{\sqrt{2}}{2} \neq \frac{1}{2}$$

We conclude that $\dfrac{\pi}{4}$ is not a solution.

Next replace θ by $\dfrac{\pi}{6}$ in the equation. The result is

$$\sin \frac{\pi}{6} = \frac{1}{2}$$

We conclude that $\dfrac{\pi}{6}$ is a solution of the given equation.

The equation given in Example 1 has other solutions besides $\theta = \dfrac{\pi}{6}$. For example, $\theta = \dfrac{5\pi}{6}$ is also a solution, as is $\theta = \dfrac{13\pi}{6}$. (You should check this for yourself.) In fact, the equation has an infinite number of solutions due to the periodicity of the sine function, as can be seen in Figure 25.

Figure 25

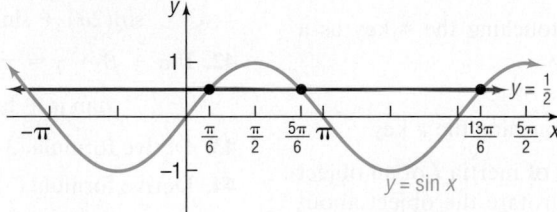

Unless the domain of the variable is restricted, we need to find *all* the solutions of a trigonometric equation. As the next example illustrates, finding all the solutions can be accomplished by first finding solutions over an interval whose length equals the period of the function and then adding multiples of that period to the solutions found. Let's look at some examples.

EXAMPLE 2 **Finding All the Solutions of a Trigonometric Equation**

Solve the equation: $\cos \theta = \dfrac{1}{2}$

Give a general formula for all the solutions. List eight of the solutions.

Solution

The period of the cosine function is 2π. In the interval $[0, 2\pi)$, there are two angles θ for which $\cos \theta = \dfrac{1}{2}$: $\theta = \dfrac{\pi}{3}$ and $\theta = \dfrac{5\pi}{3}$. See Figure 26. Because the cosine function has period 2π, all the solutions of $\cos \theta = \dfrac{1}{2}$ may be given by the general formula

$$\theta = \frac{\pi}{3} + 2k\pi \quad \text{or} \quad \theta = \frac{5\pi}{3} + 2k\pi \quad \text{k any integer}$$

Eight of the solutions are

$$\underbrace{-\frac{5\pi}{3},}_{k = -1} \underbrace{-\frac{\pi}{3},}_{} \underbrace{\frac{\pi}{3},}_{k = 0} \underbrace{\frac{5\pi}{3},}_{} \underbrace{\frac{7\pi}{3},}_{k = 1} \underbrace{\frac{11\pi}{3},}_{} \underbrace{\frac{13\pi}{3},}_{k = 2} \underbrace{\frac{17\pi}{3}}_{}$$

Figure 26

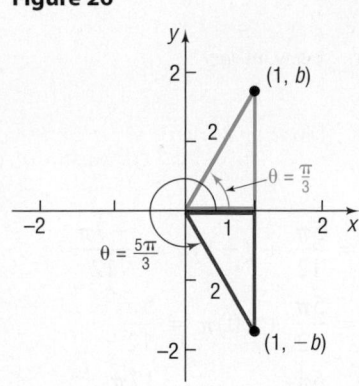

$\theta = \frac{\pi}{3}$

$\theta = \frac{5\pi}{3}$

$(1, b)$

$(1, -b)$

Figure 27

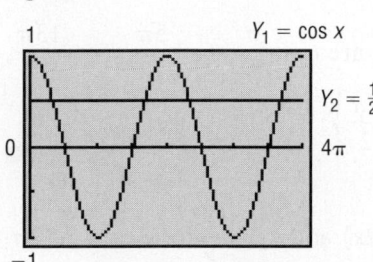

$Y_1 = \cos x$

$Y_2 = \frac{1}{2}$

4π

✓ **Check:** We can verify the solutions by graphing $Y_1 = \cos x$ and $Y_2 = \dfrac{1}{2}$ to determine where the graphs intersect. (Be sure to graph in radian mode.) See Figure 27. The graph of Y_1 intersects the graph of Y_2 at $x = 1.05 \left(\approx \dfrac{\pi}{3} \right)$, $5.24 \left(\approx \dfrac{5\pi}{3} \right)$, $7.33 \left(\approx \dfrac{7\pi}{3} \right)$, and $11.52 \left(\approx \dfrac{11\pi}{3} \right)$, rounded to two decimal places.

⟣═══ **Now Work** PROBLEM 31

In most of our work, we shall be interested only in finding solutions of trigonometric equations for $0 \le \theta < 2\pi$.

EXAMPLE 3 **Solving a Linear Trigonometric Equation**

Solve the equation: $2 \sin \theta + \sqrt{3} = 0$, $0 \le \theta < 2\pi$

Solution

We solve the equation for $\sin \theta$.

$$2 \sin \theta + \sqrt{3} = 0$$

$$2 \sin \theta = -\sqrt{3} \quad \text{Subtract } \sqrt{3} \text{ from both sides.}$$

$$\sin \theta = -\frac{\sqrt{3}}{2} \quad \text{Divide both sides by 2.}$$

In the interval $[0, 2\pi)$, there are two angles θ for which $\sin \theta = -\dfrac{\sqrt{3}}{2}$: $\theta = \dfrac{4\pi}{3}$ and $\theta = \dfrac{5\pi}{3}$. The solution set is $\left\{ \dfrac{4\pi}{3}, \dfrac{5\pi}{3} \right\}$.

⟣═══ **Now Work** PROBLEM 7

When the argument of the trigonometric function in an equation is a multiple of θ, the general formula must be used to solve the equation.

EXAMPLE 4 **Solving a Trigonometric Equation**

Solve the equation: $\sin(2\theta) = \dfrac{1}{2}$, $0 \le \theta < 2\pi$

Solution

In the interval $[0, 2\pi)$, the sine function has a value $\dfrac{1}{2}$ at $\dfrac{\pi}{6}$ and $\dfrac{5\pi}{6}$. See Figure 28 on page 652. Since the period of the sine function is 2π and the argument is 2θ in

Figure 28

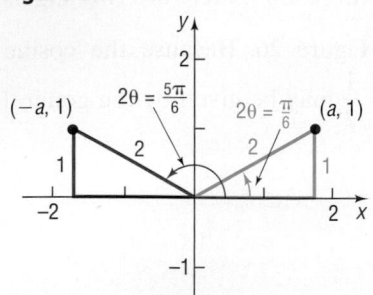

the equation $\sin(2\theta) = \dfrac{1}{2}$, we write the general formula that gives all the solutions.

$$2\theta = \frac{\pi}{6} + 2k\pi \quad \text{or} \quad 2\theta = \frac{5\pi}{6} + 2k\pi \quad \textit{k any integer}$$

$$\theta = \frac{\pi}{12} + k\pi \quad \text{or} \quad \theta = \frac{5\pi}{12} + k\pi \quad \textit{Divide by 2.}$$

Then

$$\theta = \frac{\pi}{12} + (-1)\pi = \frac{-11\pi}{12} \quad \textit{k = -1} \qquad \theta = \frac{5\pi}{12} + (-1)\pi = \frac{-7\pi}{12}$$

$$\theta = \frac{\pi}{12} + (0)\pi = \frac{\pi}{12} \quad \textit{k = 0} \qquad \theta = \frac{5\pi}{12} + (0)\pi = \frac{5\pi}{12}$$

$$\theta = \frac{\pi}{12} + (1)\pi = \frac{13\pi}{12} \quad \textit{k = 1} \qquad \theta = \frac{5\pi}{12} + (1)\pi = \frac{17\pi}{12}$$

$$\theta = \frac{\pi}{12} + (2)\pi = \frac{25\pi}{12} \quad \textit{k = 2} \qquad \theta = \frac{5\pi}{12} + (2)\pi = \frac{29\pi}{12}$$

In the interval $[0, 2\pi)$, the solutions of $\sin(2\theta) = \dfrac{1}{2}$ are $\theta = \dfrac{\pi}{12}, \theta = \dfrac{5\pi}{12}, \theta = \dfrac{13\pi}{12}$,

and $\theta = \dfrac{17\pi}{12}$. The solution set is $\left\{ \dfrac{\pi}{12}, \dfrac{5\pi}{12}, \dfrac{13\pi}{12}, \dfrac{17\pi}{12} \right\}$.

 ✓**Check:** Verify these solutions by graphing $Y_1 = \sin(2x)$ and $Y_2 = \dfrac{1}{2}$ for $0 \leq x \leq 2\pi$.

WARNING In solving a trigonometric equation for θ, $0 \leq \theta < 2\pi$, in which the argument is not θ (as in Example 4), you must write down all the solutions first and then list those that are in the interval $[0, 2\pi)$. Otherwise, solutions may be lost. For example, in solving $\sin(2\theta) = \dfrac{1}{2}$, if you merely write the solutions $2\theta = \dfrac{\pi}{6}$ and $2\theta = \dfrac{5\pi}{6}$, you will find only $\theta = \dfrac{\pi}{12}$ and $\theta = \dfrac{5\pi}{12}$ and miss the other solutions. ∎

—✏— **Now Work** PROBLEM 13

EXAMPLE 5 **Solving a Trigonometric Equation**

Solve the equation: $\tan\left(\theta - \dfrac{\pi}{2}\right) = 1, \quad 0 \leq \theta < 2\pi$

Solution The period of the tangent function is π. In the interval $[0, \pi)$, the tangent function has the value 1 when the argument is $\dfrac{\pi}{4}$. Because the argument is $\theta - \dfrac{\pi}{2}$ in the given equation, we write the general formula that gives all the solutions.

$$\theta - \frac{\pi}{2} = \frac{\pi}{4} + k\pi \quad \textit{k any integer}$$

$$\theta = \frac{3\pi}{4} + k\pi$$

In the interval $[0, 2\pi)$, $\theta = \dfrac{3\pi}{4}$ and $\theta = \dfrac{3\pi}{4} + \pi = \dfrac{7\pi}{4}$ are the only solutions.

The solution set is $\left\{ \dfrac{3\pi}{4}, \dfrac{7\pi}{4} \right\}$.

The next example illustrates how to solve trigonometric equations using a calculator. Remember that the function keys on a calculator will only give values consistent with the definition of the function.

EXAMPLE 6	Solving a Trigonometric Equation with a Calculator

Use a calculator to solve the equation $\sin\theta = 0.3, 0 \leq \theta < 2\pi$. Express any solutions in radians, rounded to two decimal places.

Solution To solve $\sin\theta = 0.3$ on a calculator, first set the mode to radians. Then use the $\boxed{\sin^{-1}}$ key to obtain

$$\theta = \sin^{-1}(0.3) \approx 0.3046927$$

Figure 29

Rounded to two decimal places, $\theta = \sin^{-1}(0.3) = 0.30$ radian. Because of the definition of $y = \sin^{-1} x$, the angle θ that we obtain is the angle $-\dfrac{\pi}{2} \leq \theta \leq \dfrac{\pi}{2}$ for which $\sin\theta = 0.3$. Another angle for which $\sin\theta = 0.3$ is $\pi - 0.30$. See Figure 29. The angle $\pi - 0.30$ is the angle in quadrant II, where $\sin\theta = 0.3$. The solutions for $\sin\theta = 0.3, 0 \leq \theta < 2\pi$, are

$$\theta = 0.30 \text{ radian} \quad \text{and} \quad \theta = \pi - 0.30 \approx 2.84 \text{ radians}$$

The solution set is $\{0.30, 2.84\}$. ∎

WARNING Example 6 illustrates that caution must be exercised when solving trigonometric equations on a calculator. Remember that the calculator supplies an angle only within the restrictions of the definition of the inverse trigonometric function. To find the remaining solutions, you must identify other quadrants, if any, in which a solution may be located. ∎

— **Now Work** PROBLEM 41

8.7 Assess Your Understanding

'Are You Prepared?' *Answers are given at the end of these exercises. If you get a wrong answer, read the pages listed in red.*

1. Solve: $3x - 5 = -x + 1$ (pp. 86–92)

2. $\sin\left(\dfrac{\pi}{4}\right) =$ _____; $\cos\left(\dfrac{8\pi}{3}\right) =$ _____ (pp. 529–532 and 540–548)

Concepts and Vocabulary

3. Two solutions of the equation $\sin\theta = \dfrac{1}{2}$ are _____ and _____.

4. All the solutions of the equation $\sin\theta = \dfrac{1}{2}$ are _____.

5. *True or False* Most trigonometric equations have unique solutions.

6. *True or False* The equation $\sin\theta = 2$ has a real solution that can be found using a calculator.

Skill Building

In Problems 7–30, solve each equation on the interval $0 \leq \theta < 2\pi$.

7. $2\sin\theta + 3 = 2$

8. $1 - \cos\theta = \dfrac{1}{2}$

9. $4\cos^2\theta = 1$

10. $\tan^2\theta = \dfrac{1}{3}$

11. $2\sin^2\theta - 1 = 0$

12. $4\cos^2\theta - 3 = 0$

13. $\sin(3\theta) = -1$

14. $\tan\dfrac{\theta}{2} = \sqrt{3}$

15. $\cos(2\theta) = -\dfrac{1}{2}$

16. $\tan(2\theta) = -1$

17. $\sec\dfrac{3\theta}{2} = -2$

18. $\cot\dfrac{2\theta}{3} = -\sqrt{3}$

19. $2\sin\theta + 1 = 0$

20. $\cos\theta + 1 = 0$

21. $\tan\theta + 1 = 0$

22. $\sqrt{3}\cot\theta + 1 = 0$

23. $4\sec\theta + 6 = -2$

24. $5\csc\theta - 3 = 2$

25. $3\sqrt{2}\cos\theta + 2 = -1$

26. $4\sin\theta + 3\sqrt{3} = \sqrt{3}$

27. $\cos\left(2\theta - \dfrac{\pi}{2}\right) = -1$

28. $\sin\left(3\theta + \dfrac{\pi}{18}\right) = 1$

29. $\tan\left(\dfrac{\theta}{2} + \dfrac{\pi}{3}\right) = 1$

30. $\cos\left(\dfrac{\theta}{3} - \dfrac{\pi}{4}\right) = \dfrac{1}{2}$

In Problems 31–40, solve each equation. Give a general formula for all the solutions. List six solutions.

31. $\sin\theta = \dfrac{1}{2}$

32. $\tan\theta = 1$

33. $\tan\theta = -\dfrac{\sqrt{3}}{3}$

34. $\cos\theta = -\dfrac{\sqrt{3}}{2}$

35. $\cos\theta = 0$

36. $\sin \theta = \dfrac{\sqrt{2}}{2}$ **37.** $\cos(2\theta) = -\dfrac{1}{2}$ **38.** $\sin(2\theta) = -1$ **39.** $\sin \dfrac{\theta}{2} = -\dfrac{\sqrt{3}}{2}$ **40.** $\tan \dfrac{\theta}{2} = -1$

In Problems 41–52, use a calculator to solve each equation on the interval $0 \le \theta < 2\pi$. Round answers to two decimal places.

41. $\sin \theta = 0.4$ **42.** $\cos \theta = 0.6$ **43.** $\tan \theta = 5$ **44.** $\cot \theta = 2$

45. $\cos \theta = -0.9$ **46.** $\sin \theta = -0.2$ **47.** $\sec \theta = -4$ **48.** $\csc \theta = -3$

49. $5 \tan \theta + 9 = 0$ **50.** $4 \cot \theta = -5$ **51.** $3 \sin \theta - 2 = 0$ **52.** $4 \cos \theta + 3 = 0$

Applications and Extensions

53. What are the x-intercepts of the graph of $f(x) = 4 \sin^2 x - 3$ on the interval $[0, 2\pi]$?

54. What are the x-intercepts of the graph of $f(x) = 2 \cos(3x) + 1$ on the interval $[0, \pi]$?

55. $f(x) = 3 \sin x$.
 (a) Find the x-intercepts of the graph of f on the interval $[-2\pi, 4\pi]$.
 (b) Graph $f(x) = 3 \sin x$ on the interval $[-2\pi, 4\pi]$.
 (c) Solve $f(x) = \dfrac{3}{2}$ on the interval $[-2\pi, 4\pi]$. What points are on the graph of f? Label these points on the graph drawn in part (b).
 (d) Use the graph drawn in part (b) along with the results of part (c) to determine the values of x such that $f(x) > \dfrac{3}{2}$ on the interval $[-2\pi, 4\pi]$.

56. $f(x) = 2 \cos x$.
 (a) Find the x-intercepts of the graph of f on the interval $[-2\pi, 4\pi]$.
 (b) Graph $f(x) = 2 \cos x$ on the interval $[-2\pi, 4\pi]$.
 (c) Solve $f(x) = -\sqrt{3}$ on the interval $[-2\pi, 4\pi]$. What points are on the graph of f? Label these points on the graph drawn in part (b).
 (d) Use the graph drawn in part (b) along with the results of part (c) to determine the values of x such that $f(x) < -\sqrt{3}$ on the interval $[-2\pi, 4\pi]$.

57. $f(x) = 4 \tan x$.
 (a) Solve $f(x) = -4$.
 (b) For what values of x is $f(x) < -4$ on the interval $\left(-\dfrac{\pi}{2}, \dfrac{\pi}{2}\right)$?

58. $f(x) = \cot x$.
 (a) Solve $f(x) = -\sqrt{3}$.
 (b) For what values of x is $f(x) > -\sqrt{3}$ on the interval $(0, \pi)$?

59. (a) Graph $f(x) = 3 \sin(2x) + 2$ and $g(x) = \dfrac{7}{2}$ on the same Cartesian plane for the interval $[0, \pi]$.
 (b) Solve $f(x) = g(x)$ on the interval $[0, \pi]$ and label the points of intersection on the graph drawn in part (b).
 (c) Solve $f(x) > g(x)$ on the interval $[0, \pi]$.
 (d) Shade the region bounded by $f(x) = 3 \sin(2x) + 2$ and $g(x) = \dfrac{7}{2}$ between the two points found in part (b) on the graph drawn in part (a).

60. (a) Graph $f(x) = 2 \cos \dfrac{x}{2} + 3$ and $g(x) = 4$ on the same Cartesian plane for the interval $[0, 4\pi]$.

60. (b) Solve $f(x) = g(x)$ on the interval $[0, 4\pi]$ and label the points of intersection on the graph drawn in part (b).
 (c) Solve $f(x) < g(x)$ on the interval $[0, 4\pi]$.
 (d) Shade the region bounded by $f(x) = 2 \cos \dfrac{x}{2} + 3$ and $g(x) = 4$ between the two points found in part (b) on the graph drawn in part (a).

61. (a) Graph $f(x) = -4 \cos x$ and $g(x) = 2 \cos x + 3$ on the same Cartesian plane for the interval $[0, 2\pi]$.
 (b) Solve $f(x) = g(x)$ on the interval $[0, 2\pi]$ and label the points of intersection on the graph drawn in part (b).
 (c) Solve $f(x) > g(x)$ on the interval $[0, 2\pi]$.
 (d) Shade the region bounded by $f(x) = -4 \cos x$ and $g(x) = 2 \cos x + 3$ between the two points found in part (b) on the graph drawn in part (a).

62. (a) Graph $f(x) = 2 \sin x$ and $g(x) = -2 \sin x + 2$ on the same Cartesian plane for the interval $[0, 2\pi]$.
 (b) Solve $f(x) = g(x)$ on the interval $[0, 2\pi]$ and label the points of intersection on the graph drawn in part (b).
 (c) Solve $f(x) > g(x)$ on the interval $[0, 2\pi]$.
 (d) Shade the region bounded by $f(x) = 2 \sin x$ and $g(x) = -2 \sin x + 2$ between the two points found in part (b) on the graph drawn in part (a).

63. The Ferris Wheel In 1893, George Ferris engineered the Ferris Wheel. It was 250 feet in diameter. If the wheel makes 1 revolution every 40 seconds, then the function

$$h(t) = 125 \sin\left(0.157t - \dfrac{\pi}{2}\right) + 125$$

represents the height h, in feet, of a seat on the wheel as a function of time t, where t is measured in seconds. The ride begins when $t = 0$.
 (a) During the first 40 seconds of the ride, at what time t is an individual on the Ferris Wheel exactly 125 feet above the ground?
 (b) During the first 80 seconds of the ride, at what time t is an individual on the Ferris Wheel exactly 250 feet above the ground?
 (c) During the first 40 seconds of the ride, over what interval of time t is an individual on the Ferris Wheel more than 125 feet above the ground?

64. Tire Rotation The P215/65R15 Cobra Radial G/T tire has a diameter of exactly 26 inches. Suppose that a car's wheel is making 2 revolutions per second (the car is traveling a little less than 10 miles per hour). Then $h(t) = 13 \sin\left(4\pi t - \dfrac{\pi}{2}\right) + 13$ represents the height h (in inches) of a point on the tire as a function of time t (in seconds). The car starts to move when $t = 0$.

(a) During the first second that the car is moving, at what time t is the point on the tire exactly 13 inches above the ground?

(b) During the first second that the car is moving, at what time t is the point on the tire exactly 6.5 inches above the ground?

(c) During the first second that the car is moving, at what time t is the point on the tire more than 13 inches above the ground?

Source: Cobra Tire

65. **Holding Pattern** An airplane is asked to stay within a holding pattern near Chicago's O'Hare International Airport. The function $d(x) = 70 \sin(0.65x) + 150$ represents the distance d, in miles, that the airplane is from the airport at time x, in minutes.

(a) When the plane enters the holding pattern, $x = 0$, how far is it from O'Hare?

(b) During the first 20 minutes after the plane enters the holding pattern, at what time x is the plane exactly 100 miles from the airport?

(c) During the first 20 minutes after the plane enters the holding pattern, at what time x is the plane more than 100 miles from the airport?

(d) While the plane is in the holding pattern, will it ever be within 70 miles of the airport? Why?

66. **Projectile Motion** A golfer hits a golf ball with an initial velocity of 100 miles per hour. The range R of the ball as a function of the angle θ to the horizontal is given by $R(\theta) = 672 \sin(2\theta)$, where R is measured in feet.

(a) At what angle θ should the ball be hit if the golfer wants the ball to travel 450 feet (150 yards)?

(b) At what angle θ should the ball be hit if the golfer wants the ball to travel 540 feet (180 yards)?

(c) At what angle θ should the ball be hit if the golfer wants the ball to travel at least 480 feet (160 yards)?

(d) Can the golfer hit the ball 720 feet (240 yards)?

 The following discussion of Snell's Law of Refraction (named after Willebrord Snell, 1580–1626) is needed for Problems 67–73. Light, sound, and other waves travel at different speeds, depending on the media (air, water, wood, and so on) through which they pass. Suppose that light travels from a point A in one medium, where its speed is v_1, to a point B in another medium, where its speed is v_2. Refer to the figure, where the angle θ_1 is called the **angle of incidence** and the angle θ_2 is the **angle of refraction**. Snell's Law, which can be proved using calculus, states that*

$$\frac{\sin \theta_1}{\sin \theta_2} = \frac{v_1}{v_2}$$

*The ratio $\dfrac{v_1}{v_2}$ is called the **index of refraction**. Some values are given in the following table.*

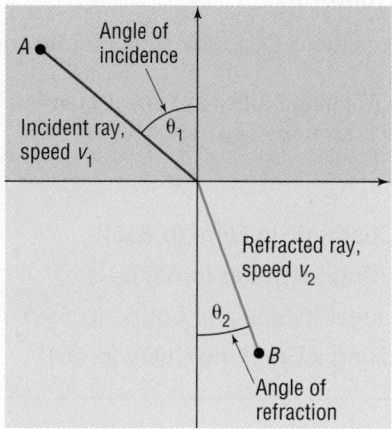

Some Indexes of Refraction	
Medium	**Index of Refraction[†]**
Water	1.33
Ethyl alcohol (20°C)	1.36
Carbon disulfide	1.63
Air (1 atm and 0°C)	1.00029
Diamond	2.42
Fused quartz	1.46
Glass, crown	1.52
Glass, dense flint	1.66
Sodium chloride	1.54

67. The index of refraction of light in passing from a vacuum into water is 1.33. If the angle of incidence is 40°, determine the angle of refraction.

68. The index of refraction of light in passing from a vacuum into dense flint glass is 1.66. If the angle of incidence is 50°, determine the angle of refraction.

69. Ptolemy, who lived in the city of Alexandria in Egypt during the second century AD, gave the measured values in the table following for the angle of incidence θ_1 and the angle of refraction θ_2 for a light beam passing from air into water. Do

these values agree with Snell's Law? If so, what index of refraction results? (These data are interesting as the oldest recorded physical measurements.) [†]

θ_1	θ_2	θ_1	θ_2
10°	8°	50°	35°0′
20°	15°30′	60°	40°30′
30°	22°30′	70°	45°30′
40°	29°0′	80°	50°0′

* Because this law was also deduced by René Descartes in France, it is also known as Descartes's Law.

[†] For light of wavelength 589 nanometers, measured with respect to a vacuum. The index with respect to air is negligibly different in most cases.

70. **Bending Light** The speed of yellow sodium light (wavelength of 589 nanometers) in a certain liquid is measured to be 1.92×10^8 meters per second. What is the index of refraction of this liquid, with respect to air, for sodium light?*

 [**Hint:** The speed of light in air is approximately 2.998×10^8 meters per second.]

71. **Bending Light** A beam of light with a wavelength of 589 nanometers traveling in air makes an angle of incidence of $40°$ on a slab of transparent material, and the refracted beam makes an angle of refraction of $26°$. Find the index of refraction of the material.*

72. **Bending Light** A light ray with a wavelength of 589 nanometers (produced by a sodium lamp) traveling through air makes an angle of incidence of $30°$ on a smooth, flat slab of crown glass. Find the angle of refraction.*

73. A light beam passes through a thick slab of material whose index of refraction is n_2. Show that the emerging beam is parallel to the incident beam.*

74. **Brewster's Law** If the angle of incidence and the angle of refraction are complementary angles, the angle of incidence is referred to as the Brewster angle θ_B. The Brewster angle is related to the index of refractions of the two media, n_1 and n_2, by the equation $n_1 \sin \theta_B = n_2 \cos \theta_B$, where n_1 is the index of refraction of the incident medium and n_2 is the index of refraction of the refractive medium. Determine the Brewster angle for a light beam traveling through water (at $20°C$) that makes an angle of incidence with a smooth, flat slab of crown glass.

* Adapted from Halliday and Resnick, *Fundamentals of Physics*, 7th ed., 2005, John Wiley & Sons.

Discussion and Writing

75. Explain in your own words how you would use your calculator to solve the equation $\cos x = -0.6, 0 \le x < 2\pi$. How would you modify your approach to solve the equation $\cot x = 5, 0 < x < 2\pi$?

'Are You Prepared?' Answers

1. $\left\{ \dfrac{3}{2} \right\}$ 2. $\dfrac{\sqrt{2}}{2}; -\dfrac{1}{2}$

8.8 Trigonometric Equations (II)

PREPARING FOR THIS SECTION *Before getting started, review the following:*

- Solving Quadratic Equations by Factoring (Section 1.2, pp. 98–99)
- The Quadratic Formula (Section 1.2, pp. 102–104)

- Solving Equations Quadratic in Form (Section 1.4, pp. 119–121)
- Using a Graphing Utility to Solve Equations (Appendix, Section 4, pp. A8–A10)

\ **Now Work** the 'Are You Prepared?' problems on page 661.

OBJECTIVES 1 Solve Trigonometric Equations Quadratic in Form (p. 656)
2 Solve Trigonometric Equations Using Identities (p. 657)
3 Solve Trigonometric Equations Linear in Sine and Cosine (p. 659)
4 Solve Trigonometric Equations Using a Graphing Utility (p. 661)

1 Solve Trigonometric Equations Quadratic in Form

In this section we continue our study of trigonometric equations. Many trigonometric equations can be solved by applying techniques that we already know, such as applying the quadratic formula (if the equation is a second-degree polynomial) or factoring.

EXAMPLE 1 | **Solving a Trigonometric Equation Quadratic in Form**

Solve the equation: $2 \sin^2 \theta - 3 \sin \theta + 1 = 0$, $0 \le \theta < 2\pi$

Solution This equation is a quadratic equation (in $\sin \theta$) that can be factored.

$$2 \sin^2 \theta - 3 \sin \theta + 1 = 0 \quad 2x^2 - 3x + 1 = 0, \quad x = \sin \theta$$
$$(2 \sin \theta - 1)(\sin \theta - 1) = 0 \quad (2x - 1)(x - 1) = 0$$

$$2 \sin \theta - 1 = 0 \quad \text{or} \quad \sin \theta - 1 = 0 \quad \text{Use the Zero-Product Property.}$$

$$\sin \theta = \frac{1}{2} \quad \text{or} \quad \sin \theta = 1$$

Solving each equation in the interval $[0, 2\pi)$, we obtain

$$\theta = \frac{\pi}{6}, \qquad \theta = \frac{5\pi}{6}, \qquad \theta = \frac{\pi}{2}$$

The solution set is $\left\{ \dfrac{\pi}{6}, \dfrac{5\pi}{6}, \dfrac{\pi}{2} \right\}$.

━━━ **Now Work** PROBLEM 7

2 Solve Trigonometric Equations Using Identities

When a trigonometric equation contains more than one trigonometric function, identities sometimes can be used to obtain an equivalent equation that contains only one trigonometric function.

EXAMPLE 2	**Solving a Trigonometric Equation Using Identities**

Solve the equation: $3 \cos \theta + 3 = 2 \sin^2 \theta, \quad 0 \le \theta < 2\pi$

Solution The equation in its present form contains a sine and a cosine. However, a form of the Pythagorean Identity can be used to transform the equation into an equivalent expression containing only cosines.

$$3 \cos \theta + 3 = 2 \sin^2 \theta$$
$$3 \cos \theta + 3 = 2(1 - \cos^2 \theta) \qquad \text{sin}^2 \theta = 1 - \cos^2 \theta$$
$$3 \cos \theta + 3 = 2 - 2 \cos^2 \theta$$
$$2 \cos^2 \theta + 3 \cos \theta + 1 = 0 \qquad \text{Quadratic in } \cos \theta$$
$$(2 \cos \theta + 1)(\cos \theta + 1) = 0 \qquad \text{Factor.}$$
$$2 \cos \theta + 1 = 0 \quad \text{or} \quad \cos \theta + 1 = 0 \quad \text{Use the Zero-Product Property.}$$
$$\cos \theta = -\frac{1}{2} \quad \text{or} \quad \cos \theta = -1$$

Solving each equation in the interval $[0, 2\pi)$, we obtain

$$\theta = \frac{2\pi}{3}, \qquad \theta = \frac{4\pi}{3}, \qquad \theta = \pi$$

The solution set is $\left\{ \dfrac{2\pi}{3}, \pi, \dfrac{4\pi}{3} \right\}$.

✓ **Check:** Graph $Y_1 = 3 \cos x + 3$ and $Y_2 = 2 \sin^2 x, 0 \le x \le 2\pi$, and find the points of intersection. How close are your approximate solutions to the exact ones found in this example?

EXAMPLE 3	**Solving a Trigonometric Equation Using Identities**

Solve the equation: $\cos(2\theta) + 3 = 5 \cos \theta, \quad 0 \le \theta < 2\pi$

Solution First, we observe that the given equation contains two cosine functions, but with different arguments, θ and 2θ. We use the Double-angle Formula $\cos(2\theta) = 2 \cos^2 \theta - 1$ to obtain an equivalent equation containing only $\cos \theta$.

$$\cos(2\theta) + 3 = 5 \cos \theta$$
$$(2 \cos^2 \theta - 1) + 3 = 5 \cos \theta \qquad \text{cos}(2\theta) = 2 \cos^2 \theta - 1$$

$$2\cos^2 \theta - 5\cos \theta + 2 = 0 \quad \text{Place in standard form.}$$

$$(\cos \theta - 2)(2\cos \theta - 1) = 0 \quad \text{Factor.}$$

$$\cos \theta = 2 \quad \text{or} \quad \cos \theta = \frac{1}{2} \quad \text{Solve by using the Zero-Product property.}$$

For any angle θ, $-1 \leq \cos \theta \leq 1$; therefore, the equation $\cos \theta = 2$ has no solution.

The solutions of $\cos \theta = \frac{1}{2}$, $0 \leq \theta < 2\pi$, are

$$\theta = \frac{\pi}{3}, \qquad \theta = \frac{5\pi}{3}$$

The solution set is $\left\{ \dfrac{\pi}{3}, \dfrac{5\pi}{3} \right\}$.

■

──── **Now Work** PROBLEM 23

EXAMPLE 4	**Solving a Trigonometric Equation Using Identities**

Solve the equation: $\cos^2 \theta + \sin \theta = 2$, $0 \leq \theta < 2\pi$

Solution · This equation involves two trigonometric functions, sine and cosine. We use a form of the Pythagorean Identity, $\sin^2 \theta + \cos^2 \theta = 1$, to rewrite the equation in terms of $\sin \theta$.

$$\cos^2 \theta + \sin \theta = 2$$

$$(1 - \sin^2 \theta) + \sin \theta = 2 \quad \cos^2 \theta = 1 - \sin^2 \theta$$

$$\sin^2 \theta - \sin \theta + 1 = 0$$

This is a quadratic equation in $\sin \theta$. The discriminant is $b^2 - 4ac = 1 - 4 = -3 < 0$. Therefore, the equation has no real solution. The solution set is the empty set, $\varnothing$.

■

✓ **Check:** Graph $Y_1 = \cos^2 x + \sin x$ and $Y_2 = 2$ to see that the two graphs never intersect, so the equation $Y_1 = Y_2$ has no real solution.

EXAMPLE 5	**Solving a Trigonometric Equation Using Identities**

Solve the equation: $\sin \theta \cos \theta = -\dfrac{1}{2}$, $0 \leq \theta < 2\pi$

Solution The left side of the given equation is in the form of the Double-angle Formula $2 \sin \theta \cos \theta = \sin(2\theta)$, except for a factor of 2. We multiply each side by 2.

$$\sin \theta \cos \theta = -\frac{1}{2}$$

$$2 \sin \theta \cos \theta = -1 \quad \text{Multiply each side by 2.}$$

$$\sin(2\theta) = -1 \quad \text{Double-angle Formula}$$

The argument here is 2θ. So we need to write all the solutions of this equation and then list those that are in the interval $[0, 2\pi)$. Because $\sin\left(\dfrac{3\pi}{2} + 2\pi k \right) = -1$, for any integer k we have

$$2\theta = \frac{3\pi}{2} + 2k\pi \quad k \text{ any integer}$$

$$\theta = \frac{3\pi}{4} + k\pi$$

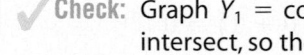

$$\underset{\substack{\uparrow \\ k = -1}}{\theta = \frac{3\pi}{4} + (-1)\pi = -\frac{\pi}{4}}, \quad \underset{\substack{\uparrow \\ k = 0}}{\theta = \frac{3\pi}{4} + (0)\pi = \frac{3\pi}{4}}, \quad \underset{\substack{\uparrow \\ k = 1}}{\theta = \frac{3\pi}{4} + (1)\pi = \frac{7\pi}{4}}, \quad \underset{\substack{\uparrow \\ k = 2}}{\theta = \frac{3\pi}{4} + (2)\pi = \frac{11\pi}{4}}$$

The solutions in the interval $[0, 2\pi)$ are

$$\theta = \frac{3\pi}{4}, \qquad \theta = \frac{7\pi}{4}$$

The solution set is $\left\{\dfrac{3\pi}{4}, \dfrac{7\pi}{4}\right\}$.

3 Solve Trigonometric Equations Linear in Sine and Cosine

Sometimes it is necessary to square both sides of an equation to obtain expressions that allow the use of identities. Remember, squaring both sides of an equation may introduce extraneous solutions. As a result, apparent solutions must be checked.

EXAMPLE 6	**Solving a Trigonometric Equation Linear in Sine and Cosine**

Solve the equation: $\sin\theta + \cos\theta = 1$, $0 \le \theta < 2\pi$

Solution A Attempts to use available identities do not lead to equations that are easy to solve. (Try it yourself.) Given the form of this equation, we decide to square each side.

$$\sin\theta + \cos\theta = 1$$

$$(\sin\theta + \cos\theta)^2 = 1 \qquad \text{Square each side.}$$

$$\sin^2\theta + 2\sin\theta\cos\theta + \cos^2\theta = 1 \qquad \text{Remove parentheses.}$$

$$2\sin\theta\cos\theta = 0 \qquad \text{\small $\sin^2\theta + \cos^2\theta = 1$}$$

$$\sin\theta\cos\theta = 0$$

Setting each factor equal to zero, we obtain

$$\sin\theta = 0 \quad \text{or} \quad \cos\theta = 0$$

The apparent solutions are

$$\theta = 0, \qquad \theta = \pi, \qquad \theta = \frac{\pi}{2}, \qquad \theta = \frac{3\pi}{2}$$

Because we squared both sides of the original equation, we must check these apparent solutions to see if any are extraneous.

$$\theta = 0: \quad \sin 0 + \cos 0 = 0 + 1 = 1 \qquad \text{\small A solution}$$

$$\theta = \pi: \quad \sin\pi + \cos\pi = 0 + (-1) = -1 \qquad \text{\small Not a solution}$$

$$\theta = \frac{\pi}{2}: \quad \sin\frac{\pi}{2} + \cos\frac{\pi}{2} = 1 + 0 = 1 \qquad \text{\small A solution}$$

$$\theta = \frac{3\pi}{2}: \quad \sin\frac{3\pi}{2} + \cos\frac{3\pi}{2} = -1 + 0 = -1 \qquad \text{\small Not a solution}$$

The values $\theta = \pi$ and $\theta = \dfrac{3\pi}{2}$ are extraneous. The solution set is $\left\{0, \dfrac{\pi}{2}\right\}$.

Solution B We start with the equation

$$\sin\theta + \cos\theta = 1$$

and divide each side by $\sqrt{2}$. (The reason for this choice will become apparent shortly.) Then

$$\frac{1}{\sqrt{2}}\sin\theta + \frac{1}{\sqrt{2}}\cos\theta = \frac{1}{\sqrt{2}}$$

The left side now resembles the formula for the sine of the sum of two angles, one of which is θ. The other angle is unknown (call it ϕ.) Then

$$\sin(\theta + \phi) = \sin\theta\cos\phi + \cos\theta\sin\phi = \frac{1}{\sqrt{2}} = \frac{\sqrt{2}}{2} \qquad (1)$$

where

$$\cos\phi = \frac{1}{\sqrt{2}} = \frac{\sqrt{2}}{2}, \qquad \sin\phi = \frac{1}{\sqrt{2}} = \frac{\sqrt{2}}{2}, \qquad 0 \le \phi < 2\pi$$

Figure 30

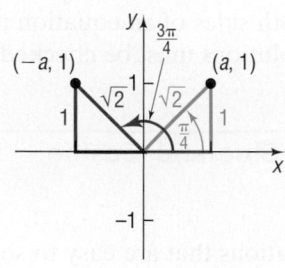

The angle ϕ is therefore $\dfrac{\pi}{4}$. As a result, equation (1) becomes

$$\sin\left(\theta + \frac{\pi}{4}\right) = \frac{\sqrt{2}}{2}$$

In the interval $[0, 2\pi)$, there are two angles whose sine is $\dfrac{\sqrt{2}}{2}$: $\dfrac{\pi}{4}$ and $\dfrac{3\pi}{4}$. See Figure 30. As a result,

$$\theta + \frac{\pi}{4} = \frac{\pi}{4} \quad \text{or} \quad \theta + \frac{\pi}{4} = \frac{3\pi}{4}$$

$$\theta = 0 \quad \text{or} \qquad \theta = \frac{\pi}{2}$$

The solution set is $\left\{0, \dfrac{\pi}{2}\right\}$.

This second method of solution can be used to solve any linear equation in the variables $\sin\theta$ and $\cos\theta$. Let's look at an example.

EXAMPLE 7 **Solving a Trigonometric Equation Linear in Sin θ and Cos θ**

Solve:

$$a\sin\theta + b\cos\theta = c \qquad (2)$$

where a, b, and c are constants and either $a \ne 0$ or $b \ne 0$.

Solution We divide each side of equation (2) by $\sqrt{a^2 + b^2}$. Then

$$\frac{a}{\sqrt{a^2 + b^2}}\sin\theta + \frac{b}{\sqrt{a^2 + b^2}}\cos\theta = \frac{c}{\sqrt{a^2 + b^2}} \qquad (3)$$

Figure 31

There is a unique angle ϕ, $0 \le \phi < 2\pi$, for which

$$\cos\phi = \frac{a}{\sqrt{a^2 + b^2}} \quad \text{and} \quad \sin\phi = \frac{b}{\sqrt{a^2 + b^2}} \qquad (4)$$

Figure 31 shows the situation for $a > 0$ and $b > 0$. Equation (3) may be written as

$$\sin\theta\cos\phi + \cos\theta\sin\phi = \frac{c}{\sqrt{a^2 + b^2}}$$

or, equivalently,

$$\sin(\theta + \phi) = \frac{c}{\sqrt{a^2 + b^2}} \qquad (5)$$

where ϕ satisfies equation (4).

If $|c| > \sqrt{a^2 + b^2}$, then $\sin(\theta + \phi) > 1$ or $\sin(\theta + \phi) < -1$, and equation (5) has no solution.

If $|c| \leq \sqrt{a^2 + b^2}$, then all the solutions of equation (5) are

$$\theta + \phi = \sin^{-1}\frac{c}{\sqrt{a^2 + b^2}} + 2k\pi \quad \text{or} \quad \theta + \phi = \pi - \sin^{-1}\frac{c}{\sqrt{a^2 + b^2}} + 2k\pi$$

Because the angle ϕ is determined by equations (4), these provide the solutions to equation (2).

Now Work PROBLEM 41

4 Solve Trigonometric Equations Using a Graphing Utility

The techniques introduced in this section apply only to certain types of trigonometric equations. Solutions for other types are usually studied in calculus, using numerical methods. In the next example, we show how a graphing utility may be used to obtain solutions.

EXAMPLE 8

Solving Trigonometric Equations Using a Graphing Utility

Solve: $5 \sin x + x = 3$

Express the solution(s) rounded to two decimal places.

Solution

Figure 32

This type of trigonometric equation cannot be solved by previous methods. A graphing utility, though, can be used here. The solution(s) of this equation is the same as the points of intersection of the graphs of $Y_1 = 5 \sin x + x$ and $Y_2 = 3$. See Figure 32.

There are three points of intersection; the x-coordinates are the solutions that we seek. Using INTERSECT, we find

$$x = 0.52, \quad x = 3.18, \quad x = 5.71$$

The solution set is $\{0.52, 3.18, 5.71\}$.

Now Work PROBLEM 53

8.8 Assess Your Understanding

'Are You Prepared?' *Answers are given at the end of these exercises. If you get a wrong answer, read the pages listed in red.*

1. Find the real solutions of $4x^2 - x - 5 = 0$. (pp. 98–99)

2. Find the real solutions of $x^2 - x - 1 = 0$. (pp. 102–109)

3. Find the real solutions of $(2x - 1)^2 - 3(2x - 1) - 4 = 0$. (pp. 119–121)

4. Use a graphing utility to solve $5x^3 - 2 = x - x^2$. Round answers to two decimal places. (pp. A8–A10)

Skill Building

In Problems 5–46, solve each equation on the interval $0 \leq \theta < 2\pi$.

5. $2 \cos^2 \theta + \cos \theta = 0$

6. $\sin^2 \theta - 1 = 0$

7. $2 \sin^2 \theta - \sin \theta - 1 = 0$

8. $2 \cos^2 \theta + \cos \theta - 1 = 0$

9. $(\tan \theta - 1)(\sec \theta - 1) = 0$

10. $(\cot \theta + 1)\left(\csc \theta - \frac{1}{2}\right) = 0$

11. $\sin^2 \theta - \cos^2 \theta = 1 + \cos \theta$

12. $\cos^2 \theta - \sin^2 \theta + \sin \theta = 0$

13. $\sin^2 \theta = 6(\cos \theta + 1)$

14. $2 \sin^2 \theta = 3(1 - \cos \theta)$

15. $\cos(2\theta) + 6 \sin^2 \theta = 4$

16. $\cos(2\theta) = 2 - 2 \sin^2 \theta$

17. $\cos \theta = \sin \theta$

18. $\cos \theta + \sin \theta = 0$

19. $\tan \theta = 2 \sin \theta$

20. $\sin(2\theta) = \cos \theta$

21. $\sin \theta = \csc \theta$

22. $\tan \theta = \cot \theta$

23. $\cos(2\theta) = \cos \theta$

24. $\sin(2\theta) \sin \theta = \cos \theta$

25. $\sin(2\theta) + \sin(4\theta) = 0$

26. $\cos(2\theta) + \cos(4\theta) = 0$

27. $\cos(4\theta) - \cos(6\theta) = 0$

28. $\sin(4\theta) - \sin(6\theta) = 0$

29. $1 + \sin\theta = 2\cos^2\theta$

30. $\sin^2\theta = 2\cos\theta + 2$

31. $2\sin^2\theta - 5\sin\theta + 3 = 0$

32. $2\cos^2\theta - 7\cos\theta - 4 = 0$

33. $3(1 - \cos\theta) = \sin^2\theta$

34. $4(1 + \sin\theta) = \cos^2\theta$

35. $\tan^2\theta = \dfrac{3}{2}\sec\theta$

36. $\csc^2\theta = \cot\theta + 1$

37. $3 - \sin\theta = \cos(2\theta)$

38. $\cos(2\theta) + 5\cos\theta + 3 = 0$

39. $\sec^2\theta + \tan\theta = 0$

40. $\sec\theta = \tan\theta + \cot\theta$

41. $\sin\theta - \sqrt{3}\cos\theta = 1$

42. $\sqrt{3}\sin\theta + \cos\theta = 1$

43. $\tan(2\theta) + 2\sin\theta = 0$

44. $\tan(2\theta) + 2\cos\theta = 0$

45. $\sin\theta + \cos\theta = \sqrt{2}$

46. $\sin\theta + \cos\theta = -\sqrt{2}$

In Problems 47–52, find the real zeros of each trigonometric function on the interval $0 \le x < 2\pi$.

47. $f(x) = 4\cos^2 x - 1$

48. $f(x) = 4\sin^2 x - 3$

49. $f(x) = \sin(2x) - \sin x$

50. $f(x) = \cos(2x) + \cos x$

51. $f(x) = \sin^2 x + 2\cos x + 2$

52. $f(x) = \cos(2x) + \sin^2 x$

In Problems 53–64, use a graphing utility to solve each equation. Express the solution(s) rounded to two decimal places.

53. $x + 5\cos x = 0$

54. $x - 4\sin x = 0$

55. $22x - 17\sin x = 3$

56. $19x + 8\cos x = 2$

57. $\sin x + \cos x = x$

58. $\sin x - \cos x = x$

59. $x^2 - 2\cos x = 0$

60. $x^2 + 3\sin x = 0$

61. $x^2 - 2\sin(2x) = 3x$

62. $x^2 = x + 3\cos(2x)$

63. $6\sin x - e^x = 2, \quad x > 0$

64. $4\cos(3x) - e^x = 1, \quad x > 0$

Applications and Extensions

65. Constructing a Rain Gutter A rain gutter is to be constructed of aluminum sheets 12 inches wide. After marking off a length of 4 inches from each edge, this length is bent up at an angle θ. See the illustration. The area A of the opening as a function of θ is given by

$$A(\theta) = 16\sin\theta(\cos\theta + 1) \quad 0° < \theta < 90°$$

|← 4 in →|← 4 in →|← 4 in →|
|← 12 in →|

(a) In calculus, you will be asked to find the angle θ that maximizes A by solving the equation

$$\cos(2\theta) + \cos\theta = 0, \quad 0° < \theta < 90°$$

Solve this equation for θ by using the Double-angle Formula.

(b) Solve the equation in part (a) for θ by writing the sum of the two cosines as a product.

(c) What is the maximum area A of the opening?

(d) Graph $A = A(\theta), 0° \le \theta \le 90°$, and find the angle θ that maximizes the area A. Also find the maximum area. Compare the results to the answers found earlier.

66. Projectile Motion An object is propelled upward at an angle θ, $45° < \theta < 90°$, to the horizontal with an initial velocity of v_0 feet per second from the base of an inclined plane that makes an angle of $45°$ with the horizontal. See the illustration. If air resistance is ignored, the distance R that the object travels up the inclined plane is given by

$$R(\theta) = \frac{v_0^2\sqrt{2}}{32}[\sin(2\theta) - \cos(2\theta) - 1]$$

(a) In calculus, you will be asked to find the angle θ that maximizes R by solving the equation

$$\sin(2\theta) + \cos(2\theta) = 0$$

Solve this equation for θ using the method of Example 7.

(b) Solve this equation for θ by dividing each side by $\cos(2\theta)$.

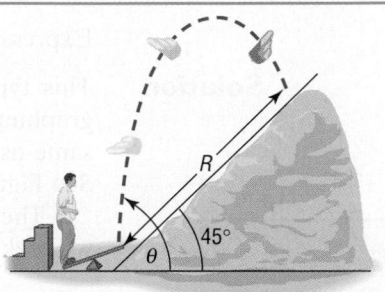

(c) What is the maximum distance R if $v_0 = 32$ feet per second?

(d) Graph $R = R(\theta), 45° \le \theta \le 90°$, and find the angle θ that maximizes the distance R. Also find the maximum distance. Use $v_0 = 32$ feet per second. Compare the results with the answers found earlier.

67. Heat Transfer In the study of heat transfer, the equation $x + \tan x = 0$ occurs. Graph $Y_1 = -x$ and $Y_2 = \tan x$ for $x \ge 0$. Conclude that there are an infinite number of points of intersection of these two graphs. Now find the first two positive solutions of $x + \tan x = 0$ rounded to two decimal places.

68. Carrying a Ladder Around a Corner Two hallways, one of width 3 feet, the other of width 4 feet, meet at a right angle. See the illustration.

(a) Show that the length L as a function of θ is

$$L(\theta) = 4\csc\theta + 3\sec\theta$$

⚠ (b) In calculus, you will be asked to find the length of the longest ladder that can turn the corner by solving the equation

$$3 \sec\theta \tan\theta - 4\csc\theta \cot\theta = 0, \quad 0° < \theta < 90°$$

Solve this equation for θ.

(c) What is the length of the longest ladder that can be carried around the corner?

(d) Graph $L = L(\theta), 0° \leq \theta \leq 90°$, and find the angle θ that minimizes the length L.

(e) Compare the result with the one found in part (b). Explain why the two answers are the same.

69. Projectile Motion The horizontal distance that a projectile will travel in the air (ignoring air resistance) is given by the equation

$$R(\theta) = \frac{v_0^2 \sin(2\theta)}{g}$$

where v_0 is the initial velocity of the projectile, θ is the angle of elevation, and g is acceleration due to gravity (9.8 meters per second squared).

(a) If you can throw a baseball with an initial speed of 34.8 meters per second, at what angle of elevation θ should you direct the throw so that the ball travels a distance of 107 meters before striking the ground?

(b) Determine the maximum distance that you can throw the ball.

(c) Graph $R = R(\theta)$, with $v_0 = 34.8$ meters per second.

(d) Verify the results obtained in parts (a) and (b) using a graphing utility.

70. Projectile Motion Refer to Problem 69.

(a) If you can throw a baseball with an initial speed of 40 meters per second, at what angle of elevation θ should you direct the throw so that the ball travels a distance of 110 meters before striking the ground?

(b) Determine the maximum distance that you can throw the ball.

(c) Graph $R = R(\theta)$, with $v_0 = 40$ meters per second.

(d) Verify the results obtained in parts (a) and (b) using a graphing utility.

'Are You Prepared?' Answers

1. $\left\{-1, \dfrac{5}{4}\right\}$ **2.** $\left\{\dfrac{1 - \sqrt{5}}{2}, \dfrac{1 + \sqrt{5}}{2}\right\}$ **3.** $\left\{0, \dfrac{5}{2}\right\}$ **4.** $\{0.76\}$

CHAPTER REVIEW

Things to Know

Definitions of the six inverse trigonometric functions

$y = \sin^{-1} x$ means $x = \sin y$ where $-1 \leq x \leq 1, \quad -\dfrac{\pi}{2} \leq y \leq \dfrac{\pi}{2}$ (p. 603)

$y = \cos^{-1} x$ means $x = \cos y$ where $-1 \leq x \leq 1, \quad 0 \leq y \leq \pi$ (p. 606)

$y = \tan^{-1} x$ means $x = \tan y$ where $-\infty < x < \infty, \quad -\dfrac{\pi}{2} < y < \dfrac{\pi}{2}$ (p. 609)

$y = \sec^{-1} x$ means $x = \sec y$ where $|x| \geq 1, \quad 0 \leq y \leq \pi, \quad y \neq \dfrac{\pi}{2}$ (p. 616)

$y = \csc^{-1} x$ means $x = \csc y$ where $|x| \geq 1, \quad -\dfrac{\pi}{2} \leq y \leq \dfrac{\pi}{2}, \quad y \neq 0$ (p. 616)

$y = \cot^{-1} x$ means $x = \cot y$ where $-\infty < x < \infty, \quad 0 < y < \pi$ (p. 616)

Sum and Difference Formulas (pp. 627, 630, and 632)

$\cos(\alpha + \beta) = \cos\alpha \cos\beta - \sin\alpha \sin\beta$ $\qquad$ $\cos(\alpha - \beta) = \cos\alpha \cos\beta + \sin\alpha \sin\beta$

$\sin(\alpha + \beta) = \sin\alpha \cos\beta + \cos\alpha \sin\beta$ $\qquad$ $\sin(\alpha - \beta) = \sin\alpha \cos\beta - \cos\alpha \sin\beta$

$\tan(\alpha + \beta) = \dfrac{\tan\alpha + \tan\beta}{1 - \tan\alpha \tan\beta}$ $\qquad$ $\tan(\alpha - \beta) = \dfrac{\tan\alpha - \tan\beta}{1 + \tan\alpha \tan\beta}$

Double-angle Formulas (pp. 637 and 638)

$\sin(2\theta) = 2\sin\theta \cos\theta$ $\qquad$ $\cos(2\theta) = \cos^2\theta - \sin^2\theta$ $\qquad$ $\tan(2\theta) = \dfrac{2\tan\theta}{1 - \tan^2\theta}$

$\cos(2\theta) = 2\cos^2\theta - 1$ $\qquad$ $\cos(2\theta) = 1 - 2\sin^2\theta$

Half-angle Formulas (pp. 640, 641 and 643)

$\sin^2\dfrac{\alpha}{2} = \dfrac{1 - \cos\alpha}{2}$ $\qquad$ $\cos^2\dfrac{\alpha}{2} = \dfrac{1 + \cos\alpha}{2}$ $\qquad$ $\tan^2\dfrac{\alpha}{2} = \dfrac{1 - \cos\alpha}{1 + \cos\alpha}$

$\sin\dfrac{\alpha}{2} = \pm\sqrt{\dfrac{1 - \cos\alpha}{2}}$ $\qquad$ $\cos\dfrac{\alpha}{2} = \pm\sqrt{\dfrac{1 + \cos\alpha}{2}}$ $\qquad$ $\tan\dfrac{\alpha}{2} = \pm\sqrt{\dfrac{1 - \cos\alpha}{1 + \cos\alpha}} = \dfrac{1 - \cos\alpha}{\sin\alpha} = \dfrac{\sin\alpha}{1 + \cos\alpha}$

where the $+$ or $-$ is determined by the quadrant of $\dfrac{\alpha}{2}$.

Product-to-Sum Formulas (p. 646)

$$\sin \alpha \sin \beta = \frac{1}{2}[\cos(\alpha - \beta) - \cos(\alpha + \beta)]$$

$$\cos \alpha \cos \beta = \frac{1}{2}[\cos(\alpha - \beta) + \cos(\alpha + \beta)]$$

$$\sin \alpha \cos \beta = \frac{1}{2}[\sin(\alpha + \beta) + \sin(\alpha - \beta)]$$

Sum-to-Product Formulas (p. 647)

$$\sin \alpha + \sin \beta = 2 \sin \frac{\alpha + \beta}{2} \cos \frac{\alpha - \beta}{2} \qquad \sin \alpha - \sin \beta = 2 \sin \frac{\alpha - \beta}{2} \cos \frac{\alpha + \beta}{2}$$

$$\cos \alpha + \cos \beta = 2 \cos \frac{\alpha + \beta}{2} \cos \frac{\alpha - \beta}{2} \qquad \cos \alpha - \cos \beta = -2 \sin \frac{\alpha + \beta}{2} \sin \frac{\alpha - \beta}{2}$$

Objectives

Section	You should be able to . . .	Review Exercises
8.1	**1** Find the exact value of inverse sine, cosine, and tangent functions (p. 603)	1–6
	2 Find an approximate value of inverse sine, cosine, and tangent functions (p. 604)	121–124
	3 Use properties of inverse functions to find exact values of certain composite functions (p. 605)	9–20
	4 Find the inverse function of a trigonometric function (p. 610)	33–36
	5 Solve equations involving inverse trigonometric functions (p. 611)	133, 134
8.2	**1** Find the exact value of expressions involving the inverse sine, cosine, and tangent functions (p. 615)	21–32
	2 Know the definition of the inverse secant, cosecant, and cotangent functions (p. 616)	7, 8, 29, 30
	3 Use a calculator to evaluate $\sec^{-1} x$, $\csc^{-1} x$, and $\cot^{-1} x$ (p. 616)	125, 126
	4 Write a trigonometric expression as an algebraic expression (p. 617)	37–40
8.3	**1** Use algebra to simplify trigonometric expressions (p. 621)	41–72
	2 Establish identities (p. 622)	41–58
8.4	**1** Use sum and difference formulas to find exact values (p. 628)	73–80, 81–90, (a)–(d), 135
	2 Use sum and difference formulas to establish identities (p. 629)	59–62
	3 Use sum and difference formulas involving inverse trigonometric functions (p. 633)	91–94
8.5	**1** Use double-angle formulas to find exact values (p. 637)	81–90(e), (f), 95, 96
	2 Use double-angle formulas to establish identities (p. 638)	63–67
	3 Use half-angle formulas to find exact values (p. 640)	81–90(g), (h), 135
8.6	**1** Express products as sums (p. 646)	68
	2 Express sums as products (p. 647)	69–72
8.7	**1** Solve equations involving a single trigonometric function (p. 650)	97–106
8.8	**1** Solve trigonometric equations quadratic in form (p. 656)	113, 114
	2 Solve trigonometric equations using identities (p. 657)	97–112, 115–118
	3 Solve trigonometric equations linear in sine and cosine (p. 659)	119, 120
	4 Solve trigonometric equations using a graphing utility (p. 661)	127–132

Review Exercises

In Problems 1–8, find the exact value of each expression. Do not use a calculator.

1. $\sin^{-1} 1$ **2.** $\cos^{-1} 0$ **3.** $\tan^{-1} 1$ **4.** $\sin^{-1}\left(-\frac{1}{2}\right)$

5. $\cos^{-1}\left(-\frac{\sqrt{3}}{2}\right)$ **6.** $\tan^{-1}\left(-\sqrt{3}\right)$ **7.** $\sec^{-1}\sqrt{2}$ **8.** $\cot^{-1}(-1)$

In Problems 9–32, find the exact value, if any, of each composite function. If there is no value, say it is "not defined." Do not use a calculator.

9. $\sin^{-1}\left(\sin\dfrac{3\pi}{8}\right)$

10. $\cos^{-1}\left(\cos\dfrac{3\pi}{4}\right)$

11. $\tan^{-1}\left(\tan\dfrac{2\pi}{3}\right)$

12. $\sin^{-1}\left[\sin\left(-\dfrac{\pi}{8}\right)\right]$

13. $\cos^{-1}\left(\cos\dfrac{15\pi}{7}\right)$

14. $\sin^{-1}\left[\sin\left(-\dfrac{8\pi}{9}\right)\right]$

15. $\sin(\sin^{-1}0.9)$

16. $\cos(\cos^{-1}0.6)$

17. $\cos[\cos^{-1}(-0.3)]$

18. $\tan[\tan^{-1}5]$

19. $\cos[\cos^{-1}(-1.6)]$

20. $\sin(\sin^{-1}1.6)$

21. $\sin^{-1}\left(\cos\dfrac{2\pi}{3}\right)$

22. $\cos^{-1}\left(\tan\dfrac{3\pi}{4}\right)$

23. $\tan^{-1}\left(\tan\dfrac{7\pi}{4}\right)$

24. $\cos^{-1}\left(\cos\dfrac{7\pi}{6}\right)$

25. $\tan\left[\sin^{-1}\left(-\dfrac{\sqrt{3}}{2}\right)\right]$

26. $\tan\left[\cos^{-1}\left(-\dfrac{1}{2}\right)\right]$

27. $\sec\left(\tan^{-1}\dfrac{\sqrt{3}}{3}\right)$

28. $\csc\left(\sin^{-1}\dfrac{\sqrt{3}}{2}\right)$

29. $\sin\left(\cot^{-1}\dfrac{3}{4}\right)$

30. $\cos\left(\csc^{-1}\dfrac{5}{3}\right)$

31. $\tan\left[\sin^{-1}\left(-\dfrac{4}{5}\right)\right]$

32. $\tan\left[\cos^{-1}\left(-\dfrac{3}{5}\right)\right]$

In Problems 33–36, find the inverse function f^{-1} of each function f. State the domain of f^{-1} and f.

33. $f(x) = 2\sin(3x)$

34. $f(x) = \tan(2x + 3) - 1$

35. $f(x) = -\cos x + 3$

36. $f(x) = 2\sin(-x + 1)$

In Problems 37–40, write each trigonometric expression as an algebraic expression in u.

37. $\cos(\sin^{-1}u)$

38. $\cos(\csc^{-1}u)$

39. $\sin(\csc^{-1}u)$

40. $\tan(\csc^{-1}u)$

In Problems 41–72, establish each identity.

41. $\tan\theta\cot\theta - \sin^2\theta = \cos^2\theta$

42. $\sin\theta\csc\theta - \sin^2\theta = \cos^2\theta$

43. $\sin^2\theta(1 + \cot^2\theta) = 1$

44. $(1 - \sin^2\theta)(1 + \tan^2\theta) = 1$

45. $5\cos^2\theta + 3\sin^2\theta = 3 + 2\cos^2\theta$

46. $4\sin^2\theta + 2\cos^2\theta = 4 - 2\cos^2\theta$

47. $\dfrac{1 - \cos\theta}{\sin\theta} + \dfrac{\sin\theta}{1 - \cos\theta} = 2\csc\theta$

48. $\dfrac{\sin\theta}{1 + \cos\theta} + \dfrac{1 + \cos\theta}{\sin\theta} = 2\csc\theta$

49. $\dfrac{\cos\theta}{\cos\theta - \sin\theta} = \dfrac{1}{1 - \tan\theta}$

50. $1 - \dfrac{\sin^2\theta}{1 + \cos\theta} = \cos\theta$

51. $\dfrac{\csc\theta}{1 + \csc\theta} = \dfrac{1 - \sin\theta}{\cos^2\theta}$

52. $\dfrac{1 + \sec\theta}{\sec\theta} = \dfrac{\sin^2\theta}{1 - \cos\theta}$

53. $\csc\theta - \sin\theta = \cos\theta\cot\theta$

54. $\dfrac{\csc\theta}{1 - \cos\theta} = \dfrac{1 + \cos\theta}{\sin^3\theta}$

55. $\dfrac{1 - \sin\theta}{\sec\theta} = \dfrac{\cos^3\theta}{1 + \sin\theta}$

56. $\dfrac{1 - \cos\theta}{1 + \cos\theta} = (\csc\theta - \cot\theta)^2$

57. $\dfrac{1 - 2\sin^2\theta}{\sin\theta\cos\theta} = \cot\theta - \tan\theta$

58. $\dfrac{(2\sin^2\theta - 1)^2}{\sin^4\theta - \cos^4\theta} = 1 - 2\cos^2\theta$

59. $\dfrac{\cos(\alpha + \beta)}{\cos\alpha\sin\beta} = \cot\beta - \tan\alpha$

60. $\dfrac{\sin(\alpha - \beta)}{\sin\alpha\cos\beta} = 1 - \cot\alpha\tan\beta$

61. $\dfrac{\cos(\alpha - \beta)}{\cos\alpha\cos\beta} = 1 + \tan\alpha\tan\beta$

62. $\dfrac{\cos(\alpha + \beta)}{\sin\alpha\cos\beta} = \cot\alpha - \tan\beta$

63. $(1 + \cos\theta)\tan\dfrac{\theta}{2} = \sin\theta$

64. $\sin\theta\tan\dfrac{\theta}{2} = 1 - \cos\theta$

65. $2\cot\theta\cot(2\theta) = \cot^2\theta - 1$

66. $2\sin(2\theta)(1 - 2\sin^2\theta) = \sin(4\theta)$

67. $1 - 8\sin^2\theta\cos^2\theta = \cos(4\theta)$

68. $\dfrac{\sin(3\theta)\cos\theta - \sin\theta\cos(3\theta)}{\sin(2\theta)} = 1$

69. $\dfrac{\sin(2\theta) + \sin(4\theta)}{\cos(2\theta) + \cos(4\theta)} = \tan(3\theta)$

70. $\dfrac{\sin(2\theta) + \sin(4\theta)}{\sin(2\theta) - \sin(4\theta)} + \dfrac{\tan(3\theta)}{\tan\theta} = 0$

71. $\dfrac{\cos(2\theta) - \cos(4\theta)}{\cos(2\theta) + \cos(4\theta)} - \tan\theta\tan(3\theta) = 0$

72. $\cos(2\theta) - \cos(10\theta) = \tan(4\theta)[\sin(2\theta) + \sin(10\theta)]$

In Problems 73–80, find the exact value of each expression.

73. $\sin 165°$

74. $\tan 105°$

75. $\cos\dfrac{5\pi}{12}$

76. $\sin\left(-\dfrac{\pi}{12}\right)$

77. $\cos 80°\cos 20° + \sin 80°\sin 20°$

78. $\sin 70°\cos 40° - \cos 70°\sin 40°$

79. $\tan\dfrac{\pi}{8}$

80. $\sin\dfrac{5\pi}{8}$

In Problems 81–90, use the information given about the angles α and β to find the exact value of:

(a) $\sin(\alpha + \beta)$

(b) $\cos(\alpha + \beta)$

(c) $\sin(\alpha - \beta)$

(d) $\tan(\alpha + \beta)$

(e) $\sin(2\alpha)$

(f) $\cos(2\beta)$

(g) $\sin\dfrac{\beta}{2}$

(h) $\cos\dfrac{\alpha}{2}$

81. $\sin\alpha = \dfrac{4}{5}, 0 < \alpha < \dfrac{\pi}{2}; \sin\beta = \dfrac{5}{13}, \dfrac{\pi}{2} < \beta < \pi$

82. $\cos\alpha = \dfrac{4}{5}, 0 < \alpha < \dfrac{\pi}{2}; \cos\beta = \dfrac{5}{13}, -\dfrac{\pi}{2} < \beta < 0$

83. $\sin\alpha = -\dfrac{3}{5}, \pi < \alpha < \dfrac{3\pi}{2}; \cos\beta = \dfrac{12}{13}, \dfrac{3\pi}{2} < \beta < 2\pi$

84. $\sin\alpha = -\dfrac{4}{5}, -\dfrac{\pi}{2} < \alpha < 0; \cos\beta = -\dfrac{5}{13}, \dfrac{\pi}{2} < \beta < \pi$

85. $\tan \alpha = \frac{3}{4}, \pi < \alpha < \frac{3\pi}{2}; \tan \beta = \frac{12}{5}, 0 < \beta < \frac{\pi}{2}$

86. $\tan \alpha = -\frac{4}{3}, \frac{\pi}{2} < \alpha < \pi; \cot \beta = \frac{12}{5}, \pi < \beta < \frac{3\pi}{2}$

87. $\sec \alpha = 2, -\frac{\pi}{2} < \alpha < 0; \sec \beta = 3, \frac{3\pi}{2} < \beta < 2\pi$

88. $\csc \alpha = 2, \frac{\pi}{2} < \alpha < \pi; \sec \beta = -3, \frac{\pi}{2} < \beta < \pi$

89. $\sin \alpha = -\frac{2}{3}, \pi < \alpha < \frac{3\pi}{2}; \cos \beta = -\frac{2}{3}, \pi < \beta < \frac{3\pi}{2}$

90. $\tan \alpha = -2, \frac{\pi}{2} < \alpha < \pi; \cot \beta = -2, \frac{\pi}{2} < \beta < \pi$

In Problems 91–96, find the exact value of each expression.

91. $\cos\left(\sin^{-1}\frac{3}{5} - \cos^{-1}\frac{1}{2}\right)$

92. $\sin\left(\cos^{-1}\frac{5}{13} - \cos^{-1}\frac{4}{5}\right)$

93. $\tan\left[\sin^{-1}\left(-\frac{1}{2}\right) - \tan^{-1}\frac{3}{4}\right]$

94. $\cos\left[\tan^{-1}(-1) + \cos^{-1}\left(-\frac{4}{5}\right)\right]$

95. $\sin\left[2\cos^{-1}\left(-\frac{3}{5}\right)\right]$

96. $\cos\left(2\tan^{-1}\frac{4}{3}\right)$

In Problems 97–120, solve each equation on the interval $0 \le \theta < 2\pi$.

97. $\cos \theta = \frac{1}{2}$

98. $\sin \theta = -\frac{\sqrt{3}}{2}$

99. $2\cos \theta + \sqrt{2} = 0$

100. $\tan \theta + \sqrt{3} = 0$

101. $\sin(2\theta) + 1 = 0$

102. $\cos(2\theta) = 0$

103. $\tan(2\theta) = 0$

104. $\sin(3\theta) = 1$

105. $\sec^2 \theta = 4$

106. $\csc^2 \theta = 1$

107. $\sin \theta = \tan \theta$

108. $\cos \theta = \sec \theta$

109. $\sin \theta + \sin(2\theta) = 0$

110. $\cos(2\theta) = \sin \theta$

111. $\sin(2\theta) - \cos \theta - 2\sin \theta + 1 = 0$

112. $\sin(2\theta) - \sin \theta - 2\cos \theta + 1 = 0$

113. $2\sin^2 \theta - 3\sin \theta + 1 = 0$

114. $2\cos^2 \theta + \cos \theta - 1 = 0$

115. $4\sin^2 \theta = 1 + 4\cos \theta$

116. $8 - 12\sin^2 \theta = 4\cos^2 \theta$

117. $\sin(2\theta) = \sqrt{2}\cos \theta$

118. $1 + \sqrt{3}\cos \theta + \cos(2\theta) = 0$

119. $\sin \theta - \cos \theta = 1$

120. $\sin \theta - \sqrt{3}\cos \theta = 2$

In Problems 121–126, use a calculator to find an approximate value for each expression, rounded to two decimal places.

121. $\sin^{-1} 0.7$

122. $\cos^{-1}\frac{4}{5}$

123. $\tan^{-1}(-2)$

124. $\cos^{-1}(-0.2)$

125. $\sec^{-1} 3$

126. $\cot^{-1}(-4)$

 In Problems 127–132, use a graphing utility to solve each equation on the interval $0 \le x \le 2\pi$. Approximate any solutions rounded to two decimal places.

127. $2x = 5\cos x$

128. $2x = 5\sin x$

129. $2\sin x + 3\cos x = 4x$

130. $3\cos x + x = \sin x$

131. $\sin x = \ln x$

132. $\sin x = e^{-x}$

In Problems 133 and 134, find the exact solution of each equation.

133. $-3\sin^{-1} x = \pi$

134. $2\cos^{-1} x + \pi = 4\cos^{-1} x$

135. Use a half-angle formula to find the exact value of $\sin 15°$. Then use a difference formula to find the exact value of $\sin 15°$. Show that the answers found are the same.

136. If you are given the value of $\cos \theta$ and want the exact value of $\cos(2\theta)$, what form of the double-angle formula for $\cos(2\theta)$ is most efficient to use?

CHAPTER TEST

In Problems 1–6, find the exact value of each expression. Express angles in radians.

1. $\sec^{-1}\left(\frac{2}{\sqrt{3}}\right)$

2. $\sin^{-1}\left(-\frac{\sqrt{2}}{2}\right)$

3. $\sin^{-1}\left(\sin\frac{11\pi}{5}\right)$

4. $\tan\left(\tan^{-1}\frac{7}{3}\right)$

5. $\cot\left(\csc^{-1}\sqrt{10}\right)$

6. $\sec\left(\cos^{-1}\left(-\frac{3}{4}\right)\right)$

In Problems 7–10, use a calculator to evaluate each expression. Express angles in radians.

7. $\sin^{-1} 0.382$

8. $\sec^{-1} 1.4$

9. $\tan^{-1} 3$

10. $\cot^{-1} 5$

In Problems 11–16 establish each identity.

11. $\dfrac{\csc \theta + \cot \theta}{\sec \theta + \tan \theta} = \dfrac{\sec \theta - \tan \theta}{\csc \theta - \cot \theta}$

12. $\sin \theta \tan \theta + \cos \theta = \sec \theta$

13. $\tan \theta + \cot \theta = 2\csc(2\theta)$

14. $\dfrac{\sin(\alpha + \beta)}{\tan \alpha + \tan \beta} = \cos \alpha \cos \beta$

15. $\sin(3\theta) = 3\sin \theta - 4\sin^3 \theta$

16. $\dfrac{\tan \theta - \cot \theta}{\tan \theta + \cot \theta} = 1 - 2\cos^2 \theta$

In Problems 17–24 use sum, difference, product or half-angle formulas to find the exact value of each expression.

17. $\cos 15°$

18. $\tan 75°$

19. $\sin\left(\dfrac{1}{2}\cos^{-1}\dfrac{3}{5}\right)$

20. $\tan\left(2\sin^{-1}\dfrac{6}{11}\right)$

21. $\cos\left(\sin^{-1}\dfrac{2}{3} + \tan^{-1}\dfrac{3}{2}\right)$

22. $\sin 75° \cos 15°$

23. $\sin 75° + \sin 15°$

24. $\cos 65° \cos 20° + \sin 65° \sin 20°$

In Problems 25–29, solve each equation on $0 \le \theta < 2\pi$.

25. $4\sin^2\theta - 3 = 0$

26. $-3\cos\left(\dfrac{\pi}{2} - \theta\right) = \tan\theta$

27. $\cos^2\theta + 2\sin\theta\cos\theta - \sin^2\theta = 0$

28. $\sin(\theta + 1) = \cos\theta$

29. $4\sin^2\theta + 7\sin\theta = 2$

30. Stage 16 of the 2004 Tour de France was a time trial from Bourg d'Oisans to L'Alpe d'Huez. The average grade (slope as a percent) for most of the 15-kilometer mountainous trek was 7.9%. What was the change in elevation from the beginning to the end of the route?

CUMULATIVE REVIEW

1. Find the real solutions, if any, of the equation
$3x^2 + x - 1 = 0$.

2. Find an equation for the line containing the points $(-2, 5)$ and $(4, -1)$. What is the distance between these points? What is their midpoint?

3. Test the equation $3x + y^2 = 9$ for symmetry with respect to the x-axis, y-axis, and origin. List the intercepts.

4. Use transformations to graph the equation $y = |x - 3| + 2$.

5. Use transformations to graph the equation $y = 3e^x - 2$.

6. Use transformations to graph the equation
$y = \cos\left(x - \dfrac{\pi}{2}\right) - 1$.

7. Sketch a graph of each of the following functions. Label at least three points on each graph. Name the inverse function of each and show its graph.
(a) $y = x^3$
(b) $y = e^x$
(c) $y = \sin x$, $\quad -\dfrac{\pi}{2} \le x \le \dfrac{\pi}{2}$
(d) $y = \cos x$, $\quad 0 \le x \le \pi$

8. If $\sin\theta = -\dfrac{1}{3}$ and $\pi < \theta < \dfrac{3\pi}{2}$, find the exact value of:
(a) $\cos\theta$
(b) $\tan\theta$
(c) $\sin(2\theta)$
(d) $\cos(2\theta)$
(e) $\sin\left(\dfrac{1}{2}\theta\right)$
(f) $\cos\left(\dfrac{1}{2}\theta\right)$

9. Find the exact value of $\cos(\tan^{-1} 2)$.

10. If $\sin\alpha = \dfrac{1}{3}, \dfrac{\pi}{2} < \alpha < \pi$, and $\cos\beta = -\dfrac{1}{3}, \pi < \beta < \dfrac{3\pi}{2}$, find the exact value of:
(a) $\cos\alpha$
(b) $\sin\beta$
(c) $\cos(2\alpha)$
(d) $\cos(\alpha + \beta)$
(e) $\sin\dfrac{\beta}{2}$

11. For the function
$f(x) = 2x^5 - x^4 - 4x^3 + 2x^2 + 2x - 1$:
(a) Find the real zeros and their multiplicity.
(b) Find the intercepts.
(c) Find the power function that the graph of f resembles for large $|x|$.
(d) Graph f using a graphing utility.
(e) Approximate the turning points, if any exist.
(f) Use the information obtained in parts (a)–(e) to sketch a graph of f by hand.
(g) Identify the intervals on which f is increasing, decreasing, or constant.

12. If $f(x) = 2x^2 + 3x + 1$ and $g(x) = x^2 + 3x + 2$, solve:
(a) $f(x) = 0$
(b) $f(x) = g(x)$
(c) $f(x) > 0$
(d) $f(x) \ge g(x)$

CHAPTER PROJECTS

I. **Waves** A stretched string that is attached at both ends, pulled in a direction perpendicular to the string, and released has motion that is described as wave motion. If we assume no friction and a length such that there are no "echoes" (that is, the wave doesn't bounce back), the transverse motion (motion perpendicular to the string) can be described by the equation

$$y = y_m \sin(kx - \omega t)$$

where y_m is the amplitude measured in meters and k and ω are constants. The height of the sound wave depends on the distance x from one endpoint of the string and on the time t,

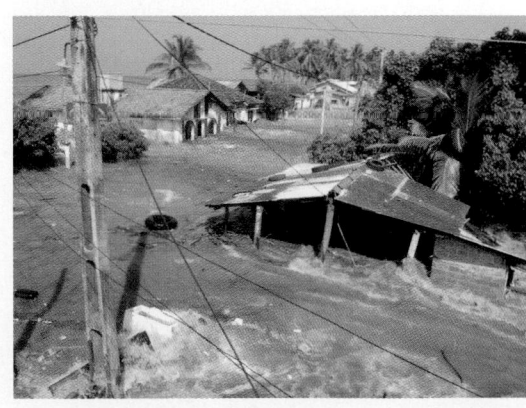

so a typical wave has horizontal and vertical motion over time.

(a) What is the amplitude of the wave
$$y = 0.00421 \sin(68.3x - 2.68t)?$$

(b) The value of ω is the angular frequency measured in radians per second. What is the angular frequency of the wave given in part (a)?

(c) The frequency f is the number of vibrations per second (hertz) made by the wave as it passes a certain point. Its value is found using the formula $f = \dfrac{\omega}{2\pi}$.
What is the frequency of the wave given in part (a)?

(d) The wavelength, λ, of a wave is the shortest distance at which the wave pattern repeats itself for a constant t.
Thus, $\lambda = \dfrac{2\pi}{k}$. What is the wavelength of the wave given in part (a)?

(e) Graph the height of the string a distance $x = 1$ meter from an endpoint.

(f) If two waves travel simultaneously along the same stretched string, the vertical displacement of the string when both waves act is $y = y_1 + y_2$, where y_1 is the vertical displacement of the first wave and y_2 is the vertical

displacement of the second wave. This result is called the Principle of Superposition and was analyzed by the French mathematician Jean Baptiste Fourier (1768–1830). When two waves travel along the same string, one wave will differ from the other wave by a phase constant ϕ. That is,

$$y_1 = y_m \sin(kx - \omega t)$$
$$y_2 = y_m \sin(kx - \omega t + \phi)$$

assuming that each wave has the same amplitude. Write $y_1 + y_2$ as a product using the Sum-to-Product Formulas.

(g) Suppose that two waves are moving in the same direction along a stretched string. The amplitude of each wave is 0.0045 meter, and the phase difference between them is 2.5 radians. The wavelength, λ, of each wave is 0.09 meter and the frequency, f, is 2.3 hertz. Find y_1, y_2, and $y_1 + y_2$.

(h) Using a graphing utility, graph y_1, y_2, and $y_1 + y_2$ on the same viewing window.

(i) Redo parts (g) and (h) with the phase difference between the waves being 0.4 radian.

(j) What effect does the phase difference have on the amplitude of $y_1 + y_2$?

The following projects are available on the Instructor's Resource Center (IRC):

II. Project at Motorola *Sending Pictures Wirelessly* The electronic transmission of pictures is made practical by image compression, mathematical methods which greatly reduce the number of bits of data used to compose the picture.

III. Calculus of Differences Finding consecutive difference quotients is called finding finite differences and is used to analyze the graph of an unknown function.

Applications of Trigonometric Functions

9

From Lewis and Clark to Landsat

For $140, you can buy a handheld Global Positioning System receiver that will gauge your latitude and longitude to within a couple of meters. But in 1804, when Meriwether Lewis and William Clark ventured across the Louisiana Territory, a state of the art positioning system consisted of an octant, a pocket chronometer, and a surveyor's compass.

But somehow, Clark—the cartographer in the group—made do. When San Francisco map collector David Rumsey took his copy of Lewis and Clark's published map of their journey, scanned it into a computer, and matched landmarks such as river junctions against corresponding features on today's maps, he found that it took only a slight amount of digital stretching and twisting to make Clark's map conform

to modern coordinates. In fact, Rumsey was able to combine Clark's depiction of his party's route to the Pacific with pages from government atlases from the 1870s and 1970s and photos from NASA Landsat satellites, creating a digital composite that documents not only a historic adventure, but also the history of mapmaking itself.

Source: Wade Roush, "From Lewis and Clark to Landsat: David Rumsey's Digital Maps Marry Past and Present," *Technology Review,* 108, no. 7 (July 2005): 26–27.

—See the Chapter Project—

A Look Back

In Chapter 7 we defined the six trigonometric functions using right triangles and then extended this definition to include general angles. In particular, we learned to evaluate the trigonometric functions. We also learned how to graph sinusoidal functions.

A Look Ahead

In this chapter, we use the trigonometric functions to solve applied problems. The first four sections deal with applications involving right triangles and *oblique triangles*, triangles that do not have a right angle. To solve problems involving oblique triangles, we will develop the Law of Sines and the Law of Cosines. We will also develop formulas for finding the area of a triangle.

The final section deals with applications of sinusoidal functions involving simple harmonic motion and damped motion.

Outline

9.1 Applications Involving Right Triangles

PREPARING FOR THIS SECTION *Before getting started, review the following:*

- Pythagorean Theorem (Section R.3, pp. 30–31)
- Trigonometric Equations (I) (Section 8.7, pp. 649–653)
- Complementary Angles (Section 7.2, pp. 523–525)

Now Work the 'Are You Prepared?' problems on page 673.

OBJECTIVES 1 Solve Right Triangles (p. 670)
2 Solve Applied Problems (p. 671)

1 Solve Right Triangles

In the discussion that follows, we will always label a right triangle so that side a is opposite angle A, side b is opposite angle B, and side c is the hypotenuse, as shown in Figure 1. **To solve a right triangle** means to find the missing lengths of its sides and the measurements of its angles. We shall follow the practice of expressing the lengths of the sides rounded to two decimal places and expressing angles in degrees rounded to one decimal place. (Be sure that your calculator is in degree mode.)

To solve a right triangle, we need to know one of the acute angles A or B and a side, or else two sides. Then we make use of the Pythagorean Theorem and the fact that the sum of the angles of a triangle is 180°. The sum of the angles A and B in a right triangle is therefore 90°.

Figure 1

THEOREM

For the right triangle shown in Figure 1, we have

$$c^2 = a^2 + b^2 \qquad A + B = 90°$$

EXAMPLE 1 | **Solving a Right Triangle**

Use Figure 2. If $b = 2$ and $A = 40°$, find a, c, and B.

Solution Since $A = 40°$ and $A + B = 90°$, we find that $B = 50°$. To find the sides a and c, we use the facts that

$$\tan 40° = \frac{a}{2} \quad \text{and} \quad \cos 40° = \frac{2}{c}$$

Now solve for a and c.

$$a = 2 \tan 40° \approx 1.68 \quad \text{and} \quad c = \frac{2}{\cos 40°} \approx 2.61$$

Figure 2

══ **Now Work** PROBLEM 9

EXAMPLE 2 | **Solving a Right Triangle**

Use Figure 3. If $a = 3$ and $b = 2$, find c, A, and B.

Solution Since $a = 3$ and $b = 2$, then, by the Pythagorean Theorem, we have

$$c^2 = a^2 + b^2 = 3^2 + 2^2 = 9 + 4 = 13$$
$$c = \sqrt{13} \approx 3.61$$

Figure 3

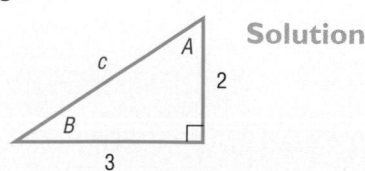

NOTE *To avoid round-off errors when using a calculator, we will store un-rounded values in memory for use in subsequent calculations.* ∎

To find angle A, we use the fact that

$$\tan A = \frac{3}{2} \quad \text{so} \quad A = \tan^{-1}\frac{3}{2}$$

Set the mode on your calculator to degrees. Then, rounded to one decimal place, we find that $A = 56.3°$. Since $A + B = 90°$, we find that $B = 33.7°$. ∎

━━━━ **Now Work** PROBLEM 19

2 Solve Applied Problems

In Section 7.3, we used right triangle trigonometry to find the lengths of unknown sides of a right triangle given the measure of an angle and the length of a side. Now that we understand the concept of inverse trigonometric functions and know how to solve trigonometric equations, we can solve applied problems that require finding the measure of an angle given the lengths of two sides in a right triangle.

| EXAMPLE 3 | Finding the Inclination of a Mountain Trail |

A straight trail leads from the Alpine Hotel, elevation 8000 feet, to a scenic overlook, elevation 11,100 feet. The length of the trail is 14,100 feet. What is the inclination (grade) of the trail? That is, what is the angle B in Figure 4?

Solution As we can see in Figure 4, we know the length of the side opposite angle B and the length of the hypotenuse. The angle B obeys the equation

Figure 4

$$\sin B = \frac{3100}{14,100}$$

Using a calculator,

$$B = \sin^{-1}\frac{3100}{14,100} \approx 12.7°$$

The inclination (grade) of the trail is approximately $12.7°$. ∎

━━━━ **Now Work** PROBLEM 25

| EXAMPLE 4 | The Gibb's Hill Lighthouse, Southampton, Bermuda |

In operation since 1846, the Gibb's Hill Lighthouse stands 117 feet high on a hill 245 feet high, so its beam of light is 362 feet above sea level. A brochure states that the light can be seen on the horizon about 26 miles distant. Verify the accuracy of this statement.

Figure 5

Solution Figure 5 illustrates the situation. The central angle θ, positioned at the center of Earth, radius 3960 miles, obeys the equation

$$\cos \theta = \frac{3960}{3960 + \dfrac{362}{5280}} \approx 0.999982687 \quad \text{1 mile = 5280 feet}$$

Solving for θ, we find

$$\theta \approx 0.33715° \approx 20.23'$$

The brochure does not indicate whether the distance is measured in nautical miles or statute miles. Let's calculate both distances.

The distance s in nautical miles (refer to Problem 114, p. 516) is the measure of the angle θ in minutes, so $s \approx 20.23$ nautical miles.

The distance s in statute miles is given by the formula $s = r\theta$, where θ is measured in radians. Then, since

$$\theta \approx 20.23' \approx 0.33715° \approx 0.00588 \text{ radian}$$

$$\uparrow \qquad\qquad \uparrow$$
$$1' = \frac{1°}{60} \qquad 1° = \frac{\pi}{180} \text{ radian}$$

we find that

$$s = r\theta \approx (3960)(0.00588) \approx 23.3 \text{ miles}$$

In either case, it would seem that the brochure overstated the distance somewhat.

Figure 6

In navigation and surveying, the **direction** or **bearing** from a point O to a point P equals the acute angle θ between the ray OP and the vertical line through O, the north–south line.

Figure 6 illustrates some bearings. Notice that the bearing from O to P_1 is denoted by the symbolism N30°E, indicating that the bearing is 30° east of north. In writing the bearing from O to P, the direction north or south always appears first, followed by an acute angle, followed by east or west. In Figure 6, the bearing from O to P_2 is S50°W, and from O to P_3 it is N70°W.

EXAMPLE 5 Finding the Bearing of an Object

In Figure 6, what is the bearing from O to an object at P_4?

Solution The acute angle between the ray OP_4 and the north–south line through O is given as 20°. The bearing from O to P_4 is S20°E.

EXAMPLE 6 Finding the Bearing of an Airplane

A Boeing 777 aircraft takes off from O'Hare Airport on runway 2 LEFT, which has a bearing of N20°E.* After flying for 1 mile, the pilot of the aircraft requests permission to turn 90° and head toward the northwest. The request is granted. After the plane goes 2 miles in this direction, what bearing should the control tower use to locate the aircraft?

Figure 7

Solution Figure 7 illustrates the situation. After flying 1 mile from the airport O (the control tower), the aircraft is at P. After turning 90° toward the northwest and flying 2 miles, the aircraft is at the point Q. In triangle OPQ, the angle θ obeys the equation

$$\tan\theta = \frac{2}{1} = 2 \quad \text{so} \quad \theta = \tan^{-1}2 \approx 63.4°$$

The acute angle between north and the ray OQ is $63.4° - 20° = 43.4°$. The bearing of the aircraft from O to Q is N43.4°W.

━ **Now Work** PROBLEM 29

* In air navigation, the term **azimuth** denotes the positive angle measured clockwise from the north (N) to a ray OP. In Figure 6, the azimuth from O to P_1 is 30°; the azimuth from O to P_2 is 230°; the azimuth from O to P_3 is 290°. In naming runways, the units digit is left off the azimuth. Runway 2 LEFT means the left runway with a direction of azimuth 20° (bearing N20°E). Runway 23 is the runway with azimuth 230° and bearing S50°W.

9.1 Assess Your Understanding

1. In a right triangle, if the length of the hypotenuse is 5 and the length of one of the other sides is 3, what is the length of the third side? (pp. 30–31)

2. *True or False* $\sin 52° = \cos 48°$. (pp. 523–525)

3. If θ is an acute angle, solve the equation $\tan \theta = \frac{1}{2}$. Express your answer in degrees, rounded to one decimal place. (pp. 649–653)

4. If θ is an acute angle, solve the equation $\sin \theta = \frac{1}{2}$. (pp. 649–653)

Concepts and Vocabulary

5. *True or False* In a right triangle, one of the angles is 90° and the sum of the other two angles is 90°.

6. In navigation or surveying, the ____ from a point O to a point P equals the acute angle θ between ray OP and the vertical line through O, the north–south line.

7. *True or False* In a right triangle, if two sides are known, we can solve the triangle.

8. *True or False* In a right triangle, if we know the two acute angles, we can solve the triangle.

Skill Building

In Problems 9–22, use the right triangle shown below. Then, using the given information, solve the triangle.

9. $b = 5$, $B = 20°$; find a, c, and A

10. $b = 4$, $B = 10°$; find a, c, and A

11. $a = 6$, $B = 40°$; find b, c, and A

12. $a = 7$, $B = 50°$; find b, c, and A

13. $b = 4$, $A = 10°$; find a, c, and B

14. $b = 6$, $A = 20°$; find a, c, and B

15. $a = 5$, $A = 25°$; find b, c, and B

16. $a = 6$, $A = 40°$; find b, c, and B

17. $c = 9$, $B = 20°$; find b, a, and A

18. $c = 10$, $A = 40°$; find b, a, and B

19. $a = 5$, $b = 3$; find c, A, and B

20. $a = 2$, $b = 8$; find c, A, and B

21. $a = 2$, $c = 5$; find b, A, and B

22. $b = 4$, $c = 6$; find a, A, and B

Applications and Extensions

23. **Geometry** The hypotenuse of a right triangle is 5 inches. If one leg is 2 inches, find the degree measure of each angle.

24. **Geometry** The hypotenuse of a right triangle is 3 feet. If one leg is 1 foot, find the degree measure of each angle.

25. **Finding the Angle of Elevation of the Sun** At 10 AM on April 26, 2006, a building 300 feet high casts a shadow 50 feet long. What is the angle of elevation of the Sun?

26. **Directing a Laser Beam** A laser beam is to be directed through a small hole in the center of a circle of radius 10 feet. The origin of the beam is 35 feet from the circle (see the figure). At what angle of elevation should the beam be aimed to ensure that it goes through the hole?

27. **Finding the Speed of a Truck** A state trooper is hidden 30 feet from a highway. One second after a truck passes, the angle θ between the highway and the line of observation from the patrol car to the truck is measured. See the illustration.

(a) If the angle measures 15°, how fast is the truck traveling? Express the answer in feet per second and in miles per hour.

(b) If the angle measures 20°, how fast is the truck traveling? Express the answer in feet per second and in miles per hour.

(c) If the speed limit is 55 miles per hour and a speeding ticket is issued for speeds of 5 miles per hour or more over the limit, for what angles should the trooper issue a ticket?

28. **Security** A security camera in a neighborhood bank is mounted on a wall 9 feet above the floor. What angle of depression should be used if the camera is to be directed to a spot 6 feet above the floor and 12 feet from the wall?

29. **Finding the Bearing of an Aircraft** A DC-9 aircraft leaves Midway Airport from runway 4 RIGHT, whose bearing is N40°E. After flying for $\frac{1}{2}$ mile, the pilot requests permission to turn 90° and head toward the southeast. The permission is granted. After the airplane goes 1 mile in this direction, what bearing should the control tower use to locate the aircraft?

30. **Finding the Bearing of a Ship** A ship leaves the port of Miami with a bearing of S80°E and a speed of 15 knots. After 1 hour, the ship turns 90° toward the south. After 2 hours, maintaining the same speed, what is the bearing to the ship from port?

31. **Niagara Falls Incline Railway** Situated between Portage Road and the Niagara Parkway directly across from the Canadian Horseshoe Falls, the Falls Incline Railway is a funicular that carries passengers up an embankment to Table Rock Observation Point. If the length of the track is 51.8 meters and the angle of inclination is 36°2′, determine the height of the embankment.

Source: www.niagaraparks.com

32. **Sears Tower** The Sears Tower in Chicago is the third tallest building in the world and is topped by a high antenna. A surveyor on the ground makes the following measurement:

1. The angle of elevation from his position to the top of the building is 34°.
2. The distance from his position to the top of the building is 2593 feet.
3. The distance from his position to the top of the antenna is 2743 feet.
 (a) How far away from the (base of the) building is the surveyor located?
 (b) How tall is the building?
 (c) What is the angle of elevation from the surveyor to the top of the antenna?
 (d) How tall is the antenna?

Source: www.infoplease.com/ce6/us/A0844218.html

33. **Chicago Skyscrapers** The angle of inclination from the base of the John Hancock Center to the top of the main structure of the Sears Tower is approximately 10.3°. If the main structure of the Sears Tower is 1451 feet tall, how far apart are the two skyscrapers? Assume the bases of the two buildings are at the same elevation.

Source: www.emporis.com

34. **Estimating the Width of the Mississippi River** A tourist at the top of the Gateway Arch (height, 630 feet) in St. Louis, Missouri observes a boat moored on the Illinois side of the Mississippi River 2070 feet directly across from the Arch. She also observes a boat moored on the Missouri side

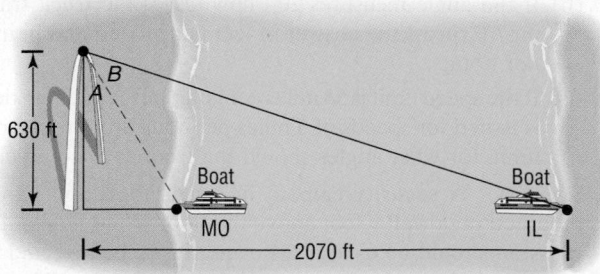

directly across from the first boat (see diagram). Given that $B = \cot^{-1}\frac{67}{55}$, estimate the width of the Mississippi River at the St. Louis riverfront.

Source: U.S. Army Corps of Engineers

35. **Finding the Pitch of a Roof** A carpenter is preparing to put a roof on a garage that is 20 feet by 40 feet by 20 feet. A steel support beam 46 feet in length is positioned in the center of the garage. To support the roof, another beam will be attached to the top of the center beam (see the figure). At what angle of elevation is the new beam? In other words, what is the pitch of the roof?

36. **Shooting Free Throws in Basketball** The eyes of a basketball player are 6 feet above the floor. The player is at the free-throw line, which is 15 feet from the center of the basket rim (see the figure). What is the angle of elevation from the player's eyes to the center of the rim?

[**Hint:** The rim is 10 feet above the floor.]

37. **Geometry** Find the value of the angle θ in degrees rounded to the nearest tenth of a degree.

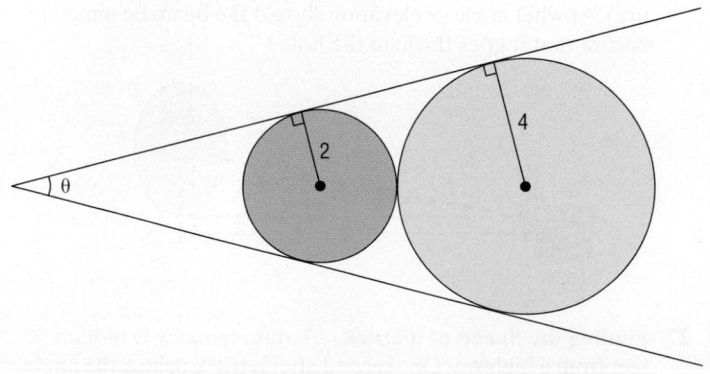

38. **Surveillance Satellites** A surveillance satellite circles Earth at a height of h miles above the surface. Suppose that d is the

distance, in miles, on the surface of Earth that can be observed from the satellite. See the illustration.

(a) Find an equation that relates the central angle θ to the height h.
(b) Find an equation that relates the observable distance d and θ.
(c) Find an equation that relates d and h.
(d) If d is to be 2500 miles, how high must the satellite orbit above Earth?
(e) If the satellite orbits at a height of 300 miles, what distance d on the surface can be observed?

Discussion and Writing

39. The Gibb's Hill Lighthouse, Southampton, Bermuda In operation since 1846, the Gibb's Hill Lighthouse stands 117 feet high on a hill 245 feet high, so its beam of light is 362 feet above sea level. A brochure states that ships 40 miles away can see the light and planes flying at 10,000 feet can see it 120 miles away. Verify the accuracy of these statements. What assumption did the brochure make about the height of the ship?

'Are You Prepared?' Answers

1. 4 **2.** False **3.** 26.6° **4.** 30°

9.2 The Law of Sines

PREPARING FOR THIS SECTION *Before getting started, review the following:*

- Trigonometric Equations (I) (Section 8.7, pp. 649–653)
- Geometry Essentials (Section R.3, pp. 30–35)
- Difference Formula for the Sine Function (Section 8.4, p. 630)

Now Work the 'Are You Prepared?' problems on page 682.

OBJECTIVES **1** Solve SAA or ASA Triangles (p. 676)
2 Solve SSA Triangles (p. 677)
3 Solve Applied Problems (p. 679)

If none of the angles of a triangle is a right angle, the triangle is called **oblique.** An oblique triangle will have either three acute angles or two acute angles and one obtuse angle (an angle between 90° and 180°). See Figure 8.

Figure 8

(a) All angles are acute **(b)** Two acute angles and one obtuse angle

Figure 9

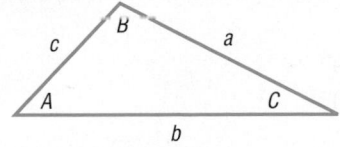

In the discussion that follows, we will always label an oblique triangle so that side a is opposite angle A, side b is opposite angle B, and side c is opposite angle C, as shown in Figure 9.

To **solve an oblique triangle** means to find the lengths of its sides and the measurements of its angles. To do this, we shall need to know the length of one side*

* The reason we need to know the length of one side is that, if we only know the angles, this will result in a family of *similar triangles*.

along with (i) two angles; (ii) one angle and one other side; or (iii) the other two sides. There are four possibilities to consider:

> **CASE 1:** One side and two angles are known (ASA or SAA).
> **CASE 2:** Two sides and the angle opposite one of them are known (SSA).
> **CASE 3:** Two sides and the included angle are known (SAS).
> **CASE 4:** Three sides are known (SSS).

Figure 10 illustrates the four cases.

Figure 10

Case 1: ASA Case 1: SAA Case 2: SSA Case 3: SAS Case 4: SSS

WARNING Oblique triangles cannot be solved using the methods of Section 9.1. Do you know why? ■

The **Law of Sines** is used to solve triangles for which Case 1 or 2 holds. Cases 3 and 4 are considered when we study the Law of Cosines in the next section.

THEOREM **Law of Sines**

For a triangle with sides a, b, c and opposite angles A, B, C, respectively,

$$\frac{\sin A}{a} = \frac{\sin B}{b} = \frac{\sin C}{c} \qquad \textbf{(1)}$$

A proof of the Law of Sines is given at the end of this section. The Law of Sines actually consists of three equalities:

$$\frac{\sin A}{a} = \frac{\sin B}{b} \qquad \frac{\sin A}{a} = \frac{\sin C}{c} \qquad \frac{\sin B}{b} = \frac{\sin C}{c}$$

Formula (1) is merely a compact way to write these three equations.

In applying the Law of Sines to solve triangles, we use the fact that the sum of the angles of any triangle equals 180°; that is,

$$A + B + C = 180° \qquad \textbf{(2)}$$

1 Solve SAA or ASA Triangles

Our first two examples show how to solve a triangle when one side and two angles are known (Case 1: SAA or ASA).

EXAMPLE 1 **Using the Law of Sines to Solve a SAA Triangle**

Solve the triangle: $A = 40°, B = 60°, a = 4$

Solution Figure 11 shows the triangle that we want to solve. The third angle C is found using equation (2).

Figure 11

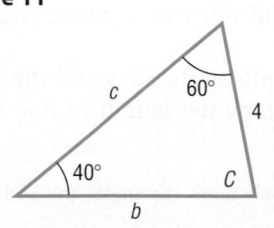

$$A + B + C = 180°$$
$$40° + 60° + C = 180°$$
$$C = 80°$$

Now we use the Law of Sines (twice) to find the unknown sides b and c.

$$\frac{\sin A}{a} = \frac{\sin B}{b} \qquad \frac{\sin A}{a} = \frac{\sin C}{c}$$

NOTE Although not a check, we can verify the reasonableness of our answer by determining if the longest side is opposite the largest angle and the shortest side is opposite the smallest angle. ∎

Because $a = 4$, $A = 40°$, $B = 60°$, and $C = 80°$, we have

$$\frac{\sin 40°}{4} = \frac{\sin 60°}{b} \qquad \frac{\sin 40°}{4} = \frac{\sin 80°}{c}$$

Solving for b and c, we find that

$$b = \frac{4 \sin 60°}{\sin 40°} \approx 5.39 \qquad c = \frac{4 \sin 80°}{\sin 40°} \approx 6.13$$

Notice in Example 1 that we found b and c by working with the given side a. This is better than finding b first and working with a rounded value of b to find c.

═══ **Now Work** PROBLEM **9**

EXAMPLE 2	**Using the Law of Sines to Solve an ASA Triangle**

Solve the triangle: $A = 35°$, $B = 15°$, $c = 5$

Solution

Figure 12 illustrates the triangle that we want to solve. Because we know two angles ($A = 35°$ and $B = 15°$), we find the third angle using equation (2).

Figure 12

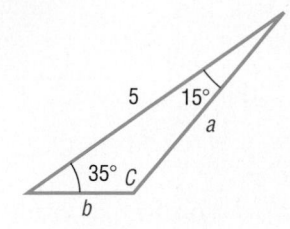

$$A + B + C = 180°$$
$$35° + 15° + C = 180°$$
$$C = 130°$$

Now we know the three angles and one side ($c = 5$) of the triangle. To find the remaining two sides a and b, we use the Law of Sines (twice).

$$\frac{\sin A}{a} = \frac{\sin C}{c} \qquad\qquad \frac{\sin B}{b} = \frac{\sin C}{c}$$

$$\frac{\sin 35°}{a} = \frac{\sin 130°}{5} \qquad\qquad \frac{\sin 15°}{b} = \frac{\sin 130°}{5}$$

$$a = \frac{5 \sin 35°}{\sin 130°} \approx 3.74 \qquad b = \frac{5 \sin 15°}{\sin 130°} \approx 1.69$$

═══ **Now Work** PROBLEM **23**

2 Solve SSA Triangles

Figure 13

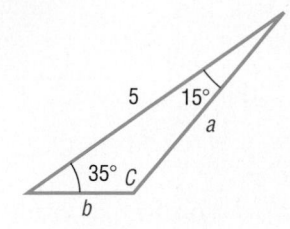

$\sin A = \frac{h}{b}$

Case 2 (SSA), which applies to triangles for which two sides and the angle opposite one of them are known, is referred to as the **ambiguous case,** because the known information may result in one triangle, two triangles, or no triangle at all. Suppose that we are given sides a and b and angle A, as illustrated in Figure 13. The key to determining the possible triangles, if any, that may be formed from the given information lies primarily with the relative size of side a and the height h and the fact that $h = b \sin A$.

No Triangle If $a < h = b \sin A$, then side a is not sufficiently long to form a triangle. See Figure 14.

One Right Triangle If $a = h = b \sin A$, then side a is just long enough to form a right triangle. See Figure 15.

Figure 14
$a < h = b \sin A$

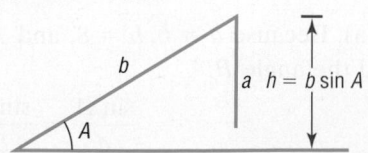

Figure 15
$a = h = b \sin A$

Two Triangles If $h = b \sin A < a$, and $a < b$ two distinct triangles can be formed from the given information. See Figure 16.

One Triangle If $a \geq b$, only one triangle can be formed. See Figure 17.

Figure 16
$b \sin A < a$ and $a < b$

Figure 17
$a \geq b$

Fortunately, we do not have to rely on an illustration or complicated relationships to draw the correct conclusion in the ambiguous case. The Law of Sines will lead us to the correct determination. Let's see how.

EXAMPLE 3 **Using the Law of Sines to Solve a SSA Triangle (One Solution)**

Solve the triangle: $a = 3, b = 2, A = 40°$

Solution See Figure 18(a). Because $a = 3, b = 2$, and $A = 40°$ are known, we use the Law of Sines to find the angle B.

Figure 18(a)

$$\frac{\sin A}{a} = \frac{\sin B}{b}$$

Then

$$\frac{\sin 40°}{3} = \frac{\sin B}{2}$$

$$\sin B = \frac{2 \sin 40°}{3} \approx 0.43$$

NOTE Here we computed B_1 by determining the value of $\sin^{-1}\left(\dfrac{2 \sin 40°}{3}\right)$. If you use the rounded value and evaluate $\sin^{-1}(0.43)$, you will obtain a slightly different result. ∎

There are two angles $B, 0° < B < 180°$, for which $\sin B \approx 0.43$.

$$B_1 \approx 25.4° \quad \text{and} \quad B_2 \approx 180° - 25.4° = 154.6°$$

The second possibility, $B_2 \approx 154.6°$, is ruled out, because $A = 40°$, making $A + B_2 \approx 194.6° > 180°$. Now, using $B_1 \approx 25.4°$, we find that

$$C = 180° - A - B_1 \approx 180° - 40° - 25.4° = 114.6°$$

The third side c may now be determined using the Law of Sines.

$$\frac{\sin A}{a} = \frac{\sin C}{c}$$

$$\frac{\sin 40°}{3} = \frac{\sin 114.6°}{c}$$

Figure 18(b)

$$c = \frac{3 \sin 114.6°}{\sin 40°} \approx 4.24$$

Figure 18(b) illustrates the solved triangle. ∎

EXAMPLE 4 **Using the Law of Sines to Solve a SSA Triangle (Two Solutions)**

Solve the triangle: $a = 6, b = 8, A = 35°$

Solution See Figure 19(a). Because $a = 6, b = 8$, and $A = 35°$ are known, we use the Law of Sines to find the angle B.

$$\frac{\sin A}{a} = \frac{\sin B}{b}$$

Figure 19(a)

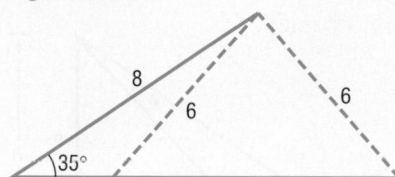

Then

$$\frac{\sin 35°}{6} = \frac{\sin B}{8}$$

$$\sin B = \frac{8 \sin 35°}{6} \approx 0.76$$

$$B_1 \approx 49.9° \quad \text{or} \quad B_2 \approx 180° - 49.9° = 130.1°$$

For both choices of B, we have $A + B < 180°$. There are two triangles, one containing the angle $B_1 \approx 49.9°$ and the other containing the angle $B_2 \approx 130.1°$. The third angle C is either

$$C_1 = 180° - A - B_1 \approx 95.1° \quad \text{or} \quad C_2 = 180° - A - B_2 \approx 14.9°$$
$$\uparrow \qquad\qquad\qquad\qquad\qquad\qquad \uparrow$$
$$\begin{array}{c} A = 35° \\ B_1 = 49.9° \end{array} \qquad\qquad\qquad\qquad \begin{array}{c} A = 35° \\ B_2 = 130.1° \end{array}$$

The third side c obeys the Law of Sines, so we have

Figure 19(b)

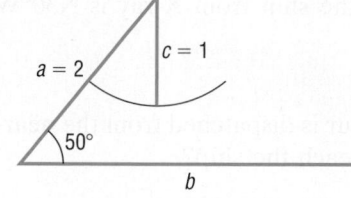

$$\frac{\sin A}{a} = \frac{\sin C_1}{c_1} \qquad\qquad\qquad \frac{\sin A}{a} = \frac{\sin C_2}{c_2}$$

$$\frac{\sin 35°}{6} = \frac{\sin 95.1°}{c_1} \qquad\qquad \frac{\sin 35°}{6} = \frac{\sin 14.9°}{c_2}$$

$$c_1 = \frac{6 \sin 95.1°}{\sin 35°} \approx 10.42 \qquad c_2 = \frac{6 \sin 14.9°}{\sin 35°} \approx 2.69$$

The two solved triangles are illustrated in Figure 19(b).

EXAMPLE 5 **Using the Law of Sines to Solve a SSA Triangle (No Solution)**

Solve the triangle: $a = 2, c = 1, C = 50°$

Solution Because $a = 2$, $c = 1$, and $C = 50°$ are known, we use the Law of Sines to find the angle A.

Figure 20

$$\frac{\sin A}{a} = \frac{\sin C}{c}$$

$$\frac{\sin A}{2} = \frac{\sin 50°}{1}$$

$$\sin A = 2 \sin 50° \approx 1.53$$

Since there is no angle A for which $\sin A > 1$, there can be no triangle with the given measurements. Figure 20 illustrates the measurements given. Notice that, no matter how we attempt to position side c, it will never touch side b to form a triangle.

Now Work PROBLEMS 25 AND 31

3 Solve Applied Problems

The Law of Sines is particularly useful for solving certain applied problems.

EXAMPLE 6 **Finding the Height of a Mountain**

To measure the height of a mountain, a surveyor takes two sightings of the peak at a distance 900 meters apart on a direct line to the mountain.* See Figure 21(a) on page 680. The first observation results in an angle of elevation of 47°, and the second results in an angle of elevation of 35°. If the transit is 2 meters high, what is the height h of the mountain?

*For simplicity, we assume that these sightings are at the same level.

Figure 21

(a) (b)

Solution Figure 21(b) shows the triangles that replicate the illustration in Figure 21(a). Since $C + 47° = 180°$, we find that $C = 133°$. Also, since $A + C + 35° = 180°$, we find that $A = 180° - 35° - C = 145° - 133° = 12°$. We use the Law of Sines to find c.

$$\frac{\sin A}{a} = \frac{\sin C}{c} \qquad A = 12°, C = 133°, a = 900$$

$$c = \frac{900 \sin 133°}{\sin 12°} \approx 3165.86$$

Using the larger right triangle, we have

$$\sin 35° = \frac{b}{c} \qquad c = 3165.86$$

$$b = 3165.86 \sin 35° \approx 1815.86 \approx 1816 \text{ meters}$$

The height of the peak from ground level is approximately $1816 + 2 = 1818$ meters. ∎

Now Work PROBLEM 39

| EXAMPLE 7 | **Rescue at Sea** |

Coast Guard Station Zulu is located 120 miles due west of Station X-ray. A ship at sea sends an SOS call that is received by each station. The call to Station Zulu indicates that the bearing of the ship from Zulu is N40°E (40° east of north). The call to Station X-ray indicates that the bearing of the ship from X-ray is N30°W (30° west of north).

(a) How far is each station from the ship?

(b) If a helicopter capable of flying 200 miles per hour is dispatched from the nearest station to the ship, how long will it take to reach the ship?

Solution (a) Figure 22 illustrates the situation. The angle C is found to be

$$C = 180° - 50° - 60° = 70°$$

Figure 22

The Law of Sines can now be used to find the two distances a and b that we seek.

$$\frac{\sin 50°}{a} = \frac{\sin 70°}{120}$$

$$a = \frac{120 \sin 50°}{\sin 70°} \approx 97.82 \text{ miles}$$

$$\frac{\sin 60°}{b} = \frac{\sin 70°}{120}$$

$$b = \frac{120 \sin 60°}{\sin 70°} \approx 110.59 \text{ miles}$$

Station Zulu is about 111 miles from the ship, and Station X-ray is about 98 miles from the ship.

(b) The time t needed for the helicopter to reach the ship from Station X-ray is found by using the formula

$$(\text{Velocity}, v)(\text{Time}, t) = \text{Distance}, a$$

Then

$$t = \frac{a}{v} = \frac{97.82}{200} \approx 0.49 \text{ hour} \approx 29 \text{ minutes}$$

It will take about 29 minutes for the helicopter to reach the ship.

Now Work PROBLEM 37

Proof of the Law of Sines To prove the Law of Sines, we construct an altitude of length h from one of the vertices of a triangle. Figure 23(a) shows h for a triangle with three acute angles, and Figure 23(b) shows h for a triangle with an obtuse angle. In each case, the altitude is drawn from the vertex at B. Using either illustration, we have

$$\sin C = \frac{h}{a}$$

from which

$$h = a \sin C \tag{3}$$

From Figure 23(a), it also follows that

$$\sin A = \frac{h}{c}$$

from which

$$h = c \sin A \tag{4}$$

From Figure 23(b), it follows that

$$\sin(180° - A) = \sin A = \frac{h}{c}$$
$$\uparrow$$
$$\sin(180° - A) = \sin 180° \cos A - \cos 180° \sin A = \sin A$$

which again gives

$$h = c \sin A$$

So, whether the triangle has three acute angles or has two acute angles and one obtuse angle, equations (3) and (4) hold. As a result, we may equate the expressions for h in equations (3) and (4) to get

$$a \sin C = c \sin A$$

from which

$$\frac{\sin A}{a} = \frac{\sin C}{c} \tag{5}$$

In a similar manner, by constructing the altitude h' from the vertex of angle A as shown in Figure 24, we can show that

$$\sin B = \frac{h'}{c} \quad \text{and} \quad \sin C = \frac{h'}{b}$$

Equating the expressions for h', we find that

$$h' = c \sin B = b \sin C$$

from which

$$\frac{\sin B}{b} = \frac{\sin C}{c} \tag{6}$$

When equations (5) and (6) are combined, we have equation (1), the Law of Sines.

Figure 23

(a)

(b)

Figure 24

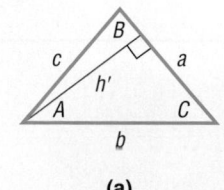

(a)

(b)

9.2 Assess Your Understanding

'Are You Prepared?' *Answers are given at the end of these exercises. If you get a wrong answer, read the pages listed in red.*

1. The difference formula for the sine function is $\sin(A - B) =$ _____. (p. 630)

2. If θ is an acute angle, solve the equation $\cos \theta = \dfrac{\sqrt{3}}{2}$. (pp. 649–653)

3. The two triangles shown are similar. Find the missing length. (pp. 30–35)

 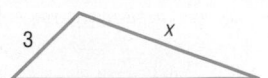

Concepts and Vocabulary

4. If none of the angles of a triangle is a right angle, the triangle is called _____.

5. For a triangle with sides a, b, c and opposite angles A, B, C, the Law of Sines states that _____.

6. *True or False* An oblique triangle in which two sides and an angle are given always results in at least one triangle.

7. *True or False* The sum of the angles of any triangle equals $180°$.

8. *True or False* The ambiguous case refers to the fact that, when two sides and the angle opposite one of them are given, sometimes the Law of Sines cannot be used.

Skill Building

In Problems 9–16, solve each triangle.

9.

10.

11.

12.

13.

14.

15.

16.

In Problems 17–24, solve each triangle.

17. $A = 40°$, $B = 20°$, $a = 2$

18. $A = 50°$, $C = 20°$, $a = 3$

19. $B = 70°$, $C = 10°$, $b = 5$

20. $A = 70°$, $B = 60°$, $c = 4$

21. $A = 110°$, $C = 30°$, $c = 3$

22. $B = 10°$, $C = 100°$, $b = 2$

23. $A = 40°$, $B = 40°$, $c = 2$

24. $B = 20°$, $C = 70°$, $a = 1$

In Problems 25–36, two sides and an angle are given. Determine whether the given information results in one triangle, two triangles, or no triangle at all. Solve any triangle(s) that results.

25. $a = 3$, $b = 2$, $A = 50°$

26. $b = 4$, $c = 3$, $B = 40°$

27. $b = 5$, $c = 3$, $B = 100°$

28. $a = 2$, $c = 1$, $A = 120°$

29. $a = 4$, $b = 5$, $A = 60°$

30. $b = 2$, $c = 3$, $B = 40°$

31. $b = 4$, $c = 6$, $B = 20°$

32. $a = 3$, $b = 7$, $A = 70°$

33. $a = 2$, $c = 1$, $C = 100°$

34. $b = 4$, $c = 5$, $B = 95°$

35. $a = 2$, $c = 1$, $C = 25°$

36. $b = 4$, $c = 5$, $B = 40°$

Applications and Extensions

37. **Rescue at Sea** Coast Guard Station Able is located 150 miles due south of Station Baker. A ship at sea sends an SOS call that is received by each station. The call to Station Able indicates that the ship is located N55°E; the call to Station Baker indicates that the ship is located S60°E.

 (a) How far is each station from the ship?

 (b) If a helicopter capable of flying 200 miles per hour is dispatched from the station nearest the ship, how long will it take to reach the ship?

38. **Surveying** Consult the figure to the right. To find the distance from the house at A to the house at B, a surveyor measures $\angle BAC$ to be 35° and then walks off a distance of

100 feet to C and measures $\angle ACB$ to be $50°$. What is the distance from A to B?

39. Finding the Length of a Ski Lift Consult the figure. To find the length of the span of a proposed ski lift from P to Q, a surveyor measures $\angle DPQ$ to be $25°$ and then walks off a distance of 1000 feet to R and measures $\angle PRQ$ to be $15°$. What is the distance from P to Q?

40. Finding the Height of a Mountain Use the illustration in Problem 39 to find the height QD of the mountain.

41. Finding the Height of an Airplane An aircraft is spotted by two observers who are 1000 feet apart. As the airplane passes over the line joining them, each observer takes a sighting of the angle of elevation to the plane, as indicated in the figure. How high is the airplane?

42. Finding the Height of the Bridge over the Royal Gorge The highest bridge in the world is the bridge over the Royal Gorge of the Arkansas River in Colorado. Sightings to the same point at water level directly under the bridge are taken from each side of the 880-foot-long bridge, as indicated in the figure. How high is the bridge?

Source: Guinness Book of World Records

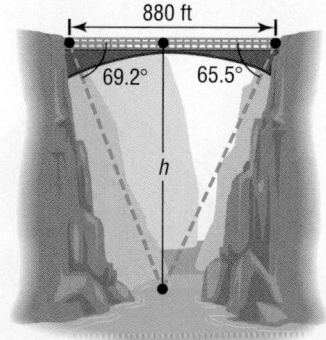

43. Landscaping Pat needs to determine the height of a tree before cutting it down to be sure that it will not fall on a nearby fence. The angle of elevation of the tree from one position on a flat path from the tree is $30°$, and from a second position 40 feet farther along this path it is $20°$. What is the height of the tree?

44. Construction A loading ramp 10 feet long that makes an angle of $18°$ with the horizontal is to be replaced by one that makes an angle of $12°$ with the horizontal. How long is the new ramp?

45. Commercial Navigation Adam must fly home to St. Louis from a business meeting in Oklahoma City. One flight option flies directly to St. Louis, a distance of about 461.1 miles. A second flight option flies first to Kansas City and then connects to St. Louis. The bearing from Oklahoma City to Kansas City is N29.6°E, and the bearing from Oklahoma City to St. Louis is N57.7°E. The bearing from St. Louis to Oklahoma City is S62.1°W, and the bearing from St. Louis to Kansas City is N79.4°W. How many more frequent flyer miles will Adam receive if he takes the connecting flight rather than the direct flight?

Source: www.landings.com

46. Time Lost due to a Navigation Error In attempting to fly from city P to city Q, an aircraft followed a course that was $10°$ in error, as indicated in the figure. After flying a distance of 50 miles, the pilot corrected the course by turning at point R and flying 70 miles farther. If the constant speed of the aircraft was 250 miles per hour, how much time was lost due to the error?

47. Finding the Lean of the Leaning Tower of Pisa The famous Leaning Tower of Pisa was originally 184.5 feet high.* At a

* On February 27, 1964, the government of Italy requested aid in preventing the tower from toppling. A multinational task force of engineers, mathematicians, and historians was assigned and met on the Azores islands to discuss stabilization methods. After over two decades of work on the subject, the tower was closed to the public in January 1990. During the time that the tower was closed the bells were removed to relieve some weight and cables were cinched around the third level and anchored several hundred meters away. Apartments and houses in the path of the tower were vacated for safety concerns. After a decade of corrective reconstruction and stabilization efforts, the tower was reopened to the public on December 15, 2001. Many methods were proposed to stabilize the tower, including the addition of 800 metric tons of lead counterweights to the raised end of the base. The final solution to correcting the lean was to remove 38 cubic meters of soil from underneath the raised end. The tower has been declared stable for at least another 300 years.

Source: http://en.wikipedia.org/wiki/Leaning_Tower_of_Pisa, page last modified June 28, 2006

distance of 123 feet from the base of the tower, the angle of elevation to the top of the tower is found to be 60°. Find ∠RPQ indicated in the figure. Also, find the perpendicular distance from R to PQ.

48. Crankshafts on Cars On a certain automobile, the crankshaft is 3 inches long and the connecting rod is 9 inches long (see the figure). At the time when ∠OPQ is 15°, how far is the piston (P) from the center (O) of the crankshaft?

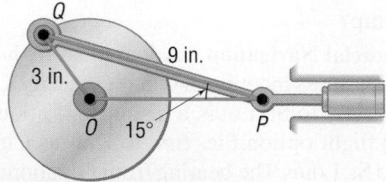

49. Constructing a Highway U.S. 41, a highway whose primary directions are north–south, is being constructed along the west coast of Florida. Near Naples, a bay obstructs the straight path of the road. Since the cost of a bridge is prohibitive, engineers decide to go around the bay. The illustration shows the path that they decide on and the measurements taken. What is the length of highway needed to go around the bay?

50. Calculating Distances at Sea The navigator of a ship at sea spots two lighthouses that she knows to be 3 miles apart along a straight seashore. She determines that the angles formed between two line-of-sight observations of the lighthouses and the line from the ship directly to shore are 15° and 35°. See the illustration.

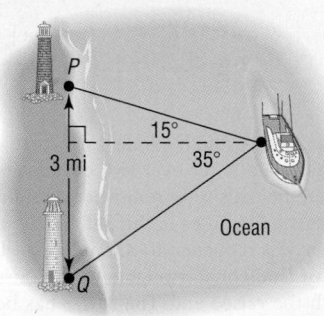

(a) How far is the ship from lighthouse P?
(b) How far is the ship from lighthouse Q?
(c) How far is the ship from shore?

51. Designing an Awning An awning that covers a sliding glass door that is 88 inches tall forms an angle of 50° with the wall. The purpose of the awning is to prevent sunlight from entering the house when the angle of elevation of the Sun is more than 65°. See the figure. Find the length L of the awning.

52. Finding Distances A forest ranger is walking on a path inclined at 5° to the horizontal directly toward a 100-foot-tall fire observation tower. The angle of elevation from the path to the top of the tower is 40°. How far is the ranger from the tower at this time?

53. Great Pyramid of Cheops One of the original Seven Wonders of the World, the Great Pyramid of Cheops was built about 2580 BC. Its original height was 480 feet 11 inches, but due to the loss of its topmost stones, it is now shorter. Find the current height of the Great Pyramid, using the information given in the illustration.

Source: Guinness Book of World Records

54. Determining the Height of an Aircraft Two sensors are spaced 700 feet apart along the approach to a small airport. When an aircraft is nearing the airport, the angle of elevation from the first sensor to the aircraft is 20°, and from the second sensor to the aircraft it is 15°. Determine how high the aircraft is at this time.

55. Mercury The distance from the Sun to Earth is approximately 149,600,000 kilometers (km). The distance from the Sun to Mercury is approximately 57,910,000 km. The **elongation angle** α is the angle formed between the line of sight from Earth to the Sun and the line of sight from Earth to Mercury. See the figure. Suppose that the elongation angle for Mercury is 15°. Use this information to find the possible distances between Earth and Mercury.

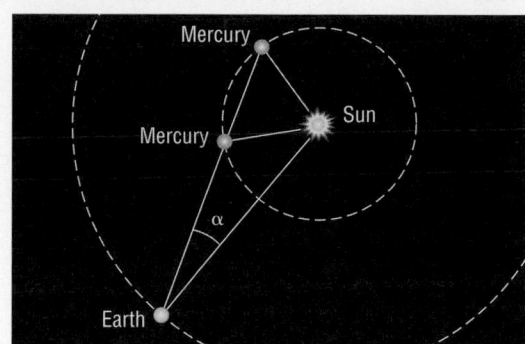

56. Venus The distance from the Sun to Earth is approximately 149,600,000 km. The distance from the Sun to Venus is approximately 108,200,000 km. The elongation angle α is the angle formed between the line of sight from Earth to the Sun and the line of sight from Earth to Venus. Suppose that the elongation angle for Venus is 10°. Use this information to find the possible distances between Earth and Venus.

57. The Original Ferris Wheel George Washington Gale Ferris, Jr. designed the original Ferris wheel for the 1893 World's Columbian Exposition in Chicago, Illinois. The wheel had 36 equally spaced cars each the size of a school bus. The distance between adjacent cars was approximately 22 feet. Determine the diameter of the wheel to the nearest foot.

Source: Carnegie Library of Pittsburgh, www.clpgh.org

58. Mollweide's Formula For any triangle, Mollweide's Formula (named after Karl Mollweide, 1774–1825) states that

$$\frac{a+b}{c} = \frac{\cos\left[\frac{1}{2}(A-B)\right]}{\sin\left(\frac{1}{2}C\right)}$$

Derive it.

[**Hint:** Use the Law of Sines and then a Sum-to-Product Formula. Notice that this formula involves all six parts of a triangle. As a result, it is sometimes used to check the solution of a triangle.]

59. Mollweide's Formula Another form of Mollweide's Formula is

$$\frac{a-b}{c} = \frac{\sin\left[\frac{1}{2}(A-B)\right]}{\cos\left(\frac{1}{2}C\right)}$$

Derive it.

60. For any triangle, derive the formula

$$a = b\cos C + c\cos B$$

[**Hint:** Use the fact that $\sin A = \sin(180° - B - C)$.]

61. Law of Tangents For any triangle, derive the Law of Tangents.

$$\frac{a-b}{a+b} = \frac{\tan\left[\frac{1}{2}(A-B)\right]}{\tan\left[\frac{1}{2}(A+B)\right]}$$

[**Hint:** Use Mollweide's Formula.]

62. Circumscribing a Triangle Show that

$$\frac{\sin A}{a} = \frac{\sin B}{b} = \frac{\sin C}{c} = \frac{1}{2r}$$

where r is the radius of the circle circumscribing the triangle PQR whose sides are $a, b,$ and $c,$ as shown in the figure.

[**Hint:** Draw the diameter PP'. Then $B = \angle PQR = \angle PP'R,$ and angle $\angle PRP' = 90°$.]

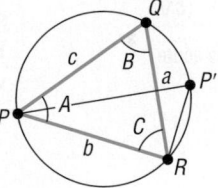

Discussion and Writing

63. Make up three problems involving oblique triangles. One should result in one triangle, the second in two triangles, and the third in no triangle.

64. What do you do first if you are asked to solve a triangle and are given one side and two angles?

65. What do you do first if you are asked to solve a triangle and are given two sides and the angle opposite one of them?

'Are You Prepared?' Answers

1. $\sin A \cos B - \cos A \sin B$ **2.** 30° **3.** $\dfrac{15}{2}$

9.3 The Law of Cosines

PREPARING FOR THIS SECTION *Before getting started, review the following:*

- Trigonometric Equations (I) (Section 8.7, pp. 649–659)
- Distance Formula (Section 2.1, p. 157)

Now Work the 'Are You Prepared?' problems on page 689.

OBJECTIVES 1 Solve SAS Triangles (p. 686)
2 Solve SSS Triangles (p. 687)
3 Solve Applied Problems (p. 688)

In the previous section, we used the Law of Sines to solve Case 1 (SAA or ASA) and Case 2 (SSA) of an oblique triangle. In this section, we derive the Law of Cosines and use it to solve the remaining cases, 3 and 4.

CASE 3: Two sides and the included angle are known (SAS).

CASE 4: Three sides are known (SSS).

THEOREM

Law of Cosines

For a triangle with sides a, b, c and opposite angles A, B, C, respectively,

$$c^2 = a^2 + b^2 - 2ab \cos C \qquad (1)$$
$$b^2 = a^2 + c^2 - 2ac \cos B \qquad (2)$$
$$a^2 = b^2 + c^2 - 2bc \cos A \qquad (3)$$

Figure 25

(a) Angle C is acute

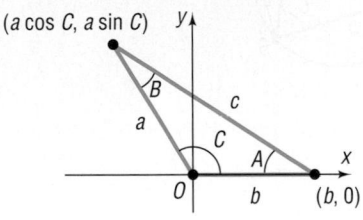

(b) Angle C is obtuse

Proof We will prove only formula (1) here. Formulas (2) and (3) may be proved using the same argument.

We begin by strategically placing a triangle on a rectangular coordinate system so that the vertex of angle C is at the origin and side b lies along the positive x-axis. Regardless of whether C is acute, as in Figure 25(a), or obtuse, as in Figure 25(b), the vertex of angle B has coordinates $(a \cos C, a \sin C)$. The vertex of angle A has coordinates $(b, 0)$.

We can now use the distance formula to compute c^2.

$$\begin{aligned}
c^2 &= (b - a \cos C)^2 + (0 - a \sin C)^2 \\
&= b^2 - 2ab \cos C + a^2 \cos^2 C + a^2 \sin^2 C \\
&= b^2 - 2ab \cos C + a^2(\cos^2 C + \sin^2 C) \\
&= a^2 + b^2 - 2ab \cos C
\end{aligned}$$ ∎

Each of formulas (1), (2), and (3) may be stated in words as follows:

THEOREM

Law of Cosines

The square of one side of a triangle equals the sum of the squares of the other two sides minus twice their product times the cosine of their included angle.

Observe that if the triangle is a right triangle (so that, say, $C = 90°$), formula (1) becomes the familiar Pythagorean Theorem: $c^2 = a^2 + b^2$. The Pythagorean Theorem is a special case of the Law of Cosines!

1 Solve SAS Triangles

Let's see how to use the Law of Cosines to solve Case 3 (SAS), which applies to triangles for which two sides and the included angle are known.

| EXAMPLE 1 | Using the Law of Cosines to Solve a SAS Triangle |

Solve the triangle: $a = 2,\quad b = 3,\quad C = 60°$

Solution See Figure 26. The Law of Cosines makes it easy to find the third side, c.

Figure 26

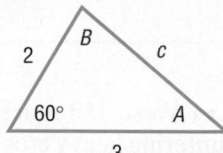

$$c^2 = a^2 + b^2 - 2ab\cos C$$
$$= 4 + 9 - 2\cdot 2\cdot 3\cdot \cos 60° \qquad {\scriptstyle a = 2, b = 3, C = 60°}$$
$$= 13 - \left(12\cdot\frac{1}{2}\right) = 7$$
$$c = \sqrt{7}$$

Side c is of length $\sqrt{7}$. To find the angles A and B, we may use either the Law of Sines or the Law of Cosines. It is preferable to use the Law of Cosines, since it will lead to an equation with one solution. Using the Law of Sines would lead to an equation with two solutions that would need to be checked to determine which solution fits the given data. We choose to use formulas (2) and (3) of the Law of Cosines to find A and B.

For A:

$$a^2 = b^2 + c^2 - 2bc\cos A$$
$$2bc\cos A = b^2 + c^2 - a^2$$
$$\cos A = \frac{b^2 + c^2 - a^2}{2bc} = \frac{9 + 7 - 4}{2\cdot 3\sqrt{7}} = \frac{12}{6\sqrt{7}} = \frac{2\sqrt{7}}{7}$$
$$A = \cos^{-1}\frac{2\sqrt{7}}{7} \approx 40.9°$$

For B:

$$b^2 = a^2 + c^2 - 2ac\cos B$$
$$\cos B = \frac{a^2 + c^2 - b^2}{2ac} = \frac{4 + 7 - 9}{4\sqrt{7}} = \frac{1}{2\sqrt{7}} = \frac{\sqrt{7}}{14}$$
$$B = \cos^{-1}\frac{\sqrt{7}}{14} \approx 79.1°$$

Notice that $A + B + C = 40.9° + 79.1° + 60° = 180°$, as required.

━━━━━ **Now Work** PROBLEM 9

2 Solve SSS Triangles

The next example illustrates how the Law of Cosines is used when three sides of a triangle are known, Case 4 (SSS).

| EXAMPLE 2 | Using the Law of Cosines to Solve a SSS Triangle |

Solve the triangle: $a = 4, b = 3, c = 6$

Solution See Figure 27. To find the angles A, B, and C we proceed as we did in the latter part of the solution to Example 1.

For A:

$$\cos A = \frac{b^2 + c^2 - a^2}{2bc} = \frac{9 + 36 - 16}{2\cdot 3\cdot 6} = \frac{29}{36}$$
$$A = \cos^{-1}\frac{29}{36} \approx 36.3°$$

Figure 27

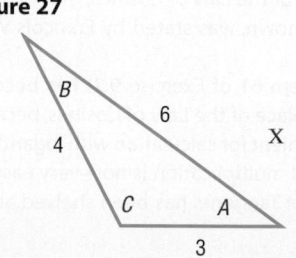

For B:

$$\cos B = \frac{a^2 + c^2 - b^2}{2ac} = \frac{16 + 36 - 9}{2\cdot 4\cdot 6} = \frac{43}{48}$$
$$B = \cos^{-1}\frac{43}{48} \approx 26.4°$$

Since we know A and B,

$$C = 180° - A - B \approx 180° - 36.3° - 26.4° = 117.3°$$

Now Work PROBLEM 15

3 Solve Applied Problems

EXAMPLE 3

Correcting a Navigational Error

A motorized sailboat leaves Naples, Florida, bound for Key West, 150 miles away. Maintaining a constant speed of 15 miles per hour, but encountering heavy crosswinds and strong currents, the crew finds, after 4 hours, that the sailboat is off course by 20°.

(a) How far is the sailboat from Key West at this time?
(b) Through what angle should the sailboat turn to correct its course?
(c) How much time has been added to the trip because of this? (Assume that the speed remains at 15 miles per hour.)

Solution

See Figure 28. With a speed of 15 miles per hour, the sailboat has gone 60 miles after 4 hours. We seek the distance x of the sailboat from Key West. We also seek the angle θ that the sailboat should turn through to correct its course.

(a) To find x, we use the Law of Cosines, since we know two sides and the included angle.

$$x^2 = 150^2 + 60^2 - 2(150)(60)\cos 20° \approx 9186.53$$
$$x \approx 95.8$$

Figure 28

The sailboat is about 96 miles from Key West.

(b) We now know three sides of the triangle, so we can use the Law of Cosines again to find the angle A opposite the side of length 150 miles.

$$150^2 = 96^2 + 60^2 - 2(96)(60)\cos A$$
$$9684 = -11{,}520 \cos A$$
$$\cos A \approx -0.8406$$
$$A \approx 147.2°$$

The sailboat should turn through an angle of

$$\theta = 180° - A \approx 180° - 147.2° = 32.8°$$

The sailboat should turn through an angle of about 33° to correct its course.

(c) The total length of the trip is now $60 + 96 = 156$ miles. The extra 6 miles will only require about 0.4 hour or 24 minutes more if the speed of 15 miles per hour is maintained.

Now Work PROBLEM 35

Historical Feature

The Law of Sines was known vaguely long before it was explicitly stated by Nasir Eddin (about AD 1250). Ptolemy (about AD 150) was aware of it in a form using a chord function instead of the sine function. But it was first clearly stated in Europe by Regiomontanus, writing in 1464.

The Law of Cosines appears first in Euclid's *Elements* (Book II), but in a well-disguised form in which squares built on the sides of triangles are added and a rectangle representing the cosine term is subtracted. It was thus known to all mathematicians because of their familiarity with Euclid's work. An early modern form of the Law of Cosines, that for finding the angle when the sides are known, was stated by François Viète (in 1593).

The Law of Tangents (see Problem 61 of Exercise 9.2) has become obsolete. In the past it was used in place of the Law of Cosines, because the Law of Cosines was very inconvenient for calculation with logarithms or slide rules. Mixing of addition and multiplication is now very easy on a calculator, however, and the Law of Tangents has been shelved along with the slide rule.

9.3 Assess Your Understanding

1. Write the formula for the distance d from $P_1 = (x_1, y_1)$ to $P_2 = (x_2, y_2)$. (p. 157)

2. If θ is an acute angle, solve the equation $\cos \theta = \dfrac{\sqrt{2}}{2}$. (pp. 649–653)

Concepts and Vocabulary

3. If three sides of a triangle are given, the Law of _____ is used to solve the triangle.

4. If one side and two angles of a triangle are given, the Law of _____ is used to solve the triangle.

5. If two sides and the included angle of a triangle are given, the Law of _____ is used to solve the triangle.

6. **True or False** Given only the three sides of a triangle, there is insufficient information to solve the triangle.

7. **True or False** Given two sides and the included angle, the first thing to do to solve the triangle is to use the Law of Sines.

8. **True or False** A special case of the Law of Cosines is the Pythagorean Theorem.

Skill Building

In Problems 9–16, solve each triangle.

9.

10.

11.

12.

13.

14.

15.

16.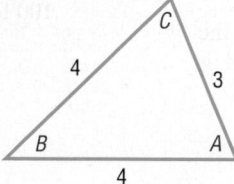

In Problems 17–32, solve each triangle.

17. $a = 3$, $b = 4$, $C = 40°$

18. $a = 2$, $c = 1$, $B = 10°$

19. $b = 1$, $c = 3$, $A = 80°$

20. $a = 6$, $b = 4$, $C = 60°$

21. $a = 3$, $c = 2$, $B = 110°$

22. $b = 4$, $c = 1$, $A = 120°$

23. $a = 2$, $b = 2$, $C = 50°$

24. $a = 3$, $c = 2$, $B = 90°$

25. $a = 12$, $b = 13$, $c = 5$

26. $a = 4$, $b = 5$, $c = 3$

27. $a = 2$, $b = 2$, $c = 2$

28. $a = 3$, $b = 3$, $c = 2$

29. $a = 5$, $b = 8$, $c = 9$

30. $a = 4$, $b = 3$, $c = 6$

31. $a = 10$, $b = 8$, $c = 5$

32. $a = 9$, $b = 7$, $c = 10$

Applications and Extensions

33. **Distance to the Green** A golfer hits an errant tee shot that lands in the rough. A marker in the center of the fairway is 150 yards from the center of the green. While standing on the marker and facing the green, the golfer turns 110° toward his ball. He then paces off 35 yards to his ball. See the figure. How far is the ball from the center of the green?

34. **Navigation** An airplane flies due north from Ft. Myers to Sarasota, a distance of 150 miles, and then turns through an angle of 50° and flies to Orlando, a distance of 100 miles. See the figure on next page.

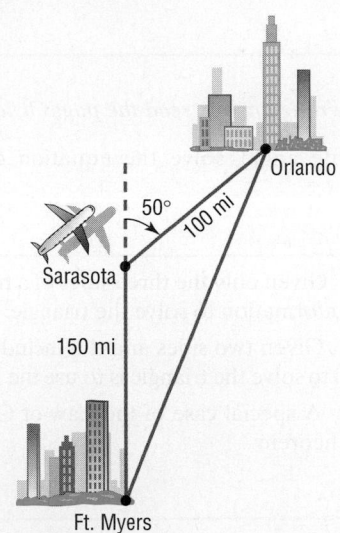

(a) How far is it directly from Ft. Myers to Orlando?

(b) What bearing should the pilot use to fly directly from Ft. Myers to Orlando?

35. **Avoiding a Tropical Storm** A cruise ship maintains an average speed of 15 knots in going from San Juan, Puerto Rico, to Barbados, West Indies, a distance of 600 nautical miles. To avoid a tropical storm, the captain heads out of San Juan in a direction of 20° off a direct heading to Barbados. The captain maintains the 15-knot speed for 10 hours, after which time the path to Barbados becomes clear of storms.

(a) Through what angle should the captain turn to head directly to Barbados?

(b) Once the turn is made, how long will it be before the ship reaches Barbados if the same 15-knot speed is maintained?

36. **Revising a Flight Plan** In attempting to fly from Chicago to Louisville, a distance of 330 miles, a pilot inadvertently took a course that was 10° in error, as indicated in the figure.

(a) If the aircraft maintains an average speed of 220 miles per hour and if the error in direction is discovered after

15 minutes, through what angle should the pilot turn to head toward Louisville?

(b) What new average speed should the pilot maintain so that the total time of the trip is 90 minutes?

37. **Major League Baseball Field** A Major League baseball diamond is actually a square 90 feet on a side. The pitching rubber is located 60.5 feet from home plate on a line joining home plate and second base.

(a) How far is it from the pitching rubber to first base?

(b) How far is it from the pitching rubber to second base?

(c) If a pitcher faces home plate, through what angle does he need to turn to face first base?

38. **Little League Baseball Field** According to Little League baseball official regulations, the diamond is a square 60 feet on a side. The pitching rubber is located 46 feet from home plate on a line joining home plate and second base.

(a) How far is it from the pitching rubber to first base?

(b) How far is it from the pitching rubber to second base?

(c) If a pitcher faces home plate, through what angle does he need to turn to face first base?

39. **Finding the Length of a Guy Wire** The height of a radio tower is 500 feet, and the ground on one side of the tower slopes upward at an angle of 10° (see the figure).

(a) How long should a guy wire be if it is to connect to the top of the tower and be secured at a point on the sloped side 100 feet from the base of the tower?

(b) How long should a second guy wire be if it is to connect to the middle of the tower and be secured at a point 100 feet from the base on the flat side?

40. **Finding the Length of a Guy Wire** See the figure below. A radio tower 500 feet high is located on the side of a hill with

an inclination to the horizontal of 5°. How long should two guy wires be if they are to connect to the top of the tower and be secured at two points 100 feet directly above and directly below the base of the tower?

41. Wrigley Field, Home of the Chicago Cubs The distance from home plate to the fence in dead center in Wrigley Field is 400 feet (see the figure). How far is it from the fence in dead center to third base?

42. Little League Baseball The distance from home plate to the fence in dead center at the Oak Lawn Little League field is 280 feet. How far is it from the fence in dead center to third base?

[**Hint:** The distance between the bases in Little League is 60 feet.]

43. Building a Swing Set Clint is building a wooden swing set for his children. Each supporting end of the swing set is to be an A-frame constructed with two 10-foot-long 4 by 4s joined at a 45° angle. To prevent the swing set from tipping over, Clint wants to secure the base of each A-frame to concrete footings. How far apart should the footings for each A-frame be?

44. Rods and Pistons Rod *OA* rotates about the fixed point *O* so that point *A* travels on a circle of radius *r*. Connected to point *A* is another rod *AB* of length *L* > 2*r*, and point *B* is connected to a piston. See the figure. Show that the distance *x* between point *O* and point *B* is given by

$$x = r \cos \theta + \sqrt{r^2 \cos^2 \theta + L^2 - r^2}$$

where θ is the angle of rotation of rod *OA*.

45. Geometry Show that the length *d* of a chord of a circle of radius *r* is given by the formula

$$d = 2r \sin \frac{\theta}{2}$$

where θ is the central angle formed by the radii to the ends of the chord. See the figure. Use this result to derive the fact that sin θ < θ, where θ > 0 is measured in radians.

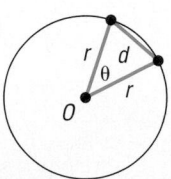

46. For any triangle, show that

$$\cos \frac{C}{2} = \sqrt{\frac{s(s-c)}{ab}}$$

where $s = \frac{1}{2}(a + b + c)$.

[**Hint:** Use a Half-angle Formula and the Law of Cosines.]

47. For any triangle show that

$$\sin \frac{C}{2} = \sqrt{\frac{(s-a)(s-b)}{ab}}$$

where $s = \frac{1}{2}(a + b + c)$.

48. Use the Law of Cosines to prove the identity

$$\frac{\cos A}{a} + \frac{\cos B}{b} + \frac{\cos C}{c} = \frac{a^2 + b^2 + c^2}{2abc}$$

Discussion and Writing

49. What do you do first if you are asked to solve a triangle and are given two sides and the included angle?

50. What do you do first if you are asked to solve a triangle and are given three sides?

51. Make up an applied problem that requires using the Law of Cosines.

52. Write down your strategy for solving an oblique triangle.

'Are You Prepared?' Answers

1. $d = \sqrt{(x_2 - x_1)^2 + (y_2 - y_1)^2}$ **2.** $\theta = 45°$

9.4 Area of a Triangle

PREPARING FOR THIS SECTION *Before getting started, review the following:*

• Geometry Essentials (Section R.3, pp. 30–35)

Now Work the 'Are You Prepared?' problem on page 694.

OBJECTIVES 1 Find the Area of SAS Triangles (p. 692)
2 Find the Area of SSS Triangles (p. 693)

In this section, we will derive several formulas for calculating the area of a triangle. The most familiar of these is the following:

THEOREM

The area K of a triangle is

$$K = \frac{1}{2}bh \tag{1}$$

where b is the base and h is an altitude drawn to that base.

Proof The derivation of this formula is rather easy once a rectangle of base b and height h is constructed around the triangle. See Figures 29 and 30.

Triangles 1 and 2 in Figure 30 are equal in area, as are triangles 3 and 4. Consequently, the area of the triangle with base b and altitude h is exactly half the area of the rectangle, which is bh.

Figure 29

Figure 30

1 Find the Area of SAS Triangles

Figure 31

If the base b and altitude h to that base are known, then we can find the area of such a triangle using formula (1). Usually, though, the information required to use formula (1) is not given. Suppose, for example, that we know two sides a and b and the included angle C. See Figure 31. Then the altitude h can be found by noting that

$$\frac{h}{a} = \sin C$$

so that

$$h = a \sin C$$

Using this fact in formula (1) produces

$$K = \frac{1}{2}bh = \frac{1}{2}b(a \sin C) = \frac{1}{2}ab \sin C$$

We now have the formula

$$K = \frac{1}{2}ab \sin C \tag{2}$$

By dropping altitudes from the other two vertices of the triangle, we obtain the following corresponding formulas:

$$K = \frac{1}{2}bc \sin A \tag{3}$$

$$K = \frac{1}{2}ac \sin B \tag{4}$$

It is easiest to remember these formulas using the following wording:

THEOREM

The area K of a triangle equals one-half the product of two of its sides times the sine of their included angle.

| EXAMPLE 1 | **Finding the Area of a SAS Triangle** |

Find the area K of the triangle for which $a = 8$, $b = 6$, and $C = 30°$.

Solution See Figure 32. We use formula (2) to get

$$K = \frac{1}{2}ab \sin C = \frac{1}{2} \cdot 8 \cdot 6 \cdot \sin 30° = 12 \text{ square units}$$

Figure 32

━━━ **Now Work** PROBLEM 5

2 Find the Area of SSS Triangles

If the three sides of a triangle are known, another formula, called **Heron's Formula** (named after Heron of Alexandria), can be used to find the area of a triangle.

THEOREM **Heron's Formula**

The area K of a triangle with sides a, b, and c is

$$K = \sqrt{s(s - a)(s - b)(s - c)} \qquad (5)$$

where $s = \frac{1}{2}(a + b + c)$.

A proof of Heron's Formula is given at the end of this section.

| EXAMPLE 2 | **Finding the Area of a SSS Triangle** |

Find the area of a triangle whose sides are 4, 5, and 7.

Solution We let $a = 4$, $b = 5$, and $c = 7$. Then

$$s = \frac{1}{2}(a + b + c) = \frac{1}{2}(4 + 5 + 7) = 8$$

Heron's Formula then gives the area K as

$$K = \sqrt{s(s - a)(s - b)(s - c)} = \sqrt{8 \cdot 4 \cdot 3 \cdot 1} = \sqrt{96} = 4\sqrt{6} \text{ square units}$$

━━━ **Now Work** PROBLEM 11

Proof of Heron's Formula The proof that we shall give uses the Law of Cosines and is quite different from the proof given by Heron.

From the Law of Cosines,

$$c^2 = a^2 + b^2 - 2ab \cos C$$

and the Half-angle Formula

$$\cos^2 \frac{C}{2} = \frac{1 + \cos C}{2}$$

we find that

$$\cos^2 \frac{C}{2} = \frac{1 + \cos C}{2} = \frac{1 + \dfrac{a^2 + b^2 - c^2}{2ab}}{2}$$

$$= \frac{a^2 + 2ab + b^2 - c^2}{4ab} = \frac{(a + b)^2 - c^2}{4ab}$$

$$= \underset{\underset{\text{Factor}}{\uparrow}}{\frac{(a + b - c)(a + b + c)}{4ab}} = \underset{\underset{\substack{a + b - c = a + b + c - 2c \\ = 2s - 2c = 2(s - c)}}{\uparrow}}{\frac{2(s - c) \cdot 2s}{4ab}} = \frac{s(s - c)}{ab} \qquad (6)$$

Similarly, using $\sin^2 \dfrac{C}{2} = \dfrac{1 - \cos C}{2}$, we find that

$$\sin^2 \frac{C}{2} = \frac{(s - a)(s - b)}{ab} \qquad (7)$$

Now we use formula (2) for the area.

$$
\begin{aligned}
K &= \frac{1}{2}ab \sin C \\
&= \frac{1}{2}ab \cdot 2 \sin \frac{C}{2} \cos \frac{C}{2} \qquad &\sin C = \sin\left[2\left(\frac{C}{2}\right)\right] = 2\sin\frac{C}{2}\cos\frac{C}{2} \\
&= ab\sqrt{\frac{(s - a)(s - b)}{ab}}\sqrt{\frac{s(s - c)}{ab}} \qquad &\text{Use equations (6) and (7).} \\
&= \sqrt{s(s - a)(s - b)(s - c)}
\end{aligned}
$$

Historical Feature

Heron's Formula (also known as *Hero's Formula*) is due to Heron of Alexandria (first century AD), who had, besides his mathematical talents, a good deal of engineering skills. In various temples his mechanical devices produced effects that seemed supernatural, and visitors presumably were thus influenced to generosity. Heron's book *Metrica*, on making such devices, has survived and was discovered in 1896 in the city of Constantinople.

Heron's Formulas for the area of a triangle caused some mild discomfort in Greek mathematics, because a product with two factors was an area, while one with three factors was a volume, but four factors seemed contradictory in Heron's time.

9.4 Assess Your Understanding

'Are You Prepared?' *Answer as given at the end of these exercises. If you get a wrong answer, read the page listed in red.*

1. The area of a triangle whose base is b and whose height is h is _____ (p. 31)

Concepts and Vocabulary

2. If three sides of a triangle are given, _____ Formula is used to find the area of the triangle.

3. *True or False* No formula exists for finding the area of a triangle when only three sides are given.

4. *True or False* Given two sides and the included angle, there is a formula that can be used to find the area of the triangle.

Skill Building

In Problems 5–12, find the area of each triangle. Round answers to two decimal places.

5.

6.

7.

8.

9.

10.

11.

12.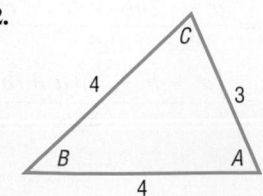

In Problems 13–24, find the area of each triangle. Round answers to two decimal places.

13. $a = 3$, $b = 4$, $C = 40°$

14. $a = 2$, $c = 1$, $B = 10°$

15. $b = 1$, $c = 3$, $A = 80°$

16. $a = 6$, $b = 4$, $C = 60°$

17. $a = 3$, $c = 2$, $B = 110°$

18. $b = 4$, $c = 1$, $A = 120°$

19. $a = 12$, $b = 13$, $c = 5$

20. $a = 4$, $b = 5$, $c = 3$

21. $a = 2$, $b = 2$, $c = 2$

22. $a = 3$, $b = 3$, $c = 2$

23. $a = 5$, $b = 8$, $c = 9$

24. $a = 4$, $b = 3$, $c = 6$

Applications and Extensions

25. Area of an ASA Triangle If two angles and the included side are given, the third angle is easy to find. Use the Law of Sines to show that the area K of a triangle with side a and angles A, B, and C is

$$K = \frac{a^2 \sin B \sin C}{2 \sin A}$$

26. Area of a Triangle Prove the two other forms of the formula given in Problem 25.

$$K = \frac{b^2 \sin A \sin C}{2 \sin B} \quad \text{and} \quad K = \frac{c^2 \sin A \sin B}{2 \sin C}$$

In Problems 27–32, use the results of Problem 25 or 26 to find the area of each triangle. Round answers to two decimal places.

27. $A = 40°$, $B = 20°$, $a = 2$

28. $A = 50°$, $C = 20°$, $a = 3$

29. $B = 70°$, $C = 10°$, $b = 5$

30. $A = 70°$, $B = 60°$, $c = 4$

31. $A = 110°$, $C = 30°$, $c = 3$

32. $B = 10°$, $C = 100°$, $b = 2$

33. Area of a Segment Find the area of the segment (shaded in blue in the figure) of a circle whose radius is 8 feet, formed by a central angle of 70°.

[**Hint:** Subtract the area of the triangle from the area of the sector to obtain the area of the segment.]

34. Area of a Segment Find the area of the segment of a circle whose radius is 5 inches, formed by a central angle of 40°.

35. Cost of a Triangular Lot The dimensions of a triangular lot are 100 feet by 50 feet by 75 feet. If the price of such land is $3 per square foot, how much does the lot cost?

36. Amount of Material to Make a Tent A cone-shaped tent is made from a circular piece of canvas 24 feet in diameter by removing a sector with central angle 100° and connecting the ends. What is the surface area of the tent?

37. Dimensions of Home Plate The dimensions of home plate at any major league baseball stadium are shown. Find the area of home plate.

38. Computing Areas Find the area of the shaded region enclosed in a semicircle of diameter 10 inches. The length of the chord PQ is 8 inches.

[**Hint:** Triangle PQR is a right triangle.]

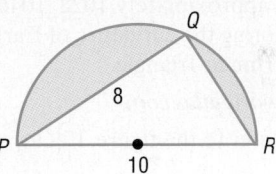

39. Geometry Consult the figure, which shows a circle of radius r with center at O. Find the area K of the shaded region as a function of the central angle θ.

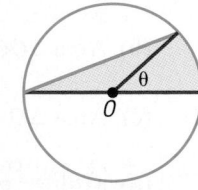

40. Approximating the Area of a Lake To approximate the area of a lake, a surveyor walks around the perimeter of the lake, taking the measurements shown in the illustration. Using this technique, what is the approximate area of the lake?

[**Hint:** Use the Law of Cosines on the three triangles shown and then find the sum of their areas.]

41. The Flatiron Building Completed in 1902 in New York City, the Flatiron Building is triangular shaped and bounded by 22nd Street, Broadway, and 5th Avenue. The building measures approximately 87 feet on the 22nd Street side, 190 feet on the Broadway side, and 173 feet on the 5th Avenue side. Approximate the ground area covered by the building.

Source: Sarah Bradford Landau and Carl W. Condit, *Rise of the New York Skyscraper: 1865–1913.* New Haven, CT: Yale University Press, 1996

42. Bermuda Triangle The Bermuda Triangle is roughly defined by Hamilton, Bermuda; San Juan, Puerto Rico; and Fort Lauderdale, Florida. The distances from Hamilton to Fort Lauderdale, Fort Lauderdale to San Juan, and San Juan to Hamilton are approximately 1028, 1046, and 965 miles respectively. Ignoring the curvature of Earth, approximate the area of the Bermuda Triangle.

Source: www.worldatlas.com

43. Geometry Refer to the figure. If $|OA| = 1$, show that:

(a) Area $\triangle OAC = \dfrac{1}{2} \sin \alpha \cos \alpha$

(b) Area $\triangle OCB = \dfrac{1}{2}|OB|^2 \sin \beta \cos \beta$

(c) Area $\triangle OAB = \dfrac{1}{2}|OB| \sin(\alpha + \beta)$

(d) $|OB| = \dfrac{\cos \alpha}{\cos \beta}$

(e) $\sin(\alpha + \beta) = \sin \alpha \cos \beta + \cos \alpha \sin \beta$

[**Hint:** Area $\triangle OAB =$ Area $\triangle OAC +$ Area $\triangle OCB$.]

44. Geometry Refer to the figure, in which a unit circle is drawn. The line segment DB is tangent to the circle and θ is acute.

(a) Express the area of $\triangle OBC$ in terms of $\sin \theta$ and $\cos \theta$.

(b) Express the area of $\triangle OBD$ in terms of $\sin \theta$ and $\cos \theta$.

(c) The area of the sector $\overset{\frown}{OBC}$ of the circle is $\dfrac{1}{2}\theta$, where θ is measured in radians. Use the results of parts (a) and (b) and the fact that

$$\text{Area } \triangle OBC < \text{Area } \overset{\frown}{OBC} < \text{Area } \triangle OBD$$

to show that

$$1 < \frac{\theta}{\sin \theta} < \frac{1}{\cos \theta}$$

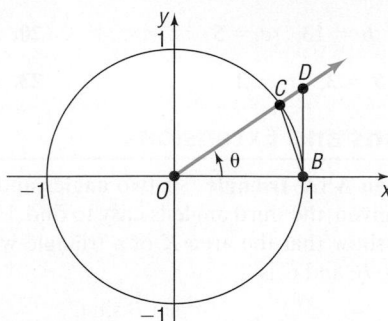

45. The Cow Problem* A cow is tethered to one corner of a square barn, 10 feet by 10 feet, with a rope 100 feet long. What is the maximum grazing area for the cow?

[**Hint:** See the illustration.]

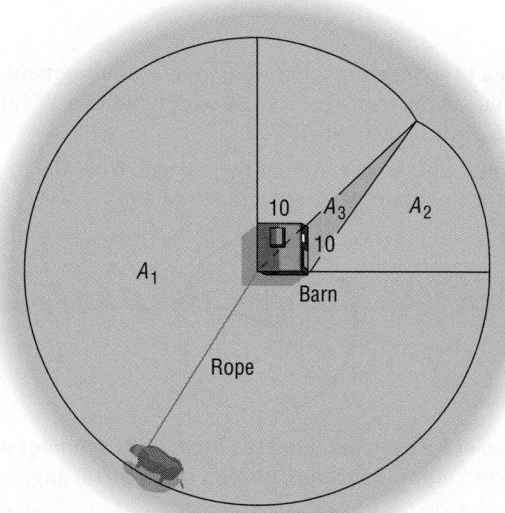

46. Another Cow Problem If the barn in Problem 45 is rectangular, 10 feet by 20 feet, what is the maximum grazing area for the cow?

47. Perfect Triangles A *perfect triangle* is one having natural number sides for which the area is numerically equal to the perimeter. Show that the triangles with the given side lengths are perfect.

(a) 9, 10, 17 (b) 6, 25, 29

Source: M. V. Bonsangue, G. E. Gannon, E. Buchman, and N. Gross, "In Search of Perfect Triangles," *Mathematics Teacher,* Vol. 92, No. 1, 1999: 56–61

48. If h_1, h_2, and h_3 are the altitudes dropped from P, Q, and R, respectively, in a triangle (see the figure), show that

$$\frac{1}{h_1} + \frac{1}{h_2} + \frac{1}{h_3} = \frac{s}{K}$$

* Suggested by Professor Teddy Koukounas of Suffolk Community College, who learned of it from an old farmer in Virginia. Solution provided by Professor Kathleen Miranda of SUNY at Old Westbury.

49. Show that a formula for the altitude h from a vertex to the opposite side a of a triangle is

$$h = \frac{a \sin B \sin C}{\sin A}$$

where K is the area of the triangle and $s = \frac{1}{2}(a + b + c)$.

[**Hint:** $h_1 = \frac{2K}{a}$.]

Inscribed Circle *For Problems 50–53, the lines that bisect each angle of a triangle meet in a single point O, and the perpendicular distance r from O to each side of the triangle is the same. The circle with center at O and radius r is called the **inscribed circle** of the triangle (see the figure).*

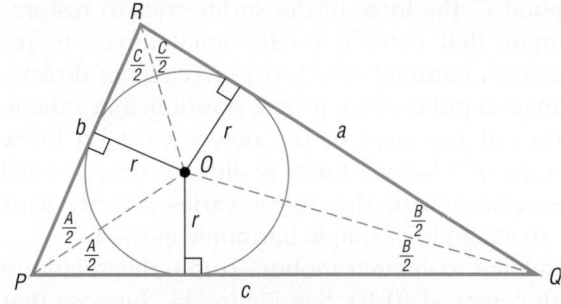

50. Apply the formula from Problem 49 to triangle OPQ to show that

$$r = \frac{c \sin \dfrac{A}{2} \sin \dfrac{B}{2}}{\cos \dfrac{C}{2}}$$

51. Use the result of Problem 50 and the result of Problems 46 and 47 in Section 9.3 to show that

$$\cot \frac{C}{2} = \frac{s - c}{r}$$

where $s = \frac{1}{2}(a + b + c)$.

52. Show that

$$\cot \frac{A}{2} + \cot \frac{B}{2} + \cot \frac{C}{2} = \frac{s}{r}$$

53. Show that the area K of triangle PQR is $K = rs$ where $s = \frac{1}{2}(a + b + c)$. Then show that

$$r = \sqrt{\frac{(s - a)(s - b)(s - c)}{s}}$$

Discussion and Writing

54. What do you do first if you are asked to find the area of a triangle and are given two sides and the included angle?

55. What do you do first if you are asked to find the area of a triangle and are given three sides?

'Are You Prepared?' Answer

1. $K = \frac{1}{2}bh$

9.5 Simple Harmonic Motion; Damped Motion; Combining Waves

PREPARING FOR THIS SECTION *Before getting started, review the following:*

- Sinusoidal Graphs (Section 7.6, pp. 563–569)

Now Work the 'Are You Prepared?' problem on page 704.

OBJECTIVES **1** Find an Equation for an Object in Simple Harmonic Motion (p. 697)

2 Analyze Simple Harmonic Motion (p. 700)

3 Analyze an Object in Damped Motion (p. 700)

4 Graph the Sum of Two Functions (p. 702)

1 Find an Equation for an Object in Simple Harmonic Motion

Many physical phenomena can be described as simple harmonic motion. Radio and television waves, light waves, sound waves, and water waves exhibit motion that is simple harmonic.

The swinging of a pendulum, the vibrations of a tuning fork, and the bobbing of a weight attached to a coiled spring are examples of vibrational motion. In this type

Vibrating tuning fork

Figure 33

Coiled spring

of motion, an object swings back and forth over the same path. In Figure 33, the point B is the **equilibrium (rest) position** of the vibrating object. The **amplitude** is the distance from the object's rest position to its point of greatest displacement (either point A or point C in Figure 33). The **period** is the time required to complete one vibration, that is, the time it takes to go from, say, point A through B to C and back to A.

Simple harmonic motion is a special kind of vibrational motion in which the acceleration a of the object is directly proportional to the negative of its displacement d from its rest position. That is, $a = -kd, k > 0$.

For example, when the mass hanging from the spring in Figure 33 is pulled down from its rest position B to the point C, the force of the spring tries to restore the mass to its rest position. Assuming that there is no frictional force* to retard the motion, the amplitude will remain constant. The force increases in direct proportion to the distance that the mass is pulled from its rest position. Since the force increases directly, the acceleration of the mass of the object must do likewise, because (by Newton's Second Law of Motion) force is directly proportional to acceleration. As a result, the acceleration of the object varies directly with its displacement, and the motion is an example of simple harmonic motion.

Simple harmonic motion is related to circular motion. To see this relationship, consider a circle of radius a, with center at $(0, 0)$. See Figure 34. Suppose that an object initially placed at $(a, 0)$ moves counterclockwise around the circle at a constant angular speed ω. Suppose further that after time t has elapsed the object is at the point $P = (x, y)$ on the circle. The angle θ, in radians, swept out by the ray $\overrightarrow{OP}$ in this time t is

$$\theta = \omega t$$

The coordinates of the point P at time t are

$$x = a \cos \theta = a \cos(\omega t)$$
$$y = a \sin \theta = a \sin(\omega t)$$

Figure 34

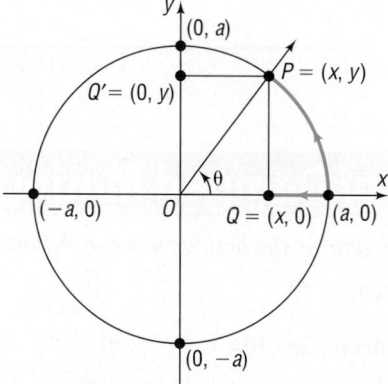

Corresponding to each position $P = (x, y)$ of the object moving about the circle, there is the point $Q = (x, 0)$, called the **projection of P on the x-axis.** As P moves around the circle at a constant rate, the point Q moves back and forth between the points $(a, 0)$ and $(-a, 0)$ along the x-axis with a motion that is simple harmonic. Similarly, for each point P there is a point $Q' = (0, y)$, called the **projection of P on the y-axis.** As P moves around the circle, the point Q' moves back and forth between the points $(0, a)$ and $(0, -a)$ on the y-axis with a motion that is simple harmonic. Simple harmonic motion can be described as the projection of constant circular motion on a coordinate axis.

* If friction is present, the amplitude will decrease with time to 0. This type of motion is an example of **damped motion,** which is discussed later in this section.

To put it another way, again consider a mass hanging from a spring where the mass is pulled down from its rest position to the point C and then released. See Figure 35(a). The graph shown in Figure 35(b) describes the displacement d of the object from its rest position as a function of time t, assuming that no frictional force is present.

Figure 35

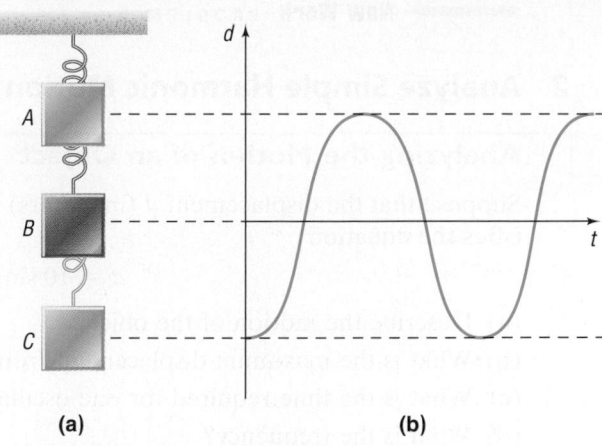

(a) (b)

THEOREM **Simple Harmonic Motion**

An object that moves on a coordinate axis so that the distance d from its rest position at time t is given by either

$$d = a \cos(\omega t) \quad \text{or} \quad d = a \sin(\omega t)$$

where a and $\omega > 0$ are constants, moves with simple harmonic motion. The motion has amplitude $|a|$ and period $\dfrac{2\pi}{\omega}$.

The **frequency** f of an object in simple harmonic motion is the number of oscillations per unit time. Since the period is the time required for one oscillation, it follows that the frequency is the reciprocal of the period; that is,

$$f = \frac{\omega}{2\pi} \qquad \omega > 0$$

EXAMPLE 1 **Finding an Equation for an Object in Harmonic Motion**

Figure 36

Rest position

$t = 0$

Suppose that an object attached to a coiled spring is pulled down a distance of 5 inches from its rest position and then released. If the time for one oscillation is 3 seconds, write an equation that relates the displacement d of the object from its rest position after time t (in seconds). Assume no friction.

Solution The motion of the object is simple harmonic. See Figure 36. When the object is released ($t = 0$), the displacement of the object from the rest position is -5 units (since the object was pulled down). Because $d = -5$ when $t = 0$, it is easier to use the cosine function*

$$d = a \cos(\omega t)$$

to describe the motion. Now the amplitude is $|-5| = 5$ and the period is 3, so

$$a = -5 \quad \text{and} \quad \frac{2\pi}{\omega} = \text{period} = 3, \quad \omega = \frac{2\pi}{3}$$

* No phase shift is required if a cosine function is used.

NOTE In the solution to Example 1, we let $a = -5$, since the initial motion is down. If the initial direction were up, we would let $a = 5$. ∎

An equation of the motion of the object is

$$d = -5 \cos\left[\frac{2\pi}{3}t\right]$$

Now Work PROBLEM 5

2 Analyze Simple Harmonic Motion

| EXAMPLE 2 | Analyzing the Motion of an Object |

Suppose that the displacement d (in meters) of an object at time t (in seconds) satisfies the equation

$$d = 10 \sin(5t)$$

(a) Describe the motion of the object.
(b) What is the maximum displacement from its resting position?
(c) What is the time required for one oscillation?
(d) What is the frequency?

Solution We observe that the given equation is of the form

$$d = a \sin(\omega t) \quad d = 10 \sin(5t)$$

where $a = 10$ and $\omega = 5$.

(a) The motion is simple harmonic.
(b) The maximum displacement of the object from its resting position is the amplitude: $|a| = 10$ meters.
(c) The time required for one oscillation is the period:

$$\text{Period} = \frac{2\pi}{\omega} = \frac{2\pi}{5} \text{ seconds}$$

(d) The frequency is the reciprocal of the period. Thus,

$$\text{Frequency} = f = \frac{5}{2\pi} \text{ oscillations per second}$$

Now Work PROBLEM 13

3 Analyze an Object in Damped Motion

Most physical phenomena are affected by friction or other resistive forces. These forces remove energy from a moving system and thereby damp its motion. For example, when a mass hanging from a spring is pulled down a distance a and released, the friction in the spring causes the distance that the mass moves from its at-rest position to decrease over time. As a result, the amplitude of any real oscillating spring or swinging pendulum decreases with time due to air resistance, friction, and so forth. See Figure 37.

Figure 37

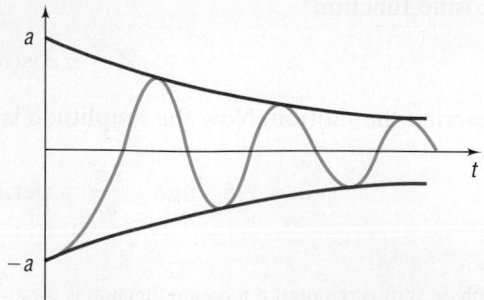

A function that describes this phenomenon maintains a sinusoidal component, but the amplitude of this component will decrease with time to account for the damping effect. In addition, the period of the oscillating component will be affected by the damping. The next result, from physics, describes damped motion.

THEOREM

Damped Motion

The displacement d of an oscillating object from its at-rest position at time t is given by

$$d(t) = ae^{-bt/(2m)}\cos\left(\sqrt{\omega^2 - \frac{b^2}{4m^2}}\,t\right)$$

where b is the **damping factor** or **damping coefficient** and m is the mass of the oscillating object. Here $|a|$ is the displacement at $t = 0$ and $\dfrac{2\pi}{\omega}$ is the period under simple harmonic motion (no damping).

Notice for $b = 0$ (zero damping) that we have the formula for simple harmonic motion with amplitude $|a|$ and period $\dfrac{2\pi}{\omega}$.

EXAMPLE 3	Analyzing a Damped Vibration Curve

Analyze the damped vibration curve

$$d(t) = e^{-t/\pi}\cos t, \quad t \geq 0$$

Solution The displacement d is the product of $y = e^{-t/\pi}$ and $y = \cos t$. Using properties of absolute value and the fact that $|\cos t| \leq 1$, we find that

$$|d(t)| = |e^{-t/\pi}\cos t| = |e^{-t/\pi}||\cos t| \leq |e^{-t/\pi}| = e^{-t/\pi}$$
$$\uparrow$$
$$e^{-t/\pi} > 0$$

As a result,

$$-e^{-t/\pi} \leq d(t) \leq e^{-t/\pi}$$

This means that the graph of d will lie between the graphs of $y = e^{-t/\pi}$ and $y = -e^{-t/\pi}$, the **bounding curves** of d.

Also, the graph of d will touch these graphs when $|\cos t| = 1$, that is, when $t = 0, \pi, 2\pi$, and so on. The x-intercepts of the graph of d occur when $\cos t = 0$, that is, at $\dfrac{\pi}{2}, \dfrac{3\pi}{2}, \dfrac{5\pi}{2}$, and so on. See Table 1.

Table 1

t	0	$\dfrac{\pi}{2}$	π	$\dfrac{3\pi}{2}$	2π
$e^{-t/\pi}$	1	$e^{-1/2}$	e^{-1}	$e^{-3/2}$	e^{-2}
$\cos t$	1	0	-1	0	1
$d(t) = e^{-t/\pi}\cos t$	1	0	$-e^{-1}$	0	e^{-2}
Point on graph of d	$(0, 1)$	$\left(\dfrac{\pi}{2}, 0\right)$	$(\pi, -e^{-1})$	$\left(\dfrac{3\pi}{2}, 0\right)$	$(2\pi, e^{-2})$

We graph $y = \cos t$, $y = e^{-t/\pi}$, $y = -e^{-t/\pi}$, and $d(t) = e^{-t/\pi} \cos t$ in Figure 38.

Figure 38

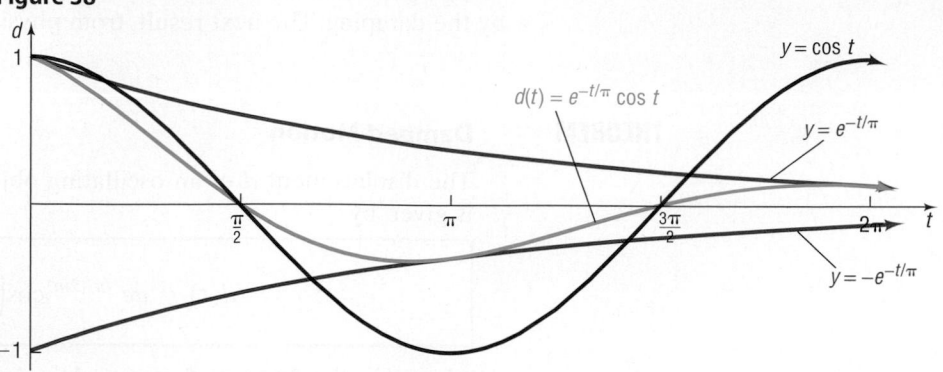

Exploration

Graph $Y_1 = e^{-x/\pi} \cos x$, along with $Y_2 = e^{-x/\pi}$, and $Y_3 = -e^{-x/\pi}$, for $0 \le x \le 2\pi$. Determine where Y_1 has its first turning point (local minimum). Compare this to where Y_1 intersects Y_3.

Result Figure 39 shows the graphs of $Y_1 = e^{-x/\pi} \cos x$, $Y_2 = e^{-x/\pi}$, and $Y_3 = -e^{-x/\pi}$. Using MINIMUM, the first turning point occurs at $x \approx 2.83$; Y_1 INTERSECTS Y_3 at $x = \pi \approx 3.14$.

Figure 39

━━━━ **Now Work** PROBLEM 21

4 Graph the Sum of Two Functions

Many physical and biological applications require the graph of the sum of two functions, such as

$$f(x) = x + \sin x \quad \text{or} \quad g(x) = \sin x + \cos(2x)$$

For example, if two tones are emitted, the sound produced is the sum of the waves produced by the two tones. See Problem 51 for an explanation of Touch-Tone phones.

To graph the sum of two (or more) functions, we can use the method of adding y-coordinates described next.

EXAMPLE 4	Graphing the Sum of Two Functions

Use the method of adding y-coordinates to graph $f(x) = x + \sin x$.

Solution First, we graph the component functions,

$$y = f_1(x) = x \qquad y = f_2(x) = \sin x$$

in the same coordinate system. See Figure 40(a). Now, select several values of x, say, $x = 0$, $x = \dfrac{\pi}{2}$, $x = \pi$, $x = \dfrac{3\pi}{2}$, and $x = 2\pi$, at which we compute $f(x) = f_1(x) + f_2(x)$. Table 2 shows the computation. We plot these points and connect them to get the graph, as shown in Figure 40(b).

Table 2

x	0	$\frac{\pi}{2}$	π	$\frac{3\pi}{2}$	2π
$y = f_1(x) = x$	0	$\frac{\pi}{2}$	π	$\frac{3\pi}{2}$	2π
$y = f_2(x) = \sin x$	0	1	0	-1	0
$f(x) = x + \sin x$	0	$\frac{\pi}{2} + 1 \approx 2.57$	π	$\frac{3\pi}{2} - 1 \approx 3.71$	2π
Point on graph of f	$(0, 0)$	$\left(\frac{\pi}{2}, 2.57\right)$	(π, π)	$\left(\frac{3\pi}{2}, 3.71\right)$	$(2\pi, 2\pi)$

Figure 40

(a) (b)

In Figure 40(b), notice that the graph of $f(x) = x + \sin x$ intersects the line $y = x$ whenever $\sin x = 0$. Also, notice that the graph of f is not periodic.

 ✓**Check:** Graph $Y_1 = x$, $Y_2 = \sin x$, and $Y_3 = x + \sin x$ and compare the result with Figure 40(b). Use INTERSECT to verify that the graphs of Y_1 and Y_3 intersect when $\sin x = 0$.

The next example shows a periodic graph.

EXAMPLE 5 **Graphing the Sum of Two Sinusoidal Functions**

Use the method of adding y-coordinates to graph

$$f(x) = \sin x + \cos(2x)$$

Solution Table 3 shows the steps for computing several points on the graph of f. Figure 41 on page 704 illustrates the graphs of the component functions, $y = f_1(x) = \sin x$ and $y = f_2(x) = \cos(2x)$, and the graph of $f(x) = \sin x + \cos(2x)$, which is shown in red.

Table 3

x	$-\frac{\pi}{2}$	0	$\frac{\pi}{2}$	π	$\frac{3\pi}{2}$	2π
$y = f_1(x) = \sin x$	-1	0	1	0	-1	0
$y = f_2(x) = \cos(2x)$	-1	1	-1	1	-1	1
$f(x) = \sin x + \cos(2x)$	-2	1	0	1	-2	1
Point on graph of f	$\left(-\frac{\pi}{2}, -2\right)$	$(0, 1)$	$\left(\frac{\pi}{2}, 0\right)$	$(\pi, 1)$	$\left(\frac{3\pi}{2}, -2\right)$	$(2\pi, 1)$

Figure 41

Notice that f is periodic, with period 2π.

 ✓ **Check:** Graph $Y_1 = \sin x$, $Y_2 = \cos(2x)$, and $Y_3 = \sin x + \cos(2x)$ and compare the result with Figure 41.

━━━ **Now Work** PROBLEM 25

9.5 Assess Your Understanding

'Are You Prepared?' *Answer given at the end of these exercises. If you get a wrong answer, read the pages listed in red.*

1. The amplitude A and period T of $f(x) = 5\sin(4x)$ are _____ and _____. (pp. 563–569)

Concepts and Vocabulary

2. The motion of an object obeys the equation $d = 4\cos(6t)$. Such motion is described as _____ _____. The number 4 is called the _____.

3. When a mass hanging from a spring is pulled down and then released, the motion is called _____ _____ if there is no fric-

tional force to retard the motion, and the motion is called _____ if there is friction.

4. *True or False* If the distance d of an object from its rest position at time t is given by a sinusoidal graph, the motion of the object is simple harmonic motion.

Skill Building

In Problems 5–8, an object attached to a coiled spring is pulled down a distance a from its rest position and then released. Assuming that the motion is simple harmonic with period T, write an equation that relates the displacement d of the object from its rest position after t seconds. Also assume that the positive direction of the motion is up.

5. $a = 5$; $T = 2$ seconds

6. $a = 10$; $T = 3$ seconds

7. $a = 6$; $T = \pi$ seconds

8. $a = 4$; $T = \dfrac{\pi}{2}$ seconds

9. Rework Problem 5 under the same conditions except that, at time $t = 0$, the object is at its resting position and moving down.

10. Rework Problem 6 under the same conditions except that, at time $t = 0$, the object is at its resting position and moving down.

11. Rework Problem 7 under the same conditions except that, at time $t = 0$, the object is at its resting position and moving down.

12. Rework Problem 8 under the same conditions except that, at time $t = 0$, the object is at its resting position and moving down.

In Problems 13–20, the displacement d (in meters) of an object at time t (in seconds) is given.
 (a) Describe the motion of the object.
 (b) What is the maximum displacement from its resting position?
 (c) What is the time required for one oscillation?
 (d) What is the frequency?

13. $d = 5\sin(3t)$

14. $d = 4\sin(2t)$

15. $d = 6\cos(\pi t)$

16. $d = 5\cos\left(\dfrac{\pi}{2}t\right)$

17. $d = -3\sin\left(\dfrac{1}{2}t\right)$

18. $d = -2\cos(2t)$

19. $d = 6 + 2\cos(2\pi t)$

20. $d = 4 + 3\sin(\pi t)$

In Problems 21–24, graph each damped vibration curve for $0 \le t \le 2\pi$.

21. $d(t) = e^{-t/\pi} \cos(2t)$ **22.** $d(t) = e^{-t/2\pi} \cos(2t)$ **23.** $d(t) = e^{-t/2\pi} \cos t$ **24.** $d(t) = e^{-t/4\pi} \cos t$

In Problems 25–32, use the method of adding y-coordinates to graph each function.

25. $f(x) = x + \cos x$ **26.** $f(x) = x + \cos(2x)$ **27.** $f(x) = x - \sin x$

28. $f(x) = x - \cos x$ **29.** $f(x) = \sin x + \cos x$ **30.** $f(x) = \sin(2x) + \cos x$

31. $g(x) = \sin x + \sin(2x)$ **32.** $g(x) = \cos(2x) + \cos x$

Applications and Extensions

In Problems 33–38, an object of mass m (in grams) attached to a coiled spring with damping factor b (in grams per second) is pulled down a distance a (in centimeters) from its rest position and then released. Assume that the positive direction of the motion is up and the period is T (in seconds) under simple harmonic motion.
 (a) Write an equation that relates the distance d of the object from its rest position after t seconds.
 (b) Graph the equation found in part (a) for 5 oscillations using a graphing utility.

33. $m = 25$, $a = 10$, $b = 0.7$, $T = 5$ **34.** $m = 20$, $a = 15$, $b = 0.75$, $T = 6$

35. $m = 30$, $a = 18$, $b = 0.6$, $T = 4$ **36.** $m = 15$, $a = 16$, $b = 0.65$, $T = 5$

37. $m = 10$, $a = 5$, $b = 0.8$, $T = 3$ **38.** $m = 10$, $a = 5$, $b = 0.7$, $T = 3$

In Problems 39–44, the distance d (in meters) of the bob of a pendulum of mass m (in kilograms) from its rest position at time t (in seconds) is given. The bob is released from the left of its rest position and represents a negative direction.
 (a) Describe the motion of the object. Be sure to give the mass and damping factor.
 (b) What is the initial displacement of the bob? That is, what is the displacement at $t = 0$?
 (c) Graph the motion using a graphing utility.
 (d) What is the displacement of the bob at the start of the second oscillation?
 (e) What happens to the displacement of the bob as time increases without bound?

39. $d = -20e^{-0.7t/40} \cos\left(\sqrt{\left(\dfrac{2\pi}{5}\right)^2 - \dfrac{0.49}{1600}} \, t \right)$ **40.** $d = -20e^{-0.8t/40} \cos\left(\sqrt{\left(\dfrac{2\pi}{5}\right)^2 - \dfrac{0.64}{1600}} \, t \right)$

41. $d = -30e^{-0.6t/80} \cos\left(\sqrt{\left(\dfrac{2\pi}{7}\right)^2 - \dfrac{0.36}{6400}} \, t \right)$ **42.** $d = -30e^{-0.5t/70} \cos\left(\sqrt{\left(\dfrac{\pi}{2}\right)^2 - \dfrac{0.25}{4900}} \, t \right)$

43. $d = -15e^{-0.9t/30} \cos\left(\sqrt{\left(\dfrac{\pi}{3}\right)^2 - \dfrac{0.81}{900}} \, t \right)$ **44.** $d = -10e^{-0.8t/50} \cos\left(\sqrt{\left(\dfrac{2\pi}{3}\right)^2 - \dfrac{0.64}{2500}} \, t \right)$

45. Loudspeaker A loudspeaker diaphragm is oscillating in simple harmonic motion described by the equation $d = a \cos(\omega t)$ with a frequency of 520 hertz (cycles per second) and a maximum displacement of 0.80 millimeter. Find ω and then determine the equation that describes the movement of the diaphragm.

46. Colossus Added to Six Flags St. Louis in 1986, the Colossus is a giant Ferris wheel. Its diameter is 165 feet, it rotates at a rate of about 1.6 revolutions per minute, and the bottom of the wheel is 15 feet above the ground. Determine an equation that relates a rider's height above the ground at time t. Assume the passenger begins the ride at the bottom of the wheel.

Source: Six Flags Theme Parks, Inc.

47. Tuning Fork The end of a tuning fork moves in simple harmonic motion described by the equation $d = a \sin(\omega t)$. If a tuning fork for the note A above middle C on an even-tempered scale (A_4, the tone by which an orchestra tunes itself) has a frequency of 440 hertz (cycles per second), find ω. If the maximum displacement of the end of the tuning fork is 0.01 millimeter, determine the equation that describes the movement of the tuning fork.

Source: David Lapp. Physics of Music and Musical Instruments. Medford, MA: Tufts University, 2003

48. Tuning Fork The end of a tuning fork moves in simple harmonic motion described by the equation $d = a \sin(\omega t)$. If a tuning fork for the note E above middle C on an even-tempered scale (E_4) has a frequency of approximately 329.63 hertz (cycles per second), find ω. If the maximum displacement of the end of the tuning fork is 0.025 millimeter, determine the equation that describes the movement of the tuning fork.

Source: David Lapp. Physics of Music and Musical Instruments. Medford, MA: Tufts University, 2003

49. Charging a Capacitor See the illustration. If a charged capacitor is connected to a coil by closing a switch, energy is transferred to the coil and then back to the capacitor in an oscillatory motion. The voltage V (in volts) across the capacitor will gradually diminish to 0 with time t (in seconds).

(a) Graph the function relating V and t:

$$V(t) = e^{-t/3}\cos(\pi t), \qquad 0 \le t \le 3$$

(b) At what times t will the graph of V touch the graph of $y = e^{-t/3}$? When does the graph of V touch the graph of $y = -e^{-t/3}$?

(c) When will the voltage V be between -0.4 and 0.4 volt?

50. The Sawtooth Curve An oscilloscope often displays a *sawtooth curve*. This curve can be approximated by sinusoidal curves of varying periods and amplitudes.

(a) Use a graphing utility to graph the following function, which can be used to approximate the sawtooth curve.

$$f(x) = \frac{1}{2}\sin(2\pi x) + \frac{1}{4}\sin(4\pi x), \qquad 0 \le x \le 4$$

(b) A better approximation to the sawtooth curve is given by

$$f(x) = \frac{1}{2}\sin(2\pi x) + \frac{1}{4}\sin(4\pi x) + \frac{1}{8}\sin(8\pi x)$$

Use a graphing utility to graph this function for $0 \le x \le 4$ and compare the result to the graph obtained in part (a).

(c) A third and even better approximation to the sawtooth curve is given by

$$f(x) = \frac{1}{2}\sin(2\pi x) + \frac{1}{4}\sin(4\pi x) + \frac{1}{8}\sin(8\pi x) + \frac{1}{16}\sin(16\pi x)$$

Use a graphing utility to graph this function for $0 \le x \le 4$ and compare the result to the graphs obtained in parts (a) and (b).

(d) What do you think the next approximation to the sawtooth curve is?

51. Touch-Tone Phones On a Touch-Tone phone, each button produces a unique sound. The sound produced is the sum of two tones, given by

$$y = \sin(2\pi l t) \quad \text{and} \quad y = \sin(2\pi h t)$$

where l and h are the low and high frequencies (cycles per second) shown in the illustration. For example, if you touch 7, the low frequency is $l = 852$ cycles per second and the high frequency is $h = 1209$ cycles per second. The sound emitted by touching 7 is

$$y = \sin[2\pi(852)t] + \sin[2\pi(1209)t]$$

Use a graphing utility to graph the sound emitted by touching 7.

Touch-Tone phone

52. Use a graphing utility to graph the sound emitted by the * key on a Touch-Tone phone. See Problem 51.

53. CBL Experiment Pendulum motion is analyzed to estimate simple harmonic motion. A plot is generated with the position of the pendulum over time. The graph is used to find a sinusoidal curve of the form $y = A\cos[B(x - C)] + D$. Determine the amplitude, period, and frequency. (Activity 16, Real-World Math with the CBL System.)

54. CBL Experiment The sound from a tuning fork is collected over time. Determine the amplitude, frequency, and period of the graph. A model of the form $y = A\cos[B(x - C)]$ is fitted to the data. (Activity 23, Real-World Math with the CBL System.)

Discussion and Writing

55. Use a graphing utility to graph the function $f(x) = \dfrac{\sin x}{x}$, $x > 0$. Based on the graph, what do you conjecture about the value of $\dfrac{\sin x}{x}$ for x close to 0?

56. Use a graphing utility to graph $y = x\sin x$, $y = x^2\sin x$, and $y = x^3\sin x$ for $x > 0$. What patterns do you observe?

57. Use a graphing utility to graph $y = \dfrac{1}{x}\sin x$, $y = \dfrac{1}{x^2}\sin x$, and $y = \dfrac{1}{x^3}\sin x$ for $x > 0$. What patterns do you observe?

58. How would you explain to a friend what simple harmonic motion is? How would you explain damped motion?

'Are You Prepared?' Answer

1. $A = 5; T = \dfrac{\pi}{2}$

CHAPTER REVIEW

Things to Know

Formulas

Law of Sines (p. 676)

$$\frac{\sin A}{a} = \frac{\sin B}{b} = \frac{\sin C}{c}$$

Law of Cosines (p. 686)

$$c^2 = a^2 + b^2 - 2ab \cos C$$

$$b^2 = a^2 + c^2 - 2ac \cos B$$

$$a^2 = b^2 + c^2 - 2bc \cos A$$

Area of a triangle (pp. 692–693)

$$K = \frac{1}{2}bh \quad K = \frac{1}{2}ab \sin C \quad K = \frac{1}{2}bc \sin A \quad K = \frac{1}{2}ac \sin B$$

$$K = \sqrt{s(s-a)(s-b)(s-c)}, \quad \text{where} \quad s = \frac{1}{2}(a+b+c)$$

Objectives

Section	You should be able to ...	Review Exercises
9.1	1 Solve right triangles (p. 670)	1–4
	2 Solve applied problems (p. 671)	35, 36, 45–47
9.2	1 Solve SAA or ASA triangles (p. 676)	5, 6, 22
	2 Solve SSA triangles (p. 677)	7–10, 12, 17, 18, 21
	3 Solve applied problems (p. 679)	37–39
9.3	1 Solve SAS triangles (p. 686)	11, 15, 16, 23, 24
	2 Solve SSS triangles (p. 687)	13, 14, 19, 20
	3 Solve applied problems (p. 688)	40, 41
9.4	1 Find the area of SAS triangles (p. 692)	25–28, 43, 44
	2 Find the area of SSS triangles (p. 693)	29–32, 42
9.5	1 Find an equation for an object in simple harmonic motion (p. 697)	48, 49
	2 Analyze simple harmonic motion (p. 700)	50–53
	3 Analyze an object in damped motion (p. 700)	54–57
	4 Graph the sum of two functions (p. 702)	58, 59

Review Exercises

In Problems 1–4, solve each triangle.

1.

2.

3.

4.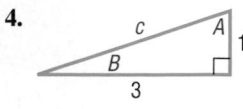

In Problems 5–24, find the remaining angle(s) and side(s) of each triangle, if it (they) exists. If no triangle exists, say "No triangle."

5. $A = 50°$, $B = 30°$, $a = 1$

6. $A = 10°$, $C = 40°$, $c = 2$

7. $A = 100°$, $a = 5$, $c = 2$

8. $a = 2$, $c = 5$, $A = 60°$

9. $a = 3$, $c = 1$, $C = 110°$

10. $a = 3$, $c = 1$, $C = 20°$

11. $a = 3$, $c = 1$, $B = 100°$

12. $a = 3$, $b = 5$, $B = 80°$

13. $a = 2$, $b = 3$, $c = 1$

14. $a = 10$, $b = 7$, $c = 8$

15. $a = 1$, $b = 3$, $C = 40°$

16. $a = 4$, $b = 1$, $C = 100°$

17. $a = 5$, $b = 3$, $A = 80°$

18. $a = 2$, $b = 3$, $A = 20°$

19. $a = 1$, $b = \frac{1}{2}$, $c = \frac{4}{3}$

20. $a = 3$, $b = 2$, $c = 2$

21. $a = 3$, $A = 10°$, $b = 4$

22. $a = 4$, $A = 20°$, $B = 100°$

23. $c = 5$, $b = 4$, $A = 70°$

24. $a = 1$, $b = 2$, $C = 60°$

In Problems 25–34, find the area of each triangle.

25. $a = 2$, $b = 3$, $C = 40°$ **26.** $b = 5$, $c = 5$, $A = 20°$ **27.** $b = 4$, $c = 10$, $A = 70°$

28. $a = 2$, $b = 1$, $C = 100°$ **29.** $a = 4$, $b = 3$, $c = 5$ **30.** $a = 10$, $b = 7$, $c = 8$

31. $a = 4$, $b = 2$, $c = 5$ **32.** $a = 3$, $b = 2$, $c = 2$

33. $A = 50°$, $B = 30°$, $a = 1$ **34.** $A = 10°$, $C = 40°$, $c = 3$

35. Finding the Grade of a Mountain Trail A straight trail with a uniform inclination leads from a hotel, elevation 5000 feet, to a lake in a valley, elevation 4100 feet. The length of the trail is 4100 feet. What is the inclination (grade) of the trail?

36. Geometry The hypotenuse of a right triangle is 12 feet. If one leg is 8 feet, find the degree measure of each angle.

37. Finding the Height of a Helicopter Two observers simultaneously measure the angle of elevation of a helicopter. One angle is measured as 25°, the other as 40° (see the figure). If the observers are 100 feet apart and the helicopter lies over the line joining them, how high is the helicopter?

38. Determining Distances at Sea Rebecca, the navigator of a ship at sea, spots two lighthouses that she knows to be 2 miles apart along a straight shoreline. She determines that the angles formed between two line-of-sight observations of the lighthouses and the line from the ship directly to shore are 12° and 30°. See the illustration.
(a) How far is the ship from lighthouse L_1?
(b) How far is the ship from lighthouse L_2?
(c) How far is the ship from shore?

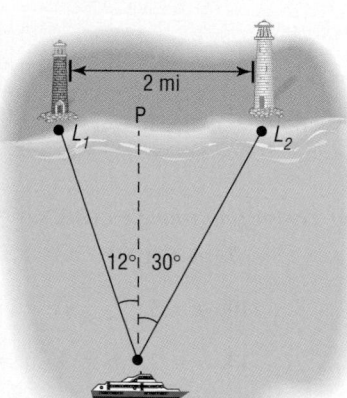

39. Constructing a Highway A highway whose primary directions are north–south is being constructed along the west coast of Florida. Near Naples, a bay obstructs the straight path of the road. Since the cost of a bridge is prohibitive, engineers decide to go around the bay. The illustration shows the path that they decide on and the measurements taken. What is the length of highway needed to go around the bay?

40. Correcting a Navigational Error A sailboat leaves St. Thomas bound for an island in the British West Indies, 200 miles away. Maintaining a constant speed of 18 miles per hour, but encountering heavy crosswinds and strong currents, the crew finds after 4 hours that the sailboat is off course by 15°.
(a) How far is the sailboat from the island at this time?
(b) Through what angle should the sailboat turn to correct its course?
(c) How much time has been added to the trip because of this? (Assume that the speed remains at 18 miles per hour.)

41. Surveying Two homes are located on opposite sides of a small hill. See the illustration. To measure the distance between them, a surveyor walks a distance of 50 feet from house P to point R, uses a transit to measure $\angle PRQ$, which is found to be 80°, and then walks to house Q, a distance of 60 feet. How far apart are the houses?

42. Approximating the Area of a Lake To approximate the area of a lake, Cindy walks around the perimeter of the lake, taking the measurements shown in the illustration. Using this technique, what is the approximate area of the lake?

[**Hint:** Use the Law of Cosines on the three triangles shown and then find the sum of their areas.]

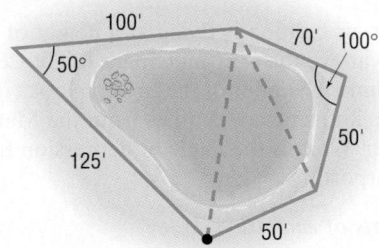

43. Calculating the Cost of Land The irregular parcel of land shown in the figure is being sold for $100 per square foot. What is the cost of this parcel?

44. Area of a Segment Find the area of the segment of a circle whose radius is 6 inches formed by a central angle of 50°.

45. Finding the Bearing of a Ship The *Majesty* leaves the Port at Boston for Bermuda with a bearing of S80°E at an average

speed of 10 knots. After 1 hour, the ship turns 90° toward the southwest. After 2 hours at an average speed of 20 knots, what is the bearing of the ship from Boston?

46. Drive Wheels of an Engine The drive wheel of an engine is 13 inches in diameter, and the pulley on the rotary pump is 5 inches in diameter. If the shafts of the drive wheel and the pulley are 2 feet apart, what length of belt is required to join them as shown in the figure?

47. Rework Problem 46 if the belt is crossed, as shown in the figure.

In Problems 48 and 49, an object attached to a coiled spring is pulled down a distance a from its rest position and then released. Assuming that the motion is simple harmonic with period T, write an equation that relates the displacement d of the object from its rest position after t seconds. Also assume that the positive direction of the motion is up.

48. $a = 3$; $T = 4$ seconds

49. $a = 5$; $T = 6$ seconds

In Problems 50–53, the distance d (in feet) that an object travels in time t (in seconds) is given.
(a) Describe the motion of the object.
(b) What is the maximum displacement from its rest position?
(c) What is the time required for one oscillation?
(d) What is the frequency?

50. $d = 6 \sin(2t)$ **51.** $d = 2 \cos(4t)$ **52.** $d = -2 \cos(\pi t)$ **53.** $d = -3 \sin\left[\dfrac{\pi}{2}t\right]$

In Problems 54 and 55, an object of mass m attached to a coiled spring with damping factor b is pulled down a distance a from its rest position and then released. Assume that the positive direction of the motion is up and the period is T under simple harmonic motion.
(a) Write an equation that relates the distance d of the object from its rest position after t seconds.
(b) Graph the equation found in part (a) for 5 oscillations.

54. $m = 40$ grams; $a = 15$ centimeters; $b = 0.75$ gram/second; $T = 5$ seconds

55. $m = 25$ grams; $a = 13$ centimeters; $b = 0.65$ gram/second; $T = 4$ seconds

In Problems 56 and 57, the distance d (in meters) of the bob of a pendulum of mass m (in kilograms) from its rest position at time t (in seconds) is given.
(a) Describe the motion of the object.
(b) What is the initial displacement of the bob? That is, what is the displacement at t = 0?
(c) Graph the motion using a graphing utility.
(d) What is the displacement of the bob at the start of the second oscillation?
(e) What happens to the displacement of the bob as time increases without bound?

56. $d = -15e^{-0.6t/40} \cos\left(\sqrt{\left(\dfrac{2\pi}{5}\right)^2 - \dfrac{0.36}{1600}}\, t\right)$ **57.** $d = -20e^{-0.5t/60} \cos\left(\sqrt{\left(\dfrac{2\pi}{3}\right)^2 - \dfrac{0.25}{3600}}\, t\right)$

In Problems 58 and 59, use the method of adding y-coordinates to graph each function.

58. $y = 2 \sin x + \cos(2x)$

59. $y = 2 \cos(2x) + \sin \dfrac{x}{2}$

CHAPTER TEST

1. A 12-foot ladder leans against a building. The top of the ladder leans against the wall 10.5 feet from the ground. What is the angle formed by the ground and the ladder?

2. A hot-air balloon is flying at a height of 600 feet and is directly above the Marshall Space Flight Center in Huntsville, Alabama. The pilot of the balloon looks down at the airport that is known to be 5 miles from the Marshall Space Flight Center. What is the angle of depression from the balloon to the airport?

In Problems 3–5, use the given information to determine the three remaining parts of each triangle.

3.

4.

5.

In Problems 6–8, solve each triangle.

6. $A = 55°$, $C = 20°$, $a = 4$

7. $a = 3$, $b = 7$, $A = 40°$

8. $a = 8$, $b = 4$, $C = 70°$

9. Find the area of the triangle described in Problem 8.

10. Find the area of the triangle described in Problem 5.

11. Find the area of the shaded region enclosed in a semicircle of diameter 8 centimeters. The length of the chord AB is 6 centimeters.

[**Hint:** Triangle ABC is a right triangle.]

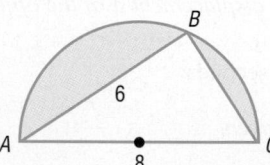

12. Find the area of the quadrilateral shown.

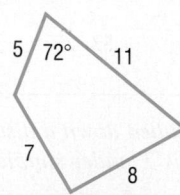

13. Madison wants to swim across Lake William from the fishing lodge (A) to the boat ramp (B), but she wants to know the

distance first. Highway 20 goes right past the boat ramp and County Road 3 goes to the lodge. The two roads intersect at point (C), 4.2 miles from the ramp and 3.5 miles from the lodge. Madison uses a transit to measure the angle of intersection of the two roads to be 32°. How far will she need to swim?

14. Given that $\triangle OAB$ is an isosceles triangle and the shaded sector is a semicircle, find the area of the entire region. Express your answer as a decimal rounded to two places.

15. The area of the triangle shown below is $54\sqrt{6}$ square units; find the lengths of the sides.

16. Logan is playing on her swing. One full swing (front to back to front) takes 6 seconds and at the peak of her swing she is at an angle of 42° with the vertical. If her swing is 5 feet long, and we ignore all resistive forces, write an equation that relates her horizontal displacement (from the rest position) after time t.

CUMULATIVE REVIEW

1. Find the real solutions, if any, of the equation $3x^2 + 1 = 4x$.

2. Find an equation for the circle with center at the point $(-5, 1)$ and radius 3. Graph this circle.

3. What is the domain of the function

$$f(x) = \sqrt{x^2 - 3x - 4}?$$

4. Graph the function $y = 3\sin(\pi x)$.

5. Graph the function $y = -2\cos(2x - \pi)$.

6. If $\tan\theta = -2$ and $\dfrac{3\pi}{2} < \theta < 2\pi$, find the exact value of:

 (a) $\sin\theta$ (b) $\cos\theta$ (c) $\sin(2\theta)$

 (d) $\cos(2\theta)$ (e) $\sin\left(\dfrac{1}{2}\theta\right)$ (f) $\cos\left(\dfrac{1}{2}\theta\right)$

7. Graph each of the following functions on the interval $[0, 4]$:

 (a) $y = e^x$ (b) $y = \sin x$

 (c) $y = e^x \sin x$ (d) $y = 2x + \sin x$

8. Sketch the graph of each of the following functions:

 (a) $y = x$ (b) $y = x^2$ (c) $y = \sqrt{x}$

 (d) $y = x^3$ (e) $y = e^x$ (f) $y = \ln x$

 (g) $y = \sin x$ (h) $y = \cos x$ (i) $y = \tan x$

9. Solve the triangle:

10. In the complex number system, solve the equation

$$3x^5 - 10x^4 + 21x^3 - 42x^2 + 36x - 8 = 0$$

11. Analyze the graph of the rational function

$$R(x) = \frac{2x^2 - 7x - 4}{x^2 + 2x - 15}$$

12. Solve $3^x = 12$. Round your answer to two decimal places.

13. Solve $\log_3(x + 8) + \log_3 x = 2$.

14. Suppose that $f(x) = 4x + 5$ and $g(x) = x^2 + 5x - 24$.

 (a) Solve $f(x) = 0$. (e) Solve $g(x) \leq 0$.

 (b) Solve $f(x) = 13$. (f) Graph $y = f(x)$.

 (c) Solve $f(x) = g(x)$. (g) Graph $y = g(x)$.

 (d) Solve $f(x) > 0$.

CHAPTER PROJECTS

I. Spherical Trigonometry When the distance between two locations on the surface of Earth is small, we can compute the distance in statutory miles. Using this assumption, we can use the Law of Sines and the Law of Cosines to approximate distances and angles. However, if you look at a globe, you notice that Earth is a sphere, so, as the distance between two points on its surface increases, the linear distance is less accurate because of curvature. Under this circumstance, we need to take into account the curvature of Earth when using the Law of Sines and the Law of Cosines.

1. Draw a spherical triangle and label each vertex by A, B, and C. Then connect each vertex by a radius to the center O of the sphere. Now, draw tangent lines to the sides a and b of the triangle that go through C. Extend the lines OA and OB to intersect the tangent lines at P and Q, respectively. See the diagram. List the plane right triangles. Determine the measures of the central angles.

2. Apply the Law of Cosines to triangles OPQ and CPQ to find two expressions for the length of PQ.

3. Subtract the expressions in part (2) from each other. Solve for the term containing $\cos c$.

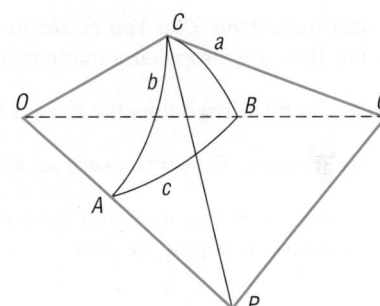

4. Use the Pythagorean Theorem to find another value for $OQ^2 - CQ^2$ and $OP^2 - CP^2$. Now solve for $\cos c$.

5. Replacing the ratios in part (4) by the cosines of the sides of the spherical triangle, you should now have the Law of Cosines for spherical triangles:

$$\cos c = \cos a \cos b + \sin a \sin b \cos C$$

Source: For the spherical Law of Cosines; see *Mathematics from the Birth of Numbers* by Jan Gullberg. W. W. Norton & Co., Publishers, 1996, pp. 491–494.

II. The Lewis and Clark Expedition Lewis and Clark followed several rivers in their trek from what is now Great Falls, Montana, to the Pacific coast. First, they went down the Missouri and Jefferson rivers from Great Falls to Lemhi, Idaho. Because the two cities are on different longitudes and different latitudes, we must account for the curvature of Earth when computing the distance that they traveled. Assume that the radius of Earth is 3960 miles.

1. Great Falls is at approximately 47.5°N and 111.3°W. Lemhi is at approximately 45.0°N and 113.5°W. (We will assume

that the rivers flow straight from Great Falls to Lemhi on the surface of Earth.) This line is called a geodesic line. Apply the Law of Cosines for a spherical triangle to find the angle between Great Falls and Lemhi. (The central angles are found by using the differences in the latitudes and longitudes of the towns. See the diagram.) Then find the length of the arc joining the two towns. (Recall $s = r\theta$.)

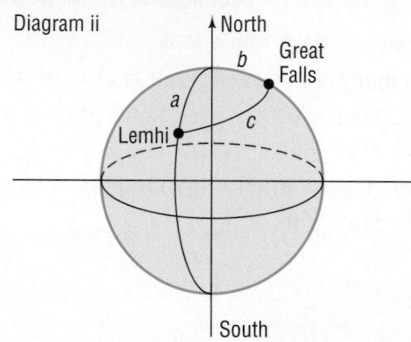

Diagram ii

2. From Lemhi, they went up the Bitteroot River and the Snake River to what is now Lewiston and Clarkston on the border of Idaho and Washington. Although this is not really a side to a triangle, we will make a side that goes from Lemhi to Lewiston and Clarkston. If Lewiston and Clarkston are at about 46.5°N 117.0°W, find the distance from Lemhi using the Law of Cosines for a spherical triangle and the arc length.

3. How far did the explorers travel just to get that far?

4. Draw a plane triangle connecting the three towns. If the distance from Lewiston to Great Falls is 282 miles and the angle at Great Falls is 42° and the angle at Lewiston is 48.5°, find the distance from Great Falls to Lemhi and from Lemhi to Lewiston. How do these distances compare with the ones computed in parts (a) and (b)?

Source: For Lewis and Clark Expedition: *American Journey: The Quest for Liberty to 1877, Texas Edition.* Prentice Hall, 1992, p. 345.

Source: For map coordinates: *National Geographic Atlas of the World,* published by National Geographic Society, 1981, pp. 74–75.

The following projects are available at the Instructor's Resource Center (IRC):

III. Project at Motorola: *How Can You Build or Analyze a Vibration Profile?* Fourier functions are not only important to analyze vibrations, but they are also what a mathematician would call interesting. Complete the project to see why.

IV. Leaning Tower of Pisa Trigonometry is used to analyze the apparent height and tilt of the Leaning Tower of Pisa.

V. Locating Lost Treasure Clever treasure seekers who know the Law of Sines are able to efficiently find a buried treasure.

VI. Jacob's Field Angles of elevation and the Law of Sines are used to determine the height of the stadium wall and the distance from home plate to the top of the wall.

Polar Coordinates; Vectors

<div style="text-align: right; font-size: 3em; font-weight: bold;">10</div>

How Do Airplanes Fly?

Have you ever watched a big jetliner lumber into position on the runway for takeoff and wonder, "How does that thing ever get off the ground?" You know it's because of the wing that it stays up in the air, but how does it really work?

When air flows around a wing, it creates lift. The way it creates lift is based on the wing's movement through the air and the air pressure created around the wing. An airplane's wing, in varying degrees depending on the type and design of the airplane, is curved over the top of the wing and straighter underneath the wing. As air hits the wing, it is "split in two," with air moving both over and under the wing. Since the top of the wing has more curve than the bottom of the wing, the air moving over the top of the wing has farther to travel, and thus must move faster than the air moving underneath the wing. The air moving over the top of the wing now exerts less air pressure on the wing than the slower-moving air under the wing. Lift is created.

The difference in air pressure is the primary force creating lift on a wing, but one other force exerted on the wing also helps to produce lift. This is the force of deflection. Air moving along the underside of the wing is deflected downward. Remember the Newtonian principle: For every action, there is an equal and opposite reaction. The air that is deflected downward (action) helps to push the wing upward (reaction), producing more lift.

These two natural forces on the wing, pressure and deflection, produce lift. The faster the wing moves through the air, the greater the forces become, and the greater the lift.

Source: Thomas Schueneman. How do airplanes fly? http://meme.essortment.com/howdoairplane_rlmi.htm, accessed August 2006. © 2002 by Pagewise, Inc. Used with permission.

—See the Chapter Project—

A Look Back, A Look Ahead

This chapter is in two parts: Polar Coordinates, Sections 10.1–10.3, and Vectors, Sections 10.4–10.5. They are independent of each other and may be covered in any order.

Sections 10.1–10.3: In Chapter 2 we introduced rectangular coordinates x and y and discussed the graph of an equation in two variables involving x and y. In Sections 10.1 and 10.2, we introduce an alternative to rectangular coordinates, polar coordinates, and discuss graphing equations that involve polar coordinates. In Section 6.3, we discussed raising a real number to a real power. In Section 10.3 we extend this idea by raising a complex number to a real power. As it turns out, polar coordinates are useful for the discussion.

Sections 10.4 and 10.5: We have seen in many chapters that often we are required to solve an equation to obtain a solution to applied problems. In the last two sections of this chapter, we develop the notion of a vector and show how it can be used to solve certain types of applied problems, particularly in physics and engineering.

Outline

10.1 Polar Coordinates

PREPARING FOR THIS SECTION *Before getting started, review the following:*

- Rectangular Coordinates (Section 2.1, p. 156)
- Definitions of the Sine and Cosine Functions (Section 7.4, pp. 540–542)

- Inverse Tangent Function (Section 8.1, pp. 608–610)
- Completing the Square (Section 1.2, pp. 99–101)

Now Work the 'Are You Prepared?' problems on page 720.

OBJECTIVES 1 Plot Points Using Polar Coordinates (p. 714)
2 Convert from Polar Coordinates to Rectangular Coordinates (p. 716)
3 Convert from Rectangular Coordinates to Polar Coordinates (p. 718)
4 Transform Equations from Polar to Rectangular Form (p. 719)

So far, we have always used a system of rectangular coordinates to plot points in the plane. Now we are ready to describe another system, called *polar coordinates*. As we shall soon see, in many instances polar coordinates offer certain advantages over rectangular coordinates.

In a rectangular coordinate system, you will recall, a point in the plane is represented by an ordered pair of numbers (x, y), where x and y equal the signed distance of the point from the y-axis and x-axis, respectively. In a polar coordinate system, we select a point, called the **pole,** and then a ray with vertex at the pole, called the **polar axis.** See Figure 1. Comparing the rectangular and polar coordinate systems, we see that the origin in rectangular coordinates coincides with the pole in polar coordinates, and the positive x-axis in rectangular coordinates coincides with the polar axis in polar coordinates.

Figure 1

1 Plot Points Using Polar Coordinates

A point P in a polar coordinate system is represented by an ordered pair of numbers (r, θ). If $r > 0$, then r is the distance of the point from the pole; θ is an angle (in degrees or radians) formed by the polar axis and a ray from the pole through the point. We call the ordered pair (r, θ) the **polar coordinates** of the point. See Figure 2.

As an example, suppose that the polar coordinates of a point P are $\left(2, \dfrac{\pi}{4}\right)$. We locate P by first drawing an angle of $\dfrac{\pi}{4}$ radian, placing its vertex at the pole and its initial side along the polar axis. Then we go out a distance of 2 units along the terminal side of the angle to reach the point P. See Figure 3.

Figure 2

Figure 3

Now Work PROBLEM 19

In using polar coordinates (r, θ), it is possible for the first entry r to be negative. When this happens, instead of the point being on the terminal side of θ, it is on the ray from the pole extending in the direction *opposite* the terminal side of θ at a distance $|r|$ units from the pole. See Figure 4 for an illustration.

For example, to plot the point $\left(-3, \dfrac{2\pi}{3}\right)$, we use the ray in the opposite direction of $\dfrac{2\pi}{3}$ and go out $|-3| = 3$ units along that ray. See Figure 5.

Figure 4

Figure 5

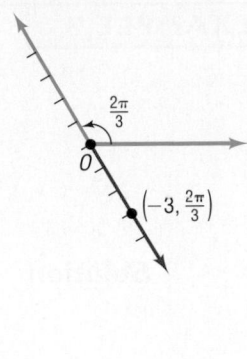

EXAMPLE 1 **Plotting Points Using Polar Coordinates**

Plot the points with the following polar coordinates:

(a) $\left(3, \dfrac{5\pi}{3}\right)$ (b) $\left(2, -\dfrac{\pi}{4}\right)$ (c) $(3, 0)$ (d) $\left(-2, \dfrac{\pi}{4}\right)$

Solution Figure 6 shows the points.

Figure 6

(a) (b) (c) (d)

Now Work PROBLEMS 11 AND 27

Recall that an angle measured counterclockwise is positive and an angle measured clockwise is negative. This convention has some interesting consequences relating to polar coordinates. Let's see what these consequences are.

EXAMPLE 2 **Finding Several Polar Coordinates of a Single Point**

Consider again the point P with polar coordinates $\left(2, \dfrac{\pi}{4}\right)$, as shown in Figure 7(a).

Because $\dfrac{\pi}{4}$, $\dfrac{9\pi}{4}$, and $-\dfrac{7\pi}{4}$ all have the same terminal side, we also could have

located this point P by using the polar coordinates $\left(2, \dfrac{9\pi}{4}\right)$ or $\left(2, -\dfrac{7\pi}{4}\right)$, as shown

in Figures 7(b) and (c). The point $\left(2, \dfrac{\pi}{4}\right)$ can also be represented by the polar co-

ordinates $\left(-2, \dfrac{5\pi}{4}\right)$. See Figure 7(d).

Figure 7

(a) (b) (c) (d)

EXAMPLE 3	Finding Other Polar Coordinates of a Given Point

Plot the point P with polar coordinates $\left(3, \dfrac{\pi}{6}\right)$, and find other polar coordinates (r, θ) of this same point for which:

(a) $r > 0$, $\quad 2\pi \le \theta < 4\pi$ (b) $r < 0$, $\quad 0 \le \theta < 2\pi$

(c) $r > 0$, $\quad -2\pi \le \theta < 0$

Solution The point $\left(3, \dfrac{\pi}{6}\right)$ is plotted in Figure 8.

Figure 8

$P = \left(3, \tfrac{\pi}{6}\right)$

$\dfrac{\pi}{6}$

0

(a) We add 1 revolution (2π radians) to the angle $\dfrac{\pi}{6}$ to get

$$P = \left(3, \frac{\pi}{6} + 2\pi\right) = \left(3, \frac{13\pi}{6}\right). \text{ See Figure 9.}$$

(b) We add $\dfrac{1}{2}$ revolution (π radians) to the angle $\dfrac{\pi}{6}$ and replace 3 by -3 to get

$$P = \left(-3, \frac{\pi}{6} + \pi\right) = \left(-3, \frac{7\pi}{6}\right). \text{ See Figure 10.}$$

(c) We subtract 2π from the angle $\dfrac{\pi}{6}$ to get $P = \left(3, \dfrac{\pi}{6} - 2\pi\right) = \left(3, -\dfrac{11\pi}{6}\right)$. See Figure 11.

Figure 9	**Figure 10**	**Figure 11**

Now Work PROBLEM 31

These examples show a major difference between rectangular coordinates and polar coordinates. In the former, each point has exactly one pair of rectangular coordinates; in the latter, a point can have infinitely many pairs of polar coordinates.

SUMMARY

A point with polar coordinates (r, θ) also can be represented by either of the following:

$$(r, \theta + 2\pi k) \quad \text{or} \quad (-r, \theta + \pi + 2\pi k) \qquad k \text{ any integer}$$

The polar coordinates of the pole are $(0, \theta)$, where θ can be any angle.

2 Convert from Polar Coordinates to Rectangular Coordinates

Sometimes we need to convert coordinates or equations in rectangular form to polar form, and vice versa. To do this, we recall that the origin in rectangular coordinates is the pole in polar coordinates and that the positive x-axis in rectangular coordinates is the polar axis in polar coordinates.

THEOREM

Conversion from Polar Coordinates to Rectangular Coordinates

If P is a point with polar coordinates (r, θ), the rectangular coordinates (x, y) of P are given by

$$x = r \cos \theta \qquad y = r \sin \theta \tag{1}$$

Figure 12

Proof Suppose that P has the polar coordinates (r, θ). We seek the rectangular coordinates (x, y) of P. Refer to Figure 12.

If $r = 0$, then, regardless of θ, the point P is the pole, for which the rectangular coordinates are $(0, 0)$. Formula (1) is valid for $r = 0$.

If $r > 0$, the point P is on the terminal side of θ, and $r = d(O, P) = \sqrt{x^2 + y^2}$. Since

$$\cos \theta = \frac{x}{r} \qquad \sin \theta = \frac{y}{r}$$

we have

$$x = r \cos \theta \qquad y = r \sin \theta$$

If $r < 0$, the point $P = (r, \theta)$ can be represented as $(-r, \pi + \theta)$, where $-r > 0$. Since

$$\cos(\pi + \theta) = -\cos \theta = \frac{x}{-r} \qquad \sin(\pi + \theta) = -\sin \theta = \frac{y}{-r}$$

we have

$$x = r \cos \theta \qquad y = r \sin \theta \qquad \blacksquare$$

| EXAMPLE 4 | **Converting from Polar Coordinates to Rectangular Coordinates** |

Find the rectangular coordinates of the points with the following polar coordinates:

(a) $\left(6, \dfrac{\pi}{6}\right)$ (b) $\left(-4, -\dfrac{\pi}{4}\right)$

Solution We use formula (1): $x = r \cos \theta$ and $y = r \sin \theta$.

Figure 13

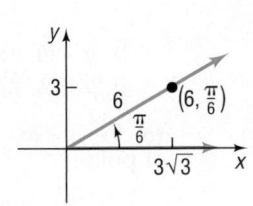

(a)

(a) Figure 13(a) shows $\left(6, \dfrac{\pi}{6}\right)$ plotted. Notice that $\left(6, \dfrac{\pi}{6}\right)$ lies in quadrant I of the rectangular coordinate system. So we expect both the x-coordinate and the y-coordinate to be positive. With $r = 6$ and $\theta = \dfrac{\pi}{6}$, we have

$$x = r \cos \theta = 6 \cos \frac{\pi}{6} = 6 \cdot \frac{\sqrt{3}}{2} = 3\sqrt{3}$$

$$y = r \sin \theta = 6 \sin \frac{\pi}{6} = 6 \cdot \frac{1}{2} = 3$$

The rectangular coordinates of the point $\left(6, \dfrac{\pi}{6}\right)$ are $\left(3\sqrt{3}, 3\right)$, which lies in quadrant I, as expected.

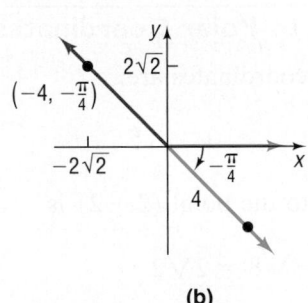

(b)

(b) Figure 13(b) shows $\left(-4, -\dfrac{\pi}{4}\right)$ plotted. Notice that $\left(-4, -\dfrac{\pi}{4}\right)$ lies in quadrant II of the rectangular coordinate system. With $r = -4$ and $\theta = -\dfrac{\pi}{4}$, we have

$$x = r \cos \theta = -4 \cos\left(-\frac{\pi}{4}\right) = -4 \cdot \frac{\sqrt{2}}{2} = -2\sqrt{2}$$

$$y = r \sin \theta = -4 \sin\left(-\frac{\pi}{4}\right) = -4\left(-\frac{\sqrt{2}}{2}\right) = 2\sqrt{2}$$

COMMENT Most calculators have the capability of converting from polar coordinates to rectangular coordinates. Consult your owner's manual for the proper key strokes. Since this procedure is often tedious, you will find that using formula (1) is faster. ∎

The rectangular coordinates of the point $\left(-4, -\dfrac{\pi}{4}\right)$ are $\left(-2\sqrt{2}, 2\sqrt{2}\right)$, which lies in quadrant II, as expected.

 ∎

Now Work PROBLEMS **39** AND **51**

3 Convert from Rectangular Coordinates to Polar Coordinates

Converting from rectangular coordinates (x, y) to polar coordinates (r, θ) is a little more complicated. Notice that we begin each example by plotting the given rectangular coordinates.

| EXAMPLE 5 | **Converting from Rectangular Coordinates to Polar Coordinates** |

Find polar coordinates of a point whose rectangular coordinates are $(0, 3)$.

Solution

Figure 14

See Figure 14. The point $(0, 3)$ lies on the y-axis a distance of 3 units from the origin (pole), so $r = 3$. A ray with vertex at the pole through $(0, 3)$ forms an angle $\theta = \dfrac{\pi}{2}$ with the polar axis. Polar coordinates for this point can be given by $\left(3, \dfrac{\pi}{2}\right)$.

COMMENT Most calculators have the capability of converting from rectangular coordinates to polar coordinates. Consult your owner's manual for the proper keystrokes. ∎

Figure 15 shows polar coordinates of points that lie on either the x-axis or the y-axis. In each illustration, $a > 0$.

Figure 15

(a) $(x, y) = (a, 0), a > 0$ **(b)** $(x, y) = (0, a), a > 0$ **(c)** $(x, y) = (-a, 0), a > 0$ **(d)** $(x, y) = (0, -a), a > 0$

Now Work PROBLEM 55

| EXAMPLE 6 | **Converting from Rectangular Coordinates to Polar Coordinates** |

Find polar coordinates of a point whose rectangular coordinates are:

(a) $(2, -2)$ (b) $\left(-1, -\sqrt{3}\right)$

Solution

Figure 16

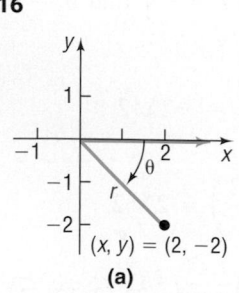

(a)

(a) See Figure 16(a). The distance r from the origin to the point $(2, -2)$ is

$$r = \sqrt{x^2 + y^2} = \sqrt{(2)^2 + (-2)^2} = \sqrt{8} = 2\sqrt{2}$$

We find θ by recalling that $\tan \theta = \dfrac{y}{x}$, so $\theta = \tan^{-1}\dfrac{y}{x}$, $-\dfrac{\pi}{2} < \theta < \dfrac{\pi}{2}$.

Since $(2, -2)$ lies in quadrant IV, we know that $-\dfrac{\pi}{2} < \theta < 0$. As a result,

$$\theta = \tan^{-1}\frac{y}{x} = \tan^{-1}\left(\frac{-2}{2}\right) = \tan^{-1}(-1) = -\frac{\pi}{4}$$

A set of polar coordinates for this point is $\left(2\sqrt{2}, -\dfrac{\pi}{4}\right)$. Other possible representations include $\left(2\sqrt{2}, \dfrac{7\pi}{4}\right)$ and $\left(-2\sqrt{2}, \dfrac{3\pi}{4}\right)$.

Figure 16

$(x, y) = (-1, -\sqrt{3})$
(b)

(b) See Figure 16(b). The distance r from the origin to the point $\left(-1, -\sqrt{3}\right)$ is

$$r = \sqrt{(-1)^2 + \left(-\sqrt{3}\right)^2} = \sqrt{4} = 2$$

To find θ, we use $\theta = \tan^{-1}\dfrac{y}{x}$, $-\dfrac{\pi}{2} < \theta < \dfrac{\pi}{2}$. Since the point $\left(-1, -\sqrt{3}\right)$ lies in quadrant III and the inverse tangent function gives an angle in quadrant I, we add π to the result to obtain an angle in quadrant III.

$$\theta = \pi + \tan^{-1}\left(\frac{-\sqrt{3}}{-1}\right) = \pi + \tan^{-1}\sqrt{3} = \pi + \frac{\pi}{3} = \frac{4\pi}{3}$$

A set of polar coordinates for this point is $\left(2, \dfrac{4\pi}{3}\right)$. Other possible representations include $\left(-2, \dfrac{\pi}{3}\right)$ and $\left(2, -\dfrac{2\pi}{3}\right)$.

Figure 17 shows how to find polar coordinates of a point that lies in a quadrant when its rectangular coordinates (x, y) are given.

Figure 17

(a) $r = \sqrt{x^2 + y^2}$
$\theta = \tan^{-1}\dfrac{y}{x}$

(b) $r = \sqrt{x^2 + y^2}$
$\theta = \pi + \tan^{-1}\dfrac{y}{x}$

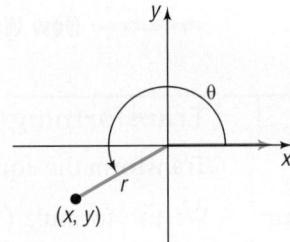

(c) $r = \sqrt{x^2 + y^2}$
$\theta = \pi + \tan^{-1}\dfrac{y}{x}$

(d) $r = \sqrt{x^2 + y^2}$
$\theta = \tan^{-1}\dfrac{y}{x}$

Based on the preceding discussion, we have the formulas

$$r^2 = x^2 + y^2 \qquad \tan\theta = \frac{y}{x} \qquad \text{if } x \neq 0 \qquad \textbf{(2)}$$

To use formula (2) effectively, follow these steps:

Steps for Converting from Rectangular to Polar Coordinates

STEP 1: Always plot the point (x, y) first, as we did in Examples 5 and 6.
STEP 2: If $x = 0$ or $y = 0$, use your illustration to find (r, θ). See Figure 15.
STEP 3: If $x \neq 0$ and $y \neq 0$, then $r = \sqrt{x^2 + y^2}$.
STEP 4: To find θ, first determine the quadrant in which the point lies.

Quadrant I or IV: $\theta = \tan^{-1}\dfrac{y}{x}$ Quadrant II or III: $\theta = \pi + \tan^{-1}\dfrac{y}{x}$

See Figure 17.

━━━ **Now Work** PROBLEM 59

4 Transform Equations from Polar to Rectangular Form

Formulas (1) and (2) may also be used to transform equations from polar form to rectangular form, and vice-versa. Two common techniques for transforming an equation from polar form to rectangular form are

1. multiplying both sides of the equation by r
2. squaring both sides of the equation.

| EXAMPLE 7 | Transforming an Equation from Polar to Rectangular Form |

Transform the equation $r = 6 \cos \theta$ from polar coordinates to rectangular coordinates, and identify the graph.

Solution If we multiply each side by r, it will be easier to apply formulas (1) and (2).

$$r = 6 \cos \theta$$
$$r^2 = 6r \cos \theta \quad \text{Multiply each side by } r.$$
$$x^2 + y^2 = 6x \quad r^2 = x^2 + y^2; x = r \cos \theta$$

This is the equation of a circle, so we proceed to complete the square to obtain the standard form of the equation.

$$x^2 + y^2 = 6x$$
$$(x^2 - 6x) + y^2 = 0 \quad \text{General form}$$
$$(x^2 - 6x + 9) + y^2 = 9 \quad \text{Complete the square in } x.$$
$$(x - 3)^2 + y^2 = 9 \quad \text{Factor.}$$

This is the standard form of the equation of a circle with center $(3, 0)$ and radius 3. ∎

➤ **Now Work** PROBLEM 75

| EXAMPLE 8 | Transforming an Equation from Rectangular to Polar Form |

Transform the equation $4xy = 9$ from rectangular coordinates to polar coordinates.

Solution We use formula (1): $x = r \cos \theta$ and $y = r \sin \theta$.

$$4xy = 9$$
$$4(r \cos \theta)(r \sin \theta) = 9 \quad x = r \cos \theta, y = r \sin \theta$$
$$4r^2 \cos \theta \sin \theta = 9$$

This is the polar form of the equation. It can be simplified as shown next:

$$2r^2(2 \sin \theta \cos \theta) = 9 \quad \text{Factor out } 2r^2.$$
$$2r^2 \sin(2\theta) = 9 \quad \text{Double-angle Formula}$$

➤ **Now Work** PROBLEM 69

10.1 Assess Your Understanding

'Are You Prepared?' *Answers are given at the end of these exercises. If you get a wrong answer, read the pages listed in red.*

1. Plot the point whose rectangular coordinates are $(3, -1)$. (p. 156)

2. To complete the square of $x^2 + 6x$, add _____. (pp. 99–101)

3. If $P = (a, b)$ is a point on the terminal side of the angle θ at a distance r from the origin, then $\sin \theta =$ _____. (pp. 540–542)

4. $\tan^{-1}(-1) =$ _____. (pp. 608–610)

Concepts and Vocabulary

5. In polar coordinates, the origin is called the _____ and the positive x-axis is referred to as the _____ _____.

6. Another representation in polar coordinates for the point $\left(2, \dfrac{\pi}{3}\right)$ is $\left(\underline{\quad}, \dfrac{4\pi}{3}\right)$.

7. The polar coordinates $\left(-2, \dfrac{\pi}{6}\right)$ are represented in rectangular coordinates by (_____, _____).

8. **True or False** The polar coordinates of a point are unique.

9. **True or False** The rectangular coordinates of a point are unique.

10. **True or False** In (r, θ), the number r can be negative.

Skill Building

In Problems 11–18, match each point in polar coordinates with either A, B, C, or D on the graph.

11. $\left(2, -\dfrac{11\pi}{6}\right)$ 12. $\left(-2, -\dfrac{\pi}{6}\right)$ 13. $\left(-2, \dfrac{\pi}{6}\right)$ 14. $\left(2, \dfrac{7\pi}{6}\right)$

15. $\left(2, \dfrac{5\pi}{6}\right)$ 16. $\left(-2, \dfrac{5\pi}{6}\right)$ 17. $\left(-2, \dfrac{7\pi}{6}\right)$ 18. $\left(2, \dfrac{11\pi}{6}\right)$

In Problems 19–30, plot each point given in polar coordinates.

19. $(3, 90°)$ 20. $(4, 270°)$ 21. $(-2, 0)$ 22. $(-3, \pi)$

23. $\left(6, \dfrac{\pi}{6}\right)$ 24. $\left(5, \dfrac{5\pi}{3}\right)$ 25. $(-2, 135°)$ 26. $(-3, 120°)$

27. $\left(-1, -\dfrac{\pi}{3}\right)$ 28. $\left(-3, -\dfrac{3\pi}{4}\right)$ 29. $(-2, -\pi)$ 30. $\left(-3, -\dfrac{\pi}{2}\right)$

In Problems 31–38, plot each point given in polar coordinates, and find other polar coordinates (r, θ) of the point for which:

 (a) $r > 0$, $-2\pi \le \theta < 0$ (b) $r < 0$, $0 \le \theta < 2\pi$ (c) $r > 0$, $2\pi \le \theta < 4\pi$

31. $\left(5, \dfrac{2\pi}{3}\right)$ 32. $\left(4, \dfrac{3\pi}{4}\right)$ 33. $(-2, 3\pi)$ 34. $(-3, 4\pi)$

35. $\left(1, \dfrac{\pi}{2}\right)$ 36. $(2, \pi)$ 37. $\left(-3, -\dfrac{\pi}{4}\right)$ 38. $\left(-2, -\dfrac{2\pi}{3}\right)$

In Problems 39–54, the polar coordinates of a point are given. Find the rectangular coordinates of each point.

39. $\left(3, \dfrac{\pi}{2}\right)$ 40. $\left(4, \dfrac{3\pi}{2}\right)$ 41. $(-2, 0)$ 42. $(-3, \pi)$

43. $(6, 150°)$ 44. $(5, 300°)$ 45. $\left(-2, \dfrac{3\pi}{4}\right)$ 46. $\left(-2, \dfrac{2\pi}{3}\right)$

47. $\left(-1, -\dfrac{\pi}{3}\right)$ 48. $\left(-3, -\dfrac{3\pi}{4}\right)$ 49. $(-2, -180°)$ 50. $(-3, -90°)$

51. $(7.5, 110°)$ 52. $(-3.1, 182°)$ 53. $(6.3, 3.8)$ 54. $(8.1, 5.2)$

In Problems 55–66, the rectangular coordinates of a point are given. Find polar coordinates for each point.

55. $(3, 0)$ 56. $(0, 2)$ 57. $(-1, 0)$ 58. $(0, -2)$

59. $(1, -1)$ 60. $(-3, 3)$ 61. $\left(\sqrt{3}, 1\right)$ 62. $\left(-2, -2\sqrt{3}\right)$

63. $(1.3, -2.1)$ 64. $(-0.8, -2.1)$ 65. $(8.3, 4.2)$ 66. $(-2.3, 0.2)$

In Problems 67–74, the letters x and y represent rectangular coordinates. Write each equation using polar coordinates (r, θ).

67. $2x^2 + 2y^2 = 3$ 68. $x^2 + y^2 = x$ 69. $x^2 = 4y$ 70. $y^2 = 2x$

71. $2xy = 1$ 72. $4x^2y = 1$ 73. $x = 4$ 74. $y = -3$

In Problems 75–82, the letters r and θ represent polar coordinates. Write each equation using rectangular coordinates (x, y).

75. $r = \cos \theta$ 76. $r = \sin \theta + 1$ 77. $r^2 = \cos \theta$ 78. $r = \sin \theta - \cos \theta$

79. $r = 2$ 80. $r = 4$ 81. $r = \dfrac{4}{1 - \cos \theta}$ 82. $r = \dfrac{3}{3 - \cos \theta}$

Applications and Extensions

83. **Chicago** In Chicago, the road system is set up like a Cartesian plane, where streets are indicated by the number of blocks they are from Madison Street and State Street. For example, Wrigley Field in Chicago is located at 1060 West Addison, which is 10 blocks west of State Street and 36 blocks north of Madison Street. Treat the intersection of Madison Street and State Street as the origin of a coordinate system, with East being the positive x-axis.

(a) Write the location of Wrigley Field using rectangular coordinates.

(b) Write the location of Wrigley Field using polar coordinates. Use the East direction for the polar axis. Express θ in degrees.

(c) U.S. Cellular Field, home of the White Sox, is located at 35th and Princeton, which is 3 blocks west of State Street and 35 blocks south of Madison. Write the location of U.S. Cellular Field using rectangular coordinates.

(d) Write the location of U.S. Cellular Field using polar coordinates. Use the East direction for the polar axis. Express θ in degrees.

84. Show that the formula for the distance d between two points $P_1 = (r_1, \theta_1)$ and $P_2 = (r_2, \theta_2)$ is

$$d = \sqrt{r_1^2 + r_2^2 - 2r_1 r_2 \cos(\theta_2 - \theta_1)}$$

City of Chicago, Illinois

Discussion and Writing

85. In converting from polar coordinates to rectangular coordinates, what formulas will you use?

86. Explain how you proceed to convert from rectangular coordinates to polar coordinates.

87. Is the street system in your town based on a rectangular coordinate system, a polar coordinate system, or some other system? Explain.

'Are You Prepared?' Answers

1. **2.** 9 **3.** $\dfrac{b}{r}$ **4.** $-\dfrac{\pi}{4}$

10.2 Polar Equations and Graphs

PREPARING FOR THIS SECTION *Before getting started, review the following:*

- Symmetry (Section 2.2, pp. 167–168)
- Circles (Section 2.4, pp. 189–193)
- Even–Odd Properties of Trigonometric Functions (Section 7.5, pp. 556–557)

- Difference Formulas for Sine and Cosine (Section 8.4, pp. 627 and 630)
- Value of the Sine and Cosine Functions at Certain Angles (Section 7.3, pp. 529–532 and Section 7.4, pp. 540–547)

Now Work the 'Are You Prepared?' problems on page 735.

OBJECTIVES **1** Graph and Identify Polar Equations by Converting to Rectangular Equations (p. 723)
 2 Test Polar Equations for Symmetry (p. 727)
 3 Graph Polar Equations by Plotting Points (p. 728)

Just as a rectangular grid may be used to plot points given by rectangular coordinates, as in Figure 18(a), we can use a grid consisting of concentric circles (with centers at the pole) and rays (with vertices at the pole) to plot points given by polar coordinates, as shown in Figure 18(b). We shall use such **polar grids** to graph *polar equations*.

Figure 18

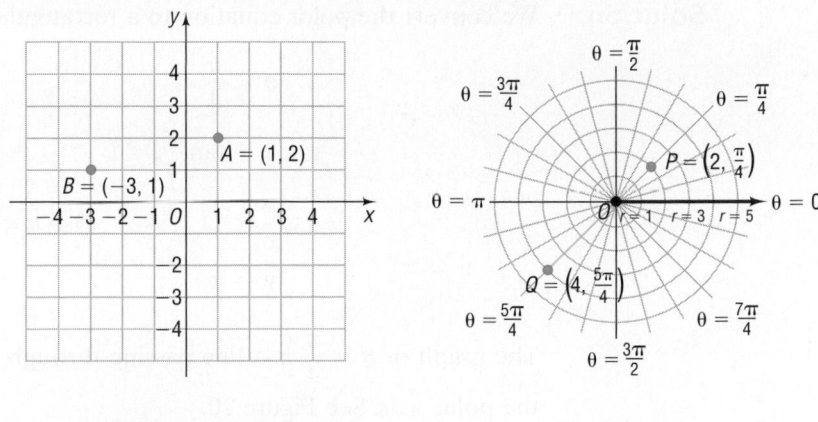

(a) Rectangular grid **(b)** Polar grid

DEFINITION

An equation whose variables are polar coordinates is called a **polar equation.** The **graph of a polar equation** consists of all points whose polar coordinates satisfy the equation.

1 Graph and Identify Polar Equations by Converting to Rectangular Equations

One method that we can use to graph a polar equation is to convert the equation to rectangular coordinates. In the discussion that follows, (x, y) represent the rectangular coordinates of a point P, and (r, θ) represent polar coordinates of the point P.

| EXAMPLE 1 | Identifying and Graphing a Polar Equation (Circle) |

Identify and graph the equation: $r = 3$

Solution We convert the polar equation to a rectangular equation.

$$r = 3$$
$$r^2 = 9 \quad \text{Square both sides.}$$
$$x^2 + y^2 = 9 \quad r^2 = x^2 + y^2$$

The graph of $r = 3$ is a circle, with center at the pole and radius 3. See Figure 19.

Figure 19
$r = 3$ or $x^2 + y^2 = 9$

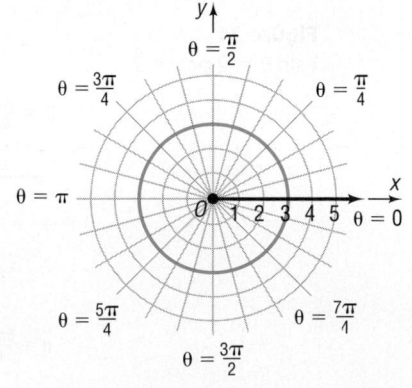

Now Work PROBLEM 13

| EXAMPLE 2 | Identifying and Graphing a Polar Equation (Line) |

Identify and graph the equation: $\theta = \dfrac{\pi}{4}$

Solution We convert the polar equation to a rectangular equation.

$$\theta = \frac{\pi}{4}$$

$$\tan \theta = \tan \frac{\pi}{4} = 1 \quad \text{Take the tangent of both sides.}$$

$$\frac{y}{x} = 1 \qquad \tan \theta = \frac{y}{x}$$

$$y = x$$

The graph of $\theta = \frac{\pi}{4}$ is a line passing through the pole making an angle of $\frac{\pi}{4}$ with the polar axis. See Figure 20.

Figure 20

$\theta = \frac{\pi}{4}$ or $y = x$

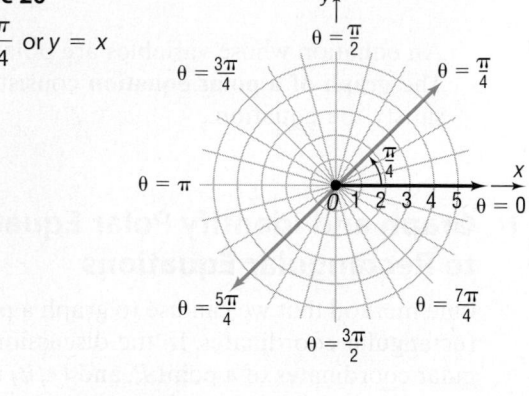

━━━ **Now Work** PROBLEM 15

EXAMPLE 3 **Identifying and Graphing a Polar Equation (Horizontal Line)**

Identify and graph the equation: $r \sin \theta = 2$

Solution Since $y = r \sin \theta$, we can write the equation as

$$y = 2$$

We conclude that the graph of $r \sin \theta = 2$ is a horizontal line 2 units above the pole. See Figure 21.

Figure 21

$r \sin \theta = 2$ or $y = 2$

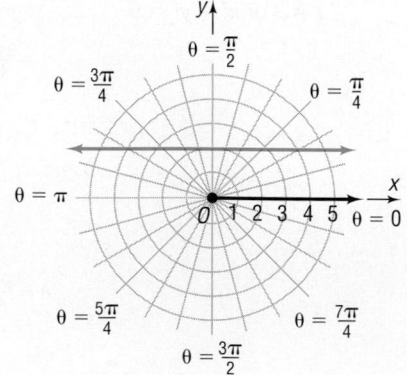

COMMENT A graphing utility can be used to graph polar equations. Read *Using a Graphing Utility to Graph a Polar Equation* in the Appendix, Section 8. Then graph $r = \dfrac{2}{\sin \theta}$ and compare the result with Figure 21. ∎

| EXAMPLE 4 | **Identifying and Graphing a Polar Equation (Vertical Line)** |

Identify and graph the equation: $r \cos \theta = -3$

Solution Since $x = r \cos \theta$, we can write the equation as

$$x = -3$$

We conclude that the graph of $r \cos \theta = -3$ is a vertical line 3 units to the left of the pole. See Figure 22.

Figure 22
$r \cos \theta = -3$ or $x = -3$

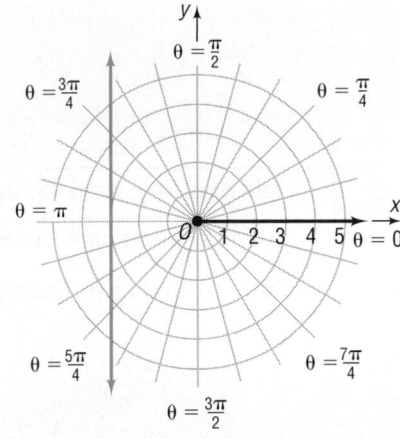

Based on Examples 3 and 4, we are led to the following results. (The proofs are left as exercises. See Problems 75 and 76.)

| THEOREM | Let a be a nonzero real number. Then the graph of the equation |

$$r \sin \theta = a$$

is a horizontal line a units above the pole if $a > 0$ and $|a|$ units below the pole if $a < 0$.
 The graph of the equation

$$r \cos \theta = a$$

is a vertical line a units to the right of the pole if $a > 0$ and $|a|$ units to the left of the pole if $a < 0$.

━━━ **Now Work** PROBLEM 19

| EXAMPLE 5 | **Identifying and Graphing a Polar Equation (Circle)** |

Identify and graph the equation: $r = 4 \sin \theta$

Solution To transform the equation to rectangular coordinates, we multiply each side by r.

$$r^2 = 4r \sin \theta$$

Now we use the facts that $r^2 = x^2 + y^2$ and $y = r \sin \theta$. Then

$$x^2 + y^2 = 4y$$
$$x^2 + (y^2 - 4y) = 0$$
$$x^2 + (y^2 - 4y + 4) = 4 \quad \text{Complete the square in } y.$$
$$x^2 + (y - 2)^2 = 4 \quad \text{Factor.}$$

This is the standard equation of a circle with center at $(0, 2)$ in rectangular coordinates and radius 2. See Figure 23.

Figure 23
$r = 4 \sin \theta$ or $x^2 + (y - 2)^2 = 4$

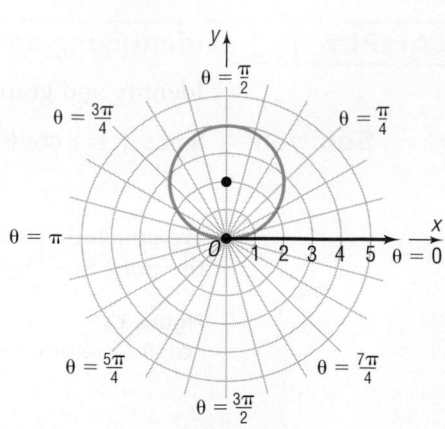

EXAMPLE 6	**Identifying and Graphing a Polar Equation (Circle)**

Identify and graph the equation: $r = -2 \cos \theta$

Solution We proceed as in Example 5.

$$r^2 = -2r \cos \theta \qquad \text{Multiply both sides by } r.$$
$$x^2 + y^2 = -2x \qquad r^2 = x^2 + y^2; \quad x = r \cos \theta$$
$$x^2 + 2x + y^2 = 0$$
$$(x^2 + 2x + 1) + y^2 = 1 \qquad \text{Complete the square in } x.$$
$$(x + 1)^2 + y^2 = 1 \qquad \text{Factor.}$$

This is the standard equation of a circle with center at $(-1, 0)$ in rectangular coordinates and radius 1. See Figure 24.

Figure 24
$r = -2 \cos \theta$ or $(x + 1)^2 + y^2 = 1$

Exploration

Using a square screen, graph $r_1 = \sin \theta, r_2 = 2 \sin \theta$, and $r_3 = 3 \sin \theta$. Do you see the pattern? Clear the screen and graph $r_1 = -\sin \theta, r_2 = -2 \sin \theta$, and $r_3 = -3 \sin \theta$. Do you see the pattern? Clear the screen and graph $r_1 = \cos \theta, r_2 = 2 \cos \theta$, and $r_3 = 3 \cos \theta$. Do you see the pattern? Clear the screen and graph $r_1 = -\cos \theta, r_2 = -2 \cos \theta$, and $r_3 = -3 \cos \theta$. Do you see the pattern?

Based on Examples 5 and 6 and the preceding Exploration, we are led to the following results. (The proofs are left as exercises. See Problems 77–80.)

THEOREM Let a be a positive real number. Then

Equation	Description
(a) $r = 2a \sin \theta$	Circle: radius a; center at $(0, a)$ in rectangular coordinates
(b) $r = -2a \sin \theta$	Circle: radius a; center at $(0, -a)$ in rectangular coordinates

(c) $r = 2a \cos \theta$ Circle: radius a; center at $(a, 0)$ in rectangular coordinates

(d) $r = -2a \cos \theta$ Circle: radius a; center at $(-a, 0)$ in rectangular coordinates

Each circle passes through the pole.

Now Work PROBLEM 21

The method of converting a polar equation to an identifiable rectangular equation to obtain the graph is not always helpful, nor is it always necessary. Usually, we set up a table that lists several points on the graph. By checking for symmetry, it may be possible to reduce the number of points needed to draw the graph.

2 Test Polar Equations for Symmetry

In polar coordinates, the points (r, θ) and $(r, -\theta)$ are symmetric with respect to the polar axis (and to the x-axis). See Figure 25(a). The points (r, θ) and $(r, \pi - \theta)$ are symmetric with respect to the line $\theta = \dfrac{\pi}{2}$ (the y-axis). See Figure 25(b). The points (r, θ) and $(-r, \theta)$ are symmetric with respect to the pole (the origin). See Figure 25(c).

Figure 25

(a) Points symmetric with respect to the polar axis

(b) Points symmetric with respect to the line $\theta = \dfrac{\pi}{2}$

(c) Points symmetric with respect to the pole

The following tests are a consequence of these observations.

THEOREM

Tests for Symmetry

Symmetry with Respect to the Polar Axis (x-Axis)

In a polar equation, replace θ by $-\theta$. If an equivalent equation results, the graph is symmetric with respect to the polar axis.

Symmetry with Respect to the Line $\theta = \dfrac{\pi}{2}$ (y-Axis)

In a polar equation, replace θ by $\pi - \theta$. If an equivalent equation results, the graph is symmetric with respect to the line $\theta = \dfrac{\pi}{2}$.

Symmetry with Respect to the Pole (Origin)

In a polar equation, replace r by $-r$. If an equivalent equation results, the graph is symmetric with respect to the pole.

The three tests for symmetry given here are *sufficient* conditions for symmetry, but they are not *necessary* conditions. That is, an equation may fail these tests and still have a graph that is symmetric with respect to the polar axis, the line $\theta = \dfrac{\pi}{2}$, or the pole. For example, the graph of $r = \sin(2\theta)$ turns out to be symmetric with respect to the polar axis, the line $\theta = \dfrac{\pi}{2}$, and the pole, but all three tests given here fail. See also Problems 81–83.

3 Graph Polar Equations by Plotting Points

EXAMPLE 7	**Graphing a Polar Equation (Cardioid)**

Graph the equation: $r = 1 - \sin\theta$

Solution We check for symmetry first.

Polar Axis: Replace θ by $-\theta$. The result is

$$r = 1 - \sin(-\theta) = 1 + \sin\theta \quad \text{sin}(-\theta) = -\sin\theta$$

The test fails, so the graph *may or may not* be symmetric with respect to the polar axis.

The Line $\theta = \dfrac{\pi}{2}$**:** Replace θ by $\pi - \theta$. The result is

$$r = 1 - \sin(\pi - \theta) = 1 - (\sin\pi\cos\theta - \cos\pi\sin\theta)$$
$$= 1 - [0\cdot\cos\theta - (-1)\sin\theta] = 1 - \sin\theta$$

The test is satisfied, so the graph is symmetric with respect to the line $\theta = \dfrac{\pi}{2}$.

The Pole: Replace r by $-r$. Then the result is $-r = 1 - \sin\theta$, so $r = -1 + \sin\theta$. The test fails, so the graph *may or may not* be symmetric with respect to the pole.

Next we identify points on the graph by assigning values to the angle θ and calculating the corresponding values of r. Due to the symmetry with respect to the line $\theta = \dfrac{\pi}{2}$, we only need to assign values to θ from $-\dfrac{\pi}{2}$ to $\dfrac{\pi}{2}$, as given in Table 1.

Now we plot the points (r, θ) from Table 1 and trace out the graph, beginning at the point $\left(2, -\dfrac{\pi}{2}\right)$ and ending at the point $\left(0, \dfrac{\pi}{2}\right)$. Then we reflect this portion of the graph about the line $\theta = \dfrac{\pi}{2}$ (the y-axis) to obtain the complete graph. See Figure 26.

Table 1

θ	$r = 1 - \sin\theta$
$-\dfrac{\pi}{2}$	$1 - (-1) = 2$
$-\dfrac{\pi}{3}$	$1 - \left(-\dfrac{\sqrt{3}}{2}\right) \approx 1.87$
$-\dfrac{\pi}{6}$	$1 - \left(-\dfrac{1}{2}\right) = \dfrac{3}{2}$
0	$1 - 0 = 1$
$\dfrac{\pi}{6}$	$1 - \dfrac{1}{2} = \dfrac{1}{2}$
$\dfrac{\pi}{3}$	$1 - \dfrac{\sqrt{3}}{2} \approx 0.13$
$\dfrac{\pi}{2}$	$1 - 1 = 0$

Figure 26

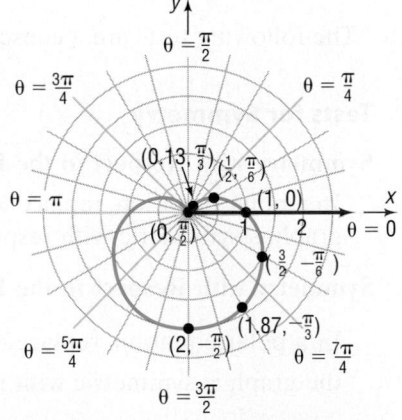

Exploration

Graph $r_1 = 1 + \sin\theta$. Clear the screen and graph $r_1 = 1 - \cos\theta$. Clear the screen and graph $r_1 = 1 + \cos\theta$. Do you see a pattern?

The curve in Figure 26 is an example of a *cardioid* (a heart-shaped curve).

DEFINITION **Cardioids** are characterized by equations of the form

$$r = a(1 + \cos\theta) \qquad r = a(1 + \sin\theta)$$
$$r = a(1 - \cos\theta) \qquad r = a(1 - \sin\theta)$$

where $a > 0$. The graph of a cardioid passes through the pole.

Now Work PROBLEM 37

EXAMPLE 8	Graphing a Polar Equation (Limaçon without an Inner Loop)

Graph the equation: $r = 3 + 2\cos\theta$

Solution

We check for symmetry first.

Polar Axis: Replace θ by $-\theta$. The result is

$$r = 3 + 2\cos(-\theta) = 3 + 2\cos\theta \qquad \cos(-\theta) = \cos\theta$$

The test is satisfied, so the graph is symmetric with respect to the polar axis.

The Line $\theta = \dfrac{\pi}{2}$: Replace θ by $\pi - \theta$. The result is

$$r = 3 + 2\cos(\pi - \theta) = 3 + 2(\cos\pi\cos\theta + \sin\pi\sin\theta)$$
$$= 3 - 2\cos\theta$$

The test fails, so the graph *may or may not* be symmetric with respect to the line $\theta = \dfrac{\pi}{2}$.

The Pole: Replace r by $-r$. The test fails, so the graph *may or may not* be symmetric with respect to the pole.

Next we identify points on the graph by assigning values to the angle θ and calculating the corresponding values of r. Due to the symmetry with respect to the polar axis, we only need to assign values to θ from 0 to π, as given in Table 2.

Now we plot the points (r, θ) from Table 2 and trace out the graph, beginning at the point $(5, 0)$ and ending at the point $(1, \pi)$. Then we reflect this portion of the graph about the polar axis (the x-axis) to obtain the complete graph. See Figure 27.

Table 2

θ	$r = 3 + 2\cos\theta$
0	$3 + 2(1) = 5$
$\dfrac{\pi}{6}$	$3 + 2\left(\dfrac{\sqrt{3}}{2}\right) \approx 4.73$
$\dfrac{\pi}{3}$	$3 + 2\left(\dfrac{1}{2}\right) = 4$
$\dfrac{\pi}{2}$	$3 + 2(0) = 3$
$\dfrac{2\pi}{3}$	$3 + 2\left(-\dfrac{1}{2}\right) = 2$
$\dfrac{5\pi}{6}$	$3 + 2\left(-\dfrac{\sqrt{3}}{2}\right) \approx 1.27$
π	$3 + 2(-1) = 1$

Figure 27

The curve in Figure 27 is an example of a *limaçon* (the French word for *snail*) *without an inner loop.*

Exploration

Graph $r_1 = 3 - 2\cos\theta$. Clear the screen and graph $r_1 = 3 + 2\sin\theta$. Clear the screen and graph $r_1 = 3 - 2\sin\theta$. Do you see a pattern?

DEFINITION

Limaçons without an inner loop are characterized by equations of the form

$$r = a + b\cos\theta, \qquad r = a + b\sin\theta$$
$$r = a - b\cos\theta, \qquad r = a - b\sin\theta$$

where $a > 0$, $b > 0$, and $a > b$. The graph of a limaçon without an inner loop does not pass through the pole.

➡ **Now Work** PROBLEM 43

| EXAMPLE 9 | Graphing a Polar Equation (Limaçon with an Inner Loop) |

Graph the equation: $r = 1 + 2\cos\theta$

Solution

First, we check for symmetry.

Polar Axis: Replace θ by $-\theta$. The result is

$$r = 1 + 2\cos(-\theta) = 1 + 2\cos\theta$$

The test is satisfied, so the graph is symmetric with respect to the polar axis.

The Line $\theta = \dfrac{\pi}{2}$: Replace θ by $\pi - \theta$. The result is

$$r = 1 + 2\cos(\pi - \theta) = 1 + 2(\cos\pi\cos\theta + \sin\pi\sin\theta)$$
$$= 1 - 2\cos\theta$$

The test fails, so the graph *may or may not* be symmetric with respect to the line $\theta = \dfrac{\pi}{2}$.

The Pole: Replace r by $-r$. The test fails, so the graph *may or may not* be symmetric with respect to the pole.

Next we identify points on the graph of $r = 1 + 2\cos\theta$ by assigning values to the angle θ and calculating the corresponding values of r. Due to the symmetry with respect to the polar axis, we only need to assign values to θ from 0 to π, as given in Table 3.

Now we plot the points (r, θ) from Table 3, beginning at $(3, 0)$ and ending at $(-1, \pi)$. See Figure 28(a). Finally, we reflect this portion of the graph about the polar axis (the x-axis) to obtain the complete graph. See Figure 28(b).

Table 3

θ	$r = 1 + 2\cos\theta$
0	$1 + 2(1) = 3$
$\dfrac{\pi}{6}$	$1 + 2\left(\dfrac{\sqrt{3}}{2}\right) \approx 2.73$
$\dfrac{\pi}{3}$	$1 + 2\left(\dfrac{1}{2}\right) = 2$
$\dfrac{\pi}{2}$	$1 + 2(0) = 1$
$\dfrac{2\pi}{3}$	$1 + 2\left(-\dfrac{1}{2}\right) = 0$
$\dfrac{5\pi}{6}$	$1 + 2\left(-\dfrac{\sqrt{3}}{2}\right) \approx -0.73$
π	$1 + 2(-1) = -1$

Figure 28

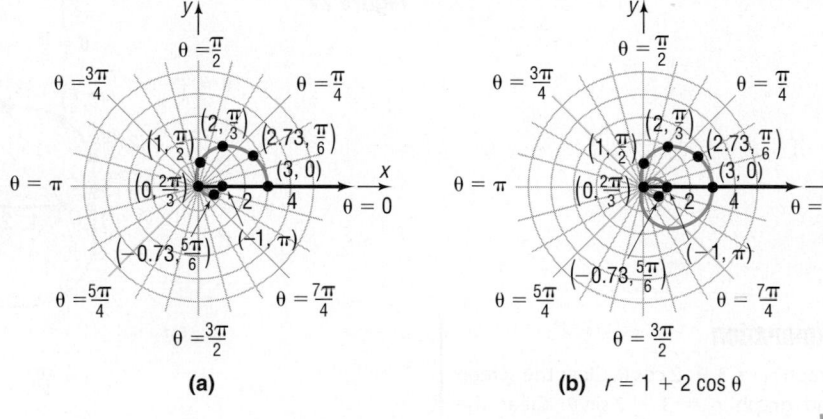

(a)

(b) $r = 1 + 2\cos\theta$

Exploration

Graph $r_1 = 1 - 2\cos\theta$. Clear the screen and graph $r_1 = 1 + 2\sin\theta$. Clear the screen and graph $r_1 = 1 - 2\sin\theta$. Do you see a pattern?

The curve in Figure 28(b) is an example of a *limaçon with an inner loop*.

DEFINITION

Limaçons with an inner loop are characterized by equations of the form

$$r = a + b\cos\theta, \qquad r = a + b\sin\theta$$
$$r = a - b\cos\theta, \qquad r = a - b\sin\theta$$

where $a > 0$, $b > 0$, and $a < b$. The graph of a limaçon with an inner loop will pass through the pole twice.

Now Work PROBLEM 45

| EXAMPLE 10 | Graphing a Polar Equation (Rose) |

Graph the equation: $r = 2 \cos(2\theta)$

Solution

We check for symmetry.

Polar Axis: If we replace θ by $-\theta$, the result is

$$r = 2 \cos[2(-\theta)] = 2 \cos(2\theta)$$

The test is satisfied, so the graph is symmetric with respect to the polar axis.

The Line $\theta = \dfrac{\pi}{2}$: If we replace θ by $\pi - \theta$, we obtain

$$r = 2 \cos[2(\pi - \theta)] = 2 \cos(2\pi - 2\theta) = 2 \cos(2\theta)$$

The test is satisfied, so the graph is symmetric with respect to the line $\theta = \dfrac{\pi}{2}$.

The Pole: Since the graph is symmetric with respect to both the polar axis and the line $\theta = \dfrac{\pi}{2}$, it must be symmetric with respect to the pole. Do you see why?

Next, we construct Table 4. Due to the symmetry with respect to the polar axis, the line $\theta = \dfrac{\pi}{2}$, and the pole, we consider only values of θ from 0 to $\dfrac{\pi}{2}$.

We plot and connect these points in Figure 29(a). Finally, because of symmetry, we reflect this portion of the graph first about the polar axis (the x-axis) and then about the line $\theta = \dfrac{\pi}{2}$ (the y-axis) to obtain the complete graph. See Figure 29(b).

Table 4

θ	$r = 2 \cos(2\theta)$
0	$2(1) = 2$
$\dfrac{\pi}{6}$	$2\left(\dfrac{1}{2}\right) = 1$
$\dfrac{\pi}{4}$	$2(0) = 0$
$\dfrac{\pi}{3}$	$2\left(-\dfrac{1}{2}\right) = -1$
$\dfrac{\pi}{2}$	$2(-1) = -2$

Figure 29

(a)

(b) $r = 2 \cos(2\theta)$

The curve in Figure 29(b) is called a *rose* with four petals.

Exploration

Graph $r_1 = 2 \cos(4\theta)$; clear the screen and graph $r_1 = 2 \cos(6\theta)$. How many petals did each of these graphs have? Clear the screen and graph, in order, each on a clear screen, $r_1 = 2 \cos(3\theta)$, $r_1 = 2 \cos(5\theta)$, and $r_1 = 2 \cos(7\theta)$. What do you notice about the number of petals?

DEFINITION

Rose curves are characterized by equations of the form

$$r = a \cos(n\theta), \qquad r = a \sin(n\theta), \qquad a \neq 0, \qquad n \geq 2$$

and have graphs that are rose shaped. If n is even, the rose has $2n$ petals; if n is odd, the rose has n petals.

══════► **Now Work** PROBLEM 49

| | EXAMPLE 11 | Graphing a Polar Equation (Lemniscate) |

Graph the equation: $r^2 = 4\sin(2\theta)$

Solution

We leave it to you to verify that the graph is symmetric with respect to the pole. Table 5 lists points on the graph for values of $\theta = 0$ through $\theta = \dfrac{\pi}{2}$. Note that there are no points on the graph for $\dfrac{\pi}{2} < \theta < \pi$ (quadrant II), since $\sin(2\theta) < 0$ for such values. The points from Table 5, where $r \geq 0$, are plotted in Figure 30(a). The remaining points on the graph are obtained by using symmetry. Figure 30(b) shows the final graph.

Table 5

θ	$r^2 = 4\sin(2\theta)$	r
0	$4(0) = 0$	0
$\dfrac{\pi}{6}$	$4\left(\dfrac{\sqrt{3}}{2}\right) = 2\sqrt{3}$	± 1.9
$\dfrac{\pi}{4}$	$4(1) = 4$	± 2
$\dfrac{\pi}{3}$	$4\left(\dfrac{\sqrt{3}}{2}\right) = 2\sqrt{3}$	± 1.9
$\dfrac{\pi}{2}$	$4(0) = 0$	0

Figure 30

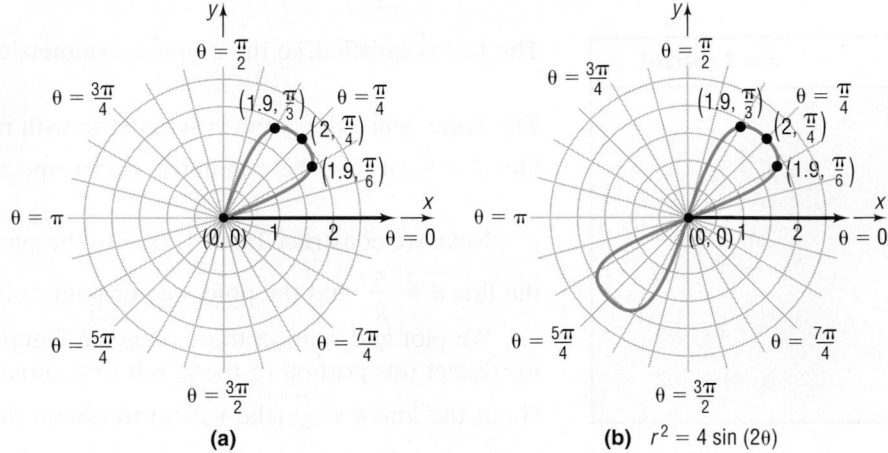

(a)

(b) $r^2 = 4\sin(2\theta)$

The curve in Figure 30(b) is an example of a *lemniscate* (from the Greek word *ribbon*).

DEFINITION

Lemniscates are characterized by equations of the form

$$r^2 = a^2\sin(2\theta), \qquad r^2 = a^2\cos(2\theta)$$

where $a \neq 0$, and have graphs that are propeller shaped.

━━━► **Now Work** PROBLEM 53

| | EXAMPLE 12 | Graphing a Polar Equation (Spiral) |

Graph the equation: $r = e^{\theta/5}$

Solution

The tests for symmetry with respect to the pole, the polar axis, and the line $\theta = \dfrac{\pi}{2}$ fail. Furthermore, there is no number θ for which $r = 0$, so the graph does not pass through the pole. We observe that r is positive for all θ, r increases as θ increases, $r \to 0$ as $\theta \to -\infty$, and $r \to \infty$ as $\theta \to \infty$. With the help of a calculator, we obtain the values in Table 6. See Figure 31 for the graph.

Table 6

θ	$r = e^{\theta/5}$
$-\dfrac{3\pi}{2}$	0.39
$-\pi$	0.53
$-\dfrac{\pi}{2}$	0.73
$-\dfrac{\pi}{4}$	0.85
0	1
$\dfrac{\pi}{4}$	1.17
$\dfrac{\pi}{2}$	1.37
π	1.87
$\dfrac{3\pi}{2}$	2.57
2π	3.51

Figure 31
$r = e^{\theta/5}$

The curve in Figure 31 is called a **logarithmic spiral,** since its equation may be written as $\theta = 5 \ln r$ and it spirals infinitely both toward the pole and away from it.

Classification of Polar Equations

The equations of some lines and circles in polar coordinates and their corresponding equations in rectangular coordinates are given in Table 7. Also included are the names and the graphs of a few of the more frequently encountered polar equations.

Table 7

Lines			
Description	Line passing through the pole making an angle α with the polar axis	Vertical line	Horizontal line
Rectangular equation	$y = (\tan \alpha)x$	$x = a$	$y = b$
Polar equation	$\theta = \alpha$	$r \cos \theta = a$	$r \sin \theta = b$
Typical graph			

(*continued*)

Circles

Description	Center at the pole, radius a	Passing through the pole, tangent to the line $\theta = \dfrac{\pi}{2}$, center on the polar axis, radius a	Passing through the pole, tangent to the polar axis, center on the line $\theta = \dfrac{\pi}{2}$, radius a
Rectangular equation	$x^2 + y^2 = a^2, \quad a > 0$	$x^2 + y^2 = \pm 2ax, \quad a > 0$	$x^2 + y^2 = \pm 2ay, \quad a > 0$
Polar equation	$r = a, \quad a > 0$	$r = \pm 2a \cos \theta, \quad a > 0$	$r = \pm 2a \sin \theta, \quad a > 0$
Typical graph			

Other Equations

Name	Cardioid	Limaçon without inner loop	Limaçon with inner loop
Polar equations	$r = a \pm a \cos \theta, \quad a > 0$ $r = a \pm a \sin \theta, \quad a > 0$	$r = a \pm b \cos \theta, \quad 0 < b < a$ $r = a \pm b \sin \theta, \quad 0 < b < a$	$r = a \pm b \cos \theta, \quad 0 < a < b$ $r = a \pm b \sin \theta, \quad 0 < a < b$
Typical graph			

Name	Lemniscate	Rose with three petals	Rose with four petals
Polar equations	$r^2 = a^2 \cos(2\theta), \quad a > 0$ $r^2 = a^2 \sin(2\theta), \quad a > 0$	$r = a \sin(3\theta), \quad a > 0$ $r = a \cos(3\theta), \quad a > 0$	$r = a \sin(2\theta), \quad a > 0$ $r = a \cos(2\theta), \quad a > 0$
Typical graph			

Sketching Quickly

If a polar equation involves only a sine (or cosine) function, you can quickly obtain a sketch of its graph by making use of Table 7, periodicity, and a short table.

EXAMPLE 13 **Sketching the Graph of a Polar Equation Quickly**

Graph the equation: $r = 2 + 2 \sin \theta$

Solution We recognize the polar equation: Its graph is a cardioid. The period of $\sin \theta$ is 2π, so we form a table using $0 \leq \theta \leq 2\pi$, compute r, plot the points (r, θ), and sketch the graph of a cardioid as θ varies from 0 to 2π. See Table 8 and Figure 32.

Table 8

Figure 32

θ	$r = 2 + 2 \sin \theta$
0	$2 + 2(0) = 2$
$\dfrac{\pi}{2}$	$2 + 2(1) = 4$
π	$2 + 2(0) = 2$
$\dfrac{3\pi}{2}$	$2 + 2(-1) = 0$
2π	$2 + 2(0) = 2$

 Calculus Comment For those of you who are planning to study calculus, a comment about one important role of polar equations is in order.

In rectangular coordinates, the equation $x^2 + y^2 = 1$, whose graph is the unit circle, is not the graph of a function. In fact, it requires two functions to obtain the graph of the unit circle:

$$y_1 = \sqrt{1 - x^2} \quad \text{Upper semicircle} \qquad y_2 = -\sqrt{1 - x^2} \quad \text{Lower semicircle}$$

In polar coordinates, the equation $r = 1$, whose graph is also the unit circle, does define a function. That is, for each choice of θ there is only one corresponding value of r, that is, $r = 1$. Since many problems in calculus require the use of functions, the opportunity to express nonfunctions in rectangular coordinates as functions in polar coordinates becomes extremely useful.

Note also that the vertical-line test for functions is valid only for equations in rectangular coordinates.

Historical Feature

Jakob Bernoulli (1654–1705)

Polar coordinates seem to have been invented by Jakob Bernoulli (1654–1705) in about 1691, although, as with most such ideas, earlier traces of the notion exist. Early users of calculus remained committed to rectangular coordinates, and polar coordinates did not become widely used until the early 1800s. Even then, it was mostly geometers who used them for describing odd curves. Finally, about the mid-1800s, applied mathematicians realized the tremendous simplification that polar coordinates make possible in the description of objects with circular or cylindrical symmetry. From then on their use became widespread.

10.2 Assess Your Understanding

'Are You Prepared?' *Answers are given at the end of these exercises. If you get a wrong answer, read the pages listed in red.*

1. If the rectangular coordinates of a point are $(4, -6)$, the point symmetric to it with respect to the origin is _____. (pp. 167–168)

2. The difference formula for cosine is $\cos(A - B) =$ _____. (p. 627)

3. The standard equation of a circle with center at $(-2, 5)$ and radius 3 is _____. (pp. 189–193)

4. Is the sine function even, odd, or neither? (pp. 556–557)

5. $\sin \dfrac{5\pi}{4} =$ _____. (pp. 540–547)

6. $\cos \dfrac{2\pi}{3} =$ _____. (pp. 540–547)

Concepts and Vocabulary

7. An equation whose variables are polar coordinates is called a(n) _____ _____.

8. Using polar coordinates (r, θ), the circle $x^2 + y^2 = 2x$ takes the form _____.

9. A polar equation is symmetric with respect to the pole if an equivalent equation results when r is replaced by _____.

10. *True or False* The tests for symmetry in polar coordinates are necessary, but not sufficient.

11. *True or False* The graph of a cardioid never passes through the pole.

12. *True or False* All polar equations have a symmetric feature.

Skill Building

In Problems 13–28, transform each polar equation to an equation in rectangular coordinates. Then identify and graph the equation.

13. $r = 4$

14. $r = 2$

15. $\theta = \dfrac{\pi}{3}$

16. $\theta = -\dfrac{\pi}{4}$

17. $r \sin \theta = 4$

18. $r \cos \theta = 4$

19. $r \cos \theta = -2$

20. $r \sin \theta = -2$

21. $r = 2 \cos \theta$

22. $r = 2 \sin \theta$

23. $r = -4 \sin \theta$

24. $r = -4 \cos \theta$

25. $r \sec \theta = 4$

26. $r \csc \theta = 8$

27. $r \csc \theta = -2$

28. $r \sec \theta = -4$

In Problems 29–36, match each of the graphs (A) through (H) to one of the following polar equations.

29. $r = 2$

30. $\theta = \dfrac{\pi}{4}$

31. $r = 2 \cos \theta$

32. $r \cos \theta = 2$

33. $r = 1 + \cos \theta$

34. $r = 2 \sin \theta$

35. $\theta = \dfrac{3\pi}{4}$

36. $r \sin \theta = 2$

(A)

(B)

(C)

(D)

(E)

(F)

(G)

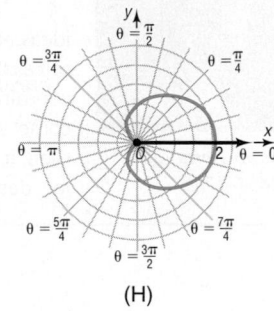

(H)

In Problems 37–60, identify and graph each polar equation.

37. $r = 2 + 2 \cos \theta$

38. $r = 1 + \sin \theta$

39. $r = 3 - 3 \sin \theta$

40. $r = 2 - 2 \cos \theta$

41. $r = 2 + \sin \theta$

42. $r = 2 - \cos \theta$

43. $r = 4 - 2 \cos \theta$

44. $r = 4 + 2 \sin \theta$

45. $r = 1 + 2 \sin \theta$

46. $r = 1 - 2 \sin \theta$

47. $r = 2 - 3 \cos \theta$

48. $r = 2 + 4 \cos \theta$

49. $r = 3 \cos(2\theta)$

50. $r = 2 \sin(3\theta)$

51. $r = 4 \sin(5\theta)$

52. $r = 3 \cos(4\theta)$

53. $r^2 = 9 \cos(2\theta)$

54. $r^2 = \sin(2\theta)$

55. $r = 2^\theta$

56. $r = 3^\theta$

57. $r = 1 - \cos \theta$

58. $r = 3 + \cos \theta$

59. $r = 1 - 3 \cos \theta$

60. $r = 4 \cos(3\theta)$

Applications and Extensions

In Problems 61–64, the polar equation for each graph is either $r = a + b\cos\theta$ or $r = a + b\sin\theta$, $a > 0, b > 0$. Select the correct equation and find the values of a and b.

61.

62.

63.

64.

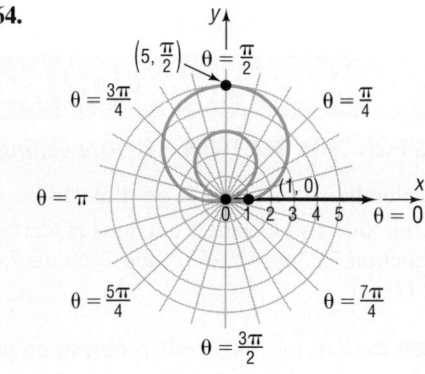

In Problems 65–74, graph each polar equation.

65. $r = \dfrac{2}{1 - \cos\theta}$ *(parabola)*

66. $r = \dfrac{2}{1 - 2\cos\theta}$ *(hyperbola)*

67. $r = \dfrac{1}{3 - 2\cos\theta}$ *(ellipse)*

68. $r = \dfrac{1}{1 - \cos\theta}$ *(parabola)*

69. $r = \theta$, $\theta \geq 0$ *(spiral of Archimedes)*

70. $r = \dfrac{3}{\theta}$ *(reciprocal spiral)*

71. $r = \csc\theta - 2$, $0 < \theta < \pi$ *(conchoid)*

72. $r = \sin\theta\tan\theta$ *(cissoid)*

73. $r = \tan\theta$, $-\dfrac{\pi}{2} < \theta < \dfrac{\pi}{2}$ *(kappa curve)*

74. $r = \cos\dfrac{\theta}{2}$

75. Show that the graph of the equation $r\sin\theta = a$ is a horizontal line a units above the pole if $a > 0$ and $|a|$ units below the pole if $a < 0$.

76. Show that the graph of the equation $r\cos\theta = a$ is a vertical line a units to the right of the pole if $a > 0$ and $|a|$ units to the left of the pole if $a < 0$.

77. Show that the graph of the equation $r = 2a\sin\theta$, $a > 0$, is a circle of radius a with center at $(0, a)$ in rectangular coordinates.

78. Show that the graph of the equation $r = -2a\sin\theta$, $a > 0$, is a circle of radius a with center at $(0, -a)$ in rectangular coordinates.

79. Show that the graph of the equation $r = 2a\cos\theta$, $a > 0$, is a circle of radius a with center at $(a, 0)$ in rectangular coordinates.

80. Show that the graph of the equation $r = -2a\cos\theta$, $a > 0$, is a circle of radius a with center at $(-a, 0)$ in rectangular coordinates.

Discussion and Writing

81. Explain why the following test for symmetry is valid: Replace r by $-r$ and θ by $-\theta$ in a polar equation. If an equivalent equation results, the graph is symmetric with respect to the line $\theta = \dfrac{\pi}{2}$ (y-axis).

 (a) Show that the test on page 727 fails for $r^2 = \cos\theta$, yet this new test works.

 (b) Show that the test on page 727 works for $r^2 = \sin\theta$, yet this new test fails.

82. Develop a new test for symmetry with respect to the pole.

 (a) Find a polar equation for which this new test fails, yet the test on page 727 works.

 (b) Find a polar equation for which the test on page 727 fails, yet the new test works.

83. Write down two different tests for symmetry with respect to the polar axis. Find examples in which one test works and the other fails. Which test do you prefer to use? Justify your answer.

'Are You Prepared?' Answers

1. $(-4, 6)$ **2.** $\cos A \cos B + \sin A \sin B$ **3.** $(x + 2)^2 + (y - 5)^2 = 9$ **4.** Odd **5.** $-\dfrac{\sqrt{2}}{2}$ **6.** $-\dfrac{1}{2}$

10.3 The Complex Plane; De Moivre's Theorem

PREPARING FOR THIS SECTION *Before getting started, review the following:*

- Complex Numbers (Section 1.3, pp. 109–114)
- Value of the Sine and Cosine Functions at Certain Angles (Section 7.3, pp. 529–532, and Sections 7.4 pp. 540–547)
- Sum and Difference Formulas for Sine and Cosine (Section 8.4, pp. 627 and 630)

Now Work the 'Are You Prepared?' problems on page 744.

OBJECTIVES 1 Convert a Complex Number from Rectangular Form to Polar Form (p. 739)

 2 Plot Points in the Complex Plane (p. 739)

 3 Find Products and Quotients of Complex Numbers in Polar Form (p. 740)

 4 Use De Moivre's Theorem (p. 741)

 5 Find Complex Roots (p. 742)

Figure 33
Complex plane

Complex numbers are discussed in Section 1.3. In that discussion, we were not prepared to give a geometric interpretation of a complex number. Now we are ready. Although we could give several interpretations, the one that follows is the easiest to understand.

 A complex number $z = x + yi$ can be interpreted geometrically as the point (x, y) in the xy-plane. Each point in the plane corresponds to a complex number and, conversely, each complex number corresponds to a point in the plane. We shall refer to the collection of such points as the **complex plane.** The x-axis will be referred to as the **real axis,** because any point that lies on the real axis is of the form $z = x + 0i = x$, a real number. The y-axis is called the **imaginary axis,** because any point that lies on it is of the form $z = 0 + yi = yi$, a pure imaginary number. See Figure 33.

DEFINITION Let $z = x + yi$ be a complex number. The **magnitude** or **modulus** of z, denoted by $|z|$, is defined as the distance from the origin to the point (x, y). That is,

$$|z| = \sqrt{x^2 + y^2} \qquad (1)$$

Figure 34

See Figure 34 for an illustration.

This definition for $|z|$ is consistent with the definition for the absolute value of a real number: If $z = x + yi$ is real, then $z = x + 0i$ and

$$|z| = \sqrt{x^2 + 0^2} = \sqrt{x^2} = |x|$$

For this reason, the magnitude of z is sometimes called the **absolute value of** z.

Recall that if $z = x + yi$ then its **conjugate,** denoted by $\bar{z}$, is $\bar{z} = x - yi$. Because $z\bar{z} = x^2 + y^2$, which is a nonnegative real number, it follows from equation (1) that the magnitude of z can be written as

$$|z| = \sqrt{z\bar{z}} \qquad (2)$$

1 Convert a Complex Number from Rectangular Form to Polar Form

When a complex number is written in the standard form $z = x + yi$, we say that it is in **rectangular,** or **Cartesian, form,** because (x, y) are the rectangular coordinates of the corresponding point in the complex plane. Suppose that (r, θ) are the polar coordinates of this point. Then

$$x = r \cos \theta, \qquad y = r \sin \theta \qquad (3)$$

DEFINITION

If $r \geq 0$ and $0 \leq \theta < 2\pi$, the complex number $z = x + yi$ may be written in **polar form** as

$$z = x + yi = (r \cos \theta) + (r \sin \theta)i = r(\cos \theta + i \sin \theta) \qquad (4)$$

Figure 35

$z = x + yi = r(\cos \theta + i \sin \theta)$,
$r \geq 0, 0 \leq \theta < 2\pi$

See Figure 35.

If $z = r(\cos \theta + i \sin \theta)$ is the polar form of a complex number, the angle θ, $0 \leq \theta < 2\pi$, is called the **argument of** z.

Also, because $r \geq 0$, we have $r = \sqrt{x^2 + y^2}$. From equation (1), it follows that the magnitude of $z = r(\cos \theta + i \sin \theta)$ is

$$|z| = r$$

2 Plot Points in the Complex Plane

EXAMPLE 1

Plotting a Point in the Complex Plane and Writing a Complex Number in Polar Form

Plot the point corresponding to $z = \sqrt{3} - i$ in the complex plane, and write an expression for z in polar form.

Solution

The point corresponding to $z = \sqrt{3} - i$ has the rectangular coordinates $(\sqrt{3}, -1)$.

The point, located in quadrant IV, is plotted in Figure 36. Because $x = \sqrt{3}$ and $y = -1$, it follows that

$$r = \sqrt{x^2 + y^2} = \sqrt{(\sqrt{3})^2 + (-1)^2} = \sqrt{4} = 2$$

Figure 36

So

$$\sin \theta = \frac{y}{r} = \frac{-1}{2}, \qquad \cos \theta = \frac{x}{r} = \frac{\sqrt{3}}{2}, \qquad 0 \leq \theta < 2\pi$$

Then $\theta = \dfrac{11\pi}{6}$ and $r = 2$, so the polar form of $z = \sqrt{3} - i$ is

$$z = r(\cos\theta + i\sin\theta) = 2\left(\cos\dfrac{11\pi}{6} + i\sin\dfrac{11\pi}{6}\right)$$

━━━ **Now Work** PROBLEM 11

| **EXAMPLE 2** | **Plotting a Point in the Complex Plane and Converting from Polar to Rectangular Form** |

Plot the point corresponding to $z = 2(\cos 30° + i\sin 30°)$ in the complex plane, and write an expression for z in rectangular form.

Figure 37

Solution To plot the complex number $z = 2(\cos 30° + i\sin 30°)$, we plot the point whose polar coordinates are $(r, \theta) = (2, 30°)$, as shown in Figure 37. In rectangular form,

$$z = 2(\cos 30° + i\sin 30°) = 2\left(\dfrac{\sqrt{3}}{2} + \dfrac{1}{2}i\right) = \sqrt{3} + i$$

━━━ **Now Work** PROBLEM 23

3 Find Products and Quotients of Complex Numbers in Polar Form

The polar form of a complex number provides an alternative method for finding products and quotients of complex numbers.

| **THEOREM** | Let $z_1 = r_1(\cos\theta_1 + i\sin\theta_1)$ and $z_2 = r_2(\cos\theta_2 + i\sin\theta_2)$ be two complex numbers. Then |

$$z_1 z_2 = r_1 r_2[\cos(\theta_1 + \theta_2) + i\sin(\theta_1 + \theta_2)] \tag{5}$$

If $z_2 \neq 0$, then

$$\dfrac{z_1}{z_2} = \dfrac{r_1}{r_2}[\cos(\theta_1 - \theta_2) + i\sin(\theta_1 - \theta_2)] \tag{6}$$

In Words

The magnitude of a complex number z is r and its argument is θ, so when

$z = r(\cos\theta + i\sin\theta)$

the magnitude of the product (quotient) of two complex numbers equals the product (quotient) of their magnitudes; the argument of the product (quotient) of two complex numbers is determined by the sum (difference) of their arguments.

Proof We will prove formula (5). The proof of formula (6) is left as an exercise (see Problem 66).

$$\begin{aligned}
z_1 z_2 &= [r_1(\cos\theta_1 + i\sin\theta_1)][r_2(\cos\theta_2 + i\sin\theta_2)] \\
&= r_1 r_2[(\cos\theta_1 + i\sin\theta_1)(\cos\theta_2 + i\sin\theta_2)] \\
&= r_1 r_2[(\cos\theta_1\cos\theta_2 - \sin\theta_1\sin\theta_2) + i(\sin\theta_1\cos\theta_2 + \cos\theta_1\sin\theta_2)] \\
&= r_1 r_2[\cos(\theta_1 + \theta_2) + i\sin(\theta_1 + \theta_2)]
\end{aligned}$$ ∎

Let's look at an example of how this theorem can be used.

| **EXAMPLE 3** | **Finding Products and Quotients of Complex Numbers in Polar Form** |

If $z = 3(\cos 20° + i\sin 20°)$ and $w = 5(\cos 100° + i\sin 100°)$, find the following (leave your answers in polar form):

(a) zw 　　　　　　　　　　　　　(b) $\dfrac{z}{w}$

Solution (a) $zw = [3(\cos 20° + i \sin 20°)][5(\cos 100° + i \sin 100°)]$

$= (3 \cdot 5)[\cos(20° + 100°) + i \sin(20° + 100°)]$

$= 15(\cos 120° + i \sin 120°)$ *Apply equation (5)*

(b) $\dfrac{z}{w} = \dfrac{3(\cos 20° + i \sin 20°)}{5(\cos 100° + i \sin 100°)}$

$= \dfrac{3}{5}[\cos(20° - 100°) + i \sin(20° - 100°)]$ *Apply equation (6)*

$= \dfrac{3}{5}[\cos(-80°) + i \sin(-80°)]$

$= \dfrac{3}{5}(\cos 280° + i \sin 280°)$ *The argument must lie between 0° and 360°.*

━━━ **Now Work** PROBLEM 33

4 Use De Moivre's Theorem

De Moivre's Theorem, stated by Abraham De Moivre (1667–1754) in 1730, but already known to many people by 1710, is important for the following reason: The fundamental processes of algebra are the four operations of addition, subtraction, multiplication, and division, together with powers and the extraction of roots. De Moivre's Theorem allows these latter fundamental algebraic operations to be applied to complex numbers.

De Moivre's Theorem, in its most basic form, is a formula for raising a complex number z to the power n, where $n \geq 1$ is a positive integer. Let's see if we can conjecture the form of the result.

Let $z = r(\cos \theta + i \sin \theta)$ be a complex number. Then, based on equation (5), we have

$n = 2$: $z^2 = r^2[\cos(2\theta) + i \sin(2\theta)]$ *Equation (5)*

$n = 3$: $z^3 = z^2 \cdot z$

$= \{r^2[\cos(2\theta) + i \sin(2\theta)]\}[r(\cos \theta + i \sin \theta)]$

$= r^3[\cos(3\theta) + i \sin(3\theta)]$ *Equation (5)*

$n = 4$: $z^4 = z^3 \cdot z$

$= \{r^3[\cos(3\theta) + i \sin(3\theta)]\}[r(\cos \theta + i \sin \theta)]$

$= r^4[\cos(4\theta) + i \sin(4\theta)]$ *Equation (5)*

Do you see the pattern?

THEOREM **De Moivre's Theorem**

If $z = r(\cos \theta + i \sin \theta)$ is a complex number, then

$$z^n = r^n[\cos(n\theta) + i \sin(n\theta)] \tag{7}$$

where $n \geq 1$ is a positive integer.

The proof of De Moivre's Theorem requires mathematical induction (which is not discussed until Section 13.4), so it is omitted here.

Let's look at some examples.

EXAMPLE 4	**Using De Moivre's Theorem**

Write $[2(\cos 20° + i \sin 20°)]^3$ in the standard form $a + bi$.

Solution

$$[2(\cos 20° + i \sin 20°)]^3 = 2^3[\cos(3 \cdot 20°) + i \sin(3 \cdot 20°)] \quad \text{Apply De Moivre's Theorem}$$

$$= 8(\cos 60° + i \sin 60°)$$

$$= 8\left(\frac{1}{2} + \frac{\sqrt{3}}{2}i\right) = 4 + 4\sqrt{3}i$$

▬▬▬▬▬▬ **Now Work** PROBLEM 41

EXAMPLE 5	**Using De Moivre's Theorem**

Write $(1 + i)^5$ in the standard form $a + bi$.

Solution

To apply De Moivre's Theorem, we must first write the complex number in polar form. Since the magnitude of $1 + i$ is $\sqrt{1^2 + 1^2} = \sqrt{2}$, we begin by writing

$$1 + i = \sqrt{2}\left(\frac{1}{\sqrt{2}} + \frac{1}{\sqrt{2}}i\right) = \sqrt{2}\left(\cos\frac{\pi}{4} + i \sin\frac{\pi}{4}\right)$$

Now

$$(1 + i)^5 = \left[\sqrt{2}\left(\cos\frac{\pi}{4} + i \sin\frac{\pi}{4}\right)\right]^5$$

$$= (\sqrt{2})^5\left[\cos\left(5 \cdot \frac{\pi}{4}\right) + i \sin\left(5 \cdot \frac{\pi}{4}\right)\right] \quad \text{Apply De Moivre's Theorem}$$

$$= 4\sqrt{2}\left(\cos\frac{5\pi}{4} + i \sin\frac{5\pi}{4}\right)$$

$$= 4\sqrt{2}\left[-\frac{1}{\sqrt{2}} + \left(-\frac{1}{\sqrt{2}}\right)i\right] = -4 - 4i$$

5 Find Complex Roots

Let w be a given complex number, and let $n \geq 2$ denote a positive integer. Any complex number z that satisfies the equation

$$z^n = w$$

is called a **complex nth root** of w. In keeping with previous usage, if $n = 2$, the solutions of the equation $z^2 = w$ are called **complex square roots** of w, and if $n = 3$, the solutions of the equation $z^3 = w$ are called **complex cube roots** of w.

THEOREM

Finding Complex Roots

Let $w = r(\cos\theta_0 + i \sin\theta_0)$ be a complex number, and let $n \geq 2$ be an integer. If $w \neq 0$, there are n distinct complex roots of w, given by the formula

$$z_k = \sqrt[n]{r}\left[\cos\left(\frac{\theta_0}{n} + \frac{2k\pi}{n}\right) + i \sin\left(\frac{\theta_0}{n} + \frac{2k\pi}{n}\right)\right] \qquad (8)$$

where $k = 0, 1, 2, \ldots, n - 1$.

Proof (Outline) We will not prove this result in its entirety. Instead, we shall show only that each z_k in equation (8) satisfies the equation $z_k^n = w$, proving that each z_k is a complex nth root of w.

$$z_k^n = \left\{ \sqrt[n]{r}\left[\cos\left(\frac{\theta_0}{n} + \frac{2k\pi}{n}\right) + i\sin\left(\frac{\theta_0}{n} + \frac{2k\pi}{n}\right)\right]\right\}^n$$

$$= (\sqrt[n]{r})^n \left\{ \cos\left[n\left(\frac{\theta_0}{n} + \frac{2k\pi}{n}\right)\right] + i\sin\left[n\left(\frac{\theta_0}{n} + \frac{2k\pi}{n}\right)\right]\right\} \quad \text{Apply De Moivre's Theorem}$$

$$= r[\cos(\theta_0 + 2k\pi) + i\sin(\theta_0 + 2k\pi)] \qquad \text{Simplify.}$$

$$= r(\cos\theta_0 + i\sin\theta_0) = w \qquad \text{The Periodic Property}$$

So each z_k, $k = 0, 1, \ldots, n-1$, is a complex nth root of w. To complete the proof, we would need to show that each z_k, $k = 0, 1, \ldots, n-1$, is, in fact, distinct and that there are no complex nth roots of w other than those given by equation (8).

EXAMPLE 6 **Finding Complex Cube Roots**

Find the complex cube roots of $-1 + \sqrt{3}i$. Leave your answers in polar form, with the argument in degrees.

Solution First, we express $-1 + \sqrt{3}i$ in polar form using degrees.

$$-1 + \sqrt{3}i = 2\left(-\frac{1}{2} + \frac{\sqrt{3}}{2}i\right) = 2(\cos 120° + i\sin 120°)$$

The three complex cube roots of $-1 + \sqrt{3}i = 2(\cos 120° + i\sin 120°)$ are

$$z_k = \sqrt[3]{2}\left[\cos\left(\frac{120°}{3} + \frac{360°k}{3}\right) + i\sin\left(\frac{120°}{3} + \frac{360°k}{3}\right)\right]$$

$$= \sqrt[3]{2}[\cos(40° + 120°k) + i\sin(40° + 120°k)] \qquad k = 0, 1, 2$$

So

$$z_0 = \sqrt[3]{2}[\cos(40° + 120°\cdot 0) + i\sin(40° + 120°\cdot 0)] = \sqrt[3]{2}(\cos 40° + i\sin 40°)$$

$$z_1 = \sqrt[3]{2}[\cos(40° + 120°\cdot 1) + i\sin(40° + 120°\cdot 1)] = \sqrt[3]{2}(\cos 160° + i\sin 160°)$$

$$z_2 = \sqrt[3]{2}[\cos(40° + 120°\cdot 2) + i\sin(40° + 120°\cdot 2)] = \sqrt[3]{2}(\cos 280° + i\sin 280°)$$

Notice that each of the three complex roots of $-1 + \sqrt{3}i$ has the same magnitude, $\sqrt[3]{2}$. This means that the points corresponding to each cube root lie the same distance from the origin; that is, the three points lie on a circle with center at the origin and radius $\sqrt[3]{2}$. Furthermore, the arguments of these cube roots are 40°, 160°, and 280°, the difference of consecutive pairs being $120° = \dfrac{360°}{3}$. This means that the three points are equally spaced on the circle, as shown in Figure 38. These results are

Figure 38

not coincidental. In fact, you are asked to show that these results hold for complex *n*th roots in Problems 63 through 65.

━━━━ **Now Work** PROBLEM 53

Historical Feature

John Wallis

The Babylonians, Greeks, and Arabs considered square roots of negative quantities to be impossible and equations with complex solutions to be unsolvable. The first hint that there was some connection between real solutions of equations and complex numbers came when Girolamo Cardano (1501–1576) and Tartaglia (1499–1557) found *real* roots of cubic equations by taking cube roots of *complex* quantities. For centuries thereafter, mathematicians worked with complex numbers without much belief in their actual existence. In 1673, John Wallis appears to have been the first to suggest the graphical representation of complex numbers, a truly significant idea that was not pursued further until about 1800. Several people, including Karl Friedrich Gauss (1777–1855), then rediscovered the idea, and graphical representation helped to establish complex numbers as equal members of the number family. In practical applications, complex numbers have found their greatest uses in the study of alternating current, where they are a commonplace tool, and in the field of subatomic physics.

Historical Problems

1. The quadratic formula will work perfectly well if the coefficients are complex numbers. Solve the following using De Moivre's Theorem where necessary.

 [**Hint:** The answers are "nice."]

 (a) $z^2 - (2 + 5i)z - 3 + 5i = 0$ (b) $z^2 - (1 + i)z - 2 - i = 0$

10.3 Assess your Understanding

'Are You Prepared?' *Answers are given at the end of these exercises. If you get a wrong answer, read the pages listed in red.*

1. The conjugate of $-4 - 3i$ is _____. (pp. 109–114)

2. The sum formula for the sine function is $\sin(A + B) =$ _____. (p. 630)

3. The sum formula for the cosine function is $\cos(A + B) =$ _____. (p. 627)

4. $\sin 120° =$ _____; $\cos 240° =$ _____. (pp. 540–547)

Concepts and Vocabulary

5. When a complex number z is written in the polar form $z = r(\cos\theta + i\sin\theta)$, the nonnegative number r is the _____ or _____ of z, and the angle θ, $0 \le \theta < 2\pi$, is the _____ of z.

6. _____ Theorem can be used to raise a complex number to a power.

7. Every non-zero complex number will have exactly _____ cube roots.

8. **True or False** De Moivre's Theorem is useful for raising a complex number to a positive integer power.

9. **True or False** Using De Moivre's Theorem, the square of a complex number will have two answers.

10. **True or False** The polar form of a complex number is unique.

Skill Building

In Problems 11–22, plot each complex number in the complex plane and write it in polar form. Express the argument in degrees.

11. $1 + i$ 12. $-1 + i$ 13. $\sqrt{3} - i$ 14. $1 - \sqrt{3}i$

15. $-3i$ 16. -2 17. $4 - 4i$ 18. $9\sqrt{3} + 9i$

19. $3 - 4i$ 20. $2 + \sqrt{3}i$ 21. $-2 + 3i$ 22. $\sqrt{5} - i$

In Problems 23–32, write each complex number in rectangular form.

23. $2(\cos 120° + i\sin 120°)$ 24. $3(\cos 210° + i\sin 210°)$ 25. $4\left(\cos\dfrac{7\pi}{4} + i\sin\dfrac{7\pi}{4}\right)$

26. $2\left(\cos\dfrac{5\pi}{6} + i\sin\dfrac{5\pi}{6}\right)$

27. $3\left(\cos\dfrac{3\pi}{2} + i\sin\dfrac{3\pi}{2}\right)$

28. $4\left(\cos\dfrac{\pi}{2} + i\sin\dfrac{\pi}{2}\right)$

29. $0.2(\cos 100° + i\sin 100°)$

30. $0.4(\cos 200° + i\sin 200°)$

31. $2\left(\cos\dfrac{\pi}{18} + i\sin\dfrac{\pi}{18}\right)$

32. $3\left(\cos\dfrac{\pi}{10} + i\sin\dfrac{\pi}{10}\right)$

In Problems 33–40, find zw and $\dfrac{z}{w}$. Leave your answers in polar form.

33. $z = 2(\cos 40° + i\sin 40°)$
$w = 4(\cos 20° + i\sin 20°)$

34. $z = \cos 120° + i\sin 120°$
$w = \cos 100° + i\sin 100°$

35. $z = 3(\cos 130° + i\sin 130°)$
$w = 4(\cos 270° + i\sin 270°)$

36. $z = 2(\cos 80° + i\sin 80°)$
$w = 6(\cos 200° + i\sin 200°)$

37. $z = 2\left(\cos\dfrac{\pi}{8} + i\sin\dfrac{\pi}{8}\right)$
$w = 2\left(\cos\dfrac{\pi}{10} + i\sin\dfrac{\pi}{10}\right)$

38. $z = 4\left(\cos\dfrac{3\pi}{8} + i\sin\dfrac{3\pi}{8}\right)$
$w = 2\left(\cos\dfrac{9\pi}{16} + i\sin\dfrac{9\pi}{16}\right)$

39. $z = 2 + 2i$
$w = \sqrt{3} - i$

40. $z = 1 - i$
$w = 1 - \sqrt{3}i$

In Problems 41–52, write each expression in the standard form a + bi.

41. $[4(\cos 40° + i\sin 40°)]^3$

42. $[3(\cos 80° + i\sin 80°)]^3$

43. $\left[2\left(\cos\dfrac{\pi}{10} + i\sin\dfrac{\pi}{10}\right)\right]^5$

44. $\left[\sqrt{2}\left(\cos\dfrac{5\pi}{16} + i\sin\dfrac{5\pi}{16}\right)\right]^4$

45. $\left[\sqrt{3}(\cos 10° + i\sin 10°)\right]^6$

46. $\left[\dfrac{1}{2}(\cos 72° + i\sin 72°)\right]^5$

47. $\left[\sqrt{5}\left(\cos\dfrac{3\pi}{16} + i\sin\dfrac{3\pi}{16}\right)\right]^4$

48. $\left[\sqrt{3}\left(\cos\dfrac{5\pi}{18} + i\sin\dfrac{5\pi}{18}\right)\right]^6$

49. $(1 - i)^5$

50. $\left(\sqrt{3} - i\right)^6$

51. $\left(\sqrt{2} - i\right)^6$

52. $\left(1 - \sqrt{5}i\right)^8$

In Problems 53–60, find all the complex roots. Leave your answers in polar form with the argument in degrees.

53. The complex cube roots of $1 + i$

54. The complex fourth roots of $\sqrt{3} - i$

55. The complex fourth roots of $4 - 4\sqrt{3}i$

56. The complex cube roots of $-8 - 8i$

57. The complex fourth roots of $-16i$

58. The complex cube roots of -8

59. The complex fifth roots of i

60. The complex fifth roots of $-i$

Applications and Extensions

61. Find the four complex fourth roots of unity (1) and plot them.

62. Find the six complex sixth roots of unity (1) and plot them.

63. Show that each complex nth root of a nonzero complex number w has the same magnitude.

64. Use the result of Problem 63 to draw the conclusion that each complex nth root lies on a circle with center at the origin. What is the radius of this circle?

65. Refer to Problem 64. Show that the complex nth roots of a nonzero complex number w are equally spaced on the circle.

66. Prove formula (6).

'Are You Prepared?' Answers

1. $-4 + 3i$ **2.** $\sin A\cos B + \cos A\sin B$ **3.** $\cos A\cos B - \sin A\sin B$ **4.** $\dfrac{\sqrt{3}}{2}; -\dfrac{1}{2}$

10.4 Vectors

Figure 39

In simple terms, a **vector** (derived from the Latin *vehere,* meaning "to carry") is a quantity that has both magnitude and direction. It is customary to represent a vector by using an arrow. The length of the arrow represents the **magnitude** of the vector, and the arrowhead indicates the **direction** of the vector.

Many quantities in physics can be represented by vectors. For example, the velocity of an aircraft can be represented by an arrow that points in the direction of movement; the length of the arrow represents speed. If the aircraft speeds up, we lengthen the arrow; if the aircraft changes direction, we introduce an arrow in the new direction. See Figure 39. Based on this representation, it is not surprising that vectors and directed line segments are somehow related.

Geometric Vectors

If P and Q are two distinct points in the xy-plane, there is exactly one line containing both P and Q [Figure 40(a)]. The points on that part of the line that joins P to Q, including P and Q, form what is called the **line segment** $\overline{PQ}$ [Figure 40(b)]. If we order the points so that they proceed from P to Q, we have a **directed line segment** from P to Q, or a **geometric vector,** which we denote by $\overrightarrow{PQ}$. In a directed line segment $\overrightarrow{PQ}$, we call P the **initial point** and Q the **terminal point,** as indicated in Figure 40(c).

Figure 40

(a) Line containing P and Q (b) Line segment $\overline{PQ}$ (c) Directed line segment $\overrightarrow{PQ}$

The magnitude of the directed line segment $\overrightarrow{PQ}$ is the distance from the point P to the point Q; that is, it is the length of the line segment. The direction of $\overrightarrow{PQ}$ is from P to Q. If a vector $\mathbf{v}$* has the same magnitude and the same direction as the directed line segment $\overrightarrow{PQ}$, we write

$$\mathbf{v} = \overrightarrow{PQ}$$

Figure 41

The vector $\mathbf{v}$ whose magnitude is 0 is called the **zero vector, 0.** The zero vector is assigned no direction.

Two vectors $\mathbf{v}$ and $\mathbf{w}$ are **equal,** written

$$\mathbf{v} = \mathbf{w}$$

if they have the same magnitude and the same direction.

For example, the three vectors shown in Figure 41 have the same magnitude and the same direction, so they are equal, even though they have different initial points

* Boldface letters will be used to denote vectors, to distinguish them from numbers. For handwritten work, an arrow is placed over the letter to signify a vector.

and different terminal points. As a result, we find it useful to think of a vector simply as an arrow, keeping in mind that two arrows (vectors) are equal if they have the same direction and the same magnitude (length).

Adding Vectors

Figure 42

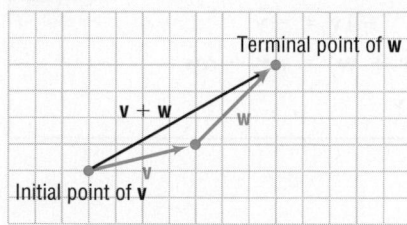

The **sum v + w** of two vectors is defined as follows: We position the vectors **v** and **w** so that the terminal point of **v** coincides with the initial point of **w**, as shown in Figure 42. The vector **v + w** is then the unique vector whose initial point coincides with the initial point of **v** and whose terminal point coincides with the terminal point of **w**.

Vector addition is **commutative.** That is, if **v** and **w** are any two vectors, then

$$\mathbf{v} + \mathbf{w} = \mathbf{w} + \mathbf{v}$$

Figure 43

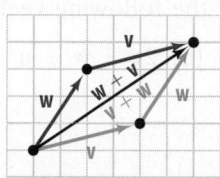

Figure 43 illustrates this fact. (Observe that the commutative property is another way of saying that opposite sides of a parallelogram are equal and parallel.)

Vector addition is also **associative.** That is, if **u**, **v**, and **w** are vectors, then

$$\mathbf{u} + (\mathbf{v} + \mathbf{w}) = (\mathbf{u} + \mathbf{v}) + \mathbf{w}$$

Figure 44

$(\mathbf{u} + \mathbf{v}) + \mathbf{w} = \mathbf{u} + (\mathbf{v} + \mathbf{w})$

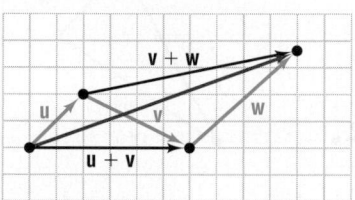

Figure 44 illustrates the associative property for vectors.

The zero vector **0** has the property that

$$\mathbf{v} + \mathbf{0} = \mathbf{0} + \mathbf{v} = \mathbf{v}$$

for any vector **v**.

If **v** is a vector, then $-\mathbf{v}$ is the vector having the same magnitude as **v**, but whose direction is opposite to **v**, as shown in Figure 45.

Furthermore,

Figure 45

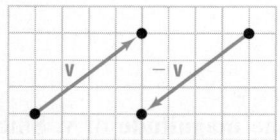

$$\mathbf{v} + (-\mathbf{v}) = \mathbf{0}$$

If **v** and **w** are two vectors, we define the **difference v − w** as

$$\mathbf{v} - \mathbf{w} = \mathbf{v} + (-\mathbf{w})$$

Figure 46 illustrates the relationships among **v**, **w**, **v + w**, and **v − w**.

Figure 46

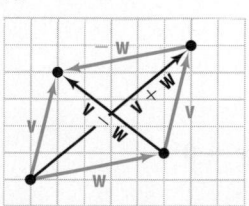

Multiplying Vectors by Numbers

When dealing with vectors, we refer to real numbers as **scalars.** Scalars are quantities that have only magnitude. Examples from physics of scalar quantities are temperature, speed, and time. We now define how to multiply a vector by a scalar.

DEFINITION If α is a scalar and **v** is a vector, the **scalar multiple** $\alpha\mathbf{v}$ is defined as follows:

1. If $\alpha > 0$, $\alpha\mathbf{v}$ is the vector whose magnitude is α times the magnitude of **v** and whose direction is the same as **v**.
2. If $\alpha < 0$, $\alpha\mathbf{v}$ is the vector whose magnitude is $|\alpha|$ times the magnitude of **v** and whose direction is opposite that of **v**.
3. If $\alpha = 0$ or if $\mathbf{v} = \mathbf{0}$, then $\alpha\mathbf{v} = \mathbf{0}$.

Figure 47

See Figure 47 for some illustrations.

For example, if **a** is the acceleration of an object of mass m due to a force **F** being exerted on it, then, by Newton's second law of motion, $\mathbf{F} = m\mathbf{a}$. Here, $m\mathbf{a}$ is the product of the scalar m and the vector **a**.

Scalar multiples have the following properties:

$$0\mathbf{v} = \mathbf{0} \qquad 1\mathbf{v} = \mathbf{v} \qquad -1\mathbf{v} = -\mathbf{v}$$
$$(\alpha + \beta)\mathbf{v} = \alpha\mathbf{v} + \beta\mathbf{v} \qquad \alpha(\mathbf{v} + \mathbf{w}) = \alpha\mathbf{v} + \alpha\mathbf{w}$$
$$\alpha(\beta\mathbf{v}) = (\alpha\beta)\mathbf{v}$$

1 Graph Vectors

EXAMPLE 1 **Graphing Vectors**

Use the vectors illustrated in Figure 48 to graph each of the following vectors:

(a) $\mathbf{v} - \mathbf{w}$ (b) $2\mathbf{v} + 3\mathbf{w}$ (c) $2\mathbf{v} - \mathbf{w} + \mathbf{u}$

Solution Figure 49 illustrates each graph.

Figure 48

Figure 49

(a) $\mathbf{v} - \mathbf{w}$ (b) $2\mathbf{v} + 3\mathbf{w}$

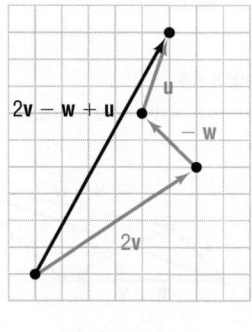

(c) $2\mathbf{v} - \mathbf{w} + \mathbf{u}$

Now Work PROBLEMS 7 AND 9

Magnitudes of Vectors

If **v** is a vector, we use the symbol $\|\mathbf{v}\|$ to represent the **magnitude** of **v**. Since $\|\mathbf{v}\|$ equals the length of a directed line segment, it follows that $\|\mathbf{v}\|$ has the following properties:

THEOREM **Properties of $\|\mathbf{v}\|$**

If **v** is a vector and if α is a scalar, then

(a) $\|\mathbf{v}\| \geq 0$ (b) $\|\mathbf{v}\| = 0$ if and only if $\mathbf{v} = \mathbf{0}$
(c) $\|-\mathbf{v}\| = \|\mathbf{v}\|$ (d) $\|\alpha\mathbf{v}\| = |\alpha|\|\mathbf{v}\|$

Property (a) is a consequence of the fact that distance is a nonnegative number. Property (b) follows, because the length of the directed line segment $\overrightarrow{PQ}$ is positive unless P and Q are the same point, in which case the length is 0. Property (c) follows because the length of the line segment $\overline{PQ}$ equals the length of the line segment $\overline{QP}$. Property (d) is a direct consequence of the definition of a scalar multiple.

DEFINITION A vector **u** for which $\|\mathbf{u}\| = 1$ is called a **unit vector.**

2 Find a Position Vector

To compute the magnitude and direction of a vector, we need an algebraic way of representing vectors.

DEFINITION

An **algebraic vector v** is represented as

$$\mathbf{v} = \langle a, b \rangle$$

where a and b are real numbers (scalars) called the **components** of the vector **v**.

Figure 50

We use a rectangular coordinate system to represent algebraic vectors in the plane. If $\mathbf{v} = \langle a, b \rangle$ is an algebraic vector whose initial point is at the origin, then **v** is called a **position vector.** See Figure 50. Notice that the terminal point of the position vector $\mathbf{v} = \langle a, b \rangle$ is $P = (a, b)$.

The next result states that any vector whose initial point is not at the origin is equal to a unique position vector.

THEOREM

Suppose that **v** is a vector with initial point $P_1 = (x_1, y_1)$, not necessarily the origin, and terminal point $P_2 = (x_2, y_2)$. If $\mathbf{v} = \overrightarrow{P_1P_2}$, then **v** is equal to the position vector

$$\mathbf{v} = \langle x_2 - x_1, y_2 - y_1 \rangle \tag{1}$$

To see why this is true, look at Figure 51.

Figure 51

$\mathbf{v} = \langle a, b \rangle = \langle x_2 - x_1, y_2 - y_1 \rangle$

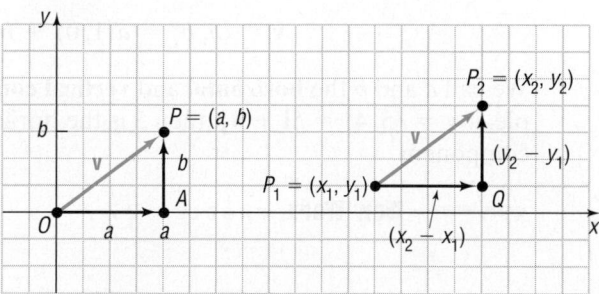

Triangle OPA and triangle P_1P_2Q are congruent. [Do you see why? The line segments have the same magnitude, so $d(O, P) = d(P_1, P_2)$; and they have the same direction, so $\angle POA = \angle P_2P_1Q$. Since the triangles are right triangles, we have angle–side–angle.] It follows that corresponding sides are equal. As a result, $x_2 - x_1 = a$ and $y_2 - y_1 = b$, so **v** may be written as

$$\mathbf{v} = \langle a, b \rangle = \langle x_2 - x_1, y_2 - y_1 \rangle$$

Because of this result, we can replace any algebraic vector by a unique position vector, and vice versa. This flexibility is one of the main reasons for the wide use of vectors.

| **EXAMPLE 2** | **Finding a Position Vector** |

Find the position vector of the vector $\mathbf{v} = \overrightarrow{P_1P_2}$ if $P_1 = (-1, 2)$ and $P_2 = (4, 6)$.

Solution By equation (1), the position vector equal to **v** is

$$\mathbf{v} = \langle 4 - (-1), 6 - 2 \rangle = \langle 5, 4 \rangle$$

See Figure 52.

Figure 52

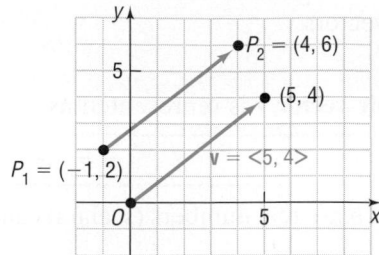

Two position vectors **v** and **w** are equal if and only if the terminal point of **v** is the same as the terminal point of **w**. This leads to the following result:

THEOREM

Equality of Vectors

Two vectors **v** and **w** are equal if and only if their corresponding components are equal. That is,

$$\text{If } \mathbf{v} = \langle a_1, b_1 \rangle \quad \text{and} \quad \mathbf{w} = \langle a_2, b_2 \rangle$$
$$\text{then} \quad \mathbf{v} = \mathbf{w} \quad \text{if and only if} \quad a_1 = a_2 \quad \text{and} \quad b_1 = b_2.$$

Figure 53

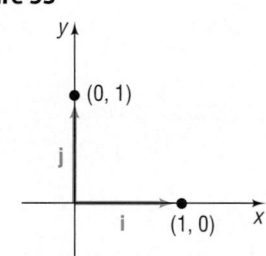

We now present an alternative representation of a vector in the plane that is common in the physical sciences. Let **i** denote the unit vector whose direction is along the positive x-axis; let **j** denote the unit vector whose direction is along the positive y-axis. Then $\mathbf{i} = \langle 1, 0 \rangle$ and $\mathbf{j} = \langle 0, 1 \rangle$, as shown in Figure 53. Any vector $\mathbf{v} = \langle a, b \rangle$ can be written using the unit vectors **i** and **j** as follows:

$$\mathbf{v} = \langle a, b \rangle = a\langle 1, 0 \rangle + b\langle 0, 1 \rangle = a\mathbf{i} + b\mathbf{j}$$

We call a and b the **horizontal** and **vertical components** of **v**, respectively. For example, if $\mathbf{v} = \langle 5, 4 \rangle = 5\mathbf{i} + 4\mathbf{j}$, then 5 is the horizontal component and 4 is the vertical component.

══ **Now Work** PROBLEM 27

3 Add and Subtract Vectors Algebraically

We define addition, subtraction, scalar multiple, and magnitude of algebraic vectors in terms of their components.

DEFINITION

Let $\mathbf{v} = a_1\mathbf{i} + b_1\mathbf{j} = \langle a_1, b_1 \rangle$ and $\mathbf{w} = a_2\mathbf{i} + b_2\mathbf{j} = \langle a_2, b_2 \rangle$ be two vectors, and let α be a scalar. Then

$$\mathbf{v} + \mathbf{w} = (a_1 + a_2)\mathbf{i} + (b_1 + b_2)\mathbf{j} = \langle a_1 + a_2, b_1 + b_2 \rangle \qquad (2)$$
$$\mathbf{v} - \mathbf{w} = (a_1 - a_2)\mathbf{i} + (b_1 - b_2)\mathbf{j} = \langle a_1 - a_2, b_1 - b_2 \rangle \qquad (3)$$
$$\alpha\mathbf{v} = (\alpha a_1)\mathbf{i} + (\alpha b_1)\mathbf{j} = \langle \alpha a_1, \alpha b_1 \rangle \qquad (4)$$
$$\|\mathbf{v}\| = \sqrt{a_1^2 + b_1^2} \qquad (5)$$

In Words

To add two vectors, add corresponding components. To subtract two vectors, subtract corresponding components.

These definitions are compatible with the geometric definitions given earlier in this section. See Figure 54.

Figure 54

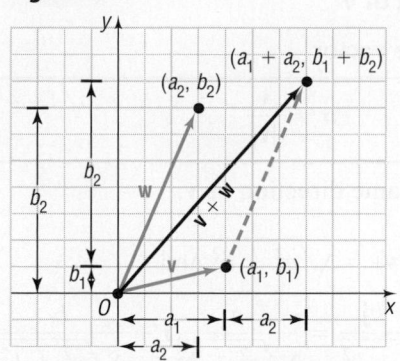

(a) Illustration of property (2)

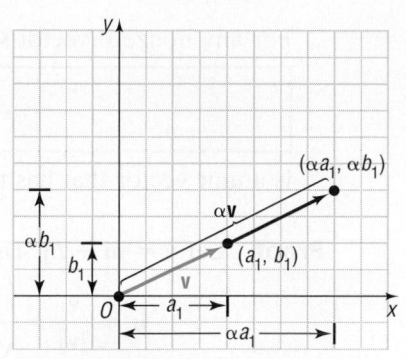

(b) Illustration of property (4), $\alpha > 0$

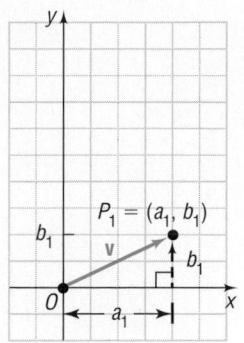

(c) Illustration of property (5):
$\| \mathbf{v} \| = $ Distance from O to P_1
$\| \mathbf{v} \| = \sqrt{a_1^2 + b_1^2}$

EXAMPLE 3	**Adding and Subtracting Vectors**

If $\mathbf{v} = 2\mathbf{i} + 3\mathbf{j} = \langle 2, 3 \rangle$ and $\mathbf{w} = 3\mathbf{i} - 4\mathbf{j} = \langle 3, -4 \rangle$, find:

(a) $\mathbf{v} + \mathbf{w}$ (b) $\mathbf{v} - \mathbf{w}$

Solution
(a) $\mathbf{v} + \mathbf{w} = (2\mathbf{i} + 3\mathbf{j}) + (3\mathbf{i} - 4\mathbf{j}) = (2 + 3)\mathbf{i} + (3 - 4)\mathbf{j} = 5\mathbf{i} - \mathbf{j}$

or

$\mathbf{v} + \mathbf{w} = \langle 2, 3 \rangle + \langle 3, -4 \rangle = \langle 2 + 3, 3 + (-4) \rangle = \langle 5, -1 \rangle$

(b) $\mathbf{v} - \mathbf{w} = (2\mathbf{i} + 3\mathbf{j}) - (3\mathbf{i} - 4\mathbf{j}) = (2 - 3)\mathbf{i} + [3 - (-4)]\mathbf{j} = -\mathbf{i} + 7\mathbf{j}$

or

$\mathbf{v} - \mathbf{w} = \langle 2, 3 \rangle - \langle 3, -4 \rangle = \langle 2 - 3, 3 - (-4) \rangle = \langle -1, 7 \rangle$

4 Find a Scalar Multiple and the Magnitude of a Vector

EXAMPLE 4	**Finding Scalar Multiples and Magnitudes**

If $\mathbf{v} = 2\mathbf{i} + 3\mathbf{j} = \langle 2, 3 \rangle$ and $\mathbf{w} = 3\mathbf{i} - 4\mathbf{j} = \langle 3, -4 \rangle$, find:

(a) $3\mathbf{v}$ (b) $2\mathbf{v} - 3\mathbf{w}$ (c) $\|\mathbf{v}\|$

Solution
(a) $3\mathbf{v} = 3(2\mathbf{i} + 3\mathbf{j}) = 6\mathbf{i} + 9\mathbf{j}$

or

$3\mathbf{v} = 3\langle 2, 3 \rangle = \langle 6, 9 \rangle$

(b) $2\mathbf{v} - 3\mathbf{w} = 2(2\mathbf{i} + 3\mathbf{j}) - 3(3\mathbf{i} - 4\mathbf{j}) = 4\mathbf{i} + 6\mathbf{j} - 9\mathbf{i} + 12\mathbf{j}$

$\qquad\qquad = -5\mathbf{i} + 18\mathbf{j}$

or

$2\mathbf{v} - 3\mathbf{w} = 2\langle 2, 3 \rangle - 3\langle 3, -4 \rangle = \langle 4, 6 \rangle - \langle 9, -12 \rangle$

$\qquad\qquad = \langle 4 - 9, 6 - (-12) \rangle = \langle -5, 18 \rangle$

(c) $\|\mathbf{v}\| = \|2\mathbf{i} + 3\mathbf{j}\| = \sqrt{2^2 + 3^2} = \sqrt{13}$

Now Work PROBLEMS **33** AND **39**

For the remainder of the section, we will express a vector $\mathbf{v}$ in the form $a\mathbf{i} + b\mathbf{j}$.

5 Find a Unit Vector

Recall that a unit vector $\mathbf{u}$ is a vector for which $\|\mathbf{u}\| = 1$. In many applications, it is useful to be able to find a unit vector $\mathbf{u}$ that has the same direction as a given vector $\mathbf{v}$.

THEOREM　**Unit Vector in the Direction of v**

For any nonzero vector **v**, the vector

$$\mathbf{u} = \frac{\mathbf{v}}{\|\mathbf{v}\|}$$

is a unit vector that has the same direction as **v**.

Proof　Let $\mathbf{v} = a\mathbf{i} + b\mathbf{j}$. Then $\|\mathbf{v}\| = \sqrt{a^2 + b^2}$ and

$$\mathbf{u} = \frac{\mathbf{v}}{\|\mathbf{v}\|} = \frac{a\mathbf{i} + b\mathbf{j}}{\sqrt{a^2 + b^2}} = \frac{a}{\sqrt{a^2 + b^2}}\mathbf{i} + \frac{b}{\sqrt{a^2 + b^2}}\mathbf{j}$$

The vector **u** is in the same direction as **v**, since $\|\mathbf{v}\| > 0$. Furthermore,

$$\|\mathbf{u}\| = \sqrt{\frac{a^2}{a^2 + b^2} + \frac{b^2}{a^2 + b^2}} = \sqrt{\frac{a^2 + b^2}{a^2 + b^2}} = 1$$

That is, **u** is a unit vector in the direction of **v**.　■

As a consequence of this theorem, if **u** is a unit vector in the same direction as a vector **v**, then **v** may be expressed as

$$\mathbf{v} = \|\mathbf{v}\|\mathbf{u} \qquad (6)$$

This way of expressing a vector is useful in many applications.

EXAMPLE 5　**Finding a Unit Vector**

Find a unit vector in the same direction as $\mathbf{v} = 4\mathbf{i} - 3\mathbf{j}$.

Solution　We find $\|\mathbf{v}\|$ first.

$$\|\mathbf{v}\| = \|4\mathbf{i} - 3\mathbf{j}\| = \sqrt{16 + 9} = 5$$

Now we multiply **v** by the scalar $\dfrac{1}{\|\mathbf{v}\|} = \dfrac{1}{5}$. A unit vector in the same direction as **v** is

$$\frac{\mathbf{v}}{\|\mathbf{v}\|} = \frac{4\mathbf{i} - 3\mathbf{j}}{5} = \frac{4}{5}\mathbf{i} - \frac{3}{5}\mathbf{j}$$

✓**Check:**　This vector is, in fact, a unit vector because

$$\left(\frac{4}{5}\right)^2 + \left(-\frac{3}{5}\right)^2 = \frac{16}{25} + \frac{9}{25} = \frac{25}{25} = 1$$

━━━ **Now Work** PROBLEM 49

6 Find a Vector from Its Direction and Magnitude

If a vector represents the speed and direction of an object, it is called a **velocity vector.** If a vector represents the direction and amount of a force acting on an object, it is called a **force vector.** In many applications, a vector is described in terms of its magnitude and direction, rather than in terms of its components. For example, a ball thrown with an initial speed of 25 miles per hour at an angle 30° to the horizontal is a velocity vector.

Suppose that we are given the magnitude $\|\mathbf{v}\|$ of a nonzero vector **v** and the angle α, $0° \le \alpha < 360°$, between **v** and **i**. To express **v** in terms of $\|\mathbf{v}\|$ and α, we first find the unit vector **u** having the same direction as **v**.

$$\mathbf{u} = \frac{\mathbf{v}}{\|\mathbf{v}\|} \quad \text{or} \quad \mathbf{v} = \|\mathbf{v}\|\mathbf{u} \qquad (7)$$

Figure 55

Look at Figure 55. The coordinates of the terminal point of **u** are $(\cos \alpha, \sin \alpha)$. Then $\mathbf{u} = \cos \alpha \mathbf{i} + \sin \alpha \mathbf{j}$ and, from (7),

$$\mathbf{v} = \|\mathbf{v}\|(\cos \alpha \, \mathbf{i} + \sin \alpha \, \mathbf{j}) \qquad (8)$$

where α is the angle between **v** and **i**.

EXAMPLE 6 **Writing a Vector When Its Magnitude and Direction Are Given**

A ball is thrown with an initial speed of 25 miles per hour in a direction that makes an angle of 30° with the positive x-axis. Express the velocity vector **v** in terms of **i** and **j**. What is the initial speed in the horizontal direction? What is the initial speed in the vertical direction?

Solution The magnitude of **v** is $\|\mathbf{v}\| = 25$ miles per hour, and the angle between the direction of **v** and **i**, the positive x-axis, is $\alpha = 30°$. By equation (8),

$$\mathbf{v} = \|\mathbf{v}\|(\cos \alpha \mathbf{i} + \sin \alpha \mathbf{j}) = 25(\cos 30°\mathbf{i} + \sin 30°\mathbf{j}) = 25\left(\frac{\sqrt{3}}{2}\mathbf{i} + \frac{1}{2}\mathbf{j}\right) = \frac{25\sqrt{3}}{2}\mathbf{i} + \frac{25}{2}\mathbf{j}$$

Figure 56

The initial speed of the ball in the horizontal direction is the horizontal component of **v**, $\dfrac{25\sqrt{3}}{2} \approx 21.65$ miles per hour. The initial speed in the vertical direction is the vertical component of **v**, $\dfrac{25}{2} = 12.5$ miles per hour. See Figure 56. ∎

Now Work PROBLEM 63

7 Analyze Objects in Static Equilibrium

Figure 57

Because forces can be represented by vectors, two forces "combine" the way that vectors "add." If $\mathbf{F}_1$ and $\mathbf{F}_2$ are two forces simultaneously acting on an object, the vector sum $\mathbf{F}_1 + \mathbf{F}_2$ is the **resultant force**. The resultant force produces the same effect on the object as that obtained when the two forces $\mathbf{F}_1$ and $\mathbf{F}_2$ act on the object. See Figure 57. An application of this concept is *static equilibrium*. An object is said to be in **static equilibrium** if (1) the object is at rest and (2) the sum of all forces acting on the object is zero, that is, if the resultant force is 0.

EXAMPLE 7 **An Object in Static Equilibrium**

Figure 58

A box of supplies that weighs 1200 pounds is suspended by two cables attached to the ceiling, as shown in Figure 58. What are the tensions in the two cables?

Solution We draw a force diagram using the vectors shown in Figure 59. The tensions in the cables are the magnitudes $\|\mathbf{F}_1\|$ and $\|\mathbf{F}_2\|$ of the force vectors $\mathbf{F}_1$ and $\mathbf{F}_2$. The magnitude of the force vector $\mathbf{F}_3$ equals 1200 pounds, the weight of the box. Now write each force vector in terms of the unit vectors **i** and **j**. For $\mathbf{F}_1$ and $\mathbf{F}_2$, we use equation (8). Remember that α is the angle between the vector and the positive x-axis.

Figure 59

$$\mathbf{F}_1 = \|\mathbf{F}_1\|(\cos 150°\mathbf{i} + \sin 150°\mathbf{j}) = \|\mathbf{F}_1\|\left(-\frac{\sqrt{3}}{2}\mathbf{i} + \frac{1}{2}\mathbf{j}\right) = -\frac{\sqrt{3}}{2}\|\mathbf{F}_1\|\mathbf{i} + \frac{1}{2}\|\mathbf{F}_1\|\mathbf{j}$$

$$\mathbf{F}_2 = \|\mathbf{F}_2\|(\cos 45°\mathbf{i} + \sin 45°\mathbf{j}) = \|\mathbf{F}_2\|\left(\frac{\sqrt{2}}{2}\mathbf{i} + \frac{\sqrt{2}}{2}\mathbf{j}\right) = \frac{\sqrt{2}}{2}\|\mathbf{F}_2\|\mathbf{i} + \frac{\sqrt{2}}{2}\|\mathbf{F}_2\|\mathbf{j}$$

$$\mathbf{F}_3 = -1200\mathbf{j}$$

For static equilibrium, the sum of the force vectors must equal zero.

$$\mathbf{F}_1 + \mathbf{F}_2 + \mathbf{F}_3 = -\frac{\sqrt{3}}{2}\|\mathbf{F}_1\|\mathbf{i} + \frac{1}{2}\|\mathbf{F}_1\|\mathbf{j} + \frac{\sqrt{2}}{2}\|\mathbf{F}_2\|\mathbf{i} + \frac{\sqrt{2}}{2}\|\mathbf{F}_2\|\mathbf{j} - 1200\mathbf{j} = \mathbf{0}$$

The **i** component and **j** component will each equal zero. This results in the two equations

$$-\frac{\sqrt{3}}{2}\|\mathbf{F}_1\| + \frac{\sqrt{2}}{2}\|\mathbf{F}_2\| = 0 \qquad \textbf{(9)}$$

$$\frac{1}{2}\|\mathbf{F}_1\| + \frac{\sqrt{2}}{2}\|\mathbf{F}_2\| - 1200 = 0 \qquad \textbf{(10)}$$

We solve equation (9) for $\|\mathbf{F}_2\|$ and obtain

$$\|\mathbf{F}_2\| = \frac{\sqrt{3}}{\sqrt{2}}\|\mathbf{F}_1\| \qquad \textbf{(11)}$$

Substituting into equation (10) and solving for $\|\mathbf{F}_1\|$, we obtain

$$\frac{1}{2}\|\mathbf{F}_1\| + \frac{\sqrt{2}}{2}\left(\frac{\sqrt{3}}{\sqrt{2}}\|\mathbf{F}_1\|\right) - 1200 = 0$$

$$\frac{1}{2}\|\mathbf{F}_1\| + \frac{\sqrt{3}}{2}\|\mathbf{F}_1\| - 1200 = 0$$

$$\frac{1 + \sqrt{3}}{2}\|\mathbf{F}_1\| = 1200$$

$$\|\mathbf{F}_1\| = \frac{2400}{1 + \sqrt{3}} \approx 878.5 \text{ pounds}$$

Substituting this value into equation (11) yields $\|\mathbf{F}_2\|$.

$$\|\mathbf{F}_2\| = \frac{\sqrt{3}}{\sqrt{2}}\|\mathbf{F}_1\| = \frac{\sqrt{3}}{\sqrt{2}} \cdot \frac{2400}{1 + \sqrt{3}} \approx 1075.9 \text{ pounds}$$

The left cable has tension of approximately 878.5 pounds and the right cable has tension of approximately 1075.9 pounds.

Now Work PROBLEM 67

Historical Feature

Josiah Gibbs
(1839–1903)

The history of vectors is surprisingly complicated for such a natural concept. In the xy-plane, complex numbers do a good job of imitating vectors. About 1840, mathematicians became interested in finding a system that would do for three dimensions what the complex numbers do for two dimensions. Hermann Grassmann (1809–1877), in Germany, and William Rowan Hamilton (1805–1865), in Ireland, both attempted to find solutions.

Hamilton's system was the *quaternions,* which are best thought of as a real number plus a vector, and do for four dimensions what complex numbers do for two dimensions. In this system the order of multiplication matters; that is, **ab ≠ ba.** Also, two products of vectors emerged, the scalar (or dot) product and the vector (or cross) product.

Grassmann's abstract style, although easily read today, was almost impenetrable during the previous century, and only a few of his ideas were appreciated. Among those few were the same scalar and vector products that Hamilton had found.

About 1880, the American physicist Josiah Willard Gibbs (1839–1903) worked out an algebra involving only the simplest concepts: the vectors and the two products. He then added some calculus, and the resulting system was simple, flexible, and well adapted to expressing a large number of physical laws. This system remains in use essentially unchanged. Hamilton's and Grassmann's more extensive systems each gave birth to much interesting mathematics, but little of this mathematics is seen at elementary levels.

10.4 Assess Your Understanding

Concepts and Vocabulary

1. A vector whose magnitude is 1 is called a(n) _____ vector.

2. The product of a vector by a number is called a(n) _____ multiple.

3. If $\mathbf{v} = a\mathbf{i} + b\mathbf{j}$, then a is called the _____ component of $\mathbf{v}$ and b is the _____ component of $\mathbf{v}$.

4. *True or False* Vectors are quantities that have magnitude and direction.

5. *True or False* Force is a physical example of a vector.

6. *True or False* Mass is a physical example of a vector.

Skill Building

In Problems 7–14, use the vectors in the figure at the right to graph each of the following vectors.

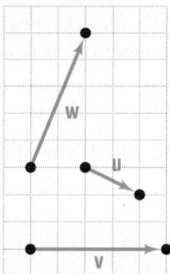

7. $\mathbf{v} + \mathbf{w}$

8. $\mathbf{u} + \mathbf{v}$

9. $3\mathbf{v}$

10. $4\mathbf{w}$

11. $\mathbf{v} - \mathbf{w}$

12. $\mathbf{u} - \mathbf{v}$

13. $3\mathbf{v} + \mathbf{u} - 2\mathbf{w}$

14. $2\mathbf{u} - 3\mathbf{v} + \mathbf{w}$

In Problems 15–22, use the figure at the right. Determine whether the given statement is true or false.

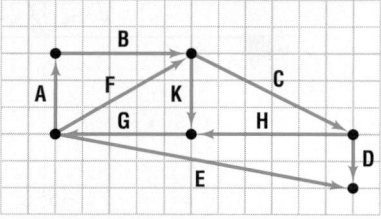

15. $\mathbf{A} + \mathbf{B} = \mathbf{F}$

16. $\mathbf{K} + \mathbf{G} = \mathbf{F}$

17. $\mathbf{C} = \mathbf{D} - \mathbf{E} + \mathbf{F}$

18. $\mathbf{G} + \mathbf{H} + \mathbf{E} = \mathbf{D}$

19. $\mathbf{E} + \mathbf{D} = \mathbf{G} + \mathbf{H}$

20. $\mathbf{H} - \mathbf{C} = \mathbf{G} - \mathbf{F}$

21. $\mathbf{A} + \mathbf{B} + \mathbf{K} + \mathbf{G} = \mathbf{0}$

22. $\mathbf{A} + \mathbf{B} + \mathbf{C} + \mathbf{H} + \mathbf{G} = \mathbf{0}$

23. If $\|\mathbf{v}\| = 4$, what is $\|3\mathbf{v}\|$?

24. If $\|\mathbf{v}\| = 2$, what is $\|-4\mathbf{v}\|$?

In Problems 25–32, the vector $\mathbf{v}$ has initial point P and terminal point Q. Write $\mathbf{v}$ in the form $a\mathbf{i} + b\mathbf{j}$; that is, find its position vector.

25. $P = (0, 0); \quad Q = (3, 4)$

26. $P = (0, 0); \quad Q = (-3, -5)$

27. $P = (3, 2); \quad Q = (5, 6)$

28. $P = (-3, 2); \quad Q = (6, 5)$

29. $P = (-2, -1); \quad Q = (6, -2)$

30. $P = (-1, 4); \quad Q = (6, 2)$

31. $P = (1, 0); \quad Q = (0, 1)$

32. $P = (1, 1); \quad Q = (2, 2)$

In Problems 33–38, find $\|\mathbf{v}\|$.

33. $\mathbf{v} = 3\mathbf{i} - 4\mathbf{j}$

34. $\mathbf{v} = -5\mathbf{i} + 12\mathbf{j}$

35. $\mathbf{v} = \mathbf{i} - \mathbf{j}$

36. $\mathbf{v} = -\mathbf{i} - \mathbf{j}$

37. $\mathbf{v} = -2\mathbf{i} + 3\mathbf{j}$

38. $\mathbf{v} = 6\mathbf{i} + 2\mathbf{j}$

In Problems 39–44, find each quantity if $\mathbf{v} = 3\mathbf{i} - 5\mathbf{j}$ and $\mathbf{w} = -2\mathbf{i} + 3\mathbf{j}$.

39. $2\mathbf{v} + 3\mathbf{w}$

40. $3\mathbf{v} - 2\mathbf{w}$

41. $\|\mathbf{v} - \mathbf{w}\|$

42. $\|\mathbf{v} + \mathbf{w}\|$

43. $\|\mathbf{v}\| - \|\mathbf{w}\|$

44. $\|\mathbf{v}\| + \|\mathbf{w}\|$

In Problems 45–50, find the unit vector in the same direction as $\mathbf{v}$.

45. $\mathbf{v} = 5\mathbf{i}$

46. $\mathbf{v} = -3\mathbf{j}$

47. $\mathbf{v} = 3\mathbf{i} - 4\mathbf{j}$

48. $\mathbf{v} = -5\mathbf{i} + 12\mathbf{j}$

49. $\mathbf{v} = \mathbf{i} - \mathbf{j}$

50. $\mathbf{v} = 2\mathbf{i} - \mathbf{j}$

51. Find a vector $\mathbf{v}$ whose magnitude is 4 and whose component in the $\mathbf{i}$ direction is twice the component in the $\mathbf{j}$ direction.

52. Find a vector $\mathbf{v}$ whose magnitude is 3 and whose component in the $\mathbf{i}$ direction is equal to the component in the $\mathbf{j}$ direction.

53. If $\mathbf{v} = 2\mathbf{i} - \mathbf{j}$ and $\mathbf{w} = x\mathbf{i} + 3\mathbf{j}$, find all numbers x for which $\|\mathbf{v} + \mathbf{w}\| = 5$.

54. If $P = (-3, 1)$ and $Q = (x, 4)$, find all numbers x such that the vector represented by $\overrightarrow{PQ}$ has length 5.

In Problems 55–60, write the vector **v** *in the form* $a\mathbf{i} + b\mathbf{j}$, *given its magnitude* $\|\mathbf{v}\|$ *and the angle* α *it makes with the positive x-axis.*

55. $\|\mathbf{v}\| = 5$, $\alpha = 60°$

56. $\|\mathbf{v}\| = 8$, $\alpha = 45°$

57. $\|\mathbf{v}\| = 14$, $\alpha = 120°$

58. $\|\mathbf{v}\| = 3$, $\alpha = 240°$

59. $\|\mathbf{v}\| = 25$, $\alpha = 330°$

60. $\|\mathbf{v}\| = 15$, $\alpha = 315°$

Applications and Extensions

61. Computer Graphics The field of computer graphics utilizes vectors to compute translations of points. For example, if the point $(-3, 2)$ is to be translated by $\mathbf{v} = \langle 5, 2\rangle$, then the new location will be $\mathbf{u}' = \mathbf{u} + \mathbf{v} = \langle -3, 2\rangle + \langle 5, 2\rangle = \langle 2, 4\rangle$. As illustrated in the figure, the point $(-3, 2)$ is translated to $(2, 4)$ by $\mathbf{v}$.

Source: Phil Dadd. Vectors and Matrices: A Primer. www.gamedev.net/reference/articles/article1832.asp

(a) Determine the new coordinates of $(3, -1)$ if it is translated by $\mathbf{v} = \langle -4, 5\rangle$.

(b) Illustrate this translation graphically.

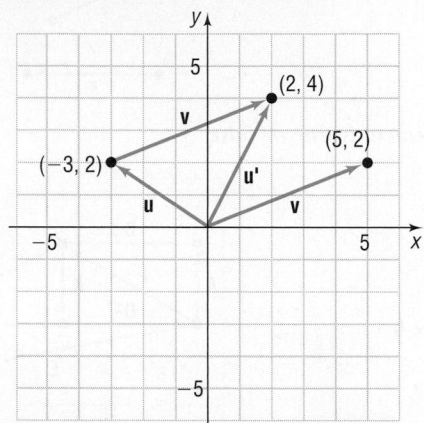

62. Computer Graphics Refer to Problem 61. The points $(-3, 0)$, $(-1, -2)$, $(3, 1)$ and $(1, 3)$ are the vertices of a parallelogram $ABCD$.

(a) Find the new vertices of a parallelogram $A'B'C'D'$ if it is translated by $\mathbf{v} = \langle 3, -2\rangle$.

(b) Find the new vertices of a parallelogram $A'B'C'D'$ if it is translated by $-\dfrac{1}{2}\mathbf{v}$.

63. Force Vectors A child pulls a wagon with a force of 40 pounds. The handle of the wagon makes an angle of 30° with the ground. Express the force vector $\mathbf{F}$ in terms of $\mathbf{i}$ and $\mathbf{j}$.

64. Force Vectors A man pushes a wheelbarrow up an incline of 20° with a force of 100 pounds. Express the force vector $\mathbf{F}$ in terms of $\mathbf{i}$ and $\mathbf{j}$.

65. Resultant Force Two forces of magnitude 40 newtons (N) and 60 N act on an object at angles of 30° and $-45°$ with the positive x-axis, as shown in the figure. Find the direction and magnitude of the resultant force; that is, find $\mathbf{F}_1 + \mathbf{F}_2$.

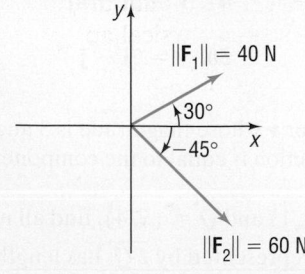

66. Resultant Force Two forces of magnitude 30 newtons (N) and 70 N act on an object at angles of 45° and 120° with the positive x-axis, as shown in the figure. Find the direction and magnitude of the resultant force; that is, find $\mathbf{F}_1 + \mathbf{F}_2$.

67. Static Equilibrium A weight of 1000 pounds is suspended from two cables as shown in the figure. What are the tensions in the two cables?

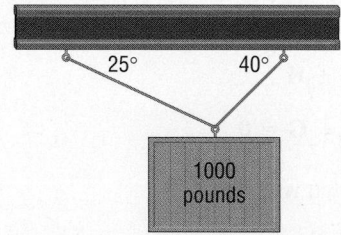

68. Static Equilibrium A weight of 800 pounds is suspended from two cables, as shown in the figure. What are the tensions in the two cables?

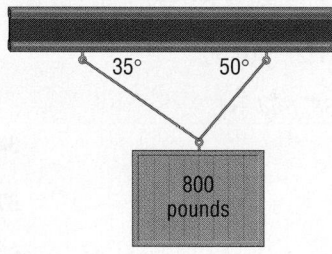

69. Static Equilibrium A tightrope walker located at a certain point deflects the rope as indicated in the figure. If the weight of the tightrope walker is 150 pounds, how much tension is in each part of the rope?

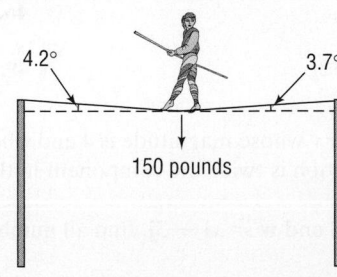

70. Static Equilibrium Repeat Problem 69 if the left angle is 3.8°, the right angle is 2.6°, and the weight of the tightrope walker is 135 pounds.

71. Truck Pull At a county fair truck pull, two pickup trucks are attached to the back end of a monster truck as illustrated in the figure. One of the pickups pulls with a force of 2000 pounds and the other pulls with a force of 3000 pounds with an angle of 45° between them. With how much force must the monster truck pull in order to remain unmoved? (**Hint:** Find the resultant force of the two trucks.)

72. Removing a Stump A farmer wishes to remove a stump from a field by pulling it out with his tractor. Having removed many stumps before, he estimates that he will need 6 tons (12,000 pounds) of force to remove the stump. However, his tractor is only capable of pulling with a force of 7000 pounds, so he asks his neighbor to help. His neighbor's tractor can pull with a force of 5500 pounds. They attach the two tractors to the stump with a 40° angle between the forces as shown in the figure.

(a) Assuming the farmer's estimate of a needed 6-ton force is correct, will the farmer be successful in removing the stump? Explain.

(b) Had the farmer arranged the tractors with a 25° angle between the forces, would he have been successful in removing the stump? Explain.

73. Static Equilibrium Show on the following graph the force needed for the object at P to be in static equilibrium.

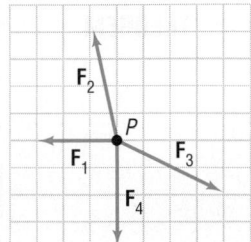

Discussion and Writing

74. Explain in your own words what a vector is. Give an example of a vector.

75. Write a brief paragraph comparing the algebra of complex numbers and the algebra of vectors.

10.5 The Dot Product

PREPARING FOR THIS SECTION *Before getting started, review the following:*

- Law of Cosines (Section 9.3, p. 686)

Now Work the 'Are You Prepared?' problem on page 763.

OBJECTIVES **1** Find the Dot Product of Two Vectors (p. 757)
2 Find the Angle between Two Vectors (p. 758)
3 Determine Whether Two Vectors Are Parallel (p. 760)
4 Determine Whether Two Vectors Are Orthogonal (p. 760)
5 Decompose a Vector into Two Orthogonal Vectors (p. 761)
6 Compute Work (p. 762)

1 Find the Dot Product of Two Vectors

The definition for a product of two vectors is somewhat unexpected. However, such a product has meaning in many geometric and physical applications.

DEFINITION If $\mathbf{v} = a_1\mathbf{i} + b_1\mathbf{j}$ and $\mathbf{w} = a_2\mathbf{i} + b_2\mathbf{j}$ are two vectors, the **dot product** $\mathbf{v} \cdot \mathbf{w}$ is defined as

$$\mathbf{v} \cdot \mathbf{w} = a_1a_2 + b_1b_2 \qquad (1)$$

EXAMPLE 1	**Finding Dot Products**

If $\mathbf{v} = 2\mathbf{i} - 3\mathbf{j}$ and $\mathbf{w} = 5\mathbf{i} + 3\mathbf{j}$, find:

(a) $\mathbf{v} \cdot \mathbf{w}$ (b) $\mathbf{w} \cdot \mathbf{v}$ (c) $\mathbf{v} \cdot \mathbf{v}$

(d) $\mathbf{w} \cdot \mathbf{w}$ (e) $\|\mathbf{v}\|$ (f) $\|\mathbf{w}\|$

Solution

(a) $\mathbf{v} \cdot \mathbf{w} = 2(5) + (-3)3 = 1$ (b) $\mathbf{w} \cdot \mathbf{v} = 5(2) + 3(-3) = 1$

(c) $\mathbf{v} \cdot \mathbf{v} = 2(2) + (-3)(-3) = 13$ (d) $\mathbf{w} \cdot \mathbf{w} = 5(5) + 3(3) = 34$

(e) $\|\mathbf{v}\| = \sqrt{2^2 + (-3)^2} = \sqrt{13}$ (f) $\|\mathbf{w}\| = \sqrt{5^2 + 3^2} = \sqrt{34}$

COMMENT A scalar multiple $\alpha\mathbf{v}$ is a vector. A dot product $\mathbf{u} \cdot \mathbf{v}$ is a scalar (real number). ∎

Since the dot product $\mathbf{v} \cdot \mathbf{w}$ of two vectors $\mathbf{v}$ and $\mathbf{w}$ is a real number (scalar), we sometimes refer to it as the **scalar product.**

The results obtained in Example 1 suggest some general properties.

THEOREM

Properties of the Dot Product

If $\mathbf{u}$, $\mathbf{v}$, and $\mathbf{w}$ are vectors, then

Commutative Property

$$\mathbf{u} \cdot \mathbf{v} = \mathbf{v} \cdot \mathbf{u} \tag{2}$$

Distributive Property

$$\mathbf{u} \cdot (\mathbf{v} + \mathbf{w}) = \mathbf{u} \cdot \mathbf{v} + \mathbf{u} \cdot \mathbf{w} \tag{3}$$

$$\mathbf{v} \cdot \mathbf{v} = \|\mathbf{v}\|^2 \tag{4}$$
$$\mathbf{0} \cdot \mathbf{v} = 0 \tag{5}$$

Proof We will prove properties (2) and (4) here and leave properties (3) and (5) as exercises (see Problems 40 and 41).

To prove property (2), we let $\mathbf{u} = a_1\mathbf{i} + b_1\mathbf{j}$ and $\mathbf{v} = a_2\mathbf{i} + b_2\mathbf{j}$. Then

$$\mathbf{u} \cdot \mathbf{v} = a_1a_2 + b_1b_2 = a_2a_1 + b_2b_1 = \mathbf{v} \cdot \mathbf{u}$$

To prove property (4), we let $\mathbf{v} = a\mathbf{i} + b\mathbf{j}$. Then

$$\mathbf{v} \cdot \mathbf{v} = a^2 + b^2 = \|\mathbf{v}\|^2 \qquad ∎$$

2 Find the Angle between Two Vectors

One use of the dot product is to calculate the angle between two vectors. We proceed as follows.

Let $\mathbf{u}$ and $\mathbf{v}$ be two vectors with the same initial point A. Then the vectors $\mathbf{u}$, $\mathbf{v}$, and $\mathbf{u} - \mathbf{v}$ form a triangle. The angle θ at vertex A of the triangle is the angle between the vectors $\mathbf{u}$ and $\mathbf{v}$. See Figure 60. We wish to find a formula for calculating the angle θ.

The sides of the triangle have lengths $\|\mathbf{v}\|$, $\|\mathbf{u}\|$, and $\|\mathbf{u} - \mathbf{v}\|$, and θ is the included angle between the sides of length $\|\mathbf{v}\|$ and $\|\mathbf{u}\|$. The Law of Cosines (Section 9.3) can be used to find the cosine of the included angle.

$$\|\mathbf{u} - \mathbf{v}\|^2 = \|\mathbf{u}\|^2 + \|\mathbf{v}\|^2 - 2\|\mathbf{u}\|\|\mathbf{v}\| \cos \theta$$

Now we use property (4) to rewrite this equation in terms of dot products.

$$(\mathbf{u} - \mathbf{v}) \cdot (\mathbf{u} - \mathbf{v}) = \mathbf{u} \cdot \mathbf{u} + \mathbf{v} \cdot \mathbf{v} - 2\|\mathbf{u}\|\|\mathbf{v}\| \cos \theta \tag{6}$$

Figure 60

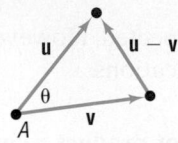

Then we apply the distributive property (3) twice on the left side of (6) to obtain

$$(\mathbf{u} - \mathbf{v}) \cdot (\mathbf{u} - \mathbf{v}) = \mathbf{u} \cdot (\mathbf{u} - \mathbf{v}) - \mathbf{v} \cdot (\mathbf{u} - \mathbf{v})$$
$$= \mathbf{u} \cdot \mathbf{u} - \mathbf{u} \cdot \mathbf{v} - \mathbf{v} \cdot \mathbf{u} + \mathbf{v} \cdot \mathbf{v}$$
$$\underset{\underset{\text{Property (2)}}{\uparrow}}{=} \mathbf{u} \cdot \mathbf{u} + \mathbf{v} \cdot \mathbf{v} - 2\,\mathbf{u} \cdot \mathbf{v} \qquad \textbf{(7)}$$

Combining equations (6) and (7), we have

$$\mathbf{u} \cdot \mathbf{u} + \mathbf{v} \cdot \mathbf{v} - 2\,\mathbf{u} \cdot \mathbf{v} = \mathbf{u} \cdot \mathbf{u} + \mathbf{v} \cdot \mathbf{v} - 2\,\|\mathbf{u}\|\|\mathbf{v}\|\cos\theta$$
$$\mathbf{u} \cdot \mathbf{v} = \|\mathbf{u}\|\|\mathbf{v}\|\cos\theta$$

We have proved the following result:

THEOREM

Angle between Vectors

If $\mathbf{u}$ and $\mathbf{v}$ are two nonzero vectors, the angle θ, $0 \le \theta \le \pi$, between $\mathbf{u}$ and $\mathbf{v}$ is determined by the formula

$$\cos\theta = \frac{\mathbf{u} \cdot \mathbf{v}}{\|\mathbf{u}\|\|\mathbf{v}\|} \qquad \textbf{(8)}$$

EXAMPLE 2 Finding the Angle θ between Two Vectors

Find the angle θ between $\mathbf{u} = 4\mathbf{i} - 3\mathbf{j}$ and $\mathbf{v} = 2\mathbf{i} + 5\mathbf{j}$.

Solution We compute the quantities $\mathbf{u} \cdot \mathbf{v}$, $\|\mathbf{u}\|$, and $\|\mathbf{v}\|$.

$$\mathbf{u} \cdot \mathbf{v} = 4(2) + (-3)(5) = -7$$
$$\|\mathbf{u}\| = \sqrt{4^2 + (-3)^2} = 5$$
$$\|\mathbf{v}\| = \sqrt{2^2 + 5^2} = \sqrt{29}$$

By formula (8), if θ is the angle between $\mathbf{u}$ and $\mathbf{v}$, then

$$\cos\theta = \frac{\mathbf{u} \cdot \mathbf{v}}{\|\mathbf{u}\|\|\mathbf{v}\|} = \frac{-7}{5\sqrt{29}} \approx -0.26$$

We find that $\theta \approx 105°$. See Figure 61.

Figure 61

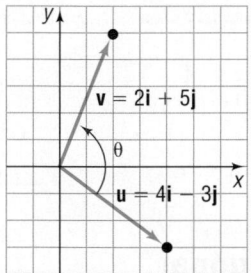

$\mathbf{v} = 2\mathbf{i} + 5\mathbf{j}$

θ

$\mathbf{u} = 4\mathbf{i} - 3\mathbf{j}$

━━ **Now Work** PROBLEMS **7(a)** AND **(b)**

EXAMPLE 3 Finding the Actual Speed and Direction of an Aircraft

A Boeing 737 aircraft maintains a constant airspeed of 500 miles per hour in the direction due south. The velocity of the jet stream is 80 miles per hour in a northeasterly direction. Find the actual speed and direction of the aircraft relative to the ground.

Solution We set up a coordinate system in which north (N) is along the positive y-axis. See Figure 62 on page 760. Let

$$\mathbf{v}_a = \text{velocity of aircraft relative to the air} = -500\mathbf{j}$$
$$\mathbf{v}_w = \text{velocity of jet stream}$$
$$\mathbf{v}_g = \text{velocity of aircraft relative to ground}$$

The velocity of the jet stream $\mathbf{v}_w$ has magnitude 80 and direction NE (northeast), so the angle between $\mathbf{v}_w$ and $\mathbf{i}$ is 45°. We express $\mathbf{v}_w$ in terms of $\mathbf{i}$ and $\mathbf{j}$ as

$$\mathbf{v}_w = 80(\cos 45°\mathbf{i} + \sin 45°\mathbf{j}) = 80\left(\frac{\sqrt{2}}{2}\mathbf{i} + \frac{\sqrt{2}}{2}\mathbf{j}\right) = 40\sqrt{2}\,(\mathbf{i} + \mathbf{j})$$

Figure 62

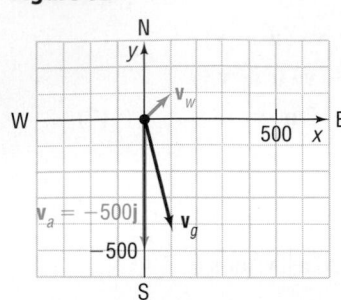

The velocity of the aircraft relative to the ground is

$$\mathbf{v}_g = \mathbf{v}_a + \mathbf{v}_w = -500\mathbf{j} + 40\sqrt{2}\,(\mathbf{i} + \mathbf{j}) = 40\sqrt{2}\mathbf{i} + \left(40\sqrt{2} - 500\right)\mathbf{j}$$

The actual speed of the aircraft is

$$\|\mathbf{v}_g\| = \sqrt{\left(40\sqrt{2}\right)^2 + \left(40\sqrt{2} - 500\right)^2} \approx 447 \text{ miles per hour}$$

The angle θ between $\mathbf{v}_g$ and the vector $\mathbf{v}_a = -500\mathbf{j}$ (the velocity of the aircraft relative to the air) is determined by the equation

$$\cos\theta = \frac{\mathbf{v}_g \cdot \mathbf{v}_a}{\|\mathbf{v}_g\|\|\mathbf{v}_a\|} \approx \frac{\left(40\sqrt{2} - 500\right)(-500)}{(447)(500)} \approx 0.9920$$

$$\theta \approx 7.3°$$

The direction of the aircraft relative to the ground is approximately S7.3°E (about 7.3° east of south).

Now Work PROBLEM 29

3 Determine Whether Two Vectors Are Parallel

Two vectors $\mathbf{v}$ and $\mathbf{w}$ are said to be **parallel** if there is a nonzero scalar α so that $\mathbf{v} = \alpha\mathbf{w}$. In this case, the angle θ between $\mathbf{v}$ and $\mathbf{w}$ is 0 or π.

EXAMPLE 4 **Determining Whether Vectors Are Parallel**

The vectors $\mathbf{v} = 3\mathbf{i} - \mathbf{j}$ and $\mathbf{w} = 6\mathbf{i} - 2\mathbf{j}$ are parallel, since $\mathbf{v} = \dfrac{1}{2}\mathbf{w}$. Furthermore, since

$$\cos\theta = \frac{\mathbf{v} \cdot \mathbf{w}}{\|\mathbf{v}\|\|\mathbf{w}\|} = \frac{18 + 2}{\sqrt{10}\,\sqrt{40}} = \frac{20}{\sqrt{400}} = 1$$

the angle θ between $\mathbf{v}$ and $\mathbf{w}$ is 0.

4 Determine Whether Two Vectors Are Orthogonal

Figure 63
v is orthogonal to **w**.

If the angle θ between two nonzero vectors $\mathbf{v}$ and $\mathbf{w}$ is $\dfrac{\pi}{2}$, the vectors $\mathbf{v}$ and $\mathbf{w}$ are called **orthogonal**.* See Figure 63.

Since $\cos\dfrac{\pi}{2} = 0$, it follows from formula (8) that if $\mathbf{v}$ and $\mathbf{w}$ are orthogonal then $\mathbf{v} \cdot \mathbf{w} = 0$.

On the other hand, if $\mathbf{v} \cdot \mathbf{w} = 0$, then either $\mathbf{v} = 0$ or $\mathbf{w} = 0$ or $\cos\theta = 0$. In the latter case, $\theta = \dfrac{\pi}{2}$, and $\mathbf{v}$ and $\mathbf{w}$ are orthogonal. If $\mathbf{v}$ or $\mathbf{w}$ is the zero vector, then, since the zero vector has no specific direction, we adopt the convention that the zero vector is orthogonal to every vector.

THEOREM

Two vectors $\mathbf{v}$ and $\mathbf{w}$ are orthogonal if and only if

$$\mathbf{v} \cdot \mathbf{w} = 0$$

Orthogonal, perpendicular, and *normal* are all terms that mean "meet at a right angle." It is customary to refer to two vectors as being *orthogonal,* two lines as being *perpendicular,* and a line and a plane or a vector and a plane as being *normal.*

| EXAMPLE 5 | Determining Whether Two Vectors Are Orthogonal |

Figure 64

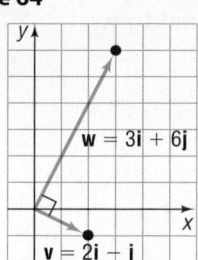

The vectors

$$\mathbf{v} = 2\mathbf{i} - \mathbf{j} \quad \text{and} \quad \mathbf{w} = 3\mathbf{i} + 6\mathbf{j}$$

are orthogonal, since

$$\mathbf{v} \cdot \mathbf{w} = 6 - 6 = 0$$

See Figure 64.

━━━━━ **Now Work** PROBLEM 7(c)

5 Decompose a Vector into Two Orthogonal Vectors

Figure 65

In many physical applications, it is necessary to find "how much" of a vector is applied in a given direction. Look at Figure 65. The force $\mathbf{F}$ due to gravity is pulling straight down (toward the center of Earth) on the block. To study the effect of gravity on the block, it is necessary to determine how much of $\mathbf{F}$ is actually pushing the block down the incline ($\mathbf{F}_1$) and how much is pressing the block against the incline ($\mathbf{F}_2$), at a right angle to the incline. Knowing the **decomposition** of $\mathbf{F}$ often will allow us to determine when friction is overcome and the block will slide down the incline.

Suppose that $\mathbf{v}$ and $\mathbf{w}$ are two nonzero vectors with the same initial point P. We seek to decompose $\mathbf{v}$ into two vectors: $\mathbf{v}_1$, which is parallel to $\mathbf{w}$, and $\mathbf{v}_2$, which is orthogonal to $\mathbf{w}$. See Figure 66(a) and (b). The vector $\mathbf{v}_1$ is called the **vector projection of v onto w.**

Figure 66

(a)

(b)

The vector $\mathbf{v}_1$ is obtained as follows: From the terminal point of $\mathbf{v}$, drop a perpendicular to the line containing $\mathbf{w}$. The vector $\mathbf{v}_1$ is the vector from P to the foot of this perpendicular. The vector $\mathbf{v}_2$ is given by $\mathbf{v}_2 = \mathbf{v} - \mathbf{v}_1$. Note that $\mathbf{v} = \mathbf{v}_1 + \mathbf{v}_2$, the vector $\mathbf{v}_1$ is parallel to $\mathbf{w}$, and the vector $\mathbf{v}_2$ is orthogonal to $\mathbf{w}$. This is the decomposition of $\mathbf{v}$ that we wanted.

Now we seek a formula for $\mathbf{v}_1$ that is based on a knowledge of the vectors $\mathbf{v}$ and $\mathbf{w}$. Since $\mathbf{v} = \mathbf{v}_1 + \mathbf{v}_2$, we have

$$\mathbf{v} \cdot \mathbf{w} = (\mathbf{v}_1 + \mathbf{v}_2) \cdot \mathbf{w} = \mathbf{v}_1 \cdot \mathbf{w} + \mathbf{v}_2 \cdot \mathbf{w} \qquad (9)$$

Since $\mathbf{v}_2$ is orthogonal to $\mathbf{w}$, we have $\mathbf{v}_2 \cdot \mathbf{w} = 0$. Since $\mathbf{v}_1$ is parallel to $\mathbf{w}$, we have $\mathbf{v}_1 = \alpha\mathbf{w}$ for some scalar α. Equation (9) can be written as

$$\mathbf{v} \cdot \mathbf{w} = \alpha\mathbf{w} \cdot \mathbf{w} = \alpha\|\mathbf{w}\|^2 \qquad \text{\footnotesize $\mathbf{v}_1 = \alpha\mathbf{w}; \mathbf{v}_2 \cdot \mathbf{w} = 0$}$$

$$\alpha = \frac{\mathbf{v} \cdot \mathbf{w}}{\|\mathbf{w}\|^2}$$

Then

$$\mathbf{v}_1 = \alpha\mathbf{w} = \frac{\mathbf{v} \cdot \mathbf{w}}{\|\mathbf{w}\|^2}\mathbf{w}$$

THEOREM If $\mathbf{v}$ and $\mathbf{w}$ are two nonzero vectors, the vector projection of $\mathbf{v}$ onto $\mathbf{w}$ is

$$\mathbf{v}_1 = \frac{\mathbf{v} \cdot \mathbf{w}}{\|\mathbf{w}\|^2}\mathbf{w} \qquad (10)$$

The decomposition of $\mathbf{v}$ into $\mathbf{v}_1$ and $\mathbf{v}_2$, where $\mathbf{v}_1$ is parallel to $\mathbf{w}$ and $\mathbf{v}_2$ is orthogonal to $\mathbf{w}$, is

$$\mathbf{v}_1 = \frac{\mathbf{v} \cdot \mathbf{w}}{\|\mathbf{w}\|^2}\mathbf{w} \qquad \mathbf{v}_2 = \mathbf{v} - \mathbf{v}_1 \qquad (11)$$

| EXAMPLE 6 | Decomposing a Vector into Two Orthogonal Vectors |

Find the vector projection of $\mathbf{v} = \mathbf{i} + 3\mathbf{j}$ onto $\mathbf{w} = \mathbf{i} + \mathbf{j}$. Decompose $\mathbf{v}$ into two vectors $\mathbf{v}_1$ and $\mathbf{v}_2$, where $\mathbf{v}_1$ is parallel to $\mathbf{w}$ and $\mathbf{v}_2$ is orthogonal to $\mathbf{w}$.

Figure 67

Solution We use formulas (10) and (11).

$$\mathbf{v}_1 = \frac{\mathbf{v} \cdot \mathbf{w}}{\|\mathbf{w}\|^2}\mathbf{w} = \frac{1+3}{\left(\sqrt{2}\right)^2}\mathbf{w} = 2\mathbf{w} = 2(\mathbf{i} + \mathbf{j})$$

$$\mathbf{v}_2 = \mathbf{v} - \mathbf{v}_1 = (\mathbf{i} + 3\mathbf{j}) - 2(\mathbf{i} + \mathbf{j}) = -\mathbf{i} + \mathbf{j}$$

See Figure 67.

 Now Work PROBLEM 19

6 Compute Work

In elementary physics, the **work** W done by a constant force $\mathbf{F}$ in moving an object from a point A to a point B is defined as

Figure 68

$$W = (\text{magnitude of force})(\text{distance}) = \|\mathbf{F}\|\|\overrightarrow{AB}\|$$

Work is commonly measured in foot-pounds or in newton-meters (joules).

In this definition, it is assumed that the force $\mathbf{F}$ is applied along the line of motion. If the constant force $\mathbf{F}$ is not along the line of motion, but, instead, is at an angle θ to the direction of the motion, as illustrated in Figure 68, the **work W done by $\mathbf{F}$** in moving an object from A to B is defined as

$$W = \mathbf{F} \cdot \overrightarrow{AB} \tag{12}$$

This definition is compatible with the force times distance definition, since

$$W = (\text{amount of force in the direction of } \overrightarrow{AB})(\text{distance})$$

$$= \|\text{projection of } \mathbf{F} \text{ on } AB\|\|\overrightarrow{AB}\| = \frac{\mathbf{F} \cdot \overrightarrow{AB}}{\|\overrightarrow{AB}\|^2}\|\overrightarrow{AB}\|\|\overrightarrow{AB}\| = \mathbf{F} \cdot \overrightarrow{AB}$$

↑
Use formula (10)

| EXAMPLE 7 | Computing Work |

Figure 69(a) shows a girl pulling a wagon with a force of 50 pounds. How much work is done in moving the wagon 100 feet if the handle makes an angle of 30° with the ground?

Figure 69

(a) **(b)**

Solution We position the vectors in a coordinate system in such a way that the wagon is moved from $(0, 0)$ to $(100, 0)$. The motion is from $A = (0, 0)$ to $B = (100, 0)$, so $\overrightarrow{AB} = 100\mathbf{i}$. The force vector $\mathbf{F}$, as shown in Figure 69(b), is

$$\mathbf{F} = 50(\cos 30°\mathbf{i} + \sin 30°\mathbf{j}) = 50\left(\frac{\sqrt{3}}{2}\mathbf{i} + \frac{1}{2}\mathbf{j}\right) = 25\left(\sqrt{3}\mathbf{i} + \mathbf{j}\right)$$

By formula (12), the work done is

$$W = \mathbf{F} \cdot \overrightarrow{AB} = 25\left(\sqrt{3}\mathbf{i} + \mathbf{j}\right) \cdot 100\mathbf{i} = 2500\sqrt{3} \text{ foot-pounds}$$

Now Work PROBLEM 25

Historical Feature

We stated in an earlier Historical Feature that complex numbers were used as vectors in the plane before the general notion of a vector was clarified. Suppose that we make the correspondence

Vector ↔ Complex number

$$a\mathbf{i} + b\mathbf{j} \leftrightarrow a + bi$$

$$c\mathbf{i} + d\mathbf{j} \leftrightarrow c + di$$

Show that

$$(a\mathbf{i} + b\mathbf{j}) \cdot (c\mathbf{i} + d\mathbf{j}) = \text{real part } [(\overline{a + bi})(c + di)]$$

This is how the dot product was found originally. The imaginary part is also interesting. It is a determinant (see Section 12.3) and represents the area of the parallelogram whose edges are the vectors. This is close to some of Hermann Grassmann's ideas and is also connected with the scalar triple product of three-dimensional vectors.

10.5 Assess Your Understanding

'Are You Prepared?' *The answer is given at the end of these exercises. If you get the wrong answer, read the page listed in red.*

1. In a triangle with sides a, b, c and angles A, B, C, the Law of Cosines states that _____. (p. 686)

Concepts and Vocabulary

2. If $\mathbf{v} \cdot \mathbf{w} = 0$, then the two vectors $\mathbf{v}$ and $\mathbf{w}$ are _____.

3. If $\mathbf{v} = 3\mathbf{w}$, then the two vectors $\mathbf{v}$ and $\mathbf{w}$ are _____.

4. *True or False* If $\mathbf{v}$ and $\mathbf{w}$ are parallel vectors, then $\mathbf{v} \cdot \mathbf{w} = 0$.

5. *True or False* Given two nonzero vectors $\mathbf{v}$ and $\mathbf{w}$, it is always possible to decompose $\mathbf{v}$ into two vectors, one parallel to $\mathbf{w}$ and the other perpendicular to $\mathbf{w}$.

6. *True or False* Work is a physical example of a vector.

Skill Building

In Problems 7–16, (a) find the dot product $\mathbf{v} \cdot \mathbf{w}$; (b) find the angle between $\mathbf{v}$ and $\mathbf{w}$; (c) state whether the vectors are parallel, orthogonal, or neither.

7. $\mathbf{v} = \mathbf{i} - \mathbf{j}$, $\mathbf{w} = \mathbf{i} + \mathbf{j}$

8. $\mathbf{v} = \mathbf{i} + \mathbf{j}$, $\mathbf{w} = -\mathbf{i} + \mathbf{j}$

9. $\mathbf{v} = 2\mathbf{i} + \mathbf{j}$, $\mathbf{w} = \mathbf{i} - 2\mathbf{j}$

10. $\mathbf{v} = 2\mathbf{i} + 2\mathbf{j}$, $\mathbf{w} = \mathbf{i} + 2\mathbf{j}$

11. $\mathbf{v} = \sqrt{3}\mathbf{i} - \mathbf{j}$, $\mathbf{w} = \mathbf{i} + \mathbf{j}$

12. $\mathbf{v} = \mathbf{i} + \sqrt{3}\mathbf{j}$, $\mathbf{w} = \mathbf{i} - \mathbf{j}$

13. $\mathbf{v} = 3\mathbf{i} + 4\mathbf{j}$, $\mathbf{w} = 4\mathbf{i} + 3\mathbf{j}$

14. $\mathbf{v} = 3\mathbf{i} - 4\mathbf{j}$, $\mathbf{w} = 4\mathbf{i} - 3\mathbf{j}$

15. $\mathbf{v} = 4\mathbf{i}$, $\mathbf{w} = \mathbf{j}$

16. $\mathbf{v} = \mathbf{i}$, $\mathbf{w} = -3\mathbf{j}$

17. Find a so that the vectors $\mathbf{v} = \mathbf{i} - a\mathbf{j}$ and $\mathbf{w} = 2\mathbf{i} + 3\mathbf{j}$ are orthogonal.

18. Find b so that the vectors $\mathbf{v} = \mathbf{i} + \mathbf{j}$ and $\mathbf{w} = \mathbf{i} + b\mathbf{j}$ are orthogonal.

In Problems 19–24, decompose $\mathbf{v}$ into two vectors $\mathbf{v}_1$ and $\mathbf{v}_2$, where $\mathbf{v}_1$ is parallel to $\mathbf{w}$ and $\mathbf{v}_2$ is orthogonal to $\mathbf{w}$.

19. $\mathbf{v} = 2\mathbf{i} - 3\mathbf{j}$, $\mathbf{w} = \mathbf{i} - \mathbf{j}$

20. $\mathbf{v} = -3\mathbf{i} + 2\mathbf{j}$, $\mathbf{w} = 2\mathbf{i} + \mathbf{j}$

21. $\mathbf{v} = \mathbf{i} - \mathbf{j}$, $\mathbf{w} = -\mathbf{i} - 2\mathbf{j}$

22. $\mathbf{v} = 2\mathbf{i} - \mathbf{j}$, $\mathbf{w} = \mathbf{i} - 2\mathbf{j}$

23. $\mathbf{v} = 3\mathbf{i} + \mathbf{j}$, $\mathbf{w} = -2\mathbf{i} - \mathbf{j}$

24. $\mathbf{v} = \mathbf{i} - 3\mathbf{j}$, $\mathbf{w} = 4\mathbf{i} - \mathbf{j}$

Applications and Extensions

25. **Computing Work** Find the work done by a force of 3 pounds acting in the direction 60° to the horizontal in moving an object 6 feet from $(0, 0)$ to $(6, 0)$.

26. **Computing Work** A wagon is pulled horizontally by exerting a force of 20 pounds on the handle at an angle of 30° with the horizontal. How much work is done in moving the wagon 100 feet?

27. **Solar Energy** The amount of energy collected by a solar panel depends on the intensity of the sun's rays and the area of the panel. Let the vector **I** represent the intensity, in watts

per square centimeter, having the direction of the sun's rays. Let the vector **A** represent the area, in square centimeters, whose direction is the orientation of a solar panel. See the figure. The total number of watts collected by the panel is given by $W = |\mathbf{I} \cdot \mathbf{A}|$.

Suppose $\mathbf{I} = \langle -0.02, -0.01 \rangle$ and $\mathbf{A} = \langle 300, 400 \rangle$.

(a) Find $\|\mathbf{I}\|$ and $\|\mathbf{A}\|$ and interpret the meaning of each.

(b) Compute W and interpret its meaning.

(c) If the solar panel is to collect the maximum number of watts, what must be true about **I** and **A**?

28. **Rainfall Measurement** Let the vector **R** represent the amount of rainfall, in inches, whose direction is the inclination of the rain to a rain gauge. Let the vector **A** represent the area, in square inches, whose direction is the orientation of the opening of the rain gauge. See the figure. The volume of rain collected in the gauge, in cubic inches, is given by $V = |\mathbf{R} \cdot \mathbf{A}|$, even when the rain falls in a slanted direction or the gauge is not perfectly vertical.

Suppose $\mathbf{R} = \langle 0.75, -1.75 \rangle$ and $\mathbf{A} = \langle 0.3, 1 \rangle$.

(a) Find $\|\mathbf{R}\|$ and $\|\mathbf{A}\|$ and interpret the meaning of each.

(b) Compute V and interpret its meaning

(c) If the gauge is to collect the maximum volume of rain, what must be true about **R** and **A**?

29. **Finding the Actual Speed and Direction of an Aircraft** A Boeing 747 jumbo jet maintains an airspeed of 550 miles per hour in a southwesterly direction. The velocity of the jet stream is a constant 80 miles per hour from the west. Find the actual speed and direction of the aircraft.

Jet stream

30. **Finding the Correct Compass Heading** The pilot of an aircraft wishes to head directly east, but is faced with a wind speed of 40 miles per hour from the northwest. If the pilot maintains an airspeed of 250 miles per hour, what compass heading should be maintained to head directly east? What is the actual speed of the aircraft?

31. **Correct Direction for Crossing a River** A river has a constant current of 3 kilometers per hour. At what angle to a boat dock should a motorboat, capable of maintaining a constant speed of 20 kilometers per hour, be headed in order to reach a point directly opposite the dock? If the river is $\frac{1}{2}$ kilometer wide, how long will it take to cross?

Direction of boat due to current

32. **Crossing a River** A small motorboat in still water maintains a speed of 20 miles per hour. In heading directly across a river (that is, perpendicular to the current) whose current is 3 miles per hour, find a vector representing the speed and direction of the motorboat. What is the true speed of the motorboat? What is its direction?

33. **Braking Load** A Toyota Sienna with a gross weight of 5300 pounds is parked on a street with a slope of 8°. See the figure. Find the force required to keep the Sienna from rolling down the hill. What is the force perpendicular to the hill?

Weight = 5300 pounds

34. **Braking Load** A Pontiac Bonneville with a gross weight of 4500 pounds is parked on a street with a slope of 10°. Find the force required to keep the Bonneville from rolling down the hill. What is the force perpendicular to the hill?

35. **Ground Speed and Direction of an Airplane** An airplane has an airspeed of 500 kilometers per hour (km/hr) bearing N45°E. The wind velocity is 60 km/hr in the direction N30°W. Find the resultant vector representing the path of the plane relative to the ground. What is the ground speed of the plane? What is its direction?

36. **Ground Speed and Direction of an Airplane** An airplane has an airspeed of 600 km/hr bearing S30°E. The wind velocity is 40 km/hr in the direction S45°E. Find the resultant vector representing the path of the plane relative to the ground. What is the ground speed of the plane? What is its direction?

37. **Ramp Angle** Billy and Timmy are using a ramp to load furniture into a truck. While rolling a 250-pound piano up the ramp, they discover that the truck is too full with other

furniture for the piano to fit. Timmy holds the piano in place on the ramp while Billy repositions other items to make room for it in the truck. If the angle of inclination of the ramp is 20°, how many pounds of force must Timmy exert to hold the piano in position?

20°

250 lb

38. Incline Angle A bulldozer exerts 1000 pounds of force to prevent a 5000-pound boulder from rolling down a hill. Determine the angle of inclination of the hill.

39. Find the acute angle that a constant unit force vector makes with the positive x-axis if the work done by the force in moving a particle from $(0, 0)$ to $(4, 0)$ equals 2.

40. Prove the distributive property:
$$\mathbf{u} \cdot (\mathbf{v} + \mathbf{w}) = \mathbf{u} \cdot \mathbf{v} + \mathbf{u} \cdot \mathbf{w}$$

41. Prove property (5), $\mathbf{0} \cdot \mathbf{v} = 0$.

42. If $\mathbf{v}$ is a unit vector and the angle between $\mathbf{v}$ and $\mathbf{i}$ is α, show that $\mathbf{v} = \cos \alpha \mathbf{i} + \sin \alpha \mathbf{j}$.

43. Suppose that $\mathbf{v}$ and $\mathbf{w}$ are unit vectors. If the angle between $\mathbf{v}$ and $\mathbf{i}$ is α and that between $\mathbf{w}$ and $\mathbf{i}$ is β, use the idea of the dot product $\mathbf{v} \cdot \mathbf{w}$ to prove that
$$\cos(\alpha - \beta) = \cos \alpha \cos \beta + \sin \alpha \sin \beta$$

44. Show that the projection of $\mathbf{v}$ onto $\mathbf{i}$ is $(\mathbf{v} \cdot \mathbf{i})\mathbf{i}$. Then show that we can always write a vector $\mathbf{v}$ as
$$\mathbf{v} = (\mathbf{v} \cdot \mathbf{i})\mathbf{i} + (\mathbf{v} \cdot \mathbf{j})\mathbf{j}$$

45. (a) If $\mathbf{u}$ and $\mathbf{v}$ have the same magnitude, show that $\mathbf{u} + \mathbf{v}$ and $\mathbf{u} - \mathbf{v}$ are orthogonal.

(b) Use this to prove that an angle inscribed in a semicircle is a right angle (see the figure).

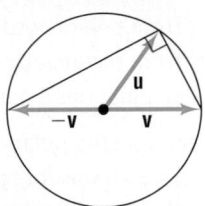

$-\mathbf{v}$ $\mathbf{u}$ $\mathbf{v}$

46. Let $\mathbf{v}$ and $\mathbf{w}$ denote two nonzero vectors. Show that the vector $\mathbf{v} - \alpha \mathbf{w}$ is orthogonal to $\mathbf{w}$ if $\alpha = \dfrac{\mathbf{v} \cdot \mathbf{w}}{\|\mathbf{w}\|^2}$.

47. Let $\mathbf{v}$ and $\mathbf{w}$ denote two nonzero vectors. Show that the vectors $\|\mathbf{w}\|\mathbf{v} + \|\mathbf{v}\|\mathbf{w}$ and $\|\mathbf{w}\|\mathbf{v} - \|\mathbf{v}\|\mathbf{w}$ are orthogonal.

48. In the definition of work given in this section, what is the work done if $\mathbf{F}$ is orthogonal to $\overrightarrow{AB}$?

49. Prove the **polarization identity**,
$$\|\mathbf{u} + \mathbf{v}\|^2 - \|\mathbf{u} - \mathbf{v}\|^2 = 4(\mathbf{u} \cdot \mathbf{v})$$

Discussion and Writing

50. Create an application different from any found in the text that requires a dot product.

'Are You Prepared?' Answer

1. $c^2 = a^2 + b^2 - 2ab \cos C$

CHAPTER REVIEW

Things to Know

Polar Coordinates (p. 714–716, 796)

Relationship between polar coordinates (r, θ) and rectangular coordinates (x, y) (pp. 716 and 769)

$x = r \cos \theta, y = r \sin \theta$

$r^2 = x^2 + y^2, \tan \theta = \dfrac{y}{x}, \quad x \neq 0$

Polar form of a complex number (p. 739)

If $z = x + yi$, then $z = r(\cos \theta + i \sin \theta)$,

where $r = |z| = \sqrt{x^2 + y^2}$, $\sin \theta = \dfrac{y}{r}$, $\cos \theta = \dfrac{x}{r}$, $0 \leq \theta < 2\pi$.

De Moivre's Theorem (p. 741)

If $z = r(\cos \theta + i \sin \theta)$, then $z^n = r^n[\cos(n\theta) + i \sin(n\theta)]$, where $n \geq 1$ is a positive integer.

nth roots of a complex number $w = r(\cos \theta_0 + i \sin \theta_0)$ (p. 742)

$\sqrt[n]{w} = \sqrt[n]{r}\left[\cos\left(\dfrac{\theta_0}{n} + \dfrac{2k\pi}{n}\right) + i \sin\left(\dfrac{\theta_0}{n} + \dfrac{2k\pi}{n}\right)\right], \quad k = 0, \dots, n-1,$

where $n \geq 2$ is an integer.

Vectors (pp. 746–748) Quantity having magnitude and direction; equivalent to a directed line segment $\overrightarrow{PQ}$

Position vector (p. 749) Vector whose initial point is at the origin

Unit vector (p. 748) Vector whose magnitude is 1

Dot product (p. 757) If $\mathbf{v} = a_1\mathbf{i} + b_1\mathbf{j}$ and $\mathbf{w} = a_2\mathbf{i} + b_2\mathbf{j}$, then $\mathbf{v} \cdot \mathbf{w} = a_1a_2 + b_1b_2$

Angle θ between two nonzero
vectors $\mathbf{u}$ and $\mathbf{v}$ (p. 759)

$$\cos \theta = \frac{\mathbf{u} \cdot \mathbf{v}}{\|\mathbf{u}\| \|\mathbf{v}\|}$$

Objectives

Section		You should be able to . . .	Review Exercises
10.1	1	Plot points using polar coordinates (p. 714)	1–6
	2	Convert from polar coordinates to rectangular coordinates (p. 716)	1–6
	3	Convert from rectangular coordinates to polar coordinates (p. 718)	7–12
	4	Transform equations from polar to rectangular form (p. 719)	13(a)–18(a)
10.2	1	Graph and identify polar equations by converting to rectangular equations (p. 723)	13(b)–18(b)
	2	Test polar equations for symmetry (p. 727)	19–24
	3	Graph polar equations by plotting points (p. 728)	19–24
10.3	1	Convert a complex number from rectangular form to polar form (p. 739)	25–28
	2	Plot points in the complex plane (p. 739)	29–34
	3	Find products and quotients of complex numbers in polar form (p. 740)	35–40
	4	Use De Moivre's Theorem (p. 741)	41–48
	5	Find complex roots (p. 742)	49–50
10.4	1	Graph vectors (p. 748)	51–54
	2	Find a position vector (p. 749)	55–58
	3	Add and subtract vectors algebraically (p. 750)	59, 60
	4	Find a scalar multiple and the magnitude of a vector (p. 751)	55–58, 61–66
	5	Find a unit vector (p. 751)	67, 68
	6	Find a vector from its direction and magnitude (p. 752)	69, 70
	7	Analyze objects in static equilibrium (p. 753)	85
10.5	1	Find the dot product of two vectors (p. 757)	71–74
	2	Find the angle between two vectors (p. 758)	71–74, 83, 84, 86
	3	Determine whether two vectors are parallel (p. 760)	75–80
	4	Determine whether two vectors are orthogonal (p. 760)	75–80
	5	Decompose a vector into two orthogonal vectors (p. 761)	81, 82
	6	Compute work (p. 762)	87

Review Exercises

In Problems 1–6, plot each point given in polar coordinates, and find its rectangular coordinates.

1. $\left(3, \dfrac{\pi}{6}\right)$ **2.** $\left(4, \dfrac{2\pi}{3}\right)$ **3.** $\left(-2, \dfrac{4\pi}{3}\right)$ **4.** $\left(-1, \dfrac{5\pi}{4}\right)$ **5.** $\left(-3, -\dfrac{\pi}{2}\right)$ **6.** $\left(-4, -\dfrac{\pi}{4}\right)$

In Problems 7–12, the rectangular coordinates of a point are given. Find two pairs of polar coordinates (r, θ) for each point, one with $r > 0$ and the other with $r < 0$. Express θ in radians.

7. $(-3, 3)$ **8.** $(1, -1)$ **9.** $(0, -2)$ **10.** $(2, 0)$ **11.** $(3, 4)$ **12.** $(-5, 12)$

In Problems 13–18, the letters r and θ represent polar coordinates. (a) Write each polar equation as an equation in rectangular coordinates (x, y). (b) Identify the equation and graph it.

13. $r = 2 \sin \theta$ **14.** $3r = \sin \theta$ **15.** $r = 5$

16. $\theta = \dfrac{\pi}{4}$ **17.** $r \cos \theta + 3r \sin \theta = 6$ **18.** $r^2 + 4r \sin \theta - 8r \cos \theta = 5$

In Problems 19–24, sketch the graph of each polar equation. Be sure to test for symmetry.

19. $r = 4 \cos \theta$ **20.** $r = 3 \sin \theta$ **21.** $r = 3 - 3 \sin \theta$

22. $r = 2 + \cos \theta$ **23.** $r = 4 - \cos \theta$ **24.** $r = 1 - 2 \sin \theta$

In Problems 25–28, write each complex number in polar form. Express each argument in degrees.

25. $-1 - i$ **26.** $-\sqrt{3} + i$ **27.** $4 - 3i$ **28.** $3 - 2i$

In Problems 29–34, write each complex number in the standard form $a + bi$ and plot each in the complex plane.

29. $2(\cos 150° + i \sin 150°)$ **30.** $3(\cos 60° + i \sin 60°)$ **31.** $3\left(\cos \dfrac{2\pi}{3} + i \sin \dfrac{2\pi}{3}\right)$

32. $4\left(\cos \dfrac{3\pi}{4} + i \sin \dfrac{3\pi}{4}\right)$ **33.** $0.1(\cos 350° + i \sin 350°)$ **34.** $0.5(\cos 160° + i \sin 160°)$

In Problems 35–40, find zw and $\dfrac{z}{w}$. Leave your answers in polar form.

35. $z = \cos 80° + i \sin 80°$
$w = \cos 50° + i \sin 50°$

36. $z = \cos 205° + i \sin 205°$
$w = \cos 85° + i \sin 85°$

37. $z = 3\left(\cos \dfrac{9\pi}{5} + i \sin \dfrac{9\pi}{5}\right)$
$w = 2\left(\cos \dfrac{\pi}{5} + i \sin \dfrac{\pi}{5}\right)$

38. $z = 2\left(\cos \dfrac{5\pi}{3} + i \sin \dfrac{5\pi}{3}\right)$
$w = 3\left(\cos \dfrac{\pi}{3} + i \sin \dfrac{\pi}{3}\right)$

39. $z = 5(\cos 10° + i \sin 10°)$
$w = \cos 355° + i \sin 355°$

40. $z = 4(\cos 50° + i \sin 50°)$
$w = \cos 340° + i \sin 340°$

In Problems 41–48, write each expression in the standard form $a + bi$.

41. $[3(\cos 20° + i \sin 20°)]^3$

42. $[2(\cos 50° + i \sin 50°)]^3$

43. $\left[\sqrt{2}\left(\cos \dfrac{5\pi}{8} + i \sin \dfrac{5\pi}{8}\right)\right]^4$

44. $\left[2\left(\cos \dfrac{5\pi}{16} + i \sin \dfrac{5\pi}{16}\right)\right]^4$

45. $\left(1 - \sqrt{3}i\right)^6$

46. $(2 - 2i)^8$

47. $(3 + 4i)^4$

48. $(1 - 2i)^4$

49. Find all the complex cube roots of 27.

50. Find all the complex fourth roots of -16.

In Problems 51–54, use the figure to graph each of the following:

51. $\mathbf{u} + \mathbf{v}$ **52.** $\mathbf{v} + \mathbf{w}$

53. $2\mathbf{u} + 3\mathbf{v}$ **54.** $5\mathbf{v} - 2\mathbf{w}$

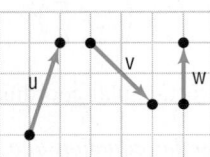

In Problems 55–58, the vector $\mathbf{v}$ is represented by the directed line segment $\overrightarrow{PQ}$. Write $\mathbf{v}$ in the form $a\mathbf{i} + b\mathbf{j}$ and find $\|\mathbf{v}\|$.

55. $P = (1, -2);$ $Q = (3, -6)$ **56.** $P = (-3, 1);$ $Q = (4, -2)$

57. $P = (0, -2);$ $Q = (-1, 1)$ **58.** $P = (3, -4);$ $Q = (-2, 0)$

In Problems 59–68, use the vectors $\mathbf{v} = -2\mathbf{i} + \mathbf{j}$ and $\mathbf{w} = 4\mathbf{i} - 3\mathbf{j}$ to find:

59. $\mathbf{v} + \mathbf{w}$ **60.** $\mathbf{v} - \mathbf{w}$ **61.** $4\mathbf{v} - 3\mathbf{w}$ **62.** $-\mathbf{v} + 2\mathbf{w}$

63. $\|\mathbf{v}\|$ **64.** $\|\mathbf{v} + \mathbf{w}\|$ **65.** $\|\mathbf{v}\| + \|\mathbf{w}\|$ **66.** $\|2\mathbf{v}\| - 3\|\mathbf{w}\|$

67. Find a unit vector in the same direction as $\mathbf{v}$.

68. Find a unit vector in the opposite direction of $\mathbf{w}$.

69. Find the vector $\mathbf{v}$ in the xy-plane with magnitude 3 if the angle between $\mathbf{v}$ and $\mathbf{i}$ is 60°.

70. Find the vector $\mathbf{v}$ in the xy-plane with magnitude 5 if the angle between $\mathbf{v}$ and $\mathbf{i}$ is 150°.

In Problems 71–74, find the dot product $\mathbf{v} \cdot \mathbf{w}$ and the angle between $\mathbf{v}$ and $\mathbf{w}$.

71. $\mathbf{v} = -2\mathbf{i} + \mathbf{j}, \quad \mathbf{w} = 4\mathbf{i} - 3\mathbf{j}$

72. $\mathbf{v} = 3\mathbf{i} - \mathbf{j}, \quad \mathbf{w} = \mathbf{i} + \mathbf{j}$

73. $\mathbf{v} = \mathbf{i} - 3\mathbf{j}, \quad \mathbf{w} = -\mathbf{i} + \mathbf{j}$

74. $\mathbf{v} = \mathbf{i} + 4\mathbf{j}, \quad \mathbf{w} = 3\mathbf{i} - 2\mathbf{j}$

In Problems 75–80, determine whether $\mathbf{v}$ and $\mathbf{w}$ are parallel, orthogonal, or neither.

75. $\mathbf{v} = 2\mathbf{i} + 3\mathbf{j}; \quad \mathbf{w} = -4\mathbf{i} - 6\mathbf{j}$

76. $\mathbf{v} = -2\mathbf{i} - \mathbf{j}; \quad \mathbf{w} = 2\mathbf{i} + \mathbf{j}$

77. $\mathbf{v} = 3\mathbf{i} - 4\mathbf{j}; \quad \mathbf{w} = -3\mathbf{i} + 4\mathbf{j}$

78. $\mathbf{v} = -2\mathbf{i} + 2\mathbf{j}; \quad \mathbf{w} = -3\mathbf{i} + 2\mathbf{j}$

79. $\mathbf{v} = 3\mathbf{i} - 2\mathbf{j}; \quad \mathbf{w} = 4\mathbf{i} + 6\mathbf{j}$

80. $\mathbf{v} = -4\mathbf{i} + 2\mathbf{j}; \quad \mathbf{w} = 2\mathbf{i} + 4\mathbf{j}$

In Problems 81–82, decompose $\mathbf{v}$ into two vectors, one parallel to $\mathbf{w}$ and the other orthogonal to $\mathbf{w}$.

81. $\mathbf{v} = 2\mathbf{i} + \mathbf{j}; \quad \mathbf{w} = -4\mathbf{i} + 3\mathbf{j}$

82. $\mathbf{v} = -3\mathbf{i} + 2\mathbf{j}; \quad \mathbf{w} = -2\mathbf{i} + \mathbf{j}$

83. Actual Speed and Direction of a Swimmer A swimmer can maintain a constant speed of 5 miles per hour. If the swimmer heads directly across a river that has a current moving at the rate of 2 miles per hour, what is the actual speed of the swimmer? (See the figure.) If the river is 1 mile wide, how far downstream will the swimmer end up from the point directly across the river from the starting point?

84. Actual Speed and Direction of an Airplane An airplane has an airspeed of 500 kilometers per hour (km/hr) in a northerly direction. The wind velocity is 60 km/hr in a south-easterly direction. Find the actual speed and direction of the plane relative to the ground.

85. Static Equilibrium A weight of 2000 pounds is suspended from two cables, as shown in the figure. What are the tensions in each cable?

86. Actual Speed and Distance of a Motorboat A small motorboat is moving at a true speed of 11 miles per hour (mi/hr) in a southerly direction. The current is known to be from the northeast at 3 mi/hr. What is the speed of the motorboat relative to the water? In what direction does the compass indicate that the boat is headed?

87. Computing Work Find the work done by a force of 5 pounds acting in the direction 60° to the horizontal in moving an object 20 feet from $(0, 0)$ to $(20, 0)$.

CHAPTER TEST

In Problems 1–3, plot each point given in polar coordinates.

1. $\left(2, \dfrac{3\pi}{4}\right)$ **2.** $\left(3, -\dfrac{\pi}{6}\right)$ **3.** $\left(-4, \dfrac{\pi}{3}\right)$

4. Convert $\left(2, 2\sqrt{3}\right)$ from rectangular coordinates to polar coordinates (r, θ), where $r > 0$ and $0 \le \theta < 2\pi$.

In Problems 5–7, convert the polar equation to a rectangular equation. Graph the equation.

5. $r = 7$ **6.** $\tan \theta = 3$ **7.** $r \sin^2 \theta + 8 \sin \theta = r$

In Problems 8 and 9, test the polar equation for symmetry with respect to the pole, the polar axis, and the line $\theta = \dfrac{\pi}{2}$.

8. $r^2 \cos \theta = 5$ **9.** $r = 5 \sin \theta \cos^2 \theta$

In Problems 10–12, perform the given operation, where $z = 2(\cos 85° + i \sin 85°)$ and $w = 3(\cos 22° + i \sin 22°)$. Write your answer in polar form.

10. zw **11.** $\dfrac{w}{z}$ **12.** w^5

13. Find all the cube roots of $-8 + 8\sqrt{3}i$. Write all answers in the form $a + bi$ and then plot them in rectangular coordinates.

In Problems 14–18, $P_1 = \left(3\sqrt{2}, 7\sqrt{2}\right)$ and $P_2 = \left(8\sqrt{2}, 2\sqrt{2}\right)$.

14. Find the position vector $\mathbf{v}$ equal to $\overrightarrow{P_1 P_2}$. **15.** Find $\|\mathbf{v}\|$.

16. Find the unit vector in the direction of $\mathbf{v}$. **17.** Find the angle between $\mathbf{v}$ and $\mathbf{i}$.

18. Decompose $\mathbf{v}$ into its vertical and horizontal components.

In Problems 19–22, $\mathbf{v}_1 = \langle 4, 6 \rangle$, $\mathbf{v}_2 = \langle -3, -6 \rangle$, $\mathbf{v}_3 = \langle -8, 4 \rangle$, and $\mathbf{v}_4 = \langle 10, 15 \rangle$.

19. Find the vector $\mathbf{v}_1 + 2\mathbf{v}_2 - \mathbf{v}_3$ **20.** Which two vectors are parallel?

21. Which two vectors are orthogonal? **22.** Find the angle between vectors $\mathbf{v}_1$ and $\mathbf{v}_2$.

23. Static Equilibrium A 1200-pound chandelier is to be suspended over a large ballroom; the chandelier will be hung on a cable whose ends will be attached to the ceiling, 16 feet apart. The chandelier will be free hanging so that the ends of the cable will make equal angles with the ceiling. If the top of the chandelier is to be 16 feet from the ceiling, what is the minimum tension the cable must be able to endure?

CUMULATIVE REVIEW

1. Find the real solutions, if any, of the equation $e^{x^2-9} = 1$.

2. Find an equation for the line containing the origin that makes an angle of $30°$ with the positive x-axis.

3. Find an equation for the circle with center at the point $(0, 1)$ and radius 3. Graph this circle.

4. What is the domain of the function $f(x) = \ln(1 - 2x)$?

5. Test the equation $x^2 + y^3 = 2x^4$ for symmetry with respect to the x-axis, the y-axis, and the origin.

6. Graph the function $y = |\ln x|$.

7. Graph the function $y = |\sin x|$.

8. Graph the function $y = \sin|x|$.

9. Find the exact value of $\sin^{-1}\left(-\dfrac{1}{2}\right)$.

10. Graph the equations $x = 3$ and $y = 4$ on the same set of rectangular coordinates.

11. Graph the equations $r = 2$ and $\theta = \dfrac{\pi}{3}$ on the same set of polar coordinates.

12. What is the amplitude and period of $y = -4\cos(\pi x)$.

CHAPTER PROJECTS

Source: www.aeromuseum.org/eduHowtoFly.html

I. **Modelling Aircraft Motion** Four aerodynamic forces act on an airplane in flight: lift, weight, thrust, and drag. While an aircraft is in flight, these four forces continuously battle each other. Weight opposes lift and drag opposes thrust. See the figure. In balanced flight at constant speed, both the lift and weight are equal and the thrust and drag are equal.

1. What will happen to the aircraft if the lift is held constant while the weight is decreased (say from burning off fuel)?

2. What will happen to the aircraft if the lift is decreased while the weight is held constant?

3. What will happen to the aircraft if the thrust is increased while the drag is held constant?

4. What will happen to the aircraft if the drag is increased while the thrust is held constant?

In 1903 the Wright brothers made the first controlled powered flight. The weight of their plane was approximately 700 pounds (lb). Newton's Second Law of motion states that force = mass × acceleration ($F = ma$). If the mass is measured in kilograms (kg) and acceleration in meters per second squared (m/s^2), then the force will be measured in newtons (N). [**Note:** $1\,\text{N} = 1\,\text{kg} \cdot \text{m/s}^2$.]

5. If $1\,\text{kg} = 2.205\,\text{lb}$, convert the weight of the Wright brother's plane to kilograms.

6. If acceleration due to gravity is $a = 9.80\,\text{m/s}^2$, determine the force due to weight on the Wright brother's plane.

7. What must be true about the lift force of the Wright brother's plane in order for it to get off the ground?

8. The weight of a fully loaded Cessna 170B is 2200 lb. What lift force is required to get this plane off the ground?

9. The maximum gross weight of a Boeing 747 is 255,000 lb. What lift force is required to get this jet off the ground?

The following projects are available at the Instructor's Resource Center (IRC):

II. **Project at Motorola** *Signal Fades Due to Interference* Complex trigonometric functions are used to assure that a cellphone has optimal reception as the user travels up and down an elevator.

III. **Compound Interest** The effect of continuously compounded interest is analyzed using polar coordinates.

IV. **Complex Equations** Analysis of complex equations illustrates the connections between complex and real equations. At times using complex equations is more efficient for proving mathematical theorems.

Analytic Geometry

11

Pluto's Unusual Orbit

Pluto is about 39 times as far from the Sun as Earth is. Its average distance from the Sun is about 3,647,240,000 miles (5,869,660,000 kilometers). Pluto travels around the Sun in an elliptical (oval-shaped) orbit. At some point in its orbit, it comes closer to the Sun than Neptune, the outermost planet. It stays inside Neptune's orbit for about 20 Earth-years. This event occurs every 248 Earth-years, which is about the same number of Earth-years it takes Pluto to travel once around the Sun. Pluto entered Neptune's orbit on January 23, 1979, and remained there until February 11, 1999.

Source: Spinrad, Hyron. "Pluto." World Book Online Reference Center. 2004. World Book, Inc. http://www.worldbookonline.com/wb/ Article?id=ar435500. Used with permission of World Book, Inc.

—See the Chapter Project—

A Look Back

In Chapter 2, we introduced rectangular coordinates and showed how geometry problems can be solved algebraically. We defined a circle geometrically, and then used the distance formula and rectangular coordinates to obtain an equation for a circle.

A Look Ahead

In this chapter we give geometric definitions for the conics and use the distance formula and rectangular coordinates to obtain their equations.

Historically, Apollonius (200 BC) was among the first to study *conics* and discover some of their interesting properties. Today, conics are still studied because of their many uses. *Paraboloids of revolution* (parabolas rotated about their axes of symmetry) are used as signal collectors (the satellite dishes used with radar and dish TV, for example), as solar energy collectors, and as reflectors (telescopes, light projection, and so on). The planets circle the Sun in approximately *elliptical* orbits. Elliptical surfaces can be used to reflect signals such as light and sound from one place to another. A third conic, the *hyperbola*, can be used to determine the location of lightning strikes.

The Greeks used the methods of Euclidean geometry to study conics. However, we shall use the more powerful methods of analytic geometry, bringing to bear both algebra and geometry, for our study of conics.

This chapter concludes with a section on equations of conics in polar coordinates and a section on plane curves and parametric equations.

Outline

11.1 Conics

1 Know the Names of the Conics

The word *conic* derives from the word *cone*, which is a geometric figure that can be constructed in the following way: Let *a* and *g* be two distinct lines that intersect at a point *V*. Keep the line *a* fixed. Now rotate the line *g* about *a* while maintaining the same angle between *a* and *g*. The collection of points swept out (generated) by the line *g* is called a **(right circular) cone.** See Figure 1. The fixed line *a* is called the **axis** of the cone; the point *V* is its **vertex;** the lines that pass through *V* and make the same angle with *a* as *g* are **generators** of the cone. Each generator is a line that lies entirely on the cone. The cone consists of two parts, called **nappes,** that intersect at the vertex.

Figure 1

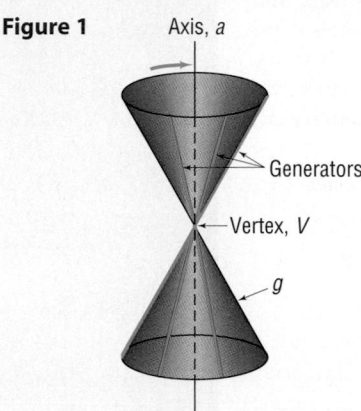

Axis, *a*

Generators

Vertex, *V*

g

Conics, an abbreviation for **conic sections,** are curves that result from the intersection of a right circular cone and a plane. The conics we shall study arise when the plane does not contain the vertex, as shown in Figure 2. These conics are **circles** when the plane is perpendicular to the axis of the cone and intersects each generator; **ellipses** when the plane is tilted slightly so that it intersects each generator, but intersects only one nappe of the cone; **parabolas** when the plane is tilted farther so that it is parallel to one (and only one) generator and intersects only one nappe of the cone; and **hyperbolas** when the plane intersects both nappes.

If the plane does contain the vertex, the intersection of the plane and the cone is a point, a line, or a pair of intersecting lines. These are usually called **degenerate conics.**

Figure 2

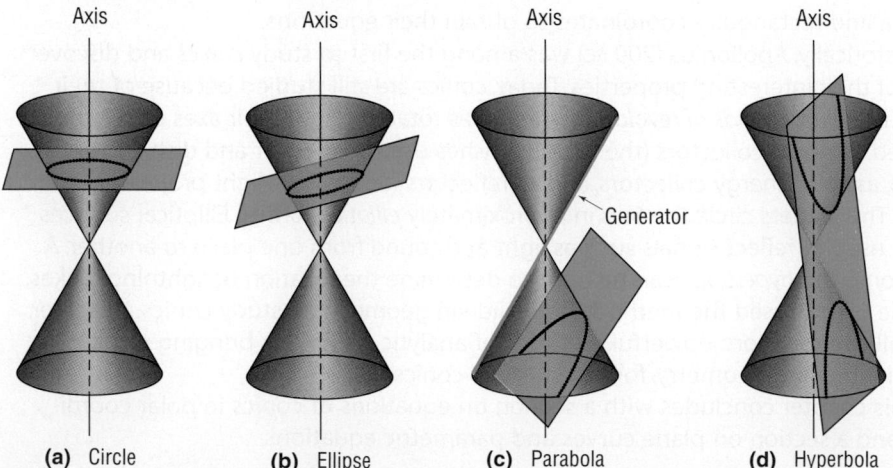

Axis Axis Axis Axis

Generator

(a) Circle **(b)** Ellipse **(c)** Parabola **(d)** Hyperbola

11.2 The Parabola

PREPARING FOR THIS SECTION *Before getting started, review the following:*

- Distance Formula (Section 2.1, p. 157)
- Symmetry (Section 2.2, pp. 167–169)
- Square Root Method (Section 1.2, pp. 98–99)

- Completing the Square (Section 1.2, pp. 99–101)
- Graphing Techniques: Transformations (Section 3.5, pp. 252–260)

Now Work the 'Are You Prepared?' problems on page 779.

OBJECTIVES 1 Analyze Parabolas with Vertex at the Origin (p. 773)
2 Analyze Parabolas with Vertex at (h, k) (p. 776)
3 Solve Applied Problems Involving Parabolas (p. 778)

We stated earlier (Section 4.3) that the graph of a quadratic function is a parabola. In this section, we begin with a geometric definition of a parabola and use it to obtain an equation.

DEFINITION

A **parabola** is the collection of all points P in the plane that are the same distance from a fixed point F as they are from a fixed line D. The point F is called the **focus** of the parabola, and the line D is its **directrix.** As a result, a parabola is the set of points P for which

$$d(F, P) = d(P, D) \qquad (1)$$

Figure 3

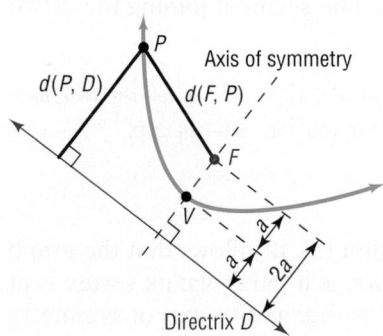

Figure 3 shows a parabola (in blue). The line through the focus F and perpendicular to the directrix D is called the **axis of symmetry** of the parabola. The point of intersection of the parabola with its axis of symmetry is called the **vertex** V.

Because the vertex V lies on the parabola, it must satisfy equation (1): $d(F, V) = d(V, D)$. The vertex is midway between the focus and the directrix. We shall let a equal the distance $d(F, V)$ from F to V. Now we are ready to derive an equation for a parabola. To do this, we use a rectangular system of coordinates positioned so that the vertex V, focus F, and directrix D of the parabola are conveniently located.

1 Analyze Parabolas with Vertex at the Origin

If we choose to locate the vertex V at the origin $(0, 0)$, we can conveniently position the focus F on either the x-axis or the y-axis. First, we consider the case where the focus F is on the positive x-axis, as shown in Figure 4. Because the distance from F to V is a, the coordinates of F will be $(a, 0)$ with $a > 0$. Similarly, because the distance from V to the directrix D is also a and, because D must be perpendicular to the x-axis (since the x-axis is the axis of symmetry), the equation of the directrix D must be $x = -a$.

Figure 4

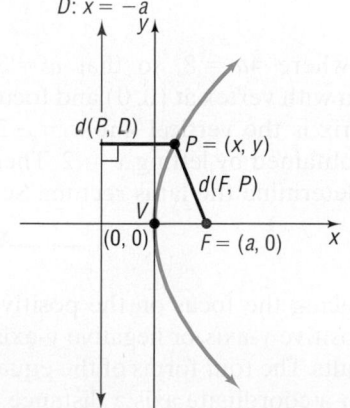

Now, if $P = (x, y)$ is any point on the parabola, P must obey equation (1):

$$d(F, P) = d(P, D)$$

So we have

$$\sqrt{(x - a)^2 + y^2} = |x + a| \qquad \text{Use the Distance Formula.}$$

$$(x - a)^2 + y^2 = (x + a)^2 \qquad \text{Square both sides.}$$

$$x^2 - 2ax + a^2 + y^2 = x^2 + 2ax + a^2 \qquad \text{Remove parentheses.}$$

$$y^2 = 4ax \qquad \text{Simplify.}$$

THEOREM

Equation of a Parabola: Vertex at (0, 0), Focus at (a, 0), a > 0

The equation of a parabola with vertex at $(0, 0)$, focus at $(a, 0)$, and directrix $x = -a, a > 0$, is

$$y^2 = 4ax \tag{2}$$

EXAMPLE 1

Finding the Equation of a Parabola and Graphing It

Find an equation of the parabola with vertex at $(0, 0)$ and focus at $(3, 0)$. Graph the equation.

Solution

Figure 5

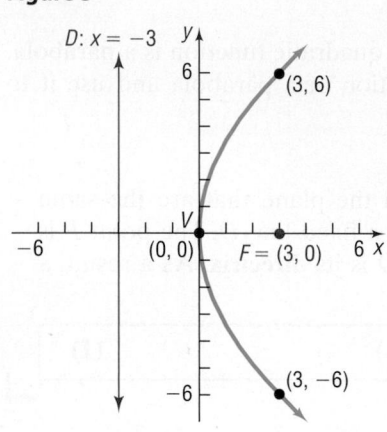

The distance from the vertex $(0, 0)$ to the focus $(3, 0)$ is $a = 3$. Based on equation (2), the equation of this parabola is

$$y^2 = 4ax$$
$$y^2 = 12x \quad a = 3$$

To graph this parabola, it is helpful to plot the points on the graph directly above or below the focus. To locate these two points, we let $x = 3$. Then

$$y^2 = 12x = 12(3) = 36$$
$$y = \pm 6 \qquad \text{Solve for } y.$$

The points on the parabola directly above or below the focus are $(3, 6)$ and $(3, -6)$. These points help in graphing the parabola because they determine the "opening." See Figure 5.

In general, the points on a parabola $y^2 = 4ax$ that lie directly above or below the focus $(a, 0)$ are each at a distance $2a$ from the focus. This follows from the fact that if $x = a$ then $y^2 = 4ax = 4a^2$, so $y = \pm 2a$. The line segment joining these two points is called the **latus rectum;** its length is $4a$.

 COMMENT To graph the parabola $y^2 = 12x$ discussed in Example 1, we need to graph the two functions $Y_1 = \sqrt{12x}$ and $Y_2 = -\sqrt{12x}$. Do this and compare what you see with Figure 5. ∎

Now Work PROBLEM 19

By reversing the steps we used to obtain equation (2), it follows that the graph of an equation of the form of equation (2), $y^2 = 4ax$, is a parabola; its vertex is at $(0, 0)$, its focus is at $(a, 0)$, its directrix is the line $x = -a$, and its axis of symmetry is the x-axis.

For the remainder of this section, the direction **"Analyze the equation"** will mean to find the vertex, focus, and directrix of the parabola and graph it.

EXAMPLE 2

Analyzing the Equation of a Parabola

Analyze the equation: $y^2 = 8x$

Figure 6

Solution

The equation $y^2 = 8x$ is of the form $y^2 = 4ax$, where $4a = 8$, so that $a = 2$. Consequently, the graph of the equation is a parabola with vertex at $(0, 0)$ and focus on the positive x-axis at $(a, 0) = (2, 0)$. The directrix is the vertical line $x = -2$. The two points that determine the latus rectum are obtained by letting $x = 2$. Then $y^2 = 16$, so $y = \pm 4$. The points $(2, -4)$ and $(2, 4)$ determine the latus rectum. See Figure 6 for the graph. ∎

Recall that we arrived at equation (2) after placing the focus on the positive x-axis. If the focus is placed on the negative x-axis, positive y-axis, or negative y-axis, a different form of the equation for the parabola results. The four forms of the equation of a parabola with vertex at $(0, 0)$ and focus on a coordinate axis a distance a

from $(0, 0)$ are given in Table 1, and their graphs are given in Figure 7. Notice that each graph is symmetric with respect to its axis of symmetry.

Table 1

Equations of a Parabola: Vertex at (0, 0); Focus on an Axis; $a > 0$				
Vertex	**Focus**	**Directrix**	**Equation**	**Description**
$(0, 0)$	$(a, 0)$	$x = -a$	$y^2 = 4ax$	Axis of symmetry is the x-axis, opens right
$(0, 0)$	$(-a, 0)$	$x = a$	$y^2 = -4ax$	Axis of symmetry is the x-axis, opens left
$(0, 0)$	$(0, a)$	$y = -a$	$x^2 = 4ay$	Axis of symmetry is the y-axis, opens up
$(0, 0)$	$(0, -a)$	$y = a$	$x^2 = -4ay$	Axis of symmetry is the y-axis, opens down

Figure 7

(a) $y^2 = 4ax$

(b) $y^2 = -4ax$

(c) $x^2 = 4ay$

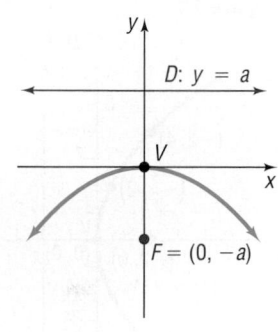

(d) $x^2 = -4ay$

EXAMPLE 3 **Analyzing the Equation of a Parabola**

Analyze the equation: $x^2 = -12y$

Figure 8

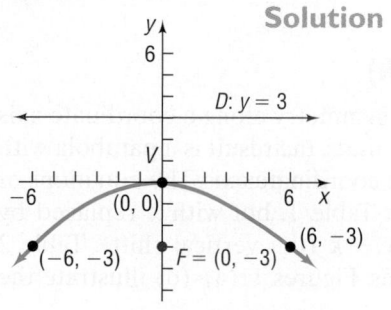

Solution The equation $x^2 = -12y$ is of the form $x^2 = -4ay$, with $a = 3$. Consequently, the graph of the equation is a parabola with vertex at $(0, 0)$, focus at $(0, -3)$, and directrix the line $y = 3$. The parabola opens down, and its axis of symmetry is the y-axis. To obtain the points defining the latus rectum, let $y = -3$. Then $x^2 = 36$, so $x = \pm 6$. The points $(-6, -3)$ and $(6, -3)$ determine the latus rectum. See Figure 8 for the graph.

✏️ **Now Work** PROBLEM 39

EXAMPLE 4 **Finding the Equation of a Parabola**

Find the equation of the parabola with focus at $(0, 4)$ and directrix the line $y = -4$. Graph the equation.

Figure 9

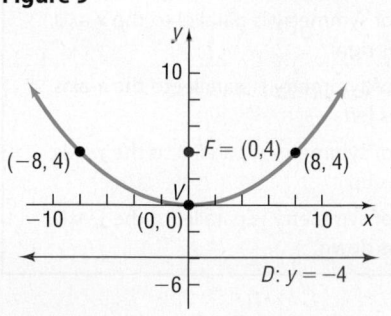

Solution A parabola whose focus is at $(0, 4)$ and whose directrix is the horizontal line $y = -4$ will have its vertex at $(0, 0)$. (Do you see why? The vertex is midway between the focus and the directrix.) Since the focus is on the positive y-axis at $(0, 4)$, the equation of this parabola is of the form $x^2 = 4ay$, with $a = 4$; that is,

$$x^2 = 4ay = 4(4)y = 16y$$
$$\uparrow$$
$$a = 4$$

The points $(8, 4)$ and $(-8, 4)$ determine the latus rectum. Figure 9 shows the graph of $x^2 = 16y$.

| EXAMPLE 5 | Finding the Equation of a Parabola |

Find the equation of a parabola with vertex at $(0, 0)$ if its axis of symmetry is the x-axis and its graph contains the point $\left(-\dfrac{1}{2}, 2\right)$. Find its focus and directrix, and graph the equation.

Solution

The vertex is at the origin, the axis of symmetry is the x-axis, and the graph contains a point in the second quadrant, so the parabola opens to the left. We see from Table 1 that the form of the equation is

$$y^2 = -4ax$$

Because the point $\left(-\dfrac{1}{2}, 2\right)$ is on the parabola, the coordinates $x = -\dfrac{1}{2}, y = 2$ must satisfy the equation. Substituting $x = -\dfrac{1}{2}$ and $y = 2$ into the equation, we find that

Figure 10

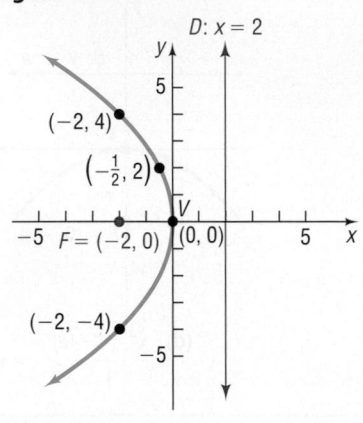

$$4 = -4a\left(-\dfrac{1}{2}\right) \qquad y^2 = -4ax;\ x = -\dfrac{1}{2}, y = 2$$

$$a = 2$$

The equation of the parabola is

$$y^2 = -4(2)x = -8x$$

The focus is at $(-2, 0)$ and the directrix is the line $x = 2$. Letting $x = -2$, we find $y^2 = 16$, so $y = \pm 4$. The points $(-2, 4)$ and $(-2, -4)$ determine the latus rectum. See Figure 10.

━━━━ **Now Work** PROBLEM 27

2 Analyze Parabolas with Vertex at (h, k)

If a parabola with vertex at the origin and axis of symmetry along a coordinate axis is shifted horizontally h units and then vertically k units, the result is a parabola with vertex at (h, k) and axis of symmetry parallel to a coordinate axis. The equations of such parabolas have the same forms as those in Table 1, but with x replaced by $x - h$ (the horizontal shift) and y replaced by $y - k$ (the vertical shift). Table 2 gives the forms of the equations of such parabolas. Figures 11(a)–(d) illustrate the graphs for $h > 0, k > 0$.

Table 2

Equations of a Parabola: Vertex at (h, k); Axis of Symmetry Parallel to a Coordinate Axis; $a > 0$				
Vertex	**Focus**	**Directrix**	**Equation**	**Description**
(h, k)	$(h + a, k)$	$x = h - a$	$(y - k)^2 = 4a(x - h)$	Axis of symmetry is parallel to the x-axis, opens right
(h, k)	$(h - a, k)$	$x = h + a$	$(y - k)^2 = -4a(x - h)$	Axis of symmetry is parallel to the x-axis, opens left
(h, k)	$(h, k + a)$	$y = k - a$	$(x - h)^2 = 4a(y - k)$	Axis of symmetry is parallel to the y-axis, opens up
(h, k)	$(h, k - a)$	$y = k + a$	$(x - h)^2 = -4a(y - k)$	Axis of symmetry is parallel to the y-axis, opens down

Figure 11

(a) $(y - k)^2 = 4a(x - h)$

(b) $(y - k)^2 = -4a(x - h)$

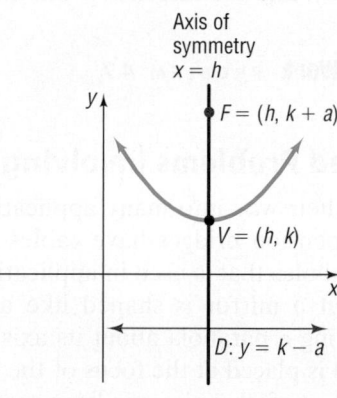

(c) $(x - h)^2 = 4a(y - k)$

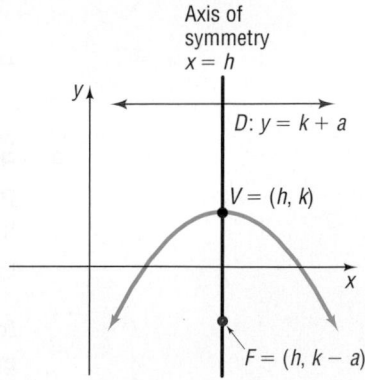

(d) $(x - h)^2 = -4a(y - k)$

| EXAMPLE 6 | Finding the Equation of a Parabola, Vertex Not at the Origin |

Find an equation of the parabola with vertex at $(-2, 3)$ and focus at $(0, 3)$. Graph the equation.

Solution

Figure 12

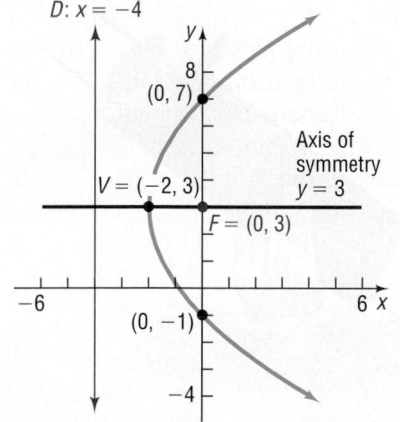

The vertex $(-2, 3)$ and focus $(0, 3)$ both lie on the horizontal line $y = 3$ (the axis of symmetry). The distance a from the vertex $(-2, 3)$ to the focus $(0, 3)$ is $a = 2$. Also, because the focus lies to the right of the vertex, we know that the parabola opens to the right. Consequently, the form of the equation is

$$(y - k)^2 = 4a(x - h)$$

where $(h, k) = (-2, 3)$ and $a = 2$. Therefore, the equation is

$$(y - 3)^2 = 4 \cdot 2[x - (-2)]$$
$$(y - 3)^2 = 8(x + 2)$$

If $x = 0$, then $(y - 3)^2 = 16$. Then $y - 3 = \pm 4$, so $y = -1$ or $y = 7$. The points $(0, -1)$ and $(0, 7)$ determine the latus rectum; the line $x = -4$ is the directrix. See Figure 12.

Now Work PROBLEM 29

Polynomial equations define parabolas whenever they involve two variables that are quadratic in one variable and linear in the other. To analyze this type of equation, we first complete the square of the variable that is quadratic.

| EXAMPLE 7 | Analyzing the Equation of a Parabola |

Analyze the equation: $x^2 + 4x - 4y = 0$

Figure 13

Axis of symmetry
$x = -2$

$F = (-2, 0)$

$V = (-2, -1)$

$D: y = -2$

Solution

To analyze the equation $x^2 + 4x - 4y = 0$, we complete the square involving the variable x.

$$x^2 + 4x - 4y = 0$$
$$x^2 + 4x = 4y \qquad \text{Isolate the terms involving x on the left side.}$$
$$x^2 + 4x + 4 = 4y + 4 \qquad \text{Complete the square on the left side.}$$
$$(x + 2)^2 = 4(y + 1) \qquad \text{Factor.}$$

This equation is of the form $(x - h)^2 = 4a(y - k)$, with $h = -2, k = -1$, and $a = 1$. The graph is a parabola with vertex at $(h, k) = (-2, -1)$ that opens up. The focus is at $(-2, 0)$, and the directrix is the line $y = -2$. See Figure 13.

━━━━━ **Now Work** PROBLEM 47

3 Solve Applied Problems Involving Parabolas

Parabolas find their way into many applications. For example, as we discussed in Section 4.4, suspension bridges have cables in the shape of a parabola. Another property of parabolas that is used in applications is their reflecting property.

Suppose that a mirror is shaped like a **paraboloid of revolution,** a surface formed by rotating a parabola about its axis of symmetry. If a light (or any other emitting source) is placed at the focus of the parabola, all the rays emanating from the light will reflect off the mirror in lines parallel to the axis of symmetry. This principle is used in the design of searchlights, flashlights, certain automobile headlights, and other such devices. See Figure 14.

Conversely, suppose that rays of light (or other signals) emanate from a distant source so that they are essentially parallel. When these rays strike the surface of a parabolic mirror whose axis of symmetry is parallel to these rays, they are reflected to a single point at the focus. This principle is used in the design of some solar energy devices, satellite dishes, and the mirrors used in some types of telescopes. See Figure 15.

Figure 14
Search Light

Rays of light

Light at focus

Figure 15
Telescope

EXAMPLE 8	Satellite Dish

A satellite dish is shaped like a paraboloid of revolution. The signals that emanate from a satellite strike the surface of the dish and are reflected to a single point, where the receiver is located. If the dish is 8 feet across at its opening and 3 feet deep at its center, at what position should the receiver be placed? That is, where is the focus?

Solution

Figure 16(a) shows the satellite dish. We draw the parabola used to form the dish on a rectangular coordinate system so that the vertex of the parabola is at the origin and its focus is on the positive y-axis. See Figure 16(b).

Figure 16

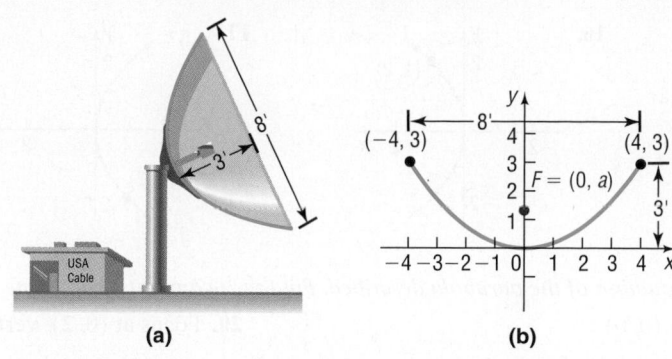

(a) (b)

The form of the equation of the parabola is

$$x^2 = 4ay$$

and its focus is at $(0, a)$. Since $(4, 3)$ is a point on the graph, we have

$$4^2 = 4a(3) \quad x = 4, y = 3$$

$$a = \frac{4}{3} \quad \text{Solve for } a.$$

The receiver should be located $1\frac{1}{3}$ feet from the base of the dish, along its axis of symmetry.

 Now Work PROBLEM 63

11.2 Assess Your Understanding

'Are You Prepared?' *Answers are given at the end of these exercises. If you get a wrong answer, read the pages listed in red.*

1. The formula for the distance d from $P_1 = (x_1, y_1)$ to $P_2 = (x_2, y_2)$ is $d =$ _____. (p. 157)

2. To complete the square of $x^2 - 4x$, add _____. (pp. 99–101)

3. Use the Square Root Method to find the real solutions of $(x + 4)^2 = 9$. (pp. 98–99)

4. The point that is symmetric with respect to the x-axis to the point $(-2, 5)$ is _____. (pp. 167–169)

5. To graph $y = (x - 3)^2 + 1$, shift the graph of $y = x^2$ to the right _____ units and then _____ 1 unit. (pp. 252–260)

Concepts and Vocabulary

6. A(n) _____ is the collection of all points in the plane such that the distance from each point to a fixed point equals its distance to a fixed line.

7. The surface formed by rotating a parabola about its axis of symmetry is called a _____ _____ _____.

8. *True or False* The vertex of a parabola is a point on the parabola that also is on its axis of symmetry.

9. *True or False* If a light is placed at the focus of a parabola, all the rays reflected off the parabola will be parallel to the axis of symmetry.

10. *True or False* The graph of a quadratic function is a parabola.

Skill Building

In Problems 11–18, the graph of a parabola is given. Match each graph to its equation.

(a) $y^2 = 4x$

(b) $x^2 = 4y$

(c) $y^2 = -4x$

(d) $x^2 = -4y$

(e) $(y - 1)^2 = 4(x - 1)$

(f) $(x + 1)^2 = 4(y + 1)$

(g) $(y - 1)^2 = 4(x \quad 1)$

(h) $(x + 1)^2 = -4(y + 1)$

11.

12.

13.

14.

15.

16.

17.

18.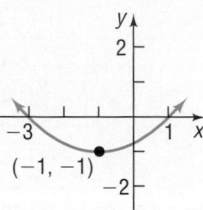

In Problems 19–36, find the equation of the parabola described. Find the two points that define the latus rectum, and graph the equation.

19. Focus at $(4, 0)$; vertex at $(0, 0)$

20. Focus at $(0, 2)$; vertex at $(0, 0)$

21. Focus at $(0, -3)$; vertex at $(0, 0)$

22. Focus at $(-4, 0)$; vertex at $(0, 0)$

23. Focus at $(-2, 0)$; directrix the line $x = 2$

24. Focus at $(0, -1)$; directrix the line $y = 1$

25. Directrix the line $y = -\dfrac{1}{2}$; vertex at $(0, 0)$

26. Directrix the line $x = -\dfrac{1}{2}$; vertex at $(0, 0)$

27. Vertex at $(0, 0)$; axis of symmetry the y-axis; containing the point $(2, 3)$

28. Vertex at $(0, 0)$; axis of symmetry the x-axis; containing the point $(2, 3)$

29. Vertex at $(2, -3)$; focus at $(2, -5)$

30. Vertex at $(4, -2)$; focus at $(6, -2)$

31. Vertex at $(-1, -2)$; focus at $(0, -2)$

32. Vertex at $(3, 0)$; focus at $(3, -2)$

33. Focus at $(-3, 4)$; directrix the line $y = 2$

34. Focus at $(2, 4)$; directrix the line $x = -4$

35. Focus at $(-3, -2)$; directrix the line $x = 1$

36. Focus at $(-4, 4)$; directrix the line $y = -2$

In Problems 37–54, find the vertex, focus, and directrix of each parabola. Graph the equation.

37. $x^2 = 4y$

38. $y^2 = 8x$

39. $y^2 = -16x$

40. $x^2 = -4y$

41. $(y - 2)^2 = 8(x + 1)$

42. $(x + 4)^2 = 16(y + 2)$

43. $(x - 3)^2 = -(y + 1)$

44. $(y + 1)^2 = -4(x - 2)$

45. $(y + 3)^2 = 8(x - 2)$

46. $(x - 2)^2 = 4(y - 3)$

47. $y^2 - 4y + 4x + 4 = 0$

48. $x^2 + 6x - 4y + 1 = 0$

49. $x^2 + 8x = 4y - 8$

50. $y^2 - 2y = 8x - 1$

51. $y^2 + 2y - x = 0$

52. $x^2 - 4x = 2y$

53. $x^2 - 4x = y + 4$

54. $y^2 + 12y = -x + 1$

In Problems 55–62, write an equation for each parabola.

55.

56.

57.

58.

59.

60.

61.

62.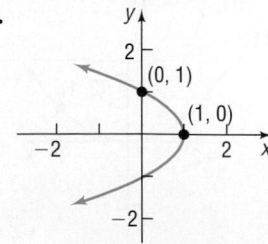

Applications and Extensions

63. Satellite Dish A satellite dish is shaped like a paraboloid of revolution. The signals that emanate from a satellite strike the surface of the dish and are reflected to a single point, where the receiver is located. If the dish is 10 feet across at its opening and 4 feet deep at its center, at what position should the receiver be placed?

64. Constructing a TV Dish A cable TV receiving dish is in the shape of a paraboloid of revolution. Find the location of the receiver, which is placed at the focus, if the dish is 6 feet across at its opening and 2 feet deep.

65. Constructing a Flashlight The reflector of a flashlight is in the shape of a paraboloid of revolution. Its diameter is 4 inches and its depth is 1 inch. How far from the vertex should the light bulb be placed so that the rays will be reflected parallel to the axis?

66. Constructing a Headlight A sealed-beam headlight is in the shape of a paraboloid of revolution. The bulb, which is placed at the focus, is 1 inch from the vertex. If the depth is to be 2 inches, what is the diameter of the headlight at its opening?

67. Suspension Bridge The cables of a suspension bridge are in the shape of a parabola, as shown in the figure. The towers supporting the cable are 600 feet apart and 80 feet high. If the cables touch the road surface midway between the towers, what is the height of the cable from the road at a point 150 feet from the center of the bridge?

68. Suspension Bridge The cables of a suspension bridge are in the shape of a parabola. The towers supporting the cable are 400 feet apart and 100 feet high. If the cables are at a height of 10 feet midway between the towers, what is the height of the cable at a point 50 feet from the center of the bridge?

69. Searchlight A searchlight is shaped like a paraboloid of revolution. If the light source is located 2 feet from the base along the axis of symmetry and the opening is 5 feet across, how deep should the searchlight be?

70. Searchlight A searchlight is shaped like a paraboloid of revolution. If the light source is located 2 feet from the base along the axis of symmetry and the depth of the searchlight is 4 feet, what should the width of the opening be?

71. Solar Heat A mirror is shaped like a paraboloid of revolution and will be used to concentrate the rays of the sun at its focus, creating a heat source. See the figure. If the mirror is 20 feet across at its opening and is 6 feet deep, where will the heat source be concentrated?

72. Reflecting Telescope A reflecting telescope contains a mirror shaped like a paraboloid of revolution. If the mirror is 4 inches across at its opening and is 3 inches deep, where will the collected light be concentrated?

73. Parabolic Arch Bridge A bridge is built in the shape of a parabolic arch. The bridge has a span of 120 feet and a maximum height of 25 feet. See the illustration. Choose a suitable rectangular coordinate system and find the height of the arch at distances of 10, 30, and 50 feet from the center.

74. Parabolic Arch Bridge A bridge is to be built in the shape of a parabolic arch and is to have a span of 100 feet. The height of the arch a distance of 40 feet from the center is to be 10 feet. Find the height of the arch at its center.

75. Gateway Arch The Gateway Arch in St. Louis is often mistaken to be parabolic is shape. In fact, it is a *catenary*, which has a more complicated formula than a parabola. The Arch is 625 feet high and 598 feet wide at its base.

(a) Find the equation of a parabola with the same dimensions. Let x equal the horizontal distance from the center of the arc.

(b) The table below gives the height of the Arch at various widths; find the corresponding heights for the parabola found in (a).

Width (ft)	Height (ft)
567	100
478	312.5
308	525

(c) Do the data support the notion that the Arch is in the shape of a parabola?

Source: Wikipedia, the free encyclopedia

76. Show that an equation of the form

$$Ax^2 + Ey = 0, \qquad A \neq 0, E \neq 0$$

is the equation of a parabola with vertex at $(0, 0)$ and axis of symmetry the y-axis. Find its focus and directrix.

77. Show that an equation of the form

$$Cy^2 + Dx = 0, \qquad C \neq 0, D \neq 0$$

is the equation of a parabola with vertex at $(0, 0)$ and axis of symmetry the x-axis. Find its focus and directrix.

78. Show that the graph of an equation of the form

$$Ax^2 + Dx + Ey + F = 0, \qquad A \neq 0$$

(a) Is a parabola if $E \neq 0$.

(b) Is a vertical line if $E = 0$ and $D^2 - 4AF = 0$.

(c) Is two vertical lines if $E = 0$ and $D^2 - 4AF > 0$.

(d) Contains no points if $E = 0$ and $D^2 - 4AF < 0$.

79. Show that the graph of an equation of the form

$$Cy^2 + Dx + Ey + F = 0, \quad C \neq 0$$

(a) Is a parabola if $D \neq 0$.
(b) Is a horizontal line if $D = 0$ and $E^2 - 4CF = 0$.
(c) Is two horizontal lines if $D = 0$ and $E^2 - 4CF > 0$.
(d) Contains no points if $D = 0$ and $E^2 - 4CF < 0$.

'Are You Prepared?' Answers

1. $d = \sqrt{(x_2 - x_1)^2 + (y_2 - y_1)^2}$ **2.** 4 **3.** $x + 4 = \pm 3; \{-7, -1\}$ **4.** $(-2, -5)$ **5.** 3; up

11.3 The Ellipse

PREPARING FOR THIS SECTION *Before getting started, review the following:*

- Distance Formula (Section 2.1, p. 157)
- Completing the Square (Section 1.2, pp. 99–101)
- Intercepts (Section 2.2, pp. 165–166)
- Symmetry (Section 2.2, pp. 167–169)
- Circles (Section 2.4, pp. 189–193)
- Graphing Techniques: Transformations (Section 3.5, pp. 252–260)

Now Work the 'Are You Prepared?' problems on page 789.

OBJECTIVES 1 Analyze Ellipses with Center at the Origin (p. 782)
2 Analyze Ellipses with Center at (h, k) (p. 786)
3 Solve Applied Problems Involving Ellipses (p. 788)

DEFINITION

An **ellipse** is the collection of all points in the plane, the sum of whose distances from two fixed points, called the **foci,** is a constant.

Figure 17

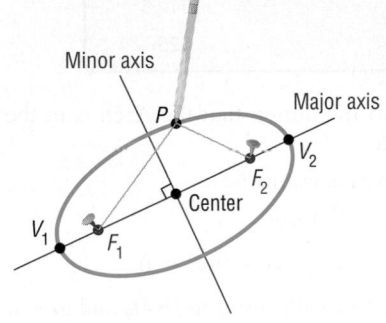

The definition contains within it a physical means for drawing an ellipse. Find a piece of string (the length of this string is the constant referred to in the definition). Then take two thumbtacks (the foci) and stick them into a piece of cardboard so that the distance between them is less than the length of the string. Now attach the ends of the string to the thumbtacks and, using the point of a pencil, pull the string taut. See Figure 17. Keeping the string taut, rotate the pencil around the two thumbtacks. The pencil traces out an ellipse, as shown in Figure 17.

In Figure 17, the foci are labeled F_1 and F_2. The line containing the foci is called the **major axis.** The midpoint of the line segment joining the foci is the **center** of the ellipse. The line through the center and perpendicular to the major axis is the **minor axis.**

The two points of intersection of the ellipse and the major axis are the **vertices,** V_1 and V_2, of the ellipse. The distance from one vertex to the other is the **length of the major axis.** The ellipse is symmetric with respect to its major axis, with respect to its minor axis, and with respect to its center.

Figure 18

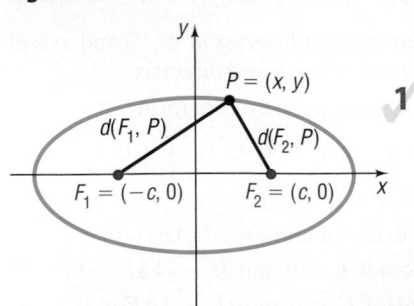

1 Analyze Ellipses with Center at the Origin

With these ideas in mind, we are now ready to find the equation of an ellipse in a rectangular coordinate system. First, we place the center of the ellipse at the origin. Second, we position the ellipse so that its major axis coincides with a coordinate axis. Suppose that the major axis coincides with the x-axis, as shown in Figure 18. If c is the distance from the center to a focus, one focus will be at $F_1 = (-c, 0)$ and the other at $F_2 = (c, 0)$. As we shall see, it is convenient to let $2a$ denote the constant

distance referred to in the definition. Then, if $P = (x, y)$ is any point on the ellipse, we have

$$d(F_1, P) + d(F_2, P) = 2a \qquad \text{Sum of the distances from } P \text{ to the foci equals a constant, } 2a.$$

$$\sqrt{(x + c)^2 + y^2} + \sqrt{(x - c)^2 + y^2} = 2a \qquad \text{Use the Distance Formula.}$$

$$\sqrt{(x + c)^2 + y^2} = 2a - \sqrt{(x - c)^2 + y^2} \qquad \text{Isolate one radical.}$$

$$(x + c)^2 + y^2 = 4a^2 - 4a\sqrt{(x - c)^2 + y^2} \qquad \text{Square both sides.}$$
$$+ (x - c)^2 + y^2$$

$$x^2 + 2cx + c^2 + y^2 = 4a^2 - 4a\sqrt{(x - c)^2 + y^2} \qquad \text{Remove parentheses.}$$
$$+ x^2 - 2cx + c^2 + y^2$$

$$4cx - 4a^2 = -4a\sqrt{(x - c)^2 + y^2} \qquad \text{Simplify; isolate the radical.}$$

$$cx - a^2 = -a\sqrt{(x - c)^2 + y^2} \qquad \text{Divide each side by 4.}$$

$$(cx - a^2)^2 = a^2[(x - c)^2 + y^2] \qquad \text{Square both sides again.}$$

$$c^2x^2 - 2a^2cx + a^4 = a^2(x^2 - 2cx + c^2 + y^2) \qquad \text{Remove parentheses.}$$

$$(c^2 - a^2)x^2 - a^2y^2 = a^2c^2 - a^4 \qquad \text{Rearrange the terms.}$$

$$(a^2 - c^2)x^2 + a^2y^2 = a^2(a^2 - c^2) \qquad \text{Multiply each side by } -1; \text{ factor } a^2 \text{ on the right side.} \quad \textbf{(1)}$$

To obtain points on the ellipse off the x-axis, it must be that $a > c$. To see why, look again at Figure 18. Then

$$d(F_1, P) + d(F_2, P) > d(F_1, F_2) \qquad \text{The sum of the lengths of two sides of a triangle is greater than the length of the third side.}$$

$$2a > 2c \qquad d(F_1, P) + d(F_2, P) = 2a, \quad d(F_1, F_2) = 2c$$

$$a > c$$

Since $a > c$, we also have $a^2 > c^2$, so $a^2 - c^2 > 0$. Let $b^2 = a^2 - c^2, b > 0$. Then $a > b$ and equation (1) can be written as

$$b^2x^2 + a^2y^2 = a^2b^2$$

$$\frac{x^2}{a^2} + \frac{y^2}{b^2} = 1 \qquad \text{Divide each side by } a^2b^2.$$

As you can verify, this equation is symmetric with respect to the x-axis, y-axis, and origin.

Because the major axis is the x-axis, we find the vertices of this ellipse by letting $y = 0$. The vertices satisfy the equation $\dfrac{x^2}{a^2} = 1$, the solutions of which are $x = \pm a$. Consequently, the vertices of this ellipse are $V_1 = (-a, 0)$ and $V_2 = (a, 0)$. The y-intercepts of the ellipse, found by letting $x = 0$, have coordinates $(0, -b)$ and $(0, b)$. These four intercepts, $(a, 0)$, $(-a, 0)$, $(0, b)$, and $(0, -b)$, are used to graph the ellipse.

Figure 19

THEOREM

Equation of an Ellipse: Center at (0, 0); Major Axis along the x-Axis

An equation of the ellipse with center at $(0, 0)$, foci at $(-c, 0)$ and $(c, 0)$, and vertices at $(-a, 0)$ and $(a, 0)$ is

$$\frac{x^2}{a^2} + \frac{y^2}{b^2} = 1, \qquad \text{where } a > b > 0 \text{ and } b^2 = a^2 - c^2 \qquad \textbf{(2)}$$

The major axis is the x-axis. See Figure 19.

Notice in Figure 19 the right triangle formed with the points $(0, 0)$, $(c, 0)$, and $(0, b)$. Because $b^2 = a^2 - c^2$ (or $b^2 + c^2 = a^2$), the distance from the focus at $(c, 0)$ to the point $(0, b)$ is a.

This can be seen another way. Look at the two right triangles in Figure 19. They are congruent. Do you see why? Because the sum of the distances from the foci to a point on the ellipse is $2a$, it follows that the distance from $(c, 0)$ to $(0, b)$ is a.

EXAMPLE 1	Finding an Equation of an Ellipse

Find an equation of the ellipse with center at the origin, one focus at $(3, 0)$, and a vertex at $(-4, 0)$. Graph the equation.

Solution

Figure 20

$$\frac{x^2}{16} + \frac{y^2}{7} = 1$$

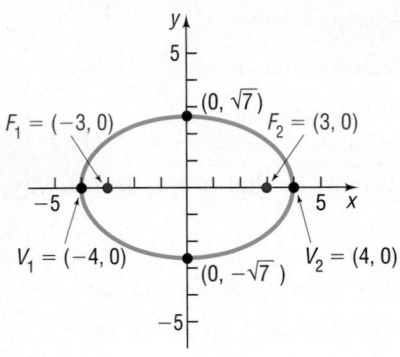

The ellipse has its center at the origin and, since the given focus and vertex lie on the x-axis, the major axis is the x-axis. The distance from the center, $(0, 0)$, to one of the foci, $(3, 0)$, is $c = 3$. The distance from the center, $(0, 0)$, to one of the vertices, $(-4, 0)$, is $a = 4$. From equation (2), it follows that

$$b^2 = a^2 - c^2 = 16 - 9 = 7$$

so an equation of the ellipse is

$$\frac{x^2}{16} + \frac{y^2}{7} = 1$$

Figure 20 shows the graph.

Notice in Figure 20 how we used the intercepts of the equation to graph the ellipse. Following this practice will make it easier for you to obtain an accurate graph of an ellipse when graphing.

 COMMENT *The intercepts of the ellipse also provide information about how to set the viewing rectangle for graphing an ellipse. To graph the ellipse*

$$\frac{x^2}{16} + \frac{y^2}{7} = 1$$

discussed in Example 1, we set the viewing rectangle using a square screen that includes the intercepts, perhaps $-4.5 \leq x \leq 4.5$, $-3 \leq y \leq 3$. Then we proceed to solve the equation for y:

$$\frac{x^2}{16} + \frac{y^2}{7} = 1$$

$$\frac{y^2}{7} = 1 - \frac{x^2}{16} \qquad \text{Subtract } \frac{x^2}{16} \text{ from each side.}$$

$$y^2 = 7\left(1 - \frac{x^2}{16}\right) \qquad \text{Multiply both sides by 7.}$$

Figure 21

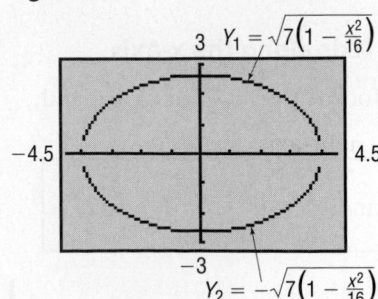

$$y = \pm\sqrt{7\left(1 - \frac{x^2}{16}\right)} \qquad \text{Take the square root of each side.}$$

Now graph the two functions

$$Y_1 = \sqrt{7\left(1 - \frac{x^2}{16}\right)} \quad \text{and} \quad Y_2 = -\sqrt{7\left(1 - \frac{x^2}{16}\right)}$$

Figure 21 shows the result.

━━━► **Now Work** PROBLEM **27**

An equation of the form of equation (2), with $a > b$, is the equation of an ellipse with center at the origin, foci on the x-axis at $(-c, 0)$ and $(c, 0)$, where $c^2 = a^2 - b^2$, and major axis along the x-axis.

For the remainder of this section, the direction **"Analyze the equation"** will mean to find the center, major axis, foci, and vertices of the ellipse and graph it.

EXAMPLE 2	Analyzing the Equation of an Ellipse

Analyze the equation: $\dfrac{x^2}{25} + \dfrac{y^2}{9} = 1$

Solution The given equation is of the form of equation (2), with $a^2 = 25$ and $b^2 = 9$. The equation is that of an ellipse with center $(0, 0)$ and major axis along the x-axis. The vertices are at $(\pm a, 0) = (\pm 5, 0)$. Because $b^2 = a^2 - c^2$, we find that

$$c^2 = a^2 - b^2 = 25 - 9 = 16$$

The foci are at $(\pm c, 0) = (\pm 4, 0)$. Figure 22 shows the graph.

Figure 22

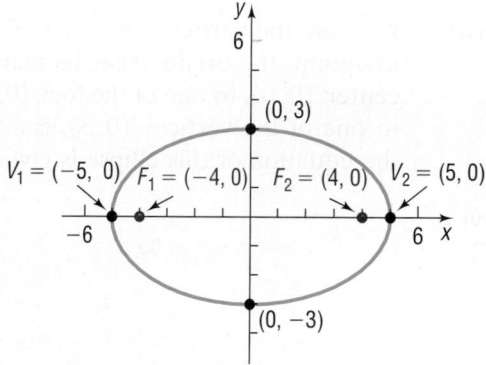

➡ **Now Work** PROBLEM 17

If the major axis of an ellipse with center at $(0, 0)$ lies on the y-axis, the foci are at $(0, -c)$ and $(0, c)$. Using the same steps as before, the definition of an ellipse leads to the following result:

THEOREM **Equation of an Ellipse: Center at (0, 0); Major Axis along the y-Axis**

An equation of the ellipse with center at $(0, 0)$, foci at $(0, -c)$ and $(0, c)$, and vertices at $(0, -a)$ and $(0, a)$ is

$$\frac{x^2}{b^2} + \frac{y^2}{a^2} = 1, \qquad \text{where } a > b > 0 \text{ and } b^2 = a^2 - c^2 \qquad \textbf{(3)}$$

The major axis is the y-axis.

Figure 23

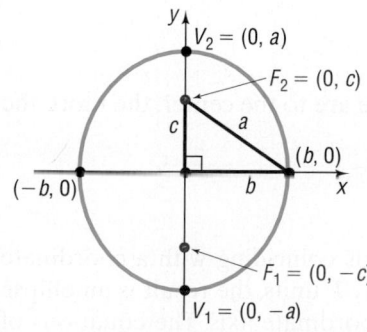

Figure 23 illustrates the graph of such an ellipse. Again, notice the right triangle with the points at $(0, 0)$, $(b, 0)$, and $(0, c)$.

Look closely at equations (2) and (3). Although they may look alike, there is a difference! In equation (2), the larger number, a^2, is in the denominator of the x^2-term, so the major axis of the ellipse is along the x-axis. In equation (3), the larger number, a^2, is in the denominator of the y^2-term, so the major axis is along the y-axis.

EXAMPLE 3	Analyzing the Equation of an Ellipse

Analyze the equation: $9x^2 + y^2 = 9$

Figure 24

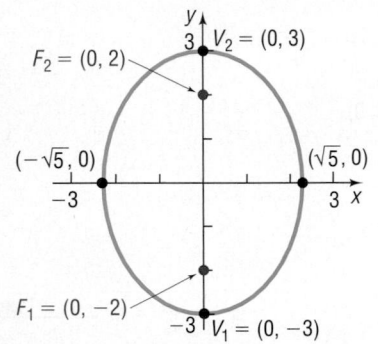

Solution

To put the equation in proper form, we divide each side by 9.

$$x^2 + \frac{y^2}{9} = 1$$

The larger number, 9, is in the denominator of the y^2-term so, based on equation (3), this is the equation of an ellipse with center at the origin and major axis along the y-axis. Also, we conclude that $a^2 = 9$, $b^2 = 1$, and $c^2 = a^2 - b^2 = 9 - 1 = 8$. The vertices are at $(0, \pm a) = (0, \pm 3)$, and the foci are at $(0, \pm c) = (0, \pm 2\sqrt{2})$. Figure 24 shows the graph.

━━ Now Work PROBLEM 21

EXAMPLE 4

Finding an Equation of an Ellipse

Find an equation of the ellipse having one focus at $(0, 2)$ and vertices at $(0, -3)$ and $(0, 3)$. Graph the equation.

Figure 25

Solution

Because the vertices are at $(0, -3)$ and $(0, 3)$, the center of this ellipse is at their midpoint, the origin. Also, its major axis lies on the y-axis. The distance from the center, $(0, 0)$, to one of the foci, $(0, 2)$, is $c = 2$. The distance from the center, $(0, 0)$, to one of the vertices, $(0, 3)$, is $a = 3$. So $b^2 = a^2 - c^2 = 9 - 4 = 5$. The form of the equation of this ellipse is given by equation (3).

$$\frac{x^2}{b^2} + \frac{y^2}{a^2} = 1$$

$$\frac{x^2}{5} + \frac{y^2}{9} = 1$$

Figure 25 shows the graph.

━━ Now Work PROBLEM 29

The circle may be considered a special kind of ellipse. To see why, let $a = b$ in equation (2) or (3). Then

$$\frac{x^2}{a^2} + \frac{y^2}{a^2} = 1$$

$$x^2 + y^2 = a^2$$

This is the equation of a circle with center at the origin and radius a. The value of c is

$$c^2 = a^2 - b^2 = 0$$
$$\underset{a = b}{\uparrow}$$

We conclude that the closer the two foci of an ellipse are to the center, the more the ellipse will look like a circle.

2 Analyze Ellipses with Center at (h, k)

If an ellipse with center at the origin and major axis coinciding with a coordinate axis is shifted horizontally h units and then vertically k units, the result is an ellipse with center at (h, k) and major axis parallel to a coordinate axis. The equations of such ellipses have the same forms as those given in equations (2) and (3), except that x is replaced by $x - h$ (the horizontal shift) and y is replaced by $y - k$ (the vertical shift). Table 3 gives the forms of the equations of such ellipses, and Figure 26 shows their graphs.

Table 3

Equations of an Ellipse: Center at (h, k); Major Axis Parallel to a Coordinate Axis				
Center	**Major Axis**	**Foci**	**Vertices**	**Equation**
(h, k)	Parallel to the x-axis	$(h + c, k)$	$(h + a, k)$	$\dfrac{(x - h)^2}{a^2} + \dfrac{(y - k)^2}{b^2} = 1$,
		$(h - c, k)$	$(h - a, k)$	$a > b$ and $b^2 = a^2 - c^2$
(h, k)	Parallel to *the* y-axis	$(h, k + c)$	$(h, k + a)$	$\dfrac{(x - h)^2}{b^2} + \dfrac{(y - k)^2}{a^2} = 1$,
		$(h, k - c)$	$(h, k - a)$	$a > b$ and $b^2 = a^2 - c^2$

Figure 26

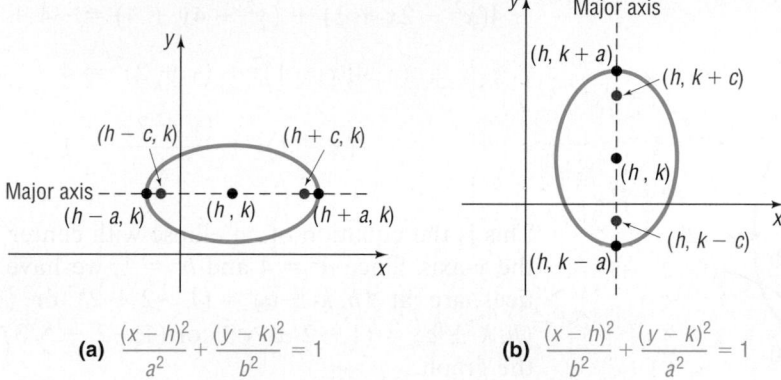

$$\textbf{(a)} \quad \frac{(x - h)^2}{a^2} + \frac{(y - k)^2}{b^2} = 1 \qquad\qquad \textbf{(b)} \quad \frac{(x - h)^2}{b^2} + \frac{(y - k)^2}{a^2} = 1$$

EXAMPLE 5 **Finding an Equation of an Ellipse, Center Not at the Origin**

Find an equation for the ellipse with center at $(2, -3)$, one focus at $(3, -3)$, and one vertex at $(5, -3)$. Graph the equation.

Solution The center is at $(h, k) = (2, -3)$, so $h = 2$ and $k = -3$. Since the center, focus, and vertex all lie on the line $y = -3$, the major axis is parallel to the x-axis. The distance from the center $(2, -3)$ to a focus $(3, -3)$ is $c = 1$; the distance from the center $(2, -3)$ to a vertex $(5, -3)$ is $a = 3$. Then $b^2 = a^2 - c^2 = 9 - 1 = 8$. The form of the equation is

$$\frac{(x - h)^2}{a^2} + \frac{(y - k)^2}{b^2} = 1, \quad \text{where } h = 2, k = -3, a = 3, b = 2\sqrt{2}$$

$$\frac{(x - 2)^2}{9} + \frac{(y + 3)^2}{8} = 1$$

To graph the equation, we use the center $(h, k) = (2, -3)$ to locate the vertices. The major axis is parallel to the x-axis, so the vertices are $a = 3$ units left and right of the center $(2, -3)$. Therefore, the vertices are

Figure 27

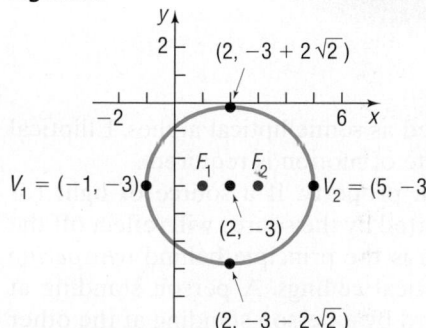

$$V_1 = (2 - 3, -3) = (-1, -3) \quad \text{and} \quad V_2 = (2 + 3, -3) = (5, -3)$$

Since $c = 1$ and the major axis is parallel to the x-axis, the foci are 1 unit left and right of the center. Therefore, the foci are

$$F_1 = (2 - 1, -3) = (1, -3) \quad \text{and} \quad F_2 = (2 + 1, -3) = (3, -3)$$

Finally, we use the value of $b = 2\sqrt{2}$ to find the two points above and below the center.

$$\left(2, -3 - 2\sqrt{2}\right) \quad \text{and} \quad \left(2, -3 + 2\sqrt{2}\right)$$

Figure 27 shows the graph.

Now Work PROBLEM 55

| EXAMPLE 6 | Analyzing the Equation of an Ellipse |

Analyze the equation: $4x^2 + y^2 - 8x + 4y + 4 = 0$

Solution We proceed to complete the squares in x and in y.

$$4x^2 + y^2 - 8x + 4y + 4 = 0$$

$$4x^2 - 8x + y^2 + 4y = -4 \qquad \text{Group like variables; place the constant on the right side.}$$

$$4(x^2 - 2x) + (y^2 + 4y) = -4 \qquad \text{Factor out 4 from the first two terms.}$$

$$4(x^2 - 2x + 1) + (y^2 + 4y + 4) = -4 + 4 + 4 \qquad \text{Complete each square.}$$

$$4(x - 1)^2 + (y + 2)^2 = 4 \qquad \text{Factor.}$$

$$(x - 1)^2 + \frac{(y + 2)^2}{4} = 1 \qquad \text{Divide each side by 4.}$$

Figure 28

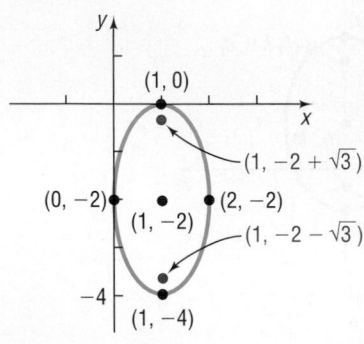

This is the equation of an ellipse with center at $(1, -2)$ and major axis parallel to the y-axis. Since $a^2 = 4$ and $b^2 = 1$, we have $c^2 = a^2 - b^2 = 4 - 1 = 3$. The vertices are at $(h, k \pm a) = (1, -2 \pm 2)$ or $(1, 0)$ and $(1, -4)$. The foci are at $(h, k \pm c) = (1, -2 \pm \sqrt{3})$ or $(1, -2 - \sqrt{3})$ and $(1, -2 + \sqrt{3})$. Figure 28 shows the graph.

■

Now Work PROBLEM 47

3 Solve Applied Problems Involving Ellipses

Ellipses are found in many applications in science and engineering. For example, the orbits of the planets around the Sun are elliptical, with the Sun's position at a focus. See Figure 29.

Figure 29

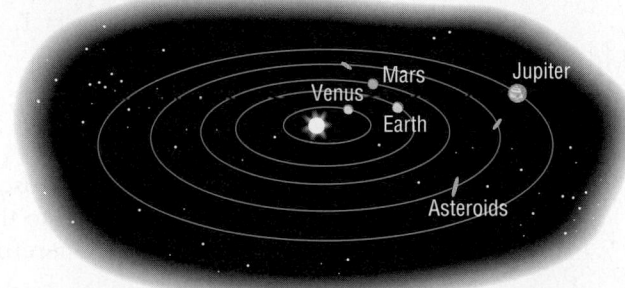

Stone and concrete bridges are often shaped as semielliptical arches. Elliptical gears are used in machinery when a variable rate of motion is required.

Ellipses also have an interesting reflection property. If a source of light (or sound) is placed at one focus, the waves transmitted by the source will reflect off the ellipse and concentrate at the other focus. This is the principle behind *whispering galleries,* which are rooms designed with elliptical ceilings. A person standing at one focus of the ellipse can whisper and be heard by a person standing at the other focus, because all the sound waves that reach the ceiling are reflected to the other person.

| EXAMPLE 7 | A Whispering Gallery |

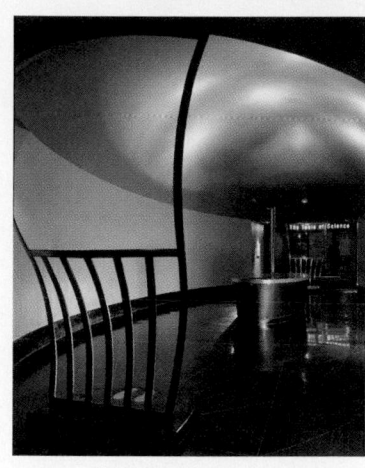

The whispering gallery in the Museum of Science and Industry in Chicago is 47.3 feet long. The distance from the center of the room to the foci is 20.3 feet. Find an equation that describes the shape of the room. How high is the room at its center?

Source: Chicago Museum of Science and Industry Web site; www.msichicago.org

Solution We set up a rectangular coordinate system so that the center of the ellipse is at the origin and the major axis is along the x-axis. The equation of the ellipse is

$$\frac{x^2}{a^2} + \frac{y^2}{b^2} = 1$$

Since the length of the room is 47.3 feet, the distance from the center of the room to each vertex (the end of the room) will be $\frac{47.3}{2} = 23.65$ feet; so $a = 23.65$ feet. The distance from the center of the room to each focus is $c = 20.3$ feet. See Figure 30. Since $b^2 = a^2 - c^2$, we find $b^2 = 23.65^2 - 20.3^2 = 147.2325$. An equation that describes the shape of the room is given by

$$\frac{x^2}{23.65^2} + \frac{y^2}{147.2325} = 1$$

The height of the room at its center is $b = \sqrt{147.2325} \approx 12.1$ feet.

Figure 30

(Graph: ellipse arch with center at origin. Points labeled: (0, 12.1) at top, 15 marked on y-axis; (−23.65, 0) and (23.65, 0) as vertices; (−20.3, 0) and (20.3, 0) as foci; −25 and 25 marked on x-axis.)

 Now Work PROBLEM 71

11.3 Assess Your Understanding

'Are You Prepared?' *Answers are given at the end of these exercises. If you get a wrong answer, read the pages listed in red.*

1. The distance d from $P_1 = (2, -5)$ to $P_2 = (4, -2)$ is $d =$ _____. (p. 157)

2. To complete the square of $x^2 - 3x$, add _____. (pp. 99–101)

3. Find the intercepts of the equation $y^2 = 16 - 4x^2$. (pp. 165–166)

4. The point that is symmetric with respect to the y-axis to the point $(-2, 5)$ is _____. (pp. 167–169)

5. To graph $y = (x + 1)^2 - 4$, shift the graph of $y = x^2$ to the (left/right) _____ unit(s) and then (up/down) _____ unit(s). (pp. 252–260)

6. The standard equation of a circle with center at $(2, -3)$ and radius 1 is _____. (pp. 189–193)

Concepts and Vocabulary

7. A(n) _____ is the collection of all points in the plane the sum of whose distances from two fixed points is a constant.

8. For an ellipse, the foci lie on a line called the _____ axis.

9. For the ellipse $\frac{x^2}{4} + \frac{y^2}{25} = 1$, the vertices are the points _____ and _____.

10. *True or False* The foci, vertices, and center of an ellipse lie on a line called the axis of symmetry.

11. *True or False* If the center of an ellipse is at the origin and the foci lie on the y-axis, the ellipse is symmetric with respect to the x-axis, the y-axis, and the origin.

12. *True or False* A circle is a certain type of ellipse.

Skill Building

In Problems 13–16, the graph of an ellipse is given. Match each graph to its equation.

(a) $\dfrac{x^2}{4} + y^2 = 1$ (b) $x^2 + \dfrac{y^2}{4} = 1$ (c) $\dfrac{x^2}{16} + \dfrac{y^2}{4} = 1$ (d) $\dfrac{x^2}{4} + \dfrac{y^2}{16} = 1$

13. **14.** **15.** **16.**

In Problems 17–26, find the vertices and foci of each ellipse. Graph each equation.

17. $\dfrac{x^2}{25} + \dfrac{y^2}{4} = 1$ **18.** $\dfrac{x^2}{9} + \dfrac{y^2}{4} = 1$ **19.** $\dfrac{x^2}{9} + \dfrac{y^2}{25} = 1$ **20.** $x^2 + \dfrac{y^2}{16} = 1$

21. $4x^2 + y^2 = 16$ **22.** $x^2 + 9y^2 = 18$ **23.** $4y^2 + x^2 = 8$ **24.** $4y^2 + 9x^2 = 36$

25. $x^2 + y^2 = 16$ **26.** $x^2 + y^2 = 4$

In Problems 27–38, find an equation for each ellipse. Graph the equation.

27. Center at $(0, 0)$; focus at $(3, 0)$; vertex at $(5, 0)$

28. Center at $(0, 0)$; focus at $(-1, 0)$; vertex at $(3, 0)$

29. Center at $(0, 0)$; focus at $(0, -4)$; vertex at $(0, 5)$

30. Center at $(0, 0)$; focus at $(0, 1)$; vertex at $(0, -2)$

31. Foci at $(\pm 2, 0)$; length of the major axis is 6

32. Foci at $(0, \pm 2)$; length of the major axis is 8

33. Focus at $(-4, 0)$; vertices at $(\pm 5, 0)$

34. Focus at $(0, -4)$; vertices at $(0, \pm 8)$

35. Foci at $(0, \pm 3)$; x-intercepts are ± 2

36. Vertices at $(\pm 4, 0)$; y-intercepts are ± 1

37. Center at $(0, 0)$; vertex at $(0, 4)$; $b = 1$

38. Vertices at $(\pm 5, 0)$; $c = 2$

In Problems 39–42, write an equation for each ellipse.

39. **40.** **41.** **42.**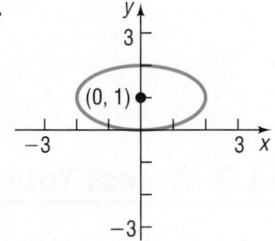

In Problems 43–54, analyze each equation; that is, find the center, foci, and vertices of each ellipse. Graph each equation.

43. $\dfrac{(x-3)^2}{4} + \dfrac{(y+1)^2}{9} = 1$ **44.** $\dfrac{(x+4)^2}{9} + \dfrac{(y+2)^2}{4} = 1$ **45.** $(x+5)^2 + 4(y-4)^2 = 16$

46. $9(x-3)^2 + (y+2)^2 = 18$ **47.** $x^2 + 4x + 4y^2 - 8y + 4 = 0$ **48.** $x^2 + 3y^2 - 12y + 9 = 0$

49. $2x^2 + 3y^2 - 8x + 6y + 5 = 0$ **50.** $4x^2 + 3y^2 + 8x - 6y = 5$ **51.** $9x^2 + 4y^2 - 18x + 16y - 11 = 0$

52. $x^2 + 9y^2 + 6x - 18y + 9 = 0$ **53.** $4x^2 + y^2 + 4y = 0$ **54.** $9x^2 + y^2 - 18x = 0$

In Problems 55–64, find an equation for each ellipse. Graph the equation.

55. Center at $(2, -2)$; vertex at $(7, -2)$; focus at $(4, -2)$

56. Center at $(-3, 1)$; vertex at $(-3, 3)$; focus at $(-3, 0)$

57. Vertices at $(4, 3)$ and $(4, 9)$; focus at $(4, 8)$

58. Foci at $(1, 2)$ and $(-3, 2)$; vertex at $(-4, 2)$

59. Foci at $(5, 1)$ and $(-1, 1)$; length of the major axis is 8

60. Vertices at $(2, 5)$ and $(2, -1)$; $c = 2$

61. Center at $(1, 2)$; focus at $(4, 2)$; contains the point $(1, 3)$

62. Center at $(1, 2)$; focus at $(1, 4)$; contains the point $(2, 2)$

63. Center at $(1, 2)$; vertex at $(4, 2)$; contains the point $(1, 3)$

64. Center at $(1, 2)$; vertex at $(1, 4)$; contains the point $(2, 2)$

In Problems 65–68, graph each function. Be sure to label all the intercepts.
[**Hint:** *Notice that each function is half an ellipse.*]

65. $f(x) = \sqrt{16 - 4x^2}$ **66.** $f(x) = \sqrt{9 - 9x^2}$ **67.** $f(x) = -\sqrt{64 - 16x^2}$ **68.** $f(x) = -\sqrt{4 - 4x^2}$

Applications and Extensions

69. Semielliptical Arch Bridge An arch in the shape of the upper half of an ellipse is used to support a bridge that is to span a river 20 meters wide. The center of the arch is 6 meters above the center of the river. See the figure. Write an equation for the ellipse in which the *x*-axis coincides with the water level and the *y*-axis passes through the center of the arch.

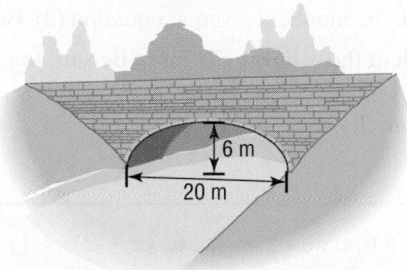

70. Semielliptical Arch Bridge The arch of a bridge is a semi-ellipse with a horizontal major axis. The span is 30 feet, and the top of the arch is 10 feet above the major axis. The roadway is horizontal and is 2 feet above the top of the arch. Find the vertical distance from the roadway to the arch at 5-foot intervals along the roadway.

71. Whispering Gallery A hall 100 feet in length is to be designed as a whispering gallery. If the foci are located 25 feet from the center, how high will the ceiling be at the center?

72. Whispering Gallery Jim, standing at one focus of a whispering gallery, is 6 feet from the nearest wall. His friend is standing at the other focus, 100 feet away. What is the length of this whispering gallery? How high is its elliptical ceiling at the center?

73. Semielliptical Arch Bridge A bridge is built in the shape of a semielliptical arch. The bridge has a span of 120 feet and a maximum height of 25 feet. Choose a suitable rectangular coordinate system and find the height of the arch at distances of 10, 30, and 50 feet from the center.

74. Semielliptical Arch Bridge A bridge is to be built in the shape of a semielliptical arch and is to have a span of 100 feet. The height of the arch, at a distance of 40 feet from the center, is to be 10 feet. Find the height of the arch at its center.

75. Racetrack Design Consult the figure. A racetrack is in the shape of an ellipse, 100 feet long and 50 feet wide. What is the width 10 feet from a vertex?

76. Semielliptical Arch Bridge An arch for a bridge over a highway is in the form of half an ellipse. The top of the arch is 20 feet above the ground level (the major axis). The highway has four lanes, each 12 feet wide; a center safety strip 8 feet wide; and two side strips, each 4 feet wide. What should the span of the bridge be (the length of its major axis) if the height 28 feet from the center is to be 13 feet?

77. Installing a Vent Pipe A homeowner is putting in a fireplace that has a 4-inch-radius vent pipe. He needs to cut an elliptical hole in his roof to accommodate the pipe. If the pitch of his roof is $\dfrac{5}{4}$, (a rise of 5, run of 4) what are the dimensions of the hole?

Source: www.pen.k12.va.us

78. Volume of a Football A football is in the shape of a **prolate spheroid,** which is simply a solid obtained by rotating an ellipse $\left(\dfrac{x^2}{a^2} + \dfrac{y^2}{b^2} = 1 \right)$ about its major axis. An inflated NFL football averages 11.125 inches in length and 28.25 inches in center circumference. If the volume of a prolate spheroid is $\dfrac{4}{3}\pi ab^2$, how much air does the football contain? (Neglect material thickness)?

Source: www.answerbag.com

In Problems 79–82, use the fact that the orbit of a planet about the Sun is an ellipse, with the Sun at one focus. The **aphelion** *of a planet is its greatest distance from the Sun, and the* **perihelion** *is its shortest distance. The* **mean distance** *of a planet from the Sun is the length of the semimajor axis of the elliptical orbit. See the illustration.*

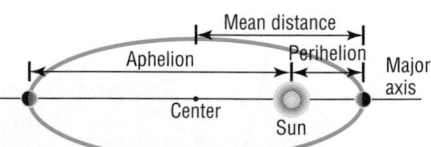

79. Earth The mean distance of Earth from the Sun is 93 million miles. If the aphelion of Earth is 94.5 million miles, what is the perihelion? Write an equation for the orbit of Earth around the Sun.

80. Mars The mean distance of Mars from the Sun is 142 million miles. If the perihelion of Mars is 128.5 million miles, what is the aphelion? Write an equation for the orbit of Mars about the Sun.

81. Jupiter The aphelion of Jupiter is 507 million miles. If the distance from the center of its elliptical orbit to the Sun is

23.2 million miles, what is the perihelion? What is the mean distance? Write an equation for the orbit of Jupiter around the Sun.

82. Pluto The perihelion of Pluto is 4551 million miles, and the distance from the center of its elliptical orbit to the Sun is 897.5 million miles. Find the aphelion of Pluto. What is the mean distance of Pluto from the Sun? Write an equation for the orbit of Pluto about the Sun.

83. Show that an equation of the form
$$Ax^2 + Cy^2 + F = 0, \qquad A \neq 0, C \neq 0, F \neq 0$$
where A and C are of the same sign and F is of opposite sign,

 (a) Is the equation of an ellipse with center at $(0,0)$ if $A \neq C$.

 (b) Is the equation of a circle with center $(0,0)$ if $A = C$.

84. Show that the graph of an equation of the form
$$Ax^2 + Cy^2 + Dx + Ey + F = 0, \qquad A \neq 0, C \neq 0$$
where A and C are of the same sign,

 (a) Is an ellipse if $\dfrac{D^2}{4A} + \dfrac{E^2}{4C} - F$ is the same sign as A.

 (b) Is a point if $\dfrac{D^2}{4A} + \dfrac{E^2}{4C} - F = 0$.

 (c) Contains no points if $\dfrac{D^2}{4A} + \dfrac{E^2}{4C} - F$ is of opposite sign to A.

Discussion and Writing

85. The **eccentricity** e of an ellipse is defined as the number $\dfrac{c}{a}$, where a and c are the numbers given in equation (2). Because $a > c$, it follows that $e < 1$. Write a brief paragraph about the general shape of each of the following ellipses. Be sure to justify your conclusions.

 (a) Eccentricity close to 0 (b) Eccentricity $= 0.5$ (c) Eccentricity close to 1

'Are You Prepared?' Answers

1. $\sqrt{13}$ **2.** $\dfrac{9}{4}$ **3.** $(-2,0), (2,0), (0,-4), (0,4)$ **4.** $(2,5)$ **5.** left; 1; down: 4 **6.** $(x-2)^2 + (y+3)^2 = 1$

11.4 The Hyperbola

PREPARING FOR THIS SECTION *Before getting started, review the following:*

- Distance Formula (Section 2.1, p. 157)
- Completing the Square (Section 1.2, pp. 99–101)
- Intercepts (Section 2.2, pp. 165–166)
- Symmetry (Section 2.2, pp. 167–169)
- Asymptotes (Section 5.2, pp. 346–352)
- Graphing Techniques: Transformations (Section 3.5, pp. 252–260)
- Square Root Method (Section 1.2, pp. 98–99)

\\ **Now Work** the 'Are You Prepared?' problems on page 801.

OBJECTIVES **1** Analyze Hyperbolas with Center at the Origin (p. 792)
 2 Find the Asymptotes of a Hyperbola (p. 797)
 3 Analyze Hyperbolas with Center at (h, k) (p. 798)
 4 Solve Applied Problems Involving Hyperbolas (p. 800)

Figure 31

DEFINITION A **hyperbola** is the collection of all points in the plane, the difference of whose distances from two fixed points, called the **foci,** is a constant.

Figure 31 illustrates a hyperbola with foci F_1 and F_2. The line containing the foci is called the **transverse axis.** The midpoint of the line segment joining the foci is the **center** of the hyperbola. The line through the center and perpendicular to the transverse axis is the **conjugate axis.** The hyperbola consists of two separate curves, called **branches,** that are symmetric with respect to the transverse axis, conjugate axis, and center. The two points of intersection of the hyperbola and the transverse axis are the **vertices,** V_1 and V_2, of the hyperbola.

1 Analyze Hyperbolas with Center at the Origin

With these ideas in mind, we are now ready to find the equation of a hyperbola in the rectangular coordinate system. First, we place the center at the origin. Next, we

Figure 32
$d(F_1, P) - d(F_2, P) = \pm 2a$

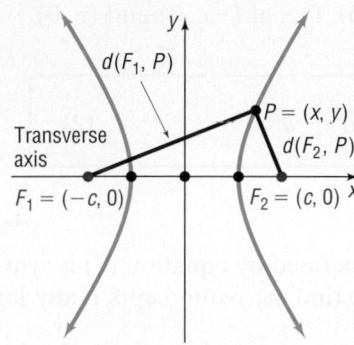

position the hyperbola so that its transverse axis coincides with a coordinate axis. Suppose that the transverse axis coincides with the x-axis, as shown in Figure 32.

If c is the distance from the center to a focus, one focus will be at $F_1 = (-c, 0)$ and the other at $F_2 = (c, 0)$. Now we let the constant difference of the distances from any point $P = (x, y)$ on the hyperbola to the foci F_1 and F_2 be denoted by $\pm 2a$. (If P is on the right branch, the $+$ sign is used; if P is on the left branch, the $-$ sign is used.) The coordinates of P must satisfy the equation

$$d(F_1, P) - d(F_2, P) = \pm 2a \qquad \text{Difference of the distances from } P \text{ to the foci equals } \pm 2a.$$

$$\sqrt{(x + c)^2 + y^2} - \sqrt{(x - c)^2 + y^2} = \pm 2a \qquad \text{Use the Distance Formula.}$$

$$\sqrt{(x + c)^2 + y^2} = \pm 2a + \sqrt{(x - c)^2 + y^2} \qquad \text{Isolate one radical.}$$

$$(x + c)^2 + y^2 = 4a^2 \pm 4a\sqrt{(x - c)^2 + y^2} \qquad \text{Square both sides.}$$
$$+ (x - c)^2 + y^2$$

Next we remove the parentheses.

$$x^2 + 2cx + c^2 + y^2 = 4a^2 \pm 4a\sqrt{(x - c)^2 + y^2} + x^2 - 2cx + c^2 + y^2$$

$$4cx - 4a^2 = \pm 4a\sqrt{(x - c)^2 + y^2} \qquad \text{Simplify; isolate the radical.}$$

$$cx - a^2 = \pm a\sqrt{(x - c)^2 + y^2} \qquad \text{Divide each side by 4.}$$

$$(cx - a^2)^2 = a^2[(x - c)^2 + y^2] \qquad \text{Square both sides.}$$

$$c^2x^2 - 2ca^2x + a^4 = a^2(x^2 - 2cx + c^2 + y^2) \qquad \text{Simplify.}$$

$$c^2x^2 + a^4 = a^2x^2 + a^2c^2 + a^2y^2 \qquad \text{Remove parentheses and simplify.}$$

$$(c^2 - a^2)x^2 - a^2y^2 = a^2c^2 - a^4 \qquad \text{Rearrange terms.}$$

$$(c^2 - a^2)x^2 - a^2y^2 = a^2(c^2 - a^2) \qquad \text{Factor } a^2 \text{ on the right side.} \quad \textbf{(1)}$$

To obtain points on the hyperbola off the x-axis, it must be that $a < c$. To see why, look again at Figure 32.

$$d(F_1, P) < d(F_2, P) + d(F_1, F_2) \qquad \text{Use triangle } F_1PF_2.$$

$$d(F_1, P) - d(F_2, P) < d(F_1, F_2)$$

$$2a < 2c \qquad \qquad \begin{array}{l} P \text{ is on the right branch, so} \\ d(F_1, P) - d(F_2, P) = 2a; \\ d(F_1, F_2) = 2c. \end{array}$$

$$a < c$$

Since $a < c$, we also have $a^2 < c^2$, so $c^2 - a^2 > 0$. Let $b^2 = c^2 - a^2, b > 0$. Then equation (1) can be written as

$$b^2x^2 - a^2y^2 = a^2b^2$$

$$\frac{x^2}{a^2} - \frac{y^2}{b^2} = 1 \qquad \text{Divide each side by } a^2b^2.$$

To find the vertices of the hyperbola defined by this equation, let $y = 0$. The vertices satisfy the equation $\dfrac{x^2}{a^2} = 1$, the solutions of which are $x = \pm a$.

Consequently, the vertices of the hyperbola are $V_1 = (-a, 0)$ and $V_2 = (a, 0)$. Notice that the distance from the center $(0, 0)$ to either vertex is a.

THEOREM

Equation of a Hyperbola: Center at (0, 0); Transverse Axis along the *x*-Axis

An equation of the hyperbola with center at $(0, 0)$, foci at $(-c, 0)$ and $(c, 0)$, and vertices at $(-a, 0)$ and $(a, 0)$ is

$$\frac{x^2}{a^2} - \frac{y^2}{b^2} = 1, \qquad \text{where } b^2 = c^2 - a^2 \qquad (2)$$

The transverse axis is the *x*-axis.

Figure 33

$\dfrac{x^2}{a^2} - \dfrac{y^2}{b^2} = 1, \quad b^2 = c^2 - a^2$

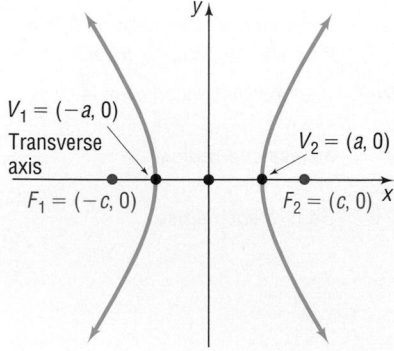

See Figure 33. As you can verify, the hyperbola defined by equation (2) is symmetric with respect to the *x*-axis, *y*-axis, and origin. To find the *y*-intercepts, if any, let $x = 0$ in equation (2). This results in the equation $\dfrac{y^2}{b^2} = -1$, which has no real solution. We conclude that the hyperbola defined by equation (2) has no *y*-intercepts. In fact, since $\dfrac{x^2}{a^2} - 1 = \dfrac{y^2}{b^2} \geq 0$, it follows that $\dfrac{x^2}{a^2} \geq 1$. There are no points on the graph for $-a < x < a$.

EXAMPLE 1

Finding and Graphing an Equation of a Hyperbola

Find an equation of the hyperbola with center at the origin, one focus at $(3, 0)$, and one vertex at $(-2, 0)$. Graph the equation.

Solution

The hyperbola has its center at the origin, and the transverse axis coincides with the *x*-axis. One focus is at $(c, 0) = (3, 0)$, so $c = 3$. One vertex is at $(-a, 0) = (-2, 0)$, so $a = 2$. From equation (2), it follows that $b^2 = c^2 - a^2 = 9 - 4 = 5$, so an equation of the hyperbola is

$$\frac{x^2}{4} - \frac{y^2}{5} = 1$$

To graph a hyperbola, it is helpful to locate and plot other points on the graph. For example, to find the points above and below the foci, we let $x = \pm 3$. Then

$$\frac{x^2}{4} - \frac{y^2}{5} = 1$$

$$\frac{(\pm 3)^2}{4} - \frac{y^2}{5} = 1 \qquad x = \pm 3$$

$$\frac{9}{4} - \frac{y^2}{5} = 1$$

$$\frac{y^2}{5} = \frac{5}{4}$$

$$y^2 = \frac{25}{4}$$

$$y = \pm \frac{5}{2}$$

Figure 34

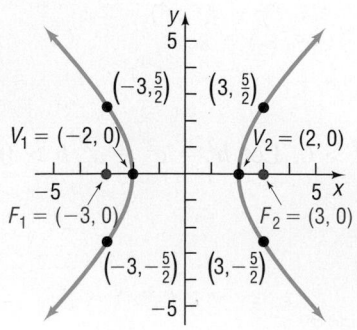

The points above and below the foci are $\left(\pm 3, \dfrac{5}{2} \right)$ and $\left(\pm 3, -\dfrac{5}{2} \right)$. These points determine the "opening" of the hyperbola. See Figure 34.

COMMENT To graph the hyperbola $\dfrac{x^2}{4} - \dfrac{y^2}{5} = 1$ discussed in Example 1, we need to graph the two functions $Y_1 = \sqrt{5}\sqrt{\dfrac{x^2}{4} - 1}$ and $Y_2 = -\sqrt{5}\sqrt{\dfrac{x^2}{4} - 1}$. Do this and compare what you see with Figure 34.

Now Work PROBLEM **17**

An equation of the form of equation (2) is the equation of a hyperbola with center at the origin, foci on the x-axis at $(-c, 0)$ and $(c, 0)$, where $c^2 = a^2 + b^2$, and transverse axis along the x-axis.

For the next two examples, the direction **"Analyze the equation"** will mean to find the center, transverse axis, vertices, and foci of the hyperbola and graph it.

EXAMPLE 2 **Analyzing the Equation of a Hyperbola**

Analyze the equation: $\dfrac{x^2}{16} - \dfrac{y^2}{4} = 1$

Solution The given equation is of the form of equation (2), with $a^2 = 16$ and $b^2 = 4$. The graph of the equation is a hyperbola with center at $(0, 0)$ and transverse axis along the x-axis. Also, we know that $c^2 = a^2 + b^2 = 16 + 4 = 20$. The vertices are at $(\pm a, 0) = (\pm 4, 0)$, and the foci are at $(\pm c, 0) = (\pm 2\sqrt{5}, 0)$.

To locate the points on the graph above and below the foci, we let $x = \pm 2\sqrt{5}$. Then

$$\frac{x^2}{16} - \frac{y^2}{4} = 1$$

$$\frac{\left(\pm 2\sqrt{5}\right)^2}{16} - \frac{y^2}{4} = 1 \qquad x = \pm 2\sqrt{5}$$

$$\frac{20}{16} - \frac{y^2}{4} = 1$$

$$\frac{5}{4} - \frac{y^2}{4} = 1$$

$$\frac{y^2}{4} = \frac{1}{4}$$

$$y = \pm 1$$

The points above and below the foci are $(\pm 2\sqrt{5}, 1)$ and $(\pm 2\sqrt{5}, -1)$. See Figure 35.

Figure 35

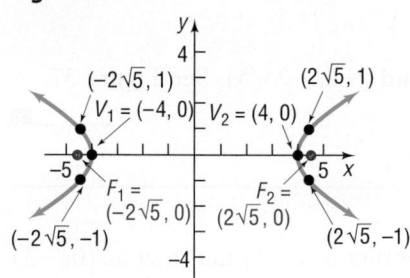

The next result gives the form of the equation of a hyperbola with center at the origin and transverse axis along the y-axis.

THEOREM **Equation of a Hyperbola: Center at (0, 0); Transverse Axis along the y-Axis**

An equation of the hyperbola with center at $(0, 0)$, foci at $(0, -c)$ and $(0, c)$, and vertices at $(0, -a)$ and $(0, a)$ is

$$\frac{y^2}{a^2} - \frac{x^2}{b^2} = 1, \qquad \text{where } b^2 = c^2 - a^2 \tag{3}$$

The transverse axis is the y-axis.

Figure 36

$\dfrac{y^2}{a^2} - \dfrac{x^2}{b^2} = 1, b^2 = c^2 - a^2$

$F_2 = (0, c)$

$V_2 = (0, a)$

$V_1 = (0, -a)$

$F_1 = (0, -c)$

Figure 36 shows the graph of a typical hyperbola defined by equation (3).

An equation of the form of equation (2), $\dfrac{x^2}{a^2} - \dfrac{y^2}{b^2} = 1$, is the equation of a hyperbola with center at the origin, foci on the x-axis at $(-c, 0)$ and $(c, 0)$, where $c^2 = a^2 + b^2$, and transverse axis along the x-axis.

An equation of the form of equation (3), $\dfrac{y^2}{a^2} - \dfrac{x^2}{b^2} = 1$, is the equation of a hyperbola with center at the origin, foci on the y-axis at $(0, -c)$ and $(0, c)$, where $c^2 = a^2 + b^2$, and transverse axis along the y-axis.

Notice the difference in the forms of equations (2) and (3). When the y^2-term is subtracted from the x^2-term, the transverse axis is along the x-axis. When the x^2-term is subtracted from the y^2-term, the transverse axis is along the y-axis.

EXAMPLE 3 Analyzing the Equation of a Hyperbola

Analyze the equation: $y^2 - 4x^2 = 4$

Solution To put the equation in proper form, we divide each side by 4:

$$\frac{y^2}{4} - x^2 = 1$$

Since the x^2-term is subtracted from the y^2-term, the equation is that of a hyperbola with center at the origin and transverse axis along the y-axis. Also, comparing the above equation to equation (3), we find $a^2 = 4$, $b^2 = 1$, and $c^2 = a^2 + b^2 = 5$. The vertices are at $(0, \pm a) = (0, \pm 2)$, and the foci are at $(0, \pm c) = (0, \pm\sqrt{5})$.

To locate other points on the graph, we let $x = \pm 2$. Then

$$y^2 - 4x^2 = 4$$
$$y^2 - 4(\pm 2)^2 = 4 \qquad x = \pm 2$$
$$y^2 - 16 = 4$$
$$y^2 = 20$$
$$y = \pm 2\sqrt{5}$$

Figure 37

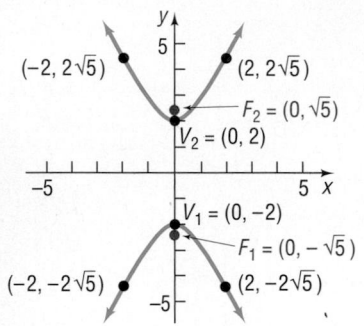

Four other points on the graph are $(\pm 2, 2\sqrt{5})$ and $(\pm 2, -2\sqrt{5})$. See Figure 37.

EXAMPLE 4 Finding an Equation of a Hyperbola

Find an equation of the hyperbola having one vertex at $(0, 2)$ and foci at $(0, -3)$ and $(0, 3)$. Graph the equation.

Figure 38 **Solution** Since the foci are at $(0, -3)$ and $(0, 3)$, the center of the hyperbola, which is at their midpoint, is the origin. Also, the transverse axis is along the y-axis. The given information also reveals that $c = 3$, $a = 2$, and $b^2 = c^2 - a^2 = 9 - 4 = 5$. The form of the equation of the hyperbola is given by equation (3):

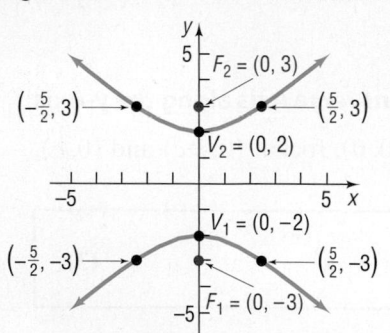

$$\frac{y^2}{a^2} - \frac{x^2}{b^2} = 1$$

$$\frac{y^2}{4} - \frac{x^2}{5} = 1$$

Let $y = \pm 3$ to obtain points on the graph across from the foci. See Figure 38.

🖉 **Now Work** PROBLEM 19

Look at the equations of the hyperbolas in Examples 2 and 4. For the hyperbola in Example 2, $a^2 = 16$ and $b^2 = 4$, so $a > b$; for the hyperbola in Example 4, $a^2 = 4$ and $b^2 = 5$, so $a < b$. We conclude that, for hyperbolas, there are no requirements involving the relative sizes of a and b. Contrast this situation to the case of an ellipse, in which the relative sizes of a and b dictate which axis is the major axis. Hyperbolas have another feature to distinguish them from ellipses and parabolas: Hyperbolas have asymptotes.

2 Find the Asymptotes of a Hyperbola

Recall from Section 5.2 that a horizontal or oblique asymptote of a graph is a line with the property that the distance from the line to points on the graph approaches 0 as $x \to -\infty$ or as $x \to \infty$. The asymptotes provide information about the end behavior of the graph of a hyperbola.

THEOREM

Asymptotes of a Hyperbola

The hyperbola $\dfrac{x^2}{a^2} - \dfrac{y^2}{b^2} = 1$ has the two oblique asymptotes

$$y = \frac{b}{a}x \quad \text{and} \quad y = -\frac{b}{a}x \qquad (4)$$

Proof We begin by solving for y in the equation of the hyperbola.

$$\frac{x^2}{a^2} - \frac{y^2}{b^2} = 1$$

$$\frac{y^2}{b^2} = \frac{x^2}{a^2} - 1$$

$$y^2 = b^2\left(\frac{x^2}{a^2} - 1\right)$$

Since $x \neq 0$, we can rearrange the right side in the form

$$y^2 = \frac{b^2 x^2}{a^2}\left(1 - \frac{a^2}{x^2}\right)$$

$$y = \pm\frac{bx}{a}\sqrt{1 - \frac{a^2}{x^2}}$$

Now, as $x \to -\infty$ or as $x \to \infty$, the term $\dfrac{a^2}{x^2}$ approaches 0, so the expression under the radical approaches 1. So, as $x \to -\infty$ or as $x \to \infty$, the value of y approaches $\pm\dfrac{bx}{a}$; that is, the graph of the hyperbola approaches the lines

$$y = -\frac{b}{a}x \quad \text{and} \quad y = \frac{b}{a}x$$

These lines are oblique asymptotes of the hyperbola. ∎

The asymptotes of a hyperbola are not part of the hyperbola, but they do serve as a guide for graphing a hyperbola. For example, suppose that we want to graph the equation

$$\frac{x^2}{a^2} - \frac{y^2}{b^2} = 1$$

We begin by plotting the vertices $(-a, 0)$ and $(a, 0)$. Then we plot the points $(0, -b)$ and $(0, b)$ and use these four points to construct a rectangle, as shown in Figure 39. The diagonals of this rectangle have slopes $\dfrac{b}{a}$ and $-\dfrac{b}{a}$, and their extensions are the asymptotes $y = \dfrac{b}{a}x$ and $y = -\dfrac{b}{a}x$ of the hyperbola. If we graph the asymptotes, we can use them to establish the "opening" of the hyperbola and avoid plotting other points.

Figure 39

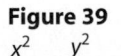

$$\frac{x^2}{a^2} - \frac{y^2}{b^2} = 1$$

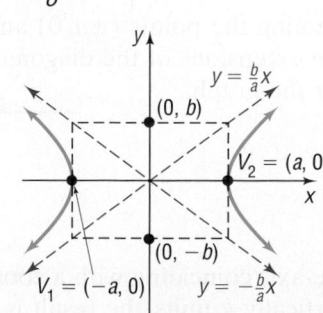

THEOREM **Asymptotes of a Hyperbola**

The hyperbola $\dfrac{y^2}{a^2} - \dfrac{x^2}{b^2} = 1$ has the two oblique asymptotes

$$y = \frac{a}{b}x \quad \text{and} \quad y = -\frac{a}{b}x \qquad\qquad (5)$$

You are asked to prove this result in Problem 74.

For the remainder of this section, the direction **"Analyze the equation"** will mean to find the center, transverse axis, vertices, foci, and asymptotes of the hyperbola and graph it.

EXAMPLE 5 **Analyzing the Equation of a Hyperbola**

Figure 40

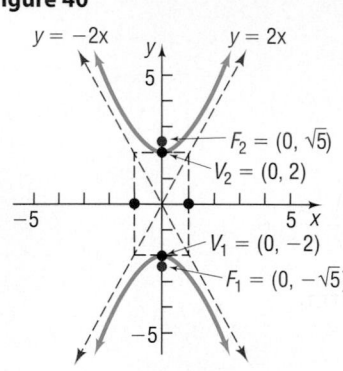

Analyze the equation: $\dfrac{y^2}{4} - x^2 = 1$

Solution Since the x^2-term is subtracted from the y^2-term, the equation is of the form of equation (3) and is a hyperbola with center at the origin and transverse axis along the y-axis. Also, comparing this equation to equation (3), we find that $a^2 = 4, b^2 = 1$, and $c^2 = a^2 + b^2 = 5$. The vertices are at $(0, \pm a) = (0, \pm 2)$, and the foci are at $(0, \pm c) = (0, \pm\sqrt{5})$. Using equation (5) with $a = 2$ and $b = 1$, the asymptotes are the lines $y = \dfrac{a}{b}x = 2x$ and $y = -\dfrac{a}{b}x = -2x$. Form the rectangle containing the points $(0, \pm a) = (0, \pm 2)$ and $(\pm b, 0) = (\pm 1, 0)$. The extensions of the diagonals of this rectangle are the asymptotes. Now graph the rectangle, the asymptotes, and the hyperbola. See Figure 40.

EXAMPLE 6 **Analyzing the Equation of a Hyperbola**

Analyze the equation: $9x^2 - 4y^2 = 36$

Solution Divide each side of the equation by 36 to put the equation in proper form.

$$\frac{x^2}{4} - \frac{y^2}{9} = 1$$

Figure 41

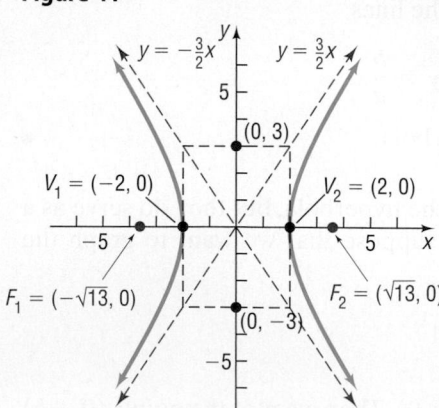

We now proceed to analyze the equation. The center of the hyperbola is the origin. Since the x^2-term is first in the equation, we know that the transverse axis is along the x-axis and the vertices and foci will lie on the x-axis. Using equation (2), we find $a^2 = 4, b^2 = 9$, and $c^2 = a^2 + b^2 = 13$. The vertices are $a = 2$ units left and right of the center at $(\pm a, 0) = (\pm 2, 0)$, the foci are $c = \sqrt{13}$ units left and right of the center at $(\pm c, 0) = (\pm\sqrt{13}, 0)$, and the asymptotes have the equations

$$y = \frac{b}{a}x = \frac{3}{2}x \quad \text{and} \quad y = -\frac{b}{a}x = -\frac{3}{2}x$$

To graph the hyperbola, form the rectangle containing the points $(\pm a, 0)$ and $(0, \pm b)$, that is, $(-2, 0), (2, 0), (0, -3)$, and $(0, 3)$. The extensions of the diagonals of this rectangle are the asymptotes. See Figure 41 for the graph.

Now Work PROBLEM 29

3 **Analyze Hyperbolas with Center at (h, k)**

If a hyperbola with center at the origin and transverse axis coinciding with a coordinate axis is shifted horizontally h units and then vertically k units, the result is a hyperbola with center at (h, k) and transverse axis parallel to a coordinate axis. The

equations of such hyperbolas have the same forms as those given in equations (2) and (3), except that x is replaced by $x - h$ (the horizontal shift) and y is replaced by $y - k$ (the vertical shift). Table 4 gives the forms of the equations of such hyperbolas. See Figure 42 for typical graphs.

Table 4

	Transverse				
Center	Axis	Foci	Vertices	Equation	Asymptotes
(h, k)	Parallel to the x-axis	$(h \pm c, k)$	$(h \pm a, k)$	$\dfrac{(x - h)^2}{a^2} - \dfrac{(y - k)^2}{b^2} = 1$, $b^2 = c^2 - a^2$	$y - k = \pm\dfrac{b}{a}(x - h)$
(h, k)	Parallel to the y-axis	$(h, k \pm c)$	$(h, k \pm a)$	$\dfrac{(y - k)^2}{a^2} - \dfrac{(x - h)^2}{b^2} = 1$, $b^2 = c^2 - a^2$	$y - k = \pm\dfrac{a}{b}(x - h)$

Equation of a Hyperbola: Center at (h, k); Transverse Axis Parallel to a Coordinate Axis

Figure 42

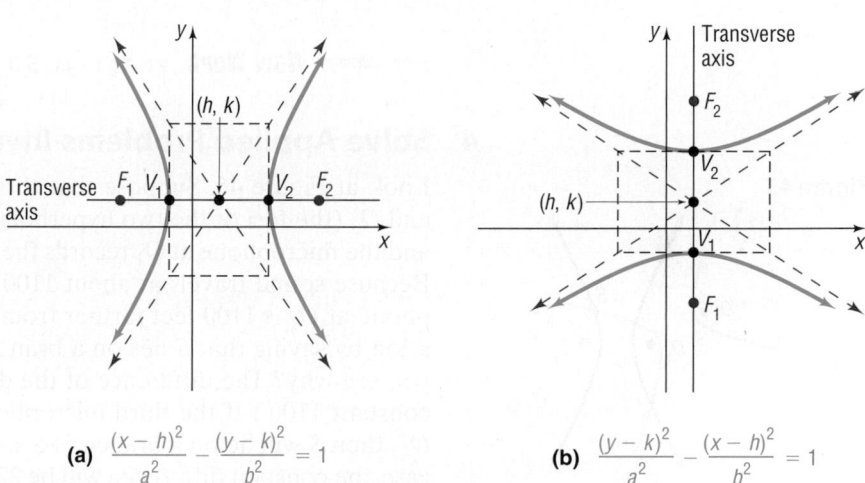

(a) $\dfrac{(x - h)^2}{a^2} - \dfrac{(y - k)^2}{b^2} = 1$

(b) $\dfrac{(y - k)^2}{a^2} - \dfrac{(x - h)^2}{b^2} = 1$

EXAMPLE 7

Finding an Equation of a Hyperbola, Center Not at the Origin

Find an equation for the hyperbola with center at $(1, -2)$, one focus at $(4, -2)$, and one vertex at $(3, -2)$. Graph the equation.

Figure 43

Solution The center is at $(h, k) = (1, -2)$, so $h = 1$ and $k = -2$. Since the center, focus, and vertex all lie on the line $y = -2$, the transverse axis is parallel to the x-axis. The distance from the center $(1, -2)$ to the focus $(4, -2)$ is $c = 3$; the distance from the center $(1, -2)$ to the vertex $(3, -2)$ is $a = 2$. Then, $b^2 = c^2 - a^2 = 9 - 4 = 5$. The equation is

$$\frac{(x - h)^2}{a^2} - \frac{(y - k)^2}{b^2} = 1$$

$$\frac{(x - 1)^2}{4} - \frac{(y + 2)^2}{5} = 1$$

See Figure 43.

Now Work PROBLEM 39

EXAMPLE 8

Analyzing the Equation of a Hyperbola

Analyze the equation: $-x^2 + 4y^2 - 2x - 16y + 11 = 0$

Solution We complete the squares in x and in y.

$$-x^2 + 4y^2 - 2x - 16y + 11 = 0$$
$$-(x^2 + 2x) + 4(y^2 - 4y) = -11 \qquad \text{Group terms.}$$
$$-(x^2 + 2x + 1) + 4(y^2 - 4y + 4) = -11 - 1 + 16 \qquad \text{Complete each square.}$$
$$-(x + 1)^2 + 4(y - 2)^2 = 4 \qquad \text{Factor.}$$
$$(y - 2)^2 - \frac{(x + 1)^2}{4} = 1 \qquad \text{Divide each side by 4.}$$

Figure 44

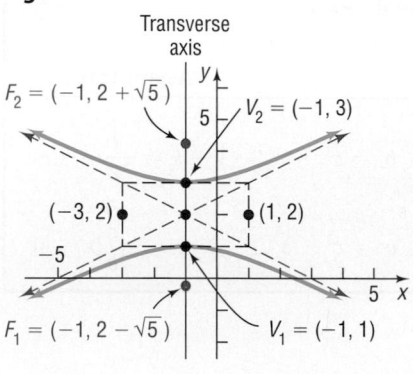

This is the equation of a hyperbola with center at $(-1, 2)$ and transverse axis parallel to the y-axis. Also, $a^2 = 1$ and $b^2 = 4$, so $c^2 = a^2 + b^2 = 5$. Since the transverse axis is parallel to the y-axis, the vertices and foci are located a and c units above and below the center, respectively. The vertices are at $(h, k \pm a) = (-1, 2 \pm 1)$, or $(-1, 1)$ and $(-1, 3)$. The foci are at $(h, k \pm c) = \left(-1, 2 \pm \sqrt{5}\right)$. The asymptotes are $y - 2 = \frac{1}{2}(x + 1)$ and $y - 2 = -\frac{1}{2}(x + 1)$. Figure 44 shows the graph.

Now Work PROBLEM 53

4 Solve Applied Problems Involving Hyperbolas

Figure 45

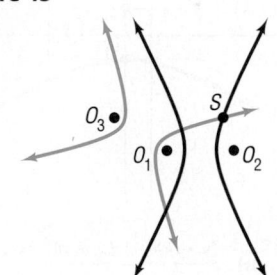

Look at Figure 45. Suppose that three microphones are located at points O_1, O_2, and O_3 (the foci of the two hyperbolas). In addition, suppose that a gun is fired at S and the microphone at O_1 records the gun shot 1 second after the microphone at O_2. Because sound travels at about 1100 feet per second, we conclude that the microphone at O_1 is 1100 feet farther from the gunshot than O_2. We can model this situation by saying that S lies on a branch of a hyperbola with foci at O_1 and O_2. (Do you see why? The difference of the distances from S to O_1 and from S to O_2 is the constant 1100.) If the third microphone at O_3 records the gunshot 2 seconds after O_1, then S will lie on a branch of a second hyperbola with foci at O_1 and O_3. In this case, the constant difference will be 2200. The intersection of the two hyperbolas will identify the location of S.

EXAMPLE 9	Lightning Strikes

Suppose that two people standing 1 mile apart both see a flash of lightning. After a period of time, the person standing at point A hears the thunder. One second later, the person standing at point B hears the thunder. If the person at B is due west of the person at A and the lightning strike is known to occur due north of the person standing at point A, where did the lightning strike?

Solution See Figure 46 in which the ordered pair (x, y) represents the location of the lightning strike. We know that sound travels at 1100 feet per second, so the person at point A is 1100 feet closer to the lightning strike than the person at point B. Since the difference of the distance from (x, y) to A and the distance from (x, y) to B is the constant 1100, the point (x, y) lies on a hyperbola whose foci are at A and B.

Figure 46

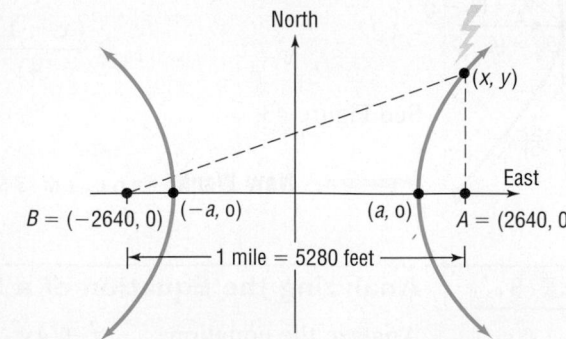

An equation of the hyperbola is

$$\frac{x^2}{a^2} - \frac{y^2}{b^2} = 1$$

where $2a = 1100$, so $a = 550$.

Because the distance between the two people is 1 mile (5280 feet) and each person is at a focus of the hyperbola, we have

$$2c = 5280$$

$$c = \frac{5280}{2} = 2640$$

Since $b^2 = c^2 - a^2 = 2640^2 - 550^2 = 6{,}667{,}100$, the equation of the hyperbola that describes the location of the lightning strike is

$$\frac{x^2}{550^2} - \frac{y^2}{6{,}667{,}100} = 1$$

Refer again to Figure 46. Since the lightning strikes due north of the individual at the point $A = (2640, 0)$, we let $x = 2640$ and solve the resulting equation.

$$\frac{2640^2}{550^2} - \frac{y^2}{6{,}667{,}100} = 1$$

$$-\frac{y^2}{6{,}667{,}100} = -22.04 \qquad \text{Subtract } \frac{2640^2}{550^2} \text{ from both sides.}$$

$$y^2 = 146{,}942{,}884 \qquad \text{Multiply both sides by } -6{,}667{,}100.$$

$$y = 12{,}122 \qquad \begin{array}{l} y > 0 \text{ since the lightning strike} \\ \text{occured in quadrant I} \end{array}$$

✓ **Check:** The difference between the distance from $(2640, 12, 122)$ to the person at the point $B = (-2640, 0)$ and the distance from $(2640, 12, 122)$ to the person at the point $A = (2640, 0)$ should be 1100. Using the distance formula, we find the difference in the distances is

$$\sqrt{[2640 - (-2640)]^2 + (12{,}122 - 0)^2} - \sqrt{(2640 - 2640)^2 + (12{,}122 - 0)^2} = 1100$$

as required.
The lightning strike is 12,122 feet north of the person standing at point A.

━━ **Now Work** PROBLEM 65

11.4 Assess Your Understanding

'Are You Prepared?' *Answers are given at the end of these exercises. If you get a wrong answer, read the pages listed in red.*

1. The distance d from $P_1 = (3, -4)$ to $P_2 = (-2, 1)$ is $d = $ _____. (p. 157)

2. To complete the square of $x^2 + 5x$, add _____. (pp. 99–101)

3. Find the intercepts of the equation $y^2 = 9 + 4x^2$. (pp. 165–166)

4. **True or False** The equation $y^2 = 9 + x^2$ is symmetric with respect to the x-axis, the y-axis, and the origin. (pp. 167–169)

5. To graph $y = (x - 5)^3 - 4$, shift the graph of $y = x^3$ to the (left/right) _____ unit(s) and then (up/down) _____ unit(s). (pp. 252–260)

6. Find the vertical asymptotes, if any, and the horizontal or oblique asymptotes, if any, of $y = \dfrac{x^2 - 9}{x^2 - 4}$. (pp. 346–352)

Concepts and Vocabulary

7. A(n) _____ is the collection of points in the plane the difference of whose distances from two fixed points is a constant.

8. For a hyperbola, the foci lie on a line called the _____ _____.

9. The asymptotes of the hyperbola $\dfrac{x^2}{4} - \dfrac{y^2}{9} = 1$ are _____ and _____.

10. *True or False* The foci of a hyperbola lie on a line called the axis of symmetry.

11. *True or False* Hyperbolas always have asymptotes.

12. *True or False* A hyperbola will never intersect its transverse axis.

Skill Building

In Problems 13–16, the graph of a hyperbola is given. Match each graph to its equation.

(a) $\dfrac{x^2}{4} - y^2 = 1$

(b) $x^2 - \dfrac{y^2}{4} = 1$

(c) $\dfrac{y^2}{4} - x^2 = 1$

(d) $y^2 - \dfrac{x^2}{4} = 1$

13.

14.

15.

16.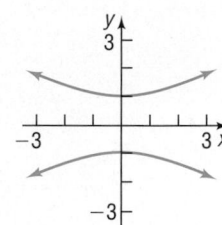

In Problems 17–26, find an equation for the hyperbola described. Graph the equation.

17. Center at $(0,0)$; focus at $(3,0)$; vertex at $(1,0)$

18. Center at $(0,0)$; focus at $(0,5)$; vertex at $(0,3)$

19. Center at $(0,0)$; focus at $(0,-6)$; vertex at $(0,4)$

20. Center at $(0,0)$; focus at $(-3,0)$; vertex at $(2,0)$

21. Foci at $(-5,0)$ and $(5,0)$; vertex at $(3,0)$

22. Focus at $(0,6)$; vertices at $(0,-2)$ and $(0,2)$

23. Vertices at $(0,-6)$ and $(0,6)$; asymptote the line $y = 2x$

24. Vertices at $(-4,0)$ and $(4,0)$; asymptote the line $y = 2x$

25. Foci at $(-4,0)$ and $(4,0)$; asymptote the line $y = -x$

26. Foci at $(0,-2)$ and $(0,2)$; asymptote the line $y = -x$

In Problems 27–34, find the center, transverse axis, vertices, foci, and asymptotes. Graph each equation.

27. $\dfrac{x^2}{25} - \dfrac{y^2}{9} = 1$

28. $\dfrac{y^2}{16} - \dfrac{x^2}{4} = 1$

29. $4x^2 - y^2 = 16$

30. $4y^2 - x^2 = 16$

31. $y^2 - 9x^2 = 9$

32. $x^2 - y^2 = 4$

33. $y^2 - x^2 = 25$

34. $2x^2 - y^2 = 4$

In Problems 35–38, write an equation for each hyperbola.

35.

36.

37.

38.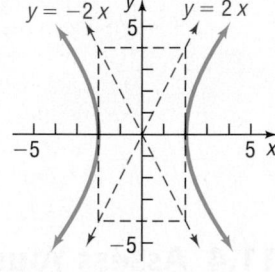

In Problems 39–46, find an equation for the hyperbola described. Graph the equation.

39. Center at $(4,-1)$; focus at $(7,-1)$; vertex at $(6,-1)$

40. Center at $(-3,1)$; focus at $(-3,6)$; vertex at $(-3,4)$

41. Center at $(-3,-4)$; focus at $(-3,-8)$; vertex at $(-3,-2)$

42. Center at $(1,4)$; focus at $(-2,4)$; vertex at $(0,4)$

43. Foci at $(3,7)$ and $(7,7)$; vertex at $(6,7)$

44. Focus at $(-4,0)$ vertices at $(-4,4)$ and $(-4,2)$

45. Vertices at $(-1,-1)$ and $(3,-1)$; asymptote the line $y + 1 = \dfrac{3}{2}(x-1)$

46. Vertices at $(1,-3)$ and $(1,1)$; asymptote the line $y + 1 = \dfrac{3}{2}(x-1)$

In Problems 47–60, find the center, transverse axis, vertices, foci, and asymptotes. Graph each equation.

47. $\dfrac{(x-2)^2}{4} - \dfrac{(y+3)^2}{9} = 1$

48. $\dfrac{(y+3)^2}{4} - \dfrac{(x-2)^2}{9} = 1$

49. $(y-2)^2 - 4(x+2)^2 = 4$

50. $(x+4)^2 - 9(y-3)^2 = 9$

51. $(x+1)^2 - (y+2)^2 = 4$

52. $(y-3)^2 - (x+2)^2 = 4$

53. $x^2 - y^2 - 2x - 2y - 1 = 0$ 54. $y^2 - x^2 - 4y + 4x - 1 = 0$ 55. $y^2 - 4x^2 - 4y - 8x - 4 = 0$

56. $2x^2 - y^2 + 4x + 4y - 4 = 0$ 57. $4x^2 - y^2 - 24x - 4y + 16 = 0$ 58. $2y^2 - x^2 + 2x + 8y + 3 = 0$

59. $y^2 - 4x^2 - 16x - 2y - 19 = 0$ 60. $x^2 - 3y^2 + 8x - 6y + 4 = 0$

In Problems 61–64, graph each function. Be sure to label any intercepts.
[**Hint:** *Notice that each function is half a hyperbola.*]

61. $f(x) = \sqrt{16 + 4x^2}$ 62. $f(x) = -\sqrt{9 + 9x^2}$ 63. $f(x) = -\sqrt{-25 + x^2}$ 64. $f(x) = \sqrt{-1 + x^2}$

Applications and Extensions

65. **Fireworks Display** Suppose that two people standing 2 miles apart both see the burst from a fireworks display. After a period of time, the first person standing at point *A* hears the burst. One second later, the second person standing at point *B* hears the burst. If the person at point *B* is due west of the person at point *A* and if the display is known to occur due north of the person at point *A*, where did the fireworks display occur?

66. **Lightning Strikes** Suppose that two people standing 1 mile apart both see a flash of lightning. After a period of time, the first person standing at point *A* hears the thunder. Two seconds later, the second person standing at point *B* hears the thunder. If the person at point *B* is due west of the person at point *A* and if the lightning strike is known to occur due north of the person standing at point *A*, where did the lightning strike?

67. **Nuclear Power Plant** Some nuclear power plants utilize "natural draft" cooling towers in the shape of a **hyperboloid,** a solid obtained by rotating a hyperbola about its conjugate axis. Suppose such a cooling tower has a base diameter of 400 feet and the diameter at its narrowest point, 360 feet above the ground, is 200 feet. If the diameter at the top of the tower is 300 feet, how tall is the tower?

 Source: Bay Area Air Quality Management District

68. **An Explosion** Two recording devices are set 2400 feet apart, with the device at point *A* to the west of the device at point *B*. At a point between the devices, 300 feet from point *B*, a small amount of explosive is detonated. The recording devices record the time until the sound reaches each. How far directly north of point *B* should a second explosion be done so that the measured time difference recorded by the devices is the same as that for the first detonation?

69. **Rutherford's Experiment** In May 1911, Ernest Rutherford published a paper in *Philosophical Magazine*. In this article, he described the motion of alpha particles as they are shot at a piece of gold foil 0.00004 cm thick. Before conducting this experiment, Rutherford expected that the alpha particles would shoot through the foil just as a bullet would shoot through snow. Instead, a small fraction of the alpha particles bounced off the foil. This led to the conclusion that the nucleus of an atom is dense, while the remainder of the atom is sparse. Only the density of the nucleus could cause the alpha particles to deviate from their path. The figure shows a diagram from Rutherford's paper that indicates that the deflected alpha particles follow the path of one branch of a hyperbola.
 (a) Find an equation of the asymptotes under this scenario.
 (b) If the vertex of the path of the alpha particles is 10 cm from the center of the hyperbola, find an equation that describes the path of the particle.

70. **Hyperbolic Mirrors** Hyperbolas have interesting reflective properties that make them useful for lenses and mirrors. For example, if a ray of light strikes a convex hyperbolic mirror on a line that would (theoretically) pass through its rear focus, it is reflected through the front focus. This property, and that of the parabola, were used to develop the *Cassegrain* telescope in 1672. The focus of the parabolic mirror and the rear focus of the hyperbolic mirror are the same point. The rays are collected by the parabolic mirror, reflected toward the (common) focus, and thus are reflected by the hyperbolic mirror through the opening to its front focus, where the eyepiece is located. If the equation of the hyperbola is $\dfrac{y^2}{9} - \dfrac{x^2}{16} = 1$ and the focal length (distance from the vertex to the focus) of the parabola is 6, find the equation of the parabola.

 Source: www.enchantedlearning.com

71. The **eccentricity** *e* of a hyperbola is defined as the number $\dfrac{c}{a}$, where *a* and *c* are the numbers given in equation (2). Because $c > a$, it follows that $e > 1$. Describe the general shape of a hyperbola whose eccentricity is close to 1. What is the shape if *e* is very large?

72. A hyperbola for which $a = b$ is called an **equilateral hyperbola.** Find the eccentricity *e* of an equilateral hyperbola.

 [**Note:** The eccentricity of a hyperbola is defined in Problem 71.]

73. Two hyperbolas that have the same set of asymptotes are called **conjugate.** Show that the hyperbolas

 $$\frac{x^2}{4} - y^2 = 1 \quad \text{and} \quad y^2 - \frac{x^2}{4} = 1$$

 are conjugate. Graph each hyperbola on the same set of coordinate axes.

74. Prove that the hyperbola

 $$\frac{y^2}{a^2} - \frac{x^2}{b^2} = 1$$

has the two oblique asymptotes

$$y = \frac{a}{b}x \quad \text{and} \quad y = -\frac{a}{b}x$$

75. Show that the graph of an equation of the form

$$Ax^2 + Cy^2 + F = 0, \quad A \neq 0, C \neq 0, F \neq 0$$

where A and C are of opposite sign, is a hyperbola with center at $(0, 0)$.

76. Show that the graph of an equation of the form

$$Ax^2 + Cy^2 + Dx + Ey + F = 0, \quad A \neq 0, C \neq 0$$

where A and C are of opposite sign,

(a) Is a hyperbola if $\dfrac{D^2}{4A} + \dfrac{E^2}{4C} - F \neq 0$.

(b) Is two intersecting lines if

$$\frac{D^2}{4A} + \frac{E^2}{4C} - F = 0$$

'Are You Prepared?' Answers

1. $5\sqrt{2}$ **2.** $\dfrac{25}{4}$ **3.** $(0, -3), (0, 3)$ **4.** True **5.** right; 5; down; 4 **6.** Vertical: $x = -2, x = 2$; horizontal: $y = 1$

11.5 Rotation of Axes; General Form of a Conic

PREPARING FOR THIS SECTION *Before getting started, review the following:*

- Sum Formulas for Sine and Cosine (Section 8.4, pp. 627 and 630)
- Half-angle Formulas for Sine and Cosine (Section 8.5, p. 640)
- Double-angle Formulas for Sine and Cosine (Section 8.5, p. 637)

Now Work the 'Are You Prepared?' problems on page 810.

OBJECTIVES 1 Identify a Conic (p. 804)
2 Use a Rotation of Axes to Transform Equations (p. 805)
3 Analyze an Equation Using a Rotation of Axes (p. 808)
4 Identify Conics without a Rotation of Axes (p. 810)

In this section, we show that the graph of a general second-degree polynomial containing two variables x and y, that is, an equation of the form

$$Ax^2 + Bxy + Cy^2 + Dx + Ey + F = 0 \tag{1}$$

where A, B, and C are not simultaneously 0, is a conic. We shall not concern ourselves here with the degenerate cases of equation (1), such as $x^2 + y^2 = 0$, whose graph is a single point $(0, 0)$; or $x^2 + 3y^2 + 3 = 0$, whose graph contains no points; or $x^2 - 4y^2 = 0$, whose graph is two lines, $x - 2y = 0$ and $x + 2y = 0$.

We begin with the case where $B = 0$. In this case, the term containing xy is not present, so equation (1) has the form

$$Ax^2 + Cy^2 + Dx + Ey + F = 0$$

where either $A \neq 0$ or $C \neq 0$.

1 Identify a Conic

We have already discussed the procedure for identifying the graph of this kind of equation; we complete the squares of the quadratic expressions in x or y, or both. Once this has been done, the conic can be identified by comparing it to one of the forms studied in Sections 11.2 through 11.4.

In fact, though, we can identify the conic directly from the equation without completing the squares.

THEOREM **Identifying Conics without Completing the Squares**

Excluding degenerate cases, the equation

$$Ax^2 + Cy^2 + Dx + Ey + F = 0 \qquad (2)$$

where A and C cannot both equal zero:

(a) Defines a parabola if $AC = 0$.

(b) Defines an ellipse (or a circle) if $AC > 0$.

(c) Defines a hyperbola if $AC < 0$.

Proof

(a) If $AC = 0$, then either $A = 0$ or $C = 0$, but not both, so the form of equation (2) is either

$$Ax^2 + Dx + Ey + F = 0, \qquad A \neq 0$$

or

$$Cy^2 + Dx + Ey + F = 0, \qquad C \neq 0$$

Using the results of Problems 78 and 79 in Exercise 11.2, it follows that, except for the degenerate cases, the equation is a parabola.

(b) If $AC > 0$, then A and C are of the same sign. Using the results of Problem 84 in Exercise 11.3, except for the degenerate cases, the equation is an ellipse if $A \neq C$ or a circle if $A = C$.

(c) If $AC < 0$, then A and C are of opposite sign. Using the results of Problem 76 in Exercise 11.4, except for the degenerate cases, the equation is a hyperbola. ∎

We will not be concerned with the degenerate cases of equation (2). However, in practice, you should be alert to the possibility of degeneracy.

EXAMPLE 1 Identifying a Conic without Completing the Squares

Identify each equation without completing the squares.

(a) $3x^2 + 6y^2 + 6x - 12y = 0$ \qquad (b) $2x^2 - 3y^2 + 6y + 4 = 0$

(c) $y^2 - 2x + 4 = 0$

Solution (a) We compare the given equation to equation (2) and conclude that $A = 3$ and $C = 6$. Since $AC = 18 > 0$, the equation is an ellipse.

(b) Here $A = 2$ and $C = -3$, so $AC = -6 < 0$. The equation is a hyperbola.

(c) Here $A = 0$ and $C = 1$, so $AC = 0$. The equation is a parabola.

◼

Now Work PROBLEM 11

Although we can now identify the type of conic represented by any equation of the form of equation (2) without completing the squares, we will still need to complete the squares if we desire additional information about the conic, such as its graph.

Now we turn our attention to equations of the form of equation (1), where $B \neq 0$. To discuss this case, we first need to investigate a new procedure: *rotation of axes*.

2 Use a Rotation of Axes to Transform Equations

In a **rotation of axes,** the origin remains fixed while the x-axis and y-axis are rotated through an angle θ to a new position; the new positions of the x-axis and the y-axis are denoted by x' and y', respectively, as shown in Figure 47(a).

Figure 47

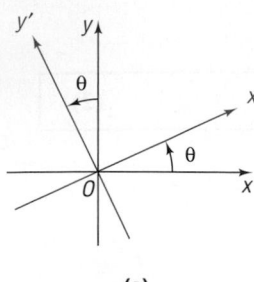

(a)

(b)

Now look at Figure 47(b). There the point P has the coordinates (x, y) relative to the xy-plane, while the same point P has coordinates (x', y') relative to the $x'y'$-plane. We seek relationships that will enable us to express x and y in terms of x', y', and θ.

As Figure 47(b) shows, r denotes the distance from the origin O to the point P, and α denotes the angle between the positive x'-axis and the ray from O through P. Then, using the definitions of sine and cosine, we have

$$x' = r \cos \alpha \qquad\qquad y' = r \sin \alpha \qquad\qquad \text{(3)}$$
$$x = r \cos(\theta + \alpha) \qquad y = r \sin(\theta + \alpha) \qquad \text{(4)}$$

Now

$$
\begin{aligned}
x &= r \cos(\theta + \alpha) \\
&= r(\cos \theta \cos \alpha - \sin \theta \sin \alpha) && \text{Apply the Sum Formula for sine.} \\
&= (r \cos \alpha)(\cos \theta) - (r \sin \alpha)(\sin \theta) \\
&= x' \cos \theta - y' \sin \theta && \text{By equation (3)}
\end{aligned}
$$

Similarly,

$$
\begin{aligned}
y &= r \sin(\theta + \alpha) \\
&= r(\sin \theta \cos \alpha + \cos \theta \sin \alpha) && \text{Apply the Sum Formula for sine.} \\
&= x' \sin \theta + y' \cos \theta && \text{By equation (3)}
\end{aligned}
$$

THEOREM

Rotation Formulas

If the x- and y-axes are rotated through an angle θ, the coordinates (x, y) of a point P relative to the xy-plane and the coordinates (x', y') of the same point relative to the new x'- and y'-axes are related by the formulas

$$x = x' \cos \theta - y' \sin \theta \qquad y = x' \sin \theta + y' \cos \theta \qquad \text{(5)}$$

EXAMPLE 2 **Rotating Axes**

Express the equation $xy = 1$ in terms of new $x'y'$-coordinates by rotating the axes through a $45°$ angle. Discuss the new equation.

Solution Let $\theta = 45°$ in equation (5). Then

$$x = x' \cos 45° - y' \sin 45° = x' \frac{\sqrt{2}}{2} - y' \frac{\sqrt{2}}{2} = \frac{\sqrt{2}}{2}(x' - y')$$

$$y = x' \sin 45° + y' \cos 45° = x' \frac{\sqrt{2}}{2} + y' \frac{\sqrt{2}}{2} = \frac{\sqrt{2}}{2}(x' + y')$$

Substituting these expressions for x and y in $xy = 1$ gives

$$\left[\frac{\sqrt{2}}{2}(x' - y') \right]\left[\frac{\sqrt{2}}{2}(x' + y') \right] = 1$$

$$\frac{1}{2}(x'^2 - y'^2) = 1$$

$$\frac{x'^2}{2} - \frac{y'^2}{2} = 1$$

Figure 48

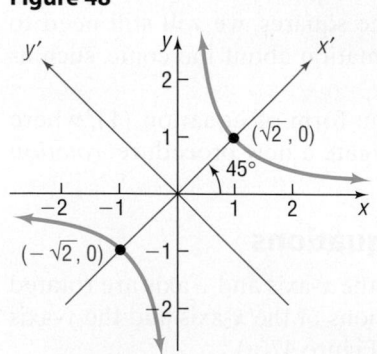

This is the equation of a hyperbola with center at $(0, 0)$ and transverse axis along the x'-axis. The vertices are at $(\pm\sqrt{2}, 0)$ on the x'-axis; the asymptotes are $y' = x'$ and $y' = -x'$ (which correspond to the original x- and y-axes). See Figure 48 for the graph.

As Example 2 illustrates, a rotation of axes through an appropriate angle can transform a second-degree equation in x and y containing an xy-term into one in x' and y' in which no $x'y'$-term appears. In fact, we will show that a rotation of axes through an appropriate angle will transform any equation of the form of equation (1) into an equation in x' and y' without an $x'y'$-term.

To find the formula for choosing an appropriate angle θ through which to rotate the axes, we begin with equation (1),

$$Ax^2 + Bxy + Cy^2 + Dx + Ey + F = 0, \qquad B \neq 0$$

Next we rotate through an angle θ using rotation formulas (5).

$$
A(x' \cos \theta - y' \sin \theta)^2 + B(x' \cos \theta - y' \sin \theta)(x' \sin \theta + y' \cos \theta)
$$
$$
+ C(x' \sin \theta + y' \cos \theta)^2 + D(x' \cos \theta - y' \sin \theta)
$$
$$
+ E(x' \sin \theta + y' \cos \theta) + F = 0
$$

By expanding and collecting like terms, we obtain

$$
(A \cos^2 \theta + B \sin \theta \cos \theta + C \sin^2 \theta)x'^2 + [B(\cos^2 \theta - \sin^2 \theta) + 2(C - A)(\sin \theta \cos \theta)]x'y'
$$
$$
+ (A \sin^2 \theta - B \sin \theta \cos \theta + C \cos^2 \theta)y'^2
$$
$$
+ (D \cos \theta + E \sin \theta)x'
$$
$$
+ (-D \sin \theta + E \cos \theta)y' + F = 0 \qquad \textbf{(6)}
$$

In equation (6), the coefficient of $x'y'$ is

$$B(\cos^2 \theta - \sin^2 \theta) + 2(C - A)(\sin \theta \cos \theta)$$

Since we want to eliminate the $x'y'$-term, we select an angle θ so that this coefficient is 0.

$$B(\cos^2 \theta - \sin^2 \theta) + 2(C - A)(\sin \theta \cos \theta) = 0$$
$$B \cos(2\theta) + (C - A) \sin(2\theta) = 0 \qquad \text{\textit{Double-angle}}$$
$$\text{\textit{Formulas}}$$
$$B \cos(2\theta) = (A - C) \sin(2\theta)$$
$$\cot(2\theta) = \frac{A - C}{B} \qquad B \neq 0$$

THEOREM To transform the equation

$$Ax^2 + Bxy + Cy^2 + Dx + Ey + F = 0 \qquad B \neq 0$$

into an equation in x' and y' without an $x'y'$-term, rotate the axes through an angle θ that satisfies the equation

$$\boxed{\cot(2\theta) = \frac{A - C}{B}} \qquad \textbf{(7)}$$

Equation (7) has an infinite number of solutions for θ. We shall adopt the convention of choosing the acute angle θ that satisfies (7). Then we have the following two possibilities:

If $\cot(2\theta) \geq 0$, then $0° < 2\theta \leq 90°$, so $0° < \theta \leq 45°$.
If $\cot(2\theta) < 0$, then $90° < 2\theta < 180°$, so $45° < \theta < 90°$.

Each of these results in a counterclockwise rotation of the axes through an acute angle θ.*

* Any rotation through an angle θ that satisfies $\cot(2\theta) = \dfrac{A - C}{B}$ will eliminate the $x'y'$-term. However, the final form of the transformed equation may be different (but equivalent), depending on the angle chosen.

WARNING Be careful if you use a calculator to solve equation (7).

1. If $\cot(2\theta) = 0$, then $2\theta = 90°$ and $\theta = 45°$.
2. If $\cot(2\theta) \neq 0$, first find $\cos(2\theta)$. Then use the inverse cosine function key(s) to obtain 2θ, $0° < 2\theta < 180°$. Finally, divide by 2 to obtain the correct acute angle θ. ∎

3 Analyze an Equation Using a Rotation of Axes

For the remainder of this section, the direction **"Analyze the equation"** will mean to transform the given equation so that it contains no xy-term and to graph the equation.

| EXAMPLE 3 | Analyzing an Equation Using a Rotation of Axes |

Analyze the equation: $x^2 + \sqrt{3}xy + 2y^2 - 10 = 0$

Solution Since an xy-term is present, we must rotate the axes. Using $A = 1$, $B = \sqrt{3}$, and $C = 2$ in equation (7), the appropriate acute angle θ through which to rotate the axes satisfies the equation

$$\cot(2\theta) = \frac{A - C}{B} = \frac{-1}{\sqrt{3}} = -\frac{\sqrt{3}}{3} \qquad 0° < 2\theta < 180°$$

Since $\cot(2\theta) = -\dfrac{\sqrt{3}}{3}$, we find $2\theta = 120°$, so $\theta = 60°$. Using $\theta = 60°$ in rotation formulas (5), we find

$$x = x'\cos 60° - y'\sin 60° = \frac{1}{2}x' - \frac{\sqrt{3}}{2}y' = \frac{1}{2}\left(x' - \sqrt{3}y'\right)$$

$$y = x'\sin 60° + y'\cos 60° = \frac{\sqrt{3}}{2}x' + \frac{1}{2}y' = \frac{1}{2}\left(\sqrt{3}x' + y'\right)$$

Substituting these values into the original equation and simplifying, we have

$$x^2 + \sqrt{3}xy + 2y^2 - 10 = 0$$

Figure 49

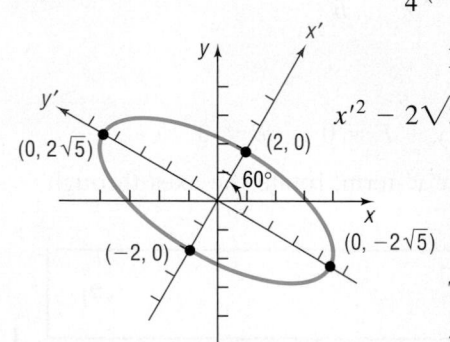

$$\frac{1}{4}\left(x' - \sqrt{3}y'\right)^2 + \sqrt{3}\left[\frac{1}{2}\left(x' - \sqrt{3}y'\right)\right]\left[\frac{1}{2}\left(\sqrt{3}x' + y'\right)\right] + 2\left[\frac{1}{4}\left(\sqrt{3}x' + y'\right)^2\right] = 10$$

Multiply both sides by 4 and expand to obtain

$$x'^2 - 2\sqrt{3}x'y' + 3y'^2 + \sqrt{3}\left(\sqrt{3}x'^2 - 2x'y' - \sqrt{3}y'^2\right) + 2\left(3x'^2 + 2\sqrt{3}x'y' + y'^2\right) = 40$$

$$10x'^2 + 2y'^2 = 40$$

$$\frac{x'^2}{4} + \frac{y'^2}{20} = 1$$

This is the equation of an ellipse with center at $(0, 0)$ and major axis along the y'-axis. The vertices are at $(0, \pm 2\sqrt{5})$ on the y'-axis. See Figure 49 for the graph. ∎

━━ **Now Work** PROBLEM 31

In Example 3, the acute angle θ through which to rotate the axes was easy to find because of the numbers that we used in the given equation. In general, the equation $\cot(2\theta) = \dfrac{A - C}{B}$ will not have such a "nice" solution. As the next example shows, we can still find the appropriate rotation formulas without using a calculator approximation by applying Half-angle Formulas.

| EXAMPLE 4 | Analyzing an Equation Using a Rotation of Axes |

Analyze the equation: $4x^2 - 4xy + y^2 + 5\sqrt{5}x + 5 = 0$

Solution Letting $A = 4$, $B = -4$, and $C = 1$ in equation (7), the appropriate angle θ through which to rotate the axes satisfies

$$\cot(2\theta) = \frac{A - C}{B} = \frac{3}{-4} = -\frac{3}{4}$$

To use rotation formulas (5), we need to know the values of $\sin\theta$ and $\cos\theta$. Since we seek an acute angle θ, we know that $\sin\theta > 0$ and $\cos\theta > 0$. We use the Half-angle Formulas in the form

$$\sin\theta = \sqrt{\frac{1 - \cos(2\theta)}{2}} \qquad \cos\theta = \sqrt{\frac{1 + \cos(2\theta)}{2}}$$

Now we need to find the value of $\cos(2\theta)$. Since $\cot(2\theta) = -\dfrac{3}{4}$, then $90° < 2\theta < 180°$ (Do you know why?), so $\cos(2\theta) = -\dfrac{3}{5}$. Then

$$\sin\theta = \sqrt{\frac{1 - \cos(2\theta)}{2}} = \sqrt{\frac{1 - \left(-\dfrac{3}{5}\right)}{2}} = \sqrt{\frac{4}{5}} = \frac{2}{\sqrt{5}} = \frac{2\sqrt{5}}{5}$$

$$\cos\theta = \sqrt{\frac{1 + \cos(2\theta)}{2}} = \sqrt{\frac{1 + \left(-\dfrac{3}{5}\right)}{2}} = \sqrt{\frac{1}{5}} = \frac{1}{\sqrt{5}} = \frac{\sqrt{5}}{5}$$

With these values, the rotation formulas (5) are

$$x = \frac{\sqrt{5}}{5}x' - \frac{2\sqrt{5}}{5}y' = \frac{\sqrt{5}}{5}(x' - 2y')$$

$$y = \frac{2\sqrt{5}}{5}x' + \frac{\sqrt{5}}{5}y' = \frac{\sqrt{5}}{5}(2x' + y')$$

Substituting these values in the original equation and simplifying, we obtain

$$4x^2 - 4xy + y^2 + 5\sqrt{5}x + 5 = 0$$

$$4\left[\frac{\sqrt{5}}{5}(x' - 2y')\right]^2 - 4\left[\frac{\sqrt{5}}{5}(x' - 2y')\right]\left[\frac{\sqrt{5}}{5}(2x' + y')\right]$$

$$+ \left[\frac{\sqrt{5}}{5}(2x' + y')\right]^2 + 5\sqrt{5}\left[\frac{\sqrt{5}}{5}(x' - 2y')\right] = -5$$

Multiply both sides by 5 and expand to obtain

$$4(x'^2 - 4x'y' + 4y'^2) - 4(2x'^2 - 3x'y' - 2y'^2)$$

$$+ 4x'^2 + 4x'y' + y'^2 + 25(x' - 2y') = -25$$

$$25y'^2 - 50y' + 25x' = -25 \qquad \text{Combine like terms.}$$

$$y'^2 - 2y' + x' = -1 \qquad \text{Divide by 25.}$$

$$y'^2 - 2y' + 1 = -x' \qquad \text{Complete the square in } y'.$$

$$(y' - 1)^2 = -x'$$

This is the equation of a parabola with vertex at $(0, 1)$ in the $x'y'$-plane. The axis of symmetry is parallel to the x'-axis. Using a calculator to solve $\sin\theta = \dfrac{2\sqrt{5}}{5}$, we find that $\theta \approx 63.4°$. See Figure 50 for the graph.

Figure 50

———— **Now Work** PROBLEM **37**

4 Identify Conics without a Rotation of Axes

Suppose that we are required only to identify (rather than analyze) an equation of the form

$$Ax^2 + Bxy + Cy^2 + Dx + Ey + F = 0, \qquad B \neq 0 \qquad \textbf{(8)}$$

If we apply the rotation formulas (5) to this equation, we obtain an equation of the form

$$A'x'^2 + B'x'y' + C'y'^2 + D'x' + E'y' + F' = 0 \qquad \textbf{(9)}$$

where A', B', C', D', E', and F' can be expressed in terms of A, B, C, D, E, F and the angle θ of rotation (see Problem 53). It can be shown that the value of $B^2 - 4AC$ in equation (8) and the value of $B'^2 - 4A'C'$ in equation (9) are equal no matter what angle θ of rotation is chosen (see Problem 55). In particular, if the angle θ of rotation satisfies equation (7), then $B' = 0$ in equation (9), and $B^2 - 4AC = -4A'C'$. Since equation (9) then has the form of equation (2),

$$A'x'^2 + C'y'^2 + D'x' + E'y' + F' = 0$$

we can identify it without completing the squares, as we did in the beginning of this section. In fact, now we can identify the conic described by any equation of the form of equation (8) without a rotation of axes.

THEOREM

Identifying Conics without a Rotation of Axes

Except for degenerate cases, the equation

$$Ax^2 + Bxy + Cy^2 + Dx + Ey + F = 0$$

(a) Defines a parabola if $B^2 - 4AC = 0$.
(b) Defines an ellipse (or a circle) if $B^2 - 4AC < 0$.
(c) Defines a hyperbola if $B^2 - 4AC > 0$.

You are asked to prove this theorem in Problem 56.

EXAMPLE 5 **Identifying a Conic without a Rotation of Axes**

Identify the equation: $8x^2 - 12xy + 17y^2 - 4\sqrt{5}x - 2\sqrt{5}y - 15 = 0$

Solution Here $A = 8, B = -12$, and $C = 17$, so $B^2 - 4AC = -400$. Since $B^2 - 4AC < 0$, the equation defines an ellipse.

Now Work PROBLEM 43

11.5 Assess Your Understanding

'Are You Prepared?' *Answers are given at the end of these exercises. If you get a wrong answer, read the pages listed in red.*

1. The sum formula for the sine function is $\sin(A + B) = $ _____. (p. 630)

2. The Double-angle Formula for the sine function is $\sin(2\theta) = $ _____. (p. 637)

3. If θ is acute, the Half-angle Formula for the sine function is $\sin\dfrac{\theta}{2} = $ _____. (p. 641)

4. If θ is acute, the Half-angle Formula for the cosine function is $\cos\dfrac{\theta}{2} = $ _____. (p. 641)

Concepts and Vocabulary

5. To transform the equation
$$Ax^2 + Bxy + Cy^2 + Dx + Ey + F = 0, \qquad B \neq 0$$
into one in x' and y' without an $x'y'$-term, rotate the axes through an acute angle θ that satisfies the equation _____.

6. Identify the conic: $x^2 - 2y^2 - x - y - 18 = 0$ _____.

7. Identify the conic: $x^2 + 2xy + 3y^2 - 2x + 4y + 10 = 0$
_____.

8. True or False The equation $ax^2 + 6y^2 - 12y = 0$ defines
an ellipse if $a > 0$.

9. True or False The equation $3x^2 + bxy + 12y^2 = 10$ de-
fines a parabola if $b = -12$.

10. True or False To eliminate the xy-term from the equation
$x^2 - 2xy + y^2 - 2x + 3y + 5 = 0$, rotate the axes through
an angle θ, where $\cot \theta = B^2 - 4AC$.

Skill Building

In Problems 11–20, identify each equation without completing the squares.

11. $x^2 + 4x + y + 3 = 0$

12. $2y^2 - 3y + 3x = 0$

13. $6x^2 + 3y^2 - 12x + 6y = 0$

14. $2x^2 + y^2 - 8x + 4y + 2 = 0$

15. $3x^2 - 2y^2 + 6x + 4 = 0$

16. $4x^2 - 3y^2 - 8x + 6y + 1 = 0$

17. $2y^2 - x^2 - y + x = 0$

18. $y^2 - 8x^2 - 2x - y = 0$

19. $x^2 + y^2 - 8x + 4y = 0$

20. $2x^2 + 2y^2 - 8x + 8y = 0$

In Problems 21–30, determine the appropriate rotation formulas to use so that the new equation contains no xy-term.

21. $x^2 + 4xy + y^2 - 3 = 0$

22. $x^2 - 4xy + y^2 - 3 = 0$

23. $5x^2 + 6xy + 5y^2 - 8 = 0$

24. $3x^2 - 10xy + 3y^2 - 32 = 0$

25. $13x^2 - 6\sqrt{3}xy + 7y^2 - 16 = 0$

26. $11x^2 + 10\sqrt{3}xy + y^2 - 4 = 0$

27. $4x^2 - 4xy + y^2 - 8\sqrt{5}x - 16\sqrt{5}y = 0$

28. $x^2 + 4xy + 4y^2 + 5\sqrt{5}y + 5 = 0$

29. $25x^2 - 36xy + 40y^2 - 12\sqrt{13}x - 8\sqrt{13}y = 0$

30. $34x^2 - 24xy + 41y^2 - 25 = 0$

*In Problems 31–42, rotate the axes so that the new equation contains no xy-term. Analyze and graph the new equation. Refer to Problems
21–30 for Problems 31–40.*

31. $x^2 + 4xy + y^2 - 3 = 0$

32. $x^2 - 4xy + y^2 - 3 = 0$

33. $5x^2 + 6xy + 5y^2 - 8 = 0$

34. $3x^2 - 10xy + 3y^2 - 32 = 0$

35. $13x^2 - 6\sqrt{3}xy + 7y^2 - 16 = 0$

36. $11x^2 + 10\sqrt{3}xy + y^2 - 4 = 0$

37. $4x^2 - 4xy + y^2 - 8\sqrt{5}x - 16\sqrt{5}y = 0$

38. $x^2 + 4xy + 4y^2 + 5\sqrt{5}y + 5 = 0$

39. $25x^2 - 36xy + 40y^2 - 12\sqrt{13}x - 8\sqrt{13}y = 0$

40. $34x^2 - 24xy + 41y^2 - 25 = 0$

41. $16x^2 + 24xy + 9y^2 - 130x + 90y = 0$

42. $16x^2 + 24xy + 9y^2 - 60x + 80y = 0$

In Problems 43–52, identify each equation without applying a rotation of axes.

43. $x^2 + 3xy - 2y^2 + 3x + 2y + 5 = 0$

44. $2x^2 - 3xy + 4y^2 + 2x + 3y - 5 = 0$

45. $x^2 - 7xy + 3y^2 - y - 10 = 0$

46. $2x^2 - 3xy + 2y^2 - 4x - 2 = 0$

47. $9x^2 + 12xy + 4y^2 - x - y = 0$

48. $10x^2 + 12xy + 4y^2 - x - y + 10 = 0$

49. $10x^2 - 12xy + 4y^2 - x - y - 10 = 0$

50. $4x^2 + 12xy + 9y^2 - x - y = 0$

51. $3x^2 - 2xy + y^2 + 4x + 2y - 1 = 0$

52. $3x^2 + 2xy + y^2 + 4x - 2y + 10 = 0$

Applications and Extensions

In Problems 53–56, apply the rotation formulas (5) to
$$Ax^2 + Bxy + Cy^2 + Dx + Ey + F = 0$$
to obtain the equation
$$A'x'^2 + B'x'y' + C'y'^2 + D'x' + E'y' + F' = 0$$

53. Express $A', B', C', D', E',$ and F' in terms of $A, B, C, D, E,$
$F,$ and the angle θ of rotation.
[**Hint:** Refer to equation (6).]

54. Show that $A + C = A' + C'$, and thus show that $A + C$ is
invariant; that is, its value does not change under a rotation
of axes.

55. Refer to Problem 54. Show that $B^2 - 4AC$ is invariant.

56. Prove that, except for degenerate cases, the equation
$$Ax^2 + Bxy + Cy^2 + Dx + Ey + F = 0$$
(a) Defines a parabola if $B^2 - 4AC = 0$.
(b) Defines an ellipse (or a circle) if $B^2 - 4AC < 0$.
(c) Defines a hyperbola if $B^2 - 4AC > 0$.

57. Use rotation formulas (5) to show that distance is invariant
under a rotation of axes. That is, show that the distance from
$P_1 = (x_1, y_1)$ to $P_2 = (x_2, y_2)$ in the xy-plane equals the dis-
tance from $P_1 = (x_1', y_1')$ to $P_2 = (x_2', y_2')$ in the $x'y'$-plane.

58. Show that the graph of the equation $x^{1/2} + y^{1/2} = a^{1/2}$ is part
of the graph of a parabola.

Discussion and Writing

59. Formulate a strategy for discussing and graphing an equa-
tion of the form
$$Ax^2 + Cy^2 + Dx + Ey + F = 0$$

60. How does your strategy change if the equation is of the fol-
lowing form?
$$Ax^2 + Bxy + Cy^2 + Dx + Ey + F = 0$$

'Are You Prepared?' Answers

1. $\sin A \cos B + \cos A \sin B$ **2.** $2 \sin \theta \cos \theta$ **3.** $\sqrt{\dfrac{1 - \cos \theta}{2}}$ **4.** $\sqrt{\dfrac{1 + \cos \theta}{2}}$

11.6 Polar Equations of Conics

PREPARING FOR THIS SECTION *Before getting started, review the following:*

- Polar Coordinates (Section 10.1, pp. 714–720)

Now Work the 'Are You Prepared?' problems on page 816.

OBJECTIVES **1** Analyze and Graph Polar Equations of Conics (p. 812)
 2 Convert the Polar Equation of a Conic to a Rectangular Equation (p. 816)

1 Analyze and Graph Polar Equations of Conics

In Sections 11.2 through 11.4, we gave separate definitions for the parabola, ellipse, and hyperbola based on geometric properties and the distance formula. In this section, we present an alternative definition that simultaneously defines all these conics. As we shall see, this approach is well suited to polar coordinate representation. (Refer to Section 10.1.)

DEFINITION

Let D denote a fixed line called the **directrix;** let F denote a fixed point called the **focus,** which is not on D; and let e be a fixed positive number called the **eccentricity.** A **conic** is the set of points P in the plane such that the ratio of the distance from F to P to the distance from D to P equals e. That is, a conic is the collection of points P for which

$$\frac{d(F, P)}{d(D, P)} = e \qquad (1)$$

If $e = 1$, the conic is a **parabola.**
If $e < 1$, the conic is an **ellipse.**
If $e > 1$, the conic is a **hyperbola.**

Observe that if $e = 1$ the definition of a parabola in equation (1) is exactly the same as the definition used earlier in Section 11.2.

In the case of an ellipse, the **major axis** is a line through the focus perpendicular to the directrix. In the case of a hyperbola, the **transverse axis** is a line through the focus perpendicular to the directrix. For both an ellipse and a hyperbola, the eccentricity e satisfies

$$e = \frac{c}{a} \qquad (2)$$

Figure 51

where c is the distance from the center to the focus and a is the distance from the center to a vertex.

Just as we did earlier using rectangular coordinates, we derive equations for the conics in polar coordinates by choosing a convenient position for the focus F and the directrix D. The focus F is positioned at the pole, and the directrix D is either parallel or perpendicular to the polar axis.

Suppose that we start with the directrix D perpendicular to the polar axis at a distance p units to the left of the pole (the focus F). See Figure 51.

If $P = (r, \theta)$ is any point on the conic, then, by equation (1),

$$\frac{d(F, P)}{d(D, P)} = e \quad \text{or} \quad d(F, P) = e \cdot d(D, P) \tag{3}$$

Now we use the point Q obtained by dropping the perpendicular from P to the polar axis to calculate $d(D, P)$.

$$d(D, P) = p + d(O, Q) = p + r \cos \theta$$

Using this expression and the fact that $d(F, P) = d(O, P) = r$ in equation (3), we get

$$
\begin{aligned}
d(F, P) &= e \cdot d(D, P) \\
r &= e(p + r \cos \theta) \\
r &= ep + er \cos \theta \\
r - er \cos \theta &= ep \\
r(1 - e \cos \theta) &= ep \\
r &= \frac{ep}{1 - e \cos \theta}
\end{aligned}
$$

THEOREM

Polar Equation of a Conic; Focus at the Pole; Directrix Perpendicular to the Polar Axis a Distance p to the Left of the Pole

The polar equation of a conic with focus at the pole and directrix perpendicular to the polar axis at a distance p to the left of the pole is

$$r = \frac{ep}{1 - e \cos \theta} \tag{4}$$

where e is the eccentricity of the conic.

EXAMPLE 1 | Analyzing and Graphing the Polar Equation of a Conic

Analyze and graph the equation: $\quad r = \dfrac{4}{2 - \cos \theta}$

Solution The given equation is not quite in the form of equation (4), since the first term in the denominator is 2 instead of 1. We divide the numerator and denominator by 2 to obtain

$$r = \frac{2}{1 - \dfrac{1}{2}\cos \theta} \qquad r = \frac{ep}{1 - e \cos \theta}$$

This equation is in the form of equation (4), with

$$e = \frac{1}{2} \quad \text{and} \quad ep = 2$$

Then

$$\frac{1}{2}p = 2, \quad \text{so} \quad p = 4$$

We conclude that the conic is an ellipse, since $e = \dfrac{1}{2} < 1$. One focus is at the pole, and the directrix is perpendicular to the polar axis, a distance of $p = 4$ units to the left of the pole. It follows that the major axis is along the polar axis. To find the vertices, we let $\theta = 0$ and $\theta = \pi$. The vertices of the ellipse are $(4, 0)$ and $\left(\dfrac{4}{3}, \pi\right)$. The midpoint of the vertices is $\left(\dfrac{4}{3}, 0\right)$ in polar coordinates. [Do you see why? The vertices $(4, 0)$ and $\left(\dfrac{4}{3}, \pi\right)$ in polar coordinates are $(4, 0)$ and $\left(-\dfrac{4}{3}, 0\right)$

in rectangular coordinates. The midpoint in rectangular coordinates is $\left(\frac{4}{3}, 0\right)$, which is also $\left(\frac{4}{3}, 0\right)$ in polar coordinates.] The center of the ellipse is $\left(\frac{4}{3}, 0\right)$.

Then $a =$ distance from the center to a vertex $= \frac{8}{3}$. Using $a = \frac{8}{3}$ and $e = \frac{1}{2}$ in equation (2), $e = \frac{c}{a}$, we find $c = ae = \frac{4}{3}$. Finally, using $a = \frac{8}{3}$ and $c = \frac{4}{3}$ in $b^2 = a^2 - c^2$, we have

$$b^2 = a^2 - c^2 = \frac{64}{9} - \frac{16}{9} = \frac{48}{9}$$

$$b = \frac{4\sqrt{3}}{3}$$

Figure 52 shows the graph.

Figure 52

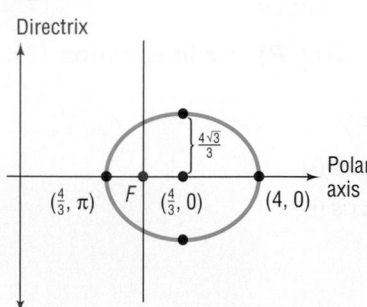

Check: In polar mode with $\theta\min = 0, \theta\max = 2\pi$, and $\theta\text{step} = \frac{\pi}{24}$, graph

$$r_1 = \frac{4}{2 - \cos\theta}$$ and compare the result with Figure 52.

Now Work PROBLEM 11

Exploration Graph $r_1 = \dfrac{4}{2 + \cos\theta}$ and compare the result with Figure 52. What do you conclude? Clear the screen and graph $r_1 = \dfrac{4}{2 - \sin\theta}$ and then $r_1 = \dfrac{4}{2 + \sin\theta}$. Compare each of these graphs with Figure 52. What do you conclude?

Equation (4) was obtained under the assumption that the directrix was perpendicular to the polar axis at a distance p units to the left of the pole. A similar derivation (see Problem 43), in which the directrix is perpendicular to the polar axis at a distance p units to the right of the pole, results in the equation

$$r = \frac{ep}{1 + e\cos\theta}$$

In Problems 44 and 45, you are asked to derive the polar equations of conics with focus at the pole and directrix parallel to the polar axis. Table 5 summarizes the polar equations of conics.

Table 5

Polar Equations of Conics (Focus at the Pole, Eccentricity e)	
Equation	**Description**
(a) $r = \dfrac{ep}{1 - e\cos\theta}$	Directrix is perpendicular to the polar axis at a distance p units to the left of the pole.
(b) $r = \dfrac{ep}{1 + e\cos\theta}$	Directrix is perpendicular to the polar axis at a distance p units to the right of the pole.
(c) $r = \dfrac{ep}{1 + e\sin\theta}$	Directrix is parallel to the polar axis at a distance p units above the pole.
(d) $r = \dfrac{ep}{1 - e\sin\theta}$	Directrix is parallel to the polar axis at a distance p units below the pole.

Eccentricity

If $e = 1$, the conic is a parabola; the axis of symmetry is perpendicular to the directrix.

If $e < 1$, the conic is an ellipse; the major axis is perpendicular to the directrix.

If $e > 1$, the conic is a hyperbola; the transverse axis is perpendicular to the directrix.

| EXAMPLE 2 | Analyzing and Graphing the Polar Equation of a Conic |

Analyze and graph the equation: $\quad r = \dfrac{6}{3 + 3\sin\theta}$

Solution To place the equation in proper form, we divide the numerator and denominator by 3 to get

$$r = \frac{2}{1 + \sin\theta}$$

Referring to Table 5, we conclude that this equation is in the form of equation (c) with

$$e = 1 \quad \text{and} \quad ep = 2$$
$$p = 2 \quad {\scriptstyle e\,=\,1}$$

Figure 53

The conic is a parabola with focus at the pole. The directrix is parallel to the polar axis at a distance 2 units above the pole; the axis of symmetry is perpendicular to the polar axis. The vertex of the parabola is at $\left(1, \dfrac{\pi}{2}\right)$. (Do you see why?)

See Figure 53 for the graph. Notice that we plotted two additional points, $(2, 0)$ and $(2, \pi)$, to assist in graphing.

 ✓ **Check:** Graph $r = \dfrac{6}{3 + 3\sin\theta}$ and compare the result with Figure 53.

━━━ **Now Work** PROBLEM 13

| EXAMPLE 3 | Analyzing and Graphing the Polar Equation of a Conic |

Analyze and graph the equation: $\quad r = \dfrac{3}{1 + 3\cos\theta}$

Solution This equation is in the form of equation (b) in Table 5. We conclude that

$$e = 3 \quad \text{and} \quad ep = 3$$
$$p = 1 \quad {\scriptstyle e\,=\,3}$$

This is the equation of a hyperbola with a focus at the pole. The directrix is perpendicular to the polar axis, 1 unit to the right of the pole. The transverse axis is along the polar axis. To find the vertices, we let $\theta = 0$ and $\theta = \pi$. The vertices are $\left(\dfrac{3}{4}, 0\right)$ and $\left(-\dfrac{3}{2}, \pi\right)$. The center, which is at the midpoint of $\left(\dfrac{3}{4}, 0\right)$ and $\left(-\dfrac{3}{2}, \pi\right)$, is $\left(\dfrac{9}{8}, 0\right)$. Then $c =$ distance from the center to a focus $= \dfrac{9}{8}$. Since $e = 3$, it follows from equation (2), $e = \dfrac{c}{a}$, that $a = \dfrac{3}{8}$. Finally, using $a = \dfrac{3}{8}$ and $c = \dfrac{9}{8}$ in $b^2 = c^2 - a^2$, we find

$$b^2 = c^2 - a^2 = \frac{81}{64} - \frac{9}{64} = \frac{72}{64} = \frac{9}{8}$$

$$b = \frac{3}{2\sqrt{2}} = \frac{3\sqrt{2}}{4}$$

Figure 54

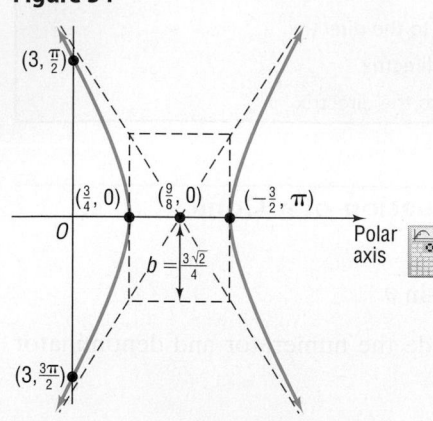

Figure 54 shows the graph. Notice that we plotted two additional points, $\left(3, \dfrac{\pi}{2}\right)$ and $\left(3, \dfrac{3\pi}{2}\right)$, on the left branch and used symmetry to obtain the right branch. The asymptotes of this hyperbola were found in the usual way by constructing the rectangle shown.

✓ **Check:** Graph $r = \dfrac{1}{1 + 3\cos\theta}$ and compare the result with Figure 54.

── **Now Work** PROBLEM 17

2 Convert the Polar Equation of a Conic to a Rectangular Equation

| EXAMPLE 4 | Converting a Polar Equation to a Rectangular Equation |

Convert the polar equation

$$r = \frac{1}{3 - 3\cos\theta}$$

to a rectangular equation.

Solution The strategy here is first to rearrange the equation and square each side before using the transformation equations.

$$r = \frac{1}{3 - 3\cos\theta}$$

$$3r - 3r\cos\theta = 1$$

$$3r = 1 + 3r\cos\theta \qquad \text{Rearrange the equation.}$$

$$9r^2 = (1 + 3r\cos\theta)^2 \qquad \text{Square each side.}$$

$$9(x^2 + y^2) = (1 + 3x)^2 \qquad x^2 + y^2 = r^2;\ x = r\cos\theta$$

$$9x^2 + 9y^2 = 9x^2 + 6x + 1$$

$$9y^2 = 6x + 1$$

This is the equation of a parabola in rectangular coordinates.

── **Now Work** PROBLEM 25

11.6 Assess Your Understanding

'Are You Prepared?' *Answers are given at the end of these exercises. If you get a wrong answer, read the pages listed in red.*

1. If (x, y) are the rectangular coordinates of a point P and (r, θ) are its polar coordinates, then $x =$ _____ and $y =$ _____. (pp. 714–720)

2. Transform the equation $r = 6\cos\theta$ from polar coordinates to rectangular coordinates. (pp. 714–720)

Concepts and Vocabulary

3. The polar equation $r = \dfrac{8}{4 - 2\sin\theta}$ is a conic whose eccentricity is _____. It is a(n) _____ whose directrix is _____ to the polar axis at a distance _____ units _____ the pole.

4. The eccentricity e of a parabola is _____, of an ellipse it is _____, and of a hyperbola it is _____.

5. **True or False** If (r, θ) are polar coordinates, the equation $r = \dfrac{2}{2 + 3\sin\theta}$ defines a hyperbola.

6. **True or False** The eccentricity of any parabola is 1.

Skill Building

In Problems 7–12, identify the conic that each polar equation represents. Also, give the position of the directrix.

7. $r = \dfrac{1}{1 + \cos\theta}$

8. $r = \dfrac{3}{1 - \sin\theta}$

9. $r = \dfrac{4}{2 - 3\sin\theta}$

10. $r = \dfrac{2}{1 + 2\cos\theta}$

11. $r = \dfrac{3}{4 - 2\cos\theta}$

12. $r = \dfrac{6}{8 + 2\sin\theta}$

In Problems 13–24, analyze each equation and graph it.

13. $r = \dfrac{1}{1 + \cos\theta}$

14. $r = \dfrac{3}{1 - \sin\theta}$

15. $r = \dfrac{8}{4 + 3\sin\theta}$

16. $r = \dfrac{10}{5 + 4\cos\theta}$

17. $r = \dfrac{9}{3 - 6\cos\theta}$

18. $r = \dfrac{12}{4 + 8\sin\theta}$

19. $r = \dfrac{8}{2 - \sin\theta}$

20. $r = \dfrac{8}{2 + 4\cos\theta}$

21. $r(3 - 2\sin\theta) = 6$

22. $r(2 - \cos\theta) = 2$

23. $r = \dfrac{6\sec\theta}{2\sec\theta - 1}$

24. $r = \dfrac{3\csc\theta}{\csc\theta - 1}$

In Problems 25–36, convert each polar equation to a rectangular equation.

25. $r = \dfrac{1}{1 + \cos\theta}$

26. $r = \dfrac{3}{1 - \sin\theta}$

27. $r = \dfrac{8}{4 + 3\sin\theta}$

28. $r = \dfrac{10}{5 + 4\cos\theta}$

29. $r = \dfrac{9}{3 - 6\cos\theta}$

30. $r = \dfrac{12}{4 + 8\sin\theta}$

31. $r = \dfrac{8}{2 - \sin\theta}$

32. $r = \dfrac{8}{2 + 4\cos\theta}$

33. $r(3 - 2\sin\theta) = 6$

34. $r(2 - \cos\theta) = 2$

35. $r = \dfrac{6\sec\theta}{2\sec\theta - 1}$

36. $r = \dfrac{3\csc\theta}{\csc\theta - 1}$

In Problems 37–42, find a polar equation for each conic. For each, a focus is at the pole.

37. $e = 1$; directrix is parallel to the polar axis 1 unit above the pole.

38. $e = 1$; directrix is parallel to the polar axis 2 units below the pole.

39. $e = \dfrac{4}{5}$; directrix is perpendicular to the polar axis 3 units to the left of the pole.

40. $e = \dfrac{2}{3}$; directrix is parallel to the polar axis 3 units above the pole.

41. $e = 6$; directrix is parallel to the polar axis 2 units below the pole.

42. $e = 5$; directrix is perpendicular to the polar axis 5 units to the right of the pole.

Applications and Extensions

43. Derive equation (b) in Table 5:
$$r = \frac{ep}{1 + e\cos\theta}$$

44. Derive equation (c) in Table 5:
$$r = \frac{ep}{1 + e\sin\theta}$$

45. Derive equation (d) in Table 5:
$$r = \frac{ep}{1 - e\sin\theta}$$

46. Orbit of Mercury The planet Mercury travels around the Sun in an elliptical orbit given approximately by
$$r = \frac{(3.442)10^7}{1 - 0.206\cos\theta}$$

where r is measured in miles and the Sun is at the pole. Find the distance from Mercury to the Sun at *aphelion* (greatest distance from the Sun) and at *perihelion* (shortest distance from the Sun). See the figure. Use the aphelion and perihelion to graph the orbit of Mercury using a graphing utility.

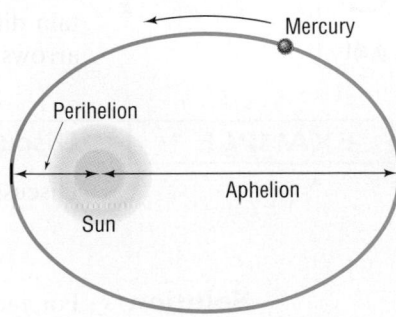

'Are You Prepared?' Answers

1. $r\cos\theta$; $r\sin\theta$

2. $x^2 + y^2 = 6x$ or $(x - 3)^2 + y^2 = 9$

11.7 Plane Curves and Parametric Equations

PREPARING FOR THIS SECTION *Before getting started, review the following:*

• Amplitude and Period of Sinusoidal Graphs (Section 7.6, pp. 564–565)

Now Work the 'Are You Prepared?' problem on page 827.

> **OBJECTIVES** 1 Graph Parametric Equations (p. 818)
> 2 Find a Rectangular Equation for a Curve Defined Parametrically (p. 819)
> 3 Use Time as a Parameter in Parametric Equations (p. 821)
> 4 Find Parametric Equations for Curves Defined by Rectangular Equations (p. 824)

Equations of the form $y = f(x)$, where f is a function, have graphs that are intersected no more than once by any vertical line. The graphs of many of the conics and certain other, more complicated, graphs do not have this characteristic. Yet each graph, like the graph of a function, is a collection of points (x, y) in the xy-plane; that is, each is a *plane curve*. In this section, we discuss another way of representing such graphs.

DEFINITION

Let $x = f(t)$ and $y = g(t)$, where f and g are two functions whose common domain is some interval I. The collection of points defined by

$$(x, y) = (f(t), g(t))$$

is called a **plane curve.** The equations

$$x = f(t) \qquad y = g(t)$$

where t is in I, are called **parametric equations** of the curve. The variable t is called a **parameter.**

1 Graph Parametric Equations

Parametric equations are particularly useful in describing movement along a curve. Suppose that a curve is defined by the parametric equations

$$x = f(t), \qquad y = g(t), \qquad a \le t \le b$$

Figure 55

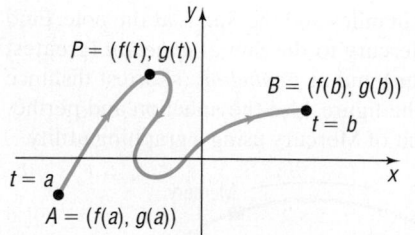

where f and g are each defined over the interval $a \le t \le b$. For a given value of t, we can find the value of $x = f(t)$ and $y = g(t)$, obtaining a point (x, y) on the curve. In fact, as t varies over the interval from $t = a$ to $t = b$, successive values of t give rise to a directed movement along the curve; that is, the curve is traced out in a certain direction by the corresponding succession of points (x, y). See Figure 55. The arrows show the direction, or **orientation,** along the curve as t varies from a to b.

EXAMPLE 1 | Discussing a Curve Defined by Parametric Equations

Discuss the curve defined by the parametric equations

$$x = 3t^2, \qquad y = 2t, \qquad -2 \le t \le 2$$

Solution For each number t, $-2 \le t \le 2$, there corresponds a number x and a number y. For example, when $t = -2$, then $x = 3(-2)^2 = 12$ and $y = 2(-2) = -4$. When $t = 0$, then $x = 0$ and $y = 0$. Indeed, we can set up a table listing various choices of the parameter t and the corresponding values for x and y, as shown in Table 6. Plotting these points and connecting them with a smooth curve leads to Figure 56, where the arrows are used to indicate the orientation.

Table 6

t	x	y	(x, y)
−2	12	−4	(12, −4)
−1	3	−2	(3, −2)
0	0	0	(0, 0)
1	3	2	(3, 2)
2	12	4	(12, 4)

Figure 56

 COMMENT Most graphing utilities have the capability of graphing parametric equations. See Section 9 in the Appendix. ∎

2 Find a Rectangular Equation for a Curve Defined Parametrically

The curve given in Example 1 should be familiar. To identify it accurately, we find the corresponding rectangular equation by eliminating the parameter t from the parametric equations given in Example 1:

$$x = 3t^2, \qquad y = 2t, \qquad -2 \le t \le 2$$

Noting that we can readily solve for t in $y = 2t$, obtaining $t = \dfrac{y}{2}$, we substitute this expression in the other equation.

$$x = 3t^2 = 3\left(\frac{y}{2}\right)^2 = \frac{3y^2}{4}$$

$$\uparrow$$
$$t = \frac{y}{2}$$

This equation, $x = \dfrac{3y^2}{4}$, is the equation of a parabola with vertex at $(0, 0)$ and axis of symmetry along the x-axis.

Exploration In FUNCtion mode, graph $x = \dfrac{3y^2}{4}\left(Y_1 = \sqrt{\dfrac{4x}{3}} \text{ and } Y_2 = -\sqrt{\dfrac{4x}{3}}\right)$ with Xmin = 0, Xmax = 15, Ymin = −5, Ymax = 5. Compare this graph with Figure 56. How do the graphs differ?

Notice that the parameterized curve discussed in Example 1 and shown in Figure 56 is only a part of the parabola $x = \dfrac{3y^2}{4}$. The graph of the rectangular equation obtained by eliminating the parameter will, in general, contain more points than the original parameterized curve. Care must therefore be taken when a parameterized curve is graphed after eliminating the parameter. Even so, the process of eliminating the parameter t of a parameterized curve to identify it accurately is sometimes a better approach than merely plotting points. However, the elimination process sometimes requires a little ingenuity.

EXAMPLE 2	Finding the Rectangular Equation of a Curve Defined Parametrically

Find the rectangular equation of the curve whose parametric equations are

$$x = a \cos t \qquad y = a \sin t$$

where $a > 0$ is a constant. Graph this curve, indicating its orientation.

Solution The presence of sines and cosines in the parametric equations suggests that we use a Pythagorean Identity. In fact, since

$$\cos t = \frac{x}{a} \qquad \sin t = \frac{y}{a}$$

we find that

$$\cos^2 t + \sin^2 t = 1$$

$$\left(\frac{x}{a}\right)^2 + \left(\frac{y}{a}\right)^2 = 1$$

$$x^2 + y^2 = a^2$$

Figure 57

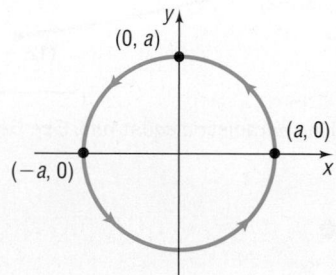

The curve is a circle with center at $(0, 0)$ and radius a. As the parameter t increases, say from $t = 0$ [the point $(a, 0)$] to $t = \dfrac{\pi}{2}$ [the point $(0, a)$] to $t = \pi$ [the point $(-a, 0)$], we see that the corresponding points are traced in a counterclockwise direction around the circle. The orientation is as indicated in Figure 57.

Now Work PROBLEMS 7 AND 19

Let's analyze the curve in Example 2 further. The domain of each parametric equation is $-\infty < t < \infty$. This means, the graph in Figure 57 is actually being repeated each time that t increases by 2π.

If we wanted the curve to consist of exactly 1 revolution in the counterclockwise direction, we could write

$$x = a \cos t, \qquad y = a \sin t, \qquad 0 \le t \le 2\pi$$

This curve starts at $t = 0$ [the point $(a, 0)$] and, proceeding counterclockwise around the circle, ends at $t = 2\pi$ [also the point $(a, 0)$].

If we wanted the curve to consist of exactly three revolutions in the counterclockwise direction, we could write

$$x = a \cos t, \qquad y = a \sin t, \qquad -2\pi \le t \le 4\pi$$

or

$$x = a \cos t, \qquad y = a \sin t, \qquad 0 \le t \le 6\pi$$

or

$$x = a \cos t, \qquad y = a \sin t, \qquad 2\pi \le t \le 8\pi$$

EXAMPLE 3 **Describing Parametric Equations**

Find rectangular equations for the following curves defined by parametric equations. Graph each curve.

(a) $x = a \cos t$, $y = a \sin t$, $0 \le t \le \pi$, $a > 0$

(b) $x = -a \sin t$, $y = -a \cos t$, $0 \le t \le \pi$, $a > 0$

Solution (a) We eliminate the parameter t using a Pythagorean Identity.

$$\cos^2 t + \sin^2 t = 1$$

$$\left(\frac{x}{a}\right)^2 + \left(\frac{y}{a}\right)^2 = 1$$

$$x^2 + y^2 = a^2$$

The curve defined by these parametric equations is a circle, with radius a and center at $(0, 0)$. The circle begins at the point $(a, 0)$, $t = 0$; passes through the point $(0, a)$, $t = \dfrac{\pi}{2}$; and ends at the point $(-a, 0)$, $t = \pi$.

Figure 58

See Figure 58. The parametric equations define an upper semicircle of radius a with a counterclockwise orientation. The rectangular equation is

$$y = \sqrt{a^2 - x^2}, \qquad -a \le x \le a$$

(b) We eliminate the parameter t using a Pythagorean Identity.

$$\sin^2 t + \cos^2 t = 1$$

$$\left(\frac{x}{-a}\right)^2 + \left(\frac{y}{-a}\right)^2 = 1$$

$$x^2 + y^2 = a^2$$

Figure 59

The curve defined by these parametric equations is a circle, with radius a and center at $(0, 0)$. The circle begins at the point $(0, -a)$, $t = 0$; passes through the point $(-a, 0)$, $t = \dfrac{\pi}{2}$; and ends at the point $(0, a)$, $t = \pi$.

See Figure 59. The parametric equations define a left semicircle of radius a with a clockwise orientation. The rectangular equation is

$$x = -\sqrt{a^2 - y^2}, \qquad -a \le y \le a$$

Seeing the Concept

Graph $x = \cos t$, $y = \sin t$, $0 \le t \le 2\pi$. Compare to Figure 57. Graph $x = \cos t$, $y = \sin t$ for $0 \le t \le \pi$. Compare to Figure 58. Graph $x = -\sin t$, $y = -\cos t$ for $0 \le t \le \pi$. Compare to Figure 59.

Example 3 illustrates the versatility of parametric equations for replacing complicated rectangular equations, while providing additional information about orientation. These characteristics make parametric equations very useful in applications, such as projectile motion.

3 Use Time as a Parameter in Parametric Equations

If we think of the parameter t as time, the parametric equations $x = f(t)$ and $y = g(t)$ of a curve specify how the x- and y-coordinates of a moving point vary with time.

For example, we can use parametric equations to describe the motion of an object, sometimes referred to as **curvilinear motion.** Using parametric equations, we can specify not only where the object travels, that is, its location (x, y), but also when it gets there, that is, the time t.

When an object is propelled upward at an inclination θ to the horizontal with initial speed v_0, the resulting motion is called **projectile motion.** See Figure 60(a).

In calculus it is shown that the parametric equations of the path of a projectile fired at an inclination θ to the horizontal, with an initial speed v_0, from a height h above the horizontal are

$$x = (v_0 \cos \theta)t \qquad y = -\frac{1}{2}gt^2 + (v_0 \sin \theta)t + h \qquad \textbf{(1)}$$

where t is the time and g is the constant acceleration due to gravity (approximately 32 ft/sec/sec or 9.8 m/sec/sec). See Figure 60(b).

Figure 60

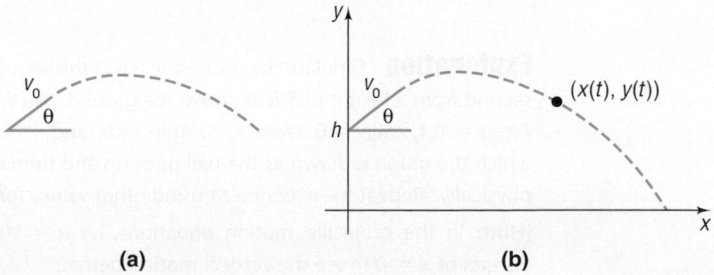

(a) (b)

| | EXAMPLE 4 | | Projectile Motion |

Figure 61

Suppose that Adam hit a golf ball with an initial velocity of 150 feet per second at an angle of 30° to the horizontal. See Figure 61.

(a) Find parametric equations that describe the position of the ball as a function of time.

(b) How long is the golf ball in the air?

(c) When is the ball at its maximum height? Determine the maximum height of the ball.

(d) Determine the horizontal distance that the ball traveled.

(e) Using a graphing utility, simulate the motion of the golf ball by simultaneously graphing the equations found in part (a).

Solution

(a) We have $v_0 = 150$ ft/sec, $\theta = 30°, h = 0$ (the ball is on the ground), and $g = 32$ ft/sec^2 (since the units are in feet and seconds). Substituting these values into equations (1), we find that

$$x = (v_0 \cos \theta)t = (150 \cos 30°)t = 75\sqrt{3}\,t$$

$$y = -\frac{1}{2}gt^2 + (v_0 \sin \theta)t + h = -\frac{1}{2}(32)t^2 + (150 \sin 30°)t + 0$$

$$= -16t^2 + 75t$$

(b) To determine the length of time that the ball is in the air, we solve the equation $y = 0$.

$$-16t^2 + 75t = 0$$

$$t(-16t + 75) = 0$$

$$t = 0 \text{ sec} \quad \text{or} \quad t = \frac{75}{16} = 4.6875 \text{ sec}$$

The ball will strike the ground after 4.6875 seconds.

(c) Notice that the height y of the ball is a quadratic function of t, so the maximum height of the ball can be found by determining the vertex of $y = -16t^2 + 75t$. The value of t at the vertex is

$$t = \frac{-b}{2a} = \frac{-75}{-32} = 2.34375 \text{ sec}$$

The ball is at its maximum height after 2.34375 seconds. The maximum height of the ball is found by evaluating the function y at $t = 2.34375$ seconds.

$$\text{Maximum height} = -16(2.34375)^2 + (75)2.34375 \approx 87.89 \text{ feet}$$

Figure 62

(d) Since the ball is in the air for 4.6875 seconds, the horizontal distance that the ball travels is

$$x = \left(75\sqrt{3}\right)4.6875 \approx 608.92 \text{ feet} \quad x = 75\sqrt{3}\,t$$

(e) We enter the equations from part (a) into a graphing utility with $T\text{min} = 0, T\text{max} = 4.7$, and $T\text{step} = 0.1$. We use ZOOM-SQUARE to avoid any distortion to the angle of elevation. See Figure 62.

Exploration Simulate the motion of a ball thrown straight up with an initial speed of 100 feet per second from a height of 5 feet above the ground. Use PARametric mode with $T\text{min} = 0, T\text{max} = 6.5, T\text{step} = 0.1, X\text{min} = 0, X\text{max} = 5, Y\text{min} = 0,$ and $Y\text{max} = 180$. What happens to the speed with which the graph is drawn as the ball goes up and then comes back down? How do you interpret this physically? Repeat the experiment using other values for $T\text{step}$. How does this affect the experiment?

[**Hint:** In the projectile motion equations, let $\theta = 90°, v_0 = 100, h = 5,$ and $g = 32$. Use $x = 3$ instead of $x = 0$ to see the vertical motion better.]

Result See Figure 63. In Figure 63(a) the ball is going up. In Figure 63(b) the ball is near its highest point. Finally, in Figure 63(c) the ball is coming back down.

Notice that, as the ball goes up, its speed decreases, until at the highest point it is zero. Then the speed increases as the ball comes back down.

Figure 63

(a) ($t \approx 0.7$) (b) ($t \approx 3$) (c) ($t \approx 4$)

── **Now Work** PROBLEM 49

A graphing utility can be used to simulate other kinds of motion as well. Let's work Example 5 from Section 1.7 again.

EXAMPLE 5 **Simulating Motion**

Tanya, who is a long distance runner, runs at an average velocity of 8 miles per hour. Two hours after Tanya leaves your house, you leave in your Honda and follow the same route. If your average velocity is 40 miles per hour, how long will it be before you catch up to Tanya? See Figure 64. Use a simulation of the two motions to verify the answer.

Figure 64

Time t

$t = 0$ 2 hr $t = 2$

$t = 2$

Solution We begin with two sets of parametric equations: one to describe Tanya's motion, the other to describe the motion of the Honda. We choose time $t = 0$ to be when Tanya leaves the house. We choose $y_1 = 2$ as Tanya's path and $y_2 = 4$ as the parallel path of the Honda in order to more easily see the two motions. The horizontal distances traversed in time t (Distance = Velocity × Time) are

$$\text{Tanya:} \quad x_1 = 8t \qquad \text{Honda:} \quad x_2 = 40(t - 2)$$

The Honda catches up to Tanya when $x_1 = x_2$.

$$8t = 40(t - 2)$$
$$8t = 40t - 80$$
$$-32t = -80$$
$$t = \frac{-80}{-32} = 2.5$$

The Honda catches up to Tanya 2.5 hours after Tanya leaves the house.

In PARametric mode with Tstep = 0.01, we simultaneously graph

$$\text{Tanya:} \quad x_1 = 8t \qquad \text{Honda:} \quad x_2 = 40(t - 2)$$
$$y_1 = 2 \qquad\qquad\qquad y_2 = 4$$

for $0 \le t \le 3$.

Figure 65 shows the relative position of Tanya and the Honda for $t = 0, t = 2, t = 2.25, t = 2.5,$ and $t = 2.75$.

Figure 65

$t = 0$ $t = 2$ $t = 2.25$

$t = 2.5$ $t = 2.75$

4 Find Parametric Equations for Curves Defined by Rectangular Equations

We now take up the question of how to find parametric equations of a given curve. If a curve is defined by the equation $y = f(x)$, where f is a function, one way of finding parametric equations is to let $x = t$. Then $y = f(t)$ and

$$x = t, \quad y = f(t), \qquad t \text{ in the domain of } f$$

are parametric equations of the curve.

EXAMPLE 6 **Finding Parametric Equations for a Curve Defined by a Rectangular Equation**

Find parametric equations for the equation $y = x^2 - 4$.

Solution Let $x = t$. Then the parametric equations are

$$x = t, \quad y = t^2 - 4, \qquad -\infty < t < \infty$$

Another less obvious approach to Example 6 is to let $x = t^3$. Then the parametric equations become

$$x = t^3, \quad y = t^6 - 4, \qquad -\infty < t < \infty$$

Care must be taken when using this approach, since the substitution for x must be a function that allows x to take on all the values stipulated by the domain of f. For example, letting $x = t^2$ so that $y = t^4 - 4$ does not result in equivalent parametric equations for $y = x^2 - 4$, since only points for which $x \geq 0$ are obtained.

━━ **Now Work** PROBLEM 33

EXAMPLE 7 **Finding Parametric Equations for an Object in Motion**

Find parametric equations for the ellipse

$$x^2 + \frac{y^2}{9} = 1$$

where the parameter t is time (in seconds) and

(a) The motion around the ellipse is clockwise, begins at the point $(0, 3)$, and requires 1 second for a complete revolution.

(b) The motion around the ellipse is counterclockwise, begins at the point $(1, 0)$, and requires 2 seconds for a complete revolution.

Figure 66

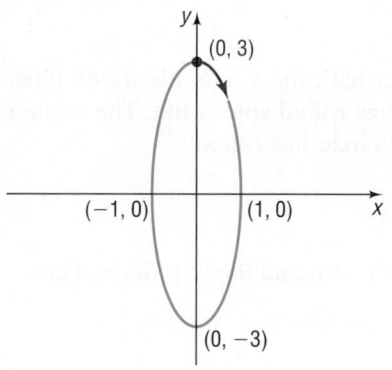

Solution (a) See Figure 66. Since the clockwise motion begins at the point $(0, 3)$, we want $x = 0$ and $y = 3$ when $t = 0$. Since the given equation is an ellipse, we begin by letting

$$x = \sin(\omega t) \qquad \frac{y}{3} = \cos(\omega t)$$

for some constant ω. These parametric equations satisfy the equation of the ellipse. Further, with this choice, when $t = 0$, we have $x = 0$ and $y = 3$.

For the motion to be clockwise, the motion will have to begin with the value of x increasing and y decreasing as t increases. This requires that $\omega > 0$. [Do you know why? If $\omega > 0$, then $x = \sin(\omega t)$ is increasing when $t > 0$ is near zero and $y = 3 \cos(\omega t)$ is decreasing when $t > 0$ is near zero.] See the red part of the graph in Figure 66.

Finally, since 1 revolution requires 1 second, the period $\dfrac{2\pi}{\omega} = 1$, so $\omega = 2\pi$.

Parametric equations that satisfy the conditions stipulated are

$$x = \sin(2\pi t), \qquad y = 3 \cos(2\pi t), \qquad 0 \le t \le 1 \qquad \textbf{(2)}$$

Figure 67

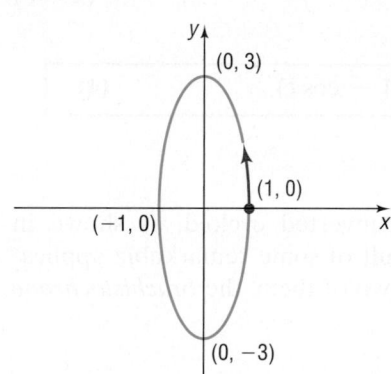

(b) See Figure 67. Since the counterclockwise motion begins at the point $(1, 0)$, we want $x = 1$ and $y = 0$ when $t = 0$. Since the given equation is an ellipse, we begin by letting

$$x = \cos(\omega t) \qquad \frac{y}{3} = \sin(\omega t)$$

for some constant ω. These parametric equations satisfy the equation of the ellipse. Further, with this choice, when $t = 0$, we have $x = 1$ and $y = 0$.

For the motion to be counterclockwise, the motion will have to begin with the value of x decreasing and y increasing as t increases. This requires that $\omega > 0$. [Do you know why?] Finally, since 1 revolution requires 2 seconds, the period is $\dfrac{2\pi}{\omega} = 2$, so $\omega = \pi$. The parametric equations that satisfy the conditions stipulated are

$$x = \cos(\pi t), \qquad y = 3 \sin(\pi t), \qquad 0 \le t \le 2 \qquad \textbf{(3)}$$

Either of equations (2) or (3) can serve as parametric equations for the ellipse $x^2 + \dfrac{y^2}{9} = 1$ given in Example 7. The direction of the motion, the beginning point, and the time for 1 revolution merely serve to help us arrive at a particular parametric representation.

Now Work PROBLEM 39

The Cycloid

Suppose that a circle of radius a rolls along a horizontal line without slipping. As the circle rolls along the line, a point P on the circle will trace out a curve called a **cycloid**. See Figure 68.

We seek parametric equations* for a cycloid. We begin with a circle of radius a and take the fixed line on which the circle rolls as the x-axis. Let the origin be one

* Any attempt to derive the rectangular equation of a cycloid would soon demonstrate how complicated the task is.

Figure 68

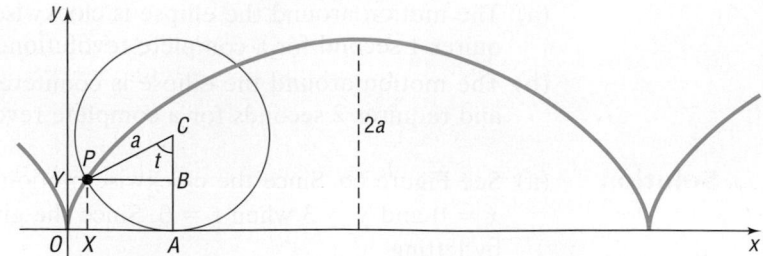

of the points at which the point P comes in contact with the x-axis. Figure 68 illustrates the position of this point P after the circle has rolled somewhat. The angle t (in radians) measures the angle through which the circle has rolled.

Since we require no slippage, it follows that

$$\text{Arc } AP = d(O, A)$$

The length of the arc AP is given by $s = r\theta$, where $r = a$ and $\theta = t$ radians. Then

$$at = d(O, A) \qquad s = r\theta, \text{ where } r = a \text{ and } \theta = t$$

The x-coordinate of the point P is

$$d(O, X) = d(O, A) - d(X, A) = at - a \sin t = a(t - \sin t)$$

The y-coordinate of the point P is equal to

$$d(O, Y) = d(A, C) - d(B, C) = a - a \cos t = a(1 - \cos t)$$

The parametric equations of the cycloid are

$$\boxed{x = a(t - \sin t) \qquad y = a(1 - \cos t) \qquad \textbf{(4)}}$$

Exploration

Graph $x = t - \sin t, y = 1 - \cos t$, $0 \le t \le 3\pi$, using your graphing utility with Tstep $= \dfrac{\pi}{36}$ and a square screen. Compare your results with Figure 68.

Applications to Mechanics

If a is negative in equation (4), we obtain an inverted cycloid, as shown in Figure 69(a). The inverted cycloid occurs as a result of some remarkable applications in the field of mechanics. We shall mention two of them: the *brachistochrone* and the *tautochrone*.*

Figure 69

(a) Inverted cycloid (b) Curve of quickest descent (c) All reach Q at the same time

The **brachistochrone** is the curve of quickest descent. If a particle is constrained to follow some path from one point A to a lower point B (not on the same vertical line) and is acted on only by gravity, the time needed to make the descent is least if the path is an inverted cycloid. See Figure 69(b). This remarkable discovery, which is attributed to many famous mathematicians (including Johann Bernoulli and Blaise Pascal), was a significant step in creating the branch of mathematics known as the *calculus of variations*.

To define the **tautochrone,** let Q be the lowest point on an inverted cycloid. If several particles placed at various positions on an inverted cycloid simultaneously begin to slide down the cycloid, they will reach the point Q at the same time, as

*In Greek, *brachistochrone* means "the shortest time" and *tautochrone* means "equal time."

Figure 70

Cycloid Cycloid

Cycloid

indicated in Figure 69(c). The tautochrone property of the cycloid was used by Christiaan Huygens (1629–1695), the Dutch mathematician, physicist, and astronomer, to construct a pendulum clock with a bob that swings along a cycloid See Figure 70. In Huygen's clock, the bob was made to swing along a cycloid by suspending the bob on a thin wire constrained by two plates shaped like cycloids. In a clock of this design, the period of the pendulum is independent of its amplitude.

11.7 Assess Your Understanding

'Are You Prepared?' *Answers are given at the end of these exercises. If you get a wrong answer, read the pages listed in red.*

1. The function $f(x) = 3 \sin(4x)$ has amplitude _____ and period _____. (pp. 564–565)

Concepts and Vocabulary

2. Let $x = f(t)$ and $y = g(t)$, where f and g are two functions whose common domain is some interval I. The collection of points defined by $(x, y) = (f(t), g(t))$ is called a(n) _____ _____. The variable t is called a(n) _____.

3. The parametric equations $x = 2 \sin t$, $y = 3 \cos t$ define a(n) _____.

4. If a circle rolls along a horizontal line without slippage, a point P on the circle will trace out a curve called a(n) _____.

5. *True or False* Parametric equations defining a curve are unique.

6. *True or False* Curves defined using parametric equations have an orientation.

Skill Building

In Problems 7–26, graph the curve whose parametric equations are given and show its orientation. Find the rectangular equation of each curve.

7. $x = 3t + 2$, $y = t + 1$; $0 \le t \le 4$

8. $x = t - 3$, $y = 2t + 4$; $0 \le t \le 2$

9. $x = t + 2$, $y = \sqrt{t}$; $t \ge 0$

10. $x = \sqrt{2t}$, $y = 4t$; $t \ge 0$

11. $x = t^2 + 4$, $y = t^2 - 4$; $-\infty < t < \infty$

12. $x = \sqrt{t} + 4$, $y = \sqrt{t} - 4$; $t \ge 0$

13. $x = 3t^2$, $y = t + 1$; $-\infty < t < \infty$

14. $x = 2t - 4$, $y = 4t^2$; $-\infty < t < \infty$

15. $x = 2e^t$, $y = 1 + e^t$; $t \ge 0$

16. $x = e^t$, $y = e^{-t}$; $t \ge 0$

17. $x = \sqrt{t}$, $y = t^{3/2}$; $t \ge 0$

18. $x = t^{3/2} + 1$, $y = \sqrt{t}$; $t \ge 0$

19. $x = 2 \cos t$, $y = 3 \sin t$; $0 \le t \le 2\pi$

20. $x = 2 \cos t$, $y = 3 \sin t$; $0 \le t \le \pi$

21. $x = 2 \cos t$, $y = 3 \sin t$; $-\pi \le t \le 0$

22. $x = 2 \cos t$, $y = \sin t$; $0 \le t \le \dfrac{\pi}{2}$

23. $x = \sec t$, $y = \tan t$; $0 \le t \le \dfrac{\pi}{4}$

24. $x = \csc t$, $y = \cot t$; $\dfrac{\pi}{4} \le t \le \dfrac{\pi}{2}$

25. $x = \sin^2 t$, $y = \cos^2 t$; $0 \le t \le 2\pi$

26. $x = t^2$, $y = \ln t$; $t > 0$

In Problems 27–34, find two different parametric equations for each rectangular equation.

27. $y = 4x - 1$

28. $y = -8x + 3$

29. $y = x^2 + 1$

30. $y = -2x^2 + 1$

31. $y = x^3$

32. $y = x^4 + 1$

33. $x = y^{3/2}$

34. $x = \sqrt{y}$

In Problems 35–38, find parametric equations that define the curve shown.

35.

36.

37.

38.

In Problems 39–42, find parametric equations for an object that moves along the ellipse $\dfrac{x^2}{4} + \dfrac{y^2}{9} = 1$ with the motion described.

39. The motion begins at $(2, 0)$, is clockwise, and requires 2 seconds for a complete revolution.

40. The motion begins at $(0, 3)$, is counterclockwise, and requires 1 second for a complete revolution.

41. The motion begins at $(0, 3)$, is clockwise, and requires 1 second for a complete revolution.

42. The motion begins at $(2, 0)$, is counterclockwise, and requires 3 seconds for a complete revolution.

In Problems 43 and 44, the parametric equations of four curves are given. Graph each of them, indicating the orientation.

43. C_1: $x = t$, $y = t^2$; $-4 \leq t \leq 4$

C_2: $x = \cos t$, $y = 1 - \sin^2 t$; $0 \leq t \leq \pi$

C_3: $x = e^t$, $y = e^{2t}$; $0 \leq t \leq \ln 4$

C_4: $x = \sqrt{t}$, $y = t$; $0 \leq t \leq 16$

44. C_1: $x = t$, $y = \sqrt{1 - t^2}$; $-1 \leq t \leq 1$

C_2: $x = \sin t$, $y = \cos t$; $0 \leq t \leq 2\pi$

C_3: $x = \cos t$, $y = \sin t$; $0 \leq t \leq 2\pi$

C_4: $x = \sqrt{1 - t^2}$, $y = t$; $-1 \leq t \leq 1$

In Problems 45–48, use a graphing utility to graph the curve defined by the given parametric equations.

45. $x = t \sin t$, $y = t \cos t$, $t > 0$

46. $x = \sin t + \cos t$, $y = \sin t - \cos t$

47. $x = 4 \sin t - 2 \sin(2t)$
$y = 4 \cos t - 2 \cos(2t)$

48. $x = 4 \sin t + 2 \sin(2t)$
$y = 4 \cos t + 2 \cos(2t)$

Applications and Extensions

49. Projectile Motion Bob throws a ball straight up with an initial speed of 50 feet per second from a height of 6 feet.

(a) Find parametric equations that describe the motion of the ball as a function of time.

(b) How long is the ball in the air?

(c) When is the ball at its maximum height? Determine the maximum height of the ball.

(d) Simulate the motion of the ball by graphing the equations found in part (a).

50. Projectile Motion Alice throws a ball straight up with an initial speed of 40 feet per second from a height of 5 feet.

(a) Find parametric equations that describe the motion of the ball as a function of time.

(b) How long is the ball in the air?

(c) When is the ball at its maximum height? Determine the maximum height of the ball.

(d) Simulate the motion of the ball by graphing the equations found in part (a).

51. Catching a Train Bill's train leaves at 8:06 AM and accelerates at the rate of 2 meters per second per second. Bill, who can run 5 meters per second, arrives at the train station 5 seconds after the train has left and runs for the train.

(a) Find parametric equations that describe the motions of the train and Bill as a function of time.

[**Hint:** The position s at time t of an object having acceleration a is $s = \dfrac{1}{2}at^2$.]

(b) Determine algebraically whether Bill will catch the train. If so, when?

(c) Simulate the motion of the train and Bill by simultaneously graphing the equations found in part (a).

52. Catching a Bus Jodi's bus leaves at 5:30 PM and accelerates at the rate of 3 meters per second per second. Jodi, who can run 5 meters per second, arrives at the bus station 2 seconds after the bus has left and runs for the bus.

(a) Find parametric equations that describe the motions of the bus and Jodi as a function of time.

[**Hint:** The position s at time t of an object having acceleration a is $s = \dfrac{1}{2}at^2$.]

(b) Determine algebraically whether Jodi will catch the bus. If so, when?

(c) Simulate the motion of the bus and Jodi by simultaneously graphing the equations found in part (a).

53. Projectile Motion Ichiro throws a baseball with an initial speed of 145 feet per second at an angle of $20°$ to the horizontal. The ball leaves Ichiro's hand at a height of 5 feet.

(a) Find parametric equations that describe the position of the ball as a function of time.

(b) How long is the ball in the air?

(c) Determine the horizontal distance that the ball traveled.

(d) When is the ball at its maximum height? Determine the maximum height of the ball.

(e) Using a graphing utility, simultaneously graph the equations found in part (a).

54. Projectile Motion Barry Bonds hit a baseball with an initial speed of 125 feet per second at an angle of $40°$ to the horizontal. The ball was hit at a height of 3 feet off the ground.

(a) Find parametric equations that describe the position of the ball as a function of time.

(b) How long is the ball in the air?

(c) Determine the horizontal distance that the ball traveled.

(d) When is the ball at its maximum height? Determine the maximum height of the ball.

(e) Using a graphing utility, simultaneously graph the equations found in part (a).

55. Projectile Motion Suppose that Adam hits a golf ball off a cliff 300 meters high with an initial speed of 40 meters per second at an angle of $45°$ to the horizontal.

(a) Find parametric equations that describe the position of the ball as a function of time.

(b) How long is the ball in the air?

(c) Determine the horizontal distance that the ball traveled.

(d) When is the ball at its maximum height? Determine the maximum height of the ball.

(e) Using a graphing utility, simultaneously graph the equations found in part (a).

56. Projectile Motion Suppose that Karla hits a golf ball off a cliff 300 meters high with an initial speed of 40 meters per second at an angle of 45° to the horizontal on the Moon (gravity on the Moon is one-sixth of that on Earth).

(a) Find parametric equations that describe the position of the ball as a function of time.

(b) How long is the ball in the air?

(c) Determine the horizontal distance that the ball traveled.

(d) When is the ball at its maximum height? Determine the maximum height of the ball.

(e) Using a graphing utility, simultaneously graph the equations found in part (a).

57. Uniform Motion A Toyota Paseo (traveling east at 40 mph) and a Pontiac Bonneville (traveling north at 30 mph) are heading toward the same intersection. The Paseo is 5 miles from the intersection when the Bonneville is 4 miles from the intersection. See the figure.

(a) Find parametric equations that describe the motion of the Paseo and Bonneville.

(b) Find a formula for the distance between the cars as a function of time.

(c) Graph the function in part (b) using a graphing utility.

(d) What is the minimum distance between the cars? When are the cars closest?

(e) Simulate the motion of the cars by simultaneously graphing the equations found in part (a).

58. Uniform Motion A Cessna (heading south at 120 mph) and a Boeing 747 (heading west at 600 mph) are flying toward the same point at the same altitude. The Cessna is 100 miles from the point where the flight patterns intersect, and the 747 is 550 miles from this intersection point. See the figure.

(a) Find parametric equations that describe the motion of the Cessna and the 747.

(b) Find a formula for the distance between the planes as a function of time.

(c) Graph the function in part (b) using a graphing utility.

(d) What is the minimum distance between the planes? When are the planes closest?

(e) Simulate the motion of the planes by simultaneously graphing the equations found in part (a).

59. The Green Monster The left field wall at Fenway Park is 310 feet from home plate; the wall itself (affectionately named The Green Monster) is 37 feet high. A batted ball must clear the wall to be a home run. Suppose a ball leaves the bat 3 feet off the ground, at an angle of 45°. Use $g = 32$ ft/sec^2 as the acceleration due to gravity and ignore any air resistance.

(a) Find parametric equations that describe the position of the ball as a function of time.

(b) What is the maximum height of the ball if it leaves the bat with a speed of 90 miles per hour? Give answer in feet.

(c) How far is the ball from home plate at its maximum height? Give answer in feet.

(d) If the ball is hit straight down the left field wall, will it clear the "Green Monster"? If it does, by how much does it clear the wall?

Source: The Boston Red Sox

60. Projectile Motion The position of a projectile fired with an initial velocity v_0 feet per second and at an angle θ to the horizontal at the end of t seconds is given by the parametric equations

$$x = (v_0 \cos \theta)t \qquad y = (v_0 \sin \theta)t - 16t^2$$

See the following illustration.

(a) Obtain the rectangular equation of the trajectory and identify the curve.

(b) Show that the projectile hits the ground ($y = 0$) when $t = \dfrac{1}{16}v_0 \sin \theta$.

(c) How far has the projectile traveled (horizontally) when it strikes the ground? In other words, find the range R.

(d) Find the time t when $x = y$. Then find the horizontal distance x and the vertical distance y traveled by the projectile in this time. Then compute $\sqrt{x^2 + y^2}$. This is the distance R, the range, that the projectile travels up a plane inclined at 45° to the horizontal ($x = y$).

See the following illustration. (See also Problem 83 in Section 8.5.)

61. Show that the parametric equations for a line passing through the points (x_1, y_1) and (x_2, y_2) are

$$x = (x_2 - x_1)t + x_1$$
$$y = (y_2 - y_1)t + y_1, \quad -\infty < t < \infty$$

What is the orientation of this line?

62. Hypocycloid The hypocycloid is a curve defined by the parametric equations

$$x(t) = \cos^3 t, \quad y(t) = \sin^3 t, \quad 0 \le t \le 2\pi$$

(a) Graph the hypocycloid using a graphing utility.

(b) Find rectangular equations of the hypocycloid.

Discussion and Writing

63. In Problem 62, we graphed the hypocycloid. Now graph the rectangular equations of the hypocycloid. Did you obtain a complete graph? If not, experiment until you do.

64. Look up the curves called *hypocycloid* and *epicycloid*. Write a report on what you find. Be sure to draw comparisons with the cycloid.

'Are You Prepared?' Answers

1. $3; \dfrac{\pi}{2}$

CHAPTER REVIEW

Things to Know

Equations

Parabola (pp. 773–778)	See Tables 1 and 2 (pp. 775 and 776).
Ellipse (pp. 782–788)	See Table 3 (p. 787).
Hyperbola (pp. 792–800)	See Table 4 (p. 799).
General equation of a conic (p. 810)	$Ax^2 + Bxy + Cy^2 + Dx + Ey + F = 0$ Parabola if $B^2 - 4AC = 0$
	Ellipse (or circle) if $B^2 - 4AC < 0$
	Hyperbola if $B^2 - 4AC > 0$
Polar equations of a conic with focus at the pole (pp. 812–816)	See Table 5 (p. 815).
Parametric equations of a curve (p. 818)	$x = f(t), y = g(t), t$ is the parameter

Definitions

Parabola (p. 773)	Set of points P in the plane for which $d(F, P) = d(P, D)$, where F is the focus and D is the directrix
Ellipse (p. 782)	Set of points P in the plane, the sum of whose distances from two fixed points (the foci) is a constant
Hyperbola (p. 792)	Set of points P in the plane, the difference of whose distances from two fixed points (the foci) is a constant
Conic in polar coordinates (p. 812)	$\dfrac{d(F, P)}{d(P, D)} = e$ Parabola if $e = 1$
	Ellipse if $e < 1$
	Hyperbola if $e > 1$

Formulas

Rotation formulas (p. 806)	$x = x' \cos \theta - y' \sin \theta$
	$y = x' \sin \theta + y' \cos \theta$
Angle θ of rotation that eliminates the $x'y'$-term (p. 807)	$\cot(2\theta) = \dfrac{A - C}{B}, \quad 0° < \theta < 90°$

Objectives

Section		You should be able to . . .	Review Exercises
11.1	1	Know the names of the conics (p. 772)	1–32
11.2	1	Analyze parabolas with vertex at the origin (p. 773)	1, 2, 21, 24
	2	Analyze parabolas with vertex at (h, k) (p. 776)	7, 11, 12, 17, 18, 27, 30
	3	Solve applied problems involving parabolas (p. 778)	77, 78
11.3	1	Analyze ellipses with center at the origin (p. 782)	5, 6, 10, 22, 25
	2	Analyze ellipses with center at (h, k) (p. 786)	14–16, 19, 28, 31
	3	Solve applied problems involving ellipses (p. 788)	79, 80
11.4	1	Analyze hyperbolas with center at the origin (p. 792)	3, 4, 8, 9, 23, 26
	2	Find the asymptotes of a hyperbola (p. 797)	3, 4, 8, 9
	3	Analyze hyperbolas with center at (h, k) (p. 798)	13, 20, 29, 32–36
	4	Solve applied problems involving hyperbolas (p. 800)	81
11.5	1	Identify a conic (p. 804)	37–40
	2	Use a rotation of axes to transform equations (p. 805)	47–52
	3	Analyze an equation using a rotation of axes (p. 808)	47–52
	4	Identify conics without a rotation of axes (p. 810)	41–46
11.6	1	Analyze and graph polar equations of conics (p. 812)	53–58
	2	Convert the polar equation of a conic to a rectangular equation (p. 816)	59–62
11.7	1	Graph parametric equations (p. 818)	63–68
	2	Find a rectangular equation for a curve defined parametrically (p. 819)	63–68
	3	Use time as a parameter in parametric equations (p. 821)	82, 83
	4	Find parametric equations for curves defined by rectangular equations (p. 824)	69–72

Review Exercises

In Problems 1–20, identify each equation. If it is a parabola, give its vertex, focus, and directrix; if it is an ellipse, give its center, vertices, and foci; if it is a hyperbola, give its center, vertices, foci, and asymptotes.

1. $y^2 = -16x$ **2.** $16x^2 = y$ **3.** $\dfrac{x^2}{25} - y^2 = 1$ **4.** $\dfrac{y^2}{25} - x^2 = 1$

5. $\dfrac{y^2}{25} + \dfrac{x^2}{16} = 1$ **6.** $\dfrac{x^2}{9} + \dfrac{y^2}{16} = 1$ **7.** $x^2 + 4y = 4$ **8.** $3y^2 - x^2 = 9$

9. $4x^2 - y^2 = 8$ **10.** $9x^2 + 4y^2 = 36$ **11.** $x^2 - 4x = 2y$ **12.** $2y^2 - 4y = x - 2$

13. $y^2 - 4y - 4x^2 + 8x = 4$ **14.** $4x^2 + y^2 + 8x - 4y + 4 = 0$ **15.** $4x^2 + 9y^2 - 16x - 18y = 11$

16. $4x^2 + 9y^2 - 16x + 18y = 11$ **17.** $4x^2 - 16x + 16y + 32 = 0$ **18.** $4y^2 + 3x - 16y + 19 = 0$

19. $9x^2 + 4y^2 - 18x + 8y = 23$ **20.** $x^2 - y^2 - 2x - 2y = 1$

In Problems 21–36, find an equation of the conic described. Graph the equation.

21. Parabola; focus at $(-2, 0)$; directrix the line $x = 2$

22. Ellipse; center at $(0, 0)$; focus at $(0, 3)$; vertex at $(0, 5)$

23. Hyperbola; center at $(0,0)$; focus at $(0,4)$; vertex at $(0,-2)$

24. Parabola; vertex at $(0, 0)$; directrix the line $y = -3$

25. Ellipse; foci at $(-3, 0)$ and $(3, 0)$; vertex at $(4, 0)$

26. Hyperbola; vertices at $(-2, 0)$ and $(2, 0)$; focus at $(4, 0)$

27. Parabola; vertex at $(2, -3)$; focus at $(2, -4)$

28. Ellipse; center at $(-1, 2)$; focus at $(0, 2)$; vertex at $(2, 2)$

29. Hyperbola; center at $(-2, -3)$; focus at $(-4, -3)$; vertex at $(-3, -3)$

30. Parabola; focus at $(3, 6)$; directrix the line $y = 8$

31. Ellipse; foci at $(-4, 2)$ and $(-4, 8)$; vertex at $(-4, 10)$

32. Hyperbola; vertices at $(-3, 3)$ and $(5, 3)$; focus at $(7, 3)$

33. Center at $(-1, 2)$; $a = 3$; $c = 4$; transverse axis parallel to the x-axis

34. Center at $(4, -2)$; $a = 1$; $c = 4$; transverse axis parallel to the y-axis

35. Vertices at $(0, 1)$ and $(6, 1)$; asymptote the line $3y + 2x = 9$

36. Vertices at $(4, 0)$ and $(4, 4)$; asymptote the line $y + 2x = 10$

In Problems 37–46, identify each conic without completing the squares and without applying a rotation of axes.

37. $y^2 + 4x + 3y - 8 = 0$

38. $2x^2 - y + 8x = 0$

39. $x^2 + 2y^2 + 4x - 8y + 2 = 0$

40. $x^2 - 8y^2 - x - 2y = 0$

41. $9x^2 - 12xy + 4y^2 + 8x + 12y = 0$

42. $4x^2 + 4xy + y^2 - 8\sqrt{5}x + 16\sqrt{5}y = 0$

43. $4x^2 + 10xy + 4y^2 - 9 = 0$

44. $4x^2 - 10xy + 4y^2 - 9 = 0$

45. $x^2 - 2xy + 3y^2 + 2x + 4y - 1 = 0$

46. $4x^2 + 12xy - 10y^2 + x + y - 10 = 0$

In Problems 47–52, rotate the axes so that the new equation contains no xy-term. Analyze and graph the new equation.

47. $2x^2 + 5xy + 2y^2 - \dfrac{9}{2} = 0$

48. $2x^2 - 5xy + 2y^2 - \dfrac{9}{2} = 0$

49. $6x^2 + 4xy + 9y^2 - 20 = 0$

50. $x^2 + 4xy + 4y^2 + 16\sqrt{5}x - 8\sqrt{5}y = 0$

51. $4x^2 - 12xy + 9y^2 + 12x + 8y = 0$

52. $9x^2 - 24xy + 16y^2 + 80x + 60y = 0$

In Problems 53–58, identify the conic that each polar equation represents and graph it.

53. $r = \dfrac{4}{1 - \cos\theta}$

54. $r = \dfrac{6}{1 + \sin\theta}$

55. $r = \dfrac{6}{2 - \sin\theta}$

56. $r = \dfrac{2}{3 + 2\cos\theta}$

57. $r = \dfrac{8}{4 + 8\cos\theta}$

58. $r = \dfrac{10}{5 + 20\sin\theta}$

In Problems 59–62, convert each polar equation to a rectangular equation.

59. $r = \dfrac{4}{1 - \cos\theta}$

60. $r = \dfrac{6}{2 - \sin\theta}$

61. $r = \dfrac{8}{4 + 8\cos\theta}$

62. $r = \dfrac{2}{3 + 2\cos\theta}$

In Problems 63–68, graph the curve whose parametric equations are given and show its orientation. Find the rectangular equation of each curve.

63. $x = 4t - 2, \quad y = 1 - t; \quad -\infty < t < \infty$

64. $x = 2t^2 + 6, \quad y = 5 - t; \quad -\infty < t < \infty$

65. $x = 3\sin t, \quad y = 4\cos t + 2; \quad 0 \le t \le 2\pi$

66. $x = \ln t, \quad y = t^3; \quad t > 0$

67. $x = \sec^2 t, \quad y = \tan^2 t; \quad 0 \le t \le \dfrac{\pi}{4}$

68. $x = t^{\frac{3}{2}}, \quad y = 2t + 4; \quad t \ge 0$

In Problems 69 and 70, find two different parametric equations for each rectangular equation.

69. $y = -2x + 4$

70. $y = 2x^2 - 8$

In Problems 71 and 72, find parametric equations for an object that moves along the ellipse $\dfrac{x^2}{16} + \dfrac{y^2}{9} = 1$ with the motion described.

71. The motion begins at $(4, 0)$, is counterclockwise, and requires 4 seconds for a complete revolution.

72. The motion begins at $(0, 3)$, is clockwise, and requires 5 seconds for a complete revolution.

73. Find an equation of the hyperbola whose foci are the vertices of the ellipse $4x^2 + 9y^2 = 36$ and whose vertices are the foci of this ellipse.

74. Find an equation of the ellipse whose foci are the vertices of the hyperbola $x^2 - 4y^2 = 16$ and whose vertices are the foci of this hyperbola.

75. Describe the collection of points in a plane so that the distance from each point to the point $(3, 0)$ is three-fourths of its distance from the line $x = \dfrac{16}{3}$.

76. Describe the collection of points in a plane so that the distance from each point to the point $(5, 0)$ is five-fourths of its distance from the line $x = \dfrac{16}{5}$.

77. Searchlight A searchlight is shaped like a paraboloid of revolution. If a light source is located 1 foot from the vertex

along the axis of symmetry and the opening is 2 feet across, how deep should the mirror be in order to reflect the light rays parallel to the axis of symmetry?

78. Parabolic Arch Bridge A bridge is built in the shape of a parabolic arch. The bridge has a span of 60 feet and a maximum height of 20 feet. Find the height of the arch at distances of 5, 10, and 20 feet from the center.

79. Semi elliptical Arch Bridge A bridge is built in the shape of a semielliptical arch. The bridge has a span of 60 feet and a maximum height of 20 feet. Find the height of the arch at distances of 5, 10, and 20 feet from the center.

80. Whispering Gallery The figure shows the specifications for an elliptical ceiling in a hall designed to be a whispering gallery. Where are the foci located in the hall?

81. Calibrating Instruments In a test of their recording devices, a team of seismologists positioned two of the devices 2000 feet apart, with the device at point A to the west of the device at point B. At a point between the devices and 200 feet from point B, a small amount of explosive was detonated and a note made of the time at which the sound reached each device. A second explosion is to be carried out at a point directly north of point B. How far north should the site of the second explosion be chosen so that the measured time difference recorded by the devices for the second detonation is the same as that recorded for the first detonation?

82. Uniform Motion Mary's train leaves at 7:15 AM and accelerates at the rate of 3 meters per second per second. Mary, who can run 6 meters per second, arrives at the train station 2 seconds after the train has left.

(a) Find parametric equations that describe the motion of the train and Mary as a function of time.
[**Hint:** The position s at time t of an object having acceleration a is $s = \dfrac{1}{2}at^2$.]

(b) Determine algebraically whether Mary will catch the train. If so, when?

(c) Simulate the motions of the train and Mary by simultaneously graphing the equations found in part (a).

83. Projectile Motion Drew Bledsoe throws a football with an initial speed of 80 feet per second at an angle of 35° to the horizontal. The ball leaves Drew Bledsoe's hand at a height of 6 feet.

(a) Find parametric equations that describe the position of the ball as a function of time.

(b) How long is the ball in the air?

(c) When is the ball at its maximum height? Determine the maximum height of the ball.

(d) Determine the horizontal distance that the ball travels.

(e) Using a graphing utility, simultaneously graph the equations found in part (a).

84. Formulate a strategy for discussing and graphing an equation of the form

$$Ax^2 + Bxy + Cy^2 + Dx + Ey + F = 0$$

CHAPTER TEST

In Problems 1–3, identify each equation. If it is a parabola, give its vertex, focus, and directrix; if an ellipse, give its center, vertices, and foci; if a hyperbola, give its center, vertices, foci, and asymptotes.

1. $\dfrac{(x+1)^2}{4} - \dfrac{y^2}{9} = 1$

2. $8y = (x-1)^2 - 4$

3. $2x^2 + 3y^2 + 4x - 6y = 13$

In Problems 4–6, find an equation of the conic described; graph the equation.

4. Parabola: focus $(-1, 4.5)$, vertex $(-1, 3)$

5. Ellipse: center $(0, 0)$, vertex $(0, -4)$, focus $(0, 3)$

6. Hyperbola: center $(2, 2)$, vertex $(2, 4)$, contains the point $\left(2 + \sqrt{10}, 5\right)$

In Problems 7–9, identify each conic without completing the square or rotating axes.

7. $2x^2 + 5xy + 3y^2 + 3x - 7 = 0$

8. $3x^2 - xy + 2y^2 + 3y + 1 = 0$

9. $x^2 - 6xy + 9y^2 + 2x - 3y - 2 = 0$

10. Given the equation $41x^2 - 24xy + 34y^2 - 25 = 0$, rotate the axes so that there is no xy-term. Analyze and graph the new equation.

11. Identify the conic represented by the polar equation $r = \dfrac{3}{1 - 2\cos\theta}$. Find the rectangular equation.

12. Graph the curve whose parametric equations are given and show its orientation. Find the rectangular equation for the curve.

$$x = 3t - 2, \quad y = 1 - \sqrt{t}, \quad 0 \le t \le 9$$

13. A parabolic reflector (paraboloid of revolution) is used by TV crews at football games to pick up the referee's announcements, quarterback signals, and so on. A microphone is placed at the focus of the parabola. If a certain reflector is 4 feet wide and 1.5 feet deep, where should the microphone be placed?

CUMULATIVE REVIEW

1. For $f(x) = -3x^2 + 5x - 2$, find

$$\dfrac{f(x+h) - f(x)}{h} \quad h \neq 0$$

2. In the complex number system, solve the equation

$$9x^4 + 33x^3 - 71x^2 - 57x - 10 = 0.$$

3. For what numbers x is $6 - x \ge x^2$?

4. (a) Find the domain and range of $y = 3^x + 2$.
(b) Find the inverse of $y = 3^x + 2$ and state its domain and range.

5. If $f(x) = \log_4(x - 2)$:
(a) Solve $f(x) = 2$.
(b) Solve $f(x) \le 2$.

6. Find an equation for each of the following graphs:

(a) Line:

(b) Circle:

(c) Ellipse:

(d) Parabola:

(e) Hyperbola:

(f) Exponential:

7. Find all the solutions of the equation $\sin(2\theta) = 0.5$.

8. Find a polar equation for the line containing the origin that makes an angle of $30°$ with the positive x-axis.

9. Find a polar equation for the circle with center at the point $(0, 4)$ and radius 4. Graph this circle.

10. What is the domain of the function $f(x) = \dfrac{3}{\sin x + \cos x}$?

11. Solve the equation $\cot(2\theta) = 1$, where $0° < \theta < 90°$.

12. Find the rectangular equation of the curve
$$\begin{cases} x = 5 \tan t \\ y = 5 \sec^2 t \end{cases} \quad -\frac{\pi}{2} < t < \frac{\pi}{2}.$$

CHAPTER PROJECTS

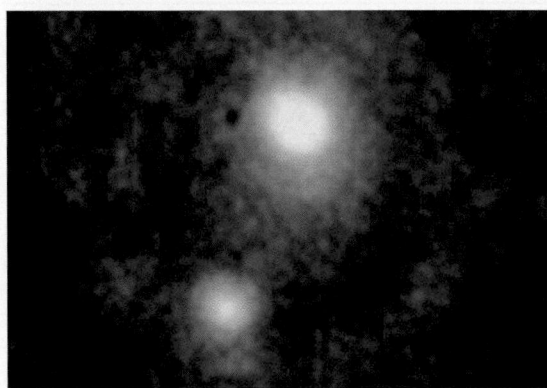

I. **The Orbits of Neptune and Pluto** The orbit of a planet about the Sun is an ellipse, with the Sun at one focus. The **aphelion** of a planet is its greatest distance from the Sun and the **perihelion** is its shortest distance. The **mean distance** of a planet from the Sun is the length of the semimajor axis of the elliptical orbit. See the illustration.

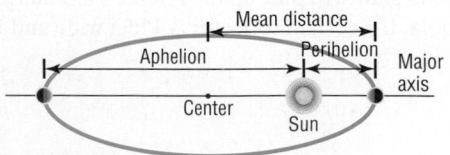

1. The aphelion of Neptune is 4532.2×10^6 km and its perihelion is 4458.0×10^6 km. Find the equation for the orbit of Neptune around the Sun.

2. The aphelion of Pluto* is 7381.2×10^6 km and its perihelion is 4445.8×10^6 km. Find the equation for the orbit of Pluto around the Sun.

3. Graph the orbits of Pluto and Neptune on a graphing utility. Knowing that the orbits of the planets intersect, what is wrong with the graphs you obtained?

4. The graphs of the orbits drawn in part 3 have the same center, but their foci lie in different locations. To see an accurate representation, the location of the Sun (a focus) needs to be the same for both graphs. This can be accomplished by shifting Pluto's orbit to the left. The shift amount is equal to Pluto's distance from the center [in the graph in part 3] to the Sun minus Neptune's distance from the center to the Sun. Find the new equation representing the orbit of Pluto.

5. Graph the equation for the orbit of Pluto found in part 4 along with the equation of the orbit of Neptune. Do you see that Pluto's orbit is sometimes inside Neptune's?

6. Find the point(s) of intersection of the two orbits.

7. Do you think two planets will ever collide?

The following projects can be found on the Instructor's Resource Center (IRC):

II. **Project at Motorola** *Distorted Deployable Space Reflector Antennas* An engineer designs an antenna that will deploy in space to collect sunlight.

III. **Constructing a Bridge over the East River** The size of ships using a river and fluctuations in water height due to tides or flooding must be considered when designing a bridge that will cross a major waterway.

IV. **Systems of Parametric Equations** Choosing an approach to use when solving a system of equations depends on the form of the system and on the domains of the equations.

*Pluto's status was reduced to a dwarf planet in September 2006.

Systems of Equations and Inequalities

<div style="text-align: right">**12**</div>

Economic Outcomes

Annual Earnings of Young Adults

For both males and females, earnings increase with education: full-time workers with at least a bachelor's degree have higher median earnings than those with less education. For example, in 2003, male college graduates earned 93 percent more than male high school completers. Females with a bachelor's or higher degree earned 91 percent more than female high school completers. Males and females who dropped out of high school earned 37 and 39 percent less, respectively, than male and female high school completers.

The median earnings of young adults who have at least a bachelor's degree declined in the 1970s relative to their counterparts who were high school completers, before increasing between 1980 and 2003. Males with a bachelor's degree or higher had earnings 19 percent higher than male high school completers in 1980 and had earnings 93 percent higher in 2003. Among females, those with at least a bachelor's degree had earnings 34 percent higher than female high school completers in 1980, compared with earnings 91 percent higher in 2003.

—See the Chapter Project—

A Look Back

In Chapters 1, 4, 5, 6, and 8, we solved various kinds of equations and inequalities involving a single variable.

A Look Ahead

In this chapter we take up the problem of solving equations and inequalities containing two or more variables. There are various ways to solve such problems:

The *method of substitution* for solving equations in several unknowns goes back to ancient times.

The *method of elimination*, although it had existed for centuries, was put into systematic order by Karl Friedrich Gauss (1777–1855) and by Camille Jordan (1838–1922).

The theory of *matrices* was developed in 1857 by Arthur Cayley (1821–1895), although only later were matrices used as we use them in this chapter. Matrices have become a very flexible instrument, useful in almost all areas of mathematics.

The method of *determinants* was invented by Takakazu Seki Kôwa (1642–1708) in 1683 in Japan and by Gottfried Wilhelm von Leibniz (1646–1716) in 1693 in Germany. *Cramer's Rule* is named after Gabriel Cramer (1704–1752) of Switzerland, who popularized the use of determinants for solving linear systems.

Section 12.5, on *partial fraction decomposition*, provides an application of systems of equations. This particular application is one that is used in integral calculus.

Section 12.8 introduces *linear programming*, a modern application of linear inequalities. This topic is particularly useful for students interested in operations research.

Outline

835

12.1 Systems of Linear Equations: Substitution and Elimination

PREPARING FOR THIS SECTION *Before getting started, review the following:*

• Linear Equations (Section 1.1, pp. 86–94) • Lines (Section 2.3, pp. 173–185)

Now Work the 'Are You Prepared?' problems on page 847.

OBJECTIVES 1 Solve Systems of Equations by Substitution (p. 838)
2 Solve Systems of Equations by Elimination (p. 840)
3 Identify Inconsistent Systems of Equations Containing Two Variables (p. 841)
4 Express the Solution of a System of Dependent Equations Containing Two Variables (p. 842)
5 Solve Systems of Three Equations Containing Three Variables (p. 843)
6 Identify Inconsistent Systems of Equations Containing Three Variables (p. 844)
7 Express the Solution of a System of Dependent Equations Containing Three Variables (p. 845)

We begin with an example.

EXAMPLE 1 | **Movie Theater Ticket Sales**

A movie theater sells tickets for $8.00 each, with seniors receiving a discount of $2.00. One evening the theater took in $3580 in revenue. If x represents the number of tickets sold at $8.00 and y the number of tickets sold at the discounted price of $6.00, write an equation that relates these variables.

Solution Each nondiscounted ticket brings in $8.00, so x tickets will bring in $8x$ dollars. Similarly, y discounted tickets bring in $6y$ dollars. Since the total brought in is $3580, we must have

$$8x + 6y = 3580$$

∎

In Example 1, suppose that we also know that 525 tickets were sold that evening. Then we have another equation relating the variables x and y:

$$x + y = 525$$

The two equations

$$8x + 6y = 3580$$
$$x + y = 525$$

form a *system* of equations.

In general, a **system of equations** is a collection of two or more equations, each containing one or more variables. Example 2 gives some illustrations of systems of equations.

EXAMPLE 2 | **Examples of Systems of Equations**

(a) $\begin{cases} 2x + y = 5 & (1) \\ -4x + 6y = -2 & (2) \end{cases}$ Two equations containing two variables, x and y

(b) $\begin{cases} x + y^2 = 5 & (1) \\ 2x + y = 4 & (2) \end{cases}$ Two equations containing two variables, x and y

$$(c) \begin{cases} x + y + z = 6 & (1) \\ 3x - 2y + 4z = 9 & (2) \\ x - y - z = 0 & (3) \end{cases}$$ Three equations containing three variables, x, y, and z

$$(d) \begin{cases} x + y + z = 5 & (1) \\ x - y \phantom{{}+ z} = 2 & (2) \end{cases}$$ Two equations containing three variables, x, y, and z

$$(e) \begin{cases} x + y + z = 6 & (1) \\ 2x \phantom{{}+ y} + 2z = 4 & (2) \\ \phantom{2x +{}} y + z = 2 & (3) \\ x \phantom{{}+ y + 2z} = 4 & (4) \end{cases}$$ Four equations containing three variables, x, y, and z

We use a brace, as shown, to remind us that we are dealing with a system of equations. We also will find it convenient to number each equation in the system.

A **solution** of a system of equations consists of values for the variables that are solutions of each equation of the system. To **solve** a system of equations means to find all solutions of the system.

For example, $x = 2$, $y = 1$ is a solution of the system in Example 2(a), because

$$\begin{cases} 2x + y = 5 & (1) \\ -4x + 6y = -2 & (2) \end{cases} \qquad \begin{cases} 2(2) + 1 = 4 + 1 = 5 \\ -4(2) + 6(1) = -8 + 6 = -2 \end{cases}$$

We may also write this solution as the ordered pair $(2, 1)$.

A solution of the system in Example 2(b) is $x = 1$, $y = 2$, because

$$\begin{cases} x + y^2 = 5 & (1) \\ 2x + y = 4 & (2) \end{cases} \qquad \begin{cases} 1 + 2^2 = 1 + 4 = 5 \\ 2(1) + 2 = 2 + 2 = 4 \end{cases}$$

Another solution of the system in Example 2(b) is $x = \dfrac{11}{4}$, $y = -\dfrac{3}{2}$, which you can check for yourself.

A solution of the system in Example 2(c) is $x = 3$, $y = 2$, $z = 1$, because

$$\begin{cases} x + y + z = 6 & (1) \\ 3x - 2y + 4z = 9 & (2) \\ x - y - z = 0 & (3) \end{cases} \qquad \begin{cases} 3 + 2 + 1 = 6 & (1) \\ 3(3) - 2(2) + 4(1) = 9 - 4 + 4 = 9 & (2) \\ 3 - 2 - 1 = 0 & (3) \end{cases}$$

We may also write this solution as the ordered triplet $(3, 2, 1)$.

Note that $x = 3$, $y = 3$, $z = 0$ is not a solution of the system in Example 2(c).

$$\begin{cases} x + y + z = 6 & (1) \\ 3x - 2y + 4z = 9 & (2) \\ x - y - z = 0 & (3) \end{cases} \qquad \begin{cases} 3 + 3 + 0 = 6 & (1) \\ 3(3) - 2(3) + 4(0) = 3 \neq 9 & (2) \\ 3 - 3 - 0 = 0 & (3) \end{cases}$$

Although $x = 3$, $y = 3$, and $z = 0$ satisfy equations (1) and (3), they do not satisfy equation (2). Any solution of the system must satisfy *each* equation of the system.

Now Work PROBLEM 9

When a system of equations has at least one solution, it is said to be **consistent;** otherwise, it is called **inconsistent.**

An equation in n variables is said to be **linear** if it is equivalent to an equation of the form

$$a_1 x_1 + a_2 x_2 + \cdots + a_n x_n = b$$

where $x_1, x_2, \ldots, x_n$ are n distinct variables, $a_1, a_2, \ldots, a_n, b$ are constants, and at least one of the a's is not 0.

Some examples of linear equations are

$$2x + 3y = 2 \qquad 5x - 2y + 3z = 10 \qquad 8x + 8y - 2z + 5w = 0$$

If each equation in a system of equations is linear, we have a **system of linear equations.** The systems in Examples 2(a), (c), (d), and (e) are linear, whereas the system in Example 2(b) is nonlinear. In this chapter we shall solve linear systems in Sections 12.1 to 12.3. We discuss nonlinear systems in Section 12.6.

We begin by discussing a system of two linear equations containing two variables. We can view the problem of solving such a system as a geometry problem. The graph of each equation in such a system is a line. So a system of two linear equations containing two variables represents a pair of lines. The lines either (1) intersect or (2) are parallel or (3) are **coincident** (that is, identical).

1. If the lines intersect, the system of equations has one solution, given by the point of intersection. The system is **consistent** and the equations are **independent.** See Figure 1(a).

2. If the lines are parallel, the system of equations has no solution, because the lines never intersect. The system is **inconsistent.** See Figure 1(b).

3. If the lines are coincident, the system of equations has infinitely many solutions, represented by the totality of points on the line. The system is **consistent** and the equations are **dependent.** See Figure 1(c).

Figure 1

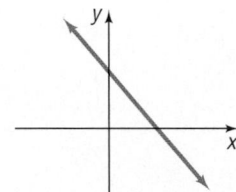

(a) Intersecting lines; system has one solution

(b) Parallel lines; system has no solution

(c) Coincident lines; system has infinitely many solutions

EXAMPLE 3	**Graphing a System of Linear Equations**

Graph the system: $\begin{cases} 2x + y = 5 & (1) \\ -4x + 6y = 12 & (2) \end{cases}$

Figure 2

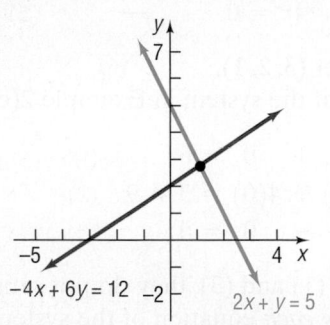

Solution Equation (1) in slope–intercept form is $y = -2x + 5$, which has slope -2 and y-intercept 5. Equation (2) in slope–intercept form is $y = \frac{2}{3}x + 2$, which has slope $\frac{2}{3}$ and y-intercept 2. Figure 2 shows their graphs. ∎

From the graph in Figure 2, we see that the lines intersect, so the system given in Example 3 is consistent. We can also use the graph as a means of approximating the solution. For this system, the solution appears to be close to the point $(1, 3)$. The actual solution, which you should verify, is $\left(\dfrac{9}{8}, \dfrac{11}{4}\right)$.

> **Seeing the Concept**
> Graph the lines $2x + y = 5$ ($Y_1 = -2x + 5$) and $-4x + 6y = 12$ $\left(Y_2 = \dfrac{2}{3}x + 2\right)$ and compare what you see with Figure 2. Use INTERSECT to verify that the point of intersection is $(1.125, 2.75)$.

1 Solve Systems of Equations by Substitution

To obtain exact solutions, we use algebraic methods. A number of methods are available to us for solving systems of linear equations algebraically. In this section, we introduce two methods: *substitution* and *elimination.* We illustrate the **method of substitution** by solving the system given in Example 3.

EXAMPLE 4	**Solving a System of Linear Equations by Substitution**

Solve: $\begin{cases} 2x + y = 5 & (1) \\ -4x + 6y = 12 & (2) \end{cases}$

Solution We solve the first equation for y, obtaining

$$2x + y = 5 \qquad (1)$$
$$y = -2x + 5 \quad \text{Subtract 2x from each side of (1).}$$

We substitute this result for y into the second equation. The result is an equation containing just the variable x, which we can then solve.

$$-4x + 6y = 12 \qquad (2)$$
$$-4x + 6(-2x + 5) = 12 \qquad \text{Substitute } y = -2x + 5 \text{ in (2).}$$
$$-4x - 12x + 30 = 12 \qquad \text{Remove parentheses.}$$
$$-16x = -18 \qquad \text{Combine like terms and subtract 30 from both sides.}$$
$$x = \frac{-18}{-16} = \frac{9}{8} \qquad \text{Divide each side by } -16 \text{ and simplify.}$$

Once we know that $x = \dfrac{9}{8}$, we can find the value of y by **back-substitution,** that is, by substituting $\dfrac{9}{8}$ for x in one of the original equations. We will use the first equation.

$$2x + y = 5 \qquad (1)$$
$$y = -2x + 5 \qquad \text{Subtract 2x from each side.}$$
$$y = -2\left(\frac{9}{8}\right) + 5 \qquad \text{Substitute } x = \frac{9}{8} \text{ in (1).}$$
$$= -\frac{9}{4} + \frac{20}{4} = \frac{11}{4}$$

The solution of the system is $x = \dfrac{9}{8} = 1.125$, $y = \dfrac{11}{4} = 2.75$. We can also write the solution as the ordered pair $\left(\dfrac{9}{8}, \dfrac{11}{4}\right)$.

✓ **Check:** $\begin{cases} 2x + y = 5: & 2\left(\dfrac{9}{8}\right) + \dfrac{11}{4} = \dfrac{9}{4} + \dfrac{11}{4} = \dfrac{20}{4} = 5 \\ -4x + 6y = 12: & -4\left(\dfrac{9}{8}\right) + 6\left(\dfrac{11}{4}\right) = -\dfrac{9}{2} + \dfrac{33}{2} = \dfrac{24}{2} = 12 \end{cases}$

The method used to solve the system in Example 4 is called **substitution.** The steps to be used are outlined next.

Steps for Solving by Substitution

STEP 1: Pick one of the equations and solve for one of the variables in terms of the remaining variables.

STEP 2: Substitute the result into the remaining equations.

STEP 3: If one equation in one variable results, solve this equation. Otherwise repeat Steps 1 and 2 until a single equation with one variable remains.

STEP 4: Find the values of the remaining variables by back-substitution.

STEP 5: Check the solution found.

Now Use Substitution to Work PROBLEM 19

2 Solve Systems of Equations by Elimination

A second method for solving a system of linear equations is the *method of elimination*. This method is usually preferred over substitution if substitution leads to fractions or if the system contains more than two variables. Elimination also provides the necessary motivation for solving systems using matrices (the subject of Section 12.2).

The idea behind the method of elimination is to replace the original system of equations by an equivalent system so that adding two of the equations eliminates a variable. The rules for obtaining equivalent equations are the same as those studied earlier. However, we may also interchange any two equations of the system and/or replace any equation in the system by the sum (or difference) of that equation and a nonzero multiple of any other equation in the system.

> **In Words**
>
> When using elimination, we want to get the coefficients of one of the variables to be negatives of one another.

Rules for Obtaining an Equivalent System of Equations

1. Interchange any two equations of the system.

2. Multiply (or divide) each side of an equation by the same nonzero constant.

3. Replace any equation in the system by the sum (or difference) of that equation and a nonzero multiple of any other equation in the system.

An example will give you the idea. As you work through the example, pay particular attention to the pattern being followed.

EXAMPLE 5 | **Solving a System of Linear Equations by Elimination**

Solve: $\begin{cases} 2x + 3y = 1 & (1) \\ -x + y = -3 & (2) \end{cases}$

Solution

We multiply each side of equation (2) by 2 so that the coefficients of x in the two equations are negatives of one another. The result is the equivalent system

$$\begin{cases} 2x + 3y = 1 & (1) \\ -2x + 2y = -6 & (2) \end{cases}$$

If we add equations (1) and (2), we obtain an equation containing just the variable y, which we can solve.

$$\begin{cases} 2x + 3y = 1 & (1) \\ \underline{-2x + 2y = -6} & (2) \\ 5y = -5 & \text{Add (1) and (2).} \\ y = -1 & \text{Solve for } y. \end{cases}$$

We back-substitute this value for y in equation (1) and simplify.

$$\begin{aligned} 2x + 3y &= 1 & (1) \\ 2x + 3(-1) &= 1 & \text{Substitute } y = -1 \text{ in (1).} \\ 2x &= 4 & \text{Simplify.} \\ x &= 2 & \text{Solve for } x. \end{aligned}$$

The solution of the original system is $x = 2$, $y = -1$ or using ordered pairs $(2, -1)$. We leave it to you to check the solution.

The procedure used in Example 5 is called the **method of elimination.** Notice the pattern of the solution. First, we eliminated the variable x from the second equation. Then we back-substituted; that is, we substituted the value found for y back into the first equation to find x.

Now Use Elimination to Work PROBLEM 19

Let's return to the movie theater example of Example 1.

| **EXAMPLE 6** | **Movie Theater Ticket Sales** |

A movie theater sells tickets for $8.00 each, with seniors receiving a discount of $2.00. One evening the theater sold 525 tickets and took in $3580 in revenue. How many of each type of ticket were sold?

Solution If x represents the number of tickets sold at $8.00 and y the number of tickets sold at the discounted price of $6.00, the given information results in the system of equations

$$\begin{cases} 8x + 6y = 3580 & (1) \\ x + y = 525 & (2) \end{cases}$$

We use the method of elimination. We choose to multiply the second equation by -6, and then add the equations.

$$\begin{cases} 8x + 6y = 3580 & (1) \\ \underline{-6x - 6y = -3150} & (2) \\ 2x = 430 & \text{Add the equations} \\ x = 215 \end{cases}$$

Now we back-substitute $x = 215$ into equation (2).

Since $x + y = 525$, then $y = 525 - x = 525 - 215 = 310$. So 215 nondiscounted tickets and 310 senior discount tickets were sold.

3 Identify Inconsistent Systems of Equations Containing Two Variables

The previous examples dealt with consistent systems of equations that had a single solution. The next two examples deal with two other possibilities that may occur, the first being a system that has no solution.

| **EXAMPLE 7** | **An Inconsistent System of Linear Equations** |

Solve: $\begin{cases} 2x + y = 5 & (1) \\ 4x + 2y = 8 & (2) \end{cases}$

Solution We choose to use the method of substitution and solve equation (1) for y.

$$2x + y = 5 \qquad (1)$$
$$y = -2x + 5 \qquad \text{Subtract } 2x \text{ from each side.}$$

Now substitute $y = -2x + 5$ for y in equation (2) and solve for x.

$$4x + 2y = 8 \qquad (2)$$
$$4x + 2(-2x + 5) = 8 \qquad \text{Substitute } y = -2x + 5 \text{ in (2).}$$
$$4x - 4x + 10 = 8 \qquad \text{Remove parentheses.}$$
$$0 = -2 \qquad \text{Subtract 10 from both sides.}$$

This equation is false. We conclude that the system has no solution and is therefore inconsistent.

Figure 3

Figure 3 illustrates the pair of lines whose equations form the system in Example 7. Notice that the graphs of the two equations are lines, each with slope -2; one has a y-intercept of 5, the other a y-intercept of 4. The lines are parallel and have no point of intersection. This geometric statement is equivalent to the algebraic statement that the system has no solution.

Seeing the Concept

Graph the lines $2x + y = 5$ ($Y_1 = -2x + 5$) and $4x + 2y = 8$ ($Y_2 = -2x + 4$) and compare what you see with Figure 3. How can you be sure that the lines are parallel?

4 Express the Solution of a System of Dependent Equations Containing Two Variables

EXAMPLE 8 | **Solving a System of Dependent Equations**

Solve: $\begin{cases} 2x + y = 4 & (1) \\ -6x - 3y = -12 & (2) \end{cases}$

Solution We choose to use the method of elimination.

$\begin{cases} 2x + y = 4 & (1) \\ -6x - 3y = -12 & (2) \end{cases}$

$\begin{cases} 6x + 3y = 12 & (1) \quad \text{Multiply each side of equation (1) by 3.} \\ -6x - 3y = -12 & (2) \end{cases}$

$\begin{cases} 6x + 3y = 12 & (1) \\ 0 = 0 & (2) \quad \text{Replace equation (2) by the sum of equations (1) and (2).} \end{cases}$

The original system is equivalent to a system containing one equation, so the equations are dependent. This means that any values of x and y for which $6x + 3y = 12$ or, equivalently, $2x + y = 4$ are solutions. For example, $x = 2$, $y = 0$; $x = 0$, $y = 4$; $x = -2$, $y = 8$; $x = 4$, $y = -4$; and so on, are solutions. There are, in fact, infinitely many values of x and y for which $2x + y = 4$, so the original system has infinitely many solutions. We will write the solution of the original system either as

$$y = -2x + 4$$

where x can be any real number, or as

$$x = -\frac{1}{2}y + 2$$

where y can be any real number.

Using ordered pairs, we write the solution as $\{(x, y) \mid y = -2x + 4, x \text{ any real number}\}$ or as $\{(x, y) \mid x = -\frac{1}{2}y + 2, y \text{ any real number}\}$.

Figure 4 illustrates the situation presented in Example 8. Notice that the graphs of the two equations are lines, each with slope -2 and each with y-intercept 4. The lines are coincident. Notice also that equation (2) in the original system is -3 times equation (1), indicating that the two equations are dependent.

For the system in Example 8, we can write down some of the infinite number of solutions by assigning values to x and then finding $y = -2x + 4$.

If $x = -1$, then $y = -2(-1) + 4 = 6$.

If $x = 0$, then $y = 4$.

If $x = 2$, then $y = 0$.

The ordered pairs $(-1, 6)$, $(0, 4)$, and $(2, 0)$ are three of the points on the line in Figure 4.

Figure 4

⌕ **Seeing the Concept**

Graph the lines $2x + y = 4$ ($Y_1 = -2x + 4$) and $-6x - 3y = -12$ ($Y_2 = -2x + 4$) and compare what you see with Figure 4. How can you be sure that the lines are coincident?

━━━━▶ **Now Work** PROBLEMS 25 AND 29

5 Solve Systems of Three Equations Containing Three Variables

Just as with a system of two linear equations containing two variables, a system of three linear equations containing three variables has either (1) exactly one solution (a consistent system with independent equations), or (2) no solution (an inconsistent system), or (3) infinitely many solutions (a consistent system with dependent equations).

We can view the problem of solving a system of three linear equations containing three variables as a geometry problem. The graph of each equation in such a system is a plane in space. A system of three linear equations containing three variables represents three planes in space. Figure 5 illustrates some of the possibilities.

Figure 5

(a) Consistent system; **(b)** Consistent system; **(c)** Inconsistent system;
one solution infinite number of solutions no solution

Recall that a **solution** to a system of equations consists of values for the variables that are solutions of each equation of the system. For example, $x = 3$, $y = -1, z = -5$ or, using ordered triplets, $(3, -1, -5)$, is a solution to the system of equations

$$\begin{cases} x + y + z = -3 & (1) & 3 + (-1) + (-5) = -3 \\ 2x - 3y + 6z = -21 & (2) & 2(3) - 3(-1) + 6(-5) = 6 + 3 - 30 = -21 \\ -3x + 5y = -14 & (3) & -3(3) + 5(-1) = -9 - 5 = -14 \end{cases}$$

because these values of the variables are solutions of each equation.

Typically, when solving a system of three linear equations containing three variables, we use the method of elimination. Recall that the idea behind the method of elimination is to form equivalent equations so that adding two of the equations eliminates a variable.

Let's see how elimination works on a system of three equations containing three variables.

━━

EXAMPLE 9 **Solving a System of Three Linear Equations with Three Variables**

Use the method of elimination to solve the system of equations.

$$\begin{cases} x + y - z = -1 & (1) \\ 4x - 3y + 2z = 16 & (2) \\ 2x - 2y - 3z = 5 & (3) \end{cases}$$

Solution For a system of three equations, we attempt to eliminate one variable at a time, using pairs of equations until an equation with a single variable remains. Our strategy for solving this system will be to use equation (1) to eliminate the variable x from equations (2) and (3). Alternatively, we could use equation (1) to eliminate either y or z from equations (2) and (3). Try one of these approaches for yourself.

We begin by multiplying each side of equation (1) by -4 and adding the result to equation (2). (Do you see why? The coefficients of x are now negatives of one another.) We also multiply equation (1) by -2 and add the result to equation (3). Notice that these two procedures result in the elimination of the variable x from equations (2) and (3).

$$\begin{array}{ll} x + y - z = -1 & (1) \\ 4x - 3y + 2z = 16 & (2) \end{array}$$
Multiply by -4

$$\begin{array}{ll} -4x - 4y + 4z = 4 & (1) \\ 4x - 3y + 2z = 16 & (2) \\ \hline -7y + 6z = 20 & \text{Add.} \end{array}$$

$$\begin{cases} x + y - z = -1 & (1) \\ -7y + 6z = 20 & (2) \\ -4y - z = 7 & (3) \end{cases}$$

$$\begin{array}{ll} x + y - z = -1 & (1) \\ 2x - 2y - 3z = 5 & (3) \end{array}$$
Multiply by -2

$$\begin{array}{ll} -2x - 2y + 2z = 2 & (1) \\ 2x - 2y - 3z = 5 & (3) \\ \hline -4y - z = 7 & \text{Add.} \end{array}$$

We now concentrate on equations (2) and (3), treating them as a system of two equations containing two variables. It is easiest to eliminate z. We multiply each side of equation (3) by 6 and add equations (2) and (3). The result is the new equation (3).

$$\begin{array}{ll} -7y + 6z = 20 & (2) \\ -4y - z = 7 & (3) \end{array}$$
Multiply by 6

$$\begin{array}{ll} -7y + 6z = 20 & (2) \\ -24y - 6z = 42 & (3) \\ \hline -31y = 62 & \text{Add.} \end{array}$$

$$\begin{cases} x + y - z = -1 & (1) \\ -7y + 6z = 20 & (2) \\ -31y = 62 & (3) \end{cases}$$

We now solve equation (3) for y by dividing both sides of the equation by -31.

$$\begin{cases} x + y - z = -1 & (1) \\ -7y + 6z = 20 & (2) \\ y = -2 & (3) \end{cases}$$

Back-substitute $y = -2$ in equation (2) and solve for z.

$$\begin{array}{ll} -7y + 6z = 20 & (2) \\ -7(-2) + 6z = 20 & \text{Substitute } y = -2 \text{ in (2).} \\ 6z = 6 & \text{Subtract 14 from both sides of the equation.} \\ z = 1 & \text{Divide both sides of the equation by 6.} \end{array}$$

Finally, we back-substitute $y = -2$ and $z = 1$ in equation (1) and solve for x.

$$\begin{array}{ll} x + y - z = -1 & (1) \\ x + (-2) - 1 = -1 & \text{Substitute } y = -2 \text{ and } z = 1 \text{ in (1).} \\ x - 3 = -1 & \text{Simplify.} \\ x = 2 & \text{Add 3 to both sides.} \end{array}$$

The solution of the original system is $x = 2$, $y = -2$, $z = 1$ or, using ordered triplets, $(2, -2, 1)$. You should check this solution. ∎

Look back over the solution given in Example 9. Note the pattern of removing one of the variables from two of the equations, followed by solving this system of two equations and two unknowns. Although which variables to remove is your choice, the methodology remains the same for all systems.

Now Work PROBLEM 43

6 Identify Inconsistent Systems of Equations Containing Three Variables

| EXAMPLE 10 | **An Inconsistent System of Linear Equations** |

Solve: $\begin{cases} 2x + y - z = -2 & (1) \\ x + 2y - z = -9 & (2) \\ x - 4y + z = 1 & (3) \end{cases}$

Solution Our strategy is the same as in Example 9. However, in this system, it seems easiest to eliminate the variable z first. Do you see why?

Multiply each side of equation (1) by -1 and add the result to equation (2). Also, add equations (2) and (3).

$$\begin{array}{rl} -2x - y + z = & 2 \quad \text{(1) Multiply by } -1. \\ x + 2y - z = & -9 \quad \text{(2)} \\ \hline -x + y \qquad = & -7 \quad \text{Add.} \end{array}$$

$$\begin{array}{rl} x + 2y - z = & -9 \quad \text{(2)} \\ x - 4y + z = & 1 \quad \text{(3)} \\ \hline 2x - 2y \qquad = & -8 \quad \text{Add.} \end{array}$$

$$\begin{cases} 2x + y - z = -2 & \text{(1)} \\ -x + y \quad\;\; = -7 & \text{(2)} \\ 2x - 2y \quad\;\; = -8 & \text{(3)} \end{cases}$$

We now concentrate on equations (2) and (3), treating them as a system of two equations containing two variables. Multiply each side of equation (2) by 2 and add the result to equation (3).

$$\begin{array}{l} -x + y = -7 \quad \text{(2) Multiply by 2.} \\ 2x - 2y = -8 \quad \text{(3)} \end{array} \qquad \begin{array}{rl} -2x + 2y = & -14 \quad \text{(2)} \\ 2x - 2y = & -8 \quad \text{(3)} \\ \hline 0 = & -22 \quad \text{Add.} \end{array} \qquad \begin{cases} 2x + y - z = -2 & \text{(1)} \\ -x + y \quad\;\; = -7 & \text{(2)} \\ 0 = -22 & \text{(3)} \end{cases}$$

Equation (3) has no solution so the system is inconsistent.

7 Express the Solution of a System of Dependent Equations Containing Three Variables

EXAMPLE 11 **Solving a System of Dependent Equations**

Solve: $\begin{cases} x - 2y - z = 8 & \text{(1)} \\ 2x - 3y + z = 23 & \text{(2)} \\ 4x - 5y + 5z = 53 & \text{(3)} \end{cases}$

Solution Our plan is to eliminate x from equations (2) and (3). Multiply each side of equation (1) by -2 and add the result to equation (2). Also, multiply each side of equation (1) by -4 and add the result to equation (3).

$$\begin{array}{l} x - 2y - z = 8 \quad \text{(1) Multiply by } -2. \\ 2x - 3y + z = 23 \quad \text{(2)} \end{array} \qquad \begin{array}{rl} -2x + 4y + 2z = & -16 \quad \text{(1)} \\ 2x - 3y + z = & 23 \quad \text{(2)} \\ \hline y + 3z = & 7 \quad \text{Add.} \end{array}$$

$$\begin{array}{l} x - 2y - z = 8 \quad \text{(1) Multiply by } -4. \\ 4x - 5y + 5z = 53 \quad \text{(3)} \end{array} \qquad \begin{array}{rl} -4x + 8y + 4z = & -32 \quad \text{(1)} \\ 4x - 5y + 5z = & 53 \quad \text{(2)} \\ \hline 3y + 9z = & 21 \quad \text{Add.} \end{array} \qquad \begin{cases} x - 2y - z = 8 & \text{(1)} \\ y + 3z = 7 & \text{(2)} \\ 3y + 9z = 21 & \text{(3)} \end{cases}$$

Treat equations (2) and (3) as a system of two equations containing two variables, and eliminate the variable y by multiplying both sides of equation (2) by -3 and adding the result to equation (3).

$$\begin{array}{l} y + 3z = 7 \quad \text{Multiply by } -3. \\ 3y + 9z = 21 \end{array} \qquad \begin{array}{rl} -3y - 9z = & -21 \\ 3y + 9z = & 21 \quad \text{Add.} \\ \hline 0 = & 0 \end{array} \qquad \begin{cases} x - 2y - z = 8 & \text{(1)} \\ y + 3z = 7 & \text{(2)} \\ 0 = 0 & \text{(3)} \end{cases}$$

The original system is equivalent to a system containing two equations, so the equations are dependent and the system has infinitely many solutions. If we solve

equation (2) for y, we can express y in terms of z as $y = -3z + 7$. Substitute this expression into equation (1) to determine x in terms of z.

$$x - 2y - z = 8 \qquad \text{(1)}$$
$$x - 2(-3z + 7) - z = 8 \qquad \text{Substitute } y = -3z + 7 \text{ in (1).}$$
$$x + 6z - 14 - z = 8 \qquad \text{Remove parentheses.}$$
$$x + 5z = 22 \qquad \text{Combine like terms.}$$
$$x = -5z + 22 \qquad \text{Solve for } x.$$

We will write the solution to the system as

$$\begin{cases} x = -5z + 22 \\ y = -3z + 7 \end{cases}$$

where z can be any real number.

This way of writing the solution makes it easier to find specific solutions of the system. To find specific solutions, choose any value of z and use the equations $x = -5z + 22$ and $y = -3z + 7$ to determine x and y. For example, if $z = 0$, then $x = 22$ and $y = 7$, and if $z = 1$, then $x = 17$ and $y = 4$.

Using ordered triplets, the solution is

$$\{(x, y, z) \mid x = -5z + 22,\ y = -3z + 7,\ z \text{ any real number}\}$$

━━━ **Now Work** PROBLEM 45

Two distinct points in the Cartesian plane determine a unique line. Given three noncollinear points, we can find the (unique) quadratic function whose graph contains these three points.

| **EXAMPLE 12** | **Curve Fitting** |

Find real numbers a, b, and c so that the graph of the quadratic function $y = ax^2 + bx + c$ contains the points $(-1, -4)$, $(1, 6)$, and $(3, 0)$.

Solution

We require that the three points satisfy the equation $y = ax^2 + bx + c$.

For the point $(-1, -4)$ we have: $-4 = a(-1)^2 + b(-1) + c \qquad -4 = a - b + c$
For the point $(1, 6)$ we have: $6 = a(1)^2 + b(1) + c \qquad 6 = a + b + c$
For the point $(3, 0)$ we have: $0 = a(3)^2 + b(3) + c \qquad 0 = 9a + 3b + c$

We wish to determine a, b, and c so that each equation is satisfied. That is, we want to solve the following system of three equations containing three variables:

$$\begin{cases} a - b + c = -4 & \text{(1)} \\ a + b + c = 6 & \text{(2)} \\ 9a + 3b + c = 0 & \text{(3)} \end{cases}$$

Solving this system of equations, we obtain $a = -2$, $b = 5$, and $c = 3$. So the quadratic function whose graph contains the points $(-1, -4)$, $(1, 6)$, and $(3, 0)$ is

$$y = -2x^2 + 5x + 3 \qquad y = ax^2 + bx + c,\ a = -2, b = 5, c = 3$$

Figure 6 shows the graph of the function along with the three points.

Figure 6

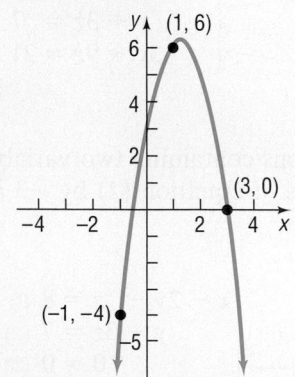

━━━ **Now Work** PROBLEM 69

12.1 Assess Your Understanding

'Are You Prepared?' *Answers are given at the end of these exercises. If you get a wrong answer, read the pages listed in red.*

1. Solve the equation: $3x + 4 = 8 - x$. (pp. 86–94)

2. (a) Graph the line: $3x + 4y = 12$.
 (b) What is the slope of a line parallel to this line? (pp. 173–185)

Concepts and Vocabulary

3. If a system of equations has no solution, it is said to be _____.

4. If a system of equations has one or more solutions, the system is said to be _____.

5. *True or False* A system of two linear equations containing two variables always has at least one solution.

6. *True or False* A solution of a system of equations consists of values for the variables that are solutions of each equation of the system.

Skill Building

In Problems 7–16, verify that the values of the variables listed are solutions of the system of equations.

7. $\begin{cases} 2x - y = 5 \\ 5x + 2y = 8 \end{cases}$
 $x = 2, y = -1; (2, -1)$

8. $\begin{cases} 3x + 2y = 2 \\ x - 7y = -30 \end{cases}$
 $x = -2, y = 4; (-2, 4)$

9. $\begin{cases} 3x - 4y = 4 \\ \frac{1}{2}x - 3y = -\frac{1}{2} \end{cases}$
 $x = 2, y = \frac{1}{2}; \left(2, \frac{1}{2}\right)$

10. $\begin{cases} 2x + \frac{1}{2}y = 0 \\ 3x - 4y = -\frac{19}{2} \end{cases}$
 $x = -\frac{1}{2}, y = 2; \left(-\frac{1}{2}, 2\right)$

11. $\begin{cases} x - y = 3 \\ \frac{1}{2}x + y = 3 \end{cases}$
 $x = 4, y = 1; (4, 1)$

12. $\begin{cases} x - y = 3 \\ -3x + y = 1 \end{cases}$
 $x = -2, y = -5; (-2, -5)$

13. $\begin{cases} 3x + 3y + 2z = 4 \\ x - y - z = 0 \\ 2y - 3z = -8 \end{cases}$
 $x = 1, y = -1, z = 2;$
 $(1, -1, 2)$

14. $\begin{cases} 4x - z = 7 \\ 8x + 5y - z = 0 \\ -x - y + 5z = 6 \end{cases}$
 $x = 2, y = -3, z = 1;$
 $(2, -3, 1)$

15. $\begin{cases} 3x + 3y + 2z = 4 \\ x - 3y + z = 10 \\ 5x - 2y - 3z = 8 \end{cases}$
 $x = 2, y = -2, z = 2; (2, -2, 2)$

16. $\begin{cases} 4x - 5z = 6 \\ 5y - z = -17 \\ -x - 6y + 5z = 24 \end{cases}$
 $x = 4, y = -3, z = 2; (4, -3, 2)$

In Problems 17–54, solve each system of equations. If the system has no solution, say that it is inconsistent.

17. $\begin{cases} x + y = 8 \\ x - y = 4 \end{cases}$

18. $\begin{cases} x + 2y = 5 \\ x + y = 3 \end{cases}$

19. $\begin{cases} 5x - y = 13 \\ 2x + 3y = 12 \end{cases}$

20. $\begin{cases} x + 3y = 5 \\ 2x - 3y = -8 \end{cases}$

21. $\begin{cases} 3x = 24 \\ x + 2y = 0 \end{cases}$

22. $\begin{cases} 4x + 5y = -3 \\ -2y = -4 \end{cases}$

23. $\begin{cases} 3x - 6y = 2 \\ 5x + 4y = 1 \end{cases}$

24. $\begin{cases} 2x + 4y = \frac{2}{3} \\ 3x - 5y = -10 \end{cases}$

25. $\begin{cases} 2x + y = 1 \\ 4x + 2y = 3 \end{cases}$

26. $\begin{cases} x - y = 5 \\ -3x + 3y = 2 \end{cases}$

27. $\begin{cases} 2x - y = 0 \\ 3x + 2y = 7 \end{cases}$

28. $\begin{cases} 3x + 3y = -1 \\ 4x + y = \frac{8}{3} \end{cases}$

29. $\begin{cases} x + 2y = 4 \\ 2x + 4y = 8 \end{cases}$

30. $\begin{cases} 3x - y = 7 \\ 9x - 3y = 21 \end{cases}$

31. $\begin{cases} 2x - 3y = -1 \\ 10x + y = 11 \end{cases}$

32. $\begin{cases} 3x - 2y = 0 \\ 5x + 10y = 4 \end{cases}$

33. $\begin{cases} 2x + 3y = 6 \\ x - y = \frac{1}{2} \end{cases}$

34. $\begin{cases} \frac{1}{2}x + y = -2 \\ x - 2y = 8 \end{cases}$

35. $\begin{cases} \frac{1}{2}x + \frac{1}{3}y = 3 \\ \frac{1}{4}x - \frac{2}{3}y = -1 \end{cases}$

36. $\begin{cases} \frac{1}{3}x - \frac{3}{2}y = -5 \\ \frac{3}{4}x + \frac{1}{3}y = 11 \end{cases}$

37. $\begin{cases} 3x - 5y = 3 \\ 15x + 5y = 21 \end{cases}$

38. $\begin{cases} 2x - y = -1 \\ x + \dfrac{1}{2}y = \dfrac{3}{2} \end{cases}$

39. $\begin{cases} \dfrac{1}{x} + \dfrac{1}{y} = 8 \\ \dfrac{3}{x} - \dfrac{5}{y} = 0 \end{cases}$

40. $\begin{cases} \dfrac{4}{x} - \dfrac{3}{y} = 0 \\ \dfrac{6}{x} + \dfrac{3}{2y} = 2 \end{cases}$

[**Hint:** Let $u = \dfrac{1}{x}$ and $v = \dfrac{1}{y}$, and solve for u and v. Then $x = \dfrac{1}{u}$ and $y = \dfrac{1}{v}$.]

41. $\begin{cases} x - y = 6 \\ 2x - 3z = 16 \\ 2y + z = 4 \end{cases}$

42. $\begin{cases} 2x + y = -4 \\ -2y + 4z = 0 \\ 3x - 2z = -11 \end{cases}$

43. $\begin{cases} x - 2y + 3z = 7 \\ 2x + y + z = 4 \\ -3x + 2y - 2z = -10 \end{cases}$

44. $\begin{cases} 2x + y - 3z = 0 \\ -2x + 2y + z = -7 \\ 3x - 4y - 3z = 7 \end{cases}$

45. $\begin{cases} x - y - z = 1 \\ 2x + 3y + z = 2 \\ 3x + 2y = 0 \end{cases}$

46. $\begin{cases} 2x - 3y - z = 0 \\ -x + 2y + z = 5 \\ 3x - 4y - z = 1 \end{cases}$

47. $\begin{cases} x - y - z = 1 \\ -x + 2y - 3z = -4 \\ 3x - 2y - 7z = 0 \end{cases}$

48. $\begin{cases} 2x - 3y - z = 0 \\ 3x + 2y + 2z = 2 \\ x + 5y + 3z = 2 \end{cases}$

49. $\begin{cases} 2x - 2y + 3z = 6 \\ 4x - 3y + 2z = 0 \\ -2x + 3y - 7z = 1 \end{cases}$

50. $\begin{cases} 3x - 2y + 2z = 6 \\ 7x - 3y + 2z = -1 \\ 2x - 3y + 4z = 0 \end{cases}$

51. $\begin{cases} x + y - z = 6 \\ 3x - 2y + z = -5 \\ x + 3y - 2z = 14 \end{cases}$

52. $\begin{cases} x - y + z = -4 \\ 2x - 3y + 4z = -15 \\ 5x + y - 2z = 12 \end{cases}$

53. $\begin{cases} x + 2y - z = -3 \\ 2x - 4y + z = -7 \\ -2x + 2y - 3z = 4 \end{cases}$

54. $\begin{cases} x + 4y - 3z = -8 \\ 3x - y + 3z = 12 \\ x + y + 6z = 1 \end{cases}$

Applications and Extensions

55. The perimeter of a rectangular floor is 90 feet. Find the dimensions of the floor if the length is twice the width.

56. The length of fence required to enclose a rectangular field is 3000 meters. What are the dimensions of the field if it is known that the difference between its length and width is 50 meters?

57. **Orbital Launches** In 2005 there was a total of 55 commercial and noncommercial orbital launches worldwide. In addition, the number of noncommercial orbital launches was one more than twice the number of commercial orbital launches. Determine the number of commercial and noncommercial orbital launches in 2005.
Source: Federal Aviation Administration

58. **Movie Theater Tickets** A movie theater charges $9.00 for adults and $7.00 for senior citizens. On a day when 325 people paid an admission, the total receipts were $2495. How many who paid were adults? How many were seniors?

59. **Mixing Nuts** A store sells cashews for $5.00 per pound and peanuts for $1.50 per pound. The manager decides to mix 30 pounds of peanuts with some cashews and sell the mixture for $3.00 per pound. How many pounds of cashews should be mixed with the peanuts so that the mixture will produce the same revenue as would selling the nuts separately?

60. **Financial Planning** A recently retired couple needs $12,000 per year to supplement their Social Security. They have $150,000 to invest to obtain this income. They have decided on two investment options: AA bonds yielding 10% per annum and a Bank Certificate yielding 5%.

(a) How much should be invested in each to realize exactly $12,000?

(b) If, after 2 years, the couple requires $14,000 per year in income, how should they reallocate their investment to achieve the new amount?

61. **Computing Wind Speed** With a tail wind, a small Piper aircraft can fly 600 miles in 3 hours. Against this same wind, the Piper can fly the same distance in 4 hours. Find the average wind speed and the average airspeed of the Piper.

62. **Computing Wind Speed** The average airspeed of a single-engine aircraft is 150 miles per hour. If the aircraft flew the same distance in 2 hours with the wind as it flew in 3 hours against the wind, what was the wind speed?

63. **Restaurant Management** A restaurant manager wants to purchase 200 sets of dishes. One design costs $25 per set, while another costs $45 per set. If she only has $7400 to spend, how many of each design should be ordered?

64. **Cost of Fast Food** One group of people purchased 10 hot dogs and 5 soft drinks at a cost of $35.00. A second bought 7 hot dogs and 4 soft drinks at a cost of $25.25. What is the cost of a single hot dog? A single soft drink?

We paid $35.00.
How much is one hot dog?
How much is one soda?

We paid $25.25.
How much is one hot dog?
How much is one soda?

65. Computing a Refund The grocery store we use does not mark prices on its goods. My wife went to this store, bought three 1-pound packages of bacon and two cartons of eggs, and paid a total of $13.45. Not knowing that she went to the store, I also went to the same store, purchased two 1-pound packages of bacon and three cartons of eggs, and paid a total of $11.45. Now we want to return two 1-pound packages of bacon and two cartons of eggs. How much will be refunded?

66. Finding the Current of a Stream Pamela requires 3 hours to swim 15 miles downstream on the Illinois River. The return trip upstream takes 5 hours. Find Pamela's average speed in still water. How fast is the current? (Assume that Pamela's speed is the same in each direction.)

67. Pharmacy A doctor's prescription calls for a daily intake containing 40 mg of vitamin C and 30 mg of vitamin D. Your pharmacy stocks two liquids that can be used: one contains 20% vitamin C and 30% vitamin D, the other 40% vitamin C and 20% vitamin D. How many milligrams of each compound should be mixed to fill the prescription?

68. Pharmacy A doctor's prescription calls for the creation of pills that contain 12 units of vitamin B_{12} and 12 units of vitamin E. Your pharmacy stocks two powders that can be used to make these pills: one contains 20% vitamin B_{12} and 30% vitamin E, the other 40% vitamin B_{12} and 20% vitamin E. How many units of each powder should be mixed in each pill?

69. Curve Fitting Find real numbers a, b, and c so that the graph of the function $y = ax^2 + bx + c$ contains the points $(-1, 4)$, $(2, 3)$, and $(0, 1)$.

70. Curve Fitting Find real numbers a, b, and c so that the graph of the function $y = ax^2 + bx + c$ contains the points $(-1, -2)$, $(1, -4)$, and $(2, 4)$.

71. IS-LM Model in Economics In economics, the IS curve is a linear equation that represents all combinations of income Y and interest rates r that maintain an equilibrium in the market for goods in the economy. The LM curve is a linear equation that represents all combinations of income Y and interest rates r that maintain an equilibrium in the market for money in the economy. In an economy, suppose the equilibrium level of income (in millions of dollars) and interest rates satisfy the system of equations

$$\begin{cases} 0.06Y - 5000r = 240 \\ 0.06Y + 6000r = 900 \end{cases}$$

Find the equilibrium level of income and interest rates.

72. IS-LM Model in Economics In economics, the IS curve is a linear equation that represents all combinations of income Y and interest rates r that maintain an equilibrium in the market for goods in the economy. The LM curve is a linear equation that represents all combinations of income Y and interest rates r that maintain an equilibrium in the market for money in the economy. In an economy, suppose the equilibrium level of income (in millions of dollars) and interest rates satisfy the system of equations

$$\begin{cases} 0.05Y - 1000r = 10 \\ 0.05Y + 800r = 100 \end{cases}$$

Find the equilibrium level of income and interest rates.

73. Electricity: Kirchhoff's Rules An application of Kirchhoff's Rules to the circuit shown results in the following system of equations:

$$\begin{cases} I_2 = I_1 + I_3 \\ 5 - 3I_1 - 5I_2 = 0 \\ 10 - 5I_2 - 7I_3 = 0 \end{cases}$$

Find the currents I_1, I_2, and I_3.

Source: Physics for Scientists & Engineers, 3rd ed., by Serway. © 1990 Brooks/Cole, a division of Thomson Learning.

74. Electricity: Kirchhoff's Rules An application of Kirchhoff's Rules to the circuit shown results in the following system of equations:

$$\begin{cases} I_3 = I_1 + I_2 \\ 8 = 4I_3 + 6I_2 \\ 8I_1 = 4 + 6I_2 \end{cases}$$

Find the currents I_1, I_2, and I_3.

Source: Physics for Scientists & Engineers, 3rd ed., by Serway. © 1990 Brooks/Cole, a division of Thomson Learning.

75. Theater Revenues A Broadway theater has 500 seats, divided into orchestra, main, and balcony seating. Orchestra seats sell for $50, main seats for $35, and balcony seats for $25. If all the seats are sold, the gross revenue to the theater is $17,100. If all the main and balcony seats are sold, but only half the orchestra seats are sold, the gross revenue is $14,600. How many are there of each kind of seat?

76. Theater Revenues A movie theater charges $8.00 for adults, $4.50 for children, and $6.00 for senior citizens. One day the theater sold 405 tickets and collected $2320 in receipts. There were twice as many children's tickets sold as adult tickets. How many adults, children, and senior citizens went to the theater that day?

77. Nutrition A dietitian wishes a patient to have a meal that has 66 grams of protein, 94.5 grams of carbohydrates, and 910 milligrams of calcium. The hospital food service tells the dietitian that the dinner for today is chicken, corn, and 2% milk. Each serving of chicken has 30 grams of protein, 35 grams of carbohydrates, and 200 milligrams of calcium. Each serving of corn has 3 grams of protein, 16 grams of carbohydrates, and 10 milligrams of calcium. Each glass of 2% milk has 9 grams of protein, 13 grams of carbohydrates, and 300 milligrams of calcium. How many servings of each food should the dietitian provide for the patient?

78. Investments Kelly has $20,000 to invest. As her financial planner, you recommend that she diversify into three investments: Treasury bills that yield 5% simple interest, Treasury bonds that yield 7% simple interest, and corporate bonds that yield 10% simple interest. Kelly wishes to earn $1390 per year in income. Also, Kelly wants her investment in Treasury bills to be $3000 more than her investment in corporate bonds. How much money should Kelly place in each investment?

79. Prices of Fast Food One group of customers bought 8 deluxe hamburgers, 6 orders of large fries, and 6 large colas for $26.10. A second group ordered 10 deluxe hamburgers, 6 large fries, and 8 large colas and paid $31.60. Is there

sufficient information to determine the price of each food item? If not, construct a table showing the various possibilities. Assume that the hamburgers cost between $1.75 and $2.25, the fries between $0.75 and $1.00, and the colas between $0.60 and $0.90.

80. Prices of Fast Food Use the information given in Problem 79. Suppose that a third group purchased 3 deluxe hamburgers, 2 large fries, and 4 large colas for $10.95. Now is there sufficient information to determine the price of each food item? If so, determine each price.

81. Painting a House Three painters, Beth, Bill, and Edie, working together, can paint the exterior of a home in 10 hours. Bill and Edie together have painted a similar house in 15 hours. One day, all three worked on this same kind of house for 4 hours, after which Edie left. Beth and Bill required 8 more hours to finish. Assuming no gain or loss in efficiency, how long should it take each person to complete such a job alone?

Discussion and Writing

82. Make up a system of three linear equations containing three variables that has:
 (a) No solution
 (b) Exactly one solution
 (c) Infinitely many solutions
 Give the three systems to a friend to solve and critique.

83. Write a brief paragraph outlining your strategy for solving a system of two linear equations containing two variables.

84. Do you prefer the method of substitution or the method of elimination for solving a system of two linear equations containing two variables? Give reasons.

'Are You Prepared?' Answers

1. $\{1\}$ **2.** (a)

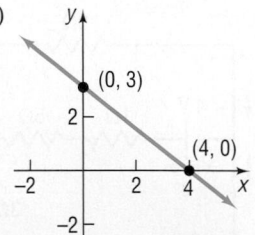

(b) $-\dfrac{3}{4}$

12.2 Systems of Linear Equations: Matrices

OBJECTIVES 1 Write the Augmented Matrix of a System of Linear Equations (p. 851)
2 Write the System of Equations from the Augmented Matrix (p. 852)
3 Perform Row Operations on a Matrix (p. 853)
4 Solve a System of Linear Equations Using Matrices (p. 854)

The systematic approach of the method of elimination for solving a system of linear equations provides another method of solution that involves a simplified notation. Consider the following system of linear equations:

$$\begin{cases} x + 4y = 14 \\ 3x - 2y = 0 \end{cases}$$

If we choose not to write the symbols used for the variables, we can represent this system as

$$\begin{bmatrix} 1 & 4 & | & 14 \\ 3 & -2 & | & 0 \end{bmatrix}$$

where it is understood that the first column represents the coefficients of the variable x, the second column the coefficients of y, and the third column the constants on the right side of the equal signs. The vertical line serves as a reminder of the equal signs. The large square brackets are used to denote a *matrix* in algebra.

DEFINITION A **matrix** is defined as a rectangular array of numbers,

$$\begin{array}{c} \\ \text{Row 1} \\ \text{Row 2} \\ \\ \text{Row } i \\ \\ \text{Row } m \end{array} \begin{bmatrix} \overset{\text{Column 1}}{a_{11}} & \overset{\text{Column 2}}{a_{12}} & \cdots & \overset{\text{Column } j}{a_{1j}} & \cdots & \overset{\text{Column } n}{a_{1n}} \\ a_{21} & a_{22} & \cdots & a_{2j} & \cdots & a_{2n} \\ \vdots & \vdots & & \vdots & & \vdots \\ a_{i1} & a_{i2} & \cdots & a_{ij} & \cdots & a_{in} \\ \vdots & \vdots & & \vdots & & \vdots \\ a_{m1} & a_{m2} & \cdots & a_{mj} & \cdots & a_{mn} \end{bmatrix} \quad \textbf{(1)}$$

Each number a_{ij} of the matrix has two indexes: the **row index** i and the **column index** j. The matrix shown in display (1) has m rows and n columns. The numbers a_{ij} are usually referred to as the **entries** of the matrix. For example, a_{23} refers to the entry in the second row, third column.

1 Write the Augmented Matrix of a System of Linear Equations

Now we will use matrix notation to represent a system of linear equations. The matrix used to represent a system of linear equations is called an **augmented matrix.** In writing the augmented matrix of a system, the variables of each equation must be on the left side of the equal sign and the constants on the right side. A variable that does not appear in an equation has a coefficient of 0.

EXAMPLE 1 Writing the Augmented Matrix of a System of Linear Equations

Write the augmented matrix of each system of equations.

(a) $\begin{cases} 3x - 4y = -6 & (1) \\ 2x - 3y = -5 & (2) \end{cases}$

(b) $\begin{cases} 2x - y + z = 0 & (1) \\ x + z - 1 = 0 & (2) \\ x + 2y - 8 = 0 & (3) \end{cases}$

Solution (a) The augmented matrix is

$$\begin{bmatrix} 3 & -4 & | & -6 \\ 2 & -3 & | & -5 \end{bmatrix}$$

(b) Care must be taken that the system be written so that the coefficients of all variables are present (if any variable is missing, its coefficient is 0). Also, all

constants must be to the right of the equal sign. We need to rearrange the given system as follows:

$$\begin{cases} 2x - y + z = 0 & \text{(1)} \\ x + z - 1 = 0 & \text{(2)} \\ x + 2y - 8 = 0 & \text{(3)} \end{cases}$$

$$\begin{cases} 2x - y + z = 0 & \text{(1)} \\ x + 0 \cdot y + z = 1 & \text{(2)} \\ x + 2y + 0 \cdot z = 8 & \text{(3)} \end{cases}$$

The augmented matrix is

$$\left[\begin{array}{ccc|c} 2 & -1 & 1 & 0 \\ 1 & 0 & 1 & 1 \\ 1 & 2 & 0 & 8 \end{array}\right]$$

If we do not include the constants to the right of the equal sign, that is, to the right of the vertical bar in the augmented matrix of a system of equations, the resulting matrix is called the **coefficient matrix** of the system. For the systems discussed in Example 1, the coefficient matrices are

$$\begin{bmatrix} 3 & -4 \\ 2 & -3 \end{bmatrix} \quad \text{and} \quad \begin{bmatrix} 2 & -1 & 1 \\ 1 & 0 & 1 \\ 1 & 2 & 0 \end{bmatrix}$$

━━━ **Now Work** PROBLEM 7

2 Write the System of Equations from the Augmented Matrix

EXAMPLE 2 **Writing the System of Linear Equations from the Augmented Matrix**

Write the system of linear equations corresponding to each augmented matrix.

(a) $\left[\begin{array}{cc|c} 5 & 2 & 13 \\ -3 & 1 & -10 \end{array}\right]$ (b) $\left[\begin{array}{ccc|c} 3 & -1 & -1 & 7 \\ 2 & 0 & 2 & 8 \\ 0 & 1 & 1 & 0 \end{array}\right]$

Solution (a) The matrix has two rows and so represents a system of two equations. The two columns to the left of the vertical bar indicate that the system has two variables. If x and y are used to denote these variables, the system of equations is

$$\begin{cases} 5x + 2y = 13 & \text{(1)} \\ -3x + y = -10 & \text{(2)} \end{cases}$$

(b) Since the augmented matrix has three rows, it represents a system of three equations. Since there are three columns to the left of the vertical bar, the system contains three variables. If x, y, and z are the three variables, the system of equations is

$$\begin{cases} 3x - y - z = 7 & \text{(1)} \\ 2x + 2z = 8 & \text{(2)} \\ y + z = 0 & \text{(3)} \end{cases}$$

3 Perform Row Operations on a Matrix

Row operations on a matrix are used to solve systems of equations when the system is written as an augmented matrix. There are three basic row operations.

> ### Row Operations
>
> 1. Interchange any two rows.
> 2. Replace a row by a nonzero multiple of that row.
> 3. Replace a row by the sum of that row and a constant nonzero multiple of some other row.

These three row operations correspond to the three rules given earlier for obtaining an equivalent system of equations. When a row operation is performed on a matrix, the resulting matrix represents a system of equations equivalent to the system represented by the original matrix.

For example, consider the augmented matrix

$$\begin{bmatrix} 1 & 2 & | & 3 \\ 4 & -1 & | & 2 \end{bmatrix}$$

Suppose that we want to apply a row operation to this matrix that results in a matrix whose entry in row 2, column 1 is a 0. The row operation to use is

Multiply each entry in row 1 by -4 and add the result
to the corresponding entries in row 2. **(2)**

If we use R_2 to represent the new entries in row 2 and we use r_1 and r_2 to represent the original entries in rows 1 and 2, respectively, we can represent the row operation in statement (2) by

$$R_2 = -4r_1 + r_2$$

Then

$$\begin{bmatrix} 1 & 2 & | & 3 \\ 4 & -1 & | & 2 \end{bmatrix} \xrightarrow[\substack{\uparrow \\ R_2 = -4r_1 + r_2}]{} \begin{bmatrix} 1 & 2 & | & 3 \\ -4(1) + 4 & -4(2) + (-1) & | & -4(3) + 2 \end{bmatrix} = \begin{bmatrix} 1 & 2 & | & 3 \\ 0 & -9 & | & -10 \end{bmatrix}$$

As desired, we now have the entry 0 in row 2, column 1.

EXAMPLE 3 **Applying a Row Operation to an Augmented Matrix**

Apply the row operation $R_2 = -3r_1 + r_2$ to the augmented matrix

$$\begin{bmatrix} 1 & -2 & | & 2 \\ 3 & -5 & | & 9 \end{bmatrix}$$

Solution The row operation $R_2 = -3r_1 + r_2$ tells us that the entries in row 2 are to be replaced by the entries obtained after multiplying each entry in row 1 by -3 and adding the result to the corresponding entries in row 2.

$$\begin{bmatrix} 1 & -2 & | & 2 \\ 3 & -5 & | & 9 \end{bmatrix} \xrightarrow[\substack{\uparrow \\ R_2 = -3r_1 + r_2}]{} \begin{bmatrix} 1 & -2 & | & 2 \\ -3(1) + 3 & (-3)(-2) + (-5) & | & -3(2) + 9 \end{bmatrix} = \begin{bmatrix} 1 & -2 & | & 2 \\ 0 & 1 & | & 3 \end{bmatrix}$$

═══ **Now Work** PROBLEM **17**

| EXAMPLE 4 | Finding a Particular Row Operation |

Find a row operation that will result in the augmented matrix

$$\begin{bmatrix} 1 & -2 & | & 2 \\ 0 & 1 & | & 3 \end{bmatrix}$$

having a 0 in row 1, column 2.

Solution We want a 0 in row 1, column 2. This result can be accomplished by multiplying row 2 by 2 and adding the result to row 1. That is, we apply the row operation $R_1 = 2r_2 + r_1$.

$$\begin{bmatrix} 1 & -2 & | & 2 \\ 0 & 1 & | & 3 \end{bmatrix} \rightarrow \begin{bmatrix} 2(0) + 1 & 2(1) + (-2) & | & 2(3) + 2 \\ 0 & 1 & | & 3 \end{bmatrix} = \begin{bmatrix} 1 & 0 & | & 8 \\ 0 & 1 & | & 3 \end{bmatrix}$$

$$\uparrow$$
$$R_1 = 2r_2 + r_1$$

∎

A word about the notation that we have introduced. A row operation such as $R_1 = 2r_2 + r_1$ changes the entries in row 1. Note also that for this type of row operation we change the entries in a given row by multiplying the entries in some other row by an appropriate nonzero number and adding the results to the original entries of the row to be changed.

4 Solve a System of Linear Equations Using Matrices

To solve a system of linear equations using matrices, we use row operations on the augmented matrix of the system to obtain a matrix that is in *row echelon form*.

DEFINITION A matrix is in **row echelon form** when the following conditions are met:

1. The entry in row 1, column 1 is a 1, and only 0's appear below it.
2. The first nonzero entry in each row after the first row is a 1, only 0's appear below it, and the 1 appears to the right of the first nonzero entry in any row above.
3. Any rows that contain all 0's to the left of the vertical bar appear at the bottom.

⌐

For example, for a system of three equations containing three variables with a unique solution, the augmented matrix is in row echelon form if it is of the form

$$\begin{bmatrix} 1 & a & b & | & d \\ 0 & 1 & c & | & e \\ 0 & 0 & 1 & | & f \end{bmatrix}$$

where a, b, c, d, e, and f are real numbers. The last row of this augmented matrix states that $z = f$. We can then determine the value of y using back-substitution with $z = f$, since row 2 represents the equation $y + cz = e$. Finally, x is determined using back-substitution again.

Two advantages of solving a system of equations by writing the augmented matrix in row echelon form are the following:

1. The process is algorithmic; that is, it consists of repetitive steps that can be programmed on a computer.
2. The process works on any system of linear equations, no matter how many equations or variables are present.

The next example shows how to write a matrix in row echelon form.

| EXAMPLE 5 | Solving a System of Linear Equations Using Matrices (Row Echelon Form) |

Solve:
$$\begin{cases} 2x + 2y & = 6 \quad (1) \\ x + y + z = 1 \quad (2) \\ 3x + 4y - z = 13 \quad (3) \end{cases}$$

Solution First, we write the augmented matrix that represents this system.

$$\begin{bmatrix} 2 & 2 & 0 & | & 6 \\ 1 & 1 & 1 & | & 1 \\ 3 & 4 & -1 & | & 13 \end{bmatrix}$$

The first step requires getting the entry 1 in row 1, column 1. An interchange of rows 1 and 2 is the easiest way to do this. [Note that this is equivalent to interchanging equations (1) and (2) of the system.]

$$\begin{bmatrix} 1 & 1 & 1 & | & 1 \\ 2 & 2 & 0 & | & 6 \\ 3 & 4 & -1 & | & 13 \end{bmatrix}$$

Next, we want a 0 in row 2, column 1 and a 0 in row 3, column 1. We use the row operations $R_2 = -2r_1 + r_2$ and $R_3 = -3r_1 + r_3$ to accomplish this. Notice that row 1 is unchanged using these row operations. Also, do you see that performing these row operations simultaneously is the same as doing one followed by the other?

$$\begin{bmatrix} 1 & 1 & 1 & | & 1 \\ 2 & 2 & 0 & | & 6 \\ 3 & 4 & -1 & | & 13 \end{bmatrix} \rightarrow \begin{bmatrix} 1 & 1 & 1 & | & 1 \\ 0 & 0 & -2 & | & 4 \\ 0 & 1 & -4 & | & 10 \end{bmatrix}$$
$$\begin{array}{c} R_2 = -2r_1 + r_2 \\ R_3 = -3r_1 + r_3 \end{array}$$

Now we want the entry 1 in row 2, column 2. Interchanging rows 2 and 3 will accomplish this.

$$\begin{bmatrix} 1 & 1 & 1 & | & 1 \\ 0 & 0 & -2 & | & 4 \\ 0 & 1 & -4 & | & 10 \end{bmatrix} \rightarrow \begin{bmatrix} 1 & 1 & 1 & | & 1 \\ 0 & 1 & -4 & | & 10 \\ 0 & 0 & -2 & | & 4 \end{bmatrix}$$

Finally, we want a 1 in row 3, column 3. To obtain it, we use the row operation $R_3 = -\dfrac{1}{2}r_3$. The result is

$$\begin{bmatrix} 1 & 1 & 1 & | & 1 \\ 0 & 1 & -4 & | & 10 \\ 0 & 0 & -2 & | & 4 \end{bmatrix} \rightarrow \begin{bmatrix} 1 & 1 & 1 & | & 1 \\ 0 & 1 & -4 & | & 10 \\ 0 & 0 & 1 & | & -2 \end{bmatrix}$$
$$R_3 = -\dfrac{1}{2}r_3$$

This matrix is the row echelon form of the augmented matrix. The third row of this matrix represents the equation $z = -2$. Using $z = -2$, we back-substitute into the equation $y - 4z = 10$ (from the second row) and obtain

$$y - 4z = 10$$
$$y - 4(-2) = 10 \quad z = -2$$
$$y = 2 \quad \textit{Solve for y.}$$

Finally, we back-substitute $y = 2$ and $z = -2$ into the equation $x + y + z = 1$ (from the first row) and obtain

$$x + y + z = 1$$
$$x + 2 + (-2) = 1 \quad \text{\small $y = 2, z = -2$}$$
$$x = 1 \quad \text{\small Solve for x.}$$

The solution of the system is $x = 1$, $y = 2$, $z = -2$ or, using ordered triplets, $(1, 2, -2)$.

The steps that we used to solve the system of linear equations in Example 5 can be summarized as follows:

Matrix Method for Solving a System of Linear Equations (Row Echelon Form)

STEP 1: Write the augmented matrix that represents the system.

STEP 2: Perform row operations that place the entry 1 in row 1, column 1.

STEP 3: Perform row operations that leave the entry 1 in row 1, column 1 unchanged, while causing 0's to appear below it in column 1.

STEP 4: Perform row operations that place the entry 1 in row 2, column 2, but leave the entries in columns to the left unchanged. If it is impossible to place a 1 in row 2, column 2, proceed to place a 1 in row 2, column 3. Once a 1 is in place, perform row operations to place 0's below it. [Place any rows that contain only 0's on the left side of the vertical bar, at the bottom of the matrix.]

STEP 5: Now repeat Step 4, placing a 1 in the next row, but one column to the right. Continue until the bottom row or the vertical bar is reached.

STEP 6: The matrix that results is the row echelon form of the augmented matrix. Analyze the system of equations corresponding to it to solve the original system.

> **In Words**
> To obtain an augmented matrix in row echelon form:
> - Add rows, exchange rows, or multiply the row by a nonzero constant
> - Work from top to bottom and left to right
> - Get 1's in the main diagonal with 0's below the 1's.

In the next example, we solve a system of linear equations using these steps.

| **EXAMPLE 6** | **Solving a System of Linear Equations Using Matrices (Row Echelon Form)** |

Solve: $\begin{cases} x - y + z = 8 & (1) \\ 2x + 3y - z = -2 & (2) \\ 3x - 2y - 9z = 9 & (3) \end{cases}$

Solution

STEP 1: The augmented matrix of the system is

$$\left[\begin{array}{rrr|r} 1 & -1 & 1 & 8 \\ 2 & 3 & -1 & -2 \\ 3 & -2 & -9 & 9 \end{array}\right]$$

STEP 2: Because the entry 1 is already present in row 1, column 1, we can go to step 3.

STEP 3: Perform the row operations $R_2 = -2r_1 + r_2$ and $R_3 = -3r_1 + r_3$. Each of these leaves the entry 1 in row 1, column 1 unchanged, while causing 0's to appear under it.

> **In Words**
> Notice we use multiples of the entry in row 1, column 1 to obtain 0's in the entries below the 1.

$$\left[\begin{array}{rrr|r} 1 & -1 & 1 & 8 \\ 2 & 3 & -1 & -2 \\ 3 & -2 & -9 & 9 \end{array}\right] \rightarrow \left[\begin{array}{rrr|r} 1 & -1 & 1 & 8 \\ 0 & 5 & -3 & -18 \\ 0 & 1 & -12 & -15 \end{array}\right]$$
$$\begin{array}{l} R_2 = -2r_1 + r_2 \\ R_3 = -3r_1 + r_3 \end{array}$$

STEP 4: The easiest way to obtain the entry 1 in row 2, column 2 without altering column 1 is to interchange rows 2 and 3. (Another way would be to multiply row 2 by $\frac{1}{5}$, but this introduces fractions).

$$\begin{bmatrix} 1 & -1 & 1 & | & 8 \\ 0 & 1 & -12 & | & -15 \\ 0 & 5 & -3 & | & -18 \end{bmatrix}$$

To get a 0 under the 1 in row 2, column 2, perform the row operation $R_3 = -5r_2 + r_3$.

$$\begin{bmatrix} 1 & -1 & 1 & | & 8 \\ 0 & 1 & -12 & | & -15 \\ 0 & 5 & -3 & | & -18 \end{bmatrix} \rightarrow \begin{bmatrix} 1 & -1 & 1 & | & 8 \\ 0 & 1 & -12 & | & -15 \\ 0 & 0 & 57 & | & 57 \end{bmatrix}$$

$$\uparrow$$
$$R_3 = -5r_2 + r_3$$

STEP 5: Continuing, we obtain a 1 in row 3, column 3 by using $R_3 = \frac{1}{57}r_3$.

$$\begin{bmatrix} 1 & -1 & 1 & | & 8 \\ 0 & 1 & -12 & | & -15 \\ 0 & 0 & 57 & | & 57 \end{bmatrix} \rightarrow \begin{bmatrix} 1 & -1 & 1 & | & 8 \\ 0 & 1 & -12 & | & -15 \\ 0 & 0 & 1 & | & 1 \end{bmatrix}$$

$$\uparrow$$
$$R_3 = \frac{1}{57}r_3$$

STEP 6: The matrix on the right is the row echelon form of the augmented matrix. The system of equations represented by the matrix in row echelon form is

$$\begin{cases} x - y + z = 8 & (1) \\ y - 12z = -15 & (2) \\ z = 1 & (3) \end{cases}$$

Using $z = 1$, we back-substitute to get

$$\begin{cases} x - y + 1 = 8 & (1) \\ y - 12(1) = -15 & (2) \end{cases} \xrightarrow{\text{Simplify.}} \begin{cases} x - y = 7 & (1) \\ y = -3 & (2) \end{cases}$$

We get $y = -3$, and back-substituting into $x - y = 7$, we find that $x = 4$. The solution of the system is $x = 4$, $y = -3$, $z = 1$ or, using ordered triplets, $(4, -3, 1)$.

Sometimes it is advantageous to write a matrix in **reduced row echelon form.** In this form, row operations are used to obtain entries that are 0 above (as well as below) the leading 1 in a row. For example, the row echelon form obtained in the solution to Example 6 is

$$\begin{bmatrix} 1 & -1 & 1 & | & 8 \\ 0 & 1 & -12 & | & -15 \\ 0 & 0 & 1 & | & 1 \end{bmatrix}$$

To write this matrix in reduced row echelon form, we proceed as follows:

$$\begin{bmatrix} 1 & -1 & 1 & | & 8 \\ 0 & 1 & -12 & | & -15 \\ 0 & 0 & 1 & | & 1 \end{bmatrix} \rightarrow \begin{bmatrix} 1 & 0 & -11 & | & -7 \\ 0 & 1 & -12 & | & -15 \\ 0 & 0 & 1 & | & 1 \end{bmatrix} \rightarrow \begin{bmatrix} 1 & 0 & 0 & | & 4 \\ 0 & 1 & 0 & | & -3 \\ 0 & 0 & 1 & | & 1 \end{bmatrix}$$

$$\uparrow$$
$$R_1 = r_2 + r_1$$

$$\uparrow$$
$$R_1 = 11r_3 + r_1$$
$$R_2 = 12r_3 + r_2$$

The matrix is now written in reduced row echelon form. The advantage of writing the matrix in this form is that the solution to the system, $x = 4$, $y = -3$, $z = 1$, is readily found, without the need to back-substitute. Another advantage will be seen in Section 12.4, where the inverse of a matrix is discussed. The methodology used to write a matrix in reduced row echelon form is called **Gauss–Jordan elimination.**

──── **Now Work** PROBLEMS 37 AND 47

The matrix method for solving a system of linear equations also identifies systems that have infinitely many solutions and systems that are inconsistent. Let's see how.

| EXAMPLE 7 | **Solving a Dependent System of Linear Equations Using Matrices** |

Solve: $\begin{cases} 6x - y - z = 4 & (1) \\ -12x + 2y + 2z = -8 & (2) \\ 5x + y - z = 3 & (3) \end{cases}$

Solution We start with the augmented matrix of the system and proceed to obtain a 1 in row 1, column 1 with 0's below.

$$\begin{bmatrix} 6 & -1 & -1 & | & 4 \\ -12 & 2 & 2 & | & -8 \\ 5 & 1 & -1 & | & 3 \end{bmatrix} \rightarrow \begin{bmatrix} 1 & -2 & 0 & | & 1 \\ -12 & 2 & 2 & | & -8 \\ 5 & 1 & -1 & | & 3 \end{bmatrix} \rightarrow \begin{bmatrix} 1 & -2 & 0 & | & 1 \\ 0 & -22 & 2 & | & 4 \\ 0 & 11 & -1 & | & -2 \end{bmatrix}$$

$$\uparrow \qquad\qquad\qquad\qquad \uparrow$$
$$R_1 = -1r_3 + r_1 \qquad\qquad R_2 = 12r_1 + r_2$$
$$R_3 = -5r_1 + r_3$$

Obtaining a 1 in row 2, column 2 without altering column 1 can be accomplished by $R_2 = -\dfrac{1}{22}r_2$ or by $R_3 = \dfrac{1}{11}r_3$ and interchanging rows 2 and 3 or by $R_2 = \dfrac{23}{11}r_3 + r_2$. We shall use the first of these.

$$\begin{bmatrix} 1 & -2 & 0 & | & 1 \\ 0 & -22 & 2 & | & 4 \\ 0 & 11 & -1 & | & -2 \end{bmatrix} \rightarrow \begin{bmatrix} 1 & -2 & 0 & | & 1 \\ 0 & 1 & -\dfrac{1}{11} & | & -\dfrac{2}{11} \\ 0 & 11 & -1 & | & -2 \end{bmatrix} \rightarrow \begin{bmatrix} 1 & -2 & 0 & | & 1 \\ 0 & 1 & -\dfrac{1}{11} & | & -\dfrac{2}{11} \\ 0 & 0 & 0 & | & 0 \end{bmatrix}$$

$$\uparrow \qquad\qquad\qquad\qquad \uparrow$$
$$R_2 = -\dfrac{1}{22}r_2 \qquad\qquad R_3 = -11r_2 + r_3$$

This matrix is in row echelon form. Because the bottom row consists entirely of 0's, the system actually consists of only two equations.

$$\begin{cases} x - 2y = 1 & (1) \\ y - \dfrac{1}{11}z = -\dfrac{2}{11} & (2) \end{cases}$$

To make it easier to write down some of the solutions, we express both x and y in terms of z.

From the second equation, $y = \dfrac{1}{11}z - \dfrac{2}{11}$. Now back-substitute this solution for y into the first equation to get

$$x = 2y + 1 = 2\left(\dfrac{1}{11}z - \dfrac{2}{11}\right) + 1 = \dfrac{2}{11}z + \dfrac{7}{11}$$

The original system is equivalent to the system

$$\begin{cases} x = \dfrac{2}{11}z + \dfrac{7}{11} & \text{(1)} \\[2mm] y = \dfrac{1}{11}z - \dfrac{2}{11} & \text{(2)} \end{cases}$$

where z can be any real number.

Let's look at the situation. The original system of three equations is equivalent to a system containing two equations. This means that any values of x, y, z that satisfy both

$$x = \dfrac{2}{11}z + \dfrac{7}{11} \quad \text{and} \quad y = \dfrac{1}{11}z - \dfrac{2}{11}$$

will be solutions. For example, $z = 0$, $x = \dfrac{7}{11}$, $y = -\dfrac{2}{11}$; $z = 1$, $x = \dfrac{9}{11}$, $y = -\dfrac{1}{11}$; and $z = -1$, $x = \dfrac{5}{11}$, $y = -\dfrac{3}{11}$ are some of the solutions of the original system. There are, in fact, infinitely many values of x, y, and z for which the two equations are satisfied. That is, the original system has infinitely many solutions. We will write the solution of the original system as

$$\begin{cases} x = \dfrac{2}{11}z + \dfrac{7}{11} \\[2mm] y = \dfrac{1}{11}z - \dfrac{2}{11} \end{cases}$$

where z can be any real number, or, using ordered triplets, as

$$\left\{ (x, y, z) \,\middle|\, x = \dfrac{2}{11}z + \dfrac{7}{11}, \ y = \dfrac{1}{11}z - \dfrac{2}{11}, \ z \text{ any real number} \right\}.$$

We can also find the solution by writing the augmented matrix in reduced row echelon form. Starting with the row echelon form, we have

$$\begin{bmatrix} 1 & -2 & 0 & \bigm| & 1 \\[2mm] 0 & 1 & -\dfrac{1}{11} & \bigm| & -\dfrac{2}{11} \\[2mm] 0 & 0 & 0 & \bigm| & 0 \end{bmatrix} \rightarrow \begin{bmatrix} 1 & 0 & -\dfrac{2}{11} & \bigm| & \dfrac{7}{11} \\[2mm] 0 & 1 & -\dfrac{1}{11} & \bigm| & -\dfrac{2}{11} \\[2mm] 0 & 0 & 0 & \bigm| & 0 \end{bmatrix}$$

$$\underset{\uparrow}{R_1 = 2r_2 + r_1}$$

The matrix on the right is in reduced row echelon form. The corresponding system of equations is

$$\begin{cases} x - \dfrac{2}{11}z = \dfrac{7}{11} & \text{(1)} \\[2mm] y - \dfrac{1}{11}z = -\dfrac{2}{11} & \text{(2)} \end{cases}$$

or, equivalently,

$$\begin{cases} x = \dfrac{2}{11}z + \dfrac{7}{11} & \text{(1)} \\[2mm] y = \dfrac{1}{11}z - \dfrac{2}{11} & \text{(2)} \end{cases}$$

where z can be any real number.

Now Work PROBLEM 53

| EXAMPLE 8 | Solving an Inconsistent System of Linear Equations Using Matrices |

$$\text{Solve: } \begin{cases} x + y + z = 6 \\ 2x - y - z = 3 \\ x + 2y + 2z = 0 \end{cases}$$

Solution We proceed as follows, beginning with the augmented matrix.

$$\begin{bmatrix} 1 & 1 & 1 & | & 6 \\ 2 & -1 & -1 & | & 3 \\ 1 & 2 & 2 & | & 0 \end{bmatrix} \rightarrow \begin{bmatrix} 1 & 1 & 1 & | & 6 \\ 0 & -3 & -3 & | & -9 \\ 0 & 1 & 1 & | & -6 \end{bmatrix} \rightarrow \begin{bmatrix} 1 & 1 & 1 & | & 6 \\ 0 & 1 & 1 & | & -6 \\ 0 & -3 & -3 & | & -9 \end{bmatrix} \rightarrow \begin{bmatrix} 1 & 1 & 1 & | & 6 \\ 0 & 1 & 1 & | & -6 \\ 0 & 0 & 0 & | & -27 \end{bmatrix}$$

$$\begin{array}{c} R_2 = -2r_1 + r_2 \\ R_3 = -1r_1 + r_3 \end{array} \qquad \text{Interchange rows 2 and 3.} \qquad R_3 = 3r_2 + r_3$$

This matrix is in row echelon form. The bottom row is equivalent to the equation

$$0x + 0y + 0z = -27$$

which has no solution. The original system is inconsistent.

Now Work PROBLEM 27

The matrix method is especially effective for systems of equations for which the number of equations and the number of variables are unequal. Here, too, such a system is either inconsistent or consistent. If it is consistent, it will have either exactly one solution or infinitely many solutions.

Let's look at a system of four equations containing three variables.

| EXAMPLE 9 | Solving a System of Linear Equations Using Matrices |

$$\text{Solve: } \begin{cases} x - 2y + z = 0 & (1) \\ 2x + 2y - 3z = -3 & (2) \\ y - z = -1 & (3) \\ -x + 4y + 2z = 13 & (4) \end{cases}$$

Solution We proceed as follows, beginning with the augmented matrix.

$$\begin{bmatrix} 1 & -2 & 1 & | & 0 \\ 2 & 2 & -3 & | & -3 \\ 0 & 1 & -1 & | & -1 \\ -1 & 4 & 2 & | & 13 \end{bmatrix} \rightarrow \begin{bmatrix} 1 & -2 & 1 & | & 0 \\ 0 & 6 & -5 & | & -3 \\ 0 & 1 & -1 & | & -1 \\ 0 & 2 & 3 & | & 13 \end{bmatrix} \rightarrow \begin{bmatrix} 1 & -2 & 1 & | & 0 \\ 0 & 1 & -1 & | & -1 \\ 0 & 6 & -5 & | & -3 \\ 0 & 2 & 3 & | & 13 \end{bmatrix}$$

$$\begin{array}{c} R_2 = -2r_1 + r_2 \\ R_4 = r_1 + r_4 \end{array} \qquad \text{Interchange rows 2 and 3.}$$

$$\rightarrow \begin{bmatrix} 1 & -2 & 1 & | & 0 \\ 0 & 1 & -1 & | & -1 \\ 0 & 0 & 1 & | & 3 \\ 0 & 0 & 5 & | & 15 \end{bmatrix} \rightarrow \begin{bmatrix} 1 & -2 & 1 & | & 0 \\ 0 & 1 & -1 & | & -1 \\ 0 & 0 & 1 & | & 3 \\ 0 & 0 & 0 & | & 0 \end{bmatrix}$$

$$\begin{array}{c} R_3 = -6r_2 + r_3 \\ R_4 = -2r_2 + r_4 \end{array} \qquad R_4 = -5r_3 + r_4$$

We could stop here, since the matrix is in row echelon form, and back-substitute $z = 3$ to find x and y. Or we can continue to obtain the reduced row echelon form.

$$\begin{bmatrix} 1 & -2 & 1 & | & 0 \\ 0 & 1 & -1 & | & -1 \\ 0 & 0 & 1 & | & 3 \\ 0 & 0 & 0 & | & 0 \end{bmatrix} \rightarrow \begin{bmatrix} 1 & 0 & -1 & | & -2 \\ 0 & 1 & -1 & | & -1 \\ 0 & 0 & 1 & | & 3 \\ 0 & 0 & 0 & | & 0 \end{bmatrix} \rightarrow \begin{bmatrix} 1 & 0 & 0 & | & 1 \\ 0 & 1 & 0 & | & 2 \\ 0 & 0 & 1 & | & 3 \\ 0 & 0 & 0 & | & 0 \end{bmatrix}$$
$$\uparrow \qquad\qquad\qquad\qquad \uparrow$$
$$R_1 = 2r_2 + r_1 \qquad\qquad R_1 = r_3 + r_1$$
$$R_2 = r_3 + r_2$$

The matrix is now in reduced row echelon form, and we can see that the solution is $x = 1$, $y = 2$, $z = 3$, or using ordered triplets, $(1, 2, 3)$.

━━━━━━━━➤ **Now Work** PROBLEM 69

EXAMPLE 10	**Penalties in the 2006 Fifa World Cup**

Italy and France combined for a total of 46 penalties during the 2006 Fifa World Cup. The penalties were a combination of fouls, yellow cards (cautions), and red cards (expulsions). There was one less red card than half the number of yellow cards and one more foul than 8 times the total number of cards. How many of each type of penalty were there during the match?

Source: fifaworldcup.com

Solution Let f, y, and r represent the number of fouls, yellow cards, and red cards, respectively. There was a total of 46 penalties, which is found by adding the number of fouls, yellow cards, and red cards. The first equation is $f + y + r = 46$. There was one less red card than half the number of yellow cards, so, the number of red cards equals $\frac{1}{2}$ the number of yellow cards, minus 1. This statement leads to the second equation: $r = \frac{1}{2}y - 1$. We also know that there was one more foul than 8 times the total number of cards, so, the number of fouls equals 1 plus the product of 8 and the sum of the number of yellow cards and the number of red cards. This statement gives the third equation: $f = 1 + 8(y + r)$.

Putting each equation in standard form, we have the following system of equations:

$$\begin{cases} f + y + r = 46 & (1) \\ \quad -\frac{1}{2}y + r = -1 & (2) \\ f - 8y - 8r = 1 & (3) \end{cases}$$

We begin with the augmented matrix and proceed as follows:

$$\begin{bmatrix} 1 & 1 & 1 & | & 46 \\ 0 & -\frac{1}{2} & 1 & | & -1 \\ 1 & -8 & -8 & | & 1 \end{bmatrix} \rightarrow \begin{bmatrix} 1 & 1 & 1 & | & 46 \\ 0 & -\frac{1}{2} & 1 & | & -1 \\ 0 & -9 & -9 & | & -45 \end{bmatrix} \rightarrow \begin{bmatrix} 1 & 1 & 1 & | & 46 \\ 0 & 1 & -2 & | & 2 \\ 0 & -9 & -9 & | & -45 \end{bmatrix}$$
$$\uparrow \qquad\qquad\qquad\qquad \uparrow$$
$$R_3 = -r_1 + r_3 \qquad\qquad R_2 = -2r_2$$

$$\rightarrow \begin{bmatrix} 1 & 1 & 1 & | & 46 \\ 0 & 1 & -2 & | & 2 \\ 0 & 0 & -27 & | & -27 \end{bmatrix} \rightarrow \begin{bmatrix} 1 & 1 & 1 & | & 46 \\ 0 & 1 & -2 & | & 2 \\ 0 & 0 & 1 & | & 1 \end{bmatrix}$$
$$\uparrow \qquad\qquad\qquad\qquad \uparrow$$
$$R_3 = 9r_2 + r_3 \qquad\qquad R_3 = -\frac{1}{27}r_3$$

The matrix is now in echelon form. The final matrix represents the system

$$\begin{cases} f + y + r = 46 & (1) \\ y - 2r = 2 & (2) \\ r = 1 & (3) \end{cases}$$

From equation (3), we determine that 1 red card was given. Back-substitute $r = 1$ into equation (2) to find that $y = 4$, so 4 yellow cards were given. Back-substitute these values into equation (1) and to find that $f = 41$, so 41 fouls were called.

■

 COMMENT Most graphing utilities have the capability to put an augmented matrix into row echelon form (ref) and also reduced row echelon form (rref). See the Appendix, Section 7, for a discussion. ■

12.2 Assess Your Understanding

Concepts and Vocabulary

1. An m by n rectangular array of numbers is called a(n) _____.

2. The matrix used to represent a system of linear equations is called a(n) _____ matrix.

3. *True or False* The augmented matrix of a system of two equations containing three variables has two rows and four columns.

4. *True or False* The matrix $\begin{bmatrix} 1 & 3 & | & -2 \\ 0 & 1 & | & 5 \\ 0 & 0 & | & 0 \end{bmatrix}$ is in row echelon form.

Skill Building

In Problems 5–16, write the augmented matrix of the given system of equations.

5. $\begin{cases} x - 5y = 5 \\ 4x + 3y = 6 \end{cases}$

6. $\begin{cases} 3x + 4y = 7 \\ 4x - 2y = 5 \end{cases}$

7. $\begin{cases} 2x + 3y - 6 = 0 \\ 4x - 6y + 2 = 0 \end{cases}$

8. $\begin{cases} 9x - y = 0 \\ 3x - y - 4 = 0 \end{cases}$

9. $\begin{cases} 0.01x - 0.03y = 0.06 \\ 0.13x + 0.10y = 0.20 \end{cases}$

10. $\begin{cases} \dfrac{4}{3}x - \dfrac{3}{2}y = \dfrac{3}{4} \\ -\dfrac{1}{4}x + \dfrac{1}{3}y = \dfrac{2}{3} \end{cases}$

11. $\begin{cases} x - y + z = 10 \\ 3x + 3y = 5 \\ x + y + 2z = 2 \end{cases}$

12. $\begin{cases} 5x - y - z = 0 \\ x + y = 5 \\ 2x - 3z = 2 \end{cases}$

13. $\begin{cases} x + y - z = 2 \\ 3x - 2y = 2 \\ 5x + 3y - z = 1 \end{cases}$

14. $\begin{cases} 2x + 3y - 4z = 0 \\ x - 5z + 2 = 0 \\ x + 2y - 3z = -2 \end{cases}$

15. $\begin{cases} x - y - z = 10 \\ 2x + y + 2z = -1 \\ -3x + 4y = 5 \\ 4x - 5y + z = 0 \end{cases}$

16. $\begin{cases} x - y + 2z - w = 5 \\ x + 3y - 4z + 2w = 2 \\ 3x - y - 5z - w = -1 \end{cases}$

In Problems 17–24, write the system of equations corresponding to each augmented matrix. Then perform each row operation on the given augmented matrix.

17. $\begin{bmatrix} 1 & -3 & | & -2 \\ 2 & -5 & | & 5 \end{bmatrix}$ $\quad R_2 = -2r_1 + r_2$

18. $\begin{bmatrix} 1 & -3 & | & -3 \\ 2 & -5 & | & -4 \end{bmatrix}$ $\quad R_2 = -2r_1 + r_2$

19. $\begin{bmatrix} 1 & -3 & 4 & | & 3 \\ 3 & -5 & 6 & | & 6 \\ -5 & 3 & 4 & | & 6 \end{bmatrix}$ (a) $R_2 = -3r_1 + r_2$
(b) $R_3 = 5r_1 + r_3$

20. $\begin{bmatrix} 1 & -3 & 3 & | & -5 \\ -4 & -5 & -3 & | & -5 \\ -3 & -2 & 4 & | & 6 \end{bmatrix}$ (a) $R_2 = 4r_1 + r_2$
(b) $R_3 = 3r_1 + r_3$

21. $\begin{bmatrix} 1 & -3 & 2 & | & -6 \\ 2 & -5 & 3 & | & -4 \\ -3 & -6 & 4 & | & 6 \end{bmatrix}$ (a) $R_2 = -2r_1 + r_2$
(b) $R_3 = 3r_1 + r_3$

22. $\begin{bmatrix} 1 & -3 & -4 & | & -6 \\ 6 & -5 & 6 & | & -6 \\ -1 & 1 & 4 & | & 6 \end{bmatrix}$ (a) $R_2 = -6r_1 + r_2$
(b) $R_3 = r_1 + r_3$

23. $\begin{bmatrix} 5 & -3 & 1 & | & -2 \\ 2 & -5 & 6 & | & -2 \\ -4 & 1 & 4 & | & 6 \end{bmatrix}$ (a) $R_1 = -2r_2 + r_1$
(b) $R_1 = r_3 + r_1$

24. $\begin{bmatrix} 4 & -3 & -1 & | & 2 \\ 3 & -5 & 2 & | & 6 \\ -3 & -6 & 4 & | & 6 \end{bmatrix}$ (a) $R_1 = -r_2 + r_1$
(b) $R_1 = r_3 + r_1$

In Problems 25–36, the reduced row echelon form of a system of linear equations is given. Write the system of equations corresponding to the given matrix. Use x, y; or x, y, z; or x_1, x_2, x_3, x_4 as variables. Determine whether the system is consistent or inconsistent. If it is consistent, give the solution.

25. $\begin{bmatrix} 1 & 0 & | & 5 \\ 0 & 1 & | & -1 \end{bmatrix}$

26. $\begin{bmatrix} 1 & 0 & | & -4 \\ 0 & 1 & | & 0 \end{bmatrix}$

27. $\begin{bmatrix} 1 & 0 & 0 & | & 1 \\ 0 & 1 & 0 & | & 2 \\ 0 & 0 & 0 & | & 3 \end{bmatrix}$

28. $\begin{bmatrix} 1 & 0 & 0 & | & 0 \\ 0 & 1 & 0 & | & 0 \\ 0 & 0 & 0 & | & 2 \end{bmatrix}$

29. $\begin{bmatrix} 1 & 0 & 2 & | & -1 \\ 0 & 1 & -4 & | & -2 \\ 0 & 0 & 0 & | & 0 \end{bmatrix}$

30. $\begin{bmatrix} 1 & 0 & 4 & | & 4 \\ 0 & 1 & 3 & | & 2 \\ 0 & 0 & 0 & | & 0 \end{bmatrix}$

31. $\begin{bmatrix} 1 & 0 & 0 & 0 & | & 1 \\ 0 & 1 & 0 & 1 & | & 2 \\ 0 & 0 & 1 & 2 & | & 3 \end{bmatrix}$

32. $\begin{bmatrix} 1 & 0 & 0 & 0 & | & 1 \\ 0 & 1 & 0 & 2 & | & 2 \\ 0 & 0 & 1 & 3 & | & 0 \end{bmatrix}$

33. $\begin{bmatrix} 1 & 0 & 0 & 4 & | & 2 \\ 0 & 1 & 1 & 3 & | & 3 \\ 0 & 0 & 0 & 0 & | & 0 \end{bmatrix}$

34. $\begin{bmatrix} 1 & 0 & 0 & 0 & | & 1 \\ 0 & 1 & 0 & 0 & | & 2 \\ 0 & 0 & 1 & 2 & | & 3 \end{bmatrix}$

35. $\begin{bmatrix} 1 & 0 & 0 & 1 & | & -2 \\ 0 & 1 & 0 & 2 & | & 2 \\ 0 & 0 & 1 & -1 & | & 0 \\ 0 & 0 & 0 & 0 & | & 0 \end{bmatrix}$

36. $\begin{bmatrix} 1 & 0 & 0 & 0 & | & 1 \\ 0 & 1 & 0 & 0 & | & 2 \\ 0 & 0 & 1 & 0 & | & 3 \\ 0 & 0 & 0 & 1 & | & 0 \end{bmatrix}$

In Problems 37–72, solve each system of equations using matrices (row operations). If the system has no solution, say that it is inconsistent.

37. $\begin{cases} x + y = 8 \\ x - y = 4 \end{cases}$

38. $\begin{cases} x + 2y = 5 \\ x + y = 3 \end{cases}$

39. $\begin{cases} 2x - 4y = -2 \\ 3x + 2y = 3 \end{cases}$

40. $\begin{cases} 3x + 3y = 3 \\ 4x + 2y = \dfrac{8}{3} \end{cases}$

41. $\begin{cases} x + 2y = 4 \\ 2x + 4y = 8 \end{cases}$

42. $\begin{cases} 3x - y = 7 \\ 9x - 3y = 21 \end{cases}$

43. $\begin{cases} 2x + 3y = 6 \\ x - y = \dfrac{1}{2} \end{cases}$

44. $\begin{cases} \dfrac{1}{2}x + y = -2 \\ x - 2y = 8 \end{cases}$

45. $\begin{cases} 3x - 5y = 3 \\ 15x + 5y = 21 \end{cases}$

46. $\begin{cases} 2x - y = -1 \\ x + \dfrac{1}{2}y = \dfrac{3}{2} \end{cases}$

47. $\begin{cases} x - y = 6 \\ 2x - 3z = 16 \\ 2y + z = 4 \end{cases}$

48. $\begin{cases} 2x + y = -4 \\ -2y + 4z = 0 \\ 3x - 2z = -11 \end{cases}$

49. $\begin{cases} x - 2y + 3z = 7 \\ 2x + y + z = 4 \\ -3x + 2y - 2z = -10 \end{cases}$

50. $\begin{cases} 2x + y - 3z = 0 \\ -2x + 2y + z = -7 \\ 3x - 4y - 3z = 7 \end{cases}$

51. $\begin{cases} 2x - 2y - 2z = 2 \\ 2x + 3y + z = 2 \\ 3x + 2y = 0 \end{cases}$

52. $\begin{cases} 2x - 3y - z = 0 \\ -x + 2y + z = 5 \\ 3x - 4y - z = 1 \end{cases}$

53. $\begin{cases} -x + y + z = -1 \\ -x + 2y - 3z = -4 \\ 3x - 2y - 7z = 0 \end{cases}$

54. $\begin{cases} 2x - 3y - z = 0 \\ 3x + 2y + 2z = 2 \\ x + 5y + 3z = 2 \end{cases}$

55. $\begin{cases} 2x - 2y + 3z = 6 \\ 4x - 3y + 2z = 0 \\ -2x + 3y - 7z = 1 \end{cases}$

56. $\begin{cases} 3x - 2y + 2z = 6 \\ 7x - 3y + 2z = -1 \\ 2x - 3y + 4z = 0 \end{cases}$

57. $\begin{cases} x + y - z = 6 \\ 3x - 2y + z = -5 \\ x + 3y - 2z = 14 \end{cases}$

58. $\begin{cases} x - y + z = -4 \\ 2x - 3y + 4z = -15 \\ 5x + y - 2z = 12 \end{cases}$

59. $\begin{cases} x + 2y - z = -3 \\ 2x - 4y + z = -7 \\ -2x + 2y - 3z = 4 \end{cases}$

60. $\begin{cases} x + 4y - 3z = -8 \\ 3x - y + 3z = 12 \\ x + y + 6z = 1 \end{cases}$

61. $\begin{cases} 3x + y - z = \dfrac{2}{3} \\ 2x - y + z = 1 \\ 4x + 2y = \dfrac{8}{3} \end{cases}$

62. $\begin{cases} x + y = 1 \\ 2x - y + z = 1 \\ x + 2y + z = \dfrac{8}{3} \end{cases}$

63. $\begin{cases} x + y + z + w = 4 \\ 2x - y + z = 0 \\ 3x + 2y + z - w = 6 \\ x - 2y - 2z + 2w = -1 \end{cases}$

64. $\begin{cases} x + y + z + w = 4 \\ -x + 2y + z = 0 \\ 2x + 3y + z - w = 6 \\ -2x + y - 2z + 2w = -1 \end{cases}$

65. $\begin{cases} x + 2y + z = 1 \\ 2x - y + 2z = 2 \\ 3x + y + 3z = 3 \end{cases}$

66. $\begin{cases} x + 2y - z = 3 \\ 2x - y + 2z = 6 \\ x - 3y + 3z = 4 \end{cases}$

67. $\begin{cases} x - y + z = 5 \\ 3x + 2y - 2z = 0 \end{cases}$

68. $\begin{cases} 2x + y - z = 4 \\ -x + y + 3z = 1 \end{cases}$

69. $\begin{cases} 2x + 3y - z = 3 \\ x - y - z = 0 \\ -x + y + z = 0 \\ x + y + 3z = 5 \end{cases}$

70. $\begin{cases} x - 3y + z = 1 \\ 2x - y - 4z = 0 \\ x - 3y + 2z = 1 \\ x - 2y = 5 \end{cases}$

71. $\begin{cases} 4x + y + z - w = 4 \\ x - y + 2z + 3w = 3 \end{cases}$

72. $\begin{cases} -4x + y = 5 \\ 2x - y + z - w = 5 \\ z + w = 4 \end{cases}$

Applications and Extensions

73. Curve Fitting Find the function $y = ax^2 + bx + c$ whose graph contains the points $(1, 2)$, $(-2, -7)$, and $(2, -3)$.

74. Curve Fitting Find the function $y = ax^2 + bx + c$ whose graph contains the points $(1, -1)$, $(3, -1)$, and $(-2, 14)$.

75. Curve Fitting Find the function $f(x) = ax^3 + bx^2 + cx + d$ for which $f(-3) = -112$, $f(-1) = -2$, $f(1) = 4$, and $f(2) = 13$.

76. Curve Fitting Find the function $f(x) = ax^3 + bx^2 + cx + d$ for which $f(-2) = -10$, $f(-1) = 3$, $f(1) = 5$, and $f(3) = 15$.

77. Nutrition A dietitian at Palos Community Hospital wants a patient to have a meal that has 78 grams of protein, 59 grams of carbohydrates, and 75 milligrams of vitamin A. The hospital food service tells the dietitian that the dinner for today is salmon steak, baked eggs, and acorn squash. Each serving of salmon steak has 30 grams of protein, 20 grams of carbohydrates, and 2 milligrams of vitamin A. Each serving of baked eggs contains 15 grams of protein, 2 grams of carbohydrates, and 20 milligrams of vitamin A. Each serving of acorn squash contains 3 grams of protein, 25 grams of carbohydrates, and 32 milligrams of vitamin A. How many servings of each food should the dietitian provide for the patient?

78. Nutrition A dietitian at General Hospital wants a patient to have a meal that has 47 grams of protein, 58 grams of carbohydrates, and 630 milligrams of calcium. The hospital food service tells the dietitian that the dinner for today is pork chops, corn on the cob, and 2% milk. Each serving of pork chops has 23 grams of protein, 0 grams of carbohydrates, and 10 milligrams of calcium. Each serving of corn on the cob contains 3 grams of protein, 16 grams of carbohydrates, and 10 milligrams of calcium. Each glass of 2% milk contains 9 grams of protein, 13 grams of carbohydrates, and 300 milligrams of calcium. How many servings of each food should the dietitian provide for the patient?

79. Financial Planning Carletta has $10,000 to invest. As her financial consultant, you recommend that she invest in Treasury bills that yield 6%, Treasury bonds that yield 7%, and corporate bonds that yield 8%. Carletta wants to have an annual income of $680, and the amount invested in corporate bonds must be half that invested in Treasury bills. Find the amount in each investment.

80. Landscaping A landscape company is hired to plant trees in three new subdivisions. The company charges the developer for each tree planted, an hourly rate to plant the trees, and a fixed delivery charge. In one subdivision it took 166 labor hours to plant 250 trees for a cost of $7520. In a second subdivision it took 124 labor hours to plant 200 trees for a cost of $5945. In the final subdivision it took 200 labor hours to plant 300 trees for a cost of $8985. Determine the cost for each tree, the hourly labor charge, and the fixed delivery charge.

Sources: gurney.com; www.bx.org

81. Production To manufacture an automobile requires painting, drying, and polishing. Epsilon Motor Company produces three types of cars: the Delta, the Beta, and the Sigma. Each Delta requires 10 hours for painting, 3 hours for drying, and 2 hours for polishing. A Beta requires 16 hours for painting, 5 hours for drying, and 3 hours for polishing, and a Sigma requires 8 hours for painting, 2 hours for drying, and 1 hour for polishing. If the company has 240 hours for painting, 69 hours for drying, and 41 hours for polishing per month, how many of each type of car are produced?

82. Production A Florida juice company completes the preparation of its products by sterilizing, filling, and labeling bottles. Each case of orange juice requires 9 minutes for sterilizing, 6 minutes for filling, and 1 minute for labeling. Each case of grapefruit juice requires 10 minutes for sterilizing, 4 minutes for filling, and 2 minutes for labeling. Each case of tomato juice requires 12 minutes for sterilizing, 4 minutes for filling, and 1 minute for labeling. If the company runs the sterilizing machine for 398 minutes, the filling machine for 164 minutes, and the labeling machine for 58 minutes, how many cases of each type of juice are prepared?

83. Electricity: Kirchhoff's Rules An application of Kirchhoff's Rules to the circuit shown results in the following system of equations:

$$\begin{cases} -4 + 8 - 2I_2 = 0 \\ 8 = 5I_4 + I_1 \\ 4 = 3I_3 + I_1 \\ I_3 + I_4 = I_1 \end{cases}$$

Find the currents I_1, I_2, I_3, and I_4.

Source: Based on Raymond Serway, *Physics,* 3rd ed. (Philadelphia: Saunders, 1990), Prob. 34, p. 790.

84. Electricity: Kirchhoff's Rules An application of Kirchhoff's Rules to the circuit shown results in the following system of equations:

$$\begin{cases} I_1 = I_3 + I_2 \\ 24 - 6I_1 - 3I_3 = 0 \\ 12 + 24 - 6I_1 - 6I_2 = 0 \end{cases}$$

Find the currents I_1, I_2, and I_3.

Source: Ibid., Prob. 38, p. 791.

85. Financial Planning Three retired couples each require an additional annual income of $2000 per year. As their financial consultant, you recommend that they invest some money in Treasury bills that yield 7%, some money in corporate bonds

that yield 9%, and some money in junk bonds that yield 11%. Prepare a table for each couple showing the various ways that their goals can be achieved:

(a) If the first couple has $20,000 to invest.

(b) If the second couple has $25,000 to invest.

(c) If the third couple has $30,000 to invest.

(d) What advice would you give each couple regarding the amount to invest and the choices available?

[**Hint:** Higher yields generally carry more risk.]

86. Financial Planning A young couple has $25,000 to invest. As their financial consultant, you recommend that they invest some money in Treasury bills that yield 7%, some money in corporate bonds that yield 9%, and some money in junk bonds that yield 11%. Prepare a table showing the various ways that this couple can achieve the following goals:

(a) The couple wants $1500 per year in income.

(b) The couple wants $2000 per year in income.

(c) The couple wants $2500 per year in income.

(d) What advice would you give this couple regarding the income that they require and the choices available?

[**Hint:** Higher yields generally carry more risk.]

87. Pharmacy A doctor's prescription calls for a daily intake of a supplement containing 40 mg of vitamin C and 30 mg of vitamin D. Your pharmacy stocks three supplements that can be used: one contains 20% vitamin C and 30% vitamin D; a second, 40% vitamin C and 20% vitamin D; and a third, 30% vitamin C and 50% vitamin D. Create a table showing the possible combinations that could be used to fill the prescription.

88. Pharmacy A doctor's prescription calls for the creation of pills that contain 12 units of vitamin B_{12} and 12 units of vitamin E. Your pharmacy stocks three powders that can be used to make these pills: one contains 20% vitamin B_{12} and 30% vitamin E; a second, 40% vitamin B_{12} and 20% vitamin E; and a third, 30% vitamin B_{12} and 40% vitamin E. Create a table showing the possible combinations of each powder that could be mixed in each pill.

Discussion and Writing

89. Write a brief paragraph or two that outline your strategy for solving a system of linear equations using matrices.

90. When solving a system of linear equations using matrices, do you prefer to place the augmented matrix in row echelon form or in reduced row echelon form? Give reasons for your choice.

91. Make up a system of three linear equations containing three variables that has:

(a) No solution

(b) Exactly one solution

(c) Infinitely many solutions

Give the three systems to a friend to solve and critique.

12.3 Systems of Linear Equations: Determinants

In the preceding section, we described a method of using matrices to solve a system of linear equations. This section deals with yet another method for solving systems of linear equations; however, it can be used only when the number of equations equals the number of variables. Although the method will work for any system (provided that the number of equations equals the number of variables), it is most often used for systems of two equations containing two variables or three equations containing three variables. This method, called *Cramer's Rule,* is based on the concept of a *determinant.*

1 Evaluate 2 by 2 Determinants

DEFINITION

If a, b, c, and d are four real numbers, the symbol

$$D = \begin{vmatrix} a & b \\ c & d \end{vmatrix}$$

is called a **2 by 2 determinant.** Its value is the number $ad - bc$; that is,

$$D = \begin{vmatrix} a & b \\ c & d \end{vmatrix} = ad - bc \qquad \textbf{(1)}$$

COMMENT A matrix is an array of numbers; it has no value. A determinant represents a number. ∎

The following device may be helpful for remembering the value of a 2 by 2 determinant:

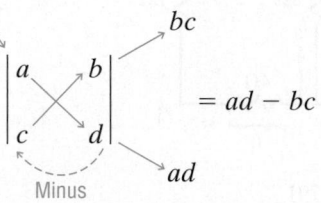

$$= ad - bc$$

EXAMPLE 1 **Evaluating a 2 × 2 Determinant**

Evaluate: $\begin{vmatrix} 3 & -2 \\ 6 & 1 \end{vmatrix}$

Solution $\begin{vmatrix} 3 & -2 \\ 6 & 1 \end{vmatrix} = (3)(1) - (6)(-2)$

$$= 3 - (-12)$$

$$= 15$$

∎

Now Work PROBLEM 7

2 Use Cramer's Rule to Solve a System of Two Equations Containing Two Variables

Let's now see the role that a 2 by 2 determinant plays in the solution of a system of two equations containing two variables. Consider the system

$$\begin{cases} ax + by = s & (1) \\ cx + dy = t & (2) \end{cases} \qquad \textbf{(2)}$$

We shall use the method of elimination to solve this system.

Provided $d \neq 0$ and $b \neq 0$, this system is equivalent to the system

$$\begin{cases} adx + bdy = sd & (1) \quad \text{Multiply by } d. \\ bcx + bdy = tb & (2) \quad \text{Multiply by } b. \end{cases}$$

Subtracting the second equation from the first equation, we get

$$(ad - bc)x + 0 \cdot y = sd - tb$$

which can be rewritten using determinant notation as

$$\begin{vmatrix} a & b \\ c & d \end{vmatrix} x = \begin{vmatrix} s & b \\ t & d \end{vmatrix}$$

If $D = \begin{vmatrix} a & b \\ c & d \end{vmatrix} = ad - bc \neq 0$, we can solve for x to get

$$x = \frac{\begin{vmatrix} s & b \\ t & d \end{vmatrix}}{\begin{vmatrix} a & b \\ c & d \end{vmatrix}} = \frac{\begin{vmatrix} s & b \\ t & d \end{vmatrix}}{D} \qquad (3)$$

Return now to the original system (2). Provided that $a \neq 0$ and $c \neq 0$, the system is equivalent to

$$\begin{cases} acx + bcy = cs & (1) \quad \text{Multiply by } c. \\ acx + ady = at & (2) \quad \text{Multiply by } a. \end{cases}$$

Subtracting the first equation from the second equation, we get

$$0 \cdot x + (ad - bc)y = at - cs$$

which can be rewritten using determinant notation as

$$\begin{vmatrix} a & b \\ c & d \end{vmatrix} y = \begin{vmatrix} a & s \\ c & t \end{vmatrix}$$

If $D = \begin{vmatrix} a & b \\ c & d \end{vmatrix} = ad - bc \neq 0$, we can solve for y to get

$$y = \frac{\begin{vmatrix} a & s \\ c & t \end{vmatrix}}{\begin{vmatrix} a & b \\ c & d \end{vmatrix}} = \frac{\begin{vmatrix} a & s \\ c & t \end{vmatrix}}{D} \qquad (4)$$

Equations (3) and (4) lead us to the following result, called **Cramer's Rule.**

THEOREM

Cramer's Rule for Two Equations Containing Two Variables

The solution to the system of equations

$$\begin{cases} ax + by = s & (1) \\ cx + dy = t & (2) \end{cases} \qquad (5)$$

is given by

$$x = \frac{\begin{vmatrix} s & b \\ t & d \end{vmatrix}}{\begin{vmatrix} a & b \\ c & d \end{vmatrix}} \qquad y = \frac{\begin{vmatrix} a & s \\ c & t \end{vmatrix}}{\begin{vmatrix} a & b \\ c & d \end{vmatrix}} \qquad (6)$$

provided that

$$D = \begin{vmatrix} a & b \\ c & d \end{vmatrix} = ad - bc \neq 0$$

In this derivation for Cramer's Rule, we assumed that none of the numbers a, b, c, and d was 0. In Problem 61 you will be asked to complete the proof under the less stringent condition that $D = ad - bc \neq 0$.

Now look carefully at the pattern in Cramer's Rule. The denominator in the solution (6) is the determinant of the coefficients of the variables.

$$\begin{cases} ax + by = s \\ cx + dy = t \end{cases} \qquad D = \begin{vmatrix} a & b \\ c & d \end{vmatrix}$$

In the solution for x, the numerator is the determinant, denoted by D_x, formed by replacing the entries in the first column (the coefficients of x) of D by the constants on the right side of the equal sign.

$$D_x = \begin{vmatrix} s & b \\ t & d \end{vmatrix}$$

In the solution for y, the numerator is the determinant, denoted by D_y, formed by replacing the entries in the second column (the coefficients of y) of D by the constants on the right side of the equal sign.

$$D_y = \begin{vmatrix} a & s \\ c & t \end{vmatrix}$$

Cramer's Rule then states that, if $D \neq 0$,

$$x = \frac{D_x}{D} \qquad y = \frac{D_y}{D} \qquad (7)$$

| EXAMPLE 2 | **Solving a System of Linear Equations Using Determinants** |

Use Cramer's Rule, if applicable, to solve the system

$$\begin{cases} 3x - 2y = 4 & (1) \\ 6x + y = 13 & (2) \end{cases}$$

Solution The determinant D of the coefficients of the variables is

$$D = \begin{vmatrix} 3 & -2 \\ 6 & 1 \end{vmatrix} = (3)(1) - (6)(-2) = 15$$

Because $D \neq 0$, Cramer's Rule (7) can be used.

$$x = \frac{D_x}{D} = \frac{\begin{vmatrix} 4 & -2 \\ 13 & 1 \end{vmatrix}}{15} \qquad\qquad y = \frac{D_y}{D} = \frac{\begin{vmatrix} 3 & 4 \\ 6 & 13 \end{vmatrix}}{15}$$

$$= \frac{(4)(1) - (13)(-2)}{15} \qquad\qquad = \frac{(3)(13) - (6)(4)}{15}$$

$$= \frac{30}{15} \qquad\qquad\qquad\qquad = \frac{15}{15}$$

$$= 2 \qquad\qquad\qquad\qquad\quad = 1$$

The solution is $x = 2$, $y = 1$, or, using ordered pairs $(2, 1)$.

In attempting to use Cramer's Rule, if the determinant D of the coefficients of the variables equals 0 (so that Cramer's Rule is not applicable), the system either is inconsistent or has infinitely many solutions.

Now Work PROBLEM 15

3 Evaluate 3 by 3 Determinants

To use Cramer's Rule to solve a system of three equations containing three variables, we need to define a 3 by 3 determinant.

A **3 by 3 determinant** is symbolized by

$$\begin{vmatrix} a_{11} & a_{12} & a_{13} \\ a_{21} & a_{22} & a_{23} \\ a_{31} & a_{32} & a_{33} \end{vmatrix} \tag{8}$$

in which $a_{11}, a_{12}, \ldots,$ are real numbers.

As with matrices, we use a double subscript to identify an entry by indicating its row and column numbers. For example, the entry a_{23} is in row 2, column 3.

The value of a 3 by 3 determinant may be defined in terms of 2 by 2 determinants by the following formula:

$$\begin{vmatrix} a_{11} & a_{12} & a_{13} \\ a_{21} & a_{22} & a_{23} \\ a_{31} & a_{32} & a_{33} \end{vmatrix} = a_{11} \underset{\substack{\uparrow \\ \text{2 by 2} \\ \text{determinant} \\ \text{left after} \\ \text{removing the row} \\ \text{and column} \\ \text{containing } a_{11}}}{\begin{vmatrix} a_{22} & a_{23} \\ a_{32} & a_{33} \end{vmatrix}} \overset{\text{Minus}}{-} a_{12} \underset{\substack{\uparrow \\ \text{2 by 2} \\ \text{determinant} \\ \text{left after} \\ \text{removing the row} \\ \text{and column} \\ \text{containing } a_{12}}}{\begin{vmatrix} a_{21} & a_{23} \\ a_{31} & a_{33} \end{vmatrix}} \overset{\text{Plus}}{+} a_{13} \underset{\substack{\uparrow \\ \text{2 by 2} \\ \text{determinant} \\ \text{left after} \\ \text{removing the row} \\ \text{and column} \\ \text{containing } a_{13}}}{\begin{vmatrix} a_{21} & a_{22} \\ a_{31} & a_{32} \end{vmatrix}} \tag{9}$$

The 2 by 2 determinants shown in formula (9) are called **minors** of the 3 by 3 determinant. For an n by n determinant, the **minor** M_{ij} of entry a_{ij} is the determinant resulting from removing the ith row and jth column.

EXAMPLE 3

Finding Minors of a 3 by 3 Determinant

For the determinant $A = \begin{vmatrix} 2 & -1 & 3 \\ -2 & 5 & 1 \\ 0 & 6 & -9 \end{vmatrix}$, find: (a) M_{12} (b) M_{23}

Solution

(a) M_{12} is the determinant that results from removing the first row and second column from A.

$$A = \begin{vmatrix} 2 & -1 & 3 \\ -2 & 5 & 1 \\ 0 & 6 & -9 \end{vmatrix} \qquad M_{12} = \begin{vmatrix} -2 & 1 \\ 0 & -9 \end{vmatrix} = (-2)(-9) - (0)(1) = 18$$

(b) M_{23} is the determinant that results from removing the second row and third column from A.

$$A = \begin{vmatrix} 2 & -1 & 3 \\ -2 & 5 & 1 \\ 0 & 6 & -9 \end{vmatrix} \qquad M_{23} = \begin{vmatrix} 2 & -1 \\ 0 & 6 \end{vmatrix} = (2)(6) - (0)(-1) = 12$$

Referring back to formula (9), we see that each element a_{ij} is multiplied by its minor, but sometimes this term is added and other times subtracted. To determine whether to add or subtract a term, we must consider the *cofactor*.

DEFINITION

For an n by n determinant A, the **cofactor** of entry a_{ij}, denoted by A_{ij}, is given by

$$A_{ij} = (-1)^{i+j}M_{ij}$$

where M_{ij} is the minor of entry a_{ij}.

The exponent of $(-1)^{i+j}$ is the sum of the row and column of the entry a_{ij}, so if $i + j$ is even, $(-1)^{i+j}$ will equal 1, and if $i + j$ is odd, $(-1)^{i+j}$ will equal -1.

To find the value of a determinant, multiply each entry in any row or column by its cofactor and sum the results. This process is referred to as **expanding across a row or column.** For example, the value of the 3 by 3 determinant in formula (9) was found by expanding across row 1.

If we choose to expand down column 2, we obtain

$$\begin{vmatrix} a_{11} & a_{12} & a_{13} \\ a_{21} & a_{22} & a_{23} \\ a_{31} & a_{32} & a_{33} \end{vmatrix} = (-1)^{1+2} a_{12} \begin{vmatrix} a_{21} & a_{23} \\ a_{31} & a_{33} \end{vmatrix} + (-1)^{2+2} a_{22} \begin{vmatrix} a_{11} & a_{13} \\ a_{31} & a_{33} \end{vmatrix} + (-1)^{3+2} a_{32} \begin{vmatrix} a_{11} & a_{13} \\ a_{21} & a_{23} \end{vmatrix}$$

Expand down column 2.

If we choose to expand across row 3, we obtain

$$\begin{vmatrix} a_{11} & a_{12} & a_{13} \\ a_{21} & a_{22} & a_{23} \\ a_{31} & a_{32} & a_{33} \end{vmatrix} = (-1)^{3+1} a_{31} \begin{vmatrix} a_{12} & a_{13} \\ a_{22} & a_{23} \end{vmatrix} + (-1)^{3+2} a_{32} \begin{vmatrix} a_{11} & a_{13} \\ a_{21} & a_{23} \end{vmatrix} + (-1)^{3+3} a_{33} \begin{vmatrix} a_{11} & a_{12} \\ a_{21} & a_{22} \end{vmatrix}$$

Expand across row 3.

It can be shown that the value of a determinant does not depend on the choice of the row or column used in the expansion. However, expanding across a row or column that has an element equal to 0 reduces the amount of work needed to compute the value of the determinant.

EXAMPLE 4	**Evaluating a 3 × 3 Determinant**

Find the value of the 3 by 3 determinant: $\begin{vmatrix} 3 & 0 & -1 \\ 4 & 6 & 2 \\ 8 & -2 & 3 \end{vmatrix}$

Solution Because of the 0, it is easier to expand across row 1 or down row 2. We choose to expand across row 1.

$$\begin{vmatrix} 3 & 0 & -1 \\ 4 & 6 & 2 \\ 8 & -2 & 3 \end{vmatrix} = (-1)^{1+1} \cdot 3 \cdot \begin{vmatrix} 6 & 2 \\ -2 & 3 \end{vmatrix} + (-1)^{1+2} \cdot 0 \cdot \begin{vmatrix} 4 & 2 \\ 8 & 3 \end{vmatrix} + (-1)^{1+3} \cdot (-1) \cdot \begin{vmatrix} 4 & 6 \\ 8 & -2 \end{vmatrix}$$

$$= 3(18 + 4) - 0 + (-1)(-8 - 48)$$

$$= 3(22) + (-1)(-56)$$

$$= 66 + 56 = 122$$

■

━━━ **Now Work** PROBLEM 11

4 Use Cramer's Rule to Solve a System of Three Equations Containing Three Variables

Consider the following system of three equations containing three variables.

$$\begin{cases} a_{11}x + a_{12}y + a_{13}z = c_1 \\ a_{21}x + a_{22}y + a_{23}z = c_2 \\ a_{31}x + a_{32}y + a_{33}z = c_3 \end{cases} \qquad \textbf{(10)}$$

It can be shown that if the determinant D of the coefficients of the variables is not 0; that is, if

$$D = \begin{vmatrix} a_{11} & a_{12} & a_{13} \\ a_{21} & a_{22} & a_{23} \\ a_{31} & a_{32} & a_{33} \end{vmatrix} \neq 0$$

the unique solution of system (10) is given by

Cramer's Rule for Three Equations Containing Three Variables

$$x = \frac{D_x}{D} \qquad y = \frac{D_y}{D} \qquad z = \frac{D_z}{D}$$

where

$$D_x = \begin{vmatrix} c_1 & a_{12} & a_{13} \\ c_2 & a_{22} & a_{23} \\ c_3 & a_{32} & a_{33} \end{vmatrix} \qquad D_y = \begin{vmatrix} a_{11} & c_1 & a_{13} \\ a_{21} & c_2 & a_{23} \\ a_{31} & c_3 & a_{33} \end{vmatrix} \qquad D_z = \begin{vmatrix} a_{11} & a_{12} & c_1 \\ a_{21} & a_{22} & c_2 \\ a_{31} & a_{32} & c_3 \end{vmatrix}$$

Do you see the similarity of this pattern and the pattern observed earlier for a system of two equations containing two variables?

| **EXAMPLE 5** | **Using Cramer's Rule** |

Use Cramer's Rule, if applicable, to solve the following system:

$$\begin{cases} 2x + y - z = 3 & (1) \\ -x + 2y + 4z = -3 & (2) \\ x - 2y - 3z = 4 & (3) \end{cases}$$

Solution The value of the determinant D of the coefficients of the variables is

$$D = \begin{vmatrix} 2 & 1 & -1 \\ -1 & 2 & 4 \\ 1 & -2 & -3 \end{vmatrix} = (-1)^{1+1} \cdot 2 \cdot \begin{vmatrix} 2 & 4 \\ -2 & -3 \end{vmatrix} + (-1)^{1+2} \cdot 1 \cdot \begin{vmatrix} -1 & 4 \\ 1 & -3 \end{vmatrix} + (-1)^{1+3}(-1) \begin{vmatrix} -1 & 2 \\ 1 & -2 \end{vmatrix}$$

$$= 2(2) - 1(-1) + (-1)(0)$$

$$= 4 + 1 = 5$$

Because $D \neq 0$, we proceed to find the values of D_x, D_y, and D_z.

$$D_x = \begin{vmatrix} 3 & 1 & -1 \\ -3 & 2 & 4 \\ 4 & -2 & -3 \end{vmatrix} = (-1)^{1+1} \cdot 3 \cdot \begin{vmatrix} 2 & 4 \\ -2 & -3 \end{vmatrix} + (-1)^{1+2} \cdot 1 \cdot \begin{vmatrix} -3 & 4 \\ 4 & -3 \end{vmatrix} + (-1)^{1+3}(-1) \begin{vmatrix} -3 & 2 \\ 4 & -2 \end{vmatrix}$$

$$= 3(2) - 1(-7) + (-1)(-2) = 15$$

$$D_y = \begin{vmatrix} 2 & 3 & -1 \\ -1 & -3 & 4 \\ 1 & 4 & -3 \end{vmatrix} = (-1)^{1+1} \cdot 2 \cdot \begin{vmatrix} -3 & 4 \\ 4 & -3 \end{vmatrix} + (-1)^{1+2} \cdot 3 \cdot \begin{vmatrix} -1 & 4 \\ 1 & -3 \end{vmatrix} + (-1)^{1+3}(-1) \begin{vmatrix} -1 & -3 \\ 1 & 4 \end{vmatrix}$$

$$= 2(-7) - 3(-1) + (-1)(-1)$$

$$= -14 + 3 + 1 = -10$$

$$D_z = \begin{vmatrix} 2 & 1 & 3 \\ -1 & 2 & -3 \\ 1 & -2 & 4 \end{vmatrix} = (-1)^{1+1} \cdot 2 \cdot \begin{vmatrix} 2 & -3 \\ -2 & 4 \end{vmatrix} + (-1)^{1+2} \cdot 1 \cdot \begin{vmatrix} -1 & -3 \\ 1 & 4 \end{vmatrix} + (-1)^{1+3} \cdot 3 \cdot \begin{vmatrix} -1 & 2 \\ 1 & -2 \end{vmatrix}$$

$$= 2(2) - 1(-1) + 3(0) = 5$$

As a result,

$$x = \frac{D_x}{D} = \frac{15}{5} = 3, \qquad y = \frac{D_y}{D} = \frac{-10}{5} = -2, \qquad z = \frac{D_z}{D} = \frac{5}{5} = 1$$

The solution is $x = 3$, $y = -2$, $z = 1$, or, using ordered triplets, $(3, -2, 1)$.

If the determinant of the coefficients of the variables of a system of three linear equations containing three variables is 0, Cramer's Rule is not applicable. In such a case, the system either is inconsistent or has infinitely many solutions. To solve such a system, use elimination or row reduction on the augmented matrix.

──── **Now Work** PROBLEM 33

5 Know Properties of Determinants

Determinants have several properties that are sometimes helpful for obtaining their value. We list some of them here.

THEOREM
The value of a determinant changes sign if any two rows (or any two columns) are interchanged. **(11)**

Proof for 2 by 2 Determinants

$$\begin{vmatrix} a & b \\ c & d \end{vmatrix} = ad - bc \quad \text{and} \quad \begin{vmatrix} c & d \\ a & b \end{vmatrix} = bc - ad = -(ad - bc) \qquad \blacksquare$$

EXAMPLE 6 **Demonstrating Theorem (11)**

$$\begin{vmatrix} 3 & 4 \\ 1 & 2 \end{vmatrix} = 6 - 4 = 2 \qquad \begin{vmatrix} 1 & 2 \\ 3 & 4 \end{vmatrix} = 4 - 6 = -2$$

THEOREM
If all the entries in any row (or any column) equal 0, the value of the determinant is 0. **(12)**

Proof Expand across the row (or down the column) containing the 0's. ■

THEOREM
If any two rows (or any two columns) of a determinant have corresponding entries that are equal, the value of the determinant is 0. **(13)**

You are asked to prove this result for a 3 by 3 determinant in which the entries in column 1 equal the entries in column 3 in Problem 64.

EXAMPLE 7 **Demonstrating Theorem (13)**

$$\begin{vmatrix} 1 & 2 & 3 \\ 1 & 2 & 3 \\ 4 & 5 & 6 \end{vmatrix} = (-1)^{1+1} \cdot 1 \cdot \begin{vmatrix} 2 & 3 \\ 5 & 6 \end{vmatrix} + (-1)^{1+2} \cdot 2 \cdot \begin{vmatrix} 1 & 3 \\ 4 & 6 \end{vmatrix} + (-1)^{1+3} \cdot 3 \cdot \begin{vmatrix} 1 & 2 \\ 4 & 5 \end{vmatrix}$$

$$= 1(-3) - 2(-6) + 3(-3) = -3 + 12 - 9 = 0$$

THEOREM
If any row (or any column) of a determinant is multiplied by a nonzero number k, the value of the determinant is also changed by a factor of k. **(14)**

You are asked to prove this result for a 3 by 3 determinant using row 2 in Problem 63.

| EXAMPLE 8 | Demonstrating Theorem (14) |

$$\begin{vmatrix} 1 & 2 \\ 4 & 6 \end{vmatrix} = 6 - 8 = -2$$

$$\begin{vmatrix} k & 2k \\ 4 & 6 \end{vmatrix} = 6k - 8k = -2k = k(-2) = k\begin{vmatrix} 1 & 2 \\ 4 & 6 \end{vmatrix}$$

THEOREM If the entries of any row (or any column) of a determinant are multiplied by a nonzero number k and the result is added to the corresponding entries of another row (or column), the value of the determinant remains unchanged.

(15)

In Problem 65, you are asked to prove this result for a 3 by 3 determinant using rows 1 and 2.

| EXAMPLE 9 | Demonstrating Theorem (15) |

$$\begin{vmatrix} 3 & 4 \\ 5 & 2 \end{vmatrix} = -14 \qquad \begin{vmatrix} 3 & 4 \\ 5 & 2 \end{vmatrix} \rightarrow \begin{vmatrix} -7 & 0 \\ 5 & 2 \end{vmatrix} = -14$$

Multiply row 2 by -2 and add to row 1.

12.3 Assess Your Understanding

Concepts and Vocabulary

1. Cramer's Rule uses _____ to solve a system of linear equations.

2. $D = \begin{vmatrix} a & b \\ c & d \end{vmatrix} =$ _____

3. *True or False* A 3 by 3 determinant can never equal 0.

4. *True or False* The value of a determinant remains unchanged if any two rows or any two columns are interchanged.

Skill Building

In Problems 5–14, find the value of each determinant.

5. $\begin{vmatrix} 3 & 1 \\ 4 & 2 \end{vmatrix}$

6. $\begin{vmatrix} 6 & 1 \\ 5 & 2 \end{vmatrix}$

7. $\begin{vmatrix} 6 & 4 \\ -1 & 3 \end{vmatrix}$

8. $\begin{vmatrix} 8 & -3 \\ 4 & 2 \end{vmatrix}$

9. $\begin{vmatrix} -3 & -1 \\ 4 & 2 \end{vmatrix}$

10. $\begin{vmatrix} -4 & 2 \\ -5 & 3 \end{vmatrix}$

11. $\begin{vmatrix} 3 & 4 & 2 \\ 1 & -1 & 5 \\ 1 & 2 & -2 \end{vmatrix}$

12. $\begin{vmatrix} 1 & 3 & -2 \\ 6 & 1 & -5 \\ 8 & 2 & 3 \end{vmatrix}$

13. $\begin{vmatrix} 4 & -1 & 2 \\ 6 & -1 & 0 \\ 1 & -3 & 4 \end{vmatrix}$

14. $\begin{vmatrix} 3 & -9 & 4 \\ 1 & 4 & 0 \\ 8 & -3 & 1 \end{vmatrix}$

In Problems 15–42, solve each system of equations using Cramer's Rule if it is applicable. If Cramer's Rule is not applicable, say so.

15. $\begin{cases} x + y = 8 \\ x - y = 4 \end{cases}$

16. $\begin{cases} x + 2y = 5 \\ x - y = 3 \end{cases}$

17. $\begin{cases} 5x - y = 13 \\ 2x + 3y = 12 \end{cases}$

18. $\begin{cases} x + 3y = 5 \\ 2x - 3y = -8 \end{cases}$

19. $\begin{cases} 3x = 24 \\ x + 2y = 0 \end{cases}$

20. $\begin{cases} 4x + 5y = -3 \\ -2y = -4 \end{cases}$

21. $\begin{cases} 3x - 6y = 24 \\ 5x + 4y = 12 \end{cases}$

22. $\begin{cases} 2x + 4y = 16 \\ 3x - 5y = -9 \end{cases}$

23. $\begin{cases} 3x - 2y = 4 \\ 6x - 4y = 0 \end{cases}$

24. $\begin{cases} -x + 2y = 5 \\ 4x - 8y = 6 \end{cases}$

25. $\begin{cases} 2x - 4y = -2 \\ 3x + 2y = 3 \end{cases}$

26. $\begin{cases} 3x + 3y = 3 \\ 4x + 2y = \dfrac{8}{3} \end{cases}$

27. $\begin{cases} 2x - 3y = -1 \\ 10x + 10y = 5 \end{cases}$

28. $\begin{cases} 3x - 2y = 0 \\ 5x + 10y = 4 \end{cases}$

29. $\begin{cases} 2x + 3y = 6 \\ x - y = \dfrac{1}{2} \end{cases}$

30. $\begin{cases} \dfrac{1}{2}x + y = -2 \\ x - 2y = 8 \end{cases}$

31. $\begin{cases} 3x - 5y = 3 \\ 15x + 5y = 21 \end{cases}$

32. $\begin{cases} 2x - y = -1 \\ x + \dfrac{1}{2}y = \dfrac{3}{2} \end{cases}$

33. $\begin{cases} x + y - z = 6 \\ 3x - 2y + z = -5 \\ x + 3y - 2z = 14 \end{cases}$

34. $\begin{cases} x - y + z = -4 \\ 2x - 3y + 4z = -15 \\ 5x + y - 2z = 12 \end{cases}$

35. $\begin{cases} x + 2y - z = -3 \\ 2x - 4y + z = -7 \\ -2x + 2y - 3z = 4 \end{cases}$

36. $\begin{cases} x + 4y - 3z = -8 \\ 3x - y + 3z = 12 \\ x + y + 6z = 1 \end{cases}$

37. $\begin{cases} x - 2y + 3z = 1 \\ 3x + y - 2z = 0 \\ 2x - 4y + 6z = 2 \end{cases}$

38. $\begin{cases} x - y + 2z = 5 \\ 3x + 2y = 4 \\ -2x + 2y - 4z = -10 \end{cases}$

39. $\begin{cases} x + 2y - z = 0 \\ 2x - 4y + z = 0 \\ -2x + 2y - 3z = 0 \end{cases}$

40. $\begin{cases} x + 4y - 3z = 0 \\ 3x - y + 3z = 0 \\ x + y + 6z = 0 \end{cases}$

41. $\begin{cases} x - 2y + 3z = 0 \\ 3x + y - 2z = 0 \\ 2x - 4y + 6z = 0 \end{cases}$

42. $\begin{cases} x - y + 2z = 0 \\ 3x + 2y = 0 \\ -2x + 2y - 4z = 0 \end{cases}$

In Problems 43–48, solve for x.

43. $\begin{vmatrix} x & x \\ 4 & 3 \end{vmatrix} = 5$

44. $\begin{vmatrix} x & 1 \\ 3 & x \end{vmatrix} = -2$

45. $\begin{vmatrix} x & 1 & 1 \\ 4 & 3 & 2 \\ -1 & 2 & 5 \end{vmatrix} = 2$

46. $\begin{vmatrix} 3 & 2 & 4 \\ 1 & x & 5 \\ 0 & 1 & -2 \end{vmatrix} = 0$

47. $\begin{vmatrix} x & 2 & 3 \\ 1 & x & 0 \\ 6 & 1 & -2 \end{vmatrix} = 7$

48. $\begin{vmatrix} x & 1 & 2 \\ 1 & x & 3 \\ 0 & 1 & 2 \end{vmatrix} = -4x$

In Problems 49–56, use properties of determinants to find the value of each determinant if it is known that

$$\begin{vmatrix} x & y & z \\ u & v & w \\ 1 & 2 & 3 \end{vmatrix} = 4$$

49. $\begin{vmatrix} 1 & 2 & 3 \\ u & v & w \\ x & y & z \end{vmatrix}$

50. $\begin{vmatrix} x & y & z \\ u & v & w \\ 2 & 4 & 6 \end{vmatrix}$

51. $\begin{vmatrix} x & y & z \\ -3 & -6 & -9 \\ u & v & w \end{vmatrix}$

52. $\begin{vmatrix} 1 & 2 & 3 \\ x-u & y-v & z-w \\ u & v & w \end{vmatrix}$

53. $\begin{vmatrix} 1 & 2 & 3 \\ x-3 & y-6 & z-9 \\ 2u & 2v & 2w \end{vmatrix}$

54. $\begin{vmatrix} x & y & z-x \\ u & v & w-u \\ 1 & 2 & 2 \end{vmatrix}$

55. $\begin{vmatrix} 1 & 2 & 3 \\ 2x & 2y & 2z \\ u-1 & v-2 & w-3 \end{vmatrix}$

56. $\begin{vmatrix} x+3 & y+6 & z+9 \\ 3u-1 & 3v-2 & 3w-3 \\ 1 & 2 & 3 \end{vmatrix}$

Applications and Extensions

57. Geometry: Equation of a Line An equation of the line containing the two points (x_1, y_1) and (x_2, y_2) may be expressed as the determinant

$$\begin{vmatrix} x & y & 1 \\ x_1 & y_1 & 1 \\ x_2 & y_2 & 1 \end{vmatrix} = 0$$

Prove this result by expanding the determinant and comparing the result to the two-point form of the equation of a line.

58. Geometry: Collinear Points Using the result obtained in Problem 57, show that three distinct points (x_1, y_1), (x_2, y_2), and (x_3, y_3) are collinear (lie on the same line) if and only if

$$\begin{vmatrix} x_1 & y_1 & 1 \\ x_2 & y_2 & 1 \\ x_3 & y_3 & 1 \end{vmatrix} = 0$$

59. Geometry: Area of a Triangle A triangle has vertices (x_1, y_1), (x_2, y_2), and (x_3, y_3). Show that the area of the triangle is given by the absolute value of D, where

$$D = \frac{1}{2}\begin{vmatrix} x_1 & x_2 & x_3 \\ y_1 & y_2 & y_3 \\ 1 & 1 & 1 \end{vmatrix}.$$ Use this formula to find the area of a triangle with vertices $(2, 3)$, $(5, 2)$, and $(6, 5)$.

60. Show that $\begin{vmatrix} x^2 & x & 1 \\ y^2 & y & 1 \\ z^2 & z & 1 \end{vmatrix} = (y - z)(x - y)(x - z)$.

61. Complete the proof of Cramer's Rule for two equations containing two variables.

[**Hint:** In system (5), page 867, if $a = 0$, then $b \neq 0$ and $c \neq 0$, since $D = -bc \neq 0$. Now show that equation (6) provides a

solution of the system when $a = 0$. There are then three remaining cases: $b = 0$, $c = 0$, and $d = 0$.]

62. Interchange columns 1 and 3 of a 3 by 3 determinant. Show that the value of the new determinant is -1 times the value of the original determinant.

63. Multiply each entry in row 2 of a 3 by 3 determinant by the number k, $k \neq 0$. Show that the value of the new determinant is k times the value of the original determinant.

64. Prove that a 3 by 3 determinant in which the entries in column 1 equal those in column 3 has the value 0.

65. Prove that, if row 2 of a 3 by 3 determinant is multiplied by k, $k \neq 0$, and the result is added to the entries in row 1, there is no change in the value of the determinant.

12.4 Matrix Algebra

OBJECTIVES **1** Find the Sum and Difference of Two Matrices (p. 876)
2 Find Scalar Multiples of a Matrix (p. 878)
3 Find the Product of Two Matrices (p. 879)
4 Find the Inverse of a Matrix (p. 884)
5 Solve a System of Linear Equations Using an Inverse Matrix (p. 887)

In Section 12.2, we defined a matrix as a rectangular array of real numbers and used an augmented matrix to represent a system of linear equations. There is, however, a branch of mathematics, called **linear algebra,** that deals with matrices in such a way that an algebra of matrices is permitted. In this section, we provide a survey of how this **matrix algebra** is developed.

Before getting started, we restate the definition of a matrix.

DEFINITION

A **matrix** is defined as a rectangular array of numbers:

$$
\begin{array}{c}
\\
\text{Row 1} \\
\text{Row 2} \\
\vdots \\
\text{Row } i \\
\vdots \\
\text{Row } m
\end{array}
\begin{array}{ccccccc}
\text{Column 1} & \text{Column 2} & & \text{Column } j & & \text{Column } n \\
\begin{bmatrix}
a_{11} & a_{12} & \cdots & a_{1j} & \cdots & a_{1n} \\
a_{21} & a_{22} & \cdots & a_{2j} & \cdots & a_{2n} \\
\vdots & \vdots & & \vdots & & \vdots \\
a_{i1} & a_{i2} & \cdots & a_{ij} & \cdots & a_{in} \\
\vdots & \vdots & & \vdots & & \vdots \\
a_{m1} & a_{m2} & \cdots & a_{mj} & \cdots & a_{mn}
\end{bmatrix}
\end{array}
$$

Each number a_{ij} of the matrix has two indexes: the **row index** i and the **column index** j. The matrix shown here has m rows and n columns. The numbers a_{ij} are usually referred to as the **entries** of the matrix. For example, a_{23} refers to the entry in the second row, third column.

Let's begin with an example that illustrates how matrices can be used to conveniently represent an array of information.

EXAMPLE 1 **Arranging Data in a Matrix**

In a survey of 900 people, the following information was obtained:

200 males	Thought federal defense spending was too high
150 males	Thought federal defense spending was too low
45 males	Had no opinion
315 females	Thought federal defense spending was too high
125 females	Thought federal defense spending was too low
65 females	Had no opinion

We can arrange these data in a rectangular array as follows:

	Too High	Too Low	No Opinion
Male	200	150	45
Female	315	125	65

or as the matrix

$$\begin{bmatrix} 200 & 150 & 45 \\ 315 & 125 & 65 \end{bmatrix}$$

This matrix has two rows (representing male and female) and three columns (representing "too high," "too low," and "no opinion").

The matrix we developed in Example 1 has 2 rows and 3 columns. In general, a matrix with m rows and n columns is called an **m by n matrix.** The matrix we developed in Example 1 is a 2 by 3 matrix and contains $2 \cdot 3 = 6$ entries. An m by n matrix will contain $m \cdot n$ entries.

If an m by n matrix has the same number of rows as columns; that is, if $m = n$, then the matrix is referred to as a **square matrix.**

EXAMPLE 2 **Examples of Matrices**

(a) $\begin{bmatrix} 5 & 0 \\ -6 & 1 \end{bmatrix}$ A 2 by 2 square matrix (b) $\begin{bmatrix} 1 & 0 & 3 \end{bmatrix}$ A 1 by 3 matrix

(c) $\begin{bmatrix} 6 & -2 & 4 \\ 4 & 3 & 5 \\ 8 & 0 & 1 \end{bmatrix}$ A 3 by 3 square matrix

1 Find the Sum and Difference of Two Matrices

We begin our discussion of matrix algebra by first defining what is meant by two matrices being equal and then defining the operations of addition and subtraction. It is important to note that these definitions require each matrix to have the same number of rows *and* the same number of columns as a condition for equality and for addition and subtraction.

We usually represent matrices by capital letters, such as A, B, and C.

DEFINITION Two matrices A and B are said to be **equal,** written as

$$A = B$$

provided that A and B have the same number of rows and the same number of columns and each entry a_{ij} in A is equal to the corresponding entry b_{ij} in B.

For example,

$$\begin{bmatrix} 2 & 1 \\ 0.5 & -1 \end{bmatrix} = \begin{bmatrix} \sqrt{4} & 1 \\ \dfrac{1}{2} & -1 \end{bmatrix} \quad \text{and} \quad \begin{bmatrix} 3 & 2 & 1 \\ 0 & 1 & -2 \end{bmatrix} = \begin{bmatrix} \sqrt{9} & \sqrt{4} & 1 \\ 0 & 1 & \sqrt[3]{-8} \end{bmatrix}$$

$$\begin{bmatrix} 4 & 1 \\ 6 & 1 \end{bmatrix} \neq \begin{bmatrix} 4 & 0 \\ 6 & 1 \end{bmatrix} \qquad \text{Because the entries in row 1, column 2 are not equal}$$

$$\begin{bmatrix} 4 & 1 & 2 \\ 6 & 1 & 2 \end{bmatrix} \neq \begin{bmatrix} 4 & 1 & 2 & 3 \\ 6 & 1 & 2 & 4 \end{bmatrix} \qquad \text{Because the matrix on the left has 3 columns and the matrix on the right has 4 columns.}$$

Suppose that A and B represent two m by n matrices. We define their **sum** $A + B$ to be the m by n matrix formed by adding the corresponding entries a_{ij} of A and b_{ij} of B. The **difference** $A - B$ is defined as the m by n matrix formed by subtracting the entries b_{ij} in B from the corresponding entries a_{ij} in A. Addition and subtraction of matrices are allowed only for matrices having the same number m of rows and the same number n of columns. For example, a 2 by 3 matrix and a 2 by 4 matrix cannot be added or subtracted.

EXAMPLE 3 **Adding and Subtracting Matrices**

Suppose that

$$A = \begin{bmatrix} 2 & 4 & 8 & -3 \\ 0 & 1 & 2 & 3 \end{bmatrix} \quad \text{and} \quad B = \begin{bmatrix} -3 & 4 & 0 & 1 \\ 6 & 8 & 2 & 0 \end{bmatrix}$$

Find: (a) $A + B$ (b) $A - B$

Solution (a) $A + B = \begin{bmatrix} 2 & 4 & 8 & -3 \\ 0 & 1 & 2 & 3 \end{bmatrix} + \begin{bmatrix} -3 & 4 & 0 & 1 \\ 6 & 8 & 2 & 0 \end{bmatrix}$

$$= \begin{bmatrix} 2 + (-3) & 4 + 4 & 8 + 0 & -3 + 1 \\ 0 + 6 & 1 + 8 & 2 + 2 & 3 + 0 \end{bmatrix} \quad \text{Add corresponding entries.}$$

$$= \begin{bmatrix} -1 & 8 & 8 & -2 \\ 6 & 9 & 4 & 3 \end{bmatrix}$$

(b) $A - B = \begin{bmatrix} 2 & 4 & 8 & -3 \\ 0 & 1 & 2 & 3 \end{bmatrix} - \begin{bmatrix} -3 & 4 & 0 & 1 \\ 6 & 8 & 2 & 0 \end{bmatrix}$

$$= \begin{bmatrix} 2 - (-3) & 4 - 4 & 8 - 0 & -3 - 1 \\ 0 - 6 & 1 - 8 & 2 - 2 & 3 - 0 \end{bmatrix} \quad \text{Subtract corresponding entries.}$$

$$= \begin{bmatrix} 5 & 0 & 8 & -4 \\ -6 & -7 & 0 & 3 \end{bmatrix}$$

Figure 7

 Seeing the Concept

Graphing utilities can make the sometimes tedious process of matrix algebra easy. In fact, most graphing calculators can handle matrices as large as 9 by 9, some even larger ones. Enter the matrices into a graphing utility. Name them [A] and [B]. Figure 7 shows the results of adding and subtracting [A] and [B].

Now Work PROBLEM 7

Many of the algebraic properties of sums of real numbers are also true for sums of matrices. Suppose that A, B, and C are m by n matrices. Then matrix addition is **commutative.** That is,

Commutative Property of Matrix Addition

$$A + B = B + A$$

Matrix addition is also **associative.** That is,

Associative Property of Matrix Addition

$$(A + B) + C = A + (B + C)$$

Although we shall not prove these results, the proofs, as the following example illustrates, are based on the commutative and associative properties for real numbers.

| EXAMPLE 4 | **Demonstrating the Commutative Property** |

$$\begin{bmatrix} 2 & 3 & -1 \\ 4 & 0 & 7 \end{bmatrix} + \begin{bmatrix} -1 & 2 & 1 \\ 5 & -3 & 4 \end{bmatrix} = \begin{bmatrix} 2 + (-1) & 3 + 2 & -1 + 1 \\ 4 + 5 & 0 + (-3) & 7 + 4 \end{bmatrix}$$

$$= \begin{bmatrix} -1 + 2 & 2 + 3 & 1 + (-1) \\ 5 + 4 & -3 + 0 & 4 + 7 \end{bmatrix}$$

$$= \begin{bmatrix} -1 & 2 & 1 \\ 5 & -3 & 4 \end{bmatrix} + \begin{bmatrix} 2 & 3 & -1 \\ 4 & 0 & 7 \end{bmatrix}$$

∎

A matrix whose entries are all equal to 0 is called a **zero matrix.** Each of the following matrices is a zero matrix.

$$\begin{bmatrix} 0 & 0 \\ 0 & 0 \end{bmatrix}$$ 2 by 2 square zero matrix $$\begin{bmatrix} 0 & 0 & 0 \\ 0 & 0 & 0 \end{bmatrix}$$ 2 by 3 zero matrix $$\begin{bmatrix} 0 & 0 & 0 \end{bmatrix}$$ 1 by 3 zero matrix

Zero matrices have properties similar to the real number 0. If A is an m by n matrix and 0 is an m by n zero matrix, then

$$A + 0 = 0 + A = A$$

In other words, the zero matrix is the additive identity in matrix algebra.

2 Find Scalar Multiples of a Matrix

We can also multiply a matrix by a real number. If k is a real number and A is an m by n matrix, the matrix kA is the m by n matrix formed by multiplying each entry a_{ij} in A by k. The number k is sometimes referred to as a **scalar,** and the matrix kA is called a **scalar multiple** of A.

| EXAMPLE 5 | **Operations Using Matrices** |

Suppose that

$$A = \begin{bmatrix} 3 & 1 & 5 \\ -2 & 0 & 6 \end{bmatrix} \qquad B = \begin{bmatrix} 4 & 1 & 0 \\ 8 & 1 & -3 \end{bmatrix} \qquad C = \begin{bmatrix} 9 & 0 \\ -3 & 6 \end{bmatrix}$$

Find: (a) $4A$ (b) $\dfrac{1}{3}C$ (c) $3A - 2B$

Solution (a) $4A = 4\begin{bmatrix} 3 & 1 & 5 \\ -2 & 0 & 6 \end{bmatrix} = \begin{bmatrix} 4 \cdot 3 & 4 \cdot 1 & 4 \cdot 5 \\ 4(-2) & 4 \cdot 0 & 4 \cdot 6 \end{bmatrix} = \begin{bmatrix} 12 & 4 & 20 \\ -8 & 0 & 24 \end{bmatrix}$

(b) $\dfrac{1}{3}C = \dfrac{1}{3}\begin{bmatrix} 9 & 0 \\ -3 & 6 \end{bmatrix} = \begin{bmatrix} \dfrac{1}{3} \cdot 9 & \dfrac{1}{3} \cdot 0 \\ \dfrac{1}{3}(-3) & \dfrac{1}{3} \cdot 6 \end{bmatrix} = \begin{bmatrix} 3 & 0 \\ -1 & 2 \end{bmatrix}$

(c) $3A - 2B = 3\begin{bmatrix} 3 & 1 & 5 \\ -2 & 0 & 6 \end{bmatrix} - 2\begin{bmatrix} 4 & 1 & 0 \\ 8 & 1 & -3 \end{bmatrix}$

$= \begin{bmatrix} 3 \cdot 3 & 3 \cdot 1 & 3 \cdot 5 \\ 3(-2) & 3 \cdot 0 & 3 \cdot 6 \end{bmatrix} - \begin{bmatrix} 2 \cdot 4 & 2 \cdot 1 & 2 \cdot 0 \\ 2 \cdot 8 & 2 \cdot 1 & 2(-3) \end{bmatrix}$

$= \begin{bmatrix} 9 & 3 & 15 \\ -6 & 0 & 18 \end{bmatrix} - \begin{bmatrix} 8 & 2 & 0 \\ 16 & 2 & -6 \end{bmatrix}$

$= \begin{bmatrix} 9 - 8 & 3 - 2 & 15 - 0 \\ -6 - 16 & 0 - 2 & 18 - (-6) \end{bmatrix}$

$= \begin{bmatrix} 1 & 1 & 15 \\ -22 & -2 & 24 \end{bmatrix}$

Check: Enter the matrices $[A]$, $[B]$, and $[C]$ into a graphing utility. Then find $4A, \frac{1}{3}C$, and $3A - 2B$.

Now Work PROBLEM 11

We list next some of the algebraic properties of scalar multiplication. Let h and k be real numbers, and let A and B be m by n matrices. Then

Properties of Scalar Multiplication

$$k(hA) = (kh)A$$
$$(k + h)A = kA + hA$$
$$k(A + B) = kA + kB$$

3 Find the Product of Two Matrices

Unlike the straightforward definition for adding two matrices, the definition for multiplying two matrices is not what we might expect. In preparation for this definition, we need the following definitions:

A **row vector** R is a 1 by n matrix

$$R = \begin{bmatrix} r_1 & r_2 & \cdots & r_n \end{bmatrix}$$

A **column vector** C is an n by 1 matrix

$$C = \begin{bmatrix} c_1 \\ c_2 \\ \vdots \\ c_n \end{bmatrix}$$

DEFINITION The **product** RC of R times C is defined as the number

$$RC = \begin{bmatrix} r_1 & r_2 \cdots r_n \end{bmatrix} \begin{bmatrix} c_1 \\ c_2 \\ \vdots \\ c_n \end{bmatrix} = r_1 c_1 + r_2 c_2 + \cdots + r_n c_n$$

Notice that a row vector and a column vector can be multiplied only if they contain the same number of entries.

EXAMPLE 6 **The Product of a Row Vector by a Column Vector**

If $R = \begin{bmatrix} 3 & -5 & 2 \end{bmatrix}$ and $C = \begin{bmatrix} 3 \\ 4 \\ -5 \end{bmatrix}$, then

$$RC = \begin{bmatrix} 3 & -5 & 2 \end{bmatrix} \begin{bmatrix} 3 \\ 4 \\ -5 \end{bmatrix} = 3 \cdot 3 + (-5)4 + 2(-5) = 9 - 20 - 10 = -21$$

∎

Let's look at an application of the product of a row vector by a column vector.

EXAMPLE 7 **Using Matrices to Compute Revenue**

A clothing store sells men's shirts for \$40, silk ties for \$20, and wool suits for \$400. Last month, the store had sales consisting of 100 shirts, 200 ties, and 50 suits. What was the total revenue due to these sales?

Solution We set up a row vector R to represent the prices of each item and a column vector C to represent the corresponding number of items sold. Then

$$\begin{array}{cc} \begin{matrix} \text{Prices} \\ \text{Shirts Ties Suits} \end{matrix} & \begin{matrix} \text{Number} \\ \text{sold} \end{matrix} \end{array}$$

$$R = \begin{bmatrix} 40 & 20 & 400 \end{bmatrix}, \qquad C = \begin{bmatrix} 100 \\ 200 \\ 50 \end{bmatrix} \begin{matrix} \text{Shirts} \\ \text{Ties} \\ \text{Suits} \end{matrix}$$

The total revenue obtained is the product RC. That is,

$$RC = \begin{bmatrix} 40 & 20 & 400 \end{bmatrix} \begin{bmatrix} 100 \\ 200 \\ 50 \end{bmatrix}$$

$$= \underbrace{40 \cdot 100}_{\text{Shirt revenue}} + \underbrace{20 \cdot 200}_{\text{Tie revenue}} + \underbrace{400 \cdot 50}_{\text{Suit revenue}} = \underbrace{\$28,000}_{\text{Total revenue}}$$

∎

The definition for multiplying two matrices is based on the definition of a row vector times a column vector.

DEFINITION Let A denote an m by r matrix and let B denote an r by n matrix. The **product** AB is defined as the m by n matrix whose entry in row i, column j is the product of the ith row of A and the jth column of B.

The definition of the product AB of two matrices A and B, in this order, requires that the number of columns of A equal the number of rows of B; otherwise, no product is defined.

An example will help to clarify the definition.

EXAMPLE 8	Multiplying Two Matrices

Find the product AB if

$$A = \begin{bmatrix} 2 & 4 & -1 \\ 5 & 8 & 0 \end{bmatrix} \quad \text{and} \quad B = \begin{bmatrix} 2 & 5 & 1 & 4 \\ 4 & 8 & 0 & 6 \\ -3 & 1 & -2 & -1 \end{bmatrix}$$

Solution First, we note that A is 2 by 3 and B is 3 by 4, so the product AB is defined and will be a 2 by 4 matrix.

Suppose that we want the entry in row 2, column 3 of AB. To find it, we find the product of the row vector from row 2 of A and the column vector from column 3 of B.

$$\begin{array}{c} \text{Column 3 of } B \\ \text{Row 2 of } A \begin{bmatrix} 1 \\ 0 \\ -2 \end{bmatrix} = 5 \cdot 1 + 8 \cdot 0 + 0(-2) = 5 \\ \begin{bmatrix} 5 & 8 & 0 \end{bmatrix} \end{array}$$

So far, we have

$$AB = \begin{bmatrix} \underline{\quad} & \underline{\quad} & \overset{\text{Column 3}}{\underset{\downarrow}{5}} & \underline{\quad} \end{bmatrix} \leftarrow \text{Row 2}$$

Now, to find the entry in row 1, column 4 of AB, we find the product of row 1 of A and column 4 of B.

$$\begin{array}{c} \text{Column 4 of } B \\ \text{Row 1 of } A \begin{bmatrix} 4 \\ 6 \\ -1 \end{bmatrix} = 2 \cdot 4 + 4 \cdot 6 + (-1)(-1) = 33 \\ \begin{bmatrix} 2 & 4 & -1 \end{bmatrix} \end{array}$$

Continuing in this fashion, we find AB.

$$AB = \begin{bmatrix} 2 & 4 & -1 \\ 5 & 8 & 0 \end{bmatrix} \begin{bmatrix} 2 & 5 & 1 & 4 \\ 4 & 8 & 0 & 6 \\ -3 & 1 & -2 & -1 \end{bmatrix}$$

$$= \begin{bmatrix} \begin{array}{l}\text{Row 1 of } A \\ \text{times} \\ \text{column 1 of } B\end{array} & \begin{array}{l}\text{Row 1 of } A \\ \text{times} \\ \text{column 2 of } B\end{array} & \begin{array}{l}\text{Row 1 of } A \\ \text{times} \\ \text{column 3 of } B\end{array} & \begin{array}{l}\text{Row 1 of } A \\ \text{times} \\ \text{column 4 of } B\end{array} \\ \\ \begin{array}{l}\text{Row 2 of } A \\ \text{times} \\ \text{column 1 of } B\end{array} & \begin{array}{l}\text{Row 2 of } A \\ \text{times} \\ \text{column 2 of } B\end{array} & \begin{array}{l}\text{Row 2 of } A \\ \text{times} \\ \text{column 3 of } B\end{array} & \begin{array}{l}\text{Row 2 of } A \\ \text{times} \\ \text{column 4 of } B\end{array} \end{bmatrix}$$

$$= \begin{bmatrix} 2 \cdot 2 + 4 \cdot 4 + (-1)(-3) & 2 \cdot 5 + 4 \cdot 8 + (-1)1 & 2 \cdot 1 + 4 \cdot 0 + (-1)(-2) & 33 \text{ (from earlier)} \\ 5 \cdot 2 + 8 \cdot 4 + 0(-3) & 5 \cdot 5 + 8 \cdot 8 + 0 \cdot 1 & 5 \text{ (from earlier)} & 5 \cdot 4 + 8 \cdot 6 + 0(-1) \end{bmatrix}$$

$$= \begin{bmatrix} 23 & 41 & 4 & 33 \\ 42 & 89 & 5 & 68 \end{bmatrix}$$

■

Now Work PROBLEM 23

Notice that for the matrices given in Example 8 the product BA is not defined, because B is 3 by 4 and A is 2 by 3.

Check: Enter the matrices [A] and [B]. Then find AB. (See what happens if you try to find BA.)

Another result that can occur when multiplying two matrices is illustrated in the next example.

EXAMPLE 9 | **Multiplying Two Matrices**

If

$$A = \begin{bmatrix} 2 & 1 & 3 \\ 1 & -1 & 0 \end{bmatrix} \quad \text{and} \quad B = \begin{bmatrix} 1 & 0 \\ 2 & 1 \\ 3 & 2 \end{bmatrix}$$

find: (a) AB (b) BA

Solution (a) $AB = \begin{bmatrix} 2 & 1 & 3 \\ 1 & -1 & 0 \end{bmatrix} \begin{bmatrix} 1 & 0 \\ 2 & 1 \\ 3 & 2 \end{bmatrix} = \begin{bmatrix} 13 & 7 \\ -1 & -1 \end{bmatrix}$

 2 by 3 3 by 2 2 by 2

(b) $BA = \begin{bmatrix} 1 & 0 \\ 2 & 1 \\ 3 & 2 \end{bmatrix} \begin{bmatrix} 2 & 1 & 3 \\ 1 & -1 & 0 \end{bmatrix} = \begin{bmatrix} 2 & 1 & 3 \\ 5 & 1 & 6 \\ 8 & 1 & 9 \end{bmatrix}$

 3 by 2 2 by 3 3 by 3

Notice in Example 9 that AB is 2 by 2 and BA is 3 by 3. It is possible for both AB and BA to be defined, yet be unequal. In fact, even if A and B are both n by n matrices so that AB and BA are each defined and n by n, AB and BA will usually be unequal.

EXAMPLE 10 | **Multiplying Two Square Matrices**

If

$$A = \begin{bmatrix} 2 & 1 \\ 0 & 4 \end{bmatrix} \quad \text{and} \quad B = \begin{bmatrix} -3 & 1 \\ 1 & 2 \end{bmatrix}$$

find: (a) AB (b) BA

Solution (a) $AB = \begin{bmatrix} 2 & 1 \\ 0 & 4 \end{bmatrix} \begin{bmatrix} -3 & 1 \\ 1 & 2 \end{bmatrix} = \begin{bmatrix} -5 & 4 \\ 4 & 8 \end{bmatrix}$

(b) $BA = \begin{bmatrix} -3 & 1 \\ 1 & 2 \end{bmatrix} \begin{bmatrix} 2 & 1 \\ 0 & 4 \end{bmatrix} = \begin{bmatrix} -6 & 1 \\ 2 & 9 \end{bmatrix}$

The preceding examples demonstrate that an important property of real numbers, the commutative property of multiplication, is not shared by matrices. In general:

THEOREM Matrix multiplication is not commutative.

Now Work PROBLEMS **13** AND **15**

Next we give two of the properties of real numbers that are shared by matrices. Assuming that each product and sum is defined, we have the following:

Associative Property of Matrix Multiplication

$$A(BC) = (AB)C$$

Distributive Property

$$A(B + C) = AB + AC$$

For an n by n square matrix, the entries located in row i, column i, $1 \leq i \leq n$, are called the **diagonal entries** or **the main diagonal**. An n by n square matrix whose diagonal entries are 1's, while all other entries are 0's, is called the **identity matrix I_n**. For example,

$$I_2 = \begin{bmatrix} 1 & 0 \\ 0 & 1 \end{bmatrix}, \qquad I_3 = \begin{bmatrix} 1 & 0 & 0 \\ 0 & 1 & 0 \\ 0 & 0 & 1 \end{bmatrix}$$

and so on.

EXAMPLE 11 **Multiplication with an Identity Matrix**

Let

$$A = \begin{bmatrix} -1 & 2 & 0 \\ 0 & 1 & 3 \end{bmatrix} \quad \text{and} \quad B = \begin{bmatrix} 3 & 2 \\ 4 & 6 \\ 5 & 2 \end{bmatrix}$$

Find: (a) AI_3 (b) I_2A (c) BI_2

Solution (a) $AI_3 = \begin{bmatrix} -1 & 2 & 0 \\ 0 & 1 & 3 \end{bmatrix} \begin{bmatrix} 1 & 0 & 0 \\ 0 & 1 & 0 \\ 0 & 0 & 1 \end{bmatrix} = \begin{bmatrix} -1 & 2 & 0 \\ 0 & 1 & 3 \end{bmatrix} = A$

(b) $I_2A = \begin{bmatrix} 1 & 0 \\ 0 & 1 \end{bmatrix} \begin{bmatrix} -1 & 2 & 0 \\ 0 & 1 & 3 \end{bmatrix} = \begin{bmatrix} -1 & 2 & 0 \\ 0 & 1 & 3 \end{bmatrix} = A$

(c) $BI_2 = \begin{bmatrix} 3 & 2 \\ 4 & 6 \\ 5 & 2 \end{bmatrix} \begin{bmatrix} 1 & 0 \\ 0 & 1 \end{bmatrix} = \begin{bmatrix} 3 & 2 \\ 4 & 6 \\ 5 & 2 \end{bmatrix} = B$

Example 11 demonstrates the following property:

Identity Property

If A is an m by n matrix, then

$$I_m A = A \quad \text{and} \quad AI_n = A$$

If A is an n by n square matrix,

$$AI_n = I_n A = A$$

An identity matrix has properties similar to those of the real number 1. In other words, the identity matrix is a multiplicative identity in matrix algebra.

4 Find the Inverse of a Matrix

DEFINITION Let A be a square n by n matrix. If there exists an n by n matrix A^{-1}, read "A inverse," for which

$$AA^{-1} = A^{-1}A = I_n$$

then A^{-1} is called the **inverse** of the matrix A.

NOTE If the determinant of A is zero, A is singular. (Refer to Section 12.3.) ■

As we shall soon see, not every square matrix has an inverse. When a matrix A does have an inverse A^{-1}, then A is said to be **nonsingular.** If a matrix A has no inverse, it is called **singular.**

EXAMPLE 12 **Multiplying a Matrix by Its Inverse**

Show that the inverse of

$$A = \begin{bmatrix} 3 & 1 \\ 2 & 1 \end{bmatrix} \quad \text{is} \quad A^{-1} = \begin{bmatrix} 1 & -1 \\ -2 & 3 \end{bmatrix}$$

Solution We need to show that $AA^{-1} = A^{-1}A = I_2$.

$$AA^{-1} = \begin{bmatrix} 3 & 1 \\ 2 & 1 \end{bmatrix}\begin{bmatrix} 1 & -1 \\ -2 & 3 \end{bmatrix} = \begin{bmatrix} 1 & 0 \\ 0 & 1 \end{bmatrix} = I_2$$

$$A^{-1}A = \begin{bmatrix} 1 & -1 \\ -2 & 3 \end{bmatrix}\begin{bmatrix} 3 & 1 \\ 2 & 1 \end{bmatrix} = \begin{bmatrix} 1 & 0 \\ 0 & 1 \end{bmatrix} = I_2$$

■

We now show one way to find the inverse of

$$A = \begin{bmatrix} 3 & 1 \\ 2 & 1 \end{bmatrix}$$

Suppose that A^{-1} is given by

$$A^{-1} = \begin{bmatrix} x & y \\ z & w \end{bmatrix} \tag{1}$$

where x, y, z, and w are four variables. Based on the definition of an inverse, if A has an inverse,

$$AA^{-1} = I_2$$

$$\begin{bmatrix} 3 & 1 \\ 2 & 1 \end{bmatrix}\begin{bmatrix} x & y \\ z & w \end{bmatrix} = \begin{bmatrix} 1 & 0 \\ 0 & 1 \end{bmatrix}$$

$$\begin{bmatrix} 3x + z & 3y + w \\ 2x + z & 2y + w \end{bmatrix} = \begin{bmatrix} 1 & 0 \\ 0 & 1 \end{bmatrix}$$

Because corresponding entries must be equal, it follows that this matrix equation is equivalent to two systems of linear equations.

$$\begin{cases} 3x + z = 1 \\ 2x + z = 0 \end{cases} \qquad \begin{cases} 3y + w = 0 \\ 2y + w = 1 \end{cases}$$

The augmented matrix of each system is

$$\begin{bmatrix} 3 & 1 & | & 1 \\ 2 & 1 & | & 0 \end{bmatrix} \quad \begin{bmatrix} 3 & 1 & | & 0 \\ 2 & 1 & | & 1 \end{bmatrix} \tag{2}$$

The usual procedure would be to transform each augmented matrix into reduced row echelon form. Notice, though, that the left sides of the augmented matrices are equal, so the same row operations (see Section 12.2) can be used to reduce each one. We find it more efficient to combine the two augmented matrices (2) into a single matrix, as shown next, and then transform it into reduced row echelon form.

$$\begin{bmatrix} 3 & 1 & | & 1 & 0 \\ 2 & 1 & | & 0 & 1 \end{bmatrix}$$

We attempt to transform the left side into an identity matrix.

$$\begin{bmatrix} 3 & 1 & | & 1 & 0 \\ 2 & 1 & | & 0 & 1 \end{bmatrix} \underset{\substack{\uparrow \\ R_1 = -1r_2 + r_1}}{\rightarrow} \begin{bmatrix} 1 & 0 & | & 1 & -1 \\ 2 & 1 & | & 0 & 1 \end{bmatrix}$$

$$\underset{\substack{\uparrow \\ R_2 = -2r_1 + r_2}}{\rightarrow} \begin{bmatrix} 1 & 0 & | & 1 & -1 \\ 0 & 1 & | & -2 & 3 \end{bmatrix} \qquad \text{(3)}$$

Matrix (3) is in reduced row echelon form.

Now we reverse the earlier step of combining the two augmented matrices in (2) and write the single matrix (3) as two augmented matrices.

$$\begin{bmatrix} 1 & 0 & | & 1 \\ 0 & 1 & | & -2 \end{bmatrix} \quad \text{and} \quad \begin{bmatrix} 1 & 0 & | & -1 \\ 0 & 1 & | & 3 \end{bmatrix}$$

We conclude from these matrices that $x = 1, z = -2$, and $y = -1, w = 3$. Substituting these values into matrix (1), we find that

$$A^{-1} = \begin{bmatrix} 1 & -1 \\ -2 & 3 \end{bmatrix}$$

Notice in display (3) that the 2 by 2 matrix to the right of the vertical bar is, in fact, the inverse of A. Also notice that the identity matrix I_2 is the matrix that appears to the left of the vertical bar. These observations and the procedures followed above will work in general.

Procedure for Finding the Inverse of a Nonsingular Matrix

To find the inverse of an n by n nonsingular matrix A, proceed as follows:

STEP 1: Form the matrix $[A|I_n]$.
STEP 2: Transform the matrix $[A|I_n]$ into reduced row echelon form.
STEP 3: The reduced row echelon form of $[A|I_n]$ will contain the identity matrix I_n on the left of the vertical bar; the n by n matrix on the right of the vertical bar is the inverse of A.

In other words, if A is nonsingular, we begin with the matrix $[A|I_n]$ and, after transforming it into reduced row echelon form, we end up with the matrix $[I_n|A^{-1}]$.

Let's look at another example.

EXAMPLE 13 Finding the Inverse of a Matrix

The matrix

$$A = \begin{bmatrix} 1 & 1 & 0 \\ -1 & 3 & 4 \\ 0 & 4 & 3 \end{bmatrix}$$

is nonsingular. Find its inverse.

Solution First, we form the matrix

$$[A|I_3] = \begin{bmatrix} 1 & 1 & 0 & | & 1 & 0 & 0 \\ -1 & 3 & 4 & | & 0 & 1 & 0 \\ 0 & 4 & 3 & | & 0 & 0 & 1 \end{bmatrix}$$

Next, we use row operations to transform $[A|I_3]$ into reduced row echelon form.

$$\begin{bmatrix} 1 & 1 & 0 & | & 1 & 0 & 0 \\ -1 & 3 & 4 & | & 0 & 1 & 0 \\ 0 & 4 & 3 & | & 0 & 0 & 1 \end{bmatrix} \rightarrow \begin{bmatrix} 1 & 1 & 0 & | & 1 & 0 & 0 \\ 0 & 4 & 4 & | & 1 & 1 & 0 \\ 0 & 4 & 3 & | & 0 & 0 & 1 \end{bmatrix} \rightarrow \begin{bmatrix} 1 & 1 & 0 & | & 1 & 0 & 0 \\ 0 & 1 & 1 & | & \frac{1}{4} & \frac{1}{4} & 0 \\ 0 & 4 & 3 & | & 0 & 0 & 1 \end{bmatrix}$$
$$\uparrow \qquad\qquad\qquad\qquad \uparrow$$
$$R_2 = r_1 + r_2 \qquad\qquad\qquad R_2 = \frac{1}{4}r_2$$

$$\rightarrow \begin{bmatrix} 1 & 0 & -1 & | & \frac{3}{4} & -\frac{1}{4} & 0 \\ 0 & 1 & 1 & | & \frac{1}{4} & \frac{1}{4} & 0 \\ 0 & 0 & -1 & | & -1 & -1 & 1 \end{bmatrix} \rightarrow \begin{bmatrix} 1 & 0 & -1 & | & \frac{3}{4} & -\frac{1}{4} & 0 \\ 0 & 1 & 1 & | & \frac{1}{4} & \frac{1}{4} & 0 \\ 0 & 0 & 1 & | & 1 & 1 & -1 \end{bmatrix}$$
$$\uparrow \qquad\qquad\qquad\qquad\qquad \uparrow$$
$$R_1 = -1r_2 + r_1 \qquad\qquad\qquad R_3 = -1r_3$$
$$R_3 = -4r_2 + r_3$$

$$\rightarrow \begin{bmatrix} 1 & 0 & 0 & | & \frac{7}{4} & \frac{3}{4} & -1 \\ 0 & 1 & 0 & | & -\frac{3}{4} & -\frac{3}{4} & 1 \\ 0 & 0 & 1 & | & 1 & 1 & -1 \end{bmatrix}$$
$$\uparrow$$
$$R_1 = r_3 + r_1$$
$$R_2 = -1r_3 + r_2$$

The matrix $[A|I_3]$ is now in reduced row echelon form, and the identity matrix I_3 is on the left of the vertical bar. The inverse of A is

$$A^{-1} = \begin{bmatrix} \frac{7}{4} & \frac{3}{4} & -1 \\ -\frac{3}{4} & -\frac{3}{4} & 1 \\ 1 & 1 & -1 \end{bmatrix}$$

You can (and should) verify that this is the correct inverse by showing that

$$AA^{-1} = A^{-1}A = I_3$$

Figure 8

✓ **Check:** Enter the matrix A into a graphing utility. Figure 8 shows A^{-1}.

Now Work PROBLEM 31

If transforming the matrix $[A|I_n]$ into reduced row echelon form does not result in the identity matrix I_n to the left of the vertical bar, A is singular and has no inverse. The next example demonstrates such a matrix.

| EXAMPLE 14 | **Showing That a Matrix Has No Inverse** |

Show that the matrix $A = \begin{bmatrix} 4 & 6 \\ 2 & 3 \end{bmatrix}$ has no inverse.

Solution Proceeding as in Example 13, we form the matrix

$$[A|I_2] = \begin{bmatrix} 4 & 6 & | & 1 & 0 \\ 2 & 3 & | & 0 & 1 \end{bmatrix}$$

Then we use row operations to transform $[A|I_2]$ into reduced row echelon form.

$$[A|I_2] = \begin{bmatrix} 4 & 6 & | & 1 & 0 \\ 2 & 3 & | & 0 & 1 \end{bmatrix} \rightarrow \begin{bmatrix} 1 & \dfrac{3}{2} & | & \dfrac{1}{4} & 0 \\ 2 & 3 & | & 0 & 1 \end{bmatrix} \rightarrow \begin{bmatrix} 1 & \dfrac{3}{2} & | & \dfrac{1}{4} & 0 \\ 0 & 0 & | & -\dfrac{1}{2} & 1 \end{bmatrix}$$

$$\uparrow \qquad\qquad\qquad \uparrow$$
$$R_1 = \frac{1}{4}r_1 \qquad\qquad R_2 = -2r_1 + r_2$$

The matrix $[A|I_2]$ is sufficiently reduced for us to see that the identity matrix cannot appear to the left of the vertical bar. We conclude that A is singular and so has no inverse.

■

 ✓ Check: Enter the matrix A. Try to find its inverse. What happens?

— **Now Work** PROBLEM 59

5 Solve a System of Linear Equations Using an Inverse Matrix

Inverse matrices can be used to solve systems of equations in which the number of equations is the same as the number of variables.

| EXAMPLE 15 | **Using the Inverse Matrix to Solve a System of Linear Equations** |

Solve the system of equations: $\begin{cases} x + y & = 3 \\ -x + 3y + 4z = -3 \\ 4y + 3z = 2 \end{cases}$

Solution If we let

$$A = \begin{bmatrix} 1 & 1 & 0 \\ -1 & 3 & 4 \\ 0 & 4 & 3 \end{bmatrix}, \qquad X = \begin{bmatrix} x \\ y \\ z \end{bmatrix}, \qquad B = \begin{bmatrix} 3 \\ -3 \\ 2 \end{bmatrix}$$

the original system of equations can be written compactly as the matrix equation

$$AX = B \qquad\qquad (4)$$

We know from Example 13 that the matrix A has the inverse A^{-1}, so we multiply each side of equation (4) by A^{-1}.

$$AX = B$$
$$A^{-1}(AX) = A^{-1}B \quad \text{Multiply both sides by } A^{-1}.$$
$$(A^{-1}A)X = A^{-1}B \quad \text{Associative Property of multiplication}$$
$$I_3X = A^{-1}B \quad \text{Definition of an inverse matrix}$$
$$X = A^{-1}B \quad \text{Property of the identity matrix} \qquad (5)$$

Now we use (5) to find $X = \begin{bmatrix} x \\ y \\ z \end{bmatrix}$.

$$X = \begin{bmatrix} x \\ y \\ z \end{bmatrix} = A^{-1}B = \begin{bmatrix} \frac{7}{4} & \frac{3}{4} & -1 \\ -\frac{3}{4} & -\frac{3}{4} & 1 \\ 1 & 1 & -1 \end{bmatrix} \begin{bmatrix} 3 \\ -3 \\ 2 \end{bmatrix} = \begin{bmatrix} 1 \\ 2 \\ -2 \end{bmatrix}$$

$\uparrow$
Example 13

The solution is $x = 1$, $y = 2$, $z = -2$, or, using triplets, $(1, 2, -2)$.

The method used in Example 15 to solve a system of equations is particularly useful when it is necessary to solve several systems of equations in which the constants appearing to the right of the equal signs change, while the coefficients of the variables on the left side remain the same. See Problems 39–58 for some illustrations. Be careful; this method can only be used if the inverse exists. If it does not exist, row reduction must be used since the system is either inconsistent or dependent.

Now Work PROBLEM **43**

Historical Feature

Arthur Cayley (1821–1895)

Matrices were invented in 1857 by Arthur Cayley (1821–1895) as a way of efficiently computing the result of substituting one linear system into another (see Historical Problem 3). The resulting system had incredible richness, in the sense that a wide variety of mathematical systems could be mimicked by the matrices. Cayley and his friend James J. Sylvester (1814–1897) spent much of the rest of their lives elaborating the theory. The torch was then passed to Georg Frobenius (1849–1917), whose deep investigations established a central place for matrices in modern mathematics. In 1924, rather to the surprise of physicists, it was found that matrices (with complex numbers in them) were exactly the right tool for describing the behavior of atomic systems. Today, matrices are used in a wide variety of applications.

Historical Problems

1. **Matrices and Complex Numbers** Frobenius emphasized in his research how matrices could be used to mimic other mathematical systems. Here, we mimic the behavior of complex numbers using matrices. Mathematicians call such a relationship an *isomorphism*.

 Complex number $\longleftrightarrow$ Matrix
 $$a + bi \longleftrightarrow \begin{bmatrix} a & b \\ -b & a \end{bmatrix}$$

 Note that the complex number can be read off the top line of the matrix. Thus,

 $$2 + 3i \longleftrightarrow \begin{bmatrix} 2 & 3 \\ -3 & 2 \end{bmatrix} \text{ and } \begin{bmatrix} 4 & -2 \\ 2 & 4 \end{bmatrix} \longleftrightarrow 4 - 2i$$

 (a) Find the matrices corresponding to $2 - 5i$ and $1 + 3i$.
 (b) Multiply the two matrices.
 (c) Find the corresponding complex number for the matrix found in part (b).
 (d) Multiply $2 - 5i$ and $1 + 3i$. The result should be the same as that found in part (c).

 The process also works for addition and subtraction. Try it for yourself.

2. Compute $(a + bi)(a - bi)$ using matrices. Interpret the result.

3. **Cayley's Definition of Matrix Multiplication** Cayley invented matrix multiplication to simplify the following problem:

 $$\begin{cases} u = ar + bs \\ v = cr + ds \end{cases} \qquad \begin{cases} x = ku + lv \\ y = mu + nv \end{cases}$$

 (a) Find x and y in terms of r and s by substituting u and v from the first system of equations into the second system of equations.
 (b) Use the result of part (a) to find the 2 by 2 matrix A in

 $$\begin{bmatrix} x \\ y \end{bmatrix} = A \begin{bmatrix} r \\ s \end{bmatrix}$$

 (c) Now look at the following way to do it. Write the equations in matrix form.

 $$\begin{bmatrix} u \\ v \end{bmatrix} = \begin{bmatrix} a & b \\ c & d \end{bmatrix} \begin{bmatrix} r \\ s \end{bmatrix} \qquad \begin{bmatrix} x \\ y \end{bmatrix} = \begin{bmatrix} k & l \\ m & n \end{bmatrix} \begin{bmatrix} u \\ v \end{bmatrix}$$

 So

 $$\begin{bmatrix} x \\ y \end{bmatrix} = \begin{bmatrix} k & l \\ m & n \end{bmatrix} \begin{bmatrix} a & b \\ c & d \end{bmatrix} \begin{bmatrix} r \\ s \end{bmatrix}$$

 Do you see how Cayley defined matrix multiplication?

12.4 Assess Your Understanding

Concepts and Vocabulary

1. A matrix B, for which $AB = I_n$, the identity matrix, is called the _____ of A.

2. A matrix that has the same number of rows as columns is called a(n) _____ matrix.

3. In the algebra of matrices, the matrix that has properties similar to the number 1 is called the _____ matrix.

4. *True or False* Every square matrix has an inverse.

5. *True or False* Matrix multiplication is commutative.

6. *True or False* Any pair of matrices can be multiplied.

Skill Building

In Problems 7–22, use the following matrices to compute the given expression.

$$A = \begin{bmatrix} 0 & 3 & -5 \\ 1 & 2 & 6 \end{bmatrix}, \quad B = \begin{bmatrix} 4 & 1 & 0 \\ -2 & 3 & -2 \end{bmatrix}, \quad C = \begin{bmatrix} 4 & 1 \\ 6 & 2 \\ -2 & 3 \end{bmatrix}$$

7. $A + B$

8. $A - B$

9. $4A$

10. $-3B$

11. $3A - 2B$

12. $2A + 4B$

13. AC

14. BC

15. CA

16. CB

17. $C(A + B)$

18. $(A + B)C$

19. $AC - 3I_2$

20. $CA + 5I_3$

21. $CA - CB$

22. $AC + BC$

In Problems 23–28, find the product.

23. $\begin{bmatrix} 2 & -2 \\ 1 & 0 \end{bmatrix} \begin{bmatrix} 2 & 1 & 4 & 6 \\ 3 & -1 & 3 & 2 \end{bmatrix}$

24. $\begin{bmatrix} 4 & 1 \\ 2 & 1 \end{bmatrix} \begin{bmatrix} -6 & 6 & 1 & 0 \\ 2 & 5 & 4 & -1 \end{bmatrix}$

25. $\begin{bmatrix} 1 & 2 & 3 \\ 0 & -1 & 4 \end{bmatrix} \begin{bmatrix} 1 & 2 \\ -1 & 0 \\ 2 & 4 \end{bmatrix}$

26. $\begin{bmatrix} 1 & -1 \\ -3 & 2 \\ 0 & 5 \end{bmatrix} \begin{bmatrix} 2 & 8 & -1 \\ 3 & 6 & 0 \end{bmatrix}$

27. $\begin{bmatrix} 1 & 0 & 1 \\ 2 & 4 & 1 \\ 3 & 6 & 1 \end{bmatrix} \begin{bmatrix} 1 & 3 \\ 6 & 2 \\ 8 & -1 \end{bmatrix}$

28. $\begin{bmatrix} 4 & -2 & 3 \\ 0 & 1 & 2 \\ -1 & 0 & 1 \end{bmatrix} \begin{bmatrix} 2 & 6 \\ 1 & -1 \\ 0 & 2 \end{bmatrix}$

In Problems 29–38, each matrix is nonsingular. Find the inverse of each matrix.

29. $\begin{bmatrix} 2 & 1 \\ 1 & 1 \end{bmatrix}$

30. $\begin{bmatrix} 3 & -1 \\ -2 & 1 \end{bmatrix}$

31. $\begin{bmatrix} 6 & 5 \\ 2 & 2 \end{bmatrix}$

32. $\begin{bmatrix} -4 & 1 \\ 6 & -2 \end{bmatrix}$

33. $\begin{bmatrix} 2 & 1 \\ a & a \end{bmatrix} \quad a \neq 0$

34. $\begin{bmatrix} b & 3 \\ b & 2 \end{bmatrix} \quad b \neq 0$

35. $\begin{bmatrix} 1 & -1 & 1 \\ 0 & -2 & 1 \\ -2 & -3 & 0 \end{bmatrix}$

36. $\begin{bmatrix} 1 & 0 & 2 \\ -1 & 2 & 3 \\ 1 & -1 & 0 \end{bmatrix}$

37. $\begin{bmatrix} 1 & 1 & 1 \\ 3 & 2 & -1 \\ 3 & 1 & 2 \end{bmatrix}$

38. $\begin{bmatrix} 3 & 3 & 1 \\ 1 & 2 & 1 \\ 2 & -1 & 1 \end{bmatrix}$

In Problems 39–58, use the inverses found in Problems 29–38 to solve each system of equations.

39. $\begin{cases} 2x + y = 8 \\ x + y = 5 \end{cases}$

40. $\begin{cases} 3x - y = 8 \\ -2x + y = 4 \end{cases}$

41. $\begin{cases} 2x + y = 0 \\ x + y = 5 \end{cases}$

42. $\begin{cases} 3x - y = 4 \\ -2x + y = 5 \end{cases}$

43. $\begin{cases} 6x + 5y = 7 \\ 2x + 2y = 2 \end{cases}$

44. $\begin{cases} -4x + y = 0 \\ 6x - 2y = 14 \end{cases}$

45. $\begin{cases} 6x + 5y = 13 \\ 2x + 2y = 5 \end{cases}$

46. $\begin{cases} -4x + y = 5 \\ 6x - 2y = -9 \end{cases}$

47. $\begin{cases} 2x + y = -3 \\ ax + ay = -a \end{cases} \quad a \neq 0$

48. $\begin{cases} bx + 3y = 2b + 3 \\ bx + 2y = 2b + 2 \end{cases} \quad b \neq 0$

49. $\begin{cases} 2x + y = \dfrac{7}{a} \\ ax + ay = 5 \end{cases} \quad a \neq 0$

50. $\begin{cases} bx + 3y = 14 \\ bx + 2y = 10 \end{cases} \quad b \neq 0$

51. $\begin{cases} x - y + z = 0 \\ -2y + z = -1 \\ -2x - 3y = -5 \end{cases}$

52. $\begin{cases} x + 2z = 6 \\ -x + 2y + 3z = -5 \\ x - y = 6 \end{cases}$

53. $\begin{cases} x - y + z = 2 \\ -2y + z = 2 \\ -2x - 3y = \dfrac{1}{2} \end{cases}$

54. $\begin{cases} x + 2z = 2 \\ -x + 2y + 3z = -\dfrac{3}{2} \\ x - y = 2 \end{cases}$

55. $\begin{cases} x + y + z = 9 \\ 3x + 2y - z = 8 \\ 3x + y + 2z = 1 \end{cases}$

56. $\begin{cases} 3x + 3y + z = 8 \\ x + 2y + z = 5 \\ 2x - y + z = 4 \end{cases}$

57. $\begin{cases} x + y + z = 2 \\ 3x + 2y - z = \dfrac{7}{3} \\ 3x + y + 2z = \dfrac{10}{3} \end{cases}$

58. $\begin{cases} 3x + 3y + z = 1 \\ x + 2y + z = 0 \\ 2x - y + z = 4 \end{cases}$

In Problems 59–64, show that each matrix has no inverse.

59. $\begin{bmatrix} 4 & 2 \\ 2 & 1 \end{bmatrix}$

60. $\begin{bmatrix} -3 & \dfrac{1}{2} \\ 6 & -1 \end{bmatrix}$

61. $\begin{bmatrix} 15 & 3 \\ 10 & 2 \end{bmatrix}$

62. $\begin{bmatrix} -3 & 0 \\ 4 & 0 \end{bmatrix}$

63. $\begin{bmatrix} -3 & 1 & -1 \\ 1 & -4 & -7 \\ 1 & 2 & 5 \end{bmatrix}$

64. $\begin{bmatrix} 1 & 1 & -3 \\ 2 & -4 & 1 \\ -5 & 7 & 1 \end{bmatrix}$

In Problems 65–68, use a graphing utility to find the inverse, if it exists, of each matrix. Round answers to two decimal places.

65. $\begin{bmatrix} 25 & 61 & -12 \\ 18 & -2 & 4 \\ 8 & 35 & 21 \end{bmatrix}$

66. $\begin{bmatrix} 18 & -3 & 4 \\ 6 & -20 & 14 \\ 10 & 25 & -15 \end{bmatrix}$

67. $\begin{bmatrix} 44 & 21 & 18 & 6 \\ -2 & 10 & 15 & 5 \\ 21 & 12 & -12 & 4 \\ -8 & -16 & 4 & 9 \end{bmatrix}$

68. $\begin{bmatrix} 16 & 22 & -3 & 5 \\ 21 & -17 & 4 & 8 \\ 2 & 8 & 27 & 20 \\ 5 & 15 & -3 & -10 \end{bmatrix}$

In Problems 69–72, use the idea behind Example 15 with a graphing utility to solve the following systems of equations. Round answers to two decimal places.

69. $\begin{cases} 25x + 61y - 12z = 10 \\ 18x - 12y + 7y = -9 \\ 3x + 4y - z = 12 \end{cases}$

70. $\begin{cases} 25x + 61y - 12z = 15 \\ 18x - 12y + 7z = -3 \\ 3x + 4y - z = 12 \end{cases}$

71. $\begin{cases} 25x + 61y - 12z = 21 \\ 18x - 12y + 7z = 7 \\ 3x + 4y - z = -2 \end{cases}$

72. $\begin{cases} 25x + 61y - 12z = 25 \\ 18x - 12y + 7z = 10 \\ 3x + 4y - z = -4 \end{cases}$

Applications and Extensions

73. **College Tuition** Nikki and Joe take classes at a community college LCCC and a local university SIUE. The number of credit hours taken and the cost per credit hour (2006–2007 academic year, tuition only) are as follows:

	LCCC	SIUE
Nikki	6	9
Joe	3	12

Cost per Credit Hour	
LCCC	$71.00
SIUE	$158.60

(a) Write a matrix A for the credit hours taken by each student and a matrix B for the cost per credit hour.

(b) Compute AB and interpret the results.

Sources: www.lc.edu, www.siue.edu

74. **School Loan Interest** Jamal and Stephanie each have school loans issued from the same two banks. The amounts borrowed and the monthly interest rates are given next (interest is compounded monthly):

	Lender 1	Lender 2
Jamal	$4000	$3000
Stephanie	$2500	$3800

Monthly Interest Rate	
Lender 1	0.011 (1.1%)
Lender 2	0.006 (0.6%)

(a) Write a matrix A for the amounts borrowed by each student and a matrix B for the monthly interest rates.

(b) Compute AB and interpret the results.

(c) Let $C = \begin{bmatrix} 1 \\ 1 \end{bmatrix}$. Compute $A(C + B)$ and interpret the results.

75. **Computing the Cost of Production** The Acme Steel Company is a producer of stainless steel and aluminum containers. On a certain day, the following stainless steel containers were manufactured: 500 with 10-gallon capacity, 350 with 5-gallon capacity, and 400 with 1-gallon capacity. On the same day, the following aluminum containers were manufactured: 700 with 10-gallon capacity, 500 with 5-gallon capacity, and 850 with 1-gallon capacity.

(a) Find a 2 by 3 matrix representing these data. Find a 3 by 2 matrix to represent the same data.

(b) If the amount of material used in the 10-gallon containers is 15 pounds, the amount used in the 5-gallon containers is 8 pounds, and the amount used in the 1-gallon containers is 3 pounds, find a 3 by 1 matrix representing the amount of material used.

(c) Multiply the 2 by 3 matrix found in part (a) and the 3 by 1 matrix found in part (b) to get a 2 by 1 matrix showing the day's usage of material.

(d) If stainless steel costs Acme $0.10 per pound and aluminum costs $0.05 per pound, find a 1 by 2 matrix representing cost.

(e) Multiply the matrices found in parts (c) and (d) to determine the total cost of the day's production.

76. **Computing Profit** Rizza Ford has two locations, one in the city and the other in the suburbs. In January, the city location sold 400 subcompacts, 250 intermediate-size cars, and 50 SUVs; in February, it sold 350 subcompacts, 100 intermediates, and 30 SUVs. At the suburban location in January, 450 subcompacts, 200 intermediates, and 140 SUVs were sold. In February, the suburban location sold 350 subcompacts, 300 intermediates, and 100 SUVs.

(a) Find 2 by 3 matrices that summarize the sales data for each location for January and February (one matrix for each month).

(b) Use matrix addition to obtain total sales for the two-month period.

(c) The profit on each kind of car is $100 per subcompact, $150 per intermediate, and $200 per SUV. Find a 3 by 1 matrix representing this profit.

(d) Multiply the matrices found in parts (b) and (c) to get a 2 by 1 matrix showing the profit at each location.

77. Cryptography One method of encryption is to use a matrix to encrypt the message and then use the corresponding inverse matrix to decode the message. The encrypted matrix, E, is obtained by multiplying the message matrix, M, by a key matrix, K. The original message can be retrieved by multiplying the encrypted matrix by the inverse of the key matrix. That is, $E = M \cdot K$ and $M = E \cdot K^{-1}$.

(a) Given the key matrix $K = \begin{bmatrix} 2 & 1 & 1 \\ 1 & 1 & 0 \\ 1 & 1 & 1 \end{bmatrix}$, find its inverse, K^{-1}. (**Note:** This key matrix is known as the Q_2^3 Fibonacci encryption matrix.)

(b) Use your result from part (a) to decode the encrypted matrix $E = \begin{bmatrix} 47 & 34 & 33 \\ 44 & 36 & 27 \\ 47 & 41 & 20 \end{bmatrix}$.

(c) Each entry in your result for part (b) represents the position of a letter in the English alphabet ($A = 1, B = 2, C = 3$, and so on). What is the original message?

Source: goldenmuseum.com

78. Economic Mobility The relative income of a child (low, medium, or high) generally depends on the relative income of the child's parents. The matrix P, given by

$$\begin{array}{cc} & \text{Parent's Income} \\ & \begin{array}{ccc} \text{L} & \text{M} & \text{H} \end{array} \\ P = & \begin{bmatrix} 0.4 & 0.2 & 0.1 \\ 0.5 & 0.6 & 0.5 \\ 0.1 & 0.2 & 0.4 \end{bmatrix} \begin{array}{l} \text{L} \\ \text{M} \\ \text{H} \end{array} \end{array} \text{Child's income}$$

is called a *left stochastic transition matrix*. For example, the entry $P_{21} = 0.5$ means that 50% of the children of low relative income parents will transition to the medium level income. The diagonal entry $P_{i,i}$ represents the percent of children who remain in the same income level as their parents. Assuming the transition matrix is valid from one generation to the next, compute and interpret P^2.

Source: Understanding Mobility in America, April 2006

79. Consider the 2 by 2 square matrix

$$A = \begin{bmatrix} a & b \\ c & d \end{bmatrix}$$

If $D = ad - bc \neq 0$, show that A is nonsingular and that

$$A^{-1} = \frac{1}{D} \begin{bmatrix} d & -b \\ -c & a \end{bmatrix}$$

Discussion and Writing

80. Create a situation different from any found in the text that can be represented by a matrix.

12.5 Partial Fraction Decomposition

PREPARING FOR THIS SECTION *Before getting started, review the following:*

- Identity (Section 1.1, p. 86)
- Proper and Improper Rational Functions (Section 5.2, p. 349)
- Factoring Polynomials (Review, Section R.5, pp. 49–55)
- Complex Zeros; Fundamental Theorem of Algebra (Section 5.6, pp. 389–393)

Now Work the 'Are You Prepared?' problems on page 897.

OBJECTIVES 1 Decompose $\dfrac{P}{Q}$, Where Q Has Only Nonrepeated Linear Factors (p. 892)

2 Decompose $\dfrac{P}{Q}$, Where Q Has Repeated Linear Factors (p. 894)

3 Decompose $\dfrac{P}{Q}$, Where Q Has a Nonrepeated Irreducible Quadratic Factor (p. 896)

4 Decompose $\dfrac{P}{Q}$, Where Q Has a Repeated Irreducible Quadratic Factor (p. 897)

Consider the problem of adding two rational expressions:

$$\frac{3}{x + 4} \quad \text{and} \quad \frac{2}{x - 3}$$

The result is

$$\frac{3}{x+4} + \frac{2}{x-3} = \frac{3(x-3) + 2(x+4)}{(x+4)(x-3)} = \frac{5x-1}{x^2+x-12}$$

The reverse procedure, of starting with the rational expression $\dfrac{5x-1}{x^2+x-12}$ and writing it as the sum (or difference) of the two simpler fractions $\dfrac{3}{x+4}$ and $\dfrac{2}{x-3}$, is referred to as **partial fraction decomposition,** and the two simpler fractions are called **partial fractions.** Decomposing a rational expression into a sum of partial fractions is important in solving certain types of calculus problems. This section presents a systematic way to decompose rational expressions.

We begin by recalling that a rational expression is the ratio of two polynomials, say, P and $Q \neq 0$. We assume that P and Q have no common factors. Recall also that a rational expression $\dfrac{P}{Q}$ is called **proper** if the degree of the polynomial in the numerator is less than the degree of the polynomial in the denominator. Otherwise, the rational expression is termed **improper.**

Because any improper rational expression can be reduced by long division to a mixed form consisting of the sum of a polynomial and a proper rational expression, we shall restrict the discussion that follows to proper rational expressions.

The partial fraction decomposition of the rational expression $\dfrac{P}{Q}$ depends on the factors of the denominator Q. Recall from Section 5.6 that any polynomial whose coefficients are real numbers can be factored (over the real numbers) into products of linear and/or irreducible quadratic factors. This means that the denominator Q of the rational expression $\dfrac{P}{Q}$ will contain only factors of one or both of the following types:

1. *Linear factors* of the form $x - a$, where a is a real number.
2. *Irreducible quadratic factors* of the form $ax^2 + bx + c$, where a, b, and c are real numbers, $a \neq 0$, and $b^2 - 4ac < 0$ (which guarantees that $ax^2 + bx + c$ cannot be written as the product of two linear factors with real coefficients).

As it turns out, there are four cases to be examined. We begin with the case for which Q has only nonrepeated linear factors.

1 Decompose $\dfrac{P}{Q}$, Where Q Has Only Nonrepeated Linear Factors

Case 1: Q has only nonrepeated linear factors.

Under the assumption that Q has only nonrepeated linear factors, the polynomial Q has the form

$$Q(x) = (x - a_1)(x - a_2) \cdots (x - a_n)$$

where none of the numbers $a_1, a_2, \ldots, a_n$ is equal. In this case, the partial fraction decomposition of $\dfrac{P}{Q}$ is of the form

$$\frac{P(x)}{Q(x)} = \frac{A_1}{x - a_1} + \frac{A_2}{x - a_2} + \cdots + \frac{A_n}{x - a_n} \qquad (1)$$

where the numbers $A_1, A_2, \ldots, A_n$ are to be determined.

We show how to find these numbers in the example that follows.

| EXAMPLE 1 | **Nonrepeated Linear Factors** |

Write the partial fraction decomposition of $\dfrac{x}{x^2 - 5x + 6}$.

Solution First, we factor the denominator,

$$x^2 - 5x + 6 = (x - 2)(x - 3)$$

and conclude that the denominator contains only nonrepeated linear factors. Then we decompose the rational expression according to equation (1):

$$\frac{x}{x^2 - 5x + 6} = \frac{A}{x - 2} + \frac{B}{x - 3} \qquad \text{(2)}$$

where A and B are to be determined. To find A and B, we clear the fractions by multiplying each side by $(x - 2)(x - 3) = x^2 - 5x + 6$. The result is

$$x = A(x - 3) + B(x - 2) \qquad \text{(3)}$$

or

$$x = (A + B)x + (-3A - 2B)$$

This equation is an identity in x. We equate the coefficients of like powers of x to get

$$\begin{cases} 1 = A + B & \text{Equate coefficients of } x: 1x = (A + B)x. \\ 0 = -3A - 2B & \text{Equate the constants: } 0 = -3A - 2B. \end{cases}$$

This system of two equations containing two variables, A and B, can be solved using whatever method you wish. Solving it, we get

$$A = -2 \qquad B = 3$$

From equation (2), the partial fraction decomposition is

$$\frac{x}{x^2 - 5x + 6} = \frac{-2}{x - 2} + \frac{3}{x - 3}$$

✓**Check:** The decomposition can be checked by adding the rational expressions.

$$\frac{-2}{x - 2} + \frac{3}{x - 3} = \frac{-2(x - 3) + 3(x - 2)}{(x - 2)(x - 3)} = \frac{x}{(x - 2)(x - 3)}$$
$$= \frac{x}{x^2 - 5x + 6}$$

The numbers to be found in the partial fraction decomposition can sometimes be found more readily by using suitable choices for x (which may include complex numbers) in the identity obtained after fractions have been cleared. In Example 1, the identity after clearing fractions is equation (3):

$$x = A(x - 3) + B(x - 2)$$

If we let $x = 2$ in this expression, the term containing B drops out, leaving $2 = A(-1)$, or $A = -2$. Similarly, if we let $x = 3$, the term containing A drops out, leaving $3 = B$. As before, $A = -2$ and $B = 3$.

━━➤ **Now Work** PROBLEM 1 3

2 Decompose $\dfrac{P}{Q}$, Where Q Has Repeated Linear Factors

Case 2: Q has repeated linear factors.

If the polynomial Q has a repeated linear factor, say $(x - a)^n, n \geq 2$ an integer, then, in the partial fraction decomposition of $\dfrac{P}{Q}$, we allow for the terms

$$\frac{A_1}{x - a} + \frac{A_2}{(x - a)^2} + \cdots + \frac{A_n}{(x - a)^n}$$

where the numbers $A_1, A_2, \ldots, A_n$ are to be determined.

EXAMPLE 2

Repeated Linear Factors

Write the partial fraction decomposition of $\dfrac{x + 2}{x^3 - 2x^2 + x}$.

Solution

First, we factor the denominator,

$$x^3 - 2x^2 + x = x(x^2 - 2x + 1) = x(x - 1)^2$$

and find that the denominator has the nonrepeated linear factor x and the repeated linear factor $(x - 1)^2$. By Case 1, we must allow for the term $\dfrac{A}{x}$ in the decomposition; and, by Case 2, we must allow for the terms $\dfrac{B}{x - 1} + \dfrac{C}{(x - 1)^2}$ in the decomposition.

We write

$$\frac{x + 2}{x^3 - 2x^2 + x} = \frac{A}{x} + \frac{B}{x - 1} + \frac{C}{(x - 1)^2} \qquad \textbf{(4)}$$

Again, we clear fractions by multiplying each side by $x^3 - 2x^2 + x = x(x - 1)^2$. The result is the identity

$$x + 2 = A(x - 1)^2 + Bx(x - 1) + Cx \qquad \textbf{(5)}$$

If we let $x = 0$ in this expression, the terms containing B and C drop out, leaving $2 = A(-1)^2$, or $A = 2$. Similarly, if we let $x = 1$, the terms containing A and B drop out, leaving $3 = C$. Then equation (5) becomes

$$x + 2 = 2(x - 1)^2 + Bx(x - 1) + 3x$$

Now let $x = 2$ (any choice other than 0 or 1 will work as well). The result is

$$4 = 2(1)^2 + B(2)(1) + 3(2)$$
$$4 = 2 + 2B + 6$$
$$2B = -4$$
$$B = -2$$

We have $A = 2$, $B = -2$, and $C = 3$.

From equation (4), the partial fraction decomposition is

$$\frac{x + 2}{x^3 - 2x^2 + x} = \frac{2}{x} + \frac{-2}{x - 1} + \frac{3}{(x - 1)^2}$$

■

EXAMPLE 3

Repeated Linear Factors

Write the partial fraction decomposition of $\dfrac{x^3 - 8}{x^2(x - 1)^3}$.

Solution The denominator contains the repeated linear factor x^2 and the repeated linear factor $(x - 1)^3$. The partial fraction decomposition takes the form

$$\frac{x^3 - 8}{x^2(x - 1)^3} = \frac{A}{x} + \frac{B}{x^2} + \frac{C}{x - 1} + \frac{D}{(x - 1)^2} + \frac{E}{(x - 1)^3} \qquad \textbf{(6)}$$

As before, we clear fractions and obtain the identity

$$x^3 - 8 = Ax(x - 1)^3 + B(x - 1)^3 + Cx^2(x - 1)^2$$
$$+ Dx^2(x - 1) + Ex^2 \qquad \textbf{(7)}$$

Let $x = 0$. (Do you see why this choice was made?) Then

$$-8 = B(-1)$$
$$B = 8$$

Now let $x = 1$ in equation (7). Then

$$-7 = E$$

Use $B = 8$ and $E = -7$ in equation (7) and collect like terms.

$$x^3 - 8 = Ax(x - 1)^3 + 8(x - 1)^3$$
$$+ Cx^2(x - 1)^2 + Dx^2(x - 1) - 7x^2$$
$$x^3 - 8 - 8(x^3 - 3x^2 + 3x - 1) + 7x^2 = Ax(x - 1)^3 + Cx^2(x - 1)^2 + Dx^2(x - 1)$$
$$-7x^3 + 31x^2 - 24x = x(x - 1)[A(x - 1)^2 + Cx(x - 1) + Dx]$$
$$x(x - 1)(-7x + 24) = x(x - 1)[A(x - 1)^2 + Cx(x - 1) + Dx]$$
$$-7x + 24 = A(x - 1)^2 + Cx(x - 1) + Dx \qquad \textbf{(8)}$$

We now work with equation (8). Let $x = 0$. Then

$$24 = A$$

Now let $x = 1$ in equation (8). Then

$$17 = D$$

Use $A = 24$ and $D = 17$ in equation (8) and collect like terms.

$$-7x + 24 = 24(x - 1)^2 + Cx(x - 1) + 17x$$

Now let $x = 2$. Then

$$-14 + 24 = 24 + C(2) + 34$$
$$-48 = 2C$$
$$-24 = C$$

We now know all the numbers A, B, C, D, and E, so, from equation (6), we have the decomposition

$$\frac{x^3 - 8}{x^2(x - 1)^3} = \frac{24}{x} + \frac{8}{x^2} + \frac{-24}{x - 1} + \frac{17}{(x - 1)^2} + \frac{-7}{(x - 1)^3}$$

━━━━━ Now Work Example 3 by solving the system of five equations containing five variables that the expansion of equation (7) leads to.

━━━━━ **Now Work** PROBLEM 19

The final two cases involve irreducible quadratic factors. A quadratic factor is irreducible if it cannot be factored into linear factors with real coefficients. A quadratic expression $ax^2 + bx + c$ is irreducible whenever $b^2 - 4ac < 0$. For example, $x^2 + x + 1$ and $x^2 + 4$ are irreducible.

3 Decompose $\dfrac{P}{Q}$, Where Q Has a Nonrepeated Irreducible Quadratic Factor

Case 3: Q contains a nonrepeated irreducible quadratic factor

If Q contains a nonrepeated irreducible quadratic factor of the form $ax^2 + bx + c$, then, in the partial fraction decomposition of $\dfrac{P}{Q}$, allow for the term

$$\frac{Ax + B}{ax^2 + bx + c}$$

where the numbers A and B are to be determined.

| EXAMPLE 4 | **Nonrepeated Irreducible Quadratic Factor** |

Write the partial fraction decomposition of $\dfrac{3x - 5}{x^3 - 1}$.

Solution We factor the denominator,

$$x^3 - 1 = (x - 1)(x^2 + x + 1)$$

and find that it has a nonrepeated linear factor $x - 1$ and a nonrepeated irreducible quadratic factor $x^2 + x + 1$. We allow for the term $\dfrac{A}{x - 1}$ by Case 1, and we allow for the term $\dfrac{Bx + C}{x^2 + x + 1}$ by Case 3. We write

$$\frac{3x - 5}{x^3 - 1} = \frac{A}{x - 1} + \frac{Bx + C}{x^2 + x + 1} \qquad (9)$$

We clear fractions by multiplying each side of equation (9) by $x^3 - 1 = (x - 1)(x^2 + x + 1)$ to obtain the identity

$$3x - 5 = A(x^2 + x + 1) + (Bx + C)(x - 1) \qquad (10)$$

Expand the identity in (10) to obtain

$$3x - 5 = (A + B)x^2 + (A - B + C)x + (A - C)$$

This identity leads to the system of equations

$$\begin{cases} A + B & = 0 & (1) \\ A - B + C & = 3 & (2) \\ A \quad - C & = -5 & (3) \end{cases}$$

The solution of this system is $A = -\dfrac{2}{3}$, $B = \dfrac{2}{3}$, $C = \dfrac{13}{3}$. Then from equation (9), we see that

$$\frac{3x - 5}{x^3 - 1} = \frac{-\dfrac{2}{3}}{x - 1} + \frac{\dfrac{2}{3}x + \dfrac{13}{3}}{x^2 + x + 1}$$

■

━━━ **Now Work Example 4 using equation (10) and assigning values to x**

━━━ **Now Work** PROBLEM 21

4 Decompose $\dfrac{P}{Q}$, Where Q Has a Repeated Irreducible Quadratic Factor

Case 4: Q contains a repeated irreducible quadratic factor.

If the polynomial Q contains a repeated irreducible quadratic factor $(ax^2 + bx + c)^n$, $n \geq 2$, n an integer, then, in the partial fraction decomposition of $\dfrac{P}{Q}$, allow for the terms

$$\frac{A_1x + B_1}{ax^2 + bx + c} + \frac{A_2x + B_2}{(ax^2 + bx + c)^2} + \cdots + \frac{A_nx + B_n}{(ax^2 + bx + c)^n}$$

where the numbers $A_1, B_1, A_2, B_2, \ldots, A_n, B_n$ are to be determined.

EXAMPLE 5 **Repeated Irreducible Quadratic Factor**

Write the partial fraction decomposition of $\dfrac{x^3 + x^2}{(x^2 + 4)^2}$.

Solution The denominator contains the repeated irreducible quadratic factor $(x^2 + 4)^2$, so we write

$$\frac{x^3 + x^2}{(x^2 + 4)^2} = \frac{Ax + B}{x^2 + 4} + \frac{Cx + D}{(x^2 + 4)^2} \qquad \textbf{(11)}$$

We clear fractions to obtain

$$x^3 + x^2 = (Ax + B)(x^2 + 4) + Cx + D$$

Collecting like terms yields the identity

$$x^3 + x^2 = Ax^3 + Bx^2 + (4A + C)x + 4B + D$$

Equating coefficients, we arrive at the system

$$\begin{cases} A = 1 \\ B = 1 \\ 4A + C = 0 \\ D + 4B = 0 \end{cases}$$

The solution is $A = 1$, $B = 1$, $C = -4$, $D = -4$. From equation (11),

$$\frac{x^3 + x^2}{(x^2 + 4)^2} = \frac{x + 1}{x^2 + 4} + \frac{-4x - 4}{(x^2 + 4)^2}$$

Now Work PROBLEM **35**

12.5 Assess Your Understanding

'Are You Prepared?' *Answers are given at the end of these exercises. If you get a wrong answer, read the pages listed in red.*

1. **True or False** The equation $(x - 1)^2 - 1 = x(x - 2)$ is an example of an identity. (p. 86)

2. **True or False** The rational expression $\dfrac{5x^2 - 1}{x^3 + 1}$ is proper. (p. 349)

3. Factor completely: $3x^4 + 6x^3 + 3x^2$. (pp. 49–55)

4. **True or False** Every polynomial with real numbers as coefficients can be factored into products of linear and/or irreducible quadratic factors. (pp. 389–393)

Skill Building

In Problems 5–12, tell whether the given rational expression is proper or improper. If improper, rewrite it as the sum of a polynomial and a proper rational expression.

5. $\dfrac{x}{x^2 - 1}$

6. $\dfrac{5x + 2}{x^3 - 1}$

7. $\dfrac{x^2 + 5}{x^2 - 4}$

8. $\dfrac{3x^2 - 2}{x^2 - 1}$

9. $\dfrac{5x^3 + 2x - 1}{x^2 - 4}$

10. $\dfrac{3x^4 + x^2 - 2}{x^3 + 8}$

11. $\dfrac{x(x - 1)}{(x + 4)(x - 3)}$

12. $\dfrac{2x(x^2 + 4)}{x^2 + 1}$

In Problems 13–46, write the partial fraction decomposition of each rational expression.

13. $\dfrac{4}{x(x - 1)}$

14. $\dfrac{3x}{(x + 2)(x - 1)}$

15. $\dfrac{1}{x(x^2 + 1)}$

16. $\dfrac{1}{(x + 1)(x^2 + 4)}$

17. $\dfrac{x}{(x - 1)(x - 2)}$

18. $\dfrac{3x}{(x + 2)(x - 4)}$

19. $\dfrac{x^2}{(x - 1)^2(x + 1)}$

20. $\dfrac{x + 1}{x^2(x - 2)}$

21. $\dfrac{1}{x^3 - 8}$

22. $\dfrac{2x + 4}{x^3 - 1}$

23. $\dfrac{x^2}{(x - 1)^2(x + 1)^2}$

24. $\dfrac{x + 1}{x^2(x - 2)^2}$

25. $\dfrac{x - 3}{(x + 2)(x + 1)^2}$

26. $\dfrac{x^2 + x}{(x + 2)(x - 1)^2}$

27. $\dfrac{x + 4}{x^2(x^2 + 4)}$

28. $\dfrac{10x^2 + 2x}{(x - 1)^2(x^2 + 2)}$

29. $\dfrac{x^2 + 2x + 3}{(x + 1)(x^2 + 2x + 4)}$

30. $\dfrac{x^2 - 11x - 18}{x(x^2 + 3x + 3)}$

31. $\dfrac{x}{(3x - 2)(2x + 1)}$

32. $\dfrac{1}{(2x + 3)(4x - 1)}$

33. $\dfrac{x}{x^2 + 2x - 3}$

34. $\dfrac{x^2 - x - 8}{(x + 1)(x^2 + 5x + 6)}$

35. $\dfrac{x^2 + 2x + 3}{(x^2 + 4)^2}$

36. $\dfrac{x^3 + 1}{(x^2 + 16)^2}$

37. $\dfrac{7x + 3}{x^3 - 2x^2 - 3x}$

38. $\dfrac{x^5 + 1}{x^6 - x^4}$

39. $\dfrac{x^2}{x^3 - 4x^2 + 5x - 2}$

40. $\dfrac{x^2 + 1}{x^3 + x^2 - 5x + 3}$

41. $\dfrac{x^3}{(x^2 + 16)^3}$

42. $\dfrac{x^2}{(x^2 + 4)^3}$

43. $\dfrac{4}{2x^2 - 5x - 3}$

44. $\dfrac{4x}{2x^2 + 3x - 2}$

45. $\dfrac{2x + 3}{x^4 - 9x^2}$

46. $\dfrac{x^2 + 9}{x^4 - 2x^2 - 8}$

'Are You Prepared?' Answers

1. True **2.** True **3.** $3x^2(x + 1)^2$ **4.** True

12.6 Systems of Nonlinear Equations

PREPARING FOR THIS SECTION *Before getting started, review the following:*

- Lines (Section 2.3, pp. 173–185)
- Circles (Section 2.4, pp. 189–193)
- Parabolas (Section 11.2, pp. 773–779)
- Ellipses (Section 11.3, pp. 782–789)
- Hyperbolas (Section 11.4, pp. 792–801)

Now Work the 'Are You Prepared?' problems on page 904.

OBJECTIVES 1 Solve a System of Nonlinear Equations Using Substitution (p. 898)
2 Solve a System of Nonlinear Equations Using Elimination (p. 900)

1 Solve a System of Nonlinear Equations Using Substitution

In Section 12.1, we observed that the solution to a system of linear equations could be found geometrically by determining the point(s) of intersection (if any) of the equations in the system. Similarly, when solving systems of nonlinear equations, the solution(s) also represents the point(s) of intersection (if any) of the graphs of the equations.

There is no general methodology for solving a system of nonlinear equations. At times substitution is best; other times, elimination is best; and sometimes neither of these methods works. Experience and a certain degree of imagination are your allies here.

Before we begin, two comments are in order.

1. If the system contains two variables and if the equations in the system are easy to graph, then graph them. By graphing each equation in the system, you can get an idea of how many solutions a system has and approximately where they are located.

2. Extraneous solutions can creep in when solving nonlinear systems, so it is imperative that all apparent solutions be checked.

| EXAMPLE 1 | Solving a System of Nonlinear Equations |

Solve the following system of equations:

$$\begin{cases} 3x - y = -2 & (1) \\ 2x^2 - y = 0 & (2) \end{cases}$$

Solution Using Substitution

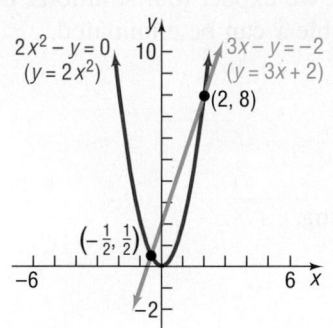

Figure 9

$2x^2 - y = 0$
$(y = 2x^2)$

$3x - y = -2$
$(y = 3x + 2)$

$(2, 8)$

$\left(-\frac{1}{2}, \frac{1}{2}\right)$

First, we notice that the system contains two variables and that we know how to graph each equation. Equation (1) is the line $y = 3x + 2$ and equation (2) is the parabola $y = 2x^2$. See Figure 9. The system apparently has two solutions.

We will use substitution to solve the system. We choose to solve equation (1) for y.

$$3x - y = -2 \qquad \text{Equation (1)}$$
$$y = 3x + 2$$

We substitute this expression for y in equation (2). The result is an equation containing just the variable x, which we can then solve.

$$2x^2 - y = 0 \qquad \text{Equation (2)}$$
$$2x^2 - (3x + 2) = 0 \qquad \text{Substitute } 3x + 2 \text{ for } y.$$
$$2x^2 - 3x - 2 = 0 \qquad \text{Remove parentheses.}$$
$$(2x + 1)(x - 2) = 0 \qquad \text{Factor.}$$
$$2x + 1 = 0 \quad \text{or} \quad x - 2 = 0 \qquad \text{Apply the Zero-Product Property.}$$
$$x = -\frac{1}{2} \quad \text{or} \quad x = 2$$

Using these values for x in $y = 3x + 2$, we find

$$y = 3\left(-\frac{1}{2}\right) + 2 = \frac{1}{2} \quad \text{or} \quad y = 3(2) + 2 = 8$$

The apparent solutions are $x = -\frac{1}{2}$, $y = \frac{1}{2}$ and $x = 2$, $y = 8$.

✓ **Check:** For $x = -\frac{1}{2}$, $y = \frac{1}{2}$,

$$\begin{cases} 3\left(-\frac{1}{2}\right) - \frac{1}{2} = -\frac{3}{2} - \frac{1}{2} = -2 & (1) \\ 2\left(-\frac{1}{2}\right)^2 - \frac{1}{2} = 2\left(\frac{1}{4}\right) - \frac{1}{2} = 0 & (2) \end{cases}$$

For $x = 2$, $y = 8$,

$$\begin{cases} 3(2) - 8 = 6 - 8 = -2 & (1) \\ 2(2)^2 - 8 = 2(4) - 8 = 0 & (2) \end{cases}$$

Each solution checks. Now we know that the graphs in Figure 9 intersect at the points $\left(-\dfrac{1}{2}, \dfrac{1}{2}\right)$ and $(2, 8)$.

 ✓ **Check:** Graph $3x - y = -2$ ($Y_1 = 3x + 2$) and $2x^2 - y = 0$ ($Y_2 = 2x^2$) and compare what you see with Figure 9. Use INTERSECT (twice) to find the two points of intersection.

 Now Work PROBLEM 15 USING SUBSTITUTION

2 Solve a System of Nonlinear Equations Using Elimination

Our next example illustrates how the method of elimination works for nonlinear systems.

| EXAMPLE 2 | **Solving a System of Nonlinear Equations** |

Solve: $\begin{cases} x^2 + y^2 = 13 & (1) \\ x^2 - y = 7 & (2) \end{cases}$

Solution Using Elimination

Equation (1) is a circle and equation (2) is the parabola $y = x^2 - 7$. We graph each equation, as shown in Figure 10. Based on the graph, we expect four solutions. By subtracting equation (2) from equation (1), the variable x can be eliminated.

Figure 10

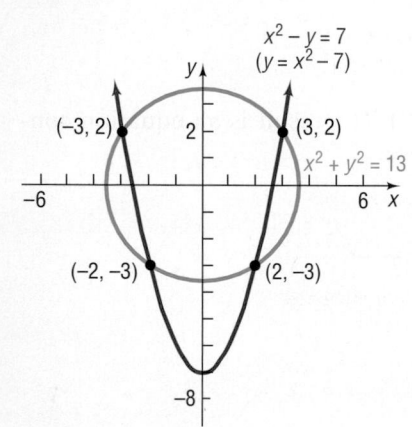

$$\begin{cases} x^2 + y^2 = 13 \\ x^2 - y = 7 \\ \hline y^2 + y = 6 \end{cases} \quad \text{Subtract}$$

This quadratic equation in y can be solved by factoring.

$$y^2 + y - 6 = 0$$
$$(y + 3)(y - 2) = 0$$
$$y = -3 \quad \text{or} \quad y = 2$$

We use these values for y in equation (2) to find x.

If $y = 2$, then $x^2 = y + 7 = 9$, so $x = 3$ or -3.

If $y = -3$, then $x^2 = y + 7 = 4$, so $x = 2$ or -2.

We have four solutions: $x = 3, y = 2$; $x = -3, y = 2$; $x = 2, y = -3$; and $x = -2, y = -3$.

You should verify that, in fact, these four solutions also satisfy equation (1), so all four are solutions of the system. The four points, $(3, 2)$, $(-3, 2)$, $(2, -3)$, and $(-2, -3)$, are the points of intersection of the graphs. Look again at Figure 10.

 ✓ **Check:** Graph $x^2 + y^2 = 13$ and $x^2 - y = 7$. (Remember that to graph $x^2 + y^2 = 13$ requires two functions: $Y_1 = \sqrt{13 - x^2}$ and $Y_2 = -\sqrt{13 - x^2}$.) Use INTERSECT to find the points of intersection.

 Now Work PROBLEM 13 USING ELIMINATION

| EXAMPLE 3 | **Solving a System of Nonlinear Equations** |

Solve: $\begin{cases} x^2 - y^2 = 4 & (1) \\ y = x^2 & (2) \end{cases}$

Figure 11

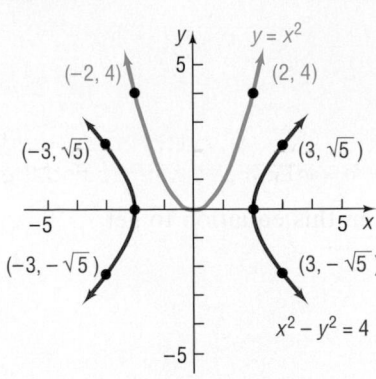

Solution Equation (1) is a hyperbola and equation (2) is a parabola. See Figure 11. It appears the system has no solution. We verify this using substitution. Replace x^2 by y in equation (1). The result is

$$x^2 - y^2 = 4 \qquad \text{Equation (1)}$$
$$y - y^2 = 4 \qquad y = x^2$$
$$y^2 - y + 4 = 0 \qquad \text{Place in standard form.}$$

This is a quadratic equation. Its discriminant is $(-1)^2 - 4 \cdot 1 \cdot 4 = 1 - 16 = -15 < 0$. The equation has no real solutions, so the system is inconsistent. The graphs of these two equations do not intersect.

■

EXAMPLE 4 | **Solving a System of Nonlinear Equations**

Solve: $\begin{cases} x^2 + x + y^2 - 3y + 2 = 0 & (1) \\ \quad x + 1 + \dfrac{y^2 - y}{x} = 0 & (2) \end{cases}$

Solution Using Elimination First, we multiply equation (2) by x to eliminate the fraction. The result is an equivalent system because x cannot be 0. [Look at equation (2) to see why.]

$$\begin{cases} x^2 + x + y^2 - 3y + 2 = 0 & (1) \\ x^2 + x + y^2 - y = 0 & (2) \quad x \neq 0 \end{cases}$$

Now subtract equation (2) from equation (1) to eliminate x. The result is

$$-2y + 2 = 0$$
$$y = 1 \qquad \text{Solve for y.}$$

To find x, we back-substitute $y = 1$ in equation (1):

$$x^2 + x + y^2 - 3y + 2 = 0 \qquad \text{Equation (1)}$$
$$x^2 + x + 1 - 3 + 2 = 0 \qquad \text{Substitute 1 for y in (1).}$$
$$x^2 + x = 0 \qquad \text{Simplify.}$$
$$x(x + 1) = 0 \qquad \text{Factor.}$$
$$x = 0 \quad \text{or} \quad x = -1 \qquad \text{Apply the Zero-Product Property.}$$

Because x cannot be 0, the value $x = 0$ is extraneous, and we discard it.

✓ **Check:** We now check $x = -1$, $y = 1$:

$$\begin{cases} (-1)^2 + (-1) + 1^2 - 3(1) + 2 = 1 - 1 + 1 - 3 + 2 = 0 & (1) \\ \quad -1 + 1 + \dfrac{1^2 - 1}{-1} = 0 + \dfrac{0}{-1} = 0 & (2) \end{cases}$$

The solution is $x = -1$, $y = 1$. The point of intersection of the graphs of the equations is $(-1, 1)$.

■

In Problem 55 you are asked to graph the equations given in Example 4. Be sure to show holes in the graph of equation (2) for $x = 0$.

 Now Work PROBLEMS **29** AND **49**

EXAMPLE 5 | **Solving a System of Nonlinear Equations**

Solve: $\begin{cases} 3xy - 2y^2 = -2 & (1) \\ 9x^2 + 4y^2 = 10 & (2) \end{cases}$

Solution We multiply equation (1) by 2 and add the result to equation (2) to eliminate the y^2 terms.

$$\begin{cases} 6xy - 4y^2 = -4 & (1) \\ 9x^2 + 4y^2 = 10 & (2) \end{cases}$$

$$\begin{array}{ll} 9x^2 + 6xy = 6 & \text{Add.} \\ 3x^2 + 2xy = 2 & \text{Divide each side by 3.} \end{array}$$

Since $x \neq 0$ (do you see why?), we can solve for y in this equation to get

$$y = \frac{2 - 3x^2}{2x} \qquad x \neq 0 \quad (3)$$

Now substitute for y in equation (2) of the system.

$$9x^2 + 4y^2 = 10 \qquad \text{Equation (2)}$$

$$9x^2 + 4\left(\frac{2 - 3x^2}{2x}\right)^2 = 10 \qquad \text{Substitute } y = \frac{2 - 3x^2}{2x} \text{ in (2).}$$

$$9x^2 + \frac{4 - 12x^2 + 9x^4}{x^2} = 10 \qquad \text{Expand and simplify}$$

$$9x^4 + 4 - 12x^2 + 9x^4 = 10x^2 \qquad \text{Multiply both sides by } x^2.$$

$$18x^4 - 22x^2 + 4 = 0 \qquad \text{Subtract } 10x^2 \text{ from both sides.}$$

$$9x^4 - 11x^2 + 2 = 0 \qquad \text{Divide both sides by 2.}$$

This quadratic equation (in x^2) can be factored:

$$(9x^2 - 2)(x^2 - 1) = 0$$

$$9x^2 - 2 = 0 \qquad \text{or} \qquad x^2 - 1 = 0$$

$$x^2 = \frac{2}{9} \qquad \text{or} \qquad x^2 = 1$$

$$x = \pm\sqrt{\frac{2}{9}} = \pm\frac{\sqrt{2}}{3} \qquad \text{or} \qquad x = \pm 1$$

To find y, we use equation (3):

If $x = \dfrac{\sqrt{2}}{3}$: $\quad y = \dfrac{2 - 3x^2}{2x} = \dfrac{2 - \dfrac{2}{3}}{2\left(\dfrac{\sqrt{2}}{3}\right)} = \dfrac{4}{2\sqrt{2}} = \sqrt{2}$

If $x = -\dfrac{\sqrt{2}}{3}$: $\quad y = \dfrac{2 - 3x^2}{2x} = \dfrac{2 - \dfrac{2}{3}}{2\left(-\dfrac{\sqrt{2}}{3}\right)} = \dfrac{4}{-2\sqrt{2}} = -\sqrt{2}$

If $x = 1$: $\quad y = \dfrac{2 - 3x^2}{2x} = \dfrac{2 - 3(1)^2}{2} = -\dfrac{1}{2}$

If $x = -1$: $\quad y = \dfrac{2 - 3x^2}{2x} = \dfrac{2 - 3(-1)^2}{-2} = \dfrac{1}{2}$

The system has four solutions: $\left(\dfrac{\sqrt{2}}{3}, \sqrt{2}\right), \left(-\dfrac{\sqrt{2}}{3}, -\sqrt{2}\right), \left(1, -\dfrac{1}{2}\right), \left(-1, \dfrac{1}{2}\right)$. Check them for yourself.

━━━━ **Now Work** PROBLEM 47

The next example illustrates an imaginative solution to a system of nonlinear equations.

EXAMPLE 6	Running a Long Distance Race

In a 50-mile race, the winner crosses the finish line 1 mile ahead of the second-place runner and 4 miles ahead of the third-place runner. Assuming that each runner maintains a constant speed throughout the race, by how many miles does the second-place runner beat the third-place runner?

3 miles — 1 mile

Solution Let v_1, v_2, and v_3 denote the speeds of the first-, second-, and third-place runners, respectively. Let t_1 and t_2 denote the times (in hours) required for the first-place runner and second-place runner to finish the race. Then we have the system of equations

$$\begin{cases} 50 = v_1 t_1 & (1) \quad \text{First-place runner goes 50 miles in } t_1 \text{ hours.} \\ 49 = v_2 t_1 & (2) \quad \text{Second-place runner goes 49 miles in } t_1 \text{ hours.} \\ 46 = v_3 t_1 & (3) \quad \text{Third-place runner goes 46 miles in } t_1 \text{ hours.} \\ 50 = v_2 t_2 & (4) \quad \text{Second-place runner goes 50 miles in } t_2 \text{ hours.} \end{cases}$$

We seek the distance d of the third-place runner from the finish at time t_2. At time t_2, the third-place runner has gone a distance of $v_3 t_2$ miles, so the distance d remaining is $50 - v_3 t_2$. Now

$$d = 50 - v_3 t_2$$

$$= 50 - v_3 \left(t_1 \cdot \frac{t_2}{t_1} \right)$$

$$= 50 - (v_3 t_1) \cdot \frac{t_2}{t_1}$$

$$= 50 - 46 \cdot \frac{\dfrac{50}{v_2}}{\dfrac{50}{v_1}} \qquad \begin{cases} \text{From (3), } v_3 t_1 = 46 \\ \text{From (4), } t_2 = \dfrac{50}{v_2} \\ \text{From (1), } t_1 = \dfrac{50}{v_1} \end{cases}$$

$$= 50 - 46 \cdot \frac{v_1}{v_2}$$

$$= 50 - 46 \cdot \frac{50}{49} \qquad \text{From the quotient of (1) and (2).}$$

$$\approx 3.06 \text{ miles}$$

Historical Feature

In the beginning of this section, we said that imagination and experience are important in solving systems of nonlinear equations. Indeed, these kinds of problems lead into some of the deepest and most difficult parts of modern mathematics. Look again at the graphs in Examples 1 and 2 of this section (Figures 9 and 10). We see that Example 1 has two solutions, and Example 2 has four solutions. We might conjecture that the number of solutions is equal to the product of the degrees of the equations involved. This conjecture was indeed made by Etienne Bezout (1730–1783), but working out the details took about 150 years. It turns out that, to arrive at the correct number of intersections, we must count not only the complex number intersections, but also those intersections that, in a certain sense, lie at infinity. For example, a parabola and a line lying on the axis of the parabola intersect at the vertex and at infinity. This topic is part of the study of algebraic geometry.

Historical Problem

A papyrus dating back to 1950 BC contains the following problem: "A given surface area of 100 units of area shall be represented as the sum of two squares whose sides are to each other as $1 : \frac{3}{4}$."

Solve for the sides by solving the system of equations

$$\begin{cases} x^2 + y^2 = 100 \\ x = \dfrac{3}{4}y \end{cases}$$

12.6 Assess Your Understanding

'Are You Prepared?' *Answers are given at the end of these exercises. If you get a wrong answer, read the pages listed in red.*

1. Graph the equation: $y = 3x + 2$ (pp. 173–185)

2. Graph the equation: $y + 4 = x^2$ (pp. 773–779)

3. Graph the equation: $y^2 = x^2 - 1$ (pp. 792–801)

4. Graph the equation: $x^2 + 4y^2 = 4$ (pp. 782–789)

Skill Building

In Problems 5–24, graph each equation of the system. Then solve the system to find the points of intersection.

5. $\begin{cases} y = x^2 + 1 \\ y = x + 1 \end{cases}$

6. $\begin{cases} y = x^2 + 1 \\ y = 4x + 1 \end{cases}$

7. $\begin{cases} y = \sqrt{36 - x^2} \\ y = 8 - x \end{cases}$

8. $\begin{cases} y = \sqrt{4 - x^2} \\ y = 2x + 4 \end{cases}$

9. $\begin{cases} y = \sqrt{x} \\ y = 2 - x \end{cases}$

10. $\begin{cases} y = \sqrt{x} \\ y = 6 - x \end{cases}$

11. $\begin{cases} x = 2y \\ x = y^2 - 2y \end{cases}$

12. $\begin{cases} y = x - 1 \\ y = x^2 - 6x + 9 \end{cases}$

13. $\begin{cases} x^2 + y^2 = 4 \\ x^2 + 2x + y^2 = 0 \end{cases}$

14. $\begin{cases} x^2 + y^2 = 8 \\ x^2 + y^2 + 4y = 0 \end{cases}$

15. $\begin{cases} y = 3x - 5 \\ x^2 + y^2 = 5 \end{cases}$

16. $\begin{cases} x^2 + y^2 = 10 \\ y = x + 2 \end{cases}$

17. $\begin{cases} x^2 + y^2 = 4 \\ y^2 - x = 4 \end{cases}$

18. $\begin{cases} x^2 + y^2 = 16 \\ x^2 - 2y = 8 \end{cases}$

19. $\begin{cases} xy = 4 \\ x^2 + y^2 = 8 \end{cases}$

20. $\begin{cases} x^2 = y \\ xy = 1 \end{cases}$

21. $\begin{cases} x^2 + y^2 = 4 \\ y = x^2 - 9 \end{cases}$

22. $\begin{cases} xy = 1 \\ y = 2x + 1 \end{cases}$

23. $\begin{cases} y = x^2 - 4 \\ y = 6x - 13 \end{cases}$

24. $\begin{cases} x^2 + y^2 = 10 \\ xy = 3 \end{cases}$

In Problems 25–54, solve each system. Use any method you wish.

25. $\begin{cases} 2x^2 + y^2 = 18 \\ xy = 4 \end{cases}$

26. $\begin{cases} x^2 - y^2 = 21 \\ x + y = 7 \end{cases}$

27. $\begin{cases} y = 2x + 1 \\ 2x^2 + y^2 = 1 \end{cases}$

28. $\begin{cases} x^2 - 4y^2 = 16 \\ 2y - x = 2 \end{cases}$

29. $\begin{cases} x + y + 1 = 0 \\ x^2 + y^2 + 6y - x = -5 \end{cases}$

30. $\begin{cases} 2x^2 - xy + y^2 = 8 \\ xy = 4 \end{cases}$

31. $\begin{cases} 4x^2 - 3xy + 9y^2 = 15 \\ 2x + 3y = 5 \end{cases}$

32. $\begin{cases} 2y^2 - 3xy + 6y + 2x + 4 = 0 \\ 2x - 3y + 4 = 0 \end{cases}$

33. $\begin{cases} x^2 - 4y^2 + 7 = 0 \\ 3x^2 + y^2 = 31 \end{cases}$

34. $\begin{cases} 3x^2 - 2y^2 + 5 = 0 \\ 2x^2 - y^2 + 2 = 0 \end{cases}$

35. $\begin{cases} 7x^2 - 3y^2 + 5 = 0 \\ 3x^2 + 5y^2 = 12 \end{cases}$

36. $\begin{cases} x^2 - 3y^2 + 1 = 0 \\ 2x^2 - 7y^2 + 5 = 0 \end{cases}$

37. $\begin{cases} x^2 + 2xy = 10 \\ 3x^2 - xy = 2 \end{cases}$

38. $\begin{cases} 5xy + 13y^2 + 36 = 0 \\ xy + 7y^2 = 6 \end{cases}$

39. $\begin{cases} 2x^2 + y^2 = 2 \\ x^2 - 2y^2 + 8 = 0 \end{cases}$

40. $\begin{cases} y^2 - x^2 + 4 = 0 \\ 2x^2 + 3y^2 = 6 \end{cases}$

41. $\begin{cases} x^2 + 2y^2 = 16 \\ 4x^2 - y^2 = 24 \end{cases}$

42. $\begin{cases} 4x^2 + 3y^2 = 4 \\ 2x^2 - 6y^2 = -3 \end{cases}$

43. $\begin{cases} \dfrac{5}{x^2} - \dfrac{2}{y^2} + 3 = 0 \\ \dfrac{3}{x^2} + \dfrac{1}{y^2} = 7 \end{cases}$

44. $\begin{cases} \dfrac{2}{x^2} - \dfrac{3}{y^2} + 1 = 0 \\ \dfrac{6}{x^2} - \dfrac{7}{y^2} + 2 = 0 \end{cases}$

45. $\begin{cases} \dfrac{1}{x^4} + \dfrac{6}{y^4} = 6 \\ \dfrac{2}{x^4} - \dfrac{2}{y^4} = 19 \end{cases}$

46. $\begin{cases} \dfrac{1}{x^4} - \dfrac{1}{y^4} = 1 \\ \dfrac{1}{x^4} + \dfrac{1}{y^4} = 4 \end{cases}$

47. $\begin{cases} x^2 - 3xy + 2y^2 = 0 \\ x^2 + xy = 6 \end{cases}$

48. $\begin{cases} x^2 - xy - 2y^2 = 0 \\ xy + x + 6 = 0 \end{cases}$

49. $\begin{cases} y^2 + y + x^2 - x - 2 = 0 \\ y + 1 + \dfrac{x - 2}{y} = 0 \end{cases}$

50. $\begin{cases} x^3 - 2x^2 + y^2 + 3y - 4 = 0 \\ x - 2 + \dfrac{y^2 - y}{x^2} = 0 \end{cases}$

51. $\begin{cases} \log_x y = 3 \\ \log_x(4y) = 5 \end{cases}$

52. $\begin{cases} \log_x(2y) = 3 \\ \log_x(4y) = 2 \end{cases}$

53. $\begin{cases} \ln x = 4 \ln y \\ \log_3 x = 2 + 2 \log_3 y \end{cases}$

54. $\begin{cases} \ln x = 5 \ln y \\ \log_2 x = 3 + 2 \log_2 y \end{cases}$

55. Graph the equations given in Example 4.

56. Graph the equations given in Problem 49.

In Problems 57–62, graph each equation and find the point(s) of intersection, if any.

57. The line $x + 2y = 0$ and
the circle $(x - 1)^2 + (y - 1)^2 = 5$

58. The line $x + 2y + 6 = 0$ and
the circle $(x + 1)^2 + (y + 1)^2 = 5$

59. The circle $(x - 1)^2 + (y + 2)^2 = 4$ and
the parabola $y^2 + 4y - x + 1 = 0$

60. The circle $(x + 2)^2 + (y - 1)^2 = 4$ and
the parabola $y^2 - 2y - x - 5 = 0$

61. $y = \dfrac{4}{x - 3}$ and the circle $x^2 - 6x + y^2 + 1 = 0$

62. $y = \dfrac{4}{x + 2}$ and the circle $x^2 + 4x + y^2 - 4 = 0$

In Problems 63–70, use a graphing utility to solve each system of equations. Express the solution(s) rounded to two decimal places.

63. $\begin{cases} y = x^{2/3} \\ y = e^{-x} \end{cases}$

64. $\begin{cases} y = x^{3/2} \\ y = e^{-x} \end{cases}$

65. $\begin{cases} x^2 + y^3 = 2 \\ x^3 y = 4 \end{cases}$

66. $\begin{cases} x^3 + y^2 = 2 \\ x^2 y = 4 \end{cases}$

67. $\begin{cases} x^4 + y^4 = 12 \\ xy^2 = 2 \end{cases}$

68. $\begin{cases} x^4 + y^4 = 6 \\ xy = 1 \end{cases}$

69. $\begin{cases} xy = 2 \\ y = \ln x \end{cases}$

70. $\begin{cases} x^2 + y^2 = 4 \\ y = \ln x \end{cases}$

Applications and Extensions

71. The difference of two numbers is 2 and the sum of their squares is 10. Find the numbers.

72. The sum of two numbers is 7 and the difference of their squares is 21. Find the numbers.

73. The product of two numbers is 4 and the sum of their squares is 8. Find the numbers.

74. The product of two numbers is 10 and the difference of their squares is 21. Find the numbers.

75. The difference of two numbers is the same as their product, and the sum of their reciprocals is 5. Find the numbers.

76. The sum of two numbers is the same as their product, and the difference of their reciprocals is 3. Find the numbers.

77. The ratio of a to b is $\dfrac{2}{3}$. The sum of a and b is 10. What is the ratio of $a + b$ to $b - a$?

78. The ratio of a to b is $4:3$. The sum of a and b is 14. What is the ratio of $a - b$ to $a + b$?

79. Geometry The perimeter of a rectangle is 16 inches and its area is 15 square inches. What are its dimensions?

80. Geometry An area of 52 square feet is to be enclosed by two squares whose sides are in the ratio of $2:3$. Find the sides of the squares.

81. Geometry Two circles have circumferences that add up to 12π centimeters and areas that add up to 20π square centimeters. Find the radius of each circle.

82. Geometry The altitude of an isosceles triangle drawn to its base is 3 centimeters, and its perimeter is 18 centimeters. Find the length of its base.

83. The Tortoise and the Hare In a 21-meter race between a tortoise and a hare, the tortoise leaves 9 minutes before the hare. The hare, by running at an average speed of 0.5 meter per hour faster than the tortoise, crosses the finish line 3 minutes before the tortoise. What are the average speeds of the tortoise and the hare?

84. Running a Race In a 1-mile race, the winner crosses the finish line 10 feet ahead of the second-place runner and 20 feet ahead of the third-place runner. Assuming that each runner maintains a constant speed throughout the race, by how many feet does the second-place runner beat the third-place runner?

85. Constructing a Box A rectangular piece of cardboard, whose area is 216 square centimeters, is made into an open box by cutting a 2-centimeter square from each corner and turning up the sides. See the figure. If the box is to have a volume of 224 cubic centimeters, what size cardboard should you start with?

86. Constructing a Cylindrical Tube A rectangular piece of cardboard, whose area is 216 square centimeters, is made into a cylindrical tube by joining together two sides of the rectangle. See the figure. If the tube is to have a volume of 224 cubic centimeters, what size cardboard should you start with?

87. Fencing A farmer has 300 feet of fence available to enclose a 4500-square-foot region in the shape of adjoining squares, with sides of length x and y. See the figure. Find x and y.

88. Bending Wire A wire 60 feet long is cut into two pieces. Is it possible to bend one piece into the shape of a square and the other into the shape of a circle so that the total area enclosed by the two pieces is 100 square feet? If this is possible, find the length of the side of the square and the radius of the circle.

89. Geometry Find formulas for the length l and width w of a rectangle in terms of its area A and perimeter P.

90. Geometry Find formulas for the base b and one of the equal sides l of an isosceles triangle in terms of its altitude h and perimeter P.

91. Descartes's Method of Equal Roots Descartes's method for finding tangents depends on the idea that, for many graphs, the tangent line at a given point is the *unique* line that intersects the graph at that point only. We will apply his method to find an equation of the tangent line to the parabola $y = x^2$ at the point $(2, 4)$. See the figure.

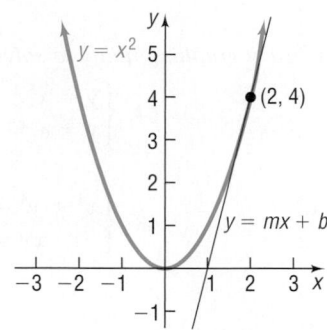

First, we know that the equation of the tangent line must be in the form $y = mx + b$. Using the fact that the point $(2, 4)$ is on the line, we can solve for b in terms of m and get the equation $y = mx + (4 - 2m)$. Now we want $(2, 4)$ to be the *unique* solution to the system

$$\begin{cases} y = x^2 \\ y = mx + 4 - 2m \end{cases}$$

From this system, we get $x^2 - mx + (2m - 4) = 0$. By using the quadratic formula, we get

$$x = \frac{m \pm \sqrt{m^2 - 4(2m - 4)}}{2}$$

To obtain a unique solution for x, the two roots must be equal; in other words, the discriminant $m^2 - 4(2m - 4)$ must be 0. Complete the work to get m, and write an equation of the tangent line.

In Problems 92–98, use Descartes's method from Problem 91 to find the equation of the line tangent to each graph at the given point.

92. $x^2 + y^2 = 10$; at $(1, 3)$

93. $y = x^2 + 2$; at $(1, 3)$

94. $x^2 + y = 5$; at $(-2, 1)$

95. $2x^2 + 3y^2 = 14$; at $(1, 2)$

96. $3x^2 + y^2 = 7$; at $(-1, 2)$

97. $x^2 - y^2 = 3$; at $(2, 1)$

98. $2y^2 - x^2 = 14$; at $(2, 3)$

99. If r_1 and r_2 are two solutions of a quadratic equation $ax^2 + bx + c = 0$, it can be shown that

$$r_1 + r_2 = -\frac{b}{a} \quad \text{and} \quad r_1 r_2 = \frac{c}{a}$$

Solve this system of equations for r_1 and r_2.

Discussion and Writing

100. A circle and a line intersect at most twice. A circle and a parabola intersect at most four times. Deduce that a circle and the graph of a polynomial of degree 3 intersect at most six times. What do you conjecture about a polynomial of degree 4? What about a polynomial of degree n? Can you explain your conclusions using an algebraic argument?

101. Suppose that you are the manager of a sheet metal shop. A customer asks you to manufacture 10,000 boxes, each box being open on top. The boxes are required to have a square base and a 9-cubic-foot capacity. You construct the boxes by cutting out a square from each corner of a square piece of sheet metal and folding along the edges.

(a) What are the dimensions of the square to be cut if the area of the square piece of sheet metal is 100 square feet?

(b) Could you make the box using a smaller piece of sheet metal? Make a list of the dimensions of the box for various pieces of sheet metal.

'Are You Prepared?' Answers

1.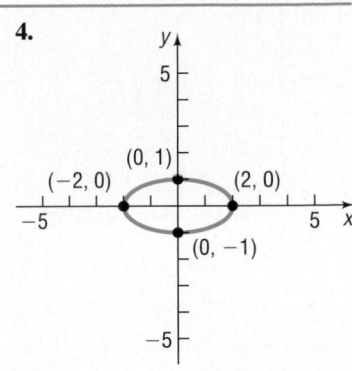

2.

3.

4.

12.7 Systems of Inequalities

PREPARING FOR THIS SECTION *Before getting started, review the following:*

- Solving Linear Inequalities (Section 1.5, pp. 124–131)
- Lines (Section 2.3, pp. 173–185)

- Circles (Section 2.4, pp. 189–193)
- Graphing Techniques: Transformations (Section 3.5, pp. 252–260)

Now Work the 'Are You Prepared?' problems on page 912.

OBJECTIVES **1** Graph an Inequality (p. 907)

2 Graph a System of Inequalities (p. 909)

In Section 1.5, we discussed inequalities in one variable. In this section, we discuss inequalities in two variables.

EXAMPLE 1 **Examples of Inequalities in Two Variables**

(a) $3x + y \leq 6$ (b) $x^2 + y^2 < 4$ (c) $y^2 > x$

1 Graph an Inequality

An inequality in two variables x and y is **satisfied** by an ordered pair (a, b) if, when x is replaced by a and y by b, a true statement results. The **graph of an inequality in two variables** x and y consists of all points (x, y) whose coordinates satisfy the inequality.

Let's look at an example.

EXAMPLE 2 **Graphing an Inequality**

Graph the linear inequality: $3x + y \leq 6$

Solution We begin by graphing the equation

$$3x + y = 6$$

formed by replacing (for now) the $\leq$ symbol with an $=$ sign. The graph of the equation is a line. See Figure 12(a) on page 908. This line is part of the graph of the inequality that we seek because the inequality is nonstrict. (Do you see why? We are seeking points for which $3x + y$ is less than *or equal to* 6.)

Figure 12

(a) $3x + y = 6$

(b) Graph of $3x + y \leq 6$

Now let's test a few randomly selected points to see whether they belong to the graph of the inequality.

	$3x + y \leq 6$	Conclusion
$(4, -1)$	$3(4) + (-1) = 11 > 6$	Does not belong to the graph
$(5, 5)$	$3(5) + 5 = 20 > 6$	Does not belong to the graph
$(-1, 2)$	$3(-1) + 2 = -1 \leq 6$	Belongs to the graph
$(-2, -2)$	$3(-2) + (-2) = -8 \leq 6$	Belongs to the graph

Look again at Figure 12(a). Notice that the two points that belong to the graph both lie on the same side of the line, and the two points that do not belong to the graph lie on the opposite side. As it turns out, this is always the case. The graph we seek consists of all points that lie on the same side of the line as $(-1, 2)$ and $(-2, -2)$ and is shown as the shaded region in Figure 12(b).

Now Work PROBLEM 15

The graph of any inequality in two variables may be obtained in a like way. First, the equation corresponding to the inequality is graphed, using dashes if the inequality is strict and solid marks if it is nonstrict. This graph, in almost every case, will separate the xy-plane into two or more regions. In each region, either all points satisfy the inequality or no points satisfy the inequality. The use of a single test point in each region is all that is required to determine whether the points of that region are part of the graph. The steps to follow are given next.

NOTE The strict inequalities are $<$ and $>$. The nonstrict inequalities are $\leq$ and $\geq$. ■

Steps for Graphing an Inequality

STEP 1: Replace the inequality symbol by an equal sign and graph the resulting equation. If the inequality is strict, use dashes; if it is nonstrict, use a solid mark. This graph separates the xy-plane into two or more regions.

STEP 2: In each region, select a test point P.

(a) If the coordinates of P satisfy the inequality, so do all the points in that region. Indicate this by shading the region.

(b) If the coordinates of P do not satisfy the inequality, none of the points in that region do.

EXAMPLE 3	**Graphing an Inequality**

Graph: $x^2 + y^2 \leq 4$

Solution First, we graph the equation $x^2 + y^2 = 4$, a circle of radius 2, center at the origin. A solid circle will be used because the inequality is not strict. We use two test points, one inside the circle, the other outside.

Figure 13

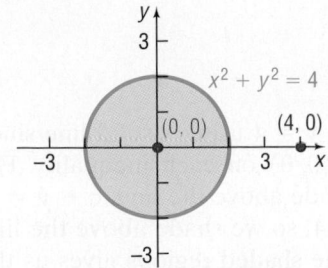

Inside $(0, 0)$: $x^2 + y^2 = 0^2 + 0^2 = 0 \le 4$ *Belongs to the graph*

Outside $(4, 0)$: $x^2 + y^2 = 4^2 + 0^2 = 16 > 4$ *Does not belong to the graph*

All the points inside and on the circle satisfy the inequality. See Figure 13.

━━━━━ **Now Work** PROBLEM 17

Linear Inequalities

Linear inequalities are inequalities in one of the forms

$$Ax + By < C \qquad Ax + By > C \qquad Ax + By \le C \qquad Ax + By \ge C$$

where A and B are not both zero.

The graph of the corresponding equation of a linear inequality is a line that separates the xy-plane into two regions, called **half-planes.** See Figure 14.

Figure 14

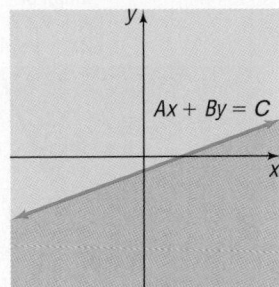

As shown, $Ax + By = C$ is the equation of the boundary line and it divides the plane into two half-planes: one for which $Ax + By < C$ and the other for which $Ax + By > C$. Because of this, for linear inequalities, only one test point is required.

EXAMPLE 4	**Graphing Linear Inequalities**

Graph:

(a) $y < 2$ (b) $y \ge 2x$

Solution (a) The graph of the equation $y = 2$ is a horizontal line and is not part of the graph of the inequality. Since $(0, 0)$ satisfies the inequality, the graph consists of the half-plane below the line $y = 2$. See Figure 15.

(b) The graph of the equation $y = 2x$ is a line and is part of the graph of the inequality. Using $(3, 0)$ as a test point, we find it does not satisfy the inequality $[0 < 2 \cdot 3]$. Points in the half-plane on the opposite side of $(3, 0)$ satisfy the inequality. See Figure 16.

Figure 15

Figure 16

 COMMENT A graphing utility can be used to graph inequalities. To see how, read Section 6 in the Appendix. ∎

━━━━━ **Now Work** PROBLEM 13

2 Graph a System of Inequalities

The **graph of a system of inequalities** in two variables x and y is the set of all points (x, y) that simultaneously satisfy *each* inequality in the system. The graph of a system of inequalities can be obtained by graphing each inequality individually and then determining where, if at all, they intersect.

EXAMPLE 5 Graphing a System of Linear Inequalities

Graph the system: $\begin{cases} x + y \geq 2 \\ 2x - y \leq 4 \end{cases}$

Solution We begin by graphing the lines $x + y = 2$ and $2x - y = 4$ using a solid line since the inequality is non-strict. We use the test point $(0, 0)$ on each inequality. For example, $(0, 0)$ does not satisfy $x + y \geq 2$, so we shade above the line $x + y = 2$. See Figure 17(a). Also, $(0, 0)$ does satisfy $2x - y \leq 4$, so we shade above the line $2x - y = 4$. See Figure 17(b). The intersection of the shaded regions gives us the result presented in Figure 17(c).

Figure 17

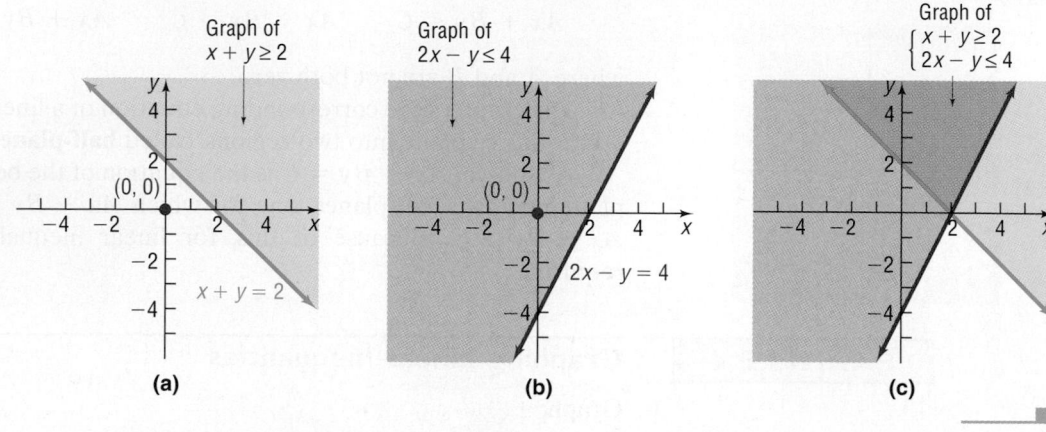

(a) (b) (c)

➤ **Now Work** PROBLEM 23

EXAMPLE 6 Graphing a System of Linear Inequalities

Graph the system: $\begin{cases} x + y \leq 2 \\ x + y \geq 0 \end{cases}$

Solution See Figure 18. The overlapping purple-shaded region between the two boundary lines is the graph of the system.

Figure 18 $x + y = 0$ $x + y = 2$

Graph of
$\begin{cases} x + y \leq 2 \\ x + y \geq 0 \end{cases}$

➤ **Now Work** PROBLEM 29

EXAMPLE 7 Graphing a System of Linear Inequalities

Graph the systems:

(a) $\begin{cases} 2x - y \geq 0 \\ 2x - y \geq 2 \end{cases}$ (b) $\begin{cases} x + 2y \leq 2 \\ x + 2y \geq 6 \end{cases}$

Solution (a) See Figure 19. The overlapping purple-shaded region is the graph of the system. Note that the graph of the system is identical to the graph of the single inequality $2x - y \geq 2$.

(b) See Figure 20. Because no overlapping region results, there are no points in the xy-plane that simultaneously satisfy each inequality. The system has no solution.

Figure 19

Graph of
$$\begin{cases} 2x - y \geq 0 \\ 2x - y \geq 2 \end{cases}$$

$2x - y = 0$

$2x - y = 2$

Figure 20

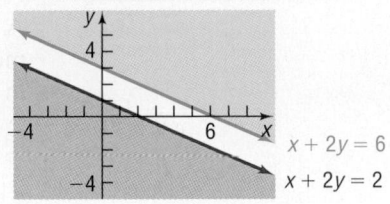

$x + 2y = 6$

$x + 2y = 2$

EXAMPLE 8 | **Graphing a System of Nonlinear Inequalities**

Graph the region below the graph of $x + y = 2$ and above the graph of $y = x^2 - 4$ by graphing the system:

$$\begin{cases} y \geq x^2 - 4 \\ x + y \leq 2 \end{cases}$$

Label all points of intersection.

Solution Figure 21 shows the graph of the region above the graph of the parabola $y = x^2 - 4$ and below the graph of the line $x + y = 2$. The points of intersection are found by solving the system of equations

$$\begin{cases} y = x^2 - 4 \\ x + y = 2 \end{cases}$$

Using substitution, we find

$$x + (x^2 - 4) = 2$$
$$x^2 + x - 6 = 0$$
$$(x + 3)(x - 2) = 0$$
$$x = -3 \quad \text{or} \quad x = 2$$

The two points of intersection are $(-3, 5)$ and $(2, 0)$.

Figure 21

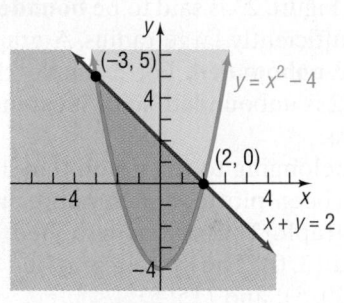

$(-3, 5)$

$y = x^2 - 4$

$(2, 0)$

$x + y = 2$

Now Work PROBLEM 37

EXAMPLE 9 | **Graphing a System of Four Linear Inequalities**

Graph the system: $$\begin{cases} x + y \geq 3 \\ 2x + y \geq 4 \\ x \geq 0 \\ y \geq 0 \end{cases}$$

Solution See Figure 22. The two inequalities $x \geq 0$ and $y \geq 0$ require the graph of the system to be in quadrant I, which is shaded light gray. We concentrate on the remaining two inequalities. The intersection of the graphs of these two inequalities and quadrant I is shown in dark purple.

Figure 22

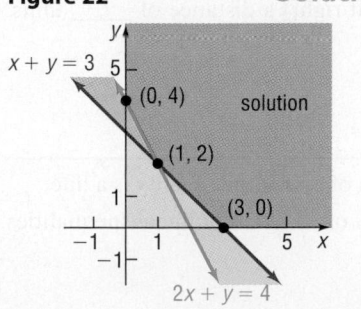

$x + y = 3$

$(0, 4)$ solution

$(1, 2)$

$(3, 0)$

$2x + y = 4$

| EXAMPLE 10 | Financial Planning |

A retired couple can invest up to $25,000. As their financial adviser, you recommend that they place at least $15,000 in Treasury bills yielding 6% and at most $5000 in corporate bonds yielding 9%.

(a) Using x to denote the amount of money invested in Treasury bills and y the amount invested in corporate bonds, write a system of linear inequalities that describes the possible amounts of each investment. We shall assume that x and y are in thousands of dollars.

(b) Graph the system.

Solution (a) The system of linear inequalities is

Figure 23

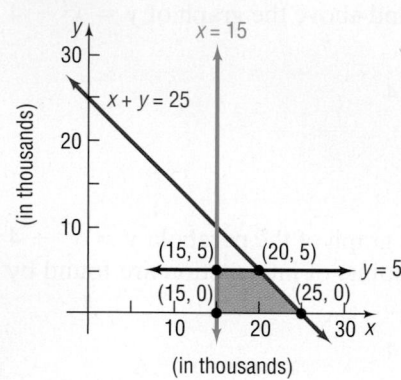

$$\begin{cases} x \geq 0 \\ y \geq 0 \\ x + y \leq 25 \\ x \geq 15 \\ y \leq 5 \end{cases}$$

x and y are nonnegative variables since they represent money invested in thousands of dollars.
The total of the two investments, $x + y$, cannot exceed $25,000.
At least $15,000 in Treasury bills
At most $5000 in corporate bonds

(b) See the shaded region in Figure 23. Note that the inequalities $x \geq 0$ and $y \geq 0$ require that the graph of the system be in quadrant I.

The graph of the system of linear inequalities in Figure 23 is said to be **bounded,** because it can be contained within some circle of sufficiently large radius. A graph that cannot be contained in any circle is said to be **unbounded.** For example, the graph of the system of linear inequalities in Figure 22 is unbounded, since it extends indefinitely in the positive x and positive y directions.

Notice in Figures 22 and 23 that those points belonging to the graph that are also points of intersection of boundary lines have been plotted. Such points are referred to as **vertices** or **corner points** of the graph. The system graphed in Figure 22 has three corner points: $(0, 4)$, $(1, 2)$, and $(3, 0)$. The system graphed in Figure 23 has four corner points: $(15, 0)$, $(25, 0)$, $(20, 5)$, and $(15, 5)$.

These ideas will be used in the next section in developing a method for solving linear programming problems, an important application of linear inequalities.

━━━━━━ **Now Work** PROBLEM 45

12.7 Assess Your Understanding

'Are You Prepared?' *Answers are given at the end of these exercises. If you get a wrong answer, read the pages listed in red.*

1. Solve the inequality: $3x + 4 < 8 - x$. (pp. 128–131)

2. Graph the equation: $3x - 2y = 6$. (pp. 173–185)

3. Graph the equation: $x^2 + y^2 = 9$. (pp. 189–193)

4. Graph the equation: $y = x^2 + 4$. (pp. 252–260)

5. *True or False* The lines $2x + y = 4$ and $4x + 2y = 0$ are parallel. (pp. 179–185)

6. The graph of $y = (x - 2)^2$ may be obtained by shifting the graph of _____ to the (left/right) a distance of _____ units. (pp. 252–260)

Concepts and Vocabulary

7. An inequality in two variables x and y is _____ by an ordered pair (a, b) if, when x is replaced by a and y by b, a true statement results.

8. The graph of a linear inequality is called a(n) _____.

9. *True or False* The graph of a linear inequality is a line.

10. *True or False* The graph of a system of linear inequalities is sometimes unbounded.

Skill Building

In Problems 11–22, graph each inequality.

11. $x \geq 0$

12. $y \geq 0$

13. $x \geq 4$

14. $y \leq 2$

15. $2x + y \geq 6$

16. $3x + 2y \leq 6$

17. $x^2 + y^2 > 1$

18. $x^2 + y^2 \leq 9$

19. $y \leq x^2 - 1$

20. $y > x^2 + 2$

21. $xy \geq 4$

22. $xy \leq 1$

In Problems 23–34, graph each system of linear inequalities.

23. $\begin{cases} x + y \leq 2 \\ 2x + y \geq 4 \end{cases}$

24. $\begin{cases} 3x - y \geq 6 \\ x + 2y \leq 2 \end{cases}$

25. $\begin{cases} 2x - y \leq 4 \\ 3x + 2y \geq -6 \end{cases}$

26. $\begin{cases} 4x - 5y \leq 0 \\ 2x - y \geq 2 \end{cases}$

27. $\begin{cases} 2x - 3y \leq 0 \\ 3x + 2y \leq 6 \end{cases}$

28. $\begin{cases} 4x - y \geq 2 \\ x + 2y \geq 2 \end{cases}$

29. $\begin{cases} x - 2y \leq 6 \\ 2x - 4y \geq 0 \end{cases}$

30. $\begin{cases} x + 4y \leq 8 \\ x + 4y \geq 4 \end{cases}$

31. $\begin{cases} 2x + y \geq -2 \\ 2x + y \geq 2 \end{cases}$

32. $\begin{cases} x - 4y \leq 4 \\ x - 4y \geq 0 \end{cases}$

33. $\begin{cases} 2x + 3y \geq 6 \\ 2x + 3y \leq 0 \end{cases}$

34. $\begin{cases} 2x + y \geq 0 \\ 2x + y \geq 2 \end{cases}$

In Problems 35–42, graph each system of inequalities.

35. $\begin{cases} x^2 + y^2 \leq 9 \\ x + y \geq 3 \end{cases}$

36. $\begin{cases} x^2 + y^2 \geq 9 \\ x + y \leq 3 \end{cases}$

37. $\begin{cases} y \geq x^2 - 4 \\ y \leq x - 2 \end{cases}$

38. $\begin{cases} y^2 \leq x \\ y \geq x \end{cases}$

39. $\begin{cases} x^2 + y^2 \leq 16 \\ y \geq x^2 - 4 \end{cases}$

40. $\begin{cases} x^2 + y^2 \leq 25 \\ y \leq x^2 - 5 \end{cases}$

41. $\begin{cases} xy \geq 4 \\ y \geq x^2 + 1 \end{cases}$

42. $\begin{cases} y + x^2 \leq 1 \\ y \geq x^2 - 1 \end{cases}$

In Problems 43–52, graph each system of linear inequalities. Tell whether the graph is bounded or unbounded, and label the corner points.

43. $\begin{cases} x \geq 0 \\ y \geq 0 \\ 2x + y \leq 6 \\ x + 2y \leq 6 \end{cases}$

44. $\begin{cases} x \geq 0 \\ y \geq 0 \\ x + y \geq 4 \\ 2x + 3y \geq 6 \end{cases}$

45. $\begin{cases} x \geq 0 \\ y \geq 0 \\ x + y \geq 2 \\ 2x + y \geq 4 \end{cases}$

46. $\begin{cases} x \geq 0 \\ y \geq 0 \\ 3x + y \leq 6 \\ 2x + y \leq 2 \end{cases}$

47. $\begin{cases} x \geq 0 \\ y \geq 0 \\ x + y \geq 2 \\ 2x + 3y \leq 12 \\ 3x + y \leq 12 \end{cases}$

48. $\begin{cases} x \geq 0 \\ y \geq 0 \\ x + y \geq 2 \\ x + y \leq 10 \\ 2x + y \leq 3 \end{cases}$

49. $\begin{cases} x \geq 0 \\ y \geq 0 \\ x + y \geq 2 \\ x + y \leq 8 \\ 2x + y \leq 10 \end{cases}$

50. $\begin{cases} x \geq 0 \\ y \geq 0 \\ x + y \geq 2 \\ x + y \leq 8 \\ x + 2y \geq 1 \end{cases}$

51. $\begin{cases} x \geq 0 \\ y \geq 0 \\ x + 2y \geq 1 \\ x + 2y \leq 10 \end{cases}$

52. $\begin{cases} x \geq 0 \\ y \geq 0 \\ x + 2y \geq 1 \\ x + 2y \leq 10 \\ x + y \geq 2 \\ x + y \leq 8 \end{cases}$

In Problems 53–56, write a system of linear inequalities that has the given graph.

53.

54.

55.

56.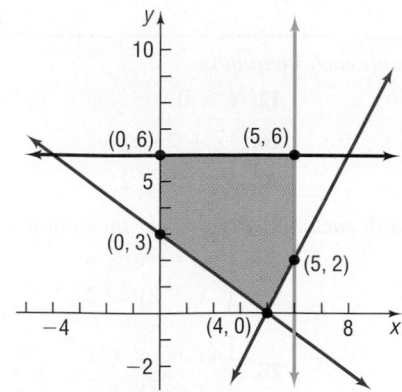

Applications and Extensions

57. Financial Planning A retired couple has up to \$50,000 to invest. As their financial adviser, you recommend that they place at least \$35,000 in Treasury bills yielding 7% and at most \$10,000 in corporate bonds yielding 10%.

(a) Using x to denote the amount of money invested in Treasury bills and y the amount invested in corporate bonds, write a system of linear inequalities that describes the possible amounts of each investment.

(b) Graph the system and label the corner points.

58. Manufacturing Trucks Mike's Toy Truck Company manufacturers two models of toy trucks, a standard model and a deluxe model. Each standard model requires 2 hours for painting and 3 hours for detail work; each deluxe model requires 3 hours for painting and 4 hours for detail work. Two painters and three detail workers are employed by the company, and each works 40 hours per week.

(a) Using x to denote the number of standard model trucks and y to denote the number of deluxe model trucks, write a system of linear inequalities that describes the possible number of each model of truck that can be manufactured in a week.

(b) Graph the system and label the corner points.

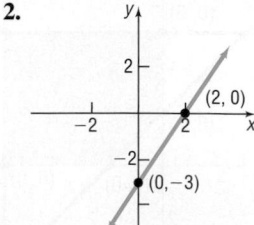

59. Blending Coffee Bill's Coffee House, a store that specializes in coffee, has available 75 pounds of A grade coffee and 120 pounds of B grade coffee. These will be blended into 1 pound packages as follows: An economy blend that contains 4 ounces of A grade coffee and 12 ounces of B grade coffee and a superior blend that contains 8 ounces of A grade coffee and 8 ounces of B grade coffee.

(a) Using x to denote the number of packages of the economy blend and y to denote the number of packages of the superior blend, write a system of linear inequalities that describes the possible number of packages of each kind of blend.

(b) Graph the system and label the corner points.

60. Mixed Nuts Nola's Nuts, a store that specializes in selling nuts, has available 90 pounds of cashews and 120 pounds of peanuts. These are to be mixed in 12-ounce packages as follows: a lower-priced package containing 8 ounces of peanuts and 4 ounces of cashews and a quality package containing 6 ounces of peanuts and 6 ounces of cashews.

(a) Use x to denote the number of lower-priced packages and use y to denote the number of quality packages. Write a system of linear inequalities that describes the possible number of each kind of package.

(b) Graph the system and label the corner points.

61. Transporting Goods A small truck can carry no more than 1600 pounds of cargo nor more than 150 cubic feet of cargo. A printer weighs 20 pounds and occupies 3 cubic feet of space. A microwave oven weighs 30 pounds and occupies 2 cubic feet of space.

(a) Using x to represent the number of microwave ovens and y to represent the number of printers, write a system of linear inequalities that describes the number of ovens and printers that can be hauled by the truck.

(b) Graph the system and label the corner points.

'Are You Prepared?' Answers

1. $\{x \mid x < 1\}$

2.

3.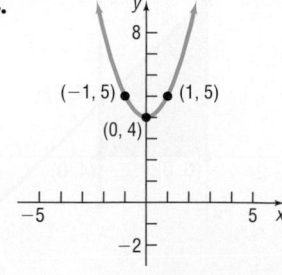

4.

5. True

6. $y = x^2$; right; 2

12.8 Linear Programming

Historically, linear programming evolved as a technique for solving problems involving resource allocation of goods and materials for the U.S. Air Force during World War II. Today, linear programming techniques are used to solve a wide variety of problems, such as optimizing airline scheduling and establishing telephone lines. Although most practical linear programming problems involve systems of several hundred linear inequalities containing several hundred variables, we will limit our discussion to problems containing only two variables, because we can solve such problems using graphing techniques.*

1 Set up a Linear Programming Problem

We begin by returning to Example 10 of the previous section.

EXAMPLE 1 Financial Planning

A retired couple has up to $25,000 to invest. As their financial adviser, you recommend that they place at least $15,000 in Treasury bills yielding 6% and at most $5000 in corporate bonds yielding 9%. How much money should be placed in each investment so that income is maximized?

The problem given in Example 1 is typical of a *linear programming problem.* The problem requires that a certain linear expression, the income, be maximized. If I represents income, x the amount invested in Treasury bills at 6%, and y the amount invested in corporate bonds at 9%, then

$$I = 0.06x + 0.09y$$

We shall assume, as before, that I, x, and y are in thousands of dollars.

The linear expression $I = 0.06x + 0.09y$ is called the **objective function.** Furthermore, the problem requires that the maximum income be achieved under certain conditions or **constraints,** each of which is a linear inequality involving the variables. (See Example 10 in Section 12.7.) The linear programming problem given in Example 1 may be restated as

$$\text{Maximize} \qquad I = 0.06x + 0.09y$$

subject to the conditions that

$$x \geq 0, \quad y \geq 0$$
$$x + y \leq 25$$
$$x \geq 15$$
$$y \leq 5$$

In general, every linear programming problem has two components:

1. A linear objective function that is to be maximized or minimized
2. A collection of linear inequalities that must be satisfied simultaneously

* The **simplex method** is a way to solve linear programming problems involving many inequalities and variables. This method was developed by George Dantzig in 1946 and is particularly well suited for computerization. In 1984, Narendra Karmarkar of Bell Laboratories discovered a way of solving large linear programming problems that improves on the simplex method.

DEFINITION

A **linear programming problem** in two variables x and y consists of maximizing (or minimizing) a linear objective function

$$z = Ax + By \qquad A \text{ and } B \text{ are real numbers, not both } 0$$

subject to certain conditions, or constraints, expressible as linear inequalities in x and y.

2 Solve a Linear Programing Problem

To maximize (or minimize) the quantity $z = Ax + By$, we need to identify points (x, y) that make the expression for z the largest (or smallest) possible. But not all points (x, y) are eligible; only those that also satisfy each linear inequality (constraint) can be used. We refer to each point (x, y) that satisfies the system of linear inequalities (the constraints) as a **feasible point.** In a linear programming problem, we seek the feasible point(s) that maximizes (or minimizes) the objective function.

Let's look again at the linear programming problem in Example 1.

EXAMPLE 2 **Analyzing a Linear Programming Problem**

Consider the linear programming problem

$$\text{Maximize} \qquad I = 0.06x + 0.09y$$

subject to the conditions that

$$x \geq 0$$
$$y \geq 0$$
$$x + y \leq 25$$
$$x \geq 15$$
$$y \leq 5$$

Graph the constraints. Then graph the objective function for $I = 0, 0.9, 1.35, 1.65,$ and 1.8.

Solution

Figure 24 shows the graph of the constraints. We superimpose on this graph the graph of the objective function for the given values of I.

For $I = 0$, the objective function is the line $0 = 0.06x + 0.09y$.

For $I = 0.9$, the objective function is the line $0.9 = 0.06x + 0.09y$.

For $I = 1.35$, the objective function is the line $1.35 = 0.06x + 0.09y$.

For $I = 1.65$, the objective function is the line $1.65 = 0.06x + 0.09y$.

For $I = 1.8$, the objective function is the line $1.8 = 0.06x + 0.09y$.

Figure 24

DEFINITION

A **solution** to a linear programming problem consists of a feasible point that maximizes (or minimizes) the objective function, together with the corresponding value of the objective function.

One condition for a linear programming problem in two variables to have a solution is that the graph of the feasible points be bounded. (Refer to page 912.)

If none of the feasible points maximizes (or minimizes) the objective function or if there are no feasible points, the linear programming problem has no solution.

Consider the linear programming problem stated in Example 2, and look again at Figure 24. The feasible points are the points that lie in the shaded region. For example, $(20, 3)$ is a feasible point, as are $(15, 5)$, $(20, 5)$, $(18, 4)$, and so on. To find the solution of the problem requires that we find a feasible point (x, y) that makes $I = 0.06x + 0.09y$ as large as possible. Notice that as I increases in value from $I = 0$ to $I = 0.9$ to $I = 1.35$ to $I = 1.65$ to $I = 1.8$ we obtain a collection of parallel lines. Further, notice that the largest value of I that can be obtained using feasible points is $I = 1.65$, which corresponds to the line $1.65 = 0.06x + 0.09y$. Any larger value of I results in a line that does not pass through any feasible points. Finally, notice that the feasible point that yields $I = 1.65$ is the point $(20, 5)$, a corner point. These observations form the basis of the following result, which we state without proof.

THEOREM

Location of the Solution of a Linear Programming Problem

If a linear programming problem has a solution, it is located at a corner point of the graph of the feasible points.

If a linear programming problem has multiple solutions, at least one of them is located at a corner point of the graph of the feasible points.

In either case, the corresponding value of the objective function is unique.

We shall not consider here linear programming problems that have no solution. As a result, we can outline the procedure for solving a linear programming problem as follows:

Procedure for Solving a Linear Programming Problem

STEP 1: Write an expression for the quantity to be maximized (or minimized). This expression is the objective function.

STEP 2: Write all the constraints as a system of linear inequalities and graph the system.

STEP 3: List the corner points of the graph of the feasible points.

STEP 4: List the corresponding values of the objective function at each corner point. The largest (or smallest) of these is the solution.

| EXAMPLE 3 | Solving a Minimum Linear Programming Problem |

Minimize the expression

$$z = 2x + 3y$$

subject to the constraints

$$y \leq 5, \qquad x \leq 6 \qquad x + y \geq 2, \qquad x \geq 0, \qquad y \geq 0$$

Solution

The objective function is $z = 2x + 3y$. We seek the smallest value of z that can occur if x and y are solutions of the system of linear inequalities

$$\begin{cases} y \leq 5 \\ x \leq 6 \\ x + y \geq 2 \\ x \geq 0 \\ y \geq 0 \end{cases}$$

The graph of this system (the set of feasible points) is shown as the shaded region in Figure 25. We have also plotted the corner points. Table 1 lists the corner points and the corresponding values of the objective function. From the table, we can see that the minimum value of z is 4, and it occurs at the point $(2, 0)$.

Figure 25

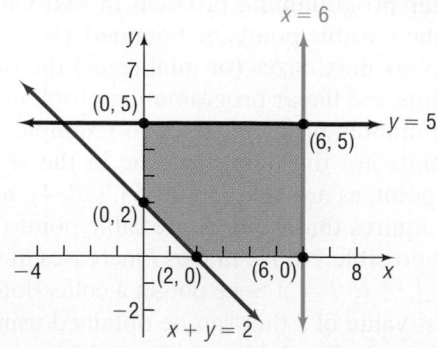

Table 1

Corner Point (x, y)	Value of the Objective Function $z = 2x + 3y$
$(0, 2)$	$z = 2(0) + 3(2) = 6$
$(0, 5)$	$z = 2(0) + 3(5) = 15$
$(6, 5)$	$z = 2(6) + 3(5) = 27$
$(6, 0)$	$z = 2(6) + 3(0) = 12$
$(2, 0)$	$z = 2(2) + 3(0) = 4$

➤ **Now Work** PROBLEMS 5 AND 11

EXAMPLE 4 **Maximizing Profit**

At the end of every month, after filling orders for its regular customers, a coffee company has some pure Colombian coffee and some special-blend coffee remaining. The practice of the company has been to package a mixture of the two coffees into 1-pound packages as follows: a low-grade mixture containing 4 ounces of Colombian coffee and 12 ounces of special-blend coffee and a high-grade mixture containing 8 ounces of Colombian and 8 ounces of special-blend coffee. A profit of $0.30 per package is made on the low-grade mixture, whereas a profit of $0.40 per package is made on the high-grade mixture. This month, 120 pounds of special-blend coffee and 100 pounds of pure Colombian coffee remain. How many packages of each mixture should be prepared to achieve a maximum profit? Assume that all packages prepared can be sold.

Solution We begin by assigning symbols for the two variables.

x = Number of packages of the low-grade mixture
y = Number of packages of the high-grade mixture

If P denotes the profit, then

$$P = \$0.30x + \$0.40y$$

This expression is the objective function. We seek to maximize P subject to certain constraints on x and y. Because x and y represent numbers of packages, the only meaningful values for x and y are nonnegative integers. So we have the two constraints

$$x \geq 0, \qquad y \geq 0 \quad \text{Nonnegative constraints}$$

We also have only so much of each type of coffee available. For example, the total amount of Colombian coffee used in the two mixtures cannot exceed 100 pounds, or 1600 ounces. Because we use 4 ounces in each low-grade package and 8 ounces in each high-grade package, we are led to the constraint

$$4x + 8y \leq 1600 \quad \text{Colombian coffee constraint}$$

Similarly, the supply of 120 pounds, or 1920 ounces, of special-blend coffee leads to the constraint

$$12x + 8y \leq 1920 \quad \text{Special-blend coffee constraint}$$

The linear programming problem may be stated as

$$\text{Maximize} \qquad P = 0.3x + 0.4y$$

subject to the constraints

$$x \geq 0, \qquad y \geq 0, \qquad 4x + 8y \leq 1600, \qquad 12x + 8y \leq 1920$$

The graph of the constraints (the feasible points) is illustrated in Figure 26. We list the corner points and evaluate the objective function at each. In Table 2, we can see that the maximum profit, $84, is achieved with 40 packages of the low-grade mixture and 180 packages of the high-grade mixture.

Figure 26

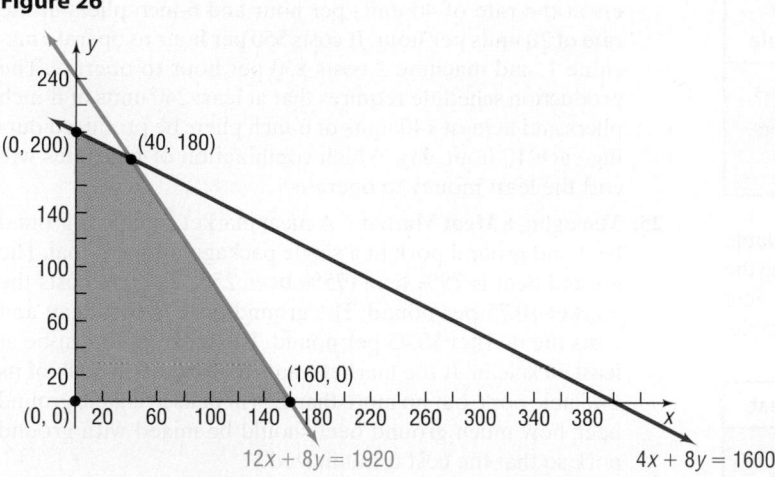

Table 2

Corner Point (x, y)	Value of Profit $P = 0.3x + 0.4y$
(0, 0)	$P = 0$
(0, 200)	$P = 0.3(0) + 0.4(200) = \80
(40, 180)	$P = 0.3(40) + 0.4(180) = \84
(160, 0)	$P = 0.3(160) + 0.4(0) = \48

━━━ **Now Work** PROBLEM 19

12.8 Assess Your Understanding

Concepts and Vocabulary

1. A linear programming problem requires that a linear expression, called the _____ _____, be maximized or minimized.

2. True or False If a linear programming problem has a solution, it is located at a corner point of the graph of the feasible points.

Skill Building

In Problems 3–8, find the maximum and minimum value of the given objective function of a linear programming problem. The figure illustrates the graph of the feasible points.

3. $z = x + y$

4. $z = 2x + 3y$

5. $z = x + 10y$

6. $z = 10x + y$

7. $z = 5x + 7y$

8. $z = 7x + 5y$

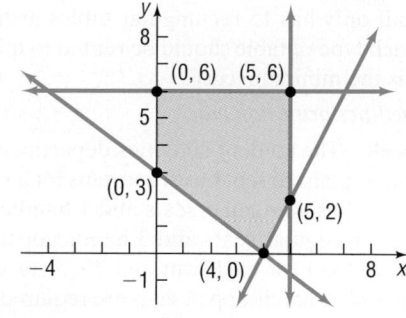

In Problems 9–18, solve each linear programming problem.

9. Maximize $z = 2x + y$ subject to $x \geq 0, \quad y \geq 0, \quad x + y \leq 6, \quad x + y \geq 1$

10. Maximize $z = x + 3y$ subject to $x \geq 0, \quad y \geq 0, \quad x + y \geq 3, \quad x \leq 5, \quad y \leq 7$

11. Minimize $z = 2x + 5y$ subject to $x \geq 0, \quad y \geq 0, \quad x + y \geq 2, \quad x \leq 5, \quad y \leq 3$

12. Minimize $z = 3x + 4y$ subject to $x \geq 0, \quad y \geq 0, \quad 2x + 3y \geq 6, \quad x + y \leq 8$

13. Maximize $z = 3x + 5y$ subject to $x \geq 0, \quad y \geq 0, \quad x + y \geq 2, \quad 2x + 3y \leq 12, \quad 3x + 2y \leq 12$

14. Maximize $z = 5x + 3y$ subject to $x \geq 0, \quad y \geq 0, \quad x + y \geq 2, \quad x + y \leq 8, \quad 2x + y \leq 10$

15. Minimize $z = 5x + 4y$ subject to $x \geq 0, \quad y \geq 0, \quad x + y \geq 2, \quad 2x + 3y \leq 12, \quad 3x + y \leq 12$

16. Minimize $z = 2x + 3y$ subject to $x \geq 0, \quad y \geq 0, \quad x + y \geq 3, \quad x + y \leq 9, \quad x + 3y \geq 6$

17. Maximize $z = 5x + 2y$ subject to $x \geq 0, \quad y \geq 0, \quad x + y \leq 10, \quad 2x + y \geq 10, \quad x + 2y \geq 10$

18. Maximize $z = 2x + 4y$ subject to $x \geq 0, \quad y \geq 0, \quad 2x + y \geq 4, \quad x + y \leq 9$

Applications and Extensions

19. Maximizing Profit A manufacturer of skis produces two types: downhill and cross-country. Use the following table to determine how many of each kind of ski should be produced to achieve a maximum profit. What is the maximum profit? What would the maximum profit be if the time available for manufacturing is increased to 48 hours?

	Downhill	Cross country	Time Available
Manufacturing time per ski	2 hours	1 hour	40 hours
Finishing time per ski	1 hour	1 hour	32 hours
Profit per ski	$70	$50	

20. Farm Management A farmer has 70 acres of land available for planting either soybeans or wheat. The cost of preparing the soil, the workdays required, and the expected profit per acre planted for each type of crop are given in the following table:

	Soybeans	Wheat
Preparation cost per acre	$60	$30
Workdays required per acre	3	4
Profit per acre	$180	$100

The farmer cannot spend more than $1800 in preparation costs nor use more than a total of 120 workdays. How many acres of each crop should be planted to maximize the profit? What is the maximum profit? What is the maximum profit if the farmer is willing to spend no more than $2400 on preparation?

21. Banquet Seating A banquet hall offers two types of tables for rent: 6-person rectangular tables at a cost of $28 each and 10-person round tables at a cost of $52 each. Kathleen would like to rent the hall for a wedding banquet and needs tables for 250 people. The room can have a maximum of 35 tables and the hall only has 15 rectangular tables available. How many of each type of table should be rented to minimize cost and what is the minimum cost?

Source: facilities.princeton.edu

22. Spring Break The student activities department of a community college plans to rent buses and vans for a spring-break trip. Each bus has 40 regular seats and 1 handicapped seat; each van has 8 regular seats and 3 handicapped seats. The rental cost is $350 for each van and $975 for each bus. If 320 regular and 36 handicapped seats are required for the trip, how many vehicles of each type should be rented to minimize cost?

Source: www.busrates.com

23. Return on Investment An investment broker is instructed by her client to invest up to $20,000, some in a junk bond yielding 9% per annum and some in Treasury bills yielding 7% per annum. The client wants to invest at least $8000 in T-bills and no more than $12,000 in the junk bond.

(a) How much should the broker recommend that the client place in each investment to maximize income if the client insists that the amount invested in T-bills must equal or exceed the amount placed in junk bonds?

(b) How much should the broker recommend that the client place in each investment to maximize income if the client insists that the amount invested in T-bills must not exceed the amount placed in junk bonds?

24. Production Scheduling In a factory, machine 1 produces 8-inch pliers at the rate of 60 units per hour and 6-inch pliers at the rate of 70 units per hour. Machine 2 produces 8-inch pliers at the rate of 40 units per hour and 6-inch pliers at the rate of 20 units per hour. It costs $50 per hour to operate machine 1, and machine 2 costs $30 per hour to operate. The production schedule requires that at least 240 units of 8-inch pliers and at least 140 units of 6-inch pliers be produced during each 10-hour day. Which combination of machines will cost the least money to operate?

25. Managing a Meat Market A meat market combines ground beef and ground pork in a single package for meat loaf. The ground beef is 75% lean (75% beef, 25% fat) and costs the market $0.75 per pound. The ground pork is 60% lean and costs the market $0.45 per pound. The meat loaf must be at least 70% lean. If the market wants to use at least 50 lb of its available pork, but no more than 200 lb of its available ground beef, how much ground beef should be mixed with ground pork so that the cost is minimized?

26. Ice Cream The Mom and Pop Ice Cream Company makes two kinds of chocolate ice cream: regular and premium. The properties of 1 gallon of each type are shown in the table:

	Regular	Premium
Flavoring	24 oz	20 oz
Milk-fat products	12 oz	20 oz
Shipping weight	5 lbs	6 lbs
Profit	$0.75	$0.90

In addition, current commitments require the company to make at least 1 gallon of premium for every 4 gallons of regular. Each day, the company has available 725 pounds of flavoring and 425 pounds of milk-fat products. If the company can ship no more than 3000 pounds of product per day, how many gallons of each type should be produced daily to maximize profit?

Source: www.scitoys.com/ingredients/ice_cream.html

27. Maximizing Profit on Ice Skates A factory manufactures two kinds of ice skates: racing skates and figure skates. The racing skates require 6 work-hours in the fabrication department, whereas the figure skates require 4 work-hours there. The racing skates require 1 work-hour in the finishing department, whereas the figure skates require 2 work-hours there. The fabricating department has available at most 120 work-hours per day, and the finishing department has no more

than 40 work-hours per day available. If the profit on each racing skate is $10 and the profit on each figure skate is $12, how many of each should be manufactured each day to maximize profit? (Assume that all skates made are sold.)

28. **Financial Planning** A retired couple has up to $50,000 to place in fixed-income securities. Their financial adviser suggests two securities to them: one is an AAA bond that yields 8% per annum; the other is a certificate of deposit (CD) that yields 4%. After careful consideration of the alternatives, the couple decides to place at most $20,000 in the AAA bond and at least $15,000 in the CD. They also instruct the financial adviser to place at least as much in the CD as in the AAA bond. How should the financial adviser proceed to maximize the return on their investment?

29. **Product Design** An entrepreneur is having a design group produce at least six samples of a new kind of fastener that he wants to market. It costs $9.00 to produce each metal fastener and $4.00 to produce each plastic fastener. He wants to have at least two of each version of the fastener and needs to have all the samples 24 hours from now. It takes 4 hours to produce each metal sample and 2 hours to produce each plastic sample. To minimize the cost of the samples, how many of each kind should the entrepreneur order? What will be the cost of the samples?

30. **Animal Nutrition** Kevin's dog Amadeus likes two kinds of canned dog food. Gourmet Dog costs 40 cents a can and has 20 units of a vitamin complex; the calorie content is 75 calories. Chow Hound costs 32 cents a can and has 35 units of vitamins and 50 calories. Kevin likes Amadeus to have at least 1175 units of vitamins a month and at least 2375 calories during the same time period. Kevin has space to store only 60 cans of dog food at a time. How much of each kind of dog food should Kevin buy each month to minimize his cost?

31. **Airline Revenue** An airline has two classes of service: first class and coach. Management's experience has been that each aircraft should have at least 8 but no more than 16 first-class seats and at least 80 but not more than 120 coach seats.

(a) If management decides that the ratio of first class to coach seats should never exceed 1:12, with how many of each type of seat should an aircraft be configured to maximize revenue?

(b) If management decides that the ratio of first class to coach seats should never exceed 1:8, with how many of each type of seat should an aircraft be configured to maximize revenue?

(c) If you were management, what would you do?
[**Hint:** Assume that the airline charges C for a coach seat and F for a first-class seat; $C > 0, F > C$.]

Discussion and Writing

32. Explain in your own words what a linear programming problem is and how it can be solved.

CHAPTER REVIEW

Things to Know

Systems of equations (pp. 836–846)

Systems with no solutions are inconsistent. Systems with a solution are consistent.

Consistent systems of linear equations have either a unique solution or an infinite number of solutions.

Matrix (p. 851) Rectangular array of numbers, called entries

Augmented matrix (p. 851)

Row operations (p. 853)

Row echelon form (p. 854)

Determinants and Cramer's Rule (pp. 866, 867, and 871)

Matrix (p. 875)

m by n matrix (p. 876)	Matrix with m rows and n columns
Identity matrix I_n (p. 883)	An n by n square matrix whose diagonal entries are 1's, while all other entries are 0's
Inverse of a matrix (pp. 884–885)	A^{-1} is the inverse of A if $AA^{-1} = A^{-1}A = I_n$
Nonsingular matrix (p. 884)	A square matrix that has an inverse

Linear programming problem (p. 916)

Maximize (or minimize) a linear objective function, $z = Ax + By$, subject to certain conditions, or constraints, expressible as linear inequalities in x and y. A feasible point (x, y) is a point that satisfies the constraints of a linear programming problem.

Location of solution (p. 917)

If a linear programming problem has a solution, it is located at a corner point of the graph of the feasible points. If a linear programming problem has multiple solutions, at least one of them is located at a corner point of the graph of the feasible points. In either case, the corresponding value of the objective function is unique.

Objectives

Section		You should be able to . . .	Review Exercises
12.1	1	Solve systems of equations by substitution (p. 838)	1–14, 101, 102, 105–107
	2	Solve systems of equations by elimination (p. 840)	1–14, 101, 102, 105–107
	3	Identify inconsistent systems of equations containing two variables (p. 841)	9, 10, 13, 98
	4	Express the solution of a system of dependent equations containing two variables (p. 842)	14, 97
	5	Solve systems of three equations containing three variables (p. 843)	15–18, 99, 100, 103
	6	Identify inconsistent systems of equations containing three variables (p. 844)	18
	7	Express the solution of a system of dependent equations containing three variables (p. 845)	17
12.2	1	Write the augmented matrix of a system of linear equations (p. 851)	35–44
	2	Write the system of equations from the augmented matrix (p. 852)	19, 20
	3	Perform row operations on a matrix (p. 853)	35–44
	4	Solve a system of linear equations using matrices (p. 854)	35–44
12.3	1	Evaluate 2 by 2 determinants (p. 866)	45, 46
	2	Use Cramer's Rule to solve a system of two equations containing two variables (p. 866)	51–54
	3	Evaluate 3 by 3 determinants (p. 869)	47–50
	4	Use Cramer's Rule to solve a system of three equations containing three variables (p. 870)	55, 56
	5	Know properties of determinants (p. 872)	57, 58
12.4	1	Find the sum and difference of two matrices (p. 876)	21, 22
	2	Find scalar multiples of a matrix (p. 878)	23, 24
	3	Find the product of two matrices (p. 879)	25–28
	4	Find the inverse of a matrix (p. 884)	29–34
	5	Solve a system of linear equations using an inverse matrix (p. 887)	35–44
12.5	1	Decompose $\dfrac{P}{Q}$, where Q has only nonrepeated linear factors (p. 892)	59, 60
	2	Decompose $\dfrac{P}{Q}$, where Q has repeated linear factors (p. 894)	61, 62
	3	Decompose $\dfrac{P}{Q}$, where Q has a nonrepeated irreducible quadratic factor (p. 896)	63, 64, 67, 68
	4	Decompose $\dfrac{P}{Q}$, where Q has a repeated irreducible quadratic factor (p. 897)	65, 66
12.6	1	Solve a system of nonlinear equations using substitution (p. 898)	69–78
	2	Solve a system of nonlinear equations using elimination (p. 900)	69–78
12.7	1	Graph an inequality (p. 907)	79–82
	2	Graph a system of inequalities (p. 909)	83–92, 104
12.8	1	Set up a linear programming problem (p. 915)	108, 109
	2	Solve a linear programming problem (p. 916)	93–96, 108, 109

Review Exercises

In Problems 1–18, solve each system of equations using the method of substitution or the method of elimination. If the system has no solution, say that it is inconsistent.

1. $\begin{cases} 2x - y = 5 \\ 5x + 2y = 8 \end{cases}$

2. $\begin{cases} 2x + 3y = 2 \\ 7x - y = 3 \end{cases}$

3. $\begin{cases} 3x - 4y = 4 \\ x - 3y = \dfrac{1}{2} \end{cases}$

4. $\begin{cases} 2x + y = 0 \\ 5x - 4y = -\dfrac{13}{2} \end{cases}$

5. $\begin{cases} x - 2y - 4 = 0 \\ 3x + 2y - 4 = 0 \end{cases}$

6. $\begin{cases} x - 3y + 5 = 0 \\ 2x + 3y - 5 = 0 \end{cases}$

7. $\begin{cases} y = 2x - 5 \\ x = 3y + 4 \end{cases}$

8. $\begin{cases} x = 5y + 2 \\ y = 5x + 2 \end{cases}$

9. $\begin{cases} x - 3y + 4 = 0 \\ \dfrac{1}{2}x - \dfrac{3}{2}y + \dfrac{4}{3} = 0 \end{cases}$ 10. $\begin{cases} x + \dfrac{1}{4}y = 2 \\ y + 4x + 2 = 0 \end{cases}$ 11. $\begin{cases} 2x + 3y - 13 = 0 \\ 3x - 2y = 0 \end{cases}$ 12. $\begin{cases} 4x + 5y = 21 \\ 5x + 6y = 42 \end{cases}$

13. $\begin{cases} 3x - 2y = 8 \\ x - \dfrac{2}{3}y = 12 \end{cases}$ 14. $\begin{cases} 2x + 5y = 10 \\ 4x + 10y = 20 \end{cases}$ 15. $\begin{cases} x + 2y - z = 6 \\ 2x - y + 3z = -13 \\ 3x - 2y + 3z = -16 \end{cases}$

16. $\begin{cases} x + 5y - z = 2 \\ 2x + y + z = 7 \\ x - y + 2z = 11 \end{cases}$ 17. $\begin{cases} 2x - 4y + z = -15 \\ x + 2y - 4z = 27 \\ 5x - 6y - 2z = -3 \end{cases}$ 18. $\begin{cases} x - 4y + 3z = 15 \\ -3x + y - 5z = -5 \\ -7x - 5y - 9z = 10 \end{cases}$

In Problems 19 and 20, write the system of equations corresponding to the given augmented matrix.

19. $\begin{bmatrix} 3 & 2 & | & 8 \\ 1 & 4 & | & -1 \end{bmatrix}$

20. $\begin{bmatrix} 1 & 2 & 5 & | & -2 \\ 5 & 0 & -3 & | & 8 \\ 2 & -1 & 0 & | & 0 \end{bmatrix}$

In Problems 21–28, use the following matrices to compute each expression.

$$A = \begin{bmatrix} 1 & 0 \\ 2 & 4 \\ -1 & 2 \end{bmatrix}, \quad B = \begin{bmatrix} 4 & -3 & 0 \\ 1 & 1 & -2 \end{bmatrix}, \quad C = \begin{bmatrix} 3 & -4 \\ 1 & 5 \\ 5 & 2 \end{bmatrix}$$

21. $A + C$ 22. $A - C$ 23. $6A$ 24. $-4B$

25. AB 26. BA 27. CB 28. BC

In Problems 29–34, find the inverse, if there is one, of each matrix. If there is not an inverse, say that the matrix is singular.

29. $\begin{bmatrix} 4 & 6 \\ 1 & 3 \end{bmatrix}$ 30. $\begin{bmatrix} -3 & 2 \\ 1 & -2 \end{bmatrix}$ 31. $\begin{bmatrix} 1 & 3 & 3 \\ 1 & 2 & 1 \\ 1 & -1 & 2 \end{bmatrix}$

32. $\begin{bmatrix} 3 & 1 & 2 \\ 3 & 2 & -1 \\ 1 & 1 & 1 \end{bmatrix}$ 33. $\begin{bmatrix} 4 & -8 \\ -1 & 2 \end{bmatrix}$ 34. $\begin{bmatrix} -3 & 1 \\ -6 & 2 \end{bmatrix}$

In Problems 35–44, solve each system of equations using matrices. If the system has no solution, say that it is inconsistent.

35. $\begin{cases} 3x - 2y = 1 \\ 10x + 10y = 5 \end{cases}$ 36. $\begin{cases} 3x + 2y = 6 \\ x - y = -\dfrac{1}{2} \end{cases}$ 37. $\begin{cases} 5x - 6y - 3z = 6 \\ 4x - 7y - 2z = -3 \\ 3x + y - 7z = 1 \end{cases}$

38. $\begin{cases} 2x + y + z = 5 \\ 4x - y - 3z = 1 \\ 8x + y - z = 5 \end{cases}$ 39. $\begin{cases} x - 2z = 1 \\ 2x + 3y = -3 \\ 4x - 3y - 4z = 3 \end{cases}$ 40. $\begin{cases} x + 2y - z = 2 \\ 2x - 2y + z = -1 \\ 6x + 4y + 3z = 5 \end{cases}$

41. $\begin{cases} x - y + z = 0 \\ x - y - 5z - 6 = 0 \\ 2x - 2y + z - 1 = 0 \end{cases}$ 42. $\begin{cases} 4x - 3y + 5z = 0 \\ 2x + 4y - 3z = 0 \\ 6x + 2y + z = 0 \end{cases}$ 43. $\begin{cases} x - y - z - t = 1 \\ 2x + y - z + 2t = 3 \\ x - 2y - 2z - 3t = 0 \\ 3x - 4y + z + 5t = -3 \end{cases}$

44. $\begin{cases} x - 3y + 3z - t = 4 \\ x + 2y - z = -3 \\ x + 3z + 2t = 3 \\ x + y + 5z = 6 \end{cases}$

In Problems 45–50, find the value of each determinant.

45. $\begin{vmatrix} 3 & 4 \\ 1 & 3 \end{vmatrix}$ 46. $\begin{vmatrix} -4 & 0 \\ 1 & 3 \end{vmatrix}$ 47. $\begin{vmatrix} 1 & 4 & 0 \\ -1 & 2 & 6 \\ 4 & 1 & 3 \end{vmatrix}$

48. $\begin{vmatrix} 2 & 3 & 10 \\ 0 & 1 & 5 \\ -1 & 2 & 3 \end{vmatrix}$ 49. $\begin{vmatrix} 2 & 1 & -3 \\ 5 & 0 & 1 \\ 2 & 6 & 0 \end{vmatrix}$ 50. $\begin{vmatrix} -2 & 1 & 0 \\ 1 & 2 & 3 \\ -1 & 4 & 2 \end{vmatrix}$

In Problems 51–56, use Cramer's Rule, if applicable, to solve each system.

51. $\begin{cases} x - 2y = 4 \\ 3x + 2y = 4 \end{cases}$

52. $\begin{cases} x - 3y = -5 \\ 2x + 3y = 5 \end{cases}$

53. $\begin{cases} 2x + 3y - 13 = 0 \\ 3x - 2y = 0 \end{cases}$

54. $\begin{cases} 3x - 4y - 12 = 0 \\ 5x + 2y + 6 = 0 \end{cases}$

55. $\begin{cases} x + 2y - z = 6 \\ 2x - y + 3z = -13 \\ 3x - 2y + 3z = -16 \end{cases}$

56. $\begin{cases} x - y + z = 8 \\ 2x + 3y - z = -2 \\ 3x - y - 9z = 9 \end{cases}$

In Problems 57 and 58, use properties of determinants to find the value of each determinant if it is known that $\begin{vmatrix} x & y \\ a & b \end{vmatrix} = 8.$

57. $\begin{vmatrix} 2x & y \\ 2a & b \end{vmatrix}$

58. $\begin{vmatrix} y & x \\ b & a \end{vmatrix}$

In Problems 59–68, write the partial fraction decomposition of each rational expression.

59. $\dfrac{6}{x(x-4)}$

60. $\dfrac{x}{(x+2)(x-3)}$

61. $\dfrac{x-4}{x^2(x-1)}$

62. $\dfrac{2x-6}{(x-2)^2(x-1)}$

63. $\dfrac{x}{(x^2+9)(x+1)}$

64. $\dfrac{3x}{(x-2)(x^2+1)}$

65. $\dfrac{x^3}{(x^2+4)^2}$

66. $\dfrac{x^3+1}{(x^2+16)^2}$

67. $\dfrac{x^2}{(x^2+1)(x^2-1)}$

68. $\dfrac{4}{(x^2+4)(x^2-1)}$

In Problems 69–78, solve each system of equations.

69. $\begin{cases} 2x + y + 3 = 0 \\ x^2 + y^2 = 5 \end{cases}$

70. $\begin{cases} x^2 + y^2 = 16 \\ 2x - y^2 = -8 \end{cases}$

71. $\begin{cases} 2xy + y^2 = 10 \\ 3y^2 - xy = 2 \end{cases}$

72. $\begin{cases} 3x^2 - y^2 = 1 \\ 7x^2 - 2y^2 - 5 = 0 \end{cases}$

73. $\begin{cases} x^2 + y^2 = 6y \\ x^2 = 3y \end{cases}$

74. $\begin{cases} 2x^2 + y^2 = 9 \\ x^2 + y^2 = 9 \end{cases}$

75. $\begin{cases} 3x^2 + 4xy + 5y^2 = 8 \\ x^2 + 3xy + 2y^2 = 0 \end{cases}$

76. $\begin{cases} 3x^2 + 2xy - 2y^2 = 6 \\ xy - 2y^2 + 4 = 0 \end{cases}$

77. $\begin{cases} x^2 - 3x + y^2 + y = -2 \\ \dfrac{x^2 - x}{y} + y + 1 = 0 \end{cases}$

78. $\begin{cases} x^2 + x + y^2 = y + 2 \\ x + 1 = \dfrac{2 - y}{x} \end{cases}$

In Problems 79–82 graph each inequality.

79. $3x + 4y \le 12$

80. $2x - 3y \ge 6$

81. $y \le x^2$

82. $x \ge y^2$

In Problems 83–88, graph each system of inequalities. Tell whether the graph is bounded or unbounded, and label the corner points.

83. $\begin{cases} -2x + y \le 2 \\ x + y \ge 2 \end{cases}$

84. $\begin{cases} x - 2y \le 6 \\ 2x + y \ge 2 \end{cases}$

85. $\begin{cases} x \ge 0 \\ y \ge 0 \\ x + y \le 4 \\ 2x + 3y \le 6 \end{cases}$

86. $\begin{cases} x \ge 0 \\ y \ge 0 \\ 3x + y \ge 6 \\ 2x + y \ge 2 \end{cases}$

87. $\begin{cases} x \ge 0 \\ y \ge 0 \\ 2x + y \le 8 \\ x + 2y \ge 2 \end{cases}$

88. $\begin{cases} x \ge 0 \\ y \ge 0 \\ 3x + y \le 9 \\ 2x + 3y \ge 6 \end{cases}$

In Problems 89–92, graph each system of inequalities.

89. $\begin{cases} x^2 + y^2 \le 16 \\ x + y \ge 2 \end{cases}$

90. $\begin{cases} y^2 \le x - 1 \\ x - y \le 3 \end{cases}$

91. $\begin{cases} y \le x^2 \\ xy \le 4 \end{cases}$

92. $\begin{cases} x^2 + y^2 \ge 1 \\ x^2 + y^2 \le 4 \end{cases}$

In Problems 93–96, solve each linear programming problem.

93. Maximize $z = 3x + 4y$ subject to $x \ge 0, y \ge 0, 3x + 2y \ge 6, x + y \le 8$

94. Maximize $z = 2x + 4y$ subject to $x \ge 0, y \ge 0, x + y \le 6, x \ge 2$

95. Minimize $z = 3x + 5y$ subject to $x \ge 0, y \ge 0, x + y \ge 1, 3x + 2y \le 12, x + 3y \le 12$

96. Minimize $z = 3x + y$ subject to $x \ge 0, y \ge 0, x \le 8, y \le 6, 2x + y \ge 4$

97. Find A so that the system of equations has infinitely many solutions.

$$\begin{cases} 2x + 5y = 5 \\ 4x + 10y = A \end{cases}$$

98. Find A so that the system in Problem 97 is inconsistent.

99. Curve Fitting Find the quadratic function $y = ax^2 + bx + c$ that passes through the three points $(0, 1)$, $(1, 0)$, and $(-2, 1)$.

100. Curve Fitting Find the general equation of the circle that passes through the three points $(0, 1)$, $(1, 0)$, and $(-2, 1)$.

[Hint: The general equation of a circle is
$x^2 + y^2 + Dx + Ey + F = 0.$]

101. Blending Coffee A coffee distributor is blending a new coffee that will cost $6.90 per pound. It will consist of a blend of $6.00 per pound coffee and $9.00 per pound coffee. What amounts of each type of coffee should be mixed to achieve the desired blend?

[**Hint:** Assume that the weight of the blended coffee is 100 pounds.]

$6.00/lb $6.90/lb $9.00/lb

102. Farming A 1000-acre farm in Illinois is used to grow corn and soybeans. The cost per acre for raising corn is $65, and the cost per acre for soybeans is $45. If $54,325 has been budgeted for costs and all the acreage is to be used, how many acres should be allocated for each crop?

103. Cookie Orders A cookie company makes three kinds of cookies, oatmeal raisin, chocolate chip, and shortbread, packaged in small, medium, and large boxes. The small box contains 1 dozen oatmeal raisin and 1 dozen chocolate chip; the medium box has 2 dozen oatmeal raisin, 1 dozen chocolate chip, and 1 dozen shortbread; the large box contains 2 dozen oatmeal raisin, 2 dozen chocolate chip, and 3 dozen shortbread. If you require exactly 15 dozen oatmeal raisin, 10 dozen chocolate chip, and 11 dozen shortbread, how many of each size box should you buy?

104. Mixed Nuts A store that specializes in selling nuts has available 72 pounds of cashews and 120 pounds of peanuts. These are to be mixed in 12-ounce packages as follows: a lower-priced package containing 8 ounces of peanuts and 4 ounces of cashews and a quality package containing 6 ounces of peanuts and 6 ounces of cashews.
(a) Use x to denote the number of lower-priced packages and use y to denote the number of quality packages. Write a system of linear inequalities that describes the possible number of each kind of package.
(b) Graph the system and label the corner points.

105. Determining the Speed of the Current of the Aguarico River On a recent trip to the Cuyabeno Wildlife Reserve in the Amazon region of Ecuador, I took a 100-kilometer trip by speedboat down the Aguarico River from Chiritza to the Flotel Orellana. As I watched the Amazon unfold, I wondered how fast the speedboat was going and how fast the current of the white-water Aguarico River was. I timed the trip downstream at 2.5 hours and the return trip at 3 hours. What were the two speeds?

106. Finding the Speed of the Jet Stream On a flight between Midway Airport in Chicago and Ft. Lauderdale, Florida, a Boeing 737 jet maintains an airspeed of 475 miles per hour. If the trip from Chicago to Ft. Lauderdale takes 2 hours, 30 minutes and the return flight takes 2 hours, 50 minutes, what is the speed of the jet stream? (Assume that the speed of the jet stream remains constant at the various altitudes of the plane and that the plane flies with the jet stream one way and against it the other way.)

107. Constant Rate Jobs If Bruce and Bryce work together for 1 hour and 20 minutes, they will finish a certain job. If Bryce and Marty work together for 1 hour and 36 minutes, the same job can be finished. If Marty and Bruce work together, they can complete this job in 2 hours and 40 minutes. How long will it take each of them working alone to finish the job?

108. Maximizing Profit on Figurines A factory manufactures two kinds of ceramic figurines: a dancing girl and a mermaid. Each requires three processes: molding, painting, and glazing. The daily labor available for molding is no more than 90 work-hours, labor available for painting does not exceed 120 work-hours, and labor available for glazing is no more than 60 work-hours. The dancing girl requires 3 work-hours for molding, 6 work-hours for painting, and 2 work-hours for glazing. The mermaid requires 3 work-hours for molding, 4 work-hours for painting, and 3 work-hours for glazing. If the profit on each figurine is $25 for dancing girls and $30 for mermaids, how many of each should be produced each day to maximize profit? If management decides to produce the number of each figurine that maximizes profit, determine which of these processes has work-hours assigned to it that are not used.

109. Minimizing Production Cost A factory produces gasoline engines and diesel engines. Each week the factory is obligated to deliver at least 20 gasoline engines and at least 15 diesel engines. Due to physical limitations, however, the factory cannot make more than 60 gasoline engines nor more than 40 diesel engines in any given week. Finally, to prevent layoffs, a total of at least 50 engines must be produced. If gasoline engines cost $450 each to produce and diesel engines cost $550 each to produce, how many of each should be produced per week to minimize the cost? What is the excess capacity of the factory; that is, how many of each kind of engine is being produced in excess of the number that the factory is obligated to deliver?

110. Describe four ways of solving a system of three linear equations containing three variables. Which method do you prefer? Why?

CHAPTER TEST

In Problems 1–4, solve each system of equations using the method of substitution or the method of elimination. If the system has no solution, say that it is inconsistent.

1. $\begin{cases} -2x + y = -7 \\ 4x + 3y = 9 \end{cases}$

2. $\begin{cases} \dfrac{1}{3}x - 2y = 1 \\ 5x - 30y = 18 \end{cases}$

3. $\begin{cases} x - y + 2z = 5 \\ 3x + 4y - z = -2 \\ 5x + 2y + 3z = 8 \end{cases}$

4. $\begin{cases} 3x + 2y - 8z = -3 \\ -x - \dfrac{2}{3}y + z = 1 \\ 6x - 3y + 15z = 8 \end{cases}$

5. Write the augmented matrix corresponding to the system of equations: $\begin{cases} 4x - 5y + z = 0 \\ -2x - y + 6 = -19. \\ x + 5y - 5z = 10 \end{cases}$

6. Write the system of equations corresponding to the augmented matrix: $\begin{bmatrix} 3 & 2 & 4 & | & -6 \\ 1 & 0 & 8 & | & 2 \\ -2 & 1 & 3 & | & -11 \end{bmatrix}$.

In Problems 7–10, use the given matrices to compute each expression.

$$A = \begin{bmatrix} 1 & -1 \\ 0 & -4 \\ 3 & 2 \end{bmatrix}, \quad B = \begin{bmatrix} 1 & -2 & 5 \\ 0 & 3 & 1 \end{bmatrix}, \quad C = \begin{bmatrix} 4 & 6 \\ 1 & -3 \\ -1 & 8 \end{bmatrix}$$

7. $2A + C$

8. $A - 3C$

9. AC

10. BA

In Problems 11 and 12, find the inverse of each nonsingular matrix.

11. $A = \begin{bmatrix} 3 & 2 \\ 5 & 4 \end{bmatrix}$

12. $B = \begin{bmatrix} 1 & -1 & 1 \\ 2 & 5 & -1 \\ 2 & 3 & 0 \end{bmatrix}$

In Problems 13–16, solve each system of equations using matrices. If the system has no solution, say that it is inconsistent.

13. $\begin{cases} 6x + 3y = 12 \\ 2x - y = -2 \end{cases}$

14. $\begin{cases} x + \dfrac{1}{4}y = 7 \\ 8x + 2y = 56 \end{cases}$

15. $\begin{cases} x + 2y + 4z = -3 \\ 2x + 7y + 15z = -12 \\ 4x + 7y + 13z = -10 \end{cases}$

16. $\begin{cases} 2x + 2y - 3z = 5 \\ x - y + 2z = 8 \\ 3x + 5y - 8z = -2 \end{cases}$

In Problems 17 and 18, find the value of each determinant.

17. $\begin{vmatrix} -2 & 5 \\ 3 & 7 \end{vmatrix}$

18. $\begin{vmatrix} 2 & -4 & 6 \\ 1 & 4 & 0 \\ -1 & 2 & -4 \end{vmatrix}$

In Problems 19 and 20, use Cramer's Rule, if possible, to solve each system.

19. $\begin{cases} 4x + 3y = -23 \\ 3x - 5y = 19 \end{cases}$

20. $\begin{cases} 4x - 3y + 2z = 15 \\ -2x + y - 3z = -15 \\ 5x - 5y + 2z = 18 \end{cases}$

In Problems 21 and 22, solve each system of equations.

21. $\begin{cases} 3x^2 + y^2 = 12 \\ y^2 = 9x \end{cases}$

22. $\begin{cases} 2y^2 - 3x^2 = 5 \\ y - x = 1 \end{cases}$

23. Graph the system of inequalities: $\begin{cases} x^2 + y^2 \le 100 \\ 4x - 3y \ge 0 \end{cases}$.

In Problems 24 and 25, write the partial fraction decomposition of each rational expression.

24. $\dfrac{3x + 7}{(x + 3)^2}$

25. $\dfrac{4x^2 - 3}{x(x^2 + 3)^2}$

26. Graph the system of inequalities. Tell whether the graph is bounded or unbounded, and label all corner points.

$$\begin{cases} x \ge 0 \\ y \ge 0 \\ x + 2y \ge 8 \\ 2x - 3y \ge 2 \end{cases}$$

27. Maximize $z = 5x + 8y$ subject to $x \ge 0, 2x + y \le 8$, and $x - 3y \le -3$.

28. Megan went clothes shopping and bought 2 pairs of flare jeans, 2 camisoles, and 4 T-shirts for $90.00. At the same store, Paige bought one pair of flare jeans and 3 T-shirts for $42.50, while Kara bought 1 pair of flare jeans, 3 camisoles, and 2 T-shirts for $62.00. Determine the price of each clothing item.

CUMULATIVE REVIEW

In Problems 1–6, solve each equation.

1. $2x^2 - x = 0$

2. $\sqrt{3x + 1} = 4$

3. $2x^3 - 3x^2 - 8x - 3 = 0$

4. $3^x = 9^{x+1}$

5. $\log_3(x - 1) + \log_3(2x + 1) = 2$

6. $3^x = e$

7. Determine whether the function $g(x) = \dfrac{2x^3}{x^4 + 1}$ is even, odd, or neither. Is the graph of g symmetric with respect to the x-axis, y-axis, or origin?

8. Find the center and radius of the circle $x^2 + y^2 - 2x + 4y - 11 = 0$. Graph the circle.

9. Graph $f(x) = 3^{x-2} + 1$ using transformations. What is the domain, range, and horizontal asymptote of f?

10. The function $f(x) = \dfrac{5}{x + 2}$ is one-to-one. Find f^{-1}. Find the domain and the range of f and the domain and the range of f^{-1}.

11. Graph each equation.

 (a) $y = 3x + 6$ (b) $x^2 + y^2 = 4$

 (c) $y = x^3$ (d) $y = \dfrac{1}{x}$

 (e) $y = \sqrt{x}$ (f) $y = e^x$

 (g) $y = \ln x$ (h) $2x^2 + 5y^2 = 1$

 (i) $x^2 - 3y^2 = 1$ (j) $x^2 - 2x - 4y + 1 = 0$

 (k) $y = \sin x$ (l) $y = 3\sin(2x)$

12. $f(x) = x^3 - 3x + 5$

 (a) Using a graphing utility, graph f and approximate the zero(s) of f.

 (b) Using a graphing utility, approximate the local maxima and local minima.

 (c) Determine the intervals on which f is increasing.

13. Solve: $2\sin^2 x = 3\cos x$

CHAPTER PROJECTS

I. Markov Chains A **Markov chain** (or process) is one in which future outcomes are determined by a current state. Future outcomes are based on probabilities. The probability of moving to a certain state depends only on the state previously occupied and does not vary with time. An example of a Markov chain is the maximum education achieved by children based on the highest education attained by their parents, where the states are (1) earned college degree, (2) high-school diploma only, (3) elementary school only. If p_{ij} is the probability of moving from state i to state j, the **transition matrix** is the $m \times m$ matrix

$$P = \begin{bmatrix} p_{11} & p_{12} & \cdots & p_{1m} \\ \vdots & \vdots & & \vdots \\ p_{m1} & p_{m2} & \cdots & p_{mm} \end{bmatrix}$$

The table represents the probabilities of the highest educational level of children based on the highest educational level of their parents. For example, the table shows that the probability p_{21} is 40% that parents with a high-school education (row 2) will have children with a college education (column 1).

Highest Educational Level of Parents	Maximum Education That Children Achieve		
	College	High School	Elementary
College	80%	18%	2%
High school	40%	50%	10%
Elementary	20%	60%	20%

1. Convert the percentages to decimals.

2. What is the transition matrix?

3. Sum across the rows. What do you notice? Why do you think that you obtained this result?

4. If P is the transition matrix of a Markov chain, the $(i, j)^{\text{th}}$ entry of P^n (nth power of P) gives the probability of passing from state i to state j in n stages. What is the probability that a grandchild of a college graduate is a college graduate?

5. What is the probability that the grandchild of a high school graduate finishes college?

6. The row vector $v^{(0)} = [0.277 \quad 0.575 \quad 0.148]$ represents the proportion of the U.S. population 25 years or older that has college, high school, and elementary school, respectively, as the highest educational level in 2004.* In a Markov chain the probability distribution $v^{(k)}$ after k stages is $v^{(k)} = v^{(0)}P^k$, where P^k is the kth power of the transition matrix. What will be the distribution of highest educational attainment of the grandchildren of the current population?

7. Calculate $P^3, P^4, P^5, \dots$. Continue until the matrix does not change. This is called the long-run distribution. What is the long-run distribution of highest educational attainment of the population?

Source: U.S. Census Bureau.

The following projects are available at the Instructor's Resource Center (IRC):

II. Project at Motorola: *Error Control Coding* The high powered engineering needed to assure that wireless communications are transmitted correctly is analyzed using matrices to control coding errors.

III. Using Matrices to Find the Line of Best Fit Have you wondered how our calculators get a line of best fit? See how to find the line by solving a matrix equation.

IV. CBL Experiment Simulate two people walking toward each other at a constant rate. Then solve the resulting system of equations to determine when and where they will meet.

Sequences; Induction; the Binomial Theorem

The Future of the World Population

New York, 24 February—World population is expected to increase by 2.6 billion over the next 45 years, from 6.5 billion today to 9.1 billion in 2050. Almost all growth will take place in the less developed regions, where today's 5.3 billion population is expected to swell to 7.8 billion in 2050. By contrast, the population of the more developed regions will remain mostly unchanged, at 1.2 billion.

Future population growth is highly dependent on the path that future fertility takes. In the medium variant, fertility declines from 2.6 children per woman today to slightly over 2 children per woman in 2050. If fertility were to remain about half a child above the levels projected in the medium variant, world population would reach 10.6 billion by 2050. A fertility path half a child below the medium would lead to a population of 7.7 billion by mid-century. That is, at the world level, continued population growth until 2050 is inevitable even if the decline of fertility accelerates.

Because of its low and declining rate of population growth, the population of developed countries as a whole is expected to remain virtually unchanged between 2005 and 2050, at about 1.2 billion. In contrast, the population of the 50 least developed countries is projected to more than double, passing from 0.8 billion in 2005 to 1.7 billion in 2050. Growth in the rest of the developing world is also projected to be robust, though less rapid, with its population rising from 4.5 billion to 6.1 billion between 2005 and 2050.

Source: U.N. Press Release POP/918, February 24, 2005

—*See the Chapter Project*—

A Look Back, A Look Ahead

This chapter may be divided into three independent parts: Sections 13.1–13.3, Section 13.4, and Section 13.5.

In Chapter 3, we defined a function and its domain, which was usually some set of real numbers. In Sections 13.1–13.3, we discuss a sequence, which is a function whose domain is the set of positive integers.

Throughout this text, where it seemed appropriate, we have given proofs of many of the results. In Section 13.4, a technique for proving theorems involving natural numbers is discussed.

In Chapter R, Section R.4, there are formulas for expanding $(x + a)^2$ and $(x + a)^3$. In Section 13.5, we discuss the Binomial Theorem, a formula for the expansion of $(x + a)^n$, where n is any positive integer.

The topics introduced in this chapter are covered in more detail in courses titled *Discrete Mathematics*. Applications of these topics can be found in the fields of computer science, engineering, business and economics, the social sciences, and the physical and biological sciences.

Outline

13.1 Sequences

PREPARING FOR THIS SECTION *Before getting started, review the following concept:*

- Functions (Section 3.1, pp. 208–219)

Now Work the 'Are You Prepared?' problems on page 936.

OBJECTIVES **1** Write the First Several Terms of a Sequence (p. 930)
 2 Write the Terms of a Sequence Defined by a Recursive Formula (p. 933)
 3 Use Summation Notation (p. 934)
 4 Find the Sum of a Sequence (p. 935)

DEFINITION A **sequence** is a function whose domain is the set of positive integers.

Because a sequence is a function, it will have a graph. In Figure 1(a), we show the graph of the function $f(x) = \dfrac{1}{x}$, $x > 0$. If all the points on this graph were removed except those whose x-coordinates are positive integers, that is, if all points were removed except $(1, 1)$, $\left(2, \dfrac{1}{2}\right)$, $\left(3, \dfrac{1}{3}\right)$, and so on, the remaining points would be the graph of the sequence $f(n) = \dfrac{1}{n}$, as shown in Figure 1(b). Notice that we use n to represent the independent variable in a sequence. This serves to remind us that n is a positive integer.

Figure 1

(a) $f(x) = \frac{1}{x}, x > 0$ **(b)** $f(n) = \frac{1}{n}, n$ a positive integer

1 Write the First Several Terms of a Sequence

A sequence is usually represented by listing its values in order. For example, the sequence whose graph is given in Figure 1(b) might be represented as

$$f(1), f(2), f(3), f(4), \ldots \quad \text{or} \quad 1, \frac{1}{2}, \frac{1}{3}, \frac{1}{4}, \ldots$$

The list never ends, as the ellipsis indicates. The numbers in this ordered list are called the **terms** of the sequence.

In dealing with sequences, we usually use subscripted letters, such as a_1, to represent the first term, a_2 for the second term, a_3 for the third term, and so on. For the sequence $f(n) = \dfrac{1}{n}$, we write

$$a_1 = f(1) = 1, \quad a_2 = f(2) = \frac{1}{2}, \quad a_3 = f(3) = \frac{1}{3}, \quad a_4 = f(4) = \frac{1}{4}, \ldots, \quad a_n = f(n) = \frac{1}{n}, \ldots$$

In other words, we usually do not use the traditional function notation $f(n)$ for sequences. For this particular sequence, we have a rule for the nth term, which is $a_n = \dfrac{1}{n}$, so it is easy to find any term of the sequence.

When a formula for the nth term (sometimes called the **general term**) of a sequence is known, rather than write out the terms of the sequence, we usually represent the entire sequence by placing braces around the formula for the nth term. For example, the sequence whose nth term is $b_n = \left(\dfrac{1}{2}\right)^n$ may be represented as

$$\{b_n\} = \left\{\left(\frac{1}{2}\right)^n\right\}$$

or by

$$b_1 = \frac{1}{2}, \quad b_2 = \frac{1}{4}, \quad b_3 = \frac{1}{8}, \dots, \quad b_n = \left(\frac{1}{2}\right)^n, \dots$$

EXAMPLE 1 | **Writing the First Several Terms of a Sequence**

Write down the first six terms of the following sequence and graph it.

$$\{a_n\} = \left\{\frac{n-1}{n}\right\}$$

Figure 2

Solution | The first six terms of the sequence are

$$a_1 = 0, \quad a_2 = \frac{1}{2}, \quad a_3 = \frac{2}{3}, \quad a_4 = \frac{3}{4}, \quad a_5 = \frac{4}{5}, \quad a_6 = \frac{5}{6}$$

See Figure 2 for the graph.

 COMMENT Graphing utilities can be used to write the terms of a sequence and graph them. Figure 3 shows the sequence given in Example 1 generated on a TI-84 Plus graphing calculator. We can see the first few terms of the sequence on the viewing window. You need to press the right arrow key to scroll right to see the remaining terms of the sequence. Figure 4 shows a graph of the sequence. Notice that the first term of the sequence is not visible since it lies on the x-axis. TRACEing the graph will allow you to see the terms of the sequence. The TABLE feature can also be used to generate the terms of the sequence. See Table 1.

Figure 3　　　　　**Figure 4**　　　　**Table 1**

Now Work PROBLEM 17

EXAMPLE 2 | **Writing the First Several Terms of a Sequence**

Write down the first six terms of the following sequence and graph it.

$$\{b_n\} = \left\{(-1)^{n+1}\left(\frac{2}{n}\right)\right\}$$

Figure 5

Solution The first six terms of the sequence are

$$b_1 = 2, \quad b_2 = -1, \quad b_3 = \frac{2}{3}, \quad b_4 = -\frac{1}{2}, \quad b_5 = \frac{2}{5}, \quad b_6 = -\frac{1}{3}$$

See Figure 5 for the graph.

Notice in the sequence $\{b_n\}$ in Example 2 that the signs of the terms **alternate.** When this occurs, we use factors such as $(-1)^{n+1}$, which equals 1 if n is odd and -1 if n is even, or $(-1)^n$, which equals -1 if n is odd and 1 if n is even.

EXAMPLE 3 **Writing the First Several Terms of a Sequence**

Write down the first six terms of the following sequence and graph it.

Figure 6

$$\{c_n\} = \left\{ \begin{array}{ll} n & \text{if } n \text{ is even} \\ \dfrac{1}{n} & \text{if } n \text{ is odd} \end{array} \right\}$$

Solution The first six terms of the sequence are

$$c_1 = 1, \quad c_2 = 2, \quad c_3 = \frac{1}{3}, \quad c_4 = 4, \quad c_5 = \frac{1}{5}, \quad c_6 = 6$$

See Figure 6 for the graph.

➤ **Now Work** PROBLEM 19

Sometimes a sequence is indicated by an observed pattern in the first few terms that makes it possible to infer the makeup of the nth term. In the example that follows, a sufficient number of terms of the sequence is given so that a natural choice for the nth term is suggested.

EXAMPLE 4 **Determining a Sequence from a Pattern**

(a) $e, \dfrac{e^2}{2}, \dfrac{e^3}{3}, \dfrac{e^4}{4}, \ldots \qquad a_n = \dfrac{e^n}{n}$

(b) $1, \dfrac{1}{3}, \dfrac{1}{9}, \dfrac{1}{27}, \ldots \qquad b_n = \dfrac{1}{3^{n-1}}$

(c) $1, 3, 5, 7, \ldots \qquad c_n = 2n - 1$

(d) $1, 4, 9, 16, 25, \ldots \qquad d_n = n^2$

(e) $1, -\dfrac{1}{2}, \dfrac{1}{3}, -\dfrac{1}{4}, \dfrac{1}{5}, \ldots \qquad e_n = (-1)^{n+1}\left(\dfrac{1}{n}\right)$

➤ **Now Work** PROBLEM 27

The Factorial Symbol

DEFINITION If $n \geq 0$ is an integer, the **factorial symbol $n!$** is defined as follows:

$$0! = 1 \qquad 1! = 1$$
$$n! = n(n-1) \cdot \cdots \cdot 3 \cdot 2 \cdot 1 \qquad \text{if } n \geq 2$$

Table 2

n	n!
0	1
1	1
2	2
3	6
4	24
5	120
6	720

For example, $2! = 2 \cdot 1 = 2$, $3! = 3 \cdot 2 \cdot 1 = 6$, $4! = 4 \cdot 3 \cdot 2 \cdot 1 = 24$, and so on. Table 2 lists the values of $n!$ for $0 \le n \le 6$.

Because

$$n! = \underbrace{n(n-1)(n-2) \cdot \cdots \cdot 3 \cdot 2 \cdot 1}_{(n-1)!}$$

we can use the formula

$$n! = n(n-1)!$$

to find successive factorials. For example, because $6! = 720$, we have

$$7! = 7 \cdot 6! = 7(720) = 5040$$

and

$$8! = 8 \cdot 7! = 8(5040) = 40{,}320$$

━━━━ **Now Work** PROBLEM 11

Exploration

Your calculator has a factorial key. Use it to see how fast factorials increase in value. Find the value of 69!. What happens when you try to find 70!? In fact, 70! is larger than 10^{100} (a googol).

2 Write the Terms of a Sequence Defined by a Recursive Formula

A second way of defining a sequence is to assign a value to the first (or the first few) term(s) and specify the nth term by a formula or equation that involves one or more of the terms preceding it. Sequences defined this way are said to be defined **recursively,** and the rule or formula is called a **recursive formula.**

EXAMPLE 5 | **Writing the Terms of a Recursively Defined Sequence**

Write down the first five terms of the following recursively defined sequence.

$$s_1 = 1, \qquad s_n = n s_{n-1}$$

Solution The first term is given as $s_1 = 1$. To get the second term, we use $n = 2$ in the formula $s_n = n s_{n-1}$ to get $s_2 = 2 s_1 = 2 \cdot 1 = 2$. To get the third term, we use $n = 3$ in the formula to get $s_3 = 3 s_2 = 3 \cdot 2 = 6$. To get a new term requires that we know the value of the preceding term. The first five terms are

$$s_1 = 1$$
$$s_2 = 2 \cdot 1 = 2$$
$$s_3 = 3 \cdot 2 = 6$$
$$s_4 = 4 \cdot 6 = 24$$
$$s_5 = 5 \cdot 24 = 120$$

Do you recognize this sequence? $s_n = n!$

EXAMPLE 6 | **Writing the Terms of a Recursively Defined Sequence**

Write down the first five terms of the following recursively defined sequence.

$$u_1 = 1, \qquad u_2 = 1, \qquad u_{n+2} = u_n + u_{n+1}$$

Solution We are given the first two terms. To get the third term requires that we know both of the previous two terms. That is,

$$u_1 = 1$$
$$u_2 = 1$$
$$u_3 = u_1 + u_2 = 1 + 1 = 2$$
$$u_4 = u_2 + u_3 = 1 + 2 = 3$$
$$u_5 = u_3 + u_4 = 2 + 3 = 5$$

The sequence defined in Example 6 is called the **Fibonacci sequence,** and the terms of this sequence are called **Fibonacci numbers.** These numbers appear in a wide variety of applications (see Problems 85–88).

Now Work PROBLEMS 35 AND 43

3 Use Summation Notation

It is often important to be able to find the sum of the first n terms of a sequence $\{a_n\}$, that is,

$$a_1 + a_2 + a_3 + \cdots + a_n$$

Rather than write down all these terms, we introduce a more concise way to express the sum, called **summation notation.** Using summation notation, we write the sum as

$$a_1 + a_2 + a_3 + \cdots + a_n = \sum_{k=1}^{n} a_k$$

The symbol Σ (the Greek letter sigma, which is an S in our alphabet) is simply an instruction to sum, or add up, the terms. The integer k is called the **index** of the sum; it tells you where to start the sum and where to end it. The expression

$$\sum_{k=1}^{n} a_k$$

is an instruction to add the terms a_k of the sequence $\{a_n\}$ starting with $k = 1$ and ending with $k = n$. We read the expression as "the sum of a_k from $k = 1$ to $k = n$."

EXAMPLE 7

Expanding Summation Notation

Write out each sum.

(a) $\displaystyle\sum_{k=1}^{n} \frac{1}{k}$ (b) $\displaystyle\sum_{k=1}^{n} k!$

Solution (a) $\displaystyle\sum_{k=1}^{n} \frac{1}{k} = 1 + \frac{1}{2} + \frac{1}{3} + \cdots + \frac{1}{n}$ (b) $\displaystyle\sum_{k=1}^{n} k! = 1! + 2! + \cdots + n!$

Now Work PROBLEM 51

EXAMPLE 8

Writing a Sum in Summation Notation

Express each sum using summation notation.

(a) $1^2 + 2^2 + 3^2 + \cdots + 9^2$ (b) $1 + \dfrac{1}{2} + \dfrac{1}{4} + \dfrac{1}{8} + \cdots + \dfrac{1}{2^{n-1}}$

Solution (a) The sum $1^2 + 2^2 + 3^2 + \cdots + 9^2$ has 9 terms, each of the form k^2, and starts at $k = 1$ and ends at $k = 9$:

$$1^2 + 2^2 + 3^2 + \cdots + 9^2 = \sum_{k=1}^{9} k^2$$

(b) The sum

$$1 + \frac{1}{2} + \frac{1}{4} + \frac{1}{8} + \cdots + \frac{1}{2^{n-1}}$$

has n terms, each of the form $\dfrac{1}{2^{k-1}}$, and starts at $k = 1$ and ends at $k = n$:

$$1 + \frac{1}{2} + \frac{1}{4} + \frac{1}{8} + \cdots + \frac{1}{2^{n-1}} = \sum_{k=1}^{n} \frac{1}{2^{k-1}}$$

The index of summation need not always begin at 1 or end at n; for example, we could have expressed the sum in Example 8(b) as

$$\sum_{k=0}^{n-1} \frac{1}{2^k} = 1 + \frac{1}{2} + \frac{1}{4} + \cdots + \frac{1}{2^{n-1}}$$

Letters other than k may be used as the index. For example,

$$\sum_{j=1}^{n} j! \quad \text{and} \quad \sum_{i=1}^{n} i!$$

each represent the same sum as the one given in Example 7(b).

━━━━━━ **Now Work** PROBLEM 61

4 Find the Sum of a Sequence

Next we list some properties of sequences using summation notation. These properties are useful for adding the terms of a sequence.

THEOREM

Properties of Sequences

If $\{a_n\}$ and $\{b_n\}$ are two sequences and c is a real number, then:

$$\sum_{k=1}^{n} (ca_k) = ca_1 + ca_2 + \cdots + ca_n = c(a_1 + a_2 + \cdots + a_n) = c \sum_{k=1}^{n} a_k \quad \textbf{(1)}$$

$$\sum_{k=1}^{n} (a_k + b_k) = \sum_{k=1}^{n} a_k + \sum_{k=1}^{n} b_k \quad \textbf{(2)}$$

$$\sum_{k=1}^{n} (a_k - b_k) = \sum_{k=1}^{n} a_k - \sum_{k=1}^{n} b_k \quad \textbf{(3)}$$

$$\sum_{k=j+1}^{n} a_k = \sum_{k=1}^{n} a_k - \sum_{k=1}^{j} a_k, \quad \text{where } 0 < j < n \quad \textbf{(4)}$$

The proof of property (1) follows from the distributive property of real numbers. The proofs of properties 2 and 3 are based on the commutative and associative properties of real numbers. Property (4) states that the sum from $j + 1$ to n equals the sum from 1 to n minus the sum from 1 to j. It can be helpful to employ this property when the index of summation begins at a number larger than 1.

Next we give some formulas for finding the sums of certain sequences.

THEOREM

Formulas for Sums of Sequences

$$\sum_{k=1}^{n} c = \underbrace{c + c + \cdots + c}_{n \text{ terms}} = cn \quad c \text{ is a real number} \quad \textbf{(5)}$$

$$\sum_{k=1}^{n} k = 1 + 2 + 3 + \cdots + n = \frac{n(n + 1)}{2} \quad \textbf{(6)}$$

$$\sum_{k=1}^{n} k^2 = 1^2 + 2^2 + 3^2 + \cdots + n^2 = \frac{n(n + 1)(2n + 1)}{6} \quad \textbf{(7)}$$

$$\sum_{k=1}^{n} k^3 = 1^3 + 2^3 + 3^3 + \cdots + n^3 = \left[\frac{n(n + 1)}{2} \right]^2 \quad \textbf{(8)}$$

The proof of formula (5) follows from the definition of summation notation. You are asked to prove formula (6) in Problem 92. The proofs of formulas (7) and (8) require mathematical induction, which is discussed in Section 13.4.

Notice the difference between formulas (5) and (6). In (5), the constant c is being summed from 1 to n, while in (6), the index of summation k is being summed from 1 to n.

EXAMPLE 9	Finding the Sum of a Sequence

Find the sum of each sequence.

(a) $\displaystyle\sum_{k=1}^{5}(3k)$ (b) $\displaystyle\sum_{k=1}^{10}(k^3+1)$

(c) $\displaystyle\sum_{k=1}^{24}(k^2-7k+2)$ (d) $\displaystyle\sum_{k=6}^{20}(4k^2)$

Solution (a) $\displaystyle\sum_{k=1}^{5}(3k)=3\sum_{k=1}^{5}k$ Property (1)

$$=3\left(\frac{5(5+1)}{2}\right)$$ Formula (6)

$$=3(15)$$

$$=45$$

(b) $\displaystyle\sum_{k=1}^{10}(k^3+1)=\sum_{k=1}^{10}k^3+\sum_{k=1}^{10}1$ Property (2)

$$=\left(\frac{10(10+1)}{2}\right)^2+1(10)$$ Formulas (8) and (5)

$$=3025+10$$

$$=3035$$

(c) $\displaystyle\sum_{k=1}^{24}(k^2-7k+2)=\sum_{k=1}^{24}k^2-\sum_{k=1}^{24}(7k)+\sum_{k=1}^{24}2$ Properties (2) and (3)

$$=\sum_{k=1}^{24}k^2-7\sum_{k=1}^{24}k+\sum_{k=1}^{24}2$$ Property (1)

$$=\frac{24(24+1)(2\cdot24+1)}{6}-7\left(\frac{24(24+1)}{2}\right)+2(24)$$ Formulas (7), (6), (5)

$$=4900-2100+48$$

$$=2848$$

(d) Notice that the index of summation starts at 6. We use property (4) as follows:

$$\sum_{k=6}^{20}(4k^2)=4\sum_{k=6}^{20}k^2=4\left[\sum_{k=1}^{20}k^2-\sum_{k=1}^{5}k^2\right]=4\left[\frac{20(21)(41)}{6}-\frac{5(6)(11)}{6}\right]$$

Property (1) Property (4) Formula (7)

$$=4[2870-55]=11{,}260$$

Now Work PROBLEM 73

13.1 Assess Your Understanding

'Are You Prepared?' *Answers are given at the end of these exercises. If you get a wrong answer, read the pages listed in red.*

1. For the function $f(x)=\dfrac{x-1}{x}$, find $f(2)$ and $f(3)$. (pp. 208–219)

2. *True or False* A function is a relation between two sets D and R so that each element x in the first set D is related to exactly one element y in the second set R. (pp. 208–219)

Concepts and Vocabulary

3. A(n) _____ is a function whose domain is the set of positive integers.

4. For the sequence $\{s_n\} = \{4n - 1\}$, the first term is $s_1 =$ _____ and the fourth term is $s_4 =$ _____.

5. $\sum_{k=1}^{4} (2k) =$ _____.

6. *True or False* Sequences are sometimes defined recursively.

7. *True or False* A sequence is a function.

8. *True or False* $\sum_{k=1}^{2} k = 3$

Skill Building

In Problems 9–14, evaluate each factorial expression.

9. $10!$

10. $9!$

11. $\dfrac{9!}{6!}$

12. $\dfrac{12!}{10!}$

13. $\dfrac{3! \, 7!}{4!}$

14. $\dfrac{5! \, 8!}{3!}$

In Problems 15–26, write down the first five terms of each sequence.

15. $\{s_n\} = \{n\}$

16. $\{s_n\} = \{n^2 + 1\}$

17. $\{a_n\} = \left\{\dfrac{n}{n + 2}\right\}$

18. $\{b_n\} = \left\{\dfrac{2n + 1}{2n}\right\}$

19. $\{c_n\} = \{(-1)^{n+1} n^2\}$

20. $\{d_n\} = \left\{(-1)^{n-1}\left(\dfrac{n}{2n - 1}\right)\right\}$

21. $\{s_n\} = \left\{\dfrac{2^n}{3^n + 1}\right\}$

22. $\{s_n\} = \left\{\left(\dfrac{4}{3}\right)^n\right\}$

23. $\{t_n\} = \left\{\dfrac{(-1)^n}{(n + 1)(n + 2)}\right\}$

24. $\{a_n\} = \left\{\dfrac{3^n}{n}\right\}$

25. $\{b_n\} = \left\{\dfrac{n}{e^n}\right\}$

26. $\{c_n\} = \left\{\dfrac{n^2}{2^n}\right\}$

In Problems 27–34, the given pattern continues. Write down the nth term of a sequence $\{a_n\}$ suggested by the pattern.

27. $\dfrac{1}{2}, \dfrac{2}{3}, \dfrac{3}{4}, \dfrac{4}{5}, \cdots$

28. $\dfrac{1}{1 \cdot 2}, \dfrac{1}{2 \cdot 3}, \dfrac{1}{3 \cdot 4}, \dfrac{1}{4 \cdot 5}, \cdots$

29. $1, \dfrac{1}{2}, \dfrac{1}{4}, \dfrac{1}{8}, \cdots$

30. $\dfrac{2}{3}, \dfrac{4}{9}, \dfrac{8}{27}, \dfrac{16}{81}, \cdots$

31. $1, -1, 1, -1, 1, -1, \ldots$

32. $1, \dfrac{1}{2}, 3, \dfrac{1}{4}, 5, \dfrac{1}{6}, 7, \dfrac{1}{8}, \ldots$

33. $1, -2, 3, -4, 5, -6, \ldots$

34. $2, -4, 6, -8, 10, \ldots$

In Problems 35–48, a sequence is defined recursively. Write down the first five terms.

35. $a_1 = 2$; $a_n = 3 + a_{n-1}$

36. $a_1 = 3$; $a_n = 4 - a_{n-1}$

37. $a_1 = -2$; $a_n = n + a_{n-1}$

38. $a_1 = 1$; $a_n = n - a_{n-1}$

39. $a_1 = 5$; $a_n = 2a_{n-1}$

40. $a_1 = 2$; $a_n = -a_{n-1}$

41. $a_1 = 3$; $a_n = \dfrac{a_{n-1}}{n}$

42. $a_1 = -2$; $a_n = n + 3a_{n-1}$

43. $a_1 = 1$; $a_2 = 2$; $a_n = a_{n-1} \cdot a_{n-2}$

44. $a_1 = -1$; $a_2 = 1$; $a_n = a_{n-2} + na_{n-1}$

45. $a_1 = A$; $a_n = a_{n-1} + d$

46. $a_1 = A$; $a_n = ra_{n-1}$, $r \neq 0$

47. $a_1 = \sqrt{2}$; $a_n = \sqrt{2 + a_{n-1}}$

48. $a_1 = \sqrt{2}$; $a_n = \sqrt{\dfrac{a_{n-1}}{2}}$

In Problems 49–58, write out each sum.

49. $\sum_{k=1}^{n} (k + 2)$

50. $\sum_{k=1}^{n} (2k + 1)$

51. $\sum_{k=1}^{n} \dfrac{k^2}{2}$

52. $\sum_{k=1}^{n} (k + 1)^2$

53. $\sum_{k=0}^{n} \dfrac{1}{3^k}$

54. $\sum_{k=0}^{n} \left(\dfrac{3}{2}\right)^k$

55. $\sum_{k=0}^{n-1} \dfrac{1}{3^{k+1}}$

56. $\sum_{k=0}^{n-1} (2k + 1)$

57. $\sum_{k=2}^{n} (-1)^k \ln k$

58. $\sum_{k=3}^{n} (-1)^{k+1} 2^k$

In Problems 59–68, express each sum using summation notation.

59. $1 + 2 + 3 + \cdots + 20$

60. $1^3 + 2^3 + 3^3 + \cdots + 8^3$

61. $\dfrac{1}{2} + \dfrac{2}{3} + \dfrac{3}{4} + \cdots + \dfrac{13}{13 + 1}$

62. $1 + 3 + 5 + 7 + \cdots + [2(12) - 1]$

63. $1 - \dfrac{1}{3} + \dfrac{1}{9} - \dfrac{1}{27} + \cdots + (-1)^6 \left(\dfrac{1}{3^6}\right)$

64. $\dfrac{2}{3} - \dfrac{4}{9} + \dfrac{8}{27} - \cdots + (-1)^{12} \left(\dfrac{2}{3}\right)^{11}$

65. $3 + \dfrac{3^2}{2} + \dfrac{3^3}{3} + \cdots + \dfrac{3^n}{n}$

66. $\dfrac{1}{e} + \dfrac{2}{e^2} + \dfrac{3}{e^3} + \cdots + \dfrac{n}{e^n}$

67. $a + (a + d) + (a + 2d) + \cdots + (a + nd)$

68. $a + ar + ar^2 + \cdots + ar^{n-1}$

In Problems 69–80, find the sum of each sequence.

69. $\displaystyle\sum_{k=1}^{40} 5$

70. $\displaystyle\sum_{k=1}^{50} 8$

71. $\displaystyle\sum_{k=1}^{40} k$

72. $\displaystyle\sum_{k=1}^{24} (-k)$

73. $\displaystyle\sum_{k=1}^{20} (5k + 3)$

74. $\displaystyle\sum_{k=1}^{26} (3k - 7)$

75. $\displaystyle\sum_{k=1}^{16} (k^2 + 4)$

76. $\displaystyle\sum_{k=0}^{14} (k^2 - 4)$

77. $\displaystyle\sum_{k=10}^{60} (2k)$

78. $\displaystyle\sum_{k=8}^{40} (-3k)$

79. $\displaystyle\sum_{k=5}^{20} k^3$

80. $\displaystyle\sum_{k=4}^{24} k^3$

Applications and Extensions

81. Credit Card Debt John has a balance of $3000 on his Discover card that charges 1% interest per month on any unpaid balance. John can afford to pay $100 toward the balance each month. His balance each month after making a $100 payment is given by the recursively defined sequence

$$B_0 = \$3000, \qquad B_n = 1.01B_{n-1} - 100$$

Determine John's balance after making the first payment. That is, determine B_1.

82. Trout Population A pond currently has 2000 trout in it. A fish hatchery decides to add an additional 20 trout each month. In addition, it is known that the trout population is growing 3% per month. The size of the population after n months is given by the recursively defined sequence

$$p_0 = 2000, \qquad p_n = 1.03p_{n-1} + 20$$

How many trout are in the pond after two months? That is, what is p_2?

83. Car Loans Phil bought a car by taking out a loan for $18,500 at 0.5% interest per month. Phil's normal monthly payment is $434.47 per month, but he decides that he can afford to pay $100 extra toward the balance each month. His balance each month is given by the recursively defined sequence

$$B_0 = \$18,500, \qquad B_n = 1.005B_{n-1} - 534.47$$

Determine Phil's balance after making the first payment. That is, determine B_1.

84. Environmental Control The Environmental Protection Agency (EPA) determines that Maple Lake has 250 tons of pollutant as a result of industrial waste and that 10% of the pollutant present is neutralized by solar oxidation every year. The EPA imposes new pollution control laws that result in 15 tons of new pollutant entering the lake each year. The amount of pollutant in the lake after n years is given by the recursively defined sequence

$$p_0 = 250, \qquad p_n = 0.9p_{n-1} + 15$$

Determine the amount of pollutant in the lake after 2 years. That is, determine p_2.

85. Growth of a Rabbit Colony A colony of rabbits begins with one pair of mature rabbits, which will produce a pair of offspring (one male, one female) each month. Assume that all rabbits mature in 1 month and produce a pair of offspring (one male, one female) after 2 months. If no rabbits

ever die, how many pairs of mature rabbits are there after 7 months?

[**Hint:** A Fibonacci sequence models this colony. Do you see why?]

86. Fibonacci Sequence Let

$$u_n = \dfrac{\left(1 + \sqrt{5}\right)^n - \left(1 - \sqrt{5}\right)^n}{2^n\sqrt{5}}$$

define the nth term of a sequence.

(a) Show that $u_1 = 1$ and $u_2 = 1$.

(b) Show that $u_{n+2} = u_{n+1} + u_n$.

(c) Draw the conclusion that $\{u_n\}$ is a Fibonacci sequence.

87. Pascal's Triangle Divide the triangular array shown (called Pascal's triangle) using diagonal lines as indicated. Find the sum of the numbers in each diagonal row. Do you recognize this sequence?

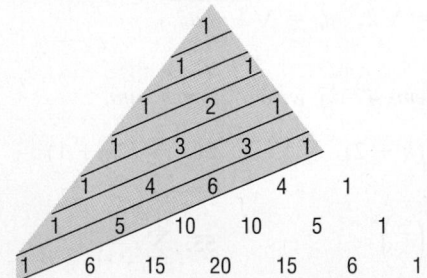

88. Fibonacci Sequence Use the result of Problem 86 to do the following problems:

(a) Write the first 11 terms of the Fibonacci sequence.

(b) Write down the first 10 terms of the ratio $\dfrac{u_{n+1}}{u_n}$.

(c) As n gets large, what number does the ratio approach? This number is referred to as the **golden ratio.** Rectangles whose sides are in this ratio were considered pleasing to

the eye by the Greeks. For example, the facade of the Parthenon was constructed using the golden ratio.

(d) Write down the first 10 terms of the ratio $\dfrac{u_n}{u_{n+1}}$.

(e) As n gets large, what number does the ratio approach? This number is also referred to as the **conjugate golden ratio.** This ratio is believed to have been used in the construction of the Great Pyramid in Egypt. The ratio equals the sum of the areas of the four face triangles divided by the total surface area of the Great Pyramid.

 89. Approximating $f(x) = e^x$ In calculus, it can be shown that

$$f(x) = e^x = \sum_{k=0}^{\infty} \frac{x^k}{k!}$$

We can approximate the value of $f(x) = e^x$ for any x using the following sum

$$f(x) = e^x \approx \sum_{k=0}^{n} \frac{x^k}{k!}$$

for some n.

(a) Approximate $f(1.3)$ with $n = 4$.

(b) Approximate $f(1.3)$ with $n = 7$.

(c) Use a calculator to approximate $f(1.3)$.

 (d) Using trial and error along with a graphing utility's SEQuence mode, determine the value of n required to approximate $f(1.3)$ correct to eight decimal places.

 90. Approximating $f(x) = e^x$ Refer to Problem 89.

(a) Approximate $f(-2.4)$ with $n = 3$.

(b) Approximate $f(-2.4)$ with $n = 6$.

(c) Use a calculator to approximate $f(-2.4)$.

(d) Using trial and error along with a graphing utility's SEQuence mode, determine the value of n required to approximate $f(-2.4)$ correct to eight decimal places.

91. Bode's Law In 1772, Johann Bode published the following formula for predicting the mean distances, in astronomical units (AU), of the planets from the sun:

$$a_1 = 0.4, \qquad \{a_n\} = \{0.4 + 0.3 \cdot 2^{n-2}\}, n \geq 2$$

where n is the number of the planet from the sun.

(a) Determine the first eight terms of this sequence.

(b) At the time of Bode's publication, the known planets were Mercury (0.39 AU), Venus (0.72 AU), Earth (1 AU),

Mars (1.52 AU), Jupiter (5.20 AU), and Saturn (9.54 AU). How do the actual distances compare to the terms of the sequence?

(c) The planet Uranus was discovered in 1781 and the asteroid Ceres was discovered in 1801. The mean orbital distances from the sun to Uranus and Ceres* are 19.2 AU and 2.77 AU, respectively. How well do these values fit within the sequence?

(d) Determine the ninth and tenth terms of Bode's sequence.

(e) The planets Neptune and Pluto* were discovered in 1846 and 1930, respectively. Their mean orbital distances from the sun are 30.07 AU and 39.44 AU, respectively. How do these actual distances compare to the terms of the sequence?

(f) On July 29, 2005, NASA announced the discovery of a tenth planet* ($n = 11$), which has temporarily been named 2003 UB313* until a permanent name is decided on. Use Bode's Law to predict the mean orbital distance of 2003 UB313 from the sun. Its actual mean distance is not yet known, but 2003 UB313 is currently about 97 astronomical units from the sun.

Sources: NASA.

92. Show that

$$1 + 2 + \cdots + (n - 1) + n = \frac{n(n + 1)}{2}$$

[**Hint:** Let

$$S = 1 + 2 + \cdots + (n - 1) + n$$
$$S = n + (n - 1) + (n - 2) + \cdots + 1$$

Add these equations. Then

$$2S = \underbrace{[1 + n] + [2 + (n - 1)] + \cdots + [n + 1]}_{n \text{ terms in brackets}}$$

Now complete the derivation.]

*Ceres, Pluto, and 2003 UB313 are now referred to as dwarf planets.

Discussion and Writing

93. Investigate various applications that lead to a Fibonacci sequence, such as art, architecture, or financial markets. Write an essay on these applications.

'Are You Prepared?' Answers

1. $f(2) = \dfrac{1}{2}; f(3) = \dfrac{2}{3}$ **2.** True

13.2 Arithmetic Sequences

OBJECTIVES 1 Determine If a Sequence Is Arithmetic (p. 940)

2 Find a Formula for an Arithmetic Sequence (p. 941)

3 Find the Sum of an Arithmetic Sequence (p. 942)

1 Determine If a Sequence Is Arithmetic

When the difference between successive terms of a sequence is always the same number, the sequence is called **arithmetic**.

DEFINITION

An **arithmetic sequence*** may be defined recursively as $a_1 = a$, $a_n - a_{n-1} = d$, or as

$$a_1 = a, \qquad a_n = a_{n-1} + d \qquad \qquad (1)$$

where $a_1 = a$ and d are real numbers. The number a is the first term, and the number d is called the **common difference.**

The terms of an arithmetic sequence with first term a_1 and common difference d follow the pattern

$$a_1, \quad a_1 + d, \quad a_1 + 2d, \quad a_1 + 3d, \dots$$

EXAMPLE 1 **Determining If a Sequence Is Arithmetic**

The sequence

$$4, \quad 6, \quad 8, \quad 10, \dots$$

is arithmetic since the difference of successive terms is 2. The first term is $a_1 = 4$, and the common difference is $d = 2$.

■

EXAMPLE 2 **Determining If a Sequence Is Arithmetic**

Show that the following sequence is arithmetic. Find the first term and the common difference.

$$\{s_n\} = \{3n + 5\}$$

Solution The first term is $s_1 = 3 \cdot 1 + 5 = 8$. The nth term and the $(n - 1)$st term of the sequence $\{s_n\}$ are

$$s_n = 3n + 5 \quad \text{and} \quad s_{n-1} = 3(n - 1) + 5 = 3n + 2$$

Their difference d is

$$d = s_n - s_{n-1} = (3n + 5) - (3n + 2) = 5 - 2 = 3$$

Since the difference of any two successive terms is the constant 3, the sequence is arithmetic and the common difference is 3.

■

EXAMPLE 3 **Determining If a Sequence Is Arithmetic**

Show that the sequence $\{t_n\} = \{4 - n\}$ is arithmetic. Find the first term and the common difference.

Solution The first term is $t_1 = 4 - 1 = 3$. The nth term and the $(n - 1)$st term are

$$t_n = 4 - n \quad \text{and} \quad t_{n-1} = 4 - (n - 1) = 5 - n$$

Their difference d is

$$d = t_n - t_{n-1} = (4 - n) - (5 - n) = 4 - 5 = -1$$

*Sometimes called an **arithmetic progression.**

Since the difference of any two successive terms is the constant -1; $\{t_n\}$ is an arithmetic sequence whose common difference is -1.

Now Work PROBLEM 5

2 Find a Formula for an Arithmetic Sequence

Suppose that a is the first term of an arithmetic sequence whose common difference is d. We seek a formula for the nth term, a_n. To see the pattern, we write down the first few terms.

$$a_1 = a$$
$$a_2 = a_1 + d = a_1 + 1 \cdot d$$
$$a_3 = a_2 + d = (a_1 + d) + d = a_1 + 2 \cdot d$$
$$a_4 = a_3 + d = (a_1 + 2 \cdot d) + d = a_1 + 3 \cdot d$$
$$a_5 = a_4 + d = (a_1 + 3 \cdot d) + d = a_1 + 4 \cdot d$$
$$\vdots$$
$$a_n = a_{n-1} + d = [a_1 + (n-2)d] + d = a_1 + (n-1)d$$

We are led to the following result:

THEOREM

nth Term of an Arithmetic Sequence

For an arithmetic sequence $\{a_n\}$ whose first term is a_1 and whose common difference is d, the nth term is determined by the formula

$$a_n = a_1 + (n-1)d \tag{2}$$

| EXAMPLE 4 | **Finding a Particular Term of an Arithmetic Sequence** |

Find the forty-first term of the arithmetic sequence: $2, 6, 10, 14, 18, \ldots$

Solution The first term of this arithmetic sequence is $a_1 = 2$, and the common difference is $d = 4$. By formula (2), the nth term is

$$a_n = 2 + (n-1)4 \qquad a_n = a_1 + (n-1)d; a_1 = 2, d = 4$$

The forty-first term is

$$a_{41} = 2 + 40 \cdot 4 = 162$$

| EXAMPLE 5 | **Finding a Recursive Formula for an Arithmetic Sequence** |

The eighth term of an arithmetic sequence is 75, and the twentieth term is 39.

(a) Find the first term and the common difference.
(b) Give a recursive formula for the sequence.
(c) What is the nth term of the sequence?

Solution (a) By formula (2), we know that $a_n = a_1 + (n-1)d$. As a result,

$$\begin{cases} a_8 = a_1 + 7d = 75 \\ a_{20} = a_1 + 19d = 39 \end{cases}$$

This is a system of two linear equations containing two variables, a_1 and d, which we can solve by elimination. Subtracting the second equation from the first, we get

$$-12d = 36$$
$$d = -3$$

With $d = -3$, we use $a_1 + 7d = 75$ and find that $a_1 = 75 - 7d = 75 - 7(-3) = 96$. The first term is $a_1 = 96$, and the common difference is $d = -3$.

(b) Using formula (1), a recursive formula for this sequence is

$$a_1 = 96, \qquad a_n = a_{n-1} - 3$$

(c) Using formula (2), a formula for the nth term of the sequence $\{a_n\}$ is

$$a_n = a_1 + (n - 1)d = 96 + (n - 1)(-3) = 99 - 3n$$

──── **Now Work** PROBLEMS 21 AND 27

3 Find the Sum of an Arithmetic Sequence

The next result gives two formulas for finding the sum of the first n terms of an arithmetic sequence.

THEOREM

Sum of the First n Terms of an Arithmetic Sequence

Let $\{a_n\}$ be an arithmetic sequence with first term a_1 and common difference d. The sum S_n of the first n terms of $\{a_n\}$ may be found in two ways:

$$S_n = a_1 + a_2 + a_3 + \cdots + a_n$$

$$= \sum_{k=1}^{n} [a_1 + (k - 1)d] = \frac{n}{2}[2a_1 + (n - 1)d] \qquad (3)$$

$$S_n = a_1 + a_2 + a_3 + \cdots + a_n$$

$$= \sum_{k=1}^{n} [a_1 + (k - 1)d] = \frac{n}{2}(a_1 + a_n) \qquad (4)$$

Proof

$$S_n = a_1 + a_2 + a_3 + \cdots + a_n \qquad \text{Sum of first } n \text{ terms}$$

$$= a_1 + (a_1 + d) + (a_1 + 2d) + \cdots + [a_1 + (n - 1)d] \qquad \text{Formula (2)}$$

$$= (a_1 + a_1 + \cdots + a_1) + [d + 2d + \cdots + (n - 1)d] \qquad \text{Rearrange terms}$$

$$\underbrace{}_{n \text{ terms}}$$

$$= na_1 + d[1 + 2 + \cdots + (n - 1)]$$

$$= na_1 + d\left[\frac{(n - 1)n}{2}\right] \qquad \text{Formula 6, Section 13.1}$$

$$= na_1 + \frac{n}{2}(n - 1)d$$

$$= \frac{n}{2}[2a_1 + (n - 1)d] \qquad \text{Factor out } \frac{n}{2}; \text{ this is Formula (3).}$$

$$= \frac{n}{2}[a_1 + a_1 + (n - 1)d]$$

$$= \frac{n}{2}(a_1 + a_n) \qquad \text{Use Formula (2); this is Formula (4).}$$

There are two ways to find the sum of the first n terms of an arithmetic sequence. Notice that formula (3) involves the first term and common difference, whereas formula (4) involves the first term and the nth term. Use whichever form is easier.

EXAMPLE 6	**Finding the Sum of an Arithmetic Sequence**

Find the sum S_n of the first n terms of the sequence $\{a_n\} = \{3n + 5\}$; that is, find

$$8 + 11 + 14 + \cdots + (3n + 5) = \sum_{k=1}^{n} (3k + 5)$$

Solution The sequence $\{a_n\} = \{3n + 5\}$ is an arithmetic sequence with first term $a_1 = 8$ and the nth term $a_n = 3n + 5$. To find the sum S_n, we use formula (4).

$$S_n = \sum_{k=1}^{n} (3k + 5) = \frac{n}{2}[8 + (3n + 5)] = \frac{n}{2}(3n + 13)$$

$$\uparrow$$
$$S_n = \frac{n}{2}(a_1 + a_n)$$

➤ **Now Work** PROBLEM 35

EXAMPLE 7	**Finding the Sum of an Arithmetic Sequence**

Find the sum: $60 + 64 + 68 + 72 + \cdots + 120$

Solution This is the sum S_n of an arithmetic sequence $\{a_n\}$ whose first term is $a_1 = 60$ and whose common difference is $d = 4$. The nth term is $a_n = 120$. We use formula (2) to find n.

$$a_n = a_1 + (n - 1)d \qquad \text{Formula (2)}$$
$$120 = 60 + (n - 1) \cdot 4 \qquad a_n = 120, a_1 = 60, d = 4$$
$$60 = 4(n - 1) \qquad \text{Simplify}$$
$$15 = n - 1 \qquad \text{Simplify}$$
$$n = 16 \qquad \text{Solve for } n$$

Now we use formula (4) to find the sum S_{16}

$$60 + 64 + 68 + \cdots + 120 = S_{16} = \frac{16}{2}(60 + 120) = 1440$$

$$\uparrow$$
$$S_n = \frac{n}{2}(a_1 + a_n)$$

➤ **Now Work** PROBLEM 39

EXAMPLE 8	**Creating a Floor Design**

A ceramic tile floor is designed in the shape of a trapezoid 20 feet wide at the base and 10 feet wide at the top. See Figure 7. The tiles, 12 inches by 12 inches, are to be placed so that each successive row contains one less tile than the preceding row. How many tiles will be required?

Figure 7

Solution The bottom row requires 20 tiles and the top row, 10 tiles. Since each successive row requires one less tile, the total number of tiles required is

$$S = 20 + 19 + 18 + \cdots + 11 + 10$$

This is the sum of an arithmetic sequence; the common difference is -1. The number of terms to be added is $n = 11$, with the first term $a_1 = 20$ and the last term $a_{11} = 10$. The sum S is

$$S = \frac{n}{2}(a_1 + a_{11}) = \frac{11}{2}(20 + 10) = 165$$

In all, 165 tiles will be required.

13.2 Assess Your Understanding

Concepts and Vocabulary

1. In a(n) _____ sequence, the difference between successive terms is a constant.

2. *True or False* In an arithmetic sequence the sum of the first and last terms equals twice the sum of all the terms.

Skill Building

In Problems 3–12, show that each sequence is arithmetic. Find the common difference and write out the first four terms.

3. $\{s_n\} = \{n + 4\}$
4. $\{s_n\} = \{n - 5\}$
5. $\{a_n\} = \{2n - 5\}$
6. $\{b_n\} = \{3n + 1\}$
7. $\{c_n\} = \{6 - 2n\}$

8. $\{a_n\} = \{4 - 2n\}$
9. $\{t_n\} = \left\{\frac{1}{2} - \frac{1}{3}n\right\}$
10. $\{t_n\} = \left\{\frac{2}{3} + \frac{n}{4}\right\}$
11. $\{s_n\} = \{\ln 3^n\}$
12. $\{s_n\} = \{e^{\ln n}\}$

In Problems 13–20, find the nth term of the arithmetic sequence $\{a_n\}$ whose initial term a and common difference d are given. What is the fifty-first term?

13. $a_1 = 2$; $d = 3$
14. $a_1 = -2$; $d = 4$
15. $a_1 = 5$; $d = -3$
16. $a_1 = 6$; $d = -2$

17. $a_1 = 0$; $d = \frac{1}{2}$
18. $a_1 = 1$; $d = -\frac{1}{3}$
19. $a_1 = \sqrt{2}$; $d = \sqrt{2}$
20. $a_1 = 0$; $d = \pi$

In Problems 21–26, find the indicated term in each arithmetic sequence.

21. 100th term of $2, 4, 6, \ldots$
22. 80th term of $-1, 1, 3, \ldots$
23. 90th term of $1, -2, -5, \ldots$

24. 80th term of $5, 0, -5, \ldots$
25. 80th term of $2, \frac{5}{2}, 3, \frac{7}{2}, \ldots$
26. 70th term of $2\sqrt{5}, 4\sqrt{5}, 6\sqrt{5}, \ldots$

In Problems 27–34, find the first term and the common difference of the arithmetic sequence described. Give a recursive formula for the sequence. Find a formula for the nth term.

27. 8th term is 8; 20th term is 44
28. 4th term is 3; 20th term is 35
29. 9th term is -5; 15th term is 31

30. 8th term is 4; 18th term is -96
31. 15th term is 0; 40th term is -50
32. 5th term is -2; 13th term is 30

33. 14th term is -1; 18th term is -9
34. 12th term is 4; 18th term is 28

In Problems 35–52, find each sum.

35. $1 + 3 + 5 + \cdots + (2n - 1)$
36. $2 + 4 + 6 + \cdots + 2n$
37. $7 + 12 + 17 + \cdots + (2 + 5n)$

38. $-1 + 3 + 7 + \cdots + (4n - 5)$
39. $2 + 4 + 6 + \cdots + 70$
40. $1 + 3 + 5 + \cdots + 59$

41. $5 + 9 + 13 + \cdots + 49$
42. $2 + 5 + 8 + \cdots + 41$
43. $73 + 78 + 83 + 88 + \cdots + 558$

44. $7 + 1 - 5 - 11 - \cdots - 299$
45. $4 + 4.5 + 5 + 5.5 + \cdots + 100$
46. $8 + 8\frac{1}{4} + 8\frac{1}{2} + 8\frac{3}{4} + 9 + \cdots + 50$

47. $\sum_{n=1}^{80}(2n - 5)$
48. $\sum_{n=1}^{90}(3 - 2n)$
49. $\sum_{n=1}^{100}\left(6 - \frac{1}{2}n\right)$
50. $\sum_{n=1}^{80}\left(\frac{1}{3}n + \frac{1}{2}\right)$

51. The sum of the first 120 terms of the sequence $14, 16, 18, 20, \ldots$.
52. The sum of the first 46 terms of the sequence $2, -1, -4, -7, \ldots$.

Applications and Extensions

53. Find x so that $x + 3, 2x + 1$, and $5x + 2$ are consecutive terms of an arithmetic sequence.

54. Find x so that $2x, 3x + 2$, and $5x + 3$ are consecutive terms of an arithmetic sequence.

55. Drury Lane Theater The Drury Lane Theater has 25 seats in the first row and 30 rows in all. Each successive row contains one additional seat. How many seats are in the theater?

56. Football Stadium The corner section of a football stadium has 15 seats in the first row and 40 rows in all. Each successive row contains two additional seats. How many seats are in this section?

57. Creating a Mosaic A mosaic is designed in the shape of an equilateral triangle, 20 feet on each side. Each tile in the mosaic is in the shape of an equilateral triangle, 12 inches to a side. The tiles are to alternate in color as shown in the illustration. How many tiles of each color will be required?

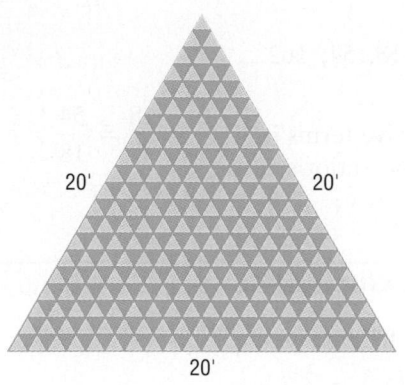

58. Constructing a Brick Staircase A brick staircase has a total of 30 steps. The bottom step requires 100 bricks. Each successive step requires two less bricks than the prior step.

(a) How many bricks are required for the top step?

(b) How many bricks are required to build the staircase?

59. Cooling Air As a parcel of air rises (for example, as it is pushed over a mountain), it cools at the *dry adiabatic lapse rate* of 5.5° F per 1000 feet until it reaches its dew point. If the ground temperature is 67° F, write a formula for the sequence of temperatures, $\{T_n\}$, of a parcel of air that has risen n thousand feet. What is the temperature of a parcel of air if it has risen 5000 feet?

Source: National Aeronautics and Space Administration

60. Citrus Ladders Ladders used by fruit pickers are typically tapered with a wide bottom for stability and a narrow top for ease of picking. Suppose the bottom rung of such a ladder is 49 inches wide and the top rung is 24 inches wide. How many rungs does the ladder have if each rung is 2.5 inches shorter than the one below it? How much material would be needed to make the rungs for the ladder described?

Source: www.stokesladders.com

61. Seats in an Amphitheater An outdoor amphitheater has 35 seats in the first row, 37 in the second row, 39 in the third row, and so on. There are 27 rows altogether. How many can the amphitheater seat?

62. Stadium Construction How many rows are in the corner section of a stadium containing 2040 seats if the first row has 10 seats and each successive row has 4 additional seats?

63. Salary Suppose that you just received a job offer with a starting salary of $35,000 per year and a guaranteed raise of $1400 per year. How many years will it take before your aggregate salary is $280,000?

[**Hint:** Your aggregate salary after 2 years is $35,000 + ($35,000 + $1400).]

Discussion and Writing

64. Make up an arithmetic sequence. Give it to a friend and ask for its twentieth term.

65. Describe the similarities and differences between arithmetic sequences and linear functions.

13.3 Geometric Sequences; Geometric Series

PREPARING FOR THIS SECTION *Before getting started, review the following:*

• Compound Interest (Section 6.7, pp. 465–472)

Now Work the 'Are You Prepared?' problems on page 954.

OBJECTIVES **1** Determine If a Sequence Is Geometric (p. 946)

2 Find a Formula for a Geometric Sequence (p. 947)

3 Find the Sum of a Geometric Sequence (p. 948)

4 Determine whether a Geometric Series Converges or Diverges (p. 949)

5 Solve Annuity Problems (p. 952)

1 Determine If a Sequence Is Geometric

When the ratio of successive terms of a sequence is always the same nonzero number, the sequence is called **geometric.**

DEFINITION

A **geometric sequence*** may be defined recursively as $a_1 = a$, $\dfrac{a_n}{a_{n-1}} = r$, or as

$$a_1 = a, \qquad a_n = ra_{n-1} \qquad\qquad \textbf{(1)}$$

where $a_1 = a$ and $r \neq 0$ are real numbers. The number a_1 is the first term, and the nonzero number r is called the **common ratio.**

The terms of a geometric sequence with first term a_1 and common ratio r follow the pattern

$$a_1, \quad a_1 r, \quad a_1 r^2, \quad a_1 r^3, \ldots$$

EXAMPLE 1 | **Determining If a Sequence Is Geometric**

The sequence

$$2, \ 6, \ 18, \ 54, \ 162, \ldots$$

is geometric since the ratio of successive terms is 3; $\left(\dfrac{6}{2} = \dfrac{18}{6} = \dfrac{54}{18} = \cdots = 3 \right)$. The first term is $a_1 = 2$, and the common ratio is 3.

EXAMPLE 2 | **Determining If a Sequence Is Geometric**

Show that the following sequence is geometric.

$$\{s_n\} = 2^{-n}$$

Find the first term and the common ratio.

Solution

The first term is $s_1 = 2^{-1} = \dfrac{1}{2}$. The nth term and the $(n - 1)$st term of the sequence $\{s_n\}$ are

$$s_n = 2^{-n} \quad \text{and} \quad s_{n-1} = 2^{-(n-1)}$$

Their ratio is

$$\frac{s_n}{s_{n-1}} = \frac{2^{-n}}{2^{-(n-1)}} = 2^{-n+(n-1)} = 2^{-1} = \frac{1}{2}$$

Because the ratio of successive terms is the nonzero constant $\dfrac{1}{2}$, the sequence $\{s_n\}$ is geometric with common ratio $\dfrac{1}{2}$.

EXAMPLE 3 | **Determining If a Sequence Is Geometric**

Show that the following sequence is geometric.

$$\{t_n\} = \{4^n\}$$

Find the first term and the common ratio.

*Sometimes called a **geometric progression.**

Solution The first term is $t_1 = 4^1 = 4$. The nth term and the $(n - 1)$st term are

$$t_n = 4^n \quad \text{and} \quad t_{n-1} = 4^{n-1}$$

Their ratio is

$$\frac{t_n}{t_{n-1}} = \frac{4^n}{4^{n-1}} = 4^{n-(n-1)} = 4$$

The sequence, $\{t_n\}$, is a geometric sequence with common ratio 4.

⬛

━━━━━▶ **Now Work** PROBLEM 11

2 Find a Formula for a Geometric Sequence

Suppose that a_1 is the first term of a geometric sequence with common ratio $r \neq 0$. We seek a formula for the nth term a_n. To see the pattern, we write down the first few terms:

$$a_1 = a_1 \cdot 1 = a_1 r^0$$
$$a_2 = ra_1 = a_1 r^1$$
$$a_3 = ra_2 = r(a_1 r) = a_1 r^2$$
$$a_4 = ra_3 = r(a_1 r^2) = a_1 r^3$$
$$a_5 = ra_4 = r(a_1 r^3) = a_1 r^4$$
$$\vdots$$
$$a_n = ra_{n-1} = r(a_1 r^{n-2}) = a_1 r^{n-1}$$

We are led to the following result:

THEOREM

nth Term of a Geometric Sequence

For a geometric sequence $\{a_n\}$ whose first term is a_1 and whose common ratio is r, the nth term is determined by the formula

$$a_n = a_1 r^{n-1} \qquad r \neq 0 \tag{2}$$

EXAMPLE 4 **Finding a Particular Term of a Geometric Sequence**

(a) Find the nth term of the geometric sequence: $10, 9, \dfrac{81}{10}, \dfrac{729}{100} \ldots$

(b) Find the ninth term of this sequence.

(c) Find a recursive formula for this sequence.

Solution (a) The first term of this geometric sequence is $a_1 = 10$ and the common ratio is $\dfrac{9}{10}$. (Use $\dfrac{9}{10}$ or $\dfrac{\frac{81}{10}}{9} = \dfrac{9}{10}$ or any two successive terms.) Then by formula (2), the nth term is

$$a_n = 10 \left(\frac{9}{10} \right)^{n-1} \qquad a_n = a_1 r^{n-1}; \, a_1 = 10, \, r = \frac{9}{10}$$

Exploration

Use a graphing utility to find the ninth term of the sequence given in Example 4. Use it to find the twentieth and fiftieth terms. Now use a graphing utility to graph the recursive formula found in Example 4(c). Conclude that the graph of the recursive formula behaves like the graph of an exponential function. How is r, the common ratio, related to a, the base of the exponential function $y = a^x$?

(b) The ninth term is

$$a_9 = 10\left(\frac{9}{10}\right)^{9-1} = 10\left(\frac{9}{10}\right)^8 \approx 4.3046721$$

(c) The first term in the sequence is 10 and the common ratio is $r = \dfrac{9}{10}$. Using formula (1), the recursive formula is $a_1 = 10$, $a_n = \dfrac{9}{10}a_{n-1}$.

Now Work PROBLEMS 33, 41, AND 49

3 Find the Sum of a Geometric Sequence

The next result gives us a formula for finding the sum of the first n terms of a geometric sequence.

THEOREM | **Sum of the First n Terms of a Geometric Sequence**

Let $\{a_n\}$ be a geometric sequence with first term a_1 and common ratio r, where $r \neq 0, r \neq 1$. The sum S_n of the first n terms of $\{a_n\}$ is

$$S_n = a_1 + a_1r + a_1r^2 + \cdots + a_1r^{n-1}$$

$$= \sum_{k=1}^{n}a_1r^{k-1} = a_1 \cdot \frac{1-r^n}{1-r} \qquad r \neq 0, 1 \qquad \text{(3)}$$

Proof The sum S_n of the first n terms of $\{a_n\} = \{a_1r^{n-1}\}$ is

$$S_n = a_1 + a_1r + \cdots + a_1r^{n-1} \qquad \text{(4)}$$

Multiply each side by r to obtain

$$rS_n = a_1r + a_1r^2 + \cdots + a_1r^n \qquad \text{(5)}$$

Now, subtract (5) from (4). The result is

$$S_n - rS_n = a_1 - a_1r^n$$
$$(1-r)S_n = a_1(1-r^n)$$

Since $r \neq 1$, we can solve for S_n.

$$S_n = a_1 \cdot \frac{1-r^n}{1-r} \qquad \blacksquare$$

EXAMPLE 5 | **Finding the Sum of the First n Terms of a Geometric Sequence**

Find the sum S_n of the first n terms of the sequence $\left\{\left(\dfrac{1}{2}\right)^n\right\}$; that is, find

$$\frac{1}{2} + \frac{1}{4} + \frac{1}{8} + \cdots + \left(\frac{1}{2}\right)^n = \sum_{k=1}^{n}\frac{1}{2}\left(\frac{1}{2}\right)^{k-1}$$

Solution The sequence $\left\{\left(\dfrac{1}{2}\right)^n\right\}$ is a geometric sequence with $a_1 = \dfrac{1}{2}$ and $r = \dfrac{1}{2}$. So we use

formula (3) to get

$$S_n = \sum_{k=1}^{n} \frac{1}{2}\left(\frac{1}{2}\right)^{k-1} = \frac{1}{2} + \frac{1}{4} + \frac{1}{8} + \cdots + \left(\frac{1}{2}\right)^n$$

$$= \frac{1}{2}\left[\frac{1 - \left(\frac{1}{2}\right)^n}{1 - \frac{1}{2}}\right] \qquad \text{Formula (3); } a_1 = \frac{1}{2},\ r = \frac{1}{2}$$

$$= \frac{1}{2}\left[\frac{1 - \left(\frac{1}{2}\right)^n}{\frac{1}{2}}\right]$$

$$= 1 - \left(\frac{1}{2}\right)^n$$

━━━━━ **Now Work** PROBLEM 55

EXAMPLE 6 **Using a Graphing Utility to Find the Sum of a Geometric Sequence**

Use a graphing utility to find the sum of the first 15 terms of the sequence $\left\{\left(\frac{1}{3}\right)^n\right\}$; that is, find

$$\frac{1}{3} + \frac{1}{9} + \frac{1}{27} + \cdots + \left(\frac{1}{3}\right)^{15} = \sum_{k=1}^{15} \frac{1}{3}\left(\frac{1}{3}\right)^{k-1}$$

Figure 8

```
sum(seq((1/3)^n,
n,1,15,1))
      .4999999652
```

Solution Figure 8 shows the result obtained using a TI-84 Plus graphing calculator. The sum of the first 15 terms of the sequence $\left\{\left(\frac{1}{3}\right)^n\right\}$ is 0.4999999652.

━━━━━ **Now Work** PROBLEM 61

4 Determine whether a Geometric Series Converges or Diverges

DEFINITION An infinite sum of the form

$$a_1 + a_1 r + a_1 r^2 + \cdots + a_1 r^{n-1} + \cdots$$

with first term a_1 and common ratio r, is called an **infinite geometric series** and is denoted by

$$\sum_{k=1}^{\infty} a_1 r^{k-1}$$

Based on formula (3), the sum S_n of the first n terms of a geometric series is

$$S_n = a_1 \cdot \frac{1 - r^n}{1 - r} = \frac{a_1}{1 - r} - \frac{a_1 r^n}{1 - r} \qquad \textbf{(6)}$$

NOTE In calculus, we use limit notation and write

$$L = \lim_{n \to \infty} S_n = \lim_{n \to \infty} \sum_{k=1}^{n} a_1 r^{k-1} = \sum_{k=1}^{\infty} a_1 r^{k-1}$$
∎

If this finite sum S_n approaches a number L as $n \to \infty$, we say the infinite geometric series $\sum_{k=1}^{\infty} a_1 r^{k-1}$ **converges.** We call L the **sum of the infinite geometric series.** and we write

$$L = \sum_{k=1}^{\infty} a_1 r^{k-1}$$

If a series does not converge, it is called a **divergent series.**

THEOREM

Convergence of an Infinite Geometric Series

If $|r| < 1$, the infinite geometric series $\sum_{k=1}^{\infty} a_1 r^{k-1}$ converges. Its sum is

$$\sum_{k=1}^{\infty} a_1 r^{k-1} = \frac{a_1}{1 - r} \tag{7}$$

Intuitive Proof Since $|r| < 1$, it follows that $|r^n|$ approaches 0 as $n \to \infty$. Then, based on formula (6), the term $\dfrac{a_1 r^n}{1 - r}$ approaches 0, so the sum S_n approaches $\dfrac{a_1}{1 - r}$ as $n \to \infty$. ∎

EXAMPLE 7 **Determining whether a Geometric Series Converges or Diverges**

Determine if the geometric series

$$\sum_{k=1}^{\infty} 2 \left(\frac{2}{3} \right)^{k-1} = 2 + \frac{4}{3} + \frac{8}{9} + \cdots$$

converges or diverges. If it converges, find its sum.

Solution Comparing $\sum_{k=1}^{\infty} 2 \left(\dfrac{2}{3} \right)^{k-1}$ to $\sum_{k=1}^{\infty} a_1 r^{k-1}$, we see that the first term is $a_1 = 2$, and the common ratio is $r = \dfrac{2}{3}$. Since $|r| < 1$, the series converges. We use formula (7) to find its sum:

$$\sum_{k=1}^{\infty} 2 \left(\frac{2}{3} \right)^{k-1} = 2 + \frac{4}{3} + \frac{8}{9} + \cdots = \frac{2}{1 - \dfrac{2}{3}} = 6$$

Now Work PROBLEM 67

EXAMPLE 8 **Repeating Decimals**

Show that the repeating decimal $0.999\ldots$ equals 1.

Solution The decimal $0.999\ldots = 0.9 + 0.09 + 0.009 + \cdots = \dfrac{9}{10} + \dfrac{9}{100} + \dfrac{9}{1000} + \cdots$ is an infinite geometric series. We will write it in the form $\sum_{k=1}^{\infty} a_1 r^{k-1}$ so we can use formula (7).

$$0.999\ldots = \frac{9}{10} + \frac{9}{100} + \frac{9}{1000} + \cdots = \sum_{k=1}^{\infty} \frac{9}{10^k} = \sum_{k=1}^{\infty} \frac{9}{10 \cdot 10^{k-1}} = \sum_{k=1}^{\infty} \frac{9}{10} \left(\frac{1}{10} \right)^{k-1}$$

Now we can compare this series to $\sum_{k=1}^{\infty} a_1 r^{k-1}$ and conclude that $a_1 = \dfrac{9}{10}$ and $r = \dfrac{1}{10}$.

Since $|r| < 1$, the series converges and its sum is

$$0.999\ldots = \frac{\dfrac{9}{10}}{1 - \dfrac{1}{10}} = \frac{\dfrac{9}{10}}{\dfrac{9}{10}} = 1$$

The repeating decimal $0.999\ldots$ equals 1.

EXAMPLE 9 **Pendulum Swings**

Figure 9

18"

Initially, a pendulum swings through an arc of 18 inches. See Figure 9. On each successive swing, the length of the arc is 0.98 of the previous length.

(a) What is the length of the arc of the 10th swing?
(b) On which swing is the length of the arc first less than 12 inches?
(c) After 15 swings, what total distance will the pendulum have swung?
(d) When it stops, what total distance will the pendulum have swung?

Solution (a) The length of the first swing is 18 inches.
The length of the second swing is 0.98(18) inches.
The length of the third swing is $0.98(0.98)(18) = 0.98^2(18)$ inches.
The length of the arc of the 10th swing is

$$(0.98)^9(18) \approx 15.007 \text{ inches}$$

(b) The length of the arc of the nth swing is $(0.98)^{n-1}(18)$. For this to be exactly 12 inches requires that

$$(0.98)^{n-1}(18) = 12$$

$$(0.98)^{n-1} = \frac{12}{18} = \frac{2}{3} \qquad \text{\small Divide both sides by 18.}$$

$$n - 1 = \log_{0.98}\left(\frac{2}{3}\right) \qquad \text{\small Express as a logarithm.}$$

$$n = 1 + \frac{\ln\left(\dfrac{2}{3}\right)}{\ln 0.98} \approx 1 + 20.07 = 21.07 \quad \text{\small Solve for n; use the Change of Base Formula.}$$

The length of the arc of the pendulum exceeds 12 inches on the 21st swing and is first less than 12 inches on the 22nd swing.

(c) After 15 swings, the pendulum will have swung the following total distance L:

$$L = \underset{\text{1st}}{18} + \underset{\text{2nd}}{0.98(18)} + \underset{\text{3rd}}{(0.98)^2(18)} + \underset{\text{4th}}{(0.98)^3(18)} + \cdots + \underset{\text{15th}}{(0.98)^{14}(18)}$$

This is the sum of a geometric sequence. The common ratio is 0.98; the first term is 18. The sum has 15 terms, so

$$L = 18 \cdot \frac{1 - 0.98^{15}}{1 - 0.98} \approx 18(13.07) \approx 235.3 \text{ inches}$$

The pendulum will have swung through 235.3 inches after 15 swings.

(d) When the pendulum stops, it will have swung the following total distance T:

$$T = 18 + 0.98(18) + (0.98)^2(18) + (0.98)^3(18) + \cdots$$

This is the sum of an infinite geometric series. The common ratio is $r = 0.98$; the first term is $a_1 = 18$. Since $|r| < 1$, the series converges. Its sum is

$$T = \frac{a_1}{1 - r} = \frac{18}{1 - 0.98} = 900$$

The pendulum will have swung a total of 900 inches when it finally stops.

━━━━━ **Now Work** PROBLEM 87

5 Solve Annuity Problems

In Section 6.7 we developed the compound interest formula that gives the future value when a fixed amount of money is deposited in an account that pays interest compounded periodically. Often, though, money is invested in small amounts at periodic intervals. An **annuity** is a sequence of equal periodic deposits. The periodic deposits may be made annually, quarterly, monthly, or daily.

When deposits are made at the same time that the interest is credited, the annuity is called **ordinary.** We will only deal with ordinary annuities here. The **amount of an annuity** is the sum of all deposits made plus all interest paid.

Suppose the interest rate that an account earns is i percent per payment period (expressed as a decimal). For example, if an account pays 12% compounded monthly (12 times a year), then $i = \dfrac{0.12}{12} = 0.01$. If an account pays 8% compounded quarterly (4 times a year), then $i = \dfrac{0.08}{4} = 0.02$.

To develop a formula for the amount of an annuity, suppose that P is deposited each payment period for n payment periods in an account that earns i percent per payment period. When the last deposit is made at the nth payment period, the first deposit of P has earned interest compounded for $n - 1$ payment periods, the second deposit of P has earned interest compounded for $n - 2$ payment periods, and so on. Table 3 shows the value of each deposit after n deposits have been made.

Table 3

Deposit	1	2	3	...	$n - 1$	n
Amount	$P(1 + i)^{n-1}$	$P(1 + i)^{n-2}$	$P(1 + i)^{n-3}$	...	$P(1 + i)$	P

The amount A of the annuity is the sum of the amounts shown in Table 3; that is,

$$A = P \cdot (1 + i)^{n-1} + P \cdot (1 + i)^{n-2} + \cdots + P \cdot (1 + i) + P$$
$$= P[1 + (1 + i) + \cdots + (1 + i)^{n-1}]$$

The expression in brackets is the sum of a geometric sequence with n terms and a common ratio of $(1 + i)$. As a result,

$$A = P[1 + (1 + i) + \cdots + (1 + i)^{n-2} + (1 + i)^{n-1}]$$
$$= P\frac{1 - (1 + i)^n}{1 - (1 + i)} = P\frac{1 - (1 + i)^n}{-i} = P\frac{(1 + i)^n - 1}{i}$$

We have established the following result:

THEOREM

Amount of an Annuity

Suppose P is the deposit in dollars made at the end of each payment period for an annuity paying i percent interest per payment period. The amount A of the annuity after n deposits is

$$A = P\frac{(1 + i)^n - 1}{i} \tag{8}$$

NOTE: In using formula (8), remember that when the nth deposit is made, the first deposit has earned interest for $n - 1$ compounding periods.

EXAMPLE 10	Determining the Amount of an Annuity

To save for retirement, Brett decides to place $2000 into an Individual Retirement Account (IRA) each year for the next 30 years. What will the value of the IRA be when Brett makes his 30th deposit? Assume that the rate of return of the IRA is 10% per annum compounded annually.

Solution This is an ordinary annuity with $n = 30$ annual deposits of $P = \$2000$. The rate of interest per payment period is $i = \dfrac{0.10}{1} = 0.10$. The amount A of the annuity after 30 deposits is

$$A = 2000\frac{(1 + 0.10)^{30} - 1}{0.10} = 2000(164.494023) = \$328,988.05$$

EXAMPLE 11	Determining the Amount of an Annuity

To save for her daughter's college education, Ms. Miranda decides to put $50 aside every month in a credit union account paying 10% interest compounded monthly. She begins this savings program when her daughter is 3 years old. How much will she have saved by the time she makes the 180th deposit? How old is her daughter at this time?

Solution This is an annuity with $P = \$50$, $n = 180$, and $i = \dfrac{0.10}{12}$. The amount A saved is

$$A = 50\frac{\left(1 + \dfrac{0.10}{12}\right)^{180} - 1}{\dfrac{0.10}{12}} = 50(414.47035) = \$20,723.52$$

Since there are 12 deposits per year, when the 180th deposit is made $\dfrac{180}{12} = 15$ years have passed and Ms. Miranda's daughter is 18 years old.

Now Work PROBLEM 91

Historical Feature

Fibonacci

Sequences are among the oldest objects of mathematical investigation, having been studied for over 3500 years. After the initial steps, however, little progress was made until about 1600.

Arithmetic and geometric sequences appear in the Rhind papyrus, a mathematical text containing 85 problems copied around 1650 BC by the Egyptian scribe Ahmes from an earlier work (see Historical Problem 1). Fibonacci (AD 1220) wrote about problems similar to those found in the Rhind papyrus, leading one to suspect that Fibonacci may have had material available that is now lost. This material would have been in the non-Euclidean Greek tradition of Heron (about AD 75) and Diophantus (about AD 250). One problem, again modified slightly, is still with us in the familiar puzzle rhyme "As I was going to St. Ives ... " (see Historical Problem 2).

The Rhind papyrus indicates that the Egyptians knew how to add up the terms of an arithmetic or geometric sequence, as did the Babylonians. The rule for summing up a geometric sequence is found in Euclid's *Elements* (Book IX, 35, 36), where, like all Euclid's algebra, it is presented in a geometric form.

Investigations of other kinds of sequences began in the 1500s, when algebra became sufficiently developed to handle the more complicated problems. The development of calculus in the 1600s added a powerful new tool, especially for finding the sum of infinite series, and the subject continues to flourish today.

(Continued)

Historical Problems

1. *Arithmetic sequence problem from the Rhind papyrus (statement modified slightly for clarity)* One hundred loaves of bread are to be divided among five people so that the amounts that they receive form an arithmetic sequence. The first two together receive one-seventh of what the last three receive. How many loaves does each receive?

 [*Partial answer:* First person receives $1\frac{2}{3}$ loaves.]

2. The following old English children's rhyme resembles one of the Rhind papyrus problems.

 > As I was going to St. Ives
 > I met a man with seven wives

 Each wife had seven sacks
 Each sack had seven cats
 Each cat had seven kits [kittens]
 Kits, cats, sacks, wives
 How many were going to St. Ives?

 (a) Assuming that the speaker and the cat fanciers met by traveling in opposite directions, what is the answer?

 (b) How many kittens are being transported?

 (c) Kits, cats, sacks, wives; how many?

13.3 Assess Your Understanding

'Are You Prepared?' *Answers are given at the end of these exercises. If you get a wrong answer, read the pages listed in red.*

1. If $1000 is invested at 4% per annum compounded semiannually, how much is in the account after two years? (pp. 465–472)

2. How much do you need to invest now at 5% per annum compounded monthly so that in 1 year you will have $10,000? (pp. 465–472)

Concepts and Vocabulary

3. In a(n) _____ sequence the ratio of successive terms is a constant.

4. If $|r| < 1$, the sum of the geometric series $\sum\limits_{k=1}^{\infty} ar^{k-1}$ is _____.

5. A sequence of equal periodic deposits is called a(n) _____.

6. *True or False* A geometric sequence may be defined recursively.

7. *True or False* In a geometric sequence the common ratio is always a positive number.

8. *True or False* For a geometric sequence with first term a_1 and common ratio r, where $r \neq 0, r \neq 1$, the sum of the first n terms is $S_n = a_1 \cdot \dfrac{1 - r^n}{1 - r}$.

Skill Building

In Problems 9–18, a geometric sequence is given. Find the common ratio and write out the first four terms.

9. $\{s_n\} = \{3^n\}$

10. $\{s_n\} = \{(-5)^n\}$

11. $\{a_n\} = \left\{-3\left(\dfrac{1}{2}\right)^n\right\}$

12. $\{b_n\} = \left\{\left(\dfrac{5}{2}\right)^n\right\}$

13. $\{c_n\} = \left\{\dfrac{2^{n-1}}{4}\right\}$

14. $\{d_n\} = \left\{\dfrac{3^n}{9}\right\}$

15. $\{e_n\} = \{2^{n/3}\}$

16. $\{f_n\} = \{3^{2n}\}$

17. $\{t_n\} = \left\{\dfrac{3^{n-1}}{2^n}\right\}$

18. $\{u_n\} = \left\{\dfrac{2^n}{3^{n-1}}\right\}$

In Problems 19–32, determine whether the given sequence is arithmetic, geometric, or neither. If the sequence is arithmetic, find the common difference; if it is geometric, find the common ratio.

19. $\{n + 2\}$

20. $\{2n - 5\}$

21. $\{4n^2\}$

22. $\{5n^2 + 1\}$

23. $\left\{3 - \dfrac{2}{3}n\right\}$

24. $\left\{8 - \dfrac{3}{4}n\right\}$

25. $1, 3, 6, 10, \ldots$

26. $2, 4, 6, 8, \ldots$

27. $\left\{\left(\dfrac{2}{3}\right)^n\right\}$

28. $\left\{\left(\dfrac{5}{4}\right)^n\right\}$

29. $-1, -2, -4, -8, \ldots$

30. $1, 1, 2, 3, 5, 8, \ldots$

31. $\{3^{n/2}\}$

32. $\{(-1)^n\}$

In Problems 33–40, find the fifth term and the nth term of the geometric sequence whose initial term a_1 and common ratio r are given.

33. $a_1 = 2;\quad r = 3$

34. $a_1 = -2;\quad r = 4$

35. $a_1 = 5;\quad r = -1$

36. $a_1 = 6;\quad r = -2$

37. $a_1 = 0;\quad r = \dfrac{1}{2}$

38. $a_1 = 1;\quad r = -\dfrac{1}{3}$

39. $a_1 = \sqrt{2};\quad r = \sqrt{2}$

40. $a_1 = 0;\quad r = \dfrac{1}{\pi}$

In Problems 41–46, find the indicated term of each geometric sequence.

41. 7th term of $1, \dfrac{1}{2}, \dfrac{1}{4}, \ldots$

42. 8th term of $1, 3, 9, \ldots$

43. 9th term of $1, -1, 1, \ldots$

44. 10th term of $-1, 2, -4, \ldots$

45. 8th term of $0.4, 0.04, 0.004, \ldots$

46. 7th term of $0.1, 1.0, 10.0, \ldots$

In Problems 47–54, find the nth term a_n of each geometric sequence. When given, r is the common ratio.

47. $7, 14, 28, 56, \ldots$

48. $5, 10, 20, 40, \ldots$

49. $-3, 1, -\dfrac{1}{3}, \dfrac{1}{9}, \ldots$

50. $4, 1, \dfrac{1}{4}, \dfrac{1}{16}, \ldots$

51. $a_6 = 243; \quad r = -3$

52. $a_2 = 7; \quad r = \dfrac{1}{3}$

53. $a_2 = 7; \quad a_4 = 1575$

54. $a_3 = \dfrac{1}{3}; \quad a_6 = \dfrac{1}{81}$

In Problems 55–60, find each sum.

55. $\dfrac{1}{4} + \dfrac{2}{4} + \dfrac{2^2}{4} + \dfrac{2^3}{4} + \cdots + \dfrac{2^{n-1}}{4}$

56. $\dfrac{3}{9} + \dfrac{3^2}{9} + \dfrac{3^3}{9} + \cdots + \dfrac{3^n}{9}$

57. $\displaystyle\sum_{k=1}^{n} \left(\dfrac{2}{3}\right)^k$

58. $\displaystyle\sum_{k=1}^{n} 4 \cdot 3^{k-1}$

59. $-1 - 2 - 4 - 8 - \cdots - (2^{n-1})$

60. $2 + \dfrac{6}{5} + \dfrac{18}{25} + \cdots + 2\left(\dfrac{3}{5}\right)^{n-1}$

For Problems 61–66, use a graphing utility to find the sum of each geometric sequence.

61. $\dfrac{1}{4} + \dfrac{2}{4} + \dfrac{2^2}{4} + \dfrac{2^3}{4} + \cdots + \dfrac{2^{14}}{4}$

62. $\dfrac{3}{9} + \dfrac{3^2}{9} + \dfrac{3^3}{9} + \cdots + \dfrac{3^{15}}{9}$

63. $\displaystyle\sum_{n=1}^{15} \left(\dfrac{2}{3}\right)^n$

64. $\displaystyle\sum_{n=1}^{15} 4 \cdot 3^{n-1}$

65. $-1 - 2 - 4 - 8 - \cdots - 2^{14}$

66. $2 + \dfrac{6}{5} + \dfrac{18}{25} + \cdots + 2\left(\dfrac{3}{5}\right)^{15}$

In Problems 67–82, determine whether each infinite geometric series converges or diverges. If it converges, find its sum.

67. $1 + \dfrac{1}{3} + \dfrac{1}{9} + \cdots$

68. $2 + \dfrac{4}{3} + \dfrac{8}{9} + \cdots$

69. $8 + 4 + 2 + \cdots$

70. $6 + 2 + \dfrac{2}{3} + \cdots$

71. $2 - \dfrac{1}{2} + \dfrac{1}{8} - \dfrac{1}{32} + \cdots$

72. $1 - \dfrac{3}{4} + \dfrac{9}{16} - \dfrac{27}{64} + \cdots$

73. $8 + 12 + 18 + 27 + \cdots$

74. $9 + 12 + 16 + \dfrac{64}{3} + \cdots$

75. $\displaystyle\sum_{k=1}^{\infty} 5\left(\dfrac{1}{4}\right)^{k-1}$

76. $\displaystyle\sum_{k=1}^{\infty} 8\left(\dfrac{1}{3}\right)^{k-1}$

77. $\displaystyle\sum_{k=1}^{\infty} \dfrac{1}{2} 3^{k-1}$

78. $\displaystyle\sum_{k=1}^{\infty} 3\left(\dfrac{3}{2}\right)^{k-1}$

79. $\displaystyle\sum_{k=1}^{\infty} 6\left(-\dfrac{2}{3}\right)^{k-1}$

80. $\displaystyle\sum_{k=1}^{\infty} 4\left(-\dfrac{1}{2}\right)^{k-1}$

81. $\displaystyle\sum_{k=1}^{\infty} 3\left(\dfrac{2}{3}\right)^k$

82. $\displaystyle\sum_{k=1}^{\infty} 2\left(\dfrac{3}{4}\right)^k$

Applications and Extensions

83. Find x so that x, $x + 2$, and $x + 3$ are consecutive terms of a geometric sequence.

84. Find x so that $x - 1$, x, and $x + 2$ are consecutive terms of a geometric sequence.

85. **Salary Increases** Suppose that you have just been hired at an annual salary of \$18,000 and expect to receive annual increases of 5%. What will your salary be when you begin your fifth year?

86. **Equipment Depreciation** A new piece of equipment cost a company \$15,000. Each year, for tax purposes, the company depreciates the value by 15%. What value should the company give the equipment after 5 years?

87. **Pendulum Swings** Initially, a pendulum swings through an arc of 2 feet. On each successive swing, the length of the arc is 0.9 of the previous length.
 (a) What is the length of the arc of the 10th swing?
 (b) On which swing is the length of the arc first less than 1 foot?
 (c) After 15 swings, what total length will the pendulum have swung?
 (d) When it stops, what total length will the pendulum have swung?

88. **Bouncing Balls** A ball is dropped from a height of 30 feet. Each time it strikes the ground, it bounces up to 0.8 of the previous height.

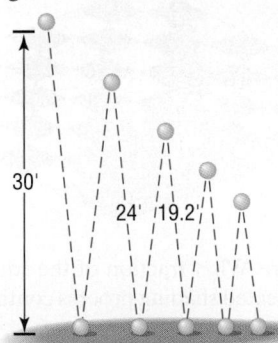

30' 24' 19.2'

 (a) What height will the ball bounce up to after it strikes the ground for the third time?
 (b) How high will it bounce after it strikes the ground for the nth time?
 (c) How many times does the ball need to strike the ground before its bounce is less than 6 inches?
 (d) What total distance does the ball travel before it stops bouncing?

89. Retirement Christine contributes $100 each month to her 401(k). What will be the value of Christine's 401(k) after the 360th deposit (30 years) if the per annum rate of return is assumed to be 12% compounded monthly?

90. Saving for a Home Jolene wants to purchase a new home. Suppose that she invests $400 per month into a mutual fund. If the per annum rate of return of the mutual fund is assumed to be 10% compounded monthly, how much will Jolene have for a down payment after the 36th deposit (3 years)?

91. Tax Sheltered Annuity Don contributes $500 at the end of each quarter to a tax sheltered annuity (TSA). What will the value of the TSA be after the 80th deposit (20 years) if the per annum rate of return is assumed to be 8% compounded quarterly?

92. Retirement Ray contributes $1000 to an Individual Retirement Account (IRA) semiannually. What will the value of the IRA be when Ray makes his 30th deposit (after 15 years) if the per annum rate of return is assumed to be 10% compounded semiannually?

93. Sinking Fund Scott and Alice want to purchase a vacation home in 10 years and need $50,000 for a down payment. How much should they place in a savings account each month if the per annum rate of return is assumed to be 6% compounded monthly?

94. Sinking Fund For a child born in 1996, a 4-year college education at a public university is projected to be $150,000. Assuming an 8% per annum rate of return compounded monthly, how much must be contributed to a college fund every month to have $150,000 in 18 years when the child begins college?

95. Grains of Wheat on a Chess Board In an old fable, a commoner who had saved the king's life was told he could ask the king for any just reward. Being a shrewd man, the commoner said, "A simple wish, sire. Place one grain of wheat on the first square of a chessboard, two grains on the second square, four grains on the third square, continuing until you have filled the board. This is all I seek." Compute the total number of grains needed to do this to see why the request, seemingly simple, could not be granted. (A chessboard consists of 8 × 8 = 64 squares.)

96. Look at the figure. What fraction of the square is eventually shaded if the indicated shading process continues indefinitely?

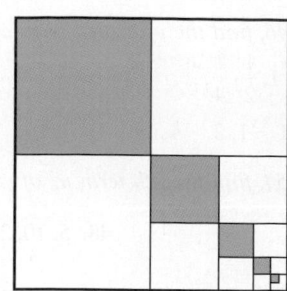

97. Multiplier Suppose that, throughout the U.S. economy, individuals spend 90% of every additional dollar that they earn. Economists would say that an individual's **marginal propensity to consume** is 0.90. For example, if Jane earns an additional dollar, she will spend $0.9(1) = \$0.90$ of it. The individual that earns $0.90 (from Jane) will spend 90% of it or $0.81. This process of spending continues and results in an infinite geometric series as follows:

$$1, 0.90, 0.90^2, 0.90^3, 0.90^4, \ldots$$

The sum of this infinite geometric series is called the **multiplier**. What is the multiplier if individuals spend 90% of every additional dollar that they earn?

98. Multiplier Refer to Problem 97. Suppose that the marginal propensity to consume throughout the U.S. economy is 0.95. What is the multiplier for the U.S. economy?

99. Stock Price One method of pricing a stock is to discount the stream of future dividends of the stock. Suppose that a stock pays $P per year in dividends and, historically, the dividend has been increased i% per year. If you desire an annual rate of return of r%, this method of pricing a stock states that the price that you should pay is the present value of an infinite stream of payments:

$$\text{Price} = P + P\frac{1+i}{1+r} + P\left(\frac{1+i}{1+r}\right)^2 + P\left(\frac{1+i}{1+r}\right)^3 + \cdots$$

The price of the stock is the sum of an infinite geometric series. Suppose that a stock pays an annual dividend of $4.00 and, historically, the dividend has been increased 3% per year. You desire an annual rate of return of 9%. What is the most you should pay for the stock?

100. Stock Price Refer to Problem 99. Suppose that a stock pays an annual dividend of $2.50 and, historically, the dividend has increased 4% per year. You desire an annual rate of return of 11%. What is the most that you should pay for the stock?

101. A Rich Man's Promise A rich man promises to give you $1000 on September 1, 2007. Each day thereafter he will give you $\dfrac{9}{10}$ of what he gave you the previous day. What is the first date on which the amount you receive is less than 1¢? How much have you received when this happens?

Discussion and Writing

102. Critical Thinking You are interviewing for a job and receive two offers:

 A: $20,000 to start, with guaranteed annual increases of 6% for the first 5 years

 B: $22,000 to start, with guaranteed annual increases of 3% for the first 5 years

Which offer is best if your goal is to be making as much as possible after 5 years? Which is best if your goal is to make as much money as possible over the contract (5 years)?

103. Critical Thinking Which of the following choices, *A* or *B*, results in more money?

A: To receive $1000 on day 1, $999 on day 2, $998 on day 3, with the process to end after 1000 days

B: To receive $1 on day 1, $2 on day 2, $4 on day 3, for 19 days

104. **Critical Thinking** You have just signed a 7-year professional football league contract with a beginning salary of $2,000,000 per year. Management gives you the following options with regard to your salary over the 7 years.

1. A bonus of $100,000 each year

2. An annual increase of 4.5% per year beginning after 1 year

3. An annual increase of $95,000 per year beginning after 1 year

Which option provides the most money over the 7-year period? Which the least? Which would you choose? Why?

105. **Critical Thinking** Suppose you were offered a job in which you would work 8 hours per day for 5 workdays per week for 1 month at hard manual labor. Your pay the first day would

be 1 penny. On the second day your pay would be two pennies; the third day 4 pennies. Your pay would double on each successive workday. There are 22 workdays in the month. There will be no sick days. If you miss a day of work, there is no pay or pay increase. How much would you get paid if you work all 22 days? How much do you get paid for the 22nd workday? What risks do you run if you take this job offer? Would you take the job?

106. Can a sequence be both arithmetic and geometric? Give reasons for your answer.

107. Make up a geometric sequence. Give it to a friend and ask for its 20th term.

108. Make up two infinite geometric series, one that has a sum and one that does not. Give them to a friend and ask for the sum of each series.

109. Describe the similarities and differences between geometric sequences and exponential functions.

'Are You Prepared?' Answers

1. $1082.43 2. $9513.28

13.4 Mathematical Induction

> **OBJECTIVE** 1 Prove Statements Using Mathematical Induction (p. 957)

1 Prove Statements Using Mathematical Induction

Mathematical induction is a method for proving that statements involving natural numbers are true for all natural numbers.*

For example, the statement "$2n$ is always an even integer" can be proved for all natural numbers n by using mathematical induction. Also, the statement "the sum of the first n positive odd integers equals n^2," that is,

$$1 + 3 + 5 + \cdots + (2n - 1) = n^2 \qquad \textbf{(1)}$$

can be proved for all natural numbers n by using mathematical induction.

Before stating the method of mathematical induction, let's try to gain a sense of the power of the method. We shall use the statement in equation (1) for this purpose by restating it for various values of $n = 1, 2, 3, \ldots$.

$n = 1$ The sum of the first positive odd integer is 1^2; $1 = 1^2$.

$n = 2$ The sum of the first 2 positive odd integers is 2^2;
 $1 + 3 = 4 = 2^2$.

$n = 3$ The sum of the first 3 positive odd integers is 3^2;
 $1 + 3 + 5 = 9 = 3^2$.

$n = 4$ The sum of the first 4 positive odd integers is 4^2;
 $1 + 3 + 5 + 7 = 16 = 4^2$.

Although from this pattern we might conjecture that statement (1) is true for any choice of n, can we really be sure that it does not fail for some choice of n? The method of proof by mathematical induction will, in fact, prove that the statement is true for all n.

* Recall that the natural numbers are the numbers 1, 2, 3, 4, In other words, the terms *natural numbers* and *positive integers* are synonymous.

THEOREM **The Principle of Mathematical Induction**

Suppose that the following two conditions are satisfied with regard to a statement about natural numbers:

CONDITION I: The statement is true for the natural number 1.

CONDITION II: If the statement is true for some natural number k, it is also true for the next natural number $k + 1$.

Then the statement is true for all natural numbers.

We shall not prove this principle. However, we can provide a physical interpretation that will help us to see why the principle works. Think of a collection of natural numbers obeying a statement as a collection of infinitely many dominoes. See Figure 10.

Figure 10

Now, suppose that we are told two facts:

1. The first domino is pushed over.

2. If one domino falls over, say the kth domino, so will the next one, the $(k + 1)$st domino.

Is it safe to conclude that *all* the dominoes fall over? The answer is yes, because if the first one falls (Condition I), the second one does also (by Condition II); and if the second one falls, so does the third (by Condition II); and so on.

Now let's prove some statements about natural numbers using mathematical induction.

EXAMPLE 1 **Using Mathematical Induction**

Show that the following statement is true for all natural numbers n.

$$1 + 3 + 5 + \cdots + (2n - 1) = n^2 \tag{2}$$

Solution We need to show first that statement (2) holds for $n = 1$. Because $1 = 1^2$, statement (2) is true for $n = 1$. Condition I holds.

Next, we need to show that Condition II holds. Suppose that we know for some k that

$$1 + 3 + \cdots + (2k - 1) = k^2 \tag{3}$$

We wish to show that, based on equation (3), statement (2) holds for $k + 1$. We look at the sum of the first $k + 1$ positive odd integers to determine whether this sum equals $(k + 1)^2$.

$$1 + 3 + \cdots + (2k - 1) + [2(k + 1) - 1] = \underbrace{[1 + 3 + \cdots + (2k - 1)]}_{= \, k^2 \text{ by equation (3)}} + (2k + 1)$$

$$= k^2 + (2k + 1)$$

$$= k^2 + 2k + 1 = (k + 1)^2$$

Conditions I and II are satisfied; by the Principle of Mathematical Induction, statement (2) is true for all natural numbers n.

EXAMPLE 2 **Using Mathematical Induction**

Show that the following statement is true for all natural numbers n.

$$2^n > n$$

Solution First, we show that the statement $2^n > n$ holds when $n = 1$. Because $2^1 = 2 > 1$, the inequality is true for $n = 1$. Condition I holds.

Next, we assume, for some natural number k, that $2^k > k$. We wish to show that the formula holds for $k + 1$; that is, we wish to show that $2^{k+1} > k + 1$. Now

$$2^{k+1} = 2 \cdot 2^k > 2 \cdot k = k + k \geq k + 1$$
$$\underset{\substack{\uparrow \\ \text{We know that} \\ 2^k > k.}}{} \qquad \underset{\substack{\uparrow \\ k \geq 1}}{}$$

If $2^k > k$, then $2^{k+1} > k + 1$, so Condition II of the Principle of Mathematical Induction is satisfied. The statement $2^n > n$ is true for all natural numbers n.

EXAMPLE 3	**Using Mathematical Induction**

Show that the following formula is true for all natural numbers n.

$$1 + 2 + 3 + \cdots + n = \frac{n(n + 1)}{2} \qquad \textbf{(4)}$$

Solution First, we show that formula (4) is true when $n = 1$. Because

$$\frac{1(1 + 1)}{2} = \frac{1(2)}{2} = 1$$

Condition I of the Principle of Mathematical Induction holds.

Next, we assume that formula (4) holds for some k, and we determine whether the formula then holds for $k + 1$. We assume that

$$1 + 2 + 3 + \cdots + k = \frac{k(k + 1)}{2} \quad \text{for some } k \qquad \textbf{(5)}$$

Now we need to show that

$$1 + 2 + 3 + \cdots + k + (k + 1) = \frac{(k + 1)[(k + 1) + 1]}{2} = \frac{(k + 1)(k + 2)}{2}$$

We do this as follows:

$$1 + 2 + 3 + \cdots + k + (k + 1) = \underbrace{[1 + 2 + 3 + \cdots + k]}_{\substack{= \frac{k(k + 1)}{2} \quad \text{by equation (5)}}} + (k + 1)$$

$$= \frac{k(k + 1)}{2} + (k + 1)$$

$$= \frac{k^2 + k + 2k + 2}{2}$$

$$= \frac{k^2 + 3k + 2}{2} = \frac{(k + 1)(k + 2)}{2}$$

Condition II also holds. As a result, formula (4) is true for all natural numbers n.

$\longrightarrow$ **Now Work** PROBLEM 1

EXAMPLE 4	**Using Mathematical Induction**

Show that $3^n - 1$ is divisible by 2 for all natural numbers n.

Solution First, we show that the statement is true when $n = 1$. Because $3^1 - 1 = 3 - 1 = 2$ is divisible by 2, the statement is true when $n = 1$. Condition I is satisfied.

Next, we assume that the statement holds for some k, and we determine whether the statement then holds for $k + 1$. We assume that $3^k - 1$ is divisible by 2 for some k. We need to show that $3^{k+1} - 1$ is divisible by 2. Now

$$3^{k+1} - 1 = 3^{k+1} - 3^k + 3^k - 1 \qquad \text{Subtract and add } 3^k.$$
$$= 3^k(3 - 1) + (3^k - 1) = 3^k \cdot 2 + (3^k - 1)$$

Because $3^k \cdot 2$ is divisible by 2 and $3^k - 1$ is divisible by 2, it follows that $3^k \cdot 2 + (3^k - 1) = 3^{k+1} - 1$ is divisible by 2. Condition II is also satisfied. As a result, the statement "$3^n - 1$ is divisible by 2" is true for all natural numbers n.

WARNING The conclusion that a statement involving natural numbers is true for all natural numbers is made only after *both* Conditions I and II of the Principle of Mathematical Induction have been satisfied. Problem 28 demonstrates a statement for which only Condition I holds, but the statement is not true for all natural numbers. Problem 29 demonstrates a statement for which only Condition II holds, but the statement is not true for any natural number. ∎

13.4 Assess Your Understanding

Skill Building

In Problems 1–22, use the Principle of Mathematical Induction to show that the given statement is true for all natural numbers n.

1. $2 + 4 + 6 + \cdots + 2n = n(n + 1)$

2. $1 + 5 + 9 + \cdots + (4n - 3) = n(2n - 1)$

3. $3 + 4 + 5 + \cdots + (n + 2) = \frac{1}{2}n(n + 5)$

4. $3 + 5 + 7 + \cdots + (2n + 1) = n(n + 2)$

5. $2 + 5 + 8 + \cdots + (3n - 1) = \frac{1}{2}n(3n + 1)$

6. $1 + 4 + 7 + \cdots + (3n - 2) = \frac{1}{2}n(3n - 1)$

7. $1 + 2 + 2^2 + \cdots + 2^{n-1} = 2^n - 1$

8. $1 + 3 + 3^2 + \cdots + 3^{n-1} = \frac{1}{2}(3^n - 1)$

9. $1 + 4 + 4^2 + \cdots + 4^{n-1} = \frac{1}{3}(4^n - 1)$

10. $1 + 5 + 5^2 + \cdots + 5^{n-1} = \frac{1}{4}(5^n - 1)$

11. $\frac{1}{1 \cdot 2} + \frac{1}{2 \cdot 3} + \frac{1}{3 \cdot 4} + \cdots + \frac{1}{n(n + 1)} = \frac{n}{n + 1}$

12. $\frac{1}{1 \cdot 3} + \frac{1}{3 \cdot 5} + \frac{1}{5 \cdot 7} + \cdots + \frac{1}{(2n - 1)(2n + 1)} = \frac{n}{2n + 1}$

13. $1^2 + 2^2 + 3^2 + \cdots + n^2 = \frac{1}{6}n(n + 1)(2n + 1)$

14. $1^3 + 2^3 + 3^3 + \cdots + n^3 = \frac{1}{4}n^2(n + 1)^2$

15. $4 + 3 + 2 + \cdots + (5 - n) = \frac{1}{2}n(9 - n)$

16. $-2 - 3 - 4 - \cdots - (n + 1) = -\frac{1}{2}n(n + 3)$

17. $1 \cdot 2 + 2 \cdot 3 + 3 \cdot 4 + \cdots + n(n + 1) = \frac{1}{3}n(n + 1)(n + 2)$

18. $1 \cdot 2 + 3 \cdot 4 + 5 \cdot 6 + \cdots + (2n - 1)(2n) = \frac{1}{3}n(n + 1)(4n - 1)$

19. $n^2 + n$ is divisible by 2.

20. $n^3 + 2n$ is divisible by 3.

21. $n^2 - n + 2$ is divisible by 2.

22. $n(n + 1)(n + 2)$ is divisible by 6.

Applications and Extensions

In Problems 23–27, prove each statement.

23. If $x > 1$, then $x^n > 1$.

24. If $0 < x < 1$, then $0 < x^n < 1$.

25. $a - b$ is a factor of $a^n - b^n$.

[**Hint:** $a^{k+1} - b^{k+1} = a(a^k - b^k) + b^k(a - b)$]

26. $a + b$ is a factor of $a^{2n+1} + b^{2n+1}$.

27. $(1 + a)^n \geq 1 + na$, for $a > 0$

28. Show that the statement "$n^2 - n + 41$ is a prime number" is true for $n = 1$, but is not true for $n = 41$.

29. Show that the formula
$$2 + 4 + 6 + \cdots + 2n = n^2 + n + 2$$
obeys Condition II of the Principle of Mathematical Induction. That is, show that if the formula is true for some k it is also true for $k + 1$. Then show that the formula is false for $n = 1$ (or for any other choice of n).

30. Use mathematical induction to prove that if $r \neq 1$ then
$$a + ar + ar^2 + \cdots + ar^{n-1} = a\frac{1 - r^n}{1 - r}$$

31. Use mathematical induction to prove that
$$a + (a + d) + (a + 2d)$$
$$+ \cdots + [a + (n - 1)d] = na + d\frac{n(n - 1)}{2}$$

32. Extended Principle of Mathematical Induction The Extended Principle of Mathematical Induction states that if Conditions I and II hold, that is,

(I) A statement is true for a natural number j.

(II) If the statement is true for some natural number $k \geq j$, then it is also true for the next natural number $k + 1$.

then the statement is true for all natural numbers $\geq j$. Use the Extended Principle of Mathematical Induction to show that the number of diagonals in a convex polygon of n sides is $\frac{1}{2}n(n - 3)$.

[Hint: Begin by showing that the result is true when $n = 4$ (Condition I).]

33. Geometry Use the Extended Principle of Mathematical Induction to show that the sum of the interior angles of a convex polygon of n sides equals $(n - 2) \cdot 180°$.

Discussion and Writing

34. How would you explain the Principle of Mathematical Induction to a friend?

13.5 The Binomial Theorem

OBJECTIVES **1** Evaluate $\binom{n}{j}$ (p. 961)

2 Use the Binomial Theorem (p. 963)

Formulas have been given for expanding $(x + a)^n$ for $n = 2$ and $n = 3$. The *Binomial Theorem** is a formula for the expansion of $(x + a)^n$ for any positive integer n. If $n = 1, 2, 3$, and 4, the expansion of $(x + a)^n$ is straightforward.

$(x + a)^1 = x + a$ Two terms, beginning with x^1 and ending with a^1

$(x + a)^2 = x^2 + 2ax + a^2$ Three terms, beginning with x^2 and ending with a^2

$(x + a)^3 = x^3 + 3ax^2 + 3a^2x + a^3$ Four terms, beginning with x^3 and ending with a^3

$(x + a)^4 = x^4 + 4ax^3 + 6a^2x^2 + 4a^3x + a^4$ Five terms, beginning with x^4 and ending with a^4

Notice that each expansion of $(x + a)^n$ begins with x^n and ends with a^n. As you read from left to right, the powers of x are decreasing by 1, while the powers of a are increasing by 1. Also, the number of terms equals $n + 1$. Notice, too, that the degree of each monomial in the expansion equals n. For example, in the expansion of $(x + a)^3$, each monomial $(x^3, 3ax^2, 3a^2x, a^3)$ is of degree 3. As a result, we might conjecture that the expansion of $(x + a)^n$ would look like this:

$$(x + a)^n = x^n + \underline{\quad} ax^{n-1} + \underline{\quad} a^2x^{n-2} + \cdots + \underline{\quad} a^{n-1}x + a^n$$

where the blanks are numbers to be found. This is, in fact, the case, as we shall see shortly.

Before we can fill in the blanks, we need to introduce the symbol $\binom{n}{j}$.

1 Evaluate $\binom{n}{j}$

COMMENT On a graphing calculator, the symbol $\binom{n}{j}$ may be denoted by tho koy $\boxed{\text{nCr}}$.

We define the symbol $\binom{n}{j}$, read "n taken j at a time," as follows:

DEFINITION If j and n are integers with $0 \leq j \leq n$, the symbol $\binom{n}{j}$ is defined as

$$\binom{n}{j} = \frac{n!}{j!(n - j)!} \tag{1}$$

* The name *binomial* is derived from the fact that $x + a$ is a binomial; that is, it contains two terms.

EXAMPLE 1

Evaluating $\binom{n}{j}$

Find:

(a) $\binom{3}{1}$ (b) $\binom{4}{2}$ (c) $\binom{8}{7}$ (d) $\binom{65}{15}$

Solution

(a) $\binom{3}{1} = \dfrac{3!}{1!(3-1)!} = \dfrac{3!}{1!\,2!} = \dfrac{3 \cdot 2 \cdot 1}{1(2 \cdot 1)} = \dfrac{6}{2} = 3$

(b) $\binom{4}{2} = \dfrac{4!}{2!(4-2)!} = \dfrac{4!}{2!\,2!} = \dfrac{4 \cdot 3 \cdot 2 \cdot 1}{(2 \cdot 1)(2 \cdot 1)} = \dfrac{24}{4} = 6$

(c) $\binom{8}{7} = \dfrac{8!}{7!(8-7)!} = \dfrac{8!}{7!\,1!} = \dfrac{8 \cdot \cancel{7!}}{\cancel{7!} \cdot 1!} = \dfrac{8}{1} = 8$

$$\uparrow$$
$$8! = 8 \cdot 7!$$

Figure 11

(d) Figure 11 shows the solution using a TI-84 Plus graphing calculator. So

$$\binom{65}{15} \approx 2.073746998 \times 10^{14}$$

◾

Now Work PROBLEM 5

Four useful formulas involving the symbol $\binom{n}{j}$ are

$$\binom{n}{0} = 1 \qquad \binom{n}{1} = n \qquad \binom{n}{n-1} = n \qquad \binom{n}{n} = 1$$

Proof

$$\binom{n}{0} = \dfrac{n!}{0!(n-0)!} = \dfrac{\cancel{n!}}{0!\,\cancel{n!}} = \dfrac{1}{1} = 1$$

$$\binom{n}{1} = \dfrac{n!}{1!(n-1)!} = \dfrac{n!}{(n-1)!} = \dfrac{n(n-1)!}{(n-1)!} = n$$

You are asked to show the remaining two formulas in Problem 45. ◾

Suppose that we arrange the values of the symbol $\binom{n}{j}$ in a triangular display, as shown next and in Figure 12.

$$\binom{0}{0}$$

$$\binom{1}{0} \quad \binom{1}{1}$$

$$\binom{2}{0} \quad \binom{2}{1} \quad \binom{2}{2}$$

$$\binom{3}{0} \quad \binom{3}{1} \quad \binom{3}{2} \quad \binom{3}{3}$$

$$\binom{4}{0} \quad \binom{4}{1} \quad \binom{4}{2} \quad \binom{4}{3} \quad \binom{4}{4}$$

$$\binom{5}{0} \quad \binom{5}{1} \quad \binom{5}{2} \quad \binom{5}{3} \quad \binom{5}{4} \quad \binom{5}{5}$$

Figure 12
Pascal triangle

This display is called the **Pascal triangle,** named after Blaise Pascal (1623–1662), a French mathematician.

The Pascal triangle has 1's down the sides. To get any other entry, add the two nearest entries in the row above it. The shaded triangles in Figure 12 illustrate this feature of the Pascal triangle. Based on this feature, the row corresponding to $n = 6$ is found as follows:

$$n = 5 \rightarrow \quad 1 \quad 5 \quad 10 \quad 10 \quad 5 \quad 1$$
$$n = 6 \rightarrow \quad 1 \quad 6 \quad 15 \quad 20 \quad 15 \quad 6 \quad 1$$

Later we shall prove that this addition always works (see the theorem on page 965).

Although the Pascal triangle provides an interesting and organized display of the symbol $\binom{n}{j}$, in practice it is not all that helpful. For example, if you wanted to know the value of $\binom{12}{5}$, you would need to produce 13 rows of the triangle before seeing the answer. It is much faster to use the definition (1).

2 Use the Binomial Theorem

Now we are ready to state the **Binomial Theorem.**

THEOREM

Binomial Theorem

Let x and a be real numbers. For any positive integer n, we have

$$(x + a)^n = \binom{n}{0}x^n + \binom{n}{1}ax^{n-1} + \cdots + \binom{n}{j}a^j x^{n-j} + \cdots + \binom{n}{n}a^n$$

$$= \sum_{j=0}^{n} \binom{n}{j} x^{n-j} a^j \tag{2}$$

Now you know why we needed to introduce the symbol $\binom{n}{j}$; these symbols are the numerical coefficients that appear in the expansion of $(x + a)^n$. Because of this, the symbol $\binom{n}{j}$ is called a **binomial coefficient.**

EXAMPLE 2

Expanding a Binomial

Use the Binomial Theorem to expand $(x + 2)^5$.

Solution In the Binomial Theorem, let $a = 2$ and $n = 5$. Then

$$(x + 2)^5 = \binom{5}{0}x^5 + \binom{5}{1}2x^4 + \binom{5}{2}2^2x^3 + \binom{5}{3}2^3x^2 + \binom{5}{4}2^4x + \binom{5}{5}2^5$$

↑
Use equation (2).

$$= 1 \cdot x^5 + 5 \cdot 2x^4 + 10 \cdot 4x^3 + 10 \cdot 8x^2 + 5 \cdot 16x + 1 \cdot 32$$

↑
Use row $n = 5$ of the Pascal triangle or formula (1) for $\binom{n}{j}$.

$$= x^5 + 10x^4 + 40x^3 + 80x^2 + 80x + 32$$

EXAMPLE 3 **Expanding a Binomial**

Expand $(2y - 3)^4$ using the Binomial Theorem.

Solution First, we rewrite the expression $(2y - 3)^4$ as $[2y + (-3)]^4$. Now we use the Binomial Theorem with $n = 4$, $x = 2y$, and $a = -3$.

$$[2y + (-3)]^4 = \binom{4}{0}(2y)^4 + \binom{4}{1}(-3)(2y)^3 + \binom{4}{2}(-3)^2(2y)^2$$

$$+ \binom{4}{3}(-3)^3(2y) + \binom{4}{4}(-3)^4$$

$$= 1 \cdot 16y^4 + 4(-3)8y^3 + 6 \cdot 9 \cdot 4y^2 + 4(-27)2y + 1 \cdot 81$$

↑
Use row $n = 4$ of the Pascal triangle or formula (1) for $\binom{n}{j}$.

$$= 16y^4 - 96y^3 + 216y^2 - 216y + 81$$

In this expansion, note that the signs alternate due to the fact that $a = -3 < 0$.

Now Work PROBLEM 21

EXAMPLE 4 **Finding a Particular Coefficient in a Binomial Expansion**

Find the coefficient of y^8 in the expansion of $(2y + 3)^{10}$.

Solution We write out the expansion using the Binomial Theorem.

$$(2y + 3)^{10} = \binom{10}{0}(2y)^{10} + \binom{10}{1}(2y)^9(3)^1 + \binom{10}{2}(2y)^8(3)^2 + \binom{10}{3}(2y)^7(3)^3$$

$$+ \binom{10}{4}(2y)^6(3)^4 + \cdots + \binom{10}{9}(2y)(3)^9 + \binom{10}{10}(3)^{10}$$

From the third term in the expansion, the coefficient of y^8 is

$$\binom{10}{2}(2)^8(3)^2 = \frac{10!}{2! \, 8!} \cdot 2^8 \cdot 9 = \frac{10 \cdot 9 \cdot 8!}{2 \cdot 8!} \cdot 2^8 \cdot 9 = 103,680$$

As this solution demonstrates, we can use the Binomial Theorem to find a particular term in an expansion without writing the entire expansion.

Based on the expansion of $(x + a)^n$, the term containing x^j is

$$\binom{n}{n-j}a^{n-j}x^j \qquad \text{(3)}$$

For example, we can solve Example 4 by using formula (3) with $n = 10, a = 3, x = 2y$, and $j = 8$. Then the term containing y^8 is

$$\binom{10}{10-8}3^{10-8}(2y)^8 = \binom{10}{2} \cdot 3^2 \cdot 2^8 \cdot y^8 = \frac{10!}{2! \, 8!} \cdot 9 \cdot 2^8 y^8$$

$$= \frac{10 \cdot 9 \cdot 8!}{2 \cdot 8!} \cdot 9 \cdot 2^8 y^8 = 103{,}680 y^8$$

EXAMPLE 5 **Finding a Particular Term in a Binomial Expansion**

Find the sixth term in the expansion of $(x + 2)^9$.

Solution A We expand using the Binomial Theorem until the sixth term is reached.

$$(x + 2)^9 = \binom{9}{0}x^9 + \binom{9}{1}x^8 \cdot 2 + \binom{9}{2}x^7 \cdot 2^2 + \binom{9}{3}x^6 \cdot 2^3 + \binom{9}{4}x^5 \cdot 2^4$$

$$+ \binom{9}{5}x^4 \cdot 2^5 + \cdots$$

The sixth term is

$$\binom{9}{5}x^4 \cdot 2^5 = \frac{9!}{5! \, 4!} \cdot x^4 \cdot 32 = 4032x^4$$

Solution B The sixth term in the expansion of $(x + 2)^9$, which has 10 terms total, contains x^4. (Do you see why?) By formula (3), the sixth term is

$$\binom{9}{9-4}2^{9-4}x^4 = \binom{9}{5}2^5 x^4 = \frac{9!}{5! \, 4!} \cdot 32x^4 = 4032x^4$$

 Now Work PROBLEMS 29 AND 35

Next we show that the *triangular addition* feature of the Pascal triangle illustrated in Figure 12 always works.

THEOREM If n and j are integers with $1 \le j \le n$, then

$$\binom{n}{j-1} + \binom{n}{j} = \binom{n+1}{j} \qquad \text{(4)}$$

Proof

$$\binom{n}{j-1} + \binom{n}{j} = \frac{n!}{(j-1)![n-(j-1)]!} + \frac{n!}{j!(n-j)!}$$

$$= \frac{n!}{(j-1)!(n-j+1)!} + \frac{n!}{j!(n-j)!}$$

$$= \frac{jn!}{j(j-1)!(n-j+1)!} + \frac{(n-j+1)n!}{j!(n-j+1)(n-j)!}$$

Multiply the first term by $\dfrac{j}{j}$ and the second term by $\dfrac{n-j+1}{n-j+1}$.

$$= \frac{jn!}{j!(n-j+1)!} + \frac{(n-j+1)n!}{j!(n-j+1)!}$$

Now the denominators are equal.

$$= \frac{jn! + (n-j+1)n!}{j!(n-j+1)!}$$

$$= \frac{n!(j+n-j+1)}{j!(n-j+1)!}$$

$$= \frac{n!(n+1)}{j!(n-j+1)!} = \frac{(n+1)!}{j![(n+1)-j]!} = \binom{n+1}{j}$$ ∎

Historical Feature

*Omar Khayyám
(1050–1123)*

The case $n = 2$ of the Binomial Theorem, $(a + b)^2$, was known to Euclid in 300 BC, but the general law seems to have been discovered by the Persian mathematician and astronomer Omar Khayyám (1050–1123), who is also well known as the author of the *Rubáiyát*, a collection of four-line poems making observations on the human condition. Omar Khayyám did not state the Binomial Theorem explicitly, but he claimed to have a method for extracting third, fourth, fifth roots, and so on. A little study shows that one must know the Binomial Theorem to create such a method.

The heart of the Binomial Theorem is the formula for the numerical coefficients, and, as we saw, they can be written in a symmetric triangular form. The Pascal triangle appears first in the books of Yang Hui (about 1270) and Chu Shih-chieh (1303). Pascal's name is attached to the triangle because of the many applications he made of it, especially to counting and probability. In establishing these results, he was one of the earliest users of mathematical induction.

Many people worked on the proof of the Binomial Theorem, which was finally completed for all n (including complex numbers) by Niels Abel (1802–1829).

13.5 Assess Your Understanding

Concepts and Vocabulary

1. The _____ _____ is a triangular display of the binomial coefficients.

2. $\binom{6}{2} =$ _____

3. *True or False* $\binom{n}{j} = \frac{j!}{(n-j)!\, n!}$

4. The _____ _____ can be used to expand expressions like $(2x + 3)^6$.

Skill Building

In Problems 5–16, evaluate each expression.

5. $\binom{5}{3}$ **6.** $\binom{7}{3}$ **7.** $\binom{7}{5}$ **8.** $\binom{9}{7}$

9. $\binom{50}{49}$ **10.** $\binom{100}{98}$ **11.** $\binom{1000}{1000}$ **12.** $\binom{1000}{0}$

13. $\binom{55}{23}$ **14.** $\binom{60}{20}$ **15.** $\binom{47}{25}$ **16.** $\binom{37}{19}$

In Problems 17–28, expand each expression using the Binomial Theorem.

17. $(x + 1)^5$ **18.** $(x - 1)^5$ **19.** $(x - 2)^6$ **20.** $(x + 3)^5$

21. $(3x + 1)^4$ **22.** $(2x + 3)^5$ **23.** $(x^2 + y^2)^5$ **24.** $(x^2 - y^2)^6$

25. $\left(\sqrt{x} + \sqrt{2}\right)^6$ **26.** $\left(\sqrt{x} - \sqrt{3}\right)^4$ **27.** $(ax + by)^5$ **28.** $(ax - by)^4$

In Problems 29–42, use the Binomial Theorem to find the indicated coefficient or term.

29. The coefficient of x^6 in the expansion of $(x + 3)^{10}$

30. The coefficient of x^3 in the expansion of $(x - 3)^{10}$

31. The coefficient of x^7 in the expansion of $(2x - 1)^{12}$

32. The coefficient of x^3 in the expansion of $(2x + 1)^{12}$

33. The coefficient of x^7 in the expansion of $(2x + 3)^9$

34. The coefficient of x^2 in the expansion of $(2x - 3)^9$

35. The fifth term in the expansion of $(x + 3)^7$

36. The third term in the expansion of $(x - 3)^7$

37. The third term in the expansion of $(3x - 2)^9$

38. The sixth term in the expansion of $(3x + 2)^8$

39. The coefficient of x^0 in the expansion of $\left(x^2 + \dfrac{1}{x} \right)^{12}$

40. The coefficient of x^0 in the expansion of $\left(x - \dfrac{1}{x^2} \right)^9$

41. The coefficient of x^4 in the expansion of $\left(x - \dfrac{2}{\sqrt{x}} \right)^{10}$

42. The coefficient of x^2 in the expansion of $\left(\sqrt{x} + \dfrac{3}{\sqrt{x}} \right)^8$

Applications and Extensions

43. Use the Binomial Theorem to find the numerical value of $(1.001)^5$ correct to five decimal places.
[**Hint:** $(1.001)^5 = (1 + 10^{-3})^5$]

44. Use the Binomial Theorem to find the numerical value of $(0.998)^6$ correct to five decimal places.

45. Show that $\dbinom{n}{n-1} = n$ and $\dbinom{n}{n} = 1$.

46. Show that if n and j are integers with $0 \le j \le n$ then
$$\binom{n}{j} = \binom{n}{n-j}$$
Conclude that the Pascal triangle is symmetric with respect to a vertical line drawn from the topmost entry.

47. If n is a positive integer, show that
$$\binom{n}{0} + \binom{n}{1} + \cdots + \binom{n}{n} = 2^n$$
[**Hint:** $2^n = (1 + 1)^n$; now use the Binomial Theorem.]

48. If n is a positive integer, show that
$$\binom{n}{0} - \binom{n}{1} + \binom{n}{2} - \cdots + (-1)^n \binom{n}{n} = 0$$

49. $\dbinom{5}{0}\left(\dfrac{1}{4}\right)^5 + \dbinom{5}{1}\left(\dfrac{1}{4}\right)^4\left(\dfrac{3}{4}\right) + \dbinom{5}{2}\left(\dfrac{1}{4}\right)^3\left(\dfrac{3}{4}\right)^2$
$+ \dbinom{5}{3}\left(\dfrac{1}{4}\right)^2\left(\dfrac{3}{4}\right)^3 + \dbinom{5}{4}\left(\dfrac{1}{4}\right)\left(\dfrac{3}{4}\right)^4 + \dbinom{5}{5}\left(\dfrac{3}{4}\right)^5 = ?$

50. Stirling's Formula An approximation for $n!$, when n is large, is given by
$$n! \approx \sqrt{2n\pi}\left(\dfrac{n}{e}\right)^n\left(1 + \dfrac{1}{12n - 1}\right)$$
Calculate 12!, 20!, and 25! on your calculator. Then use Stirling's formula to approximate 12!, 20!, and 25!.

CHAPTER REVIEW

Things to Know

Sequence (p. 930)	A function whose domain is the set of positive integers.
Factorials (p. 932)	$0! = 1$, $1! = 1$, $n! = n(n - 1) \cdot \cdots \cdot 3 \cdot 2 \cdot 1$ if $n \ge 2$ is an integer
Arithmetic sequence (pp. 940 and 941)	$a_1 = a$, $a_n = a_{n-1} + d$, where $a_1 = a = $ first term, $d = $ common difference $a_n = a_1 + (n - 1)d$
Sum of the first n terms of an arithmetic sequence (p. 942)	$S_n = \dfrac{n}{2}[2a_1 + (n - 1)d] = \dfrac{n}{2}(a_1 + a_n)$
Geometric sequence (pp. 946 and 947)	$a_1 = a$, $a_n = ra_{n-1}$, where $a_1 = a = $ first term, $r = $ common ratio $a_n = a_1 r^{n-1}$, $r \ne 0$
Sum of the first n terms of a geometric sequence (p. 948)	$S_n = a_1 \dfrac{1 - r^n}{1 - r}$, $r \ne 0, 1$
Infinite geometric series (p. 949)	$a_1 + a_1 r + \cdots + a_1 r^{n-1} + \cdots = \displaystyle\sum_{k=1}^{\infty} a_1 r^{k-1}$
Sum of a convergent infinite geometric series (p. 950)	If $\|r\| < 1$, $\displaystyle\sum_{k=1}^{\infty} a_1 r^{k-1} = \dfrac{a_1}{1 - r}$
Principle of Mathematical Induction (p. 958)	Suppose the following two conditions are satisfied. Condition I: The statement is true for the natural number 1. Condition II: If the statement is true for some natural number k, it is also true for $k + 1$. Then the statement is true for all natural numbers.

Binomial coefficient (p. 961)	$\dbinom{n}{j} = \dfrac{n!}{j!(n-j)!}$
Pascal triangle (p. 963)	See Figure 12.
Binomial Theorem (p. 963)	$(x+a)^n = \dbinom{n}{0}x^n + \dbinom{n}{1}ax^{n-1} + \cdots + \dbinom{n}{j}a^j x^{n-j} + \cdots + \dbinom{n}{n}a^n = \displaystyle\sum_{j=0}^{n}\dbinom{n}{j}x^{n-j}a^j$

Objectives

Section	You should be able to . . .	Review Exercises
13.1	1 Write the first several terms of a sequence (p. 930)	1–4
	2 Write the terms of a sequence defined by a recursive formula (p. 933)	5–8
	3 Use summation notation (p. 934)	9–12
	4 Find the sum of a sequence (p. 935)	25, 26
13.2	1 Determine if a sequence is arithmetic (p. 940)	13–24
	2 Find a formula for an arithmetic sequence (p. 941)	31, 32, 35, 37–40, 65
	3 Find the sum of an arithmetic sequence (p. 942)	13, 14, 19, 20, 27, 28, 66
13.3	1 Determine if a sequence is geometric (p. 946)	13–24
	2 Find a formula for a geometric sequence (p. 947)	22, 33, 34, 36, 70
	3 Find the sum of a geometric sequence (p. 948)	17, 18, 21, 22, 29, 30, 67(a)–(c), 70
	4 Determine whether a geometric series converges or diverges (p. 949)	41–48, 67(d)
	5 Solve annuity problems (p. 952)	68, 69
13.4	1 Prove statements using mathematical induction (p. 957)	49–54
13.5	1 Evaluate $\dbinom{n}{j}$ (p. 961)	55, 56
	2 Use the Binomial Theorem (p. 963)	57–64

Review Exercises

In Problems 1–8, write down the first five terms of each sequence.

1. $\{a_n\} = \left\{(-1)^n\left(\dfrac{n+3}{n+2}\right)\right\}$ **2.** $\{b_n\} = \{(-1)^{n+1}(2n+3)\}$ **3.** $\{c_n\} = \left\{\dfrac{2^n}{n^2}\right\}$ **4.** $\{d_n\} = \left\{\dfrac{e^n}{n}\right\}$

5. $a_1 = 3;\quad a_n = \dfrac{2}{3}a_{n-1}$ **6.** $a_1 = 4;\quad a_n = -\dfrac{1}{4}a_{n-1}$ **7.** $a_1 = 2;\quad a_n = 2 - a_{n-1}$ **8.** $a_1 = -3;\quad a_n = 4 + a_{n-1}$

In Problems 9 and 10, write out each sum.

9. $\displaystyle\sum_{k=1}^{4}(4k+2)$ **10.** $\displaystyle\sum_{k=1}^{3}(3-k^2)$

In Problems 11 and 12, express each sum using summation notation.

11. $1 - \dfrac{1}{2} + \dfrac{1}{3} - \dfrac{1}{4} + \cdots + \dfrac{1}{13}$ **12.** $2 + \dfrac{2^2}{3} + \dfrac{2^3}{3^2} + \cdots + \dfrac{2^{n+1}}{3^n}$

In Problems 13–24, determine whether the given sequence is arithmetic, geometric, or neither. If the sequence is arithmetic, find the common difference and the sum of the first n terms. If the sequence is geometric, find the common ratio and the sum of the first n terms.

13. $\{a_n\} = \{n+5\}$ **14.** $\{b_n\} = \{4n+3\}$ **15.** $\{c_n\} = \{2n^3\}$ **16.** $\{d_n\} = \{2n^2 - 1\}$

17. $\{s_n\} = \{2^{3n}\}$ **18.** $\{u_n\} = \{3^{2n}\}$ **19.** $0, 4, 8, 12, \ldots$ **20.** $1, -3, -7, -11, \ldots$

21. $3, \dfrac{3}{2}, \dfrac{3}{4}, \dfrac{3}{8}, \dfrac{3}{16}, \ldots$ **22.** $5, -\dfrac{5}{3}, \dfrac{5}{9}, -\dfrac{5}{27}, \dfrac{5}{81}, \ldots$ **23.** $\dfrac{2}{3}, \dfrac{3}{4}, \dfrac{4}{5}, \dfrac{5}{6}, \ldots$ **24.** $\dfrac{3}{2}, \dfrac{5}{4}, \dfrac{7}{6}, \dfrac{9}{8}, \dfrac{11}{10}, \ldots$

In Problems 25–30, find each sum.

25. $\displaystyle\sum_{k=1}^{50}(3k)$

26. $\displaystyle\sum_{k=1}^{30}k^2$

27. $\displaystyle\sum_{k=1}^{30}(3k-9)$

28. $\displaystyle\sum_{k=1}^{40}(-2k+8)$

29. $\displaystyle\sum_{k=1}^{7}\left(\frac{1}{3}\right)^k$

30. $\displaystyle\sum_{k=1}^{10}(-2)^k$

In Problems 31–36, find the indicated term in each sequence. [**Hint:** Find the general term first.]

31. 9th term of $3, 7, 11, 15, \ldots$

32. 8th term of $1, -1, -3, -5, \ldots$

33. 11th term of $1, \dfrac{1}{10}, \dfrac{1}{100}, \ldots$

34. 11th term of $1, 2, 4, 8, \ldots$

35. 9th term of $\sqrt{2}, 2\sqrt{2}, 3\sqrt{2}, \ldots$

36. 9th term of $\sqrt{2}, 2, 2^{3/2}, \ldots$

In Problems 37–40, find a general formula for each arithmetic sequence.

37. 7th term is 31; 20th term is 96

38. 8th term is -20; 17th term is -47

39. 10th term is 0; 18th term is 8

40. 12th term is 30; 22nd term is 50

In Problems 41–48, determine whether each infinite geometric series converges or diverges. If it converges, find its sum.

41. $3 + 1 + \dfrac{1}{3} + \dfrac{1}{9} + \cdots$

42. $2 + 1 + \dfrac{1}{2} + \dfrac{1}{4} + \cdots$

43. $2 - 1 + \dfrac{1}{2} - \dfrac{1}{4} + \cdots$

44. $6 - 4 + \dfrac{8}{3} - \dfrac{16}{9} + \cdots$

45. $\dfrac{1}{2} + \dfrac{3}{4} + \dfrac{9}{8} + \cdots$

46. $\displaystyle\sum_{k=1}^{\infty}5\cdot\left(-\frac{5}{4}\right)^{k-1}$

47. $\displaystyle\sum_{k=1}^{\infty}4\left(\frac{1}{2}\right)^{k-1}$

48. $\displaystyle\sum_{k=1}^{\infty}3\left(-\frac{3}{4}\right)^{k-1}$

In Problems 49–54, use the Principle of Mathematical Induction to show that the given statement is true for all natural numbers.

49. $3 + 6 + 9 + \cdots + 3n = \dfrac{3n}{2}(n+1)$

50. $2 + 6 + 10 + \cdots + (4n-2) = 2n^2$

51. $2 + 6 + 18 + \cdots + 2\cdot3^{n-1} = 3^n - 1$

52. $3 + 6 + 12 + \cdots + 3\cdot2^{n-1} = 3(2^n - 1)$

53. $1^2 + 4^2 + 7^2 + \cdots + (3n-2)^2 = \dfrac{1}{2}n(6n^2 - 3n - 1)$

54. $1\cdot3 + 2\cdot4 + 3\cdot5 + \cdots + n(n+2) = \dfrac{n}{6}(n+1)(2n+7)$

In Problems 55 and 56, evaluate each binomial coefficient.

55. $\dbinom{5}{2}$

56. $\dbinom{8}{6}$

In Problems 57–60, expand each expression using the Binomial Theorem.

57. $(x+2)^5$

58. $(x-3)^4$

59. $(2x+3)^5$

60. $(3x-4)^4$

61. Find the coefficient of x^7 in the expansion of $(x+2)^9$.

62. Find the coefficient of x^3 in the expansion of $(x-3)^8$.

63. Find the coefficient of x^2 in the expansion of $(2x+1)^7$.

64. Find the coefficient of x^6 in the expansion of $(2x+1)^8$.

65. Constructing a Brick Staircase A brick staircase has a total of 25 steps. The bottom step requires 80 bricks. Each successive step requires three less bricks than the prior step.
(a) How many bricks are required for the top step?
(b) How many bricks are required to build the staircase?

66. Creating a Floor Design A mosaic tile floor is designed in the shape of a trapezoid 30 feet wide at the base and 15 feet wide at the top. The tiles, 12 inches by 12 inches, are to be placed so that each successive row contains one less tile than the row below. How many tiles will be required?

67. Bouncing Balls A ball is dropped from a height of 20 feet. Each time it strikes the ground, it bounces up to three-quarters of the previous height.
(a) What height will the ball bounce up to after it strikes the ground for the third time?

(b) How high will it bounce after it strikes the ground for the *n*th time?
(c) How many times does the ball need to strike the ground before its bounce is less than 6 inches?
(d) What total distance does the ball travel before it stops bouncing?

68. Retirement Planning Chris gets paid once a month and contributes $200 each pay period into his 401(k). If Chris plans on retiring in 20 years, what will be the value of his 401(k) if the per annum rate of return of the 401(k) is 10% compounded monthly?

69. Retirement Planning Jacky contributes $500 every quarter to an IRA. If Jacky plans on retiring in 30 years, what will be the value of the IRA if the per annum rate of return of the IRA is 8% compounded quarterly?

70. Salary Increases Your friend has just been hired at an annual salary of $20,000. If she expects to receive annual increases of 4%, what will be her salary as she begins her fifth year?

CHAPTER TEST

In Problems 1 and 2, write down the first five terms of each sequence.

1. $\{s_n\} = \left\{\dfrac{n^2 - 1}{n + 8}\right\}$

2. $a_1 = 4, a_n = 3a_{n-1} + 2$

In Problems 3 and 4, write out each sum. Evaluate each sum.

3. $\displaystyle\sum_{k=1}^{3} (-1)^{k+1}\left(\dfrac{k+1}{k^2}\right)$

4. $\displaystyle\sum_{k=1}^{4}\left[\left(\dfrac{2}{3}\right)^k - k\right]$

5. Write the following sum using summation notation.
$$-\dfrac{2}{5} + \dfrac{3}{6} - \dfrac{4}{7} + \cdots + \dfrac{11}{14}$$

In Problems 6–11, determine whether the given sequence is arithmetic, geometric, or neither. If the sequence is arithmetic, find the common difference and the sum of the first n terms. If the sequence is geometric, find the common ratio and the sum of the first n terms.

6. $6, 12, 36, 144, \ldots$

7. $\left\{-\dfrac{1}{2} \cdot 4^n\right\}$

8. $-2, -10, -18, -26, \ldots$

9. $\left\{-\dfrac{n}{2} + 7\right\}$

10. $25, 10, 4, \dfrac{8}{5}, \ldots$

11. $\left\{\dfrac{2n - 3}{2n + 1}\right\}$

12. Determine whether the infinite geometric series
$$256 - 64 + 16 - 4 + \cdots$$
converges or diverges. If it converges, find its sum.

13. Expand $(3m + 2)^5$ using the Binomial Theorem.

14. Use the Principle of Mathematical Induction to show that the given statement is true for all natural numbers.
$$\left(1 + \dfrac{1}{1}\right)\left(1 + \dfrac{1}{2}\right)\left(1 + \dfrac{1}{3}\right)\cdots\left(1 + \dfrac{1}{n}\right) = n + 1$$

15. A 2004 Dodge Durango sold for $31,000. If the vehicle loses 15% of its value each year, how much will it be worth after 10 years?

16. A weightlifter begins his routine by benching 100 pounds and increases the weight by 30 pounds for each set. If he does 10 repetitions in each set, what is the total weight lifted after 5 sets?

CUMULATIVE REVIEW

1. Find all the solutions, real and complex, of the equation
$$|x^2| = 9$$

2. (a) Graph the circle $x^2 + y^2 = 100$ and the parabola $y = 3x^2$.

 (b) Solve the system of equations: $\begin{cases} x^2 + y^2 = 100 \\ y = 3x^2 \end{cases}$

 (c) Where do the circle and the parabola intersect?

3. Solve the equation $2e^x = 5$.

4. Find an equation of the line with slope 5 and x-intercept 2.

5. Find the standard equation of the circle whose center is the point $(-1, 2)$ if $(3, 5)$ is a point on the circle.

6. $f(x) = \dfrac{3x}{x - 2}$, $g(x) = 2x + 1$

 Find:

 (a) $(f \circ g)(2)$ (b) $(g \circ f)(4)$ (c) $(f \circ g)(x)$

 (d) The domain of $(f \circ g)(x)$

 (e) $(g \circ f)(x)$ (f) The domain of $(g \circ f)(x)$

 (g) The function g^{-1} and its domain

 (h) The function f^{-1} and its domain

7. Find the equation of an ellipse with center at the origin, a focus at $(0, 3)$, and a vertex at $(0, 4)$.

8. Find the equation of a parabola with vertex at $(-1, 2)$ and focus at $(-1, 3)$.

9. Find the polar equation of a circle with center at $(0, 4)$ that passes through the pole. What is the rectangular equation?

10. Solve the equation
$$2\sin^2 x - \sin x - 3 = 0, \quad 0 \le x < 2\pi$$

11. Find the exact value of $\cos^{-1}(-0.5)$.

12. If $\sin\theta = \dfrac{1}{4}$ and θ is in the second quadrant, find:

 (a) $\cos\theta$ (b) $\tan\theta$

 (c) $\sin(2\theta)$ (d) $\cos(2\theta)$

 (e) $\sin\left(\dfrac{1}{2}\theta\right)$

CHAPTER PROJECTS

I. Population Growth The size of the population of the United States essentially depends on its current population, the birth and death rates of the population, and immigration. Suppose that b represents the birth rate of the U.S. population and d represents its death rate. Then $r = b - d$ represents the growth rate of the population, where r varies from year to year. The U.S. population after n years can be modeled using the recursive function

$$p_n = (1 + r)p_{n-1} + I$$

where I represents net immigration into the United States.

1. Using data from the National Center for Health Statistics *www.fedstats.gov*, determine the birth and death rates for all races for the most recent year that data are available. Birth rates are given as the number of live births per 1000 population, while death rates are given as the number of deaths per 100,000 population. Each must be computed as the number of births (deaths) per individual. For example, in 2000, the birth rate was 14.7 per 1000 and the death rate was 873.1 per 100,000, so $b = \dfrac{14.7}{1000} = 0.0147$, while $d = \dfrac{873.1}{100,000} = 0.008731$.

 Next, using data from the Immigration and Naturalization Service *www.fedstats.gov*, determine the net immigration to the United States for the same year used to obtain b and d in Problem 1.

2. Determine the value of r, the growth rate of the population.

3. Find a recursive formula for the population of the United States.

4. Use the recursive formula to predict the population of the United States in the following year. In other words, if data are available up to the year 2005, predict the U.S. population in 2006.

5. Compare your prediction to actual data.

6. Repeat Problems 1–5 for Uganda using the CIA World Factbook at *www.cia.gov/cia/publications/factbook/index.html* (in 2000, the birth rate was 48.04 per 1000 and the death rate was 18.44 per 1000).

7. Do your results for the United States (a developed country) and Uganda (a developing country) seem in line with the article in the chapter opener? Explain.

8. Do you think the recursive formula found in Problem 3 will be useful in predicting future populations? Why or why not?

The following projects are available at the Instructor's Resource Center (IRC):

II. Project at Motorola *Digital Wireless Communication* Cell phones take speech and change it into digital code using only zeros and ones. See how the code length can be modeled using a mathematical sequence.

III. Economics Economists use the current price of a good and a recursive model to predict future consumer demand and to determine future production.

IV. Standardized Tests Many tests of intelligence, aptitude, and achievement contain questions asking for the terms of a mathematical sequence.

Counting and Probability

14

Deal or No Deal

By LYNN ELBER, AP Television Writer—LOS ANGELES—The promise of an easy million bucks, a stage crowded with sexy models and the smoothly calibrated charm of host **Howie Mandel** made "Deal or No Deal" an unexpected hit in television's December dead zone. Based on a series that debuted in Holland in 2002 and became an international hit, "Deal or No Deal" is about luck and playing the odds. Contestants are faced with 26 briefcases held by 26 models, each case with a hidden value ranging from a penny to the top prize that will escalate by week's end to $3 million. As the game progresses and cases are eliminated, a contestant weighs the chance of snaring a big prize against lesser but still tempting offers made by the show's "bank," represented by an anonymous, silhouetted figure.

Source: Adapted from *Lynn Elber, "'Deal or No Deal' back with bigger prizes," Associated Press, February 24, 2006. © 2006 Associated Press.*

—See the Chapter Project—

A Look Back

We first introduced sets in Chapter R. We have been using sets to represent solutions of equations and inequalities and to represent the domain and range of functions.

A Look Ahead

Here we discuss methods for counting the number of elements in a set and the role of sets in probability.

14.1 Counting

PREPARING FOR THIS SECTION *Before getting started, review the following:*

- Sets (Review, Section R.1, pp. 1–3)

Now Work the 'Are You Prepared?' problems on page 978.

OBJECTIVES 1 Find All the Subsets of a Set (p. 974)
2 Count the Number of Elements in a Set (p. 974)
3 Solve Counting Problems Using the Multiplication Principle (p. 976)

Counting plays a major role in many diverse areas, such as probability, statistics, and computer science; counting techniques are a part of a branch of mathematics called **combinatorics.**

1 Find All the Subsets of a Set

We begin by reviewing the ways that two sets can be compared.

If two sets A and B have precisely the same elements, we say that A and B are **equal** and write $A = B$.

If each element of a set A is also an element of a set B, we say that A is a **subset** of B and write $A \subseteq B$.

If $A \subseteq B$ and $A \neq B$, we say that A is a **proper subset** of B and write $A \subset B$.

If $A \subseteq B$, every element in set A is also in set B, but B may or may not have additional elements. If $A \subset B$, every element in A is also in B, and B has at least one element not found in A.

Finally, we agree that the empty set is a subset of every set; that is,

$$\emptyset \subseteq A \qquad \text{for any set } A$$

EXAMPLE 1 **Finding All the Subsets of a Set**

Write down all the subsets of the set $\{a, b, c\}$.

Solution To organize our work, we write down all the subsets with no elements, then those with one element, then those with two elements, and finally those with three elements. These will give us all the subsets. Do you see why?

0 Elements	1 Element	2 Elements	3 Elements
$\emptyset$	$\{a\}, \{b\}, \{c\}$	$\{a, b\}, \{b, c\}, \{a, c\}$	$\{a, b, c\}$

Now Work PROBLEM 7

2 Count the Number of Elements in a Set

As you count the number of students in a classroom or the number of pennies in your pocket, what you are really doing is matching, on a one-to-one basis, each object to be counted with the set of counting numbers, $1, 2, 3, \ldots, n$, for some number n. If a set A matched up in this fashion with the set $\{1, 2, \ldots, 25\}$, you would conclude that there are 25 elements in the set A. We use the notation $n(A) = 25$ to indicate that there are 25 elements in the set A.

Because the empty set has no elements, we write

$$n(\emptyset) = 0$$

In Words
We use the notation $n(A)$ to mean the number of elements in set A.

If the number of elements in a set is a nonnegative integer, we say that the set is **finite.** Otherwise, it is **infinite.** We shall concern ourselves only with finite sets.

Look again at Example 1. A set with 3 elements has $2^3 = 8$ subsets. This result can be generalized.

If A is a set with n elements, A has 2^n subsets.

For example, the set $\{a, b, c, d, e\}$ has $2^5 = 32$ subsets.

EXAMPLE 2 **Analyzing Survey Data**

In a survey of 100 college students, 35 were registered in College Algebra, 52 were registered in Computer Science I, and 18 were registered in both courses.

(a) How many students were registered in College Algebra or Computer Science I?

(b) How many were registered in neither course?

Solution (a) First, let A = set of students in College Algebra

B = set of students in Computer Science I

Then the given information tells us that

$$n(A) = 35, \qquad n(B) = 52, \qquad n(A \cap B) = 18,$$

Figure 1

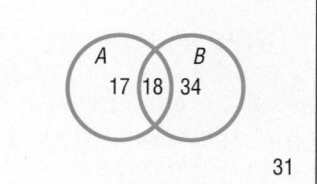

Refer to Figure 1. Since $n(A \cap B) = 18$, we know that the common part of the circles representing set A and set B has 18 elements. In addition, we know that the remaining portion of the circle representing set A will have $35 - 18 = 17$ elements. Similarly, we know that the remaining portion of the circle representing set B has $52 - 18 = 34$ elements. We conclude that $17 + 18 + 34 = 69$ students were registered in College Algebra or Computer Science I.

(b) Since 100 students were surveyed, it follows that $100 - 69 = 31$ were registered in neither course.

∎

━━━ **Now Work** PROBLEMS **15** AND **25**

The solution to Example 2 contains the basis for a general counting formula. If we count the elements in each of two sets A and B, we necessarily count twice any elements that are in both A and B, that is, those elements in $A \cap B$. To count correctly the elements that are in A or B, that is, to find $n(A \cup B)$, we need to subtract those in $A \cap B$ from $n(A) + n(B)$.

THEOREM **Counting Formula**

If A and B are finite sets,

$$n(A \cup B) = n(A) + n(B) - n(A \cap B) \tag{1}$$

Refer back to Example 2. Using (1), we have

$$n(A \cup B) = n(A) + n(B) - n(A \cap B)$$
$$= 35 + 52 - 18$$
$$= 69$$

There are 69 students registered in College Algebra or Computer Science I.

A special case of the counting formula (1) occurs if A and B have no elements in common. In this case, $A \cap B = \emptyset$, so $n(A \cap B) = 0$.

THEOREM **Addition Principle of Counting**

If two sets A and B have no elements in common, that is,

$$\text{if } A \cap B = \emptyset, \text{ then } n(A \cup B) = n(A) + n(B) \tag{2}$$

We can generalize formula (2).

THEOREM **General Addition Principle of Counting**

If, for n sets $A_1, A_2, \ldots, A_n$, no two have elements in common,

$$n(A_1 \cup A_2 \cup \cdots \cup A_n) = n(A_1) + n(A_2) + \cdots + n(A_n) \qquad \textbf{(3)}$$

EXAMPLE 3 **Counting**

In the year 2004, U.S. universities awarded 42,155 doctoral degrees. Table 1 lists the number of doctorates conferred by broad fields of study.

Table 1

Broad Field of Study	Number of Doctorates
Physical sciences	6049
Engineering	5776
Life sciences	8819
Social sciences	6795
Humanities	5467
Education	6635
Professional/other fields	2614

Source: NSF/NIH/USED/NEH/USDA/NASA, *2004 Survey of Earned Doctorates*

(a) How many doctorates were awarded by U.S. universities in physical sciences or life sciences?

(b) How many doctorates were awarded by U.S. universities in physical sciences, life sciences, or engineering?

Solution Let A represent the set of physical science doctorates, B represent the set of life science doctorates, and C represent the set of engineering doctorates. No two of the sets $A, B,$ and C have elements in common since a single degree cannot be classified into more than one broad field of study. Then

$$n(A) = 6049 \qquad n(B) = 8819 \qquad n(C) = 5776$$

(a) Using formula (2), we have

$$n(A \cup B) = n(A) + n(B) = 6049 + 8819 = 14{,}868$$

There were 14,868 doctorates awarded in physical sciences or life sciences.

(b) Using formula (3), we have

$$n(A \cup B \cup C) = n(A) + n(B) + n(C) = 6049 + 8819 + 5776 = 20{,}644$$

There were 20,644 doctorates awarded in physical sciences, life sciences, or engineering.

■

Now Work PROBLEM 29

3 Solve Counting Problems Using the Multiplication Principle

We begin with an example that illustrates the multiplication principle.

| EXAMPLE 4 | Counting the Number of Possible Meals |

The fixed-price dinner at Mabenka Restaurant provides the following choices:

Appetizer: soup or salad

Entree: baked chicken, broiled beef patty, baby beef liver, or roast beef au jus

Dessert: ice cream or cheese cake

How many different meals can be ordered?

Solution Ordering such a meal requires three separate decisions:

Choose an Appetizer **Choose an Entree** **Choose a Dessert**

2 choices 4 choices 2 choices

Look at the **tree diagram** in Figure 2. We see that, for each choice of appetizer, there are 4 choices of entrees. And for each of these $2 \cdot 4 = 8$ choices, there are 2 choices for dessert. A total of

$$2 \cdot 4 \cdot 2 = 16$$

different meals can be ordered.

Figure 2

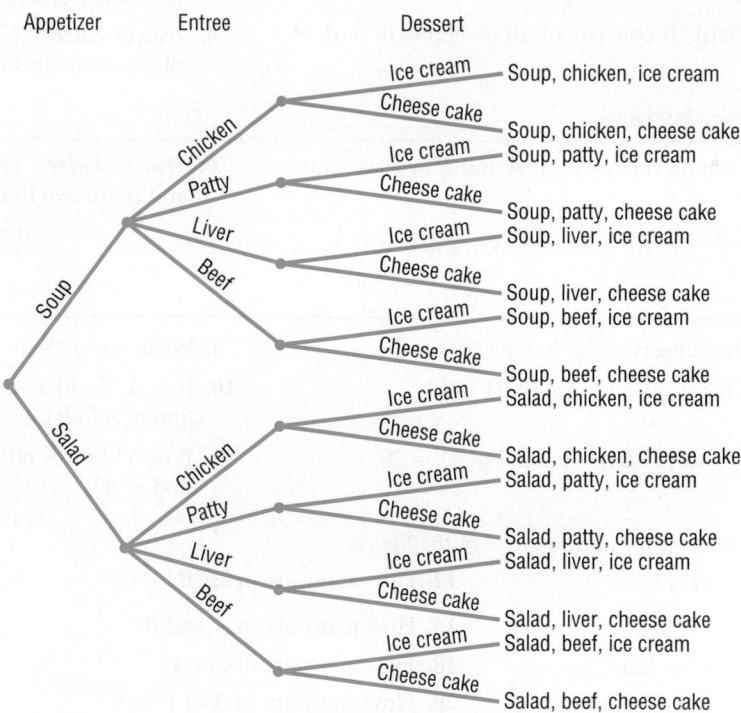

Appetizer Entree Dessert

Example 4 demonstrates a general principle of counting.

THEOREM **Multiplication Principle of Counting**

If a task consists of a sequence of choices in which there are p selections for the first choice, q selections for the second choice, r selections for the third choice, and so on, the task of making these selections can be done in

$$p \cdot q \cdot r \cdots$$

different ways.

EXAMPLE 5	Forming Codes

How many two-symbol code words can be formed if the first symbol is an upper-case letter and the second symbol is a digit?

Solution It sometimes helps to begin by listing some of the possibilities. The code consists of an uppercase letter followed by a digit, so some possibilities are A1, A2, B3, X0, and so on. The task consists of making two selections: the first selection requires choosing an uppercase letter (26 choices) and the second task requires choosing a digit (10 choices). By the Multiplication Principle, there are

$$26 \cdot 10 = 260$$

different codewords of the type described.

Now Work PROBLEM 21

14.1 Assess Your Understanding

'Are You Prepared?' *Answers are given at the end of these exercises. If you get a wrong answer, read the pages listed in red.*

1. The _____ of A and B consists of all elements in either A or B or both. (p. 2)

2. The _____ of A with B consists of all elements in both A and B. (p. 2)

3. *True or False* The intersection of two sets is always a subset of their union. (p. 2)

4. *True or False* If A is a set, the complement of A is the set of all the elements in the universal set that are not in A. (p. 3)

Concepts and Vocabulary

5. The Counting Formula states that if A and B are two finite sets then

$$n(A \cup B) = \underline{\hspace{2cm}}.$$

6. *True or False* The Multiplication Principle states that if A and B are two finite sets then

$$n(A \cup B) = n(A) \cdot n(B)$$

Skill Building

7. Write down all the subsets of $\{a, b, c, d\}$.

8. Write down all the subsets of $\{a, b, c, d, e\}$.

9. If $n(A) = 15, n(B) = 20$, and $n(A \cap B) = 10$, find $n(A \cup B)$.

10. If $n(A) = 30, n(B) = 40$, and $n(A \cup B) = 45$, find $n(A \cap B)$.

11. If $n(A \cup B) = 50, n(A \cap B) = 10$, and $n(B) = 20$, find $n(A)$.

12. If $n(A \cup B) = 60, n(A \cap B) = 40$, and $n(A) = n(B)$, find $n(A)$.

In Problems 13–20, use the information given in the figure.

13. How many are in set A?

14. How many are in set B?

15. How many are in A or B?

16. How many are in A and B?

17. How many are in A but not C?

18. How many are not in A?

19. How many are in A and B and C?

20. How many are in A or B or C?

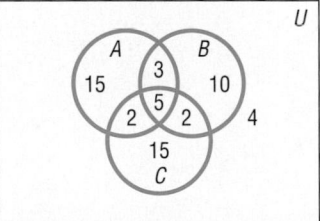

Applications and Extensions

21. **Shirts and Ties** A man has 5 shirts and 3 ties. How many different shirt and tie arrangements can he wear?

22. **Blouses and Skirts** A woman has 5 blouses and 8 skirts. How many different outfits can she wear?

23. **Four-digit Numbers** How many four-digit numbers can be formed using the digits 0, 1, 2, 3, 4, 5, 6, 7, 8, and 9 if the first digit cannot be 0? Repeated digits are allowed.

24. **Five-digit Numbers** How many five-digit numbers can be formed using the digits 0, 1, 2, 3, 4, 5, 6, 7, 8, and 9 if the first digit cannot be 0 or 1? Repeated digits are allowed.

25. **Analyzing Survey Data** In a consumer survey of 500 people, 200 indicated that they would be buying a major appliance within the next month, 150 indicated that they would buy a car, and 25 said that they would purchase both a major appliance and a car. How many will purchase neither? How many will purchase only a car?

26. **Analyzing Survey Data** In a student survey, 200 indicated that they would attend Summer Session I and 150 indicated Summer Session II. If 75 students plan to attend both summer sessions and 275 indicated that they would attend neither session, how many students participated in the survey?

27. Analyzing Survey Data In a survey of 100 investors in the stock market,

 50 owned shares in IBM

 40 owned shares in AT&T

 45 owned shares in GE

 20 owned shares in both IBM and GE

 15 owned shares in both AT&T and GE

 20 owned shares in both IBM and AT&T

 5 owned shares in all three

(a) How many of the investors surveyed did not have shares in any of the three companies?

(b) How many owned just IBM shares?

(c) How many owned just GE shares?

(d) How many owned neither IBM nor GE?

(e) How many owned either IBM or AT&T but no GE?

28. Classifying Blood Types Human blood is classified as either Rh+ or Rh−. Blood is also classified by type: A, if it contains an A antigen but not a B antigen; B, if it contains a B antigen but not an A antigen; AB, if it contains both A and B antigens; and O, if it contains neither antigen. Draw a Venn diagram illustrating the various blood types. Based on this classification, how many different kinds of blood are there?

29. Demographics The following data represent the marital status of males 18 years old and older in 2004.

(a) Determine the number of males 18 years old and older who are widowed or divorced.

(b) Determine the number of males 18 years old and older who are married, widowed, or divorced.

Marital Status	Number (in thousands)
Married	62,486
Widowed	2,643
Divorced	8,954
Never married	29,561

Source: Current Population Survey

30. Demographics The following data represent the marital status of females 18 years old and older in 2004.

(a) Determine the number of females 18 years old and older who are widowed or divorced.

Marital Status	Number (in thousands)
Married	64,829
Widowed	11,140
Divorced	12,803
Never married	23,654

Source: Current Population Survey

(b) Determine the number of females 18 years old and older who are married, widowed, or divorced.

31. Stock Portfolios As a financial planner, you are asked to select one stock each from the following groups: 8 DOW stocks, 15 NASDAQ stocks, and 4 global stocks. How many different portfolios are possible?

Discussion and Writing

32. Make up a problem different from any found in the text that requires the addition principle of counting to solve. Give it to a friend to solve and critique.

33. Investigate the notion of counting as it relates to infinite sets. Write an essay on your findings.

'Are You Prepared?' Answers

1. union **2.** intersection **3.** True **4.** True

14.2 Permutations and Combinations

PREPARING FOR THIS SECTION *Before getting started, review the following:*

- Factorial (Section 13.1, pp. 931–933).

Now Work the 'Are You Prepared?' problems on page 985.

OBJECTIVES **1** Solve Counting Problems Using Permutations Involving *n* Distinct Objects (p. 979)

 2 Solve Counting Problems Using Combinations (p. 982)

 3 Solve Counting Problems Using Permutations Involving *n* Nondistinct Objects (p. 984)

1 Solve Counting Problems Using Permutations Involving *n* Distinct Objects

We begin with a definition.

DEFINITION A **permutation** is an ordered arrangement of *r* objects chosen from *n* objects.

We discuss three types of permutations:

1. The n objects are distinct (different), and repetition is allowed in the selection of r of them. [Distinct, with repetition]

2. The n objects are distinct (different), and repetition is not allowed in the selection of r of them, where $r \leq n$. [Distinct, without repetition]

3. The n objects are not distinct, and we use all of them in the arrangement. [Not distinct]

We take up the first two types here and deal with the third type at the end of this section.

The first type of permutation is handled using the Multiplication Principle.

| EXAMPLE 1 | Counting Airport Codes [Permutation: Distinct, with Repetition] |

The International Airline Transportation Association (IATA) assigns three-letter codes to represent airport locations. For example, the airport code for Ft. Lauderdale, Florida, is FLL. Notice that repetition is allowed in forming this code. How many airport codes are possible?

Solution We are choosing 3 letters from 26 letters and arranging them in order. In the ordered arrangement a letter may be repeated. This is an example of a permutation with repetition in which 3 objects are chosen from 26 distinct objects.

The task of counting the number of such arrangements consists of making three selections. Each selection requires choosing a letter of the alphabet (26 choices). By the Multiplication Principle, there are

$$26 \cdot 26 \cdot 26 = 26^3 = 17{,}576$$

possible airport codes.

The solution given to Example 1 can be generalized.

THEOREM **Permutations: Distinct Objects with Repetition**

The number of ordered arrangements of r objects chosen from n objects, in which the n objects are distinct and repetition is allowed, is n^r.

➝ **Now Work** PROBLEM 33

We begin the discussion of permutations in which the objects are distinct and repetition is not allowed with an example.

| EXAMPLE 2 | Forming Codes [Permutation: Distinct, without Repetition] |

Suppose that we wish to establish a three-letter code using any of the 26 uppercase letters of the alphabet, but we require that no letter be used more than once. How many different three-letter codes are there?

Solution Some of the possibilities are ABC, ABD, ABZ, ACB, CBA, and so on. The task consists of making three selections. The first selection requires choosing from 26 letters. Because no letter can be used more than once, the second selection requires choosing from 25 letters. The third selection requires choosing from 24 letters. (Do you see why?) By the Multiplication Principle, there are

$$26 \cdot 25 \cdot 24 = 15{,}600$$

different three-letter codes with no letter repeated.

For the second type of permutation, we introduce the following notation.

> The notation $P(n, r)$ represents the number of ordered arrangements of r objects chosen from n distinct objects, where $r \leq n$ and repetition is not allowed.

For example, the question posed in Example 2 asks for the number of ways that the 26 letters of the alphabet can be arranged in order using three nonrepeated letters. The answer is

$$P(26, 3) = 26 \cdot 25 \cdot 24 = 15,600$$

EXAMPLE 3	Lining Up People

In how many ways can 5 people be lined up?

Solution The 5 people are distinct. Once a person is in line, that person will not be repeated elsewhere in the line; and, in lining up people, order is important. We have a permutation of 5 objects taken 5 at a time. We can line up 5 people in

$$P(5, 5) = \underbrace{5 \cdot 4 \cdot 3 \cdot 2 \cdot 1}_{5 \text{ factors}} = 120 \text{ ways}$$

■

Now Work PROBLEM 35

To arrive at a formula for $P(n, r)$, we note that the task of obtaining an ordered arrangement of n objects in which only $r \leq n$ of them are used, without repeating any of them, requires making r selections. For the first selection, there are n choices; for the second selection, there are $n - 1$ choices; for the third selection, there are $n - 2$ choices; ...; for the rth selection, there are $n - (r - 1)$ choices. By the Multiplication Principle, we have

$$\overset{\text{1st} \quad \text{2nd} \quad \text{3rd} \qquad \qquad r\text{th}}{P(n, r) = n \cdot (n - 1) \cdot (n - 2) \cdot \cdots \cdot [n - (r - 1)]}$$
$$= n \cdot (n - 1) \cdot (n - 2) \cdot \cdots \cdot (n - r + 1)$$

This formula for $P(n, r)$ can be compactly written using factorial notation.*

$$P(n, r) = n \cdot (n - 1) \cdot (n - 2) \cdot \cdots \cdot (n - r + 1)$$
$$= n \cdot (n - 1) \cdot (n - 2) \cdot \cdots \cdot (n - r + 1) \cdot \frac{(n - r) \cdot \cdots \cdot 3 \cdot 2 \cdot 1}{(n - r) \cdot \cdots \cdot 3 \cdot 2 \cdot 1} = \frac{n!}{(n - r)!}$$

THEOREM **Permutations of r Objects Chosen from n Distinct Objects without Repetition**

The number of arrangements of n objects using $r \leq n$ of them, in which

1. the n objects are distinct,
2. once an object is used it cannot be repeated, and
3. order is important,

is given by the formula

$$P(n, r) = \frac{n!}{(n - r)!} \tag{1}$$

*Recall that $0! = 1$, $1! = 1$, $2! = 2 \cdot 1, \ldots, n! = n(n - 1) \cdot \cdots \cdot 3 \cdot 2 \cdot 1$.

EXAMPLE 4	Computing Permutations

Evaluate: (a) $P(7, 3)$ (b) $P(6, 1)$ (c) $P(52, 5)$

Solution We shall work parts (a) and (b) in two ways.

(a) $P(7, 3) = \underbrace{7 \cdot 6 \cdot 5}_{3 \text{ factors}} = 210$

or

$$P(7, 3) = \frac{7!}{(7 - 3)!} = \frac{7!}{4!} = \frac{7 \cdot 6 \cdot 5 \cdot \cancel{4!}}{\cancel{4!}} = 210$$

(b) $P(6, 1) = \underbrace{6}_{1 \text{ factor}} = 6$

or

$$P(6, 1) = \frac{6!}{(6 - 1)!} = \frac{6!}{5!} = \frac{6 \cdot \cancel{5!}}{\cancel{5!}} = 6$$

Figure 3

(c) Figure 3 shows the solution using a TI-84 Plus graphing calculator. So
$$P(52, 5) = 311,875,200$$

Now Work PROBLEM 7

EXAMPLE 5	The Birthday Problem

All we know about Shannon, Patrick, and Ryan is that they have different birthdays. If we listed all the possible ways this could occur, how many would there be? Assume that there are 365 days in a year.

Solution This is an example of a permutation in which 3 birthdays are selected from a possible 365 days, and no birthday may repeat itself. The number of ways that this can occur is

$$P(365, 3) = \frac{365!}{(365 - 3)!} = \frac{365 \cdot 364 \cdot 363 \cdot \cancel{362!}}{\cancel{362!}} = 365 \cdot 364 \cdot 363 = 48,228,180$$

There are 48,228,180 ways in a group of three people for each to have a different birthday.

Now Work PROBLEM 47

2 Solve Counting Problems Using Combinations

In a permutation, order is important. For example, the arrangements ABC, CAB, BAC, ... are considered different arrangements of the letters A, B, and C. In many situations, though, order is unimportant. For example, in the card game of poker, the order in which the cards are received does not matter; it is the *combination* of the cards that matters.

DEFINITION A **combination** is an arrangement, without regard to order, of r objects selected from n distinct objects without repetition, where $r \leq n$. The notation $C(n, r)$ represents the number of combinations of n distinct objects using r of them.

EXAMPLE 6	Listing Combinations

List all the combinations of the 4 objects a, b, c, d taken 2 at a time. What is $C(4, 2)$?

Solution One combination of a, b, c, d taken 2 at a time is

$$ab$$

We exclude *ba* from the list because order is not important in a combination (this means that we do not distinguish *ab* from *ba*). The list of all combinations of *a*, *b*, *c*, *d* taken 2 at a time is

$$ab, \quad ac, \quad ad, \quad bc, \quad bd, \quad cd$$

so

$$C(4, 2) = 6$$

We can find a formula for $C(n, r)$ by noting that the only difference between a permutation of type 2 (distinct, without repetition) and a combination is that we disregard order in combinations. To determine $C(n, r)$, we need only eliminate from the formula for $P(n, r)$ the number of permutations that were simply rearrangements of a given set of *r* objects. This can be determined from the formula for $P(n, r)$ by calculating $P(r, r) = r!$. So, if we divide $P(n, r)$ by $r!$, we will have the desired formula for $C(n, r)$:

$$C(n, r) = \frac{P(n, r)}{r!} = \frac{\dfrac{n!}{(n - r)!}}{\underset{\uparrow}{r!}} = \frac{n!}{(n - r)!r!}$$

<p style="text-align:center;">Use formula (1).</p>

We have proved the following result:

THEOREM

Number of Combinations of *n* Distinct Objects Taken *r* at a Time

The number of arrangements of *n* objects using $r \leq n$ of them, in which

1. the *n* objects are distinct,
2. once an object is used, it cannot be repeated, and
3. order is not important,

is given by the formula

$$C(n, r) = \frac{n!}{(n - r)!r!} \tag{2}$$

Based on formula (2), we discover that the symbol $C(n, r)$ and the symbol $\dbinom{n}{r}$ for the binomial coefficients are, in fact, the same. The Pascal triangle (see Section 13.5) can be used to find the value of $C(n, r)$. However, because it is more practical and convenient, we will use formula (2) instead.

EXAMPLE 7 **Using Formula (2)**

Use formula (2) to find the value of each expression.

(a) $C(3, 1)$ (b) $C(6, 3)$ (c) $C(n, n)$ (d) $C(n, 0)$ (e) $C(52, 5)$

Solution (a) $C(3, 1) = \dfrac{3!}{(3 - 1)!1!} = \dfrac{3!}{2!1!} = \dfrac{3 \cdot 2 \cdot 1}{2 \cdot 1 \cdot 1} = 3$

(b) $C(6, 3) = \dfrac{6!}{(6 - 3)!3!} = \dfrac{6 \cdot 5 \cdot 4 \cdot 3!}{3! \cdot 3!} = \dfrac{6 \cdot 5 \cdot 4}{6} = 20$

(c) $C(n, n) = \dfrac{n!}{(n - n)!n!} = \dfrac{n!}{0!n!} = \dfrac{1}{1} = 1$

Figure 4

(d) $C(n, 0) = \dfrac{n!}{(n-0)!0!} = \dfrac{\cancel{n!}}{\cancel{n!}0!} = \dfrac{1}{1} = 1$

(e) Figure 4 shows the solution using a TI-84 Plus graphing calculator. So

$$C(52, 5) = 2,598,960$$

—— **Now Work** PROBLEM 15

EXAMPLE 8 | **Forming Committees**

How many different committees of 3 people can be formed from a pool of 7 people?

Solution The 7 people are distinct. More important, though, is the observation that the order of being selected for a committee is not significant. The problem asks for the number of combinations of 7 objects taken 3 at a time.

$$C(7, 3) = \frac{7!}{4!3!} = \frac{7 \cdot 6 \cdot 5 \cdot \cancel{4!}}{\cancel{4!}3!} = \frac{7 \cdot \cancel{6} \cdot 5}{\cancel{6}} = 35$$

Thirty-five different committees can be formed.

EXAMPLE 9 | **Forming Committees**

In how many ways can a committee consisting of 2 faculty members and 3 students be formed if 6 faculty members and 10 students are eligible to serve on the committee?

Solution The problem can be separated into two parts: the number of ways that the faculty members can be chosen, $C(6, 2)$, and the number of ways that the student members can be chosen, $C(10, 3)$. By the Multiplication Principle, the committee can be formed in

$$C(6, 2) \cdot C(10, 3) = \frac{6!}{4!2!} \cdot \frac{10!}{7!3!} = \frac{6 \cdot 5 \cdot \cancel{4!}}{\cancel{4!}2!} \cdot \frac{10 \cdot 9 \cdot 8 \cdot \cancel{7!}}{\cancel{7!}3!}$$

$$= \frac{30}{2} \cdot \frac{720}{6} = 1800 \text{ ways}$$

—— **Now Work** PROBLEM 49

3 Solve Counting Problems Using Permutations Involving *n* Nondistinct Objects

We begin with an example.

EXAMPLE 10 | **Forming Different Words**

How many different words (real or imaginary) can be formed using all the letters in the word REARRANGE?

Solution Each word formed will have 9 letters: 3 R's, 2 A's, 2 E's, 1 N, and 1 G. To construct each word, we need to fill in 9 positions with the 9 letters:

$$\underline{1} \ \underline{2} \ \underline{3} \ \underline{4} \ \underline{5} \ \underline{6} \ \underline{7} \ \underline{8} \ \underline{9}$$

The process of forming a word consists of five tasks:

Task 1: Choose the positions for the 3 R's.
Task 2: Choose the positions for the 2 A's.
Task 3: Choose the positions for the 2 E's.

Task 4: Choose the position for the 1 N.

Task 5: Choose the position for the 1 G.

Task 1 can be done in $C(9, 3)$ ways. There then remain 6 positions to be filled, so Task 2 can be done in $C(6, 2)$ ways. There remain 4 positions to be filled, so Task 3 can be done in $C(4, 2)$ ways. There remain 2 positions to be filled, so Task 4 can be done in $C(2, 1)$ ways. The last position can be filled in $C(1, 1)$ way. Using the Multiplication Principle, the number of possible words that can be formed is

$$C(9, 3) \cdot C(6, 2) \cdot C(4, 2) \cdot C(2, 1) \cdot C(1, 1) = \frac{9!}{3! \cdot 6!} \cdot \frac{6!}{2! \cdot 4!} \cdot \frac{4!}{2! \cdot 2!} \cdot \frac{2!}{1! \cdot 1!} \cdot \frac{1!}{0! \cdot 1!}$$

$$= \frac{9!}{3! \cdot 2! \cdot 2! \cdot 1! \cdot 1!} = 15,120$$

15,120 possible words can be formed.

The form of the expression before the answer to Example 10 is suggestive of a general result. Had the letters in REARRANGE each been different, there would have been $P(9, 9) = 9!$ possible words formed. This is the numerator of the answer. The presence of 3 R's, 2 A's, and 2 E's reduces the number of different words, as the entries in the denominator illustrate. We are led to the following result:

THEOREM

Permutations Involving n Objects That Are Not Distinct

The number of permutations of n objects of which n_1 are of one kind, n_2 are of a second kind, ..., and n_k are of a kth kind is given by

$$\frac{n!}{n_1! \cdot n_2! \cdot \cdots \cdot n_k!} \tag{3}$$

where $n = n_1 + n_2 + \cdots + n_k$.

EXAMPLE 11 **Arranging Flags**

How many different vertical arrangements are there of 8 flags if 4 are white, 3 are blue, and 1 is red?

Solution We seek the number of permutations of 8 objects, of which 4 are of one kind, 3 are of a second kind, and 1 is of a third kind. Using formula (3), we find that there are

$$\frac{8!}{4! \cdot 3! \cdot 1!} = \frac{8 \cdot 7 \cdot 6 \cdot 5 \cdot 4!}{4! \cdot 3! \cdot 1!} = 280 \text{ different arrangements}$$

━━━▶ **Now Work** PROBLEM 51

14.2 Assess Your Understanding

'Are You Prepared?' *Answers are given at the end of these exercises. If you get a wrong answer, read the pages listed in red.*

1. $0! = $ _____; $1! = $ _____. (p. 932)

2. *True or False* $n! = \dfrac{(n+1)!}{n}$. (pp. 932–933)

Concepts and Vocabulary

3. A(n) _____ is an ordered arrangement of r objects chosen from n objects.

4. A(n) _____ is an arrangement of r objects chosen from n distinct objects, without repetition and without regard to order.

5. *True or False* In a combination problem, order is not important.

6. *True or False* In some permutation problems, once an object is used, it cannot be repeated.

Skill Building

In Problems 7–14, find the value of each permutation.

7. $P(6, 2)$ **8.** $P(7, 2)$ **9.** $P(4, 4)$ **10.** $P(8, 8)$

11. $P(7, 0)$ **12.** $P(9, 0)$ **13.** $P(8, 4)$ **14.** $P(8, 3)$

In Problems 15–22, use formula (2) to find the value of each combination.

15. $C(8, 2)$ **16.** $C(8, 6)$ **17.** $C(7, 4)$ **18.** $C(6, 2)$

19. $C(15, 15)$ **20.** $C(18, 1)$ **21.** $C(26, 13)$ **22.** $C(18, 9)$

Applications and Extensions

23. List all the ordered arrangements of 5 objects $a, b, c, d,$ and e choosing 3 at a time without repetition. What is $P(5, 3)$?

24. List all the ordered arrangements of 5 objects $a, b, c, d,$ and e choosing 2 at a time without repetition. What is $P(5, 2)$?

25. List all the ordered arrangements of 4 objects 1, 2, 3, and 4 choosing 3 at a time without repetition. What is $P(4, 3)$?

26. List all the ordered arrangements of 6 objects 1, 2, 3, 4, 5, and 6 choosing 3 at a time without repetition. What is $P(6, 3)$?

27. List all the combinations of 5 objects $a, b, c, d,$ and e taken 3 at a time. What is $C(5, 3)$?

28. List all the combinations of 5 objects $a, b, c, d,$ and e taken 2 at a time. What is $C(5, 2)$?

29. List all the combinations of 4 objects 1, 2, 3, and 4 taken 3 at a time. What is $C(4, 3)$?

30. List all the combinations of 6 objects 1, 2, 3, 4, 5, and 6 taken 3 at a time. What is $C(6, 3)$?

31. Forming Codes How many two-letter codes can be formed using the letters $A, B, C,$ and D? Repeated letters are allowed.

32. Forming Codes How many two-letter codes can be formed using the letters $A, B, C, D,$ and E? Repeated letters are allowed.

33. Forming Numbers How many three-digit numbers can be formed using the digits 0 and 1? Repeated digits are allowed.

34. Forming Numbers How many three-digit numbers can be formed using the digits 0, 1, 2, 3, 4, 5, 6, 7, 8, and 9? Repeated digits are allowed.

35. Lining Up People In how many ways can 4 people be lined up?

36. Stacking Boxes In how many ways can 5 different boxes be stacked?

37. Forming Codes How many different three-letter codes are there if only the letters $A, B, C, D,$ and E can be used and no letter can be used more than once?

38. Forming Codes How many different four-letter codes are there if only the letters $A, B, C, D, E,$ and F can be used and no letter can be used more than once?

39. Stocks on the NYSE Companies whose stocks are listed on the New York Stock Exchange (NYSE) have their company name represented by either 1, 2, or 3 letters (repetition of letters is allowed). What is the maximum number of companies that can be listed on the NYSE?

40. Stocks on the NASDAQ Companies whose stocks are listed on the NASDAQ stock exchange have their company name represented by either 4 or 5 letters (repetition of letters is allowed). What is the maximum number of companies that can be listed on the NASDAQ?

41. Establishing Committees In how many ways can a committee of 4 students be formed from a pool of 7 students?

42. Establishing Committees In how many ways can a committee of 3 professors be formed from a department having 8 professors?

43. Possible Answers on a True/False Test How many arrangements of answers are possible for a true/false test with 10 questions?

44. Possible Answers on a Multiple-choice Test How many arrangements of answers are possible in a multiple-choice test with 5 questions, each of which has 4 possible answers?

45. Arranging Books Five different mathematics books are to be arranged on a student's desk. How many arrangements are possible?

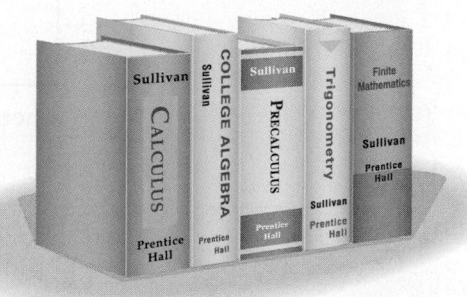

46. Forming License Plate Numbers How many different license plate numbers can be made using 2 letters followed by 4 digits selected from the digits 0 through 9, if

 (a) letters and digits may be repeated?

 (b) letters may be repeated, but digits may not be repeated?

 (c) neither letters nor digits may be repeated?

47. Birthday Problem In how many ways can 2 people each have different birthdays? Assume that there are 365 days in a year.

48. Birthday Problem In how many ways can 5 people each have different birthdays? Assume that there are 365 days in a year.

49. Forming a Committee A student dance committee is to be formed consisting of 2 boys and 3 girls. If the membership is to be chosen from 4 boys and 8 girls, how many different committees are possible?

50. Forming a Committee The student relations committee of a college consists of 2 administrators, 3 faculty members, and 5 students. Four administrators, 8 faculty members, and 20 students are eligible to serve. How many different committees are possible?

51. Forming Words How many different 9-letter words (real or imaginary) can be formed from the letters in the word ECONOMICS?

52. Forming Words How many different 11-letter words (real or imaginary) can be formed from the letters in the word MATHEMATICS?

53. Selecting Objects An urn contains 7 white balls and 3 red balls. Three balls are selected. In how many ways can the 3 balls be drawn from the total of 10 balls:

(a) If 2 balls are white and 1 is red?

(b) If all 3 balls are white?

(c) If all 3 balls are red?

54. Selecting Objects An urn contains 15 red balls and 10 white balls. Five balls are selected. In how many ways can the 5 balls be drawn from the total of 25 balls:

(a) If all 5 balls are red?

(b) If 3 balls are red and 2 are white?

(c) If at least 4 are red balls?

55. Senate Committees The U.S. Senate has 100 members. Suppose that it is desired to place each senator on exactly 1 of 7 possible committees. The first committee has 22 members, the second has 13, the third has 10, the fourth has 5, the fifth has 16, and the sixth and seventh have 17 apiece. In how many ways can these committees be formed?

56. Football Teams A defensive football squad consists of 25 players. Of these, 10 are linemen, 10 are linebackers, and 5 are safeties. How many different teams of 5 linemen, 3 linebackers, and 3 safeties can be formed?

57. Baseball In the American Baseball League, a designated hitter may be used. How many batting orders is it possible for a manager to use? (There are 9 regular players on a team.)

58. Baseball In the National Baseball League, the pitcher usually bats ninth. If this is the case, how many batting orders is it possible for a manager to use?

59. Baseball Teams A baseball team has 15 members. Four of the players are pitchers, and the remaining 11 members can play any position. How many different teams of 9 players can be formed?

60. World Series In the World Series the American League team (A) and the National League team (N) play until one team wins four games. If the sequence of winners is designated by letters (for example, $NAAAA$ means that the National League team won the first game and the American League won the next four), how many different sequences are possible?

61. Basketball Teams A basketball team has 6 players who play guard (2 of 5 starting positions). How many different teams are possible, assuming that the remaining 3 positions are filled and it is not possible to distinguish a left guard from a right guard?

62. Basketball Teams On a basketball team of 12 players, 2 only play center, 3 only play guard, and the rest play forward (5 players on a team: 2 forwards, 2 guards, and 1 center). How many different teams are possible, assuming that it is not possible to distinguish left and right guards and left and right forwards?

63. Combination Locks A combination lock displays 50 numbers. To open it, you turn to a number, then rotate clockwise to a second number, and then counterclockwise to the third number.

(a) How many different lock combinations are there?

(b) Comment on the description of such a lock as a *combination* lock.

Discussion and Writing

64. Create a problem different from any found in the text that requires a permutation to solve. Give it to a friend to solve and critique.

65. Create a problem different from any found in the text that requires a combination to solve. Give it to a friend to solve and critique.

66. Explain the difference between a permutation and a combination. Give an example to illustrate your explanation.

'Are You Prepared?' Answers

1. 1; 1

2. False

14.3 Probability

Probability is an area of mathematics that deals with experiments that yield random results, yet admit a certain regularity. Such experiments do not always produce the same result or outcome, so the result of any one observation is not predictable. However, the results of the experiment over a long period do produce regular patterns that enable us to predict with remarkable accuracy.

EXAMPLE 1	Tossing a Fair Coin

In tossing a fair coin, we know that the outcome is either a head or a tail. On any particular throw, we cannot predict what will happen, but, if we toss the coin many times, we observe that the number of times that a head comes up is approximately equal to the number of times that we get a tail. It seems reasonable, therefore, to assign a probability of $\frac{1}{2}$ that a head comes up and a probability of $\frac{1}{2}$ that a tail comes up.

1 Construct Probability Models

The discussion in Example 1 constitutes the construction of a **probability model** for the experiment of tossing a fair coin once. A probability model has two components: a sample space and an assignment of probabilities. A **sample space** S is a set whose elements represent all the possibilities that can occur as a result of the experiment. Each element of S is called an **outcome.** To each outcome, we assign a number, called the **probability** of that outcome, which has two properties:

1. The probability assigned to each outcome is nonnegative.

2. The sum of all the probabilities equals 1.

DEFINITION A **probability model** with the sample space

$$S = \{e_1, e_2, \ldots, e_n\}$$

where $e_1, e_2, \ldots, e_n$ are the possible outcomes and $P(e_1), P(e_2), \ldots, P(e_n)$ are the respective probabilities of these outcomes, requires that

$$P(e_1) \geq 0, P(e_2) \geq 0, \ldots, P(e_n) \geq 0 \tag{1}$$

$$\sum_{i=1}^{n} P(e_i) = P(e_1) + P(e_2) + \cdots + P(e_n) = 1 \tag{2}$$

EXAMPLE 2	Determining Probability Models

In a bag of M&Ms,™ the candies are colored red, green, blue, brown, yellow, and orange. Suppose that a candy is drawn from the bag and the color is recorded. The sample space of this experiment is {red, green, blue, brown, yellow, orange}. Determine which of the following are probability models.

(a)

Outcome	Probability
red	0.3
green	0.15
blue	0
brown	0.15
yellow	0.2
orange	0.2

(b)

Outcome	Probability
red	0.1
green	0.1
blue	0.1
brown	0.4
yellow	0.2
orange	0.3

(c)

Outcome	Probability
red	0.3
green	−0.3
blue	0.2
brown	0.4
yellow	0.2
orange	0.2

(d)

Outcome	Probability
red	0
green	0
blue	0
brown	0
yellow	1
orange	0

Solution

(a) This model is a probability model since all the outcomes have probabilities that are nonnegative and the sum of the probabilities is 1.

(b) This model is not a probability model because the sum of the probabilities is not 1.

(c) This model is not a probability model because $P(\text{green})$ is less than 0. Recall, all probabilities must be nonnegative.

(d) This model is a probability model because all the outcomes have probabilities that are nonnegative, and the sum of the probabilities is 1. Notice that $P(\text{yellow}) = 1$, meaning that this outcome will occur with 100% certainty each time that the experiment is repeated. This means that the bag of M&Ms™ has only yellow candies.

 Now Work PROBLEM 7

EXAMPLE 3	**Constructing a Probability Model**

An experiment consists of rolling a fair die once. A die is a cube with each face having either 1, 2, 3, 4, 5, or 6 dots on it. See Figure 5. Construct a probability model for this experiment.

Figure 5

Solution A sample space S consists of all the possibilities that can occur. Because rolling the die will result in one of six faces showing, the sample space S consists of

$$S = \{1, 2, 3, 4, 5, 6\}$$

Because the die is fair, one face is no more likely to occur than another. As a result, our assignment of probabilities is

$$P(1) = \frac{1}{6} \qquad P(2) = \frac{1}{6}$$

$$P(3) = \frac{1}{6} \qquad P(4) = \frac{1}{6}$$

$$P(5) = \frac{1}{6} \qquad P(6) = \frac{1}{6}$$

Now suppose that a die is loaded (weighted) so that the probability assignments are

$$P(1) = 0, \quad P(2) = 0, \quad P(3) = \frac{1}{3}, \quad P(4) = \frac{2}{3}, \quad P(5) = 0, \quad P(6) = 0$$

This assignment would be made if the die were loaded so that only a 3 or 4 could occur and the 4 is twice as likely as the 3 to occur. This assignment is consistent with the definition, since each assignment is nonnegative and the sum of all the probability assignments equals 1.

━━ **Now Work** PROBLEM 23

EXAMPLE 4	**Constructing a Probability Model**

An experiment consists of tossing a coin. The coin is weighted so that heads (H) is three times as likely to occur as tails (T). Construct a probability model for this experiment.

Solution The sample space S is $S = \{H, T\}$. If x denotes the probability that a tail occurs,

$$P(T) = x \quad \text{and} \quad P(H) = 3x$$

Since the sum of the probabilities of the possible outcomes must equal 1, we have

$$P(T) + P(H) = x + 3x = 1$$
$$4x = 1$$
$$x = \frac{1}{4}$$

We assign the probabilities

$$P(T) = \frac{1}{4}, \qquad P(H) = \frac{3}{4}$$

━━ **Now Work** PROBLEM 27

In working with probability models, the term **event** is used to describe a set of possible outcomes of the experiment. An event E is some subset of the sample space S. The **probability of an event** $E, E \neq \varnothing$, denoted by $P(E)$, is defined as the sum of the probabilities of the outcomes in E. We can also think of the probability of an event E as the likelihood that the event E occurs. If $E = \varnothing$, then $P(E) = 0$; if $E = S$, then $P(E) = P(S) = 1$.

2 Compute Probabilities of Equally Likely Outcomes

When the same probability is assigned to each outcome of the sample space, the experiment is said to have **equally likely outcomes.**

THEOREM **Probability for Equally Likely Outcomes**

If an experiment has n equally likely outcomes and if the number of ways that an event E can occur is m, then the probability of E is

$$P(E) = \frac{\text{Number of ways that } E \text{ can occur}}{\text{Number of possible outcomes}} = \frac{m}{n} \qquad (3)$$

If S is the sample space of this experiment,

$$P(E) = \frac{n(E)}{n(S)} \qquad (4)$$

EXAMPLE 5	Calculating Probabilities of Events Involving Equally Likely Outcomes

Calculate the probability that in a 3-child family there are 2 boys and 1 girl. Assume equally likely outcomes.

Solution

We begin by constructing a tree diagram to help in listing the possible outcomes of the experiment. See Figure 6, where B stands for boy and G for girl. The sample space S of this experiment is

$$S = \{BBB, BBG, BGB, BGG, GBB, GBG, GGB, GGG\}$$

so $n(S) = 8$.

We wish to know the probability of the event E: "having two boys and one girl." From Figure 6, we conclude that $E = \{BBG, BGB, GBB\}$, so $n(E) = 3$. Since the outcomes are equally likely, the probability of E is

$$P(E) = \frac{n(E)}{n(S)} = \frac{3}{8}$$

Figure 6

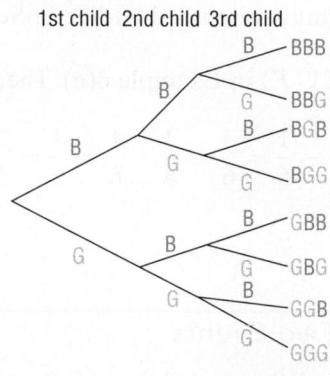

1st child 2nd child 3rd child

Now Work PROBLEM 37

So far, we have calculated probabilities of single events. We will now compute probabilities of multiple events, called **compound probabilities.**

EXAMPLE 6	Computing Compound Probabilities

Consider the experiment of rolling a single fair die. Let E represent the event "roll an odd number," and let F represent the event "roll a 1 or 2."

(a) Write the event E and F. What is $n(E \cap F)$?
(b) Write the event E or F. What is $n(E \cup F)$?
(c) Compute $P(E)$. Compute $P(F)$.
(d) Compute $P(E \cap F)$.
(e) Compute $P(E \cup F)$.

Solution

The sample space S of the experiment is $\{1, 2, 3, 4, 5, 6\}$, so $n(S) = 6$. Since the die is fair, the outcomes are equally likely. The event E: "roll an odd number" is $\{1, 3, 5\}$, and the event F: "roll a 1 or 2" is $\{1, 2\}$, so $n(E) = 3$ and $n(F) = 2$.

(a) The word *and* in probability means the intersection of two events. The event E and F is
$$E \cap F = \{1, 3, 5\} \cap \{1, 2\} = \{1\} \qquad n(E \cap F) = 1$$

(b) The word *or* in probability means the union of the two events. The event E or F is
$$E \cup F = \{1, 3, 5\} \cup \{1, 2\} = \{1, 2, 3, 5\} \qquad n(E \cup F) = 4$$

(c) We use formula (4). Then
$$P(E) = \frac{n(E)}{n(S)} = \frac{3}{6} = \frac{1}{2} \qquad P(F) = \frac{n(F)}{n(S)} = \frac{2}{6} = \frac{1}{3}$$

(d) $P(E \cap F) = \dfrac{n(E \cap F)}{n(S)} = \dfrac{1}{6}$

(e) $P(E \cup F) = \dfrac{n(E \cup F)}{n(S)} = \dfrac{4}{6} = \dfrac{2}{3}$

3 Find Probabilities of the Union of Two Events

The next formula can be used to find the probability of the union of two events.

THEOREM

For any two events E and F,

$$P(E \cup F) = P(E) + P(F) - P(E \cap F) \qquad (5)$$

This result is a consequence of the counting formula discussed earlier in Section 14.1.

For example, we can use formula (5) to find $P(E \cup F)$ in Example 6(e). Then

$$P(E \cup F) = P(E) + P(F) - P(E \cap F) = \frac{1}{2} + \frac{1}{3} - \frac{1}{6} = \frac{3}{6} + \frac{2}{6} - \frac{1}{6} = \frac{4}{6} = \frac{2}{3}$$

as before.

EXAMPLE 7 **Computing Probabilities of the Union of Two Events**

If $P(E) = 0.2$, $P(F) = 0.3$, and $P(E \cap F) = 0.1$, find the probability of E or F. That is, find $P(E \cup F)$.

Solution We use formula (5).

$$\text{Probability of } E \text{ or } F = P(E \cup F) = P(E) + P(F) - P(E \cap F)$$
$$= 0.2 + 0.3 - 0.1 = 0.4$$

A Venn diagram can sometimes be used to obtain probabilities. To construct a Venn diagram representing the information in Example 7, we draw two sets E and F. We begin with the fact that $P(E \cap F) = 0.1$. See Figure 7(a). Then, since $P(E) = 0.2$ and $P(F) = 0.3$, we fill in E with $0.2 - 0.1 = 0.1$ and F with $0.3 - 0.1 = 0.2$. See Figure 7(b). Since $P(S) = 1$, we complete the diagram by inserting $1 - (0.1 + 0.1 + 0.2) = 0.6$ outside the circles. See Figure 7(c). Now it is easy to see, for example, that the probability of F, but not E, is 0.2. Also, the probability of neither E nor F is 0.6.

Figure 7

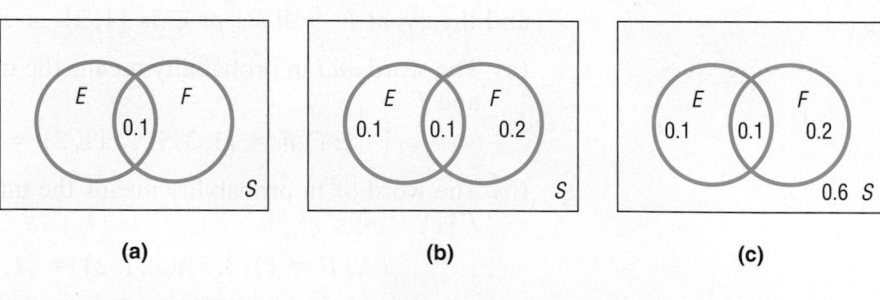

(a) (b) (c)

Now Work PROBLEM 45

If events E and F are disjoint so that $E \cap F = \varnothing$, we say they are **mutually exclusive.** In this case, $P(E \cap F) = 0$, and formula (5) takes the following form:

THEOREM

Mutually Exclusive Events

If E and F are mutually exclusive events,

$$P(E \cup F) = P(E) + P(F) \qquad (6)$$

EXAMPLE 8	Computing Probabilities of the Union of Two Mutually Exclusive Events

If $P(E) = 0.4$ and $P(F) = 0.25$, and E and F are mutually exclusive, find $P(E \cup F)$.

Solution Since E and F are mutually exclusive, we use formula (6).

$$P(E \cup F) = P(E) + P(F) = 0.4 + 0.25 = 0.65$$

➟ **Now Work** PROBLEM 47

4 Use the Complement Rule to Find Probabilities

Recall, if A is a set, the complement of A, denoted $\overline{A}$, is the set of all elements in the universal set U not in A. We similarly define the complement of an event.

DEFINITION **Complement of an Event**

Let S denote the sample space of an experiment, and let E denote an event. The **complement of E,** denoted $\overline{E}$, is the set of all outcomes in the sample space S that are not outcomes in the event E.

The complement of an event E, that is, $\overline{E}$, in a sample space S has the following two properties:

$$E \cap \overline{E} = \varnothing, \qquad E \cup \overline{E} = S$$

Since E and $\overline{E}$ are mutually exclusive, it follows from (6) that

$$P(E \cup \overline{E}) = P(S) = 1, \qquad P(E) + P(\overline{E}) = 1, \qquad P(\overline{E}) = 1 - P(E)$$

We have the following result:

THEOREM **Computing Probabilities of Complementary Events**

If E represents any event and $\overline{E}$ represents the complement of E, then

$$P(\overline{E}) = 1 - P(E) \tag{7}$$

EXAMPLE 9	Computing Probabilities Using Complements

On the local news the weather reporter stated that the probability of rain tomorrow is 40%. What is the probability that it will not rain?

Solution The complement of the event "rain" is "no rain."

$$P(\text{no rain}) = 1 - P(\text{rain}) = 1 - 0.4 = 0.6$$

There is a 60% chance of no rain tomorrow.

➟ **Now Work** PROBLEM 51

EXAMPLE 10	Birthday Problem

What is the probability that in a group of 10 people at least 2 people have the same birthday? Assume that there are 365 days in a year.

Solution We assume that a person is as likely to be born on one day as another, so we have equally likely outcomes.

We first determine the number of outcomes in the sample space S. There are 365 possibilities for each person's birthday. Since there are 10 people in the group, there are 365^{10} possibilities for the birthdays. [For one person in the group, there are 365 days on which his or her birthday can fall; for two people, there are $(365)(365) = 365^2$ pairs of days; and, in general, using the Multiplication Principle, for n people there are 365^n possibilities.] So

$$n(S) = 365^{10}$$

We wish to find the probability of the event E: "at least two people have the same birthday." It is difficult to count the elements in this set; it is much easier to count the elements of the complementary event $\overline{E}$: "no two people have the same birthday."

We find $n(\overline{E})$ as follows: Choose one person at random. There are 365 possibilities for his or her birthday. Choose a second person. There are 364 possibilities for this birthday, if no two people are to have the same birthday. Choose a third person. There are 363 possibilities left for this birthday. Finally, we arrive at the tenth person. There are 356 possibilities left for this birthday. By the Multiplication Principle, the total number of possibilities is

$$n(\overline{E}) = 365 \cdot 364 \cdot 363 \cdot \ \cdots \ \cdot 356$$

The probability of the event $\overline{E}$ is

$$P(\overline{E}) = \frac{n(\overline{E})}{n(S)} = \frac{365 \cdot 364 \cdot 363 \cdot \ \cdots \ \cdot 356}{365^{10}} \approx 0.883$$

The probability of two or more people in a group of 10 people having the same birthday is then

$$P(E) = 1 - P(\overline{E}) \approx 1 - 0.883 = 0.117$$

The birthday problem can be solved for any group size. The following table gives the probabilities for two or more people having the same birthday for various group sizes. Notice that the probability is greater than $\frac{1}{2}$ for any group of 23 or more people.

	Number of People															
	5	10	15	20	21	22	23	24	25	30	40	50	60	70	80	90
Probability That Two or More Have the Same Birthday	0.027	0.117	0.253	0.411	0.444	0.476	0.507	0.538	0.569	0.706	0.891	0.970	0.994	0.99916	0.99991	0.99999

Now Work PROBLEM 71

Historical Feature

Blaise Pascal
(1623–1662)

Set theory, counting, and probability first took form as a systematic theory in an exchange of letters (1654) between Pierre de Fermat (1601–1665) and Blaise Pascal (1623–1662). They discussed the problem of how to divide the stakes in a game that is interrupted before completion, knowing how many points each player needs to win. Fermat solved the problem by list-

ing all possibilities and counting the favorable ones, whereas Pascal made use of the triangle that now bears his name. As mentioned in the text, the entries in Pascal's triangle are equivalent to $C(n, r)$. This recognition of the role of $C(n, r)$ in counting is the foundation of all further developments.

The first book on probability, the work of Christiaan Huygens (1629–1695), appeared in 1657. In it, the notion of mathematical expectation is explored. This allows the calculation of the profit or loss that a

gambler might expect, knowing the probabilities involved in the game (see the Historical Problems that follow).

Although Girolamo Cardano (1501–1576) wrote a treatise on probability, it was not published until 1663 in Cardano's collected works, and this was too late to have any effect on the early development of the theory.

In 1713, the posthumously published *Ars Conjectandi* of Jakob Bernoulli (1654–1705) gave the theory the form it would have until 1900. Recently, both combinatorics (counting) and probability have undergone rapid development due to the use of computers.

A final comment about notation. The notations $C(n, r)$ and $P(n, r)$ are variants of a form of notation developed in England after 1830. The

notation $\binom{n}{r}$ for $C(n, r)$ goes back to Leonhard Euler (1707–1783), but is now losing ground because it has no clearly related symbolism of the same type for permutations. The set symbols $\cup$ and $\cap$ were introduced by Giuseppe Peano (1858–1932) in 1888 in a slightly different context. The inclusion symbol $\subset$ was introduced by E. Schroeder (1841–1902) about 1890. The treatment of set theory in the text is due to George Boole (1815–1864), who wrote $A + B$ for $A \cup B$ and AB for $A \cap B$ (statisticians still use AB for $A \cap B$).

Historical Problems

1. *The Problem Discussed by Fermat and Pascal* A game between two equally skilled players, *A* and *B*, is interrupted when *A* needs 2 points to win and *B* needs 3 points. In what proportion would the stakes be divided?

 (a) *Fermat's solution* List all possible outcomes that can occur as a result of four more plays. The probabilities for *A* to win and *B* to win then determine how the stakes should be divided.

 (b) *Pascal's solution* Use combinations to determine the number of ways that the 2 points needed for *A* to win could occur in four plays. Then use combinations to determine the number of ways that the 3 points needed for *B* to win could occur. This is trickier than it looks, since *A* can win with 2 points in either two plays, three plays, or four plays. Compute the probabilities and compare with the results in part (a).

2. *Huygens's Mathematical Expectation* In a game with *n* possible outcomes with probabilities $p_1, p_2, \ldots, p_n$, suppose that the *net* winnings are $w_1, w_2, \ldots, w_n$, respectively. Then the mathematical expectation is

$$E = p_1 w_1 + p_2 w_2 + \cdots + p_n w_n$$

The number *E* represents the profit or loss per game in the long run. The following problems are a modification of those of Huygens.

 (a) A fair die is tossed. A gambler wins \$3 if he throws a 6 and \$6 if he throws a 5. What is his expectation?
 [**Hint:** $w_1 = w_2 = w_3 = w_4 = 0$]

 (b) A gambler plays the same game as in part (a), but now the gambler must pay \$1 to play. This means that $w_5 = \$5, w_6 = \2, and $w_1 = w_2 = w_3 = w_4 = -\1 What is the expectation?

14.3 Assess Your Understanding

Concepts and Vocabulary

1. When the same probability is assigned to each outcome of a sample space, the experiment is said to have _____ _____ outcomes.

2. The _____ of an event *E* is the set of all outcomes in the sample space *S* that are not outcomes in the event *E*.

3. *True or False* The probability of an event can never equal 0.

4. *True or False* In a probability model, the sum of all probabilities is 1.

Skill Building

5. In a probability model, which of the following numbers could be the probability of an outcome?

$$0, \quad 0.01, \quad 0.35, \quad -0.4, \quad 1, \quad 1.4$$

6. In a probability model, which of the following numbers could be the probability of an outcome?

$$1.5, \quad \frac{1}{2}, \quad \frac{3}{4}, \quad \frac{2}{3}, \quad 0, \quad -\frac{1}{4}$$

7. Determine whether the following is a probability model.

Outcome	Probability
1	0.2
2	0.3
3	0.1
4	0.4

8. Determine whether the following is a probability model.

Outcome	Probability
Jim	0.4
Bob	0.3
Faye	0.1
Patricia	0.2

9. Determine whether the following is a probability model.

Outcome	Probability
Linda	0.3
Jean	0.2
Grant	0.1
Ron	0.3

10. Determine whether the following is a probability model.

Outcome	Probability
Lanny	0.3
Joanne	0.2
Nelson	0.1
Rich	0.5
Judy	−0.1

In Problems 11–16, construct a probability model for each experiment.

11. Tossing a fair coin twice

12. Tossing two fair coins once

13. Tossing two fair coins, then a fair die

14. Tossing a fair coin, a fair die, and then a fair coin

15. Tossing three fair coins once

16. Tossing one fair coin three times

In Problems 17–22, use the following spinners to construct a probability model for each experiment.

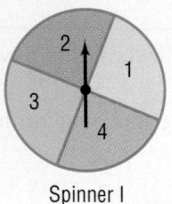
Spinner I
(4 equal areas)

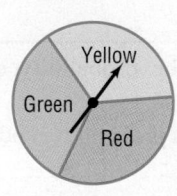
Spinner II
(3 equal areas)

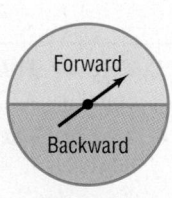
Spinner III
(2 equal areas)

17. Spin spinner I, then spinner II. What is the probability of getting a 2 or a 4, followed by Red?

18. Spin spinner III, then spinner II. What is the probability of getting Forward, followed by Yellow or Green?

19. Spin spinner I, then II, then III. What is the probability of getting a 1, followed by Red or Green, followed by Backward?

20. Spin spinner II, then I, then III. What is the probability of getting Yellow, followed by a 2 or a 4, followed by Forward?

21. Spin spinner I twice, then spinner II. What is the probability of getting a 2, followed by a 2 or a 4, followed by Red or Green?

22. Spin spinner III, then spinner I twice. What is the probability of getting Forward, followed by a 1 or a 3, followed by a 2 or a 4?

In Problems 23–26, consider the experiment of tossing a coin twice. The table lists six possible assignments of probabilities for this experiment. Using this table, answer the following questions.

Assignments	Sample Space			
	HH	**HT**	**TH**	**TT**
A	$\frac{1}{4}$	$\frac{1}{4}$	$\frac{1}{4}$	$\frac{1}{4}$
B	0	0	0	1
C	$\frac{3}{16}$	$\frac{5}{16}$	$\frac{5}{16}$	$\frac{3}{16}$
D	$\frac{1}{2}$	$\frac{1}{2}$	$-\frac{1}{2}$	$\frac{1}{2}$
E	$\frac{1}{4}$	$\frac{1}{4}$	$\frac{1}{4}$	$\frac{1}{8}$
F	$\frac{1}{9}$	$\frac{2}{9}$	$\frac{2}{9}$	$\frac{4}{9}$

23. Which of the assignments of probabilities is(are) consistent with the definition of a probability model?

24. Which of the assignments of probabilities should be used if the coin is known to be fair?

25. Which of the assignments of probabilities should be used if the coin is known to always come up tails?

26. Which of the assignments of probabilities should be used if tails is twice as likely as heads to occur?

27. Assigning Probabilities A coin is weighted so that heads is four times as likely as tails to occur. What probability should we assign to heads? to tails?

28. Assigning Probabilities A coin is weighted so that tails is twice as likely as heads to occur. What probability should we assign to heads? to tails?

29. Assigning Probabilities A die is weighted so that an odd-numbered face is twice as likely to occur as an even-numbered face. What probability should we assign to each face?

30. Assigning Probabilities A die is weighted so that a six cannot appear. The other faces occur with the same probability. What probability should we assign to each face?

For Problems 31–34, the sample space is $S = \{1, 2, 3, 4, 5, 6, 7, 8, 9, 10\}$. Suppose that the outcomes are equally likely.

31. Compute the probability of the event $E = \{1, 2, 3\}$.

32. Compute the probability of the event $F = \{3, 5, 9, 10\}$.

33. Compute the probability of the event E: "an even number."

34. Compute the probability of the event F: "an odd number."

For Problems 35 and 36, an urn contains 5 white marbles, 10 green marbles, 8 yellow marbles, and 7 black marbles.

35. If one marble is selected, determine the probability that it is white.

36. If one marble is selected, determine the probability that it is black.

In Problems 37–40, assume equally likely outcomes.

37. Determine the probability of having 3 boys in a 3-child family.

38. Determine the probability of having 3 girls in a 3-child family.

39. Determine the probability of having 1 girl and 3 boys in a 4-child family.

40. Determine the probability of having 2 girls and 2 boys in a 4-child family.

For Problems 41–44, two fair dice are rolled.

41. Determine the probability that the sum of the two dice is 7.

42. Determine the probability that the sum of the two dice is 11.

43. Determine the probability that the sum of the two dice is 3.

44. Determine the probability that the sum of the two dice is 12.

In Problems 45–48, find the probability of the indicated event if $P(A) = 0.25$ and $P(B) = 0.45$.

45. $P(A \cup B)$ if $P(A \cap B) = 0.15$

46. $P(A \cap B)$ if $P(A \cup B) = 0.6$

47. $P(A \cup B)$ if A, B are mutually exclusive

48. $P(A \cap B)$ if A, B are mutually exclusive

49. If $P(A) = 0.60$, $P(A \cup B) = 0.85$, and $P(A \cap B) = 0.05$, find $P(B)$.

50. If $P(B) = 0.30$, $P(A \cup B) = 0.65$, and $P(A \cap B) = 0.15$, find $P(A)$.

51. Automobile Theft According to the Insurance Information Institute, in 2004 there was a 13% probability that an automobile theft in the United States would be cleared by arrests. If an automobile theft case is randomly selected, what is the probability that it was not cleared by an arrest?

52. Pet Ownership According to the American Pet Products Manufacturers Association's *2005–2006 National Pet Owners Survey,* there is a 63% probability that a U.S. household owns a pet. If a U.S. household is randomly selected, what is the probability that it does not own a pet?

53. Cat Ownership According to the American Pet Products Manufacturers Association's *2005–2006 National Pet Owners Survey,* in 2004 there was a 34% probability that a U.S. pet owner owned a cat. If a U.S. pet owner is randomly selected, what is the probability that he or she does not own a cat?

54. Doctorate Degrees According to the National Science Foundation, in 2004 there was a 13.7% probability that a doctoral degree awarded at a U.S. university was awarded in engineering. If a 2004 U.S. doctoral recipient is randomly selected, what is the probability that his or her degree was not in engineering?

55. Online Gambling According to a Harris poll (January 12–17, 2006), 5% of U.S. adults admitted to having spent money gambling online. If a U.S. adult is selected at random, what is the probability that he or she has never spent any money gambling online?

56. Girl Scout Cookies According to the Girl Scouts of America, in March 2006, 9% of all Girl Scout cookies sold are shortbread/trefoils. If a box of Girl Scout cookies is selected at random, what is the probability that it is not shortbread/trefoils?

For Problems 57–60, a golf ball is selected at random from a container. If the container has 9 white balls, 8 green balls, and 3 orange balls, find the probability of each event.

57. The golf ball is white or green.

58. The golf ball is white or orange.

59. The golf ball is not white.

60. The golf ball is not green.

61. On the "Price is Right" there is a game in which a bag is filled with 3 strike chips and 5 numbers. Let's say that the numbers in the bag are 0, 1, 3, 6, and 9. What is the probability of selecting a strike chip or the number 1?

62. Another game on the "Price is Right" requires the contestant to spin a wheel with numbers 5, 10, 15, 20, . . . , 100. What is the probability that the contestant spins 100 or 30?

Problems 63–66, are based on a consumer survey of annual incomes in 100 households. The following table gives the data.

Income	$0–9999	$10,000–19,999	$20,000–29,999	$30,000–39,999	$40,000 or more
Number of households	5	35	30	20	10

63. What is the probability that a household has an annual income of $30,000 or more?

64. What is the probability that a household has an annual income between $10,000 and $29,999, inclusive?

65. What is the probability that a household has an annual income of less than $20,000?

66. What is the probability that a household has an annual income of $20,000 or more?

67. Surveys In a survey about the number of TV sets in a house, the following probability table was constructed:

Number of TV sets	0	1	2	3	4 or more
Probability	0.05	0.24	0.33	0.21	0.17

Find the probability of a house having:

(a) 1 or 2 TV sets (b) 1 or more TV sets

(c) 3 or fewer TV sets (d) 3 or more TV sets

(e) Fewer than 2 TV sets (f) Fewer than 1 TV set

(g) 1, 2, or 3 TV sets (h) 2 or more TV sets

68. Checkout Lines Through observation, it has been determined that the probability for a given number of people waiting in line at the "5 items or less" checkout register of a supermarket is as follows:

Number waiting in line	0	1	2	3	4 or more
Probability	0.10	0.15	0.20	0.24	0.31

Find the probability of:

(a) At most 2 people in line

(b) At least 2 people in line

(c) At least 1 person in line

69. In a certain Algebra and Trigonometry class, there are 18 freshmen and 15 sophomores. Of the 18 freshmen, 10 are male, and of the 15 sophomores, 8 are male. Find the probability that a randomly selected student is:

(a) A freshman or female

(b) A sophomore or male

70. The faculty of the mathematics department at Joliet Junior College is composed of 4 females and 9 males. Of the 4 females, 2 are under the age of 40, and of the males 3 are under age 40. Find the probability that a randomly selected faculty member is:

(a) Female or under age 40

(b) Male or over age 40

71. **Birthday Problem** What is the probability that at least 2 people have the same birthday in a group of 12 people? Assume that there are 365 days in a year.

72. **Birthday Problem** What is the probability that at least 2 people have the same birthday in a group of 35 people? Assume that there are 365 days in a year.

73. **Winning a Lottery** In a certain lottery, there are ten balls, numbered 1, 2, 3, 4, 5, 6, 7, 8, 9, 10. Of these, five are drawn in order. If you pick five numbers that match those drawn in the correct order, you win $1,000,000. What is the probability of winning such a lottery?

CHAPTER REVIEW

Things to Know

Counting formula (p. 975)	$n(A \cup B) = n(A) + n(B) - n(A \cap B)$
Addition Principle of Counting (p. 975)	If $A \cap B = \varnothing$, then $n(A \cup B) = n(A) + n(B)$.
Multiplication Principle of Counting (p. 977)	If a task consists of a sequence of choices in which there are p selections for the first choice, q selections for the second choice, and so on, the task of making these selections can be done in $p \cdot q \cdot \cdots$ different ways.
Permutation (p. 979)	An ordered arrangement of r objects chosen from n objects
Number of permutations: Distinct, with repetition (p. 980)	n^r The n objects are distinct (different), and repetition is allowed in the selection of r of them.
Number of permutations: Distinct, without repetition (p. 981)	$P(n, r) = n(n - 1) \cdot \cdots \cdot [n - (r - 1)] = \dfrac{n!}{(n - r)!}$ The n objects are distinct (different), and repetition is not allowed in the selection of r of them, where $r \le n$.
Combinations (p. 982)	An arrangement, without regard to order, of r objects selected from n distinct objects, where $r \le n$
Number of combinations (p. 983)	$C(n, r) = \dfrac{P(n, r)}{r!} = \dfrac{n!}{(n - r)! r!}$
Number of permutations: Not distinct (p. 985)	$\dfrac{n!}{n_1! n_2! \cdots n_k!}$ The number of permutations of n objects of which n_1 are of one kind, n_2 are of a second kind, ..., and n_k are of a kth kind, where $n = n_1 + n_2 + \cdots + n_k$
Sample space (p. 988)	Set whose elements represent the possible outcomes that can occur as a result of an experiment
Probability (p. 988)	A nonnegative number assigned to each outcome of a sample space; the sum of all the probabilities of the outcomes equals 1.
Probability for equally likely outcomes (p. 990)	$P(E) = \dfrac{n(E)}{n(S)}$ The same probability is assigned to each outcome.
Probability of the union of two events (p. 992)	$P(E \cup F) = P(E) + P(F) - P(E \cap F)$
Probability of the complement of an event (p. 993)	$P(\overline{E}) = 1 - P(E)$

Objectives

Section	You should be able to . . .	Review Exercises
14.1	**1** Find all the subsets of a set (p. 974)	1, 2
	2 Count the number of elements in a set (p. 974)	3–10
	3 Solve counting problems using the Multiplication Principle (p. 976)	15–18
14.2	**1** Solve counting problems using permutations involving n distinct objects (p. 979)	11, 12, 19, 20, 24–28, 33(a)
	2 Solve counting problems using combinations (p. 982)	13, 14, 21–23, 31, 32
	3 Solve counting problems using permutations involving n nondistinct objects (p. 984)	29, 30
14.3	**1** Construct probability models (p. 988)	33(b)
	2 Compute probabilities of equally likely outcomes (p. 990)	33(b), 34(a), 35(a), 36–39
	3 Find probabilities of the union of two events (p. 992)	40
	4 Use the Complement Rule to find probabilities (p. 993)	33(c), 34(b), 35(b), 36

Review Exercises

1. Write down all the subsets of the set {Dave, Joanne, Erica}.

2. Write down all the subsets of the set {Green, Blue, Red}.

3. If $n(A) = 8, n(B) = 12$, and $n(A \cap B) = 3$, find $n(A \cup B)$.

4. If $n(A) = 12, n(A \cup B) = 30$, and $n(A \cap B) = 6$, find $n(B)$.

In Problems 5–10, use the information supplied in the figure.

5. How many are in A?

6. How many are in A or B?

7. How many are in A and C?

8. How many are not in B?

9. How many are in neither A nor C?

10. How many are in B but not in C?

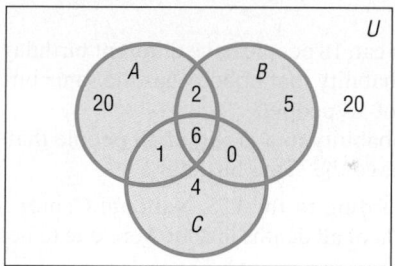

In Problems 11–14, compute the given expression.

11. $P(8, 3)$

12. $P(7, 3)$

13. $C(8, 3)$

14. $C(7, 3)$

15. Stocking a Store A clothing store sells pure wool and polyester–wool suits. Each suit comes in 3 colors and 10 sizes. How many suits are required for a complete assortment?

16. Wiring In connecting a certain electrical device, 5 wires are to be connected to 5 different terminals. How many different wirings are possible if 1 wire is connected to each terminal?

17. Baseball On a given day, the American Baseball League schedules 7 games. How many different outcomes are possible, assuming that each game is played to completion?

18. Baseball On a given day, the National Baseball League schedules 6 games. How many different outcomes are possible, assuming that each game is played to completion?

19. Choosing Seats If 4 people enter a bus having 9 vacant seats, in how many ways can they be seated?

20. Arranging Letters How many different arrangements are there of the letters in the word ROSE?

21. Choosing a Team In how many ways can a squad of 4 relay runners be chosen from a track team of 8 runners?

22. Writing a Test A professor has 10 similar problems to put on a test that has 3 problems. How many different tests can she design?

23. Baseball In how many ways can 2 teams from 14 teams in the American League be chosen without regard to which team is at home?

24. Arranging Books on a Shelf There are 5 different French books and 5 different Spanish books. How many ways are there to arrange them on a shelf if:
(a) Books of the same language must be grouped together, French on the left, Spanish on the right?
(b) French and Spanish books must alternate in the grouping, beginning with a French book?

25. Telephone Numbers Using the digits $0, 1, 2, \ldots, 9$, how many 7-digit numbers can be formed if the first digit cannot be 0 or 9 and if the last digit is greater than or equal to 2 and less than or equal to 3? Repeated digits are allowed.

26. Home Choices A contractor constructs homes with 5 different choices of exterior finish, 3 different roof arrangements, and 4 different window designs. How many different types of homes can be built?

27. License Plate Possibilities A license plate consists of 1 letter, excluding O and I, followed by a 4-digit number that cannot have a 0 in the lead position. How many different plates are possible?

28. Binary Codes Using the digits 0 and 1, how many different numbers consisting of 8 digits can be formed?

29. Forming Different Words How many different words, real or imaginary, can be formed using all the letters in the word MISSING?

30. Arranging Flags How many different vertical arrangements are there of 10 flags if 4 are white, 3 are blue, 2 are green, and 1 is red?

31. Forming Committees A group of 9 people is going to be formed into committees of 4, 3, and 2 people. How many committees can be formed if:
(a) A person can serve on any number of committees?
(b) No person can serve on more than one committee?

32. Forming Committees A group consists of 5 men and 8 women. A committee of 4 is to be formed from this group, and policy dictates that at least 1 woman be on this committee.
(a) How many committees can be formed that contain exactly 1 man?
(b) How many committees can be formed that contain exactly 2 women?
(c) How many committees can be formed that contain at least 1 man?

33. Birthday Problem For this problem, assume that a year has 365 days.
(a) How many ways can 18 people have different birthdays?
(b) What is the probability that nobody has the same birthday in a group of 18 people?
(c) What is the probability in a group of 18 people that at least 2 people have the same birthday?

34. Death Rates According to the U.S. National Center for Health Statistics, 29% of all deaths in 2001 were due to heart disease.
(a) What is the probability that a randomly selected death in 2001 was due to heart disease?
(b) What is the probability that a randomly selected death in 2001 was not due to heart disease?

35. Unemployment According to the U.S. Bureau of Labor Statistics, 5.8% of the U.S. labor force was unemployed in 2002.
(a) What is the probability that a randomly selected member of the U.S. labor force was unemployed in 2002?
(b) What is the probability that a randomly selected member of the U.S. labor force was not unemployed in 2002?

36. From a box containing three 40-watt bulbs, six 60-watt bulbs, and eleven 75-watt bulbs, a bulb is drawn at random. What is the probability that the bulb is 40 watts? What is the probability that it is not a 75-watt bulb?

37. You have four $1 bills, three $5 bills, and two $10 bills in your wallet. If you pick a bill at random, what is the probability that it will be a $1 bill?

38. Each letter in the word ROSE is written on an index card and the cards are shuffled. When the cards are dealt, what is the probability that they spell the word ROSE?

39. Each of the numbers, 1, 2, ..., 100 is written on an index card and the cards are shuffled. If a card is selected at random, what is the probability that the number on the card is divisible by 5? What is the probability that the card selected is either a 1 or names a prime number?

40. At the Milex tune-up and brake repair shop, the manager has found that a car will require a tune-up with a probability of 0.6, a brake job with a probability of 0.1, and both with a probability of 0.02.
(a) What is the probability that a car requires either a tune-up or a brake job?
(b) What is the probability that a car requires a tune-up but not a brake job?
(c) What is the probability that a car requires neither a tune-up nor a brake job?

CHAPTER TEST

In Problems 1–4, a survey of 70 college freshmen asked whether students planned to take biology, chemistry, or physics during their first year. Use the diagram to answer each question.

1. How many of the surveyed students plan to take physics during their first year?

2. How many of the surveyed students do not plan to take biology, chemistry, or physics during their first year?

3. How many of the surveyed students plan to take only biology and chemistry during their first year?

4. How many of the surveyed students plan to take physics or chemistry during their first year?

In Problems 5–7, compute the given expression.

5. 7! **6.** $P(10, 6)$ **7.** $C(11, 5)$

8. M&M's® offers customers the opportunity to create their own color mix of candy. There are 21 colors to choose from, and customers are allowed to select up to 6 different colors. How many different color mixes are possible, assuming that no color is selected more than once?

9. How many distinct 8-letter words (real or imaginary) can be formed from the letters in the word REDEEMED?

10. In horse racing, an exacta bet requires the bettor to pick the first two horses in the exact order. If there are 8 horses in a race, in how many ways could you make an exacta bet?

11. On February 20, 2004, the Ohio Bureau of Motor Vehicles unveiled the state's new license plate format. The plate consists of three letters (A–Z) followed by 4 digits (0–9). Assume that all letters and digits may be used except that the third letter cannot be O, I, or Z. If repetitions are allowed, how many different plates are possible?

12. Kiersten applies for admission to the University of Southern California (USC) and Florida State University (FSU). She estimates that she has a 60% chance of being admitted to USC, a 70% chance of being admitted to FSU, and a 35% chance of being admitted to both universities.
 (a) What is the probability that she will be admitted to either USC or FSU?
 (b) What is the probability that she will not be admitted to FSU?

13. A cooler contains 8 bottles of Pepsi, 5 bottles of Coke, 4 bottles of Mountain Dew, and 3 bottles of IBC.

 (a) What is the probability that a bottle chosen at random is Coke?
 (b) What is the probability that a bottle chosen at random is either Pepsi or IBC?

14. A study on the age distribution of a community college gave the following table:

Age	17 and under	18–20	21–24	25–34	35–64	65 and over
Probability	0.03	???	0.23	0.29	0.25	0.01

What must be the probability that a randomly selected student at the college is between 18 and 20 years old?

15. Powerball is a multistate lottery in which 5 white balls from a drum with 53 balls and 1 red ball from a drum with 42 red balls are selected. For a $1 ticket, players get one chance at winning the jackpot by matching all 6 numbers. What is the probability of selecting the winning numbers on a $1 play?

16. If you roll a die five times, what is the probability that you obtain exactly 2 fours?

CUMULATIVE REVIEW

1. Solve: $3x^2 - 2x = -1$

2. Graph $f(x) = x^2 + 4x - 5$ by determining whether the graph opens up or down and by finding the vertex, axis of symmetry, and intercepts.

3. Graph $f(x) = 2(x + 1)^2 - 4$ using transformations.

4. Solve: $|x - 4| \leq 0.01$

5. Find the complex zeros of
$$f(x) = 5x^4 - 9x^3 - 7x^2 - 31x - 6$$

6. Graph $g(x) = 3^{x-1} + 5$ using transformations. Determine the domain, the range, and horizontal asymptote of g.

7. What is the exact value of $\log_3 9$?

8. Solve: $\log_2(3x - 2) + \log_2 x = 4$

9. Solve the system: $\begin{cases} x - 2y + z = 15 \\ 3x + y - 3z = -8 \\ -2x + 4y - z = -27 \end{cases}$

10. What is the 33rd term in the sequence $-3, 1, 5, 9, \ldots$? What is the sum of the first 20 terms?

11. Graph: $y = 3\sin(2x + \pi)$

12. Solve the following triangle and determine its area.

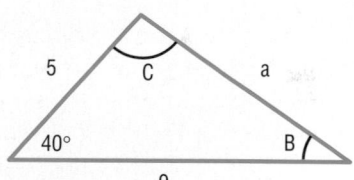

CHAPTER PROJECTS

I. **The Monty Hall Game** The *Monty Hall Game,* based on a segment from the game show *Let's Make a Deal,* is a classic probability problem that continues to stir debate. A more recent game show, *Deal or No Deal* (see the chapter opening vignette) has often been compared to the classic *Monty Hall Game.*

(a) Research the *Monty Hall Game* and *Deal or No Deal.*

(b) In the *Monty Hall Game,* what is the probability that a contestant wins if she does not switch? What is the probability of winning if she does switch? Perform a simulation to estimate the probabilities. Do the values agree with your research?

(c) Suppose the game *Deal or No Deal* is played with only three suitcases. Explain why this game is not the same as the *Monty Hall Game*.

(d) Suppose the game *Deal or No Deal* is played with 26 suitcases and the contestant is not allowed to switch at the end. What is the probability that the contestant will win the grand prize?

(e) Suppose the game *Deal or No Deal* is played with 26 suitcases and the contestant is allowed to switch at the end. Perform a simulation to estimate the probability that the contestant will win the grand prize if he does not switch at the end.

(f) Repeat part (e), but assume the contestant will always switch at the end.

The following projects are available at the Instructor's Resource Center (IRC):

II. Project at Motorola *Probability of Error Digital Wireless Communications* Transmission errors in digital communications can often be detected by adding an extra digit of code to each transmitted signal. Investigate the probability of identifying an erroneous code using this simple coding method.

III. Surveys Polling (or taking a survey) is big business in the United States. Take and analyze a survey; then consider why different pollsters might get different results.

IV. Law of Large Numbers The probability an event occurs, such as a head in a coin toss, is the proportion of heads you expect in the long run. A simulation is used to show that the more times a coin is flipped, the closer the probability that half of the outcomes is head.

V. Simulation Electronic simulation of an experiment is often an economical way to investigate a theoretical probability. Develop a theory without leaving your desk.

Graphing Utilities

Outline

1 The Viewing Rectangle

All graphing utilities, that is, all graphing calculators and all computer software graphing packages, graph equations by plotting points on a screen. The screen itself actually consists of small rectangles, called **pixels.** The more pixels the screen has, the better the resolution. Most graphing calculators have 2048 pixels per square inch; most computer screens have 4096 to 8192 pixels per square inch. When a point to be plotted lies inside a pixel, the pixel is turned on (lights up). The graph of an equation is a collection of pixels. Figure 1 shows how the graph of $y = 2x$ looks on a TI-84 Plus graphing calculator.

The screen of a graphing utility will display the coordinate axes of a rectangular coordinate system. However, you must set the scale on each axis. You must also include the smallest and largest values of x and y that you want included in the graph. This is called **setting the viewing rectangle** or **viewing window.** Figure 2 illustrates a typical viewing window.

To select the viewing window, we must give values to the following expressions:

Xmin: the smallest value of x

Xmax: the largest value of x

Xscl: the number of units per tick mark on the x-axis

Ymin: the smallest value of y

Ymax: the largest value of y

Yscl: the number of units per tick mark on the y-axis

Figure 3 illustrates these settings and their relation to the Cartesian coordinate system.

Figure 1
$y = 2x$

Figure 2

Figure 3

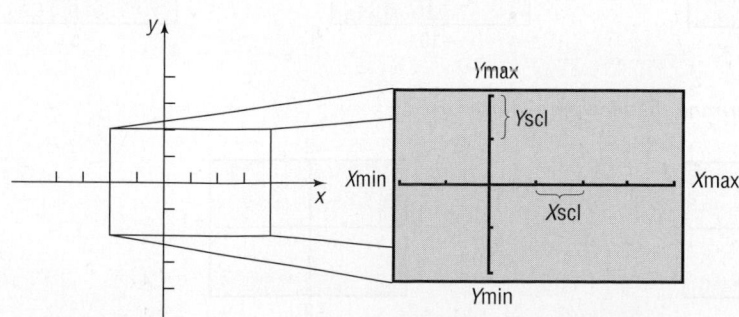

If the scale used on each axis is known, we can determine the minimum and maximum values of x and y shown on the screen by counting the tick marks. Look again at Figure 2. For a scale of 1 on each axis, the minimum and maximum values of x are -10 and 10, respectively; the minimum and maximum values of y are also -10 and 10. If the scale is 2 on each axis, then the minimum and maximum values of x are -20 and 20, respectively; and the minimum and maximum values of y are -20 and 20, respectively.

Conversely, if we know the minimum and maximum values of x and y, we can determine the scales being used by counting the tick marks displayed. We shall follow the practice of showing the minimum and maximum values of x and y in our illustrations so that you will know how the viewing window was set. See Figure 4.

Figure 4

$$\text{means} \quad \begin{aligned} X\text{min} &= -3 & Y\text{min} &= -4 \\ X\text{max} &= 3 & Y\text{max} &= 4 \\ X\text{scl} &= 1 & Y\text{scl} &= 2 \end{aligned}$$

| EXAMPLE 1 | **Finding the Coordinates of a Point Shown on a Graphing Utility Screen** |

Find the coordinates of the point shown in Figure 5. Assume that the coordinates are integers.

Figure 5

Solution First we note that the viewing window used in Figure 5 is

$$\begin{aligned} X\text{min} &= -3 & Y\text{min} &= -4 \\ X\text{max} &= 3 & Y\text{max} &= 4 \\ X\text{scl} &= 1 & Y\text{scl} &= 2 \end{aligned}$$

The point shown is 2 tick units to the left on the horizontal axis (scale = 1) and 1 tick up on the vertical axis (scale = 2). The coordinates of the point shown are $(-2, 2)$.

$\blacksquare$

1 Exercises

In Problems 1–4, determine the coordinates of the points shown. Write down the quadrant in which each point lies. Assume that the coordinates are integers.

1. **2.** **3.** **4.**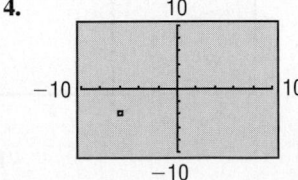

In Problems 5–10, determine the viewing window used.

5. **6.** **7.**

8. **9.** **10.**

In Problems 11–16, select a setting so that each of the given points will lie within the viewing rectangle.

11. $(-10, 5), (3, -2), (4, -1)$ **12.** $(5, 0), (6, 8), (-2, -3)$ **13.** $(40, 20), (-20, -80), (10, 40)$

14. $(-80, 60), (20, -30), (-20, -40)$ **15.** $(0, 0), (100, 5), (5, 150)$ **16.** $(0, -1), (100, 50), (-10, 30)$

In Problems 17–20, find the length of the line segment. Assume that the endpoints of each line segment have integer coordinates.

17. **18.** **19.** **20.**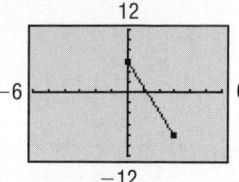

2 Using a Graphing Utility to Graph Equations

From Examples 2 and 3 in Chapter 2, Section 2.2, we see that a graph can be obtained by plotting points in a rectangular coordinate system and connecting them. Graphing utilities perform these same steps when graphing an equation. For example, the TI-84 Plus determines 95 evenly spaced input values,* starting at Xmin and ending at Xmax, uses the equation to determine the output values, plots these points on the screen, and finally (if in the connected mode) draws a line between consecutive points.

To graph an equation in two variables x and y using a graphing utility requires that the equation be written in the form $y = \{expression\ in\ x\}$. If the original equation is not in this form, replace it by equivalent equations until the form $y = \{expression\ in\ x\}$ is obtained. In general, there are four ways to obtain equivalent equations.

> ## Procedures That Result in Equivalent Equations
>
> **1.** Interchange the two sides of the equation:
> $$\text{Replace} \quad 3x + 5 = y \quad \text{by} \quad y = 3x + 5$$
>
> **2.** Simplify the sides of the equation by combining like terms, eliminating parentheses, and so on:
> $$\text{Replace} \quad (2y + 2) + 6 = 2x + 5(x + 1)$$
> $$\text{by} \quad 2y + 8 = 7x + 5$$
>
> **3.** Add or subtract the same expression on both sides of the equation:
> $$\text{Replace} \quad y + 3x - 5 = 4$$
> $$\text{by} \quad y + 3x - 5 + 5 = 4 + 5$$
>
> **4.** Multiply or divide both sides of the equation by the same nonzero expression:
> $$\text{Replace} \quad 3y = 6 - 2x$$
> $$\text{by} \quad \frac{1}{3} \cdot 3y = \frac{1}{3}(6 - 2x)$$

* These input values depend on the values of Xmin and Xmax. For example, if Xmin $= -10$ and Xmax $= 10$, then the first input value will be -10 and the next input value will be $-10 + \dfrac{10 - (-10)}{94} = -9.7872$, and so on.

| EXAMPLE 1 | Expressing an Equation in the Form $y = \{\text{expression in } x\}$ |

Solve for y: $2y + 3x - 5 = 4$

Solution We replace the original equation by a succession of equivalent equations.

$$2y + 3x - 5 = 4$$
$$2y + 3x - 5 + 5 = 4 + 5 \qquad \text{Add 5 to both sides.}$$
$$2y + 3x = 9 \qquad \text{Simplify.}$$
$$2y + 3x - 3x = 9 - 3x \qquad \text{Subtract } 3x \text{ from both sides.}$$
$$2y = 9 - 3x \qquad \text{Simplify.}$$
$$\frac{2y}{2} = \frac{9 - 3x}{2} \qquad \text{Divide both sides by 2.}$$
$$y = \frac{9 - 3x}{2} \qquad \text{Simplify.}$$

Now we are ready to graph equations using a graphing utility. Most graphing utilities require the following steps:

Steps for Graphing an Equation Using a Graphing Utility

STEP 1: Solve the equation for y in terms of x.

STEP 2: Get into the graphing mode of your graphing utility. The screen will usually display $Y_1 = $, prompting you to enter the expression involving x that you found in Step 1. (Consult your manual for the correct way to enter the expression; for example, $y = x^2$ might be entered as $x^{\wedge}2$ or as $x*x$ or as $x \, x^Y \, 2$).

STEP 3: Select the viewing window. Without prior knowledge about the behavior of the graph of the equation, it is common to select the **standard viewing window*** initially. In this text the standard viewing window is

$$X\text{min} = -10 \qquad Y\text{min} = -10$$
$$X\text{max} = 10 \qquad Y\text{max} = 10$$
$$X\text{scl} = 1 \qquad Y\text{scl} = 1$$

STEP 4: Graph.

STEP 5: Adjust the viewing window until a complete graph is obtained.

| EXAMPLE 2 | Graphing an Equation on a Graphing Utility |

Graph the equation: $6x^2 + 3y = 36$

Solution **STEP 1:** We solve for y in terms of x.

$$6x^2 + 3y = 36$$
$$3y = -6x^2 + 36 \qquad \text{Subtract } 6x^2 \text{ from both sides of the equation.}$$
$$y = -2x^2 + 12 \qquad \text{Divide both sides of the equation by 3 and simplify.}$$

STEP 2: From the $Y_1 = $ screen, enter the expression $-2x^2 + 12$ after the prompt.

STEP 3: Set the viewing window to the standard viewing window.

STEP 4: Graph. The screen should look like Figure 6.

* Some graphing utilities have a ZOOM-STANDARD feature that automatically sets the viewing window to the standard viewing window and graphs the equation.

Figure 6

Figure 7

Figure 8

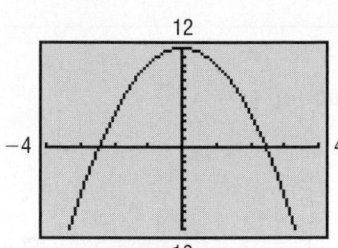

STEP 5: The graph of $y = -2x^2 + 12$ is not complete. The value of Ymax must be increased so that the top portion of the graph is visible. After increasing the value of Ymax to 12, we obtain the graph in Figure 7. The graph is now complete.

Look again at Figure 7. Although a complete graph is shown, the graph might be improved by adjusting the values of Xmin and Xmax. Figure 8 shows the graph of $y = -2x^2 + 12$ using Xmin $= -4$ and Xmax $= 4$. Do you think this is a better choice for the viewing window?

| EXAMPLE 3 | **Creating a Table and Graphing an Equation** |

Create a table and graph the equation: $y = x^3$

Solution Most graphing utilities have the capability of creating a table of values for an equation. (Check your manual to see if your graphing utility has this capability.) Table 1 illustrates a table of values for $y = x^3$ on a TI-84 Plus. See Figure 9 for the graph.

Table 1

X	Y1
-3	-27
-2	-8
-1	-1
0	0
1	1
2	8
3	27

Y1◼X^3

Figure 9

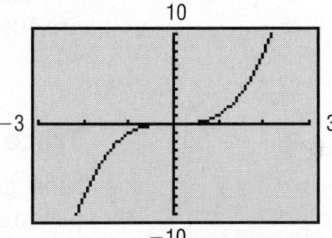

2 Exercises

In Problems 1–16, graph each equation using the following viewing windows:

(a) Xmin $= -5$	(b) Xmin $= -10$	(c) Xmin $= -10$	(d) Xmin $= -5$
Xmax $= 5$	Xmax $= 10$	Xmax $= 10$	Xmax $= 5$
Xscl $= 1$	Xscl $= 1$	Xscl $= 2$	Xscl $= 1$
Ymin $= -4$	Ymin $= -8$	Ymin $= -8$	Ymin $= -20$
Ymax $= 4$	Ymax $= 8$	Ymax $= 8$	Ymax $= 20$
Yscl $= 1$	Yscl $= 1$	Yscl $= 2$	Yscl $= 5$

1. $y = x + 2$ **2.** $y = x - 2$ **3.** $y = -x + 2$ **4.** $y = -x - 2$

5. $y = 2x + 2$ **6.** $y = 2x - 2$ **7.** $y = -2x + 2$ **8.** $y = -2x - 2$

9. $y = x^2 + 2$ **10.** $y = x^2 - 2$ **11.** $y = -x^2 + 2$ **12.** $y = -x^2 - 2$

13. $3x + 2y = 6$ **14.** $3x - 2y = 6$ **15.** $-3x + 2y = 6$ **16.** $-3x - 2y = 6$

17–32. For each of the preceding equations 1–16, create a table, $-3 \leq x \leq 3$, and list points on the graph.

3 Using a Graphing Utility to Locate Intercepts and Check for Symmetry

Value and Zero (or Root)

Most graphing utilities have an eVALUEate feature that, given a value of x, determines the value of y for an equation. We can use this feature to evaluate an equation at $x = 0$ to determine the y-intercept. Most graphing utilities also have a ZERO (or ROOT) feature that can be used to determine the x-intercept(s) of an equation.

EXAMPLE 1	Finding Intercepts Using a Graphing Utility

Use a graphing utility to find the intercepts of the equation $y = x^3 - 8$.

Solution Figure 10(a) shows the graph of $y = x^3 - 8$.

Figure 10

(a) (b) (c)

The eVALUEate feature of a TI-84 Plus graphing calculator accepts as input a value of x and determines the value of y. If we let $x = 0$, we find that the y-intercept is -8. See Figure 10(b).

The ZERO feature of a TI-84 Plus is used to find the x-intercept(s). See Figure 10(c). The x-intercept is 2.

Trace

Most graphing utilities allow you to move from point to point along the graph, displaying on the screen the coordinates of each point. This feature is called TRACE.

EXAMPLE 2	Using TRACE to Locate Intercepts

Graph the equation $y = x^3 - 8$. Use TRACE to locate the intercepts.

Solution Figure 11 shows the graph of $y = x^3 - 8$.

Figure 11

Activate the TRACE feature. As you move the cursor along the graph, you will see the coordinates of each point displayed. When the cursor is on the y-axis, we find that the y-intercept is -8. See Figure 12.

Figure 12

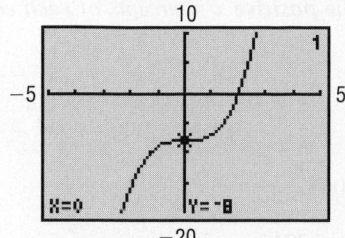

Continue moving the cursor along the graph. Just before you get to the x-axis, the display will look like the one in Figure 13(a). (Due to differences in graphing utilities, your display may be slightly different from the one shown here.)

Figure 13

(a)

(b)

In Figure 13(a), the negative value of the y-coordinate indicates that we are still below the x-axis. The next position of the cursor is shown in Figure 13(b). The positive value of the y-coordinate indicates that we are now above the x-axis. This means that between these two points the x-axis was crossed. The x-intercept lies between 1.9148936 and 2.0212766.

EXAMPLE 3

Graphing the Equation $y = \dfrac{1}{x}$

Graph the equation: $y = \dfrac{1}{x}$

With the viewing window set as

$$X\min = -3 \qquad Y\min = -4$$
$$X\max = 3 \qquad Y\max = 4$$
$$X\text{scl} = 1 \qquad Y\text{scl} = 1$$

use TRACE to infer information about intercepts and symmetry.

Figure 14

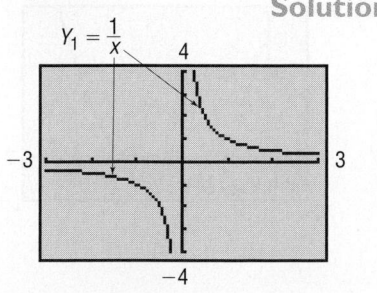

Solution Figure 14 illustrates the graph. We infer from the graph that there are no intercepts; we may also infer that symmetry with respect to the origin is a possibility. The TRACE feature on a graphing utility can provide further evidence of symmetry with respect to the origin. Using TRACE, we observe that for any ordered pair (x, y) the ordered pair $(-x, -y)$ is also a point on the graph. For example, the points $(0.95744681, 1.0444444)$ and $(-0.95744681, -1.0444444)$ both lie on the graph.

3 Exercises

In Problems 1–6, use ZERO (or ROOT) to approximate the smaller of the two x-intercepts of each equation. Express the answer rounded to two decimal places.

1. $y = x^2 + 4x + 2$ **2.** $y = x^2 + 4x - 3$ **3.** $y = 2x^2 + 4x + 1$

4. $y = 3x^2 + 5x + 1$ **5.** $y = 2x^2 - 3x - 1$ **6.** $y = 2x^2 - 4x - 1$

*In Problems 7–14, use ZERO (or ROOT) to approximate the **positive** x-intercepts of each equation. Express each answer rounded to two decimal places.*

7. $y = x^3 + 3.2x^2 - 16.83x - 5.31$

8. $y = x^3 + 3.2x^2 - 7.25x - 6.3$

9. $y = x^4 - 1.4x^3 - 33.71x^2 + 23.94x + 292.41$

10. $y = x^4 + 1.2x^3 - 7.46x^2 - 4.692x + 15.2881$

11. $y = \pi x^3 - (8.88\pi + 1)x^2 - (42.066\pi - 8.88)x + 42.066$

12. $y = \pi x^3 - (5.63\pi + 2)x^2 - (108.392\pi - 11.26)x + 216.784$

13. $y = x^3 + 19.5x^2 - 1021x + 1000.5$

14. $y = x^3 + 14.2x^2 - 4.8x - 12.4$

4 Using a Graphing Utility to Solve Equations

For many equations, there are no algebraic techniques that lead to a solution. For such equations, a graphing utility can often be used to investigate possible solutions. When a graphing utility is used to solve an equation, usually *approximate* solutions are obtained. Unless otherwise stated, we shall follow the practice of giving approximate solutions *rounded to two decimal places*.

The ZERO (or ROOT) feature of a graphing utility can be used to find the solutions of an equation when one side of the equation is 0. In using this feature to solve equations, we make use of the fact that the x-intercepts (or zeros) of the graph of an equation are found by letting $y = 0$ and solving the equation for x. Solving an equation for x when one side of the equation is 0 is equivalent to finding where the graph of the corresponding equation crosses or touches the x-axis.

EXAMPLE 1 **Using ZERO (or ROOT) to Approximate Solutions of an Equation**

Find the solution(s) of the equation $x^2 - 6x + 7 = 0$. Round answers to two decimal places.

Solution The solutions of the equation $x^2 - 6x + 7 = 0$ are the same as the x-intercepts of the graph of $Y_1 = x^2 - 6x + 7$. We begin by graphing Y_1. See Figure 15(a).

From the graph there appear to be two x-intercepts (solutions to the equation): one between 1 and 2, the other between 4 and 5.

Figure 15

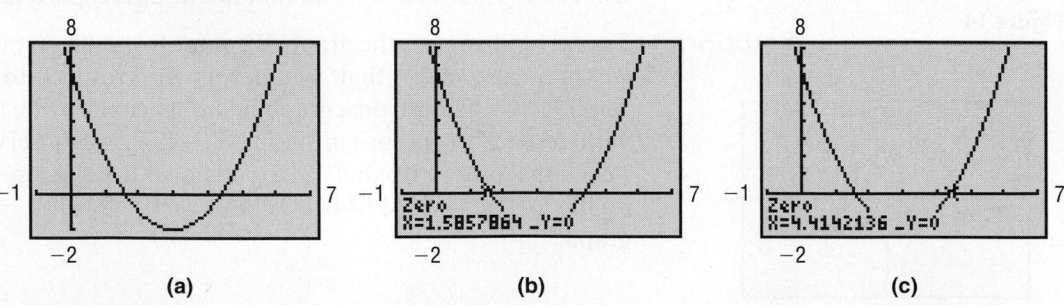

(a) (b) (c)

Using the ZERO (or ROOT) feature of our graphing utility, we determine that the x-intercepts, and so the solutions to the equation, are $x = 1.59$ and $x = 4.41$, rounded to two decimal places. See Figures 15(b) and (c).

A second method for solving equations using a graphing utility involves the INTERSECT feature of the graphing utility. This feature is used most effectively when one side of the equation is not 0.

| EXAMPLE 2 | Using INTERSECT to Approximate Solutions of an Equation |

Find the solution(s) to the equation $3(x - 2) = 5(x - 1)$. Round answers to two decimal places.

Solution We begin by graphing each side of the equation as follows: graph $Y_1 = 3(x - 2)$ and $Y_2 = 5(x - 1)$. See Figure 16(a).

Figure 16

(a) (b)

Figure 17

At the point of intersection of the graphs, the value of the y-coordinate is the same. We conclude that the x-coordinate of the point of intersection represents the solution to the equation. Do you see why? The INTERSECT feature on a graphing utility determines the point of intersection of the graphs. Using this feature, we find that the graphs intersect at $(-0.5, -7.5)$. See Figure 16(b). The solution of the equation is therefore $x = -0.5$.

✓ **Check:** We can verify our solution by evaluating each side of the equation with -0.5 STOred in x. See Figure 17. Since the left side of the equation equals the right side of the equation, the solution checks.

SUMMARY

The steps to follow for approximating solutions of equations are given next.

Steps for Approximating Solutions of Equations Using ZERO (or ROOT)

STEP 1: Write the equation in the form {*expression in x*} = 0.

STEP 2: Graph $Y_1 = $ {*expression in x*}.

Be sure that the graph is complete. That is, be sure that all the intercepts are shown on the screen.

STEP 3: Use ZERO (or ROOT) to determine each x-intercept of the graph.

Steps for Approximating Solutions of Equations Using INTERSECT

STEP 1: Graph $Y_1 = $ {*expression in x on the left side of the equation*}.

Graph $Y_2 = $ {*expression in x on the right side of the equation*}.

STEP 2: Use INTERSECT to determine each x-coordinate of the point(s) of intersection, if any.

Be sure that the graphs are complete. That is, be sure that all the points of intersection are shown on the screen.

| EXAMPLE 3 | Solving a Radical Equation |

Find the real solutions of the equation $\sqrt[3]{2x - 4} - 2 = 0$.

Solution Figure 18 shows the graph of the equation $Y_1 = \sqrt[3]{2x - 4} - 2$. From the graph, we see one x-intercept near 6. Using ZERO (or ROOT), we find that the x-intercept is 6. The only solution is $x = 6$.

Figure 18

5 Square Screens

Figure 19

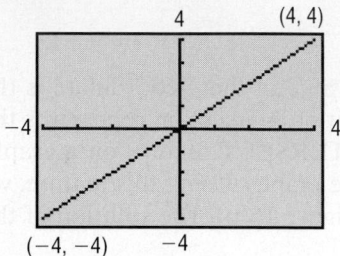

Most graphing utilities have a rectangular screen. Because of this, using the same settings for both x and y will result in a distorted view. For example, Figure 19 shows the graph of the line $y = x$ connecting the points $(-4, -4)$ and $(4, 4)$.

We expect the line to bisect the first and third quadrants, but it doesn't. We need to adjust the selections for Xmin, Xmax, Ymin, and Ymax so that a **square screen** results. On most graphing utilities, this is accomplished by setting the ratio of x to y at $3 : 2$.* For example, if

$$X\text{min} = -6 \qquad Y\text{min} = -4$$
$$X\text{max} = 6 \qquad Y\text{max} = 4$$

then the ratio of x to y is

$$\frac{X\text{max} - X\text{min}}{Y\text{max} - Y\text{min}} = \frac{6 - (-6)}{4 - (-4)} = \frac{12}{8} = \frac{3}{2}$$

for a ratio of $3 : 2$, resulting in a square screen.

EXAMPLE 1 **Examples of Viewing Rectangles That Result in Square Screens**

(a) Xmin = -3 (b) Xmin = -6 (c) Xmin = -6
 Xmax = 3 Xmax = 6 Xmax = 6
 Xscl = 1 Xscl = 1 Xscl = 1
 Ymin = -2 Ymin = -4 Ymin = -4
 Ymax = 2 Ymax = 4 Ymax = 4
 Yscl = 1 Yscl = 1 Yscl = 2

Figure 20

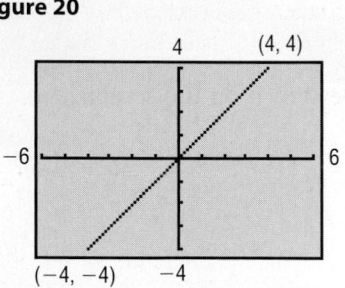

Figure 20 shows the graph of the line $y = x$ on a square screen using the viewing rectangle given in part (b). Notice that the line now bisects the first and third quadrants. Compare this illustration to Figure 19.

5 Exercises

In Problems 1–8, determine which of the given viewing rectangles result(s) in a square screen.

1. Xmin = -3 Ymin = -2 **2.** Xmin = -5 Ymin = -4 **3.** Xmin = 0 Ymin = -2
 Xmax = 3 Ymax = 2 Xmax = 5 Ymax = 4 Xmax = 9 Ymax = 4
 Xscl = 2 Yscl = 2 Xscl = 1 Yscl = 1 Xscl = 3 Yscl = 2

* Some graphing utilities have a built-in function that automatically squares the screen. For example, the TI-85 has a ZSQR function that does this. Some graphing utilities require a ratio other than $3 : 2$ to square the screen. For example, the HP 48G requires the ratio of x to y to be $2 : 1$ for a square screen. Consult your manual.

4. Xmin $= -6$ Ymin $= -4$
 Xmax $= 6$ Ymax $= 4$
 Xscl $= 1$ Yscl $= 2$

5. Xmin $= -6$ Ymin $= -2$
 Xmax $= 6$ Ymax $= 2$
 Xscl $= 1$ Yscl $= 0.5$

6. Xmin $= -6$ Ymin $= -4$
 Xmax $= 6$ Ymax $= 4$
 Xscl $= 2$ Yscl $= 1$

7. Xmin $= 0$ Ymin $= -2$
 Xmax $= 9$ Ymax $= 4$
 Xscl $= 1$ Yscl $= 1$

8. Xmin $= -6$ Ymin $= -4$
 Xmax $= 6$ Ymax $= 4$
 Xscl $= 2$ Yscl $= 2$

9. If Xmin $= -4$, Xmax $= 8$, and Xscl $= 1$, how should Ymin, Ymax, and Yscl be selected so that the viewing rectangle contains the point $(4, 8)$ and the screen is square?

10. If Xmin $= -6$, Xmax $= 12$, and Xscl $= 2$, how should Ymin, Ymax, and Yscl be selected so that the viewing rectangle contains the point $(4, 8)$ and the screen is square?

6 Using a Graphing Utility to Graph Inequalities

It is easiest to begin with an example.

| EXAMPLE 1 | Graphing an Inequality Using a Graphing Utility |

Use a graphing utility to graph: $3x + y - 6 \leq 0$

Solution We begin by graphing the equation $3x + y - 6 = 0$ ($Y_1 = -3x + 6$). See Figure 21.

Figure 21

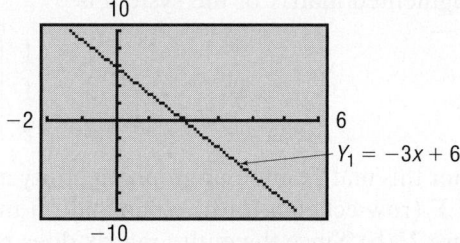

As with graphing by hand, we need to test points selected from each region and determine whether they satisfy the inequality. To test the point $(-1, 2)$, for example, enter $3(-1) + 2 - 6 \leq 0$. See Figure 22(a). The 1 that appears indicates that the statement entered (the inequality) is true. When the point $(5, 5)$ is tested, a 0 appears, indicating that the statement entered is false. Thus, $(-1, 2)$ is a part of the graph of the inequality and $(5, 5)$ is not. Figure 22(b) shows the graph of the inequality on a TI-84 Plus.*

Figure 22

(a) (b)

The steps to follow to graph an inequality using a graphing utility are given next.

* Consult your owner's manual for shading techniques.

Steps for Graphing an Inequality Using a Graphing Utility

STEP 1: Replace the inequality symbol by an equal sign, solve the equation for y, and graph the equation.

STEP 2: In each region, select a test point P and determine if the coordinates of P satisfy the inequality.

(a) If the test point satisfies the inequality, then so do all the points in the region. Indicate this by using the graphing utility to shade the region.

(b) If the coordinates of P do not satisfy the inequality, then none of the points in that region do.

7 Using a Graphing Utility to Solve Systems of Linear Equations

Most graphing utilities have the capability to put the augmented matrix of a system of linear equations in row echelon form. The next example, Example 6 from Section 12.2, demonstrates this feature using a TI-84 Plus graphing calculator.

EXAMPLE 1 Solving a System of Linear Equations Using a Graphing Utility

Solve: $\begin{cases} x - y + z = 8 & (1) \\ 2x + 3y - z = -2 & (2) \\ 3x - 2y - 9z = 9 & (3) \end{cases}$

Solution The augmented matrix of the system is

$$\begin{bmatrix} 1 & -1 & 1 & \bigm| & 8 \\ 2 & 3 & -1 & \bigm| & -2 \\ 3 & -2 & -9 & \bigm| & 9 \end{bmatrix}$$

We enter this matrix into our graphing utility and name it A. See Figure 23(a). Using the REF (row echelon form) command on matrix A, we obtain the results shown in Figure 23(b). Since the entire matrix does not fit on the screen, we need to scroll right to see the rest of it. See Figure 23(c).

Figure 23

(a)　　　　　　　(b)　　　　　　　(c)

The system of equations represented by the matrix in row echelon form is

$$\begin{bmatrix} 1 & -\dfrac{2}{3} & -3 & \bigm| & 3 \\ 0 & 1 & \dfrac{15}{13} & \bigm| & -\dfrac{24}{13} \\ 0 & 0 & 1 & \bigm| & 1 \end{bmatrix} \quad \begin{cases} x - \dfrac{2}{3}y - 3z = 3 & (1) \\ y + \dfrac{15}{13}z = -\dfrac{24}{13} & (2) \\ z = 1 & (3) \end{cases}$$

Using $z = 1$, we back-substitute to get

$$\begin{cases} x - \dfrac{2}{3}y - 3(1) = 3 & (1) \\ y + \dfrac{15}{13}(1) = -\dfrac{24}{13} & (2) \end{cases} \xrightarrow[\text{Simplify.}]{} \begin{cases} x - \dfrac{2}{3}y = 6 & (1) \\ y = \dfrac{-39}{13} = -3 & (2) \end{cases}$$

Solving the second equation for y, we find that $y = -3$. Back-substituting $y = -3$ into $x - \dfrac{2}{3}y = 6$, we find that $x = 4$. The solution of the system is $x = 4$, $y = -3$, $z = 1$.

■

Notice that the row echelon form of the augmented matrix using the graphing utility differs from the row echelon form found in Example 6 of Section 12.2, yet both matrices provide the same solution! This is because the two solutions used different row operations to obtain the row echelon form. In all likelihood, the two solutions parted ways in Step 4 of the algebraic solution, where we avoided introducing fractions by interchanging rows 2 and 3.

Most graphing utilities also have the ability to put a matrix in reduced row echelon form. Figure 24 shows the reduced row echelon form of the augmented matrix from Example 1 using the RREF command on a TI-84 Plus graphing calculator. Using this command, we see that the solution of the system is $x = 4$, $y = -3$, $z = 1$.

Figure 24

8 Using a Graphing Utility to Graph a Polar Equation

Most graphing utilities require the following steps to obtain the graph of a polar equation. Be sure to be in POLar mode.

Graphing a Polar Equation Using a Graphing Utility

STEP 1: Set the mode to POLar. Solve the equation for r in terms of θ.

STEP 2: Select the viewing rectangle in polar mode. Besides setting Xmin, Xmax, Xscl, and so forth, the viewing rectangle in polar mode requires setting the minimum and maximum values for θ and an increment setting for θ (θstep). In addition, a square screen and radian measure should be used.

STEP 3: Enter the expression involving θ that you found in Step 1. (Consult your manual for the correct way to enter the expression.)

STEP 4: Graph.

EXAMPLE 1 Graphing a Polar Equation Using a Graphing Utility

Use a graphing utility to graph the polar equation $r \sin \theta = 2$.

Solution **STEP 1:** We solve the equation for r in terms of θ.

$$r \sin \theta = 2$$

$$r = \frac{2}{\sin \theta}$$

STEP 2: From the POLar mode, select the viewing rectangle. We will use the one given next.

θmin $= 0$	Xmin $= -9$	Ymin $= -6$
θmax $= 2\pi$	Xmax $= 9$	Ymax $= 6$
θstep $= \dfrac{\pi}{24}$	Xscl $= 1$	Yscl $= 1$

θstep determines the number of points that the graphing utility will plot. For example, if θstep is $\dfrac{\pi}{24}$, the graphing utility will evaluate r at $\theta = 0(\theta\text{min}), \dfrac{\pi}{24}, \dfrac{2\pi}{24}, \dfrac{3\pi}{24}$, and so forth, up to $2\pi(\theta\text{max})$. The smaller that

Figure 25

θstep is, the more points that the graphing utility will plot. The student is encouraged to experiment with different values for θmin, θmax, and θstep to see how the graph is affected.

STEP 3: Enter the expression $\dfrac{2}{\sin \theta}$ after the prompt $r_1 =$.

STEP 4: Graph.

The graph is shown in Figure 25.

9 Using a Graphing Utility to Graph Parametric Equations

Most graphing utilities have the capability of graphing parametric equations. The following steps are usually required to obtain the graph of parametric equations. Check your owner's manual to see how yours works.

Graphing Parametric Equations Using a Graphing Utility

STEP 1: Set the mode to PARametric. Enter $x(t)$ and $y(t)$.

STEP 2: Select the viewing window. In addition to setting Xmin, Xmax, Xscl, and so on, the viewing window in parametric mode requires setting minimum and maximum values for the parameter t and an increment setting for t (Tstep).

STEP 3: Graph.

EXAMPLE 1

Graphing a Curve Defined by Parametric Equations Using a Graphing Utility

Graph the curve defined by the parametric equations

$$x = 3t^2, \qquad y = 2t, \qquad -2 \le t \le 2$$

Solution

STEP 1: Enter the equations $x(t) = 3t^2$, $y(t) = 2t$ with the graphing utility in PARametric mode.

STEP 2: Select the viewing window. The interval is $-2 \le t \le 2$, so we select the following square viewing window:

$$
\begin{array}{lll}
T\text{min} = -2 & X\text{min} = 0 & Y\text{min} = -5 \\
T\text{max} = 2 & X\text{max} = 15 & Y\text{max} = 5 \\
T\text{step} = 0.1 & X\text{scl} = 1 & Y\text{scl} = 1
\end{array}
$$

We choose Tmin $= -2$ and Tmax $= 2$ because $-2 \le t \le 2$. Finally, the choice for Tstep will determine the number of points that the graphing utility will plot. For example, with Tstep at 0.1, the graphing utility will evaluate x and y at $t = -2, -1.9, -1.8$, and so on. The smaller the Tstep is, the more points the graphing utility will plot. The reader is encouraged to experiment with different values of Tstep to see how the graph is affected.

STEP 3: Graph. Notice the direction in which the graph is drawn. This direction shows the orientation of the curve.

The graph shown in Figure 26 is complete.

Figure 26

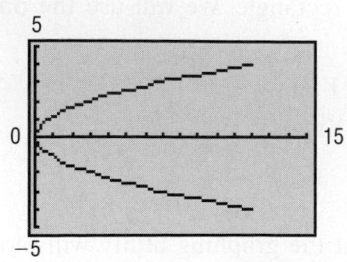

Exploration

Graph the following parametric equations using a graphing utility with $Xmin = 0$, $Xmax = 15$, $Ymin = -5$, $Ymax = 5$, and $Tstep = 0.1$.

1. $x = \dfrac{3t^2}{4}$, $\quad y = t$, $\quad -4 \le t \le 4$

2. $x = 3t^2 + 12t + 12$, $\quad y = 2t + 4$, $\quad -4 \le t \le 0$

3. $x = 3t^{2/3}$, $\quad y = 2\sqrt[3]{t}$, $\quad -8 \le t \le 8$

Compare these graphs to the graph in Figure 26. Conclude that parametric equations defining a curve are not unique; that is, different parametric equations can represent the same graph.

Exploration

In FUNCtion mode, graph $x = \dfrac{3y^2}{4}$ $\left(Y_1 = \sqrt{\dfrac{4x}{3}} \text{ and } Y_2 = -\sqrt{\dfrac{4x}{3}} \right)$ with $Xmin = 0$, $Xmax = 15$, $Ymin = -5$, $Ymax = 5$. Compare this graph with Figure 26. Why do the graphs differ?

Answers

CHAPTER R Review

Historical Problems (page 15)

1. (a) 1,20 **(b)** 2,50 **2. (a)** $\dfrac{7}{3} = 2.333\ldots$ **(b)** $\dfrac{39}{8} = 4.875$ **(c)** $\dfrac{84{,}823}{27{,}000} = 3.141592592\ldots$

R.1 Assess Your Understanding (page 15)

1. rational **2.** 31 **3.** Distributive **4.** $5(x + 3) = 6$ **5.** T **6.** F **7.** F **8.** T **9.** $\{1, 2, 3, 4, 5, 6, 7, 8, 9\}$ **11.** $\{4\}$

13. $\{1, 3, 4, 6\}$ **15.** $\{0, 2, 6, 7, 8\}$ **17.** $\{0, 1, 2, 3, 5, 6, 7, 8, 9\}$ **19.** $\{0, 1, 2, 3, 5, 6, 7, 8, 9\}$ **21. (a)** $\{2, 5\}$ **(b)** $\{-6, 2, 5\}$

(c) $\left\{-6, \dfrac{1}{2}, -1.333\ldots, 2, 5\right\}$ **(d)** $\{\pi\}$ **(e)** $\left\{-6, \dfrac{1}{2}, -1.333\ldots, \pi, 2, 5\right\}$ **23. (a)** $\{1\}$ **(b)** $\{0, 1\}$ **(c)** $\left\{0, 1, \dfrac{1}{2}, \dfrac{1}{3}, \dfrac{1}{4}\right\}$ **(d)** None

(e) $\left\{0, 1, \dfrac{1}{2}, \dfrac{1}{3}, \dfrac{1}{4}\right\}$ **25. (a)** None **(b)** None **(c)** None **(d)** $\left\{\sqrt{2}, \pi, \sqrt{2} + 1, \pi + \dfrac{1}{2}\right\}$ **(e)** $\left\{\sqrt{2}, \pi, \sqrt{2} + 1, \pi + \dfrac{1}{2}\right\}$

27. (a) 18.953 **(b)** 18.952 **29. (a)** 28.653 **(b)** 28.653 **31. (a)** 0.063 **(b)** 0.062 **33. (a)** 9.999 **(b)** 9.998 **35. (a)** 0.429 **(b)** 0.428

37. (a) 34.733 **(b)** 34.733 **39.** $3 + 2 = 5$ **41.** $x + 2 = 3 \cdot 4$ **43.** $3y = 1 + 2$ **45.** $x - 2 = 6$ **47.** $\dfrac{x}{2} = 6$ **49.** 7 **51.** 6 **53.** 1

55. $\dfrac{13}{3}$ **57.** -11 **59.** 11 **61.** -4 **63.** 1 **65.** 6 **67.** $\dfrac{2}{7}$ **69.** $\dfrac{4}{45}$ **71.** $\dfrac{23}{20}$ **73.** $\dfrac{79}{30}$ **75.** $\dfrac{13}{36}$ **77.** $-\dfrac{16}{45}$ **79.** $\dfrac{1}{60}$ **81.** $\dfrac{15}{22}$ **83.** 1 **85.** $\dfrac{15}{8}$

87. $6x + 24$ **89.** $x^2 - 4x$ **91.** $\dfrac{3}{2}x - 1$ **93.** $x^2 + 6x + 8$ **95.** $x^2 - x - 2$ **97.** $x^2 - 10x + 16$ **99.** $2x + 3x = (2 + 3)x = 5x$

101. $2(3 \cdot 4) = 2 \cdot 12 = 24; (2 \cdot 3) \cdot (2 \cdot 4) = 6 \cdot 8 = 48$ **103.** No; $2 - 3 \neq 3 - 2$ **105.** No; $\dfrac{2}{3} \neq \dfrac{3}{2}$ **107.** Symmetric Property
109. No; no

R.2 Assess Your Understanding (page 26)

1. variable **2.** origin **3.** strict **4.** base; exponent or power **5.** 1.2345678×10^3 **6.** T **7.** T **8.** F **9.** F **10.** F
11. **13.** $>$ **15.** $>$ **17.** $>$ **19.** $=$ **21.** $<$ **23.** $x > 0$ **25.** $x < 2$ **27.** $x \leq 1$

29. **31.** **33.** 1 **35.** 2 **37.** 6 **39.** 4 **41.** -28 **43.** $\dfrac{4}{5}$ **45.** 0 **47.** 1 **49.** 5 **51.** 1 **53.** 22

55. 2 **57.** $x = 0$ **59.** $x = 3$ **61.** None **63.** $x = 0, x = 1, x = -1$ **65.** $\{x \,|\, x \neq 5\}$ **67.** $\{x \,|\, x \neq -4\}$ **69.** $0°C$ **71.** $25°C$ **73.** 16

75. $\dfrac{1}{16}$ **77.** $\dfrac{1}{9}$ **79.** 9 **81.** 5 **83.** 4 **85.** $64x^6$ **87.** $\dfrac{x^4}{y^2}$ **89.** $\dfrac{x}{y}$ **91.** $-\dfrac{8x^3 z}{9y}$ **93.** $\dfrac{16x^2}{9y^2}$ **95.** -4 **97.** 5 **99.** 4 **101.** 2 **103.** $\sqrt{5}$

105. $\dfrac{1}{2}$ **107.** $10; 0$ **109.** 81 **111.** 304,006.671 **113.** 0.004 **115.** 481.890 **117.** 0.000 **119.** 4.542×10^2 **121.** 1.3×10^{-2}

123. 3.2155×10^4 **125.** 4.23×10^{-4} **127.** 61,500 **129.** 0.001214 **131.** 110,000,000 **133.** 0.081 **135.** $A = lw$ **137.** $C = \pi d$

139. $A = \dfrac{\sqrt{3}}{4}x^2$ **141.** $V = \dfrac{4}{3}\pi r^3$ **143.** $V = x^3$ **145. (a)** \$6000 **(b)** \$8000 **147.** $|x - 4| \geq 6$ **149. (a)** $2 \leq 5$ **(b)** $6 > 5$

151. (a) Yes **(b)** No **153.** 400,000,000 m **155.** 0.0000005 m **157.** 5×10^{-4} in. **159.** 5.865696×10^{12} mi **161.** No; $\dfrac{1}{3}$ is larger;
0.000333... **163.** No

R.3 Assess Your Understanding (page 36)

1. right; hypotenuse **2.** $A = \dfrac{1}{2}bh$ **3.** $C = 2\pi r$ **4.** similar **5.** T **6.** T **7.** F **8.** T **9.** T **10.** F **11.** 13 **13.** 26 **15.** 25 **17.** Right
triangle; 5 **19.** Not a right triangle **21.** Right triangle; 25 **23.** Not a right triangle **25.** 8 in.2 **27.** 4 in.2 **29.** $A = 25\pi$ m^2; $C = 10\pi$ m

31. $V = 224$ ft^3; $S = 232$ ft^2 **33.** $V = \dfrac{256}{3}\pi$ cm^3; $S = 64\pi$ cm^2 **35.** $V = 648\pi$ in.3; $S = 306\pi$ in.2 **37.** π square units **39.** 2π square units

41. $x = 4$ units; $A = 90°$; $B = 60°$; $C = 30°$ **43.** $x = 67.5$ units; $A = 60°$; $B = 95°$; $C = 25°$ **45.** About 16.8 ft

47. 64 ft^2 **49.** $24 + 2\pi \approx 30.28$ ft^2; $16 + 2\pi \approx 22.28$ ft **51.** 160 paces **53.** About 5.477 mi **55.** From 100 ft: 12.2 mi; From 150 ft: 15.0 mi

R.4 Assess Your Understanding (page 47)

1. 4; 3 **2.** $x^4 - 16$ **3.** $x^3 - 8$ **4.** F **5.** T **6.** F **7.** Monomial; variable: x; coefficient: 2; degree: 3 **9.** Not a monomial; the exponent
of the variable is not a nonnegative integer **11.** Monomial; variables: x, y; coefficient: -2; degree: 3 **13.** Not a monomial; the exponent
of one of the variables is not a nonnegative integer **15.** Not a monomial; it has more than one term **17.** Yes; 2 **19.** Yes; 0

21. No; the variable of one of the terms is not a nonnegative integer **23.** Yes; 3 **25.** No; the polynomial of the denominator has a degree greater than 0 **27.** $x^2 + 7x + 2$ **29.** $x^3 - 4x^2 + 9x + 7$ **31.** $6x^5 + 5x^4 + 3x^2 + x$ **33.** $7x^2 - x - 7$ **35.** $-2x^3 + 18x^2 - 18$
37. $2x^2 - 4x + 6$ **39.** $15y^2 - 27y + 30$ **41.** $x^3 + x^2 - 4x$ **43.** $-8x^5 - 10x^2$ **45.** $x^3 + 3x^2 - 2x - 4$ **47.** $x^2 + 6x + 8$ **49.** $2x^2 + 9x + 10$
51. $x^2 - 2x - 8$ **53.** $x^2 - 5x + 6$ **55.** $2x^2 - x - 6$ **57.** $-2x^2 + 11x - 12$ **59.** $2x^2 + 8x + 8$ **61.** $x^2 - xy - 2y^2$
63. $-6x^2 - 13xy - 6y^2$ **65.** $x^2 - 49$ **67.** $4x^2 - 9$ **69.** $x^2 + 8x + 16$ **71.** $x^2 - 8x + 16$ **73.** $9x^2 - 16$ **75.** $4x^2 - 12x + 9$
77. $x^2 - y^2$ **79.** $9x^2 - y^2$ **81.** $x^2 + 2xy + y^2$ **83.** $x^2 - 4xy + 4y^2$ **85.** $x^3 - 6x^2 + 12x - 8$ **87.** $8x^3 + 12x^2 + 6x + 1$
89. $4x^2 - 11x + 23$; remainder -45 **91.** $4x - 3$; remainder $x + 1$ **93.** $5x^2 - 13$; remainder $x + 27$ **95.** $2x^2$; remainder $-x^2 + x + 1$
97. $x^2 - 2x + \dfrac{1}{2}$; remainder $\dfrac{5}{2}x + \dfrac{1}{2}$ **99.** $-4x^2 - 3x - 3$; remainder -7 **101.** $x^2 - x - 1$; remainder $2x + 2$ **103.** $x^2 + ax + a^2$;
remainder 0

R.5 Assess Your Understanding (page 56)

1. $3x(x - 2)(x + 2)$ **2.** Prime **3.** T **4.** F **5.** $3(x + 2)$ **7.** $a(x^2 + 1)$ **9.** $x(x^2 + x + 1)$ **11.** $2x(x - 1)$ **13.** $3xy(x - 2y + 4)$
15. $(x + 1)(x - 1)$ **17.** $(2x + 1)(2x - 1)$ **19.** $(x + 4)(x - 4)$ **21.** $(5x + 2)(5x - 2)$ **23.** $(x + 1)^2$ **25.** $(x + 2)^2$ **27.** $(x - 5)^2$
29. $(2x + 1)^2$ **31.** $(4x + 1)^2$ **33.** $(x - 3)(x^2 + 3x + 9)$ **35.** $(x + 3)(x^2 - 3x + 9)$ **37.** $(2x + 3)(4x^2 - 6x + 9)$
39. $(x + 2)(x + 3)$ **41.** $(x + 6)(x + 1)$ **43.** $(x + 5)(x + 2)$ **45.** $(x - 8)(x - 2)$ **47.** $(x - 8)(x + 1)$ **49.** $(x + 8)(x - 1)$
51. $(x + 2)(2x + 3)$ **53.** $(x - 2)(2x + 1)$ **55.** $(2x + 3)(3x + 2)$ **57.** $(3x + 1)(x + 1)$ **59.** $(z + 1)(2z + 3)$ **61.** $(x + 2)(3x - 4)$
63. $(x - 2)(3x + 4)$ **65.** $(x + 4)(3x + 2)$ **67.** $(x + 4)(3x - 2)$ **69.** $(x + 6)(x - 6)$ **71.** $2(1 + 2x)(1 - 2x)$ **73.** $(x + 1)(x + 10)$
75. $(x - 7)(x - 3)$ **77.** $4(x^2 - 2x + 8)$ **79.** Prime **81.** $-(x - 5)(x + 3)$ **83.** $3(x + 2)(x - 6)$ **85.** $y^2(y + 5)(y + 6)$
87. $(2x + 3)^2$ **89.** $2(3x + 1)(x + 1)$ **91.** $(x - 3)(x + 3)(x^2 + 9)$ **93.** $(x - 1)^2(x^2 + x + 1)^2$ **95.** $x^5(x - 1)(x + 1)$
97. $(4x + 3)^2$ **99.** $-(4x - 5)(4x + 1)$ **101.** $(2y - 5)(2y - 3)$ **103.** $-(3x - 1)(3x + 1)(x^2 + 1)$ **105.** $(x + 3)(x - 6)$
107. $(x + 2)(x - 3)$ **109.** $(3x - 5)(9x^2 - 3x + 7)$ **111.** $(x + 5)(3x + 11)$ **113.** $(x - 1)(x + 1)(x + 2)$
115. $(x - 1)(x + 1)(x^2 - x + 1)$ **117.** $2(3x + 4)(9x + 13)$ **119.** $2x(3x + 5)$ **121.** $5(x + 3)(x - 2)^2(x + 1)$
123. $3(4x - 3)(4x - 1)$ **125.** $6(3x - 5)(2x + 1)^2(5x - 4)$ **127.** The possibilities are $(x \pm 1)(x \pm 4) = x^2 \pm 5x + 4$ or
$(x \pm 2)(x \pm 2) = x^2 \pm 4x + 4$, none of which equals $x^2 + 4$.

R.6 Assess Your Understanding (page 60)

1. quotient; divisor; remainder **2.** $-3\overline{)2\ 0\ -5\ 1}$ **3.** T **4.** T **5.** $x^2 + x + 4$; remainder 12 **7.** $3x^2 + 11x + 32$; remainder 99
9. $x^4 - 3x^3 + 5x^2 - 15x + 46$; remainder -138 **11.** $4x^5 + 4x^4 + x^3 + x^2 + 2x + 2$; remainder 7 **13.** $0.1x^2 - 0.11x + 0.321$;
remainder -0.3531 **15.** $x^4 + x^3 + x^2 + x + 1$; remainder 0 **17.** No **19.** Yes **21.** Yes **23.** No **25.** Yes **27.** -9

R.7 Assess Your Understanding (page 69)

1. lowest terms **2.** least common multiple **3.** T **4.** F **5.** $\dfrac{3}{x - 3}$ **7.** $\dfrac{x}{3}$ **9.** $\dfrac{4x}{2x - 1}$ **11.** $\dfrac{y + 5}{2(y + 1)}$ **13.** $\dfrac{x + 5}{x - 1}$ **15.** $-(x + 7)$
17. $\dfrac{3}{5x(x - 2)}$ **19.** $\dfrac{2x(x^2 + 4x + 16)}{x + 4}$ **21.** $\dfrac{8}{3x}$ **23.** $\dfrac{x - 3}{x + 7}$ **25.** $\dfrac{4x}{(x - 2)(x - 3)}$ **27.** $\dfrac{4}{5(x - 1)}$ **29.** $-\dfrac{(x - 4)^2}{4x}$ **31.** $\dfrac{(x + 3)^2}{(x - 3)^2}$
33. $\dfrac{(x - 4)(x + 3)}{(x - 1)(2x + 1)}$ **35.** $\dfrac{x + 5}{2}$ **37.** $\dfrac{(x - 2)(x + 2)}{2x - 3}$ **39.** $\dfrac{3x - 2}{x - 3}$ **41.** $\dfrac{x + 9}{2x - 1}$ **43.** $\dfrac{4 - x}{x - 2}$ **45.** $\dfrac{2(x + 5)}{(x - 1)(x + 2)}$
47. $\dfrac{3x^2 - 2x - 3}{(x + 1)(x - 1)}$ **49.** $\dfrac{-(11x + 2)}{(x + 2)(x - 2)}$ **51.** $\dfrac{2(x^2 - 2)}{x(x - 2)(x + 2)}$ **53.** $(x - 2)(x + 2)(x + 1)$ **55.** $x(x - 1)(x + 1)$ **57.** $x^3(2x - 1)^2$
59. $x(x - 1)^2(x + 1)(x^2 + x + 1)$ **61.** $\dfrac{5x}{(x - 6)(x - 1)(x + 4)}$ **63.** $\dfrac{2(2x^2 + 5x - 2)}{(x - 2)(x + 2)(x + 3)}$ **65.** $\dfrac{5x + 1}{(x - 1)^2(x + 1)^2}$
67. $\dfrac{-x^2 + 3x + 13}{(x - 2)(x + 1)(x + 4)}$ **69.** $\dfrac{x^3 - 2x^2 + 4x + 3}{x^2(x + 1)(x - 1)}$ **71.** $\dfrac{-1}{x(x + h)}$ **73.** $\dfrac{x + 1}{x - 1}$ **75.** $\dfrac{(x - 1)(x + 1)}{2x(2x + 1)}$ **77.** $\dfrac{2(5x - 1)}{(x - 2)(x + 1)^2}$
79. $\dfrac{-2x(x^2 - 2)}{(x + 2)(x^2 - x - 3)}$ **81.** $\dfrac{-1}{x - 1}$ **83.** $\dfrac{3x - 1}{2x + 1}$ **85.** $\dfrac{19}{(3x - 5)^2}$ **87.** $\dfrac{(x + 1)(x - 1)}{(x^2 + 1)^2}$ **89.** $\dfrac{x(3x + 2)}{(3x + 1)^2}$ **91.** $-\dfrac{(x + 3)(3x - 1)}{(x^2 + 1)^2}$
93. $f = \dfrac{R_1 \cdot R_2}{(n - 1)(R_1 + R_2)}$; $\dfrac{2}{15}$ m

R.8 Assess Your Understanding (page 77)

3. index **4.** T **5.** cube root **6.** F **7.** 3 **9.** -2 **11.** $2\sqrt{2}$ **13.** $-2x\sqrt[3]{x}$ **15.** x^3y^2 **17.** x^2y **19.** $6\sqrt{x}$ **21.** $6x\sqrt{x}$ **23.** $15\sqrt[3]{3}$
25. $12\sqrt{3}$ **27.** $7\sqrt{2}$ **29.** $\sqrt{2}$ **31.** $2\sqrt{3}$ **33.** $-\sqrt[3]{2}$ **35.** $x - 2\sqrt{x} + 1$ **37.** $(2x - 1)\sqrt[3]{2x}$ **39.** $(2x - 15)\sqrt{2x}$
41. $-(x + 5y)\sqrt[3]{2xy}$ **43.** $\dfrac{\sqrt{2}}{2}$ **45.** $-\dfrac{\sqrt{15}}{5}$ **47.** $\dfrac{\left(5 + \sqrt{2}\right)\sqrt{3}}{23}$ **49.** $\dfrac{8\sqrt{5} - 19}{41}$ **51.** $\dfrac{5\sqrt[3]{4}}{2}$ **53.** $\dfrac{2x + h - 2\sqrt{x^2 + xh}}{h}$
55. 4 **57.** -3 **59.** 64 **61.** $\dfrac{1}{27}$ **63.** $\dfrac{27\sqrt{2}}{32}$ **65.** $\dfrac{27\sqrt{2}}{32}$ **67.** $x^{7/12}$ **69.** xy^2 **71.** $x^{2/3}y$ **73.** $\dfrac{8x^{5/4}}{y^{3/4}}$ **75.** $\dfrac{3x + 2}{(1 + x)^{1/2}}$ **77.** $\dfrac{x(3x^2 + 2)}{(x^2 + 1)^{1/2}}$

79. $\dfrac{22x + 5}{10\sqrt{x} - 5\sqrt{4x + 3}}$ **81.** $\dfrac{2 + x}{2(1 + x)^{3/2}}$ **83.** $\dfrac{4 - x}{(x + 4)^{3/2}}$ **85.** $\dfrac{1}{x^2(x^2 - 1)^{1/2}}$ **87.** $\dfrac{1 - 3x^2}{2\sqrt{x}(1 + x^2)^2}$ **89.** $\dfrac{1}{2}(5x + 2)(x + 1)^{1/2}$

91. $2x^{1/2}(3x - 4)(x + 1)$ **93.** $(x^2 + 4)^{1/3}(11x^2 + 12)$ **95.** $(3x + 5)^{1/3}(2x + 3)^{1/2}(17x + 27)$ **97.** $\dfrac{3(x + 2)}{2x^{1/2}}$ **99.** 1.41 **101.** 1.59

103. 4.89 **105.** 2.15 **107. (a)** 15,660.4 gal **(b)** 390.7 gal **109.** $2\sqrt{2}\pi \approx 8.89$ sec **111.** $\dfrac{\pi\sqrt{3}}{6} \approx 0.91$ sec

Review Exercises (page 81)

1. $\{1, 3, 5, 6, 7, 8\}$ **3.** $\{3, 7\}$ **5.** $\{2, 4, 6, 8, 9\}$ **7.** $\{1, 2, 4, 5, 6, 9\}$ **9. (a)** None **(b)** $\{-10\}$ **(c)** $\left\{-10, 0.65, 1.343434\ldots, \dfrac{1}{9}\right\}$

(d) $\{\sqrt{7}\}$ **(e)** $\left\{-10, 0.65, 1.343434\ldots, \sqrt{7}, \dfrac{1}{9}\right\}$ **11.** 14 **13.** $\dfrac{3}{4}$ **15.** 3 **17.** $4x - 12$ **19.** **21.** 5 **23.** -10

25. -49 **27.** 5 **29.** $\{x | x \neq 6\}$ **31.** $x = 3\sqrt{2}$ units; $A = 90°$; $B = 45°$; $C = 45°$ **33.** 5.0625 **35.** Coefficients: $3, 4, -2, 0, 5, -12$;
degree: 5 **37.** $2x^4 - 2x^3 + x^2 + 5x + 3$ **39.** $6x^2 - 7xy - 5y^2$ **41.** $16x^2 - 1$ **43.** $x^3 - 7x - 6$ **45.** $3x^2 + 8x + 25$; remainder 79
47. $-3x^2 + 4$; remainder -2 **49.** $x^4 - x^3 + x^2 - x + 1$; remainder 0 **51.** $(x + 7)(x - 2)$ **53.** $(3x + 2)(2x - 3)$ **55.** $3(x + 2)(x - 7)$

57. $(2x + 1)(4x^2 - 2x + 1)$ **59.** $(2x + 3)(x - 1)(x + 1)$ **61.** $(5x - 2)(5x + 2)$ **63.** Prime **65.** $(x + 4)^2$ **67.** $\dfrac{2x + 7}{x - 2}$

69. $\dfrac{3(3x - 1)}{(x + 3)(3x + 1)}$ **71.** $\dfrac{4x}{(x + 1)(x - 1)}$ **73.** $\dfrac{x^2 + 17x + 2}{(x - 2)(x + 2)^2}$ **75.** $\dfrac{x - 1}{x - 3}$ **77.** $4\sqrt{2}$ **79.** $-2\sqrt[3]{2}$ **81.** $2\sqrt{2}$ **83.** $\dfrac{1}{x^5 y}$ **85.** $\dfrac{125}{x^2 y}$

87. $\dfrac{3x}{5y^2}$ **89.** $3xy^4\sqrt[3]{x}$ **91.** $\dfrac{4\sqrt{5}}{5}$ **93.** $-2(1 + \sqrt{2})$ **95.** $-\dfrac{3 + \sqrt{5}}{2}$ **97.** $\dfrac{2(1 + x^2)}{(2 + x^2)^{1/2}}$ **99.** $\dfrac{x(3x + 16)}{2(x + 4)^{3/2}}$ **101.** $\dfrac{1}{x^2\sqrt{x^2 - 1}}$

103. $(x^2 + 4)^{1/3}(11x^2 + 12)$ **105.** 2.81421906×10^8 **107.** Yes **109.** \$0.35 per share **111.** $16\pi \approx 50.27$ ft^2; $10\pi \approx 31.42$ ft

Chapter Test (page 84)

1. (a) $\{7\}$ **(b)** $\{0, 7\}$ **(c)** $\left\{0, 1.2, 7, \dfrac{1}{2}\right\}$ **(d)** $\left\{\sqrt{2}, \pi\right\}$ **(e)** $\left\{0, 1.2, \sqrt{2}, 7, \dfrac{1}{2}, \pi\right\}$ **2. (a)** -16 **(b)** 18 **(c)** 5 **(d)** -174

3. $x = \dfrac{3}{2}$; $A = 40°$; $B = 95°$; $C = 45°$ **4. (a)** $-8x^3 + 11x^2 - 14x + 11$ **(b)** $-10x^2 + 19x - 6$

5. (a) $(x - 4)(x - 2)$ **(b)** $(2x - 5)(2x + 5)$ **(c)** $(3x + 1)(2x - 7)$ **6.** $x^2 - x + 6$; remainder 2

7. (a) $3\sqrt{3}$ **(b)** -2 **(c)** $\dfrac{x}{y^4}$ **(d)** $\dfrac{64x^2}{y}$ **8.** $\dfrac{15 - 3\sqrt{3}}{22}$ **9.** **10.** $a = 0$; $b = 0$

11. 216 square feet; 84 feet of fence

CHAPTER 1 Equations and Inequalities

1.1 Assess Your Understanding (page 94)

4. F **5.** identity **6.** linear; first degree **7.** F **8.** T

9. $\{3\}$ **11.** $\{-5\}$ **13.** $\left\{\dfrac{3}{2}\right\}$ **15.** $\left\{\dfrac{5}{4}\right\}$ **17.** $\{-2\}$ **19.** $\{3\}$ **21.** $\{-1\}$ **23.** $\{-2\}$ **25.** $\{-18\}$ **27.** $\{-4\}$ **29.** $\left\{-\dfrac{3}{4}\right\}$ **31.** $\{-20\}$

33. $\{2\}$ **35.** $\{0.5\}$ **37.** $\left\{\dfrac{29}{10}\right\}$ **39.** $\{2\}$ **41.** $\{8\}$ **43.** $\{2\}$ **45.** $\{-1\}$ **47.** $\{3\}$ **49.** No solution **51.** No solution **53.** $\{-6\}$

55. $\{34\}$ **57.** $\left\{-\dfrac{20}{39}\right\}$ **59.** $\{-1\}$ **61.** $\left\{-\dfrac{11}{6}\right\}$ **63.** $\{-6\}$ **65.** $\{5.91\}$ **67.** $\{0.41\}$ **69.** $x = \dfrac{b + c}{a}$ **71.** $x = \dfrac{abc}{a + b}$ **73.** $x = a^2$

75. $a = 3$ **77.** $R = \dfrac{R_1 R_2}{R_1 + R_2}$ **79.** $R = \dfrac{mv^2}{F}$ **81.** $r = \dfrac{S - a}{S}$ **83.** \$11,500 will be invested in bonds and \$8500 in CDs. **85.** Yahoo! was
used for 1.52 billion searches; Google was used for 2.05 billion searches. **87.** The regular hourly rate is \$8.50. **89.** Brooke needs a score of 85.
91. The original price was \$500,000; purchasing the model saves \$75,000. **93.** The bookstore paid \$68.15 for the book. **95.** There were 2187
adults. **97.** The length is 19 ft; the width is 11 ft. **99.** There were about 694.06 million Internet users worldwide.

Historical Problems (page 106)

1. The area of each shaded square is 9, so the larger square will have area $85 + 4(9) = 121$. The area of the larger square is also given by the
expression $(x + 6)^2$, so $(x + 6)^2 = 121$. Taking the positive square root of each side, $x + 6 = 11$ or $x = 5$.
2. Let $z = -6$, so $z^2 + 12z - 85 = -121$. We get the equation $u^2 - 121 = 0$ or $u^2 = 121$. Thus $u = \pm 11$, so $x = \pm 11 - 6$. $x = -17$ or $x = 5$.

3.
$$\left(x + \frac{b}{2a}\right)^2 = \left(\frac{\sqrt{b^2 - 4ac}}{2a}\right)^2$$
$$\left(x + \frac{b}{2a}\right)^2 - \left(\frac{\sqrt{b^2 - 4ac}}{2a}\right)^2 = 0$$
$$\left(x + \frac{b}{2a} - \frac{\sqrt{b^2 - 4ac}}{2a}\right)\left(x + \frac{b}{2a} + \frac{\sqrt{b^2 - 4ac}}{2a}\right) = 0$$
$$\left(x + \frac{b - \sqrt{b^2 - 4ac}}{2a}\right)\left(x + \frac{b + \sqrt{b^2 - 4ac}}{2a}\right) = 0$$
$$x = \frac{-b + \sqrt{b^2 - 4ac}}{2a} \text{ or } x = \frac{-b - \sqrt{b^2 - 4ac}}{2a}$$

1.2 Assess Your Understanding (page 106)

5. add; $\left(\frac{5}{2}\right)^2 = \frac{25}{4}$ **6.** discriminant; negative **7.** F **8.** F **9.** $\{0, 9\}$ **11.** $\{-5, 5\}$ **13.** $\{-3, 2\}$ **15.** $\left\{-\frac{1}{2}, 3\right\}$ **17.** $\{-4, 4\}$ **19.** $\{2, 6\}$

21. $\left\{\frac{3}{2}\right\}$ **23.** $\left\{-\frac{2}{3}, \frac{3}{2}\right\}$ **25.** $\left\{-\frac{2}{3}, \frac{3}{2}\right\}$ **27.** $\left\{-\frac{3}{4}, 2\right\}$ **29.** $\{-5, 5\}$ **31.** $\{-1, 3\}$ **33.** $\{-3, 0\}$ **35.** 16 **37.** $\frac{1}{16}$ **39.** $\frac{1}{9}$ **41.** $\{-7, 3\}$

43. $\left\{-\frac{1}{4}, \frac{3}{4}\right\}$ **45.** $\left\{\frac{-1 - \sqrt{7}}{6}, \frac{-1 + \sqrt{7}}{6}\right\}$ **47.** $\{2 - \sqrt{2}, 2 + \sqrt{2}\}$ **49.** $\{2 - \sqrt{5}, 2 + \sqrt{5}\}$ **51.** $\left\{1, \frac{3}{2}\right\}$ **53.** No real solution

55. $\left\{\frac{-1 - \sqrt{5}}{4}, \frac{-1 + \sqrt{5}}{4}\right\}$ **57.** $\left\{0, \frac{9}{4}\right\}$ **59.** $\left\{\frac{1}{3}\right\}$ **61.** $\left\{-\frac{2}{3}, 1\right\}$ **63.** $\left\{\frac{3 - \sqrt{29}}{10}, \frac{3 + \sqrt{29}}{10}\right\}$ **65.** $\left\{\frac{-2 - \sqrt{10}}{2}, \frac{-2 + \sqrt{10}}{2}\right\}$

67. $\left\{\frac{1 - \sqrt{33}}{8}, \frac{1 + \sqrt{33}}{8}\right\}$ **69.** $\left\{\frac{9 - \sqrt{73}}{2}, \frac{9 + \sqrt{73}}{2}\right\}$ **71.** $\{0.63, 3.47\}$ **73.** $\{-2.80, 1.07\}$ **75.** $\{-0.85, 1.17\}$ **77.** $\{-8.16, -0.22\}$

79. $\{-\sqrt{5}, \sqrt{5}\}$ **81.** $\left\{\frac{1}{4}\right\}$ **83.** $\left\{-\frac{3}{5}, \frac{5}{2}\right\}$ **85.** $\left\{-\frac{1}{2}, \frac{2}{3}\right\}$ **87.** $\left\{\frac{-\sqrt{2} + 2}{2}, \frac{-\sqrt{2} - 2}{2}\right\}$ **89.** $\left\{\frac{-1 - \sqrt{17}}{2}, \frac{-1 + \sqrt{17}}{2}\right\}$ **91.** $\{5\}$
93. No real solution **95.** Repeated real solution **97.** Two unequal real solutions **99.** Academic year 2009–2010 **101.** The dimensions are 11 ft by 13 ft.
103. The dimensions are 5 m by 8 m. **105.** The dimensions should be 4 ft by 4 ft. **107. (a)** The ball strikes the ground after 6 sec. **(b)** The ball passes the top of the building on its way down after 5 sec. **109.** The dimensions should be 11.55 cm by 6.55 cm by 3 cm. **111.** The border will be 2.71 ft wide.
113. The border will be 2.56 ft wide. **115.** The screen of a 37-inch TV in 4 : 3 format has an area of 657.12 square inches; the screen of a 37-inch TV in 16 : 9 format has an area of 584.97 square inches. The traditional TV has a larger screen. **117.** 36 consecutive integers must be added.
119. $\frac{-b + \sqrt{b^2 - 4ac}}{2a} + \frac{-b - \sqrt{b^2 - 4ac}}{2a} = \frac{-2b}{2a} = -\frac{b}{a}$ **121.** $k = \frac{1}{2}$ or $k = -\frac{1}{2}$
123. $ax^2 + bx + c = 0, x = \frac{-b \pm \sqrt{b^2 - 4ac}}{2a}; ax^2 - bx + c = 0, x = \frac{b \pm \sqrt{(-b)^2 - 4ac}}{2a} = \frac{b \pm \sqrt{b^2 - 4ac}}{2a} = -\frac{-b \pm \sqrt{b^2 - 4ac}}{2a}$ **125.** (b)

1.3 Assess Your Understanding (page 116)

4. real; imaginary; imaginary unit **5.** $\{-2i, 2i\}$ **6.** F **7.** T **8.** F **9.** $8 + 5i$ **11.** $-7 + 6i$ **13.** $-6 - 11i$ **15.** $6 - 18i$ **17.** $6 + 4i$

19. $10 - 5i$ **21.** 37 **23.** $\frac{6}{5} + \frac{8}{5}i$ **25.** $1 - 2i$ **27.** $\frac{5}{2} - \frac{7}{2}i$ **29.** $-\frac{1}{2} + \frac{\sqrt{3}}{2}i$ **31.** $2i$ **33.** $-i$ **35.** i **37.** -6 **39.** $-10i$ **41.** $-2 + 2i$

43. 0 **45.** 0 **47.** $2i$ **49.** $5i$ **51.** $5i$ **53.** $\{-2i, 2i\}$ **55.** $\{-4, 4\}$ **57.** $\{3 - 2i, 3 + 2i\}$ **59.** $\{3 - i, 3 + i\}$ **61.** $\left\{\frac{1}{4} - \frac{1}{4}i, \frac{1}{4} + \frac{1}{4}i\right\}$

63. $\left\{\frac{1}{5} - \frac{2}{5}i, \frac{1}{5} + \frac{2}{5}i\right\}$ **65.** $\left\{-\frac{1}{2} - \frac{\sqrt{3}}{2}i, -\frac{1}{2} + \frac{\sqrt{3}}{2}i\right\}$ **67.** $\{2, -1 - \sqrt{3}i, -1 + \sqrt{3}i\}$ **69.** $\{-2, 2, -2i, 2i\}$ **71.** $\{-3i, -2i, 2i, 3i\}$

73. Two complex solutions that are conjugates of each other **75.** Two unequal real solutions **77.** A repeated real solution **79.** $2 - 3i$ **81.** 6

83. 25 **85.** $2 + 3i$ ohms **87.** $z + \bar{z} = (a + bi) + (a - bi) = 2a; z - \bar{z} = (a + bi) - (a - bi) = 2bi$

89. $\overline{z + w} = \overline{(a + bi) + (c + di)} = \overline{(a + c) + (b + d)i} = (a + c) - (b + d)i = (a - bi) + (c - di) = \bar{z} + \bar{w}$

1.4 Assess Your Understanding (page 122)

4. extraneous **5.** quadratic in form **6.** T **7.** $\{1\}$ **9.** No real solution **11.** $\{-13\}$ **13.** $\{4\}$ **15.** $\{-1\}$ **17.** $\{0, 64\}$ **19.** $\{3\}$ **21.** $\{2\}$

23. $\left\{-\frac{8}{5}\right\}$ **25.** $\{8\}$ **27.** $\{-1, 3\}$ **29.** $\{1, 5\}$ **31.** $\{1\}$ **33.** $\{5\}$ **35.** $\{2\}$ **37.** $\{-4, 4\}$ **39.** $\{0, 3\}$ **41.** $\{-2, -1, 1, 2\}$ **43.** $\{-1, 1\}$

45. $\{-2, 1\}$ **47.** $\{-6, -5\}$ **49.** $\left\{-\frac{1}{3}\right\}$ **51.** $\left\{-\frac{3}{2}, 2\right\}$ **53.** $\left\{0, \frac{1}{16}\right\}$ **55.** $\{16\}$ **57.** $\{1\}$ **59.** $\left\{\left(\frac{9 - \sqrt{17}}{8}\right)^4, \left(\frac{9 + \sqrt{17}}{8}\right)^4\right\}$

61. $\{\sqrt{2}, \sqrt{3}\}$ **63.** $\{-4, 1\}$ **65.** $\left\{-2, -\frac{1}{2}\right\}$ **67.** $\left\{-\frac{3}{2}, \frac{1}{3}\right\}$ **69.** $\left\{-\frac{1}{8}, 27\right\}$ **71.** $\left\{-2, -\frac{4}{5}\right\}$ **73.** $\{-3, 0, 3\}$ **75.** $\left\{0, \frac{3}{4}\right\}$ **77.** $\{-5, 0, 4\}$

79. $\{-1, 1\}$ **81.** $\{-2, 2, 3\}$ **83.** $\left\{-2, \frac{1}{2}, 2\right\}$ **85.** $\left\{\frac{2}{5}\right\}$ **87.** $\left\{0, \frac{5}{2}, 3\right\}$ **89.** $\{0.34, 11.66\}$ **91.** $\{-1.03, 1.03\}$ **93.** $\{-1.85, 0.17\}$ **95.** $\{1.5, 5\}$

97. The depth of the well is 229.94 ft. **99.** 220.7 ft

1.5 Assess Your Understanding *(page 132)*

3. negative **4.** closed interval **5.** Multiplication Properties **6.** T **7.** T **8.** T **9.** F **10.** T **11.** $[0, 2]; 0 \le x \le 2$ **13.** $[2, \infty); x \ge 2$ **15.** $[0, 3);$
$0 \le x < 3$ **17. (a)** $6 < 8$ **(b)** $-2 < 0$ **(c)** $9 < 15$ **(d)** $-6 > -10$ **19. (a)** $7 > 0$ **(b)** $-1 > -8$ **(c)** $12 > -9$ **(d)** $-8 < 6$ **21. (a)** $2x + 4 < 5$
(b) $2x - 4 < -3$ **(c)** $6x + 3 < 6$ **(d)** $-4x - 2 > -4$

23. $[0, 4]$

25. $[4, 6)$

27. $[4, \infty)$

29. $(-\infty, -4)$

31. $2 \le x \le 5$

33. $-3 < x < -2$

35. $x \ge 4$

37. $x < -3$

39. $<$ **41.** $>$ **43.** $\ge$ **45.** $<$ **47.** $\le$ **49.** $>$ **51.** $\ge$

53. $\{x|x < 4\}$ or $(-\infty, 4)$

55. $\{x|x \ge -1\}$ or $[-1, \infty)$

57. $\{x|x > 3\}$ or $(3, \infty)$

59. $\{x|x \ge 2\}$ or $[2, \infty)$

61. $\{x|x > -7\}$ or $(-7, \infty)$

63. $\left\{x\middle|x \le \dfrac{2}{3}\right\}$ or $\left(-\infty, \dfrac{2}{3}\right]$

65. $\{x|x < -20\}$ or $(-\infty, -20)$

67. $\left\{x\middle|x \ge \dfrac{4}{3}\right\}$ or $\left[\dfrac{4}{3}, \infty\right)$

69. $\{x|3 \le x \le 5\}$ or $[3, 5]$

71. $\left\{x\middle|\dfrac{2}{3} \le x \le 3\right\}$ or $\left[\dfrac{2}{3}, 3\right]$

73. $\left\{x\middle|-\dfrac{11}{2} < x < \dfrac{1}{2}\right\}$ or $\left(-\dfrac{11}{2}, \dfrac{1}{2}\right)$

75. $\{x|-6 < x < 0\}$ or $(-6, 0)$

77. $\{x|x < -5\}$ or $(-\infty, -5)$

79. $\{x|x \ge -1\}$ or $[-1, \infty)$

81. $\left\{x\middle|\dfrac{1}{2} \le x < \dfrac{5}{4}\right\}$ or $\left[\dfrac{1}{2}, \dfrac{5}{4}\right)$

83. $\left\{x\middle|x < -\dfrac{1}{2}\right\}$ or $\left(-\infty, -\dfrac{1}{2}\right)$

85. $\left\{x\middle|x > \dfrac{10}{3}\right\}$ or $\left(\dfrac{10}{3}, \infty\right)$

87. $\{x|x > 3\}$ or $(3, \infty)$

89. $a = 3, b = 5$ **91.** $a = -12, b = -8$ **93.** $a = 3, b = 11$ **95.** $a = \dfrac{1}{4}, b = 1$ **97.** $a = 4, b = 16$ **99.** $\{x|x \ge -2\}$ **101.** $21 < \text{Age} < 30$

103. (a) Male ≥ 79.66 years **(b)** Female ≥ 83.58 years **(c)** A female can expect to live 3.92 years longer. **105.** The agent's commission ranges
from \$45,000 to \$95,000, inclusive. As a percent of selling price, the commission ranges from 5% to 8.6%, inclusive. **107.** The amount withheld
varies from \$98.30 to \$148.30, inclusive. **109.** The usage varies from 675.41 kW · hr to 2500.91 kW · hr, inclusive. **111.** The dealer's cost varies
from \$15,254.24 to \$16,071.43, inclusive. **113. (a)** You need at least a 74 on the fifth test. **(b)** You need at least a 77 on the fifth test.

115. $\dfrac{a+b}{2} - a = \dfrac{a+b-2a}{2} = \dfrac{b-a}{2} > 0$; therefore, $a < \dfrac{a+b}{2}$. $b - \dfrac{a+b}{2} = \dfrac{2b-a-b}{2} = \dfrac{b-a}{2} > 0$; therefore, $b > \dfrac{a+b}{2}$

117. $(\sqrt{ab})^2 - a^2 = ab - a^2 = a(b-a) > 0$; thus $(\sqrt{ab})^2 > a^2$ and $\sqrt{ab} > a$.
$b^2 - (\sqrt{ab})^2 = b^2 - ab = b(b-a) > 0$; thus $b^2 > (\sqrt{ab})^2$ and $b > \sqrt{ab}$.

119. $h - a = \dfrac{2ab}{a+b} - a = \dfrac{ab-a^2}{a+b} = \dfrac{a(b-a)}{a+b} > 0$; thus $h > a$. $b - h = b - \dfrac{2ab}{a+b} = \dfrac{b^2-ab}{a+b} = \dfrac{b(b-a)}{a+b} > 0$; thus $h < b$.

121. Since $0 < a < b$, then $a - b < 0$ and $\dfrac{a-b}{ab} < 0$. So $\dfrac{a}{ab} - \dfrac{b}{ab} < 0$, or $\dfrac{1}{b} - \dfrac{1}{a} < 0$. Therefore, $\dfrac{1}{b} < \dfrac{1}{a}$. And $0 < \dfrac{1}{b}$ because $b > 0$.

1.6 Assess Your Understanding *(page 137)*

3. $\{-5, 5\}$ **4.** $\{x|-5 < x < 5\}$ **5.** T **6.** T **7.** $\{-3, 3\}$ **9.** $\{-4, 1\}$ **11.** $\left\{-1, \dfrac{3}{2}\right\}$ **13.** $\{-4, 4\}$ **15.** $\{2\}$ **17.** $\left\{-\dfrac{27}{2}, \dfrac{27}{2}\right\}$

19. $\left\{-\dfrac{36}{5}, \dfrac{24}{5}\right\}$ **21.** No solution **23.** $\left\{-\dfrac{1}{2}, \dfrac{1}{2}\right\}$ **25.** $\{3, 3\}$ **27.** $\{-1, 3\}$ **29.** $\{-2, -1, 0, 1\}$ **31.** $\left\{\dfrac{8}{7}, 4\right\}$ **33.** $\left\{-\dfrac{1}{2}, 0\right\}$

35. $\{x|-4 < x < 4\}; (-4, 4)$

37. $\{x|x < -4 \text{ or } x > 4\}; (-\infty, -4) \cup (4, \infty)$

39. $\{x|1 < x < 3\}; (1, 3)$

41. $\left\{t\middle|-\dfrac{2}{3} \le t \le 2\right\}; \left[-\dfrac{2}{3}, 2\right]$

43. $\left\{x\middle|x \le \dfrac{1}{2} \text{ or } x \ge \dfrac{5}{2}\right\}; \left(-\infty, \dfrac{1}{2}\right] \cup \left[\dfrac{5}{2}, \infty\right)$

45. $\left\{x\middle|-1 < x < \dfrac{3}{2}\right\}; \left(-1, \dfrac{3}{2}\right)$

47. $\{x|x < -1 \text{ or } x > 2\}; (-\infty, -1) \text{ or } (2, \infty)$ **49.** No solution **51.** $\left\{x \middle| x < -\dfrac{3}{2} \text{ or } x > \dfrac{3}{2}\right\}; \left(-\infty, -\dfrac{3}{2}\right) \cup \left(\dfrac{3}{2}, \infty\right)$

53. $\{x|-1 \le x \le 2\}; [-1, 2]$ **55.** No solution **57.** All real numbers; $(-\infty, \infty)$

59. $\left\{x \middle| -\dfrac{9}{4} < x < \dfrac{3}{4}\right\}; \left(-\dfrac{9}{4}, \dfrac{3}{4}\right)$ **61.** All real numbers; $(-\infty, \infty)$

63. $|x - 98.6| \ge 1.5; x \le 97.1 \text{ or } x \ge 100.1$ **65.** $|x - 13.4| < 1.35$; between 12.05 and 14.75 books per year are read.
67. $|x - 3| < \dfrac{1}{2}; \dfrac{5}{2} < x < \dfrac{7}{2}$ **69.** $|x + 3| > 2; x < -5 \text{ or } x > -1$ **71.** $a = 2, b = 8$ **73.** $a = -15, b = -7$ **75.** $a = -1, b = -\dfrac{1}{15}$
77. $b - a = (\sqrt{b} - \sqrt{a})(\sqrt{b} + \sqrt{a})$. Since $\sqrt{b} - \sqrt{a} > 0, \sqrt{a} > 0, \sqrt{b} > 0$, then $b - a > 0$, so $a < b$.
79. $(a + b)^2 = a^2 + 2ab + b^2 \le |a|^2 + 2|a||b| + |b|^2 = (|a| + |b|)^2$; thus, $|a + b| \le |a| + |b|$.
81. $x^2 - a < 0; (x - \sqrt{a})(x + \sqrt{a}) < 0$; therefore, $-\sqrt{a} < x < \sqrt{a}$. **83.** $\{x|-1 < x < 1\}$ **85.** $\{x|x \le -3 \text{ or } x \ge 3\}$
87. $\{x|-4 \le x \le 4\}$ **89.** $\{x|x < -2 \text{ or } x > 2\}$ **91.** $\{-1, 5\}$

1.7 Assess Your Understanding (page 145)

1. mathematical modeling **2.** interest **3.** uniform motion **4.** F **5.** T **6.** $100 - x$ **7.** $A = \pi r^2; r = $ radius, $A = $ area **9.** $A = s^2; A = $ area,
$s = $ length of a side **11.** $F = ma; F = $ force, $m = $ mass, $a = $ acceleration **13.** $W = Fd; W = $ work, $F = $ force, $d = $ distance **15.** $C = 150x$;
$C = $ total variable cost, $x = $ number of dishwashers **17.** Invest \$31,250 in bonds and \$18,750 in CDs. **19.** \$11,600 was loaned out at 8%.
21. Mix 75 lb of Earl Grey tea with 25 lb of Orange Pekoe tea. **23.** Mix 160 lb of cashews with the almonds. **25.** The speed of the current is 2.286
mi/hr. **27.** The speed of the current is 5 mi/hr. **29.** Karen walked at 4.05 ft/sec. **31.** A doubles tennis court is 78 feet long and 36 feet wide.
33. Working together, it takes 12 min. **35. (a)** The dimensions are 10 ft by 5 ft. **(b)** The area is 50 sq ft. **(c)** The dimensions would be 7.5 ft by 7.5 ft.
(d) The area would be 56.25 sq ft. **37.** The defensive back catches up to the tight end at the tight end's 45-yd line. **39.** Add $\dfrac{2}{3}$ gal of water.
41. Evaporate 10.67 oz of water. **43.** 40 g of 12-karat gold should be mixed with 20 g of pure gold. **45.** Mike passes Dan $\dfrac{1}{3}$ mile from the start, 2 min
from the time Mike started to race. **47.** Start the auxiliary pump at 9:45 AM. **49.** The tub will fill in 1 hr. **51.** Run: 12 miles; bicycle: 75 miles.
53. Lewis would beat Burke by 16.75 m. **55.** Set the original price at \$40. At 50% off, there will be no profit. **59.** The tail wind was 91.47 knots.

Review Exercises (page 150)

1. $\{-18\}$ **3.** $\{6\}$ **5.** $\left\{\dfrac{1}{5}\right\}$ **7.** $\{6\}$ **9.** No real solution **11.** $\left\{\dfrac{11}{8}\right\}$ **13.** $\left\{-2, \dfrac{3}{2}\right\}$ **15.** $\left\{\dfrac{1 - \sqrt{13}}{4}, \dfrac{1 + \sqrt{13}}{4}\right\}$ **17.** $\{-3, 3\}$

19. No real solution **21.** $\{-2, -1, 1, 2\}$ **23.** $\{2\}$ **25.** $\left\{\dfrac{13}{2}\right\}$ **27.** $\left\{\dfrac{\sqrt{5}}{2}\right\}$ **29.** $\left\{\dfrac{9}{4}\right\}$ **31.** $\left\{-1, \dfrac{1}{2}\right\}$ **33.** $\left\{\dfrac{m}{1 - n}, \dfrac{m}{1 + n}\right\}$

35. $\left\{-\dfrac{9b}{5a}, \dfrac{2b}{a}\right\}$ **37.** $\left\{-\dfrac{9}{5}\right\}$ **39.** $\{-5, 2\}$ **41.** $\left\{-\dfrac{5}{3}, 3\right\}$ **43.** $\left\{0, \dfrac{3}{2}\right\}$ **45.** $\left\{-\dfrac{5}{2}, -2, 2\right\}$

47. $\{x|x \ge 14\}; [14, \infty)$ **49.** $\left\{x \middle| -\dfrac{31}{2} \le x \le \dfrac{33}{2}\right\}; \left[-\dfrac{31}{2}, \dfrac{33}{2}\right]$ **51.** $\{x|-23 < x < -7\}; (-23, -7)$

53. $\left\{x \middle| -\dfrac{3}{2} < x < -\dfrac{7}{6}\right\}; \left(-\dfrac{3}{2}, -\dfrac{7}{6}\right)$ **55.** $\{x|x \le -2 \text{ or } x \ge 7\}; (-\infty, -2] \cup [7, \infty)$ **57.** $\left\{x \middle| 0 \le x \le \dfrac{4}{3}\right\}; \left[0, \dfrac{4}{3}\right]$

59. $\left\{x \middle| x < -1 \text{ or } x > \dfrac{7}{3}\right\}; (-\infty, -1) \cup \left(\dfrac{7}{3}, \infty\right)$ **61.** 9 **63.** $\dfrac{4}{9}$ **65.** $4 + 7i$ **67.** $-3 + 2i$ **69.** $\dfrac{9}{10} - \dfrac{3}{10}i$ **71.** -1 **73.** $-46 + 9i$

75. $\left\{-\dfrac{1}{2} - \dfrac{\sqrt{3}}{2}i, -\dfrac{1}{2} + \dfrac{\sqrt{3}}{2}i\right\}$ **77.** $\left\{\dfrac{-1 - \sqrt{17}}{4}, \dfrac{-1 + \sqrt{17}}{4}\right\}$ **79.** $\left\{\dfrac{1}{2} - \dfrac{\sqrt{11}}{2}i, \dfrac{1}{2} + \dfrac{\sqrt{11}}{2}i\right\}$ **81.** $\left\{\dfrac{1}{2} - \dfrac{\sqrt{23}}{2}i, \dfrac{1}{2} + \dfrac{\sqrt{23}}{2}i\right\}$

83. $p = 2l + 2w$ **85.** The interest is \$630. **87.** The storm is 3300 ft away. **89.** The search plane can go as far as 616 miles. **91.** The helicopter
will reach the life raft in a little less than 1 hr 35 min. **93.** The Metra commuter averages 30 mi/hr; the Amtrak averages 80 mi/hr. **95.** It takes

Clarissa 10 days by herself. **97.** Add 256 oz of water. **99.** 5 cm and 12 cm **101.** Mix 90 cm³ of 15% HCl with the 60 cm³ of 40% of HCl to obtain 150 cm³ of 25% HCl. **103.** It will take the smaller pump 2 hr. **105.** Scott receives $400,000, Alice receives $300,000, and Tricia receives $200,000. **107.** It would take the older copier 180 min or 3 hr. **109.** The freight train is 190.67 feet long.

Chapter Test *(page 152)*

1. $\left\{\dfrac{5}{2}\right\}$ **2.** $\{-2, 3\}$ **3.** $\{-2, 2\}$ **4.** $\left\{\dfrac{9}{2}\right\}$ **5.** $\{0, 3\}$ **6.** $\left\{-2, -\dfrac{2}{3}, 2\right\}$ **7.** No real solution

8. $\left[-\dfrac{2}{3}, \dfrac{16}{3}\right]$ **9.** $\left(-4, \dfrac{4}{3}\right)$ **10.** $(-\infty, -1] \cup [6, \infty)$

11. $-\dfrac{3}{5} - \dfrac{1}{5}i$ **12.** $\left\{\dfrac{1}{2} - i, \dfrac{1}{2} + i\right\}$ **13.** Add $6\dfrac{2}{3}$ lb of $8/lb coffee to get $26\dfrac{2}{3}$ lb of $5/lb coffee.

CHAPTER 2 Graphs

2.1 Assess Your Understanding *(page 160)*

5. abscissa; ordinate **6.** quadrants **7.** midpoint **8.** F **9.** F **10.** T

11. (a) Quadrant II **(b)** *x*-axis
(c) Quadrant III **(d)** Quadrant I
(e) *y*-axis **(f)** Quadrant IV

13. The points will be on a vertical line that is 2 units to the right of the *y*-axis.

15. $\sqrt{5}$
17. $\sqrt{10}$
19. $2\sqrt{17}$
21. $\sqrt{85}$
23. $\sqrt{53}$
25. $\sqrt{6.89} \approx 2.62$
27. $\sqrt{a^2 + b^2}$

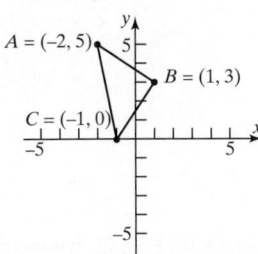

29. $d(A, B) = \sqrt{13}$
$d(B, C) = \sqrt{13}$
$d(A, C) = \sqrt{26}$
$\left(\sqrt{13}\right)^2 + \left(\sqrt{13}\right)^2 = \left(\sqrt{26}\right)^2$
Area $= \dfrac{13}{2}$ square units

31. $d(A, B) = \sqrt{130}$
$d(B, C) = \sqrt{26}$
$d(A, C) = 2\sqrt{26}$
$\left(\sqrt{26}\right)^2 + \left(2\sqrt{26}\right)^2 = \left(\sqrt{130}\right)^2$
Area $= 26$ square units

33. $d(A, B) = 4$
$d(B, C) = \sqrt{41}$
$d(A, C) = 5$
$4^2 + 5^2 = \left(\sqrt{41}\right)^2$
Area $= 10$ square units

35. $(4, 0)$ **37.** $\left(\dfrac{3}{2}, 1\right)$ **39.** $(5, -1)$ **41.** $(1.05, 0.7)$ **43.** $\left(\dfrac{a}{2}, \dfrac{b}{2}\right)$ **45.** $(2, 2); (2, -4)$ **47.** $(0, 0); (8, 0)$ **49.** $(1, 2)$ **51.** $\sqrt{17}; 2\sqrt{5}; \sqrt{29}$

53. $\left(\dfrac{s}{2}, \dfrac{s}{2}\right)$ **55.** $d(P_1, P_2) = 6; d(P_2, P_3) = 4; d(P_1, P_3) = 2\sqrt{13}$; right triangle

57. $d(P_1, P_2) = 2\sqrt{17}; d(P_2, P_3) = \sqrt{34}; d(P_1, P_3) = \sqrt{34}$; isosceles right triangle **59.** $90\sqrt{2} \approx 127.28$ ft
61. (a) $(90, 0), (90, 90), (0, 90)$ **(b)** $5\sqrt{2161} \approx 232.43$ ft **(c)** $30\sqrt{149} \approx 366.20$ ft
63. $d = 50t$ mi **65. (a)** $(2.65, 1.6)$ **(b)** Approximately 1.285 units

2.2 Assess Your Understanding *(page 171)*

3. intercepts **4.** $y = 0$ **5.** *y*-axis **6.** 4 **7.** $(-3, 4)$ **8.** T **9.** F **10.** F **11.** $(0, 0)$ is on the graph. **13.** $(0, 3)$ is on the graph.
15. $(0, 2)$ and $\left(\sqrt{2}, \sqrt{2}\right)$ are on the graph.

17. $(-2, 0), (0, 2)$

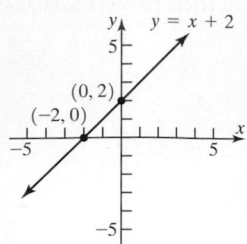

19. $(-4, 0), (0, 8)$

21. $(-1, 0), (1, 0), (0, -1)$

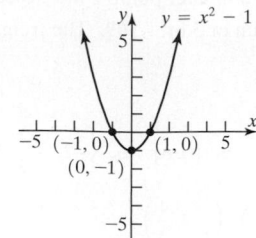

23. $(-2, 0), (2, 0), (0, 4)$

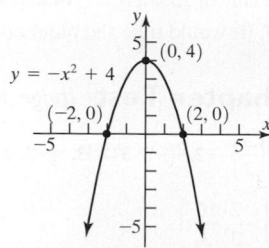

25. $(3, 0), (0, 2)$

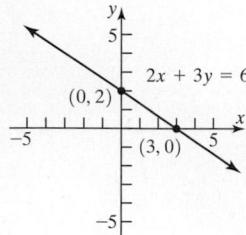

27. $(-2, 0), (2, 0), (0, 9)$

29.

31.

33.

35.

37.

39. (a) $(-1, 0), (1, 0)$
(b) Symmetric with respect to the x-axis, the y-axis, and the origin

41. (a) $\left(-\dfrac{\pi}{2}, 0\right), (0, 1), \left(\dfrac{\pi}{2}, 0\right)$
(b) Symmetric with respect to the y-axis

43. (a) $(0, 0)$
(b) Symmetric with respect to the x-axis

45. (a) $(-2, 0), (0, 0), (2, 0)$
(b) Symmetric with respect to the origin

47. (a) $(-1, 0), (0, -1), (1, 0)$
(b) Symmetric with respect to the y-axis

49. (a) No intercepts
(b) Symmetric with respect to the origin

51.

53.

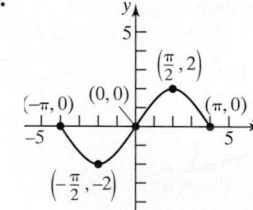

55. $(-4, 0), (0, -2), (0, 2)$; symmetric with respect to the x-axis **57.** $(0, 0)$; symmetric with respect to the origin **59.** $(0, 9), (3, 0), (-3, 0)$; symmetric with respect to the y-axis **61.** $(-2, 0), (2, 0), (0, -3), (0, 3)$; symmetric with respect to the x-axis, y-axis, and origin **63.** $(0, -27), (3, 0)$; no symmetry

65. $(0, -4), (4, 0), (-1, 0)$; no symmetry **67.** $(0, 0)$; symmetric with respect to the origin **69.** $(0, 0)$; symmetric with respect to the origin

71.

73.

75. $b = 13$

77. $a = -4$ or $a = 1$

79. $(-1, -2)$

81. 4

83. (a) $(0, 0), (2, 0), (0, 1), (0, -1)$ **(b)** x-axis symmetry

85. (a) $y = \sqrt{x^2}$ and $y = |x|$ have the same graph. **(b)** $\sqrt{x^2} = |x|$ **(c)** $x \geq 0$ for $y = (\sqrt{x})^2$, while x can be any real number for $y = x$.
(d) $y \geq 0$ for $y = \sqrt{x^2}$

2.3 Assess Your Understanding *(page 185)*

1. undefined; 0 **2.** 3; 2 **3.** $y = b$; y-intercept **4.** T **5.** F **6.** T **7.** $m_1 = m_2$; y-intercepts; $m_1 m_2 = -1$ **8.** 2 **9.** $-\dfrac{1}{2}$ **10.** False

11. (a) Slope $= \dfrac{1}{2}$

 (b) If x increases by 2 units, y will increase by 1 unit.

13. (a) Slope $= -\dfrac{1}{3}$

 (b) If x increases by 3 units, y will decrease by 1 unit.

15. Slope $= -\dfrac{3}{2}$

17. Slope $= -\dfrac{1}{2}$

19. Slope $= 0$

21. Slope undefined

23.

25.

27.

29.

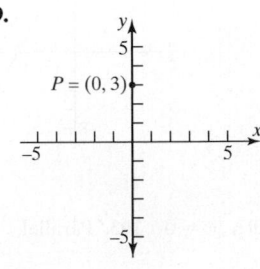

31. $(2, 6)$; $(3, 10)$; $(4, 14)$ **33.** $(4, -7)$; $(6, -10)$; $(8, -13)$ **35.** $(-1, -5)$; $(0, -7)$; $(1, -9)$ **37.** $x - 2y = 0$ or $y = \dfrac{1}{2}x$

39. $x + y = 2$ or $y = -x + 2$ **41.** $2x - y = 3$ or $y = 2x - 3$ **43.** $x + 2y = 5$ or $y = -\dfrac{1}{2}x + \dfrac{5}{2}$ **45.** $3x - y = -9$ or $y = 3x + 9$

47. $2x + 3y = -1$ or $y = -\dfrac{2}{3}x - \dfrac{1}{3}$ **49.** $x - 2y = -5$ or $y = \dfrac{1}{2}x + \dfrac{5}{2}$ **51.** $3x + y = 3$ or $y = -3x + 3$ **53.** $x - 2y = 2$ or $y = \dfrac{1}{2}x - 1$

55. $x = 2$; no slope–intercept form **57.** $y = 2$ **59.** $2x - y = -4$ or $y = 2x + 4$ **61.** $2x - y = 0$ or $y = 2x$ **63.** $x = 4$; no slope–intercept form

65. $2x + y = 0$ or $y = -2x$ **67.** $x - 2y = -3$ or $y = \dfrac{1}{2}x + \dfrac{3}{2}$ **69.** $y = 4$

71. Slope $= 2$; y-intercept $= 3$ **73.** Slope $= 2$; y-intercept $= -2$ **75.** Slope $= \dfrac{1}{2}$; y-intercept $= 2$ **77.** Slope $= -\dfrac{1}{2}$; y-intercept $= 2$

 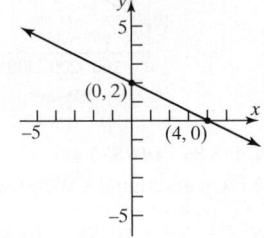

79. Slope $= \dfrac{2}{3}$; y-intercept $= -2$ **81.** Slope $= -1$; y-intercept $= 1$ **83.** Slope undefined; no y-intercept **85.** Slope $= 0$; y-intercept $= 5$

 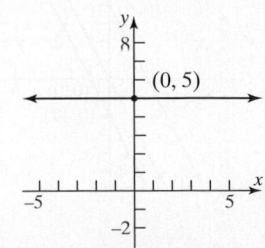

87. Slope $= 1$; y-intercept $= 0$ **89.** Slope $= \dfrac{3}{2}$; y-intercept $= 0$ **91. (a)** x-intercept: 3; y-intercept: 2 **93. (a)** x-intercept: -10; y-intercept: 8

(b) **(b)**

 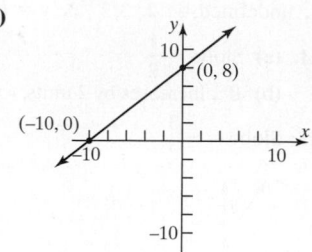

95. (a) x-intercept: 3; y-intercept: $\dfrac{21}{2}$ **97. (a)** x-intercept: 2; y-intercept: 3 **99. (a)** x-intercept: 5; y-intercept: -2

(b) **(b)** **(b)**

 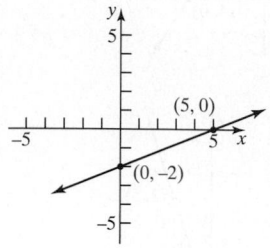

101. $y = 0$ **103.** Parallel **105.** Neither **107.** $x - y = -2$ or $y = x + 2$ **109.** $x + 3y = 3$ or $y = -\dfrac{1}{3}x + 1$

111. $P_1 = (-2, 5)$, $P_2 = (1, 3)$, $m_1 = -\dfrac{2}{3}$; $P_2 = (1, 3)$, $P_3 = (-1, 0)$, $m_2 = \dfrac{3}{2}$; because $m_1 m_2 = -1$, the lines are perpendicular and the points $(-2, 5)$, $(1, 3)$, and $(-1, 0)$ are the vertices of a right triangle; thus, the points P_1, P_2, and P_3 are the vertices of a right triangle.

113. $P_1 = (-1, 0)$, $P_2 = (2, 3)$, $m = 1$; $P_3 = (1, -2)$, $P_4 = (4, 1)$, $m = 1$; $P_1 = (-1, 0)$, $P_3 = (1, -2)$, $m = -1$; $P_2 = (2, 3)$, $P_4 = (4, 1)$, $m = -1$; opposite sides are parallel, and adjacent sides are perpendicular; the points are the vertices of a rectangle.

115. $C = 0.20x + 29$; $51.00; $75.00 **117.** $C = 0.53x + 1,070,000$

119. (a) $C = 0.08275x + 7.58, 0 \le x \le 400$

(b)

(c) $15.86 **(d)** $32.41
(e) Each additional k W-hr used adds $0.08275 to the bill.

121. $^{\circ}\text{C} = \dfrac{5}{9}(^{\circ}\text{F} - 32)$; approximately 21.1°C

123. (a) $y = -\dfrac{2}{25}x + 30$ **(b)** x-intercept: 375; The ramp meets the floor 375 in. (31.25 ft) from the base of the platform.
(c) The ramp does not meet design requirements. It has a run of 31.25 ft long. **(d)** The only slope possible for the ramp to comply with the requirement is for it to drop 1 in. for every 12-in. run.

125. (a) $A = \dfrac{1}{5}x + 20,000$ **(b)** $80,000 **(c)** Each additional box sold requires an additional $0.20 in advertising.

127. All have the same slope, 2; the lines are parallel.

129. (b), (c), (e), (g) **131.** (c) **137.** No; no **139.** They are the same line.
141. Yes, if the y-intercept is 0.

2.4 Assess Your Understanding *(page 193)*

3. F **4.** radius **5.** T **6.** F **7.** Center $(2, 1)$; radius $= 2$; $(x - 2)^2 + (y - 1)^2 = 4$ **9.** Center $\left(\frac{5}{2}, 2\right)$; radius $= \frac{3}{2}$; $\left(x - \frac{5}{2}\right)^2 + (y - 2)^2 = \frac{9}{4}$

11. $x^2 + y^2 = 4$;
$x^2 + y^2 - 4 = 0$

13. $x^2 + (y - 2)^2 = 4$;
$x^2 + y^2 - 4y = 0$

15. $(x - 4)^2 + (y + 3)^2 = 25$;
$x^2 + y^2 - 8x + 6y = 0$

17. $(x + 2)^2 + (y - 1)^2 = 16$;
$x^2 + y^2 + 4x - 2y - 11 = 0$

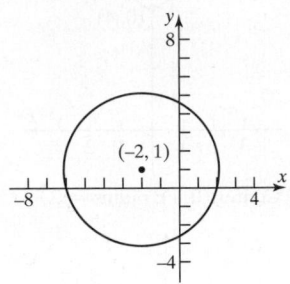

19. $\left(x - \frac{1}{2}\right)^2 + y^2 = \frac{1}{4}$;
$x^2 + y^2 - x = 0$

21. **(a)** $(h, k) = (0, 0)$; $r = 2$
(b)

23. **(a)** $(h, k) = (3, 0)$; $r = 2$
(b)

25. **(a)** $(h, k) = (1, 2)$; $r = 3$
(b)

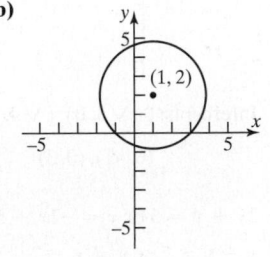

(c) $(\pm 2, 0)$; $(0, \pm 2)$

(c) $(1, 0)$; $(5, 0)$

(c) $\left(1 \pm \sqrt{5}, 0\right)$; $\left(0, 2 \pm 2\sqrt{2}\right)$

27. **(a)** $(h, k) = (-2, 2)$; $r = 3$
(b)

29. **(a)** $(h, k) = \left(\frac{1}{2}, -1\right)$; $r = \frac{1}{2}$
(b)

31. **(a)** $(h, k) = (3, -2)$; $r = 5$
(b)

33. **(a)** $(h, k) = (-2, 0)$; $r = 2$
(b)

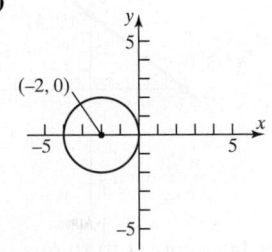

(c) $\left(-2 \pm \sqrt{5}, 0\right)$; $\left(0, 2 \pm \sqrt{5}\right)$ **(c)** $(0, -1)$ **(c)** $\left(3 \pm \sqrt{21}, 0\right)$; $(0, -6)$, $(0, 2)$ **(c)** $(0, 0)$, $(-4, 0)$

35. $x^2 + y^2 = 13$ **37.** $(x - 2)^2 + (y - 3)^2 = 9$ **39.** $(x + 1)^2 + (y - 3)^2 = 5$ **41.** $(x + 1)^2 + (y - 3)^2 = 1$ **43.** (c) **45.** (b) **47.** 18 units2
49. $x^2 + (y - 139)^2 = 15{,}625$ **51.** $x^2 + y^2 + 2x + 4y - 4168.16 = 0$ **53.** $\sqrt{2}x + 4y - 9\sqrt{2} = 0$ **55.** $(1, 0)$ **57.** $y = 2$ **59.** (b), (c), (e), (g)

2.5 Assess Your Understanding *(page 199)*

1. $y = kx$ **2.** F **3.** $y = \frac{1}{5}x$ **5.** $A = \pi x^2$ **7.** $F = \frac{250}{d^2}$ **9.** $z = \frac{1}{5}(x^2 + y^2)$ **11.** $M = \frac{9d^2}{2\sqrt{x}}$ **13.** $T^2 = \frac{8a^3}{d^2}$ **15.** $V = \frac{4\pi}{3}r^3$ **17.** $A = \frac{1}{2}bh$

19. $F = 6.67 \times 10^{-11}\left(\frac{mM}{d^2}\right)$ **21.** $p = 0.00649B$; \$941.05 **23.** 144 ft; 2 sec **25.** 2.25 **27.** $R = 3.95g$; \$41.48 **29.** **(a)** $D = \frac{429}{p}$ **(b)** 143 bags

31. 450 cm^3 **33.** 124.76 lb **35.** $V = \pi r^2 h$ **37.** 54.86 lb **39.** $\sqrt[3]{6} \approx 1.82$ in. **41.** 2812.5 joules **43.** 384 psi

Review Exercises *(page 202)*

1. **(a)** $2\sqrt{5}$ **(b)** $(2, 1)$ **(c)** $\frac{1}{2}$ **(d)** For each run of 2, there is a rise of 1.

3. **(a)** 5 **(b)** $\left(-\frac{1}{2}, 1\right)$ **(c)** $-\frac{4}{3}$ **(d)** For each run of 3, there is a rise of -4. **5.** **(a)** 12 **(b)** $(4, 2)$ **(c)** undefined **(d)** no change in x

7.

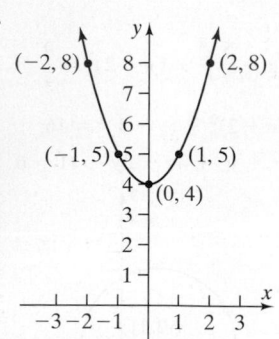

9. $(0, 0)$; symmetric with respect to the x-axis
11. $(\pm 4, 0), (0, \pm 2)$; symmetric with respect to the x-axis, y-axis, and origin
13. $(0, 1)$; symmetric with respect to the y-axis
15. $(0, 0), (-1, 0), (0, -2)$; no symmetry
17. $(x + 2)^2 + (y - 3)^2 = 16$ **19.** $(x + 1)^2 + (y + 2)^2 = 1$

21. Center $(0, 1)$; radius $= 2$

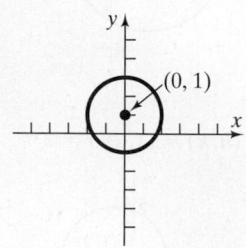

Intercepts: $(-\sqrt{3}, 0), (\sqrt{3}, 0),$
$(0, -1), (0, 3)$

23. Center $(1, -2)$; radius $= 3$

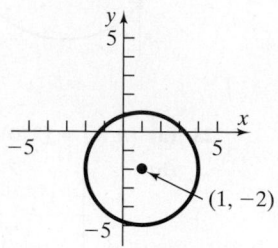

Intercepts: $(1 - \sqrt{5}, 0), (1 + \sqrt{5}, 0),$
$(0, -2 - 2\sqrt{2}), (0, -2 + 2\sqrt{2})$

25. Center $(1, -2)$; radius $= \sqrt{5}$

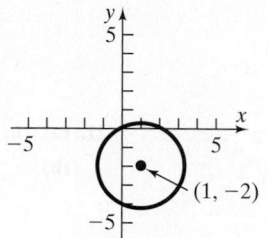

Intercepts: $(0, 0), (2, 0), (0, -4)$

27. $2x + y = 5$ or $y = -2x + 5$ **29.** $x = -3$; no slope–intercept form **31.** $x + 5y = -10$ or $y = -\dfrac{1}{5}x - 2$ **33.** $2x - 3y = -19$ or $y = \dfrac{2}{3}x + \dfrac{19}{3}$
35. $x - y = 7$ or $y = x - 7$

37. Slope $= \dfrac{4}{5}$; y-intercept $= 4$

39. Slope $= \dfrac{3}{2}$; y-intercept $= \dfrac{1}{2}$

41. Intercepts: $(6, 0), (0, -4)$

43. Intercepts: $(4, 0), (0, 6)$

45.

47.

49. **(a)** $d(A, B) = 2\sqrt{5}; d(B, C) = \sqrt{145}; d(A, C) = 5\sqrt{5}; [d(A, B)]^2 + [d(A, C)]^2 = (2\sqrt{5})^2 + (5\sqrt{5})^2 = 20 + 125 = 145 = [d(B, C)]^2$

(b) Slope from A to B is -2; slope from A to C is $\dfrac{1}{2}$. Since $(-2)\left(\dfrac{1}{2}\right) = -1$, the lines are perpendicular.

51. Slope from A to B is -1; slope from A to C is -1. **53.** $R = 3.59g$; $\$40.21$ **55.** $a \approx 36$ million miles

Chapter Test (page 204)

1. $d = 2\sqrt{13}$ **2.** $(2, 1)$ **3.** **(a)** $m = -\dfrac{2}{3}$ **(b)** For every 3-unit change in x, y will change by -2 units.

4.

5.

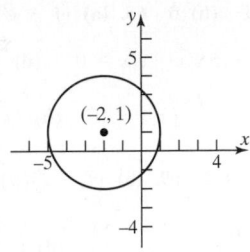

6. Intercepts: $(-3, 0), (3, 0), (0, 9)$; symmetric with respect to the y-axis

7. $y = -2x + 2$

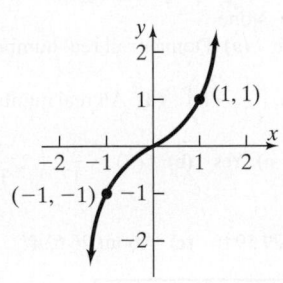

8. $x^2 + y^2 - 8x + 6y = 0$

9. Center: $(-2, 1)$; radius: 3

10. Parallel line: $y = -\dfrac{2}{3}x - \dfrac{1}{3}$; perpendicular line: $y = \dfrac{3}{2}x + 3$ **11.** 14.69 ohms

Cumulative Review (page 204)

1. $\left\{\dfrac{5}{3}\right\}$ **2.** $\{-3, 4\}$ **3.** $\left\{-\dfrac{1}{2}, 3\right\}$ **4.** $\{1 - \sqrt{3}, 1 + \sqrt{3}\}$ **5.** No real solution **6.** $\{4\}$ **7.** $\{1, 3\}$ **8.** $\{-2 - 2\sqrt{2}, -2 + 2\sqrt{2}\}$

9. $\{-3i, 3i\}$ **10.** $\{1 - 2i, 1 + 2i\}$ **11.** $\{x | x \le 5\}$ or $(\infty, 5]$; **12.** $\{x | -5 < x < 1\}$ or $(-5, 1)$;

13. $\{x | 1 \le x \le 3\}$ or $[1, 3]$; **14.** $\{x | x < -5 \text{ or } x > 1\}$ or $(\infty, -5) \cup (1, \infty)$;

15. $5\sqrt{2}$; $\left(\dfrac{3}{2}, \dfrac{1}{2}\right)$ **16.** (a), (b)

17.

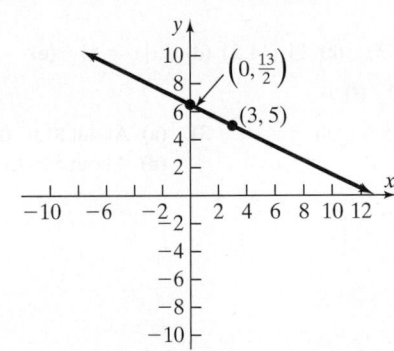

18. $y = -2x + 2$

19. $y = -\dfrac{1}{2}x + \dfrac{13}{2}$;

20.

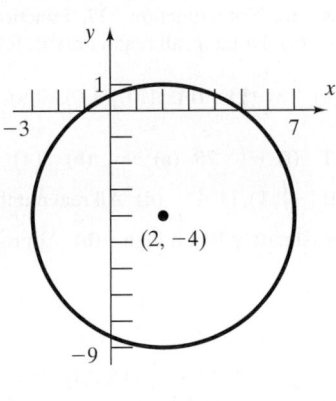

CHAPTER 3 Functions and Their Graphs

3.1 Assess Your Understanding (page 219)

5. independent; dependent **6.** range **7.** $[0, 5]$ **8.** $\ne$; f; g **9.** $(g - f)(x)$ **10.** F **11.** T **12.** T **13.** F **14.** F

15. Function; Domain: {Elvis, Colleen, Kaleigh, Marissa}, Range: {January 8, March 15, September 17} **17.** Not a function **19.** Not a function

21. Function; Domain: $\{1, 2, 3, 4\}$; Range: $\{3\}$ **23.** Not a function **25.** Function; Domain: $\{-2, -1, 0, 1\}$, Range: $\{0, 1, 4\}$ **27.** Function

29. Function **31.** Not a function **33.** Not a function **35.** Function **37.** Not a function **39.** (a) -4 (b) 1 (c) -3 (d) $3x^2 - 2x - 4$

(e) $-3x^2 - 2x + 4$ (f) $3x^2 + 8x + 1$ (g) $12x^2 + 4x - 4$ (h) $3x^2 + 6xh + 3h^2 + 2x + 2h - 4$

41. (a) 0 **(b)** $\dfrac{1}{2}$ **(c)** $-\dfrac{1}{2}$ **(d)** $\dfrac{-x}{x^2+1}$ **(e)** $\dfrac{-x}{x^2+1}$ **(f)** $\dfrac{x+1}{x^2+2x+2}$ **(g)** $\dfrac{2x}{4x^2+1}$ **(h)** $\dfrac{x+h}{x^2+2xh+h^2+1}$

43. (a) 4 **(b)** 5 **(c)** 5 **(d)** $|x|+4$ **(e)** $-|x|-4$ **(f)** $|x+1|+4$ **(g)** $2|x|+4$ **(h)** $|x+h|+4$

45. (a) $-\dfrac{1}{5}$ **(b)** $-\dfrac{3}{2}$ **(c)** $\dfrac{1}{8}$ **(d)** $\dfrac{2x-1}{3x+5}$ **(e)** $\dfrac{-2x-1}{3x-5}$ **(f)** $\dfrac{2x+3}{3x-2}$ **(g)** $\dfrac{4x+1}{6x-5}$ **(h)** $\dfrac{2x+2h+1}{3x+3h-5}$ **47.** All real numbers

49. All real numbers **51.** $\{x\,|\,x \neq -4, x \neq 4\}$ **53.** $\{x\,|\,x \neq 0\}$ **55.** $\{x\,|\,x \geq 4\}$ **57.** $\{x\,|\,x > 9\}$ **59.** $\{x\,|\,x > 1\}$

61. (a) $(f+g)(x) = 5x+1$; All real numbers **(b)** $(f-g)(x) = x+7$; All real numbers **(c)** $(f \cdot g)(x) = 6x^2 - x - 12$; All real numbers
(d) $\left(\dfrac{f}{g}\right)(x) = \dfrac{3x+4}{2x-3};\ \left\{x\,\middle|\,x \neq \dfrac{3}{2}\right\}$ **(e)** 16 **(f)** 11 **(g)** 10 **(h)** -7 **63. (a)** $(f+g)(x) = 2x^2 + x - 1$; All real numbers

(b) $(f-g)(x) = -2x^2 + x - 1$; All real numbers **(c)** $(f \cdot g)(x) = 2x^3 - 2x^2$; All real numbers **(d)** $\left(\dfrac{f}{g}\right)(x) = \dfrac{x-1}{2x^2};\ \{x\,|\,x \neq 0\}$
(e) 20 **(f)** -29 **(g)** 8 **(h)** 0 **65. (a)** $(f+g)(x) = \sqrt{x} + 3x - 5;\ \{x\,|\,x \geq 0\}$ **(b)** $(f-g)(x) = \sqrt{x} - 3x + 5;\ \{x\,|\,x \geq 0\}$
(c) $(f \cdot g)(x) = 3x\sqrt{x} - 5\sqrt{x};\ \{x\,|\,x \geq 0\}$ **(d)** $\left(\dfrac{f}{g}\right)(x) = \dfrac{\sqrt{x}}{3x-5}\ \left\{x\,\middle|\,x \geq 0, x \neq \dfrac{5}{3}\right\}$ **(e)** $\sqrt{3}+4$ **(f)** -5 **(g)** $\sqrt{2}$ **(h)** $-\dfrac{1}{2}$

67. (a) $(f+g)(x) = 1 + \dfrac{2}{x};\ \{x\,|\,x \neq 0\}$ **(b)** $(f-g)(x) = 1;\ \{x\,|\,x \neq 0\}$ **(c)** $(f \cdot g)(x) = \dfrac{1}{x} + \dfrac{1}{x^2};\ \{x\,|\,x \neq 0\}$ **(d)** $\left(\dfrac{f}{g}\right)(x) = x+1;\ \{x\,|\,x \neq 0\}$
(e) $\dfrac{5}{3}$ **(f)** 1 **(g)** $\dfrac{3}{4}$ **(h)** 2 **69. (a)** $(f+g)(x) = \dfrac{6x+3}{3x-2};\ \left\{x\,\middle|\,x \neq \dfrac{2}{3}\right\}$ **(b)** $(f-g)(x) = \dfrac{-2x+3}{3x-2};\ \left\{x\,\middle|\,x \neq \dfrac{2}{3}\right\}$
(c) $(f \cdot g)(x) = \dfrac{8x^2+12x}{(3x-2)^2};\ \left\{x\,\middle|\,x \neq \dfrac{2}{3}\right\}$ **(d)** $\left(\dfrac{f}{g}\right)(x) = \dfrac{2x+3}{4x};\ \left\{x\,\middle|\,x \neq 0; x \neq \dfrac{2}{3}\right\}$ **(e)** 3 **(f)** $-\dfrac{1}{2}$ **(g)** $\dfrac{7}{2}$ **(h)** $\dfrac{5}{4}$ **71.** $g(x) = 5 - \dfrac{7}{2}x$

73. 4 **75.** $2x + h - 1$ **77.** $6x + 3h - 2$ **79.** $3x^2 + 3xh + h^2$ **81.** $A = -\dfrac{7}{2}$ **83.** $A = -4$ **85.** $A = 8$; undefined at $x = 3$ **87.** $A(x) = \dfrac{1}{2}x^2$

89. $G(x) = 10x$ **91. (a)** P is the dependent variable; a is the independent variable. **(b)** $P(20) = 197.34$ million; In 2005, there are 197.34 million
people 20 years of age or older. **(c)** $P(0) = 290.580$ million; In 2005, there are 290.580 million people. **93. (a)** 15.1 m, 14.071 m, 12.944 m, 11.719 m
(b) 1.01 sec, 1.43 sec, 1.75 sec **(c)** 2.02 sec **95. (a)** \$222 **(b)** \$225 **(c)** \$220 **(d)** \$230 **97.** $R(x) = \dfrac{L(x)}{P(x)}$ **99.** $H(x) = P(x) \cdot I(x)$

101. (a) $P(x) = -0.05x^3 + 0.8x^2 + 155x - 500$ **(b)** $P(15) = \$1836.25$ **(c)** When 15 hundred cellphones are sold, the profit is \$1836.25.

103. Only $h(x) = 2x$

3.2 Assess Your Understanding *(page 226)*

3. vertical **4.** $5; -3$ **5.** $a = -2$ **6.** F **7.** F **8.** T **9. (a)** $f(0) = 3; f(-6) = -3$ **(b)** $f(6) = 0; f(11) = 1$ **(c)** Positive **(d)** Negative
 (e) $-3, 6$, and 10 **(f)** $-3 < x < 6; 10 < x \leq 11$ **(g)** $\{x\,|\,-6 \leq x \leq 11\}$ **(h)** $\{y\,|\,-3 \leq y \leq 4\}$ **(i)** $-3, 6, 10$ **(j)** 3 **(k)** 3 times **(l)** Once
 (m) $0, 4$ **(n)** $-5, 8$ **11.** Not a function **13.** Function **(a)** Domain: $\{x\,|\,-\pi \leq x \leq \pi\}$; Range: $\{y\,|\,-1 \leq y \leq 1\}$ **(b)** $\left(-\dfrac{\pi}{2}, 0\right), \left(\dfrac{\pi}{2}, 0\right), (0, 1)$
 (c) y-axis **15.** Not a function **17.** Function **(a)** Domain: $\{x\,|\,x > 0\}$; Range: all real numbers **(b)** $(1, 0)$ **(c)** None
19. Function **(a)** Domain: all real numbers; Range: $\{y\,|\,y \leq 2\}$ **(b)** $(-3, 0), (3, 0), (0, 2)$ **(c)** y-axis **21.** Function **(a)** Domain: all real numbers;

Range: $\{y\,|\,y \geq -3\}$ **(b)** $(1, 0), (3, 0), (0, 9)$ **(c)** None **23. (a)** Yes **(b)** $f(-2) = 9; (-2, 9)$ **(c)** $0, \dfrac{1}{2}; (0, -1), \left(\dfrac{1}{2}, -1\right)$ **(d)** All real numbers

(e) $-\dfrac{1}{2}, 1$ **(f)** -1 **25. (a)** No **(b)** $f(4) = -3; (4, -3)$ **(c)** $14; (14, 2)$ **(d)** $\{x\,|\,x \neq 6\}$ **(e)** -2 **(f)** $-\dfrac{1}{3}$ **27. a)** Yes **(b)** $f(2) = \dfrac{8}{17}; \left(2, \dfrac{8}{17}\right)$
(c) $-1, 1; (-1, 1), (1, 1)$ **(d)** All real numbers **(e)** 0 **(f)** 0

29. (a) Approximately 10.4 ft high **(b)** Approximately 9.9 ft high **31. (a)** About 81.07 ft **(b)** About 129.59 ft **(c)** About 26.63 ft
(c) **(d)** About 528.13 ft **(e)**

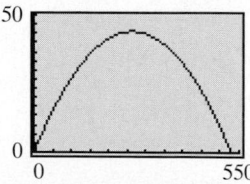

(d) The ball will not go through the hoop; $h(15) \approx 8.4$ ft. **(f)** About 115.07 ft and 413.05 ft **(g)** 275 ft; maximum height
 If $v = 30$ ft/sec, $h(15) = 10$ ft. shown in the table is 131.8 ft **(h)** 264 ft

33. (a) **(b)** **35. (a)** 3 **(b)** -2 **(c)** -1 **(d)** 1 **(e)** 2 **(f)** $-\dfrac{1}{3}$

37. The x-intercepts can number anywhere from 0 to
infinitely many. There is at most one y-intercept.

39. (a) III **(b)** IV **(c)** I **(d)** V **(e)** II

(c) 600 mi/hr

41.

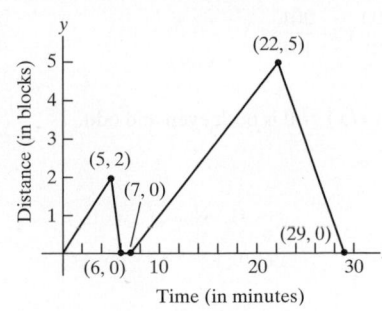

43. (a) 2 hr elapsed during which Kevin was between 0 and 3 mi from home.
 (b) 0.5 hr elapsed during which Kevin was 3 mi from home.
 (c) 0.3 hr elapsed during which Kevin was between 0 and 3 mi from home.
 (d) 0.2 hr elapsed during which Kevin was 0 mi from home.
 (e) 0.9 hr elapsed during which Kevin was between 0 and 2.8 mi from home.
 (f) 0.3 hr elapsed during which Kevin was 2.8 mi from home.
 (g) 1.1 hr elapsed during which Kevin was between 0 and 2.8 mi from home.
 (h) 3 mi **(i)** 2 times

45. No points whose x-coordinate is 5 or whose y-coordinate is 0 can be on the graph.

3.3 Assess Your Understanding *(page 238)*

6. increasing **7.** even; odd **8.** T **9.** T **10.** F **11.** Yes **13.** No **15.** $(-8, -2); (0, 2); (5, \infty)$ **17.** Yes; 10 **19.** $-2, 2; 6, 10$
21. (a) $(-2, 0), (0, 3), (2, 0)$ **(b)** Domain: $\{x | -4 \le x \le 4\}$ or $[-4, 4]$; Range: $\{y | 0 \le y \le 3\}$ or $[0, 3]$
 (c) Increasing on $(-2, 0)$ and $(2, 4)$; Decreasing on $(-4, -2)$ and $(0, 2)$ **(d)** Even
23. (a) $(0, 1)$ **(b)** Domain: all real numbers; Range: $\{y | y > 0\}$ or $(0, \infty)$ **(c)** Increasing on $(-\infty, \infty)$ **(d)** Neither
25. (a) $(-\pi, 0), (0, 0), (\pi, 0)$ **(b)** Domain: $\{x | -\pi \le x \le \pi\}$ or $[-\pi, \pi]$; Range: $\{y | -1 \le y \le 1\}$ or $[-1, 1]$
 (c) Increasing on $\left(-\dfrac{\pi}{2}, \dfrac{\pi}{2}\right)$; Decreasing on $\left(-\pi, -\dfrac{\pi}{2}\right)$ and $\left(\dfrac{\pi}{2}, \pi\right)$ **(d)** Odd
27. (a) $\left(0, \dfrac{1}{2}\right), \left(\dfrac{1}{3}, 0\right), \left(\dfrac{5}{2}, 0\right)$ **(b)** Domain: $\{x | -3 \le x \le 3\}$ or $[-3, 3]$; Range: $\{y | -1 \le y \le 2\}$ or $[-1, 2]$
 (c) Increasing on $(2, 3)$; Decreasing on $(-1, 1)$; Constant on $(-3, -1)$ and $(1, 2)$ **(d)** Neither **29. (a)** $0; 3$ **(b)** $-2, 2; 0, 0$
31. (a) $\dfrac{\pi}{2}; 1$ **(b)** $-\dfrac{\pi}{2}; -1$ **33.** Odd **35.** Even **37.** Odd **39.** Neither **41.** Even **43.** Odd

45.

Increasing: $(-2, -1), (1, 2)$
Decreasing: $(-1, 1)$
Local maximum: $(-1, 4)$
Local minimum: $(1, 0)$

47.

Increasing: $(-2, -0.77), (0.77, 2)$
Decreasing: $(-0.77, 0.77)$
Local maximum: $(-0.77, 0.19)$
Local minimum: $(0.77, -0.19)$

49.

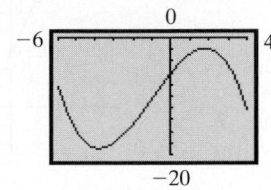

Increasing: $(-3.77, 1.77)$
Decreasing: $(-6, -3.77), (1.77, 4)$
Local maximum: $(1.77, -1.91)$
Local minimum: $(-3.77, -18.89)$

51.

Increasing: $(-1.87, 0), (0.97, 2)$
Decreasing: $(-3, -1.87), (0, 0.97)$
Local maximum: $(0, 3)$
Local minima: $(-1.87, 0.95), (0.97, 2.65)$

53. (a) -4 **(b)** -8 **(c)** -10 **55. (a)** 17 **(b)** -1 **(c)** 11 **57. (a)** 5 **(b)** $y = 5x - 2$ **59. (a)** -1; **(b)** $y = -x$ **61. (a)** 4 **(b)** $y = 4x - 8$

63. (a) $V(x) = x(24 - 2x)^2$
 (b) 972 in.3 **(c)** 160 in.3
 (d) V is largest when $x = 4$.

65. (a)

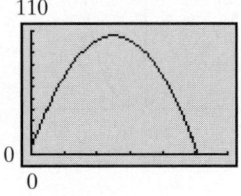

 (b) 2.5 sec
 (c) 106 ft

67. (a)

 (b) 10 riding lawn mowers/hr
 (c) \$239/mower

69. (a) On average, the population is increasing at a rate of 0.036 g/hr from 0 to 2.5 hrs.
 (b) On average, from 4.5 to 6 hrs, the population is increasing at a rate of 0.1 g/hr. **(c)** The average rate of change is increasing over time.
71. (a) 1 **(b)** 0.5 **(c)** 0.1 **(d)** 0.01 **(e)** 0.001
 (f)

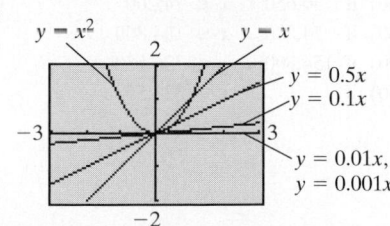

73. (a) 2 **(b)** 2; 2; 2; 2
 (c) $y = 2x + 5$
 (d)

75. (a) $2x + h + 2$
 (b) 4.5; 4.1; 4.01; 4
 (c) $y = 4.01x - 1.01$
 (d)

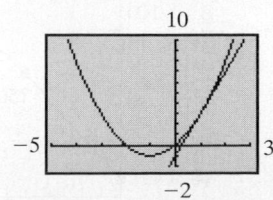

 (g) They are getting closer to the tangent line at $(0, 0)$.
 (h) They are getting closer to 0.

77. (a) $4x + 2h - 3$ **(b)** $2; 1.2; 1.02; 1$
(c) $y = 1.02x - 1.02$
(d)

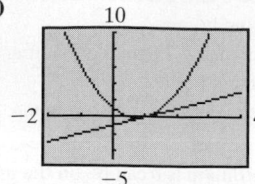

79. (a) $-\dfrac{1}{(x + h)x}$ **(b)** $-\dfrac{2}{3}; -\dfrac{10}{11}; -\dfrac{100}{101}; -1$ **(c)** $y = -\dfrac{100}{101}x + \dfrac{201}{101}$
(d)

83. At most one
85. Yes; the function $f(x) = 0$ is both even and odd.

3.4 Assess Your Understanding (page 248)

4. $(-\infty, 0)$ **5.** piecewise-defined **6.** T **7.** F **8.** F **9.** C **11.** E **13.** B **15.** F

17.

19.

21.

23.

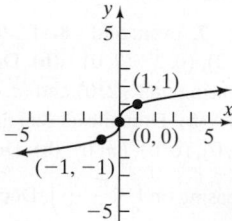

25. (a) 4 **(b)** 2 **(c)** 5 **27. (a)** -4 **(b)** -2 **(c)** 0 **(d)** 25
29. (a) All real numbers
(b) $(0, 1)$
(c)

31. (a) All real numbers
(b) $(0, 3)$
(c)

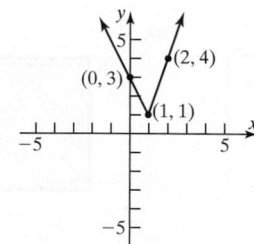

33. (a) $\{x | x \geq -2\}; [-2, \infty)$
(b) $(0, 3), (2, 0)$
(c)

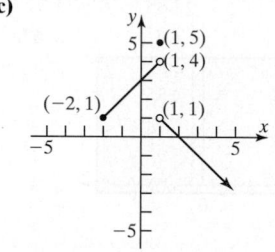

(d) $\{y | y \neq 0\}; (-\infty, 0) \cup (0, \infty)$
(d) $\{y | y \geq 1\}; [1, \infty)$
(d) $\{y | y < 4, y = 5\}; (-\infty, 4) \cup \{5\}$

35. (a) All real numbers
(b) $(-1, 0), (0, 0)$
(c)

37. (a) $\{x | x \geq -2, x \neq 0\}; [-2, 0) \cup (0, \infty)$
(b) No intercepts
(c)

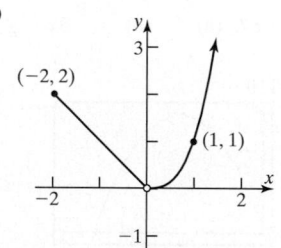

39. (a) All real numbers
(b) $(x, 0)$ for $0 \leq x < 1$
(c)

(d) All real numbers
(d) $\{y | y > 0\}; (0, \infty)$
(d) Set of even integers

41. $f(x) = \begin{cases} -x & \text{if } -1 \leq x \leq 0 \\ \frac{1}{2}x & \text{if } 0 < x \leq 2 \end{cases}$ (Other answers are possible.)

43. $f(x) = \begin{cases} -x & \text{if } x \leq 0 \\ -x + 2 & \text{if } 0 < x \leq 2 \end{cases}$ (Other answers are possible.)

45. (a) 2 **(b)** 3 **(c)** -4 **47. (a)** \$35 **(b)** \$61 **(c)** \$35.40
49. (a) \$67.43
(b) \$477.04
(c) $C = \begin{cases} 1.15955x + 9.45 & \text{if } 0 \leq x \leq 50 \\ 0.91025x + 21.915 & \text{if } x > 50 \end{cases}$
(d)

51. For schedule X:

$$f(x) = \begin{cases} 0.10x & \text{if} & 0 < x \leq 7550 \\ 755 + 0.15(x - 7550) & \text{if} & 7550 < x \leq 30{,}650 \\ 4220 + 0.25(x - 30{,}650) & \text{if} & 30{,}650 < x \leq 74{,}200 \\ 15{,}107.50 + 0.28(x - 74{,}200) & \text{if} & 74{,}200 < x \leq 154{,}800 \\ 37{,}675.50 + 0.33(x - 154{,}800) & \text{if} & 154{,}800 < x \leq 336{,}550 \\ 97{,}653.00 + 0.35(x - 336{,}550) & \text{if} & x > 336{,}550 \end{cases}$$

53. (a)

(b) $C = 50 + 0.4(x - 100)$

(c) $C = 170 + 0.25(x - 400)$

55. $f(x) = \begin{cases} x & \text{if } 0 \le x < 10 \\ 10 & \text{if } 10 \le x < 500 \\ 30 & \text{if } 500 \le x < 1000 \\ 50 & \text{if } 1000 \le x < 1500 \\ 70 & \text{if } 1500 \le x \end{cases}$

57. (a) $10°C$ **(b)** $4°C$ **(c)** $-3°C$ **(d)** $-4°C$ **(e)** The wind chill is equal to the air temperature.
(f) At wind speed greater than 20 m/sec, the wind chill factor depends only on the air temperature.

59.

$C(x) = \begin{cases} 0.39 & 0 < x \le 1 \\ 0.63 & 1 < x \le 2 \\ 0.87 & 2 < x \le 3 \\ 1.11 & 3 < x \le 4 \\ 1.35 & 4 < x \le 5 \\ 1.59 & 5 < x \le 6 \\ 1.83 & 6 < x \le 7 \\ 2.07 & 7 < x \le 8 \\ 2.31 & 8 < x \le 9 \\ 2.55 & 9 < x \le 10 \\ 2.79 & 10 < x \le 11 \\ 3.03 & 11 < x \le 12 \\ 3.27 & 12 < x \le 13 \end{cases}$

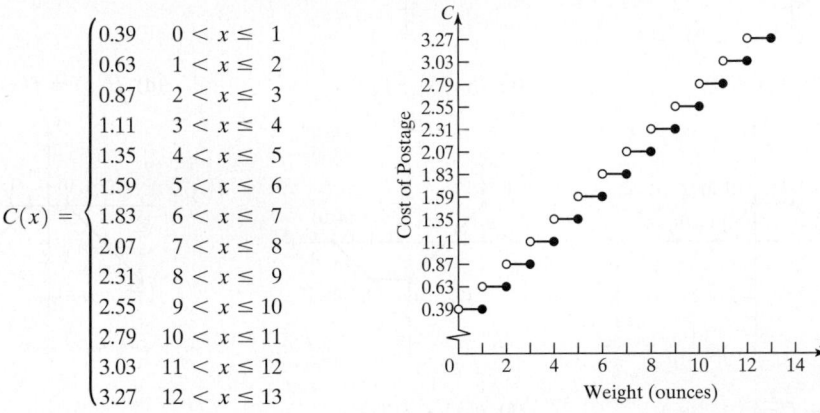

61. Each graph is that of $y = x^2$, but shifted horizontally. If $y = (x - k)^2, k > 0$, the shift is right k units; if $y = (x + k)^2, k > 0$, the shift is left k units.
63. The graph of $y = -f(x)$ is the reflection about the x-axis of the graph of $y = f(x)$. **65.** Yes. The graph of $y = (x - 1)^3 + 2$ is the graph of
$y = x^3$ shifted right 1 unit and up 2 units. **67.** They all have the same general shape. All three go through the points $(-1, -1), (0, 0),$ and $(1, 1)$.
As the exponent increases, the steepness of the curve increases (except near $x = 0$).

3.5 Assess Your Understanding (page 261)

1. horizontal; right **2.** y **3.** $-5; -2; 2$ **4.** T **5.** F **6.** T **7.** B **9.** H **11.** I **13.** L **15.** F **17.** G **19.** $y = (x - 4)^3$ **21.** $y = x^3 + 4$

23. $y = -x^3$ **25.** $y = 4x^3$ **27.** (1) $y = \sqrt{x} + 2$; (2) $y = -(\sqrt{x} + 2)$; (3) $y = -(\sqrt{-x} + 2)$
29. (1) $y = -\sqrt{x}$; (2) $y = -\sqrt{x} + 2$; (3) $y = -\sqrt{x + 3} + 2$ **31.** (c) **33.** (c)

35.

37.

39.

41.

43.

45.

47.

49.

51.

53.

55.

57.

59.

61.

63.

65. (a) $F(x) = f(x) + 3$

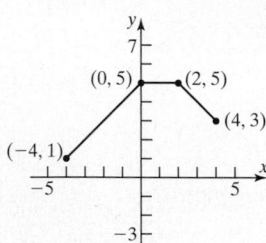

(b) $G(x) = f(x + 2)$

(c) $P(x) = -f(x)$

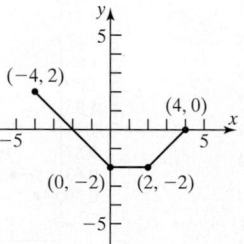

(d) $H(x) = f(x + 1) - 2$

(e) $Q(x) = \frac{1}{2}f(x)$

(f) $g(x) = f(-x)$

(g) $h(x) = f(2x)$

67. (a) $F(x) = f(x) + 3$

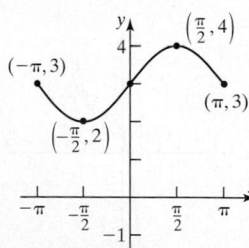

(b) $G(x) = f(x + 2)$

(c) $P(x) = -f(x)$

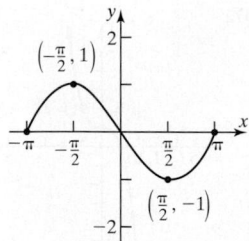

(d) $H(x) = f(x + 1) - 2$

(e) $Q(x) = \frac{1}{2}f(x)$

(f) $g(x) = f(-x)$

(g) $h(x) = f(2x)$

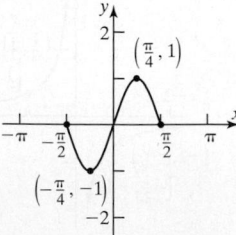

69. (a) -7 and 1 **(b)** -3 and 5 **(c)** -5 and 3 **(d)** -3 and 5 **71. (a)** $(-3, 3)$ **(b)** $(4, 10)$ **(c)** Decreasing on $(-1, 5)$ **(d)** Decreasing on $(-5, 1)$

73. (a)

(b)

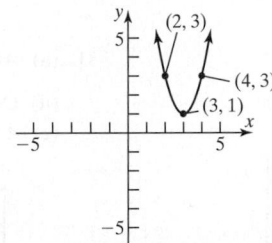

75. $f(x) = (x + 1)^2 - 1$

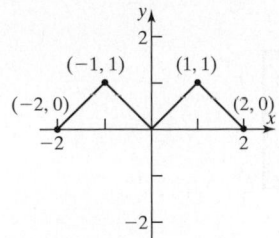

77. $f(x) = (x - 4)^2 - 15$

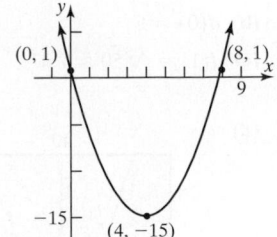

79. $f(x) = \left(x + \dfrac{1}{2}\right)^2 + \dfrac{3}{4}$

81. $f(x) = 2(x - 3)^2 + 1$

83. $f(x) = -3(x + 2)^2 - 5$

85.

87. (a) $72°F; 65°F$

(b) The temperature decreases by $2°$ to $70°F$ during the day and $63°F$ overnight.

(c) The time at which the temperature adjusts between the daytime and overnight settings is moved to 1 hr sooner. It begins warming up at 5:00 AM instead of 6:00 AM, and it begins cooling down at 8:00 PM instead of 9:00 PM.

89.

91. (a)

(b) 10% tax

(c) Y_1 is the graph of $p(x)$ shifted down vertically 10,000 units.
Y_2 is the graph of $p(x)$ vertically compressed by a factor of 0.9.

(d) 10% tax

3.6 Assess Your Understanding (page 267)

1. (a) $d(x) = \sqrt{x^4 - 15x^2 + 64}$

 (b) $d(0) = 8$

 (c) $d(1) = \sqrt{50} \approx 7.07$

 (d)

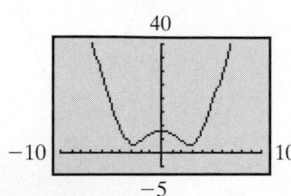

 (e) d is smallest when $x \approx -2.74$
 or $x \approx 2.74$.

3. (a) $d(x) = \sqrt{x^2 - x + 1}$

 (b)

 (c) d is smallest when $x = 0.5$.

5. $A(x) = \dfrac{1}{2}x^4$

7. (a) $A(x) = x(16 - x^2)$

 (b) Domain: $\{x \mid 0 < x < 4\}$

 (c) The area is largest when $x \approx 2.31$.

9. (a) $A(x) = 4x\sqrt{4 - x^2}$

 (c) A is largest when $x \approx 1.41$.

 (b) $p(x) = 4x + 4\sqrt{4 - x^2}$

 (d) p is largest when $x \approx 1.41$.

11. (a) $A(x) = x^2 + \dfrac{25 - 20x + 4x^2}{\pi}$

 (d) Domain: $\{x \mid 0 < x < 2.5\}$

 (c) A is smallest when $x \approx 1.40$ m.

13. (a) $C(x) = x$ **(b)** $A(x) = \dfrac{x^2}{4\pi}$ **15. (a)** $A(r) = 2r^2$ **(b)** $p(r) = 6r$ **17.** $A(x) = \left(\dfrac{\pi}{3} - \dfrac{\sqrt{3}}{4}\right)x^2$

19. (a) $d(t) = \sqrt{2500t^2 - 360t + 13}$

 (b) d is smallest when $t \approx 0.07$ hr.

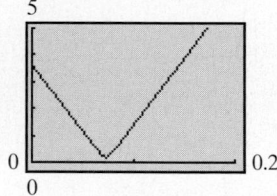

21. $V(r) = \dfrac{\pi H(R - r)r^2}{R}$

23. (a) $T(x) = \dfrac{12 - x}{5} + \dfrac{\sqrt{x^2 + 4}}{3}$

 (b) $\{x \mid 0 \le x \le 12\}$

 (c) 3.09 hr

 (d) 3.55 hr

Review Exercises (page 271)

1. Function; domain $\{-1, 2, 4\}$, range $\{0, 3\}$ **3. (a)** 2 **(b)** -2 **(c)** $-\dfrac{3x}{x^2 - 1}$ **(d)** $-\dfrac{3x}{x^2 - 1}$ **(e)** $\dfrac{3(x - 2)}{x^2 - 4x + 3}$ **(f)** $\dfrac{6x}{4x^2 - 1}$ **5. (a)** 0 **(b)** 0

(c) $\sqrt{x^2 - 4}$ **(d)** $-\sqrt{x^2 - 4}$ **(e)** $\sqrt{x^2 - 4x}$ **(f)** $2\sqrt{x^2 - 1}$ **7. (a)** 0 **(b)** 0 **(c)** $\dfrac{x^2 - 4}{x^2}$ **(d)** $-\dfrac{x^2 - 4}{x^2}$ **(e)** $\dfrac{x(x - 4)}{(x - 2)^2}$ **(f)** $\dfrac{x^2 - 1}{x^2}$

9. $\{x \mid x \ne -3, x \ne 3\}$ **11.** $\{x \mid x \le 2\}$ **13.** $\{x \mid x > 0\}$ **15.** $\{x \mid x \ne -3, x \ne 1\}$

17. $(f + g)(x) = 2x + 3$; Domain: all real numbers
 $(f - g)(x) = -4x + 1$; Domain: all real numbers
 $(f \cdot g)(x) = -3x^2 + 5x + 2$; Domain: all real numbers
 $\left(\dfrac{f}{g}\right)(x) = \dfrac{2 - x}{3x + 1}$; Domain: $\left\{x \mid x \ne -\dfrac{1}{3}\right\}$

19. $(f + g)(x) = 3x^2 + 4x + 1$; Domain: all real numbers
 $(f - g)(x) = 3x^2 - 2x + 1$; Domain: all real numbers
 $(f \cdot g)(x) = 9x^3 + 3x^2 + 3x$; Domain: all real numbers
 $\left(\dfrac{f}{g}\right)(x) = \dfrac{3x^2 + x + 1}{3x}$; Domain: $\{x \mid x \ne 0\}$

21. $(f + g)(x) = \dfrac{x^2 + 2x - 1}{x(x - 1)}$; Domain: $\{x \mid x \ne 0, 1\}$

 $(f - g)(x) = \dfrac{x^2 + 1}{x(x - 1)}$; Domain: $\{x \mid x \ne 0, 1\}$

 $(f \cdot g)(x) = \dfrac{x + 1}{x(x - 1)}$; Domain: $\{x \mid x \ne 0, 1\}$

 $\left(\dfrac{f}{g}\right)(x) = \dfrac{x(x + 1)}{x - 1}$; Domain: $\{x \mid x \ne 0, 1\}$

23. $-4x + 1 - 2h$

25. (a) Domain: $\{x | -4 \leq x \leq 3\}$; Range: $\{y | -3 \leq y \leq 3\}$ **(b)** (0, 0) **(c)** -1 **(d)** -4 **(e)** $\{x | 0 < x \leq 3\}$

(f) **(g)** **(h)**

27. (a) Domain: $\{x | -4 \leq x \leq 4\}$ or $[-4, 4]$
 Range: $\{y | -3 \leq y \leq 1\}$ or $[-3, 1]$
(b) Increasing on $(-4, -1)$ and $(3, 4)$;
 Decreasing on $(-1, 3)$
(c) Local maximum is 1 and occurs at $x = -1$.
 Local minimum is -3 and occurs at $x = 3$.
(d) No symmetry
(e) Neither
(f) x-intercepts: $-2, 0, 4$; y-intercept: 0

29. Odd **31.** Even
33. Neither **35.** Odd

37.

Local maximum: $(-0.91, 4.04)$
Local minimum: $(0.91, -2.04)$
Increasing: $(-3, -0.91)$; $(0.91, 3)$
Decreasing: $(-0.91, 0.91)$

39.

Local maximum: $(0.41, 1.53)$
Local minima: $(-0.34, 0.54)$; $(1.80, -3.56)$
Increasing: $(-0.34, 0.41)$; $(1.80, 3)$
Decreasing: $(-2, -0.34)$; $(0.41, 1.80)$

41. (a) 23 **(b)** 7 **(c)** 47
43. -5 **45.** -17
47. No **49.** Yes

51.

53.

55.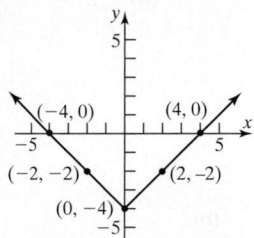

Intercepts: $(-4, 0)$, $(4, 0)$, $(0, -4)$
Domain: all real numbers
Range: $\{y | y \geq -4\}$ or $[-4, \infty)$

57.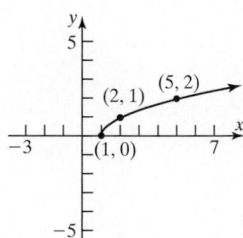

Intercept: $(0, 0)$
Domain: all real numbers
Range: $\{y | y \leq 0\}$ or $(-\infty, 0]$

59.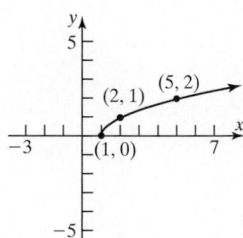

Intercept: $(1, 0)$

Domain: $\{x | x \geq 1\}$ or $[1, \infty)$

Range: $\{y | y \geq 0\}$ or $[0, \infty)$

61.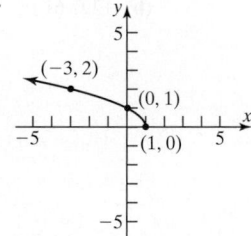

Intercepts: $(0, 1)$, $(1, 0)$

Domain: $\{x | x \leq 1\}$ or $(-\infty, 1]$

Range: $\{y | y \geq 0\}$ or $[0, \infty)$

63.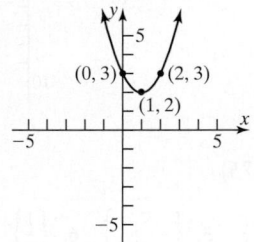

Intercept: $(0, 3)$

Domain: all real numbers

Range: $\{y | y \geq 2\}$ or $[2, \infty)$

65.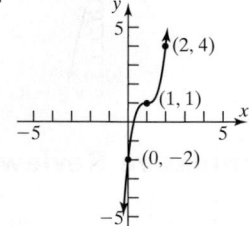

Intercepts: $(0, -2)$,

$\left(1 - \dfrac{\sqrt[3]{9}}{3}, 0\right)$ or about $(0.3, 0)$

Domain: all real numbers
Range: all real numbers

67. (a) $\{x | x > -2\}$ or $(-2, \infty)$ **(b)** $(0, 0)$
(c)

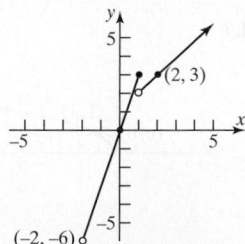

(d) $\{y | y > -6\}$ or $(-6, \infty)$

69. (a) $\{x | x \geq -4\}$ or $[-4, \infty)$ **(b)** $(0, 1)$
(c)

(d) $\{y | -4 \leq y < 0$ or $y > 0\}$
 or $[-4, 0) \cup (0, \infty)$

71. $A = 11$ **73.** If the radius doubles, the volume of the new sphere is 8 times as large as the original sphere, and the surface area of the new sphere is 4 times as large as the original sphere.

75. (a) $A = 2\pi r^2 + \dfrac{200}{r}$

(b) 123.22 ft^2 **(c)** 150.53 ft^2

(d) 197.08 ft^2

(e)

A is smallest when $r \approx 2.52$ ft.

Chapter Test (page 274)

1. (a) Function; domain: $\{2, 4, 6, 8\}$; range: $\{5, 6, 7, 8\}$ **(b)** Not a function **(c)** Not a function **(d)** Function; domain; all real numbers; range: $\{y | y \ge 2\}$ **2.** Domain: $\left\{x | x \le \dfrac{4}{5}\right\}$; $f(-1) = 3$ **3.** Domain: $\{x | x \ne -2\}$; $g(-1) = 1$ **4.** Domain: $\{x | x \ne -9, x \ne 4\}$; $h(-1) = \dfrac{1}{8}$

5. (a) Domain: $\{x | -5 \le x \le 5\}$; range: $\{y | -3 \le y \le 3\}$ **(b)** $(0, 2), (-2, 0)$, and $(2, 0)$ **(c)** $f(1) = 3$ **(d)** $x = -5$ and $x = 3$
(e) $\{x | -5 \le x < -2 \text{ or } 2 < x \le 5\}$ or $[-5, -2) \cup (2, 5]$
6. Local maxima: $f(-0.85) \approx -0.86$; $f(2.35) \approx 15.55$; Local minima: $f(0) = -2$; the function is increasing on the intervals $(-5, -0.85)$ and $(0, 2.35)$ and decreasing on the intervals $(-0.85, 0)$ and $(2.35, 5)$.

7. (a)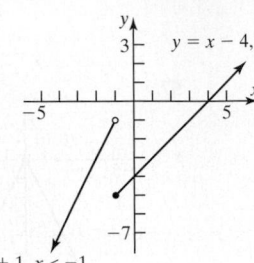

(b) $(0, -4), (4, 0)$
(c) $g(-5) = -9$
(d) $g(2) = -2$

8. 19
9. (a) $(f - g) = 2x^2 - 3x + 3$
(b) $(f \cdot g) = 6x^3 - 4x^2 + 3x - 2$
(c) $f(x + h) - f(x) = 4xh + 2h^2$

10. (a) **(b)**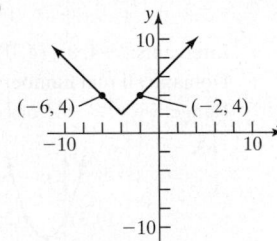

11. (a) 8.67% occurring in 1997 $(x \approx 5)$
(b) The model predicts that the interest rate will be -10.343%. This is not reasonable.

12. (a) $V(x) = \dfrac{x^2}{8} - \dfrac{5x}{4} + \dfrac{\pi x^2}{64}$

(b) 1297.61 ft^3

Cumulative Review (page 275)

1. $\{6\}$ **2.** $\left\{0, \dfrac{1}{3}\right\}$ **3.** $\{-1, 9\}$ **4.** $\left\{\dfrac{1}{3}, \dfrac{1}{2}\right\}$ **5.** $\left\{-\dfrac{7}{2}, \dfrac{1}{2}\right\}$ **6.** $\left\{\dfrac{1}{2}\right\}$

7. $\left\{x | x < -\dfrac{4}{3}\right\}$; $\left(-\infty, -\dfrac{4}{3}\right)$ **8.** $\{x | 1 < x < 4\}$; $(1, 4)$ **9.** $\left\{x | x \le -2 \text{ or } x \ge \dfrac{3}{2}\right\}$; $(-\infty, -2] \cup \left[\dfrac{3}{2}, \infty\right)$

10. (a) distance: $\sqrt{29}$ **(b)** midpoint: $\left(\dfrac{1}{2}, -4\right)$ **(c)** slope: $-\dfrac{2}{5}$

11. **12.** **13.** **14.**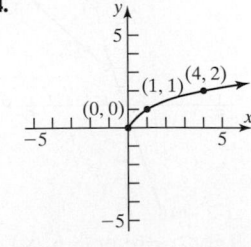

15. Intercepts: $(0, -3), (-2, 0), (2, 0)$; symmetry with respect to the y-axis **16.** $y = \frac{1}{2}x + 5$

17.

18.

19.
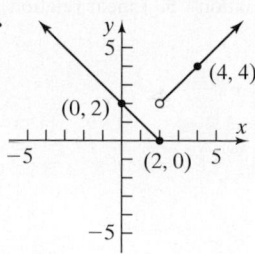

CHAPTER 4 Linear and Quadratic Functions

4.1 Assess Your Understanding (page 284)

7. slope; y-intercept **8.** $-4; 3$ **9.** positive **10.** T **11.** F **12.** F

13. (a) $m = 2; b = 3$
(b)

(c) 2 **(d)** Increasing

15. (a) $m = -3; b = 4$
(b)

(c) -3 **(d)** Decreasing

17. (a) $m = \frac{1}{4}; b = -3$
(b)
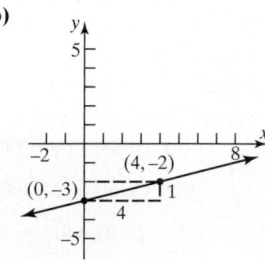
(c) $\frac{1}{4}$ **(d)** Increasing

19. (a) $m = 0; b = 4$
(b)
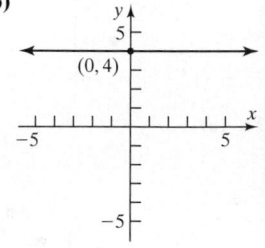
(c) 0 **(d)** Constant

21. Linear; -3
23. Nonlinear
25. Nonlinear
27. Linear; 0

29. (a) $\frac{1}{4}$ **(b)** $\left\{ x \middle| x > \frac{1}{4} \right\}$ or $\left(\frac{1}{4}, \infty \right)$
(c) 1 **(d)** $\{ x | x \le 1 \}$ or $(-\infty, 1]$
(e)
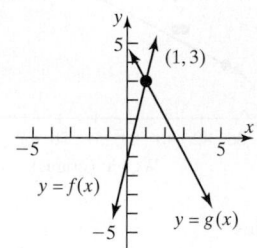

31. (a) 40 **(b)** 88 **(c)** -40 **(d)** $\{ x | x > 40 \}$ or $(40, \infty)$ **(e)** $\{ x | x \le 88 \}$ or $(-\infty, 88]$
(f) $\{ x | -40 < x < 88 \}$ or $(-40, 88)$
33. (a) -4 **(b)** $\{ x | x < -4 \}$ or $(-\infty, -4)$
35. (a) -6 **(b)** $\{ x | -6 \le x < 5 \}$ or $[-6, 5)$
37. (a) \$45 **(b)** 180 mi **(c)** 260 mi **39. (a)** \$778.22 **(b)** 2006 **(c)** 2012
41. (a) \$16; 600 T-shirts **(b)** $0 < p < 16$ **(c)** The price will increase
43. (a) $\{ x | 7300 \le x \le 29{,}700 \}$ **(b)** \$2335
(c) The independent variable is adjusted gross income, x.
The dependent variable is the tall bill, T.
(d)
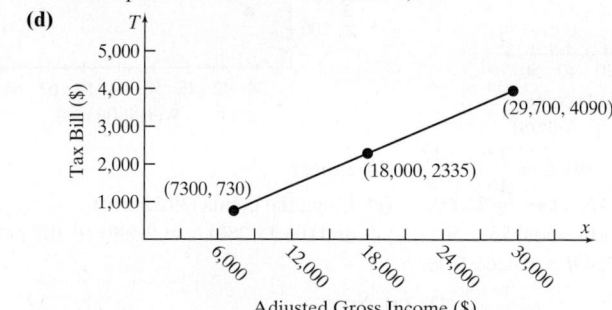
(e) \$21,500
45. (a) $x = 5000$ **(b)** $x > 5000$

47. (a) $V(x) = -1000x + 3000$
(b)

(c) \$1000 **(d)** After 1 year

49. (a) $C(x) = 90x + 1800$
(b)

(c) \$3060 **(d)** 22 bicycles

51. (a) $C(x) = 0.07x + 29$
(b) \$36.70; \$45.10
53. (d), (e)
55. $b = 0; m = 0$

4.2 Assess Your Understanding (page 290)

1. scatter diagram **2.** T **3.** Linear relation **5.** Linear relation **7.** Nonlinear relation

9. (a) **(c)** **(e)**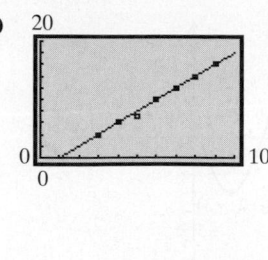

(b) Answers will vary. Using (3, 4) and (9, 16), $y = 2x - 2$. **(d)** $y = 2.0357x - 2.3571$

11. (a) **(b)** Answers will vary. Using $(-2, -4)$ and $(2, 5)$, $y = \frac{9}{4}x + \frac{1}{2}$. **(c)** **(d)** $y = 2.2x + 1.2$ **(e)**

13. (a) **(b)** Answers will vary. Using $(-20, 100)$ and $(-10, 140)$, $y = 4x + 180$. **(c)** 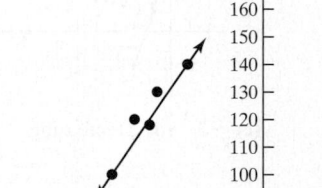 **(d)** $y = 3.8613x + 180.2920$ **(e)**

15. (a) **17. (a)** **(d)**

(b) Using (18, 13) and (50, 39), $C = \frac{13}{16}I - \frac{13}{8}$. **(b)** Linear **(e)** 269 calories

(c) As disposable income increases by $1000, consumption increases by about $812.50. **(c)** Using the points (39.52, 210) and (66.45, 280), $y = 2.599x + 107.274$. **(f)** If the weight of a candy bar is increased by 1 gram, the number of calories will increase by 2.599.

(d) $32,500 **(e)** $C = 0.7549I + 0.6266; 0.987$

19. (a) **21. (a)** No **(b)** **23.**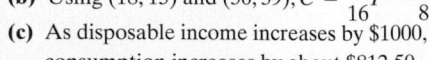

(b) $C(H) = 0.3734H + 7.3268$

(c) If height increases by 1 inch, head circumference increases by about 0.3734 inch.

(d) About 17.0 inches

(e) About 26.98 inches

(c) $D = -1.3355p + 86.1974; -0.949$

(d) If the price increases $1, the quantity sold per day decreases by about 1.34 pairs of jeans.

(e) $D(p) = -1.3355p + 86.1974$ **(f)** $\{p \mid 0 < p \le 64\}$ **(g)** About 49 pairs

The data do not follow a linear pattern.

25. No linear relation

4.3 Assess Your Understanding *(page 302)*

5. parabola **6.** axis or axis of symmetry **7.** $-\dfrac{b}{2a}$ **8.** T **9.** T **10.** T **11.** C **13.** F **15.** G **17.** H

19.

21.

23.

25.

27. $f(x) = (x + 2)^2 - 2$

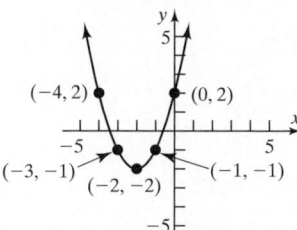

29. $f(x) = 2(x - 1)^2 - 1$

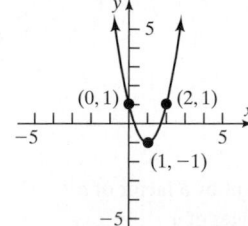

31. $f(x) = -(x + 1)^2 + 1$

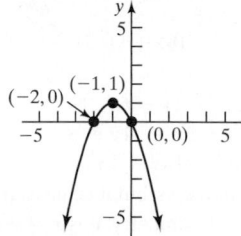

33. $f(x) = \dfrac{1}{2}(x + 1)^2 - \dfrac{3}{2}$

35. (a)

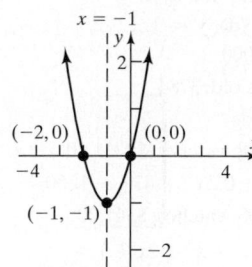

(b) Domain: $(-\infty, \infty)$
Range: $[-1, \infty)$
(c) Decreasing: $(-\infty, -1)$
Increasing: $(-1, \infty)$

37. (a)

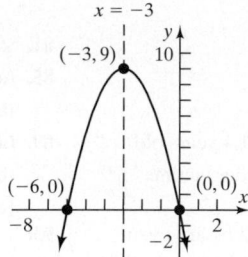

(b) Domain: $(-\infty, \infty)$
Range: $(-\infty, 9]$
(c) Increasing: $(-\infty, -3)$
Decreasing: $(-3, \infty)$

39. (a)

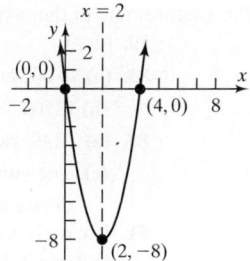

(b) Domain: $(-\infty, \infty)$
Range: $[-8, \infty)$
(c) Decreasing: $(-\infty, 2)$
Increasing: $(2, \infty)$

41. (a)

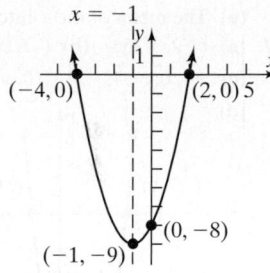

(b) Domain: $(-\infty, \infty)$
Range: $[-9, \infty)$
(c) Decreasing: $(-\infty, -1)$
Increasing: $(-1, \infty)$

43. (a)

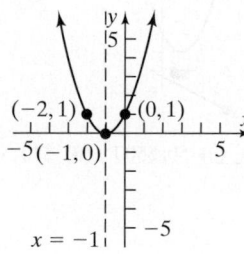

(b) Domain: $(-\infty, \infty)$
Range: $[0, \infty)$
(c) Decreasing: $(-\infty, -1)$
Increasing: $(-1, \infty)$

45. (a)

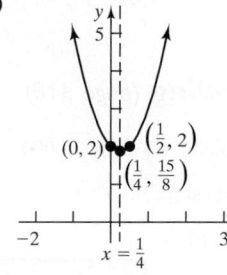

(b) Domain: $(-\infty, \infty)$
Range: $\left[\dfrac{15}{8}, \infty\right)$
(c) Decreasing: $\left(-\infty, \dfrac{1}{4}\right)$
Increasing: $\left(\dfrac{1}{4}, \infty\right)$

47. (a)

(b) Domain: $(-\infty, \infty)$
Range: $\left(-\infty, -\dfrac{5}{2}\right]$
(c) Increasing: $\left(-\infty, \dfrac{1}{2}\right)$
Decreasing: $\left(\dfrac{1}{2}, \infty\right)$

49. (a)

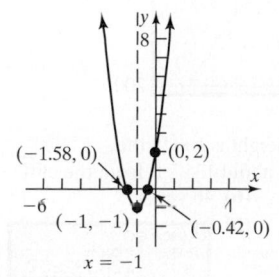

(b) Domain: $(-\infty, \infty)$
Range: $[-1, \infty)$
(c) Decreasing: $(-\infty, -1)$
Increasing: $(-1, \infty)$

51. (a)

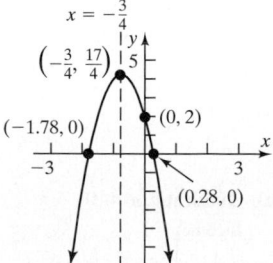

(b) Domain: $(-\infty, \infty)$
Range: $\left(-\infty, \dfrac{17}{4}\right]$
(c) Increasing: $\left(-\infty, -\dfrac{3}{4}\right)$
Decreasing: $\left(-\dfrac{3}{4}, \infty\right)$

53. $f(x) = (x + 1)^2 - 2 = x^2 + 2x - 1$ **55.** $f(x) = -(x + 3)^2 + 5 = -x^2 - 6x - 4$ **57.** $f(x) = 2(x - 1)^2 - 3 = 2x^2 - 4x - 1$
59. Minimum value; -18 **61.** Minimum value; -21 **63.** Maximum value; 21 **65.** Maximum value; 13 **67.** $a = 6, b = 0, c = 2$
69. (a), (c), (d) **71. (a), (c), (d)** **73. (a), (c), (d)**

 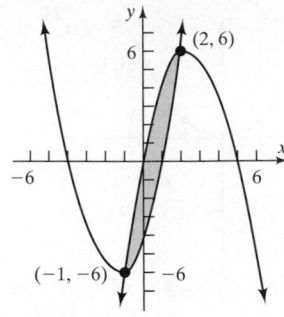

(b) $\{-1, 3\}$ **(b)** $\{-1, 3\}$ **(b)** $\{-1, 2\}$

75. (a) $a = 1: f(x) = (x + 3)(x - 1) = x^2 + 2x - 3$
$a = 2: f(x) = 2(x + 3)(x - 1) = 2x^2 + 4x - 6$
$a = -2: f(x) = -2(x + 3)(x - 1) = -2x^2 - 4x + 6$
$a = 5: f(x) = 5(x + 3)(x - 1) = 5x^2 + 10x - 15$
(b) The value of a does not affect the x-intercepts, but it changes the y-intercept by a factor of a.
(c) The value of a does not affect the axis of symmetry. It is $x = -1$ for all values of a.
(d) The value of a does not affect the x-coordinate of the vertex. However, the y-coordinate of the vertex is multiplied by a.
(e) The mean of the x-intercepts is the x-coordinate of the vertex.

77. (a) $(-2, -25)$ **(b)** $(-7, 0), (3, 0)$ **79.** $(2, 2)$ **81.** $500; \$1,000,000$
(c) $-4, 0; (-4, -21), (0, -21)$ **83. (a)** 70,000 mp3 players **85. (a)** 37.4 years old; 3784
(d) **(b)** $\$2500$ **(b)** Decreasing
 87. (a) 2149 victims **(b)** 31.4 years old **89. (a)** 187 or 188 watches; $\$7031.20$
 (c) The number of murder victims **(b)** $P(x) = -0.2x^2 + 43x - 1750$
 decreases. **(c)** 107 or 108 watches; $\$561.20$
 91. If x is even, then ax^2 and bx are even **93.**
 and $ax^2 + bx$ is even, which means that
 $ax^2 + bx + c$ is odd. If x is odd, then $y = x^2 + 2x + 1$
 ax^2 and bx are odd and $ax^2 + bx$ is $y = x^2 + 2x$
 even, which means that $ax^2 + bx + c$ is $y = x^2 + 2x - 3$
 odd. In either case, $f(x)$ is odd.
 95. $b^2 - 4ac < 0$ **97.** No

4.4 Assess Your Understanding (page 310)

3. (a) $R(x) = -\frac{1}{6}x^2 + 100x$ **(b)** $\$13,333.33$ **(c)** $300; \$15,000$ **(d)** $\$50$ **5. (a)** $R(x) = -\frac{1}{5}x^2 + 20x$ **(b)** $\$255$ **(c)** $50; \$500$ **(d)** $\$10$

7. (a) $A(w) = -w^2 + 200w$ **(b)** A is largest when $w = 100$ yd. **(c)** 10,000 yd^2 **9.** 2,000,000 m^2

11. (a) $\frac{625}{16} \approx 39$ ft **(b)** $\frac{7025}{32} \approx 219.5$ ft **(d)** **13.** 18.75 m **15.** 3 in.
(c) About 170 ft 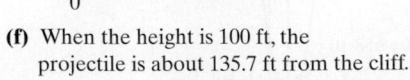 **17.** $\frac{750}{\pi}$ m by 375 m

 19. (a) 171 ft **(b)** 49 mph

 21. $x = \frac{a}{2}$ **23.** $\frac{38}{3}$ **25.** $\frac{248}{3}$

(f) When the height is 100 ft, the
projectile is about 135.7 ft from the cliff.

27. (a) Quadratic, $a < 0$ **(b)** 46 years **(c)** $\$33,528$
 (d) **(e)**

29. (a) Quadratic, $a < 0$

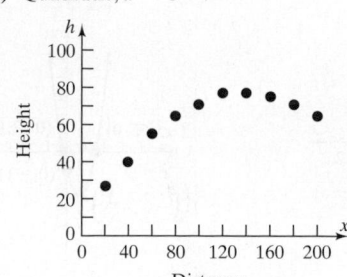

(b) About 139.2 ft

(c) About 77.4 ft

(d)

QuadReg
y=ax²+bx+c
a=-.0037121212
b=1.031818182
c=5.666666667

(e)

4.5 Assess Your Understanding (page 316)

3. (a) $\{x \mid x < -2 \text{ or } x > 2\}$; $(-\infty, -2)$ or $(2, \infty)$ **(b)** $\{x \mid -2 \le x \le 2\}$; $[-2, 2]$
5. (a) $\{x \mid -2 \le x \le 1\}$; $[-2, 1]$ **(b)** $\{x \mid x < -2 \text{ or } x > 1\}$; $(-\infty, -2)$ or $(1, \infty)$
7. $\{x \mid -2 < x < 5\}$; $(-2, 5)$ **9.** $\{x \mid x < 0 \text{ or } x > 4\}$; $(-\infty, 0)$ or $(4, \infty)$ **11.** $\{x \mid -3 < x < 3\}$; $(-3, 3)$

13. $\{x \mid x < -4 \text{ or } x > 3\}$; $(-\infty, -4)$ or $(3, \infty)$ **15.** $\left\{x \mid -\dfrac{1}{2} < x < 3\right\}$; $\left(-\dfrac{1}{2}, 3\right)$ **17.** $\{x \mid x < -1 \text{ or } x > 8\}$; $(-\infty, -1)$ or $(8, \infty)$

19. No real solution **21.** $\left\{x \mid x < -\dfrac{2}{3} \text{ or } x > \dfrac{3}{2}\right\}$; $\left(-\infty, -\dfrac{2}{3}\right)$ or $\left(\dfrac{3}{2}, \infty\right)$ **23.** $\{x \mid x \le -4 \text{ or } x \ge 4\}$; $(-\infty, -4]$ or $[4, \infty)$

25. (a) $\{-1, 1\}$ **(b)** $\{-1\}$ **(c)** $\{-1, 4\}$ **(d)** $\{x \mid x < -1 \text{ or } x > 1\}$; $(-\infty, -1)$ or $(1, \infty)$ **(e)** $\{x \mid x \le -1\}$ or $(-\infty, -1]$
(f) $\{x \mid x < -1 \text{ or } x > 4\}$; $(-\infty, -1)$ or $(4, \infty)$ **(g)** $\left\{x \mid x \le -\sqrt{2} \text{ or } x \ge \sqrt{2}\right\}$; $\left(-\infty, -\sqrt{2}\right]$ or $\left[\sqrt{2}, \infty\right)$

27. (a) $\{-1, 1\}$ **(b)** $\left\{-\dfrac{1}{4}\right\}$ **(c)** $\{-4, 0\}$ **(d)** $\{x \mid -1 < x < 1\}$; $(-1, 1)$ **(e)** $\left\{x \mid x \le -\dfrac{1}{4}\right\}$ or $\left(-\infty, -\dfrac{1}{4}\right]$ **(f)** $\{x \mid -4 < x < 0\}$; $(-4, 0)$ **(g)** $\{0\}$

29. (a) $\{-2, 2\}$ **(b)** $\{-2, 2\}$ **(c)** $\{-2, 2\}$ **(d)** $\{x \mid x < -2 \text{ or } x > 2\}$; $(-\infty, -2)$ or $(2, \infty)$ **(e)** $\{x \mid x \le -2 \text{ or } x \ge 2\}$; $(-\infty, -2]$ or $[2, \infty)$
(f) $\{x \mid x < -2 \text{ or } x > 2\}$; $(-\infty, -2)$ or $(2, \infty)$ **(g)** $\left\{x \mid x \le -\sqrt{5} \text{ or } x \ge \sqrt{5}\right\}$; $\left(-\infty, -\sqrt{5}\right]$ or $\left[\sqrt{5}, \infty\right)$ **31. (a)** $\{-1, 2\}$ **(b)** $\{-2, 1\}$ **(c)** $\{0\}$
(d) $\{x \mid x < -1 \text{ or } x > 2\}$; $(-\infty, -1)$ or $(2, \infty)$ **(e)** $\{x \mid -2 \le x \le 1\}$; $[-2, 1]$ **(f)** $\{x \mid x < 0\}$; $(-\infty, 0)$

(g) $\left\{x \mid x \le \dfrac{1 - \sqrt{13}}{2} \text{ or } x \ge \dfrac{1 + \sqrt{13}}{2}\right\}$; $\left(-\infty, \dfrac{1 - \sqrt{13}}{2}\right]$ or $\left[\dfrac{1 + \sqrt{13}}{2}, \infty\right)$ **33. (a)** 5 sec **(b)** The ball is more than 96 ft above the ground
from time t between 2 and 3 sec, $2 < t < 3$. **35. (a)** \$0, \$1000 **(b)** The revenue is more than \$800,000 for prices between \$276.39 and \$723.61,
\$276.39 $< p <$ \$723.61. **37. (a)** $\{c \mid 0.112 < c < 81.907\}$; $(0.112, 81.907)$ **(b)** It is possible to hit a target 75 km away if $c = 0.651$ or $c = 1.536$.

Review Exercises (page 318)

1. (a) $m = 2; b = -5$

(b)

(c) Increasing

3. (a) $m = \dfrac{4}{5}; b = -6$

(b)

(c) Increasing

5. (a) $m = 0; b = 4$

(b)

(c) Constant

7. Linear; Slope: 5

9.

11.

13.

15. (a)

(b) Domain: $(-\infty, \infty)$
Range: $[2, \infty)$
(c) Decreasing: $(-\infty, 2)$
Increasing: $(2, \infty)$

17. (a)

19. (a)

21. (a)

23. (a)

(b) Domain: $(-\infty, \infty)$

Range: $[-16, \infty)$

(c) Decreasing: $(-\infty, 0)$

Increasing: $(0, \infty)$

(b) Domain: $(-\infty, \infty)$

Range: $(-\infty, 1]$

(c) Increasing: $\left(-\infty, \frac{1}{2}\right)$

Decreasing: $\left(\frac{1}{2}, \infty\right)$

(b) Domain: $(-\infty, \infty)$

Range: $\left[\frac{1}{2}, \infty\right)$

(c) Decreasing: $\left(-\infty, -\frac{1}{3}\right)$

Increasing: $\left(-\frac{1}{3}, \infty\right)$

(b) Domain: $(-\infty, \infty)$

Range: $\left[-\frac{7}{3}, \infty\right)$

(c) Decreasing: $\left(-\infty, -\frac{2}{3}\right)$

Increasing: $\left(-\frac{2}{3}, \infty\right)$

25. Minimum value; 1 **27.** Maximum value; 12 **29.** Maximum value; 16 **31.** $\{x | -8 < x < 2\}$; $(-8, -2)$

33. $\left\{ x | x \le -\frac{1}{3} \text{ or } x \ge 5 \right\}$; $\left(-\infty, -\frac{1}{3}\right]$ or $[5, \infty)$ **35. (a)** Company A: $C(x) = 0.06x + 7$; Company B: $C(x) = 0.08x$ **(b)** 350 min

(c) $0 \le x < 350$ **37. (a)** $R(x) = -\frac{1}{10}x^2 + 150x$ **(b)** \$14,000 **(c)** 750; \$56,250 **(d)** \$75 **39.** 4,166,666.7m²

41. (a) 63 clubs **(b)** \$151.90 **43.** 25 units²

45. (a)

(b) Yes; the two variables appear to be linearly related.

(c) $y = 1.3902x + 1.1140$

(d) 37.95 mm

Chapter Test *(page 320)*

1. (a) Slope: -4; y-intercept: 3

(b) Decreasing

(c)

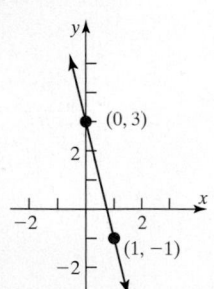

2. $\left(-\frac{4}{3}, 0\right), (2, 0), (0, -8)$

3. $\left(\frac{2 - \sqrt{6}}{2}, 0\right), \left(\frac{2 + \sqrt{6}}{2}, 0\right), (0, 1)$

4. $\{-1, 3\}$

5.

6. (a) Opens up

(b) $(2, -8)$

(c) $x = 2$

(d) x-intercepts: $\frac{6 - 2\sqrt{6}}{3}, \frac{6 + 2\sqrt{6}}{3}$; y-intercept: 4

(e)

7. Maximum value; 21

8. $\{x | x \le 4 \text{ or } x \ge 6\}$; $(-\infty, 4]$ or $[6, \infty)$

9. (a) $C(m) = 0.15m + 129.50$

(b) \$258.50

(c) 562 miles

10. (a) 1997, about 8.67%

(b) About -10.34; No

Cumulative Review Exercises *(page 321)*

1. $5\sqrt{2}; \dfrac{3}{2}, \dfrac{1}{2}$ **2.** $(-2, -1)$ and $(2, 3)$ are on the graph.

3. $\left\{x \,\middle|\, x \geq -\dfrac{3}{5}\right\}$ or $\left[-\dfrac{3}{5}, \infty\right)$

4. $y = -2x + 2$

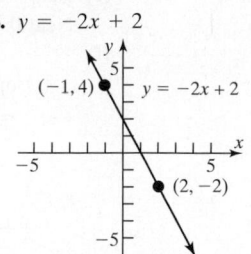

5. $y = -\dfrac{1}{2}x + \dfrac{13}{2}$

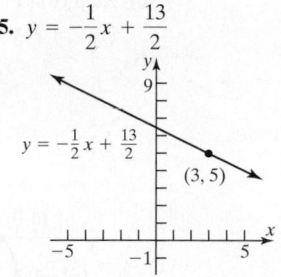

6. $(x - 2)^2 + (y + 4)^2 = 25$

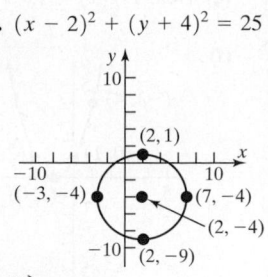

7. Yes **8. (a)** -3 **(b)** $x^2 - 4x - 2$ **(c)** $x^2 + 4x + 1$ **(d)** $-x^2 + 4x - 1$ **(e)** $x^2 - 3$ **(f)** $2x + h - 4$ **9.** $\left\{z \,\middle|\, z \neq \dfrac{7}{6}\right\}$

10. Yes **11. (a)** No **(b)** -1; $(-2, -1)$ is on the graph. **(c)** -8; $(-8, 2)$ is on the graph. **12.** Neither **13.** Local maximum is 5.30 and occurs at $x = -1.29$. Local minimum is -3.30 and occurs at $x = 1.29$. Increasing: $(-4, -1.29)$ and $(1.29, 4)$ Decreasing: $(-1.29, 1.29)$

14. (a) -4 **(b)** $\{x \,|\, x > -4\}$ or $(-4, \infty)$

15. (a) Domain: $\{x \,|\, -4 \leq x \leq 4\}$; Range: $\{y \,|\, -1 \leq y \leq 3\}$ **(b)** $(-1, 0), (0, -1), (1, 0)$ **(c)** y-axis **(d)** 1 **(e)** -4 and 4 **(f)** $\{x \,|\, -1 < x < 1\}$

(g)

(h)

(i)

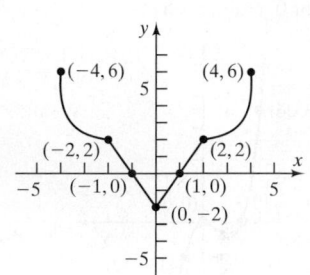

(j) Even **(k)** $(0, 4)$

CHAPTER 5 Polynomial and Rational Functions

5.1 Assess Your Understanding *(page 339)*

5. smooth; continuous **6.** zero or root **7.** touches **8.** T **9.** F **10.** F **11.** Yes; degree 3 **13.** Yes; degree 2 **15.** No; x is raised to the -1 power.

17. No; x is raised to the $\dfrac{3}{2}$ power. **19.** Yes; degree 4 **21.** Yes; degree 4

23.

25.

27.

29.

31.

33.

35.

37. $f(x) = x^3 - 3x^2 - x + 3$ for $a = 1$
39. $f(x) = x^3 - x^2 - 12x$ for $a = 1$
41. $f(x) = x^4 - 15x^2 + 10x + 24$ for $a = 1$
43. $f(x) = x^3 - 5x^2 + 3x + 9$ for $a = 1$

45. (a) 7, multiplicity 1; -3, multiplicity 2 **(b)** Graph touches the x-axis at -3 and crosses it at 7. **(c)** Near -3: $f(x) \approx -30(x + 3)^2$; Near 7: $f(x) \approx 300(x - 7)$ **(d)** 2 **(e)** $y = 3x^3$ **47. (a)** 2, multiplicity 3 **(b)** Graph crosses the x-axis at 2 **(c)** Near 2: $f(x) \approx 20(x - 2)^3$

(d) 4 **(e)** $y = 4x^5$ **49. (a)** $-\dfrac{1}{2}$, multiplicity 2 **(b)** Graph touches the x-axis at $-\dfrac{1}{2}$. **(c)** Near $-\dfrac{1}{2}$: $f(x) \approx -36.125\left(x + \dfrac{1}{2}\right)^2$

(d) 5 **(e)** $y = -2x^6$. **51. (a)** 5, multiplicity 3; -4, multiplicity 2 **(b)** Graph touches the x-axis at -4 and crosses it at 5.

(c) Near -4: $f(x) \approx -729(x + 4)^2$; Near 5: $f(x) \approx 81(x - 5)^3$ **(d)** 4 **(e)** $y = x^5$ **53. (a)** No real zeros **(b)** Graph neither crosses nor touches the x-axis. **(c)** No real zeros **(d)** 5 **(e)** $y = 3x^6$ **55. (a)** 0, multiplicity 2; $-\sqrt{2}, \sqrt{2}$, multiplicity 1 **(b)** Graph touches the x-axis at 0 and crosses at $-\sqrt{2}$ and $\sqrt{2}$. **(c)** Near $-\sqrt{2}$: $f(x) \approx 11.31(x + \sqrt{2})$; Near 0: $f(x) \approx 4x^2$; Near $\sqrt{2}$: $f(x) \approx -11.31(x - \sqrt{2})$ **(d)** 3 **(e)** $y = -2x^4$

57. Could be; zeros: $-1, 1, 2$; least degree is 3. **59.** Cannot be the graph of a polynomial; gap at $x = -1$ **61.** c, e, f **63.** c, e

65. (a) x-intercept: 1; y-intercept: 1
(b) Touches at 1
(c) $y = x^2$ **(d)** 1
(e) Near 1: $f(x) = (x - 1)^2$
(f)

67. (a) x-intercepts: 0, 3; y-intercept: 0
(b) Touches at 0; crosses at 3
(c) $y = x^3$ **(d)** 2
(e) Near 0: $f(x) \approx -3x^2$; Near 3: $f(x) \approx 9(x - 3)$
(f)

69. (a) x-intercepts: -4, 0; y-intercept: 0
(b) Crosses at -4, 0
(c) $y = 6x^4$ **(d)** 3
(e) Near -4: $f(x) \approx -384(x + 4)$; Near 0: $f(x) \approx 24x^3$
(f)

71. (a) x-intercepts: -2, 0; y-intercept: 0
(b) Crosses at -2; touches at 0
(c) $y = -4x^3$ **(d)** 2
(e) Near -2: $f(x) \approx -16(x + 2)$; Near 0: $f(x) \approx -8x^2$
(f)

73. (a) x-intercepts: -4, 1, 2; y-intercept: 8
(b) Crosses at -4, 1, 2
(c) $y = x^3$ **(d)** 2
(e) Near -4: $f(x) \approx 30(x + 4)$; Near 1: $f(x) \approx -5(x - 1)$; Near 2: $f(x) \approx 6(x - 2)$
(f)

75. $f(x) = 4x - x^3 = -x(x^2 - 4) = -x(x + 2)(x - 2)$
(a) x-intercepts: -2, 0, 2; y-intercept: 0
(b) Crosses at -2, 0, and 2
(c) $y = -x^3$ **(d)** 2
(e) Near -2: $f(x) \approx -8(x + 2)$; Near 0: $f(x) \approx 4x$; Near 2: $f(x) \approx -8(x - 2)$
(f)

77. (a) x-intercepts: -2, 0, 2; y-intercept: 0
(b) Crosses at -2, 2; touches at 0
(c) $y = x^4$ **(d)** 3
(e) Near -2: $f(x) \approx -16(x + 2)$; Near 0: $f(x) \approx -4x^2$; Near 2: $f(x) \approx 16(x - 2)$
(f)

79. (a) x-intercepts: -2, 2; y-intercept: 16
(b) Touches at -2, 2
(c) $y = x^4$ **(d)** 3
(e) Near -2: $f(x) \approx 16(x + 2)^2$; Near 2: $f(x) \approx 16(x - 2)^2$
(f)

81. (a) x-intercepts: -1, 1, 3; y-intercept: -3
(b) Crosses at -1, 3; touches at 1
(c) $y = x^4$ **(d)** 3
(e) Near -1: $f(x) \approx -16(x + 1)$; Near 1: $f(x) \approx -4(x - 1)^2$; Near 3: $f(x) \approx 16(x - 3)$
(f)

83. (a) x-intercepts: -2, 4; y-intercept: 64
(b) Touches at -2 and 4
(c) $y = x^4$ **(d)** 3
(e) Near -2: $f(x) \approx 36(x + 2)^2$; Near 4: $f(x) \approx 36(x - 4)^2$
(f)

85. (a) x-intercepts: 0, 2; y-intercept: 0
(b) Touches at 0; crosses at 2
(c) $y = x^5$ **(d)** 4
(e) Near 0: $f(x) \approx -6x^2$; Near 2: $f(x) \approx 28(x - 2)$
(f)

87. (a) x-intercepts: -1, 0, 1; y-intercept: 0
(b) Crosses at 1; touches at -1 and 0
(c) $y = -x^5$ **(d)** 4
(e) Near -1: $f(x) \approx 2(x + 1)^2$; Near 0: $f(x) \approx x^2$; Near 1: $f(x) \approx -4(x - 1)$
(f)

89. (a) Degree 3; $y = x^3$

(b)

(c) x-intercepts: $-1.26, -0.20, 1.26$;
y-intercept: -0.31752

(d)

Above on $(-1.26, -0.20)$ and $(1.26, \infty)$;
below on $(-\infty, -1.26)$ and $(-0.20, 1.26)$

(e) Local maximum at $(-0.80, 0.57)$;
local minimum at $(0.66, -0.99)$

(f)

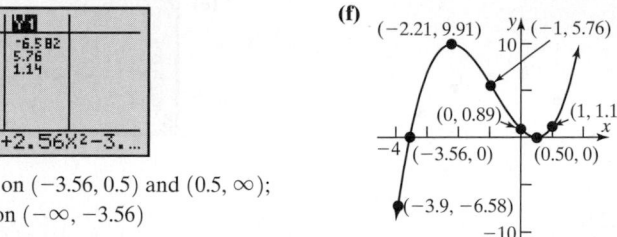

(g) Domain: all real numbers; range: all real numbers

(h) Increasing on $(-\infty, -0.80)$ and $(0.66, \infty)$; decreasing on $(-0.80, 0.66)$

91. (a) Degree 3; $y = x^3$

(b)

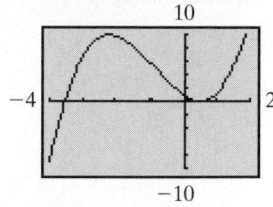

(c) x-intercepts: $-3.56, 0.50$; y-intercept: 0.89

(d)

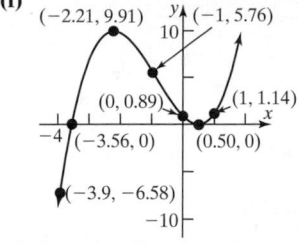

Above on $(-3.56, 0.5)$ and $(0.5, \infty)$;
below on $(-\infty, -3.56)$

(e) Local maximum at $(-2.21, 9.91)$; local minimum at $(0.50, 0)$

(f)

(g) Domain: all real numbers; range: all real numbers

(h) Increasing on $(-\infty, -2.21)$ and $(0.50, \infty)$; decreasing on $(-2.21, 0.50)$

93. (a) Degree 4; $y = x^4$

(b)

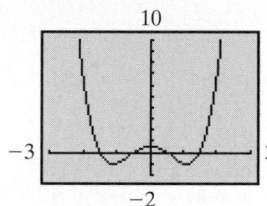

(c) x-intercepts: $-1.5, -0.5, 0.5, 1.5$; y-intercept: 0.5625

(d)

Above on $(-\infty, -1.5)$, $(-0.5, 0.5)$,
and $(1.5, \infty)$; below on $(-1.5, -0.5)$
and $(0.5, 1.5)$

(e) Local minima at $(-1.12, -1)$ and $(1.12, -1)$; local maximum at $(0, 0.5625)$

(f)

(g) Domain: all real numbers; range: $\{y | y >= -1\}$

(h) Increasing on $(-1.12, 0)$ and $(1.12, \infty)$; decreasing on $(-\infty, -1.12)$ and $(0, 1.12)$

95. (a) Degree 4; $y = 2x^4$

(b)

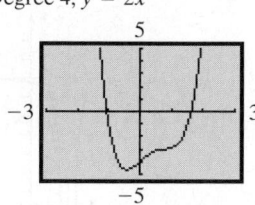

(c) x-intercepts: $-1.07, 1.62$; y-intercept: -4

(d)

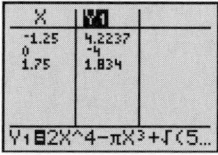

Above on $(-\infty, -1.07)$ and
$(1.62, \infty)$; below on $(-1.07, 1.62)$

(e) Local minimum at $(-0.42, -4.64)$

(f)

(g) Domain: all real numbers; range: $\{y | y >= -4.64\}$

(h) Increasing on $(-0.42, \infty)$; decreasing on $(-\infty, -0.42)$

97. (a) Degree 5; $y = -2x^5$

(b)

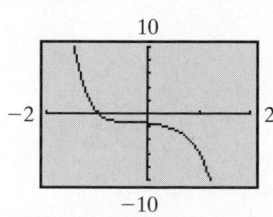

(c) x-intercept: -0.98; y-intercept: $-\sqrt{2}$

(d)

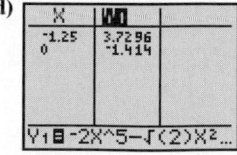

Above on $(-\infty, -0.98)$; below on
$(-0.98, \infty)$

(e) None

(f)

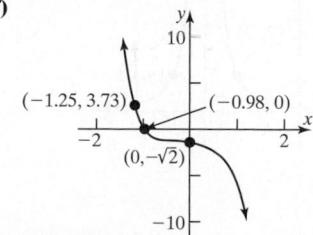

(g) Domain: all real numbers; range: all real numbers

(h) Decreasing on $(-\infty, \infty)$

99. (a)

The relationship appears to cubic.

(b) Between 6 and 7 major hurricanes

(c)
```
CubicReg
y=ax³+bx²+cx+d
a=.1590909091
b=-2.32034632
c=9.33008658
d=-2.214285714
```

(d) 12.5

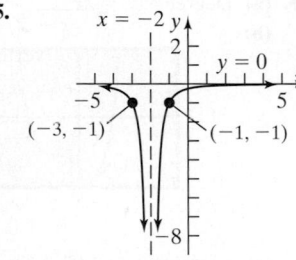

(e) Approximately 10 major hurricanes

101. (a)

The relation appears to be cubic.

(b) Average rate of change: 2.27° /hr.
(c) Average rate of change: 0° /hr.
(d) The predicted temperature at 5 PM is 64°F.

(e)
```
CubicReg
y=ax³+bx²+cx+d
a=-.0102974186
b=.3173761424
c=-1.374242424
d=45.39285714
R²=.9584628573
```

(g) The y-intercept, 45.4°, is the predicted temperature at midnight.

(f)

107. (a)−(d)

5.2 Assess Your Understanding (page 352)

5. $y = 1$ **6.** $x = -1$ **7.** Proper **8.** F **9.** T **10.** T **11.** All real numbers except 3; $\{x | x \neq 3\}$

13. All real numbers except 2 and −4; $\{x | x \neq 2, x \neq -4\}$ **15.** All real numbers except $-\dfrac{1}{2}$ and 3; $\left\{ x \middle| x \neq -\dfrac{1}{2}, x \neq 3 \right\}$

17. All real numbers except 2; $\{x | x \neq 2\}$ **19.** All real numbers **21.** All real numbers except −3 and 3; $\{x | x \neq -3, x \neq 3\}$

23. (a) Domain: $\{x | x \neq 2\}$; Range: $\{y | y \neq 1\}$ **(b)** $(0, 0)$ **(c)** $y = 1$ **(d)** $x = 2$ **(e)** None

25. (a) Domain: $\{x | x \neq 0\}$; Range: all real numbers **(b)** $(-1, 0), (1, 0)$ **(c)** None **(d)** $x = 0$ **(e)** $y = 2x$

27. (a) Domain: $\{x | x \neq -2, x \neq 2\}$; Range: $\{y | y \leq 0, y > 1\}$ **(b)** $(0, 0)$ **(c)** $y = 1$ **(d)** $x = -2, x = 2$ **(e)** None

29.

31.

33.

35.

37.

39.

41. Vertical asymptote: $x = -4$; horizontal asymptote: $y = 3$
43. Vertical asymptote: $x = 3$; oblique asymptote: $y = x + 5$
45. Vertical asymptotes: $x = 1, x = -1$; horizontal asymptote: $y = 0$

47. Vertical asymptote: $x = 0$; horizontal asymptote: $y = 0$ **49.** Vertical asymptote: $x = 0$; oblique asymptote: $y = 3x$

51. Vertical asymptote: $x = 0$; oblique asymptote: $y = -(x + 1)$ **53. (a)** 9.8208 m/sec² **(b)** 9.8195 m/sec² **(c)** 9.7936 m/sec² **(d)** h-axis **(e)** ∅

55. (a)

(b) Horizontal: $R_{tot} = 10$; as the resistance of R_2 increases without bound, the total resistance approaches 10 ohms, the resistance R_1.

(c) $R_1 \approx 103.5$ ohms

5.3 Assess Your Understanding (page 366)

2. in lowest terms **3.** F **4.** F **5.** F **6.** T

7. 1. Domain: $\{x \mid x \neq 0, x \neq -4\}$; no y-intercept **2.** R is in lowest terms; x-intercept: -1 **3.** R is in lowest terms; vertical asymptotes: $x = 0, x = -4$

4. Horizontal asymptote: $y = 0$, intersected at $(-1, 0)$

5.

Interval	$(-\infty, -4)$	$(-4, -1)$	$(-1, 0)$	$(0, \infty)$
Number Chosen	-5	-2	$-\frac{1}{2}$	1
Value of R	$R(-5) = -\frac{4}{5}$	$R(-2) = \frac{1}{4}$	$R\left(-\frac{1}{2}\right) = -\frac{2}{7}$	$R(1) = \frac{2}{5}$
Location of Graph	Below x-axis	Above x-axis	Below x-axis	Above x-axis
Point on Graph	$\left(-5, -\frac{4}{5}\right)$	$\left(-2, \frac{1}{4}\right)$	$\left(-\frac{1}{2}, -\frac{2}{7}\right)$	$\left(1, \frac{2}{5}\right)$

6–7.

9. 1. $R(x) = \dfrac{3(x+1)}{2(x+2)}$; domain: $\{x \mid x \neq -2\}$; y-intercept: $\dfrac{3}{4}$ **2.** R is in lowest terms; x-intercept: -1

3. R is in lowest terms; vertical asymptote: $x = -2$ **4.** Horizontal asymptote: $y = \dfrac{3}{2}$, not intersected

5.

Interval	$(-\infty, -2)$	$(-2, -1)$	$(-1, \infty)$
Number Chosen	-3	$-\frac{3}{2}$	0
Value of R	$R(-3) = 3$	$R\left(-\frac{3}{2}\right) = -\frac{3}{2}$	$R(0) = \frac{3}{4}$
Location of Graph	Above x-axis	Below x-axis	Above x-axis
Point on Graph	$(-3, 3)$	$\left(-\frac{3}{2}, -\frac{3}{2}\right)$	$\left(0, \frac{3}{4}\right)$

6–7.

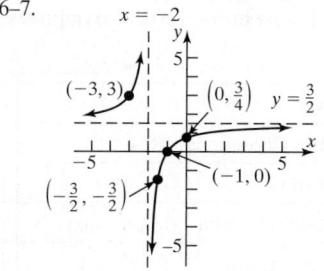

11. 1. $R(x) = \dfrac{3}{(x+2)(x-2)}$; domain: $\{x \mid x \neq -2, x \neq 2\}$; y-intercept: $-\dfrac{3}{4}$ **2.** R is in lowest terms; no x-intercept

3. R is in lowest terms; vertical asymptotes: $x = 2, x = -2$ **4.** Horizontal asymptote: $y = 0$, not intersected

5.

Interval	$(-\infty, -2)$	$(-2, 2)$	$(2, \infty)$
Number Chosen	-3	0	3
Value of R	$R(-3) = \frac{3}{5}$	$R(0) = -\frac{3}{4}$	$R(3) = \frac{3}{5}$
Location of Graph	Above x-axis	Below x-axis	Above x-axis
Point on Graph	$\left(-3, \frac{3}{5}\right)$	$\left(0, -\frac{3}{4}\right)$	$\left(3, \frac{3}{5}\right)$

6–7.

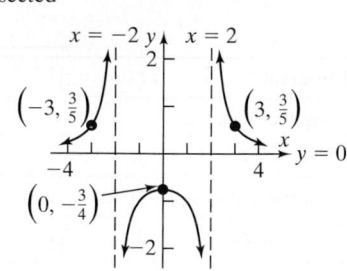

13. 1. $P(x) = \dfrac{x^4 + x^2 + 1}{(x+1)(x-1)}$; domain: $\{x \mid x \neq -1, x \neq 1\}$; y-intercept: -1 **2.** P is in lowest terms; no x-intercept

3. P is in lowest terms; vertical asymptotes: $x = -1, x = 1$ **4.** No horizontal or oblique asymptotes

5.

Interval	$(-\infty, -1)$	$(-1, 1)$	$(1, \infty)$
Number Chosen	-2	0	2
Value of P	$P(-2) = 7$	$P(0) = -1$	$P(2) = 7$
Location of Graph	Above x-axis	Below x-axis	Above x-axis
Point on Graph	$(-2, 7)$	$(0, -1)$	$(2, 7)$

6–7.

15. 1. $H(x) = \dfrac{(x-1)(x^2+x+1)}{(x+3)(x-3)}$; domain: $\{x \mid x \neq -3, x \neq 3\}$; y-intercept: $\dfrac{1}{9}$ 2. H is in lowest terms; x-intercept: 1

3. H is in lowest terms; vertical asymptotes: $x = 3$, $x = -3$ 4. Oblique asymptote: $y = x$, intersected at $\left(\dfrac{1}{9}, \dfrac{1}{9}\right)$

5.

Interval	$(-\infty, -3)$	$(-3, 1)$	$(1, 3)$	$(3, \infty)$
Number Chosen	-5	0	2	5
Value of H	$H(-5) \approx -7.9$	$H(0) = \frac{1}{9}$	$H(2) = -1.4$	$H(5) = 7.75$
Location of Graph	Below x-axis	Above x-axis	Below x-axis	Above x-axis
Point on Graph	$(-5, -7.9)$	$\left(0, \frac{1}{9}\right)$	$(2, -1.4)$	$(5, 7.75)$

6–7.

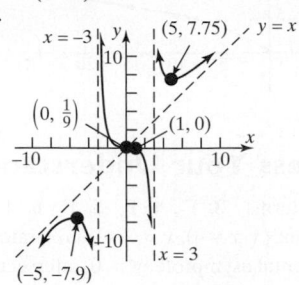

17. 1. $R(x) = \dfrac{x^2}{(x+3)(x-2)}$; domain: $\{x \neq -3, x \neq 2\}$; y-intercept: 0 2. R is in lowest terms; x-intercept: 0

3. R is in lowest terms; vertical asymptotes: $x = 2$, $x = -3$ 4. Horizontal asymptote: $y = 1$, intersected at $(6, 1)$

5.

Interval	$(-\infty, -3)$	$(-3, 0)$	$(0, 2)$	$(2, \infty)$
Number Chosen	-6	-1	1	3
Value of R	$R(-6) = \frac{3}{2}$	$R(-1) = -\frac{1}{6}$	$R(1) = -\frac{1}{4}$	$R(3) = \frac{3}{2}$
Location of Graph	Above x-axis	Below x-axis	Below x-axis	Above x-axis
Point on Graph	$\left(-6, \frac{3}{2}\right)$	$\left(-1, -\frac{1}{6}\right)$	$\left(1, -\frac{1}{4}\right)$	$\left(3, \frac{3}{2}\right)$

6–7.

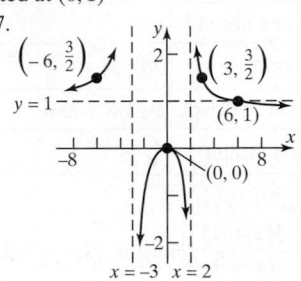

19. 1. $G(x) = \dfrac{x}{(x+2)(x-2)}$; domain: $\{x \mid x \neq -2, x \neq 2\}$; y-intercept: 0 2. G is in lowest terms; x-intercept: 0

3. G is in lowest terms; vertical asymptotes: $x = -2$, $x = 2$ 4. Horizontal asymptote: $y = 0$, intersected at $(0, 0)$

5.

Interval	$(-\infty, -2)$	$(-2, 0)$	$(0, 2)$	$(2, \infty)$
Number Chosen	-3	-1	1	3
Value of G	$G(-3) = -\frac{3}{5}$	$G(-1) = \frac{1}{3}$	$G(1) = -\frac{1}{3}$	$G(3) = \frac{3}{5}$
Location of Graph	Below x-axis	Above x-axis	Below x-axis	Above x-axis
Point on Graph	$\left(-3, -\frac{3}{5}\right)$	$\left(-1, \frac{1}{3}\right)$	$\left(1, -\frac{1}{3}\right)$	$\left(3, \frac{3}{5}\right)$

6–7.

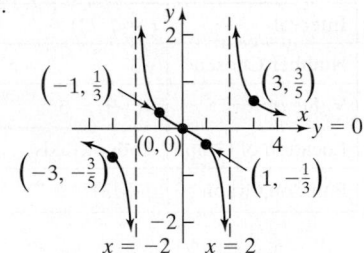

21. 1. $R(x) = \dfrac{3}{(x-1)(x+2)(x-2)}$; domain: $\{x \mid x \neq 1, x \neq -2, x \neq 2\}$; y-intercept: $\dfrac{3}{4}$ 2. R is in lowest terms; no x-intercept

3. R is in lowest terms; vertical asymptotes: $x = -2$, $x = 1$, $x = 2$ 4. Horizontal asymptote: $y = 0$, not intersected

5.

Interval	$(-\infty, -2)$	$(-2, 1)$	$(1, 2)$	$(2, \infty)$
Number Chosen	-3	0	$\frac{3}{2}$	3
Value of R	$R(-3) = -\frac{3}{20}$	$R(0) = \left(\frac{3}{4}\right)$	$R\left(\frac{3}{2}\right) = -\frac{24}{7}$	$R(3) = \frac{3}{10}$
Location of Graph	Below x-axis	Above x-axis	Below x-axis	Above x-axis
Point on Graph	$\left(-3, -\frac{3}{20}\right)$	$\left(0, \frac{3}{4}\right)$	$\left(\frac{3}{2}, -\frac{24}{7}\right)$	$\left(3, \frac{3}{10}\right)$

6–7.

23. 1. $H(x) = \dfrac{(x+1)(x-1)}{(x^2+4)(x+2)(x-2)}$; domain: $\{x \mid x \neq -2, x \neq 2\}$; y-intercept: $\dfrac{1}{16}$

2. H is in lowest terms; x-intercepts: $-1, 1$

3. H is in lowest terms; vertical asymptotes: $x = -2$, $x = 2$

4. Horizontal asymptote: $y = 0$, intersected at $(-1, 0)$ and $(1, 0)$

5.

Interval	$(-\infty, -2)$	$(-2, -1)$	$(-1, 1)$	$(1, 2)$	$(2, \infty)$
Number Chosen	-3	-1.5	0	1.5	3
Value of H	$H(-3) \approx 0.12$	$H(-1.5) \approx -0.11$	$H(0) = \frac{1}{16}$	$H(1.5) \approx -0.11$	$H(3) \approx 0.12$
Location of Graph	Above x-axis	Below x-axis	Above x-axis	Below x-axis	Above x-axis
Point on Graph	$(-3, 0.12)$	$(-1.5, -0.11)$	$\left(0, \frac{1}{16}\right)$	$(1.5, -0.11)$	$(3, 0.12)$

6–7.

25. 1. $F(x) = \dfrac{(x+1)(x-4)}{x+2}$; domain: $\{x | x \neq -2\}$; y-intercept: -2 2. F is in lowest terms; x-intercepts: $-1, 4$

 3. F is in lowest terms; vertical asymptote: $x = -2$ 4. Oblique asymptote: $y = x - 5$, not intersected

 5.

Interval	$(-\infty, -2)$	$(-2, -1)$	$(-1, 4)$	$(4, \infty)$
Number Chosen	-3	-1.5	0	5
Value of F	$F(-3) = -14$	$F(-1.5) = 5.5$	$F(0) = -2$	$F(5) \approx 0.86$
Location of Graph	Below x-axis	Above x-axis	Below x-axis	Above x-axis
Point on Graph	$(-3, -14)$	$(-1.5, 5.5)$	$(0, -2)$	$(5, 0.86)$

6–7.

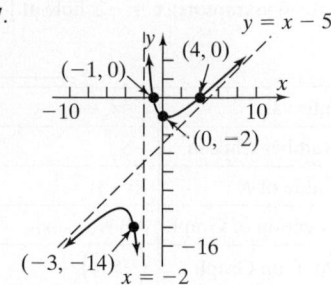

27. 1. $R(x) = \dfrac{(x+4)(x-3)}{x-4}$; domain: $\{x | x \neq 4\}$; y-intercept: 3 2. R is in lowest terms; x-intercepts: $-4, 3$

 3. R is in lowest terms; vertical asymptote: $x = 4$ 4. Oblique asymptote: $y = x + 5$, not intersected

 5.

Interval	$(-\infty, -4)$	$(-4, 3)$	$(3, 4)$	$(4, \infty)$
Number Chosen	-5	0	3.5	5
Value of R	$R(-5) = -\frac{8}{9}$	$R(0) = 3$	$R(3.5) = -7.5$	$R(5) = 18$
Location of Graph	Below x-axis	Above x-axis	Below x-axis	Above x-axis
Point on Graph	$(-5, -\frac{8}{9})$	$(0, 3)$	$(3.5, -7.5)$	$(5, 18)$

6–7.

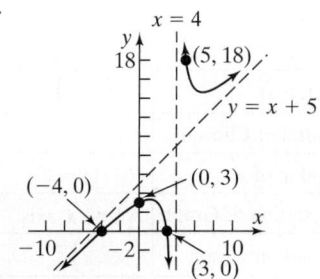

29. 1. $F(x) = \dfrac{(x+4)(x-3)}{x+2}$; domain: $\{x | x \neq -2\}$; y-intercept: -6 2. F is in lowest terms; x-intercepts: $-4, 3$

 3. F is in lowest terms; vertical asymptote: $x = -2$ 4. Oblique asymptote: $y = x - 1$, not intersected

 5.

Interval	$(-\infty, -4)$	$(-4, -2)$	$(-2, 3)$	$(3, \infty)$
Number Chosen	-5	-3	0	4
Value of F	$F(-5) = -\frac{8}{3}$	$F(-3) = 6$	$F(0) = -6$	$F(4) = \frac{4}{3}$
Location of Graph	Below x-axis	Above x-axis	Below x-axis	Above x-axis
Point on Graph	$\left(-5, -\frac{8}{3}\right)$	$(-3, 6)$	$(0, -6)$	$\left(4, \frac{4}{3}\right)$

6–7.

31. 1. Domain: $\{x | x \neq -3\}$; y-intercept: 0 2. x-intercepts: $0, 1$ 3. vertical asymptote: $x = -3$

 4. Horizontal asymptote: $y = 1$, not intersected

 5.

Interval	$(-\infty, -3)$	$(-3, 0)$	$(0, 1)$	$(1, \infty)$
Number Chosen	-4	-1	$\frac{1}{2}$	2
Value of R	$R(-4) = 100$	$R(-1) = -0.5$	$R\left(\frac{1}{2}\right) \approx 0.003$	$R(2) = 0.016$
Location of Graph	Above x-axis	Below x-axis	Above x-axis	Above x-axis
Point on Graph	$(-4, 100)$	$(-1, -0.5)$	$\left(\frac{1}{2}, 0.003\right)$	$(2, 0.016)$

 6 7.

Enlarged view

33. 1. $R(x) = \dfrac{(x+4)(x-3)}{(x+2)(x-3)}$; domain: $\{x \mid x \neq -2, x \neq 3\}$; y-intercept: 2 2. In lowest terms, $R(x) = \dfrac{x+4}{x+2}$; x-intercept: -4

3. Vertical asymptote: $x = -2$; hole at $\left(3, \dfrac{7}{5}\right)$ 4. Horizontal asymptote: $y = 1$, not intersected

5.

Interval	$(-\infty, -4)$	$(-4, -2)$	$(-2, 3)$	$(3, \infty)$
Number Chosen	-5	-3	0	4
Value of R	$R(-5) = \frac{1}{3}$	$R(-3) = -1$	$R(0) = 2$	$R(4) = \frac{4}{3}$
Location of Graph	Above x-axis	Below x-axis	Above x-axis	Above x-axis
Point on Graph	$\left(-5, \frac{1}{3}\right)$	$(-3, -1)$	$(0, 2)$	$\left(4, \frac{4}{3}\right)$

6–7.
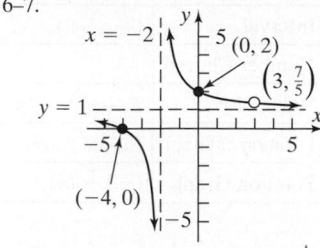

35. 1. $R(x) = \dfrac{(3x+1)(2x-3)}{(x-2)(2x-3)}$; domain: $\left\{x \mid x \neq \dfrac{3}{2}, x \neq 2\right\}$; y-intercept: $-\dfrac{1}{2}$ 2. In lowest terms, $R(x) = \dfrac{3x+1}{x-2}$; x-intercept: $-\dfrac{1}{3}$

3. Vertical asymptote: $x = 2$; hole at $\left(\dfrac{3}{2}, -11\right)$ 4. Horizontal asymptote: $y = 3$, not intersected

5.

Interval	$\left(-\infty, -\frac{1}{3}\right)$	$\left(-\frac{1}{3}, \frac{3}{2}\right)$	$\left(\frac{3}{2}, 2\right)$	$(2, \infty)$
Number Chosen	-1	0	1.7	6
Value of R	$R(-1) = \frac{2}{3}$	$R(0) = -\frac{1}{2}$	$R(1.7) \approx -20.3$	$R(6) = 4.75$
Location of Graph	Above x-axis	Below x-axis	Below x-axis	Above x-axis
Point on Graph	$\left(-1, \frac{2}{3}\right)$	$\left(0, -\frac{1}{2}\right)$	$(1.7, -20.3)$	$(6, 4.75)$

6–7.

37. 1. $R(x) = \dfrac{(x+3)(x+2)}{x+3}$; domain: $\{x \mid x \neq -3\}$; y-intercept: 2 2. In lowest terms, $R(x) = x + 2$; x-intercept: -2

3. Vertical asymptote: none; hole at $(-3, -1)$ 4. Oblique asymptote: $y = x + 2$ intersected at all points except $x = -3$

5.

Interval	$(-\infty, -3)$	$(-3, -2)$	$(-2, \infty)$
Number Chosen	-4	-2.5	0
Value of R	$R(-4) = -2$	$R(-2.5) = -\frac{1}{2}$	$R(0) = 2$
Location of Graph	Below x-axis	Below x-axis	Above x-axis
Point on Graph	$(-4, -2)$	$\left(-2.5, -\frac{1}{2}\right)$	$(0, 2)$

6–7.
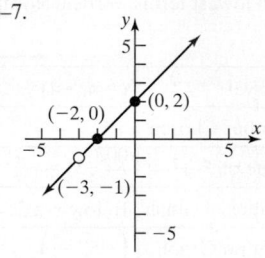

39. 1. $f(x) = \dfrac{x^2 + 1}{x}$; domain: $\{x \mid x \neq 0\}$; no y-intercepts 2. f is in lowest terms; no x-intercepts

3. f is in lowest terms; vertical asymptote: $x = 0$ 4. Oblique asymptotes: $y = x$, not intersected

5.

Interval	$(-\infty, 0)$	$(0, \infty)$
Number Chosen	-1	1
Value of f	$f(-1) = -2$	$f(1) = 2$
Location of Graph	Below x-axis	Above x-axis
Point on Graph	$(-1, -2)$	$(1, 2)$

6–7.

41. 1. $f(x) = \dfrac{x^3 + 1}{x} = \dfrac{(x+1)(x^2 - x + 1)}{x}$; domain: $\{x \mid x \neq 0\}$; no y-intercepts 2. f is in lowest terms; x-intercept: -1

3. f is in lowest terms; vertical asymptote: $x = 0$ 4. No horizontal or oblique asymptotes

5.

Interval	$(-\infty, -1)$	$(-1, 0)$	$(0, \infty)$
Number Chosen	-2	$-\frac{1}{2}$	1
Value of f	$f(-2) = 3.5$	$f\left(-\frac{1}{2}\right) = -1.75$	$f(1) = 2$
Location of Graph	Above x-axis	Below x-axis	Above x-axis
Point on Graph	$(-2, 3.5)$	$\left(-\frac{1}{2}, -1.75\right)$	$(1, 2)$

6–7.

43. 1. $f(x) = \dfrac{x^4 + 1}{x^3}$; domain: $\{x | x \neq 0\}$; no y-intercepts 2. f is in lowest terms; no x-intercepts

3. f is in lowest terms; vertical asymptote: $x = 0$ 4. Oblique asymptote: $y = x$, not intersected

5.

6–7.

Interval	$(-\infty, 0)$	$(0, \infty)$
Number Chosen	-1	1
Value of f	$f(-1) = -2$	$f(1) = 2$
Location of Graph	Below x-axis	Above x-axis
Point on Graph	$(-1, -2)$	$(1, 2)$

45. One possibility: $R(x) = \dfrac{x^2}{x^2 - 4}$ **47.** One possibility: $R(x) = \dfrac{(x-1)(x-3)\left(x^2 + \frac{4}{3}\right)}{(x+1)^2(x-2)^2}$

49. (a) t-axis; $C(t) \to 0$

(b) 0.4

(c) 0.71 hr after injection

51. (a) $C(x) = 16x + \dfrac{5000}{x} + 100$

(b) $x > 0$

(c)

(d) Approximately 17.7 ft by 56.6 ft (longer side parallel to river)

53. (a) $S(x) = 2x^2 + \dfrac{40{,}000}{x}$

(b) 10,000

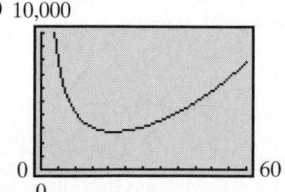

(c) 2784.95 in.2
(d) 21.54 in. $\times$ 21.54 in. $\times$ 21.54 in.
(e) To minimize the cost of materials needed for construction

55. (a) $C(r) = 12\pi r^2 + \dfrac{4000}{r}$

57. No. Each function is a quotient of polynomials, but it is not written in lowest terms. Each function is undefined for $x = 1$; each graph has a hole at $x = 1$.

(b) 6000

The cost is smallest when $r = 3.76$ cm.

5.4 Assess Your Understanding (page 373)

2. F **3.** $\{x | x < -2\}$; $(-\infty, -2)$ **5.** $\{x | x > 4\}$; $(4, \infty)$ **7.** $\{x | x \leq -3 \text{ or } 0 \leq x \leq 3\}$; $(-\infty, -3] \cup [0,3]$

9. $\{x | -4 < x < 0 \text{ or } x > 0\}$; $(-4, 0) \cup (0, \infty)$ **11.** $\{x | x \geq 1\}$; $[1, \infty)$ **13.** $\{x | x \leq 1 \text{ or } 2 \leq x \leq 3\}$; $(-\infty, 1] \text{ or } [2,3]$

15. $\{x | -1 < x < 0 \text{ or } x > 3\}$; $(-1, 0) \cup (3, \infty)$ **17.** $\{x | x < -1 \text{ or } x > 1\}$; $(-\infty, -1) \cup (1, \infty)$ **19.** $\{x | x < -1 \text{ or } r > 1\}$; $(-\infty, -1) \cup (1, \infty)$

21. $\{x | x < -1 \text{ or } x > 1\}$; $(-\infty, -1) \cup (1, \infty)$ **23.** $\{x | x \leq -1 \text{ or } 0 < x \leq 1\}$; $(-\infty, -1] \cup (0,1]$ **25.** $\{x | x < -1 \text{ or } x > 1\}$; $(-\infty, -1) \cup (1, \infty)$

27. $\left\{ x \middle| x < -\dfrac{2}{3} \text{ or } 0 < x < \dfrac{3}{2} \right\}$; $\left(-\infty, -\dfrac{2}{3}\right) \cup \left(0, \dfrac{3}{2}\right)$ **29.** $\{x | x < 2\}$; $(-\infty, 2)$ **31.** $\{x | -2 < x \leq 9\}$; $(-2, 9]$

33. $\{x | x < 2 \text{ or } 3 < x < 5\}$; $(-\infty, 2) \cup (3,5)$ **35.** $\{x | x < -3 \text{ or } -1 < x < 1 \text{ or } x > 2\}$; $(-\infty, -3) \cup; (-1, 1) \cup (2, \infty)$

37. $\{x | x < -5 \text{ or } -4 \leq x \leq -3 \text{ or } x = 0 \text{ or } x > 1\}$; $(-\infty, -5) \cup [-4, -3] \cup \{0\} \cup (1, \infty)$ **39.** $\left\{ x \middle| -\dfrac{1}{2} < x < 1 \text{ or } x > 3 \right\}$; $\left(-\dfrac{1}{2}, 1\right) \cup (3, \infty)$

41. $\{x | x > 4\}$; $(4, \infty)$ **43.** $\{x | x \leq -2 \text{ or } x \geq 2\}$; $(-\infty, -2] \text{ or } [2, \infty)$ **45.** $\{x | x < -4 \text{ or } x \geq 2\}$; $(-\infty, -4) \text{ or } [2, \infty)$

47.

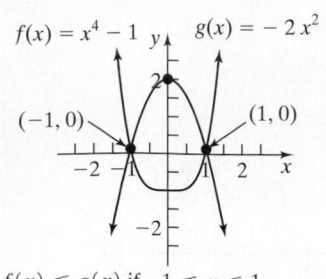

$f(x) = x^4 - 1$ $g(x) = -2x^2 + 2$

$f(x) \le g(x)$ if $-1 \le x \le 1$

49.

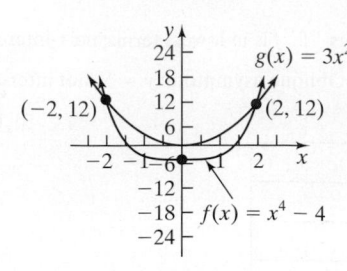

$g(x) = 3x^2$

$f(x) = x^4 - 4$

$f(x) \le g(x)$ if $-2 \le x \le 2$

51. Produce at least 250 bicycles
53. **(a)** The stretch is less than 39 ft.
 (b) The ledge should be at least 84 ft above the ground for a 150 lb jumper.

Historical Problems *(page 386)*

1.
$$\left(x - \frac{b}{3}\right)^3 + b\left(x - \frac{b}{3}\right)^2 + c\left(x - \frac{b}{3}\right) + d = 0$$
$$x^3 - bx^2 + \frac{b^2 x}{3} - \frac{b^3}{27} + bx^2 - \frac{2b^2 x}{3} + \frac{b^3}{9} + cx - \frac{bc}{3} + d = 0$$
$$x^3 + \left(c - \frac{b^2}{3}\right)x + \left(\frac{2b^3}{27} - \frac{bc}{3} + d\right) = 0$$

Let $p = c - \dfrac{b^2}{3}$ and $q = \dfrac{2b^3}{27} - \dfrac{bc}{3} + d$. Then $x^3 + px + q = 0$.

2.
$$(H + K)^3 + p(H + K) + q = 0$$
$$H^3 + 3H^2 K + 3HK^2 + K^3 + pH + pK + q = 0$$

Let $3HK = -p$.

$$H^3 - pH - pK + K^3 + pH + pK + q = 0, \quad H^3 + K^3 = -q$$

3.
$$3HK = -p$$
$$K = -\frac{p}{3H}$$
$$H^3 + \left(-\frac{p}{3H}\right)^3 = -q$$
$$H^3 - \frac{p^3}{27H^3} = -q$$
$$27H^6 - p^3 = -27qH^3$$
$$27H^6 + 27qH^3 - p^3 = 0$$
$$H^3 = \frac{-27q \pm \sqrt{(27q)^2 - 4(27)(-p^3)}}{2 \cdot 27}$$
$$H^3 = \frac{-q}{2} \pm \sqrt{\frac{27^2 q^2}{2^2 (27^2)} + \frac{4(27)p^3}{2^2(27^2)}}$$
$$H^3 = \frac{-q}{2} \pm \sqrt{\frac{q^2}{4} + \frac{p^3}{27}}$$
$$H = \sqrt[3]{\frac{-q}{2} + \sqrt{\frac{q^2}{4} + \frac{p^3}{27}}}$$

Choose the positive root for now.

4. $H^3 + K^3 = -q$
$$K^3 = -q - H^3$$
$$K^3 = -q - \left[\frac{-q}{2} + \sqrt{\frac{q^2}{4} + \frac{p^3}{27}}\right]$$
$$K^3 = \frac{-q}{2} - \sqrt{\frac{q^2}{4} + \frac{p^3}{27}}$$
$$K = \sqrt[3]{\frac{-q}{2} - \sqrt{\frac{q^2}{4} + \frac{p^3}{27}}}$$

5. $x = H + K$
$$x = \sqrt[3]{\frac{-q}{2} + \sqrt{\frac{q^2}{4} + \frac{p^3}{27}}} + \sqrt[3]{\frac{-q}{2} - \sqrt{\frac{q^2}{4} + \frac{p^3}{27}}} \quad \text{(Note that had we}$$

used the negative root in 3 the result would be the same.)

6. $x = 3$ **7.** $x = 2$ **8.** $x = 2$

5.5 Assess Your Understanding *(page 386)*

5. Remainder; Dividend **6.** $f(c)$ **7.** -4 **8.** F **9.** F **10.** T **11.** $R = f(2) = 8$; no **13.** $R = f(2) = 0$; yes **15.** $R = f(-3) = 0$; yes

17. $R = f(-4) = 1$; no **19.** $R = f\left(\dfrac{1}{2}\right) = 0$; yes **21.** 7; 3 or 1 positive; 2 or 0 negative **23.** 6; 2 or 0 positive; 2 or 0 negative

25. 3; 2 or 0 positive; 1 negative **27.** 4; 2 or 0 positive; 2 or 0 negative **29.** 5; 0 positive; 3 or 1 negative **31.** 6; 1 positive; 1 negative

33. $\pm 1, \pm\dfrac{1}{3}$ **35.** $\pm 1, \pm 3$ **37.** $\pm 1, \pm 2, \pm\dfrac{1}{4}, \pm\dfrac{1}{2}$ **39.** $\pm 1, \pm 3, \pm 9, \pm\dfrac{1}{2}, \pm\dfrac{1}{3}, \pm\dfrac{1}{6}, \pm\dfrac{3}{2}, \pm\dfrac{9}{2}$ **41.** $\pm 1, \pm 2, \pm 3, \pm 4, \pm 6, \pm 12, \pm\dfrac{1}{2}, \pm\dfrac{3}{2}$

43. $\pm 1, \pm 2, \pm 4, \pm 5, \pm 10, \pm 20, \pm\dfrac{1}{2}, \pm\dfrac{5}{2}, \pm\dfrac{1}{3}, \pm\dfrac{2}{3}, \pm\dfrac{4}{3}, \pm\dfrac{5}{3}, \pm\dfrac{10}{3}, \pm\dfrac{20}{3}, \pm\dfrac{1}{6}, \pm\dfrac{5}{6}$ **45.** $-3, -1, 2; f(x) = (x + 3)(x + 1)(x - 2)$

47. $\dfrac{1}{2}; f(x) = 2\left(x - \dfrac{1}{2}\right)(x^2 + 1)$ **49.** $2, \sqrt{5}, -\sqrt{5}; f(x) = 2(x - 2)(x - \sqrt{5})(x + \sqrt{5})$

51. $-1, 0.5, \sqrt{3}, -\sqrt{3}; f(x) = 2(x + 1)\left(x - \dfrac{1}{2}\right)(x - \sqrt{3})(x + \sqrt{3})$ **53.** 1, multiplicity 2; $-2, -1; f(x) = (x + 2)(x + 1)(x - 1)^2$

55. $-1, -0.25; f(x) = 4(x + 1)\left(x + \dfrac{1}{4}\right)(x^2 + 2)$ **57.** $\{-1, 2\}$ **59.** $\left\{\dfrac{2}{3}, -1 + \sqrt{2}, -1 - \sqrt{2}\right\}$ **61.** $\left\{\dfrac{1}{3}, \sqrt{5}, -\sqrt{5}\right\}$ **63.** $\{-3, -2\}$

65. $\left\{-\dfrac{1}{3}\right\}$ **67.** $\left\{\dfrac{1}{2}, 2, 5\right\}$

69. y-intercept: -6
x-intercepts: $-3, -1, 2$

71.

73.

75.

77.

79.

81. 5 **83.** 2 **85.** 5 **87.** $\dfrac{3}{2}$ **89.** $f(0) = -1; f(1) = 10$

91. $f(-5) = -58; f(-4) = 2$

93. $f(1.4) = -0.17536; f(1.5) = 1.40625$ **95.** 0.21

97. -4.04 **99.** 1.15 **101.** 2.53 **103.** $k = 5$ **105.** -7

107. If $f(x) = x^n - c^n$, then $f(c) = c^n - c^n = 0$, so $x - c$ is a factor of f. **109.** 5 **111.** 7 in.

113. All the potential rational zeros are integers, so r is either an integer or is not a rational zero (and is therefore irrational).

117. No, by the Rational Zeros Theorem, $\dfrac{1}{3}$ is not a potential rational zero. **119.** No, by the Rational Zeros Theorem, $\dfrac{2}{3}$ is not a potential rational zero.

5.6 Assess Your Understanding (page 393)

3. one **4.** $3 - 4i$ **5.** T **6.** F **7.** $4 + i$ **9.** $-i, 1 - i$ **11.** $-i, -2i$ **13.** $-i$ **15.** $2 - i, -3 + i$
17. $f(x) = x^4 - 14x^3 + 77x^2 - 200x + 208; a = 1$ **19.** $f(x) = x^5 - 4x^4 + 7x^3 - 8x^2 + 6x - 4; a = 1$
21. $f(x) = x^4 - 6x^3 + 10x^2 - 6x + 9; a = 1$ **23.** $-2i, 4$ **25.** $2i, -3, \dfrac{1}{2}$ **27.** $3 + 2i, -2, 5$ **29.** $4i, -\sqrt{11}, \sqrt{11}, -\dfrac{2}{3}$

31. $1, -\dfrac{1}{2} - \dfrac{\sqrt{3}}{2}i, -\dfrac{1}{2} + \dfrac{\sqrt{3}}{2}i; f(x) = (x - 1)\left(x + \dfrac{1}{2} + \dfrac{\sqrt{3}}{2}i\right)\left(x + \dfrac{1}{2} - \dfrac{\sqrt{3}}{2}i\right)$ **33.** $2, 3 - 2i, 3 + 2i; f(x) = (x - 2)(x - 3 + 2i)(x - 3 - 2i)$

35. $-i, i, -2i, 2i; f(x) = (x + i)(x - i)(x + 2i)(x - 2i)$ **37.** $-5i, 5i, -3, 1; f(x) = (x + 5i)(x - 5i)(x + 3)(x - 1)$ **39.** $-4, \dfrac{1}{3}, 2 - 3i, 2 + 3i;$

$f(x) = 3(x + 4)\left(x - \dfrac{1}{3}\right)(x - 2 + 3i)(x - 2 - 3i)$ **41.** Zeros that are complex numbers must occur in conjugate pairs; or a polynomial with real

coefficients of odd degree must have at least one real zero. **43.** If the remaining zero were a complex number, its conjugate would also be a zero, creating a polynomial of degree 5.

Review Exercises (page 396)

1. Polynomial of degree 5 **3.** Not a polynomial
5.

7.

9.

11. (a) x-intercepts: $-4, -2, 0$; y-intercept: 0
(b) Crosses at $-4, -2, 0$
(c) $y = x^3$ **(d)** 2
(e) Near -4: $f(x) \approx 8(x + 4)$
Near -2: $f(x) \approx -4(x + 2)$
Near 0: $f(x) \approx 8x$
(f)

13. (a) x-intercepts: $-4, 2$; y-intercept: 16
(b) Crosses at -4; touches at 2
(c) $y = x^3$ **(d)** 2
(e) Near -4: $f(x) \approx 36(x + 4)$
Near 2: $f(x) \approx 6(x - 2)^2$
(f)

15. (a) $f(x) = -2x^2(x - 2)$
x-intercepts: 0, 2; y-intercept: 0
(b) Touches at 0; crosses at 2
(c) $y = -2x^3$ **(d)** 2
(e) Near 0: $f(x) \approx 4x^2$
Near 2: $f(x) \approx -8(x - 2)$
(f)

17. **(a)** x-intercepts: $-3, -1, 1$; y-intercept: 3 **(f)**

(b) Crosses at $-3, -1$; touches at 1

(c) $y = x^4$ **(d)** 3

(e) Near -3: $f(x) \approx -32(x + 3)$

Near -1: $f(x) \approx 8(x + 1)$;

Near 1: $f(x) \approx 8(x - 1)^2$

19. Domain: $\{x \mid x \neq -3, x \neq 3\}$; horizontal asymptote: $y = 0$; vertical asymptotes: $x = -3, x = 3$ **21.** Domain: $\{x \mid x \neq -2\}$; horizontal asymptote: $y = 1$; vertical asymptote: $x = -2$

23. 1. $R(x) = \dfrac{2(x - 3)}{x}$; domain: $\{x \mid x \neq 0\}$; no y-intercept 2. R is in lowest terms; x-intercept: 3

3. R is in lowest terms; vertical asymptote: $x = 0$ 4. Horizontal asymptote: $y = 2$; not intersected

5.

Interval	$(-\infty, 0)$	$(0, 3)$	$(3, \infty)$
Number Chosen	-2	1	4
Value of R	$R(-2) = 5$	$R(1) = -4$	$R(4) = \frac{1}{2}$
Location of Graph	Above x-axis	Below x-axis	Above x-axis
Point on Graph	$(-2, 5)$	$(1, -4)$	$\left(4, \frac{1}{2}\right)$

6–7.

25. 1. Domain: $\{x \mid x \neq 0, x \neq 2\}$; no y-intercept 2. H is in lowest terms; x-intercept: -2 3. H is in lowest terms; vertical asymptotes: $x = 0, x = 2$
4. Horizontal asymptote: $y = 0$; intersected at $(-2, 0)$

5.

Interval	$(-\infty, -2)$	$(-2, 0)$	$(0, 2)$	$(2, \infty)$
Number Chosen	-3	-1	1	3
Value of H	$H(-3) = -\frac{1}{15}$	$H(-1) = \frac{1}{3}$	$H(1) = -3$	$H(3) = \frac{5}{3}$
Location of Graph	Below x-axis	Above x-axis	Below x-axis	Above x-axis
Point on Graph	$\left(-3, -\frac{1}{15}\right)$	$\left(-1, \frac{1}{3}\right)$	$(1, -3)$	$\left(3, \frac{5}{3}\right)$

6–7.

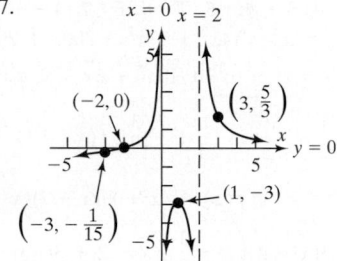

27. 1. $R(x) = \dfrac{(x + 3)(x - 2)}{(x - 3)(x + 2)}$; domain: $\{x \mid x \neq -2, x \neq 3\}$; y-intercept: 1

2. R is in lowest terms; x-intercepts: $-3, 2$

3. R is in lowest terms; asymptote: $x = -2, x = 3$

4. Horizontal asymptote: $y = 1$; intersected at $(0, 1)$

5.

Interval	$(-\infty, -3)$	$(-3, -2)$	$(-2, 2)$	$(2, 3)$	$(3, \infty)$
Number Chosen	-4	-2.5	0	2.5	4
Value of R	$R(-4) = 0.43$	$R(-2.5) = -0.82$	$R(0) = 1$	$R(2.5) = -1.22$	$R(4) = \frac{7}{3}$
Location of Graph	Above x-axis	Below x-axis	Above x-axis	Below x-axis	Above x-axis
Point on Graph	$(-4, 0.43)$	$(-2.5, -0.82)$	$(0, 1)$	$(2.5, -1.22)$	$\left(4, \frac{7}{3}\right)$

6–7.

29. 1. $F(x) = \dfrac{x^3}{(x + 2)(x - 2)}$; domain: $\{x \mid x \neq -2, x \neq 2\}$; y-intercept: 0 2. F is in lowest terms; x-intercept: 0

3. F is in lowest terms; vertical asymptotes: $x = -2, x = 2$ 4. Oblique asymptote: $y = x$; intersected at $(0, 0)$

5.

Interval	$(-\infty, -2)$	$(-2, 0)$	$(0, 2)$	$(2, \infty)$
Number Chosen	-3	-1	1	3
Value of F	$F(-3) = -\frac{27}{5}$	$F(-1) = \frac{1}{3}$	$F(1) = -\frac{1}{3}$	$F(3) = \frac{27}{5}$
Location of Graph	Below x-axis	Above x-axis	Below x-axis	Above x-axis
Point on Graph	$\left(-3, -\frac{27}{5}\right)$	$\left(-1, \frac{1}{3}\right)$	$\left(1, -\frac{1}{3}\right)$	$\left(3, \frac{27}{5}\right)$

6–7.

31. 1. Domain: $\{x \mid x \neq 1\}$; y-intercept: 0 2. R is in lowest terms; x-intercept: 0 3. R is in lowest terms; vertical asymptote: $x = 1$
4. No oblique or horizontal asymptote

5.

Interval	$(-\infty, 0)$	$(0, 1)$	$(1, \infty)$
Number Chosen	-2	$\frac{1}{2}$	2
Value of R	$R(-2) = \frac{32}{9}$	$R\left(\frac{1}{2}\right) = \frac{1}{2}$	$R(2) = 32$
Location of Graph	Above x-axis	Above x-axis	Above x-axis
Point on Graph	$\left(-2, \frac{32}{9}\right)$	$\left(\frac{1}{2}, \frac{1}{2}\right)$	$(2, 32)$

6–7.

33. 1. $G(x) = \dfrac{(x+2)(x-2)}{(x+1)(x-2)}$; domain: $\{x \mid x \neq -1, x \neq 2\}$; y-intercept: 2 2. In lowest terms, $G(x) = \dfrac{x+2}{x+1}$; x-intercept: -2

3. Vertical asymptote: $x = -1$; hole at $\left(2, \dfrac{4}{3}\right)$ 4. Horizontal asymptote: $y = 1$, not intersected

5.

Interval	$(-\infty, -2)$	$(-2, -1)$	$(-1, 2)$	$(2, \infty)$
Number Chosen	-3	-1.5	0	3
Value of G	$G(-3) = \frac{1}{2}$	$G(-1.5) = -1$	$G(0) = 2$	$G(3) = 1.25$
Location of Graph	Above x-axis	Below x-axis	Above x-axis	Above x-axis
Point on Graph	$\left(-3, \frac{1}{2}\right)$	$(-1.5, -1)$	$(0, 2)$	$(3, 1.25)$

6–7.

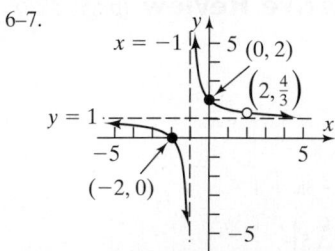

35. $\{x \mid x < -2 \text{ or } -1 < x < 2\}$; $(-\infty, -2) \cup (-1, 2)$

37. $\{x \mid -3 < x \leq 3\}$; $(-3, 3]$

39. $\{x \mid x < 1 \text{ or } x > 2\}$; $(-\infty, 1) \cup (2, \infty)$

41. $\{x \mid 1 \leq x \leq 2 \text{ or } x > 3\}$; $[1, 2] \cup (3, \infty)$

43. $\{x \mid x < -4 \text{ or } 2 < x < 4 \text{ or } x > 6\}$; $(-\infty, -4) \cup (2, 4) \cup (6, \infty)$

45. $R = 10$; g is not a factor of f. **47.** $R = 0$; g is a factor of f. **49.** $f(4) = 47{,}105$

51. 4, 2, or 0 positive; 2 or 0 negative **53.** $\pm 1, \pm 3, \pm\frac{1}{2}, \pm\frac{3}{2}, \pm\frac{1}{3}, \pm\frac{1}{4}, \pm\frac{3}{4}, \pm\frac{1}{6}, \pm\frac{1}{12}$ **55.** $-2, 1, 4$; $f(x) = (x+2)(x-1)(x-4)$

57. $\dfrac{1}{2}$, multiplicity 2; -2; $f(x) = 4\left(x - \dfrac{1}{2}\right)^2 (x+2)$ **59.** 2, multiplicity 2; **61.** $\{-3, 2\}$ **63.** $\left\{-3, -1, -\dfrac{1}{2}, 1\right\}$ **65.** 5 **67.** $\dfrac{37}{2}$

69. $f(0) = -1$; $f(1) = 1$ **71.** $f(0) = -1$; $f(1) = 1$ **73.** 1.52 **75.** 0.93 **77.** $4 - i$; $f(x) = x^3 - 14x^2 + 65x - 102$

79. $-i, 1 - i$; $f(x) = x^4 - 2x^3 + 3x^2 - 2x + 2$ **81.** $-2, 1, 4$; $f(x) = (x+2)(x-1)(x-4)$ **83.** $-2, \dfrac{1}{2}$ (multiplicity 2); $f(x) = 4(x+2)\left(x - \dfrac{1}{2}\right)^2$

85. 2 (multiplicity 2), $-\sqrt{5}i, \sqrt{5}i$; $f(x) = \left(x + \sqrt{5}i\right)\left(x - \sqrt{5}i\right)(x-2)^2$

87. $-3, 2, -\dfrac{\sqrt{2}}{2}i, \dfrac{\sqrt{2}}{2}i$; $f(x) = 2(x+3)(x-2)\left(x + \dfrac{\sqrt{2}}{2}i\right)\left(x - \dfrac{\sqrt{2}}{2}i\right)$

89. (a) $A(r) = 2\pi r^2 + \dfrac{500}{r}$ (b) 223.22 cm^2 (c) 257.08 cm^2 (d)

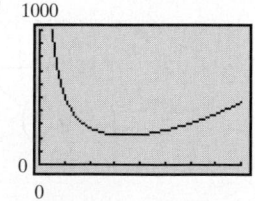

A is smallest when $r \approx 3.41$ cm.

Chapter Test (page 398)

1.

2. (a) 3 (b) Every zero of g lies between -15 and 15.
(c) $\dfrac{p}{q}$: $\pm\dfrac{1}{2}, \pm 1, \pm\dfrac{3}{2}, \pm\dfrac{5}{2}, \pm 3, \pm 5, \pm\dfrac{15}{2}, \pm 15$
(d) $-5, -\dfrac{1}{2}, 3$; $g(x) = (x+5)(2x+1)(x-3)$
(e) y-intercept: -15; x-intercepts: $-5, -\dfrac{1}{2}, 3$
(f) Crosses at $-5, -\dfrac{1}{2}, 3$ (g) $y = 2x^3$
(h) Near -5: $g(x) \approx 72(x+5)$
Near $-\dfrac{1}{2}$: $g(x) \approx -\dfrac{63}{4}(2x+1)$
Near 3: $g(x) \approx 56(x-3)$

(i)

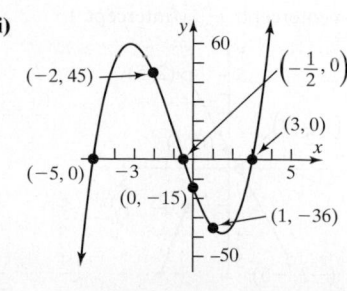

3. $4, -5i, 5i$ **4.** $\left\{1, \dfrac{5 - \sqrt{61}}{6}, \dfrac{5 + \sqrt{61}}{6}\right\}$ **5.** Domain: $\{x \mid x \neq -10, x \neq 4\}$; asymptotes: $x = -10, y = 2$

6. Domain: $\{x \mid x \neq -1\}$; asymptotes: $x = -1, y = x + 1$

7.

8. Answers may vary. One possibility is $f(x) = x^4 - 4x^3 - 2x^2 + 20x$.

9. Answers may vary. One possibility is $r(x) = \dfrac{2(x - 9)(x - 1)}{(x - 4)(x - 9)}$.

10. $f(0) = 8; f(4) = -36$

Since $f(0) = 8 > 0$ and $f(4) = -36 < 0$, the Intermediate Value Theorem guarantees that there is at least one real zero between 0 and 4.

11. $\{x \mid x < 3 \text{ or } x > 8\}$; $(-\infty, 3) \cup (8, \infty)$

Cumulative Review (page 398)

1. $\sqrt{26}$

2. $\{x \mid x \leq 0 \text{ or } x \geq 1\}$; $(-\infty, 0]$ or $[1, \infty)$

3. $\{x \mid -1 < x < 4\}$; $(-1, 4)$

4. $f(x) = -3x + 1$

5. $y = 2x - 1$

6.

7. Not a function; 3 has two images. **8.** $\{0, 2, 4\}$ **9.** $\left\{x \mid x \geq \dfrac{3}{2}\right\}$; $\left[\dfrac{3}{2}, \infty\right)$

10. Center: $(-2, 1)$; radius: 3

11. x-intercepts: $-3, 0, 3$; y-intercept: 0; symmetric with respect to the origin

12. $y = -\dfrac{2}{3}x + \dfrac{17}{3}$

13. Not a function; it fails the Vertical Line Test.

14. (a) 22 **(b)** $x^2 - 5x - 2$ **(c)** $-x^2 - 5x + 2$ **(d)** $9x^2 + 15x - 2$ **(e)** $2x + h + 5$

15. (a) $\{x \mid x \neq 1\}$ **(b)** No; $(2, 7)$ is on the graph. **(c)** 4; $(3, 4)$ is on the graph. **(d)** $\dfrac{7}{4}$; $\left(\dfrac{7}{4}, 9\right)$ is on the graph.

16.

17.

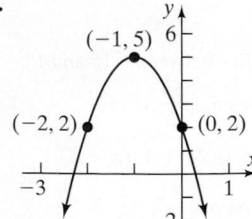

18. 6; $y = 6x - 1$

19. (a) x-intercepts: $-5, -1, 5$; y-intercept: -3

(b) No symmetry

(c) Neither

(d) Increasing: $(-\infty, -3)$ and $(2, \infty)$; decreasing: $(-3, 2)$

(e) Local maximum is 5 and occurs at $x = -3$.

(f) Local minimum is -6 and occurs at $x = 2$.

20. Odd

21. (a) Domain: $\{x \mid x > -3\}$ or $(-3, \infty)$

(b) x-intercept: $-\dfrac{1}{2}$; y-intercept: 1

(c)

22.

23. (a) $(f + g)(x) = x^2 - 9x - 6$; domain: all real numbers

(b) $\left(\dfrac{f}{g}\right)(x) = \dfrac{x^2 - 5x + 1}{-4x - 7}$; domain: $\left\{x \mid x \neq -\dfrac{7}{4}\right\}$

24. (a) $R(x) = -\dfrac{1}{10}x^2 + 150x$

(b) \$14,000 **(c)** 750; \$56,250 **(d)** \$75

(d) Range: $\{y \mid y < 5\}$ or $(-\infty, 5)$

CHAPTER 6 Exponential and Logarithmic Functions

6.1 Assess Your Understanding (page 407)

4. $(g \circ f)(x)$ **5.** F **6.** F **7. (a)** -1 **(b)** -1 **(c)** 8 **(d)** 0 **(e)** 8 **(f)** -7 **9. (a)** 4 **(b)** 5 **(c)** -1 **(d)** -2 **11. (a)** 98 **(b)** 49 **(c)** 4

(d) 4 **13. (a)** 97 **(b)** $-\dfrac{163}{2}$ **(c)** 1 **(d)** $-\dfrac{3}{2}$ **15. (a)** $2\sqrt{2}$ **(b)** $2\sqrt{2}$ **(c)** 1 **(d)** 0 **17. (a)** $\dfrac{1}{17}$ **(b)** $\dfrac{1}{5}$ **(c)** 1 **(d)** $\dfrac{1}{2}$ **19. (a)** $\dfrac{3}{\sqrt[3]{4}+1}$

(b) 1 **(c)** $\dfrac{6}{5}$ **(d)** 0 **21.** $\{x | x \neq 0, x \neq 2\}$ **23.** $\{x | x \neq -4, x \neq 0\}$ **25.** $\left\{ x \middle| x \geq -\dfrac{3}{2} \right\}$ **27.** $\{x | x \geq 1\}$ **29. (a)** $(f \circ g)(x) = 6x + 3$;

all real numbers **(b)** $(g \circ f)(x) = 6x + 9$; all real numbers **(c)** $(f \circ f)(x) = 4x + 9$; all real numbers **(d)** $(g \circ g)(x) = 9x$; all real numbers

31. (a) $(f \circ g)(x) = 3x^2 + 1$; all real numbers **(b)** $(g \circ f)(x) = 9x^2 + 6x + 1$; all real numbers **(c)** $(f \circ f)(x) = 9x + 4$; all real numbers

(d) $(g \circ g)(x) = x^4$; all real numbers **33. (a)** $(f \circ g)(x) = x^4 + 8x^2 + 16$; all real numbers **(b)** $(g \circ f)(x) = x^4 + 4$; all real numbers

(c) $(f \circ f)(x) = x^4$; all real numbers **(d)** $(g \circ g)(x) = x^4 + 8x^2 + 20$; all real numbers **35. (a)** $(f \circ g)(x) = \dfrac{3x}{2 - x}$; $\{x | x \neq 0, x \neq 2\}$

(b) $(g \circ f)(x) = \dfrac{2(x - 1)}{3}$; $\{x | x \neq 1\}$ **(c)** $(f \circ f)(x) = \dfrac{3(x - 1)}{4 - x}$; $\{x | x \neq 1, x \neq 4\}$ **(d)** $(g \circ g)(x) = x$; $\{x | x \neq 0\}$

37. (a) $(f \circ g)(x) = \dfrac{4}{4 + x}$; $\{x | x \neq -4, x \neq 0\}$ **(b)** $(g \circ f)(x) = \dfrac{-4(x - 1)}{x}$; $\{x | x \neq 0, x \neq 1\}$ **(c)** $(f \circ f)(x) = x$; $\{x | x \neq 1\}$

(d) $(g \circ g)(x) = x$; $\{x | x \neq 0\}$ **39. (a)** $(f \circ g)(x) = \sqrt{2x + 3}$; $\left\{ x \middle| x \geq -\dfrac{3}{2} \right\}$ **(b)** $(g \circ f)(x) = 2\sqrt{x} + 3$; $\{x | x \geq 0\}$

(c) $(f \circ f)(x) = \sqrt[4]{x}$; $\{x | x \geq 0\}$ **(d)** $(g \circ g)(x) = 4x + 9$; all real numbers **41. (a)** $(f \circ g)(x) = x$; $\{x | x \geq 1\}$

(b) $(g \circ f)(x) = |x|$; all real numbers **(c)** $(f \circ f)(x) = x^4 + 2x^2 + 2$; all real numbers **(d)** $(g \circ g)(x) = \sqrt{\sqrt{x - 1} - 1}$; $\{x | x \geq 2\}$

43. (a) $(f \circ g)(x) = -\dfrac{4x - 17}{2x - 1}$; $\left\{ x \middle| x \neq 3; \ x \neq \dfrac{1}{2} \right\}$ **(b)** $(g \circ f)(x) = -\dfrac{3x - 3}{2x + 8}$; $\{x | x \neq -4; \ x \neq -1\}$

(c) $(f \circ f)(x) = -\dfrac{2x + 5}{x - 2}$; $\{x | x \neq -1; \ x \neq 2\}$ **(d)** $(g \circ g)(x) = -\dfrac{3x - 4}{2x - 11}$; $\left\{ x \middle| x \neq \dfrac{11}{2}; \ x \neq 3 \right\}$

45. $(f \circ g)(x) = f(g(x)) = f\left(\dfrac{1}{2}x\right) = 2\left(\dfrac{1}{2}x\right) = x$; $(g \circ f)(x) = g(f(x)) = g(2x) = \dfrac{1}{2}(2x) = x$

47. $(f \circ g)(x) = f(g(x)) = f(\sqrt[3]{x}) = (\sqrt[3]{x})^3 = x$; $(g \circ f)(x) = g(f(x)) = g(x^3) = \sqrt[3]{x^3} = x$

49. $(f \circ g)(x) = f(g(x)) = f\left(\dfrac{1}{2}(x + 6)\right) = 2\left[\dfrac{1}{2}(x + 6)\right] - 6 = x + 6 - 6 = x$;

$(g \circ f)(x) = g(f(x)) = g(2x - 6) = \dfrac{1}{2}(2x - 6 + 6) = \dfrac{1}{2}(2x) = x$

51. $(f \circ g)(x) = f(g(x)) = f\left(\dfrac{1}{a}(x - b)\right) = a\left[\dfrac{1}{a}(x - b)\right] + b = x$; $(g \circ f)(x) = g(f(x)) = g(ax + b) = \dfrac{1}{a}(ax + b - b) = x$

53. $f(x) = x^4$; $g(x) = 2x + 3$ (Other answers are possible.) **55.** $f(x) = \sqrt{x}$; $g(x) = x^2 + 1$ (Other answers are possible.)

57. $f(x) = |x|$; $g(x) = 2x + 1$ (Other answers are possible.) **59.** $(f \circ g)(x) = 11$; $(g \circ f)(x) = 2$ **61.** $-3, 3$ **63. (a)** $(f \circ g)(x) = acx + ad + b$

(b) $(g \circ f)(x) = acx + bc + d$ **(c)** The domains of both $f \circ g$ and $g \circ f$ are all real numbers **(d)** $f \circ g = g \circ f$ when $ad + b = bc + d$

65. $S(t) = \dfrac{16}{9}\pi t^6$ **67.** $C(t) = 15{,}000 + 800{,}000t - 40{,}000t^2$ **69.** $C(p) = \dfrac{2\sqrt{100 - p}}{25} + 600, 0 \leq p \leq 100$ **71.** $V(r) = 2\pi r^3$

73. (a) $f(x) = 0.8382x$ **(b)** $g(x) = 140.9687x$ **(c)** $g(f(x)) = g(0.8382x) = 118.15996x$ **(d)** $118{,}159.96$ yen

75. f and g are odd functions, so $f(-x) = -f(x)$ and $g(-x) = -g(x)$. Then $(f \circ g)(-x) = f(g(-x)) = f(-g(x)) = -f(g(x)) = -(f \circ g)(x)$. So $f \circ g$ is also odd.

6.2 Assess Your Understanding (page 419)

4. one-to-one **5.** $y = x$ **6.** $[4, \infty)$ **7.** F **8.** T **9.** one-to-one **11.** Not one-to-one **13.** Not one-to-one **15.** one-to-one **17.** one-to-one **19.** Not one-to-one **21.** one-to-one

23.

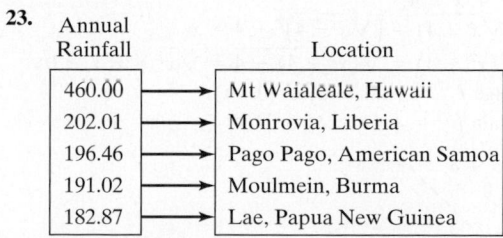

Annual Rainfall → Location

Domain: $\{460.00, 202.01, 196.46, 191.02, 182.87\}$
Range: $\{\text{Mt Waialeale, Monrovia, Pago Pago, Moulmein, Lae}\}$

25.

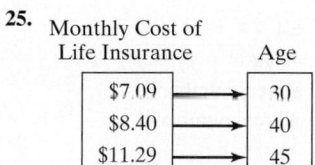

Monthly Cost of Life Insurance → Age

Domain: $\{\$7.09, \$8.04, \$11.29\}$
Range: $\{30, 40, 45\}$

27. $\{(5, -3), (9, -2), (2, -1), (11, 0), (-5, 1)\}$
Domain: $\{5, 9, 2, 11, -5\}$
Range: $\{-3, -2, -1, 0, 1\}$

29. $\{(1, -2), (2, -3), (0, -10), (9, 1), (4, 2)\}$
Domain: $\{1, 2, 0, 9, 4\}$
Range: $\{-2, -3, -10, 1, 2\}$

31. $f(g(x)) = f\left(\dfrac{1}{3}(x - 4)\right) = 3\left[\dfrac{1}{3}(x - 4)\right] + 4$

$= (x - 4) + 4 = x$

$g(f(x)) = g(3x + 4) = \dfrac{1}{3}[(3x + 4) - 4] = \dfrac{1}{3}(3x) = x$

33. $f(g(x)) = f\left(\dfrac{x}{4} + 2\right) = 4\left[\dfrac{x}{4} + 2\right] - 8 = (x + 8) - 8 = x$

$g(f(x)) = g(4x - 8) = \dfrac{4x - 8}{4} + 2 = (x - 2) + 2 = x$

35. $f(g(x)) = f\left(\sqrt[3]{x + 8}\right) = \left(\sqrt[3]{x + 8}\right)^3 - 8 = (x + 8) - 8 = x$

$g(f(x)) = g(x^3 - 8) = \sqrt[3]{(x^3 - 8) + 8} = \sqrt[3]{x^3} = x$

37. $f(g(x)) = f\left(\dfrac{1}{x}\right) = \dfrac{1}{\left(\dfrac{1}{x}\right)} = x; g(f(x)) = g\left(\dfrac{1}{x}\right) = \dfrac{1}{\left(\dfrac{1}{x}\right)} = x$

39. $f(g(x)) = f\left(\dfrac{4x - 3}{2 - x}\right) = \dfrac{2\left(\dfrac{4x - 3}{2 - x}\right) + 3}{\dfrac{4x - 3}{2 - x} + 4}$

$= \dfrac{2(4x - 3) + 3(2 - x)}{4x - 3 + 4(2 - x)} = \dfrac{5x}{5} = x$

$g(f(x)) = g\left(\dfrac{2x + 3}{x + 4}\right) = \dfrac{4\left(\dfrac{2x + 3}{x + 4}\right) - 3}{2 - \dfrac{2x + 3}{x + 4}}$

$= \dfrac{4(2x + 3) - 3(x + 4)}{2(x + 4) - (2x + 3)} = \dfrac{5x}{5} = x$

41.

43.

45.

47. $f^{-1}(x) = \dfrac{1}{3}x$

$f(f^{-1}(x)) = f\left(\dfrac{1}{3}x\right) = 3\left(\dfrac{1}{3}x\right) = x$

$f^{-1}(f(x)) = f^{-1}(3x) = \dfrac{1}{3}(3x) = x$

Domain f = range f^{-1} = all real numbers
Range f = domain f^{-1} = all real numbers

49. $f^{-1}(x) = \dfrac{x}{4} - \dfrac{1}{2}$

$f(f^{-1}(x)) = f\left(\dfrac{x}{4} - \dfrac{1}{2}\right) = 4\left(\dfrac{x}{4} - \dfrac{1}{2}\right) + 2 = (x - 2) + 2 = x$

$f^{-1}(f(x)) = f^{-1}(4x + 2) = \dfrac{4x + 2}{4} - \dfrac{1}{2} = \left(x + \dfrac{1}{2}\right) - \dfrac{1}{2} = x$

Domain f = range f^{-1} = all real numbers
Range f = domain f^{-1} = all real numbers

51. $f^{-1}(x) = \sqrt[3]{x + 1}$
$f(f^{-1}(x)) = f\left(\sqrt[3]{x + 1}\right) = \left(\sqrt[3]{x + 1}\right)^3 - 1 = x$
$f^{-1}(f(x)) = f^{-1}(x^3 - 1) = \sqrt[3]{(x^3 - 1) + 1} = x$
Domain f = range f^{-1} = all real numbers
Range f = domain f^{-1} = all real numbers

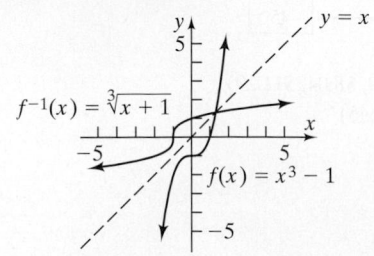

53. $f^{-1}(x) = \sqrt{x - 4}, x \geq 4$
$f(f^{-1}(x)) = f\left(\sqrt{x - 4}\right) = \left(\sqrt{x - 4}\right)^2 + 4 = x$
$f^{-1}(f(x)) = f^{-1}(x^2 + 4) = \sqrt{(x^2 + 4) - 4} = \sqrt{x^2} = x, x \geq 0$
Domain f = range f^{-1} = $\{x | x \geq 0\}$ or $[0, \infty)$
Range f = domain f^{-1} = $\{x | x \geq 4\}$ or $[4, \infty)$

55. $f^{-1}(x) = \dfrac{4}{x}$

$$f(f^{-1}(x)) = f\left(\dfrac{4}{x}\right) = \dfrac{4}{\left(\dfrac{4}{x}\right)} = x$$

$$f^{-1}(f(x)) = f^{-1}\left(\dfrac{4}{x}\right) = \dfrac{4}{\left(\dfrac{4}{x}\right)} = x$$

Domain f = range f^{-1} = all real numbers except 0
Range f = domain f^{-1} = all real numbers except 0

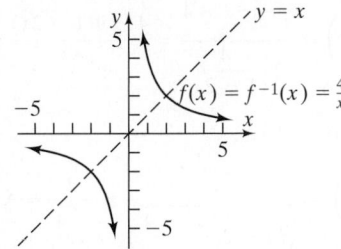

57. $f^{-1}(x) = \dfrac{2x + 1}{x}$

$$f(f^{-1}(x)) = f\left(\dfrac{2x + 1}{x}\right) = \dfrac{1}{\dfrac{2x + 1}{x} - 2} = \dfrac{x}{(2x + 1) - 2x} = x$$

$$f^{-1}(f(x)) = f^{-1}\left(\dfrac{1}{x - 2}\right) = \dfrac{2\left(\dfrac{1}{x - 2}\right) + 1}{\dfrac{1}{x - 2}} = \dfrac{2 + (x - 2)}{1} = x$$

Domain f = range f^{-1} = all real numbers except 2
Range f = domain f^{-1} = all real numbers except 0

59. $f^{-1}(x) = \dfrac{2 - 3x}{x}$

$$f(f^{-1}(x)) = f\left(\dfrac{2 - 3x}{x}\right) = \dfrac{2}{3 + \dfrac{2 - 3x}{x}} = \dfrac{2x}{3x + 2 - 3x} = \dfrac{2x}{2} = x$$

$$f^{-1}(f(x)) = f^{-1}\left(\dfrac{2}{3 + x}\right) = \dfrac{2 - 3\left(\dfrac{2}{3 + x}\right)}{\dfrac{2}{3 + x}} = \dfrac{2(3 + x) - 3 \cdot 2}{2} = \dfrac{2x}{2} = x$$

Domain f = all real numbers except -3
Range f = domain f^{-1} = all real numbers except 0

61. $f^{-1}(x) = \dfrac{-2x}{x - 3}$

$$f(f^{-1}(x)) = f\left(\dfrac{-2x}{x - 3}\right) = \dfrac{3\left(\dfrac{-2x}{x - 3}\right)}{\dfrac{-2x}{x - 3} + 2} = \dfrac{3(-2x)}{-2x + 2(x - 3)} = \dfrac{-6x}{-6} = x$$

$$f^{-1}(f(x)) = f^{-1}\left(\dfrac{3x}{x + 2}\right) = \dfrac{-2\left(\dfrac{3x}{x + 2}\right)}{\dfrac{3x}{x + 2} - 3} = \dfrac{-2(3x)}{3x - 3(x + 2)} = \dfrac{-6x}{-6} = x$$

Domain f = all real numbers except -2
Range f = domain f^{-1} = all real numbers except 3

63. $f^{-1}(x) = \dfrac{x}{3x - 2}$

$$f(f^{-1}(x)) = f\left(\dfrac{x}{3x - 2}\right) = \dfrac{2\left(\dfrac{x}{3x - 2}\right)}{3\left(\dfrac{x}{3x - 2}\right) - 1} = \dfrac{2x}{3x - (3x - 2)} = \dfrac{2x}{2} = x$$

$$f^{-1}(f(x)) = f^{-1}\left(\dfrac{2x}{3x - 1}\right) = \dfrac{\dfrac{2x}{3x - 1}}{3\left(\dfrac{2x}{3x - 1}\right) - 2} = \dfrac{2x}{6x - 2(3x - 1)} = \dfrac{2x}{2} = x$$

Domain f = all real numbers except $\dfrac{1}{3}$

Range f = domain f^{-1} = all real numbers except $\dfrac{2}{3}$

65. $f^{-1}(x) = \dfrac{3x + 4}{2x - 3}$

$f(f^{-1}(x)) = f\left(\dfrac{3x + 4}{2x - 3}\right) = \dfrac{3\left(\dfrac{3x + 4}{2x - 3}\right) + 4}{2\left(\dfrac{3x + 4}{2x - 3}\right) - 3}$

$= \dfrac{3(3x + 4) + 4(2x - 3)}{2(3x + 4) - 3(2x - 3)}$

$= \dfrac{17x}{17} = x$

$f^{-1}(f(x)) = f^{-1}\left(\dfrac{3x + 4}{2x - 3}\right) = \dfrac{3\left(\dfrac{3x + 4}{2x - 3}\right) + 4}{2\left(\dfrac{3x + 4}{2x - 3}\right) - 3}$

$= \dfrac{3(3x + 4) + 4(2x - 3)}{2(3x + 4) - 3(2x - 3)} = \dfrac{17x}{17} = x$

Domain f = all real numbers except $\dfrac{3}{2}$

Range f = domain f^{-1} = all real numbers except $\dfrac{3}{2}$

67. $f^{-1}(x) = \dfrac{-2x + 3}{x - 2}$

$f(f^{-1}(x)) = f\left(\dfrac{-2x + 3}{x - 2}\right) = \dfrac{2\left(\dfrac{-2x + 3}{x - 2}\right) + 3}{\dfrac{-2x + 3}{x - 2} + 2}$

$= \dfrac{2(-2x + 3) + 3(x - 2)}{-2x + 3 + 2(x - 2)} = \dfrac{-x}{-1} = x$

$f^{-1}(f(x)) = f^{-1}\left(\dfrac{2x + 3}{x + 2}\right) = \dfrac{-2\left(\dfrac{2x + 3}{x + 2}\right) + 3}{\dfrac{2x + 3}{x + 2} - 2}$

$= \dfrac{-2(2x + 3) + 3(x + 2)}{2x + 3 - 2(x + 2)} = \dfrac{-x}{-1} = x$

Domain f = all real numbers except -2

Range f = domain f^{-1} = all real numbers except 2

69. $f^{-1}(x) = \dfrac{2}{\sqrt{1 - 2x}}$

$f(f^{-1}(x)) = f\left(\dfrac{2}{\sqrt{1 - 2x}}\right) = \dfrac{\dfrac{4}{1 - 2x} - 4}{2 \cdot \dfrac{4}{1 - 2x}} = \dfrac{4 - 4(1 - 2x)}{2 \cdot 4}$

$= \dfrac{8x}{8} = x$

$f^{-1}(f(x)) = f^{-1}\left(\dfrac{x^2 - 4}{2x^2}\right) = \dfrac{2}{\sqrt{1 - 2\left(\dfrac{x^2 - 4}{2x^2}\right)}} = \dfrac{2}{\sqrt{\dfrac{4}{x^2}}} = \sqrt{x^2}$

$= x$, since $x > 0$

Domain f = $\{x | x > 0\}$ or $(0, \infty)$

Range f = domain f^{-1} = $\left\{x \Big| x < \dfrac{1}{2}\right\}$ or $\left(-\infty, \dfrac{1}{2}\right)$

71. (a) 0 **(b)** 2 **(c)** 0 **(d)** 1 **73.** 7 **75.** Domain of f^{-1}: $[-2, \infty)$; range of f^{-1}: $[5, \infty)$ **77.** Domain of g^{-1}: $[0, \infty)$; range of g^{-1}: all real numbers
79. Increasing on the interval $(f(0), f(5))$ **81.** $f^{-1}(x) = \dfrac{1}{m}(x - b), m \neq 0$ **83.** Quadrant I **85.** Possible answer: $f(x) = |x|, x \geq 0$, is
one-to-one; $f^{-1}(x) = x, x \geq 0$

87. (a) $r(d) = \dfrac{d + 90.39}{6.97}$

(b) $r(d(r)) = \dfrac{6.97r - 90.39 + 90.39}{6.97} = \dfrac{6.97r}{6.97} = r$

$d(r(d)) = 6.97\left(\dfrac{d + 90.39}{6.97}\right) - 90.39 = d + 90.39 - 90.39 = d$

(c) 56 miles per hour

89. (a) 77.6 kg

(b) $h(W) = \dfrac{W - 50}{2.3} + 60 = \dfrac{W + 88}{2.3}$

(d) 73 inches

(c) $h(W(h)) = \dfrac{50 + 2.3(h - 60) + 88}{2.3} = \dfrac{2.3h}{2.3} = h$

$W(h(W)) = 50 + 2.3\left(\dfrac{W + 88}{2.3} - 60\right)$

$= 50 + W + 88 - 138 = W$

91. (a) $\{g | 30,650 \leq g \leq 74,200\}$

(b) $\{T | 4220 \leq T \leq 15,107.50\}$

(c) $g(T) = \dfrac{T - 4220}{0.25} + 30,650$

Domain: $\{T | 4220 \leq T \leq 15,107.50\}$
Range: $\{g | 30,650 \leq g \leq 74,200\}$

93. (a) t represents time, so $t \geq 0$.

(b) $t(H) = \sqrt{\dfrac{H - 100}{-4.9}} = \sqrt{\dfrac{100 - H}{4.9}}$

(c) 2.02 seconds

95. $f^{-1}(x) = \dfrac{-dx + b}{cx - a}; f = f^{-1}$ if $a = -d$

99. No

6.3 Assess Your Understanding *(page 432)*

6. $\left(-1, \dfrac{1}{a}\right)$, $(0, 1)$, $(1, a)$ **7.** 1 **8.** 4 **9.** F **10.** F **11. (a)** 11.212 **(b)** 11.587 **(c)** 11.664 **(d)** 11.665 **13. (a)** 8.815 **(b)** 8.821 **(c)** 8.824
(d) 8.825 **15. (a)** 21.217 **(b)** 22.217 **(c)** 22.440 **(d)** 22.459 **17.** 3.320 **19.** 0.427 **21.** Not exponential **23.** Exponential; $a = 4$ **25.** Exponential;
$a = 2$ **27.** Not exponential **29.** B **31.** D **33.** A **35.** E

37.

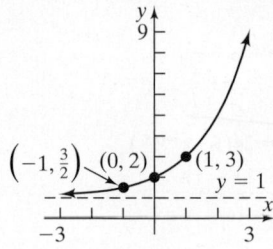

Domain: All real numbers
Range: $\{y|y > 1\}$ or $(1, \infty)$
Horizontal asymptote: $y = 1$

39.

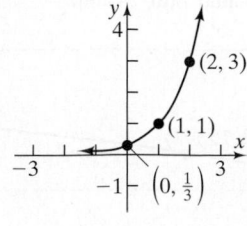

Domain: All real numbers
Range: $\{y|y > 0\}$ or $(0, \infty)$
Horizontal asymptote: $y = 0$

41.

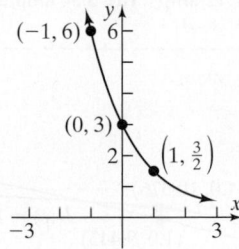

Domain: All real numbers
Range: $\{y|y > 0\}$ or $(0, \infty)$
Horizontal asymptote: $y = 0$

43.

Domain: All real numbers
Range: $\{y|y > -2\}$ or $(-2, \infty)$
Horizontal asymptote: $y = -2$

45.

Domain: All real numbers
Range: $\{y|y > 2\}$ or $(2, \infty)$
Horizontal asymptote: $y = 2$

47.

Domain: All real numbers
Range: $\{y|y > 2\}$ or $(2, \infty)$
Horizontal asymptote: $y = 2$

49.

Domain: All real numbers
Range: $\{y|y > 0\}$ or $(0, \infty)$
Horizontal asymptote: $y = 0$

51.

Domain: All real numbers
Range: $\{y|y > 0\}$ or $(0, \infty)$
Horizontal asymptote: $y = 0$

53.

Domain: All real numbers
Range: $\{y|y < 5\}$ or $(-\infty, 5)$
Horizontal asymptote: $y = 5$

55.

Domain: All real numbers
Range: $\{y|y < 2\}$ or $(-\infty, 2)$
Horizontal asymptote: $y = 2$

57. $\{3\}$ **59.** $\{-4\}$ **61.** $\{2\}$ **63.** $\left\{\dfrac{3}{2}\right\}$ **65.** $\left\{-\sqrt{2}, 0, \sqrt{2}\right\}$ **67.** $\{6\}$

69. $\{-1, 7\}$ **71.** $\{-4, 2\}$ **73.** $\{-4\}$ **75.** $\{1, 2\}$

77. (a) $16; (4, 16)$ **(b)** $-4; \left(-4, \dfrac{1}{16}\right)$ **79. (a)** $\dfrac{9}{4}; \left(-1, \dfrac{9}{4}\right)$

(b) $3; (3, 66)$ **81. (a)** $10; (-2, 10)$ **(b)** $3; \left(3, -\dfrac{13}{8}\right)$

83. $\dfrac{1}{49}$ **85.** $\dfrac{1}{4}$ **87.** $f(x) = 3^x$ **89.** $f(x) = -6^x$

91.

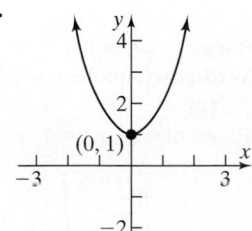

Domain: $(-\infty, \infty)$
Range: $[1, \infty)$
Intercept: $(0, 1)$

93.

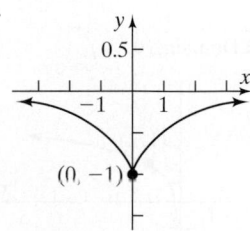

Domain: $(-\infty, \infty)$
Range: $[-1, 0)$
Intercept: $(0, -1)$

95. (a) 74% **(b)** 47% **97. (a)** \$12,123 **(b)** \$6443 **99.** 3.35 mg; 0.45 mg

101. (a) 0.632 **(b)** 0.982 **(c)** 1

(d)

(e) About 7 min

103. (a) 0.0516 **(b)** 0.0888

105. (a) 70.95% **(b)** 72.62%

(c) 100%

107. (a) 5.41 amp, 7.59 amp, 10.38 amp **(b)** 12 amp **(d)** 3.34 amp, 5.31 amp, 9.44 amp **(e)** 24 amp

(c), (f)

109. $n = 4: 2.7083$; $n = 6: 2.7181$; $n = 8: 2.7182788$; $n = 10: 2.7182818$

111. $\dfrac{f(x + h) - f(x)}{h} = \dfrac{a^{x+h} - a^x}{h} = \dfrac{a^x a^h - a^x}{h} = \dfrac{a^x(a^h - 1)}{h}$ **113.** $f(-x) = a^{-x} = \dfrac{1}{a^x} = \dfrac{1}{f(x)}$

115. (a) $f(-x) = \dfrac{1}{2}(e^{-x} - e^{-(-x)}) = \dfrac{1}{2}(e^{-x} - e^x) = -\dfrac{1}{2}(e^x - e^{-x}) = -f(x)$ **117.** $f(1) = 5, f(2) = 17, f(3) = 257, f(4) = 65,537$

$f(5) = 4,294,967,297 = 641 \times 6,700,417$

(b)

6.4 Assess Your Understanding *(page 446)*

4. $\{x | x > 0\}$ or $(0, \infty)$ **5.** $\left(\dfrac{1}{a}, -1\right), (1, 0), (a, 1)$ **6.** 1 **7.** F **8.** T **9.** $2 = \log_3 9$ **11.** $2 = \log_a 1.6$ **13.** $x = \log_2 7.2$ **15.** $x = \ln 8$

17. $2^3 = 8$ **19.** $a^6 = 3$ **21.** $3^x = 2$ **23.** $e^x = 4$ **25.** 0 **27.** 2 **29.** -4 **31.** $\dfrac{1}{2}$ **33.** 4 **35.** $\dfrac{1}{2}$ **37.** $\{x | x > 3\}$; $(3, \infty)$ **39.** All real numbers

except 0; $\{x | x \neq 0\}$ **41.** $\{x | x > 10\}$; $(10, \infty)$ **43.** $\{x | x > -1\}$; $(-1, \infty)$ **45.** $\{x | x < -1$ or $x > 0\}$; $(-\infty, -1) \cup (0, \infty)$

47. $\{x | x \geq 1\}$; $[1, \infty)$ **49.** 0.511 **51.** 30.099 **53.** 2.303 **55.** -53.991 **57.** $\sqrt{2}$

59.

61.

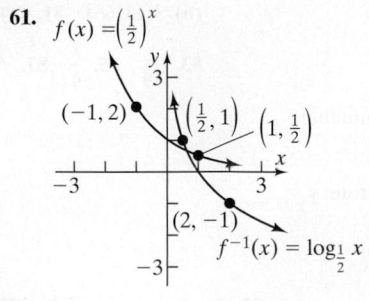

63. B
65. D
67. A
69. E

71. (a) Domain: $(-4, \infty)$ **(c)** Range: $(-\infty, \infty)$ **73. (a)** Domain: $(0, \infty)$ **(c)** Range: $(-\infty, \infty)$
 (b) Vertical asymptote: $x = -4$ **(b)** Vertical asymptote: $x = 0$
 (d) $f^{-1}(x) = e^x - 4$ **(d)** $f^{-1}(x) = e^{x-2}$
 (e) Range of f: $(-\infty, \infty)$ **(e)** Range of f: $(-\infty, \infty)$
 (f) **(f)**

75. **(a)** Domain: $(0, \infty)$

(b)

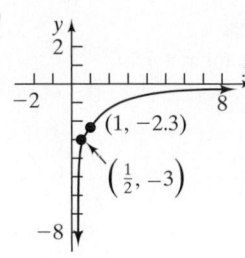

(c) Range: $(-\infty, \infty)$
Vertical asymptote: $x = 0$

(d) $f^{-1}(x) = \dfrac{1}{2}e^{x+3}$
(e) Range of f: $(-\infty, \infty)$
(f)

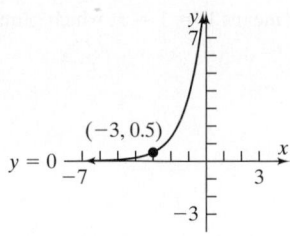

77. **(a)** Domain: $(4, \infty)$

(b)

(c) Range: $(-\infty, \infty)$
Vertical asymptote: $x = 4$

(d) $f^{-1}(x) = 10^{x-2} + 4$
(e) Range of f: $(-\infty, \infty)$
(f)

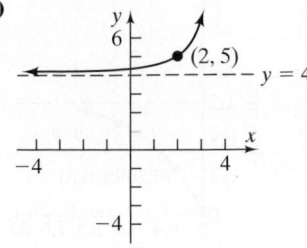

79. **(a)** Domain: $(0, \infty)$

(b)

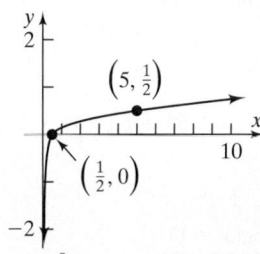

(c) Range: $(-\infty, \infty)$
Vertical asymptote: $x = 0$

(d) $f^{-1}(x) = \dfrac{1}{2} \cdot 10^{2x}$
(e) Range of f: $(-\infty, \infty)$
(f)

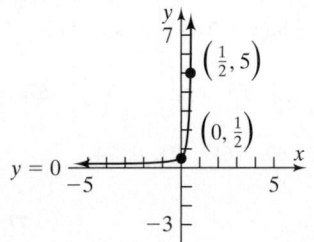

81. **(a)** Domain: $(-2, \infty)$

(b)

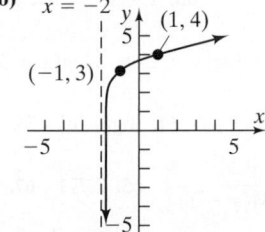

(c) Range: $(-\infty, \infty)$
Vertical asymptote: $x = -2$

(d) $f^{-1}(x) = 3^{x-3} - 2$
(e) Range of f: $(-\infty, \infty)$
(f)

83. **(a)** Domain: $(-\infty, \infty)$

(b)

(c) Range: $(-3, \infty)$
Horizontal asymptote: $y = -3$

(d) $f^{-1}(x) = \ln(x+3) - 2$
(e) Range of f: $(-3, \infty)$
(f)

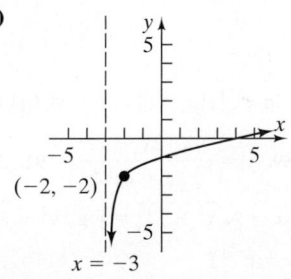

85. **(a)** Domain: $(-\infty, \infty)$

(b)

(c) Range: $(4, \infty)$
Horizontal asymptote: $y = 4$

(d) $f^{-1}(x) = 3\log_2(x-4)$
(e) Range of f: $(4, \infty)$
(f)

87. $\{9\}$ **89.** $\left\{\dfrac{7}{2}\right\}$ **91.** $\{2\}$ **93.** $\{5\}$ **95.** $\{3\}$ **97.** $\{2\}$ **99.** $\left\{\dfrac{\ln 10}{3}\right\}$ **101.** $\left\{\dfrac{\ln 8 - 5}{2}\right\}$ **103.** $\{-2\sqrt{2}, 2\sqrt{2}\}$ **105.** $\{-1\}$ **107.** $\left\{5\ln\dfrac{7}{5}\right\}$

109. $\left\{2 - \log\dfrac{5}{2}\right\}$ **111. (a)** $\left\{x \mid x > -\dfrac{1}{2}\right\}; \left(-\dfrac{1}{2}, \infty\right)$ **(b)** $4; (40, 4)$ **(c)** $4; (4, 2)$

113.

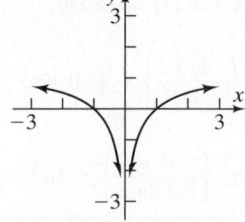

Domain: $\{x \mid x \neq 0\}$
Range: $(-\infty, \infty)$
Intercepts: $(-1, 0), (1, 0)$

115.

Domain: $\{x \mid x > 0\}$
Range: $\{y \mid y \geq 0\}$
Intercept: $(1, 0)$

117. (a) 1 **(b)** 2 **(c)** 3 **(d)** It increases **(e)** 0.000316 **(f)** 3.981×10^{-8}
119. (a) 5.97 km **(b)** 0.90 km **121. (a)** 6.93 min **(b)** 16.09 min **(c)** No, since $F(t)$ can never equal 1 **123.** $h \approx 2.29$, so the time between injections is about 2 hr, 17 min.

125. 0.2695 sec
0.8959 sec

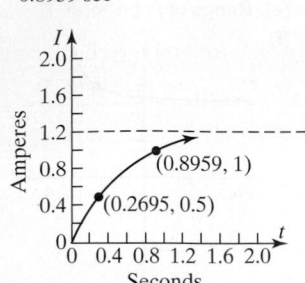

127. 50 decibels (dB)
129. 90 dB
131. 8.1 **133. (a)** $k = 11.216$ **(b)** 6.73 **(c)** 0.41% **(d)** 0.14%
135. Because $y = \log_1 x$ means $1^y = 1 = x$, which cannot be true for $x \neq 1$

6.5 Assess Your Understanding (page 457)

1. sum **2.** 7 **3.** $r \log_a M$ **4.** F **5.** F **6.** T **7.** 71 **9.** -4 **11.** 7 **13.** 1 **15.** 1 **17.** 2 **19.** $\dfrac{5}{4}$ **21.** 4 **23.** $a + b$ **25.** $b - a$ **27.** $3a$

29. $\dfrac{1}{5}(a + b)$ **31.** $2 + \log_5 x$ **33.** $3 \log_2 z$ **35.** $1 + \ln x$ **37.** $\ln x + x$ **39.** $2 \log_a u + 3 \log_a v$ **41.** $2 \ln x + \dfrac{1}{2}\ln(1 - x)$

43. $3 \log_2 x - \log_2(x - 3)$ **45.** $\log x + \log(x + 2) - 2 \log(x + 3)$ **47.** $\dfrac{1}{3}\ln(x - 2) + \dfrac{1}{3}\ln(x + 1) - \dfrac{2}{3}\ln(x + 4)$

49. $\ln 5 + \ln x + \dfrac{1}{2}\ln(1 + 3x) - 3 \ln(x - 4)$ **51.** $\log_5 u^3 v^4$ **53.** $\log_3\left(\dfrac{1}{x^{5/2}}\right)$ **55.** $\log_4\left[\dfrac{x - 1}{(x + 1)^4}\right]$ **57.** $-2 \ln(x - 1)$ **59.** $\log_2[x(3x - 2)^4]$

61. $\log_a\left(\dfrac{25x^6}{\sqrt{2x + 3}}\right)$ **63.** $\log_2\left[\dfrac{(x + 1)^2}{(x + 3)(x - 1)}\right]$ **65.** 2.771 **67.** -3.880 **69.** 5.615 **71.** 0.874

73. $y = \dfrac{\log x}{\log 4}$

75. $y = \dfrac{\log(x + 2)}{\log 2}$

77. $y = \dfrac{\log(x + 1)}{\log(x - 1)}$

79. (a) $(f \circ g)(x) = x$; $\{x | x \text{ is any real number}\}$ or $(-\infty, \infty)$
(b) $(g \circ f)(x) = x$; $\{x | x > 0\}$ or $(0, \infty)$ **(c)** 5 **(d)** $(f \circ h)(x) = \ln x^2$; $\{x | x \neq 0\}$ or $(-\infty, 0) \cup (0, \infty)$ **(e)** 2

81. $y = Cx$ **83.** $y = Cx(x + 1)$ **85.** $y = Ce^{3x}$ **87.** $y = Ce^{-4x} + 3$ **89.** $y = \dfrac{\sqrt[3]{C}(2x + 1)^{1/6}}{(x + 4)^{1/9}}$ **91.** 3 **93.** 1

95. $\log_a\left(x + \sqrt{x^2 - 1}\right) + \log_a\left(x - \sqrt{x^2 - 1}\right) = \log_a\left[\left(x + \sqrt{x^2 - 1}\right)\left(x - \sqrt{x^2 - 1}\right)\right] = \log_a[x^2 - (x^2 - 1)] = \log_a 1 = 0$

97. $\ln(1 + e^{2x}) = \ln[e^{2x}(e^{-2x} + 1)] = \ln e^{2x} + \ln(e^{-2x} + 1) = 2x + \ln(1 + e^{-2x})$

99. $y = f(x) = \log_a x$; $a^y = x$ implies $a^y = \left(\dfrac{1}{a}\right)^{-y} = x$, so $-y = \log_{1/a} x = -f(x)$.

101. $f(x) = \log_a x$; $f\left(\dfrac{1}{x}\right) = \log_a \dfrac{1}{x} = \log_a 1 - \log_a x = -f(x)$

103. $\log_a \dfrac{M}{N} = \log_a(M \cdot N^{-1}) = \log_a M + \log_a N^{-1} = \log_a M - \log_a N$, since $a^{\log_a N^{-1}} = N^{-1}$ implies $a^{-\log_a N^{-1}} = N$; i.e., $\log_a N = -\log_a N^{-1}$.

6.6 Assess Your Understanding (page 463)

5. $\{16\}$ **7.** $\left\{\dfrac{16}{5}\right\}$ **9.** $\{6\}$ **11.** $\{16\}$ **13.** $\left\{\dfrac{1}{3}\right\}$ **15.** $\{3\}$ **17.** $\{5\}$ **19.** $\left\{\dfrac{21}{8}\right\}$ **21.** $\{-6\}$ **23.** $\{-2\}$ **25.** $\left\{-1 + \sqrt{1 + e^4}\right\} \approx \{6.456\}$

27. $\left\{\dfrac{-5 + 3\sqrt{5}}{2}\right\} \approx \{0.854\}$ **29.** $\{2\}$ **31.** $\left\{\dfrac{9}{2}\right\}$ **33.** $\{8\}$ **35.** $\{\log_2 10\} = \left\{\dfrac{\ln 10}{\ln 2}\right\} \approx \{3.322\}$ **37.** $\{-\log_8 1.2\} = \left\{-\dfrac{\ln 1.2}{\ln 8}\right\} \approx \{-0.088\}$

39. $\left\{\dfrac{1}{3}\log_2 \dfrac{8}{5}\right\} = \left\{\dfrac{\ln \dfrac{8}{5}}{3 \ln 2}\right\} \approx \{0.226\}$ **41.** $\left\{\dfrac{\ln 3}{2 \ln 3 + \ln 4}\right\} \approx \{0.307\}$ **43.** $\left\{\dfrac{\ln 7}{\ln 0.6 + \ln 7}\right\} \approx \{1.356\}$ **45.** $\{0\}$ **47.** $\left\{\dfrac{\ln \pi}{1 + \ln \pi}\right\} \approx \{0.534\}$

49. $\left\{\dfrac{\ln 3}{\ln 2}\right\} \approx \{1.585\}$ **51.** $\{0\}$ **53.** $\left\{\log_4\left(-2 + \sqrt{7}\right)\right\} \approx \{-0.315\}$ **55.** $\{\log_5 4\} \approx \{0.861\}$ **57.** No real solution **59.** $\{\log_4 5\} \approx \{1.161\}$

61. $\{2.79\}$ **63.** $\{-0.57\}$ **65.** $\{-0.70\}$ **67.** $\{0.57\}$ **69.** $\{0.39, 1.00\}$ **71.** $\{1.32\}$ **73.** $\{1.31\}$ **75.** $\{1\}$ **77.** $\{16\}$ **79.** $\left\{-1, \dfrac{2}{3}\right\}$ **81.** $\{0\}$

83. $\left\{\ln\left(2 + \sqrt{5}\right)\right\} \approx \{1.444\}$ **85.** $\left\{e^{\frac{\ln 5 \cdot \ln 3}{\ln 15}}\right\} \approx \{1.921\}$ **87. (a)** $\{5\}, (5, 3)$ **(b)** $\{5\}, (5, 4)$ **(c)** $\{1\}$, yes, at $(1, 2)$ **(d)** $\{5\}$ **(e)** $\left\{-\dfrac{1}{11}\right\}$

89. (a)

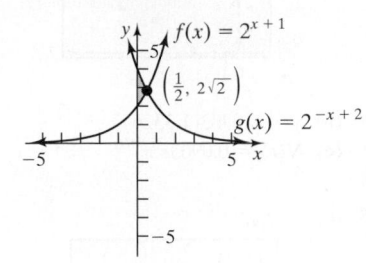

(b) $(0.710, 6.541)$
(c) $\{x | x > 0.710\}$ or $(0.710, \infty)$

91. (a), (b)

(c) $(\log_3 10, 10)$

93. (a), (b)

(c) $\left(\dfrac{1}{2}, 2\sqrt{2}\right)$

95. (a)

(b) $\{x | x < 2\}$ or $(-\infty, 2)$

97. (a) 2010 **(b)** 2027 **99. (a)** After 2.4 yr **(b)** After 6.5 yr **(c)** After 10 yr

6.7 Assess Your Understanding *(page 472)*

3. $108.29 **5.** $609.50 **7.** $697.09 **9.** $12.46 **11.** $125.23 **13.** $88.72 **15.** $860.72 **17.** $554.09 **19.** $59.71 **21.** $361.93 **23.** $6\dfrac{1}{4}\%$ compounded annually **25.** 9% compounded monthly **27.** 5.095% **29.** 5.127% **31.** 25.992% **33.** 24.573% **35. (a)** About 8.69 yr **(b)** About 8.66 yr

37. 6.823% **39.** 5.09 yr; 5.07 yr **41.** 15.27 yr or 15 yr, 3 mo **43.** $104,335 **45.** $12,910.62 **47.** About $30.17 per share or $3017 **49.** Not quite. Jim will have $1057.60. The second bank gives a better deal, since Jim will have $1060.62 after 1 yr. **51.** Will has $11,632.73; Henry has $10,947.89.

53. (a) $49,581 **(b)** $33,235 **55.** Approximately $602 billion **57.** $940.90 **59.** 2.53% **61.** 34.31 yr **63. (a)** $1364.62 **(b)** $1353.35

65. $4631.93

67. (a) 6.12 yr

(b) 18.45 yr

(c) $mP = P\left(1 + \dfrac{r}{n}\right)^{nt}$

$m = \left(1 + \dfrac{r}{n}\right)^{nt}$

$\ln m = \ln\left(1 + \dfrac{r}{n}\right)^{nt} = nt \ln\left(1 + \dfrac{r}{n}\right)$

$t = \dfrac{\ln m}{n \ln\left(1 + \dfrac{r}{n}\right)}$

69. (a) 2.51% **(b)** In 2022 or after 27 yr

71. 22.7 yr

6.8 Assess Your Understanding *(page 484)*

1. (a) 500 insects **(b)** $0.02 = 2\%$ per day **(c)** About 611 insects **(d)** After about 23.5 days **(e)** After about 34.7 days **3. (a)** $-0.0244 = -2.44\%$ per year **(b)** About 391.7 g **(c)** After about 9.1 yr **(d)** 28.4 yr **5. (a)** $N(t) = N_0 e^{kt}$ **(b)** 5832 **(c)** 3.9 days **7. (a)** $N(t) = N_0 e^{kt}$ **(b)** 25,198
9. 9.797 g **11.** 9727 yr ago **13. (a)** 5:18 PM **(b)** About 14.3 min **(c)** The temperature of the pan approaches 70°F. **15.** 18.63°C; 25.1°C
17. 1.7 ppm; 7.17 days or 172 hr **19.** 0.26 M, 6.58 ln or 395 min **21.** 26.6 days **23. (a)** 90% **(b)** 12.86% **(c)** 40.70% **(d)** 2012 **25. (a)** 1000 g
(b) 43.9% per hour **(c)** 30 g **(d)** 616.6 g **(e)** After 9.85 hr **(f)** About 7.9 hr
27. (a) 9.23×10^{-3}, or about 0 **(d)** 57.91°, 43.99°, 30.07°
(b) 0.81, or about 1
(c) 5.01, or about 5

6.9 Assess Your Understanding *(page 491)*

1. (a)

(d)

3. (a)

(d)

(b) $y = 0.0903(1.3326)^x$
(c) $N(t) = 0.0903e^{0.2871t}$

(e) 0.67
(f) After about 7.37 hr

(b) $y = 100.326(0.8769)^x$
(c) $A(t) = 100.326e^{-0.1314t}$

(e) 5.3 weeks **(f)** 0.14 g
(g) After about 12.3 weeks

5. (a)

(b)

$y = 751.470(0.95188)^x$

(c) $A(t) = 751.470e^{-0.04932t}$

(d)

(e) Approximately 359 billion cigarettes
(f) In 2019 or about 24 yr after 1995

7. (a)

(b) $y = 32,741.02 - 6070.96 \ln x$ **(c)**

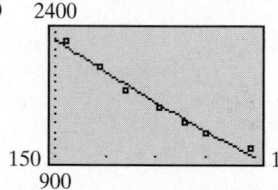

(d) Approximately 168 computers

9. (a)

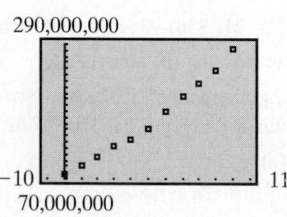

(b) $y = \dfrac{799,475,916.5}{1 + 9.1968e^{-0.0160x}}$

(c)

(d) 799,475,917
(e) Approximately 292,251,184
(f) 2007

11. (a)

(b)

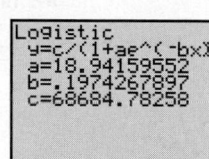

$y = \dfrac{68,684.8}{1 + 18.9416e^{-0.1974x}}$

(c)

(d) The carrying capacity is about 68,685,000 subscribers.

(e) In 2015 there will be a predicted 68,505,000 subscribers.

Review Exercises *(page 496)*

1. (a) -26 **(b)** -241 **(c)** 16 **(d)** -1 **3. (a)** $\sqrt{11}$ **(b)** 1 **(c)** $\sqrt{\sqrt{6} + 2}$ **(d)** 19 **5. (a)** e^4 **(b)** $3e^{-2} - 2$ **(c)** e^{e^4} **(d)** -17

7. $(f \circ g)(x) = 1 - 3x$, all real numbers; $(g \circ f)(x) = 7 - 3x$, all real numbers;
$(f \circ f)(x) = x$, all real numbers; $(g \circ g)(x) = 9x + 4$, all real numbers

9. $(f \circ g)(x) = 27x^2 + 3|x| + 1$, all real numbers; $(g \circ f)(x) = 3|3x^2 + x + 1|$, all real numbers;
$(f \circ f)(x) = 3(3x^2 + x + 1)^2 + 3x^2 + x + 2$, all real numbers; $(g \circ g)(x) = 9|x|$, all real numbers

11. $(f \circ g)(x) = \dfrac{1 + x}{1 - x}, \{x | x \neq 0, x \neq 1\}; (g \circ f)(x) = \dfrac{x - 1}{x + 1}, \{x | x \neq -1, x \neq 1\}; (f \circ f)(x) = x, \{x | x \neq 1\}; (g \circ g)(x) = x, \{x | x \neq 0\}$

13. (a) One-to-one **(b)** $\{(2, 1), (5, 3), (8, 5), (10, 6)\}$

15.

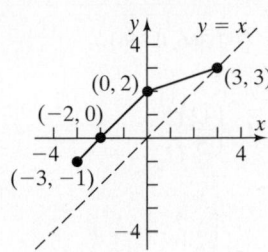

17. $f^{-1}(x) = \dfrac{2x + 3}{5x - 2}$

$$f(f^{-1}(x)) = \dfrac{2\left(\dfrac{2x + 3}{5x - 2}\right) + 3}{5\left(\dfrac{2x + 3}{5x - 2}\right) - 2} = x$$

$$f^{-1}(f(x)) = \dfrac{2\left(\dfrac{2x + 3}{5x - 2}\right) + 3}{5\left(\dfrac{2x + 3}{5x - 2}\right) - 2} = x$$

Domain of f = range of f^{-1} = all real numbers except $\dfrac{2}{5}$

Range of f = domain of f^{-1} = all real numbers except $\dfrac{2}{5}$

19. $f^{-1}(x) = \dfrac{x + 1}{x}$

$$f(f^{-1}(x)) = \dfrac{1}{\dfrac{x + 1}{x} - 1} = x$$

$$f^{-1}(f(x)) = \dfrac{\dfrac{1}{x - 1} + 1}{\dfrac{1}{x - 1}} = x$$

Domain of f = range of f^{-1} = all real numbers except 1

Range of f = domain of f^{-1} = all real numbers except 0

21. $f^{-1}(x) = \dfrac{27}{x^3}$

$$f(f^{-1}(x)) = \dfrac{3}{\left(\dfrac{27}{x^3}\right)^{1/3}} = x$$

$$f^{-1}(f(x)) = \dfrac{27}{\left(\dfrac{3}{x^{1/3}}\right)^3} = x$$

Domain of f = range of f^{-1} = all real numbers except 0

Range of f = domain of f^{-1} = all real numbers except 0

23. (a) 81 **(b)** 2 **(c)** $\dfrac{1}{9}$ **(d)** -3 **25.** $\log_5 z = 2$ **27.** $5^{13} = u$ **29.** $\left\{x \,\middle|\, x > \dfrac{2}{3}\right\}; \left(\dfrac{2}{3}, \infty\right)$ **31.** $\{x \mid x < 1 \text{ or } x > 2\}; (-\infty, 1) \cup (2, \infty)$ **33.** -3

35. $\sqrt{2}$ **37.** 0.4 **39.** $\log_3 u + 2\log_3 v - \log_3 w$ **41.** $2 \log x + \dfrac{1}{2}\log(x^3 + 1)$ **43.** $\ln x + \dfrac{1}{3}\ln(x^2 + 1) - \ln(x - 3)$ **45.** $\dfrac{25}{4}\log_4 x$ **47.** $-2\ln(x + 1)$

49. $\log\left(\dfrac{4x^3}{[(x + 3)(x - 2)]^{1/2}}\right)$ **51.** 2.124 **53.**

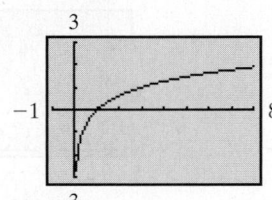

55. (a) Domain of f: $(-\infty, \infty)$

(b)

(c) Range of f: $(0, \infty)$

Horizontal asymptote: $y = 0$

(d) $f^{-1}(x) = 3 + \log_2 x$

(e) Range of f: $(0, \infty)$

(f)

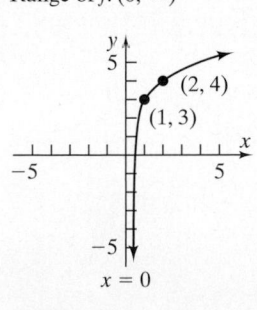

57. (a) Domain of f: $(-\infty, \infty)$

(b)

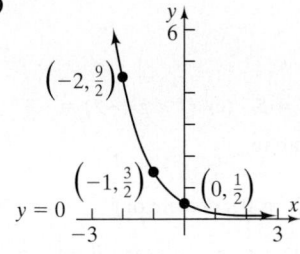

(c) Range of f: $(0, \infty)$

Horizontal asymptote: $y = 0$

(d) $f^{-1}(x) = -\log_3(2x)$

(e) Range of f: $(0, \infty)$

(f)

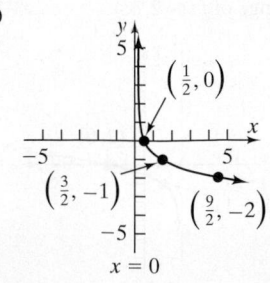

59. (a) Domain of f: $(-\infty, \infty)$

(b)

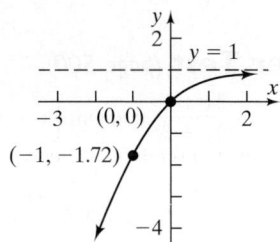

(c) Range of f: $(-\infty, 1)$

Horizontal asymptote: $y = 1$

(d) $f^{-1}(x) = -\ln(1 - x)$

(e) Range of f: $(-\infty, 1)$

(f)

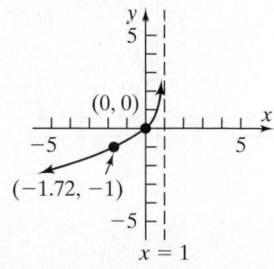

61. (a) Domain of f: $(-3, \infty)$
(b)

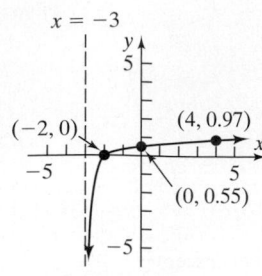

(c) Range of $(-\infty, \infty)$
Vertical asymptote: $x = -3$

(d) $f^{-1}(x) = e^{2x} - 3$

(e) Range of f: $(-\infty, \infty)$

(f)

63. $\left\{\dfrac{1}{4}\right\}$ **65.** $\left\{\dfrac{-1-\sqrt{3}}{2}, \dfrac{-1+\sqrt{3}}{2}\right\} \approx \{-1.366, 0.366\}$

67. $\left\{\dfrac{1}{4}\right\}$ **69.** $\left\{\dfrac{2\ln 3}{\ln 5 - \ln 3}\right\} \approx \{4.301\}$ **71.** $\left\{\dfrac{12}{5}\right\}$

73. $\{83\}$ **75.** $\left\{\dfrac{1}{2}, -3\right\}$ **77.** $\{-1\}$

79. $\{1 - \ln 5\} \approx \{-0.609\}$

81. $\left\{\log_3\left(-2 + \sqrt{7}\right)\right\} = \left\{\dfrac{\ln\left(-2+\sqrt{7}\right)}{\ln 3}\right\} \approx \{-0.398\}$

83. (a), (e)

(b) $\{3\}$, $(6, 3)$ **(c)** $\{10\}$, $(10, 4)$ **(d)** $\left\{x \,\middle|\, x > \dfrac{5}{2}\right\}$ or $\left(\dfrac{5}{2}, \infty\right)$

(e) $f^{-1}(x) = 2^{x-1} + 2$

85. 3229.5 m **87. (a)** 37.3 W **(b)** 6.9 dB **89. (a)** 9.85 yr **(b)** 4.27 yr **91.** \$41,668.97 **93.** 24,203 yr ago **95.** 7,237,249,063 **97.** \$483.67 billion

99. (a) 165.1

(c) 165.1

(b) $y = 165.73(0.9951)^x$

(d) Approximately 83 sec

101. (a)

(c)

(b) $C = \dfrac{46.93}{1 + 21.273e^{-0.7306t}}$

(d) About 47 people; 50 people

(e) 2.4 days; during the hour 10 of day 3

(f) 9.5 days

Chapter Test *(page 500)*

1. (a) $f \circ g = \dfrac{2x + 7}{2x + 3}$; domain: $\left\{x \,\middle|\, x \neq -\dfrac{3}{2}\right\}$ **(b)** $(g \circ f)(-2) = 5$ **(c)** $(f \circ g)(-2) = -3$

2. (a) The function is not one-to-one. **(b)** The function is one-to-one.

3. $f^{-1}(x) = \dfrac{2 + 5x}{3x}$; domain of $f = \left\{x \,\middle|\, x \neq \dfrac{5}{3}\right\}$, range of $f = \{y \,|\, y \neq 0\}$; domain of $f^{-1} = \{x \,|\, x \neq 0\}$; range of $f^{-1} = \left\{y \,\middle|\, y \neq \dfrac{5}{3}\right\}$

4. The point $(-5, 3)$ must be on the graph of f^{-1}. **5.** $x = 5$ **6.** $b = 4$ **7.** $x = 625$ **8.** $e^3 + 2 \approx 22.086$ **9.** $\log 20 \approx 1.301$

10. $\log_3 21 = \dfrac{\ln 21}{\ln 3} \approx 2.771$ **11.** $\ln 133 \approx 4.890$

12. (a) Domain of f: $\{x \,|\, -\infty < x < \infty\}$ or $(-\infty, \infty)$
(b)

(c) Range of f: $\{y \,|\, y > -2\}$ or $(-2, \infty)$ Horizontal asymptote: $y = -2$

(d) $f^{-1}(x) = \log_4(x + 2) - 1$

(e) Range of f: $(-2, \infty)$

(f)

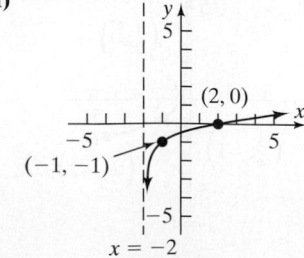

13. (a) Domain of f: $\{x|x > 2\}$ or $(2, \infty)$ **(d)** $f^{-1}(x) = 5^{1-x} + 2$ **14.** $\{1\}$ **15.** $\{91\}$ **16.** $\{-\ln 2\} \approx \{-0.693\}$

(b)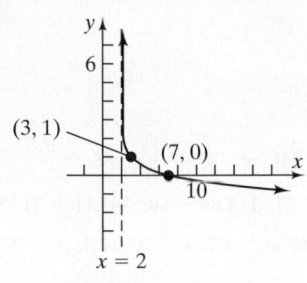

(e) Range of f: $(-\infty, \infty)$

(f)

17. $\left\{\dfrac{1 - \sqrt{13}}{2}, \dfrac{1 + \sqrt{13}}{2}\right\} \approx \{-1.303, 2.303\}$

18. $\left\{\dfrac{3 \ln 7}{1 - \ln 7}\right\} \approx \{-6.172\}$

19. $\{2\sqrt{6}\} \approx \{4.899\}$

20. $2 + 3\log_2 x - \log_2(x - 6) - \log_2(x + 3)$

21. About 250.39 days

22. (a) \$1033.82 **(b)** \$963.42 **(c)** 11.9 yr

23. (a) About 83 dB

(c) Range of f: $\{y|-\infty < y < \infty\}$
or $(-\infty, \infty)$; Vertical symptote: $x = 2$

(b) The pain threshold will be exceeded
if 31,623 people shouted at the
same time.

Cumulative Review (page 501)

1. Yes; no **2. (a)** 10 **(b)** $2x^2 + 3x + 1$ **(c)** $2x^2 + 4xh + 2h^2 - 3x - 3h + 1$ **3.** $\left(\dfrac{1}{2}, \dfrac{\sqrt{3}}{2}\right)$ is on the graph. **4.** $\{-26\}$

5.

6. (a)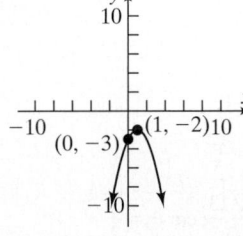

(b) $\{x|-\infty < x < \infty\}$

7. $f(x) = 2(x - 4)^2 - 8 = 2x^2 - 16x + 24$

8.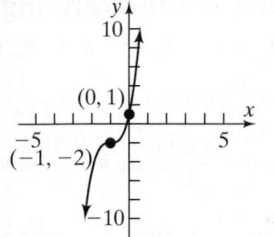

9. $f(g(x)) = \dfrac{4}{(x - 3)^2} + 2$; domain: $\{x|x \neq 3\}$; 3

10. (a) Zeros: $-4, -\dfrac{1}{4}, 2$

(b) x-intercepts: $-4, -\dfrac{1}{4}, 2$; y-intercept: -8

(c) Local maximum of 60.75 occurs at $x = -2.5$
Local minimum of -25 occurs at $x = 1$

(d)

11. (a), (c)

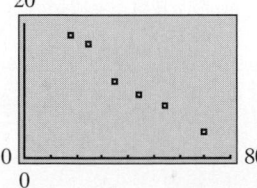

Domain g = range g^{-1} = $(-\infty, \infty)$
Range g = domain g^{-1} = $(2, \infty)$
(b) $g^{-1}(x) = \log_3(x - 2)$

12. $\left\{-\dfrac{3}{2}\right\}$ **13.** $\{2\}$

14. (a) $\{-1\}$
(b) $\{x|x > -1\}$ or $(-1, \infty)$
(c) $\{25\}$

15. (a) **(b)** Answers will vary.

CHAPTER 7 Trigonometric Functions

7.1 Assess Your Understanding (page 513)

3. Standard position **4.** $r\theta; \dfrac{1}{2}r^2\theta$ **5.** $\dfrac{s}{t}; \dfrac{\theta}{t}$ **6.** F **7.** T **8.** T **9.** T **10.** F

11. **13.** **15.**

17. **19.** **21.**

23. $40.17°$ **25.** $1.03°$ **27.** $9.15°$ **29.** $40°19'12''$ **31.** $18°15'18''$ **33.** $19°59'24''$ **35.** $\dfrac{\pi}{6}$ **37.** $\dfrac{4\pi}{3}$ **39.** $-\dfrac{\pi}{3}$ **41.** π **43.** $-\dfrac{3\pi}{4}$ **45.** $-\dfrac{\pi}{2}$

47. $60°$ **49.** $-225°$ **51.** $90°$ **53.** $15°$ **55.** $-90°$ **57.** $-30°$ **59.** 0.30 **61.** -0.70 **63.** 2.18 **65.** $179.91°$ **67.** $114.59°$ **69.** $362.11°$ **71.** 5 m

73. 6 ft **75.** 0.6 radian **77.** $\dfrac{\pi}{3} \approx 1.047$ in. **79.** 25 m^2 **81.** $2\sqrt{3} \approx 3.464$ ft **83.** 0.24 radian **85.** $\dfrac{\pi}{3} \approx 1.047$ in.2 **87.** $s = 2.094$ ft; $A = 2.094$ ft^2

89. $s = 14.661$ yd; $A = 87.965$ yd^2 **91.** $3\pi \approx 9.42$ in; $5\pi \approx 15.71$ in. **93.** $2\pi \approx 6.28$ m^2 **95.** $\dfrac{675\pi}{2} \approx 1060.29$ ft^2

97. $\omega = \dfrac{1}{60}$ radian/s; $v = \dfrac{1}{12}$ cm/s **99.** Approximately 452.5 rpm **101.** Approximately 359 mi **103.** Approximately 898 mi/h

105. Approximately 2292 mi/h **107.** $\dfrac{3}{4}$ rpm **109.** Approximately 2.86 mi/h **111.** Approximately 31.47 rpm **113.** Approximately 1037 mi/h

115. Radius ≈ 3979 mi; circumference $\approx 25{,}000$ mi **117.** $v_1 = r_1\omega_1,\ v_2 = r_2\omega_2$, and $v_1 = v_2$, so $r_1\omega_1 = r_2\omega_2 \Rightarrow \dfrac{r_1}{r_2} = \dfrac{\omega_2}{\omega_1}$.

7.2 Assess Your Understanding *(page 525)*

3. complementary **4.** cosine **5.** $62°$ **6.** 1 **7.** T **8.** F **9.** T **10.** F

11. $\sin\theta = \dfrac{5}{13};\ \cos\theta = \dfrac{12}{13};\ \tan\theta = \dfrac{5}{12};\ \csc\theta = \dfrac{13}{5};\ \sec\theta = \dfrac{13}{12};\ \cot\theta = \dfrac{12}{5}$

13. $\sin\theta = \dfrac{2\sqrt{13}}{13};\ \cos\theta = \dfrac{3\sqrt{13}}{13};\ \tan\theta = \dfrac{2}{3};\ \csc\theta = \dfrac{\sqrt{13}}{2};\ \sec\theta = \dfrac{\sqrt{13}}{3};\ \cot\theta = \dfrac{3}{2}$

15. $\sin\theta = \dfrac{\sqrt{3}}{2};\ \cos\theta = \dfrac{1}{2};\ \tan\theta = \sqrt{3};\ \csc\theta = \dfrac{2\sqrt{3}}{3};\ \sec\theta = 2;\ \cot\theta = \dfrac{\sqrt{3}}{3}$ **17.** $\sin\theta = \dfrac{\sqrt{6}}{3};\ \cos\theta = \dfrac{\sqrt{3}}{3};\ \tan\theta = \sqrt{2};\ \csc\theta = \dfrac{\sqrt{6}}{2};$

$\sec\theta = \sqrt{3};\ \cot\theta = \dfrac{\sqrt{2}}{2}$ **19.** $\sin\theta = \dfrac{\sqrt{5}}{5};\ \cos\theta = \dfrac{2\sqrt{5}}{5};\ \tan\theta = \dfrac{1}{2};\ \csc\theta = \sqrt{5};\ \sec\theta = \dfrac{\sqrt{5}}{2};\ \cot\theta = 2$

21. $\tan\theta = \dfrac{\sqrt{3}}{3};\ \csc\theta = 2;\ \sec\theta = \dfrac{2\sqrt{3}}{3};\ \cot\theta = \sqrt{3}$ **23.** $\tan\theta = \dfrac{2\sqrt{5}}{5};\ \csc\theta = \dfrac{3}{2};\ \sec\theta = \dfrac{3\sqrt{5}}{5};\ \cot\theta = \dfrac{\sqrt{5}}{2}$

25. $\cos\theta = \dfrac{\sqrt{2}}{2};\ \tan\theta = 1;\ \csc\theta = \sqrt{2};\ \sec\theta = \sqrt{2};\ \cot\theta = 1$ **27.** $\sin\theta = \dfrac{2\sqrt{2}}{3};\ \tan\theta = 2\sqrt{2};\ \csc\theta = \dfrac{3\sqrt{2}}{4};\ \sec\theta = 3;\ \cot\theta = \dfrac{\sqrt{2}}{4}$

29. $\sin\theta = \dfrac{\sqrt{5}}{5};\ \cos\theta = \dfrac{2\sqrt{5}}{5};\ \csc\theta = \sqrt{5};\ \sec\theta = \dfrac{\sqrt{5}}{2};\ \cot\theta = 2$ **31.** $\sin\theta = \dfrac{2\sqrt{2}}{3};\ \cos\theta = \dfrac{1}{3};\ \tan\theta = 2\sqrt{2};\ \csc\theta = \dfrac{3\sqrt{2}}{4};\ \cot\theta = \dfrac{\sqrt{2}}{4}$

33. $\sin\theta = \dfrac{\sqrt{6}}{3};\ \cos\theta = \dfrac{\sqrt{3}}{3};\ \csc\theta = \dfrac{\sqrt{6}}{2};\ \sec\theta = \sqrt{3};\ \cot\theta = \dfrac{\sqrt{2}}{2}$ **35.** $\sin\theta = \dfrac{1}{2};\ \cos\theta = \dfrac{\sqrt{3}}{2};\ \tan\theta = \dfrac{\sqrt{3}}{3};\ \sec\theta = \dfrac{2\sqrt{3}}{3};\ \cot\theta = \sqrt{3}$

37. 1 **39.** 1 **41.** 0 **43.** 0 **45.** 1 **47.** 0 **49.** 0 **51.** 1 **53.** 1 **55. (a)** $\dfrac{1}{2}$ **(b)** $\dfrac{3}{4}$ **(c)** 2 **(d)** 2 **57. (a)** 17 **(b)** $\dfrac{1}{4}$ **(c)** 4 **(d)** $\dfrac{17}{16}$

59. (a) $\dfrac{1}{4}$ **(b)** 15 **(c)** 4 **(d)** $\dfrac{16}{15}$ **61. (a)** 0.78 **(b)** 0.79 **(c)** 1.27 **(d)** 1.27 **(e)** 1.61 **(f)** 0.78 **(g)** 0.62 **(h)** 1.27 **63.** 0.6 **65.** $20°$

67. (a) 10 min

(b) 20 min

(c) $T(\theta) = 5\left(1 - \dfrac{1}{3\tan\theta} + \dfrac{1}{\sin\theta}\right)$

(d) Approximately 15.8 min

(e) Approximately 10.4 min

(f)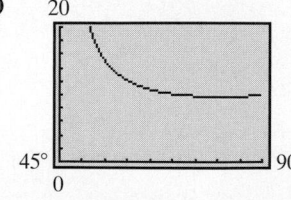

$70.5°$; 177 ft, 9.7 min

69. (a) $Z = 200\sqrt{13} \approx 721.1$ ohms

(b) $\tan\phi = \dfrac{2}{3};\ \sin\phi = \dfrac{2\sqrt{13}}{13};\ \cos\phi = \dfrac{3\sqrt{13}}{13};\ \cot\phi = \dfrac{3}{2};\ \csc\phi = \dfrac{\sqrt{13}}{2};\ \sec\phi = \dfrac{\sqrt{13}}{3}$

71. (a) $|OA| = |OC| = 1$; angle OAC = angle OCA; angle OAC + angle OAC + $180° - \theta = 180°$; angle $OAC = \dfrac{\theta}{2}$

(b) $\sin\theta = \dfrac{|CD|}{|OC|} = |CD|;\ \cos\theta = \dfrac{|OD|}{|OC|} = |OD|$ **(c)** $\tan\dfrac{\theta}{2} = \dfrac{|CD|}{|AD|} = \dfrac{\sin\theta}{1 + |OD|} = \dfrac{\sin\theta}{1 + \cos\theta}$

73. $h = x\tan\theta$ and $h = (1 - x)\tan n\theta$; thus, $x\tan\theta = (1 - x)\tan n\theta$, so $x = \dfrac{\tan n\theta}{\tan\theta + \tan n\theta}$.

75. (a) Area $\triangle OAC = \frac{1}{2}|AC|\,|OC| = \frac{1}{2}\cdot\frac{|AC|}{1}\cdot\frac{|OC|}{1} = \frac{1}{2}\sin\alpha\cos\alpha$ **(b)** Area $\triangle OCB = \frac{1}{2}|BC|\,|OC| = \frac{1}{2}|OB|^2\frac{|BC|}{|OB|}\cdot\frac{|OC|}{|OB|} = \frac{1}{2}|OB|^2\sin\beta\cos\beta$

(c) Area $\triangle OAB = \frac{1}{2}|BD|\,|OA| = \frac{1}{2}|OB|\frac{|BD|}{|OB|} = \frac{1}{2}|OB|\sin(\alpha+\beta)$ **(d)** $\dfrac{\cos\alpha}{\cos\beta} = \dfrac{1}{\dfrac{|OC|}{|OB|}} = |OB|$

(e) Area $\triangle OAB = $ area $\triangle OAC + $ area $\triangle OCB$

$$\frac{1}{2}|OB|\sin(\alpha+\beta) = \frac{1}{2}\sin\alpha\cos\alpha + \frac{1}{2}|OB|^2\sin\beta\cos\beta$$

$$\sin(\alpha+\beta) = \frac{\sin\alpha\cos\alpha + |OB|^2\sin\beta\cos\beta}{|OB|}$$

$$\sin(\alpha+\beta) = \frac{\sin\alpha(|OB|\cos\beta) + |OB|^2\sin\beta\left(\dfrac{\cos\alpha}{|OB|}\right)}{|OB|}$$

$$\sin(\alpha+\beta) = \sin\alpha\cos\beta + \cos\alpha\sin\beta$$

77. $\sin\alpha = \tan\alpha\,\cos\alpha = \cos\beta\cos\alpha$
$\qquad = \cos\beta\tan\beta = \sin\beta;$
$$\sin^2\alpha + \cos^2\alpha = 1$$
$$\sin^2\alpha + \tan^2\beta = 1$$
$$\sin^2\alpha + \frac{\sin^2\beta}{\cos^2\beta} = 1$$
$$\sin^2\alpha + \frac{\sin^2\alpha}{1-\sin^2\alpha} = 1$$
$$\sin^2\alpha - \sin^4\alpha + \sin^2\alpha = 1 - \sin^2\alpha$$
$$\sin^4\alpha - 3\sin^2\alpha + 1 = 0$$
$$\sin^2\alpha = \frac{3\pm\sqrt{5}}{2}$$
$$\sin^2\alpha = \frac{3-\sqrt{5}}{2}\quad \sin\alpha = \sqrt{\frac{3-\sqrt{5}}{2}}$$

79. Since $a^2 + b^2 = c^2, a > 0, b > 0,$
then $0 < a^2 < c^2$ or $0 < a < c.$
Thus $0 < \dfrac{a}{c} < 1$ and $0 < \cos\theta < 1$
or $\dfrac{\cos\theta}{\cos\theta} < \dfrac{1}{\cos\theta}$; therefore, $\sec\theta > 1.$

7.3 Assess Your Understanding (page 536)

1. $\dfrac{3}{2}$ **2.** 0.91 **3.** T **4.** F **5.** $\sin 45° = \dfrac{\sqrt{2}}{2}$; $\cos 45° = \dfrac{\sqrt{2}}{2}$; $\tan 45° = 1$; $\csc 45° = \sqrt{2}$; $\sec 45° = \sqrt{2}$; $\cot 45° = 1$ **7.** $\dfrac{\sqrt{3}}{2}$ **9.** $\dfrac{1}{2}$ **11.** $\dfrac{3}{4}$

13. $\sqrt{3}$ **15.** $\dfrac{\sqrt{3}}{4}$ **17.** $\sqrt{2}$ **19.** 2 **21.** $\sqrt{2}+\dfrac{4\sqrt{3}}{3}$ **23.** $-\dfrac{8}{3}$ **25.** $\dfrac{1}{2}$ **27.** 0 **29.** 0.47 **31.** 0.38 **33.** 1.33 **35.** 0.31 **37.** 3.73 **39.** 1.04

41. 0.84 **43.** 0.02 **45.** 0.31 **47.** $R \approx 310.56$ ft; $H \approx 77.64$ ft **49.** $R \approx 19{,}541.95$ m; $H \approx 2278.14$ m **51. (a)** 1.20 s **(b)** 1.12 s **(c)** 1.20 s

53. (a) $T(\theta) = 1 + \dfrac{2}{3\sin\theta} - \dfrac{1}{4\tan\theta}$ **55.** 4.59 in.; 6.55 in. **57. (a)** 5.52 in. or 11.83 in. **59.** 70.02 ft **61.** 985.91 ft

(b) 1.9 h; 0.57 h **(c)** 1.69 h; 0.75 h **63.** 137.37m **65.** 1978.09 ft **67.** 60.27 ft **69.** 203.52 ft **71.** 554.52 ft **73.** 3.83 mi

(d) 1.63 h; 0.86 h **(e)** 1.67 h **(f)** 2.75 h **75.** The white ball should hit the top cushion 4.125 ft from the upper left corner.

(g)

67.98°; 1.62 h; 0.9 h

77.

θ	0.5	0.4	0.2	0.1	0.01	0.001	0.0001	0.00001
$f(\theta) = \dfrac{\sin\theta}{\theta}$	0.9589	0.9735	0.9933	0.9983	1.0000	1.0000	1.0000	1.0000

$\dfrac{\sin\theta}{\theta}$ approaches 1 as $\theta \to 0.$

79. 1 **81.** $\dfrac{\sqrt{2}}{2}$

7.4 Assess Your Understanding (page 548)

1. tangent; cotangent **2.** coterminal **3.** 60° **4.** F **5.** T **6.** T **7.** 60° **8.** I, IV **9.** $\dfrac{\pi}{2},\dfrac{3\pi}{2}$ **10.** $\dfrac{\pi}{3}$ **11.** $\sin\theta = \dfrac{4}{5}$; $\cos\theta = -\dfrac{3}{5}$;

$\tan\theta = -\dfrac{4}{3}$; $\csc\theta = \dfrac{5}{4}$; $\sec\theta = -\dfrac{5}{3}$; $\cot\theta = -\dfrac{3}{4}$ **13.** $\sin\theta = -\dfrac{3\sqrt{13}}{13}$; $\cos\theta = \dfrac{2\sqrt{13}}{13}$; $\tan\theta = -\dfrac{3}{2}$; $\csc\theta = -\dfrac{\sqrt{13}}{3}$; $\sec\theta = \dfrac{\sqrt{13}}{2}$; $\cot\theta = -\dfrac{2}{3}$

15. $\sin\theta = -\dfrac{\sqrt{2}}{2}$; $\cos\theta = -\dfrac{\sqrt{2}}{2}$; $\tan\theta = 1$; $\csc\theta = -\sqrt{2}$; $\sec\theta = -\sqrt{2}$; $\cot\theta = 1$

17. $\sin\theta = \dfrac{1}{2}$; $\cos\theta = \dfrac{\sqrt{3}}{2}$; $\tan\theta = \dfrac{\sqrt{3}}{3}$; $\csc\theta = 2$; $\sec\theta = \dfrac{2\sqrt{3}}{3}$; $\cot\theta = \sqrt{3}$

19. $\sin\theta = -\dfrac{\sqrt{2}}{2}$; $\cos\theta = \dfrac{\sqrt{2}}{2}$; $\tan\theta = -1$; $\csc\theta = -\sqrt{2}$; $\sec\theta = \sqrt{2}$; $\cot\theta = -1$ **21.** $\dfrac{\sqrt{2}}{2}$ **23.** 1 **25.** 1 **27.** $\sqrt{3}$

29. $\dfrac{\sqrt{2}}{2}$ **31.** 0 **33.** II **35.** IV **37.** IV **39.** III **41.** 30° **43.** 60° **45.** 30° **47.** $\dfrac{\pi}{4}$ **49.** $\dfrac{\pi}{3}$ **51.** 45° **53.** $\dfrac{\pi}{3}$ **55.** 80° **57.** $\dfrac{\pi}{4}$ **59.** $\dfrac{1}{2}$

61. $\dfrac{\sqrt{2}}{2}$ **63.** $\dfrac{1}{2}$ **65.** $\dfrac{\sqrt{2}}{2}$ **67.** -2 **69.** $-\sqrt{3}$ **71.** $\dfrac{\sqrt{2}}{2}$ **73.** $\sqrt{3}$ **75.** $-\dfrac{\sqrt{2}}{2}$ **77.** $-\dfrac{\sqrt{3}}{2}$ **79.** $-\sqrt{3}$ **81.** $\sqrt{2}$ **83.** 0 **85.** 0 **87.** -1

89. $\cos\theta = -\dfrac{5}{13}$; $\tan\theta = -\dfrac{12}{5}$; $\csc\theta = \dfrac{13}{12}$; $\sec\theta = -\dfrac{13}{5}$; $\cot\theta = -\dfrac{5}{12}$ **91.** $\sin\theta = -\dfrac{3}{5}$; $\tan\theta = \dfrac{3}{4}$; $\csc\theta = -\dfrac{5}{3}$; $\sec\theta = -\dfrac{5}{4}$; $\cot\theta = \dfrac{4}{3}$

93. $\cos\theta = -\dfrac{12}{13}$; $\tan\theta = \dfrac{5}{12}$; $\csc\theta = \dfrac{13}{5}$; $\sec\theta = -\dfrac{13}{12}$; $\cot\theta = -\dfrac{12}{5}$ **95.** $\sin\theta = -\dfrac{2\sqrt{2}}{3}$; $\tan\theta = 2\sqrt{2}$; $\csc\theta = -\dfrac{3\sqrt{2}}{4}$; $\sec\theta = -3$; $\cot\theta = \dfrac{\sqrt{2}}{4}$

97. $\cos\theta = -\dfrac{\sqrt{5}}{3}$; $\tan\theta = -\dfrac{2\sqrt{5}}{5}$; $\csc\theta = \dfrac{3}{2}$; $\sec\theta = -\dfrac{3\sqrt{5}}{5}$; $\cot\theta = -\dfrac{\sqrt{5}}{2}$

99. $\sin\theta = -\dfrac{\sqrt{3}}{2}$; $\cos\theta = \dfrac{1}{2}$; $\tan\theta = -\sqrt{3}$; $\csc\theta = -\dfrac{2\sqrt{3}}{3}$; $\cot\theta = -\dfrac{\sqrt{3}}{3}$

101. $\sin\theta = -\dfrac{3}{5}$; $\cos\theta = \dfrac{4}{5}$; $\csc\theta = -\dfrac{5}{3}$; $\sec\theta = -\dfrac{5}{4}$; $\cot\theta = \dfrac{4}{3}$ **103.** $\sin\theta = \dfrac{\sqrt{10}}{10}$; $\cos\theta = -\dfrac{3\sqrt{10}}{10}$; $\csc\theta = \sqrt{10}$; $\sec\theta = -\dfrac{\sqrt{10}}{3}$; $\cot\theta = -3$

105. $\sin\theta = -\dfrac{1}{2}$; $\cos\theta = -\dfrac{\sqrt{3}}{2}$; $\tan\theta = \dfrac{\sqrt{3}}{3}$; $\sec\theta = -\dfrac{2\sqrt{3}}{3}$; $\cot\theta = \sqrt{3}$ **107.** 0 **109.** -0.2 **111.** 3 **113.** -5 **115.** 0

117. (a) Approximately 16.6 ft **(b)** **(c)** 67.5°

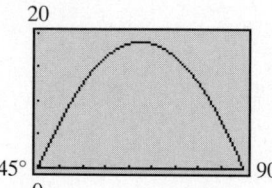

7.5 Assess Your Understanding (page 557)

4. 2π; π **5.** All real numbers, except odd multiples of $\dfrac{\pi}{2}$ **6.** All real numbers from -1 to 1 inclusive **7.** -0.2; 0.2 **8.** T

9. $\sin t = -\dfrac{1}{2}$; $\cos t = \dfrac{\sqrt{3}}{2}$; $\tan t = -\dfrac{\sqrt{3}}{3}$; $\csc t = -2$; $\sec t = \dfrac{2\sqrt{3}}{3}$; $\cot t = -\sqrt{3}$

11. $\sin t = -\dfrac{\sqrt{2}}{2}$; $\cos t = -\dfrac{\sqrt{2}}{2}$; $\tan t = 1$; $\csc t = -\sqrt{2}$; $\sec t = -\sqrt{2}$; $\cot t = 1$

13. $\sin t = \dfrac{2}{3}$; $\cos t = \dfrac{\sqrt{5}}{3}$; $\tan t = \dfrac{2\sqrt{5}}{5}$; $\csc t = \dfrac{3}{2}$; $\sec t = \dfrac{3\sqrt{5}}{5}$; $\cot t = \dfrac{\sqrt{5}}{2}$

15. $\sin\theta = -\dfrac{4}{5}$; $\cos\theta = \dfrac{3}{5}$; $\tan\theta = -\dfrac{4}{3}$; $\csc\theta = -\dfrac{5}{4}$; $\sec\theta = \dfrac{5}{3}$; $\cot\theta = -\dfrac{3}{4}$

17. $\sin\theta = \dfrac{3\sqrt{13}}{13}$; $\cos\theta = -\dfrac{2\sqrt{13}}{13}$; $\tan\theta = -\dfrac{3}{2}$; $\csc\theta = \dfrac{\sqrt{13}}{3}$; $\sec\theta = -\dfrac{\sqrt{13}}{2}$; $\cot\theta = -\dfrac{2}{3}$

19. $\sin\theta = -\dfrac{\sqrt{2}}{2}$; $\cos\theta = -\dfrac{\sqrt{2}}{2}$; $\tan\theta = 1$; $\csc\theta = -\sqrt{2}$; $\sec\theta = -\sqrt{2}$; $\cot\theta = 1$ **21.** $\dfrac{\sqrt{2}}{2}$ **23.** 1 **25.** 1 **27.** $\sqrt{3}$ **29.** $\dfrac{\sqrt{2}}{2}$ **31.** 0

33. $\sqrt{2}$ **35.** $\dfrac{\sqrt{3}}{3}$ **37.** $-\dfrac{\sqrt{3}}{2}$ **39.** $-\dfrac{\sqrt{3}}{3}$ **41.** 2 **43.** -1 **45.** -1 **47.** $\dfrac{\sqrt{2}}{2}$ **49.** 0 **51.** $-\sqrt{2}$ **53.** $\dfrac{2\sqrt{3}}{3}$ **55.** -1 **57.** -2

59. $\dfrac{2-\sqrt{2}}{2}$ **61.** All real numbers **63.** At odd multiples of $\dfrac{\pi}{2}$ **65.** At odd multiples of $\dfrac{\pi}{2}$ **67.** $[-1, 1]$ **69.** $(-\infty, \infty)$ **71.** $(-\infty, -1]$ or $[1, \infty)$

73. Odd; yes; origin **75.** Odd; yes; origin **77.** Even; yes; y-axis **79.** 0.9 **81.** 9 **83. (a)** $-\dfrac{1}{3}$ **(b)** 1 **85. (a)** -2 **(b)** 6 **87. (a)** -4 **(b)** -12

89. (a) Using graph: $\sin 1 \approx 0.8$, $\cos 1 \approx 0.5$, $\tan 1 \approx 1.6$, $\csc 1 \approx 1.3$, $\sec 1 \approx 2$, $\cot 1 \approx 0.6$;
 Using calculator: $\sin 1 \approx 0.8$, $\cos 1 \approx 0.5$, $\tan 1 \approx 1.6$, $\csc 1 \approx 1.2$, $\sec 1 \approx 1.9$, $\cot 1 \approx 0.6$

 (b) Using graph: $\sin 5.1 \approx -0.9$, $\cos 5.1 \approx 0.4$, $\tan 5.1 \approx -2.3$, $\csc 5.1 \approx -1.1$, $\sec 5.1 \approx 2.5$, $\cot 5.1 \approx -0.4$;
 Using calculator: $\sin 5.1 \approx -0.9$, $\cos 5.1 \approx 0.4$, $\tan 5.1 \approx -2.4$, $\csc 5.1 \approx -1.1$, $\sec 5.1 \approx 2.6$, $\cot 5.1 \approx -0.4$

91. Let $P = (x, y)$ be the point on the unit circle that corresponds to t. Consider the equation $\tan t = \dfrac{y}{x} = a$. Then $y = ax$.

But $x^2 + y^2 = 1$, so $x^2 + a^2x^2 = 1$. Thus $x = \pm\dfrac{1}{\sqrt{1+a^2}}$ and $y = \pm\dfrac{a}{\sqrt{1+a^2}}$; that is, for any real number a, there is a point $P = (x, y)$ on the

unit circle for which $\tan t = a$. In other words, $-\infty < \tan t < \infty$, and the range of the tangent function is the set of all real numbers.

93. Suppose that there is a number $p, 0 < p < 2\pi$, for which $\sin(\theta + p) = \sin\theta$ for all θ. If $\theta = 0$, then $\sin(0 + p) = \sin p = \sin 0 = 0$;

so $p = \pi$. If $\theta = \dfrac{\pi}{2}$, then $\sin\left(\dfrac{\pi}{2} + p\right) = \sin\left(\dfrac{\pi}{2}\right)$. But $p = \pi$. Thus $\sin\left(\dfrac{3\pi}{2}\right) = -1 = \sin\left(\dfrac{\pi}{2}\right) = 1$. This is impossible. Therefore, the

smallest positive number p for which $\sin(\theta + p) = \sin\theta$ for all θ is $p = 2\pi$.

95. $\sec\theta = \dfrac{1}{\cos\theta}$; since $\cos\theta$ has period 2π, so does $\sec\theta$.

97. If $P = (a, b)$ is the point on the unit circle corresponding to θ, then $Q = (-a, -b)$ is the point on the unit circle corresponding to $\theta + \pi$.

Thus $\tan(\theta + p) = \dfrac{(-b)}{(-a)} = \dfrac{b}{a} = \tan\theta$. If there exists a number $p, 0 < p < \pi$, for which $\tan(\theta + p) = \tan\theta$ for all θ, then, if $\theta = 0$,

$\tan(p) = \tan 0 = 0$. But this means that p is a multiple of π. Since no multiple of π exists in the interval $(0, \pi)$, this is a impossible. Therefore,

the fundamental period of $f(\theta) = \tan\theta$ is π. **99.** $m = \dfrac{\sin\theta - 0}{\cos\theta - 0} = \dfrac{\sin\theta}{\cos\theta} = \tan\theta$

7.6 Assess Your Understanding *(page 570)*

3. $1; \dfrac{\pi}{2} + 2\pi k, k$ any integer **4.** $3; \pi$ **5.** $3; \dfrac{\pi}{3}$ **6.** T **7.** F **8.** T **9.** 0 **11.** $-\dfrac{\pi}{2} < x < \dfrac{\pi}{2}$ **13.** 1 **15.** $0, \pi, 2\pi$

17. $\sin x = 1$ for $x = -\dfrac{3\pi}{2}, \dfrac{\pi}{2}$; $\sin x = -1$ for $x = -\dfrac{\pi}{2}, \dfrac{3\pi}{2}$ **19.** Amplitude = 2; period = 2π **21.** Amplitude = 4; period = π

23. Amplitude = 6; period = 2 **25.** Amplitude = $\dfrac{1}{2}$; period = $\dfrac{4\pi}{3}$ **27.** Amplitude = $\dfrac{5}{3}$; period = 3 **29.** F **31.** A **33.** H **35.** C **37.** J

39. A **41.** B

43.

45.

47.

49.

51.

53.

55.

57.

59.

61.

63.

65.
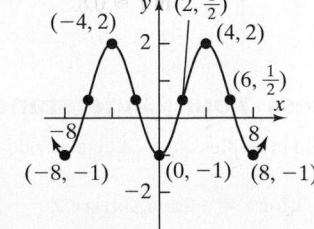

67. $y = \pm 3 \sin(2x)$ **69.** $y = \pm 3 \sin(\pi x)$ **71.** $y = 5 \cos\left(\dfrac{\pi}{4}x\right)$ **73.** $y = -3 \cos\left(\dfrac{1}{2}x\right)$ **75.** $y = \dfrac{3}{4}\sin(2\pi x)$ **77.** $y = -\sin\left(\dfrac{3}{2}x\right)$

79. $y = -\cos\left(\dfrac{4\pi}{3}x\right) + 1$ **81.** $y = 3 \sin\left(\dfrac{\pi}{2}x\right)$ **83.** $y = -4 \cos(3x)$ **85.** $\dfrac{2}{\pi}$ **87.** $\dfrac{\sqrt{2}}{\pi}$

89. $(f \circ g)(x) = \sin(4x)$ **91.** $(f \circ g)(x) = -2 \cos x$ **93.** Period $= \dfrac{1}{30}$ s; amplitude $= 220$ amp

$(g \circ f)(x) = 4 \sin x$ $(g \circ f)(x) = \cos(-2x)$

95. (a) Amplitude $= 220$ V; period $= \dfrac{1}{60}$ s **(c)** $I(t) = 22 \sin(120\pi t)$ **(d)** Amplitude $= 22$ amp; period $= \dfrac{1}{60}$ s

(b), (e)

97. (a) $P(t) = \dfrac{[V_0 \sin(2\pi f t)]^2}{R} = \dfrac{V_0^2}{R}\sin^2(2\pi f t)$ **(b)** Since the graph of P has amplitude $\dfrac{V_0^2}{2R}$ and period $\dfrac{1}{2f}$ and is of the form $y = A \cos(\omega t) + B$, then $A = -\dfrac{V_0^2}{2R}$ and $B = \dfrac{V_0^2}{2R}$. Since $\dfrac{1}{2f} = \dfrac{2\pi}{\omega}$, then $\omega = 4\pi f$. Therefore, $P(t) = -\dfrac{V_0^2}{2R}\cos(4\pi f t) + \dfrac{V_0^2}{2R} = \dfrac{V_0^2}{2R}[1 - \cos(4\pi f t)]$.

99. (a) Physical potential: $\omega = \dfrac{2\pi}{23}$; emotional potential: $\omega = \dfrac{\pi}{14}$; intellectual potential: $\omega = \dfrac{2\pi}{33}$

(b)

(c) No

(d) Physical potential peaks at 15 days after 20th birthday. Emotional potential is 50% at 17 days, with a maximum at 10 days and a minimum at 24 days. Intellectual potential starts fairly high, drops to a minimum at 13 days, and rises to a maximum at 29 days.

101.

7.7 Assess Your Understanding *(Page 579)*

3. origin; odd multiples of $\dfrac{\pi}{2}$ **4.** y-axis; odd multiples of $\dfrac{\pi}{2}$ **5.** $y = \cos x$ **6.** T **7.** 0 **9.** 1

11. $\sec x = 1$ for $x = -2\pi, 0, 2\pi$; $\sec x = -1$ for $x = -\pi, \pi$ **13.** $-\dfrac{3\pi}{2}, -\dfrac{\pi}{2}, \dfrac{\pi}{2}, \dfrac{3\pi}{2}$ **15.** $-\dfrac{3\pi}{2}, -\dfrac{\pi}{2}, \dfrac{\pi}{2}, \dfrac{3\pi}{2}$

17.

19.

21.

23.

25.

27.

29.

31.

33.

35.

37.

39.

41. $\dfrac{2\sqrt{3}}{\pi}$ **43.** $\dfrac{6\sqrt{3}}{\pi}$

45. $(f \circ g)(x) = \tan(4x)$ $(g \circ f)(x) = 4\tan x$ **47.** $(f \circ g)(x) = -2\cot x$ $(g \circ f)(x) = \cot(-2x)$

49. (a) $L(\theta) = \dfrac{3}{\cos \theta} + \dfrac{4}{\sin \theta}$

$= 3 \sec \theta + 4 \csc \theta$

(b)

(c) 0.83 **(d)** 9.86 ft

51.

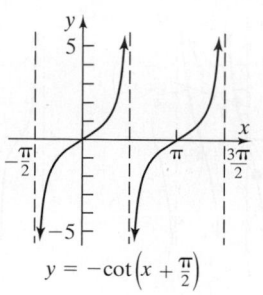

$y = \tan x$

$y = -\cot\left(x + \frac{\pi}{2}\right)$

7.8 Assess Your Understanding *(page 589)*

1. phase shift **2.** F

3. Amplitude $= 4$

Period $= \pi$

Phase shift $= \dfrac{\pi}{2}$

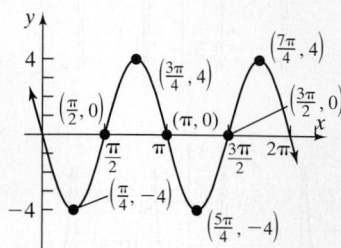

5. Amplitude $= 2$

Period $= \dfrac{2\pi}{3}$

Phase shift $= -\dfrac{\pi}{6}$

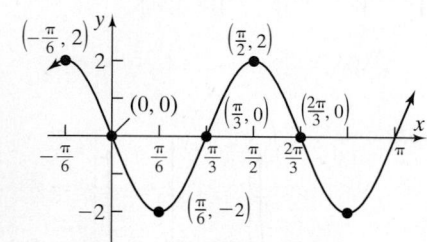

7. Amplitude $= 3$

Period $= \pi$

Phase shift $= -\dfrac{\pi}{4}$

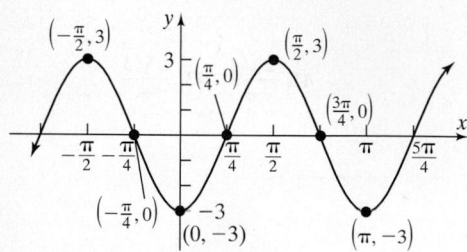

9. Amplitude $= 4$

Period $= 2$

Phase shift $= -\dfrac{2}{\pi}$

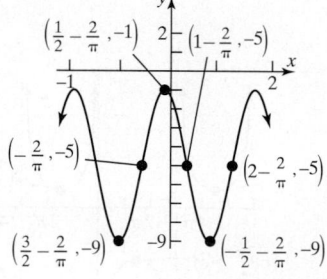

11. Amplitude $= 3$

Period $= 2$

Phase shift $= \dfrac{2}{\pi}$

13. Amplitude $= 3$

Period $= \pi$

Phase shift $= \dfrac{\pi}{4}$

15. $y = 2 \sin\left[2\left(x - \dfrac{1}{2}\right)\right]$ or $y = 2 \sin(2x - 1)$

17. $y = 3 \sin\left[\dfrac{2}{3}\left(x + \dfrac{1}{3}\right)\right]$ or $y = 3 \sin\left(\dfrac{2}{3}x + \dfrac{2}{9}\right)$

19.

21.

23.

25.

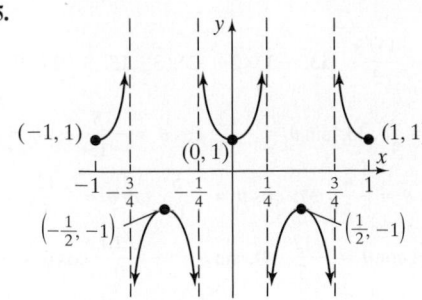

27. Period $= \dfrac{1}{15}$ s; amplitude

$= 120$ amp; phase shift $= \dfrac{1}{90}$ s

29. (a)

(c)

(e)

(b) $y = 15.9 \sin\left(\dfrac{\pi}{6}x - \dfrac{2\pi}{3}\right) + 40.1$

(d) $y = 15.62 \sin(0.517x - 2.096) + 40.377$

31. (a)

(c)

(e)

(b) $y = 24.95 \sin\left(\dfrac{\pi}{6}x - \dfrac{2\pi}{3}\right) + 50.45$

(d) $y = 25.693 \sin(0.476x - 1.814) + 49.854$

33. (a) 4:08 PM **(b)** $y = 4.4 \sin\left[\dfrac{4\pi}{25}(x - 0.5083)\right] + 3.8$ or **(c)**

$$y = 4.4 \sin\left[\dfrac{4\pi}{25}x - 0.2555\right] + 3.8$$

(d) 8.2 ft

35. (a) $y = 1.61 \sin\left(\dfrac{2\pi}{365}x - 1.39\right) + 12.14$

(b) 12.42 h

(c)

37. (a) $y = 6.975 \sin\left(\dfrac{2\pi}{365}x - 1.39\right) + 12.445$

(b) 13.67 h

(c)

(d) The actual hours of sunlight on April 1, 2005 was 12.43 hours. This is very close to the predicted amount of 12.42 hours.

(d) The actual hours of sunlight on April 1, 2005 was 13.43 hours. This is close to the predicted amount of 13.67 hours.

Review Exercises (page 594)

1. $\dfrac{3\pi}{4}$ 3. $\dfrac{\pi}{10}$ 5. $135°$ 7. $-450°$ 9. $\dfrac{1}{2}$ 11. $\dfrac{3\sqrt{2}}{2} - \dfrac{4\sqrt{3}}{3}$ 13. $-3\sqrt{2} - 2\sqrt{3}$ 15. 3 17. 0 19. 0 21. 1 23. 1 25. 1 27. -1 29. 1

31. $\cos\theta = \dfrac{3}{5}$; $\tan\theta = \dfrac{4}{3}$; $\csc\theta = \dfrac{5}{4}$; $\sec\theta = \dfrac{5}{3}$; $\cot\theta = \dfrac{3}{4}$ 33. $\sin\theta = -\dfrac{12}{13}$; $\cos\theta = -\dfrac{5}{13}$; $\csc\theta = -\dfrac{13}{12}$; $\sec\theta = -\dfrac{13}{5}$; $\cot\theta = \dfrac{5}{12}$

35. $\sin\theta = \dfrac{3}{5}$; $\cos\theta = -\dfrac{4}{5}$; $\tan\theta = -\dfrac{3}{4}$; $\csc\theta = \dfrac{5}{3}$; $\cot\theta = -\dfrac{4}{3}$ 37. $\cos\theta = -\dfrac{5}{13}$; $\tan\theta = -\dfrac{12}{5}$; $\csc\theta = \dfrac{13}{12}$; $\sec\theta = -\dfrac{13}{5}$; $\cot\theta = -\dfrac{5}{12}$

39. $\cos\theta = \dfrac{12}{13}$; $\tan\theta = -\dfrac{5}{12}$; $\csc\theta = -\dfrac{13}{5}$; $\sec\theta = \dfrac{13}{12}$; $\cot\theta = -\dfrac{12}{5}$ 41. $\sin\theta = -\dfrac{\sqrt{10}}{10}$; $\cos\theta = -\dfrac{3\sqrt{10}}{10}$; $\csc\theta = -\sqrt{10}$; $\sec\theta = -\dfrac{\sqrt{10}}{3}$; $\cot\theta = 3$

43. $\sin\theta = -\dfrac{2\sqrt{2}}{3}$; $\cos\theta = \dfrac{1}{3}$; $\tan\theta = -2\sqrt{2}$; $\csc\theta = -\dfrac{3\sqrt{2}}{4}$; $\cot\theta = -\dfrac{\sqrt{2}}{4}$

45. $\sin\theta = \dfrac{\sqrt{5}}{5}$; $\cos\theta = -\dfrac{2\sqrt{5}}{5}$; $\tan\theta = -\dfrac{1}{2}$; $\csc\theta = \sqrt{5}$; $\sec\theta = -\dfrac{\sqrt{5}}{2}$

47.

49.

51.

53.

55.

57.

59.

61.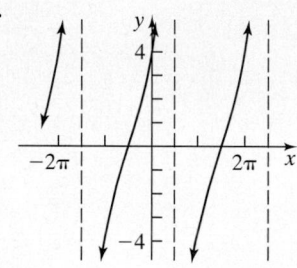

63. Amplitude = 4; period = 2π

65. Amplitude = 8; period = 4

67. Amplitude = 4

Period = $\dfrac{2\pi}{3}$

Phase shift = 0

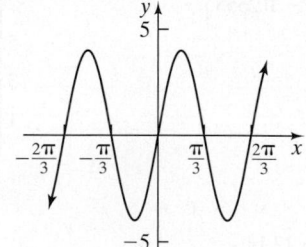

69. Amplitude = 2

Period = π

Phase shift = $\dfrac{\pi}{2}$

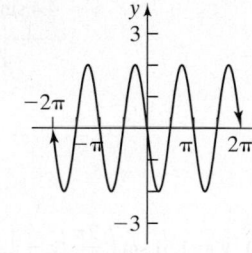

71. Amplitude = $\dfrac{1}{2}$

Period = $\dfrac{4\pi}{3}$

Phase shift = $\dfrac{2\pi}{3}$

73. Amplitude = $\dfrac{2}{3}$

Period = 2

Phase shift = $\dfrac{6}{\pi}$

75. $y = 5\cos\dfrac{x}{4}$ **77.** $y = -6\cos\left(\dfrac{\pi}{4}x\right)$ **79.** $\sin\theta = \dfrac{5}{13}$; $\cos\theta = \dfrac{12}{13}$; $\tan\theta = \dfrac{5}{12}$; $\csc\theta = \dfrac{13}{5}$; $\sec\theta = \dfrac{13}{12}$; $\cot\theta = \dfrac{12}{5}$

81. $\sin\theta = -\dfrac{4}{5}$; $\cos\theta = \dfrac{3}{5}$; $\tan\theta = -\dfrac{4}{3}$; $\csc\theta = -\dfrac{5}{4}$; $\sec\theta = \dfrac{5}{3}$; $\cot\theta = -\dfrac{3}{4}$ **83.** $\dfrac{\pi}{5}$ **85.** Domain: all real numbers except odd multiples of $\dfrac{\pi}{2}$; range: all real

numbers ≤ -1 or ≥ 1 **87.** $\dfrac{\pi}{3} \approx 1.047$ ft; $\dfrac{\pi}{3} \approx 1.047$ ft^2 **89.** Approximately 114.59 rev/h **91.** 0.1 rev/s $= \dfrac{\pi}{5}$ radian/s **93.** 839.10 ft **95.** 23.32 ft **97.** 2.15 mi

99. (a) $\dfrac{1}{15}$ s **(b)** 220 amp **101. (a)** $y = 2.455\sin\left(\dfrac{2\pi}{365}x - 1.39\right) + 12.175$

(c) $-\dfrac{1}{180}$ s **(d)**

(b) 12.61 h **(c)**

Chapter Test *(page 598)*

1. $\dfrac{13\pi}{9}$ **2.** $-\dfrac{20\pi}{9}$ **3.** $\dfrac{13\pi}{180}$ **4.** $-22.5°$ **5.** $810°$ **6.** $135°$ **7.** $\dfrac{1}{2}$ **8.** 0 **9.** $-\dfrac{1}{2}$ **10.** $-\dfrac{\sqrt{3}}{3}$ **11.** 2 **12.** $\dfrac{3(1-\sqrt{2})}{2}$ **13.** 0.292 **14.** 0.309

15. -1.524 **16.** 2.747 **17.**

	$\sin\theta$	$\cos\theta$	$\tan\theta$	$\sec\theta$	$\csc\theta$	$\cot\theta$
θ in QI	$+$	$+$	$+$	$+$	$+$	$+$
θ in QII	$+$	$-$	$-$	$-$	$+$	$-$
θ in QIII	$-$	$-$	$+$	$-$	$-$	$+$
θ in QIV	$-$	$+$	$-$	$+$	$-$	$-$

18. $-\dfrac{3}{5}$

19. $\cos\theta = -\dfrac{2\sqrt{6}}{7}$; $\tan\theta = -\dfrac{5\sqrt{6}}{12}$; $\csc\theta = \dfrac{7}{5}$; $\sec\theta = -\dfrac{7\sqrt{6}}{12}$; $\cot\theta = -\dfrac{2\sqrt{6}}{5}$ **20.** $\sin\theta = -\dfrac{\sqrt{5}}{3}$; $\tan\theta = -\dfrac{\sqrt{5}}{2}$; $\csc\theta = -\dfrac{3\sqrt{5}}{5}$; $\sec\theta = \dfrac{3}{2}$;

$\cot\theta = -\dfrac{2\sqrt{5}}{5}$ **21.** $\sin\theta = \dfrac{12}{13}$; $\cos\theta = -\dfrac{5}{13}$; $\csc\theta = \dfrac{13}{12}$; $\sec\theta = -\dfrac{13}{5}$; $\cot\theta = -\dfrac{5}{12}$ **22.** $\dfrac{7\sqrt{53}}{53}$ **23.** $-\dfrac{5\sqrt{146}}{146}$ **24.** $-\dfrac{1}{2}$

25.

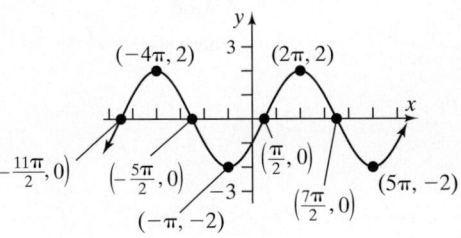

26.

$y = \tan\left(-x + \dfrac{\pi}{4}\right) + 2$

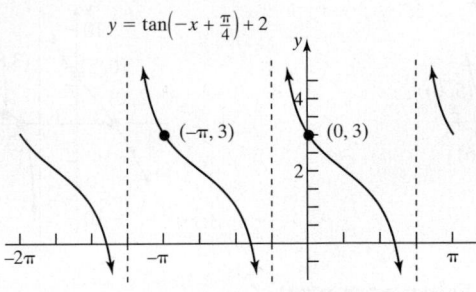

27. $y = -3\sin\left(3x + \dfrac{3\pi}{4}\right)$

28. 78.93 ft^2 **29.** 143.5 rpm **30.** The ship is about 838 ft from the statue. **31.** The building is 122.4 ft tall.

Cumulative Review *(page 598)*

1. $\left\{-1, \dfrac{1}{2}\right\}$ **2.** $y - 5 = -3(x + 2)$ or $y = -3x - 1$ **3.** $x^2 + (y + 2)^2 = 16$

4. A line; slope $\dfrac{2}{3}$; intercepts $(6, 0)$ and $(0, -4)$ **5.** A circle; center $(1, -2)$; radius 3 **6.**

7. (a)

(b)

(c)

(d)

(e)

(f)

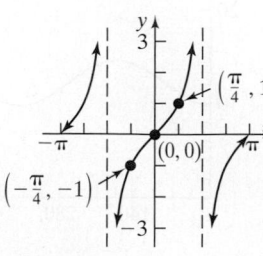

8. $f^{-1}(x) = \frac{1}{3}(x + 2)$

10.

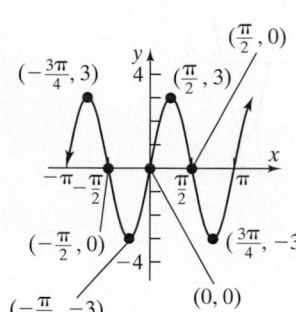

9. -2

11. $3 - \frac{3\sqrt{3}}{2}$

12. $y = 2(3^x)$

13. $y = 3 \cos\left(\frac{\pi}{6}x\right)$

14. (a) $f(x) = -3x - 3$;
 $m = -3$; $(-1, 0), (0, -3)$

(b) $f(x) = (x - 1)^2 - 6$; $(0, -5)$,
 $\left(-\sqrt{6} + 1, 0\right), \left(\sqrt{6} + 1, 0\right)$

(c) We have that $y = 3$ when $x = -2$ and $y = -6$ when $x = 1$. Both points satisfy $y = ae^x$. Therefore, for $(-2, 3)$ we have $3 = ae^{-2}$, which implies that $a = 3e^2$. But for $(1, -6)$ we have $-6 = ae^1$, which implies that $a = -6e^{-1}$. Therefore, there is no exponential function $y = ae^x$ that contains $(-2, 3)$ and $(1, -6)$.

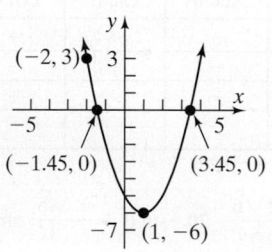

15. (a) $f(x) = \frac{1}{6}(x + 2)(x - 3)(x - 5)$

(b) $R(x) = -\frac{(x + 2)(x - 3)(x - 5)}{3(x - 2)}$

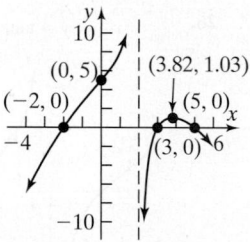

CHAPTER 8 Analytic Trigonometry

8.1 Assess Your Understanding *(page 611)*

7. $x = \sin y$ **8.** $\frac{\pi}{2}$ **9.** $\frac{\pi}{5}$ **10.** F **11.** T **12.** T **13.** 0 **15.** $-\frac{\pi}{2}$ **17.** 0 **19.** $\frac{\pi}{4}$ **21.** $\frac{\pi}{3}$ **23.** $\frac{5\pi}{6}$ **25.** 0.10 **27.** 1.37 **29.** 0.51 **31.** -0.38

33. -0.12 **35.** 1.08 **37.** $\frac{4\pi}{5}$ **39.** $-\frac{3\pi}{8}$ **41.** $-\frac{\pi}{8}$ **43.** $-\frac{\pi}{5}$ **45.** $\frac{1}{4}$ **47.** 4 **49.** Not defined **51.** π

53. $f^{-1}(x) = \sin^{-1}\frac{x - 2}{5}$ **55.** $f^{-1}(x) = \frac{1}{3}\cos^{-1}\left(-\frac{x}{2}\right)$ **57.** $f^{-1}(x) = -\tan^{-1}(x + 3) - 1$ **59.** $f^{-1}(x) = \frac{1}{2}\left[\sin^{-1}\left(\frac{x}{3}\right) - 1\right]$

Domain of f: $(-\infty, \infty)$ Domain of f: $(-\infty, \infty)$ Domain of f: $x \neq \frac{(2k + 1)\pi}{2} - 1$; k an integer Domain of f: $(-\infty, \infty)$

Domain of f^{-1}: $[-3, 7]$ Domain of f^{-1}: $[-2, 2]$ Domain of f^{-1}: $(-\infty, \infty)$ Domain of f^{-1}: $[-3, 3]$

61. $\left\{\frac{\sqrt{2}}{2}\right\}$ **63.** $\left\{-\frac{1}{4}\right\}$ **65.** $\left\{\sqrt{3}\right\}$ **67.** $\{-1\}$ **69. (a)** 13.92 hr or 13 hr, 55 min **(b)** 12 hr **(c)** 13.85 hr or 13 hr, 51 min

71. (a) 13.3 hr or 13 hr, 18 min **(b)** 12 hr **(c)** 13.26 hr or 13 hr, 15 min **73. (a)** 12 hr **(b)** 12 hr **(c)** 12 hr **(d)** It is 12 hr. **75.** 3.35 min

77. (a) $\frac{\pi}{3}$ square units **(b)** $\frac{5\pi}{12}$ square units **79.** 4251 mi

8.2 Assess Your Understanding (page 618)

4. $x = \sec y;\ \geq 1;\ 0;\ \pi$ **5.** $\dfrac{\sqrt{2}}{2}$ **6.** F **7.** T **8.** T **9.** $\dfrac{\sqrt{2}}{2}$ **11.** $-\dfrac{\sqrt{3}}{3}$ **13.** 2 **15.** $\sqrt{2}$ **17.** $-\dfrac{\sqrt{2}}{2}$ **19.** $\dfrac{2\sqrt{3}}{3}$ **21.** $\dfrac{3\pi}{4}$ **23.** $\dfrac{\pi}{6}$ **25.** $\dfrac{\sqrt{2}}{4}$

27. $\dfrac{\sqrt{5}}{2}$ **29.** $-\dfrac{\sqrt{14}}{2}$ **31.** $-\dfrac{3\sqrt{10}}{10}$ **33.** $\sqrt{5}$ **35.** $-\dfrac{\pi}{4}$ **37.** $\dfrac{\pi}{6}$ **39.** $-\dfrac{\pi}{2}$ **41.** $\dfrac{\pi}{6}$ **43.** $\dfrac{2\pi}{3}$ **45.** 1.32 **47.** 0.46 **49.** -0.34 **51.** 2.72 **53.** -0.73

55. 2.55 **57.** $\dfrac{1}{\sqrt{1+u^2}}$ **59.** $\dfrac{u}{\sqrt{1-u^2}}$ **61.** $\dfrac{\sqrt{u^2-1}}{|u|}$ **63.** $\dfrac{\sqrt{u^2-1}}{u}$ **65.** $\dfrac{1}{u}$ **67.** $\dfrac{5}{13}$ **69.** $\dfrac{3\pi}{4}$ **71.** $-\dfrac{3}{4}$ **73.** $\dfrac{5}{13}$ **75.** $\dfrac{\pi}{6}$ **77.** $-\sqrt{15}$

79. (a) $\theta = 31.89°$
 (b) 54.64 ft in diameter
 (c) 37.96 ft high

81. (a) $\theta = 22.3°$
 (b) $v_0 = 2940.23$ ft/sec

83.

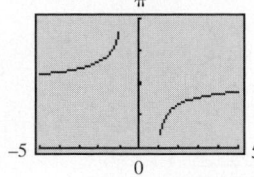

8.3 Assess Your Understanding (page 624)

3. identity; conditional **4.** -1 **5.** 0 **6.** T **7.** F **8.** T **9.** $\dfrac{1}{\cos\theta}$ **11.** $\dfrac{1+\sin\theta}{\cos\theta}$ **13.** $\dfrac{1}{\sin\theta\cos\theta}$ **15.** 2 **17.** $\dfrac{3\sin\theta + 1}{\sin\theta + 1}$

19. $\csc\theta \cdot \cos\theta = \dfrac{1}{\sin\theta} \cdot \cos\theta = \dfrac{\cos\theta}{\sin\theta} = \cot\theta$ **21.** $1 + \tan^2(-\theta) = 1 + (-\tan\theta)^2 = 1 + \tan^2\theta = \sec^2\theta$

23. $\cos\theta(\tan\theta + \cot\theta) = \cos\theta\left(\dfrac{\sin\theta}{\cos\theta} + \dfrac{\cos\theta}{\sin\theta}\right) = \cos\theta\left(\dfrac{\sin^2\theta + \cos^2\theta}{\cos\theta\sin\theta}\right) = \cos\theta\left(\dfrac{1}{\cos\theta\sin\theta}\right) = \dfrac{1}{\sin\theta} = \csc\theta$

25. $\tan u \cot u - \cos^2 u = \dfrac{\sin u}{\cos u} \cdot \dfrac{\cos u}{\sin u} - \cos^2 u = 1 - \cos^2 u = \sin^2 u$

27. $(\sec\theta - 1)(\sec\theta + 1) = \sec^2\theta - 1 = \tan^2\theta$ **29.** $(\sec\theta + \tan\theta)(\sec\theta - \tan\theta) = \sec^2\theta - \tan^2\theta = 1$

31. $\cos^2\theta(1 + \tan^2\theta) = \cos^2\theta + \cos^2\theta\tan^2\theta = \cos^2\theta + \cos^2\theta \cdot \dfrac{\sin^2\theta}{\cos^2\theta} = \cos^2\theta + \sin^2\theta = 1$

33. $(\sin\theta + \cos\theta)^2 + (\sin\theta - \cos\theta)^2 = \sin^2\theta + 2\sin\theta\cos\theta + \cos^2\theta + \sin^2\theta - 2\sin\theta\cos\theta + \cos^2\theta$

$$= \sin^2\theta + \cos^2\theta + \sin^2\theta + \cos^2\theta = 1 + 1 = 2$$

35. $\sec^4\theta - \sec^2\theta = \sec^2\theta(\sec^2\theta - 1) = (1 + \tan^2\theta)\tan^2\theta = \tan^4\theta + \tan^2\theta$

37. $\sec u - \tan u = \dfrac{1}{\cos u} - \dfrac{\sin u}{\cos u} = \dfrac{1 - \sin u}{\cos u} \cdot \dfrac{1 + \sin u}{1 + \sin u} = \dfrac{1 - \sin^2 u}{\cos u(1 + \sin u)} = \dfrac{\cos^2 u}{\cos u(1 + \sin u)} = \dfrac{\cos u}{1 + \sin u}$

39. $3\sin^2\theta + 4\cos^2\theta = 3\sin^2\theta + 3\cos^2\theta + \cos^2\theta = 3(\sin^2\theta + \cos^2\theta) + \cos^2\theta = 3 + \cos^2\theta$

41. $1 - \dfrac{\cos^2\theta}{1 + \sin\theta} = 1 - \dfrac{1 - \sin^2\theta}{1 + \sin\theta} = 1 - \dfrac{(1 + \sin\theta)(1 - \sin\theta)}{1 + \sin\theta} = 1 - (1 - \sin\theta) = \sin\theta$

43. $\dfrac{1 + \tan v}{1 - \tan v} = \dfrac{1 + \dfrac{1}{\cot v}}{1 - \dfrac{1}{\cot v}} = \dfrac{\dfrac{\cot v + 1}{\cot v}}{\dfrac{\cot v - 1}{\cot v}} = \dfrac{\cot v + 1}{\cot v - 1}$ **45.** $\dfrac{\sec\theta}{\csc\theta} + \dfrac{\sin\theta}{\cos\theta} = \dfrac{\dfrac{1}{\cos\theta}}{\dfrac{1}{\sin\theta}} + \tan\theta = \dfrac{\sin\theta}{\cos\theta} + \tan\theta = \tan\theta + \tan\theta = 2\tan\theta$

47. $\dfrac{1 + \sin\theta}{1 - \sin\theta} = \dfrac{1 + \dfrac{1}{\csc\theta}}{1 - \dfrac{1}{\csc\theta}} = \dfrac{\dfrac{\csc\theta + 1}{\csc\theta}}{\dfrac{\csc\theta - 1}{\csc\theta}} = \dfrac{\csc\theta + 1}{\csc\theta - 1}$

49. $\dfrac{1 - \sin v}{\cos v} + \dfrac{\cos v}{1 - \sin v} = \dfrac{(1 - \sin v)^2 + \cos^2 v}{\cos v(1 - \sin v)} = \dfrac{1 - 2\sin v + \sin^2 v + \cos^2 v}{\cos v(1 - \sin v)} = \dfrac{2 - 2\sin v}{\cos v(1 - \sin v)} = \dfrac{2(1 - \sin v)}{\cos v(1 - \sin v)} = \dfrac{2}{\cos v} = 2\sec v$

51. $\dfrac{\sin\theta}{\sin\theta - \cos\theta} = \dfrac{1}{\dfrac{\sin\theta - \cos\theta}{\sin\theta}} = \dfrac{1}{1 - \dfrac{\cos\theta}{\sin\theta}} = \dfrac{1}{1 - \cot\theta}$

53. $(\sec\theta - \tan\theta)^2 = \sec^2\theta - 2\sec\theta\tan\theta + \tan^2\theta = \dfrac{1}{\cos^2\theta} - \dfrac{2\sin\theta}{\cos^2\theta} + \dfrac{\sin^2\theta}{\cos^2\theta} = \dfrac{1 - 2\sin\theta + \sin^2\theta}{\cos^2\theta} = \dfrac{(1 - \sin\theta)^2}{1 - \sin^2\theta} = \dfrac{(1 - \sin\theta)^2}{(1 - \sin\theta)(1 + \sin\theta)}$

$$= \dfrac{1 - \sin\theta}{1 + \sin\theta}$$

55. $\dfrac{\cos\theta}{1 - \tan\theta} + \dfrac{\sin\theta}{1 - \cot\theta} = \dfrac{\cos\theta}{1 - \dfrac{\sin\theta}{\cos\theta}} + \dfrac{\sin\theta}{1 - \dfrac{\cos\theta}{\sin\theta}} = \dfrac{\cos\theta}{\dfrac{\cos\theta - \sin\theta}{\cos\theta}} + \dfrac{\sin\theta}{\dfrac{\sin\theta - \cos\theta}{\sin\theta}} = \dfrac{\cos^2\theta}{\cos\theta - \sin\theta} + \dfrac{\sin^2\theta}{\sin\theta - \cos\theta}$

$$= \dfrac{\cos^2\theta - \sin^2\theta}{\cos\theta - \sin\theta} = \dfrac{(\cos\theta - \sin\theta)(\cos\theta + \sin\theta)}{\cos\theta - \sin\theta} = \sin\theta + \cos\theta$$

57. $\tan\theta + \dfrac{\cos\theta}{1+\sin\theta} = \dfrac{\sin\theta}{\cos\theta} + \dfrac{\cos\theta}{1+\sin\theta} = \dfrac{\sin\theta(1+\sin\theta)+\cos^2\theta}{\cos\theta(1+\sin\theta)} = \dfrac{\sin\theta+\sin^2\theta+\cos^2\theta}{\cos\theta(1+\sin\theta)} = \dfrac{\sin\theta+1}{\cos\theta(1+\sin\theta)} = \dfrac{1}{\cos\theta} = \sec\theta$

59. $\dfrac{\tan\theta+\sec\theta-1}{\tan\theta-\sec\theta+1} = \dfrac{\tan\theta+(\sec\theta-1)}{\tan\theta-(\sec\theta-1)} \cdot \dfrac{\tan\theta+(\sec\theta-1)}{\tan\theta+(\sec\theta-1)} = \dfrac{\tan^2\theta+2\tan\theta(\sec\theta-1)+\sec^2\theta-2\sec\theta+1}{\tan^2\theta-(\sec^2\theta-2\sec\theta+1)}$

$= \dfrac{\sec^2\theta-1+2\tan\theta(\sec\theta-1)+\sec^2\theta-2\sec\theta+1}{\sec^2\theta-1-\sec^2\theta+2\sec\theta-1} = \dfrac{2\sec^2\theta-2\sec\theta+2\tan\theta(\sec\theta-1)}{-2+2\sec\theta}$

$= \dfrac{2\sec\theta(\sec\theta-1)+2\tan\theta(\sec\theta-1)}{2(\sec\theta-1)} = \dfrac{2(\sec\theta-1)(\sec\theta+\tan\theta)}{2(\sec\theta-1)} = \tan\theta+\sec\theta$

61. $\dfrac{\tan\theta-\cot\theta}{\tan\theta+\cot\theta} = \dfrac{\dfrac{\sin\theta}{\cos\theta}-\dfrac{\cos\theta}{\sin\theta}}{\dfrac{\sin\theta}{\cos\theta}+\dfrac{\cos\theta}{\sin\theta}} = \dfrac{\dfrac{\sin^2\theta-\cos^2\theta}{\cos\theta\sin\theta}}{\dfrac{\sin^2\theta+\cos^2\theta}{\cos\theta\sin\theta}} = \dfrac{\sin^2\theta-\cos^2\theta}{1} = \sin^2\theta-\cos^2\theta$

63. $\dfrac{\tan u-\cot u}{\tan u+\cot u}+1 = \dfrac{\dfrac{\sin u}{\cos u}-\dfrac{\cos u}{\sin u}}{\dfrac{\sin u}{\cos u}+\dfrac{\cos u}{\sin u}}+1 = \dfrac{\dfrac{\sin^2 u-\cos^2 u}{\cos u\sin u}}{\dfrac{\sin^2 u+\cos^2 u}{\cos u\sin u}}+1 = \sin^2 u-\cos^2 u+1 = \sin^2 u+(1-\cos^2 u) = 2\sin^2 u$

65. $\dfrac{\sec\theta+\tan\theta}{\cot\theta+\cos\theta} = \dfrac{\dfrac{1}{\cos\theta}+\dfrac{\sin\theta}{\cos\theta}}{\dfrac{\cos\theta}{\sin\theta}+\cos\theta} = \dfrac{\dfrac{1+\sin\theta}{\cos\theta}}{\dfrac{\cos\theta+\cos\theta\sin\theta}{\sin\theta}} = \dfrac{1+\sin\theta}{\cos\theta} \cdot \dfrac{\sin\theta}{\cos\theta(1+\sin\theta)} = \dfrac{\sin\theta}{\cos\theta}\cdot\dfrac{1}{\cos\theta} = \tan\theta\sec\theta$

67. $\dfrac{1-\tan^2\theta}{1+\tan^2\theta}+1 = \dfrac{1-\tan^2\theta}{\sec^2\theta}+1 = \dfrac{1}{\sec^2\theta}-\dfrac{\tan^2\theta}{\sec^2\theta}+1 = \cos^2\theta-\dfrac{\dfrac{\sin^2\theta}{\cos^2\theta}}{\dfrac{1}{\cos^2\theta}}+1 = \cos^2\theta-\sin^2\theta+1 = \cos^2\theta+(1-\sin^2\theta) = 2\cos^2\theta$

69. $\dfrac{\sec\theta-\csc\theta}{\sec\theta\csc\theta} = \dfrac{\sec\theta}{\sec\theta\csc\theta}-\dfrac{\csc\theta}{\sec\theta\csc\theta} = \dfrac{1}{\csc\theta}-\dfrac{1}{\sec\theta} = \sin\theta-\cos\theta$

71. $\sec\theta-\cos\theta = \dfrac{1}{\cos\theta}-\cos\theta = \dfrac{1-\cos^2\theta}{\cos\theta} = \dfrac{\sin^2\theta}{\cos\theta} = \sin\theta\cdot\dfrac{\sin\theta}{\cos\theta} = \sin\theta\tan\theta$

73. $\dfrac{1}{1-\sin\theta}+\dfrac{1}{1+\sin\theta} = \dfrac{1+\sin\theta+1-\sin\theta}{(1+\sin\theta)(1-\sin\theta)} = \dfrac{2}{1-\sin^2\theta} = \dfrac{2}{\cos^2\theta} = 2\sec^2\theta$

75. $\dfrac{\sec\theta}{1-\sin\theta} = \dfrac{\sec\theta}{1-\sin\theta}\cdot\dfrac{1+\sin\theta}{1+\sin\theta} = \dfrac{\sec\theta(1+\sin\theta)}{1-\sin^2\theta} = \dfrac{\sec\theta(1+\sin\theta)}{\cos^2\theta} = \dfrac{1+\sin\theta}{\cos^3\theta}$

77. $\dfrac{(\sec v-\tan v)^2+1}{\csc v(\sec v-\tan v)} = \dfrac{\sec^2 v-2\sec v\tan v+\tan^2 v+1}{\dfrac{1}{\sin v}\left(\dfrac{1}{\cos v}-\dfrac{\sin v}{\cos v}\right)} = \dfrac{2\sec^2 v-2\sec v\tan v}{\dfrac{1}{\sin v}\left(\dfrac{1-\sin v}{\cos v}\right)} = \dfrac{\dfrac{2}{\cos^2 v}-\dfrac{2\sin v}{\cos^2 v}}{\dfrac{1-\sin v}{\sin v\cos v}} = \dfrac{2-2\sin v}{\cos^2 v}\cdot\dfrac{\sin v\cos v}{1-\sin v}$

$= \dfrac{2(1-\sin v)}{\cos v}\cdot\dfrac{\sin v}{1-\sin v} = \dfrac{2\sin v}{\cos v} = 2\tan v$

79. $\dfrac{\sin\theta+\cos\theta}{\cos\theta}-\dfrac{\sin\theta-\cos\theta}{\sin\theta} = \dfrac{\sin\theta}{\cos\theta}+1-1+\dfrac{\cos\theta}{\sin\theta} = \dfrac{\sin^2\theta+\cos^2\theta}{\cos\theta\sin\theta} = \dfrac{1}{\cos\theta\sin\theta} = \sec\theta\csc\theta$

81. $\dfrac{\sin^3\theta+\cos^3\theta}{\sin\theta+\cos\theta} = \dfrac{(\sin\theta+\cos\theta)(\sin^2\theta-\sin\theta\cos\theta+\cos^2\theta)}{\sin\theta+\cos\theta} = \sin^2\theta+\cos^2\theta-\sin\theta\cos\theta = 1-\sin\theta\cos\theta$

83. $\dfrac{\cos^2\theta-\sin^2\theta}{1-\tan^2\theta} = \dfrac{\cos^2\theta-\sin^2\theta}{1-\dfrac{\sin^2\theta}{\cos^2\theta}} = \dfrac{\cos^2\theta-\sin^2\theta}{\dfrac{\cos^2\theta-\sin^2\theta}{\cos^2\theta}} = \cos^2\theta$

85. $\dfrac{(2\cos^2\theta-1)^2}{\cos^4\theta-\sin^4\theta} = \dfrac{[2\cos^2\theta-(\sin^2\theta+\cos^2\theta)]^2}{(\cos^2\theta-\sin^2\theta)(\cos^2\theta+\sin^2\theta)} = \dfrac{(\cos^2\theta-\sin^2\theta)^2}{\cos^2\theta-\sin^2\theta} = \cos^2\theta-\sin^2\theta = (1-\sin^2\theta)-\sin^2\theta = 1-2\sin^2\theta$

87. $\dfrac{1+\sin\theta+\cos\theta}{1+\sin\theta-\cos\theta} = \dfrac{(1+\sin\theta)+\cos\theta}{(1+\sin\theta)-\cos\theta}\cdot\dfrac{(1+\sin\theta)+\cos\theta}{(1+\sin\theta)+\cos\theta} = \dfrac{1+2\sin\theta+\sin^2\theta+2(1+\sin\theta)\cos\theta+\cos^2\theta}{1+2\sin\theta+\sin^2\theta-\cos^2\theta}$

$= \dfrac{1+2\sin\theta+\sin^2\theta+2(1+\sin\theta)(\cos\theta)+(1-\sin^2\theta)}{1+2\sin\theta+\sin^2\theta-(1-\sin^2\theta)} = \dfrac{2+2\sin\theta+2(1+\sin\theta)(\cos\theta)}{2\sin\theta+2\sin^2\theta}$

$= \dfrac{2(1+\sin\theta)+2(1+\sin\theta)(\cos\theta)}{2\sin\theta(1+\sin\theta)} = \dfrac{2(1+\sin\theta)(1+\cos\theta)}{2\sin\theta(1+\sin\theta)} = \dfrac{1+\cos\theta}{\sin\theta}$

89. $(a\sin\theta+b\cos\theta)^2+(a\cos\theta-b\sin\theta)^2 = a^2\sin^2\theta+2ab\sin\theta\cos\theta+b^2\cos^2\theta+a^2\cos^2\theta-2ab\sin\theta\cos\theta+b^2\sin^2\theta$

$= a^2(\sin^2\theta+\cos^2\theta)+b^2(\cos^2\theta+\sin^2\theta) = a^2+b^2$

91. $\dfrac{\tan\alpha + \tan\beta}{\cot\alpha + \cot\beta} = \dfrac{\tan\alpha + \tan\beta}{\dfrac{1}{\tan\alpha} + \dfrac{1}{\tan\beta}} = \dfrac{\tan\alpha + \tan\beta}{\dfrac{\tan\beta + \tan\alpha}{\tan\alpha\tan\beta}} = (\tan\alpha + \tan\beta)\cdot\dfrac{\tan\alpha\tan\beta}{\tan\alpha + \tan\beta} = \tan\alpha\tan\beta$

93. $(\sin\alpha + \cos\beta)^2 + (\cos\beta + \sin\alpha)(\cos\beta - \sin\alpha) = (\sin^2\alpha + 2\sin\alpha\cos\beta + \cos^2\beta) + (\cos^2\beta - \sin^2\alpha)$

$\quad = 2\cos^2\beta + 2\sin\alpha\cos\beta = 2\cos\beta(\cos\beta + \sin\alpha)$

95. $\ln|\sec\theta| = \ln|\cos\theta|^{-1} = -\ln|\cos\theta|$ **97.** $\ln|1 + \cos\theta| + \ln|1 - \cos\theta| = \ln(|1 + \cos\theta||1 - \cos\theta|) = \ln|1 - \cos^2\theta| = \ln|\sin^2\theta| = 2\ln|\sin\theta|$

99. $g(x) = \sec x - \cos x = \dfrac{1}{\cos x} - \cos x = \dfrac{1}{\cos x} - \dfrac{\cos^2 x}{\cos x} = \dfrac{1 - \cos^2 x}{\cos x} = \dfrac{\sin^2 x}{\cos x} = \sin x\cdot\dfrac{\sin x}{\cos x} = \sin x\cdot\tan x = f(x)$

101. $f(\theta) = \dfrac{1 - \sin\theta}{\cos\theta} - \dfrac{\cos\theta}{1 + \sin\theta} = \dfrac{1 - \sin\theta}{\cos\theta}\cdot\dfrac{1 + \sin\theta}{1 + \sin\theta} - \dfrac{\cos\theta}{1 + \sin\theta}\cdot\dfrac{\cos\theta}{\cos\theta} = \dfrac{1 - \sin^2\theta}{\cos\theta(1 + \sin\theta)} - \dfrac{\cos^2\theta}{\cos\theta(1 + \sin\theta)}$

$\quad = \dfrac{\cos^2\theta}{\cos\theta(1 + \sin\theta)} - \dfrac{\cos^2\theta}{\cos\theta(1 + \sin\theta)} = 0 = g(\theta)$

103. $1200\sec\theta\,(2\sec^2\theta - 1) = 1200\dfrac{1}{\cos\theta}\left(\dfrac{2}{\cos^2\theta} - 1\right) = 1200\dfrac{1}{\cos\theta}\left(\dfrac{2}{\cos^2\theta} - \dfrac{\cos^2\theta}{\cos^2\theta}\right) = 1200\dfrac{1}{\cos\theta}\left(\dfrac{2 - \cos^2\theta}{\cos^2\theta}\right) = \dfrac{1200\,(1 + 1 - \cos^2\theta)}{\cos^3\theta}$

$\quad = \dfrac{1200\,(1 + \sin^2\theta)}{\cos^3\theta}$

8.4 Assess Your Understanding (page 634)

4. – **5.** – **6.** F **7.** F **8.** F **9.** $\dfrac{1}{4}\left(\sqrt6 + \sqrt2\right)$ **11.** $\dfrac{1}{4}\left(\sqrt2 - \sqrt6\right)$ **13.** $-\dfrac{1}{4}\left(\sqrt2 + \sqrt6\right)$ **15.** $2 - \sqrt3$

17. $-\dfrac{1}{4}\left(\sqrt6 + \sqrt2\right)$ **19.** $\sqrt6 - \sqrt2$ **21.** $\dfrac{1}{2}$ **23.** 0 **25.** 1 **27.** -1 **29.** $\dfrac{1}{2}$ **31. (a)** $\dfrac{2\sqrt5}{25}$ **(b)** $\dfrac{11\sqrt5}{25}$ **(c)** $\dfrac{2\sqrt5}{5}$ **(d)** 2

33. (a) $\dfrac{4 - 3\sqrt3}{10}$ **(b)** $\dfrac{-3 - 4\sqrt3}{10}$ **(c)** $\dfrac{4 + 3\sqrt3}{10}$ **(d)** $\dfrac{25\sqrt3 + 48}{39}$ **35. (a)** $-\dfrac{5 + 12\sqrt3}{26}$ **(b)** $\dfrac{12 - 5\sqrt3}{26}$ **(c)** $\dfrac{-5 + 12\sqrt3}{26}$ **(d)** $\dfrac{-240 + 169\sqrt3}{69}$

37. (a) $-\dfrac{2\sqrt2}{3}$ **(b)** $\dfrac{-2\sqrt2 + \sqrt3}{6}$ **(c)** $\dfrac{-2\sqrt2 + \sqrt3}{6}$ **(d)** $\dfrac{9 - 4\sqrt2}{7}$ **39.** $\dfrac{1 - 2\sqrt6}{6}$ **41.** $\dfrac{\sqrt3 - 2\sqrt2}{6}$ **43.** $\dfrac{8\sqrt2 - 9\sqrt3}{5}$

45. $\sin\left(\dfrac{\pi}{2} + \theta\right) = \sin\dfrac{\pi}{2}\cos\theta + \cos\dfrac{\pi}{2}\sin\theta = 1\cdot\cos\theta + 0\cdot\sin\theta = \cos\theta$

47. $\sin(\pi - \theta) = \sin\pi\cos\theta - \cos\pi\sin\theta = 0\cdot\cos\theta - (-1)\sin\theta = \sin\theta$

49. $\sin(\pi + \theta) = \sin\pi\cos\theta + \cos\pi\sin\theta = 0\cdot\cos\theta + (-1)\sin\theta = -\sin\theta$

51. $\tan(\pi - \theta) = \dfrac{\tan\pi - \tan\theta}{1 + \tan\pi\tan\theta} = \dfrac{0 - \tan\theta}{1 + 0\cdot\tan\theta} = -\tan\theta$ **53.** $\sin\left(\dfrac{3\pi}{2} + \theta\right) = \sin\dfrac{3\pi}{2}\cos\theta + \cos\dfrac{3\pi}{2}\sin\theta = (-1)\cos\theta + 0\cdot\sin\theta = -\cos\theta$

55. $\sin(\alpha + \beta) + \sin(\alpha - \beta) = \sin\alpha\cos\beta + \cos\alpha\sin\beta + \sin\alpha\cos\beta - \cos\alpha\sin\beta = 2\sin\alpha\cos\beta$

57. $\dfrac{\sin(\alpha + \beta)}{\sin\alpha\cos\beta} = \dfrac{\sin\alpha\cos\beta + \cos\alpha\sin\beta}{\sin\alpha\cos\beta} = \dfrac{\sin\alpha\cos\beta}{\sin\alpha\cos\beta} + \dfrac{\cos\alpha\sin\beta}{\sin\alpha\cos\beta} = 1 + \cot\alpha\tan\beta$

59. $\dfrac{\cos(\alpha + \beta)}{\cos\alpha\cos\beta} = \dfrac{\cos\alpha\cos\beta - \sin\alpha\sin\beta}{\cos\alpha\cos\beta} = \dfrac{\cos\alpha\cos\beta}{\cos\alpha\cos\beta} - \dfrac{\sin\alpha\sin\beta}{\cos\alpha\cos\beta} = 1 - \tan\alpha\tan\beta$

61. $\dfrac{\sin(\alpha + \beta)}{\sin(\alpha - \beta)} = \dfrac{\sin\alpha\cos\beta + \cos\alpha\sin\beta}{\sin\alpha\cos\beta - \cos\alpha\sin\beta} = \dfrac{\dfrac{\sin\alpha\cos\beta + \cos\alpha\sin\beta}{\cos\alpha\cos\beta}}{\dfrac{\sin\alpha\cos\beta - \cos\alpha\sin\beta}{\cos\alpha\cos\beta}} = \dfrac{\dfrac{\sin\alpha\cos\beta}{\cos\alpha\cos\beta} + \dfrac{\cos\alpha\sin\beta}{\cos\alpha\cos\beta}}{\dfrac{\sin\alpha\cos\beta}{\cos\alpha\cos\beta} - \dfrac{\cos\alpha\sin\beta}{\cos\alpha\cos\beta}} = \dfrac{\tan\alpha + \tan\beta}{\tan\alpha - \tan\beta}$

63. $\cot(\alpha + \beta) = \dfrac{\cos(\alpha + \beta)}{\sin(\alpha + \beta)} = \dfrac{\cos\alpha\cos\beta - \sin\alpha\sin\beta}{\sin\alpha\cos\beta + \cos\alpha\sin\beta} = \dfrac{\dfrac{\cos\alpha\cos\beta - \sin\alpha\sin\beta}{\sin\alpha\sin\beta}}{\dfrac{\sin\alpha\cos\beta + \cos\alpha\sin\beta}{\sin\alpha\sin\beta}} = \dfrac{\dfrac{\cos\alpha\cos\beta}{\sin\alpha\sin\beta} - \dfrac{\sin\alpha\sin\beta}{\sin\alpha\sin\beta}}{\dfrac{\sin\alpha\cos\beta}{\sin\alpha\sin\beta} + \dfrac{\cos\alpha\sin\beta}{\sin\alpha\sin\beta}} = \dfrac{\cot\alpha\cot\beta - 1}{\cot\beta + \cot\alpha}$

65. $\sec(\alpha + \beta) = \dfrac{1}{\cos(\alpha + \beta)} = \dfrac{1}{\cos\alpha\cos\beta - \sin\alpha\sin\beta} = \dfrac{\dfrac{1}{\sin\alpha\sin\beta}}{\dfrac{\cos\alpha\cos\beta - \sin\alpha\sin\beta}{\sin\alpha\sin\beta}} = \dfrac{\dfrac{1}{\sin\alpha}\cdot\dfrac{1}{\sin\beta}}{\dfrac{\cos\alpha\cos\beta}{\sin\alpha\sin\beta} - \dfrac{\sin\alpha\sin\beta}{\sin\alpha\sin\beta}} = \dfrac{\csc\alpha\csc\beta}{\cot\alpha\cot\beta - 1}$

67. $\sin(\alpha - \beta)\sin(\alpha + \beta) = (\sin\alpha\cos\beta - \cos\alpha\sin\beta)(\sin\alpha\cos\beta + \cos\alpha\sin\beta) = \sin^2\alpha\cos^2\beta - \cos^2\alpha\sin^2\beta$

$\quad = (\sin^2\alpha)(1 - \sin^2\beta) - (1 - \sin^2\alpha)(\sin^2\beta) = \sin^2\alpha - \sin^2\beta$

69. $\sin(\theta + k\pi) = \sin\theta\cos k\pi + \cos\theta\sin k\pi = (\sin\theta)(-1)^k + (\cos\theta)(0) = (-1)^k\sin\theta$, k any integer

71. $\dfrac{\sqrt{3}}{2}$ **73.** $-\dfrac{24}{25}$ **75.** $-\dfrac{33}{65}$ **77.** $\dfrac{63}{65}$ **79.** $\dfrac{48 + 25\sqrt{3}}{39}$ **81.** $\dfrac{4}{3}$ **83.** $u\sqrt{1 - v^2} - v\sqrt{1 - u^2}$: $-1 \le u \le 1$; $-1 \le v \le 1$

85. $\dfrac{u\sqrt{1 - v^2} - v}{\sqrt{1 + u^2}}$: $-\infty < u < \infty$; $-1 \le v \le 1$ **87.** $\dfrac{uv - \sqrt{1 - u^2}\sqrt{1 - v^2}}{v\sqrt{1 - u^2} + u\sqrt{1 - v^2}}$: $-1 \le u \le 1$; $-1 \le v \le 1$

89. Let $\alpha = \sin^{-1} v$ and $\beta = \cos^{-1} v$. Then $\sin\alpha = \cos\beta = v$, and since $\sin\alpha = \cos\left(\dfrac{\pi}{2} - \alpha\right)$, $\cos\left(\dfrac{\pi}{2} - \alpha\right) = \cos\beta$.

If $v \ge 0$, then $0 \le \alpha \le \dfrac{\pi}{2}$, so $\left(\dfrac{\pi}{2} - \alpha\right)$ and β both lie on $\left[0, \dfrac{\pi}{2}\right]$. If $v < 0$, then $-\dfrac{\pi}{2} \le \alpha < 0$, so $\left(\dfrac{\pi}{2} - \alpha\right)$ and β both lie on $\left(\dfrac{\pi}{2}, \pi\right]$.

Either way, $\cos\left(\dfrac{\pi}{2} - \alpha\right) = \cos\beta$ implies $\dfrac{\pi}{2} - \alpha = \beta$, or $\alpha + \beta = \dfrac{\pi}{2}$.

91. Let $\alpha = \tan^{-1}\dfrac{1}{v}$ and $\beta = \tan^{-1} v$. Because $v \ne 0$, $\alpha, \beta \ne 0$. Then $\tan\alpha = \dfrac{1}{v} = \dfrac{1}{\tan\beta} = \cot\beta$, and since

$\tan\alpha = \cot\left(\dfrac{\pi}{2} - \alpha\right)$, $\cot\left(\dfrac{\pi}{2} - \alpha\right) = \cot\beta$. Because $v > 0$, $0 < \alpha < \dfrac{\pi}{2}$, and so $\left(\dfrac{\pi}{2} - \alpha\right)$ and β both lie on $\left(0, \dfrac{\pi}{2}\right)$.

Then $\cot\left(\dfrac{\pi}{2} - \alpha\right) = \cot\beta$ implies $\dfrac{\pi}{2} - \alpha = \beta$, or $\alpha = \dfrac{\pi}{2} - \beta$.

93. $\sin(\sin^{-1} v + \cos^{-1} v) = \sin(\sin^{-1} v)\cos(\cos^{-1} v) + \cos(\sin^{-1} v)\sin(\cos^{-1} v) = (v)(v) + \sqrt{1 - v^2}\sqrt{1 - v^2} = v^2 + 1 - v^2 = 1$

95. $\dfrac{\sin(x + h) - \sin x}{h} = \dfrac{\sin x\cos h + \cos x\sin h - \sin x}{h} = \dfrac{\cos x\sin h - \sin x(1 - \cos h)}{h} = \cos x \cdot \dfrac{\sin h}{h} - \sin x \cdot \dfrac{1 - \cos h}{h}$

97. (a) $\tan(\tan^{-1}1 + \tan^{-1}2 + \tan^{-1}3) = \tan((\tan^{-1}1 + \tan^{-1}2) + \tan^{-1}3) = \dfrac{\tan(\tan^{-1}1 + \tan^{-1}2) + \tan(\tan^{-1}3)}{1 - \tan(\tan^{-1}1 + \tan^{-1}2)\tan(\tan^{-1}3)}$

$= \dfrac{\dfrac{\tan(\tan^{-1}1) + \tan(\tan^{-1}2)}{1 - \tan(\tan^{-1}1)\tan(\tan^{-1}2)} + 3}{1 - \dfrac{\tan(\tan^{-1}1) + \tan(\tan^{-1}2)}{1 - \tan(\tan^{-1}1)\tan(\tan^{-1}2)} \cdot 3} = \dfrac{\dfrac{1 + 2}{1 - 1\cdot 2} + 3}{1 - \dfrac{1 + 2}{1 - 1\cdot 2} \cdot 3} = \dfrac{\dfrac{3}{-1} + 3}{1 - \dfrac{3}{-1} \cdot 3} = \dfrac{-3 + 3}{1 + 9} = \dfrac{0}{10} = 0$

(b) From the definition of the inverse tangent function $0 < \tan^{-1}1 < \dfrac{\pi}{2}$, $0 < \tan^{-1}2 < \dfrac{\pi}{2}$, and $0 < \tan^{-1}3 < \dfrac{\pi}{2}$, so $0 < \tan^{-1}1 + \tan^{-1}2 + \tan^{-1}3 < \dfrac{3\pi}{2}$.

On the interval $\left(0, \dfrac{3\pi}{2}\right)$, $\tan\theta = 0$ if and only if $\theta = \pi$. Therefore, from part (a), $\tan^{-1}1 + \tan^{-1}2 + \tan^{-1}3 = \pi$.

99. $\tan\theta = \tan(\theta_2 - \theta_1) = \dfrac{\tan\theta_2 - \tan\theta_1}{1 + \tan\theta_1\tan\theta_2} = \dfrac{m_2 - m_1}{1 + m_1 m_2}$

101. $2\cot(\alpha - \beta) = \dfrac{2}{\tan(\alpha - \beta)} = 2\left(\dfrac{1 + \tan\alpha\tan\beta}{\tan\alpha - \tan\beta}\right) = 2\left(\dfrac{1 + (x + 1)(x - 1)}{(x + 1) - (x - 1)}\right) = 2\left(\dfrac{1 + x^2 - 1}{x + 1 - x + 1}\right) = \dfrac{2x^2}{2} = x^2$

103. $\tan\dfrac{\pi}{2}$ is not defined; $\tan\left(\dfrac{\pi}{2} - \theta\right) = \dfrac{\sin\left(\dfrac{\pi}{2} - \theta\right)}{\cos\left(\dfrac{\pi}{2} - \theta\right)} = \dfrac{\cos\theta}{\sin\theta} = \cot\theta$.

8.5 Assess Your Understanding (page 643)

1. $\sin^2\theta$; $2\cos^2\theta$; $2\sin^2\theta$ **2.** $1 - \cos\theta$ **3.** $\sin\theta$ **4.** T **5.** F **6.** F **7. (a)** $\dfrac{24}{25}$ **(b)** $\dfrac{7}{25}$ **(c)** $\dfrac{\sqrt{10}}{10}$ **(d)** $\dfrac{3\sqrt{10}}{10}$

9. (a) $\dfrac{24}{25}$ **(b)** $-\dfrac{7}{25}$ **(c)** $\dfrac{2\sqrt{5}}{5}$ **(d)** $-\dfrac{\sqrt{5}}{5}$ **11. (a)** $-\dfrac{2\sqrt{2}}{3}$ **(b)** $\dfrac{1}{3}$ **(c)** $\sqrt{\dfrac{3 + \sqrt{6}}{6}}$ **(d)** $\sqrt{\dfrac{3 - \sqrt{6}}{6}}$

13. (a) $\dfrac{4\sqrt{2}}{9}$ **(b)** $-\dfrac{7}{9}$ **(c)** $\dfrac{\sqrt{3}}{3}$ **(d)** $\dfrac{\sqrt{6}}{3}$ **15. (a)** $-\dfrac{4}{5}$ **(b)** $\dfrac{3}{5}$ **(c)** $\sqrt{\dfrac{5 + 2\sqrt{5}}{10}}$ **(d)** $\sqrt{\dfrac{5 - 2\sqrt{5}}{10}}$

17. (a) $-\dfrac{3}{5}$ **(b)** $-\dfrac{4}{5}$ **(c)** $\dfrac{1}{2}\sqrt{\dfrac{10 - \sqrt{10}}{5}}$ **(d)** $-\dfrac{1}{2}\sqrt{\dfrac{10 + \sqrt{10}}{5}}$ **19.** $\dfrac{\sqrt{2 - \sqrt{2}}}{2}$ **21.** $1 - \sqrt{2}$ **23.** $-\dfrac{\sqrt{2 + \sqrt{3}}}{2}$

25. $\dfrac{2}{\sqrt{2 + \sqrt{2}}} = (2 - \sqrt{2})\sqrt{2 + \sqrt{2}}$ **27.** $-\dfrac{\sqrt{2 - \sqrt{2}}}{2}$ **29.** $-\dfrac{4}{5}$ **31.** $\dfrac{\sqrt{10(5 - \sqrt{5})}}{10}$ **33.** $\dfrac{4}{3}$ **35.** $-\dfrac{7}{8}$ **37.** $\dfrac{\sqrt{10}}{4}$ **39.** $-\dfrac{\sqrt{15}}{3}$

41. $\sin^4\theta = (\sin^2\theta)^2 = \left(\dfrac{1 - \cos(2\theta)}{2}\right)^2 = \dfrac{1}{4}[1 - 2\cos(2\theta) + \cos^2(2\theta)] = \dfrac{1}{4} - \dfrac{1}{2}\cos(2\theta) + \dfrac{1}{4}\cos^2(2\theta)$

$= \dfrac{1}{4} - \dfrac{1}{2}\cos(2\theta) + \dfrac{1}{4}\left(\dfrac{1 + \cos(4\theta)}{2}\right) = \dfrac{1}{4} - \dfrac{1}{2}\cos(2\theta) + \dfrac{1}{8} + \dfrac{1}{8}\cos(4\theta) = \dfrac{3}{8} - \dfrac{1}{2}\cos(2\theta) + \dfrac{1}{8}\cos(4\theta)$

43. $\cos(3\theta) = 4\cos^3\theta - 3\cos\theta$ **45.** $\sin(5\theta) = 16\sin^5\theta - 20\sin^3\theta + 5\sin\theta$ **47.** $\cos^4\theta - \sin^4\theta = (\cos^2\theta + \sin^2\theta)(\cos^2\theta - \sin^2\theta) = \cos(2\theta)$

49. $\cot(2\theta) = \dfrac{1}{\tan(2\theta)} = \dfrac{1 - \tan^2\theta}{2\tan\theta} = \dfrac{1 - \dfrac{1}{\cot^2\theta}}{2\left(\dfrac{1}{\cot\theta}\right)} = \dfrac{\dfrac{\cot^2\theta - 1}{\cot^2\theta}}{\dfrac{2}{\cot\theta}} = \dfrac{\cot^2\theta - 1}{\cot^2\theta} \cdot \dfrac{\cot\theta}{2} = \dfrac{\cot^2\theta - 1}{2\cot\theta}$

51. $\sec(2\theta) = \dfrac{1}{\cos(2\theta)} = \dfrac{1}{2\cos^2\theta - 1} = \dfrac{1}{\dfrac{2}{\sec^2\theta} - 1} = \dfrac{1}{\dfrac{2 - \sec^2\theta}{\sec^2\theta}} = \dfrac{\sec^2\theta}{2 - \sec^2\theta}$ **53.** $\cos^2(2u) - \sin^2(2u) = \cos[2(2u)] = \cos(4u)$

55. $\dfrac{\cos(2\theta)}{1 + \sin(2\theta)} = \dfrac{\cos^2\theta - \sin^2\theta}{1 + 2\sin\theta\cos\theta} = \dfrac{(\cos\theta - \sin\theta)(\cos\theta + \sin\theta)}{\sin^2\theta + \cos^2\theta + 2\sin\theta\cos\theta} = \dfrac{(\cos\theta - \sin\theta)(\cos\theta + \sin\theta)}{(\sin\theta + \cos\theta)(\sin\theta + \cos\theta)} = \dfrac{\cos\theta - \sin\theta}{\cos\theta + \sin\theta}$

$= \dfrac{\dfrac{\cos\theta - \sin\theta}{\sin\theta}}{\dfrac{\cos\theta + \sin\theta}{\sin\theta}} = \dfrac{\dfrac{\cos\theta}{\sin\theta} - \dfrac{\sin\theta}{\sin\theta}}{\dfrac{\cos\theta}{\sin\theta} + \dfrac{\sin\theta}{\sin\theta}} = \dfrac{\cot\theta - 1}{\cot\theta + 1}$

57. $\sec^2\dfrac{\theta}{2} = \dfrac{1}{\cos^2\left(\dfrac{\theta}{2}\right)} = \dfrac{1}{\dfrac{1 + \cos\theta}{2}} = \dfrac{2}{1 + \cos\theta}$

59. $\cot^2\dfrac{v}{2} = \dfrac{1}{\tan^2\left(\dfrac{v}{2}\right)} = \dfrac{1}{\dfrac{1 - \cos v}{1 + \cos v}} = \dfrac{1 + \cos v}{1 - \cos v} = \dfrac{1 + \dfrac{1}{\sec v}}{1 - \dfrac{1}{\sec v}} = \dfrac{\dfrac{\sec v + 1}{\sec v}}{\dfrac{\sec v - 1}{\sec v}} = \dfrac{\sec v + 1}{\sec v} \cdot \dfrac{\sec v}{\sec v - 1} = \dfrac{\sec v + 1}{\sec v - 1}$

61. $\dfrac{1 - \tan^2\left(\dfrac{\theta}{2}\right)}{1 + \tan^2\left(\dfrac{\theta}{2}\right)} = \dfrac{1 - \dfrac{1 - \cos\theta}{1 + \cos\theta}}{1 + \dfrac{1 - \cos\theta}{1 + \cos\theta}} = \dfrac{\dfrac{1 + \cos\theta - (1 - \cos\theta)}{1 + \cos\theta}}{\dfrac{1 + \cos\theta + 1 - \cos\theta}{1 + \cos\theta}} = \dfrac{2\cos\theta}{1 + \cos\theta} \cdot \dfrac{1 + \cos\theta}{2} = \cos\theta$

63. $\dfrac{\sin(3\theta)}{\sin\theta} - \dfrac{\cos(3\theta)}{\cos\theta} = \dfrac{\sin(3\theta)\cos\theta - \cos(3\theta)\sin\theta}{\sin\theta\cos\theta} = \dfrac{\sin(3\theta - \theta)}{\dfrac{1}{2}(2\sin\theta\cos\theta)} = \dfrac{2\sin(2\theta)}{\sin(2\theta)} = 2$

65. $\tan(3\theta) = \tan(\theta + 2\theta) = \dfrac{\tan\theta + \tan(2\theta)}{1 - \tan\theta\tan(2\theta)} = \dfrac{\tan\theta + \dfrac{2\tan\theta}{1 - \tan^2\theta}}{1 - \dfrac{\tan\theta(2\tan\theta)}{1 - \tan^2\theta}} = \dfrac{\tan\theta - \tan^3\theta + 2\tan\theta}{1 - \tan^2\theta - 2\tan^2\theta} = \dfrac{3\tan\theta - \tan^3\theta}{1 - 3\tan^2\theta}$

67. $\dfrac{1}{2}(\ln|1 - \cos(2\theta)| - \ln 2) = \ln\left(\dfrac{|1 - \cos(2\theta)|}{2}\right)^{1/2} = \ln|\sin^2\theta|^{1/2} = \ln|\sin\theta|$ **69.** $\dfrac{\sqrt{3}}{2}$ **71.** $\dfrac{7}{25}$ **73.** $\dfrac{24}{7}$ **75.** $\dfrac{24}{25}$ **77.** $\dfrac{1}{5}$ **79.** $\dfrac{25}{7}$

81. (a) $W = 2D(\csc\theta - \cot\theta) = 2D\left(\dfrac{1}{\sin\theta} - \dfrac{\cos\theta}{\sin\theta}\right) = 2D\dfrac{1 - \cos\theta}{\sin\theta} = 2D\tan\dfrac{\theta}{2}$ **(b)** $\theta = 24.45°$

83. (a) $R = \dfrac{v_0^2\sqrt{2}}{16}\cos\theta(\sin\theta - \cos\theta)$ **(b)** **(c)** $\theta = 67.5°$ makes R largest.

$= \dfrac{v_0^2\sqrt{2}}{32}(2\cos\theta\sin\theta - 2\cos^2\theta)$

$= \dfrac{v_0^2\sqrt{2}}{32}[\sin(2\theta) - \cos(2\theta) - 1]$

85. $A = \dfrac{1}{2}h(\text{base}) = h\left(\dfrac{1}{2}\text{base}\right) = s\cos\dfrac{\theta}{2} \cdot s\sin\dfrac{\theta}{2} = \dfrac{1}{2}s^2\sin\theta$ **87.** $\sin(2\theta) = \dfrac{4x}{4 + x^2}$ **89.** $-\dfrac{1}{4}$

91. $\dfrac{2z}{1 + z^2} = \dfrac{2\tan\left(\dfrac{\alpha}{2}\right)}{1 + \tan^2\left(\dfrac{\alpha}{2}\right)} = \dfrac{2\tan\left(\dfrac{\alpha}{2}\right)}{\sec^2\left(\dfrac{\alpha}{2}\right)} = \dfrac{\dfrac{2\sin\left(\dfrac{\alpha}{2}\right)}{\cos\left(\dfrac{\alpha}{2}\right)}}{\dfrac{1}{\cos^2\left(\dfrac{\alpha}{2}\right)}} = 2\sin\left(\dfrac{\alpha}{2}\right)\cos\left(\dfrac{\alpha}{2}\right) = \sin\left(2 \cdot \dfrac{\alpha}{2}\right) = \sin\alpha$

93.

95. $\sin\dfrac{\pi}{24} = \dfrac{\sqrt{2}}{4}\sqrt{4 - \sqrt{6} - \sqrt{2}};$

$\cos\dfrac{\pi}{24} = \dfrac{\sqrt{2}}{4}\sqrt{4 + \sqrt{6} + \sqrt{2}}$

97. $\sin^3\theta + \sin^3(\theta + 120°) + \sin^3(\theta + 240°) = \sin^3\theta + (\sin\theta\cos 120° + \cos\theta\sin 120°)^3 + (\sin\theta\cos 240° + \cos\theta\sin 240°)^3$

$= \sin^3\theta + \left(-\dfrac{1}{2}\sin\theta + \dfrac{\sqrt{3}}{2}\cos\theta\right)^3 + \left(-\dfrac{1}{2}\sin\theta - \dfrac{\sqrt{3}}{2}\cos\theta\right)^3$

$= \sin^3\theta + \dfrac{1}{8}\left(3\sqrt{3}\cos^3\theta - 9\cos^2\theta\sin\theta + 3\sqrt{3}\cos\theta\sin^2\theta - \sin^3\theta\right) - \dfrac{1}{8}\left(\sin^3\theta + 3\sqrt{3}\sin^2\theta\cos\theta + 9\sin\theta\cos^2\theta + 3\sqrt{3}\cos^3\theta\right)$

$= \dfrac{3}{4}\sin^3\theta - \dfrac{9}{4}\cos^2\theta\sin\theta = \dfrac{3}{4}[\sin^3\theta - 3\sin\theta(1 - \sin^2\theta)] = \dfrac{3}{4}(4\sin^3\theta - 3\sin\theta) = -\dfrac{3}{4}\sin(3\theta)$ (from Example 2)

8.6 Assess Your Understanding (page 648)

1. $\dfrac{1}{2}[\cos(2\theta) - \cos(6\theta)]$ **3.** $\dfrac{1}{2}[\sin(6\theta) + \sin(2\theta)]$ **5.** $\dfrac{1}{2}[\cos(2\theta) + \cos(8\theta)]$ **7.** $\dfrac{1}{2}[\cos\theta - \cos(3\theta)]$ **9.** $\dfrac{1}{2}[\sin(2\theta) + \sin\theta]$ **11.** $2\sin\theta\cos(3\theta)$

13. $2\cos(3\theta)\cos\theta$ **15.** $2\sin(2\theta)\cos\theta$ **17.** $2\sin\theta\sin\dfrac{\theta}{2}$ **19.** $\dfrac{\sin\theta + \sin(3\theta)}{2\sin(2\theta)} = \dfrac{2\sin(2\theta)\cos\theta}{2\sin(2\theta)} = \cos\theta$

21. $\dfrac{\sin(4\theta) + \sin(2\theta)}{\cos(4\theta) + \cos(2\theta)} = \dfrac{2\sin(3\theta)\cos\theta}{2\cos(3\theta)\cos\theta} = \dfrac{\sin(3\theta)}{\cos(3\theta)} = \tan(3\theta)$ **23.** $\dfrac{\cos\theta - \cos(3\theta)}{\sin\theta + \sin(3\theta)} = \dfrac{2\sin(2\theta)\sin\theta}{2\sin(2\theta)\cos\theta} = \dfrac{\sin\theta}{\cos\theta} = \tan\theta$

25. $\sin\theta[\sin\theta + \sin(3\theta)] = \sin\theta[2\sin(2\theta)\cos\theta] = \cos\theta[2\sin(2\theta)\sin\theta] = \cos\theta\left[2\cdot\dfrac{1}{2}[\cos\theta - \cos(3\theta)]\right] = \cos\theta[\cos\theta - \cos(3\theta)]$

27. $\dfrac{\sin(4\theta) + \sin(8\theta)}{\cos(4\theta) + \cos(8\theta)} = \dfrac{2\sin(6\theta)\cos(2\theta)}{2\cos(6\theta)\cos(2\theta)} = \dfrac{\sin(6\theta)}{\cos(6\theta)} = \tan(6\theta)$

29. $\dfrac{\sin(4\theta) + \sin(8\theta)}{\sin(4\theta) - \sin(8\theta)} = \dfrac{2\sin(6\theta)\cos(-2\theta)}{2\sin(-2\theta)\cos(6\theta)} = \dfrac{\sin(6\theta)}{\cos(6\theta)}\cdot\dfrac{\cos(2\theta)}{-\sin(2\theta)} = \tan(6\theta)[-\cot(2\theta)] = -\dfrac{\tan(6\theta)}{\tan(2\theta)}$

31. $\dfrac{\sin\alpha + \sin\beta}{\sin\alpha - \sin\beta} = \dfrac{2\sin\dfrac{\alpha+\beta}{2}\cos\dfrac{\alpha-\beta}{2}}{2\sin\dfrac{\alpha-\beta}{2}\cos\dfrac{\alpha+\beta}{2}} = \dfrac{\sin\dfrac{\alpha+\beta}{2}}{\cos\dfrac{\alpha+\beta}{2}}\cdot\dfrac{\cos\dfrac{\alpha-\beta}{2}}{\sin\dfrac{\alpha-\beta}{2}} = \tan\dfrac{\alpha+\beta}{2}\cot\dfrac{\alpha-\beta}{2}$

33. $\dfrac{\sin\alpha + \sin\beta}{\cos\alpha + \cos\beta} = \dfrac{2\sin\dfrac{\alpha+\beta}{2}\cos\dfrac{\alpha-\beta}{2}}{2\cos\dfrac{\alpha+\beta}{2}\cos\dfrac{\alpha-\beta}{2}} = \dfrac{\sin\dfrac{\alpha+\beta}{2}}{\cos\dfrac{\alpha+\beta}{2}} = \tan\dfrac{\alpha+\beta}{2}$

35. $1 + \cos(2\theta) + \cos(4\theta) + \cos(6\theta)$

$= [1 + \cos(6\theta)] + [\cos(2\theta) + \cos(4\theta)]$

$= 2\cos^2(3\theta) + 2\cos(3\theta)\cos(-\theta)$

$= 2\cos(3\theta)[\cos(3\theta) + \cos\theta]$

$= 2\cos(3\theta)[2\cos(2\theta)\cos\theta]$

$= 4\cos\theta\cos(2\theta)\cos(3\theta)$

37. **(a)** $y = 2\sin(2061\pi t)\cos(357\pi t)$ **(b)** $y_{max} = 2$

(c)

39. $I_u = I_x\cos^2\theta + I_y\sin^2\theta - 2I_{xy}\sin\theta\cos\theta = I_x\cos^2\theta + I_y\sin^2\theta - I_{xy}\sin 2\theta$

$= I_x\left(\dfrac{\cos 2\theta + 1}{2}\right) + I_y\left(\dfrac{1 - \cos 2\theta}{2}\right) - I_{xy}\sin 2\theta$

$= \dfrac{I_x}{2}\cos 2\theta + \dfrac{I_x}{2} + \dfrac{I_y}{2} - \dfrac{I_y}{2}\cos 2\theta - I_{xy}\sin 2\theta$

$= \dfrac{I_x + I_y}{2} + \dfrac{I_x - I_y}{2}\cos 2\theta - I_{xy}\sin 2\theta$

$I_v = I_x\sin^2\theta + I_y\cos^2\theta + 2I_{xy}\sin\theta\cos\theta = I_x\left(\dfrac{1 - \cos 2\theta}{2}\right) + I_y\left(\dfrac{\cos 2\theta + 1}{2}\right) + I_{xy}\sin 2\theta$

$= \dfrac{I_x}{2} - \dfrac{I_x}{2}\cos 2\theta + \dfrac{I_y}{2}\cos 2\theta + \dfrac{I_y}{2} + I_{xy}\sin 2\theta$

$= \dfrac{I_x + I_y}{2} - \dfrac{I_x - I_y}{2}\cos 2\theta + I_{xy}\sin 2\theta$

41. $\sin(2\alpha) + \sin(2\beta) + \sin(2\gamma) = 2\sin(\alpha + \beta)\cos(\alpha - \beta) + \sin(2\gamma) = 2\sin(\alpha + \beta)\cos(\alpha - \beta) + 2\sin\gamma\cos\gamma$

$= 2\sin(\pi - \gamma)\cos(\alpha - \beta) + 2\sin\gamma\cos\gamma = 2\sin\gamma\cos(\alpha - \beta) + 2\sin\gamma\cos\gamma = 2\sin\gamma[\cos(\alpha - \beta) + \cos\gamma]$

$= 2\sin\gamma\left(2\cos\dfrac{\alpha - \beta + \gamma}{2}\cos\dfrac{\alpha - \beta - \gamma}{2}\right) = 4\sin\gamma\cos\dfrac{\pi - 2\beta}{2}\cos\dfrac{2\alpha - \pi}{2} = 4\sin\gamma\cos\left(\dfrac{\pi}{2} - \beta\right)\cos\left(\alpha - \dfrac{\pi}{2}\right)$

$= 4\sin\gamma\sin\beta\sin\alpha$

43.

$$\sin(\alpha - \beta) = \sin \alpha \cos \beta - \cos \alpha \sin \beta$$
$$\sin(\alpha + \beta) = \sin \alpha \cos \beta + \cos \alpha \sin \beta$$
$$\sin(\alpha - \beta) + \sin(\alpha + \beta) = 2 \sin \alpha \cos \beta$$
$$\sin \alpha \cos \beta = \frac{1}{2}[\sin(\alpha + \beta) + \sin(\alpha - \beta)]$$

45. $2 \cos \dfrac{\alpha + \beta}{2} \cos \dfrac{\alpha - \beta}{2} = 2 \cdot \dfrac{1}{2}\left[\cos\left(\dfrac{\alpha + \beta}{2} + \dfrac{\alpha - \beta}{2}\right) + \cos\left(\dfrac{\alpha + \beta}{2} - \dfrac{\alpha - \beta}{2}\right)\right] = \cos \dfrac{2\alpha}{2} + \cos \dfrac{2\beta}{2} = \cos \alpha + \cos \beta$

8.7 Assess Your Understanding (page 653)

3. $\dfrac{\pi}{6}; \dfrac{5\pi}{6}$ **4.** $\left\{\theta \middle| \theta = \dfrac{\pi}{6} + 2\pi k, \theta = \dfrac{5\pi}{6} + 2\pi k, k \text{ any integer}\right\}$ **5.** F **6.** F **7.** $\left\{\dfrac{7\pi}{6}, \dfrac{11\pi}{6}\right\}$ **9.** $\left\{\dfrac{\pi}{3}, \dfrac{2\pi}{3}, \dfrac{4\pi}{3}, \dfrac{5\pi}{3}\right\}$ **11.** $\left\{\dfrac{\pi}{4}, \dfrac{3\pi}{4}, \dfrac{5\pi}{4}, \dfrac{7\pi}{4}\right\}$

13. $\left\{\dfrac{\pi}{2}, \dfrac{7\pi}{6}, \dfrac{11\pi}{6}\right\}$ **15.** $\left\{\dfrac{\pi}{3}, \dfrac{2\pi}{3}, \dfrac{4\pi}{3}, \dfrac{5\pi}{3}\right\}$ **17.** $\left\{\dfrac{4\pi}{9}, \dfrac{8\pi}{9}, \dfrac{16\pi}{9}\right\}$ **19.** $\left\{\dfrac{7\pi}{6}, \dfrac{11\pi}{6}\right\}$ **21.** $\left\{\dfrac{3\pi}{4}, \dfrac{7\pi}{4}\right\}$ **23.** $\left\{\dfrac{2\pi}{3}, \dfrac{4\pi}{3}\right\}$

25. $\left\{\dfrac{3\pi}{4}, \dfrac{5\pi}{4}\right\}$ **27.** $\left\{\dfrac{3\pi}{4}, \dfrac{7\pi}{4}\right\}$ **29.** $\left\{\dfrac{11\pi}{6}\right\}$ **31.** $\left\{\theta \middle| \theta = \dfrac{\pi}{6} + 2k\pi, \theta = \dfrac{5\pi}{6} + 2k\pi\right\}; \dfrac{\pi}{6}, \dfrac{5\pi}{6}, \dfrac{13\pi}{6}, \dfrac{17\pi}{6}, \dfrac{25\pi}{6}, \dfrac{29\pi}{6}$

33. $\left\{\theta \middle| \theta = \dfrac{5\pi}{6} + k\pi\right\}; \dfrac{5\pi}{6}, \dfrac{11\pi}{6}, \dfrac{17\pi}{6}, \dfrac{23\pi}{6}, \dfrac{29\pi}{6}, \dfrac{35\pi}{6}$ **35.** $\left\{\theta \middle| \theta = \dfrac{\pi}{2} + 2k\pi, \theta = \dfrac{3\pi}{2} + 2k\pi\right\}; \dfrac{\pi}{2}, \dfrac{3\pi}{2}, \dfrac{5\pi}{2}, \dfrac{7\pi}{2}, \dfrac{9\pi}{2}, \dfrac{11\pi}{2}$

37. $\left\{\theta \middle| \theta = \dfrac{\pi}{3} + k\pi, \theta = \dfrac{2\pi}{3} + k\pi\right\}; \dfrac{\pi}{3}, \dfrac{2\pi}{3}, \dfrac{4\pi}{3}, \dfrac{5\pi}{3}, \dfrac{7\pi}{3}, \dfrac{8\pi}{3}$ **39.** $\left\{\theta \middle| \theta = \dfrac{8\pi}{3} + 4k\pi, \theta = \dfrac{10\pi}{3} + 4k\pi\right\}; \dfrac{8\pi}{3}, \dfrac{10\pi}{3}, \dfrac{20\pi}{3}, \dfrac{22\pi}{3}, \dfrac{32\pi}{3}, \dfrac{34\pi}{3}$

41. $\{0.41, 2.73\}$ **43.** $\{1.37, 4.51\}$ **45.** $\{2.69, 3.59\}$ **47.** $\{1.82, 4.46\}$ **49.** $\{2.08, 5.22\}$ **51.** $\{0.73, 2.41\}$ **53.** $\dfrac{\pi}{3}, \dfrac{2\pi}{3}, \dfrac{4\pi}{3}, \dfrac{5\pi}{3}$

55. (a) $-2\pi, -\pi, 0, \pi, 2\pi, 3\pi, 4\pi$

(b)

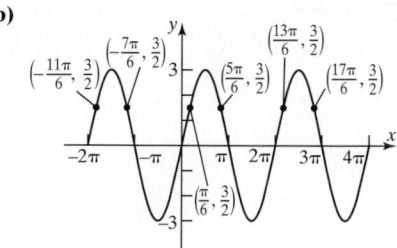

(c) $\left\{-\dfrac{11\pi}{6}, -\dfrac{7\pi}{6}, \dfrac{\pi}{6}, \dfrac{5\pi}{6}, \dfrac{13\pi}{6}, \dfrac{17\pi}{6}\right\}$

(d) $\left\{x \middle| -\dfrac{11\pi}{6} < x < -\dfrac{7\pi}{6} \text{ or } \dfrac{\pi}{6} < x < \dfrac{5\pi}{6} \text{ or } \dfrac{13\pi}{6} < x < \dfrac{17\pi}{6}\right\}$

57. (a) $\left\{x \middle| x = -\dfrac{\pi}{4} + k\pi\right\}$ **(b)** $-\dfrac{\pi}{2} < x < -\dfrac{\pi}{4}$ or $\left(-\dfrac{\pi}{2}, -\dfrac{\pi}{4}\right)$

59. (a), (d)

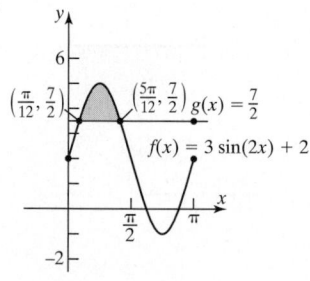

(b) $\left\{\dfrac{\pi}{12}, \dfrac{5\pi}{12}\right\}$

(c) $\left\{x \middle| \dfrac{\pi}{12} < x < \dfrac{5\pi}{12}\right\}$ or $\left(\dfrac{\pi}{12}, \dfrac{5\pi}{12}\right)$

61. (a), (d)

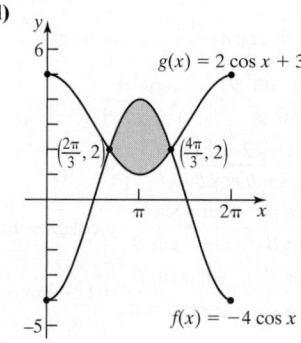

(b) $\left\{\dfrac{2\pi}{3}, \dfrac{4\pi}{3}\right\}$

(c) $\left\{x \middle| \dfrac{2\pi}{3} < x < \dfrac{4\pi}{3}\right\}$ or $\left(\dfrac{2\pi}{3}, \dfrac{4\pi}{3}\right)$

63. (a) 10 sec; 30 sec **(b)** 20 sec; 60 sec

(c) $10 < x < 30$ or $(10, 30)$

65. (a) 150 mi **(b)** 6.06, 8.44, 15.72, 18.11 min

(c) Before 6.06 min, between 8.44 and 15.72 min, and after 18.11 min **(d)** No

67. 28.90° **69.** Yes; it varies from 1.25 to 1.34.

71. 1.47

73. If θ is the original angle of incidence and ϕ is the angle of refraction, then $\dfrac{\sin \theta}{\sin \phi} = n_2$.

The angle of incidence of the emerging beam is also ϕ, and the index of refraction is $\dfrac{1}{n_2}$.

Thus, θ is the angle of refraction of the emerging beam.

8.8 Assess Your Understanding (page 661)

5. $\left\{\dfrac{\pi}{2}, \dfrac{2\pi}{3}, \dfrac{4\pi}{3}, \dfrac{3\pi}{2}\right\}$ **7.** $\left\{\dfrac{\pi}{2}, \dfrac{7\pi}{6}, \dfrac{11\pi}{6}\right\}$ **9.** $\left\{0, \dfrac{\pi}{4}, \dfrac{5\pi}{4}\right\}$ **11.** $\left\{\dfrac{\pi}{2}, \dfrac{2\pi}{3}, \dfrac{4\pi}{3}, \dfrac{3\pi}{2}\right\}$ **13.** $\{\pi\}$ **15.** $\left\{\dfrac{\pi}{3}, \dfrac{2\pi}{3}, \dfrac{4\pi}{3}, \dfrac{5\pi}{3}\right\}$ **17.** $\left\{\dfrac{\pi}{4}, \dfrac{5\pi}{4}\right\}$ **19.** $\left\{0, \dfrac{\pi}{3}, \pi, \dfrac{5\pi}{3}\right\}$

21. $\left\{\dfrac{\pi}{2}, \dfrac{3\pi}{2}\right\}$ **23.** $\left\{0, \dfrac{2\pi}{3}, \dfrac{4\pi}{3}\right\}$ **25.** $\left\{0, \dfrac{\pi}{3}, \dfrac{\pi}{2}, \dfrac{2\pi}{3}, \pi, \dfrac{4\pi}{3}, \dfrac{3\pi}{2}, \dfrac{5\pi}{3}\right\}$ **27.** $\left\{0, \dfrac{\pi}{5}, \dfrac{2\pi}{5}, \dfrac{3\pi}{5}, \dfrac{4\pi}{5}, \pi, \dfrac{6\pi}{5}, \dfrac{7\pi}{5}, \dfrac{8\pi}{5}, \dfrac{9\pi}{5}\right\}$ **29.** $\left\{\dfrac{\pi}{6}, \dfrac{5\pi}{6}, \dfrac{3\pi}{2}\right\}$ **31.** $\left\{\dfrac{\pi}{2}\right\}$

33. $\{0\}$ **35.** $\left\{\dfrac{\pi}{3}, \dfrac{5\pi}{3}\right\}$ **37.** No real solution **39.** No real solution **41.** $\left\{\dfrac{\pi}{2}, \dfrac{7\pi}{6}\right\}$ **43.** $\left\{0, \dfrac{\pi}{3}, \pi, \dfrac{5\pi}{3}\right\}$ **45.** $\left\{\dfrac{\pi}{4}\right\}$ **47.** $\dfrac{\pi}{3}, \dfrac{2\pi}{3}, \dfrac{4\pi}{3}, \dfrac{5\pi}{3}$

49. $0, \dfrac{\pi}{3}, \pi, \dfrac{5\pi}{3}$ **51.** π **53.** $-1.31, 1.98, 3.84$ **55.** 0.52 **57.** 1.26 **59.** $-1.02, 1.02$ **61.** $0, 2.15$ **63.** $0.76, 1.35$

65. (a) $60°$ **(b)** $60°$ **67.** $2.03, 4.91$ **69. (a)** $30°, 60°$

(c) $A(60°) = 12\sqrt{3}$ in.2 **(b)** 123.6 m

(d) 21 **(c)** 130

$\theta_{\max} = 60°$
Maximum area $= 20.78$ in.2

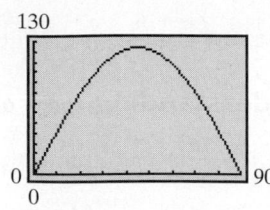

Review Exercises (page 664)

1. $\dfrac{\pi}{2}$ **3.** $\dfrac{\pi}{4}$ **5.** $\dfrac{5\pi}{6}$ **7.** $\dfrac{\pi}{4}$ **9.** $\dfrac{3\pi}{8}$ **11.** $-\dfrac{\pi}{3}$ **13.** $\dfrac{\pi}{7}$ **15.** 0.9 **17.** -0.3 **19.** Not defined **21.** $-\dfrac{\pi}{6}$ **23.** $-\dfrac{\pi}{4}$ **25.** $-\sqrt{3}$ **27.** $\dfrac{2\sqrt{3}}{3}$

29. $\dfrac{4}{5}$ **31.** $-\dfrac{4}{3}$ **33.** $f^{-1}(x) = \dfrac{1}{3}\sin^{-1}\left(\dfrac{x}{2}\right)$; Domain of f: $(-\infty, \infty)$; Domain of f^{-1}: $[-2, 2]$

35. $f^{-1}(x) = \cos^{-1}(3 - x)$; Domain of f: $(-\infty, \infty)$; Domain of f^{-1}: $[2, 4]$ **37.** $\sqrt{1 - u^2}$ **39.** $\dfrac{1}{u}$ **41.** $\tan\theta\cot\theta - \sin^2\theta = 1 - \sin^2\theta = \cos^2\theta$

43. $\sin^2\theta(1 + \cot^2\theta) = \sin^2\theta\csc^2\theta = 1$ **45.** $5\cos^2\theta + 3\sin^2\theta = 2\cos^2\theta + 3(\cos^2\theta + \sin^2\theta) = 3 + 2\cos^2\theta$

47. $\dfrac{1 - \cos\theta}{\sin\theta} + \dfrac{\sin\theta}{1 - \cos\theta} = \dfrac{(1 - \cos\theta)^2 + \sin^2\theta}{\sin\theta(1 - \cos\theta)} = \dfrac{1 - 2\cos\theta + \cos^2\theta + \sin^2\theta}{\sin\theta(1 - \cos\theta)} = \dfrac{2(1 - \cos\theta)}{\sin\theta(1 - \cos\theta)} = 2\csc\theta$

49. $\dfrac{\cos\theta}{\cos\theta - \sin\theta} = \dfrac{\cos\theta}{\dfrac{\cos\theta - \sin\theta}{\cos\theta}} = \dfrac{1}{1 - \dfrac{\sin\theta}{\cos\theta}} = \dfrac{1}{1 - \tan\theta}$

51. $\dfrac{\csc\theta}{1 + \csc\theta} = \dfrac{\dfrac{1}{\sin\theta}}{1 + \dfrac{1}{\sin\theta}} = \dfrac{1}{1 + \sin\theta} = \dfrac{1}{1 + \sin\theta} \cdot \dfrac{1 - \sin\theta}{1 - \sin\theta} = \dfrac{1 - \sin\theta}{1 - \sin^2\theta} = \dfrac{1 - \sin\theta}{\cos^2\theta}$

53. $\csc\theta - \sin\theta = \dfrac{1}{\sin\theta} - \sin\theta = \dfrac{1 - \sin^2\theta}{\sin\theta} = \dfrac{\cos^2\theta}{\sin\theta} = \cos\theta \cdot \dfrac{\cos\theta}{\sin\theta} = \cos\theta\cot\theta$

55. $\dfrac{1 - \sin\theta}{\sec\theta} = \cos\theta(1 - \sin\theta) \cdot \dfrac{1 + \sin\theta}{1 + \sin\theta} = \dfrac{\cos\theta(1 - \sin^2\theta)}{1 + \sin\theta} = \dfrac{\cos^3\theta}{1 + \sin\theta}$

57. $\cot\theta - \tan\theta = \dfrac{\cos\theta}{\sin\theta} - \dfrac{\sin\theta}{\cos\theta} = \dfrac{\cos^2\theta - \sin^2\theta}{\sin\theta\cos\theta} = \dfrac{1 - 2\sin^2\theta}{\sin\theta\cos\theta}$

59. $\dfrac{\cos(\alpha + \beta)}{\cos\alpha\sin\beta} = \dfrac{\cos\alpha\cos\beta - \sin\alpha\sin\beta}{\cos\alpha\sin\beta} = \dfrac{\cos\alpha\cos\beta}{\cos\alpha\sin\beta} - \dfrac{\sin\alpha\sin\beta}{\cos\alpha\sin\beta} = \cot\beta - \tan\alpha$

61. $\dfrac{\cos(\alpha - \beta)}{\cos\alpha\cos\beta} = \dfrac{\cos\alpha\cos\beta + \sin\alpha\sin\beta}{\cos\alpha\cos\beta} = \dfrac{\cos\alpha\cos\beta}{\cos\alpha\cos\beta} + \dfrac{\sin\alpha\sin\beta}{\cos\alpha\cos\beta} = 1 + \tan\alpha\tan\beta$

63. $(1 + \cos\theta)\left(\tan\dfrac{\theta}{2}\right) = (1 + \cos\theta) \cdot \dfrac{\sin\theta}{1 + \cos\theta} = \sin\theta$

65. $2\cot\theta\cot 2\theta = 2\left(\dfrac{\cos\theta}{\sin\theta}\right)\left(\dfrac{\cos 2\theta}{\sin 2\theta}\right) = \dfrac{2\cos\theta(\cos^2\theta - \sin^2\theta)}{2\sin^2\theta\cos\theta} = \dfrac{\cos^2\theta - \sin^2\theta}{\sin^2\theta} = \cot^2\theta - 1$

67. $1 - 8\sin^2\theta\cos^2\theta = 1 - 2(2\sin\theta\cos\theta)^2 = 1 - 2\sin^2(2\theta) = \cos(4\theta)$ **69.** $\dfrac{\sin(2\theta) + \sin(4\theta)}{\cos(2\theta) + \cos(4\theta)} = \dfrac{2\sin(3\theta)\cos(-\theta)}{2\cos(3\theta)\cos(-\theta)} = \tan(3\theta)$

71. $\dfrac{\cos(2\theta) - \cos(4\theta)}{\cos(2\theta) + \cos(4\theta)} - \tan\theta\tan(3\theta) = \dfrac{-2\sin(3\theta)\sin(-\theta)}{2\cos(3\theta)\cos(-\theta)} - \tan\theta\tan(3\theta) = \tan(3\theta)\tan\theta - \tan\theta\tan(3\theta) = 0$

73. $\dfrac{1}{4}(\sqrt{6} - \sqrt{2})$ **75.** $\dfrac{1}{4}(\sqrt{6} - \sqrt{2})$ **77.** $\dfrac{1}{2}$ **79.** $\sqrt{2} - 1$ **81. (a)** $-\dfrac{33}{65}$ **(b)** $-\dfrac{56}{65}$ **(c)** $-\dfrac{63}{65}$ **(d)** $\dfrac{33}{56}$ **(e)** $\dfrac{24}{25}$ **(f)** $\dfrac{119}{169}$ **(g)** $\dfrac{5\sqrt{26}}{26}$

(h) $\dfrac{2\sqrt{5}}{5}$ **83. (a)** $-\dfrac{16}{65}$ **(b)** $-\dfrac{63}{65}$ **(c)** $-\dfrac{56}{65}$ **(d)** $\dfrac{16}{63}$ **(e)** $\dfrac{24}{25}$ **(f)** $\dfrac{119}{169}$ **(g)** $\dfrac{\sqrt{26}}{26}$ **(h)** $-\dfrac{\sqrt{10}}{10}$ **85. (a)** $-\dfrac{63}{65}$ **(b)** $\dfrac{16}{65}$ **(c)** $\dfrac{33}{65}$ **(d)** $-\dfrac{63}{16}$

(e) $\dfrac{24}{25}$ **(f)** $-\dfrac{119}{169}$ **(g)** $\dfrac{2\sqrt{13}}{13}$ **(h)** $-\dfrac{\sqrt{10}}{10}$ **87. (a)** $\dfrac{-\sqrt{3} - 2\sqrt{2}}{6}$ **(b)** $\dfrac{1 - 2\sqrt{6}}{6}$ **(c)** $\dfrac{-\sqrt{3} + 2\sqrt{2}}{6}$ **(d)** $\dfrac{8\sqrt{2} + 9\sqrt{3}}{23}$ **(e)** $-\dfrac{\sqrt{3}}{2}$

(f) $-\dfrac{7}{9}$ **(g)** $\dfrac{\sqrt{3}}{3}$ **(h)** $\dfrac{\sqrt{3}}{2}$ **89. (a)** 1 **(b)** 0 **(c)** $-\dfrac{1}{9}$ **(d)** Not defined **(e)** $\dfrac{4\sqrt{5}}{9}$ **(f)** $-\dfrac{1}{9}$ **(g)** $\dfrac{\sqrt{30}}{6}$ **(h)** $-\dfrac{\sqrt{6}\sqrt{3} - \sqrt{5}}{6}$ **91.** $\dfrac{4 + 3\sqrt{3}}{10}$

93. $-\dfrac{48 + 25\sqrt{3}}{39}$ **95.** $-\dfrac{24}{25}$ **97.** $\left\{\dfrac{\pi}{3}, \dfrac{5\pi}{3}\right\}$ **99.** $\left\{\dfrac{3\pi}{4}, \dfrac{5\pi}{4}\right\}$ **101.** $\left\{\dfrac{3\pi}{4}, \dfrac{7\pi}{4}\right\}$ **103.** $\left\{0, \dfrac{\pi}{2}, \pi, \dfrac{3\pi}{2}\right\}$ **105.** $\left\{\dfrac{\pi}{3}, \dfrac{2\pi}{3}, \dfrac{4\pi}{3}, \dfrac{5\pi}{3}\right\}$ **107.** $\{0, \pi\}$

109. $\left\{0, \dfrac{2\pi}{3}, \pi, \dfrac{4\pi}{3}\right\}$ **111.** $\left\{0, \dfrac{\pi}{6}, \dfrac{5\pi}{6}\right\}$ **113.** $\left\{\dfrac{\pi}{6}, \dfrac{\pi}{2}, \dfrac{5\pi}{6}\right\}$ **115.** $\left\{\dfrac{\pi}{3}, \dfrac{5\pi}{3}\right\}$ **117.** $\left\{\dfrac{\pi}{4}, \dfrac{\pi}{2}, \dfrac{3\pi}{4}, \dfrac{3\pi}{2}\right\}$ **119.** $\left\{\dfrac{\pi}{2}, \pi\right\}$ **121.** 0.78 **123.** −1.11

125. 1.23 **127.** $\{1.11\}$ **129.** $\{0.87\}$ **131.** $\{2.22\}$ **133.** $\left\{-\dfrac{\sqrt{3}}{2}\right\}$

Chapter Test *(page 666)*

1. $\dfrac{\pi}{6}$ **2.** $-\dfrac{\pi}{4}$ **3.** $\dfrac{\pi}{5}$ **4.** $\dfrac{7}{3}$ **5.** 3 **6.** $-\dfrac{4}{3}$ **7.** ≈ 0.392 **8.** ≈ 0.775 **9.** ≈ 1.249 **10.** ≈ 0.197

11. $\dfrac{(\csc\theta + \cot\theta)}{(\sec\theta + \tan\theta)} = \dfrac{(\csc\theta + \cot\theta)}{(\sec\theta + \tan\theta)} \cdot \dfrac{(\csc\theta - \cot\theta)}{(\csc\theta - \cot\theta)} = \dfrac{(\csc^2\theta - \cot^2\theta)}{(\sec\theta + \tan\theta)(\csc\theta - \cot\theta)} = \dfrac{1}{(\sec\theta + \tan\theta)(\csc\theta - \cot\theta)}$

$= \dfrac{1}{(\sec\theta + \tan\theta)(\csc\theta - \cot\theta)} \cdot \dfrac{\sec\theta - \tan\theta}{\sec\theta - \tan\theta} = \dfrac{\sec\theta - \tan\theta}{(\sec^2\theta - \tan^2\theta)(\csc\theta - \cot\theta)} = \dfrac{\sec\theta - \tan\theta}{\csc\theta - \cot\theta}$

12. $\sin\theta \tan\theta + \cos\theta = \sin\theta \cdot \dfrac{\sin\theta}{\cos\theta} + \cos\theta = \dfrac{\sin^2\theta}{\cos\theta} + \dfrac{\cos^2\theta}{\cos\theta} = \dfrac{\sin^2\theta + \cos^2\theta}{\cos\theta} = \dfrac{1}{\cos\theta} = \sec\theta$

13. $\tan\theta + \cot\theta = \dfrac{\sin\theta}{\cos\theta} + \dfrac{\cos\theta}{\sin\theta} = \dfrac{\sin^2\theta}{\sin\theta\cos\theta} + \dfrac{\cos^2\theta}{\sin\theta\cos\theta} = \dfrac{\sin^2\theta + \cos^2\theta}{\sin\theta\cos\theta} = \dfrac{1}{\sin\theta\cos\theta} = \dfrac{2}{2\sin\theta\cos\theta} = \dfrac{2}{\sin(2\theta)} = 2\csc(2\theta)$

14. $\dfrac{\sin(\alpha + \beta)}{\tan\alpha + \tan\beta} = \dfrac{\sin\alpha\cos\beta + \cos\alpha\sin\beta}{\dfrac{\sin\alpha}{\cos\alpha} + \dfrac{\sin\beta}{\cos\beta}} = \dfrac{\sin\alpha\cos\beta + \cos\alpha\sin\beta}{\dfrac{\sin\alpha\cos\beta}{\cos\alpha\cos\beta} + \dfrac{\cos\alpha\sin\beta}{\cos\alpha\cos\beta}} = \dfrac{\sin\alpha\cos\beta + \cos\alpha\sin\beta}{\dfrac{\sin\alpha\cos\beta + \cos\alpha\sin\beta}{\cos\alpha\cos\beta}}$

$= \dfrac{\sin\alpha\cos\beta + \cos\alpha\sin\beta}{1} \cdot \dfrac{\cos\alpha\cos\beta}{\sin\alpha\cos\beta + \cos\alpha\sin\beta} = \cos\alpha\cos\beta$

15. $\sin(3\theta) = \sin(\theta + 2\theta) = \sin\theta\cos(2\theta) + \cos\theta\sin(2\theta) = \sin\theta \cdot (\cos^2\theta - \sin^2\theta) + \cos\theta \cdot 2\sin\theta\cos\theta = \sin\theta\cos^2\theta - \sin^3\theta + 2\sin\theta\cos^2\theta$

$= 3\sin\theta\cos^2\theta - \sin^3\theta = 3\sin\theta(1 - \sin^2\theta) - \sin^3\theta = 3\sin\theta - 3\sin^3\theta - \sin^3\theta = 3\sin\theta - 4\sin^3\theta$

16. $\dfrac{(\tan\theta - \cot\theta)}{(\tan\theta + \cot\theta)} = \dfrac{\dfrac{\sin\theta}{\cos\theta} - \dfrac{\cos\theta}{\sin\theta}}{\dfrac{\sin\theta}{\cos\theta} + \dfrac{\cos\theta}{\sin\theta}} = \dfrac{\dfrac{\sin^2\theta - \cos^2\theta}{\sin\theta\cos\theta}}{\dfrac{\sin^2\theta + \cos^2\theta}{\sin\theta\cos\theta}} = \dfrac{\sin^2\theta - \cos^2\theta}{\sin^2\theta + \cos^2\theta} = \dfrac{-\cos(2\theta)}{1} = -(2\cos^2\theta - 1) = 1 - 2\cos^2\theta$

17. $\dfrac{\sqrt{2}}{4}(\sqrt{3} + 1)$ **18.** $2 + \sqrt{3}$ **19.** $\dfrac{\sqrt{5}}{5}$ **20.** $\dfrac{12\sqrt{85}}{49}$ **21.** $\dfrac{2\sqrt{13}(\sqrt{5} - 3)}{39}$ **22.** $\dfrac{2 + \sqrt{3}}{4}$ **23.** $\dfrac{\sqrt{6}}{2}$ **24.** $\dfrac{\sqrt{2}}{2}$

25. $\left\{\dfrac{\pi}{3}, \dfrac{2\pi}{3}, \dfrac{4\pi}{3}, \dfrac{5\pi}{3}\right\}$ **26.** $\{0, 1.911, \pi, 4.373\}$ **27.** $\left\{\dfrac{3\pi}{8}, \dfrac{7\pi}{8}, \dfrac{11\pi}{8}, \dfrac{15\pi}{8}\right\}$ **28.** $\{0.285, 3.427\}$ **29.** $\{0.253, 2.889\}$

30. The change in elevation during the time trial was about 1.18 km.

Cumulative Review *(page 667)*

1. $\left\{\dfrac{-1 - \sqrt{13}}{6}, \dfrac{-1 + \sqrt{13}}{6}\right\}$ **2.** $y + 1 = -1(x - 4)$ or $x + y = 3; 6\sqrt{2}; (1, 2)$ **3.** *x*-axis symmetry; $(0, -3), (0, 3), (3, 0)$

4.

5.

6.

7. (a)

(b)

(c)

(d)

8. (a) $-\dfrac{2\sqrt{2}}{3}$ **(b)** $\dfrac{\sqrt{2}}{4}$ **(c)** $\dfrac{4\sqrt{2}}{9}$ **(d)** $\dfrac{7}{9}$ **(e)** $\sqrt{\dfrac{3+2\sqrt{2}}{6}}$ **(f)** $-\sqrt{\dfrac{3-2\sqrt{2}}{6}}$

9. $\dfrac{\sqrt{5}}{5}$ **10. (a)** $-\dfrac{2\sqrt{2}}{3}$ **(b)** $-\dfrac{2\sqrt{2}}{3}$ **(c)** $\dfrac{7}{9}$ **(d)** $\dfrac{4\sqrt{2}}{9}$ **(e)** $\dfrac{\sqrt{6}}{3}$

11. (a) $f(x) = (2x-1)(x-1)^2(x+1)^2; \dfrac{1}{2}$ multiplicity 1; 1 and -1 multiplicity 2 **(b)** $(0,-1); \left(\dfrac{1}{2},0\right); (-1,0); (1,0)$ **(c)** $y = 2x^5$

(d)

(f)

12. (a) $\left\{-1, -\dfrac{1}{2}\right\}$ **(b)** $\{-1, 1\}$ **(c)** $(-\infty, -1) \cup \left(-\dfrac{1}{2}, \infty\right)$ **(d)** $(-\infty, -1] \cup [1, \infty)$

(e) Minima $(-0.29, -1.33), (1, 0)$
Maxima $(-1, 0), (0.69, 0.10)$

(g) Increasing: $(-\infty, -1), (-0.29, 0.69), (1, \infty)$
Decreasing: $(-1, -0.29), (0.69, 1)$

CHAPTER 9 Applications of Trigonometric Functions

9.1 Assess Your Understanding (page 673)

5. T **6.** direction or bearing **7.** T **8.** F **9.** $a \approx 13.74, c \approx 14.62, A = 70°$ **11.** $b \approx 5.03, c \approx 7.83, A = 50°$ **13.** $a \approx 0.71, c \approx 4.06, B = 80°$
15. $b \approx 10.72, c \approx 11.83, B = 65°$ **17.** $b \approx 3.08, a \approx 8.46, A = 70°$ **19.** $c \approx 5.83, A \approx 59.0°, B \approx 31.0°$ **21.** $b \approx 4.58, A \approx 23.6°, B \approx 66.4°$
23. $23.6°$ and $66.4°$ **25.** $80.5°$ **27. (a)** 111.96 ft/sec or 76.3 mi/hr **(b)** 82.42 ft/sec or 56.2 mi/hr **(c)** Under $18.8°$ **29.** S76.6°E
31. The embankment is 30.5 m high. **33.** The buildings are 7984 ft apart. **35.** $69.0°$ **37.** $38.9°$

9.2 Assess Your Understanding (page 682)

4. oblique **5.** $\dfrac{\sin A}{a} = \dfrac{\sin B}{b} = \dfrac{\sin C}{c}$ **6.** F **7.** T **8.** F **9.** $a \approx 3.23, b \approx 3.55, A = 40°$ **11.** $a \approx 3.25, c \approx 4.23, B = 45°$
13. $C = 95°, c \approx 9.86, a \approx 6.36$ **15.** $A = 40°, a = 2, c \approx 3.06$ **17.** $C = 120°, b \approx 1.06, c \approx 2.69$ **19.** $A = 100°, a \approx 5.24, c \approx 0.92$
21. $B = 40°, a \approx 5.64, b \approx 3.86$ **23.** $C = 100°, a \approx 1.31, b \approx 1.31$ **25.** One triangle; $B \approx 30.7°, C \approx 99.3°, c \approx 3.86$ **27.** One triangle;
$C \approx 36.2°, A \approx 43.8°, a \approx 3.51$ **29.** No triangle **31.** Two triangles; $C_1 \approx 30.9°, A_1 \approx 129.1°, a_1 \approx 9.07$ or $C_2 \approx 149.1°, A_2 \approx 10.9°, a_2 \approx 2.20$
33. No triangle **35.** Two triangles; $A_1 \approx 57.7°, B_1 \approx 97.3°, b_1 \approx 2.35$ or $A_2 \approx 122.3°, B_2 \approx 32.7°, b_2 \approx 1.28$ **37. (a)** Station Able is about 143.33 mi
from the ship: Station Baker is about 135.58 mi from the ship. **(b)** Approximately 41 min **39.** 1490.48 ft **41.** 381.69 ft **43.** The tree is 39.4 ft high.
45. Adam receives 88.3 more frequent flyer miles. **47.** $84.7°; 183.72$ ft **49.** 2.64 mi **51.** 38.5 in. **53.** 449.36 ft **55.** 187,600,000 km or 101,440,000 km
57. The diameter is 252 ft.

59. $\dfrac{a-b}{c} = \dfrac{a}{c} - \dfrac{b}{c} = \dfrac{\sin A}{\sin C} - \dfrac{\sin B}{\sin C} = \dfrac{\sin A - \sin B}{\sin C} = \dfrac{2\sin\left(\dfrac{A-B}{2}\right)\cos\left(\dfrac{A+B}{2}\right)}{2\sin\dfrac{C}{2}\cos\dfrac{C}{2}} = \dfrac{\sin\left(\dfrac{A-B}{2}\right)\cos\left(\dfrac{\pi}{2}-\dfrac{C}{2}\right)}{\sin\dfrac{C}{2}\cos\dfrac{C}{2}} = \dfrac{\sin\left(\dfrac{A-B}{2}\right)}{\cos\dfrac{C}{2}}$

61. $\dfrac{a-b}{a+b} = \dfrac{\dfrac{a-b}{c}}{\dfrac{a+b}{c}} = \dfrac{\dfrac{\cos\dfrac{C}{2}}{\cos\left[\dfrac{1}{2}(A-B)\right]}}{\dfrac{\cos\left[\dfrac{1}{2}(A-B)\right]}{\sin\dfrac{C}{2}}} = \dfrac{\tan\left[\dfrac{1}{2}(A-B)\right]}{\cot\dfrac{C}{2}} = \dfrac{\tan\left[\dfrac{1}{2}(A-B)\right]}{\tan\left(\dfrac{\pi}{2}-\dfrac{C}{2}\right)} = \dfrac{\tan\left[\dfrac{1}{2}(A-B)\right]}{\tan\left[\dfrac{1}{2}(A+B)\right]}$

9.3 Assess Your Understanding (page 689)

3. Cosines **4.** Sines **5.** Cosines **6.** F **7.** F **8.** T **9.** $b \approx 2.95, A \approx 28.7°, C \approx 106.3°$ **11.** $c \approx 3.75, A \approx 32.1°, B \approx 52.9°$
13. $A \approx 48.5°, B \approx 38.6°, C \approx 92.9°$ **15.** $A \approx 127.2°, B \approx 32.1°, C \approx 20.7°$ **17.** $c \approx 2.57, A \approx 48.6°, B \approx 91.4°$

19. $a \approx 2.99$, $B \approx 19.2°$, $C \approx 80.8°$ **21.** $b \approx 4.14$, $A \approx 43.0°$, $C \approx 27.0°$ **23.** $c \approx 1.69$, $A = 65.0°$, $B = 65.0°$ **25.** $A \approx 67.4°$, $B = 90°$, $C \approx 22.6°$

27. $A = 60°$, $B = 60°$, $C = 60°$ **29.** $A \approx 33.6°$, $B \approx 62.2°$, $C \approx 84.3°$ **31.** $A \approx 97.9°$, $B \approx 52.4°$, $C \approx 29.7°$ **33.** 165 yd **35. (a)** 26.4° **(b)** 30.8 hr

37. (a) 63.7 ft **(b)** 66.8 ft **(c)** 92.8° **39. (a)** 492.6 ft **(b)** 269.3 ft **41.** 342.3 ft **43.** The footings should be 7.65 ft apart.

45. Suppose $0 < \theta < \pi$. Then by the Law of Cosines: $d^2 = r^2 + r^2 - 2r^2 \cos\theta = 4r^2\left(\dfrac{1 - \cos\theta}{2}\right) \Rightarrow d = 2r\sqrt{\dfrac{1 - \cos\theta}{2}} = 2r\sin\dfrac{\theta}{2}$. Since for any

angle in $(0, \pi)$, d is strictly less than the length of the arc subtended by θ, that is, $d < r\theta$, then $2r\sin\dfrac{\theta}{2} < r\theta$, or $2\sin\dfrac{\theta}{2} < \theta$. Since $\cos\dfrac{\theta}{2} < 1$, then, for

$0 < \theta < \pi$, $\sin\theta = 2\sin\dfrac{\theta}{2}\cos\dfrac{\theta}{2} < 2\sin\dfrac{\theta}{2} < \theta$. If $\theta \ge \pi$, then since $\sin\theta \le 1$, $\sin\theta < \theta$. Thus $\sin\theta < \theta$ for all $\theta > 0$.

47. $\sin\dfrac{C}{2} = \sqrt{\dfrac{1 - \cos C}{2}} = \sqrt{\dfrac{1 - \dfrac{a^2 + b^2 - c^2}{2ab}}{2}} = \sqrt{\dfrac{2ab - a^2 - b^2 + c^2}{4ab}} = \sqrt{\dfrac{c^2 - (a - b)^2}{4ab}} = \sqrt{\dfrac{(c + a - b)(c + b - a)}{4ab}}$

$= \sqrt{\dfrac{(2s - 2b)(2s - 2a)}{4ab}} = \sqrt{\dfrac{(s - a)(s - b)}{ab}}$

9.4 Assess Your Understanding (page 694)

2. Heron's **3.** F **4.** T **5.** 2.83 **7.** 2.99 **9.** 14.98 **11.** 9.56 **13.** 3.86 **15.** 1.48 **17.** 2.82 **19.** 30 **21.** 1.73 **23.** 19.90

25. $K = \dfrac{1}{2}ab\sin C = \dfrac{1}{2}a\sin C\left(\dfrac{a\sin B}{\sin A}\right) = \dfrac{a^2\sin B\sin C}{2\sin A}$ **27.** 0.92 **29.** 2.27 **31.** 5.44 **33.** 9.03 sq ft **35.** \$5446.38

37. The area of home plate is about 216.5 in². **39.** $K = \dfrac{1}{2}r^2(\theta + \sin\theta)$ **41.** The ground area is 7517.4 ft².

43. (a) Area $\triangle OAC = \dfrac{1}{2}|OC||AC| = \dfrac{1}{2}\cdot\dfrac{|OC|}{1}\cdot\dfrac{|AC|}{1} = \dfrac{1}{2}\sin\alpha\cos\alpha$

(b) Area $\triangle OCB = \dfrac{1}{2}|BC||OC| = \dfrac{1}{2}|OB|^2\dfrac{|BC|}{|OB|}\cdot\dfrac{|OC|}{|OB|} = \dfrac{1}{2}|OB|^2\sin\beta\cos\beta$

(c) Area $\triangle OAB = \dfrac{1}{2}|BD||OA| = \dfrac{1}{2}|OB|\dfrac{|BD|}{|OB|} = \dfrac{1}{2}|OB|\sin(\alpha + \beta)$

(d) $\dfrac{\cos\alpha}{\cos\beta} = \dfrac{\dfrac{|OC|}{1}}{\dfrac{|OC|}{|OB|}} = |OB|$

(e) Area $\triangle OAB = $ Area $\triangle OAC + $ Area $\triangle OCB$

$\dfrac{1}{2}|OB|\sin(\alpha + \beta) = \dfrac{1}{2}\sin\alpha\cos\alpha + \dfrac{1}{2}|OB|^2\sin\beta\cos\beta$

$\sin(\alpha + \beta) = \dfrac{1}{|OB|}\sin\alpha\cos\alpha + |OB|\sin\beta\cos\beta$

$\sin(\alpha + \beta) = \dfrac{\cos\beta}{\cos\alpha}\sin\alpha\cos\alpha + \dfrac{\cos\alpha}{\cos\beta}\sin\beta\cos\beta$

$\sin(\alpha + \beta) = \sin\alpha\cos\beta + \cos\alpha\sin\beta$

45. 31,145 ft² **47. (a)** The perimeter and area are both 36. **(b)** The perimeter and area are both 60.

49. $K = \dfrac{1}{2}ah = \dfrac{1}{2}ab\sin C \Rightarrow h = b\sin C = \dfrac{a\sin B\sin C}{\sin A}$

51. $\cot\dfrac{C}{2} = \dfrac{\cos\dfrac{C}{2}}{\sin\dfrac{C}{2}} = \dfrac{\dfrac{c\sin\dfrac{A}{2}\sin\dfrac{B}{2}}{r}}{\sqrt{\dfrac{(s - a)(s - b)}{ab}}} = \dfrac{c\sqrt{\dfrac{(s - b)(s - c)}{bc}}\sqrt{\dfrac{(s - a)(s - c)}{ac}}}{r\sqrt{\dfrac{(s - a)(s - b)}{ab}}} = \dfrac{c}{r}\sqrt{\dfrac{(s - c)^2}{c^2}} = \dfrac{s - c}{r}$

53. $K = $ area of triangle $QOR + $ area of $ROP + $ area of $POQ = \dfrac{1}{2}ar + \dfrac{1}{2}br + \dfrac{1}{2}cr = r\dfrac{1}{2}(a + b + c) = rs$, so

$r = \dfrac{K}{s} = \dfrac{\sqrt{s(s - a)(s - b)(s - c)}}{s} = \sqrt{\dfrac{(s - a)(s - b)(s - c)}{s}}$.

9.5 Assess Your Understanding (page 704)

2. Simple harmonic; amplitude **3.** Simple harmonic; damped **4.** T **5.** $d = -5\cos(\pi t)$ **7.** $d = -6\cos(2t)$ **9.** $d = -5\sin(\pi t)$

11. $d = -6\sin(2t)$ **13. (a)** Simple harmonic **(b)** 5 m **(c)** $\dfrac{2\pi}{3}$ sec **(d)** $\dfrac{3}{2\pi}$ oscillation/sec **15. (a)** Simple harmonic **(b)** 6 m **(c)** 2 sec

(d) $\dfrac{1}{2}$ oscillation/sec **17. (a)** Simple harmonic **(b)** 3 m **(c)** 4π sec **(d)** $\dfrac{1}{4\pi}$ oscillation/sec **19. (a)** Simple harmonic **(b)** 2 m **(c)** 1 sec

(d) 1 oscillation/sec

21.

23.

25.

27.

29.

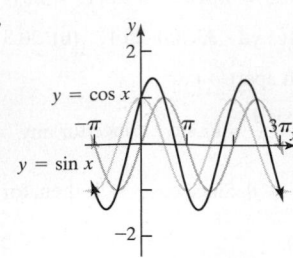

$y = \cos x$

$y = \sin x$

31.

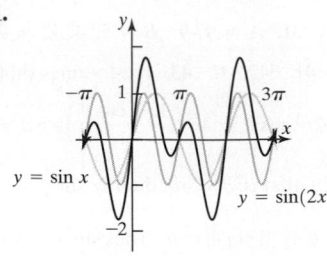

$y = \sin x$

$y = \sin(2x)$

33. (a) $d = -10e^{-0.7t/50}\cos\left(\sqrt{\dfrac{4\pi^2}{25} - \dfrac{0.49}{2500}}\,t\right)$

(b)

35. (a) $d = -18e^{-0.6t/60}\cos\left(\sqrt{\dfrac{\pi^2}{4} - \dfrac{0.36}{3600}}\,t\right)$

(b)

37. (a) $d = -5e^{-0.8t/20}\cos\left(\sqrt{\dfrac{4\pi^2}{9} - \dfrac{0.64}{400}}\,t\right)$

(b)

39. (a) The motion is damped.
The bob has mass $m = 20$ kg
with a damping factor of 0.7 kg/sec.

(b) 20 m leftward

(c)

(d) 18.33 m leftward **(e)** $d \to 0$

41. (a) The motion is damped.
The bob has mass $m = 40$ kg
with a damping factor of 0.6 kg/sec.

(b) 30 m leftward

(c)

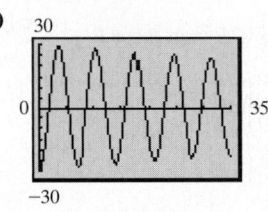

(d) 28.47 m leftward **(e)** $d \to 0$

43. (a) The motion is damped.
The bob has mass $m = 15$ kg
with a damping factor of 0.9 kg/sec.

(b) 15 m leftward

(c)

(d) 12.53 m leftward **(e)** $d \to 0$

45. $\omega = 1040\pi$;
$d = 0.80\cos(1040\pi t)$

47. $\omega = 880\pi$;
$d = 0.01\sin(880\pi t)$

49. (a)

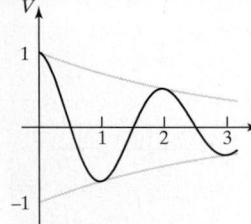

(b) At $t = 0, 2$; at $t = 1, t = 3$
(c) During the approximate intervals
$0.35 < t < 0.67$, $1.29 < t < 1.75$,
and $2.19 < t \le 3$

51.

55.

57. $y = \dfrac{1}{x}\sin x$

$y = \dfrac{1}{x^2}\sin x$

$y = \dfrac{1}{x^3}\sin x$

Review Exercises (page 707)

1. $A = 70°, b \approx 3.42, a \approx 9.40$ **3.** $a \approx 4.58, A \approx 66.4°, B \approx 23.6°$ **5.** $C = 100°, b \approx 0.65, c \approx 1.29$ **7.** $B \approx 56.8°, C \approx 23.2°, b \approx 4.25$
9. No triangle **11.** $b \approx 3.32, A \approx 62.8°, C \approx 17.2°$ **13.** No triangle **15.** $c \approx 2.32, A \approx 16.1°, B \approx 123.9°$ **17.** $B \approx 36.2°, C \approx 63.8°, c \approx 4.55$
19. $A \approx 39.6°, B \approx 18.6°, C \approx 121.9°$ **21.** Two triangles: $B_1 \approx 13.4°, C_1 \approx 156.6°, c_1 \approx 6.86$ or $B_2 \approx 166.6°, C_2 \approx 3.4°, c_2 \approx 1.02$ **23.** $a \approx 5.23$,
$B \approx 46.0°, C \approx 64.0°$ **25.** 1.93 **27.** 18.79 **29.** 6 **31.** 3.80 **33.** 0.32 **35.** 12.7° **37.** 29.97 ft **39.** 6.22 mi **41.** 71.12 ft **43.** \$222,983.51

45. S4.0°E **47.** 79.69 in. **49.** $d = -5\cos\left(\dfrac{\pi}{3}t\right)$ **51. (a)** Simple harmonic **(b)** 2 ft **(c)** $\dfrac{\pi}{2}$ sec **(d)** $\dfrac{2}{\pi}$ oscillation/sec **53. (a)** Simple harmonic

(b) 3 ft **(c)** 4 sec **(d)** $\dfrac{1}{4}$ oscillation/sec

55. (a) $d = -13e^{-0.65t/50} \cos\left(\sqrt{\dfrac{\pi^2}{4} - \dfrac{0.4225}{2500}} \, t\right)$

(b)

57. (a) The motion is damped. The bob has mass $m = 30$ kg with a damping factor of 0.5 kg/sec.

(b) 20 m leftward

(c)

(d) 19.51 m leftward

(e) $d \to 0$

59.

Chapter Test (page 710)

1. $61.0°$ **2.** $1.3°$ **3.** $a = 15.88, B \approx 57.5°, C \approx 70.5°$ **4.** $b \approx 6.85, C = 117°, c \approx 16.30$ **5.** $A \approx 52.4°, B \approx 29.7°, C \approx 97.9°$

6. $b \approx 4.72, c \approx 1.67, B = 105°$ **7.** No triangle **8.** $c \approx 7.62, \quad A \approx 80.5°, B \approx 29.5°$ **9.** 15.04 square units **10.** 19.81 square units **11.** The area of the shaded region is 9.26 cm². **12.** 54.15 square units **13.** Madison will have to swim about 2.23 miles. **14.** 12.63 square units **15.** The lengths of the sides are 15, 18, and 21. **16.** $d = 5(\sin 42°)\sin\left(\dfrac{\pi t}{3}\right)$ or $d \approx 3.346 \sin\left(\dfrac{\pi t}{3}\right)$

Cumulative Review (page 711)

1. $\left\{\dfrac{1}{3}, 1\right\}$ **2.** $(x + 5)^2 + (y - 1)^2 = 9$ **3.** $\{x \mid x \leq -1 \text{ or } x \geq 4\}$ **4.** **5.**

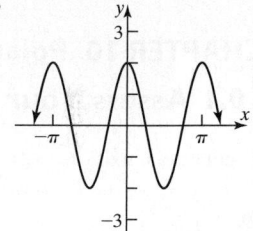

6. (a) $-\dfrac{2\sqrt{5}}{5}$ **(b)** $\dfrac{\sqrt{5}}{5}$ **(c)** $-\dfrac{4}{5}$ **(d)** $-\dfrac{3}{5}$ **(e)** $\sqrt{\dfrac{5 - \sqrt{5}}{10}}$ **(f)** $-\sqrt{\dfrac{5 + \sqrt{5}}{10}}$

7. (a) **(b)** **(c)** **(d)**

8. (a) **(b)** **(c)** **(d)**

(e) **(f)** **(g)** **(h)** **(i)**

9. Two triangles: $A_1 \approx 59.0°$, $B_1 \approx 81.0°$, $b_1 \approx 23.05$ or $A_2 \approx 121.0°$, $B_2 \approx 19.0°$, $b_2 \approx 7.59$ **10.** $\left\{-2i, 2i, \dfrac{1}{3}, 1, 2\right\}$

11. $R(x) = \dfrac{(2x + 1)(x - 4)}{(x + 5)(x - 3)}$; domain: $\{x \mid x \neq -5, x \neq 3\}$

Intercepts: $\left(-\dfrac{1}{2}, 0\right), (4, 0), \left(0, \dfrac{4}{15}\right)$

No symmetry

Vertical asymptotes: $x = -5$, $x = 3$

Horizontal asymptote: $y = 2$

Intersects: $\left(\dfrac{26}{11}, 2\right)$

12. $\{2.26\}$ **13.** $\{1\}$ **14. (a)** $\left\{-\dfrac{5}{4}\right\}$ **(b)** $\{2\}$ **(c)** $\left\{\dfrac{-1 - 3\sqrt{13}}{2}, \dfrac{-1 + 3\sqrt{13}}{2}\right\}$ **(d)** $\left\{x \mid x > -\dfrac{5}{4}\right\}$ or $\left(-\dfrac{5}{4}, \infty\right)$

(e) $\left\{x \mid -8 \leq x \leq 3\right\}$ or $[-8, 3]$ **(f)**

(g)

CHAPTER 10 Polar Coordinates; Vectors

10.1 Assess Your Understanding (page 720)

5. pole; polar axis **6.** -2 **7.** $\left(-\sqrt{3}, -1\right)$ **8.** F **9.** T **10.** T **11.** A **13.** C **15.** B **17.** A

19.

21.

23.

25.

27.

29.

31.

(a) $\left(5, -\dfrac{4\pi}{3}\right)$

(b) $\left(-5, \dfrac{5\pi}{3}\right)$

(c) $\left(5, \dfrac{8\pi}{3}\right)$

33.

(a) $(2, -2\pi)$

(b) $(-2, \pi)$

(c) $(2, 2\pi)$

35.

(a) $\left(1, -\dfrac{3\pi}{2}\right)$

(b) $\left(-1, \dfrac{3\pi}{2}\right)$

(c) $\left(1, \dfrac{5\pi}{2}\right)$

37.

(a) $\left(3, -\dfrac{5\pi}{4}\right)$

(b) $\left(-3, \dfrac{7\pi}{4}\right)$

(c) $\left(3, \dfrac{11\pi}{4}\right)$

39. $(0, 3)$ **41.** $(-2, 0)$ **43.** $\left(-3\sqrt{3}, 3\right)$ **45.** $\left(\sqrt{2}, -\sqrt{2}\right)$ **47.** $\left(-\dfrac{1}{2}, \dfrac{\sqrt{3}}{2}\right)$ **49.** $(2, 0)$ **51.** $(-2.57, 7.05)$ **53.** $(-4.98, -3.85)$ **55.** $(3, 0)$

57. $(1, \pi)$ **59.** $\left(\sqrt{2}, -\dfrac{\pi}{4}\right)$ **61.** $\left(2, \dfrac{\pi}{6}\right)$ **63.** $(2.47, -1.02)$ **65.** $(9.30, 0.47)$ **67.** $r^2 = \dfrac{3}{2}$ or $r = \dfrac{\sqrt{6}}{2}$ **69.** $r^2 \cos^2 \theta - 4r \sin \theta = 0$

71. $r^2 \sin 2\theta = 1$ **73.** $r \cos \theta = 4$ **75.** $x^2 + y^2 - x = 0$ or $\left(x - \dfrac{1}{2}\right)^2 + y^2 = \dfrac{1}{4}$ **77.** $(x^2 + y^2)^{3/2} - x = 0$ **79.** $x^2 + y^2 = 4$ **81.** $y^2 = 8(x + 2)$

83. (a) $(-10, 36)$ **(b)** $\left(2\sqrt{349}, 180° + \tan^{-1}\left(-\dfrac{18}{5}\right)\right) \approx (37.36, 105.5°)$ **(c)** $(-3, -35)$ **(d)** $\left(\sqrt{1234}, 180° + \tan^{-1}\left(\dfrac{35}{3}\right)\right) \approx (35.13, 265.1°)$

10.2 Assess Your Understanding (page 735)

7. polar equation **8.** $r = 2 \cos \theta$ **9.** $-r$ **10.** F **11.** F **12.** F

13. $x^2 + y^2 = 16$; circle, radius 4, center at pole

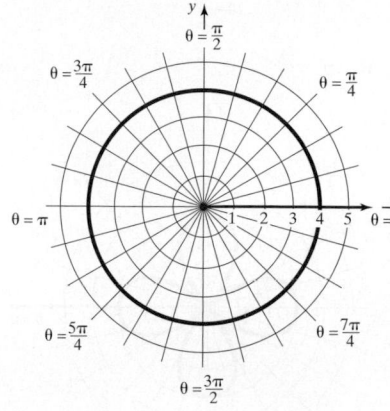

15. $y = \sqrt{3}x$; line through pole, making an angle of $\dfrac{\pi}{3}$ with polar axis

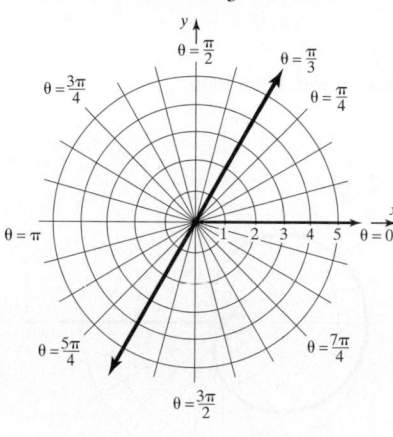

17. $y = 4$; horizontal line 4 units above the pole

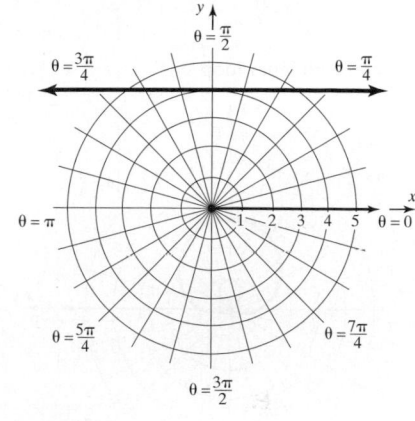

19. $x = -2$; vertical line 2 units to the left of the pole

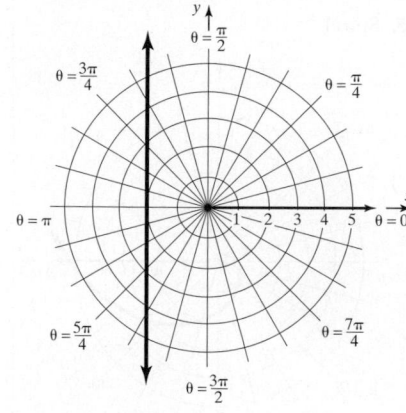

21. $(x - 1)^2 + y^2 = 1$; circle, radius 1, center $(1, 0)$ in rectangular coordinates

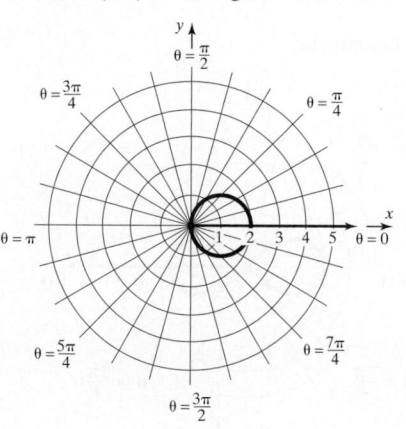

23. $x^2 + (y + 2)^2 = 4$; circle, radius 2, center at $(0, -2)$ in rectangular coordinates

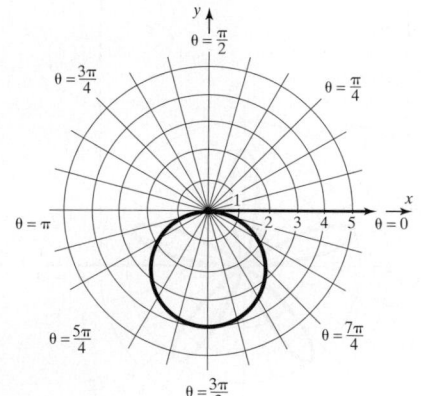

25. $(x - 2)^2 + y^2 = 4$; circle, radius 2, center at $(2, 0)$ in rectangular coordinates

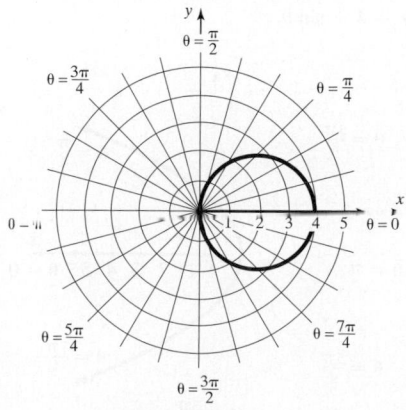

27. $x^2 + (y + 1)^2 = 1$; circle, radius 1, center at $(0, -1)$ in rectangular coordinates

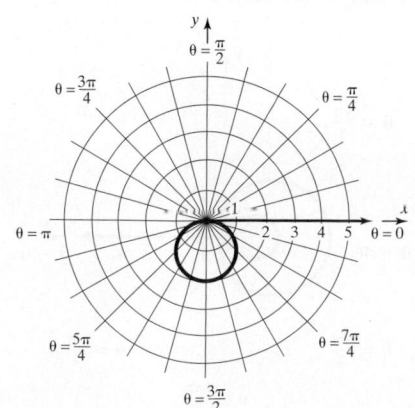

29. E **31.** F **33.** H **35.** D
37. Cardioid

39. Cardioid

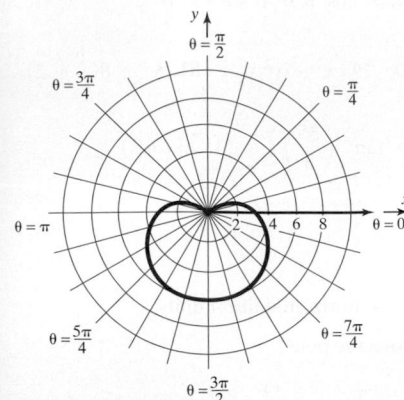

41. Limaçon without inner loop

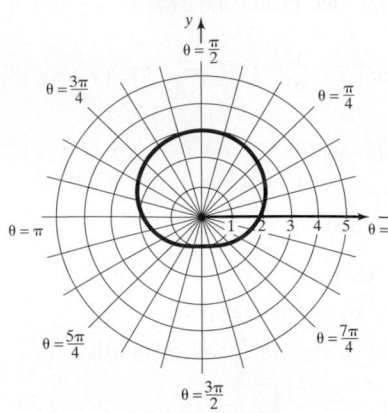

43. Limaçon without inner loop

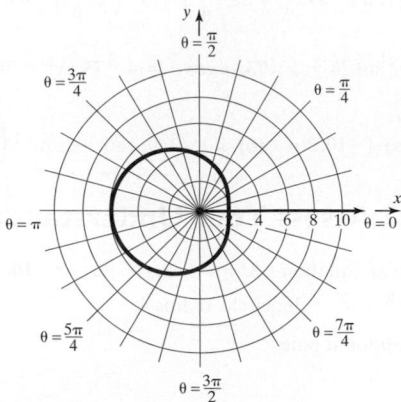

45. Limaçon with inner loop

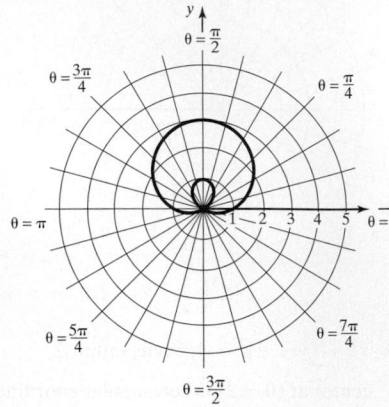

47. Limaçon with inner loop

49. Rose

51. Rose

53. Lemniscate

55. Spiral

57. Cardioid

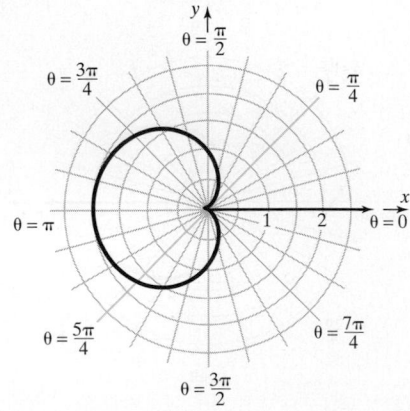

59. Limaçon with inner loop

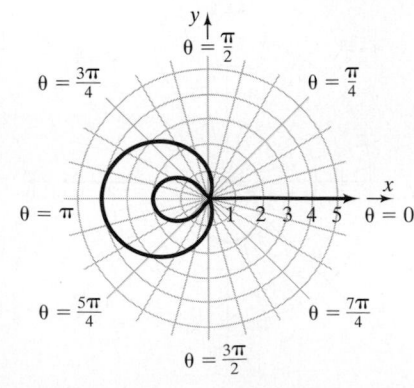

61. $r = 3 + 3\cos\theta$

63. $r = 4 + \sin\theta$

65.

67.

69.

71.

73.

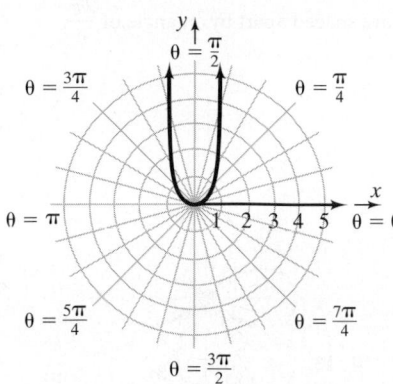

75. $r \sin \theta = a$
$y = a$

77.
$$r = 2a \sin \theta$$
$$r^2 = 2ar \sin \theta$$
$$x^2 + y^2 = 2ay$$
$$x^2 + y^2 - 2ay = 0$$
$$x^2 + (y - a)^2 = a^2$$
Circle, radius a, center at $(0, a)$
in rectangular coordinates

79.
$$r = 2a \cos \theta$$
$$r^2 = 2ar \cos \theta$$
$$x^2 + y^2 = 2ax$$
$$x^2 - 2ax + y^2 = 0$$
$$(x - a)^2 + y^2 = a^2$$
Circle, radius a, center at $(a, 0)$
in rectangular coordinates

81. (a) $r^2 = \cos \theta: r^2 = \cos(\pi - \theta)$
$r^2 = -\cos \theta$
Not equivalent; test fails.
$(-r)^2 = \cos(-\theta)$
$r^2 = \cos \theta$
New test works.

(b) $r^2 = \sin \theta: r^2 = \sin(\pi - \theta)$
$r^2 = \sin \theta$
Test works.
$(-r)^2 = \sin(-\theta)$
$r^2 = -\sin \theta$
Not equivalent; new test fails.

Historical Problems *(page 744)*

1. (a) $1 + 4i, 1 + i$ **(b)** $-1, 2 + i$

10.3 Assess Your Understanding *(page 744)*

5. magnitude; modulus; argument **6.** De Moivre's **7.** three **8.** T **9.** F **10.** T

11.

$\sqrt{2}(\cos 45° + i \sin 45°)$

13.

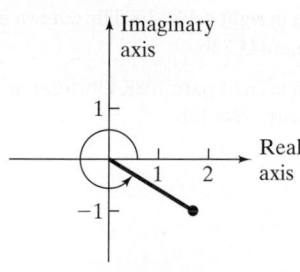

$2(\cos 330° + i \sin 330°)$

15.

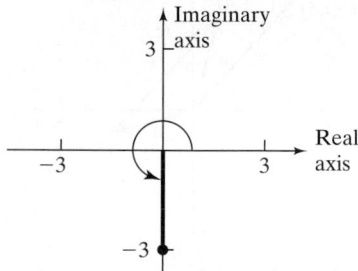

$3(\cos 270° + i \sin 270°)$

17.

$4\sqrt{2}(\cos 315° + i \sin 315°)$

19.

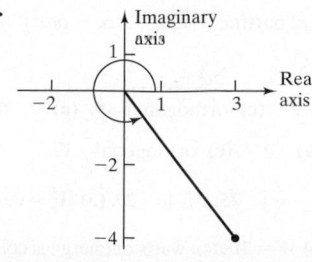

$5(\cos 306.9° + i \sin 306.9°)$

21.

$\sqrt{13}(\cos 123.7° + i \sin 123.7°)$

23. $-1 + \sqrt{3}i$ **25.** $2\sqrt{2} - 2\sqrt{2}i$ **27.** $-3i$ **29.** $-0.035 + 0.197i$ **31.** $1.970 + 0.347i$

33. $zw = 8(\cos 60° + i \sin 60°); \dfrac{z}{w} = \dfrac{1}{2}(\cos 20° + i \sin 20°)$ **35.** $zw = 12(\cos 40° + i \sin 40°); \dfrac{z}{w} = \dfrac{3}{4}(\cos 220° + i \sin 220°)$

37. $zw = 4\left(\cos \dfrac{9\pi}{40} + i \sin \dfrac{9\pi}{40}\right); \dfrac{z}{w} = \cos \dfrac{\pi}{40} + i \sin \dfrac{\pi}{40}$ **39.** $zw = 4\sqrt{2}(\cos 15° + i \sin 15°); \dfrac{z}{w} = \sqrt{2}(\cos 75° + i \sin 75°)$ **41.** $-32 + 32\sqrt{3}i$

43. $32i$ **45.** $\dfrac{27}{2} + \dfrac{27\sqrt{3}}{2}i$ **47.** $-\dfrac{25\sqrt{2}}{2} + \dfrac{25\sqrt{2}}{2}i$ **49.** $-4 + 4i$ **51.** $-23 + 14.142i$

53. $\sqrt[6]{2}(\cos 15° + i \sin 15°), \sqrt[6]{2}(\cos 135° + i \sin 135°), \sqrt[6]{2}(\cos 255° + i \sin 255°)$

55. $\sqrt[4]{8}(\cos 75° + i \sin 75°), \sqrt[4]{8}(\cos 165° + i \sin 165°), \sqrt[4]{8}(\cos 255° + i \sin 255°), \sqrt[4]{8}(\cos 345° + i \sin 345°)$

57. $2(\cos 67.5° + i \sin 67.5°), 2(\cos 157.5° + i \sin 157.5°), 2(\cos 247.5° + i \sin 247.5°), 2(\cos 337.5° + i \sin 337.5°)$

59. $\cos 18° + i \sin 18°, \cos 90° + i \sin 90°, \cos 162° + i \sin 162°, \cos 234° + i \sin 234°, \cos 306° + i \sin 306°$

61. $1, i, -1, -i$

63. Look at formula (8);
$|z_k| = \sqrt[n]{r}$ for all k.

65. Look at formula (8).
The z_k are spaced apart by an angle of $\dfrac{2\pi}{n}$.

10.4 Assess Your Understanding *(page 755)*

1. unit **2.** scalar **3.** horizontal; vertical **4.** T **5.** T **6.** F

7. **9.** **11.** **13.**

15. T **17.** F **19.** F **21.** T **23.** 12 **25.** $v = 3i + 4j$ **27.** $v = 2i + 4j$ **29.** $v = 8i - j$ **31.** $v = -i + j$ **33.** 5 **35.** $\sqrt{2}$ **37.** $\sqrt{13}$ **39.** $-j$

41. $\sqrt{89}$ **43.** $\sqrt{34} - \sqrt{13}$ **45.** i **47.** $\dfrac{3}{5}i - \dfrac{4}{5}j$ **49.** $\dfrac{\sqrt{2}}{2}i - \dfrac{\sqrt{2}}{2}j$ **51.** $v = \dfrac{8\sqrt{5}}{5}i + \dfrac{4\sqrt{5}}{5}j$ or $v = -\dfrac{8\sqrt{5}}{5}i - \dfrac{4\sqrt{5}}{5}j$

53. $\left\{-2 + \sqrt{21}, -2 - \sqrt{21}\right\}$ **55.** $v = \dfrac{5}{2}i + \dfrac{5\sqrt{3}}{2}j$ **57.** $v = -7i + 7\sqrt{3}j$ **59.** $v = \dfrac{25\sqrt{3}}{2}i - \dfrac{25}{2}j$

61. **(a)** $(-1, 4)$
(b)

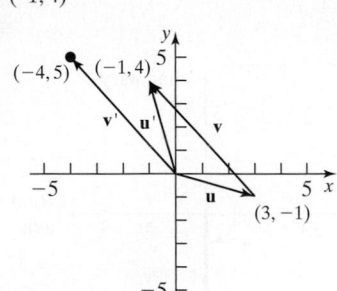

63. $F = 20\sqrt{3}i + 20j$

65. $F = \left(20\sqrt{3} + 30\sqrt{2}\right)i + \left(20 - 30\sqrt{2}\right)j$

67. Tension in right cable: 1000 lb; tension in
left cable: 845.2 lb

69. Tension in right part: 1088.4 lb; tension
in left part: 1089.1 lb

71. The truck must pull with a force
of 4635. 2 lb.

73.

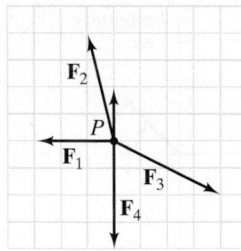

Historical Problem *(page 763)*

$(ai + bj) \cdot (ci + dj) = ac + bd$

Real part $[(\overline{a + bi})(c + di)] = $ real part$[(a - bi)(c + di)] = $ real part$[ac + adi - bci - bdi^2] = ac + bd$

10.5 Assess Your Understanding *(page 763)*

2. orthogonal **3.** parallel **4.** F **5.** T **6.** F **7. (a)** 0 **(b)** 90° **(c)** orthogonal **9. (a)** 0 **(b)** 90° **(c)** orthogonal **11. (a)** $\sqrt{3} - 1$ **(b)** 75°
(c) neither **13. (a)** 24 **(b)** 16.3° **(c)** neither **15. (a)** 0 **(b)** 90° **(c)** orthogonal **17.** $\dfrac{2}{3}$ **19.** $v_1 = \dfrac{5}{2}i - \dfrac{5}{2}j, v_2 = -\dfrac{1}{2}i - \dfrac{1}{2}j$

21. $v_1 = -\dfrac{1}{5}i - \dfrac{2}{5}j, v_2 = \dfrac{6}{5}i - \dfrac{3}{5}j$ **23.** $v_1 = \dfrac{14}{5}i + \dfrac{7}{5}j, v_2 = \dfrac{1}{5}i - \dfrac{2}{5}j$ **25.** 9 ft-lb **27. (a)** $\|I\| = 0.022$; the intensity of the sun's rays is approximately

0.022 W/cm². $\|A\| = 500$; the area of the solar panel is 500 cm². **(b)** $W = 10$; ten watts of energy is collected. **(c)** Vectors I and A should be parallel with
the solar panels facing the sun. **29.** 496.7 mi/hr; 38.5° west of south **31.** 8.6° off direct heading across the current, upstream; 1.52 min

33. Force required to keep Sienna from rolling down the hill: 737.6 lb; force perpendicular to the hill: 5248.4 lb

35. $\mathbf{v} = \left(250\sqrt{2} - 30\right)\mathbf{i} + \left(250\sqrt{2} + 30\sqrt{3}\right)\mathbf{j}$; 518.8 km/hr; N38.6°E **37.** Timmy must exert 85.5 lb. **39.** 60° **41.** Let $\mathbf{v} = a\mathbf{i} + b\mathbf{j}$. Then

$\mathbf{0} \cdot \mathbf{v} = 0a + 0b = 0$. **43.** $\mathbf{v} = \cos\alpha\mathbf{i} + \sin\alpha\mathbf{j}, 0 \le \alpha \le \pi$; $\mathbf{w} = \cos\beta\mathbf{i} + \sin\beta\mathbf{j}, 0 \le \beta \le \pi$. If θ is the angle between $\mathbf{v}$ and $\mathbf{w}$, then $\mathbf{v} \cdot \mathbf{w} = \cos\theta$,

since $\|\mathbf{v}\| = 1$ and $\|\mathbf{w}\| = 1$. Now $\theta = \alpha - \beta$ or $\theta = \beta - \alpha$. Since the cosine function is even, $\mathbf{v} \cdot \mathbf{w} = \cos(\alpha - \beta)$.

Also, $\mathbf{v} \cdot \mathbf{w} = \cos\alpha\cos\beta + \sin\alpha\sin\beta$. So, $\cos(\alpha - \beta) = \cos\alpha\cos\beta + \sin\alpha\sin\beta$.

45. (a) If $\mathbf{u} = a_1\mathbf{i} + b_1\mathbf{j}$ and $\mathbf{v} = a_2\mathbf{i} + b_2\mathbf{j}$, then, since $\|\mathbf{u}\| = \|\mathbf{v}\|, a_1^2 + b_1^2 = \|\mathbf{u}\|^2 = \|\mathbf{v}\|^2 = a_2^2 + b_2^2$, and

$(\mathbf{u} + \mathbf{v}) \cdot (\mathbf{u} - \mathbf{v}) = (a_1 + a_2)(a_1 - a_2) + (b_1 + b_2)(b_1 - b_2) = (a_1^2 + b_1^2) - (a_2^2 + b_2^2) = 0$.

(b) The legs of the angle can be made to correspond to vectors $\mathbf{u} + \mathbf{v}$ and $\mathbf{u} - \mathbf{v}$.

47. $(\|\mathbf{w}\|\mathbf{v} + \|\mathbf{v}\|\mathbf{w}) \cdot (\|\mathbf{w}\|\mathbf{v} - \|\mathbf{v}\|\mathbf{w}) = \|\mathbf{w}\|^2\mathbf{v} \cdot \mathbf{v} - \|\mathbf{w}\|\|\mathbf{v}\|\mathbf{v} \cdot \mathbf{w} + \|\mathbf{v}\|\|\mathbf{w}\|\mathbf{w} \cdot \mathbf{v} - \|\mathbf{v}\|^2\mathbf{w} \cdot \mathbf{w} = \|\mathbf{w}\|^2\mathbf{v} \cdot \mathbf{v} - \|\mathbf{v}\|^2\mathbf{w} \cdot \mathbf{w} = \|\mathbf{w}\|^2\|\mathbf{v}\|^2 - \|\mathbf{v}\|^2\|\mathbf{w}\|^2 = 0$

49. $\|\mathbf{u} + \mathbf{v}\|^2 - \|\mathbf{u} - \mathbf{v}\|^2 = (\mathbf{u} + \mathbf{v}) \cdot (\mathbf{u} + \mathbf{v}) - (\mathbf{u} - \mathbf{v}) \cdot (\mathbf{u} - \mathbf{v}) = (\mathbf{u} \cdot \mathbf{u} + \mathbf{u} \cdot \mathbf{v} + \mathbf{v} \cdot \mathbf{u} + \mathbf{v} \cdot \mathbf{v}) - (\mathbf{u} \cdot \mathbf{u} - \mathbf{u} \cdot \mathbf{v} - \mathbf{v} \cdot \mathbf{u} + \mathbf{v} \cdot \mathbf{v})$

$= 2(\mathbf{u} \cdot \mathbf{v}) + 2(\mathbf{v} \cdot \mathbf{u}) = 4(\mathbf{u} \cdot \mathbf{v})$

Review Exercises *(page 766)*

1. $\left(\dfrac{3\sqrt{3}}{2}, \dfrac{3}{2}\right)$

3. $\left(1, \sqrt{3}\right)$

5. $(0, 3)$

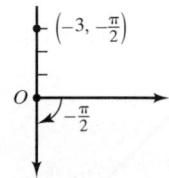

7. $\left(3\sqrt{2}, \dfrac{3\pi}{4}\right), \left(-3\sqrt{2}, -\dfrac{\pi}{4}\right)$ **9.** $\left(2, -\dfrac{\pi}{2}\right), \left(-2, \dfrac{\pi}{2}\right)$ **11.** $(5, 0.93), (-5, 4.07)$

13. (a) $x^2 + (y - 1)^2 = 1$ **(b)** Circle, radius 1, **15. (a)** $x^2 + y^2 = 25$ **(b)** Circle, radius 5, **17. (a)** $x + 3y = 6$ **(b)** Line through $(6, 0)$ and $(0, 2)$

center $(0, 1)$ in rectangular coordinates center at pole in rectangular coordinates

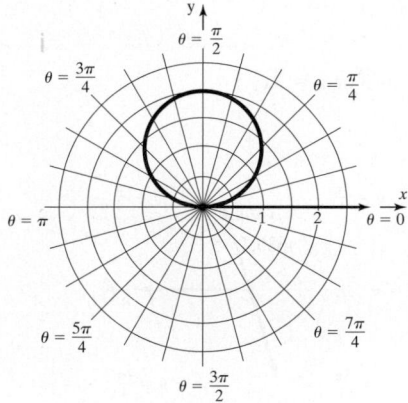

19. Circle; radius 2, center at $(2, 0)$ in **21.** Cardioid; symmetric with respect **23.** Limaçon without inner loop;

rectangular coordinates; symmetric to the line $\theta = \dfrac{\pi}{2}$ symmetric with respect to the polar axis

with respect to the polar axis

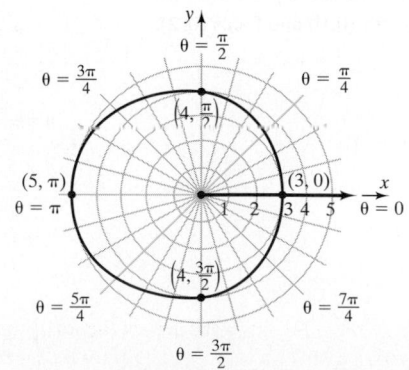

25. $\sqrt{2}(\cos 225° + i \sin 225°)$ **27.** $5(\cos 323.1° + i \sin 323.1°)$

29. $-\sqrt{3} + i$ **31.** $-\dfrac{3}{2} + \dfrac{3\sqrt{3}}{2}i$ **33.** $0.10 - 0.02i$

 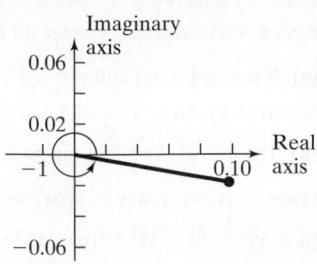

35. $zw = \cos 130° + i \sin 130°; \dfrac{z}{w} = \cos 30° + i \sin 30°$ **37.** $zw = 6(\cos 0 + i \sin 0) = 6; \dfrac{z}{w} = \dfrac{3}{2}\left(\cos\dfrac{8\pi}{5} + i \sin\dfrac{8\pi}{5}\right)$

39. $zw = 5(\cos 5° + i \sin 5°); \dfrac{z}{w} = 5(\cos 15° + i \sin 15°)$ **41.** $\dfrac{27}{2} + \dfrac{27\sqrt{3}}{2}i$ **43.** $4i$ **45.** 64 **47.** $-527 - 336i$

49. $3, 3(\cos 120° + i \sin 120°), 3(\cos 240° + i \sin 240°)$ or $3, -\dfrac{3}{2} + \dfrac{3\sqrt{3}}{2}i, -\dfrac{3}{2} - \dfrac{3\sqrt{3}}{2}i$

51. **53.**

55. $\mathbf{v} = 2\mathbf{i} - 4\mathbf{j}; \|\mathbf{v}\| = 2\sqrt{5}$ **57.** $\mathbf{v} = -\mathbf{i} + 3\mathbf{j}; \|\mathbf{v}\| = \sqrt{10}$

59. $2\mathbf{i} - 2\mathbf{j}$ **61.** $-20\mathbf{i} + 13\mathbf{j}$ **63.** $\sqrt{5}$ **65.** $\sqrt{5} + 5 \approx 7.24$ **67.** $-\dfrac{2\sqrt{5}}{5}\mathbf{i} + \dfrac{\sqrt{5}}{5}\mathbf{j}$

69. $\mathbf{v} = \dfrac{3}{2}\mathbf{i} + \dfrac{3\sqrt{3}}{2}\mathbf{j}$ **71.** $\mathbf{v} \cdot \mathbf{w} = -11; \theta \approx 169.7°$ **73.** $\mathbf{v} \cdot \mathbf{w} = -4; \theta \approx 153.4°$

75. Parallel **77.** Parallel **79.** Orthogonal **81.** $\mathbf{v}_1 = \dfrac{4}{5}\mathbf{i} - \dfrac{3}{5}\mathbf{j}; \mathbf{v}_2 = \dfrac{6}{5}\mathbf{i} + \dfrac{8}{5}\mathbf{j}$

83. $\sqrt{29} \approx 5.39$ mi/hr; 0.4 mi **85.** Left cable: 1843.21 lb; right cable: 1630.41 lb

87. 50 ft-lb

Chapter Test (page 768)

1.–3. 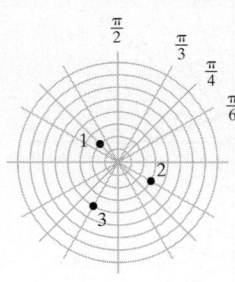 **4.** $\left(4, \dfrac{\pi}{3}\right)$ **5.** $x^2 + y^2 = 49$ **6.** $\dfrac{y}{x} = 3$ or $y = 3x$

 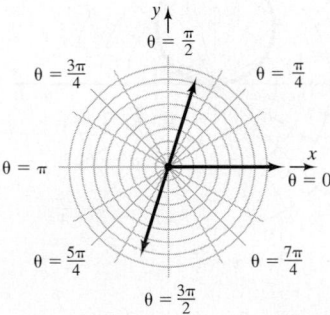

7. $8y = x^2$ or $4(2)y = x^2$

The graph is a parabola with vertex $(0, 0)$ and focus $(0, 2)$.

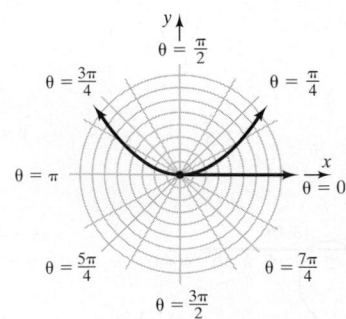

8. $r^2 \cos \theta = 5$ is symmetric about the pole, the polar axis, and the line $\theta = \dfrac{\pi}{2}$.

9. $r = 5 \sin \theta \cos^2 \theta$ is symmetric about the line $\theta = \dfrac{\pi}{2}$. The tests for symmetry about the pole and the polar axis fail, so the graph of $r = 5 \sin \theta \cos^2 \theta$ may or may not be symmetric about the pole or polar axis.

10. $zw = 6(\cos 107° + i \sin 107°)$

11. $\dfrac{w}{z} = \dfrac{3}{2}(\cos 297° + i \sin 297°)$

12. $w^5 = 243(\cos 110° + i \sin 110°)$

13. $z_1 = 2\sqrt[3]{2}(\cos 160° + i \sin 160°)$

14. $\mathbf{v} = \langle 5\sqrt{2}, -5\sqrt{2} \rangle$ **15.** $\|\mathbf{v}\| = 10$

16. $\mathbf{u} = \dfrac{\mathbf{v}}{\|\mathbf{v}\|} = \left\langle \dfrac{\sqrt{2}}{2}, -\dfrac{\sqrt{2}}{2} \right\rangle$ **17.** $315°$ off the positive x-axis

18. $\mathbf{v} = 5\sqrt{2}\mathbf{i} - 5\sqrt{2}\mathbf{j}$ **19.** $\mathbf{v}_1 + 2\mathbf{v}_2 - \mathbf{v}_3 = \langle 6, -10 \rangle$

20. Vectors $\mathbf{v}_1$ and $\mathbf{v}_4$ are parallel.

21. Vectors $\mathbf{v}_2$ and $\mathbf{v}_3$ are orthogonal. **22.** $172.87°$

23. The cable must be able to endure a tension of approximately 670.82 lb.

Cumulative Review (page 769)

1. $\{-3, 3\}$

2. $y = \dfrac{\sqrt{3}}{3}x$

3. $x^2 + (y - 1)^2 = 9$

4. $\left\{ x \mid x < \dfrac{1}{2} \right\}$ or $\left(-\infty, \dfrac{1}{2} \right)$

5. Symmetry with respect to the y-axis

6.

7.

8.

9. $-\dfrac{\pi}{6}$

10.

11.

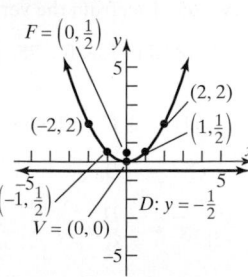

12. Amplitude: 4; period: 2

CHAPTER 11 Analytic Geometry

11.2 Assess Your Understanding (page 779)

6. parabola **7.** paraboloid of revolution **8.** T **9.** T **10.** T **11.** B **13.** E **15.** H **17.** C

19. $y^2 = 16x$

21. $x^2 = -12y$

23. $y^2 = -8x$

25. $x^2 = 2y$

27. $x^2 = \dfrac{4}{3}y$

29. $(x - 2)^2 = -8(y + 3)$

31. $(y + 2)^2 = 4(x + 1)$

33. $(x + 3)^2 = 4(y - 3)$

35. $(y + 2)^2 = -8(x + 1)$

37. Vertex: $(0, 0)$; focus: $(0, 1)$; directrix: $y = -1$

39. Vertex: $(0, 0)$; focus: $(-4, 0)$; directrix: $x = 4$

41. Vertex: $(-1, 2)$; focus: $(1, 2)$; directrix: $x = -3$

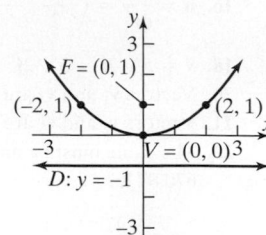

43. Vertex: $(3, -1)$; focus: $\left(3, -\dfrac{5}{4}\right)$; directrix: $y = -\dfrac{3}{4}$

45. Vertex: $(2, -3)$; focus: $(4, -3)$; directrix: $x = 0$

47. Vertex: $(0, 2)$; focus: $(-1, 2)$; directrix: $x = 1$

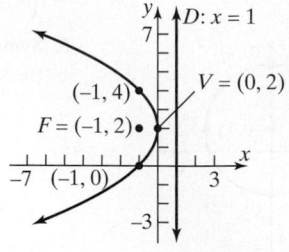

49. Vertex: $(-4, -2)$; focus: $(-4, -1)$; directrix: $y = -3$

51. Vertex: $(-1, -1)$; focus: $\left(-\dfrac{3}{4}, -1\right)$; directrix: $x = -\dfrac{5}{4}$

53. Vertex: $(2, -8)$; focus: $\left(2, -\dfrac{31}{4}\right)$; directrix: $y = -\dfrac{33}{4}$

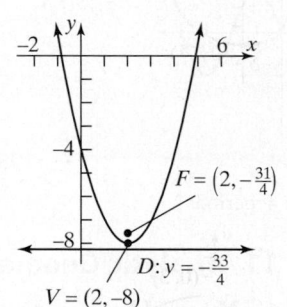

55. $(y - 1)^2 = x$ **57.** $(y - 1)^2 = -(x - 2)$ **59.** $x^2 = 4(y - 1)$ **61.** $y^2 = \dfrac{1}{2}(x + 2)$ **63.** 1.5625 ft from the base of the dish, along the axis of symmetry **65.** 1 in. from the vertex, along the axis of symmetry **67.** 20 ft **69.** 0.78125 ft **71.** 4.17 ft from the base, along the axis of symmetry

73. 24.31 ft, 18.75 ft, 7.64 ft **75. (a)** $y = -\dfrac{625}{(299)^2}x^2 + 625$ **(b)** 567 ft: 63.12 ft; 478 ft: 225.67 ft; 308 ft: 459.2 ft **(c)** No

77. $Cy^2 + Dx = 0, C \ne 0, D \ne 0$

$$Cy^2 = -Dx$$

$$y^2 = -\dfrac{D}{C}x$$

This is the equation of a parabola with vertex at $(0, 0)$ and axis of symmetry the x-axis.

The focus is $\left(-\dfrac{D}{4C}, 0\right)$; the directrix is the line $x = \dfrac{D}{4C}$. The parabola opens to the right if $-\dfrac{D}{C} > 0$ and to the left if $-\dfrac{D}{C} < 0$.

79. $Cy^2 + Dx + Ey + F = 0, C \ne 0$

$$Cy^2 + Ey = -Dx - F$$

$$y^2 + \dfrac{E}{C}y = -\dfrac{D}{C}x - \dfrac{F}{C}$$

$$\left(y + \dfrac{E}{2C}\right)^2 = -\dfrac{D}{C}x - \dfrac{F}{C} + \dfrac{E^2}{4C^2}$$

$$\left(y + \dfrac{E}{2C}\right)^2 = -\dfrac{D}{C}x + \dfrac{E^2 - 4CF}{4C^2}$$

(a) If $D \ne 0$, then the equation may be written as

$$\left(y + \dfrac{E}{2C}\right)^2 = -\dfrac{D}{C}\left(x - \dfrac{E^2 - 4CF}{4CD}\right).$$

This is the equation of a parabola with vertex at $\left(\dfrac{E^2 - 4CF}{4CD}, -\dfrac{E}{2C}\right)$ and axis of symmetry parallel to the x-axis.

(b)–(d) If $D = 0$, the graph of the equation contains no points if $E^2 - 4CF < 0$, is a single horizontal line if $E^2 - 4CF = 0$, and is two horizontal lines if $E^2 - 4CF > 0$.

11.3 Assess Your Understanding (page 789)

7. ellipse **8.** major **9.** $(0, -5); (0, 5)$ **10.** F **11.** T **12.** T **13.** C **15.** B

17. Vertices: $(-5, 0), (5, 0)$

Foci: $\left(-\sqrt{21}, 0\right), \left(\sqrt{21}, 0\right)$

19. Vertices: $(0, -5), (0, 5)$

Foci: $(0, -4), (0, 4)$

21. $\dfrac{x^2}{4} + \dfrac{y^2}{16} = 1$

Vertices: $(0, -4), (0, 4)$

Foci: $\left(0, -2\sqrt{3}\right), \left(0, 2\sqrt{3}\right)$

23. $\dfrac{x^2}{8} + \dfrac{y^2}{2} = 1$

Vertices: $\left(-2\sqrt{2}, 0\right), \left(2\sqrt{2}, 0\right)$

Foci: $\left(-\sqrt{6}, 0\right), \left(\sqrt{6}, 0\right)$

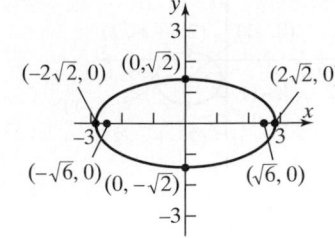

25. $\dfrac{x^2}{16} + \dfrac{y^2}{16} = 1$

Vertices: $(-4, 0), (4, 0), (0, -4), (0, 4)$

Focus: $(0, 0)$

27. $\dfrac{x^2}{25} + \dfrac{y^2}{16} = 1$

29. $\dfrac{x^2}{9} + \dfrac{y^2}{25} = 1$

31. $\dfrac{x^2}{9} + \dfrac{y^2}{5} = 1$

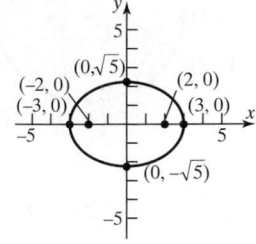

33. $\dfrac{x^2}{25} + \dfrac{y^2}{9} = 1$

35. $\dfrac{x^2}{4} + \dfrac{y^2}{13} = 1$

37. $x^2 + \dfrac{y^2}{16} = 1$

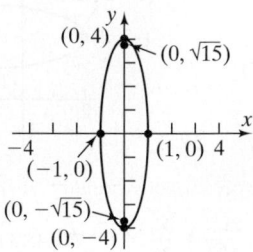

39. $\dfrac{(x + 1)^2}{4} + (y - 1)^2 = 1$

41. $(x - 1)^2 + \dfrac{y^2}{4} = 1$

43. Center: $(3, -1)$; vertices: $(3, -4), (3, 2)$;

foci: $\left(3, -1 - \sqrt{5}\right), \left(3, -1 + \sqrt{5}\right)$

45. $\dfrac{(x + 5)^2}{16} + \dfrac{(y - 4)^2}{4} = 1$

Center: $(-5, 4)$; vertices: $(-9, 4), (-1, 4)$;

foci: $\left(-5 - 2\sqrt{3}, 4\right), \left(-5 + 2\sqrt{3}, 4\right)$

47. $\dfrac{(x + 2)^2}{4} + (y - 1)^2 = 1$

Center: $(-2, 1)$; vertices: $(-4, 1), (0, 1)$;

foci: $\left(-2 - \sqrt{3}, 1\right), \left(-2 + \sqrt{3}, 1\right)$

49. $\dfrac{(x-2)^2}{3} + \dfrac{(y+1)^2}{2} = 1$

Center: $(2, -1)$; vertices: $\left(2 - \sqrt{3}, -1\right)$;

$\left(2 + \sqrt{3}, -1\right)$; foci: $(1, -1), (3, -1)$

51. $\dfrac{(x-1)^2}{4} + \dfrac{(y+2)^2}{9} = 1$

Center: $(1, -2)$; vertices: $(1, -5), (1, 1)$;

foci: $\left(1, -2 - \sqrt{5}\right), \left(1, -2 + \sqrt{5}\right)$

53. $x^2 + \dfrac{(y+2)^2}{4} = 1$

Center: $(0, -2)$; vertices: $(0, -4), (0, 0)$;

foci: $\left(0, -2 - \sqrt{3}\right), \left(0, -2 + \sqrt{3}\right)$

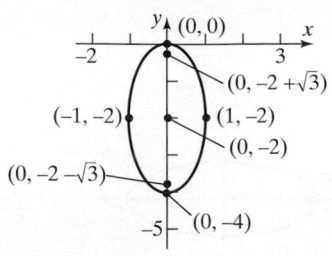

55. $\dfrac{(x-2)^2}{25} + \dfrac{(y+2)^2}{21} = 1$

57. $\dfrac{(x-4)^2}{5} + \dfrac{(y-6)^2}{9} = 1$

59. $\dfrac{(x-2)^2}{16} + \dfrac{(y-1)^2}{7} = 1$

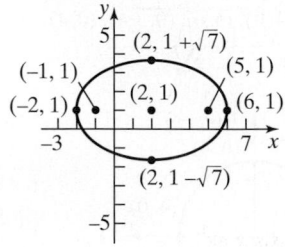

61. $\dfrac{(x-1)^2}{10} + (y-2)^2 = 1$

63. $\dfrac{(x-1)^2}{9} + (y-2)^2 = 1$

65.

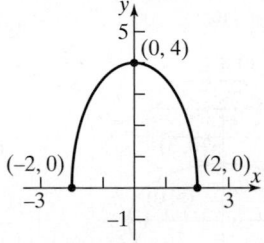

67.

69. $\dfrac{x^2}{100} + \dfrac{y^2}{36} = 1$ **71.** 43.3 ft **73.** 24.65 ft, 21.65 ft, 13.82 ft **75.** 30 ft **77.** The elliptical hole will have a

major axis of length $2\sqrt{41}$ in. and a minor axis of length 8 in. **79.** 91.5 million mi; $\dfrac{x^2}{(93)^2} + \dfrac{y^2}{8646.75} = 1$

81. Perihelion: 460.6 million mi; mean distance: 483.8 million mi; $\dfrac{x^2}{(483.8)^2} + \dfrac{y^2}{233,524.2} = 1$

83. (a) $Ax^2 + Cy^2 + F = 0$

$Ax^2 + Cy^2 = -F$

If A and C are of the same sign and F is of opposite sign, then the equation takes the form

$\dfrac{x^2}{\left(-\dfrac{F}{A}\right)} + \dfrac{y^2}{\left(-\dfrac{F}{C}\right)} = 1$, where $-\dfrac{F}{A}$ and $-\dfrac{F}{C}$ are positive. This is the equation of an ellipse with center at $(0,0)$.

(b) If $A = C$, the equation may be written as $x^2 + y^2 = -\dfrac{F}{A}$.

This is the equation of a circle with center at $(0,0)$ and radius equal to $\sqrt{-\dfrac{F}{A}}$.

11.4 Assess Your Understanding (page 801)

7. hyperbola **8.** transverse axis

9. $y = \frac{3}{2}x$; $y = -\frac{3}{2}x$ **10.** F **11.** T

12. F **13.** B **15.** A

17. $x^2 - \dfrac{y^2}{8} = 1$

19. $\dfrac{y^2}{16} - \dfrac{x^2}{20} = 1$

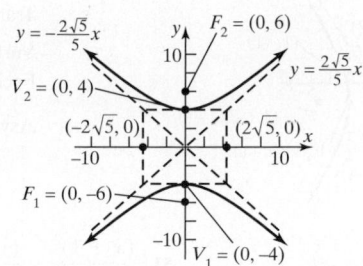

21. $\dfrac{x^2}{9} - \dfrac{y^2}{16} = 1$

23. $\dfrac{y^2}{36} - \dfrac{x^2}{9} = 1$

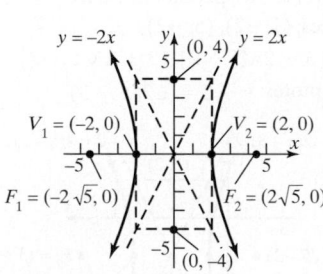

25. $\dfrac{x^2}{8} - \dfrac{y^2}{8} = 1$

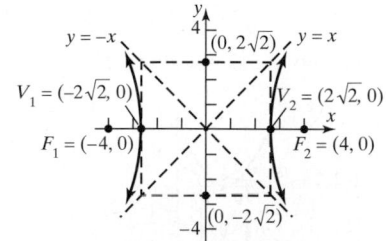

27. $\dfrac{x^2}{25} - \dfrac{y^2}{9} = 1$

Center: $(0,0)$
Transverse axis: x-axis
Vertices: $(-5, 0)$, $(5, 0)$
Foci: $\left(-\sqrt{34}, 0\right)$, $\left(\sqrt{34}, 0\right)$
Asymptotes: $y = \pm\dfrac{3}{5}x$

29. $\dfrac{x^2}{4} - \dfrac{y^2}{16} = 1$

Center: $(0,0)$
Transverse axis: x-axis
Vertices: $(-2, 0)$, $(2, 0)$
Foci: $\left(-2\sqrt{5}, 0\right)$, $\left(2\sqrt{5}, 0\right)$
Asymptotes: $y = \pm 2x$

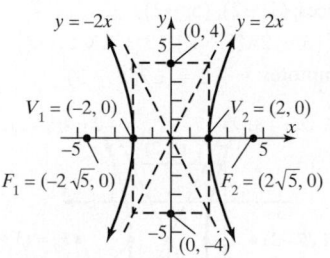

31. $\dfrac{y^2}{9} - x^2 = 1$

Center: $(0,0)$
Transverse axis: y-axis
Vertices: $(0, -3)$, $(0, 3)$
Foci: $\left(0, -\sqrt{10}\right)$, $\left(0, \sqrt{10}\right)$
Asymptotes: $y = \pm 3x$

33. $\dfrac{y^2}{25} - \dfrac{x^2}{25} = 1$

Center: $(0,0)$
Transverse axis: y-axis
Vertices: $(0, -5)$, $(0, 5)$
Foci: $\left(0, -5\sqrt{2}\right)$, $\left(0, 5\sqrt{2}\right)$
Asymptotes: $y = \pm x$

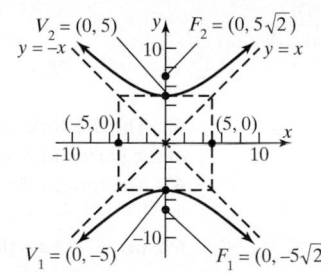

35. $x^2 - y^2 = 1$

37. $\dfrac{y^2}{36} - \dfrac{x^2}{9} = 1$

39. $\dfrac{(x-4)^2}{4} - \dfrac{(y+1)^2}{5} = 1$

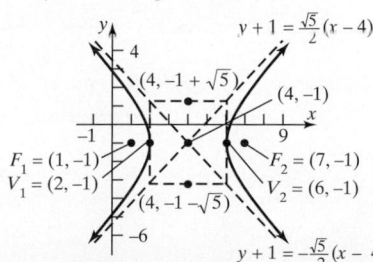

41. $\dfrac{(y+4)^2}{4} - \dfrac{(x+3)^2}{12} = 1$

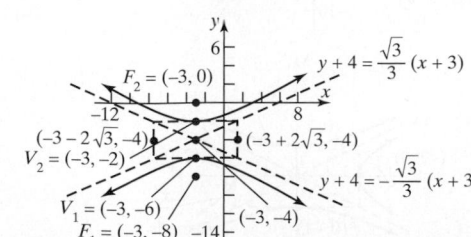

43. $(x-5)^2 - \dfrac{(y-7)^2}{3} = 1$

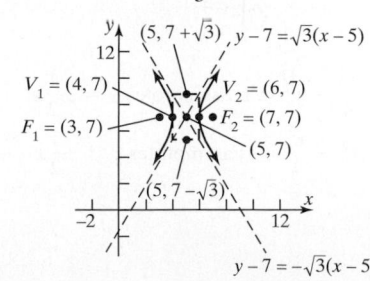

45. $\dfrac{(x-1)^2}{4} - \dfrac{(y+1)^2}{9} = 1$

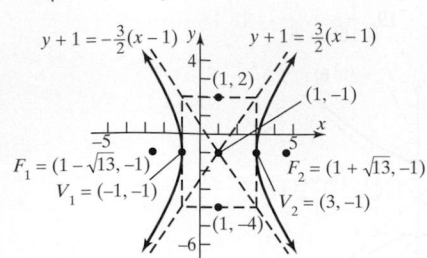

47. $\dfrac{(x-2)^2}{4} - \dfrac{(y+3)^2}{9} = 1$

Center: $(2, -3)$

Transverse axis: parallel to x-axis

Vertices: $(0, -3), (4, -3)$

Foci: $\left(2 - \sqrt{13}, -3\right), \left(2 + \sqrt{13}, -3\right)$

Asymptotes: $y + 3 = \pm\dfrac{3}{2}(x - 2)$

49. $\dfrac{(y-2)^2}{4} - (x+2)^2 = 1$

Center: $(-2, 2)$

Transverse axis: parallel to y-axis

Vertices: $(-2, 0), (-2, 4)$

Foci: $\left(-2, 2 - \sqrt{5}\right), \left(-2, 2 + \sqrt{5}\right)$

Asymptotes: $y - 2 = \pm 2(x + 2)$

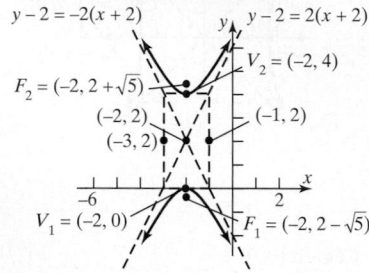

51. $\dfrac{(x+1)^2}{4} - \dfrac{(y+2)^2}{4} = 1$

Center: $(-1, -2)$

Transverse axis: parallel to x-axis

Vertices: $(-3, -2), (1, -2)$

Foci: $\left(-1 - 2\sqrt{2}, -2\right), \left(-1 + 2\sqrt{2}, -2\right)$

Asymptotes: $y + 2 = \pm(x + 1)$

53. $(x-1)^2 - (y+1)^2 = 1$

Center: $(1, -1)$

Transverse axis: parallel to x-axis

Vertices: $(0, -1), (2, -1)$

Foci: $\left(1 - \sqrt{2}, -1\right), \left(1 + \sqrt{2}, -1\right)$

Asymptotes: $y + 1 = \pm(x - 1)$

55. $\dfrac{(y-2)^2}{4} - (x+1)^2 = 1$

Center: $(-1, 2)$

Transverse axis: parallel to y-axis

Vertices: $(-1, 0), (-1, 4)$

Foci: $\left(-1, 2 - \sqrt{5}\right), \left(-1, 2 + \sqrt{5}\right)$

Asymptotes: $y - 2 = \pm 2(x + 1)$

57. $\dfrac{(x-3)^2}{4} - \dfrac{(y+2)^2}{16} = 1$

Center: $(3, -2)$

Transverse axis: parallel to x-axis

Vertices: $(1, -2), (5, -2)$

Foci: $\left(3 - 2\sqrt{5}, -2\right), \left(3 + 2\sqrt{5}, -2\right)$

Asymptotes: $y + 2 = \pm 2(x - 3)$

59. $\dfrac{(y-1)^2}{4} - (x+2)^2 = 1$

Center: $(-2, 1)$

Transverse axis: parallel to y-axis

Vertices: $(-2, -1), (-2, 3)$

Foci: $\left(-2, 1 - \sqrt{5}\right), \left(-2, 1 + \sqrt{5}\right)$

Asymptotes: $y - 1 = \pm 2(x + 2)$

61.

63.

65. The fireworks display is 50,138 ft north of the person at point A.

67. The tower is 592.4 ft tall.

69. (a) $y = \pm x$ (b) $\dfrac{x^2}{100} - \dfrac{y^2}{100} = 1, x \geq 0$

73. $\dfrac{x^2}{4} - y^2 = 1$; asymptotes $y = \pm\dfrac{1}{2}x$;

$y^2 - \dfrac{x^2}{4} = 1$; asymptotes $y = \pm\dfrac{1}{2}x$

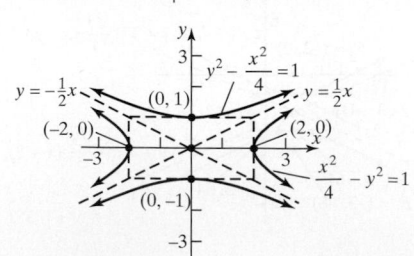

75. $Ax^2 + Cy^2 + F = 0$ If A and C are of opposite sign and $F \neq 0$, this equation may be written as $\dfrac{x^2}{\left(-\dfrac{F}{A}\right)} + \dfrac{y^2}{\left(-\dfrac{F}{C}\right)} = 1$,

$Ax^2 + Cy^2 = -F$ where $-\dfrac{F}{A}$ and $-\dfrac{F}{C}$ are opposite in sign. This is the equation of a hyperbola with center $(0, 0)$.

The transverse axis is the x-axis if $-\dfrac{F}{A} > 0$; the transverse axis is the y-axis if $-\dfrac{F}{A} < 0$.

11.5 Assess Your Understanding *(page 810)*

5. $\cot(2\theta) = \dfrac{A - C}{B}$ **6.** Hyperbola **7.** Ellipse **8.** T **9.** T **10.** F **11.** Parabola **13.** Ellipse **15.** Hyperbola **17.** Hyperbola **19.** Circle

21. $x = \dfrac{\sqrt{2}}{2}(x' - y'),\ y = \dfrac{\sqrt{2}}{2}(x' + y')$ **23.** $x = \dfrac{\sqrt{2}}{2}(x' - y'),\ y = \dfrac{\sqrt{2}}{2}(x' + y')$ **25.** $x = \dfrac{1}{2}(x' - \sqrt{3}y'),\ y = \dfrac{1}{2}(\sqrt{3}x' + y')$

27. $x = \dfrac{\sqrt{5}}{5}(x' - 2y'),\ y = \dfrac{\sqrt{5}}{5}(2x' + y')$ **29.** $x = \dfrac{\sqrt{13}}{13}(3x' - 2y'),\ y = \dfrac{\sqrt{13}}{13}(2x' + 3y')$

31. $\theta = 45°$ (see Problem 21)

$x'^2 - \dfrac{y'^2}{3} = 1$

Hyperbola

Center at origin

Transverse axis is the x'-axis.

Vertices at $(\pm 1, 0)$

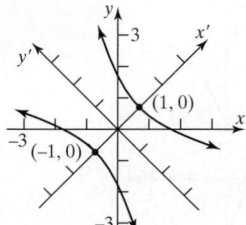

33. $\theta = 45°$ (see Problem 23)

$x'^2 + \dfrac{y'^2}{4} = 1$

Ellipse

Center at $(0, 0)$

Major axis is the y'-axis.

Vertices at $(0, \pm 2)$

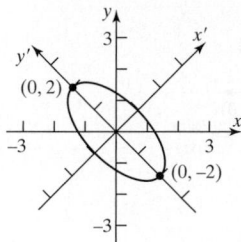

35. $\theta = 60°$ (see Problem 25)

$\dfrac{x'^2}{4} + y'^2 = 1$

Ellipse

Center at $(0, 0)$

Major axis is the x'-axis.

Vertices at $(\pm 2, 0)$

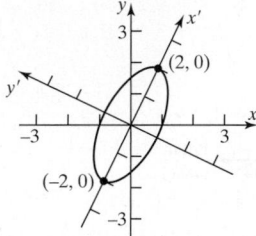

37. $\theta \approx 63°$ (see Problem 27)

$y'^2 = 8x'$

Parabola

Vertex at $(0, 0)$

Focus at $(2, 0)$

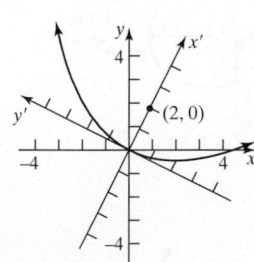

39. $\theta \approx 34°$ (see Problem 29)

$\dfrac{(x' - 2)^2}{4} + y'^2 = 1$

Ellipse

Center at $(2, 0)$

Major axis is the x'-axis.

Vertices at $(4, 0)$ and $(0, 0)$

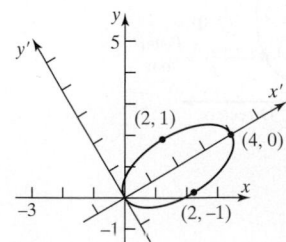

41. $\cot(2\theta) = \dfrac{7}{24}$

$\theta = \sin^{-1}\left(\dfrac{3}{5}\right) \approx 37°$

$(x' - 1)^2 = -6\left(y' - \dfrac{1}{6}\right)$

Parabola

Vertex at $\left(1, \dfrac{1}{6}\right)$

Focus at $\left(1, -\dfrac{4}{3}\right)$

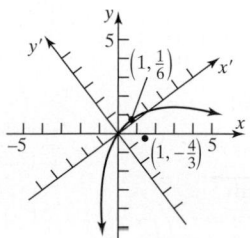

43. Hyperbola **45.** Hyperbola **47.** Parabola **49.** Ellipse **51.** Ellipse

53. Refer to equation (6): $A' = A\cos^2\theta + B\sin\theta\cos\theta + C\sin^2\theta$
$B' = B(\cos^2\theta - \sin^2\theta) + 2(C - A)(\sin\theta\cos\theta)$
$C' = A\sin^2\theta - B\sin\theta\cos\theta + C\cos^2\theta$
$D' = D\cos\theta + E\sin\theta$
$E' = -D\sin\theta + E\cos\theta$
$F' = F$

55. Use Problem 53 to find $B'^2 - 4A'C'$. After much cancellation, $B'^2 - 4A'C' = B^2 - 4AC$.

57. The distance between P_1 and P_2 in the $x'y'$-plane equals $\sqrt{(x_2' - x_1')^2 + (y_2' - y_1')^2}$.

Assuming that $x' = x \cos \theta - y \sin \theta$ and $y' = x \sin \theta + y \cos \theta$, then

$$(x_2' - x_1')^2 = (x_2 \cos \theta - y_2 \sin \theta - x_1 \cos \theta + y_1 \sin \theta)^2$$

$$= \cos^2 \theta (x_2 - x_1)^2 - 2 \sin \theta \cos \theta (x_2 - x_1)(y_2 - y_1) + \sin^2 \theta (y_2 - y_1)^2, \text{ and}$$

$$(y_2' - y_1')^2 = (x_2 \sin \theta + y_2 \cos \theta - x_1 \sin \theta - y_1 \cos \theta)^2 = \sin^2 \theta (x_2 - x_1)^2 + 2 \sin \theta \cos \theta (x_2 - x_1)(y_2 - y_1) + \cos^2 \theta (y_2 - y_1)^2.$$

Therefore, $(x_2' - x_1')^2 + (y_2' - y_1')^2 = \cos^2 \theta (x_2 - x_1)^2 + \sin^2 \theta (x_2 - x_1)^2 + \sin^2 \theta (y_2 - y_1)^2 + \cos^2 \theta (y_2 - y_1)^2$

$$= (x_2 - x_1)^2 (\cos^2 \theta + \sin^2 \theta) + (y_2 - y_1)^2 (\sin^2 \theta + \cos^2 \theta) = (x_2 - x_1)^2 + (y_2 - y_1)^2.$$

11.6 Assess Your Understanding *(page 816)*

3. $\frac{1}{2}$; ellipse; parallel; 4; below **4.** 1; <1; >1 **5.** T **6.** T **7.** Parabola; directrix is perpendicular to the polar axis 1 unit to

the right of the pole. **9.** Hyperbola; directrix is parallel to the polar axis $\frac{4}{3}$ units below the pole.

11. Ellipse; directrix is perpendicular to the polar axis $\frac{3}{2}$ units to the left of the pole.

13. Parabola; directrix is perpendicular to the polar axis 1 unit to the right of the pole; vertex is at $\left(\frac{1}{2}, 0\right)$.

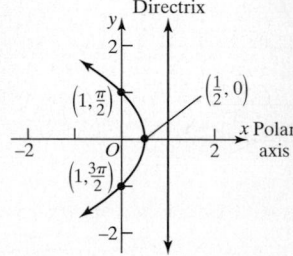

15. Ellipse; directrix is parallel to the polar axis $\frac{8}{3}$ units above the pole; vertices are at $\left(\frac{8}{7}, \frac{\pi}{2}\right)$ and $\left(8, \frac{3\pi}{2}\right)$.

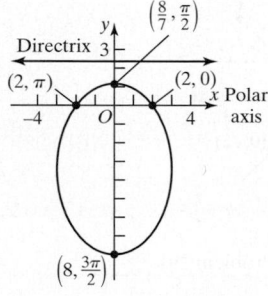

17. Hyperbola; directrix is perpendicular to the polar axis $\frac{3}{2}$ units to the left of the pole; vertices are at $(-3, 0)$ and $(1, \pi)$.

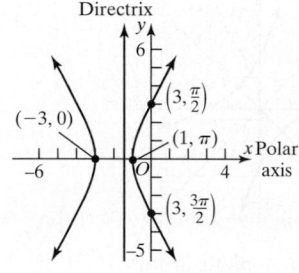

19. Ellipse; directrix is parallel to the polar axis 8 units below the pole; vertices are at $\left(8, \frac{\pi}{2}\right)$ and $\left(\frac{8}{3}, \frac{3\pi}{2}\right)$.

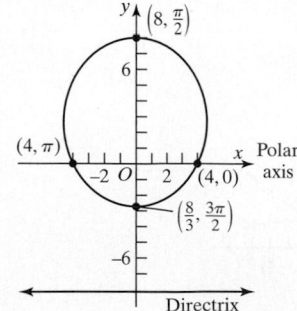

21. Ellipse; directrix is parallel to the polar axis 3 units below the pole; vertices are at $\left(6, \frac{\pi}{2}\right)$ and $\left(\frac{6}{5}, \frac{3\pi}{2}\right)$.

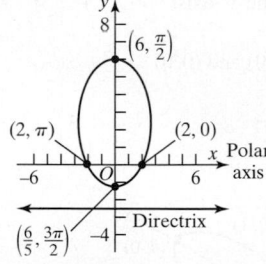

23. Ellipse; directrix is perpendicular to the polar axis 6 units to the left of the pole; vertices are at $(6, 0)$ and $(2, \pi)$.

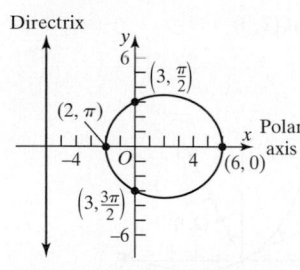

25. $y^2 + 2x - 1 = 0$ **27.** $16x^2 + 7y^2 + 48y - 64 = 0$ **29.** $3x^2 - y^2 + 12x + 9 = 0$ **31.** $4x^2 + 3y^2 - 16y - 64 = 0$

33. $9x^2 + 5y^2 - 24y - 36 = 0$ **35.** $3x^2 + 4y^2 - 12x - 36 = 0$ **37.** $r = \dfrac{1}{1 + \sin \theta}$ **39.** $r = \dfrac{12}{5 - 4 \cos \theta}$ **41.** $r = \dfrac{12}{1 - 6 \sin \theta}$

43. Use $d(D, P) = p - r \cos \theta$ in the derivation of equation (a) in Table 5.

45. Use $d(D, P) = p + r \sin \theta$ in the derivation of equation (a) in Table 5.

11.7 Assess Your Understanding (page 827)

2. plane curve; parameter 3. ellipse 4. cycloid 5. F 6. T

7.

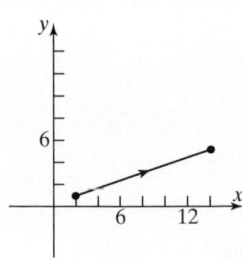

$x - 3y + 1 = 0$

9.

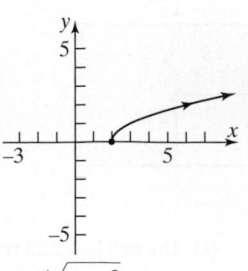

$y = \sqrt{x - 2}$

11.

$x = y + 8$

13.

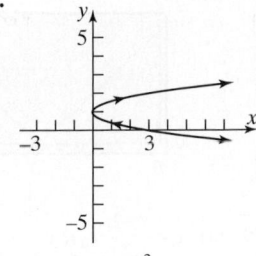

$x = 3(y - 1)^2$

15.

$2y = 2 + x$

17.

$y = x^3$

19.

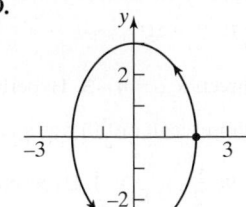

$\dfrac{x^2}{4} + \dfrac{y^2}{9} = 1$

21.

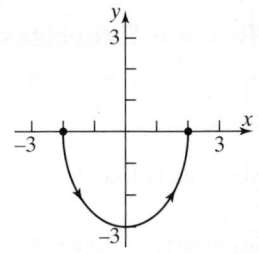

$\dfrac{x^2}{4} + \dfrac{y^2}{9} = 1$

23.

$x^2 - y^2 = 1$

25.

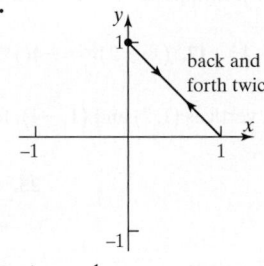

back and forth twice

$x + y = 1$

27. $x = t$ or $x = \dfrac{t + 1}{4}$
 $y = 4t - 1$ $y = t$

29. $x = t$ or $x = t^3$
 $y = t^2 + 1$ $y = t^6 + 1$

31. $x = t$ or $x = \sqrt[3]{t}$
 $y = t^3$ $y = t$

33. $x = t$ or $x = t^3$
 $y = t^{2/3}$ $y = t^2, t \geq 0$

35. $x = t + 2, y = t, 0 \leq t \leq 5$

37. $x = 3 \cos t, y = 2 \sin t, 0 \leq t \leq 2\pi$ 39. $x = 2 \cos(\pi t), y = -3 \sin(\pi t), 0 \leq t \leq 2$ 41. $x = 2 \sin(2\pi t), y = 3 \cos(2\pi t), 0 \leq t \leq 1$

43.

45.

47.

51. (a) Train: $x_1 = t^2, y_1 = 1$;

49. (a) $x = 3$
 $y = -16t^2 + 50t + 6$
(b) 3.24 sec
(c) 1.5625 sec; 45.0625 ft
(d) 50

51. (a) Train: $x_1 = t^2, y_1 = 1$;
 Bill: $x_2 = 5(t - 5), y_2 = 3$
(b) Bill won't catch the train.
(c)

$t = 9.5$

53. (a) $x = (145 \cos 20°)t$
 $y = -16t^2 + (145 \sin 20°)t + 5$
(b) 3.197 sec (c) 435.61 ft
(d) 1.55 sec; 43.43 ft
(e)

55. (a) $x = (40 \cos 45°)t$
 $y = -4.9t^2 + (40 \sin 45°)t + 300$
(b) 11.226 sec (c) 317.5 m
(d) 2.886 sec; 340.8 m
(e)

57. (a) Paseo: $x = 40t - 5, y = 0$; Bonneville: $x = 0, y = 30t - 4$ **(d)** 0.2 mi; 7.68 min
 (b) $d = \sqrt{(40t - 5)^2 + (30t - 4)^2}$ **(e)** Turn axes off to see the graph:
 (c)

59. (a) $x = \dfrac{\sqrt{2}}{2} v_0 t, y = -16t^2 + \dfrac{\sqrt{2}}{2} v_0 t + 3$ **(b)** Maximum height is 139.1 ft. **(c)** The ball is 272.25 ft from home plate.

(d) Yes, the ball will clear the wall by about 99.5 ft. **61.** The orientation is from (x_1, y_1) to (x_2, y_2).

Review Exercises (page 831)

1. Parabola; vertex $(0, 0)$, focus $(-4, 0)$, directrix $x = 4$ **3.** Hyperbola; center $(0, 0)$, vertices $(5, 0)$ and $(-5, 0)$, foci $\left(\sqrt{26}, 0\right)$ and $\left(-\sqrt{26}, 0\right)$,

asymptotes $y = \dfrac{1}{5}x$ and $y = -\dfrac{1}{5}x$ **5.** Ellipse; center $(0, 0)$, vertices $(0, 5)$ and $(0, -5)$, foci $(0, 3)$ and $(0, -3)$ **7.** $x^2 = -4(y - 1)$: Parabola;

vertex $(0, 1)$, focus $(0, 0)$, directrix $y = 2$ **9.** $\dfrac{x^2}{2} - \dfrac{y^2}{8} = 1$: Hyperbola; center $(0, 0)$, vertices $\left(\sqrt{2}, 0\right)$ and $\left(-\sqrt{2}, 0\right)$, foci $\left(\sqrt{10}, 0\right)$ and $\left(-\sqrt{10}, 0\right)$,

asymptotes $y = 2x$ and $y = -2x$ **11.** $(x - 2)^2 = 2(y + 2)$: Parabola; vertex $(2, -2)$, focus $\left(2, -\dfrac{3}{2}\right)$, directrix $y = -\dfrac{5}{2}$ **13.** $\dfrac{(y - 2)^2}{4} - (x - 1)^2 = 1$:

Hyperbola; center $(1, 2)$, vertices $(1, 4)$ and $(1, 0)$, foci $\left(1, 2 + \sqrt{5}\right)$ and $\left(1, 2 - \sqrt{5}\right)$, asymptotes $y - 2 = \pm 2(x - 1)$ **15.** $\dfrac{(x - 2)^2}{9} + \dfrac{(y - 1)^2}{4} = 1$:

Ellipse; center $(2, 1)$, vertices $(5, 1)$ and $(-1, 1)$, foci $\left(2 + \sqrt{5}, 1\right)$ and $\left(2 - \sqrt{5}, 1\right)$ **17.** $(x - 2)^2 = -4(y + 1)$: Parabola; vertex $(2, -1)$, focus

$(2, -2)$, directrix $y = 0$ **19.** $\dfrac{(x - 1)^2}{4} + \dfrac{(y + 1)^2}{9} = 1$: Ellipse; center $(1, -1)$, vertices $(1, 2)$ and $(1, -4)$, foci $\left(1, -1 + \sqrt{5}\right)$ and $\left(1, -1 - \sqrt{5}\right)$

21. $y^2 = -8x$

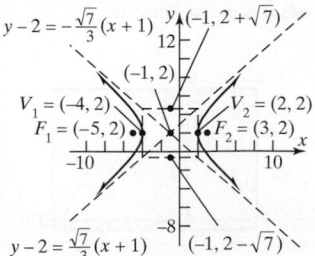

23. $\dfrac{y^2}{4} - \dfrac{x^2}{12} = 1$

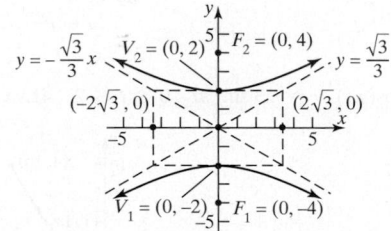

25. $\dfrac{x^2}{16} + \dfrac{y^2}{7} = 1$

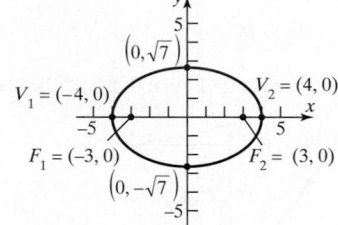

27. $(x - 2)^2 = -4(y + 3)$

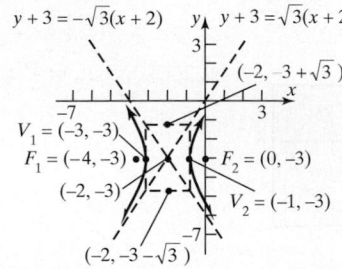

29. $(x + 2)^2 - \dfrac{(y + 3)^2}{3} = 1$

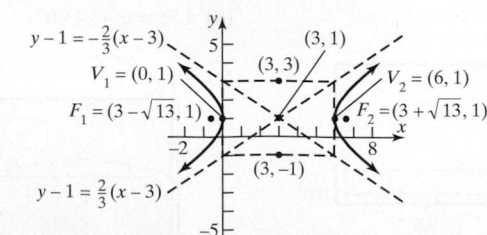

31. $\dfrac{(x + 4)^2}{16} + \dfrac{(y - 5)^2}{25} = 1$

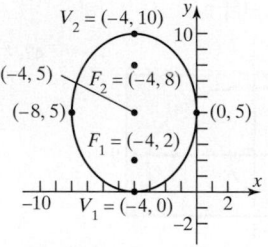

33. $\dfrac{(x + 1)^2}{9} - \dfrac{(y - 2)^2}{7} = 1$

35. $\dfrac{(x - 3)^2}{9} - \dfrac{(y - 1)^2}{4} = 1$

37. Parabola
39. Ellipse
41. Parabola
43. Hyperbola
45. Ellipse

47. $x'^2 - \dfrac{y'^2}{9} = 1$

Hyperbola

Center at the origin

Transverse axis the x'-axis

Vertices at $(\pm 1, 0)$

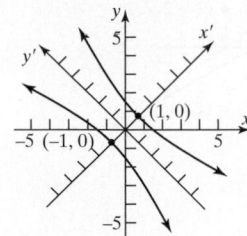

49. $\dfrac{x'^2}{2} + \dfrac{y'^2}{4} = 1$

Ellipse

Center at origin

Major axis the y'-axis

Vertices at $(0, \pm 2)$

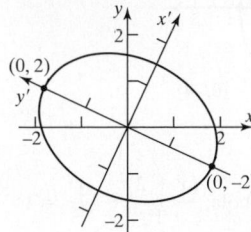

51. $y'^2 = -\dfrac{4\sqrt{13}}{13} x'$

Parabola

Vertex at the origin

Focus on the x'-axis at $\left(-\dfrac{\sqrt{13}}{13}, 0 \right)$

53. Parabola; directrix is perpendicular to the polar axis 4 units to the left of the pole; vertex is $(2, \pi)$.

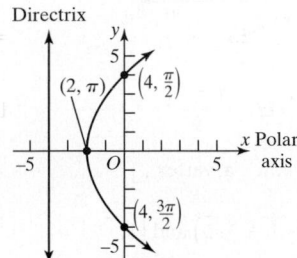

55. Ellipse; directrix is parallel to the polar axis 6 units below the pole; vertices are $\left(6, \dfrac{\pi}{2} \right)$ and $\left(2, \dfrac{3\pi}{2} \right)$.

57. Hyperbola; directrix is perpendicular to the polar axis 1 unit to the right of the pole; vertices are $\left(\dfrac{2}{3}, 0 \right)$ and $(-2, \pi)$.

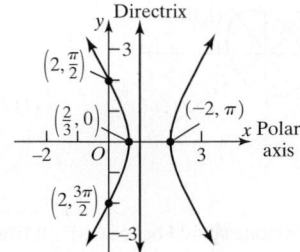

59. $y^2 - 8x - 16 = 0$

61. $3x^2 - y^2 - 8x + 4 = 0$

63.

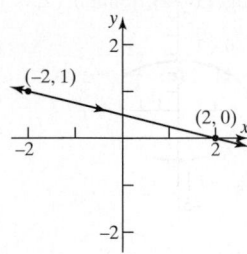

$x + 4y = 2$

65.

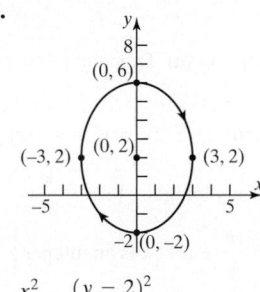

$\dfrac{x^2}{9} + \dfrac{(y-2)^2}{16} = 1$

67.

$1 + y = x$

69. $x = t, y = -2t + 4, -\infty < t < \infty$ $x = \dfrac{t-4}{-2}, y = t, -\infty < t < \infty$ **71.** $x = 4\cos\left(\dfrac{\pi}{2} t \right), y = 3\sin\left(\dfrac{\pi}{2} t \right), 0 \le t \le 4$

73. $\dfrac{x^2}{5} - \dfrac{y^2}{4} = 1$ **75.** The ellipse $\dfrac{x^2}{16} + \dfrac{y^2}{7} = 1$ **77.** $\dfrac{1}{4}$ ft or 3 in. **79.** 19.72 ft, 18.86 ft, 14.91 ft **81.** 450 ft

83. (a) $x = (80\cos 35°)t$

$y = -16t^2 + (80\sin 35°)t + 6$

(b) 2.9932 sec **(c)** 1.4339 sec; 38.9 ft

(d) 196.15 ft **(e)**

Chapter Test (page 833)

1. Hyperbola; center: $(-1, 0)$; vertices: $(-3, 0)$ and $(1, 0)$; foci: $\left(-1 - \sqrt{13}, 0 \right)$ and $\left(-1 + \sqrt{13}, 0 \right)$; asymptotes: $y = -\dfrac{3}{2}(x + 1)$ and $y = \dfrac{3}{2}(x + 1)$

2. Parabola; vertex: $\left(1, -\dfrac{1}{2} \right)$; focus: $\left(1, \dfrac{3}{2} \right)$; directrix: $y = -\dfrac{5}{2}$

3. Ellipse; center: $(-1, 1)$; foci: $\left(-1 - \sqrt{3}, 1 \right)$ and $\left(-1 + \sqrt{3}, 1 \right)$; vertices: $(-4, 1)$ and $(2, 1)$

4. $(x + 1)^2 = 6(y - 3)$

5. $\dfrac{x^2}{7} + \dfrac{y^2}{16} = 1$

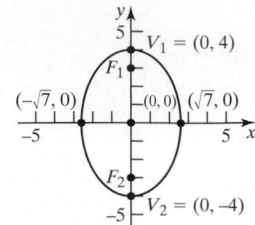

6. $\dfrac{(y - 2)^2}{4} - \dfrac{(x - 2)^2}{8} = 1$

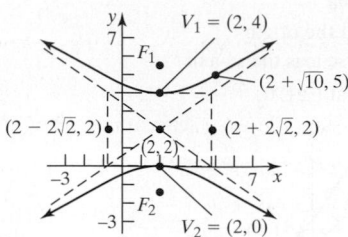

7. Hyperbola **8.** Ellipse **9.** Parabola

10. $x'^2 + 2y'^2 = 1$; This is the equation of an ellipse with center at $(0, 0)$ in the $x'y'$-plane. The vertices axis of symmetry.

11. Hyperbola; $\dfrac{(x + 2)^2}{1} - \dfrac{y^2}{3} = 1$

12. $y = 1 - \sqrt{\dfrac{x + 2}{3}}$

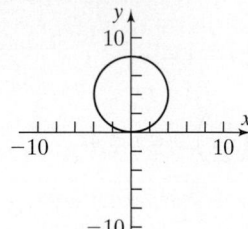

13. The microphone should be located $\dfrac{2}{3}$ ft from the base of the reflector, along its axis of symmetry.

Cumulative Review *(page 833)*

1. $-6x + 5 - 3h$ **2.** $\left\{-5, -\dfrac{1}{3}, 2\right\}$ **3.** $\{x \mid -3 \le x \le 2\}$ or $[-3, 2]$ **4.** **(a)** Domain: $(-\infty, \infty)$; range: $(2, \infty)$ **(b)** $y = \log_3 (x - 2)$; domain: $(2, \infty)$;

range: $(-\infty, \infty)$ **5.** **(a)** $\{18\}$ **(b)** $(2, 18)$ **6.** **(a)** $y = 2x - 2$ **(b)** $(x - 2)^2 + y^2 = 4$ **(c)** $\dfrac{x^2}{9} + \dfrac{y^2}{4} = 1$ **(d)** $y = 2(x - 1)^2$ **(e)** $\dfrac{y^2}{1} - \dfrac{x^2}{3} = 1$

(f) $y = 4^x$ **7.** $\theta = \dfrac{\pi}{12} \pm \pi k, k$ is any integer; $\theta = \dfrac{5\pi}{12} \pm \pi k, k$ is any integer **8.** $\theta = \dfrac{\pi}{6}$

9. $r = 8 \sin \theta$ **10.** $\left\{x \mid x \ne \dfrac{3\pi}{4} \pm \pi k, k \text{ is an integer}\right\}$ **11.** $\{22.5°\}$ **12.** $y = \dfrac{x^2}{5} + 5$

CHAPTER 12 Systems of Equations and Inequalities

12.1 Assess Your Understanding *(page 847)*

3. inconsistent **4.** consistent **5.** F **6.** T

7. $\begin{cases} 2(2) - (-1) = 5 \\ 5(2) + 2(-1) = 8 \end{cases}$ **9.** $\begin{cases} 3(2) - 4\left(\dfrac{1}{2}\right) = 4 \\ \dfrac{1}{2}(2) - 3\left(\dfrac{1}{2}\right) = -\dfrac{1}{2} \end{cases}$ **11.** $\begin{cases} 4 - 1 = 3 \\ \dfrac{1}{2}(4) + 1 = 3 \end{cases}$ **13.** $\begin{cases} 3(1) + 3(-1) + 2(2) = 4 \\ 1 - (-1) - 2 = 0 \\ 2(-1) - 3(2) = -8 \end{cases}$ **15.** $\begin{cases} 3(2) + 3(-2) + 2(2) = 41 \\ 3(2) - 3(-2) + 2(2) = 10 \\ 5(2) - 2(-2) - 3(2) = 81 \end{cases}$

17. $x = 6, y = 2; (6, 2)$ **19.** $x = 3, y = 2; (3, 2)$ **21.** $x = 8, y = -4; (8, -4)$ **23.** $x = \dfrac{1}{3}, y = -\dfrac{1}{6}; \left(\dfrac{1}{3}, -\dfrac{1}{6}\right)$ **25.** Inconsistent

27. $x = 1, y = 2; (1, 2)$ **29.** $\{(x, y) \mid x = 4 - 2y, y \text{ is any real number}\}$ or $\left\{(x, y) \mid y = \dfrac{4 - x}{2}, x \text{ is any real number}\right\}$

31. $x = 1, y = 1; (1, 1)$ **33.** $x = \dfrac{3}{2}, y = 1; \left(\dfrac{3}{2}, 1\right)$ **35.** $x = 4, y = 3; (4, 3)$ **37.** $x = \dfrac{4}{3}, y = \dfrac{1}{5}; \left(\dfrac{4}{3}, \dfrac{1}{5}\right)$ **39.** $x = \dfrac{1}{5}, y = \dfrac{1}{3}; \left(\dfrac{1}{5}, \dfrac{1}{3}\right)$

41. $x = 8, y = 2, z = 0; (8, 2, 0)$ **43.** $x = 2, y = -1, z = 1; (2, -1, 1)$ **45.** Inconsistent

47. $\{(x, y, z) \mid x = 5z - 2, y = 4z - 3; z \text{ is any real number}\}$ **49.** Inconsistent **51.** $x = 1, y = 3, z = -2; (1, 3, -2)$

53. $x = -3, y = \dfrac{1}{2}, z = 1; \left(-3, \dfrac{1}{2}, 1\right)$ **55.** Length 30 ft; width 15 ft **57.** There were 18 commercial launches and 37 noncommercial launches in 2005.

59. 22.5 lb **61.** Average wind speed 25 mph; average airspeed 175 mph **63.** 80 \$25 sets and 120 \$45 sets **65.** \$9.96 **67.** Mix 50 mg of first compound

with 75 mg of second. **69.** $a = \dfrac{4}{3}, b = -\dfrac{5}{3}, c = 1$ **71.** $y = 9000, r = 0.06$ **73.** $I_1 = \dfrac{10}{71}, I_2 = \dfrac{65}{71}, I_3 = \dfrac{55}{71}$ **75.** 100 orchestra, 210 main, and

190 balcony seats **77.** 1.5 chicken, 1 corn, 2 milk

79. If x = price of hamburgers, y = price of fries,

and z = price of colas, then

$x = 2.75 - z, y = \dfrac{41}{60} + \dfrac{1}{3}z, \$0.60 \le z \le \$0.90.$

There is not sufficient information:

x	\$2.13	\$2.01	\$1.86
y	\$0.89	\$0.93	\$0.98
z	\$0.62	\$0.74	\$0.89

81. It will take Beth 30 hr, Bill 24 hr, and Edie 40 hr.

12.2 Assess Your Understanding (page 862)

1. matrix **2.** augmented **3.** T **4.** T

5. $\begin{bmatrix} 1 & -5 & 5 \\ 4 & 3 & 6 \end{bmatrix}$ **7.** $\begin{bmatrix} 2 & 3 & 6 \\ 4 & -6 & -2 \end{bmatrix}$ **9.** $\begin{bmatrix} 0.01 & -0.03 & 0.06 \\ 0.13 & 0.10 & 0.20 \end{bmatrix}$ **11.** $\begin{bmatrix} 1 & -1 & 1 & 10 \\ 3 & 3 & 0 & 5 \\ 1 & 1 & 2 & 2 \end{bmatrix}$ **13.** $\begin{bmatrix} 1 & 1 & -1 & 2 \\ 3 & -2 & 0 & 2 \\ 5 & 3 & -1 & 1 \end{bmatrix}$ **15.** $\begin{bmatrix} 1 & -1 & -1 & 10 \\ 2 & 1 & 2 & -1 \\ -3 & 4 & 0 & 5 \\ 4 & -5 & 1 & 0 \end{bmatrix}$

17. $\begin{cases} x - 3y = -2 & (1) \\ 2x - 5y = 5 & (2) \end{cases}$ $\begin{bmatrix} 1 & -3 & -2 \\ 0 & 1 & 9 \end{bmatrix}$ **19.** $\begin{cases} x - 3y + 4z = 3 & (1) \\ 3x - 5y + 6z = 6 & (2) \\ -5x + 3y + 4z = 6 & (3) \end{cases}$ **(a)** $\begin{bmatrix} 1 & -3 & 4 & 3 \\ 0 & 4 & -6 & -3 \\ -5 & 3 & 4 & 6 \end{bmatrix}$ **(b)** $\begin{bmatrix} 1 & -3 & 4 & 3 \\ 3 & -5 & 6 & 6 \\ 0 & -12 & 24 & 21 \end{bmatrix}$

21. $\begin{cases} x - 3y + 2z = -6 & (1) \\ 2x - 5y + 3z = -4 & (2) \\ -3x - 6y + 4z = 6 & (3) \end{cases}$ **(a)** $\begin{bmatrix} 1 & -3 & 2 & -6 \\ 0 & 1 & -1 & 8 \\ -3 & -6 & 4 & 6 \end{bmatrix}$ **(b)** $\begin{bmatrix} 1 & -3 & 2 & -6 \\ 2 & -5 & 3 & -4 \\ 0 & -15 & 10 & -12 \end{bmatrix}$

23. $\begin{cases} 5x - 3y + z = -2 & (1) \\ 2x - 5y + 6z = -2 & (2) \\ -4x + y + 4z = 6 & (3) \end{cases}$ **(a)** $\begin{bmatrix} 1 & 7 & -11 & 2 \\ 2 & -5 & 6 & -2 \\ -4 & 1 & 4 & 6 \end{bmatrix}$ **(b)** $\begin{bmatrix} 1 & -2 & 5 & 4 \\ 2 & -5 & 6 & -2 \\ -4 & 1 & 4 & 6 \end{bmatrix}$ **25.** $\begin{cases} x = 5 \\ y = -1 \end{cases}$ **27.** $\begin{cases} x = 1 \\ y = 2 \\ 0 = 3 \end{cases}$

Consistent: $x = 5, y = -1$ or $(5, -1)$ Inconsistent

29. $\begin{cases} x + 2z = -1 \\ y - 4z = -2 \\ 0 = 0 \end{cases}$ **31.** $\begin{cases} x_1 = 1 \\ x_2 + x_4 = 2 \\ x_3 + 2x_4 = 3 \end{cases}$ **33.** $\begin{cases} x_1 + 4x_4 = 2 \\ x_2 + x_3 + 3x_4 = 3 \\ 0 = 0 \end{cases}$ **35.** $\begin{cases} x_1 + x_4 = -2 \\ x_2 + 2x_4 = 2 \\ x_3 - x_4 = 0 \end{cases}$

Consistent: Consistent: Consistent: Consistent:

$\begin{cases} x = -1 - 2z \\ y = -2 + 4z \\ z \text{ is any real number or} \end{cases}$ $\begin{cases} x_1 = 1, x_2 = 2 - x_4 \\ x_3 = 3 - 2x_4 \\ x_4 \text{ is any real number or} \end{cases}$ $\begin{cases} x_1 = 2 - 4x_4 \\ x_2 = 3 - x_3 - 3x_4 \\ x_3, x_4 \text{ are any real numbers or} \end{cases}$ $\begin{cases} x_1 = -2 - x_4 \\ x_2 = 2 - 2x_4 \\ x_3 = x_4 \\ x_4 \text{ is any real number or} \end{cases}$

$\{(x, y, z) \mid x = -1 - 2z,$ $\{(x_1, x_2, x_3, x_4) \mid x_1 = 1,$ $\{(x_1, x_2, x_3, x_4) \mid x_1 = 2 - 4x_4,$ $\{(x_1, x_2, x_3, x_4) \mid x_1 = -2 - x_4,$

$y = -2 + 4z, z \text{ is any real}$ $x_2 = 2 - x_4, x_3 = 3 - 2x_4,$ $x_2 = 3 - x_4 - 3x_4, x_3, x_4 \text{ are}$ $x_2 = 2 - 2x_4, x_3 = x_4, x_4 \text{ is any}$

number\} $x_4 \text{ is any real number}\}$ any real numbers\} real number\}

37. $x = 6, y = 2; (6, 2)$ **39.** $x = \dfrac{1}{2}, y = \dfrac{3}{4}; \left(\dfrac{1}{2}, \dfrac{3}{4}\right)$ **41.** $x = 4 - 2y, y \text{ is any real number}; \{(x, y) \mid x = 4 - 2y, y \text{ is any real number}\}$

43. $x = \dfrac{3}{2}, y = 1; \left(\dfrac{3}{2}, 1\right)$ **45.** $x = \dfrac{4}{3}, y = \dfrac{1}{5}; \left(\dfrac{4}{3}, \dfrac{1}{5}\right)$ **47.** $x = 8, y = 2, z = 0; (8, 2, 0)$ **49.** $x = 2, y = -1, z = 1; (2, -1, 1)$ **51.** Inconsistent

53. $x = 5z - 2, y = 4z - 3, \text{ where } z \text{ is any real number}; \{(x, y, z) \mid x = 5z - 2, y = 4x - 3, z \text{ is any real number}\}$ **55.** Inconsistent

57. $x = 1, y = 3, z = -2; (1, 3, -2)$ **59.** $x = -3, y = \dfrac{1}{2}, z = 1; \left(-3, \dfrac{1}{2}, 1\right)$ **61.** $x = \dfrac{1}{3}, y = \dfrac{2}{3}, z = 1; \left(\dfrac{1}{3}, \dfrac{2}{3}, 1\right)$ **63.** $x = 1, y = 2, z = 0, w = 1;$

$(1, 2, 0, 1)$ **65.** $y = 0, z = 1 - x, x \text{ is any real number or } \{(x, y, z) \mid y = 0, z = 1 - x, x \text{ is any real number}\}$ **67.** $x = 2, y = z - 3, z \text{ is any}$

real number or $\{(x, y, z) \mid x = 2, y = z - 3, z \text{ is any real number}\}$ **69.** $x = \dfrac{13}{9}, y = \dfrac{7}{18}, z = \dfrac{19}{18}; \left(\dfrac{13}{9}, \dfrac{7}{18}, \dfrac{19}{18}\right)$

71. $x = \dfrac{7}{5} - \dfrac{3}{5}z - \dfrac{2}{5}w, y = -\dfrac{8}{5} + \dfrac{7}{5}z + \dfrac{13}{5}w, \text{ where } z \text{ and } w \text{ are any real numbers or } \left\{(x, y, z, w) \mid x = \dfrac{7}{5} - \dfrac{3}{5}z - \dfrac{2}{5}w, y = -\dfrac{8}{5} + \dfrac{7}{5}z + \dfrac{13}{5}w,\right.$

$z \text{ and } w \text{ are any real numbers}\}$ **73.** $y = -2x^2 + x + 3$ **75.** $f(x) = 3x^3 - 4x^2 + 5$ **77.** 1.5 salmon steak, 2 baked eggs, 1 acorn squash

79. \$4000 in Treasury bills, \$4000 in Treasury bonds, \$2000 in corporate bonds **81.** 8 Deltas, 5 Betas, 10 Sigmas **83.** $I_1 = \dfrac{44}{23}, I_2 = 2, I_3 = \dfrac{16}{23}, I_4 = \dfrac{28}{23}$

85. (a)

Amount Invested At		
7%	**9%**	**11%**
0	10,000	10,000
1000	8000	11,000
2000	6000	12,000
3000	4000	13,000
4000	2000	14,000
5000	0	15,000

(b)

Amount Invested At		
7%	**9%**	**11%**
12,500	12,500	0
14,500	8500	2000
16,500	4500	4000
18,750	0	6250

(c) All the money invested at 7% provides $2100, more than what is required.

87.

First Liquid	Second Liquid	Third Liquid
50 mg	75 mg	0 mg
36 mg	76 mg	8 mg
22 mg	77 mg	16 mg
8 mg	78 mg	24 mg

12.3 Assess Your Understanding (page 873)

1. determinants **2.** $ad - bc$ **3.** F **4.** F **5.** 2 **7.** 22 **9.** -2 **11.** 10 **13.** -26 **15.** $x = 6, y = 2; (6, 2)$ **17.** $x = 3, y = 2; (3, 2)$

19. $x = 8, y = -4; (8, -4)$ **21.** $x = 4, y = -2; (4, -2)$ **23.** Not applicable **25.** $x = \dfrac{1}{2}, y = \dfrac{3}{4}; \left(\dfrac{1}{2}, \dfrac{3}{4}\right)$ **27.** $x = \dfrac{1}{10}, y = \dfrac{2}{5}; \left(\dfrac{1}{10}, \dfrac{2}{5}\right)$

29. $x = \dfrac{3}{2}, y = 1; \left(\dfrac{3}{2}, 1\right)$ **31.** $x = \dfrac{4}{3}, y = \dfrac{1}{5}; \left(\dfrac{4}{3}, \dfrac{1}{5}\right)$ **33.** $x = 1, y = 3, z = -2; (1, 3, -2)$ **35.** $x = -3, y = \dfrac{1}{2}, z = 1; \left(-3, \dfrac{1}{2}, 1\right)$

37. Not applicable **39.** $x = 0, y = 0, z = 0; (0, 0, 0)$ **41.** Not applicable **43.** -5 **45.** $\dfrac{13}{11}$ **47.** 0 or -9 **49.** -4 **51.** 12 **53.** 8 **55.** 8

57. $(y_1 - y_2)x - (x_1 - x_2)y + (x_1 y_2 - x_2 y_1) = 0$

$$(y_1 - y_2)x + (x_2 - x_1)y = x_2 y_1 - x_1 y_2$$

$$(x_2 - x_1)y - (x_2 - x_1)y_1 = (y_2 - y_1)x + x_2 y_1 - x_1 y_2 - (x_2 - x_1)y_1$$

$$(x_2 - x_1)(y - y_1) = (y_2 - y_1)x - (y_2 - y_1)x_1$$

$$y - y_1 = \frac{y_2 - y_1}{x_2 - x_1}(x - x_1)$$

59. The triangle has an area of 5 square units.

61. If $a = 0$, we have
$$by = s$$
$$cx + dy = t$$
Thus, $y = \dfrac{s}{b}$ and $x = \dfrac{t - dy}{c}$
$$= \dfrac{tb - ds}{bc}$$

Using Cramer's Rule, we get
$$x = \frac{sd - tb}{-bc} = \frac{tb - sd}{bc}$$
$$y = \frac{-sc}{-bc} = \frac{s}{b}$$

If $b = 0$, we have
$$ax = s$$
$$cx + dy = t$$
Since $D = ad \neq 0$, then
$a \neq 0$ and $d \neq 0$.
Thus, $x = \dfrac{s}{a}$ and
$$y = \frac{t - cx}{d} = \frac{ta - cs}{ad}$$

Using Cramer's Rule, we get
$$x = \frac{sd}{ad} = \frac{s}{a}$$
$$y = \frac{ta - cs}{ad}$$

If $c = 0$, we have
$$ax + by = s$$
$$dy = t$$
Since $D = ad \neq 0$, then
$a \neq 0$ and $d \neq 0$.
Thus, $y = \dfrac{t}{d}$ and $x = \dfrac{s - by}{a}$
$$= \dfrac{sd - bt}{ad}$$

Using Cramer's Rule, we get
$$x = \frac{sd - bt}{ad}$$
$$y = \frac{at}{ad} = \frac{t}{d}$$

If $d = 0$, we have
$$ax + by = s$$
$$cx = t$$
Since $D = -bc \neq 0$, then
$b \neq 0$ and $c \neq 0$.
Thus, $x = \dfrac{t}{c}$ and $y = \dfrac{s - ax}{b}$
$$= \dfrac{sc - at}{bc}$$

Using Cramer's Rule, we get
$$x = \frac{-tb}{-bc} = \frac{t}{c}$$
$$y = \frac{at - sc}{-bc} = \frac{sc - at}{bc}$$

63.
$$\begin{vmatrix} a_{11} & a_{12} & a_{13} \\ ka_{21} & ka_{22} & ka_{23} \\ a_{31} & a_{32} & a_{33} \end{vmatrix} = -ka_{21}(a_{12}a_{33} - a_{32}a_{13}) + ka_{22}(a_{11}a_{33} - a_{31}a_{13}) - ka_{23}(a_{11}a_{32} - a_{31}a_{12})$$

$$= k[-a_{21}(a_{12}a_{33} - a_{32}a_{13}) + a_{22}(a_{11}a_{33} - a_{31}a_{13}) - a_{23}(a_{11}a_{32} - a_{31}a_{12})] = k\begin{vmatrix} a_{11} & a_{12} & a_{13} \\ a_{21} & a_{22} & a_{23} \\ a_{31} & a_{32} & a_{33} \end{vmatrix}$$

65.
$$\begin{vmatrix} a_{11} + ka_{21} & a_{12} + ka_{22} & a_{13} + ka_{23} \\ a_{21} & a_{22} & a_{23} \\ a_{31} & a_{32} & a_{33} \end{vmatrix} = (a_{11} + ka_{21})(a_{22}a_{33} - a_{32}a_{23}) - (a_{12} + ka_{22})(a_{21}a_{33} - a_{31}a_{23}) + (a_{13} + ka_{23})(a_{21}a_{32} - a_{31}a_{22})$$

$$= a_{11}a_{22}a_{33} - a_{11}a_{32}a_{23} + \overline{ka_{21}a_{22}a_{33}} - \overline{ka_{21}a_{32}a_{23}} - a_{12}a_{21}a_{33} + a_{12}a_{31}a_{23}$$
$$\quad - \overline{ka_{22}a_{21}a_{33}} + \overline{ka_{22}a_{31}a_{23}} + a_{13}a_{21}a_{32} - a_{13}a_{31}a_{22} + \overline{ka_{23}a_{21}a_{32}} - \overline{ka_{23}a_{31}a_{22}}$$
$$= a_{11}a_{22}a_{33} - a_{11}a_{32}a_{23} - a_{12}a_{21}a_{33} + a_{12}a_{31}a_{23} + a_{13}a_{21}a_{32} - a_{13}a_{31}a_{22}$$
$$= a_{11}(a_{22}a_{33} - a_{32}a_{23}) - a_{12}(a_{21}a_{33} - a_{31}a_{23}) + a_{13}(a_{21}a_{32} - a_{31}a_{22})$$
$$= \begin{vmatrix} a_{11} & a_{12} & a_{13} \\ a_{21} & a_{22} & a_{23} \\ a_{31} & a_{32} & a_{33} \end{vmatrix}$$

Historical Problems (page 888)

1. (a) $2 - 5i \longleftrightarrow \begin{bmatrix} 2 & -5 \\ 5 & 2 \end{bmatrix}$, $1 + 3i \longleftrightarrow \begin{bmatrix} 1 & 3 \\ -3 & 1 \end{bmatrix}$ **(b)** $\begin{bmatrix} 2 & -5 \\ 5 & 2 \end{bmatrix}\begin{bmatrix} 1 & 3 \\ -3 & 1 \end{bmatrix} = \begin{bmatrix} 17 & 1 \\ -1 & 17 \end{bmatrix}$ **(c)** $17 + i$ **(d)** $17 + i$

2. $\begin{bmatrix} a & b \\ -b & a \end{bmatrix}\begin{bmatrix} a & -b \\ b & a \end{bmatrix} = \begin{bmatrix} a^2+b^2 & 0 \\ 0 & b^2+a^2 \end{bmatrix}$; the product is a real number.

3. (a) $x = k(ar + bs) + l(cr + ds) = r(ka + lc) + s(kb + ld)$ **(b)** $A = \begin{bmatrix} ka + lc & kb + ld \\ ma + nc & mb + nd \end{bmatrix}$
$\quad\quad y = m(ar + bs) + n(cr + ds) = r(ma + nc) + s(mb + nd)$

12.4 Assess Your Understanding (page 889)

1. inverse **2.** square **3.** identity **4.** F **5.** F **6.** F

7. $\begin{bmatrix} 4 & 4 & -5 \\ -1 & 5 & 4 \end{bmatrix}$ **9.** $\begin{bmatrix} 0 & 12 & -20 \\ 4 & 8 & 24 \end{bmatrix}$ **11.** $\begin{bmatrix} -8 & 7 & -15 \\ 7 & 0 & 22 \end{bmatrix}$ **13.** $\begin{bmatrix} 28 & -9 \\ 4 & 23 \end{bmatrix}$ **15.** $\begin{bmatrix} 1 & 14 & -14 \\ 2 & 22 & -18 \\ 3 & 0 & 28 \end{bmatrix}$ **17.** $\begin{bmatrix} 15 & 21 & -16 \\ 22 & 34 & -22 \\ -11 & 7 & 22 \end{bmatrix}$ **19.** $\begin{bmatrix} 25 & -9 \\ 4 & 20 \end{bmatrix}$

21. $\begin{bmatrix} -13 & 7 & -12 \\ -18 & 10 & -14 \\ 17 & -7 & 34 \end{bmatrix}$ **23.** $\begin{bmatrix} -2 & 4 & 2 & 8 \\ 2 & 1 & 4 & 6 \end{bmatrix}$ **25.** $\begin{bmatrix} 5 & 14 \\ 9 & 16 \end{bmatrix}$ **27.** $\begin{bmatrix} 9 & 2 \\ 34 & 13 \\ 47 & 20 \end{bmatrix}$ **29.** $\begin{bmatrix} 1 & -1 \\ -1 & 2 \end{bmatrix}$ **31.** $\begin{bmatrix} 1 & -\dfrac{5}{2} \\ -1 & 3 \end{bmatrix}$ **33.** $\begin{bmatrix} 1 & -\dfrac{1}{a} \\ -1 & \dfrac{2}{a} \end{bmatrix}$

35. $\begin{bmatrix} 3 & -3 & 1 \\ -2 & 2 & -1 \\ -4 & 5 & -2 \end{bmatrix}$ **37.** $\begin{bmatrix} -\dfrac{5}{7} & \dfrac{1}{7} & \dfrac{3}{7} \\ \dfrac{9}{7} & \dfrac{1}{7} & -\dfrac{4}{7} \\ \dfrac{3}{7} & -\dfrac{2}{7} & \dfrac{1}{7} \end{bmatrix}$ **39.** $x = 3, y = 2$ or $(3, 2)$ **41.** $x = -5, y = 10$ or $(-5, 10)$ **43.** $x = 2, y = -1$ or $(2, -1)$

45. $x = \dfrac{1}{2}, y = 2$ or $\left(\dfrac{1}{2}, 2\right)$ **47.** $x = -2, y = 1$ or $(-2, 1)$ **49.** $x = \dfrac{2}{a}, y = \dfrac{3}{a}$ or $\left(\dfrac{2}{a}, \dfrac{3}{a}\right)$ **51.** $x = -2, y = 3, z = 5$ or $(-2, 3, 5)$

53. $x = \dfrac{1}{2}, y = -\dfrac{1}{2}, z = 1$ or $\left(\dfrac{1}{2}, -\dfrac{1}{2}, 1\right)$ **55.** $x = -\dfrac{34}{7}, y = \dfrac{85}{7}, z = \dfrac{12}{7}$ or $\left(-\dfrac{34}{7}, \dfrac{85}{7}, \dfrac{12}{7}\right)$ **57.** $x = \dfrac{1}{3}, y = 1, z = \dfrac{2}{3}$ or $\left(\dfrac{1}{3}, 1, \dfrac{2}{3}\right)$

59. $\left[\begin{array}{cc|cc} 4 & 2 & 1 & 0 \\ 2 & 1 & 0 & 1 \end{array}\right] \rightarrow \left[\begin{array}{cc|cc} 1 & \frac{1}{2} & \frac{1}{4} & 0 \\ 2 & 1 & 0 & 1 \end{array}\right] \rightarrow \left[\begin{array}{cc|cc} 1 & \frac{1}{2} & \frac{1}{4} & 0 \\ 0 & 0 & -\frac{1}{2} & 1 \end{array}\right]$ **61.** $\left[\begin{array}{cc|cc} 15 & 3 & 1 & 0 \\ 10 & 2 & 0 & 1 \end{array}\right] \rightarrow \left[\begin{array}{cc|cc} 1 & \frac{1}{5} & \frac{1}{15} & 0 \\ 10 & 2 & 0 & 1 \end{array}\right] \rightarrow \left[\begin{array}{cc|cc} 1 & \frac{1}{5} & \frac{1}{15} & 0 \\ 0 & 0 & -\frac{2}{3} & 1 \end{array}\right]$

63. $\left[\begin{array}{ccc|ccc} -3 & 1 & -1 & 1 & 0 & 0 \\ 1 & -4 & -7 & 0 & 1 & 0 \\ 1 & 2 & 5 & 0 & 0 & 1 \end{array}\right] \rightarrow \left[\begin{array}{ccc|ccc} 1 & 2 & 5 & 0 & 0 & 1 \\ 1 & -4 & -7 & 0 & 1 & 0 \\ -3 & 1 & -1 & 1 & 0 & 0 \end{array}\right] \rightarrow \left[\begin{array}{ccc|ccc} 1 & 2 & 5 & 0 & 0 & 1 \\ 0 & -6 & -12 & 0 & 1 & -1 \\ 0 & 7 & 14 & 1 & 0 & 3 \end{array}\right]$

$\rightarrow \left[\begin{array}{ccc|ccc} 1 & 2 & 5 & 0 & 0 & 1 \\ 0 & 1 & 2 & 0 & -\frac{1}{6} & \frac{1}{6} \\ 0 & 1 & 2 & \frac{1}{7} & 0 & \frac{3}{7} \end{array}\right] \rightarrow \left[\begin{array}{ccc|ccc} 1 & 2 & 5 & 0 & 0 & 1 \\ 0 & 1 & 2 & 0 & -\frac{1}{6} & \frac{1}{6} \\ 0 & 0 & 0 & \frac{1}{7} & \frac{1}{6} & \frac{11}{42} \end{array}\right]$ **65.** $\begin{bmatrix} 0.01 & 0.05 & -0.01 \\ 0.01 & -0.02 & 0.01 \\ -0.02 & 0.01 & 0.03 \end{bmatrix}$ **67.** $\begin{bmatrix} 0.02 & -0.04 & -0.01 & 0.01 \\ -0.02 & 0.05 & 0.03 & -0.03 \\ 0.02 & 0.01 & -0.04 & 0.00 \\ -0.02 & 0.06 & 0.07 & 0.06 \end{bmatrix}$

69. $x = 4.57, y = -6.44, z = -24.07$ or $(4.57, -6.44, -24.07)$ **71.** $x = -1.19, y = 2.46, z = 8.27$ or $(-1.19, 2.46, 8.27)$

73. (a) $A = \begin{bmatrix} 6 & 9 \\ 3 & 12 \end{bmatrix}; B = \begin{bmatrix} 71.00 \\ 158.6 \end{bmatrix}$ **(b)** $AB = \begin{bmatrix} 1853.40 \\ 2116.20 \end{bmatrix}$; Nikki's total tuition is \$1853.40, and Joe's total tuition is \$2116.20.

75. (a) $\begin{bmatrix} 500 & 350 & 400 \\ 700 & 500 & 850 \end{bmatrix}$; $\begin{bmatrix} 500 & 700 \\ 350 & 500 \\ 400 & 850 \end{bmatrix}$ **(b)** $\begin{bmatrix} 15 \\ 8 \\ 3 \end{bmatrix}$ **(c)** $\begin{bmatrix} 11{,}500 \\ 17{,}050 \end{bmatrix}$ **(d)** $[0.10 \quad 0.05]$ **(e)** $2002.50

77. (a) $K^{-1} = \begin{bmatrix} 1 & 0 & -1 \\ -1 & 1 & 1 \\ 0 & -1 & 1 \end{bmatrix}$ **(b)** $M = \begin{bmatrix} 13 & 1 & 20 \\ 8 & 9 & 19 \\ 6 & 21 & 14 \end{bmatrix}$ **(c)** Math is fun.

79. If $D = ad - bc \neq 0$, then $a \neq 0$ and $d \neq 0$, or $b \neq 0$ and $c \neq 0$. Assuming the former,

$$\left[\begin{array}{cc|cc} a & b & 1 & 0 \\ c & d & 0 & 1 \end{array}\right] \to \left[\begin{array}{cc|cc} 1 & \dfrac{b}{a} & \dfrac{1}{a} & 0 \\ c & d & 0 & 1 \end{array}\right] \to \left[\begin{array}{cc|cc} 1 & \dfrac{b}{a} & \dfrac{1}{a} & 0 \\ 0 & \dfrac{D}{a} & -\dfrac{c}{a} & 1 \end{array}\right] \to \left[\begin{array}{cc|cc} 1 & \dfrac{b}{a} & \dfrac{1}{a} & 0 \\ 0 & 1 & -\dfrac{c}{D} & \dfrac{a}{D} \end{array}\right] \to \left[\begin{array}{cc|cc} 1 & 0 & \dfrac{d}{D} & -\dfrac{b}{d} \\ 0 & 1 & -\dfrac{c}{D} & \dfrac{a}{D} \end{array}\right]$$

$$R_1 = \frac{1}{a} r_1 \qquad R_2 = -c r_1 + r_2 \qquad R_2 = \frac{a}{D} r_2 \qquad R_1 = -\frac{b}{a} r_2 + r_1$$

12.5 Assess Your Understanding (page 897)

5. Proper **7.** Improper; $1 + \dfrac{9}{x^2 - 4}$ **9.** Improper; $5x + \dfrac{22x - 1}{x^2 - 4}$ **11.** Improper; $1 + \dfrac{-2(x-6)}{(x+4)(x-3)}$ **13.** $\dfrac{-4}{x} + \dfrac{4}{x-1}$ **15.** $\dfrac{1}{x} + \dfrac{-x}{x^2+1}$

17. $\dfrac{-1}{x-1} + \dfrac{2}{x-2}$ **19.** $\dfrac{\frac{1}{4}}{x+1} + \dfrac{\frac{3}{4}}{x-1} + \dfrac{\frac{1}{2}}{(x-1)^2}$ **21.** $\dfrac{\frac{1}{12}}{x-2} + \dfrac{-\frac{1}{12}(x+4)}{x^2+2x+4}$ **23.** $\dfrac{\frac{1}{4}}{x-1} + \dfrac{\frac{1}{4}}{(x-1)^2} + \dfrac{-\frac{1}{4}}{x+1} + \dfrac{\frac{1}{4}}{(x+1)^2}$

25. $\dfrac{-5}{x+2} + \dfrac{5}{x+1} + \dfrac{-4}{(x+1)^2}$ **27.** $\dfrac{\frac{1}{4}}{x} + \dfrac{1}{x^2} + \dfrac{-\frac{1}{4}(x+4)}{x^2+4}$ **29.** $\dfrac{\frac{2}{3}}{x+1} + \dfrac{\frac{1}{3}(x+1)}{x^2+2x+4}$ **31.** $\dfrac{\frac{2}{7}}{3x-2} + \dfrac{\frac{1}{7}}{2x+1}$ **33.** $\dfrac{\frac{3}{4}}{x+3} + \dfrac{\frac{1}{4}}{x-1}$

35. $\dfrac{1}{x^2+4} + \dfrac{2x-1}{(x^2+4)^2}$ **37.** $\dfrac{-1}{x} + \dfrac{2}{x-3} + \dfrac{-1}{x+1}$ **39.** $\dfrac{4}{x-2} + \dfrac{-3}{x-1} + \dfrac{-1}{(x-1)^2}$ **41.** $\dfrac{x}{(x^2+16)^2} + \dfrac{-16x}{(x^2+16)^3}$

43. $\dfrac{-\frac{8}{7}}{2x+1} + \dfrac{\frac{4}{7}}{x-3}$ **45.** $\dfrac{\frac{2}{9}}{x} + \dfrac{\frac{1}{3}}{x^2} + \dfrac{\frac{1}{6}}{x-3} + \dfrac{\frac{1}{18}}{x-3}$

Historical Problem (page 903)

$x = 6$ units, $y = 8$ units

12.6 Assess Your Understanding (page 904)

5.

7.

9.

11.

13.

15.

17.

19.
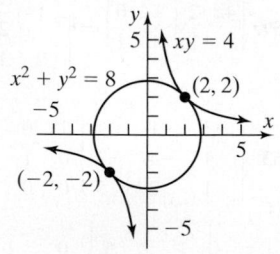

21. No points of intersection

23.

25. $x = 1, y = 4; x = -1, y = -4; x = 2\sqrt{2}, y = \sqrt{2}; x = -2\sqrt{2}, y = -\sqrt{2}$ or $(1, 4), (-1, -4), \left(2\sqrt{2}, \sqrt{2}\right), \left(-2\sqrt{2}, -\sqrt{2}\right)$

27. $x = 0, y = 1; x = -\dfrac{2}{3}, y = -\dfrac{1}{3}$ or $(0, 1), \left(-\dfrac{2}{3}, -\dfrac{1}{3}\right)$ **29.** $x = 0, y = -1; x = \dfrac{5}{2}, y = -\dfrac{7}{2}$ or $(0, -1), \left(\dfrac{5}{2}, -\dfrac{7}{2}\right)$

31. $x = 2, y = \dfrac{1}{3}; x = \dfrac{1}{2}, y = \dfrac{4}{3}$ or $\left(2, \dfrac{1}{3}\right), \left(\dfrac{1}{2}, \dfrac{4}{3}\right)$ **33.** $x = 3, y = 2; x = 3, y = -2; x = -3, y = 2; x = -3, y = -2$ or

$(3, 2), (3, -2), (-3, 2), (-3, -2)$ **35.** $x = \dfrac{1}{2}, y = \dfrac{3}{2}; x = \dfrac{1}{2}, y = -\dfrac{3}{2}; x = -\dfrac{1}{2}, y = \dfrac{3}{2}; x = -\dfrac{1}{2}, y = -\dfrac{3}{2}$ or

$\left(\dfrac{1}{2}, \dfrac{3}{2}\right), \left(\dfrac{1}{2}, -\dfrac{3}{2}\right), \left(-\dfrac{1}{2}, \dfrac{3}{2}\right), \left(-\dfrac{1}{2}, -\dfrac{3}{2}\right)$ **37.** $x = \sqrt{2}, y = 2\sqrt{2}; x = -\sqrt{2}, y = -2\sqrt{2}$ or $\left(\sqrt{2}, 2\sqrt{2}\right), \left(-\sqrt{2}, -2\sqrt{2}\right)$

39. No real solution exists. **41.** $x = \dfrac{8}{3}, y = \dfrac{2\sqrt{10}}{3}; x = -\dfrac{8}{3}, y = \dfrac{2\sqrt{10}}{3}; x = \dfrac{8}{3}, y = -\dfrac{2\sqrt{10}}{3}; x = -\dfrac{8}{3}, y = -\dfrac{2\sqrt{10}}{3}$ or

$\left(\dfrac{8}{3}, \dfrac{2\sqrt{10}}{3}\right), \left(-\dfrac{8}{3}, \dfrac{2\sqrt{10}}{3}\right), \left(\dfrac{8}{3}, -\dfrac{2\sqrt{10}}{3}\right), \left(-\dfrac{8}{3}, -\dfrac{2\sqrt{10}}{3}\right)$ **43.** $x = 1, y = \dfrac{1}{2}; x = -1, y = \dfrac{1}{2}; x = 1, y = -\dfrac{1}{2}; x = -1, y = -\dfrac{1}{2}$ or

$\left(1, \dfrac{1}{2}\right), \left(-1, \dfrac{1}{2}\right), \left(1, -\dfrac{1}{2}\right), \left(-1, -\dfrac{1}{2}\right)$ **45.** No real solution exists. **47.** $x = \sqrt{3}, y = \sqrt{3}; x = -\sqrt{3}, y = -\sqrt{3}; x = 2, y = 1; x = -2, y = -1$ or

$\left(\sqrt{3}, \sqrt{3}\right), \left(-\sqrt{3}, -\sqrt{3}\right), (2, 1), (-2, -1)$ **49.** $x = 0, y = -2; x = 0, y = 1; x = 2, y = -1$ or $(0, -2), (0, 1), (2, -1)$

51. $x = 2, y = 8$ or $(2, 8)$ **53.** $x = 81, y = 3$ or $(81, 3)$

55.

57.

59.

61.

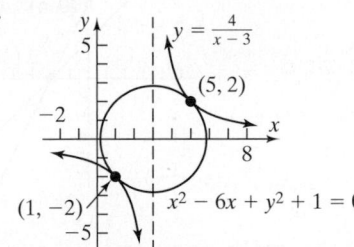

63. $x = 0.48, y = 0.62$ **65.** $x = -1.65, y = -0.89$ **67.** $x = 0.58, y = 1.86; x = 1.81, y = 1.05; x = 0.58, y = -1.86; x = 1.81, y = -1.05$

69. $x = 2.35, y = 0.85$ **71.** 3 and 1; −3 and −1 **73.** 2 and 2; −2 and −2 **75.** $\dfrac{1}{2}$ and $\dfrac{1}{3}$ **77.** 5 **79.** 5 in. by 3 in. **81.** 2 cm and 4 cm

83. tortoise: 7 m/hr, hare: $7\dfrac{1}{2}$ m/hr **85.** 12 cm by 18 cm **87.** $x = 60$ ft; $y = 30$ ft **89.** $l = \dfrac{P + \sqrt{P^2 - 16A}}{4}; w = \dfrac{P - \sqrt{P^2 - 16A}}{4}$

91. $y = 4x - 4$ **93.** $y = 2x + 1$ **95.** $y = -\dfrac{1}{3}x + \dfrac{7}{3}$ **97.** $y = 2x - 3$ **99.** $r_1 = \dfrac{-b + \sqrt{b^2 - 4ac}}{2a}; r_2 = \dfrac{-b - \sqrt{b^2 - 4ac}}{2a}$

101. (a) 4.274 ft by 4.274 ft or 0.093 ft by 0.093 ft

12.7 Assess Your Understanding (page 912)

7. satisfied **8.** half-plane **9.** F **10.** T

11.

13.

15.

17.

19.

21.

23.

25.

27.

29.

31.

33. No solution

35.

37.

39.

41.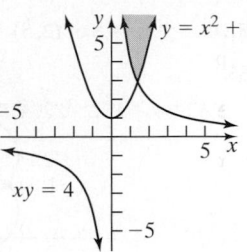

43. Bounded; corner points

$(0, 0), (3, 0), (2, 2), (0, 3)$

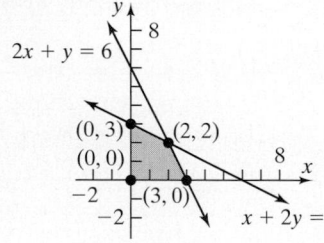

45. Unbounded; corner points

$(2, 0), (0, 4)$

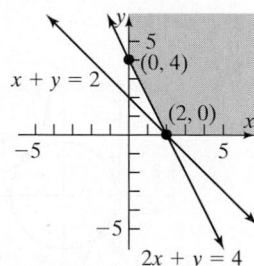

47. Bounded; corner points $(2, 0), (4, 0),$

$\left(\dfrac{24}{7}, \dfrac{12}{7}\right), (0, 4), (0, 2)$

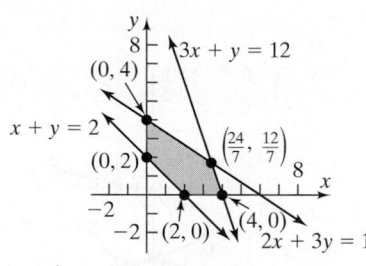

49. Bounded; corner points $(2, 0), (5, 0),$
$(2, 6), (0, 8), (0, 2)$

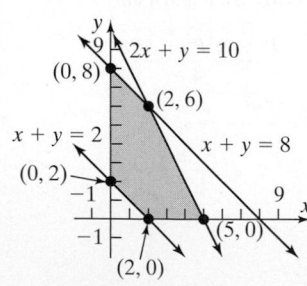

51. Bounded; corner points

$(1, 0), (10, 0), (0, 5), \left(0, \dfrac{1}{2}\right)$

53. $\begin{cases} x \le 4 \\ x + y \le 6 \\ x \ge 0 \\ y \ge 0 \end{cases}$

55. $\begin{cases} x \le 20 \\ y \ge 15 \\ x + y \le 50 \\ x - y \le 0 \\ x \ge 0 \end{cases}$

57. (a) $\begin{cases} x + y \le 50{,}000 \\ x \ge 35{,}000 \\ y \le 10{,}000 \\ x \ge 0 \\ y \ge 0 \end{cases}$

(b)

59. (a) $\begin{cases} x \ge 0 \\ y \ge 0 \\ x + 2y \le 300 \\ 3x + 2y \le 480 \end{cases}$

(b)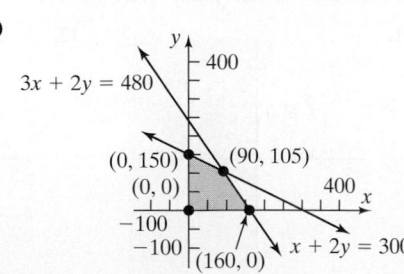

61. (a) $\begin{cases} 3x + 2y \le 160 \\ 2x + 3y \le 150 \\ \quad x \ge 0 \\ \quad y \ge 0 \end{cases}$ **(b)**

12.8 Assess Your Understanding (page 919)

1. objective function **2.** T **3.** Maximum value is 11; minimum value is 3. **5.** Maximum value is 65; minimum value is 4.
7. Maximum value is 67; minimum value is 20. **9.** The maximum value of z is 12, and it occurs at the point $(6, 0)$.
11. The minimum value of z is 4, and it occurs at the point $(2, 0)$. **13.** The maximum value of z is 20, and it occurs at the point $(0, 4)$.
15. The minimum value of z is 8, and it occurs at the point $(0, 2)$. **17.** The maximum value of z is 50, and it occurs at the point $(10, 0)$.
19. 8 downhill, 24 cross-country; $1760; $1920 **21.** Rent 15 rectangular tables and 16 round tables for a minimum cost of $1252.00.
23. (a) $10,000 in a junk bond and $10,000 in Treasury bills **(b)** $12,000 in junk bonds and $8000 in Treasury bills **25.** 100 lb of ground beef and 50 lb
of pork **27.** 10 racing skates, 15 figure skates **29.** 2 metal samples, 4 plastic samples; $34 **31. (a)** 10 first class, 120 coach **(b)** 15 first class, 120 coach

Review Exercises (page 922)

1. $x = 2, y = -1$ or $(2, -1)$ **3.** $x = 2, y = \frac{1}{2}$ or $\left(2, \frac{1}{2}\right)$ **5.** $x = 2, y = -1$ or $(2, -1)$ **7.** $x = \frac{11}{5}, y = -\frac{3}{5}$ or $\left(\frac{11}{5}, -\frac{3}{5}\right)$ **9.** Inconsistent

11. $x = 2, y = 3$ or $(2, 3)$ **13.** Inconsistent **15.** $x = -1, y = 2, z = -3$ or $(-1, 2, -3)$ **17.** $x = \frac{7}{4}z + \frac{39}{4}, y = \frac{9}{8}z + \frac{69}{8}$, z is any real number or

$\left\{(x, y, z,) \mid x = \frac{7}{4}z + \frac{39}{4}, y = \frac{9}{8}z + \frac{69}{8}, z \text{ is any real number}\right\}$ **19.** $\begin{cases} 3x + 2y = 8 \\ x + 4y = -1 \end{cases}$ **21.** $\begin{bmatrix} 4 & -4 \\ 3 & 9 \\ 4 & 4 \end{bmatrix}$ **23.** $\begin{bmatrix} 6 & 0 \\ 12 & 24 \\ -6 & 12 \end{bmatrix}$ **25.** $\begin{bmatrix} 4 & -3 & 0 \\ 12 & -2 & -8 \\ -2 & 5 & -4 \end{bmatrix}$

27. $\begin{bmatrix} 8 & -13 & 8 \\ 9 & 2 & -10 \\ 22 & -13 & -4 \end{bmatrix}$ **29.** $\begin{bmatrix} \frac{1}{2} & -1 \\ -\frac{1}{6} & \frac{2}{3} \end{bmatrix}$ **31.** $\begin{bmatrix} -\frac{5}{7} & \frac{9}{7} & \frac{3}{7} \\ \frac{1}{7} & \frac{1}{7} & -\frac{2}{7} \\ \frac{3}{7} & -\frac{4}{7} & \frac{1}{7} \end{bmatrix}$ **33.** Singular **35.** $x = \frac{2}{5}, y = \frac{1}{10}$ or $\left(\frac{2}{5}, \frac{1}{10}\right)$

37. $x = 9, y = \frac{13}{3}, z = \frac{13}{3}$ or $\left(9, \frac{13}{3}, \frac{13}{3}\right)$ **39.** $x = -\frac{1}{2}, y = -\frac{2}{3}, z = -\frac{3}{4}$ or $\left(-\frac{1}{2}, -\frac{2}{3}, -\frac{3}{4}\right)$ **41.** $z = -1, x = y + 1$, y is any real number or

$\left\{(x, y, z) \mid x = y + 1, z = -1, y \text{ is any real number}\right\}$ **43.** $x = 4, y = 2, z = 3, t = -2$ or $(4, 2, 3, -2)$ **45.** 5 **47.** 108 **49.** -100

51. $x = 2, y = -1$ or $(2, -1)$ **53.** $x = 2, y = 3$ or $(2, 3)$ **55.** $x = -1, y = 2, z = -3$ or $(-1, 2, -3)$ **57.** 16 **59.** $\dfrac{-\frac{3}{2}}{x} + \dfrac{\frac{3}{2}}{x - 4}$

61. $\dfrac{-3}{x - 1} + \dfrac{3}{x} + \dfrac{4}{x^2}$ **63.** $\dfrac{-\frac{1}{10}}{x + 1} + \dfrac{\frac{1}{10}x + \frac{9}{10}}{x^2 + 9}$ **65.** $\dfrac{x}{x^2 + 4} + \dfrac{-4x}{(x^2 + 4)^2}$ **67.** $\dfrac{\frac{1}{2}}{x^2 + 1} + \dfrac{\frac{1}{4}}{x - 1} + \dfrac{-\frac{1}{4}}{x + 1}$ **69.** $x = -\frac{2}{5}, y = -\frac{11}{5}; x = -2, y = 1$ or

$\left(-\frac{2}{5}, -\frac{11}{5}\right), (-2, 1)$ **71.** $x = 2\sqrt{2}, y = \sqrt{2}; x = -2\sqrt{2}, y = -\sqrt{2}$ or $\left(2\sqrt{2}, \sqrt{2}\right), \left(-2\sqrt{2}, -\sqrt{2}\right)$ **73.** $x = 0, y = 0; x = -3, y = 3;$

$x = 3, y = 3$ or $(0, 0), (-3, 3), (3, 3)$ **75.** $x = \sqrt{2}, y = -\sqrt{2}; x = -\sqrt{2}, y = \sqrt{2}; x = \frac{4}{3}\sqrt{2}, y = -\frac{2}{3}\sqrt{2}; x = -\frac{4}{3}\sqrt{2}, y = \frac{2}{3}\sqrt{2}$ or

$\left(\sqrt{2}, -\sqrt{2}\right), \left(-\sqrt{2}, \sqrt{2}\right), \left(\frac{4}{3}\sqrt{2}, -\frac{2}{3}\sqrt{2}\right), \left(-\frac{4}{3}\sqrt{2}, \frac{2}{3}\sqrt{2}\right)$ **77.** $x = 1, y = -1$ or $(1, -1)$

79.

81.

83. Unbounded

85. Bounded

87. Bounded

89.

91.

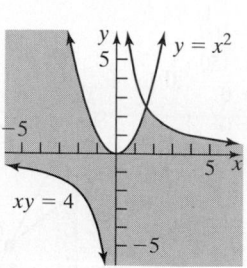

93. The maximum value is 32 when $x = 0$ and $y = 8$. **95.** The minimum value is 3 when $x = 1$ and $y = 0$. **97.** 10 **99.** $y = -\dfrac{1}{3}x^2 - \dfrac{2}{3}x + 1$

101. 70 pounds of $6.00 coffee and 30 pounds of $9.00 coffee **103.** 1 small, 5 medium, 2 large **105.** Speedboat: 36.67 km/hr; Aguarico River: 3.33 km/hr
107. Bruce: 4 hours; Bryce: 2 hours; Marty: 8 hours **109.** 35 gasoline engines, 15 diesel engines; 15 gasoline engines, 0 diesel engines

Chapter Test (page 925)

1. $x = 3, y = -1$ **2.** Inconsistent. **3.** $x = -z + \dfrac{18}{7}, y = z - \dfrac{17}{7}$, where z can be any real number. **4.** $x = \dfrac{1}{3}, y = -2, z = 0$

5. $\begin{bmatrix} 4 & -5 & 1 & | & 0 \\ -2 & -1 & 6 & | & -19 \\ 1 & 5 & -5 & | & 10 \end{bmatrix}$ **6.** $\begin{cases} 3x + 2y + 4z = -6 \\ 1x + 0y + 8z = 2 \\ -2x + 1y + 3z = -11 \end{cases}$ or $\begin{cases} 3x + 2y + 4z = -6 \\ x + 8z = 2 \\ -2x + y + 3z = -11 \end{cases}$ **7.** $\begin{bmatrix} 6 & 4 \\ 1 & -11 \\ 5 & 12 \end{bmatrix}$ **8.** $\begin{bmatrix} -11 & -19 \\ -3 & 5 \\ 6 & -22 \end{bmatrix}$

9. The operation cannot be performed. **10.** $\begin{bmatrix} 16 & 17 \\ 3 & -10 \end{bmatrix}$ **11.** $\begin{bmatrix} 2 & -1 \\ -\dfrac{5}{2} & \dfrac{3}{2} \end{bmatrix}$ **12.** $\begin{bmatrix} 3 & 3 & -4 \\ -2 & -2 & 3 \\ -4 & -5 & 7 \end{bmatrix}$ **13.** $x = \dfrac{1}{2}, y = 3$ **14.** The system is dependent

and therefore has an infinite number of solutions. Any ordered pair satisfying the equation $x = -\dfrac{1}{4}y + 7$ or $y = -4x + 28$ is a solution to the system.

15. $x = 1, y = -2, z = 0$ **16.** Inconsistent **17.** -29 **18.** -12 **19.** $x = -2, y = -5$ **20.** $x = 1, y = -1, z = 4$ **21.** $(1, -3)$ and $(1, 3)$

22. $(3, 4)$ and $(1, 2)$ **23.**

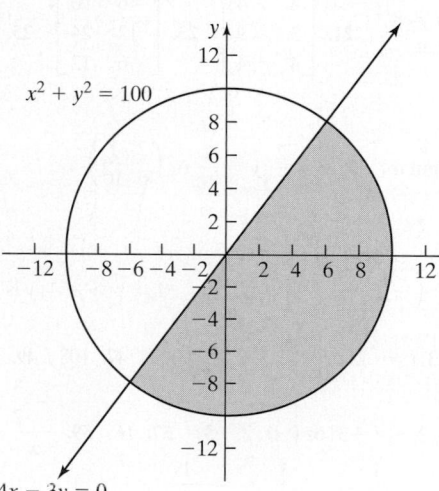

24. $\dfrac{3}{x + 3} + \dfrac{-2}{(x + 3)^2}$

25. $\dfrac{-\dfrac{1}{3}}{x} + \dfrac{\dfrac{1}{3}x}{(x^2 + 3)} + \dfrac{5x}{(x^2 + 3)^2}$

26. The graph is unbounded. The corner points are $(4, 2)$ and $(8, 0)$:

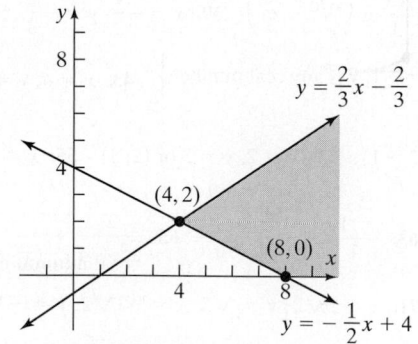

27. The maximum value of z is 64, and it occurs at the point $(0, 8)$.

28. Flare jeans cost $24.50, camisoles cost $8.50, and T-shirts cost $6.00.

Cumulative Review (page 926)

1. $\left\{ 0, \dfrac{1}{2} \right\}$ **2.** $\{5\}$ **3.** $\left\{ -1, -\dfrac{1}{2}, 3 \right\}$ **4.** $\{-2\}$ **5.** $\left\{ \dfrac{5}{2} \right\}$ **6.** $\left\{ \dfrac{1}{\ln 3} \right\}$ **7.** Odd; symmetric with respect to origin

8. Center: $(1, -2)$; radius $= 4$

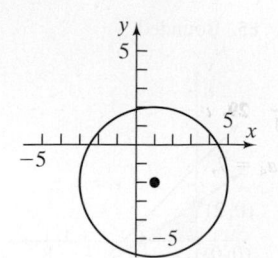

9. Domain: all real numbers
Range: $\{y | y > 1\}$
Horizontal asymptote: $y = 1$

10. $f^{-1}(x) = \dfrac{5}{x} - 2$
Domain of f: $\{x | x \neq -2\}$
Range of f: $\{y | y \neq 0\}$
Domain of f^{-1}: $\{x | x \neq 0\}$
Range of f^{-1}: $\{y | y \neq -2\}$

11. (a)

(b)

(c)

(d)

(e)

(f)

(g)

(h)

(i)

(j)

(k)

(l)

12. (a) -2.28 **(b)** Local maximum of 7 at $x = -1$; local minimum of 3 at $x = 1$ **(c)** $(-\infty, -1), (1, \infty)$ **13.** $\{x \mid x = \pm \dfrac{\pi}{3} + 2\pi k, k \text{ any integer}\}$

CHAPTER 13 Sequences; Induction; The Binomial Theorem

13.1 Assess Your Understanding (page 936)

3. sequence **4.** $3; 15$ **5.** 20 **6.** T **7.** T **8.** T **9.** 3,628,800 **11.** 504 **13.** 1260 **15.** $s_1 = 1, s_2 = 2, s_3 = 3, s_4 = 4, s_5 = 5$

17. $a_1 = \dfrac{1}{3}, a_2 = \dfrac{1}{2}, a_3 = \dfrac{3}{5}, a_4 = \dfrac{2}{3}, a_5 = \dfrac{5}{7}$ **19.** $c_1 = 1, c_2 = -4, c_3 = 9, c_4 = -16, c_5 = 25$ **21.** $s_1 = \dfrac{1}{2}, s_2 = \dfrac{2}{5}, s_3 = \dfrac{2}{7}, s_4 = \dfrac{8}{41}, s_5 = \dfrac{8}{61}$

23. $t_1 = -\dfrac{1}{6}, t_2 = \dfrac{1}{12}, t_3 = -\dfrac{1}{20}, t_4 = \dfrac{1}{30}, t_5 = -\dfrac{1}{42}$ **25.** $b_1 = \dfrac{1}{e}, b_2 = \dfrac{2}{e^2}, b_3 = \dfrac{3}{e^3}, b_4 = \dfrac{4}{e^4}, b_5 = \dfrac{5}{e^5}$ **27.** $a_n = \dfrac{n}{n+1}$ **29.** $a_n = \dfrac{1}{2^{n-1}}$

31. $a_n = (-1)^{n+1}$ **33.** $a_n = (-1)^{n+1}n$ **35.** $a_1 = 2, a_2 = 5, a_3 = 8, a_4 = 11, a_5 = 14$ **37.** $a_1 = -2, a_2 = 0, a_3 = 3, a_4 = 7, a_5 = 12$

39. $a_1 = 5, a_2 = 10, a_3 = 20, a_4 = 40, a_5 = 80$ **41.** $a_1 = 3, a_2 = \dfrac{3}{2}, a_3 = \dfrac{1}{2}, a_4 = \dfrac{1}{8}, a_5 = \dfrac{1}{40}$ **43.** $a_1 = 1, a_2 = 2, a_3 = 2, a_4 = 4, a_5 = 8$

45. $a_1 = A, a_2 = A + d, a_3 = A + 2d, a_4 = A + 3d, a_5 = A + 4d$

47. $a_1 = \sqrt{2}, a_2 = \sqrt{2 + \sqrt{2}}, a_3 = \sqrt{2 + \sqrt{2 + \sqrt{2}}},$

$a_4 = \sqrt{2 + \sqrt{2 + \sqrt{2 + \sqrt{2}}}},$

$a_5 = \sqrt{2 + \sqrt{2 + \sqrt{2 + \sqrt{2 + \sqrt{2}}}}}$

49. $3 + 4 + \cdots + (n + 2)$

51. $\dfrac{1}{2} + 2 + \dfrac{9}{2} + \cdots + \dfrac{n^2}{2}$

53. $1 + \dfrac{1}{3} + \dfrac{1}{9} + \cdots + \dfrac{1}{3^n}$

55. $\dfrac{1}{3} + \dfrac{1}{9} + \cdots + \dfrac{1}{3^n}$ **57.** $\ln 2 - \ln 3 + \ln 4 - \cdots + (-1)^n \ln n$ **59.** $\displaystyle\sum_{k=1}^{20} k$ **61.** $\displaystyle\sum_{k=1}^{13} \dfrac{k}{k+1}$ **63.** $\displaystyle\sum_{k=0}^{6} (-1)^k \left(\dfrac{1}{3^k}\right)$ **65.** $\displaystyle\sum_{k=1}^{n} \dfrac{3^k}{k}$

67. $\displaystyle\sum_{k=0}^{n} (a + kd)$ or $\displaystyle\sum_{k=1}^{n+1} [a + (k-1)d]$ **69.** 200 **71.** 820 **73.** 1110 **75.** 1560 **77.** 3570 **79.** 44,000 **81.** \$2930 **83.** \$18,058.03

85. 21 pairs **87.** Fibonacci sequence **89.** (a) 3.630170833 (b) 3.669060828 (c) 3.669296668 (d) 12

91. (a) $a_1 = 0.4$; $a_2 = 0.7$; $a_3 = 1$; $a_4 = 1.6$; $a_5 = 2.8$; $a_6 = 5.2$; $a_7 = 10$; $a_8 = 19.6$

(b) Except for term 5, which has no match, Bode's formula provides excellent approximations for the mean distances of the planets from the sun.

(c) The mean distance of Ceres from the Sun is approximated by $a_5 = 2.8$; and that of Uranus is $a_8 = 19.6$.

(d) $a_9 = 38.8$; $a_{10} = 77.2$

(e) Pluto's distance is approximated by a_9, but no term approximates Neptune's mean distance from the Sun.

(f) According to Bode's Law, the mean orbital distance of 2003 UB 313 will be 154 AU from the sun.

13.2 Assess Your Understanding *(page 944)*

1. arithmetic **2.** F **3.** $s_n - s_{n-1} = (n+4) - [(n-1)+4] = n + 4 - (n+3) = n + 4 - n - 3 = 1$, a constant; $d = 1$; $s_1 = 5, s_2 = 6, s_3 = 7, s_4 = 8$

5. $a_n - a_{n-1} = (2n - 5) - [2(n-1) - 5] = 2n - 5 - (2n - 2 - 5) = 2n - 5 - (2n - 7) = 2n - 5 - 2n + 7 = 2$, a constant;

$d = 2$; $a_1 = -3, a_2 = -1, a_3 = 1, a_4 = 3$

7. $c_n - c_{n-1} = (6 - 2n) - [6 - 2(n-1)] = 6 - 2n - (6 - 2n + 2) = 6 - 2n - (8 - 2n) = 6 - 2n - 8 + 2n = -2$, a constant;

$d = -2$; $c_1 = 4, c_2 = 2, c_3 = 0, c_4 = -2$

9. $t_n - t_{n-1} = \left(\dfrac{1}{2} - \dfrac{1}{3}n\right) - \left[\dfrac{1}{2} - \dfrac{1}{3}(n-1)\right] = \dfrac{1}{2} - \dfrac{1}{3}n - \left(\dfrac{1}{2} - \dfrac{1}{3}n + \dfrac{1}{3}\right) = \dfrac{1}{2} - \dfrac{1}{3}n - \left(\dfrac{5}{6} - \dfrac{1}{3}n\right) = \dfrac{1}{2} - \dfrac{1}{3}n - \dfrac{5}{6} + \dfrac{1}{3}n = -\dfrac{1}{3}$, a constant;

$d = -\dfrac{1}{3}$; $t_1 = \dfrac{1}{6}, t_2 = -\dfrac{1}{6}, t_3 = -\dfrac{1}{2}, t_4 = -\dfrac{5}{6}$

11. $s_n - s_{n-1} = \ln 3^n - \ln 3^{n-1} = n \ln 3 - (n-1) \ln 3 = n \ln 3 - (n \ln 3 - \ln 3) = n \ln 3 - n \ln 3 + \ln 3 = \ln 3$, a constant;

$d = \ln 3$; $s_1 = \ln 3, s_2 = 2 \ln 3, s_3 = 3 \ln 3, s_4 = 4 \ln 3$

13. $a_n = 3n - 1$; $a_{51} = 152$ **15.** $a_n = 8 - 3n$; $a_{51} = -145$ **17.** $a_n = \dfrac{1}{2}(n-1)$; $a_{51} = 25$ **19.** $a_n = \sqrt{2}n$; $a_{51} = 51\sqrt{2}$ **21.** 200

23. -266 **25.** $\dfrac{83}{2}$ **27.** $a_1 = -13$; $d = 3$; $a_n = a_{n-1} + 3$; $a_n = -16 + 3n$

29. $a_1 = -53$; $d = 6$; $a_n = a_{n-1} + 6$; $a_n = -59 + 6n$ **31.** $a_1 = 28$; $d = -2$; $a_n = a_{n-1} - 2$; $a_n = 30 - 2n$

33. $a_1 = 25$; $d = -2$; $a_n = a_{n-1} - 2$; $a_n = 27 - 2n$ **35.** n^2 **37.** $\dfrac{n}{2}(9 + 5n)$ **39.** 1260 **41.** 324 **43.** 30,919 **45.** 10,036

47. 6080 **49.** -1925 **51.** 15,960 **53.** $-\dfrac{3}{2}$ **55.** 1185 seats **57.** 210 of beige and 190 blue

59. $\{T_n\} = \{-5.5n + 67\}$; $T_5 = 39.5°F$ **61.** The amphitheater has 1647 seats. **63.** 8 yr

Historical Problems *(page 954)*

1. $1\dfrac{2}{3}$ loaves, $10\dfrac{5}{6}$ loaves, 20 loaves, $29\dfrac{1}{6}$ loaves, $38\dfrac{1}{3}$ loaves **2.** (a) 1 person (b) 2401 kittens (c) 2800

13.3 Assess Your Understanding *(page 954)*

3. geometric **4.** $\dfrac{a}{1-r}$ **5.** annuity **6.** T **7.** F **8.** T **9.** $r = 3$; $s_1 = 3, s_2 = 9, s_3 = 27, s_4 = 81$

11. $r = \dfrac{1}{2}$; $a_1 = -\dfrac{3}{2}, a_2 = -\dfrac{3}{4}, a_3 = -\dfrac{3}{8}, a_4 = -\dfrac{3}{16}$ **13.** $r = 2$; $c_1 = \dfrac{1}{4}, c_2 = \dfrac{1}{2}, c_3 = 1, c_4 = 2$ **15.** $r = 2^{1/3}$; $e_1 = 2^{1/3}, e_2 = 2^{2/3}, e_3 = 2, e_4 = 2^{4/3}$

17. $r = \dfrac{3}{2}$; $t_1 = \dfrac{1}{2}, t_2 = \dfrac{3}{4}, t_3 = \dfrac{9}{8}, t_4 = \dfrac{27}{16}$ **19.** Arithmetic; $d = 1$ **21.** Neither **23.** Arithmetic; $d = -\dfrac{2}{3}$ **25.** Neither **27.** Geometric; $r = \dfrac{2}{3}$

29. Geometric; $r = 2$ **31.** Geometric; $r = 3^{1/2}$ **33.** $a_5 = 162$; $a_n = 2 \cdot 3^{n-1}$ **35.** $a_5 = 5$; $a_n = 5 \cdot (-1)^{n-1}$ **37.** $a_5 = 0$; $a_n = 0$

39. $a_5 = 4\sqrt{2}$; $a_n = (\sqrt{2})^n$ **41.** $a_7 = \dfrac{1}{64}$ **43.** $a_9 = 1$ **45.** $a_8 = 0.00000004$ **47.** $a_n = 7 \cdot 2^{n-1}$ **49.** $a_n = -3 \cdot \left(-\dfrac{1}{3}\right)^{n-1} = \left(-\dfrac{1}{3}\right)^{n-2}$

51. $a_n = -(-3)^{n-1}$ **53.** $a_n = \dfrac{7}{15}(15)^{n-1} = 7 \cdot 15^{n-2}$ **55.** $-\dfrac{1}{4}(1 - 2^n)$ **57.** $2\left[1 - \left(\dfrac{2}{3}\right)^n\right]$ **59.** $1 - 2^n$

61.

63.

65.

67. Converges; $\dfrac{3}{2}$ **69.** Converges; 16 **71.** Converges; $\dfrac{8}{5}$ **73.** Diverges **75.** Converges; $\dfrac{20}{3}$ **77.** Diverges **79.** Converges; $\dfrac{18}{5}$

81. Converges; 6 **83.** -4 **85.** \$21,879.11 **87.** **(a)** 0.775 ft **(b)** 8th **(c)** 15.88 ft **(d)** 20 ft **89.** \$349,496.41 **91.** \$96,885.98

93. \$305.10 **95.** 1.845×10^{19} **97.** 10 **99.** \$72.67 per share **101.** December 20, 2007; \$9999.92

103. Option B results in more money (\$524.287 versus \$500,500). **105.** Total pay: \$41,943.03; pay on day 22: \$20,971.52

13.4 Assess Your Understanding (page 960)

1. (I) $n = 1$: $2(1) = 2$ and $1(1 + 1) = 2$
(II) If $2 + 4 + 6 + \cdots + 2k = k(k + 1)$, then $2 + 4 + 6 + \cdots + 2k + 2(k + 1) = (2 + 4 + 6 + \cdots + 2k) + 2(k + 1)$
$= k(k + 1) + 2(k + 1) = k^2 + 3k + 2 = (k + 1)(k + 2) = (k + 1)[(k + 1) + 1]$.

3. (I) $n = 1$: $1 + 2 = 3$ and $\dfrac{1}{2}(1)(1 + 5) = \dfrac{1}{2}(6) = 3$

(II) If $3 + 4 + 5 + \cdots + (k + 2) = \dfrac{1}{2}k(k + 5)$, then $3 + 4 + 5 + \cdots + (k + 2) + [(k + 1) + 2]$

$= [3 + 4 + 5 + \cdots + (k + 2)] + (k + 3) = \dfrac{1}{2}k(k + 5) + k + 3 = \dfrac{1}{2}(k^2 + 7k + 6) = \dfrac{1}{2}(k + 1)(k + 6) = \dfrac{1}{2}(k + 1)[(k + 1) + 5]$.

5. (I) $n = 1$: $3(1) - 1 = 2$ and $\dfrac{1}{2}(1)[3(1) + 1] = \dfrac{1}{2}(4) = 2$

(II) If $2 + 5 + 8 + \cdots + (3k - 1) = \dfrac{1}{2}k(3k + 1)$, then $2 + 5 + 8 + \cdots + (3k - 1) + [3(k + 1) - 1]$

$= [2 + 5 + 8 + \cdots + (3k - 1)] + (3k + 2) = \dfrac{1}{2}k(3k + 1) + (3k + 2) = \dfrac{1}{2}(3k^2 + 7k + 4) = \dfrac{1}{2}(k + 1)(3k + 4)$

$= \dfrac{1}{2}(k + 1)[3(k + 1) + 1]$.

7. (I) $n = 1$: $2^{1-1} = 1$ and $2^1 - 1 = 1$
(II) If $1 + 2 + 2^2 + \cdots + 2^{k-1} = 2^k - 1$, then $1 + 2 + 2^2 + \cdots + 2^{k-1} + 2^{(k+1)-1} = (1 + 2 + 2^2 + \cdots + 2^{k-1}) + 2^k$
$= 2^k - 1 + 2^k = 2(2^k) - 1 = 2^{k+1} - 1$.

9. (I) $n = 1$: $4^{1-1} = 1$ and $\dfrac{1}{3}(4^1 - 1) = \dfrac{1}{3}(3) = 1$

(II) If $1 + 4 + 4^2 + \cdots + 4^{k-1} = \dfrac{1}{3}(4^k - 1)$, then $1 + 4 + 4^2 + \cdots + 4^{k-1} + 4^{(k+1)-1} = (1 + 4 + 4^2 + \cdots + 4^{k-1}) + 4^k$

$= \dfrac{1}{3}(4^k - 1) + 4^k = \dfrac{1}{3}[4^k - 1 + 3(4^k)] = \dfrac{1}{3}[4(4^k) - 1] = \dfrac{1}{3}(4^{k+1} - 1)$.

11. (I) $n = 1$: $\dfrac{1}{1 \cdot 2} = \dfrac{1}{2}$ and $\dfrac{1}{1 + 1} = \dfrac{1}{2}$

(II) If $\dfrac{1}{1 \cdot 2} + \dfrac{1}{2 \cdot 3} + \dfrac{1}{3 \cdot 4} + \cdots + \dfrac{1}{k(k + 1)} = \dfrac{k}{k + 1}$, then $\dfrac{1}{1 \cdot 2} + \dfrac{1}{2 \cdot 3} + \dfrac{1}{3 \cdot 4} + \cdots + \dfrac{1}{k(k + 1)} + \dfrac{1}{(k + 1)[(k + 1) + 1]}$

$= \left[\dfrac{1}{1 \cdot 2} + \dfrac{1}{2 \cdot 3} + \dfrac{1}{3 \cdot 4} + \cdots + \dfrac{1}{k(k + 1)}\right] + \dfrac{1}{(k + 1)(k + 2)} = \dfrac{k}{k + 1} + \dfrac{1}{(k + 1)(k + 2)} = \dfrac{k(k + 2) + 1}{(k + 1)(k + 2)}$

$= \dfrac{k^2 + 2k + 1}{(k + 1)(k + 2)} = \dfrac{(k + 1)^2}{(k + 1)(k + 2)} = \dfrac{k + 1}{k + 2} = \dfrac{k + 1}{(k + 1) + 1}$.

13. (I) $n = 1$: $1^2 = 1$ and $\dfrac{1}{6} \cdot 1 \cdot 2 \cdot 3 = 1$

(II) If $1^2 + 2^2 + 3^2 + \cdots + k^2 = \dfrac{1}{6}k(k + 1)(2k + 1)$, then $1^2 + 2^2 + 3^2 + \cdots + k^2 + (k + 1)^2$

$= (1^2 + 2^2 + 3^2 + \cdots + k^2) + (k + 1)^2 = \dfrac{1}{6}k(k + 1)(2k + 1) + (k + 1)^2 = \dfrac{1}{6}(2k^3 + 9k^2 + 13k + 6)$

$= \dfrac{1}{6}(k + 1)(k + 2)(2k + 3) = \dfrac{1}{6}(k + 1)[(k + 1) + 1][2(k + 1) + 1]$.

15. (I) $n = 1$: $5 - 1 = 4$ and $\dfrac{1}{2}(1)(9 - 1) = \dfrac{1}{2} \cdot 8 = 4$

(II) If $4 + 3 + 2 + \cdots + (5 - k) = \dfrac{1}{2}k(9 - k)$, then $4 + 3 + 2 + \cdots + (5 - k) + [5 - (k + 1)]$

$= [4 + 3 + 2 + \cdots + (5 - k)] + 4 - k = \dfrac{1}{2}k(9 - k) + 4 - k = \dfrac{1}{2}(9k - k^2 + 8 - 2k) = \dfrac{1}{2}(-k^2 + 7k + 8)$

$= \dfrac{1}{2}(k + 1)(8 - k) = \dfrac{1}{2}(k + 1)[9 - (k + 1)]$.

17. (I) $n = 1: 1 \cdot (1 + 1) = 2$ and $\frac{1}{3} \cdot 1 \cdot 2 \cdot 3 = 2$

(II) If $1 \cdot 2 + 2 \cdot 3 + 3 \cdot 4 + \cdots + k(k + 1) = \frac{1}{3}k(k + 1)(k + 2)$, then $1 \cdot 2 + 2 \cdot 3 + 3 \cdot 4 + \cdots + k(k + 1)$

$+ (k + 1)[(k + 1) + 1] = [1 \cdot 2 + 2 \cdot 3 + 3 \cdot 4 + \cdots + k(k + 1)] + (k + 1)(k + 2)$

$= \frac{1}{3}k(k + 1)(k + 2) + \frac{1}{3} \cdot 3(k + 1)(k + 2) = \frac{1}{3}(k + 1)(k + 2)(k + 3) = \frac{1}{3}(k + 1)[(k + 1) + 1][(k + 1) + 2]$.

19. (I) $n = 1: 1^2 + 1 = 2$, which is divisible by 2.

(II) If $k^2 + k$ is divisible by 2, then $(k + 1)^2 + (k + 1) = k^2 + 2k + 1 + k + 1 = (k^2 + k) + 2k + 2$. Since $k^2 + k$ is divisible by 2 and $2k + 2$ is divisible by 2, $(k + 1)^2 + (k + 1)$ is divisible by 2.

21. (I) $n = 1: 1^2 - 1 + 2 = 2$ which is divisible by 2.

(II) If $k^2 - k + 2$ is divisible by 2, then $(k + 1)^2 - (k + 1) + 2 = k^2 + 2k + 1 - k - 1 + 2 = (k^2 - k + 2) + 2k$. Since $k^2 - k + 2$ is divisible by 2 and $2k$ is divisible by 2, $(k + 1)^2 - (k + 1) + 2$ is divisible by 2.

23. (I) $n = 1:$ If $x > 1$, then $x^1 = x > 1$.

(II) Assume, for an arbitrary natural number k, that if $x > 1$ then $x^k > 1$. Multiply both sides of the inequality $x^k > 1$ by x. If $x > 1$, then $x^{k+1} > x > 1$.

25. (I) $n = 1: a - b$ is a factor of $a^1 - b^1 = a - b$.

(II) If $a - b$ is a factor of $a^k - b^k$, then $a^{k+1} - b^{k+1} = a(a^k - b^k) + b^k(a - b)$.

Since $a - b$ is a factor of $a^k - b^k$ and $a - b$ is a factor of $a - b$, then $a - b$ is a factor of $a^{k+1} - b^{k+1}$.

27. (I) $n = 1: (1 + a)^1 = 1 + a \geq 1 + 1 \cdot a$

(II) Assume that there is an integer k for which the inequality holds. So $(1 + a)^k \geq 1 + ka$. We need to show that $(1 + a)^{k+1} \geq 1 + (k + 1)a$. $(1 + a)^{k+1} = (1 + a)^k (1 + a) \geq (1 + ka)(1 + a) = 1 + ka^2 + a + ka = 1 + (k + 1)a + ka^2 \geq 1 + (k + 1)a$.

29. If $2 + 4 + 6 + \cdots + 2k = k^2 + k + 2$, then $2 + 4 + 6 + \cdots + 2k + 2(k + 1)$

$= (2 + 4 + 6 + \cdots + 2k) + 2k + 2 = k^2 + k + 2 + 2k + 2 = k^2 + 3k + 4 = (k^2 + 2k + 1) + (k + 1) + 2$

$= (k + 1)^2 + (k + 1) + 2.$

But $2 \cdot 1 = 2$ and $1^2 + 1 + 2 = 4$. The fact is that $2 + 4 + 6 + \cdots + 2n = n^2 + n$, not $n^2 + n + 2$ (Problem 1).

31. (I) $n = 1: [a + (1 - 1)d] = a$ and $1 \cdot a + d\frac{1 \cdot (1 - 1)}{2} = a$.

(II) If $a + (a + d) + (a + 2d) + \cdots + [a + (k - 1)d] = ka + d\frac{k(k - 1)}{2}$, then

$a + (a + d) + (a + 2d) + \cdots + [a + (k - 1)d] + [a + ((k + 1) - 1)d] = ka + d\frac{k(k - 1)}{2} + a + kd$

$= (k + 1)a + d\frac{k(k - 1) + 2k}{2} = (k + 1)a + d\frac{(k + 1)(k)}{2} = (k + 1)a + d\frac{(k + 1)[(k + 1) - 1]}{2}$.

33. (I) $n = 3:$ The sum of the angles of a triangle is $(3 - 2) \cdot 180° = 180°$.

(II) Assume for some $k \geq 3$ that the sum of the angles of a convex polygon of k sides is $(k - 2) \cdot 180°$. A convex polygon of $k + 1$ sides consists of a convex polygon of k sides plus a triangle (see the illustration). The sum of the angles is $(k - 2) \cdot 180° + 180° = (k - 1) \cdot 180° = [(k + 1) - 2] \cdot 180°$.

k sides

$k + 1$ sides

13.5 Assess Your Understanding *(page 966)*

1. Pascal triangle **2.** 15 **3.** F **4.** Binomial Theorem **5.** 10 **7.** 21 **9.** 50 **11.** 1 **13.** $\approx 1.8664 \times 10^{15}$ **15.** $\approx 1.4834 \times 10^{13}$

17. $x^5 + 5x^4 + 10x^3 + 10x^2 + 5x + 1$ **19.** $x^6 - 12x^5 + 60x^4 - 160x^3 + 240x^2 - 192x + 64$ **21.** $81x^4 + 108x^3 + 54x^2 + 12x + 1$

23. $x^{10} + 5x^8y^2 + 10x^6y^4 + 10x^4y^6 + 5x^2y^8 + y^{10}$ **25.** $x^3 + 6\sqrt{2}x^{5/2} + 30x^2 + 40\sqrt{2}x^{3/2} + 60x + 24\sqrt{2}x^{1/2} + 8$

27. $a^5x^5 + 5a^4bx^4y + 10a^3b^2x^3y^2 + 10a^2b^3x^2y^3 + 5ab^4xy^4 + b^5y^5$ **29.** 17,010 **31.** $-101,376$ **33.** 41,472 **35.** $2835x^3$

37. $314,928x^7$ **39.** 495 **41.** 3360 **43.** 1.00501

45. $\binom{n}{n - 1} = \frac{n!}{(n - 1)! [n - (n - 1)]!} = \frac{n!}{(n - 1)! 1!} = \frac{n \cdot (n - 1)!}{(n - 1)!} = n; \binom{n}{n} = \frac{n!}{n! (n - n)!} = \frac{n!}{n!0!} = \frac{n!}{n!} = 1$

47. $2^n = (1 + 1)^n = \binom{n}{0}1^n + \binom{n}{1}(1)^{n-1}(1) + \cdots + \binom{n}{n}1^n = \binom{n}{0} + \binom{n}{1} + \cdots + \binom{n}{n}$ **49.** 1

Review Exercises *(page 968)*

1. $a_1 = -\frac{4}{3}, a_2 = \frac{5}{4}, a_3 = -\frac{6}{5}, a_4 = \frac{7}{6}, a_5 = -\frac{8}{7}$ **3.** $c_1 = 2, c_2 = 1, c_3 = \frac{8}{9}, c_4 = 1, c_5 = \frac{32}{25}$ **5.** $a_1 = 3, a_2 = 2, a_3 = \frac{4}{3}, a_4 = \frac{8}{9}, a_5 = \frac{16}{27}$

7. $a_1 = 2, a_2 = 0, a_3 = 2, a_4 = 0, a_5 = 2$ **9.** $6 + 10 + 14 + 18 = 48$ **11.** $\sum_{k=1}^{13}(-1)^{k+1}\frac{1}{k}$ **13.** Arithmetic; $d = 1; S_n = \frac{n}{2}(n + 11)$

15. Neither **17.** Geometric; $r = 8$; $S_n = \dfrac{8}{7}(8^n - 1)$ **19.** Arithmetic; $d = 4$; $S_n = 2n(n - 1)$

21. Geometric; $r = \dfrac{1}{2}$; $S_n = 6\left[1 - \left(\dfrac{1}{2}\right)^n\right]$ **23.** Neither **25.** 3825 **27.** 1125 **29.** $\dfrac{1093}{2187} \approx 0.49977$ **31.** 35 **33.** $\dfrac{1}{10^{10}}$ **35.** $9\sqrt{2}$

37. $\{a_n\} = \{5n - 4\}$ **39.** $\{a_n\} = \{n - 10\}$ **41.** Converges; $\dfrac{9}{2}$ **43.** Converges; $\dfrac{4}{3}$ **45.** Diverges **47.** Converges; 8

49. (I) $n = 1: 3 \cdot 1 = 3$ and $\dfrac{3 \cdot 1}{2}(1 + 1) = 3$

(II) If $3 + 6 + 9 + \cdots + 3k = \dfrac{3k}{2}(k + 1)$, then $3 + 6 + 9 + \cdots + 3k + 3(k + 1) = (3 + 6 + 9 + \cdots + 3k) + (3k + 3)$

$= \dfrac{3k}{2}(k + 1) + (3k + 3) = \dfrac{3k^2}{2} + \dfrac{3k}{2} + \dfrac{6k}{2} + \dfrac{6}{2} = \dfrac{3}{2}(k^2 + 3k + 2) = \dfrac{3}{2}(k + 1)(k + 2) = \dfrac{3(k + 1)}{2}[(k + 1) + 1]$.

51. (I) $n = 1: 2 \cdot 3^{1-1} = 2$ and $3^1 - 1 = 2$

(II) If $2 + 6 + 18 + \cdots + 2 \cdot 3^{k-1} = 3^k - 1$, then $2 + 6 + 18 + \cdots + 2 \cdot 3^{k-1} + 2 \cdot 3^{(k+1)-1} = (2 + 6 + 18 + \cdots + 2 \cdot 3^{k-1}) + 2 \cdot 3^k$

$= 3^k - 1 + 2 \cdot 3^k = 3 \cdot 3^k - 1 = 3^{k+1} - 1$.

53. (I) $n = 1: (3 \cdot 1 - 2)^2 = 1$ and $\dfrac{1}{2} \cdot 1 \cdot [6(1)^2 - 3(1) - 1] = 1$

(II) If $1^2 + 4^2 + 7^2 + \cdots + (3k - 2)^2 = \dfrac{1}{2}k(6k^2 - 3k - 1)$, then $1^2 + 4^2 + 7^2 + \cdots + (3k - 2)^2 + [3(k + 1) - 2]^2$

$= [1^2 + 4^2 + 7^2 + \cdots + (3k - 2)^2] + (3k + 1)^2 = \dfrac{1}{2}k(6k^2 - 3k - 1) + (3k + 1)^2 = \dfrac{1}{2}(6k^3 - 3k^2 - k) + (9k^2 + 6k + 1)$

$= \dfrac{1}{2}(6k^3 + 15k^2 + 11k + 2) = \dfrac{1}{2}(k + 1)(6k^2 + 9k + 2) = \dfrac{1}{2}(k + 1)[6(k + 1)^2 - 3(k + 1) - 1]$.

55. 10 **57.** $x^5 + 10x^4 + 40x^3 + 80x^2 + 80x + 32$ **59.** $32x^5 + 240x^4 + 720x^3 + 1080x^2 + 810x + 243$ **61.** 144 **63.** 84

65. (a) 8 bricks **(b)** 1100 bricks **67. (a)** $20\left(\dfrac{3}{4}\right)^3 = \dfrac{135}{16}$ ft **(b)** $20\left(\dfrac{3}{4}\right)^n$ ft **(c)** 13 times **(d)** 140 ft **69.** \$244,129.08

Chapter Test *(page 970)*

1. $0, \dfrac{3}{10}, \dfrac{8}{11}, \dfrac{5}{4}, \dfrac{24}{13}$ **2.** $4, 14, 44, 134, 404$ **3.** $2 - \dfrac{3}{4} + \dfrac{4}{9} = \dfrac{61}{36}$ **4.** $-\dfrac{1}{3} - \dfrac{14}{9} - \dfrac{73}{27} - \dfrac{308}{81} = -\dfrac{680}{81}$ **5.** $\displaystyle\sum_{k=1}^{10}(-1)^k\left(\dfrac{k + 1}{k + 4}\right)$ **6.** Neither

7. Geometric; $r = 4$; $S_n = \dfrac{2}{3}(1 - 4^n)$ **8.** Arithmetic; $d = -8$; $S_n = n(2 - 4n)$ **9.** Arithmetic; $d = -\dfrac{1}{2}$; $S_n = \dfrac{n}{4}(27 - n)$

10. Geometric; $r = \dfrac{2}{5}$; $S_n = \dfrac{125}{3}\left[1 - \left(\dfrac{2}{5}\right)^n\right]$ **11.** Neither **12.** Converges; $\dfrac{1024}{5}$ **13.** $243m^5 + 810m^4 + 1080m^3 + 720m^2 + 240m + 32$

14. First we show that the statement holds for $n = 1$. $\left(1 + \dfrac{1}{1}\right) = 1 + 1 = 2$. The equality is true for $n = 1$, so Condition I holds. Next we assume that

$\left(1 + \dfrac{1}{1}\right)\left(1 + \dfrac{1}{2}\right)\left(1 + \dfrac{1}{3}\right)\cdots\left(1 + \dfrac{1}{n}\right) = n + 1$ is true for some k, and we determine whether the formula then holds for $k + 1$. We assume that

$\left(1 + \dfrac{1}{1}\right)\left(1 + \dfrac{1}{2}\right)\left(1 + \dfrac{1}{3}\right)\cdots\left(1 + \dfrac{1}{k}\right) = k + 1$. Now we need to show that $\left(1 + \dfrac{1}{1}\right)\left(1 + \dfrac{1}{2}\right)\left(1 + \dfrac{1}{3}\right)\cdots\left(1 + \dfrac{1}{k}\right)\left(1 + \dfrac{1}{k + 1}\right)$

$= (k + 1) + 1 = k + 2$. We do this as follows:

$\left(1 + \dfrac{1}{1}\right)\left(1 + \dfrac{1}{2}\right)\left(1 + \dfrac{1}{3}\right)\cdots\left(1 + \dfrac{1}{k}\right)\left(1 + \dfrac{1}{k + 1}\right) = \left[\left(1 + \dfrac{1}{1}\right)\left(1 + \dfrac{1}{2}\right)\left(1 + \dfrac{1}{3}\right)\cdots\left(1 + \dfrac{1}{k}\right)\right]\left(1 + \dfrac{1}{k + 1}\right) = (k + 1)$

$\left(1 + \dfrac{1}{k + 1}\right)$ (induction assumption) $= (k + 1) \cdot 1 + (k + 1) \cdot \dfrac{1}{k + 1} = k + 1 + 1 = k + 2$

Condition II also holds. Thus, the formula holds true for all natural numbers.

15. After 10 years, the Durango will be worth \$6103.11. **16.** The weightlifter will have lifted a total of 8000 pounds after 5 sets.

Cumulative Review *(page 970)*

1. $\{-3, 3, -3i, 3i\}$ **2. (a)**

(b) $\left\{\left(\sqrt{\dfrac{-1 + \sqrt{3601}}{18}}, \dfrac{-1 + \sqrt{3601}}{6}\right), \left(-\sqrt{\dfrac{-1 + \sqrt{3601}}{18}}, \dfrac{-1 + \sqrt{3601}}{6}\right)\right\}$

(c) The circle and the parabola intersect at

$\left(\sqrt{\dfrac{-1 + \sqrt{3601}}{18}}, \dfrac{-1 + \sqrt{3601}}{6}\right), \left(-\sqrt{\dfrac{-1 + \sqrt{3601}}{18}}, \dfrac{-1 + \sqrt{3601}}{6}\right)$.

3. $\left\{\ln\left(\dfrac{5}{2}\right)\right\}$ **4.** $y = 5x - 10$ **5.** $x^2 + y^2 + 2x - 4y - 20 = 0$

6. (a) 5 **(b)** 13 **(c)** $\dfrac{6x + 3}{2x - 1}$ **(d)** $\left\{x \middle| x \neq \dfrac{1}{2}\right\}$ **(e)** $\dfrac{7x - 2}{x - 2}$ **(f)** $\{x | x \neq 2\}$ **(g)** $g^{-1}(x) = \dfrac{1}{2}(x-1)$; all reals **(h)** $f^{-1}(x) = \dfrac{2x}{x - 3}$; $\{x | x \neq 3\}$

7. $\dfrac{x^2}{7} + \dfrac{y^2}{16} = 1$ **8.** $(x + 1)^2 = 4(y - 2)$ **9.** $r = 8\sin\theta$; $x^2 + (y - 4)^2 = 16$ **10.** $\left\{\dfrac{3\pi}{2}\right\}$ **11.** $\dfrac{2\pi}{3}$

12. (a) $-\dfrac{\sqrt{15}}{4}$ **(b)** $-\dfrac{\sqrt{15}}{15}$ **(c)** $-\dfrac{\sqrt{15}}{8}$ **(d)** $\dfrac{7}{8}$ **(e)** $\sqrt{\dfrac{1 + \dfrac{\sqrt{15}}{4}}{2}} = \dfrac{\sqrt{4 + \sqrt{15}}}{2\sqrt{2}}$

CHAPTER 14 Counting and Probability

14.1 Assess Your Understanding (page 978)

5. $n(A) + n(B) - n(A \cap B)$ **6.** F **7.** $\emptyset$, $\{a\}$, $\{b\}$, $\{c\}$, $\{d\}$, $\{a, b\}$, $\{a, c\}$, $\{a, d\}$, $\{b, c\}$, $\{b, d\}$, $\{c, d\}$, $\{a, b, c\}$, $\{b, c, d\}$, $\{a, c, d\}$, $\{a, b, d\}$, $\{a, b, c, d\}$ **9.** 25 **11.** 40 **13.** 25 **15.** 37 **17.** 18 **19.** 5 **21.** 15 different arrangements **23.** 9000 numbers **25.** 175; 125
27. (a) 15 **(b)** 15 **(c)** 15 **(d)** 25 **(e)** 40 **29. (a)** 11,597 thousand **(b)** 74.083 thousand **31.** 480 portfolios.

14.2 Assess Your Understanding (page 985)

3. permutation **4.** combination **5.** T **6.** T **7.** 30 **9.** 24 **11.** 1 **13.** 1680 **15.** 28 **17.** 35 **19.** 1 **21.** 10,400,600 **23.** $\{abc, abd, abe, acb,$
$acd, ace, adb, adc, ade, aeb, aec, aed, bac, bad, bae, bca, bcd, bce, bda, bdc, bde, bea, bec, bed, cab, cad, cae, cba, cbd, cbe, cda, cdb, cde, cea, ceb, ced,$
$dab, dac, dae, dba, dbc, dbe, dca, dcb, dce, dea, deb, dec, eab, eac, ead, eba, ebc, ebd, eca, ecb, ecd, eda, edb, edc\}$; 60 **25.** $\{123, 124, 132, 134, 142, 143,$
$213, 214, 231, 234, 241, 243, 312, 314, 321, 324, 341, 342, 412, 413, 421, 423, 431, 432\}$; 24 **27.** $\{abc, abd, abe, acd, ace, ade, bcd, bce, bde, cde\}$; 10
29. $\{123, 124, 134, 234\}$; 4 **31.** 16 **33.** 8 **35.** 24 **37.** 60 **39.** 18,278 **41.** 35 **43.** 1024 **45.** 120 **47.** 132,860 **49.** 336 **51.** 90,720
53. (a) 63 **(b)** 35 **(c)** 1 **55.** 1.157×10^{76} **57.** 362,880 **59.** 660 **61.** 15 **63. (a)** 125,000; 117,600 **(b)** A better name for a *combination* lock
would be a *permutation* lock because the order of the numbers matters.

Historical Problems (page 995)

1. (a) $\{AAAA, AAAB, AABA, AABB, ABAA, ABAB, ABBA, ABBB, BAAA, BAAB, BABA, BABB, BBAA, BBAB, BBBA, BBBB\}$

(b) $P(A \text{ wins}) = \dfrac{C(4, 2) + C(4, 3) + C(4, 4)}{2^4} = \dfrac{6 + 4 + 1}{16} = \dfrac{11}{16}$; $P(B \text{ wins}) = \dfrac{C(4, 3) + C(4, 4)}{2^4} = \dfrac{4 + 1}{16} = \dfrac{5}{16}$

2. (a) $\$\dfrac{3}{2} = \1.50 **(b)** $\$\dfrac{1}{2} = \0.50

14.3 Assess Your Understanding (page 995)

1. equally likely **2.** complement **3.** F **4.** T **5.** 0, 0.01, 0.35, 1 **7.** Probability model **9.** Not a probability model
11. $S = \{HH, HT, TH, TT\}$; $P(HH) = \dfrac{1}{4}$, $P(HT) = \dfrac{1}{4}$, $P(TH) = \dfrac{1}{4}$, $P(TT) = \dfrac{1}{4}$ **13.** $S = \{HH1, HH2, HH3, HH4, HH5, HH6, HT1, HT2, HT3,$

$HT4, HT5, HT6, TH1, TH2, TH3, TH4, TH5, TH6, TT1, TT2, TT3, TT4, TT5, TT6\}$; each outcome has the probability of $\dfrac{1}{24}$.

15. $S = \{HHH, HHT, HTH, HTT, THH, THT, TTH, TTT\}$; each outcome has the probability of $\dfrac{1}{8}$.

17. $S = \{$ 1 Yellow, 1 Red, 1 Green, 2 Yellow, 2 Red, 2 Green, 3 Yellow, 3 Red, 3 Green, 4 Yellow, 4 Red, 4 Green$\}$; each outcome has the probability of

$\dfrac{1}{12}$; thus, $P(2 \text{ Red}) + P(4 \text{ Red}) = \dfrac{1}{12} + \dfrac{1}{12} = \dfrac{1}{6}$.

19. $S = \{$1 Yellow Forward, 1 Yellow Backward, 1 Red Forward, 1 Red Backward, 1 Green Forward, 1 Green Backward, 2 Yellow Forward, 2 Yellow
Backward, 2 Red Forward, 2 Red Backward, 2 Green Forward, 2 Green Backward, 3 Yellow Forward, 3 Yellow Backward, 3 Red Forward, 3 Red
Backward, 3 Green Forward, 3 Green Backward, 4 Yellow Forward, 4 Yellow Backward, 4 Red Forward, 4 Red Backward, 4 Green Forward, 4 Green

Backward$\}$; each outcome has the probability of $\dfrac{1}{24}$; thus, $P(1 \text{ Red Backward}) + P(1 \text{ Green Backward}) = \dfrac{1}{24} + \dfrac{1}{24} = \dfrac{1}{12}$.

21. $S = \{$11 Red, 11 Yellow, 11 Green, 12 Red, 12 Yellow, 12 Green, 13 Red, 13 Yellow, 13 Green, 14 Red, 14 Yellow, 14 Green, 21 Red, 21 Yellow, 21 Green,
22 Red, 22 Yellow, 22 Green, 23 Red, 23 Yellow, 23 Green, 24 Red, 24 Yellow, 24 Green, 31 Red, 31 Yellow, 31 Green, 32 Red, 32 Yellow, 32 Green, 33 Red,
33 Yellow, 33 Green, 34 Red, 34 Yellow, 34 Green, 41 Red, 41 Yellow, 41 Green, 42 Red, 42 Yellow, 42 Green, 43 Red, 43 Yellow, 43 Green, 44 Red,

44 Yellow, 44 Green$\}$; each outcome has the probability of $\dfrac{1}{48}$; thus, $E = \{22 \text{ Red}, 22 \text{ Green}, 24 \text{ Red}, 24 \text{ Green}\}$; $P(E) = \dfrac{n(E)}{n(S)} = \dfrac{4}{48} = \dfrac{1}{12}$.

23. A, B, C, F **25.** B **27.** $P(H) = \dfrac{4}{5}$; $P(T) = \dfrac{1}{5}$ **29.** $P(1) = P(3) = P(5) = \dfrac{2}{9}$; $P(2) = P(4) = P(6) = \dfrac{1}{9}$ **31.** $\dfrac{3}{10}$ **33.** $\dfrac{1}{2}$ **35.** $\dfrac{1}{6}$ **37.** $\dfrac{1}{8}$

39. $\dfrac{1}{4}$ **41.** $\dfrac{1}{6}$ **43.** $\dfrac{1}{18}$ **45.** 0.55 **47.** 0.70 **49.** 0.30 **51.** 0.87 **53.** 0.66 **55.** 0.95 **57.** $\dfrac{17}{20}$ **59.** $\dfrac{11}{20}$ **61.** $\dfrac{1}{2}$ **63.** $\dfrac{3}{10}$ **65.** $\dfrac{2}{5}$

67. (a) 0.57 **(b)** 0.95 **(c)** 0.83 **(d)** 0.38 **(e)** 0.29 **(f)** 0.05 **(g)** 0.78 **(h)** 0.71 **69. (a)** $\dfrac{25}{33}$ **(b)** $\dfrac{25}{33}$ **71.** 0.167 **73.** 0.000033069

Review Exercises *(page 999)*

1. ∅, {Dave}, {Joanne}, {Erica}, {Dave, Joanne}, {Dave, Erica}, {Joanne, Erica}, {Dave, Joanne, Erica} **3.** 17 **5.** 29 **7.** 7 **9.** 25 **11.** 336
13. 56 **15.** 60 **17.** 128 **19.** 3024 **21.** 1680 **23.** 91 **25.** 1,600,000 **27.** 216,000 **29.** 1260 **31. (a)** 381,024 **(b)** 1260 **33. (a)** $8.634628387 \times 10^{45}$
(b) 0.6531 **(c)** 0.3469 **35. (a)** 0.058 **(b)** 0.942 **37.** $\dfrac{4}{9}$ **39.** 0.2; 0.26

Chapter Test *(page 1000)*

1. 22 **2.** 3 **3.** 8 **4.** 45 **5.** 5040 **6.** 151,200 **7.** 462 **8.** There are 54,264 ways to choose 6 colors from the 21 available colors. **9.** There are 840
distinct arrangements of the letters in the word REDEEMED. **10.** There are 56 different exacta bets for an 8-horse race. **11.** There are 155,480,000
possible license plates using the new format. **12. (a)** 0.95 **(b)** 0.30 **13. (a)** 0.25 **(b)** 0.55 **14.** 0.19
15. $P(\text{win on } \$1 \text{ play}) = \dfrac{1}{120{,}526{,}770} \approx 0.0000000083$ **16.** $P(\text{exactly 2 fours}) = \dfrac{1250}{7776} \approx 0.1608$

Cumulative Review *(page 1001)*

1. $\left\{ \dfrac{1}{3} - \dfrac{\sqrt{2}}{3}i, \dfrac{1}{3} + \dfrac{\sqrt{2}}{3}i \right\}$ **2.** **3.** **4.** $\{x \mid 3.99 \le x \le 4.01\}$ or $[3.99, 4.01]$

5. $\left\{ -\dfrac{1}{2} + \dfrac{\sqrt{7}}{2}i, -\dfrac{1}{2} - \dfrac{\sqrt{7}}{2}i, -\dfrac{1}{5}, 3 \right\}$ **6.** **7.** 2 **8.** $\left\{ \dfrac{8}{3} \right\}$ **9.** $x = 2, y = -5, z = 3$ **10.** 125; 700

11. **12.**

Domain: all real numbers
Range: $\{y \mid y > 5\}$
Horizontal asymptote: $y = 5$

Area ≈ 14.46 square units

APPENDIX Graphing Utilities

1 Exercises *(page A2)*

1. $(-1, 4)$; II **3.** $(3, 1)$; I **5.** $X\text{min} = -6, X\text{max} = 6, X\text{scl} = 2, Y\text{min} = -4, Y\text{max} = 4, Y\text{scl} = 2$
7. $X\text{min} = -6, X\text{max} = 6, X\text{scl} = 2, Y\text{min} = -1, Y\text{max} = 3, Y\text{scl} = 1$ **9.** $X\text{min} = 3, X\text{max} = 9, X\text{scl} = 1, Y\text{min} = 2, Y\text{max} = 10, Y\text{scl} = 2$
11. $X\text{min} = -11, X\text{max} = 5, X\text{scl} = 1, Y\text{min} = -3, Y\text{max} = 6, Y\text{scl} = 1$
13. $X\text{min} = -30, X\text{max} = 50, X\text{scl} = 10, Y\text{min} = -90, Y\text{max} = 50, Y\text{scl} = 10$
15. $X\text{min} = -10, X\text{max} = 110, X\text{scl} = 10, Y\text{min} = -10, Y\text{max} = 160, Y\text{scl} = 10$ **17.** $4\sqrt{10}$ **19.** $2\sqrt{65}$

2 Exercises *(page A5)*

1. (a) **(b)** **(c)** **(d)**

3. (a) **(b)** **(c)** **(d)**

5. (a) (b) (c) (d)

7. (a) (b) (c) (d)

9. (a) (b) (c) (d)

11. (a) (b) (c) (d)

13. (a) (b) (c) (d)

15. (a) (b) (c) (d)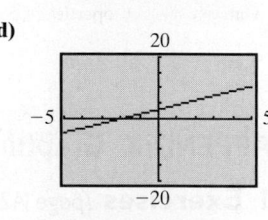

17. $Y_1 = X+2$ **19.** $Y_1 = -X+2$ **21.** $Y_1 = 2X+2$ **23.** $Y_1 = -2X+2$

25. $Y_1 = X^2+2$ **27.** $Y_1 = -X^2+2$ **29.** $Y_1 = -(3/2)X+3$ **31.** $Y_1 = (3/2)X+3$

3 Exercises (page A7)

1. -3.41 **3.** -1.71 **5.** -0.28 **7.** 3.00 **9.** 4.50 **11.** $0.32, 12.30$ **13.** $1.00, 23.00$

5 Exercises (page A10)

1. Yes **3.** Yes **5.** No **7.** Yes **9.** Answers may vary. A possible answer is $Y\min = 4$, $Y\max = 12$, and $Y\mathrm{scl} = 1$.

Index